Fundamentals of Industrial Hygiene

Third Edition

Barbara A. Plog, MPH, CIH, CSP
Editor

George S. Benjamin, MD, FACS
Technical Adviser

Maureen A. Kerwin, MPH
Technical Adviser

National Safety Council

Editor: Barbara A. Plog, MPH, CIH, CSP
Technical Advisers: George S. Benjamin, MD, FACS
Maureen A. Kerwin, MPH
Project Editor: Jodey B. Schonfeld
Interior Design and Composition: North Coast Associates
Cover Design: Russell Schneck Design

Printed in the United States of America
93 5

Library of Congress Cataloging in Publication Data
National Safety Council
Fundamentals of Industrial Hygiene
International Standard Book Number: 0–87912–082–7
Library of Congress Catalog Card Number: 87–60256
10M293

Product Number: 15133–0000

Occupational Safety and Health Series

The National Safety Council's OCCUPATIONAL SAFETY AND HEALTH SERIES is composed of four volumes written to help readers establish and maintain safety and health programs. The latest information on establishing priorities, collecting and analyzing data to help identify problems, and developing methods and procedures to reduce or eliminate illness and accidents, thus mitigating injury and minimizing economic loss resulting from accidents, is contained in all volumes in the series:

ACCIDENT PREVENTION MANUAL FOR INDUSTRIAL OPERATIONS (2-Volume Set)
Administration and Programs
Engineering and Technology
FUNDAMENTALS OF INDUSTRIAL HYGIENE
INTRODUCTION TO OCCUPATIONAL HEALTH AND SAFETY

Other hardcover books published by the Council include:
MOTOR FLEET SAFETY MANUAL
SUPERVISORS GUIDE TO HUMAN RELATIONS
SUPERVISORS SAFETY MANUAL

Contents

Foreword to the Third Edition

FUNDAMENTALS OF INDUSTRIAL HYGIENE is unique among the books on occupational health and safety. As an introductory text, *FUNDAMENTALS OF INDUSTRIAL HYGIENE* is unparalleled in its completeness and ease of use. Seasoned professionals can obtain a quick update by a review of selected chapters on specific topics. The information provided in the tables, appendices, and index make *FUNDAMENTALS OF INDUSTRIAL HYGIENE* an essential and economical reference for all occupational health and safety professionals. It is understandable why *FUNDAMENTALS OF INDUSTRIAL HYGIENE* has become the best-selling book on the subject.

On a personal note, *FUNDAMENTALS OF INDUSTRIAL HYGIENE* has been an integral part of my career in industrial hygiene. It was the discovery of a copy of the first edition in a used bookstore, which caused me to become interested in the subject. The second edition has served me as a textbook in graduate school and a reference in my professional practice. As a contributor to the third edition, I have had the opportunity to give back to this book some of what it has given to me. If you have any interest in the subject, don't wait until the third edition of *FUNDAMENTALS OF INDUSTRIAL HYGIENE* is available in used bookstores. Everyone with an interest in occupational health and safety will find value in this book.

Theodore J. Hogan, PhD, CIH
Chairman, Occupational Health Hazards Committee
National Safety Council

Preface

Much has occurred in the field of industrial hygiene since the second edition of *FUNDAMENTALS OF INDUSTRIAL HYGIENE* was published.

The recent passage of the federal Hazard Communication Standard (*CFR* 1910.1200) once again has thrust the profession of industrial hygiene into the center of the public stage. This Standard mandates hazard determination, labels and warnings, material safety data sheets, information and training programs, and a written hazard communication program for employees who may be exposed to hazardous substances during the course of their work. This piece of legislation has been called the most significant occupational health and safety legislation since the Williams Steiger Occupational Safety and Health Act of 1970 created the Occupational Safety and Health Administration (OSHA) and the National Institute for Occupational Safety and Health (NIOSH)—the federal safety and health apparatus—and mandated the provision of a safe and healthy working environment for all working women and men in the United States.

Initially, *CFR* 1910.1200 covered the basic manufacturing sector only, represented by Standard Industrial Classification (SIC) codes 20–39; however, OSHA issued a Final Standard on August 24, 1987, expanding coverage to the remainder of U.S. industry. The OSHA took this action to comply with a court ruling arising from litigation initiated by the AFL-CIO and the United Steelworkers of America. The unions had asked the courts to direct OSHA to broaden its coverage of the Standard to include employees in the nonmanufacturing sector. These newly covered employers in the nonmanufacturing sector must be in compliance with the Standard by May 24, 1988.

New federal asbestos regulations for general industry and construction (*CFR* 1910.1001 and *CFR* 1926.58, respectively) became effective in 1987 as well. The regulations reduced the Permissible Exposure Limit to 0.2 fibers per cubic centimeter and mandated an array of work methods, hazard controls, medical and air monitoring, and respiratory protection practices to be followed. This new legislation, combined with intensive Environmental Protection Agency (EPA) activity and regulations on asbestos in schools, has generated extensive asbestos abatement activity nationwide, adding to the high visibility of and demand for industrial hygiene services.

In addition to these developments, government actions in the area of hazardous wastes are impacting the industrial hygiene profession. The 1986 reauthorization by Congress of the 1980 Comprehensive Environmental Response, Compensation and Liability Act (CERCLA) creating Superfund has mandated extensive new requirements on risk assessment, underground storage tanks, community right-to-know, and training. Under Title III of Superfund, companies presently covered under the Hazard Communication Standard are required to submit copies of the material safety data sheets for their hazardous chemicals to the state emergency response commission, the local emergency planning committee, and the local fire department. Covered facilities must also submit emergency and hazardous chemical inventory forms to the same state and local authorities.

The 1986 Superfund Amendments and Reauthorization Act (SARA) also directed OSHA to issue regulations protecting the health and safety of hazardous waste workers (including those involved in emergency response operations). New OSHA regulations were issued and became effective the same date, December 19, 1986, and require measures in areas of hazard analysis, protection plans, work practices, information and training, medical surveillance, hazard controls, personal protective equipment, exposure limits, decontamination procedures, and emergency response.

Thus, the third edition of *FUNDAMENTALS OF INDUSTRIAL HYGIENE* comes at a time of intense federal regulatory activity that has profoundly affected the scope and the practice of industrial hygiene. Revisions of existing chapters and exciting new additions to the book fully address this activity. Part Seven—Governmental Regulations and Their Impact, contains a greatly revised and expanded chapter on governmental regulations and an original new chapter on the History of the Occupational Safety and Health Administration. This new offering details the creation of OSHA and its 17-year history—the history of an agency shaped, perhaps more than any other, by the political winds of change.

Other new additions to the third edition include completely new chapters on: (1) Ergonomics, an area that bridges the domain of the industrial hygienist and the safety professional and one that has seen vast new application in these fields and on (2) Computerizing an Industrial Hygiene Program, an arena in which personal computers have revolutionized how industrial hygienists manage data.

In addition, there are new sections on Acquired Immune Deficiency Syndrome (AIDS), Legionnaire's Disease, advances in industrial hygiene instrumentation, new training programs for OSHA compliance officers, the Hearing Conservation Amendment, and much more.

The primary purpose of this book is to provide a reference for those who have either an interest in or a direct responsibility for the recognition, evaluation, and control of occupational health hazards. Thus, it is intended to be of use to industrial hygienists, industrial hygiene students, physicians, nurses, safety personnel from labor and industry, labor organizations, public service groups, government agencies, and manufacturers. Others who may find this reference helpful include consultants, architects, lawyers, and allied professional personnel who work with those engaged in business, industry, and agriculture. It is hoped that this book will be of use to those responsible for planning and carrying out programs to minimize occupational health hazards.

An understanding of the fundamentals of industrial hygiene is very important to anyone involved in environmental, community, or occupational health. This manual, then, should be of help in defining the magnitude and extent of an industrial hygiene problem; it should help the reader decide when expert help is needed.

The *FUNDAMENTALS OF INDUSTRIAL HYGIENE* is also intended to be used either as a self-instructional text or as a text for an industrial hygiene fundamentals course, such as the one offered by the National Safety Council, various colleges and universities, and professional organizations.

The increase in the number and complexity of substances found in the workplace—substances that may spill over into the community environment—makes imperative the dissemination, as efficiently and conveniently as possible, of certain basic information relating to occupational health hazards and resultant occupational diseases.

The book is organized into seven parts; each can stand alone as a reference source. For that reason, we have permitted a certain amount of redundancy.

Part One introduces the subject areas to be covered, in an overview of the fundamentals of industrial hygiene.

Part Two includes chapters on the fundamental aspects of the anatomy, physiology, hazards, and pathology of the lungs, skin, ears, and eyes. This background lays the groundwork for understanding how these organ systems interrelate and function.

Part Three is concerned with the recognition of specific environmental factors or stresses. The chemical substances, physical agents, and biological and ergonomic hazards present in the workplace are covered. Anticipation of these hazards is hopefully the result.

Part Four describes methods and techniques of evaluating the hazard. Included is one of the more important aspects of an industrial hygiene program—the methods used to evaluate the extent of exposure to harmful chemical and physical agents. Basic information is given on the various types of instruments available to measure these stresses and on how to use the instruments properly to obtain valid measurements.

Part Five deals with the control of environmental hazards. Although industrial hygiene problems vary, the basic principles of health hazards control, problem-solving techniques, and the examples of engineering control measures given here are general enough to have wide application. To augment the basics, specific information is covered in the chapters on industrial ventilation.

Part Six is directed specifically to persons responsible for conducting and organizing industrial hygiene programs. The fundamental concepts of the roles of the industrial hygienist, the nurse, the safety professional, and the physician, in implementing a successful program, are discussed in detail. Particular attention is paid to a discussion of the practice of industrial hygiene in the public and private sectors and to a description of the professional certification of industrial hygienists. Computerizing an industrial hygiene program is discussed in one of the book's new chapters.

Part Seven contains up-to-date information on governmental regulations and their impact upon the practice of industrial hygiene.

Appendix A provides sources of help. One of the most difficult parts of getting any project started is finding sources of help and information. For this reason, we have included a completely updated and comprehensive annotated bibliography, and a listing of professional and service organizations, governmental agencies, training courses, and audiovisual materials.

Other appendices include Threshold Limit Values, Permissible Exposure Limits, information on specific chemical hazards, a review of mathematics, instructions on conversion of units, and a glossary of terms used in industrial hygiene, occupational health, and pollution control. An extensive index is included to assist the reader in locating information in this text.

Because this manual will be revised periodically, contributions and comments from readers are welcome.

Barbara A. Plog, MPH, CIH, CSP
Editor

PART I

History and Development

Overview of Industrial Hygiene

by Julian B. Olishifski, MS, PE, CSP
Revised by
Barbara A. Plog, MPH, CIH, CSP

INDUSTRIAL HYGIENE

Industrial hygiene is primarily concerned with the control of occupational health hazards that arise as a result of or during work. Industrial hygiene has been defined as "that science and art devoted to the anticipation, recognition, evaluation, and control of those environmental factors or stresses arising in or from the workplace, which may cause sickness, impaired health and well-being, or significant discomfort among workers or among the citizens of the community."

According to the American Academy of Industrial Hygiene's Code of Ethics for the professional, practice of industrial hygiene, the primary responsibility of the industrial hygienist is as follows:

(In general)

- to protect the health of employees
- to maintain an objective attitude toward the recognition, evaluation and control of health hazards regardless of external influences, realizing that the health and welfare of workers and others may depend upon the industrial hygienist's professional judgment
- to counsel employees regarding the health hazards and necessary precautions to avoid adverse health effects

(To employers)

- to respect confidences, advise honestly, and report findings and recommendations accurately
- to act responsibly in the application of industrial hygiene principles toward the attainment of healthful working environments
- to hold responsibilities to the employer or client subservient to the ultimate responsibility to protect the health of employees.

The industrial hygienist, although basically trained in engineering, physics, chemistry, or biology, has acquired by postgraduate study and/or experience a knowledge of the effects upon health of chemical and physical agents under various levels of exposure. The industrial hygienist is involved with the monitoring and analytical methods required to detect the extent of exposure, and the engineering and other methods used for hazard control.

The industrial hygienist recognizes that environmental stresses may endanger life and health, accelerate the aging process, or cause significant discomfort.

Evaluation of the magnitude of the environmental factors and stresses arising in or from the workplace is done by the industrial hygienist, aided by training, experience, and quantitative measurement of the chemical, physical, ergonomic, or biological stresses. The industrial hygienist can thus give an expert opinion as to the degree of risk posed by the environmental stresses.

Industrial hygiene includes the development of corrective measures in order to control health hazards by either reducing or eliminating the exposure. These control procedures may include such measures as the substitution of harmful or toxic materials with less dangerous ones, changing of work processes to eliminate or minimize work exposure, installation of exhaust ventilation systems, "good housekeeping" (including adequate methods of disposing of wastes), and the provision of proper personal protective equipment.

Basically, an effective industrial hygiene program would consist of the application of knowledge to the anticipation and recognition of health hazards arising out of work operations and processes, evaluation and measurement of the magnitude of the hazard—based on past experience and study, and control of the hazard.

This introductory chapter on fundamental concepts and the remaining chapters in this book are organized along the lines of the definitions just given.

Occupational health hazards may mean (1) conditions that cause legally compensable illnesses, or it may mean (2) any conditions in the workplace that impair the health of employees enough to make them lose time from work or to cause significant discomfort. Both are undesirable. Both are preventable. Their correction is properly a responsibility of management.

The safety professional and the industrial hygienist

Most safety professionals are already involved in some aspects of industrial hygiene. They study work operations, look for potential hazards, and make recommendations to minimize these hazards. The industrial hygienist, through specialized study and training, has the expertise to deal with these complex problems. If the safety professional carries on the day-to-day safety functions involving immediate decisions, he or she must know when and where to get help on industrial hygiene problems. (For more details, see Chapter 25, The Safety Professional.)

After the industrial hygienist surveys the plant, makes recommendations, and suggests certain control measures, it may become the safety professional's responsibility to see that the control measures are being applied and followed. Or such responsibility may be vested in an individual whose education and training is in the combined disciplines of safety and health.

Figure 1-1. The company or plant physician may have to depend on the skills, techniques, and knowledge of the health and safety professional to provide insight into an employee's job. The health and safety professional serves as a link between the health hazards of manufacturing operations of a company and its medical department or physician.

The occupational physician and the industrial hygienist

The occupational health program requires the services of the professional disciplines of the occupational physician, the occupational nurse, and the industrial hygienist, each supported by ancillary safety and health professionals, including industrial toxicologists and health physicists.

The occupational physician has acquired, by graduate training and/or industrial experience, extensive knowledge of cause-effect relationships of the chemical and physical agents, the signs and symptoms of chronic and acute exposures, and the treatment of adverse effects. (For more details, see Chapter 26, The Occupational Physician.)

The industrial hygienist effectively provides information about the manufacturing operations of a company to its medical department. The physician depends on the skills, techniques, and knowledge of the industrial hygienist to provide insight into the health background of an employee's job (Figure 1-1). In many cases, it is extremely difficult to differentiate between the symptoms of occupational and nonoccupational disease. The industrial hygienist, by pointing out the work operations and their associated hazards, enables the physician to correlate the employee's condition and complaints with the known potential job health hazards.

The physician uses the information provided by the industrial hygienist on the hazards present in the industrial environment to:

- determine employee response to the work environment
- correlate employee complaints with potential hazard areas
- undertake special biochemical tests to determine if normal bodily functions have been impaired
- provide the employee medical guidance on general health problems in relation to the physical requirements of the job
- through physical examinations, select workers for job assignments where preexisting conditions will not be aggravated nor will the worker's presence endanger the health and safety of others.

Management, the employee, and the industrial hygienist

To create an optimal occupational health program within a company, support from top management is critical. This support must pervade all levels of management, and be thorough and ongoing. Both timely execution of written directives and management example are key elements.

The employee plays a major role in the occupational health program. Employees are excellent sources of information on work processes and procedures and the perceived hazards of their daily operations. The industrial hygienist will benefit from this source of information and often obtain innovative suggestions for controlling hazards.

FEDERAL REGULATIONS

Federal regulations require an expansion in the scope of health programs, particularly in the area of medical examinations and environmental health tests, to protect the health of the worker.

The goal of an occupational health program is the maintenance and promotion of employee health and well-being by protecting employees from undesirable health effects that can result from inadequately controlled equipment, processes, materials, products, and wastes. (See Chapter 28, The Industrial Hygiene Program, for more details.)

The extent of an employer's industrial hygiene or occupational health program depends largely upon the nature of the operations, the materials and equipment used, and the type of hazards the operations may generate.

Occupational Safety and Health Act

An overview of the Occupational Safety and Health Act (OSHAct) is presented in this section. For more details, however, see Chapters 30 and 31, respectively, Governmental Regulations and The History of the Occupational Safety and Health Administration.

The declared congressional purpose of the OSHAct of 1970 is to "assure so far as possible every working man and woman in the nation safe and healthful working conditions and to preserve our human resources." The federal government is authorized to develop and set mandatory occupational safety and health standards applicable to any business affecting interstate commerce. The responsibility for promulgating and enforcing occupational safety and health standards rests with the Department of Labor.

The OSHAct of 1970 has brought a restructuring of programs and activities relating to safeguarding the health of the worker. Uniform occupational health regulations now apply to all businesses engaged in commerce, regardless of their locations within the jurisdiction. Threshold Limit Values for the year 1968 have been incorporated into the regulations and have the effect of law. (See Appendix B-2, Permissible Exposure Levels.)

The 1970 OSHAct has clearly defined procedures for promulgating regulations, conducting investigations for compliance, and handling availability of exposure data on workers, keeping of records, and other problems.

Nearly every employer is required to implement some element of an industrial hygiene or occupational health or hazard communication program, to be responsive to OSHA and the OSHAct and its health regulations or those of another regulatory agency. For example, the Mine Safety and Health Administration (MSHA) is charged with enforcing MSHA health regulations in all underground or surface mines.

National Institute for Occupational Safety and Health

The OSHAct established a National Institute for Occupational Safety and Health (NIOSH) to conduct research on the health effects of exposures in the work environment, to develop criteria for dealing with toxic materials and harmful agents, including safe levels of exposure, to train an adequate supply of professional personnel to meet the purposes of the OSHAct, and, in general, to conduct research and assistance programs for improving protection and maintenance of worker health.

Also authorized by the OSHAct are NIOSH programs for medical examinations and tests in the workplace as may be necessary to determine, for the purpose of research, the incidence of occupational illness and the susceptibility of employees to such illnesses. These examinations may also be at government expense. Another NIOSH function is annual publication of a list of all known toxic substances and the concentration at which toxicity is known to occur.

The NIOSH publishes industrywide studies on chronic or low-level exposure to a variety of industrial materials, processes, and stresses, which can affect the potential for illness and disease or loss of functional capacity in aging adults. The NIOSH is also responsible for conducting research on which new standards can be based.

Health standards promulgation

The OSHAct establishes a set of criteria that the employer will use in protecting employees against health hazards and harmful materials. Section 6(b)(5) of the act requires:

> The Secretary of Labor in promulgating standards dealing with toxic materials or harmful physical agents under this subsection of the OSHAct, shall set the standard, which most adequately assures to the extent feasible, on the basis of the best available evidence, that no employee will suffer material impairment of health or functional capacity even if such employee has regular exposure to the hazard dealt with by such standard for the period of his [sic] working life.

Basically, this establishes guidelines to make certain that such standards contain all of the substantive data that relate to the issue or subject covered by the standards so that the employer and employees can better understand compliance requirements.

Rules on use of labels and other forms of warning necessary to alert employees of all hazards to which they might be exposed are among requirements in the OSHAct. Employees should be aware of potential relevant symptoms of exposure and appropriate emergency treatment. Proper conditions and precautions should also be familiar to employees.

When appropriate, such standards (*1*) must also prescribe suitable protection equipment and control or technological procedures to be used in connection with such hazards and (2) must provide for monitoring or measuring employee exposure at such locations and intervals; both are to be performed as necessary for protection of employees.

Where appropriate, any such standard shall also prescribe the type and frequency of medical examinations or other tests that shall be made available, by the employer or at no cost to employees exposed to such hazards in order to determine most effectively whether the health of such employees is adversely affected by such exposures.

Also, Section 8(c)(3) of the OSHAct, requires OSHA to issue regulations requiring employers to maintain accurate records of employee exposures to potentially toxic materials or harmful physical agents that are required to be monitored or measured, provide employees or their representatives with an opportunity to observe such monitoring or measuring and to have access to these records, and make appropriate provisions for each employee or former employee to have access to records indicating the employee's exposure to toxic or harmful physical agents. (For more details, see Chapter 28, The Industrial Hygiene Program.)

In addition, the standard's requirements stipulate that each employer shall promptly notify any employee who has been or is being exposed to toxic materials or harmful physical agents in concentrations or at levels that exceed those prescribed by any standard, and (the employer) must inform any employee who is being thus exposed of the corrective action being taken.

The employer should be aware that the OSHAct regulations covering occupational health also deal with environmental controls. These regulations deal with air contaminants (gases, fumes, vapors, dusts, and mists), noise, and ionizing radiation. Employers whose operations deal with any of these exposures must become familiar with industrial hygiene concepts and controls.

ENVIRONMENTAL FACTORS OR STRESSES

The various environmental factors or stresses that can cause sickness, impaired health, or significant discomfort in workers can be classified as chemical, physical, biological, or ergonomic.

Chemical hazards. These arise from excessive airborne concentrations of mists, vapors, gases, or solids that are in the form of dusts or fumes. In addition to the hazard of inhalation, many of these materials may act as skin irritants or may be toxic by absorption through the skin.

Physical hazards. These include excessive levels of electromagnetic and ionizing radiations, noise, vibration, and extremes of temperature and pressure.

Ergonomic hazards. These include improperly designed tools or work areas. Improper lifting or reaching, poor visual conditions, or repeated motions in an awkward position can result in accidents or illnesses in the occupational environment. Designing the tools and the job to be done to fit the worker should be of prime importance. Intelligent application of engineering and biomechanical principles is required to eliminate hazards of this kind.

Biological hazards. These include insects, molds, fungi, and bacterial contamination (sanitation and housekeeping items as potable water, removal of industrial waste and sewage, food handling, and personal cleanliness.) (Biological and chemical hazards overlap.)

Exposure to many of the harmful stresses or hazards listed can produce an immediate response due to the intensity of the hazard; or the response can result from longer exposure at a lower intensity.

The occupational hazards in the work environment include thermal, electromagnetic, and mechanical energies as well as chemical agents that adversely affect bodily functions.

In certain life situations and occupations, depending upon the duration and severity of exposure, the work environment can produce significant subjective responses or strain. The energies and agents responsible for these effects are referred to as environmental stresses. An employee is most often exposed to an intricate interplay of many stresses, not to a single environmental stress.

Chemical hazards

The majority of the occupational health hazards arise from inhaling chemical agents in the form of vapors, gases, dusts, fumes, and mists, or by skin contact with these materials. The degree of risk of handling a given substance depends on the magnitude and duration of exposure. (See Chapter 16, Evaluation, for more details.)

To recognize occupational factors or stresses, a health and safety professional must first know about the chemicals used as raw materials and the nature of the products and by-products manufactured.

This sometimes requires great effort. The required information can be obtained from the Material Safety Data Sheet (MSDS) (Figure 1-2) that must be supplied by the chemical manufacturer or importer to the purchaser for all hazardous materials under the Hazard Communication Standard. The MSDS is a summary of the important health, safety, and toxicological information on the chemical or the mixture ingredients. Other stipulations of the Hazard Communication Standard require that all containers of hazardous substances in the workplace be labeled with appropriate warning and identification labels. See Chapters 30, Governmental Regulations, and 31, History of the Occupational Safety and Health Administration, for further discussion of the Hazard Communication Standard.

If the MSDS or the label does not give complete information but only trade names, it may be necessary to contact the manufacturer of the chemicals to obtain this information.

Many industrial materials such as resins and polymers are relatively inert and nontoxic under normal conditions of use, but when heated or machined, they may decompose to form highly toxic by-products. Information concerning these types of hazardous products and by-products must also be included in the company's Hazard Communication Program.

Breathing of some materials can irritate the upper respiratory tract or the terminal passages of the lungs and the air sacs, depending upon the solubility of the material. Contact of irritants with the skin surface can produce various kinds of dermatitis.

The presence of excessive amounts of biologically inert gases can dilute the atmospheric oxygen below the level required to maintain the normal blood saturation value for oxygen and disturb cellular processes. Other gases and vapors can prevent the blood from carrying oxygen to the tissues or intefere with its transfer from the blood to the tissue, thus producing chemical asphyxia or suffocation. Carbon monoxide and hydrogen cyanide are examples of chemical asphyxiants.

Some substances may affect the central nervous system and brain to produce narcosis and/or anaesthesia. In varying degrees, many solvents have these effects. Substances are often classified according to the major reaction that they produce, as asphyxiants, systemic toxins, pneumoconiosis-producing agents, carcinogens, irritant gases, and so on.

Solvents. This section discusses some general hazards arising from the use of solvents; a more detailed description is given in Chapter 6, Solvents.

Solvent vapors enter the body mainly by inhalation, although some skin absorption can occur. The vapors are absorbed from the lungs into the blood, and are distributed mainly to tissues with a high content of fat and lipids, such as the central nervous system, liver, and bone marrow. Solvents include aliphatic and aromatic hydrocarbons, alcohols, aldehydes, ketones, chlorinated hydrocarbons, and carbon disulfide.

Occupational exposure can occur in many different processes, such as the degreasing of metals in the machine industry, the extraction of fats or oils in the chemical or food industry, in dry cleaning, painting, and in the plastics industry, including the viscose-rayon industry.

The widespread industrial use of solvents presents a major problem to the industrial hygienist, the safety professional, and

Material Safety Data Sheet May be used to comply with OSHA's Hazard Communication Standard, 29 CFR 1910.1200. Standard must be consulted for specific requirements.	**U.S. Department of Labor** Occupational Safety and Health Administration (Non-Mandatory Form) Form Approved OMB No. 1218-0072 
IDENTITY *(As Used on Label and List)*	*Note: Blank spaces are not permitted. If any item is not applicable, or no information is available, the space must be marked to indicate that.*

Section I

Manufacturer's Name	Emergency Telephone Number
Address *(Number, Street, City, State, and ZIP Code)*	Telephone Number for Information
	Date Prepared
	Signature of Preparer *(optional)*

Section II — Hazardous Ingredients/Identity Information

Hazardous Components (Specific Chemical Identity; Common Name(s))	OSHA PEL	ACGIH TLV	Other Limits Recommended	% *(optional)*

Section III — Physical/Chemical Characteristics

Boiling Point		Specific Gravity (H_2O = 1)	
Vapor Pressure (mm Hg.)		Melting Point	
Vapor Density (AIR = 1)		Evaporation Rate (Butyl Acetate = 1)	
Solubility in Water			
Appearance and Odor			

Section IV — Fire and Explosion Hazard Data

Flash Point (Method Used)	Flammable Limits	LEL	UEL
Extinguishing Media			
Special Fire Fighting Procedures			
Unusual Fire and Explosion Hazards			

(Reproduce locally) OSHA 174, Sept. 1985

Figure 1-2. Material Safety Data Sheet—its format meets the requirements of the federal Hazard Communication Standard. (Continued)

Section V — Reactivity Data

Stability	Unstable		Conditions to Avoid
	Stable		

Incompatibility (*Materials to Avoid*)

Hazardous Decomposition or Byproducts

Hazardous Polymerization	May Occur		Conditions to Avoid
	Will Not Occur		

Section VI — Health Hazard Data

Route(s) of Entry: Inhalation? Skin? Ingestion?

Health Hazards (*Acute and Chronic*)

Carcinogenicity: NTP? IARC Monographs? OSHA Regulated?

Signs and Symptoms of Exposure

Medical Conditions
Generally Aggravated by Exposure

Emergency and First Aid Procedures

Section VII — Precautions for Safe Handling and Use

Steps to Be Taken in Case Material Is Released or Spilled

Waste Disposal Method

Precautions to Be Taken in Handling and Storing

Other Precautions

Section VIII — Control Measures

Respiratory Protection (*Specify Type*)

Ventilation	Local Exhaust	Special
	Mechanical (*General*)	Other

Protective Gloves	Eye Protection

Other Protective Clothing or Equipment

Work/Hygienic Practices

Page 2

☆ USGPO 1986-491-529/45775

Figure 1-2. (Concluded)

others charged with the responsibility for maintaining a safe, healthful working environment. Getting the job done using solvents without hazard to employees or property depends upon the proper selection, application, handling, and control of solvents and an understanding of their properties.

A good working knowledge of the physical properties, nomenclature, and effects of exposure is absolutely necessary in making a proper assessment of a solvent exposure. Nomenclature can be misleading. For example, "benzine" is sometimes mistakenly referred to as "benzene," a completely different solvent. Some commercial grades of benzine may contain benzene as a contaminant.

Use the information on the MSDA (Figure 1-2) or the manufacturer's label for the specific name and composition of the solvents involved. (See details in Chapter 6, Solvents.)

The severity of a hazard in the use of organic solvents depends on the following factors.

1. how the solvent is used
2. type of job operation which determines how the workers are exposed
3. work pattern
4. duration of exposure
5. operating temperature
6. exposed liquid surface
7. ventilation rates
8. evaporation rate of solvent
9. pattern of airflow
10. concentration of vapor in workroom air
11. housekeeping

The solvent hazard therefore is determined not only by the toxicity of the solvent itself but by the conditions of its use—who, what, how, where, and how long.

The health and safety professional can obtain much valuable information by observing the manner in which health hazards are generated, the number of people involved, and the control measures in use.

After the list of chemicals and physical conditions to which employees are exposed has been prepared, determine which of the chemicals or agents result in hazardous exposures and need further study.

Dangerous materials are those chemicals that may, under specific circumstances, cause injury to persons or damage to property because of reactivity, instability, spontaneous decomposition, flammability, or volatility. Under this definition, we will consider substances, mixtures, or compounds that are explosive, corrosive, flammable, or toxic.

Explosives are those substances, mixtures, or compounds capable of entering into a combustion reaction so rapidly and violently as to cause an explosion.

Corrosives are capable of destroying living tissue and have a destructive effect on other substances, particularly on combustible materials; this effect can result in a fire or explosion.

Flammable liquids are those liquids with a flash point of 38 C (100 F) or less, although those with higher flash points can be both combustible and dangerous.

Toxic chemicals are those gases, liquids, or solids, which through their chemical properties can produce injurious or lethal effects upon contact with body cells.

Oxidizing materials are those chemicals that will decompose readily under certain conditions to yield oxygen. They may cause a fire in contact with combustible materials, can react violently with water, and when involved in a fire can react violently.

Dangerous gases are those gases that can cause lethal or injurious effects and damage to property by their toxic, corrosive, flammable, or explosive physical and chemical properties.

When they are used, storage of dangerous chemicals should be limited to one day's supply, consistent with the safe and efficient operation of the process. The storage should also comply with applicable local laws and ordinances. An approved storehouse should be provided for the main supply of hazardous materials.

For hazardous materials information, MSDSs can be consulted to elicit toxicological information. The information is useful to the medical, purchasing, managerial, engineering, and health and safety departments in setting guidelines for safe use of these materials in industry. This information is also very helpful in the event of an emergency. The information should include those materials actually in use, with those that may be contemplated for early future use. Possibly the best and earliest source of information concerning such materials is from the purchasing agent. Thus, a close liaison should be set up between the purchasing agent and health and safety personnel so that early information will be provided concerning materials in use, and those to be ordered, and to ensure that MSDSs are received and reviewed for all hazardous substances.

Toxicity v. hazard. The toxicity of a material is not synonymous with its quality for being a health hazard. Toxicity is the capacity of a material to produce injury or harm. Hazard is the possibility that exposure to a material will cause injury when a specific quantity is used under certain conditions.

The key elements to be considered when evaluating a health hazard are as follows:

- How much of the material must be in contact with a body cell and for how long to produce injury?
- What is the probability that the material will be absorbed or come in contact with body cells?
- What is the rate of generation of airborne contaminants?
- What control measures are in use?

The effects of exposure to a substance depend on dose, rate, physical state of the substance, temperature, site of absorption, diet, and general state of a person's health.

Physical hazards

Problems relating to such things as noise, temperature extremes, ionizing radiation, nonionizing radiation, and pressure extremes are physical stresses. It is important that the employer, supervisor, and those responsible for safety and health are alert to these hazards because of the possible immediate or cumulative effects on the health of the employees in the workplace.

Noise (unwanted sound) is a form of vibration conducted through solids, liquids, or gases. The effects of noise on humans include the following:

- psychological effects (noise can startle, annoy, and disrupt concentration, sleep, or relaxation)
- interference with communication by speech, and, as a consequence, interference with job performance and safety
- physiological effects (noise-induced loss of hearing, or aural pain when the exposure is severe).

Damage risk criteria. If the ear is subjected to high levels of noise for a sufficient period of time, some loss of hearing may occur.

A number of factors can influence the effect of the noise exposure. These include the following:

- variation in individual susceptibility
- the total energy of the sound
- the frequency distribution of the sound
- other characteristics of the noise exposure, such as whether it is continuous, intermittent, or made up of a series of impacts
- the total daily duration of exposure
- the length of employment in the noise environment.

Because of the complex relationships of noise and exposure time to threshold shift (reduction in hearing level) and the many contributory causes, establishing criteria for protecting workers against hearing loss is difficult. However, criteria have been developed to protect against hearing loss in the speech-frequency range. These criteria are known as the Threshold Limit Values for Noise. (See Chapter 9, Industrial Noise, and Appendix B-1, Threshold Limit Values, for more details.)

There are three nontechnical rules-of-thumb to determine if the work area has excessive noise levels:

- If it is necessary to speak very loudly or shout directly into the ear of a person in order to be understood, it is possible that the exposure limit for noise is being exceeded. Conversation becomes difficult when the noise level exceeds 70 decibels (dBA).
- If employees say that they have heard noises and ringing noises in their ears at the end of the workday, they may be exposed to too much noise.
- If employees complain that the sounds of speech or music seem muffled after leaving work, but that their hearing is fairly clear in the morning when they return to work, they may be exposed to noise levels that cause a partial temporary loss of hearing, which can become permanent upon repeated exposure.

Permissible levels. The criteria for hearing conservation, required by OSHAct in 29 *CFR* 1910.95, establishes the permissible levels of harmful noise to which an employee may be subjected. The permissible decibel levels and hours (duration per day) are specified. For example, a noise of 90 dBA is permissible for 8 hours; 95 dBA, for 4 hours, etc. (See Chapter 9, Industrial Noise, for more details.)

Basically, the regulations stipulate that when employees are subjected to sound that exceeds the permissible limits, feasible adminstrative or engineering controls shall be utilized. If such controls fail to reduce sound exposure within permissible levels, personal protective equipment must be provided and used to reduce sound levels to within the permissible levels. Exposure to impulsive or impact noise should not exceed 140 dBA peak sound pressure level.

According to the Hearing Conservation Amendment to 29 *CFR* 1910.95, in all cases when the sound levels exceed 85 dBA on an 8-hour time-weighed average (TWA), a continuing, effective hearing conservation program shall be administered. The Hearing Conservation Amendment specifies the essential elements of a Hearing Conservation Program. (See Chapter 9, Industrial Noise, for a discussion of noise and OSHA noise regulations.)

Administering a hearing conservation program goes beyond the wearing of earplugs or earmuffs. Such programs can be complex, and professional guidance is essential for establishing programs that will be responsive to the need. Valid noise exposure information, correlated with audiometric tests results are needed to help both health and safety and medical personnel to make intelligent decisions about hearing conservation programs.

The effectiveness of a hearing conservation program depends upon the cooperation of employers, employees, and others concerned. Management's responsibility in this type of program includes noise measurements, initiation of noise control measures, provision of hearing protection equipment where it is required, audiometric testing of employees to measure their hearing levels (thresholds), and information and training programs for employees.

The employee's responsibility is to properly use the protective equipment provided by management, and to observe any rules or regulations on the use of equipment, to minimize the noise exposure.

Extremes of temperature. Probably the most elementary factor of environmental control is control of the thermal environment in which people work. Extremes of temperature affect the amount of work that people can do and the manner in which they do it. In industry, the problem is more often high temperatures rather than low temperatures. (More details on this subject are given in Chapter 12, Temperature Extremes.)

The body continuously produces heat through its metabolic processes. Since the body processes are designed to operate only within a very narrow range of temperature, the body must dissipate this heat as rapidly as it is produced if it is to function efficiently. A sensitive and rapidly acting set of temperature-sensing devices in the body must also control the rates of its temperature-regulating processes. (This mechanism is described in Chapter 3, The Skin.)

Heat stress is a common problem as are the problems presented by a very cold environment. Evaluation of heat stress by interpreting information relating the physiology of a person to the physical aspects of the environment is not simple or easy. Considerably more is involved than simply taking a number of air-temperature measurements and making decisions on the basis of this information.

One question that must be asked is, are people merely uncomfortable or are the conditions such that continued exposure will cause the body temperature to fall below or rise above safe limits. This makes it difficult for an individual, armed with a clipboard full of data, to interpret how another individual actually feels or is adversely affected.

People function efficiently only in a very narrow body temperature range, a "core" temperature measured deep inside the body, not on the skin or at body extremities. Fluctuations in core temperatures exceeding about 2 F below, or 3 F above, the normal core temperature of 37.6 C (99.6 F), which is 37 C mouth temperature (98.6 F, mouth temperature), impair performance markedly. If this five-degree range is exceeded, a health hazard exists.

The body attempts to counteract the effects of high temperature by increasing the heart rate. The capillaries in the skin also dilate to bring more blood to the surface so that the rate of cooling is increased. Sweating is an important factor in cooling the body.

Heatstroke is caused by exposure to an environment in which the body is unable to cool itself sufficiently, thus the body temperature rises rapidly. Heatstroke is a much more serious condition than heat cramps or heat exhaustion (described next). An important predisposing factor is excessive physical exertion or moderate exertion in extreme heat conditions. The method of control is to reduce the temperature of the surroundings or to increase the ability of the body to cool itself, so that body temperature does not rise. In heatstroke, sweating may cease and the body temperature can quickly rise to fatal levels. It is critical to undertake emergency cooling of the body even while medical help is on the way. Studies show that the higher the body temperature upon admission to emergency rooms, the higher the death rate. Heatstroke is a life-threatening medical emergency.

Heat cramps can result from exposure to high temperature for a relatively long time, particularly if accompanied by heavy exertion, with excessive loss of salt and moisture from the body. Even if the moisture is replaced by drinking plenty of water, an excessive loss of salt can provide heat cramps or heat exhaustion.

Heat exhaustion can also result from physical exertion in a hot environment. Its signs are a mildly elevated temperature, pallor, weak pulse, dizziness, profuse sweating, and cool moist skin.

Environmental measurements. In many heat stress studies, the variables commonly measured are work energy metabolism (often estimated rather than measured), air movement, air temperature, humidity, and radiant heat. See Chapter 12, Temperature Extremes, for illustrations and more details.

Air movement is measured with some type of anemometer and the air temperature with a thermometer, that is often referred to as a "dry bulb" thermometer.

Humidity or the moisture content of the air, is generally measured with a psychrometer, which gives both dry bulb and wet bulb temperatures. Using these temperatures and referring to a psychrometric chart, the relative humidity can be established.

The term "wet bulb" is commonly used to describe the temperature obtained by having a wet wick over the mercury-well bulb of an ordinary thermometer. Evaporation of moisture in the wick, to the extent that the moisture content of the surrounding air permits, cools the thermometer to a temperature below that registered by the "dry bulb." The combined readings of the dry bulb and wet bulb thermometers are then used to calculate percent relative humidity, absolute moisture content of the air, and water vapor pressure.

Radiant heat is a form of electromagnetic energy similar to light but of longer wavelength. Radiant heat (from such sources as red-hot metal, open flames, and the sun) has no appreciable heating effect on the air it passes through, but its energy is absorbed by any object it strikes, thus heating the person, wall, machine, or whatever object it falls upon. Protection requires placing opaque shields or screens between the person and the radiating surface.

An ordinary dry bulb thermometer alone will not measure radiant heat. If, however, the ordinary thermometer bulb is fixed in the center of a metal toilet float that has been painted dull black, and the top of the thermometer stem protrudes outside through a one-hole cork or rubber stopper, radiant heat can be measured by the heat absorbed in this sphere. This device is known as a globe thermometer.

Heat loss. Conduction becomes an important means of heat loss when the body is in contact with a good cooling agent, such as water. For this reason, when people are immersed in cold water, they become chilled much more rapidly and effectively than when exposed to air of the same temperature.

Air movement cools the body by convection, the moving air removes the air film or the saturated air (which is formed very rapidly by evaporation of sweat) and replaces it with a fresh air layer, capable of accepting more moisture from the skin.

Heat stress indices. The methods commonly used to estimate heat stress relate various physiological and environmental variables and end up with one number that then serves as a guide for evaluating stress. For example, the "effective temperature index" combines air temperature (dry bulb), humidity (wet bulb), and air movement to produce a single index called an "effective temperature."

Another index is the wet bulb globe temperature (WBGT). The numerical value of the WBGT Index is calculated by the following equations.

Indoors or outdoors with no solar loads:

$$WBGT = 0.7\ WB + 0.3\ GT$$

Outdoors with solar load:

$$WBGT = 0.7\ WB + 0.2\ GT + 0.1\ DB$$

WB = natural wet bulb temperature
GT = globe temperature
DB = dry bulb temperature.

In the NIOSH *Criteria Document on Hot Environments* (see Bibliography), NIOSH states that when impermeable clothing is worn, the WBGT should not be used, because evaporative cooling would be limited. The WBGT combines the effect of humidity and air movement, air temperature and radiation, and air temperature. It has been successfully used for environmental heat stress monitoring at military camps to control heat stress casualties. The measurements are few and easy to make; the instrumentation is simple, inexpensive, and rugged, and the calculations are straightforward. It is also the index used in the American Conference of Governmental Industrial Hygienists' (ACGIH) *Threshold Limit Value and Biological Exposure Indices Booklet* (see Appendix B-1). The ACGIH recommends TLVs for continuous work in hot environments as well as when 25, 50, or 75 percent of each working hour is at rest. Regulating allowable exposure time in the heat is a viable alternative technique for permitting necessary work to continue under heat-stress conditions that would be intolerable for continuous exposure. The NIOSH criteria document also contains a complete recommended heat stress control program including work practices.

Work practices include acclimatization periods, work and rest regimens, distribution of work load with time, regular breaks of a minimum of one per hour, provision for water intake, protective clothing, and application of engineering controls. Experience has shown that workers do not stand a hot job very well at first, but develop tolerance rapidly through acclimatization and acquire full endurance in a week to a month. (For more details, see Chapter 12, Temperature Extremes, and the NIOSH criteria document.)

Cold stress. Generally, the answer to a cold work area is to supply heat where possible, except for areas that must be cold, such as food storage areas.

General hypothermia is an acute problem resulting from prolonged cold exposure and heat loss. If an individual becomes fatigued during physical activity, they will be more prone to heat loss, and as exhaustion approaches, sudden vasodilation (blood vessel dilation) occurs with resultant rapid loss of heat.

Cold stress is proportional to the total thermal gradient between the skin and the environment because this gradient determines the rate of heat loss from the body by radiation and convection. When vasoconstriction (blood vessel constriction) is no longer adequate to maintain body heat balance, shivering becomes an important mechanism for increasing body temperature by causing metabolic heat production to increase to several times the resting rate.

General physical activity acts to increase metabolic heat. With clothing providing the proper insulation to minimize heat loss, a satisfactory microclimate can be maintained. Only exposed body surfaces are then likely to be excessively chilled and frostbitten. If clothing becomes wet either from contact with water or due to sweating during intensive physical work, its cold-insulating property will be greatly diminished.

Frostbite occurs when there is actual freezing of the tissues. Theoretically, the freezing point of the skin is about −1 C (30 F); however, with increasing wind velocity, heat loss is greater and frostbite will occur more rapidly. Once started, freezing progresses rapidly. For example, if the wind velocity reaches 20 mph, exposed flesh can freeze within about 1 minute at −10 C (14 F). Furthermore, if the skin comes in direct contact with objects whose surface temperature is below freezing point, frostbite can develop at the point of contact in spite of warm environmental temperatures. Air movement is more important in cold environments than in hot because the combined effect of wind and temperature can produce a condition called "windchill." The windchill index should be consulted by everyone facing exposure to low temperature and strong winds.

Ionizing radiation. A brief description of ionizing radiation hazards is given in this section; for complete description, read Chapter 10, Ionizing Radiation.

To understand a little about ionization, recall that the human body is made up of various chemical compounds which are in turn composed of molecules and atoms. Each atom has a nucleus with its own outer system of electrons.

When ionization of body tissues occurs, some of the electrons surrounding the atoms are forcibly ejected from their orbits. The greater the intensity of the ionizing radiation, the more ions will be created, and the more physical damage will be done to the cells.

Light consisting of electromagnetic radiation from the sun that strikes the surface of the earth is very similar to x-rays and gamma-radiation; it differs only in wavelength and energy content. (See description in Chapter 11, Nonionizing Radiation.) However, the energy level of sunlight at the earth's surface is too low to disturb orbital electrons and consequently sunlight is not referred to as ionizing, even though it has enough energy to cause severe skin burns over a period of time.

The exact mechanism of the manner in which ionization affects body cells and tissue is complex. At the risk of oversimplifying some basic physical principles and ignoring others, the purpose of this section is to present enough information so the health and safety professional will realize the problems involved and know when to call upon health physicists or radiation safety experts for help.

There are at least three basic factors that must be considered in such an approach to radiation safety:

- Radioactive materials emit energy that can damage living tissue.
- There are different kinds of radioactivity presenting different kinds of radiation safety problems. The types of ionizing radiation with which we will be concerned are alpha-, beta-, x-(x-ray), and gamma-radiation, and neutrons.
- Radioactive materials can be hazardous in two different ways. Certain materials can be hazardous even when located some distance away from the body. These are external hazards. Other types are hazardous only when they get inside the body by virtue of breathing, eating, or through broken skin. These are referred to as internal radiation hazards.

Instruments are available for evaluating possible radiation hazards. Meters or other devices are used for measuring radiation levels and doses.

Kinds of radioactivity. The five kinds of radioactivity that are of concern are alpha, beta, x-ray, gamma, and neutron. The first four are the most important since neutron sources usually are not used in ordinary manufacturing operations.

Of the five types of radiation mentioned, alpha particles are the least penetrating. They will not penetrate thin barriers. For example, paper, cellophane, and skin will stop alpha particles.

Beta-radiation has considerably more penetrating power than alpha radiation. A quarter of an inch of aluminum can stop the more energetic betas. As far as x-rays are concerned, virtually everyone is familiar with their penetrating ability and the fact that a barrier, such as concrete or lead, is required to stop them.

Gamma-rays are, for all practical purposes, the same as x-rays and require the same kinds of heavy shielding materials.

Neutrons are very penetrating and have characteristics that make it necessary to employ shielding materials of high hydrogen atom content rather than the use of mass alone.

Although the type of radiation from one radioactive material may be the same as that emitted by several other different radioactive materials, there may be a wide variation in energies.

The amount of energy a particular kind of radioactive material possesses is defined in terms of MeV (million electron volts); the greater the number of MeV, the greater the energy. Each radioactive material emits its own particular kind(s) of radiation with energy measured in terms of MeV.

External v. internal hazards. Radioactive materials that emit x-rays, gamma-rays, or neutrons are external hazards. In other words, such materials can be located some distance from the body and emit radiation that will produce ionization (and thus damage) as it passes through the body. Control by limiting exposure time, working at a safe distance, use of barriers or shielding, or a combination of all three is required for adequate protection against external radiation hazards.

As long as a radioactive material that emits only alpha particles remains outside the body, it will not cause trouble. Internally, it is a hazard because the ionizing ability of alpha particles at very short distances in soft tissue makes them a veritable bulldozer. Once inside the body—in the lungs, stomach, as an open wound, for example—there is no thick layer of skin to serve as a barrier and damage results. Alpha-emitting radioactive materials that will concentrate as persisting deposits in specific parts of the body are considered very hazardous.

Beta-emitters are generally considered to be an internal hazard although they also can be classed as an external hazard because they can produce burns when in contact with the skin. They require the same precautions as do alpha-emitters if there is a chance they can become airborne. In addition, some shielding may be required.

Measuring ionizing radiation. Many types of meters are used to measure various kinds of ionizing radiation. But these meters are useless unless they are accurately calibrated for the type of radiation they are designed to measure.

Meters with very thin windows in the probes can be used to check for alpha-radiation. Geiger-Muller and ionization chamber-type instruments are used for measuring beta-, gamma-, and x-radiation. Special types of meters are available for measuring neutrons.

Devices are available that will measure accumulated amounts (doses) of radiation. Film badges are used as dosimeters to record the amount of radiation received from beta-, x-ray, or gamma-radiation while special badges are available to record neutron radiation.

Film badges are worn by an individual continuously during each monitor period, and depending upon how they are worn, they will allow an estimate of an accumulated dose of radiation to the whole body or to just a part of the body, such as a hand or arm.

Alpha-radiation cannot be measured with film badges because the alpha particles will not penetrate the paper that must be used over the film emulsion to exclude light. (For more details on measurement and governmental regulations for ionizing radiation, see Chapter 10, Ionizing Radiation.)

Nonionizing radiation is a form of electromagnetic radiation with varying effects on the body, depending largely on the particular wavelength of the radiation involved. In the following paragraphs, in approximate order of decreasing wavelength and increasing frequency, are some hazards associated with different regions of the nonionizing electromagnetic radiation spectrum. Nonionizing radiation is covered in detail by OSHAct regulations 29 *CFR* 1910.97, and in Chapter 11, Nonionizing Radiation.

Low frequency. The longer wavelengths, including powerline transmission frequencies, broadcast radio, and shortwave radio, can produce general heating of the body. The health hazard from these radiations is very small, however, since it is unlikely that they would be found in intensities great enough to cause significant effect. An exception can be found very close to powerful radio transmitter aerials.

Microwaves have wavelengths of 3 m–3 mm (100-100,000 megahertz, MHz). They are found in radar, communications, some types of cooking, and diathermy applications. Microwave intensities may be sufficient to cause significiant heating of tissues.

The effect is related to wavelength, power intensity, and time of exposure. Generally, the longer wavelengths will produce a greater penetration and temperature rise in deeper tissues than the shorter wavelengths. However, for a given power intensity, there is less subjective awareness to the heat from longer wavelengths than there is to the heat from shorter wavelengths, because of the absorption of the longer wavelength radiation beneath the body's surface.

An intolerable rise in body temperature, as well as localized damage to specific organs, can result from an exposure of sufficient intensity and time. In addition, flammable gases and vapors can ignite when they are inside metallic objects located in a microwave beam.

Infrared radiation does not penetrate below the superficial layer of the skin so that its only effect is to heat the skin and the tissues immediately below it. Except for thermal burns, the health hazard upon exposure to low level conventional infrared radiation sources is negligible. (For information on possible damage to the eye, consult Chapter 11.)

Visible radiation, which is about midway in the electromagnetic spectrum, is important because it can affect both the quality and accuracy of work. Good lighting conditions generally result in increased product quality with less spoilage and increased production.

Lighting should be bright enough for easy and efficient sight, and directed so that it does not create glare. Ilumination levels and brightness ratios recommended for manufacturing and service industries are published by the Illuminating Engineering Society. The OSHAct has not, to date, adopted the ANSI standard RP7-1983, *Practice for Industrial Lighting,* per se. It has, however, adopted many other ANSI Standards in its "general Industry Standards" (29 *CFR* 1910) and ANSI A11.1 is adopted by reference. Such adoptions have the force and effect of law, where such standards apply.

One of the most objectionable features of lighting is glare—brightness within the field of vision that causes discomfort or interferes with seeing. The brightness can be caused by either direct or reflected light. To prevent glare, the source of light should be kept well above the line of vision, or shielded with opaque or translucent material.

Almost as bad is an area of excessively high brightness in the visual field. A higher reflecting white paper in the center of a dark, nonreflecting surface, or a brightly illuminated control handle on a dark or dirty machine are two examples.

To prevent such conditions, keep surfaces uniformly light or dark with little difference in surface reflectivity. Color contrasts are acceptable, however.

Although it is generally best to provide even, shadow-free light, some jobs require contrast lighting. In these cases, keep the general (or background) light well diffused and glareless and add a supplementary source of light that will cast shadows where needed.

Ultraviolet radiation in industry can be found around electrical arcs, and such arcs should be shielded by materials opaque to the ultraviolet. The fact that a material can be opaque to ultraviolet has no relation to its opacity to other parts of the spectrum. Ordinary window glass, for instance, is almost completely opaque to the ultraviolet in sunlight although transparent to the visible wavelengths. A piece of plastic dyed a deep red-violet may be almost entirely opaque in the visible part of the spectrum and transparent in the near-ultraviolet.

Electric welding arcs and germicidal lamps are the most common strong producers of ultraviolet radiation in industry. The ordinary fluorescent lamp generates a good deal of ultraviolet inside the bulb, but it is essentially all absorbed by the bulb and its coating.

The most common exposure to ultraviolet radiation is from direct sunlight, and a familiar result of overexposure—one that is known to all sunbathers—is sunburn. Most everyone is also familiar with certain compounds and lotions that reduce the effects of the sun's rays, but many are unaware that some industrial materials, such as cresols, make the skin especially sensitive to ultraviolet rays. So much so that after having been exposed to cresols, even a short exposure in the sun usually results in a severe sunburn.

Lasers emit beams of coherent radiation of a single color or wavelength and frequency, in contrast to conventional light sources, that produce random, disordered light wave mixtures of various frequencies. The laser (an acronym that stands for Light Amplification by Stimulated Emission of Radiation) is made up of light waves that are nearly parallel to each other, all traveling in the same direction. Atoms are 'pumped' full of energy, and when they are stimulated to fall to a lower energy level, they give off radiation that is directed to produce the coherent laser beam. (See Chapter 11, Nonionizing Radiation, for more details.)

The maser, the laser's predecessor, emits microwaves instead of light. Some companies call their lasers "optical masers." Since the laser is highly collimated (has a small divergence angle), it can have a large energy density in a narrow beam. Direct viewing of the laser source or its reflections should be avoided. The work area should contain no reflective surface (such as mirrors or highly polished furniture) for even a reflected laser beam can be hazardous. Suitable shielding to contain the laser beam should be provided. The OSHAct covers protection against laser hazards in it construction regulations.

Biological effects. The eye is the organ that is most vulnerable to injury induced by laser energy. The reason for this is the ability of the cornea and lens to focus the parallel laser beam on a small spot on the retina.

The fact that infrared radiation of certain lasers may not be visible to the naked eye contributes to the potential hazard.

Lasers generating in the ultraviolet range of the electromagnetic spectrum can produce corneal burns rather than retinal damage, because of the way the eye handles ultraviolet light. (See Chapter 11, Nonionizing Radiation.)

Other factors that have a bearing on the degree of eye injury induced by laser light are as follows:

- the pupil size—the smaller the pupil diameter, the less the amount of laser energy permitted to the retina
- the ability of the cornea and lens to focus the incident light on the retina
- the distance from the source of energy to the retina
- the energy and wavelength of the laser
- the pigmentation of the eye of the subject
- the place of the retina where the light is focused
- the divergence of the laser light
- the presence of scattering media in the light path.

A discussion of laser beam characteristics and protective eyewear can be found in Chapter 11.

Extremes of pressure. It has been recognized from the beginning of caisson work (work performed in a watertight structure) that men working under pressures greater than at a normal atmospheric one, are subject to various ills connected with the job. Hyperbaric (greater than normal pressures) environments are also encountered by divers operating under water, whether by holding the breath while diving, breathing from a self-contained underwater breathing apparatus (SCUBA), or by breathing gas mixtures supplied by compression from the surface.

Occupational exposures occur in caisson or tunneling operations, where a compressed gas environment is used to exclude water or mud and to provide support for structures. Man can withstand large pressures providing air has free access to lungs, sinuses and the middle ear. Unequal distribution of pressure can result in barotrauma (tissue damage resulting from expansion or contraction of gas spaces found within or adjacent to the body, which can occur either during compression [descent] or during decompression [ascent]).

The teeth, sinuses, and ears are frequently affected by pressure differentials. For example, gas spaces that can be present adjacent to tooth roots or fillings may be compressed during descent. Fluid or tissue forced into these spaces can cause pain either during descent or ascent. Sinus blockage caused by occlusion of the sinus aperture by inflamed nasal mucosa prevents equalization of pressures.

Under some conditions of work at high pressure, the concentration of carbon dioxide in the atmosphere can be considerably increased so that the carbon dioxide will act as a narcotic. Keeping the oxygen concentration high minimizes this condition, but does not prevent it. The procedure is useful where the carbon dioxide concentration cannot be kept at a proper level.

Decompression sickness, commonly known as the "bends," results from the release of nitrogen bubbles into the circulation and tissues during decompression. If the bubbles lodge at the joints and under muscles, they cause severe cramps. To prevent this, decompression is carried out slowly and by stages so that the nitrogen can be eliminated slowly and without the formation of bubbles.

Deep-sea divers are supplied with a mixture of helium and oxygen for breathing, and because helium is an inert diluent and less soluble in blood and tissue than is nitrogen, it presents a less formidable decompression problem.

One of the most common troubles encountered by workers under compressed air is pain and congestion in the ears from inability to ventilate the middle ear properly during compression and decompression. As a result, many workers subjected to increased air pressures suffer from temporary hearing loss; some have permanent hearing loss. This damage is believed to be caused by obstruction of the Eustachian tubes, which prevents proper equalization of pressure from the throat to the middle ear.

The effects of reduced pressure on the worker are much the same as the effects of decompression from a high pressure. If pressure is reduced too rapidly, decompression sickness and ear disturbances similar to the diver's conditions can result.

Ergonomic hazards

"Ergonomics" literally means the customs, habits, and laws of work. According to the International Labor Office it is ". . .The application of human biological science in conjunction with the engineering sciences to achieve the optimum mutual adjustment of man [sic] and his [sic] work, the benefits being measured in terms of human efficiency and well-being." The topic of ergonomics is covered briefly here. (For more details, see Chapter 13, Ergonomics.)

The ergonomics approach goes beyond productivity, health, and safety. It includes consideration of the total physiological and psychological demands of the job upon the worker.

In the broad sense, the benefits that can be expected from designing work systems to minimize physical stress on workers are as follows:

- more efficient operation
- fewer accidents
- lower cost of operation
- reduced training time
- more effective use of personnel.

The human body can endure considerable discomfort and stress and can perform many awkward and unnatural movements—for a limited period of time. However, when awkward conditions or motions are continued for prolonged periods, the physiological limitations of the worker can be exceeded. To ensure a continued high level of performance, work systems must be tailored to human capacities and limitations.

Ergonomics considers the physiological and psychological stresses of the task. The task should not require excessive muscular effort, considering the age, sex, and state of health of the worker. The job should not be so easy that boredom and inattention lead to unnecessary errors, scrap material, and accidents. Ergonomic stresses can impair the health and efficiency of the worker just as significantly as the other more commonly recognized environmental stresses.

The task of the design engineer and health and safety professional is to find the happy blend between "easy" and "difficult" jobs. In any human-machine system, there are tasks that are better performed by people than by machine—and, conversely, tasks that are better handled by machines.

Ergonomics deals with the interactions between humans and such traditional environmental elements as atmospheric contaminants, heat, light, sound, and all tools and equipment pertaining to the workplace. The modern concept is that people are to be considered the monitoring link of a human-machine environment system.

In any activity, a person receives and processes information, and then acts upon it. The receptor function occurs largely through the sense organs of the eyes and the ear, but information can also be conveyed through the senses of smell, touch, or sensations of heat or cold. This information is conveyed to the central mechanism of the brain and spinal cord, where the information is processed to arrive at a decision. This can involve the integration of the information, which has already been stored in the brain, and decisions can vary from automatic responses to those involving a high degree of reasoning and logic.

Having received the information and processed it, the individual then takes action (control) as a result of the decision, usually through muscular activity based on the skeletal framework of the body. When an individual's activity involves the operation of a piece of equipment, the person will often form part of a "closed-loop servosystem" displaying many of the feedback characteristics of such a system. The person usually forms that part of the system that makes decisions and thus has a fundamental part to play in the efficiency of the system.

Biomechanics—physical demands. Biomechanics can be a very effective tool in preventing excessive work stress. Biomechanics means the mechanics of biological organisms. It deals with the functioning of the structural elements of the body and the effects of external and internal forces on the various parts of the body.

Cumulative effects of excessive ergonomic stress on the worker can, in a rather insidious and subtle manner, result in physical illnesses and injuries, such as "trigger finger," tenosynovitis, and bursitis.

Cases of excessive fatigue and discomfort are, in many cases, forerunners of soreness and pain. By exerting a strong distracting influence on a worker, these stresses can render the worker more prone to major accidents. Discomfort and fatigue tend to make the worker less capable of maintaining the proper vigilance for the safe performance of the task.

Some of the principles of biomechanics can be illustrated by considering different parts of the human anatomy, for example, the hand.

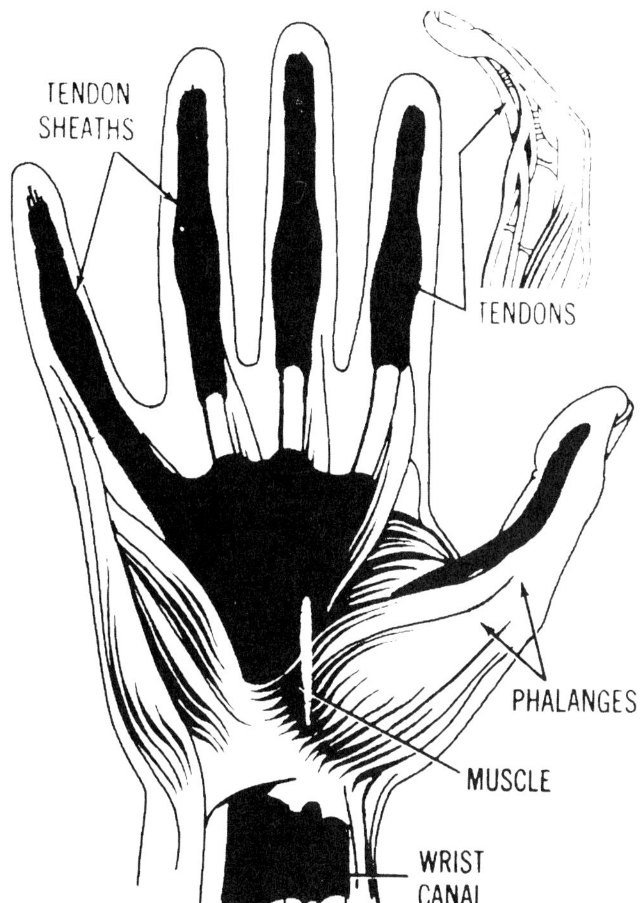

Figure 1-3. Diagram of hand anatomy.

Hand anatomy. The flexing action in the fingers is controlled by tendons attached to muscles in the forearm. The tendons, which run in lubricated sheaths, enter the hand through a tunnel in the wrist formed by bones and ligaments and continue on to point of attachment to the different segments, or phalanges, of the fingers (Figure 1-3).

When the wrist is bent toward the little finger side, the tendons tend to bunch up on one side of the tunnel through which they enter the hand. If an excessive amount of force is continuously applied with the fingers while the wrist is flexed, or if the flexing motion is repeated rapidly over a long period of time, the resulting friction can produce inflammation of the tendon sheaths, or tenosynovitis.

The palm of the hand, which contains a network of nerves and blood vessels, should never be used as a hammer nor subjected to continued firm pressure. Repetitive or prolonged pressure on the nerves and blood vessels in this area can result in pain either in the palm itself or at any point along the nerve pathways up through the arm and shoulder. Other parts of the body, elbow joints, shoulders, etc., can become painful for similar reasons.

Mechanical vibration. A condition known to stonecutters as "dead fingers" or "white fingers" (Raynaud's phenomenon) occurs mainly in the fingers of the hand used to guide the cutting tool. The circulation in this hand becomes impaired, and when exposed to cold the fingers become white and without sensation, as though mildly frostbitten. The white appearance usually disappears when the fingers are warmed for some time, but a few cases are sufficiently disabling that the victims are forced to seek other types of work. In many instances both hands are affected.

The condition has been observed in a number of other occupations involving the use of vibrating tools, such as the air hammers used for scarfing metal surfaces, the air chisels for chipping castings in the metal trades, and the chain saws used in forestry. The injury is caused by vibrating the fingers as they grip the tools to guide them in performing their tasks. The related damage to blood vessels can progress to nearly complete obstruction of the vessels' capacity for bloodflow.

Prevention should be directed at reducing the vibrational energy transferred to the fingers (perhaps by the use of padding) and by changing energy and the frequency of the vibration. Low frequencies, 25-75 hertz, are more damaging than higher frequencies.

Lifting. The injuries resulting from manual handling of objects and materials are especially prominent in the total of compensable injuries. This is obviously a problem for considerable concern to the health and safety professional, and represents an area where the biomechanical data relating to lifting and carrying can be applied in the work layout and design of jobs, which require handling of materials. (For more details, see Chapter 13, Ergonomics, and the NIOSH *Guide to Manual Lifting.*)

The relevant data concerning lifting can be classified into task, human, and environmental variables.

- Task variables
  1. location of object to be lifted
  2. size of the object to be lifted
  3. height from which and to which the object is lifted
  4. frequency of lift
  5. weight of object
  6. working position
- Human variables
  1. sex of worker
  2. age of worker
  3. training of worker
  4. physical fitness or conditioning of worker
  5. body dimensions, such as height of the worker
- Environmental variables
  1. extremes of temperature
  2. humidity
  3. air contaminants.

Static work. Another very fatiguing situation encountered in industry, which unfortunately is frequently and easily overlooked, is static, or isometric, work. Since very little outward movement occurs, it seems that no muscular effort is involved. Often, however, there is more muscular fatigue being generated than if some outward movement occurred. A cramped working posture, for example, is a substantial source of static muscular loading.

In general, maintaining any set of muscles in a rigid, unsupported position for long periods of time results in muscular strain. The blood supply to the contracted muscle is diminished, a local deficiency of oxygen can occur, and waste products accumulate. Alternating static and dynamic work, or providing support for partial relaxation of the member involved, will alleviate this problem.

The need for armrests is usually indicated in two types of situations. One is the case just mentioned—to relieve the isometric muscular work involved in holding the arm in a fixed, unsupported position for long periods of time. The second case is where the arm is pressed against a hard surface such as the edge of a bench or machine. The pressure on the soft tissues overlaying the bones can cause bruises and pain. Padded arm-rests have solved numerous problems of both types (see Figure 1-4).

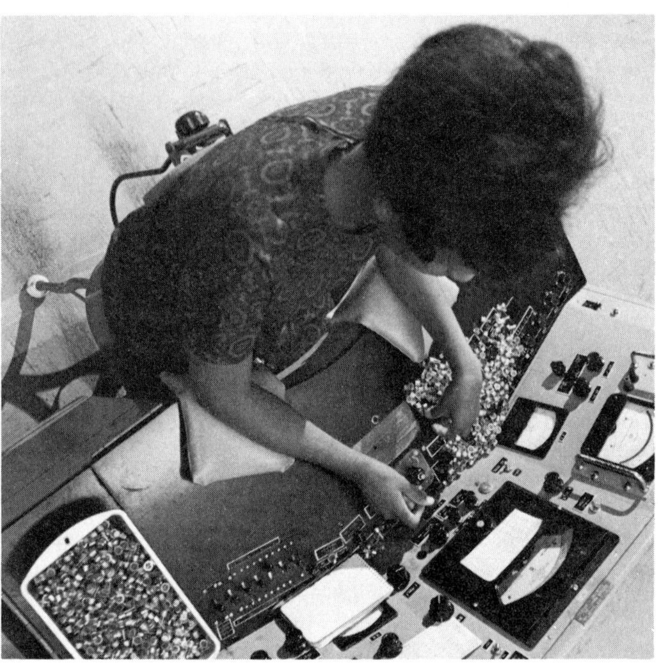

Figure 1-4. Worker uses pads to keep her forearm off the sharp table edge.

Workplace design. Relating the physical characteristics and capabilities of the worker to the design of equipment and to the layout of the workplace is another key ergonomic concept. When this is done, the result should be an increase in efficiency, a decrease in human error, and a consequent reduction in accident frequency. However, several different types of information are needed; namely, a description of the job, an understanding of the kinds of equipment that will be used, a description of the kinds of people who will use the equipment, and, finally, the biological characteristics of these people.

In general, the first three items—job, equipment, and users—can be defined both accurately and easily. The biological characteristics of the users, however, can often be determined satisfactorily only from special surveys that yield descriptive data on human body size and biomechanical abilities and limitations.

Anthropometric data. The anthropometric data consists of various heights, lengths, and breadths used to establish the minimum clearances and spatial accommodations, and the functional arm, leg, and body movements that are made by the worker during the performance of the task.

Behavioral aspects—mental demands. One important aspect of industrial machine design directly related to the safety and productivity of the worker concerns the design of displays and controls. Displays are one of the most common types of operator input; the others include direct sensing and verbal or visual commands. Displays tell the operator what the machine is doing and how it is performing. Problems of display design are primarily related to the human senses.

A machine operator can successfully control equipment only to the extent that the operator receives clear, unambiguous information when needed, on all pertinent aspects of the task. Accidents, or operational errors, often occur because a worker has misinterpreted, or was unable to obtain information from displays, concerning the functioning of the equipment. Displays are usually visual, though they also can be auditory (for example, a warning bell rather than a warning light), especially when there is danger of overloading the visual sensory channels.

Design of controls. An operator must decide upon the proper course of action and manipulate controls to produce any desired change in the machine's performance. The efficiency and effectiveness—that is, the safety—with which controls can be operated depends upon the extent to which information on the dynamics of human movement (or biomechanics) has been incorporated in their design. This is particularly true whenever controls must be operated at high speed, against large resistances, with great precision, or over long periods of time.

Controls should be designed so that rapid, accurate settings easily can be made without undue fatigue, thereby avoiding many accidents and operational errors. Since there is a wide variety of machine controls, ranging from the simple on-off action of pushbuttons to very complex mechanisms, advance analysis of the task requirements must be made. On the basis of considerable experimental evidence, it is possible to recommend the most appropriate control and its desirable range of operation.

In general, from the standpoint of human engineering, the mechanical design of equipment must be compatible with the biological and psychological characteristics of the operator. The effectiveness of the human-machine combination can be greatly enhanced by treating the operator and the equipment as a unified system. Thus, the instruments should be considered as extensions of the operator's nervous and perceptual systems, the controls as extensions of the hands, and the feet as simple tools. Any control that is difficult to reach or operate, any instrument dial that has poor legibility, any seat that induces poor posture or discomfort, or any obstruction of vision can contribute directly to an accident or illness.

Biological hazards

Biological hazards are any virus, bacteria, fungus, parasite, or any living organism that can cause a disease in human beings. They can be a part of the total environment or associated with certain occupations.

Diseases transmitted from animals to humans are common—infectious and parasitic diseases can also result from exposure to contaminated water, insects, or infected people. (See Chapter 14, Biological Hazards, for more details.)

Industrial sanitation. The requirements for sanitation and personal facilities are covered in the OSHAct safety and health regulations 29 *CFR* 1910, Subpart J—General Environmental Controls. The OSHAct's regulations for carcinogens require special personal health and sanitary facilities for employees working with potentially carcinogenic materials.

Water supply. Potable water should be provided in workplaces, when needed for drinking and personal washing, for cooking, washing of foods, cooking, or eating utensils, washing of food preparation premises, and personal service rooms.

Drinking fountain surfaces must be constructed of materials impervious to water and not subject to oxidation. The nozzle of the fountain must be located to prevent the return of water in the jet or bowl to the nozzle orifice. A guard over the nozzle shall prevent contact with the nozzle by the mouth or nose of persons using the drinking fountain.

Potable drinking water dispensers must be designed, constructed, and served so that sanitary conditions are maintained, capable of being closed, and equipped with a tap. Ice that comes in contact with drinking water shall be made of potable water and maintained in a sanitary condition. Standing water in cooling towers and other air-moving systems should be monitored for *Legionnella* bacteria (see Chapter 14, Biological Hazards, for details).

Outlets for nonpotable water, such as water for industrial or firefighting purposes, shall be posted or otherwise marked in a manner that indicates clearly that the water is unsafe and is not to be used as drinking water. Nonpotable water systems or systems carrying any other nonpotable substance should be constructed so as to prevent backflow or backsiphonage.

HARMFUL AGENTS—ROUTE OF ENTRY

In order for a harmful agent to exert its toxic effect it must come into contact with a body cell, and must enter the body via:

inhalation
skin absorption
ingestion.

Chemical compounds in the form of liquids, gases, mists, dusts, fumes, and vapors can cause problems by inhalation (breathing), absorption (through direct contact with the skin), or ingestion (eating or drinking).

Inhalation

Inhalation involves those airborne contaminants that can be inhaled directly into the lungs and can be physically classified as gases, vapors, and particulate matter that includes dusts, fumes, smokes, aerosols, and mists.

Inhalation, as a route of entry, is particularly important because of the rapidity with which a toxic material can be absorbed in the lungs, pass into the bloodstream, and reach the brain. Inhalation is the major route of entry for hazardous chemicals in the work environment.

Absorption

Absorption through the skin can occur quite rapidly if the skin is cut or abraded. Intact skin, however, offers a reasonably good barrier to chemicals. Unfortunately, there are many compounds that can be absorbed through intact skin.

Some substances are absorbed by way of the openings for hair follicles and others dissolve in the fats and oils of the skin, such as organic lead compounds, many nitro compounds, and organic phosphate pesticides. Compounds that are good solvents for fats (such as toluene and xylene) also can cause problems by being absorbed through the skin.

Many organic compounds, such as TNT, cyanides, and most aromatic amines, amides, and phenols, can produce systemic poisoning by direct contact with the skin.

Ingestion

In the workplace, people can unknowingly eat or drink harmful chemicals. Toxic compounds are capable of being absorbed from the gastrointestinal tract into the blood. Lead oxide can cause serious problems if people working with this material are allowed to eat or smoke in work areas. In this situation, careful and thorough washing is required both before eating and at the end of every shift.

Inhaled toxic dusts can also be ingested in amounts that may cause trouble. If the toxic dust swallowed with food or saliva is not soluble in digestive fluids, it is eliminated directly through the intestinal tract. Toxic materials that are readily soluble in digestive fluids can be absorbed into the blood from the digestive system.

It is important to study all routes of entry when evaluating the work environment—candy bars or lunches in work area, solvents being used to clean work clothing and hands, in addition to airborne contaminants in working areas. (For more details, see Chapter 15, Toxicology.)

TYPES OF AIRBORNE CONTAMINANTS

There are precise meanings of certain words commonly used in industrial hygiene. These must be used correctly in order to (1) understand the requirements of OSHAct regulations; (2) effectively communicate with other workers in this area, and; (3) intelligently prepare purchase orders to procure health services and personal protective equipment. For example, a fume respirator is worthless as protection against gases or vapors. Too frequently, terms (such as gases, vapors, fumes, and mists) are used interchangeably. Each term has a definite meaning and describes a certain state of matter.

States of matter

Dusts. These are solid particles generated by handling, crushing, grinding, rapid impact, detonation, and decrepitation (breaking apart by heating) of organic or inorganic materials, such as rock, ore, metal, coal, wood, and grain.

Dust is a term used in industry to describe airborne solid particles that range in size from 0.1–25 micrometers (μm). One micrometer = 0.0001 centimeter or 1/25,400 inch. Dusts more than 5μm in size usually will not remain airborne long enough to present an inhalation problem (see Chapter 7, Particulates).

Dust can enter the air from various sources; for instance, when a dusty material is handled, such as when lead oxide is dumped into a mixer or when talc is dusted on a product. When solid materials are reduced to small sizes in processes such as grinding, crushing, blasting, shaking, and drilling the mechanical action of the grinding or shaking device supplies energy to disperse the dust.

Evaluating dust exposures properly requires knowledge of the chemical composition, particle size, dust concentration in air, how it is dispersed, and many other factors described here. Although in the case of gases, the concentration that reaches the alveolar sacs will be nearly like the concentration in the air breathed, this is not the case for aerosols or dust particles. Large particles, more than 10 μm aerodynamic diameter, can be deposited through gravity and impaction in large ducts before they reach the very small sacs (alveoli). Only the smaller particles will reach the alveoli. (See Chapter 2, The Lungs, for more details.)

Excepting some fibrous materials, dust particles must usually be smaller than 5 μm in order to penetrate to the alveoli or inner recess of the lungs.

A person with normal eyesight can detect dust particles as small as 50 μm in diameter. Smaller airborne particles can be detected individually by the naked eye only when strong light is reflected from them. Particles of dust of respirable size (less than 10 μm) cannot be seen without the aid of a microscope, but they may be perceived as a "haze."

Most industrial dusts consist of particles that vary widely in size, with the small particles greatly outnumbering the large ones. Consequently (with few exceptions), when dust is noticeable in the air near a dusty operation, probably more invisible dust particles than visible ones are present. A process that produces dust fine enough to remain suspended in the air long enough to be breathed should be regarded as hazardous until it can be proved safe.

There is no simple one-to-one relationship between the concentration of an atmospheric contaminant and duration of exposure and the rate of dosage by the hazardous agent to the critical site within the body. For a given magnitude of atmospheric exposure to a potentially toxic particulate contaminant, the resulting hazard can range from an insignificant level to one of great danger, depending upon the toxicity of the material, the size of the inhaled particles, and other factors that determine their fate in the respiratory system.

Fumes. These are formed when the material from a volatilized solid condenses in cool air. The solid particles that are formed make up a fume that is extremely fine—usually less than 1.0 μm in diameter. In most cases, the hot vapor reacts with the air to form an oxide. Gases and vapors are not fumes, although the terms are often mistakenly used interchangeably.

Welding, metalizing, and other operations involving vapors from molten metals may produce fumes; these may be harmful under certain conditions. Arc welding volatilizes metal vapor that condenses—as the metal or its oxide—in the air around the arc. In addition, the rod coating is partially volatilized. These fumes, because they are extremely fine, are readily inhaled.

Other toxic fumes—such as those formed when welding structures that have been painted with lead-based paints, or when welding galvanized metal—can produce severe symptoms of toxicity rather rapidly, unless fumes are controlled with good, local exhaust ventilation, or the welder is protected by respiratory protective equipment.

Fortunately, most soldering operations do not require temperatures high enough to volatilize an appreciable amount of lead. However, the lead in molten solder pots is oxidized by contact with air at the surface. If this oxide, often called dross, is mechanically dispersed into the air, it can produce a severe lead poisoning hazard.

In operations when lead dust may be present in air, such as soldering or lead battery-making, preventing occupational poisoning is largely a matter of scrupulously clean housekeeping to prevent the lead oxide from becoming dispersed into the air. It is customary to enclose melting pots, dross boxes, and similar operations, and to ventilate them adequately to control the hazard.

Smoke. This consists of carbon or soot particles less than 0.1 μm in size, and results from the incomplete combustion of carbonaceous materials such as coal or oil. Smoke generally contains droplets as well as dry particles. Tobacco, for instance, produces a wet smoke composed of minute tarry droplets.

Aerosols. These are liquid droplets or solid particles of fine enough particle size to remain dispersed in air for a prolonged period of time.

Mists. These are suspended liquid droplets generated by condensation of liquids from the vapor back to the liquid state or by breaking up a liquid into a dispersed state, such as by splashing, forming, or atomizing. The term mist is applied to a finely divided liquid suspended in the atmosphere. Examples are the oil mist produced during cutting and grinding operations, acid mists from electroplating, acid or alkali mists from pickling operations, paint spray mist in painting operations, and the condensation of water vapor to form a fog or rain.

Gases. These are formless fluids that expand to occupy the space or enclosure in which they are confined. Gases are a state of matter in which the molecules are unrestricted by cohesive forces. Examples are arc-welding gases, internal combustion engine exhaust gases, and air.

Vapors. These are the volatile form of substances that are normally in the solid or liquid state at room temperature and pressure. Evaporation is the process by which a liquid is changed into the vapor state and mixed with the surrounding atmosphere. Solvents with low boiling points volatilize readily at room temperature.

In addition to the definitions concerning states of matter that find daily usage in the vocabulary of the industrial hygienist, other terms used to describe degree of exposure include the following:

ppm—parts of vapor or gases per million parts of contaminated air by volume at room temperature and pressure

mppcf—millions of particles of a particulate per cubic foot of air

mg/m^3—milligrams of a substance per cubic meter of air.

The health and safety professional recognizes that air contaminants exist as a gas, dust, fume, mist, or vapor in the workroom air. In evaluating the degree of exposure, the measured concentration of the air contaminant is compared to limits or exposure guidelines that appear in the published standards on levels of exposure (see Appendix B-1).

Respiratory hazards

Airborne chemical agents that enter the lungs can pass directly into the bloodstream and be carried to other parts of the body. The respiratory system consists of organs contributing to normal respiration or breathing. Strictly speaking, it includes the nose, mouth, upper throat, larynx, trachea, and bronchi, which are all air passages or airways and the lungs where oxygen is passed into the blood and carbon dioxide is given off. Finally, it includes the diaphragm and the muscles of the chest, which perform the normal respiratory movements of inspiration and expiration (see Chapter 2, The Lungs).

All living cells of the body are engaged in a series of chemical processes—the name given to the sum total of these processes is metabolism. In the course of its metabolism, each cell consumes oxygen and produces carbon dioxide as a waste substance.

Respiratory hazards can be broken down into two main groups:

1. oxygen deficiency: the oxygen concentration (or partial pressure of oxygen) is below that level considered safe for human exposure
2. air that contains harmful or toxic contaminants.

Oxygen-deficient atmospheres. Each living cell in the body requires a constant supply of oxygen. Some cells are more dependent on a continuing oxygen supply than others. Some cells in the brain and nervous system can be injured or die after 4–6 minutes without oxygen. These cells, if destroyed, cannot be regenerated or replaced, and permanent changes and impaired functioning of the brain can result from such damage. Other cells in the body are not as critically dependent on an oxygen supply since they can be replaced.

Normal air at sea level contains approximately 21 percent oxygen and 79 percent nitrogen and other inert gases. At sea level and normal barometric pressure (760 millimeters of mercury [mm Hg] or 101.3 kPa) the partial pressure of oxygen would be 21 percent of 760 mm, or 160 mm. The partial pressure of nitrogen and inert gases would be 600 mm (79 percent of 760 mm).

At higher altitudes or under conditions of reduced barometric pressure, the relative proportions of oxygen and nitrogen would remain the same, but the partial pressure of each gas would be decreased. The partial pressure of oxygen at the alveolar surface of the lung is critical, because it determines the rate of oxygen diffusion through the moist lung tissue membranes.

Deficiency of oxygen in the atmosphere of confined spaces can be a problem in industry. For this reason, the oxygen content of any tank or other confined space (as well as the levels of any toxic contaminants) should be measured before entry is made. Instruments such as the oxygen analyzer are commercially available for this purpose. See Chapters 18, Air-Sampling Instruments, and 19, Direct-Reading Gas and Vapor Monitors, for more details.

The first physiological signs of an oxygen deficiency (anoxia) are an increased rate and depth of breathing. Oxygen concentrations of less than 16 percent by volume cause dizziness, rapid heartbeat, and headache. A worker should never enter or remain in areas where tests have indicated such concentrations unless wearing some sort of supplied-air or self-contained respiratory equipment. (See Chapter 23, Respiratory Protective Equipment, for more details.)

Oxygen-deficient atmospheres can cause an inability to move and a semiconscious lack of concern about the imminence of death. In cases of abrupt entry into areas containing little or no oxygen, the person usually has no warning symptoms, immediately loses consciousness, and has no recollection of the incident if rescued in time to be revived. The senses cannot be relied upon to alert or warn a person of atmospheres deficient in oxygen.

Oxygen-deficient atmospheres can occur in tanks, vats, holds of ships, silos, mines, or in areas where the air may be diluted or displaced by asphyxiating levels of gases or vapors, or where the oxygen may have been consumed by chemical or biological reactions.

Ordinary jobs involving maintenance and repair of systems for storing and transporting fluids or entering tanks or tunnels for cleaning and repairs are controlled almost entirely by the immediate supervisor, and, accordingly, that person should be particularly knowledgeable of all rules and precautions to assure the safety of persons required to work in such atmospheres. Safeguards should be meticulously adhered to.

As an example, there should be a standard operating procedure for entering tanks. Such procedures should be consistent with OSHAct regulations and augmented by in-house procedures, which may enhance the basic OSHAct rules. The American National Standards Institute (ANSI) lists confined space procedures in its Respiratory Protection Standard and NIOSH has also issued guidelines for work in confined spaces including a *Criteria Document for Working in Confined Spaces.* (See Bibliography at the end of this chapter.) Even if a tank is empty, it may have been closed for some time and developed an oxygen deficiency through chemical reactions of residues left in the tank. It may be unsafe to enter without proper respiratory protection.

The hazard of airborne contaminants. Inhaling harmful materials can irritate the upper respiratory tract and lung tissue, or the terminal passages of the lungs and the air sacs, depending upon the solubility of the material.

Inhalation of biologically inert gases can dilute the atmospheric oxygen below the normal blood saturation value and disturb cellular processes. Other gases and vapors may prevent the blood from carrying oxygen to the tissues or interfere with its transfer from the blood to the tissue, producing chemical asphyxia.

Inhaled contaminants that adversely affect the lungs fall into three general categories:

1. aerosols (particulates), which, when deposited in the lungs, can produce either rapid local tissue damage, some slower tissue reactions, eventual disease, or only physical plugging
2. toxic vapors and gases that produce adverse reaction in the tissue of the lungs
3. some toxic aerosols or gases that do not affect the lung tissue locally but (1) are passed from the lungs into the bloodstream, where they are carried to other body organs, or (2) have adverse effects on the oxygen-carrying capacity of the blood cells.

An example of the first type (aerosols) is silica dust, which causes fibrotic growth (scar tissue) in the lungs. Other harmful aerosols are fungi found in sugar cane residues, producing bagassosis.

An example of the second type (toxic gases) is hydrogen fluoride, a gas that directly affects lung tissue. It is a primary irritant of mucous membranes, even causing chemical burns. Inhalation of this gas will cause pulmonary edema and direct interference with the gas transfer function of the alveolar lining.

An example of the third type is carbon monoxide, a toxic gas passed into the bloodstream without essentially harming the lung. The carbon monoxide passes through the alveolar walls into the blood, where it ties up the hemoglobin so that it cannot accept oxygen, thus causing oxygen starvation. Cyanide gas has another effect—it prevents enzymatic utilization of molecular oxygen by cells.

Sometimes several types of lung hazards occur simultaneously. In mining operations, for example, explosives release oxides of nitrogen. These impair the bronchial clearance mechanism, so that coal dust (of the particle sizes associated with the explosions) is not efficiently cleansed from the lungs.

If a compound is very soluble—such as ammonia, sulfuric acid, or hydrochloric acid—it is rapidly absorbed in the upper respiratory tract and during the initial phases of exposure does not penetrate deeply into the lungs. Consequently the nose and throat become very irritated.

Compounds that are insoluble in body fluids cause considerably less throat irritation than the soluble ones, but can penetrate deeply into the lungs. Thus, a very serious hazard can be present and not be recognized immediately because of a lack of warning that the local irritation would otherwise provide. Examples of such compounds (gases) are nitrogen dioxide and ozone. The immediate danger from these compounds in high concentrations is acute lung irritation or, possibly later, chemical pneumonia.

There are numerous chemical compounds that do not follow the general solubility rule. Such compounds are not very soluble in water and yet are very irritating to the eyes and respiratory tract. They also can cause lung damage and even death under certain conditions.

THRESHOLD LIMIT VALUES

Threshold Limit Values (TLVs) are exposure guidelines that have been established for airborne concentrations of many chemical compounds. The health and safety professional or other responsible person involved in discussing problems with the employee should understand something about TLVs and the terminology in which their concentrations are expressed. (See Chapter 15, Toxicology, and Appendix B-1 for more details.)

Threshold Limit Values refer to airborne concentrations of substances, and it is believed represent conditions under which nearly all workers may be repeatedly exposed, day after day, without adverse effect. Control of the work environment is based on the assumption that for each substance there is some safe or tolerable level of exposure below which no significant adverse effect occurs. These tolerable levels are called Threshold Limit Values. The copyrighted trademark, "Threshold Limit Value" refers to limits published by ACGIH. The TLVs are reviewed and updated annually to reflect the most current information on the effects of each substance assigned a TLV. (See Appendix B-1 and the Bibliography at the end of this chapter.)

The data for establishing TLVs comes from animal studies, human studies, and industrial experience, and the limit may be selected for several reasons. As mentioned earlier in this chapter, the TLV can be based on the fact that a substance is very irritating to the majority of people exposed, or other substances may be asphyxiants. Still other reasons for establishing a TLV for a given substance include the fact that certain chemical compounds are anesthetic, fibrogenic, can cause allergic reactions, or malignancies. Some additional TLVs have been established because exposure above a certain airborne concentration is a nuisance.

The amount and nature of the information available for establishing a TLV varies from substance to substance; consequently, the precision of the estimated TLV continues to be subject to revision. The latest documentation for that substance should be consulted to assess the present data available for a given substance.

Because individual susceptibility varies widely, an occasional exposure of an individual at (or even below) the threshold limit may not prevent discomfort or aggravation of a preexisting condition, or occupational illness. In addition to the TLVs set for chemical compounds, there are limits for physical agents, such as noise, microwaves, and heat stress (see Chapters 9, Industrial Noise, and 11, Nonionizing Radiation, and Appendix B-1).

The ACGIH periodically publishes a documentation of TLVs in which it gives the data and information upon which the TLV for each substance is based. This documentation can be used to provide health and safety professionals with insight to aid professional judgment when applying the TLVs.

The most current edition of the ACGIH *Threshold Limit Values and Biological Exposure Indices* should be used. When referring to an ACGIH TLV, the year of publication should always preface the value, such as "the 1987 TLV for nitric oxide was 25 ppm." Note that the TLVs are not mandatory federal or state employee exposure standards, and the term TLV should not be used for standards published by OSHA or any agency except the ACGIH.

Three categories of Threshold Limit Values are specified as follows:

Time-Weighted Average (TLV-TWA). This is the time-weighted average concentration for a normal 8-hour workday or 40-hour workweek, to which nearly all workers may be repeatedly exposed, day after day, without adverse effect.

Short-Term Exposure Limit (TLV-STEL). This is the maximal concentration to which workers can be exposed for a period of up to 15 minutes continuously without suffering from any of the following:

1. irritation
2. chronic or irreversible tissue change
3. narcosis of sufficient degree to increase the likelihood of accidental injury, impair self-rescue, or materially reduce work efficiency.

A STEL is a 15-minute TWA exposure that should not be exceeded at any time during a workday, even if the 8-hour TWA is within the TLV. Exposures at the STEL should not be longer than 15 minutes and should not be repeated more than four times daily. There should be at least 60 minutes between successive exposures at the STEL.

The TLV-STEL is not a separate, independent exposure limit, it supplements the TWA limit when there are recognized acute effects from a substance that has primarily chronic effects. The STELs are recommended only when toxic effects in humans or animals have been reported from high short-term exposures.

Note: none of the limits mentioned here, especially the TWA-STEL, should be used as engineering design criteria.

Ceiling (TLV-C). This is the concentration that should not be exceeded during any part of the working exposure. To assess a TLV-C, the conventional industrial hygiene practice is to sample during a 15-minute period, except for those substances that can cause immediate irritation with exceedingly short exposures.

For some substances (for example, irritant gases), only one category, the TLV-C, may be relevant. For other substances, two or three categories may be relevant, depending on their physiological action. If any one of these three TLVs is exceeded, a potential hazard from that substance is presumed to exist.

Limits based on physical irritation should be considered no less binding than those based on physical impairment. Increasing evidence shows that physical irritation can initiate, promote, or accelerate physical impairment via interaction with other chemical or biological agents.

The amount by which threshold limits can be exceeded for short periods without injury to health depends on many factors, such as the nature of the contaminant, whether very high concentrations, even for a short period, produce acute poisoning, whether the effects are cumulative, the frequency with which high concentrations occur, and the duration of such periods. All factors must be considered when deciding whether a hazardous condition exists.

Skin notation

Nearly 25 percent of the substances in the TLV list are followed by the designation "Skin." This refers to potential exposure through the cutaneous route usually by direct contact with the substance. Vehicles such as certain solvents can alter skin absorption. This designation is intended to suggest appropriate measures for the prevention of cutaneous absorption.

Mixtures

Special consideration should be given in assessing the health hazards that can be associated with exposure to mixtures of two or more substances.

Federal Occupational Safety and Health Standards

The first compilation of the health and safety standards promulgated by the Department of Labor's OSHA in 1970 was derived from the then-existing federal standards and national consensus standards. Thus, many of the 1968 TLVs established by the ACGIH became federal standards or Permissible Exposure Limits (PELs). Also, certain workplace quality standards known as maximal acceptable concentrations of the American National Standards Institute (ANSI) were incorporated as federal health standards in 29 *CFR* 1910.1000 (Table Z-2) as national consensus standards.

In adopting the ACGIH TLVs, OSHA also adopted the concept of the TWA for a workday. In general:

$$\text{TWA} = \frac{C_aT_a + C_bT_b + \ldots C_nT_n}{8}$$

where:
T_a = the time of the first exposure period during the shift
C_a = the concentration of contaminant in period "a"
T_b = another time period during the shift
C_b = the concentration during period "b"
T_n = the nth or final time period in the shift
C_n = the concentration during period "n."

This simply provides a summation throughout the workday of the product of the concentrations and the time periods for those concentrations encountered in each time interval and averaged over an 8-hour standard workday.

EVALUATION

Evaluation can be defined as the decision-making process resulting in an opinion on the degree of health hazard that exists from chemical or physical agents from industrial operations. The basic approach to controlling occupational disease consists of evaluating the potential hazard and controlling the specific hazard by suitable industrial hygiene techniques. (See Chapters 16, Evaluation, and 17, Methods of Evaluation, for more details.)

Evaluation involves judging the magnitude of the chemical, physical, biological, or ergonomic stresses. Deciding if a health hazard exists is based on comparing environmental measurements with hygienic guides, TLVs, OSHA PELS, and/or reports in the literature.

Evaluation, in the broad sense, also includes determining the levels of physical and chemical agents arising out of a process to study the related work procedures and to determine the effectiveness of a given piece of equipment used to control the hazards from that process.

Recognizing industrial health hazards involves knowledge and understanding of the several types of workplace environmental stresses and the effect(s) of these stresses upon the health of the worker. Control involves the reduction of environmental stresses to values that the worker can tolerate without impairment of health or productivity. Measuring and quantitating environmental stress are the essential ingredients for modern industrial hygiene, and are instrumental in conserving the health and well-being of workers.

Basic hazard-recognition procedures

There is a basic, systematic procedure for recognizing and evaluating environmental health hazards, which includes the following questions:

What is produced?
What raw material is used?
What materials are added in the process?
What equipment is involved?
What is the cycle of operations?
What operational procedures are used?
Is there a written procedure for the safe handling and storage of materials?
What about dust control, cleanup after spills, and waste disposal?
Are the ventilating and exhaust systems adequate?
Does the plant layout minimize exposure?
Is the plant well equipped with safety appliances such as showers, masks, respirators, and emergency eyewash fountains?
Are safe operating procedures outlined and enforced?
Is a complete hazard communication program that meets state or federal OSHA requirements in effect?

Understand the industrial process well enough to see where contaminants are released. For each process perform the following:

- For each contaminant find the OSHA PEL or other safe exposure guidelines based on the toxicological effect of the material.
- Determine the actual level of exposure to harmful physical agents.
- Determine the number of employees exposed and for how long.
- Identify the chemicals and contaminants in the process.
- Determine the level of airborne contaminants using air-sampling techniques.
- Calculate the resulting daily average and peak exposures from the air-sampling results and employee exposure times.
- Compare the calculated exposures with OSHA standards, the TLV listing published by the ACGIH, the hygienic guides, or other toxicological recommendations.

All of the above will be discussed in detail in the following chapters.

Information required

Detailed information should be obtained regarding what hazardous materials are used within a plant, type of job operation, how the workers are exposed, work pattern, levels of air contamination, duration of exposure, what control measures are used, and other pertinent information. The hazard potential of the material is determined not only by its inherent toxicity, but also by the conditions of use—who uses what, where, and how long?

To recognize hazardous environmental factors or stresses, a health and safety professional must first know the raw materials used and the nature of the products and by-products manufactured. Consult MSDSs for the substances.

Any person responsible for maintaining a safe, healthful work environment should be thoroughly acquainted with the concentrations of harmful materials or energies that may be encountered in the industrial environment for which they are responsible.

If a plant is going to handle a hazardous material, it is necessary for the health and safety professional to consider all the unexpected events that can occur, and what precautions are required in case of an accident, to prevent or control atmospheric release of a toxic material.

After these considerations have been studied and proper countermeasures installed, operating and maintenance personnel must be taught the proper operation of the health and safety control measures. Only in this way can personnel be made aware of the possible hazards that could exist and the needs for certain built-in safety features.

The operating and maintenance people should set up a routine procedure (at frequent, stated intervals) for testing the emergency industrial hygiene and safety provisions that are not used in normal, ordinary plant or process operations.

Degree of hazard

The degree of hazard from exposure to harmful environmental factors or stresses would depend on the following:

- nature of the material or energy involved
- intensity of the exposure
- duration of the exposure.

The key elements to be considered when evaluating a health hazard are: (1) how much of the material in contact with body cells is required to produce injury, (2) the probability of the material being absorbed by the body to result in an injury, (3) rate that airborne contaminant is generated, (4) total time of contact, and (5) control measures in use.

Medical aspects. The medical staff, the industrial hygienist, and other health professionals must be aware that certain occupations present hazards to the worker. Many regulations provide for a minimum medical surveillance program and specify certain tests and procedures necessary for the control of that exposure (see Chapter 26, The Occupational Physician, for more details).

The industrial physician should be familiar with all jobs, materials, and processes that are used. An occasional inspection trip will help keep the physician abreast of what is going on in these areas. This visual inventory of hazards helps the physician suggest to the health and safety professional ways to protect employees from actual or potentially harmful environments.

This basic knowledge also aids the industrial physician when recommending placement of employees in jobs best suited to their physical capabilities.

Air-sampling

The importance of the sampling location, the proper time to sample, and the number of samples to be taken during the course of an investigation of the work environment cannot be overstressed.

Although this procedure might appear to be a routine, mechanical job, actually, it is an art requiring detailed knowledge of the sampling equipment and its shortcomings; where and when to sample; and how to weigh the many factors that can influence the sample results—such as ambient temperature, season of the year, unusual problems or upsets in plant operations, and interferences from other contaminants. The sample must usually be taken in the breathing zone of a particular employee (see Figure 1-5).

Figure 1-5. Portable pump with intake positioned to collect continuous samples from the breathing zone of a worker.

The air volume sampled must be sufficient to permit a representative determination of the contaminant to properly compare the result with the TLV or PEL. The sampling period must usually be such as to give a direct measure of the average full-shift exposure of the employees concerned. The sample must be sealed and identified if it is to be shipped to a laboratory so that it will be possible to identify positively the time and place of sampling and the individual who took the sample.

Area samples, taken by setting the sampling equipment in a fixed position in the work area, are useful as an index of general contamination. However, the actual exposure of the employee at the point of generation of the contaminant can have an exposure greater than is indicated by an area sample.

To meet the requirement of establishing the TWA concentrations, the sampling method and time periods should be chosen to average out fluctuations that commonly occur in a day's work. If there are wide fluctuations in concentration, the long-term samples should be supplemented by samples designed to catch the peaks separately.

If the exposure being measured is from a continuous operation, it is necessary to follow the particular operator through two cycles of operation, or through the full shift if operations follow a random pattern during the day. For operations of this sort, it is particularly important to find out what the workers do when the equipment is down for maintenance or process change. Such periods are frequently also periods of maximum exposure.

As an example of the very small concentrations involved, the hygienist commonly samples and measures substances in the air of the working environment in concentrations ranging from 1–100 ppm. Some idea of the magnitude of these concentrations can be appreciated when one realizes that one inch in 16 miles is one part per million; one cent in $10,000, one ounce of salt in 62,500 pounds of sugar, one ounce of oil in 7,812.5 gallons of water—all represent one part per million.

OCCUPATIONAL SKIN DISEASES

Some general observations on dermatitis are given in this chapter, but more detailed information is given in Chapter 8, Industrial Dermatoses. Occupational dermatoses can be caused by organic substances, such as formaldehyde, and solvents or inorganic materials, such as acids and alkalis, and chromium and nickel compounds. Skin irritants are usually either liquids or dusts.

Types

There are two general types of dermatitis—primary irritation and sensitization.

Primary irritation dermatitis. Nearly all persons suffer primary irritation dermatitis from mechanical agents, such as friction, from physical agents, such as heat or cold, and from chemical agents, such as acids, alkalis, irritant gases, and vapors. Brief contact with a high concentration of a primary irritant or prolonged exposure to a low concentration causes inflammation. Allergy is not a factor in these conditions.

Sensitization dermatitis. This type results from an allergic reaction to a given substance. The sensitivity becomes established during the induction period, which may be a few days to a few months. After the sensitivity is established, exposure to even a small amount of the sensitizing material is likely to produce a severe reaction.

Some substances can produce both primary irritation dermatitis and sensitization dermatitis. Among them are organic solvents, chromic acid, and epoxy resin systems.

Causes

Occupational dermatitis can be caused by chemical, mechanical, physical, and biological agents, and plant poisons.

Chemical agents are the predominant causes of dermatitis in manufacturing industries. Cutting oils and similar substances are significant because the oil dermatitis that they cause is probably of greater interest to industrial concerns than is any other type of dermatitis.

Detergents and solvents remove the natural oils from the skin or react with the oils of the skin to increase susceptibility to reactions from chemicals that ordinarily do not affect the skin. Materials that remove the natural oils include alkalis, soap, and turpentine.

Dessicators, hygroscopic agents, and anhydrides take water out of the skin and generate heat. Examples are sulfur dioxide and trioxide, phosphorus pentoxide, strong acids, such as sulfuric, and strong alkalis, such as potash.

Protein precipitants tend to coagulate the outer layers of the skin. They include all the heavy metallic salts and those that form alkaline albuminates on combining with the skin, such as mercuric and ferric chloride. Alcohol, tannic acid, formaldehyde, picric acid, phenol, and intense ultraviolet rays are other examples of protein-precipatating agents.

Oxidizers unite with hydrogen and liberate nascent oxygen on the skin. Such materials include nitrates, chlorine, iodine, bromine, hypochlorites, ferric chloride, hydrogen peroxide, chromic acid, permanganates, and ozone.

Solvents extract essential skin constituents. Examples are ketones, aliphatic and aromatic hydrocarbons, halogenated hydrocarbons, ethers, esters, and certain nitro compounds.

Allergic or anaphylactic proteins stimulate the production of antibodies that cause skin reactions in sensitive persons. The sources of these antigens are usually cereals, flour, and pollens, but can include feathers, scales, flesh, fur, and other emanations.

Mechanical causes of skin irritation include friction, pressure, and trauma, which may become infected with either bacteria or fungi.

Physical agents leading to occupational dermatitis include heat, cold, sunlight, x-rays, ionizing radiation, and electricity. The x-rays and other ionizing radiation can cause dermatitis, severe burns, and even cancer. Prolonged exposure to sunlight produces skin changes.

Biological agents causing dermatitis can be bacterial, fungal, or parasitic. Boils and folliculitis caused by staphylococci and streptococci, and general infection from occupational wounds are probably best known among the bacterial skin infections. These can be occupationally induced infections.

Fungi cause athlete's foot and other types of dermatitis among kitchen workers, bakers and fruit handlers; fur, hide, and wool handlers or sorters; barbers; and horticulturists. Parasites cause grain itch and often occur among handlers of grains and straws, and particularly among farmers, laborers, miners, fruit handlers, and horticulturists.

Plant poisons causing dermatitis are produced by several hundred species of plants. The best known are poison ivy, poison oak, and poison sumac. Dermatitis from these three sources can result from bodily contact with any part of the plant, exposure of any part of the body to smoke from the burning plant, or contact with clothing or other objects previously exposed to the plant.

Physical examinations

Preplacement examinations help identify those especially susceptible to skin irritations. The examining physician should be given detailed information on the type of work for which the applicant is being considered. If the work involves exposure to skin irritants, the physician should determine if the prospective employee has deficiencies or characteristics likely to predispose him/her to dermatitis (see Chapter 26, The Occupational Physician, for more details).

Preventive measures

Before introducing new or different chemicals in an established process, possible dermatitis hazards should be carefully considered. Once anticipated, suitable engineering controls should be devised and built into the processes to avoid these hazards.

The type, number, and amounts of skin irritants used in various industrial processes affect the degree of control that can be readily obtained, but the primary objective in every case should be to eliminate skin contact as completely as possible. The preventive measures discussed in Chapter 20, Methods of Control, can be adapted to control industrial dermatitis.

CONTROL METHODS

The type and extent of control methods depend upon physical, chemical, and toxic properties of the air contaminant or energy involved, evaluation made of the exposure, and the operation that produces the energy or disperses the contaminant. The extensive controls needed for lead oxide dust, for example, would not be needed for limestone dust, since much more limestone dust is permissible.

General methods of controlling harmful environmental factors or stresses include the following:

- substituting a less harmful material for one that is hazardous to health
- changing or altering a process to minimize worker contact
- isolating or enclosing a process or work operation to reduce the number of persons exposed
- wet methods to reduce generation of dust in operations, such as mining and quarrying
- local exhaust at the generating and dispersing points of contaminants
- general or dilution ventilation with clean air to provide a healthful atmosphere
- personal protective devices, such as special clothing, eye, and respiratory protection
- good housekeeping, including cleanliness of the workplace, waste disposal, adequate washing and eating facilities, healthful drinking water, and control of insects and rodents
- special control methods for specific hazards, such as reduction of exposure time, film badges, and similar monitoring devices, continuous sampling with preset alarms
- medical programs to detect intake of toxic materials
- training and education to supplement engineering controls.

(These topics are discussed in detail in Chapters 20 and 23.)

General methods for controlling environmental factors or stresses include the use of engineering controls, administrative controls, and personal protective equipment. Alternatives that

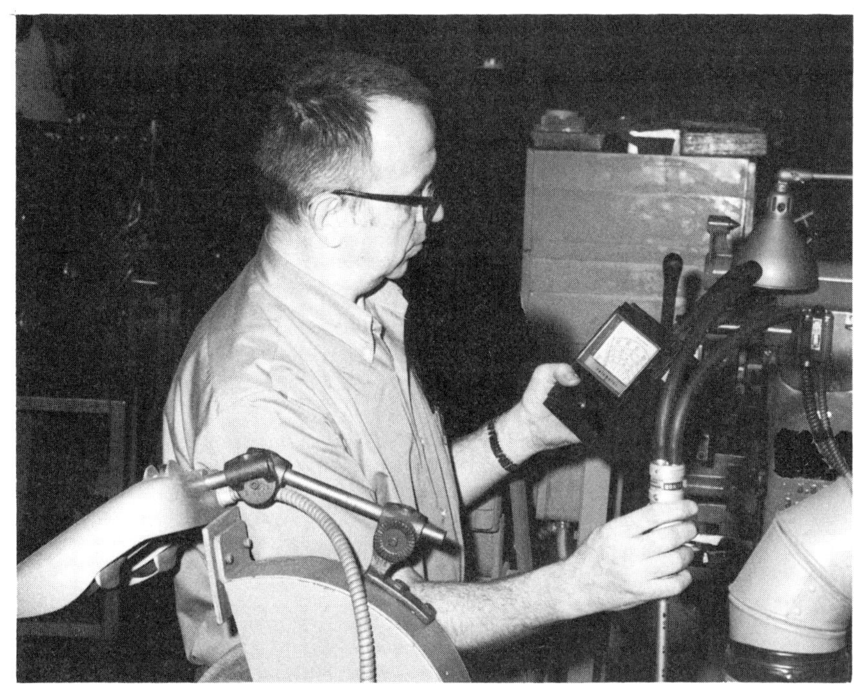

Figure 1-6. Industrial hygienist checks for adequate ventilation rates.

can be applied individually or with another to achieve the desired level of protection exist within each category of these control methods.

Engineering controls

Substituting or replacing a toxic material with a harmless one is a very practical method of eliminating an industrial health hazard. In many cases a solvent with a lower order of toxicity or flammability can be substituted for a more hazardous one. In a solvent substitution, it is always advisable to experiment on a small scale before making the new solvent part of the operation or process.

A change in process often offers an ideal chance to improve working conditions as well as quality and production. In some cases, a process can be modified to reduce the exposure to a dust or fume and thus markedly reduce the hazard. Brush painting or dipping instead of spray painting minimizes the concentration of airborne contaminants from toxic pigments. Structural bolts in place of riveting, steam-cleaning instead of vapor degreasing of parts, and airless spraying techniques and electrostatic devices to replace hand-spraying are examples of a change of process. When buying individual machines, the need for accessory ventilation, noise and vibration suppression, and heat control should be considered before purchase.

Noisy operations can be isolated from the people nearby by a physical barrier (such as an acoustic box to contain noise from a whining blower or a rip saw). Isolation is particularly useful for limited operations requiring relatively few workers or where control by any other method is not feasible.

Enclosing the process or equipment is a desirable method of control, because it can minimize escape of the contaminant into the workroom atmosphere, whether noise, fumes, dust, etc. Examples of this type of control are glove box enclosures and abrasive shot blast machines for cleaning castings.

In the chemical industry, isolating hazardous processes in closed systems is a widespread practice. The use of a closed system is one reason why the manufacture of toxic substances can be less hazardous than their use.

Dust hazards frequently can be minimized or greatly reduced by spraying water at the source of dust dispersion. "Wetting down" is one of the simplest methods for dust control. However, its effectiveness depends upon proper wetting of the dust and keeping it moist. To be effective, the addition of a wetting agent to the water and proper and timely disposal of the wetted dust before it dries out and is redispersed may be necessary.

Ventilation

The major use of exhaust ventilation for contaminant control is to prevent health hazards from airborne materials. The OSHA has ventilation standards for abrasive blasting, grinding, polishing and buffing operations, spray finishing operations, and open-surface tanks (Figure 1-6). For more details, see Chapters 21, Industrial Ventilation, and 22, General Ventilation.

A local exhaust system traps and removes the air contaminant near the generating source, which usually makes this method much more effective than general ventilation (Figure 1-7). Local exhaust ventilation should be used, therefore, when exposures to the contaminant cannot be controlled by substitution, changing the process, isolation, or enclosure. Even though a process has been isolated, it still may require a local exhaust system.

General or dilution ventilation—removing and adding air to dilute the concentration of a contaminant to below hazardous levels—uses natural or forced air movement through open doors, windows, roof ventilators, and chimneys. General exhaust fans can be mounted in roofs, walls, or windows (see Chapters 21 and 22 for more details).

Consideration must be given to providing replacement air, especially during winter months. Dilution ventilation is feasible only if the quantity of air contaminant is not excessive, and particularly if the contaminant is released at a substantial distance from the worker's breathing zone. General ventilation should not be used where there is a major, localized source of contami-

Figure 1-7. Control of harmful fumes and gases close to the source, such as this exhaust hood over the die head of plastics extruder, is effective and reduces the necessity of larger air movement required for general ventilation.

nation (especially highly toxic dusts and fumes). A local exhaust system is more effective in such cases.

Air conditioning does not substitute for air cleaning. Air conditioning is mainly concerned with control of air temperature and humidity and can be accomplished by systems that accomplish little or no air cleaning. An air-conditioning system usually uses an air washer as part of the means to accomplish temperature and humidity control. However, these air washers are not designed as efficient air cleaners and should not be considered as such.

Processes in which materials are crushed, ground, or transported are potential sources of dust dispersion, and should be controlled either by wet methods or enclosed and ventilated by local exhaust ventilation. Points where conveyors are loaded or discharged, transfer points along the conveying sytem, and heads or boots of elevators should be enclosed as well as ventilated. (For more details, see Chapter 21, Industrial Ventilation.)

Personal protective equipment

When not feasible to render the working environment completely safe, it may be necessary to protect the worker from that environment by personal protective equipment. This is considered to be a secondary control method to engineering and administrative controls and should be considered as a last resort.

Where not possible to enclose or isolate the process or equipment, ventilation or other control measures should be provided; where there are short exposures to hazardous concentrations of contaminants; where unavoidable spills may occur—personal protective equipment must be provided and used.

Personal protective devices have one serious drawback—they do nothing to reduce or eliminate the hazard. They interpose a barrier between worker and hazard; if the barrier fails, immediate exposure is the result. The supervisor must be constantly alert to make sure that required protective equipment is worn by those workers who need supplementary protection as may be required by OSHA standards.

Administrative controls

When exposure cannot be reduced to permissible levels through engineering controls, as in the case of air contaminants or noise, an effort should be made to limit the employee's exposure through administrative controls.

Examples of some administrative controls are as follows:

- arranging work schedules and the related duration of exposures so that employees are minimally exposed to health hazards
- transferring employees who have reached their upper permissible limits of exposure to an environment where no further additional exposure will be experienced.

Where exposure levels exceed the PEL for one worker in one day, the job can be assigned to two, three, or as many workers needed to keep each one's duration of exposure within the PEL. In the case of noise, other possibilities may involve intermittent use of noisy equipment.

Administrative controls must only be designed by knowledgeable health and safety professionals, and used cautiously and judiciously. They are not as satisfactory as installing engineering controls and have been criticized by some as a means of spreading exposures instead of reducing or eliminating the exposure.

Good housekeeping plays a key role in occupational health protection. Basically, it is another tool for preventing dispersion of dangerous contaminants and for maintaining a safe and healthful working condition required by law. Immediate cleanup of any spills or toxic material, by workers wearing proper protective equipment, is a very important control measure. Good housekeeping is also essential where solvents are stored, handled, and used. Leaking containers or spigots should be fixed immediately, and spills cleaned promptly. All solvent-soaked rags or absorbents should be placed in airtight metal receptacles and removed daily (see Chapter 6, Solvents, for more details).

It is impossible to have an effective occupational health program without good maintenance and housekeeping—workers should be informed about the need for these controls. Proper training and education is a vital element for successful implementation of any control effort, and is required by law as part of a complete federal or state OSHA Hazard Communication Program.

SOURCES OF HELP

Specialized help is available from a number of sources. Every supplier of products or services is likely to have competent professional staff who can provide technical assistance or guidance. Many insurance companies that carry workers' compensation insurance provide industrial hygiene consultation services, just as they provide periodic safety inspections.

Professional consultants and privately owned laboratories are available on a fee basis for concentrated studies of a specific problem or for a plantwide or companywide survey, which can be undertaken to identify and catalog individual environmental exposures. Lists of certified analytical laboratories and industrial hygiene consultants are available from the American Industrial Hygiene Association.

Many states have excellent industrial hygiene departments that can provide consultation on a specific problem. Appendix A, Sources of Help, contains names and addresses of state and provincial health and hygiene agencies. The NIOSH has a Technical Information Center that can provide information on specific problems. Scientific and technical societies that can help with problems are listed in Appendix A. Some provide consultation services to nonmembers; they all have much accessible technical information. A list of organizations concerned with industrial hygiene is included in Appendix A.

SUMMARY

No matter what health hazards are encounterd, the approach of the industrial hygienist is essentially the same. Using methods relevant to the problem, they secure qualitative and quantitative estimates of the extent of hazard. This data is then compared with the recommended exposure guidelines. If a situation hazardous to life or health is shown, recommendations for correction are made. The industrial hygienist's recommendations place particular emphasis on effectiveness of control, cost, and ease of maintenance of the control measures.

Recognition ▪ evaluation ▪ control—these are the fundamental concepts of providing all workers with a healthy working environment.

BIBLIOGRAPHY

American Conference of Governmental Industrial Hygienists (ACGIH). *Documentation of Threshold Limit Values,* 3rd ed. Cincinnati: ACGIH, 1971.

______. *Sampling Instruments Manual,* 2nd ed. Cincinnati: ACGIH, 1972.

______. *Threshold Limit Values and Biological Exposure Indices.* Cincinnati: ACGIH, published annually.

______, and Committee on Industrial Ventilation. *Industrial Ventilation: A Manual of Recommended Practice,* 19th ed. Lansing, MI: ACGIH, 1986.

American Industrial Hygiene Association (AIHA). *Engineering Field Reference Manual.* Akron, OH: AIHA, 1984.

______. *Ergonomics Guide Series—Manual Lifting.* Akron, OH: AIHA, 1970.

______. *Hygienic Guide Series.* Akron, OH: AIHA, published periodically.

______. *Industrial Noise Manual,* 4th ed. Akron, OH: AIHA, 1986.

______. *Respiratory Protection Program: A Manual on Guidelines.* Akron, OH: AIHA, 1980.

American National Standards Institute, 1430 Broadway, New York, NY 10017.
Standards:
–*Practices for Respiratory Protection,* Z88.2-1980.
–*Practices for Respiratory Equipment during Fumigation,* Z88.3-1983.
–*Practices for Respiratory Protection for the Fire Service,* Z88.5-1981.
–*Physical Qualifications for Respirator Use,* Z88.6-1984.

Burgess, W. A. *Recognition of Health Hazards in Industry: A Review of Materials and Processes.* New York: John Wiley & Sons (Wiley-Interscience Publ.), 1981.

Cralley, L. J., and Cralley, L. V., series eds. *Industrial Hygiene Aspects of Plant Operations:* Vol. 1: *Process Flows;* Mutchler, J. F., ed. Vol. 2: *Unit Operations and Product Fabrication;* Caplan, K. J., ed. Vol. 3: *Selection, Layout, and Building Design.* New York: MacMillan Publ. Co., 1986.

Damon, F. A. Organizing for ergonomics. *American Industrial Hygiene Association Journal,* 12 (Nov. 1967):6.

Grandjean, E. *Fitting the Task to the Man: An Ergonomic Approach.* London: Taylor & Francis Ltd., 1980.

Gosselin, R. E., et al. *Clinical Toxicology of Commercial Products: Acute Poisoning,* 5th ed. Baltimore: The Williams & Wilkins Co., 1984.

Key, M. M., et al., eds. *Occupational Diseases—A Guide to Their Recognition,* rev. ed. Washington, DC: USHHS (U.S. Government Printing Office), 1977.

LaDou, J., ed. *Introduction to Occupational Health and Safety.* Chicago: National Safety Council, 1986.

Levy, Barry S., and Wegman, D. H. *Occupational Health: Recognizing and Preventing Work Related Disease.* Boston: Little, Brown & Co., 1983.

McDermott, H. J. *Handbook of Ventilation for Contaminant Control,* 2nd ed. Stoneham, MA: Butterworth, 1985.

Merchant, J. A., Boehlecke, B. A., Taylor, G., Eds. *Occupational Respiratory Diseases.* Cincinnati: NIOSH Publications Dissemination, 1986.

Murrell, K. F. *Ergonomics: Man in His Working Environment.* New York: Methuen, Inc., 1980.

National Institute for Occupational Safety and Health (NIOSH), USDHHS, Division of Safety Research. *A Worker's Guide to Confined Spaces.* Morgantown, WV: NIOSH.

______. *Work Practices Guide for Manual Lifting,* NIOSH Pub. No. 81-122. Cincinnati: NIOSH Publications Dissemination, March 1981.

______. *Criteria for a Recommended Standard: Working in Confined Spaces,* NIOSH Pub. No. 80-106. Cincinnati: NIOSH Publications Dissemination, 1979.

______. *Criteria for a Recommended Standard, Occupational Exposure to Hot Environments,* revised criteria, NIOSH Publ. No. 86-113. Cincinnati: NIOSH Publications Dissemination, 1986.

National Safety Council. *Accident Prevention Manual for Industrial Operations,* 9th ed: Vol. 1: *Administration and Programs;* Vol. 2: *Engineering and Technology.* Chicago: National Safety Council, 1988.

National Safety Council, 444 N. Michigan Ave., Chicago IL 60611. Industrial Data Sheets, latest index.

Patty, F. A., ed. *Industrial Hygiene and Toxicology:* vol. 1, 3rd ed., 1978; vol. 2A, 3rd ed., 1980; vol. 2B, 3rd ed., 1981; vol. 2C, 3rd ed., 1981; vol. 3A, 2nd ed., 1985; vol. 3B, 1985.

Procter, N. H., and Hughes, J. P. *Chemical Hazards of the Workplace.* Philadelphia: J. B. Lippincott Co., 1978.

Protecting Workers' Lives: A Safety and Health Guide for Unions. Chicago: National Safety Council, 1983.

Schilling, R. S., ed. *Occupational Health Practice,* 2nd ed. Stoneham, MA: Butterworth, 1981.

Snook, S. H., and Irvine, C. H. Maximum acceptable weight of lift. *American Industrial Hygiene Association Journal* 28(July-August 1967):4.

Tichauer, E. R. Ergonomics: The state of the art. *American Industrial Hygiene Association Journal.*

Wasserman, D. E., and Taylor, W., eds. *Proceedings of the International Occupational Hand-Arm Vibration Conference,* National Institute for Occupational Safety and Health, Cincinnati, April 1977.

Woodson, W. E. *Human Factors Reference Guide for Process Plants.* New York: McGraw-Hill, 1986.

Zenz, C. *Occupational Medicine—Principles and Practical Applications,* 2nd ed. Chicago: Year Book Medical Publ., 1988.

PART II

Anatomy, Physiology, and Pathology

2

The Lungs

by Julian B. Olishifski, MS, PE, CSP
Revised by George S. Benjamin, MD, FACS

THE MATERIAL IN THIS CHAPTER on human respiration is intended primarily for engineers and health and safety professionals who must evaluate and control industrial health hazards.

Establishing an effective industrial hygiene program calls for an understanding of the anatomy and physiology of the human respiratory system. The respiratory system presents a quick and direct avenue of entry for toxic materials into the body because of its intimate association with the circulatory system and the constant need to oxygenate human tissue cells. Anything affecting the respiratory system affects the entire human organism—whether it is insufficient oxygen or contaminated air.

Humans can survive for weeks without food, for days without water—but only a few minutes without air. Air must reach the lungs almost constantly so oxygen can be extracted and distributed via the blood to every body cell. The life-giving component of air is oxygen, which constitutes a little less than one-fifth of its volume.

All living cells of the body are engaged in a series of chemical processes. The total of these processes is metabolism. In the course of the body's metabolism, each cell consumes oxygen and produces carbon dioxide as a waste substance.

Each living cell in the body requires a constant supply of oxygen. Some cells, however, need a more constant oxygen supply than others—cells in the brain and heart may die after 4–6 minutes without oxygen. These cells can never be replaced, and permanent changes result from such damage. Other cells in the body are not so critically dependent on an oxygen supply, because they are replaceable.

Thus, the respiratory system by which oxygen is delivered to the body and carbon dioxide removed is a very important part of the body. The respiratory system consists of all the organs of the body that contribute to normal respiration or breathing. Strictly speaking, it includes the nose, mouth, upper throat, larynx, trachea, and bronchi, which are all air passages or airways. It includes the lungs, where oxygen is passed into the blood and carbon dioxide is given off. Finally, it includes the diaphragm and the muscles of the chest, which permit normal respiratory movements (Figure 2-1).

ANATOMY

Nose

The nose consists of an external and an internal portion. The external portion of the nose protrudes from the face and is highly variable in shape. The upper part of this triangular structure is held in a fixed position by the supporting nasal bones that form the bridge of the nose. The lower portion is movable because of its pliable framework of fibrous tissue, cartilage, and skin.

The internal portion of the nose lies within the skull between the base of the cranium and the roof of the mouth, and is in front of the nasopharynx (the upper extension of the throat). The skull bones that enter into the formation of the nose include the frontal, the sphenoid, the ethmoid, the nasal, the maxillary, the lacrimal, the vomer, and the palatine and inferior conchae.

The nasal septum is a narrow partition that divides the nose into right and left nasal cavities. In some people the nasal septum is markedly deflected to one side causing the affected nasal cavity to be almost completely obstructed; this condition is called deviated nasal septum.

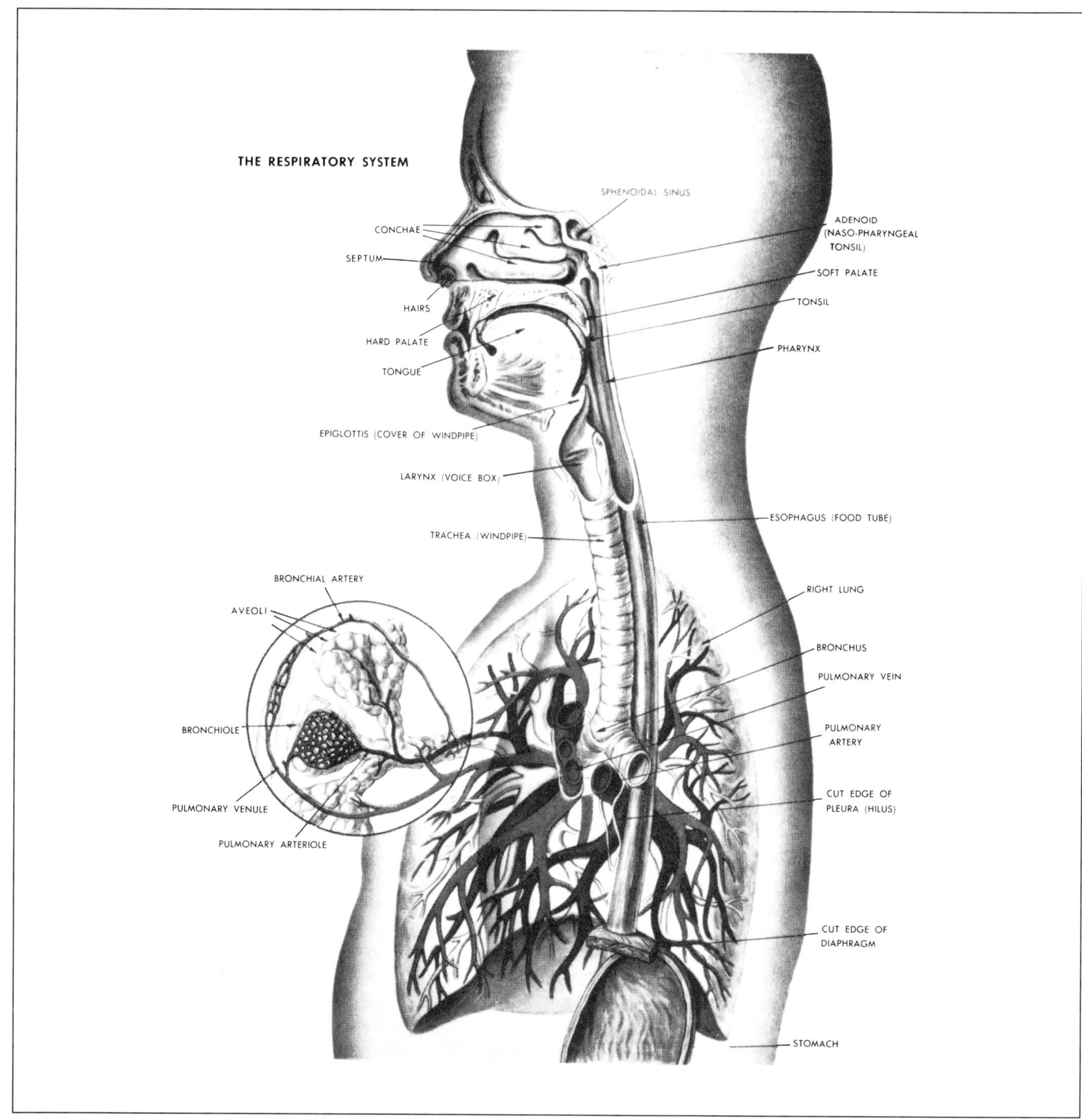

Figure 2-1. Schematic drawing of respiratory system. (Reprinted with permission from *The Wonderful Human Machine.* Chicago, American Medical Association, 1971)

The nasal cavities are open to the outside through the anterior nares (or nostrils); toward the rear, they open into the nasopharynx by means of the posterior nares or conchae. The vestibule of each cavity is the dilated portion just inside the nostril. Toward the front, the vestibule is lined with skin and presents a ring of coarse hairs which serve to trap dust particles. Toward the rear, the lining of the vestibule changes from skin to a highly vascular ciliated mucous membrane, called the nasal mucosa, which lines the rest of the nasal cavity.

Extending into the nasal cavity from the base of the skull are large nerve filaments which are part of the sense organ for smell. From these filaments, information on odors is relayed to the olfactory nerve which goes to the brain.

Turbinates. Near the middle of the nasal cavity, and on both sides of the septum, are a series of scroll-like bones called the conchae, or turbinates. The purpose of the turbinates is to increase the amount of tissue surface within the nose so that incoming air has a greater opportunity to be conditioned before it continues on its way to the lungs.

Respiration begins with the nose, which is specially designed for the purpose, although there are times when you breathe through the mouth as well. When you perform any vigorous activity, and begin to puff and pant, you are breathing rapidly through the mouth to provide the blood with the extra oxygen needed.

However, the mouth is not designed for breathing. You may have noticed this on cold days when you make a deliberate effort to keep your mouth tightly closed, because if you take air in through the mouth you can feel its coldness. Cold air passing through the mouth has no chance to become properly warmed. But cold as the air may be, you can breathe comfortably through the nose.

Air enters through the nares or nostrils, passes through a web of nasal hairs, and flows posteriorly toward the nasopharynx. The air is warmed and moistened in its passage and partially depleted of particles. Some particles are removed by impaction on the nasal hairs and at bends in the air path, and others by sedimentation.

In mouth breathing, some particles are deposited, primarily by impaction, in the oral cavity and at the back of the throat. These particles are rapidly passed to the esophagus by swallowing.

Mucus. The surfaces of the turbinates, like the rest of the interior walls of the nose, are covered with mucous membranes. These membranes secrete a fluid called mucus. The film of mucus is produced continuously and drains slowly into the throat. The mucus gives up heat and moisture to incoming air and serves as a trap for bacteria and dust in the air. It also helps dilute any irritating substances in the air.

The common cold involves an inflammation of the mucous membrane of the nose. It is characterized by an acute congestion of the mucous membrane and increased secretion of mucus. It is difficult to breath through the nose because of the swelling of the mucous membrane and the accumulated secretions clogging the air passageway.

In cold weather, the membranes can increase the flow of mucus. If the atmosphere is unusually dry, as in an improperly heated building, the mucus may lose its moisture too rapidly and the membrane may become dry and irritated.

Cilia. In addition to the mucus, the membrane is coated with cilia, or hairlike filaments, that move in coordinated waves to propel mucus and trapped particles toward the nostrils. The millions of cilia lining the nasal cavity help the mucus clean the incoming air. When breathing through the mouth, the protective benefits of the cilia and mucus are lost.

In summary, the nose serves not only as a passageway for air going to and from the lungs but also as an air conditioner and as the sense organ for smell. The importance of breathing through the nose is obvious as it serves to moisten and filter and warm or cool the air that is on its way to the lungs (Figure 2-2).

Pharynx

From the nasal cavity, air moves into the pharynx, or throat. Seven tubes enter the pharynx: the two from the nasal cavity, the eustachian tubes which lead to the ears, the mouth cavity, the opening of the esophagus, and the opening of the windpipe.

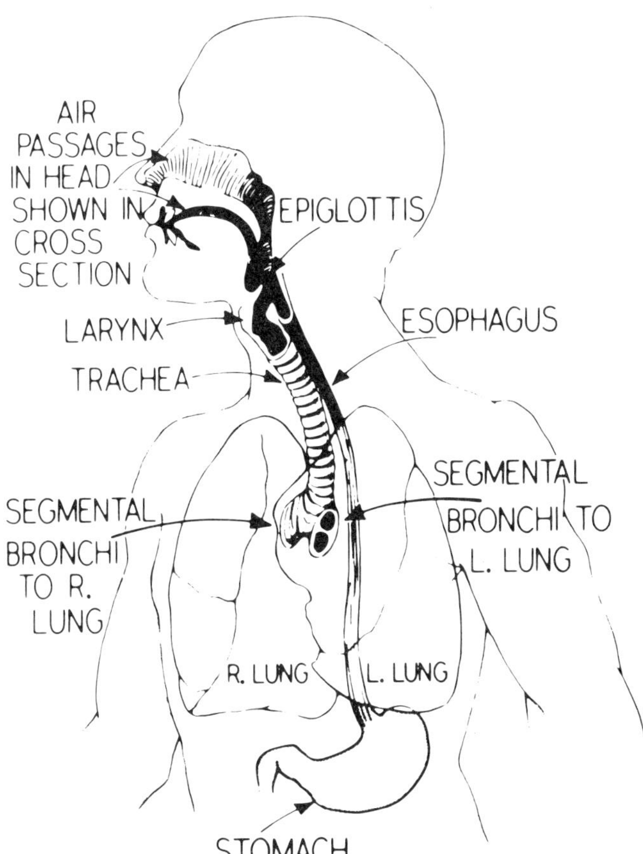

Figure 2-2. Parts of the human respiratory system are shown. Air enters through the mouth and nose, passes down the trachea and into the lungs. (Reprinted with permission from *Emergency Care and Transportation of the Sick and Injured,* American Academy of Orthopaedic Surgeons)

The pharynx or throat is a tubular passageway attached to the base of the skull and extending downward behind the nasal cavity, the mouth, and the larynx to continue as the esophagus, food tube. Its walls are composed of skeletal muscle, and the lining consists of mucous membrane. The nasal passage joins the food canal just behind the mouth. The union of the two passageways at this point makes it possible to breathe with reasonable comfort through the mouth when the nasal passages are blocked because of a cold or allergy.

The nasopharynx is the superior portion of the pharyngeal cavity; it lies behind the nasal cavities and above the level of the soft palate. The function of this portion of the pharynx is purely respiratory, and its ciliated mucosal lining is continuous with that of the nasal cavities. The inflammation of the throat that often accompanies a cold causes the mucous membrane of the region to overproduce mucus. The situation is made worse by mucus entering the throat from the inflamed nasal passages above (postnasal drip). The spasmodic coughing that accompanies a cold is the body's attempt to get rid of the mucus.

At the bottom of the throat are two passageways; the esophagus behind and the trachea in front. Food and liquids entering the pharynx pass into the esophagus, which carries them to the stomach. Air and other gases enter the trachea to go to the lung.

Epiglottis. Guarding the opening of the trachea is a thin, leaf-

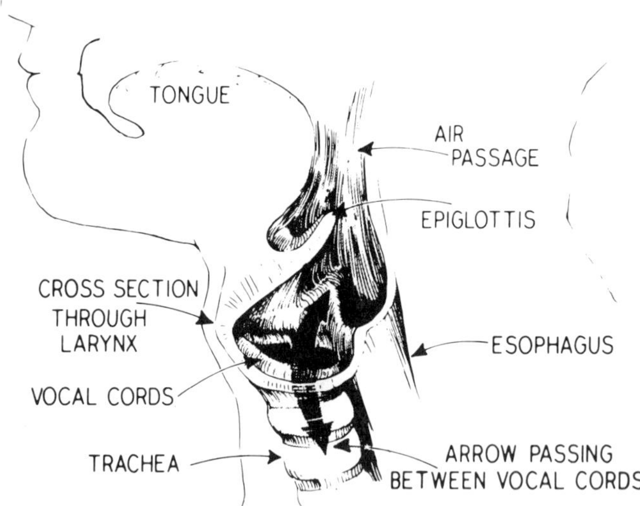

Figure 2-3. Illustrates the anatomy of the neck, showing the epiglottis, larynx, vocal cords, trachea, and the esophagus. (Reprinted with permission from *Emergency Care and Transportation of the Sick and Injured,* American Academy of Orthopaedic Surgeons)

shaped structure, called the epiglottis (Figure 2-3). This structure helps food glide from the mouth to the esophagus.

Everyone is aware that swallowing food and breathing cannot take place at the same time—without danger of choking. But nature has devised a way for food and air to use the same general opening, the pharynx, with only an occasional mix-up.

The incoming air travels through the nasal cavity and through the larynx by crossing over the path used by food on its way to the stomach. Similarly, food crosses over the route of air. But, when food is swallowed, the larynx rises against the base of the tongue to help seal the opening.

If food accidentally starts down the wrong way, into the lungs rather than the stomach, there are explosive protests from the lungs. Any contact of a sizable liquid or solid particle with the trachea sets off a cough, an explosive expulsion of air that will blow it out again. A cough can be a very powerful force. A slight breathing in, closing of the glottis, buildup of pressure, and a sudden release of the trapped air are involved. Also, stimulation of the larynx can cause spasm of the vocal cords with total obstruction of breathing.

Normally, swallowing blocks off the glottis, halts breathing briefly, and assures correct division of air and food. However, an unconscious person may lack this automatic response, and if a drink is given, it may proceed straight into the lungs.

The diaphragm is sometimes subject to periodic spasms of contraction that enlarge the lung cavities and lead to a quick inrush of air. The vocal cords come together to stop the flow, and the air, so suddenly set into motion and so suddenly stopped, makes the sharp noise called the hiccup. Hiccups may be due to indigestion, overloaded stomach, irritation under the surface of the diaphragm, imbibing too much alcohol, or many other possible causes, including heart attacks.

Larynx

The larynx, or voice box, serves as a passageway for air between the pharynx and the trachea. It lies in the midline of the neck, below the hyoid bone and in front of the laryngopharynx. The unique structure of the larynx enables it to function somewhat like a valve on guard duty at the entrance to the windpipe, controlling air flow and preventing anything but air from entering the lower air passages. Exhalation of air through the larynx is controlled by voluntary muscles and so enables the larynx to become the organ of voice.

The larynx is a triangular box composed of nine cartilages that are joined together by ligaments and controlled by skeletal muscles. The larynx is lined with ciliated mucous membrane (except the vocal folds), and the cilia move particles upward to the pharynx.

Vocal cords. The larynx, or voice box, is at the top of the windpipe, or trachea, which takes air to the lungs. But while incoming air passes through the boxlike larynx, it is actually air expelled from the lungs that makes voice sounds. In the front of the larynx, two folds of membranes, the vocal cords, are attached and held by tiny cartilages. Muscles attached to the cartilages move the vocal cords, which are made to vibrate by air exhaled from the lungs.

During ordinary breathing, the vocal cords are held toward the walls of the larynx so that air can pass without being obstructed. During speech, the vocal cords swing over the center of the tube and muscles contract to tense the vocal cords.

Speech. Sounds are created as air is forced past the vocal cords, making them vibrate. These vibrations make the sound. These vibrations can be felt, if the fingers are placed lightly on the larynx (Adam's apple) while speaking.

The vibrations are carried through the air upward into the pharynx, mouth, nasal cavities, and sinuses, which serve as resonating chambers. The greater the force and amount of air from the lungs, the louder the voice. Pitch differences result from variations in the tension on the cords. The larger the larynx and the longer the cords, the deeper the voice. The average man's vocal cords are about 1.9 cm (0.75 in.) long. Shorter vocal cords give women higher-pitched voices. Words and other understandable sounds are formed by the tongue and muscles of the mouth.

Infections of the throat and nasal passages alter the shape of the resonating chambers and change the voice, roughening it so that it sounds hoarse. When the membranes of the larynx themselves are affected (laryngitis), speech may be reduced to a whisper. In whispering, the vocal cords are not involved and use is made of tissue folds, sometimes called false vocal cords, that lie just above the vocal cords themselves.

Trachea

The windpipe, or trachea, is a tube about 11.5 cm (4.5 in.) long and 2.5 cm (1 in.) in diameter, extending from the bottom of the larynx through the neck and into the chest cavity. At its lower end it divides into two tubes, the right and left bronchi. The esophagus, which carries food to the stomach, is immediately behind the trachea.

Rings of cartilage hold the trachea and bronchi open. If the head is tilted back, the tube can be felt as the fingers run down the front of the neck. The ridges produced by the alternation of cartilage and fibrous tissue are also felt, giving the tube a feeling of roughness.

The windpipe wall is lined with mucous membrane, and there are many hairlike cilia fanning upward toward the throat, moving dust particles that have been caught in the sticky membrane, thus preventing them from reaching the lungs.

The path of the esophagus, which carries food to the stomach, runs immediately behind that of the trachea. At the point behind the middle of the breastbone, where the aorta arches away from the heart, the trachea divides into two branches—the right and the left bronchi.

Respiratory infections such as colds and sore throats may sometimes extend down into the trachea; they are then called tracheitis. Inflammation of the walls of these passages causes harsh breathing and deep cough.

Bronchi

The trachea divides into the right and left main stem bronchi under the sternum (breastbone) approximately where the second and third ribs connect to the sternum. Each bronchus enters the lung of its own side through the hilus (meaning an opening through which vessels, nerves, etc., enter or leave an organ).

The right main stem bronchus is wider and shorter than the left. Its direction is almost identical with that of the trachea. This is why most aspirated material enters the right lung.

Each bronchus leads to a separate lung, and in doing so divides and subdivides into continually smaller, finer, and more numerous tubes, something in the fashion of the branches of a tree—the whole structure is sometimes called the bronchial tree. In the larger branches there also is stiffening by rings of cartilage, but as the branches get smaller the cartilage becomes reduced to small plates and finally disappears.

The smaller branches of the bronchial tree, bronchioles, are a possible source of another sort of unpleasantness. The fine subdivisions of the air passages are lined by circular muscles, which through contraction or relaxation can alter the diameter, thus helping to control the flow of air through the lungs. Sometimes, as a result of infection or an allergic reaction to some foreign substance, there is a spasmodic contraction of the small muscles and a swelling of the mucous membrane of the bronchioles. The air passages narrow and airflow is reduced.

Lungs

There are two lungs, one on each side of the thoracic cage (Figure 2-4). The lungs are suspended within the thoracic cage by the trachea, by the arteries and veins running to and from the heart, and by pulmonary ligaments.

The lungs extend from the collarbone to the diaphragm, one on the right side of the body and one on the left. Taken together, they fill almost all of the thoracic cavity. The two lungs are not quite mirror images of each other. The right lung, which is slightly the larger of the two, is partially divided into three lobes, the left lung is divided only into two.

The mediastinum is the compartment between the left and right lung. It contains the heart, great vessels (aorta, vena cava, pulmonary veins, and arteries), nerves, trachea, main stem bronchi, and esophagus.

Pleura. The lungs are covered by a double membrane. One, the pleural membrane, lies over the lungs; the other lines the chest cavity (Figure 2-5). They are separated by a thin layer of fluid that, during breathing, prevents the two membranes from rubbing against each other. Inflammation of the pleura can cause roughness and irritation, the condition called pleurisy.

The potential intrapleural space is between the two pleural layers. This space has a negative atmospheric pressure. An introduction of air between the pleural layers (pneumothorax) would decrease or disrupt this negative pressure and the lung would partially or totally collapse.

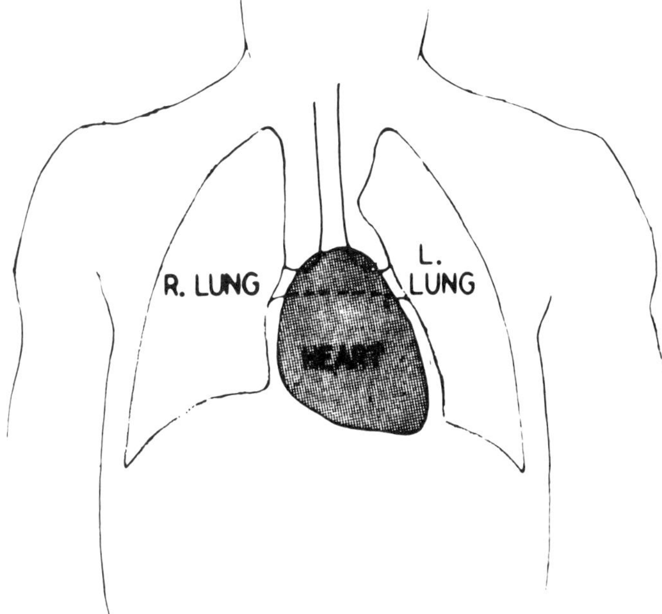

Figure 2-4. The relative size and spatial relationship of the human heart and lungs. (Reprinted with permission from *Emergency Care and Transportation of the Sick and Injured,* American Academy of Orthopaedic Surgeons)

The urge to collapse is counteracted by a pull in the opposite direction. The lung surface clings tenaciously to the chest wall, not by physical bonds, but from the negative pressure of the intrapleural space. Normally, this negative pressure acts somewhat like a suction cup to pull the lung against the chest wall and keep it expanded.

Alveoli. Within a lung, the bronchi divide and subdivide, becoming smaller and smaller, until the branches reach a very fine size, which are called bronchioles. The respiratory bronchioles lead into several ducts, each duct ends in a cluster of air sacs, which resemble a tiny bunch of grapes. The air sacs end in tiny air sacs called alveoli (Figure 2-6).

The walls of the alveoli are two cells thick and oxygen can pass freely across those thin membranes. It can pass freely in both directions, of course, but the blood coming to the lungs has a lower partial pressure of oxygen than inspired air, so the net exchange is from the lungs to the bloodstream.

The human respiratory branches successively from the trachea to some 25–100 million branches. These branches terminate in some 300 million air sacs, or alveoli. The cross section of the trachea is about 2 cm^2, and the combined cross sections of the alveolar ducts, which handle about the same quantity of air, are about 8,000 cm^2 (8 sq ft).

The respiratory surface in the lungs ranges from about 28 m^2 (300 sq ft) at rest to about 93 m^2 (1,000 sq ft) at deepest inspiration. The membrane separating the alveolar air space from circulating blood may be only one or two cells thick. In the course of an 8-hour day of moderate work, a human breathes about 8.5 m^3 (300 cu ft) of air. Contrast the forced

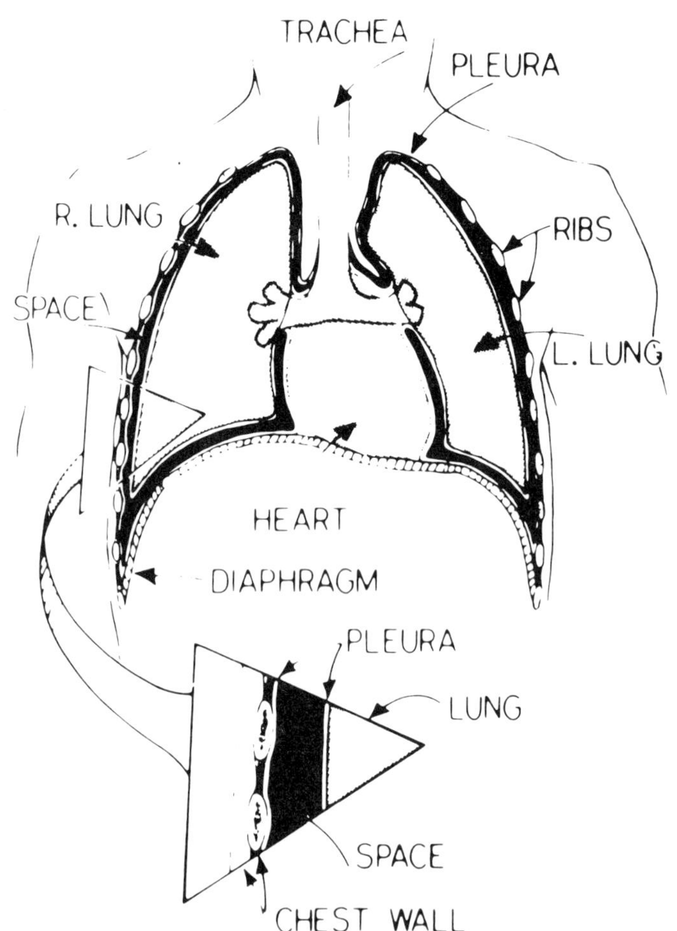

Figure 2-5. The lungs, pleura, pleural space. Inset shows the chest wall relationships. (Reprinted with permission from *Emergency Care and Transportation of the Sick and Injured,* American Academy of Orthopaedic Surgeons)

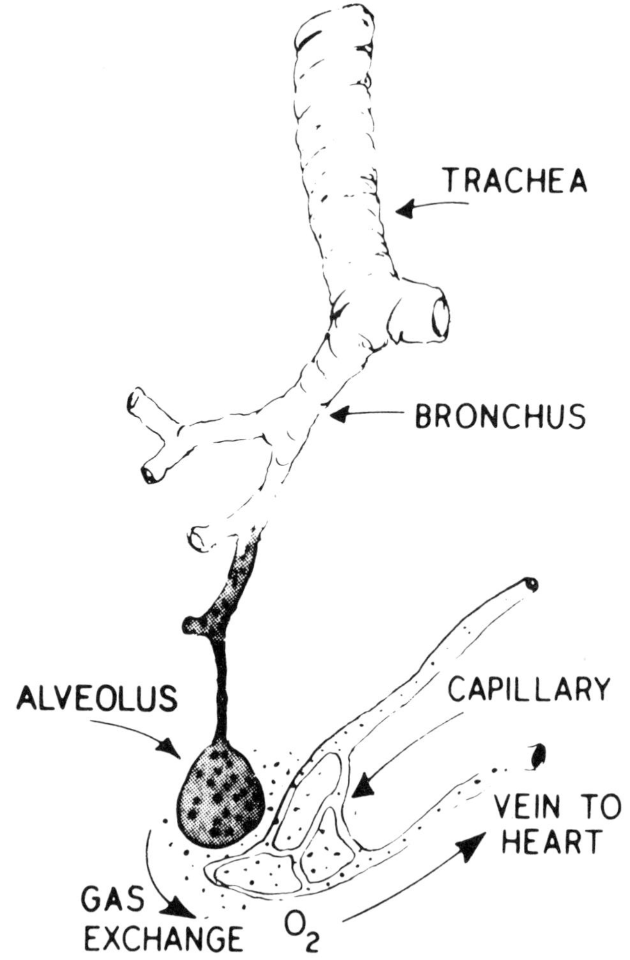

Figure 2-6. The branching characteristic of the trachea into smaller airways ending in an alveolus is shown. (Reprinted with permission from *Emergency Care and Transportation of the Sick and Injured,* American Academy of Orthopaedic Surgeons)

ventilation exposure of the large delicate lung surface with the ambient air exposure of the skin, which has some 1.9 m^2 (20 sq ft) of surface and a thickness measured in millimeters. It is evident that the lungs represent by far the most extensive and intimate contact of the body with the ambient atmosphere.

The respiratory tract, with its successive branches and tortuous passageways, is a highly efficient dust collector. Essentially all particles entering the respiratory system larger than 4 or 5 micrometers (μm) are deposited in it. About half of those of 1-μm size appear to be deposited and the other half exhaled. The sites of deposition in the system are different for various sizes. Discussion of dust deposition in the respiratory system is simplified by the concept of equivalent size of particles. The equivalent size of a particle is the diameter of a unit density sphere, which has the same terminal falling velocity in still air as does the particle.

Particles greater than 2.5 or 3 μm equivalent size are deposited, for the most part, in the upper respiratory system, that is, the nasal cavity, the trachea, the bronchial tubes, and other air passages; whereas particles 2 μm in equivalent size are deposited about equally in the upper respiratory system and in the alveolar or pulmonary air spaces. Particles about 1 μm in size are deposited more efficiently in the alveolar spaces than elsewhere; essentially none are collected in the upper respiratory system. For more details, see Chapter 7, Particulates.

RESPIRATION

The process through which the body combines oxygen with food substances, and thus produces energy, is called metabolism (Figure 2-7). Respiration refers to the tissue enzyme oxidation processes that use oxygen and produce carbon dioxide. More generally this term designates the phases of oxygen supply and carbon dioxide removal. The following outline shows the general subdivision of the overall process:

1. breathing—movement of chest/lung complex to ventilate the alveoli
2. external respiration—exchange of gas (oxygen and carbon dioxide) between lung (alveolar) air and blood
3. internal respiration—exchange of gas between tissue blood and the tissue cells
4. true respiration—ultimate utilization of oxygen by the cells with the coincident release of carbon dioxide.

To a biochemist, respiration refers to the enzymatic processes

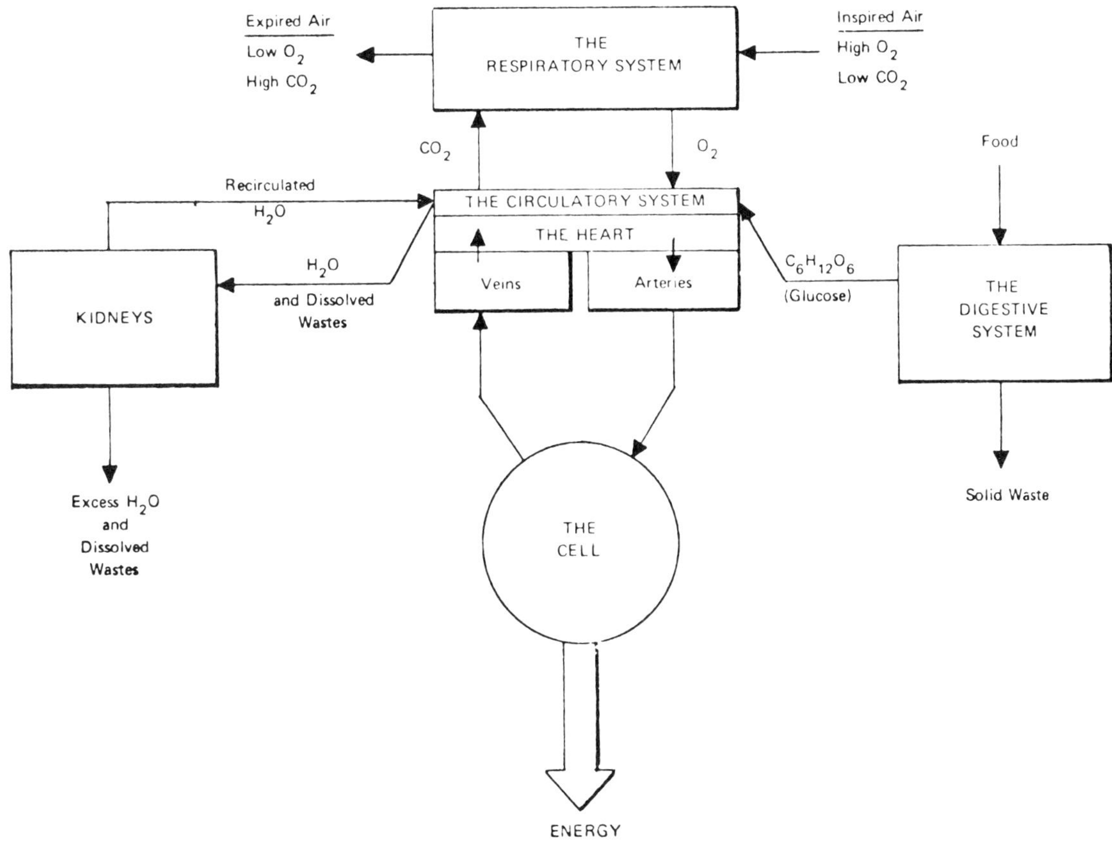

Figure 2-7. The conversion of food into energy (the metabolic process) is illustrated. (Reprinted from *A Guide to Industrial Respiratory Protection,* NIOSH Publication No. 76-189)

in the tissues that use oxygen and produce carbon dioxide. The blood contains a chemical that is part protein and part iron pigment, called hemoglobin. The hemoglobin binds oxygen when the blood flows through regions where oxygen is plentiful—as in the alveoli and releases it to tissues that are consuming oxygen. Similarly, the carbon dioxide produced when the body cells burn their fuel is dissolved in the bloodstream as it flows through the tissues where carbon dioxide is plentiful, and is released in the lungs where carbon dioxide is comparatively scarce.

Carbon dioxide is always present in the atmosphere, but the proportion of carbon dioxide in air exhaled from the lungs is 100 times greater. The proportion of water vapor in air exhaled from the lungs is about 10 times greater than that of the normal atmosphere. Everyone has no doubt noticed the moisture that accumulates on a glass window when the nose and mouth are close to it. Breath appears as a white cloud on cold days when the low temperature of the air causes the exhaled water vapor to condense.

Gas exchange

Gases diffuse rapidly from areas of higher to lower concentrations. In the body the concentration of oxygen is higher in alveolar air than it is in the blood coming to the lungs from the right ventricle; therefore, oxygen diffuses into the blood from the alveolar air. On the other hand, the concentration of oxygen is low in the cells of the body tissues and in tissue fluid; therefore, oxygen diffuses from the blood in the capillaries into the tissue fluid and on into cells.

If there is a pressure difference across a permeable membrane like that separating the alveoli from the pulmonary capillaries, gas molecules pass from the high- to the low-pressure region until the pressures are equalized (Figure 2-8).

The concentration of carbon dioxide in the tissue cells and the tissue fluid is higher than it is in the blood in the capillaries. Therefore, carbon dioxide diffuses from tissue cells and tissue fluid into the blood. The concentration of carbon dioxide is higher in blood coming to the lungs from the right ventricle than it is in alveolar air; therefore, it diffuses from blood in pulmonary capillaries into the alveolar air.

On entering the bloodstream, both oxygen and carbon dioxide immediately go into simple physical solution in the plasma. However, since the plasma can hold only a small amount of gas in solution, most of the oxygen and the carbon dioxide quickly enter into chemical combinations with other blood constituents.

Oxygen tension. Only a small amount of oxygen is carried in solution in the plasma. However, it is this oxygen that exerts tension or pressure and is available for immediate diffusion when blood reaches the systemic capillaries (Figure 2-9). The remaining oxygen in the blood is combined with hemoglobin in the red blood cells to form oxyhemoglobin ($HHbO_2$). This oxygen is given up readily by the hemoglobin whenever the oxygen tension of the plasma decreases, so that as fast as oxygen diffuses from the plasma in tissue capillaries it is replenished by more from the oxyhemoglobin. Hemoglobin that has given up its load of oxygen is called reduced hemoglobin (HHb) (Figure 2-10). See Chapter 15, Industrial Toxicology, for more details.

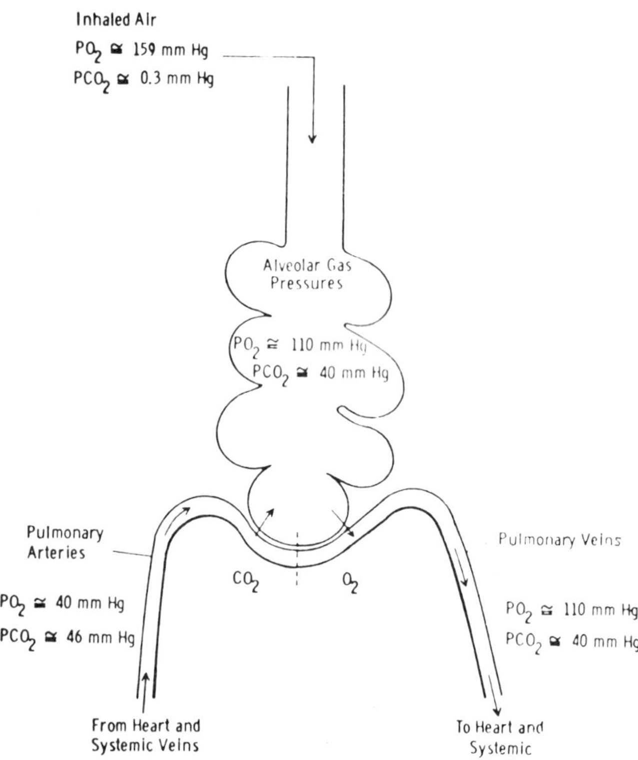

Figure 2-8. Partial pressures of various gases involved in the gas exchange in the lungs are shown. (Reprinted from *A Guide to Industrial Respiratory Protection,* NIOSH Publication No. 76-189)

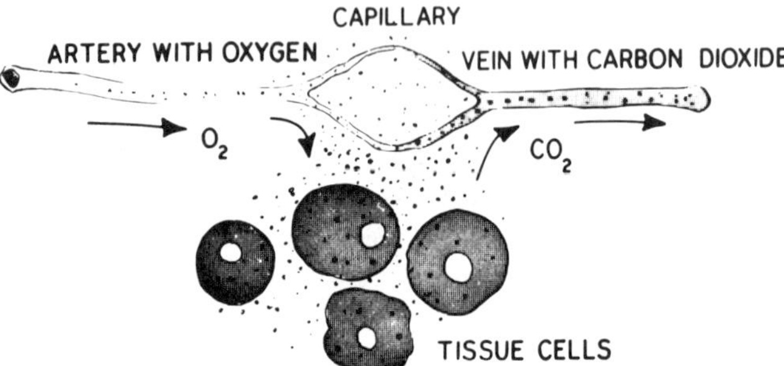

Figure 2-9. Exchange of oxygen and carbon dioxide between blood vessels, capillaries, and tissue cells is shown. Oxygen passes from the blood to the capillaries to the tissue cells. Carbon dioxide passes from the tissue cells to the capillaries and into the blood. (Reprinted with permission from *Emergency Care and Transportation of the Sick and Injured,* American Academy of Orthopaedic Surgeons)

In most people during routine activities, the depth and rate of breathing movements are regulated for the maintenance of carbon dioxide in the arterial blood. Oxygen want can be regulating, but only when the oxygen content of the inspired gases is reduced to nearly half that in air at sea level. Oxygen partial pressure, except in some unusual circumstances, should always be high enough so that the breathing is regulated by the body requirements for carbon dioxide.

The oxygen content of lung air is determined by the oxygen content of the inspired gases, the flushing of the lungs required for carbon dioxide regulation, and the rate of oxygen uptake by the blood as it passes through the lungs.

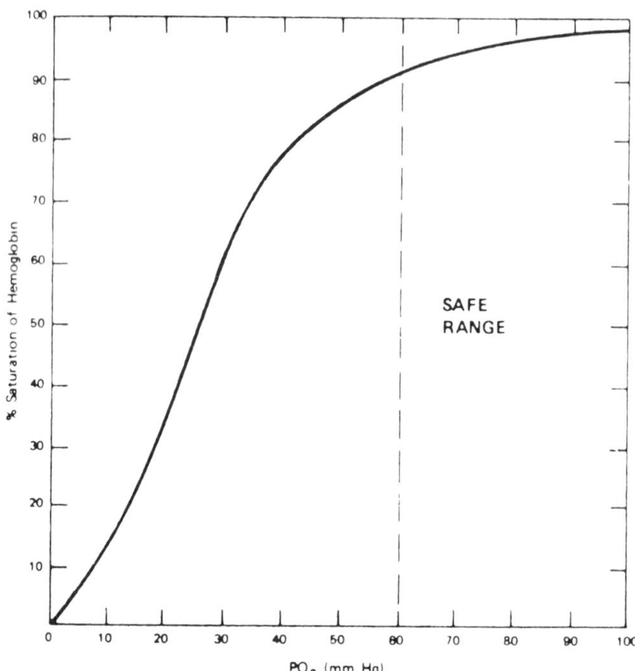

Figure 2-10. Percent saturation of hemoglobin with oxygen at various partial pressures is shown here in the hemoglobin saturation curve. (Reprinted from *A Guide to Industrial Respiratory Protection,* NIOSH Publication No. 76-189)

Mechanics of breathing

Breathing is the act of taking fresh air into and expelling stale air from the lungs. Breathing is accomplished by changes in the size of the chest cavity. Twelve pairs of ribs surround and guard the lungs. They are joined to the spine at the back and curve around the chest to form a cage. In front, the top seven pairs are connected to the breastbone. The next three pairs are connected to the rib above. The last two pairs, unconnected in front, are called floating ribs. The entire cage is flexible and can be expanded readily by special muscles. The rib cage forms the wall of the chest; the dome-shaped diaphragm forms the floor of the chest cavity. The diaphragm is attached to the breastbone in front, the spinal column in back, and the lower ribs on the sides.

Pressure changes

The basic principle underlying the movement of any gas is that it travels from an area of higher pressure to an area of lower pressure, or from a point of greater concentration of molecules to a point of lower concentration. This principle applies not only to the flow of air into and out of the lungs but also to the diffusion of oxygen and carbon dioxide through alveolar and capillary membranes. The respiratory muscles and the elasticity of the lungs make the necessary changes in the pressure gradient possible, so that air first flows into the air passages, and then is expelled.

Atmospheric pressure is that pressure exerted against all parts of the body by the surrounding air. It averages 760 mm of mercury (Hg) at sea level. Any pressure that falls below atmospheric pressure is called a negative pressure and represents a partial vacuum.

Intrapulmonic pressure is the pressure of air within the bronchial tree and the alveoli. During each respiratory cycle this pressure fluctuates below and above atmospheric pressure as air moves into and out of the lungs. Intrapulmonic pressure is below atmospheric pressure during inspiration, equal to atmospheric pressure at the end of inspiration, above atmospheric pressure during expiration, and again equal to atmospheric pressure at the end of expiration.

This series of changes in intrapulmonic pressure is repeated with each respiratory cycle. Whenever the size of the thoracic cavity remains constant for a few seconds, or in a position of rest, the intrapulmonic pressure is equal to atmospheric pressure.

Lungs have no way of filling themselves. Movement of the thoracic cage and the diaphragm permits air to enter the lungs. The thoracic cage is a semirigid bony case enclosed by muscle and skin. The diaphragm is a muscular partition separating the chest and abdominal cavities.

The chest cage can be compared to a bellows. The ribs maintain the shape of the chest bellows. The opening of the chest bellows is through the trachea. Air moves through the trachea to and from the lung to fill and empty the air sacs (Figure 2-11). When a bellows is opened, the volume it can hold is increased, causing a slight vacuum. This lowers the air pressure inside the bellows, and causes the higher air pressure outside the bellows to drive air through the opening, thereby filling the bellows.

When the air pressure inside equals the pressure outside, air stops moving into the bellows. Air will move from a high-pressure area to a low-pressure area until the pressure in both areas is equal. Therefore, as the bellows is closed, the pressure inside becomes higher than outside and air is expelled (Figure 2-11).

During inspiration (inhaling), the diaphragm and rib muscles contract. When the diaphragm contracts, it moves downward and enlarges the thoracic cavity from top to bottom. When the rib muscles contract, they raise the ribs. This enlarges the chest cavity (bellows) in all dimensions. This enlargement of the thoracic cavity reduces the pressure within the chest. The action is identical to that of opening a bellows. Air rushes into the lungs. Take a deep breath to see how the chest increases in size. This is the active muscular part of breathing.

During expiration (exhaling), the diaphragm and the rib muscles relax. As these muscles relax, the chest cavity is decreased in size in all dimensions. As the chest cavity decreases in size, the air in the lungs is pressed into a smaller space, the pressure is increased, and air is pushed out through the trachea. Decrease in size of the chest cavity after relaxation is accomplished largely by action of elastic tissue in the lung, which stretches for inhalation and recoils after muscular relaxation.

Control of breathing

Breathing is controlled by a series of respiratory centers in the nervous system. One center is in the medulla, the part of the brain at the top of the spinal cord (Figure 2-12).

Respiratory center. Nerve impulses originating in the motor areas of the cerebral cortex and traveling to the respiratory center

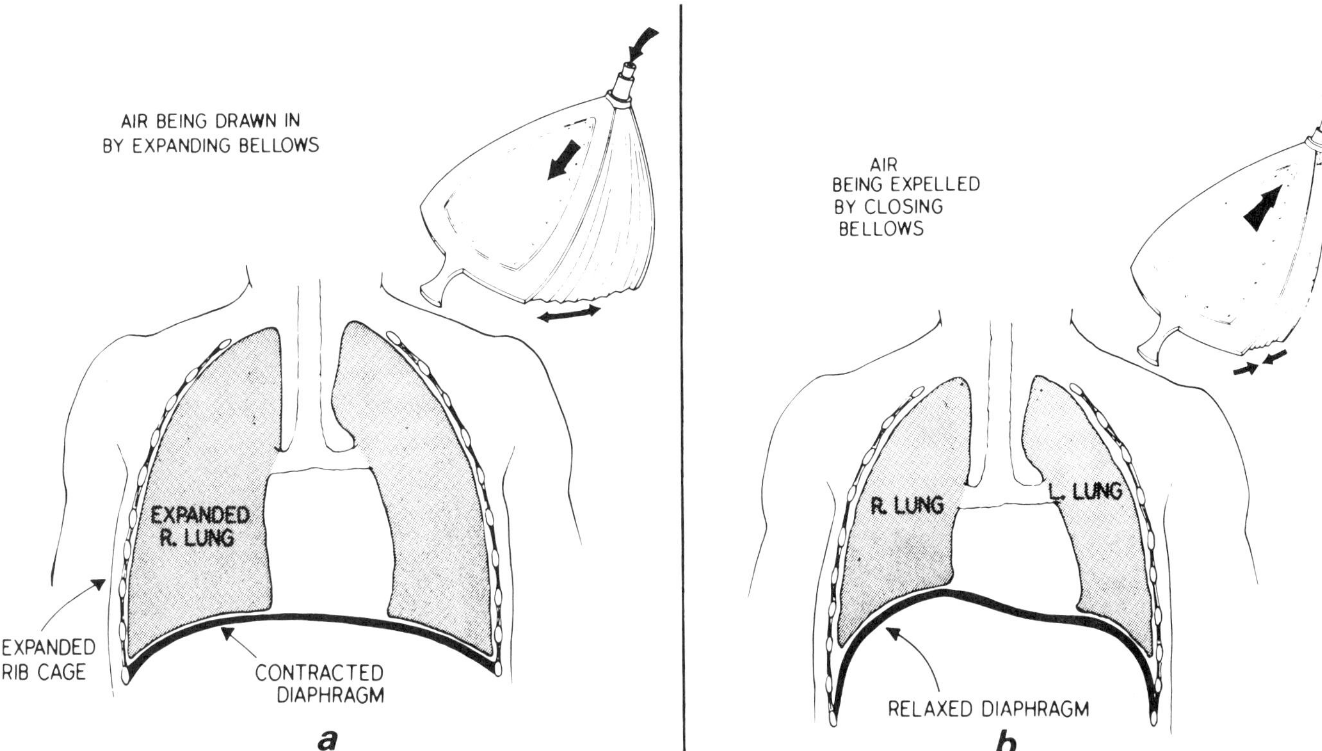

Figure 2-11. Inhalation is similar to the act of air entering a bellows. It occurs when the diaphragm contracts and the ribs expand. Exhalation is similar to the act of air leaving a bellows. It occurs when the diaphragm and ribs relax. (Reprinted with permission from *Emergency Care and Transportation of the Sick and Injured,* American Academy of Orthopaedic Surgeons)

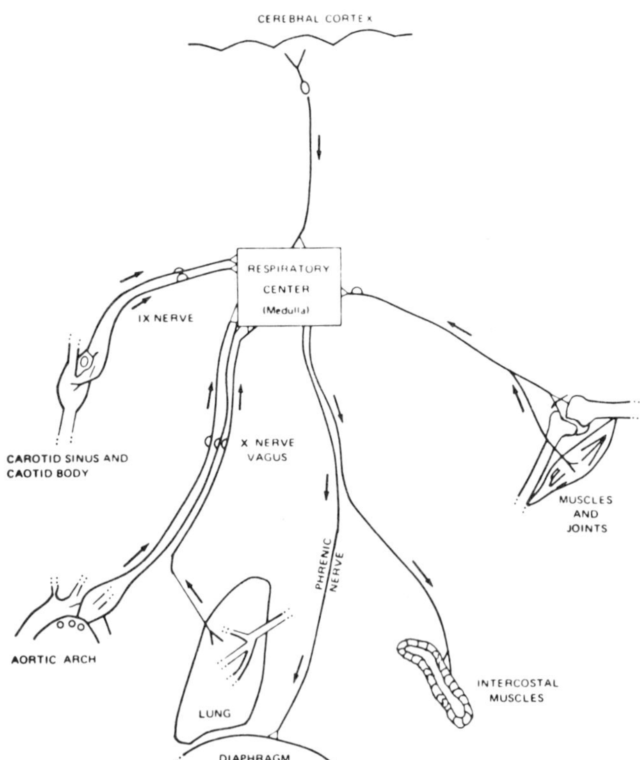

Figure 2-12. Normal rhythmic breathing is controlled by the requirement to ventilate the lungs to remove carbon dioxide as fast as it is produced by metabolic activity. The factors effective in controlling breathing are illustrated schematically. (Reprinted from *Bioastronautics Data Book.* National Aeronautics and Space Administration)

enable us to consciously alter the rate and the depth of breathing. For example, during the act of speaking or singing, breath control is very important.

You can hold your breath voluntarily for a short period of time. However, voluntary control is limited, and the respiratory center will ignore messages from the cortex when it is necessary to meet the body's basic needs.

Carbon dioxide. Breathing action can be triggered by the centers when there is an increase in the amount of carbon dioxide in the blood, or when there is a drop in the oxygen level of the blood.

If you hold your breath, carbon dioxide accumulates in the blood until, finally, it so strongly stimulates the respiratory control center of the brain that you are forced to breathe again. The length of time the breath can be held varies from 25–75 seconds; some people can hold their breath even longer.

Rate. Even in a relaxed state, you breathe in and out 10–14 times a minute, with each breath lasting 4–6 seconds. In the space of a minute, 4.3–5.7 liters (L) (9–12 pt) of air are taken in. The fact is that the body has small reserves of oxygen; all of it is consumed within less than half a minute after the start of vigorous exertion. And with such exertion, the need for air increases many times so that the breathing rate may speed up to one breath per second and a total intake of 120 L (31 gal) of air per minute.

In a normal day, you breathe some 12,491 L (3,300 gal) of air—enough to occupy a space about 8 ft × 8 ft × 8 ft—and, in a lifetime, a prodigious quantity will be consumed, enough to occupy 368,120 m^3 (13 million cu ft) of space.

Lung volumes and capacities

For descriptive convenience the total capacity of the lung at full inspiration is divided into several functional subdivisions. These are illustrated in Figure 2-13.

The four primary lung volumes that do not overlap are as follows:

1. tidal volume (TV)—the volume of gas inspired or expired during each respiratory cycle;
2. inspiratory reserve volume (IRV)—the maximal volume that can be forcibly expired following a normal inspiration (from the end-inspiratory position);
3. expiratory reserve volume (ERV)—the maximum amount of air that can be forcibly expired following a normal expiration;
4. residual volume (RV)—the amount of air remaining in the lungs following a maximum expiratory effort.

Each of the four following capacities includes two or more of the primary volumes:

1. total lung capacity (TLC)—the sum of all four of the primary lung volumes;
2. inspiratory capacity (IC)—the maximum volume by which the lung can be increased by a maximum inspiratory effort from midposition;
3. vital capacity (VC)—the maximum amount of air that can be exhaled from the lungs after a maximum inspiration. It is the sum of the inspiratory reserve volume, tidal volume, and expiratory reserve volume;
4. functional residual capacity (FRC)—the normal volume at the end of passive exhalation; that is, the gas volume that normally remains in the lung and functions as the residual capacity.

In an ordinary inhalation the first air to enter the lungs is the air that was in the bronchi, throat, and nose—air that had left the lungs in the previous expiration but had not been pushed out as far as the outside world. Then, after an inspiration is complete, some of the fresh air that entered through the nostrils remains in the air passages; here it is useless, and is expired

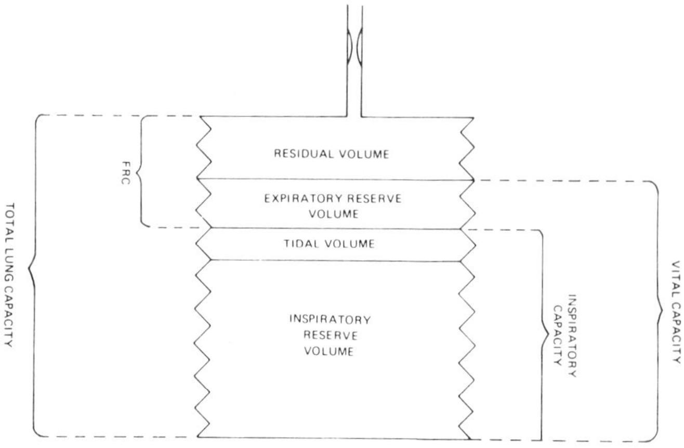

Figure 2-13. Inspiratory capacity and tidal capacity. (Reprinted from *Bioastronautics Data Book.* National Aeronautics and Space Administration)

again before ever it can get to the lungs. The dead space represents the air passages between the nostrils and lungs. Fresh air actually entering the lungs with each breath may amount to no more than 350 m^3 (21.4 $in.^3$). This represents only 1/18 of the lungs' total capacity and is called the tidal volume.

The partial replacement of the air within the lungs (alveolar air) by the shallow breathing we normally engage in is sufficient for ordinary purposes. We are quite capable of also taking a deep drag of air as well, forcing far more into the lungs than would ordinarily enter. After about 500 cm^3 (30.5 $in.^3$) of air have been inhaled in a normal quiet breath, an additional 2,500 cm^3 (153 $in.^3$) can be sucked in. On the other hand, you can force 42.7 $in.^3$ (700 cm^3) of additional air out of the lungs after an ordinary quiet expiration is completed. By forcing all possible air out of the lungs and then drawing in the deepest possible breath, well over 4,000 cm^3 (1.7 $ft.^3$) of new air can be brought into the lungs in one breath. This is the vital capacity.

Even with the utmost straining, the lungs cannot be completely emptied of air. After the last bubble has been forced out, about 1,200 cm^3 (73 $in.^3$) remain. This is the residual volume and is a measure of the necessary inefficiency of the lungs.

Vital lung capacity is measured by inhaling as deeply as possible and blowing as much as possible into a spirometer. The quantity of expelled air varies with body size and age. A medium-sized man may have a vital capacity of 4,000–4,500 cm^3 (1.7–1.9 ft^3) between the ages of 20–40 years. However, as the elasticity of tissues decreases with age, the vital capacity diminishes and may be as much as 20 percent less at age 60 and 40 percent less at age 75.

Spirometry means measurement of air—the ventilatory capacity of the lungs. The spirometer achieves this by measuring volumes of air and relating them to time.

Change in the ability to move air into and out of the lungs in a normal manner results in what is called either obstructive or restrictive ventilatory defect, or a combination of the two. In obstructive bronchopulmonary disease (for example, reduced forced expiratory flow [FEF] and forced expiratory volume in 1 second [FEV_1]) there is reduction of airflow rates and prolongation of expiration.

Forced vital capacity (FVC) is the maximal volume of air which can be exhaled forcefully after a maximal inspiration. For all practical purposes, the VC without forced effort and the FVC are identical in most people.

Regarding the early detection of pneumoconiosis, the FVC is of variable use. In asbestosis, the FVC has been regarded as the most sensitive indicator of early disease and is frequently impaired before there are radiographic abnormalities. Conversely, x-ray changes may be evident in the silicotic while the FVC is still normal.

Forced expiratory volume in one second (FEV_1) is that volume of air which can be forcibly expelled during the first second of expiration.

Forced expiratory flow during the middle half of the FVC ($FEF_{25\text{-}75\%}$) can be defined as the average rate of flow during the middle two quarters of the forced expiratory effort. Compared to the FEV, it is more sensitive in detecting early airways obstruction and tends to reflect changes in airways less than 2 mm in diameter. Airborne substances are thought to exert their initial deleterious effects in these smaller bronchi and bronchioles. Smoking one cigarette can lower the FEF for several hours.

HAZARDS

Now, let's look at some of the unhealthy conditions to which the lung is subject, the associated terminology, and some typical hazardous substances.

The membrane lining of the nasal passages can be affected by a number of causes. The resultant condition is called rhinitis. Inflammation in the larynx is called laryngitis, that of the bronchial tubes, bronchitis. Constriction of the tube muscles, in response to irritation, allergy, or other stimulus is called asthma.

In the lung sacs, a number of conditions can develop:

- Atelectasis means incomplete expansion. It refers to imperfect expansion of the lungs at birth or partial collapse of the lung after birth. The latter is caused by occlusion of a bronchus, perhaps by a plug of heavy mucus, with subsequent absorption of the air, or by external compression, as from a pleural effusion or a tumor. The atelectatic portion of the lung will pass blood through without adding oxygen or removing carbon dioxide.
- Emphysema. The term emphysema derives from Greek words meaning overinflated. The overinflated structures are microscopic air sacs of the lungs (alveoli). Tiny bronchioles through which air flows to and from the air sacs have muscle fibers in their walls. These structures can become hypertrophied and lose elasticity. Then air flows into the air sacs easily but cannot flow out easily because of the narrowed diameter of bronchioles. The patient can breathe in, but cannot breathe out efficiently; this leaves too much stale air in his lungs. As pressure builds up in the air cells, their thin walls are stretched to the point of rupture, so several air spaces communicate and the area of surfaces where gas exchange takes place is decreased.
- Pleurisy is caused when the outer lung lining (the visceral pleura) and the chest cavity's inner lining (the parietal pleura) lose their lubricating properties. The resultant friction causes irritation and pain. The thin, glistening layer of pleura which is inseparably bound to the lung has no pain fibers, but the opposing pleura is richly supplied. Normally the pleural layers glide over each other on a thin film of lubricating fluid. Disease may cause the pleura to become inflamed and adherent, or fluids may accumulate in the interspace, separating the layers.
- Pneumonitis applies to any inflammation of the lung. It is essentially equivalent to the term pneumonia, the latter being reserved for certain types of pneumonitis, usually infectious.
- Bronchitis refers to inflammation of the lining of the bronchial tubes.
- Pneumoconiosis (dusty lung) is a general word for various pulmonary manifestations of dust inhalation, whether the dust is harmful or not. Two common forms of pneumoconiosis are silicosis and asbestosis. The typical pathological condition in harmful pneumoconiosis is the existence of fibrotic (stringy) tissue in the alveolar sacs, or at lymph nodes in the lungs. This fibrotic tissue, caused by some dust particles, reduces the efficiency of the lungs by making them less resilient, and by reducing the effective working surface for gaseous exchange. A simple benign pneumoconiosis with little disturbance of pulmonary function can also predispose to such diseases as pneumonia and tuberculosis.

Air contaminants

The fate of the air contaminant once it reaches the deep portion of the lung (alveoli) depends on its solubility and reactivity. The more soluble, reactive substances may evoke acute inflammatory reactions and pulmonary edema. Most of the particles that reach the deep lung (alveoli) are engulfed by macrophages that migrate proximally to the airways and are either expectorated or swallowed or may enter the interstitial tissues. Once in the deep lung, however, chemical components in particles or in the vapor state are capable of being absorbed into the bloodstream.

Inhaled contaminants that adversely affect the lungs fall into three general categories:

- aerosols and dusts, which, when deposited in the lungs, may produce either tissue damage, tissue reaction, disease, or physical obstruction
- toxic gases that produce adverse reaction in the tissue of the lungs themselves
- toxic aerosols or gases that do not affect the lung tissue, but are passed from the lung into the bloodstream, where they are carried to other body organs or have adverse effects on the oxygen-carrying capacity of the bloodstream itself (Chapter 15, Industrial Toxicology).

An example of the first type is asbestos fiber, which causes fibrotic growth in the alveolar tissue, narrowing the ducts or limiting the effective area of the alveolar lining. Other harmful aerosols are certain fungi found in sugar cane residues, producing bagassosis.

Potential health hazards from dust occur on three levels. (1) The inhalation of sufficient quantities of dust, regardless of its chemical composition, can cause a person to choke or cough. It can accumulate in the lungs and the pores of the skin. (2) Depending upon its chemical composition, the dust can cause allergic or sensitization reaction in the respiratory tract or on the skin. (3) Depending upon both its size and chemical composition, the dust can, by physical irritation or chemical action, damage the vital internal tissues.

Fibrosis can be produced by certain insoluble and relatively inert fibrous and nonfibrous solid particulates found in industry. It is now thought that one of the prerequisites for particulate-induced bronchogenic carcinoma may be the insolubility of the particulate in the fluids and tissues of the respiratory tract, which thereby allows requisite residence time in the lung for tumor induction.

Some of the highly reactive industrial gases and vapors of low solubility can produce an immediate irritation and inflammation of the respiratory tract and pulmonary edema. Prolonged or continued exposure to these gases and vapors can lead to chronic inflammatory or neoplastic changes or to fibrosis of the lung.

Hydrogen fluoride is a gas that directly affects lung tissue. It is a primary irritant of mucous membranes, causing chemical burns. Inhalation of this gas will cause pulmonary edema, and direct interference with the gas transfer function of the alveoli lining.

Very soluble gases, such as sulfur dioxide and ammonia, seldom proceed much farther down the respiratory tract than the bronchi. Less soluble gases, such as nitrogen dioxide, phosgene, and ozone, reach the deeper recesses of the respiratory tract, affecting mainly the bronchioles and the adjacent alveolar spaces, where they may produce pulmonary edema within a few hours.

Carbon monoxide (CO) is a toxic gas that is passed via the lungs into the bloodstream, but does not damage the lungs. The carbon monoxide passes through the alveolar walls into the blood, where it ties up red corpuscles so they cannot accept oxygen, thus causing oxygen starvation.

Many of the metal oxides of submicron particle size (called fume) produce both immediate and long-term effects; the latter can occur in organs and tissues remote from the site of entry. For example, cadmium oxide fume inhaled at concentrations well above the Threshold Limit Value (TLV) may produce immediate pulmonary edema that can be fatal; in addition, inhalation for many years of the fume at concentrations of a few multiples of the TLV can result in eventual renal injury and pulmonary emphysema.

Individual susceptibility to respiratory toxins is difficult to assess. In the occupational setting, workers exposed to the same environment for equal periods of time may develop different degrees of pulmonary disease. This can be due to the variation of the rate of clearance from the lung, the effect of cigarette smoking, coexistent pulmonary disease, and genetic factors.

NATURAL DEFENSES

The respiratory system has a rather complete set of mechanisms for shrugging off insults—the warming and humidifying effects of the nasal and throat passages (as defenses against very cold or overly dry air), the mucous lining, and the mechanical valves in the throat.

Because the mucous lining plays an important role in the cleansing of aerosols from the lungs, it deserves closer inspection. Cells in the windpipe and bronchi produce mucus that is constantly being carried toward the mouth by tiny hairlike projections, called cilia, waving in synchrony. This moving blanket acts as a vehicle to carry foreign substances up and out of the system to the throat, where they can be expectorated or swallowed.

In a healthy lung, aerosols that get into a bronchiole can be carried back out of the system in a matter of hours. Given adequate recovery time (about 16 hours) after an 8-hour exposure to dust, the healthy lung can thus cleanse itself.

Other defense mechanisms include muscular contraction of the bronchial tubes upon irritation—this reaction restricts the airflow and thus minimizes intake of the irritating substance—and the cough and sneeze, which tend to rid the upper respiratory tract of irritants.

Defenses of the alveoli

Thus far we have only discussed the defenses of the airways leading to the alveoli. In general, only very fine particles and gases will reach the alveolar sacs. The larger the particle, the sooner it will be deposited through impaction or gravity on the lining of the airway tubes leading to the sacs.

In the case of gases, the concentration that reaches the alveolar sacs will be nearly the same as the concentration in the air breathed. With aerosols, this is not the case. Large particles, more than 10 μ, will be deposited long before they reach the sacs, through gravity and impaction. Only the smaller particles will reach the alveoli. In the sacs, Brownian movement of the particles results in deposition by diffusion.

Since the very small aerosol particles are the only ones likely to reach the alveoli in great quantities, and since the alveoli are the most important area in the lungs, it is clear that the minute aerosols are potentially more harmful than larger aerosols. What happens to the very small suspended particles and the gases that do reach the alveolar sacs?

First, since the air in the sacs is nearly quiescent (because the sacs are dead ends), the majority of the aerosols will be deposited. The deposition may be through diffusion (these aerosols act almost like gases), or gravitational settling.

Particles may fall prey to the mobile phagocyte cells, which are the type of white blood cells capable of ingesting particles. Once laden with foreign matter, these cells can:

- migrate to the bronchioles, where the mucous lining carries them out of the system
- pass through the alveolar membrane into the lymph vessels associated with the blood capillaries
- be destroyed (if the contaminant is cytotoxic) and break up, releasing the particles into the alveolar sac.

If the aerosol is not removed by these means, it can form a deposit in the sac. Such deposits may or may not acutely affect the health of the lungs.

All of the defense mechanisms are subject to some deterioration and slowing down with age or ill health. Thus, an older worker's lungs will not cleanse themselves as quickly or efficiently as those of a younger person. Also, some contaminants may impede the defense mechanisms themselves, increasing the rate of retention of the contaminant in the lungs.

AMA GUIDES FOR EVALUATING IMPAIRMENT

This section on determining the percent impairment is adapted from the American Medical Association's (AMA) *Guides to the Evaluation of Permanent Impairment* and is included to assist health and safety professionals in interpreting and understanding medical reports of workers' compensation cases.

The AMA publication assists physicians in evaluating permanent impairment of the respiratory system and the effect that such impairment has on a person's ability to perform the activities of daily life. Permanent impairment of the respiratory system is not necessarily a static condition. A changing process can be present, so that it may be desirable to reevaluate the patient's impairment at appropriate intervals.

The measurable degree of dysfunction of the respiratory system does not necessarily parallel either the extent and severity of the anatomic changes of the lungs or the patient's own account of difficulties in carrying out the activities of daily life. Among the reasons for this phenomenon are the large pulmonary reserves normally present; the existence of disease of other systems, particularly the cardiovascular system; the wide variation in certain physiological measurements in normal individuals; and the patient's emotional response to respiratory disease or injury.

Many tests of pulmonary function have value and interest as guides to therapy and prognosis. For most patients, however, most of these are neither practical nor necessary for the assignment to a particular class of impairment. Judicious interpretation of the results of tests of ventilatory function combined with the clinical impression gained from weighing all the information gathered should permit a physician to place the patient in the proper class of impairment.

Rating of impairment

The classification of respiratory impairment is based primarily on (1) the degree of dyspnea (shortness of breath), and (2) the degree of impairment of ventilatory function, the easiest aspect of pulmonary function to measure. There are considerations and symptoms other than dyspnea that can contribute to the impairment in patients with disease of the respiratory system. Such factors can modify the patient's ability to perform the activities of daily life, but these are often difficult to evaluate. They include, for example, malaise, fatigability, wheezing, excessive cough, or other factors associated with chronic lung infection.

Procedures useful in evaluating impairment of the respiratory system include but are not limited to: (1) complete history and physical examination with special reference to cardiopulmonary symptoms and signs; (2) chest roentgenography (posteroanterior, PA) in full inspiration, lateral, and other procedures as indicated; (3) hematocrit and/or hemoglobin determination; (4) electrocardiogram; (5) performance of the following tests of ventilation (terminology defined in Table 2-A): (a) 1-second forced expiratory volume, (b) forced vital capacity; (6) other pulmonary function tests as indicated, such as blood gas studies and diffusion studies.

The AMA guides include an outline for a history and physical examination useful in evaluating impairment of the respiratory system and a discussion of the roentgenographic examination.

Dyspnea

Dyspnea, or shortness of breath, is a subjective finding. It is a consciousness of effort in breathing, which, in the patient's opinion, is at a higher level of exertion than it should be. The term is also used to describe labored breathing. In recording abnormalities of breathing, the physician should indicate whether there is an increased rate of respiration (tachypnea) or an increased depth and rate of respiration (hyperpnea).

The ability to accurately describe the complaint of shortness of breath varies from patient to patient, depending on intelligence, facility with the language, preoccupation with health, and/or the patient's opinion as to what is normal for a person of the same age and particular activities. A sedentary person can have considerable impairment of respiratory function before being aware of exertional dyspnea, although he may have already abandoned hobbies requiring vigorous activity. On the other hand, an athlete might be disturbed by a slight decrease in this breathing reserve. Consequently, the description of dyspnea may not correlate well with the results of tests of pulmonary function.

There are many causes of dyspnea, but the mechanism usually involved is the increased work of breathing, whether due to decreased breathing capacity or increased ventilatory requirements. Less commonly, dyspnea is due to circulatory impairment or decreased oxygen-carrying capacity of the blood, resulting in hypoxemia.

The most common form of decreased breathing capacity is obstructive in nature; that is, there is increased resistance to the flow of air as in chronic bronchitis, chronic obstructive pulmonary emphysema, or chronic bronchial asthma with airway resistance. Decreased breathing capacity is called restrictive if there is decreased compliance (increased stiffness) of the lungs or thorax due to chest-wall weakness or atrophy, thoracoplasty, air or fluid in the pleural space, or pulmonary fibrosis. Patients who have a decreased amount of functioning lung tissue, as after pulmonary resection, can also be placed in this group. Frequently, both obstructive and restrictive ventilatory defects are present in the same patient.

Table 2-A. Terminology of Certain Ventilatory Measurements

Terms Used	*Symbol*	*Description*	*Remarks*
Vital capacity	VC	The largest volume of air measured on complete expiration after the deepest inspiration without forced or rapid effort.	Test *not* recommended for rating purposes, since result may be normal in those with severe respiratory impairment.
Forced vital capacity	FVC	The vital capacity performed with expiration as forceful and rapid as possible.	Formerly called timed vital capacity.
One-second forced vital capacity	FEV_1	Volume of air exhaled during the performance of a forced vital capacity in the first second.	
One-second forced expiratory volume expressed as a percentage of FVC	$\frac{FEV_1}{FVC} \times 100$	The observed FEV_1 expressed as a percentage of the observed FVC.	This value normally should exceed 70 percent. A lower value indicates the presence of some degree of obstructive airway disease.
Maximal voluntary ventilation	MVV	Volume of air which a subject can breathe with voluntary maximal effort for a given time (10-15 sec, if possible, equated to 1 min).	Formerly called maximum breathing capacity—may be useful in evaluating ability to work with respirator.

(Reprinted with permission from the American Medical Association [AMA]). *Guides to the Evaluation of Permanent Impairment,* 1st ed. Chicago: AMA, 1971)

The important aspects of pulmonary function as applicable to dyspnea include (1) ventilation of the lungs with air, including its distribution to the alveoli; (2) diffusion of oxygen and carbon dioxide across the alveolar capillary membrane; and (3) perfusion of the alveolar capillaries by the pulmonary circulation. All of these functions can be impaired in varying degrees in some disorders, such as chronic obstructive emphysema.

Tests of pulmonary function

Ventilation. The tests of ventilatory function have certain limitations:

1. They require maximal voluntary effort by the patient, who for various reasons may be unable or reluctant to perform the tests as well as ventilatory capacity permits. For example, the performance may be affected by the patient's lack of understanding of the test; state of physical training; fear of cough, chest pain, hemoptysis, or worsening of dyspnea; motivation and cooperation; the effects of other illness, particularly heart disease; and the effects of certain temporary factors on the particular day of the test, such as the presence of a respiratory infection or bronchospasm.
2. The results of these tests vary considerably among normal people of the same sex, age, and height.
3. Infrequently, significant impairment of respiratory function can exist even though the patient can perform the tests of ventilatory function normally; that is, the bellows action of the lungs and thorax is normal, but there are abnormalities of pulmonary circulation and/or gas exchange that give rise to the impairment and necessitate other procedures for evaluation.

Various types of apparatus are available that give a permanent record and that readily permit measurement of the FEV_1 and the FVC, and determination of the maximal voluntary ventilation (MVV). These tests can be understood by patients after a short explanation and instruction period, but most patients will need to be encouraged to put forth their best effort. The FEV_1 and FVC should each be administered three times, with the best test result determined as most representative of the patient's ability. The test should not be considered valid unless the best two curves agree within 5 percent. The MVV is a fatiguing test, requiring considerable muscular effort on the part of the patient who must breathe as deeply and as rapidly as possible for 10–15 sec, and for this reason, the better of two attempts (rather than three) should be accepted as more representative of the patient's ability. The degree of permanent impairment of patients known to have severe impairment of ventilation can usually be established without the MVV test.

If the forced expiratory volume test is interpreted as showing airflow obstruction, the test might be repeated 5–10 min after the patient has inhaled a nebulized bronchodilator. If there is at least 15 percent improvement in the performance of the test, the possible reversibility of the airway obstruction, and, incidentally, the presumed efficiency of bronchodilator therapy are established. However, the best results of tests before bronchodilation should be used in determining the degree of impairment.

Results of tests of ventilatory function should be expressed in terms of liters or liters per minute, and also as a percentage of the predicted normal. The FVC as a percentage of the predicted normal is taken as a measure of restriction impairment, but the ratio of actual FEV_1 to actual FVC is considered a better measure of obstructive impairment than is the value of measured FEV_1 as a percentage of predicted FEV_1. Tables for predicting the "normal" value of FVC, FEV_2, and FVC/FEV_1 are given in the AMA *Guides to the Evaluation of Permanent Impairment.*

Diffusion studies, determination of exercise capacity, and arterial blood-gas determinations are useful when a patient's symptoms do not correlate well with spirometric studies. The diffusing capacity of carbon monoxide (single breath D_{CO}) is available in most pulmonary function laboratories. It detects interference with passage of gases across the alveolar membrane such as may occur in interstitial fibrosis. Tables of normal values are given in the AMA guides, but many factors not related to pulmonary impairment can affect this measurement.

Quantitative exercise capacity measurements can be done using a treadmill or stationary bicycle but such testing can be hazardous to individuals in poor health.

Determinations of partial pressures of oxygen and carbon dioxide in arterial blood, particularly before and after exercise, can be useful in certain cases. These measurements require arterial puncture, thus, are not suitable for routine evaluation.

There are other measurements of pulmonary function available in specialized laboratories. These are not regarded as being sufficiently standardized for evaluation of impairment.

SUMMARY

The nose. This external organ is lined by an extensive mucous membrane that warms, moistens, and filters the air passing through. It is the organ of smell and gives resonance to the voice.

The pharynx. Placed at the back of the nose, mouth, and larynx, this cylindrical tube allows passage of food and air.

The larynx. This anterior structure in the neck is the voice box. With its cartilaginous walls it is held open during inspiration and expiration.

The trachea and bronchi. These airways are lined with ciliated mucous membrane and have rings of cartilage to maintain patency. At midsternal level the trachea divides into two bronchi, one going to each lung. The left bronchus is longer and more horizontal than the right to accommodate the heart; consequently inhaled foreign bodies find their way more easily into the right bronchus. These structures are the main sensory area for the initiation of the cough reflex. Their ciliated linings sweep mucus upward to the throat.

The lungs. These two spongy cone-shaped organs occupy the major portion of the thoracic cavity. The space between them is called the mediastinum, and contains the heart, blood vessels, and all tubes passing to and from the abdomen. The lungs are made up of ever-branching bronchioles; at the end of each there is an alveolar duct, from which many alveoli open rather like a balloon and in the shape of a bunch of grapes.

The alveoli. These have a rich blood supply from the pulmonary arteries, the blood and the air being in close contact, thus permitting the interchange of oxygen into the blood and carbon dioxide into the air. The bronchioles, alveoli, and blood vessels are supported on elastic connective tissue, which, with lymphatic vessels and glands, and nerves, form the substance of the lungs.

The vital capacity of the lungs is 3–4 L of air, but only half a liter is changed with each quiet respiration. As well as the gaseous exchange in the lungs, heat and moisture are lost from the body.

There are spontaneous inspiratory nervous impulses arising from a center in the brain stem. This center is influenced by stimuli from many chemical and mechanical receptors. The stimulus for expiration is of nervous origin and arises from stretching of the nerve endings in the alveolar wall; this cuts out the impulses that produced inspiration and by elastic recoil and relaxation of muscle, expiration is produced.

Human lungs are size-selective dust collectors. Only relatively small particles, generally those less than 5 μ in diameter, reach the alveolar spaces. Such small particles move with the air currents; they settle very slowly in still air, and even when thrown into the air with high velocity they travel through the air only a short distance.

The lungs have a very large surface, 28–92 m^2 (300–1,000 sq ft) of very delicate tissue. This surface is exposed to contaminants in the air breathed. The lungs have good defenses against particulates; when unimpaired, these clearance mechanisms remove about 99 percent of the insoluble dust deposited in the lungs.

BIBLIOGRAPHY

American Academy of Orthopaedic Surgeons. Establishing an effective respiratory protection program—The respiratory system. *National Safety News,* June 1971.

American Medical Association (AMA). *Guides to the Evaluation of Permanent Impairment,* 2nd ed. Chicago: AMA, 1984.

Bateman, H.E., and Mason, R.M. *Applied Anatomy and Physiology and the Speech and Hearing Mechanism.* Springfield, IL: Charles C. Thomas Publ., 1984.

Johnstone, R.T., and Miller, S.E. *Occupational Diseases and Industrial Medicine.* Ann Arbor, MI: 1960 Books on Demand, University Microfilms, 1960.

Mackler, P.T., and Mead, J., eds. *Handbook of Physiology: Mechanics of Breathing.* Vol 3, parts 1, 2. Bethesda, MD: American Physiological Society, 1968.

Morgan, E.J. Establishing an effective respiratory protection program. *National Safety News,* December 1971.

National Institute for Occupational Safety and Health (NIOSH). *A Guide to Industrial Respiratory Protection.* Springfield, VA: National Technical Information Service.

Zenz, C., ed. *Occupational Medicine—Principles and Practical Applications,* 2nd ed. Chicago: Year Book Medical Publ., 1988.

3

The Skin

by Julian B. Olishifski, MS, PE, CSP
Revised by James S. Taylor, MD

THE SKIN IS THE LARGEST ORGAN of the body, comprising about 1.8 m^2 (2,880 sq in., or 19 sq ft) of surface area and about 15 percent of total body weight. Skin is a tough flexible cover and is the first body barrier to come into contact with a wide variety of industrial hazards. The skin is subjected to attack from heat, cold, moisture, radiation, bacteria, fungi, and penetrating objects. A health and safety professional should have a basic understanding of the anatomy, physiology, and defense mechanisms of the skin, before recommending proper control measures.

The skin performs a number of important functions. Among these are protecting the body (1) against invasion by microorganisms (fungi, bacteria, etc.), (2) against injury to vital internal organs, (3) against the rays of the sun, and (4) against the loss of moisture. The skin also serves as an organ of sensory perception. One square centimeter of skin has about 3.4 meters of nerves (1 sq in. of skin has about 72 ft of nerves). It also contains hundreds of pain receptors, plus pressure, heat, and cold receptors (Figure 3-1).

Temperature regulation is yet another job performed by the skin. One square inch of skin also contains about 4.6 m (15 ft) of blood vessels, which dilate (widen) when the body needs to lose heat or constrict (narrow) when the body must reduce the amount of heat loss through the skin.

When the surrounding air is comparatively warm, the skin is cooled by evaporation of moisture excreted by the sweat glands. There are between 2 and 5 million sweat glands over the surface of the body excluding mucous membranes. The greatest concentration of sweat glands occurs on the palms of the hands and the soles of the feet. Their function depends on an intact nerve supply. Thermoregulatory sweating is controlled by a heat regulator in the brain. Emotions stimulate sweating primarily on the palms and soles.

The surface of skin may look smooth, but if examined under a magnifying glass, countless ridges and valleys can be seen, in which the many small openings of pores, hair follicles, and sweat glands are found. There are also different patterns of skin texture on the palm of the hand, in contrast to the back of the hand (Figure 3-2). The skin generally is soft and flexible, and, in young people, more elastic.

A number of predisposing factors interact to determine the degree to which a person's skin responds to chemical, physical, and biological insults. These include type of skin (pigmentation, dryness, amount of hair), age, sex, season, previous skin diseases, allergies, and personal hygiene.

A worker's skin is very vulnerable to occupational hazards. Surveys indicate that dermatological conditions other than injuries accounted for almost 40 percent of all occupational diseases reported to the Bureau of Labor Statistics in 1983. Occupational skin disease is underreported and results in considerable lost time from work.

Although most occupational skin disorders are treated by primary care and occupational physicians, dermatologists are often consulted. Dermatology is the branch of medicine concerned with the diagnosis, treatment, and prevention of diseases of the skin. Some dermatologists have had special training in occupational skin disorders. The most common occupational afflictions of the skin are discussed in Chapter 8, Industrial Dermatoses.

Some disorders that are visible in the skin do not arise primarily in the skin but in other organs. Thus, the skin is an early warning system, and its examination is very important in physical diagnosis, occasionally furnishing the first clue to identification of serious systemic diseases.

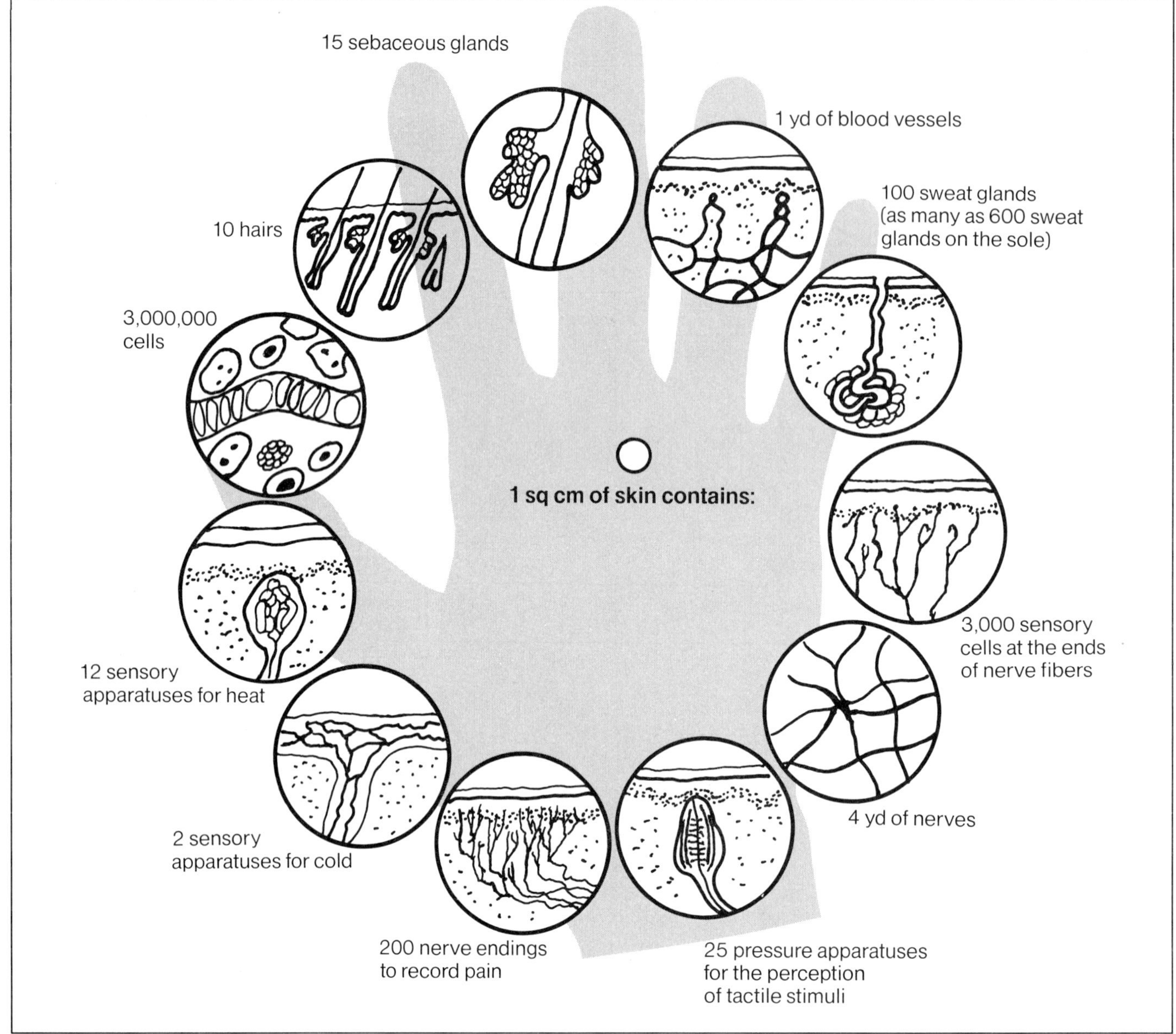

Figure 3-1. Structures in the skin. (Adapted from the American Medical Association [AMA]). *Today's Health Guide.* Chicago: AMA, 1965)

ANATOMY

Three distinct layers of tissue make up the skin: from the surface downward, they are the epidermis, the dermis, and the subcutaneous layer.

The thickness of the skin varies from 0.5 mm on the eyelid (the dermis is thinnest here) to 3 or 4 mm on the palms of the hands and soles of the feet (the epidermis is thickest here).

In some parts of the body, there are as many as 60 layers of cells, for example, the palms of the hands and the soles of the feet; in some areas it is thin, notably in the skin folds—the axillae (armpits), under the breasts, the groin, and between the fingers and toes.

Epidermis

The top layer of the epidermis is composed of dead cells variously called the horny layer, stratum corneum, or keratin layer. This layer stands up fairly well against chemical attack—with the notable exception of alkali. It serves as the chief rate-limiting barrier against absorption of water and aqueous solutions, but offers little protection against lipid-soluble materials (such as organic solvents) or gases.

The stratum corneum, or horny layer gradually flakes off, or soaks off when wet. The horny layer is constantly being replaced by cells pushed toward the surface as new cells are formed in the deeper germinative layer of the epidermis. This regenerative and sloughing characteristic serves to some extent as a protection against chemicals and microorganisms.

Brisk rubbing with a towel peels off little rolls of material composed of dead outer skin cells that never will be missed. This incessant shedding of flaky material goes mostly unnoticed, unless, for instance, a person has dandruff or must peel off insensate skin patches after a sunburn.

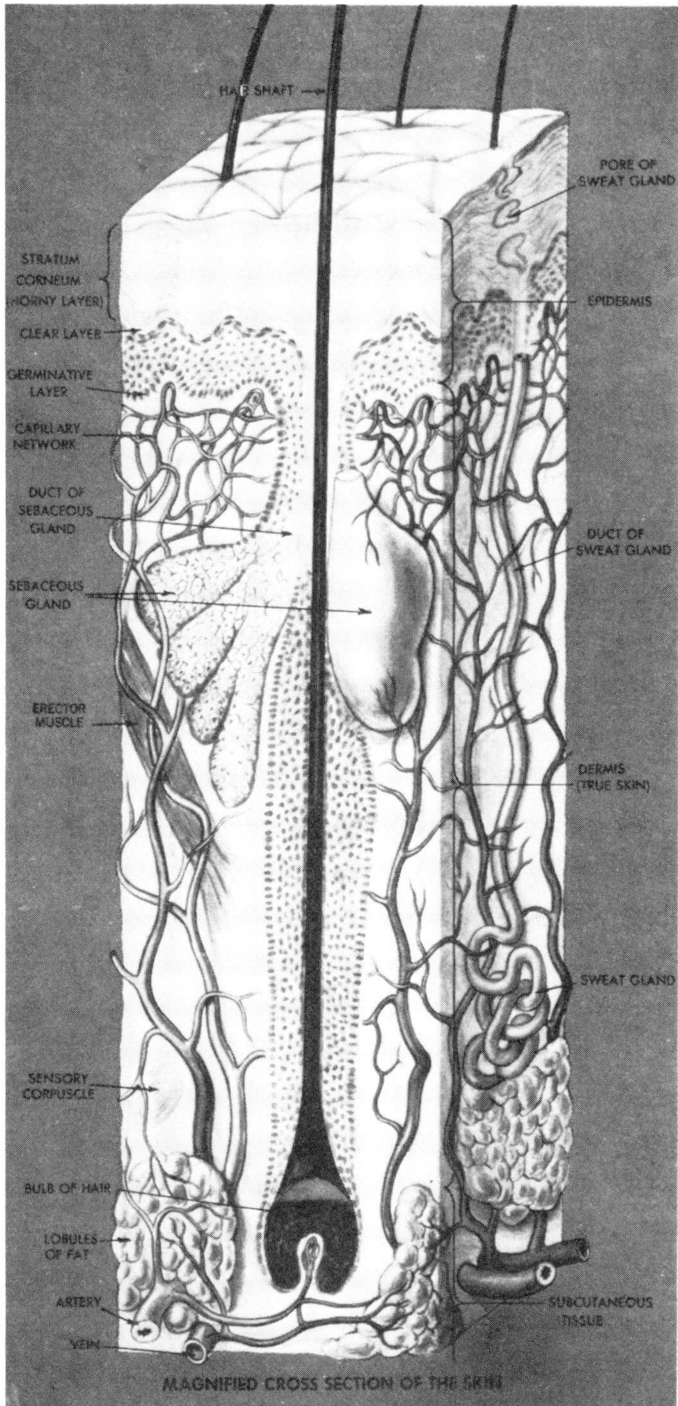

Figure 3-2. Magnified cross section of the skin. (Reprinted with permission from the AMA. *Today's Health Guide.* Chicago: AMA, 1965.)

There are four cell types in the epidermis:

1. Keratinocytes, which make up the bulk of the epidermis, form from below and move up to become dead horny cells.
2. Melanocytes or pigment-forming cells synthesize melanin (pigment) granules, which are then transferred to keratinocytes. It is the amount of melanin in keratinocytes that determines the degree of pigmentation of skin and hair. The absolute number of melanocytes in human skin is the same for all races. Differences in coloration among races result in differences in the number, size, degree of pigment formation, distribution, and rate of degradation of pigment granules within keratinocytes. Melanin proliferates under stimulus of certain wavelengths of sunlight and becomes visible as suntan or freckles. Moles are growths that also contain melanin. Some people with little or no pigment in their skin have albinism, an inherited abnormality in which melanin (pigment) production is decreased. Vitiligo is a more common disorder, in which loss of melanocytes also results in areas of cutaneous pigment loss. Some chemicals, for example, phenolic germicides, can destroy pigment after occupational or environmental exposure.
3. Langerhans' cells, located in the mid-epidermis, account for 4 percent of all epidermal cells and play an important role in various immune processes, especially allergic contact dermatitis.
4. Merkel cells function as slowly adapting receptors of the touch sensation.

Although the epidermis is an active tissue it is not richly supplied with blood, because blood vessels are absent. However, it is bathed in lymph, a fluid derived from the blood.

The epidermis is thin enough so that the nerve endings in the dermis are sufficiently close to the surface to supply the fine sense of touch. Some of this sensation is lost where areas of the skin are chronically subjected to friction, resulting in subsequent thickening of the epidermis to provide protection in the form of a callus. Thus, the soles of the feet are commonly callused among those who habitually walk barefoot, as are the palms of the hands of those who do heavy work.

Dermis

Beneath the epidermis is the dermis, also called the corium, which is 15–40 times thicker than the epidermis, depending on location. It contains connective tissue composed of collagen elastic fibers and ground substance, is strong and elastic, and is the part of animal skins that makes leather when tanned. It is laced with blood vessels, nerve fibers, receptor organs for sensations of touch, pain, heat, and cold, contains muscular elements, hair follicles, and oil and sweat glands (Figure 3-2).

The dermis is tough and resilient, and is the main natural protection against trauma. When injured, it can form new tissue, a scar, to repair itself.

A layer of tiny cone-shaped objects called papillae is at the top of the dermis. There are perhaps 150 million papillae scattered over the body. They are more numerous in areas such as the fingertips, where the skin appears to be more sensitive. Nerve fibers and special nerve endings are found in many of the papillae. As a result, the sense of touch is best developed in areas where papillae with nerve endings occur in the greatest numbers.

The papillary layer fits snugly against the outer layer of skin, the epidermis, which has ridges corresponding to those of the papillae. The ridges prevent the various skin layers from slipping.

The ridges on the surfaces of the fingertips form whirls, loops, and arches, which we call fingerprints; dermatoglyphics is the study of the patterns of the ridges of the skin and of the fingers, palms, toes, and soles. Similar ridges appear on the soles of the feet. Because it is unlikely that two persons will have exactly the same pattern of ridges, fingerprints are used by the police to identify persons, and footprints are used by hospitals to identify babies.

The larger component of the dermis (reticular dermis) extends from the base of the papillary dermis to the subcutaneous fat. Muscle fibers are commonly seen in the reticular dermis on the face and neck.

Subcutaneous layer

Beneath the dermis is a layer of subcutaneous tissue with fatty and resilient elements that helps to cushion and insulate the skin above it. The distinguishing feature of the subcutaneous layer is the presence of fat. Also present are the lower parts of some eccrine and apocrine sweat glands and hairs as well as hairs, nerves, blood and lymphatic vessels and cells, and fibrous partitions composed of collagen, elastic tissue, and reticulum. It links the dermis with tissue covering the muscles and bones.

Absorption of subcutaneous fat and softer parts of the skin removes bouncy supporting material, and since the external skin does not have the capability to shrink at the same rate, it tends to collapse and become enfolded in wrinkles.

Glands in the skin

Two main types of glands are located in the dermis. One, already mentioned, is the sweat gland. Under the microscope it appears as a tightly coiled tube deep in the dermis with a corkscrewlike tubule that rises through the epidermis to the surface of the skin.

The second type is the sebaceous or oil gland, which usually occurs in or near a hair follicle. Sebaceous glands are located in all parts of the skin except on the palms and soles. They are particularly numerous on the face and scalp.

Sweat glands

Sweat glands excrete a fluid known as sweat, or perspiration. The working or secreting parts of the glands are intricately coiled little tubules in the dermis. There are two kinds of sweat glands, producing different kinds of sweat.

Apocrine. One kind, called apocrine sweat, is not very important physiologically but has some social significance. Apocrine sweat glands open into hair follicles and are limited to a few regions of the body, particularly the underarm and genital areas. Apocrine sweat is sterile when excreted but decomposes when contaminated by bacteria from the skin surface resulting in a strong and characteristic odor. The purpose of the many cosmetic underarm preparations is to remove these bacteria or block gland excretion.

Eccrine. The other kind of sweat, called eccrine, is of towering importance to our comfort, if not our lives. Multitudes of eccrine sweat glands are present almost everywhere in the skin, except the lips and certain other areas. They are crowded in largest numbers into skin of the palms, soles, and forehead.

Eccrine sweat is little more than extremely dilute salt water. Its function is to help the body to dissipate excessive internal heat by evaporation from the surface of the skin.

Sebaceous glands

There are many oil-secreting or sebaceous glands in the skin. They are distributed over almost the entire body, and are largest in regions of the forehead, face, neck, and chest—typically, the areas involved in common acne, a condition with which cell-clogged oil glands are associated. The primary function of this oil substance, or sebum, is lubrication—in this case, of the hair-shaft and horny surface layers of the skin. A certain amount of natural skin oil is necessary to keep skin and hair soft and pliable.

A strap of internal, plain, involuntary muscle tissue, the arrectore pilorum or raiser of hair is inserted into the lower portion of the hair follicle below the sebaceous glands and originating in the connective tissue of the upper dermis. When we have goose flesh these muscles are active in an attempt to produce heat for us. The raised hairs trap a layer of air, which, being a bad conductor of heat, limits heat loss from the skin.

Blood vessels

The skin is richly supplied with small blood vessels. The blood supply in the skin accounts for the reddening of sunburn and the coloration of the fingers beneath the nails. Engorgement of the blood vessels accounts for the reddening of the skin when we blush.

Vascular birthmarks, for example, hemangiomas, strawberry marks, and port wine stains, get their coloration from unusually large numbers of tiny blood vessels concentrated in a small area of the skin.

Hair

Hair and nails are a modified form of skin cells containing keratin as the major structural material. Keratin is produced by the same processes that change living cells into dead, horny cells of the outermost skin layer. However, hair and nails are almost total keratin.

With the exception of the palms and soles, hair follicles populate the entire cutaneous surface, though in many areas they are so inconspicuous or vestigial that they are never noticed.

Several kinds of hair are part of the skin. They range in texture from the soft, almost invisible hair on the forehead to the long hairs of the scalp and the short, stiff hairs on the eyelids, the eyelashes.

The hair follicles develop as downgrowths of the layers of the skin. The hair then grows outward from the bottom of the follicle. Thus, the body hair actually is a special form of the skin itself.

Each hair has a root, which is anchored at the bottom of a microscopic shaft called the follicle. It also has a shaft, which extends past the top of the follicle. The follicle enters the epidermis and passes deep into the dermal layer at an angle. The follicles of long hairs can extend into the subcutaneous layer. Oil, or sebaceous, glands empty into the follicle. At the root of the hair is a cone-shaped papilla that is similar to the peglike papillae that underlie the ridges of the fingers, palms, and soles.

The shaft of the hair is covered with tiny, overlapping scales. An inner layer of cells in the hair shaft contains pigment that gives the hair its color. Curly hair appears flattened when seen in cross section under a microscope. Straight hairs appear round or oval in cross section.

Most hair tips project from the skin at a slant. Minute muscles attached to follicle structures have the fascinating potentiality of making the hair stand on end. This phenomenon is best observed in an angry cat. (The closest that most humans come to it is the experience popularly known as goose pimples, already discussed.)

Both the hair follicles and the sweat glands serve as routes for substances to enter the body through the skin. Physicians sometimes use this absorptive ability of the skin in administering certain drugs—some chemicals placed on the skin can be detected in the saliva a few minutes later. In the workplace the skin is a potential portal of entry for a number of hazardous chemicals.

Nails

The fingernails and toenails, like hair, are a specialized form of the skin. The fully developed nail then overlays a modified part of the dermis that is the bed of the nail. The base portion of the nail is covered by epidermis.

Nails are essentially the same in structure as hairs, except that nails are flat, hard plates. The living part of a nail lies in the matrix back of the half-moon or lunula. If the dead nail plate, which constitutes most of the visible part of the nail, is destroyed by injury, a new nail will grow if the matrix is intact.

The rate of growth of nails varies and depends upon such factors as the age of the person and the season of the year. Nails grow faster in young people and during the summer months.

PHYSIOLOGY AND FUNCTIONS

The skin performs a number of important functions. Among these are protecting the body against invasion of bacteria, against injury to vital internal organs, against the rays of the sun, and against the loss of moisture.

Temperature regulation

For a discussion of the role of the skin in heat regulation of the body, see Chapter 12, Temperature Extremes.

Sweat

Sweat is constantly being produced, usually in proportion to the temperature of the environment. This results in the need to lose heat by evaporation, which is more effective than simple radiation. In cool, dry weather the amount of sweat produced is relatively small and the rate of evaporation can keep up with it. The skin remains dry to the touch, and a person is not aware of sweating. This is insensible perspiration.

When heat production of the body is increased, or when the ambient temperature is unusually high, the sweat glands produce more perspiration. The rate of production can then outstrip the rate of evaporation, particularly if humidity is high, since the rate of evaporation declines with the rise in humidity. Perspiration then collects on the body in visible drops and we are conscious of sweating. However, heat is lost only when the sweat evaporates. By far the most efficient means of body-cooling is evaporation of water—sweat. This process goes on even though the skin can seem to be dry.

One way of increasing the rate at which water is evaporated from the body is to breathe rapidly and carry quantities of air from the moist surfaces of the mouth, throat, and lungs. Humans cannot do this in comfort, but it is the chief method of cooling available to dogs; in warm weather, dogs will sit with mouth open, tongue extended, and pant.

Ultraviolet light

Skin protects not only against mechanical shocks, but also against various forms of ultraviolet (UV) light. (See Chapter 11, Nonionizing Radiation, for a discussion of the forms.) Most animals are protected from sunlight by scales, hair, and feathers, which effectively absorb the sun's rays without harm to themselves.

Humans only have the relatively thin layer of epidermis as protection from the sun's UV rays. Ultraviolet light energy produces chemical changes within the skin's cells; the effects vary with the time of the year, the geographic area, and the hour of the day. Generally, after initial exposure to summer sun at midday, skin shows reddening or erythema, which may not appear for several hours. If the dose of sunlight is intense, the erythema may be followed by blistering and peeling of the outer layer of epidermal cells (Figure 3-3).

If the erythema is not severe, it fades in a few days and the skin gradually acquires a tan coloration (suntan). The tan color is produced by darkening of existing pigment (immediate pigment darkening) and by increase in pigment-formation. When skin is exposed to the sun, it is believed that melanin pigment moves toward the surface of the skin and is replaced by new melanin in the lower cell layer. Along with pigmentation increase, the stratum corneum thickens to furnish additional protection against solar radiation injury. One or two weeks may be required to develop a suntan by moderate daily doses of sunlight; and the tan begins to fade if occasional exposure to sunlight is not continued.

Solar UV radiation can induce actinic (solar) degeneration and skin cancer. Photoaging and natural chronologic aging are different entities. Sun damage is not an acceleration of age-dependent changes. For example, actinic degeneration does not occur in healthy protected skin, even of very old persons.

As protection against UV light, human skin is equipped with the capacity to form pigment (melanin), which absorbs UV light without harm to itself, and thus acts as a protective umbrella over the regions beneath.

There is now preliminary evidence that the immune system of humans is affected by UV radiation and that environmental sources of radiation can have similar effects, for example, contact photoallergy.

One beneficial normal effect of UV radiation on skin is the photochemistry that leads to the production of vitamin D_3. In most industrial countries sufficient vitamin D is added to food to meet normal daily requirements.

Skin absorption

A waxy type of mixture composed of sebum, breakdown products of keratin, and sweat, called the surface lipid film, coats the outer surface of the keratin layer. There is no evidence that this normal coating has any barrier function.

The epidermis and more particularly the stratum corneum or outer horny layer serve as the major permeability barrier impeding the entry of foreign chemicals into the body. Overall, the skin is selectively permeable—more impermeable than permeable—and shows regional variation in absorptive capacity. Absorption of materials through the skin markedly increases when the continuity of the skin is disrupted by an abrasion, laceration, or a puncture. The hair follicles and sweat ducts are now believed to play only a minor role in skin absorption. They act as diffusion shunts, that is, relatively easy pathways through the skin for certain substances such as ions, polyfunctional polar compounds, and very large molecules that move across the stratum corneum very slowly. After this initial phase, however, most of the percutaneous absorption of all substances takes place

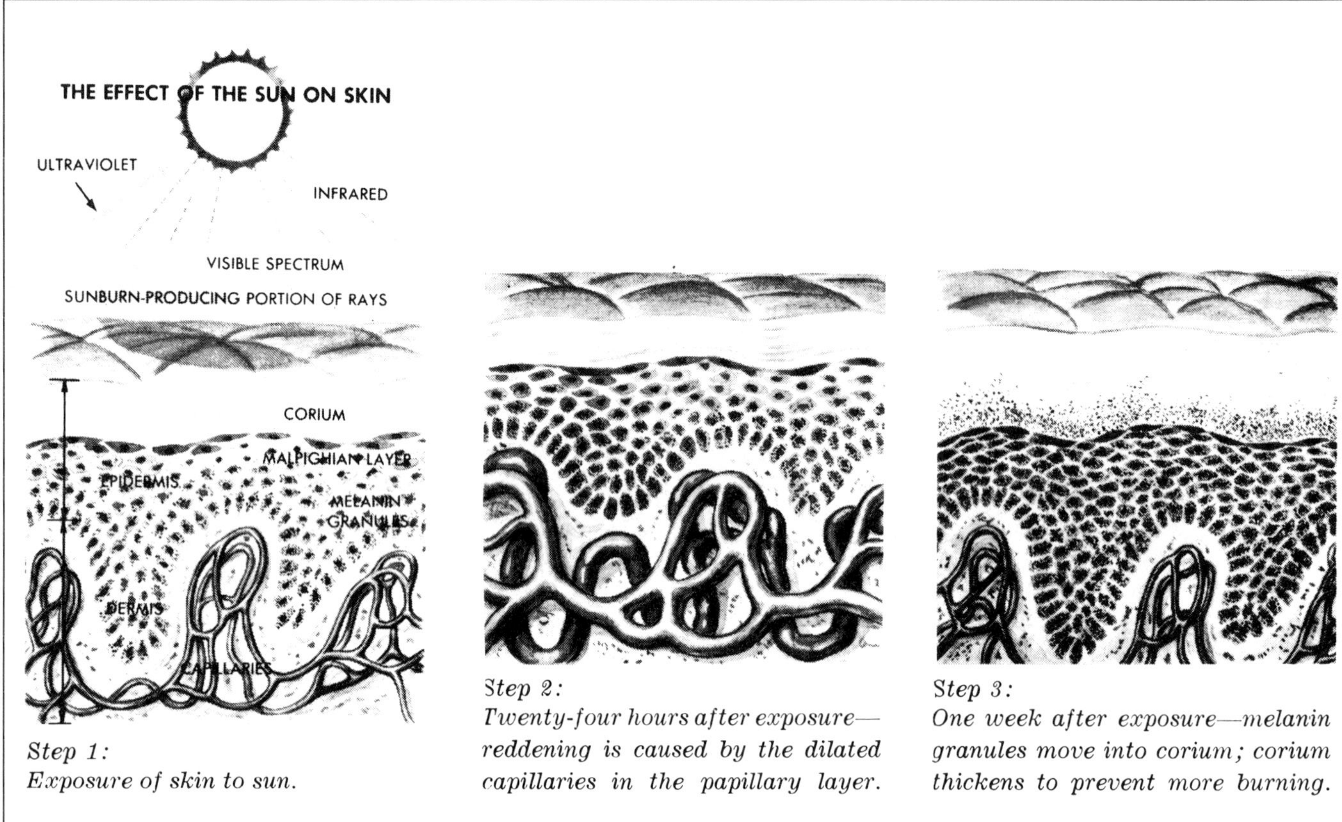

Figure 3-3. The effect of the sun on skin. (Reprinted with permission from the AMA. *Today's Health Guide.* Chicago: AMA, 1965)

across the stratum corneum, which has a much greater surface area than that of the hair follicles and sweat ducts. Absorption of fat-soluble chemicals and oils can also occur via the hair follicle.

DIRECT CAUSES OF OCCUPATIONAL SKIN DISEASE

There are unlimited substances and conditions capable of inducing a skin disorder in the workplace. Each year new causes are reported and most can be classified under one of the following five broad headings:

- chemical
- mechanical
- physical
- biological
- botanical.

Chemical

Organic and inorganic chemicals are the predominant causes of dermatoses in the work environment (Table 3-A). They constitute a never-ending list, because each year the chemical spectrum gains additional agents capable of injuring the skin. Chemical agents may be divided into two groups—primary irritants and sensitizers.

Primary irritants. These are likely to affect most people; some actually affect everyone. These agents react on contact. The reaction alters the chemistry of the skin by dissolving a portion of it, by precipitating the protein of the cells, or by some other chemical reaction. The result can range from tissue destruction (chemical burn) to inflammation (dermatitis), depending on the strength of the agent and the duration of the exposure.

Primary irritants damage skin because they have an innate chemical capacity to do so. Many primary irritants are water-soluble and, thereby, actively able to react with certain tissues within the skin. Even the water-insoluble compounds, which comprise many of the solvents, react with the lipid (fatty) elements within skin. The precise mechanism of primary irritation on the skin is not known, but some useful generalizations serve as indices to explain the activity of groups of materials in the irritant category. About 80 percent of all occupational dermatoses are caused by primary irritants. Dermatitis caused by a primary irritant is referred to as irritant contact dermatitis, because the skin irritation is normally confined to the area of direct contact.

Most inorganic and organic acids act as primary irritants. Certain inorganic alkalis, such as ammonium hydroxide, calcium chloride, sodium carbonate, and sodium hydroxide, are skin irritants. Organic alkalis, particularly amines, also are active irritants. Metallic salts, especially arsenicals, chromates, mercurials, nickel sulphate, and zinc chloride, severely irritate the skin. Organic solvents include many substances, such as chlorinated hydrocarbons, petroleum-based compounds, ketones, alcohols, and terpenes, among others, which irritate the skin because of their solvent qualities (Figure 3-4).

Keratin solvents. All of the alkalis, organic and inorganic,

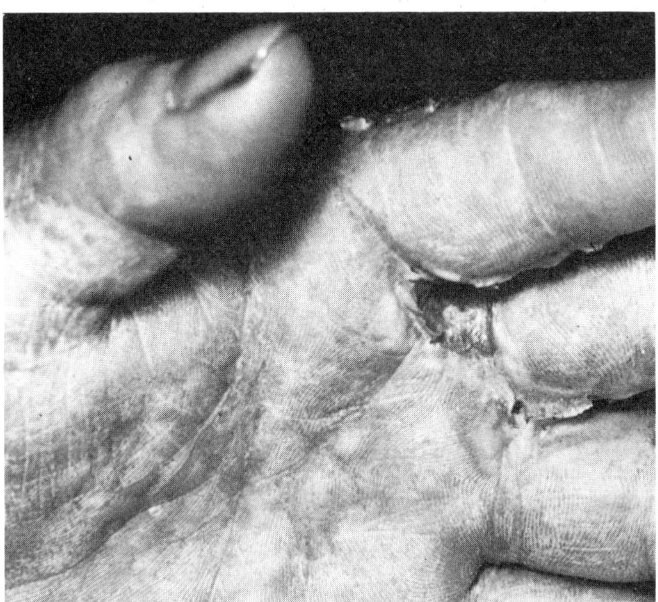

Figure 3-4. Eczematous dermatitis is a form of contact dermatitis caused by working with organic solvents. It is one of the most prevalent types of dermatitis.

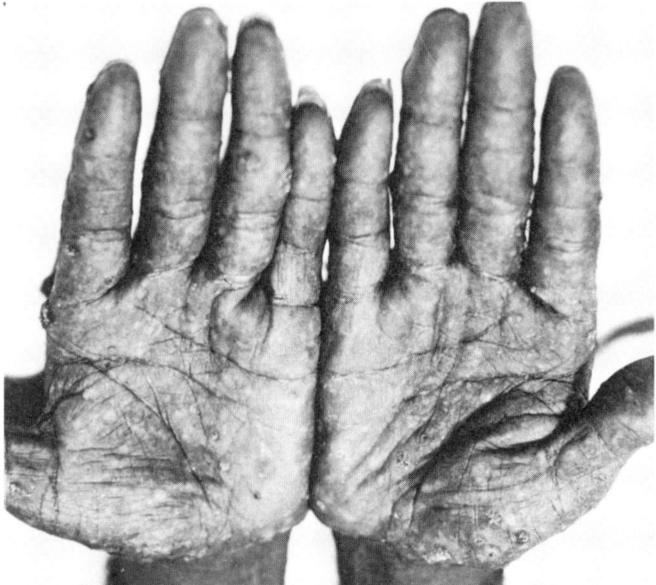

Figure 3-5. Nodules in the keratin layer of the skin, which may result from repeated exposure to certain tars or coal tar derivatives.

injure the keratin layer when concentration and exposure time are adequate. These agents soften the keratin cells and succeed in removing many of them. At the same time, they bring about considerable water loss from this layer with resulting dry, cracked skin. This prepares the way for secondary infection and also, at times, for the development of allergic sensitization.

Keratin stimulants. Several chemicals stimulate the skin so that it undertakes growth patterns that can lead to tumor or cancer formation. Certain petroleum products, a number of the coal tar-based materials, arsenic, and some of the chlorinated hydrocarbons can stimulate the epidermal cells to produce these effects (Figure 3-5).

Fats and oil solvents. Just as organic solvents dissolve oily and greasy industrial soils, they remove the skin's surface lipids and disturb the keratin layer of cells so that they can no longer maintain their water-holding capacity. Workers exposed each day to the action of the organic solvents develop exceedingly dry and cracked skin.

Protein precipitants. Several of the heavy metal salts precipitate protein and denature it. The salts of arsenic, chromium, mercury, and zinc are best known for this action.

Reducers. In sufficient concentrations, salicylic acid, oxalic acid, urea, as well as other substances, can actually reduce the keratin layer so that it is no longer protective, and an occupational dermatosis results.

Sensitizers. Some primary skin irritants also sensitize. Certain irritants sensitize a person so that a dermatitis develops from a very low, nonirritating concentration of a compound that previously could have been handled without any problem.

Some chemical and many plant substances as well as biological agents are classified as sensitizers. Initial skin contact with them may not produce irritation, but after repeated or extended exposure some people develop an allergic reaction termed allergic contact dermatitis. Clinically, allergic contact dermatitis is often indistinguishable from irritant contact dermatitis (see Chapter 8, Industrial Dermatoses, for further discussion of allergic contact dermatitis and patch testing).

Examples of substances that are both irritants and allergens include turpentine, formaldehyde, chromic acid, and epoxy resin components. Common sensitizers are plant oleoresins, for example, poison ivy, epoxy resins, azo dyes, certain spices, certain metals, such as nickel and chromium, and topical medicaments, for example, neomycin.

Other chemicals can sensitize the skin to light. Known as photosensitizers, these chemicals include coal tar and pitch derivatives, fluorescent dyes, salicylanilides, hexachlorophene, and some plants.

Mechanical

Trauma at work can be mild, moderate, or severe and occur as a single or repeated event. Friction results in the formation of a blister or callus, pressure in thickening and color change, sharp objects in laceration, and external force in bruising, being punctured, or being torn off. A commonly cited example is fibrous glass, which can cause irritation, itching, and scratching. Secondary infection may complicate blisters, calluses, or breaks in the skin.

Physical

Physical agents such as heat, cold, and radiation can cause occupational dermatoses. For example, high temperatures cause perspiration and softening of the outer horny layer of the skin. This can lead to miliaria or heat rash that is common among workers exposed to hot humid weather, electric furnaces, hot metals, and other sources of heat.

High temperatures can also cause systemic symptoms and signs, such as heat cramps, heat exhaustion, and even heat stroke. Burns can result from electric shock, sources of ionizing radiation, molten metals and glass, and solvents or detergents being used at elevated temperatures.

Exposures to low temperatures can cause frostbite and result

Table 3-A. Selected Chemical Causes of Skin Disorders

Chemical	*Primary Irritants*	*Sensitizers*	*Selected Skin Manifestations (Some also have important systemic effects on other organs.)*	*Selected Occupations, Trades, or Processes Where Exposure Can Occur*
ACIDS				
Acetic	X	?	Dermatitis and ulceration	Manufacturing acetate rayon, textile printing and dyeing, vinyl plastic makers
Carbolic (Phenol)	X		Corrosive action on skin, local anesthetic effect	Carbolic acid makers, disinfectant manufacturing, dye makers, pharmaceutical workers, plastic manufacturing
Chromic	X	X	Ulcers ("chrome holes") on skin, inflammation and perforation of nasal septum	Platers, manufacturing organic chemicals and dyestuffs
Cresylic	X		Corrosive to skin, local anesthetic effect	Manufacturing disinfectants, coal tar pitch workers, foundry workers
Formic	X		Severe irritation with blisters and ulcerations	Rubber and laundry workers, mordanters, cellulose formate makers, airplane dope makers
Hydrochloric	X		Irritation and ulceration of skin	Bleachers, picklers (metals), refiners (metals), tinners, chemical manufacturing
Hydrofluoric	X		Severe chemical burn with blisters, erosion, or ulceration	Enamel manufacturing, etchers, hydrofluoric acid makers, flurochemical workers
Lactic	X		Ulceration (if strong solutions are used)	Adhesives, plastics, textiles
Nitric	X		Severe skin burns and ulcers	Nitric acid workers, electroplaters metal cleaners, acid dippers, nitrators, dye makers
Oxalic	X		Severe corrosive action on skin, cyanosis (bluish discoloration), and brittleness of nails	Tannery workers, blueprint paper makers, oxalic acid makers
Picric	X	X	Erythema, dermatitis, scaling, yellow discoloration of skin and hair	Explosives workers, picric acid makers, dyers and dye makers, tannery workers
Sulfuric	X		Corrosive action on skin, severe inflammation of mucous membranes	Nitrators, picklers (metals), dippers, chemical manufacturing
ALKALIS				
Calcium cyanamide	X		Irritation and ulceration	Fertilizer makers, agricultural workers, nitrogen compound makers
Calcium oxide	X		Dermatitis, burns, or ulceration	Lime workers, manufacturing of calcium, salts, glass, and fertilizer
Potassium hydroxide	X		Severe corrosion of skin, deep-seated persistent ulcers, loss of fingernails	Potassium hydroxide makers, electroplaters, paper, soap, and printing ink makers
Sodium hydroxide	X		Severe corrosion of skin, deep-seated persistent ulcers, loss of fingernails	Sodium hydroxide makers, bleachers, soap and dye makers, petroleum refiners, mercerizers, plastic manufacturing
Sodium silicate	X		Corrosion of skin, ulcers on fingers	Cements, water softeners, detergents
Sodium or potassium cyanide	X		Blisters, ulcers	Electroplaters, case hardening, extraction of gold
Trisodium phosphate	X		Blisters, ulcers	Photographic developers, leather tanning, industrial cleaning detergents
SALTS OR ELEMENTS				
Antimony and its compounds	X	?	Irritation and lichenoid eruptions of skin	Antimony extractors, glass and rubber mixers, manufacturing of various alloys, fireworks, and aniline colors
Arsenic and its compounds	X	X	Spotty pigmentation of skin, perforation of nasal septum, skin cancer, keratoses especially on palms and soles, dermatitis, pustules	Leather workers, manufacturing insecticides, glass industry, agriculture, pesticides, tanning, taxidermy, alloy, lubricating oils
Barium and its compounds	X		Irritation of skin	Barium carbonate, fireworks, textile dyes, and paint makers
Bromine and its compounds	X		Irritation, vesicles & ulceration; acne	Bromine extractors, bromine salts makers, dye and drug makers, photographic trades
Chromium and its compounds	X	X	Pitlike ulcers (chrome holes) on skin, perforation of nasal septum, dermatitis	Chromium platers, dye industry workers, chrome manufacturing, leather tanners

(Continued)

Table 3-A. Selected Chemical Causes of Skin Disorders (continued)

Chemical	*Primary Irritants*	*Sensitizers*	*Selected Skin Manifestations (Some also have important systemic effects on other organs.)*	*Selected Occupations, Trades, or Processes Where Exposure Can Occur*
SALTS OR ELEMENTS (continued)				
Mercury and its compounds	X	X	Corrosion and irritation of skin, dermatitis	Explosives manufacturing, silver and gold extractors, manufacturing electrical appliances and scientific equipment, hat making
Nickel salts	X	X	Folliculitis, dermatitis	Nickel platers, alloy makers
Sodium and certain of its compounds	X		Burns and ulceration	Bleaching: detergent, paper, glass, tetraethyl lead manufacturing
Zinc chloride	X	?	Ulcers of skin and nasal septum	Manufacturing chemicals, dyestuffs, paper, disinfectants
SOLVENTS				
Acetone	X		Dry (defatted) skin	Spray painters, celluloid industry, artificial silk and leather workers, acetylene workers, lacquer and varnish makers, garage mechanics
Benzene and its homologues (toluene and xylene)	X		Dry (defatted) skin	Chemical and rubber manufacturing
Carbon disulfide	X	X	Dry (defatted) irritated skin	Extraction of oils, fats, and a wide range of other materials, manufacture of rayon, rubber, rubber cements, germicides, and other chemicals
Trichloroethylene	X	?	Dermatitis	Degreasers, chemical intermediates
Turpentine	X	X	Dermatitis	Painters, furniture polishers, lacquerers, artists
SOME DYE INTERMEDIATES				
Dinitrobenzene	X		Yellow discoloration of skin, hair, and eyes	Dye manufacturing
Nitro and nitroso compounds	X	X	Dermatitis	Dye manufacturing
Phenyl hydrazine	X	X	Severe chemical burns, dermatitis	Dye and pharmaceutical manufacturing
PETROLEUM AND COAL—TAR DERIVATIVES				
Petroleum oils	X		Dermatitis, folliculitis	Petroleum workers, machinists, mechanics
Pitch and asphalt	X		Dermatitis, folliculitis, keratoses, skin cancer	Manufacturing pitch and asphalt, roofers
Tar (coal)	X	X	Dermatitis, folliculitis, skin cancer, eye inflammation (keratitis)	Tar manufacturing: manufacturing roofing paper and pitch; road building and repairing
DYES (e.g., paraphenylenediamine)	X		Contact dermatitis (erythema, blisters, edema)	Dye workers, cosmetologists
RUBBER ACCELERATORS AND ANTIOXIDANTS				
Mercaptobenzothiazole, tetramethylthiuram disulfide, diethylthiourea, and para phenylenediamine	X		Contact dermatitis (erythema, blisters, edema)	Rubber workers, such as compound mixers and calender and mill operators; fabricators of rubber products
SOAPS AND SOAP POWDERS	X		Dermatitis, dry skin, paronychia (inflammation around fingernails)	Soap manufacturing, dishwashers, soda fountain clerks, maintenance workers—all associated with wet work
INSECTICIDES				
Arsenic	X		See above under salts or elements	Manufacturing and applying insecticides
Pentachlorophenols	X	?	Dermatitis, chloracne	Pesticide and wood preservative
Creosote	X	X	Dermatitis, folliculitis, keratoses, hyperpigmentation, skin cancer	Manufacturing wood preservatives, railroad ties, coal tar lamp black and pitch workers
Fluorides	X		Severe burns, dermatitis	Manufacturing insecticides, enamel manufacturing
Phenylmercury compounds	X	X	Dermatitis	Manufacturing and applying fungicides and disinfectants
Pyrethrum		X	Dermatitis	Manufacturing and applying insecticides

(Continued)

Table 3-A. Selected Chemical Causes of Skin Disorders (concluded)

Chemical	*Primary Irritants*	*Sensitizers*	*Selected Skin Manifestations (Some also have important systemic effects on other organs.)*	*Selected Occupations, Trades, or Processes Where Exposure Can Occur*
INSECTICIDES (continued)				
Rotenone	X		Dermatitis	Manufacturing and applying insecticides
RESINS (NATURAL)*				
Cashew nut oils		X	Severe poison ivylike dermatitis	Handlers of unprocessed cashew nuts, varnish
Rosin		X	Dermatitis	Adhesive and paper mill workers, dentists, rubber industry
Shellac		X	Dermatitis	Coating, cosmetics
Synthetic resins such as phenol-formaldehyde, urea-formaldehyde, epoxy, vinyl, polyurethane, polyester, acrylate, cellulose esters	X	X	Dermatitis	Plastic workers, varnish makers

* The skin reactions from this group of chemicals in some instances are due to the essential composition of the synthetic resin, but in other cases are due to the presence of added compounds such as plasticizers and other modifying agents.

Chemical	*Primary Irritants*	*Sensitizers*	*Selected Skin Manifestations*	*Selected Occupations*
EXPLOSIVES				
Nitrates, mercury fulminate, tetryl, lead azide, TNT, nitroglycerin	X	X	Severe irritation, dermatitis, skin discoloration	Explosives manufacturing, shell loading
METAL WORKING FLUIDS				
Cutting oils	X		Oil acne (folliculitis), rare dermatitis	Machinists
Coolants—synthetic and semisynthetic	X	X	Dermatitis	Machinists
OTHER				
Isocyanates such as TDI, MDI, HDI	X	X	Dermatitis	Polyurethane makers, adhesive workers, organic chemical synthesizers
Vinyl chloride	X		Dermatitis, acro-osteolysis	Polyvinyl resin, rubber and organic chemical makers
Formaldehyde	X	X	Dermatitis	Undertakers, biologists, textile workers

in permanent damage to blood vessels. The ears, nose, fingers, and toes are the most often frostbitten. Electric utility and telephone line workers, highway maintenance workers, farmers, fishermen, police officers, letter carriers, and other outdoor employees are most often affected.

Sunlight is the greatest source of skin-damaging radiation and, consequently, is a source of danger to construction workers, farmers, fishermen, foresters, and all others who are outdoors for extended periods of time. The most serious effect on the skin is skin cancer.

Increasing numbers of people come into casual or prolonged contact with artificial UV light sources, such as molten metals and glass, welding operations, and the plasma torch. A newer light source is found in operations using the laser. Because laser beams can injure the skin, eye, and other biological tissue, it is important that the appropriate protective devices are used.

Ionizing radiation sources include the following:

- alpha-radiation, though not injurious to skin, is dangerous if radioactive substances, for example, plutonium, are inhaled or ingested.
- beta-radiation can injure skin by contact and the body in general if the substances, for example, phosphorus-32, are inhaled or ingested.
- gamma-radiation and x-rays are well-known skin (radiodermatitis and skin cancer) and systemic (internal) hazards when sufficient exposure occurs. Radiodermatitis is characterized by dry skin, hair loss, telangiectasia, spiderlike angiomas, and hyperkeratosis. Skin cancer may ultimately develop (see Chapter 10, Ionizing Radiation, for more information).

Biological

Bacteria, viruses, fungi, and parasites can produce cutaneous and/or systemic disease of occupational origin. Animal breeders, agricultural workers, bakers, culinary employees, florists, horticulturists, laboratory technicians, and tannery workers are among those at greater risk of developing infections. Examples include anthrax in hide processors, yeast infections of the nail in dishwashers, bartenders, and others engaged in wet work, and animal ringworm in farmers and veterinarians. Parasitic mites are common inhabitants of grain and other foodstuffs and attack those handling such materials, for example, grocers, truckers, longshoremen, and farmers. Outdoor workers, such as bricklayers and plumbers in southeastern states, risk contracting animal hookworm via larvae deposited by infected animals in sandy soil. When these substances are known to be connected with work, all necessary precautions for preventing disease must be exercised. A common type of skin infection seen among workers is caused by staphylococci invading the site of a previous wound.

Botanical

Many plants and woods, of which poison ivy and poison oak are the most common, can cause contact dermatitis. Irritant contact dermatitis can also occur from some plants, and although the chemical identity of many of the toxins is not known, the allergen or irritant agent occurs in the leaves, stems, flowers, bark, or other part of the plant. Other plants, such as wild parsnip and diseased celery (pink-rot) are photosensitizers. With woods, dermatitis occurs especially when they are being sandpapered, polished, and cut. Fomites can carry and transmit these allergens, and they can also be disseminated by the smoke from burning.

DEFENSE MECHANISMS

In conclusion, anyone who works is a candidate for contracting an occupational skin disease, yet most workers are not affected by such disorders, because the skin is a primary organ of defense. The skin is able to perform its many defense functions because of its location, structure, and physiological activity.

Look now at the specific defenses the skin has in terms of its protection against typical industrial hazards.

1. Against bacteria—the skin has the defense of being a naturally dry terrain (except in places like armpits and the groin, and during abnormal sweating), and has a normal contingent of bacteria that tends to destroy pathogenic bacteria. Free fatty acids in the surface oil also can have some antibacterial value. The immune defenses of the skin also defend against infections.
2. Against sunlight—the skin has two defenses: an increase in pigmentation, and thickening of the stratum corneum.
3. Against primary irritants—the natural defenses are the buffering action of surface components such as amino acids, lactic acid, and lactate derived from the sweat gland and dermal metabolism and from carbon dioxide diffusing from within to the skin surface. Sweat can act as a diluent to decrease the effect of water-soluble toxins. Conversely, it enhances hydration and maceration of the barrier, thereby promoting percutaneous absorption.
4. Against injury—the skin's resilience, especially of the dermis, provides a measure of resistance to forceful impact. The cutaneous nerves also provide information about the state of the external environment, for example, touch and temperature.
5. Against excessive increase or decrease in body heat—the body's thermoregulatory mechanisms operate controlling activity of sweat glands and blood vessels.
6. Against the absorption of chemicals through the skin—this is where the most important function is performed. The skin is a flexible body envelope and the epidermal barrier, especially the stratum corneum, provides a significant blockade against water loss from the body and penetration of the skin by chemical agents. See Chapter 8, Industrial Dermatoses, for more information on the subject and for selected information from the American Medical Association's *Guides to Evaluation of Permanent Impairment,* for use in interpreting and understanding medical reports of workers' compensation cases.

CONCLUSION

Causative factors relating to occupational dermatoses are complex, and workers' particular skin reactions to specific substances or combinations of substances are also complex and highly individual. Accurate diagnosis of a particular employee's skin problem is best left to his/her personal physician acting in concert with the plant medical personnel and a dermatologist. Further details of the diagnosis, treatment, and prevention of occupational skin diseases are described in Chapter 8.

BIBLIOGRAPHY

Adams, R.M. *Occupational Skin Diseases.* New York: Grune & Stratton, 1983.

American Medical Association (AMA) Advisory Committee on Occupational Dermatoses of the Council on Industrial Health. *Occupational Dermatoses: A Series of Five Reports.* Chicago: AMA, 1959.

Cooley, D.G., ed. *Family Medical Guide.* New York: Better Homes and Gardens Books, 1973.

Fisher, A.A. *Contact Dermatitis.* Philadelphia: Lea & Febiger, 1986.

Fitzpatrick, T.B., et al. *Dermatology in General Medicine.* New York: McGraw Hill, 3rd ed., 1987.

Goldsmith, L.A., ed. *Biochemistry and Physiology of the Skin.* New York: Oxford, 1983.

Hurley, H.J. Permeability of the skin. In *Dermatology,* edited by Moschella, S.L., and Hurley, H.J. Philadelphia: Saunders, 1985.

ILO Encyclopedia of Occupational Health and Safety, 3rd ed. Geneva, Switzerland: International Labor Organization, 1983.

Industrial dermatitis: Part 1: The skin. *National Safety News* 112(October 1975):59-62.

Industrial dermatitis: Part 2: Primary irritation. *National Safety News* 112(November 1975):107-112.

Industrial dermatitis: Part 3: Sensitization dermatitis. *National Safety News* 112(December 1975):65-68.

Jakubovic, H.R., and Ackerman, A.B. Structure and Function of Skin. In *Dermatology,* 2nd ed., edited by Moschella, S.L., and Hurley, H.J. Philadelphia: Saunders, 1985.

Key, M.M., et al., ed. *Occupational Diseases—A Guide to Their Recognition.* DHEW NIOSH Publication No. 77-181, 1977.

Litt, J.Z. *Your Skin and How to Live in It.* Cleveland: Corinthian Press, 1980.

Lubowe, I.I. *New Hope for Your Skin.* New York: Pocket Books, Inc., 1966.

Mitchell, J.A., and Rook, A. *Botanical Dermatology.* Vancouver: J.A. Mitchell, 1979.

National Safety Council, 444 N. Michigan Ave., Chicago, IL 60611.
Industrial Data Sheet:
–*Poison Ivy, Poison Oak, and Poison Sumac,* No. 304.

Report of the Advisory Committee on Cutaneous Hazards to the Assistant Secretary of Labor, OSHA. Occupational Safety and Health Administration, U.S. Department of Labor, December 19, 1978.

Rook, A., et al. *Textbook of Dermatology,* 4th ed. Oxford: Blackwell, 1986.

Schwartz, L., Tulipan, L., and Birmingham, D.J. *Occupational Diseases of the Skin,* 3rd ed. Philadelphia: Lea & Febiger, 1957.

Taylor, J.S. "Occupational Dermatoses." In *Clinical Medicine for the Occupational Physician,* edited by Alderman, M.H., and Hanley, M.J. New York: Dekker, 1982, pp 299–344.

Taylor, J.S., Parrish, J.A., and Blank, I.H. Environmental reactions to chemical, physical, and biological agents. *Journal of the American Academy of Dermatology* 11(1984):1007–1021.

The number one occupational illness: Dermatitis. *National Safety News* 105(June 1972): 38-43.

Toxic plants. *National Safety News* 105(June 1972):45-49.

The Ears

by Julian B. Olishifski, MS, PE, CSP
Revised by George S. Benjamin, MD, FACS

INTRODUCTION

THE ANATOMY OF THE HUMAN EAR is very complex, and its minute size and protective encasement in hard, dense bone further complicate scientific study of this delicate organ. Many aspects of ear function are unknown, particularly those involving the inner ear and the pathways leading to the brain. For safety and health professionals, this chapter presents an overview of the structure of the auditory mechanism, how it seems to work, and the common causes of impairment of this sensitive organ.

The auditory mechanism enables us to hear sound. In air, sound is defined as variations of pressure above and below the ambient atmospheric pressure. These air pressure fluctuations, or sound waves, vary in intensity, harmonic content, frequency, and direction. The word sound also indicates the sensation experienced when pressure fluctuations strike the ear. (Chapter 9, Industrial Noise, covers this subject in detail.)

Through the ear, sound waves can be detected within a range of 20–20,000 Hz and then are converted into electrical impulses that are transmitted to the brain for interpretation. The brain performs this function by receiving the sensory information as a series of electrical impulses and sending or shunting these impulses from one brain cell to another.

The organ of hearing is divided into three parts—the external, the middle, and the inner ear (Figure 4-1).

ANATOMY

External ear

The external ear is divided into two sections—the portion attached to the outer surface of the head, called the auricle or pinna, and the external auditory canal.

The pinna. This is the most visible part of the ear, a delicately folded cartilaginous structure with a few small muscles, covered by subcutaneous tissue and skin. The pinna functions as a funnel to collect sound waves. Many animals can voluntarily control the ear muscles and aim the pinnae to enhance the collection of sound waves.

The external auditory canal. This canal, or meatus, is a skin-lined pouch about 3.8 cm (1.5 in.) long, supported in its outer third by the cartilage of the pinna and in its inner two thirds by bone of the skull. At its innermost end lies the tympanic membrane or eardrum that separates the external from the middle ear.

The small hairs or vibrissae and ceruminal glands that secrete a waxy substance called cerumen are located in the skin of the outer third of the ear canal. The hairs serve a protective function by filtering out particulate matter and other large foreign bits of debris. Cerumen, both sticky and bactericidal, prevents smaller particles from entering the ear canal and keeps the canal healthy and free of infection.

Middle ear

The middle ear is the space or cavity, about 1-2 mL in volume and lying between the eardrum and the bony wall of the inner ear (Figure 4-2). The middle ear is lined with mucous membrane essentially the same as that lining the mouth. The ossicles, which are the smallest bones in the body, are located within the middle ear cavity (Figure 4-3). The ossicles connect the

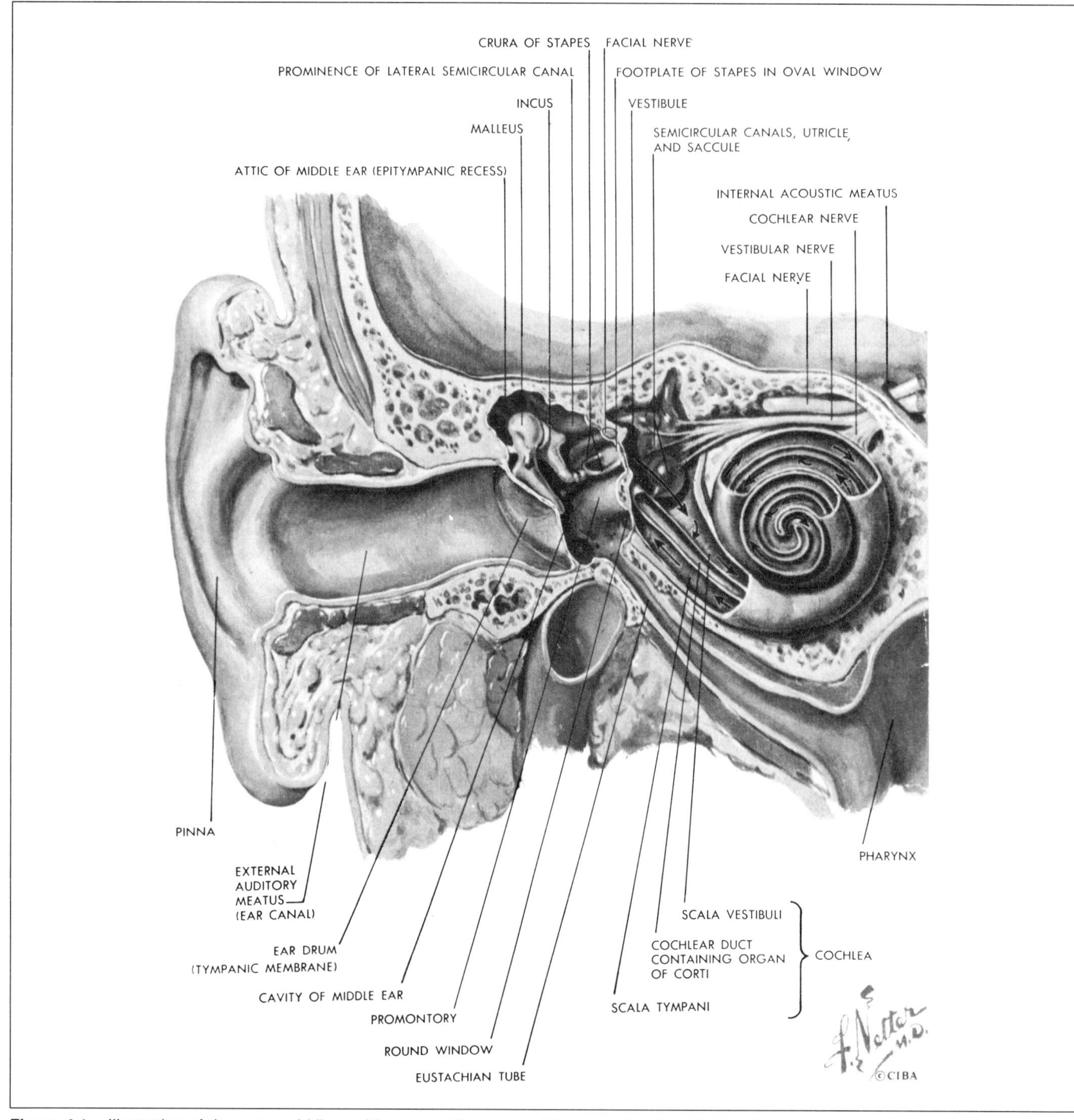

Figure 4-1. Illustration of the outer, middle, and inner ear. (Reprinted with permission from Netter, F.H. *Clinical Symposia.* CIBA Pharmaceutical Co.)

eardrum to an opening in the wall of the inner ear called the oval window.

Picture the middle ear space as a cube:

1. The outer wall is formed by the eardrum.
2. The inner wall is the bony partition separating the inner ear from the middle ear. The round and oval windows fit into this wall and include the only two movable barriers between the middle and inner ear.
3. The front wall opens into the eustachian tube.
4. The back wall opens into the mastoid air cells.
5. The roof separates the middle ear from the temporal lobe of the brain.
6. The floor separates the middle ear from the jugular vein and the internal carotid artery, which lies high in the neck.

The sound-conducting mechanism in the middle ear includes the eardrum and three ossicles (bones) that are supported by ligaments and moved by two muscles.

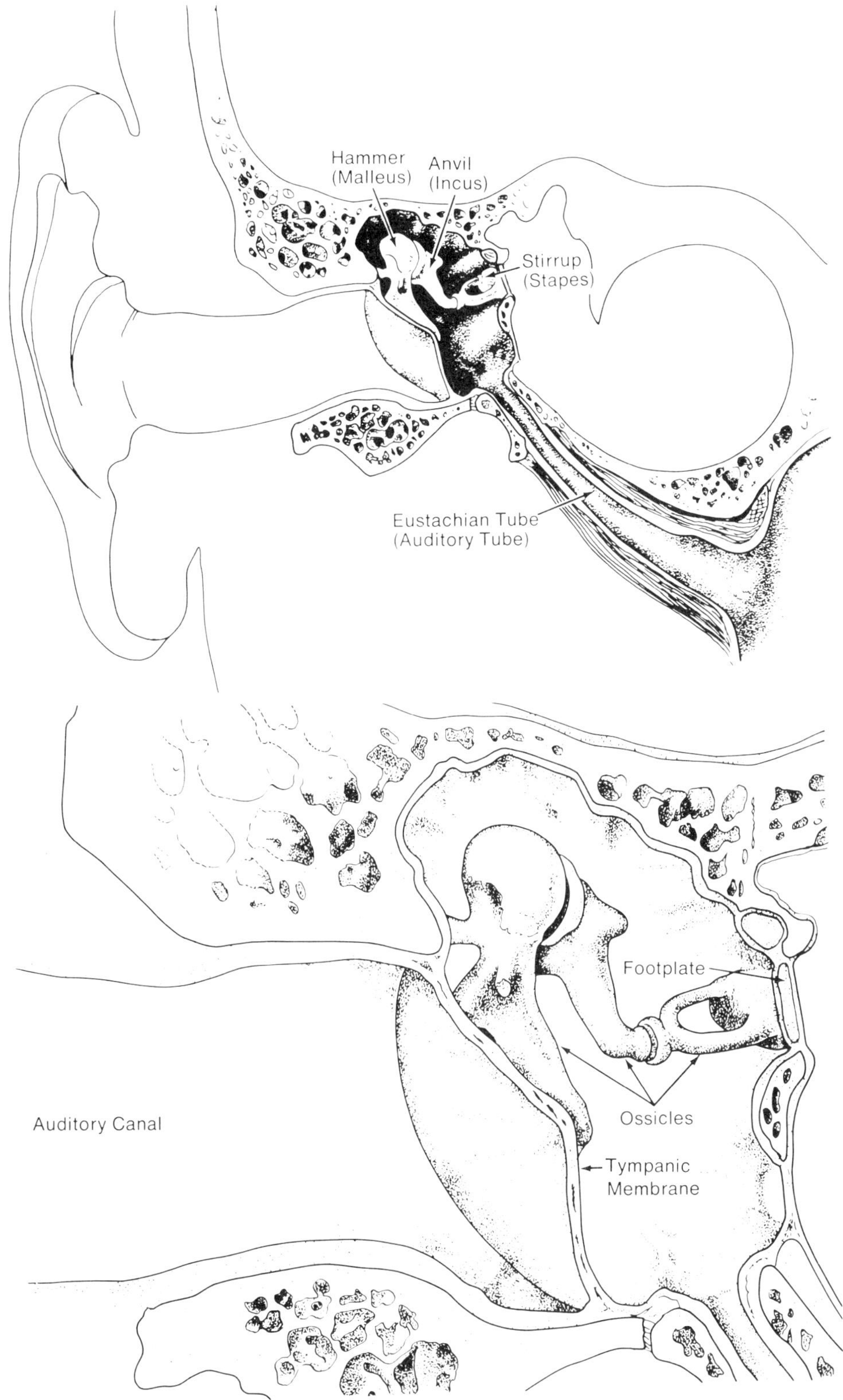

Figure 4-2. The middle ear is contained within the temporal bone and is made up of the eustachian tube, the middle ear space, and the mastoid air cell system. (Adapted from Figure 4-1)

Figure 4-3. The ossicles, located within the middle ear cavity, link the eardrum to an opening in the wall of the inner ear (the oval window). (Adapted from Figure 4-1)

The eustachian tube. This serves to equalize the pressure in the middle ear with the external atmospheric pressure. It opens during swallowing and yawning. It tends to remain closed when the pressure is increasing, as during rapid descent in an airplane. In some persons after frequent infection, the tube is closed permanently by adhesions, and their hearing is impaired greatly. Whenever the pressure is unequal on the two sides of the eardrum, it is not free to vibrate in response to sound waves.

The eardrum. This membrane separates the external ear canal from the middle ear. It consists of an inner layer of mucous membrane and a middle layer of fibrous tissue. It is spiderweb in form, with radial and circular fibers for structural support.

Ossicular chain. The ossicles, which together are called the ossicular chain, are the malleus, incus, and the stapes.

- The malleus or hammer is fastened to the eardrum by the handle. The head lies in the upper area of the middle ear cavity and is connected to the incus.
- The incus, also called the anvil, is the second ossicle and has a long projection that runs downward and joins the stapes.
- The stapes, also called the stirrup, lies at almost right angles to the long axis of the incus. The two branches of the stapes, anterior and posterior, end in the footplate that fits into the oval window.

When the handle of the hammer is set into motion by movement of the eardrum, the action is transferred mechanically through the ossicular chain to the oval window.

The oval window and the round window. These are located on the inner wall of the middle ear. The round window is covered by a very thin membrane that moves out as the footplate in the oval window moves in. As the action is reversed and the footplate in the oval window is pulled out, the round window membrane moves inward.

Mastoid air cell system. On the back wall of the middle ear space is an opening that extends into the mastoid. This opening resembles a honeycomb of spaces filled with air. The mucous membrane lining the middle ear is continuous with that of the pharynx and the mastoid air cells; thus it is possible for infection to travel along the mucous membrane from the nose or the throat to the middle ear and to the mastoid air cells.

Inner ear

The inner ear contains the receptors for hearing and for position sense. It consists of a bony labyrinth which contains a membranous labyrinth.

The bony labyrinth consists of a series of tiny canals and cavities hollowed out of the petrous portion of the temporal bone. They contain a watery fluid called perilymph. There are three bony divisions: the cochlea, the vestibule, and the semicircular canals (Figure 4-4).

The cochlea. This is shaped like a snail shell (see Figure 4-5). It is a bony tube winding around a central pillar of bone called the modiolus. The membranous cochlear duct resembles a lopsided triangle as it lies within the bony cochlea. It extends like a shelf across the bony canal and is attached to the sides. The portion of the bony canal above the cochlear duct is called the scala (stairway) vestibuli, and the portion below is called the scala tympani. They are continuous with each other through a tiny opening (helicotrema) in the apical end of the cochlea.

The scala vestibuli begins beneath the footplate of the stapes and is separated from the scala media below it by Reissner's membrane. The scala vestibuli continues to the helicotrema where it joins the scala tympani.

The scala tympani lies below the scala media, separated from it by the basilar membrane, and ends at the round window. The oval and round windows lie at opposite ends of the perilymphatic space of the cochlea (Figure 4-6).

Organ of Corti. The cochlear duct is connected with the inner wall of the bony canal by the osseous spiral lamina, and with the outer wall by the spiral ligament. The roof of the cochlear duct is thin and is called Reissner's membrane. The floor is composed of the basilar membrane. Resting on this basilar membrane is the spiral organ of Corti, the essential receptor end organ for hearing (Figure 4-5). It is a very complicated structure, consisting of a supporting framework on which rest the hair cells.

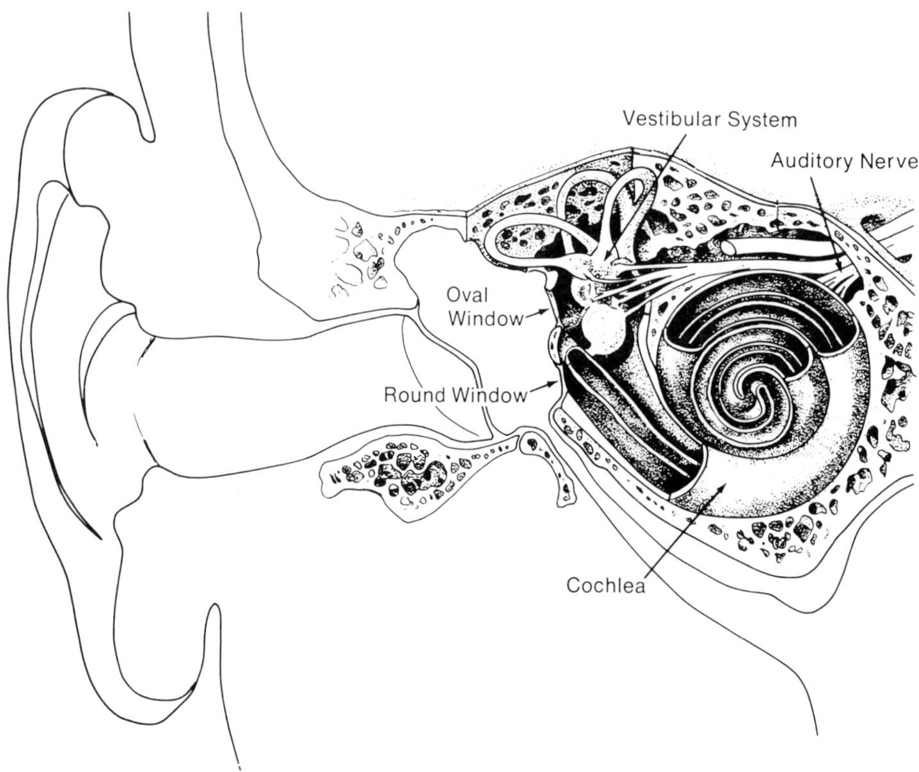

Figure 4-4. The major components of the inner ear, including the vestibular system and the cochlea. (Adapted from Figure 4-1)

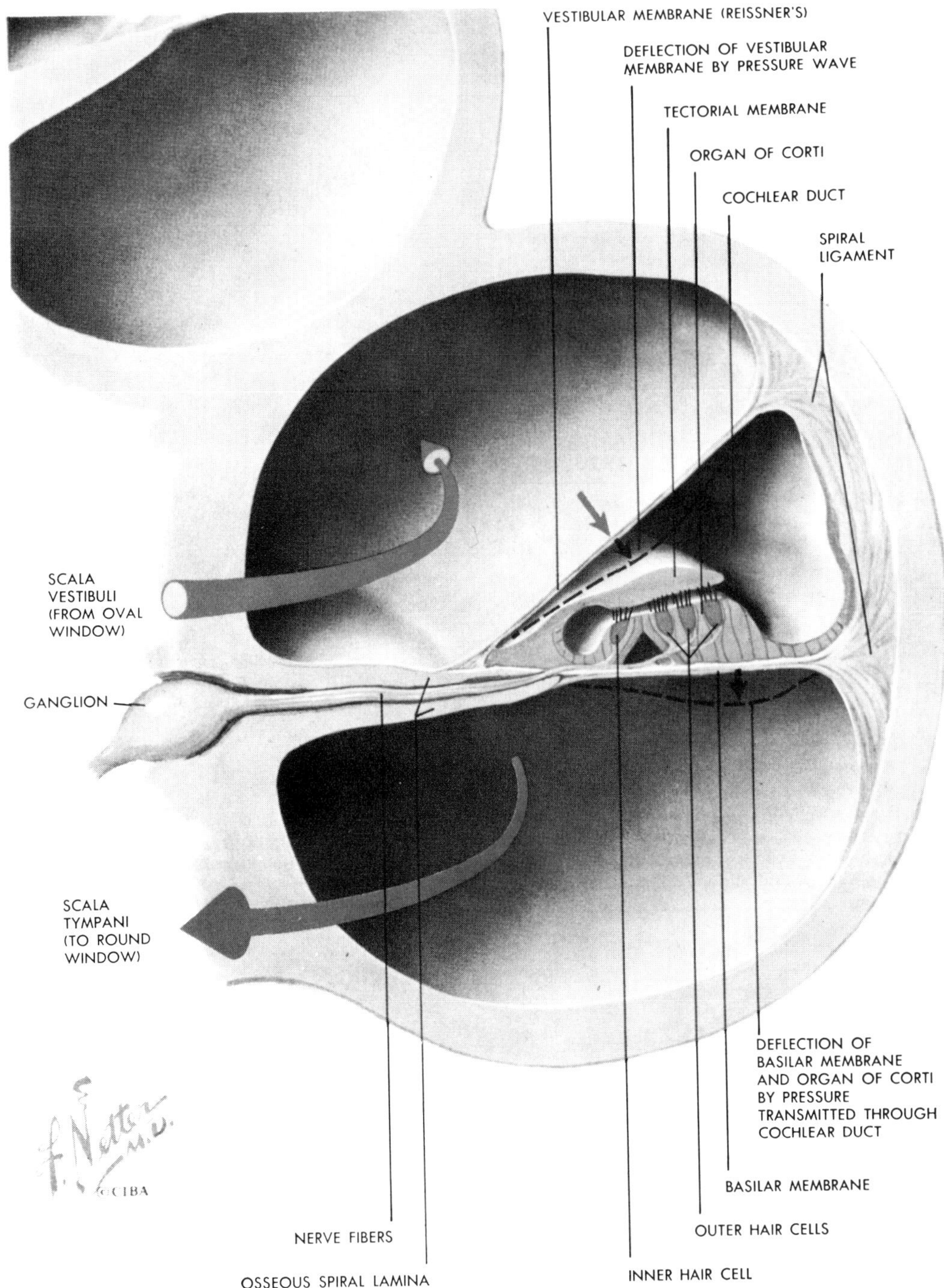

Figure 4-5. This schematic diagram depicts the transmission of sound across the cochlear duct, stimulating the hair cells. (Reprinted with permission from Netter, F.H. *Clinical Symposia.* CIBA Pharmaceutical Co.)

PHYSIOLOGY

External ear

The function of the outer ear in the hearing process is relatively simple—the external portion of the ear collects sound waves from the air and funnels them into the ear canal where they are transported to the eardrum. The collected sound waves cause the eardrum to move back and forth in a vibrating mechanical motion that is passed on to the bones of the middle ear (Figure 4-6).

Middle ear

The primary function of the middle ear in the hearing process is to transfer sound energy from the outer to the inner ear. As the eardrum vibrates, it transfers its motion to the attached hammer (malleus). Since the bones of the ossicular chain are connected to one another, the movements of the hammer are passed on to the anvil, and finally to the stirrup embedded in the oval window.

As the stirrup moves back and forth in a rocking motion, it

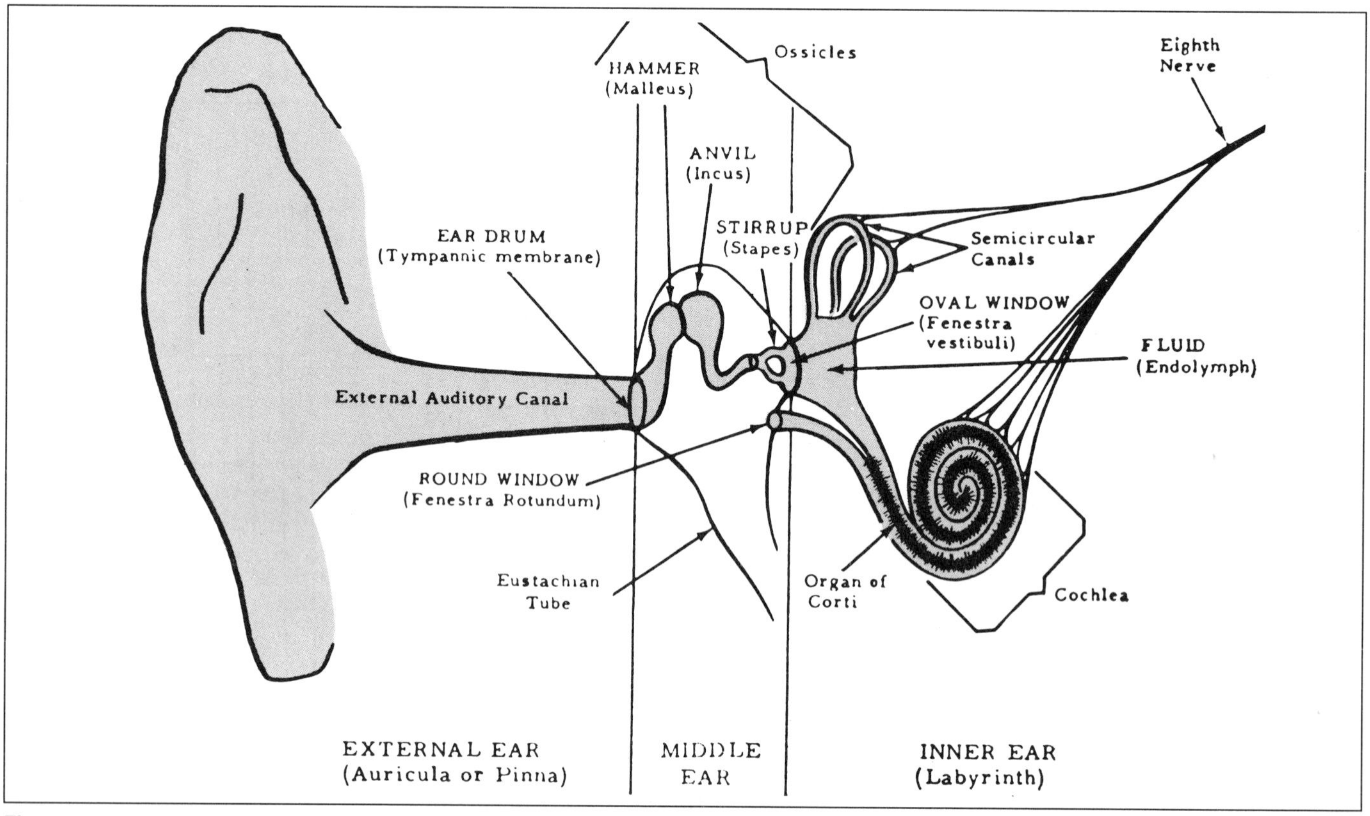

Figure 4-6. How the ear hears—wave motions in the air set up sympathetic vibrations that are transmitted by the eardrum and the three bones in the middle ear to the fluid-filled chamber of the inner ear. In the process, the relatively large, but feeble, air-induced vibrations of the eardrum are converted to much smaller, but more powerful mechanical vibrations by the three ossicles, and finally into fluid vibrations. The wave motion in the fluid is sensed by the nerves in the cochlea, which transmit neural messages to the brain. (Reprinted with permission from the American Foundrymen's Society)

passes the vibrations on into the inner ear through the oval window. Thus, the mechanical motion of the eardrum is effectively transmitted through the middle ear and into the fluid of the inner ear.

Amplification. The sound-conducting mechanism also amplifies sound by two main mechanisms. First, the large surface area of the drum compared to the small surface area of the base of the stapes (footplate) results in a hydraulic effect. The eardrum has about 25 times as much surface area as the oval window. All of the sound pressure collected on the eardrum is transmitted through the ossicular chain and is concentrated on the much smaller area of the oval window. This produces a significant increase in pressure (Figure 4-7).

The bones of the ossicular chain are arranged in such a way that they act as a series of levers. The long arms are nearest the eardrum, and the shorter arms are toward the oval window. The fulcrums are located where the individual bones meet. A small pressure on the long arm of the lever produces a much stronger pressure on the shorter arm. Since the longer arm is attached to the eardrum and the shorter arm is attached to the oval window, the ossicular chain acts as an amplifier of sound pressure. The magnification effect of the entire sound conducting mechanism is about 22-to-1.

Two tiny muscles attach to the ossicular chain, the stapedius to the neck of the stapes bone and the tensor tympani to the malleus. Loud sounds cause these muscles to contract, which

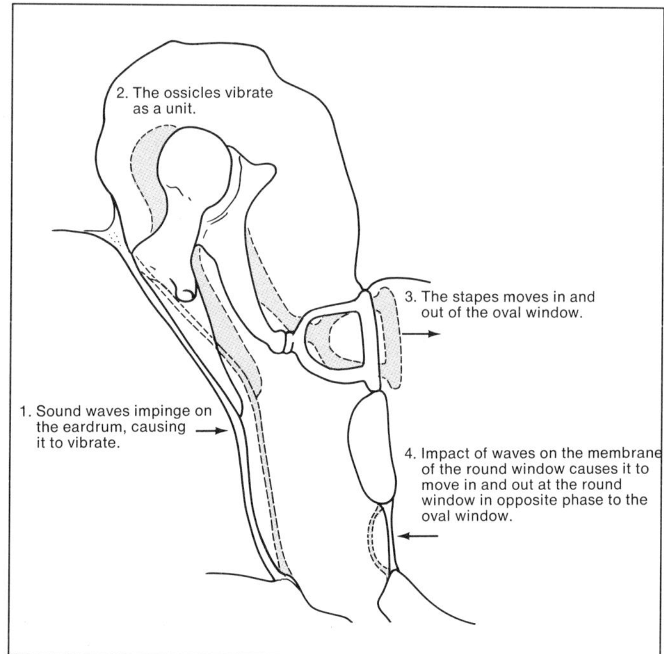

Figure 4-7. The eardrum has about 25 times as much surface area as the oval window. All of the sound energy collected on the eardrum is transmitted through the ossicular chain to the smaller area of the oval window.

stiffens and diminishes or damps the movement of the ossicular chain.

Inner ear

The major components of the inner ear include the vestibular receptive system and the cochlea, housed within the compact temporal bone called the osseous labyrinth. It is filled with fluid (perilymph) in which a tubular membrane (membranous labyrinth) floats. The membranous labyrinth is filled with a fluid of a slightly different chemical composition called endolymph. Endolymph bathes the balance receptors and the hearing organ (organ of Corti) located within the membranous labyrinth.

The back portion of the inner ear consists of the three semicircular canals positioned at right angles to one another, each containing a single balance receptor. The front section (cochlea) is shaped like a snail shell, coiled two and one-half times around its own axis; it houses the organ of Corti. These two regions are separated by the vestibule that contains two additional receptors for balance.

Vestibular system. Our sense of balance is not dependent on hearing, but on organs of equilibrium. Near the cochlea are three semicircular canals lying in planes at right angles to each other. The canals contain fluid that responds to movements and, over intricate nerve pathways to the brain, gives information about positions of the body (Figure 4-8).

The vestibular branch of the acoustic nerve transmits impulses to the cerebral cortex, and we recognize the position of our head in space as it relates to the pull of gravity.

If you are rotated in a chair, the endolymph in the semicircular canals is set into motion and stimulates the hair cells. A sensation of vertigo or dizziness occurs, together with a peculiar movement of the eyes called nystagmus. Nystagmus consists of a rapid movement of the eyes in one direction and a slow movement in the opposite direction; they appear to oscillate. Nausea may occur. We call these rotation sensations motion sickness.

The cochlea. This is a tubular bony structure lined with a membrane containing thousands of feathery hair cells tuned to vibrate to different sound frequencies. Nerve endings are contained in a complex, slightly elevated structure over the floor of the tube forming the cochlea. This area (organ of Corti) is the center of the sense of hearing.

Vibrations of the stapedial footplate set into motion the fluids of the inner ear. As the basilar membrane is displaced, a shearing movement occurs on the tectorial surface that drags the hair cells attached to the nerve endings. This sets up electrical impulses that are appropriately coded and transmitted to the brain via the auditory (cochlear) nerve (Figure 4-9).

The nerve endings in the cochlea are sensitive to different frequencies. Those sensitive to high frequencies are located at the large base end of the cochlea near the oval and round windows. The nerve endings that respond to low frequencies are located at the small end of the cochlea.

PATHOLOGY

Although the human ear is subject to a number of types of disorders that can cause hearing loss, the major occupational hazard is excessive unwanted sound (noise).

However, there are many nonjob-related causes of hearing loss. Some 25 percent of newly employed workers come to their jobs with some degree of preexisting hearing loss. (See Chapter 9, Industrial Noise, for details.)

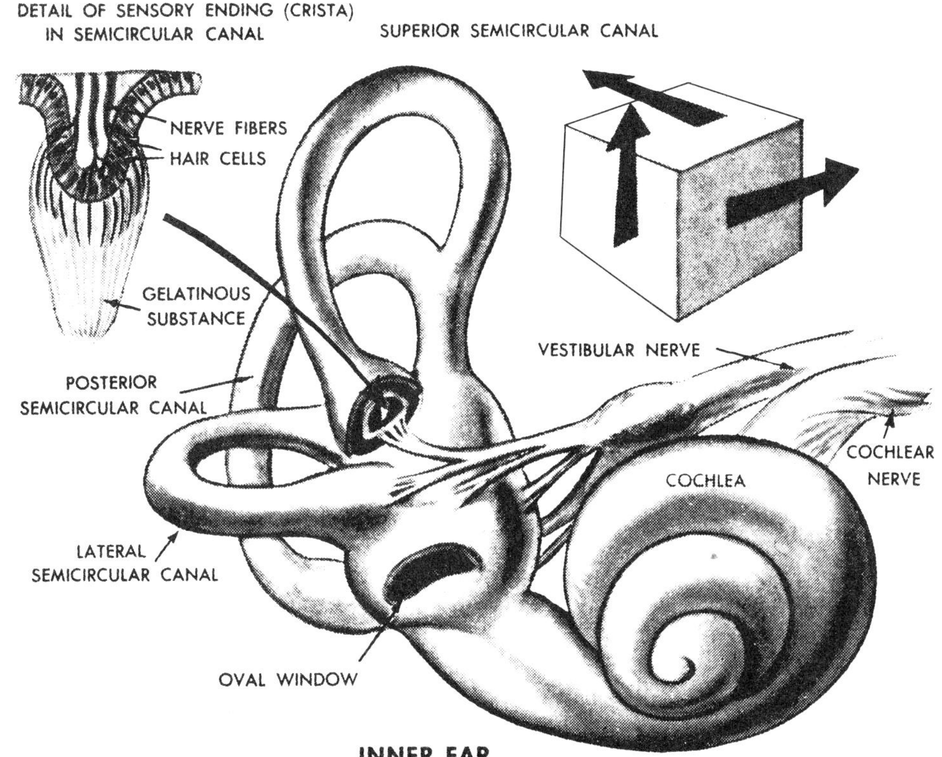

Figure 4-8. The three subdivisions of the inner ear—the cochlea, the vestibule, and the semicircular canals. (Reprinted with permission from the American Medical Association [AMA]. *The Wonderful Human Machine.* Chicago: AMA, 1971)

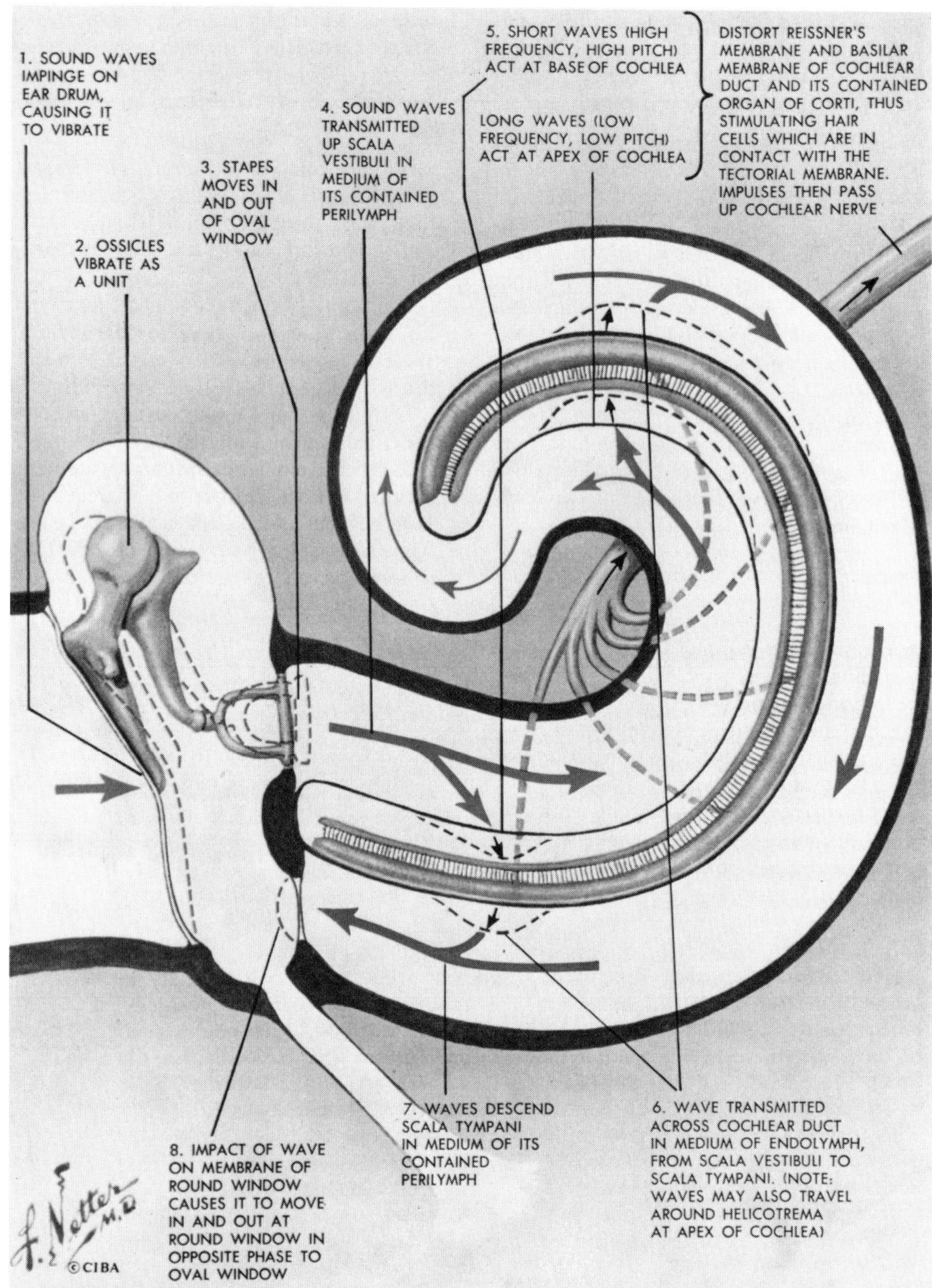

Figure 4-9. The mechanism for transmission of sound vibrations from the eardrum through the cochlea. (Reprinted with permission from Netter, F.H. *Clinical Symposia.* CIBA Pharmaceutical Co.)

Hearing impairments that were not induced by noise can arise from:

- physical blockage of the auditory canals with excessive wax, foreign bodies, etc.
- traumatic damage—such as punctured eardrums or displacement of the ossicles
- disease damage—childhood diseases, for example, smallpox infections of the inner ear; degenerative diseases; tumors, etc.
- hereditary or prenatal damages
- drug-induced damages—such as from use of streptomycin, quinine
- presbycusis—natural reduced hearing sensitivity due to aging.

Off-the-job noise exposure to loud music, motorcycles, snowmobiles, private airplanes, etc., can be a significant cause of hearing loss. All health and safety professionals responsible for hearing tests should know about the effects of these nonoccupational noise exposures, in order to avoid confusing nonoccupational with occupational causes. Under the workers' compensation laws of many states, a hearing loss may be compensable, if it is due to any occupational exposure.

External ear

The external ear, because of its prominence and its thin tight skin, is especially subject to sunburn and frostbite. Thus, it often needs to be protected from the elements. Injured cartilage is replaced by fibrous tissue; and repeated injuries will result in the cauliflower-shaped ear seen on many boxers. Disorders of the auricle include congenital malformations (in which the cartilage is misshapen) and protruding or lop ears, both of which may be surgically corrected. The auricle is the most common site in the ear for malignancies. Dermatitis and infection are common in this area.

Ear canal

The ear canal leading to the middle ear is normally kept healthy by wax (cerumen) which protects it from drying and scaling or from damage by water while swimming or bathing.

Foreign objects accidentally lodged in the ear can be dangerous and should be removed only by a physician. A live insect in the ear canal can be especially annoying or painful. If this happens, drop light mineral oil into the canal, to suffocate and quiet the insect until it can be removed.

The external ear canal is prone to infection because of its increased skin temperature and humidity. In this area, bacterial infections are most common; while fungus (otomycosis) and viral causes are less frequent. Skin disorders (dermatitis) are also frequent ear canal problems. Generalized skin disorders, usually of the scalp, may extend to the outer ear.

An abnormal narrowing of the ear canal is called stenosis and may be caused by infection or injury. Tumors are rare in this area, the most common are benign bony masses.

Normally, ear canals are self-cleansing but occasionally a failure of this mechanism results in wax impaction. The use of cotton-tipped swabs for cleaning purposes tends to pack wax into the ear canal. Also, swabbing will stimulate excess production of wax. However, the external ear canal must be almost totally occluded (blocked) before attenuation of sound occurs. If there is an opening in the ear canal as small as the diameter of one red blood cell, sound can get through to the eardrum and hearing can be maintained. Consequently, before any loss in hearing can occur from ear wax, there must be a considerable quantity involved. Wax impaction should preferably be cleaned by an otologist because of the risks of injury and infection inherent in the cleaning process.

Eardrum

Infections localized to the eardrum are rare and when they do occur, are caused by viruses, such as the virus of shingles (herpes zoster). However, the eardrum is frequently included in infections of the external auditory canal or the middle ear.

Perforations are most frequently caused by infections or by injuries. A blow to the ear, compressing air in the external auditory canal can cause sufficient force to rupture the eardrum. Sudden pressure changes, such as occur in scuba diving or explosions, cause perforations. Perforations of the eardrum can be repaired by surgical grafting procedures, but most heal spontaneously.

Eustachian tube

Retractions of the eardrum are caused by poor middle ear ventilation due to eustachian tube dysfunction.

Failure of the eustachian tube to ventilate results in a vacuum in the middle ear space, which in turn causes one of two pathological events to occur: (1) it pulls fluid into the middle ear, resulting in a condition called nonsuppurative otitis media (noninfectious inflammation of the middle ear); or (2) it pulls the eardrum inward, interferes with mobility, and causes hearing loss. Eventually, the fluid organizes into fibrous tissues, adhesions form about the ossicles, and hearing loss is permanent.

The opposite condition, which is uncommon, is a patent eustachian tube in which the tube constantly remains open. However, if it does occur, the results are an annoying symptom of hearing one's own voice and breath sounds (autophonia) in the involved ear.

Middle ear

The middle ear space is prone to infectious diseases, especially in childhood years. These are predominantly bacterial in origin and called suppurative otitis media. Since the middle ear space connects with the mastoid air cell system, infection may easily spread to this area (mastoiditis). Before the days of antibiotics, these were serious, often life-threatening, problems due to the risk of spread of infection to the brain or major vessels surrounding the ear. While less likely to occur today, these dangers still exist.

Congenital deformities of the middle ear are not too common and are usually associated with structural abnormalities of the sound conducting system. Tumors in the middle ear are rare.

Ossicles. There are only two ways that disease adversely affects hearing via the ossicular chain: (1) by fixation, so that the chain cannot vibrate or vibrates inefficiently, and (2) by interruption, producing a gap in the chain. Fixation may occur as a result of developmental errors, adhesions, or scars from old middle ear infections or bone diseases that affect this area.

Otosclerosis is a prime example of fixation and is the most common cause of progressive conductive hearing loss seen in this country. It usually begins in early adult life. Interruptions are usually caused by middle ear infections, cholesteatomas, or head injuries.

It is important to realize that if disease is confined to the outer ear or middle ear or both, the resulting hearing loss will be conductive. Conductive losses are losses for loudness, not for clarity. Most of these disorders causing conductive losses are medically or surgically correctable. Some are dangerous and some are progressive. Therefore, all cases should have the benefit of otologic evaluation and care.

Inner ear

Disorders of the inner ear result in a sensory hearing loss. Inner ear hearing losses may or may not have associated losses for clarity. Congenital sensory hearing losses (inner ear hearing losses present at the time of birth) may be of many different types. Some of these are inherited and are called familial sensory hearing losses. Inflammatory disorders include suppurative labyrinthitis, a very rare condition; viral labyrinthitis caused by such organisms as mumps, measles, and viruses;

and toxic labyrinthitis which relates to inner ear dysfunction, caused by hormonal (hypothyroidism) and allergic factors, and to drugs that have an adverse effect on the hearing mechanism.

Concussions of the inner ear may also occur and usually cause an incomplete sensory loss with some, if not complete, recovery of hearing. Hearing acuity, like visual acuity, tends to decrease with age after the fourth decade. This loss, called presbycusis, is of the same sensorineural type, and may be due to the noise trauma of everyday life. Tumors affecting the inner ear are rare and usually do not originate within the cochlea but affect the inner ear by way of extension to this structure.

Meniere's disease affects both parts of the inner ear (hearing and balance) and its cause is unknown. It is characterized by episodic dizziness, often severe and associated with nausea and vomiting, fluctuant sensory hearing loss that is generally progressive, noise in the ear or tinnitus, and a peculiar sensation of fullness in the involved ear.

Tinnitus (head noises)

Tinnitus is not a disease, but is frequently a feature of various ear diseases—or of no detectable disease at all. For instance, a person with a perforated eardrum may have tinnitus which disappears when the drum heals. Tinnitus is a common symptom of otosclerosis and many other middle ear conditions or infections, and persons with impaired hearing often complain of it. But tinnitus can arise anywhere along the nervous pathways of hearing and the tissues serving them, and it is exceedingly difficult, and often impossible, to pinpoint the site of trouble with precision. It may be associated with hearing loss at a specific frequency.

Head noises are hissing, ringing, whistling, whizzing, roaring, booming sounds but the kind most frequently described by patients is "like steam coming out of a kettle" or "foghorn," or "bells." The patient with tinnitus usually works at peace in noisy surroundings which drown out his head noises, and the condition may be most annoying at night when the room is quiet.

Sometimes head noises (a more serviceable word than tinnitus) can be cured by so simple a measure as removing hard masses of wax from the ears. Other local causes may be found. Clicking and ticking noises may be produced by contraction of small muscles, or in the eustachian tube by sticky mucoid material which can snap like a rubber band when the tube opens and closes. Drugs such as quinine, aspirin, and alcohol may cause tinnitus.

THE HEARING PROCESS

The outer ear serves to funnel and conduct sound vibrations to the eardrum through the ear canal. The eardrum vibrates in response to the sound waves that strike it. This vibratory movement, in turn, is transmitted to the chain of three tiny bones in the middle ear. These small bones, the ossicles, conduct the sound vibration across the air-filled middle ear cavity to a fluid in the delicate inner ear. The vibrating action of the ossicle creates waves in the inner ear fluid that stimulate microscopic hair cells. The stimulation of these hair cells generates nerve impulses, which pass along the auditory nerve to the brain for interpretation.

The outer and middle sections of the ear serve to conduct sound energy to the deeper structures. Therefore, the outer and middle ear, collectively, serve as the conductive hearing mechanism. In contrast, the deeper structures, including the inner ear and the auditory nerve, are referred to as the sensorineural mechanism. The terms conductive and sensorineural are used to describe two major types of hearing impairment.

As long as the hearing mechanism functions normally, the ear has the ability to detect sounds of minute intensity, and at the same time to tolerate sounds of great intensity. The loudest sound the normal ear can tolerate is more than one hundred million (10^8) times more powerful than the faintest sound the ear can detect (see Figure 4-10). Further, a young listener with normal hearing can detect sounds across a very wide frequency range—from very low pitched sounds of 20 Hz to very high-pitched sounds of 20,000 Hz.

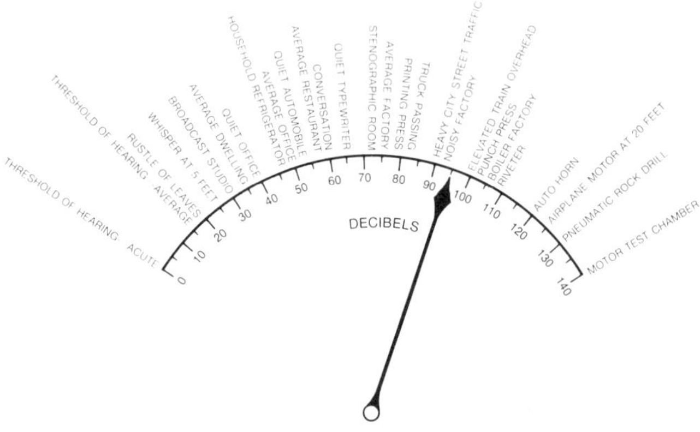

Figure 4-10. Typical sound levels associated with various activities.

Although nature has surrounded the delicate ear mechanism with hard, protective bone, any portion of the ear can become impaired. The part of the hearing mechanism affected and the extent of damage has a direct bearing on the type of hearing loss that results.

Audiometer

An audiometer is a frequency-controlled, audio-signal generator. It produces pure tones at various frequencies and intensities for use in measuring hearing sensitivity. When hearing thresholds are measured, essentially it is a person's ability to hear the simplest form of sound (called pure tones) that is being measured.

The audiometer was developed to provide an electronic pure-tone sound stimulus similar to the tuning fork. In one respect the audiometer is superior to the tuning fork; intensities can be controlled much more accurately and, therefore, the results can be more carefully quantified.

Audiogram

The audiometer is used to test hearing by finding the minimum intensity levels at which a person is able to distinguish various sounds. The results are recorded on a standard chart—the audiogram.

Frequency. Shown across the top of the audiogram in Figure

4-11 are several numbers (125–8,000 Hz). These numbers represent the frequency or pitch of sounds, expressed in hertz (Hz). The lower numbers (125, 250) to the left represent low-pitch sound. For example, a 250-Hz tone sounds like middle C on a piano. The tones become progressively higher in pitch as one moves to the right across to the higher numbers. A 4,000-Hz tone sounds much like a piccolo hitting a high note.

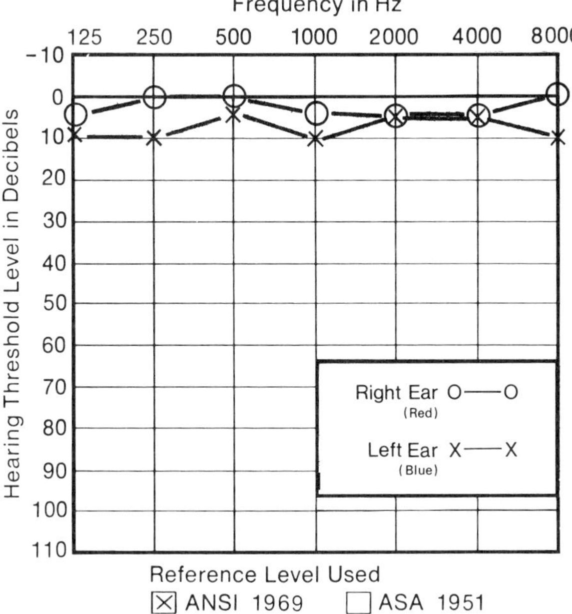

Figure 4-11. A typical manual audiogram showing hearing thresholds within the normal range.

Intensity. In contrast, the numbers on the left side of the audiogram in Figure 4-11 indicate the intensity or loudness of the sound, which is measured in decibels (dBA). The smaller the number, the fainter the sound. When measuring a person's hearing, the level is established at each test frequency where the sound can just barely be heard. This level is called the threshold of hearing.

In audiometry, the further a person's threshold is below the zero line of the audiogram, the greater is the loss of hearing. Common practice is to record the pure tone thresholds by air conduction for the right ear on the audiogram as red circles. In contrast, a blue X is recorded to reflect each threshold for the left ear.

If a sound must be made louder than 25 dBA in the speech-important frequencies for a person to detect its presence, the thresholds begin to fall into the range of hearing impairment. Thus, the more intense (louder) the sound from the audiometer must be for a person to hear it, the greater is that individual's hearing loss. However, as long as hearing is normal, or nearly normal, across the speech frequency range (500–2,000 Hz) the person should have little difficulty hearing speech in ordinary listening situations.

The American Academy of Ophthalmology and Otolaryngology has recommended that audiograms be drawn to a scale like the illustration shown in Figure 4-11. For every 20-dBA interval measured along one side and for one octave measured across the top (250–500 Hz, for example), there will be a perfect square. The reason for this recommended scale is that the appearance or pattern of a hearing loss can be altered a good deal by changing the dimensions of an audiogram. If the proportions of the audiogram are different from standard dimensions, a person's hearing loss may look quite different than if it were plotted on the standard audiogram format. Customarily, audiograms are scaled in 10-dBA steps. Obviously, if a person has a threshold of 55 dBA, it is plotted on the appropriate frequency line at the halfway point between 50 and 60 dBA.

Hearing losses plotted on a chart produce a profile of a person's hearing. A trained person can review an audiogram to determine the type and degree of hearing loss, and can estimate the difficulty in communication this loss will cause.

Calibration. Hearing level for a pure tone is the number of decibels that the listener's threshold of hearing is as compared to the standard audiometric zero for that frequency. It is the reading on the hearing threshold level (hearing loss) dial of an audiometer that is calibrated according to American National Standard Institute (ANSI) standard *Specifications for Audiometers* (S 3.6-1969 R1973). Prior to the adoption of this standard, audiometers were calibrated according to American Standard Z24.5-1951. There are still in use some audiometers calibrated to the older values (ASA-1951). For this reason, each audiogram or record of an audiometric test should include a specific notation as to whether it is based on the current ANSI-1969 or the ASA-1951 reference levels.

HEARING LOSS

A steady loss of hearing acuity occurs as we grow older. The normal young ear can hear tones within a range of 20 Hz—the lowest bass note of a piano—up to high-pitched sounds of 20,000 Hz. People in their sixties are lucky to hear normal level sounds at 12,000 Hz. This hearing loss is greater for the high-frequency sounds and must be considered normal since it happens to practically everybody as the years roll on.

A slight loss of perception of high-pitched sounds muffles some of the shrillness of the world. But hearing impairment severe enough to make ordinary conversation difficult or impossible to understand is quite another matter. The pitch of human speech ranges between 300 and 4,000 Hz. These are the frequencies most vital for communication. Inability to hear well within this range is a serious personal and social handicap. Hard-of-hearing persons are often blamed unfairly for being crotchety, cantankerous, rude, and suspicious, but in reality, they may not answer questions because they have not heard them.

Bone-conduction tests

Until now, only one of the two ways that sound reaches the inner ear has been discussed—that is by air conduction where sound travels from the outer ear through the bones of the middle ear into the fluid of the inner ear (Figure 4-12). However, there is another way to introduce sound and to measure hearing. This is by bone conduction, where sound travels directly to the inner ear via the bones of the skull; a route that by-passes the outer and middle ear. Bone-conduction audiometry is rarely performed by the industrial audiometric technician; however, a basic understanding of bone-conduction tests can help when interpreting audiograms.

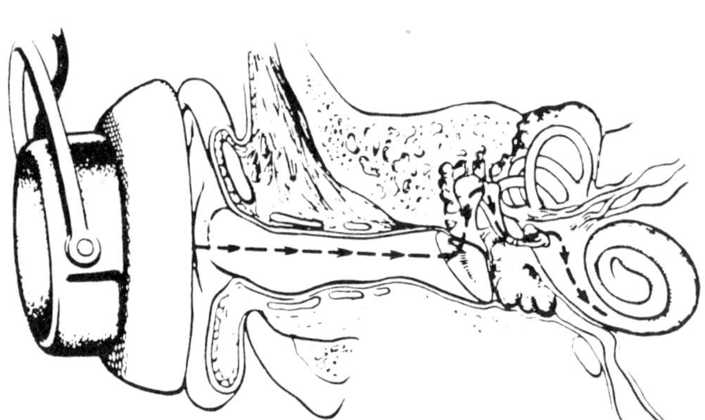

Figure 4-12. An air-conduction earphone is depicted—note that the earphone is placed directly over the external ear canal, and the sound waves are conducted (by air) to the eardrum and through the middle ear to the inner ear.

During such a test, a bone vibrator is placed on the mastoid bone behind the auricle (outer ear), and it is held in place by a headband. It is, in fact, a unit that vibrates the skull (Figure 4-13).

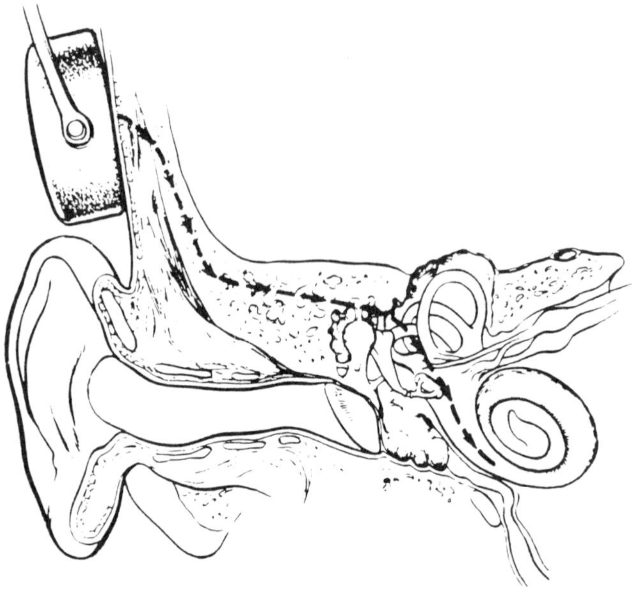

Figure 4-13. Sound can be transmitted directly to the inner ear through the bones of the skull using a bone-conduction vibrator placed on the mastoid bone behind the outer ear. The broken line (with arrows) shows the path taken by the sound waves through the bony areas of the head to the inner ear.

Obviously, bone has more resistance to vibration than the air column in the outer ear canal. As a result, it takes a good deal more intensity for a listener to detect sound from the bone vibrator than from an earphone. This increase in sound output is built into the audiometer when it is manufactured and calibrated at the factory.

Types of hearing loss

Conductive. When test results show that a person has depressed hearing by air conduction but normal hearing by bone conduction, the presence of a conductive hearing loss is indicated (Figure 4-14). In other words, the conductive mechanism is impaired in some way since the normal bone conduction responses indicate that the deeper structures of the ear are intact.

Sensorineural. If, however, the person hears just as poorly by bone conduction as by air conduction, then the hearing loss can be due to damage in the deep structures of the ear. (No matter how sound is presented to the sensorineural mechanism, it is met by an insufficient receiver in the inner ear.) This then would indicate a sensorineural loss (Figure 4-15).

Mixed hearing loss. A third type of hearing loss is a combination of conductive and sensorineural. This is referred to as a mixed hearing loss. If the person has a mixed type of impairment, hearing loss will show on both types of tests.

A conductive loss is due simply to some impairment of sound transmission before it reaches the inner ear. A conductive impairment, then, is one that results from some interference with the function of the outer or the middle ear.

Any blockage—usually ear wax or infection—of the outer ear that results in a loss of sound energy being conducted to the middle ear can cause a conductive hearing loss. Similarly, any impairment in the sound transmission system of the middle ear can cause a conductive hearing loss. Of course, such a loss could also be due to malfunction in both the outer and the middle ear.

In contrast, a hearing impairment that involves only the inner ear or the auditory nerve is classified as a sensorineural impairment. (Sensori refers to the sense organ in the inner ear, neural refers to the nerve fibers.) A sensorineural loss can involve either an impairment of the cochlea, the auditory nerve, or both.

It is virtually impossible to tell from an audiogram whether the damage is in the inner ear or in the auditory nerve, which is the transmission line to the brain. For this reason the loss is labeled sensorineural, because the specific area of damage cannot be determined from audiometric findings.

In the past, two different terms were also used to describe sensorineural losses. These terms were nerve deafness and perceptive deafness. These are losing their popularity, however.

Of course, it is possible for a person to have something wrong with both the conductive hearing mechanism and the sensorineural hearing mechanism. In fact, the damage does not even have to occur at the same time. For example, a person may have a sensorineural loss and then later develop a middle ear infection, which would then produce a conductive hearing loss. If the person had both types of hearing loss, it would be called mixed hearing loss, as discussed earlier.

Certain drugs such as quinine and streptomycin can inflict permanent injury upon the auditory nerve. Sensorineural hearing loss is a rare, but permanent, complication of mumps. Blows and skull fractures can damage ear structures. Advanced infections of the middle ear are less common since the advent of antibiotics, but are still a significant cause of hearing impairment. Tendencies to some ear defects seem to run in families. This is true of otosclerosis, which has a hereditary component.

EFFECTS OF NOISE EXPOSURE

Effects of noise on humans can be classified in various ways. For example, the effects can be treated in the context of health or medical problems owing to their underlying biological basis.

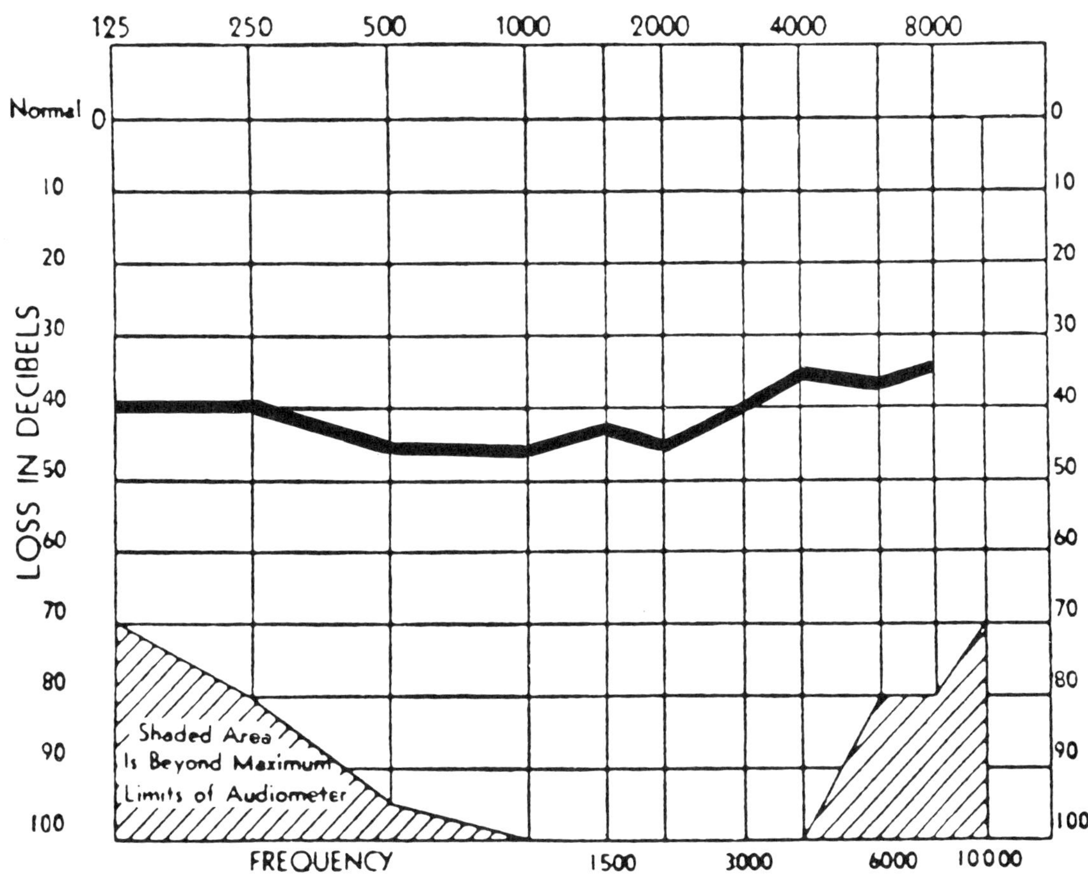

Figure 4-14. Audiogram shows conductive or middle ear hearing loss. Characterized by a relatively flat curve, this loss is not caused by noise exposure.

Noise-induced hearing loss involves damage to the structure of the hearing organ. In contrast speech interference and annoyance can be considered to be nonbiological problems because they involve no pathology.

The ear is especially adapted and most responsive to the pressure changes caused by airborne sounds or noise. The outer and middle ear structures are rarely damaged by exposure to intense sound energy, although explosive sounds or blasts can rupture the eardrum and possibly dislodge the ossicular chain. More commonly, excessive exposure produces hearing loss that involves injury to the hair cells of the inner ear.

Temporary threshold shift

Temporary threshold shift (TTS) of the hearing level can be produced by a brief exposure to high-level sound. Temporary threshold shift is greatest immediately after exposure to excessive noise and progressively diminishes with increasing rest time, as the ear recovers from the apparent noise overstimulation. A noise capable of causing significant TTS with brief exposures is probably capable of causing a significant permanent threshold shift (PTS) upon prolonged or recurrent exposures.

Permanent threshold shift

Permanent threshold shift resembles TTS except that the recovery of hearing is less than total. Important variables in the development of temporary and permanent hearing threshold changes include the following:

1. Sound level—sound levels must exceed 60–80 dBA before the typical person will experience TTS.
2. Frequency distribution of sound—sounds having most of their energy in the speech frequencies are more potent in causing a threshold shift than are sounds having most of their energy below the speech frequencies.
3. Duration of sound—the longer the sound lasts, the greater the amount of threshold shift.
4. Temporal distribution of sound exposure—the number and length of quiet periods between periods of sound influences the potentiality of threshold shift.
5. Individual differences in tolerance of sound—these may vary greatly among individuals.
6. Type of sound: steady-state, intermittent, impulse, or impact—the tolerance to peak sound pressure is greatly reduced by increasing the rise time and/or burst duration of the sound.

Noise-induced hearing loss

When a person is first exposed to hazardous levels of noise, the initial change usually observed is a loss of hearing in the higher frequency range, usually a dip or a notch at about 4,000 Hz. After a rest period away from the noise, the hearing usually returns to its former level. For practical purposes, a rest period of 14 hours or so away from the noise is adequate to return the threshold to previous levels.

Permanent damage from noise is generally classified as noise-

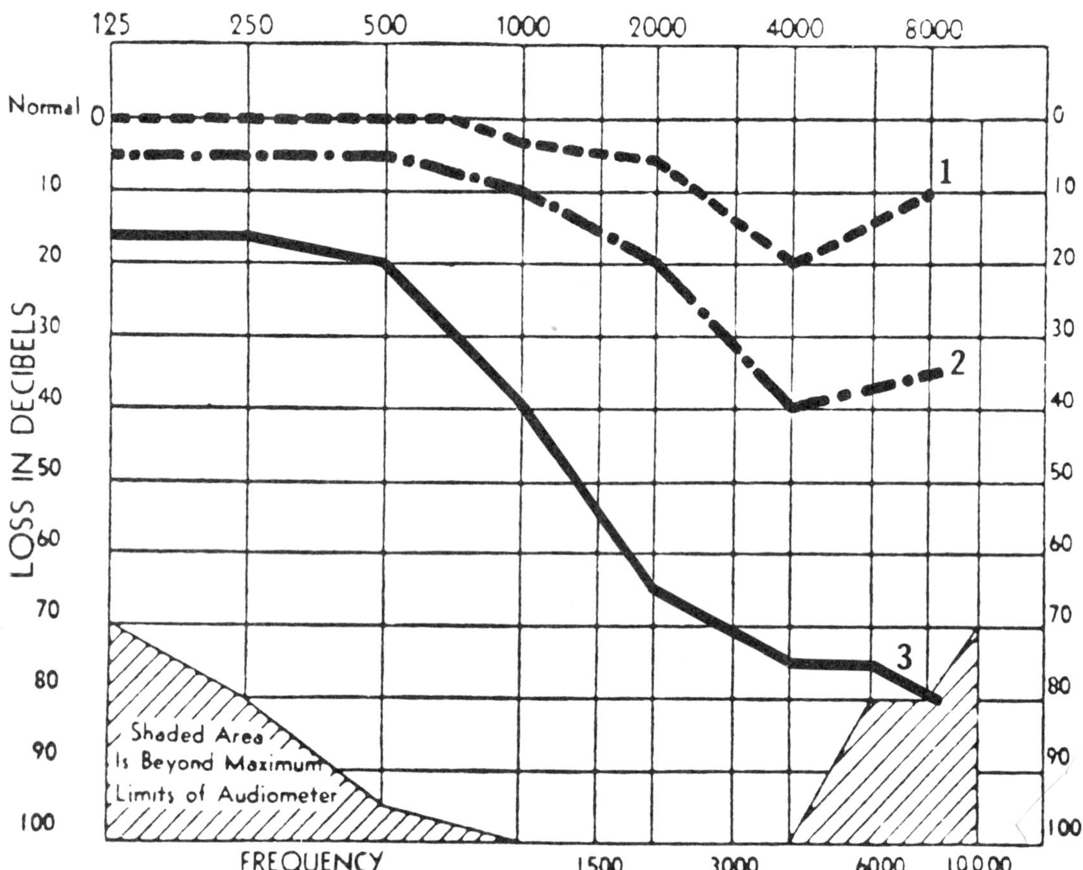

Figure 4-15. Sensorineural hearing losses of the kind produced by noise or other causes. Curve 1 = early; curve 2 = intermediate; curve 3 = advanced. (As shown, curve 3 might include some involvement of presbycusis.)

induced hearing loss or acoustic trauma, depending upon the nature of exposure. The long-term cumulative effects of repeated and prolonged hazardous noise exposure result in permanent pathologic changes in the cochlea and irreversible threshold shifts in the hearing acuity. This is referred to as noise-induced hearing loss. It is usually represented audiometrically by a notch at 4,000 Hz. Because the hearing loss, however, does not necessarily stop here, further exposure may result in a deepening and widening of the notch. When the hearing loss involves the speech frequency range, considerable difficulty in hearing conversational speech develops.

The effect of noise on hearing depends on the amount and characteristics of the noise as well as the duration of exposure. In some instances, employment for a few hours or days in a noisy industrial environment or exposure to a single sound of damaging intensity may suffice to produce a permanent hearing loss. This is often referred to as acoustic trauma. Yet, others working in the same occupational noise atmosphere for many years retain normal hearing acuity and are unaffected. The major deterioration of hearing, however, occurs during the initial 5–10 years of employment in a noise-risk environment.

Noise is a pervasive, insidious cause of hearing loss. It causes no particular pain unless it is as loud as a rifle blast. The ears have considerable comeback power from temporary brief exposure to noise and ordinarily recover overnight. However, prolonged exposure to intense noise gradually damages the inner ear.

Susceptibility to noise-induced hearing loss varies greatly from one individual to another. Beyond certain levels of extremely high intensity, it is generally agreed that all individuals are susceptible—provided the exposure is long enough. Generally speaking, tests of temporary threshold shift throw no specific light on the susceptibility of individuals to permanent threshold damage.

COMMUNICATION PROBLEMS

The communicative problem of persons with a noise-induced hearing loss is very frustrating to them and is easily misunderstood by family and friends. This problem causes a great deal of inconsistent auditory behavior—the person appears to hear very well at some times and very poorly at others, and is thus often accused of not paying attention.

It is helpful to understand the kind of communicative problem imposed by a hearing loss caused by substantial exposure to noise.

Hearing v. understanding

There are two important characteristics of normal hearing—the ability to hear sounds as loud as they truly are, and the ability to hear sounds with complete clarity. It is important that the distinction between these two characteristics of hearing be understood.

Loudness

If Figure 4-16 (left) is held at arm's length, the printing in the center is obvious but almost impossible to read. A close look

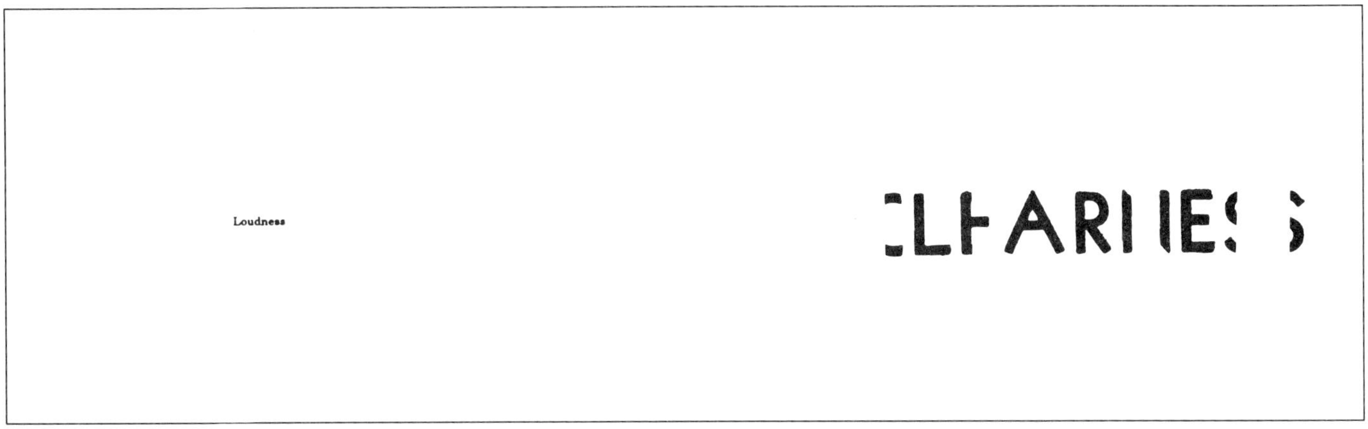

Figure 4-16. Both sides of this illustration represent, by analogy, two important dimensions of normal hearing—the ability to hear sounds as loudly as they really are, and the ability to hear sounds with complete clarity. *Left,* it is almost impossible to read the word, LOUDNESS; this is comparable, visually, to not being able to hear faint sounds. Moving the figure closer makes it easier to read, just as increasing the volume makes it easier to hear. *Right,* no matter how closely the other illustration is held, it is difficult to interpret. The word is not clear, because some important parts of the letters are missing. This visually shows a hearing difficulty caused by a loss in the ability to distinguish between various sounds (the word is CLEARNESS).

confirms that it is indeed a word and an even closer look reveals that the word is LOUDNESS. This example illustrates a very common problem associated with hearing loss—the inability to hear soft sounds. If the sounds are made louder, a person with only a loss of hearing sensitivity will have much less difficulty.

Clarity

Now, if Figure 4-16 (right) is held at arm's length, there is no difficulty in seeing everything within the box. But can the word be read? If not, the reason is that too much of the word is missing, even though it is large enough. No matter how closely you look, it is difficult to read because some of the important parts are missing. This illustrates a problem in loss of clarity or an inability to distinguish between the various sounds in the language. (The word, incidentally, is CLEARNESS.)

If damage is sustained to any portion of the outer and middle ear, the primary result is an inability to hear soft speech. Clarity, however, is preserved. The key to clarity in hearing is held by the inner ear mechanism and the nerve fibers that carry the message to the brain.

If the inner ear or auditory nerve is damaged, not only a loss in the loudness of sounds can be expected, but, in many cases, a loss in clarity as well. In such dual problems associated with sensorineural hearing loss, speech seems muffled or fuzzy no matter how loud it is. In these cases, the major problem is trouble in understanding what is being said.

Someone with a hearing loss from noise exposure often has this kind of problem. The tiny hair cells that respond to specific speech sounds may be so severely damaged that they cannot react when the vibrations from the outside strike them. At the same time, hair cells for other speech sounds may be functioning normally. Someone with this hearing impairment misses parts of words and parts of sentences and often misunderstands what is being said. This can be a very subtle problem. In fact, persons with this loss often are not aware of the reason that they do not hear everything.

To summarize, persons with a high-frequency, noise-induced loss can hear speech, but may not understand what is being said.

Speech sounds

Characteristic speech sounds can be related to the two principal kinds of speech—vowels and consonants.

The vowel sounds—located in the lower frequencies—are the more powerful speech sounds. Therefore, vowels carry the energy for speech. In contrast, the consonant sounds—located in the higher frequencies—are the keys to distinguishing one word from another, especially if the words sound alike. This is the heart of the communicative problem of persons with a noise-induced (high-frequency) hearing loss. They cannot distinguish the difference between similar words, such as stop and shop.

It is quite easy to miss a key sound in a word. In turn, this could change the meaning of a key word in a sentence. As a result, the entire sentence might be misunderstood.

Persons with a high-frequency hearing loss often get along fairly well in quiet listening situations. But as soon as they are in a place where there is a lot of background noise, such as in traffic or on the job, it becomes difficult to communicate through hearing alone. For example, if a speaker continued talking in the presence of a typical background sound, a listener should hear quite well.

But, if a listener developed a hearing loss for all speech sounds above 1,000 Hz, a marked loss in communication would be noticed immediately. In other words, so long as it is quiet, a person can use good hearing below the midpitch range to an advantage. Unfortunately, most noises around us interfere with the low-pitched speech sounds and cause us to miss many of these sounds as they occur in words and sentences. Thus, the hearing loss caused by the noise and the hearing loss from damage to the ear result in a greater hearing problem than occurs with just one of these conditions.

Hearing aids

Although there is no medical or surgical cure for sensorineural deafness, using auditory rehabilitation methods, much can be done to help someone compensate for a hearing handicap and lead a normal life, with minimal effect on personality or social and economic status.

Fitting of a proper type of hearing aid, when indicated, is a vital part of an auditory rehabilitation program. However, a patient must be psychologically prepared to accept a hearing aid. Many are reluctant to use hearing aids and never use them, or use them ineffectively. Before recommending a hearing aid, determine whether the patient will be helped by it enough to justify purchasing one. This is particularly important in cases of sensorineural hearing loss, in which the problem is more one of discrimination than of amplification.

In persons whose hearing losses resulted from noise exposure, one of the most important benefits that a hearing aid provides is to enable the individual to hear what is already heard, but with greater ease—it minimizes the severe stress of listening. Although comprehension may not be enhanced, there nevertheless may be great benefit from the device, because it relieves tension, fatigue, and some complications of hearing impairment.

Patients who seek early medical attention for hearing loss are wise in many respects. If the condition can be helped by medical or surgical means, there is a better chance of being helped. If a hearing aid is necessary, the sooner it is acquired, the less dramatic environmental noises will be when audible again—ranging from the barking of dogs and the crying of babies to more melodious sounds.

Once purchased, thousands of hearing aids, given too brief and halfhearted a trial, are relegated to a desk drawer. Overlong postponement in acquiring the aid is sometimes a factor, or a patient expects to hear normally with a hearing aid, when the condition of his hearing organs makes such a result impossible. The physician, audiologist, and the hearing aid dealer should make it clear to the patient that a hearing aid is not a substitute for a normal ear, especially in a patient with sensorineural deafness. One common cause of disappointment with a hearing aid is patients' expectations for invisible or inconspicuous aid, when what they should be looking for is an aid that will enable them to understand conversation with maximum effectiveness.

AMA GUIDES FOR EVALUATING IMPAIRMENT

This section on determining the percent impairment is adapted from the American Medical Association's (AMA) *Guides to Evaluation of Permanent Impairment,* and is included to assist health and safety professionals in interpreting and understanding medical reports of workers' compensation cases.

This section includes criteria for use in evaluating permanent impairment resulting from the principal dysfunctions of the ear, and thereby determine the corresponding percentage of permanent impairment in terms of the whole person.

Although the ear and related structures each have multiple functions, some of which are closely allied, permanent impairment usually results from a clinically established deviation from normal in one or both of the following functions: hearing and equilibrium.

It is recognized that in certain infrequent cases other dysfunctions or disturbances may occur. If such a dysfunction or disturbance produces permanent impairment, a value that is based on its severity and importance and consistent with established values should be assigned by the examining physician. This assigned value is then combined with any other impairment values in the case under consideration.

As in other types of impairments, competent evaluation presupposes that the necessary equipment for testing the various functions is available to the physician. It should be emphasized, however, that impairment of normal function should be determined without the aid of any prosthetic device.

Furthermore, the physician should be able to explain and substantiate the conclusions regarding the patient's status. When oral or written reports are required, the physician should clearly indicate that the information supplied represents the findings and conclusions as of the date evaluation was made. Such reports should include a history, findings by both physical and laboratory examinations, consultations, diagnoses, classification (when used), and the resulting impairment value. In all cases, special emphasis is expected on observations of an objective nature. When these are not readily available, sound clinical judgment must be substituted.

The functions of the ear are hearing and equilibrium, which are considered separately in this chapter. The criteria for evaluating hearing impairment are relatively specific. On the other hand, it has been necessary to provide rather general criteria for disturbances of equilibrium.

Such disturbances of the ear as chronic otorrhea and otalgia are not measurable and, therefore, a value based on its severity and importance and consistent with established values should be assigned by the physician.

Deformities of the auricle and cosmetic defects that do not alter function are not considered. Tinnitus per se does not result in permanent impairment.

Binaural hearing impairment must be used in determining impairment of the whole person. The American Medical Association has pubished *Guides to the Evaluation of Permanent Impairment,* which include a formula for binaural hearing that is based on the hearing levels of each ear tested separately. The *Guides* rate the hearing loss of one ear to binaural loss by a ratio of 1-to-5. Because many state agencies have adopted the use of the AMA formula for determining hearing impairment, it is important that the industrial hygienist or occupational physician understand them

OSHA HEARING CONSERVATION PROGRAM (29 *CFR* 1910.95)

On April 7, 1983, the Hearing Conservation Amendment (HCA), included in the *Federal Register 29 CFR* 1910.95, became effective. The Occupational Health and Safety Administration (OSHA) Noise Regulation provides specifics on the content of required hearing conservation programs. The regulation covers:

- noise monitoring
- audiometric testing program
- definition of standard (permanent) threshold shift
- employee follow-up and referral
- hearing protection
- employee training
- record-keeping

This amendment requires that all OSHA-covered workers are to be included in hearing conservation programs if they are exposed at or above an 8-hour TWA of 85 dBA or more. Chapter 9 contains a complete discussion of the requirements of the HCA. Unfortunately, however, there are a number of persons who will develop threshold shifts on prolonged exposure to levels at least as low as 80 dBA.

The *Noise and Hearing Conservation Manual* published by the American Industrial Hygiene Association contains a chapter on workers' compensation laws as they relate to hearing loss (see Bibliography). Existing compensation laws do not, however, take into account the effects of high-frequency loss on the ability to discriminate speech sounds, especially against noisy backgrounds.

BIBLIOGRAPHY

Alberti, P.W., ed. *Personal Hearing Protection in Industry.* New York: Raven Press, 1982.

American Medical Association (AMA), Committee on Rating of Mental and Physical Impairment. *Guides to the Evaluation of Permanent Impairment,* 2nd ed. Chicago: AMA, 1984.

American Industrial Hygiene Association. *Noise and Hearing Conservation Manual.* Akron, OH: AIHA, 1986.

Bateman, H.E., and Mason, R.M. *Applied Anatomy and Physiology of the Speech and Hearing Mechanism.* Springfield, IL: Charles C. Thomas Publ., 1984.

Davis, H., and Silverman, S.R., eds. *Hearing and Deafness,* 4th ed. New York: Holt, Rinehart and Winston, 1978.

Henderson, D., Hamernik, R.P., Dosanjh, D.S., and Mills, J.H. *Effects of Noise on Hearing.* New York: Raven Press, 1976.

Hughes, G.B., ed. *Textbook of Clinical Otology.* New York: Thieme, Inc., 1985.

Newby, H.A., and Popelka, G.R. *Audiology,* 5th ed. Englewood Cliffs, N.J.: Prentice-Hall, 1985.

Paparella, M., and Schumrick, D.A., eds. *Otolaryngology,* vol. 2, *The Ear.* Philadelphia, PA: W.B. Saunders Co., 1980.

Sataloff, J., and Michael, P. *Hearing Conservation.* Springfield, IL: Charles C. Thomas Publ., 1973.

Schuknecht, H.F. *Pathology of the Ear.* Cambridge, MA: Harvard University Press, 1974.

The Eyes

by Julian B. Olishifski, MS, PE, CSP
Revised by George S. Benjamin, MD, FACS

The eye may be the organ most vulnerable to occupational injuries. Although the eye has some natural defenses, they do not compare with the healing properties of the skin, the automatic cleansing abilities of the lungs, or the recuperative powers of the ear. Consequently the eye is at greater risk, and eye and face protection is an area of major importance in the practice of occupational health.

The demands placed upon the eye in modern workplaces and practices, such as prolonged viewing at close distances or prolonged viewing at distances neither near nor far, are great. Both of these conditions can cause acute and chronic eye fatigue and visual discomfort. There are other dangers from hazardous substances that can be absorbed into the eye system or from machinery, which, if guarding mechanisms should fail, can propel objects capable of causing traumatic injury to the eye.

The eye, with its remarkable ability to translate radiant light energy into neural impulses, which are transmitted to the visual cortex of the brain, is certainly one of the most valued organs. Protection of the sensory organs of this complex system should be a high priority in every occupational health and safety program.

ANATOMY

A look at the structure of the human eye and how it can be affected by industrial hazards emphasizes the need for sound eye protection programs. The eyeball is housed in a cone of cushioning fatty tissue that insulates it from the skull's bony eye socket. The skull has brow and cheek ridges projecting in front of the eyeball. It is composed of specialized tissue—tissue that does not react to injury like other body tissue (Figure 5-1).

Eyeball

The eyeball consists of three coats, or layers, of tissue surrounding the transparent internal structures. There is an external fibrous layer, a middle vascular layer, and an inner layer of nerve tissue.

The outermost fibrous layer of the eyeball consists of the sclera and the cornea. The sclera, also called the white of the eye, is composed of dense fibrous tissue and is the outer, protective and supporting layer of the eyeball. The sclera is modified from a white, opaque membrane to the transparent cornea. The cornea is composed of a dense fibrous connective tissue and has no blood vessels. The cornea must be transparent to let light through to the receptors within the eyeball.

The middle vascular layer of the eyeball is heavily pigmented and contains many blood vessels that help nourish other tissues.

The nerve layer, or retina, is the third and innermost layer of the eyeball. Toward the rear, the retina is continuous with the optic nerve; and toward the front, it ends a short distance behind the ciliary body in a wavy border called the ora serrata. The retina is composed of two parts, the outer part is pigmented and attached to the choroid layer, and the inner part, consisting of nerve tissue.

The front of the eyeball is protected by a smooth, transparent layer of tissue called the conjunctiva. A similar membrane covers the inner surfaces of the eyelids. The eyelids also contain dozens of tiny tarsal glands that secrete an oil to lubricate the surfaces of the eyeball and eyelids. Still further protection is provided by the lacrimal gland, located at the outer edge of

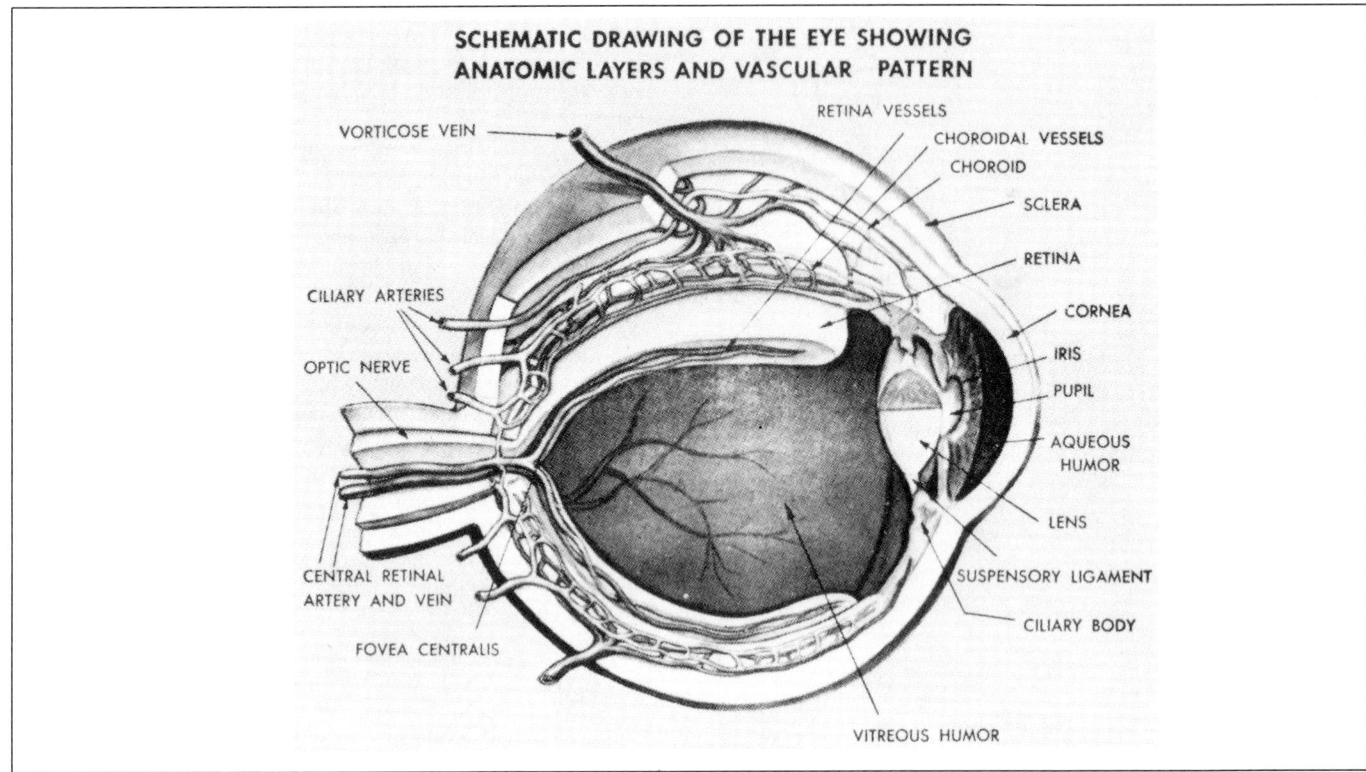

Figure 5-1. Schematic drawing of the eye showing anatomic layers and vascular pattern. (Reprinted with permission from the American Medical Association [AMA]. *The Wonderful Human Machine.* Chicago: AMA, 1971)

the eyesocket. It secretes tears to clean the protective membrane and keep it moist (Figure 5-2).

The region between the cornea and the lens is filled with a salty, clear fluid known as the aqueous humor. The eyeball behind the lens is filled with a jellylike substance called the vitreous humor.

Light rays enter the transparent cornea, and are refracted at the curved interface between air and the fluid bathing the cornea. After passing through the cornea and the clear liquid, the aqueous humor contained in the anterior chamber, the bundle of rays is restricted by a circular variable aperture, the pupil. Its size is changed through action of the muscles of the iris.

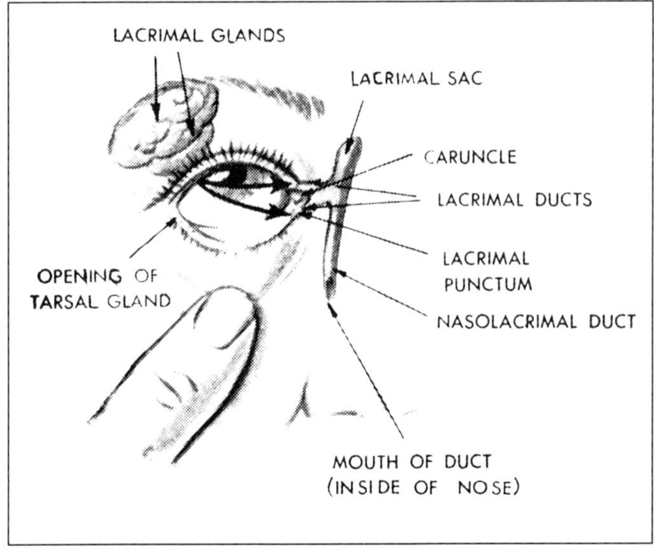

Figure 5-2. Illustrations of the eye and tear ducts. (Reprinted with permission from the AMA. *The Wonderful Machine.* Chicago: AMA, 1971)

The light rays are further refracted by passage through the lens, traversing the clear, jellylike vitreous humor of the posterior chamber, so that, in a properly focused eye, a sharp image is formed on the retina. Scattering of light within the eye is minimized by a darkly pigmented layer of tissue underlying the retina, called the choroid. The choroid contains an extremely rich blood supply, which seems to dissipate the heat resulting from absorbed light energy. The shape of the eyeball is maintained by its enclosure in an elastic capsule, the sclera, and by the fluids within that are maintained at positive pressure.

The lens is attached by suspensory ligaments to the ciliary body, a muscular organ further attached to the sclera. The ciliary body muscles alter the lens shape to fine focus the incoming light beam. Ordinarily, these muscles are active only when looking at objects closer than 6.1 m (20 ft). Consequently, when doing close work, it is restful to pause occasionally and look out a window into the distance. Many complaints of eye fatigue are really complaints about tired ciliary muscles. With increasing age the lens gradually loses some of its accommodative powers, and no amount of ciliary effort can replace holding a book at arm's length to read it.

The pigmented iris, overlying the lens, is a muscular structure, designed to expand or contract and thus regulate the amount of light entering the eye. The circular aperture formed by the iris is called the pupil.

The aqueous and vitreous humor, and other eye tissues, are composed primarily of water, thus their absorption characteristics are similar to those of water.

Retina

The retina, a thin membrane lining the rear of the eye, contains the light-sensitive cells. These cells are of two functionally discrete types, called rods and cones. They get their name from the rodlike and conelike shapes seen when the layer is viewed under a microscope. The rods are more sensitive to light than the cones; the cones are sensitive to colors.

There are more rod cells than cones, each eye has about 120 million rods and only 6 million cones. The rods are incapable of color discrimination, since they contain a single photosensitive pigment. There are fewer cone cells, and they are less sensitive to low levels of luminance. There are three types of cones in the human eye, each contains a different photopigment with peak response to a particular part of the visible spectrum. Thus, by differential transmission of nerve impulses upon stimulation, the cones are able to encode information about the spectral content of the image so that the observer experiences the sensation of color.

Binocular vision

Binocular vision refers to vision with two eyes. The advantages of binocular vision are (1) a larger visual field and (b) a perception of depth or steroscopic vision. There is a slight difference in the images on the two retinas; there is a right-eyed picture on the right retina and a left-eyed picture on the left retina. It is as if the same landscape were photographed twice, with the camera in two positions a slight distance apart. The two images blend in consciousness and give us an impression of depth or solidity.

EYE PROBLEMS

Before discussing eye problems, a few definitions of the specialists involved in this field are necessary.

Specialists

Ophthalmologist. An ophthalmologist (oculist) is a doctor of medicine, who is licensed to practice all branches of medicine, and specializes in the examination of the eye and its related structures and in the prevention, diagnosis, and medical and surgical treatment of eye defects and diseases. An ophthalmologist also prescribes whatever medication and correction is required, including eyeglasses and contact lenses.

Education and training qualify an ophthalmologist to relate findings observed in an examination of the eye to those diseases in other parts and systems of the body, which may have an effect on the eye. Almost every medical doctor of this type has passed American Board examinations and is qualified as an expert in this field.

Optometrist. An optometrist has the education, training, and licensure necessary to examine eyes for abnormal visual problems not due to disease. Since an optometrist is not a physician, this specialist cannot prescribe drugs, but can prescribe, fit, and supply eyeglasses and contact lenses. During an optometric examination, if there is suspicion of a defect or disease requiring medical or surgical treatment, the patient should be referred to an appropriate medical source for treatment.

Optician. An optician is a person who manufactures eyeglasses at the request of an ophthalmologist or an optometrist. Most states license opticians only after a period of training and require them to follow standards for measuring frames and grinding lenses for eyeglasses. In addition, some opticians fit contact lenses following the prescription of an opthalmologist or optometrist.

Examining instruments

The vision tests used in industry are screening tests, and are used to detect various substandard elements of vision. They are not diagnostic. They do not tell anyone why a person may see better with the right eye than the left; the test merely states what the acuity rating was at that time.

Snellen chart

The most common industrial test for distance acuity is the Snellen wall chart in its several variations. The Snellen test chart consists of block letters in diminishing sizes so at various distances the appropriate letter will subtend a visual angle of 5 minutes at the nodal point of the eye. Thus the top large letter will appear to be the same size when it is 61 m (200 ft) away as will the standard at 6.1 m (20 ft).

The distance of 6.1 m (20 ft) is considered to be infinity. This means that the rays of light coming from an illuminated object are parallel; they neither diverge nor converge. If the object is closer than 6.1 m (20 ft), the light rays diverge and must be made parallel by action of the lens within the eye or by the addition of a supplementary lens held in front of the eye, otherwise they will not come to a sharp focus on the area of central visual acuity of the retina.

The bending of the parallel rays to converge to a focus on the retina is done by the cornea and the lens of the eye. Looking at an object at 6.1 m (20 ft) or more, the normal lens is relaxed into its usual biconvex shape. The parallel rays of light that are bent (refracted) by the cornea and lens cross at the nodal point of the eye (about 7 mm behind the cornea) and continuing their straight course fall upon the retina, forming an inverted image.

It is important to check several factors. The distance from the chart to the individual being tested needs to be 6.1 m (20 ft), or 3 m (10 ft) if a mirror and reversed chart are used. The lighting should be uniform, its source not visible to the person being tested. The chart should be clean. Finally, the tester must be trained to hold the cover correctly over the eye not being tested, to vary the order of lines and letters, and to be alert for any unusual factors. It is important to separately test and record the vision of each eye, and the vision when reading the chart with both eyes.

Satisfactory vision at a distance does not ensure adequate near-point vision, making it important to recognize near-point abnormalities. Many industrial work situations involve near-point seeing even though they are not confined to near-point work.

Industrial vision testing should be done with standardized and foolproof tests to detect and screen out substandard visual functions. Accuracy is vital, because, in some jobs, workers with visual defects can put themselves and others at risk. For instance, one has to depend on the accuracy of the depth perception test and the test record when choosing a crane operator. It is in the best interests of safety that an individual wholly or greatly lacking in depth perception (stereopsis) or a person with monocular vision should not be deliberately selected to operate a crane or an electric truck.

Admittedly, there are accident-free records for one-eyed operators. By trial and error, a crane operator with this problem substitutes other guides—stationary machinery, windows, pillars.

A competent examination of the eyes requires the use of a number of special examining instruments.

- An ophthalmoscope permits study of the interior structures and the slit-lamp microscope allows study of the structures in the anterior through high magnification.
- A tonometer is used to measure the pressures within the eyeball.
- A perimeter is used to map the limits of the fields of vision.
- A gonioscope views the angle of the chamber where the outflow drainage apparatus of the eye is found.
- A retinoscope can detect a refractive aberration, for example, myopia and astigmatism or modifications of the same. The instrument can be fitted with test lenses to determine which lenses will improve the vision (Figure 5-3).

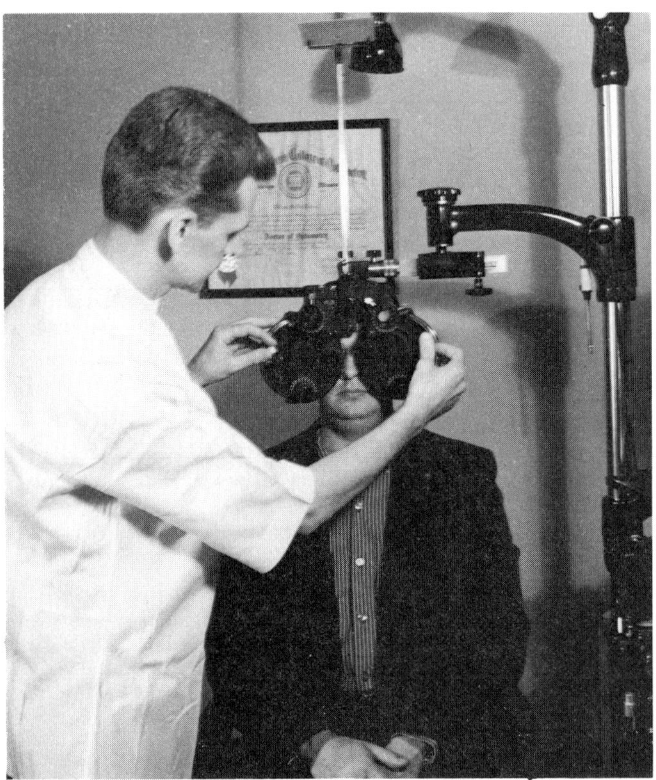

Figure 5-3. A retinoscope is used with a set of test lenses to find which ones aid vision.

- Vision screening device—in a typical device, the test slides are set in two drums rotated by a crank on the outside of the instrument and are not visible except through the viewing box. Tests in the distance drum are at the optical equivalent of 8 m distance. Tests in the near drum are at a downward angle and at the optical equivalent of 0.32 m (13 in.) distance from the instrument lenses. Adapters are available for mid-range, 0.75 m (30 in.), screening. This distance is critical when testing those involved in prolonged use of video display terminals (VDTs) (Figure 5-4). (Video display terminals are discussed later in this chapter.)

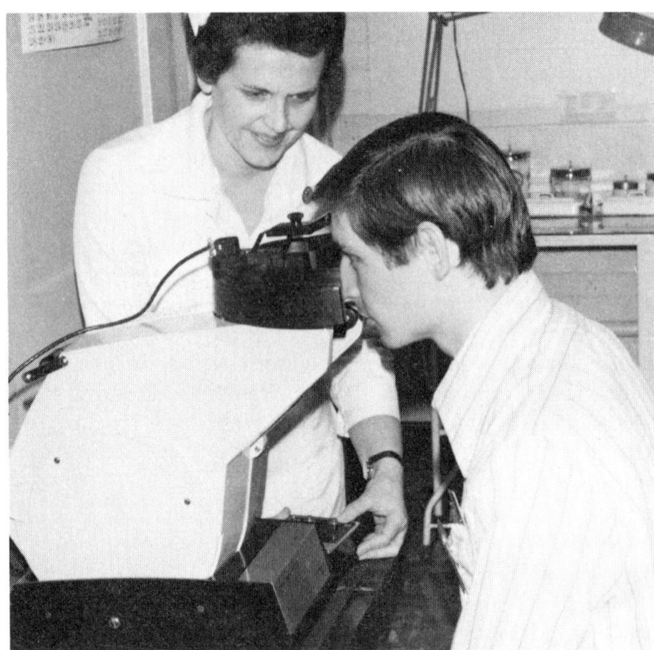

Figure 5-4. Vision screening device tests for acuity, depth, color, and vertical and lateral balance between the two eyes at various distances. (Reprinted with permission of Titmus Optical, Inc.)

Eye defects

At the time of this writing, according to the American Foundation for the Blind, 223 of every 100,000 people are legally blind. Eye diseases and defects affecting the eyes are the leading causes of blindness; but frequently, blindness is the result of misunderstanding, neglect, and delay in seeking aid. Many potentially blinding disorders can be prevented, arrested, or even cured with prompt attention. In addition, about 40 percent of the population wears glasses, indicating that the vision of almost one of every two people is imperfect.

Three common eye defects—farsightedness, nearsightedness, and astigmatism—are the results of simple optical aberrations in the eye.

Farsightedness. When the eyeball is too short from front to back, the light rays come to a focus beyond the retina. Light rays coming from a distant object may reach their focus at the retina, so that distant vision is good, but near vision is blurred. The treatment is to wear a convex lens that will converge the light rays from near objects so that they will be brought to a focus on the retina (Figure 5-5).

The closer an object is brought to the eye, the more convex the human eye lens must become in order to focus it. Through aging, the human lens loses its elasticity and its power of thickening, so that by the time we are in our forties it may seem as if our arms are too short to read a book. This condition is called presbyopia, and is overcome by wearing convex lenses, often in a bifocal. The strength of these lenses may need to be increased as we age.

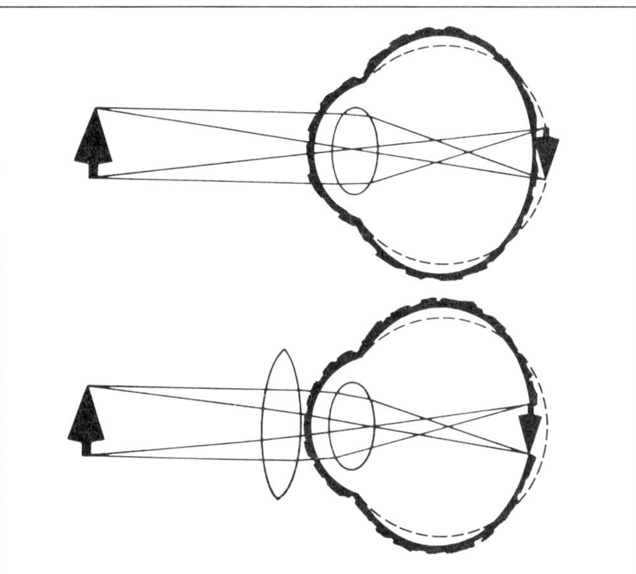

Figure 5-5. In farsightedness, or hyperopia, the image of an object is focused behind the retina of an eyeball that is too short. A convex lens brings the light rays into focus on the retina. Although hyperopic persons may be able to see things sharply by thickening the lens of the eye (accommodation), this involves effort of inner muscles of the eye and may cause eye fatigue.

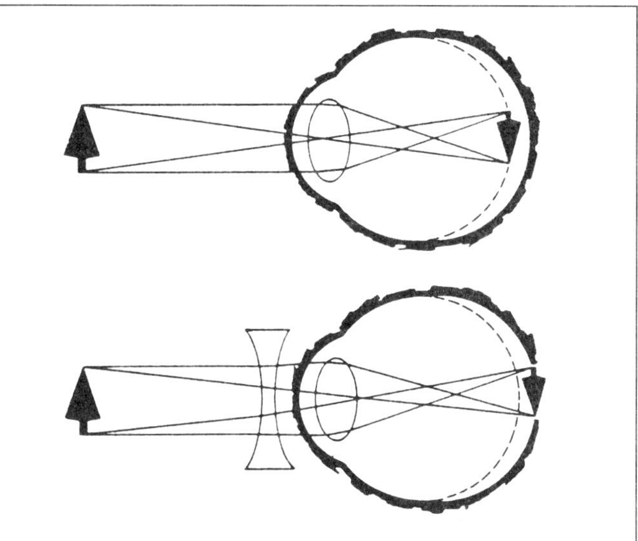

Figure 5-6. In nearsightedness, or myopia, the image of an object (unless it is held close to the eyes) falls in front of the retina instead of upon it, and the object is seen indistinctly. The condition is corrected by placing a concave lens of proper curvature to bring the image into focus on the retina. (Reprinted with permission from Cooley, D.G., ed. *Family Medical Guide.* New York: Better Homes and Gardens Books, 1973)

Nearsightedness. If the eyeball is too long from front to back, as it is in nearsightedness, or myopia, the image of an object 6.1 m (20 ft) or more away falls somewhere in front of the retina, and can only be sharply focused by looking through a concave lens, which diverges the rays coming from the object. By bringing the object near enough to the eyes, a myopic person can get a good focus on the retina (Figure 5-6).

Astigmatism. If the curvature of the cornea is irregular so some rays of light are bent more in one diameter than in another, the resultant image is blurred because, if one part of the ray is focused the other part is not. This is something like the distortion produced by a wavy pane of glass, and is called astigmatism. It is corrected by using a lens that bends the rays of light in only one diameter (axis). This lens is called a cylindrical lens and it can be turned in the trial frame to its proper axis to even up the focusing of the rays of light in all parts (Figure 5-7).

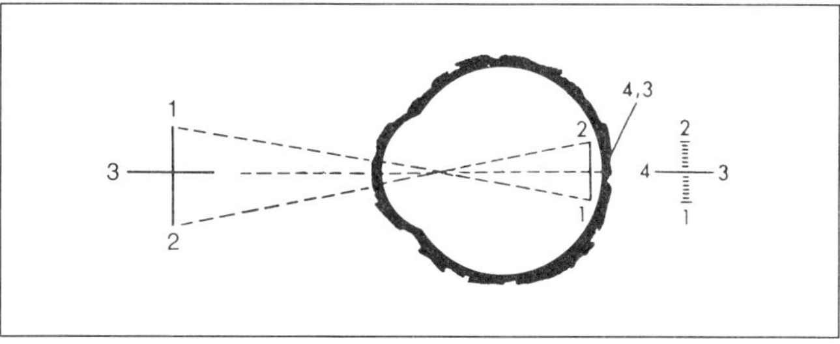

Figure 5-7. Astigmatism resulting from irregular curvature of the cornea is something like distortion produced by a wavy pane of glass. Drawing shows light rays (3,4) in sharp focus on retina, with light rays (1,2) focused in front of the retina, resulting in a blurred image. Small diagram on the right shows horizontal image in focus, vertical image out of focus. A cylindrical lens in the proper axis to bring light rays to even focus corrects astigmatism.

Eyeglasses

The purpose of wearing eyeglasses is to help focus the rays of light onto the retina. The eye is a receiving organ only; the image on the retina is carried into the brain and interpreted.

Glasses cannot change the eye or produce any disease even if they are badly fitted.

A prescription for glasses may look something like this: + 2.0 D^1 ◯+ 0.50 cyl ax 90.

The sign + indicates a convex lens suitable for a farsighted person. The sign − indicates a concave lens for a myopic person. The D is an abbreviation for diopter, which indicates the strength or power of the lens. A diopter is a unit of measurement of the refractive or light-bending power of a lens. The normal human lens in its relaxed biconvex shape has a power of about 10 D. The symbol ◯ means combined with.

Thus, the prescription in the example means the optician grinds a convex spherical lens of 2.0 D combined with a convex cylindrical lens of half a diopter (0.50) situated vertically (axis 90 degrees). Lens prescriptions may look strange, but opticians everywhere know what they mean.

Incidentally, it is a good idea to carry your lens prescription when traveling in the event your glasses are lost or smashed. A spare set of glasses is good insurance, particularly for persons who would be seriously handicapped without them.

VISUAL PERFORMANCE

Normal visual performance involves a number of interdependent discriminations made in response to the visual environment and mediated by the visual system.

Visual acuity

There are many definitions of the term, visual acuity; all, however, incorporate the concept of detail resolution. Many test patterns have been used to measure acuity, from single dots to twin stars, gratings, broken rings, checkerboards, and letters. There is no general agreement on the choice of a test, and results from the different patterns are often at odds. The Snellen Letter Test is probably the most familiar, and is widely used, despite the fact that it is a test of letter recognition rather than of retinal resolution.

The most satisfactory expression for visual acuity is the amount of critical detail that can just be discriminated.

Some important variables affecting visual acuity are:

Luminance. The level of adaptation of the eye has a profound effect upon visual acuity.

Position in the field. At photopic levels, acuity is best at the fovea, and drops off as the retinal periphery is approached, owing to receptor population differences. Nocturnal acuity is quite poor, with essential blindness at the fovea and best resolution appearing in the periphery where rod packing is densest.

Duration. When the pattern is exposed for only a short time, measured acuity diminishes.

Contrast. Visual acuity decreases as the contrast between pattern and background is diminished. The form of this relationship depends upon the adapting luminance.

Dark adaptation

Optimal visual discrimination under conditions of very low light can be made only if the eyes are adapted to the level of the prevailing light, or even lower. If a fully light-adapted eye is suddenly plunged into darkness, its initial sensitivity is poor. With time, however, sensitivity begins to increase as a result of photochemical regeneration, certain functional neural changes, and, to a much smaller degree, enlargement of the pupil of the eye. If the eye remains in total darkness for 30–60 minutes, the adaptation process will be nearly complete and the sensitivity of the eye in those parts of the retina where both rods and cones are present will have increased by a factor of 10,000 for white light.

There are several important operational consequences of the dark adaptation process and its properties:

- Best performance on a task at low light requires that the eye is preadapted to an appropriately low level for sufficient time to obtain maximum sensitivity.
- Since the rods are more sensitive than the cones at low light levels, detection capability will occur best on those parts of the retina where rods abound (10–30 degrees from the fovea), and averted vision is required for optimal performance.
- Since the rods are relatively insensitive to extreme red wavelengths, dark adaptation will proceed if the observer wears red goggles or if the illumination provided is a very deep red, for example, in a spacecraft or photographic darkroom. By this means it is possible to continue to use the high-acuity capability of the central fovea at elevated luminance levels for reading instruments and such, while the adaptation process goes on; although vision naturally will be monochromatic in this case.
- Because the two eyes are essentially independent in adaptation, it is possible to maintain dark adaptation in one, for example, by means of an eye patch, while the other is used at high light levels.

Depth perception

Depth can be estimated by an experienced observer through use of various cues. Some of these cues are provided by the nature of the scene of interest; others are inherent in the observer.

In cases where only internal cues to distance are available, that is when objects of interest are of unknown size and shape, it is evident that the observer must depend upon his stereoscopic acuity, accommodation and convergence, and, where possible, movement parallax. Accommodation is the only effective cue to distances at ranges of a meter or less, and even here it is inaccurate. Convergence alone is a somewhat more useful cue, but is limited to a range out to about 20 m. Stereoscopic acuity, however, provides a powerful cue to distance.

EYE DISORDERS

Conjunctivitis

Various types of conjunctivitis, or inflammation of the mucous membrane, can develop beneath the eyelids. In such cases, the eye becomes scratchy and red and has a discharge. Most often the cause is bacteria or viruses.

A fairly common type of infection, especially in adults, is caused by the herpes simplex virus on the anterior surface of the eyeball. This can lead to blurred vision, scarring, and permanent damage to vision. It may affect one or both eyes and quite often recurs. Topical steroids often used for conjunctivitis will greatly aggravate the condition. Therefore, any signs suggesting inflammation near the cornea immediately should be referred to a physician who is experienced in treating eye disorders.

Inflammation of the interior eye is common in adults. One of the most common areas of infection is the uveal tract, which is the middle coat of the eye. Inflammation of this type damages the retina, the lens, and the cornea. Uveal inflammations are quite often associated with diseases of the joint, lung, or intestinal tract. A search must be made for disease elsewhere in the body that might be the cause of the eye problem. When found, the primary cause should be treated. The eye should also be treated to prevent damage to vision.

Glaucoma

Glaucoma is a leading cause of blindness. The most common form develops when the fluid that normally fills the eyeball, the aqueous humor, fails to drain properly. Ordinarily, the fluid is continuously produced within the eye, and excess amounts drain off through a small duct near the iris. But aging, infection, injuries, congenital defects, and other causes can constrict or block the drain. Fluid pressure then builds up and the pressure, if great and of long duration, may damage the optic nerve.

In the acute type of glaucoma, vision may dim suddenly, the eyeball becomes painful, and the victim feels quite ill. But the insidious type of glaucoma causes no pain, injuring vision very slowly. Sometimes it may make itself known by the appearance of colored rings and halos about bright objects or by dimming of side vision.

Much can be done to preserve vision in most cases, provided glaucoma is diagnosed in time. Surgical enlargement of the drainage ducts can be used, or medicine may be effective in constricting the pupil enough to allow the canals to open by themselves.

Cataracts

Cataracts are opaque spots that form on the lens and impair vision of many elderly and some younger people. There is, unfortunately, no primary preventive method known. Cataracts can be removed surgically at any time and at practically any age. Depending on the condition of the lens, the retina, and other factors, occasionally a cataract can be treated without surgery.

Excessive brightness

Good sunglasses can protect the eyes in bright sunlight. Poor ones compound or create problems. Do not wear glasses with scratches and irregularities. Some glasses are too lightly tinted to do much good; good glasses reduce the invisible as well as the visible light (this will be discussed later in this chapter under Irradiation burns).

The best sunglasses have ground and polished lenses and are worth the investment. If a person uses regular glasses, it is worthwhile to have a pair of sunglasses ground to the prescription rather than use possibly inferior sunglasses over a regular correction.

Night blindness

Inability to see well or at all in dim light can mean something is wrong not only with the eye but with the entire visual system. Night blindness, as it is called, is a threat to safety, particularly on the highway, because a driver may have 20/20 vision and not realize that his vision is somewhat impaired at night. The condition produces no discernible change in eye tissues, so it cannot be diagnosed unless a patient tells the physician of difficulty in reading road signs or picking out objects at night. It is not normal to have trouble seeing in dim light since sufficient accommodation occurs in two or three minutes.

Eyestrain

Eyestrain can result from a need for eyeglasses or from using corrective glasses with the wrong correction. Eye muscle strain may also result from unfavorable conditions such as improper lighting while reading or doing close work. To avoid strain when reading, do not face the light; it should come from behind and to the side. Be sure light bulbs are strong enough (75–100 watts). Hold the book or paper about 0.4 m (16–18 in.) away and slightly below eye level. Avoid glare, occasionally rest the eyes by shifting focus and looking off into the distance.

Nystagmus

Nystagmus, involuntary movement of the eyeballs, may occur among those workers who, for extended periods, subject their eyes to abnormal and unaccustomed movements. Complaints of objects dancing before the eyes, headaches, dizziness, and general fatigue are associated symptoms; all can clear up quickly if a change of work is made. The involuntary movements of the eyeball characteristic of nystagmus exceptionally can be induced by occupational causes affecting the eyes through the central nervous system or by some extraneous cause. The most prevalent form of occupational nystagmus is seen in miners.

PHYSICAL HAZARDS

The eye is subject to several kinds of physical injury—blows from blunt objects, cuts from sharp objects, and damage from foreign bodies.

Blows from objects

A blow from a blunt object can produce direct pressure on the eyeball, or, if the object delivering the blow seals the rim of the bony orbit on impact, it can exert hydraulic pressure. Such blows may cause contusion of the iris, lens, retina, or even optic nerve. Violent blows might result in rupture of the entire globe or fracture of the thin lower plate of the bony orbit with entrapment of the eye muscles.

Contusions may result in serious, irreversible injury if not treated promptly and adequately. Hemorrhaging releases blood, which can be toxic to eye tissues, and physical dislocations of lens, retina, and other parts are unlikely to repair themselves. Lacerations of the cornea, lid, or conjunctiva can be caused by any sharp object—from the corner of a piece of typing paper to a knife.

Corneal lacerations and abrasions

Corneal lacerations, if full-thickness, may allow the aqueous solution behind the cornea to gush out—until the iris, which is like wet tissue paper, is pulled toward the laceration and plugs the wound. The iris can be put back in place, the laceration sutured, and the eye may be nearly as good as new. More common corneal injuries are scrapes or abrasions that do not penetrate to the chamber behind the cornea. Such abrasions are very painful, but will heal within several days if treated properly. If they are too deep or allowed to become infected, scars that interfere with vision occur.

Lacerations of the lid will heal, but the scar tissue may pull the lid into an unnatural position. In addition to cosmetic deformity, there is a chance of damage to the eye due to the lids not closing completely or from lashes turning in against the eyeball. Vertical lacerations are more serious in this respect.

So the eye, being highly specialized tissue, is more likely to suffer permanent damage from injury than, for example, a finger.

This does not mean the eye has no natural defenses. The bony ridges of the skull protect the eyeball from traumatic injury caused by massive impact. A baseball, for instance, is too big to crush the eyeball—it will be stopped by the bony orbit.

The cushioning layers of conjunctiva and muscle around the eyeball tend to absorb impact. The fact that the eyeball can be displaced in its socket also is a defense against injury. Related to this is the fact that the optic nerve is extra long to allow displacement of the ball without rupture of the nerve.

Blink reflex

The eye is most subject to attack at the corneal surface. Here, the eye is equipped with an automatic wiper and washer combination. The washers are the lacrimal glands; the wiper is the blinking action.

The teary blink washes foreign bodies from the corneal or conjunctival surfaces before they can become embedded. The triggering mechanism is irritation.

The reflex blink also can act like a door to shut out a foreign object heading for the eye—if the eye can see it coming and it isn't coming too fast. It is apparent that protective equipment for the eye is apparatus to improve or extend these natural defenses—the bony ridge, the blink, and the tear glands. These natural defenses might be adequate to protect against light and small foreign objects, and to wash away small quantities of mildly toxic liquids that might get into the eye, but they are no match for industrial eye hazards such as small high-speed particles or caustic powders and liquids.

Foreign bodies

Invasion by a foreign body is the most common type of physical injury to the eye. Not all foreign bodies, however, affect the eye in the same way.

Foreign bodies affecting the conjunctiva are not usually very serious. They may result in redness and discomfort, but not vision damage. Bodies on the conjunctiva, however, can be transferred to the cornea and become embedded if a person rubs his eye. So even with minor irritations of the conjunctiva a trip to the nurse is advisable. If there is obvious irritation and no object can be found, it is advisable to immediately see a physician.

Some industrial eye injuries may appear trivial, but can become serious due to complications. The most common complication is infection, which can cause delayed healing and corneal scarring. The infection can be carried into the intraocular tissues by a foreign substance, and the bacteria can originate either from sources outside of the eye or from pathogenic organisms already present on the lids, conjunctiva, or in lacrimal apparatus.

What effects can be expected from corneal foreign bodies?

- Pain—because the cornea is heavily endowed with nerves, an object sitting on the surface of the cornea will constantly stimulate the nerves.
- Infection—can be carried by the foreign particle as bacteria or fungi, or from fingers used to rub the eye. This used to be much more common; antibiotics have greatly reduced the problem.
- Scarring—corneal tissue will heal, but the scars are optically imperfect, and may obscure vision.

What can be expected from intraocular foreign bodies?

- Infection—is much less of a problem with low-speed, low-mass particles. This is because the speed of small metallic particles often creates enough heat to effect sterilization. Wood particles, however, will not heat up and, if they penetrate the eye, can cause dangerous infection. This usually leads to a marked reduction in vision.
- Damage—depending on its angle, point of entry, and speed, an intraocular particle may cause traumatic damage to the cornea, iris, lens, or retina—individually or together. Damage to the lens is especially serious, because it is not supplied with blood, and is slow to heal. Also, any damage to the lens can act as a catalyst for protein coagulation, resulting in opacity and loss of vision.

Pure copper particles can cause serious damage to the eye, because the toxic copper molecules become deposited in the lens, cornea, and iris (chalcosis). Copper alloys do not seem to have any toxic effects.

Pain cannot be counted on to alert the worker that there is a foreign body in his or her eye. The cornea is very sensitive, but if the object has penetrated into the eyeball, there may be no acute pain.

Thermal burns

Heat can destroy eye and eyelid tissue just as it does any other body tissue. But eye tissues may not recover as well as skin and muscle from such trauma. The lids are more likely to be involved in burns than the eye itself, since involuntary closing of the eye is an almost certain response to excessive heat.

Irradiation burns

Damage mechanisms. Light in sufficient amounts may damage eye tissue, ranging from barely detectable impairment to gross lesions. The principal damage depends on the tissue involved and the energy of the incident light photon. Far-infrared light usually effects damage through a general increase in tissue temperature, whereas far-ultraviolet light generally causes specific photochemical reactions.

Damage to lens cells may not be apparent for some time after insult because of the low level of metabolic activity. Low-degree damage is evidenced by vision clouding or cataract and usually is not reversible. Should recovery occur, it is a slow process. Lens damage may be cumulative because dead cells cannot be eliminated from the lens capsule, causing a progressive loss of visual acuity.

Retinal damage can take a number of forms. Generally, the neural components of the retina, such as the photoreceptors, may regenerate when slightly injured, but usually degenerate when extensively injured.

Common industrial radiation wavelengths that can harm the eye include:

Ultraviolet. Exposures to ultraviolet (UV) light usually occur near welding operations. Harmful effects of visible and UV radiation are seen in welding, particularly in electric arc welding. These include welder's flash, an acute inflammation of the cornea and conjunctiva known as acute keratoconjunctivitis, which develops in about 6 hours after even a momentary exposure to the arc-light. The welder rarely is involved, being too close to the arc to look at it without an eyeshield. But welders' helpers and other bystanders frequently suffer from exposure.

Infrared. Unlike UV, infrared (IR) radiations pass easily through the cornea and their energy is absorbed by the lens and retina. With automation of metals operations, eye damage from IR radiation is not as common as it once was.

Visible light. Various combinations of light sources, exposure durations, and experimental animals have been used to determine the threshold level of light capable of producing a visible retinal lesion. Unfortunately, the various experiments recorded were not systematically designed or standardized, causing numerous gaps and inconsistencies in the reports. Such parameters as pulse duration and irradiated spot area or diameter on the retina are notably lacking. Methods of measurement often are unstated, making comparisons and appraisals of accuracy difficult.

Many models have been proposed to explain the production of visible lesions on the retina from exposure to laser light. Most of the models consider thermal injury to be the only cause of damage. (See Chapter 11, Nonionizing Radiation, for more details.)

Chemical hazards

The effects of accidental contamination of the eye with chemicals varies from minor irritation to complete loss of vision. In addition to accidental splashing, some mists, vapors, and gases produce eye irritation, either acute or chronic. In some instances, a chemical that does no damage to the eye can be sufficiently absorbed and cause systemic poisoning.

Exposure to irritant chemicals provokes acute inflammation of the cornea (acute keratitis) with pinpoint vacuoles (holes) of the cornea, which rapidly break down into erosions. Some industrial chemicals irritate the mucous membrane, stimulating lacrimation (excessive watering of the eyes). Other results may be discoloration of the conjunctiva, disturbances of vision, double vision from paralysis of the eye muscles, optic atrophy and temporary or permanent blindness.

Chemical burns. Because caustics are much more injurious to the eyes than acids, the medical prognosis of caustic burns always has to be guarded. An eye might not look too bad on the first day after exposure to a caustic, but later it may deteriorate markedly. This is in contrast to acid burns, when the initial appearance is a good indication of the ultimate damage.

The reason for this is strong acids tend to precipitate a protein barrier that prevents further penetration into the tissue. The alkalis do not do this; they continue to soak into the tissue as long as they are allowed to remain in the eye.

The ultimate result of a chemical burn may be a scar on the cornea. If this is not in front of the window in the iris, vision may not be greatly hampered. If the scar is superficial, a corneal transplant can alleviate burn damage. Densely scarred corneal tissue cannot be repaired by transplants, contrary to popular belief.

When the chemical penetrates the anterior chamber of the eye, the condition is called iritis (irritation caused by bathing the iris with the chemical agent). Glaucoma may be a complication of chemical iritis.

Evaluating eye hazards

It does not take special training or engineering skills to identify most eye hazards. When people handle acids or caustics, when airborne particles of dust, wood, metal, or stone occur, or when blows from blunt objects are likely—eye protection is necessary.

Workers directly involved with operations producing these hazards are usually included in protective equipment programs, but often workers on the perimeter of eye-hazardous operations are left unprotected—with costly results. Who has not heard of pieces of broken metal tools that found an eye after traveling great distances from the drill press, or of the welder's helper who failed to turn his or her back a few times and got painful corneal burns?

Even when work only occasionally brings an employee near eye hazards, the safest policy is to encourage eye protection all day long. The outdated technique of hanging a pair of community goggles near the grinding wheel is an example of the "eye-hazard job" approach to eye protection. The jobs that involved eye hazards were identified and eye protection was required only when the worker actually was doing the job.

The eye-hazard area concept is a better approach. It provides eye protection equipment for workers, neighboring workers, supervisors, and visitors. An eye-hazard area is an area where the continuous or intermittent work being performed can cause an eye injury to anyone in the area. With proper enforcement and designation of areas, this approach is the most effective way to prevent eye injuries.

First aid

What should an employee do if foreign material does get into the eye?

Propelled object injuries require immediate medical attention. Even for foreign bodies on the corneal surface, self-help should be discouraged—removal of such particles is a job for a trained medical staff member.

Chemical splash injuries require a different approach. Here the extent of permanent damage depends almost entirely on how the victim reacts. If the victim of a concentrated caustic splash gets quickly to an eyewash fountain and properly irrigates the eye for at least 15 minutes, and promptly receives expert medical attention, the chances are good for a clear cornea or, at most, minimal damage.

Such irrigation should be with plain water from standard eyewash fountains, emergency showers, hoses, or any other available sources. Water for eye irrigation should be clean, and within certain temperature limits for comfort. (Tests have shown that 33 C (112 F) is about the upper threshold limit for comfort, but colder water, even ice water, apparently causes no harm and is not uncomfortable enough to discourage irrigation.)

ANSI Z358.1–1981

The American National Standard Institute (ANSI) standard Z358.1–1981 covers the design and function of eyewash fountains; the water should meet potable standards. It has been noted that acanthamoebae capable of infecting traumatized eyes may be present in potable water. No cases have been directly attributed to the presence of these organisms in eyewash stations, but it seems prudent to follow the ANSI recommendation of a weekly systemic flushing. At least three minutes of flushing will significantly reduce the number of organisms.

Portable units are intended for brief irrigation of an injury. A full 15-minute flushing of the injured eye at a stationary station should follow. It has been suggested that water for a portable station be treated with calcium hypochlorite to 25 ppm free chlorine to eliminate acanthamoebae.

Some industrial medical units use sterile water for irrigation. Use of water substitutes, such as neutralizing solutions, boric acid solutions, and mineral oil, is discouraged by nearly all industrial ophthalmologists, because in many instances, such preparations can cause eye damage greater than if no irrigation were used at all.

PROTECTIVE EQUIPMENT

All eye-protection equipment is designed to enhance one or more of the eye's natural defenses. Chipper's cup goggles extend the bony ridge protecting the eye socket and provide an auxiliary, more penetration-resistant, cornea. Chemical splash goggles are better than a blinking eyelid.

There is a tremendous variety of eye protection available, from throwaway visitor's eye shields to trifocal prescription safety spectacles and from welder's helmets to clip-on, antiglare lenses. But the classic safety glasses, with or without sideshields, are probably adequate for 90 percent of general industrial work.

The requirement for proper eye protection should be impartially enforced to give maximum protection for the degree of hazard involved. On certain jobs, 100 percent eye protection must be insisted upon.

Protection of the eyes and face from injury by physical and chemical agents or by radiation is vital in any occupational safety program. Eye-protective devices must be considered as optical instruments, and should be carefully selected, fitted, and used.

Unfortunately, the very term safety glasses can be confusing. A Food and Drug Administration (FDA) ruling, which became effective January 1, 1972, requires that all prescription eyeglass and sunglass lenses be impact-resistant. However, such lenses are not the equivalent of industrial-quality safety lenses, and they should not be used in an industrial environment where protection is mandatory.

ANSI Z87.1-1979

Only safety eyewear that meets or exceeds the requirements of ANSI standard Z87.1–1979, *Practice for Occupational and Educational Eye and Face Protection* (referenced in OSHAct Regulations), is approved for full-time use by industrial workers.

The Z87 standard specifies that industrial safety lenses are at least 3 mm thick and capable of withstanding impact from a 1-in. diameter steel ball dropped 1.3 m (50 in.). The FDA ruling does not mention lens thickness, and requires that a smaller, 16-mm (5/8 in.), diameter steel ball be used to verify impact-resistance. To pass this test, the lens cannot become chipped or displaced from the frame.

Impact protection

Three types of equipment are used to protect eyes from flying particles—spectacles with impact-resistant lenses, flexible or cushion-fitting goggles, and chipping goggles.

Spectacles. Spectacles without sideshields should be used for limited hazards requiring only frontal protection. Where side as well as frontal protection is required, the spectacles must have sideshields. Full-cup sideshields are designed to restrict side entry of flying particles. Semifold or flatfold sideshields can be used where lateral protection only is required. Snap-on and clip-on sideshield types are not acceptable unless they are secured (Figure 5-8).

Flexible-fitting goggles. These should have a wholly flexible frame forming the lens holder. Cushion-fitting goggles should have a rigid plastic frame with a separate, cushioned surface on the facial contact area (Figure 5-9). Both flexible and cushion goggles usually have a single plastic lens. These goggles are designed to give the eyes frontal and side protection from flying particles. Most models will fit over ordinary ophthalmic spectacles.

Chipping goggles. These have contour-shaped rigid plastic eyecups, come in two styles—one for persons who do not wear eyeglasses, and one to fit over corrective glasses. Chipping goggles should be used where maximum protection from flying particles is needed.

If lenses will be exposed to pitting from grinding wheel sparks, a transparent and durable coating can be applied to them.

Eye protection for welding

In addition to damage from physical and chemical agents, the eyes are subject to the effects of radiant energies. Ultraviolet,

Figure 5-8. Full-cup sideshields are designed to restrict the entry of flying objects from the side of the wearer.

Figure 5-9. Flexible-fitting goggles should have a flexible frame forming the lens holder.

visible, and infrared bands of the spectrum are all able to produce harmful effects upon the eyes, and therefore require special attention.

Welding processes emit radiations in three spectral bands. Depending upon the flux used and the size and temperature of the pool of melted metal, welding processes will emit UV, visible, and IR radiation—the proportion of the energy emitted in the visible range increases as the temperature rises.

All welding presents problems, mostly in the control of IR and visible radiations. Heavy gas welding and cutting operations, and arc cutting and welding exceeding 30 amperes, present additional problems in control of UV.

Welders can choose the shade of lenses they prefer within one or two shade numbers. Following are shades commonly used:

- Shades No. 1.5–No. 3.0 are intended to protect against glare from snow, ice, and reflecting surfaces; and against stray flashes and reflected radiation from cutting and welding operations in the immediate vicinity. These shades also are suggested for goggles or spectacles with sideshields worn under helmets in arc-welding operations, particularly gas-shielded arc-welding operations.
- Shade No. 4, the same as shades 1.5–3.0, but for greater radiation intensity.

For welding, cutting, brazing, or soldering operations, the guide for the selection of proper shade numbers of filter lenses or windows is given in ANSI Z87.1, *Eye and Face Protection.* (For more details, see Chapter 11, Nonionizing Radiation.)

Laser beam protection

No one type of glass offers protection from all laser wavelengths. Consequently, most laser-using firms do not depend on safety glasses to protect an employee's eyes from laser burns. Some point out that laser goggles or glasses might give a false sense of security, tempting the wearer to be exposed to unnecessary hazards (Figure 5-10).

Nevertheless, researchers and laser technicians do frequently need eye protection. Both spectacles and goggles are available—and glass for protection against nearly all the known lasers can be ordered specially from eyewear manufacturers. Typically, the eyewear will have maximum attenuation at a specific laser wavelength; protection falls off rather rapidly at other wavelengths. (For more details, again refer to Chapter 11, Nonionizing Radiation.)

Video display terminals (VDTs)

Much energy and concern have been focused on health problems associated with the use of VDTs. There is no doubt that use of these devices can cause increased eye fatigue and visual discomfort. Factors involved in visual discomfort are poor contrast between the characters and background; high contrast between the screen and other surfaces, such as the documents; and glare from and flicker on the screen. Long periods of eye fixation and refractive errors are also significant contributors.

Frequently, the displays are at a distance of about 76 cm (30 in.) from the eye of the operator. Special corrective lenses may be required, because the distance is neither near nor far vision.

Regulations for the design and operation of VDT workstations have been issued by some countries, but most authorities oppose rigid rules. Because of the variety of visual problems and work practices of operators, proper design should allow for flexibility in the placement of the screen, keyboard, source documents, and work surfaces. Ambient lighting must be adjustable; bright backgrounds, such as windows, should be eliminated; and appropriate rest periods are indicated. (See Chapter 13, Ergonomics, for additional information.)

Figure 5-10. One type of safety glasses suitable for protection against laser beams.

Plastic v. glass lenses

When making a decision between plastic and glass lenses, there are a number of things to consider:

- Both can pass impact tests when of certain formulation and thickness.
- Glass has a slightly lower resistance than plastic to breakage from sharp objects.
- Tests show that plastic lenses have more favorable resistance to small objects moving at high rates of speed than glass lenses.
- Abrasion resistance, while not good with plastic, is improved when plastics are coated.
- Plastics are resistant to hot materials. Hot metal invariably shatters glass but not plastic. Hot metal also tends to adhere to glass.
- Plastics generally show surface reaction to some chemicals, but satisfactorily stop splashes and protect the eyes.
- Whereas fogging occurs on both glass and plastic, it usually takes longer for plastic to fog. Plastic goggles are available with a hydrophylic coating, which tends to prevent fogging. There are also double-lens plastic goggles that operate on the thermopane principle and suppress fogging to a large extent.

Sunglasses

Safety sunglasses for workers whose duties keep them outdoors is an understandable practice, but this is not so, where millions who labor indoors are concerned. No tinted lenses of any kind should be worn indoors, unless specifically required because of excessive glare or eye-hazardous radiation. There may be some employees whose eyesight would benefit from lightly tinted lenses because of a particular eye condition, but not very many.

Phototropic lenses. Phototropic or photochromic lenses automatically change tint from light to dark and back again, depending upon their exposure to UV light. The convenience of sunglasses with variable-tint lenses is obvious, even though such lenses will not react indoors, in a car, or anywhere else that UV light cannot reach.

The ANSI Z87 standard recommends a variety of fixed-density tinted lenses for use in specific job situations involving radiation harmful to vision. Each tint is assigned an individual shade number, which is inscribed on the front surface of each such lens. The current Z87 standard makes no mention of phototropic lenses; this fact alone should give pause to health and safety specialists who might be tempted to switch rather than fight pressure to shift a work force immediately into phototropic lenses.

Such lenses may have a future for outdoor use; for instance, by telephone line and brush crews, and gas-line transmission workers. However, until safety eyewear manufacturers are willing—or able—to certify phototropic lenses as being fully in compliance with the Z87 standard, industry would be well advised to play it safe and stay within ANSI guidelines.

Contact lenses

Contact lenses have certain advantages in many applications outside the industrial environment. However, with rare exceptions, contact lenses have few practical applications for people employed in manufacturing and other types of industrial firms.

Contact lenses do not provide eye protection in the industrial sense; their use without appropriate eye or face protective devices of industrial quality should not be permitted in a hazardous environment, according to the National Society for the Prevention of Blindness and ANSI standard Z87.1-1979.

Many contact lens wearers are employed in industry; a great many others will accept such work in the future. Thus, the significance and possible effects of an industrial environment on the eyesight of workers who wear contact lenses are of particular concern to the health and safety professional.

It is advisable that a worker who uses contact lenses also has a pair of prescription safety glasses available. Accidental displacement or loss of a contact lens may occur without warning, thereby causing immediate incapacitation by sudden change of vision, excessive tearing, light sensitivity, and involuntary squeezing together of the eyelids.

Another important factor often overlooked when contact lenses are worn by an industrial worker is spectacle blur. The vision of either eye may be blurred when contact lenses are removed. Although the change in visual acuity is not extreme, the blurred vision may persist for as long as an hour.

Comfort and fit

To be comfortable, eye-protective equipment must be properly fitted. Corrective spectacles should be fitted only by members of the ophthalmic profession. However, a technician can be trained to fit, adjust, and maintain eye-protective equipment; each worker, of course, should be taught the proper care of the device being used.

To give the widest possible field of vision, goggles should be close to the eyes, but the eyelashes should not touch the lenses (Figure 5-11).

Various defogging materials are available. Before a selection is made, test to determine the most effective type for a specific application.

In areas where goggles or other types of eye protection are extensively used, goggle-cleaning stations should be conveniently located, along with defogging materials, wiping tissues, and a waste receptacle.

VISION CONSERVATION PROGRAM

There are four steps in a vision conservation program:

1. the environmental survey
2. vision screening program
3. remedial program
4. professional fitting and follow-up procedure.

Environmental survey

The environment should be surveyed by persons qualified in industrial vision. The survey should access: (1) the likelihood of injury and potential severity of injury from the worker's job operation; (2) the potential for injury from adjacent operations; and (3) the optimum visual acuity requirements for fast, safe, efficient operations.

The environmental survey includes illumination measurements and recommendations for improvements to make the workplace safer. Often, simply cleaning existing lighting can increase illumination 100 percent.

Job working distances and viewing angles should be measured so eye doctors will have the necessary information to pre-

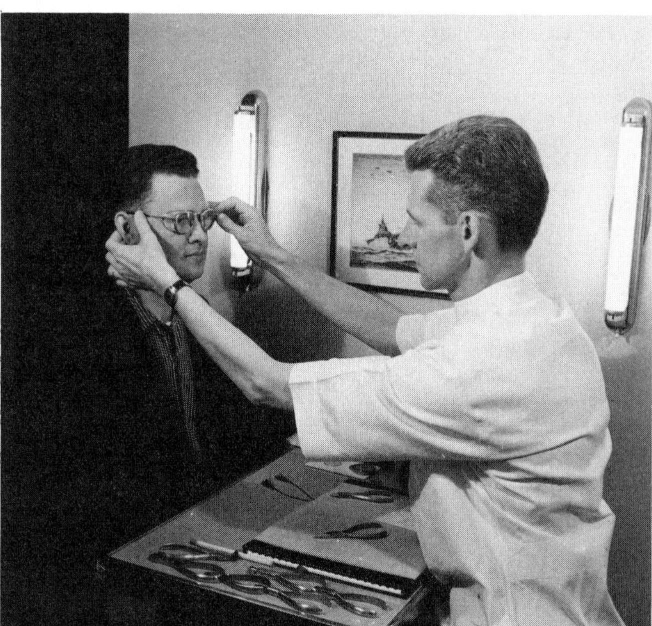

Figure 5-11. Safety glasses should be properly fitted and adjusted.

scribe lens strengths affording optimum comfort, efficiency, safety, and ergonomic advantage (Figure 5-12).

Each workstation should be free of toxic or corrosive materials, and employees instructed in the correct use of eyewash facilities. Eyewash fountains should be examined to make certain they work properly and provide an even flow of water.

All environmental factors influencing an employee's visual performance and safety should be written in a visual job description, in terms understandable to the eye doctors who will be caring for employees with deficient vision.

Vision screening program

The next step, after working conditions and visual requirements are known, is to determine the visual status of the work force. Reliable vision screening instruments are available for use by nurses or trained technicians.

These instruments test the visual acuity in each eye separately, both eyes together, both at near point (usually working distance) as well as at far point (distant vision), plus binocular coordination (the ability to make the two eyes work together as a team).

Additional specific tests—such as color vision, field of vision, glaucoma testing, and depth perception—can be added when a need is indicated.

Upon completion of this second step, the visual requirements of an employee's job and individual visual capability are known.

Often an employee actually performs many jobs. In these instances, recommendations are made after the job most frequently performed is compared with the one most visually demanding.

Remedial program

Each employee is told the result of his or her vision screening; persons showing a deficiency are referred to the eye doctor of their choice. But this time, an employee goes to the doctor with the prescription form for safety glasses, and a copy of a written description of the visual aspect of the job or duties at work. The job description includes recommendations concerning the type of prescription that will make the worker more comfortable and more efficient.

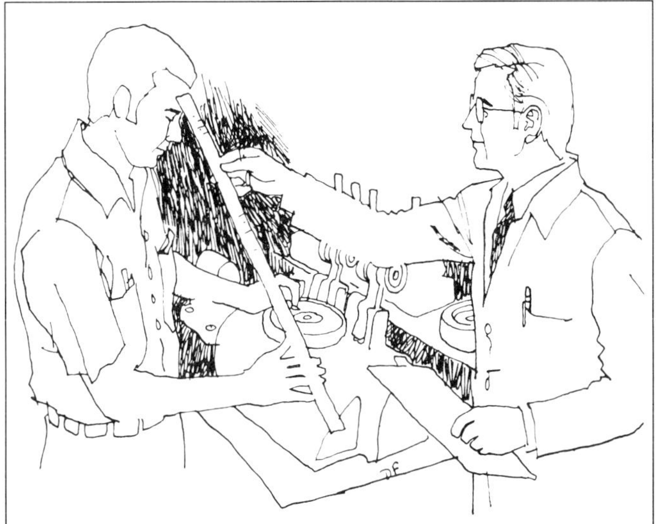

Figure 5-12. Measuring the distance from the eye to the work plane is essential for proper safety vision prescriptions.

Professional fitting

The final step in a vision conservation program is professionally fitting the protective-corrective safety eyewear to the employee. Proper fitting and sizing is essential and cannot be correctly done by the tool crib attendant or purchasing agent. Workers come in various sizes and shapes as do safety glasses. Proper measurement and fitting can be the difference between a successful and unsuccessful program.

A perfect prescription is valueless if the frame hurts so much that it cannot be worn. This fitting procedure is of equal or greater importance to those employees required to wear nonprescription safety glasses. These employees usually are the most difficult to comfortably fit. They are not accustomed to having anything in front of their eyes, or feeling weight on their nose or ears.

Considerable care must be exercised in the proper care and fitting of this type employee. This adjustment and alignment procedure should be provided on a continuing basis.

Human vision is not static; it is constantly changing. Employees, especially those who are over age 40, should be advised that gradual, sometimes unnoticed, changes in their vision may affect their safety and efficiency.

Continued testing must be instituted as a health-safety department policy. If a health specialist is not employed at the plant, arrangements for this service can be made with your local health care community. The local academy of ophthalmology or optometry usually will be able to provide trained personnel.

Guidelines

The following guidelines should be a part of every plant vision conservation program:

- Make it a 100 percent program, include everyone. Employees will accept it more readily, and it will be easier to administer. Promote it well in advance; get union cooperation.
- Make certain that safety eyewear is properly fitted. A few jobs (welding, labs, lasers) will require special types. Optical companies will assist in correct fitting and explaining maintenance.
- Include stations for eye first aid and for cleaning lenses.
- Control eye hazards at the source; install safety glass guards on machines to prevent flying chips or splashing liquids; install enclosures to control fine dusts, mists, or vapors.
- Make sure all areas have adequate lighting, are free from glare, and are painted in colors that emphasize depth perception and highlight potential hazards.
- Post signs advising eye protection—ALL PERSONNEL AND VISITORS MUST WEAR PROTECTIVE EYEWEAR.
- A preplacement eye examination should be given to all new employees. Periodic follow-up examinations should be scheduled, especially for those employees over 40.

AMA GUIDES FOR EVALUATING IMPAIRMENT

This section is adapted from the American Medical Association's (AMA) *Guides to Evaluation of Permanent Impairment.* It is included in this chapter to assist health and safety professionals in interpreting and understanding medical reports of workers' compensation cases.

This section includes criteria for use in evaluating permanent impairment of the visual system. It was originally adapted from a report prepared by a committee of the Council on Industrial Health in 1955 and entitled Estimation of Loss of Visual Efficiency. That and previous reports from the American Medical Association have been the basis for rating under numerous programs for the disabled.

Table 5-A. Visual Acuity Notations With Corresponding Percentages of Loss of Central Vision for Distance

Snellen Notations			
English	*Metric 6*	*Metric 4*	*% Loss*
20/15	6/5	4/3	0
20/20	6/5	4/4	0
20/25	6/7.5	4/5	5
20/30	6/10	4/6	10
20/40	6/12	4/8	15
20/50	6/15	4/10	25
20/60	6/20	4/12	35
20/70	6/22	4/14	40
20/80	6/24	4/16	45
20/100	6/30	4/20	50
20/125	6/38	4/25	60
20/150	6/50	4/30	70
20/200	6/60	4/40	80
20/300	6/90	4/60	85
20/400	6/120	4/80	90
20/800	6/240	4/160	95

For Near

Near Snellen		Revised Jaeger Standard	American point-type	% Loss
Inches	*Centimeters*			
14/14	35/35	1	3	0
14/18	35/45	2	4	0
14/21	35/53	3	5	5
14/24	35/60	4	6	7
14/28	35/70	5	7	10
14/35	35/88	6	8	50
14/40	35/100	7	9	55
14/45	35/113	8	10	60
14/60	35/150	9	11	80
14/70	35/175	10	12	85
14/80	35/200	11	13	87
14/88	35/220	12	14	90
14/112	35/280	13	21	95
14/140	35/350	14	23	98

(Reprinted with permission from American Medical Association [AMA]. *Guides to the Evaluation of Permanent Impairment,* 2nd ed. Chicago: AMA, 1984)

This guide provides a simplified method for determining permanent visual impairment and its effect on an individual's ability to perform the activities of daily living. Diminished visual ability is expressed as a percentage of impairment of the visual system. Diminished ability of the individual is expressed as a percentage of impairment of the whole person.

Visual impairment in varying degrees occurs in the presence of a deviation from normal in one or more functions of the eye, including (2) corrected visual acuity for distance and near, (2) visual fields, and (3) ocular motility with absence of diplopia. Evaluation of visual impairment is based on these three functions. Although they are not equally important, vision is imperfect without their coordinated function.

Other ocular functions and disturbances are considered to the extent they are reflected in one or more of the three coordinated functions. Such ocular disturbances include a slight paralysis of movement (paresis of accommodation), a paralysis of the sphincter of the iris (iridoplagia), distorted image (metamorphosia), inversion of the eyelid causing the lashes to rub against the eyeball (entropion), eversion of the eyelid (ectropion), a persistent flow of tears from excessive secretion or impeded outflow (epiphora), and a condition where the eyelids cannot be entirely closed (lagophthalmia).

To the extent that any ocular disturbance causes impairment not reflected in visual acuity, visual fields, or ocular motility without double vision (diplopia), it must be evaluated by the physician as to the degree of impairment and be combined with the impairment of the visual system as determined in the calculations which follow. In such circumstances, a physician will be guided by the value relative to the measurable functions on which evaluation is based in this guide, as well as by the value assigned to other ocular impairments in the individual case. Deformities of the orbit and cosmetic defects which do not alter ocular function are not considered.

The following equipment is necessary to test the functions of the eyes:

1. Visual acuity test charts for (1) distance: for purposes of standardization the Snellen test chart with block letters or numbers, the illiterate E chart, or Landolt's broken-ring chart are desirable; (2) near: there are many such charts available, using a similar Snellen notation for inches or Jaeger print or point type notation.
2. A standard perimeter with radius of 330 mm and a tangent screen.
3. Refraction equipment.

Criteria and methods for evaluation

Central visual acuity. Illumination of the test chart should be at least 5 foot-candles (50 lux) and the chart or reflecting surface should not be dirty or discolored from age. The test distances should be 6 m (20 ft) for distance and 36 cm (14 in.) for near vision.

Central vision should be measured and recorded for distance and near with and without correction. The use of contact lenses might further improve vision reduced by irregular astigmatism from corneal injury or disease. However, the practical difficulties of fitting and expense and tolerance of such lenses, in addition to the fact that they are sometimes medically contraindicated, are sufficiently important at present to recommend regular ophthalmic lenses be used to obtain the best corrected vision. In the absence of contraindications, contact lenses are acceptable.

Visual acuity for distance should be recorded in the notation of a fraction in which (1) the numerator is the test distance in feet or meters; and (2) the denominator is the distance at which the smallest letter discriminated by the patient would subtend 5 minutes of arc (the distance at which an eye with 20/20 vision would see that letter).

This fractional designation is purely a convenient form of notation and does not imply percentage of visual acuity. A similar Snellen notation with use of inches or a comparable Jaeger

Table 5-B—Loss of Central Vision* in Percentage

Snellen Rating for Distance in Feet	*Approximate Snellen Rating for Near in Inches*													
	14/14	*14/18*	*14/21*	*14/24*	*14/28*	*14/35*	*14/40*	*14/45*	*14/60*	*14/70*	*14/80*	*14/88*	*14/112*	*14/140*
20/15	0	0	3	4	5	25	27	30	40	43	44	45	48	49
	50	50	52	52	53	63	64	65	70	72	72	73	74	75
20/20	0	0	3	4	5	25	27	30	40	43	44	46	48	49
	50	50	52	52	53	63	64	65	70	72	72	73	74	75
20/25	3	3	5	6	8	28	30	33	43	45	46	48	50	52
	52	52	53	53	54	64	65	67	72	73	73	74	75	76
20/30	5	5	8	9	10	30	32	35	45	48	49	50	53	54
	53	53	54	54	55	65	66	68	73	74	74	75	76	77
20/40	8	8	10	11	13	33	35	38	48	50	51	53	55	57
	54	54	55	56	57	67	68	69	74	75	76	77	78	79
20/50	13	13	15	16	18	38	40	43	53	55	56	58	60	62
	57	57	58	58	59	69	70	72	77	78	78	79	80	81
20/60	16	16	18	20	22	41	44	46	56	59	60	61	64	65
	58	58	59	60	61	70	72	73	78	79	80	81	82	83
20/80	20	20	23	24	25	45	47	50	60	63	64	65	68	69
	60	60	62	62	63	73	74	75	80	82	82	83	84	85
20/100	25	25	28	29	30	50	52	55	65	68	69	70	73	74
	63	63	64	64	65	75	76	78	83	84	84	85	87	87
20/125	30	30	33	34	35	55	57	60	70	73	74	75	78	79
	65	65	67	67	68	78	79	80	85	87	87	88	89	90
20/150	34	34	37	38	39	59	61	64	74	77	78	79	82	83
	67	67	68	69	70	80	81	82	87	88	89	90	91	92
20/200	40	40	43	44	45	65	67	70	80	83	84	85	88	89
	70	70	72	72	73	83	84	85	90	91	92	93	94	95
20/300	43	43	45	46	48	68	70	73	83	85	86	88	90	92
	72	72	73	73	74	84	85	87	91	93	93	94	95	96
20/400	45	45	48	49	50	70	72	75	85	88	89	90	93	94
	73	73	74	74	75	85	86	88	93	94	94	95	97	97
20/800	48	48	50	51	53	73	75	78	88	90	91	93	95	97
	74	74	75	76	77	87	88	89	94	95	96	97	98	99

*Upper figure = % loss of central vision without allowance for monocular aphakia; lower figure = % loss of central vision with allowance for monocular aphakia. (Reprinted with permission from AMA. *Guides to the Evaluation of Permanent Impairment,* 2nd ed. Chicago: AMA, 1984)

or point-type notation can be used in designating visual acuity for near.

The visual acuity notations for distance and near with corresponding percentages of loss of central vision which appear in Table 5-A are included solely to indicate the basic values used in developing Table 5-B.

Simple addition of two percentages of loss corresponding to appropriate notations for distance and near does not provide the true percentage loss of central vision. In accordance with accepted principles, true loss of central vision is the mean of the two percentages (Table 5-B).

Aphakia, absence of the lens, is considered an additional visual handicap and, if present, is assigned a value of 50 percent decrease in the remaining corrected central vision (Table 5-B).

To determine loss of central vision in one eye:

1. Measure and record central visual acuity for distance and near with and without corrective lenses, either conventional or contacts.
2. Consult Table 5-B for corresponding loss of central vision depending on the presence of monocular aphakia (absence of the crystaline lens).

EXAMPLE: Without allowance for monocular aphakia: 14/56 for near and 20/200 for distance produces 80 percent loss of central vision. With allowance for monocular aphakia (applicable to corrected vision only): 14/56 for near and 20/200 for distance produces 90 percent loss of central vision.

Visual fields. The extent of the visual field is determined by the usual perimetric methods with a white target which subtends a 0.5-degree angle (a 3-mm white disk at a distance of 330 mm) under illumination of not less than 7 footcandles. A 6/330 white disk should be used for aphakia. The test object is brought from the periphery to the seeing area. At least two peripheral fields should be obtained which agree within 15 degrees in each meridian. The reliability of the patient's responses should be noted. The result is plotted on an ordinary visual field chart on each of the eight 45-degree principal meridians. For more complete information including perimetric and other methods of calculating percent loss of visual field, consult Chapter 6 in the AMA's *Guides to the Evaluation of Permanent Impairment.*

Ocular motility. Unless diplopia (double vision) is present within 30 degrees of the center of fixation, it rarely causes significant visual loss except on looking downward. The extent of the diplopia in the various directions of gaze is determined on the perimeter of 330 mm or on a bowl perimeter 30 cm from the patient's eyes in each of the 45-degree meridians, with use of a small test light and without the addition of colored lenses or correcting prisms.

BIBLIOGRAPHY

American Medical Association (AMA). *Guides to the Evaluation of Permanent Impairment,* 2nd ed. Chicago: AMA, 1984.

AMA. *The Wonderful Human Machine.* Chicago: AMA, 1971.

Brant, W.M. *Toxicology of the Eye.* Springfield, IL: Charles C. Thomas Publ., 1974.

Dawson, H. *Physiology of the Eye,* 4th ed. New York: Academic Press, 1980.

Dawson, H. *The Eye,* 3rd ed. New York: Academic Press, 1984.

Fraunfelder, F.T., and Roy, F.H. *Current Occular Therapy.* Philadelphia, PA: W.B. Saunders Co., 1980.

Grandjean, E. *Ergonomics and Health in Modern Offices.* Philadelphia, PA: Taylor & Francis, 1984.

Ironside, K.S., Lyle, M.M., and Tyndall, R.L. *The Presence of Acanthamoebae in Portable and Stationary Eyewash Stations,* in press.

Zenz, Carl, ed. *Occupational Medicine—Principles and Practical Applications,* 2nd ed. Chicago: Year Book Medical Publ., 1988.

PART III

Recognition of Hazards

Solvents

by Donald R. McFee, ScD, CIH, PE, CSP
Peter Zavon, CIH

A POTENTIAL THREAT TO THE HEALTH, productivity, and efficiency of workers in most occupations and industries (as well as nonoccupational environments) is their exposure to organic solvents. No one fully comprehends the total effect, yet all of us are exposed and we all are affected.

Exposures to solvents occur throughout life, from conception to death. For example, organic solvent vapors inhaled by a mother can reach the fetus. The elderly often spend their last days in a hospital where the odors of solvents and disinfectants often prevail. Exposures also occur in the course of daily living—ranging from the inhalation of vapors from a newspaper freshly off the press to the intake of a cleaning solvent by all routes of exposure, either at home or at work. Effects from the exposure may range from a simple objection to an odor, to death at high concentrations. In between, there is a whole spectrum of effects.

WHY ARE SOLVENTS SO IMPORTANT?

Solvents convert substances into a form suitable for a particular use. Solvents are so significant because many substances are most useful when in solution.

Organic solvents are used in the home as dry cleaning agents, paint thinners, and spot removers; in the office as typewriter key cleaners, desktop cleaners, and wax removers; in commercial laundries as dry cleaning liquids; on the farm as pesticides; and in laboratories as chemical reagents, and drying, cleaning, and liquid extraction agents. Many consumer products packaged in cans and drums contain mixtures of organic solvents.

Because of the near-infinite number of combinations possible for the variables involved—hundreds of different solvents, degree of concentration, duration of exposure, combined effects with other solvents, and health and age of an exposed person—general statements about effects of solvents on a person are difficult to make. The problem lies not so much in the effect itself, but rather in determining which effects are harmful and at what level.

When an exposure exceeds certain threshold levels, many of the effects are harmful, and a person's health and functioning may be impaired. In some cases the effects are irreversible and damage can be permanent.

As with many safety and health measures, we often do not use the necessary controls; too often there is more contact between the solvent and the skin than an exposed person realizes. Sometimes the local ventilation in the breathing zone is not adequate. Frequent fires and explosions attest to inadequate ventilation—concentrations that can cause fire and explosion are far greater than those levels that have toxicological effects.

As a rule, even with good general ventilation, common cleaning solvents have vapor pressures that produce concentrations on the order of 100–1,000 parts per million (ppm) in the breathing zone of the user (Figure 6-1). Very toxic solvents should be used only in a hood with local exhaust ventilation.

The Threshold Limit Value (TLV) is that concentration by volume in air under which it is believed that nearly all workers may be repeatedly exposed day after day without adverse effects. A list of TLVs is published yearly by the American Conference of Governmental Industrial Hygienists (see Appendix B-1).

The TLV should not be confused with the permissible exposure level (PEL), set forth in the regulations promulgated by the Occupational Safety and Health Administration (OSHA)

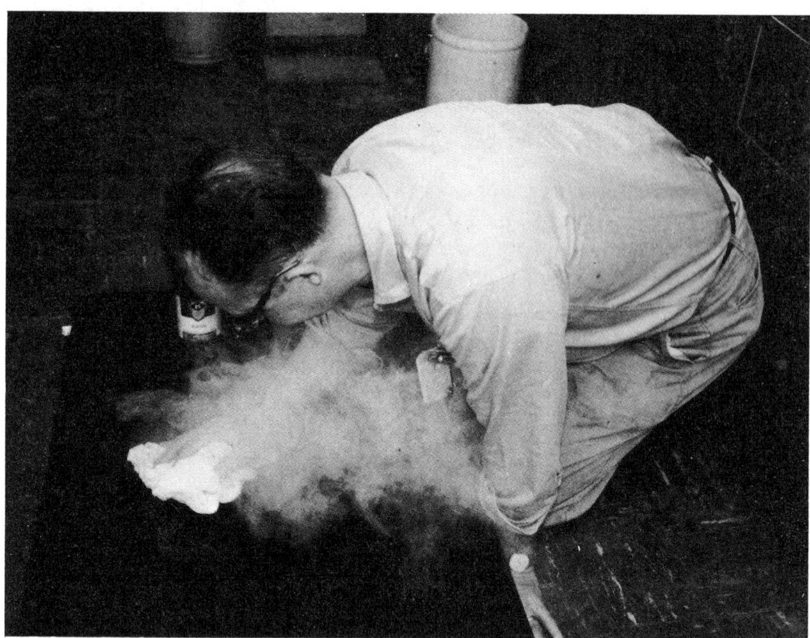

Figure 6-1. Despite the good general ventilation in this room (about 15 changes per hour), the breathing zone of this technician, who is using cleaning rags, has a vapor buildup. There should be a positive movement of air away from him. (Courtesy Argonne National Laboratory)

(see Appendix B-2). Most of the OSHA levels were adopted initially from the 1968 TLV list. The OSHA standards are a part of the law and are not updated annually. The TLV list is updated annually; therefore, consult the latest listing for guidance. Some OSHA standards, however, do contain more information on solvents than the single TLV, and this information should be used.

Classifications

The term solvent means material used to dissolve another material, and it includes aqueous systems as well as nonaqueous systems.

Aqueous systems are those based on water. Examples are aqueous solutions of acids, alkalis, detergents, and other materials. In general, aqueous systems have low vapor pressures; thus, the potential hazard by inhalation and subsequent systemic toxicity is not great.

A solution can be defined as a mixture of two or more substances. A solution has uniform chemical and physical properties. It can also be defined as "a system whose [*sic*] component parts are two or more molecular species, there being no boundary surfaces between these parts larger than molecules." There are two components to every solution—the solvent and the solute. As a matter of convenience, we designate the excessive part of a solution as the solvent; the solute is the lesser component. Thus, we have a gaseous solution when a substance is dissolved in a gas, a liquid solution when a substance is dissolved in a liquid, and a solid solution when a substance is dissolved in a solid.

However, for simplicity, in this chapter the term solvent will include only organic liquids commonly used to dissolve other organic materials. Organic solvents include materials such as naphtha, mineral spirits, turpentine, benzine, benzene, alcohol, perchloroethylene, and trichloroethylene.

Organic chemistry is the chemistry of the compounds of carbon. The carbon atom can form single, double, and triple bonds to other carbon atoms and to atoms of other elements. The bonds are covalent and have definite directions in space. A molecular chain (or skeleton) consists of a line of carbon atoms with branches of carbon atoms, or functional groups. These functional groups may contain oxygen (O), nitrogen (N), phosphorus (P), and sulfur (S). (See the periodic classification of the elements in Appendix C, Chemical Hazards.)

A functional group in an organic molecule is a region in the molecule where reactions can take place. Double and triple bonds and the presence of atoms other than carbon make up typical functional groups.

Organic compounds are named according to the number of carbon atoms in the basic skeletal chain. The location of the functional groups is designated by the number of the carbon atom to which it is attached. Carbon skeletons with straight chains of carbon with all carbon-hydrogen bonds filled with a hydrogen atom are referred to as alkanes. These are further identified as aliphatic or paraffin hydrocarbons. Those forming rings are identified by the prefix "cyclic" or "cyclo." One specific stable ring structure contains six carbon atoms with three double bonds. Skeletons containing this ring are identified by the term aromatic. (See Table 6-A for molecular configurations.)

Isomers are molecules that have the same basic skeletal structure with the same number and kinds of atoms. But the atoms are arranged differently; this imparts different physical and chemical properties to them.

The common organic solvents can be classified as aliphatic, cyclic, aromatic, halogenated hydrocarbons, ketones, esters, alcohols, and ethers. Each class has a characteristic molecular structure as shown in Table 6-A.

A good working knowledge of the nomenclature, the characteristic molecular structure, and the different toxicities is helpful in making a proper assessment of a solvent problem.

Nomenclature itself can often be misleading to the health and safety professional. For example, benzine and benzene are different solvents that are often confused and that have greatly different toxic effects. Some commercial grades of benzine may contain benzene as a contaminant that must be reckoned with. Benzene is a solvent that has been shown to cause leukemia—a type of cancer.

Even a scientifically trained user often has only a vague and sometimes completely erroneous knowledge of the particular solvent preparation in use. It is good practice to verify the specific name and composition of the solvents involved with direct evidence from the label, from the manufacturer, or from the laboratory. Only after verification of name and composition should one attempt to evaluate the potential effect or hazard of a solvent.

Manufacturers now are required by governmental regulations to provide information on the composition of their trade name materials. The minimum information is that contained in the Material Safety Data Sheet (Figure 6-2). (See Chapter 30 for a discussion of these regulations.)

There is usually little value in manufacturers withholding such information. Adequate information on composition can be determined by a competent chemist in a properly equipped laboratory. Furthermore, a manufacturer or distributor can provide a solvent tailored to meet the needs of the user far more economically than can the user.

Hawley's *Condensed Chemical Dictionary,* Windholz's *The Merck Index,* Gleason's *Clinical Toxicology of Commercial Products,* and the NFPA *Flash Point Index of Trade Name Liquids* provide general information and descriptions of many solvents, including trade name materials. These are helpful references for classifying and understanding the composition of a solvent (See Bibliography for more information.)

EFFECTS

Physiological effects

The physiological effects of different solvents are far too complex and variable to be discussed in detail here; however, certain generalizations can be made.

Aqueous systems. These are known for their irritant effects following prolonged exposure. Contact dermatitis from aqueous solutions is quite common—witness the ubiquitous "dishpan hands." Excessive levels of mists in the air (resulting from heating, agitation, and spraying) can cause throat irritation and bronchitis. Many other effects and hazards are possible if there are reactions between the chemicals involved and the container. Halley (see References) cites a number of examples. As a rule, aqueous systems, because of their low vapor pressure and ease of control, are not a problem, but they cannot be dismissed as being without potential hazard.

Organic solvents. These present a different type of problem. Vapor pressures are usually higher, and the potential for inhalation of toxic quantities is much greater. Some of the effects are generalized in the following paragraphs. For detailed information on specific organic solvents, consult *Patty's Industrial Hygiene and Toxicology,* Gerarde's *Toxicology and Biochemistry of Aromatic Hydrocarbons,* Browning's *Toxicity* and *Metabolism of Industrial Solvents,* the AIHA *Hygienic Guide Series* and the NIOSH *Registry of Toxic Effects of Chemical Substances* or the NIOSH criteria document on the subject solvent (if available) (see Bibliography).

All organic solvents affect the central nervous system to some extent, because they act as depressants and anesthetics. They also cause other effects. Depending upon the degree of exposure and the solvent involved, these effects may range from mild narcosis to death from respiratory arrest.

Aliphatic hydrocarbons. The aliphatic compounds take their name from the Greek word, *aliphe,* meaning fat, since fats are derivatives of this class of hydrocarbons (Table 6-A).

Aliphatic hydrocarbons are further classified as alkanes, alkenes, cycloalkanes, cycloalkenes, acetylenes, and arenes. Petroleum and natural gas are the most important sources of alkanes, alkenes, and cycloalkanes. Coal tar is an important source of arenes. High molecular weight alkanes are broken down (cracked) catalytically to increase the yield of gasoline from petroleum. Ethylene (H_2CCH_2) is an important by-product of cracking and is used to make plastics and ethanol (C_2H_5OH).

The saturated aliphatic hydrocarbons, known as alkanes or paraffins, are those with all bond positions satisfied (or saturated) by bonding with hydrogen. The paraffin series has been assigned the characteristic "-ane" suffix for naming those compounds, for example, isobutane, 2-methylpentane and 2,2-dimethylpentane. They are as inert biochemically as they are chemically. Even as air pollutants, they are among the least reactive and do not pose a significant problem. The paraffins are good solvents for natural rubber. They act primarily as depressants to the central nervous system.

A relatively high level is required for toxic effects. The TLVs generally range from 100 ppm and higher. The exception is normal hexane. Repeated exposures to excessive concentrations may cause peripheral neuritis. The TLV is 50 ppm. Other isomers have not been found to cause this.

The unsaturated aliphatic hydrocarbons, the alkenes (also referred to as olefins) and the alkynes, with double and triple bonds respectively, are similarly inert in the body. They are, however, more chemically reactive than the saturated hydrocarbons. As air pollutants, they are reactive and create a control problem. The primary problem with the aliphatics is dermatitis.

Crude oil (petroleum) is mainly a very complex mixture of aliphatic compounds. It contains alkanes, alkenes, cycloalkenes, and arenes, as well as small amounts of nitrogen and sulfur compounds, which vary depending on the source.

Petroleum is separated into mixtures of hydrocarbons by fractional distillation. Gasoline is the fraction of petroleum boiling between room temperature and 200 C and is mainly made up of C_5 to C_{11} hydrocarbons, with C_8 predominating. It has been estimated that there are as many as 500 different hydrocarbons in gasoline alone. About 150 of them have been separated and identified.

Cyclic hydrocarbons. The cyclic hydrocarbons act much in the same manner as the aliphatics, still they are not quite as inert. A significant percentage of cyclic hydrocarbons are metabolized to compounds with a low order of toxicity.

The unsaturated cyclic hydrocarbons generally are more irritating than the saturated forms.

Aromatic hydrocarbons. These get their names from aroma, meaning pleasant odor. The molecules are usually characterized by one or more six-carbon rings. This classification once served to distinguish pertroleum and coal-tar hydrocarbon solvents. Now, however, aromatics are derived from both sources.

Benzene and other aromatics do not undergo the addition reactions shown by alkenes and alkynes, but do undergo aromatic substitution reactions in which an atom or group of atoms

Table 6-A. Major Classes of Common Organic Solvents*

Aliphatic Hydrocarbons (Acyclic)

Straight or branched chains of carbon and hydrogen.

```
    H   H   H   H   H   H
    |   |   |   |   |   |
H - C - C - C - C - C - C - H
    |   |   |   |   |   |
    H   H   H   H   H   H
```

n-Hexane—50 ppm/150 mm Hg
Hexane isomers—500 ppm
Octane—300 ppm

Cyclic Hydrocarbons (Cycloparaffins, naphthenes)

Ring structure saturated and unsaturated with hydrogen.

```
          H   H
          |   |
          C - C
        / |   | \
       /  H   H  \
  H - C - H     H - C - H
       \  H   H  /
        \ |   | /
          C - C
          |   |
          H   H
```

Cyclohexane—300 ppm/104 mm Hg
Turpentine—100 ppm/156-169 F
(Turpentines are mixtures primarily of the unsaturated cyclic hydrocarbons and pinene)

Nitro-Hydrocarbons

Contains an NO_2 group.

```
    H   H     O
    |   |    /
H - C - C - N
    |   |    \
    H   H     O
```

Nitroethane—100 ppm/25 mm Hg

Aromatic Hydrocarbons

Contain a 6-carbon ring structure with one hydrogen per carbon bound by energy from several resonant forms.

```
          H   H
          |   |
          C - C
        //     \\
  H - C          C - H
        \      /
          C = C
          |   |
          H   H
```

**†Benzene—10 ppm/95 mm Hg
Toluene—100 ppm/30 mm Hg
Xylene—100 ppm/10 mm Hg

Halogenated Hydrocarbons

A halogen atom has replaced one or more of hydrogen atoms on the hydrocarbon.

```
       Cl
       |
  Cl - C - Cl
       |
       Cl
```

Tetrachloromethane—5 ppm/113 mm Hg
(carbon tetrachloride)
1, 1, 1, trichloroethane—350 ppm/127 mm (methyl chloroform)
Trichlorotrifluoroethane—1,000 ppm/331 mm (fluorocarbon No. 113)

*Synonyms are given in parentheses. A description of the characteristic group, a schematic of a typical molecular structure, the names of the structure that is shown (underlined), the names of other typical solvents in the class, the TLV (in parts per million by volume), and either a distillation range in degrees F, or the vapor pressure in millimeters of mercury at 25 C (or other temperature, as noted) are given under each heading in the table.
**Typical calculated value. See latest TLV list for determining procedure.
†Leukemogenic agent.

replaces one of the hydrogen atoms on the ring. Aromatic substitution reactions of benzene produce a wide variety of useful products (see Table 6-A).

The aromatic hydrocarbon, benzene, is notorious for its effect on the blood-forming tissues of the bone marrow. Gerarde (see Bibliography) has shown that, in animals, injury may result from a single exposure. Benzene is now indicated as a leukemogenic agent. This has reduced the extent of its use in solvents. Toxic levels of benzene are easily absorbed through the skin and by inhalation. Benzene should not be used for cleaning processes or for any process requiring skin contact or where the concentration in the air is not controlled by proper ventilation. The 1986-87 TLV for benzene is 10 ppm. However, NIOSH recommends that the PEL be reduced to 0.1 ppm, averaged over an 8-hour period, and that benzene be regulated as an occupational carcinogen.

The aromatic hydrocarbons, in general, are local irritants and vasodilators that cause severe pulmonary and vascular injury when absorbed in sufficient concentrations. They also are potent narcotics. The primary problem with the common aromatic solvents other than benzene are dermatitis and effects on the central nervous system.

Halogenated hydrocarbons. The term halogen is applied to five elements—fluorine, chlorine, bromine, iodine, and astatine. The halogens are a remarkable family of elements, marked by their great chemical activity and unique properties. Stability, nonflammability, and a wide range of solvency are but a few of the characteristics imparted by their application (Table 6-A).

The effects of the halogenated hydrocarbons vary considerably with the number and type of halogen atoms present in the

Table 6-A. Major Classes of Common Organic Solvents* (continued)

Esters

Formed by interaction of an organic acid with an alcohol.

```
    H      O
    |     //
H - C - C      H   H
    |     \    |   |
    H      O - C - C - H
               |   |
               H   H
```

Ethyl acetate—400 ppm/100 mm Hg
Amyl acetate—100 ppm/5 mm Hg

Ketones

Contain the double bonded carbonyl group, C=O, with 2 hydrocarbon groups on the carbon.

```
    H      O
    |     //
H - C - C   H   H
    |     \ |   |
    H       C - C - H
            |   |
            H   H
```

Methyl ethyl ketone—200 ppm/100 mm
Acetone—750 ppm/226 mm

Aldehydes

Contain the double-bonded carbonyl group, C=0, with only one hydrocarbon group on the carbon.

```
    H      O
    |     //
H - C - C
    |     \
    H      H
```

Acetaldehyde—100 ppm/740 mm @ 20 C

Alcohols

Contain a single –OH group

```
    H
    |
H - C - OH
    |
    H
```

Methanol—200 ppm/125 mm Hg (methyl alcohol, wood alcohol).
Ethanol—1,000 ppm/50 mm (ethyl alcohol, grain alcohol).
Propanol—200 ppm/20.8 mm

Ethers

Contain the C—O—C linkage.

```
    H   H       H   H
    |   |       |   |
H - C - C - O - C - C - H
    |   |       |   |
    H   H       H   H
```

Ethyl ether—400 ppm/440 mm Hg
Isopropyl ether—250 ppm/120 mm
Ethylene glycol monomethyl ether—5 ppm/9.7 mm (methyl Cellosolve)
Ethylene glycol monoethyl ether—50 ppm/5.3 mm (Cellosolve)

Glycols

Contain double –OH groups.

```
        H   H
        |   |
H - O - C - C - O - H
        |   |
        H   H
```

Ethylene glycol
(1,2-ethanediol)—50 ppm ceiling/0.06 mm @ 20 C

molecule. Carbon tetrachloride at one end of the scale is highly toxic, acting acutely by injury to the kidneys, the liver, the central nervous system, and the gastrointestinal tract. The 1986-87 TLV for carbon tetrachloride is 5 ppm; however, NIOSH recommends that the PEL be reduced to 2 ppm averaged over a 1-hour period and that the chemical be regulated as an occupational carcinogen.

Chronic exposure to carbon tetrachloride also damages the liver and kidneys and is suspected of causing liver cancer. Carbon tetrachloride has become the classical liver toxicant for use in studies on the effects of damage to the liver. As with benzene, this solvent should not be used for open cleaning processes where there is skin contact or where the concentration in the breathing zone may exceed recommended levels.

Replacing some of the chlorine atoms with fluorine as in trichlorotrifluoroethane produces a compound with a low level of toxicity. Its present TLV is 1,000 ppm as a ceiling limit. The depressant effect on the central nervous system and cardiac arrhythmia occur at concentrations much greater than the TLV. Being nonflammable and of low toxicity, it is a good general solvent to substitute for the more hazardous materials.

Material Safety Data Sheet May be used to comply with OSHA's Hazard Communication Standard, 29 CFR 1910.1200. Standard must be consulted for specific requirements.	**U.S. Department of Labor** Occupational Safety and Health Administration (Non-Mandatory Form) Form Approved OMB No. 1218-0072
IDENTITY *(As Used on Label and List)*	*Note: Blank spaces are not permitted. If any item is not applicable, or no information is available, the space must be marked to indicate that.*

Section I

Manufacturer's Name	Emergency Telephone Number
Address *(Number, Street, City, State, and ZIP Code)*	Telephone Number for Information
	Date Prepared
	Signature of Preparer *(optional)*

Section II — Hazardous Ingredients/Identity Information

Hazardous Components (Specific Chemical Identity; Common Name(s))	OSHA PEL	ACGIH TLV	Other Limits Recommended	% *(optional)*

Section III — Physical/Chemical Characteristics

Boiling Point		Specific Gravity (H_2O = 1)	
Vapor Pressure (mm Hg.)		Melting Point	
Vapor Density (AIR = 1)		Evaporation Rate (Butyl Acetate = 1)	
Solubility in Water			
Appearance and Odor			

Section IV — Fire and Explosion Hazard Data

Flash Point (Method Used)	Flammable Limits	LEL	UEL
Extinguishing Media			
Special Fire Fighting Procedures			
Unusual Fire and Explosion Hazards			

(Reproduce locally) OSHA 174, Sept. 1985

Figure 6-2. Material Safety Data Sheet—this form, or an equivalent, is used by manufacturers and distributors to provide health and safety information on solvents (Continued).

Section V — Reactivity Data

Stability	Unstable		Conditions to Avoid
	Stable		

Incompatibility (*Materials to Avoid*)

Hazardous Decomposition or Byproducts

Hazardous Polymerization	May Occur		Conditions to Avoid
	Will Not Occur		

Section VI — Health Hazard Data

Route(s) of Entry:	Inhalation?	Skin?	Ingestion?

Health Hazards *(Acute and Chronic)*

Carcinogenicity:	NTP?	IARC Monographs?	OSHA Regulated?

Signs and Symptoms of Exposure

Medical Conditions
Generally Aggravated by Exposure

Emergency and First Aid Procedures

Section VII — Precautions for Safe Handling and Use

Steps to Be Taken in Case Material Is Released or Spilled

Waste Disposal Method

Precautions to Be Taken in Handling and Storing

Other Precautions

Section VIII — Control Measures

Respiratory Protection *(Specify Type)*

Ventilation	Local Exhaust	Special
	Mechanical *(General)*	Other

Protective Gloves	Eye Protection

Other Protective Clothing or Equipment

Work/Hygienic Practices

Page 2

☆ USGPO 1986-491-529/45775

Figure 6-2. (Concluded)

The chlorinated hydrocarbons, in general, are more toxic than the common fluorinated hydrocarbon solvents. Specific effects and toxicities vary widely, but the most common effects from the chlorinated hydrocarbons of intermediate toxicity (trichloroethylene, for example) are the depressant effect on the central nervous system, dermatitis, and injury to the liver. Personality changes also have been noted.

In addition, the chlorinated hydrocarbons, especially trichloroethylene, are noted for their synergistic effects with alcohol. These include the flushed, red face and significant personality changes. This must be taken into account when evaluating industrial exposure since a significant portion of the working population may also ingest alcohol. Results of recent studies indicate that perchloroethylene should be treated as a suspect carcinogen until more information is available.

Nitro-hydrocarbons. These vary in their toxicological effects, depending on whether the hydrocarbon is a paraffin or an aromatic. Nitroparaffins are known more for their irritant effects accompanied by nausea; effects on the central nervous system and liver become significant during acute exposures; 2-nitropropane is listed as a suspect carcinogen, but with an assigned TLV and low carcinogenic potency. The nitroaromatics (like nitrobenzene) are much more acutely hazardous. They cause the formation of methomoglobin and act on the central nervous system, the liver, and other organs.

Oxygen-containing functional groups. These are found in alcohols (–OH), aldehydes (–CHO), and ketones ($-\overset{\overset{O}{\|}}{C}-$), as well as in carboxylic acids (–COOH) and their esters (–COOR) and anhydrides ($-\overset{\overset{O}{\|}}{C}-O-\overset{\overset{O}{\|}}{C}-$).

Esters. These are produced by the esterification of an acid with an alcohol. The particular properties of the esters are, therefore, partly determined by the parent alcohol. The esters are good solvents for surface coatings.

The esters are noted for their irritating effects to exposed skin surfaces and to the respiratory tract. They also are potent anesthetics. Cumulative effects of the common esters used as solvents are not significant except for those conditions resulting from irritation.

Ketones. These have become increasingly important solvents for acetate rayon and vinyl resin coatings. Ketones are stable solvents with high dilution ratios for hydrocarbon diluents. They are freely miscible with most lacquer solvents and diluents, and their compatibility with lacquer ingredients gives good blush resistance. Generally, ketones are good solvents for cellulose esters and ethers and many natural and synthetic resins.

The common ketones generally exert a narcotic-type action. All are irritating to the eyes, nose, and throat and, for this reason, high concentrations are not usually tolerated. Methyl ethyl ketone in conjunction with toluene or xylene has been reported to elicit vertigo and nausea. Lower tolerable concentrations may impair judgment and thereby create secondary hazards. The lower saturated aliphatic ketones are rapidly excreted and for this reason cause only minor systemic effects. Methyl *n*-butyl ketone received widespread attention during the 1970s, when it was pinpointed as the etiological agent producing a high incidence of peripheral neuritis in one working population.

The alcohols. One of the most important classes of industrial solvents are characterized by the presence of a hydroxyl group (–OH). The saturated alcohols are widely used as solvents. The alcohols are formed by the replacement of one or more hydrogen atoms by one or more hydroxyl groups.

These polar compounds are classified on the basis of both the number of hydroxyl groups, and on the nature of the radicals attached to the hydroxyl groups. The monohydric alcohols, which contain one hydroxyl group, are known simply as alcohols; dihydric alcohols have two hydroxyl groups, and are known as glycols; trihydric alcohols have three hydroxyl groups and are called glycerols or polyols.

The alcohols are noted for their effect on the central nervous system and the liver but they vary widely in their degree of toxicity.

Methanol (H_3COH) and ethanol (H_3CCH_2OH) are the two most important industrial alcohols. Methanol is made by catalytic hydrogenation of carbon monoxide and may one day replace gasoline and natural gas as a fuel because it can be made from coal. Ethanol is made by fermentation of starch (or other carbohydrate) and by hydration of ethene.

Methanol is responsible for causing several industrial fatalities and injuries, notably, impairment of vision and injury of the optic nerve. Methanol slowly produces toxic metabolites. For this reason, its chronic toxicity is greater than that of ethanol.

Ethanol is used industrially in a denatured form. It is quickly metabolized in the body and largely converted to carbon dioxide, and is the least toxic of the alcohols. Any toxicity it causes can be more related to the denaturants. The undesirable effects of ethanol primarily are related to its excessive use, which affects the drinker's physical safety, and which can synergize the effects of other solvents or medications.

Alcohol is a depressant, not a stimulant. Medically, alcohol depresses the central nervous system, slowing down the activity of the brain and spinal cord. A large enough dose of alcohol can sedate the brain to a point when involuntary functions like breathing are lost. Death can be the result.

The commonly accepted concentration of blood alcohol concentration (BAC) that results in intoxication is 0.10 (100 mg/100 mL of blood, or 10 mg/210 L of breath). Death usually occurs at BACs of 0.50 and above. Several states consider a driver intoxicated if the BAC is 0.08. The impairment threshold is 0.05 BAC. Once a person's BAC rises above 0.05, they are impaired and involvement in accidents becomes more likely.

Propanol is metabolized to toxic by-products and is more toxic internally than ethanol but less toxic than the higher homologues. Except for removing fat from the skin, neither isopropanol nor *n*-propanol are considered as industrial hazards.

Ethers. These are made up of two hydrocarbon groups held together by an oxygen atom. They are made by combining two molecules of the corresponding alcohol. Compared with alcohols, ethers are characterized by their greater volatility, lower solubility in water, and higher solvent power for oils, fats, and greases. Because of their stability and ease of recovery, ethers are widely used for extraction purposes. Mixtures of the lower alkyl ethers with alcohols make efficient solvents for cellulose esters. The so-called epoxides (cyclic ethers) differ from other ethers, which are chemically inert, by being highly active chemically because of their unstable three-membered rings. Since the epoxides react with the labile (unstable) hydrogen atom of water, alcohols, amines, and similar substances, they form a wide range of industrially important compounds.

The primary reaction to the saturated and unsaturated alkyl ethers, such as ethyl and divinyl ether, is anesthesia, as well as irritation of the mucous membranes. However, the greatest safety hazard with these ethers is their tendency to form explosive peroxides. Take care to preclude this. Halogenated ethers generally are highly toxic and the reader should refer to the more comprehensive references for information on these materials (see Bibliography, especially Clayton and Clayton, E. Browning, and the Hygienic Guides).

The glycols. The glycols, like Cellosolves and the Carbitols, are colorless liquids of mild odor. They are miscible with most liquids and owe this wide solubility to the presence of the hydroxyl, the ether, and alkyl groups in the molecule. The glycol dialkyl ethers are pure ethers with mild and pleasant odor. They are better solvents for resins and oils than are the monoethers. As a rule these compounds are more volatile than the monoethers with the same boiling point.

The glycol ethers exert their effects upon the brain, the blood, and the kidneys. Of these, ethylene glycol monomethyl ether is the most toxic. It is rapidly absorbed through the skin and is noted for elicitation of neurological symptoms, including changes in personality. It also may affect the reproductive system in both men and women. Ethylene glycol monoethyl ether, on the other hand, is definitely less toxic than either ethylene glycol monomethyl ether or ethylene glycol monobutyl ether. There are few reports of industrial disease or injury resulting from the use of ethylene glycol monoethyl ether.

The glycols have low vapor pressures, and thus, have not been an industrial problem from the viewpoint of inhalation. Those intoxications that have occurred usually resulted from ingestion; the kidney was the main organ affected.

Due to the low vapor pressures, inhalation exposures are not likely unless the material is heated or sprayed. In such cases the mists and vapors are irritating.

The aldehydes. These are well known for skin and mucosal irritation and for their action on the central nervous system. Dermatitis from the aldehydes is common. The aldehydes also are characterized by their sensitizing properties. Allergic responses are common.

This information indicates only the general toxicological effects, which can be used to assist in determining hazard potential and establishing a frame of reference. Actual effects from a specific solvent or from a given mixture of solvents may vary considerably, as indicated in the introductory paragraphs. In each classification there seems to be one or more specific solvents that are more hazardous than other homologues.

Hazard potential

The toxicological effects alone are not adequate to assess the hazard potential of a solvent. The vapor pressure, ventilation, and manner of usage will determine the concentration in air, and thus the amount of material available to produce an effect.

The vapor/hazard ratio number. This is one approach toward a numerical comparison of potential hazard under a given set of conditions. This number is the ratio of the equilibrium vapor concentration at 25 C (77 F) to the TLV (ppm/ppm)—the lower the ratio, the lower the potential hazard. As an example, hexone (methyl isobutyl ketone, MIBK) with a TLV of 50 ppm might be judged potentially more hazardous than 2-butanone (methyl ethyl ketone, MEK), which has a TLV of 200 ppm, if judged on the basis of the TLV only. When the potential for vaporization is taken into account it becomes apparent that hexone (MIBK) should have a lower hazard potential. The vapor hazard ratio for hexone (MIBK) is only 186, whereas the vapor hazard ratio for 2-butanone (MEK) is 625 (Table 6-B).

Other factors. Other factors need to be taken into consideration. For example, handling procedures and type of clothing will determine the degree of skin contact and absorption. Even the degree of a user's respect for the hazard potential can be a decisive factor.

Ignition temperature, flash point, and other factors determining the potential for fire and explosion also must be considered. Although concentrations that are safe from a toxicological viewpoint are much lower than the lower flammable limits of flammable solvents, concentrations at potential points of ignition may be far higher than concentrations in the user's breathing zone.

Evaluation of hazard potential requires assessing the consequences of exposure, the degree of exposure, and all factors contributing to the exposure.

Fire and explosion

If ventilation is adequate from a toxicological viewpoint, the potential for fire and explosion will be minimized. As stated in the previous paragraph, localized concentrations at openings to the container, in depressions, in pits, pockets, at vent openings, and at the liquid surface may be within the flammable range.

The potential for fire and explosion can be minimized by using nonflammable solvents and solvents with flash points greater than 60 C (140 F). However, the nonflammable halogenated hydrocarbons decompose when subjected to high temperatures and give off toxic and corrosive decomposition products (such as phosgene, hydrochloric acid, and hydrofluoric acid); so they cannot be used freely in the presence of flames, electrical equipment with open arcs, or other high-temperature sources.

If flammable solvents with flash points lower than about 60 C must be used, proper precautions must be taken. Eliminate sources of ignition, such as flames, sparks, high temperatures, and smoking. Properly bond and ground equipment for handling flammable solvents and install in accordance with the national and state electrical codes for such installations. Adequately train workers in fire protective measures.

Minimum legal standards for flammable solvents in general industry are covered by the Occupational Safety and Health Administration regulations §1910.106 and local and state fire codes in the area of application. The OSHA standards for specific industries supplement these requirements. See also NFPA Standard No. 30, *Flammable and Combustible Liquids Code.*

Additional information on flammable liquids has been extracted from the National Safety Council's Industrial Data Sheet No. 532, *Flammable and Combustible Liquids in Small*

Table 6-B. Organic Liquids Arranged in Order of Vapor Hazard

Substance	Vapor Hazard[a]	Threshold Limit (ppm by volume)[b]	B.P. (deg C)[c]
Acrylonitrile[d]	112,000	2	78.9
Carbon disulfide	46,000	10	46.3
Butylamine	34,000	5	77.8
Diethylamine	31,500	10	55.5
Carbon tetrachloride[d]	28,340	5	76.8
Chloroform[d]	24,850	10	61.2
Allyl alcohol	16,450	2	96.6
Tetranitromethane	15,800	1	126.0
Ethylene dichloride (1,2 Dichloroethane)	11,600	10	83.7
1,1,2,2-Tetrachloroethane	8,420	1	146.3
Methylene chloride	4,320	100	40.1
Hexane (n-hexane)	4,100	50	69.0
Ethyl formate	3,160	100	54.0
Methyl cellosolve (2-Methoxyethanol)	3,150	5	124-25
Ethyl bromide	3,030	200	38.4
Trichloroethylene	2,000	50	87.2
Acetic acid	1,970	10	118.1
Dioxane (diethylene dioxide)	1,960	25	101.0
Ethylene diamine	1,710	10	55.5
Cellosolve (1-Ethoxy-ethanol)	1,640	5	135.1
Benzyl chloride	1,580	1	179.4
1,1-Dichloroethane	1,450	200	57.3
Ethyl ether	1,380	400	34.6
Methyl acetate	1,380	200	57.1
Acetic anahydride	1,340	5	139.6
Hexanone (Methyl butyl ketone)	1,000	5	128.0
Propylene dichloride (1,2-Dichloropropane)	910	75	96.8
2-Nitropropane[d]	896	25	120.3
Mesityl oxide	893	15	130.0
Dimethylaniline (N-Dimethylaniline)	870	5	193.0
Cellosolve acetate (ethoxethyl acetate)	840	5	156.3
Propyl ether (Isopropyl ether)	840	250	69.0
Methyl alcohol	820	200	64.7
Methyl cellosolve acetate (Ethylene glycol monomethyl ether acetate)	815	5	144.5
Pentane	750	600	36.3
2-Butanone (MEK)	625	200	79.6
Methylal (Dimethoxymethane)	526	1,000	42-3
1,1,1 Trichloroethane (Methyl chloroform)	489	350	74.1
Perchloroethylene (Tetrachloroethylene)[d]	474	50	120.8
Nitrobenzene	474	1	210.9
Nitromethane	434	100	101.0
Cyclohexane	427	300	80-81
Propyl acetate	390	200	88.4
Acetone	387	750	56.5
Toluene	368	100	110.8
Aniline	330	2	184.4
Ethyl acetate	303	400	77.1
Dichloroethyl ether	288	5	178.5
Propyl alcohol	280	200	82.5
Methyl isobutyl carbinol (Methyl amyl alcohol)	263	25	131.8
Nitroethane	263	100	114.8
Cyclohexanone	226	25	155-6
Amyl alcohol (Isoamyl alcohol)	220	100	132.0
Chlorobenzene (Monochlorobenzene)	210	75	132.1
Styrene monomer (Phenyl ethylene)	194	50	145-6
Hexone (Methyl isobutyl ketone)	186	50	117-19
Cresol (all isomers)	184	5	191-203
Butyl alcohol (n-Butanol)	184	50	117.0
O-Toluidine[d]	165	2	199.7
Heptane (n-Heptane)	151	400	98.4
Methylcyclohexane	151	400	101.0
Diisobutyl ketone	142	25	168.1
Nitrotoluene	138	2	222-38
Phenol	132	5	181.4
Isophorone	130	5	215.0
Ethyl benzene	126	100	136.2
Pentanone (Methyl propyl ketone)	112	200	95
Butyl acetate (n-Butyl acetate)	105	150	127
Xylene	100	100	139-44
Methylcyclohexanone	94	50	162-70
Ethyl alcohol (Ethanol)	76	1000	78.4
p-Tertiary butyl toluene	72	10	200
Turpentine	66	100	120-80
Octane	57	300	99-125
Amyl acetate	53	100	142.0
Cyclohexanol	47	50	160-61
Stoddard solvent	35	100	150-90
Methylcyclohexanol	14	50	165-75
Phenylhydrazine	11	5	243.5
O-Dichlorobenzene	9	50	179
Diacetone alcohol (4-Hydroxy-4-methyl-pentanone-2)	8	50	167.9

NOTES:

[a]Ratio (ppm/ppm) of equilibrium vapor concentration at 25 C to the Threshold Limit Value, computed from vapor pressure data in references 1 and 2.
[b]From ACGIH Threshold Limit Values for 1985–86.
[c]Boiling point at 760 mm Hg (references 1, 2, and 3). Observed boiling points in mixtures may be lower due to formation of azeotropes.
[d]Suspect carcinogen TLV may be lowered. Consult latest TLV list.

References:
[1]Mellan, I. *Industrial Solvents.* Reinhold Publishing Corp., New York, 1939.
[2]Jordan, T.E. *Vapor Pressure of Organic Compounds.* Interscience Publishers, Inc., New York, 1954.
[3]Lange, N.A. *Handbook of Chemistry,* 6th ed. Handbook Publishers, Inc., Sandusky, Ohio.

Containers, and is condensed at the end of this chapter for the reader's convenience. More detailed information may be found in the data sheet itself and in the NFPA *Fire Protection Handbook.*

Air pollution

Solvents may become hazardous to the public in the form of air pollutants when released outdoors. Solvent hydrocarbons are important compounds in the formulation of photochemical smog. In the presence of sunlight, they react with atomic oxygen and ozone to produce aldehydes, acids, nitrates, and a whole series of other irritant and noxious compounds.

The greatest portion of hydrocarbons contributing to air pollution originates from automobiles, but a significant amount also comes from the tons of solvents that are exhausted daily from industrial cleaning and surface-coating processes.

Some solvents are more reactive to sunlight and contribute heavily to the smog problem. The use of such solvents is being curtailed in more and more areas, especially the large cities. Other solvents are less reactive and are exempt from stringent control. The following order of photochemical reactivity is listed as a guide.

- olefins (unsaturated open-chain hydrocarbons containing one or more double bonds) are more reactive
- aromatics (except benzene)
- branched ketones, including methyl isobutyl ketone
- chlorinated ethylenes, including trichloroethylene (except perchloroethylene)
- normal ketones (for example methyl ethyl ketone)
- alcohols and aldehydes
- branched paraffins
- cyclic paraffins
- normal paraffins
- benzene, acetone, perchloroethylene, and the saturated halogenated hydrocarbons.

There are differing opinions as to the exact order of reactivity and many solvents have yet to be tested. The trend is toward the development and use of nonreactive solvent blends.

Upper atmosphere effects. In addition to the smog-related materials discussed previously, fluorocarbons, such as trichlorotrifluoroethane and related materials, catalyze the destruction of ozone in the upper atmosphere. Although the extent of this reaction is not well established, their production and use has been reduced. Should the destruction of ozone by fluorocarbons and other materials prove to be significant, the amount of solar ultraviolet radiation reaching the earth's surface may increase. This would impair agricultural production and increase the incidence of skin cancer.

HEALTH AND SAFETY PROCEDURES

Responsibility of health and safety personnel

Personnel concerned with health and safety should recognize that the use of solvents can be a major threat to health and that controls are usually necessary to prevent detrimental physiological effects.

Surveys or searches should be made for evidence of disease. Dermatitis, unusual behavior, coughing, or complaints of irritation, headache, and ill feeling are all outward signs of potential disease that warrants further investigation. Such forms of evidence can be found by an experienced person. Positive findings serve to justify the effort and provide convincing evidence for educating personnel to the need for corrective actions.

Note conditions and practices that contribute to excessive exposure and call them to the attention of responsible personnel. Train users to handle solvents properly to prevent injurious exposures. Set guidelines to direct operating personnel in the selection, use, and handling of solvents. Prohibit general use of highly toxic solvents, highly flammable solvents, or solvents that are extremely hazardous, unless special evaluation or authorization is obtained.

Finally, provide technical assistance to help the user select the least hazardous solvents, design and obtain proper ventilation, eliminate the risk of fire, eliminate skin contact, and evaluate those situations when workers might be exposed to excessive levels.

Selection of solvents

One of the most effective means of controlling solvents is to use the least hazardous solvent. By simply substituting a less toxic or less volatile solvent, one can minimize or, sometimes, eliminate a hazard. The fact that a certain solvent has been specified does not mean that it is the only one or even the best one for a particular use. Usually the solvent specified is the most familiar one.

This fact is more apparent if one compares the TLVs and the vapor pressures or distillation ranges of different solvents in each class. Xylene is a good cleaning agent that can usually be substituted for benzene, for example. If an aromatic hydrocarbon is not required, then it may be replaced with less toxic aliphatic mineral spirits. The potential for fire can be minimized by the introduction of nonflammable trifluorotrichloroethane or trichloroethane.

The best all-around solvent is water. It is nontoxic, nonflammable, and (with the proper additives) it forms an aqueous solvent system that is good for many organic materials. For the cleanup of inorganic soils, aqueous solvent systems are still the best. The disadvantages are corrosiveness of many aqueous solutions and long drying time of the water, but with the proper equipment even these can be controlled.

The aliphatic hydrocarbons are good for dissolving nonpolar organic soils such as oils and lubricants. The aliphatics, however, are not effective cleaners for dissolving or removing many tenacious inorganic materials.

The aromatic hydrocarbons are especially effective on resins and polymeric materials. In between the aromatic and aliphatic hydrocarbons in solvent power are the cyclic hydrocarbons. The halogenated hydrocarbons are effective solvents for a wide range of both nonpolar and semipolar compounds.

The nitrohydrocarbons have not been used to any large extent as cleaning agents. Their greatest use has been as solvents for esters, resins, waxes, paints, and the like. Because the ketones, alcohols, esters, ethers, aldehydes, and glycols are more soluble in water than are the other classes, they are good solvents for the more polar compounds. These solvents are often used as cleaning agents alone or combined with one or more other solvents, especially water. They are useful as solvents for paints, varnishes, and plastics. The publication *Contamination Control* (see Bibliography) is a good source of information for the effectiveness of the different solvents against various soils.

Remember that for nearly every process there is an effective solvent or solvent blend that has low toxicity and low flammability. For example, several companies have switched to water containing an alkaline cleaner as a replacement for naphtha and other such organic solvents for cleaning hydraulic tubing, tanks, and other containers. Inhibited 1,1,1-trichloroethane has replaced carbon tetrachloride as a household spot remover.

As a guide, the following suggestions might be used:

- Use an aqueous (water) solution if possible.
- If water is not suitable, use a so-called "safety solvent," but make sure there is adequate ventilation. Safety solvents include such compounds as inhibited 1,1,1-trichloroethane, the aliphatic hydrocarbons with high flash points, and the fluorinated hydrocarbons.
- Solvents that are more toxic than the safety solvents are to be used only with properly engineered local exhaust systems. Solvents, such as trichloroethylene, toluene, and ethylene dichloride, are in this category.
- Highly toxic or highly flammable solvents, like benzene, carbon tetrachloride, and gasoline, should be prohibited as general cleaning solvents.

Definite dividends may be noted as a result of promoting this policy. The number of employees who might have exposures exceeding the TLV can be reduced significantly. The number of small fires resulting from the use of flammable bench solvents also can be reduced.

Enclosure and ventilation

The major portal of entry for solvents into the body is the lungs (Chapter 2, The Lungs). The lungs have a surface area of about 55–75 m² (85,000–115,000 sq in.); much of this area is permeated with thin-walled capillaries. Solvents in the breathing zone are drawn into the lungs during breathing, quickly absorbed into the bloodstream, and distributed to other parts of the body.

The first and most effective way of preventing the inhalation of solvents is to keep them out of the breathing zone. This is done by using closed systems and local exhaust ventilation.

Open vessels should be covered except when in use. Systems should be designed to prevent leakage and spillage and to collect and contain the solvent in the event of a leak or spill. For open and closed system design parameters, see Cralley.

Ventilation must be considered for any process using solvents. Even storage areas require adequate general ventilation to prevent accumulation and buildup of flammable or toxic concentrations (Chapters 21 and 22 on industrial and general ventilation).

Storage of flammable and toxic solvents should only be in refrigerators designed or prepared for that use. They should be posted as to having this feature. Such refrigerators have had their ignition sources removed. Refrigerators used for storage of food and beverages should not be used for any other purpose.

Local exhaust ventilation is necessary to capture the vapors at their point of origin and thus prevent excessive concentrations in the breathing zone of the worker. If general ventilation is good, for solvents of low toxicity, a simple pedestal fan blowing the vapors away from the user's breathing zone will often be sufficient. If a highly toxic solvent is being used, or if general ventilation is poor, a local exhaust system is necessary to remove the vapors.

All control measures should maintain concentrations of solvents in the breathing zone well below those action levels, above which the administrative and medical control measures specified by OSHA must be implemented. Present trends in worker's compensation insurance and federal regulations justify designs that are well on the safe side.

Ventilation systems are a topic in themselves and the reader should refer to Chapters 21 and 22, and to the ACGIH *Industrial Ventilation Manual* (see Bibliography).

Remember that the local exhaust ventilation system of removing vapors at their point of origin is usually the most satisfactory means of control.

Respirators

Do not use respirators as the only means of protection against solvents because there are too many factors that limit their use. They can be used as emergency or backup protection. Respiratory protective equipment, especially the air-purifying type, is limited by leakage around the mask edges, surface contamination, impaired efficiency with use, and need for adequate oxygen. Unless it is correctly used and properly cared for, a respirator may present a greater danger to an employee than not having any protection at all. Too often, such equipment gives a false sense of security and the wearer becomes careless and may be exposed to highly hazardous levels. Respirators should be controlled through a program that provides for proper selection, fitting, testing, and maintenance under the surveillance of competent personnel. Such a program is mandatory under present federal occupational safety and health standards. See Chapter 23, Respiratory Protective Equipment.

Air-supplied respirators or air-purifying respirators that are listed as approved by NIOSH for use against organic vapors should provide adequate protection for most solvents. Make sure, however, that the level of solvent vapors in the air does not exceed the protective factor of the respirator. Air-purifying respirators should not be used for operations where the solvent is air-sprayed, unless there is supplementary mechanical ventilation. *Respiratory Protection—A Manual and Guideline* is an excellent guide for such equipment, see Chapter 23 for more details.

Skin contact and protection

Another major route of entry for solvents is through the skin. Dermatitis is the leading industrial disease, and solvents are second only to cutting oils and lubricants in causing this disease. See Chapter 3, The Skin, and Chapter 8, Industrial Dermatoses.

Skin contact occurs through direct immersion, splashing, spilling, contact with solvent-soaked clothing, improper gloves, and contact with solvent-wet objects. Some solvents like benzene, carbon tetrachloride, and methyl alcohol can be absorbed in amounts great enough to cause physiological injury to organs other than the skin.

The most effective way and often the only way to prevent harm is to keep the solvent from the skin. This can be done with systems that eliminate the need for contact by using mechanical handling devices, such as tongs and baskets, and by using impermeable protective clothing, such as aprons, face shields, and gloves.

The use of gloves requires caution. A common mistake is to recommend rubber or neoprene gloves for use as hand protection against a solvent, regardless of the kind of solvent in use. Many solvents can quickly penetrate rubber or neoprene gloves and come in contact with the skin, creating favorable conditions for injury.

The permeability of gloves to certain solvents is really the most important characteristic to consider when selecting gloves for solvent protection. Manufacturers of solvents include permeability information with their product, often in the form of permeability tables. They suggest appropriate glove materials for particular solvents. The abrasion resistance of solvents is also given in tables—this information is often more widely available. Take care not to confuse permeability and abrasion resistance when considering the type of gloves to use with certain solvents.

Solvent permeability measurements should be made on the complete glove if the effect of weak or thin spots is to be detected. A rough comparison of the permeability of gloves plus an indication of some of the other characteristics can easily be made by turning the gloves inside out, filling them three-fourths full of solvent, sealing the cuff, and measuring the loss of weight, the stretch, and other parameters.

More precise methods for measuring glove permeability are available, but require the use of an analytical laboratory. A standard method has been published by the American Society for Testing and Materials (Test Method F739-81).

The time required for a solvent to penetrate a glove is affected by the glove's thickness and its composition. In some cases, the time to break through can be as brief as 5 minutes. For example, benzene broke through a 0.03-mm polyethylene glove in five minutes. Conversely, the same glove material had a 2-hour breakthrough time when placed against butyl acetate.

Note that gloves made of the same material and nominal thickness, but from different manufacturers, may have significantly different breakthrough times. This phenomenon may be due to differences in formulation of glove materials or manufacturing procedures used.

Examples of the degree of penetration through gloves made of various polymers are shown in Tables 6-C and 6-D, in 0.05-, 4-, and 8-hour periods of exposure.

Neoprene is good for protection against most common oils, aliphatic hydrocarbons, and for certain other solvents, but is not satisfactory for use against the aromatic hydrocarbons, the halogenated hydrocarbons, the ketones, and many other solvents. Natural rubber is not effective against these solvents, either.

Polyvinyl alcohol gloves provide adequate protection against the aromatic and chlorinated hydrocarbons, but they must be kept away from water, acetone, and other solvents miscible in water to prevent deterioration. Butyl rubber gloves often provide a suitable compromise when polyvinyl alcohol cannot be used.

Regular periodic cleaning and drying of gloves is as important as using the proper type. Keep an alternate pair of gloves available for use while the cleaned pair is being well aired and dried. When the gloves may become soiled with hard-to-remove hazardous materials, such as insecticides or epoxy resins, it is often better to discard the glove than it is to try to clean it. In some situations, gloves have to be replaced after only a few minutes' work. If the outside of any glove becomes thoroughly wetted, remove it promptly.

Disposable gloves are useful for light laboratory or assembly work, but are too easily torn or punctured for heavier work. There is no set recommendation for the use of gloves; what works well for one group of workers may not work for another. In many cases, a certain amount of trial and error is required.

Barrier creams are the *least* effective way of protecting skin. Barrier creams are not a substitute for gloves, except when there is only occasional and minor contact with a solvent, or around rotating machinery when gloves cannot be worn because of the catching hazard. Barrier creams are *not* as effective as an impervious glove. (See Chapter 8 for more information on barrier creams.)

Good personal hygiene is important whenever solvents are used. Remove spills and splashes immediately with soap and water. This includes showering and replacing solvent-soaked or splattered clothing with clean clothing immediately and as often as necessary.

Evaluation

A prime question regarding any process using a volatile solvent is whether the concentration of the solvent in the air exceeds acceptable levels. Getting the answer to this is not as difficult as it may seem and often relies on common sense.

A knowledge of the solvent, its properties, and the process in which it is used should give the investigator some idea of the potential hazards.

If a solvent of high-hazard potential is being used, if the equipment and ventilation system is poorly designed, or if performance of the system is questioned, then there is a greater probability of physiologic injury, and immediate action should be taken to evaluate and reduce the hazard before it becomes a problem.

Conduct the evaluation in an approved manner (see Clayton in References) by an industrial hygienist or a person experienced in air-sampling techniques. (See Chapters 16, 17, and 18 for details; the following paragraphs summarize the procedures.)

Samples can be collected in the field and returned to the laboratory for analysis or—and this is the trend—they may be collected and analyzed on the spot with direct-reading instrumentation. In this way, a much greater number of samples and much more information can be obtained and evaluated immediately. Recorders can be connected to direct-reading meters to obtain a continuous recording of the meter reading. Peak concentrations that elicit subjective complaints, which from an integrated toxicological viewpoint may be insignificant, become apparent. Such concentrations are likely to be missed with grab samples and integrated samples that are returned to the laboratory. Peak concentrations are especially important when the vapor is an irritant or is highly odorous, or if a subjective complaint is involved. The concentration above the norm must be reduced to achieve satisfaction.

Because direct-reading field instruments often require considerable laboratory backup in the way of maintenance, testing, and calibration, they are not as expedient as "the reading of a dial" might seem. The cost for a field evaluation may be much more than the cost of a laboratory analysis—but this is offset because much more information is obtained.

Direct-reading instruments

Many direct-reading field instruments are available; some are especially useful if the composition of the solvent is well known or if a large amount of sampling is required to evaluate a hazard.

Indicator tube. This is the simplest and is quite useful; however, the results can be very misleading unless the user is expe-

Table 6-C. Extent of Solvent Penetration Through Glove Materials After 0.5 Hour

	Glove Material					
Solvent	*Natural Rubber (0.4 mm)*	*Neoprene (0.4 mm)*	*Neoprene + Natural Rubber (0.5 mm)*	*Nitrile (0.4 mm)*	*PVC (0.2 mm)*	*PVA (0.4 mm)*
Carbon tetrachloride	D	D	D	A	D	A
Chloroform	D	D	D	D	D	A
Methylene chloride	D	D	D	D	D	A
Methyl iodide	D	D	D	D	D	A
1,1,2,2-Tetrachloroethane	D	D	D	C	D	A
1,1,2-Trichloroethane	D	D	D	D	D	A
Perchloroethylene	D	D	D	A	D	A
Methanol	A	A	A	A	B	D
Ethanol	A	A	A	A	B	C
2-Propanol	A	A	A	A	B	B
n-Butanol	A	A	B	A	B	A
Benzene	D	D	D	C	D	A
Toluene	D	D	D	C	D	A
Aniline	A	A	A	C	D	A
Phenol (10% water)	B	A	B	A	B	C
Acetone	B	C	B	D	D	A
Methyl ethyl ketone	C	D	C	D	D	A
Tetrahydrofuran	D	D	D	D	D	A
Dimethyl sulfoxide	A	A	A	A	B	D
Dimethyl formamide	B	A	A	C	D	D
Pyridine	C	C	C	D	D	C
Dioxane	B	B	B	A	D	A
n-Hexane	C	A	C	A	D	A
Water (^3H_2O)	A	A	A	A	A	D

Key: A = <0.1%, B = 0.1-1%, C = 1-10%, D = >10%.

Table 6-D. Extent of Solvent Penetration Through Glove Materials After 4 and 8 Hours

	Glove Material					
Solvent	*Natural Rubber (0.4 mm)*	*Neoprene (0.4 mm)*	*Neoprene + Natural Rubber (0.5 mm)*	*Nitrile (0.4 mm)*	*PVC (0.2 mm)*	*PVA (0.4 mm)*
Carbon tetrachloride				Ad		Aa
Chloroform						B
Methylene chloride						B
Methyl iodide						B
1,1,2,2-Tetrachloroethane						Ab
1,1,2-Trichloroethane						Ab
Perchloroethylene			C	C		Aa
Methanol	B	B	B	C		
Ethanol	B	Ab	B	Ab		
2-Propanol	B	Aa	B	Aa		
n-Butanol	C	Aa		Aa		C
Benzene						Ab
Toluene						Aa
Aniline	C	D	C			B
Phenol (10% water)		Ac		D		
Acetone						C
Methyl ethyl ketone						B
Tetrahydrofuran						C
Dimethyl sulfoxide	Ab	B	B	B		
Dimethyl formamide		C	C			
Pyridine						
Dioxane				D		C
n-Hexane		C		Aa		Aa
Water (^3H_2O)	Aa	Aa	Aa	Ab	B	

Key: Upper case letters are used for 4-hour data, lower case letters for 8-hour data.
A, a = <0.1%, B, b = 0.1-1%, C, c = 1-10%, D, d = >10%.

(Tables 6-C and 6-D are reprinted with permission from Sansone, E.B., and Tewari, B. The permeability of laboratory gloves to selected solvents. *American Industrial Hygiene Association Journal* [February 1978] 39:169-174)

rienced in interpreting the response and there are no interfering components. Proper calibration is important.

Some types are made for single samples. In effect, these are only grab samples that indicate conditions at the moment of sampling. Tubes made for sampling with pumps during a long period of time, which provide integrated sample results, are also available.

Indicator tubes have an accuracy of $\pm$25–50 percent. They are used as general indicators of current conditions, but should not be the only instrument used when highly accurate results are required.

Diffusion badge. The diffusion badge is available from several manufacturers. The badge is a small unit that clips to a person's lapel and relies on diffusion to sample the material of interest. The diffusion badge uses neither pump nor batteries. Some diffusion badges can be read directly while in the field, others must be sent to a laboratory for analysis.

The combustible gas meter. Although this is useful with flammable solvents having high TLVs, it must be calibrated, is not specific, and is not sufficiently accurate for concentrations of solvents much below 100 ppm. An additional drawback to the meter is that its response depends upon the concentration of oxygen and other gases present in the atmosphere being tested.

The halide meter. This is useful for direct measurement of halogenated hydrocarbons. Its disadvantage, or advantage as the case may be, is the lack of specificity for distinguishing among the different halogenated hydrocarbons.

Portable flame ionization meter. This meter gives fairly straight-line responses that are highly reproducible. Lack of specificity makes the instrument more versatile since it can be used to measure the concentration of nearly any of the common solvents in the air, especially if some other means is available to determine the qualitative composition of the vapors. Some models have gas chromatograph attachments that might help identify vapor composition.

The portable flame ionization meter has proven to be a most useful and versatile instrument. It is extremely sensitive and measures to 1 ppm or less.

For reliable results, carefully calibrate meters. An even more accurate and detailed record of conditions can be obtained by coupling a recorder to such devices.

The oxidant meter. This meter, preceded by a suitable pyrolyzer, will detect and measure certain halogenated hydrocarbons—methyl iodide and carbon tetrachloride—at levels in the parts-per-hundred-million range (Figure 6-3).

Portable gas chromatograph. This method is also available for field use (Figure 6-4). Several models are equipped with built-in microcomputers, which aid in their use and increase versatility.

Portable infrared (IR) meter. This IR meter is available for use in the field. Such meters are widely used because their applications are more specific than those of the flame ionization meter, yet they respond to a wide range of materials. Most IR meters must be carefully calibrated if reliable results are needed.

The photoionization meter. This meter is now the vogue. This method of detection is more specific and the level of sensitivity generally is an order of magnitude lower than that of the flame ionization meter.

Figure 6-3. A microcoulomb detector and recorder, commonly called an oxidant meter or ozone meter. (Courtesy Mast Development Co.)

Samples

Samples collected in the field must be returned to the laboratory when direct-reading field equipment is not available. Laboratory analyses are difficult, because of two reasons: (1) the complex nature of solvents, and (2) usually the more troublesome sample must be returned to the laboratory. Therefore, collect samples in the manner and form that are most suitable for the laboratory involved.

Grab samples. The simplest collection devices are grab-sampling devices like evacuated flasks or plastic bags. Bags made of Saran, Mylar, or Tedlar plastics are the least permeable to solvent vapors (Figure 6-5).

Larger time-weighted average samples. These can be collected by drawing the air through a collecting solution or through a solid absorbentlike silica gel, charcoal, or a molecular sieve.

Solid adsorbents. These are quite popular because they occupy a small space and the solvent can be driven off with heat or a suitable eluant, permitting analysis without interference from the presence of collecting solution.

Charcoal sampling tubes are available to collect solvent vapor samples. Collection procedures are standardized. Laboratories that are certified by the Laboratory Accreditation Committee of the American Industrial Hygiene Association participate in an interlaboratory quality-control testing program for solvent vapors collected on charcoal tubes (Figure 6-6). Samples collected on charcoal tubes and analyzed by these labs are more likely to give reliable results. Samples collected in this manner yield time-weighted average values that are directly comparable to the TLVs and the OSHA standards.

Diffusion badges for the collection of samples. These are quickly becoming standard sampling methods. See previous comments on these badges. The badges are especially useful in confirming nondetectable levels. For more specific results, the devices can be qualified by simultaneously collected samples collected on conventional tubes.

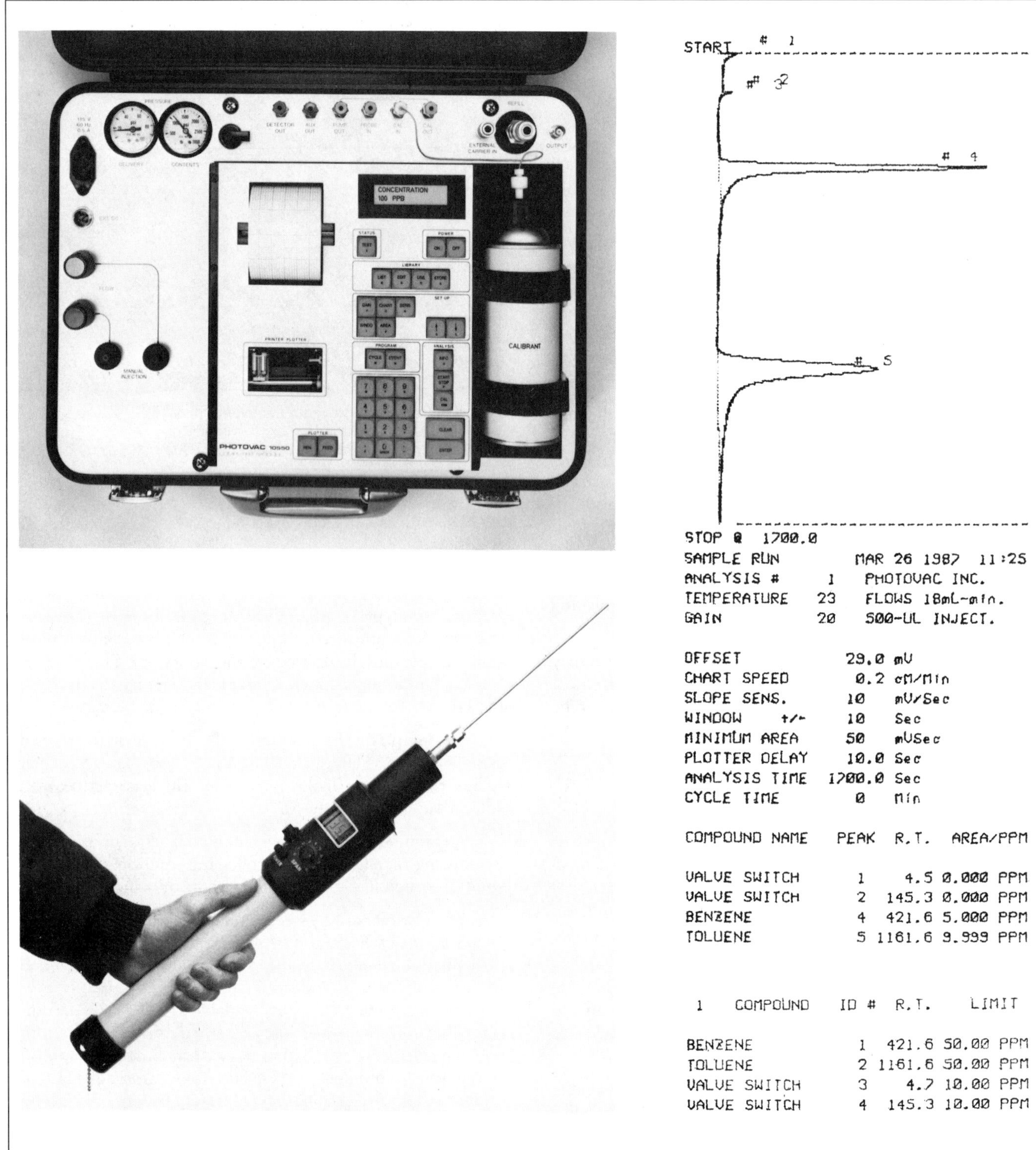

Figure 6-4. Portable gas chromatographs and analyzers for trace-level measurements of volatile organics in air, soil, and water. A sample print-out is shown. (Courtesy Photovac International, Inc.)

Laboratory instruments. All sorts of instruments as well as a number of analytical methods are available in the laboratory. But instruments are rapidly taking over from traditional chemical procedures in the laboratory as they have already done in the field. Among these instruments are the gas chromatograph, the ultraviolet (UV) spectrophotometer, the IR spectrophotometer, the mass spectrometer, and the electron-capture detector.

- The gas chromatograph is probably the most useful of these because it can separate closely related components of the solvents. With proper calibration and operation, it can be highly specific.
- The UV spectrophotometer is used for the aromatic hydrocarbons when interferences are known to be low.

Figure 6-5. Mylar bag makes a good grab-sampling device. The bag is filled by holding the top ring and letting the weight of the bottom pull it open. (Courtesy Argonne National Laboratory)

- The IR spectrophotometer is widely used for laboratory analysis of solvents in air. It offers the advantage of being able to identify certain functional chemical groups and provides greater specificity in the presence of other vapors.
- The mass spectrometer has become a highly useful tool for the identification of unknowns. It is often coupled with a gas chromatograph for separation and quantification of the different components. Such instrumentation allows industrial hygienists to evaluate and identify most solvent vapors.
- The electron-capture detector is especially suited for low levels of halogenated hydrocarbons.

There are also a number of properties and tests that are helpful indicators and screening devices to identify solvents in even the most austere laboratories. They include density, flash point, boiling range, miscibility, and the Beilstein test for halogenated and aromatic hydrocarbons. More detailed methods are described by Clayton.

BIBLIOGRAPHY

American Conference of Industrial Hygienists (ACGIH). *Threshold Limit Values and Biological Exposure Indices for 1986-1987.* Cincinnati: ACGIH, 1986.

——. Committee on Industrial Ventilation. *Industrial Ventilation—A Manual of Recommended Practice,* 19th ed. Lansing, MI: ACGIH, 1986.

American Industrial Hygiene Association (AIHA). *Heating and Cooling for Man in Industry,* 2nd ed. Akron, OH: AIHA, 1975.

——. *Respiratory Protection—A Manual and Guideline.* Akron, OH: AIHA, 1980.

——. *Hygienic Guide Series* on specific materials. Akron, OH: AIHA.

Browning, E. *Toxicity and Metabolism of Industrial Solvents.* New York: Elsevier Publishing.

Clayton, G.D., and Clayton, F.E., eds. *Patty's Industrial Hygiene and Toxicology,* 3rd ed., vol. 2A(1981), 2B(1981), 2C(1982). New York: John Wiley & Sons, 1981, 1982.

Gerarde, H.W. Toxicological studies on hydrocarbons: III. The biochemorphology of phenylalkanes and phenylalkenes. *Archives of Industrial Health* 19 (1959):403.

Toxicology and Biochemistry of Aromatic Hydrocarbons. New York: Elsevier Publishing Co., 1960.

Gleason, M. N., Gosslin, R.E., and Hodge, H.C. *Clinical Toxicology of Commercial Products,* 5th ed. Baltimore: The Williams & Wilkins Co., 1981.

Hamming, W.J. "Photochemical Reactivity of Solvents." Paper No. 670809 presented at the October 2-6, 1969, Aeronautic and Space Engineering and Manufacturing meeting, sponsored by the Society of Automotive Engineers, Inc.

International Committee of Contamination Control Societies: *Proceedings of the 4th International Symposium on Contamination Control,* September 1978, Washington, D.C. Mt. Prospect, IL: Institute of Environmental Sciences (formerly, American Association for Contamination Control).

Lunche, R.G., et al. L.A.'s rule 66 nips air pollution due to solvents. *SAE Journal* 76(2)(Nov. 1968):25.

McFee, D., and Garrison, R.P. "Process Characteristics—Open Systems." In *Industrial Hygiene Aspects of Plant Operations,* vol. 3, edited by L.V. Cralley and L. J. Cralley. New York: MacMillan Publishing Co., 1986.

McFee, D.R. How well do gloves protect hands against solvents. *Journal of the American Society of Safety Engineers,* May 1964.

National Fire Protection Association, Batterymarch Park, Quincy, MA 02269.

Fire Protection Handbook, latest edition.

–*Flammable and Combustible Liquids Code,* NFPA No. 30.

–*National Electrical Code,* NFPA No. 70.

National Institute for Occupational Safety and Health, Cincinnati, OH 45226.

——. *Criteria Documents* (listing available).

——. *Occupational Diseases: A Guide to Their Recognition,* Revised Edition, 1977. Publication Number DHEW(NIOSH) 77-181.

——. *Registry of Toxic Effects of Chemical Substances,* 1981-82.

——. *1983 Supplement to the Registry of Toxic Effects of Chemical Substances.*

——. *The Industrial Environment—Its Evaluation and Control.*

National Safety Council, 444 N. Michigan Ave., Chicago, IL 60611. Industrial Data Sheet.

–*Flammable and Combustible Liquids in Small Containers,* No. 532.

Nelson, G.O., et al. Glove permeation by organic solvents. *American Industrial Hygiene Association Journal,* 42(March 1981):217-225.

Sansone, E.B., and Tewari, Y.B. The permeability of laboratory gloves to selected solvents. *American Industrial Hygiene Association Journal,* 39(February 1978):169-174.

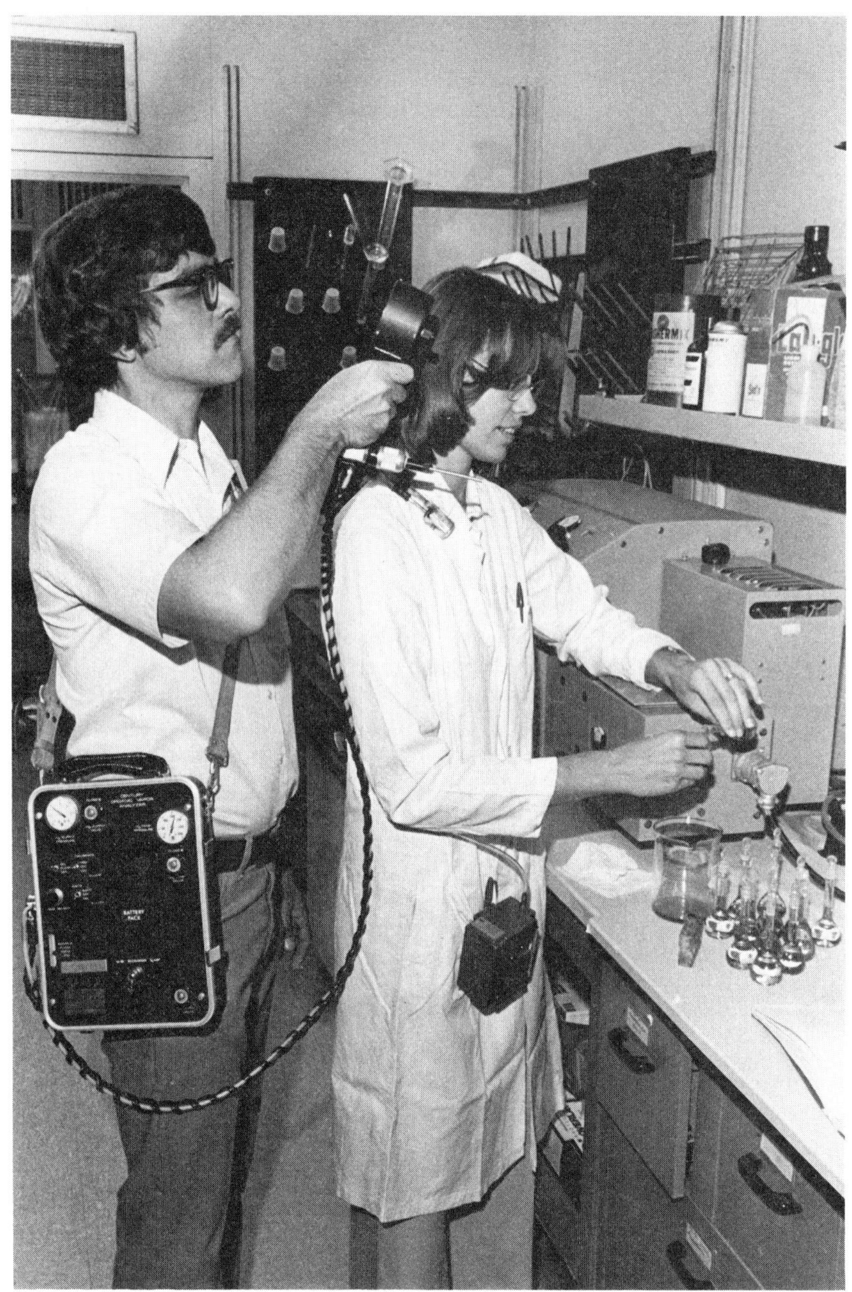

Figure 6-6. This laboratory worker is wearing a battery-powered sampling pump connected to a charcoal sampling tube on her collar. The tube will be analyzed to determine average solvent exposure. The portable monitor is being used to continuously monitor total organic vapors in the breathing zone, to determine which operations contribute most to solvent concentrations in air. (Courtesy Argonne National Laboratory)

Sansone, E.B., and Tewari, Y.B. Differences in the extent of solvent penetration through natural rubber and nitrile gloves from various manufacturers. *American Industrial Hygiene Association Journal* 41(July 1980):527.

Sax, N. I. and Lewis, R. J. *Hawley's Condensed Chemical Dictionary,* 11th ed. New York: Van Nostrand Reinhold Company, 1987.

VanDolah, R.W., et al. Flame propagation, extinguishment, and environmental effects on combustion. *Fire Technology* 1(2)(May 1965):138-145.

Williams, J.R. Permeation of glove materials by physiologically harmful chemicals. *American Industrial Hygiene Association Journal* 40(October 1979):877-882.

Windholz, M., ed. *The Merck Index,* 10th ed. Rahway, NJ: Merck & Co., Inc., 1983.

ADDENDUM—FLAMMABLE SOLVENT HAZARDS AND CONTROLS

This section was adapted from the National Safety Council Data Sheet No. 532, *Flammable and Combustible Liquids in Small Containers.*

Small container is defined as one having a capacity not exceeding 19 L (5 gal) nor used for flammable liquids under pressure. The safety features of protective devices, used on safety containers, are also incorporated into the design of dispensing and venting devices for larger containers, such as drums. Drums are frequently associated with dispensing into small containers and as receptacles of waste liquids from small containers. Because of this close relationship, many paragraphs contain comments and recommendations that relate to drums (not exceeding 227 L [60 gal]).

Flammable liquids (those with flash points below 37.8 C [100 F] in this text) are distinguished from *combustible liquids* defined in the *Flammable and Combustible Liquids Code,* NFPA No. 30, to be liquids having flash points at or above 37.8 C (100 F).

Hazards

Flammable and combustible liquids are volatile by nature, and it is their vapors combined with air, not the liquids themselves, that ignite and burn. The rate of evaporation increases with temperature. In many cases, therefore, as the temperature is increased, a flammable liquid becomes more hazardous because of the increased rate at which its vapors are evolved.

Flammable liquids vaporize and may form flammable mixtures in either open or closed containers, when leaks or spills occur and when heated. The degree of danger is determined largely by the flash point of the liquid, the concentration of the vapor-air mix, the evaporation rate, dispersion characteristics, and the possibility of a source of ignition that affords heat energy sufficient to ignite the mixture.

Vapor from a flammable liquid is usually invisible and may be difficult to detect unless a combustible-gas indicator is used. Because, in most cases, the vapor is heavier than air, it can settle to the floor or other lower levels, such as basements and the bottom of elevator shafts or stairways (Figure 6-7). However, the effects of air currents, heating, window ventilation, and the like, can be more significant in vapor travel.

Figure 6-7. Vapor from a flammable liquid ordinarily seeks the lowest level in air.

Frequently, investigations of fires involving flammable liquids indicate the cause as ignition of a vapor trail or vapor cloud, and a resultant flashback, at a considerable distance from the source of the vapor.

Numerous sources of ignition exist in many places and it is often difficult to eliminate or control them. Some common sources of ignition are open flames, smoking materials, hot surfaces, and sparks resulting from welding or cutting, operation of electrical equipment, and static electricity.

The danger of fire and explosion presented by flammable liquids can usually be eliminated or minimized by strict observance of safe storing, dispensing, and handling procedures. Ways to eliminate hazard associated with flammable liquid vapors include process modifications that substantially reduce the areas of exposed liquids, or that substitute a nonflammable or less flammable material, or a liquid with a higher flash point. Vapors must be controlled by confinement, local exhaust removal, or area ventilation so they do not accumulate to form a flammable mixture. An explosion-proof exhaust system should be used in areas where high concentrations of flammable solvents occur.

Classes of flammable and combustible liquids

Liquids are divided into two general categories by the National Fire Protection Association (NFPA 30, *Flammable and Combustible Liquids Code;* the flash points and flammable ranges of specific flammable and combustible liquids are given in the NFPA 325M, *Fire Hazard Properties of Flammable Liquids, Gases and Volatile Solids*) and are defined as follows:

- **Flammable liquid.** A liquid having a flash point below 37.8 C (100 F) and having a vapor pressure not exceeding 40 pounds per square inch (absolute) at 37.8 C shall be known as a Class I liquid.

  Class I liquids shall be subdivided as follows:

  —Class IA shall include those having flash points below 22.8 C (73 F) and having a boiling point below 37.8 C.

  —Class IB shall include those having flash points below 22.8 C and having a boiling point at or above 37.8 C.

  —Class IC shall include those having flash points at or above 22.8 C and below 37.8 C.

- **Combustible liquid.** A liquid having a flash point at or above 37.8 C.

  Combustible liquids shall be subdivided as follows:

  —Class II liquids shall include those having flash points at or above 37.8 and below 60 C (140 F).

  —Class IIIA liquids shall include those having flash points at or above 60 C and below 93.4 C (200 F).

  —Class IIIB liquids shall include those having flash points at or above 93.4 C.

Flash points

The flash point of a liquid is the lowest temperature at which it gives off enough vapor to form an ignitible mixture with the air near the surface of the liquid or within a vessel that is capable of flame propagation away from the source of ignition. Some evaporation takes place below the flash point but not in sufficient quantities to cause an ignitible mixture.

The most commonly used device for determining flash points is the Tag (Tagliabue) Closed Tester for testing liquids with flash points below 93.4 C and viscosities below 45 SUS at 37.8 C (ASTM D-56, *Test for Flash Point by Tag Closed Tester,* ANSI, ASTM D56-82, American Society for Testing and Materials, Philadelphia).

The Pensky-Martens Closed Tester is considered most accurate for testing liquids having flash points about 93.4 C and viscosities equal to or greater than 45 SUS. (*Tests for Flash Point by Pensky-Martens Closed Tester,* ANSI-ASTM D93-80 gives the procedure.)

As an alternate to closed-cup test methods, the Setaflash Closed Tester may be used for aviation turbine fuels (*Test for Flash Point of Aviation Turbine Fuels by Setaflash Closed Tester,* ANSI/ASTM D3243-77) and for paints, enamels, lacquers, varnishes, and related products or components with flash points between 0 C (32 F) and 145 C (230 F), see *Test for Flash Point of Liquids by Setaflash Closed Tester,* ASTM D3278-78. (ASTM D-3278).

The Cleveland Open-Cup Tester is commonly used for petroleum products, except fuel oils having an open-cup flash below 79 C (175 F) (described in *Test for Flash and Fire Points by Cleveland Open Cut,* ANSI/ASTM D92-78).

Another type is the Tag Open-Cup Apparatus, *Test for Flash Point of Liquids by Tag Open-Cup Apparatus,* ASTM D 1310-80, which is sometimes used for low-flash liquids to make tests representative of conditions in open tanks and for labeling and transportation purposes.

Open-cup flash points represent conditions with the liquid in the open air and are generally higher (10 to 20 percent) than closed-cup flash point figures for the same substance. When open-cup flash point figures are given, they are usually identified by the initials "OC."

Fire point

The fire point of a liquid represents the lowest temperature at which vapors evolve fast enough to support continuous combustion. The fire point temperature is usually about five degrees F above the flash point temperature.

Flammable range

A prominent factor in rating the fire hazard of a flammable liquid is its flammable range, sometimes referred to as the "explosive range." For each flammable liquid there is a minimum concentration of its vapor, in air, below which propagation of flame (propagation of flame is the self-sustaining spread of flame through the body of the flammable vapor-air mixture after introduction of the source of ignition; a vapor-air mixture at or below its lower explosive limit may burn at the point of ignition without propagating) does not occur on contact with a source of ignition because the mixture is too *lean.* There is also a maximum concentration of vapor, in air, above which propagation of flame does not occur because the mixture is too *rich.* The boundary line mixtures of vapor with air that, if ignited, will just propagate flame are known as the lower and upper flammable (explosive) limits, and are usually expressed in terms of percentage by volume of vapor, in air.

The flammable (or explosive) range includes all the concentration of a vapor in air between the lower flammable limit (LFL) and the upper flammable limit (UFL). The lower flammable limit is of particular importance because, if this percentage is small, it will take only a small amount of the liquid vaporized in air to form an ignitible mixture.

It also should be noted that if the concentration of vapor in the vapor-air mixture is above the upper flammable limit, introduction of air (by ventilation or other means) will produce a mixture within the flammable range before a safe concentration of vapor below the lower flammable limit can be reached.

Control of conditions

Controlled conditions are essential for safe handling of flammable liquids, regardless of the quantities involved. Therefore, the problems associated with each flammable liquid to be used should be analyzed to determine the extent of flammability and health hazards so that appropriate control measures can be taken. To control these hazards, one must consider the characteristics of the specific liquid, the amounts of vapor involved, potential ignition sources, the kinds of operations, usage temperature, ventilation, and type of building construction.

Because the relative degree of exposure varies widely, the judgment of a competent individual should be obtained to determine the necessity of safeguarding electrical equipment, ventilation requirements, the need for eliminating sparks, open flames, and other sources of ignition, safe materials handling procedures, proper grounding procedures, and other factors promoting the maintenance of a safe environment. (See *National Electrical Code,* NFPA 70—Chapter 5, Article 500.)

PROTECTIVE DEVICES FOR CONTAINERS

Various devices have been developed to protect flammable liquid containers against fire and explosion, and thus make safe handling and storing of these substances possible. They include self-closing covers or valves, pressure- and vacuum-relief devices, and flame arresters.

Self-closing covers

Self-closing covers retard the evaporation of liquids in storage and, by excluding outside air, minimize the chance of fire. Generally, self-closing covers on flammable liquid containers are classified into three types according to their method of closing: gravity, spring action, and combination gravity and spring action actuated by a fusible-link mechanism.

The spring-action cover is most commonly used on portable safety containers (Figure 6-8). It serves three vital safety functions: (1) it provides over-pressure relief, (2) it seals the container against leakage, and (3) it minimizes evaporation and escape of vapors.

The spring-action cap or cover is held open by hand pressure during dispensing or filling operations. When hand pressure is removed, the cover closes automatically to form a tight seal around the opening. The seal should be so effective that no significant leakage will occur even if the container is completely inverted.

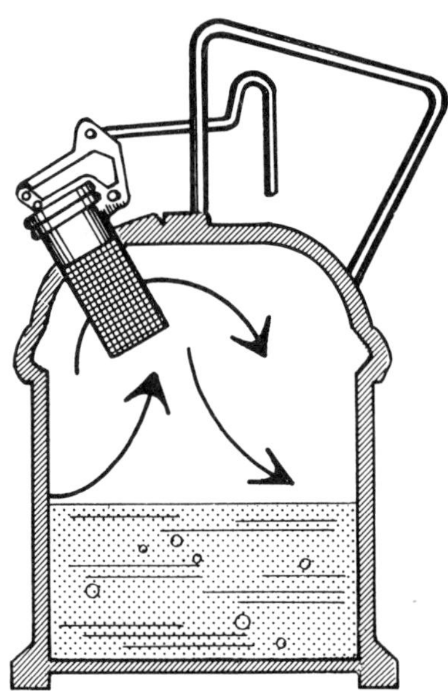

Figure 6-8. A properly designed spring-action cover will afford relief of excess vapor pressure within a portable safety container for flammable liquids. While closed, such a cover will allow no appreciable leakage of the liquid itself, even if the container is inverted.

Gravity-type covers are generally used on small parts washer tanks and dip tanks. Oily waste containers and some large tanks may also use gravity-type covers that are opened by depressing a treadle bar.

The combination-type cover has a fusible-link mechanism for safeguarding open vessels in which washing and cleaning operations are performed, and for protectively enclosing the contents of trash cans and drums. The fusible link, which holds the cover open, melts at a relatively low temperature, about 71 C (160 F). If a fire occurs in or near the container, the fusible link is designed to melt and allow the cover to fall by means of a combination of spring action and gravity. The effect is one of quickly smothering a fire inside the container or preventing a fire outside of it from making entry. Again, nonflammable or combustible liquids should be substituted for flammable liquids whenever possible in cleaning operations.

Pressure-relief devices

Because flammable liquids are volatile, it may be ncessary to consider venting for relief of vapor pressure. Pressure from expanding vapor can build up within a container if it is exposed to excessive temperature, such as from a space heating system, a fire, or another source of heat.

If a means for relieving this pressure is not provided, the pressure may become great enough to rupture the container, frequently allowing ignition of vapors and spreading fire over the surrounding area. A pressure-relief device also will prevent excessive buildup of pressure within a container that otherwise could cause vapors to spew into the face or on the clothing of a person opening the vessel.

The spring-action cover used to seal a portable safety container in itself provides pressure relief. As pressure builds up inside the container (Figure 6-8), it forces the cover to rise against the spring just enough to permit the vapors to escape. The cover then closes automatically to form a tight seal. It is a misconception that a portable safety container offers protection at its immediate exterior area. Because of the potential of vapor release, safety containers must not be stored in compact enclosures. Ventilation is necessary to dilute and disperse the vapors as they are released.

The safety bung vent for storage drums operates in a similar manner. Internal pressure forces a valve to rise against a coil spring so that the vapors can escape. After the internal pressure has been relieved, the valve closes automatically to a tight seal.

Vacuum-relief devices

Vents for vacuum relief are necessary for both functional and safety reasons. A liquid can be dispensed in a continuous flow from a so-called closed container only if the space vacated by the liquid is vented to counteract formation of a vacuum. Vaccum relief also prevents collapse of flammable liquid containers due to sudden cooling.

Vacuum relief is no problem in a portable safety container, if air can enter freely through a large pour opening and fill the space vacated by the liquid being dispensed. Vacuum-relief vents normally are built into containers that have small dispensing apertures.

Vaccum relief for storage drums often is combined wiln the pressure relief vent. Manual vacuum relief is accomplished by loosening or lifting the bung vent to open a port, which allows air to enter the drum. For automatic vacuum relief, a second valve integral with the pressure relief valve is forced open by atmospheric pressure as liquid is withdrawn from the drum. When the pressure in the drum is equal to atmospheric pressure, the automatic vacuum-relief valve closes to make a tight seal.

Flame arresters

A flame arrester, when installed on a flammable liquid container vent or opening, prevents temporary propagation of a flame into the container. Its primary function is to absorb and dissipate heat, and thus prevent vapors within the container from being set aflame by an external source of ignition. The flame arrester screen should fit properly and be held in place securely so that its protective features are not defeated.

A metal having a high heat capacity is used to fabricate a flame arrester with sufficient surface area to absorb heat. The design should be one that will permit free passage of flammable liquids, vapors, and air.

Flame arresters are typically constructed of either one or more layers of metal screen or perforate metal. In the form of a double-wall cylinder, they are provided for the dispensing and filling openings of portable safety containers. Flame arresters are also incorporated in the design of dispensing, filling, and venting accessories for drums, parts washer tanks, plunger cans, and other specialized containers.

The principal safety feature afforded by the flame arrester is protection against propagation of a flame into a container in the event a flammable air/vapor mixture develops in the container during filling or dispensing operations. The protection afforded by a flame arrester is also applicable to containers that may remain open for extended periods, for example, a parts washer, a drain can, or a plunger can.

Drums

Many industrial plants and commercial establishments purchase flammable liquids in 208-L (55 gal) drums and from them fill small containers 19 L (5 gal) or less in capacity, for routine use. Frequently, several drums may be on hand at one time containing the same or different liquids. Some of the drums, containing a reserve supply, may remain sealed until the open drum is empty.

Many users assume that it is safe to store sealed drums exactly as they are received. To be safe for storage or dispensing, a drum must be protected against fire and explosion. Full drums should be stored and protected from direct rays of the sun and other sources of heat.

A pressure- and vacuum-relief vent should be installed in a drum containing flammable liquid if there is any chance that it will be exposed to the direct sunlight or, in any other way, subjected to considerable variations in temperature. If a drum leaks, or is otherwise damaged, its contents should be immediately transferred to an approved container(s) that is clean or that previously held the same liquid.

Each drum should be checked for proper contents labeling. It is important that this label remain clearly legible to avoid confusing the contents with other flammable, combustible, or nonflammable liquids, and also to facilitate safe disposal.

Portable safety cans

Probably the most common piece of equipment for handling small quantities of a flammable liquid is the portable safety can, ranging in size from 0.5 L (1 pt) to 19 L. Safety cans are made

Figure 6-9. Laboratory use of flammable liquids often imposes a requirement for chemical purity in addition to the need for safety. Both demands are served by using recognized safety containers that are made of reaction-resistant metal or that are coated internally with a chemically impervious material. The portable containers, the tilt-type dispenser, and the flammable liquid waste can are designed with full safety features.

in numerous styles with dispensing valves, pouring spouts, or dispensing hoses. There are special safety containers made for viscous liquids, such as rubber cement and heavy oils.

Only containers that are listed or labeled by a nationally recognized testing laboratory should be considered acceptable for handling flammable liquids. (Underwriters Laboratories Inc., *Standards for Safety,* "Metal Safety Cans-UL 30" and "Non-Metallic Safety Cans for Petroleum Products-UL 1313"; Factory Mutual Approval Guide "Safety Containers and Filling, Supply, and Disposal Containers—Class Number 6051 and 6052"; and American National Standard *Plastic Containers [Jerry Cans] for Petroleum Products,* D3435-80.) Containers must be approved or listed specifically for a particular purpose, whether it is storing, carrying, dispensing, or end-use.

Some flammable liquids are often purchased in small light-weight containers with screw caps, such as one-gallon cans. The most typical are rectangular containers, commonly referred to as "Type F Cans," that range in size from 0.5 L to 3.8 L (1 gal). The design of these containers is not covered by any standard. Although designed as a shipping container, once opened, the container may be stored for a lengthy time period with occasional dispensing of the contents. When empty, such containers should not be refilled. These containers do not provide protection for the storing, carrying, or dispensing of flammable liquids that is comparable to safety containers. It is, therefore, recommended that flammable liquids be transferred from shipping containers to safety containers when conditions of storage or handling justify the need for a greater degree of safety.

Containers should be handled carefully to prevent accidental rupture and discharge of contents. If there is any sign of vapor or liquid leakage from a shipping container, if the container cap or cover is lost, or if there is evidence of significant mechanical damage to it, the contents should generally be transferred immediately. The transfer should be made to a clean, undamaged container of the same type or one that is better, which can be adequately sealed. Flammable liquids should be transferred in an area that is adequately ventilated, has adequate electrical grounding facilities, and is free from all sources of ignition.

Frequently, in laboratories, offices, and industrial and commercial establishments, common shipping containers, such as glass or plastic bottles, or light-weight metal cans are used for storing, dispensing, and carrying small quantities of flammable liquids. Such containers may provide little or no protection against fire and explosion, and should be stored in cabinets, away from heat or potential sources of ignition, and tightly sealed. Hand transporting of chemicals in glass bottles should be done only in plastic or rubber containers equipped with handles. Attempts to dispense materials from them often result in spillage; therefore, a funnel or pouring spout should be used and dispensing should be conducted in ventilated locations free from sources of ignition.

In special cases where chemical purity must be maintained, storage in glass or plastic containers is permissible but may require keeping such containers in approved cabinets or in areas protected by automatic extinguishing systems. The problem of preserving chemical purity may be solved in many cases by the use of safety cans that are made of stainless steel or other relatively impervious metal (Figure 6-9) or that are lined with a reaction-resistant material.

Unlisted or unlabeled containers

Containers that have not been listed or labeled by a nationally recognized testing laboratory are available for purchase, storing, and handling flammable liquids. Many of them are essentially a storage container that has been modified to incorporate a vent and pouring spout. If the screw cap is lost or misplaced, the vent left open, or the pouring spout left on, flammable vapors can escape and/or a liquid spill occur if the container is knocked over. Containers of this type are generally used for storing and handling fuel for gasoline-powered equipment, heaters, stoves, and lanterns. Frequently, they are subjected to storage and handling conditions that would be less hazardous if a portable safety can were used. Typical deficiencies found on unlisted or unlabeled containers are:

1. no self-closing cover
2. no flame arrester
3. light-weight metal (punctures easily)
4. weak seams (ruptures easily)
5. light-weight handle (breaks easily)
6. no vapor space (permits over-filling)
7. poor stability (tips easily).

Identifying containers

Accidental mixing of flammable liquids should be avoided and unauthorized mixing should be prohibited. The hazards of both flammability and toxicity could be increased by such mixing.

To help prevent mixing one liquid with another and to reduce the chance that one liquid will be mistaken for another, containers should be clearly marked or identified as to contents. They should be kept clean at all times so markings can be easily seen.

For all flammable liquids, the label should read:

DANGER—FLAMMABLE—KEEP AWAY FROM HEAT, SPARKS, AND OPEN FLAMES. KEEP CLOSED WHEN NOT IN USE.

The federal government, the various states, and some municipalities have laws relating to the labeling of potentially harmful substances. (See labeling requirements discussion under the OSHA Hazard Communication Standard in Chapter 30.) Most of these laws provide for warning labels identifying the chemicals falling into certain hazard classifications and generally define the types of labels to be provided. The Department of Transportation (DOT) also has labeling requirements, but they frequently apply only to interstate and not to intrastate shipments. Thus, hazardous solvents sometimes can be shipped within a state without DOT labeling. The American National Standards Institute has a labeling guide, ANSI Z129.1, and some companies provide their own labeling procedures. Whether or not labeling laws apply, each container of a flammable or combustible liquid always should be labeled in a manner that will identify its contents and indicate the hazard potential of the solvent and the safe practices required in its use. (From *Handbook of Industrial Organic Solvents,* published by the Alliance of American Insurers.)

Dispensing

Dispensing should be done by an individual from only one drum at a time and all dispensing of one material should be completed before dispensing of another material is begun.

There are two recognized devices commonly used for dispensing small quantities of flammable liquids from drums—transfer pumps and drum faucets. Only listed or labeled pumps or faucets should be used. Drums containing flammable liquids should never be pressurized, even slightly, to provide automatic dispensing.

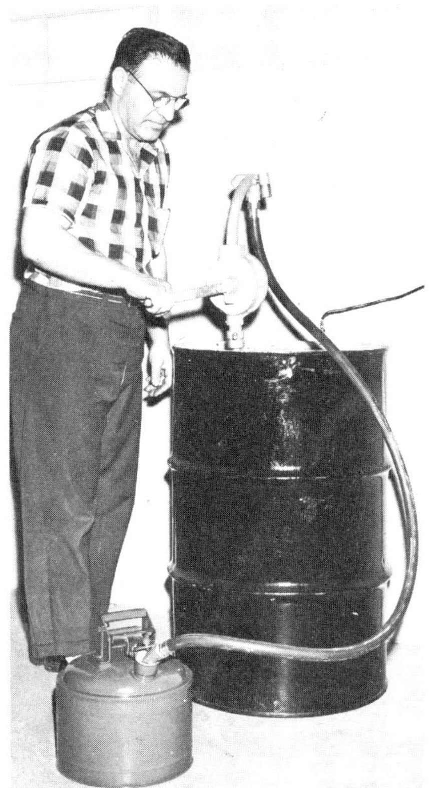

Figure 6-10. Rotary-type transfer pump threaded into the end bung of a 208-L (55 gal) drum. Flammable liquid is dispensed from drum to portable safety container via conductive hose that is designed to bond drum to container during transfer.

Of the two devices, the safer one is the hand-operated rotary type transfer pump (Figure 6-10), which is generally installed in the end bung opening of a drum mounted in the vertical position. When the pump handle is cranked, the liquid is pulled up through a suction tube and discharged through a hose or nozzle. It is essential that an electrical bond be maintained between the dispensing and receiving containers during the entire transfer operation.

Drum transfer pumps (1) permit overflow and spillage, (2) can be reversed to siphon off excess in case of overfilling, and (3) can be equipped with a drip return so that excess can drain back into the drum.

The hand-operated self-closing drum faucet is installed in the horizontal position (Figure 6-11). Liquid can be dispensed only while hand pressure is applied on the faucet handle or lever. When hand pressure is released, the faucet valve automatically shuts off the gravity flow of the liquid. No practice should be permitted and no device should be used that would prevent the faucet from closing automatically when hand pressure is removed. Flexible metal or conductive rubber hoses may be installed on the faucets to facilitate dispensing into receiving containers that have small openings.

The faucet method of dispensing also requires bonding of the vessels and the installation of a drum vent.

Figure 6-11. A hand-operated self-closing metal faucet is threaded into the end bung opening of this 208-L (55 gal) drum mounted horizontally on a wheeled rack. The worker is preparing to dispense flammable liquid from the drum to a portable safety container by means of a flexible metal hose. By keeping the hose nozzle in contact with exposed metal of the container spout, aided by spring pressure of the spout cover, the worker follows procedure to establish an electrical bond between the drum and safety can. If for any reason a bond cannot be maintained in this way during the entire transfer, the worker should use a flexible conductor with suitable clamps to bond the vessels.

As general practice, safety cans should be employed to receive flammable liquids from drums. Before the receiving container is used, it should be checked for correct contents labeling, installation, and proper operation of the required protective devices, and it should be in overall good condition.

In some instances, stations for dispensing small amounts of flammable liquids are set up at locations convenient to end-use areas in plants and other establishments (Figure 6-12). A limited number of safety supply cans are provided at such stations only during work hours so that the stations may serve to minimize handling and carrying of flammable liquids through the plant. Such areas should be clearly designated as "flammable liquid locations" to promote maximum control over sources of ignition.

Bonding and grounding

The action of transferring a liquid from one container to another may produce voltage potentials that can result in static electrical sparks capable of igniting flammable vapors. Because static charges can generate from this source, it is important to bond flammable liquid dispensing and receiving containers together before pouring. It is also important for large containers, such as drums, to be connected to an adequate electrical ground when they are used as a dispensing or receiving vessel.

To be effective, all grounding and bonding connections must be metal to metal. Therefore, all dirt, paint, rust, or corrosion should be removed from points of contact before such connections are made.

A bond or ground connector should be composed of suitable conductive materials having adequate mechanical strength, corrosion resistance, and flexibility for the service intended. Because the bond or ground does not need to have low resistance, nearly any conductor size will be satisfactory from an electrical standpoint. However, to assure adequate mechanical strength, size 3.3 or 2.6 mm (No. 8 or No. 10 AWG) wire is the minimum size that should be used. Solid conductors are satisfactory for fixed connections. Flexible conductors (stranded or braided ribbon wire) should be used for bonds that are to be connected and disconnected frequently.

Conductors may be uninsulated conductors that allow easy detection of defects by visual inspections. If the conductor is insulated for mechanical protection, it should be checked for continuity at regular intervals, depending on experience. Permanent connections may be made with screw-type ground clamps, brazing, welding, or other suitable means. Temporary connections may be made with spring (battery-type) clamps, magnetic connectors, or other special clamps that provide metal-to-metal contact.

To ground a drum from which a flammable liquid is to be dispensed, one end of a conductor should be attached to the rim of the drum with a screw clamp and the other end of the conductor should be connected to a known ground with a sturdy, bolt-on clamp or other equivalent means.

In many cases, the common practice of placing the nozzle of the dispensing container in contact with the opening of the

Figure 6-12. A dispensing station for withdrawal of small amounts of flammable liquids at a location convenient to end-use area. The worker is preparing to insert tip of dispensing faucet in spout of small container, so that spout's self-closing cover firmly grips faucet and establishes electrical bond between the vessels during transfer of the liquid.

receiving container is used to establish an electrical bond between the vessels. This method is generally not satisfactory because a good electrical contact is difficult to establish and maintain during the entire dispensing operation.

A drip return pan attached to the dispensing pump may provide effective bonding when the receiving container is placed on it preliminary to and during the transfer. Likewise, a recognized dispensing hose having a metal connector and nozzle joined by a sealed-in conductor may establish a continuous bond if the nozzle is kept in contact with exposed metal of the receiving container during the entire transfer.

However, prevention of static spark while transferring a flammable liquid from one container to another is best accomplished by prior bonding of the containers with a wire having suitable clamps (Figure 6-13).

Ventilation

Where flammable liquids are used, adequate local exhaust or general ventilation should be provided at the place of exposure (see *Industrial Ventilation,* Committee on Industrial Ventilation, American Conference of Governmental Industrial Hygienists). Requirements for removal or dilution of vapors should be set up for both the health hazard and the flammability hazard.

Heating a flammable or combustible liquid above its flash point involves considerable risk. If possible, operations requiring the use of a flammable liquid and the application of heat should be isolated. An exhaust hood that conducts the vapor safely outside the building is recommended. The exhaust fan should operate continuously whenever the liquid temperature is above its flash point.

Storage

Rooms where drums containing flammable liquids are stored, and where the liquids are dispensed, should be considered potentially hazardous areas, but consistent application of basic fire and accident prevention rules should eliminate or mitigate most hazards. Special attention should be given to the location of the room, its arrangement, its ventilation, the equipment used, and storage procedures.

Because the most effective way to control the hazard is to isolate it, the storage and dispensing room should be located in a special building separated from other buildings. If this arrangement is not feasible and the room must be located inside the establishment, it should be constructed of walls, floor, and ceiling having at least two-hour fire-resistance and Class B fire doors as minimum protection. The room should be used for the storage of flammable and combustible liquids only; that is, storage of other materials in it should be banned.

A storage and dispensing room located inside a building should be protected by strategically located portable fire extinguishers (preferably outside the room but near an entrance to it) designed for use on Class B fires. Additional protection may be achieved with an approved installation of an automatic extinguishing system that complies with the appropriate NFPA standard. The room should have both high- and low-level ventilation to the outside, special trapped floor drainage facilities, and explosion relief in the form of pressure-opening windows or a weak wall, providing 0.09 m^2 (1 sq ft) of venting for every 1.42 m^3 (50 cu ft) of room volume. For Class I liquids, electrical equipment should conform to the requirements of NFPA 70, *National Electrical Code,* and NFPA 30, *Flammable and Combustible Liquids Code,* National Fire Protection Association for Class I hazardous locations.

Design features, storage limitations, and other precautionary measures should meet the specifications given in the NFPA *Flammable and Combustible Liquids Code.* In addition, local fire ordinances must be consulted.

After provision for venting has been made, drums should be stored on metal racks with end bung openings toward an aisle and side bung openings on top. The drums, as well as the racks, should be grounded. Any spillage should be cleaned up immediately and leaking valves replaced.

Storage cabinets (see Figure 6-14) that are approved for holding containers of flammable liquids should be installed near the areas where the liquids are used. Such cabinets not only facilitate safe storage but also conserve production time because workers need not make frequent trips to remote storage areas whenever the limited quantities permitted at the point of use are exhausted. Individual containers should not exceed 19 L in capacity, and a total of not over 454 L (120 gal) should be stored in any one cabinet. Storage cabinets should conform to the specifications as outlined in the NFPA *Flammable and Combustible Liquids Code.*

Flammable liquids should not be stored in domestic-type refrigerators unless they are properly modified to remove ignition sources. Special refrigerators for this service are listed by Underwriters Laboratories Inc.

In a building used for public assembly, such as a school, church, or theater, not more than 38 L (10 gal) of Class I and Class II liquids should be stored in a single fire area outside of a storage area unless the liquids are in safety cans. A smaller

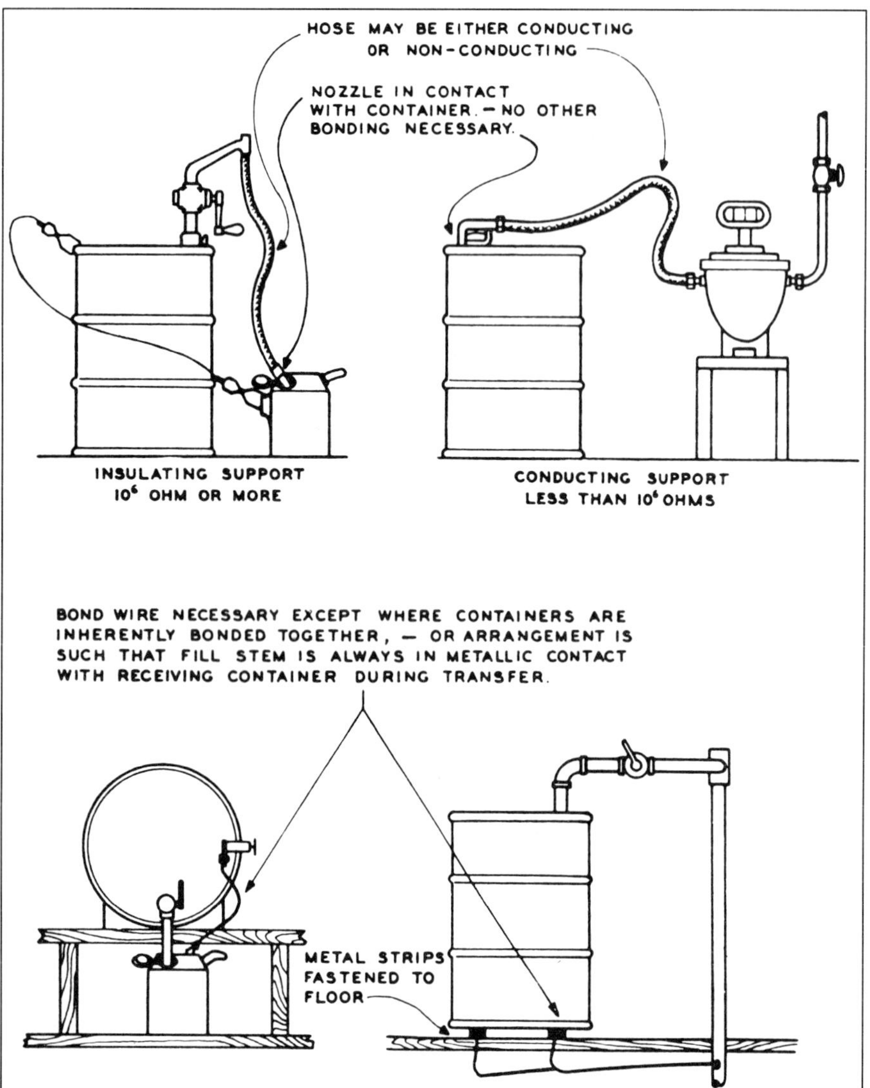

Figure 6-13. Bonding during container filling.

quantity limitation may be imposed by the authority having jurisdiction or greater protection may be required where unusual hazards to life or property are involved. The maximum permitted in a single fire area in safety cans outside of a separate inside storage area or in a storage cabinet is 95 L (25 gal) of Class I and Class II liquids combined. Generally, a safety can will provide the best protection, but another suitable container may be used if necessary precautions are observed. The quantity should be restricted to a minimum, and the place of storage should be kept at least 3 m (10 ft) away from a stairway, elevator, or exit unless guarded by a fire-resistive partition. The containers should not be exposed to any source of heat.

Waste disposal

In mercantile occupancies, retail stores, and other related areas accessible to the public, storage of Class I and Class II shall be limited to quantities needed for display and merchandising purposes and shall not exceed 81.4 liters per m^2 (2 gal/sq ft) of gross floor area. For more details, see NFPA 30.

Cloth, paper, and other solid wastes that have been soaked with a flammable or combustible liquid should be placed in listed or labeled disposal containers. Such containers are made of metal and are equipped with self-closing covers. Disposal containers should be clearly labeled as to the type of waste they are intended to receive.

The containers should not be overfilled and should be emptied at the end of each working day and their contents removed to a safe location and burned (if permitted). Incinerators can be used to handle solid waste, or waste can be burned at a safe location away from the plant or establishment under close surveillance and with suitable fire protection equipment for emergency use close at hand. In all cases, applicable governmental regulations must be closely adhered to.

Most laws and ordinances strictly prohibit pouring flammable and combustible liquids into sinks or floor drains that connect with sanitary or storm sewer facilities. Standard 14-L or 208-L (30-55 gal) drums or smaller containers should be used for waste flammable liquid disposal, and should be emptied periodically. They should be safeguarded against fire according to all applicable provisions specified for dispensing containers. The liquids should be disposed of by burning in a solvent burner or by being delivered to a waste-liquids collection agency. In many cities,

Figure 6-14. Metal storage cabinet for flammable liquid containers of maximum 19-L (5 gal) capacity each. (Courtesy The Protectoseal Co.)

there are firms that collect waste liquids. In some instances, the original supplier will pick up used liquids for reclaiming and resale.

Transporting flammable liquids

Flammable liquids cannot be transported safely in safety cans when the containers are placed in the closed compartment of a vehicle. Because safety cans incorporate pressure relief features in their design, changes in temperature and atmospheric pressure might cause vapors to be released from the container.

In a confined space in a vehicle, such as a closed compartment of a truck, vapors venting from a safety can might ultimately reach flammable proportions. A spark, for instance, from the vehicle's electrical system or other source of ignition, could ignite a flammable mixture under such conditions. Therefore, the practice of carrying a spare container of gasoline in the interior of a vehicle should be prohibited.

If gasoline or other flammable liquids must be transported on a regular basis, only vehicles suitably equipped or modified for such transport should be used. Provisions for safe temporary transport of small quantities of flammable liquids (usually one to five gallons) should include: use of rugged pressure-resistant and non-venting containers, storage during transport in a well-ventilated location, and elimination of potential ignition sources.

Fire extinguishment

Fire extinguishing methods for flammable or combustible liquid fires are:

1. exclusion of air by foam or by other smothering techniques (Special foam agents are available for fires involving flammable liquids that are miscible in water, such as alcohols, acetone, ether, and amyl acetates.)
2. cooling below the fire point by water spray or water fog
3. reducing the oxygen content of air with carbon dioxide (CO_2), or other inert gas
4. interrupting the chemical chain reaction of the flame with dry chemical agents or halogenated agents
5. shutting off the flow of flammable liquids feeding the fire.

Wherever flammable liquids are handled or used, enough portable fire extinguishers (dry chemical, carbon dioxide, AFFF, or Halon types) for use on Class B fires should be readily available to protect the specific area (see NFPA 10, *Portable Fire Extinguishers,* National Fire Protection Association). All empoyees should know where they are located and how to use them.

Personnel training

Safe practices on the part of employees who handle flammable liquids are essential in the prevention of fire and explosion. Before being permitted to undertake jobs that require the use of such liquids, workers should be fully instructed in the characteristics, hazards, and methods of control. Supervisors should make frequent checks to assure that the required safe practices are being followed consistently.

Among basic safe practices employees should be trained to observe are the following:

1. Use only listed or labeled containers in good condition and keep them closed when they are not in use.
2. Never use a container for any liquid other than that for which it is intended and so marked.
3. Keep only that quantity of liquid needed during the shift at the job site; at the end of the shift, return any unused liquid to designated storage area.
4. Clean up liquid spills immediately; dispose cleanup rags into listed or labeled containers; never use sawdust to absorb a spill.
5. Never smoke, use open flames, or create sparks where there is a possibility of igniting a flammable or combustible liquid.
6. Check bonding and grounding connections for electrical continuity.

Only specially trained personnel should be assigned to work in storage and dispensing rooms. Periodic checks should be made for adherence to the prescribed safe practices, and, when necessary, refresher training should be given.

Employees should be kept constantly aware that flammable liquids are hazardous and must be handled with particular care. Posters and signs are among the means that can be used effectively. NO SMOKING signs should be posted conspicuously in buildings and areas where smoking is prohibited because of the presence of flammable liquids.

Particulates

by Edwin L. Alpaugh
Revised by Theodore J. Hogan, PhD, CIH

THE AIR WE BREATHE contains particulates in the form of dust, and a portion of that dust is retained in the lungs (although not everyone inhales enough dust to cause illness). Pneumoconiosis—a tongue-twisting term of Greek derivation—means "dusty lung." Fume, another form of a particulate, can also cause pneumoconiosis. Fume is a form of particulate matter, differing from dust only in the way it is generated and in its particle size. Dust normally involves a wide range of particle sizes that are the result of some mechanical action, such as crushing or grinding. A fume consists of extremely small particles, less than a micrometer, or micron, in diameter, and is generated by processes such as combustion, condensation, and sublimation.

Some fibrosis is present in almost every lung. Because of the greater total exposure, the lungs of older people might reasonably be expected to exhibit a greater degree of fibrosis than would the lungs of young people. The term pneumoconiosis is restricted to cases in which fibrosis results from an exposure to inorganic dusts.

One of the first symptoms of pneumoconiosis is shortness of breath. Shortness of breath can have many causes. Consequently, a diagnosis of pneumoconiosis should not be made solely on the basis of shortness of breath or any other single symptom.

Except for skin diseases, most occupational diseases are caused by inhalation of certain materials used in a work area. This is understandable because lung tissue is by far the most efficient medium the body possesses for capturing and absorbing airborne contaminants. One reason is that the surface area of lung tissue averages 55–75 m^2 (590–920 sq ft). The surface area of the skin, by comparison, averages about 2 m^2 (22 sq ft). Consequently, dusts that are soluble in body fluids and that reach the lungs can be absorbed and eventually pass directly into the blood stream. Other dusts that are not soluble may stay in the lungs and cause local or irritant damaging action.

Toxic and irritant dusts can also be ingested in amounts that may cause trouble. If the toxic dust is swallowed with food or saliva and is not soluble in body fluids, it is eliminated directly through the intestinal tract. Toxic materials that are readily soluble in body fluids, such as lead oxide dust or lead fume, can, if swallowed, be absorbed by the digestive system and picked up by the blood.

However, both the ingestion of and skin contact with particulates are of relatively minor importance in industrial dust poisoning. Inhalation is the primary cause of severe poisoning cases.

CRITICAL EXPOSURE FACTORS

Health problems associated with the various kinds of dust exposures are influenced by four critical factors. These are as follows:

- the type of dust involved;
- the length of exposure time (possibly in years);
- the concentration of airborne dusts in the breathing zone of those exposed; and
- the size of the dust particles present in the breathing zone.

Each of the factors mentioned can be critical. However, the factors are so interrelated that each must be considered in evaluating dust exposures. For example, an airborne dust of a potentially toxic material will not cause pulmonary illness if its particle size is too large to gain access to the lungs, if it is present in low concentrations, or if exposure time is very short. Thus the importance of any one factor, as it affects health, must be evaluated in terms of the other three.

When dust exposure measurements are determined, these four factors must also be taken into account. A "total dust" measurement provides very little information about the health consequences of the exposure. However, exposure measurement techniques can be specific enough to identify the type and size of the particulate matter as well as the overall dust concentration.

Because an understanding of the four critical factors is important to anyone concerned with the recognition, evaluation, and control of industrial dust problems, each will be discussed in some detail.

Type of dust involved

Industrial dusts can be classified in two very broad categories—organic and inorganic. Organic dusts originate from plant and animal material or from synthetic material. In general, naturally occurring organic dusts tend to produce allergic responses after acute and chronic exposures. Synthetic organic dusts can produce irritation and allergic responses as well as a whole range of local effects (such as dermatitis) or systemic toxic effects (such as liver damage), depending on the particular chemical involved. Inorganic dusts can be classified as metallic or nonmetallic. Nonmetallic dusts that contain silica are further classified as crystalline or amorphous. Figure 7-1 shows the types of dust associated with each category.

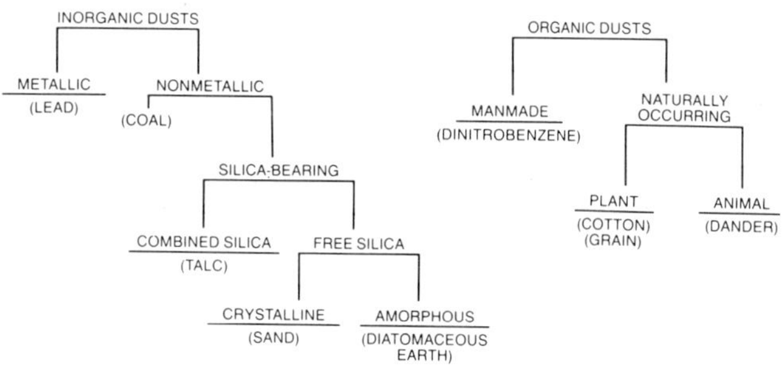

Figure 7-1. Classification for sampling and evaluating respirable dusts. (Adapted from the Bureau of Mines Circular 8503, Feb. 1971)

Inorganic metallic dusts can produce local dermatitis and sensitization (such as nickel itch) and systemic toxicity, particularly to the kidneys, blood, and central nervous system (Table 7-A). Inorganic silica-bearing dusts are normally not an acute hazard either to the skin or systemically. However, dusts containing crystalline, or "free," silica can cause pneumoconiosis as a result of chronic exposure.

Length of exposure

Pneumoconioses, such as silicosis, asbestosis, and coal workers pneumoconiosis, normally become disabling only after several years of dust exposure. Toxic metal dusts, such as lead and manganese, can cause problems after a shorter exposure time (from several days to several weeks), depending on how much of the toxic metal dust is absorbed in a specified period of time. A few hours of overexposure to some metal fumes can cause metal fume fever, a transient illness that is similar to the flu. Dusts that can cause allergic reactions or that can be very irritating may require only a brief exposure time at relatively low concentrations to cause serious problems. Air-sampling is needed to evaluate actual exposure times; sampling can be a simple process or it can be difficult, depending on the mobility of the person exposed and the fluctuations in exposure patterns of his or her job.

Dust concentration

Another critical factor in evaluating an exposure to dust is the actual concentration of dust present in the breathing zone. For many years industrial hygienists have been guided by the dust concentration figures established as Threshold Limit Values (TLVs) by the American Conference of Governmental Industrial Hygienists (ACGIH) and by the Permissible Exposure Limits (PELs) for dust in the workplace established by the Occupational Safety and Health Act (OSHA). (See Appendix B-1 for more information on TLVs.)

Airborne particulate matter, except for asbestos fibers, is usually measured by collecting particles on a preweighed filter. A known volume of air is then pulled through the filter, the filter is weighed, and the increase in weight, if any, is reported in terms of so many milligrams or micrograms of dust per cubic meter of air sampled. Additional analytical techniques can be used to identify the types of materials present in the sampled air. For example, the types and concentrations of metals present in a fume sample can be determined in the laboratory using an atomic-absorption spectrophotometer. Asbestos dust is also collected on a filter, but only the number of fibers greater than 5 μm in length are counted by microscopic examination.

For many years, the dust particles collected (usually in a liquid medium in an impinger) were actually counted, and the concentrations were reported as so many million particles per cubic foot (mppcf) of air sampled. Despite the fact that much empirical data are available relating the results of dust counts to potential health problems, the gravimetric procedure is almost universally used. This is because true breathing-zone samples covering an 8-hour exposure period are much easier to obtain with a filter than with a liquid-filled impinger. Also weighing a filter is quicker and less tedious than counting dust particles with a microscope.

Particle size

The last (but equally important) factor required in the evaluation of dust exposure is the actual size of airborne particles (Figure 7-2). By using a size-selective device (such as a cyclone) ahead of a filter at a specific airflow sampling rate, it is possible to collect respirable-sized particles (less than 10 μm) on the filter. The same end is achieved in dust counting procedures by simply not counting particles larger than 10 μm in diameter.

When a solid is broken into finely divided particles, its surface area is increased many times. For example, when one solid cubic centimeter (0.061 in.) of quartz is crushed into one-micrometer cubes, 10^{12}, or one trillion, particles with a total surface area of 6 m^2 (9,300 sq in.) are yielded, as compared with 6 cm^2 (0.930 sq in.) in area for the original cube.

When a solid is broken into finely divided particles, the volume occupied by the mass is also increased because of the voids between the particles. A dust concentration of 50 mppcf of air,

Table 7-A. Selected Toxic Dusts and Fumes

Substance	*Description and Effects*	*Threshold Limit in Milligrams per Cubic Meter of Air**
Antimony	Often associated with lead and arsenic. Hazardous from inhalation and ingestion. Soluble salts may cause dermatitis. Antimony trioxide may cause lung cancer.	0.5
Arsenic	Silvery brittle crystalline metal. Hazardous from inhalation and ingestion. Usually encountered as arsenic trioxide. Arsenic trioxide production is associated with increased cancer rates in humans.	0.2
Barium (soluble compounds)	Soluble barium chloride and sulfide are toxic when taken orally.	0.5
Beryllium	Light weight gray metal. The metal, low-fired oxides, soluble salts, and some alloys are toxic by inhalation. Suspected carcinogen.	0.002
Cadmium oxide fume	Used in some silver solders and as a metal coating. Very high acute fume exposures can be fatal.	0.05 Ceiling
Chromic acid and Chromates	Red, brown, or black crystals. Caustic action on mucous membranes or skin. Some water-insoluble chrome VI compounds are carcinogens.	0.05
Cyanide (as CN)	Nonvolatile cyanides are ingestion hazards. Cyanides inhibit tissue oxidation when inhaled and cause death.	5.0 (skin†)
Dinitrobenzene	Yellowish crystal. Hazardous as a result of skin absorption, inhalation, and ingestion.	1.0 (skin†)
Fluorides	Inorganic fluorides are highly irritant and toxic.	2.5
Hydroquinone	Colorless hexagonal crystals. Contact with the skin may cause sensitization and irritation. Excessive exposure to dust may cause corneal injury.	2.0
Iron oxide fume	Major sources are cutting and welding.	5.0
Lead	Lead fumes and lead compounds cause poisoning after prolonged exposure. Most important means of entry into body is by inhalation. Skin absorption is of significance only from such organic compounds as lead tetraethyl.	0.15
Lead arsenate	White crystals. Highly toxic.	0.15
Magnesium oxide fume	White powder. Inhalation of freshly generated fume may cause metal fume fever.	10.0
Manganese	Silver gray metal. Hazardous if fumes or dust are inhaled.	5.0 Ceiling
Pentachlorophenol	Dark-colored flakes. Harmful dust. Emits toxic fumes when heated.	0.5 (skin**)
Phosphorus (yellow)	Poisonous mainly on inhalation. Severe burn hazard from skin contact.	0.1
Picric acid	Yellow crystals or liquid. Explosive—particularly metallic salts. Toxic fumes on decomposition.	0.1 (skin†)
Selenium compounds	Toxicity varies somewhat according to the solubility of the specific compound. Often causes contact dermatitis.	0.2
Sodium hydroxide	White, deliquescent pieces or lumps. Has severe action on all body tissue.	2.0 Ceiling
Tellurium	Similar to selenium chemically and in physiological effects.	0.1
Titanium dioxide	White to black powder. Considered in the nuisance category.	10.0
Trinitrotoluene	Colorless to yellow monoclinic crystals. Emits toxic fumes of oxides of nitrogen when heated to decomposition. Highly poisonous explosive.	0.5 Ceiling
Uranium	Highly toxic—particulary to kidneys—and a radiation hazard that requires special consideration. Soluble and insoluble compounds as U (natural).	0.2
Vanadium pentoxide (respirable dust and fume)	Yellow to red crystals. Acts chiefly as an irritant to the conjunctiva and respiratory tract.	0.05
Zinc oxide fume	Amorphous white or yellow powder. The powder is essentially nontoxic, but freshly generated fume may cause metal fume fever.	5.0
Zirconium compounds	Most compounds are insoluble and have low toxicity.	5.0

*These Threshold Limit Values were adopted by the American Conference of Governmental Industrial Hygienists in 1986.

†The word "skin" in this table indicates that the substance can penetrate the skin to contribute to the exposure.

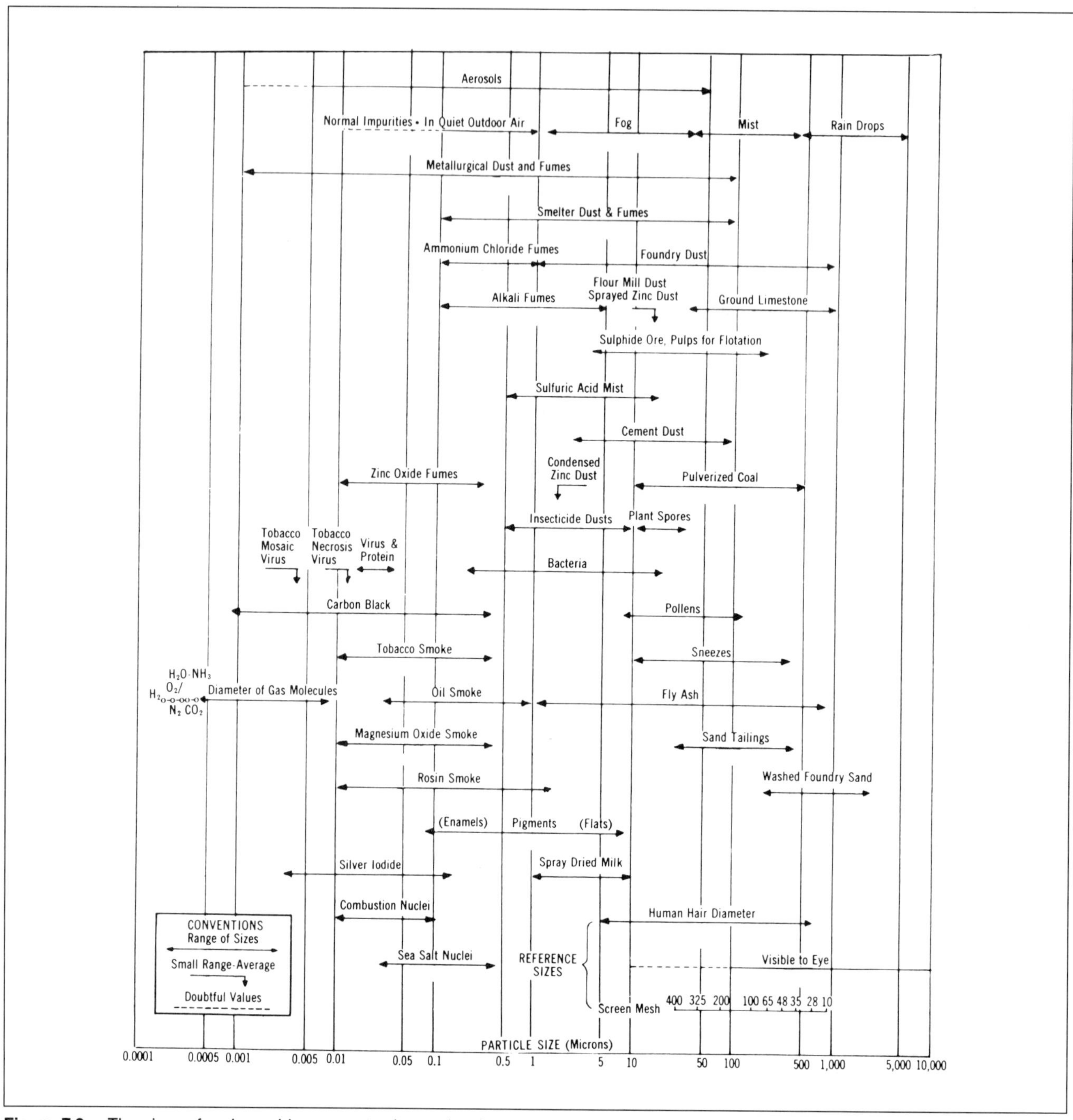

Figure 7-2. The sizes of various airborne contaminants (1 micron=1 micrometer). (Courtesy MSA)

that results when one cubic centimeter of material is reduced to particles one cubic micrometer in size, will occupy an air space of 560 m³ (20,000 cu ft).

A person with normal eyesight can detect individual dust particles as small as 50 μm in diameter. Smaller airborne particles can be detected individually by the naked eye only when strong light is reflected from them. Dust of respirable size (below 10 μm) cannot be seen as individual particles without the aid of a microscope. However, high concentrations of suspended small particles may be perceived as a haze or have the appearance of smoke.

Most industrial dusts consist of particles that vary widely in size; the small particles greatly outnumber the larger ones. Consequently, with few exceptions, when dust is noticeable in the air around an operation, more invisible dust particles than visible ones are probably present.

Dust in the air may or may not have the same composition as its parent material. The factors that determine composition are the particle size and density of each of the several components in the original mixture and the hardness of the materials. Hard materials resist the pulverizing action of a mechanical device.

Although dust particles are, of course, subject to the force of gravity, their settling rate through still air varies according to their size, density, and shape. Microscopically small particles settle more slowly than do larger particles because of the former's relatively minor density and because of the influence of Brownian movement. Mineral particles larger than 10 μm settle relatively quickly (Table 7-B). As Table 7-B demonstrates, respirable particles released into the air can remain airborne for many hours. If the rate of dust generation is constant, the dust level (and, therefore, exposure) will increase throughout the workday unless ventilation or other methods of control are effectively used.

Table 7-B. Estimated Settling Rates for Silica Dust in Still Air

Size in Micrometers	*Time to Fall 1 Foot (0.30 m) (minutes)*
0.25	590.0
0.50	187.0
1.00	54.0
2.00	14.5
5.00	2.5

With the exception of such fibrous materials as asbestos, dust particles must usually be smaller than 5 μm in order to enter the alveoli or inner recesses of the lungs. Although a few particles up to 10 μm in size may occasionally enter the lungs, nearly all the larger particles become trapped in the nasal passages, throat, larynx, trachea, and bronchi, from which they are expectorated or swallowed into the digestive tract.

Ragweed pollen, which varies from 18–25 μm in diameter, can cause hay fever as a result of its action in the upper respiratory system. This type of allergenic dust, as well as bacterial and irritant dusts, can cause difficulty even when the airborne particles are large.

When dust-laden air is inhaled, some of the larger particles are trapped by the hairs in the nose. Other dust particles are removed from the air as it passes over the most mucous membranes of the nose, throat, and other portions of the upper respiratory system.

The bronchi and other respiratory passages are covered with a large number of tiny, hairlike cilia or microscopic whiplashes, which aid in the removal of dust trapped on these moist surfaces. The cilia, all bending in one direction, make a fast stroke toward the mouth and a slower stroke away from the mouth. This action tends to push mucus and deposited dust upward to the mouth so that the particles can be expectorated or swallowed.

BIOLOGICAL REACTION

Because there are many different types of dust, fumes, and mists, the biological reaction caused by exposure to any one of them will depend on the type. A reaction may include any of the following:

Lung diseases. These are caused by the body's reaction to an accumulation of dust in the lungs. These diseases include fibrosis (scar tissue formation), bronchitis (the overproduction of mucus), asthma (the constriction of the bronchial tubes), and cancer. Restriction of lung function places an additional burden on the right side of the heart, which tries to pump more blood to the lungs to maintain an adequate oxygen supply. This additional strain can cause permanent heart damage.

Systemic reactions. These are caused when the blood absorbs inorganic toxic dusts composed of such elements as lead, manganese, cadmium, and mercury, and by absorbing certain organic compounds.

Metal fume fever. This results from the inhalation of finely divided and freshly generated fumes, of zinc, magnesium, copper, or their oxides. The inhalation of aluminum, antimony, cadmium, copper, iron, manganese, nickel, selenium, silver, and tin have also been reported to cause metal fume fever.

Allergic and sensitization reactions. These may be caused by inhalation of or skin contact with such materials as organic dusts from flour and grains and some woods and dusts of a few organic and inorganic chemicals.

Bacterial and fungus infections. These occur from inhalation of dusts containing active organisms, such as wool or fur dust containing anthrax spores or wood bark, or grain dust containing parasitic fungi.

Irritation of the nose and throat. This is caused by acid, alkali, or other irritating dusts or mists. Some dusts such as soluble chromate dusts may cause ulceration of nasal passages or even lung cancer.

Damage to internal tissues. This may result from the inhalation of radioactive materials, such as radium and its products, and from the inhalation of other particulate radioisotopes that emit highly ionizing radiation.

Threshold Limit Values

The TLVs of mineral dusts and toxic dusts considered permissible for exposures 8 hours per day, 5 days per week for most workers over a working lifetime have been published by the ACGIH. These values have been established as a result of evidence obtained by many groups in industry and from the results of laboratory studies performed on animals. They are reviewed annually and changed as necessary.

These values are set only as guidelines and are not to be considered absolute values. Evidence indicates that occupational disease will not occur if exposures are kept below these levels. On the other hand, occupational disease is likely to develop in some people if the recommended levels are consistently exceeded.

The currently recommended threshold limits of specific dusts can be found in the most recently published ACGIH list. (See Appendix B-1, for the 1986–87 TLV booklet.)

No one knows the exact concentration at which a particular person will start to develop silicosis, asbestosis, or lead poisoning. With some toxic dusts, however, extensive experience has revealed that the present threshold limits are fairly reliable.

As more experience is gained, the original threshold limits can be modified. In some cases, it has been necessary to lower the limits, because new findings showed effects at previously established levels.

Silicosis

Silicon dioxide is a term used generally when referring to amorphous silica (noncrystalline), crystallized silica such as sand (quartz), and silicates such as clay (aluminum silicate). Only the crystalline (free silica) material found in quartz, tridymite, cristobalite, and a few other nonsilicate materials causes silicosis. The crystalline structure in tridymite and cristobalite differs from that in quartz. For that reason, tridymite and cristobalite are more potent than quartz in causing silicosis.

Silicosis can be classified as a lung disease caused by the inhalation of free silica dust. Silicosis is common in industries and occupations where the crystalline form of free silica dust is present, such as in foundries, glass manufacturing plants, granite cutting operations, and where mining and tunneling in quartz rock are performed. Silicosis is found throughout the world, and in the past it has had many names, such as miner's asthma, grinder's consumption, miner's phthisis, potter's rot, and stonemason's disease. All these names, however, describe the same disease that is caused by dust from the crystalline form of free silica—usually quartz.

Crystallized silicon dioxide (SiO_2) most commonly occurs as sand, but it is also widely distributed in hard rocks and minerals. The percentage of crystalline SiO_2 in a dust mixture is the usual basis for evaluating the hazards associated with breathing in the mixture (Table 7-C).

Table 7-C. Crystalline Silicon Dioxide in Various Materials

Material	*Normal Range Percent SiO_2*
Foundry molding sand	50–90
Pottery ware body	15–25
Brick and tile compositions	10–35
Buffing wheel dressings	0–60
Road rock	0–80
Limestone (agricultural)	0–3
Feldspar	12–25
Clay	0–40
Mica	0–10
Talc	0–5
Slate and shale	5–15

To calculate the OSHA PEL using the formula under Mineral Dusts for crystalline silica (quartz) in 29 *CFR* 1910.1000, Table Z-2, proceed as follows:

$$\frac{10 \text{ mg/m}^3}{\% \text{ respirable quartz} + 2} = \text{PEL for respirable crystalline silica (quartz)}$$

For example, a sample of respirable dust contains 5.5 percent SiO_2 (quartz). Substitute 5.5 percent in this formula.

$$\text{PEL} = \frac{10 \text{ mg/m}^3}{5.5 + 2}$$

$$= \frac{10}{7.5} = 1.3 \text{ mg/m}^3$$

Therefore, levels of respirable quartz dust in the air that are above 1.3 mg/m^3 exceed the permissible exposure level. The TLV time-weighted average for crystalline silica is 0.1 mg/m^3 of the respirable quartz dust in the air. The formula for the silica TLV used to be the same as the standard OSHA formula. The common practice of analyzing dust samples for the mass and percentage of free silica has eliminated the need for the TLV formula. The PEL established by the OSHA and the TLV established by the ACGIH are virtually the same. This can be seen by substituting 100 percent in the PEL formula. This yields a PEL of 0.1 mg/m^3 for respirable dust composed of 100 percent quartz.

Influence of factors

Silicosis has manifested itself after widely different periods of exposure to silica dust. Apparently, development of the disease depends upon:

- the amount and kind of dust inhaled;
- the percentage of free silica in the dust;
- the form of the silica;
- the size of the particles inhaled;
- the duration of the exposure;
- the powers of resistance of the individual concerned; and
- the presence or absence of a complicating process (such as infection).

Action of silica on the lungs

Over the years, many theories have been advanced to explain why the crystalline form of free silica acts as it does in the lungs. These theories have been based on the hardness of the material and the effect of sharp edges, solubility phenomena, electrochemical action of the crystals, and immunological reactions. Free silica particles have a toxic effect on macrophage cells, which try to engulf and remove foreign matter from the lungs. Free silica causes macrophage cells to break and release material into the lung tissue, which can cause fibrosis.

Some dust moves out of the air spaces into other portions of the lung and, at the several points in the lung where silica dust is deposited and accumulates, a fibrous tissue develops and grows around the particle. This fibrous tissue is not as elastic as normal lung tissue and does not permit the ready passage of oxygen and carbon dioxide. Also, as it proliferates, the fibrous tissue reduces the amount of normal lung tissue. As a result, the available functional volume of the lung is reduced.

In some advanced cases, the fibrous tissue will slow down or even prevent the diffusion of oxygen from the lung to the blood in the capillaries, and the blood in the area will not be completely oxygenated. The fibrous tissue can also obliterate the blood vessels or reduce the flow of blood so that ultimately the lungs will not readily oxygenate sufficient blood for the body's needs. Then when the body's oxygen demand is increased by exertion, the individual will experience shortness of breath.

In very severe cases, fibrous tissue can so hinder the flow of blood in vessels of the lung that the heart enlarges in an effort to pump more blood. Serious enlargement of the heart is called cor pulmonale. Death can result from the cardiopulmonary effects of chronic silicosis.

EVALUATION OF EXPOSURE

This subsection is taken from the National Institute for Occupational Safety and Health (NIOSH) Criteria Document no. 75-120, *Occupational Exposure to Crystalline Silica* (see Bibliography).

The clinical signs of silicosis are not unique. Symptoms may be progressive with continued exposure to quantities of dust containing free silica, with advancing age, and with continued smoking.

Pulmonary symptomatology usually begins insidiously.

Symptoms include presence of cough, dyspnea, wheezing, and repeated nonspecific chest illnesses. Impairment of pulmonary function may be progressive. In individual cases, there may be little or no decrement when simple discrete nodular silicosis is present. However, when nodulations become larger or when conglomeration occurs, recognizable cardiopulmonary impairment tends to occur.

As is true of any of the pneumoconioses, the various progression stages of the silicotic lesions are related to the degree of exposure to free silica (exposure concentration), the duration of exposure, and the duration of time that the retained dust is permitted to react with the lung tissue. Because there are very few symptoms, very little is known about the early lesions resulting from moderately high exposure to free silica. Occasionally, exposures to very high concentrations occur in short periods of time in occupations such as sandblasting and tunneling. In these cases of acute or rapidly developing silicosis, there can be severe respiratory symptoms, and even death may occur. Roentgenographic examination of the lungs usually reveals a different pattern than that displayed by typical silicotic nodulation.

Other chemical or biological factors can influence the rate of reaction of free silica with the tissue and can create problems in diagnosis. One of the most frequent complications is the occurrence of tuberculosis with silicosis. This disease is called silicotuberculosis or tuberculosilicosis. This disease develops because silica is toxic to lung macrophage cells, which are important agents in the body's defense against tuberculosis.

The most common criteria used in diagnosing silicosis (and other occupational respiratory diseases) are the results obtained from pulmonary function tests, chest roentgenograms, and occupational exposure histories. Pulmonary function tests are objective indicators of respiratory dysfunction. However, there is no pulmonary function test specific for silicosis. A chest roentgenogram is a moderately good indicator of the degree to which tissue reacts to exposure to free silica. Unfortunately, several other disease entities can produce nearly the same roentgenographic pattern as free silica. Hamlin (1946) has found more than 20 conditions or diseases that cannot be differentiated from silicosis by roentgenographic study alone. In some cases, such as those involving dust particles of iron, tin, and barium, nodular densities with no fibrosis are produced by aggregates of particles alone. A history of exposure to free silica must be demonstrated before the other two criteria (pulmonary function and chest roentgenogram) can be used in making a diagnosis of silicosis.

Action of silicates

Silicates contain silicon and oxygen combined with other elements to form a complex molecule. Analyses of minerals are sometimes reported as percentages of oxides, which may include SiO_2, Al_2O_2, K_2O, and Fe_2O_3 (respectively silicon dioxide, aluminum oxide, potassium oxide, and ferric oxide). The silicon dioxide reported in such chemical analyses is the total of both the free silica (if present) and the silica present in the mineral. Such analyses are not reliable indications of the silicosis potential of the mineral.

Uncombined or free silica (quartz) is the most significant factor in industrial dust exposure. To properly evaluate an exposure, the percentage of uncombined silica must be determined using roentgenographic diffraction analyses or other special analytical chemical procedures.

With the exception of asbestos and some talcs, the silicate dusts do not ordinarily cause a serious disabling lung condition such as those produced by free silica. Much higher levels of silicate dusts can be tolerated.

In many industries, people have worked with silicate dusts that contained no free silica and did not develop a disability or nodulation in the lungs. A roentgenogram may show shadows indicating dust deposits in the lungs, but the pneumoconiosis is essentially harmless.

Disabling pneumoconioses due to exposure to abnormally high concentrations of mica, tremolite talc, and kaolin dusts have been described in the literature. The clinical signs for these silicate dusts are not the same as they are for free silica. The body does not have an adequate defense against indiscriminate amounts of dust; consequently, although specific symptoms have not been described for many mineral dusts, the general experience would indicate that dust levels should be kept below the most current TLVs.

Development of PELs and TLVs

The determination of hygienic exposure values for dust containing free silica has been based on the quantitative concept that the magnitude of the toxicity is proportional to the concentration of free silica in the dust. When the magnitude of toxicity is represented by an exposure limit, then the limit is inversely proportional to the percentage of free silica in the dust. The threshold limit can be expressed in millions of particles per cubic foot (mppcf) as derived from a particle count of the dust-laden environment and the following general particle count formula:

$$\text{TLV} = \frac{\text{K mppcf}}{\%\ \text{SiO}_2}$$

One of the first recommended "upper limits" for quartz-bearing industrial dusts was that suggested by Russell in 1937 for the Vermont granite industry. Based on studies in that industry, a limit of 10 mppcf for dust containing 25–35 percent quartz was recommended.

The TLVs for chemical substances and physical agents in the workroom are guides that have been adopted by the ACGIH for use in the control of occupational hazards. (The 1986–87 TLVs are given in Appendix B-1, Tables of Threshold Limit Values.)

The threshold value for quartz was first published in 1946; it was originally called a maximum allowable concentration (MAC) value and followed the pattern suggested by the particle count formula just described. However, only three ranges of free silica (quartz) content were considered, as shown in Table 7-D.

Table 7-D. Maximum Allowable Concentration Value of Free Silica Quartz

Range of SiO_2 (percent)	*MAC (mppcf)*
Silica–High (>50% free SiO_2)	5
Silica–Medium (5–50% free SiO_2)	20
Silica–Low (<5% free SiO_2)	50

Review of the early studies conducted by the Public Health

Service and others have suggested that the results of the engineering and medical studies were reasonably consistent with values calculated from the count formula using a factor designated K, equal to 250, and by adding a constant 5 to the percentage of free silica in the denominator. This formula was published by the ACGIH in 1962.

$$TLV = \frac{250 \text{ mppcf}}{\% \ SiO_2 + 5}$$

To make the TLV consistent with a 1970 revision of the TLV for nuisance dusts, the numerator K was raised to 300 and the constant 5 was raised to 10 in the denominator. Particle counting is very seldom used now in the calculation of industrial hygiene dust measurements.

Prior to the 1970 revision of the count formula, a respirable dust concentration formula utilizing a respirable mass measurements of dust was introduced:

$$TLV = \frac{10 \text{ mg/m}^3}{\% \text{ respirable free silica} + 2}$$

This formula was adopted by the OSHA and is still used today to measure PELs. The formula is based on the collection of dust by size-selective sampling devices. These instruments collect a fraction of the dust that is capable of penetrating to the gas-exchange portion of the lung where long-term retention of dust occurs. The concentration of airborne free silica in this fraction should relate closely to the degree of health hazard. As with the count formula, a constant is added to the denominator to prevent excessively high respirable dust concentrations when the fraction of free silica in the dust is low. The constant 2 limits the concentration of respirable dust with less than 1 percent free silica to 5 mg/m^3. Using the formula, the TLV for pure silica would be 0.1 mg/m^3. Because industrial hygiene laboratories are equipped to perform free silica analyses and these tests are commonly performed on each sample collected, the TLV for pure respirable free silica has been established at 0.1 mg/m^3, regardless of the total dust concentration.

In addition to quartz, other forms of free silica have been assigned a specific TLV. These values are based on experimental data or on industrial experience that indicated a need for individual identification.

Cristobalite (above 5 percent). This free silica was originally listed in 1960 with a TLV of 5 mppcf, based on (1) studies in the diatomite industry, (2) analogy with the TLV for silica, and (3) experimental studies using animals. In 1968, the TLV was reduced to one half the value obtained from either the count or mass formula for quartz, following a review of existing documentation and information produced by the TLV Committee. This information suggested that the limit of 5 mppcf for cristobalite did not allow a sufficient safety factor for the prevention of pneumoconiosis. The current TLV for respirable cristobalite is 0.05 mg/m^3.

Tridymite. This was also assigned one half the quartz value based on animal toxicity data. When tridymite dust was administered by intratracheal injection into the lungs of rats, evidence indicated that it was a more active form of free silica than quartz. Analogy was also made with cristobalite.

Fused silica dust. Although insufficient industrial experience was available to indicate the degree of hazard presented by fused silica dust, the same limit as that required by the quartz formulae was adopted in 1969. Intratracheal injection studies with rats indicated that fused silica was considerably less active than quartz. Respirable fused silica has a TLV of 0.1 mg/m^3.

Tripoli and silica flour. These were added to the TLV list in 1972, with the recommendation that the standard for these materials be derived using the respirable mass formula for quartz. Documentation for inclusion of tripoli on the list came from the study by McCord, et al. (1943), in which tissue proliferation was induced by direct intraperitoneal implantation of tripoli dust in rats and guinea pigs. The tissue proliferation was similar to that produced by quartz. Silica flour was included on the list as a result of a study that showed that silica flour has a significant fibrogenic potential because of its fine particle size.

The 1968 ACGIH recommended TLVs for quartz have been adopted by the U.S. Department of Labor under the Walsh-Healey Public Contracts Acts regulations (41 *CFR* 50.204). The TLVs have also been adopted by the U.S. Department of Interior under the Metal and Nonmetallic Health and Safety Act (Sec 6, 80 Stat. 774; 30 USC 725).

The Federal Coal Mine Health and Safety Act of 1969 (Public Law 91.973) provides that the Secretary of Health and Human Services prescribe a formula for determining the applicable standard for coal mines where quartz amounts to more than 5 percent. Such a formula has been published using 0.1 mg/m^3 of respirable quartz as a basis. (See 30 *CFR* Part 70.101, published in the *Federal Register,* vol. 36, p. 4941, dated March 16, 1971; and 30 *CFR* Part 71.10, published in the 37 *Federal Register* 6368, dated March 28, 1972). The law states that (1) The 8-hour average concentration per shift of respirable coal dust must not exceed 2 mg/m^3, and (2) If the respirable dust contains more than 5 percent quartz, the 8-hour average concentration cannot exceed the result obtained when the percentage of quartz is divided into the number 10. The answer is expressed in terms of mg/m^3. For example, if the dust contained 100 percent quartz, the allowable dust limit would be 0.1 mg/m^3.

$$\frac{10}{100\%} = \frac{0.1 \text{ mg}}{\text{m}^3} = \frac{100 \ \mu\text{g}}{\text{m}^3}$$

The present federal standard for free silica is an 8-hour time-weighted average based on the 1968 ACGIH TLV of 10 mg/m^3 ÷ percent SiO_2+2 for respirable quartz. One half this amount has been established as the limit for cristobalite and tridymite. (See 29 *CFR* Part 1910.93, published in the 37 *Federal Register* 22139, dated October 18, 1972.)

ASBESTOS

Another kind of pneumoconiosis, which involves specific lung changes, is called absestosis; it is caused by the inhalation of asbestos dust.

Asbestos is a generic term used to describe a number of naturally occurring, fibrous, hydrated mineral silicates that differ in chemical composition. These may be divided into two mineral groups: (1) Pyroxenes, which include chrysotile (3 $MgO \cdot 2SiO_2 \cdot 2H_2O$)—the type most widely used in U.S. industry; and (2) amphiboles, including amosite ([FeMg]SiO_3), crocidolite ($NaFe[SiO_3]_2{\cdot}H_3O$), tremolite ($Ca_2Mg_5Si_8O_{22}[OH]_2$), anthophyllite ($(MgFe)_7Si_8O_{22}[OH]_2]$), and actinolite ($CaO{\cdot}3[MgFe]O{\cdot}SiO_2$). Asbestos fibers are generally characterized by high tensile strength, flexibility, heat and chemical resistance, and favorable frictional properties. Certain grades of asbestos can be carded, spun, and woven, while others can be laid and pressed to form paper or used for structural reinforcement of materials such as cement, plastic, and asphalt.

Chrysotile (white asbestos). This is the fibrous form of the mineral serpentine. It is the most common variety of asbestos and is widely distributed geographically. The largest deposits are in Canada, Russia, and Rhodesia.

Crocidolite (blue asbestos). This is another important, although more specialized, form of asbestos. It is the fibrous form of riebeckite, and has fine, resilient fibers of a characteristic blue color:

Amosite. This is the fibrous variety of the mineral grunerite, a ferrous magnesium silicate mined only in South Africa. Amosite can be readily broken down into long, somewhat harsh fibers that range in color from a brownish-yellow to almost white, depending on the quality.

Anthophyllite. This is magnesium silicate of somewhat variable composition that has rather fragile brownish or off-white fibers.

Tremolite. This calcium magnesium silicate, is often a major component involved in the production of industrial and commercial talc.

Actinolite, a calcium magnesium iron silicate, is rarely used in industry.

Effects of asbestos exposure

Asbestos in its several commercial forms has been shown to be associated with the development of a variety of disease entities. These include:

Asbestosis. A diffuse, interstitial, nonmalignant scarring of the lungs;

Bronchogenic carcinoma. A malignancy of the lining of the lung's air passages;

Mesothelioma. A diffuse malignancy of the lining of the chest cavity (pleural mesothelioma) or of the lining of the abdomen (peritoneal mesothelioma);

Cancer of the stomach, colon, and rectum. The association between absestos and these forms of cancer has not yet been well established.

In its advanced stages, asbestos exposure is evident by characteristic manifestations on x-ray films, by restrictive pulmonary function, or by clinical signs, such as finger clubbing or rales (dry, cracking sounds within the lung). Its most important symptom is dyspnea, or undue shortness of breath. The disease resulting from asbestos exposure may be progressive, even in the absence of futher exposure; the inhaled fibers trapped within the lung continue their biological action. In its severe forms, death results from the inability of the body to obtain enough oxygen or from the heart's failure to pump blood through the scarred lungs.

Mesotheliomas. Mesotheliomas are diffuse and spread rapidly throughout the cavity of origin. They have yet to be successfully cured by any types of treatment, including chemotherapy, radiation, or surgery. Death usually results within a year of diagnosis. In the general population, mesothelioma is very rare. It may account for one death in several hundred thousand in the absence of an environmental or occupational asbestos exposure.

Once established, the other asbestos-associated cancers differ little from those that occur in the general population, although there may be variations in the location of the primary site. Appropriate treatment and prognosis follow the usual pattern of the particular tumor.

Lung cancer, pleural mesothelioma, and peritoneal mesothelioma usually do not become clinically evident until more than 20 years after onset of exposure. This time lag is now widely recognized. While some of these cancers may appear during the second decade following onset of occupational exposure, peak incidence is often not noted until 30 or more years later. This is true both when regular, long-term asbestos exposure is involved, and also following short-term, brief, or intermittent exposures. While variations in the time of occurrence may depend on the intensity and duration of exposure (with heavier exposure often being associated with shorter latency periods), variations among individual cases make it impossible to predict the latency period for the risk of any particular worker.

Asbestos regulations

Because of the severe health consequences associated with asbestos, exposure to asbestos is strictly regulated by OSHA. Industries that manufacture products containing asbestos and vehicle repair facilities where asbestos may be encountered are regulated under 29 *CFR* 1910.1001 (revision issued June 20, 1986). Besides limiting asbestos exposures to 0.2 fibers per cubic centimeter of air (f/cc) over an 8-hour workday, specific requirements are provided in the standard on work practices, engineering controls, air monitoring, medical monitoring, and employee education. A separate but very similar OSHA standard applies to the construction industry and other industries engaged in the handling or removal of asbestos used in building materials (including insulation). The following summary of the Asbestos OSHA Construction Industry Standard (29 *CFR* 1926.58) was prepared by David E. Owen, B.S., M.S.P.H., Director of Occupational Health, North Carolina Power and Light Company, Raleigh, NC, and is used with permission.

A summary of 29 *CRF* 1926.58 as amended June 20, 1986, Asbestos

(a) Construction work is defined to include (1) removal or encapsulation of asbestos material; (2) construction, alteration, repair, maintenance, or removal of structures that contain asbestos; (3) emergency cleanup of asbestos material; and (4) transportation, disposal, and storage of containers of asbestos on the site or location where construction activities are performed.

(b) The 8-hour Time Weighted Average (TWA) PEL is 0.2 f/cc. No short-term exposure limit is established.

(c) At multiemployer work sites, the employer performing asbestos work shall notify other employers of the work and the requirements pertaining to regulated areas (enclosures) of asbestos work.

(d) Except for "small-scale, short-duration operations" (jobs requiring less than 1 hour to complete), negative-pressure enclosures shall be established, where feasible, for removal, demolition, or renovation operations. A "competent person" [with specific abilities described in paragraphs (b) and (e)] shall supervise the entire operation to assure that the requirements of the standard are met. This person shall be trained by a course of study equivalent to that offered by the Environmental Protection Agency (EPA) Asbestos Training Centers.

(e) Initial employee exposure monitoring shall be performed at the beginning of each asbestos job with two exceptions: (1) when the employer has demonstrated that employee exposure will not exceed the action level (0.1 f/cc) or (2) when the employer has performed exposure monitoring on very similar jobs. Periodic employee exposure monitoring shall be performed each day of the job for each employee in the enclosure except when the employees are using supplied-air, positive-pressure respirators.

(f) Engineering controls of employee exposure include HEPA-filtered exhaust ventilation systems, HEPA-filtered vacuum cleaners, wetting methods, prohibition of high-speed saws and compressed air, and other generally recognized methods. These controls shall be used, as feasible, to reduce employee exposure to the lowest attainable levels. Employee rotation is prohibited as a method to achieve compliance with the permissible exposure limit.

(g) Only high-efficiency air-purifying respirators or those offering greater protection may be used against asbestos. At concentrations less than 2.0 f/cc, half-facepiece air-purifying respirators with high-efficiency filters may be used. At concentrations less than 20 f/cc, full-facepiece air-purifying respirators with high-efficiency filters may be used. Powered-air-purifying respirators and supplied-air respirators are specified for higher concentrations. Semiannual fit testing is required for employees using negative-pressure (air purifying) respirators. The preamble specifically excludes the use of disposable respirators. (Some state or local regulations stipulate that Type C air-supplying respirators must be used.)

(h) Full-body protective clothing is required for employees exposed to airborne concentrations of asbestos greater than 0.2 f/cc. The preamble states that this full-body clothing is required, regardless of high temperatures and prohibits employees' taking contaminated clothing home.

(i) Clean areas for changing out of protective clothing are required except when employees are performing "small-scale, short-duration" operations when employees are required to vacuum their clothing with HEPA-filtered vacuum cleaners prior to leaving the work area.

A decontamination area shall be established for removal, demolition, or renovation operations. This area shall consist of: (1) an equipment room for removing protective clothing (but not respirators), (2) a shower where respirators may be removed, and (3) a clean change room.

The shower shall be located, if feasible, adjacent to the work area enclosure equipment room. When this is not feasible, the shower may be located away from the work area and employees shall either HEPA-filtered vacuum clean their clothing or remove their contaminated clothing and don clean work clothes prior to proceeding to the shower.

(j) Work area enclosures shall be posted with warning signs. Containers of asbestos waste shall be labeled with warning signs. The required signs differ in wording from those currently in use. Employees shall be trained prior to their initial assignment to an asbestos job and at least annually thereafter.

(k) When vacuum cleaners are used, they shall be HEPA-filtered. All waste shall be disposed of in sealed, labeled, impermeable containers or bags.

(l) A medical surveillance program shall be offered to employees who are (1) exposed to airborne concentrations of asbestos above the action level for more than 30 days per year or (2) who are required by this standard to use negative-pressure respirators. Persons who conduct spirometry shall be specifically trained in this procedure.

The medical surveillance shall be offered initially and annually thereafter, or more frequently at the discretion of the examining physician. The medical surveillance shall include spirometry, chest x-ray films, and a health history questionnaire (shown in the OSHA standard). Chest x-ray (films) shall be interpreted by the ILO classification system.

The employer shall provide the examining physician with extensive information on the employee's duties, asbestos exposure, and personal protection equipment the employee uses. The physician shall provide the employer with a written opinion stating (1) whether any conditions were found that would place the employee at "increased risk of material health impairment" from exposure to asbestos, (2) recommendations of limitations on the employee's use of personal protection equipment, and (3) that the employee has been informed of these results.

(m) Records on objective data for exempted operations, exposure monitoring, medical surveillance, and training shall be retained in accordance with 29 *CFR* 1910.20.

(n) The requirements of paragraphs (c) through (n) become effective January 16, 1987 (210 days after the standard's publication in the *Federal Register*).

(o) Appendices on air sampling (A), respirator fit testing (C), medical surveillance health history questionnaire (D), and x-ray interpretation (E) are mandatory.

(p) Appendix G, Work Practices and Engineering Controls for Small-Scale Short-Duration Asbestos Renovation and Maintenance Activities, is not mandatory.

Before any asbestos work in the construction industry is begun, it is essential to consult the OSHA standard for its detailed explanation of effective work practices. Appendix F of 29 *CFR* 1926.58 describes work practices and engineering controls for major asbestos removal, renovation, and demolition projects. Included are instructions on how to build containments around the work area, descriptions of necessary equipment, removal methods that minimize dust, cleanup of the work area, and proper disposal methods. Appendix G of 29 *CFR* 1926.58 describes work practices and engineering controls for small-scale, short-duration asbestos jobs, such as glove bag work (Figure 7-3). It also describes methods for safely conducting cleaning and maintenance activities in a building with asbestos-containing materials.

One widespread source of asbestos exposure is in the area of automotive brake and clutch repair operations. A common practice in the past was to blow out the asbestos dust that had been deposited due to brake or clutch wear with compressed air. This can create exposures well in excess of the 0.2 f/cc limit.

Two very effective control methods are available for removing this dust in ways that will easily reduce exposures well below the OSHA action level (0.1 f/cc) and thereby avoid some of the more burdensome aspects of the regulation. The two methods, the Enclosed Cylinder HEPA Vacuum System and the Spray Can Solvent Cleaner Method, are described in detail in Appendix F of 29 *CFR* 1910.1001. The inexpensive but effective spray can solvent method is described by OSHA in the following way:

> The mechanic should begin spraying the asbestos-contaminated parts with the solvent at a sufficient distance to ensure that the asbestos particles are not dislodged by the velocity of the solvent spray. After the asbestos particles are thoroughly wetted, the spray may be brought closer to the parts and the parts may be sprayed as necessary to remove grease and other material. The automotive parts sprayed with the mist are then wiped with a rag, which must then be disposed of appropriately. Rags should be placed in a labelled plastic bag or other container while they are still wet. This ensures that the asbestos fibers will not become airborne after the brake and clutch parts have been cleaned. (If cleanup rags are laundered rather than disposed of, they must be washed using methods appropriate for the laundering of asbestos-contaminated materials.) (Source: *Federal Register* vol. 51, no. 119, June 20, 1986. 29 *CFR* 1910.1001, Appendix F.)

Spray can solvent brake cleaners are available at any auto supply store.

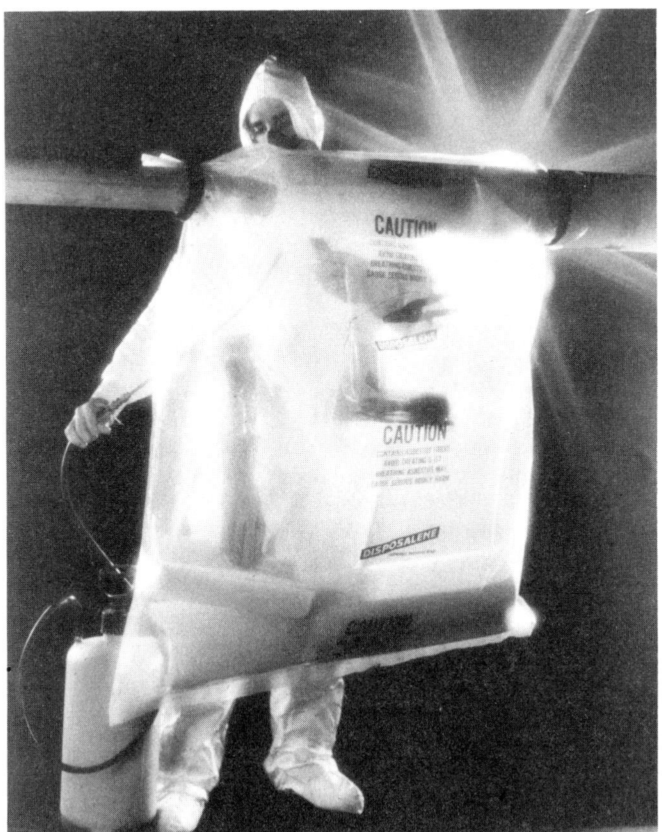

Figure 7-3. Disposable glove bag makes asbestos removal from pipes easy and safe. (Courtesy Omni Sales and Manufacturing)

BERYLLIUM INTOXICATION

Beryllium intoxication is a severe systemic disease that can result from the inhalation of dust or fumes from metallic beryllium, beryllium oxide, or soluble beryllium compounds.

There are two forms of the disease. One is an acute form of chemical pneumonitis with cough, pain, difficulty in breathing, cyanosis, and loss of weight. The other is the chronic type, known as berylliosis, in which there may be loss of appetite and weight, weakness, cough, extreme difficulty in breathing, cyanosis, and cardiac failure. Formerly, mortality was high in cases of chronic beryllium intoxication, and many of those who survived suffered from pulmonary distress.

Individual susceptibility apparently is an important factor in the development of the disease. In many instances severe symptoms can develop in one employee, while other employees doing the same work may show no signs of disability.

Beryllium intoxication has never been found in individuals who mine or handle ore only; there is also no evidence of intoxication as a result of the ingestion of beryllium oxide, beryllium metal, or any of the beryllium alloys. Only the inhalation of beryllium-bearing dusts or fumes produces the systemic disease. Accordingly, control of such dusts and fumes at or below concentrations specified by the ACGIH TLVs should be recognized as a basic protective measure.

When the soluble salts of beryllium, especially beryllium fluoride, come in contact with cuts or abrasions on the skin, deep ulcers may form that heal very slowly. Complete surgical excision of the ulcer is sometimes required before the wound can heal.

Chronic beryllium disease (berylliosis)

The clinical nature of chronic beryllium disease differs from acute pneumoconiosis in that the former often develops several years after beryllium exposure. A number of case histories have revealed a delay ranging from 5–10 years between the last beryllium exposure and the appearance of detectable evidence of disease. In some cases, a delay of 20 years or more has occurred. Further, chronic illness has been characterized as a systemic disease, which is prolonged in duration and commonly progressive in severity despite cessation of exposure.

Chronic beryllium disease results from inhalation of beryllium particulates. This disease is characterized by granulomas in the lungs, skin, and other organs. Symptoms include cough, chest pain, and general weakness. Pulmonary dysfunction and systemic effects, such as heart enlargement (cor pulmonale, leading to cardiac failure), enlargements of the liver and spleen, cyanosis, and the appearance of kidney stones also characterize the chronic illness. Beryllium has also been shown to cause lung cancer in rats and rhesus monkeys.

The present OSHA standard prescribes an 8-hour time-weighted average of 2.0 $\mu g/m^3$ with a ceiling concentration 5.0 $\mu g/m^3$. In addition, the present standard allows a peak concentration above the acceptable ceiling concentration for an 8-hour shift of 25 $\mu g/m^3$, for a maximum duration of 30 minutes. The current TLV is an 8-hour TWA of 2.0$\mu g/m^3$, and has been classified as an industrial substance suspect of carcinogenic potential in humans.

BLACK LUNG

Black lung is the name given to all lung diseases associated with chronic overexposure to coal dust. These diseases include chronic bronchitis, silicosis, and coal workers' pneumoconiosis. A coal miner may have all three diseases at the same time. The exact cause of chronic bronchitis among coal workers is not known, although cigarette smoking may be an aggravating factor. Coal dust (particularly from anthracite coal) can contain free silica, and exposure to such coal dust can cause silicosis.

Exposure to coal dust that does not contain free silica was once thought to be harmless. But since the 1940s, it has been shown that even coal dust with minimal free silica content can cause the fibrotic lung disease known as coal workers' penumoconiosis (CWP). The disease mechanism of CWP is not well understood, and the symptoms are hard to distinguish from other lung diseases. Individuals with early stages of CWP may have no symptoms that can be directly related to fibrotic changes. Instead, any respiratory deficiencies are often due to the chronic bronchitis associated with coal dust exposure.

Varying degrees of fibrotic changes and symptoms can occur with CWP. In a small number of those exposed, CWP can develop from simple fibrosis to progressive massive fibrosis. This is a condition in which small, discrete fibrotic nodules have conglomerated, resulting in a severe restriction of lung capacity.

Exposure to coal dust in mines is regulated by the Mine Safety and Health Administration (MSHA). The MSHA standard for respirable coal dust is 2 mg/m^3 over an 8-hour work shift.

Black lung in miners is a federally compensable occupational disease under the 1969 Federal Coal Mine Safety and Health Act. Current payments to miners with black lung diseases total 1.7 billion dollars annually.

Coal dust exposure away from mining operations is regulated by OSHA. The OSHA PEL for coal dust that contains less than 5 percent free silica is 2.4 milligrams of respirable dust per cubic meter (mg/m^3) of air. If the free silica content amounts to more than 5 percent, the limit becomes mg/m^3 respirable dust = 10 ÷ percent free silica + 2.

MISCELLANEOUS PNEUMOCONIOSES

Even though most dust is classified as harmless, excessive amounts of it can irritate the walls of the respiratory system. Mica dust and kaolin dust are two good examples of dusts that ordinarily are considered benign but, in excessive amounts, can cause a troublesome pneumoconiosis.

Mica pneumoconiosis

Mica pneumoconiosis has been observed in grinding operations where mica dust was present but free silica was not. There were marked changes in x-ray films obtained of the lungs, and some disability following exposure occurred. The cases occurred where the dust exposures were massive over many years.

Kaolinosis

Kaolinosis has been described as a condition induced by inhalation of dust released in the grinding and handling of kaolin (china clay). In the area where the cases occurred, dust levels of several hundred million particles per cubic foot of air were common.

Bauxite pneumoconiosis

Also known as Shaver's disease, bauxite pneumoconiosis has been found only in workers exposed to fumes containing aluminum oxide and minute or ultramicroscopic silica particles arising from smelting bauxite in the manufacture of corundum, an impure form of aluminum oxide that may contain small amounts of aluminum silicate. The disease definitely does not occur from the use of corundum grinding wheels or from exposure to other forms of aluminum oxide.

Some pneumoconioses may show marked shadows on an x-ray film, which, without the necessary information on the exposure of the individual, may be alarming in a general x-ray screening program. Clinical examination, however, often discloses no disability or symptoms. These shadows are frequently encountered when the dusts contain atoms of a relatively high molecular weight because the heavier atoms are fairly opaque to x-rays. Insoluble barium dusts and tin oxide dusts, for example, can show very marked shadows on x-ray films without producing signs of significant pathology. (Barium dust that is soluble in the body fluids, however, can produce a toxic reaction.)

Siderosis

Iron oxide, particularly excessive fumes from welding operations, may produce siderosis with a pigmentation of the lungs (black in welders' lungs and red in the lungs of iron ore miners) usually without causing disability. Some individuals with siderosis may have symptoms of chronic bronchitis and shortness of breath. The shadows produced by iron oxide in x-ray films of the lungs are somewhat similar to the shadows produced by silicosis. Because of this similarity, differential diagnosis is often difficult, and heavy exposures to iron oxide dust and fumes may lead to medicolegal problems.

Miscellaneous dusts

Limestone, marble, lime, gypsum, and Portland cement dusts apparently have minimal effects even after long exposures. These materials have little or no free silica. Also, many silicates and other minerals have not caused impairment in individuals who have inhaled the dusts, and the resulting pneumoconioses have generally been classed as benign. Some cements can contain diatomaceous earth, which can be converted to cristobalite silica with high heat. Other cements may contain asbestos. It is important to consult the material safety data sheet of any dusty material to determine the potential health hazards.

Allergic reactions

When they are in the form of dust, a large number of materials may cause various allergic reactions in susceptible individuals. Such agents include certain animal products, foods, drugs, and chemicals. Western red cedar dust, proteolytic enzyme detergents, and ethanolamines (found in some aluminum solder fluxes) can cause allergic reactions. The allergic reactions, which may be quite severe, usually involve the skin, respiratory system, and gastrointestinal system. Occasionally, two or more systems are involved. Some of the allergic reactions are dermatitis, hay fever, asthma, and hives.

Usually the victim is subjected to a series of exposures without any reactions; during this time, sensitization is built up. These exposures may occur continuously for years. Then, at the end of the "incubation period," which varies with each individual, a reaction is produced.

Two factors are required for a true allergic reaction to occur:

- a history of prior exposure to the material involved (sometimes this is not known even by the affected employees);
- a "challenge dose" of the material, which provokes the allergic reaction.

Continuous exposure may act as "desensitizing doses" and, under these conditions, an allergic individual may work without incident for long periods of time. However, reexposure following removal from the sensitizing material (such as after a vacation) causes an allergic response to recur. Unfortunately, a reaction known as "allergic alveolitis" develops in some individuals subjected to continuous exposure. This condition can be progressive and can develop into a severely disabling obstructive disease.

Medical and engineering recommendations intended to prevent allergic reactions are based on the prevention of exposure by means of personal protective equipment, ventilation methods, or by the removal of sensitized individuals from the exposure.

Bacteria and fungi

The possibility of lung infections from the inhalation of bacteria and fungi exists in several industries. Pulmonary anthrax as a result of the inhalation of dust that contains anthrax spores has occurred among persons engaged in handling wool and crushing bones of infected animals.

In addition to causing infections in individuals, a specific sensitization to mold spores may develop. This sensitization is similar in nature to an allergic reaction, except that the alveoli (air sacs) of the lungs instead of the bronchial tubes are constricted. A few hours after a sensitized individual is exposed, he or she experiences flulike symptoms as well as shortness of breath. The total lung capacity is reduced. Although these symptoms are initially reversible, repeated exposures can result in irreversible symptoms and eventually respiratory or cardiac failure.

Fungi (molds) growing on grain have been found in the sputum of workers who shovel the grain and are believed to be the cause of respiratory disorders. Likewise, fungi found in sugar can residues (bagasse) are believed to be part of the cause of bagassosis. Fungal spores found under the bark of some trees have been blamed for respiratory difficulties among employees who strip the bark from dry logs.

Although the incidence of occupationally related bacterial and fungal infections and sensitizations is relatively low, the respiratory effects can be troublesome, and in the case of pulmonary anthrax, even fatal. The basic methods of control are the same as those for controlling the pneumoconiosis-producing dusts, but to these methods, sterilization and disinfection must be added.

Measurement of airborne dust concentrations

To evaluate dust exposure, it is first necessary to determine the composition of the dust that remains suspended in the air that workers breathe.

Operations that involve the crushing, grinding, or polishing of minerals or mineral mixtures frequently do not produce airborne dusts that have the same compositon that the material being worked on has. Of primary concern is the percentage of crystalline silicon dioxide that is suspended with other dusts in the air.

In many mineral mixtures, crystalline silicon dioxide is harder to crush than is the remainder of the mixture. Thus, in most crushing and grinding operations, airborne dust is produced with a lower percentage of crystalline silicon dioxide than is found in the original mineral mixture. Occasionally, such as when the grinding operation involves a great deal of impact force, the situation is reversed. For example, measurements obtained in coal breaker houses (where coal is prepared for combustion in industrial boilers) often show that silica levels are higher in the respirable coal dust than in the original coal.

In pulverized materials, crystalline silicon dioxide is frequently found in larger particle sizes than are the softer components; when this is the case, subsequent handling may result in a higher percentage of the softer components in the airborne dust.

Therefore, it is necessary to actually obtain a sample of airborne dust and to analyze it if the composition of the dust is to be used in evaluating the hazard associated with breathing the dust.

The extremely small size of the dust—0–5 μm in diameter—makes chemical methods of analysis, which depend on the differential solubility of the various components, too unreliable to be used alone. The x-ray diffraction and infrared analysis techniques are usually preferred. These require equipment and procedures that are not commonly available in chemical laboratories. Table 7-C indicates the normal range of crystalline silicon dioxide determined by analysis of many minerals and industrial compositions.

The industrial hygienist measures the size of airborne dust particles in units known as micrometers or microns. (A micrometer or micron is one thousandth of a millimeter or approximately one twenty-five thousandth of an inch.)

As mentioned earlier in this chapter, when the standard technique is used to evaluate dust concentrations, particles larger than 10 μm are not considered because particles this size or larger do not remain suspended in still air for appreciable lengths of time, nor do they penetrate deeply into the lung if they are breathed. Larger particles seldom reach the lungs because they are filtered out in the nose and throat. The physiological effects of extremely small dust particles, however, are uncertain.

When air samples are collected in the immediate vicinity of a dust-producing operation, larger particles, which have not yet had time to settle from the air, may be collected. If a large number of these particles appear in the dust sample, the effect of their presence may have to be evaluated separately.

Dust measurement

The following subsection has been gleaned from the NIOSH Criteria Document 75-120, *Occupational Exposure to Crystalline Silica* (see Bibliography).

To evaluate either the relative hazard to health posed by dusts or the effectiveness of dust control measures, one must have a method for determining the extent of the dustiness. Ideally, the methods employed should be as closely related to the health hazard as possible. In addition to determining the percentage of free silica, the method should measure any other portion of the dust that can cause silicosis; that is, dust that penetrates and is retained in the pulmonary, nonciliated regions of the lungs.

Through the years, many collection methods have been used in the determination of dustiness, but only the basic methods will be briefly discussed here.

Count procedures. The concern of industrial hygienists over the years has been to measure that fraction of a dust that can cause pneumoconiosis. Since it has been recognized that only dust particles smaller than approximately 5 μm in aerodynamic diameter are deposited and retained in the lung, methods were sought to measure the concentration of these tiny dust particles. Microscopic counting of dust collected by impingement has long been used for this purpose.

Dust counting as an index of dust concentration, and consequently of workers' exposure, has been performed in South

Africa using the konimeter, and in Australia using a jet dust sampler. In the United Kingdom, thermal precipitation has been frequently used for dust collection. In the United States, the Greenburg-Smith and midget impingers have been commonly employed.

In these investigations, the lower limit of dust size included in counts was determined by the procedure of was implicit in the counting procedure employed. In impinger counts, where a 10 × 16-mm objective lens was used with light-field counting, the usual lower limit of particles seen was approximately 1.0 μm in diameter. Others have used dark-field illumination techniques with which it is possible to see particles as small as 0.1 μm in diameter.

Because of differences in sampling techniques and in the instruments used, comparisons of dust concentration with silicosis prevalence in different parts of the world are difficult. This indicates that if dust concentrations are measured by a counting procedure, the procedure employed should follow a standardized method to minimize differences.

Although counting methods are inefficient and yield variable results, they clearly showed the effectiveness of dust-control measures. Later, as part of their efforts to further reduce silicosis, researchers also turned to improving dust measurement methods. Because the prevalence of silicosis in the 1920s and earlier was severe, more effort has since been devoted to improving dust conditions than to refining and developing methods of dust sampling and measurement.

"Total" mass concentration methods. The simplest method of measuring dust concentration is to determine the total weight of dust collected in a given volume of air. The "total" mass, however, is determined to a considerable extent by the large dust particles, which cannot penetrate to the pulmonary spaces and cause silicosis. The proportion of dust that is small enough to penetrate into the pulmonary spaces ("respirable" dust) is extremely variable; it ranges in industrial dust clouds from as little as 5 percent to more than 50 percent (by weight).

Thus, the "total" dust concentration by weight is not a reliable index of "respirable" dust concentrations, nor is it an index of a silicosis hazard.

Respirable mass size-selection measurement (personal sampling). When evaluating a silicosis hazard, the method now generally preferred is personal (breathing zone) respirable mass sampling. Dust collection devices now available for this method of sampling also provide a means for a size-frequency analysis of the collected dust. A traditional method for such an analysis has been to collect a sample on a membrane filter and examine it by high-powered optical microscopy (about 1,000 ×), supplemented, perhaps, by electron microscopy.

The dust sample is size-separated by the design and flow characteristics of the sampling device. Such equipment includes impactors, centrifugal and gravitational separators, and a range of miniature cyclones. In addition to particle-size separation, these instruments are also capable of collecting a quantity of dust sufficient for an analysis of the free silica content of the dust.

Respirable mass samples are preferably taken over a full 8-hour shift. However, multiple, shorter period samples (over a 2– 4-hour period) may be collected during an individual's full-shift exposure period. The samples are then pooled for analytical purposes, and the average respirable mass concentration for free silica is calculated for a full-shift. The recommended equipment and the method for collecting dust containing respirable free silica are presented in Figure 7-4.

Toxic dusts and fumes

Systemic reactions are caused by the inhalation of toxic dusts and fumes from various elements and their compounds. All metallic fumes are irritating, especially when freshly generated. Industrially important metals and their compounds that can have a toxic effect when the dust or fume is inhaled include arsenic, antimony, cadmium, chromium, lead, manganese, mercury, selenium, tellurium, thallium, uranium, and a few others.

The effect of some metals, such as magnesium and zinc, appears to be transient. Although the dusts and fumes from metals with low toxicity do not require as much caution as the dusts and fumes from highly toxic metals, the former should not be neglected or disregarded. Metals with low toxicity are more easily controlled because greater amounts can be tolerated, but their dusts and fumes should be kept at reasonable levels since excessive amounts of any of them can be harmful (Table 7-A).

Lead poisoning

Lead dust and fumes can present a severe hazard to those who are overexposed to them. Unlike many other metals that are necessary nutrients at low concentrations in food, lead serves no biological function. Inhaled or, to a lesser extent, ingested, lead is readily absorbed and distributed throughout the body. Repeated exposures can cause a gradual accumulation of lead, particularly in the bones. Symptoms of chronic overexposure include anxiety, weakness, headaches, tremors, excessive tiredness, and other indicators of nervous system damage. Anemia, kidney damage, and reproductive defects in both men and women (sterility, miscarriages, birth defects, and others) can also be caused by lead.

To prevent such health hazards, OSHA has developed an exposure standard for lead (29 *CFR* 1910.1025). As one of OSHA's completed standards, it not only specifies an 8-hour PEL for lead ($50 \mu g/m^3$), but also establishes requirements for air testing, medical monitoring, respiratory protection, protective clothing, engineering controls, employee training, and hygiene facilities and practices. If there is a potential for exposure, the Lead Standard should be consulted. It provides an excellent summary of the potential hazards involved as well as the preventive measures that can be taken.

Metal fume fever

Metal fume fever is an acute condition of short duration caused by a brief high exposure to the freshly generated fumes of metals, such as zinc or magnesium, or their oxides. Symptoms appear from 4–12 hours after exposure and include fever and shaking chills. Recovery is usually complete within one day, and ordinarily the employee can return to the same job without suffering a recurrence of the symptoms. Oddly enough, daily exposure confers immunity—but not permanently. If the daily exposure is interrupted for a period, such as over a long 3-day weekend, subsequent exposure will result in a recurrence of the symptoms. The severity of the attack depends on the length of the interruption.

Heavy concentrations of fumes cause metal fume fever. Zinc oxide fume is the most common source, but cases caused by the

Figure 7-4. Respirable Dust Sampling

A. Prepare cyclone assembly and mount sampling apparatus on employee.

1. Loosen set screw on back of cyclone and remove rubber stopper (B).
2. Identify preweighed polyvinylchloride (PVC) filter cassette by number or letter.
3. Remove end plugs from filter cassette (save) and insert cassette (filter on top side) into cyclone (A).
4. Tighten set screw (make sure that cassette inlet and outlet are seated into each holder of cyclone—upper and lower [A]).
5. Attach pump to worker (B,C).
6. Connect sample hose to cyclone outlet. *NOTE: Be careful to keep cyclone in an upright position. This will prevent dust in the bottom of the cyclone from falling into filter cassette.*
7. Clip cyclone assembly to worker's collar, allowing cyclone to hang free (A,B).

B. Calibrate sampling train

1. Connect free end of sample hose to lower port of rotameter (C).
2. Connect upper port of rotameter to pump inlet.
3. Determine rotameter float elevation from calibration chart of rotameter (C). *NOTE: Flow rate should be 1.8 L/min.*
4. Hold rotameter vertically, start pump, and adjust slotted flow valve on pump with screwdriver until float elevation (reading center of ball) determined in step B-3 is established (C).
5. *Remove* rotameter and connect pump directly to cyclone (B).

C. Collect sample

1. Record start time.
2. Collect sample as detailed in attached sampling procedure, while *periodically checking flow rate* by repeating steps B-1 through B-5. (Not required with constant-flow pumps.)
3. Stop pump and record time.
4. Remove filter cassette from cycle and insert plugs. (*Caution:* Do not invert cyclone while cassette is in place.)

(Adapted with permission from NATLSCO Environmental Sciences)

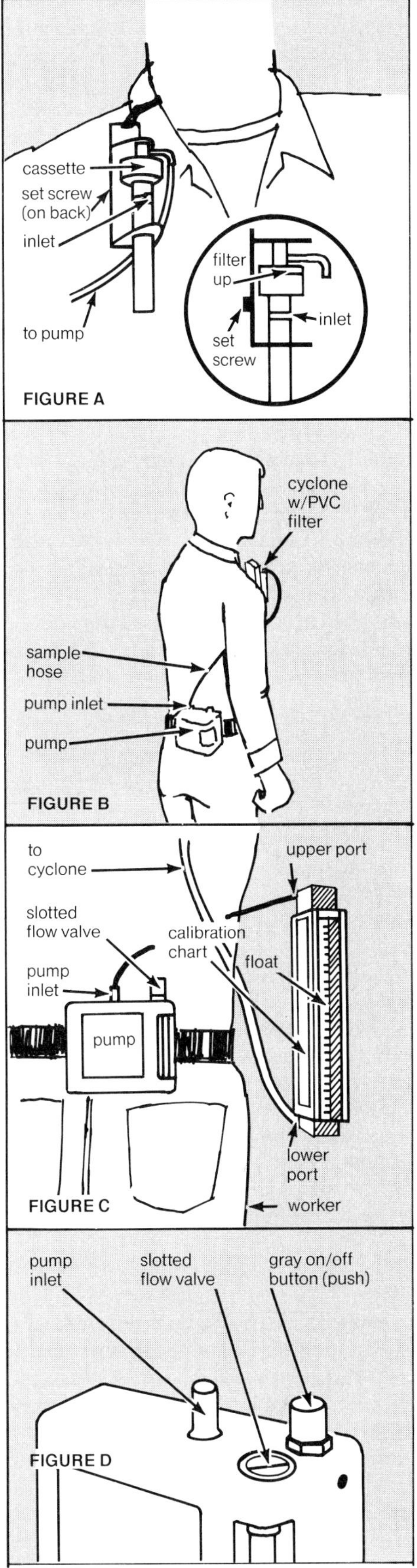

Figure 7-4. Collecting dust containing respirable free silica.

inhalation of fumes from magnesium oxide, copper oxide, and other metallic oxides have also been reported. The condition does not occur from the handling of these oxides in powder form. Apparently, the condition only results from the inhalation of extremely fine particles that have been freshly formed as fume (nascent fume.) Nickel, mercury, and other metals may also produce a fever as a result of the toxic effects of the element.

Welding fumes

The American Welding Society has published an extensive study called *Fumes and Gases in the Welding Environment* (see Bibliography) in which are identified the possible constituents of welding fumes.

Welding fumes cannot be classified simply. The composition and quantity of welding fumes depend on the alloy being welded and the process and electrodes being used. Reliable analysis of fumes cannot be made without considering the nature of the welding process and the system being examined. Reactive metals and alloys such as aluminum and titanium are arc welded in a protective, inert atmosphere such as argon. Although these arcs create relatively little fume, they do produce an intense radiation that can produce ozone.

Similar processes are used to arc weld steels, which also produces a relatively low level of fumes. Ferrous alloys are also arc welded in oxidizing environments that generate considerable fume and that can produce carbon monoxide. Such fumes generally are composed of discrete particles of amorphous slags containing iron, manganese, silicon, and other metallic constituents depending on the alloy system involved.

Chromium and nickel compounds are found in fumes when stainless steels are arc welded. Some coated and flux-cored electrodes are formulated with fluorides, and the fumes associated with them can contain significantly more fluorides than oxides.

Because of these factors, arc-welding fumes frequently must be tested for the presence of likely individual constituents to determine whether specific TLVs have been exceeded. For example, stainless steel metal fumes can contain a particularly hazardous form of chrome compound known as hexavalent chrome (sometimes known as Cr^{VI}). Some hexavalent chrome compounds (particularly the water-insoluble ones) are carcinogenic. The TLVs for chrome and chrome compounds vary depending on the type of compound present. It is important when arranging for an analysis of a chrome-containing fume that a test for both hexavalent as well as total chrome be obtained.

Conclusions based on total fume concentration are generally adequate if no toxic elements are present in welding rods, metal, or metal coatings, and conditions are not conducive to the formation of toxic gases.

Radioactive dusts

A radioactive contaminant may pose a chemical toxicity hazard in addition to an ionizing radiation exposure, and it may be present as a gas, dust, fume, or mist. For example, uranium can be a radioactive hazard and can cause kidney damage similar to that produced by other heavy metals such as lead.

Radioactive contaminants taken into the body may be deposited in various organs where they constitute sources of internal radiation. The chemical characteristics of the radioactive contaminant or isotope determine the organ in which it will be deposited. If a radioisotope has been deposited in the body, the internal exposure is regarded as continuous until the isotope is lost as a result of radiological or biological decay.

Since radioisotopes are selectively deposited in individual organs, they may cause only local irradiation. Solubility and particle size determine how much of the active material will gain access to and remain in the blood stream and in various organs.

If radioactive airborne contamination is known to be present, control measures are essential to prevent inhalation. Access to such areas must be restricted by law to admit only trained personnel equipped with radiation-monitoring dosimeters. As a minimal precaution, the use of respirators with high efficiency (HEPA) filters approved for radioactive particulate matter is required.

If the presence of contamination is unknown but suspected, determine whether or not the airborne concentrations of the radioisotope are below the limits established by the Nuclear Regulatory Commission (10 *CFR* 20, Appendix B).

Good housekeeping, personal hygiene, and good operating techniques are much more important in the safe handling of radioactive materials than in the handling of most other materials used in industry.

Engineering controls for radioactive dusts are similar to those for other dusts and depend primarily on capture at the point of generation. The difference lies in the fact that controls for radioactive dusts must be extremely efficient. Exposure limits for radioactive particulate matter are very low, and in some cases 100 percent efficiency in capture and retention is required. (Additional details are given in Chapter 10, Ionizing Radiation.)

Nuisance dusts

Even though a dust may be considered generally innocuous and not be recognized as the direct cause of a serious pathological condition, its level should be kept as low as possible. A concentration of 10 mg/m^3 of total dust (or 5 mg/m^3 respirable dust) that contains less than 1 percent silicon dioxide is suggested as the threshold limit for a number of nuisance dusts. Examples of particulate material generally considered to be nuisance dusts are listed in the ACGIH TLV booklet. When good engineering practices are employed, this level will not be exceeded. Any reduction below this level, however, may increase the comfort of employees and improve plant housekeeping.

Dust-control methods

The methods used to control industrial dust exposure are numerous and varied, and their application involves extensive technical knowledge.

The handling of dust-laden air in ventilating processes involves principles that must take into account the weight of the dust in the air and the separation of the dust from the air before it is exhausted outside the building. This can require the use of expensive and complicated equipment.

Exhausting the air without dust removal can create a dust load in the outside area that may contaminate the plant on reentry or may create neighborhood problems. (See Chapters 21 and 22 on industrial and general ventilation for a description of some of the standard methods of dust control; these include enclosure, local exhaust, general ventilation with dilution, moisture control, and the use of personal protective devices.)

The surest and most positive dust-control method is a total enclosure of the dust-producing process that includes an exhaust of the enclosure to maintain a negative pressure within the entire enclosure. This is frequently impractical, although certain pieces of equipment can be enclosed. In other instances, a total enclosure with only feed or hopper openings can be used with a sufficient exhaust system to make sure air moves into these openings at all times.

Large equipment that does not require constant attention can be enclosed in buildings with separate exhausts; workers can wear personal protective equipment when they must enter these buildings. Consideration should be given to prevent dust explosions in such operations.

Local exhaust

Local exhaust ventilation is frequently used at points of high dust production; when it can be combined with a hooded enclosure, such ventilation can be quite effective. In these instances, the enclosure should be designed with as small an opening as practical so that workers can stand outside of the hood. Generally, hooded enclosures that do not extend to the floor or the work bench, or those with more than one open side, require large amounts of air to be effective.

The ventilated enclosure of buffing and grinding wheels illustrates an application of hood design that must take into account the mechanical generation of dust (Figure 7-5). The ventilation of the wheel enclosure would be totally ineffective in collecting the high-speed particles that emerge from the point of grinding contact if they were not directed into the exhaust system (Figure 7-6). From a practical standpoint, no amount of air movement supplied by the suction applied to the hood could deflect the high-speed particles. The dust generated by the wheel is, in effect, collected by a hood designed so that mechanically generated particles are projected into it.

Exhaust slots must sometimes be used because of operational requirements. In these applications, the slots will not be completely effective if the dust is generated more than one foot from the slot or if the method of generation involves a mechanical agitation of the dust. Slots will be somewhat more effective if they are set in baffle plates or in other applications that direct air collection. Generally, exhaust slots require high suction pressures, which are expensive to maintain. In some applications of exhaust slots, the push-pull principle of supplying air from one side and directing it toward the slot on the opposite side may be helpful. However, in this application, the balance of air volumes and the dispersing effects of supplied air must be carefully considered.

General ventilation with dilution

In instances in which the sources of dust generation are numerous, widely distributed (not individually intense) general or dilution ventilation may be the best solution. But such a method can be both expensive and ineffective if exhaust locations are not properly placed, and if the ventilation is applied without adequately balancing the exhaust with well-placed, heated supplied air.

All air that is exhausted from a plant must be supplied from some source. Because the incoming air usually reaches plant air temperatures before it is exhausted, it is normally more economical—and almost always more effective—to supply the working area with heated makeup air by mechanical means than it is to rely on plant leakage to make up the supply of exhausted air.

For some reason, this principle seems to be the least understood of all factors involved in industrial ventilation. Heat and fuel are not conserved by exhausting the warmest air in the plant from the roof of the building and by expecting makeup air to enter the plant through windows and doors at "no cost." Even in warm weather, when it is desirable to remove warm air from the upper strata for comfort control, a greater degree of comfort will be attained if outside air is supplied at the level (position) of the worker. (See Chapters 21 and 22 for more information on ventilation.)

Figure 7-5. Typical in-plant installations show flexible ducts exhausting dust and gases from machinery section.

Recirculation of air

Recirculation of "clean air" from a dust collector for the purpose of conserving energy requires careful consideration. The PEL for nuisance dust (that is, dust containing less than 1 percent quartz) is 10 mg/m^3. Even a very well-maintained fabric filter dust collector operating under ideal conditions cannot remove all dust particles; that is, it cannot operate at 100 percent efficiency.

The percentage of respirable quartz determines the numerical value of the PEL according to the following formula:

$$\text{PEL} = \frac{10 \text{ mg/m}^3}{\% \text{ quartz} + 2}$$

Therefore, when a quartz level of 1 percent is present, the PEL is 3.33 mg/m^3. At a quartz level of 2 percent, it drops to 2.5 mg/m^3. Compare these figures with a "clean air" discharge from

Figure 7-6. Because the swing-frame grinder is used in many positions, local exhaust system must be adjustable. Flexible duct A permits movement of exhaust hood B as needed. (Courtesy American Foundrymen's Society)

a dust collector and you will understand the potential problems associated with recirculating air from a dust collector.

When recirculation is being performed, remember that gases such as carbon monoxide are still present at the same concentration as when they entered the collector.

Moisture control

When pulverized materials are handled, the amount of dust that will be dispersed by any given mechanical operation will vary with the material's moisture content. Wet drilling and grinding are typical examples of this application that involve excessive moisture. Foundry molding in which moist sand is used is an example of moisture control required by the process. In other applications, careful control of permissible moisture may well reduce dust generation by as much as 75 percent.

Respirator usage

Since most dusts are hazardous to the lungs, respirators are a common method of primary or secondary protection. Respirators are appropriate as a primary control during intermittent maintenance or cleaning activities when fixed engineering controls may not be feasible. Respirators can also be used as a supplement to good engineering and work practice controls for dusts to increase employee protection and comfort.

To be effective, respirators must be carefully matched to the type of particulate hazard present. The critical exposure factors that determine the degree of hazard (type of dust, length of exposure, dust concentration, and particle size) are also used to determine the type of respirator that should be employed. For example, metal fumes are extremely small particles. A respirator approved by the NIOSH for dusts and mists will not protect against these fumes. Only respirators approved by the NIOSH for fumes will be capable of filtering out these small particles (see Chapter 23, Respiratory Protective Equipment).

Very high concentrations of dust can present both a respiratory and eye irritation hazard. Full-face respirators, which cover the eyes, nose, mouth, and chin, will protect the eyes as well as provide a higher degree of respiratory protection than a half-mask because a full-face respirator maintains a better fit.

It is important to consider how long an individual is required to wear a respirator. Wearing a respirator for prolonged periods can be uncomfortable as well as physically stressful. If exposure time is prolonged and the respirator provided is not very comfortable, the wearer may decide to forego wearing the respirator or remove it for brief periods. Repeated brief breakings of the respirator-to-face seal can greatly increase the wearer's overall exposure. This can be a critical problem when exposure to highly toxic dusts is involved. Reducing the respirator wearing time through job redesign, allowing the wearer to select a comfortable as well as adequate respirator, instructing the wearer on the importance of continuously maintaining an adequate respirator-to-face seal, and monitoring of respirator use by supervisors are ways to increase worker compliance.

The OSHA regulations for some particulate matter (particularly lead and asbestos) have specific requirements on respirators. These regulations describe what types of respirators are appropriate for various levels of exposure, as well as how and when respirators are permitted to be used as part of a comprehensive approach to controlling exposures. These regulations must be consulted before respirators are used. Respirator use is also covered under the general OSHA respirator provision in 29 *CFR* 1910.134. These regulations apply to the use of all respirators and specify mandatory requirements in a minimally acceptable respirator program.

COMBUSTIBLE DUSTS

Finely divided combustible airborne materials can be hazardous. Deposits of combustible dusts on beams, machinery, and other surfaces are subject to flash fires. When combustible dusts suspended in air are ignited, they can cause severe explosions. The hazard of any given dust is related to the ease with which it is ignited and the severity of the ensuing explosion. Most of the material in this section is adapted from the *Fire Protection Handbook* (see Bibliography).

A dust cloud might be formed by a sudden disturbance resulting from repair work, a small gas ignition, or some other cause. Or the cloud might be an inherent part of the manufacturing process, as is the case within a grinding mill or a pneumatic conveying system. The ignition sensitivity of a dust cloud depends on the ignition temperature and the minimum energy required for ignition.

Safety and health professionals should be familiar with both dust explosion hazards and health hazards. The TLV for total dust is 10 mg/m^3. Since the explosive limit for many dusts is around 0.1 ounce per cubic foot of air, or 100,000 mg/m^3, the minimum explosive concentration is roughly 10,000 times the health limit.

In a dust cloud just as in a gas-air mixture, ignition of one part of the cloud will be propagated throughout the entire mixture and develop into an explosion only when the concentration of the dust in the air is between certain lower and upper explosive limits. The minimum explosive concentration, or lower limit of explosibility, is that concentration at which there is just barely enough dust in the air to propagate flame throughout the cloud after ignition at a localized point (Table 7-E).

A dust explosion is essentially the very rapid combustion of a dust cloud or suspension of dust in air, during which heat is generated at a very high rate. Any combustible solid material in finely divided form might produce a dust explosion if it is thrown into suspension in air and ignited. The conditions necessary for the explosion are a sufficiently dense dust cloud, adequate oxygen or air to support combustion, and an ignition source intense enough and in contact with the dust long enough to raise the temperature of part of the dust mixture to the ignition point.

Dust explosions usually occur as a series. Frequently, the initial explosion is small but intense enough to jar dust from beams, ledges, and other resting places. The initial explosion may even rupture small pieces of equipment within buildings, such as dust collectors or bins. This creates a much larger dust cloud through which a second explosion can propagate. It is not unusual to have a series of explosions propagating from room to room or from building to building.

Factors to be considered

A dust cloud will ignite if the particle size, dust concentration, impurities present, oxygen concentration, and the strength of the source of ignition are within certain limits. Table 7-E gives the maximum explosion pressure, the average and maximum rate of pressure rise, ignition temperature, minimum ignition energy of a dust cloud, minimum explosion concentration, and the limiting oxygen concentration in a spark ignition chamber.

Particle size. The smaller the size of the dust particle, the easier it is to ignite the dust cloud. The exposed surface area of a unit weight of material increases as the particle size decreases. The particle size also has an effect on the rate of pressure rise. At a given weight concentration of dust, a coarse dust will show a lower rate of pressure rise than a fine dust. The lower explosive limit concentration, ignition temperature, and the energy necessary for ignition will decrease as dust particle size decreases.

The likelihood that an explosive dust cloud will form depends on the dispersibility of the particles and on their ability to remain in suspension for any length of time. The fineness of dust particles has a vital effect on explosibility. Fine dust can be thrown into suspension more readily, it disperses more uniformly, and it remains in suspension longer than coarse dust; therefore, chances are greater that explosions will occur from fine dust.

Concentration. It is customary to express the concentration figures in terms of weight per unit volume, though, if the particle size distribution of the sample is not known, this expression is meaningless. The minimum explosive concentration is expressed in ounces per cubic foot of air. It is interesting to note: 1 ounce per cubic foot=one gram per liter=1,000 milligrams per liter=1,000,000 milligrams per cubic meter. The reason it works out that way is because: one ounce=28.35 grams and one cubic foot=28.32 liters.

Inert material. The presence of an inert solid reduces the combustibility of the airborne dust. The amount of inert dust necessary to prevent an explosion is usually considerably higher than concentrations that would normally be found.

Moisture in dust particles raises the ignition temperature of the dust because heat is absorbed during heating and vaporization of the moisture. The moisture in the air surrounding a dust particle has no significant effect on the course of an explosion once ignition has occurred. There is, however, a direct relationship between moisture content and minimum energy required for ignition, minimum explosive concentration, maximum pressure, and maximum rate of pressure rise.

Oxygen concentration. The percentage of oxygen in the atmosphere affects the flammability of a dust cloud. In an atmosphere of 100 percent oxygen, dust clouds ignite at lower temperatures, the minimum ignition energy is smaller, and the minimum explosive concentration is lower than in air. When the partial pressure of oxygen is decreased, the energy required for ignition is reduced, ignition temperature increases, and the maximum explosion pressure decreases. The type of inert gas used as the diluent in the reduction of the oxygen concentration also has an effect on the combustibility of the dust.

TABLE 7-E
EXPLOSION CHARACTERISTICS OF VARIOUS DUSTS

Type of dust	Ignition temp of dust cloud (deg C)	Min spark energy required for ignition of dust cloud (millijoules)	Min explosive concentration (oz per 1,000 cu ft)	Max explosion pressure (psi)	Rates of pressure rise (psi per sec)		Limiting oxygen percentage to prevent ignition of dust clouds by electric sparks
					Avg	Max	
Metal powders							
Aluminum, atomized	640	15	40	90	3,500	10,000+	7
Aluminum, milled	550	...	45	70	2,000	4,250	
Aluminum, stamped	550	10	35	100	10,000	10,000+	4
Boron (85% B, 8% Mg)	470	60	135	90	900	2,500	
Iron, carbonyl	320	20	105	50	1,500	7,000	10
Iron, electrolytic	320	240	200	45	500	1,000	13
Iron, hydrogen reduced	315	80	120	45	800	1,750	13
Magnesium, atomized	600	120	10	80	2,000	5,250	3
Magnesium, milled	520	40	20	95	3,000	10,000+	†
Magnesium, stamped	520	20	20	80	3,400	10,000+	†
Manganese	450	80	125	50	1,300	2,750	15
Silicon	775	80	100	105	2,000	10,000	13
Thorium	270	5	75	50	1,400	3,250	†
Thorium hydride	260	3	80	60	2,100	6,750	6
Tin	630	160	190	35	500	1,250	16
Titanium	330	10	45	80	3,400	10,000+	†
Titanium hydride	440	60	70	95	3,800	10,000+	13
Uranium	*	45	60	55	1,600	3,500	†
Uranium hydride	*	5	60	45	2,900	6,250	0.5
Vanadium	500	60	220	35‡	200	300	13
Zinc	600	650	480	50	600	1,750	10
Zirconium	*	5	40	65	800	8,750	†
Zirconium hydride	350	60	85	60	2,400	8,750	11
Aluminum-cobalt alloy (60-40)	950	100	100	80	2,500	8,500	
Aluminum-copper alloy (50-50)	950	100	100	70	800	2,500	
Aluminum-nickel alloy (60-40)	960	80	190	80	2,600	10,000+	
Calcium-silicon alloy	540	220	600	75	400	10,000+	
Dowmetal	430	80	20	85	3,600	10,000+	†
Ferromanganese (1.4% C)	450	80	130	45	1,400	4,250	
Ferrosilicon (80% Si)	860	280	400	90	1,500	3,600	19
Ferrotitanium, low-carbon	370	80	140	55	2,200	9,500	13
Magnesium-aluminum alloy (50-50)	535	80	50	90	4,000	10,000+	†
Plastics							
Allyl alcohol resin	500	20	35	105	2,800	10,000+	
Butadiene-styrene resin	440	60	25	80	1,400	4,000	
Cellulose acetates	320	10	25	110	2,800	6,750	11
Cellulose propionate	460	45	25	105	1,600	4,750	
Coumarone-indene resin	520	10	15	85	2,800	8,500	14
Dimethyl terephthalate	570	20	30	90	3,100	10,000	
Gums (arabic, copal, etc.)	360	30	30	95	1,500	5,000	14
Lignin resin	450	20	40	80	1,700	4,750	17
Methyl cellulose	360	20	30	100	1,900	6,000	
Methyl methacrylate	440	15	20	100	500	1,750	14
Phenolic resins	460	10	25	80	1,700	6,000	14
Pine-rosin base resin	440	...	55	80	1,900	7,500	
Polyacrylonitrile	500	20	25	90	2,000	5,000	
Polyamide	500	20	30	90	1,800	7,000	
Polyster resin-glass fiber mixture (65–35)	440	50	45	85	2,200	6,000	
Polyether alcohol resin	460	160	45	65	500	1,000	
Polyethylene resin	410	30	20	80	1,500	3,500	13
Polyethylene terephthalate	500	35	40	90	1,600	7,500	13
Polystyrene	490	15	15	90	2,400	7,000	14
Polyvinyl acetate resin	520	120	35	75	1,200	3,000	
Rubber, synthetic hard	320	30	30	95	1,100	3,000	15
Shellac	400	10	20	75	1,400	3,500	14
Styrene-maleic anhydride copolymer	470	20	30	80	2,300	9,500	
Urea resin	450	80	70	85	800	2,000	17
Vinyl butyral resin	390	10	20	60‡	500	1,000	14
Vinyl chloride-acrylonitrile polymer	530	15	35	85	1,700	4,500	
Vinyl copolymer resin	500	60	100	40	200	500	
Agricultural products							
Alfalfa	460	320	100	65	500	1,000	
Cellucotton	440	60	50	100	900	3,000	
Cinnamon	440	40	40	115	1,400	4,000	
Citrus peel, dehydrated	490	45	60	100	1,200	3,000	
Clover seed	470	80	60	60	400	1,000	15
Cocoa	420	100	45	62‡	550	1,200	
Coffee	410	160	85	50	150	250	13
Corncob meal	400	60	30	120	1,200	3,750	
Cornstarch	380	30	40	110	2,200	6,750	10
Cotton seed	470	80	55	90	800	2,500	15
Dextrin, corn	400	40	40	105	1,800	7,000	

TABLE 7-E *(Concluded)*

EXPLOSION CHARACTERISTICS OF VARIOUS DUSTS

Type of dust	*Ignition temp of dust cloud (deg C)*	*Min spark energy required for ignition of dust cloud (millijoules)*	*Min explosive concentration (oz per 1,000 cu ft)*	*Max explosion pressure (psi)*	*Rates of pressure rise (psi per sec)*		*Limiting oxygen percentage to prevent ignition of dust cloud by electric sparks*
					Avg	*Max*	
Furfural residue	440	40	40	105	1,400	4,000	
Grain dust	430	30	55	95	1,000	2,750	
Guar seed	500	60	40	105	1,400	4,750	
Lycopodium	480	40	25	85	2,300	7,000	13
Nut shells	420	50	30	105	1,900	4,000	
Onion, dehydrated	410	...	130	60‡	400	1,250	
Pea, dehydrated	560	40	50	100	2,100	6,000	
Pectin	420	35	75	110	1,800	8,000	
Potato starch	440	25	45	95	2,300	8,000	
Pyrethrum	480	80	100	80	600	1,500	
Rice	440	40	45	95	1,000	2,750	
Soybean	520	50	35	100	1,200	3,250	15
Sugar	350	30	35	90	1,600	5,000	
Tung	440	240	70	110	1,400	3,500	
Wheat dust	470	50	70	105	1,500	3,500	
Wheat flour	380	50	50	95	1,200	3,750	
Yeast	520	50	50	105	1,000	2,500	
Miscellaneous							
Adipic acid	550	70	35	75	1,200	2,750	
Aluminum stearate	400	15	15	95	1,200	4,750	
Aspirin	660	25	35	85	2,000	10,000+	
Bark dust (Douglas-fir)	540	40	30	90	2,900	9,500	
Beryllium acetate	620	100	80	80	600	2,000	17
Calcium lignin sulfonic acid	590	100	160	80	600	2,000	
Carbon, activated	660	...	...	40‡	200	300	
Casein, rennet	520	60	45	65	400	1,000	17
Cellulose	480	80	55	100	1,100	2,750	13
Charcoal (pine wood)	620	...	...	40‡	200	250	
Coal, low volatile	635	...	...	45	300	600	
Coal, medium volatile	605	120	120	60	300	600	18
Coal, high volatile (Pgh. seam)	610	60	55	85	800	2,250	16
Coal, subbituminous	455	60	45	95	1,200	3,000	
Cork	470	45	35	100	2,000	5,500	
Diazoaminobenzine	550	20	15	90	2,900	10,000+	
Dinitro-ortho-cresol	440	80	25	55‡	1,300	2,250	15
Diphenyl	650	60	35	55	400	1,500	
Gilsonite	560	25	20	90	1,200	3,750	
Hexamethylenetetramine	410	10	15	100	2,400	10,000+	14
Lactalbumin	570	50	40	90	900	2,750	13
Lignite	440	60	45	90	800	2,750	15
Liver protein	520	45	45	80	800	2,250	
Napalm	450	40	20	85	1,000	3,000	12
Paraformaldehyde	410	20	40	100	2,500	10,000+	
Peat, sphagnum	460	50	45	85	900	2,250	
Pentaerythritol	450	10	30	90	1,700	9,500	14
Phenothiazine	540	...	15	80	1,400	4,250	16
Phthalic anhydride	650	15	15	70	1,300	4,250	14
Phytosterol	330	10	25	75	1,500	8,000	
Pitch, coal tar (58% vol. matter)	710	20	35	95	1,900	6,000	15
Procaine penicillin	450	...	25	50‡	1,000	2,000	
Rubber, crude, hard	350	50	25	80	1,200	3,800	15
Secobartital sodium	520	95	105	55	250	500	
Soap	430	60	45	85	600	1,750	
Sodium alkylarylsulfonate	540	...	130	75‡	400	1,250	
Sodium benzoate	560	80	55	85	1,800	10,000+	
Sodium carboxymethyl cellulose	350	560	150	60	300	600	
Sorbic acid	470	15	25	90	3,000	10,000+	
Sulfur	190	15	35	80	1,700	4,750	11
Vitamin B_2	500	80	105	80	1,000	2,250	
Wood flour	430	20	40	110	1,600	5,500	17

From Standard Handbook for Mechanical Engineers, *rev. 7th ed. Edited by T. Baumeister. (Copyright 1967. McGraw-Hill Book Co.) Used by permission.*

When uranium, uranium hydride, and zirconium were dispersed into air at room temperature, the dust clouds ignited under some conditions.

†The oxygen reduction data in this table are based on tests made in air-CO_2 mixtures. Dust clouds of thorium, titanium, uranium, zirconium, Dowmetal, and certain magnesium and magnesium-alloy powders ignited in pure CO_2.

‡Pressure and rates of pressure rise for these dusts were measured by an older testing technique.

(Reprinted with permission from Baumeister, T, ed. *Standard Handbook for Mechanical Engineers,* rev. 7th ed. New York: McGraw Hill Book Co., 1967)

Ignition source. Dust clouds that have been in contact with an ignition source for some time can be ignited at lower temperatures than dust clouds that have instantaneous contact with the heat source.

Rate of pressure rise. The rate of pressure rise may be defined as the ratio of the increase in explosion pressure to the time interval of that increase. It is the most important single factor in evaluating the hazard of a dust and principally determines the degree of an explosion's destructiveness. The rate of pressure rise is also an important consideration in the design of explosion vents, since it largely determines the size of the vent.

When a dust explosion occurs, gaseous products are generally formed and heat is released, which raises the temperature of the air in the enclosure. Since gases expand when heated, destructive pressures will be exerted on the surrounding enclosure unless enough vent area is provided to release the hot gases before dangerous pressures are reached.

Prevention of dust explosions. Explosions can be prevented by the application of dust-tight construction, isolation of combustible dusts from sources of ignition, limitation of unauthorized use of flame- or spark-producing equipment, introduction of an inert gas, an adequate exhaust system, and good housekeeping.

BIBLIOGRAPHY

American Conference of Governmental Industrial Hygienists (ACGIH). *Air-Sampling Instruments for Evaluation of Atmospheric Contaminants,* 6th ed. Cincinnati: ACGIH, 1983.

—— *Threshold Limit Values and Biological Exposure Indices for 1986–1987.* Cincinnati: ACGIH, 1986.

—— Committee on Industrial Ventilation. *Industrial Ventilation—A Manual of Recommended Practice,* 19th ed. Lansing, MI: ACGIH, 1986.

American Welding Society (AWS). *Fumes and Gases in the Welding Environment.* Miami, FL: AWS, 1979.

Cralley, L., and Cralley, L. *Patty's Industrial Hygiene Toxicology: The Work Environment,* Vol. 3A, 2nd ed. New York: Interscience Publications, John Wiley & Sons, Inc., 1985.

Key, M.M., et al. *Occupational Diseases—A Guide to Their Recognition,* rev. ed. (DHEW Publication No. 77-181). Washington, DC: U.S. Government Printing Office, 1977.

Kusnetz, S., and Hutchinson M.K. *A Guide to the Work-Relatedness of Disease* (DHEW [NIOSH] Publication No. 79-116). Washington, DC: U.S. Government Printing Office.

Schwab, R.F. "Dusts." In *The Fire Protection Handbook,* 16th ed., edited by A.E. Cote and J.L. Linville. Quincy, MA: National Fire Protection Association, 1986.

U.S. Department of Health, Education, and Welfare. *Criteria for a Recommended Standard for Occupational Exposure to Crystalline Silica* (HEW [NIOSH] Publication No. 75-120), 1974.

Zenz, C., ed. *Occupational Medicine—Principles and Practical Applications.* 2nd ed. Chicago: Year Book Medical Publ., 1988.

Industrial Dermatoses

by Larry L. Hipp, MD
Revised by James S. Taylor, MD

INCIDENCE, PREVALENCE, AND CAUSES OF OCCUPATIONAL SKIN DISEASE

Skin diseases caused by substances or conditions in the workplace are the most frequently encountered occupational illness. Occupational skin diseases can occur in workers of all ages, in any work setting, and cause a great deal of illness, personal misery, and reduced productivity. Although the frequency of occupational skin disease often parallels the level of hygiene practiced by employers, occupational skin diseases are largely preventable. Many consider this type of disease trivial and insignificant, but occupational skin disorders can result in complex impairment. Data compiled by the Bureau of Labor Statistics (BLS) for 1983 indicate approximately 98 percent (4,748,000) of all occupational disorders are injuries and a little more than 2 percent (106,000) are diseases. Because large surface areas of skin are often directly exposed to the environment, the skin is particularly vulnerable to occupational insults. While complete data on the extent and cost of dermatological injuries are not available, the National Institute for Occupational Safety and Health (NIOSH) estimated in 1986 that skin injuries may account for 23–35 percent of all injuries. An estimated 1–1.65 million dermatological injuries may occur annually, with an estimated annual rate of skin injury of 1.4–2.2 per 100 full-time workers. The highest percentage is due to lacerations and punctures (82 percent) followed by burns (chemical and other—14 percent) (Table 8-A).

Table 8-A. Occupational Dermatological Injuries*—United States, 1983

Type of Injury	*No.*	*(%)*
Lacerations and punctures	253,141	(82.3)
Burns (nonchemical)	36,477	(11.9)
Abrasions	10,576	(3.4)
Burns (chemical)	6,828	(2.2)
Cold injuries	566	(0.2)
Radiation injuries	135	(0.04)
Total	307,723	(100.0)

*Reported by the Supplementary Data System of the Bureau of Labor Statistics from 29 participating states.

In the mid-1950s, skin disorders other than injuries accounted for 50–70 percent of all occupational diseases. This figure has been gradually decreasing and was 37 percent or 39,540 cases in 1983. The NIOSH attributes the decline of skin diseases during the past 30 years to a continuing trend toward automation, enclosure of industrial process, and educational efforts. Despite these figures, dermatitis is still the major cause of reported occupational disease in the United States. National data indicate as many as 20–25 percent of all occupational skin disease can involve lost time from work, with an average of 11 days lost per case involving lost workdays. California and South Carolina have reported similar data based on workers' compensation claims. The results of two studies show a serious underreporting of occupational disease of all types, which may mean the true incidence is anywhere from 10–50 times greater than that reported by the U.S. BLS.

The NIOSH has included work-related dermatological conditions on its list of 10 leading work-related diseases and injuries in the United States (Table 8-B). Reasons include the fact that 10–15 percent of requests that NIOSH receives for health hazard evaluation involve skin complaints, and because the economical impact of dermatological conditions is substantial. The annual cost resulting from lost worker productivity, medical care, and disability payments can range between $222 million and $1 billion.

Table 8-B. The 10 Leading Work-Related Diseases and Injuries—United States, 1982*†

Occupational lung diseases: asbestosis, byssinosis, silicosis, coal workers' pneumoconiosis, lung cancer, occupational asthma

Musculoskeletal injuries: disorders of the back, trunk, upper extremity, neck, lower extremity, traumatically induced Raynaud's phenomenon

Occupational cancers (other than lung): leukemia, mesothelioma, cancers of the bladder, nose, and liver

Severe occupational traumatic injuries: amputations, fractures, eye loss, lacerations, and traumatic deaths

Cardiovascular diseases: hypertension, coronary artery disease, acute myocardial infarction

Disorders of reproduction: infertility, spontaneous abortion, teratogenesis

Neurotoxic disorders: peripheral neuropathy, toxic encephalitis, psychoses, extreme personality changes (exposure-related)

Noise-induced loss of hearing

Dermatological conditions: dermatoses, burns (scaldings), chemical burns, contusions (abrasions)

Psychological disorders: neuroses, personality disorders, alcoholism, drug dependency

*The conditions listed under each category are to be viewed as selected examples, not comprehensive definitions of the category, and are *not* in order of incidence or importance.
†NIOSH has recently developed a suggested list of the 10 leading work-related diseases and injuries (Table 8-B). Three criteria were used to develop the list: (1) the frequency of occurrence of the disease or injury; (2) its severity in the individual case; and (3) its amenability to prevention.

Tables 8-C–8-F provide other statistical data on occupational skin disorders. Table 8-C gives incidence of occupational dermatoses (disease) by industry group for the United States in 1984. The highest incidence rate was in agriculture—28.5 cases per 10,000 full-time workers. Table 8-D lists causal agents for occupational dermatitis for the United States for 1976, based on limited data from the U.S. Labor Department. Botanical agents (plants, trees, vegetables) and chemicals lead the list. Table 8-E lists data on the U.S. industries with the highest risk for dermatitis combining all risk components. Poultry dressing plants had the greatest number of lost workdays. Table 8-F lists the 10 most hazardous industrial processes for skin disorders as determined by the experience of the OSHA Advisory Committee on Cutaneous Hazards in 1978.

A dermatosis is any abnormal condition of the skin, ranging from the mildest redness, itching, or scaling to an eczematous (superficial inflammation), ulcerative (ulcer-forming), acneiform (resembling acne), pigmentary (abnormal skin color), granulomatous (tumorlike mass, nodule), or neoplastic (new, abnormal tissue growth) disorder. An occupational dermatosis includes any skin abnormality resulting directly from, or aggravated by, the work environment. Dermatitis is a more limited term, referring to any inflammation of the skin, such as contact dermatitis or cement dermatitis.

Table 8-C. Cases and Incidence Rate of Occupational Dermatological Conditions in a Segment of Workers, by Major Industrial Divisions—United States, 1984*

Industrial Division	*No.*	*Incidence Rate†*
Agriculture/forestry/fishing	2,233	28.5
Manufacturing	23,017	12.3
Construction	2,456	6.6
Services	7,973	5.0
Transportation/utilities	2,114	4.3
Mining	393	4.0
Wholesale/retail trade	3,770	2.1
Finance/insurance/real estate	563	1.1

*Bureau of Labor Statistics Annual Survey.
†Per 10,000 full-time workers (2,000 employment hours/full-time worker/year).

Table 8-D. Causal Agents in Dermatitis: No. of Cases of Dermatitis by Source of Illness*

Agent	*No. Cases*	*% Total*
Plants, trees, vegetables	3,161	27.0
Chemicals	2,877	24.6
Miscellaneous	2,670	
Acids	95	
Alkalis	47	
Metallics	36	
Halogenated	29	
Soap and detergents	958	8.1
Coal and petroleum products	806	6.9
Miscellaneous	437	
Lubricating oils, grease	186	
Naphtha solvents	122	
Gasoline, liquid hydrocarbons	27	
Crude, fuel oil	24	
Hydrocarbon gases	10	
Food products	431	
Infectious, parasitic agents	375	3.2
Glass items	286	2.4
Mineral items (nonmetallic)	154	1.3
Metal items	114	1.0
Plastic items	125	1.1
Animal products	117	1.0
Liquids	95	0.8
Textile items	85	0.7
Drugs and medicines	84	0.7
Machines	68	0.6
Clothing	68	0.6
Animals, insects	57	0.5
Wood items	49	0.5
Hand tools (power and nonpower)	34	0.3
Flame, fire, smoke	21	0.2
Rubber products	3	—

*1976 Supplementary Data System (SDS) data—U.S. BLS. From the Report of the Advisory Committee on Cutaneous Hazards to the Assistant Secretary of Labor, OSHA, Dec. 19, 1978.

Anatomy, physiology, and direct causes

The anatomy and physiology of the skin as well as the direct causes of occupational skin disorders are discussed in detail in Chapter 3, The Skin.

Table 8-E. Highest Risk Industries for Occupational Skin Diseases or Disorders in Terms of Their Risk Components*

Hazard Rank	*Industry*	*I Incidence*	×	*P Population*	×	*S Severity*	×	*D Duration*	=	*R Lost Workdays*
1	Poultry dressing plants	16.4		89.8		0.30		10.0		4,418
2	Meat packing plants	7.2		164.3		0.31		4.3		1,577
3	Fabricated rubber products	5.5		103.2		0.22		11.5		1,436
4	Leather tanning and finishing	21.2		22.9		0.34		8.3		1,370
5	Ophthalmic goods	8.5		38.0		0.52		8.3		1,394
6	Plating and polishing	8.3		61.4		0.28		9.0		1,284
7	Frozen fruits and vegetables	7.2		43.2		0.31		12.1		1,167
8	Internal combustion engines	5.5		75.7		0.27		8.8		984
9	Canned and cured seafoods	5.6		19.7		0.36		23.7		941
10	Carburetors, pistons, rings, valves	7.0		29.4		0.24		17.9		884
11	Chemical preparations	8.3		36.7		0.23		12.3		861
12	Boat building and repairing	11.1		48.0		0.22		7.4		867
13	Fresh or frozen packaged fish	10.6		28.5		0.31		8.6		805
14	Paints and allied products	7.0		66.6		0.21		7.6		744
15	Electronic components	6.3		137.0		0.13		10.4		1,167
16	Noncurrent-carrying wiring devices	5.9		22.1		0.41		14.3		764
17	Screw machine products	8.2		43.4		0.17		11.4		690
18	Farm labor and management services	8.9		51.6		0.22		6.5		657
19	Beet sugar	9.0		15.9		0.52		8.4		625
20	Small arms	5.9		15.5		0.20		33.8		618
21	Horticultural specialties	8.1		33.0		0.21		9.5		533
22	Concrete products	5.6		63.1		0.21		7.0		519
23	Landscape and horticulture	9.2		104.2		0.14		3.4		456
24	Sporting and athletic goods	8.1		61.1		0.20		4.5		445
25	Power-driven hand tools	6.1		28.3		0.13		19.1		429
26	Electronic capacitors & resistors	15.2		30.6		0.20		13.9		1,293
27	Polishes and sanitation goods	8.4		28.9		0.23		7.7		430
28	Abrasive products	9.3		24.3		0.16		11.3		409
29	Ammunitions, except small arms	6.0		24.5		0.18		15.1		400
30	Silverware and plated ware	7.4		11.0		0.39		12.6		400
31	Storage batteries	7.5		25.3		0.25		0.2		9.49
32	Asbestos products	7.0		20.8		0.31		0.2		9.02
33	Nonferrous foundries	5.5		18.4		0.25		13.1		331
34	Cyclic crudes and intermediates	5.6		33.0		0.17		10.2		320
35	Poultry and egg processing	10.2		13.9		0.29		6.8		280

*Incidence based on cases/1,000; population values are in thousands; severity calculated as lost-workday cases/cases; duration calculated as lost workdays/lost-workday cases; risk=total lost workdays. (From the Report of the Advisory Committee on Cutaneous Hazards to the Assistant Secretary of Labor, OSHA, Dec. 19, 1978)

Table 8-F. Most Hazardous Industrial Processes for Skin Disorders*

Machine tool operations using cutting oils and coolants
Plastics manufacturing
Rubber manufacturing
Food processing
Leather tanning and finishing
Agriculture
Metal plating and cleaning
Construction
Printing
Forest products

*As determined by OSHA Standards Advisory Committee on Cutaneous Hazards, 1978.

PREDISPOSING FACTORS

In classifying and determining the severity of occupational dermatoses, consider a number of factors: the particular situation of exposure, the potential effects of the environmental agent, duration of exposure and extent of bodily exposure, and an agent's chemical stability and potential for being absorbed by or affecting the barrier layer of the skin. Other variables include preexisting skin disease or exposure to more than one agent.

Indirect or predisposing factors leading to the development of occupational dermatoses are generally associated with race, age, sex, skin type, perspiration, season of the year, personal hygiene, and allergy.

Racial characteristics

Redhead, blond, blue-eyed, light-complexioned persons are very susceptible to the acute and chronic effects of sunlight. The opposite is true of people with dark skin. Although darker-skinned people are believed to experience less reaction when handling certain chemicals (such as tar and pitch, which react with sunlight), dark skin is not universally resistant to chemical constituents in the industrial environment. Blacks are more susceptible to keloid (thickened scar) formation.

Age and experience

Younger, inexperienced, and possibly inadequately trained workers have a higher prevalence of occupational dermatoses than older workers. However, older workers can be more prone to chronic skin irritation, because their skin is generally drier.

Skin type

Workers with naturally dry skin cannot tolerate the action of solvents and detergents as well as persons with oily skin (Figure 8-1). However, those with oily skin can be predisposed to develop folliculitis and acne induced by cutting oils (Figure 8-2).

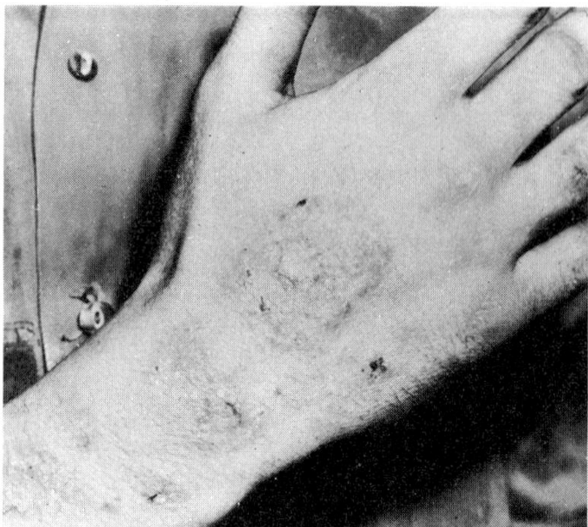

Figure 8-1. Cleaning hands with a strong petroleum solvent instead of with a good industrial cleanser resulted in this case of dermatitis.

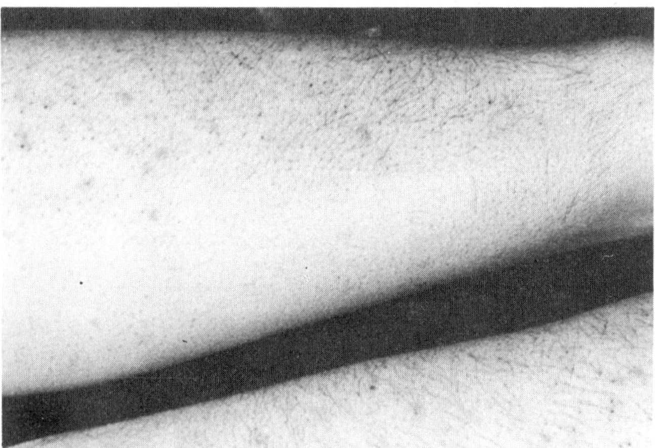

Figure 8-2. Examples of acneiform disorder shown on this worker's forearm, is frequently caused by exposure to cutting oils. Lack of splash guards and poor personal hygiene can be factors.

Perspiration

Hyperhidrosis (increased sweating) can produce maceration (softening and resultant separation) of skin already mildly irritated by rubbing (armpits, groin, etc.) in body areas favoring chemical, fungal, and bacterial action. Some materials, such as caustics, soda ash, and slaked lime become irritants in solution. Sweating, however, can also serve a protective function by diluting the toxic substances.

Sex

Because the incidence of nickel sensitivity (allergy) is much greater in women (due to ear-piercing), they are more susceptible to developing dermatitis in certain situations, for example, when handling coins (cashier) or when in contact with nickel salts, and metal alloys.

Seasons and humidity

Occupational dermatoses are more common in warm weather, when workers wear less clothing and are more likely to come in contact with external irritants. Excessive perspiration with resulting skin damage is also more common in warm weather. When a work area is hot, workers may not use protective clothing. Warm weather also means that many workers have greater exposure to sunlight, poisonous plants, and insects, the effects of which may or may not be related to the job.

Winter brings chapping from exposure to cold and wind. Heated rooms usually are low in relative humidity, which causes skin to lose moisture. Large-scale outbreaks of dermatitis in some factories has been traced to nothing more than low humidity. Clothing can keep dust particles and mechanical irritants in close contact with the skin. Infrequent bathing and changing of clothing can increase the incidence of skin irritation. (See Chapter 12, Temperature Extremes, for more information.)

Hereditary allergy (atopy)

This is the genetic tendency toward the development of atopic dermatitis, asthma, and hay fever. Atopic persons are predisposed to developing dermatitis on the basis of reduced skin threshold to chemical irritants, inherent dry skin, dysfunctional sweating, and a high skin colonization rate with the bacterium, *Staphylococcus aureus.* In an adult atopic person, the hand is the main location for dermatitis. Because contact dermatitis in nonatopic persons also frequently involves the hands, the extent that atopy influences the overall problem of contact dermatitis is not known.

Personal hygiene

This is believed to be a major factor causing occupational skin disorders. Unwashed skin covered with unwashed and unchanged clothes prolongs contact with chemicals. Responsibility for maintaining clean skin is shared by employer and employee, alike. Thus, adequate facilities for maintaining personal cleanliness should be provided in every place of employment. Educating workers in the preventive aspects of personal hygiene is imperative. Excessive skin cleansing with harsh agents can produce an irritant contact dermatitis or aggravate preexisting dermatitis.

Preexisting skin diseases

Other forms of skin irritation (eczema), such as nonoccupational contact dermatitis, palmar psoriasis, and lichen planus, can be aggravated by chemicals in the work environment. Ultraviolet light-sensitive disease, such as lupus erythematosus, and cold-induced disease, such as Raynaud's phenomenon, can be aggravated and precipitated by sunlight and cold exposure, respectively.

CLASSIFICATION OF OCCUPATIONAL SKIN DISEASE

Skin disorders are relatively easy to recognize, because they are visible. However, accurate diagnosis and classification of disease type and its relationship to employment usually require a high level of clinical skill and expertise. The varied nature of skin responses causing occupational skin disorders takes several forms. The appearance and pattern of the dermatosis infrequently indicates the provoking substance definitively, but can provide clues to the class of materials involved. Diagnosis depends on appearance and location and most importantly on the history. Preexisting skin disorders, adverse effects of treatment, and secondary infections add to the difficulty in diagnosis. The following grouping composes most occupational dermatoses.

Contact dermatitis

Contact dermatitis is the most frequent cause of occupational skin disease, accounting for most reported cases. Two types are generally recognized: irritant and allergic. Approximately 80 percent of all cases of occupational contact dermatitis result from primary irritants and 20 percent from allergic sensitizers. Both are difficult to differentiate clinically, since each can appear as an acute or chronic eczematous dermatitis. The acute form is erythematous (increased redness), vesicular (small blisters) to bullous (large vesicles), edematous (swollen), and oozing, and of short duration, lasting days or weeks. The chronic form is lichenified (thickened skin), scaly, and fissured, and usually lasts weeks, months, or years. Itching is usually a major symptom.

Contact dermatitis most often occurs on the hands, wrists, and forearms, although any area can be affected. Dusts, vapors, and mists can affect the exposed areas, including forehead, eyelids, face, ears, and neck, and frequently collect in areas where the body bends, for example, under the collar, and at the tops of shoes. The palms and soles are partially protected by a thick stratum corneum. The scalp tends to be protected by the hair; but the male genitalia are commonly affected, as an irritant is often transferred by the hands. Contact dermatitis also localizes under rings, and between fingers and toes and other cutaneous areas that rub together.

Irritant contact dermatitis. A primary skin irritant is a substance causing damage at the site of contact, because of its direct chemical or physical action on the skin. Irritants are generally divided into *strong (absolute)* and *marginal* types. Strong (absolute) irritants include strong acids, alkalis, aromatic amines, phosphorous, ethylene oxide, riot-control agents, and metallic salts, and produce an observable effect within minutes. In contrast, marginal irritants, such as soap and water, detergents, solvents, and oils can require days before clinical changes appear. Marginal irritants cause most cases of occupational irritant dermatitis and are a major skin problem in the workplace. (See Chapter 3, The Skin, for further discussion and lists of irritants and sensitizers).

Important factors to consider in irritant dermatitis are the nature of the substance: pH, solubility, physical state, concentration, duration of contact, and host and environmental factors. Despite the frequency of irritant dermatitis, much is unknown about the precise mechanisms of how irritants disturb the skin. Several points merit emphasis:

1. Contact dermatitis can occur from contact with several marginal irritants, the effects of which are cumulative.
2. Cumulative irritant contact dermatitis can lead to skin fatigue, a condition in which even mild substances can irritate the skin, or to "hardening," in which the skin eventually accommodates repeated exposure to an offending agent.
3. The clinical and histological differentiation of irritant and allergic contact dermatitis is often difficult or impossible.
4. Constant exposure to irritants impairs the barrier function of the skin and allows penetration of potential allergens.
5. Irritant and allergic contact dermatitis frequently coexists in the same patient.

Allergic contact dermatitis. A variety of industrial chemicals are potential contact allergens. The incidence of allergic contact dermatitis varies depending on the nature of the materials handled, predisposing factors, and the ability of the physician to accurately use and interpret patch tests. Allergic contact dermatitis, in contrast with primary irritation, is a form of cell-mediated, antigen-antibody immune reaction. Sensitizing agents differ from primary irritants in their mechanism of action and their effect on the skin. Unless they are concomitant irritants, most sensitizers do not produce a skin reaction on first contact. Following this sensitization phase of 1 week or longer, further contact with the same or a cross-reacting substance on the same or other parts of the body results in an acute dermatitis (elicitation phase).

Other essential points about allergic contact dermatitis include:

1. As a general rule, a key difference between primary irritation and allergic contact dermatitis is that an irritant usually affects many workers, whereas a sensitizer generally affects few. Exceptions exist with potent sensitizers, such as poison oak oleoresin or epoxy resin and components.
2. Differentiation of marginal irritants from skin allergens can be extremely difficult. Marginal irritants may require repeated or prolonged exposure before a dermatitis appears.
3. Allergic contact dermatitis may not develop for months or years after exposure to an agent.
4. Many skin sensitizers are also primary irritants, for example, chromates, nickel salts, epoxy resin hardeners.
5. However, sensitization (allergy) can be produced or maintained by allergens in minute amounts and in concentrations insufficient to irritate the nonallergic skin, for example, nickel, chromates, formaldehyde, turpentine.
6. Cross-sensitivity is an important phenomenon in which a worker sensitized to one chemical will also react to one or more closely related chemicals. A number of examples exist: Rhus antigens: poison oak, ivy, sumac, Japanese lacquer, mango, and cashew nutshell oil; aromatic amines: *p*-phenylenediamine, procaine, benzocaine, and *p*-aminobenzoic acid (sunscreens); perfume or flavoring agents: balsum of Peru, benzoin, cinnamates, and vanilla.
7. Systemic eczematization is a widespread, eczematous, contactlike dermatitis that can result from oral parenteral (intravenous or intramuscular) administration of an allergen to which a worker is sensitized topically (for example, sulfonamides and thiazide diuretics with *p*-phenylenediamine and benzocaine-containing topical anesthetics).
8. Patch testing is used to differentiate allergic contact dermatitis from irritant dermatitis.

The most frequent contact sensitizers in the general population have been determined from clinical experience and published studies on the prevalence of positive patch tests in patients evaluated for possible contact dermatitis. Major sensitizers include:

1. rhus (poison oak, ivy, or sumac),
2. *p*-phenylenediamine,
3. nickel compounds,

4. rubber compounds
5. ethylenediamine
6. topical medicaments ("caines," antihistamines, and antibiotics).

Selected additional industrial allergens include:

1. chromates
2. plastics (especially epoxy and acrylic resins)
3. formaldehyde
4. mercury compounds
5. cobalt compounds.

Photosensitivity

Photosensitivity is the capacity of an organ or organism or certain chemicals and plants to be stimulated to activity by light or to react to light. Two types are generally recognized, phototoxicity and photoallergy. Phototoxicity, like primary irritation, can affect anyone, although darkly pigmented individuals are more resistant. Photoallergens, like contact allergens, involve immune mechanisms and affect fewer people.

Industrial sources of photosensitivity can be hidden or obscure and require careful epidemiological and clinical investigation, including photopatch testing. An example is phototoxicity from *p*-aminobenzoic acid used in the manufacture of ultraviolet-cured inks. Medical personnel may be occupationally exposed to photosensitizing drugs. Other workers who can have contact with topical photosensitizers include outdoor and field workers (photosensitizers in plants and chemicals); machinists (antimicrobials in metal working fluids); pharmaceutical workers (drugs, dyes, and fragrances); oil field, road construction, and coal tar workers (tars, pitch, and other hydrocarbons).

Occupational acne

Occupational acne results from contact with petroleum and its derivatives, coal tar products, or certain halogenated aromatic hydrocarbons (Table 8-G). The eruption can be mild, involving localized, exposed, or covered areas of the body, or severe and generalized, with acne involving almost every follicular orifice. Acne caused by exposure to chlorine compounds (chloracne), in addition to being a difficult cosmetic and therapeutic problem, is of considerable concern because it is caused by highly toxic chemicals.

Occupational acne is seen most commonly in workers exposed to cutting oils in the machine tool trades. The insoluble (straight) oils are the most frequent cause (Figure 8-2). Oil acne typically starts as comedones and an inflammatory folliculitis affecting the tops of the hands and extensor surfaces of the forearms. However, covered areas of the body (thighs, lower abdomen, and buttocks) can be affected by contact with oil-saturated clothing. Although the lesions are commonly called "oil boils," they almost never develop from bacteria present in the oils.

Any form of occupational acne or preexisting or coexisting acne vulgaris (nonoccupational) can be aggravated by heat (acne tropicalis and aestivalis); constant friction (acne mechanica), with acne localized to the forehead (hard hat), waist (belt), etc; excessive scrubbing with harsh soaps (acne detergicans); cosmetics (acne cosmetica); pomade and vaseline (pomade acne); and topical corticosteroids (steroid rosacea). Acneiform eruptions from systemic medication containing bromides, iodides, and corticosteroids, and the syndrome of senile or solar comedones on the face are also to be considered in the differential diagnoses.

Table 8-G. Some Causes of Occupational Acne

Petroleum and its derivatives (e.g., crude oil and fractious cutting oils)
Coal tar products (e.g., coal tar oils, pitch, creosote)
Halogenated Aromatic Compounds (chloracnegens)
Polyhalogenated naphthalenes
Polyhalogenated biphenyls (e.g., PCBs, PBBs)
Polyhalogenated dibenzofurans
Contaminants of polychlorophenol compounds
especially herbicides (2,4,5-T and pentachlorophenol) and herbicide intermediates (trichlorophenols, e.g., dioxin)
Contaminants of 3,4-dichloroaniline and related herbicides
(Propanil and Methazole) (e.g., azo- and azoxybenzenes)

Coal tar oils, creosote, and pitch can produce extensive acne in coal tar plant workers, roofers, and road maintenance and construction workers. Comedones are typical of this form of acne. Phototoxic reactions involving both the skin and eye (keratoconjunctivitis) can complicate the picture and produce coal tar melanosis and exacerbations of the acne. Pitch kerotoses and acanthomas (precancerous and cancerous skin lesions) can develop later.

Certain halogenated aromatic chemicals are the most potent acnegens and are among the most toxic environmental compounds known. These chemicals can produce chloracne, a type of acne often refractory to therapy, which may be accompanied by systemic toxicity. Chloracne represents one of the most sensitive indicators of biological response to these chemicals, and serves as a marker of the medical and environmental impact of contamination of technical-grade chemicals with potentially highly toxic intermediates.

Pigmentary abnormalities

Industrially related pigmentary abnormalities can occur after exposure to certain chemical, physical, and biological agents. They not only can be difficult cosmetic problems, but they also can indicate exposure to potentially systemic toxins. Differentiation from various nonoccupational, genetic, metabolic, endocrine, inflammatory, and neoplastic pigmentary conditions is necessary.

Hyperpigmentation. Hyperpigmentation (skin darkening) can follow almost any dermatitis as a postinflammatory event. Chemical photosensitizers (tar, pitch, plant, and drug photosensitizers), physical agents (ultraviolet light, thermal, and ionizing radiation), and trauma (chronic itching) are common causes. Exposure to certain chemicals (arsenic, acnegenic aromatic hydrocarbons) can also cause hyperpigmentation.

Hypopigmentation. Pigment loss also can be postinflammatory. An example of hypopigmentation due to vitiligo is shown in Figure 8-3. Physical or chemical damage to the skin from thermal, ultraviolet, radiation, or chemical burns not only may cause loss of pigment, but also often leads to scarring. These changes usually pose no diagnostic problem.

However, pigment loss from certain chemical exposures can be difficult to differentiate from idiopathic vitiligo (a patchy loss of pigment from otherwise healthy skin). Occupational leukoderma (white skin) of this type was first reported from exposure to hydroquinone or its monobenzyl ether (agerite alba derivative). Monobenzyl ether of hydroquinone was once used

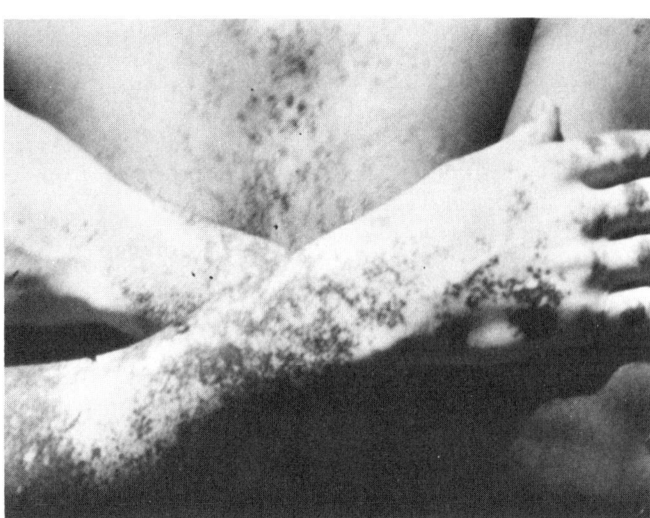

Figure 8-3. A pigmentary change in the skin (hypopigmentation) was caused by vitiligo.

in the rubber manufacturing process for industrial gloves, etc. During the past 20 years a number of phenolic compounds, principally congeners of hydroquinone, have caused leukoderma among exposed workers. They include germicidal disinfectants and antioxidants found in a variety of sources (hospital and industrial cleaners, metal working fluids, oils, latex glues, inks, paints, and plastic resins). Table 8-H is a list of some of these compounds. These chemicals interfere with melanin biosynthesis and/or destruction. Hands and forearms are usually affected, although covered parts can also be affected, possibly from ingestion or inhalation of the chemicals.

Table 8-H. Some Chemicals Producing Occupational Leukoderma

Monobenzyl ether of hydroquinone
Hydroquinone
P-Tertiary amyl phenol
P-Tertiary butyl phenol
P-Tertiary butyl catechol
Alkyl phenols
Other phenolic compounds

Sweat-induced reactions, including miliaria and intertrigo

Miliaria (prickly heat or heat rash) results from obstruction of sweat ducts and is an inflammatory reaction to retained extravasated sweat. It is a common reaction of persons who sweat profusely while exposed to heat. The lesions consist of pinpoint- to pinhead-sized papules and vesicles on the chest, back, submammary, inguinal, and axillary folds.

Intertrigo is a maceration that occurs on apposing skin surfaces and is a scaling, erythematous eruption. Superimposed yeast or superficial fungal infection can also be present. Obesity and heat exposure are aggravating factors.

Cutaneous tumors

Neoplastic changes of the skin are classified as benign lesions, precancers, or cancers. Benign viral warts (verrucae vulgaris) can occur more frequently among workers in certain occupations associated with wet work (for example, butchers). Keratoacanthomas can be occupationally associated with exposure to sunlight or contact with various tars, pitch, and oils (pitch warts and acanthomas). Although classed as benign lesions, keratoacanthomas can be extremely difficult to differentiate clinically and pathologically from squamous cell carcinoma. Pitch and tar warts (keratoses) and acanthomas can be premalignant lesions.

Excessive exposure to sunlight is the most common cause of precancers and cancers in human skin. Additionally, inorganic arsenic compounds, polycyclic aromatic hydrocarbon compounds associated with asphalt, paraffins, coal tars, oils (creosote, shale, hydrogenated, petroleum, insoluble cutting, and mineral), and ionizing radiation can cause cancer of the skin and other organs. Precancerous actinic (pertaining to rays of light that produce chemical effects) keratoses appear in sun-exposed areas, can be extensive in workers with outdoor jobs (utility line workers, farmers, construction workers, ranchers, fishermen, sailors, etc.), and can progress to squamous cell and basal cell carcinomas. Such workers frequently have other stigmata of sun exposure from solar degeneration of collagen, including hyperpigmentation, thin and wrinkled skin, and telangiectasia (a spiderlike growth composed of blood or lymph vessels). Epidemiological studies show that sunlight can also be a factor in the increased incidence of malignant melanoma. Skin biopsy is absolutely essential for the diagnosis of all types of skin cancer.

Ulcerations

Tissue injury on a skin or mucous membrane surface can result in erythema, blisters, or pustules, which may result in necrosis and ulceration. This can be caused by trauma, thermal or chemical burns, and cutaneous infection, and a number of chemicals, including certain chromium, beryllium, nickel, and platinum salts, calcium oxide, calcium arsenate, calcium nitrate, and strong acids. Cutaneous tumors can also ulcerate. Self-inflicted or unintentionally produced skin disorders commonly appear as ulcerations.

Granulomas

These represent chronic, indolent areas of inflammation and can be localized or generalized. Scar formation often results. Causes include a variety of bacterial (anthrax), mycological (sporotrichosis), viral (herpes simplex), parasitical (protothecosis), and botanical (thorns) sources. Other causes include minerals (silica, beryllium, zirconium), bone, chitin, and grease.

Alopecia

Alopecia (absence of hair from the skin areas where it is normally present) has many causes: trauma, cutaneous and systemic disease, drugs, chemicals, and other physical factors, including ionizing radiation. Industrially caused hair loss is rare and the differential diagnosis is long. Chemicals or medications can cause extensive hair shedding by precipitating telogen (resting hair) development, directly poisoning the anagen (growing) hair, or acting in other unknown ways. Other alopecia-producing chemicals include thallium (rodent poison) and boric acid. Medications, primarily cancer chemotherapeutic agents, can precipitate anagen hair loss (immediate loss). Drugs capable of causing telogen hair loss (delayed loss) include oral contraceptives, anticoagulants, propanolol, and thallium.

Nail disease

Chronic inflammation of the folds of tissue surrounding the fingernail (paronychia) with associated nail dystrophy is a frequent occupational disorder associated with wet work (bartenders, maintenance workers, kitchen workers). These are commonly associated with *Candida* species, *Pseudomonas* species, other bacteria, and dermatophyte fungi.

Nail discoloration can occur from exposure to chemicals, such as bichromates (accompanied by nail dystrophy), formaldehyde, certain amines, picric acid, nicotine, mercury, resorcinol, or iodochlorhydroxyquin (Vioform).

Nail dystrophy can also accompany exposure to a number of chemicals, especially solvents; it is also caused by trauma (occupational marks) in certain occupations (including weaving, the fur industry). Nail dystrophy can also be secondary to Raynaud's phenomenon, in cases of vibratory trauma, and in acro-osteolysis.

Systemic intoxication

A number of chemicals with or without direct toxic effect on the skin itself can be absorbed through it and cause systemic intoxication; the severity depends on the amount absorbed. A partial list of substances and their systemic effects includes: aniline (red blood cells and methemoglobinemia); benzidine (carcinoma of urinary bladder); carbon disulfide (nervous system, psychological disturbances, and atherosclerosis); carbon tetrachloride (central nervous system (CNS) depression, hepatotoxicity, and nephrotoxicity); dioxane (CNS depression, suspected carcinogen); ethylene glycol ethers (CNS depression, pulmonary edema, hepatotoxicity and nephrotoxicity); halogenated naphthalenes, diphenyls, and dioxins (neurotoxicity and hepatotoxicity, altered metabolism); methyl butyl ketone (CNS depression and peripheral neuritis); organophosphate pesticides (inhibition of enzyme cholinesterase with cardiovascular, gastrointestinal, neuromuscular, and pulmonary toxicity); tetrachloroethylene (CNS depression, suspected carcinogen); and toluene (CNS depression).

BURNS

Because all burns have essentially the same features, they are usually classified by degree according to depth of injury—first, second, and third degree.

The main types are as follows:

- explosion burns, usually affecting exposed areas (hands, face)
- steam burns, often superficial on exposed areas (more serious if with eye or respiratory contact)
- hot-water burns, frequently leading to blistering depending on water temperature—more severe if victim wearing heavy, permeable clothing
- molten-metal burns, frequently affecting lower limbs, often extremely deep with metal incrusted in skin
- hot-solid burns, normally not extensive, can be very deep
- flame burns, almost always deep, often extensive, with the type of clothing being a major factor in severity
- electricity and radiant energy burns, almost always severe, often with complications; ordinary clothing little protection.

Nature of chemical burns

Burns caused by chemicals are similar to those caused by heat. In fact, some chemicals, such as sodium hydroxide, cause not only chemical burns because of their caustic action, but also thermal burns because of the heat that can be generated when they react with moisture in the skin. After patients with chemical burns have been given emergency first aid, their treatment is the same as that for patients with thermal burns.

Both thermal burns and chemical burns destroy body tissue. But some chemicals will continue to cause damage until reaction with body tissue is complete or until the chemical is washed away by prolonged flushing with water. Strong alkalies penetrate tissue deeply, and strong acids corrode tissue with a characteristic stain.

Many concentrated chemical solutions have an affinity for water. When they come in contact with body tissue, they withdraw water from it so rapidly that the original chemical composition of the tissue (and hence the tissue itself) is destroyed. In fact, a strong caustic may dissolve even dehydrated animal tissue. The more concentrated the solution, the more rapid is the destruction.

Sulfuric, nitric, and hydrofluoric acids are the most corrosive of the inorganic acids—even more corrosive than hydrochloric acid.

Some chemicals, such as phenol, are doubly hazardous. In addition to being highly corrosive, they are poisonous when absorbed through the skin. The severity of chemical burns depends upon the following factors.

- corrosiveness of the chemical
- concentration of the chemical
- temperature of the chemical or its solution
- duration of the contact.

The first three factors are set by the very nature of the chemical and the requirements of the process in which it is used. The fourth factor, however, duration of the contact, can be controlled by the proper first-aid treatment administered without delay.

Classification of burns

Burns are commonly classified as first, second, or third degree. Some believe that second-degree burns should be further classified as superficial or deep dermal. However, for purposes of this chapter, the common classifications of first, second, and third degree are adequate and are described as follows:

First-degree burns. These are characterized by redness and heat accompanied by itching, burning, and considerable pain. Only the outer layer of the epidermis is involved.

Second-degree burns. These are highly painful and involve deeper portions of the epidermis and the upper layer of dermis. Generally, the skin is mottled red with a moist surface, and blisters are formed. Such burns are easily infected.

Third-degree burns. These are very severe forms of injury, involving loss of skin and deeper subcutaneous tissue. They are pearly-white or charred in appearance and the surface is dry. They are not exceedingly painful at first, because nerve endings are usually impaired or destroyed.

Special types of burns. Cement and hydrofluoric acid deserve special mention. Recently, there have been many reports of severe burns from kneeling in wet cement or from wet coment becoming trapped inside boots. Pressure and occlusion are important factors, as well as the need to work fast with premixed cement, which can encourage prolonged contact. Hydrofluoric acid is one of the strongest acids known and is widely used in industry. Hydrofluoric acid burns are characterized by intense pain, often delayed, and progressive deep tissue destruction (necrosis).

Immediate treatment with topical magnesium sulfate or benzalkonium chloride and calcium gluconate gels and injections is recommended.

Complications of burns

The dangers to life that result from extensive burns are infection (which causes most burn complications), loss of body fluid (plasma or lymph from the blood), and subsequent shock. Finally, the functional, cosmetic, and psychological sequelae may require the full attention of a rehabilitation team.

AMA GUIDES FOR EVALUATING IMPAIRMENT

This section on the determination of the percent of impairment is included in this chapter to assist safety and health professionals interpret medical reports of workers' compensation cases; it is taken from the American Medical Association's (AMA) *Guides to the Evaluation of Permanent Impairment* (see Bibliography).

The main objective of the *Guides* is to define as precisely as possible the meaning of medical and nonmedical statements made by physicians about persons whose health is impaired, and the ways in which these statements are understood and used. The purpose of the *Guides* is to provide clinically sound and reproducible criteria for rating permanent impairment for each organ system of the body. In particular, you are referred to the Preface and the two Appendices on Reports and Glossary of Terms.

Although the *AMA Guides* are meant to apply to evaluation of disability under public and private disability systems and under a number of jurisdictions, some state workers' compensations systems have adopted them. The *Guides* emphasize the distinction between a patient's *medical impairment,* which is an alteration of health status assessed by medical means, and the patient's *disability,* which is an alteration of the patient's capacity to meet personal, social, or occupational demands, or meet the statutory or regulatory requirements, which are assessed by nonmedical means. The *Guides* do not address causation, that is, when a particular disorder is or is not considered occupationally related. They also do not deal with temporary impairment and disability. Individual state workers' compensation statutes and other references should be consulted for further information. They do attempt to bring objectivity to an area of great subjectivity.

The chapter in the *Guides* on "The Skin" provides criteria for evaluating the effect that permanent impairment of the skin and its appendages has on a person's ability to perform the activities of daily living, including the occupation.

An established deviation from the normal functions of the skin can result in an anatomic or functional abnormality or loss, and thus constitute a permanent impairment.

Permanent impairment of the skin is any anatomic or functional abnormality or loss, including an acquired immunological capacity to react to an antigen (allergen) that persists following medical treatment and rehabilitation, and after a reasonable time has elapsed to permit optimal regeneration and other physiologic adjustments to occur. The degree of permanent impairment of the skin may not be static. Therefore, findings should be subject to review, and the patient's impairment should be reevaluated at appropriate intervals.

Evaluation of impairment is usually possible through the exercise of sound clinical judgment based on a detailed medical history, a thorough physical examination, and the judicious use of diagnostic procedures. Laboratory aids include procedures such as patch, open, scratch, intracutaneous, and serological tests for allergy, Wood's light examinations and cultures and scrapings for bacteria, fungi, and viruses, and biopsies.

Reports

Before a physician gives any written or oral statement of conclusions, a full account of the findings and observations should be made, including information on pathogenic and etiological diagnoses as well as information as to any physical and/or chemical agents that should be avoided to lessen the manifestations of the impairment. The physician should explain and substantiate the conclusions about the patient's impairment; and record the date on which the findings and conclusions were determined.

Criteria and methods of evaluation

In evaluation of permanent impairment resulting from a skin disorder, the actual functional loss is the prime consideration, rather than the extent of cosmetic or cutaneous involvement.

Impairments of other body systems can be associated with a skin impairment, e.g., behavioral or psychiatric problems; restriction of motion or ankylosis of joints; respiratory, cardiovascular, endocrine, or gastrointestinal disorders. When there is permanent impairment in more than one body system, the degree of impairment for each should be separately evaluated and combined to determine the permanent impairment of the whole person.

Pruritus

Pruritus is frequently associated with cutaneous disorders. It is a subjective, unpleasant sensation that provokes the desire to scratch or rub. The sensation is closely related to pain, in that it is mediated by pain receptors and pain fibers when they are weakly stimulated. However, the itching sensation may be intolerable. Like pain, it can be defined as a unique complex made up of afferent stimuli interacting with the emotional or affective state of the person and modified by that individual's past experience and present state of mind.

The sensation of pruritus has two elements, peripheral neural stimulation and CNS reaction, which are extremely variable in makeup and in time. The first element can vary from total absence of sensation to an awareness of stimuli as either usual or unusual sensations. The second element is also variable and is modified by the person's state of attentiveness, experience, motivation at the moment, and stimuli, such as exercise, sweating and changes in temperature.

In evaluating pruritus associated with skin disorders, the physician should consider (1) how the pruritus interferes with the person's performance of the activities of daily living, including occupation; and (2) to what extent the description of the pruritus is supported by objective skin findings, such as lichenification, excoriation, or hyperpigmentation. Subjective complaints of itching that cannot be substantiated objectively may require specialized referral.

Disfigurement

Disfigurement is an altered or abnormal appearance. This may

Table 8-I. American Medical Association's Permanent Impairment Classification for Skin Disease

Class 1 0%-5% Impairment	Class 2 10%-20% Impairment	Class 3 25%-50% Impairment	Class 4 55%-80% Impairment	Class 5 85%-95% Impairment
A patient belongs in Class 1 when signs or symptoms of skin disorder are present **and** with treatment, there is no limitation, or minimal limitation, in the performance of the activities of daily living, although exposure to certain physical or chemical agents might increase limitation temporarily.	A patient belongs in Class 2 when signs and symptoms of skin disorder are present **and** intermittent treatment is required **and** there is limitation in the performance of some of the activities of daily living.	A patient belongs in Class 3 when signs and symptoms of skin disorder are present **and** continuous treatment is required **and** there is limitation in the performance of many activities of daily living.	A patient belongs in Class 4 when signs and symptoms of skin disorder are present **and** continuous treatment is required, which may include periodic confinement at home or other domicile **and** there is limitation in the performance of many of the activities of daily living.	A patient belongs in Class 5 when signs and symptoms of skin disorder are present **and** continuous treatment is required, which necessitates confinement at home or other domicile **and** there is severe limitation in the performance of activities of daily living.

be an alteration of color, shape, or structure, or a combination of these. Disfigurement can be a residual of injury or disease, or it can accompany a recurrent or ongoing disorder. Examples include a giant pigmented nevi, nevus flammeus, cavernous hemangioma, and alterations in pigmentation.

With disfigurement there is usually no loss of body function and little or no effect on the activities of daily living. Disfigurement can produce either social rejection or impairment of self-image, with self-imposed isolation, life-style alteration or other behavioral changes. If, however, impairment due to disfigurement does exist, it is usually manifested by a change in behavior such as the person's withdrawal from society. Then, it should be evaluated in accordance with the criteria set forth in the chapter on mental and behavioral disorders in the *Guides.*

In some patients with altered pigmentation there can be loss of body function and interference in the activities of daily living, which should be evaluated in accordance with the criteria described in the following paragraphs.

The description of disfigurement is enhanced by good color photographs showing multiple views of the defect. The probable duration and permanency of the altered appearance should be stated.

The possibility of improvement in the altered appearance through medical or surgical therapy, and the extent to which the alteration can be concealed cosmetically, such as with hair pieces, wigs, or cosmetics, should be described in writing and should be depicted with photographs if possible.

Scars

Scars are cutaneous abnormalities that result from the healing of burned, traumatized, or diseased tissue, and they represent a special type of disfigurement. Scars should be described by giving their dimensions in centimeters and by describing their shape, color, anatomical location, and evidence of ulceration; their depression or elevation, which relates to whether they are atrophic or hypertrophic; their texture, which relates to whether they are soft and pliable or hard and indurated, thin or thick and smooth or rough; and their attachment, if any, to underlying bone, joints, muscles or other tissues. Good color photography with multiple views of the defect enhances the description of scars.

The tendency of a scar to disfigure should be considered in evaluating whether impairment is permanent, or whether the scar can be changed, made less visible, or concealed. Function can be restored without improving appearance, and appearance can be improved without altering anatomical or physiological function. Assignment of a percentage of impairment because of behavioral changes related to a scar should be done according to the criteria set forth in the chapter on mental and behavioral disorders.

If a scar involves the loss of sweat gland function, hair growth, nail growth, or pigment formation, the effect of such loss on performance of the activities of daily living should be evaluated. Furthermore, any loss of function due to sensory deficit, pain, or discomfort in area of the scar should be evaluated according to the criteria in the chapter on the nervous system. Loss of function due to limited motion in the area of the scar should be evaluated according to criteria in the chapter on the extremities, spine and pelvis, or, if chest wall excursion is limited, in the chapter on the respiratory system.

Patch testing—performance, interpretation, and relevance

Patch testing is not a substitute for an adequately detailed history. Nevertheless, when properly performed and interpreted, patch tests can make a significant contribution to the diagnosis and management of contact dermatoses.

The physician must be aware that patch testing can yield false-positive and false-negative results. Selecting the proper concentration of the suspected chemical, the proper vehicle, the proper site of application, and the proper type of patch is critical in assuring validity of the procedure. Making such selections and determining the relevance of the test results require considerable skill and experience.

A positive or negative patch test result should not be accepted at face value until the details of the testing procedures have been evaluated. While appropriate test concentrations and vehicles have been established for many common sensitizers, for most chemicals in existence there are *no* established vehicle and concentration standards. Further details about patch testing and its pitfalls are discussed in standard tests.

Manifestation of skin disorders can be influenced by physical and/or chemical agents that a patient may encounter. While the avoidance of these irritant agents might alleviate the manifestations of the skin disorder or even necessitate a change in occupation (disability), nevertheless any presence of a skin disorder should be recognized and evaluated in accordance with the criteria outlined here (Table 8-I).

CONTROL

Dermatoses caused by substances or conditions present in the work environment are largely preventable, but only through the combined effort of management and workers. This type of combined effort is best demonstrated in large industrial firms.

There are two major approaches to the control of occupational diseases, in general, or dermatoses, in particular: (1) environmental control measures and (2) personal hygiene methods. In both cases, the key word is cleanliness, environmental or personal.

Environment

Environmental cleanliness includes good housekeeping (discussed later in this section). Its primary function in preventing industrial dermatitis (and other industrial diseases) is to reduce the possibility of contact with the offending agent.

Planning. Proper design of equipment during plant construction is of great importance in the reduction of dermatitis and other industrial health problems. Ventilation must meet the requirements of the particular industry.

Provisions must be made for the safe handling of irritant chemicals. Pumps, valves, pipes, fittings, and the like must be maintained to eliminate (insofar as is possible) the contact of irritants with workers. Empty drums or bags used to transport incoming materials of a hazardous nature should be properly disposed of to prevent accidental exposure. Containers being readied for shipment should be filled in a manner that will prevent contact with workers, and also left clean so that truckers, warehouse workers, and others will not accidentally contact a harmful material on contaminated surfaces. Containers with harmful materials should be labeled with proper precautions.

Process control. Before any new process or work procedure is introduced and before adoption of any new or different substances in an established process, an industrial hygienist or chemist should carefully consider every aspect of the operation for possible or known dermatitis hazards—including those that can be caused by trace impurities. Analyzing work procedures and processes often requires specialized equipment and techniques.

Once the potential dermatitis-causing factors have been determined, suitable engineering controls can be instituted and built into the work processes or operations.

The best way of controlling dermatitis is to prevent skin contact with offending substances—if there is no exposure, there will be no dermatitis. Unfortunately, this is more easily said than done.

Basically, operations should be planned and engineered to assure minimal worker contact with any irritants. When possible, chemicals of low toxicity and irritant potential should be substituted. Enclosure guards and mechanical handling facilities may be necessary when an operation involves highly corrosive materials. Operations that give off dust, fumes, or vapors need suitable exhaust ventilation to minimize exposure.

Selection of materials. Much can be done to minimize hazardous conditions through the selection of materials. Dry sodium and potassium hydroxide, for example, are now available in virtually dust-free forms or, for many uses, they may be purchased as a solution and handled with pumps. Other products, too, may be available as prills (beads), pellets, granules, or solutions that will do an adequate job and reduce the dust hazard. Concentrated solutions are also finding favor, not only for safety, but for economy in handling.

Some compounds can be successfully used when the percentage of the irritant in the compound is reduced. In other cases, a less irritating or nonirritating material can be substituted for a chemical that is a skin irritant. The supplier should be asked to provide a closely related and generally satisfactory substitute for the irritating material.

Good housekeeping. Environmental cleanliness is nothing more than "good housekeeping," and it is maintained by frequently cleaning floors, walls, ceilings, windows, and machinery. Good housekeeping work is usually performed by a special maintenance group who are given direct responsibility for maintenance cleaning. In order to be effective, cleaning should be part of a plan and should be accomplished on schedule. The necessary equipment and materials to do the most effective job possible in a reasonable amount of time should be assigned, and housekeeping workers should be trained so that they perform their operations efficiently and safely.

Environmental cleanliness is important to maintain good morale, reduce contact dermatitis, and set an example for workers. Floors, walls, ceilings and light fixtures should be cleaned regularly in order to maintain the best possible conditions in the plant. (The requirements of Part 1910 of the Occupational Safety and Health Standards, Section 1910.141, "Sanitation," contain details on housekeeping, waste disposal, vermin control, water supply, toilet facilities, and washing facilities, change rooms, consumption of food and beverages on premises, and food handling.) As pointed out in the OSHA standards, washrooms, showers, toilets, and locker rooms should be kept clean and sanitary.

Many types of cleaners are available—from simple cleaning agents to complex formulations. These come in solid, liquid, or paste, and contain cleaners and sanitizing agents using synthetic detergents, soaps, and alkaline salts in combinations. Some mixtures include sanitizing agents to help prevent the spread of bacteria, fungi, and other biological agents.

Environmental cleanliness and good housekeeping are also beneficial, because they set an example for the workers and encourage personal cleanliness.

Personal cleanliness

The importance of personal cleanliness cannot be too strongly emphasized in the prevention and control of occupational dermatosis. When investigating contact dermatitis, one should also consider the possibility of irritants contacted at home or with a hobby.

Prevention of contact. When plant and process design cannot eliminate all contact with irritants, personal protective equipment must be used. Included are gloves, gauntlets, aprons, and boots made of a material that is impervious to the particular substance. These, along with goggles, afford sufficient protection in most cases. Gauntlets, aprons, boots, and gloves are available in disposable types. However, they are more subject to tears than heavier safety gear. Other gear may provide insulation against heat or light. Whatever type of personal protective equipment is required for a job, it should be carefully maintained. When the equipment becomes worn, it should be replaced.

If a worker is to minimize contact with harmful agents, the worker must have access to facilities for washing hands and be furnished other means of keeping clean at work. It is up to the

Figure 8-4. Wash facilities should be located conveniently and should be adequate for all needs.

company to provide adequate washing facilities and good cleansing materials and education on the need for good hygiene practices). Washbasins must be well designed, conveniently located, and kept clean; otherwise they will be used infrequently, if at all. The farther a worker must walk to clean up the skin, the less likelihood there is of doing so. Inconveniently located washbasins invite such undesirable practices as washing with more easily available solvents, mineral oils, or industrial detergents, none of which is intended for skin cleansing. For workers to keep their skin reasonably free of injurious agents, they must use washing facilities at least three times a day—during work (before eating, drinking, smoking, or using the restroom), before lunch, after lunch, and before leaving the plant (see Figure 8-4).

Those who work with toxic chemicals and radioactive substances must be required to take a shower after their work shift and change their clothing. Workers should be instructed in specific procedures for cleanliness. They should be told where, how, and when to wash, should be given sufficient time to wash, advised that they will be rated on this part of their job performance, and should be informed of the possible health hazards involved.

For many exposures, frequent washing alone proves a successful preventive, particularly when the dermatitis is caused by plugging of the pores, as from dust. In all cases, however, the use of large quantities of water on the skin following exposure to irritants is necessary. Safety showers and eyewash fountains should be available, and flushing should continue for at least 15 minutes.

It may be advisable in some instances to use neutralizing solutions after a thorough flushing with water. However, since some neutralizing solutions are themselves irritants, they should be used only upon the advice of a physician.

The question of the type of soap to use is important. Even a generally good soap can cause irritation on certain types of skins. For example, harsh mineral abrasives can cause dermatitis on many individuals.

The choice of a good soap may in some cases involve technical considerations, which are better left to the medical department or other qualified department than to lay persons. The basic requirements of industrial skin cleansers are as follows:

1. They should remove industrial soil quickly and efficiently.
2. They should not harmfully dehydrate, abrade, or irritate the skin by normal application.
3. They should flow easily through dispensers.
4. They should be adequately preserved against microbial contamination.

Additional desirable qualities include the following:

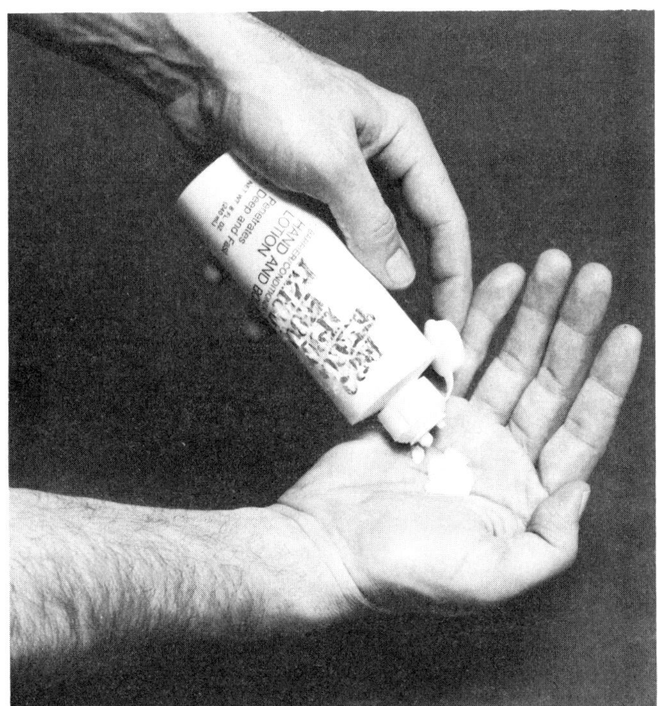

Figure 8-5. Barrier cream is applied by an employee before starting work, and is washed off before lunch; it is reapplied after lunch, and washed off before leaving work. A barrier cream is the least effective way of protecting the skin.

1. They should have aesthetic appeal (color and odor).
2. They should have good foaming qualities.

A number of cases of industrial dermatitis are reported to be caused not by substances used in the workplace, but by cleansing materials used to remove those substances. A worker may be inclined to wash the hands with those cleaning agents which are most available and work the fastest—these are often dermatitis-producing solvents.

To combat this practice, the installation in work-area washing places (as well as in regular plant washing facilities) of soap-dispensing units containing properly selected cleansing agents has proved to be a valuable measure. Such units should be placed in convenient locations, and enough of them should be provided to accommodate all employees who are exposed to skin irritants. Where soap-dispensing units are furnished, workers should be required to use them.

Barrier creams. A barrier cream is the least effective way of protecting skin (Figure 8-5). However, there are instances when a protective cream may be used for preventing contact with harmful agents, for example, if the face cannot be covered by a shield, or gloves cannot be worn (Figure 8-6). Several manufacturers compound a variety of products, each designed for a certain type of protective purpose. Thus, there are barrier creams for protecting against dry substances and those that protect against wet materials. Using a barrier cream to protect against a solvent is not as effective as using an impervious glove; however, there are compounds that offer some protection against solvents, providing the creams are applied with sufficient frequency.

Barrier creams and lotions should be used to supplement, but not to replace, personal protective equipment. Protective barrier agents should be applied to clean skin. When skin becomes soiled, both the barrier and any soil should be washed off and the cream reapplied.

Three main types of barrier creams and lotions are available.

Vanishing cream usually contains soap and emollients that coat the skin and cover the pores to make subsequent cleanup easier.

Water-repellent type leaves a thin film of water-repellent substance, such as lanolin, beeswax, petrolatum, or silicone on the skin, and helps to prevent ready contact with water-soluble irritants, such as acids, alkalies and certain metallic acids. Remember, however, that the protection may not be complete, especially when the barrier has been on the skin for some time. Alkaline cleaning solutions tend to emulsify and remove the barrier rapidly, thus leaving the skin unprotected.

Solvent-repellent types contain ingredients that repel oil and solvent. Lanolin has some oil-repellent and water-repellent properties, and can be used as an ingredient. There are two types of solvent-repellent barrier preparations, one leaving ointment film and the other, a dry, oil-repellent film. Sodium alginates, methyl cellulose, sodium silicate, and tragacanth are commonly used. Lanolin offers some protection against oils as well as water.

Special types. In addition to these three main types, a number of specialized barriers have been developed. Creams and lotions containing ultraviolet screening agents are used to help prevent overexposure to sun or other ultraviolet sources. Others have been developed to afford protection from such diverse irritants as insects and gunfire backflash.

Figure 8-6. Most glove manufacturers provide a hand protection counseling service. They assess hazards and match them with gloves or other devices. It is up to the employer to educate employees to use the proper protection for the job performed.

Table 8-J. Choosing the Right Glove

This table shows the relative resistance ratings of various glove materials to some industrial solutions. **NOTE:** The purpose of gloves is to eliminate or reduce skin exposure to chemical substances. NEVER IMMERSE the hands, even with gloves rated E (excellent).

KEY TO CHARTS: E = excellent; G = good; F = fair; P = poor (Ratings are subject to variation, depending on formulation, thickness, and whether the material is supported by fabric).

The listings were taken from various glove manufacturers and NIOSH, and are ONLY A GENERAL GUIDE. When selecting gloves for any application, contact the manufacturer, giving as much detailed information as possible, according to the following:

1. Ability of glove to resist penetration of the chemical, thus ensuring the protection of the wearer
2. Chemical composition of the solution
3. Degree of concentration
4. Abrasive effects of materials being handled
5. Temperature conditions
6. Time cycle of use
7. Specify in purchase order what materials are to be handled
8. Cost

Chemical Resistance Chart

Glove Material	*mineral acids*	*organic acids*	*caustics*	*alcohols*	*aromatic solvents*	*petroleum solvents*	*ketonic solvents*	*chlorinated solvents*
	Hydrochloric (38%)	*Acetic*	*Sodium Hydroxide (50%)*	*Methanol*	*Toluene*	*Naphtha*	*Methyl Ethyl Ketone*	*Perchlor-ethylene*
Natural Rubber	G	E	E	E	NR	F	E	NR
Neoprene	E	E	E	E	F	E	G	P
Polyvinyl Chloride	G	G	G	G	P	F	NR	NR
Polyvinyl Alcohol	P	NR	NR	F	E	E	F	E
NBR	E	G	E	G	G	E	F	G

Miscellaneous

Glove Material	*Lacquer Thinner*	*Benzene*	*Formaldehyde*	*Ethyl Acetate*	*Vegetable Oils*	*Animal Fats*	*Turpentine*	*Phenol*
Natural Rubber	F	NR	E	F	F	P	F	F
Neoprene	G	P	E	G	G	E	G	E
Polyvinyl Chloride	F	P	E	P	F	G	G	G
Polyvinyl Alcohol	E	E	P	P	E	E	E	F
NBR	G	G	E	G	E	E	E	G

Physical Performance Chart*

Coating	*Abrasion Resistance*	*Cut Resistance*	*Puncture Resistance*	*Heat Resistance*	*Flexibility*	*Ozone Resistance*	*Tear Resistance*	*Relative Cost*
Natural Rubber	E	E	E	F	E	P	E	Medium
Neoprene	E	E	G	G	G	E	G	Medium
Chlorinated Polyethylene (CPE)	E	G	G	G	G	E	G	Low
Butyl Rubber	F	G	G	E	G	E	G	High
Polyvinyl Chloride	G	P	G	P	F	E	G	Low
Polyvinyl Alcohol	F	F	F	G	P	E	G	Very High
Polyethylene	F	F	P	F	G	F	F	Low
Nitrile Rubber	E	E	E	G	E	F	G	Medium
Nitrile Rubber/Polyvinyl Chloride (Nitrile PVC)	G	G	G	F	G	E	G	Medium
Polyurethane	E	G	G	G	E	G	G	High
Styrene-butadiene Rubber (SBR)	E	G	F	G	G	F	F	Low
Viton	G	G	G	G	G	E	G	Very High

*Grip/slip is related to glove surface and is enhanced when the glove surface is rough.
Dexterity/tactility is related to glove thickness and decreases as the glove thickness increases.

Protective equipment

Clothing. Whenever irritating chemicals are likely to contaminate clothing, care must be taken to provide clean clothing at least daily. Because cases have been cited where workers' families have developed contact dermatitis or chloracne from clothing worn home from the job, it is advisable that clothing worn on the job not be worn at home. Clothing contaminated with irritants should always be thoroughly laundered before wearing it again. Clothing contaminated on the job should be changed at once.

Protective clothing. Sometimes handling irritant materials cannot be avoided; in this situation, protective clothing is a good barrier against industrial irritants. OSHA requirements for protective clothing are described in Subpart I, Section 1910.132, General Requirements for Personal Protective Equipment. Other protective clothing requirements appear in standards covering specific hazards.

All workers do not have to wear protective clothing, but for those whose jobs require it, good quality clothing should be obtained. Manufacturers provide a large selection of protective garments made of rubber, plastic film, leather, or cotton or synthetic fiber that are designed for specific purposes. For example, there is clothing that protects against acids, alkalis, extreme exposures of heat, cold, moisture, oils, and the like. When such garments must be worn, management should purchase them and enforce their use. Management should make sure that the clothing is serviced and laundered often enough to keep it protective. If work clothes are laundered at home, there is a chance that contamination of family wearing apparel with chemicals, glass fiber, or other dusts can occur.

Closely woven fabrics also protect against irritating dust. Gloves and aprons of impervious materials (such as rubber or plastic) will protect against liquids, vapors, and fumes (Table 8-J). Natural rubber gloves, aprons, boots, and sleeves are impervious to water-soluble irritants, but soon deteriorate if exposed to strong alkali and certain solvents.

Synthetic rubber, such as neoprene and many of the newer plastics, are more resistant to alkalis and solvents than is natural rubber; however, some materials are adversely affected by chlorinated hydrocarbon solvents. The protection used should be based on the particular solvents that are used. For workers who wear rubber or plastic impervious to the irritating agent, the irritant will eventually penetrate and be trapped next to the skin, causing repeated exposure every time the garment is worn.

Disposable paper and plastic garments can also be used for some tasks. Garments are also necessary in sterile areas to keep products from being contaminated (Figure 8-7).

Fabrics. Fabrics without coatings are generally unsuitable as protective clothing for toxic chemical exposures since they are all more or less permeable, and have other weaknesses when resisting chemical onslaughts. Cotton and rayon, for instance, are degraded by acids; wool is degraded by alkalis.

Safety. It is imperative that all safety gear is worn only when safely possible. Protective clothing, especially gloves, can be caught in moving machinery, resulting in serious injury.

Figure 8-7. Protective clothing is donned over work clothes to prevent contamination of both the worker and, in sterile rooms, the product.

Responsibility for control

Top management, the safety department, the purchasing department, the medical department or company physician, the supervisors, and the workers all have specific responsibilities for the prevention of industrial skin diseases and the control of exposure to skin irritants.

To control or eliminate dermatitis that could occur in an area, management should first recognize the scope of the problem and then delegate authority for action to the proper persons. When it is necessary to have more than one department work on phases of dermatitis control or elimination, the activities of those departments should be coordinated. Periodic reports on the status of the dermatitis problem within the organization should be made to management by its delegated representatives.

The industrial hygienist (or persons doing this type of work, such as the safety professional, safety committee members, the nurse, or the industrial hygienist) should gather information on dermatitis hazards of materials used in the plant and should disseminate this information among supervisors and other operating personnel. The industrial hygienist should make periodic surveys to check for exposure to skin irritants and should suggest means to correct any hazards found.

Case examples of control

The following examples of the use of controls to reduce or eliminate occupational dermatoses was taken from the 1978 Report of the OSHA Advisory Committee on Cutaneous Hazards.

Powered epoxy spraying operation. A manufacturer of household washing machines began using an epoxy material as a finished surface on its products. The epoxy material came in powdered form and was sprayed on the parts to be assembled, which were then baked in an oven to form an extremely hard surface. The spraying was done automatically, inside a booth. The parts passed through the booth hanging from an overhead conveyor. Overspray was exhausted out the bottom of the booth and into a barrel; some overspray remained on the inside walls of the booth. The only worker in the area during the spraying was an operator who sat inside an enclosed control booth and thus was not exposed to the epoxy powder.

On the midnight shift, however, when production was stopped, a cleanup crew entered the area to perform a number of duties:

1. They used air hoses to blow out the overspray that had accumulated on the inside walls of the spray booth.
2. They dumped barrels of exhausted overspray back into the supply system for reuse.
3. They swept floors and other surfaces outside the booth to clean some spray that had escaped the booth.

The powder was very fine and the slightest turbulence caused it to become airborne, and, consequently, a great concentration of epoxy dust was in the air. The cleanup crew was equipped with disposable respirators, hair covers, boots and complete coveralls. Despite the personal protection, several members of the cleanup crew broke out in rashes after the spraying had been performed for a few weeks.

The problem was solved, after an investigation, by changing the overspray exhaust system to return the overspray directly into the supply system, thus eliminating one major source of dust. Using a vacuum system rather than sweeping or air hoses eliminated the other sources of dust. No cases of dermatitis recurred.

Machining operations. Exposure to cutting fluids in machining operations constitutes one of the major causes of industrial dermatitis. Controls that have virtually eliminated dermatitis have been instituted in many machining operations. For example, in one well-controlled plant that produces diesel engines, over 2,000 workers on two shifts operating approximately 1,000 machines had not a single case of recordable occupational dermatitis in 1977, in contrast with some poorly controlled operations in which roughly 30% of the work force have skin problems.

Control programs put into effect included:

1. careful identification, by generic name, of all ingredients in the cutting fluids used
2. programs to keep the coolant free of tramp oil, foreign particles and dirt through the use of effective filters and redesigning the coolant flow system to eliminate "eddies" and "backwaters" of coolant
3. daily programs to monitor coolant characteristics, such as pH, bacteria count, etc.
4. daily programs, such as hosing down, to keep machinery clean
5. redesigning spray application to minimize coolant splash and spray
6. using splash goggles and curtains
7. use of local exhaust systems and oil collectors to reduce airborne oil mist
8. use of abundant quantities of shop rags
9. provision of paid wash time to allow operators to keep clean.

Experience shows that when the coolant is well controlled and measures are taken to reduce the amount of coolant splashed on the worker, the rate of dermatitis is reduced.

Rubber manufacturing. Improvements in rubber manufacturing operations have reduced problems with skin disease in those plants where the improvements were made. These improvements have taken many forms:

1. improved methods of material handling to reduce the amount of skin contact with rubber and related chemicals
2. substitution of known skin sensitizers by other, less hazardous chemicals, such as the replacement of isopropyl-phenyl-paraphenyl-diamine (IPPD), used as an antioxidant in tires, with other less toxic derivatives of paraphenylene diamine
3. improved methods of mixing rubber chemicals to reduce exposure to a wide variety of known skin irritants and sensitizers: such improvements have included preblending chemicals, using exhaust-ventilated mixing booths, and automation of the mixing process
4. in one plant, an air-conditioned isolation booth was installed for a worker who was strongly sensitized to an antiozonent.

Chemical manufacturing. A major producer of industrial chemicals has instituted a wide variety of controls that have resulted in a reduced rate of dermatitis. Their program includes:

1. extensive use of self-contained systems to handle chemicals to eliminate worker exposure to dermatitis-producing substances; operations were designed with a goal of zero emissions
2. mechanization of material-handling systems to eliminate worker exposure to chemicals
3. emphasis on good plant housekeeping
4. use of wipe testing to check equipment surfaces for films of toxic materials
5. adoption of extensive employee education programs to inform employees of the risks of chemicals
6. implementation of programs of personal hygiene, which, in the case of one particularly hazardous material, included three daily showers for the exposed employee
7. for handling liquid chemicals, use of seal-less pumps and, where leaks cannot be permanently sealed, use of local exhaust systems; grouping of all pumps in one central area for better control; and scaling of floors around the pumps
8. preparation of educational materials to be supplied to purchasers of chemicals, including proper controls for the materials.

Physical examinations

Preplacement examinations will help identify those people who may be especially susceptible to skin irritations. The physician in charge of the examination should be provided with detailed information regarding the type of work for which a person is being considered.

Routine use of preplacement patch tests to determine sensitivity to various material is not recommended. Patch tests will not reveal whether new workers will become sensitized to certain materials and develop dermatitis, but will only tell whether people who have previously worked on similar jobs are, or are not, sensitized to the chemicals with which they had worked.

The industrial physician has the primary responsibility for

determining whether or not an applicant may be predisposed to skin irritations and for recommending suitable placement on the basis of these findings. Nevertheless, considerable responsibility also may fall to the safety and personnel departments, supervisors, industrial hygienists, and other persons functionally responsible for accident prevention work and control of industrial diseases.

Care should be taken in restricting persons who are not specifically sensitive to the agents involved in the job just because of a history of skin trouble, unless there is active skin disease at the time of placement. In many cases, the physician is limited to simply counseling the person about risk.

DIAGNOSIS

Anyone who works can develop a skin disorder, but not all skin disorders occurring in the workplace are occupational. Arriving at the correct diagnosis is not generally difficult, but it is more than a routine exercise. The following criteria are generally used.

Appearance of the lesion. The dermatosis should fall into one of the accepted clinical types with respect to its morphological appearance.

Sites of involvement. Common sites are the hands, wrists, and forearms, but other areas can be affected. Widespread dermatitis can indicate heavy exposure to dust, because of inadequate protective clothing or poor hygiene habits.

History and course of the disease. A thorough and pertinent clinical history is the most important aspect in diagnosis of occupational dermatoses. This includes a description of the eruption, response to therapy, medical history and review of systems, a detailed work history (description of present and past jobs, moonlighting, preventive measures, cleansers, and barrier creams), and a detailed description of nonoccupational exposures. The behavior of the eruption on weekends, vacation, and sick leave can be very helpful in assessing the occupational component.

Ancillary diagnostic tests. When indicated use laboratory tests for detecting skin infections. These may include direct microscopic examination and bacterial and fungal cultures, skin biopsy for histopathological diagnoses, and patch tests to detect any occupational or nonoccupational allergens including photosensitizers. Patch tests were discussed previously in this chapter.

Treatment. Therapy of occupational skin disorders is essentially no different from that of the same nonoccupational disorder. Additionally, two key factors are often overlooked for a worker to have clear enough skin to return to employment; (1) identifying the cause of the disease, and (2) preventing a recurrence. Refer patients with skin disorders not responding to initial treatment for specialty evaluation.

MONITORING AND CONTROL TECHNOLOGY

Current sampling procedures are often difficult to apply to prevention of occupational skin disease. Some exceptions include use of wipe samples for chemical analysis, sampling cotton socks when shoe contamination is suspected, and sampling air inside a suit when the air is contaminated. Air levels of dusts and chemicals may have some limited application. The use of fluorescent tracers is currently being evaluated. Color-indicator soaps used to be employed for tetryl used in munitions plants and for mercury exposure. A recent study established methods to determine relative benefits of equipment, such as gloves and clothing, to protect skin against styrene in a reinforced plastics plant. Another study dealt with the biological surveillance of workers exposed to dimethylformamide and the influence of skin protection on its percutaneous absorption. Standardized techniques have improved measurement of the effectiveness of protective material, such as gloves and clothing against carcinogens and polychlorinated biphenyls.

BIBLIOGRAPHY

Adams, R.M. *Occupational Skin Disease.* New York: Grune & Stratton, 1983.

American Medical Association (AMA). *Guides to Evaluation of Permanent Impairment,* 2nd ed. Chicago: AMA, 1984.

Berardinelli, S. Chemical protective gloves. *Dermatologic Clinics,* in press, 1988.

Birmingham, D.J. "Occupational Dermatoses." In *Patty's Industrial Hygiene and Toxicology,* vol 1, *General Principles,* 3rd ed., edited by G.D. Clayton and F.E. Clayton. New York: Wiley Interscience, 1978.

Birmingham, D.J., ed. *The Prevention of Occupational Skin Diseases.* New York: The Soap and Detergent Association, 1981.

Cohen, S.R. Risk factors in occupational skin disease. In *Occupational and Industrial Dermatology,* 2nd ed., edited by H.I. Maibach. Chicago: Year Book Medical Publ, 1987.

Fisher, A.A. *Contact Dermatitis.* Philadelphia: Lea & Febiger, 1986.

Industrial dermatitis: Part 1: The skin. *National Safety News* Reprint 112 (1975):59-62.

Industrial dermatitis: Part 2: Primary irritation. *National Safety News* Reprint 112 (1975):107-112.

Industrial dermatitis: Part 3: Sensitization dermatitis. *National Safety News* Reprint 112 (1975):165-168.

Morbidity and Mortality Weekly Reports, Jan 21, 1983, and Sept 5, 1986. Atlanta: Centers for Disease Control.

National Safety Council, 444 N. Michigan Ave., Chicago, IL 60611. Industrial Data Sheet:
–*Chemical Burns,* I-510, Rev. 1981.

Report to the Advisory Committee on Cutaneous Hazards to the Assistant Secretary of Labor, OSHA. Occupational Safety and Health Administration, U.S. Department of Labor, Dec. 19, 1978.

Rook, A., et al. *Textbook of Dermatology,* 4th ed. Oxford: Blackwell, 1986.

Samitz, M.H. "Assessment of Cutaneous Impairment and Disability." In *Occupational and Industrial Dermatology,* 2nd ed., edited by H.I. Maibach. Chicago: Year Book Medical Publ, 1987.

Taylor, J. S. "Occupational Dermatoses." In *Clinical Medicine for the Occupational Physician,* edited by M.H. Alderman and M.J. Hanley. New York: Dekker, 1982.

——. "The Pilosebaceous Unit." In *Occupational and Industrial Dermatology,* 2nd ed., edited by H.I. Maibach. Chicago: Year Book Medical Publ, 1987.

Taylor, J.S., Parrish, J.A., and Blank, I.H. Environmental reactions to chemical, physical, and biological events. *Journal of the American Academy of Dermatology* 11(1984):1007-1021.

Industrial Noise

by Julian B. Olishifski, MS, PE, CSP
Revised by
John J. Standard, MS, MPH, CIH, CSP

The sounds of industry, growing in volume over the years, have heralded not only technical and economic progress, but also the threat of an ever-increasing incidence of hearing loss and other noise-related hazards to exposed employees. Noise is not a new hazard. Indeed, noise-induced hearing loss was observed centuries ago. In 1700, Ramazzini is "De Morbis Artificium Diatriba" described how workers who hammer copper "have their ears so injured by that perpetual din. . . that workers of this class become hard of hearing, and if they grow old at this work, completely deaf." Before the Industrial Revolution, however, comparatively few people were exposed to high levels of noise in the workplace. The advent of steam power during the Industrial Revolution first brought general attention to noise as an occupational hazard. Workers who fabricated steam boilers were found to develop hearing loss in such numbers that the malady was dubbed "boilermakers disease." The increasing mechanization that has occurred in all industries and in most trades has since aggravated the noise problem. Noise levels within the workplace, particularly those maintained in mechanized industries, are likely to be more intense and sustained than any noise levels experienced outside the workplace. The recognition, evaluation, and control of industrial noise hazards are introduced in this chapter. Basically, this involves: (1) assessing the extent of the noise problem; (2) setting objectives for a noise abatement program; (3) controlling exposure to excessive noise; and (4) monitoring the hearing of exposed employees.

COMPENSATION ASPECTS

The growing interest of employers in controlling industrial noise exposures has been stimulated by the trend toward covering hearing losses under state worker's compensation laws. Compensation laws that cover loss of hearing due to noise exposure have been enacted in many states; compensation is being awarded in other states even though hearing loss is not specifically defined in many compensation laws.

Occupational hearing loss can be defined as "a hearing impairment of one or both ears, partial or complete, arising in, or during the course of, but as the result of one's employment." It includes acoustic trauma as well as noise-induced hearing loss.

Acoustic trauma denotes injury to the sensorineural elements of the inner ear. Acoustic trauma is produced by one or a few exposures to sudden intense acoustic forms of energy resulting from blasts and explosions or from direct trauma to the head or ear. The worker should be able to relate the onset of hearing loss to one single incident. For details on ear anatomy, see Chapter 4, The Ears.

Noise-induced hearing loss, on the other hand, describes the cumulative permanent loss of hearing—always of the sensorineural type—that develops over months or years of hazardous noise exposure.

Noise-induced hearing loss usually affects both ears equally in the extent and degree of loss. It should also be kept in mind that the onset of hearing loss, its progression, its permanency, as well as the characteristics of the audiograms obtained, vary depending on whether the injury is a noise-induced hearing loss or acoustic trauma.

To establish a diagnosis of noise-induced hearing loss and a causal relationship to employment, the physician considers the following factors:

Table 9-A. Noise Exposures Above 90 dBA in Manufacturing

Code	Number of Plants in Sample	Total Number of Employees in Sample	Number Located in Areas 90 dBA and Above	Percent of Work Force Exposed	Total Work Force	Number Projected to Be Located in Areas 90 dBA and Over
Textile mill products	23	12,764	5,634	44.1	963,300	424,815
Petroleum and coal products	16	20,493	5,875	28.6	192,800	55,140
Lumber and wood products	14	5,654	1,460	25.8	601,000	155,058
Food and kindred products	17	23,690	5,959	25.1	1,898,600	476,549
Furniture and fixtures	11	10,374	1,849	17.8	465,400	82,841
Fabricated metal products	56	41,371	7,079	17.1	1,335,000	228,285
Stone, clay, and glass products	5	2,502	416	16.6	643,800	106,870
Primary metal industries	51	71,208	11,001	15.4	1,190,000	183,260
Rubber and plastic products	4	7,671	1,105	14.4	589,500	84,888
Transportation equipment	46	199,212	23,445	11.7	1,705,500	199,543
Electrical equipment and supplies	7	8,790	973	11.0	1,778,100	195,591
Chemicals and allied products	8	3,081	324	10.5	1,014,400	106,512
Apparel and other textile products	1	50	5	10.0	1,353,100	*
Paper and allied products	21	14,997	1,385	9.2	687,400	63,420
Ordnance and accessories	12	39,403	3,480	8.8	93,900	17,063
Instruments and related products	6	3,254	193	5.9	433,800	25,594
Machinery except electrical	38	25,016	1,144	4.5	1,768,000	79,560
Printing and publishing	5	5,597	237	4.2	1,085,900	45,607
TOTAL	341†	495,127	71,564	14.5	17,799,500	2,530,596

* Insufficient data for projection.
† 2,709 questionnaires were sent to the manufacturing industries listed: 1,550 were returned, and 341 of these respondents answered this question.
(Reprinted from NIOSH *Criteria Document*)

1. the employee's history of hearing loss—onset and progress;
2. the employee's occupational history, type of work, years of employment, etc.;
3. the results of the employee's otological examination;
4. the results of audiological and hearing studies performed (preplacement, periodic, and termination); and
5. the ruling out of nonindustrial causes of hearing loss.

It has been estimated that 1.7 million workers in the U.S. between 50 and 59 years of age have compensable noise-induced hearing loss. Assuming that only 10 percent of these workers file for compensation and that the average claim amounts to \$3,000, the potential cost to industry could exceed \$500 million.

Estimates show that 14 percent of the working population is employed in jobs where the noise level is in excess of 90 dBA (Table 9-A). At present, no test can predict which individuals will incur a hearing loss. If enough people are placed in an environment where the predominant noise level exceeds 90 dBA for a sufficient period of time, some individuals will incur a hearing impairment greater than that due to prebycusis (loss of hearing due to aging). The number of workers subjected to noise hazards exceeds those exposed to any other significant occupational hazard.

The development of the audiometer in recent years provided an easily reproducible means of measuring the status of an individual's hearing with appreciable accuracy. Partial hearing losses are easily measurable by using commercially available audiometers.

Noise exposure regulations

The federal regulation of occupational noise exposure started with the rules issued under the authority of the Walsh-Healey Public Contracts Act. These rules require that occupational noise exposure must be reasonably controlled to minimize fatigue and the probability of accidents. The federal occupational noise exposure regulations were originally written to apply only to contractors under the Walsh-Healey Public Contracts Act and the McNamara-O'Hara Service Contracts Act. Under the Williams-Steiger Occupational Safety and Health Act of 1970, the Bureau of Labor Standards was replaced by the Occupational Safety and Health Administration (OSHA).

The National Institute for Occupational Safety and Health (NIOSH) was established within the Department of Health and Human Services (formerly the Department of Health, Education, and Welfare) by the Occupational Safety and Health Act of 1970 to conduct research and to recommend new occupational safety and health standards. These recommendations are transmitted to the Department of Labor, which is responsible for the final setting, promulgation, and enforcement of the standards.

In 1972, NIOSH provided the Department of Labor with a document called *Criteria for a Recommended Standard: Occupational Exposure to Noise.* Subsequently, the assistant secretary of labor determined that a standard advisory committee on noise should be formed. The purpose of this OSHA Advisory Committee was to obtain and evaluate additional recommendations from labor, management, government, and independent experts. The committee considered written and oral comments directed to it by interested parties. The committee then transmitted its recommendations for a revised standard to the OSHA on December 20, 1973.

In 1974, OSHA published a proposed standard in the *Federal Register* that limited an employee's exposure level to 90 dBA, calculated as an 8-hour, time-weighted average (TWA). The NIOSH commented on OSHA's proposed standard, stating that there was a need for reducing the 8-hour exposure level to 85 dBA. However, NIOSH was unable to recommend a specific future date after which the 85-dBA noise level should become mandatory for all industries. Sufficient data were not available to demonstrate the technological feasibility of this level.

The Environmental Protection Agency (EPA) reviewed the OSHA proposal and recommended that the limit should not exceed 85 dBA. The EPA reviewed the proposed noise standard and recommended that additional studies be undertaken to explore the efficacy of reducing the permissible level still further at some future date. The proposed revisions to the OSHA rules for occupational noise exposure were published in the *Federal Register* on October 24, 1974.

Criteria and standards

Although the words criteria and standards are synonymous in professional usage (both are a measure by which something is judged), in the environmental sense, criteria are a recitation of factual information regarding cause-and-effect relationships and do not indicate what is desired or desirable. Data demonstrate that noise evokes a significant subjective physiological and psychological response pattern. However, the mere recitation of data, no matter how scientifically valid, is not acceptable to the courts, affected industries, or to the public.

The basis of environmental standards must be identified, and adequate statements for the reasons behind the decisions must be provided. Therefore, the general process of developing standards must include a credible chain of reasoning leading to the conclusion that the proposed standards are the best available, given the accumulation of data upon which these standards are based.

In the standards promulgation process, it must be clearly stated how and why the particular standard was selected. It is also necessary to demonstrate what alternatives were developed, including all information relevant to cost, impact, projected results, and so on. The way in which these alternatives were individually examined and then tested against others to produce the decision in the form of a proposed standard should be clearly shown.

A regulatory program usually implies some form of legislation. Legislation that regulates noise is nothing new in this country—many states and major cities have general and qualitative laws against excessive or unwarranted noise. But such laws are very difficult to enforce, because determining whether a violation has occurred is a matter of subjective judgment, and enforcement officials are reticent to enforce qualitative laws.

Noise regulations can and should be objective, quantitative laws. Criteria are well established for those noise levels that are objectionable to a majority of people. Equipment and measurement procedures for determining when noise standards are exceeded are readily available, and their reliability is also well established.

The regulation of environmental noise levels entails consideration of the extremely complex interrelationship among scientific, engineering, technical, social, economic, and political factors. In addition to the skills and knowledge of scientists and engineers, those of lawyers, economists, social scientists, industrial engineers, and political scientists must be utilized. More important, the views of those most affected by these noise standards—those for whom protection is being provided and those whose practices are to be regulated—should be brought into the standards-making process.

Another requirement for effective noise legislation is that it be consistent with current technology. Legislation must also be flexible, since further reductions of the noise level in the future may be desirable. The noise regulations should begin by a gradual phasing out of old and noisy devices that create the peak disturbances. Newer, quieter devices can then be phased in without imposing too great a burden on the manufacturers or operators of these devices.

Noise control requires public awareness of both the problem and of the potentials for solving it. Research and development have an important role to play in seeking new and improved ways of controlling noise. But what is primarily needed is an application, on a broad scale, of the knowledge that is already available.

PROPERTIES OF SOUND

Sound can be defined as any pressure variation (in air, water, or some other medium) that the human ear can detect. The number of pressure variations per second is called the frequency of the sound, which was formerly measured in cycles per second but is now expressed in hertz (Hz), which represents cycles per second.

Subjectively, sound may be defined as "a stimulus that produces a sensory response in the brain." The perception of sound resulting in the sensation called hearing is the principal sensory response; however, under certain conditions, additional subjective sensations, ranging from pressure in the chest cavity to actual pain in the ears, can be produced (see Chapter 4, The Ears). There are certain effects produced by sounds that appear to be universally undesirable for all people. These effects include:

1. the masking of wanted sounds, particularly speech;
2. auditory fatigue and damage to hearing;
3. excessive loudness; and
4. annoyance.

Noise

Noise usually is sound that bears no information and whose intensity usually varies randomly in time. The word *noise* is often used to mean "sound that is unwanted by the listener" because it is unpleasant. Noise interferes with the perception of wanted sound and is likely to be physiologically harmful.

Noise does not necessarily have any particular physical characteristics that distinguish it from wanted sound. No instrument can distinguish between a sound and a noise—only human reaction can.

A variety of methods have been devised that can be used to relate objective physical measurements of sound to subjective human perception. The purpose of this section is to outline the factors involved in the objective physical properties of sound and to summarize the important subjective aspects of sound.

The term *sound* is usually applied to the form of energy that produces a sensation perceived by the sense of hearing in humans, while *vibration* usually refers to the nonaudible acoustic phenomena that are recognized by the tactile experience of touch or feeling. However, there is no essential physical difference between the sonic and vibratory forms of sound energy.

The generation and propagation of sound are easily visualized by means of a simple model. Consider a plate suspended in midair (Figure 9-1). When struck, the plate vibrates rapidly back and forth. As the plate travels in either direction, it compresses the air, causing a slight increase in pressure. When the plate reverses direction, it leaves a partial vacuum, or rarefac-

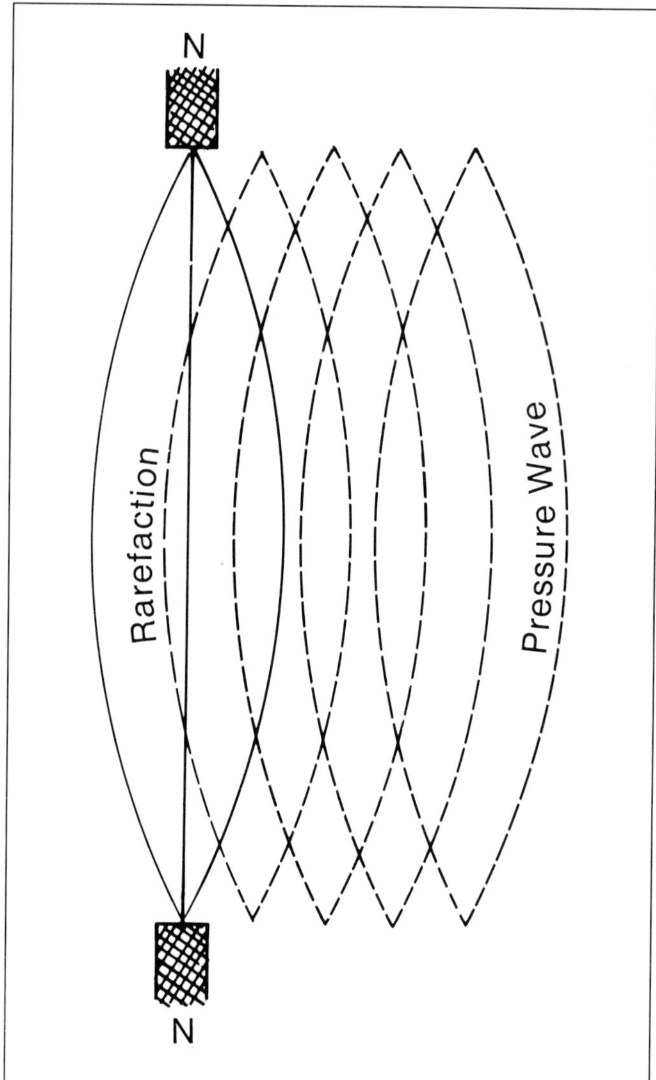

Figure 9-1. As the vibrating plate moves back and forth, it compresses the air in the direction of its motions. When it reverses direction, it produces a partial vacuum, or rarefaction, imparting energy, which radiates away from the plate as sound to the air.

tion, of the air. These alternate compressions and rarefactions cause small but repeated fluctuations in the atmospheric pressure that extend outward from the plate. When these pressure variations strike an eardrum, they cause it to vibrate in response to the slight changes in atmospheric pressure. The disturbance of the eardrum is translated into a neural sensation in the inner ear and is carried to the brain where it is interpreted as sound (Figure 9-2).

Sound is invariably produced by vibratory motion of some sort. The sounding body must act on some medium to produce vibrations that are characteristic of sound. Any type of vibration may be a source of sound, but by definition, only longitudinal vibration of the conducting medium is a sound wave.

Sound waves

Sound waves are a particular form of a general class of waves known as elastic waves. Sound waves can only occur in media

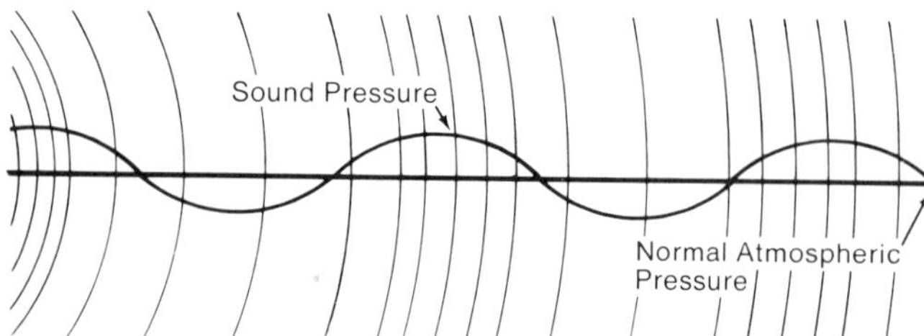

Figure 9-2. Air is an elastic medium and behaves as if it were a succession of adjoining particles. The resulting motion of the medium is known as wave motion, and the instantaneous form of the disturbance is called a sound wave.

that have the properties of mass (inertia) and elasticity. Because air possesses both inertia and elasticity, a sound wave can be propagated in air. One sound wave may have three times the frequency and one third the amplitude of another sound wave. However, if both waves cross their respective zero positions in the same direction at the same time, they are said to be in phase (Figure 9-3).

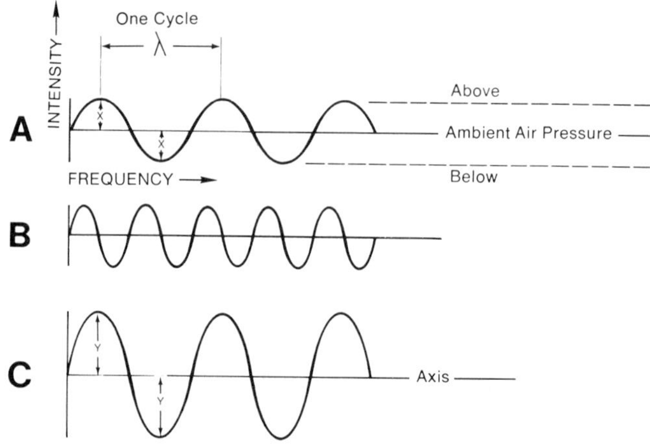

Figure 9-3. The curves shown are pictorial representations of sound waves. Frequency is related to pitch, and intensity is related to the loudness of the sound. Curve B represents a sound that would be higher in pitch than a sound represented by curve A, because the variations in air pressure, as represented by a point on the curve, cross the axis more often. The intensity (loudness) of a sound can be shown by the height of the curve; the sound represented by curve C is louder than the sound represented by curve A (distance Y is greater than distance X).

Frequency

Frequency is the number of times per second that a point on the sound source is displaced from its position of equilibrium, rebounds through the equilibrium position to a maximum displacement opposite in direction to the initial displacement, and then returns to its equilibrium position. In other words, frequency is the number of times per second a vibrating body traces one complete cycle of motion (Figure 9-4). The time required for each cycle is known as the period of the wave and is simply the reciprocal of the frequency. The description of frequency, which used to be "cycles per second," is now Hertz, abbreviated Hz.

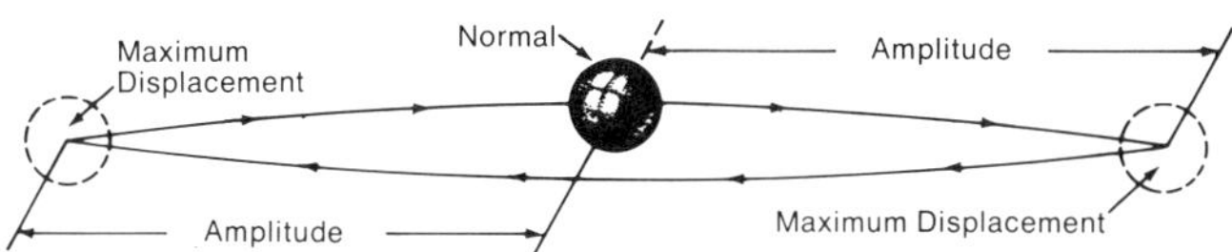

Figure 9-4. Relative positions of an air molecule during one complete cycle of motion.

Frequency is perceived as pitch. The audible range of frequencies for humans with good hearing is between 20 Hz and 20,000 Hz. Most everyday sounds contain a mixture of frequencies generated by a variety of sources. A sound's frequency composition is referred to as its spectrum. The frequency spectrum can be a determining factor in the level of annoyance caused by noise; high-frequency noise generally is more annoying than low-frequency noise. Also, narrow frequency bands or pure tones (single frequencies) can be somewhat more harmful to hearing than broadband noise.

Wavelength

Wavelength is the distance measured between two analogous points on two successive parts of a wave. In other words, wavelength is the distance that a sound wave travels in one cycle. The Greek letter lambda (λ) is used to express wavelength, and it is measured in feet or meters (Figure 9-3-A).

Wavelength is an important property of sound. For example, sound waves that have a wavelength that is much larger than the size of an obstacle are little affected by the presence of that obstacle; the sound waves will bend around it. This bending of the sound around obstacles is called diffraction.

If the wavelength of the sound is small in comparison with the size of the obstacle (such a wavelength is generated by high-frequency sounds), the sound will be reflected or scattered in many directions, and the obstacle will cast a shadow. Actually, some sound is diffracted into the shadow, and there is significant reflection of the sound. As a consequence of diffraction, a wall is of little use as a shield against low-frequency sound (long wavelength), but it can be an effective barrier against high-frequency sound (short wavelength) (Figure 9-5).

Velocity

The velocity with which the analogous pressure points on successive parts of the wave pass a given point is called the speed of sound. The speed of sound is always equal to the product of the wavelength and the frequency:

$$c = f\lambda$$

In the formula, c=speed of sound (feet or meters per second); f= frequency of sound (Hertz); and λ=wavelength (feet or meters).

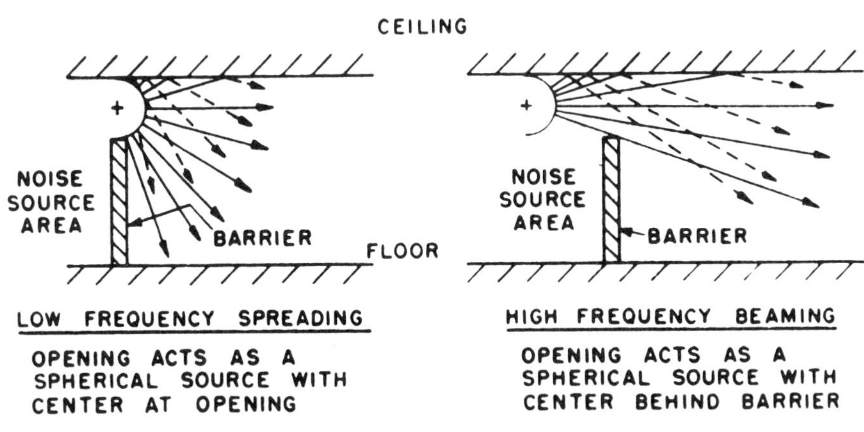

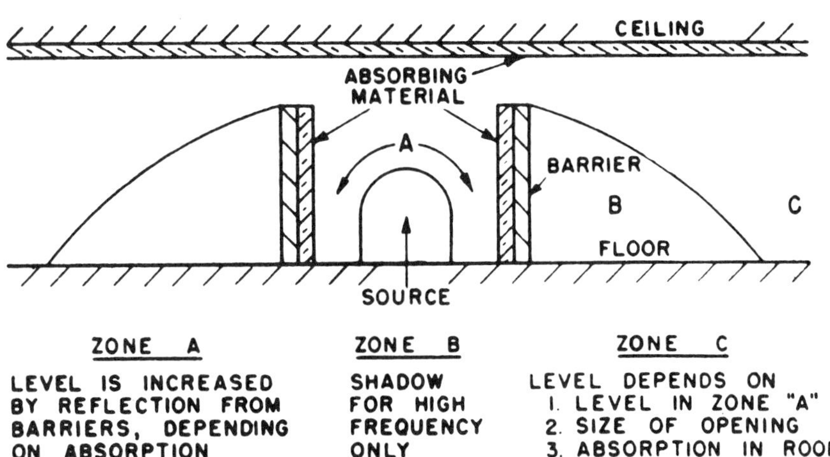

Figure 9-5. The effects of a barrier as a shield to contain noise of low or high frequency.

The speed with which the sound disturbance spreads depends on the mass and elastic properties of the medium. In air at 72 F, the speed of sound is about 344 m/sec (1,130 ft/sec). Its effects are commonly observed in echoes and in the apparent delay between a flash of lightning and the accompanying thunder.

In a homogeneous medium, the speed of sound is independent of frequency; that is, in such a medium, sounds of all frequencies travel at the same speed. When traveling through different media, the speed of sound increases as the medium becomes more dense and less compressible. For example, the speed of sound is approximately 1,433 m/sec in water, 3,962 m/sec in wood, and 5,029 m/sec in steel. Sound, therefore, may be transmitted through many media before it is eventually transmitted through air to the ears of the receiver.

Sound pressure

Sound is a slight, rapid variation in atmospheric pressure caused by some disturbance or agitation of the air. The sounds of normal conversation amount to sound pressure of only a few millionths of a pound per square inch yet can be easily heard because of the remarkable sensitivity of the human ear. The noises that can damage our hearing amount to only a few thousandths of a pound per square inch.

Most common sounds consist of a rapid, irregular series of positive pressure disturbances (compressions) and negative pressure disturbances (rarefactions) measured against the equilibrium pressure value. If we were to measure the mean value of a sound pressure disturbance, we would find it would be zero, because there are as many positive compressions as negative rarefactions. Thus, the mean value of sound pressure is not a useful measurement. We must look for a measurement that permits the effects of rarefactions to be added to (rather than subtracted from) the effects of compressions.

The root-mean-square (rms) sound pressure is one such measurement. The rms sound pressure is obtained by squaring the value of the sound pressure disturbance at each instant of time. The squared values are then added and averaged over the given time. The rms sound pressure is the square root of this time average. Since the squaring operation converts all the negative sound pressures to positive squared values, the rms sound pressure is a useful, nonzero measurement of the magnitude of the sound wave. The units used to measure sound pressure are: micropascals (μPa); newtons per square meter (N/m^2); microbars (μbar); and dynes per square centimeter (d/cm^2). Relations among these units are as follows: $1\ \mu\text{bar} = 1\ d/cm^2 = 0.1\ N/m^2 = 0.1$ Pa.

Decibels and sound pressure level

Even though the weakest sound pressure perceived as sound is a small quantity, the *range* of sound pressure perceived as sound is extremely large. The weakest sound that can be heard by a person with very good hearing in an extremely quiet location is known as the threshold of hearing. At a reference tone of 1,000 Hz, the threshold of hearing is equal to a sound pressure of 20 μPa (0.0002 μbar). The threshold of pain, or the greatest sound pressure that can be perceived without pain, is approximately 10 million times greater. It is therefore more convenient to use a *relative* scale of sound pressure rather than an absolute scale.

For this purpose, the bel, a unit of measure in electrical-communication engineering, is used. The decibel (dB) is the preferred unit for measuring sound. One decibel is one-tenth of a bel and is the minimum difference in loudness that is usually perceptible. The decibel is used to describe a level of quantities that is proportional to sound power. The general relationship for the decibel is:

$$L = 10 \log q/q_0 \quad [\text{dB}]$$

In the formula, L=level; q=the quantity whose level is being obtained; q_0=the reference level of a like quantity; and log= logarithm to the base 10.

In free progressive plain and spherical sound waves, the mean-square pressure is proportional to sound power. We may therefore define a quantity, called the sound pressure level, that describes the sound pressure in a sound field:

$$L_p = 10 \log p^2/p_0^2 = 20 \log p/p_0 \quad [\text{dB}]$$

In this formula, L_p = the sound pressure level; p = rms sound pressure; p_0 = a reference sound pressure; and log = logarithm to the base 10.

The use of the term "level" indicates that the given quantity has a certain level above a certain reference quantity. The reference pressure (p_0) sound is the threshold of hearing for an "average" person at a reference tone of 1,000 Hz, and has a value of 20 μPa.

The entire range of audible sound pressure (for individuals with normal hearing, a range of more than 10 million to one) can be compressed into a practical scale of sound pressure levels from 0–140 dB (Table 9-B).

Table 9-B. Levels of Some Common Sounds

Sound Pressure, P N/m² (Pascal)	*Sound Pressure Level, L_p dB re 20 μN/m² (μPascal)*	*Sound Source*
100,000 (1 bar)	194	Saturn rocket
20,000.0	180	
	170	
2,000.0	160	Ram jet
	150	Turbo jet
200.0	140	Threshold of pain
	135	
	130	Pipe organ
20.0	120	Riveter, chipper
	110	Punch press
2.0	100	Passing truck
	90	Factory
.2	80	Noisy office
	70	
.02	60	Conversational speech
	50	Private office
.002	40	Average residence
	30	Recording studio
.0002	20	Whisper
	10	Threshold of good hearing
.00002	0	Threshold of excellent youthful hearing

The instrument used to measure sound pressure level is called a sound level meter and consists of a microphone, attenuator, amplifier, and indicating meter. The sound level meter is calibrated in decibels to directly indicate sound pressure level.

Because a logarithmic scale is used, a small increase in decibels represents a large increase in sound energy. Technically, each increase of 3 dB represents a doubling of sound energy, an increase of 10 dB represents a tenfold increase, and a 20-dB increase represents a 100-fold increase in sound energy.

It is important to note that the decibel scale of measurement is not unique to the description of sound pressure level. By definition, the decibel is a dimensionless unit that is related to the logarithm of the ratio of a measured quantity to a reference quantity. The decibel has no meaning unless a reference quantity is specified. Because of the mathematical properties of the logarithmic function, the decibel scale can compress data involving entities of large and small magnitude into a relative scale involving small numbers. The decibel is commonly used to describe levels of acoustic intensity, acoustic power, hearing thresholds, electrical voltage, electrical current, electrical power, and so forth, as well as sound pressure levels.

Loudness

Although loudness depends primarily on sound pressure, it is also affected by frequency. (Pitch is closely related to frequency.) The reason for this is that the human ear is more sensitive to high-frequency sounds than it is to low-frequency sounds.

The upper limit of frequency at which airborne sounds can be heard depends primarily on the condition of a person's hearing and on the intensity of the sound. For young adults, this upper limit is usually quoted as being somewhere between 16,000 and 20,000 Hz. For most practical purposes, the actual figure is not important. It is important, however, to realize that most people lose sensitivity for the higher-frequency sounds as they grow older (presbycusis).

The complete hearing process seems to consist of a number of separate processes that, in themselves, are fairly complicated. No simple relationship exists between the physical measurement of a sound pressure level and the human perception of the sound. One pure tone may sound louder compared with another pure tone, even though the measured sound pressure level is the same in both cases.

Sound pressure levels, therefore, are only a part of the story and can be deceiving. The fundamental problem is that the quantities to be measured must include a person's reaction to the sound—a reaction that may be determined by such varied factors as the state of the person's health, characteristics of the sound, and attitude toward the person or device that generates the sound. In the course of time, various loudness level rating methods have been suggested, and a number of different criteria for tolerable noise levels have been proposed.

A complete physical description of sound must include its frequency spectrum, its overall sound pressure level, and the variation of both of these quantities with time. Loudness is the human subjective response to sound pressure and intensity. At any given frequency, loudness varies directly as sound pressure and intensity vary, but not in a simple, straight-line manner.

The physical characteristics of a sound as measured by an instrument and the "noisiness" of a sound as a subjective characteristic may bear little relationship one to the other. A sound level meter cannot distinguish between a pleasant sound and an unpleasant one. A human reaction is required to differentiate between a pleasant sound and a noise. Loudness is not merely a question of sound pressure level. A sound that has a constant sound pressure can be made to appear quieter or louder by changing its frequency.

Equal-loudness contours. Experiments designed to determine the response of the human ear to sound were reported by Fletcher and Munson in 1933. A reference tone and a test tone were presented alternately to the test subjects (young men), who were asked to adjust the level of the test tone until it sounded as loud to them as the reference tone (1,000 Hz). The results of these experiments yielded the familiar Fletcher-Munson, or equal-loudness, contours (Figure 9-6).

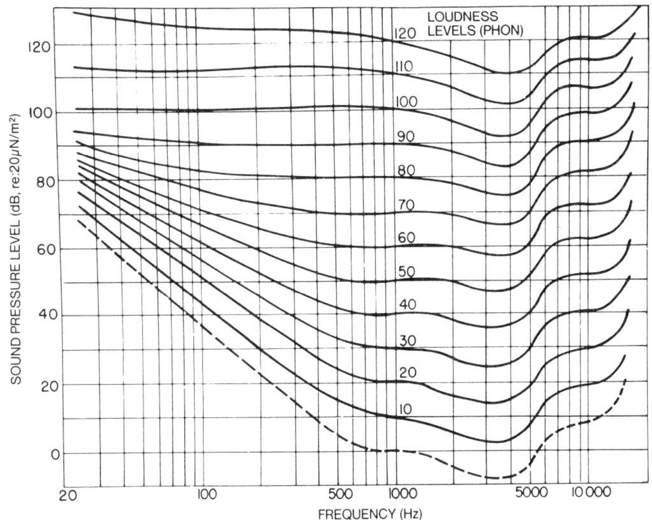

Figure 9-6. Free-field equal-loudness contours of pure tones. Because the human ear is more sensitive to the higher frequencies of sound, changing the frequency of a sound changes its relative loudness. These are also called Fletcher-Munson contours. (Adapted from the *Handbook of Noise Measurement,* 9th ed. GenRad, Inc., 1980)

The contours represent the sound pressure level necessary at each frequency to produce the same loudness response in the average listener. The nonlinearity of the ear is represented by the changing contour shapes as the sound pressure level is increased, a phenomenon that is particularly noticeable at low frequencies. The lower, dashed curve indicates the threshold of hearing, which represents the sound pressure level necessary to trigger the sensation of hearing in the average listener. The actual threshold varies as much as ±10 dB among healthy individuals.

Sound pressure weighting. It would seem relatively simple to build an electronic circuit whose sensitivity varied with frequency in the same way as the human ear. This has in fact been done and has resulted in three different internationally standardized characteristics termed weighting networks "A," "B," and "C." The A-network was designed to approximate the equal loudness curves at low sound pressure levels, the B-network was designed for medium sound pressure levels, and the C-network was designed for high levels.

The weighting networks are the sound level meter's means

of responding to some frequencies more than to others. The very low frequencies are discriminated against (attenuated) quite severely by the A-network, moderately filtered out by the B-network, and hardly attenuated at all by the C-network (Figure 9-7). Therefore, if the measured sound level of a noise is much higher on C-weighting than on A-weighting, much of the noise energy is probably of low frequency.

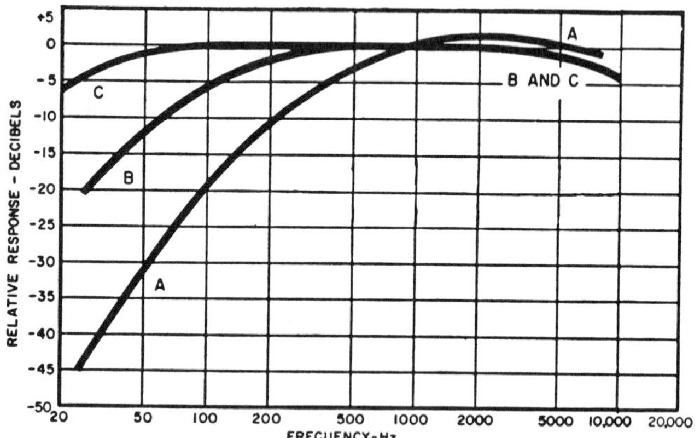

Figure 9-7. Frequency-response attenuation characteristics for the A-, B-, and C-weighting networks.

The A-weighted sound level measurement has become popular in the assessment of the overall noise hazard since this level is thought to provide a rating of industrial broadband noises that indicates the injurious effects such noise has on the human ear.

As a result of its simplicity in rating the hazard to hearing, the A-weighted sound level has been adopted as the measurement for assessing noise exposure by the American Conference of Governmental Industrial Hygienists (ACGIH). The A-weighted sound level as the preferred unit of measurement was also adopted by the U.S. Department of Labor as part of the Occupational Safety and Health Standards. The A-weighted sound levels have also been shown to provide reasonably good assessments of speech interference and community disturbance conditions and have been adopted for these purposes (Figure 9-8).

A-weighted sound levels as single-number ratings have shown excellent agreement in some cases between the A-level and its subjective effects. However, in other cases, relatively wide discrepancies occur, particularly in relation to high-level, narrow-band noise or pure tones that greatly exceed broadband noises. The most consistent results are obtained when the noises that are being compared are similar in character.

OCCUPATIONAL DAMAGE-RISK CRITERIA

The purpose of damage-risk criteria is to define maximum permissible noise levels during given periods which, if not exceeded, would result in acceptable small changes in the hearing levels of exposed employees over a working lifetime. The acceptability of a particular noise level is a function of many variables.

Increasing attention is being given by regulatory agencies and industrial and labor groups to the effects of noise exposures on employees; therefore, equitable, reliable, and practical damage-risk noise criteria are needed.

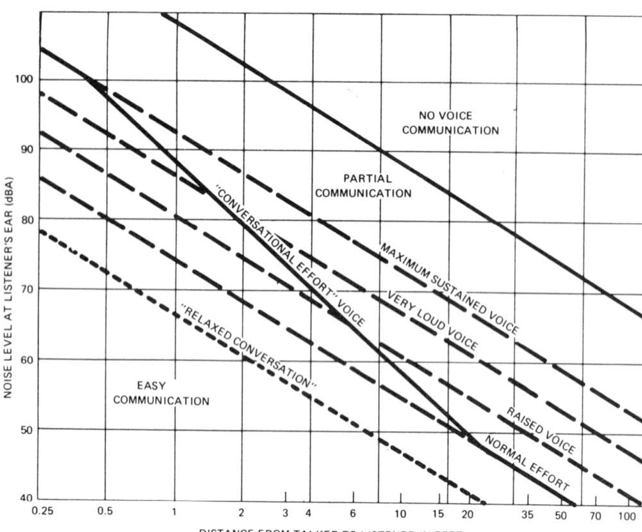

Figure 9-8. Distance at which ordinary speech can be understood (as a function of the A-weighted sound levels of the masking noise in an outdoor environment). (Reprinted from *Public Health and Welfare Criteria of Noise*, July 27, 1973, U.S. Environmental Protection Agency)

A criterion is a standard, rule, or test by which a judgment can be formed. A criterion for establishing levels for damage-risk noise requires one or more standards for judgment. Damage-risk criteria can be developed once standards are selected by which the effects of occupational noise exposure on employees can be judged.

Hearing ability

Tests for evaluating the ability to hear speech have been developed. These tests generally fall into two classes: (1) those that measure the hearing threshold or the ability to hear very faint speech sounds, and (2) those that measure discrimination, or the ability to understand speech (see Chapter 4, The Ears).

Ideally, hearing impairment should be evaluated in terms of an individual's ability (or inability) to hear everyday speech under everyday conditions. The ability of an individual to hear sentences and to repeat them correctly in a quiet environment is considered to be satisfactory evidence of adequate hearing ability. Hearing tests using pure tones are extensively employed to monitor the status of a person's hearing and the possible progression of a hearing loss. A person's ability to hear pure tones is related to the hearing of speech.

A person who works in a noisy environment should have his or her hearing checked periodically to determine if the noise exposure is producing a detrimental effect on hearing. The noise-induced hearing losses that can be measured by pure-tone audiometry are those threshold shifts that constitute a departure from a specified baseline. This baseline, or normal hearing level, has been defined as "the average hearing threshold of a group of young people who have no history of previous exposure to intense noise and no otological malfunction."

AAOO—AMA guide. After years of studying the various methods and procedures used to determine hearing damage, the Committee on Conservation of Hearing of the American Academy of Ophthalmology and Otolaryngology (AAOO) issued a report that was published by the American Medical Association (AMA) called the *Guide for the Evaluation of Hearing Impairment* (see Chapter 4). The report recommended that the average hearing level for pure tones at 500 Hz, 1,000 Hz, and 2,000 Hz be used as an indirect measure of the probable ability to hear everyday speech. If the average monaural hearing level at 500, 1,000, and 2,000 Hz is 25 dB or less, the AMA report stated that no impairment in the ability to hear everyday speech usually occurs under everyday conditions (see Chapter 4, The Ears).

Risk factors

If the ear is subjected to high levels of noise for a sufficient period of time, some loss of hearing may occur. There are many factors that affect the degree and extent of hearing loss: the intensity of the noise (sound pressure level); the type of noise (frequency spectrum); the period of exposure each day (duty cycle per day); the total work duration (years of employment); individual susceptibility; the age of the worker; coexisting hearing loss and ear disease; the character of the surroundings in which the noise is produced; the distance from the source; and the position of the ears with respect to sound waves. The first four are the most important factors, and they are referred to as *noise exposure* factors. Thus, it is not only necessary to know how much noise is present, but also the kind of noise present and its duration.

Because of the complex relationship of noise and exposure time to threshold shift (reduction in hearing level) and its many possible contributory causes, the establishment of criteria designed to protect workers against hearing loss required many years to develop.

A relatively recent effort to establish the basis for reliable noise criteria was accomplished by the Intersociety Committee on Guidelines for Noise Exposure Control. A significant part of their report is shown graphically in Figure 9-9. These curves in Figure 9-9 relate the incidence of significant hearing loss to age and the magnitude of noise exposure over a working lifetime.

Without attempting to explain the full significance of the graph, it can be stated that 20 percent of the general population between age 50 and 59 will experience hearing losses without any exposure to industrial noise. Groups of workers exposed to steady-state industrial noises over a working lifetime will show a greater increase in the incidence of hearing loss.

For example, exposure to steady-state noise at 90 dB on the A-scale of the sound level meter (90 dBA) will result in significant hearing losses to 27 percent of the exposed group. If the working lifetime exposure is 95 dBA, 36 percent of the group will show significant hearing loss.

Essentially, this graph supplies industry and other interested groups with information from which the risk of developing compensable hearing loss among groups of workers exposed to noises of different magnitudes can be predicted.

Analysis of noise exposure

The critical factors in the analysis of noise exposures are (1) the A-weighted sound level, (2) the frequency composition, or spectrum, of the noise, and (3) the duration and distribution of noise exposure during a typical workday.

It is currently believed that any exposure of unprotected ears to sound levels in excess of 115 dBA is hazardous and should

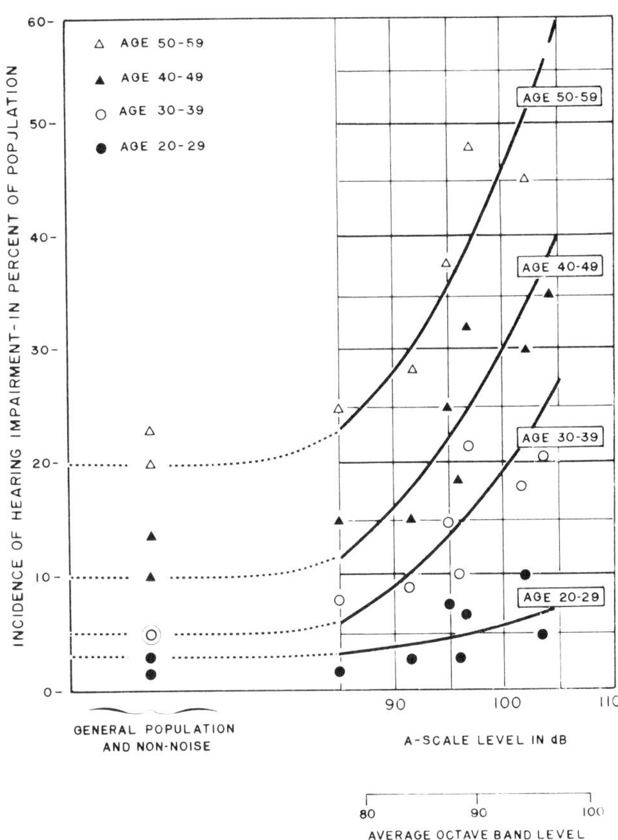

Figure 9-9. The incidence of hearing impairment in the general population and in selected populations by age groups and by occupational noise exposure. (Reprinted with permission from the *American Industrial Hygiene Association Journal*)

be avoided. Exposure to sound levels below 70 dBA can be assumed to be safe and will not produce any permanent hearing loss. The majority of industrial noise exposures fall within this 45-dBA range; thus, additional information is required, such as the type of noise and duration of exposure.

It would be very helpful to know the predominant frequencies present and the contributions from each of the frequency bands that make up the overall level. It is currently believed that noise energy with predominant frequencies above 500 Hz has a greater potential for causing hearing loss than noise energy concentrated in the low-frequency regions. It is also believed that noises that have a sharp peak in a narrow-frequency band (such as a pure tone) present a greater hazard to hearing than those noises of equal energy levels that have a continuous distribution of energy across a broad-frequency range.

The incidence of noise-induced hearing losses is directly related to total exposure time. In addition, it is believed that intermittent exposures are far less damaging to the ear than are continuous exposures, even if the sound pressure levels for the intermittent exposures are considerably higher than those during continuous exposures. The rest periods between noise exposures allow the ear to recuperate.

At present, the deleterious effects of noise exposure and the energy content of the noise cannot be directly equated. For example, doubling the energy content does not produce twice the hearing loss. In general, however, the greater the total energy

Table 9-C. Acceptable Exposures to Noise in dB(A) as a Function of the Number of Occurrences per Day

Daily Duration		*Number of Times the Noise Occurs Per Day*						
Hours	*Min*	*1*	*3*	*7*	*15*	*35*	*75*	*160 up*
8		90	90	90	90	90	90	90
6		91	93	96	98	97	95	94
4		92	95	99	102	104	102	100
2		95	99	102	106	109	114	
1		98	103	107	110	115		
	30	101	106	110	115			
	15	105	110	115				
	8	109	115					
	4	113						

This table summarizes the results of TTS studies that may be used to estimate the effect of intermittency of noise exposures on risk of hearing impairment. The information in the table may be approximated by the simple rule that for each halving of daily exposure time the noise level may be increased by 5 dB without increasing the hazard of hearing impairment. To use the table, select the column headed by the number of times the noise occurs per day, read down to the average sound level of the noise, and locate directly to the left in the first column the total duration of noise permitted for any 24-hour period. It is permissible to interpolate if necessary. Noise levels are in dBA. (Adapted from Intersociety Committee Report [1970] "Guidelines for Noise Exposure Control," *Journal of Occupational Medicine*, July 1970, Vol. 12, No. 7)

content of the noise, the shorter the time of exposure required to produce the same amount of hearing loss. However, the exact relation between time and energy is not known.

Another factor that should be considered in the analysis of noise exposures is the type of noise. For instance, impact noise is noise generated by drop hammers and punch presses, and steady-state noise is noise generated by turbines and fans. Impact noise is a sharp burst of sound; therefore, sophisticated instrumentation is necessary to determine the peak levels for this type of noise. Additional research needs to be done to fully define the effects of impact noise on the ears (Table 9-C).

The total noise exposure during a person's normal working lifetime must be known to arrive at a vaild judgment of how noise will affect that person's hearing. Instruments, such as noise dosimeters, can be used to determine the exposure pattern of a particular individual. Instruments, such as sound level meters, can be used to determine the noise exposure at a given instant in time (that is, during the time the test is being taken). An exposure pattern can be established using a series of such tests and the work history of the individual.

SOUND MEASURING INSTRUMENTS

There is a wide assortment of equipment available for noise measurements, including sound survey meters, sound level meters, octave-band analyzers, narrowband analyzers, tape and graphic level recorders, impact sound level meters, and equipment for calibrating these instruments.

For most noise problems encountered in industry, the sound level meter and octave-band analyzers provide ample information (Figure 9-10).

Sound level meters

The basic instrument used to measure sound pressure variations in air is the sound level meter. This instrument contains a microphone, an amplifier with a calibrated attenuator, a set of frequency-response networks (weighting networks), and an indicating meter (Figure 9-11).

The sound level meter is a sensitive electronic voltmeter that measures the electrical signal emitted from a microphone, which is ordinarily attached to the instrument. The alternating electrical signal emitted from the microphone is amplified suffi-

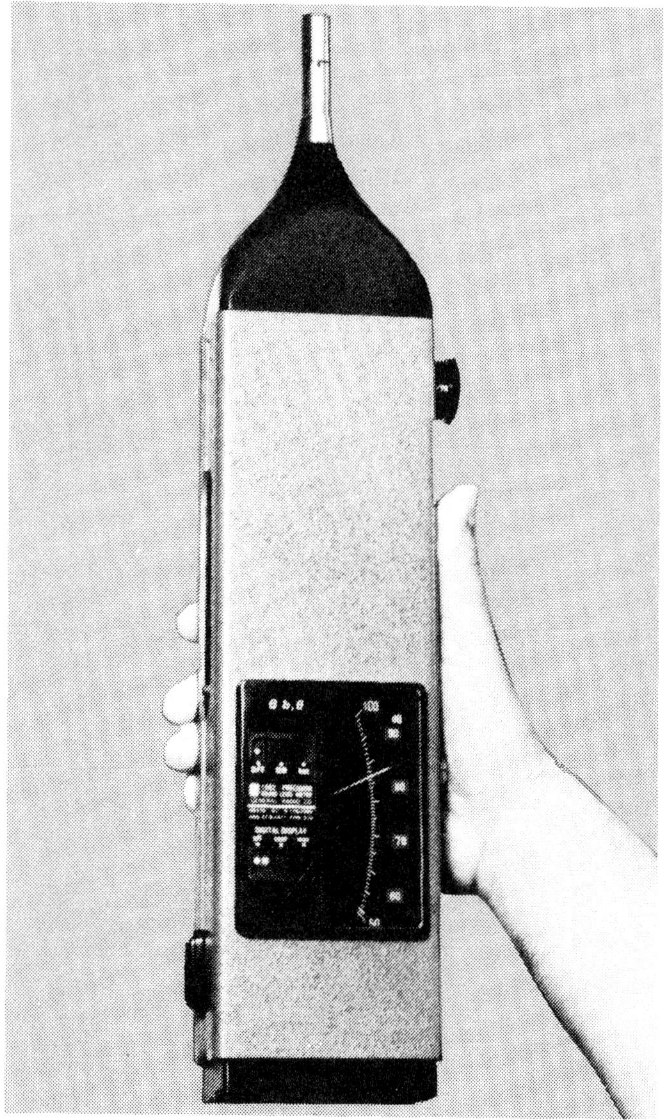

Figure 9-10. The muiltipurpose instrument shown here can be used as sound level meter, octave-band analyzer, and impact/impulse noise meter. (Courtesy GenRad)

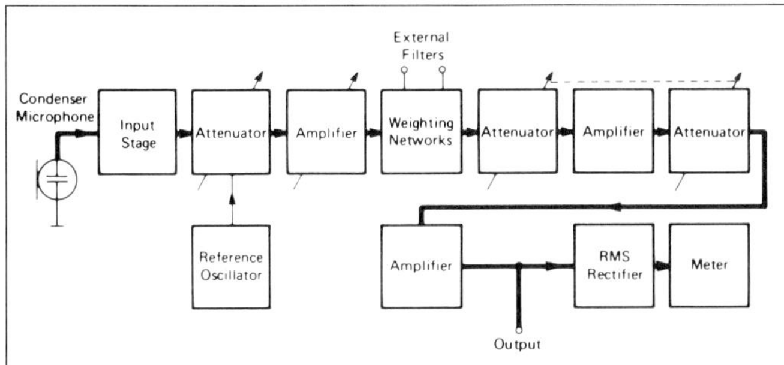

Figure 9-11. Schematic diagram of a sound level meter with auxiliary output.

ciently so that, after conversion to direct current by means of a rectifier, the signal can deflect a needle on an indicating meter. An attenuator controls the overall amplification of the instrument. The response v. frequency characteristics of the amplified signal are controlled by the weighting networks.

Some sound level meters have a measurement range of about 40–140 dB (re20μN/m^2) without the aid of special accessory equipment. Special microphones permit measurement of lower or of considerably higher sound levels. An amplifier able to register the electrical output signal of the microphone is usually provided with the sound level meter for hookup to other instruments for recording and analysis. The sound level meter is basically designed to be a device for field use and as such should be reliable, rugged, reasonably stable under battery operation, and lightweight.

Microphone. The microphone responds to sound pressure variations and produces an electrical signal that is processed by the sound level meter.

Amplifier. The amplifier in a sound level meter must have a high available gain so that it can measure the low-voltage signal from a microphone in a quiet location. It should have a wide frequency range, usually on the order of 20–20,000 Hz. The range of greatest interest in noise measurements is 50–6,000 Hz. The inherent electronic noise floor and hum level of the amplifier must be low.

Attenuators. Sound level meters are used for measuring sounds that differ greatly in level. A small portion of this range is covered by the relative deflection of the needle on the indicating meter. The rest of the range is covered by an adjustable attenuator, which is an electrical resistance network inserted in the amplifier to produce known ranges of signal level. To simplify use, it is customary to have the attenuator adjustable in steps of 10 dB.

In some instances, the attenuator may be split into sections among various amplifying stages to improve the signal-to-noise ratio of the instrument and to limit the dynamic range.

Weighting networks. The sound level meter response at various frequencies can be controlled by electrical weighting networks. The response curves for these particular networks have been established in the American National Standards Institute's publication, ANSI S1.41-1971 (R1976). C-weighting is intended to be applied to a uniform response over the frequency range from 25–8,000 Hz. Changes in the electronic circuit are sometimes made to compensate for the response of particular microphones, so that the net response is uniform (flat) within the tolerance allowed by the standards. The C-weighting network is generally used when the sound level meter supplies a signal to an auxiliary instrument for a more detailed analysis. (The weighting networks are shown in Figure 9-7.) The A-weighting network is used to determine compliance with the OSHA standard.

Metering system. After the electrical signal from the microphone is amplified and sent through the attenuators and weighting networks, the signal is used to drive a metering circuit. This metering circuit displays a value that is proportional to the electrical signal applied to it. The ANSI S1.4-1971 (R1976) standard on sound level meters specifies that the rms value of the signal should be indicated. This requirement corresponds to adding up the different components of the sound wave on an energy basis. When measuring sound, the rms value is a useful indication of the general energy content.

Since the indicating needle on the dial of a sound level meter cannot follow rapidly changing variations in sound pressure, a running average of the rectified output of the metering circuit is shown. The average time (or response speed) is determined by the meter ballistics and by response circuits (chosen using a switch).

Octave-band analyzers

For many industrial noise problems, it is necessary to use some type of analyzer to determine where the noise energy lies in the frequency spectrum. This is especially true if engineering control of noise problems is planned, because industrial noise is made up of various sound intensities at various frequencies (Figure 9-12).

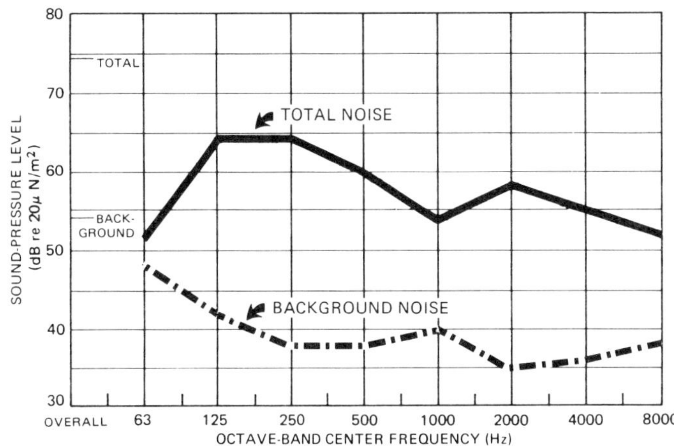

Figure 9-12. Data obtained from an octave-band analysis of a noise source showing the total noise and the background noise levels when the noise source is not in operation.

The identification of pure tone components, when present, is an extremely useful diagnostic tool for finding and quieting the noise source.

Some noise sources have a well-defined frequency content. For example, the hum generated by a fan or blower is usually centered at the blade-passage frequency, which is the product of the number of blades of the fan multiplied by the speed (revolutions per second) of the fan. Electrical current transformers usually hum in frequencies that are multiples of 60 Hz. Positive-displacement pumps have a sound pressure distribution that is directly related to the pressure pulses on either the inlet or the outlet of the pump.

Noise resulting from the discharge of steam or air-pressure relief valves has a frequency peak that is related to the pressure in the system and the diameter of the restriction preceding the discharge to the atmosphere. A peak energy content in any single octave band would provide information as to the predominant frequency of a particular noise source.

Sound analyzers that can be used to make measurements in third-octave and tenth-octave bands are also available. The narrower the band for analysis, the more sharply defined the data.

If a noisy machine is to be used in a room, the acoustical characteristics of the room and the machine's radiated sound power levels as a function of frequency must be known so that sound pressure levels produced by the machine can be estimated.

Noise dosimeters

In many work environments, it may not be adequate to measure noise exposure at a fixed location for the duration of a work shift. The worker may move about to several locations in the course of his or her duties or perform a variety of operations during the day, each generating different noise levels. The practical way to measure the noise exposure in these circumstances is with a noise-exposure monitor, or dosimeter, that can be worn by the worker and that moves with the worker during the day. The noise dosimeter records the noise energy to which the worker was exposed during the workshift.

Depending on the relative positions of the microphone, the employee, and the noise source, the sound level measured by the dosimeter may differ from that measured on a sound level meter.

SOUND SURVEYS

Sound measurement falls into two broad categories: source measurement and ambient-noise measurement. Source measurements involve the collection of acoustical data for the purpose of determining the characteristics of noise radiated by a source. The source may be a single piece of equipment or may involve a combination of equipment or systems. For example, an electric motor or an entire plant may be considered a noise source.

Ambient-noise measurements range from studying a single sound level to making a detailed analysis showing hundreds of components of a complex vibration. The number of measurements taken and the type of instruments needed depend on the information that is required. If compliance with a certain noise specification must be checked, the particular measurement required is reasonably clear. Only some guidance as to the selection of instruments and their use is needed. But if the goal is to reduce the noise produced by industrial operations in general, the situation is more complex, and careful attention to the acoustic environment is essential.

Measurement of the noise field may require using different types of sound-level measuring instruments. These measurements must be repeated as changes in noise-producing equipment or operating procedures occur.

The use of the dBA scale for preliminary noise measurement greatly simplifies the collection of sound level survey data. Detailed sound level survey and octave-band analysis data are necessary to provide sufficient information so that the proper remedial measures in noise control procedures can be determined. Calibration checks of the instruments should be made before, during, and after the sound level survey.

Source measurements

Source measurements frequently are made in the presence of noise created by other sources that form the background- or ambient-noise level. Although it is not always possible to make a clear distinction between source and ambient-noise measurement, it is important to understand that source measurements describe the characteristics of a particular sound source, while ambient-noise measurements describe the characteristics of a sound field due largely to unspecified or unknown sources.

A uniform, standard reporting procedure should be established to ensure that sufficient data are collected in a proper form for subsequent analysis. To be effective, this standard reporting procedure should include detailed descriptions of the techniques of measurement position, operating conditions, instrument calibration, exposure time, amplitude patterns, and other important variables.

Several forms have been devised to record data obtained during a screening survey. Use of these forms will facilitate the recording of pertinent information that will be extremely useful if more detailed studies are conducted later. An employee noise exposure survey is conducted by measuring noise levels at each workstation that an employee occupies throughout the day or by acquiring a sufficient sampling of data at each workstation so that the exposure of an employee while at that workstation can be evaluated. Workstations that pose particular noise exposure hazards can be readily identified using measured sound level contours if these are obtained in adequate detail.

In many industrial situations, however, it is extremely difficult to accurately evaluate the noise exposure to which a specific worker is subjected. This is due in part to the fact that the noise level to which the stationary worker is exposed throughout the working day fluctuates, making it difficult to evaluate compliance or noncompliance with the OSHA regulations. The problem can be due also to the fact that the worker's job may require that he or she spend time in areas where the noise levels vary from very low to very high.

Because of the fluctuating nature of many industrial noise levels, it would not be accurate or meaningful to use a single sound-level meter reading to estimate the daily TWA noise level.

Preliminary noise survey

A hearing conservation program should start with a preliminary plant-wide noise level survey using appropriate sound-level measuring equipment to locate operations or areas where workers may be exposed to hazardous noise levels.

Those conducting the survey will have to decide whether to purchase sound-level measuring equipment and train personnel to use it or whether the work should be contracted to an outside firm. The extent of the noise problem, the size of the plant, and the nature of the work will affect this decision. In most plants, noise surveys are conducted by a qualified engineer, an audiologist, an industrial hygienist, or a health and safety professional.

A noise survey should be carried out at work areas where it is difficult to communicate in normal tones. A noise survey should also be performed if people, after being exposed to high noise levels during their workshift, notice that speech and other sounds are muffled for several hours or if they develop ringing in the ears.

As a general guideline for conducting a noise survey, the information recorded should be sufficient to allow another individual to take the report, use the same equipment, locate the various measurement locations, and finally reproduce the measured and/or recorded data.

The preliminary noise survey normally does not define the noise environment in depth and therefore should not be used to determine employee exposure time and other details. The preliminary noise survey simply supplies sufficient data that can be used to determine if a potential noise problem exists and to indicate to what extent it may exist.

Detailed noise survey

From the preliminary noise survey, it is relatively easy to determine specific locations that require more detailed study and attention. A detailed noise study should then be made at each of these locations to determine the employee's TWA exposure.

The purposes of a detailed noise survey are to: (1) obtain specific information on the noise levels existing at each employee's work station; (2) develop guidelines for establishing engineering and/or administrative controls; (3) define areas where hearing protection will be required; and (4) determine those work areas where audiometric testing of employees is desirable and/or required. In addition, detailed noise survey data can be used to develop engineering control policies and procedures and to determine whether specific company, state, or federal requirements have been complied with.

An effective hearing conservation program always starts with the question, "Does a noise problem exist?" The answer must not simply be based on the subjective feeling that the problem exists but on the results of a careful technical definition of the problem. Answers to the following questions must be obtained:

- How noisy is each work area?
- What equipment or process is generating the noise?
- Which employees are exposed to the noise?
- How long are they exposed?

First-line supervisors can provide basic job function information concerning the duration of operation, the types of noise-producing equipment in work areas, and the percentage of time the worker spends in each of the areas. Production records can be examined, and on-site evaluation can provide information as to the extent of the noise problem.

The noise survey should be made using a general purpose sound level meter that meets standards set by ANSI S1.4-1971 (R1976), *Specifications for Sound Level Meters, Type 1 or 2.* The sound level meter should be set for A-scale slow response.

Measurements of noise exposure should be taken at approximate ear level (Figure 9-13). No worker should be exposed to steady-state or interrupted steady-state sound levels that exceed the maximum listed in the current noise standard. Other information should include the name of the individual making the noise survey as well as the date, location of measurement, and time the measurement was made. The serial numbers of the sound level meters and the date of calibration are also essential for compliance records.

The noise survey procedure is a three-step process.

Step 1—Area Measurements. Using a sound level meter set for A-scale slow response, the regularly occurring maximum noise level and the regularly occurring minimum noise level are recorded at the center of each work area. (For measurement purposes, the size of the work area should be limited to 93 m^2 [1,000 sq ft] or smaller.) If the maximum sound level does not exceed 80 dBA, it can be assumed that all employees in this work area are working in an environment with a satisfactory noise level. If the noise levels measured at the center of the work area fall between 80 and 92 dBA, then more information is needed.

Figure 9-13. Take sound level measurements near an employee's workstation.

Step 2—Workstation measurements. To evaluate the noise exposure for people working in locations where measurements at the center of the work area range between 80 and 92 dBA, measurements should be made at each employee's normal workstation. If the level varies on a regular basis, both the maximum and minimum levels should be recorded. If the noise level never goes below 90 dBA, an unsatisfactory noise exposure is indicated. If the measured level is never greater than 85 dBA, the noise exposure to which the employee is subjected can be regarded as satisfactory.

Step 3—Exposure duration. At workstations where the regularly occurring noise varies above and below the 85-dBA level, further analysis is needed.

If an employee performs varying work patterns in different work areas, it is necessary to ascertain the sound level and duration of noise exposure within each work area. Such information may be obtained by consulting the employee or the employee's supervisor, or by visual monitoring. A briefing/debriefing approach for a particular day's activities may also be used. This is performed by requesting each employee to keep a general work area/time log of his or her daily activities. The employee is then debriefed at the end of the work period to ensure that sufficient information was logged. In many cases, it may be desirable for an employee to wear a noise dosimeter that records daily exposure in terms of the current OSHA requirements.

The procedure for determining an employee's daily noise exposure rating is discussed in the section that follows.

General classes of noise exposure

There are three general classes into which occupational noise exposures may be grouped:

- continuous noise;
- intermittent noise; and
- impact-type noise.

Continuous noise. Continuous noise is normally defined as "broadband noise of approximately constant level and spectrum, to which an employee is exposed for a period of eight hours per day, 40 hours per week." A large number of industrial operations fit into this class of noise exposure. Most damage-risk criteria are written for this type of noise exposure because it is the easiest to define in terms of amplitude, frequency content, and duration.

The OSHA Noise Standard, 29 *CFR* 1910.95(a) and (b), establishes permissible employee noise exposures in terms of duration in hours per day at various sound levels. The Standard requires that the employer reduce employee exposures to the allowable level by use of feasible engineering or administrative controls. The standard defines the permissible exposure level (PEL) as that noise dose that would result from a continuous 8-hour exposure to a sound level of 90 dBA. This is a dose of 100 percent. Doses for other exposures that are either continuous or fluctuating in level are computed relative to the PEL based on a 5-dBA trading relationship between noise level and exposure time (Table 9-D). Every 5-dBA increase in noise level cuts the allowable exposure time in half. This is known as a 5-dBA doubling rate.

Table 9-D. Permissible Noise Exposures

Duration Per Day (hours)	*Sound Level, Slow Response (dBA)*
8	90
6	92
4	95
3	97
2	100
1½	102
1	105
½	110
¼ or less	115

Note: When the daily noise exposure is composed of two or more periods of noise exposure of different levels, their combined effect should be considered, rather than the individual effect of each. If the sum of the following fractions: $C_1/T_1 + C_2/T_2 \ldots C_n/T_n$ exceeds 100 percent, then the mixed exposure should be considered to exceed the limit value. C_n indicates the total time of exposure at a specified noise level, and T_n indicates the total time of exposure permitted at that level.

When employees are exposed to different noise levels during the day, the mixed exposure (E_m) must be calculated by using the following formula:

$$E_m = \frac{C_1}{T_1} + \frac{C_2}{T_2} + \frac{C_3}{T_3} + \ldots \frac{C_n}{T_n}$$

In this formula, C_n equals the amount of time an employee was exposed to noise at a specific level; T_n equals the amount of time the employee may be permitted to be exposed to that level.

If the sum of the fractions equals or exceeds 1, then "the mixed exposure is considered to exceed the limit value." Daily noise dose (D) is an expression of E_m in percentage, e.g., $E_m = 1$, is equivalent to a noise dose of 100 percent. Noise levels below 90 dBA are not considered in the calculation of daily noise dose.

Example 1. An employee is exposed to the following noise levels during the workday:

85 dBA for 3.75 hours
90 dBA for 2 hours
95 dBA for 2 hours
110 dBA for 0.25 hours

Thus, the daily noise dose is as follows:

$$D = 100\left[\frac{3.75}{\text{no limit}} \quad \text{or} \quad 0 + \frac{2}{8} + \frac{2}{4} + \frac{0.25}{0.50}\right] = 125\%$$

Since the dose exceeds 100 percent, the employee received an excessive exposure during the workday.

The permissible exposures given in Table 9-D are based on the presence of continuous noise rather than intermittent or impact-type noise. By definition, "if the variations in noise level involve maxima at intervals of one second or less, it is considered to be continuous."

Example 2. A drill runs for 15 seconds and is off 0.5 second between operations. This noise is rated at its "on" level for an entire 8-hour day. The noise generated by the drill would be "safe" only if the level were 90 dBA or less.

Further interpretation of the Standard indicates that exposure above 115 dBA is not permissible for any length of time.

Intermittent noise. Exposure to intermittent noise may be defined as "exposure to a given broadband sound-pressure level several times during a normal working day." The inspector or plant supervisor who periodically makes trips from a relatively quiet office into noisy production areas may be subject to this type of noise. Criteria established for this type of noise exposure are shown in Table 9-C.

With steady noises, it is sufficient to record the A-weighted sound level attained by the noise. With noises that are not steady, such as impulsive noises, impact noises, and the like, the temporal character of the noise requires additional specification. Both the short-term and long-term variations of the noise must be described. Nonsteady noise exposure measurements are most easily made using dosimeters.

Impact-type noise. Impact-type noise is "a sharp burst of sound," and sophisticated instrumentation is necessary to determine the peak levels for this type of noise. Noise types other than steady ones are often encountered. In general, sounds repeated more than once per second may be considered as steady. Impulsive or impact noise, such as that made by hammer blows or explosions, is generally less than one-half second in duration and does not repeat more often than once per second. Employees should not be exposed to impulsive or impact noise that exceeds a peak sound pressure level of 140 dB.

Brief, intermittent sounds that occur at intervals greater than 1 second should be evaluated in terms of level and total duration during an 8-hour day. If the noise level exhibits peaks at intervals of 1 second or less, the noise may be considered continuous, and the highest value should be used in determining exposure in accordance with Table 9-D.

Individual impulse and impact sounds can be characterized in terms of their rise time, peak sound level, and pulse duration. The rate and number of the impact sounds that constitute an exposure period are factors in judging the possible hazards of these types of sound.

It is not easy to accurately measure the sound levels of rapidly varying staccato noises. Sound level meters do not follow sudden peaks in sound pressure, and they may systematically distort and misrepresent the true sound pressure levels reached by that noise. This is particularly true when measuring impact noise—for example, the noise made by a drop forge.

NOISE CONTROL PROGRAMS

The degree of noise reduction required is determined by comparing the measured levels with acceptable noise levels. The next step is to consider various noise control measures, such as making alterations in engineering design, limiting the time of exposure, or using personal protective devices to achieve the desired level of reduction.

Every noise problem can be broken down into three parts: (1) a source that radiates sound energy, (2) a path along which the sound energy travels, and (3) a receiver, such as the human ear (Figure 9-14). The "system" approach to noise problem analysis and control will assist in understanding both the problem and the changes that will be necessary for noise reduction. If each part of the system—source, path, and receiver—is examined in detail, the overall problem will be greatly simplified. To help translate these principles into practical terms, specific examples of controlling industrial noise exposure are outlined in this section.

Source

The most desirable method of controlling a noise problem is to minimize the noise at the source. This generally means modifying existing equipment and structures or possibly introducing noise reduction measures at the design stage of new machinery and equipment.

Noise path

Because the desired amount of noise reduction cannot always

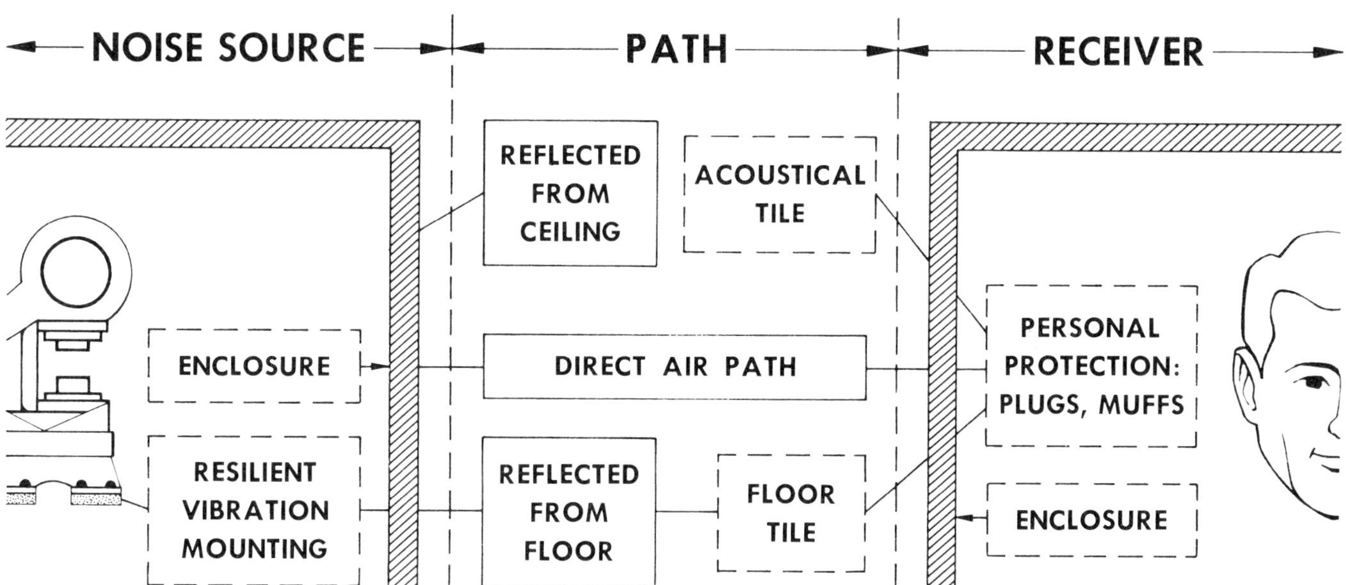

REDUCTION OF NOISE AT SOURCE BY:

1. Acoustical design
   a. Decrease energy for driving vibrating system.
   b. Change coupling between this energy and acoustical radiating system.
   c. Change structure so less sound is radiated.
2. Substitution with less noisy equipment.
3. Change in method of processing.

REDUCTION OF NOISE BY CHANGES IN PATH:

1. Increase distance between source and receiver.
2. Acoustical treatment of ceiling, walls and floor to absorb sound and reduce reverberation.
3. Enclosure of noise source.

REDUCTION OF NOISE AT RECEIVER BY:

1. Personal protection.
2. Enclosures — isolating the worker.
3. Rotation of personnel to reduce exposure time.
4. Changing job schedules.

Figure 9-14. Every noise problem can be broken down into three component parts: (1) a source that radiates sound energy; (2) a path along which the sound energy travels; (3) a receiver, such as the human ear.

be achieved by control at the source, modification at the noise path and at the receiver must also be considered.

Noise reduction along the path can be accomplished in many ways: by shielding or enclosing the source, by increasing the distance between the source and the receiver, or by placing a shield between the source and the receiver. Noise can be reduced along the path by means of baffles and enclosures placed over noise-producing equipment to minimize the transmission of noise to areas occupied by employees. Use of acoustical material on walls, ceilings, and floors to absorb sound waves and to reduce reverberations can result in significant noise reduction.

Noise produced by a source travels outward in all directions. If all of the walls, the floor, and the ceiling are hard reflecting surfaces, all the sound is reflected again and again. The sound level measured at any point in the room is the sum of the sound radiated directly by the source plus all the reflected sounds. Practically all industrial machine installations are located in such environments. These locations are known as semireverberant. Noise measured around a machine in a semireverberant location is the sum of two components: the noise radiated directly by the machine and the noise reflected from the walls, floor, and ceiling.

Close to the machine, most of the noise is radiated directly by the machine. Close to the walls, the reflected component may be predominant. Sound absorption materials applied to the walls and ceiling can reduce the reflected noise but will have no effect on the noise directly radiated by the source.

Enclosures

In many cases, the purpose of an acoustic enclosure is to prevent noise from getting inside. Soundproof booths for machine operators and audiometric testing booths for testing the hearing of employees are examples of such enclosures. More often, however, an enclosure is placed around a noise source to prevent noise from getting outside. Enclosures are normally lined with sound absorption material to decrease internal sound-pressure buildup.

Noise can best be prevented from entering or leaving an enclosure by sealing all outlets. In extreme cases, double structures can be used. Special treatment, including the use of steel and lead panels, is available to prevent noise leakage in certain cases. Gaskets around doors also can reduce noise transmission from one space to another.

Control measures

Noise control can often be designed into equipment so that little or no compromise in the design goals is required. Noise control undertaken on existing equipment is usually more difficult. Engineering control of industrial noise problems requires the skill of individuals who are highly proficient in this field.

Noise control strategies require careful objective analysis on both a practical and economic basis. Complete redesign requires that product and equipment designers immediately consider noise level a primary product or equipment specification in the design of all new products. Full replacement of all products or equipment would eventually take place depending on the service life of each. Many designers feel this approach will minimize cost increases associated with noise control measures. Existing products or equipment modifications would require manufacturers to modify or replace existing products and equipment to lower the noise levels of noisy equipment.

The existing equipment within any plant was probably selected because it was economical and efficient. However, careful acoustical design can result in quieter equipment that would even be more economical to operate than noisier equipment. Examples of noise control measures applied at the source include: the substitution of quieter machines, the use of vibration-isolation mountings, and the reduction of the external surface areas of the vibrating parts as much as possible. Machines mounted directly on floors and walls may cause them to vibrate, resulting in sound radiation. The proper use of machine mounting isolates the machines and reduces the transmission of vibrations to the floors and walls.

Although substitution of less noisy machines may have limited application, there are certain areas in which substitution has a potentially wider application. Examples include using "squeeze" type equipment in the place of drop hammers, welding in place of riveting, and instituting the chemical cleaning of metal rather than high-speed polishing and grinding.

Engineering. When starting a noise reduction program, it is most desirable to apply engineering principles that are designed to reduce noise levels. The application of known noise control principles can usually reduce any noise to any desired degree. However, economical considerations, and/or operational necessities may make some applications impractical.

Engineering controls are procedures other than administrative or personal protection procedures that reduce the sound level either at the source or within the hearing zone of the workers. The following are examples of engineering principles that can be applied to reduce noise levels.

1. Maintenance:
   a. replacement or adjustment or worn, loose, or unbalanced parts of machines;
   b. lubrication of machine parts and use of cutting oils;
   c. use of properly shaped and sharpened cutting tools.
2. Substitution of machines:
   a. larger, slower machines for smaller, faster ones;
   b. step dies for single-operation dies;
   c. presses for hammers;
   d. rotating shears for square shears;
   e. hydraulic presses for mechanical presses;
   f. belt drives for gears.
3. Substitution of processes:
   a. compression riveting for impact riveting;
   b. welding for riveting;
   c. hot working for cold working;
   d. pressing for rolling or forging.
4. The driving force of vibrating surfaces may be reduced by:
   a. reducing the forces;
   b. minimizing rotational speed;
   c. isolating.
5. The response of vibrating surfaces may be reduced by:
   a. damping;
   b. additional support;
   c. increasing the stiffness of the material;
   d. increasing the mass of vibrating members;
   e. changing the size to change resonance frequency.
6. The sound radiation from the vibrating surfaces can be reduced by:
   a. reducing the radiating area;
   b. reducing overall size;

   c. perforating surfaces.
7. Reduce the sound transmission through solids by using:
   a. flexible mountings;
   b. flexible sections in pipe runs;
   c. flexible-shaft couplings;
   d. fabric sections in ducts;
   e. resilient flooring.
8. Reduce the sound produced by gas flow by:
   a. using intake and exhaust mufflers;
   b. using fan blades designed to reduce turbulence;
   c. using large, low-speed fans instead of smaller, high speed fans;
   d. reducing the velocity of fluid flow (air);
   e. increasing the cross section of streams;
   f. reducing the pressure;
   g. reducing air turbulence.
9. Reduce noise by reducing its transmission through air by:
   a. using sound absorptive material on walls and ceiling in work areas;
   b. using sound barriers and sound absorption along the transmission path;
   c. the complete enclosure of individual machines;
   d. using baffles;
   e. confining high-noise machines to insulated rooms.
10. Isolate the operator by providing a relatively soundproof booth for the operator or attendant.

Some of the noise control measures described here can be executed quite inexpensively by plant personnel. Other controls require considerable expense and highly specialized technical knowledge to obtain the required results. The services of competent acoustical engineers should be contracted when planning and carrying out engineering noise control programs.

The possibility that excessive plant noise levels exist should be considered at the planning stage. Vendors supplying machinery and equipment should be advised that specified low noise levels will be considered in the selection process. Suppliers should be asked to provide information on the noise levels of currently available equipment. The inclusion of noise specifications in purchase orders has been used successfully to obtain quiet equipment. If purchasers of industrial equipment demand quieter machines, designers will give more consideration to the problem of noise control.

It is not enough to specify that the sound pressure level of a particular machine at the operator's station shall be 90 dBA or less. As mentioned earlier in this chapter, an increase of 3 dB represents a doubling of sound energy. If another identical machine is placed nearby, the sound level produced by the two machines can be 93 dBA at the operator's station.

To estimate the effect of a given machine on the total work environment, it is necessary to know the sound power that the machine produces. If there is no operator's work station in the machine's immediate vicinity, the sound power specifications may be sufficient; if, however, there is an operator in the near-sound field, more information is generally needed.

Objectionable noise levels produced as a by-product of manufacturing operations are found in almost every industry. Practical noise control measures are not easy to develop, and few ready-made solutions are available. Unfortunately, a standard technique or procedure that can be applied to all or even most situations cannot be presented here. The same machine, process, or noise source in two different locations may present two entirely different problems and may need to be solved in two entirely different ways.

Noise control techniques are being incorporated into products during the design stage. Machine tool buyers represent one group who currently specify maximum noise levels in their purchase orders. Equipment can be designed with lower noise levels, but performance tradeoffs involving weight, size, power consumption, and perhaps increased maintenance costs may be necessary. These new, quieter products will probably weigh more, be bigger and bulkier, cost more, and be more difficult and expensive to service and maintain.

To attain quieter products, the engineer must be prepared to trade off, to some degree, many of the design goals that have been achieved in response to market demands. However, lightweight, low cost, portable machines that are easily operated and simply maintained should not be cast aside lightly. Price increases may be inevitable, and the cost/benefit relationship should be examined in each case. In addition to paying higher original equipment costs, the user (both as a consumer and as a taxpayer) will also bear the burden of increased direct and indirect costs.

The success of a noise reduction project usually depends on the ingenuity with which basic noise control measures can be applied without decreasing the maximum use and accessibility of the machine or other noise source that is being quieted.

Administrative controls. There are many operations in which the exposure of employees to noise can be controlled administratively, that is production schedules can simply be changed or jobs can be rotated so that exposure times are reduced. This includes such measures as transferring employees from a job location with a high noise level to a job location with a lower one if this procedure would make the employee's daily noise exposure acceptable.

Administrative controls also refer to scheduling machine operating times so as to reduce the number of workers exposed to noise. For example, if an operation is performed during only one 8-hour day per week and the operator is overexposed on that one day, it might be possible to reduce the operation to one-half day (4 hours, 2 days per week). The employee then may not be overexposed.

Employees who are particularly susceptible to noise can be transferred and allowed to work in a less noisy area. Transferring employees is a limited control measure since personnel problems can be caused due to loss of seniority and prestige and lower productivity and pay.

Administrative controls may also entail any administrative decision that results in lower noise exposure. This includes complying with purchase agreements that specify maximum noise levels at the operator's position.

The sound level specification that is made part of the purchasing agreement must be more than just some general compliance statement, such as "Must meet the requirements of the OSHA." It is important to realize that OSHA sets allowable noise limits relative to the exposure of the people involved. OSHA does not relate directly to noise-generating equipment. The OSHA Noise Standard is not intended to be used as an equipment design specification and thus cannot be used as such.

Personal hearing protection. Pending the application of engineering control measures, employee exposure to noise can be reduced by the mandatory use of hearing-protective devices.

LET'S REVIEW THE FACTS

1. It is necessary for employees in certain noisy areas to wear ear protectors.
2. Prolonged exposure to excessive noise can harm the delicate hearing mechanism.
3. Ear protectors such as ear plugs or ear muffs will reduce the noise before it reaches the ear drum.
4. Your job assignment will determine whether you should wear ear plugs (inserts) or muffs (covers).
5. Speech and warning signals can be fully heard with ear protectors in noisy shop areas.

WEAR YOUR EAR PROTECTORS

1. The nurse will fit them and instruct you how to wear them.
2. Wear them for short periods to start and gradually increase the wearing time. After a few days you will be able to wear them all day with minimum discomfort.

Suggested Wearing Time Schedule

	A.M.	P.M.
1st day	= 30 minutes	— 1 hour
2nd day	= 1 hour	— 1 hour
3rd day	= 2 hours	— 2 hours
4th day	= 3 hours	— 3 hours
5th day	= all day — all day thereafter	

3. If after five days the ear protectors feel uncomfortable, come in and see the nurse in the Company hospital.
4. Ear protectors should be replaced when they become worn, stiff or lose their shape.
5. If ear protectors are misplaced, a new pair should be obtained without delay.
6. Never put soiled ear plugs into your ears. Wash the ear plugs at least once a day with soap and water.
7. With proper care, ear plugs should last for several months and ear muffs should last for several years.

OTHER POINTS TO REMEMBER

1. The best ear protector is the one that is properly fitted and worn.
2. Good protection depends on a snug fit. A small leak can destroy the effectiveness of the protection.
3. Ear plugs tend to work loose as a result of talking or chewing, and they must be re-seated from time to time during the working day.
4. If ear plugs are kept clean, skin irritations and other reactions should not occur.

YOUR HEARING IS PRICELESS

PROTECT IT

Figure 9-15. An example of a card distributed to all company employees who are required to wear some form of hearing protection. The card highlights the care and use of the device.

Occupational noise regulations require that whenever employees are exposed to excessive noise levels, feasible administrative or engineering controls should be used to reduce these levels. When these control measures cannot be completely accomplished and/or while such controls are being initiated, personnel should be protected from the effects of excessive noise levels. Such protection can, in most cases, be provided by wearing suitable hearing-protective devices.

Once management has decided that hearing protectors should be worn, the success of such a program depends largely on the method of initiation used and on the proper indoctrination of supervisory personnel and workers. Supervisors should set an example by wearing their hearing protectors when they go into noisy areas.

Some companies have found it very helpful to meet with employees or their representatives to thoroughly review the contemplated protection program and reach an understanding of the various problems involved. These include reviewing work areas where hearing protection will be provided or required and complying with state and federal regulations that require the use of hearing-protective devices. A short explanation concerning the purpose and benefits to be derived through the wearing of hearing protectors is essential (Figure 9-15).

Personal hearing-protective devices are acoustical barriers that reduce the amount of sound energy transmitted through the ear canal to receptors in the inner ear.

The sound attenuation (reduction) capability of a hearing-protective device (in decibels) is the difference in the measured hearing level threshold of an observer wearing hearing protectors (test threshold) and the measured hearing threshold when the observer's ears are uncovered (reference threshold).

Inserts or muffs are hearing-protective devices that are in common use today. The insert-type protector attenuates noise by plugging the external ear canal, while the muff-type protector encloses the auricle of the ear to provide an acoustical seal. The effectiveness of hearing-protective devices depends on several factors that are related to the manner in which the sound energy is transmitted through or around the device. Figure 9-16 shows four pathways by which sound can reach the inner ear when hearing-protective devices are worn: (1) seal leaks, (2) material leaks, (3) hearing-protective device vibration, and (4) conduction through bone and tissue.

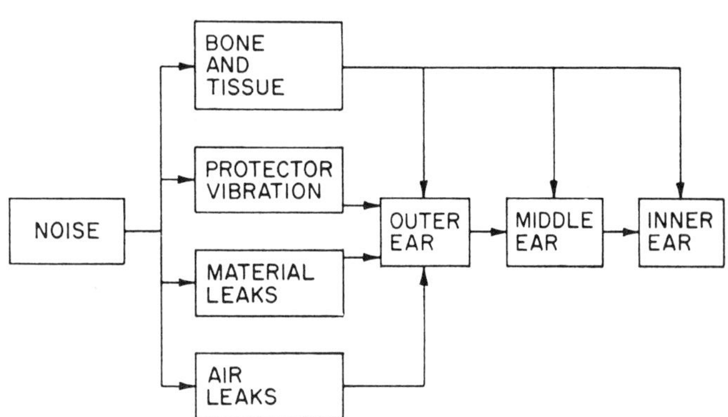

Figure 9-16. When a person is wearing a hearing protector, sound reaches the inner ear by different methods.

Seal leaks. For maximum protection, the device must form a virtually airtight seal against the ear canal or the side of the head. Inserts must accurately fit the contours of the ear canal, and muffs must accurately fit the areas surrounding the external ear. Small air leaks in the seal between the hearing protector and the skin can significantly reduce the low-frequency sound attenuation or permit a greater proportion of the low-frequency sounds to pass through. As the air leak becomes larger, attenuation lessens at all frequencies.

Material leaks. A second possible transmission pathway for sound is directly through the material of the hearing-protective device. Thus, the hearing-protective device may attenuate or prevent the passage of most of the sound energy, but some sound is still allowed to pass through.

Hearing-protective device vibration. A third pathway for sound to be transmitted to the inner ear is possible when the hearing-protective device itself is set into vibration in response to exposure to external sound energy.

Due to the flexibility of the flesh in the ear canal, earplugs can vibrate in a pistonlike manner within the ear canal. This limits their low-frequency attenuation. Likewise, an earmuff cannot be attached to the head in a totally rigid manner. Its cup will vibrate against the head like a mass/spring system. The muff's effective stiffness will be governed by the flexibility of the muff cushion and the flesh surrounding the ear, as well as by the air volume entrapped under the cup.

Bone conduction. If the ear canal were completely closed so that no sound entered the ear by this path, some sound energy could still reach the inner ear by means of bone conduction. However, the sound reaching the inner ear by such means would be about 50 dB below the level of air-conducted sound through the open ear canal. It is obvious, therefore, that no matter how the ear canal is blocked, the hearing-protective device will be bypassed by the bone-conduction pathway through the skull. A perfect hearing-protective device cannot provide more than 50 dB of effective sound attenuation.

When a hearing-protective device is properly sized and carefully fitted and adjusted for optimum performance on a laboratory subject, air leaks are minimized and material leaks, hearing-protective device vibration, and bone conduction are the primary sound transmission paths. In the workplace, however, this is usually not the case; sound transmission through air leaks is often the primary pathway.

All hearing protectors must be properly fitted when they are initially dispensed. Comfort, motivation, and training are also very important factors to consider if hearing protectors are to be successfully used.

Hearing-protective devices. Personal hearing-protective equipment can be divided into four classifications:

1. enclosures (entire head);
2. aural inserts;
3. superaural protectors; and
4. circumaural protectors.

ENCLOSURES. As the name implies, the enclosure-type hearing-protective device is incorporated in equipment that entirely envelops the head. A typical example is the helmet worn by an astronaut. In this case, attenuation at the ear is achieved through the acoustical properties of the helmet.

The maximum amount that a hearing protector can reduce the sound reaching the ear is from about 35 dB at 250 Hz to about 50 dB at the higher frequencies. By wearing hearing protectors and then adding a helmet that encloses the head, an additional 10-dB reduction of sound transmission to the ears can be achieved.

Helmets can be used to support earmuffs or earphones and cover the bony portion of the head in an attempt to reduce bone-conducted sound. Helmets are particularly suited for use in extremely high-noise level areas and where protection of the head is needed against bumps or missiles. With good design and careful fitting of the seal between the edges of the helmet and the skin of the face and neck, 5–10 dB of sound attenuation can be obtained beyond that already provided by the earmuffs or earphones within the helmet. This approach to protection against excess noise is practical only in very special applications. Cost as well as bulk normally preclude the use of helmet-type hearing protectors in a general industrial hearing conservation program.

AURAL INSERT PROTECTORS. Aural insert hearing-protective devices are normally referred to as inserts or earplugs. This type of protector is generally inexpensive, but the service life is limited, ranging from single-time use to several months. Insert-type protectors or plugs are supplied in many different configurations and are made from such materials as rubber, plastics, fine glass down, foam, and wax-impregnated cotton. The pliable materials used in these aural inserts are quite soft, and there is little danger of injury resulting from accidentally forcing the plug against the tender lining of the ear canal.

It is desirable to have the employee's ears examined by qualified medical personnel before earplugs are fitted. Occasionally, the physical shape of the ear canal precludes the use of insert-type protectors. There is also the possibility that the ear canal may be filled with hardened wax. If wax (cerumen) is a problem, the wax should be removed by qualified personnel. In some cases, the skin of the ear may be sensitive to a particular earplug material, and earplugs that do not cause an allergic response should be recommended.

Aural insert-type hearing protectors fall into three broad categories or general classifications: (1) the formable type; (2) the custom-molded type; and (3) the premolded type.

Formable protectors. Formable types of hearing-protective devices can provide good attenuation and fit all ears. Many of the formable types are designed for a one-time use only and are then thrown away. Materials from which these disposable plugs are made include very fine glass fiber (quite often referred to as Swedish wool), wax-impregnated cotton, expandable plastic, and foam.

These materials are generally rolled into a conical shape before being inserted into the ear. However, while adequate instruction must be given to emphasize the importance of a snug fit, the user must be careful not to push the material so far into the ear canal that it has to be removed by medical personnel.

Another type of material is a plasticlike substance similar in consistency to putty. The preparation of this material requires that the individual take a quantity of it and mold or form it so that it can be inserted into the ear canal. The user should be shown the correct method of forming the material. In addition, the user must be cautioned to have clean hands when forming the material and placing it in the ear. If the hands are dirty, foreign material can get into the ear canal.

Custom-molded protectors. Formable hearing-protective devices in this category are, as the name implies, custom molded for the individual user. Generally, two or more materials (packaged separately) are mixed together to form a compound that resembles soft rubber when set. For use as a hearing-protective device, the mixture is carefully placed into the outer ear with some portion of it in the ear canal, in the manner prescribed by the manufacturer. As the material sets, it molds itself to the shape of the individual ear and external ear canal. In some cases, the materials are premixed and come in a tube from which they can be injected into the ear.

Premolded aural insert protector. Premolded insert protectors are quite often referred to as prefabricated because they are usually made in large quantities in a multiple-cavity mold. The materials of construction range from soft silicone or rubber to other plastics.

There are two versions of the premolded insert protector. One is known as the universal fit type. In this type, the plug is designed to fit a wide variety of ear canal shapes and sizes. The other type of premolded protector is supplied in several different sizes to assure a good fit. The design of the plug is important. For example, the smooth bullet-shaped plug is very comfortable and provides adequate attenuation in straight ear canals; however, its performance falls off sharply in many irregularly shaped canals.

The use of premolded insert-type protectors requires proper fitting by trained personnel. In many instances, the right and left ear canals are not the same size. For this reason, properly trained personnel must prescribe the correct protector size for each ear canal. Sizing devices are available to aid in the proper fitting.

The premolded type of earplug has a number of disadvantages that limit its practical acceptability. To be effective, it has to fit snugly and, for some users, this is uncomfortable. Because the plug must fit tightly and because many people have irregular shaped ear canals, an incorrect size of plug may be selected, or the plug may not be inserted far enough, and a good fit is not obtained.

Some premolded insert protectors may shrink and become hard. This is caused primarily by ear wax (present in all ear canals). The wax extracts the plasticizer from some plug materials, causing the hardening and possible shrinkage of the plug. The degree of hardening and shrinkage of the plug varies from one individual to another depending on such factors as temperature, duration of use, and the personal hygiene of the user. Regular cleaning of the protectors with mild soap and water prolongs their useful life. To keep the plugs clean and free from contamination, most manufacturers provide a carrying case for storing the plugs when not in use.

SUPERAURAL PROTECTORS. Hearing-protective devices in this category seal the external opening of the ear canal to achieve sound attenuation. A soft, rubberlike material is held in place by a very light band or head suspension. The tension of the band holds the superaural device against the external opening of the ear canal.

CIRCUMAURAL PROTECTORS. Circumaural hearing-protective devices—usually called earmuffs—consist essentially of two cup- or dome-shaped devices that fit over the entire external ear, including the lobe, and the seal against the side of the head with a suitable cushion or pad. In general, the ear cups are made of a molded rigid plastic and are lined with a cell-type foam material. The size and shape of the ear cup vary from one manufacturer to another (Figure 9-17).

The cups are generally held in place by a spring-loaded suspension assembly or headband. The force applied is directly related to the degree of attenuation desired. The width, circumference, and material of the earmuff cushion resting against the head must be considered to maintain a proper balance of performance and comfort. To provide a good acoustical seal, the required width of the contact surface depends to a large degree on the material used in the cushion. The cup with the smallest possible circumference that will accommodate the largest ear lobes should be chosen. A slight pressure on the lobe can become painful in time, so it is very important to select a muff dome that is large enough.

The earmuffs currently on the market are supplied with replaceable ear seals or cushions that may be filled with foam, liquid, or air—the foam-filled type is the most common. The outer covering of these seals is vinyl or a similar thermoplastic material. Human perspiration tends to extract the plasticizer from the seal material, which results in an eventual stiffening of the seals. For this reason, the seals require periodic replacement; the frequency of replacement depends on the conditions of exposure.

Figure 9-17. Hearing protection should be worn by workers at those operations where the noise source cannot be controlled by engineering methods.

Selection of protector. The attenuation characteristics of a particular hearing protector must be considered before it is used for a specific application (Figure 9-18). As part of a well-planned hearing-conservation program, characteristics of the noise levels in various areas should be known. From these data and from the attenuation information available from manufacturers, it can be determined whether a given device is suitable for the intended application. Consideration must be given to the work area where the individual must use the hearing-protective device. For example, a large-volume earmuff would not be practical for an individual who must work in confined areas where there is very little head clearance. In such instances, a very small or flat ear cup or insert-type protector would be more practical.

When using muff-type protectors in special hazard areas (such as power-generating stations where there are electrical hazards), it may be desirable to use nonconductive suspension systems in connection with muff-type protectors. Also, if other personal protective equipment, such as safety hats or safety spectacles, must be worn, the degree of hearing protection required must not be compromised. The efficiency of muff-type protectors is reduced when they are worn over the frames of eye-protective

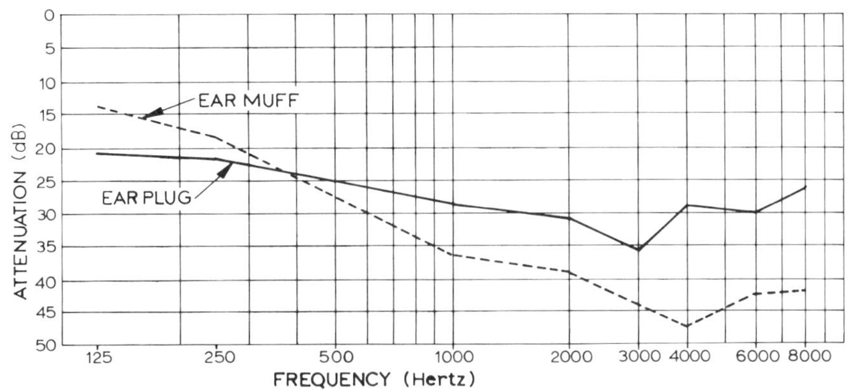

Figure 9-18. Comparison of the attenuation properties of a molded-type earplug and an earmuff protector. Note that the earplug offers greater attenuation of the lower frequencies, while the earmuff is better at the higher frequencies.

devices. The reduction in noise attenuation will depend on the type of glasses being worn as well as on the size and shape of the individual wearer's head. When eye protection is required, it is recommended that cable-type temples be used. This type has the smallest possible opening between the seal and the head.

When selecting a hearing-protective device, one should also consider the frequency of exposure to excessive noise. If exposure is relatively infrequent (once a day or once a week), an insert or plug device will probably satisfy the requirement. On the other hand, if the noise exposure is relatively frequent and the employee must wear the protective device for an extended period of time, the muff-type protector might be preferable. If the noise exposures are intermittent, the muff-type protector is probably more desirable since it is somewhat more difficult to remove and reinsert earplugs.

When determining the suitability of a hearing-protective device for a given application, the manufacturer's reported test data must be examined carefully. It is necessary to correlate that information with the specific noise exposure involved. The attenuation characteristics of the individual hearing-protective devices are compared at different frequencies.

The most convenient method by which to gauge the adequacy of a hearing protector's attenuation capacity is by checking its Noise Reduction Rating (NRR). The NRR was developed by the EPA. According to the EPA regulation, the NRR must be printed on the hearing protector package. The NRR is then correlated with an individual worker's noise environment to assess the adequacy of the attenuation characteristics of a particular hearing-protective device. Appendix B of 29 *CFR* 1910.95 describes methods of using the NRR to determine whether a particular hearing-protective device provides adequate protection within a given exposure environment.

INDUSTRIAL AUDIOMETRY

Industrial audiometry is an important part of the hearing conservation program. Briefly, the objectives to accomplish are as follows.

1. Obtain a baseline audiogram that indicates an individual's hearing ability at the time of the preplacement examination.
2. Provide a record of an employee's hearing acuity.
3. Check the effectiveness of noise control measures by measuring the hearing thresholds of exposed employees.
4. Record significant hearing threshold shifts in exposed employees during the course of their employment.
5. Comply with government regulations.

An audiometer is required to help assess an individual's hearing ability. An audiometer is an electronic instrument that converts electrical energy into sound energy in precisely variable amounts. It should meet the standards set forth in ANSI S3.6-1969 (R1973), *Specifications for Audiometers.*

An audiometer consists of an oscillator, which produces pure tone sounds at predetermined frequencies, an attenuator, which controls the intensity of the sound or tone produced, a presenter switch, and earphones, through which the person whose hearing is being tested hears the tone.

Threshold audiometry

Threshold audiometry is used to determine an employee's auditory threshold for a given stimulus. Measurements of hearing are made to determine hearing acuity and the presence of abnormal function in the ear. Before hearing can be described as abnormal, a reference point, or normal value, must be designated.

The quantity that is of interest, however, is not the sound pressure level of the normal hearing threshold, but rather the magnitude of departure from a standard reference threshold. Levels that depart from the norm can then be read directly from the setting of the audiometer attenuator dial.

Hearing threshold levels are those intensities at specific frequencies at which a sound or a tone can just be heard. The term "air conduction" describes the air path by which the test sounds generated at the earphones are conducted through air to stimulate the eardrum.

The record of measured hearing thresholds is called a threshold audiogram. Audiometric tests can also be recorded in the form of audiograms on which are plotted both sound intensity (in decibels) and frequency (in hertz). A sample audiogram is shown in Figure 9-19.

Who should be examined

Preplacement hearing threshold tests should be taken by all job applicants, not just those who are to work in noisy areas. This establishes a baseline hearing threshold for each employee for future comparison. Preplacement hearing tests are essential if a company is to protect itself from liability for preexisting hearing loss incurred elsewhere. If an employee is hired with hearing damage and he or she is subsequently exposed to high noise levels, the company may be liable for all the employee's hearing loss—unless it can be proved that the employee had a preexisting hearing loss when hired. In some states, the most recent employer is liable for all compensable hearing loss, regardless of past exposures.

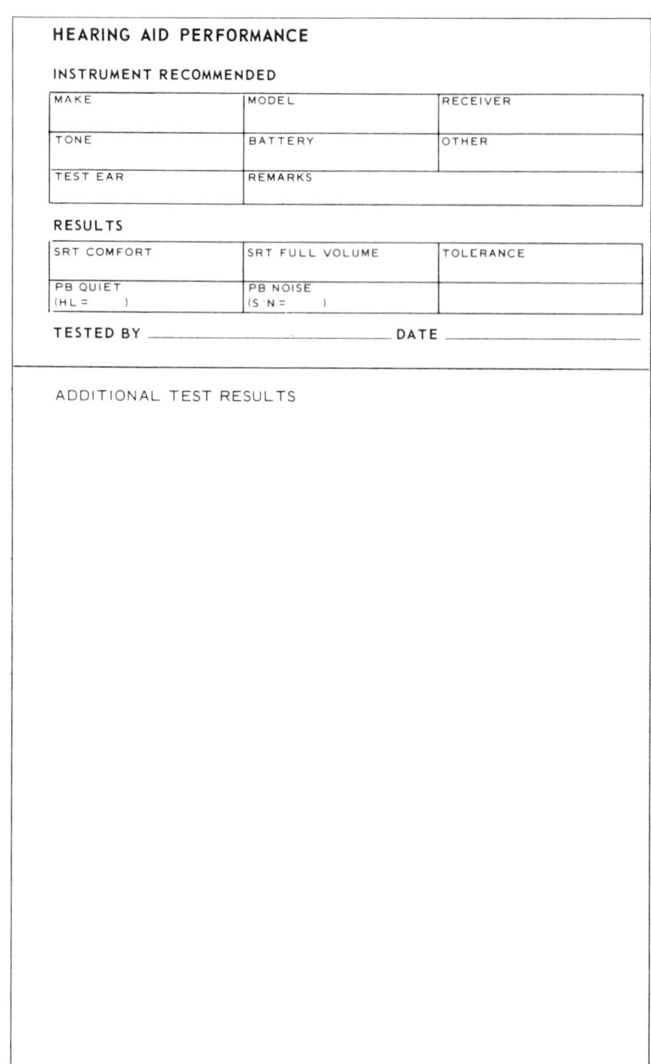

HEARING AID PERFORMANCE

INSTRUMENT RECOMMENDED

MAKE	MODEL	RECEIVER
TONE	BATTERY	OTHER
TEST EAR	REMARKS	

RESULTS

SRT COMFORT	SRT FULL VOLUME	TOLERANCE
PB QUIET (HL =)	PB NOISE (S/N =)	

TESTED BY ________ DATE ________

ADDITIONAL TEST RESULTS

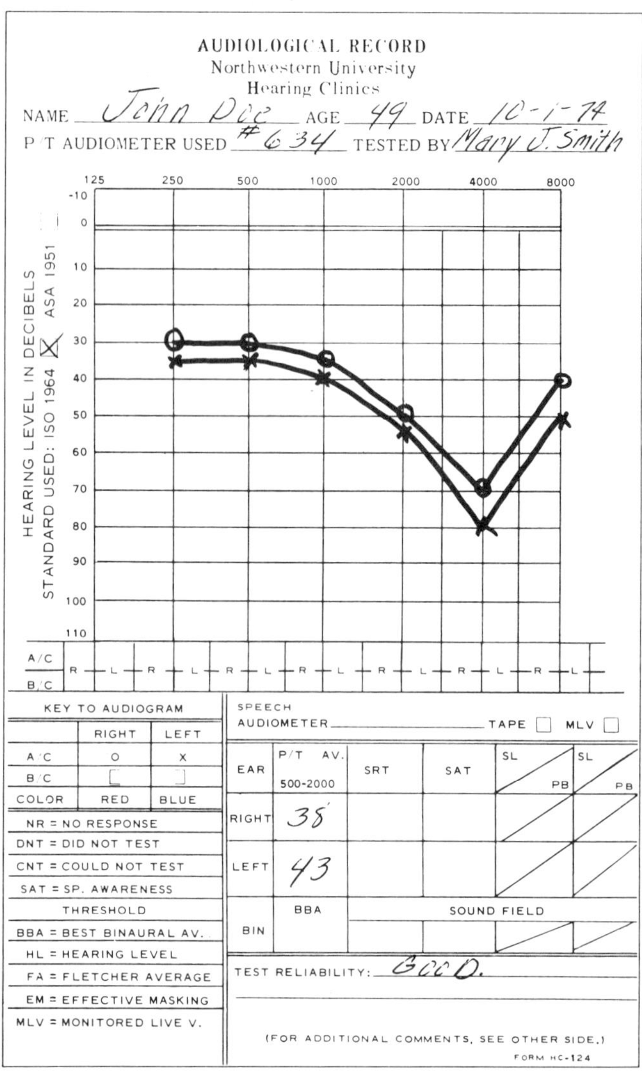

AUDIOLOGICAL RECORD
Northwestern University
Hearing Clinics

NAME John Doe AGE 49 DATE 10-1-74

P/T AUDIOMETER USED #634 TESTED BY Mary J. Smith

KEY TO AUDIOGRAM	RIGHT	LEFT
A/C	O	X
B/C	[	]
COLOR	RED	BLUE

NR = NO RESPONSE
DNT = DID NOT TEST
CNT = COULD NOT TEST
SAT = SP. AWARENESS THRESHOLD
BBA = BEST BINAURAL AV.
HL = HEARING LEVEL
FA = FLETCHER AVERAGE
EM = EFFECTIVE MASKING
MLV = MONITORED LIVE V.

SPEECH AUDIOMETER ________ TAPE ☐ MLV ☐

EAR	P/T AV. 500-2000	SRT	SAT	SL / PB	SL / PB
RIGHT	38				
LEFT	43				
BIN	BBA	SOUND FIELD			

TEST RELIABILITY: Good.

(FOR ADDITIONAL COMMENTS, SEE OTHER SIDE.)

FORM HC-124

Figure 9-19. This audiogram shows the initial effects of exposure to excessive noise. Note the decided notch at the 4,000-Hz frequency. (Courtesy Northwestern University Hearing Clinics)

Periodic follow-up hearing tests should be administered to persons stationed in areas where noise exposures exceed permissible levels.

The schedule for periodic follow-up hearing tests depends largely on the employee's noise exposure. Assuming that a record of the worker's hearing status was established at the time of employment or placement, the first reexamination should be made 9–12 months after placement. If no significant threshold shifts relative to the preplacement audiogram are noted, subsequent follow-up tests can be administered at yearly intervals. If the noise exposure is relatively low, the interval between follow-up tests can be increased. However, this decision should be based on conditions of exposure and the results of previous audiograms and clinical examinations.

Noise exposure is by no means the only reason the results of an individual's audiogram may change. When a change in hearing status is confirmed, the cause must be determined. Improper placement of earphones and excessive ambient noise in the test room would certainly affect audiogram results. Physiological changes due the the employee's state of health and age would also affect audiogram results. The individual's motivation and his or her attitude toward the test can affect performance.

The industrial audiometric program can identify persons who experience hearing threshold changes that are not related to noise exposure. These workers should be referred to their family physician for diagnosis and treatment. However, when threshold shifts related to noise exposure are identified, the subsequent procedure should be followed.

1. Check the fit of the hearing-protective device, if one is worn by the worker.
2. Repeat or initiate educational sessions to encourage the employee to wear a hearing-protective device, if it has not been worn.
3. Investigate the noise levels in the work area, particularly if a previous sound level survey failed to reveal noise hazards.

Noise exposure information, correlated with audiometric test results, is needed to make intelligent decisions about a firm's hearing conservation program. If all hearing tests and medical opinion point to a progressive deterioration of an individual's hearing, the health and safety professional should provide and enforce the use of hearing-protection equipment and/or recommend that the individual's exposure to excessive noise be controlled.

Conclusions about the general noise environment should not be based on changes in the hearing of a single individual because the variation in individual susceptibility to noise is broad. Conclusions can, however, be drawn from the average changes—or lack of them—in a group of employees exposed to the same noise environment.

Effective programs

An effective industrial audiometric program should include consideration of the following components:

- medical surveillance;
- qualified personnel;
- suitable test environment;
- calibrated equipment; and
- adequate records.

Medical surveillance. Medical surveillance is essential in a hearing-testing program so that the program can fulfill its dual purpose of detecting hearing loss and providing valid records for compensation claims. Although many smaller companies do not have a medical department, they can satisfy the general medical surveillance requirement by using part-time medical consultants.

Noise-susceptible workers are those employees who suffer handicapping hearing losses more quickly than do their colleagues under equivalent noise exposures. These workers constitute the group from whom compensation claims are most likely to arise and for whom the risk of hearing damage is likely to be greatest.

During the preplacement examination, the applicant should provide a detailed history covering his or her prior occupational experience and a personal record of illnesses and injuries. For applicants who will work in noisy environments, the history should include noise exposures in previous jobs as well as in the military service. The medical phase of the history should detail frequency of earache, ear discharge, ear injury, surgery (ear or mastoid), head injury with unconsciousness, ringing in the ears, hearing loss in the immediate family, the use of drugs, and history of allergy and toxic exposures. A standard form can be devised for this purpose.

Qualified personnel. Audiometric tests should be administered by a qualified individual, such as a specially trained nurse, an audiologist, or a certified industrial audiometric technician or occupational hearing conservationist. A certified industrial audiometric technician is an individual who has satisfactorily completed a course of training that meets, as a minimum, the guidelines established by the Intersociety Committee on Audiometric Technician Training or who has been certified by the Council for Accreditation in Occupational Hearing Conservation.

The industrial audiometric technician's duties are to perform baseline and periodic pure-tone air-conduction threshold tests. Systematic supervision and encouragement of the industrial audiometric technician by the physician, audiologist, or other qualified person in charge of the audiometric program is recommended to maintain the high motivation required for good audiometric testing. The supervision should include periodic review of the testing procedures used by the audiometric technician to make sure that they conform to established procedures.

Suitable test environment. Hearing measurements must be made in a test room or booth that conforms to the requirements established by ANSI S3.1-1960 (R1971), *Criteria for Background Noise in Audiometer Rooms.* It must be sufficiently quiet within the enclosure so that external noises do not interfere with the employee's perception of the test sounds. This usually requires a special, sound-treated enclosure (Figure 9-20).

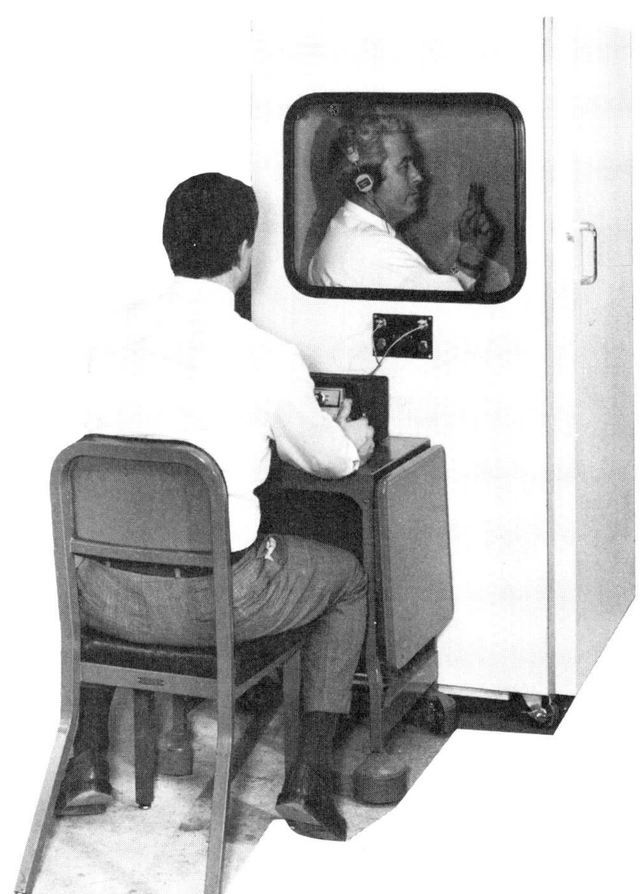

Figure 9-20. The raised finger of the subject's hand indicates that he is, in all probability, hearing the test tone. The technician must decide whether this was an accurate or inaccurate response. (Courtesy Eckel Industries, Inc.)

Hearing-testing rooms should be located away from outside walls, elevators, and locations with heating and plumbing noises. If the background noise levels in the test area do not exceed the sound levels allowed by the standard, the background noise will not affect the hearing test results. The hearing test booth or room may be either a prefabricated unit or one that is built on the premises. Doors, gaskets, and other parts of the room or booth that may deteriorate, warp, or crack should be carefully inspected periodically.

In addition to proper acoustical standards, the booth or room should allow for ease of access and egress and be provided with good, comfortable ventilation and lighting. The audiometric technician should be able to sit outside the room or booth but be able to see the interior of the room through a window.

To select the proper room, it is necessary to conduct a noise survey at the proposed test location. Noise levels at each test frequency should be measured and recorded using an octave-band sound level meter. The audiometric booth selected must have sufficient noise attenuation so that the background noise levels present at each test frequency are reduced and do not exceed the maximum permissible background levels listed in the current ANSI standard.

Calibrated equipment. Limited range, pure-tone audiometers must conform to the current ANSI standard listed in *Specifications for Audiometers.* Two basic types of audiometers are available: the automatic recording audiometers and the manually operated audiometer.

The audiometer should be subjected to a biological check each day before the instrument is used. The biological check should be made by testing the hearing of a person whose hearing threshold is known and stable. The check should include the movement and bending of cord and wire, knob turning, switch actuating, and button pushing to make sure that no sounds are produced in the earphone other than the test tones.

An exhaustive electronic calibration of the audiometric test instrument should be made annually by a repair and calibration facility that has the specialized equipment and skilled technical personnel necessary for this work. A certificate of calibration should be kept with the audiometer at all times.

Adequate records. The medical form used in audiometric testing programs should include all basic data related to the hearing evaluation. Hearing threshold values, noise exposure history, and pertinent medical history information should be accurately recorded each time an employee's hearing is tested. The employee should be identified by name, social security number, sex, and age. Additional information, such as the date and time of test (day of week, time of day), conditions of test, and the name of the examiner should also be included.

Audiometric test records should be kept for at least the duration of employment. The records could become the basis for a settlement of a hearing loss claim.

Periodic audiograms are a profile of the employee's hearing acuity. Any change from the results of previous audiograms should be explored. One reason for a hearing loss may be that the employee's hearing protectors are inadequate or improperly worn.

The audiometric testing program should be both practical and feasible. In small companies, where the total number of employees to be tested is few in number, it would be impractical to purchase a booth and audiometer. It would be more economical to consider a mobile audiometer testing service or to refer these few employees to a local, properly equipped and staffed hearing center or to a qualified physician or audiologist for an audiometric examination.

Audiometric testing is an integral part of a comprehensive hearing conservation program. The OSHA Hearing Conservation Standard discussed in the following section details specific requirements for audiometric testing.

HEARING CONSERVATION PROGRAMS

An effective hearing conservation program prevents hearing impairment as a result of noise exposure on the job. In terms of existing workers' compensation laws, an effective hearing conservation program is one that limits the amount of compensable hearing loss in the frequency range over which normal hearing is necessary for communication. It should be noted that "compensable" loss at present does not include frequencies over 4,000 Hz, although such loss will impair enjoyment of sound and may interfere with speech discrimination. In compliance with the OSHA requirements, an effective hearing conservation program must be instituted if any employee's noise exposure exceeds current limits as defined in the OSHA Noise Exposure Standard 29 *CFR* 1910.95.

In October 1974, OSHA began revising the Occupational Noise Exposure Standard. After years of collecting oral and written public testimony, which resulted in an unwieldy public record of almost 40,000 pages, OSHA promulgated revisions for the Noise Standard (46 *FR* 4078) in January 1981. These revisions were followed by deferrals, stays (46 *FR* 42622), further revisions, further public hearings, and a multiplicity of lawsuits, all of which culminated in the promulgation of a hearing conservation amendment (48 *FR* 9738) on March 8, 1983, with an effective date of April 7, 1983.

Background

It is estimated by OSHA (46 *FR* 4078) that there are 2.9 million workers in American production industries who experience 8-hour noise exposures in excess of 90 dBA. An additional 2.3 million experience exposure levels in excess of 85 dBA. The Hearing Conservation Amendment (HCA) applies to all those 5.2 million employees except for those in oil and gas well drilling and servicing industries, which are specifically exempted. Additionally, the Amendment does not apply to those engaged in construction or agriculture, although a Construction Industry Noise Standard exists (29 *CFR* 1926.52 and 1926.101). This Standard is essentially identical to paragraphs (a) and (b) of the General Industry Noise Standard.

The Occupational Noise Standard

Prior to promulgation of the HCA, the existing Noise Standard (29 *CFR* 1910.95(a) and (b)) established a permissible noise exposure level of 90 dBA for 8 hours and required the employer to reduce exposure to that level by use of feasible engineering and administrative controls. In all cases in which sound levels exceeded the permissible exposure, regardless of the use of hearing-protective devices, "a continuing, effective hearing conservation program" was required. However, the details of such a program were never mandated. Paragraphs (c) through (p) of the HCA replaced paragraph (b)(3) of 29 *CFR* 1910.95 and supplemented OSHA's definition of an "effective hearing conservation program."

Summary of the HCA

All employees whose noise exposures equal or exceed an 8-hour TWA of 85 dBA must be included in a hearing conservation program comprised of five basic components: exposure monitoring, audiometric testing, hearing protection, employee training, and record-keeping. Note that although the 8-hour TWA permissible exposure remains 90 dBA, a hearing conservation program becomes mandatory at an 8-hour TWA exposure of 85 dBA.

The following summary briefly discusses the required components of the hearing conservation program.

Monitoring. The HCA requires employers to monitor noise exposure levels in a manner that will accurately identify employees who are exposed at or above an 8-hour TWA exposure of 85 dBA. The exposure measurement must include all noise within an 80–130-dBA range. The requirement is performance-oriented and allows employers to choose the monitoring method that best suits each situation.

Employees are entitled to observe monitoring procedures and, in addition, they must be notified of the results of exposure monitoring. However, the method used to notify employees is left to the discretion of the employer.

Employers must remonitor workers' exposures whenever changes in exposures are sufficient to require new hearing protectors or whenever employees not previously included in the program because they were not exposed to an 8-hour TWA of 85 dBA are included in the program.

Instruments used for monitoring employee exposures must be calibrated to ensure that the measurements are accurate. Since calibration procedures are unique to specific instruments, employers should follow the manufacturer's instructions to determine when and how extensively to calibrate.

Audiometric testing. Audiometric testing not only monitors employee hearing acuity over time but also provides an opportunity for employers to educate employees about their hearing and the need to protect it. The audiometric testing program includes obtaining baseline audiograms and annual audiograms and initiating training and follow-up procedures. The audiometric testing program should indicate whether hearing loss is being prevented by the employer's hearing conservation program. Audiometric testing must be made available to all employees who have average exposure levels of 85 dBA. A professional (audiologist, otolaryngologist, or physician) must be responsible for the program, but he or she does not have to be present when a qualified technician is actually conducting the testing. Professional responsibilities include overseeing the program and the work of the technicians, reviewing problem audiograms, and determining whether referral is necessary. Either a professional or a trained technician may conduct audiometric testing. In addition to administering audiometric tests, the tester (or the supervising professional) is also responsible for ensuring that the tests are conducted in an appropriate test environment, for seeing that the audiometer works properly, for reviewing audiograms for standard threshold shifts (as defined in the HCA), and for identifying audiograms that require further evaluation by a professional.

Audiograms. There are two types of audiograms required in the hearing conservation program: baseline and annual audiograms. The baseline audiogram is the reference audiogram against which future audiograms are compared. Baseline audiograms must be provided within 6 months of an employee's first exposure at or above a TWA of 85 dBA. When employers obtain audiograms in mobile test vans, baseline audiograms must be completed within 1 year after an employee's first exposure to workplace noise at or above a TWA of 85 dBA. Additionally, when mobile vans are used and employers are allowed to delay baseline testing for up to a year, those employees exposed to levels of 85 dBA or more must be issued and fitted with hearing protectors 6 months after exposure. The hearing protectors are to be worn until the baseline audiogram is obtained. Baseline audiograms taken before the effective date of the amendment are acceptable as baselines in the program if the professional supervisor determines that the audiogram is valid. The annual audiogram must be conducted within 1 year of the baseline. It is important to test hearing on an annual basis to identify changes in hearing acuity so that protective follow-up measures can be initiated before hearing loss progresses.

Audiogram evaluation. Annual audiograms must be routinely compared to baseline audiograms to determine whether the audiogram is accurate and whether the employee has lost hearing ability; that is, to determine whether a standard threshold shift, or STS, has occurred. An effective program depends on a uniform definition of an STS. An STS is defined in the amendment as an average shift (or loss) in either ear of 10 dB or more at the 2,000-, 3,000-, and 4,000-Hz frequencies. A method of determining an STS by computing an average was chosen because it diminishes the number of persons identified as having an STS who are later shown not to have had a significant change in hearing ability.

If an STS is identified, the employee must be fitted or refitted with adequate hearing protectors, shown how to use them, and required to wear them. In addition, employees must be notified within 21 days from the time the determination is made that their audiometric test results indicate an STS. Some employees with an STS may need to be referred for further testing if the professional determines that their test results are questionable or if they have an ear problem of a medical nature caused or aggravated by wearing hearing protectors. If the suspected medical problem is not thought to be related to wearing protectors, employees must merely be informed that they should see a physician. If subsequent audiometric tests show that the STS identified on a previous audiogram is not persistent, employees exposed to an average level of 90 dBA may discontinue wearing hearing protectors.

A subsequent audiogram may be substituted for the original baseline audiogram if the professional supervising the program determines that the employee has experienced a persistent STS. The substituted audiogram becomes known as the revised baseline audiogram. This substitution will ensure that the same shift is not repeatedly identified. The professional may also decide to revise the baseline audiogram after an improvement in hearing has occurred, which will ensure that the baseline reflects actual thresholds as much as possible. When a baseline audiogram is revised, the employer must, of course, also retain the original audiogram. To obtain valid audiograms, audiometers must be used, maintained, and calibrated according to specifications detailed in appendices C and E of the Standard.

Hearing protectors. Hearing protectors must be made available to all workers exposed at or above a TWA of 85 dBA. This requirement will ensure that employees have access to protectors before they experience a loss in hearing. When baseline audiograms are delayed because it is inconvenient for mobile test vans to visit the workplace more than once a year, protectors must be worn by employees for any period exceeding 6 months from the time they are first exposed to 8-hour average noise levels of 85 dBA or above until their baseline audiograms are obtained. The use of hearing protectors is also mandatory for employees who have experienced threshold shifts, since these workers are particularly susceptible to noise.

With the help of a person who is trained in fitting hearing protectors, employees should decide which size and type protector is most suitable for their working environment. The protector selected should be comfortable to wear and offer sufficient attenuation to prevent hearing loss. Employees must be shown how to use and care for their protectors, and they must be supervised on the job to ensure that they contine to wear them correctly.

Hearing protectors must provide adequate attenuation in each employee's work environment. The employer must reevaluate the suitability of the employee's present protector whenever there is a change in working conditions that may render the hearing protector inadequate. If workplace noise levels increase, employees must be given more effective protectors. The protector must reduce the level of exposure to at least 90 dBA or 85 dBA when an STS has occurred.

Training. Employee training is important because when workers understand the hearing conservation program's requirements and why it is necessary to protect their hearing, they will be better motivated to actively participate in the program. Employees will be more willing to cooperate by wearing their protectors and by undergoing audiometric tests. Employees exposed to TWAs of 85 dBA and above must be trained at least annually in the following: the effects of noise; the purpose, advantages, disadvantages, and attenuation characteristics of various types of hearing protectors; the selection, fitting, and care of protectors; and the purpose and procedures of audiometric testing. Training does not have to be accomplished in one session. The program may be structured in any format, and different parts may be conducted by different individuals as long as the required topics are covered. For example, audiometric procedures could be discussed immediately prior to audiometric testing. The training requirements are such that employees must be reminded on a yearly basis that noise is hazardous to hearing, and that they can prevent damage by wearing a hearing protector, where appropriate, and by participating in audiometric testing.

Record-keeping. Noise exposure measurement records must be kept for 2 years. Records of audiometric test results must be maintained for the duration of the affected employee's employment. Audiometric test records must include the name and job classification of the employee, the date the test was performed, the examiner's name, the date of acoustic or exhaustive calibration, measurements of the background sound pressure levels in audiometric test rooms, and the employee's most recent noise exposure measurement.

SUMMARY

Because industrial noise problems are extremely complex, there is no one "standard" program that is applicable to all situations. In view of developments in the workers' compensation field, however, it behooves industry to consider and evaluate its noise problems and to take steps toward the establishment of effective hearing conservation procedures. The OSHA regulations require the control of noise exposures, employee protection against the effects of noise exposures, and the initiation of comprehensive and effective hearing conservation programs.

As outlined in this chapter, an effective hearing conservation program consists of:

- noise measurement and analysis;
- engineering control of noise exceeding permissible levels;
- hearing protection for those employees working in areas where noise cannot be feasibly controlled;
- audiometric examinations for all employees;
- employee training; and
- record-keeping.

The effectiveness of a hearing conservation program depends on the cooperation of employers, supervisors, employees, and others concerned. Management's responsibility in this type of program includes taking noise measurements, initiating noise control measures, undertaking the audiometric testing of employees, providing hearing-protective equipment where it is required, enforcing the use of such protective equipment with sound policies and by example, and informing employees of the benefits to be derived from a hearing conservation program.

It is the employee's responsibility to make proper use of the protective equipment provided by management. It is also the employee's responsibility to observe any rules or regulations in the use of equipment designed to minimize noise exposure.

Detailed references to noise, its management, effects, and control can be found in a great many books and periodicals. For those companies needing assistance in establishing hearing conservation programs, consultation services are available in a number of professional areas through private consultation, insurance, and governmental groups.

BIBLIOGRAPHY

Much of the material in this chapter is adapted from material that originally appeared in *Industrial Noise and Hearing Conservation* by J. B. Olishifski and E. R. Harford, published by the National Safety Council, 1975.

Acoustics Handbook. Palo Alto, CA: Hewlett-Packard Co., 1968.

Alberti, P. W., ed. *Personal Hearing Protection in Industry.* New York: Raven Press, 1982.

American Academy of Ophthalmology and Otolaryngology (AAOO). *Guide for Conservation of Hearing.* Rochester, MN: AAOO, 1970.

American Industrial Hygiene Association (AIHA). *Noise and Hearing Conservation Manual,* 4th ed. Akron, OH: AIHA, 1986.

American National Standards Institute, New York, NY. *Criteria for Background Noise in the Audiometer Room,* S3.1-1960 (R1971). *Method for the Measurement of Real-Ear Protection of Hearing Protectors and Physical Attenuation of Earmuffs,* S3.19-1974 (R1979).

Harris, C. M., ed. *Handbook of Noise Control,* 2nd ed. New York: McGraw-Hill Book Co., 1979.

Jones, R. S. *Noise and Vibration Control in Buildings.* New York: McGraw-Hill Book Co., 1984.

Kryter, K. D. *The Effects of Noise on Man,* 2nd ed. New York: Academic Press, Inc., 1985.

Michael, P. L. Ear protectors—their usefulness and limitations. *Archives of Environmental Health* 10(1965):612-618.

National Institute for Occupational Safety and Health. *Criteria for a Recommended Standard—Occupational Exposure to Noise.* Washington, DC: GPO, 1972, HSM 73-1101.

Newby, H. A., and Popelka, G. R. *Audiology,* 5th ed. Englewood Cliffs, NJ: Prentice-Hall, Inc., 1985.

Occupational Safety and Health Administration. *Occupational Noise Exposure and Hearing Conservation Amendment.*

Federal Register 46(11) (1981):4,078-4,181.
Federal Register 46(162) (1981):42,622-42,639.
Federal Register 48(46) (1983):9,738-9,783.
Olishifski, J. B., and Harford, E. R. *Industrial Noise and Hearing Conservation.* Chicago: National Safety Council, 1975.
Petersen, A. P. G. *Handbook of Noise Measurement,* 9th ed. Concord, MA: GenRad, Inc., 1980.
Sataloff, J., and Michael, P. *Hearing Conservation.* Springfield, IL: Charles C. Thomas Publ., 1973.
U. S. Department of Health, Education, and Welfare, National Institute for Occupational Safety and Health, Division of Technical Services, Cincinnati, Ohio.
——*Compendium of Materials for Noise Control,* Pub. No. 75-165.
——*Occupational Noise and Hearing: 1968-1972,* Pub. No. 74-116.
Ward, W. D., and Fricke, F. E., eds. *Noise as a Public Hazard.* Washington, DC: American Speech and Hearing Association, 1969.

1910.95—OCCUPATIONAL NOISE EXPOSURE

(a) Protection against the effects of noise exposure shall be provided when the sound levels exceed those shown in Table G–16 when measured on the A scale of a standard sound level meter at slow response. When noise levels are determined by octave band analysis, the equivalent A-weighted sound level may be determined as follows:

TABLE G–16—PERMISSIBLE NOISE EXPOSURES [1]

Duration per day, hours	*Sound level dBA slow response*
8	90
6	92
4	95
3	97
2	100
1½	102
1	105
½	110
¼ or less	115

[1]When the daily noise exposure is composed of two or more periods of noise exposure of different levels, their combined effect should be considered, rather than the individual effect of each. If the sum of the following fractions: $C_1/T_1 + C_2/T_2 \ldots C_n/T_n$ exceeds unity, then, the mixed exposure should be considered to exceed the limit value. C_n indicates the total time of exposure at a specified noise level, and T_n indicates the total time of exposure permitted at that level.

Figure G-9

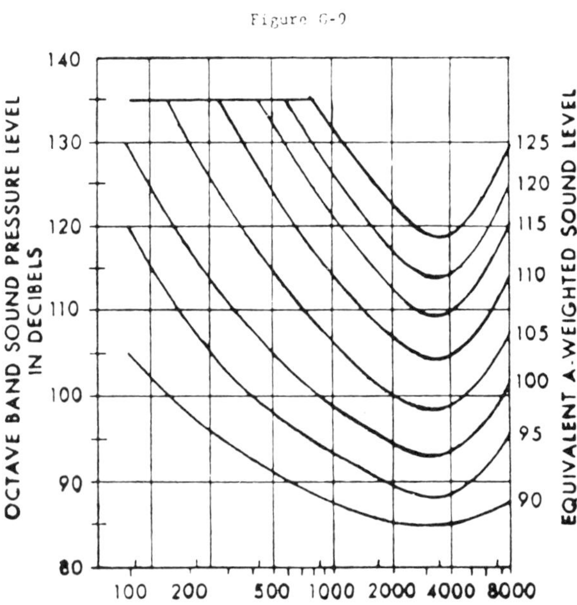

Equivalent sound level contours. Octave band sound pressure levels may be converted to the equivalent A-weighted sound level by plotting them on this graph and noting the A-weighted sound level corresponding to the point of highest penetration into the sound level contours. This equivalent A-weighted sound level, which may differ from the actual A-weighted sound level of the noise, is used to determine exposure limits from Table G–16.

(b)

(1) When employees are subjected to sound exceeding those listed in Table G-16, feasible administrative or engineering controls shall be utilized. If such controls fail to reduce sound levels within the levels of Table G-16, personal protective equipment shall be provided and used to reduce sound levels within the levels of the table.

Exposure to impulsive or impact noise should not exceed 140 dB peak sound pressure level.

(2) If the variations in noise level involve maxima at intervals of 1 second or less, it is to be considered continuous.

(c) Hearing conservation program.

(1) The employer shall administer a continuing, effective hearing conservation program, as described in paragraphs (c) through (o) of this section, whenever employee noise exposures equal or exceed an 8-hour time-weighted average sound level (TWA) of 85 decibels measured on the A scale (slow response) or, equivalently, a dose of fifty percent. For purposes of the hearing conservation program, employee noise exposures shall be computed in accordance with Appendix A and Table G-16a, and without regard to any attenuation provided by the use of personal protective equipment.

(2) For purposes of paragraphs (c) through (n) of this section, an 8-hour time-weighted average of 85 decibels or a dose of fifty percent shall also be referred to as the action level.

(d) Monitoring.

(1) When information indicates that any employee's exposure may equal or exceed an 8-

hour time-weighted average of 85 decibels, the employer shall develop and implement a monitoring program.

(i) The sampling strategy shall be designed to identify employees for inclusion in the hearing conservation program and to enable the proper selection of hearing protectors.

(ii) Where circumstances such as high worker mobility, significant variations in sound level, or a significant component of impulse noise make area monitoring generally inappropriate, the employer shall use representative personal sampling to comply with the monitoring requirements of this paragraph unless the employer can show that area sampling produces equivalent results.

(2)

(i) All continuous, intermittent and impulsive sound levels from 80 decibels to 130 decibels shall be integrated into the noise measurements.

(ii) Instruments used to measure employee noise exposure shall be calibrated to ensure measurement accuracy.

(3) Monitoring shall be repeated whenever a change in production, process, equipment or controls increases noise exposures to the extent that:

(i) Additional employees may be exposed at or above the action level; or

(ii) The attenuation provided by hearing protectors being used by employees may be rendered inadequate to meet the requirements of paragraph (j) of this section.

(e) Employee notification. The employer shall notify each employee exposed at or above an 8-hour time-weighted average of 85 decibels of the results of the monitoring.

(f) Observation of monitoring. The employer shall provide affected employees or their representatives with an opportunity to observe any noise measurements conducted pursuant to this section.

(g) Audiometric testing program.

(1) The employer shall establish and maintain an audiometric testing program as provided in this paragraph by making audiometric testing available to all employees whose exposures equal or exceed an 8-hour time-weighted average of 85 decibels.

(2) The program shall be provided at no cost to employees.

(3) Audiometric tests shall be performed by a licensed or certified audiologist, otolaryngologist, or other physician, or by a technician who is certified by the Council of Accreditation in Occupational Hearing Conservation, or who has satisfactorily demonstrated competence in administering audiometric examinations, obtaining valid audiograms, and properly using, maintaining and checking calibration and proper functioning of the audiometers being used. A technician who operates microprocessor audiometers does not need to be certified. A technician who performs audiometric tests must be responsible to an audiologist, otolaryngologist or physician.

(4) All audiograms obtained pursuant to this section shall meet the requirements of Appendix C: *Audiometric Measuring Instruments*.

(5) Baseline audiogram.

(i) Within 6 months of an employee's first exposure at or above the action level, the employer shall establish a valid baseline audiogram against which subsequent audiograms can be compared.

(ii) Mobile test van exception. Where mobile test vans are used to meet the audiometric testing obligations, the employer shall obtain a valid baseline audiogram within 1 year of an employee's first exposure at or above the action level. Where baseline audiograms are obtained more than 6 months after the employee's first exposure at or above the action level, employees shall wear hearing protectors for any period exceeding six months after first exposure until the baseline audiogram is obtained.

(iii) Testing to establish a baseline audiogram shall be preceded by at least 14 hours without exposure to workplace noise. Hearing protectors may be used as a substitute for the requirement that baseline audiograms be preceded by 14 hours without exposure to workplace noise.

(iv) The employer shall notify employees of the need to avoid high levels of non-occupational noise exposure during the 14-hour period immediately preceding the audiometric examination.

(6) Annual audiogram. At least annually after obtaining the baseline audiogram, the employer shall obtain a new audiogram for each employee exposed at or above an 8-hour time-weighted average of 85 decibels.

(7) Evaluation of audiogram.

(i) Each employee's annual audiogram shall be compared to that employee's baseline audiogram to determine if the audiogram is valid and if a standard threshold shift as defined in paragraph (g)(10) of this section has occurred. This comparison may be done by a technician.

(ii) If the annual audiogram shows that an employee has suffered a standard threshold shift, the employer may obtain a retest within 30 days and consider the results of the retest as the annual audiogram.

(iii) The audiologist, otolaryngologist, or physician shall review problem audiograms and shall determine whether there is a need for further evaluation. The employer shall provide to the person performing this evaluation the following information:

(a) A copy of the requirements for hearing conservation as set forth in paragraphs (c) through (n) of this section;

(b) The baseline audiogram and most recent audiogram of the employee to be evaluated;

(c) Measurements of background sound pressure levels in the audiometric test room as required in Appendix D: Audiometric Test Rooms.

(d) Records of audiometer calibrations required by paragraph (h)(5) of this section.

(8) Follow-up procedures.

(i) If a comparison of the annual audiogram to the baseline audiogram indicates a standard threshold shift as defined in paragraph (g)(10) of this section has occurred, the employee shall be informed of this fact in writing, within 21 days of the determination.

(ii) Unless a physician determines that the standard threshold shift is not work related or aggravated by occupational noise exposure, the employer shall ensure that the following steps are taken when a standard threshold shift occurs:

(a) Employees not using hearing protectors shall be fitted with hearing protectors, trained in their use and care, and required to use them.

(b) Employees already using hearing protectors shall be refitted and retained in the use of hearing protectors and provided with hearing protectors offering greater attenuation if necessary.

(c) The employee shall be referred for a clinical audiological evaluation or an otological examination, as appropriate, if additional testing is necessary or if the employer suspects that a medical pathology of the ear is caused or aggravated by the wearing of hearing protectors.

(d) The employee is informed of the need for an otological examination if a medical pathology of the ear that is unrelated to the use of hearing protectors is suspected.

(iii) If subsequent audiometric testing of an employee whose exposure to noise is less than an 8-hour TWA of 90 decibels indicates that a standard threshold shift is not persistent, the employer:

(a) Shall inform the employee of the new audiometric interpretation; and

(b) May discontinue the required use of hearing protectors for that employee.

(9) Revised baseline. An annual audiogram may be substituted for the baseline audiogram when, in the judgment of the audiologist, otolaryngologist or physician who is evaluating the audiogram:

(i) The standard threshold shift revealed by the audiogram is persistent; or

(ii) The hearing threshold shown in the annual audiogram indicates significant improvement over the baseline audiogram.

(10) Standard threshold shift.

(i) As used in this section, a standard threshold shift is a change in hearing threshold relative to the baseline audiogram of an average of 10 dB or more at 2000, 3000, and 4000 Hz in either ear.

(ii) In determining whether a standard threshold shift has occurred, allowance may be made for the contribution of aging (presbycusis) to the change in hearing level by correcting the annual audiogram according to the procedure described in Appendix F: *Calculation and Application of Age Correction to Audiograms.*

(h) Audiometric test requirements.

(1) Audiometric tests shall be pure tone, air conduction, hearing threshold examinations, with test frequencies including as a minimum 500, 1000, 2000, 3000, 4000, and 6000 Hz. Tests at each frequency shall be taken separately for each ear.

(2) Audiometric tests shall be conducted with audiometers (including microprocessor audiometers) that meet the specifications of, and are maintained and used in accordance with, American National Standard Specification for Audiometers, S3.6-1969.

(3) Pulsed-tone and self-recording audiometers, if used, shall meet the requirements specified in Appendix C: *Audiometric Measuring Instruments.*

(4) Audiometric examinations shall be administered in a room meeting the requirements listed in Appendix D: *Audiometric Test Rooms.*

(5) Audiometer calibration.

(i) The functional operation of the audiometer shall be checked before each day's use by testing a person with known, stable hearing thresholds, and by listening to the audiometer's output to make sure that the output is free from distorted or unwanted sounds. Deviations of 10 decibels or greater require an acoustic calibration.

(ii) Audiometer calibration shall be checked acoustically at least annually in accordance with Appendix E: *Acoustic Calibration of Audiometers.* Test frequencies below 500 Hz and above 6000 Hz may be omitted from this check. Deviations of 15 decibels or greater require an exhaustive calibration.

(iii) An exhaustive calibration shall be performed at least every two years in accordance with sections 4.1.2; 4.1.3.; 4.1.4.3; 4.2; 4.4.1; 4.4.2; 4.4.3; and 4.5 of the American National Standard Specification for Audiometers, S3.6-1969. Test frequencies below 500 Hz and above 6000 Hz may be omitted from this calibration.

(i) Hearing protectors.

(1) Employers shall make hearing protectors available to all employees exposed to an 8-hour time-weighted average of 85 decibels or greater at no cost to the employees. Hearing protectors shall be replaced as necessary.

(2) Employers shall ensure that hearing protectors are worn:

(i) By an employee who is required by paragraph (b)(1) of this section to wear personal protective equipment; and

(ii) By any employee who is exposed to an 8-hour time-weighted average of 85 decibels or greater, and who:

(a) Has not yet had a baseline audiogram established pursuant to paragraph (g)(5)(ii); or

(b) Has experienced a standard threshold shift.

(3) Employees shall be given the opportunity to select their hearing protectors from a variety of suitable hearing protectors provided by the employer.

(4) The employer shall provide training in the use and care of all hearing protectors provided to employees.

(5) The employer shall ensure proper initial fitting and supervise the correct use of all hearing protectors.

(j) Hearing protector attenuation.

(1) The employer shall evaluate hearing protector attenuation for the specific noise environments in which the protector will be used. The employer shall use one of the evaluation methods described in Appendix B: *Methods for Estimating the Adequacy of Hearing Protection Attenuation.*

(2) Hearing protectors must attenuate employee exposure at least to an 8-hour time-weighted average of 90 decibels as required by paragraph (b) of this section.

(3) For employees who have experienced a standard threshold shift, hearing protectors must attenuate employee exposure to an 8-hour time-weighted average of 85 decibels or below.

(4) The adequacy of hearing protector attenuation shall be re-evaluated whenever employee noise exposures increase to the extent that the hearing protectors provided may no longer provide adequate attenuation. The employee shall provide more effective hearing protectors where necessary.

(k) Training program.

(1) The employer shall institute a training program for all employees who are exposed to noise at or above an 8-hour time-weighted average of 85 decibels, and shall ensure employee participation in such program.

(2) The training program shall be repeated annually for each employee included in the hearing conservation program. Information provided in the training program shall be updated to be consistent with changes in protective equipment and work processes.

(3) The employer shall ensure that each employee is informed of the following:

(i) The effects of noise on hearing;

(ii) The purpose of hearing protectors, the advantages, disadvantages, and attenuation of various types, and instructions on selection, fitting, use, and care; and

(iii) The purpose of audiometric testing, and an explanation of the test procedures.

(l) Access to information and training materials.

(1) The employer shall make available to affected employees or their representatives copies of this standard and shall also post a copy in the workplace.

(2) The employer shall provide to affected employees any informational materials pertaining to the standard that are supplied to the employer by the Assistant Secretary.

(3) The employer shall provide, upon request, all materials related to the employer's training and education program pertaining to this standard to the Assistant Secretary and the Director.

(m) Recordkeeping.

(1) Exposure measurements. The employer shall maintain an accurate record of all employee exposure measurements required by paragraph (d) of this section.

(2) Audiometric tests.

(i) The employer shall retain all employee audiometric test records obtained pursuant to paragraph (g) of this section:

(ii) This record shall include:

(a) Name and job classification of the employee;

(b) Date of the audiogram;

(c) The examiner's name;

(d) Date of the last acoustic or exhaustive calibration of the audiometer; and

(e) Employee's most recent noise exposure assessment.

(f) The employer shall maintain accurate records of the measurements of the background sound pressure levels in audiometric test rooms.

(3) Record retention. The employer shall retain records required in this paragraph (m) for at least the following periods.

(i) Noise exposure measurement records shall be retained for two years.

(ii) Audiometric test records shall be retained for the duration of the affected employee's employment.

(4) Access to records. All records required by this section shall be provided upon request to employees, former employees, representatives designated by the individual employee, and the Assistant Secretary. The provisions of 29 CFR 1910.20(a)–(e) and (g)–(i) apply to access to records under this section.

(5) Transfer of records. If the employer ceases to do business, the employer shall transfer to the successor employer all records required to be maintained by this section, and the successor employer shall retain them for the remainder of the period prescribed in paragraph (m)(3) of this section.

(n) Appendices.

(1) Appendices A, B, C, D, and E to this section are incorporated as part of this section and the contents of these Appendices are mandatory.

(2) Appendices F and G to this section are informational and are not intended to create any additional obligations not otherwise imposed or to detract from any existing obligations.

(o) Exemptions. Paragraphs (c) through (n) of this section shall not apply to employers engaged in oil and gas well drilling and servicing operations.

(p) Startup date. Baseline audiograms required by paragraph (g) of this section shall be completed by March 1, 1984.

APPENDIX A: NOISE EXPOSURE COMPUTATION

This Appendix is Mandatory

I. Computation of Employee Noise Exposure

(1) Noise dose is computed using Table G-16a as follows:

(i) When the sound level, L, is constant over the entire work shift, the noise dose, D, in percent, is given by: $D=100\ C/T$ where C is the total length of the work day, in hours, and T is the reference duration corresponding to the measured sound level, L, as given in Table G-16a or by the formula shown as a footnote to that table.

(ii) When the workshift noise exposure is composed of two or more periods of noise at different levels, the total noise dose over the work day is given by:
$D=100\ (C_1/T_1+C_2/T_2+\ldots+C_n/T_n)$,
where C_n indicates the total time of exposure at a specific noise level, and T_n indicates the reference duration for that level as given by Table G-16a.

(2) The eight-hour time-weighted average sound level (TWA), in decibels, may be computed from the dose, in percent, by means of the formula: $TWA=16.61\ \log_{10}(D/100)+90$. For an eight-hour workshift with the noise level constant over the entire shift, the TWA is equal to the measured sound level.

(3) A table relating dose and TWA is given in Section II.

TABLE G-16A

A-weighted sound level, L (decibel)	Reference duration, T (hour)
80	32
81	27.9
82	24.3
83	21.1
84	16.4
85	16
86	13.9
87	12.1
88	10.6
89	9.2
90	8
91	7.0
92	6.1
93	5.3
94	4.6
95	4
96	3.5
97	3.0
98	2.6
99	2.3
100	2
101	1.7
102	1.5
103	1.3
104	1.1
105	1
106	0.87
107	0.76
108	0.66
109	0.57
110	0.5
111	0.44
112	0.38
113	0.33
114	0.29
115	0.25
116	*0.22*
117	*0.19*
118	*0.16*
119	*0.14*
120	*0.125*
121	*0.11*
122	*0.095*
123	*0.082*
124	*0.072*
125	*0.063*
126	*0.054*
127	*0.047*
128	*0.041*
129	*0.036*
130	*0.031*

In the above table the reference duration, T, is computed by

$$T=\frac{8}{2^{(L-90)/5}}$$

where L is the measured A-weighted sound level.

II. Conversion Between "Dose" and "8-Hour Time-Weighted Average" Sound Level

Compliance with paragraphs (c)-(r) of this regulation is determined by the amount of exposure to noise in the workplace. The amount of such exposure is usually measured with an audiodosimeter which gives a readout in terms of "dose." In order to better understand the requirements of the amendment, dosimeter readings can be converted to an "8-hour time-weighted average sound level." (TWA).

In order to convert the reading of a dosimeter into TWA, see Table A-1, below. This table applies to dosimeters that are set by the manufacturer to calculate dose or percent exposure according to the relationships in Table G-16a. So, for example, a dose of 91 percent over an eight hour day results in a TWA of 89.3 dB, and, a dose of 50 percent corresponds to a TWA of 85 dB.

If the dose as read on the dosimeter is less than or greater than the values found in Table A-1, the TWA may be calculated by using the formula: $TWA = 16.61 \log_{10} (D/100) + 90$ where TWA=8-hour time-weighted average sound level and D=accumulated dose in percent exposure.

Table A-1.—Conversion From "Percent Noise Exposure" or "Dose" to "8-Hour Time-Weighted Average Sound Level" (TWA)

Dose or percent noise exposure	TWA
10	73.4
15	76.3
20	78.4
25	80.0
30	81.3
35	82.4
40	83.4
45	84.2
50	85.0
55	85.7
60	86.3
65	86.9
70	87.4
75	87.9
80	88.4
81	88.5
82	88.6
83	88.7
84	88.7
85	88.8
86	88.9
87	89.0
88	89.1
89	89.2
90	89.2
91	89.3
92	89.4
93	89.5
94	89.6
95	89.6
96	89.7
97	89.8
98	89.9
99	89.9
100	90.0
101	90.1
102	90.1
103	90.2
104	90.3
105	90.4
106	90.4
107	90.5
108	90.6
109	90.6

Table A-1.—Conversions From "Percent Noise Exposure" or "Dose" to "8-Hour Time-Weighted Average Sound level" (TWA)—Continued

Dose or percent noise exposure	TWA
110	90.7
111	90.8
112	90.8
113	90.9
114	90.9
115	91.1
116	91.1
117	91.1
118	91.2
119	91.3
120	91.3
125	91.6
130	91.9
135	92.2
140	92.4
145	92.7
150	92.9
155	93.2
160	93.4
165	93.6
170	93.8
175	94.0
180	94.2
185	94.4
190	94.6
195	94.8
200	95.0
210	95.4
220	95.7
230	96.0
240	96.3
250	96.6
260	96.9
270	97.2
280	97.4
290	97.7
300	97.9
310	98.2
320	98.4
330	98.6
340	98.8
350	99.0
360	99.2
370	99.4
380	99.6
390	99.8
400	100.0
410	100.2
420	100.4
430	100.5
440	100.7
450	100.8
460	101.0
470	101.2
480	101.3
490	101.5
500	101.6
510	101.8
520	101.9
530	102.0
540	102.2
550	102.3
560	102.4
570	102.6
580	102.7

Table A-1.—Conversions From "Percent Noise Exposure" or "Dose" to "8-Hour Time-Weighted Average Sound level" (TWA)—Continued

Dose or percent noise exposure	TWA
590	102.8
600	102.9
610	103.0
620	103.2
630	103.3
640	103.4
650	103.5
660	103.6
670	103.7
680	103.8
690	103.9
700	104.0
710	104.1
720	104.2
730	104.3
740	104.4
750	104.5
760	104.6
770	104.7
780	104.8
790	104.9
800	105.0
810	105.1
820	105.2
830	105.3
840	105.4
850	105.4
860	105.5
870	105.6
880	105.7
890	105.8
900	105.8
910	105.9
920	106.0
930	106.1
940	106.2
950	106.2
960	106.3
970	106.4
980	106.5
990	106.5
999	106.6

APPENDIX B: METHODS FOR ESTIMATING THE ADEQUACY OF HEARING PROTECTOR ATTENUATION

This Appendix is Mandatory

For employees who have experienced a significant threshold shift, hearing protector attenuation must be sufficient to reduce employee exposure to a TWA of 85 dB. Employers must select one of the following methods by which to estimate the adequacy of hearing protector attenuation.

The most convenient method is the Noise Reduction Rating (NRR) developed by the Environmental Protection Agency (EPA). According to EPA regulation, the NRR must be shown on the hearing protector package. The NRR is then related to an individual worker's noise environment in order to assess the adequacy of the attenuation of a given hearing protector. This Appendix describes four methods of using the NRR to determine whether a particular hearing protector provides adequate protection within a given exposure environment. Selection among the four procedures is dependent upon the employer's noise measuring instruments.

Instead of using the NRR, employers may evaluate the adequacy of hearing protector attenuation by using one of the three methods developed by the National Institute for Occupational Safety and Health (NIOSH), which are described in the "List of Personal Hearing Protectors and Attenuation Data," HEW Publication No. 76-120, 1975, pages 21-37. These methods are known as NIOSH methods #1, #2 and #3. The NRR described below is a simplification of NIOSH method #2. The most complex method is NIOSH method #1, which is probably the most accurate method since it uses the largest amount of spectral information from the individual employee's noise environment. As in the case of the NRR method described below, if one of the NIOSH methods is used, the selected method must be applied to an individual's noise environment to assess the adequacy of the attenuation. Employers should be careful to take a sufficient number of measurements in order to achieve a representative sample for each time segment.

Note.—The employer must remember that calculated attenuation values reflect realistic values only to the extent that the protectors are properly fitted and worn.

When using the NRR to assess hearing protector adequacy, one of the following methods must be used:

(I) When using a dosimeter that is capable of C-weighted measurements:

(A) Obtain the employee's C-weighted dose for the entire workshift, and convert to TWA (see Appendix A, II).

(B) Subtract the NRR from the C-weighted TWA to obtain the estimated A-weighted TWA under the ear protector.

(II) When using a dosimeter that is not capable of C-weighted measurements, the following method may be used:

(A) Convert the A-weighted dose to TWA (see Appendix A).

(B) Subtract 7 dB from the NRR.

(C) Subtract the remainder from the A-weighted TWA to obtain the estimated A-weighted TWA under the ear protector.

(III) When using a sound level meter set to the A-weighting network:

(A) Obtain the employee's A-weighted TWA.

(B) Subtract 7 dB from the NRR, and subtract the remainder from the A-weighted TWA to obtain the estimated A-weighted TWA under the ear protector.

(IV) When using a sound level meter set on the C-weighting network:

(A) Obtain a representative sample of the C-weighted sound levels in the employee's environment.

(B) Subtract the NRR from the C-weighted average sound level to obtain the estimated A-weighted TWA under the ear protector.

(V) When using area monitoring procedures and a sound level meter set to the A-weighing network.

(A) Obtain a representative sound level for the area in question.

(B) Subtract 7 dB from the NRR and subtract the remainder from the A-weighted sound level for that area.

(vi) When using area monitoring procedures and a sound level meter set to the C-weighting network:

(A) Obtain a representative sound level for the area in question.

(B) Subtract the NRR from the C-weighted sound level for that area.

APPENDIX C: AUDIOMETRIC MEASURING INSTRUMENTS

This Appendix Is Mandatory

1. In the event that pulsed-tone audiometers are used, they shall have a tone on-time of at least 200 milliseconds.

2. Self-recording audiometers shall comply with the following requirements:

(A) The chart upon which the audiogram is traced shall have lines at positions corresponding to all multiples of 10 dB hearing level within the intensity range spanned by the audiometer. The lines shall be equally spaced and shall be separated by at least ¼ inch. Additional increments are optional. The audiogram pen tracings shall not exceed 2 dB in width.

(B) It shall be possible to set the stylus manually at the 10-dB increment lines for calibration purposes.

(C) The slewing rate for the audiometer attenuator shall not be more than 6 dB/sec except that an initial slewing rate greater than 6 dB/sec is permitted at the beginning of each new test frequency, but only until the second subject response.

(D) The audiometer shall remain at each required test frequency for 30 seconds (±3 seconds). The audiogram shall be clearly marked at each change of frequency and the actual frequency change of the audiometer shall not deviate from the frequency boundaries marked on the audiogram by more than ±3 seconds.

(E) It must be possible at each test frequency to place a horizontal line segment parallel to the time axis on the audiogram, such that the audiometric tracing crosses the line segment at least six times at that test frequency. At each test frequency the threshold shall be the average of the midpoints of the tracing excursions.

APPENDIX D: AUDIOMETRIC TEST ROOMS

This Appendix Is Mandatory

Rooms used for audiometric testing shall not have background sound pressure levels exceeding those in Table D-1 when measured by equipment conforming at least to the Type 2 requirements of American National Standard Specification for Sound Level Meters, S1.4-1971 (R1976), and to the Class II requirements of American National Standard Specification for Octave, Half-Octave, and Third-Octave Band Filter Sets, S1.11-1971 (R1976).

Table D-1.—Maximum Allowable Octave-Band Sound Pressure Levels for Audiometric Test Rooms

Octave-band center frequency (Hz)	500	1000	2000	4000	8000
Sound pressure level (dB)	40	40	47	57	62

APPENDIX E: ACOUSTIC CALIBRATION OF AUDIOMETERS

This Appendix Is Mandatory

Audiometer calibration shall be checked acoustically, at least annually, according to the procedures described in this Appendix. The equipment necessary to perform these measurments is a sound level meter, octave-band filter set, and a National Bureau of Standards 9A coupler. In making these measurements, the accuracy of the calibrating equipment shall be sufficient to determine that the audiometer is within the tolerances permitted by American Standard Specification for Audiometers, S3.6-1969.

(1) Sound Pressure Output Check

A. Place the earphone coupler over the microphone of the sound level meter and place the earphone on the coupler.

B. Set the audiometer's hearing threshold level (HTL) dial to 70 dB.

C. Measure the sound pressure level of the tones that each test frequency from 500 Hz through 6000 Hz for each earphone.

D. At each frequency the readout on the sound level meter should correspond to the levels in Table E-1 or Table E-2, as appropriate, for the type of earphone, in the column entitled "sound level meter reading."

(2) Linearity Check

A. With the earphone in place, set the frequency to 1000 Hz and the HTL dial on the audiometer to 70 dB.

B. Measure the sound levels in the coupler at each 10-dB decrement from 70 dB to 10 dB, noting the sound level meter reading at each setting.

C. For each 10-dB decrement on the audiometer the sound level meter should indicate a corresponding 10 dB decrease.

D. This measurement may be made electrically with a voltmeter connected to the earphone terminals.

(3) Tolerances

When any of the measured sound levels deviate from the levels in Table E-1 or Table E-2 by ±3 dB at any test frequency between 500 and 3000 Hz, 4 dB at 4000 Hz, or 5 dB at 6000 Hz, an exhaustive calibration is advised. An exhaustive calibration is required if the deviations are greater than 10 dB at any test frequency.

Table E-1.—Reference Threshold Levels for Telephonics—TDH-39 Earphones

Frequency, Hz	Reference threshold level for TDH-39 ear-phones, dB	Sound level meter reading, dB
500	11.5	81.5
1000	7	77
2000	9	79
3000	10	80
4000	9.5	79.5
6000	15.5	85.5

Table E-2.—Reference Threshold Levels for Telephonics—TDH-49 Earphones

Frequency, Hz	Reference threshold level for TDH-49 ear-phones, dB	Sound level meter reading, dB
500	13.5	83.5
1000	7.5	77.5
2000	11	81.0
3000	9.5	79.5
4000	10.5	80.5
6000	13.5	83.5

APPENDIX F: CALCULATIONS AND APPLICATION OF AGE CORRECTIONS TO AUDIOGRAMS

This Appendix is Non-Mandatory

In determining whether a standard threshold shift has occurred, allowance may be made for the contribution of aging to the change in hearing level by adjusting the most recent audiogram. If the employer chooses to adjust the audiogram, the employer shall follow the procedure described below. This procedure and the age correction tables were developed by the National Institute for Occupational Safety and Health in the criteria document entitled "Criteria for a Recommended Standard ... Occpational Exposure to Noise," ((HSM)-11001).

For each audiometric test frequency;

(i) Determine from Tables F-1 or F-2 the age correction values for the employee by:

(A) Finding the age at which the most recent audiogram was taken and recording the corresponding values of age corrections at 1000 Hz through 6000 Hz;

(B) Finding the age at which the baseline audiogram was taken and recording the corresponding values of age corrections at 1000 Hz through 6000 Hz.

(ii) Subtract the values found in step (i)(A) from the value found in step (i)(B).

(iii) The differences calculated in step (ii) represented that portion of the change in hearing that may be due to aging.

Example: Employee is a 32-year-old male. The audiometric history for his right ear is shown in decibels below.

Employee's age	Audiometric test frequency (Hz) 1000	2000	3000	4000	6000
26	10	5	5	10	5
*27	0	0	0	5	5
28	0	0	0	10	5
29	5	0	5	15	5
30	0	5	10	20	10
31	5	10	20	15	15
*32	5	10	10	25	20

The audiogram at age 27 is considered the baseline since it shows the best hearing threshold levels. Asterisks have been used to identify the baseline and most recent audiogram. A threshold shift of 20 dB exists at 4000 Hz between the audiograms taken at ages 27 and 32.

(The threshold shift is computed by subtracting the hearing threshold at age 27, which was 5, from the hearing threshold at age 32, which is 25). A retest audiogram has confirmed this shift. The contribution of aging to this change in hearing may be estimated in the following manner:

Go to Table F-1 and find the age correction values (in dB) for 4000 Hz at age 27 and age 32.

	Frequency (Hz) 1000	2000	3000	4000	6000
Age 32	6	5	7	10	14
Age 27	5	4	6	7	11
Difference	1	1	1	3	3

The difference represents the amount of hearing loss that may be attributed to aging in the time period between the baseline audiogram and the most recent audiogram. In this example, the difference at 4000 Hz is 3 dB. This value is subtracted from the hearing level at 4000 Hz, which in the most recent audiogram is 25, yielding 22 after adjustment. Then the hearing threshold in the baseline audiogram at 4000 Hz (5) is subtracted from the adjusted annual audiogram hearing threshold at 4000 Hz (22). Thus the age-corrected threshold shift would be 17 dB (as opposed to a threshold shift of 20 dB without age correction).

Table F-1.—Age Correction Values In Decibels For Males

Years	Audiometric Test Frequencies (Hz) 1000	2000	3000	4000	6000
20 or younger	5	3	4	5	8
21	5	3	4	5	8
22	5	3	4	5	8
23	5	3	4	6	9
24	5	3	5	6	9
25	5	3	5	7	10
26	5	4	5	7	10
27	5	4	6	7	11
28	6	4	6	8	11
29	6	4	6	8	12
30	6	4	6	9	12
31	6	4	7	9	13
32	6	5	7	10	14
33	6	5	7	10	14
34	6	5	8	11	15
35	7	5	8	11	15
36	7	5	9	12	16
37	7	6	9	12	17
38	7	6	9	13	17
39	7	6	10	14	18
40	7	6	10	14	19
41	7	6	10	14	20
42	8	7	11	16	20
43	8	7	12	16	21
44	8	7	12	17	22
45	8	7	13	18	23
46	8	8	13	19	24
47	8	8	14	19	24
48	9	8	14	20	25
49	9	9	15	21	26
50	9	9	16	22	27
51	9	9	16	23	28
52	9	10	17	24	29
53	9	10	18	25	30
54	10	10	18	26	31

Table F-1.—Age Correction Values In Decibels For Males—Continued

Years	Audiometric Test Frequencies (Hz) 1000	2000	3000	4000	6000
55	10	11	19	27	32
56	10	11	20	28	34
57	10	11	21	29	35
58	10	12	22	31	36
59	11	12	22	32	37
60 or older	11	13	23	33	38

Table F-2.—Age Correction Values In Decibels For Females

Years	Audiometric Test Frequencies (Hz) 1000	2000	3000	4000	6000
20 or younger	7	4	3	3	6
21	7	4	4	3	6
22	7	4	4	4	6
23	7	5	4	4	7
24	7	5	4	4	7
25	8	5	4	4	7
26	8	5	5	4	8
27	8	5	5	5	8

Table F-2.—Age Correction Values In Decibels For Females —Continued

Years	Audiometric Test Frequencies (Hz) 1000	2000	3000	4000	6000
28	8	5	5	5	8
29	8	5	5	5	9
30	8	6	5	5	9
31	8	6	6	5	9
32	9	6	6	6	10
33	9	6	6	6	10
34	9	6	6	6	10
35	9	6	7	7	11
36	9	7	7	7	11
37	9	7	7	7	12
38	10	7	7	7	12
39	10	7	8	8	12
40	10	7	8	8	13
41	10	8	8	8	13
42	10	8	9	9	13
43	11	8	9	9	14
44	11	8	9	9	14
45	11	8	10	10	15
46	11	9	10	10	15
47	11	9	10	11	16
48	12	9	11	11	16
49	12	9	11	11	16
50	12	10	11	12	17
51	12	10	12	12	17
52	12	10	12	13	18
53	13	10	13	13	18
54	13	11	13	14	19
55	13	11	14	14	19
56	13	11	14	15	20
57	13	11	15	15	20
58	14	12	15	16	21
59	14	12	16	16	21
60 or older	14	12	16	17	22

APPENDIX G: MONITORING NOISE LEVELS NON-MANDATORY INFORMATIONAL APPENDIX

This appendix provides information to help employers comply with the noise monitoring obligations that are part of the hearing conservation amendment.

What is the purpose of noise monitoring?

This revised amendment requires that employees be placed in a hearing conservation program if they are exposed to average noise levels of 85 dB or greater during an 8 hour workday. In order to determine if exposures are at or above this level, it may be necessary to measure or monitor the actual noise levels in the workplace and to estimate the noise exposure or "dose" received by employees during the workday.

When is it necessary to implement a noise monitoring program?

It is not necessary for every employer to measure workplace noise. Noise monitoring or measuring must be conducted only when exposures are at or above 85 dB. Factors which suggest that noise exposures in the workplace may be at this level include employee complaints about the loudness of noise, indications that employees are losing their hearing, or noisy conditions which make normal conversation difficult. The employer should also consider any information available regarding noise emitted from specific machines. In addition, actual workplace noise measurements can suggest whether or not a monitoring program should be initiated.

How is noise measured?

Basically, there are two different instruments to measure noise exposures: the sound level meter and the dosimeter. A sound level meter is a device that measures the intensity of sound at a given moment. Since sound level meters provide a measure of sound intensity at only one point in time, it is generally necessary to take a number of measurements at different times during the day to estimate noise exposure over a workday. If noise levels fluctuate, the amount of time noise remains at each of the various measured levels must be determined.

To estimate employee noise exposures with a sound level meter it is also generally necessary to take several measurements at different locations within the workplace. After appropriate sound level meter readings are obtained, people sometimes draw "maps" of the sound levels within different areas of the workplace. By using a sound level "map" and information on employee locations throughout the day, estimates of individual exposure levels can be developed. This measurement method is generally referred to as *area* noise monitoring.

A dosimeter is like a sound level meter except that it stores sound level measurements and integrates these measurements over time, providing an average noise exposure reading for a given period of time, such as an 8-hour workday. With a dosimeter, a microphone is attached to the employee's clothing and the exposure measurement is simply read at the end of the desired time period. A reader may be used to read-out the dosimeter's measurements. Since the dosimeter is worn by the employee, it measures noise levels in those locations in which the employee travels. A sound level meter can also be positioned within the immediate vicinity of the exposed worker to obtain an individual exposure estimate. Such procedures are generally referred to as *personal* noise monitoring.

Area monitoring can be used to estimate noise exposure when the noise levels are relatively constant and employees are not mobile. In workplaces where employees move about in different areas or where the noise intensity tends to fluctuate over time, noise exposure is generally more accurately estimated by the personal monitoring approach.

In situations where personal monitoring is appropriate, proper positioning of the microphone is necessary to obtain accurate measurements. With a dosimeter, the microphone is generally located on the shoulder and remains in that position for the entire workday. With a sound level meter, the microphone is stationed near the employee's head, and the instrument is usually held by an individual who follows the employee as he or she moves about.

Manufacturer's instructions, contained in dosimeter and sound level meter operating manuals, should be followed for calibration and maintenance. To ensure accurate results, it is considered good professional practice to calibrate instruments before and after each use.

How often is it necessary to monitor noise levels?

The amendment requires that when there are significant changes in machinery or production processes that may result in increased noise levels, remonitoring must be conducted to determine whether additional employees need to be included in the hearing conservation program. Many companies choose to remonitor periodically (once every year or two) to ensure that all exposed employees are included in their hearing conservation programs.

Where can equipment and technical advice be obtained?

Noise monitoring equipment may be either purchased or rented. Sound level meters cost about $500 to $1,000, while dosimeters range in price from about $750 to $1,500. Smaller companies may find it more economical to rent equipment rather than to purchase it. Names of equipment suppliers may be found in the telephone book (Yellow Pages) under headings such as: "Safety Equipment," "Industrial Hygiene," or "Engineers-Acoustical." In addition to providing information on obtaining noise monitoring equipment, many companies and individuals included under such listings can provide professional advice on how to conduct a valid noise monitoring program. Some audiological testing firms and industrial hygiene firms also provide noise monitoring services. Universities with audiology, industrial hygiene, or acoustical engineering departments may also provide information or may be able to help employers meet their obligations under this amendment.

Free, on-site assistance may be obtained from OSHA-supported state and private consultation organizations. These safety and health consultative entities generally give priority to the needs of small businesses. See the attached directory for a listing of organizations to contact for aid.

OSHA Onsite Consultation Project Directory

State	Office and address	Contact
Alabama	Alabama Consultation Program, P.O. Box 6005, University, Alabama 35486	(205) 348-7136, Mr. William Weems, Director.
Alaska	State of Alaska, Department of Labor, Occupational Safety & Health, 3301 Eagle St., Pouch 7-022, Anchorage, Alaska 99510.	(907) 276-5013, Mr. Stan Godboe, Project Manager [illegible]
American Samoa	Service not yet available.	
Arizona	Consultation and Training, Arizona Division of Occupational Safety and Health, P.O. Box 19070, 1624 W. Adams, Phoenix, Ariz. 85005.	(602) 255-5795, Mr. Thomas Ramsley, Manager.
Arkansas	OSHA Consultation, Arkansas Department of Labor, 1022 High St., Little Rock, Ark. 72202	(501) 371-2992, Mr. George Smith, Project Director
California	CAL/OSHA Consultation Service, 2nd Floor, 525 Golden Gate Avenue, San Francisco, Calif. 94102.	(415) 557-2870, Mr. Emmett Jones, Chief.
Colorado	Occupational Safety & Health Section, Colorado State University, Institute of Rural Environmental Health, 110 Veterinary Science Building, Fort Collins, Colo. 80523.	(303) 491-6151, Dr. Roy M. Buchan, Project Director
Connecticut	Division of Occupational Safety & Health, Connecticut Department of Labor, 200 Folly Brook Boulevard, Wethersfield, Conn. 06109.	(203) 566-4550, Mr. Leo Alix, Director.
Delaware	Delaware Department of Labor, Division of Industrial Affairs, 820 North French Street, 6th Floor, Wilmington, Del. 19801.	(302) 571-3908, Mr. Bruno Salvadori, Director
District of Columbia	Occupational Safety & Health Division, District of Columbia, Department Employment Services, Office of Labor Standards, 2900 Newton Street NE., Washington, D.C. 20018.	(202) 832-1230, Mr. Lorenzo M. White, Acting Associate Director.
Florida	Department of Labor & Employment Security, Bureau of Industrial Safety and Health, LaFayette Building, Room 204, 2551 Executive Center Circle West, Tallahassee, Fla. 32301.	(904) 488-3044, Mr. John C. Glenn, Administrator.
Georgia	Economic Development Division, Technology and Development Laboratory, Engineering Experiment Station, Georgia Institute of Technology, Atlanta, Ga. 30332.	(404) 894-3806, Mr. William C. Howard, Assistant to Director. Mr. James Burson, Project Manager.

OSHA ONSITE CONSULTATION PROJECT DIRECTORY—Continued

State	Office and address	Contact
Guam	Department of Labor, Government of Guam, 23548 Guam Main Facility, Agana, Guam 96921	(671) 772-6291, Joe R. San Agustin, Director.
Hawaii	Education and Information Branch, Division of Occupational Safety and Health, Suite 910, 677 Ala Moana, Honolulu, Hawaii 96813.	(808) 548-2511, Mr. Don Alper, Manager (Air Mail).
Idaho	OSHA Onsite Consultation Program, Boise State University, Community and Environmental Health, 1910 University Drive, Boise, Idaho 83725.	(208) 385-3929, Dr. Eldon Edmundson, Director.
Illinois	Division of Industrial Services, Dept. of Commerce and Community Affairs, 310 S. Michigan Avenue, 10 Floor, Chicago, Ill. 60601.	(800) 972-4140/4216 (Toll-free in State), (312) 793-3270, Mr. Stan Czwinski, Assistant Director.
Iowa	Bureau of Labor, 307 E. Seventh Street, Des Moines, Iowa 50319	(515) 281-3606, Mr. Allen J. Meier, Commissioner
Indiana	Bureau of Safety, Education and Training, Indiana Division of Labor, 1013 State Office Building, Indianapolis, Indiana 46204.	(317) 633-5845, Mr. Harold Mills, Director.
Kansas	Kansas Dept. of Human Resources, 401 Topeka Ave., Topeka, Kans. 66603	(913) 296-4086, Ms. Jerry Abbott, Secretary.
Kentucky	Education and Training, Occupational Safety and Health, Kentucky Department of Labor, 127 Building, 127 South, Frankfort, Ky. 40601.	(502) 564-6895, Mr. Larry Potter, Director.
Louisiana	No services available as yet (Pending FY 83).	
Maine	Division of Industrial Safety, Maine Dept. of Labor, Labor Station 45, State Office Building, Augusta, Maine 04333.	(207) 289-3331, Mr. Lester Wood, Director.
Maryland	Consultation Services, Division of Labor & Industry, 501 St. Paul Place, Baltimore, Maryland 21202	(301) 659-4210, Ms. Ileana O'Brien, Project Manager, 7(c)(1) Agreement.
Massachusetts	Division of Industrial Safety, Massachusetts Department of Labor and Industries, 100 Cambridge Street, Boston, Massachusetts 02202.	(617) 727-3567, Mr. Edward Noseworthy, Project Director
Michigan (Health)	Special Programs Section, Division of Occupational Health, Michigan Dept. of Public Health, 3500 N. Logan, Lansing, Mich. 48909.	(517) 373-1410, Mr. Irving Davis, Chief
Michigan (Safety)	Safety Education & Training Division Bureau of Safety and Regulation, Michigan Department of Labor, 7150 Harris Drive, Box 30015, Lansing, Michigan 48909.	(517) 322-1809, Mr. Alan Harvie, Chief.
Minnesota	Training and Education Unit, Department of Labor and Industry, 5th Floor, 444 Lafayette Road, St. Paul, Minn. 55101.	(612) 296-2973, Mr. Timothy Tierney, Project Manager
Mississippi	Division of Occupational Safety and Health, Mississippi State Board of Health, P.O. Box 1700, Jackson, Mississippi 39205.	(601) 982-6315, Mr. Henry L. Laird, Director.
Missouri	Missouri Department of Labor and Industrial Relations, 722 Jefferson Street, Jefferson City, Missouri 65101	1-(800) 392-0208, (314) 751-3403, Ms. Paula Smith, Mr. Jim Brake.
Montana	Montana Bureau of Safety & Health, Division of Workers Compensation, 815 Front Street, Helena, Montana 59601.	(406) 449-3402, Mr. Ed Gatzemeier, Chief.
Nebraska	Nebraska Department of Labor, State House Station, State Capitol, P.O. Box 94600, Lincoln, Nebraska 68509.	475-8451 Ext. 258, Mr. Joseph Carroll, Commissioner
Nevada	Department of Occupational Safety and Health, Nevada Industrial Commission, 515 E. Muffer Street, Carson City, Nev. 89714.	(702) 885-5240, Mr. Allen Traenkner, Director.
New Hampshire	For information contact	Office of Consultation Programs, Room N3472 200 Constitution Avenue, N.W. Washington, D.C. 20210, Phone: (202) 523-8985.
New Jersey	New Jersey Department of Labor and Industry Division of Work Place Standards, CN-054, Trenton, New Jersey 08625.	(609) 292-2313, FTS-8-477-2313, Mr. William Clark, Assistant Commissioner.
New Mexico	OSHA Consultation, Health and Environment Department, Environmental Improvement Division, Occupational Health & Safety Section, 4215 Montgomery Boulevard, NE., Albuquerque, New Mexico 87109.	(505) 842-3387, Mr. Albert M. Stevens, Project Manager.
New York	Division of Safety and Health, New York State Department of Labor, 2 World Trade Center, Room 6995, New York, New York 10047.	(212) 488-7746/7, Mr. Joseph Alleva, Project Manager, DOSH.
North Carolina	Consultation Services, North Carolina Department of Labor, 4 West Edenton Street, Raleigh, N.C. 27601.	(919) 733-4885, Mr. David Pierce, Director.
North Dakota	Division of Environmental Research, Department of Health, Missouri Office Building, 1200 Missouri Avenue, Bismarck, N. Dak. 58505.	(701) 224-2348, Mr. Jay Crawford, Director.
Ohio	Department of Industrial Relations, Division of Onsite Consultation, P.O. Box 825, 2323 5th Avenue, Columbus, Ohio 43216.	(800) 282-1425 (Toll-free in State), (614) 466-7485, Mr. Andrew Doehrel, Project Manager.
Oklahoma	OSHA Division, Oklahoma Department of Labor, State Capitol, Suite 118, Oklahoma City, Okla. 73105.	(405) 521-2461, Mr. Charles W. McGlon, Director.
Oregon	Consultative Section, Department of Workers' Compensation, Accident Prevention Division, Room 102, Building 1, 2110 Front Street NE., Salem, Oregon 97310.	(503) 378-2890, Mr. Jack Buckland, Supervisor.
Pennsylvania	For information contact	Office of Consultation Programs, Room N3472, 200 Constitution Avenue NW., Washington, D.C. 20210, Phone: (202) 523-8985.
Puerto Rico	Occupational Safety & Health, Puerto Rico Department of Labor and Human Resources, 505 Munoz Rivera Ave., 21st Floor, Hato Rey, Puerto Rico 00919.	(809) 754-2134, Mr. John Cinque, Assistant Secretary, (Air Mail).
Rhode Island	Division of Occupational Health, Rhode Island Department of Health, The Cannon Building, 206 Health Department Building, Providence, R.I. 02903.	(401) 277-2438, Mr. James E. Hickey, Chief.
South Carolina	Consultation and Monitoring, South Carolina Department of Labor, P.O. Box 11329, Columbia, S.C. 29211.	(803) 758-8921, Mr. Robert Peck, Director, 7(c)(1), Project.
South Dakota	South Dakota Consultation Program, South Dakota State University, S.T.A.T.E.-Engineering Extension, 201 Pugsley Center-SDSO, Brookings, S. Dak. 57007.	(605) 688-4101, Mr. James Caglan, Director.
Tennessee	OSHA Consultative Services, Tennessee Department of Labor, 2nd Floor, 501 Union Building, Nashville, Tennessee 37219.	(615) 741-2793, Mr. L. H. Craig Director.
Texas	Division of Occupational Safety and State Safety Engineer, Texas Department of Health and Resources, 1100 West 49th Street, Austin, Texas 78756.	(512) 458-7287, Mr. Walter G. Martin, P.E. Director.
Trust Territories	Service not yet available.	
Utah	Utah Job Safety and Health Consultation Service, Suite 4004, Crane Building, 307 West 200 South, Salt Lake City, Utah 84101.	(801) 533-7927/8/9, Mr. H. M. Bergeson, Project Director.
Vermont	Division of Occupational Safety and Health, Vermont Department of Labor and Industry, 118 State Street, Montpelier, Vt. 05602.	(802) 828-2765, Mr. Robert Mcleod, Project Director.
Virginia	Department of Labor and Industry, P.O. Box 12064, 205 N. 4th Street, Richmond, Va. 23241	(804) 786-5875, Mr. Robert Beard, Commissioner.
Virgin Islands	Division of Occupational Safety and Health, Virgin Islands Department of Labor, Lagoon Street, Room 207, Frederiksted, Virgin Islands 00840.	(809) 772-1315, Mr. Louis Llanos, Deputy Director-DOSH.
Washington	Department of Labor and Industry, P.O. Box 207, Olympia, Wash. 98504	(206) 753-6500, Mr. James Sullivan, Assistant Director.
West Virginia	West Virginia Department of Labor, Room 451B, State Capitol, 1900 Washington Street, Charleston, W. Va. 25305.	FTS 8-885-7890, Mr. Lawrence Barker, Commissioner.
Wisconsin (Health)	Section of Occupational Health, Department of Health and Social Services, P.O. Box 309, Madison, Wisconsin 53701.	(608) 266-0417, Ms. Patricia Natzke, Acting Chief.
Wisconsin (Safety)	Division of Safety and Buildings, Department of Industry, Labor and Human Relations, 1570 E. Moreland Blvd., Waukesha, Wis. 53186.	(414) 544-8686, Mr. Richard Michalski, Supervisor.
Wyoming	Wyoming Occupational Health and Safety Department, 200 East 8th Avenue, Cheyenne, Wyo. 82002.	(307) 777-7786, Mr. Donald Owsley, Health and Safety Administrator.

APPENDIX H: AVAILABILITY OF REFERENCED DOCUMENTS

Paragraphs (c) through (o) of 29 CFR 1910.95 and the accompanying appendices contain provisions which incorporate publications by reference. Generally, the publications provide criteria for instruments to be used in monitoring and audiometric testing. These criteria are intended to be mandatory when so indicated in the applicable paragraphs of Section 1910.95 and appendices.

It should be noted that OSHA does not require that employers purchase a copy of the referenced publications. Employers, however, may desire to obtain a copy of the referenced publications for their own information.

The designation of the paragraph of the standard in which the referenced publications appear, the titles of the publications, and the availability of the publication are as follows:

Paragraph designation	Referenced publication	Available from—
Appendix B	"List of Personal Hearing Protectors and Attenuation Data," HEW Pub. No. 76-120, 1975, NTIS-PB267461.	National Technical Information Service, Port Royal Road, Springfield, VA 22161.
Appendix D	"Specification for Sound Level Meters," S1.4-1971 (R1976).	American National Standards Institute, Inc., 1430 Broadway, New York, NY 10018.
§1910.95(k)(2), appendix E	"Specifications for Audiometers," S3.6-1969.	American National Standards Institute, Inc., 1430 Broadway, New York, NY 10018.
Appendix D	"Specification for Octave, Half-Octave and Third-Octave Band Filter Sets," S1.11-1971 (R1976).	Back Numbers Department, Dept. STD, American Institute of Physics, 333 E. 45 St., New York, NY 10017; American National Standards Institute, Inc., 1430 Broadway, New York, NY 10018.

The referenced publications (or a microfiche of the publications) are available for review at many universities and public libraries throughout the country. These publications may also be examined at the OSHA Technical Data Center, Room N2439, United States Department of Labor, 200 Constitution Avenue, NW., Washington, D.C. 20210, (202) 523-9700 or at any OSHA Regional Office (see telephone directories under United States Government—Labor Department).

APPENDIX I: DEFINITIONS

These definitions apply to the following terms as used in paragraphs (c) through (n) of 29 CFR 1910.95.

Action level—An 8-hour time-weighted average of 85 decibels measured on the A-scale, slow response, or equivalently, a dose of fifty percent.

Audiogram—A chart, graph, or table resulting from an audiometric test showing an individual's hearing threshold levels as a function of frequency.

Audiologist—A professional, specializing in the study and rehabilitation of hearing, who is certified by the American Speech-Language-Hearing Association or licensed by a state board of examiners.

Baseline audiogram—The audiogram against which future audiograms are compared.

Criterion sound level—A sound level of 90 decibels.

Decibel (dB)—Unit of measurement of sound level.

Hertz (Hz)—Unit of measurement of frequency, numerically equal to cycles per second.

Medical pathology—A disorder or disease. For purposes of this regulation, a condition or disease affecting the ear, which should be treated by a physician specialist.

Noise dose—The ratio, expressed as a percentage, of (1) the time integral, over a stated time or event, of the 0.6 power of the measured SLOW exponential time-averaged, squared A-weighted sound pressure and (2) the product of the criterion duration (8 hours) and the 0.6 power of the squared sound pressure corresponding to the criterion sound level (90 dB).

Noise dosimeter—An instrument that integrates a function of sound pressure over a period of time in such a manner that it directly indicates a noise dose.

Otolaryngologist—A physician specializing in diagnosis and treatment of disorders of the ear, nose and throat.

Representative exposure—Measurements of an employee's noise dose or 8-hour time-weighted average sound level that the employers deem to be representative of the exposures of other employees in the workplace.

Sound level—Ten times the common logarithm of the ratio of the square of the measured A-weighted sound pressure to the square of the standard reference pressure of 20 micropascals. Unit: decibels (dB). For use with this regulation, SLOW time response, in accordance with ANSI S1.4-1971 (R1976), is required.

South level meter—An instrument for the measurement of sound level.

Time-weighted average sound level—That sound level, which if constant over an 8-hour exposure, would result in the same noise dose as is measured.

10

Ionizing Radiation

by C. Lyle Cheever, MS, MBA

Basic concepts of ionizing radiation and safe handling of radioactive materials are presented in this chapter. The aim is to provide the framework within which the safety of ionizing radiation conditions can be evaluated. It is directed to the health and safety professional who needs a basic understanding of radiation safety. The health and safety professional should also know where to get consultation and technical help on specific radiation problems.

Radiation safety should be part of a company's total health and safety program. The introduction of radiation devices or radioactive materials will likely call for radiation safety reviews, engineering studies, and plant modifications. Traffic patterns for pedestrians and mobile equipment may need to be altered to minimize the spread of radioactive materials in the event of an accident.

In the absence of a health physicist or radiation control officer, the health and safety professional may be responsible for directing safety efforts to control radiation hazards. The health and safety professional should have general knowledge of the nature of radiation, the detection of radiation, Permissible Exposure Limits (PELs), biological effects of radiation, monitoring techniques, procedures, and control measures. The health and safety professional should be able to use approved radiation detectors and interpret readings, and see that plant personnel are properly advised of radiation hazards and safe procedures to be followed.

The health and safety professional should ensure that all radiation installations are properly reviewed and assure compliance with federal, state, local, and company regulations. They must also be familiar with medical and emergency procedures to be followed and see that plans are in order.

The control of radiation exposures to personnel requires effective accident prevention techniques. Radiation hazards are to some extent prevented with common safety measures. In other cases, the health and safety professional will recognize the need to consult a health physicist in order to obtain expert guidance. Some radiation control operations may require facilities and equipment which may entail extraordinary expense. It is the responsibility of the health and safety professional to see that the proper actions are taken.

Basic information on the characteristics of radiation, standards for exposure limitation, safety factors, and control measures are presented here. As an aid to further study, consult the Bibliography at the end of this chapter.

IONIZING RADIATION TERMS

In order to discuss intelligently the health and safety aspects of ionizing radiation, an understanding of some basic terminology is needed. Brief definitions of some important terms are given here. Figure 10-1 illustrates the process of radioactive disintegration.

Activity. The number of nuclear disintegrations occuring in a given quantity of material per unit time.

Alpha particle. (alpha-radiation, α). An alpha particle is made up of two neutrons and two protons giving it a unit charge of plus-two. It is emitted from the nucleus of radioactive atoms and causes high-density ionization. Alpha particles transfer their energy in a very short distance and are readily shielded by a piece of paper or the dead layer of skin. Alpha-radioactivity is therefore primarily an internal radiation hazard.

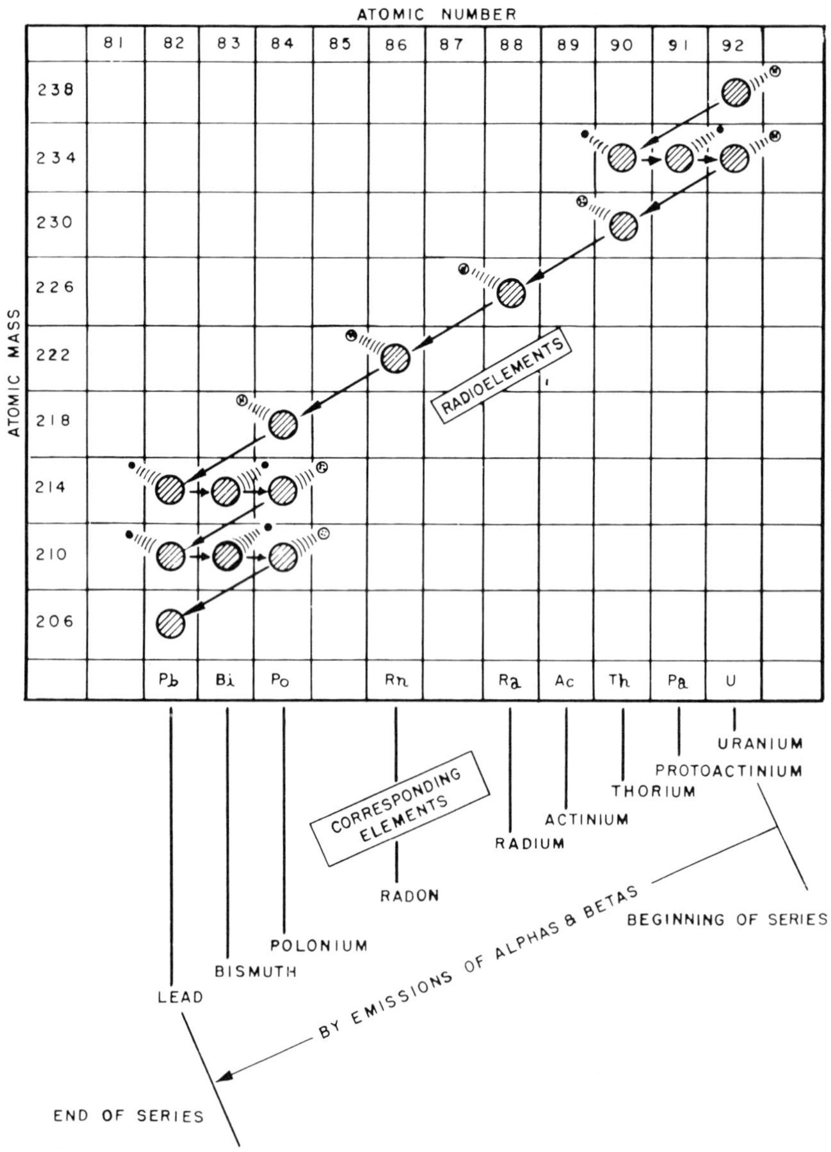

Figure 10-1. The radioactive disintegration scheme of uranium-238, by emissions of α and β particles. (Reprinted with permission from *Atomic Radiation,* RCA Service Co., Inc., 1957)

Annihilation. The process by which a negative electron and a positive electron, called a positron, combine and disappear with emission of electromagnetic radiation.

Atomic number. The atomic number is the number of protons (positively charged particles) in the nucleus of an atom. Each element has a different atomic number. The atomic number of hydrogen is 1, that of oxygen 8, iron 26, lead 82, uranium 92. The atomic number is also called the charge number and is usually denoted by "Z."

Atomic weight. The atomic weight is approximately the sum of the number of protons and neutrons found in the nucleus of an atom. This sum is also called the mass number. The atomic weight of oxygen, for example, is approximately 16, with most oxygen atoms containing 8 protons and 8 neutrons.

Background radiation. The radiation coming from sources other than the radioactive material to be measured is termed "background radiation." Background radiation is primarily a result of cosmic rays which constantly bombard the earth from outer space.

Becquerel (Bq). One disintegration per second (dps). This unit provides a measure of the rate of radioactive disintegration. There are 3.7×10^{10} becquerels per curie of radioactivity.

Beta particle. (beta-radiation, β). Beta particles are small electrically charged particles emitted from the nucleus of radioactive atoms. They are identical to electrons and have a negative electrical charge of one. Beta particles are emitted with various kinetic energies. They pose an internal exposure hazard and are often penetrating enough to cause skin burns.

Bremsstrahlung. The electromagnetic radiation associated with the deceleration of charged particles. The term can also be applied to electromagnetic radiation produced by acceleration of charged particles.

Compton effect. The glancing collision of a gamma photon with an orbital electron. The gamma photon gives up part of its energy to the electron, ejecting the electron from its orbit.

Controlled area. A specified area in which exposure of personnel to radiation or radioactive material is controlled. Controlled areas should be under the supervision of a person who has knowledge of and responsibility for applying the appropriate radiation protection practices.

Counter. A device for counting nuclear disintegrations so as to measure the amount of radioactivity. The electronic signal announcing disintegration is called a count.

Curie. A measure of the rate at which a radioactive material emits particles. One curie corresponds to 3.7×10^{10} disintegrations per second.

Disintegration. When a radioactive atom disintegrates, it emits a particle from its nucleus. What remains is a different element. When an atom of polonium disintegrates, it ejects an alpha particle and changes to a lead atom by this process (Figure 10-1).

Dose. A general term denoting the quantity of radiation or energy absorbed in a specified mass. For special purposes, its meaning should be appropriately stated, e.g., absorbed dose.

Dosimeter (dose meter). An instrument used to determine the radiation dose a person has received.

Electron. A minute atomic particle possessing the smallest amount of negative electric charge (−1). Orbital electrons rotate around the nucleus of an atom. The mass of an electron is only about 1/1,820 the mass of a proton or neutron.

Electron volt (eV). A small unit of energy—the amount of energy that an electron gains when it is acted upon by one volt. Radioactive materials emit radiation that may have energies of up to several million electron-volts, MeV. Gamma-ray energies from radioisotopes can be 4 MeV or higher. Some are emitted at relatively low energies and are correspondingly less hazardous.

Element. All atoms of an element contain a definite number of protons and therefore have the same atomic number. Various isotopes of an element are due to different numbers of neutrons in the nucleus. However, electrical charge and chemical properties of the various isotopes of an element are alike.

Film badge. A piece of masked photographic film worn like a badge for personal monitoring of radiation exposure. It is darkened by penetrating radiation, and radiation exposure can be checked by developing and interpreting the film. The type of masking depends on the type of radiation to be measured.

Gamma-rays (Gamma-radiation, γ). Gamma-rays are electromagnetic photons emitted from the nuclei of radioactive atoms. They are highly penetrating and present an external radiation exposure hazard.

Gray (Gy). Unit of absorbed radiation dose equal to one joule of absorbed energy per kilogram of matter.

Half-life. A means of classifying the rate of decay of radioisotopes according to the time it takes them to lose half their strength (intensity). Half-lives range from fractions of a second to billions of years. Cobalt-60, for example, has a half-life of 5.3 years.

Half-value layer. The thickness of a specified substance which, when introduced into the path of a given beam of radiation, reduces the value of a specified radiation quantity by one-half. It is sometimes expressed in terms of mass per unit area.

Ion. An atom or molecule that carries either a positive or negative electrical charge.

Ionizing radiation. Electromagnetic or particulate radiation capable of producing ions, directly or indirectly, by interaction with matter. In biological systems, such radiation must have a photon energy greater than 10 electron volts. This excludes most of the ultraviolet bands and all longer wave lengths.

Ionization chamber. A basic counting device to measure radioactivity.

Isotope. Nuclei which have the same atomic number. Isotopes of a given element contain the same number of protons but a different number of neutrons. Uranium-238 contains 92 protons and 146 neutrons while the isotope U-235 contains 92 protons and 143 neutrons. Thus the atomic weight (atomic mass) of U-238 is 3 higher than that of U-235.

Moderator. A material used to slow neutrons, such as used in a reactor. Slow neutrons are particularly effective in causing fission. Neutrons are slowed down when they collide with atoms of light elements such as hydrogen, deuterium, and carbon, three common moderators.

Molecule. The smallest unit of a compound or element as it exists in nature. A water molecule consists of two hydrogen atoms combined with one oxygen atom. Hence the well-known formula, H_2O. The element oxygen exists as diatomic molecules, O_2.

Neutron. An atomic particle. The neutron weighs about the same as the proton. As its name implies, the neutron has no electric charge. Neutrons make effective atomic projectiles for the bombardment of nuclei. Neutrons can also present unique external exposure hazards to personnel.

Nucleus. The inner core of the atom. The nucleus consists of neutrons and protons tightly bound together.

Pair production. The conversion of a gamma-ray into a pair of particles—an electron and a positron. This is an example of direct conversion of energy into matter, and is quantified by Einstein's famous formula: $E = Mc^2$; Energy = Mass × Velocity of Light (squared).

Photoelectric effect. Occurs when an electron is ejected from the orbit of an atom by a photon that imparts all of its energy to the electron.

Photon. A bundle (quantum) of electromagnetic radiation. x-rays, gamma-rays, and visible light and radio waves.

Plutonium. A man-made heavy element that undergoes fission under the impact of neutrons. It is a useful fuel in nuclear reactors. Plutonium can be produced by capture of slow neutrons in uranium. It is a highly hazardous alpha emitter.

Proton. An elementary particle found in the atomic nucleus. Its positive charge of one is opposite to that of the electron.

Radioactivity. The emission of very fast atomic particles and/or rays by nuclei. Some elements are naturally radioactive while others become radioactive after bombardment with neutrons or other particles. The three major forms of radioactivity are alpha (α), beta (β), and gamma (γ), named for the first three letters of the Greek alphabet.

Radioisotope. A radioactive isotope of an element. A radioisotope can be produced by placing material in a nuclear reactor and bombarding it with neutrons. Many fission products are radioisotopes. Radioisotopes are sometimes used as tracers, as energy sources for chemical processing or food pasteurization, or as energy sources for nuclear batteries.

Radium. One of the earliest known naturally radioactive ele-

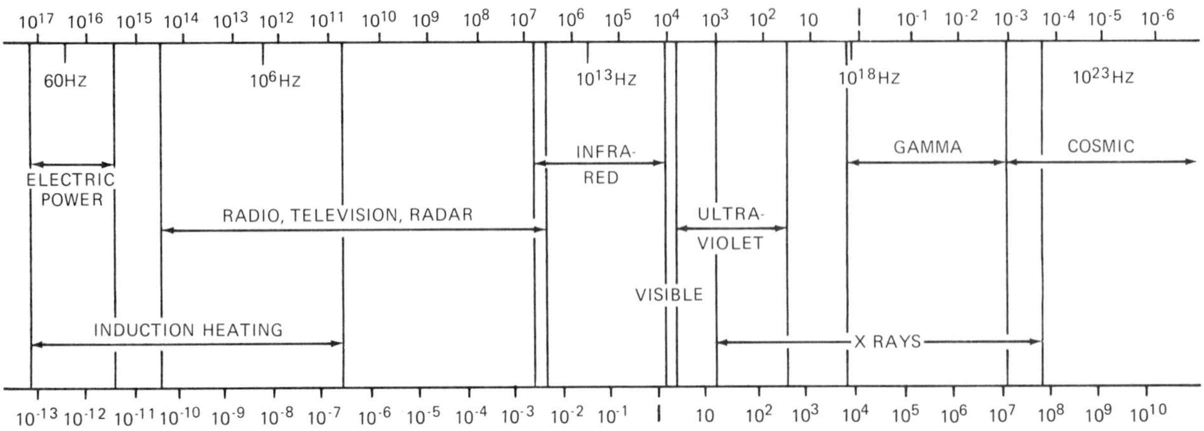

Figure 10-2. Electromagnetic spectrum illustrates energy and wavelength of the various categories of electromagnetic radiation.

ments, radium, is far more radioactive than uranium and is found in the same ores. It is a highly hazardous alpha emitter.

Roentgen (R). The amount of x- or gamma-radiation that produces ionization resulting in one electrostatic unit of charge in one cubic centimeter of dry air at standard conditions.

Roentgen equivalent man (rem). A unit of absorbed energy times a quality factor for the relative biological effect of the particular radiation as compared to gamma-radiation. Personnel exposure limits may be expressed in rem.

Scintillation counter. A radiation counting device that operates by means of tiny flashes of light (scintillations) particles produce when they strike certain crystals or liquids.

Shielding. A barrier that protects workers from harmful radiations released by radioactive materials. Lead bricks, dense concrete, water, and earth are examples of materials used for shielding.

Sievert (Sv). Unit of absorbed radiation dose times the quality factor of the radiation as compared to gamma-radiation. It is equal to the Gray times the quality factor and is equivalent to 100 rem.

Strontium-90. An isotope of strontium having a mass number of 90. Strontium-90 is an important fission product. It has a half-life of 25 years and is a highly hazardous beta emitter.

Tracer. A radioisotope mixed with a stable material. Radioisotopes enable scientists to trace chemical and physical changes in materials. Tracers are widely used in science, industry and agriculture. For example, when radioactive phosphorus is mixed with a chemical fertilizer, the uptake of radioactive phosphorus from fertilized soil can be measured in the plants as they grow.

Tritium. Often called hydrogen-three. Tritium is an extra-heavy hydrogen whose nucleus contains two neutrons and one proton. It is radioactive as a beta emitter.

Uranium. A heavy metal, the two principal natural isotopes of which are U-238 and U-235. U-235 has the only readily fissionable nucleus that occurs in appreciable quantities in nature, hence its importance as a nuclear reactor fuel. Only one part in 140 of natural uranium is U-235.

X-ray. Highly penetrating electromagnetic radiation similar to gamma-rays. The x-rays are produced by electron bombardment of target materials. They are commonly used to produce shadow pictures (roentgenograms) of the denser portions of objects.

TYPES OF IONIZING RADIATION

Radiation is a form of energy. Familiar forms of radiation energy include light (a form of radiation we can see) and infrared (a form of radiation we can feel as heat). Radio and television waves are forms of radiation that we can neither see nor feel. The relationship between the various categories of electromagnetic radiation is shown in Figure 10-2.

The x-rays and gamma-rays overlap and occupy a common range in the electromagnetic spectrum. The x-radiation is produced in the orbiting electron portion of the atom or from free electrons while gamma-radiation is produced in the nucleus. The x-rays generally are machine-produced and gamma-rays are emitted spontaneously from radioactive materials.

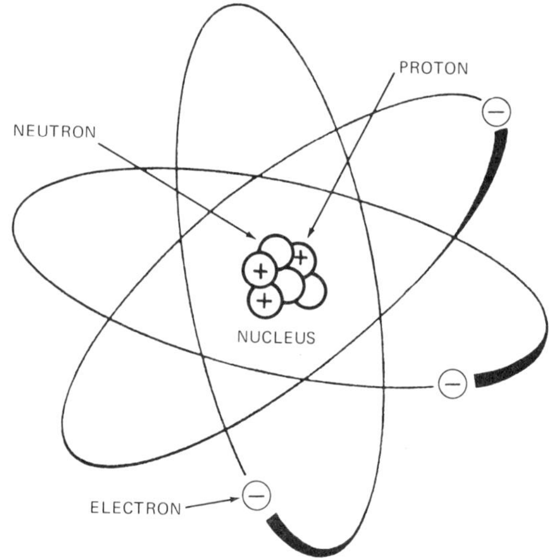

Figure 10-3. Basic model of the atom. The atom illustrated here is lithium-6.

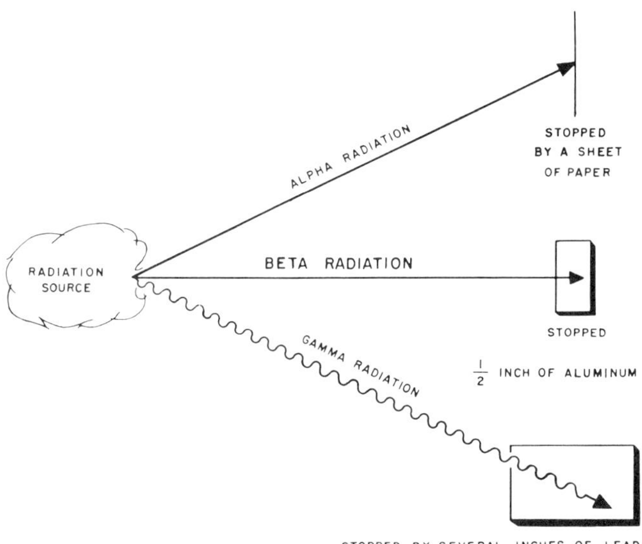

Figure 10-4. Relative penetrating power of alpha-, beta-, and gamma-radiation. (Reprinted with permission from *Atomic Radiation,* RCA Service Co., Inc., 1957)

All matter is composed of atoms and each atom has two basic parts—a heavy core or nucleus containing positively charged particles called protons and neutral particles called neutrons and relatively lightweight negatively charged particles called electrons that spin around the nucleus (Figure 10-3). Neutrons and protons were once considered basic particles. However, they have been found to be composed of even smaller particles.

Ionization is an energy transfer process that changes the normal electrical balance in an atom. If a normal atom (electrically neutral) were to lose one of its orbiting electrons (one negative charge) the atom would no longer be neutral. It would have more positive charges than negative charges making it a positive ion. The electron that was removed would be called a free electron. If it were attached to another atom, that atom would be a negative ion. The positive and negative ions thus produced are known as an ion pair.

The term nuclear radiation describes all forms of radiation energy that originate in the nucleus of a radioactive atom. In addition to the gamma-rays, fast moving particles may also be emitted from radioactive atoms.

Some materials are naturally radioactive and other materials can be made radioactive in a nuclear reactor or accelerator. Some nonradioactive atoms can be converted to radioactive atoms when an extra neuron is captured by a nucleus. The radioactive atom is unstable because of the extra energy that the neutron added to the nucleus. The excited or radioactive atoms get rid of their excess energy and return to a stable state by emitting subatomic particles and gamma-rays from the nucleus. The most important particles are alpha particles, beta particles, and neutrons.

The hazardous properties of radioactive materials are usually thought of in terms of nuclear radiation. All types of radiation share the common property of being absorbed and transferring energy to the absorbing body.

The most commonly encountered types of ionizing radiation are alpha, beta, and neutron particles and x-ray or gamma-electromagnetic-radiation. Other types of ionizing radiation are encountered in specialized facilities.

Alpha particles

Alpha particles originate in the nuclei of radioactive atoms during the process of disintegration. These particles consist of a cluster of two protons and two neutrons (giving a mass number of 4, being structurally the same as the nucleus of a helium atom). Alpha particles, on slowing down, combine with electrons from the material through which they are passing and thus become helium atoms.

The mass and electrical charge characterize the hazardous properties of alpha particles. They have a positive charge of two units and interact electrically with human tissues and other matter. Alpha particles range in energy to over 7 MeV. Because they produce dense ionization along their path in a material, they travel only a short distance. Their range is at most about 10 cm (4 in.) in air. They are stopped by a film of water, a sheet of paper, or other paper-thin material (Figure 10-4).

Detecting alpha particles with a radiation survey meter requires that the instrument probe be held close to the source and that the window on the probe be very thin and designed for alpha detection.

Alpha particles are produced by elements with a high atomic number ("Z") (Figure 10-1). Alpha emitters are hazardous when taken into the body. Because they are chemically similar to calcium in their action within the body, some alpha emitters become part of the structure of bone, remaining there for long periods of time. As they disintegrate, they emit alpha particles, which can damage tissue. Other alpha emitters are not bone seekers but concentrate in body organs such as kidney, liver, lungs, and spleen.

If the alpha-emitting material is kept outside the body, little damage results since generally alpha particles cannot penetrate the outermost dead layer of skin. Alpha emitters are considered to be internal radiation hazards.

Beta particles

Beta particles are electrons ejected from the nuclei of radioactive atoms by disintegration. They have a negative charge of one unit and cause the disintegrating atom to change to an element of a higher atomic number. Thus, strontium-90 changes to yttrium-90 on distintegration with ejection of a beta particle (electron) from the nucleus. The atomic number of the atom, (the number of protons in the nucleus) is changed from 38 to 39 by the beta particle emission. (A neutron, by losing an electron, has become a proton.)

Beta particles with the same mass as the electron but with a positive unit charge are called *positrons* and can also be ejected from nuclei by disintegration. However, a positron particle is readily annihilated by combination with an electron, yielding gamma-radiation.

Beta particles do not penetrate to the depth that x-rays or gamma-radiations of similar energy do (Figure 10-4). Their maximum range in wood is about 4 cm (1.5 in.) and they can penetrate into the human body from 0.2 to 1.3 cm (0.1–0.5 in.).

Skin burns may result from excessively high doses of beta radiation, since it requires only about 70 keV of energy for a beta particle to penetrate the dead layer of skin. Beta emitters are internal radiation hazards when taken into the body.

Beta particles have a broad distribution of energies ranging from near zero up to the maximum value specific for the particular radionuclide. The range in air of the beta particles emitted from one radionuclide may be up to 15 cm (6 in.) maximum, while the range from another radionuclide may be up to 18 m (60 ft).

Beta particles, or high-energy electrons, are emitted from a wide range of light and heavy radioactive elements. Beta particles are much smaller than alpha particles and may have a velocity approaching the velocity of light. The energy level of beta particles can be 4 MeV or higher. The energy range is similar to the energy range characteristic of gamma-rays.

The electron has a negative electrical charge which tends to limit its penetration range somewhat. Beta particles are relatively more hazardous externally than alpha particles because the penetration power is greater.

When the beta particle is slowed down or stopped, secondary x-radiation known as bremsstrahlung may be produced.

Light metals like aluminum are preferred for shielding beta particles because they produce less bremsstrahlung radiation. Common beta-radiations have ranges in air less than 9 m (30 ft). Depending on their energy, they will be stopped by the walls of a room or by a sheet of aluminum 1.3 cm (0.5 in.) thick.

Neutrons

Neutrons are not encountered as commonly as alpha and beta particles. The neutron particle, as its name implies, has no electrical charge. Neutrons exist within the nuclei of all atoms except those of the lightest isotope of hydrogen.

Neutrons are released upon disintegration of certain radioactive materials (the fissionable isotopes). Neutrons can have long or short ranges in air depending on their kinetic energy, which in turn depends on the method of neutron production. Furthermore, the range of neutrons depends on the characteristics of the material through which they pass, the property of the atoms of this material to interact with the neutrons, and finally, the kind of collisions that occur. In human tissue, the average distance for absorption of neutrons varies from about 0.6 cm (0.25 in.) to several centimeters (inches) depending on the neutron energy.

Very high-energy neutrons collide with atoms of a material through which they are passing and often break up these nuclei into high-energy fragments. The neutron itself is unstable and emits a beta particle as it decays to a proton. Interaction with the atoms of the material through which neutrons pass is the main way neutrons are removed from a beam.

Absorption of neutrons results when neutrons are deflected by, or collide with nuclei. They collide repeatedly and are deflected and slowed down. The loss of energy by the neutrons that suffer these collisions leads to an increased probability of absorption by a nucleus. When an absorption occurs, the process is called neutron capture.

In the human body, most of the captures that occur take place in hydrogen or nitrogen atoms. The subsequent result is that the nucleus of the atom is in an excited state. That is, an excess of energy is available. An atom will exist in this state only for a short period of time and will return to the ground (unexcited) state by releasing the excess energy. In the process, a proton, gamma ray, beta particle, or alpha particle may be emitted, depending on what type of atom captures the neutron. Since these secondary emissions will produce damage in tissue, the task of determining neutron dose is difficult. The health hazard which neutrons present arises then from the fact that they cause the release of secondary radiation.

Human exposures to neutrons may occur around reactors, accelerators, and sources designed to produce neutrons. Determination of neutron exposures must be made by highly skilled persons using specialized equipment. A harmful dose is dependent not only on the abundance of the neutrons, but also on their energy distribution.

X-radiation

The x-radiation is commonly thought of as electromagnetic radiation produced by an x-ray machine. When high-speed electrons are suddenly slowed down on striking a target, they lose energy in the form of x-radiation. In an x-ray machine, the voltage across the electrodes of the vacuum tube determines the energy of the electrons. The energy of the electrons principally determines the wavelength and penetrating quality of the resulting x-rays.

The character of the x-radiation is also affected by the composition of the target material inside the x-ray tube. That is, the wavelength of a portion of the x-radiation is influenced by the kind of material composing the target. Since the electrons strike and interact at various speeds, the x-ray beam has a variety of wavelengths and energies. The energy of an x-ray is inversely proportional to its wavelength. The more energy an x-ray possesses, the shorter its wavelength.

The extent of penetration of x-rays depends on wavelength and the material being irradiated. The x-rays of short wavelength are called "hard" and will penetrate several centimeters of steel. The x-rays of long or "soft" wavelengths are less penetrating. This power to penetrate through matter is termed the "quality." The "intensity" is the energy flux density. The physical properties of a beam of x- or gamma-rays are often summarized in the two concepts of intensity and quality.

The range of penetration may be expressed in terms of "half value layer." This is the thickness of material which will reduce the incident radiation by one-half. The half-value layers for x-radiation may range to several centimeters of concrete (Table 10-A).

Table 10-A. Half-Value Layer for Five Shielding Materials

Material	*Cobalt-60*	*Celsium-137*
Lead	1.24 cm	0.64 cm
Copper	2.10 cm	1.65 cm
Iron	2.21 cm	1.73 cm
Zinc	2.67 cm	2.06 cm
Concrete	6.60 cm	5.33 cm

1 cm = 0.394 in.

Gamma-radiation

Gamma-radiation is similar to x-radiation in that it is also an electromagnetic and ionizing radiation. In fact, it is identical to x-radiation except for its source being the nucleus of an atom. x-radiation refers to electromagnetic radiation which originates outside of the nucleus.

Radioactive materials, by definition, spontaneously emit one or more characteristic radiations that may include gamma-radiation. The radiation comes from an excited or unstable nucleus of an atom.

A gamma-ray emitted from a given radionuclide has a fixed

Table 10-B. Isotopes Commonly Available—Listed By Increasing Half-Life

Half-Life	*Element and Symbol*	*Atomic Number*	*Mass Number*	*Gamma-Radiation Energy (MeV)*
88 days	Sulfur (S)	16	35	none
115 days	Tantalum (Ta)	73	182	0.068, .10, .15, .22, 1.12, 1.19, 1.22
120 days	Selenium (Se)	34	75	0.12, .14, .26, .28, .40
130 days	Thulium (Tm)	69	170	0.084
138 days	Polonium (Po)	84	210	0.80
165 days	Calcium (Ca)	20	45	none
245 days	Zinc (Zn)	30	65	1.12
270 days	Cobalt (Co)	27	57	0.12, .13
253 days	Silver (Ag)	47	110	0.66, .68, .71, .76, .81, .89, .94, 1.39
284 days	Cerium (Ce)	58	144	0.08, .134
303 days	Manganese (Mn)	25	54	0.84
367 days	Ruthenium (Ru)	44	106	none
1.81 years	Europium (Eu)	63	155	0.09, .11
2.05 years	Cesium (Cs)	55	134	0.57, .60, .79
2.6 years	Promethium (Pm)	61	147	none
2.6 years	Sodium (Na)	11	22	1.277
2.7 years	Antimony (Sb)	51	125	0.18, .43, .46, .60, .64
2.6 years	Iron (Fe)	26	55	none
3.8 years	Thallium (Tl)	81	204	none
5.27 years	Cobalt (Co)	27	60	1.3, 1.12
12.46 years	Hydrogen (H)	1	3	none
12 years	Europium (Eu)	63	152	0.12, .24, .34, .78, .96, 1.09, 1.11, 1.41
16 years	Europium (Eu)	63	154	0.123, .23, .59, .72, .87, 1.00, 1.28
28.1 years	Strontium (Sr)	38	90	none
21 years	Lead (Pb)	82	210	0.047
30 years	Cesium (Cs)	55	137	0.661
92 years	Nickel (Ni)	28	63	none
1602 years	Radium (Ra)	88	226	0.186
5730 years	Carbon (C)	6	14	none
2.12×10^5 years	Technetium (Tc)	43	99	none
3.1×10^5 years	Chlorine (Cl)	17	36	none

(Reprinted from *Radiological Health Handbook,* Revised Edition, January 1970. U.S. Dept. of HHS, Rockville, MD)

energy specific to that radionuclide (Table 10-B). Gamma-rays from various radionuclides cover a wide range of wavelengths or energies. They present an external exposure problem due to their deep penetration. The half-value layer of shielding for 1.0 million electron-volts (MeV) gamma- or x-radiation is slightly more than 1.3 cm (0.5 in.) of steel.

BIOLOGICAL EFFECTS OF RADIATION

The human body, apparently, can tolerate a certain amount of exposure to ionizing radiation without impairing its overall functions. We are continuously exposed to ionizing radiation from natural sources such as cosmic radiation from outer space and from radioactive materials in the earth and materials around us and in us. This "background radiation" is part of our normal environment; we have evolved under its effect and we are continuously exposed to it.

One fundamental property of ionizing radiation is that when it passes into or through a material, it transfers energy by the ionization process. The intensity of radiation to which a material is subjected is referred to as the radiation field. The term dose is generally used to express a measure of radiation that a body or other material absorbs when exposed in a radiation field.

If the incoming and outgoing energies are almost identical in nature and amount, then little energy is transferred and the dose is small.

External radiation sources (those sources which are located outside of the body) present an entirely different set of conditions than radionuclides which have entered the body. Once inside the body, radionuclides are absorbed, metabolized, and distributed throughout the tissues and organs according to their chemical properties. Their effects on organs or tissues depend on the type and energy of the radiation and residence time within the body.

The effects of irradiation on living systems may be studied by

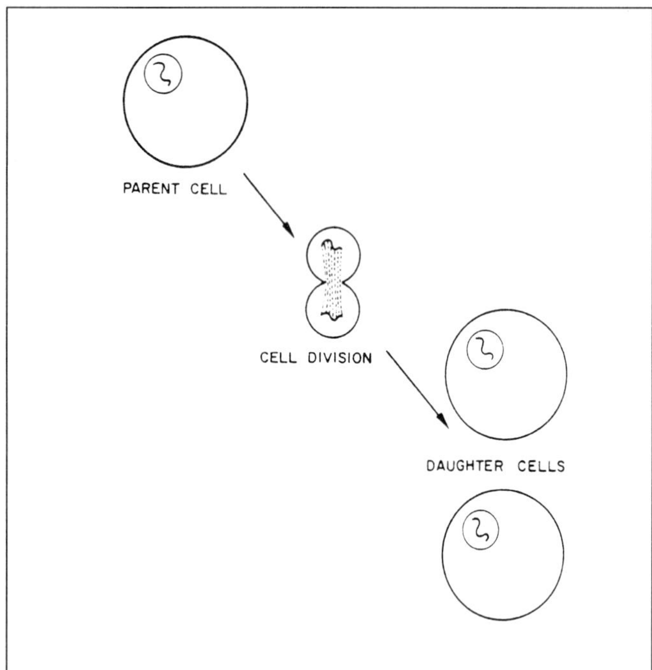

Figure 10-5. Normal cell division.

looking for effects on the living cells, for changes in the biochemical reactions, for evidence of production of disease, or for changes in life or normal growth patterns (Figures 10-5 and 10-6). Interpretation of such findings is not simple. Extensive research in this field is continually being carried out in many laboratories.

The effect of ionizing radiation in living tissue is generally assumed to be almost entirely due to the ionization process that destroys the capacity of reproduction or division in some of the cells or causes mutation in others (Figure 10-6). The human body is a complex chemical machine that constantly produces new cells to replace those that have died or been damaged. The body has a tremendous capability for repairing cell damage. Therefore, our survival depends on the body's ability to keep cell damage within the body's repair capabilities.

Ionization strips electrons from atoms and breaks their chemical bonds with other atoms. A simple molecular structure, such as water, will recombine after ionization. This, however, is not the case in a complicated living cell. Here, ionization may give many possible atomic recombinations. The rupture of a few bonds in the elaborate structure of the molecules of the living cell may have profound effects.

Types of injuries

The accumulation of knowledge of the effects of radiation on humans was begun in 1895 when Roentgen discovered the x-ray. Radiation from uranium was discovered by Becquerel in 1896. Alpha and beta particles and gamma-rays were identified shortly thereafter, and the neutron was discovered in 1930. Injurious effects were experienced by these early workers

There are two points of view for consideration of the injurious effects of ionizing radiation. These are the somatic effects (i.e., injury to individuals) and the genetic effects (i.e., changes passed on to future generations).

The degree of injury inflicted on an individual by radiation exposure depends on such factors as the total dose, the rate at which the dose is received, the kind of radiation, and the body part receiving it.

Tissues, such as the bone marrow which contains the blood-forming cells, the lining of the digestive tract, and some cells of the skin, are more sensitive to radiation that those of bone, muscle, and nerve.

In general, if the total amount of radiation is received slowly over a long period of time, a larger dose is required to produce the same degree of damage than if the same total is received over a short period. If the rate at which the dose is received is low enough, the body recovery processes may be able to keep abreast of the slight damage.

Some relatively small doses have no effect if given only once, but may shorten the life span and will produce abnormalities if continued long enough. The time between the exposure and the first signs of radiation damage is called the "latent period." The larger the dose, the shorter the latent period.

The various tissues and organs of the body are not affected equally by equal irradiation. Their responses vary considerably. For radiation protection purposes it is essential that the dose to the most sensitive organs be given primary consideration.

Over the years, allowable radiation levels have been consistently reduced as researchers have obtained more information on the effects of radiation exposure and as judgments involving acceptable risks have become more conservative. As radiation was applied to humans for healing purposes, effects such as

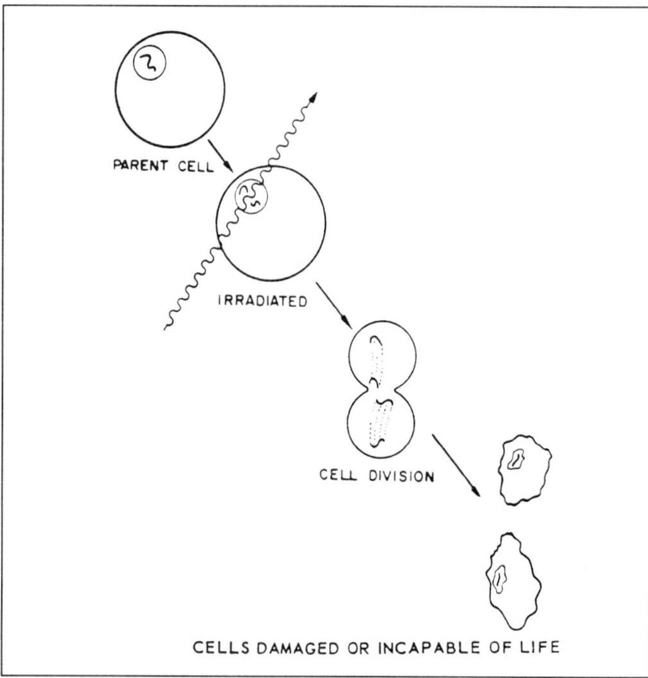

Figure 10-6. Damage or death of cells after division of irradiated parent cell. (Reprinted with permission from *Atomic Radiation,* RCA Service Co., Inc., 1957)

skin redness, dermatitis, and skin cancer were noted, as were hair loss and eye inflammation. It was found that the incidence of cancer and of certain blood diseases was higher among

Table 10-C. Maximum Permissible Dose Equivalent for Occupational Exposure

Combined whole body occupational exposure	
Prospective annual limit	5 rems in any 1 yr
Retrospective annual limit	10-15 rems in any 1 yr
Long-term accumulation	(N-18)x5 rems, where N is age in yr
Skin	15 rems in any 1 yr
Hands	75 rems in any 1 yr (25/qtr)
Forearms	30 rems in any 1 yr (10/qtr)
Other organs, tissues and organ systems	15 rems in any 1 yr (5/qtr)
Fertile women (with respect to fetus)	0.5 rem in gestation period
Dose limits for the public, or occasionally exposed individuals:	
Individual or occasional	0.5 rem in any 1 yr
Students	0.1 rem in any 1 yr
Population dose limits	
Genetic	0.17 rem average per yr
Somatic	0.17 rem average per yr
Emergency dose limits—Life saving:	
Individual (older than 45, if possible)	100 rems
Hands and forearms	200 rems, additional (300 rems, total)
Emergency dose limits—Less urgent:	
Individual	25 rems
Hands and forearms	100 rems, total
Family of radioactive patients:	
Individual (under age 45)	0.5 rems in any 1 yr
Individual (over age 45)	5 rems in any 1 yr

(Reprinted from NCRP Publication No. 43, *Review of the Current State of Radiation Protection Philosophy,* 1975)

radiologists than would be expected for a chance distribution in the population studied.

The pool of health experience data was obtained from: (1) early radiation workers, (2) medical personnel who routinely administered radiation for diagnostic and therapeutic purposes, (3) patients who were treated with radiation, and finally, (4) a group of workers who painted dials with luminous paints containing radium.

With this data, it became apparent that exposure to ionizing radiation was associated with a higher-than-normal incidence of certain diseases, such as skin, lung, and other cancers, of bone damage, and of cataracts. It appeared also that there was some evidence of life-shortening.

Geneticists consider the population as a whole rather than particular individuals. Their concern is the effect of radiation on future generations. Radiation damage to human reproductive materials can be transmitted to succeeding generations.

Genetic damage may also be caused by disease or toxic chemicals as well as by ionizing radiation. Birth defects are usually the result of defective genetic materials, but it is not possible at this time to determine what damaged the genetic material.

Relating dosage to damage

Extensive work has been done in an attempt to relate radiation dose to resulting damage. A great deal of laboratory work was also carried out with various biological systems (both plants and animals) in order to learn more about the conditions of irradiation, the effects it produced, and the relative effectiveness of the kinds and energies of radiation.

From all these studies came the basis for determining "maximum permissible levels" of exposure. The maximum permissible levels denote the radiation dose that can be tolerated with little chance of later development of adverse effects.

STANDARDS AND GUIDES

Maximum permissible levels of external and internal radiation have been published by the National Council on Radiation Protection (NCRP). (Bibliography, NCRP, *Maximum Permissible Body Burdens and Maximum Permissible Concentrations of Radionuclides in Air and in Water for Occupational Exposure,* Handbook 22; and *Review of Current State of Radiation Protection Philosophy,* Handbook 43.) Limits are shown in Table 10-C.

Guides for maximum permissible external exposure to ionizing radiation, etc., have also been published by the International Commission on Radiological Protection (ICRP).

The Federal Radiation Council has recommended "radiation protection guide" values for external radiation and " radioactivity concentration guides" for radioactive materials found in the air and water. The recommendations concerning protection of workers exposed to ionizing radiation have recently been revised.

Work can often be accomplished at no real inconvenience at levels far below those recommended by the Federal Radiation Council. The accepted practice is to keep radiation exposure as low as reasonably achievable.

Present limiting values are given separately for protection against different types of effects on health and apply to the sum of doses from external and internal sources of radiation. For cancer and genetic effects, the limiting value is specified in terms of a derived quantity called the effective dose equivalent. The effective dose equivalent received in any year by an adult worker should not exceed 5 rem. For other health effects, the limiting values are specified in terms of the dose equivalent to specifc organs and tissues. Occupational dose equivalents to individuals under the age of 18 should be limited to one-tenth of the limiting values.

There is disagreement on the magnitude, extent, and cause of effects, if any, at exposure levels below the guide levels. Much of the discussion is in the area of theory, rather than about measured cause and demonstrated effect in human beings.

Due to the sensitivity of a fetus to radiation damage, the NCRP has recommended a dose limit to the fetus of 0.5 rem during the pregnancy. The reason for the lower limit of exposure is that rapidly dividing cells are more susceptible to radiation damage than are mature cells. Women of childbearing age who may be exposed to radiation should be informed of the need to protect the fetus from excessive or unnecessary radiation exposure.

In the event that workplace exposures exceed Permissible Exposure Limits, action should be taken to reduce the exposures to within the guidelines. If this is not practicable, it will be necessary to transfer the female worker out of the radiation exposure area for the duration of the pregnancy.

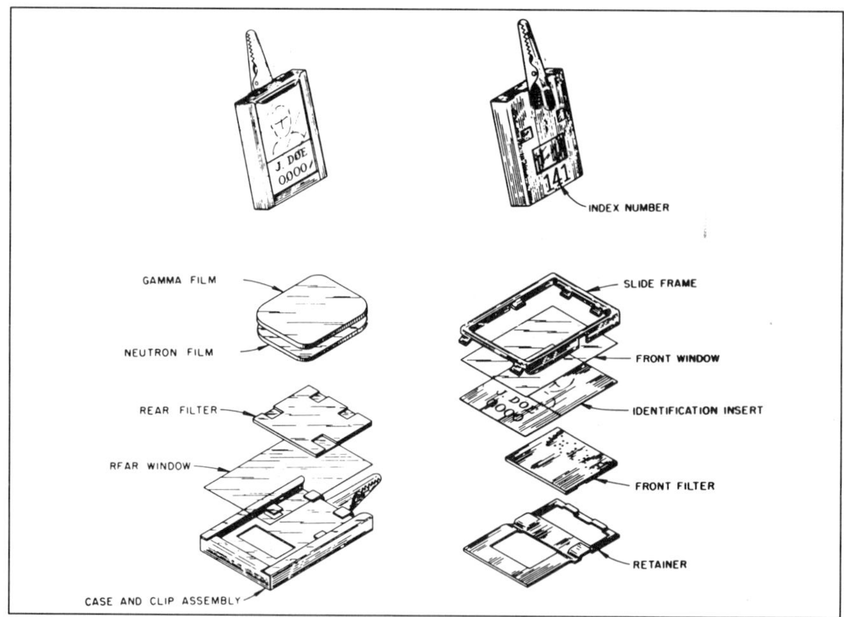

Figure 10-7. Exploded view of radiation film badge. Front view, at upper left.

Radiation dose limits, of whatever official organization or government body that has authority with respect to the user, should be considered as upper limits; the objective should be to keep exposures as low as possible (Table 10-C).

Individuals have a whole body dose limit of 3 rem during any calendar quarter and 5 rem per year for occupational exposure. Accumulated occupational dose is not to exceed 5(N-18) where N is the individual's age. It is especially important that any employee under the age of 18 not receive a dose in any calendar quarter in excess of 10 percent of the limits in Table 10-C.

For further explanation of radiation protection standards and guides, refer to the Bibliography.

MONITORING INSTRUMENTS

There are a variety of detectors and readout devices used for radiation monitoring or measurement. None is universally applicable and selection of the most appropriate detector or detectors for each radiation measurement (or type of measurement) becomes a matter of great importance.

Radiation monitoring involves the routine or special measurement of radiation fields in the vicinity of a radiation source, measurement of surface contamination, and measurement of airborne radioactivity. Such monitoring procedures are sometimes referred to as radiation surveys.

Film badge

The film badge, worn on the outer clothing, is an example of a personal radiation monitor for gamma-, x-, and beta-radiation. It consists of a small piece of photographic film wrapped in an opaque cover and supported with a metal backing. It can be pinned to the clothing or worn as a ring. Radiation interacts with the silver atoms in the photographic film to affect (expose) the film the same way as light rays do. The badge is removed at regular intervals and the film developed. The amount of darkening of a film is then compared to a control film that was not exposed to radiation during the same time period to determine the amount of radiation exposure. Figure 10-7 shows a typical film badge arrangement for monitoring personnel; the film badge provides a permanent record of dosages.

Thermoluminescence detectors

Thermoluminescence detectors (TLD) have come into widespread use for radiation exposure monitoring for gamma-, x-, and beta-radiation. These dosimeters can be worn by the person as body badges or finger rings. Most commonly they are small chips of lithium fluoride. A major advantage of the TLDs is that for x- and gamma-radiation, they are essentially energy independent from 20 keV up. The absorbed ionizing radiation energy displaces electrons from their ground state (valence state). The electrons are trapped in a metastable state but can be returned to the ground state by heating. When electrons return to the ground state, light is emitted. A TLD readout instrument is used for precise control of heating and of measuring the light emitted from the chip. The amount of light released is related to the absorbed radiation dose and, in turn, to the radiation exposure to the individual. Since the stored energy is released on readout, the readings cannot be repeated. It is therefore common practice to include two or more chips in a dosimeter.

The TLDs have also been used for monitoring of neutron radiation exposures. In this application, a thorough knowledge of the neutron energy spectrum is needed.

Pocket dosimeter

The pocket dosimeter is a direct-reading portable unit shaped like a pen with a pocket clip. A typical unit is illustrated in Figure 10-8. It is generally used to measure x- and gamma-radiation (although may respond to beta-radiation).

The dosimeter consists basically of a quartz fiber, a scale, a lens to observe the movement of the fiber across the scale, and an ionization chamber. The fiber is charged electrostatically until it reaches zero on the scale. Then as the dosimeter is exposed to radiation, some of the air atoms in the chamber become ionized. This allows the static electricity charge to leak from the quartz fiber in direct relationship to the amount of

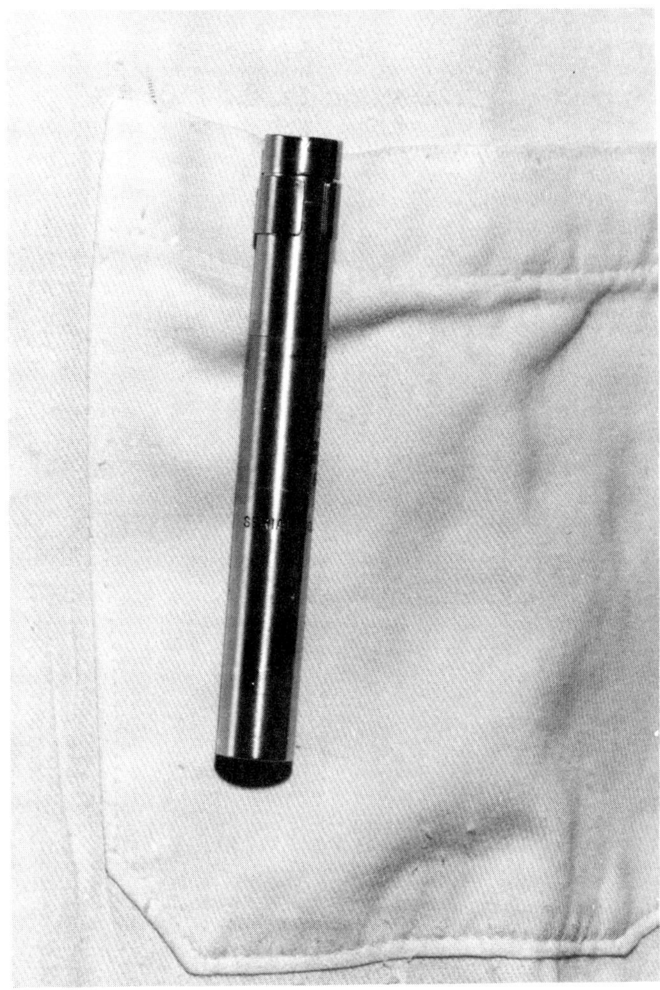

Figure 10-8. Pencil-type dosimeter indicates amount of x- and gamma-radiation exposure. (Courtesy Argonne National Laboratory)

radiation present. As the charge leaks away, the fiber deflects to some new position on the scale that indicates the amount of radiation exposure.

The main advantage of the pocket dosimeter is that it allows the individual to determine the radiation dose while working with radiation, rather than waiting until after periodic processing of a film badge or thermoluminescent dosimeter.

Other dosimeters

Personal electronic alarm dosimeters are now available to monitor the presence of x- and gamma-radiation. These units, usually Geiger-Mueller tubes, have an automatic audible alarm if significant exposure rates are encountered. These units can also indicate total integrated exposure rates at any time or provide a digital readout.

Ionization chambers

Radiation can be measured very conveniently and accurately by measuring the ionization in a small volume of air. If two plates or electrodes with an electrical potential between them are placed in a container filled with air, an ionization chamber will be formed. If the ionization chamber is exposed to a beam of

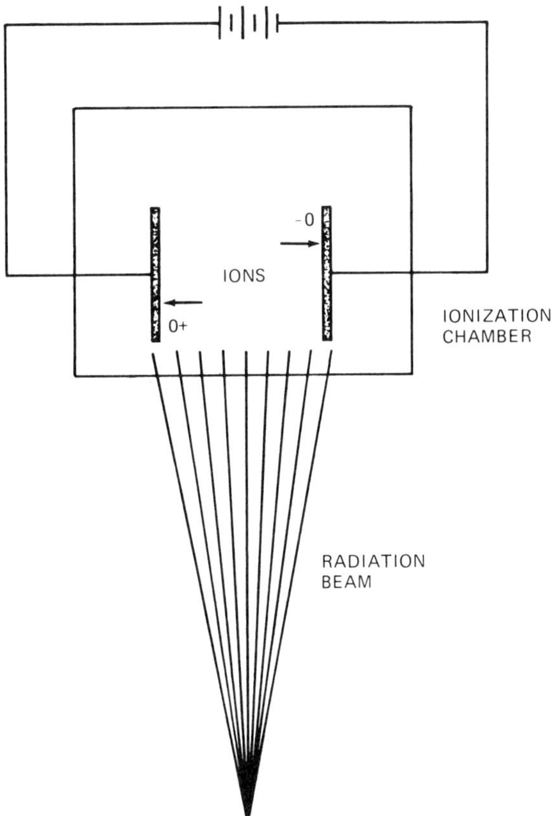

Figure 10-9. Diagram of an ionization chamber.

radiation, a current will flow in the circuit because the electrons that are knocked out of the air atoms by the radiation will be attracted by the positive electrode. In other words, ionized air becomes "conductive" (Figure 10-9).

The ionization chamber measures ionization directly and is energy independent. These units can measure gamma-, x-, beta-, and, if the window is thin enough, alpha-radiation. It is a very useful and popular tool for radiation safety work. Use of an ionization chamber instrument for gamma-radiation measurements is illustrated in Figure 10-10.

Geiger-Mueller counters

A Geiger-Mueller counter is used for beta-, gamma-, and x-radiation survey measurements because it is capable of detecting very small amounts of radiation. It is especially sensitive to beta-radiation. It uses an ionization chamber but it is filled with a special gas and has a greater voltage supplied between its electrodes. A very small number of ions are needed to put it into discharge. Electrons are freed by the initial ionization process and they acquire enough extra energy by the applied voltage to create more ions. This instrument does not give a uniform response for different radiation energy levels and is accurate only for the type of radiation energy for which it is calibrated. A radiation survey with a Geiger-Mueller counter is shown in Figure 10-11.

Calibration

Usually, the calibration of radiation meters is a laboratory procedure to be carried out by qualified experts. Under certain cir-

Figure 10-10. Gamma-radiation measurements with ionization chamber instrument. (Courtesy Argonne National Laboratory)

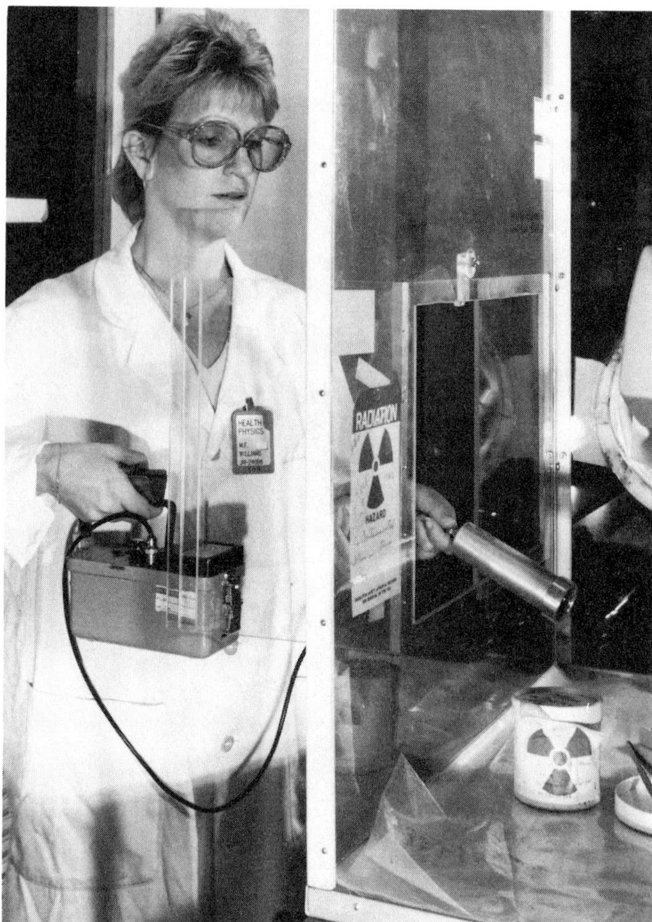

Figure 10-11. Use of a Geiger-Mueller counter. (Courtesy Argonne National Laboratory)

cumstances, however, it is possible and permissible to calibrate meters by comparing a radiation measuring instrument with a standard radiation source of known output.

Consult up-to-date manuals and become familiar with regulations concerning the use and calibration of radiation personnel monitoring devices.

BASIC SAFETY FACTORS

For external radiation exposure hazards, the basic protection measures are associated with (1) time, (2) distance, and (3) shielding.

Time

The longer the exposure, the greater the chance for radiation injury. Because there is a direct relationship between exposure dose and duration of exposure, reducing the exposure time by one-half reduces the dose received by one half.

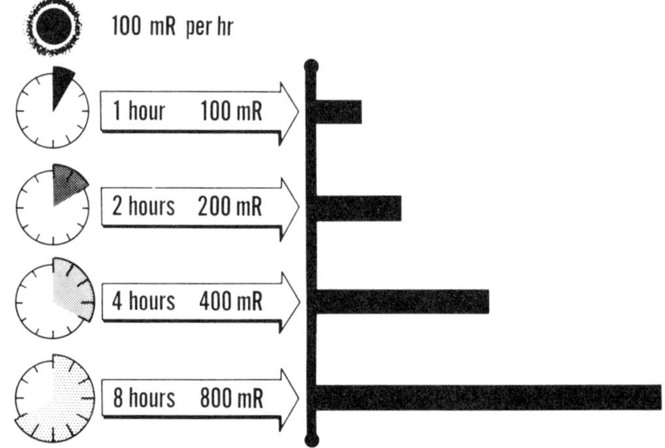

Figure 10-12. The effect of time on radiation exposure is easy to understand. If we are in an area when the radiation level from penetrating x- or gamma-radiations is 100 mr/hr, then in 1 hr, we would get 100 mr. If we stayed 2 hr, we would get 200 mr, and so on. Note: mr=millirems. (Reprinted from U. S. Nuclear Regulatory Commission, *Living with Radiation—Fundamentals*)

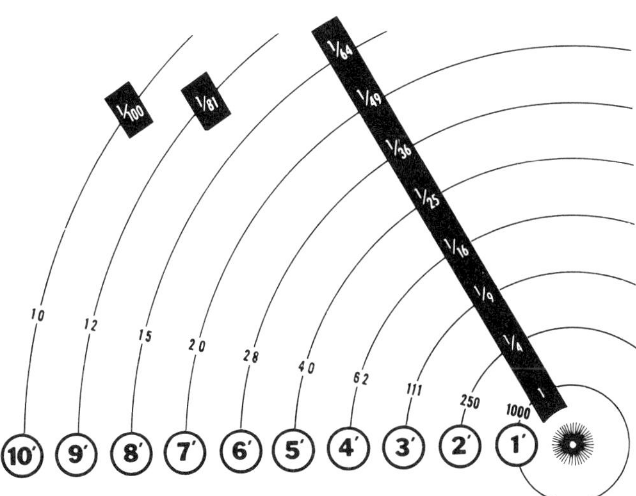

Figure 10-13. The effect of distance on radiation exposure follows the inverse square law—the intensity of radiation falls off by the square of the distance from the source. If we had a point source of radiation giving off 1,000 units of penetrating external radiation at 1 ft (or any other unit of distance), we would receive only one-fourth as much or $(\frac{1}{2})^2$, if we double our distance. If we triple our distance, we reduce the dose to 1/9 or $(\frac{1}{3})^2$. (Reprinted with permission from *Safe Handling of Radioisotopes in Industrial Radiography,* Picker X-Ray Corp., 1962)

A common practice is to reduce the exposure time and thus the exposure. From the knowledge of the dose rate in a given location and the maximum dose which is acceptable for the time period under consideration, the maximum acceptable exposure time can be calculated. If the exposure rate is 2.5 millirems/hour (mr/hr), 40 hours will result in 100 mr. If the exposure rate is 10 times higher, then the time must be reduced to one-tenth (to 4 hr) for the same dose (Figure 10-12).

In practice, the dose received in accomplishing a task can be spread over several employees so that no one person will exceed the guide levels. Only necessary exposures should be planned for a work task. Exposure rate is measured with one of several types of instruments. Direct-reading dosimeters are available, as are those that emit chirping sounds at a preset exposure rate.

Distance

As a rough approximation, the inverse square law can be applied to determine the change in external penetrating radiation exposure with change in distance from a radiation source. Figure 10-13 indicates that by doubling the distance from the source of radiation the exposure would be decreased to $(\frac{1}{2})^2$ or to one-fourth of the original amount. By increasing the distance from 2–20 m, the exposure would be decreased to $(2/20)^2$ or to 1 percent of the original amount. While the inverse square law can be used as an approximation, it should be recognized that it applies only to a point source in free space where there is no scattering of radiation. In practical applications, the radiation source may not be equivalent to a point source and surrounding surfaces may reflect some radiation so that free space does not apply. However, in most instances, the approximation will be adequate.

Table 10-D. Gamma Emitters and Radiation Levels At Various Distances from the Source*

Isotope	*0.3 m*	*0.6 m*	*1.2 m*	*2.4 m*	*3.0 m*
Cobalt-60	14.5	3.6	0.9	0.23	0.145
Radium-226	9.0	2.3	0.6	0.14	0.09
Cesium-137	4.2	1.1	0.26	0.07	0.042
Iridium-192	5.9	1.5	0.4	0.09	0.059
Thulium-170	0.027	0.007	0.002	0.0004	0.00027

*Distance in m; 3.28 ft = 1 m
(Reprinted from *Safe Handling of Radioisotopes in Industrial Radiography,* Picker X-Ray Corp., 1962)

Maintaining a safe distance is especially critical for employees who must work near inadequately shielded sources of radiation. This applies to both portable and nonportable source types (Table 10-D). In such cases, a radiation survey should be made with an appropriate survey meter by a qualified health and safety professional to establish minimum safe distances that workers must abide by.

Work involving penetrating types of radiation should be performed where permanent barriers and shielding protect workers from harmful exposure. However, it is possible that certain operations may occasion some exposure of employees. In these cases, all unnecessary exposure to radiation should be avoided, even to the extent of barring workers, including technicians, from active areas during exposure time, such as during use of x-ray equipment.

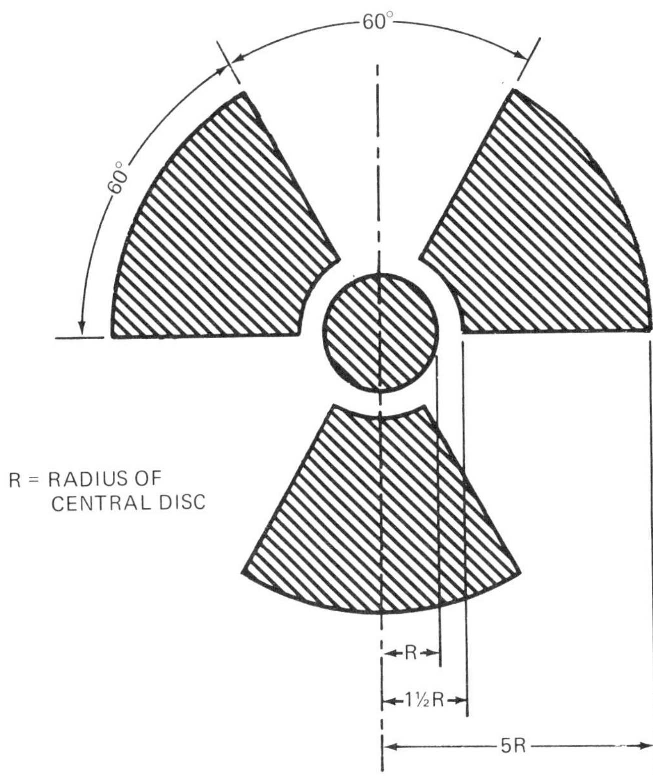

Figure 10-14. Standard symbol for radiation warning signs. (Courtesy American National Standards Institute)

Table 10-E. Approximate Tenth- and Half-Value Thickness for Shielding Radiographic Sources

Radiographic Source	Tenth- and Half-Value Thicknesses (cm)					
	Lead		Iron		Concrete	
	1/10	1/2	1/10	1/2	1/10	1/2
Co-60	4.11	1.24	7.36	2.21	22.9	6.90
Ra-226	4.70	1.42	7.70	2.31	24.4	7.37
Cs-137	2.13	0.64	5.72	1.73	18.0	5.33
Ir-192	1.63	0.48	*	*	15.7	4.83

* = No data.
Notes: The thicknesses for tenth- and half-value layers provide shielding protection from the scattered radiation resulting from deflection of the primary gamma rays within the shield well as protection for primary radiation from the source. The tenth-value layers were determined from the reduction factor *v.* shield thickness curves. The tenth-value thicknesses were taken as one-third the thickness of shielding material necessary to give a reduction factor of 1,000. The half-value thickness is equal to the tenth-value thickness divided by 3.32.
1 cm = 0.394 in.
Density of concrete assumed to be 2.35 g/cc (147 lb/cu ft).
(Reprinted from the Nuclear Regulatory Commission Publication U-2967)

A safe distance is the distance nonoperating workers must maintain from the radiation source in order to receive no more exposure than that specified in the NCRP *Radiation Protection Guides,* even if personnel were to remain at that distance continually. The distances specified for a given job may not necessarily remain constant if the radiation source in use is modified.

Hazardous areas indicated by the protection survey must be barricaded or roped off to form a restricted section which must not be entered by nonoperating workers or bystanders. Large signs bearing the standard radiation symbol with proper wording should be posted (Figure 10-14).

Shielding

Shielding is commonly used to protect against radiation from radioactive sources. The more mass that is placed between a source and a person, the less radiation the person will receive (Figure 10-15). For a high-density material, such as lead, the barrier thickness required for a given attenuation of x- or gamma-radiation is less than it is for a less dense material, such as packed earth (Table 10-E).

Shielding of neutrons requires a different approach than does shielding of x- or gamma-radiation. Neutrons transfer energy to, and are shielded most effectively by, light nuclei (hydrogen atoms are the most effective). Therefore, water or other materials rich in hydrogen content are used as shields for neutrons. Carbon atoms, such as in graphite, are also often applied for neutron shielding.

Shielding can take many forms. These include cladding on radioactive material, containers with heavy walls and covers for radioactive sources, cells with thick high-density concrete walls having viewing windows filled with high-density transparent liquid for remote handling of high-level gamma emitters, and a deep layer of water for shielding against gamma-radiation from spent nuclear reactor fuel. Shielding calculations are often highly technical and require the services of an expert in this area.

Tables of half-value layers are given in radiation handbooks and typical values for two radioactive materials, cobalt-60 and cesium-137, are given in Table 10-A.

Table 10-A states that a 6.5 cm (2.6 in.) thickness of concrete will reduce the gamma-radiation coming from any cobalt-60 source to a factor of ½. If the gamma emitter is cesium-137, then the half-value layer for concrete becomes 5.3 cm (2.1 in.). The gamma-radiation from cesium-137 is lower in energy (not as many MeVs) than the gamma-radiation from cobalt-60. Figure 10-15 illustrates how additional mass can reduce radiation levels. The example used in Figure 10-15 is a cobalt-60 source that gives a meter reading of 0.5 roentgen per hour (500 milliroentgen per hour) at a distance of 0.9 m (3 ft).

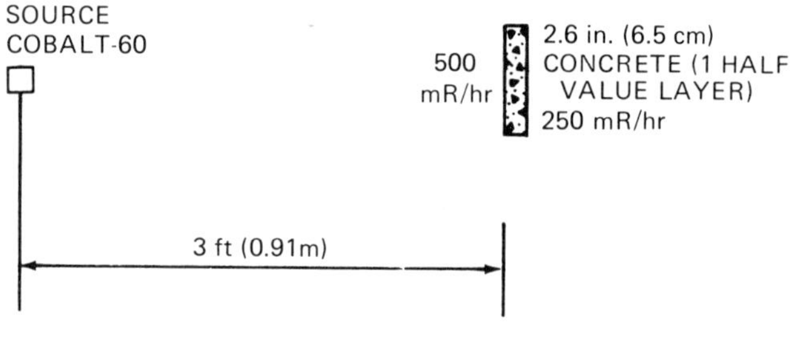

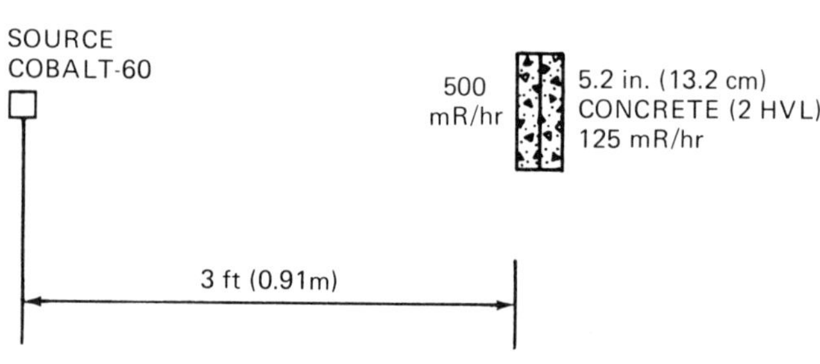

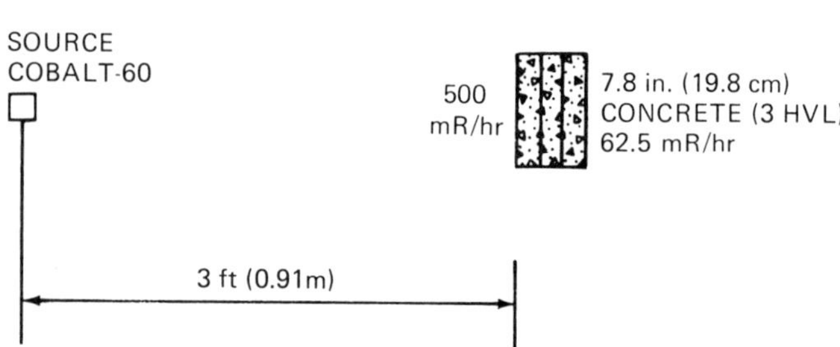

Figure 10-15. Illustration shows attenuation of cobalt-60 gamma-radiation by half-value layers of concrete.

A formula used for gamma emitters is:

R/hr at 1 ft=(6)(C)(E)(f)

This relationship states that the number of roentgens per hour measured at a distance of 0.3 m (1 ft) from a source is approximately equal to six times the curie strength (C) of the source, times the energy (E) of the gamma-radiation, times the fractional yield (f) of the gamma-radiation per disintegration. It cannot be used for estimating beta-radiation levels.

Given the name of the radioactive source and its quantity (or activity) in curies, the value for its energy in MeVs can be obtained from handbooks. For example, Table 10-B gives the energy of radiation from several radioactive isotopes. The fractional yield (f) of the gamma-radiation per disintegration can be found in the *Radiological Health Handbook,* U.S. Department of Health and Human Services, PHS, FDA Bureau of Radiological Health, Rockville, MD.

Two cautions are required for proper application of this formula:

- Some radioactive materials, such as cobalt-60, emit more than one gamma, each with different energies. The sum of the energy of the total emissions must be used. This information will be found in handbooks.
- Terms must be consistent. If the source activity is given in millicuries or microcuries it should be converted to curies. If millicuries are used in the source term, then the answer is in milliroentgens.

Example: What radiation reading would be expected at a distance of 1 foot from an unshielded 100-millicurie cobalt-60 source?

R/hr at 1 ft = (6)(C)(E)(f)

Given: C = 100 millicuries (0.1 curie)
E = 1.2 MeV + 1.3 MeV (from handbook tables)
F = 1 (from handbook tables)

Answer:
R/hr at 1 ft = 6(0.1) (1.2 + 1.3)(1) = 1.5

For a 1-millicurie cesium-137 source?

Answer: R/hr at 1 ft = 6(0.001)(0.66)(0.9) = 0.0036, which would be the same as 3.6 milliroentgens (mR) per hour at 1 ft (0.3 m).

The above examples illustrate (1) the use of the formula in calculating dose rates, and (2) show how the application of the formula can assist in the interpretation of fractional amounts of curies.

The fact that a 1-millicurie source of cesium-137 is going to be used may not mean too much to a health and safety professional who is not sure just how much 1 millicurie is, or if it will be a big problem. However, by calculating the exposure rate of 3.4 milliroentgens per hour at 1 foot, and then applying the distance rule of decreasing radiation (1 divided by the distance squared—the inverse square rule) it becomes apparent that 1.2 m (4 ft) away from the source the radiation level would be 0.21 mR/hr, which is negligible.

Table 10-D lists a number of gamma emitters and their radiation levels measured at various distances from a 1-curie source. The numbers were obtained by using the previous formula and then applying the inverse square law for distances of other than one foot.

Knowing existing radiation levels at various distances from various radioactive sources, allows use of control procedures to ensure that cumulative doses will not be in excess of the maximum permissible dose (MPD).

As a final example of how MPD figures can be used, assume that a 0.5 curie cobalt-60 source is involved in a fire. The radiation meters have been damaged and there is reason to believe that the source container is not functioning as an effective shield. Until someone with a good meter can conduct a survey and determine how severe the problem is, at what distance should the area be fenced off?

First—Calculate R/hr at 1 ft for a 0.5 curie cobalt-60 source. R/hr at 1 ft =6(0.5)(2.5)(1)=7.5 or 7,500 milliroentgen per hour at one foot.
Second—At a MPD of 100 milliroentgens per week, for a 40-hour week, the dose rate per hour should not exceed 100/40 or 2.5 mR/hr.
Third—Use the inverse square law as follows:

$$? \text{ ft} = \sqrt{\frac{\text{mR/hr at 1 ft}}{2.5 \text{ mR/hr}}}$$

The distance in feet to place the barricade equals the square root of the quotient (milliroentgen per hour at one foot divided by 2.5 milliroentgen per hour).

$$\text{Thus: } ? \text{ ft} = \sqrt{\frac{7{,}500}{2.5}} = \sqrt{3{,}000} = 54.8$$

Workers should not be allowed 17 m (55 ft) from the source location. (All examples assume that exposures are occupational to workers with accurately maintained radiation exposure records.)

CONTROL PROGRAMS

A radiation health and safety program should establish safe working procedures, detect and measure radiation, make surveys, be concerned with decontamination and disposal, laboratory and other special services, and record-keeping.

Ionizing radiation cannot be felt, seen, heard, tasted, or smelled. However, unknown radioactive isotopes can be measured and identified with instruments, and therefore can be adequately controlled.

A basic concept in radiation protection practice is the establishment of a "controlled area." Areas where radiation dose rates may be excessive can be guarded during exposure times by suitable methods, such as the erection of barriers and warning signs (Figure 10-14), stationing of attendants to keep personnel out of restricted localities, and, in extreme cases, complete closing of the areas. Safe exposure times and safe practices can be established through measurement and control, and by learning from past experience.

Work with radiation can and should be so planned and managed that radiation exposures of employees and of the general public are kept to a minimum. Controls are needed to prevent release of radioactive materials that can result in exposures above guide levels.

Potential avenues of exposure to the public are contaminated air or water, waste materials, or employees unknowingly leaving the place of work with contamination on their persons, clothing, or shoes.

Table 10-F. Classification of Isotopes According to Relative Radiotoxicity Per Unit Activity
The isotopes in each class are listed in order of increasing atomic number

CLASS 1 (very high toxicity)	Sr-90 + Y-90, *Pb-210 + Bi-210 (Ra D + E), Po-210, At-211, Ra-226 + per cent *daughter products, Ac-227, *U-233, Pu-239, *Am-241, Cm-242.
CLASS 2 (high toxicity)	Ca-45, *Fe-59, Sr-89, Y-91, Ru-106 + *Rh-106, *I-131, *Ba-140 + La-140, Ce-144 + *Pr-144, Sm-151, *Eu-154; *Tm-170, *Th-234 + *Pa-234, *natural uranium.
CLASS 3 (moderate toxicity)	*Na-22, *Na-24, P-32, S-35, Cl-36, *K-42, *Sc-46, Sc-47, *Sc-48, *V-48, *Mn-52, *Mn-54, *Mn-56, Fe-55, *Co-58, *Co-60, Ni-59, *Cu-64, *Zn-65, *Ga-72, *As-74, *As-76, *Br-82, *Rb-86, *Zr-95 - *Nb-95, *Nb-95, *Mo-99, Tc-98, *Rh-105, Pd-103 - Rh-103, *Ag-105, Ag-11, Cd-109 - *Ag-109, *Sn-113, *Te-127, *Te-129, *I-132, Cs-137 - *Ba-137, *La-140, Pr-143, Pm-147, *Ho-166, *Lu-177, *Ta-182, *W-181, *Re-183, *Ir-190, *Ir-192, Pt-191, *Pt-193, *Au-198, *Au-199, Tl-200, Tl-202, Tl-204, *Pb-203.
CLASS 4 (Slight toxicity)	H-3, *Be-7, C-14, F-18, *Cr-51, Ge-71, *Tl-201.

*Gamma-emitters
(Reprinted from *Safe Handling of Radionuclides,* Safety Series No. 1, International Atomic Energy Agency, Vienna.)

Consider the sources of radiation

One must always consider the amount and kind of radiation sources used in a contemplated operation. Suitable radiation safety operating standards can then be designed. Special circumstances can greatly affect the safety requirements.

Table 10-F lists a number of radioactive materials according to their toxicity. Table 10-G classifies the laboratory or work area required for radioactive materials of differing toxicity.

Frequently, radionuclides are used so that measurements of their radiation will disclose useful information. One example is the employment of radioactive sources to effect quantitative or dimensional control of industrial materials. For example, radionuclides can measure fluid-flow rates or thickness of semi-finished or finished products (Figure 10-16).

Figure 10-16. Radiation level near beta-ray gauge used to measure thickness of steel strip is checked with a Geiger-Mueller counter.

Figure 10-17. Pocket dosimeter and film badge should be worn during all work periods. (Courtesy Applied Health Physics)

Sealed sources. Sealed sources with widely varying amounts of radioactivity are available. A high intensity source may require so much shielding that it will not be considered portable. Shielding, control devices, and procedures should be designed by someone experienced in such work. Keeping external exposures to personnel under control can be accomplished with the aid of alarms, interlocks, strict control of access, and thorough monitoring. (Bibliography, NCRP Handbook 30, *Safe Handling of Radioactive Materials.*)

Sources of low intensity should only require keeping track of its presence.

Sources of intermediate intensity are more troublesome. They are portable and in some cases may require so little shielding that they can easily be moved by hand. Persons operating a portable source must be depended on to follow the necessary precau-

Table 10-G. Laboratory or Work Area Required for Isotopes of Increasing Radiotoxicity

Radiotoxicity of isotopes	Minimum significant quantity	Type of Laboratory or Work Area Required: Type C Good Chemical Laboratory	Type B Radioisotope Laboratory	Type A High-Level Laboratory
Very high	0.1 μc	10 μc or less	10 μc—10 mc	10 mc or more
High	1.0 μc	100 μc or less	100 μc—100 mc	100 mc or more
Moderate	10 μc	1 mc or less	1 mc—1 c	1 c or more
Slight	100 μc	10 mc or less	10 mc—10 c	10 c or more

Procedure		Modifying factor
Storage (stock solutions)	X	100
Very simple wet operations	X	10
Nromal chemical operations	X	1
Complex wet operations with risk of spills	X	0.1
Simple dry operations	X	0.1
Dry and dusty operations	X	0.01

NOTE: Modifying factors should be applied to the quantities shown in the last three columns as follows, according to the complexity of the procedures to be followed. Factors are suggested only because due regard must be paid to the circumstances affecting individual cases.

(Reprinted from *Safe Handling of Radionuclides,* Safety Series No. 1, Atomic Energy Agency, Vienna)

tions, including the use of film badges and dosimeters (Figures 10-7 and 10-17) to measure their own exposure. They must operate the source so that other persons are not accidently irradiated. (Bibliography, NSC Industrial Data Sheet 461, *Beta-Particle Sealed Sources*—rev. 1981).

Some sealed sources are subject to breakage and spillage, and even welded seals have been known to fail. The hazard can be severe and great care should be taken in handling sealed sources. Even if no known accident has occurred, such sources should be tested for leaks by standardized procedure at scheduled intervals by an experienced technician.

Radiation-producing machines. For many years, radiation-producing machines have been in use and proper operation of x-ray machines has been described in many publications. Remote actuation and shielding are common ways to control personnel exposures. If an accelerator is used, radiation safety personnel with special training or experience are required to assure proper installation and operation (Bibliography, NCRP Handbooks 33, 35, 49, and 51).

Portable x-ray units present many of the same kinds of radiation safety problems as nonportable sources of corresponding intensity. Special problems related to the portability and lack of stationary shields must also be considered.

Radioisotopes. Use of radioisotopes in the laboratory may encompass a wide range of hazards. Hazards will depend on the quantities and types of radioisotopes used as well as the kinds of operations to be performed.

Radioisotopes must be considered individually when hazards are being analyzed because their effect on human health varies with the type of radiation emitted, the biological result produced, the process or work being conducted, and the safe practices being employed by those conducting the work. No general statements can be made until specific information is gathered and analyzed. It may be necessary to consult an expert in the field.

Table 10-F groups various radioisotopes according to their relative radiotoxicity per unit activity in accordance with the International Atomic Energy Agency's (IAEA) publication, *Safe Handling of Radioisotopes.*

The IAEA explains the hazard classification for unsealed radioactive sources as follows: "Hazards arising out of the handling of unsealed sources depend on factors such as the types of compounds in which these isotopes appear, the specific activity, the volatility, the complexity of the procedures involved, and of the relative doses of radiation to the critical organs and tissues, if an accident should occur that gives rise to skin penetration, inhalation, or ingestion." Broad classifications for the radiotoxicity of various isotopes and their significant amounts are given in Tables 10-F and -G.

Spent fuel elements from a nuclear power reactor may contain on the order of a million curies, depending on the length of time they have been irradiated in the reactor and the interval of time between their removal from the reactor and measurement of their activity. Shielding is required for storage and shipment of spent fuel materials and other high-level-gamma-radiation sources.

Radioactive metals. Radioactive metals vary greatly in degree of hazard. Some have such low radiation rates that they may be held in the hand (if they are solids that are not flaking or dusting). A piece of normal uranium or an alloy of this material, for example, may be safely handled without personal protection for a few hours per week. It is good practice, nevertheless, to wear gloves when handling such materials, because metals (like uranium) can oxidize and develop a flaky surface. Metal cladding or a paint-type surface coating may be needed to avoid loose contamination. Respiratory protection may also be needed.

Certain radioactive metals have such high surface dose rates that they must bc handled only with remote control devices. If a solid radioactive metal has a loose contaminating layer of radioactivity on its surface, it will need to be handled in a ventilated enclosure equipped with high-efficiency exhaust filter. In each case, the radiation level of the particular material and its composition must be known before safe handling practices can be determined.

Instrument measurement of the dose rate from the material will help determine the most practical control by means of

Figure 10-18. Typical glove-box type of hood permits rigid control of conditions when radioactive metals are worked. (Courtesy Argonne Natonal Laboratory)

exposure time, distance, and shielding. A review of process and handling history, visual inspections, and smear and air sampling is useful for indicating the probability of contamination incidents and the control measures needed.

Uranium metal will burn under certain conditions and, along with some alloys used in chemical operations, may involve a chemical explosion hazard. Casting of uranium metal must be done in a vacuum. It is necessary to use an inert gas, rather than air, to break the vacuum. Wastes and scraps must be carefully handled from the time they are generated until they are ultimately disposed of, not only for reasons of safety, but also because of their high monetary value.

Cutting of uranium with abrasive cutoff wheels requires special local exhaust ventilation. Because chips, turnings, and finely divided uranium burn spontaneously under some conditions, it is best to convert uranium into the oxide daily as the scrap is accumulated. Uranium-contaminated waste requires packaging that meets U.S. Department of Transportation and disposal site criteria.

Criticality. Criticality is the fissioning or breaking apart of nuclei with emission of neutrons at a rate faster than neutrons are absorbed or lost from the system. This leads to a highly hazardous instantaneous burst of neutrons with associated high level gamma radiation. Whenever amounts of fissionable radioactive materials such as uranium-235 and plutonium-239 necessitate criticality controls, an expert in this area of radiation safety must be involved.

A number of fatalities have resulted from criticality incidents and personnel who are even at considerable distances from the source may receive severe radiation exposure. Stringent controls are needed over amounts of and movement of fissionable materials. Immediate local sensing of a criticality and sounding of audible alarms to trigger quick evacuation is another precaution employed for this hazard.

Plutonium. Plutonium is a highly hazardous alpha emitter, which must be handled or processed under rigidly controlled conditions. Enclosures, some of which are called glove boxes or dry boxes (Figure 10-18), must be carefully designed and are relatively expensive. The glove boxes must be maintained at a negative pressure and may need an inert atmosphere, such as argon or nitrogen to avoid fires. Plutonium is pyrophoric under certain conditions. Plutonium-239 is fissionable and presents a criticality control problem where sufficient amounts are present (>300 gm).

Glove-box exhaust gas is typically filtered by two or more high-efficiency particulate air (HEPA) filters in series. Emergency power for exhaust ventilation and alarms for loss of negative pressure in the boxes are used as system safety features.

Since glove-box gloves may develop pinholes, cracks, tears, or punctures, workers wear protective gloves. Workers should monitor their gloved hands periodically and monitor and remove protective clothing on leaving the work area.

Consider operational factors

Ascertain the required level of radiation protection and evaluate the problems that may arise by analyzing the radiation work in terms of the following factors:

- area involved (in square meters or feet), number of rooms and buildings
- number of employees exposed to radiation and their locations
- chemical and physical states of the radioactive material and the nature of its use
- incidents that might occur and their possible locations
- nonradiation hazards involved
- nature of the probable radiation exposure or release of radioactive material
  - a controlled or supervised release (as in a disposal operation)
  - an accidental release or exposure that cannot be sensed by warning alarm signals generated by radiation-detection instruments
  - a violent release of dust, droplets, gases, or vapors through fire, temperature increase, or explosion
  - a contact spread of contamination as the result of its adherence to other material or of spillage
- inherent danger of the material because of its internal or external effect on humans. This danger depends on the isotopic identification and chemical form of the material (Bibliography, NBS Handbooks 59 and 69).
- probability of detection of harmful situation by routine surveys or monitoring. Current knowledge of conditions is essential to determine the acceptability of the risks involved.
- possible effects of accidents on operations, such as interruption of production, loss of occupancy of space, loss of equipment, and cost of cleanup.

Consider employee exposure potential

The industrial accidents which can produce radioactive contamination are usually no more difficult to control than the common types of industrial accidents, but there is an added, compelling reason to control them—the danger of intake of radioactive materials by personnel.

Monitors for detection of radioactive contamination of personnel are shown in Figures 10-19 and 10-20. It is important that the appropriate monitoring instruments are available and are properly maintained and calibrated.

External hazards. Under ordinary operating conditions and barring accidents, the level of exposure to external radiation may be known through continuous monitoring instruments, dosimeters, or from previous measurements of radiation fields for identical operations. The problem is then reduced to limiting the rate of exposure or the length of exposure time. Measurement of the exposure by instruments will indicate what advantage is to be obtained by installing additional shielding or by having personnel work further from the source.

Under emergency conditions with high-exposure levels, it is advisable to rotate personnel so as to prevent exposure of any one individual above guide levels. Emergency exposure

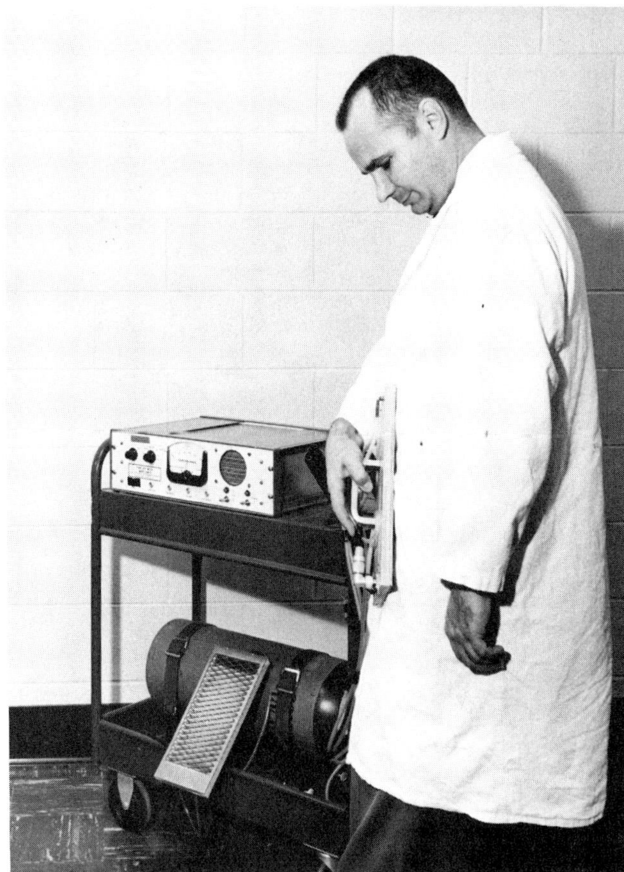

Figure 10-19. Mobile personnel monitor for hands, feet, and clothing of personnel working with radioactive materials.

levels for extreme conditions have been published in NCRP Report No. 39, *Basic Radiation Protection Criteria*. These are essentially once-in-a-lifetime allowances for handling an extremely serious situation.

Occasionally, an unexpectedly high reading will appear on a personal dosimeter. While the reading may be invalid for any one of a number of reasons, an investigation will be needed to determine the validity of the reading and indicate the corrections needed to prevent recurrences.

Internal hazards. Inhalation of radioactive materials is the most frequent route of entry into the body. Routine air-sampling may detect the concentration, and a constant air-monitoring system may be designed to sound an alarm if hazardous levels are detected. Also, employees may recognize conditions that in the past have released materials; thus they can act promptly to minimize exposure.

In the event that a spill of radioactive materials is detected by immediate recognition, hand and foot monitoring, or through air or room surface monitoring, prompt action is needed. If there is a potential inhalation hazard, personnel should evacuate to the nearest safe location. Workers' further movement should be restricted until a monitoring survey check can be made to determine the extent of radioactive contamination. Quick arrangements should be made to restrict exit from and entry into the building wing, or laboratory, until there is a determination that radioactivity has not

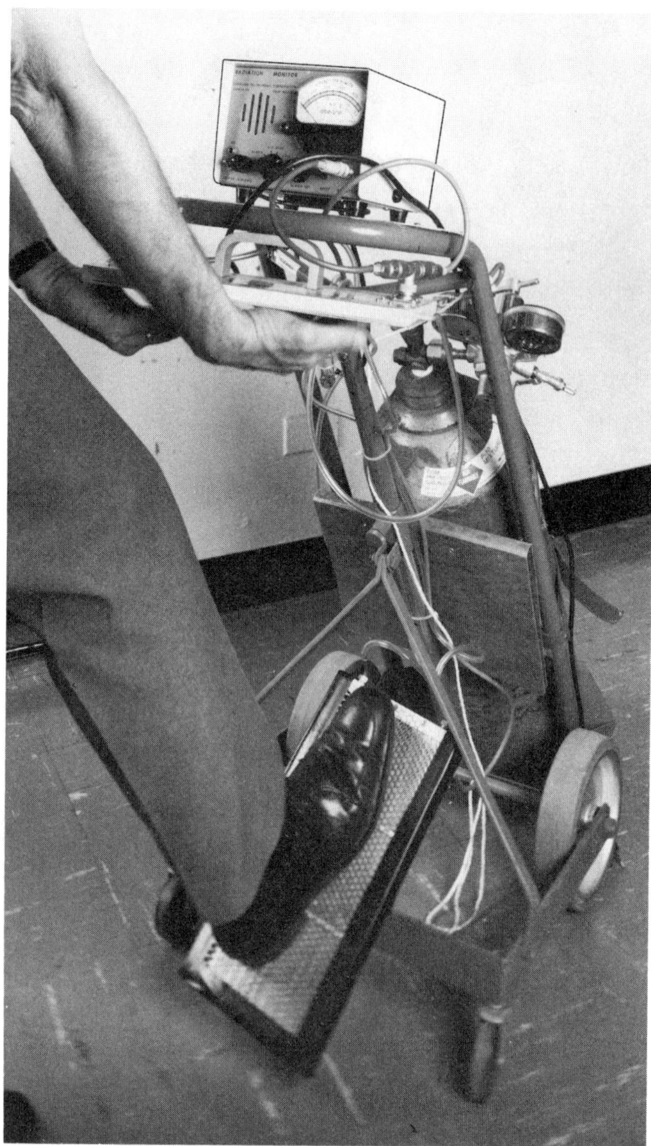

Figure 10-20. Portable hand and foot monitor for detecting radioactivity. (Courtesy Argonne National Laboratory)

and will not be tracked or carried out to other areas. Corridors should be roped off or posted to control movement until personnel-, surface-, and air-monitoring shows that contamination is controlled.

It is important to isolate and clean up a spill as soon as possible. Personnel conducting the monitoring and cleanup should wear appropriate protective equipment and clothing. The incident can be secured as soon as there is assurance that the potential hazards have been controlled.

A less common route by which radioactive materials enter the body—in rare cases this can be more serious than through the lungs—is through broken skin. Radioactive materials can readily gain entry through a minute cut or abrasion, enter the bloodstream, and thence be dispersed throughout the body. The result can be serious if there is a high concentration of the radioisotope at the skin break. If the workers have been appropriately informed, the injured person will be aware of the possible danger and can request a radiation survey of the wound area.

By and large, absorption through the intact skin is not a significant route of entry although tritiated water vapor is absorbed rapidly through unbroken skin. However, it is important to scrub up immediately on finding significant skin contamination. This minimizes the possibility of entry through inhalation or a subsequent skin break, and prevents the spread of the radioactive materials.

Analysis of the radionuclide content of the body fluids and excreta and whole body counting are important for after-the-fact detection of intake of radioactivity. When the presence of radionuclides within the body is detected, or an accidental intake of radionuclides becomes known, steps can be taken to determine the cause and prevent recurrence. This serves as a backup for environmental monitoring.

Records

Some diseases that may be caused by exposure to radiation or radioactive materials can occur after many years of exposure. Retention of suitable records relating to exposures and working conditions is desirable and is usually required.

SUMMARY

It is impossible to discuss all of the important aspects of radiation safety in a "fundamentals manual" such as this. Nevertheless, health and safety professionals will have a good background for further study and work in the field of radiation safety if the professional:

1. treats radiation and radioactive materials with respect because the potential hazards are recognized;
2. recognizes the two distinct types of exposure hazards involved—external and internal;
3. has a concept of the various types of ionizing radiation;
4. knows that calibrated monitoring instruments appropriate for the specific type of ionizing radiation must be used for measurements;
5. recognizes the importance of basic exposure control measures including time, distance, and shielding; and
6. knows where to look for information that will establish guides and limits to follow.

BIBLIOGRAPHY

American Industrial Hygiene Association, 475 Wolf Ledges Parkway, Akron, OH 44311.

Respiratory Protection Program—A Manual and Guideline, 1980.

American National Standards Institute, 1430 Broadway, New York, NY 10018.

Standards:

Administrative Practices in Radiation Monitoring (A Guide for Management), ANSI N13.2, 1969.

Classification of Sealed Radioactive Sources, ANSI N5.10, 1968.

Facilities Handling Radioactive Materials, Recommended Fire Protection Practice for, ANSI/NFPA No. 801, 1986.

Film Badge Performance, Criteria for, ANSI N13.7, 1983.

Guide for Writing Operating Manuals for Radioactive Materials Packaging, ANSI N679, 1976.

Immediate Evacuation Signal for Use in Industrial Installations Where Radiation Exposure May Occur, ANSI N2.3, 1979.

Inspection and Test Specifications for Direct Reading and Indirect Reading Quartz Fiber Pocket Dosimeters, ANSI N322, 1975.

Leakage Tests on Packages for Shipment of Radioactive Materials (issued for trial use and comments), ANSI N14.5, 1977.

On-site Instrumentation for Continuously Monitoring Radioactivity in Effluents, Specification and Performance of, ANSI N42.18, 1980.

Occupational Radiation Exposure Records Systems, Practice for, ANSI N13.6, 1966 (R1972).

Packaging and Transporting Radioactive Materials, Administrative Guide for, ANSI N14, 10.1, 1973.

Personnel Neutron Dosimeters (Neutron Energies Less Than 20 MeV), ANSI N319, 1976.

Radiation Protection Instrumentation Test and Calibration, ANSI N323, 1978 (3).

Radiation Symbol, ANSI N21.1, 1969.

Radioactive Waste Categories, Definition of, ANSI N5.8, 1967.

Radioactively Contaminated Biological Materials, Packaging and Transportation of, ANSI N14.3, 1973.

Sampling Airborne Radioactive Materials in Nuclear Facilities, Guide to, ANSI N13.1, 1969.

Shipping Packages for Type A Quantities in Radioactive Materials, Guide to Design and Use of, ANSI N14.7, 1975.

Applied Radiation and Isotopes, Part A. Elmsford, NY: Pergamon Press.

A Review of the Department of Transportation (DOT) Regulations for Transportation of Radioactive Materials. Washington, DC: U.S. DOT, Materials Transportation Bureau, Office of Hazardous Materials Operations, 1976.

Cameron, J. R., Suntharalingam, and Kenney *Thermoluminescent Dosimetry.* Madison, WI: University of Wisconsin Press, 1968.

Cember, H. *Introduction to Health Physics.* Elmsford, NY: Pergamon Press, 1983.

Code of Federal Regulations (CFR). Title 10, Chapter 1—Energy. Part 20. Standards for Protection Against Radiation. Washington, DC: Nuclear Regulatory Commission (NRC), Superintendent of Documents, 1985.

CFR 49. Transportation: Part 100-199. Revised as of October 1, 1984. Washington, DC: U.S. Government Printing Office, 1984.

Cralley, L. J., and Cralley, L. V. "Evaluation of Exposure to Ionizing Radiation." In *Patty's Industrial Hygiene and Toxicology.* New York: John Wiley & Sons, 1979.

Everything You Always Wanted to Know About Shipping High Level Nuclear Wastes. Wash-1264. U.S. Atomic Energy Commission. Washington, DC: U.S. Superintendent of Documents, 1974.

Federal Register (FR) 52—2822. Jan. 27, 1987. Interagency Recommendations for Protecting Workers From Exposure to Ionizing Radiation. Washington, DC: U.S. Government Printing Office, 1987.

General Principles of Monitoring for Radiation Protection of Workers. ICRP Publication No. 12. Elmsford, NY: Pergamon Press, 1969.

Hazardous Materials Regulations of the DOT. Tariff No. BOE-6000-E. Effective June 14, 1985, Thomas A. Phemister, Agent, 1920 L St., N.W., Washington, DC, 1985.

Health Physics, Official Journal of the Health Physics Society. Elmsford NY: Pergamon Press.

Hendee, W. R. *Medical Radiation Physics.* Chicago: Yearbook Medical Publ Inc., 1979.

Kocher, D. C. "Radioactive Decay Data Tables," DOE/TIC-11026, Technical Information Center, U.S. D.O.E., 1981.

Lanzl, L. H., Pingel, J. H., Rust, J. H., eds. *Radiation Accidents and Emergencies in Medicine, Research, and Industry.* Springfield, IL: Charles C. Thomas Publ, 1965.

Martin, A., and Harbison, S. A. *An Introduction to Radiation Protection.* London, England: Chapman and Hall, Methuen, Inc., 1980.

Medical Radionuclides: Radiation Dose and Effects. Symposium proceedings Oak Ridge, CONF-691212. Springfield, VA: Clearinghouse for Federal Scientific and Technical Information, U.S. Dept of Commerce, 1970.

Morgan, K. Z., and Turner, J. E. *Principles of Radiation Protection.* New York: John Wiley & Sons, 1973.

National Council on Radiation Protection, 7910 Woodmont Avenue, Suite 1016, Bethesda, MD 20814.

Alpha-Emitting Particles in Lungs, No. 46, 1975.

Basic Radiation Protection Criteria, No. 39, 1971.

Cesium-137 From the Environment to Man: Metabolism and Dose, No. 52, 1977.

Control and Removal of Radioactive Contamination in Laboratories, No. 8, 1951.

Dental X-Ray Protection, No. 35, 1970.

Environmental Radiation Measurements, No. 50, 1976.

General Concepts for the Dosimetry of Internally Deposited Radionuclides, No. 84, 1985.

A Handbook of Radioactivity Measurements Procedures, No. 58, 1985.

Instrumentation and Monitoring Methods for Radiation Protection, No. 57, 1978.

Krypton-85 in the Atmosphere—Accumulation, Biological Significance, and Control Technology, No. 44, 1975.

Management of Persons Accidentally Contaminated with Radionuclides, No. 65, 1980.

Maximum Permissible Body Burdens and Maximum Permissible Concentrations of Radionuclides in Air and in Water for Occupational Exposure, No. 22, 1959. (Includes Addendum I issued in August 1963.)

Measurement of Absorbed Dose of Neutrons and of Mixtures of Neutrons and Gamma Rays, No. 25, 1961.

Measurement of Neutron Flux and Spectra for Physical and Biological Applications, No. 23, 1960.

Medical Radiation Exposure of Pregnant and Potentially Pregnant Women, No. 54, 1977.

Medical X-Ray and Gamma-Ray Protection for Energies Up to 10 MeV Equipment Design and Use, No. 33, 1968.

Natural Background Radiation in the United States, No. 45, 1975.

Operational Radiation Safety Program, No. 59, 1978.

Operational Radiation Safety—Training, No. 71, 1983.

Precautions in the Management of Patients Who Have Received Therapeutic Amounts of Radionuclides, No. 37, 1970.

Protection Against Neutron Radiation, No. 38, 1971.

Protection Against Radiation from Brachytherapy Sources, No. 40, 1972.
Protection of the Thyroid Gland in the Event of Releases of Radioiodine, No. 55, 1977.
Radiation Exposure From Consumer Products and Miscellaneous Sources, No. 56, 1977.
Radiation Protection Design Guidelines for 0.1-100 MeV Particle Accelerator Facilities, No. 51, 1977.
Radiation Protection for Medical and Allied Health Personnel, No. 48, 1976.
Radiation Protection in Educational Institutions, No. 32, 1966.
Radiation Protection in Veterinary Medicine, No. 36, 1970.
Radioactive Waste Disposal in the Ocean, No. 16, 1954.
Radiological Factors Affecting Decision-Making in a Nuclear Attack, No. 42, 1974.
Recommendations for the Disposal of Carbon-14 Wastes, No. 12, 1953.
Recommendations for Waste Disposal of Phosphorus-32 and Iodine-131 for Medical Users, No. 9, 1951.
Review of NCRP Radiation Dose Limit for Embryo and Fetus in Occupationally Exposed Women, No. 53, 1977.
Review of the Current State of Radiation Protection Philosophy, No. 43, 1975.
Safe Handling of Radioactive Materials, No. 30, 1964.
Specification of Gamma-Ray Brachytherapy Sources, No. 41, 1974.
Stopping Powers for Use with Cavity Chambers, No. 27, 1961.
Structural Shielding Design and Evaluation for Medical Use X-Rays and Gamma Rays of Energies up to 10 MeV, No. 49, 1976.
Tritium Measurement Techniques, No. 47, 1976.

National Safety Council, 444 North Michigan Avenue, Chicago, IL 60611.
Beta-Particle Sealed Sources, Industrial Data Sheet, 461.
Sources of Information on Nuclear Energy and Energy Related Activities, Industrial Data Sheet, 685.

Nuclear Instruments and Methods in Physics Research Section. Amsterdam, Holland: Elsevier Science Publishers.

Nuclear Safety. Washington, DC: U.S. Government Printing Office.

Nuclear Science and Engineering, American Nuclear Society. Available from Academic Press, Inc., 111 Fifth Ave., New York, NY.

Purrington, R. G., and Patterson, H. W. *Handling Radiation Emergencies.* Boston, MA: National Fire Protection Agency, 1977.

Safety Series, International Atomic Energy Agency and the World Health Organization, Unipub, Inc., P. O. Box 443, New York, NY 10016.
Basic Factors for the Treatment and Disposal of Radioactive Wastes, No. 24, 1967.
Basic Requirements for Personnel Monitoring, No. 14, 1980.
Basic Safety Standards for Radiation Protection, No. 9, 1982.
Manual on Safety Aspects of the Design and Equipment of Hot Laboratories, No. 30, 1969.
Medical Supervision of Radiation Workers, No. 25, 1968.
Planning for the Handling of Radiation Accidents, No. 32, 1969.
Radiation Protection Procedures, No. 38, 1973.
Regulations for the Safe Transport of Radioactive Materials, No. 6, 1985.
Respirators and Protective Clothing, No. 22, 1967.
Safe Handling of Radionuclides, No. 1, 1973.
Safe Use of Radioactive Tracers in Industrial Processes, No. 40, 1974.

Shapiro, J. *Radiation Protection, A Guide for Scientists and Physicians.* Cambridge, MA: Harvard University Press, 1981.

Shleien, B., and Terpilak, M. S. *The Health Physics and Radiological Handbook.* Olney, MD: Nuclear Lectern Associates, Inc., 1984.

Stewart, D.C. *Data for Radioactive Waste Management and Nuclear Applications.* New York: John Wiley & Sons, 1985.

Wade, J. E., and Cunningham, G. E. *Radiation Monitoring, A Programmed Instruction Book.* Oak Ridge, TN: Division of Technical Information Extension, 1967.

11

Nonionizing Radiation

by Edward J. Largent
Julian Olishifski, MS, PE, CSP
Revised by Larry E. Anderson, PhD

THE NATURE, PROPERTIES, AND EFFECTS of nonionizing radiation, including lasers, are discussed. Power transmission and radio frequencies (including radar and microwave), infrared (IR), visible light, and ultraviolet (UV) regions of the electromagnetic spectrum are commonly considered to be nonionizing. When considering the entire spectrum of electromagnetic radiation, it is apparent that the nonionizing portions include those radiations of longer wavelengths and lower frequencies. These relationships are shown in Figure 11-1.

DEFINITION OF TERMS

In its simplest form, electromagnetic radiation consists of vibrating electric waves moving through space accompanied by a vibrating magnetic field showing the characteristics of wave motion. A moving object or mass can lose its kinetic energy when it collides with a stationary target. The transfer of energy by wave motion on impact with a surface is less obvious. For example, if a stone is thrown into a pool of quiet water, an ever-widening circular wave spreads over the surface from the point of impact. A cork floating on the surface bobs up and down as the wave passes, but does not move outward from the point of impact, as it would if there was an actual outward flow of water. Waves can be produced by vibrations other than those of liquid particles on the surface of water. The electric current in the transmitter antenna circuit of a radio station sets up a magnetic field and an electric field in the region around it. As the current oscillates, these fields continually build up and collapse and set up electromagnetic waves that spread outward from the antenna. These radio waves are not transmitted by the motion of air particles but by changes in the magnetic and electrical conditions of space.

There are three basic ways in which electromagnetic waves may differ: (1) in strength, that is, in the intensity of the electromagnetic forces, (2) in frequency, which is the number of times they vibrate or number of complete cycles they perform in each second, and (3) in wavelength, which is the shortest distance between consecutive similar points on the wave train. These relationships are shown in Figure 11-2.

Electromagnetic radiation can also act like discrete particles (or quanta) of radiation, each quantum (bundle of energy) having a definite value of energy and momentum. The longer the wavelength, the lower the quantum energy.

When a sufficiently energetic quantum of energy enters a substance, it may act on a neutral atom or molecule with a force large enough to remove an electron from an atom, the resulting atom is said to be ionized. As the energy level of the incident quanta is decreased, ionization ceases but inelastic collisions between the incident quanta and atoms occur and give rise to the emission of characteristic nonionizing radiation. The nonionizing region of the electromagnetic spectrum is that portion in which the energy of the incident quanta is insufficient, under normal circumstances, to dislodge orbital electrons in body tissues of humans or test animals, and produce ion pairs. As the energy (and frequency) that is inelastically imparted to an atom decreases, the wavelength of the emitted radiation grows longer.

The sun's energy that reaches the earth is transmitted by electromagnetic waves. If a narrow beam of sunlight is allowed to pass through a prism and then project upon a surface, a pattern of very vivid and intense colors will be observed; called a "spectrum," this visible light segment ranges from red at one

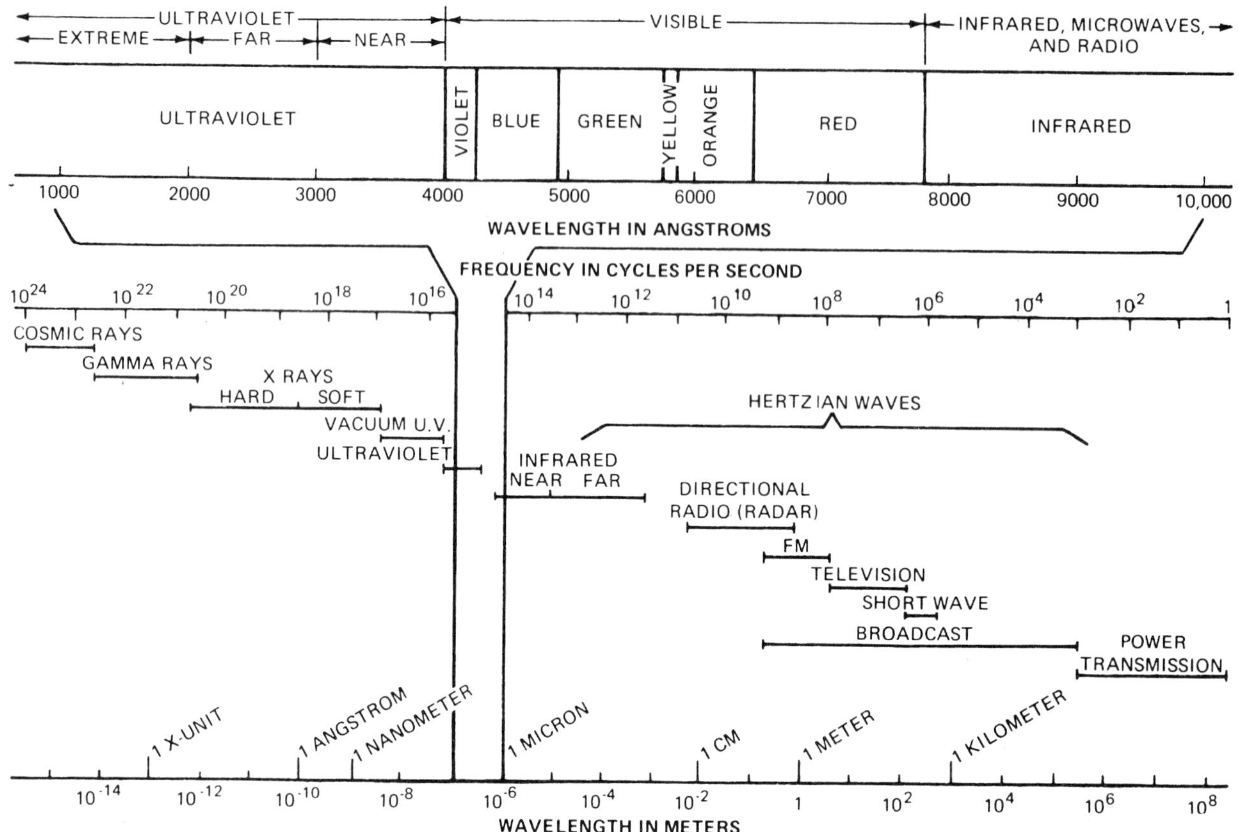

Figure 11-1. The electromagnetic spectrum, encompassing the ionizing radiations and the nonionizing radiations (expanded portion and right). Top portion expands spectrum between 10^{-7} and 10^{-8} m. Note: cycles per second (cps) = hertz (Hz).

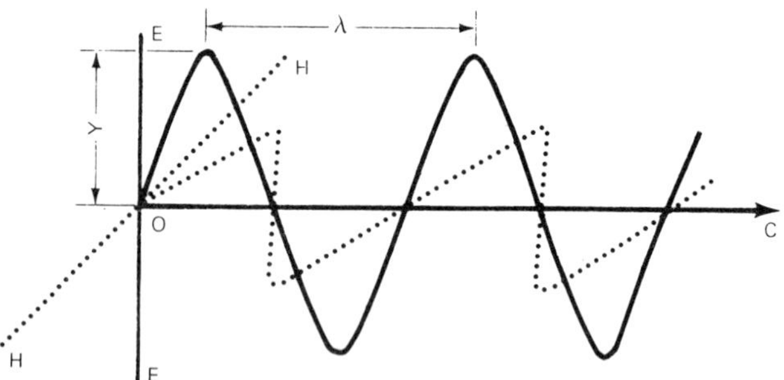

Figure 11-2. An electromagnetic wave: the electric field (*E*) oscillates in the plane of the page, at right angles to the magnetic field (*H*), which oscillates in a plane perpendicular to the page. Both are at right angles to the velocity (*C*). Field strength is represented as amplitude of the wave (*Y*); frequency is characterized by the number of complete cycles per second; and wavelength is represented by λ.

end through orange, yellow, green, blue, and indigo to violet at the other extreme. If the bulb of a thermometer is placed in the violet portion, and then moved slowly to the red portion, it will show a rise in temperature. If the thermometer is placed in the dark space immediately beyond the red portion of the spectrum, it will show a still higher temperature. This portion of the electromagnetic spectrum is known as the *infrared* (IR); the region at the other end of the spectrum (in the dark portion beyond the violet) is called the *ultraviolet* (UV).

There is no sharp subdividing line between the IR, visible, and UV regions—they all manifest electromagnetic radiation, differing from each other only in frequency, wavelength, or energy level. It is convenient, however, to separate these regions into distinct groups because of the nature of the physical and biological effects that are produced. The wavelength limits of these regions are defined in Figure 11-1. Ordinary electric current found in homes and offices is 60 hertz (Hz) (or cycles per second) and has a wavelength between 1 and 2 million meters.

All electromagnetic radiation, though it may vary widely in wavelength and frequency, has a common origin—in moving electric charges. These may arise in many different ways involving different atomic or molecular actions.

The longer electric, magnetic, and radio waves can be produced by oscillating electric circuits. Most IR radiations are obtained as radiation from hot bodies and are commonly known as heat waves. Infrared waves are emitted from the rotations and vibrations of the atoms making up the hot body. Visible light is emitted as the temperature of the hot body is raised; some visible light is also produced by electron transitions. Visible and UV light also results when an electric current is passed through a gas. The UV frequencies are due to electronic excitations of atoms and molecules. As the energy of excitation increases there is an overlap into the ionizing portion of the electromagnetic spectrum. High-speed electrons impinging upon heavy-metal targets can produce x-rays. As the energy of these high-speed electrons is increased, the radiation frequencies increase and overlap into the gamma-ray region.

Physical units of measurement

It is necessary to describe electromagnetic radiation both qualitatively and quantitatively.

Qualitative description. Electromagnetic waves can be described qualitatively in terms of wavelength or the frequency of vibration. The frequency and wavelength of electromagnetic radiation are related by the equation:

$$c = f\lambda$$

where

c = the velocity of light (3×10^{10} cm per sec) in free space

f = the frequency of vibration, in vibrations or cycles per second

λ = the wavelength, in centimeters.

The entire electromagnetic spectrum is roughly divisible into two broad regions.

- The upper region (shorter wavelength) is of particular concern to the physicists and physical scientists, who describe radiation in terms of wavelength (angstroms [A], centimeters [cm], microns [μ], millimeters [mm], and nanometers [nm]).
- The lower region (longer wavelengths) has been explored by the communications scientists and engineers, who prefer to describe electromagnetic radiation in terms of frequency (hertz [Hz], megahertz [MHz], cycles [c], kilocycles [kc], megacycles [Mc], and gigacycles [Gc]). The dimensions of these terms are shown in Table 11-A.

Table 11-A. Physical Units

Unit	*Symbol*	*Equivalent*
	WAVELENGTH	
angstrom	Å	10^{-8} cm
centimeter	cm	1 cm
micrometer	μm	10^{-4} cm
nanometer	nm	10^{-7} cm
	FREQUENCY	
hertz	Hz	1 cps
kilocycle	kc	1,000 cps
megacycle	Mc	10^6 cps
gigacycle	Gc	10^9 cps

Quantitative description. The total amount or quantity of radiation may be described in physical or biological units. The watt-second, the calorie, and the erg are physical units used to measure the total quantity of energy or work. The international standard unit is now the joule (see Appendix D). The watt, the calorie per second, or the erg per second are the physical units used to measure the time rate at which radiation is emitted. The intensity or energy density is expressed in terms of the energy incident upon a unit area or absorbed in a unit volume (watts per square centimeter or watts per cubic centimeter).

A radiating body (such as the sun or a pool of molten metal) emits light and heat radiation having a mixture of wavelengths—these can be represented by a curve of the type shown in Figure 11-3. The relative intensity varies with the wavelength.

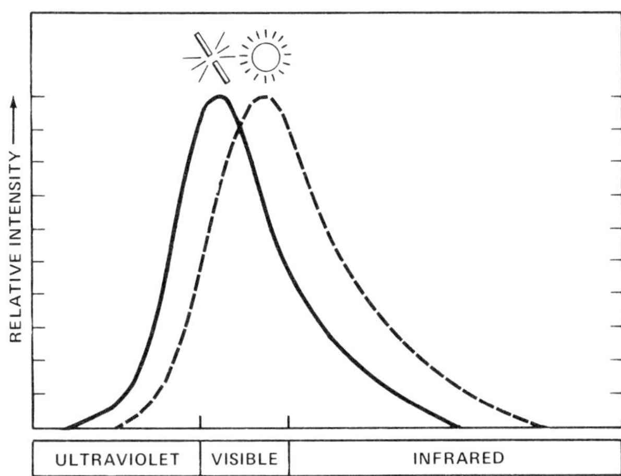

Figure 11-3. Energy distribution in the radiation from a welding arc (solid line) and the sun (broken line).

The information in Figure 11-3 can be characterized by two numbers—a wavelength corresponding to the peak of the energy curve, and a brightness figure corresponding to the energy radiated per square centimeter of hot surface. The height of the curve at any point represents the relative intensity of radiation at that wavelength. The total area under the curve represents the intensity being emitted. The intensity of electromagnetic radiation varies inversely as the square of the distance from the source. This discussion assumes that the distance is large and that there is no absorption by the intervening medium.

Biological units of response

The whole range of electromagnetic radiation may be classified according to the biological effects produced upon exposure. The UV, visible, and IR radiations have long been recognized as significant to health. More recently, radiations in the radio frequency band and even extremely low frequencies of power transmission systems have also been shown to cause effects in biological systems. Visible light is the most readily perceived. Any dark-colored glass can reduce the intensity of the visible light to a comfortable level. However, it can have little effect on the UV and IR radiations, which the eyes do not perceive. These radiations, as well as microwave and lower frequencies, fail to produce the sensation of vision, but have other very active biological properties.

Studies have shown that the human eye is not equally sensitive to all wavelengths of light. Figure 11-4 shows the relative sensation of brightness produced upon the average eye by equal amounts of energy throughout the visible spectrum. A given amount of energy of 555-nm wavelength falling upon the retina produces a greater sensation of brightness than an equal amount of energy at longer or shorter wavelengths.

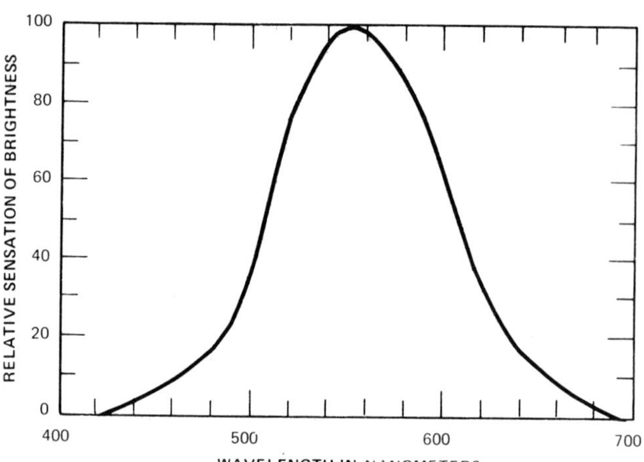

Figure 11-4. Relative sensation of brightness curve for the average human eye.

Another very important difference for biological systems between the regions of the electromagnetic spectrum is the depth of penetration of the incident radiation. Values for the depth of penetration into skin of significant amounts of different kinds of incident radiation are shown in Table 11-B. Electromagnetic radiation incident on the human body will be partially absorbed by the skin; however, some microwave radiation passes into the subcutaneous fat and into the underlying muscular tissues. The strong differences between the electrical properties of subcutaneous fat, skin, and muscle tissues are responsible for fairly complicated reflection phenomena which are generated at the tissue interfaces.

Table 11-B. Depth of Penetration of Nonionizing Radiation Into the Human Skin

Spectral Region	*Range of Penetration (mm)*
Far UV (1,800–2,900 Å)	0.01–0.1
Near UV (2,900–3,900 Å)	0.1–1
Visible spectrum (3,900–7,600 Å)	1.0–10.0
Near IR (7,600–15,000 Å)	10.0–1.0
Far IR (15,000–150,000 Å)	1.0–0.05

Microwaves—at low frequencies, most of the body-absorbed radiant energy is converted into heat in the more deeply placed body tissues. At frequencies in excess of 3,000 MHz most of the energy received by the body converts into heat in the skin. Between about 900 and 3,000 MHz, the relative heat load on skin, subcutaneous fat, and muscle tissue can vary considerably. (180 nm = 1,800 Å)

(Reprinted with permission from W. W. Coblentz, *JAMA* 123:378 [1946])

Thermal effects are produced in the skin upon exposure to energy in the IR and FM-TV-radio region. Photochemical effects may occur upon absorption of energy in the UV and visible regions of the electromagnetic spectrum. It is sometimes convenient to evaluate radiant energy in terms of the effect produced rather than in terms of the radiant energy involved. These effects may or may not be reversible, depending on many factors, such as the surface or organ irradiated, the intensity or energy density of the incident radiation, frequency, duration of exposure, and environmental temperature and humidity.

Hazard Control Act of 1968. Hazards from electromagnetic radiations attracted increasing government concern for the protection of the public, ultimately leading to the passage of the Radiation Control Act of 1968. Subsequently, regulations were promulgated by the U.S. Department of Health, Education and Welfare's (HEW) (now the Department of Health and Human Services—DHHS) Bureau of Radiation Health (BRH).

The "Radiation Control for Health and Safety Act of 1968" amends the Public Health Service Act to include control of "electronic product radiation." This includes all radiation—whether it be ionizing, nonionizing, electromagnetic, particulate, sonic, ultrasonic—resulting from the operation of an "electronic circuit." Among other things, the bill:

1. authorizes the U.S. DHHS to set performance standards to control radiation from electronic products.
2. directs DHHS to consult with industry, various associations, government agencies, and labor in prescribing standards, and to consult with a technical committee prior to prescribing standards.
3. authorizes limited in-plant inspection by DHHS based on findings as to safety.
4. requires DHHS to review and evaluate industry testing programs.
5. requires manufacturers to affix a certification of conformance to applicable standards pursuant to testing programs not disapproved by DHHS.
6. requires a manufacturer to repair a defect which relates to safe use, replace the product or refund the money, subject to DHHS regulations.
7. requires that manufacturers provide notice to dealers, distributors, first purchasers, and subsequent transferees holding warrantees against defects relating to safety or of noncompliance with standards.
8. provides for civil penalties and injunction relief.

ULTRAVIOLET RADIATION

Ultraviolet radiation occupies the region of the electromagnetic spectrum between visible light and x-rays (Figure 11-1). We know that UV radiation can initiate photochemical reactions in the skin. This process occurs when vitamin D_3 is produced. (Vitamin D_3 helps prevent rickets.) The full impact of UV radiation on human health is difficult to quantify—there have even been reports that exposure to artificially produced UV radiation has decreased the incidence of infectious diseases (Ronge). Alternatively, large doses of UV radiation have acute destructive effects on the skin and eyes. The UV spectrum has been subdivided into three regions:

1. Near: 400–300 nm
2. Far: 300–200 nm
3. Vacuum: 200–4 nm (Figure 11-5).

Biological effects resulting from exposure to UV radiation can also be used to classify various portions of the UV spectrum. The region of the electromagnetic spectrum between 400 and 300 nm, the so-called "black light" region, is called *UV-A*. In this range, fluorescence can be induced in many substances. The UV-A exposure is also the range that is responsible for pigmentation of the skin, or suntan.

The region between 320 and 280 nm is called *UV-B*, the erythemal region. Most of the biologically active and potentially harmful UV from natural sources falls within this range.

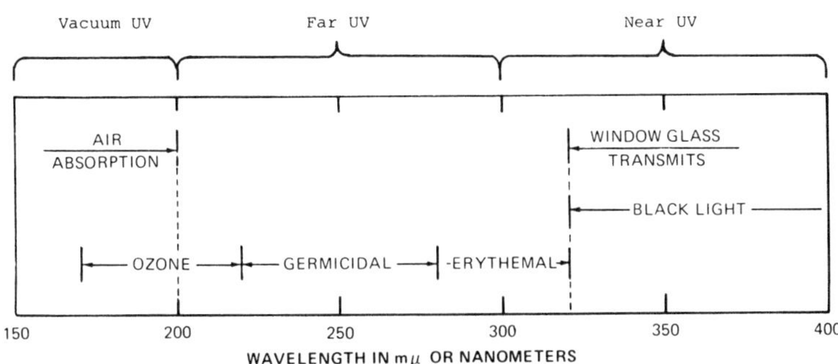

Figure 11-5. The ultraviolet (UV) spectrum.

The UV radiation in this region of the spectrum is absorbed by the cornea of the eye, the first tissue it meets, and there is no immediate effect. Later, after several hours, discomfort and the familiar sensation of sand in the eye develops. Inflammation of the cornea with tiny lesions is termed keratitis. Most UV radiation is absorbed by the cornea, but UV cataracts have been described. The human retina contains rods capable of responding to UV, but the cornea, lens, and vitreous humor are essentially opaque to these wavelengths. See Chapter 5, The Eyes, for more information.

The UV radiation can also produce sunburn of the skin, which passes into photopigmentation or suntan. Severe overexposure can cause very painful reddening of the skin and fluid-containing blisters. Photosensitizing agents have action spectra that are frequently in the UV range. Many plants (such as figs, limes, and parsnips) contain photosensitizing chemicals. Persons exposed to these agents become abnormally sensitive to skin exposure to UV radiation.

The region between 280 and 220 nm is called *UV-C* region, and is noted for its bactericidal or germicidal effect. The UV-C region occurs in the light of germicidal lamps and some welding arcs, but not in sunlight at the surface of the earth.

The region between 220 and 170 nm is the most efficient wavelength range for the production of ozone. The UV radiation in this wavelength range is strongly absorbed by air. The UV-C radiation must be studied in a vacuum or in a gas that does not absorb it.

By far the most common exposure to UV radiation is from direct sunlight. Solar irradiation exhibits a broad spectrum of UV radiation, both in intensity and wavelength. Fortunately, the surface of the earth is shielded by the atmosphere; otherwise, solar UV radiation would probably be lethal to most living organisms on the earth. Many factors affect the actual dose of UV received by humans, including: altitude above sea level, geographical latitude, time of year, presence or absence of dust and pollution. Practically no UV radiation from the sun with wavelengths below 290 nm reaches the surface of the earth, primarily due to atmospheric filtering.

The risk of developing skin cancer is directly related to the intensity and duration of exposure to sunlight (see Chapter 3, The Skin, and Chapter 8, Industrial Dermatoses). Even a short exposure in the late afternoon when the sun is low can produce a severe sunburn. The UV radiation from the sun also increases the cutaneous effects of some industrial irritants and photosensitizing chemicals. After exposure to photosensitizing compounds, such as coal tar or creosols, the skin is exceptionally sensitive to the UV rays of the sun. Some studies show increases in the rate of skin cancer in these cases. There are also compounds that minimize the effect of UV rays. Some of these are used in certain protective creams or sunblocking lotions.

The most important commercial application of UV radiation is in the production of visible light from fluorescent lamps. A fluorescent lamp bulb consists of a phosphor-coated glass tube that contains a small amount of mercury vapor. An electrical discharge through the mercury vapor generates UV radiation, which is absorbed by the phosphor coating on the inside of the tube. The phosphor fluoresces, converting the shorter wavelengths into longer wavelength (visible) energy. Incandescent and fluorescent lamps, used for general lighting, emit little or no UV radiation and are generally considered harmless. Although an ordinary fluorescent lamp generates a good deal of UV radiation inside the bulb, it is essentially absorbed in the glass bulb and its fluorescent coating.

Electric welding arcs and germicidal lamps are the most common high-level sources of UV radiation in industry. Industrial applications of "black light" include blueprinting, laundry mark identification, and instrument panel illumination. It is also used in advertising, entertainment, crime detection, photoengraving, and in sterilizing air, water, and food. Ultraviolet radiation has also been used in prevention and cure of rickets, killing of bacteria and molds, and in other therapeutic applications.

Among the several processes involving exposure to UV, electric arc-welding probably affects more workers indoors than any of the other processes. The development by workers of corneal and conjunctival irritation, or "flash" burn, also known as welder's flash, is caused by UV. The severity of the flash burn is dependent on several factors, (1) duration of exposure, (2) the various wavelengths of UV produced in various welding arcs, and (3) the energy level of the luminance and radiance during welding. All of this is made more complex by the several kinds of welds that are produced on differing base metals and in which differing filler rods are arced onto or into the base metal, along with the several sizes of welding rods used with the needed high or low amperage in the arc. Varying the distance of the rod from the base metal adds to the variations in the amounts of UV from the arc. These separate factors working together determine how far an unprotected worker must stay from the arc to avoid eye burn and the nature of the eye protection needed to prevent burning the welder's eyes as the work is performed.

In their special study of nonionizing radiation protection, W. J. Marshall et al (1977) describe several factors relating to measurements made under many different welding conditions. These include (1) base metal used, (2) welding rod diameters, and (3) the current levels used. Typical results using mild steel as a base metal; 1/16 in., 3/32 in., and 1/8 in. welding rods; and current levels of 50 amperes (amp), 100 amp, 150 amp, 200 amp, and

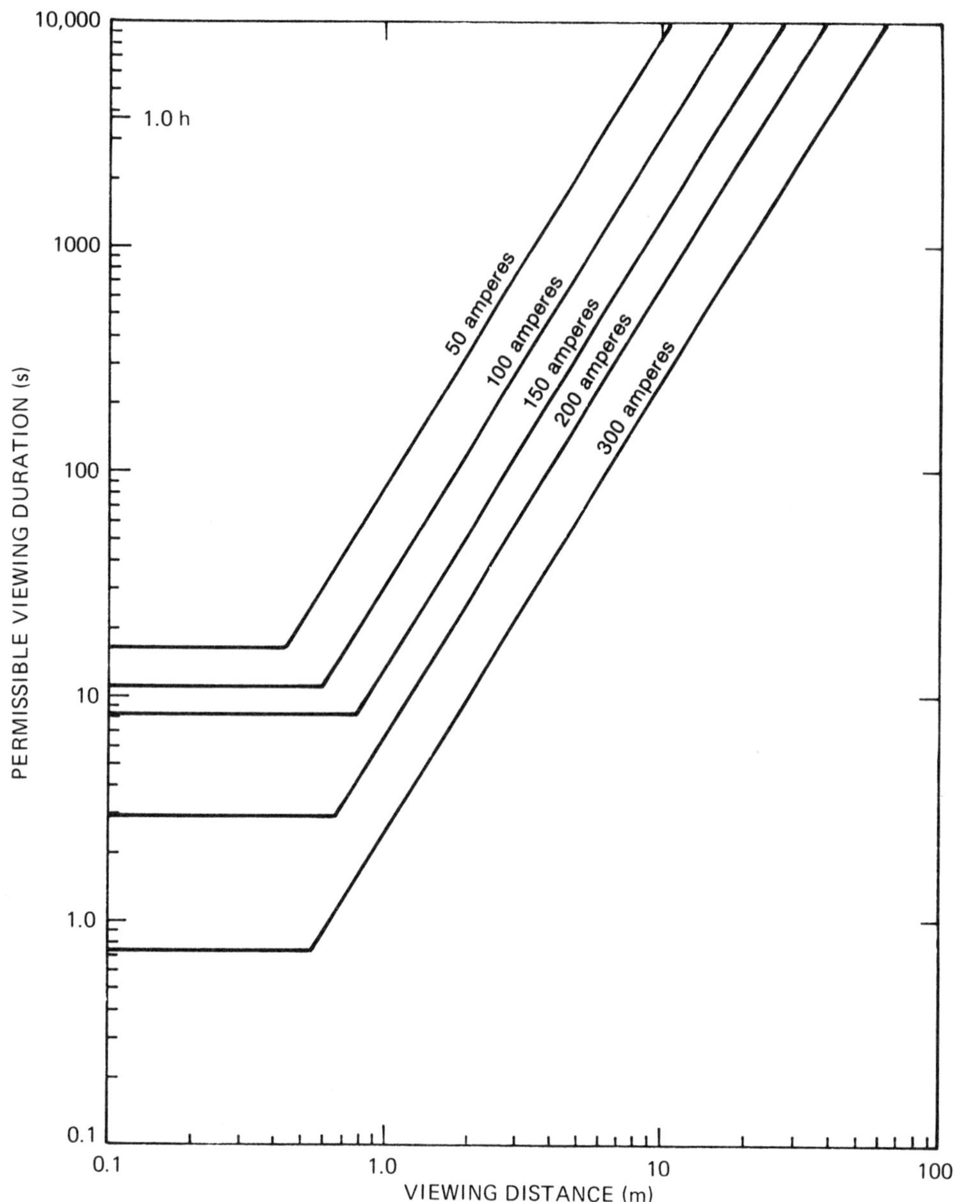

Figure 11-6. Recommended maximum exposure duration (in seconds) as a function of viewing distance (in meters). (Reprinted from *Nonionizing Radiation Protection,* U.S. Army Environmental Hygiene Agency)

250 amp show clearly that the UV radiating from the arc and blue light from the arc increased as the amperage was increased. Also, as the amperage increased the safe exposure time for workers (unprotected) *decreased* and the safe viewing distance *increased* (Figure 11-6). The report indicates that luminance and radiance of an arc increases in a relation proportional to the square of the amperage.

Variations of measured irradiance found by varying the type and size of electrodes (welding rods) are shown in Table 11-C. Again, increasing the amperage level increases the UV (irradiance) but equal amperage levels with different welding rods may result in lower irradiance levels.

Table 11-C. Effect of Varying Type of Electrode and Amperage on Measured Irradiance from Electrode Welding

Electrodes	*Amperage (Amp)*	*Irradiance (W/cm^2)*
No. 6011 Hobart, 0.625 cm	380	2.11×10^{-4}
	350	1.76×10^{-4}
No. 6011 Hobart, 0.78 cm	550	6.4×10^{-4}
	500	6.0×10^{-4}
No. 7014 Hobart, 0.78 cm	550	4.3×10^{-4}
	500	2.1×10^{-4}
No. 7024 Hobart, 0.625 cm	400	3.43×10^{-5}
	350	2.62×10^{-5}
No. 7027 A1 0.55 cm	340	2.67×10^{-4}
No. 7018 A1 0.39 cm	225	9.97×10^{-5}

(Reprinted with permission from Emmett, E. A., and Horstman, S. W., *Journal of Occupational Medicine* 18:41, 1976)

Transmission of ultraviolet radiation

Opacity of glass in relation to UV radiation has no relation to opacity in the visible part of the spectrum. Ordinary window glass, for instance, is almost completely opaque to UV in ordinary sunlight, although transparent to the visible. Such commercially available home window glass about 0.1 in. thick is

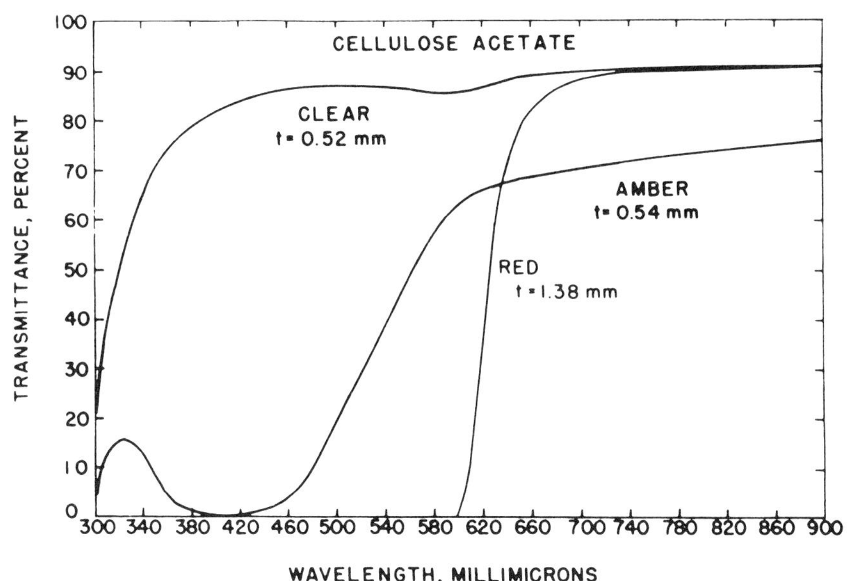

Figure 11-7. Spectral transmittances for representative samples of cellulose acetate in the UV, visible, and near-IR regions of the spectrum. (Reprinted with permission from ANSI, *The Spectral-Transmissive Properties of Plastics for Use in Eye Protection*, 1955)

practically opaque to UV radiation of wavelengths shorter than 300 nm (or 3,000 Å), and will provide sufficient protection for the eyes and skin against UV radiation from ordinary UV sources. If extremely intense sources of UV radiation and high intensities of visible and IR radiation are present, common ordinary glass may not offer adequate protection. The low transmission of UV radiation through window glass is usually attributed to the presence of traces of iron oxide. Other types of glass, developed especially for bactericidal lamps, will transmit an appreciable quantity of UV radiation. A piece of plastic dyed a deep red-violet may be almost entirely opaque in the visible part of the spectrum and transparent in the near ultraviolet. The transmission curves for a number of organic plastic materials are given in *Spectral-Transmissive Properties of Plastics for Use in Eye Protection,* published by the American National Standards Institute (1955). Sample curves are shown in Figure 11-7.

In applying the information obtained from these curves, a distinction should be made between the transmission of brand name materials and commercial resins. Commercial resins may contain ultraviolet stabilizers or dyes introduced to improve certain physical or chemical properties. In order to specify the optical property of a plastic, it is necessary not only to designate the brand name, but also its particular chemical formulation.

Health effects

Functionally, two groups of health effects can be described, those related to acute, direct exposure and those that appear over protracted periods in which there is increased risk to health. Acute effects due to direct UV exposure vary with dose and due to the limited penetration of UV into tissue, the principal short-term effects are restricted to the skin and eyes.

Dermatological. Immediate changes in the skin occur in a sequence of events including quick darkening of cellular pigment, the occurrence of erythema (sunburn), production and migration of melanin granules (suntanning); and changes in cell growth in the epidermis. The skin pigment, melanin, is directly and immediately affected by UV irradiation. Its reaction spectrum, with a wavelength maximum of 360 nm, extends to slightly greater than 400 nm.

Erythema, also occurring soon after initiation of UV exposure, is a vascular reaction that consists of vasodilation and augmented bloodflow. The extent and severity of this reaction is highly variable, depending on the skin properties (transmission of UV through the epidermal layers) and the wavelengths of UV received (Farr and Diffey, 1985). In lightly pigmented Caucasian skin, the minimal dose that will provoke erythema is about 200 J/m^2 for wavelengths between 250 and 300 nm; between 300 and 330 nm there is a sharp increase in the required effective dose and between 330 and 400 nm the level is approximately 10^5 J/m^2 (Parrish et al, 1982).

"Suntan" occurs over a slightly longer period than erythema, however the process has a similar action spectrum. This increase in pigmentation occurs both because of increased production of pigment granules and the spread of granules more uniformly throughout epidermal cells.

The final short-term effect on the skin is a change in the growth pattern of epidermal cells. After short-term slowing of growth there is an increase that gives rise to hyperplasia in the epidermis with a concomitant shedding of cells from the skin surface.

Ocular. Although a small amount of UV may not produce permanent injury to the eyes, the only safe procedure to follow is to exclude completely all harmful levels of exposure. The elimination of harmful UV radiation must be given attention, because it can do great injury to the eyes without discomfort during exposure. The potential for strong UV sources (for example, lasers) to cause irreversible damage to ocular structures is of particular concern.

It is only after the damage has been done (some 4–6 hours later) that the effects can begin to appear in the form of conjunctival irritation. Effects on the eye have been previously mentioned and consist of photokeratitis and conjunctivitis. These clinical effects are caused primarily by UV-B and UV-C irradiation.

Late effects may also occur after prolonged exposure to UV sources, natural or synthetic. These can consist of decreased elasticity in the skin giving the appearance of premature aging, or certain types of cataracts may increase. The dose-response relationships of these phenomena are unknown. Of greater

Table 11-D. Penetration of UV Radiation into Human Epidermis, at Several Wavelengths and at Several Depths*

←Depth (μm)	Layer	254	270	280	290	297	313	365	436	546	Layer	←Depth (μm)
0		100	100	100	100	100	100	100	100	100		0
10	STRATUM CORNEUM	42	30	28	30	50	67	80	85	89	STRATUM CORNEUM	10
20	STRATUM CORNEUM	18	9.1	7.9	14	25	44	64	72	80	STRATUM CORNEUM	20
30	MALPIGHIAN LAYER	5.0	2.4	2.5	6.5	15	33	50	62	72	MALPIGHIAN LAYER	30
40	MALPIGHIAN LAYER	1.4	0.65	0.78	3.0	9.6	24	39	54	65	MALPIGHIAN LAYER	40
50	MALPIGHIAN LAYER	.39	.17	.25	1.3	6.0	18	31	46	59	MALPIGHIAN LAYER	50
60	MALPIGHIAN LAYER	.11	.047	.077	.60	3.8	13	24	40	53	MALPIGHIAN LAYER	60
70	MALPIGHIAN LAYER	.030	.012	.024	.27	2.4	9.5	19	34	48	MALPIGHIAN LAYER	70

Wavelength (nm) →

* The data apply to collimated, perpendicular irradiation and are expressed as percent of incident irradiance (from Bruls et al, 1984).

concern is the increased risk of cancer particularly of the skin, in people exposed to UV. These effects are a function of dose, generally, over an extended period.

Threshold Limit Values for ultraviolet radiation

The biological effect upon exposure to UV radiation is dependent upon the intensity and energy distribution of the source. Ultraviolet radiation to the skin is absorbed primarily in the epidermis; shorter wavelengths are more strongly absorbed. The same is true for the eye; most of the radiation is absorbed by the cornea (Table 11-D).

Many sources of UV radiation (Figure 11-3) emit varying amounts of energy at different wavelengths. If the energy distribution curve of the source is known, the equivalent erythemally weighted intensity may be calculated. The energy value in each band or wavelength region is multiplied by the percent relative erythemal effectiveness (Figure 11-8) to obtain the equivalent intensity of 296.7 nm radiation. The sum of these equivalent intensities is called the erythemally weighted UV energy of that source. The Council on Physical Medicine of the American Medical Association issued the following criteria for safe exposure to radiant energy from germicidal lamps:

> The total intensity of ultraviolet radiation diffusely reflected from the walls and fixture and emanating directly from the lamp, incident on the occupant for 7 hours or less, should not exceed 0.5 μW/cm² and, for continuous exposure (24 hours a day), should not exceed 0.1 μW/cm² of wavelength 2537A (253.7 nm).

Germicidal lamps emit almost monochromatic radiation in the 253.7-nm region which is only 50 percent as effective as the 296.7-nm band in producing erythema. With the recognition that many parameters affect threshold values, I. Matelsky, in his chapter in *Industrial Hygiene Highlights* (Cralley, 1968) suggests that the following threshold doses, weighted on the basis of their action spectra, be considered:

1. Minimum erythemal dose for previously nonexposed skin—2×10^4 to 2.5×10^4 μW sec/cm² of erythemally weighted UV.
2. Minimum erythemal dose for previously exposed skin—2.5×10^4 to 3.5×10^4 W sec/cm² of keratinically weighted UV.
3. Minimum keratitic dose—1.5×10^3 W sec/cm² of keratinically weighted UV.

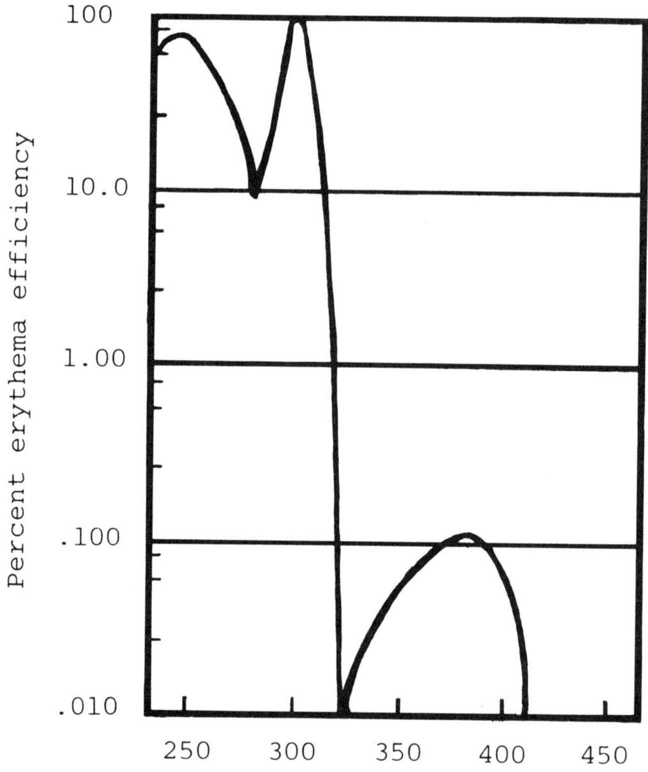

Figure 11-8. Erythema action spectrum for human skin. Note logarithmic vertical axis. (Adapted with permission from Hausser, K. W., and Vahle, W. "Sunburn and Suntanning." In *The Biological Effects of Ultraviolet Radiation with Emphasis on the Skin.* Edited by F. Urbach. Oxford: Pergamon Press, 1969, pp. 3-21)

These values are based on the standard erythemal curve, and may be very conservative if the more recently suggested effectiveness data are applied.

The 1986-1987 TLV for exposure to UV light is divided into two parts: (1) for unprotected workers to UV in the range of 320–400 nm, the total irradiance on the skin or eyes should not exceed 1 mW/cm² for periods greater than 10^3 seconds (approximately 16 minutes) and for exposure times less than 10^3 seconds, should not exceed 1 J/cm²; (2) for shorter wavelengths (200–315 nm) TLV formulae for broad band sources and tables in the *Threshold Limit Values and Biological Exposure Indices for 1986-1987* (Appendix B-1) should be used.

If the UV energy is from a narrow band or monochromatic source, TLVs for a daily 8-hour period can be read directly from the curve in Figure 11-9. This curve was adapted from the 1972 TLV as shown in the NIOSH document, *Criteria for a Recommended Standard: Occupational Exposure to Ultraviolet Radiation* (PB-21-4268). This document is an excellent source of additional information about the injurious effects of UV and for work practices for controlling exposures to UV, including protective measures to be used by exposed employees. Additional potential hazards from high voltage, ozone, oxides of nitrogen, initiation of explosions and the hazard from UV related to converting chlorinated hydrocarbon vapors to phosgene are also discussed in the NIOSH UV document.

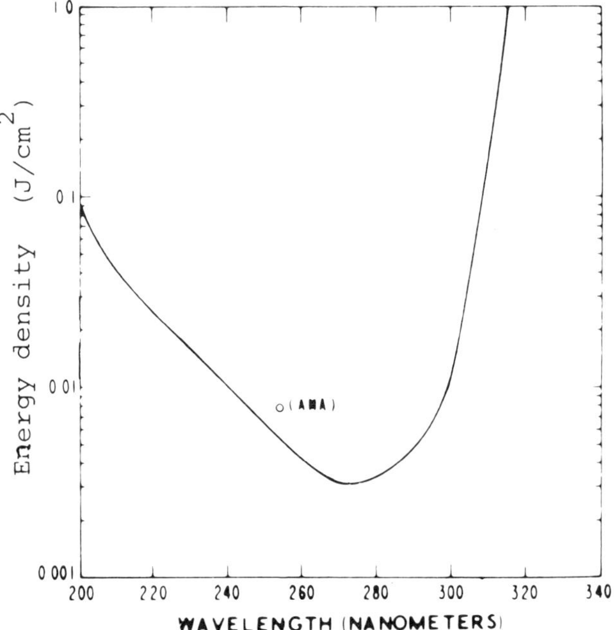

Figure 11-9. Threshold Limit Values for UV radiation. (Reprinted with permission from the ACGIH. *Threshold Limit Values and Biological Exposure Indices for 1986-1987.* Cincinnati: ACGIH, 1986—also see Appendix B-1)

VISIBLE ENERGY

The visible energy portion of the electromagnetic spectrum occupies the region between 400 nm and 750 nm, Figure 11-1. Exposure of the human eye to high brightness levels evokes a number of physiological responses—adaptation, pupillary reflex, partial or full lid closure, and shading of the eyes are protective mechanisms to prevent excessive brightness from being projected into the retina.

The subjective feeling of visual comfort is perhaps the most important criterion to be used in setting safe exposure levels to noncoherent, polychromatic visible light. The danger of retinal injury peaks in the blue light range, 425–450 nm (see the TLV in Appendix B-1).

Intense visible radiation is emitted from the sun, artificial light sources, arc-welding processes, and highly incandescent bodies. The introduction of highly collimated, monochromatic, coherent radiation produced by lasers can cause retinal burns. (See Lasers, later in this chapter.)

Units of measurement of light

The intensity of visible radiation is measured in units of candles. The rate of flow of light, referred to as luminous flux, is measured in lumens. One lumen is the flux on one square foot of a sphere, one foot in radius with a light source of one candle at the center, and radiating uniformly in all directions. Footcandles refer to a unit of illumination, which is the direct measure of the visible radiation falling on a surface. Footlamberts represent the unit measure of the physical brightness of any surface emitting or reflecting visible radiation. For example, if a light having 100 footcandles is incident on a 100 percent reflecting white surface, the physical brightness of the surface would be 100 footlamberts.

Industrial lighting

Well-balanced levels of illumination are essential in establishing safe working conditions. Industrial lighting involves a wide variety of seeing tasks, operating conditions, and economic considerations. Visual tasks may involve objects being extremely small or very large, dark or light, opaque, transparent, or translucent; and may be handled on glossy or rough surfaces.

Some less tangible factors associated with poor illumination are important contributing causes of industrial accidents. These can include direct glare, reflected glare from the work, and dark shadows which may lead to excessive visual fatigue. Visual fatigue, itself, may be a causative factor in industrial accidents. Accidents may also be caused by delayed eye adaptation when coming from bright surroundings into darker ones. Some accidents which have been attributed to an individual's carelessness can, in fact, be traced to difficulty in seeing, involving one or more of these contributing causes.

The purpose of industrial lighting is to help provide a safe working environment, to provide efficient and comfortable seeing, and to reduce losses in visual performance. It is important, therefore, to analyze the several factors which contribute to seeing; that is, the task, the environment, and the lighting.

Recommended lighting levels

The following material is adapted from the ANSI/IES standard RP7-1983, *Practice for Industrial Lighting.*

In general, one sees by reflection, transmission, and silhouette. *Silhouette* seeing involves the detection of the presence of an object and its contour because its darker outline is revealed by a contrast against lighted surroundings. *Transmission* concerns the revealing of details through the variation and transmission of white light, or the changing of color, through materials that are susceptible to penetration. By far, the most common method of seeing is by *reflected light* where light and dark areas or details are revealed by difference in reflection. Highlights and shadows, actually the degree of reflected light, portray three-dimensional configuration or contour. The three types of light reflection are shown in Figure 11-10.

The visibility of a task or object is determined by its size, contrast, time of viewing, and brightness. Each of these factors is sufficiently dependent upon the others that a deficiency in one

THREE TYPES OF LIGHT REFLECTION

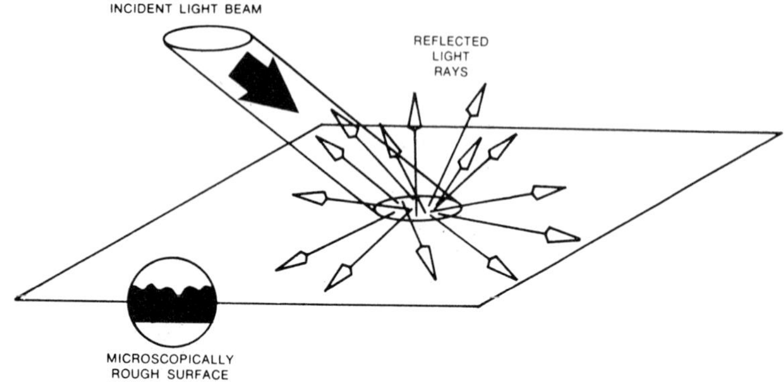

DIFFUSE REFLECTION

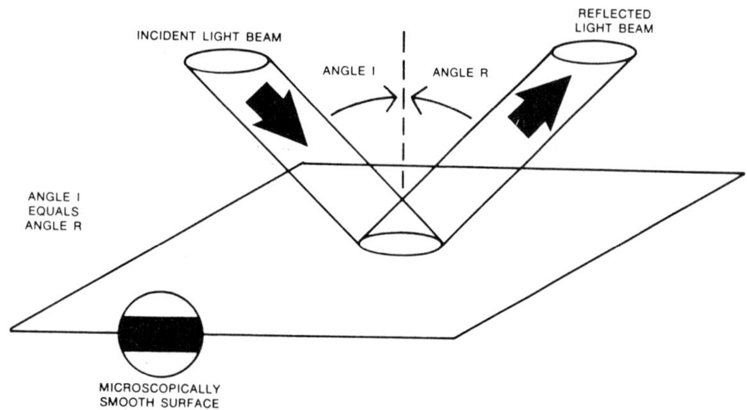

MIRROR REFLECTION

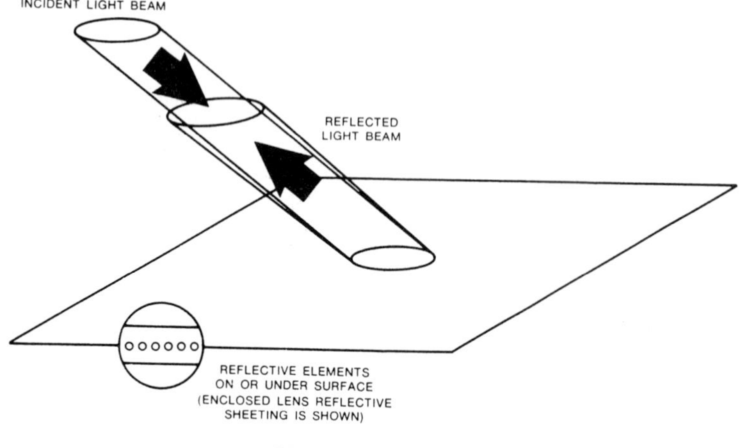

RETRO REFLECTION

Figure 11-10. Three types of light reflection. (Courtesy 3-M Co.)

may be compensated within limits by augmenting one or more of the others. As size increases, visibility increases, and up to a certain point seeing becomes easier. If an object is small, in order to see it, a person should use more light, hold the object closer to his/her eyes, or even use a magnifier.

To be readily visible, each detail of an object must differ in brightness or color from the surrounding background. If discrimination is dependent solely on brightness differences, visibility is at a maximum when the contrast of details with the background is the greatest. Therefore, within the limits available to industry, the contrast of seeing tasks (that is, between the object of regard and its immediate background) should be made as high as possible. Where it is impractical to provide good contrast conditions, high levels of illumination help compensate for the poor contrast.

If low brightness prevails, it will take a relatively longer time to accomplish the seeing task. By increasing the brightness, the time required for seeing will be shortened. The factors of size and contrast are inherent in the task itself and, within limits, time of viewing may be considered in this same category. Therefore, in general, brightness is important as the one controllable factor. Brightness resulting from the light on the task and its surroundings in the visual field may be controlled within wide limits by varying the amount and distribution of light.

Factors of good illumination

Reduced to simplest terms, the factors to be considered can be summed up under the headings of *quantity,* the amount of illumination that produces brightness on the task and surroundings, and *quality,* which pertains to the distribution of brightness in a visual environment and includes the color of light, its direction, diffusion, degree of glare, and the like.

Quantity. The desirable quantity of light for any particular installation depends primarily upon the work that is to be done. The degree of accuracy required, the fineness of detail to be observed, the color and the reflectance of the work, as well as the immediate surroundings, materially affect the brightness requirements that will produce optimum seeing conditions.

The lighting recommendations listed in Table 11-E are based on the characteristics of the visual tasks and the visual performance requirements of young adults with normal eyes. Values are given in accordance with the difficulty of the various seeing tasks. The highest illumination levels are listed for tasks involving discrimation of fine detail, low contrast, and prolonged work periods, such as detailed assembly and fine layout and bench work. The more casual or intermittent tasks having high contrast are assigned lower values. Where filter-type safety goggles are worn that materially reduce the light reaching the eye, the level of illumination should be increased in accordance with the absorption of the goggles.

Quality of illumination pertains to the distribution of brightness in the visual environment. The term is used in a positive sense and implies that all brightnesses contribute favorably to visual performance, visual comfort, ease of seeing, safety, and esthetics for the specific visual task involved. Glare, diffusion, direction, uniformity, color, brightness, and brightness ratio (Table 11-F) all have a significant effect on visibility and the ability to see easily, accurately, and quickly. Certain seeing tasks, such as discernment of fine details, require much more careful analysis and higher quality illumination than others. Areas where the seeing tasks are severe and performed over long periods of time require much higher quality than where seeing tasks are casual or of relatively short duration. Glare can be defined as any brightness within the field of vision where such characteristic would cause discomfort, annoyance, interference with vision, or eye fatigue.

An electromagnetic radiation survey was conducted by Moss, et al (1978) on several models and types of video display terminals (VDTs), typically used in the communication field. The values obtained for x-ray, UV, visible, IR, and radiofrequency levels were considerably below accepted Threshold Limit Values or standards indicating that currently VDTs of the types surveyed do not present an occupational ocular radiation hazard.

The OSHA has found little data that relate the lowest permissible limits of illumination to accident rates. Hence, OSHA standards for illumination are somewhat sparse.

The ACGIH TLVs for light and near-IR radiation are recommended for light sources that exceed 1 candela (cd)/cm^2. As noted earlier, blue light requires special emphasis, particularly in the range of 425–450 nm.

Densely colored filters may be needed for protecting the eyes of exposed workers. This is especially true in the case of arc welders in relation to the amperage levels of the arc being used. This is emphasized in Figure 11-11.

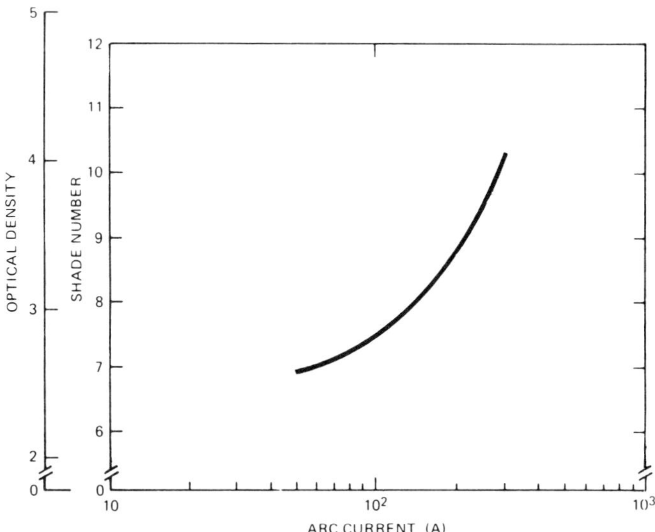

Figure 11-11. Shade number or optical density necessary for eye protection against blue light as a function of arc current (in amperes). (Reprinted from *Nonionizing Radiation Protection,* U.S. Army Environmental Hygiene Agency)

INFRARED RADIATION

It is generally considered that the IR region of the electromagnetic spectrum extends from the visible red light region (750 nm) to the 0.3-cm wavelength of microwaves (Figure 11-12). Exposures to IR radiation can occur from any surface which is at a higher temperature than the receiver. Infrared radiation can be used for any heating application where the principal

Table 11-E. Illuminance Categories and Values for Indoor Various Work Activities

Type of Activity	*Illuminance Category*	*Ranges of Illuminances*		*Reference Work Plane*
		Lux	*Footcandles*	
Public space with dark surroundings	A	20–30–50	2–3–5	
Simple orientation for short temporary visits	B	50–75–100	5–7.5–10	General lighting throughout spaces
Working spaces where visual tasks are only occasionally performed	C	100–150–200	10–15–20	
Performance of visual tasks of high contrast or large size	D	200–300–500	20–30–50	
Performance of visual tasks of medium contrast or small size	E	500–750–1,000	50–75–100	
Performance of visual tasks of low contrast or very small size	F	1,000–1,500–2,000	100–150–200	Illuminance on task
Performance of visual tasks of low contrast and very small size over a prolonged period	G	2,000–3,000–5,000	200–300–500	
Performance of very prolonged and exacting visual tasks	H	5,000–7,500–10,000	500–750–1,000	
Performance of very special tasks of extremely low contast and small size	I	10,000–15,000–20,000	1,000–1,500–2,000	Illuminance on task, obtained by a combination of general and local (supplementary lighting)

Commercial, Institutional, Residential, and Public Assembly Interiors

Area/Activity	*Illuminance Category*
Conference rooms	D
Conferring	
Critical seeing (refer to individual task)	
Assembly	
Simple	D
Moderately difficult	E
Difficult	F
Very difficult	G
Exacting	H
Candy-making	
Box department	D
Chocolate department	D
Husking, winnowing, fat extraction, crushing and refining, feeding	
Bean-cleaning, sorting, dipping, packing, wrapping	D
Milling	E
Cream-making	
Mixing, cooking, molding	D
Gum drops and jellied forms	D
Hand decorating	D
Hard candy	
Mixing, cooling, molding	D
Die cutting & sorting	E
Kiss-making and wrapping	E
Foundries	
Annealing (furnaces)	D
Cleaning	D
Core making	
Fine	F
Medium	E
Grinding and chipping	F
Inspection	
Fine	G
Medium	F
Molding	
Medium	F
Large	E
Pouring	E
Sorting	E
Cupola	C
Shakeout	D
Garages—service	
Repairs	E
Active traffic areas	C
Write-up	D
Inspection	
Simple	D
Moderately difficult	E
Difficult	F
Very difficult	G
Exacting	H
Machine shops	
Rough bench or machine work	D
Medium bench or machine work, ordinary automatic machines, rough grinding, medium buffing, and polishing	E
Fine bench or machine work, fine automatic machines, medium grinding, fine buffing, and polishing	G
Extra-fine bench or machine work, grinding, fine work	H
Paint shops	
Dipping, simple spraying, firing	D
Rubbing, ordinary hand painting and finishing art, stencil, and special spraying	
Fine hand painting and finishing	E
Extra-fine hand painting and finishing	G
Service spaces (see also, Storage rooms)	
Stairways, corridors	B
Elevators, freight & passenger	B
Toilets & washrooms	C
Sheet metal works	
Miscellaneous machines, ordinary bench work	E
Presses, shears, stamps, spinning, medium bench work	E
Punches	E
Tin place inspection, galvanized	F
Scribing	F
Storage rooms or warehouses	
Inactive	B
Active	
Rough, bulky items	C
Small items	D
Welding	
Orientation	D
Precision manual arc-welding	H

(Adapted with permission from the Illuminating Engineering Society of North America. Kaufman, J. E., ed. *IES Lighting Handbook: 1981 Application Volume.* Baltimore: Waverly Press, 1981)

Table 11-F. Recommended Maximum Brightness Ratios

	Environmental Classification		
	A	B	C
1. Between tasks and adjacent darker surroundings	3–1	3–1	5–1
2. Between tasks and adjacent lighter surroundings	1–3	1–3	1–5
3. Between tasks and more remote darker surfaces	10–1	20–1	*
4. Between tasks and more remote lighter surfaces	1–10	1–20	*
5. Between luminaires (or windows, skylights, etc.) and surfaces adjacent to them	20–1	*	*
6. Anywhere within normal field of view	40–1	*	*

*Brightness ratio control not practical.

A—Interior areas where reflectances of entire space can be controlled in line with recommendations for optimum seeing conditions.
B—Areas where reflectances of immediate work area can be controlled, but control of remote surroundings is limited.
C—Areas (indoors and outdoor) where it is completely impractical to control reflectances and difficult to alter environmental conditions.

product surfaces can be arranged for exposure to the heat sources. Transfer of energy or heat occurs whenever radiant energy emitted by one body is absorbed by another. The electromagnetic spectrum wavelengths longer than those of visible energy (750 nm) and shorter than those of radar waves are used for radiant heating. The best energy absorption of white, pastel-colored, and translucent products is obtained by using wavelength emissions longer than 2,500 nm. The majority of dark-pigmented and oxide-coated materials will readily absorb wavelength emissions from 750–9,000 nm.

Water vapor and visible aerosols, such as steam, readily absorb the longer IR wavelengths.

Typical industrial applications for IR include: (1) drying and baking of paints, varnishes, enamels, adhesives, printer's ink, and other protective coatings; (2) heating of metal parts for shrink fit assembly, forging, thermal aging, brazing, radiation testing, and conditioning surfaces for application of adhesives and welding, (3) dehydrating of textiles, paper, leather, meat, vegetables, pottery ware, and sand molds; and (4) spot and localized heating for any desired objective.

Rapid rates of heating can be provided in relatively cold surroundings by controlling the intensity of radiant energy, absorption characteristics of the exposed surfaces, and rate of heat loss to the surroundings.

Infrared radiation is perceptible as a sensation of warmth on the skin. The increase in tissue temperature upon exposure to IR radiation depends upon the wavelength, the total amount of energy delivered to the tissue, and the length of exposure. The IR radiation in the far wavelength region of 5,000 nm–0.3 cm is completely absorbed in the surface layers of the skin. Exposure to IR radiation in the region between 750–1,500 nm can cause acute skin burn and increased persistent skin pigmentation.

This short wavelength region of the IR is capable of causing injuries to the cornea, iris, retina, and lens of the eye. Excessive exposure of the eyes to luminous radiation, mainly visible and IR radiation, from furnaces and similar hot bodies, has been known for many years to produce "glass blower's cataract" or "heat cataract." This condition is an opacity of the rear surface of the lens in the eye.

Safe infrared exposure levels

The general effect of IR radiation on the eyes is summed up by I. Matelsky (see Cralley, 1968) as follows:

> The available data indicate ". . . that acute ocular damage from the incandescent hot bodies found in industry can occur with energy densities between 4–8 Ws/cm² (1–2 cal/cm²) incident upon the cornea. The ocular tissues involved would depend upon the wavelengths which are absorbed. As these relate to threshold phenomena, it would appear that a maximum permissible dose of 0.4–0.8

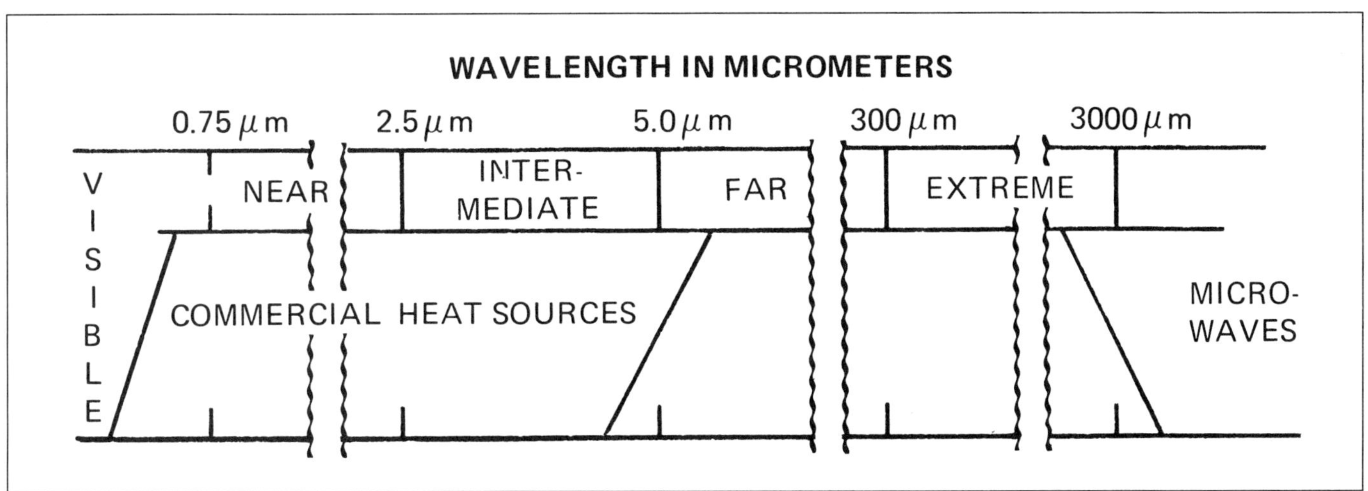

Figure 11-12. Infrared spectrum.

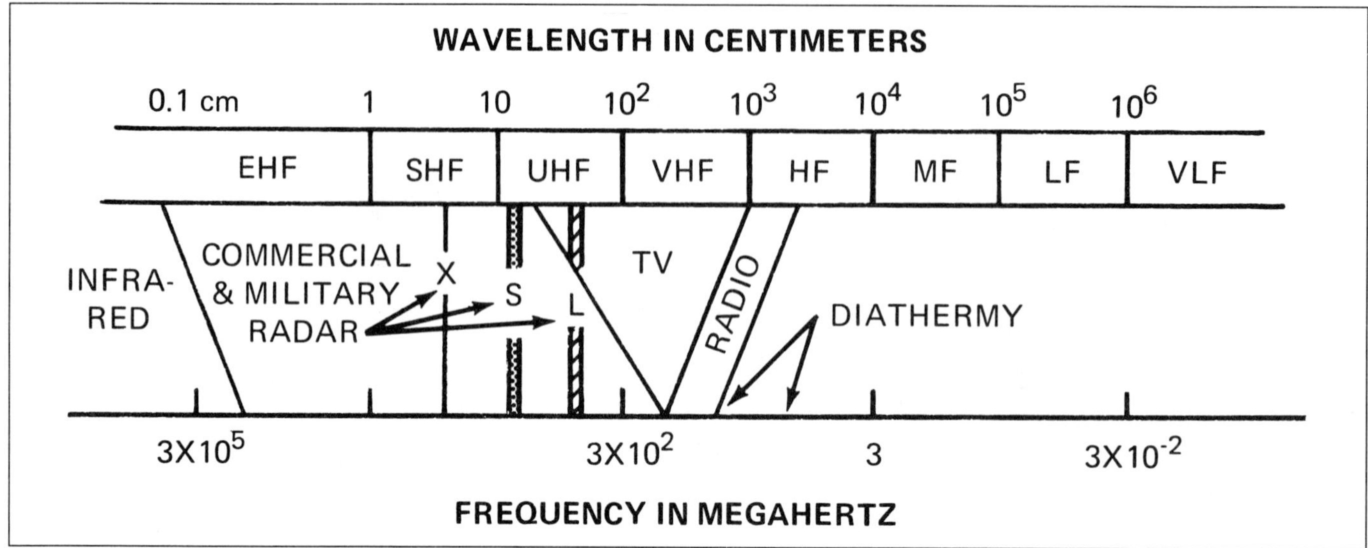

Figure 11-13. Microwave spectrum.

Ws/cm² (0.1-0.2 cal/cm²) could limit the occurrence of these acute effects. A further reduction by a factor of ten should prevent the more chronic effects on the intraocular tissues.

The ACGIH TLV for IR radiation is combined with that for visible light (see Appendix B-1). To avoid possible delayed effects upon the lens of the eye, exposure to infrared wavelengths longer than 770 nm should be limited to 10 mW/cm². Van Pelt et al (1973), in a review of biological thresholds for radiation, notes some uncertainty about the correctness of choosing 1,400 nm for separating the near from the far infrared ranges. Special protective eyewear and face shields should be considered for protecting exposed workers to these radiations.

MICROWAVES AND RADIO WAVES

Radiation produced by molecular vibration in crystals or other solid bodies is usually described by the generated wave frequency. Within the broad spectrum of radio frequencies the term microwave refers to electromagnetic radiation extending from approximately 10–300,000 MHz (Figure 11-13). This form of radiation is normally propagated into the atmosphere from antennas associated with television transmitters, FM transmitters, and radar transmitters. Microwave energy sources are also used in medical applications (diathermy devices), microwave ovens, freeze drying and wood gluing operations. For industrial, scientific and medical uses, seven I-S-M microwave frequencies have been assigned: 13.56 MHz, 27.12 MHz, 40.68 MHz, 915 MHz, 2450 MHz, 5,800 MHz, and 22,125 MHz.

Microwave radiation may be categorized into continuous wave (CW) at a single frequency, or it may be amplitude-, frequency-, or pulse-modulated. In typical radar applications, pulses of extremely short duration are emitted at a repetition frequency of a few hundred pulses per second. Peak power densities of pulsed microwaves will generally be more than 1,000 times greater than for continuous wave at the same average power.

Power intensities for microwaves are given in units of watts per square centimeter. Areas having a power intensity of over 0.01 W/cm² should be avoided. In such areas, dummy loads should be used to absorb the energy output while equipment is being operated or tested. If a dummy load cannot be used, adjacent populated areas should be protected by adequate shielding.

Interaction with biological systems

Microwave radiation can be transmitted, reflected, or absorbed upon striking a target following the same general principles as with other forms of electromagnetic radiation. Upon interaction with a biological system, radio or microwave frequency fields induce electric and magnetic fields within that system. Energy is subsequently transferred from the field to the medium with several resultant effects; attenuation of the field, heating of the medium or tissue, as well as possible nonthermal effects. One of the major thrusts in research conducted to date has been to define and evaluate the interaction of radio frequency energy with biological systems. A dose rate can be described as the specific absorption ratio (SAR), that is used to normalize the rate of radio frequency energy input into the tissue. Distribution of such energy is dependent upon many exposure factors including intensity, frequency, body size, shape and electrical properties, polarization, and the presence of reflecting surfaces. Further, the deposit of energy can be described by local SAR or lumped as whole-body SAR. However, data specifying such distributions currently are too incomplete to directly explain observed biological effects.

The primary effect upon the human body when exposed to microwave energy appears to be thermal. In general, the higher the frequency, the lower the potential health hazard. Microwave radiation greater than 3,000 MHz is readily absorbed within the skin. Frequencies less than 3,000 MHz can penetrate the outer layers of the skin and be absorbed in the underlying tissues. Serious damage can occur when underlying tissues, which have little sensation to temperature, are exposed to microwave fequencies of 3,000 MHz or less. The maximum SAR in humans occurs at about 70 MHz (30 MHz for radio frequency-grounded human).

An intolerable rise in body temperature, as well as localized damage, can result from exposure to microwaves of sufficient intensity and time.

A core temperature rise of up to 2 C/hr is estimated to occur in humans under a whole-body SAR of 1–4 W/kg at 70 MHz. Such heating of biological tissue by high level radio frequency energy could result in all of the known biological effects of heating: e.g., burns, cataracts, possible detrimental effects on reproduction, and even death. However, there continues to be unresolved discussion when evaluating biological effects, as to whether the temporal pattern of energy input is the critical factor or whether total energy input is more important.

In addition to thermal effects, other electromagnetic effects of microwaves have been investigated. One result is expansion of the recommended limits of exposure to include free-space electric and magnetic field strengths (see the microwave TLV in Appendix B-1). The exact biological effects of microwave radiation at low levels are not known. These effects often involve millimeter wavelengths and modulated or pulsed radio frequency fields. Several hypotheses have been formulated but the fundamental mechanisms of interaction are not understood and there is insufficient information to set widely accepted low level exposure limits.

Microwave heating applications

There has been considerable interest in the safe operation of microwave ovens for heating or cooking food. Microwave energy, a very convenient source of heat, has clear advantages over other heat sources in certain applications. It is clean, flexible, and reacts instantly to control. Microwave heating eliminates combustion products or convection heating from being added to the working environment. Furthermore, the great ease with which microwave energy converts to heat results in very high heating rates.

Most nonmetallic materials are translucent to microwave energy. Microwave fields penetrate for a considerable distance, as determined by the generated frequency and properties of the materials. These fields influence the motion of electrons in the material being heated as they interact with their environment at the atomic and molecular levels. The heavier atoms are not able to react as freely as the electrons with the high-frequency field. The electrons move more or less freely in the material to an extent determined by the dielectric constant and loss tangent. The motion of the electrons which relates to the dielectric constant is elastic and results essentially in stored energy. That electronic motion, which exceeds the elastic limits of the constraint imposed by the environment, results in conversion of the microwave energy to heat through hysteresis.

The frequency selected for microwave cooking ovens is usually 915 or 2,450 MHz. Currently, the choice appears to depend on the ability to produce effective generators and ovens in which energy transfer is satisfactory. In special circumstances, the lower freqency might well be favored because of greater penetration. For general-use ovens, the higher frequency has seen much greater development, but no valid reason appears to exist which would make 915 MHz equipment inadvisable.

According to E.C. Okress (1968), the complete microwave oven consists of the following eight major components:

1. the power supply that adapts line power to the generator requirement.
2. the generator or power tube that converts the power supply into microwave energy.
3. the transmission section or waveguide for energy propagation to the oven proper.
4. coupling devices that permit the transfer of energy to the load.
5. distributing devices that deliver the energy in a uniform interaction pattern.
6. the cavity or oven itself that is a resonant structure for efficient energy transfer.
7. energy sealing or trapping structure to prevent the escape of stray radiation.
8. operating controls and safety devices for the selection of cooking conditions and the protection of the operator.

W.A.G. Voss (1970) reports that microwave leakage can occur from three principal sources, (1) the generator or power tube, (2) the doors to the oven, and (3) entry, exit, and airports. If they are properly designed and installed, flanges and connectors seldom give rise to leakage, but even hairline cracks in applicators and waveguide piping can be dangerous. Misuse, such as the removing of interlocks, is the greatest hazard.

To be safe, the tube and waveguide launching section must be shielded at all times. In view of the air-cooling requirements, the most practical approach is to use perforated stainless steel or aluminum sheet placed within a generator's casing. A finely perforated sheet is also used for this purpose on doors as well as for windows (viewing screens) in oven doors. A typical screen on an oven door may be perfectly satisfactory when monitored by a power meter (most often less than 1 mW/cm^2). It must be remembered, however, that the screen is a form of diffraction grating and the leakage will be affected by the presence or absence of a dielectric material. Thus, even with leakage levels measured at less than 1 mW/cm^2, it is still possible that a wet finger may draw a spark. Consequently, the present trend is to avoid the use of screen doors and, instead, to use a solid metal or to install a combined glass-face and wavetrap viewing area.

Applicator doors cause peripheral leakage which must be controlled. Therefore, the interlock must, in all designs, be an integral part of the door. The microwave circuit must disconnect before the door opens to the point where leakage or sparking can begin to occur. Sparking, a danger in itself, causes metal deterioration and eventually a severe leakage. Food deposits, in particular fat and oil, further aggravate this situation and there is a strong case to be made for continuous smooth metal enclosures which cannot fill with foreign material. This type of enclosure can be easily cleaned and it is unaffected by normal wear. The interlock pin must be concealed and should preferably operate the main power line disconnect except for the filament circuit as well as the microwave circuit. Lever-pin types and magnetic locks are strongly preferred by designers for these reasons.

The design and operation of the microwave oven door is critical to the safe operation of the oven. The load should be visible through the door while heating, but the door should be safe and durable in use and prevent stray radiation from being emitted. A tight closure with a metal-to-metal conductive seal will serve this purpose. It should stay in position with a durable spring-loaded hinge or slides for the useful life of the oven.

As noted earlier, a Radiation Control Safety and Health Act was enacted in 1968. According to a BRH Bulletin (volume V, July 6, 1971), manufacturers of microwave equipment were improving their products by greatly reducing the amount of microwave leakage. Eure et al (1972) believed their data (Table 11-G) indicated significant reductions in microwave leakage from ovens being manufactured over a period of only 2 years. (Micro-

Table 11-G. Summary of Conveyorized, Industrial Microwave Heating Systems

Type of process	Output power in kW*	Frequency in MHz	Maximum leakage in mW/cm²	Source of maximum leakage	Max. leakage at conveyor slot in mW/cm²	Maximum potential occupational exposure at eye level in mW/cm²
1. Pharmaceutical drying	10	2450	190.0	Loose cleaning door	20.0	1.0
2. Food processing research	10	2450	70.0	Conveyor slot	70.0	3.0
3. Finish drying potato chips	60	915	20.0	Conveyor slot	20.0	1.0
4. Opening oysters	6	2450	25.0	Conveyor slot	25.0	1.0
5. Filament drying	2	2450	25.0	Conveyor slot	25.0	1.0
6. Film drying	5	2450	50.0	Conveyor slot	50.0	1.0
7. Film drying	30	915	2.0	Conveyor slot	2.0	1.0
8. Precooking chicken	120	2450	70.0	Waveguide feed	5.0	1.0
9. Food processing research	5	2450	1.0	Conveyor slot	1.0	1.0
10. Opening oysters	5	2450	60.0	Loose door hinge	8.0	1.0
11. Precooking onion rings	4.5	2450	<1.0	—	<1.0	1.0
12. Rapid heating of school lunches	15	2450	1.0	Conveyor slot	1.0	1.0
13. Thawing frozen food	25	915	1.0	Conveyor slot	1.0	1.0
14. Donut proofing	5	2450	30.0	Warped cleaning door	2.0	1.0
15. Macaroni drying	30	915	+200.0	Burned cleaning door	2.0	4.0

*Home microwave ovens generally are in 0.4 to 0.75 kW (400–750 W) range. Adapted from Eure, Nicolls, and Elder, *AJPH*, 62 (No. 12): 1573, 1972.

wave oven inspection guidelines are given in the Addendum to this chapter.)

Radio-frequency heaters

Radio- or high-frequency electrical heating equipment can be used in a wide variety of jobs in industry. Metalworking plants use it for hardening gear teeth, cutting tools, and bearing surfaces, and for annealing, soldering, and brazing. Woodworking plants use it in bonding plywood, laminating, and general gluing. The food industry makes use of high-frequency heating for sterilizing containers and killing bacteria in foods. Other uses include molding plastics, curing and vulcanizing rubber, thermosealing, and setting twist in textile materials.

The high-frequency generator has three components. The first is the rectifier unit, where ordinary single- and three-phase electric power enters and is converted into direct current power for the second component, the electronic generator or oscillator, which transforms the direct current into radio frequency power and delivers it to the third component, and work circuit. The frequency of the output power of the radio frequency generator varies for different applications from 200 kilohertz to the megahertz range, depending upon the specific heating problem. In the work circuit, the high-frequency power is used to heat material which is placed in or near a coil of copper tubing or between the electrodes of a capacitor.

There are two types of high-frequency heaters—induction and dielectric.

Induction heaters are generally used for annealing, forging, brazing, or soldering conductive materials. The metal is brought near an induction coil, usually consisting of turns of copper tubing, electrically connected to a source of high-frequency power. Alternating magnetic fields of 50 Hz up to 10 kHz (occasionally up to a few MHz) induce eddy currents in the metal to be handled. It undergoes heating as it tends to resist the flow of induced high-frequency current. Generally, the metal becomes hot on the outside first, then progressively toward the center. Typically, exposures of the whole body are quite low, although magnetic flux densities can be quite high for localized exposures (for example, up to 25 μT [microteslas] for the hands of an operator).

Dielectric heaters are used for nonconducting, dielectric materials like rubber, wood, plastics, and leather. The load to be heated is placed between two plates of the capacitor; these are called electrodes and are connected to a source of high-frequency power, generally at one of the ISM frequencies 13.56, 27.12, and 40.68 MHz. In the high-frequency field, the molecules in the material are affected in such a manner that they become agitated and cause the material to heat due to molecular friction.

Clearly, operator exposure is a major concern with radio frequency heaters. Although exposures are usually quite localized, they can occur wherein workers may absorb considerable amounts of stray radio frequency energy. Therefore, heating devices should be well shielded to divert the radio frequency energy.

POWER FREQUENCIES

Extremely low-frequency fields

Potential health and safety issues associated with power frequency transmission and electrical devices are generally of two types. The more obvious are the hazards related to shocks and currents possible from high voltage lines and equipment. The other factors are related to extremely low-frequency (ELF) radiations. In the range of ELF, due to the rapidly increasing and varied uses of electric power in industrial society, the likelihood and level of exposures for biological systems has increased by orders of magnitude over the past century (WHO, in press).

Electric fields with frequencies in the ELF range result predominantly from synthetic sources. The strongest ELF fields to which humans are normally exposed, outside of a few specific occupational settings, are those produced by electric-power generation, transmission, and distribution systems.

Unlike electric fields, magnetic fields in the ELF range often occur as a result of natural phenomena, such as thunderstorms and solar activity. Such fields are generally of low strength; however, during intense magnetic storms, these fields can reach intensities of about 0.5 μT. Of greater importance in the context of possible biological effects are the numerous ELF magnetic fields arising from synthetic sources. In the lowest intensity range, generally less than 0.3 μT, are fields found in the home and in office environments (e.g., near video display equipment) (Stuckly et al, 1983). Magnetic fields from ELF communications and power transmission systems are somewhat higher and can approach a level of up to about 15 μT.

Considerably higher flux densities can occur in the immediate proximity of industrial processes using large induction motors or heating devices. Lovsund and co-workers (1982) documented magnetic fields of from 8–70 μT in the steel industry in Sweden. Recently, significant developments in specific areas of medical care have allowed the use of pulsed magnetic fields for various diagnostic and treatment procedures. Flux densities (field strengths) from these new techniques may range from 1–10 mT (milliteslas).

Interaction with biological systems

This section will discuss how ELF electric and magnetic fields interact with living tissues. (It is taken in part from "Electric and Magnetic Fields at Extremely Low Frequencies: Interactions with Biological Systems" by L.E. Anderson and W.T. Kaune, in *Non-ionizing Radiation Protection,* WHO, in press.)

External electric fields act directly on the surface of the body of an exposed human or animal and they also induce electric currents and fields inside the body.

The electric field acting on the surface of the body of a human or animal is enhanced over most of the body surface relative to the unperturbed electric field. It is well known that this field can be perceived. The fields produced inside the body are greatly attenuated; however, they still may be involved in the occurrence of biological effects. A generally accepted mechanism of interaction of ELF electric fields with biological tissues is the direct stimulation of excitable cells. It accounts for the ability of humans and animals to perceive electric currents in their bodies and for the possibility that they can experience a shock. At the cellular level, this mechanism consists of the induction of potentials across the membranes of excitable cells sufficient to generate action potentials.

In contrast to electric-field exposure, the bodies of humans, animals, and other living organisms cause almost no perturbation in an ELF magnetic field to which they are exposed. This is true because: (1) Excluding a few highly specialized tissues that contain magnetite, living tissues contain no magnetic materials and therefore have magnetic properties almost identical to those of air. (2) The modification in the applied magnetic field, due to the secondary magnetic fields produced by currents induced in the body of the subject, is small.

Very little theoretical or experimental work has actually been carried out on the coupling of magnetic fields to living organisms. However, alternating magnetic fields do induce electric fields inside the bodies of exposed humans and animals. As indicated above, external alternating electric fields also induce electric fields inside bodies. The distributions of the fields induced by these two types of coupling are different, but at the level of the cell there would appear to be no fundamental difference.

Biological effects

Numerous studies have been initiated to determine to what extent an electrical environment containing electric or magnetic fields of 50 or 60 Hz poses a health hazard to living organisms (Grandolfo, 1985; Graves, 1985). The biological effects reported in many of the experiments have not yet confirmed any pathological effects, even after prolonged exposures to high-strength (100 kV/m) fields and high-intensity magnetic (10-mT) fields. Areas in which effects have been demonstrated appear to be associated primarily with the nervous system particularly in circadian timing mechanisms and perception of the fields (L.E. Anderson, 1985). In addition, in several instances where unconfirmed or controversial data exist (e.g., changes in brain chemistry and morphology, alterations in reproduction and development), observed effects may or may not be due to the fields. It is not yet known whether confirmed or putative effects are due to a direct interaction of the electric field with tissue or to an indirect interaction, e.g., a physiological response due to detection and/or sensory stimulation by the field. The nature of the physical mechanisms involved in field-induced effects is obscure, and such knowledge is one of the urgent goals of current research.

Results to date have demonstrated various biological effects in specific species exposed in the laboratory to a wide range of field strengths. The extrapolation of specific effects that occurred under controlled laboratory conditions to a general assessment of the health risk of a human population exposed to electric and/or magnetic fields is very tenuous.

In human studies conducted to date, principal sources of information on effects of ELF fields on humans are surveys of utility workers and people living in the vicinity of high-voltage lines, a few laboratory and clinical studies, and several epidemiological studies. Human data from these studies, including reported effects on cancer promotion, congenital malformations, reproductive performance and general health, although somewhat suggestive of adverse health effects, are not conclusive at this time. The value of many of the human studies to date has been compromised by one or more of three serious problems: (1) small sample sizes with extremely limited statistical power; (2) failure to obtain quantitative data on levels and durations of exposure; and (3) uncertainty about what constitutes an appropriate control group. While these difficulties do not necessarily invalidate the results of such studies, it is important to recognize such problems when evaluating the results.

Protective measures

Protection from electric-field exposure can be relatively easily achieved using shielding. At ELF frequencies, virtually any conducting surface will provide substantial electric-field shielding. One practical approach for personnel working in high field strength regions is to provide them with clothes that are electrically conductive. This practice is used commonly in electric utility industry by line workers who work on high-voltage transmission lines using "bare-hand" techniques. The other method to obtain protection from electric-field exposure is to limit the access of individuals to areas where electric field strengths are large.

Unfortunately, there is no practical way to shield against ELF magnetic-field exposure in virtually all cases. Thus, the only practical protective method is to limit exposure, either by limiting access of personnel to areas where magnetic fields are strong or by limiting to safe levels all magnetic fields to which humans could be exposed.

The potential for capacitive discharges and contact currents must also be addressed when discussing protective measures. In this case one must insure that all objects in an electric field have equal potentials. In theory, this can be accomplished by electrically grounding all conducting objects exposed to elec-

Figure 11-14. Horn-reflector antenna used in space communications. (Courtesy Bell Labs)

tric fields. For large, stationary objects this is a practical method, but for mobile objects (e.g., a worker), this technique may not be so useful.

Capacitive discharges and contact currents can be prevented in particular situations by electrical shielding of the affected objects. Finally, protection can be obtained by limiting access of humans and other life forms to areas where electric field strengths are large.

RADAR HAZARDS

Although in the radio frequency range, radar represents an important special microwave application, which is discussed in this separate section.

Radar, an acronym made up of the initial letters of the compound term "radio detection and ranging," is that group of radio detecting instruments that operate on the principle of microwave radiation echoing in a wavelength range from several meters to several millimeters. The comparable frequencies are of the order of 100–100,000 MHz.

Basically, the radar unit consists of a transmitter and receiver, usually operating through a common antenna (Figure 11-14). A pulse of energy is emitted by the transmitter to be picked up as an echo signal by the receiver. The signal thus received is converted by a display or sounding device into usable information.

The average power output range of radar units may vary from a fraction of a watt to hundreds of kilowatts, and is being extended toward the megawatt region. Peak pulse levels of power vary from a few watts to megawatts.

The health, electrical, and fire hazards involved in the handling and use of radar sets include the following:

1. x-radiation from high-voltage tubes
2. radioactivity from radioactive activators used in certain radar switching tubes
3. thermal effects of electromagnetic radiation on the body or parts of the body
4. toxicological hazards of gas fills contained in certain waveguides
5. electrical hazards connected with high-voltage equipment
6. fire hazards of flammable gases, fumes, vapors, explosives, and other highly combustible materials
7. materials handling hazards, particularly with respect to portable equipment and during installation and replacement, relocation, or removal of fixed equipment.

Precautions

Basically, the same interactions occur between radar sources and biological systems as that previously discussed for microwaves. Excessive temperature rise resulting from exposure to radar energy can cause damage if the body absorbs more energy than it is capable of dissipating.

The amount of heating produced in the body depends primarily upon the field strength and duration of exposure, but is also affected by the frequency of the radar unit, the proportion of fat and muscle in the body, and the relation of the body to other objects. There is no reason to believe that any frequency of low *average* power is harmless when the *pulsed* field strength is high. Conversely, there is no reason to believe that any CW frequency is harmful when field strength is low.

The point of exposure of a person to radar energies is usually near the front of the antenna, within its beam, or at the open end of the waveguide. The waveguide structure itself is usually sufficiently enclosed to eliminate the hazard except at the open end or at leaky joints.

Radar units such as those used to measure traffic speed or to map weather present no significant hazard unless they are viewed from directly in front of the antenna, while the unit is operating, and at a distance of a few feet. Larger units, such as search and warning types, however, may produce hazardous field strengths and should be checked before any person must work in front of the antenna.

Radar workers should at no time look directly into a radar beam from a high-energy unit. (In this instance, any unit producing a field strength at the point of observation in excess of 0.01 W/cm² is considered a high-energy unit.) Moreover, they should view the interior of microwave tubes, waveguides, and similar equipment only through a remote viewing device such as a periscope or telescope, unless microwave operation has ceased. Before the viewing device is used on high-energy units, the eyepiece area should be checked with appropriate instruments to assure the user that radar energy is not being deflected or focused at the eyepiece.

Radar outputs from high-energy units not fed into an antenna should be absorbed by an enclosed load. If an enclosed load is not used, the output of the waveguide should be passed into the atmosphere in such a direction or at such a distance that it will not produce a field greater than 0.01 W/cm² in the body of any person or animal in the vicinity.

Radar beams that must be discharged into a closed area should be absorbed by a microwave absorber, of which several commercial types are available. The reflected wave unit must be carefully measured so that if additional absorbers are required, they can be provided. If the intensity of the beam exceeds the stated limit, the entire area through which the beam passes must be roped off, posted, or both, to keep persons out.

Personnel who work in or around high-power radar antennas or radar test equipment should be adequately supervised and instructed to minimize the exposure received. They should work at as great a distance from the beam as practical and should not expose themselves to it unless necessary and then as infrequently and briefly as possible.

Ignition hazard

Radar equipment can cause ignition of flammable materials by inductive heating of steel, other metals, or metal wools, or by induced sparks produced across a small gap between metal elements in an irradiated area in which an explosive fuel/air mixture is present.

Radar equipment can set off such items as photoflash bulbs at a considerable distance and this in turn can ignite other substances. Because of its filamentary structure, the flashbulb is more sensitive to radio frequency than are many other devices. Properly packed photoflash bulbs present no hazard, but defectively packed bulbs or loose bulbs left in or near material that has a low ignition temperature may ignite such material or other flammable substances.

Medical control

Before employees are assigned to work in or around radar equipment, they should undergo a medical evaluation, including complete medical and occupational history, physical examination, and complete blood count. A complete eye examination, including slit-lamp examination, should be made. Whether or not the person has metal plates, pins, or other metal implants in his/her body should be included in the written history.

Personnel working with or near microwave exposures should receive periodic physical examinations in accordance with a definite schedule.

A number of studies have examined ocular lens effects associated with radar exposure data. Although some lens opacities were observed, currently, no data would support a conclusion that low-level, chronic radar exposure induces cataracts in humans. In a survey of 705 microwave workers and 118 laser workers in three Army installations, researchers Hathaway et al (1977) found no evidence of work-related ocular ill effects. See also Cleary (1980) and Elder (1984).

For normal environmental conditions and for incident electromagnetic energy of frequencies from 300 MHz–300 GHz (300,000 megahertz), the radiation protection guide is 10 mW/cm² (milliwatt per square centimeter) as averaged over any possible 0.1-hour period. This means the following:

Power Density: 10 mW/cm² for periods of 0.1 hour or more
Energy Density: 1 mWh/cm² (milliwatt hour per square centimeter) during any 0.1-hour period

The 1986-1987 TLV for radio frequency and microwave radiation covers the frequency range 10 kHz–300 GHz. The TLV chart shown in Appendix B-1 limits the average whole-body SAR to 0.4 W/kg in any 6-minute (0.1 hr) period for 3 MHz–300 GHz. Between 10 kHz and 3 MHz, the average whole-body SAR is still limited to 0.4 W/kg, but the plateau at 100 mW/cm² was set to protect against shock and burn hazards.

LASERS

Just about all present-day lasers are potential eye hazards. All produce extremely high-intensity light radiation of a single wavelength (or a narrow band of wavelengths) depending on the material used for light amplification. The selection of a proper protective filtering lens for laser operators depends on the wavelengths involved, and on the optical density needed to prevent damage to the retina of the eye.

This section will briefly discuss the history, technical details, and use of the laser; it is taken from "Laser Beam Precautions," by W. McCollough (1968).

History

The name laser is an acronym for light amplification by stimulated emission of radiation. The original concept was invented by Dr. Charles Townes in 1955. In 1958 he and Dr. Arthur Schawlou gave a paper on how to construct an optical maser. The name maser is an acronym for microwave amplification by stimulated emission of radiation. An optical maser is a laser. In 1960 Dr. Theodore Maiman, a scientist then working with the Hughes Aircraft Company, gave the world's first demonstration of a laser beam. In 1964, Dr. Charles Townes won a Nobel prize for his work on the theory of lasers.

Technical details

The light that comes from a conventional light source radiates in all directions. Light waves of varying lengths reinforce or cancel each other. Because light waves are nondirectional and of

Table 11-H. Typical Laser Medium, Type, and Operating Wavelengths

Medium	*Operating Wavelength (angstroms)*	*Operation*	*Type*	*Typical Power*
Ruby (chromium in Al_2O_3)	6,943Å	pulsed	solid-state	100,000 watts
Neodymium (Nd in calcium tungstate yttrium aluminum garnet, or glass)	10,600Å	continuous	solid-state	3 watts
Helium—neon	6,328Å	continuous	gas	0.2 watts*
Argon	4579Å–5145Å	continuous	gas	2 watts
(Carbon dioxide+nitrogen+helium)	106,000Å	continuous	gas	10–300 watts
Gallium arsenide	8,400Å	continuous	semiconductor	2 watts

*Pulsed gas lasers have been developed also, and produce power in the kilowatt range.

NOTE:
The wavelength of laser light is determined by the active medium and the mirrors used.
Lasers are classified according to the active element—crystal, or glass (solid state lasers), gas, semiconductor, or liquid.
Highest energy levels achieved by lasers so far (in the 100,000-watt range) have come from solid-state devices, such as those using ruby crystals or neodymium-doped glass as the active medium.

varying wavelengths, it is said to be incoherent. Light from a laser beam however (Figure 11-15) vibrates in a single plane, travels in only one direction and is monochromatic or the same wavelength, so it is known as coherent light.

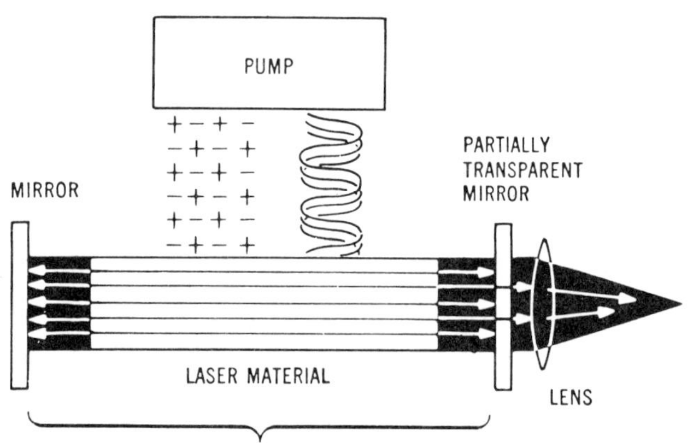

Figure 11-15. Ordinary incandescent sources emit their light in all directions, at different frequencies, and at various amplitudes. The eye interprets light from these different wavelength mixtures as white or yellow-white. Laser light, on the other hand, is coherent, monochromatic, and highly collimated (concentration of light in a narrow beam for a long distance and in the same plane of polarization). The resulting light beams have extremely high intensity and energy.

To obtain a directional beam from an incoherent conventional source of light, it must be focused, and even then it soon spreads its beam again. Ordinary artificial light sources are usually heat units such as incandescent lamps, and arc lights. In a neon tube the glass walls remain cool, but the electrons discharged inside the tube activate the neon atoms within the glass tube and accelerate them to levels normally found at high temperatures. The atoms of the source material are brought to a continuous excited state and, as they fall back to normal, lose energy through heat and visible light.

All electromagnetic energy radiates in fixed wavelengths or frequencies. All electromagnetic frequencies travel at the speed of light (186,000 miles per second or 3×10^{10} cm/sec). The higher the frequency, the higher the energy of the radiation.

Laser beams are not limited to the frequency of visible light. A laser unit produces only one wavelength or frequency, but laser units can be designed over a wide range of frequencies. One of the most powerful beams uses a carbon dioxide laser and shoots out a continuous beam of intensely hot but invisible infrared radiation. Some other lasers operate in the ultraviolet frequencies.

Types of beams

A laser consists of an active medium in a resonant optical cavity, acted upon by some source of excitation energy. There are three types of laser beam-generating active media: (1) the solid state, of which the ruby crystal is the most common type; (2) the gaseous state, of which the helium-neon is the most common; and (3) semiconductor or injection-type (Table 11-H).

Ruby crystal. The most common laser is the ruby crystal. The synthetic ruby crystal is made from aluminum oxide in which some chromium atoms replace some of the aluminum. The ruby crystal is surrounded by a flashtube energized by a capacitor. One end of the ruby crystal rod is silvered, the other end is partly silvered. When the flashtube flashes a burst of light, it excites some of the chromium atoms, raising them to a higher metastable energy level. The excited atoms drop back to their stable energy level, giving off a photon of light energy. The photons oscillate between the reflecting ends of the ruby rod, triggering other excited chromium atoms to give up a photon, so that a whole stream of photons bursts through the partial mirror in one swift pulse of coherent light. The entire sequence takes only a few thousandths of a second.

Atoms can be raised to the excited state by injecting into the system electromagnetic energy at a wavelength different than the stimulating wavelength. This is known as *pumping* (Figure 11-15).

The gaseous laser beam uses a similar principle to the ruby crystal laser beam, but uses a different energy source and a different method of pumping. In a helium-neon gas laser the proper

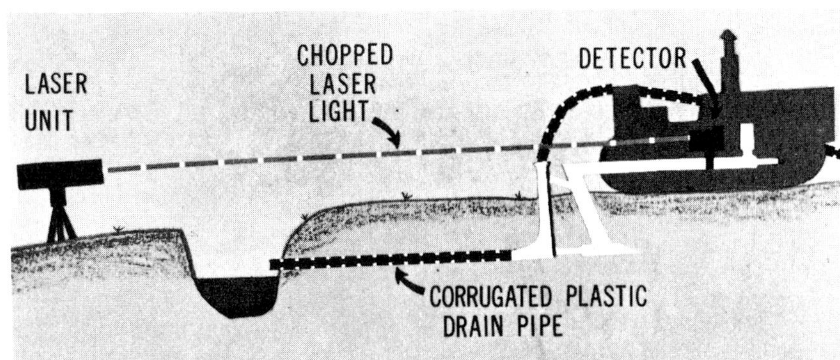

Figure 11-16. Helium-neon laser guides used in the laying of plastic drain pipe. The detector unit is mounted on the side of the tractor. The tractor operator faces away from the light source.

mixture of helium and neon is required to provide an active medium. A radio frequency or direct current excites the helium atoms. The helium atoms excite the neon atoms to a higher energy level and the neon atoms return to the normal state, giving off a photon. The output beam is stimulated by repeated passes between the reflecting end mirrors, triggering more photons. This gives off a continuous stream of laser light known as continuous wave. Peak power in laser beams achieved by CW is lower than that which can be achieved in solid state pulse lasers.

The injection laser uses a tiny semiconductor crystal which has a desirable lattice structure of its atoms. A very small quantity of light of a desired frequency will stimulate the recombination of an electron which has been excited. This will give off laser energy.

Injection lasers can approach power levels that could be dangerous to tissues other than just the eye.

The laser beam travels in parallel lines and does not spread out as ordinary light does. The energy content of the laser is therefore confined to a small diameter. A few feet from the output end of an argon gas laser, generating 18 watts of continuous wave energy, the power is 2,000 times that received on the surface of the earth from the sun. In contrast, the output from a 20 milliwatt continuous wave, helium-neon laser which emits a wavelength of 632.8 nm (a bright red light) at a few feet would not be felt on the surface of the skin. Exposure to 150 mW could result in third-degree skin burns.

If an unfocused 20 mW wave is in a beam diameter of one centimeter, it is an inconsequential amount; but if the beam is focused to an area of one square millimeter, the power density would then be 2 watts/cm^2 and would be enough to present a serious potential hazard.

A ruby laser generally emits in short bursts of energy. Usual output energy levels are 200–500 J (a joule equals one watt-second) in pulse widths of 175–350 milliseconds. The peak power obtained from 500 joules in a 350-millisecond burst is 1421 watts over an area of ⅝-in. (16 mm) diameter. If this is focused to one square millimeter, it can pierce holes in a diamond in a fraction of a second. The Q-switching laser uses a means to shorten the pulsewidths to a millionth of a second or less. Five joules of energy at a billionth of a second pulsewidth is 500,000,000 watts per second. Laser beams, indeed, can have a very high energy level.

Uses. The laser beam, because of its collimated character, has wide application in the areas of military, industry and communications. It can be employed in such diverse uses as precision measurements, guidance systems, photography, metal work, symbol recognition, and medicine. Knowing some of the uses of laser beams will help personnel and the public to be aware of situations where to expect to find the equipment being operated (Table 11-I).

Table 11-I. Typical Laser Applications

Area of use	*Personnel involved*
Business offices	Office workers
Retail establishments	Sales clerks
Communications	Communication engineers
Construction	Surveyors, equipment operators, sewer pipe installers, ceiling installers
Dentistry	Researchers, technicians
Geodesy	Aircraft pilots, surveyors
Holography	Researchers, photographers, artists
Materials processing	Processing engineers and machine technicians
Medicine	Dermatologists, ophthalmologists, surgeons, medical technologists, surgical assistants
Military	Researchers, troops during maneuvers, aircraft pilots

(Reprinted from: L. Goldman et al. "Optical Radiation, With Particular Reference to Lasers." In: *Non-ionizing Radiation Protection,* edited by M. Senss. Copenhagen: World Health Organization, 1982)

The type of laser that has the greatest use in the construction industry has been the helium-neon (He-Ne) gas laser. Its highly collimated beam has been used to project a reference line for construction equipment in such operations as dredging, tunneling, pipe laying (Figure 11-16) and bridge building and marine construction. One of the greatest requirements for such a reference line exists in operations over large bodies of water where position references are most difficult to establish. Several manufacturers provide small He-Ne lasers complete with transit mounts and collimating optics for this purpose. These lasers generally have a power output of 1–10 mW. Most small He-Ne lasers have a beam diameter of 1–3 mm, which is expanded by collimating optics to 20–30 mm (approximately 1 in.), thereby increasing the collimation of the beam. Some collimation systems have been designed to provide a fanned-shaped beam so that a reference plan is produced rather than a line. In some applications, the beam is directed at another reference point, such as a target card.

The small He-Ne gas laser has also been used for highly pre-

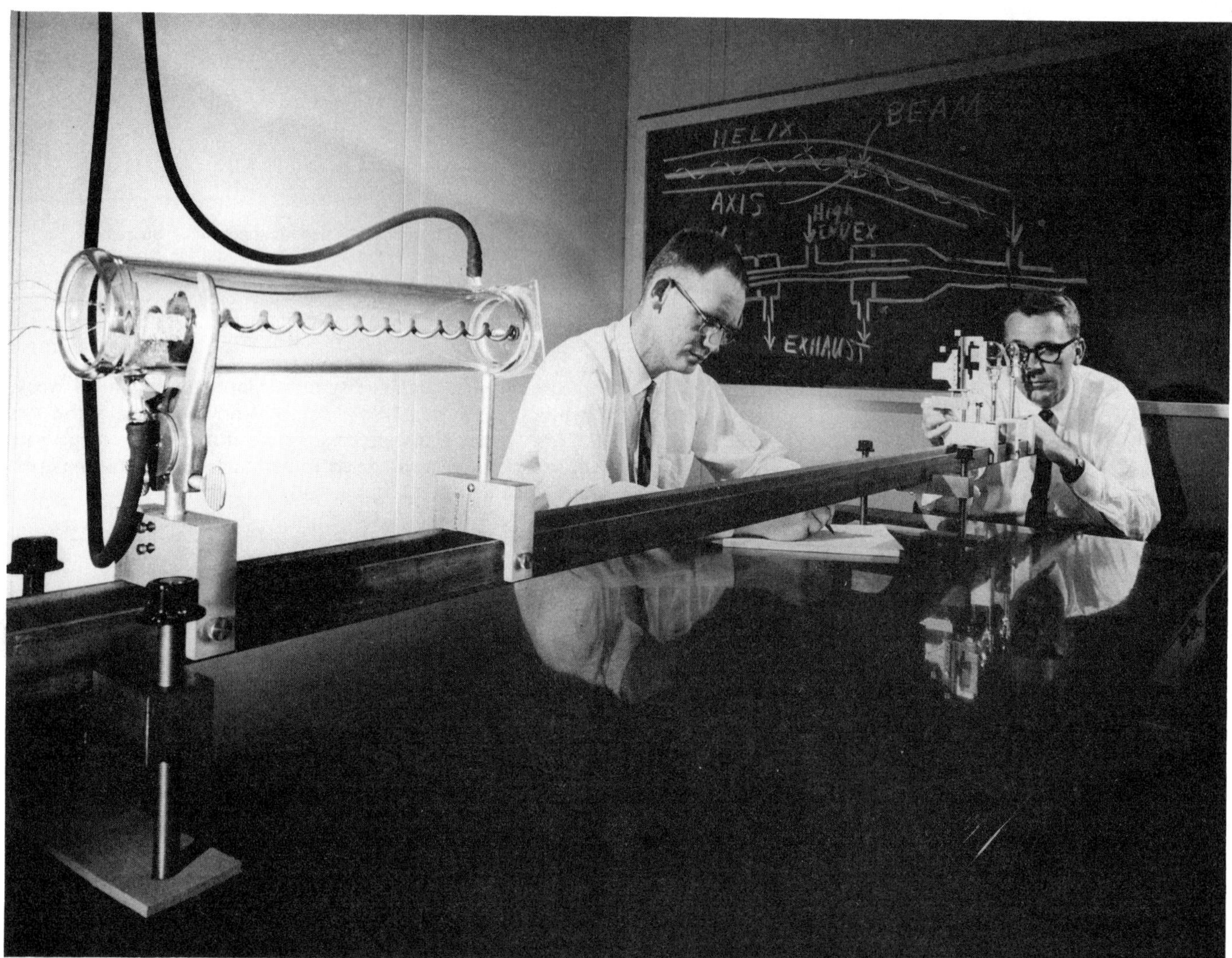

Figure 11-17. A vintage (1964) photograph showing Drs. Dwight Berreman and Andrew Hutson testing the experimental helical gas convections lens, which uses temperature-produced variations in the refractive index of a gas to guide light. (In 1964 the lens showed promise for use in guiding laser beams for use in long-distance communications.) (Courtesy Bell Labs)

cise distance measuring in surveying. The U.S. Coast and Geodetic Survey presently uses laser geodimeters. In another application, the laser beam is used for welding or micromachining fine parts. The laser photocoagulator is used by some surgeons to repair torn retinas. Laser surgery makes bloodless surgery possible. A limited-beam laser has been used to kill malignant skin tissue, remove birthmarks, or burn away warts. A laser microprobe focuses through a microscope onto an object. When a portion of the test material is vaporized, a spectrograph can then read the light from the material which is converted into a spectrum whose lines can be physically identified and used to measure the chemical constituents.

The laser beam can be used to transmit communication signals. This is rapidly becoming the most extensive use of the laser beam. A single laser beam, theoretically, can carry as many messages as all the radio communication channels now in existence. The main problem is that no light beam will penetrate fog, rain, or snow very well. To be useful for earthbound communications it may be necessary to enclose the beam within tubes. Figure 11-17 shows work being done in the early 1960s, which was instrumental in the development of laser use in long distance communications.

The hologram is a three-dimensional image that hangs in midair. One laser beam is used to illuminate the subject and bounce from it to the photographic plate. A second beam is split off from the first and bounced to the plate from a mirror. The two beams produce alternating black and white lines known as interference patterns. When a laser beam is passed through the developed plate, an image of the original object photographed will hover ghostlike some distance from the plate. Even the back of the object being photographed can be seen.

Laser beams will crumble rock and may be used in the future for drilling tunnels in rock. Lasers are also being used more in the practice of medicine. They are now used to treat the eye, skin, and internal organs, and in surgical applications requiring extreme precision. In dentistry, lasers are used for various diagnostic techniques, as well as enamel scaling and bridgework.

Hazards

This section will discuss exposure criteria, hazards, and hazard-

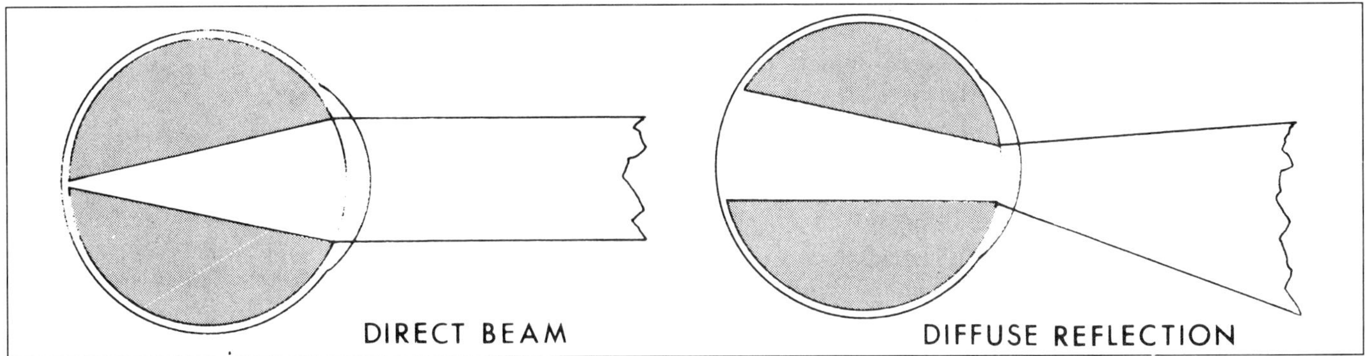

Figure 11-18. Observing laser light. Left: Parallel rays of a laser can be focused to a point image by the eye. Right: Rays from an extended source (as from a conventional lamp, or rays from a diffuse reflection of a laser beam—see Figure 11-20) produce a sizable (and less dangerous) image at the retina.

control. (It is taken, in part, from "The Amazing Laser," by D. H. Sliney, 1968.)

What characteristics of a laser make it a hazard? It seems strange to many people that a 0.1-watt laser is considered a potential ocular hazard, while a 100-watt light bulb is not. The principal reason for this is that the laser can be effectively a point source of great brightness close to the source and the light is emitted in a narrow beam, whereas conventional means of illumination are extended sources, which emit light waves in all directions and, hence, are considerably less bright. The laser also has characteristics of coherence and monochromaticity which in themselves do not contribute to the hazardous aspect.

Figure 11-18 shows how the parallel rays of a laser can be focused to a point image while rays from an extended source as from a conventional lamp (or rays from a diffuse reflection of a laser beam) produce a sizable image at the retina. Light from a laser entering the eye is concentrated 100,000 times at the retina. Because of this focusing effect, the eye (Figure 11-19) is by far the organ of the body most susceptible to damage. Hence, injury to the skin is seldom of concern except in dealing with very high-powered lasers.

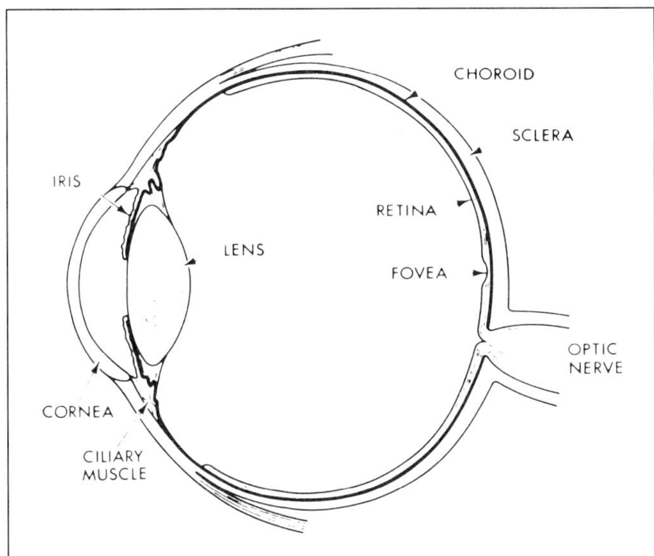

Figure 11-19. A horizontal cross-section of the human eye. Length from cornea to cone layer of retina is about 24 mm. Thickness of choroid is about 0.05 mm; thickness of the sclera, about 1.0 mm.

The hazards associated with laser operations fall into two categories—those from the laser itself and those from associated equipment.

The laser. The solid-state lasers present the greatest hazards, because of their high achievable peak powers. Gas and injection lasers, although not as powerful as the solid-state lasers, are capable of producing power outputs that can cause permanent eye damage and skin burns if proper safety practices and procedures are not followed.

The primary hazard associated with laser operations is the laser beam. This beam is capable of inflicting serious injury to personnel, in particular to the unprotected eye. Experience has shown that specular reflections may approach direct beam intensities. These reflections are difficult to predict and can make off-axis viewing just as dangerous as on-axis (direct) viewing.

The primary physical trauma associated with a laser lesion is usually thermal in nature, but thermoacoustic transion pressure waves can also produce important biological effects. Optical breakdown of biological molecules can occur as well as photochemical activation of molecules in laser-induced exposure effects.

Secondary effects must also be considered in laser use. Some of these may actually be a greater hazard to health and safety than the laser. These effects are not discussed in detail in this chapter, but include the following: possible electric shock from the high voltages associated with laser-producing equipment, volitization or vaporization of target materials with attendant inhalation hazards, ozone produced as a result of electrical discharges or ionization of the air surrounding high-power laser beams, UV and IR radiation, along with the brilliant white flashes in xenon solid-state lasers. Another danger is the buildup of high pressures of the gases in the flash lamp when it is fired, thus posing an explosion hazard. Cryogenic gases such as liquid nitrogen and liquid helium are sometimes used to cool the crystal (ruby, neodymium, etc.) in solid-state lasers. These liquified gases are capable of producing skin burns upon contact. If these gases leak into a closed room, they are capable of replacing oxygen in the atmosphere. This latter condition can produce an oxygen-deficient atmosphere. Flammable solvents and materials may be associated with a particular laser operation and are capable of being ignited by a laser beam. (See Chapter 17, Methods of Evaluation.)

Exposure criteria. A number of groups have developed recommended safe ocular exposure levels for laser radiation (subsequently referred to as safe levels). See Appendix B-1.

Since the threshold doses presented in most biological research literature are generally presented in terms of retinal intensities, early standards proposed safe levels at the retina. Such levels created serious problems for many users who were not familiar enough with physiological optics to extrapolate retinal doses to the exterior (cornea) of the eye. Safety factors ranging from 2–100 have been used by different groups in arriving at safe levels from the same biological data. Opinion varies as to whether safety levels should be based upon cell damage detected by biomicroscopic methods, by clearly evident damage to the retina determined ophthalmoscopically, or by detectable functional loss of vision. It may well be that there exists a significant difference between acute and chronic effects of laser exposure. In general, it has been the practice on the part of those who have proposed safe levels to be conservative. Except where lasers are used in an outdoor environment over long ranges, conservative levels have not resulted in signficant operational restrictions. Lasers have been classified according to their output parameters, based on IEC, USFDA, and ANSI standards. The classifications go from non-risk to high-risk, and are described in some detail at the end of the chapter.

Reflections. The foregoing exposure criteria assume that the incident laser radiation consists of a parallel beam and that the eye is focused at infinity. This would be true for intrabeam viewing within the direct beam or for a specularly reflected beam; however, most reflections which individuals view are diffuse in nature. A safe retinal irradiance can readily be related to a safe surface brightness measured at the diffuse reflector. The U.S. Army Environmental Hygiene Agency has been using the following surface "brightness" values as the upper limits in judging safe operations: 0.07J/cm^2 for Q-switched lasers, 0.9J/cm^2 for non-Q-switched lasers, and 2.5 W/cm^2 for continuous wave lasers measured at the surface. It is essential to understand the significance of these values. If a surface brightness occurring from pulsed laser illumination is viewed, it is equally as hazardous for all viewing angles or distances provided the reflection appears as an extended source. On the other hand, the likelihood of an observer's eye being within the direct or a specularly reflected beam in any practical situation is remote. When viewing an extended source of laser rays, the energy received by the eye is an image over a large area of the retina. Energy from a point source focuses at a point on the retina and can be very likely to produce a burn. These two entities are addressed separately in the laser TLVs.

In this regard, there are, probably, two principal reasons for the low incidence of laser injury in the past; namely, the low probability of a person receiving a well-collimated specular reflection in most operations, and the fact that the rule regarding the wearing of protective eyewear is seldom disregarded when using lasers capable of producing hazardous diffuse reflections.

Hazards of small He-Ne lasers. The output power of present He-Ne lasers is not sufficient to cause injury to the skin. A potential hazard exists to the eye, however, since the beam may be focused to a small spot on the retina. This hazard applies only if the eye is located within the primary beam or a beam created by reflection from a flat specular surface. Reflections of the beam from any diffuse surface are safe to view as long as the beam power density is below 2.5 W/cm^2. Present experimental and theoretical research indicates that retinal burns may result from exposure to beam intensities within the order of 1 mW/cm^2 and above. Because of effects other than retinal burns which may be of significance from viewing the beam for long periods of time, it is presently considered wise to limit ocular exposure to intensities of one μW/cm^2 or less. Average beam intensities above one mW/cm^2 exist only to a range of approximately 50 to 100 meters or not at all for most construction lasers. Thus, at these close ranges, the potential eye hazard may be significant even for short exposures. Although an occasional accidental exposure to the CW laser beams at levels below 1 mW/cm^2 would not be expected to produce a burn, repeated exposure at levels between 10^{-6} and 10^{-3} W/cm^2 is undesirable. Personnel would not be expected to stare into the beam in any case, since 10^{-6} W/cm^2 looks quite as dazzling during daylight, as 10^{-8} W/cm^2 does at night. Thus, the protective mechanisms of the eye may be expected to prevent individuals from staring into high-intensity beams. This protective feature of the eye is not helpful, however, for lasers emitting in the IR because the beam is not visible; nor is it helpful for pulsed lasers.

Optically aided viewing. An important concept to understand is the effect of viewing laser light by optical instruments, such as binoculars or telescopes. The laser light arriving at the eye after passing through a telescope or binocular may be concentrated by as much as the square of the magnification of the instrument. Thus, viewing the direct beam of a laser through 7×50 binoculars could increase the intensity level at the eye by as much as 7^2 or 49 times. This applies only if the beam is viewed directly or by specular (mirrorlike) reflection, but not if viewed by diffuse reflection. There is no additional hazard in viewing diffuse reflection with binoculars or telescopes. If personnel with binoculars are allowed to be in the direct beam at ranges normally considered safe for viewing by the unaided eye, then a single "safe range" cannot be defined.

Operational aspects

The application of the hazard criteria already discussed depends heavily upon the nature of the laser operation. For example, in one tunnel boring operation the laser beam passed above occupied areas and was directed at a photodetector at the boring machine which then gave the machine operator information as to whether he was on course. In this type of situation, no problems would be expected to arise.

On the other hand, some applications call for equipment operators to align their equipment visually by looking directly into the beam. This is not considered wise unless the average beam intensity is down to approximately 10^{-6} W/cm^2 (one μW/cm^2) for occupational exposure at night or five times this value in bright sunlight. At ranges where beam intensities are much greater than one μW/cm^2, protective eyewear may be used to reduce the intensity to a satisfactory level while still rendering the beam visible. Alternatively, the beam may be safely viewed by diffuse reflection. A He-Ne laser beam is generally visible by diffuse reflection from a white card in daylight at approximately one mW/cm^2, or 0.1 mW/cm^2 if the card is shaded. Retroreflective sheets will render lower levels visible if the viewer is near the beam axis. Goggles that cut out ambient light but transmit the laser beam are also available for rendering lower intensities of light visible; however, such goggles must not be mistaken for protective eyewear.

Reflections encountered in the field. Some applications, such as the laser geodimeter, utilize a reflected beam from a corner-cube retroreflector, which is directed back along the beam path to the geodimeter. The reflected beam may be observed through the finder telescope. The retroreflector is typically used only at distances greater than a kilometer in order to enhance the reflected beam. For close ranges, retroreflective "Scotchlite" coating is utilized. Reflections from the "Scotchlite" are not hazardous from present geodimeters. Reflections from natural surfaces, which are reasonably diffuse, are not potentially hazardous.

However, reflections from specular surfaces, such as flat glass or flat surfaces of still ponds, may be potentially hazardous if these surfaces are quite near the laser. For instance, directing the beam through a glass window from within a building would create a reflected beam with an intensity of approximately 8 percent of the primary beam (or more, depending upon incident angle and polarization of the beam). If the polished surface is curved, safe levels are achieved at shorter ranges (see Figure 11-20).

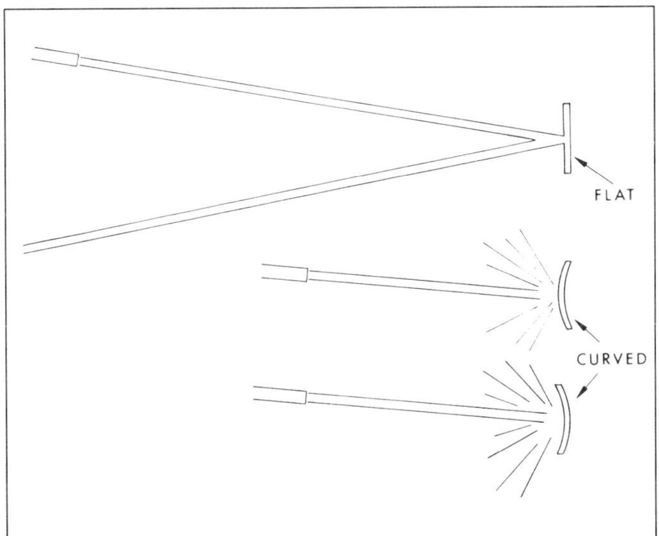

Figure 11-20. Reflections from specular (mirrorlike) surfaces (top) can be potentially hazardous if these surfaces are quite near the laser. If the polished surfaces are curved (bottom), safe levels are achieved at shorter ranges.

Laser hazard controls

Hazard controls should be designed to minimize the opportunity for ocular exposure to the direct laser beam and specular (mirrorlike) reflections. The controls should be reasonable, while not hampering the operation or creating new hazards.

Users often misunderstand the different orders of magnitudes of intensity levels found in the operational environment and the understanding of the probability of potential accidental exposures. A program of educating personnel concerned with laser hazards is an essential part of the total hazard control effort. This must be supplemented by continuing on-the-job supervision. Accessible emission limits (AELs) have been established for each class of laser. Those limits are based on many criteria: wavelength, exposure duration, CW or pulsed beams with information describing average power output, peak power, pulse duration, and repetition frequency, laser-source radiance or integrated radiance. The laser safety problem must be presented in perspective with other hazards encountered on an everyday basis.

Environmental controls may differ widely depending on whether the laser is used in a laboratory or out-of-doors. Backstops and shields to exclude the beam from occupied areas are commonly used both in and out-of-doors. Well-illuminated laboratories and limited-access rooms are important environmental controls. The prevention of unsafe acts by personnel may be achieved by the use of physical barriers; by the application of administrative procedures (education and training) and through careful supervision. The use of protective eyewear is a major control which would be mandatory when a serious risk of injury to the eye exists.

In general, it appears that present laser equipment may be operated safely by trained operating personnel without undue restrictions. However, several procedures are required to minimize long term low-level exposures. The following guidelines are based upon this conclusion.

- Lasers should not be left unattended during operation. Beam shutters or caps should be utilized, or the laser turned off when laser transmission is not actually required.
- Personnel who work with laser units should be instructed in the potential eye hazards and the importance of limiting unnecessary exposure. Personnel occupationally exposed to laser light should receive preplacement, periodic, and final eye examinations.
- A warning sign should be attached to laser equipment in a conspicuous location indicating the potential eye hazard associated with the laser and warning against looking into the primary beam and at specular reflections. Such a warning sign might read:

DANGER! LASER LIGHT
DO NOT LOOK INTO PRIMARY LASER BEAM
DO NOT AIM LASER AT FLAT GLASS
OR MIRROR SURFACES
AIM ONLY AT REFLECTIONS SUPPLIED WITH UNIT

- The use of corner-cube retroreflectors should be avoided at close ranges if the reflected beam is to be observed. Diffuse or retroreflective card targets are recommended for short ranges.
- The use of binoculars or aiming telescopes should not be used to view the direct beam, or a reflected beam from mirrors or corner-cube retroreflectors unless the beam intensities are greatly below safe levels. If necessary, a filter having a sufficient optical density is placed in the optical path of the telescope for such situations, or adequate laser-protective eyewear is worn by the operator.
- For ranges where beam intensities are significantly above one $\mu W/cm^2$, personnel receiving the beam should be provided with protective eyewear with a typical optical density of three. Protective eyewear should be considered necessary for personnel at the laser itself only if strong specular reflections are expected, if the laser is so situated that personnel can walk into the emerging beam at eye level, or when viewing strong retroreflections through the telescopes. If protective eyewear is required, it should be labeled as to the optical density at the appropriate laser wavelength or as to the laser equipment for which it was designed. Personnel who must wear protective eyewear should remain in good communication with the laser operator to make sure that eyewear is worn during laser operation (Figure 11-21).

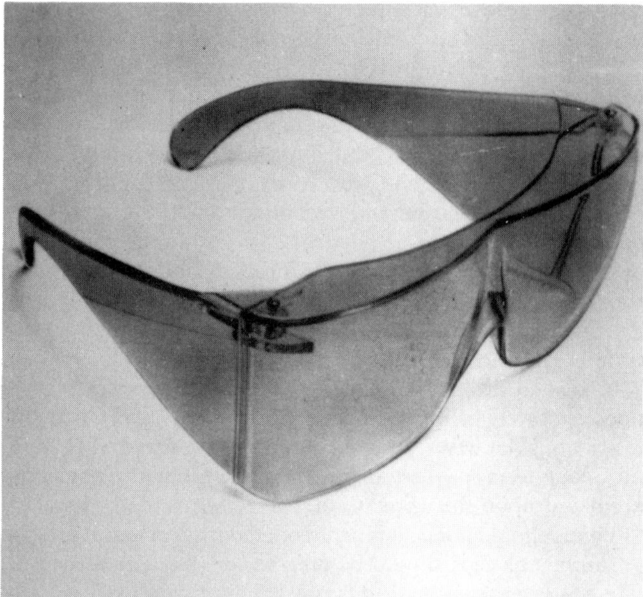

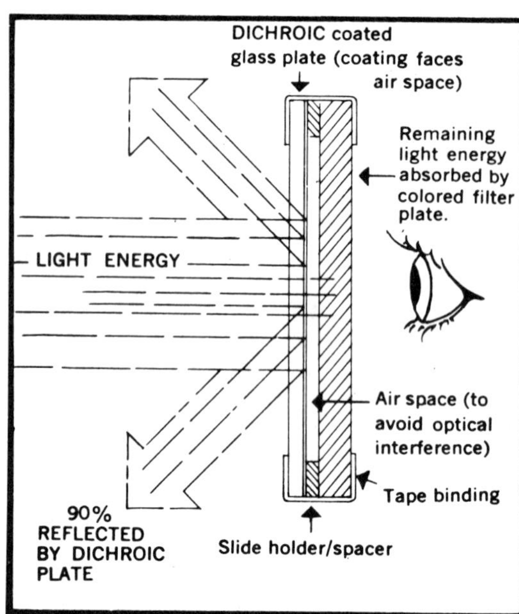

Figure 11-21. No one type of glass offers protection from all laser wavelengths. Consequently, many firms do not depend upon safety goggles to protect from laser burns, because goggles might give a false sense of security and tempt the wearer to be exposed to unnecessary hazards. Nevertheless, laser users do frequently use eye protection. Left: Laser glasses (Courtesy of Glendale Protective Technologies). Right: Antilaser eyeshield with diagram to show how it works.

- During the alignment and setup procedures, care should be taken to avoid aiming the laser into potentially occupied areas. Prior to use in heavily occupied areas, the alignment of the beam with the pointing telescope or aiming equipment should be checked if the instrument is so equipped.
- If the beam is directed through a glass window, the beam should pass perpendicularly to the plane of the glass, or protective eyewear should be required for personnel in the vicinity of the window.
- Stable mounts for the laser are important so that beam traverse can be readily controlled.
- Reflections from rain, snow, dust, and other particulate matter are not of concern unless the beam intensity is above 2.5 $\mu W/cm^2$ (seldom if ever the case with construction lasers).
- Despite the potential hazards, the laser beam can be used safely if the proper procedures and necessary precautions are followed. Only hightly trained persons should be permitted to work in an area where they could come into direct contact with a laser beam. The work area should be brightly lighted to prevent dilation of the pupils. All surfaces in laser area must be nonreflective. Work areas should be monitored regularly for ozone or other potential contaminants and stray radiation.
- Any combustible solvent or material should be stored in proper containers, and shielded from the laser beam or induced electric spark. When necessary to view work being done by a laser beam, it should be viewed through remote means such as closed-circuit television.
- All potential electrical hazards should be safeguarded and proper procedures used. Approved signs, and audible and/or visual signals should be used to indicate when a laser is operating.
- A thorough medical examination should be given to all people working with or near laser beams. This examination should be given before employment, at least annually, and upon completion of employment. The medical examination should include a complete ophthalmologic and dermatologic examination as well as a complete blood count and urinalysis and a review of the person's medical history. Persons with the following medical conditions should not be permitted to work with or near laser beam units:

1. any eye disease or defect that could be aggravated by a laser beam
2. any skin problem, particularly skin malignancies
3. chronic pulmonary or cardiovascular disease
4. chronic emotional or mental illness
5. hypothyroidism or diabetes
6. pregnancy.

These guidelines are intended only for the small He-Ne lasers discussed earlier in this section. Other guidelines would apply to other types.

For guidance on appropriate eye protection for work with lasers, see Chapter 5, The Eyes, and Chapter 20, Methods of Control.

The TLVs for laser radiation are a family of values that are to be calculated for each set of exposure conditions of wavelengths of the lasers, duration of exposure, size of the laser beam (narrow band or extended source) and the type of beam—pulsed or continuous wave. These values are derived from a series of figures and tables included in the TLV list (see Appendix B-1). Some readers will find it helpful to consult also *A Guide for Control of Laser Hazards,* listed in the Bibliography.

Methods of evaluation of laser hazards are included in Chapter 17, Methods of Evaluation.

Safety rules and recommendations

This material on safety rules and recommendations is adapted from rules and regulations of the U.S. Army's Environmental Hygiene Agency (D. H. Sliney and J. K. Franks, 1977), and is reprinted with permission from *Laser Focus Buyers' Guide,* January 1977).

This section emphasizes hazards from laser radiation, which are confined to the eye and, to a smaller extent, the skin. Electrical hazards so far have proven more serious, and procedures for handling high voltages safely are given at the end of this section.

Radiation hazards. Effects of radiation at various wavelengths on various parts of the eye are shown in Figure 11-22. Actinic ultraviolet, at wavelengths of 200–315 nm, is absorbed at the cornea; these wavelengths are responsible for "welder's flash" or photokeratitis. Actinic-UV radiation also produces "sunburn" or erythema of the skin. Near-UV radiation between 315 and 400 nm is absorbed in the lens and may contribute to some forms of cataract. At high irradiances, these wavelengths also produce "long-wave" erythema of the skin.

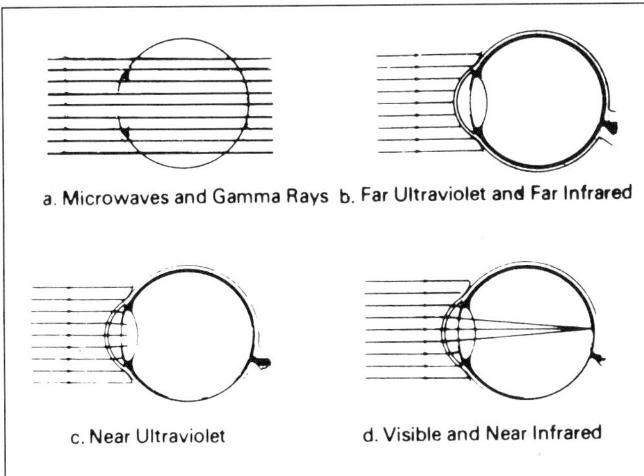

Figure 11-22. Penetration of the eye by different wavelengths of electromagnetic radiation. (Reprinted from *Safety Rules and Recommendations,* U.S. Army Environmental Hygiene Agency)

Radiation at visible, 400–780 nm, and near-infrared, 780–1,400 nm, wavelengths is transmitted through the ocular media with little loss and usually is focused to a spot on the retina 10–20 micrometers in diameter. Such focusing can cause intensities high enough to damage the retina; for that reason 400–1,400 nm is termed the ocular-hazard region.

Although far-IR radiation with wavelengths of 3 μm to 1 mm is absorbed in the front surface of the eye, some middle-IR radiation between 1.4 and 3 μm penetrates deeper and may even contribute to "glass-blower's cataract." Extensive exposure to near-infrared radiation also may contribute to such cataracts.

Few serious eye injuries due to lasers have been reported since the appearance of commercial devices. The accident rate is not that low because the ocular exposure limits are overly conservative; they are not. Instead, the possibility of accidental exposure of the eye to a collimated beam is extremely remote if a few rudimentary commonsense precautions are followed.

In only a few cases, primarily in materials-working, is there a large probability of accidental exposure. If a diffuse reflection is so bright that it is hazardous, a viewer is susceptible to injury if the source is close enough to be viewed as an extended source rather than a point source. The probability of one's eye being positioned within a narrow beam of light at the right moment and of the eye being relaxed and oriented along the beam axis obviously is infinitesimal compared with the high probability of viewing a hazardous diffuse reflection. A second case of susceptible exposure occurs if the nature of the laser operation requires an individual to fix his/her eye upon a source of direct or reflect laser radiation; this is the case with many materials-working operations, which require operators to use their central vision—the fovea—constantly to view the source of reflected laser radiation.

Ruby and neodymium laser irradiances capable of causing surface ablation in any material are almost always greater than safe exposure limits for viewing diffuse reflections. Furthermore, thse irradiances are orders of magnitude above limits for viewing a specular, or mirrorlike, reflection. There is risk, therefore, in viewing any materials-processing operation with a laser emitting at 400–1,400 nm at a distance sufficiently close to observe a diffuse reflection as an extended source.

Classification of laser hazards. A hazard-classification scheme was first presented in the ANSI standard Z-136.1-1986 on the safe use of lasers. The Bureau of Radiological Health's classification scheme, which took effect August 2, 1976 to govern the manufacture of laser devices, is described as follows; this is a slight modification of the ANSI scheme.

Class I, or exempt lasers or systems, are those that cannot under normal operating conditions emit a hazardous level of optical radiation. No warning label or control measure is required for this class.

Class II, or low-power laser devices, are those visible lasers that do not have enough output power to injure a person accidentally, but which may produce retinal injury when stared at for a long period. The BRH requires the manufacturer to affix a "caution" label like the one shown in Figure 11-23.

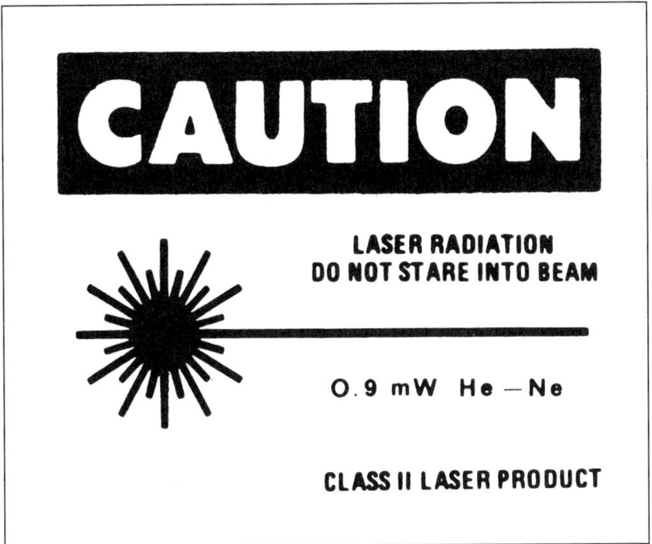

Figure 11-23. BRH label for Class II helium-neon laser. (Figures 11-23–11-26 were reprinted with permission from *Safety Rules and Recommendations,* U.S. Army Environmental Hygiene Agency)

Class IIIa covers visible lasers that cannot injure the unaided eye of a person with a normal aversion response to a bright light, but may cause injury when the energy is collected and put into the eye, as with binoculars. For Class IIIa, BRH will require a label like one shown in Figure 11-24.

Figure 11-24. BRH label for Class IIIa helium-neon laser.

Class IIIb consists of lasers which can produce accidental injury if viewed directly. The danger from such a laser is the direct or specularly reflected beam. A warning label for such a laser is shown in Figure 11-25.

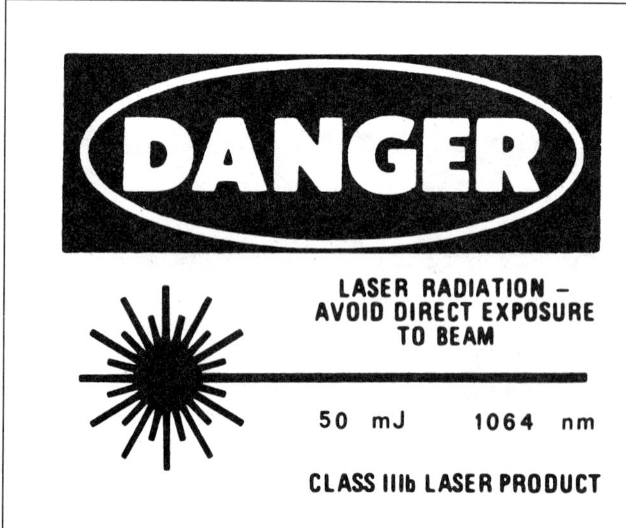

Figure 11-25. BRH label for Class IIIb neodymium laser.

Class IV includes lasers which not only produce a hazardous direct or specularly reflected beam but also can be a fire hazard or produce a hazardous diffuse reflection. A Class IV laser must have the BRH warning label shown in Figure 11-26.

Figure 11-26. BRH label for Class IV carbon dioxide laser.

The ANSI Z-136.1 standard on the safe use of lasers published in September 1976 increased the permissible exposure limits for the eye for durations greater than 10 seconds to wavelengths of 550-1,400 nm; exposure limits at blue wavelengths remain unchanged. Correction factors giving the permissible exposure limits as a function of wavelength and exposure duration are given in ANSI standard *Safe Use of Lasers.* The ANSI also eliminated Class V to make its classification system compatible with BRHs.

BRH has recognized an exemption from parts of its safety code for laser products designed for combat or combat-training operations of the Department of Defense. Laser systems exempted under this ruling include:

- lasers used in tactical systems such as laser rangefinders, target designators, and target illuminators
- tactical aircraft-display devices and eye-pointing indicators for pilots or gunners
- field-training lasers such as gallium-arsenide direct-fire simulators
- field military laser-communications transmitters
- classified military laser systems.

Other laser systems operated by the Department of Defense are not exempted from the BRH code. These include:

- industrial laser products such as container-making systems, laser welders, laser drillers, and laser machine-alignment systems
- indoor (classroom) laser-training devices including holographic and other displays
- laser geodimeters, engineering laser-distance meters, and construction lasers which are also used in civilian applications
- lasers used in basic research and in scientific instrumentation
- lasers used in office machines and merchandising equipment
- medical lasers.

Military lasers which are exempted from the BRH regulations must prominently carry the following label:

CAUTION

This electronic system has been exempted from FDA radiation safety performance standards prescribed in the *Code of Federal Regulations,* title 21, chapter 1, subchapter J, pursuant to exemption No. 76EL-01DOD issued on

26 July, 1976. This product should not be used without adequate protective devices or procedures.

In addition, the Defense Department must submit an annual report to BRH summarizing the types and numbers of laser products or lasers procured and disposed of under this exemption. Manufacturers of exempted laser systems must report to BRH any "accidental radiation occurrences," as for nonexempted lasers.

Guidelines to prevent electric shock

1. General precautions

- Avoid wearing rings, metallic watchbands and other metallic objects.
- When possible, use only one hand in working on a circuit or control device.
- Never handle electrical equipment when hands, feet or body are wet or perspiring or when standing on a wet floor.
- With high voltages, regard all floors as conductive and grounded unless covered with well-maintained dry rubber matting of a type suitable for electrical work.
- Learn rescue procedures for helping victims of apparent electrocution: Kill the circuit; remove the victim with a nonconductor if he is still in contact with the energized circuit; initiate mouth-to-mouth respiration immediately and continue until relieved by a physician; have someone call for emergency aid.

2. Precautions with high-power lasers

- Provide fault-current-limiting devices such as fuses or resistors, capable of clearing or dissipating total energy, and emergency shutoff switches. In some research laboratories these are incorporated in the laboratory bench wiring.
- Provide protection against projectiles that may be produced during faults by the use of suitable enclosures and barriers.
- Provide enclosures designed to prevent accidental contact with terminals, cables or exposed electrical contacts. Provide a grounded metal enclosure that is locked and/or interlocked.
- Prevent or contain fires by keeping combustible material away from capacitors.
- Automatically dump (deenergize), or crowbar, capacitors before opening any access door.
- When feasible, wait 24 hours before working on circuits involving high-energy capacitors.
- Provide a sufficiently short discharge time constant in the grounding system.
- Check that each capacitor is discharged, shorted and grounded before allowing access to capacitor area.
- Provide reliable grounding, shorting and interlocking.
- Install crossbars, grounding switches, cables and other safety devices to withstand the mechanical forces that could exist when faults occur or crossbar currents flow.
- Provide suitable warning devices, such as signs and lights.
- Place shorting straps at each capacitor during maintenance while capacitors are in storage.
- Provide manual grounding equipment that has the connecting cable visible for its entire length.
- Supply such safety devices as safety glasses, rubber gloves and insulating mats.
- Provide metering, control and auxiliary circuits that are protected from possible high potentials even during fault conditions.
- Inspect routinely for deformed or leaky capacitor containers.
- Provide a grounding stick that has a discharge resistor at its contact point, an insulated ground cable (transparent insulation preferred), and a grounding cable permanently attached to ground. Such a grounding stick should not be used to ground an entire large bank of capacitors. Large-capacity shorting bars, with resistors, should be part of the stationary equipment. Final assurance of discharge should be accomplished with a solid-conducting ground rod.

An extensive bibliography on laser safety is included in the Sliney and Franks (1977) article (Bibliography).

The Occupational Safety and Health Administration does not have a regulation on laser safety for general industry. The construction standards (29 *CFR,* 1926) have specific requirements for use of qualified personnel, eye protection, posters, and exposure limits.

BIBLIOGRAPHY

American Conference of Governmental Industrial Hygienists (ACGIH). *A Guide for Control of Laser Hazards.* Cincinnati: ACGIH, 1976.

American National Standards Institute, 1430 Broadway, New York, NY 10018.

Electromagnetic Fields, Safety Levels with Respect to Human Exposure to Radio Frequency, ANSI C95.1-1982.

Practice for Industrial Lighting, ANSI/IES RP7-1983.

Radio Frequency Radiation Hazard Warning Symbol, ANSI C95.2-1982.

Recommended Practice for Measurement of Hazardous RF and Microwave Electromagnetic Fields, ANSI C95.5-1981.

Safe Use of Lasers, Z136.1-1986.

Techniques and Instrumentation for the Measurement of Potentially Hazardous Electromagnetic Radiation at Microwave Frequencies, ANSI C95.3-1973 (R1979)

Anderson, L. E. "Interaction of ELF Electric and Magnetic Fields with Neural and Neuroendocrine Systems." In *Biological and Human Health Effects of Extremely Low Frequency Electromagnetic Fields.* Arlington, VA: American Institute of Biological Sciences, 1985.

Anderson, L. E., and Kanne, W. T. "Electric and Magnetic Fields at Extremely Low Frequencies: Interactions with Biological Systems." In *Non-ionizing Radiation Protection,* edited by M. J. Seuss. Copenhagen: World Health Organization (WHO), in press.

ASA Subcommittee on Transmissive Properties of Plastics. *The Spectral-Transmissive Properties of Plastics.* New York: American National Standards Institute, 1955.

Brotherton, M. *Masers and Lasers: How They Work, What They Do.* New York: McGraw-Hill Book Co., 1964.

Bruls, W. A. G., et al. Transmission of human epidermis and stratum corneum as a function of thickness in the ultraviolet and visible wavelengths. *Photochemistry and Photobiology* 40(1984):485-494.

Cleary, S. F. Microwave cataractogenesis. *Proceedings IEEE* 68(1980):49-55.

Cralley, L. V., ed. *Industrial Hygiene Highlights,* vol. 1. Pittsburgh: Industrial Hygiene Foundation of America, Inc., 1968.

Elder, J. A. "Special Senses." In *Biological Effects of Radiofrequency Radiation,* edited by J. A. Elder and D. F. Cahill. U.S. Environmental Protection Agency Publication No. EPA-600/8-83-026 (NTIS PB-85120848), 1984, pp. 6-1-6-9.

Electronic Engineering Association (EEA). A General Guide to the Safe Use of Lasers. London, UK: EEA, September, 1966.

Emmett, E. A., and Horstman, S. W. Factors influencing the output of ultraviolet radiations during welding. *Journal of Occupational Medicine* 18(1976):41.

Eure, J. A., Nicholls, J. W., and Elder, R. L. Radiation exposure from industrial microwave applications. *American Journal of Public Health* 62(1972):1573.

Farr, P. M., and Diffey, B. L. The erythemal response of human skin to ultraviolet radiation. *Journal of Investigative Dermatology* 84 (1985): 449-450.

Grandolfo, M. Michaelson, S. M., Rindi, A., eds. *Biological Effects and Dosimetry of Static and ELF Electromagnetic Fields.* New York: Plenum Press, 1985.

Graves, H. B., ed. *Biological and Human Health Effects of Extremely Low Frequency Electromagnetic Fields.* Arlington, VA: American Institute of Biological Sciences, 1985.

Hathaway, J. A., Stern, N., Soles, E. M., et al. Ocular medical surveillance on microwave and laser workers. *Journal of Occupational Medicine* 19(1977):683.

International Electrotechnical Commission (IEC). *Radiation Safety of Laser Products, Equipment Classification and User's Guide.* Geneva: Publ. W.S. 825 IEC, 1984.

Kaufman, J. E., ed. *IES Lighting Handbook: Application Volume.* New York: The Illuminating Engineering Society, 1981.

Key, M. M., Henschel, A. F., Butler, J., eds. *Occupational Diseases: A Guide to Their Recognition,* DHHS, PHS Publication No. 1097. Washington, DC: U.S. Government Printing Office, 1977.

Koller, L. R. *Ultraviolet Radiation,* 2nd ed. New York: John Wiley & Sons, Inc., 1965.

Lapp, R. E., and Andrews, H. L. *Nuclear Radiation Physics.* New York: Prentice-Hall, Inc., 1948.

Lovsund, P., Oberg, P. A., and Nilsson, S. B. G. ELF magnetic fields in electrosteel and welding industries. *Radio Science* 17(1982):359-389.

Luckiesh, M. *Applications of Germicidal, Erythemal and Infrared Energy.* New York: D. Van Nostrand Co., 1946.

Marshall, W. J., Sliney, D. H., Lyon, T. L., et al. Nonionizing Radiation Protection Special Study No. 42-0312-77, Evaluation of the Potential Retinal Hazards from Optical Radiation Generated by Electric Welding and Cutting Arcs. Aberdeen Proving Ground, MD: U.S. Army Environmental Hygiene Agency, 1977.

McCullough, W. Laser beam precautions. *National Safety Congress Transactions* 13(1968):15-18.

Moss, C. E., Murray, W. E., Parr, W. H., et al. *An Electromagnetic Radiation Survey of Selected Video Display Terminals.* Cincinnati: NIOSH (DHHS), 1978.

Okress, E. C., ed. *Microwave Power Engineering,* vol. 2., *Applications.* New York: Academic Press, 1968.

Parrish, J. A., et al. Erythema and melanogenesis action spectra of normal human skin. *Photochemistry and Photobiology* 36(198):187-191.

Powell, C. H., and Hosey, A. D., eds. *The Industrial Environment...Its Evaluation and Control: Syllabus,* DHHS, PHS Publication No. 614. Washington, DC: U.S. Government Printing Office, Rev. 1965.

Sliney, D. H. The amazing laser. *National Safety Congress Transactions* 8(1968):38-42.

Sliney, D. H., and Franks, J. K. "Safety Rules and Recommendations." In *Laser Focus Buyers' Guide,* January 1977.

Sperling, H. G., ed. *Laser Eye Effects.* A Report of the Armed Forces NRC Committee on Vision, Washington, DC, April 1968.

Stair, R. *Special-Transmissive Properties and Use of Eye-Protective Glasses.* National Bureau of Standards Circular No. 471. Washington, DC: U.S. Government Printing Office, October 8, 1948.

Stuchly, M. A., Leeuyer, D. W., and Mann, R. O. Extremely low frequency electromagnetic emissions from video display terminals, and other devices. *Health Physics* 45(1983): 713.

U.S. Atomic Energy Commission, Nevada Operations Office. *Recommendations of the Ad Hoc Laser Committee, Standards for Laser Safety,* USAEC Nevada Operations Office, Las Vegas, October 1967.

U.S. Department of Health, Education, and Welfare. BRH Bulletin. Vol. 5, No. 16, Rockville, MD: Bureau of Radiological Health, July 1971.

——— Regulations for the Administration and Enforcement of the Radiation Control Health and Safety Act of 1968. HEW Publication (FDA) 76-8035. Rockville, MD: Food and Drug Administration, 1976.

——— *Criteria for a Recommended Standard. . .Occupations Exposure to Ultraviolet Radiations.* NIOSH No. PB-21-4268, NTIS. Springfield, VA.

U.S. Food and Drug Administration. Laser Products Performance Standard, 21 *CFR* 1910, Rockville, MD, 1974.

Van Pelt, W. F., Payne, W. R., and Peterson, R. W. *A Review of Selected Bioeffects Thresholds for Various Spectral Ranges of Light.* BRH Bulletin. Rockville, MD: DHHS, Food and Drug Administration, 1973.

Voss, W. A. G. "Advances in the Use of Microwave Power." DHHS, PHS, Seminar Paper No. 008. February 1970.

Weston, B. A. *Laser Systems—Code of Practice.* London, UK: The Ministry of Aviation, November 1965.

World Health Organization (1984): Environmental Health Criteria 35: *Extremely Low Frequency (ELF) Fields.* Edited by M. J. Seuss. Geneva: WHO.

ADDENDUM

Microwave oven inspection guidelines

The material is adapted from USAEHA-RL *Technical Guide,* June 1974.

Hazards. The cooking process in microwave ovens is accomplished with microwave energy generated by a magnetron tube. This energy is the same type as that emitted by radar units. The hazards involved are due to the leakage of microwave energy and are usually confined to the area surrounding the oven door.

Leakage from microwave ovens is usually caused by a worn door seal, faulty door safety interlocks, and (on ovens with a viewing area) by the seal around the faceplate or screen. Failure of door interlocks to shut off the oven when the door is opened could expose personnel in the vicinity of the oven to microwave radiation levels a hundredfold above safe levels.

The use of metal cooking containers or aluminum foil wrapped food in these ovens also increases leakage levels. If for any reason the oven must be operated without food, to check its operation, a small container of water should be placed in the oven as a load. This load would simulate the operating condition the oven would normally be subjected to. It is possible to operate the oven in a no-load condition, but this is inadvisable since damage to the microwave generating source could occur if operated empty over an extended period of time. When operated under such no-load conditions, the leakage would be of a higher level since none of the energy is absorbed.

The performance standard for microwave ovens was amended by adding requirements that specific safety instructions be included in use and service manuals, and that all microwave ovens bear certain specified warning labels. The instructions must contain the following wording: "Precautions to avoid possible exposure to excessive microewave energy: (1) Do not attempt to operate this oven with the door open . . . (2) Do not allow soil or residue to accumulate on sealing surfaces. (3) Do not operate the oven if it is damaged. It is particularly important that the oven door close properly and that there is no damage to the door (bent); hinges and latches (broken or loosened); door seals and sealing surfaces. (4) The oven should not be adjusted or repaired by anyone except properly qualified service personnel."

Visual inspections. Although it is impossible to determine the exact amount a microwave cooking oven is leaking, without suitable instrumentation, there are certain visual checks which can be made.

As almost all instances of leakage are around the door area, the following visual checks should be made:

1. Check for loose or bent door hinges, and screws missing from hinges.
2. Sprung, warped, or misaligned doors.
3. Faulty interlocks; for example, oven should not be operable with its door slightly open.
4. Worn, missing or damaged seals around the door or viewing area.
5. Check for pitted and burnt spots around the periphery of the door closure area. This is usually caused by arcing as a result of grease buildup around the door. Ovens should be checked at frequent intervals to eliminate this arcing which causes an increase in leakage levels.
6. Check to see if personnel are using metal cooking vessels, or aluminum foil, as reflections from metal objects can also increase leakage.
7. Check to see that ovens are not being operated empty. If for any reason the oven must be operated without food, to check its operation, such as interlocks, a small bowl or beaker of water should be used to simulate a normal load. Appropriate signs indicating the potential hazard to personnel should be posted.

Inventory. A suggested oven survey sheet, which can be used for maintaining an inventory of ovens at an installation, should include the following information:

- Installation
- Oven location, building number, and using activity
- Presently in use or under procurement—if under procurement, give estimated date of installation
- Oven manufacturer
- Model number
- Serial number, if available
- Operating frequency of magnetron—this should be either 915 MHz or 2450 MHz (report other frequencies, if found)
- Number of magnetron tubes
- Name of individual responsible for oven
- Activity providing maintenance for oven
- Name of individual completing this form.

12

Temperature Extremes

by Edwin L. Alpaugh
Revised by Theodore J. Hogan, PhD, CIH

INTRODUCTION

Thermal comfort

A COMFORTABLE ENVIRONMENT is the result of simultaneous control of temperature, humidity, and air distribution within the workers' vicinity. This set of factors includes mean radiant temperature as well as the air temperature. Air circulates around the worker. The worker also has radiant heat exchange with the surrounding surfaces (i.e., glass and inside walls). Air is brought into motion within a given space either thermally or by mechanical forces. In choosing optimal conditions for comfort, knowledge of the energy expended during the course of routine physical activities is necessary, since body heat production increases in proportion to exercise intensity.

Thermal comfort is a function of many variables, including the season of the year, dry and wet bulb temperatures, and cultural practices and habits. A chart of comfort zones such as that shown in Figure 12-1, although valid for people in the United States, is not necessarily valid for any other country. People in England, for instance, would find the zone listed as "comfortable" in Figure 12-1 too warm for comfort, because their homes are kept at lower temperatures than those in the United States. Nevertheless, describing comfort zones is necessary for the proper design and operation of heating and air conditioning systems.

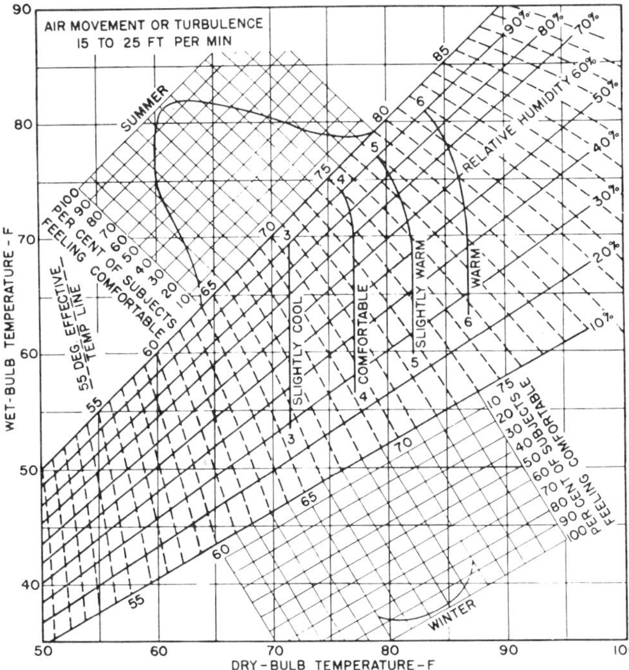

Figure 12-1. Comfort chart for still air. The winter comfort line is limited to rooms heated by central, convection-type systems (such as homes and offices) where occupants fully adapt to the artificial air conditions and not to locations where exposure is less than three hours. Optimum summer comfort line pertains to northern portion of the United States and southern Canada at elevations less than 300 m (1,000 ft) above sea level. (Reprinted with permission from ASHRAE *Handbook*)

Evaluation of the information relating the physiology of a person to the physical aspects of the environment is not a simple task. Considerably more is involved than simply taking a number of air temperature measurements and making decisions on the basis of that information.

Whenever temperature differences exist between two or more bodies, heat can be transferred. Net heat transfer is always from the body (or object) of higher temperature to the body of lower temperature and occurs by one or more of the following mechanisms.

- Conduction. The transfer of heat from one point to another within a body, or from one body to another when both bodies are in physical contact. While conduction can be a localized source of discomfort from direct physical contact with a hot or cold surface, it normally is not a significant factor to total heat stress.
- Convection. The transfer of heat from one place to another by moving gas or liquid. Natural convection results from differences in density caused by temperature differences. Thus warm air is less dense than cool air. Warm air rises relative to the cool air, and vice versa.
- Radiation. The process by which energy, electromagnetic (visible and infrared), is transmitted through space without the presence or movement of matter in or through this space.

There are two sources of heat that are important to anyone working in a hot environment: (1) internally generated metabolic heat and (2) externally imposed environmental heat.

Metabolic heat is a by-product of the chemical processes that occur within cells, tissues, and organs. Environmental heat is important because it influences the rate at which body heat can be exchanged with the environment and consequently the ease with which the body can regulate and maintain a normal temperature.

The net heat exchange between a person and the ambient environment can be expressed by:

$$H = M \pm R \pm C - E$$

where: H = body heat storage load
M = metabolic heat gain
R = radiant or infrared heat load
C = convection heat load
E = evaporative heat loss

The body tries to maintain a balance between the heat gained by work, radiant and converted heat imposed on the body, and the heat lost by sweating (evaporation). This balance can be expressed by:

$$H + E = M \pm R \pm C$$

Ideally, the change in body heat content should be zero. If this balance cannot be maintained by evaporation, then heat can build up in the body, causing a rise in internal temperature.

Metabolic heat gain (M) is composed of the basal or resting metabolism that provides the energy necessary to keep the body functioning, as well as the working metabolism that provides the energy necessary for the body to accomplish specific tasks. Metabolism can only add heat to the body; therefore, M is always positive. However, the more efficiently muscular work can be done, the less heat will be generated while accomplishing a given task. The process of improving muscular efficiency is "training."

Radiant heat load (R) is energy in the form of wavelengths that are transformed into heat when they strike an object. Whether the human body emits or receives radiant energy depends on the temperature of the body, and of the surrounding objects. Thus, R can be either negative or positive.

Convective heat load (C) is the amount of heat energy transferred between the skin and air. Human skin temperature varies over a wide range, 25–35 C (77–95 F). Air temperature in excess of skin temperature will warm the body; air temperatures less than skin temperature will cause the body to be cooled.

The evaporative heat loss from the body is signified by E. Vaporization of perspiration reduces body heat and, therefore, its value is always negative. The use of fans and blowers to increase E is a common method of cooling workers.

Conductive heat load (D) is the heat energy transferred between parts of the body and other objects when they are in direct contact. Normally, this term is insignificant and can be disregarded except in special cases, such as in swimming and diving.

Regulation of body temperature is an important physiological function. The ease with which it can be successfully accomplished is determined by an individual's ambient environment, including the air temperature, humidity, air movement, and radiant energy exchange with the surroundings.

At times, people must work in situations where there are extremes of cold or hot temperatures. Although more studies have been made on the evaluation of the stress from hot environments than that from cold, both are important and will be discussed in this chapter.

COLD ENVIRONMENTS

Many thousands of people work exposed to cold temperatures in freezer plants, meat-packing houses, cold storage facilities, farming in northern areas, cattle ranching, lumbering, and other outdoor activities. Because humans are homoiotherm (warm-blooded animals) they must maintain their body heat. If properly protected, they can work efficiently in both natural and manmade frigid climates.

How the body handles cold

(*Note:* Much of the material in this subsection was adapted from the National Safety Council's *Pocket Guide to Cold Stress.* See Bibliography for details.)

The human body is designed to function best at a rectal temperature of 38–39 C (99–100 F). The body maintains this temperature by gaining heat from food and muscular work, or by losing it through radiation and sweating. The body's first physiological defense against cold is constriction of the blood vessels of the skin and/or shivering.

In considering temperature control, it is customary to think of the body as having two main parts—a shell and a core. The shell includes millions of ultrafine blood vessels (capillaries), nerves, muscles, and fat. The heart, lungs, brain, kidneys, and other internal organs make up the core.

Cold first affects the skin, cooling the blood in the peripheral capillaries. A complex of signals from the skin and core are integrated in a portion of the brain called the hypothalamus. The hypothalamus regulates many basic body functions, including body temperature. The hypothalamus works like a thermostat, making adjustments as needed to maintain a normal temperature. When a chill signal is received, the hypothalamus begins two processes—one to conserve heat already in the body, the other to generate new heat.

Heat conservation is accomplished by causing the blood vessels in the shell to constrict, which reduces the heat loss from the surface of the skin, and making the shell an insulator. This constriction also inhibits the function of the sweat glands, preventing heat loss by evaporation.

Glucose is produced to provide additional fuel. Glucose causes the heart to beat faster, sending oxygen and glucose-rich blood to the tissues where needed.

Involuntary shivering begins in an attempt to produce heat by rapid contractions of the muscles, much as heat is generated by strenuous activity. Shivering raises the body's metabolic rate.

If someone becomes fatigued during physical activity, they will be more prone to heat loss. As exhaustion approaches, sudden enlargement of the blood vessels can occur, with a resulting rapid loss of heat.

The frequency of accidents seems to be higher in cold environments. Nerve impulses are slowed, we react sluggishly, fumble with our hands and become clumsy. There are also safety problems common to cold environments. They include ice, snow blindness, reflections from snow, and the possibility of burns from contact with cold metal surfaces.

Cold disorders

Cold injury is classified as either localized, as in frostbite, frostnip, or chillblain; or generalized, as in hypothermia.

The main factors contributing to cold injury are exposure to humidity and high winds, contact with wetness or metal, inadequate clothing, age and general health. Physical conditions that worsen the effects of cold include allergies, vascular disease, excessive smoking and drinking, and specific drugs and medicines.

Hypothermia. Air temperature alone is not enough to judge the cold hazard of a particular environment. Most cases of hypothermia develop in air temperatures between 2–10 C (30–50 F). However, by the time you consider a factor such as the windchill, the effective temperature could be significantly lower.

The first symptoms of hypothermia are uncontrollable shivering and the sensation of cold; the heartbeat slows and sometimes becomes irregular, the pulse weakens and the blood pressure changes.

Severe shaking or rigid muscles are caused by bursts of body energy and changes in the body chemistry. Uncontrollable fits of shivering, vague or slow slurred speech, memory lapses, incoherence and drowsiness are some of the symptoms that can occur. Other symptoms that can be seen before complete collapse are cool skin, slow, irregular breathing, low blood pressure, apparent exhaustion, and fatigue after rest.

As the core temperature (38–39 C [100.4–102.2 F]) drops into into the mid-range, the victim can become listless, confused, and make little or no effort to keep warm. Pain in the extremities can be the first warning of dangerous exposure to cold. Severe shivering must be taken as a sign of danger. At about 29 C (85 F), serious problems can develop because of significant drops in blood pressure, pulse rate, and respiration. In some cases, the victim may die. (The core temperatures given here are rectal temperatures.)

When someone becomes fatigued during physical activity, they become more susceptible to heat loss. As exhaustion approaches, the body's ability to contract blood vessels diminishes; blood circulation occurs closer to the surface of the skin; and rapid loss of heat and cooling begins.

Sedative drugs and alcohol increase the risk of hypothermia. Sedative drugs interfere with the transmission of impulses to the brain. Alcohol dilates the blood vessels near the skin surface, which increases heat loss and lowers body temperature.

Blood vessel abnormalities

Certain blood vessel abnormalities can be associated with increased cold sensitivity. They include Raynaud's phenomenon and acrocyanosis.

Raynaud's phenomenon. This is a condition that refers to the blanching of the distal portion of the digits. Numbness, itching, tingling, or a burning sensation may occur during intermittent attacks. The reaction is triggered by cooling of the skin. Raynaud's phenomenom is associated with a number of diseases including systemic scleroderma, pulmonary hypertension, multiple sclerosis, and an idiopathic form called Raynaud's disease. Of great importance in industry is the association of this phenomenon with the use of vibrating hand tools in a condition sometimes called white finger disease. Persistent cold sensitivity, ulceration, and amputations can occur in severe cases.

Acrocyanosis. This is a relatively benign condition in which the hands and/or feet acquire a slightly blue, purple, or greyish coloring. The condition is caused by exposure to cold that reduces the level of hemoglobin in the blood.

Thromboangiitis obliterans. This is one of the many disabling diseases that can result from tobacco use. Inflammation and fibrosis of connective tissue surrounding medium-sized arteries and veins and their walls result in blockage of the arteries. Gangrene of the affected limb often requires amputation.

Workers suffering from blood vessel abnormalities should take special precautions to avoid chilling. Some people develop sensitivity reactions when exposed to cold. Some elderly and very young persons have an impaired ability to sense cold. They may fail to respond quickly enough by increasing room temperature and adding clothing. A disabled person may not be able to leave a cold environment. Increased cold sensitivity can follow injury to an extremity—particularly frostbite or crushing injuries.

Frostbite

Frostbite can occur without hypothermia when the extremities do not receive sufficient heat from central body stores. This can occur because of inadequate circulation and/or because of inadequate insulation. Frostbite occurs when there is freezing of the fluids around the cells of the body tissues. This freezing is from exposure to extremely low temperatures. The condition results in damage to and loss of tissue. The most vulnerable parts of the body are the nose, cheeks, ears, fingers, and toes.

Damage from frostbite can affect either the outer layers of skin only, or it can include tissue beneath these outer layers. Damage from frostbite can be serious; scarring, tissue death, and amputation are all possibilities, as is permanent loss of movement in the affected parts. However, skin and nails that slough off can grow back.

The freezing point of the skin is about –1 C (30 F). As wind velocity increases, heat loss is greater and frostbite will occur more rapidly. If skin should come in contact with objects colder than freezing, frostbite may develop at the point of contact, even in a warm environment.

There are three degrees of frostbite: first degree, which is freezing without blistering or peeling; second degree, which is freezing with blistering or peeling; and third degree, which is freezing with death of skin tissues and possibly of the deeper tissues.

Symptoms of frostbite include the following:

1. the skin changes color to white or grayish-yellow, progresses to reddish-violet, and finally turns black as the tissue dies;
2. pain may be felt at first, but subsides;
3. blisters may appear;
4. the affected part is cold and numb.

When frostbite of the outer layer of skin occurs, the skin has a waxy or whitish look and is firm to the touch (the tissue underneath is still resilient). In cases of deep frostbite, the tissues are cold, pale, and solid. Injury is severe.

The first symptom of frostbite is usually an uncomfortable sensation of coldness, followed by numbness. There may be a tingling, stinging or aching feeling, or even cramping pains. The victim is often unaware of the frostbite until someone else observes the symptoms.

Trench foot. This condition may be caused by long, continuous exposure to cold without freezing, combined with persistent dampness or actual immersion in water. Edema (swelling), tingling, itching, and severe pains occur, and may be followed by blistering, death of skin tissue, and ulceration. When other areas of the body are affected, the condition is known as chilblains.

Frostnip. This occurs when the face or extremities are exposed to a cold wind, causing the skin to turn white.

Windchill index

Air temperature alone is not sufficient to judge the cold hazard of a particular environment. Heat loss from convection is probably the greatest and most deceptive factor in loss of body heat. When the air in a given environment is –1C (30 F), the body will feel cool. Given the same temperature and a wind of 40 km/h (25 mph), the air will feel bitterly cold. In essence, the wind blows away the thin layer of air that acts as an insulator between the skin and the outside air temperature.

Indices for evaluating cold environments include Threshold Limit Values (TLVs) for cold stress and the windchill index (Table 12-A). The TLVs for cold stress have been determined and should be reviewed for application to a particular work environment (see Appendix B-1). The cold stress TLVs are based on the windchill index. The windchill index is probably the best known and the most used of cold stress indices. All of the cold stress indices have limitations like those for heat stress; however, under the right conditions, the information provided can be beneficial in evaluating the cold environment.

The windchill factor is the cooling effect of any combination of temperature and wind velocity or air movement. The windchill index should be consulted by everyone facing exposure to low temperatures and wind. Note that windchill temperatures have no significance other than that expressed—the effect on the body. Although the windchill temperature can be below the freezing point of water, it will not freeze unless the air temperature is also below the freezing point.

The windchill index does not take into account the following: (1) the body part exposed to cold, (2) the level of activity with its effect on body heat production, or (3) the amount of clothing worn (Figure 12–2). Figure 12–2 shows the impor-

Table 12-A. Windchill Index

	Actual Thermometer Reading (F)									
	50	40	30	20	10	0	–10	–20	–30	–40
Wind speed in mph	*Equivalent Temperature (F)*									
calm	50	40	30	20	10	0	–10	–20	–30	–40
5	48	37	27	16	6	–5	–15	–26	–36	–47
10	40	28	16	4	–9	–21	–33	–46	–58	–70
15	36	22	9	–5	–18	–36	–45	–58	–72	–85
20	32	18	4	–10	–25	–39	–53	–67	–82	–96
25	30	16	0	–15	–29	–44	–59	–74	–88	–104
30	28	13	–2	–18	–33	–48	–63	–79	–94	–109
35	27	11	–4	–20	–35	–49	–67	–82	–98	–113
40	26	10	–6	–21	–37	–53	–69	–85	–100	–116
Over 40 mph (little added effect)	*Little Danger (for properly clothed person)*				*Increasing Danger (Danger from freezing of exposed flesh)*			*Great Danger*		

The human body senses "cold" as a result of both air temperature and wind velocity. Cooling of exposed flesh increases rapidly as the wind velocity goes up. Frostbite can occur at relatively mild temperatures if wind penetrates the body insulation. For example, when the actual air temperature of the wind is 4.4 C (40 F) and its velocity is 48 km/h (30 mph), the exposed skin would perceive this situation as an equivalent still air temperature of –11 C (13 F).

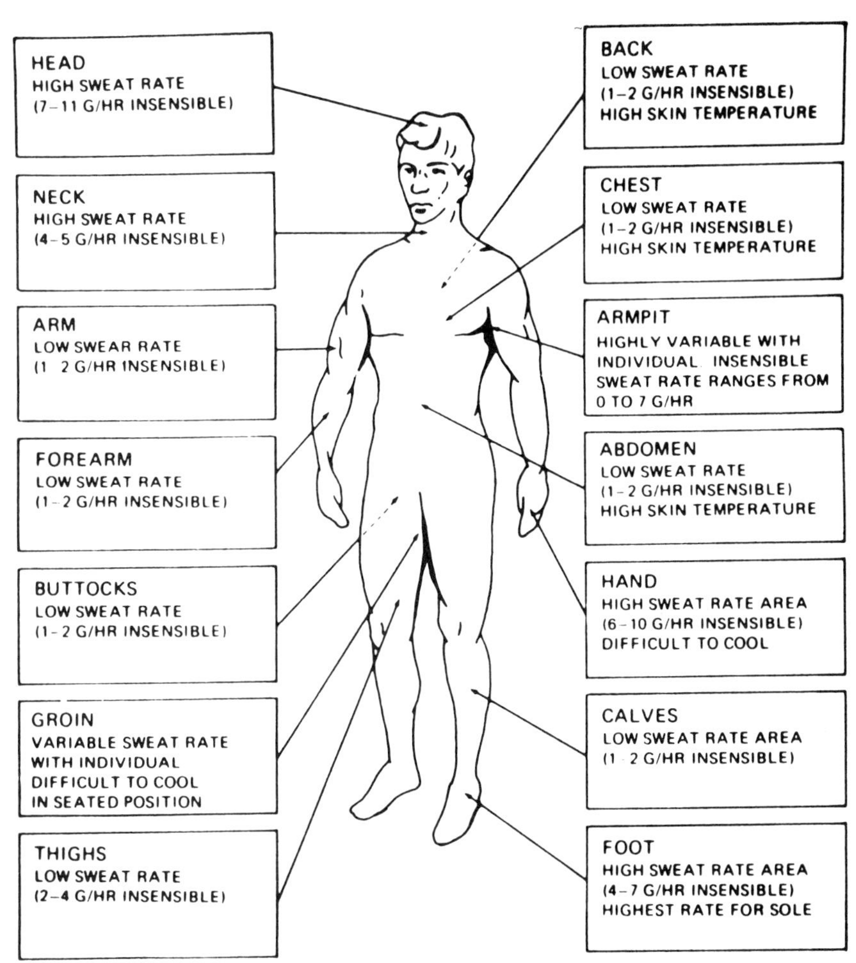

Figure 12-2. Regional cooling requirements of the human body in air at sea level at rest. (Adapted from NASA *Bioastronautics Data Book,* 2nd ed. Washington, DC: U.S. Government Printing Office, 1973)

tance of keeping the head, neck, and hands covered. The skin is the natural barrier of the body for heat and cold. Thus, skin temperature is an important factor in body thermal control.

Preventing cold stress

In preventing cold stress the health and safety professional must consider factors relating both to the individual and the environment.

Acclimatization, water and salt replacement, medical screening, continuing medical supervision, proper work clothing, and training and education will contribute to the prevention of cold stress and injury related to working in a cold environment.

Control of the environment involves engineering controls, work practices, work-rest schedules, environmental monitoring, and consideration of the windchill temperature.

Acclimatization. Some degree of acclimatization may be achieved in cold environments. With sufficient exposure to cold, the body does undergo some changes that increase comfort and reduce the risk of cold injury. However, these physiological changes are usually minor and require repeated uncomfortably cold exposures to induce them. People who are physically unfit, older, obese, taking medications or using alcohol or drugs may not acclimatize readily.

Dehydration. Working in cold areas causes significant water losses through the skin and lungs as a result of the dryness of the air. Increased fluid intake is essential to prevent dehydration, which affects the flow of blood to the extremities and increases the risk of cold injury.

Warm, sweet, caffeine-free, nonalcoholic drinks and soup should be available at the worksite for fluid replacement and caloric energy.

Salt. The body needs a certain amount of salt and other electrolytes to function properly. However, using salt tablets is not recommended. Salt tablets cause stomach irritation, which may include nausea and vomiting. A normal, balanced diet should take care of salt needs. Anyone with high blood pressure or who is on a restricted sodium diet should consult a physician for advice on salt intake.

Diet. It is important for people who work in cold environments to eat a well-balanced diet. Restricted diets can deprive the body of elements needed to withstand cold stress.

Control measures. Continuous exposure of skin should not be permitted when the windchill factor results in an equivalent temperature of −32 C (−26 F). Workers exposed to air temperatures of 2 C (35.6 F) or lower who become immersed in water or whose clothing gets wet should be given dry clothing immediately and treated for hypothermia.

Engineering controls. The following are some of the ways engineering controls can be used to reduce the stress of a cold environment;

- General or spot heating should be used to increase temperature at the workplace.
- If fine work is to be performed with bare hands for 10 or 20 minutes or more, special provisions should be made to keep the worker's hands warm. Warm air jets, radiant heaters, or contact warm plates can be used.
- The work area should be shielded if the air velocity at the job site is increased by wind, draft, or ventilating equipment.
- The air velocity in refrigerated rooms should be minimized as much as possible, and should not exceed 1 m/sec (200 fpm) at the job site.
- At temperatures below −1 C (30 F), metal handles of tools and control bars should be covered with thermal insulating material.
- Unprotected metal chair seats should not be used.
- When necessary, equipment and processes should be substituted, isolated, relocated, or redesigned to reduce cold stress at the worksite.
- Power tools, hoists, cranes or lifting aids should be used to reduce the metabolic workload.
- Heated warming shelters such as tents and cabins should be made available if work is performed continuously in an equivalent chill temperature of −7 C (20 F) or below. Workers should be encouraged to use the shelters regularly.

Engineering control of cold stress can be very complex, and often depends more on ingenuity than on standard methods.

Administrative work practice controls. These controls include work practices or rules designed to reduce the total cold-stress burden. Some of these include:

- a work-rest schedule to reduce the peak of cold stress;
- enforcing scheduled rest breaks;
- enforcing frequent intake of warm, sweet, caffeine-free, nonalcoholic drinks or soup;
- scheduling the coldest work for the warmest part of the day;
- moving work to warmer areas whenever possible;
- assigning extra workers to highly demanding tasks;
- allowing workers to pace themselves, and take extra work breaks when needed;
- making relief workers available for workers who need a break;
- teaching workers the basic principles of preventing cold stress and emergency response to cold stress;
- maintaining protective supervision or a buddy system for those who work at −12 C (10 F) or below;
- allowing new employees time to adjust to conditions before they work full-time in cold environments;
- arranging work to minimize sitting still or standing for long periods of time;
- reorganizing work procedures so as much of a job as possible is performed in a warm environment. This will reduce the amount of work that must be done in a cold environment;
- including the weight and bulkiness of clothing when estimating work performance requirements and weights to be lifted.

Special considerations. Older workers, or workers with circulatory problems need to be extra careful in the cold. Additional insulating clothing and reduced exposure time should be considered for these workers. Obese and chronically ill people need to make a special effort to follow preventive measures. Sufficient sleep and good nutrition are important for maintaining a high level of tolerance to cold. If possible, the most stressful tasks should be performed during the warmer parts of the day. Double shifts and overtime should be avoided. Rest periods should be extended to cope with increases in cold stress.

A worker should immediately go to warm shelter if any of the following symptoms are spotted: the onset of heavy shivering, frostnip, the feeling of excessive fatigue, drowsiness, and/or euphoria. The outer layer of clothing should be removed when entering a heated shelter. If possible, a change of dry work clothing should be provided to prevent workers from returning to work with wet clothing. If this is not feasible, the remaining clothing should be loosened to permit sweat to evaporate.

Alcohol should not be consumed while in the warmer environment. Anyone on medications such as blood pressure control or water pills should consult a physician about possible side effects from cold stress. It is strongly recommended that workers suffering from diseases or taking medication that interferes with normal body temperature regulation, or that reduces tolerance of cold, not be permitted to work in temperatures of −1 C (30 F) or below.

It may be advisable for workers to weigh themselves at the beginning and end of the workday to check for weight loss that might occur from progressive dehydration.

A control program for cold stress. A control program for cold stress in industry should include the following elements:

- *Medical supervision of workers;* including preplacement physicals that evaluate fitness, weight, the cardiovascular system, and other conditions that might make worker susceptible to cold stress. Medical evaluation during and after cold illnesses and a medical release for returning to work should be required.
- *Employee orientation and training* on cold stress, cold-induced illnesses and their symptoms, water and salt replacement, proper clothing, work practices and emergency first aid procedures.
- *Work-rest regimens,* with heated rest areas and enforced rest breaks.
- *Scheduled drink breaks* for recommended fluids.
- *Environmental monitoring,* using the air temperature and wind speed indices to determine wind chill and adjust work-rest schedules accordingly.
- *Reduction of cold stress* through engineering and administrative controls, and the use of personal protective equipment.

Personal protective equipment and clothing. Personal protective equipment and protective clothing is essential. The correct clothing depends on the specific cold stress situation. It is important to preserve the air space between the body and the outer layer of clothing in order to retain body heat. The more air pockets each layer of clothing has, the better the insulation. However, the insulating effect is negated if the clothing interferes with the evaporation of sweat, or if the skin or clothing is wet.

The most important parts of the body to protect are the feet, hands, head, and face. Hands and feet are the farthest from the heart, and become cooled most easily. Keeping the head covered is important, because as much as 40 percent of body heat can be lost when the head is exposed.

Clothing made of thin cotton fabric is ideal—it helps evaporate sweat by picking it up and bringing it to the surface. Loosely fitted clothing also aids sweat evaporation. Tightly fitted clothing of synthetic fabrics interferes with evaporation.

Recommended clothing may include: a cotton t-shirt and shorts or underpants under cotton and wool thermal underwear.

Two-piece long underwear is preferred, because the top can be removed and put back on as needed. Socks with high wool content are best. When two pairs of socks are worn, the inside pair should be smaller, and be made of cotton. If necessary, wool socks can also double as mittens. Wool or thermal trousers (either quilted or specially lined) are preferred. Belts can constrict and reduce circulation. Suspenders may be preferred. You will need extra room for trousers to fit over long underwear. Trousers should be lapped over boot tops to keep out snow or water. For heavy work, a felt-lined, rubber-bottomed, leather-topped boot with a removable felt insole is preferred. Boots should be waterproofed and socks changed when they become sweat soaked. Air insole cushions and felt liners should be used with chemical and/or water-resistant boots. The best foot protection is provided by insulated boots sealed inside and outside by vapor barriers. Either a wool shirt or a wool sweater over a cotton shirt should be worn. Size-graduated shirts and sweaters can be worn in layers. An anorak or snorkel coat or arctic parka should fit loosely, with a drawstring at the waist. Sleeves should fit snugly. The hood prevents the escape of warm air from around the neck, and can be extended past the face to create a frost tunnel, which warms the air for breathing. A wool knit cap provides the best head protection. When a hard hat is worn, a liner should be used. Wool mittens are more efficient insulators than gloves; they can be worn over gloves for extra warmth.

A face mask or scarf is vital when working in a cold wind. A ski mask with eye openings gives better visibility than a snorkel hood. Face protectors must be removed periodically, so the worker can be checked for signs of frostbite.

Workers should wear several layers of clothing instead of a single heavy outer garment. In addition to offering better insulation, layers of clothing can be removed as needed to keep the worker from overheating. The outer layer should be windproof and waterproof. Body heat is lost quickly if the protective layer isn't windproof.

All clothing and equipment must be properly fitted and worn to avoid interfering with the circulation. Thermal-type masks and respirators are available for those bothered by breathing very cold air. Full-facepiece respirators must have separate respirator channels to prevent fogging and frosting of the facepiece. Double-layered goggles with foam padding around the edges are effective in extremely cold conditions.

Liquids conduct heat better than air, and have a greater capacity for heat than air. For example, a spill of cold gasoline on skin can freeze the tissue quickly. That is why it is a good idea to wear chemical-resistant gloves (such as neoprene gloves with cotton inserts) for chemical handling operations. Workers handling chemicals with permeable-type gloves should always keep extra gloves available should one pair become contaminated.

Gloves should be used by workers if manual dexterity is not required, or if the air temperature falls below 16 C (60 F) for sedentary work, 4 C (40 F) for light work, and 7 C (20 F) for moderate work. Mittens should be used instead of gloves if the air temperature is 18 C (0 F) or less.

Avoid the following: Dirty or greasy clothing loses much of its insulation value. Air pockets in dirty clothes are crushed or filled, and heat can escape more easily. Denim is not a good protective fabric. It is relatively loosely woven, which allows water to penetrate and wind to blow away body heat that should be trapped between the body and clothing. Any interference with the circulation of the blood reduces the amount of heat delivered to the extremities.

Outdoor construction workers, as well as mountaineers, hikers, and farmers are likely victims of hypothermia if proper precautions to ward off exposure and exhaustion caused by cool (not necessarily freezing) temperatures, wind, and rain are not taken. When stranded during a storm in a vehicle, it is better to stay with the vehicle. The engine can furnish heat, while the vehicle itself can act as a shelter from outside elements. Care should be taken, however, to prevent a buildup of carbon monoxide gas in the closed vehicle. For additional warmth, insulation can be taken from vehicle seats and stuffed in clothing. If travel is in areas where storms occur frequently, emergency supplies should be carried to meet any weather conditions. Survival depends on a clear understanding of the situation. The rules of survival are the same, no matter what the emergency: think clearly, be properly prepared, and keep a weather eye open. These are the best safeguards against hypothermia.

HOT ENVIRONMENTS

Problems of heat stress are more common throughout industry than those presented by a very cold environment. As with cold stress problems, however, the health and safety professional must thoroughly understand all aspects of heat stress and become familiar with control methods and programs for workers exposed to extremes of heat.

Heat stress. Heat stress is the aggregate of environmental and physical work factors that constitute the total heat load imposed on the body. The environmental factors of heat stress include air temperature, radiant heat exchange, air movement, and water vapor pressure. Physical work contributes to the total heat stress of the job by producing metabolic heat in the body in proportion to the intensity of the work. Clothing also affects the heat stress.

Heat strain. Heat strain is the series of physiological responses to heat stress. These responses reflect the degree of heat stress. When the strain is excessive for the exposed individual, a feeling of discomfort or distress may result, and, finally, a heat disorder may ensue. The severity of strain will depend not only on the magnitude of the prevailing stress, but also on the age, physical fitness, degree of acclimatization, and dehydration of the worker. Figure 12-3 summarizes the human response to heat buildup.

Heat disorders. A variety of heat disorders can be distinguished clinically when individuals are exposed to excessive heat. These disorders range from simple postural heat syncope (fainting) to the complexities of heatstroke. Heat disorders are interrelated and seldom occur as distinct entities. A common feature in all heat-related disorders (except simple postural heat syncope) is some degree of elevated body temperature that may then be complicated by deficits of body water. The prognosis depends on the absolute level of the elevated body temperature, promptness of treatment to lower the body temperature, and extent of deficiency or imbalance of fluids or electrolytes. A summary of classification, clinical features, prevention, and treatment of heat illnesses is presented in Table 12-B.

Effects of heat

The central nervous system. The hypothalamus of the brain is

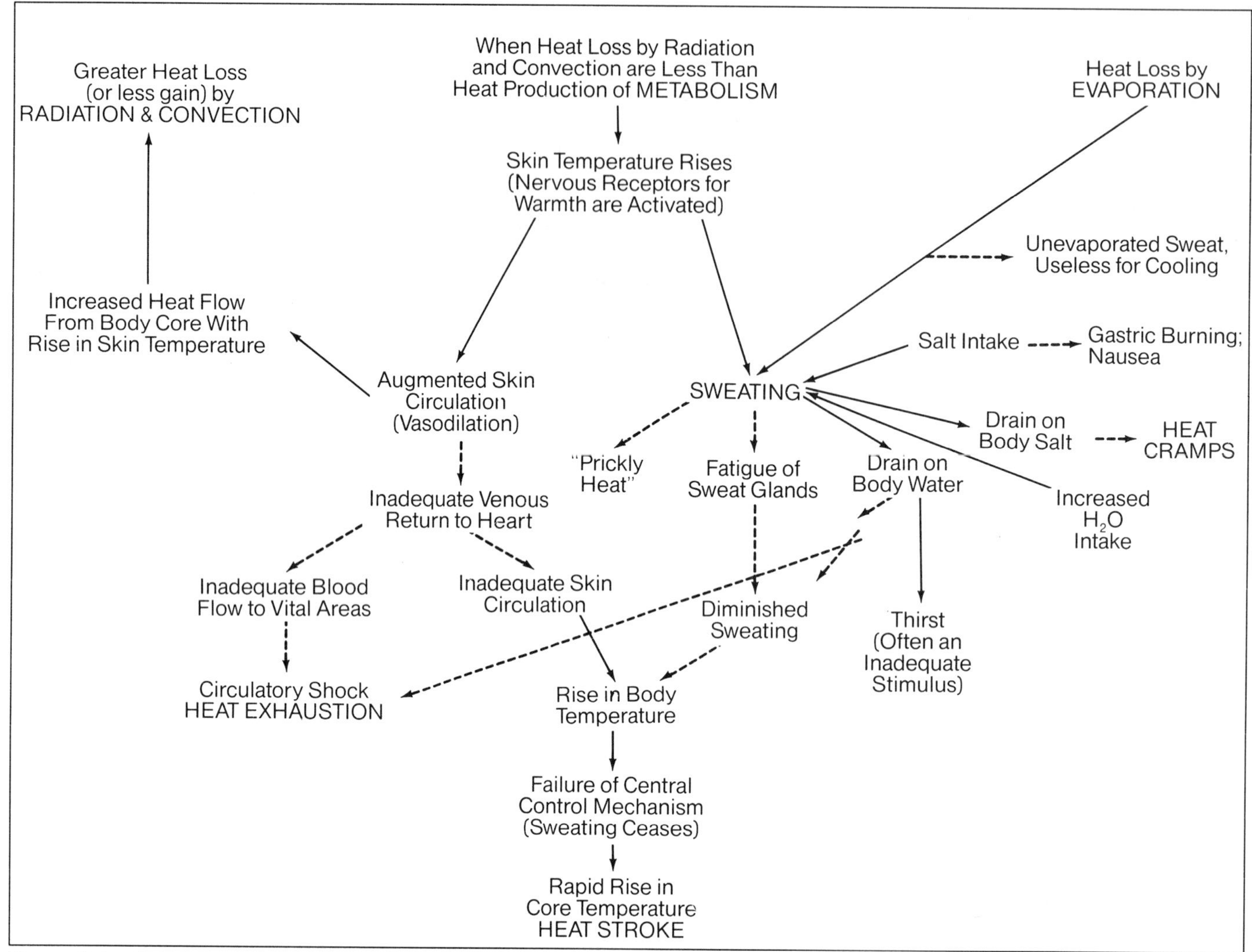

Figure 12-3. Human response to heat buildup. (Reprinted from DHHS, PHS Publication No. 614)

considered to be the central nervous system structure that acts as the primary seat of control. It is believed that the hypothalamus provides a set point and acts as a thermostat that initiates action to heat or cool the body if the core temperature of the body differs from the normal body temperature. For example, in situations where the set point temperature is exceeded, blood vessels near the skin surface are dilated and sweating is increased. This causes heat to be moved from the core of the body to the skin surface where it can be dissipated by air convection over the skin and evaporation of sweat.

Muscular activity and work capacity. While heat can be added to the body from external sources, muscular work can be a major contributor to (metabolic) heat gain. Muscles require an adequate supply of oxygen to perform without fatigue. The measure of the body's ability to supply this oxygen is called the aerobic capacity.

The proportion of maximal aerobic capacity (VO_2 max) needed to do a specific job is important for several reasons. First, the cardiovascular system must respond with an increased cardiac output. At levels of work of up to about 40 percent, VO_2 max is brought about by an increase in both stroke volume and heart rate. When maximum stroke volume is reached, additional increases in cardiac output can be achieved solely by increased heart rate (which itself has a maximum value). Further complexities arise when high work intensities are sustained for long periods, particularly when work is carried out in hot environments. Second, muscular activity is associated with an increase in muscle temperature, which then is associated with an increase in core temperature, with attendant influences on the thermoregulatory controls. Third, at high levels of exercise even in a temperate environment, the oxygen supply to the tissues may be insufficient to meet the oxygen needs of the working muscles completely.

In warmer conditions, an adequate supply of oxygen to the tissues may become a problem even at moderate work intensities due to competition for blood distribution between the working muscle and the skin. Because of the lack of oxygen, the working muscles must then begin to draw on their anaerobic reserves, deriving energy from the oxidation of glycogen ("animal starch") in the muscles. That event leads to the accumulation of lactic acid, which can be associated with the development of muscular fatigue.

Currently, recommendations for an acceptable proportion of VO_2 max for daily industrial work vary from 30–40 percent of the VO_2 max, which in comfortably cool surroundings, is

Table 12-B. Classification, Medical Aspects, and Prevention of Heat Illness

Category and Clinical Features	*Predisposing Factors*	*Underlying Physiological Disturbance*	*Treatment*	*Prevention*
1. Temperature Regulation Heatstroke				
Heatstroke: (1) Hot dry skin usually red, mottled or cyanotic; (2) Rectal temperature 40.5 C (104 F) and over; (3) confusion, loss of consciousness, convulsions, rectal temperature continues to rise; fatal if treatment delayed	(1) Sustained exertion in heat by unacclimatized workers; (2) Lack of physical fitness and obesity; (3) Recent alcohol intake; (4) Dehydration; (5) Individual susceptibility; and (6) Chronic cardiovascular disease	Failure of the central drive for sweating (cause unknown) leading to loss of evaporative cooling and an uncontrolled accelerating rise in t_{re}, there may be partial rather than complete failure of sweating	Immediate and rapid cooling by immersion in chilled water with massage or by wrapping in wet sheet with vigorous fanning with cool dry air, avoid overcooling, treat shock if present	Medical screening of workers, selection based on health and physical fitness, acclimatization for 5–7 days by graded work and heat exposure, monitoring workers during sustained work in severe heat
2. Circulatory Hypostasis Heat Syncope				
Fainting while standing erect and immobile in heat	Lack of acclimatization	Pooling of blood in dilated vessels of skin and lower parts of body	Remove to cooler area, rest recumbent position, recovery prompt and complete	Acclimatization, intermittent activity to assist venous return to heart
3. Water and/or Salt Depletion (a) Heat Exhaustion				
(1) Fatigue, nausea, headache, giddiness; (2) Skin clammy and moist; complexion pale, muddy, or hectic flush; (3) May faint on standing with rapid thready pulse and low blood pressure; (4) Oral temperature normal or low but rectal temperature, usually elevated (37.5–38.5 C) (99.5–101.3 F); water restriction type: urine volume small, highly concentrated; salt restriction type: urine less concentrated, chlorides less than 3g/L	(1) Sustained exertion in heat; (2) Lack of acclimatization; and (3) Failure to replace water lost in sweat	(1) Dehydration from deficiency of water; (2) Depletion of circulating blood volume; (3) Circulatory strain from competing demands for blood flow to skin and to active muscles	Remove to cooler environment, rest recumbent position, administer fluids by mouth, keep at rest until urine volume indicates that water balances have been restored	Acclimatize workers using a breaking-in schedule for 5–7 days, supplement dietary salt only during acclimatization, ample drinking water to be available at all times and to be taken frequently during work day
(b) Heat Cramps				
Painful spasms of muscles used during work (arms, legs, or abdominal); onset during or after work hours	(1) Heavy sweating during hot work; (2) Drinking large volumes of water without replacing salt loss	Loss of body salt in sweat, water intake dilutes electrolytes, water enters muscles, causing spasm	Salted liquids by mouth, or more prompt relief by I-V infusion	Adequate salt intake with meals; in unacclimatized workers supplement salt intake at meals
4. Skin Eruptions (a) Heat Rash (miliaria rubra; "prickly heat")				
Profuse tiny raised red vesicles (blister-like) on affected areas pricking sensations during heat exposure	Unrelieved exposure to humid heat with skin continuously wet with unevaporated sweat	Plugging of sweat gland ducts with retention of sweat and inflammatory reaction	Mild drying lotions, skin cleanliness to prevent infection	Cool sleeping quarters to allow skin to dry between heat exposures
(b) Anhidrotic Heat Exhaustion (miliaria profunda)				
Extensive areas of skin which do not sweat on heat exposure, but present gooseflesh appearance, which subsides with cool environments; associated with incapacitation in heat	Weeks or months of constant exposure to climatic heat with previous history of extensive heat rash and sunburn	Skin trauma (heat rash; sunburn) causes sweat retention deep in skin, reduced evaporative cooling causes heat intolerance	No effective treatment available for anhidrotic areas of skin, recovery of sweating occurs gradually on return to cooler climate	Treat heat rash and avoid further skin trauma by sunburn, periodic relief from sustained heat
5. Behavioral Disorders (a) Heat Fatigue—Transient				
Impaired performance of skilled sensorimotor, mental, or vigilance tasks, in heat	Performance decrement greater in unacclimatized and unskilled worker	Discomfort and physiologic strain	Not indicated unless accompanied by other heat illness	Acclimatization and training for work in the heat
(b) Heat Fatigue—Chronic				
Reduced performance capacity, lowering of self-imposed standards of social behavior (e.g., alcoholic over-indulgence), inability to concentrate, etc.	Workers at risk come from temperate climates, for long residence in tropical latitudes	Psychosocial stresses probably as important as heat stress, may involve hormonal imbalance but no positive evidence	Medical treatment for serious cases, speedy relief of symptoms on returning home	Orientation on life in hot regions (customs, climate, living conditions, etc.)

(Reprinted from DHHS *Criteria for a Recommended Standard . . . Occupational Exposure to Hot Environments—Revised Criteria 1986*. Washington, DC: U.S. Government Printing Office, 1986)

associated with rectal temperatures of, respectively, 37.4 C and 37.7 C (99.3–99.9 F), while work at 50 percent VO_2 max yields a rectal temperature of 38 C (100.4 F) in the absence of heat stress. The World Health Organization's (WHO) recommended limit for deep body temperature under conditions of prolonged daily work and heat is 38 C.

The heart cannot provide enough cardiac output to meet the peak needs of all organ systems or the need for dissipation of body heat. The increase in blood supply to the active muscles is assured by the action of locally produced vasodilator substances in inactive vascular beds. There is a progressive vasoconstriction with the severity of the exercise. This is particularly important in the vascular bed in the digestive organs where venoconstriction also permits the return of blood sequestered in its large venous bed, allowing up to 1 liter (L) of blood to be added to the circulating volume.

The sweating mechanism. Sweat glands are found in abundance in the outer layers of the skin. They are stimulated by nerves to secrete a hypotonic watery solution onto the surface of the skin. In industrial settings, sweat production at rates of about 1 L/hr has been recorded frequently. Sweat represents an important source of cooling if it is evaporated. Large losses of water by sweat also pose a potential threat to successful thermoregulation. A progressive depletion of body water content occurs if water lost is not replaced. A lack of water by itself affects thermoregulation and results in a rise in core temperature.

The rate of evaporation of sweat is controlled by the amount of humidity in the air and the air velocity. Hot environments with increasing humidity limit the amount of sweat that can be evaporated. Sweat that can not be evaporated drips from the skin and does not result in any heat loss from the body.

An important constituent of sweat is salt or sodium chloride. In most circumstances, a salt deficit does not readily occur, because the normal diet provides 8–14 gm/day. Salt supplementation of the normal diet is rarely required except possibly for heat-unacclimatized individuals during the first 2 or 3 days of heat exposure. By the end of the third day of heat exposure, a significant amount of heat acclimatization will have occurred, with a resulting decrease in salt loss in the sweat and urine, and a decrease in the salt requirement. In view of the high incidence of elevated blood pressure in the U.S. worker population and the relatively high salt content of the average U.S. diet, even in those who watch salt intake, recommending increased salt intake is probably not warranted. Salt tablets can irritate the stomach and should not be used.

Water and electrolyte balance. It is imperative to replace the water lost in sweat. It is common for workers to lose 6–8 qt of sweat during a working shift in hot industries. Progressive loss of water results in a lower sweat production and a corresponding increase in body temperature. This can be a dangerous situation for the individual.

Sweat lost in such quantities is often difficult to replace and it is not uncommon for individuals to register a water deficit of 2–3 percent or greater of the body weight at the end of a shift. Because the normal thirst mechanism is not sensitive enough to ensure a sufficient water intake, every effort should be made to encourage individuals to drink water or low sodium noncarbonated beverages. So-called electrolyte-balanced solutions advertised for heat stress relief may not provide any extra benefit over plain water, except that workers may be more willing to drink them. Whatever fluid is chosen, it should be as palatable as possible and served at a temperature of 10–15 C (50–60 F). Small quantities taken at frequent intervals, about 7 oz (150–200 mL) every 15–20 minutes, is a more effective regimen for practical fluid replacement than the intake of 7 oz or more once an hour.

Other related factors. Age, gender, and obesity influence sweat gland function. The aging process results in a more sluggish response of the sweat glands, resulting in a less effective control of body temperature. Gender also plays a slight role in heat tolerance. Purely on a basis of a lower aerobic capacity, the average woman, similar to a small man, is at a disadvantage when she has to perform the same job as an average size man. When they work at similar proportions of their VO_2 max, women perform similarly to men. Obesity predisposes individuals to heat disorders. Greater expenditures of energy are required to perform a given amount of work due to the additional weight to be carried. In addition, body surface to body weight ratio becomes less favorable for heat dissipation. More important is the lower physical fitness, decreased maximum work capacity and cardiovascular capacity frequently associated with obesity.

Alcohol has been commonly associated with the occurrence of heatstroke. The ingestion of alcohol prior to or during work in the heat should not be permitted, as alcohol reduces heat tolerance and increases the risk of heat illnesses.

Many drugs prescribed for therapeutic purposes can interfere with thermoregulation. Almost any drug that affects central nervous system activity, cardiovascular reserve (e.g., beta blockers), or body hydration (such as diuretics), could potentially affect heat tolerance. Thus a worker who requires therapeutic medications should be under the supervision of a physician who understands the potential ramifications of drugs on heat tolerance.

It has long been recognized that individuals suffering from degenerative diseases of the cardiovascular system and other diseases such as diabetes or simple malnutrition are in extra jeopardy when they are exposed to heat, and when stress is imposed on the cardiovascular system.

Acclimatization

Work in heat produces a phenomenon called acclimatization. In a heat-stressful situation, a person acclimatized to heat will have a lower heart rate, a lower body temperature, a higher sweat rate, and a more dilute (containing less salt) sweat than a person who is not acclimatized at the start of exposure to excessive heat. Acclimatization, then, is another way of reducing heat strain.

New employees and employees returning from illness or vacation must be given adequate time to acclimate to hot-working conditions.

Both physical labor and heat stress are required to initiate the body changes that result in acclimatization. Working in the heat for about 2 hours per day for a week or two will result in essentially complete acclimatization to that work-stress combination. To achieve acclimatization, the usual practice is to maintain the environmental conditions at relatively constant levels and to increase gradually (over a week) the amount of work done. Working more than 2 hours per day in the heat will not speed acclimatization, nor hinder it. Lack of adequate water or salt, however, will reduce the speed of acclimatization.

Once attained, acclimatization is lost slowly—traces will remain even 2 or 3 months following the last work in the heat. A measurable amount, however, can be lost in a few days. Workers in the heat may well experience slightly more discomfort on Monday than they did on Friday doing the same job, because of the acclimatization they have lost. If removed from exposure to excessive heat stress for as long as a week, an employee may need another period of acclimatization prior to resuming a full workload. See also the section, Control of heat stress.

Measurement of heat stress

Heat stress is caused by the interaction of a number of environmental factors and metabolic heat production (Figure 12-4). Therefore, measurements of heat stress cannot be easily reduced to a single instrument or number. The environmental factors of heat stress include: air temperature and movement, water vapor pressure, and radiant heat. Physical work contributes to total heat stress by producing metabolic heat. Clothing will also alter the amount of heat stress experienced by the worker. Heat stress evaluations should take into account all of these factors in order to provide a realistic picture of the heat stress experienced by workers. Such comprehensive information will also be necessary to select from among the various heat stress control measures.

Environmental heat measurement

Humidity, the amount of water vapor within a given space is commonly measured as the relative humidity (RH). That is, the percentage of moisture in the air relative to the amount it could hold if saturated at the same temperature. Humidity is important as a temperature-dependent expression of the actual water vapor pressure. Water vapor pressure is the key climatic factor affecting heat exchange between the body and environment by evaporation. The higher the water vapor pressure, the lower will be the evaporative heat loss.

Water vapor pressure (p_a) is the pressure at which a vapor can accumulate above its liquid if the vapor is kept in confinement, and the temperature is held constant. The standard international (SI) units for water vapor pressure are millimeters of mercury (mmHg). For calculating heat loss by evaporation of sweat, the ambient water vapor pressure must be used. The lower the ambient water vapor pressure, the higher will be the rate of evaporative heat loss.

Water vapor pressure is most commonly determined from a psychrometric chart. The psychrometric chart is the graphical representation for the relationships among the relative humidity (rh), the dew point temperature (t_{dp}), the dry bulb temperature (DBT), wet bulb temperature (WBT), and vapor pressure (p_a). By knowing any two of these five climatic factors, the other three can be obtained from the psychrometric chart (Figure 12-5).

Dew point temperature is the temperature at which air, on being cooled, becomes saturated and moisture begins to be deposited from it. For example, a glass of ice water will condense moisture on its outside surface if the dewpoint of the air around the glass is equal to or greater than the temperature of the ice water.

Dry bulb temperature is the temperature of air as registered by a thermal sensor such as an ordinary mercury-in-glass thermometer shielded from direct radiant energy sources (Figure 12-4).

Wet bulb temperature is registered by a thermometer whose bulb is covered with a wetted wick, effectively shielded from radiation, and exposed to a current of rapidly moving air. It can be determined using a powered (aspirated) or a sling psychrometer, which is whirled by hand to produce the required air velocity.

Psychrometric charts (Figure 12-5) are based on an air speed of about 4.6 m/sec (15 ft/sec) that, in most environments, produces maximum wet bulb depression. More rapid air speeds cause excessive drying of the wick. When water evaporates, it requires about 80 calories per gram; this cools the thermometer.

The natural wet bulb temperature is obtained by a wetted sensor such as wet wick over a mercury-in-glass thermometer that is exposed to natural air movement unshielded from radiation (see Figure 12-4).

A highly absorbent woven cotton wick should cover the thermometer bulb and at least one inch of the thermometer stem above the bulb. One inch of wetted wick should be exposed to the air above the top of the reservoir. The wick should be wet to the top at all times with pure distilled water. The water should be replaced when dirty.

Under unusually hot or dry conditions, excessive drying may occur and special provision may be necessary, such as an auxiliary water supply or manual wetting. If the air is saturated with water vapor, no cooling will take place and the wet bulb and dry bulb temperatures will be identical. If the air is not saturated, the cooling (or wet bulb depression) will be in propor-

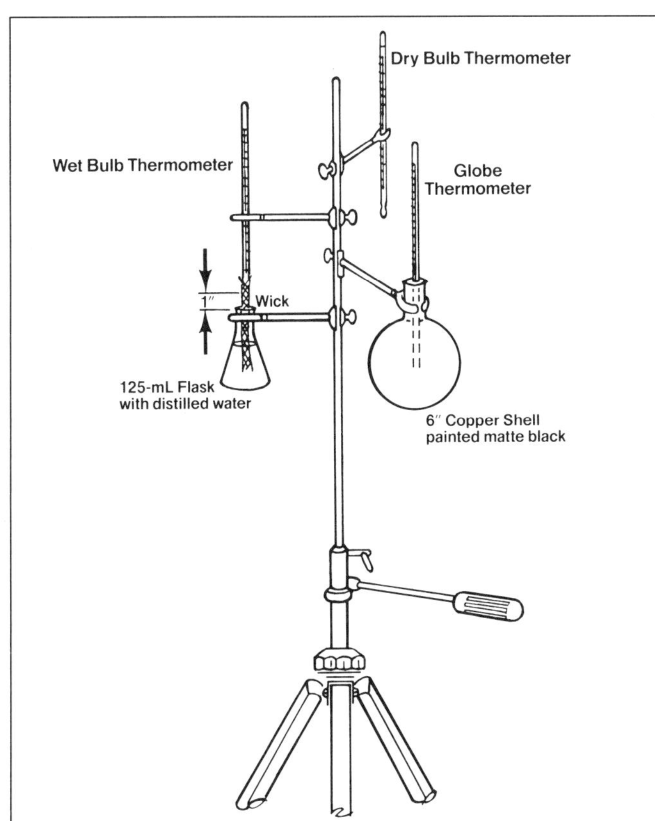

Figure 12-4. A suggested instrument arrangement for environmental measurements of heat stress to workers.

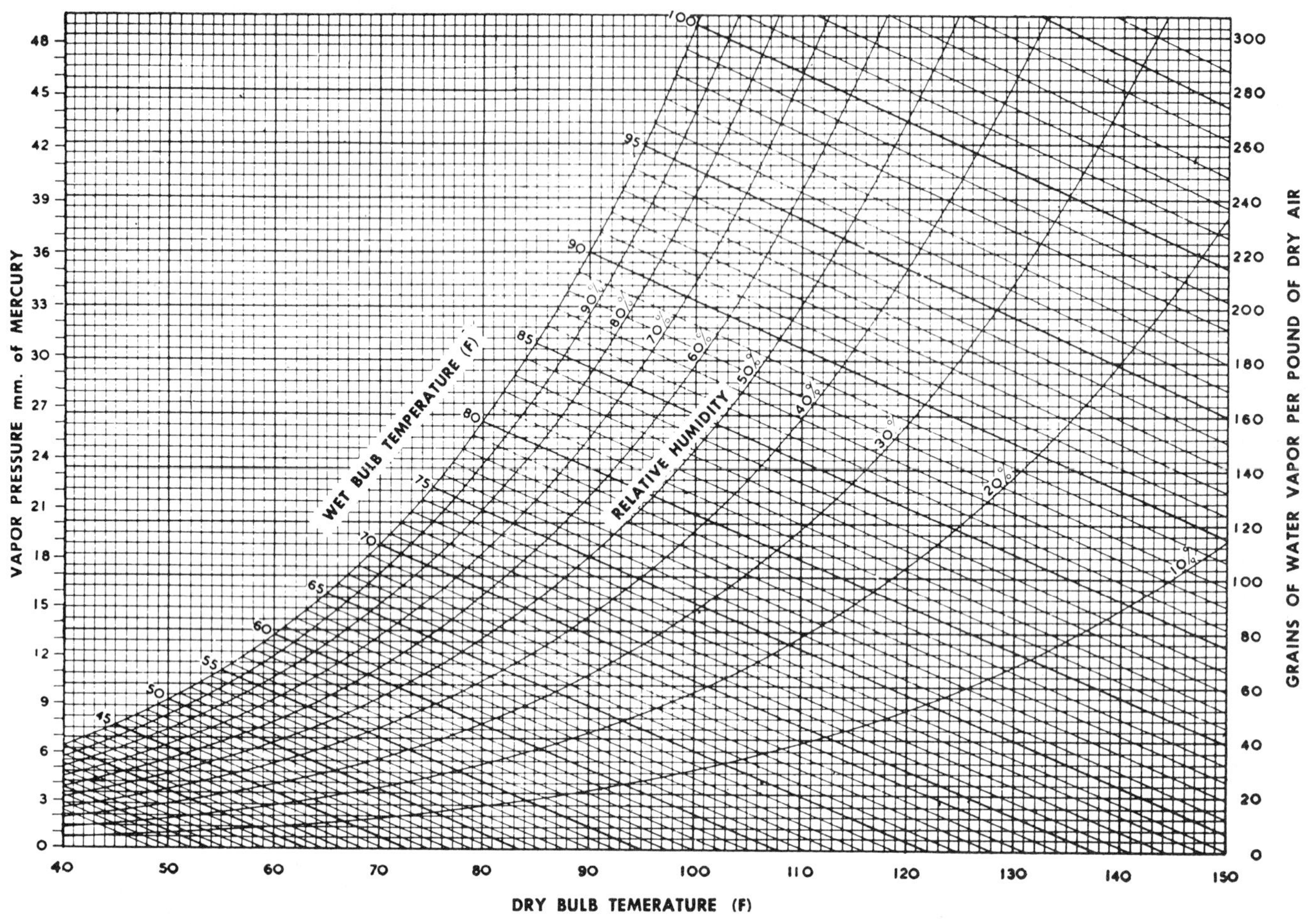

Figure 12-5. Psychrometric chart. (Reprinted with permission from ACGIH *Industrial Ventilation Manual*, 19th ed. Lansing, MI: ACGIH, 1986)

tion to the rate of water evaporation at a constant speed of air movement.

Globe temperature. Radiant heat is a form of electromagnetic energy similar to visible light but of longer wavelength (Chapter 11, Nonionizing Radiation). Radiant heat from such sources as hot metal, open flames, and the sun has little heating effect on the air it passes through. Its energy is absorbed by any object it strikes, thus heating the person, wall, or whatever solid object it falls upon (see Figure 12-4).

Thermal radiation is the transfer of heat through space between one object and another by electromagnetic wave motion. Heat will be exchanged by radiation between the surface of the body and all of the surfaces in its surroundings where the temperatures differ from its own. The intensity of the energy emitted from a surface by radiation increases as the fourth power of its absolute temperature. The intensity is usually diminished below the theoretical maximum. However, by the physical nature of the surface, the relative effect is known as emissivity.

The rate of heat transfer depends on the temperatures and emissivities of the surfaces involved, and not on the temperature of the air. Radiation can contribute substantially to heat stress imposed on the worker. Globe thermometers are used to evaluate this factor.

The globe thermometer is a thin-wall, blackened (flat matte black) copper sphere 15 cm (6 in.) in diameter, with a temperature-sensing device at its center. The temperature attained by the globe thermometer depends on the transfer of radiant energy between the surrounding surfaces and the convective heat exchange with the ambient air. In turn, this depends on ambient air speed and temperature. When using a globe thermometer, approximately 20 minutes must be allowed before the temperature can be read.

Other types of instruments can be used to obtain the globe temperature provided they yield results equivalent to those obtained with a 15-cm globe. The equilibration time of other instruments may differ from that of a 15-cm globe thermometer (see Figure 12-6).

Air velocity. The speed of air movement at the workplace makes a major contribution to evaporative and convective heat exchange by humans and can be measured by various types of anemometers. This measurement is usually expressed in units of feet per minute (meters per second). Air velocity is difficult to measure because air motion is not usually steady, nor is it usually unidirectional. For this reason, air velocity must be measured by a device such as a hot-wire thermometer or heated thermocouple, thermistor, or a thermocouple anemometer not

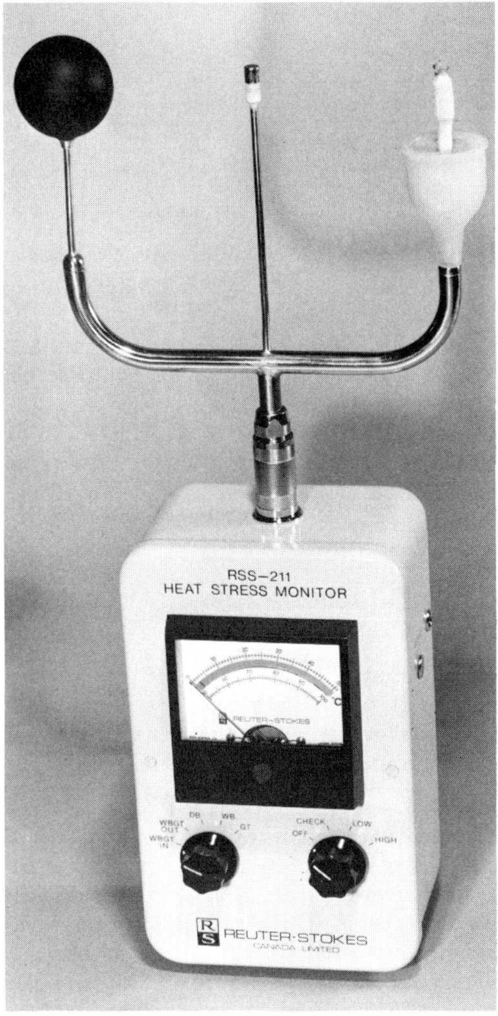

Figure 12-6. A portable instrument that measures the wetbulb globe temperature. (Courtesy Reuter-Stokes, Inc.)

Figure 12-7. A thermocouple anemometer air meter can be used to measure air velocity under various industrial conditions.

sensitive to the direction of air movement (see Figure 12-7). (Common swinging-vane or propeller-type air-velocity indicators should not be used except in very special cases of unidirectional air movement.) The velocity of air usually varies. Therefore, time interval being measured must be averaged.

Location of thermal sensors. Thermal-sensing instruments should be located at the workstation so that the actual conditions of heat exposure are measured. A person's body is a shield and therefore measurements would have to be made at the position of a worker but with the worker not there. Where measurements must be made at an occupied workstation, an effort should be made to evaluate the shielding effect of the body on thermal radiation and on air movement as that effect relates to the measurements.

When impossible or impractical to measure the environmental conditions when the worker is in the zone, an estimate of the heat exposure at that workstation may be obtained by any of the following methods:

- Immediately after the worker leaves the zone, set up the instruments in the zone where the employee has been working. This measurement of environmental conditions immediately after the worker leaves the zone is generally an adequate estimate of the worker's actual conditions. However, this will not give a reliable estimate if environmental conditions change rapidly. In some cases, it will be necessary to ask the worker to maintain the heat source(s) in the same condition(s) as it was (they were) when the task was performed.
- In those zones where the worker spends a substantial amount of time, measurements should be taken periodically. Once per hour or once per half-hour may be adequate. However, in those zones where the worker spends only a few minutes each shift, it is necessary to take only two or three environmental readings each work shift.
- Where the employee moves through a large area, several zones may be involved. It is permissible, in this case, to estimate the worker's exposure from environmental data for zones that have heat levels similar to those the employee passed through, provided the time taken to walk through the area was short.

The heat stress survey should emphasize detailed measurements on a few work areas that expose workers to the highest temperatures.

During days of data collection, it is important to obtain outdoor temperatures (psychrometric wet and dry bulb). Also record qualitative information on cloud cover and wind velocity for later possible use to predict conditions at task sites.

Metabolic heat measurement

The energy cost of an activity as measured by the metabolic heat (m) is a major element in the heat-exchange balance between the human body and the environment. Estimates of metabolic heat for use in assessing muscular workload and

Table 12-C. Estimates of Energy Metabolism (M) of Various Types of Activity

	Activity	Metabolic Rate, M			
		Btu/hr	Watts (W)	kcal/hr	kcal/min
	Sleeping	250	73	63	1.05
	Sitting quietly	400	117	100	1.75
LIGHT WORK	Sitting, moderate arm and trunk movements (e.g., desk work, typing)	450–550	130–160	113–140	1.8–2.3
	Sitting, moderate arm and leg movements (e.g., playing organ, driving car in traffic)	550–650	160–190	140–160	2.3–2.7
	Standing, light work at machine or bench, mostly arms	550–650	160–190	140–160	2.3–2.7
MODERATE WORK	Sitting, heavy arm and leg movement	650–800	190–235	165–200	2.8–3.3
	Standing, light work at machine or bench, some walking about	650–750	190–220	165–190	2.8–3.2
	Standing, moderate work at machine or bench, some walking about	750–1,000	220–290	190–250	3.2–4.2
	Walking about, with moderate lifting or pushing	1,000–1,400	290–410	250–350	4.2–5.8
HEAVY WORK	Intermittent heavy lifting, pushing or pulling (e.g., pick and shovel work)	1,500–2,000	440–590	380–500	6.3–8.3
	Hardest sustained work	2,000–2,400	590–700	500–600	8.3–10.0

Note: Values apply for a 70-kg (154 lb) man, and do not include rest pauses.

human heat regulation are obtained from tabulated descriptions of energy cost for typical work tasks and activities.

To evaluate the average energy requirements over an extended period of time for industrial tasks, it is necessary to divide the task into its basic activities and subactivities. The metabolic heat of each activity or subactivity is measured or estimated. A time weighted average (TWA) for the energy required for the task can then be obtained (Tables 12-B and 12-C).

Heat stress indices. In the past 50 years, several methods have been devised to assess and/or predict the level of heat stress and/or strain a worker might experience when working at hot industrial jobs. Some are based on the measurement of a single environmental factor (wet bulb). While others incorporate all of the important environmental factors (dry bulb, wet bulb, mean radiant temperatures, and air velocity). For all of the indices, either the level of metabolic heat production is directly incorporated into the index or the acceptable level of index values varies as a function of metabolic heat production. The following discusses the measurements required, advantages and disadvantages, and applicability to routine industrial use of some of the more frequently used heat-stress/heat strain indices.

Dry bulb temperature. The dry bulb temperature is commonly used for estimating comfort conditions for sedentary people wearing conventional indoor clothing. With light air movement and relative humidity of 20–60 percent, air temperatures of 22–25.5 C (71.6–77.9 F) are considered comfortable by most people. If work intensity is increased to moderate or heavy work, the comfort air temperature is decreased about 1.7 C (3 F) for each 25 kcal (100 Btu or 29 w) increase in the hourly metabolic heat production. Dry bulb temperature is easily measured.

Wet bulb temperature. The psychrometric wet bulb temperature may be an appropriate index for assessing heat stress under conditions where radiant temperature and air velocity are not large factors. For normally clothed individuals at low air velocities, a wet bulb temperature of about 30 C (86 F) is the upper limit for unimpaired performance on sedentary tasks and 28 C (82.4 F) is the upper limit for moderate levels of physical work. As the wet bulb temperature increases above these threshold values, performance deteriorates and the number of accidents increase.

Effective temperature. The effective temperature (ET) is a sensory index (single-number value) of the degree of warmth a person, wearing very light clothing and engaged in light activity, would experience on exposure to different combinations of air temperature, humidity, and air movement. Graphs such as that in Figure 12-8 illustrate how the ET is calculated. A corrected effective temperature (CET) is available for use where radiant heat is present and where the person may be fully clad.

The ET (or CET) does not take metabolic heat production into account. It is based only on sensations of comfort or discomfort. The ET (or CET) does not usually give a true indication of heat stress where moderate or heavy work is being performed or where the environment is very hot or humid. The ET is used more for comfort than for heat stress evaluations.

Heat stress evaluations are most often performed with the wet bulb globe temperature (WBGT) and/or the Belding-Hatch heat stress index (HSI) in ways outlined well by Horvath and Jensen, 1976 (see Bibliography).

The WBGT index is used because it is easy to determine and has the official sanction of the ACGIH and of the National Institute for Occupational Safety and Health (NIOSH) (see Addendum, "TLVs for Heat Stress"). Despite the fact that it does not correlate well with heat strain, the WBGT has been used for many years by the military with good results.

The WBGT index requires knowledge of the wet bulb temperature, the globe temperature, and the dry bulb air temperature. The WBGT is calculated for indoor exposure, or outdoor exposure with no solar load:

$$WBGT = 0.7t_{wb} + 0.3t_g$$

For outdoor sunlit exposure:

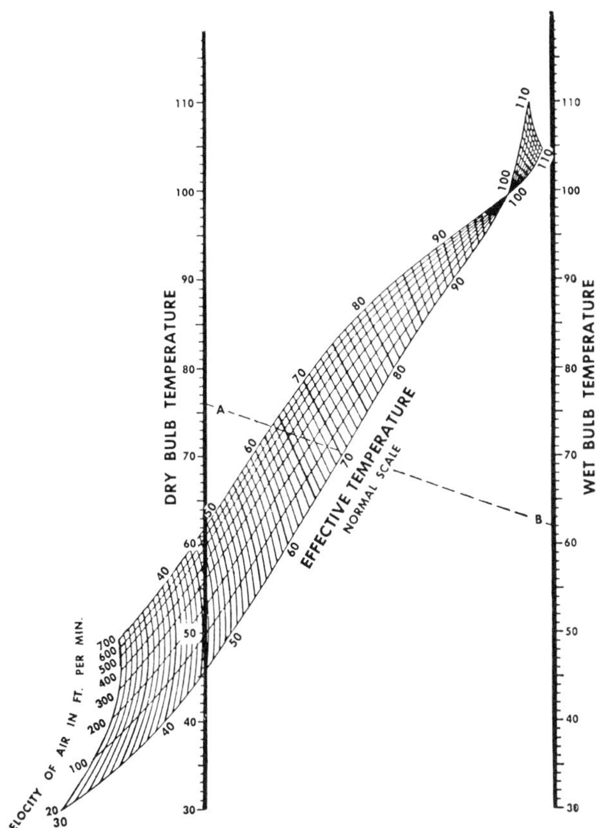

Figure 12-8. Effective temperature (ET) index is a sensory index of the degree of warmth a person experiences. The ET combines air temperature, humidity, and air movement (applicable to persons at rest and normally clothed). (Reprinted with permission from ASHRAE *Handbook*)

$$WBGT=0.7t_{wb}+0.2t_g+0.1t_a$$

where: t_{wb} =wet bulb temperature
t_g =globe temperature
t_a =dry bulb air temperature

Instrumentation to determine the WBGT index should always be located so that the readings obtained will be truly representative of the environmental conditions to which the worker is exposed.

Sensors should be at least the mean height of the worker, and due consideration should be given to the location of radiation sources and the direction of air movement. Sufficient time after setting up the instruments must be allowed for stabilization (equilibration) before readings can be obtained.

Calculating the WGBT. Where the employee is continuously exposed to a hot environment, the environmental heat exposure is considered as a series of hourly TWAs. Where the employee's exposure is intermittent (interrupted at least each 15 minutes by breaks spent in cools areas), time-weighting should be performed for periods of 2 hours.

For jobs in which heat exposure and work effort are intermittent, the TWA should be derived by recording the time spent at each task, including rest periods, and the corresponding times spent in hot locations and in cooler locations during recovery.

The 2-hour TWA is calculated by the following equation:

Average WBGT=

$$\frac{(WBGT_1)x(t_1)+(WBGT_2)xt_2+\ldots+(WBGT_n)x(t_n)}{(t_1)+(t_2)+\ldots+(t_n)}$$

Where $WBGT_1$, $WBGT_2$, and $WBGT_n$ are measured values of WBGT for the various work and rest intervals during the total time period, t_1, t_2, and t_n are the duration of the respective intervals in minutes.

Heat exposures are time-weighted for 2-hour periods during the shift. With a complete record of time spent on individual tasks and of time spent at different work locations during the shift, it is possible to derive time-weighting for shorter or longer intervals. If physiological monitoring immediately following a short period of strenuous work in hot locations gives evidence of excessive strain, time-weighting for a shorter interval (30 or 60 minutes) may be indicated.

The Threshold Limit Value (TLV) for heat stress combines three basic parameters; including: (1) metabolic demands of the task, (2) an index of severity of the environment (WBGT), and (3) percentage of time that the individual may be permitted to perform the task. The philosophy applied in the TLV is that environmental stress should not create a rise in deep body temperature above that in response to the work itself. (See Appendix B-1, Heat Stress.)

For example, a task requiring light to moderate work of 200 kcal/hr could be performed continuously in environments up to a WBGT of 30 C, but only 25 percent of the time at a WBGT of 32.2 C (90 F).

A heavier task, say 400 kcal/hr, could be performed continuously in an environment where the WBGT was up to about 25 C (77 F). This recognizes the role of metabolic heat production in the heat balance equation.

One offshoot of the interest in the WBGT by official agencies in the United States has been the development, by Mutchler, Malzahn, Vecchio, and Soule, 1975, of a method for predicting the WBGT by means of correlations between inside environmental conditions and outdoor weather conditions. Such a prediction requires a short-term environmental study at each worksite to estimate regression constants but, subsequently, good estimates of the WBGT can be made even from weather forecasts (see Bibliography).

Next to the dry bulb air temperature and the wet bulb temperature, the wet globe thermometer (botsball) is the simplest, most easily read, and most portable of the environmental measuring devices. The wet globe thermometer consists of a hollow 7.6-cm (3 in.) copper sphere covered by a black cloth, which is kept at 100 percent wetness from a water reservoir (Figure 12-9). The sensing element of a thermometer is located at the inside center of the copper sphere. The temperature inside the sphere is read on a dial at the end of the stem. Heat exchange by convection, radiation, and evaporation are integrated into a single instrument reading.

Many studies have indicated a good correlation between the wet globe temperature (WGT) and the WBGT. A simple approximation of the relationship is WBGT=WGT+2 C for conditions of moderate radiant heat and humidity. These approximations

Table 12-D. Evaluation of Index of Heat Stress

Index of Heat Stress	*Physiological and Hygienic Implications of 8-Hr. Exposures to Various Heat Stress*
−20 −10	Mild cold strain. This condition frequently exists in areas where people recover from exposure to heat.
0	No thermal strain.
+10 20 30	Mild to moderate heat strain. Where a job involves higher intellectual functions, dexterity, or alertness, subtle to substantial decrements in performance may be expected. In performance of heavy physical work, little decrement expected unless ability of individuals to perform such work under no thermal stress is marginal.
40 50 60	Severe heat strain, involving a threat to health unless people are physically fit. Break-in period required for those not previously acclimatized. Some decrement in performance of physical work is to be expected. Medical selection of personnel desirable because these conditions are unsuitable for those with cardiovascular or respiratory impairment or with chronic dermatitis. These working conditions are also unsuitable for activities requiring sustained mental effort.
70 80 90	Very severe heat strain. Only a small percentage of the population may be expected to qualify for this work. Personnel should be selected (a) by medical examination, and (b) by trial on the job (after acclimatization). Special measures are needed to assure adequate water and salt intake. Amelioration of working conditions by any feasible means is highly desirable, and may be expected to decrease the health hazard while increasing efficiency on the job. Slight "indisposition" which in most jobs would be insufficient to affect performance may render workers unfit for this exposure.
100	The maximum strain tolerated daily by fit, acclimatized young men.

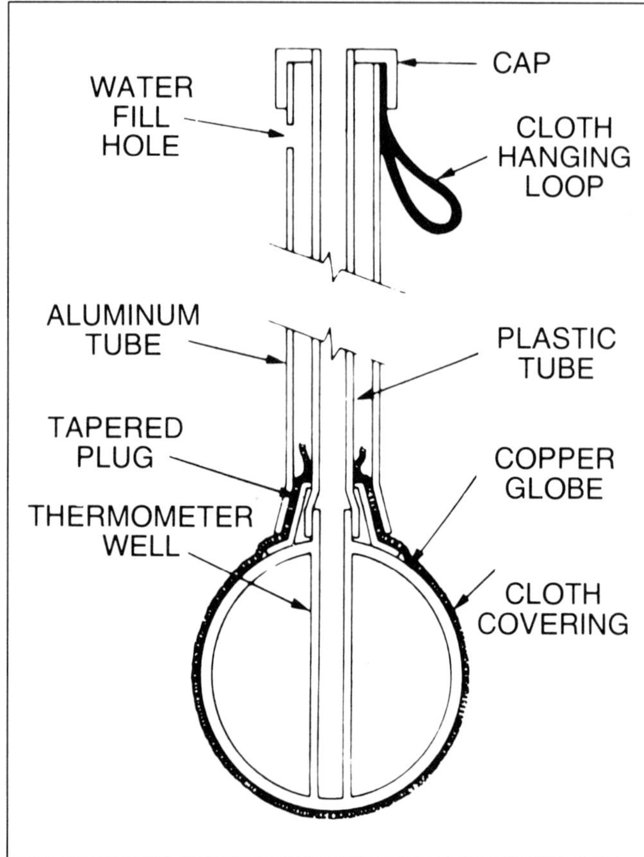

Figure 12-9. A cutaway view of wet globe thermometer plus a unit ready for use. This unit measures the environmental heat load by exchanging heat with the surroundings through conduction, convection, evaporation, and radiation. (Courtesy Howard Engineering Co.)

are probably adequate for general monitoring in industry. If the WGT shows high values, it should be followed with WBGT or other detailed measurements. The WGT, although good for screening and monitoring, does not yield data that will point to the causes and controls of heat stress in a particular environment. However, the color coded WGT display dial provides a simple and rapid indicator of the level of heat stress.

The heat stress index (HSI) is not particularly well correlated with the resulting heat strain, but its determination results in considerably more knowledge about the environment than does use of the WBGT. Use of the HSI can be expected to give more and better clues about the potential efficacy of possible control measures than can use of the WBGT.

Calculating the HSI. To calculate the HSI, measurements of the air temperature (t_a), the psychrometric wet bulb temperature (t_{wb}), the globe temperature (t_g), and the air velocity (V) are necessary for each jobsite, as is an estimate of the rate of energy expenditure (metabolic rate, M) of workers at that site. Once measurements have been made, rates of heat exchange between the worker and the environment by convection (C) and radiation (R) are calculated, and with M, are used to estimate the amount of sweating required to stay at equilibrium (E_{req}) (Figure 12-10 and Table 12-C).

Among other assumptions embodied in the HSI is that of a constant skin temperature of 35 C (95 F). This assumption is one most likely to lead to the greatest errors, and in many hot work situations will cause the stress to be overestimated. Nevertheless, calculations of R, C, E_{req}, and E_{max} and estimations of M lead to a knowledge of the relative contribution of each, and hence, may well suggest possible means of solving the problem.

The effect of various contemplated solutions can be predicted (increasing air velocity by means of pedestal fans, for instance, or installing a reflective shield between workers and a hot surface) on each of the exchange rates and on the HSI. Thus, the relative effect of each potential change can be known prior to its application. This cannot be done with the WBGT.

When the HSI equals 100, then the required evaporative capacity is just equal to the maximum evaporative capacity. Therefore, efforts should be made to keep the HSI below 100. When the HSI is greater than 100, however, the difference between E_{req} and E_{max} can be used to calculate the consequences of heat exposure under these conditions (Table 12-D).

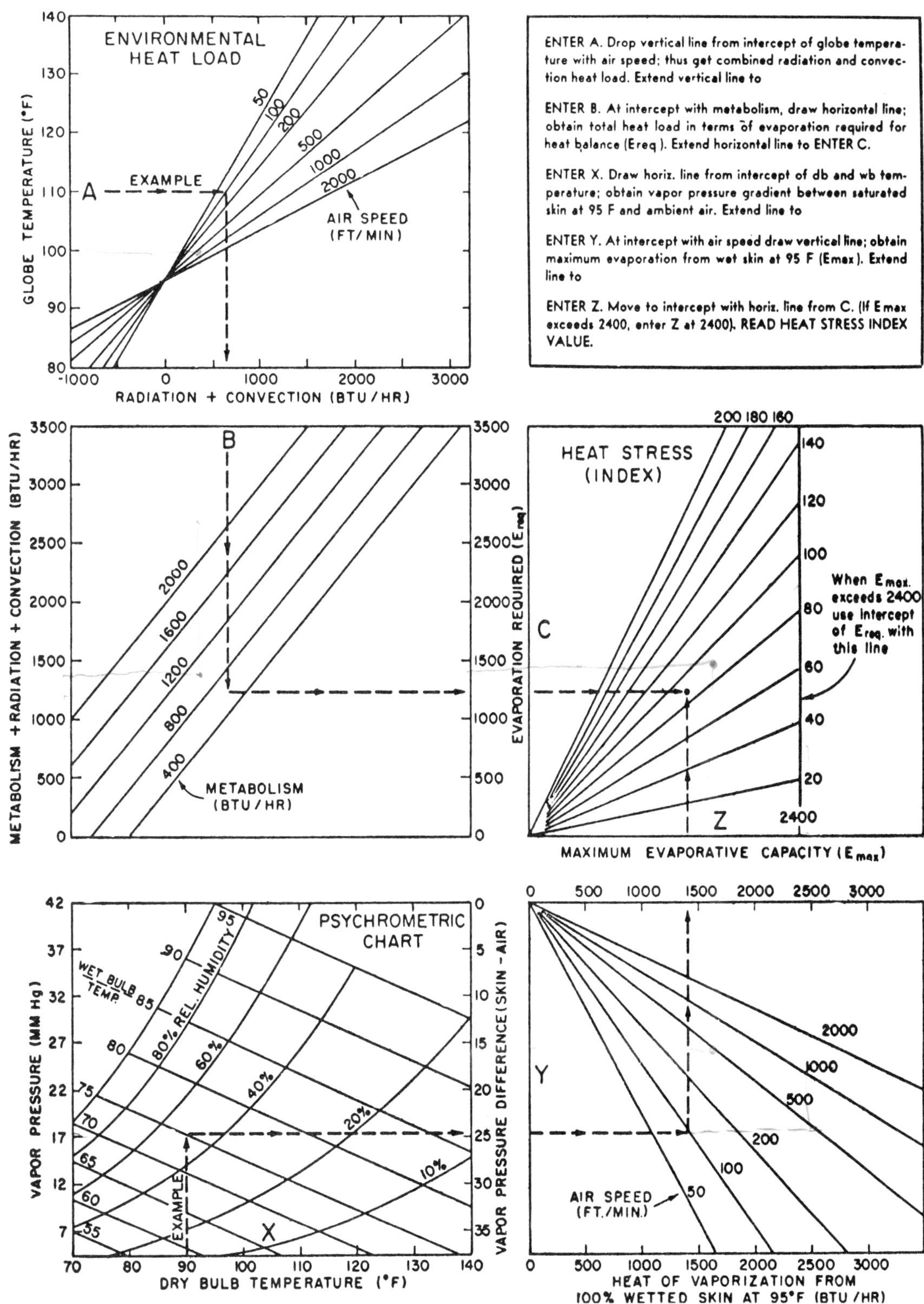

Figure 12-10. The Belding-Hatch Index calculations for the various components of heat stress. Use the nomogram as follows: (1) Enter A (left). Drop vertical line from intercept of globe temperature with air speed, thus getting combined radiation and convection heat load. Extend vertical line to enter B. At intercept with "Metabolism," draw horizontal line to obtain total heat load in terms of evaporation required for heat balance (E_{req}). Extend horizontal line to enter C. (2) Next, enter X (lower left). Draw horizontal line from intercept of dry- and wet-bulb temperatures. Obtain vapor pressure gradient between saturated skin at 95 F and ambient air. Extend line to right to enter Y. At intercept with air speed, draw vertical line; obtain maximum evaporation from wet skin (E_{max}). Extend line to enter Z. Move to intercept with horizontal line from C. If E_{max} exceeds 2,400, enter Z at 2,400. Read Heat Stress Index Value.

The rate of body heating is, for the average 70-kg (155 lb) man, about one degree Celcius per 33-watt hour absorbed (two degrees fahrenheit for each Btu absorbed) so that, knowing the rate of heat absorption, the maximum time in that environment for a specified body temperature rise can be estimated.

Control of heat stress

From a review of the heat balance equation ($H=M\pm C\pm R-E$), total heat stress can be reduced only by modifying one or more of the following factors: metabolic heat production, heat exchange by convection, heat exchange by radiation, or heat exchange by evaporation. Environmental heat load (C, R, and E) can be modified by engineering controls (e.g., ventilation, air conditioning, screening, insulation, and modification of process or operation) and protective clothing and equipment. Metabolic heat production can be modified by work practices and application of labor-reducing devices. Table 12-E is a checklist of heat stress control.

Table 12-E. Checklist for Controlling Heat Stress

Item	*Actions for Consideration*
Controls	
M, Body heat production of task	Reduce physical demands of the work; powered assistance for heavy tasks.
R, Radiative load	Interpose line-of-sight barrier; furnace wall insulation, metallic reflecting screen, heat reflective clothing, cover exposed parts of body.
C, Convective load	If air temperature is above 35 C (95 F); reduce air temperature, reduce air speed across skin, wear clothing. If air temperature is below 35 C (95 F); increase air speed across skin and reduce clothing.
E_{Max}, Maximum evaporative cooling by sweating	Increase by: decreasing humidity, increasing air speed Decrease clothing
Work practices	Shorten duration of each exposure; more frequent short exposures better than fewer long exposures. Schedule very hot jobs in cooler part of day when possible.
Exposure limit	Self-limiting, based on formal indoctrination of workers and supervisors on signs and symptoms of overstrain.
Recovery	Air-conditioned space nearby.
Personal protection R, C, and E_{max}	Cooled air, cooled fluid, or ice cooled conditioned clothing Reflective clothing or aprons
Other considerations	Determine by medical evaluation, primarily of cardiovascular status Careful break-in of unacclimatized workers Water intake at frequent intervals to prevent hypohydration Fatigue or mild illness not related to the job may temporarily contraindicate exposure (e.g., low-grade infection, diarrhea, sleepless night, alcohol ingestion)
Heat wave	Introduce heat alert program

(Reprinted from DHHS *Criteria for a Recommended Standard. . . Occupational Exposure to Hot Environments—Revised Criteria 1986.* Washington, DC: U.S. Government Printing Office, 1986)

Engineering controls. The environmental factors that can be modified by engineering procedures are those involved in convective, radiative, and evaporative heat control. The following discusses each of these.

Convective heat control. As stated earlier, the environmental variables concerned with convective heat exchange between the worker and the ambient environment are dry bulb air temperature (t_a) and the speed of air movement (V_a). When air temperature is higher than the mean skin temperature (t_{sk} of 35 C or 95 F), heat is gained by convection. The rate of heat gain is dependent on temperature differential ($t_a - t_{sk}$) and air velocity (V_a). Where t_a is below t_{sk}, heat is lost from the body; the rate of loss is dependent on $t_a - t_{sk}$ and air velocity.

Engineering approaches to enhancing convective heat exchange are limited to modifying air temperature and air movement. When t_a is less than t_{sk}, increasing air movement across the skin by increasing either general or local ventilation will increase the rate of body heat loss. When t_a exceeds t_{sk} (convective heat gain), t_a should be reduced by bringing in cooler outside air or by evaporative or refrigerative cooling of the air. As long as t_a exceeds t_{sk}, air speed should be reduced to levels that will still permit sweat to evaporate freely but will reduce convective heat gain (Table 12-E). The effect of air speed on convective heat exchange is a 0.6 root function of air speed. Spot cooling (t_a less than t_{sk}) of the individual worker can be an effective approach to controlling convective heat exchange, especially in large workshops where the cost of cooling the entire space would be prohibitive. However, spot coolers or blowers may interfere with the ventilation systems required to control toxic chemical agents.

Radiant heat control. Radiant heat exchange between the worker and hot equipment, production processes, and walls that surround the worker is a fourth power function of the difference between skin temperature (t_{sk}) and the temperature of hot objects that "see" the worker (t_r). The only engineering approach to control radiant heat gain is to reduce t_r or to shield the worker from the radiant heat source.

To reduce t_r would require (1) lowering the process temperature which is usually not compatible with the temperature requirements of the manufacturing processes; (2) relocating, insulating, or cooling the heat source; (3) placing line-of-sight radiant reflective shielding between the heat source and the worker; or (4) changing the emissivity of the hot surface by coating it. Of these alternatives, radiant reflective shielding is generally the easiest to install and the least expensive. Radiant reflective shielding can reduce the radiant heat load by as much as 80–85 percent. Some ingenuity may be required in placing the shielding so that it does not create an obstacle for the worker. Remotely operated tongs, metal chain screens, or air or hydraulically activated doors opened only as needed are some approaches that may be used.

Evaporative heat control. Heat is lost from the body when sweat is evaporated from the skin surface. The rate and amount of evaporation is a function of the speed of air movement over the skin and the difference between the water vapor pressure of the air (p_a) at ambient temperature and the water vapor

vapor pressure of the wetted skin assuming a skin temperature of 34–35 C (93.2–95 F). At any air-to-skin vapor pressure gradient, the evaporation increases as a 0.6 root function of increased air movement. Evaporative heat loss at low air velocities can be greatly increased by improving ventilation (increasing air velocity). At high air velocities (2.5 m/sec or 500 fpm); an additional increase will be ineffective except when clothing interferes with air movement over the skin.

Engineering controls of evaporative cooling can therefore assume two forms: (1) increase air movement or (2) decrease ambient water vapor pressure. Of these, increased air movement with fans or blowers is often the simplest and usually the least expensive approach to increasing the rate of evaporative heat loss. Ambient water vapor pressure reduction usually requires air-conditioning equipment (cooling compressors). In some cases the installation of air conditioning, particularly spot air conditioning, may be less expensive than the installation of increased ventilation because of the lower airflow involved. The vapor pressure of the worksite air is usually at least equal to that of the outside ambient air, except when all incoming and recirculated air is humidity controlled by absorbing or condensing the moisture from the air (e.g., by air conditioning). In addition to the ambient air as a source of water vapor, water vapor may be added from the manufacturing processes as steam, leaks from steam valves and steam lines, and evaporation of water from wet floors. Eliminating these additional sources of water vapor can help reduce the overall vapor pressure in the air and thereby increase evaporative heat loss by facilitating the rate of evaporation of sweat from the skin.

Work and hygienic practices and administrative controls. Situations exist in industries where the complete control of heat stress by the application of engineering controls may be technologically impossible or impractical, where the level of environmental heat stress can be unpredictable and variable (as seasonal heat waves), and where the exposure time can vary with the task and with unforeseen critical events. Where engineering controls of the heat stress are not practical or complete, other solutions must be sought to keep the level of total heat stress on the worker within limits which will not be accompanied by an increased risk of heat illnesses.

The application of preventive practices frequently can be an alternative or complementary approach to engineering techniques for controlling heat stress. Preventive practices include: (1) limiting or modifying the duration of exposure time; (2) reducing the metabolic component of the total heat load; (3) enhancing the heat tolerance of the worker by heat acclimatization and physical conditioning; (4) training workers in safety and health procedures for work in hot environments; and (5) medical screening of workers to discern individuals with low heat tolerance and/or physical fitness.

Limiting exposure time and/or temperature. There are several ways to control the daily length of time and temperature to which a worker is exposed in heat stress conditions.

- When possible, schedule hot jobs for the cooler part of the day (early morning, late afternoon, or night shift).
- Schedule routine maintenance and repair work in hot areas for the cooler seasons of the year.
- Alter rest-work regimen to permit more rest time.
- Provide cool areas for rest and recovery.
- Add extra personnel to reduce exposure time for each member of the crew.
- Permit freedom to interrupt work when a worker feels extreme heat discomfort.
- Increase water intake of workers on the job.
- Adjust schedule when possible so that hot operations are not performed at the same time and place as other operations that require the presence of workers, e.g., maintenance and cleanup while tapping a furnace.

Reducing metabolic heat load. In most industrial work situations, metabolic heat is not a major factor of total heat load. However, because metabolic heat represents an extra load on the circulatory system, it can be a critical component in high heat exposures. Metabolic heat production can be reduced usually by not more than 200 kcal/hr (800 Btu/hr) by:

- mechanization of the physical components of the job,
- reduction of work time (reduce workday, increase rest time, restrict double shifting),
- an increased work force.

Enhancing tolerance to heat. Stimulating acclimatization can significantly increase heat tolerance. There is, however, a wide difference in the ability of people to adapt to heat.

- A properly designed and applied heat-acclimatization program will dramatically increase the ability of workers to work at a hot job and will decrease the risk of heat-related illnesses and unsafe acts. Heat acclimatization can usually be induced in 5 to 7 days of exposure at the hot job. For workers with previous experience with the job, the acclimatization regimen should be exposure for 50 percent on day 1, 60 percent on day 2, 80 percent on day 3, and 100 percent on day 4. For new workers the schedule should be 20 percent on day 1 and a 20 percent increase on each additional day.
- Physical fitness will enhance heat tolerance for both heat-acclimatized and unacclimatized workers. The time required to develop heat acclimatization in unfit individuals is about 50 percent greater than in the physically fit.
- To ensure that water lost in sweat and urine is replaced during the work day, an adequate water supply and intake are essential for heat tolerance and prevention of heat induced illnesses. Liquids should be taken at least hourly.
- Electrolyte balance in the body fluids must be maintained to prevent some of the heat-induced illnesses. For heat-unacclimatized workers on a restricted salt diet, additional salting of food, with a physician's concurrence, during the first 2 days of heat exposure may be required to replace salt lost in sweat. The acclimatized worker loses relatively little salt in sweat; therefore, salt supplementation of the normal American diet is usually not required.

Health and safety training. Prevention of serious sequelae from heat-induced illnesses is dependent on early recognition of the signs and symptoms of impending heat illnesses and initiation of first aid and/or corrective procedures at the earliest possible moment.

- Supervisors and other personnel should be trained in recognizing the signs and symptoms of the various types of heat-induced illnesses, e.g., heat cramps, heat exhaustion, heat rash, and heatstroke, and in administering first-aid procedures (Table 12-B).

- All personnel exposed to heat should receive basic instruction on the causes and recognition of the various heat illnesses and personal care procedures that should be exercised to minimize risk.
- All personnel who use heat protective clothing and equipment should be instructed in their proper care and use.
- All personnel working in hot areas should be instructed on the effects of nonoccupational factors (drugs, alcohol and obesity) on tolerance to occupational heat stress.
- A buddy system which depends on recognition of the early signs and symptoms of heat illnesses should be initiated. Each worker and supervisor is assigned the responsibility for observing, at periodic intervals, one or more fellow workers to determine whether any of the early symptoms of a developing heat illness are present. If a worker exhibits signs and symptoms indicative of an impending heat illness, the worker should be sent to the dispensary or first-aid station for more complete evaluation of the situation. Workers on hot jobs where the heat stress exceeds the WBGT should be observed by a fellow worker or supervisor. Contingency plans for treatment, e.g., cool rest area and transportation to hospital, should be in place.

Screening for heat intolerance. The ability to tolerate heat stress varies widely even between individuals within a group of normal healthy individuals with similar heat exposure experiences. One way to reduce risk of incurring heat illnesses and disorders within a heat-exposed work force is to reduce or eliminate the exposure to heat stress of the heat-intolerant individuals. Identification of heat-intolerant individuals without the need for performing a strenuous, time-consuming heat-tolerance test would be basic to any such screening process.

Data from laboratory and field studies indicate that individuals with low physical work capacity are more likely to develop higher body temperatures than are individuals with high physical work capacity when exposed to equal work in high temperatures. None of the individuals with a maximum work capacity of 2.5 L of oxygen per minute (L/min) or above were heat intolerant, while 63 percent of those with VO_2max below 2.5 L/min were heat intolerant. It has also been shown that heat-acclimatized individuals with a VO_2max less than 2.5 L/min had a 5 percent risk of reaching heatstroke levels of body temperature (40 C or 104 F) while those with a VO_2max above 2.5 L/min had only a 0.05 percent risk.

Because tolerance to physical work in a hot environment is related to physical work capacity, heat tolerance might be predictable from physical fitness tests. However, such tests have not as yet been proven to have predictive validity for use in hot industries.

Medical screening for heat intolerance in otherwise healthy normal workers should include a history of any previous incident of heat illness. Workers who have experienced a heat illness may be less heat tolerant.

Emergency Prevention Heat-Alert Program. In plants where heat illnesses and disorders occur primarily during hot spells in the summer, a Heat-Alert Program (HAP) should be established. Such programs differ somewhat from one plant to another but the purpose is always to take advantage of the weather forecast of the National Weather Service. If a hot spell is predicted for the next day or days, a state of Heat Alert is declared to ensure that measures to prevent heat casualties will be strictly observed. Although this sounds simple and straightforward, it requires the cooperation of the administrative staff; the maintenance and operative work force; and the medical, industrial safety, and health departments. An effective HAP is described below.

Each year, early in the spring, Heat-Alert Committee consisting of an industrial physician or nurse, industrial hygienist, health and safety engineer, operation engineer, and a high ranking manager should be established. Once established, this committee should:

- Arrange a training course for all involved in the HAP, dealing with procedures to follow in the event a Heat Alert is declared. Special emphasis is given to the prevention and early recognition of heat illnesses and first-aid procedures.
- By memorandum, instruct the supervisors to:
  1. Reverse winterization of the plant, open windows, doors, skylights, and vents according to instructions for greatest ventilating efficiency at places where high air movement is needed.
  2. Check drinking fountains, fans, and air conditioners to ensure they are functional, that the necessary maintenance and repair is performed, that these facilities are regularly rechecked, and that workers know how to use them.
- Ascertain that in the medical department, as well as at job sites, all facilities required to give first aid in case of a heat illness are in a state of readiness.
- Establish criteria for the declaration of a Heat Alert; for example, a Heat Alert would be declared if the area weather forecast for the next day predicts a maximum air temperature of 35 C (95 F) or above or a maximum of 32 C (90 F)

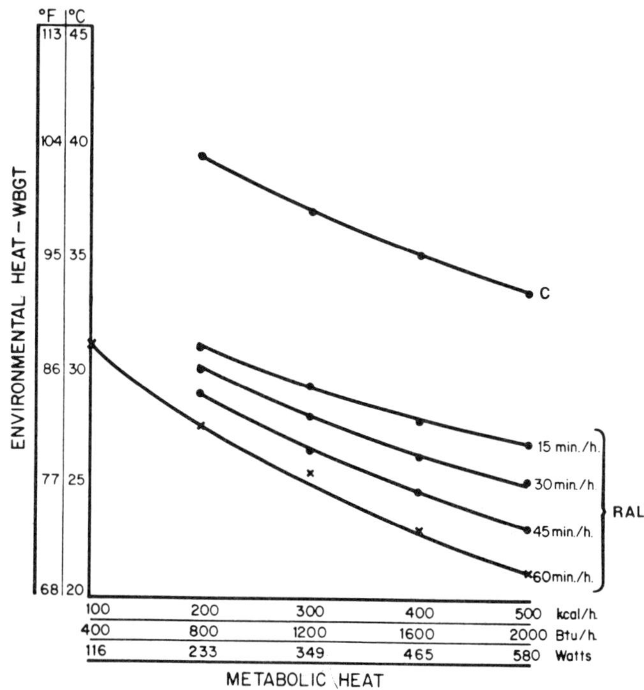

Figure 12-11. Recommended Heat-Stress Alert Limits Heat—Unacclimatized Workers. Key: C=ceiling limit; RAL=recommended alert limit. Values shown are those for an "average" worker of 70 kg (154 lb) body weight and 1.8 m² (19.4 ft²) body surface. (Reprinted from NIOSH Criteria Document)

if the predicted maximum is 5 C (9 F) above the maximum reached in any of the preceding 3 days (Figure 12-11).

Procedures to be followed during the state of Heat Alert are as follows:

- Postpone tasks that are not urgent (preventive maintenance involving high activity or heat exposure) until the hot spell is over.
- Increase the number of workers to reduce each worker's heat exposure. Introduce new workers gradually to allow acclimatization (follow heat-acclimatization procedure).
- Increase rest allowances. Let workers recover in air-conditioned rest places.
- Turn off heat sources that are not absolutely necessary.
- Require workers to drink water in small amounts frequently to prevent excessive dehydration, to weigh themselves before and after the shift, and to be sure to drink enough water to maintain body weight.
- Monitor the environmental heat at the job sites and resting places.
- Check workers' oral temperature during their most severe heat-exposure period.
- Exercise additional caution on the first day of a shift change to make sure that workers are not overexposed to heat. They may have lost some of their acclimatization over the weekend and during days off.
- Send workers who show signs of a heat disorder, even a minor one, to the medical department. The physician's permission to return to work must be given in writing.
- Restrict overtime work.

Auxiliary body cooling and protective clothing. When unacceptable levels of heat-stress occur, there generally are only four approaches to a solution: (1) modify the worker by heat acclimatization; (2) modify the clothing or equipment; (3) modify the work; or (4) modify the environment. To do everything possible to improve human tolerance would require that the individuals should be fully heat acclimated, should have good training in the use of and practice in wearing the protective clothing, should be in good physical condition, and should be encouraged to drink as much water as necessary to compensate for sweat water loss.

If heat acclimatization and physical fitness enhancement are not enough to alleviate heat stress and reduce the risk of heat illnesses, only the latter three solutions are left to deal with the problem. It may be possible to redesign ventilation systems to avoid interior humidity and temperature buildup. These may not completely solve the heat stress problem.

When air temperature is above 35 C (95 F) with an RH of 75–85 percent or when there is an intense radiant heat source, a suitable, and in some ways more functional, approach is to modify the clothing to include some form of auxiliary body cooling. Even mobile individuals can be provided some form of auxiliary cooling for limited periods of time. A properly designed system will reduce heat stress, conserve large amounts of drinking water which would otherwise be required, and allow unimpaired performance across a wide range of climatic factors. A seated individual will rarely require more than 100 W (86 kcal/hr or 344 Btu/hr) of auxiliary cooling. The most active individuals require not more than 400 W (345 kcal/hr or 1,380 Btu/hr) unless working at a level where physical exhaustion per se would limit the duration of work.

Personal protective equipment. Clothing worn by workers is extremely important in controlling heat stress. Loose-fitting clothing should be used in highly humid areas. Workers must know when more or less clothing is needed. With medium radiant heat loads, the amount of exposed skin must be minimal. Jobs with high radiant heat loads often demand that reflective garments be worn. Extreme radiant and convective heat exposure may require special insulation or even mechanically cooled suits (Figure 12-12).

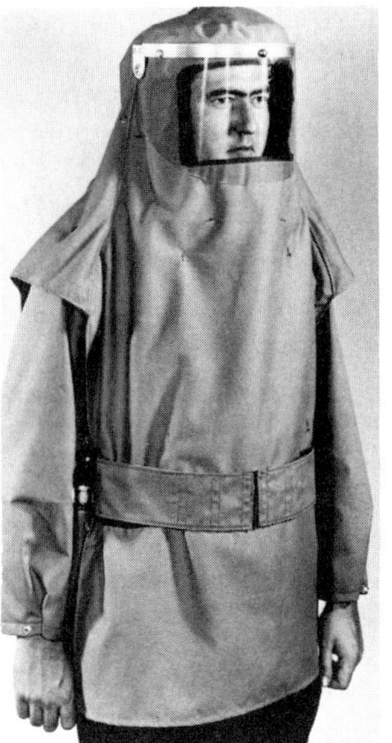

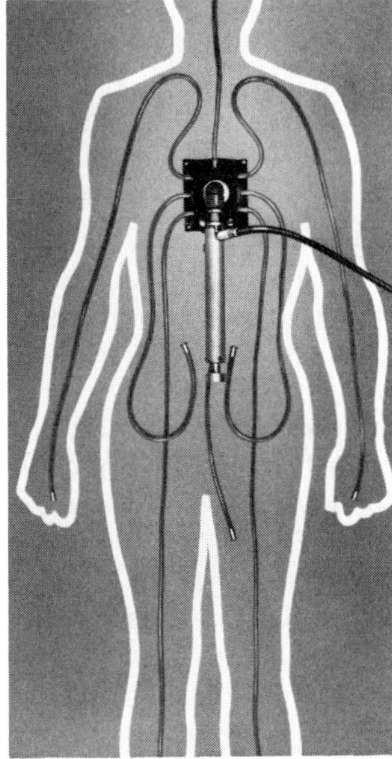

Figure 12-12. This is how the vortex tube is used to provide "personal air conditioning."

Personal protective equipment for heat stress is ordinarily used only when a person must remain in a very hot environment long enough to cause unacceptably high heat strain without protection. Examples range from driving a race car, to the descent of astronauts from orbit, the repair of firebrick in a furnace, and to the control of oil well fires.

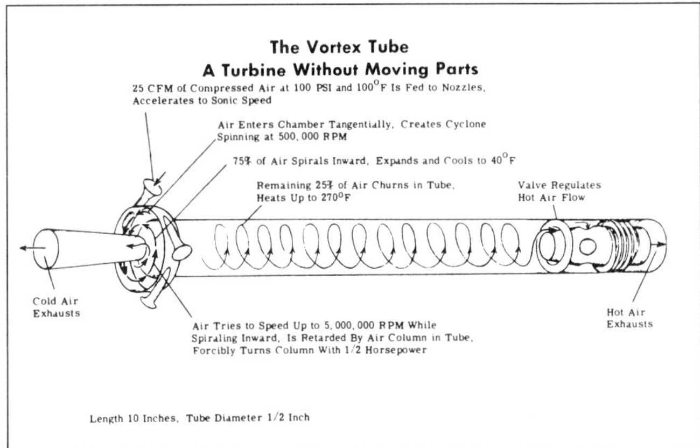

Figure 12-13. Schematic of "vortex tube" shows how it is used to provide effective personal protection against heat. (Courtesy Fulton Cryogenics)

Protective equipment adequate for the job varies from simple head cooling to essentially complete isolation of the worker from the environment in a "space" suit. The cooling medium can be air (especially cooled air, perhaps from a vortex tube, Figures 12-12 and 12-13), water, or even wet or dry ice.

Water-cooled garments. Water-cooled garments include (1) a water-cooled hood that provides cooling to the head, (2) a water-cooled vest that provides cooling to the head and torso, (3) a short, water-cooled undergarment which provides cooling to the torso, arms, and legs, and (4) a long, water-cooled undergarment which provides cooling to the head, torso, arms, and legs. None of these water-cooled systems provide cooling to the hands and feet.

Water-cooled garments and headgear require a battery driven circulating pump and container where the circulating fluid is cooled by the ice. The weight of the batteries, container, and pump will limit the amount of ice that can be carried. The amount of ice available will determine the effective use time of the water-cooled garment.

The range of cooling provided by each of the water-cooled garments versus the cooling water inlet temperature has been studied. The rate of increase in cooling, with decrease in cooling water inlet temperature, is 3.1 W/degrees C for the water-cooled cap with water-cooled vest, 17.6 W/degrees C for the short water-cooled undergarment, and 25.8 W/degrees C for the long water-cooled undergarments. A "comfortable" cooling water inlet temperature of 20 C (68 F) should provide 46 W of cooling using the water-cooled cap; 66 W using the water-cooled vest; 112 W using the water-cooled cap with water-cooled vest; 264 W using the short water-cooled undergarment; and 387 W using the long water-cooled undergarment.

Air-cooled garments. Air-cooled suits and/or hoods that distribute cooling air next to the skin are available. The total heat exchange from a completely sweat wetted skin when cooling air is supplied to the air-cooled suit is a function of cooling air temperature and cooling airflow rate. Both the total heat exchanges and the cooling power increase with cooling airflow rate and decrease with increasing cooling air inlet temperature.

For an air inlet temperature of 10 C (50 F) at 20 percent relative humidity and a flow rate of 10 ft^3/min (0.28 m^3/min), the total heat exchanges over the body surface would be 233 W in a 29.4 C (84.9 F) 85 percent relative humidity environment and 180 W in a 51.7 C (125.1 F) at 25 percent relative humidity environment. Increasing the cooling air inlet temperature to 21 C (69.8 F) at 10 percent relative humidity would reduce the total heat exchanges to 148 W and 211 W, respectively. Either air inlet temperature easily provides 100 W of cooling.

The use of a vortex tube as a source of cooled air for body cooling is applicable in many hot industrial situations (Figures 12-12 and 12-13). The vortex tube, which is attached to the worker, requires a constant source of compressed air supplied through an air hose. The hose connecting the vortex tube to the compressed air source limits the area within which the worker can operate. However, unless mobility of the worker is required, the vortex tube, even though noisy, should be considered as a simple cooled air source.

Ice packet vest. Available ice packet vests may contain as many as 72 ice packets; each packet has a surface area of approximately 64 cm^2 and contains about 46 gm of water. These ice packets are generally secured to the vest by tape. The cooling provided by each individual ice packet will vary with time and with its contact pressure with the body surface, plus any heating effect of the clothing and hot environment; thus, the environmental conditions have an effect on both the cooling provided and the duration of time this cooling is provided. Solid carbon dioxide in plastic packets can be used instead of ice packets in some models.

BIBLIOGRAPHY

American Conference of Governmental Industrial Hygienists (ACGIH). *Industrial Ventilation—A Manual of Recommended Practice,* 19th ed. Lansing, MI: ACGIH, 1986.

———. *Threshold Limit Values and Biological Exposure Indices for 1986–1987.* Cincinnati: ACGIH, 1986.

American Industrial Hygiene Association (AIHA). *Heating and Cooling for Man in Industry.* Akron, OH: AIHA, latest edition.

American Society of Heating, Refrigerating and Air Conditioning Engineers (ASHRAE). Handbooks published annually. Atlanta: ASHRAE, latest edition.

Belding, H.S., and Hatch, T.F. Index for evaluating heat stress in terms of resulting physiological strains. *Journal of the American Society of Heating and Ventilating Engineers,* Heating, Piping, and Air Conditioning Section, 27 (August 1955): 129–135.

Belding, H.S., Hertig, B.A., and Riedesel, M.L. Laboratory simulation of a hot industrial job to find effective heat stress and resulting physiologic strain. *American Industrial Hygiene Association Journal* (February 1960):25–31.

Bernard, T.E., Kenney, W.L., and Balint, L. *Heat-Stress Management Program for Nuclear Power Plants.* Palo Alto, CA: Electric Power Research Institute, 1986.

Brouha, L.A. *Physiology in Industry,* 2nd ed. New York: Pergamon Press, 1967.

Brouha, L.A. *Protecting the Worker in Hot Environments.* Pittsburgh: Industrial Hygiene Foundation, Mellon Institute, November 1965.

Burton, A.C., and Edholm, O.C. *Man in a Cold Environment.* London: Arnold, 1955.

Carlson, L.D. Human tolerance to cold. *Journal of Occupational Medicine* (March 1960):129–131.

Cold and its effect on the worker. *Occupational Health Bulletin,* Vol. 15, No. 11. Occupational Health Division, Department of National Health and Welfare, Ottawa, Ontario, Canada.

Cold Injuries. U.S. Army Aviation Digest, Department of the Army. Washington, DC: Superintendent of Documents, January 1964, pp. 27–29.

Criteria for a Recommended Standard: Occupational Exposure to Hot Environments, Revised Criteria, 1986. Washington, DC: DHHS, U.S. PHS, CDC, NIOSH, Division of Standards Development and Technology Transfer, April 1986.

The effective distribution of supply air to workers. *Michigan's Occupational Health,* Vol. 10, No. 4. Lansing, MI, Department of Health, 1965.

Fuller, F.H., and Brouha, L.A. New engineering methods for evaluating the job environment. *American Society of Heating, Refrigerating and Air Conditioning Engineers Journal* (February 1966):39–52.

Goldsmith, R., and Lewis, H.E. Polar expeditions as human laboratories. *Journal of Occupational Medicine* (March 1960):118–122.

Haines, G.F., Jr., and Hatch, T.F. Industrial heat exposures—Evaluation and control. *Heating and Ventilating* (November 1952):93–104.

Hatch, T.F. "Assessment of heat stress." In *Temperature, Its Measurement and Control in Science and Industry,* vol. 3, part 3., edited by C. M. Herzfeld. Melbourne, FL: Robert E. Krieger Publ. Co., Inc., 1972.

Hertig, B.A., and Belding, H.S. "Evaluation and Control of Heat Hazards." In *Temperature, Its Measurement and Control in Science and Industry,* vol. 3, part 3, edited by C. M. Herzfeld. Melbourne, FL: Robert E. Krieger Publ. Co., Inc., 1972.

Horvath, S.M. and Jensen, R.C., ed. *Occupational Exposures to Hot Environments.* National Institute for Occupational Safety and Health, U.S. HEW (now DHHS), January 1976.

Horvath, S.M., ed. *Cold Injuries.* New York: Josiah Macy, Jr., Foundation, 1960.

Hot Environments. Cincinnati: National Institute for Occupational Safety and Health (NIOSH), U.S. Department of Health and Human Services (DHHS), 1985.

Houghton, F.C., and Yaglou, C.P. Determination of the Comfort Zone. *Journal of the American Society of Heating and Ventilating Engineers,* 29:515.

Jensen, R.C., and Heins, D.A. *Relationships Between Several Prominent Heat Stress Indices,* DHHS (NIOSH) Publication No. 77-109. Cincinnati: NIOSH, Division of Biomedical and Behavioral Science, October 1976.

Kuhlemeier, K.V., and Miller, J.M. *Assessment of Deep Body Temperature of Women in Hot Jobs.* Cincinnati: NIOSH, DHHS, July 1977.

Leithead, C.S., and Lind, A.R. *Heat Stress and Heat Disorders.* Philadelphia: F.A. Davis Co., 1964.

Minard, D. Prevention of heat casualties in marine corps recruits. *Military Medicine* 126:261-272.

Morris, Desmond. *The Naked Ape.* New York: McGraw-Hill Book Co., 1967.

Mutchler, J.E., Malzahn, D.D., Vecchio, J.L., and Soule, R.D., eds. *An Improved Method for Monitoring Heat Stress Levels in the Workplace.* Cincinnati: NIOSH, DHHS, May 1975.

National Safety Council (NSC), 444 N. Michigan Ave., Chicago, IL 60611. Industrial Data Sheet: *Cold Room Testing of Gasoline and Diesel Engines,* No. 456, Rev. 1985.

——. *Pocket Guide to Cold Stress.* Chicago: NSC, 1985.

——. *Pocket Guide to Heat Stress.* Chicago: NSC, 1985.

Peters, W.R., Darilek, G.T., and Herzig, F.X. *Survey of Environmental Parameter for a Personal Heat Stress Monitor.* Cincinnati: NIOSH, DHHS, March 1976.

Peterson, J.E. Experimental evaluation of heat stress indices. *American Industrial Hygiene Association Journal* (May-June, 1970):305-317.

Redmond, C.K. Emes, J.L., Mazumdars, S., Magee, P.C., and Kamon, E. *Mortality of Steel Workers Employed in Hot Jobs.* Cincinnati: NIOSH, DHHS, August 1977.

Science and Technical Information Office, National Aeronautics and Space Administration, *Bioastronautics Data Book,* 2nd ed., Washington, DC: Government Printing Office, 1973.

Schooley, J.F., ed. *Temperature: Its Measurement and Control in Science and Industry.* Proceedings of the 6th International Symposium, Washington, DC, March 15, 1982.

Sellers, E.A. Cold and its influence on the worker. *Journal of Occupational Medicine* (March 1960):115-117.

Turl, L.H. Clothing for cold conditions. *Journal of Occupational Medicine* (March 1960):123-128.

13

Ergonomics

by Karl H. E. Kroemer, PhD

NOTE. This chapter relies much on Dr. Bruce A. Hertig's text for the second edition, 1979.

BACKGROUND

The term "ergonomics" was coined in 1950 by a group of physical, biological, and psychological scientists and engineers in the United Kingdom to describe interdisciplinary activities that were designed to solve problems created by wartime technology. The term is derived from the Greek roots *ergon,* which is related to work and strength, and *nomos,* indicating law or rule.

In North America, the human factor was also recognized as important although often disregarded in military human-machine systems during World War II. Behavioral scientists, anthropometrists, engineers, and others working in this emerging scientific discipline in 1957 decided to refer to their new field as "human factors," rather than ergonomics. Thus, essentially the same field of research and engineering applications is called human factors and human engineering in the United States and ergonomics throughout the rest of the world. In this text, the terms will be used synonymously.

A formal definition is as follows: "Ergonomics is the study of human characteristics for the appropriate design of the living and work environment." Ergonomic researchers strive to learn about human characteristics (capabilities, limitations, motivations, and desires) so that this knowledge can be used to adapt a human-made environment to the people involved. This knowledge may affect complex technical systems or work tasks, equipment and workstations, or the tools and utensils used at work or during leisure times. Hence, ergonomics is human-centered, transdisciplinary, and application oriented.

The goal of ergonomics/human factors ranges from making work safe and human and increasing human efficiency to creating human well-being. The Committee on Human Factors of the National Research Council (1983) stated: "Human factors engineering can be defined as the application of scientific principles, methods, and data drawn from a variety of disciplines to the development of engineering systems in which people play a significant role. Successful application is measured by improved productivity, efficiency, safety, and acceptance of the resultant system design. The disciplines that may be applied to a particular problem include psychology, cognitive science, physiology, biomechanics, applied physical anthropology, and industrial and systems engineering. The systems range from the use of a simple tool by a consumer to multiperson sociotechnical systems. They typically include both technological and human components.

"Human factors specialists. . .are united by a singular perspective on the system design process: that design begins with an understanding of the user's role in overall system performance and that systems exist to serve their users, whether they are consumers, system operators, production workers, or maintenance crews. This user-oriented design philosophy acknowledges human variability as a design parameter. The resultant designs incorporate features that take advantage of unique human capabilities as well as build in safeguards to avoid or reduce the impact of unpredictable human error."

Figure 13-1 shows how ergonomics interacts with related applied disciplines and sciences. Among the primary foundations of ergonomics are the biological sciences, particularly anatomy and physiology. Leonardo da Vinci in the 16th century, Giovanni Alfonso Borelli in the 17th century, Lavoisier, Amar,

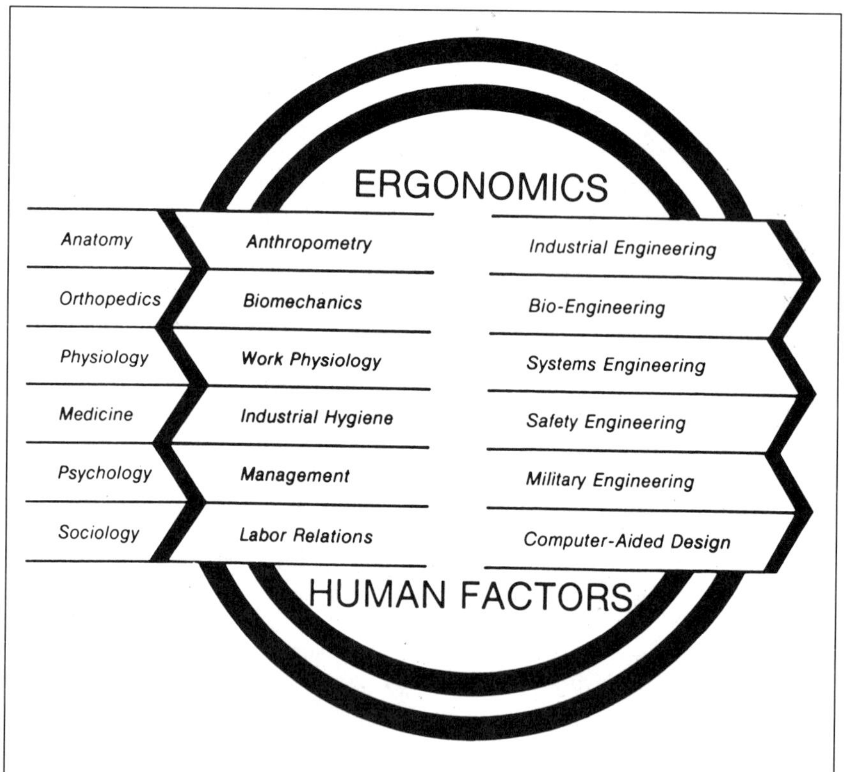

Figure 13-1. Origins and applications of ergonomics/human factors.

Rubner, Johannson, and many others in the 19th and early 20th centuries contributed ideas, concepts, theories, and practical data to forward the understanding of the role of the human body in a work environment. Among the social and behavioral sciences, anthropologists, psychologists, and sociologists have contributed toward modeling and understanding the human role in societal and technological systems, including management theories. Among the engineering disciplines, industrial engineers (using, for example, the groundwork laid by Frederick Taylor and the Gilbreths), mechanical engineers and more recently computer engineers, with their new application fields of safety, military, and space, are the major contributors to and users of ergonomic knowledge.

Ergonomic specialists have developed their own theories, methods, techniques, and tools to perform scientific research. Industrial hygienists, work physiologists, industrial psychologists, and industrial engineers are among the primary users of ergonomics. A typical modern application is computer-aided design (CAD) which incorporates the systematic consideration of human attributes translated in the form of engineering anthropometric and biomechanical data. Health and safety at work, the quality of life while at work, and, perhaps, participatory management are some well-known programmatic aspects of ergonomics.

There are three levels at which ergonomic knowledge can be used. Tolerable conditions do not pose known dangers to human life or health. Acceptable conditions are those upon which (according to the current scientific knowledge and under given sociological, technological, and organizational circumstances) the people involved can voluntarily agree. Optimal conditions are so well adapted to human characteristics, capabilities, and desires that physical, mental, and social well-being is achieved. The aim of human factors/ergonomics applications is to achieve ease and efficiency at work.

WORK PHYSIOLOGY

People perform widely differing tasks in daily work situations. These tasks must be matched with human capabilities to avoid "underloading," in which human capabilities are not utilized properly, as well as "overloading," which may cause the employee to break down and suffer reduced performance capability or even permanent damage. Work physiologists evaluate the capacities and limitations of the worker for performing physical work; they also determine human tolerance to stresses produced by the physical environment.

Capacity for physical work

An individual's physical tolerance of physical work is usually determined by the capacity of his or her respiratory and cardiovascular systems to deliver oxygen to the working muscles and to metabolize chemically stored energy. Maximum oxygen uptake is often used to describe the upper limit of this capacity. If a person is pushed beyond this limit in an emergency situation, the additional energy required is provided by anaerobic processes. The energy stores thus depleted (the oxygen debt) must be replenished following cessation of the emergency.

Tolerance times for maximal efforts are measured in minutes or even in seconds for a sprint runner. In the modern industrial setting, maximal effort may be required for brief periods, such as when an employee must heave heavy loads onto a handtruck. However, during an 8-hour shift, the average energy required usually falls well below human peak capacity.

The biochemical processes that transform foodstuffs into energy available for work are quite complex; they involve a series of aerobic and anaerobic steps. Nevertheless, measurement of the volume of oxygen consumed provides a relatively simple overall index of energy consumption and hence of the energy demands of work. Utilization of 1 liter (L) of oxygen yields approximately 5 kilocalories (kcal). To put oxygen consumption and energy demands into proper perspective, consider the abilities of trained athletes who may reach maximal oxygen uptake capacities of up to 6 L/min. Aside from a person's physique, age and gender influence the oxygen intake capacity substantially. Men who are 20 years of age have an average maximal capacity of 3–3.5 L/min; women of the same age have an average capacity of 2.3–2.8 L/min. At age 60, the capacities diminish to about 2.2–2.5 L/min for men, and 1.8–2.0 L/min for women. As with most physiological characteristics, there is considerable individual variability.

The ability to move oxygen from the air to the active muscle can be improved through physical training by up to 20 percent. Unfortunately, the efficiency with which humans convert oxygen to energy is quite low—only about 5 percent or less of the maximal oxygen uptake capacity is converted to energy in daily activity. Hence, translation of oxygen consumption figures into maximal energy production (i.e., energy that is exerted in the use of the work equipment) yields values of about 900–1,000 kcal/hr for the industrial population. However, most modern jobs require such effort not over 1 hour but over several hours, perhaps during the entire workshift. Of course, an individual's capacity also depends on other "central" functions (e.g., of the circulatory and cardiac systems) and on "local" capacities (e.g., of muscles). Readers who wish to have more detailed information on the assessment of physical work capacity may want to consult the texts by Astrand and Rodahl (1986), Grandjean (1980), Kroemer, Kroemer, and Kroemer-Elbert (1986), and the *Ergonomics Guides* published by the American Industrial Hygiene Association (AIHA).

Energy cost of work

Typically, the heaviest work that a young, fit man can sustain for prolonged periods is about 500 kcal/hr. Among the general population, this figure is somewhat lower: 400–425 kcal/hr. These figures are equivalent to about 40 percent of the maximal uptake capability.

Industrial jobs seldom demand such a high-energy expenditure over the course of a workday. Rest pauses, fetching tools, mopping the brow, and receiving instruction all tend to reduce the average energy expenditure considerably. Table 13-A lists several typical activities and their average metabolic costs; resting values are included for reference. The given values must be adjusted according to one's body weight; the table applies to a man of 70 kg (154 lb).

When intermittent tasks are performed, the average expenditure may be calculated using the following formula:

$$\overline{M} = (M_1t_1 + M_2t_2 + \ldots M_nt_n)t^{-1}$$

In this formula, $\overline{M}$ is the total metabolic energy cost; M_1, M_2, and so on are the metabolic costs of individual tasks; t_1, t_2, and so on indicate the duration of the individual task; and t is the total elapsed time.

Heart rate at work

There is a close interaction between the human circulatory and metabolic systems. Nutrients and oxygen must be brought to the working muscles and metabolic by-products removed from them to ensure proper functioning. Therefore, heart rate (which is a primary indicator of circulatory functions) and oxygen consumption (representing the metabolic processes taking place in the body) have a linear and reliable relationship in the range between light and heavy work. (However, when very light work loads or very heavy ones are being handled, that relationship may not be reliable. It is also not reliable under severe environmental conditions or when workers are under mental stress.) Given such a linear relationship, one can often simply substitute heart rate measurements for a measurement of a metabolic process such as oxygen consumption. This is a very attractive shortcut, since heart rate measurements can be performed rather easily.

The simplest technique for heart rate assessment is to palpate an artery, often in the wrist. The measurer counts the number of heartbeats over a given period of time—such as 15 seconds—and then calculates the average heart rate per minute. More refined methods utilize various plethysmographic techniques, which rely on the deformation of tissue due to changes that result when the imbedded blood vessels fill with blood. Such techniques range from mechanically measuring the change in the volume of tissues to using photoelectric techniques that react to changes in the transmissibility of light caused by the blood filling. Such changes occur in the ear lobule, for example. More expensive techniques rely on electric signals generated by the nervous control systems that control heart rate. When using this technique, electrodes are usually placed on the patient's chest.

Use of heart rate measurement has one major advantage over oxygen measurement as an indicator of metabolic processes: heart rate reacts faster to work demands and therefore more easily indicates quick changes in body functions due to changes in work requirements. More information on heart rate assessment can be obtained from Astrand and Rodahl (1986); Eastman Kodak Company (1983, 1986), and Kroemer, Kroemer, and Kroemer-Elbert (1986).

Table 13-A. Metabolic Energy Costs of Several Typical Activities

Activity	*kcal/hr*	*Btu/hr*
Resting, prone	80–90	320–360
Resting, seated	95–100	375–397
Standing, at ease	100–110	397–440
Drafting	105	415
Light assembly (bench work)	105	415
Medium assembly	160	635
Driving automobile	170	675
Walking, casual	175–225	695–900
Sheet metal work	180	715
Machining	185	730
Rock drilling	225–550	900–2,170
Mixing cement	275	1070
Walking on job	290–400	1,150–1,570
Pushing wheelbarrow	300–400	1,170–1,570
Shoveling	235–525	930–2,070
Chopping with axe	400–1,400	1,570–5,550
Climbing stairs	450–775	1,770–3,070
Slag removal	630–750	2,500–2,970

Values are for a male worker of 70 kg (154 lb). (Reprinted with permission from "Ergonomics guide to assessment of metabolic and cardiac costs of physical work." *American Industrial Hygiene Association Journal,* 32 [1971]:560-564)

Matching people and their work

Obviously, it is important to match human capabilities with the related requirements of a given job. If the job demands equal the worker's capabilities or if they exceed them, the person will be under much strain and may not be able to perform the task. Hence, various stress tests, which are administered by a physician, have been developed to assess an individual's capability to perform physically demanding work. Bicycle ergometers, treadmills, or steps are used to simulate stressful demands. The reactions of the individual in terms of oxygen consumption, heart rate, or blood pressure are used to assess that person's ability to withstand and exceed such demands. However, the examining physician needs to know what the actual work demands are. This often requires a special investigation of the actual work requirements. The industrial hygienist may be called upon to help in the assessment of the existing work demands.

Work classification. The work demands listed in Table 13-B are rated from light to extremely heavy in terms of energy expenditure per minute, and the relative heart rate in beats per minute is also given. Light work is associated with rather small energy expenditures and is accompanied by a heart rate of approximately 90 beats/min. At this level of work, the energy needs of the working muscles are supplied by oxygen available in the blood and by glycogen in the muscle. There is no buildup of lactic acid or other metabolic by-products that would limit a person's ability to continue such work.

Table 13-B. Classification of Light to Heavy Work According to Energy Expenditure and Heart Rate

Classification	*Total Energy Expenditure (kcal/min)*	*Heart Rate (beats/min)*
Light work	2.5	90 or less
Medium work	5	100
Heavy work	7.5	120
Very heavy work	10	140
Extremely heavy work	15	160 or more

At medium work, which is associated with about 100 heart beats/min, the oxygen required by the working muscles is still covered, and lactic acid developed initially at the beginning of the work period is resynthesized to glycogen during the activity.

In heavy work, during which the heart rate is about 120 beats/min, the oxygen required is still supplied if the person is physically capable to do such work and specifically trained in this job. However, the lactic acid concentration produced during the initial phase of the work is not reduced but remains high until the end of the work period. The concentration returns to normal levels after cessation of the work.

In the course of light, medium, and, if the person is capable and trained, even heavy work, the metabolic and other physiological functions can attain a steady-state condition during the work period. This indicates that all physiological functions can meet the demands and will remain essentially constant throughout the duration of the effort.

However, no steady state exists in the course of very heavy work, during which the heart rate level attains or exceeds 140 beats/min. In this case, the original oxygen deficit incurred during the early phase of the work increases throughout the duration of the effort and metabolic by-products accumulate, making intermittent rest periods necessary or even forcing the person to stop this effort completely. At even higher energy expenditures, which are associated with heart rates of 160 beats/min or more, the lactic acid concentration in the blood and the oxygen deficit achieve such magnitudes that frequent rest periods are needed. Even highly trained and capable persons are usually unable to perform such a demanding job throughout a full working day.

Hence, energy requirements or heart rate allow one to judge whether a job is energetically easy or hard. Of course, such labels as light, medium, or heavy reflect judgments of physiological events (and of their underlying job demands) that rely very much on the current socioeconomic concept of what is comfortable, acceptable, permissible, difficult, or excessive.

Rating the perceived effort

We are all able to perceive the strain generated in the body by a given work task and can make absolute and relative judgments about this perceived effort. A worker can assess and rate the relationship between the physical stimulus (that is, the work performed) and the perceived sensation associated with this stimulus. The more subjectively demanding a work load is, the higher the perceived effort.

In the 19th century, E. H. Weber (1834) and G. T. Fechner (1860) developed models of the relationships between physical stimulus and the perceptual sensation of that stimulus, i.e., the "psychophysical correlate." Weber suggested that the "just noticeable difference ΔI" that can be perceived increases with the absolute magnitude of the physical stimulus I:

$$\Delta I = \alpha * I$$

In the formula, α is a constant.

Fechner related the magnitude of the "perceived sensation P" to the magnitude of the stimulus in the following way:

$$P = \beta + \delta * \log I$$

In the formula, β and δ are constants.

In the 1950s, S. S. Stevens at Harvard and G. Ekman in Sweden introduced ratio scales, which assume a zero point and equidistant scale values. These scales have since been used to describe the relationships between the perceived intensity and the physically measured intensity of a stimulus in a variety of sensory modalities (e.g., related to sound, lighting, and climate) as follows:

$$P = \delta * I^n$$

In the formula, δ is a constant, and n ranges from 0.5–4, depending on the modality.

Since about 1960, Borg and his co-workers have modified these relationships to take deviations from previous assumptions (such as zero point and equidistance) into account and to describe the perception of different kinds of physical efforts (Borg, 1982). Borg's "general function" is as follows:

$$P = a + c(I - b)^n$$

In the formula, the constant a represents "the basic conceptual noise" (normally less than 10 percent of I), and the constant b indicates the starting point of the curve; c is a conversion factor that depends on the type of effort.

Ratio scales indicate proportions between percentages but

do not indicate absolute intensity levels; they neither allow intermodal comparisons nor comparisons between intensities perceived by different individuals. Borg has tried to overcome this problem by assuming that the subjective range and intensity level are about the same for each subject at the level of maximum intensity. In 1960, this led to the development of a category scale for rating perceived exertions (RPE). The scale ranges from 6 to 20 and matches heart rates from 60–200 beats/min. A new category begins at every second number.

The 1960 Borg RPE	*Borg's 1985 Modification**
6	6—no exertion at all
7—very, very light	7
8	8—extremely light
9—very light	9—very light
10	10
11—fairly light	11—light
12	12
13—somewhat hard	13—somewhat hard
14	14
15—hard	15—hard (heavy)
16	16
17—very hard	17—very hard
18	18
19—very, very hard	19—extremely hard
20	20—maximal exertion

*Copyrighted by Borg

In 1980, Borg proposed a category scale with ratio properties that could yield ratios, levels, and allow comparisons but still retain the same correlation (R about .88) with heart rate as the RPE scale.

The Borg General Scale (1980)

0 —nothing at all
0.5—extremely weak (just noticeable)
1 —very weak
2 —weak
3 —moderate
4 —somewhat strong
5 —strong
6
7 —very strong
8
9
10 —extremely strong (almost maximal)

The following exemplifies how the Borg scales are used. While the subject looks at the rating scale, the person administering the test says, "I will not ask you to specify the feeling but do select a number that most accurately corresponds to your perception of (experimenter specifies symptom). If you don't feel anything, for example, if there is no (symptom), you answer zero—nothing at all. If you start feeling something that is only just about noticeable, you answer 0.5—extremely weak, just noticeable. If you have an extremely strong feeling of, for instance (symptom), you answer 10—extremely strong, almost maximal. This is the absolute strongest you have ever experienced. The more you feel, the stronger the feeling, the higher the number you choose.

"Keep in mind that there are no wrong numbers. Be honest. Do not overestimate or underestimate your ratings. Do not think of any other sensation than the one I ask you about. Do you have any questions?"

Let the subject get well acquainted with the rating scale before the test. During the test, let the subject do the ratings toward the end of every work period, i.e., about 30 seconds before stopping or changing the work load. If the test must be stopped before the scheduled end of the work period, let the subject rate the feeling at the moment of stoppage.

Work/rest cycles

In the event that a task demands more of the worker than can be sustained, rest pauses must be taken. A general principle governing the schedule of work/rest cycles is to break up excessively heavy work into bouts of work that are as short as is practical for the task at hand. Frequent short rest periods reduce cumulative fatigue better than a few long breaks. The worst procedure is to let the worker go home early, exhausted.

A formula has been used to estimate the percentage of time that should be allotted to rest:

$$T_{rest}\,(\%) = \frac{M_{max} - M}{M_{rest} - M} \times 100$$

In the formula, T_{rest} is the percentage of rest time; M_{max} is the upper limit of the metabolic cost for sustained work; M is the metabolic cost of the task; and M_{rest} represents the resting (sitting) metabolism.

For example, suppose that M_{max} equals 350 kcal/hr; an average value for M_{rest} is 100 kcal/hr. Then assume that the task requires 525 kcal/hr, which is obviously too high. Apply these values to the formula as follows:

$$\text{rest time} = \frac{350\text{–}525}{100\text{–}525} \times 100 = \frac{-175}{-425} \times 100 = 41\%$$

Thus, rest pauses should be scheduled to last a total of 41 percent (24 min) of the hour.

As an alternative to the idle rest pause, one may consider intermingling light tasks with heavy tasks. To calculate the proportion of time that should be allocated to the two tasks, consider a heavy task that requires 500 kcal/hr interrupted by a light task that requires 250 kcal/hr. Again, assume that M_{max} equals 350 kcal/hr.

$$\text{light task time} = \frac{350\text{–}500}{250\text{–}500} \times 100 = \frac{-150}{-250} \times 100 = 60\%$$

Following this regimen, the hard work should consume 40 percent of the time, the light task 60 percent. The light, secondary work task thus actually constitutes rest time from the heavy, primary task. Sharpening tools or walking to get material and other interruptions can provide productive respites from heavy work.

Fatigue

Fatigue is an overexertion phenomenon that leads to a temporary decrease in physical performance. It is often associated with a buildup of lactic acid in the body, which can be metabolized to carbon dioxide and water during reduced activity, or rest.

Subjective feelings of lowered motivation and deteriorated mental and physical activities may result from fatigue. Fatigue may occur together with monotony, a sensation associated with the lack of stimuli. Fatigue-induced low performance can be completely restored to its full level by rest.

The subjective sensation of fatigue is feeling tired. When tired, a person has reduced capability and desire for either physical or mental work, and feels heavy and sluggish. The sensation of fatigue has a protective function similar to hunger and thirst. Feeling fatigued forces one to avoid further stress and allows recovery to take place.

The most important factors that produce fatigue are: physical work intensity (static and dynamic work); illness, pain, lack of rest (sleep), and poor eating habits; and psychological factors—worry, conflict, and possibly monotony. Hence, many different sources may be responsible for the sensation and the state of fatigue. In everyday experience, fatigue is often an accumulation of the effects stemming from various sources. It is evident that if the rhythm of life is to proceed normally, the possibilities of recovery should compensate for the strain which the organism undergoes daily.

Severe, continuous daily fatigue eventually leads to chronic fatigue. Not only is the feeling of tiredness intensified and continuous after work, but occasionally a person feels tired before beginning work. The following signs signal chronic fatigue:

- increased irritability—intolerance, antisocial behavior;
- tendency to depression—unmotivated worries;
- general weakness in drive and a dislike for work;
- nonspecific physical complaints—headaches, giddiness, palpitations, rapid breathing, loss of appetite, indigestion, or insomnia.

When physical complaints of this kind appear, the condition can be called "clinical fatigue." As a result of this condition, absences from work increase in frequency and duration, workers need longer rest periods, and they may show an increased susceptibility to illness. It is often difficult to separate mental or emotional events from physical causes of fatigue. In clinical fatigue, one is hardly able to distinguish between cause and effect. A negative attitude toward one's work, superiors, or place of work can just as well be a cause of clinical fatigue as the result of it.

ENGINEERING ANTHROPOMETRY AND BIOMECHANICS

Engineering anthropometry and biomechanics are among the newer scientific and application fields triggered and furthered by ergonomics. These fields describe the physical dimensions and properties of the body. Their origins can be traced to work by Quetelet, Braune, Fischer, Harless, Broca, and others working in the 19th century. In the 20th century, standardized research techniques were developed (Martin), and data were systematically applied to human-machine interfaces (Hooton, McFarland, Hertzberg). Information about the stress resistance of bones, joints, and tissues became critical; e.g., in research on automobile crashes or crew ejections from fast-flying aircraft. Combining such anthropometric and biomechanical information with physiological, psychological, and engineering aspects has added new dimensions of human factors engineering to classical physiological and behavioral information during the last few decades.

Engineering anthropometry

Anthropometry literally means "measuring the human." The human body is traditionally measured in terms of heights, breadths, depths, and distances—all straight-line, point-to-point measurements between landmarks on the body and/or reference surfaces. Also, curvatures and circumferences following contours are measured. For such measurements, the body is placed in defined upright straight postures with body segments at 180° or 90° with respect to each other. For example, the subject is required to "stand erect; heels together, buttocks, shoulder blades, and back of head touching the wall; arms vertical (or extended straight forward), fingers straight. . ." This is similar to the so-called anatomical position. When measurements are taken on a seated subject, the (flat and horizontal) surfaces of seat and foot support are so arranged that the thighs are horizontal, the lower legs are vertical, and the feet are flat on their support. The subject is nude, or nearly so, and unshod.

Figure 13-2 shows reference planes and descriptive terms often used in anthropometry. Figures 13-3 and 13-4 show important anatomical landmarks of the human body. The NASA-Webb *Anthropometric Sourcebook* (1978) contains much information on measurement techniques in general and on military anthropometric data in particular. The book by Kroemer, Kroemer, and Kroemer-Elbert (1986) condenses and updates this information.

Body dimensions are commonly measured using anthropometers, calipers, tapes, and a scale. Nearly all currently available information has been gathered in this manner. Emerging tech-

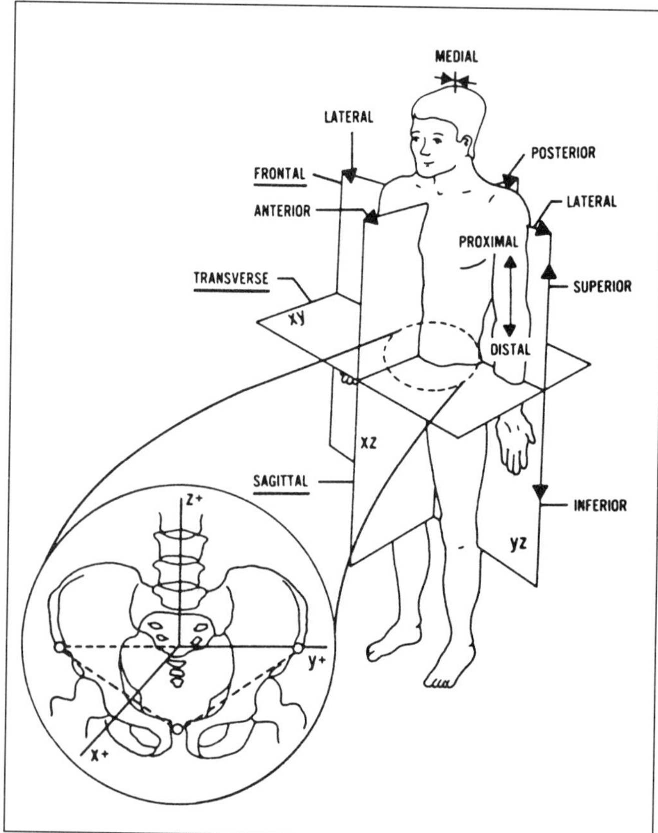

Figure 13-2. Reference planes and terms used in engineering anthropometry. (Adapted from NASA-Webb, 1978)

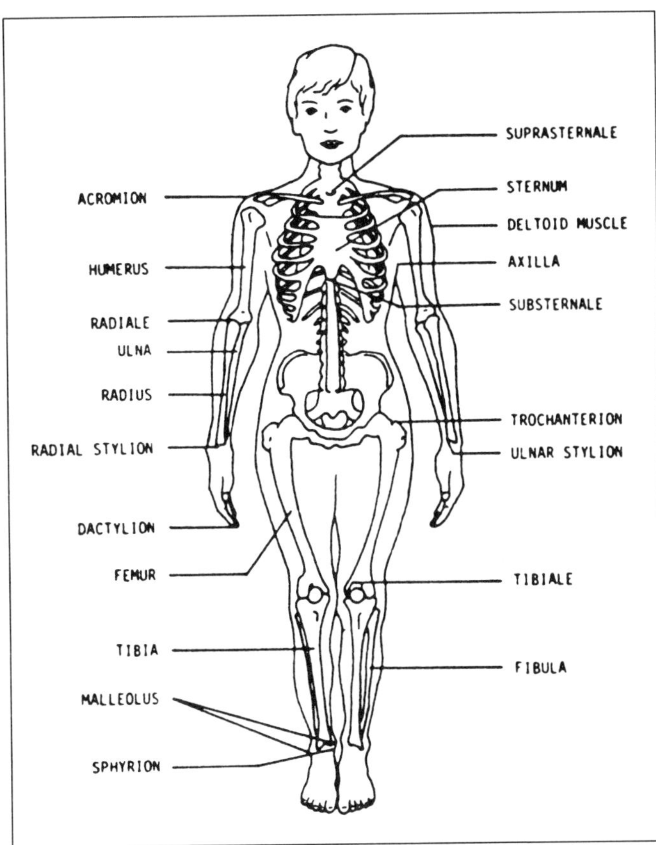

Figure 13-3. Landmarks on the human body in the frontal view. (Reprinted with permission from NASA-Webb, 1978)

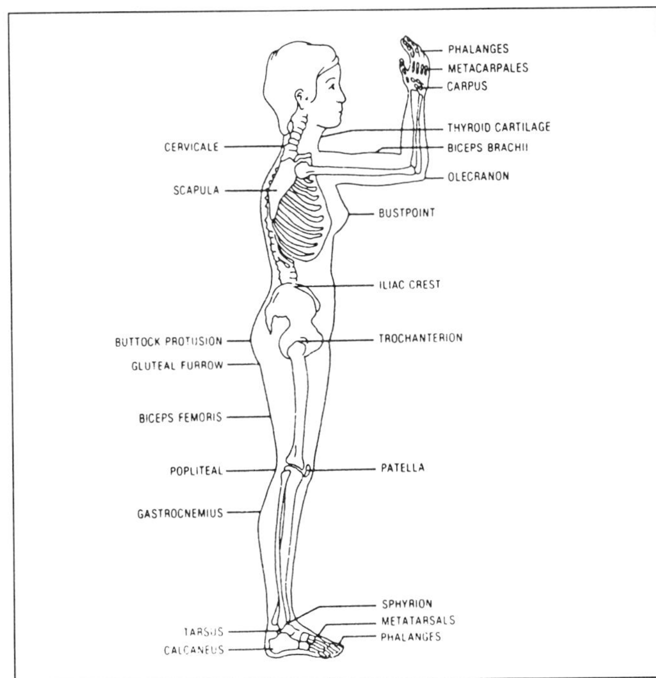

Figure 13-4. Landmarks on the human body in the lateral view. (Reprinted with permission from NASA-Webb, 1978)

niques, particularly stereophotogrammetry, are expected to result in new data compilations. The existing anthropometric information has been gathered while the subjects assume highly stylized and standardized postures, which are quite different from the body positions assumed while working, particularly when the worker is moving around. Hence, current data must be interpreted by the ergonomic designer for practical applications.

It is of interest to note that body-building typologies (somatotypes), such as those developed by Kretschmer, Sheldon, or Heath-Carter, have neither proved to be suitable for engineering anthropometry, nor are reliable predictors of attitudes, physical capabilities, or limitations relative to performance in industrial systems. Hence, somatotyping is of little or no value for engineers or managers.

Civilian dimensions. The anthropometric literature abounds with data on military personnel, while reliable and complete information on civilians is rather scarce. Hence, many of the anthropometric data applied to industrial populations were obtained from young, healthy, and (mostly) male soldiers. No large, complete, and reliable surveys of the U.S. civilian population have been performed during the last decade. Hence, physical anthropologists working with the (amply available) military data and the (scarce) civilian data have developed a compilation of estimated civilian body dimensions, which is presented in Table 13-C. The table shows the 5th, 50th, and 95th percentile dimensions as well as the standard deviation. Assuming a normal distribution of the data, the 50th percentile coincides with the average (mean) value.

Following the example of military requirements, it is common practice to design equipment, work spaces, and tools so that they fit the small body (usually that of women in the 5th percentile) as well as to the large body (usually that of men in the 95th percentile). These percentile values correspond to a standard deviation of ±1.65 around the mean. If other percentile points need to be calculated, Table 13-D can be used.

Designing for percentiles. Percentiles are very convenient to determine exactly which percentage of a known population is fitted by a design range. For example, the adjustment range of a work seat is from the lowest seat height of 35.5 cm (14.0 in.) to its highest setting of 48.8 cm (19.2 in.). This range accommodates the popliteal height (a measurement of lower leg length) from the woman in the 5th percentile to the man in the 95th percentile. One should then add heel height of approximately 2 cm to this range. Table 13-C shows that a single fixed seat height of 41 cm (16 in.) corresponds to approximately the 50th percentile popliteal height of a mixed male-female population; thus, this seat height will be too high for about half the people and too low for all the others. Designing for the average really fits nobody.

The ergonomically false concept of the average person needs to be abandoned because no one is average in several or all body dimensions. A person who is average in weight is not usually average in stature, leg length, or in arm circumference. The correct approach is to select the specific anthropometric dimension(s) to be fitted and to establish (for each dimension) the design range (such as from the 5th to the 95th percentile) so that proper fit will be achieved. Reference books by Eastman Kodak Company (1983, 1986) and Kroemer, Kroemer, and Kroemer-Elbert (1986) provide guidance for such ergonomic design.

Table 13-C. U.S. Civilian Body Dimensions, Female/Male, for Ages 20 to 60 Years,* in cm.

Dimensions	*Percentiles* 5th	50th	95th	*Standard Deviation*
Heights				
Stature (height)	149.5/161.8	160.5/173.6	171.3/184.4	6.6/6.9
Eye height	138.3/151.1	148.9/162.4	159.3/172.7	6.4/6.6*
Shoulder (acromion) height	121.1/132.3	131.1/142.8	141.9/152.4	6.1/6.1*
Elbow height	93.6/100.0	101.2/109.9	108.8/119.0	4.6/5.8
Knuckle height	64.3/69.8	70.2/75.4	75.9/80.4	3.5/3.2
Height, sitting	78.6/84.2	85.0/90.6	90.7/96.7	3.5/3.7
Eye height, sitting	67.5/72.6	73.3/78.6	78.5/84.4	3.3/3.6*
Shoulder height, sitting	49.2/52.7	55.7/59.4	61.7/65.8	3.8/4.0*
Elbow rest height, sitting	18.1/19.0	23.3/24.3	28.1/29.4	2.9/3.0
Knee height, sitting	45.2/49.3	49.8/54.3	54.5/59.3	2.7/2.9
Popliteal height, sitting	35.5/39.2	39.8/44.2	44.3/48.8	2.6/2.8
Thigh clearance height	10.6/11.4	13.7/14.4	17.5/17.7	1.8/1.7
Depths				
Chest depth	21.4/21.4	24.2/24.2	29.7/27.6	2.5/1.9*
Elbow-fingertip distance	38.5/44.1	42.1/47.9	56.0/51.4	2.2/2.2
Buttock-knee distance, sitting	51.8/54.0	56.9/59.4	62.5/64.2	3.1/3.0
Buttock-popliteal distance, sitting	43.0/44.2	48.1/49.5	53.5/54.8	3.1/3.0
Forward reach, functional	64.0/76.3	71.0/82.5	79.0/88.3	4.5/5.0
Breadths				
Elbow-to-elbow breadth	31.5/35.0	38.4/41.7	49.1/50.6	5.4/4.6
Hip breadth, sitting	31.2/30.8	36.4/35.4	43.7/40.6	3.7/2.8
Head Dimensions				
Head breadth	13.6/14.4	14.54/15.42	15.5/16.4	.57/.59
Head circumference	52.3/53.8	54.9/56.8	57.7/59.3	1.63/1.68
Interpupillary distance	5.1/5.5	5.83/6.20	6.5/6.8	.44/.39
Hand Dimensions				
Hand length	16.4/17.6	17.95/19.05	19.8/20.6	1.04/.93
Breadth, metacarpal	7.0/8.2	7.66/8.88	8.4/9.8	.41/.47
Circumference, metacarpal	16.9/19.9	18.36/21.55	19.9/23.5	.89/1.09
Thickness, meta III	2.5/2.4	2.77/2.76	3.1/3.1	.18/.21
Digit 1: breadth, interphalangeal	1.7/2.1	1.98/2.29	2.1/2.5	.12/.21
Crotch-tip length	4.7/5.1	5.36/5.88	6.1/6.6	.44/.45
Digit 2: breadth, distal joint	1.4/1.7	1.55/1.85	1.7/2.0	.10/.12
Crotch-tip length	6.1/6.8	6.88/7.52	7.8/8.2	.52/.46
Digit 3: breadth, distal joint	1.4/1.7	1.53/1.85	1.7/2.0	.09/.12
Crotch-tip length	7.0/7.8	7.77/8.53	8.7/9.5	.51/.51
Digit 4: breadth, distal joint	1.3/1.6	1.42/1.70	1.6/1.9	.09/.11
Crotch-tip length	6.5/7.4	7.29/7.99	8.2/8.9	.53/.47
Digit 5: breadth, distal joint	1.2/1.4	1.32/1.57	1.5/1.8	.09/.12
Crotch-tip length	4.8/5.4	5.44/6.08	6.2/6.99	.44/.47
Foot dimensions				
Foot length	22.3/24.8	24.1/26.9	26.2/29.0	1.19/1.28
Foot breadth	8.1/9.0	8.84/9.79	9.7/10.7	.50/.53
Lateral malleolus height	5.8/6.2	6.78/7.03	7.8/8.0	.59/.54
Weight (kg)	46.2/56.2	61.1/74.0	89.9/97.1	13.8/12.6

(Adapted from Dr. J. T. McConville, Anthropology Research Project, Yellow Springs, OH 45387, and Dr. K. W. Kennedy, USAF-AAMRL-HEG, OH 45433)

* Estimated by Kroemer in 1981.

Population changes. Throughout the last 50 years, increases in certain body dimensions have been observed. The increase in stature is well-known; many children grow taller than their parents. This increase has been in the range of about 1 cm per decade. There is reason to believe that this increase will level off in the near future. A much more pronounced increase, however, has occurred in body weight, with increments of 2 or 3 kg per decade. Whether such an increase in mass has been affected by nutrition and exercise trends ("slim and fit") remains to be seen.

Altogether, such secular developments of body dimensions are rather small and slow. Hence, for most engineers, the changes

Table 13-D. Calculation of Percentiles

Percentile p		Central Percentage Included in the Range	
$x_i = \bar{x} - kS$ (below mean)	$x_j = \bar{x} + kS$ (above mean)	x_{pi} to x_{pj}	k
0.5	99.5	99	2.576
1	99	98	2.326
2	98	96	2.06
2.5	97.5	95	1.96
3	97	94	1.88
5	95	90	1.65
10	90	80	1.28
15	85	70	1.04
16.5	83.5	67	1.00
20	80	60	0.84
25	75	50	0.67
37.5	62.5	25	0.32
50	50	0	0

in body data should have little practical consequences for the design of tools, equipment, workstations, or work clothes, since virtually none are designed to be used over many decades. However, one may have to consider increased ranges of variability in certain dimensions—and this is best expressed by percentile values.

The working population has been changing in age, health, strength, and composition. For example, the U.S. work force today has many more women in occupations that were dominated by men just a few decades ago. It is estimated that in the mid-1990s, two thirds of all U.S. workers may be women. Occupations have changed drastically; computers and service industries are pulling people from traditional workshops and industries. Traditional blue collar workers are becoming fewer and fewer. Life expectancy has increased in the United States since 1900 by 25 years; thus, people in 1980 could expect to live to 73. In 2020, approximately 18 percent of all Americans will be 65 years and older. Thus, the number of elderly workers in the U.S. work force will increase within the very near future. With the current low birth rate, a reduction in the U.S. population within a few decades would take place, but immigration is expected to keep the number of U.S. citizens from decreasing. However, an immigration, for example, of relatively short South Americans can have a pronounced effect on the anthropometry of the working population in certain geographical regions.

Hence, the local anthropometric conditions that suit workers on the shop floor, in the office, or users and operators of equipment can be quite different from national statistics.

Biomechanics

Biomechanics explains characteristics of the human body in mechanical terms. More than 300 years ago, the father of biomechanics, Giovanni Alfonso Borelli, described a model of the human body that consisted of links (bones) joined in their articulations and powered by muscles bridging the joints. This "stick person," refined and embellished with mass properties and material characteristics, still underlies most current biomechanical models of the human body. About 100 years ago, Harless determined the masses of body segments; Braune and Fischer investigated the interactions between mass distribution, body posture, and external forces applied to the body; and von Meyer discussed the body's statics and mechanics. Since then, biomechanical research has addressed, for example, responses of the body to vibrations and impacts, functions and strain properties of the spinal column, and human motion and strength characteristics.

When treating the human body as a mechanical system, many gross simplifications are necessary, and mental functions, for example, are disregarded. However, within these limitations and simplifications, a large body of useful biomechanical information is already available, and this scientific and engineering field is developing rapidly. (See Chaffin and Andersson, 1984; Pope, Frymoyer, and Andersson, 1984; Kroemer, Kroemer, and Kroemer-Elbert, 1986.)

Muscle strength. Assessment of human muscle strength is a biomechanical procedure. This assessment utilizes Newton's Second and Third Laws, which state that force is proportional to mass times acceleration, and that each action is opposed by an equivalent reaction. Since human muscle strength is not measurable at the muscle in vivo and in situ, human strength is measured as the amount of force applied to an external measuring instrument. Inside the body, muscular force vectors develop torque (also called moment) around the body joint bridged by the muscle. Torque is the product of force and its lever arm to the body joint; the direction of the force is measured at a right angle to its lever arm. In kinesiology, the lever arm is often called the mechanical advantage. Figure 13-5 uses the example of elbow flexion to illustrate these relationships. The primary flexing muscle (biceps brachii) exerts a force (M′) at the forearm at its lever arm (m) about the elbow joint. This generages a torque: T=mM′. Since, by definition, a right angle must exist between lever arm and force vector, the lever arm is smaller when the elbow angle is wide open or acute than when the elbow angle is a right angle. Figure 13-6 represents the same condition in more detail. The actual direction of the muscle force vector M usually differs from that of its vector P, which is perpendicular to the lever arm, m.

The correct unit used to express force measurement is the Newton (N). A 1-lb force is approximately 4.45 N, and 1-kg force (kg_f) equals 9.81 N. The pound, ounce, and gram are not

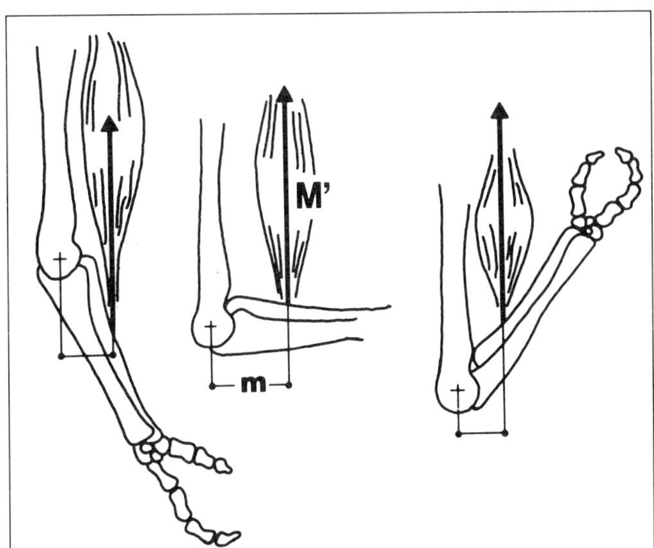

Figure 13-5. Changing lever arm (m) of the muscle force (M′) with varying elbow angle. (Reprinted with permission from Kroemer, Kroemer, and Kroemer-Elbert, 1986)

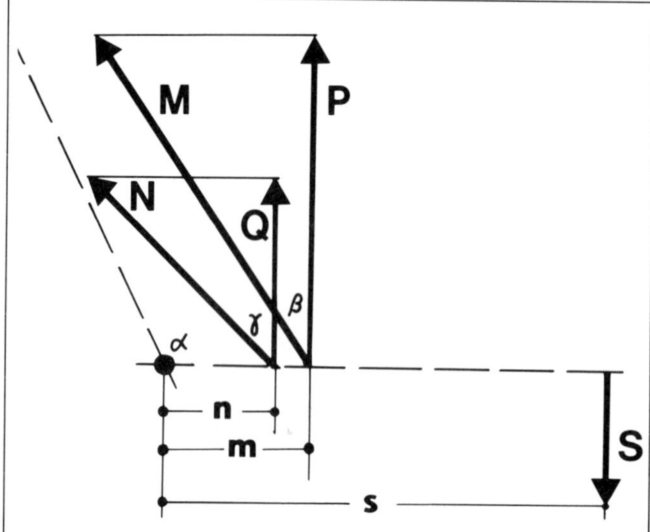

Figure 13-6. Interaction between several muscles and an external force. (Reprinted with permission from Kroemer, Kroemer, and Kroemer-Elbert, 1986)

force units but mass units. According to Newton's Second Law, force equals mass times acceleration (F=mg); hence, mass (m) exerts a proportional force, or weight, as long as the gravitational acceleration (g) applies.

Force (as well as torque) has vector qualities, which means that force must be described not only in terms of magnitude, but also by direction and by the line of force application and the point of its application.

Figure 13-6 shows that the actual direction of the muscle force vector M differs from that of the vector P, which is perpendicular to the lever arm, m. With the angle β between P and M, the relationship between the two vectors is:

$$P = M \cos \beta$$

The angle β itself varies with the elbow angle α, hence:

$$\beta = f(\alpha)$$

The torque T around the elbow joint generated by the muscle force M becomes:

$$T = mP$$

This torque is counteracted, in the balanced condition, by an equally large but opposing torque generated, for example, by a force vector S pulling down on the hand. With its lever arm, s (and assuming a right angle between S and M), S establishes equilibrium if:

$$sS = mP = mM \cos \beta$$

By measuring the external force S when the lever arms s and m as well as the angle β are known, one can calculate the muscle force vector M as follows:

$$M = sSm^{-1} (\cos \beta)^{-1}$$

In reality, the conditions are even more complex; for example, several muscles are usually involved in torque development around the elbow.

The principles of vector algebra can be applied to a chain of body segments; for example, pushing or lifting with the hands generates torques (forces) from the wrist to the elbow joint, to the shoulder, down the spinal column, and transmitted across knee joints and ankles to the ground where the forces are generated that counteract the hand forces. Such kinematic chain models have been used to assess the weak links of the human body: these are often found in the lower part of the spinal column where back injuries are so frequent.

Assessment of muscle strength

Voluntary muscle strength is of much practical interest since it moves the segments of the body and generates energy that is exerted on outside objects when one performs work. There are over 200 skeletal muscles in the body. Many actually consist of bundles of muscles, each of which is wrapped—as is the total muscle—in connective tissue in which nerves and blood vessels are embedded. The sheaths of this connective tissue influence the mechanical properties of the muscle; at the end of the muscle, the tissues combine to form tendons that connect the ends of the muscle to bones.

The only active action a muscle can take is to contract. Elongation is brought about by external forces that lengthen the muscle. Contraction is actually brought about by fine structures of the muscle, called filaments, that slide along each other. The nervous control for muscular contraction is provided by the neuromuscular system, which carries signals from the brain forward to the muscle and also provides feedback. The efferent signals that arrive at the motor units of the muscle are observable in an electromyogram (EMG).

Figure 13-7 shows a simple model of the generation of muscular strength (Kroemer, Kroemer, and Kroemer-Elbert, 1986). Signals E travel from the central nervous system (CNS) along the efferent pathways to the muscle. Here, they generate contraction, which, modified by the existing mechanical conditions, generates the strength that is applied to a measuring instrument (or a hand tool). Three feedback loops are shown. The first one indicating reflexes; the second, transmitting sensations of touch, pressure, and other kinesthetic signals; the third reporting sound and vision events related to the strength exerted.

Considering the feed-forward section of the model, it becomes apparent that, with current technology, no suitable means exist to measure the executive program or the subroutines in the CNS, or the effects of will or motivation on the signals generated. The efferent excitation impulses (E) that travel along the motor nerves to the muscles are observable in an EMG but are difficult to interpret. The resulting contraction activities at the muscles can be qualitatively observed but not quantitatively measured in humans. No instruments are available at this time that can directly measure the tensions within the muscle in situ. It is difficult and often practically impossible to record and control the mechanical conditions within the body; e.g., the lever arms of the tendon attachments or the pull angles of muscles with respect to the bones to which they are attached. However, the mechanical conditions outside the body can be observed and controlled; e.g., the kind of and location of the coupling between the body and the measuring device; the direction of exerted force or torque; the time history of this exertion; the

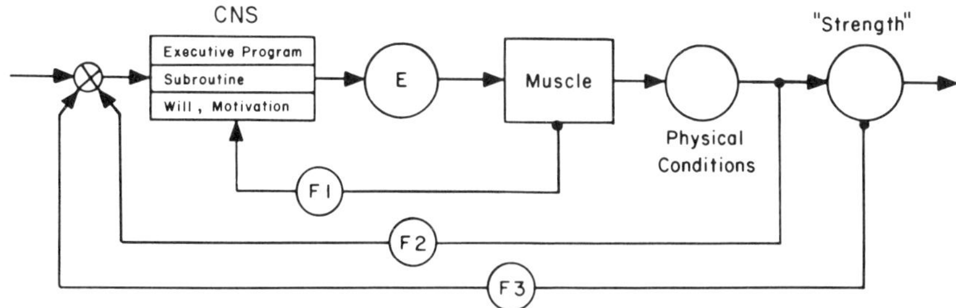

Figure 13-7. Model of generation and control of muscle strength exertion. (Reprinted with permission from Kroemer, Kroemer, and Kroemer-Elbert, 1986)

positioning and support of the body; and environmental conditions, such as temperature and humidity.

Hence, only the output of this complex system, called muscle strength, can be defined and measured. Since strength has vector qualities, it must be recorded not only in magnitude, but also in direction, point of application, and time history.

Measuring techniques. According to Newton's Second Law (force equals mass times acceleration, or torque equals moment of inertia times rotational acceleration), the measurer first has to decide whether or not acceleration shall be present.

If there is no acceleration, there will be no change in speed. If speed is set to zero, then adjacent body segments will not move with respect to each other. Hence, the length of the muscle(s) spanning the joint remains constant. Physically, this means that the measurement of muscle strength will be performed in the static condition. Biologically, this condition is called "isometric," meaning constant muscle length. Thus, in this case, the terms static and isometric are factually synonymous. Measurement of static strength is straightforward and involves only simple instrumentation. It is for this reason that almost all information on muscle strength currently available reflects isometric exertions.

If velocity is not zero but is constant, the condition is called "isokinetic," meaning constant motion. Measurement devices that provide a constant angular velocity around a given body articulation are commercially available. During their constant-speed phase (but neither at the beginning nor at the end of the motion), these provide a defined condition for which the exerted strength can be recorded. Only the angular velocity is controlled by the experimenter, while the amount of strength actually exerted at any moment remains under control of the subject.

If there is acceleration present, meaning that the velocity is not constant but variable, strength is exerted under dynamic conditions that need to be defined and controlled depending on the circumstances selected. Such experimental control is likely to be very difficult and often impractical. Typical examples of dynamic conditions are feats of strength at sports events, which can neither be easily measured nor controlled since they are highly specific to the situation and the person.

One technique used to control dynamic conditions is to let the subject exercise isoinertially, that is, using constant masses (weights). To assess strength, particularly lifting capability, weights are usually increased from test to test until one can determine the largest mass that the subject can move or lift (see the section on manual material handling in this chapter).

The strength test protocol. After selecting the type of strength test to be administered and the measurement techniques and measurement devices to be used, an experimental protocol must be devised. This includes selecting subjects, eliciting data, and guaranteeing the subjects' protection; controlling the experimental conditions; using, calibrating, and maintaining the measurement devices; and (usually) avoiding negative effects of training and fatigue. Regarding the selection of subjects, care must be taken that the subjects participating in the tests are in fact a representative sample of the population for which data are to be gathered. Regarding the management of the experimental conditions, the control over motivational aspects is particularly difficult. It is widely accepted (outside sports) that the experimenter should not exhort or encourage the subject to perform (Caldwell et al, 1974). The so-called Caldwell Regimen reads (in excerpts and revised to reflect newer developments and findings) as follows:

DEFINITION: Static strength is the capacity to produce torque or force by a maximal voluntary isometric muscular exertion. Strength has vector qualities and therefore should be described by magnitude and direction.

1. Static strength is measured according to the following conditions.
   (a) Static strength is assessed during a steady exertion sustained for 4 seconds.
   (b) The transient periods of about 1 second each, before and after the steady exertion, are disregarded.
   (c) The strength datum is the mean score recorded during the first 3 seconds of the steady exertion.
2. The subject should be made aware of the following.
   (a) The subject should be informed about the test purpose and procedures.
   (b) Instructions to the subject should be kept factual and not include emotional appeals.
   (c) The subject should be instructed to increase to maximal exertion (without jerking) in about 1 second and to maintain this effort during a 4-second count.
   (d) Inform the subject during the test session about his or her general performance in qualitative, noncomparative, positive terms. Do not give instantaneous feedback during the exertion.
   (e) Rewards, goal setting, competition, spectators, fear, noise, and so on can affect the subject's motivation and performance and, therefore, should be avoided.
3. The minimal rest period between related efforts should be 2 minutes—more, if symptoms of fatigue are apparent.
4. Describe the conditions existing during strength testing:
   (a) body parts and muscles chiefly used
   (b) body position
   (c) body support/reaction force available
   (d) coupling of the subject to the measuring device (to describe location of the strength vector)
   (e) strength measuring and recording device

5. Subject description
   (a) population and sample selection
   (b) current health status: a medical examination/questionnaire is recommended
   (c) sex
   (d) age
   (e) anthropometry (at least height and weight)
   (f) training related to the strength testing
6. Data reporting
   (a) mean (median, mode)
   (b) standard deviation
   (c) skewness
   (d) minimum and maximum values
   (e) sample size

While this regimen applies to static (isometric) strength testing, a similar procedure should be applied for isokinetic measurement. The time limitation should be deleted, but the strength exerted during the constant motion phase must be subjectively maximal throughout that whole phase of the procedure. The intention is to avoid peaks of force and to attempt a continuous exertion of maximal strength throughout. It is therefore proposed that no more than a ± 10 percent deviation from the average strength exerted during the constant motion phase be allowed during isokinetic testing as is practiced in isometric strength assessments.

MANUAL MATERIAL HANDLING: LIFTING, LOWERING, PUSHING, PULLING, CARRYING

We all handle material daily. We lift, hold, carry, push, pull, and lower while moving, packing, and storing objects. The objects may be soft or solid, bulky or small, smooth or with corners and edges; the objects may be bags, boxes, or containers that come with or without handles. We may handle material occasionally or repeatedly. We may handle material during leisure activities as part of the paid work. Manual handling involves lifting light or heavy objects. Heavy loads pose additional strain on the body owing to their weight or bulk or lack of handles. But even lightweight and small objects can strain us because we have to stretch, move, bend, or straighten out body parts, using fingers, arms, trunk, and legs.

Manual material handling is one of the most frequent causes of injury—and the most severe—in today's U.S. industry. The direct and indirect costs are enormous, and the human suffering associated with, for example, low back injuries, immeasurable.

The "four keys" of manual material handling ergonomics have been described by Kroemer (1984) as follows:

Key #1: Initial layout or improvement of *facilities* contributes essentially towards safe and efficient material transfer. What process is selected and, accordingly, how the flow of material is organized and designed in detail, determines the involvement of people and how they need to handle material.

Key #2: *Job design* determines the stress imposed on the worker by the work. Initially, the engineer must decide whether to assign certain tasks to a person or a machine. Furthermore, the layout of the task, the kind of material handling motions to be performed, the organization of work and rest periods, and many other engineering and managerial techniques determine whether a job is well-designed or not; whether it is safe, efficient, and agreeable for the operator.

Key #3: Selection, use, and improvement of *equipment,* machines, and tools, strongly affect material handling requirements. Human engineering princples, e.g., concerning operator space requirements, control design, visibility, and color and sign coding, must be considered.

Key #4: This concerns *people* as material handlers, particularly with regard to body size, strength, and energy capabilities. People are the king pins of manual material handling; they are critical because they supervise, control, operate, drive, and actually handle material. If people are not needed in the system, then it should be automated. If they are needed, the system must be designed for them.

These four keys allow a systematic analysis of the material handling problems. In fact, optimizing each of these keys (with the aid of seminars or workshops) can lead to dramatic improvements. Many of the strains and risks caused by manual materials handling at the workplace (in industry while manufacturing, assembling, or warehousing; at the grocery wholesaler or retail supermarket; or in hospitals) can be reduced or avoided by proper design of the facilities, appropriate selection and use of equipment, and intelligent job design.

A systematic way to evaluate material handling is as follows:

- Describe the whole material movement process from receiving to distribution.
- Break down the whole process into its separate functions.
- Within these functions, chart and tabulate the activities to determine manual materials handling details.
- Allocate tasks between humans and machines and determine job requirements if people must handle the material.

In this procedure, several principles of material movement are important. The "unit size" principle is of particular interest. This principle states that one can either increase the quantity (size, weight) of the unit load so that equipment use becomes feasible and appropriate for the movement of material—this is called the "big unit" principle. Or one may reduce the size and weight of the load so that one operator can safely handle the material—this is called the "small unit" principle.

If all opportunities to automate or mechanize the movement of material have been exhausted, some material handling may have to be assigned to people. In this case, job requirements must be established that will not overload the person or pose possible hazards. One must organize the task, establish job procedures, and determine details to enable the operator to perform the work safely and efficiently. Here are some guidelines:

- If people must move material, make sure the movement is predominantly carried out in the horizontal plane. Have people push and pull, rather than lift or lower, and avoid severe bending of the body.
- If material is delivered to the workplace, make sure it is placed at knuckle (hip) height.
- If people must lift or lower material, have them do so between knuckle height and shoulder height. Lifting and lowering below knuckle height or above shoulder height are most likely to result in overexertion injuries.
- If lifting and lowering must be done by people, make sure these activities occur close to and in front of the body. If

the worker must bend forward or, worse, twist the body sideways, overexertion injuries are most likely.

- If people must move material, make sure the material is light, compact, and safe to grasp. A light object will strain the spinal column and muscles less than heavy objects. Compact material can be held more closely to the body than a bulky object. A solid object with good handles is more safely held and more easily moved than pliable material.
- If people must handle material, make sure it does not have sharp edges, corners, or pinch points.
- If material is delivered in bins or containers, make sure it can be easily removed and that the operator does not have to dive into the container to reach the material.

The body as energy source

The human body must maintain an energy balance between the external demands produced by work and the work environment and the capacity of the internal body functions to produce energy. The body is an energy factory that converts chemical energy derived from food into externally useful energy. The final stages of this process take place at skeletal muscles. These use oxygen transported from the lungs by the blood. Also, the blood removes by-products generated in the energy conversion, such as carbon dioxide, water and heat. Cabon dioxide is dissipated in the lungs where, at the same time, oxygen is absorbed into the blood. Although heat and water are dissipated to a small extent through the lungs, dissipation mainly occurs through the skin (sweat). The blood circulation system is powered by the heart.

Thus, the pulmonary system (lungs), the circulatory system (heart and blood vessels), and the metabolic system (energy conversion) establish central limitations of a person's ability to perform strenuous work. However, in today's workplaces, work demands do not usually tax one's central capabilities.

A person's capability for labor is limited also by muscular strength, by the ability for movement in body joints (e.g., in the knees), or by the spinal column (e.g., by a weak back). These are local limitations of the force or work that a person can exert. Such local limitations often establish the upper limits for performance capability. For example, one may simply lack the strength to lift an object because the hands are too far extended in front of the body. The mechanical advantages at which muscles must work often determine one's ability to perform a given lifting job.

While handling material, the force exerted with the hands must be transmitted through the whole body, that is across wrists, elbows, shoulders, trunk, hips, knees, ankles, and feet to the floor. In this chain of force factors, the weakest link determines the capability of the whole body to do the job. A particularly weak link in this chain is the spinal column, particularly at the low back. Muscular strains, painful displacements in the facet joints, or deformations of the intervertebral disks often limit a person's ability for handling material.

Matching people with their tasks

Three major approaches can ensure that physical work, particularly manual material handling, is performed safely and efficiently.

1. Design tasks and equipment to improve the ease and efficiency of manual material handling.
2. This reduces the requirements for training of persons in safe and efficient procedures.
3. Select persons who are capable of performing the labor.

Obviously, these approaches are closely related. Only persons who possess the basic physical capabilities to do the labor can be trained for the job. Both selection and training can be carried out only if the specific job demands are known. Hence, knowledge of job requirements is basic to all three approaches. On the other hand, job design can be executed properly only if human capabilities and limitations are known. Information on body dimensions (Table 13-C) and on work capabilities (see the sections on physiology and biomechanics) is basic for ergonomic design.

Training for safe lifting practices

Numerous attempts have been made to train material handlers to do their work, particularly lifting, in a safe manner. Unfortunately, hopes for significant and lasting reductions of overexertion injuries through the use of training have been generally disappointing. There are several reasons for the disappointing results:

- If the job requirements are stressful, "doctoring the system" through behavioral modification will not eliminate the inherent risk. Designing a safe job is basically better than training people to behave safely in an unsafe job.
- People tend to revert to previous habits and customs if practices to replace previous ones are not reinforced and refreshed periodically.
- Emergency situations, the unusual case, the sudden quick movement, increased body weight, or impaired physical wellbeing may overly strain the body, since training does not include these conditions.

Thus, unfortunately, training for safe material handling (which is not limited to lifting) should not be expected to really solve the problem. On the other hand, if properly applied and periodically reinforced, training should help to alleviate some aspects of the basic problem.

The idea of training workers in safe and proper manual materials handling techniques has been propagated for many years. Originally it was advocated that one lift with a straight back and to unbend knees while lifting. However, the frequency and intensity of back injuries was not reduced during the last 40 years while this lifting method has been taught. Biomechanical and physiological research has shown that leg muscles used in this lifting technique do not always have the needed strength. Also, awkward and stressful postures may be assumed if this technique is applied when the object is bulky, for example. Hence, the straight back/bent knees action evolved into the "kinetic" lift, in which the back is kept mostly straight and the knees are bent, but the positions of the feet, chin, arms, hands, and torso are prescribed. Another variant is the "free style" lift, which, however, may be better for male (but not female) workers than the straight back/bent knee technique (Garg and Saxena, 1985). It appears that no single lifting method is best for all situations (Andersson and Chaffin, 1986).

Training of proper lifting techniques is an unsettled issue. It is unclear what exactly should be taught, who should be taught, how and how often a technique should be taught. This uncertainty concerns both the objectives and methods, as well as the expected results. Claims about the effectiveness of one technique or another are frequent but are usually unsupported by convincing evidence.

A thorough review of the existing literature indicates that the issue of training for the prevention of back injuries in manual materials handling is confused at best. In fact, training may not be effective in injury prevention, or its effect may be so uncertain and inconsistent that money and effort paid for training programs might be better spent on research and implementation of techniques for worker selection and ergonomic job design. Nevertheless, according to the National Institute for Occupational Safety and Health (NIOSH), "The importance of training in manual materials handling in reducing hazard is generally accepted. The lacking ingredient is largely a definition of what the training should be and how this early experience can be given to a new worker without harm. The value of any training program is open to question as there appear to have been no controlled studies showing a consequent drop in the manual material handling (MMH) accident rate or the back injury rate. Yet so long as it is a legal duty for employers to provide such training or for as long as the employer is liable to a claim of negligence for failing to train workers in safe methods of MMH, the practice is likely to continue despite the lack of evidence to support it. Meanwhile, it may be worth considering what improvements can be made to existing training techniques." (NIOSH, 1981, p. 99.)

Currently it appears that two major training approaches are most likely to be successful. One involves training in awareness and attitude through information on the physics involved in manual materials handling and on the related biomechanical and physiological events going on in one's body. The other approach is the improvement of individual physical fitness through exercise and warm-ups (which of course also influences awareness and attitude, though indirectly).

Rules for lifting. There are no comprehensive and sure-fire rules for "safe" lifting. Manual material handling is a very complex combination of moving body segments, changing joint angles, tightening muscles, and loading the spinal column. The following DOs and DO NOTs apply, however:

- DO design manual lifting and lowering out of the task and workplace. If it nevertheless must be done by a worker, perform it between knuckle and shoulder height.
- DO be in good physical shape. If you are not used to lifting and vigorous exercise, do not attempt to do difficult lifting or lowering tasks.
- DO think before acting. Place material conveniently within reach. Have handling aids available. Make sure sufficient space is cleared.
- DO get a good grip on the load. Test the weight before trying to move it. If it is too bulky or heavy, get a mechanical lifting aid or somebody else to help, or both.
- DO get the load close to the body. Place the feet close to the load. Stand in a stable position with the feet pointing in the direction of movement. Lift mostly by straightening the legs.
- DO NOT twist the back or bend sideways.
- DO NOT lift or lower awkwardly.
- DO NOT hesitate to get mechanical help or help from another person.
- DO NOT lift or lower with the arms extended.
- DO NOT continue heaving when the load is too heavy.

Personnel selection for material handling

Selecting persons who are unlikely to suffer an overexertion injury is one of the three methods to reduce the risk of musculoskeletal disorders in manual materials handling. The purpose of screening is to place only those individuals on strenuous jobs who can do them safely. Such screening may be done before employment (although this may have legal ramifications pertaining to federal equal opportunity regulations), before placement on a new job, or on occasion of routine examinations during employment. The basic premise is that the risk of overexertion injury for manual material handling decreases as the handler's capability to perform such activity increases. This means that the test should be designed so that it allows the test administrator to match a person's capabilities for manual materials handling to the actual demands of the job. This matching process requires that the test administrator know quantitatively both the job requirements and the related capabilities to be tested.

Scientists usually rely on the development and use of *models.* A model is an abstract (mathematical-physical) system that obeys specific rules and conditions and whose behavior is used to understand the real system (in this case, the worker-task) to which it is analogous in certain respects (e.g., in physiological, biomechanical, psychophysical, or other traits). A model usually represents a theory. Without proper models, reliable and suitable methods cannot be developed. A *method* is a systematic, orderly way of arranging thoughts and executing actions. A *technique* is the specific, practical manner in which actions are done; it implements the methods that are derived from models. Test techniques involve specific procedures and instruments used to obtain measurements with respect to the subject's capability to perform manual material handling activities.

Many models have been developed to describe central and local limitations just discussed. In the following section, these models are simplified and categorized by major disciplines for convenience.

Anatomical/anthropometric models. Since Alfonso Giovanni Borelli's "De Motu Animalium" (about 1680), many similar attempts have been made to simplify the human body into links (representing the long bones and vertebrae), connected at articulations (joints), and powered by muscle engines that span one or two articulations and move the links. Most engineering models follow this basic stick-man system.

Physiological models. Physiological models primarily provide information on oxygen consumpton (in L/min or L/kg/min) as a measure of energy conversion and expenditure (in kcal/min or kJ/min), and on the loading of the circulatory system (in heartbeats/min). These primarily reflect central functions.

Orthopedic models. Orthopedic models concentrate on musculoskeletal functions, which are often related to deformities, diseases, and loss of limbs. Of particular interest are the body joints, including the spinal column, often in connection with surgical or rehabilitory treatments or in connection with protheses. Thus, these models commonly address local functions.

Biomechanical models. During the last 100 years, many models have been developed (usually following the Borelli concept) in an effort to explain the behavior of the body in mechanical terms. Hence, these models rely primarily on anatomical, anthropometric, and orthopedic paradigms and inputs, with much use of computer technology in recent years. Such models usually reflect local functions.

Statistical models. In contrast to the more normative models just discussed, many statistical models have been developed on the basis of empirical data. These models try to describe the combined effects of parametric inputs into a model that is essentially defined by these parameters. Typical examples are multiple-regression models, which predict lift capacity from observed anthropometric, biomechanical, and muscle-strength parameters. The predictive power of these models is usually expressed by the square of the (combined) correlation coefficient, R^2, between the involved variables.

Psychophysical models. The psychophysical concept suggests that human capabilities are synergistically determined by bodily and mental functions and the person's ability to rate the perceived strain and to control one's actions accordingly. The psychophysical approach to the understanding of central and local functions has been successfully employed in the past to describe the effects of sound, climate, illumination, vibration, and physical exertion. (It is of considerable interest to note that the assessment of maximal voluntary muscle strength is implicit to nearly all models and is in fact essential for anatomical, orthopedic, and biomechanical models. However, its measurement depends on the voluntary cooperation [motivation] of the subject. Since there exists at present no feasible technique to assess true [structural] muscle strength, all models rely [by concept or input] in this respect on the psychophysical approach.)

Assessment methods

Based on models such as those just described, various methods have been developed to assess an individual's capabilities for performing specified handling tasks. Such assessments rest on a foundation of general epidemiological information or on more specific etiological information.

The *medical examination* primarily screens out the medically unfit on the basis of their physiological and orthopedic traits. Unless specific job requirements are known to the physician, the physician must test the person's capabilities against generic job demands. The *physiological examination* is often combined with medical testing. It identifies individual limitations in central capabilities; e.g., of pulmonary, circulatory, or metabolic functions. The *biomechanical examination* addresses mechanical functions of the body, primarily of the musculoskeletal type; e.g., load-bearing capacities of the spine or muscle strength exertable in certain postures or motions. *The psychophysical examination* addresses all (local or central) functions strained in the test. Thus, this examination may include all or many of the systems checked via medical, physiological, or biomechanical methods. It filters the strain experienced through the sensation of the subject. In tests of maximal voluntary exertions, the subject decides how much strain is acceptable under the given conditions; e.g., how large an exertion will in fact be performed voluntarily.

Screening techniques

Several screening techniques exist that (should) serve to select persons who are able to perform defined materials handling activities with no or acceptably small risks of overexertion injuries; i.e., persons whose capabilities match the job demands with a safety margin. Primarily, these tests differ in the techniques used to generate the external stresses that strain the (local and central) function capabilities to be measured.

Static techniques. Static techniques require that the subject exert isometric (or constant-length) muscle strength against an external measuring instrument. Since muscle length does not change, there is no displacement of body segments involved; hence, no time derivatives of displacement exist. This establishes a mechanically and physiologically simple case that allows the straightforward measurement of isometric muscle-strength capability screening techniques (Ayoub, Selan, and Jiang, 1984; Chaffin, 1981; NIOSH, 1981).

Dynamic techniques. Dynamic techniques appear more relevant to actual material handling activities, but they are also more complex because an infinite number of possible displacements and time derivatives (velocity, acceleration, jerking) exists. Hence, current dynamic testing techniques employ either one of two ways to generate dynamic tests stresses:

In the *isokinetic* (or constant-motion) technique, the subject moves the limb (or trunk) at constant angular velocity about a specific joint (usually the knee, hip, shoulder, or elbow) while exerting maximal voluntary torque. The test equipment used is designed so that the torque actually exerted can be monitored continuously throughout the angular displacement (Marras, King, and Joynt, 1984; Kamon, Kiser, and Pytel, 1982).

The *isoinertial* (or constant-mass) technique requires that the subject move a constant mass (weight) between two defined points. The maximal load that can actually be lifted (or lowered, carried, or held) by the subject is the measurement of that person's capability, while the forces or torques actually exerted depend (according to Newton's Second Law) on mass and acceleration. This rather realistic technique has been employed in many tests (Ayoub et al, 1980; Ayoub, Selan, and Jiang, 1984; Kroemer, 1983; McDaniel, Skandis, and Madole, 1983; Snook, 1978).

The main advantage of the static (isometric) technique is its conceptual simplicity: putting a person in a few frozen positions and measuring the force which the person can generate in one given direction over a period of just a few seconds is indeed easily understood. This allows the use of rather simple instruments and permits relatively easy control of the experimental condition. Unfortunately, static strength testing has been shown to have rather low predictive power for dynamic tasks (R^2 below 0.5).

The advantage of suitable dynamic measuring is the similarity to actual material handling, which is usually performed in motion, not motionless. The difficulty of dynamic measuring lies in the fact that the movement of body parts can generate a variety of dynamic conditions. Simplifying this by reducing velocity, the first derivative of displacement, to a constant is characteristic of isokinetic measurement techniques. After measuring isokinetic strengths around all body joints, one can synthesize the total capability of the body. Unfortunately, the pleasure reaped by the analytical scientist is the accountant's nightmare (when the price tag for the needed equipment is received). If less sophisticated equipment is used, on the other hand, the analytical scientist is deprived of much significant information.

The oldest and most practical lifting test is to actually lift a load. To have somebody simply "try it" is certainly a realistic test of lifting capability, but such a test does not provide analytical information about the contributions of individual muscles around body joints. It is, however, easily understood, easily executed, and easily controlled. It is safe within limits, reliable, and valid and has been used to test millions of military recruits (McDaniel, Skandis, and Madole, 1983). The lift test equipment consists essentially of two upright channels that guide a carriage to which weights are attached at the rear. The subject attempts to lift the carriage by its handles, not knowing what weight is attached. The "LIFTEST" regimen (Kroemer, 1983, 1985) determines the weight that a subject can lift (up to 45.5 kg) within 10 minutes. The equipment itself is inexpensive and sturdy.

As in all areas of developing scientific and applied knowledge, the state of the art changes, often rapidly. With respect to testing for lift strength, static testing was current just a few years ago but is already becoming outdated. The new isokinetic or isoinertial testing techniques are the most advanced approaches now available. Hence, while one should be cautious about applying either of these methods exclusively in the preplacement examination, other physiological, biomechanical, psychological, and medical assessments should be used as deemed appropriate.

A major problem in all current techniques of lift-strength testing is that of validity—the predictive power of the tests for "real life" (true working) conditions. Note that the test techniques discussed here assess only lifting ability, not the ability to lower, push, pull, hold, or carry, which were listed in the NIOSH lifting guide (1981) as components of manual materials handling. Static testing techniques need to be thoroughly updated or replaced to assess the dynamic components of manual work and its demands on the dynamic capabilities and limitations of the body. Figure 13-8 schematically shows engineering interventions applied to solve the problems associated with manual material handling. The main intent is to eliminate, or at least to reduce, the overexertion injury risks to material handlers.

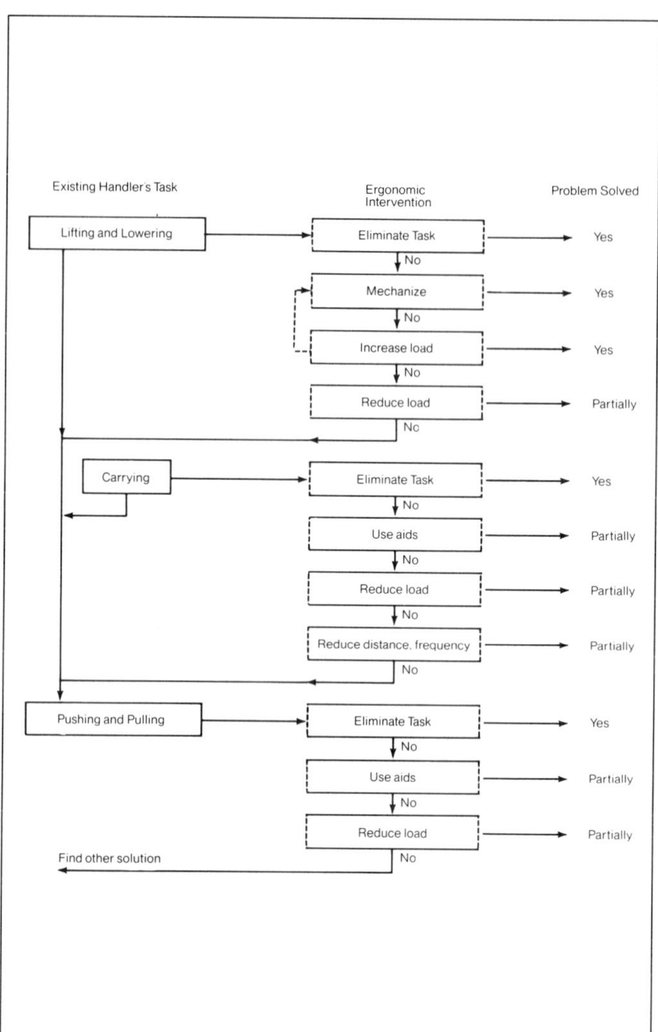

Figure 13-8. Ergonomic interventions in manual material handling.

Permissible loads (weights) for manual material handling

The tables of weights used only a few years ago that supposedly indicated the loads that could be lifted by men, women, and children are no longer legitimate or suitable. Using epidemiological, medical, physiological, biomechanical, and other approaches discussed earlier, new knowledge with regard to human capabilities in the area of manual materials handling has been gained and incorporated in various guidelines. However, although these are much more suitable than the old weight tables, the new guidelines are still based on assumptions and approaches that need refinement and further evaluation.

Limits for lifting. The NIOSH's 1981 *Work Practices Guide for Manual Lifting* describes specific weights that may be lifted. The NIOSH recommendations apply only to:

1. smooth lifting (slow and without sudden movements—NOT to lowering, pushing, pulling, holding, carrying, and so on);
2. two-handed symmetrical lifting in the sagittal plane (directly in front of the body; no twisting during the lift);
3. a load of moderate width (76 cm [30 in.] or less);
4. unrestricted lifting posture;
5. good couplings (hands with handles, shoes with floor surfaces);
6. favorable physical environment;
7. a minimal amount of physical work (such as holding, carrying, pushing, pulling, lowering) that is performed in addition to lifting; in fact so minimal that the individual may be assumed at rest when not lifting; and
8. material handlers who are physically fit and accustomed to labor.

Finally, the recommended limits do not include safety factors commonly used by engineers to ensure that unexpected conditions are accommodated.

With these preconditions, the NIOSH guide provides a formula to calculate an action limit (AL) for a load that "over 75 percent of women and over 99 percent of men could lift." Obviously, more persons could perform lifting activities below the AL, while above the AL, the frequency and severity of injuries would increase. The AL is calculated using the following formula:

$$\text{AL (kg)} = 40(15/H)(1-.004|V-75|)(.7+7.5/D)(1-F/F_{max})$$
(in metric units)

or

$$\text{AL (lb)} = 90(6/H)(1-.01|V-30|)(.7+3/D)(1-F/F_{max})$$
(in U.S. units)

In the formula, H equals a horizontal hand location (expressed in centimeters or inches) in front of the midpoint between the ankles at the origin of lift; V is the vertical location (expressed in centimeters or inches) at the origin of lift; D is the vertical travel distance (expressed in centimeters or inches) between the origin and the destination of lift; F is the average frequency of lift (in lifts/min); and F_{max} is the maximum frequency that can be sustained.

For purposes of the NIOSH guide, these variables are assumed to have the following limits.

- H is between 15 cm (6 in.) and 80 cm (32 in.). Objects cannot, in general, be closer than 15 cm without interfering with the body. Objects positioned more than 80 cm away cannot be reached by many people.
- V is between 0 cm and 175 cm (70 in.). This represents the range of vertical reach for most people.
- D is between 25 cm (10 in.) and (220-V) cm [(80-V) in.]. When the travel distance is less than 25 cm, let D equal 25.
- F is between .2 (1 lift every 5 min) and F_{max}. When lifting less than once per 5 min, let F equal 0. For other F_{max} values, see the following.

	Average Vertical Location in cm (in.)	
Period	*V>75 (30) Standing*	*V<75 (30) Stooped*
1 hour	18 lifts/min	15 lifts/min
8 hours	15 lifts/min	12 lifts/min

Note that when $F=F_{max}$, AL becomes zero, a quirk of the equation.

Discussion and examples

The equations for the AL represent a multiplicative factor weighting for each task variable.

The H factor, for horizontal location, represents the distance at which the load handles are positioned in front of the body. If H=15 cm, the factor (15/H) is 1, and no adjustment for horizontal location is necessary. If H=75 cm, the factor 15/75 is .20. Thus, the AL is reduced from 40 to 40(.20), i.e., 8 kg.

The V factor, for vertical location, involves the absolute deviation of the height of the load handles from 75 cm (the approximate knuckle height). Thus, at floor height, V=75:

$$\text{V factor} = (1-.004|V-75|) = 1-.004(0) = 1$$

If V = 15: $\text{V factor} = (1-.004|15-75|) = 1-.004(60) = .76$

Likewise, if V = 135: $\text{V factor} = (1-.004|135-75|) = 1-.004(60) = .76$

The D factor, for the vertical travel distance, ranges between 1 and .74 as D varies between 0 and 200 cm.

$$\text{D factor} = (.7 + 7.5/D)$$

If D=0, set D=25 (the minimum allowed value). Then:

$$\text{D factor} = (.7 + 7.5/25) = 1.0$$

If D=200 (the maximal height considered in the NIOSH guide):

$$\text{D factor} = (.7 + 7.5/200) = .74$$

The F factor, for frequency of lift, is more complicated. If the lifting originates below 75 cm and is performed continuously throughout the day, $F_{max}=12$. If the observed frequency is 6 lifts/min (F=6), then:

$$\text{F factor} = (1-F/F_{max}) = 1-6/12 = .5$$

The effective weight that can be lifted is thus cut in half due to the frequency with which lifting is required.

The NIOSH guide also provides a maximum permissible limit (MPL), which is set at three times the AL. At the MPL, only about 25 percent of men and less than 1 percent of women workers can lift the load. As loads become larger, even fewer persons can perform the work.

The AL and MPL establish three regions:

- Region below the AL—in this region, nominal risks are present to most industrial workers.
- Region between the AL and the MPL—these conditions are unacceptable without administrative and/or engineering controls. Such controls include:
  1. selection of persons capable of performing the job;
  2. training of persons for safe job performance;
  3. reduction of job requirements in terms of amount of load, location of load, frequency of lifting, and so on.
- Region above the MPL—these conditions are unacceptable and require engineering controls. Such controls include:
  1. automation of the job;
  2. mechanization of the job;
  3. reduction of physical requirements, i.e., reduction in the amount of the load, in the frequency that lifting is to be performed, and so on.

EXAMPLE #1. If continuous lifting occurs below knuckle height, and H=20 cm, V=40 cm, D=100 cm, and loads are lifted at a rate of 6/min, then:

$$AL = 40(15/20)(1-.004|40-75|)(.7+7.5/100)(1-6/12) = 40(.75)(.86)(.78)(.50) = 10 \text{ kg}$$

EXAMPLE #2. Suppose the load actually weighed 35 kg. In this case, one engineering control might be to reduce the frequency of lifting from 6 to 1/min. This would increase the frequency factor from .50 to .92 and consequently the AL to 18.4 kg and the MPL to 55.2 kg.

The relative weights of each factor, in this case .75, .86, .78, and .50, indicate which factors should be changed for highest effectiveness. In this case, frequency of lifting is the biggest discounted factor (.50) and should receive first consideration.

Reducing the frequency to 1 lift/min would lower the stressfulness of the job to within the level of acceptable administrative controls. In this case, 35 kg is between the AL (18.4) and the MPL (55.2). However, this should not preclude further engineering controls. It is important to realize that the job still cannot be safely performed by most women or by the majority of men. Further reductions in the frequency (F factor=.92) would be ineffective, since this factor can only be increased to 1.0.

Table 13-E. Maximal Acceptable Lift Force (kg_{force})

	(a) Width	(b) Distance	(c) Percentage	*Floor Level to Knuckle Height* — *One Lift Every*							
				5 sec	9 sec	14 sec	1 min	2 min	5 min	30 min	8 hr
Men	18	51	90	10	14	15	18	20	25	27	30
			75	14	18	20	24	26	32	36	38
			50	17	23	26	30	32	40	44	48
Women	18	51	90	9	11	13	14	15	20	22	23
			75	10	13	15	17	18	23	25	27
			50	12	16	17	19	21	26	29	31

	(a) Width	(b) Distance	(c) Percentage	*Knuckle Height to Shoulder Height* — *One Lift Every*							
				5 sec	9 sec	14 sec	1 min	2 min	5 min	30 min	8 hr
Men	18	51	90	11	14	16	16	17	20	21	22
			75	13	18	20	21	22	26	27	29
			50	16	21	24	26	27	32	34	36
Women	18	51	90	8	11	11	11	12	14	15	16
			75	9	12	12	13	14	17	18	19
			50	10	14	14	15	16	19	20	21

	(a) Width	(b) Distance	(c) Percentage	*Shoulder Height to Arm Reach* — *One Lift Every*							
				5 sec	9 sec	14 sec	1 min	2 min	5 min	30 min	8 hr
Men	18	51	90	8	11	13	15	16	18	20	21
			75	10	15	16	19	20	24	25	27
			50	13	18	20	24	25	30	32	34
Women	18	51	90	6	9	10	11	11	13	14	15
			75	6	10	11	12	12	15	16	17
			50	7	11	13	13	14	16	17	18

(a) Handles in front of the operator (cm); (b) Vertical distance of lifting (cm); (c) Acceptable to 50, 75, or 90 percent of industrial workers.

Table 13-F. Maximal Acceptable Lowering Forces (kg_{force})

	(a) Width	(b) Distance	(c) Percentage	*Knuckle Height to Floor Level* — *One Lowering Every*							
				5 sec	9 sec	14 sec	1 min	2 min	5 min	30 min	8 hr
Men	18	51	90	11	14	16	21	21	29	32	34
			75	14	19	22	27	29	37	41	44
			50	18	24	27	34	37	47	51	55
Women	18	51	90	8	13	13	16	17	21	23	25
			75	9	14	16	18	19	24	26	28
			50	11	16	18	20	21	27	30	32

	(a) Width	(b) Distance	(c) Percentage	*Shoulder Height to Knuckle Height* — *One Lowering Every*							
				5 sec	9 sec	14 sec	1 min	2 min	5 min	30 min	8 hr
Men	18	51	90	12	13	14	16	17	20	22	23
			75	15	16	18	21	22	26	28	30
			50	19	21	23	26	28	33	35	37
Women	18	51	90	11	11	11	12	13	15	16	17
			75	13	13	13	14	15	17	18	19
			50	15	15	15	16	17	20	21	22

	(a) Width	(b) Distance	(c) Percentage	*Arm Reach to Shoulder Height* — *One Lowering Every*							
				5 sec	9 sec	14 sec	1 min	2 min	5 min	30 min	8 hr
Men	18	51	90	8	10	11	13	14	16	17	18
			75	11	13	15	17	18	21	23	24
			50	14	16	18	21	22	26	28	30
Women	18	51	90	8	10	10	11	11	13	14	15
			75	10	12	12	12	13	15	16	17
			50	11	13	13	14	14	17	18	19

(a) Handles in front of the operator (cm); (b) Vertical distance of lowering (cm); (c) Acceptable to 50, 75, or 90 percent of industrial workers.

The load probably cannot be brought appreciably closer to the body (H=15 cm v. H=20 cm).

The best engineering solution at this point would be to reduce the load weight. Halving the weight (from 35 kg to 18 kg) would be one solution to bring the task within the capabilities of most people (18 kg is less than the 18.4-kg AL). An equally acceptable solution would be to increase the load drastically, say to 100 kg and provide mechanical lifting aids, thus precluding manual handling and relieving the worker of lifting altogether.

Limits for lifting, lowering, pushing, pulling, and carrying. In 1978, Snook published extensive tables of the loads and forces that male and female workers were able to manage in manual materials handling. A number of prerequisites apply as follows.

- Two-handed symmetrical materials handling is performed in the sagittal plane (directly in front of the body; no twisting during the activity).
- The load is of moderate width (76 cm [30 in.] or less).
- Working postures are unrestricted.
- Good couplings (hands with handles, shoes with floor surfaces) are present.
- Favorable physical environments are provided.
- Only a minimal amount of other physical work activities is required.
- Materials handlers are physically fit and accustomed to labor.

Tables 13-E through 13-I are exerpted from Snook's more detailed lists. (The original tables should be consulted for more information.) The main difference between the presentation of Snook's recommendations and the NIOSH guidelines is that Snook's tables show loads that are acceptable for either female or male materials handlers; specifically, loads acceptable to either 50, 75, or 90 percent of either gender. Furthermore, the tables are subdivided into three different height areas: floor to knuckle heights, between knuckle and shoulder heights, and shoulder to overhead-reach heights.

The data do not indicate individual capacity limits; rather, they represent the opinions of a sample of experienced materials handlers with regard to what loads they would handle

Table 13-G. Maximal Acceptable Push Forces (kg_{force})

One 2.1 m Push Every

	(a) Percentage	(b) Height	6 sec	12 sec	1 min	2 min	5 min	30 min	8 hr
Initial Forces									
Men	64	90	17	21	22	24	29	31	33
		75	23	27	30	31	38	40	43
		50	30	34	37	39	47	51	54
Women	57	90		16	18	18	22	24	26
		75		19	21	22	27	29	31
		50		22	25	26	32	34	37
Sustained Forces									
Men	64	90	8	11	11	12	15	17	18
		75	12	16	18	19	24	27	29
		50	17	22	26	27	34	38	40
Women	57	90		11	11	12	15	16	18
		75		13	15	16	20	22	23
		50		16	19	20	25	27	29

One 30.5 m Push Every

	(a) Percentage	(b) Height	1 min	2 min	5 min	30 min	8 hr
Initial Forces							
Men	64	90	15	17	20	21	23
		75	18	21	24	26	27
		50	21	25	28	31	32
Women	57	90	12	13	15	17	18
		75	14	15	18	19	20
		50	15	18	20	22	23
Sustained Forces							
Men	64	90	9	10	12	13	14
		75	11	12	14	15	17
		50	13	15	17	19	20
Women	57	90	9	9	11	12	13
		75	10	10	13	14	15
		50	11	12	14	16	17

(a) Height of hands above floor (cm); (b) Acceptable to 50, 75, or 90 percent of industrial workers.

Table 13-H. Maximal Acceptable Pull Forces (kg_{force})

One 2.1 m Pull Every

	(a) Percentage	(b) Height	6 sec	12 sec	1 min	2 min	5 min	30 min	8 hr
Initial Forces									
Men	64	90	20	24	28	29	32	33	35
		75	25	29	33	34	38	40	42
		50	31	35	39	41	45	47	49
Women	57	90		19	22	22	25	26	27
		75		24	26	27	30	31	32
		50		29	30	31	35	36	38
Sustained Forces									
Men	64	90	11	13	16	17	18	19	20
		75	14	17	21	22	24	25	26
		50	16	22	27	27	30	31	32
Women	57	90		11	13	13	15	16	16
		75		14	17	17	19	20	21
		50		18	20	20	23	24	25

(a) Height of hands above floor (cm); (b) Acceptable to 50, 75, or 90 percent of industrial workers.

Table 13-I. Maximal Acceptable Carry Forces (kg_{force})

One 2.1 m Carry Every

	(a) Percentage	(b) Height	6 sec	12 sec	1 min	2 min	5 min	30 min	8 hr
Men	79	90	13	17	21	22	26	28	30
		75	18	23	28	30	35	37	40
		50	23	29	36	38	45	48	50
Women	72	90	13	14	15	16	19	20	21
		75	15	16	18	19	22	23	25
		50	17	19	21	22	26	27	29

(a) Height of hands above floor (cm); (b) Acceptable to 50, 75, or 90 percent of industrial workers.

willingly and without overexertion. If the actual loads exceed the table values, administrative or engineering controls should be applied. Snook believes that industrial back injuries could be reduced by about one third if the loads above the values acceptable to 75 percent of the materials handlers could be eliminated.

Table 13-E presents the maximum acceptable weights (in kg) for lifting. Maximum weight depends on several working conditions: the area of lift (height) in front of the body where the lifting action is performed; the number of lifts per time unit; and the percentage of the working population to which this weight is acceptable. The table applies to objects whose handles (and centers of mass) are about 18 cm (7 in.) in front of the body and which are lifted vertically through a distance of 51 cm (about 20 in.). If the object must be lifted farther away from the body or if the lifting height is varied, the weights must be adjusted.

Similarly, Table 13-F presents weights that are acceptable to men and women when lowering objects.

Tables 13-G and 13-H contain data on the maximum acceptable push and pull forces. These depend on the length of push or pull distance and on the number of exertions per time unit. As a function of the differing body dimensions, the handle is somewhat higher for men than for women (64 cm, or 25 in., versus 57 cm, or 22.5 in.). A distinction is made in both these tables between the initial force needed to set the object into motion (the breakaway force) and the sustained force necessary to maintain the movement.

Table 13-I lists the weights that were acceptable for carrying a load over a distance of 2.1 m (6.9 ft) at given frequencies.

The complete tables by Snook should be consulted for conditions different from those listed. Snook's tables are very simple to use: one just selects the proper line and column and finds the appropriate number. No calculations are necessary. A few examples follow.

EXAMPLE #1. Assume a task requires a male operator to lift a box weighing 24 kg (53 lb) from the floor to a table (at knuckle height) once a minute. Table 13-E shows that 75 percent of the male population can do this task.

EXAMPLE #2. If the weight were reduced to 18 kg (40 lb), the table shows that: (1) 90 percent of the male workers would be willing to lift the object once a minute; (2) 75 percent would be willing to lift this mass at 9-second intervals; and (3) 75 percent of female workers would be willing to lift this mass once every 2 minutes.

EXAMPLE #3. What if the mass were 30 kg (66 lb)? We would expect that 50 percent of male workers would lift it once a minute, while female workers would lift it only a few times a shift.

Comparing the NIOSH guide with Snook's recommendations. Of course, it is interesting to compare the recommendations established by the NIOSH with those proposed by Snook. Only those recommendations relative to lifting can be compared, however, since the NIOSH guide does not cover lowering, pushing, pulling, or carrying. In general, the ALs recommended by the NIOSH are quite similar to those recommended by Snook for the 90th percentile of male workers, although in some cases, particularly at extreme working conditions, substantial discrepancies between the two sets of recommendations exist. However, in general, there is reasonable congruence, which is not surprising since Snook's data on lifting were part of the basic information used to develop the NIOSH guidelines. The main value of the recommendations by the NIOSH and Snook lies in the information that they provide regarding what can be presumed to be suitable and acceptable or unacceptable conditions for manual materials handling.

For new systems, the data are initial planning guides for material handling conditions that can either be performed by people or that should be assigned to machines. In following these guides, the designer is able to avoid potentially injurious job requirements and can establish both safe and economical conditions.

For the evaluation of existing material handling systems, actually existing job requirements can be compared with the table data to seek out those task demands that are likely to exceed human capabilities. Then the working conditions should be changed by engineering intervention (i.e., automation, mechanization, or lowering of demands) and/or managerial intervention (i.e., worker selection and training). Clearly, the engineering intervention is preferable because it eliminates the source of the risk.

HAND TOOLS

Many hand tools are really extensions of the hand. Pliers, for example, amplify the hand's strength, protect sensitive tissues, and extend the hand's reach. Other tools are able to do things that the hand cannot—such as soldering—but are yet held in and directed by the hand. The specific capabilities of a hand tool are often counteracted by an inappropriate design that, for instance, forces the wrist to be held bent or generates pressure points between the handle and the hand. Too often design efforts have been concentrated on the working end of the tool rather than on how it interfaces with the hand. Hence, many of our regular hand tools are acceptable only when used occasionally and need to be redesigned if they are used frequently over long periods of time. Figure 13-9 shows a typical example of straight-nose pliers that require an acute bend in the wrist. Putting the bend in the tool, not in the wrist, and fitting the tool to the form and functions of the hand are among the primary ergonomic tool design axioms.

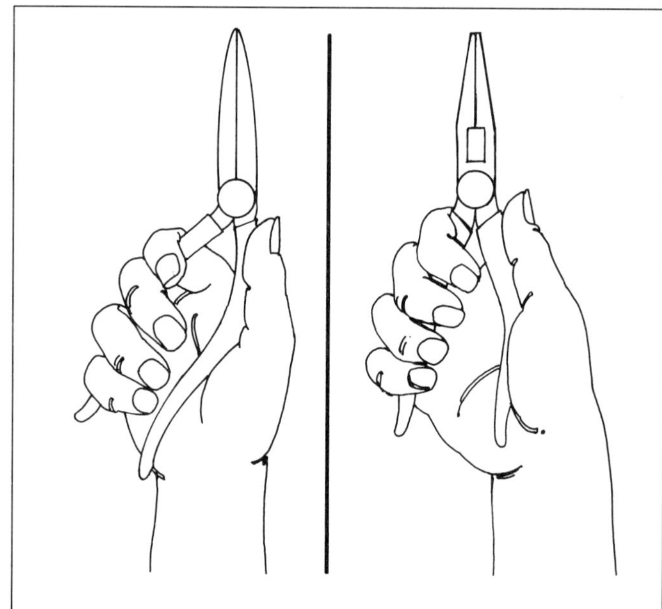

Figure 13-9. Redesigning straight-nose pliers to avoid bending the wrist and to fit the palm and fingers facilitates their industrial use.

There are many interactions between the hand and the tool. Figure 13-10 presents different couplings between hand and handle (Kroemer, 1986). Most hand tools require couplings of the kinds listed in the second half of the figure (6 through 10). Unfortunately, not much reliable information about the forces

1. Finger Touch: One finger touches an object without holding it.

2. Palm Touch: Some part of the inner surface of the hand touches the object without holding it.

3. Finger Palmar Grip (Hook Grip): One finger or several fingers hook(s) onto a ridge, or handle. This type of finger action is used where thumb counterforce is not needed.

4. Thumb-Fingertip Grip (Tip Grip): The thumb tip opposes one fingertip.

5. Thumb-Finger Palmar Grip (Pinch or Plier Grip): Thumb pad opposes the palmar pad of one finger (or the pads of several fingers) near the tips. This grip evolves easily from coupling #4.

6. Thumb-Forefinger Side Grip (Lateral Grip or Side Pinch): Thumb opposes the (radial) side of the forefinger.

7. Thumb-Two-Finger Grip (Writing Grip): Thumb and two fingers (often forefinger and index finger) oppose each other at or near the tips.

8. Thumb-Fingertips Enclosure (Disk Grip): Thumb pad and the pads of three or four fingers oppose each other near the tips (object grasped does not touch the palm). This grip evolves easily from coupling #7.

9. Finger-Palm Enclosure (Collet Enclosure): Most, or all, of the inner surface of the hand is in contact with the object while enclosing it.

10. Power Grasp: The total inner hand surface is grasping the (often cylindrical) handle which runs parallel to the knuckles and generally protrudes on one or both sides from the hand.

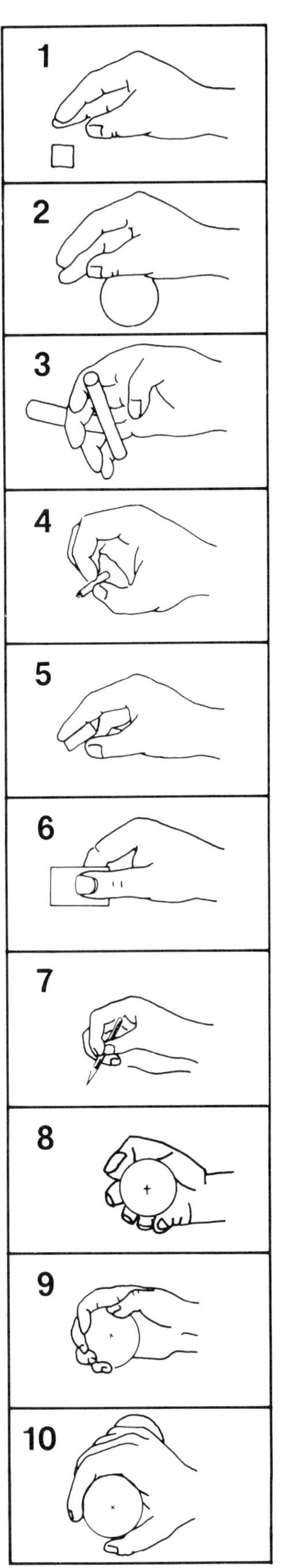

Figure 13-10. Couplings between hand and handle. (Adapted from Kroemer, 1986)

and torques exertable in these configurations is available; more related strength data need to be provided through research. Some estimated hand-strength data are given in Table 13-J.

Table 13-J. Estimated Hand-Force Capabilities for Men

Coupling #	Force (kg) 5th Percentile	50th Percentile	95th Percentile
1			
2			
3			
4	5.9	9.5	13.2
5	5.9	10.0	14.1
6	6.8	10.5	14.1
7			
8			
9			
10	34.0	45.5	56.8

(Adapted with permission from Van Cott and Kinkade, 1972)
Note: To estimate hand-force capabilities for women, multiply these values by two-thirds. More strength data are currently (1987) compiled by the author.

The shape of the handle size, both in its surface contour and its cross section, can have strong effects on a person's ability to manipulate the tool. Often the main force is generated perpendicularly to the handle surface (such as in closing pliers), but in addition, a force perpendicular to that direction must be exerted to perform the job (such as in pulling with the pliers). Hence, both the cross-sectional and longitudinal shape of the tool must be considered. Furthermore, the presence of grease or dirt between the hand and handle or the wearing of gloves can have profound effects on the coupling.

Several recent publications address the problem of handle design: Cochran and Riley (1986); Drury (1980); Greenberg and Chaffin (1979); Konz (1986); and Mital and Sanghavi (1986). Given the various tasks and uses of hand tools, only a few general guidelines can be provided; details have to be decided according to the given conditions:

- In manipulating the hand tool, the wrist should stay at or near a straight position with respect to the forearm; that is, be neither rotated nor bent. This often requires that the working side of the hand tool be at an oblique angle with the handle. (However, this may make the tool useful only for certain tasks.)
- The handle should be of such cross-sectional size (the diameter) that the hand nearly circles the handle, with not more than about 1.3 cm (.5 in.) between the fingertips and the thumb side. This means that the largest distance between two opposing sides of the handle (the diameter, if circular) should be between 2.5 and 6.4 cm (1.0–2.5 in.).
- The shape of the handle, in cross section, depends on the task to be performed; that is, on the motions involved in opening and closing the handle and on the magnitude of force or torque (moment) to be developed. In many cases, elliptical shapes or rectangular ones with well-rounded edges are advantageous if twisting (torquing or turning such as with screwdrivers) must be performed. However, more circular cross sections are preferred if the tool must be grasped in many different manners.
- The handle should easily accommodate the length of the hand in contact with it; for example, a knife handle should be at least as long as the hand enclosing it. The contour of the handle can follow the contour of the inside of the hand enfolding the handle. However, strongly formfitting the handle shape might prevent people with different hand sizes or people who grasp the handle in an unusual way from handling the tool comfortably.
- Pressure points should be avoided. These are often present if the form of the handle has many shape components, such as pronounced indentations for the fingers or sharp edges or contours.
- Rough surfaces of the handle might be uncomfortable for sensitive hands but can counteract the effects of grease that make the handle slippery.
- Flanges at the end of the handle can guide the hand to the correct position and prevent the hand from slipping off the handle. (For more information see particularly Woodson, 1981.)

Improperly designed hand tools, particularly when combined with ill-conceived workplaces and job procedures, can cause acute injuries to the musculoskeletal system of the hand-arm complex. Often, the repetition of biomechanical insults, each in itself insignificant, can lead to cumulative trauma disorders. Repetitive trauma to the upper extremity is a major cause of lost work in many hand-intensive industries. Most of these can be averted easily by paying attention to the following recommendations.

- Avoid repetitive or sustained exertions, particularly if they are accompanied by deviations from a straight wrist and/or by forceful exertions.
- Keep the elbow at the side of the body, the forearm semi-pronated, and the wrist straight.
- Use tools with handles of appropriate size and shape, as already discussed.
- Round off all edges and sharp corners on the hand tool or at the workstation with which the worker might come in contact.
- Avoid cooling the hand either through subjection to a cold environment or by strong air movement (particularly important with air-powered tools) or by contact with energy-conducting handles (e.g., metal).
- Ensure that gloves worn actually help the activity but do not hinder motion or enforce awkward wrist positions.
- Minimize vibrations of hand tools.

WORKSTATION DESIGN

The goal in designing a workstation is to promote ease and efficiency (the goals of ergonomics) for the person working. Productivity will suffer if the operator is uncomfortable or if the layout of the workstation or the job procedures are awkward. Productivity relative to both quality and quantity will be enhanced if the operator is comfortable physiologically and psychologically and if the layout of the workstation is conducive to performing the task well. Keeping this in mind, it is advisable to try to establish an ideal workstation and work environment first and to make concessions to practical limitations only if absolutely necessary.

General principles

Six general rules govern the design for a specific workplace:

1. Plan the ideal, then the practical.
2. Plan the whole, then the detail.

3. Plan the work process and equipment around the system requirements.
4. Plan the workplace layout around the process and equipment.
5. Plan the final enclosure around the workplace layout.
6. Use mockups to evaluate alternative solutions and to check the final design.

Obviously, a number of design aspects must primarily be considered:

- clearance for the operator's body for entrance and egress (including emergency exit); suitable body posture at work (including space for changing positions); operation of controls and equipment (without bumping elbows, knees, or head); and the avoidance of excessive forces or inadvertent operation of controls;
- operation of tools and controls by hand or foot (handles, switches, and so on); seat adjustment; and emergency items (stop button, flashlight, survival equipment);
- visual field and information both inside the workstation (displays and control settings) and outside (road, machine being controlled, and so on); and visual contact with co-workers;
- auditory information, such as oral communication with other workers, signals (including warning signals), and sounds from equipment (engine under load, cutting tool, and so on).

More detailed design guidelines depend on the special workstation, on the specific work task, and on the environment. Such guidelines can be found, for example, in books by Boff, Kaufman, and Thomas (1986); the Eastman Kodak Company (1983, 1986); Konz (1983); Sanders and McCormick (1987); Salvendy (1987); Van Cott and Kinkade (1972); and Woodson (1981). *Military Standards 759* and *1472* also provide a wealth of ergonomic human factors engineering information, as do, of course, other standards and regulations published by the American National Standard Institute (ANSI).

Standing v. sitting

Whether the operator should stand or sit at the workstation depends on several factors: the mobility required, the forces needed, the size of the work piece, and the required precision. The advantages of standing over sitting include more mobility, more arm strength available, less front-to-rear room required, no seat required, and greater latitude in workstation design. The advantages of sitting are that pedals can be operated with the foot more effectively (more strongly, and with more precision), it is less fatiguing to maintain the sitting posture (if a good seat is available), and manipulation and vision may be more precise. Unless the specific work task or the environment or conditions strictly demand either sitting or standing, provisions should be made to allow the operator to sit or stand at will. Figure 13-11 shows "stand-seats," which can offer a useful compromise. (More information on seat design is contained in the section in this chapter on computer workstations.)

WORKPLACE DESIGN

The most basic requirement for a workplace is that it must accommodate the person working in it. Specifically, this means that the work space for the hands should be between hip and chest height in front of the body—lower locations are preferred for heavy manual work, and higher locations are preferred for tasks that require close visual observation. Contours of reach envelopes indicate the maximum distances at which objectives can be manipulated or placed. Figure 13-12 shows an example of such reach capabilities. (See NASA-Webb 1978 for more data.)

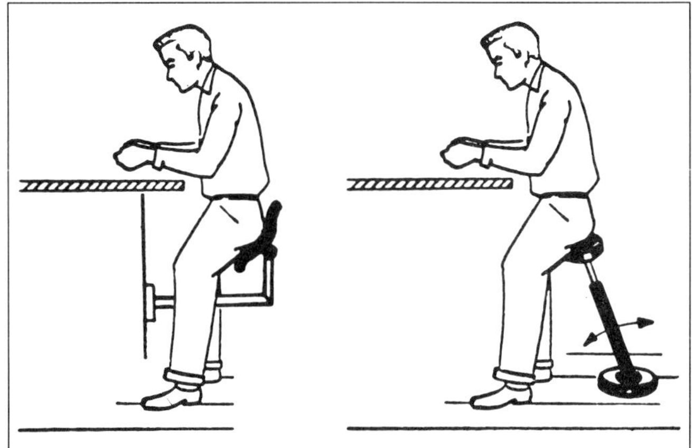

Figure 13-11. "Stand-seats." (Courtesy *American Industrial Hygiene Association Journal*)

Work objects should be located close to the front edge of the work surface so that the worker does not have to bend over and lean across the surface to grasp items. To allow the person to be close to the front edge of the work surface, sufficient room must be provided so that thighs, knees, and toes can be placed somewhat under the work surface if the work is performed while standing. For sitting operators, much more leg room must be provided under the bench, table, or desk. If foot controls are used, additional room for foot and leg motions may be needed. Pedals that must be operated continuously or frequently normally require a seated operator because if a person operated them while standing, the body weight would have to be supported on one foot.

Visual displays, such as computer screens, counters, or dials and signal lights on instruments are preferably placed in front of the body and below eye level so that the line of sight (which aligns the eyes with the visual target) is declined 10° to 40° below the horizontal level. Table 13-K lists general principles for workstation design.

Work space dimensions

Work space dimensions can be grouped in three basic categories: minimal, maximal, and adjustable dimensions. *Minimal* work space dimensions provide clearance for the worker in the working space; e.g., open leg room or clearance for ingress and egress in walkways and doors. While many dimensions, such as the open leg space under a desk, are usually determined using 95th percentiles values from anthropometric tables (such as those provided in Table 13-C), larger values need to be considered for other, more critical clearances. For example: if the height of the door frame were at the 95th percentile value, at least 5 percent of all users would bump their heads. *Maximal* work space dimensions permit smaller workers to use the equipment. This is ensured by selecting work space dimensions over which a small person can reach or by establishing control forces that are small enough so that even a weak person can operate the equipment. Usually, the 5th percentile value of the relevant body attribute is used for this design.

Adjustable dimensions permit the operator to modify the work environment and equipment so that it conforms to that

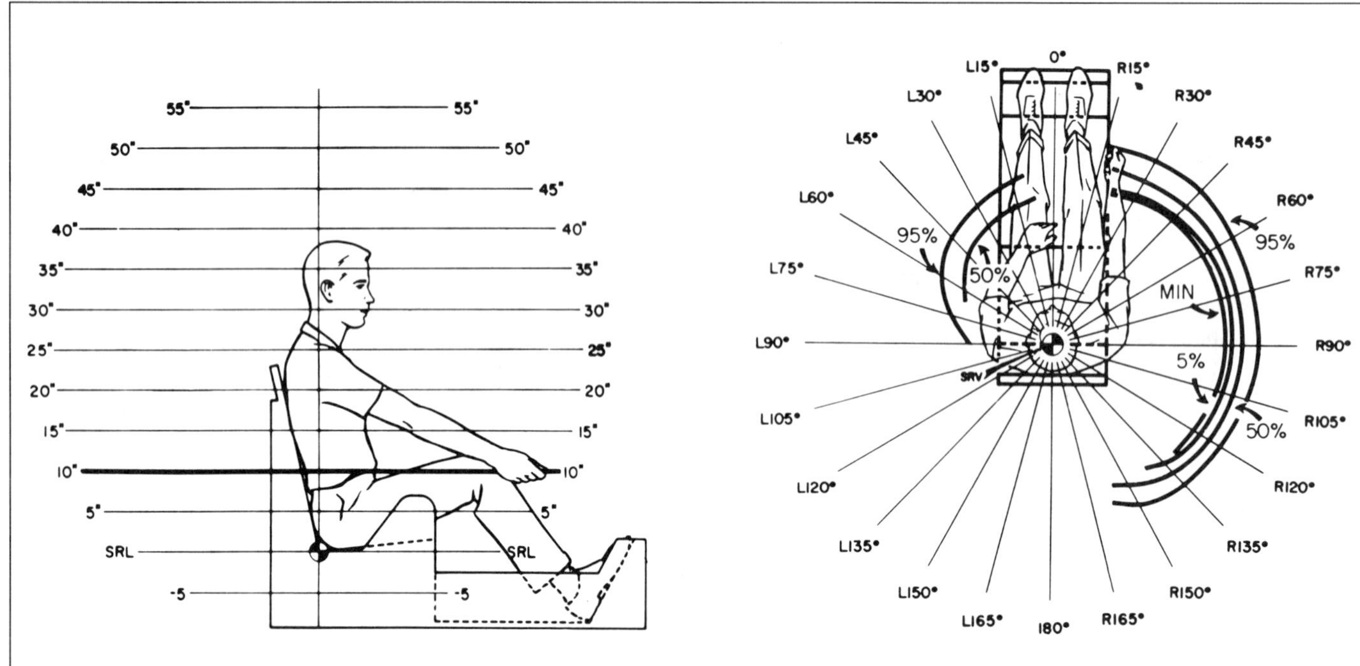

Figure 13-12. Grasping reach, in inches, to a horizontal plane 25 in. above the seat reference point. Right arm. (1 in. = 2.54 cm) (Reprinted with permission from Damon, Stoudt, and McFarland, *Human Body in Equipment Design)*

individual's particular set of anthropometric characteristics as well as to subjective preferences. A six-way adjustable seat in a truck or airplane is an example of the proper adjustment capabilities available to individual operators. Adjustable dimensions are particularly important when optimal performance with minimum effort is necessary to accomplish the work task.

Approximate dimensions for industrial workplaces are presented in Table 13-L and in Figures 13-13 and 13-14. Of course, it is most desirable to allow the worker as much free space as desired to position the body and the workpiece; however, space limitations or structural requirements might limit the available space. Still, severely limiting the leg room is not suitable since it forces the operator to assume a fixed posture.

The requirements for an industrial work seat are essentially the same as those for an office chair (see the section in this chapter on computer workplaces). However, the industrial work seat is probably somewhat more rugged—including soil-resistant upholstery. As shown in Figure 13-15, the industrial work seat must be adjustable in seat pan height between 38–51 cm (15–20 in.); the front edge should be well rounded to avoid pressure to the underside of the thighs. A backrest should be provided, if the work activities allow it. The backrest should be adjustable

Table 13-K. Ergonomic Guidelines for Workplace Design

1. In the design of the facility, assure a proper match between the facility and the operator to avoid static efforts, such as holding a work piece or hand tool. Static (isometric) muscle tension is inefficient and leads to rapid fatigue.
2. The design of the task and the design of the workplace are interrelated. The work system should be designed to prevent overloading the muscular system. Forces necessary for dynamic activities should be kept to less than 30% of the maximal forces the muscles are capable of generating. Occasionally, forces of up to 50% are acceptable when maintained for only short durations (approximately 5 minutes or less). If static effort is unavoidable, the muscular load should be kept quite low—less than 15% of the maximal muscle force.
3. Aim for the best mechanical advantage in the design of the task. Use postures for the limbs and body that provide the best lever arms for the muscles used. This avoids muscle overload.
4. Foot controls can be used by the seated operator. They are not recommended for continuous use by a standing operator because of the imbalanced posture imposed on the operator. If a pedal must be used by the standing operator, it should be operable with either foot. Avoid hard floors for the standing operator; a soft floor mat is recommended, if feasible.
5. Maintain a proper sitting height, which is usually achieved when the thighs are about horizontal, the lower legs vertical, and the feet flat on the floor. Use adjustable chairs and, if needed, footrests. When adjusting the chair, make sure that:
   a. elbows are at proper height in relation to work surface height;
   b. the footrest is adjusted to prevent pressure at undersides of the thighs;
   c. the backrest is large enough to be leaned against, at least for a break; and
   d. special seating devices are used if the task warrants them.
6. Permit change of posture—static posture causes problems in tissue compression, nerve irritation, and circulation. The operator should be able to change his or her posture frequently to avoid fatigue. Ideally, the operator should be able to alternate between sitting and standing; therefore, a workplace that can be used by either a sitting or standing operator is recommended.
7. In designing the facility, accommodate the large operator first and give that operator enough space. Then provide adjustments and support so that the smaller operator fits into the work space. For standing work, the work surface should be designed to accommodate the taller operator; use platforms to elevate shorter operators. (But watch out for stumbles and falls!) For reach, design to accommodate the shorter operator.
8. Instruct and train the operator to use good working postures whether sitting or standing, working with machines and tools, lifting or loading, or pushing or pulling loads.

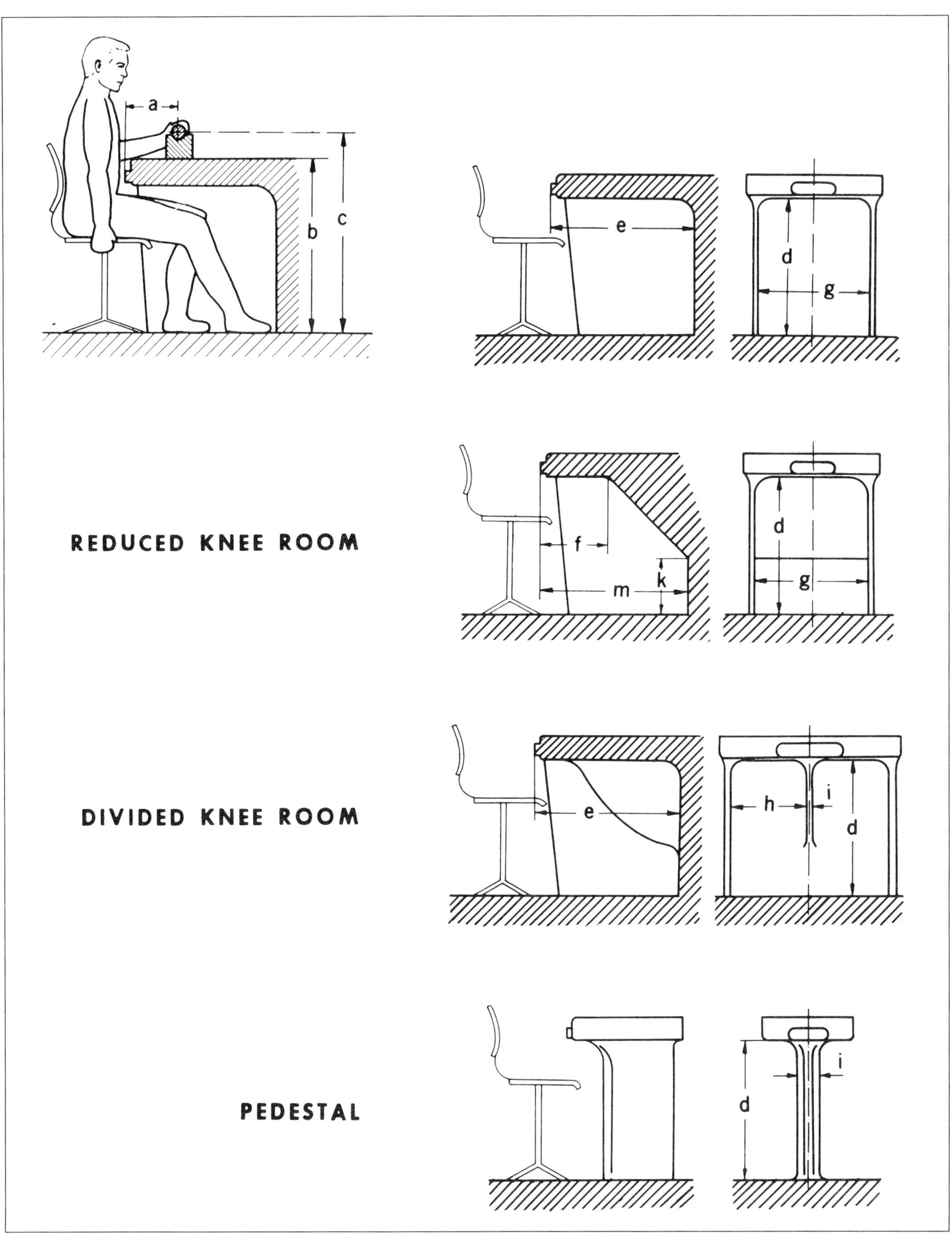

Figure 13-13. Workplace dimensions for sitting operation.

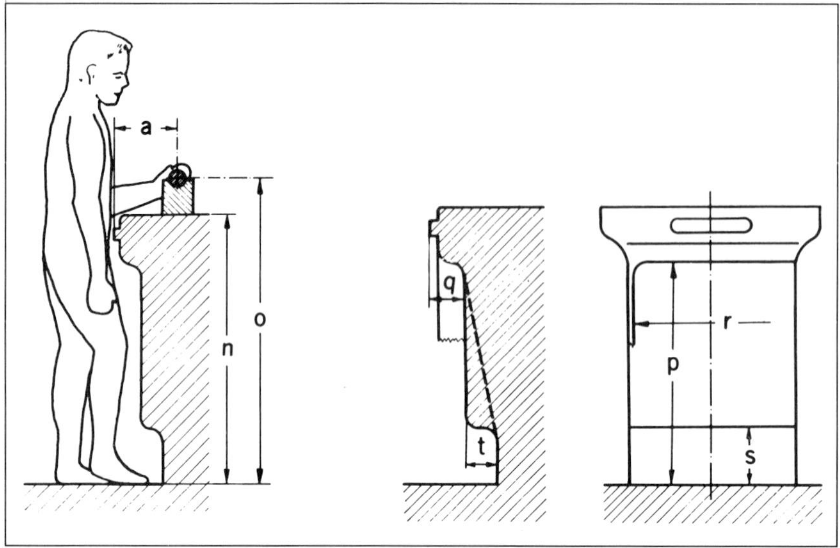

Figure 13-14. Workplace dimensions for standing operation.

Table 13-L. Approximate Dimensions (cm) of Industrial Workplaces for Sitting or Standing Operators

	Sitting		*Standing*
		Outside Dimensions	
a	15.2-25.4	a	15.2-38.1
b	50.8-71.1 if adjustable 71.1 (if fixed)	n	88.9 (approximately)
c	63.5-101.6	0	96.5-121.9
		Leg Room	
d	48.3 minimum	p	81.3 (approximately)
e	63.5 or more	q	10.2 (approximately)
f	30.5 minimum if "e" is impossible	—	—
g	63.5 or more	r	63.5 or more
h*	30.5 (20.3 minimum)	—	—
i*	2.5 at front edge	—	—
	7.6, at 12.7 from front edge	—	—
	15.2, at 25.4 from front edge	—	—
		Foot Room	
k	25.4 or more	s	20.3 (approximately)
m	63.5 or more	t	12.7 (approximately)
g	63.5 or more	r	63.5 or more
h*	30.5 (20.3 minimum)	—	—

Note: Letters correspond to measurements illustrated in Figures 13-13 and 13-14. For alternate sitting and standing, combine all appropriate dimensions.
*Avoid divider rib or pedestal.

in height and in distance from the front edge of the seat pan. To allow free mobility of the arms and shoulder blades, it probably should not extend up to the neck; however, the larger the backrest, the better it can be used for relaxation during a break from the work activities. The backrest should have a protrusion or pad at lumbar height, just like an office chair.

Objects that must be seen and observed (displays, signal lights, controls, dials, keyboards, written documents, and so forth) should be placed within the worker's visual field. The more important ones and those that must be read exactly should be placed within the preferred viewing area. Figure 13-16 and Table 13-M identify these areas.

OFFICE (COMPUTER) WORKSTATIONS

Complaints related to posture and vision are, by far, the most frequent health problems voiced by computer operators (see also, Chapter 5, The Eyes, and Chapter 11, Nonionizing Radiation). Musculoskeletal pain and discomfort, eye strain, and fatigue constitute at least half of the health problems, in some surveys up to 80 percent of all subjective and objective symptoms in North America and Europe. Apparently, some of these complaints are related. Difficulties in viewing (focusing distance, angle of gaze, and so on), together with straining curvatures of the spinal column (particularly in the neck and lumbar region), joined by fatiguing postures of shoulders and arms result

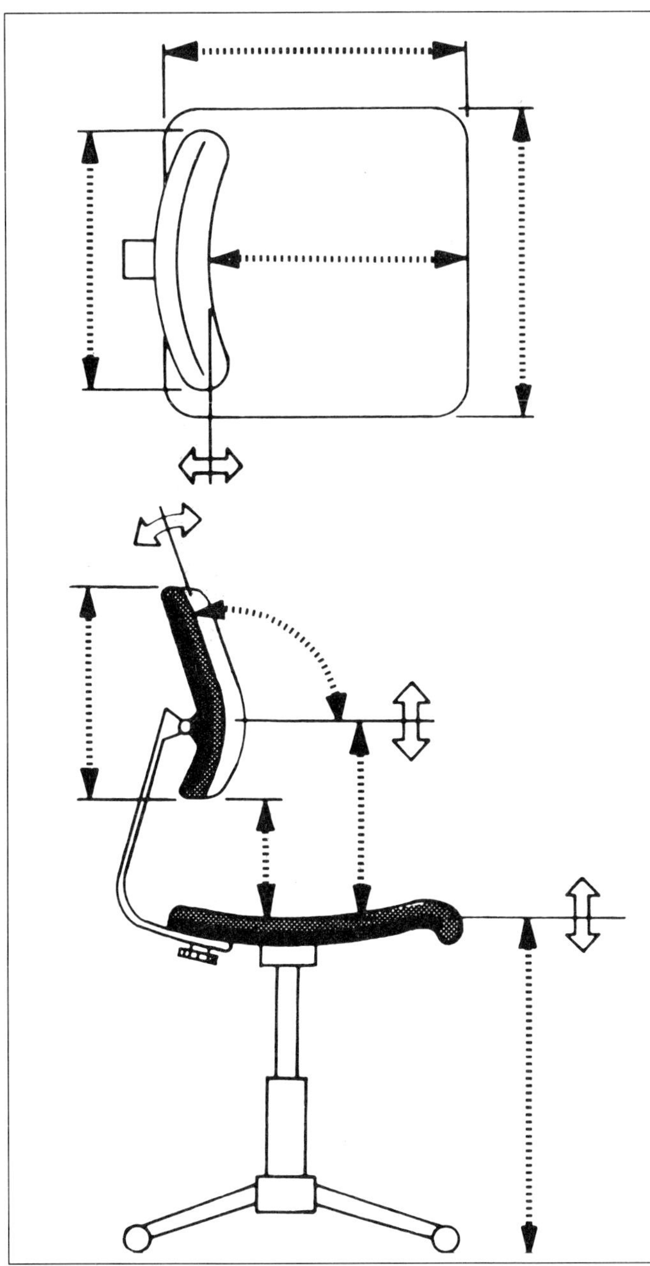

Figure 13-15. Main design features (dimensions and requirements) of a standard work seat. (Courtesy of the *American Industrial Hygiene Association Journal)*

in a stress-strain combination in which causes and effects intermix, alternate, and build on each other.

The postural problems appear to be largely caused by improperly designed and poorly arranged workstation furniture. The questionable convention of sitting straight at the desk has been carried over to the computer workplaces. Even the ANSI publication on visual display terminal workstations (draft of July 1986) advocates an "upright, or near straight" posture for deriving recommended furniture dimensions, although it acknowledges that other working postures exist—that may be more subjectively comfortable, hence individually preferred.

Successful ergonomic design of the office workstation depends on proper consideration of several interrelated aspects, which are sketched in Figure 13-17. A person perfoming a work task will assume related work postures and perform certain work activities. These are influenced by the workstation conditions, including furniture, other equipment, and the environment. All need to fit the person and the task. Also, the existing workstation components strongly affect work postures assumed by the person as well as the actual work activities. The ergonomist, of course, would prefer that desired work postures and appropriate activities determine workstation design.

The actual work posture affects the physical and the psychological well-being, while the work activities determine the output of the person working with the equipment. Feeling well, both physically and about one's performance, affects health, work attitudes, and work output. Of course, these interactive relationships are not static but vary with time.

This very brief discussion of the many and multileveled relationships among many work variables is meant to emphasize the need for carefully designing the workstation—particularly furniture, equipment, and lighting—so that the desired results of well-being and high performance are achieved.

Work task

The now proverbial office revolution has been largely brought about by the use of computers. Hence, the tasks associated with certain office job categories have changed dramatically in some cases. In many offices, "typing" is now performed on a word processor rather than on a traditional typewriter. This new technology allows the user to control the layout of texts and graphics and eliminates the frustrating and time-consuming job of retyping large chunks of material to incorporate relatively minor changes. Hence, different and more complex skills, more control functions, and more responsibilities are required from a person doing text processing than from a traditional typist.

The interactions between the user and the computer establish special requirements on visual and motor capabilities. The eyes scan source documents (for input into the computer) and the display screen on the monitor (either to obtain information or to receive feedback about material being transmitted to the computer system). The fingers input information to the system via keyboard, mouse, joystick and rotary controls, lightpen, touch panels, trackball, and so on. Voice communication with the computer, both as input and output, is also rapidly being developed.

The intensity with which the visual and motor communication links are used identifies task characteristics. Table 13-N lists visual and motor links between the user and the computer. Such listing can be used to identify the priority needs for an ergonomic design of these links.

Positioning the body in relation to the computer

The user interfaces with computers through several sensors (mostly vision, audition, and tactation) and responds to the received stimuli by motor outputs. Within the body, the nervous system transmits signals to the brain, which makes decisions about the input, and initiates and controls output activities.

The majority sensory reception occurs through the eyes, as they fix their sight on the monitor, on the source document, and even on the keys. (Complex computer keyboards require visual search and indentification of keys; "blind" touchtyping is often not feasible.) Of course, handwriting requires visual

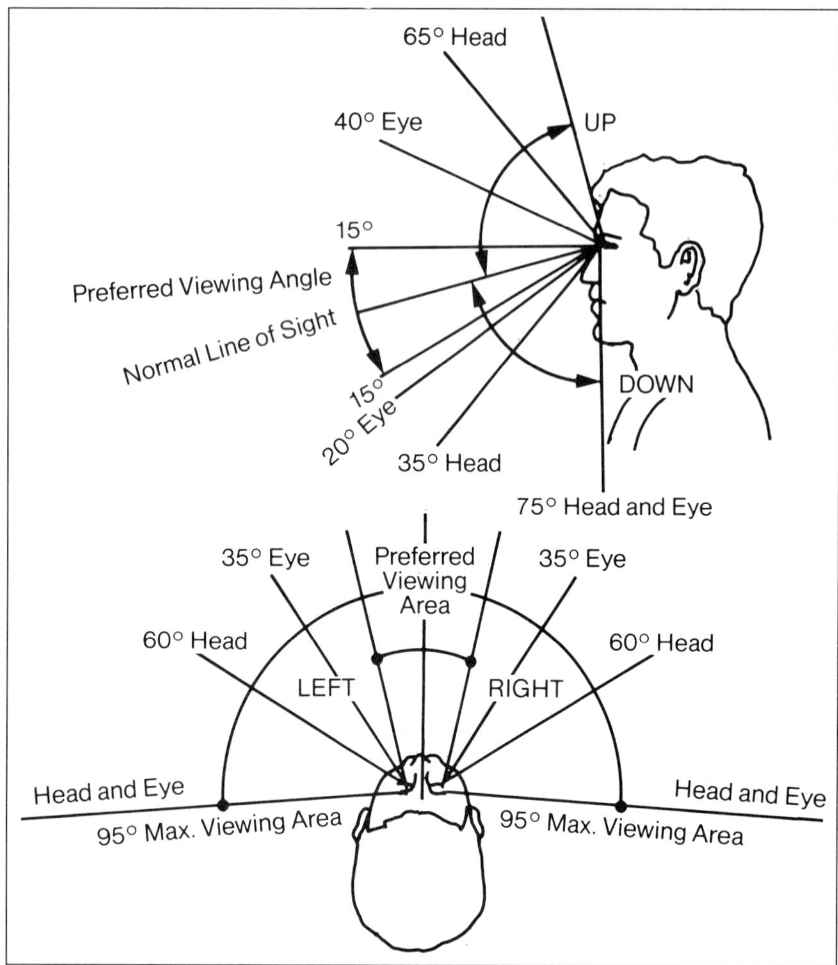

Figure 13-16. Visual field. (Reprinted from *Military Standard 759A,* 1981.)

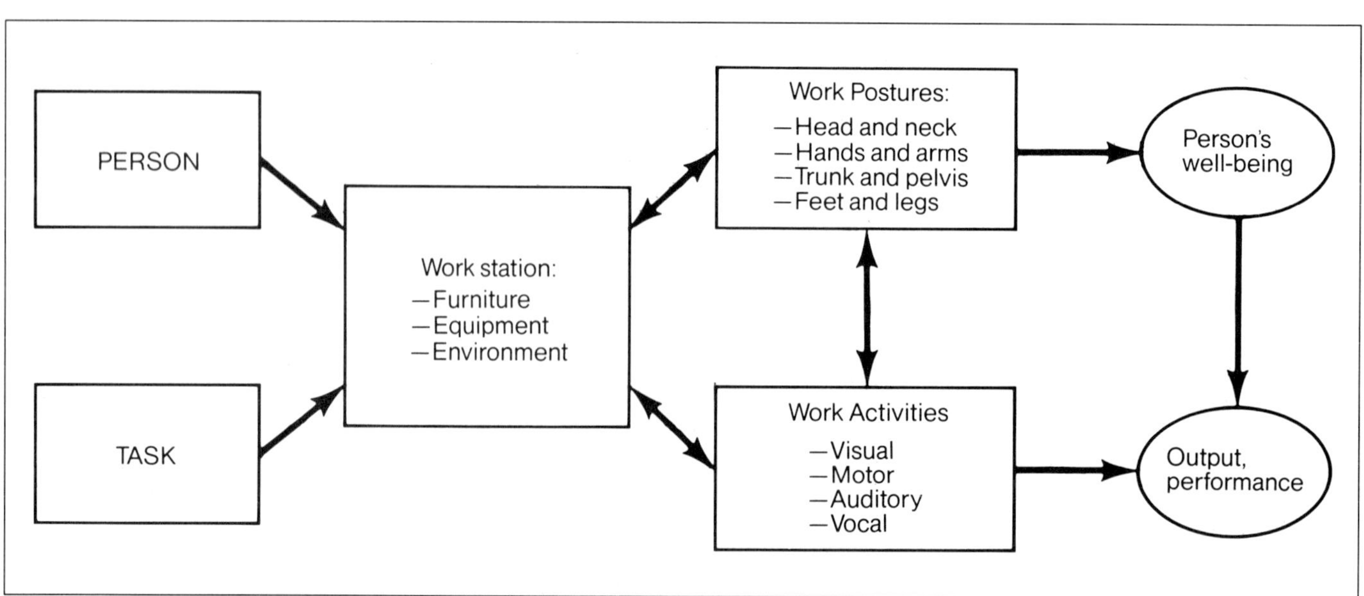

Figure 13-17. Interactions among workstation design, work postures, and work activities and their effect on the computer operator's well-being and performance.

Table 13-M. Preferred Line of Sight and Viewing Area

Direction of the Normal Line of Sight	Preferred	Maximal Deviations from the Preferred Angle: Only Eye Rotation	Only Head Rotation	Combined Eye and Head Rotation
Right or Left, each	15°	35°	60°	95°
Up	15°	40°	65°	90°
Down to distant target	15°	20°	35°	85°
to close target, (e.g., computer monitor)	10–30°	25°	35°	85°

Table 13-N. Human Linking with the Computer in Different Work Tasks

Task	*Motor Input Requirements*	*Visual Requirements*	*Work Interruptions*
Data entry	High (keyboard)	High (source & screen)	Few
Data acquisition	Medium (keyboard)	High (screen)	Varies
Word processing	High (keyboard)	High to medium (source & screen)	Few
Interactive communication	Medium	Medium (screen)	Varies
CAD	Low	High (screen & source)	Frequent

(Adapted from National Research Council, 1983)

control, as does the search for items on the work surface. Though individual preferences for the angle of the line of sight and for focusing distance vary (preference often depends on visual deficiencies and their correction by spectacles or contact lenses), the position of the person's eyes is relatively fixed with respect to the visual target. This eye fixation has rather stringent consequences for neck and trunk postures.

The hands are the major output interface between the user and the computer; the hands operate keys and other devices such as the mouse, trackball, joystick, lightpen, or touch panel. If these controls are fixed within the workstation, the operator has no choice but to keep the hands at this location. (Such controls include a fixed keyboard, a fixed touch panel, or a fixed pad for mouse movement). If such input devices are movable within the workstation area, the location of the hand with respect to the body is also more variable. However, with current technology, controls and hands are essentially in front of the body and between shoulder and lap height. Again, this has consequences for the body posture.

Foot controls are not used frequently with current designs, but occasionally. These controls obviously require that the foot be positioned over them, thereby limiting body posture.

The ear is another input channel for the operator, and the mouth is an output channel. However, sound or speech signals (not yet generally used by current computer systems) usually do not restrict the location of the head to any specified location, since acoustic signals travel through the air or can be transmitted through speakers or through phones attached to the head.

Table 13-O lists primary input and output modalities between the operator and the computer and also indicates how the modalities affect the relative positioning of the body and the computer.

Table 13-O. Links Between Human and Computer Influencing Work Posture

Input to the Operator	*Output from the Operator*	*Requirement of Locating the Operator Relative to the Computer*
Eyes	—	High
—	Hands, feet	Medium to high
Ears	—	Low
—	Mouth	Low

Healthy work postures

A generally accepted tenet has been that an upright trunk and neck is part of a healthy posture. For decades, this idea has been used—usually together with presuming right angles at hips, knees and ankles—to design office chairs and other furniture. However, are there physiological or orthopedic reasons compelling enough to make people sit straight since, if left alone, hardly anyone chooses this posture?

Sitting upright means that, from the lateral view, the spinal column forms an S-curve that shows slight forward bends (lordosis) in the neck and in the low back region, and a slight rearward bulge (kyphosis) in the chest region. While there appears to be no reason to doubt that this is —in the current evolutionary condition of civilized humans—a normal and therefore desirable posture, it needs to be discussed how such a posture should be achieved, supported or even enforced.

Staffel (1884) and seven decades later Schlegel proposed a forward-declining set surface. A seat pan design with a "Schneider Wedge" on its rear edge was popular for about a decade in Europe (for a review see Kroemer, 1971). More recently, Mandal (1982) and Congleton, Ayoub, and Smith (1985) were again promoting seat surfaces that slope downward and forward. The underlying idea is that the desired lumbar lordosis be achieved by opening the hip angle to more than 90 percent and by rotating the pelvis forward. To prevent the buttocks from sliding off the forward-declined seat, the seat surface may be saddle-shaped to fit the human underside (according to Congleton, et al), or the downward-forward thrust can be counteracted by either bearing down on the feet (according to Mandal) or by propping knees or upper shins on special pads (a balans chair). Chaffin and Andersson (1984) called this posture "semi-sitting."

Another way to bring about lordosis of the lumbar region is to push that section of the back and spinal column forward with a specially designed backrest. The "Akerblom pad" of the seat back upholstery or the inflatable lumbar cushion incorporated in the seat back construction of some car and airplane seats are examples of this design feature.

Of course, one can shape the total backrest. Apparently independent of the other, Ridder (1959) in the United States and Grandjean (1963) in Switzerland found rather similar backrest shapes to be acceptable by experimental subjects. In essence, these shapes follow the curvature of the rear side of the human body: at the bottom, concave to accept the buttocks; above the seat, slightly convex to fill in the lumbar lordosis and then nearly straight but declined backwards to support the thoracic area; at the top, again convex to follow the neck lordosis. This shape has been used successfully for seats in automobiles, aircraft, passenger trains, cars, and for easy chairs. In the traditional office, these "first class" chairs were available to managers, while more lowly persons had to make do with simpler designs that ranged down to the miserable small board attached to the so-called secretarial chair. (Extensive bibliographies and reviews of recommendations for seat designs encompassing the last three decades can be found in the publications by Grandjean, 1986; Kroemer, 1987; Lueder, 1983; and Wilson, Corlett; and Manenica, 1986.)

Biomechanical actions on the spine

The positions of the lower spine, of the pelvis, and of the thighs are not independent but, rather, influence each other; the positions of the upper trunk, the head, and arms affect the total spine. The interactions involve both the skeleton and muscles. When a persons stands, the upper surface of the lowest part of the spine, the sacrum, is severely inclined forward, thus providing a downward-slanted support basis (at the L5-S1 interface) for the lumbar section of the spine. This helps to achieve lumbar lordosis. As the pelvis, to which the sacrum is attached, rolls backward (such as when one sits upright on a flat surface), the S1-L5 interface becomes horizontal or may even be slightly slanted backwards. This brings about flattening or even kyphosis of the lumbar spine (Figure 13-18).

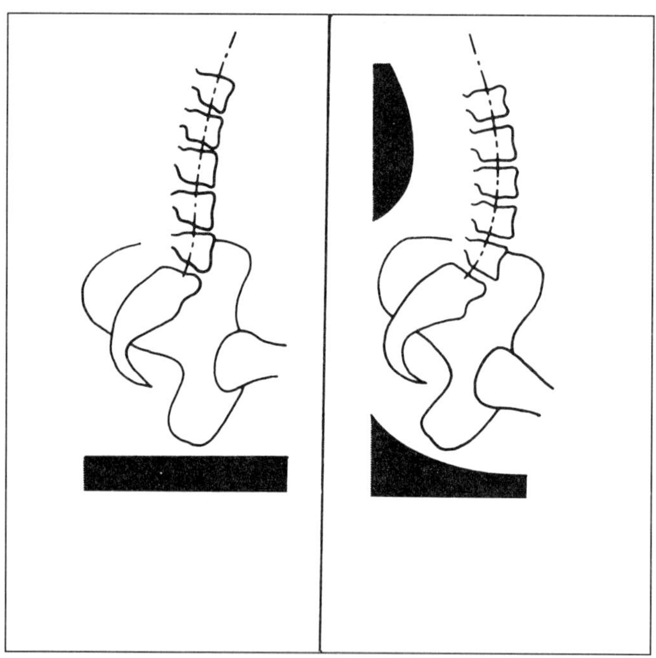

Figure 13-18. Postures of the lumbar spine of the seated operator. Kyphosis (left); lordosis (right). (Courtesy of the *American Industrial Hygiene Association Journal)*

An important muscular connection is established by the hamstring muscles, which run from the back of the calves posteriorly along the thighs and are attached to the pelvis. Thus, they cross both the knee and hip joints and, particularly if the thigh is elevated (small hip angle) and the lower legs are brought forward (large knee angle), enforce a rear rotation of the pelvis that is usually accompanied by flattening or even kyphotic bending of the lumbar spine.

Physiological and mechanical events associated with these postures have been recorded under a variety of conditions and by many researchers. Radiographic studies have established that the rotation of the pelvis and the curvature of the spinal column are closely associated. When a person sits on a flat surface with the hip angle at approximately 90°, the pelvis naturally rotates backward, and the lumbar spine flattens. If desired, such flattening can be avoided by muscle tension, by the design of the backrest, or by providing a seat surface that mechanically tilts the pelvis forward. Lumbar lordosis can also be achieved by opening the hip angle; for example, by thrusting the feet forward and the knees down. This is helped by elevating the seat pan and declining it; that is, making the front section lower than the rear part.

Disk compression and trunk muscle activities are related because the stability of the vertebrae stacked on top on each other is secured by the contraction of muscles that generate essentially vertical forces in the trunk (primarily m. latissimus dorsi, m. erector spinae, m. obliquus internus and externus abdominis, m. rectus abdominis). Directly or indirectly, these muscles pull down on the spine and keep the vertebrae aligned on top of each other. Each vertebra rests upon its lower one cushioned by the spinal disk between their main bodies and supported lateroposteriorly in the two facet joints of the articulation processes. Since the downward pull of the muscles generates disk and facet-joint compression forces (in response to upper body weight and external forces), there must be a relationship between trunk muscle activities and disk pressures. These strains are reduced when the trunk, neck, head, and arm weight are at least partially supported by a suitable backrest (Chaffin and Andersson, 1984; Grandjean, 1986; Kroemer, 1987; and Kroemer, Kroemer, and Kroemer-Elbert, 1986).

Experimental studies

In addition to empirical studies, such as those conducted by Ridder and Grandjean, many analytical experiments have been performed to measure the physiological responses of the human body to certain postures. Lundervold (1951) was among the first to record and interpret electromyographic activities of the upper body muscles associated with defined seated positions. Based on these early studies, many EMG recordings have been obtained (summarized by Chaffin and Andersson [1984], Winkel and Bendix [1986], and by Soderberg et al [1986]). As may be expected, varying activities were found in the muscles that stabilize the body, particularly in the trunk. However, involvement of muscles in the hip and lower trunk regions seems to be rather unimportant for regular seated postures; the observed EMG activities indicate very low demands on muscular capabilities, typically well below one-tenth of the maximal contraction capabilities. (However, muscular strains in the low back area could be important given unusual conditions and postures, such as leaning over the desk at which one is seated to lift an object such as a computer monitor with extended arms.)

Furthermore, the interpretation of these weak EMG signals is controversial, since one cannot necessarily assume that little muscle use (i.e., flat EMG signals) should be preferred over more extensive muscle use. "Dynamic sitting" is desired by some physiologists and biomechanics to obtain suitable muscle tone and training; to obtain orthostatic tolerance and electrolyte and fluid balance (Grieco, 1986; Kilbom, 1986); to benefit intervertebral disk metabolism (Hansson, 1986); and is believed to benefit macro- and microcirculatory aspects (Winkel, 1986), including blood pooling (Thompson, Yates, and Franzen, 1986). In light

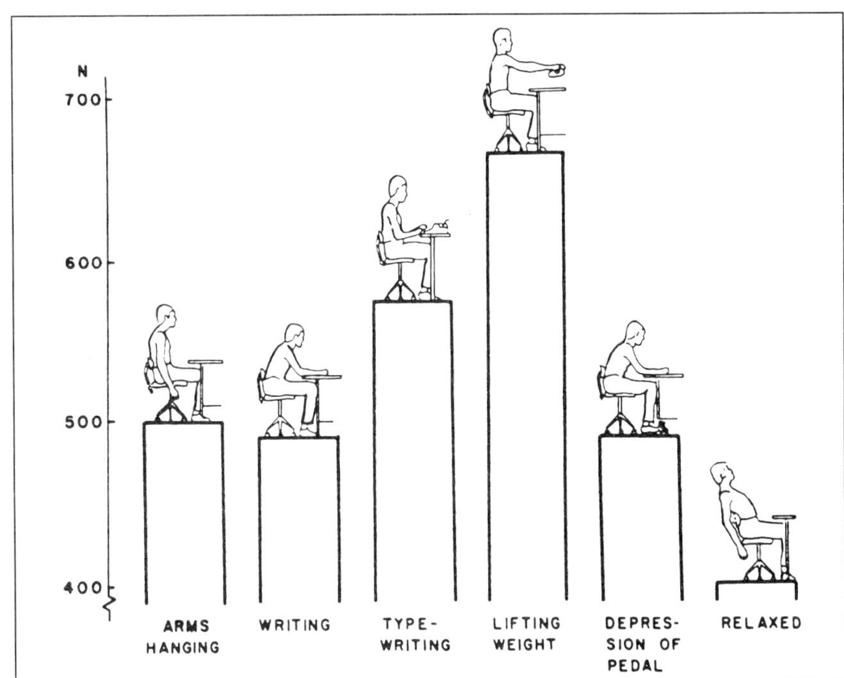

Figure 13-19. Forces in the third lumbar disk when sitting at a desk on an office chair with a small lumbar backrest. (Reprinted with permission from Chaffin and Andersson, 1984)

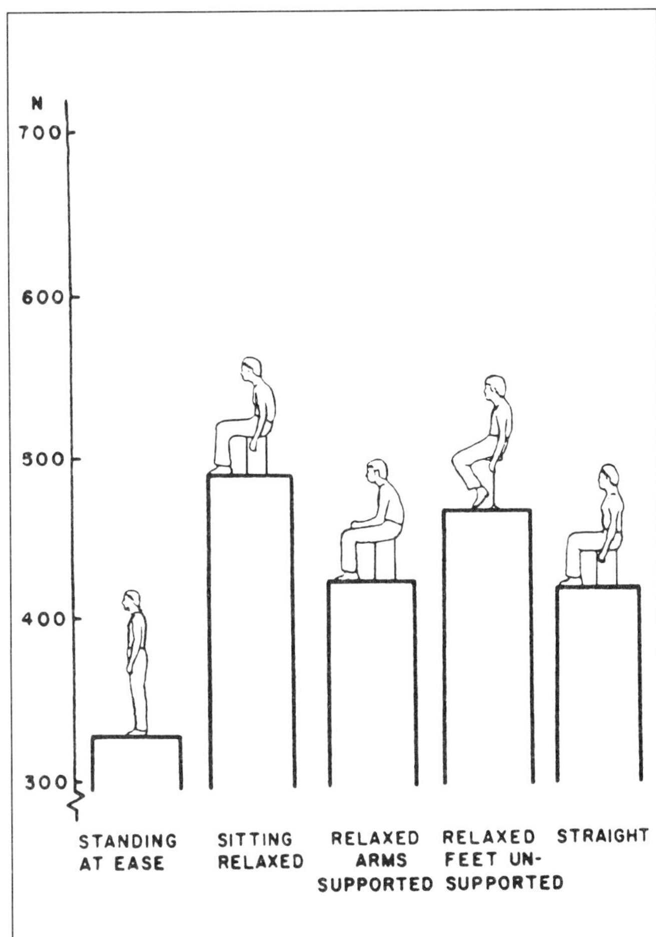

Figure 13-20. Forces in the third lumbar disk when standing or when sitting on a stool without a backrest. (Reprinted with permission from Chaffin and Andersson, 1984)

of these considerations, bursts of muscular activities while sitting should be encouraged. Maintenance of the same posture over long periods of time, and thus, continued muscle tension and spinal compression (even at the low levels just mentioned), becomes uncomfortable and should be avoided by the introduction of rest periods or physical activities and exercises.

Tension and pain in the neck area are among the most frequently mentioned health complaints of computer operators. In contrast to the events in the lumbar region, EMG activities in the neck and shoulders are often considerably higher than the 10 percent or less level reported for lower trunk muscles and often must be maintained over considerable periods of time while the head is kept in a fixed position relative to the visual object. Intensity, frequency and the length of time that such muscle contractions are maintained can generate intense discomfort, pain, and related musculosketetal health complaints that may persist over long periods of time.

Other analytical studies have addressed the pressure in the intervertebral disks, dependent on trunk posture. The most famous experiments are those performed in the 1970s in Scandinavia during which pressure transducers were pushed into spinal disks. (A thorough compilation and review was conducted by Chaffin and Andersson, 1984). These experiments showed that the amount of intradisk force in the lumbar region depended on trunk posture and support (Figures 13-19 and 13-20).

When standing at ease, the forces in the lumbar spine are in the neighborhood of 330 N. This force increases by about 100 N when one sits on a stool without a backrest. It makes little difference if one sits erect with the arms hanging or relaxed with the lower arms resting on the thighs. Sitting relaxed but letting the arms hang down increases the internal force to nearly 500 N. Thus, there is a significant increase in the spinal compression force in the lumbar region when one changes from a standing posture to a sitting one, but the differences among several sitting postures are not very pronounced.

About the same force values are obtained when one sits on

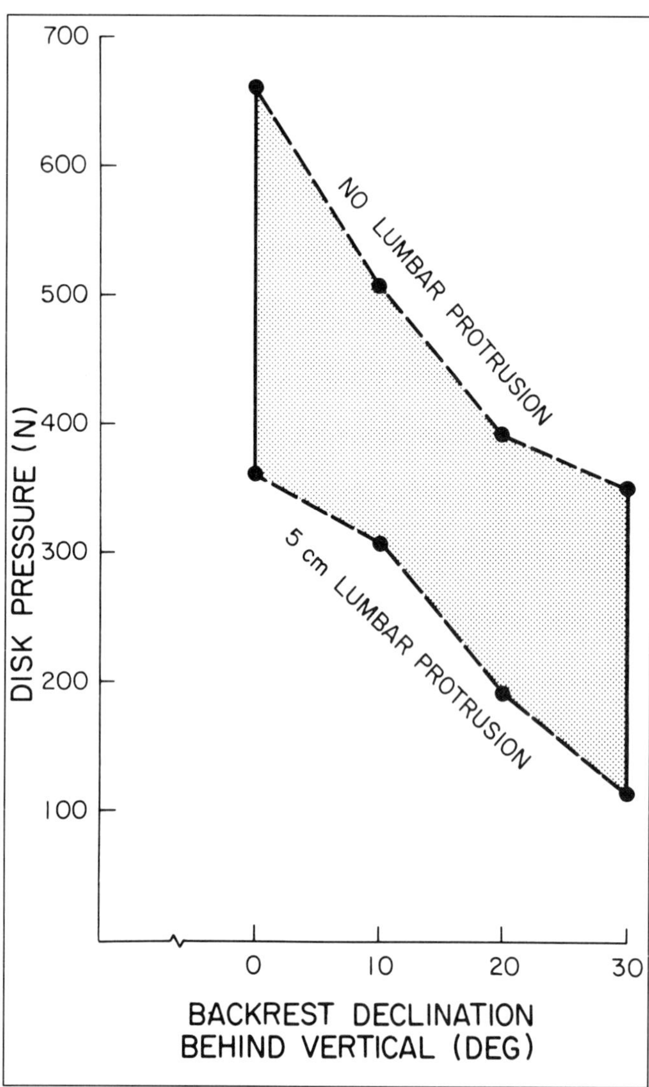

Figure 13-21. Disk forces (L3/4) depending on backrest angle and lumbar pad size. (Adapted from Chaffin and Andersson, 1984)

an office chair with a small lumbar support. Sitting with the arms hanging, writing with the arms resting on a table, and activating a pedal, all result in forces of around 500 N. The spinal forces are increased by typing, when the forearms and hands must be lifted to keyboard height. A further increase is seen when a weight is lifted in the hands with the arms extended forward (see Eklund, Corlett, and Johnsson, 1983). None of these postures makes use of the backrest. However, if one leans back over the small backrest and lets the arms hang down, the internal compression forces are reduced to approximately 400 N.

Figure 13-21 shows the effect of backrest use even more dramatically. When the backrest is upright, it cannot support the body, and rather high disk forces may occur. When the straight backrest is declined behind the vertical position, dramatic decreases in internal force are caused because part of the upper body weight is transmitted to the backrest and thus does not rest on the spinal column. An even more pronounced effect can be brought about by making the backrest protrude toward the lumbar lordosis. A protrusion of 5 cm nearly cuts the internal disk forces associated with a flat backrest in half; protrusions of 1-4 cm into the lumbar region produce intermediate effects.

These experimental results yield three important findings. The first is that sitting down from a standing position increases disk pressure by one third to one half. The second is that there are no dramatic disk pressure differences among sitting straight, sitting relaxed, or sitting with supported arms if there is no backrest or only a small lumbar board. The third finding is that the use of a suitably designed backrest brings about disk pressures that are as low as those measured in a standing person. (Certainly, these findings do not at all support the theory that sitting upright—as opposed to sitting relaxed or leaning back—reduces disk pressure.)

If the backrest consists of only a small lumbar board, it is nearly worthless unless the person is draped by leaning backward over it (Figure 13-19). A large backrest is also nearly useless when upright but highly beneficial when leaned back behind the vertical position, where it can support a large portion of the weight of the upper trunk, head, and arms. Its positive effects are dramatically enhanced if it is shaped to bring about the S-curve of the spinal column, particularly lumbar lordosis.

Relaxed leaning against a declined backrest is the least stressful sitting posture. This is a condition that is often freely chosen by persons working in an office: ". . . an impression which many observers have already perceived when visiting offices or workshops with VDT workstations: Most of the operators do not maintain an upright trunk posture. . . . In fact, the great majority of the operators lean backwards even if the chairs are not suitable for such a posture." (Grandjean, Hunting, and Nishiyama, 1984, pp. 100-101.)

This observation indicates the problems associated with trying to find objective measures that reflect the complex, holistic, subjective, and variable with-time-and-attitude feeling of comfort and well-being. To arrive at comprehensive criteria for judging the adequacy, acceptance, and comfort of design measures, some scientists use subjective evaluations. This psychophysical approach has been developed into a scientific tool that complements objective measures. (Life and Pheasant, 1984; Bhatnager, Drury, and Schiro, 1985; Drury and Francher, 1985.)

ERGONOMIC DESIGN OF OFFICE WORKSTATIONS

Several variables combine to determine the ergonomics of office workstations. These include psychological and attitudinal as well as organizational conditions. Another important variable is the physical environment, which includes climate, illumination, and general facility and work space design (ANSI, 1986; Grandjean, 1986; Koffler Office Systems Ergonomics Reports, 1985, 1986; Kroemer, 1984; Kroemer 1985).

Major influences are produced by the components of the work equipment actually used; for example, by data entry devices, such as keyboards of various designs, trackball, mouse, touchpad, lightpen, and joystick controls. Highly specific, even unusual work tasks and conditions may prevail, and individual preferences may lead to rather unconventional solutions. However, the large majority of regular workstations consist of a display unit, a data entry unit, support(s) on which these rest, a chair, and an operator. These are the system components that the following ergonomic design recommendations take into consideration.

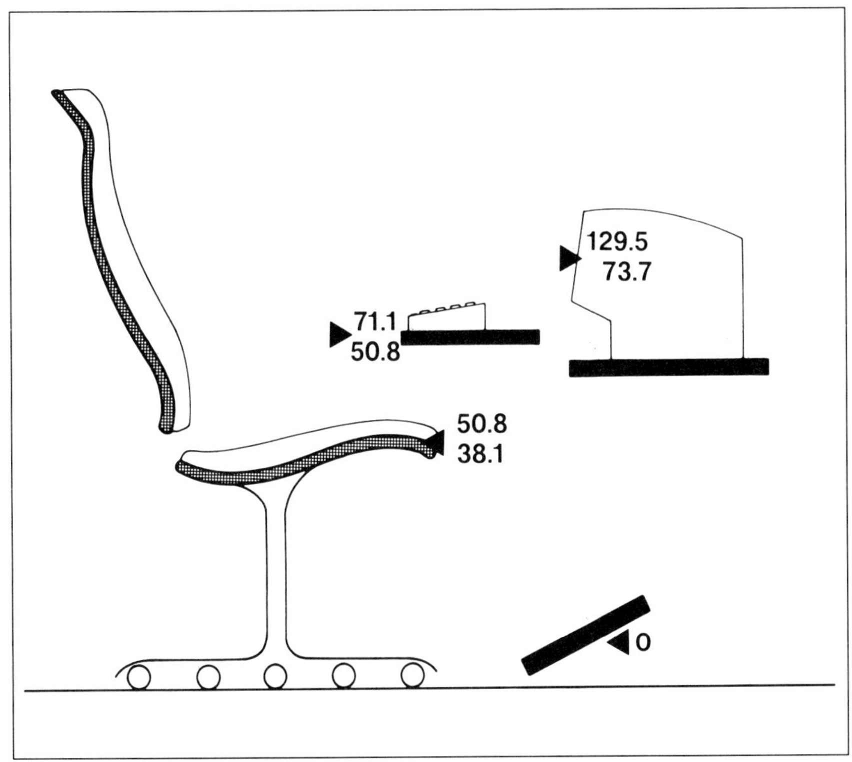

Figure 13-22. Height dimensions of a computer workstation (cm). (Reprinted with permission from Kroemer, 1985)

The operator is the most important component in the system since she or he drives the output. Hence, the operator must be accommodated first: the design of the workplace components should fit all operators, but allow many variations in working postures that may be individually quite different. The fiction of "one healthy upright posture, good for everybody, anytime" used to determine the design must be abolished.

Among the first steps in designing office furniture is the establishment of the main clearance and external dimensions, which can be derived from body measurements. There is sufficient information available to provide an approximate anthropometric overview of the U.S. civilian population. Table 13-C presents the best currently available information for U.S. female and male civilians.

Certain body dimensions of the operator determine related dimensions of the workstation as follows:

- Eye height primarily determines the location (distance from the eye and height) of the visual target: monitor, source document, notepad, keyboard.
- Elbow height and forearm length are related to the location of such motor activities as keying and writing with a pen and to the operation of such hand controls as keys, mouse, trackball, and so on.
- Knee height and thigh thickness determine the needed height of the leg room underneath tables or support surfaces.
- The forward protrusions of the knees (buttock-knee depth) and of the foot determine the needed depth of the leg room under the equipment.
- Thigh breadth determines the minimal width for the open leg room and of the width of the seat.
- Buttock-popliteal depth determines the depth of the seat pan.
- Lower leg length (popliteal height) determines the height of the seat.

Of course, other dimensions, such as functional reach, also have significant effects on workstation design dimensions. In some cases, the application of the anthropometric information to the design task is straightforward; for example, the seat pan depth must be shorter than the buttock-popliteal length to avoid uncomfortable pressure of the seat front on the soft tissues behind the knee.

Three main strategies can be pursued to determine major equipment dimensions. The first is to make the seat height, the support heights for the equipment (primarily the keyboard and other working surfaces), and the support height for the display adjustable. The second strategy assumes that the support height must be fixed (as table heights in traditional offices usually are) but that seat height and display height are adjustable. The third strategy assumes that the seat height is fixed but that the support and display heights are adjustable. (The detailed results of the three design strategies were published by Kroemer in 1985.) The common procedure is to accommodate 90 percent of the civilian population (Table 13-C); that is, only persons smaller than the 5th female percentile and larger than the 95th male percentile are excluded.

In the first ergonomic strategy for workstation design, seat height is determined from the popliteal height, adding 2 cm for heels. This results in the seat height above the floor to be adjustable from 38.1 cm (15 in.) to 50.8 cm (20 in.). Thigh thickness is then added for the necessary clearance height underneath the support structure, and 2 cm are added for the support structure. This adds up to total support surface heights of 50.8 cm (20 in.) to 71.1 cm (28 in.). The next step is to determine eye height above the (just determined) seat pan. Next, the center height of the display is determined using values for the preferred viewing distance and the preferred angle of sight (Hill and Kroemer, 1986). Accordingly, the height of the center of the computer display should be between 73.7 cm (29 in.) and 129.5 cm (51 in.) above the floor. A footrest is not needed in this design approach. Figure 13-22 illustrates these height adjustments.

Furniture designed to these dimensions provides fit for nearly all computer users but does not assume or require certain postures, such as an upright trunk or horizontal forearms. In fact, such ergonomic furniture allows the user freedom to sit almost any way he or she likes: from bending forward to leaning back; holding the legs in any posture within the leg room provided; and using either conventional seats, ergonomic chairs with large backrests, or forward-declining seat surfaces for semi-sitting. This, of course, assumes that other clearance heights, widths and depths, and work surface dimensions are ergonomically suitable as well; e.g., as listed in the ANSI Draft Standard (1986). Note that in this draft, the lowest possible adjustable seat height is 41 cm (16 in.), which is not low enough to accommodate persons with short lower legs—mostly women—without having to use footrests.

Such furniture permits the use of any data entry device, such as keyboards, trackballs, mousepads and so on. (Note, however, that support surfaces that are thicker than the 2 cm assumed in the aforementioned calculations would require the user to lift the hands and elbows, which would render the working posture more strained and uncomfortable.) The furniture accommodates any display units because it specifies only the height of the center of the screen and does not make limiting assumptions about the unit's specific enclosure and housing designs.

Ergonomic design of the seat is of primary importance for reducing physiological and biomechanical stresses on the body when seated, providing a wide range of adjustments and postures to suit the individual, and promoting well-being and performance. Given these far-ranging goals and the wide personal variability, it is clearly impossible to recommend one particular seat design. In fact, it is obvious that various designs with varying features are needed. Hence, only some basic features and dimensions can be mentioned here.

The height of the *seat surface* should be adjustable in the range of 37 cm (15 in.) to 51 cm (20 in.). The seat surface should be between approximately 38 cm (15 in.) and 43 cm (17 in.) deep, and at least 45 cm (18 in.) wide. These recommendations apply to a conventional seat whose pan is horizontal and inclined or declined by only a few degrees. Less conventional designs, such as those employed for semi-sitting, will probably deviate from these measurements. The seat surface should be comfortably but firmly upholstered to distribute pressure and to allow various sitting postures. Particular attention should be paid to the front of the seat surface, which must not generate undue pressure to the thighs of the seated person.

The *backrest* should provide a large and well-formed surface to support the back and neck. At its lowest part, the backrest must provide room for the buttocks but should have a slight protrusion (which is not to exceed 5 cm and which is preferably adjustable) to fit the lumbar concavity of the body. The height of this lumbar pad should be adjustable between 15 cm (6 in.) and 23 cm (9 in.) above the seat pan. Above the lumbar pad, the backrest can be nearly flat but must be able to be inclined from a near upright position to 20 or possibly 30 degrees behind the vertical position. At the backrest's upper part, the surface should follow the concave form of the neck. This cervical pad must be adjustable to heights between 50 cm (20 in.) and 70 cm (28 in.) above the seat surface and should be adjustable in its protrusion to allow individual variation. The width of the backrest is not critical a long as it is at least 30 cm (12 in.). Figure 13-22 illustrates the major features of a suitable backrest.

Obviously, many seat dimensions need to be adjustable to suit body form and the preferences of the individual. It has been shown in the laboratory and through practical use that *adjustment features* will be employed if they are easy and natural but, if not, they will be disregarded. Whether adjustment features should be coupled (for example, backrest tilt can be linked with seat-pan tilt) is a matter of preference and convenience. The same holds true for armrests, which may or may not be deemed desirable; however, the data illustrated in Figures 13-19 and 13-20 indicate that propping the arm on a support can reduce the compression load on the spinal column.

With regard to designs for semi-sitting, diverse opinions have been voiced, ranging from those who find the position to be highly comfortable to those who reject it. Few scientific studies have been published on the subject; Drury and Francher (1985) found no general advantages of a balans-type seat over a more conventional one. It appears that, as emphasized before, individual preferences in working postures vary widely.

A major design concern is to provide the opportunity and means to *change body posture* frequently during the work period. Maintaining a particular posture, even if it is comfortable in the beginning, becomes stressful as time passes. Changes in posture are necessary, best facilitated by brief periods of physical activity. To permit position changes for the hands/arms and eyes, the input device (e.g., a keyboard) should be movable within the work space. Also, one should be able to adjust the display screen to various heights (and angles), which requires an easily adjustable, possibly motor or spring-driven suspension system of the support surface.

If vision must be focused on the screen as well as on the source document and on the keyboard, all *visual targets* should be located close to each other—at the same distance from the eyes and in about the same direction of gaze. If the visual targets are spaced apart in direction or distance, the eye must be continuously redirected and refocused while sweeping from one target to another. This is particularly critical if reading glasses (or any other eye-correction lenses) are worn because these are usually shaped for a special focusing distance and for an assumed direction of sight. Often, the computer screen is arranged too high, forcing the operator (particularly when reading glasses are worn) to tilt the neck severely backward. This position causes muscle tension and generates strain on the cervical part of the spinal column. This strain, in turn, regularly leads to complaints about headaches and pains in the neck and shoulder region. Similar postural complaints are often voiced by persons who have to hold their arms and hands in stressful positions that are caused, for example, when the keyboard is improperly placed. Most postural complaints can be avoided by proper ergonomic design, adjustment, and use of the work equipment.

All components of the workstation must fit each other, and each must suit the operator. This requires easy adjustability. Figure 13-23 demonstrates various adjustment features that allow the user to match seat height (S) with the height of support of the input devices, or table (T)—possibly while using a footrest (F)—and to match eye position with the monitor display (D) resting on its support (M).

Another way to change working postures is to allow the computer operator, at his or her own choosing, to work for some

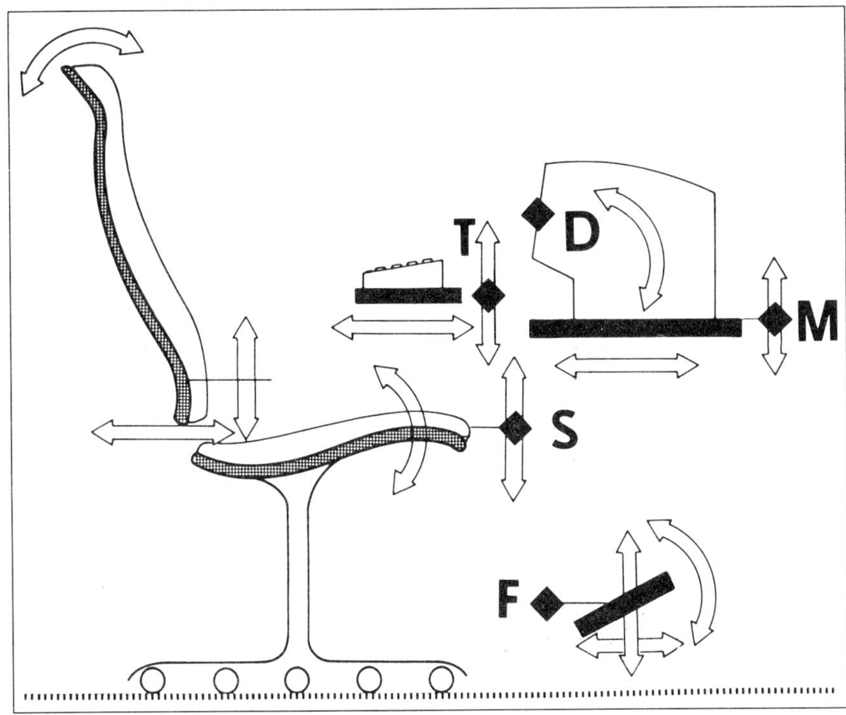

Figure 13-23. Adjustment features of a VDT office workstation. Key: S=seat height; T=table; F=footrest; D=monitor; M=support. (Reprinted with permission from Kroemer, 1985)

period of time while standing up. A stand-up workstation should be adjustable so that the input device is approximately at elbow height when standing; i.e., between 90 cm (35 in.) and 120 cm (48 in.). As in the sit-down work place, the display unit should be located close to the other visual targets. In workplaces designed for standing operation, a footrest positioned at about two-thirds the knee height (approximately 30 cm, or 12 in.) is often provided so that the operator can prop one foot up on it temporarily. This causes changes in pelvis rotation and in spine curvature.

These recommendations for the design and use of furniture assume flexibility in work organization and management attitudes. Ergonomic design of the computer workstation and its proper use can facilitate interactions and can promote the worker's well-being and high performance.

CONTROLS

Controls, operated by hand or foot, transmit inputs to a machine. Much research has been performed on controls and displays; in fact, in human factors engineering, the period after World War II is often called the "knobs and dials era." Hence, this topic is rather well researched, and summaries of the finds can be found in Van Cott and Kinkade (1972), Woodson (1981), and McCormick and Sanders (1982); Military Standards have established clear design guidelines. General rules for the selection of hand controls are supplied in the following section.

Selection of controls

Controls shall be selected for their functional usefulness and distributed so that none of the operator's limbs will be overburdened.

Continuous controls. Continuous controls shall be selected if control operation is required anywhere within the range of the control, and no setting in any given position is required.

Detent controls. Detent controls shall be selected if control operation in discrete steps is required.

Standard practice. Unless other solutions have been demonstrated to be better, the following rules apply:

1. Two-dimensional vehicle steering should be operated by a steering wheel.
2. Three-dimensional steering should be operated by a joystick or by combining levers, wheel, and pedals.
3. Primary vehicle braking should be controlled by pedal(s).
4. Primary vehicle acceleration should be controlled by a pedal or lever.
5. Compound speed should be controlled by lever.
6. Transmission gear selection should be controlled by lever or by legend switch.
7. Valves should be controlled by round knobs or T-handles.
8. Selection of one (of two or more) operating modes can be by toggle switch, push button, bar knob, rocker switch, lever, or legend switch.

Table 13-P provides general guidelines for the selection of controls for special functions. Tables 13-Q and 13-R help in the selection of controls depending on the actuation forces required.

Consistency of movement. Controls shall be selected so that the direction of the control movement is consistent with the response movement of the associated machine. The machine can be a vehicle, equipment, a component, an accessory, and so on. Table 13-S lists such control-movement stereotypes.

Control actuation force. The force (or torque) that must be applied by the operator for the actuation of the control shall be kept as low as feasible. Minimal forces shall be required particularly if the control will be operated often. If the operator is subjected to jerks and vibrations, it is better to stabilize the operator (with the help of supporting arms and wrists) than

Table 13-P. Selection of Hand Controls (General)

Function	Keylock	Toggle Switch	Push Button	Bar Knob	Round Knob	Thumbwheel Discrete	Thumbwheel Continuous	Crank	Rocker Switch	Lever	Joystick or Ball	Legend Switch
Select ON/OFF	1	1	1	3	—	—	—	—	1	—	—	1
Select ON/STANDBY/OFF	—	2	1 several	1	—	—	—	—	—	1	—	1
Select OFF/PRIME MODE/SECONDARY MODE	—	3	2 several	1	—	—	—	—	—	1	—	1
Select one or several related functions	—	2	1	—	—	—	—	—	2	—	—	—
Select one of three or more discrete alternatives	—	—	—	1	—	—	—	—	—	—	—	—
Select operating condition	—	1	1	2	—	—	—	—	1	1	—	1
ENGAGE or DISENGAGE mechanical function	—	—	—	—	—	—	—	—	—	1	—	—
Select one of mutually exclusive functions	—	—	1	—	—	—	—	—	—	—	—	1
Set value on scale	—	—	—	—	1	—	2	3	—	3	3	
Select value in discrete steps	—	—	1	1	—	1	—	—	—	—	—	—

Note: 1 = most preferred; 3 = least preferred.

Table 13-Q. Selection of Controls if Large Hand Force Is Needed*

Positions	Controls
2 discrete positions	Foot push button Hand push button Detent lever
3 to 24 discrete positions	Detent lever Bar knob
Continuous setting	Handwheel Lever Joystick Crank

*Above approximately 20 N or 20 Ncm

Table 13-R. Selection of Controls If Small Hand Force Is Needed

Positions	Controls
2 discrete points	Keylock Toggle switch Push button Rocker switch Legend switch
3 discrete positions	Toggle switch Bar knob Legend switch
4 to 24 discrete positions	Bar knob
Continuous setting	Round knob Joystick Continuous thumbwheel Crank Lever
Continuous slewing and fine adjustments	Crank Round knob

*Below approximately 20 N or 20 Ncm

to increase the control resistance to prevent uncontrolled or inadvertent actuation.

Multidimensional operation. If a machine is capable of motion in more than two dimensions, exceptions to the above recommendations may be made to ensure the consistency of anticipated responses (e.g., the forward motion of a directional control causes a boom to descend rather than to move forward). When several controls are combined in one control device, caution shall be exercised to avoid conflicts, such that, for example, control motion to the right is compatible with a clockwise roll, right turn, and direct movement to the right.

Operator-control orientation. Controls shall be oriented with respect to the operator.

Control-effect relationships. Control-effect relationships shall be apparent through design considerations similarity, proximity, grouping, coding, framing, labeling, and similar techniques.

Time lag. The time lag between the response of a system to a control input shall be kept to a minimum and shall be consistent with safe and efficient system operation.

Arrangement and grouping

Location of primary controls. The most important and frequently used controls (particularly rotary controls and those requiring fine settings) shall have the most favorable position with respect to ease of operation and reaching.

Grouping. All controls that have sequential relations, that involve a particular function or that are operated together, shall be arranged in functional groups (together with their associated displays); e.g., drive train, boom movement, and auxiliary equipment. Controls and displays within functional groups shall be arranged according to operational sequence and function.

Sequential operation. If sequential operations follow a fixed pattern, controls shall be arranged to facilitate operation; that is, in left-to-right pattern (preferred) or from top-to-bottom as on a printed page.

Consistency. The arrangement of functionally identical or

Table 13-S. Control Movement Stereotypes

Function	*Direction of Control Movement*											
	Up	*Right*	*Forward*	*Clockwise*	*Press* Squeeze*	*Down*	*Left*	*Rearward*	*Back*	*Counterclockwise*	*Pull***	*Push***
On	1	1	1	1	2	—	—	—	—	—	1	—
Off	—	—	—	—	—	1	2	2	—	1	—	2
Right	—	1	—	2	—	—	—	—	—	—	—	—
Left	—	—	—	—	—	—	1	—	2	—	—	—
Raise	1	—	—	—	—	—	—	2	—	—	—	—
Lower	—	—	2	—	—	1	—	—	—	—	—	—
Retract	2	—	—	—	—	—	—	1	—	—	2	—
Extend	—	—	1	—	—	2	—	—	—	—	—	2
Increase	2	2	1	2	—	—	—	—	—	—	—	—
Decrease	—	—	—	—	—	2	2	1	—	2	—	—
Open Valve	—	—	—	—	—	—	—	—	—	1	—	—
Close Valve	—	—	—	1	—	—	—	—	—	—	—	—

Note: 1=most preferred; 2=less preferred.
* With trigger-type control
**With push-pull switch

similar controls shall be consistent from panel to panel throughout the system, vehicle, unit, component, or equipment.

Spacing. Table 13-T offers recommendations for minimum spacing between controls. More information on spacing is provided in the following sections on the specific controls.

Detent Controls

Keylock

APPLICATION. Keylocks (also called key-operated switches) are used to prevent unauthorized machine operation. Keylocks usually control the on and off positions and ignition functions.

DESIGN RECOMMENDATIONS. Design recommendations are given in Figure 13-24 and Table 13-U. Other recommendations are as follows:

1. Keys with teeth on both edges (preferred) should fit the lock with either side up or forward.
2. Keys with a single row of teeth should be inserted into the lock with the teeth pointing up or forward.
3. Locks should be oriented so that the key's vertical position is the off position.
4. Operators should normally not be able to remove the key from the lock unless the switch is turned off.
5. The on and off positions should be labeled.

Bar knob

APPLICATION. Bar knobs (also called rotary selector switches) should be used for discrete functions when two, three, or more detented positions are required.

SHAPE. Knobs shall be bar-shaped with parallel sides, and the index end shall be tapered to a point.

DESIGN RECOMMENDATIONS. Design recommendations are illustrated in Figure 13-25 and listed in Table 13-V.

Discrete thumbwheel

APPLICATION. Thumbwheels for discrete settings may be used if the function requires a compact input device for discrete steps.

DESIGN RECOMMENDATIONS. Design recommendations are illustrated in Figure 13-26 and listed in Table 13-W.

Push button

APPLICATION. Push buttons should be used for single switch-

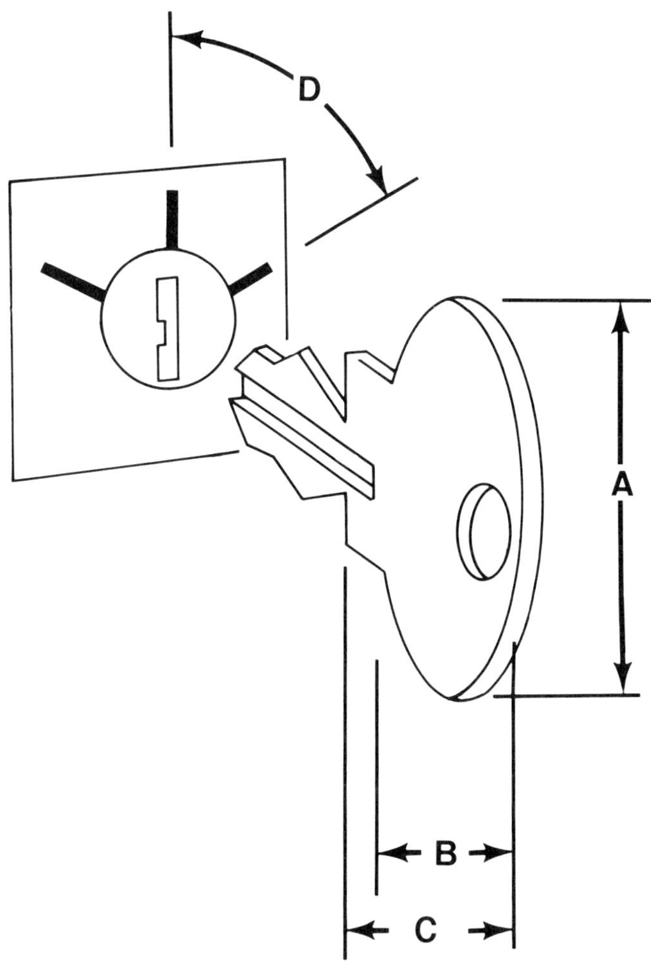

Figure 13-24. Keylock.

Table 13-T. Minimum Separation Distances for Controls

	Toggle Switches	*Push Buttons**	*Round Knob*	*Bar Knob*	*Discrete Thumbwheel*
Toggle switches	See Figure 13-29	13 mm	19 mm	19 mm	13 mm
Push buttons*	13 mm	See Figure 13-27	13 mm	13 mm	13 mm
Round knob	19 mm	13 mm	See Figure 13-31	25 mm	19 mm
Bar knob	19 mm	13 mm	25 mm	See Figure 13-25	19 mm
Discrete thumbwheel	13 mm	13 mm	19 mm	19 mm	See Figure 13-26

Note: All values are for one hand operation. Distances are measured from edge to edge of pairs of controls at their closest point of approach.
* If not separated by barriers.

Table 13-U. Dimensions of a Keylock

	A Height (mm)	*B Width (mm)*	*C Protrusion (mm)*	*D Displacement (degrees)*	*S* Separation (mm)*	*R† Resistance (Nm)*
Minimum	13	13	20	45	25	0.1
Preferred	—	—	—	—	—	—
Maximum	75	38	—	90	25	0.7

Note: Letters A-D correspond to measurements illustrated in Figure 13-24.
* Between closest edges of two adjacent keys.
† Control should "snap" into detent position and not be able to stop between detents.

Table 13-V. Dimensions of a Bar Knob

						S Separation	
	L Length (mm)	*W Width (mm)*	*H Height (mm)*	*R† Resistance (Nm)*	*A Displacement (degrees)*	*One Hand, Random Operation (mm)*	*Two Hands, Simultaneous Operation (mm)*
Minimum	25 38*	13	16	0.1	15 30**	25 38*	75 100*
Preferred	—	—	—	—	—	—	—
Maximum	100	25	75	0.7	90	50 63*	150 175*

Note: Letters L, W, H, and A correspond to measurements illustrated in Figure 13-25.
* If operator wears gloves.
† High resistance with large bar knob only. Control should snap into detent position and not be able to stop between detents.
** For blind positioning.

Table 13-W. Dimensions of a Discrete Thumbwheel

	D Diameter (mm)	*W Width (mm)*	*L Through Distance (mm)*	*H Through Depth (mm)*	*S Separation Side-by-Side (mm)*	*R* Resistance (N)*
Minimum	38	3	11	3	10	0.2
Preferred	—	—	—	—	—	—
Maximum	65	—	19	13	—	0.6

Note: Letters D, W, L, H, and S correspond to measurements illustrated in Figure 13-26.
* Control should snap into detent position and not be able to stop between detents.

ing between two conditions, for entry of a discrete control order, or for release of a locking system (e.g., of a parking brake). Push buttons can be used for momentary contact or for sustained contact.

Design recommendations. Design recommendations are given in Figure 13-27 and listed in Table 13-X.

Shape. The push button surface should normally be concave (indented) to fit the finger. When this is impractical, the surface shall provide a high degree of frictional resistance to prevent slipping. (The surface should be convex for operation with the palm of the hand.)

Positive indication. A positive indication of control activation shall be provided (e.g., snap feel, audible click, or integral light).

Legend switch

Application. Legend switches are particularly suited to display qualitative information on status of an important machine that requires the operator's attention and action.

Design recommendations. Design recommendations are given in Figure 13-28 and listed in Table 13-Y.

Legend switches should be located within a 30°-cone along the operator's line of sight.

Table 13-X. Dimensions of a Push Button

	D Diameter or Width of Square			*R Resistance*					*S Separation*			
									Single Finger			
	Fingertip (mm)	*Thumb (mm)*	*Palm of Hand (mm)*	*Little Finger (N)*	*Other Finger (N)*	*Thumb (N)*	*Palm of Hand (N)*	*A** Displacement (mm)*	*Single Operation (mm)*	*Sequential Operation (mm)*	*Different Fingers (mm)*	*Palm or Thumb (mm)*
Minimum	10 13*	19	25	0.25	0.25	1.1	1.7	3.2 16	13 25*	6	6	25
Preferred	—	—	—	—	—	—	—	—	50	13	13	150
Maximum	19	—	—	1.5	11.1	16.7	22.2	6.5 20*	—	—	—	—

Note: Letters D, A, and S correspond to measurements illustrated in Figure 13-27.
* If operator wears gloves.
** Depressed button shall stick out at least 2.5 mm.

Table 13-Y. Dimensions of a Legend Switch

	W Width (mm)	*A* Displacement (mm)	B_W Barrier Width (mm)	B_D Barrier Depth (mm)	*R* Resistance (N)
Minimum	19	3	3	5	0.28
Preferred	—	—	—	—	—
Maximum	38	6	6	6	11

Note: Letters W, A, B_w, and B_d correspond to measurements illustrated in Figure 13-28.

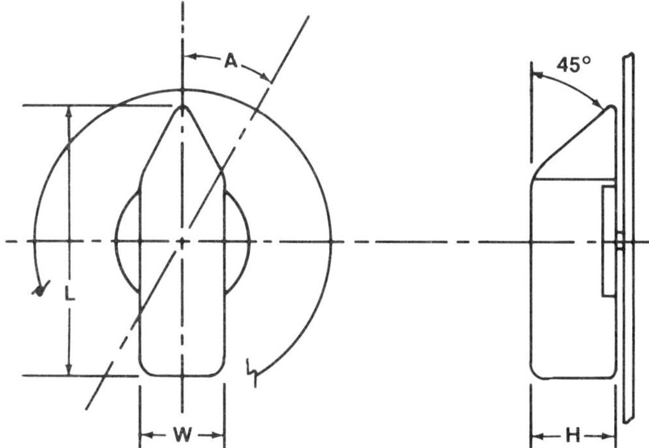

Figure 13-25. Bar knob.

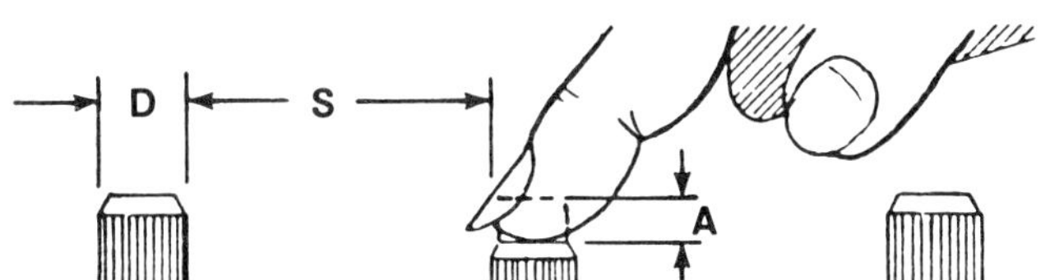

Figure 13-27. Push button.

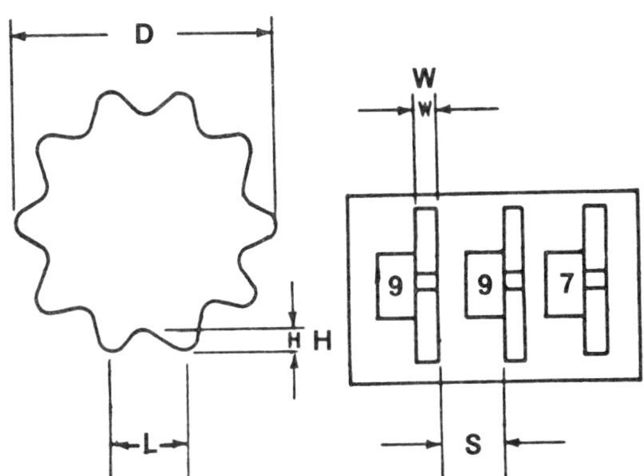

Figure 13-26. Discrete thumbwheel.

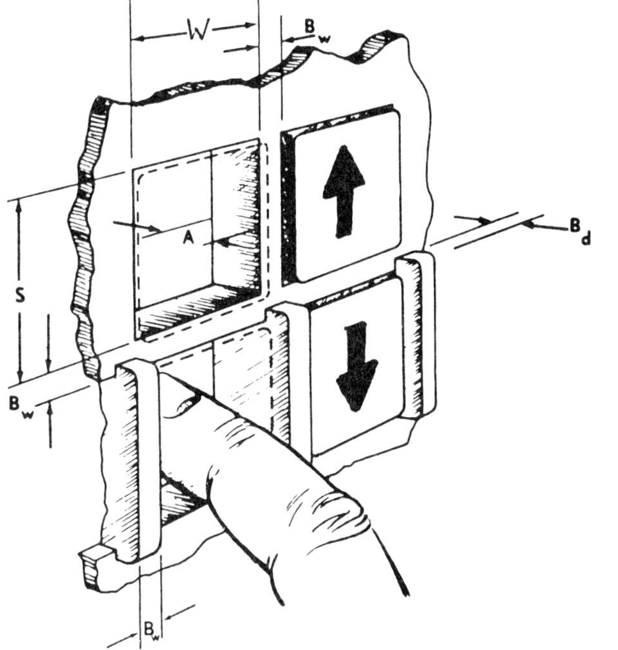

Figure 13-28. Legend switch.

Toggle switch

Application. Toggle switches may be used if two discrete positions are required. Toggle switches with three positions shall be used only where the use of a bar knob, legend switch, array of push buttons, etc., is not feasible.

Design recommendations. Design recommendations are given in Figure 13-29 and listed in Table 13-Z.

Orientation. Toggle switches should be so oriented that the handle moves in a vertical plane, with "off" in the down position. Horizontal actuation shall be employed only if compatibility with the controlled function or equipment location is desired.

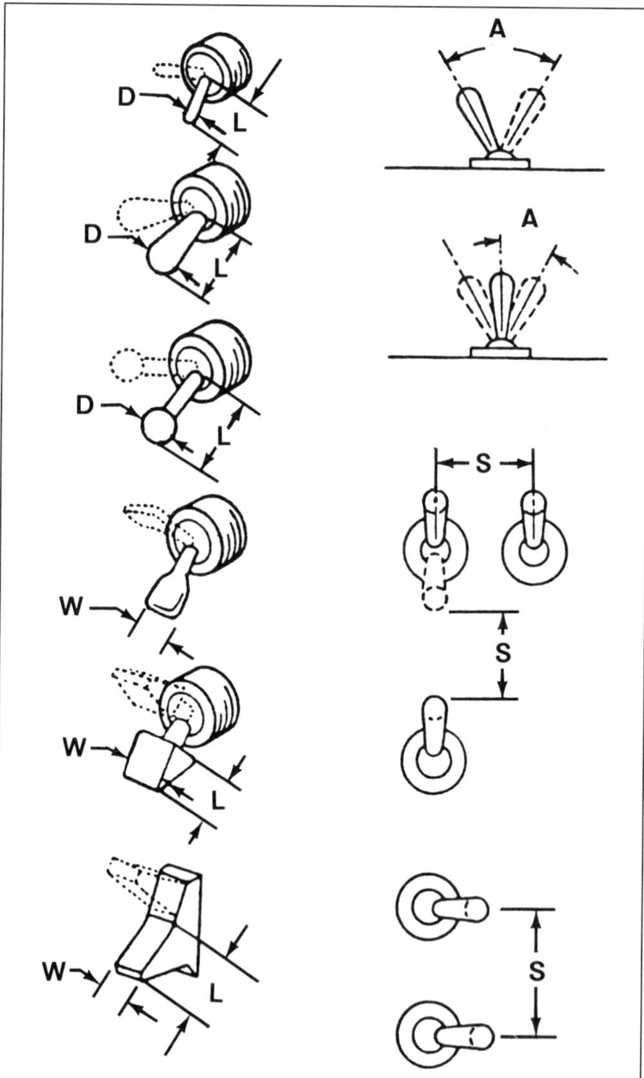

Figure 13-29. Toggle switch.

Rocker switch

Application. Rocker switches may be used if two discrete positions are required. Rocker switches protrude less from the panel than do toggle switches.

Design recommendations. Design recommendations are given in Figure 13-30 and listed in Table 13-AA.

Orientation. Rocker switches should be so oriented that the handle moves in a vertical plane, with "off" in the down position. Horizontal actuation shall be employed only if compatibility with the controlled function or equipment location is desired.

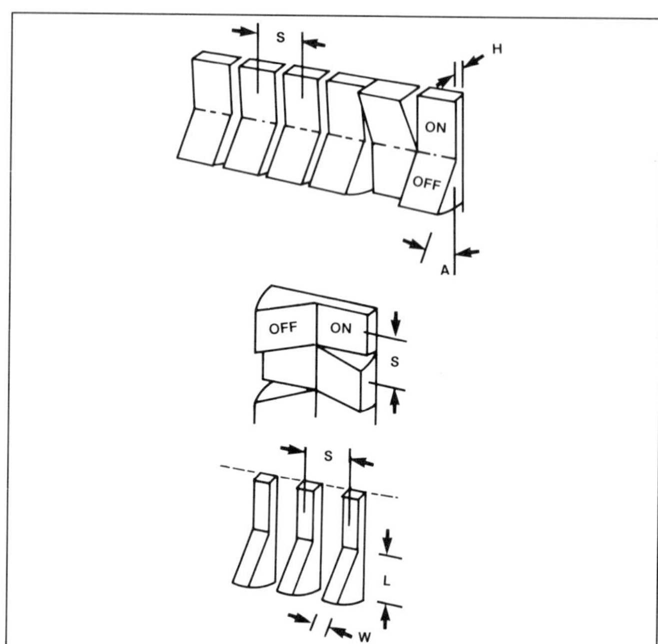

Figure 13-30. Rocker switch. Narrow width (bottom) is especially desirable for tactile definition when gloves are worn.

Continuous controls

Knob

Application. Knobs should be used when little force is required and when precise adjustments of a continuous variable are required. If positions must be distinguished, an index line on the knob should point to markers on the panel.

Design recommendations. Design recommendations are given in Figure 13-31 and Table 13-BB. Within the range specified the latter, knob size is relatively unimportant—provided the resistance is low, and the knob can be easily grasped and manipulated. When panel space is extremely limited, knobs should approximate the minimum values and should have resistance as low as possible without permitting the setting to be changed by vibration or by merely touching the control.

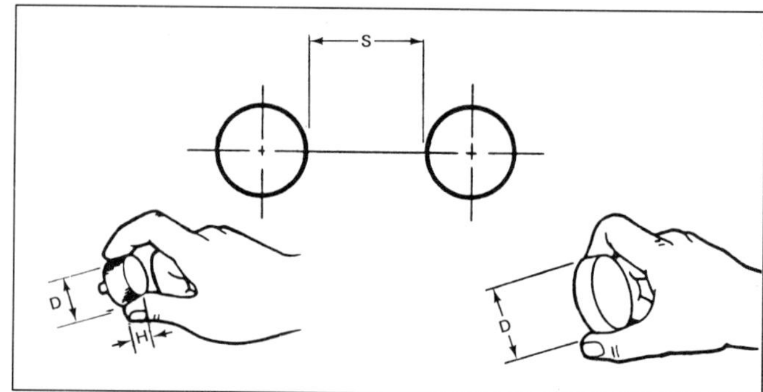

Figure 13-31. Knob.

Table 13-Z. Dimensions of a Toggle Switch

	L Arm Length (mm)	D or W Tip Diameter or Width (mm)	R† Resistance (N)	A Displacement: 2 Positions (degrees)	A Displacement: 3 Positions (degrees)	S Separation: Horizontal Array Vertical Operation, Single Finger: Random Operation (mm)	S Separation: Horizontal Array Vertical Operation, Single Finger: Sequential Operation (mm)	S Separation: Horizontal Array Vertical Operation, Several Fingers Simultaneous Operation (mm)	S Separation: Vertical Array Horizontal Operation (mm)	S Separation: Toward Each Other Tip-to-Tip (mm)
Minimum	9.5 38*	3	2.8	25	18	19	13	19 32*	25 38*	25
Preferred	—	—	4.5	—	25	50	25	28	—	—
Maximum	50	25	11	120	60	—	—	—	—	—

Note: Letters L, D, W, A, and S correspond to measurements illustrated in Figure 13-29.
* If operator wears gloves.
† Control should snap into detent position and not be able to stop between detents.

Table 13-AA. Dimensions of a Rocker Switch

	W Width (mm)	L Length (mm)	A (degrees)	H Depressed (mm)	S Separation Center to Center (mm)	R Resistance (N)
Minimum	6.5	13	30	2.5	19 32*	2.8
Preferred	—	—	—	—	—	—
Maximum	—	—	—	—	—	11.1

Note: Letters W, L, A, H, and S correspond to measurements illustrated in Figure 13-30.
* If operator wears gloves.

Table 13-BB. Dimensions of a Knob

	H Height (mm)	D Diameter: Fingertip Grip (mm)	D Diameter: Thumb and Finger Grasp (mm)	T Torque: Up to 25 mm in Diameter (Nm)	T Torque: Over 25 mm in Diameter (Nm)	S Separation: One Hand (mm)	S Separation: Two Hands Simultaneously (mm)
Minimum	13	10	25	—	—	25	50
Preferred	—	—	—	—	—	50	125
Maximum	25	100	75	0.03	0.04	—	—

Note: Letters H, D, and S correspond to measurements illustrated in Figure 13-31.

Table 13-CC. Dimensions of a Crank

	Operated by Finger and Wrist Movement (Resistance below 22 N)					Operated by Arm Movement (Resistance above 22 N)				
	L Length (mm)	D Diameter (mm)	R Turning Radius: Below 100 RPM (mm)	R Turning Radius: Above 100 RPM (mm)	S Separation (mm)	L Length (mm)	D Diameter (mm)	R Turning Radius: Below 100 RPM (mm)	R Turning Radius: Above 100 RPM (mm)	S Separation (mm)
Minimum	25	9.5	38	13	75	75	25	190	125	75
Preferred	38	13	75	58	—	95	25	—	—	—
Maximum	75	16	125	115	—	—	38	510	230	—

Note: Letters correspond to measurements illustrated in Figure 13-32.

KNOB STYLE. Unless otherwise specified, control knobs shall conform to the guidelines established in *Military Standards, MIL-STD* 1348 or *MIL-HDBK-759*.

Crank

APPLICATION. Cranks should be used primarily if the control must be rotated many times. For tasks involving large slewing movements as well as small, fine adjustments, a crank handle may be mounted on a knob or handwheel.

GRIP HANDLE. The crank's grip handle shall be designed so that it turns freely around its shaft.

DESIGN RECOMMENDATIONS. Design recommendations are given in Figure 13-32 and Table 13-CC.

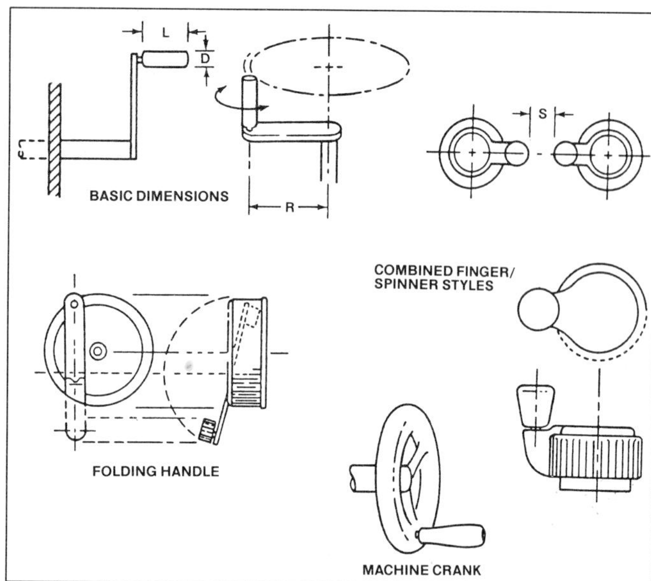

Figure 13-32. Crank.

Handwheel

APPLICATION. Handwheels that are designed for nominal two-handed operation should be used when the breakout or rotation forces are too large to be overcome with a one-hand control—provided that two hands will be available for this task.

KNURLING. Knurling or indentation shall be built into a handwheel to facilitate operator grasp.

SPINNING HANDLE. When large displacements must be made rapidly, a spinner handle may be attached to the handwheel when this is not overruled by safety considerations.

DESIGN RECOMMENDATIONS. Design recommendations are given in Figure 13-33 and Table 13-DD.

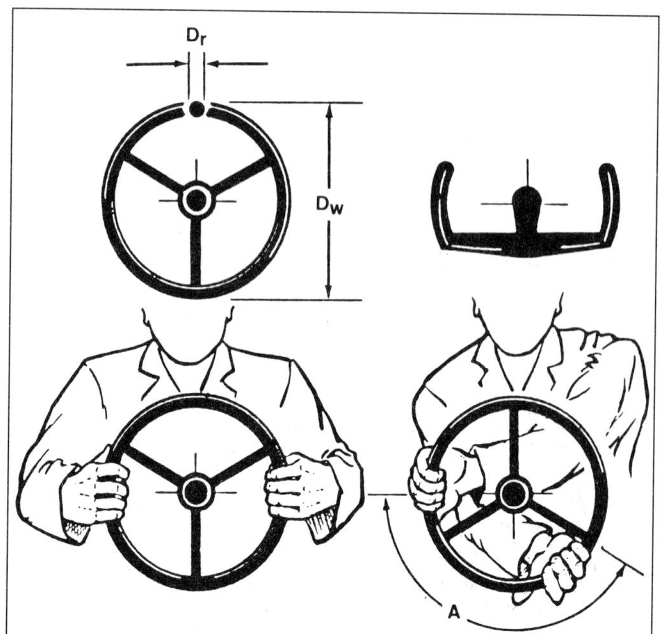

Figure 13-33. Handwheel.

Lever

APPLICATION. Lever-type controls may be used when large force or displacement is required at the control and/or when multidimensional movements are required.

DESIGN RECOMMENDATIONS. Design recommendations are given in Figure 13-34 and Table 13-EE.

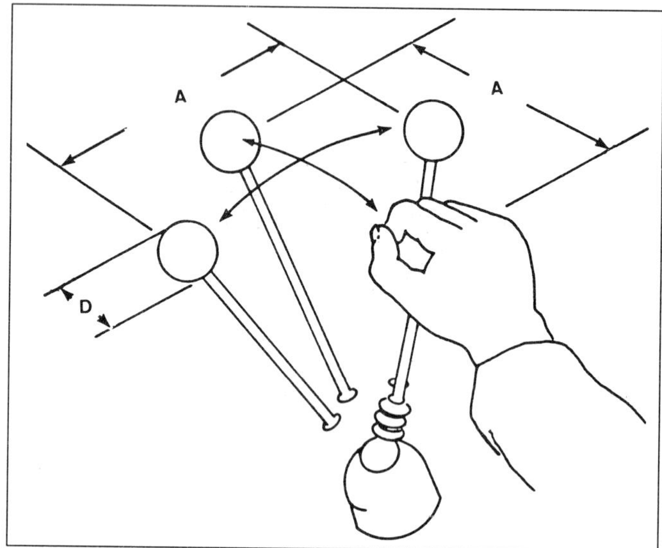

Figure 13-34. Lever.

LIMB SUPPORT. When levers are used to make fine or continuous adjustments, support shall be provided for the appropriate limb segment as follows:

- for large hand movements: elbow support;
- for small hand movements: forearm support;
- for finger movements: wrist support.

CODING. When several levers are grouped in proximity to each other, the lever handles shall be coded.

LABELING. When practicable, all levers shall be labeled with regard to function and direction of motion.

ELASTIC RESISTANCE. For joystick controls, elastic resistance that increases with displacement may be used to improve "stick feel."

HIGH-FORCE LEVERS. For occasional or emergency use, high-force levers may be used. They shall be designed to be either pulled up or pulled back toward the shoulder, with an elbow angle of 150° (±30°). The force required for operation shall not exceed 190 N. The handle diameter shall be from 25–38 mm (1–1.5 in.), and its length shall be at least 100 mm (3.9 in.). Displacement should not exceed 125 mm (5.2 in.). Clearance behind the handle and along the sides of the path of the handle shall be at least 65 mm (2.5 in.). The lever may have a thumb-button release at the hand or a clip-type release.

Continuous thumbwheel

APPLICATIONS. Thumbwheels for continuous adjustments may be used as an alternative to round knobs if a compact thumbwheel is beneficial.

DESIGN RECOMMENDATIONS. Design recommendations are given in Figure 13-35 and Table 13-FF.

Table 13-DD. Dimensions of a Handwheel

	D_w Wheel Diameter With Powersteering (mm)	D_w Wheel Diameter Without Powersteering (mm)	D_r Rim Diameter (mm)	Tilt From Vertical (degrees)	R Resistance (N)	A Displacement Both Hands on Wheel (degrees)
Minimum	355	400	19	30 Light Vehicle	20	—
Preferred	—	—	—	—	—	—
Maximum	400	510	32	45 Heavy Vehicle	220	120

Note: Letters D_w, D_r, and A correspond to measurements illustrated in Figure 13-33.

Table 13-EE. Dimensions of a Lever

	D Diameter		R Resistance				A Displacement		S Separation	
			Fore-Aft		Left-Right					
	Finger Grip (mm)	Hand Grip (mm)	One Hand (N)	Two Hands (N)	One Hand (N)	Two Hands (N)	Fore-Aft (mm)	Left-Right (mm)	One Hand (mm)	Two Hands (mm)
Minimum	13	32	9	9	9	9	—	—	50*	75
Preferred	—	—	—	—	—	—	—	—	—	—
Maximum	75	75	135	220	90	135	360	970	100*	125

Note: Letters A and D correspond to measurements illustrated in Figure 13-34.
* About 25 mm if one hand usually operates two adjacent levers simultaneously.

Table 13-FF. Design Recommendations for a Continuous Thumbwheel

	E Rim Exposure (mm)	W Width (mm)	S Separation Side-by-Side (mm)	S Separation Head-to-Foot (mm)	R Resistance (N)
Minimum	—	—	25 38*	50 75*	—
Preferred	25	3.2	—	—	—
Maximum	100	23	—	—	3.3†

Note: Letters E, W, and S correspond to measurements illustrated in Figure 13-35.
* If operator wears gloves.
† To minimize danger of inadvertent operation.

Coding

Methods and requirements. The selection of a coding mode (e.g., size or color) for a particular application shall be determined by the relative advantages and disadvantages for each type of coding. If coding is selected for the purpose of differentiating among controls, application of the code shall be uniform throughout the system. Table 13-GG lists the advantages and disadvantages of several coding types.

Location coding. Controls associated with similar functions should be in the same relative location from panel to panel.

Size coding. No more than three different sizes of controls shall be used in coding controls for discrimination by absolute size. Controls used for performing the same function on different items or equipment shall be the same size.

Shape coding. Control shapes shall be both visually and tactually identifiable and shall be free of sharp edges. (Note that a company may select a small number of shapes and sizes for internal standardization.)

Color coding. Controls may be black (17038) or gray (26231) (identified in *FED-STD-595*). If color coding is required, the following colors shall be selected: red (11105, 21105, 31105); green (14187); orange-yellow (13538, 23538, 33538); or white (17875, 27875, 37875). Use blue (15123 or 25123) only if an additional color is absolutely necessary.

Prevention of accidental activation

Location and design. Controls shall be designed and located so that they are not susceptible to being moved accidentally. Particular attention shall be given to critical controls whose inadvertent operation might cause injury to persons, damage to the machine, or degradation of system functions.

Methods. For situations in which controls must be protected from accidental activation, one or more of the following methods shall be used as applicable.

1. Locate and orient the controls so that the operator is not likely to strike or move them accidentally in the normal sequence of control movements.

Table 13-GG. Advantages and Disadvantages of Various Types of Coding

	Type of Coding					
Advantages	*Location*	*Shape*	*Size*	*Mode of Operation*	*Labeling*	*Color*
Improves visual identification	X	X	X		X	X
Improves nonvisual identification (tactual and kinesthetic)	X	X	X	X		
Helps standardization	X	X	X	X	X	X
Aids identification under low levels of illumination and colored lighting	X	X	X	X	(When transilluminated)	(When transilluminated)
May aid in identifying control position (setting)		X		X	X	
Requires little (if any) training; is not subject to forgetting					X	
Disadvantages						
May require extra space	X	X	X	X	X	
Affects manipulation of the control (ease of use)	X	X	X	X		
Limited in number of available coding categories	X	X	X	X		X
May be less effective if operator wears gloves		X	X	X		
Controls must be viewed (i.e., must be within visual areas and with adequate illumination present)					X	X

2. Recess, shield, or otherwise surround the controls by physical barriers. The control shall be entirely contained within the envelope described by the recess or barrier.
3. Cover or guard the controls. (Do not use safety or lock wire.)
4. Provide the controls with interlocks so that extra movement (e.g., a side movement out of a detent position or that provided by a pull-to-engage clutch) or the prior operation of a related or locking control is required.
5. Provide the controls with resistance (i.e., viscous or coulomb friction, spring-loading, or inertia) so that definite or sustained effort is required for actuation.
6. Provide the controls with a lock to prevent the control from passing though a position without delay when strict sequential activation is necessary (i.e., the control is moved only to the next position and is then delayed).
7. Design controls for operation by rotary action.

Dead man controls. Dead man controls, which will result in system shutdown and returns the system to a noncritical operating state when force is removed, shall be utilized when a critical system condition is produced.

Examples for guarding controls against accidental activation are shown in Figures 13-36 and 13-37.

LIGHT SIGNALS

A red signal shall be used to alert an operator that the system or any portion of the system is inoperative or that a successful mission is not possible until appropriate corrective or override action is taken. Examples of indicators which should be coded red are those that display such information as "no-go," "error," "failure," "malfunction," and so on.

A flashing red signal shall be used only to denote emergency conditions that require immediate operator action or to avert impending personnel injury, equipment damage, or both.

A yellow signal shall be used to advise an operator that a marginal condition exists. Yellow shall also be used to alert the operator to situations for which caution, rechecking, or unexpected delay is necessary.

A green signal shall be used to indicate that the monitored equipment is in satisfactory condition and that it is all right to proceed (e.g., green displays such information as "go ahead," "in tolerance," "ready," "function activated," "power on," and so on.

A white signal shall be used to indicate system conditions that do not have right or wrong implications, such as alternative functions (e.g., "rear steering on") or transitory conditions (e.g., "fan on"), provided such indication does not imply the success or failure of the operation.

A blue signal may be used as an advisory light but preferential use of blue should be avoided.

Table 13-HH lists the recommended dimensions and suitable functions of simple indicator lights.

DISPLAYS

The proper selection, design, and layout of displays ensure that the operator is provided with the necessary information. Such displays may be visual (scales, counters, lights), auditory (bells, horns), or tactile (shaped knobs, Braille writing). Selection of the proper indicator depends on the information that must be provided to the operator and on the information clutter that may already be present. More information on selection of the proper display may be found in the literature: Sanders and McCormick (1987), Woodson (1981), or in *Military Standards.* Overall guidelines are provided in the following.

- Display only the information that is essential for adequate job performance.

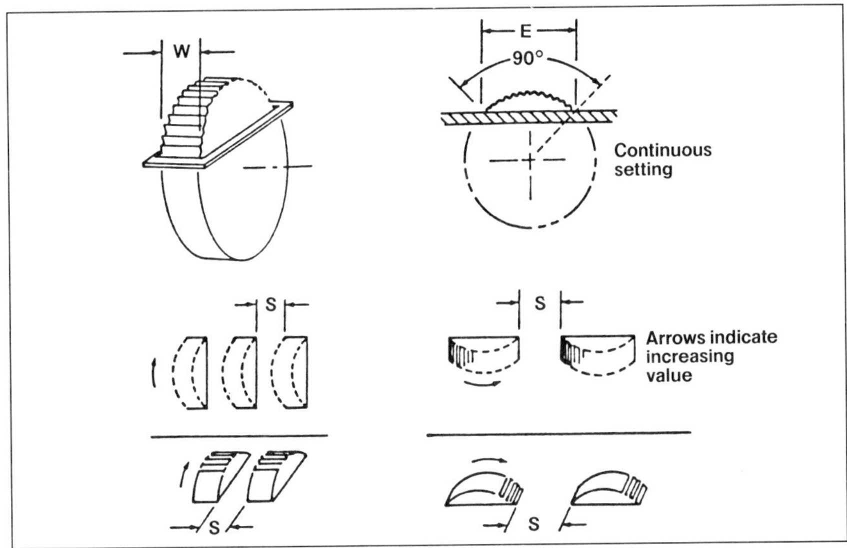

Figure 13-35. Continuous thumbwheel.

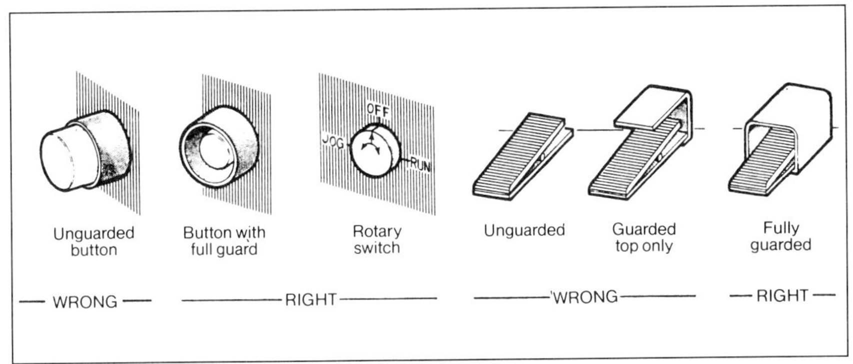

Figure 13-36. Design of controls for avoiding accidental activation. (Reprinted with permission from *Machine Design,* June 23, 1977.)

Table 13-HH. Coding of Simple Indicator Lights

	Color			
Size/Type	*Red*	*Yellow*	*Green*	*White*
13 mm diameter or smaller/steady	Malfunction; action stopped; failure; stop action	Delay; check; recheck	Go ahead; in tolerance; acceptable; ready	Functional or physical position; action in progress
25 mm diameter or larger/steady	Master summation (system or subsystem)	Extreme caution (impending danger)	Master summation (system or subsystem)	
25 mm diameter or larger/flashing	Emergency condition (impending personnel or equipment disaster)			

- Display information only as accurately as the operator's decisions and control actions require.
- Present information in such a way that a failure or malfunction of the display will be immediately obvious.
- Present information in the most direct, simple, understandable, and usable form possible.
- Arrange displays so that the operator can locate and identify them easily without unnecessary searching.
- Group displays functionally or sequentially so that the operator can use them easily.
- Make sure that all displays are properly illuminated, coded, and labeled according to function.

Generally, there are three basic types of visual displays:

1. The check display, which indicates whether or not a given condition exists. A green light to indicate normal functioning is an example of a check display.
2. The quantitative display, which indicates an exact numerical value that must be read. A clock is an example of a quantitative display.
3. The qualitative display, which indicates the approximate value or status of a changing variable or its trend of change. In fact, many quantitative indicators can be changed to qualitative ones; for instance, the numerical display of temperature can be changed to indicate simply "too cold," "acceptable," or "too hot."

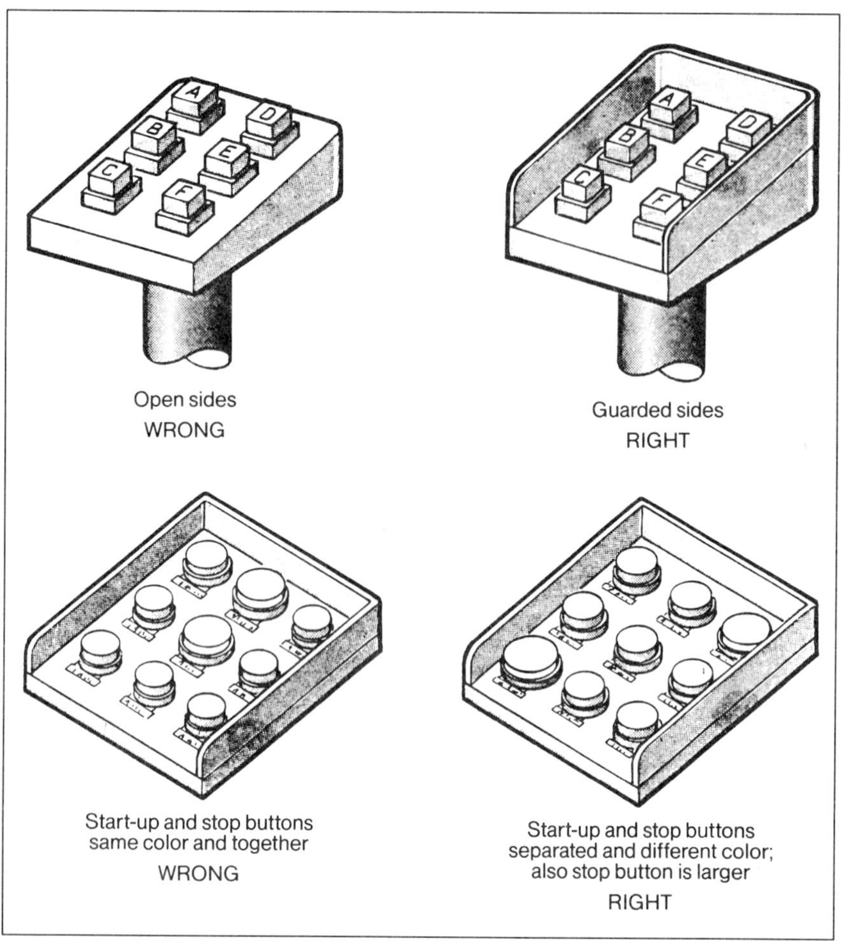

Figure 13-37. Design of push button enclosures and arrangements.

Table 13-II. Characteristics of Displays

Use	*Scalar Indicators*			
	Moving Pointer	*Moving Pointer*	*Counters*	*Pictorial Displays*
Quantitative information	Good Difficult to read while pointer is in motion	Fair Difficult to read while scale is in motion	Good Minimum time and error for exact numerical value, but difficult to read when moving	Fair Direction of motion/scale relations sometimes conflict, causing ambiguity in interpretation
Qualitative information	Good Location of pointer easy; numbers and scale need not be read; position changes easily detected	Poor Difficult to judge direction and magnitude of deviation without reading numbers and scale	Poor Numbers must be read; position changes not easily detected	Good Easily associated with real world situation
Setting	Good Simple and direct relation of motion of pointer to motion of setting knob; position change aids monitoring	Fair Relation to motion of setting knob may be ambiguous; no pointer position change to aid monitoring; not readable during rapid setting	Good Most accurate monitoring of numerical setting; relation to motion of setting knob less direct than for moving pointer; not readable during rapid setting	Good Control-display relationship easy to observe
Tracking	Good Pointer position readily controlled and monitored; simplest relation to manual control motion.	Fair No position changes to aid monitoring; relation to control motion somewhat ambiguous	Poor No gross position changes to aid monitoring	Good Same as above
Difference estimation	Good Easy to calculate positively or negatively by scanning scale	Fair Subject to reversal errors	Poor Requires mental calculation	Good Easy to calculate either quantitatively or qualitatively by visual inspection
General	Requires largest exposed and illuminated area on panel; scale length limited unless multiple pointers used	Saves panel space; only small section of scale need be exposed and illuminated; use of tape allows long scale	Most economical of space and illumination; scale length limited only by available number of digit positions	Picture/symbols need to be carefully designed and pretested

(Adapted from *Military Standard 759A,* 1981)

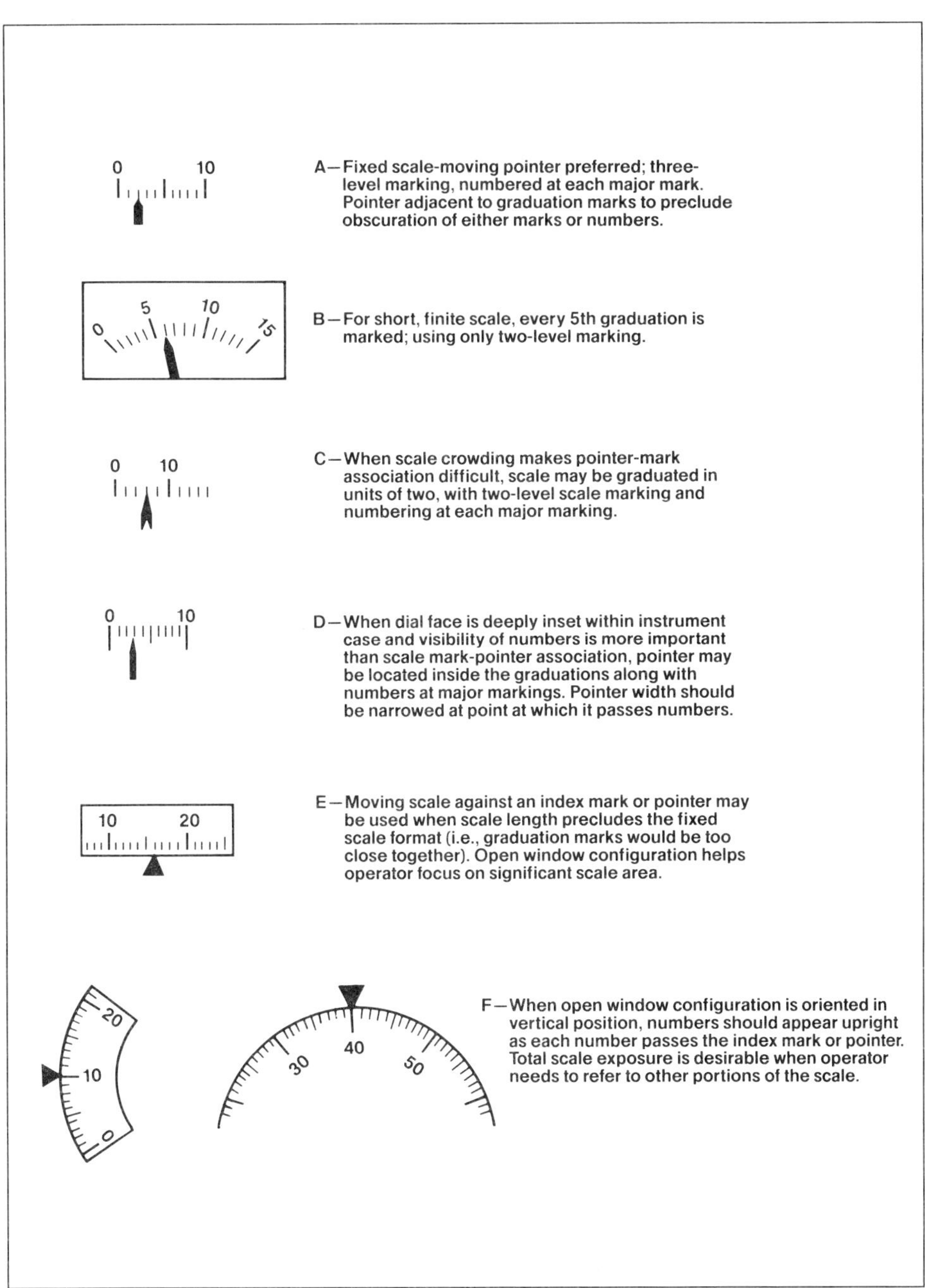

Figure 13-38. Scale graduation, pointer position, and scale numbering alternatives. (Reprinted from *Military Standard 759A,* 1981)

Table 13-II compares different kinds of displays with respect to their ability to provide information in the space needed. For a quantitative display, it is usually preferable to use an instrument with a moving pointer and a fixed scale. The scale may be circular, curved (in an arc), horizontal-straight, or vertical-straight. The other type of quantitative indicator, with a fixed pointer and a moving scale, should be used only in special cases.

Graduation and numbering is often less than optimal, hindering correct readings. Figure 13-38 provides information about correct scale graduation, numbering, and pointer positioning.

Numerals should be located outside the scale markings so that they cannot be obscured by the pointer. The pointer should ride against the other side of the scale, with its tip just short of the markings. The markings should show divisions only as fine as the operator must read. Recommended dimensions for the scale marks are presented in Figure 13-39. These dimensions are suitable even for low illumination.

Figures 13-40 and 13-41 show examples of quantitative displays in which color or form codes indicate operating conditions.

Displays should be oriented within the viewing area of the operator, with their surfaces perpendicular to the line of sight. Figure 13-16 and Table 13-M indicate the visual field and the preferred viewing area.

A group of pointer instruments can be arranged so that all pointers are aligned under normal conditions. If one of the instruments indicates a deviation from the normal condition, the displacement of the pointer from the aligned configuration will be particularly obvious. Figure 13-42 shows examples of arrangements of pointer instruments for rapid check-reading.

In many cases, instrument settings are set by controls that,

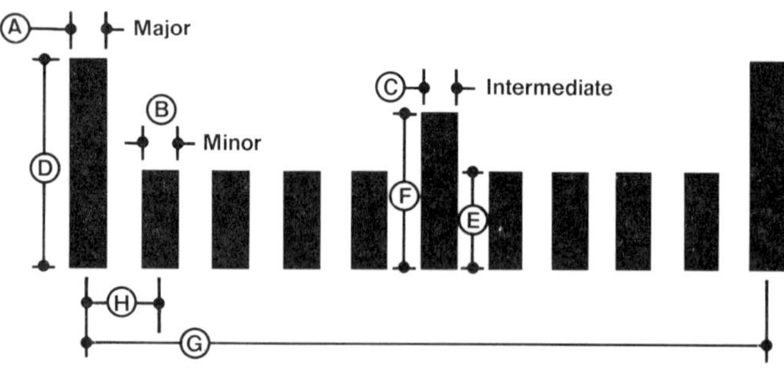

Dimension (in mm)	Viewing Distance (in mm) 710	910	1525
A (Major index width)	0.89	1.14	1.90
B (Minor index width)	0.64	0.81	1.37
C (Intermediate index width)	0.76	0.99	1.63
D (Major index height)	5.59	7.19	12.00
E (Minor index height)	2.54	3.28	5.44
F (Intermediate index height)	4.06	5.23	8.71
G (Major index separation between midpoints)	17.80	22.90	38.00
H (Minor index separation between midpoints)	1.78	2.29	3.81

Minimum scale dimensions suitable even for low illumination (1-3.4 cd/m²)

Figure 13-39. (Reprinted from *Military Standard 759A,* 1981)

for convenience and to avoid error, should be located in a suitable position with respect to the instrument. Figure 13-43 shows several examples of usable control-display arrangements.

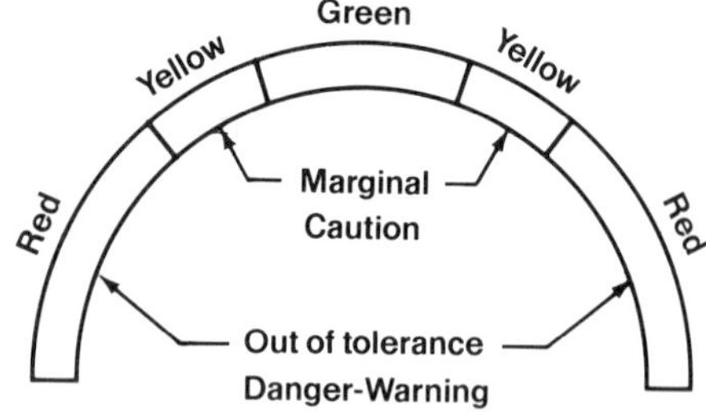

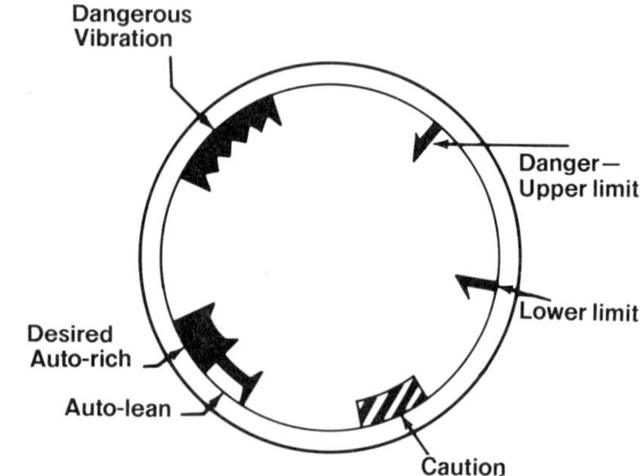

Figure 13-41. Shape and color coding of operating conditions on quantitative displays. (Reprinted from *Military Standard 759A,* 1981)

LABELING

Controls, displays, and any other items of equipment that must be located, identified, read, or manipulated, shall be appropriately and clearly labeled to permit rapid and accurate performance. No label will be required on equipment or controls whose use is obvious to the user.

Label characteristics. Labeling characteristics shall be determined by such factors as:

- the accuracy of identification required;
- the time available for recognition or other responses;

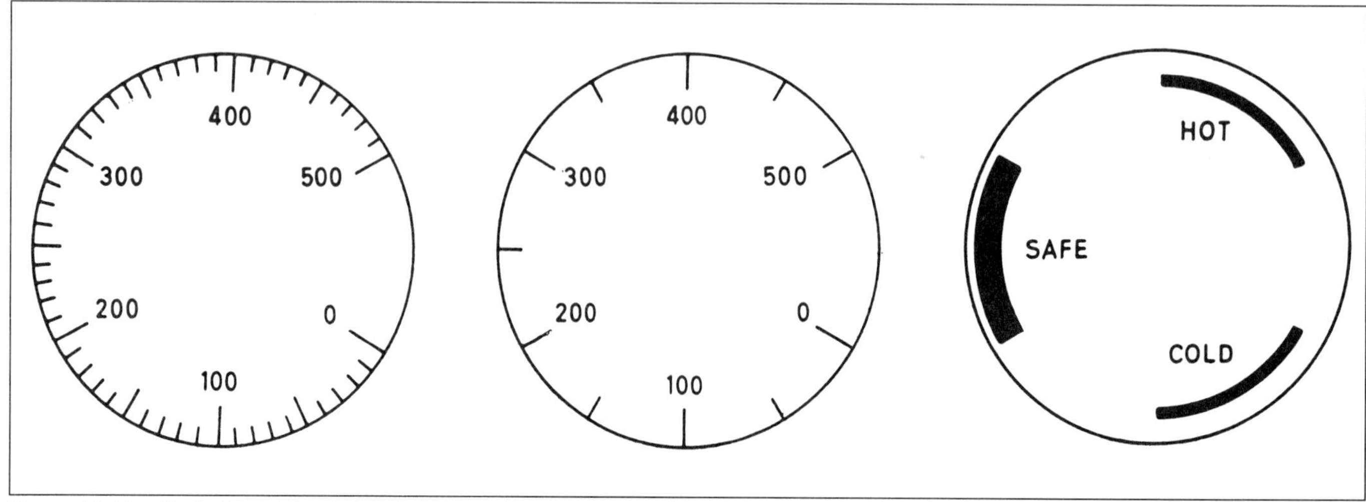

Figure 13-40. Replacing scale markings with code markings.

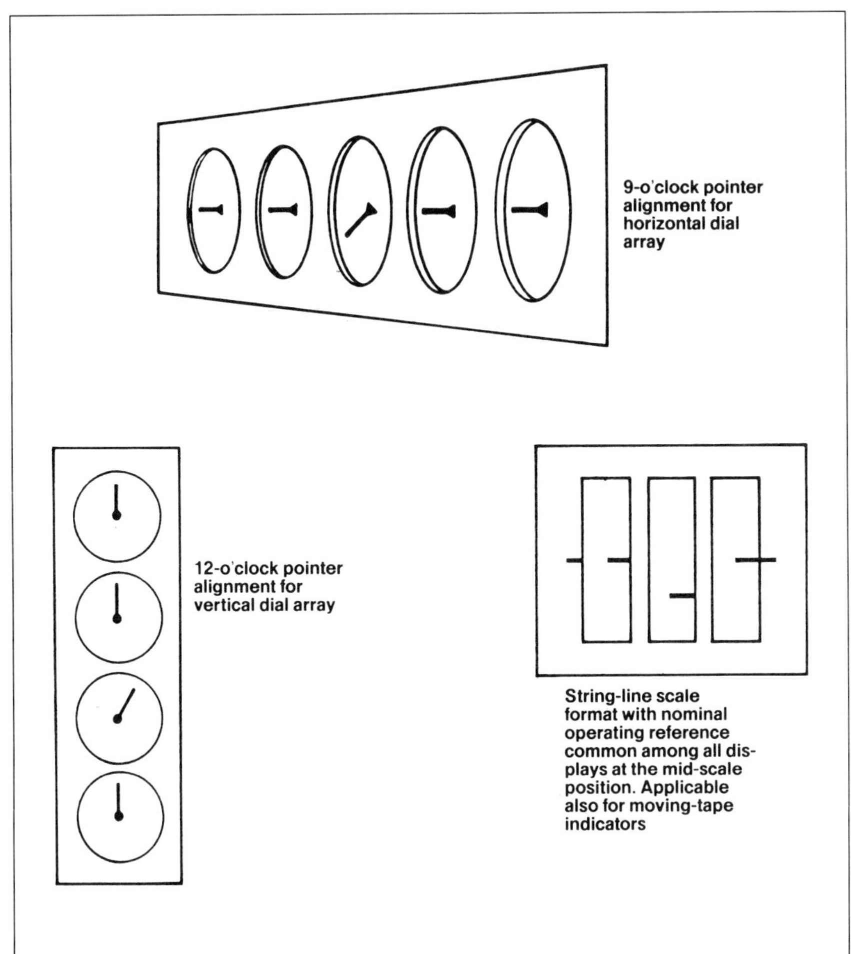

Figure 13-42. Aligned pointers for rapid check-reading. (Reprinted from *Military Standard 759A,* 1981)

- the distance at which the labels must be read;
- the illumination level and color characteristics of the illuminant;
- the critical nature of the function labeled;
- the consistency of label design within and between systems.

Orientation. Labels and the information printed thereon should be oriented horizontally so that the labels may be read quickly and easily from left to right.

Location. Labels shall be placed on or very near the items they identify so as to eliminate confusion with other items and labels.

Standardization. Placement of labels shall be consistent throughout the equipment and system.

Equipment functions. Labels should primarily describe the functions of equipment items. Secondarily, the engineering characteristics or nomenclature may be described.

Abbreviations. Standard abbreviations shall be selected. If a new abbreviation is required, its meaning shall be obvious to the intended reader. Capital letters shall be used. Periods shall be omitted except when needed to preclude misinterpretation. The same abbreviation shall be used for all tenses and for both singular and plural forms of a word.

Brevity. Labels shall be as concise as possible without distorting the intended meaning or information and shall be unambiguous. Redundancy shall be minimized. If the general function is obvious, only the specific function shall be identified (e.g., "frequency" as opposed to "frequency factor").

Familiarity. Words shall be chosen on the basis of operator familiarity whenever possible, provided the words express exactly what is intended. Brevity shall not be stressed if the results will be unfamiliar to operating personnel. Common, meaningful symbols (e.g., % and +) may be used as necessary.

Visibility and legibility. Labels and placards shall be designed to be read easily and acurately at the anticipated operational reading distances, within a vibration/motion environment, and at the particular illumination levels. The following factors must be taken into consideration: contrast between the lettering and its immediate background; the height, width, stroke width, spacing, and style of letters; and the specular reflection of the background, cover, or other components.

SUMMARY

The industrial hygienist is largely responsible for the health and well-being of the industrial employee. Of course, this concern must be combined with the company's necessity to produce, both in quantity and in quality. Fortunately, ergonomic/human factors recommendations usually bring about, directly or

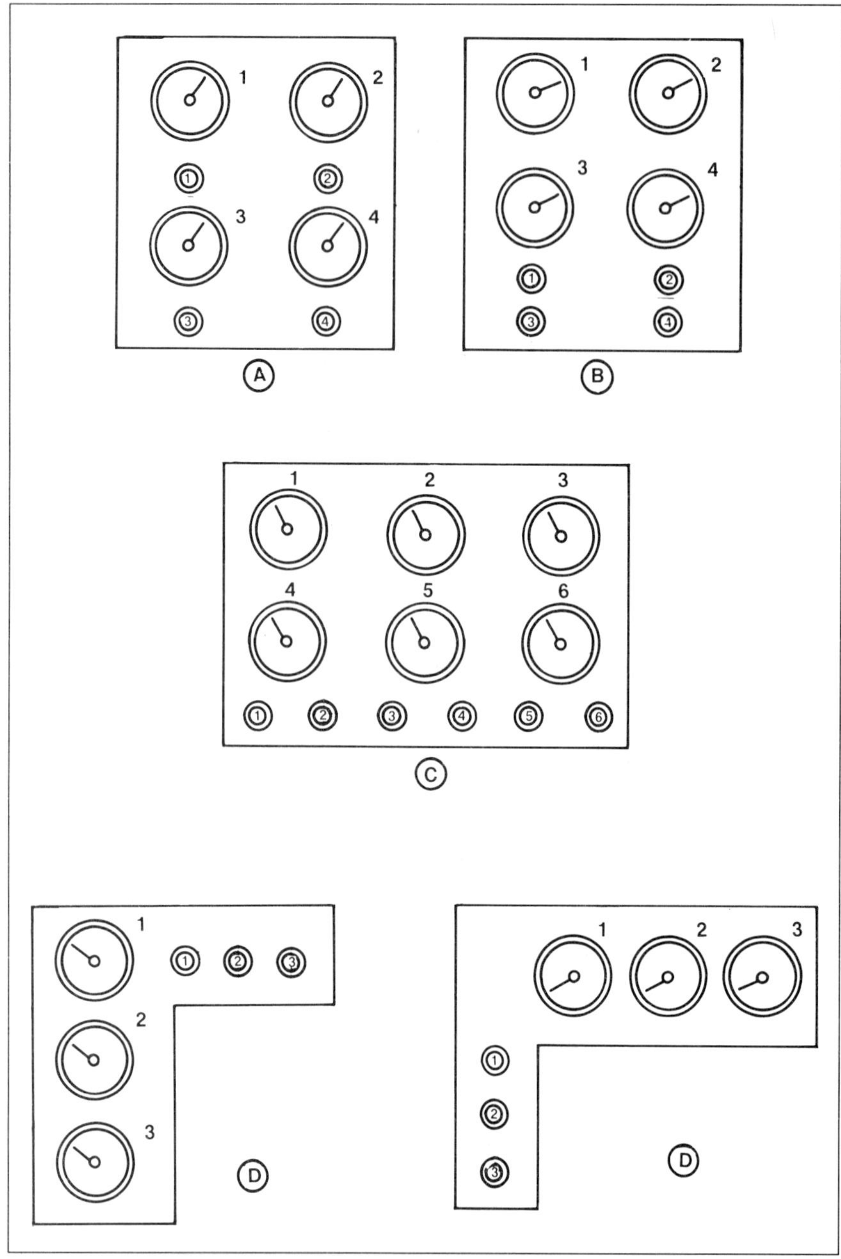

Figure 13-43. Control-display relationships. (Reprinted with permission from *Military Standard 759A,* 1981)

indirectly, improved job performance together with increased safety, health, and well-being of the worker. In recent years, both management and employee representatives, including unions, have cooperated in using ergonomics to increase the ease and efficiency with which work is carried out.

BIBLIOGRAPHY

American Industrial Hygiene Assoc., 475 Wolf Ledges Pkwy., Akron, OH 44311.

—*Ergonomics Guides,* including Kroemer, K.H.E. *Ergonomics of VDT Workplaces,* 1984.

American National Standard for Human Factors Engineering of Visual Display Terminal Workstation. Revised Review Draft, July 1986. Santa Monica, Calif.: The Human Factors Society, 1986. ANSI Draft Standard HFS 100.

Andersson, C.K., and Chaffin, D.B. A biomechanical evaluation of five lifting techniques. *Applied Ergonomics.* 17 (1986):2-8.

Astrand, P.O., and Rodahl, K. *Textbook of Work Physiology,* 3rd ed. New York: McGraw-Hill, 1986.

Ayoub, M.M., Mital, A., Asfour, S. S., et al. Review, evaluation, and comparison of models for predicting lifting capacity. *Human Factors* 22 (1980):257-269.

Ayoub, M.M., Selan, J.L., and Jiang, B.C. *A Mini-Guide for Manual Materials Handling.* Lubbock, TX: Institute of Ergonomics Research, Texas Tech University, 1984.

Bhatnager, V., Drury, C.G., and Schiro, S.G. Posture, postural discomfort, and performance. *Human Factors* 27 (1985): 189-199.

Boff, K.R., Kaufman, L., and Thomas, J.P., eds. *Handbook of Perception and Human Performance* (2 vol). New York: John Wiley & Sons, 1986.

Borg, G.A.V. Psychophysical bases of perceived exertion. *Medicine and Science in Sports and Exercise* 14 (1982):377-381.

Caldwell, L.S., Chaffin, D. B., Dukes-Dobos, F. N., et al. A proposed standard procedure for static muscle strength testing. *American Industrial Hygiene Association Journal* 35 (1974):201-206.

Chaffin, D.B. Functional assessment for heavy physical labor. *Occupational Health and Safety* 50, 24, 27, 32, 64, (1981).

Chaffin, D.B., and Andersson, G.B.J. *Occupational Biomechanics*. New York: John Wiley & Sons, 1984.

Cochran, D.J., and Riley, M.W. The effects of handle shape and size on exerted forces. *Human Factors* 28 (1986):253-265.

Congleton, J.J., Ayoub, M.M., and Smith, J.L. The design and evaluation of the neutral posture chair for surgeons. *Human Factors* 27 (1985):589-600.

Drury, C.G. Handles for manual material handling. *Applied Ergonomics* 11 (1980):35-42.

Drury, C.G., and Francher, M. Evaluation of a forward sloping chair. *Applied Ergonomics* 16 (1985):41-47.

Eastman Kodak Company. *Ergonomic Design for People at Work,* vol. 1. Belmont, CA: Lifetime Learning Publications, 1983.

Eastman Kodak Company. *Ergonomic Design for People at Work,* vol. 2, New York: Van Nostrand Reinhold, 1986.

Eklund, J.A.E., Corlett, E.N., and Johnson, F. A method for measuring the load imposed on the back of a sitting person. *Ergonomics* 26 (1983):1063-1076.

Garg, A., and Saxena, U. Physiological stresses in warehouse operations with special reference to lifting technique and gender: A case study. *American Industrial Hygiene Association Journal* 46 (1985):53-59.

Grandjean, E. *Ergonomics in Computerized Offices.* Philadelphia: Taylor and Francis, 1986.

Grandjean, E. *Fitting the Task to the Man.* London: Taylor and Francis, 1980.

Grandjean, E. *Physiologische Arbeitsgestaltung.* Munich: Otto, 1963.

Grandjean, E., Huenting, W., and Nishiyama, K. Preferred VDT workstation settings, body posture and physical impairment. *Applied Ergonomics* 15(1984):99-104.

Greenberg, L., and Chaffin, D.B. *Workers and their Tools.* Midland, MI: Pendall, 1979.

Grieco, A. Sitting posture: An old problem and a new one. *Ergonomics* 29 (1986):345-362.

Hansson, T. "Prolonged Sitting and the Back." In *Proceedings of the Conference "Work with Display Units,"* Stockholm, May 12-15, 1986. Stockholm: Swedish National Board of Occupational Safety and Health, 1986, pp. 491-492.

Hill, S.G., and Kroemer, K.H.E. Preferred declination of the line of sight. *Human Factors* 28 (1986):127-134.

Kamon, E., Kiser, D., and Pytel, J. Dynamic and static lifting capacity and muscular strength of steelmill workers. *American Industrial Hygiene Association Journal* 43 (1982): 853-857.

Kilbom, A. "Physiological Effects of Extreme Physical Inactivity." In *Proceedings of the Conference "Work with Display Units"* Stockholm, May 12-15, 1986. Stockholm: Swedish National Board of Occupational Safety and Health, 1986, pp. 486-489.

Koffler Office Systems Ergonomics Reports. Several reports by the Koffler Group. Santa Monica, CA: Author, 1985, 1986.

Konz, S. Bent hammer handles. *Human Factors* 28 (1986): 317-323.

Konz, S. *Work Design: Industrial Ergonomics,* 2nd ed. Columbus, OH: Grid, 1983.

Kroemer, K.H.E. An isoinertial technique to assess individual lifting capability. *Human Factors* 25 (1983):493-506.

Kroemer, K.H.E. Coupling the hand with the handle. *Human Factors* 28 (1986):337-339.

Kroemer, K.H.E. *Development of LIFTEST, a Dynamic Technique to Assess the Individual Capability to Lift Material.* Blacksburg, Va.: Ergonomics Laboratory, IEOR, Virginia Tech, 1982. Final Report, NIOSH Contract 210-79-0041.

Kroemer, K.H.E. Engineering Anthropometry. In Salvendy, G., ed. *Handbook of Human Factors/Ergonomics.* New York: John Wiley & Sons, 1987, pp. 154-168.

Kroemer, K.H.E. Engineering Anthropometry: Designing the Work Place to Fit the Human. In *Proceedings of the Annual Conference,* American Institute of Industrial Engineers, Detroit, May 17-20, 1981. Norcross, GA: AIIE, 1981, pp. 119-126.

Kroemer, K.H.E. *Ergonomics Manual for Manual Material Handling,* 2nd rev. ed. Blacksburg, VA: Ergonomics Laboratory, IEOR, Virginia Tech, 1984.

Kroemer, K.H.E. Office Ergonomics: Work Station Dimensions. In Alexander, D.C., and Pulat, B.M., eds. *Industrial Ergonomics.* Norcross, GA: Institute of Industrial Engineers, 1985, pp. 187-201.

Kroemer, K. H. E. Seating in plant and office. *American Industrial Hygiene Association Journal* 32(1971):633-652.

Kroemer, K.H.E. Testing individual capability to lift material: Repeatability of a dynamic test compared with static testing. *Journal of Safety Research* 16 (1985):1-7.

Kroemer, K.H.E. VDT Workstation Design. In Helander, M.G., ed. *Handbook of Human Computer Interaction.* New York: Elsevier Publ., 1987.

Kroemer, K.H.E., Kroemer, H.J., and Kroemer-Elbert, K.E. *Engineering Physiology: Physiologic Bases of Ergonomics.* New York: Elsevier Publ., 1986.

Life, M.A., and Pheasant, S.T. An integrated approach to the study of posture in keyboard operation. *Applied Ergonomics* 15 (1984):83-90.

Lueder, R.K. Seat comfort: A review of the construct in the office environment. *Human Factors* 26 (1983):339-345.

Lundervold, A. Electromyographic investigations of position and manner of working in typewriting. *Acta Physiologica Scandinavic* 24 (1951):84 (suppl.).

Mandal, A.C. The correct height of school furniture. *Human Factors* 24 (1982):257-269.

Marras, W.S., King, A.I., and Joynt, R.L. Measurement of loads on the lumbar spine under isometric and isokinetic conditions. *Spine* 9(2) (1984):176-188.

McDaniel, J.W., Skandis, R.J., and Madole, S.W. *Weight Lifting Capabilities of Air Force Basic Trainees.* Wright-Patterson AFB, Ohio: USAF Aerospace Medical Research Laboratory, 1983. AFAMRL-TR-83-0001.

Military Standards (such as 759 and 1472). Available from the Commander, U.S. Army Missile Command, DRSMI-RSD, Redstone Arsenal, AL 35898.

Mital, A., and Sanghavai, N. Comparison of maximum volitional torque exertion capabilities of males and females using common hand tools. *Human Factors* 28 (1986):283-294.

NASA-Webb. *Anthropometric Sourcebook* (3 vol). Houston: NASA, 1978. NASA Ref. Pub. No. 1024.

National Institute for Occupational Safety and Health. *Work Practices Guide for Manual Lifting.* Washington, DC: U.S. Government Printing Office, 1981. DHHS Pub. No. 81-122.

National Research Council. *Video Displays, Work, and Vision.* Washington, DC: National Academy Press, 1983.

National Research Council, Committee on Human Factors. *Research Needs for Human Factors.* Washington, DC: National Academy Press, 1983, pp. 2-3.

Pope, M.H., Frymoyer, J.W., and Andersson, G. *Occupational Low Back Pain.* New York: Praeger, 1984.

Ridder, C.A. *Basic Design Measurements for Sitting.* Fayetteville, AR: University of Arkansas, 1959. Bulletin 616, Agricultural Experiment Station.

Salvendy, G., ed. *Handbook of Human Factors.* New York: John Wiley & Sons, 1987.

Sanders, M. S., and McCormick, E.J. *Human Factors in Engineering and Design,* 6th ed. New York: McGraw-Hill, 1987.

Snook, S.H. The design of manual handling tasks. *Ergonomics* 21 (1978):963-985.

Soderberg, G.L., Blanco, M. K., Cosentino, K. A., et al. An EMG analysis of posterior trunk musculature during flat and anteriorly inclined sitting. *Human Factors* 28 (1986):483-491.

Thompson, F.J., Yates, B.J., and Frazen, O.G. "Blood Pooling in Leg Skeletal Muscles Prevented by a 'New' Venopressor Reflex Mechanism." In *Proceedings of the Conference "Work with Display Units,"* Stockholm, May 12-15, 1986, Swedish National Board of Occupational Safety and Health, 1986, pp. 493-496.

Van Cott, H.P., and Kinkade, R.G., eds. *Human Engineering Guide to Equipment Design,* rev. ed. Washington, DC: U.S. Government Printing Office, 1972.

Wilson, J., Corlett, N., and Manenica, I., eds. *Ergonomics of Working Postures.* Philadelphia: Taylor and Francis, 1986.

Winkel, J. "Macro- and Micro-Circulatory Changes during Prolonged Sedentary Work and the Need for Lower Limit Values for Leg Activity." In *Proceedings of the Conference "Work with Display Units,"* Stockholm, May 12-15, 1986. Stockholm: Swedish National Board of Occupational Safety and Health, 1986, pp. 497-500.

Winkel, J., and Bendix, T. Muscular performance during seated work evaluated by two different EMG Methods. *European Journal of Applied Physiology* 55 (1986):167-173.

Woodson, W.E. *Human Factors Design Handbook.* New York: McGraw-Hill, 1981.

14

Biological Hazards

by Alvin L. Miller PhD, CIH
Cynthia S. Volk

Biohazard is a combination of the words biological hazard and refers to plants, animals, or their products that may present a potential risk to the health and well-being of humans or animals. Biohazards can affect humans either directly through illness or indirectly through disruption of the environment. Infectious biological agents constitute five types of infections: bacterial, viral, rickettsial, and, to a lesser degree, fungal and parasitic infections. Biological hazards, including identification and classification schemes according to risk and occupational exposures, are discussed in this chapter.

Biohazards can be unique to a particular occupational group, or may threaten the general public, as the common cold virus does. While exposure to biohazards may seem obvious in an occupation such as nursing or medical research, many other occupations such as laboratory work, farming, and the handling of animal products (slaughterhouses and meat packing operations) also may pose a threat to employees due to possible exposure to infectious agents.

Biohazards can be transmitted to a person through inhalation, injection, ingestion, or contact with the skin. The combination of the number of organisms in the environment, the virulence of these organisms and the resistance of the individual ultimately determines whether or not the person will actually contract the disease. The effects of a biological agent are further compounded by the presence of concomitant physical and/or chemical stressors in the environment. For example, the incidence and severity of respiratory infections can be enhanced by the presence of irritant gases in the air. After exposure to nitrogen dioxide, animals have been found to show a greater susceptibility to pneumonia (Erlich and Henry, 1968). Thus, it is important to consider not only the biological agents, which pose a hazard to the occupational worker, but also to realize that exposure to other environmental stressors may result in an additive or synergistic effect.

Until recently, little information pertaining to biohazards was found in the literature on industrial hygiene. Published work on biological agents was, for the most part, related to research and/or laboratory uses of these substances. With the development and commercialization of biotechnology or genetically engineered organisms and investigations into office-related illness or "sick building syndrome," much more information has been generated as to real or potential risk. This is especially pertinent in the case of viral oncology (tumor) research in which the risk of cancer induction in humans from exposure to viruses is difficult to estimate, particularly as the consequences may not be shown for many years. Preventive measures cannot be delayed until appropriate scientific knowledge is available to accurately assess risk.

Many potentially oncogenic (tumor-producing), infectious, or toxic biological agents are encountered in the occupational environment, and the health and safety professional should apply control measures to minimize the hazard. Employers must maintain an awareness in protecting workers from potentially dangerous biohazards.

Before such work is undertaken, management should determine the potential hazards involved and the precautions to be undertaken. A written plan should be developed. Program and support staff should be fully informed and adequately trained. Accidentally created hazards should be anticipated with an emergency plan, and written instructions should be developed for handling them. If work requires contact with a known pathogen for which an effective vaccine is available, employees should be immunized, as appropriate. An effective placement and medical surveillance program is an essential element in any biohazard control program.

Each employee also has a clear responsibility for biohazards control, as adherence to established, safe procedures will increase the level of an employee's own protection as well as that of co-workers. Employees should report any condition or action that may create a biohazard or potential exposure to a biohazard.

BIOHAZARD CONTROL COMMITTEE

Difficult health and safety decisions arise from pursuits into areas of possible risk where information about actual risk is either scarce or nonexistent. The formation of a committee on biohazard control has been suggested to aid professionals in attempting to set up guidelines for regulating these questionable biological agents. Specifically, at this time the issue of recombinant deoxyribonucleic acid (DNA) research indicates the need for concern by safety and occupational health professionals to reduce hazards arising out of this activity.

Experimentation involving the transfer of genetic information called DNA from one species to another is an example of recombinant DNA research. In addition to providing an important new means for the study of genetics, it also provides for the possible development of plant species, which are more productive and more resistant to disease, and for production of important new chemicals such as insulin, growth hormones, and hepatitis B virus vaccine by microorganisms.

Genetic engineering includes direct techniques for modifying genetic material of living organisms, such as recombinant DNA and ribonucleic acid (RNA) modification and novel techniques, such as cell fusion and gene splicing.

The possible need for control of such experiments was pointed out initially by the scientists who were conducting them. As a result, an elaborate set of guidelines was prepared, which includes precise instructions as to the organisms that can be used and the type of physical containment necessary to assure health and safety (HEW, NIH, *Fed. Reg* 42). These guidelines were adapted by NIH (National Institutes of Health) for all projects funded by this agency.

The NIH "Recombinant DNA Research Guidelines" require that a biohazards committee performs the following functions:

- serve as a technical resource to investigators;
- develop a biohazards safety manual;
- certify that applications for research projects include provisions for the various facilities and safety procedures necessary for the safe conduct of that research.

A biohazard control committee should be composed of persons skilled in safety, industrial hygiene, microbiology, infection control, epidemiology, or other scientific fields pertinent to biohazard control. It should be the responsibility of this committee to establish, maintain, and update the policies and regulations for the control of biohazards. Specifically, the committee should perform the following functions:

- provide technical advice on biohazards to the various departments;
- coordinate and review activities of the biological research programs;
- periodically inform management and the supervisory staff on the status of biohazards control;
- review the inventory of biological materials at the facility, devoting particular attention to the evaluation of new biohazardous materials;
- identify areas where new biological hazards may exist;
- recommend policies and procedures or other technical information for biological risk assessment and reduction of biohazards.

In April 1984 the Cabinet Council on Natural Resources and the Environment (later reorganized to the Cabinet Council on Domestic Policy) formed a Working Group on Biotechnology made up of representatives from federal departments, commissions, and agencies and chaired by the President's Office of Science and Technology Policy. A "Proposal for a Coordinated Framework for Regulation of Biotechnology; Notice" was published for comment in the *Federal Register* on December 31, 1984, pages 50,856–50,907. This notice contained a matrix outlining laws, regulations, and guidelines applicable to biotechnology, policy statements of the agencies involved, a proposed scientific advisory mechanism, and a glossary of terms. The Occupational Safety and Health Administration (OSHA) published its proposed guidelines in the *Federal Register* on April 12, 1985, pages 14,468–14,469.

Following review of comments, a "Coordinated Framework for Regulation of Biotechnology; Announcement of Policy and Notice for Public Comment" was published in the *Federal Register* on June 26, 1986, pages 23,302–23,393. This policy keeps the NIH Recombinant DNA Advisory Committee intact to review applications for biotechnology projects, and creates a Biotechnology Science Coordinating Committee to assess consistency of scientific policy and scientific reviews of applications.

The reader is referred to these publications for a comprehensive review of the current status of knowledge regarding biotechnology.

Specific instructions for management of microbiological laboratories can be found in the CDC publication, *Biosafety in Microbiological and Biomedical Laboratories,* USDHHS Pub. No. 84-8395.

CLASSIFICATION OF BIOHAZARDOUS AGENTS

The identification and classification of biohazards are important tools that health and safety professionals, supervisors, and employees use to decide on the appropriate safeguards to be used to prevent infection. Two points to remember are: (1) any accident involving biohazardous materials can result in infection, and (2) when working with biological agents or materials of which epidemiology and etiology are unknown or not completely understood, it must be assumed that the material presents a biohazard.

In general, the biohazards of established infectious disease agents are known or can be estimated. In cancer research, including work in viral oncology, there is the principal risk of accidental contamination or cross-contamination with a resultant uncertain risk of health impairment to the investigator.

The United States Public Health Service (USPHS) and the United States Department of Agriculture (USDA) have participated in the formulation of a Standard Classification for Evaluating the Hazards associated with a number of Biohazardous Agents (HEW, PHS, CDC, 1974). This Standard was intended to provide a method for defining minimal safety conditions. The material in this section is adapted from this Standard.

The Standard (HEW, PHS, CDC, 1974) defines four classes of biohazardous agents with each higher number representing increased hazard. A fifth class is composed of animal pathogens that are excluded from the United States by law. The Standard provides only for the minimum health and safety conditions considered necessary. The nature of the work and the physical state and quantity of agent in use can often increase the biohazard to an extent that it should be considered even more hazardous. The five classes of agents are as follows:

Class 1—Agents of no or minimal hazard (under ordinary conditions of handling) that can be handled safely without special apparatus or equipment, using techniques generally acceptable for nonpathogenic materials—Class 1 includes all bacterial, fungal, viral, rickettsial, chlamydial, and parasitic agents not included in higher classes.

Class 2—Agents of ordinary potential hazard—this class includes agents that may produce disease of varying degrees of severity through accidental inoculation, injection, or other means of cutaneous penetration, but which can usually be adequately and safely contained by ordinary laboratory techniques.

Distribution of Class 2 biohazardous material should be limited to facilities with employees whose levels of competency are equal to or greater than those expected in a college department of microbiology.

Class 3—Agents involving special hazard, or agents derived from outside the United States that require a USDA permit for importation unless they are specified for higher classification—this class includes pathogens that require special conditions for containment.

Distribution of Class 3 biohazardous material should be only to facilities with employees whose levels of competency are equal to or greater than those expected in a college department of microbiology, and who also have had special training in handling dangerous agents and are supervised by competent scientists.

Conditions for containment include:

1. There must be a controlled access facility: suite or room separated from the activities of individuals not engaged in handling Class 3 agents and apart from the general traffic pattern of the rest of the building or laboratory.
2. Negative air pressure must be maintained at the site of work in an approved biological safety cabinet. Air can be recirculated only after it has been adequately decontaminated through high-efficiency filters.
3. Animal experiments, including cage sterilization, refuse handling, and disposal of animals are conducted with a level of precaution equivalent to conditions required for laboratory experiments.
4. Personnel at risk are immunized against agents for which immune prophylaxis is available.

Class 4—Agents that require the most stringent conditions for containment because they are extremely hazardous to personnel or can cause serious epidemic disease—this class includes Class 3 agents from outside the United States when they are used in entomological experiments and/or when other entomological experiments are conducted in the same laboratory area.

Distribution of Class 4 biohazardous material is restricted to facilities with employees whose levels of competency are equal to or greater than one would expect in a college department of microbiology, who have had special training in handling dangerous pathogens, and who are supervised by competent scientists.

Conditions for containment include all those required for Class 3 agents and the following:

1. Work areas should be in a facility that is in effect a separate building, or that are separated from other work areas by effective airlocks.
2. If the work area is not in a separate building, the entire area used for Class 4 agents should have a separate exhaust system and be under negative air pressure relative to other areas of the building. Exhaust air is decontaminated by filtration through high-efficiency filters or by some other suitable process. Class 4 agents must be manipulated only in safety cabinets equipped with absolute filters.
3. Access to work areas should be restricted to immunized individuals or to those under specific medical surveillance.
4. Protective clothing must be worn and decontaminated.
5. When an agent is used in entomological experiments, the windows, walls, floor, ceiling, and airlock of the work area must be insect-proof, and pyrethrum insecticide, or a suitable insect-killing device must be available in the airlock.

Class 5—Foreign animal pathogens that are excluded from the United States by law or whose entry is restricted by the USDA administrative policy—see Tables 14-A and 14-B for summaries of the recommended biosafety levels for work with infectious agents and infected vertebrate animals.

Bacterial agents

Bacteria are simple, one-celled organisms that are visible only under the microscope and multiply by simple division or fission into two parts. Types of bacteria include the cocci, which are round and (when magnified) look like a string of beads; the bacilli, which are rod-shaped; and the spirilla, which resemble tiny corkscrews. Some bacteria are pathogenic (disease-causing), others are harmless, and some are even useful (as in the process of fermentation).

Bacterial infections of an occupational nature can be caused by neglected minor wounds and abrasions in which the integrity of the skin surface is broken. These infections are frequently caused by mixed bacterial infections, but chief among the offending organisms are staphylococci and streptococci. Food poisoning rarely is an occupational disease, but in many cases, the worker acts as the contaminating agent of otherwise pure food. The three primary types of contamination are the bacteria of the salmonella group, *Clostridium perfringens,* and *Staphylococcus aureus.* In the case of *C. perfringens* and *S. aureus,* the disease is caused by a toxin, not actual infection. Individuals subjected to mass feeding techniques such as in the military, in prisons, and (in certain instances) in the workplace are at highest risk. Bacterial agents are classified as follows:

Class 1—All bacterial agents not included in higher classes;

Class 2—Examples: *Clostridium botulinum; Escherichia coli*—all enteropathogenic serotypes; salmonellae—all species and serotypes;

Class 3—Examples: brucellae—all species.

Table 14-A. Summary of Recommended Biosafety Levels for Infectious Agents

Biosafety Level	Practices and Techniques	Safety Equipment	Facilities
1	Standard microbiological practices	None: primary containment provided by adherence to standard laboratory practices during open bench operations.	Basic
2	Level 1 practices plus: Laboratory costs; decontamination of all infectious wastes; limited access; protective gloves and biohazard warning signs as indicated.	Partial containment equipment (i.e., Class I or II Biological Safety Cabinets) used to conduct mechanical and manipulative procedures that have high aerosol potential that may increase the risk of exposure to personnel.	Basic
3	Level 2 practices plus: Special laboratory clothing; controlled access.	Partial containment equipment used for all manipulations of infectious material.	Containment
4	Level 3 practices plus: Entrance through change room where street clothing is removed and laboratory clothing is put on; shower on exit; all wastes are decontaminated on exit from the facility.	Maximum containment equipment (i.e., Class III biological safety cabinet or partial containment equipment in combination with full-body, air-supplied, positive-pressure personnel suit) used for all procedures and activities.	Maximum containment

Reprinted from *Biosafety in Microbiological and Biomedical Laboratories,* USDHHS Pub. No. 84-8395.

Table 14-B. Summary of Recommended Biosafety Levels for Activities in Which Experimentally or Naturally Infected Vertebrate Animals Are Used

Biosafety Level	Practices and Techniques	Safety Equipment	Facilities
1	Standard animal care and management practices.	None	Basic
2	Laboratory costs; decontamination of all infectious wastes and of animal cages prior to washing; limited access; protective gloves and hazard warning signs as indicated.	Partial containment equipment and/or personal protective devices used for activities and manipulations of agents or infected animals that produce aerosols.	Basic
3	Level 2 practices plus: Special laboratory clothing; controlled access.	Partial containment equipment and/or personal protective devices used for all activities and manipulations of agents or infected animals.	Containment
4	Level 3 practices plus: Entrance through clothes change room where street clothing is removed and laboratory clothing is put on; shower on exit; all wastes are decontaminated before removal from the facility.	Maximum containment equipment (i.e., Class III biological safety cabinet or partial containment equipment in combination with full-body, air-supplied positive-pressure personnel suit) used for all procedures and activities.	Maximum containment

(Reprinted from *Biosafety in Microbiological and Biomedical Laboratories,* USDHHS Pub. No. 84-8395.)

Rickettsial and chlamydial agents

Rickettsiae are a group of coccoid or rod-shaped microorganisms, bacterial in nature but smaller in size. They are obligate parasites (meaning they depend upon their host to provide the factors they need for growth and reproduction), and cannot survive without this association. Rickettsiae have an intracellular existence, which means they can survive only within living cells and are unable to function in any other environment. These microbes are associated with and transmitted to man through bloodsucking arthropods, such as fleas, ticks, and lice. They are occasionally airborne. Rickettsiae are responsible for such diseases as typhus and Rocky Mountain spotted fever.

Chlamydiae are also obligate parasites, bacterial in nature and even smaller than rickettsiae. They are also intracellular microorganisms and are distinguished from rickettsiae by their smaller size and increasingly complex method of reproduction. Chlamydiae are usually transmitted through the air, and invade the body through the human respiratory system. Chlamydiae occur as two species; both are pathogenic to man. The primary source of human infection is from birds.

Class 1—All rickettsial and chlamydial agents not included in higher classes.

Class 2—Examples: *Lymphogranuloma venereum* agent, transmitted by sexual contact.

Class 3—Examples: psittacosis—ornithosis—trachoma group of agents; rickettsiae—all species (except *Vole rickettsiae* when used for transmission or animal inoculation experiments).

Viral agents

Viruses are a group of noncellular, parasitic pathogens which are much smaller than bacteria, rickettsiae, or chlamydiae. They are, in fact, the smallest organism known, and because of their submicroscopic size can only be seen by using an electron microscope. They are believed to be living organisms or chemical entities bordering between the living and nonliving. Viruses are obligate parasites in that they must be associated with a cell to function; they are incapable of growth or reproduction apart from living cells.

Viral diseases likely to be encountered occupationally include animal respiratory viruses, poxviruses, enteroviruses, and arboviruses. Infections can be acquired from the vector or from the handling of animals or animal products in agriculture. Laboratory-acquired infections can result from working with the agent, from accidents, from animals, from clinical or autopsy specimens, from aerosols, or from glassware. Viral transmission can occur among patients and staff of hospitals. Viral agents are classified as follows:

Class 1—All viral agents not included in higher classes.

Class 2—Examples: Adenoviruses—human—all types; influenza viruses—all types except A/PR8/34 (Class 1); measles virus; mumps virus; polioviruses—all strains (except rabies street virus which is classified as Class 3 when inoculated in carnivores); rubella virus; yellow fever virus, 17 vaccine strain.

Class 3—Examples: rabies street virus (see above); yellow fever virus—wild, when used in vitro.

Class 4—Examples: tickborne encephalitis virus complex; yellow fever virus—wild, when used for transmission or animal inoculation experiments.

Fungal agents

Fungi are now classified in the kingdom, Protista, along with algae. There are as many as 100,000 species, which differ from algae in being devoid of chlorophyll or other pigment capable of photosynthesis. They are not able to synthesize protein or other organic material from simple compounds, and are therefore parasitic (grow in or on the living body of a host) or saprophytic (grow on dead plant or animal matter). The incidence of fungal disease of an occupational nature is low and mainly confined to farmers, outdoor workers, and animal raisers. Diagnosis of fungal diseases is made by microscopic identification of the fungus with cultural confirmation. Fungal diseases can be roughly classified by systemic, subcutaneous, superficial, or hypersensitivity effects. Occasionally, the subcutaneous fungal diseases spread systemically. Hypersensitivity reactions are due to fungal antigens inhaled with dusts during agricultural or other activities. These usually involve pneumonitis (inflammation of the lungs) with asthmalike symptoms.

Fungi vary in size from the familiar mushroom to the microscopic form, which causes conditions like ringworm or athlete's foot. Fungal agents are classified as follows:

Class 1—All fungal agents not included in higher classes.

Class 2—Examples: actinomycetes; *Blastomyces dermatitidis.*

Class 3—Examples: *Histoplasma capsulatum.*

Parasitic agents

Although microbes such as bacteria and viruses can be parasitic, when we speak of parasitic agents we usually do not mean microbes, but refer to a plant or animal organism parasite. A parasite lives in or on another organism from whom it derives some advantage but to whose welfare it contributes nothing. A tapeworm living in the intestine is an example of a parasitic agent.

Parasitic infections of occupational significance are caused by protozoa, helminths, and arthropods. Diseases caused by protozoa include malaria, amebiasis, leishmaniasis, trypanosomiasis, and a variety of less common blood and gastrointestinal infections. There have been reports of serious eye infections caused by the *Acanthamoeba* species (see Chapter 5, The Eyes). Helminthic diseases include schistosomiasis, creeping eruption, and hookworm. Arthropods, such as mites and chiggers, can cause dermatoses and act as vectors or hosts for other nonarthropod parasites. Certain occupational groups are at great risk of contracting a parasitic disease because of their exposure to vectors carrying a parasitic disease, their direct contact with the infective form of a parasite, and indirectly their presence in areas where conditions are crowded or sanitation and hygiene are inadequate. A parasitic disease is any disease resulting from the invasion of the body by parasitic agents. The outcome of the host-parasite relationship depends upon both the virulence of the parasite and the resistance of the host. Parasitic agents fall into three classes.

Class 1—All parasitic agents not included in higher classes.

Class 2—Examples: *Toxoplasma gondii; Trichinella spiralis.*

Class 3—Examples: *Schistosoma mansoni.*

OCCUPATIONAL EXPOSURE TO BIOHAZARDS

A comprehensive review of the various occupations in which biological agents may be encountered is found in Donald Hunter's *Diseases of Occupations,* 1975, from which the list of infectious diseases (Table 14-C) is adapted. It provides numerous examples of occupations with potential biological hazards.

Laboratory research

The biological research laboratory is probably the most obvious workplace in which employees are subjected to biohazards, simply because the work requires handling and manipulation of biohazardous agents. However, since organisms have applications beyond the scope of the laboratory, such as control of oil spills, use as pesticides, and control of certain chemicals in hazardous waste sites, remarks made here about laboratories apply to all workplaces where such organisms are used. As explained in the National Institutes of Health's *Biohazards Safety Guide,* (NIH), 1974, "When such biological agents are capable of replication, biohazard control is needed to prevent accidents and infections in laboratory workers and to assure the validity of investigative work by preventing undesirable cross-contamination."

The immediate goal of biohazard safety is to prevent illnesses in the worker. This can be extremely difficult because in some cases the results may not be clinically manifested for a long period. While the reported number of acquired infections represents only a fraction of those that have actually occurred (G. B. Phillips, 1971), the number has increased despite advances in containment techniques because of the increased volume of microbiological research and the broadened spectrum of infectious bioagents presently under investigation. It is also essential to guard against the airborne spread of infectious material to other sites and the overall community environment.

There are several possible routes of transmission of these hazardous bioagents within the laboratory (see Table 14-D). In a 1965 study of the biological laboratory at Fort Detrick (Frederick, MD), of 1,218 accidents occurring between 1959 and 1963, less than 20 percent of the recorded infections were caused by recognized and recorded accidents (G. B. Phillips, 1965). Unsafe conditions resulted in approximately 10 percent of the cases,

Table 14-C. Selected Infectious Diseases and Some Occupations Where Agents May Be Encountered

Agent/Disease	*Occupation*
Colds, influenza, scarlet fever, diphtheria, smallpox	May be contracted anywhere.
Tuberculosis	Silica workers, people exposed to heat and organic dusts, and medical personnel.
Typhoid, diphtheria, streptococcal sore throat, bacteriaemia primary chancre of the finger, poliomyelitis, etc.	Physicians and nurses having patient contacts, pathologists working with specimens and cadavers, and laboratory and research workers.
Hepatitis B	Health care workers in hospitals and public institutions who come in contact with persons with infectious diseases, perform renal dialysis and renal transplant, or have contact with blood or excreta of patients having viral hepatitis.
Anthrax	Animal handlers and handlers of carcasses, skins, hides, or hair of infected animals, including wool carpet processors and handlers.
Glanders	Pathologists, horse trainers, veterinary surgeons, or anyone having contact with sick horses or mules.
Tetanus	Handling jute (spores in soil mixed with jute) or anyone having contact with manure.
Ringworm (in horses, cattle, deer, pigs, cats, dogs, birds)	Pet shop salesmen, stockmen, breeders of cats and dogs, and others who handle these animals.
Brucellosis (undulant fever)	Workers with cattle and pigs, beef and pork handlers in slaughterhouses.
Tularemia	Those who handle small rodents, hunters (skinning rabbits).
Q fever	Slaughterhouse workers, farmers, and veterinary surgeons (cows, pigs, sheep).
Swineherd's disease (virus among pigs)	Farmers
Louping-Ill	Laboratory workers, shepherds, farmers, veterinary officers, and slaughterhouse workers (those killing and skinning sheep).
Orf (virus in sheep)	Slaughterhouse workers, farmers, shepherds, sheep shearers, butchers, those who cook sheep.
Psittacosis (in parrots, parakeets, pigeons, ducks, turkeys, chickens, grouse, pheasants, canaries and linnets)	Pet shop personnel, gardeners, housewives, veterinary surgeons, and research workers.
Weil's disease (bacillus in rats)	Coal miners, bargemen, slaughterhouse workers, sewer workers, milkers.
Erysipeloid	Fishermen, fish handlers, sealers, and whalers.
Hookworm	Miners, agricultural laborers, planters of sugar, tobacco, tea, rice and cotton, and brick and tunnel workers.
Rabies (e.g., dogs, bats, rats, pigs, cats)	Veterinarians, letter carriers, laboratory research workers, agricultural workers.

(Reprinted with permission from D. Hunter *The Diseases of Occupations.* London, England: English Universities Press, Ltd., 1975)

Table 14-D. Sources of Viral and Rickettsial Laboratory Infections

Source	*Infection*	
	Viral	*Rickettsial*
Known accidents	59	4
Clinical specimens	11	9
Autopsy	1	3
Handling discarded glassware	7	1
Contact with infected animals or ectoparasites	45	3
Aerosols	34	26
Worked with the agent	64	25
Not indicated	3	10

(Reprinted from *Survey of Viral and Rickettsial Infections Contracted in the Laboratory* National Cancer Institute, National Institutes of Health, Bethesda, MD, p. 33)

leaving between 70 and 75 percent of the infections caused by hazardous work practices; many occurred without realization or recognition due to lack of proper employee training (G. B. Phillips, 1965). Table 14-E provides additional information about laboratory-acquired infections.

Table 14-E. Frequency Rates for Laboratory Infections

Laboratory	*Infection Rates per Million Working-Hours*	*Source*
European Laboratory, 1944-1959	50.0	Personal communication
TB Labs., Canada (except Quebec and Manitoba)	14.0	Merger, 1957
Research Institutes, 1930-1950	4.1	Sulkin and Pike, 1951
Hospital clinical labs., 1953	1.0	Bureau Labor Statistics, 1958
Public Health labs., 1930-1950	0.35	Sulkin and Pike, 1951
Hospital labs., 1930-1950	0.30	Sulkin and Pike, 1951
Biologic manufacturers, 1930-1950	0.25	Sulkin and Pike, 1951
Agricultural and veterinary schools and experimental stations, 1930-1950	0.25	Sulkin and Pike, 1951
Colleges and medical schools, 1930-1950	0.15	Sulkin and Pike, 1951
Clinical labs., 1930-1950	0.10	Sulkin and Pike, 1951

(Reprinted with permission from G. B. Phillips. *Causal Factors in Microbiological Laboratory Accidents* [PhD. dissertation, New York University, 1965])

Much exposure is due to improper work practices, which can be due to the lack of employees' knowledge as to the danger of the substances with which they are dealing. Although the degree of hazard depends mainly upon the etiological agent and the conditions of its use, it is essential that employees are instructed in and understand the hazards to which they are exposed, and the proper work practices to follow to reduce those hazards, each time they work with a biohazardous substance. If the etiology is not known or not completely understood, it must be assumed that the material presents a hazard; therefore, all possible precautions must be taken to prevent latent effects.

The OSHA has proposed a standard for occupational exposure to toxic substances in laboratories that includes, among other requirements, setting forth procedures, work practices, and employee training programs for toxic substances in the laboratory. Consult federal OSHA regulations for more information.

Another major cause of laboratory infections is through the spread of microorganisms as aerosols during routine bacteriological and virological procedures, such as grinding tissues, decanting, and using centrifuges and sonicators (Table 14-F). Employees must be instructed in the proper technique for conducting routine laboratory procedures, since much of microbiological safety is based upon proper work practices.

Table 14-F. Aerosols from Common Laboratory Procedures

Technique	*Average No. of Clumps of Organisms Recovered from Air During Operation*
Pipetting 10 mL culture into 1000 mL broth	2.4
Drop of culture falling 12 in. (30 cm) onto	
Stainless steel	49.0
Painted wood	43.0
Hand towel wet with 5 percent phenol	4.0
Re-suspending centrifuged cells with pipette	4.5
Blowing out last drop from pipette	3.8
Shattering tube during centrifuging	1,183.0
Inserting hot loop into broth culture	8.7
Streaking agar plates	0.2
Withdrawing syringe and needle from vaccine bottle	16.0
Injecting 10 guinea pigs	16.0
Making dilutions with syringe and needle	2.3
Using syringe and needle for intranasal inoculation of mice	27.0
Harvesting allantoic fluid from 5 eggs	5.6

(Reprinted from *Survey of Viral and Rickettsial Infections Contracted in the Laboratory*, National Cancer Institute, National Institutes of Health, Bethesda, MD, p. 29)

Specific occupational biohazards are recognized in the field of animal laboratory research, especially the hazard of disease transmission from animals to humans. Animals from natural habitats may have been exposed to infectious disease or indigenous latent viruses, and thus pose a potential threat to workers. Laboratory animals or animal tissues injected with specific bioagents can carry latent infectious agents to which humans are susceptible.

Airborne aerosol transmission of bioagents can also occur when working with contaminated animal tissues, organs, and blood or other body fluids. In attempting to protect personnel involved in animal experimentation, it is essential to control disease transmission both by pre-exposure immunization to known diseases (such as rabies), and by careful surveillance of exposure to zoonoses, animal bites, scratches, and unusual illnesses, either on the part of the animal or employee.

Hospitals

For many years hospitals have maintained the image of being the ultimate in cleanliness. Visible cleanliness, however, does not assure microbiological cleanliness; a hospital has many potential biological hazards that can pose a threat to employees, patients, and visitors. (A hospital has many chemical hazards, such as anesthetic waste gases, antineoplastic drugs, ethylene oxide, and cleaning chemicals, and physical hazards such as ionizing radiation, that also must be evaluated by the industrial hygienist.) The main biohazards within the hospital setting are bacterial infections (such as streptococcus or staphylococcus), and viral infections (such as hepatitis B or rubella).

Anyone entering a hospital, whether patient, visitor, or staff member, can be the carrier or be exposed to an unrecognized biohazardous agent. Usually, hospital employees take the greatest risk from exposure to these biohazards, especially those who work in laundry, housekeeping, and the laboratory. Hospital-acquired infections (known as nosocomial infections) are sometimes responsible for increasing the length of patient stays.

The primary transmission route of hospital-acquired infections seems to be through person-to-person contact. Because of the great risk to both staff and patients from undiagnosed cases, surveillance of the health status of the staff, although costly and time-consuming, is necessary to ensure control of the spread of infectious agents within the hospital. In the following sections, those hospital personnel at greatest risk and possible control measures are discussed.

Laundry. Employees responsible for the collection, cleaning, and distribution of linen are constantly exposed to pathogenic microorganisms. Exposure results from the handling of linen from infected patients—linen fouled by wound drainage and body wastes.

To avoid exposure, the following procedures should be used as a guide:

1. All linen should be placed in plastic bags at the bedside, rather than be carried through the halls to collection bags.
2. Laundry bags should be color coded in order to alert laundry staff to potential hazards.
3. All linen must be carefully bagged, and not placed carelessly in a laundry cart or closet.
4. When the laundry reaches the washing facility, the contents of the bags should be emptied directly into the washing machines.

Employees responsible for sorting and folding clean linen can also be sources of infection as a result of poor personal hygiene. Thorough handwashing and the use of rubber gloves are essential basic infection-control methods.

Laundry personnel also can suffer from contact dermatitis from using detergents and other laundry chemicals. Laundry personnel may be exposed to aerosols that arise from poor seals around the washer and dryer doors and exhaust systems. These aerosols may contain microorganisms that can pose a threat to employees.

Housekeeping. Employees assigned to the hospital housekeeping staff face many biohazardous exposures. They are possibly the single, highest-risk group in the hospital. Contact with discarded contaminated disposable equipment and general cleaning responsibilities increase the risk of infection. The widespread use of disposables, especially those used in intravenous administration and blood collection, are mainly responsible for spreading infections. Contaminated hypodermic needles and intravenous catheters can present hazards. All needles must be discarded in specially marked containers.

Dry mopping or sweeping the floors does not remove many microorganisms, nor is it designed to. Most sweeping or mopping operations push dust and other material from area to area. When mops are improperly treated, dust is dispersed back into the air. Because dust frequently serves as a collector of microorganisms, when so dispersed it may cause respiratory infections.

Vacuum cleaners are most effective in dust removal if the filter bags are clean. A two-filter bag system has been recommended, with an inner collector bag and an outer bag. Damaged bags will not effectively trap dirt and can release dangerous aerosols.

Many detergents and disinfectants lose their antimicrobial properties when left as water solutions in open containers. If a solution has been stored in a mop bucket for some time, it is conceivable that the floor can be more contaminated after the mopping than before. Water solutions of detergents should be used as soon as possible after mixing, and always within 24 hours.

Central supply. The increased use of disposables has greatly reduced the exposure of the central supply staff to infectious agents. The most serious problem in this department is cleaning surgical instruments. Although these instruments are often rinsed or dumped into a disinfectant solution before being removed from the operating theater, they may still be contaminated. Grossly contaminated material should be sterilized in an autoclave before any handling or rinsing. (An autoclave is an apparatus that sterilizes using steam under pressure.) Soaking is not an effective cleaning method; mechanical scrubbing action is more efficient, and it is during scrubbing that exposure to biohazards is the greatest. Direct injection of microorganisms is possible if the skin is punctured with dirty instruments, or if the skin has lesions which come into contact with contaminated instruments. Ultrasonic cleaners and high-pressure water streams, although effective for removing dried blood and tissue, can produce bacteria laden aerosols.

Once the instruments are cleaned, they must be separated from used equipment even though sterilization is to follow. Sterilization cannot clean—it only produces microbe-free surfaces. Sterile packs can become sources of infection when they are stored for long periods or when the integrity of the external packaging has been compromised. All sterile packs must be inspected thoroughly before use.

Nursing. Nursing employees have direct contact with patients and exposure to infection is always present. Staff involved in patient care risk direct transmission of infection from the patient, and can in turn also be a source of infection to the patient. The nursing staff can spread infection from patient to patient, to other staff members, or from patients to visitors.

All nursing personnel must dispose of contaminated equipment properly so that no hazard exists for others. Hands should be thoroughly washed after visiting each patient to minimize the chance of spreading an organism. Isolation gowns, masks, and caps must be worn where indicated, and removed before entering clean areas such as rest areas and lunch rooms.

(Both nurses and pharmacists can be involved in preparing and administering antineoplastic drugs, which are used for cancer treatment. These chemicals are highly toxic; some have been shown to be carcinogenic, mutagenic, teratogenic, fetotoxic, or hepatotoxic. See the DHHS/CDC publication, *Biosafety in Microbiological and Biomedical Laboratories,* HHS No. 84-8395; The American Society of Hospital Pharmacists (4630 Montgomery Ave., Bethesda, MD 20814), *Safe Handling of Cytotoxic Drugs—Study Guide;* and the OSHA guidelines, *Work Practice Guidelines for Personnel Dealing with Cytotoxic (Antineoplastic) Drugs,* OSHA Instruction Pub. 8-1.1, Jan. 29, 1986, Appendix A, Office of Occupational Medicine, for more details on proper handling procedures.

Pharmacy. The risk of infection spread by or to the pharmacy personnel has increased with the new responsibilities assumed by the pharmacy staff. In many hospitals satellite pharmacies on patient floors attend to patient pharmaceutical needs. Pharmacists can attend rounds and consult with patients and their families. With this greater amount of involvement in direct patient care, more pharmacists are being exposed to biohazards.

Many hospitals employ pharmacists in intravenous (IV) admixture programs. These programs are designed to reduce nonsterile addition of drugs to IV solutions. In IV admixture programs, drugs are added to IV fluids, using aseptic techniques inside a laminar flow biological safety cabinet with HEPA filters. Although these filters are designed to remove more than 99 percent of particles with diameters larger than 0.3 micrometers, they still do not guarantee sterility. The hoods are open on one side to allow access to the work area, and in order to minimize contamination, all work must be performed 15 cm (6 in.) inside the edge of the hood. In addition, the bottles or bags of solution, as well as the ampules and vials of drugs, are only sterile on the inside; therefore, the outside surfaces of the containers should be cleaned before use.

Pharmacists involved in IV fluid therapy teams also undergo biohazardous exposure due to direct patient contact and contact with used hypodermic equipment. Intravenous teams are responsible for starting the procedure, the placement of IV catheters, and (in some hospitals) the development of therapy regimens.

Physicians. Currently, in the day-to-day care of hospitalized patients, physicians are at lower risk of infection than most other employees.

Dietary. The story of "Typhoid Mary" is probably the best known example of the spread of infection to the dietary staff. One employee with a gastrointestinal infection can cause infection throughout an entire hospital. Ill employees should be encouraged to stay at home, and to visit the health service before returning to the job.

Staff involved in food preparation are faced with possible infection from raw foods, such as salmonella and botulism, which can result from contact with raw fish, meat, and some vegetables. Foods should be thoroughly washed before preparation, and once washed must be separated from unwashed foods to prevent contamination. All employees must wash their hands before beginning each food preparation.

Staphylococcal food poisoning is produced by an enterotoxin that develops as the staphylococcus grows in the food product. Onset of the symptoms usually occurs after three hours, but can vary from two to six, depending on the quantity of the toxin ingested. Symptoms are increased salivation, nausea and vomiting with retching, abdominal pain and cramps, and watery diarrhea.

Primarily, prevention consists of proper handling of food products, with clean hands and garments; and the handler should not have skin lesions. Refrigeration of the food products at a proper level will prevent the growth of the bacteria. Adequate cooking of foods will kill the salmonella organisms. Once the enterotoxin of the staphylococcus is produced in the food, no amount of cooking, freezing, or any other known method will remove it from the food.

Personal hygiene. All hospital employees must be instructed in proper hand washing techniques. Facilities must be available and kept clean; unlean sinks and lavatories discourage their use. Visitors should be encouraged to wash their hands after visiting a patient to reduce the chance of spreading infection.

All employees must wash their hands after using the toilet and before eating, drinking, or smoking. Additionally, nursing and other patient-care staff must wash their hands before and after changing a dressing on a patient.

Sterilization and disinfection. Disinfection and sterilization are not synonymous operations. Disinfection frees from infection but does not imply complete destruction of all harmful microorganisms; one can selectively disinfect for specific infectious agents while leaving other microorganisms intact. Sterilization, on the other hand, does mean the absolute destruction of all living organisms.

All equipment used to sterilize or disinfect must be properly maintained and operated. Employees handling such equipment should be thoroughly trained in its operation. A sterilizer with a leaking seal cannot effectively sterilize. An employee who opens a sterilizer before it has cooled runs the risk of compromising sterility, as well as being scalded.

Nondestructive sterilization is often required for plastic syringes, plastic tubing for replacing arteries and veins, and even for controlling the growth of organisms in spices. This is often accomplished using gases, such as ethylene oxide, which is highly toxic and carcinogenic. Special care must be taken in protecting workers (see Bibliography for more information).

All cleaning equipment such as mops, sponges, dust cloths and brooms should be cleaned after each use.

Ventilation systems should be separated to prevent the spread of airborne diseases through the system. Unauthorized personnel must be barred from isolation areas, laboratories, and other areas where their presence may put them or others at risk. Infection-control procedures should be included in the hospital policy manual, and an infection control committee should be established. An epidemiologist should be available to help evaluate and control hospital infection outbreaks.

Immunization. An effective immunization is now available for hepatitis B and rubella. Rubella immunization is recommended for most personnel in the health care setting. Individuals likely to come in contact with blood or other bodily fluids should receive the hepatitis B vaccine.

Agriculture

Occupational exposure to biohazards also occurs in agriculture. Human health is affected by the health of the animals in close contact. The sources of biohazards are the infectious diseases that both humans and animals share. Kelley Donham (1974) points out that there are four types of relationships in terms of disease transmission between humans and animals:

1. Diseases of vertebrate animals transmissible to humans and other animals (zoonoses)
2. Diseases of humans transmissible to other animals (anthropozoonoses)
3. Diseases of vertebrate animals chiefly transmissible to humans (zooanthroponoses)
4. Diseases transmissible to humans from the environment (animals being the source of environmental contamination)

Zoonoses. These fine distinctions, however, are not often made and most frequently all diseases transmitted to humans from animals are referred to as zoonoses. Donham (1974) conducted a study of those diseases most commonly found to be transmissible and categorized 48 of them as to whether they posed definite significance as a hazard, questionable significance as a hazard, and doubtful (no relationship found) significance as a hazard (Table 14-G).

The zoonoses consist of a variety of viral, bacterial, rickettsial, fungal, protozoal, and helminthic diseases. Among the most important in the United States are anthrax, brucellosis, tetanus, encephalitis, leptospiroses, Q fever, rabies, salmonellosis, and tularemia. The infectious agents enter the body by inhalation, ingestion, or through the skin and mucous membranes. Infections can be transmitted to humans by direct contact with contaminated animal flesh or body fluids (urine, feces, blood, saliva), and indirectly through contamination of the workplace by the inhalation of spore-contaminated dusts. Mosquitoes, flies, fleas, and other insect vectors also play important roles in the complex chain of transmission of those zoonoses that may be garbageborne, waterborne, or milkborne.

Farmers are prime targets because of their occupational dealings with animals. Note that contamination of the environment by an infected animal tends to spread the disease to other animals. Veterinarians run a great risk of zoonoses from accidental inoculation when vaccinating animals. Improper slaughtering and disposal of carcasses and entrails of infected animals, some of which are eaten by other animals, are important in the spread of disease.

Tetanus. Individuals whose occupations include the hazard of injury, usually of a penetrating or crush type wound, are at highest risk from tetanus. Farmers and ranchers who work around domestic animals and soil are also at risk.

Anthrax. Certification of imported hides, hair, and wool as anthrax-free by the exporting country has helped to reduce the incidence of anthrax in the United States. The chief preventive measure for high-risk industrial workers is immunization. According to the CDC, this has practically eliminated the incidence of human anthrax in the United States. In recent years,

Table 14-G. The Occupational Significance to Agricultural Workers of Selected Diseases Common to Animals and Humans*

Definite Risk	*Questionable Risk*	*Doubtful Risk*
1. Brucellosis	1. Pseudotuberculosis	1. Cowpox
2. Leptospirosis	2. Psittacosis/ornithosis	2. Taeniasis
3. Toxoplasmosis	3. Vibriosis	3. Dermatophilosis
4. Rabies	4. Salmonellosis	4. Trichinosis
5. Tetanus	5. Listeriosis	5. Pneumoccal infections
6. Anthrax	6. Tuberculosis	6. Babesiosis
7. Erysipeloid	7. Streptococcal infections (pharyngitis, erysipelas)	7. Fascioliasis
8. Q fever	8. Influenza and parainfluenza	8. Giardiasis
9. Histoplasmosis	9. Plague	
10. Blastomycosis	10. Cryptococcosis	
11. Ringworm	11. Nematodiasis (cutaneous larval migrans, visceral larval migrans, and strongyloidiases)	
12. Equine encephalitis	12. Colorado tick fever	
13. Newcastle disease	13. Pneumococcal infections	
14. Pseudocowpox	14. Encephalomyocarditis	
15. Vesicular stomatitis	15. Dirofiliariasis	
16. Contagious ecthyma		
17. Staphylococcal infections		
18. Enchinococcosis		
19. Collibacillosis		
20. Tularemia		
21. Acariasis		
22. Pasteurellosis		
23. Rocky Mountain spotted fever		
24. Spotrotrichosis		
25. Balantidiasis		

*Risk for agricultural workers in comparison to the general rural and general urban population.

the CDC has had reports of only a few cases per year; the last death from anthrax that was reported was in 1976. Improved personal hygiene of workers, protective clothing, ventilation, and housekeeping controls in the plants are also valuable in control of the disease. Vaccination of animals in areas where specific animal diseases are present and strict adherence to laws regarding animals that have contracted or died of anthrax have helped reduce agricultural incidence.

Brucellosis. Control of brucellosis in humans is contingent upon control of the disease in animals. In a given locality, the incidence in humans is often the index of the effectiveness of disease-control in the domestic animal population. Awareness of the disease by the workers and by doctors who see patients involved in occupations where exposure to infected animals is possible is very important. In the workplace, hygienic practices and proper attention to minor cuts and scratches, especially on the hands and forearm, can help control the disease.

Tularemia. Tularemia is a disease of rodents, resembling plague, which is transmitted by bites of flies, fleas, ticks, and lice; it may be acquired by humans through handling of infected animals or by airborne transmission from infected humans in the laboratory. Generally, the route of entry is through small cuts and scratches on the hands from the lesions of the infected animals. This is particularly true of hunters who dress wild rabbits and cooks who handle these wild rabbits to ready them for the table. Humans may also become infected from the bites of infected insects; from bites of animals carrying the organism in their mouth as a result of feeding on infected carcasses; by eating infected meat that has been insufficiently cooked; or by drinking contaminated water. Using rubber gloves when handling carcasses of wild animals, using safety hoods during laboratory experimentation, avoiding arthropods, flies, mosquitos, and wood ticks when working in endemic areas, not drinking raw water in endemic areas, and thorough cooking of wild game may help to control the disease.

Dermatophytoses. Dermatophytoses are caused by fungi. Farmers, animal handlers, pet and hide handlers, wool sorters, cattle ranchers, athletes, lifeguards, gymnasium employees, and animal laboratory workers are at high risk. The main reservoir of these superficial fungi is in humans and animals. Ringworm of the feet or athlete's foot developes in areas of maceration between the toes. Ringworm of the nails is generally due to spread from ringworm on the feet or hands. Ringworm of the groin occurs in the folds of the upper inner thigh where chafing and irritation are common. Ringworm of the body or ringworm of the hand are usually due to contact with animals, contaminated fomites, or other human beings with similar conditions. Ringworm of the scalp can can be due to contact with animals, but most generally it is transmitted from human to human. Prevention depends on recognition of the disease in animals and proper handling techniques with these animals. Sterilization and proper laundering of towels and general cleanliness in showers and dressing rooms of gymnasiums and swimming pools are also necessary. Education for personal hygiene is also effective control.

Hypersensitivity. Inhalation of fungus and actinomycetes spores causes hypersensitivity diseases. Repeated exposure sensitizes the individual to the spores (protein sensitization), and the disease state recurs on subsequent exposure. Farmers who handle hay in confined areas, saw mill operators, mushroom workers, sugar cane workers, cork workers, workers exposed to redwood processing, and workers handling other agricultural products on an industrial basis such as seeds, textile fibers, wood, and gum are at high risk.

Unfortunately, there is not much data relating to the extent of agriculturally related diseases, and this gives a false picture as to prevalence of the problem. This is partially due to improper diagnoses, either because the examining physician is unaware of the disease, or because of a failure to associate the agricultural occupation with the disease. The paucity of data on the extent of agriculturally related disease may also be due to the lack of compliance with state occupational disease reporting laws.

BIOHAZARD CONTROL PROGRAMS

Because of the potential dangers of biohazards, good biohazard control is the first and most important line of defense. The major emphasis should be aimed directly at the source of contamination. Control practices usually include a health surveillance program, standardized work procedures, and environmental control procedures (especially in reference to safety equipment). The following guidelines are adapted from both *Biohazards Safety Guide* (NIH, 1974) and *Biohazards in Biological Research* (Hellman, et al, 1973). For more details, the reader is urged to consult these references.

Employee health program

Preplacement physical examinations should be given to all new employees to establish a baseline reference. Periodic physical examinations should be scheduled as part of a surveillance program. Persons currently employed or transferred to the biohazardous area should be urged to regularly have physical examinations. This applies to all persons handling potentially oncogenic, biological, or toxic chemical materials and to those employees who clean laboratory glassware, handle or care for experimental animals or their tissues, or perform janitorial duties.

Management must consider the health status of the employees working with biohazards. It may be undesirable to employ persons with certain physical conditions for work with specific types of infectious agents. These, of course, are medical decisions.

Vaccination of exposed personnel is recommended when a satisfactory immunogenic preparation is available. The efficiency of vaccines for workers is evaluated on the basis of their effectiveness in preventing disease in the general population. Two possible pitfalls to this line of thinking should be mentioned. The first is that the worker may be exposed to infectious microorganisms at a higher dose level than would be expected from the normal route of exposure. Secondly, this exposure may be by a different route from that normally expected; one example would be respiratory infection from the tularemia or anthrax organism.

Consequently, vaccination is especially important for those working with biohazards. Good immunity is conferred after vaccination against smallpox, tetanus, yellow fever, botulism, hepatitis B, and diphtheria. Other vaccines, such as those for psittacosis, Q fever, and Rift Valley fever, have been or are being tried experimentally with varying degrees of succes.

Pregnant women working in oncogenic virus laboratories should be counseled at the earliest possible time as to the potential risk involved in their continued presence in the laboratory. The option of remaining in the laboratory or transferring to a nonvirological unit should be explained by the supervisor or department head.

Another vital part of a health surveillance program includes keeping permanent records of individual work assignments, especially in the case of laboratory workers.

Laboratory safety and health program

Management should establish and implement an environmental control and personnel safety and health program that should include biological safety. There should be a written safety policy and a safety manual available to all laboratory personnel. The general practices and procedures that directly affect laboratory safety and contamination control should be described. Employee training to recognize and avoid hazards is an essential element of any safety and health program. Much of the material in this section on biohazard controls is adapted from *Biohazards Safety Guide* (NIH, 1974). The CDC publication *Biosafety in Microbiological and Biomedical Laboratories* (DHHS 84-8395) details the precautions to be followed for specific infectious agents.

Routine laboratory safety and health procedures have application to the control of biohazards; however, there are certain additional precautions and procedures that must be followed when dealing with potential biohazardous agents. The following sections summarize general precautionary procedures for biohazard control and should be applied appropriately depending on the risk involved.

In general, no one should be admitted to a biohazardous area unless working specifically within that area. This particularly applies to controlled access at all times for children and pregnant women. Social visiting of personnel from one biohazardous area to another should be strictly forbidden.

Eating, drinking, smoking, and gum-chewing should not be permitted in rooms where work with infectious agents is in progress. Employees who work with biohazardous material must be sure to wash and disinfect their hands thoroughly both before and after eating or smoking. An effort should also be made to keep all surplus equipment and extraneous materials (such as scientific journals) out of biohazardous areas.

All employees and visitors in laboratories working with biohazardous agents are required to wear protective clothing according to the level of risk involved. Protective laboratory clothing should not be worn outside the work area.

The universal biohazard symbol required by the United States Department of Labor, Occupational Safety and Health Administration (OSHA) as shown in Figure 14-1, should be used to identify all restricted biohazardous areas. A posted sign displaying the symbol should indicate the name of the investigator responsible for the project, the nature of the hazard, and any special procedures or precautions required.

All laboratory rooms containing infectious substances should be designated as separate infectious disease work areas and prominently marked with the biohazard warning symbol. Cabinets and containers, which hold infectious biological agents, should also be marked with the symbol and should further be labeled indicating cleaning instructions such as INFECTIOUS—TO BE AUTOCLAVED. This will ensure proper sterilization so that no cross-contamination can occur, and will also distinguish between equipment to be used with infectious material and equipment to be used only with noninfectious material. All cultures should specifically be labeled with the name of the agent, the researchers' names, and the date.

All incubators, refrigerators, and similar storage spaces for biohazardous materials should bear a biohazard symbol as

BIOHAZARD

RESTRICTED AREA

AREA:	BLDG.		ROOM	RESPONSIBLE INVESTIGATOR
	INST./DIV.	LAB./BR.		LAB./BR. CHIEF

HAZARD OR RESTRICTION:

SPECIAL PROCEDURES OR PRECAUTIONS:

ADMIT: ☐ STAFF ☐ GUARD ☐ CLEANER ☐ MAINTENANCE ☐ EMERGENCY PERSONNEL

NOTICE	CALL OR SEE	BLDG.	ROOM	NIH PHONE	HOME PHONE
FOR ENTRY OR ADVICE					
IN CASE OF EMERGENCY					
IN CASE OF EMERGENCY					

NIH 645-4 (1-69)

Figure 14-1. The biohazard symbol—here, it is being used on a sign for restricted areas. Indicate the name of the person to be contacted, the hazard or restriction, and the special procedures or precautions needed.

shown in Figure 14-1. Refrigerators should be cleaned and checked periodically for any containers that may have been borken during storage (Figure 14-2). When equipment is no longer being used to store infectious material, the biohazard markings should be removed and the equipment decontaminated before recycling or disposal.

Floors, laboratory benches, and other surfaces in infectious agent areas should be disinfected with a suitable germicide. Adequate precautions must be taken to reduce the hazards of aerosol generating processes such as centrifugation, lyophilization, sonication, grinding, and similar procedures; and the surrounding area should be disinfected after completion.

All contaminated wastes must be decontaminated before disposal. Dry contaminated wastes should be collected in impermeable sealed bags before removal to the autoclave or incinerator. Tissue culture and other virus-containing liquid wastes should be decontaminated, either chemically or by heat, before being discharged to the community sanitary sewer system. All ani-

Figure 14-2. Technician wearing an air-supplied suit and hood checks for any broken containers. (Courtesy CDC)

mal wastes must be incinerated or autoclaved before other disposal.

Equipment known or suspected of being faulty should not be operated. When maintenance workers enter a biohazardous area, the supervisor must be sure that all necessary precautions are understood by the individuals, and that any unnecessary hazards in the area are secured or decontaminated. All equipment to be serviced must be rendered free of infectious organisms before being serviced. Potentially contaminated supplies and equipment must be decontaminated and removed from the work area. Maintenance workers' tools must also be decontaminated before removal from high-risk areas.

Sterilization can be accomplished using several different methods and types of equipment. To assure that sterilization is being accomplished, procedures and equipment must be certified before original use, and periodically checked. Certification is accomplished using instrumentation, chemical indicators, and biological indicators to verify that appropriate parameters are satisfactory and that resistant microorganisms placed in the sterilization equipment are destroyed.

Animal care and handling

Depending upon the environmental stability and infectiousness of the pathogens involved, and the method of animal testing, various degrees of isolation for experimentally infected animals are required. In a complete containment system animals are housed within gastight cabinets. The next level of containment would be provided by closed, ventilated cages.

Doors to animal-containment rooms should be kept closed at all times, and access restricted only to authorized personnel. Floors, walls, and cage racks should be washed with a disinfectant frequently, and a container of disinfectant should be kept in each room for general decontamination purposes.

All cages should be marked to indicate those animals that have been inoculated with infectious substances. After use, cages should be sterilized by autoclaving and all refuse, bowls, and watering devices should remain in the cage until after this decontamination process is performed.

Careful handling procedures should be used to minimize the dissemination of dust from both the cage and the animal. Every effort should be made to restrain the animal in avoiding accidents resulting in dissemination of infectious material. Heavy gloves should be worn by animal caretakers at all times when feeding, watering, or removing infected animals. When animals are injected or otherwise subjected to pathogenic material, laboratory workers should also wear surgeons' gloves. Under no circumstances should bare hands be placed in the cage at any time.

Necropsy of infected animals should be performed in ventilated safety cabinets and rubber gloves and surgeons' gowns should be worn by all personnel involved. The animal should be pinned or fastened down in a metal tray and the fur wet with a suitable disinfectant. When the procedure is completed, all potentially contaminated surfaces should be disinfected. Grossly contaminated gloves should be dipped in disinfectant before removal from the hands, and then sterilized. Dead animals should be placed in leak-proof containers and thoroughly autoclaved before removal and incineration.

Biological safety cabinets

Many of the routine operations that are performed in microbiological laboratories can release infectious aerosols. Biological safety cabinets protect laboratory workers from exposure to aerosols by controlling contamination at its source. These devices should be used for all procedures involving infectious substances, such as unsealing flasks and bottles, pipetting, mixing solutions, inoculating and/or performing postmortems on animals, grinding or blending substances, and operating sonicators and centrifuges. The importance of primary containment is emphasized by the fact that the majority of reported laboratory-acquired infections occur among workers who directly handle infectious materials.

As a primary containment device, the biological safety cabinet can be designed to provide near absolute or partial containment. Absolute containment cabinets are capable of preventing both aerosol and contact exposure, while partial containment cabinets are designed only to reduce exposures to accidentally created aerosols. The selection of the most effective cabinet type depends upon (1) the relative hazard of the microbiological agent and (2) the potential of the procedure involved to produce infectious aerosols.

Class 1 cabinet. This is a partial containment cabinet with a limited front opening. Because of this limitation in providing both personnel and product protection, it is recommended only for routine procedures using agents of minimal infectivity or risk. This type of safety cabinet uses negative pressure, solid wall barriers, and filtered air to contain or prevent airborne contamination.

The Class 1 cabinet has an open face, and the inward flow of room air into the device provides personnel safety and can prevent the escape of aerosols. The air velocity through the opening should be between 0.38–0.51 m/sec (75–100 feet per minute [fpm]. The exhaust air should be high-efficiency particulate air (HEPA)-filtered and then exhausted from the laboratory. The HEPA filter has an efficiency of 99.97 percent for particulates of at least 0.3 micrometers in diameter.

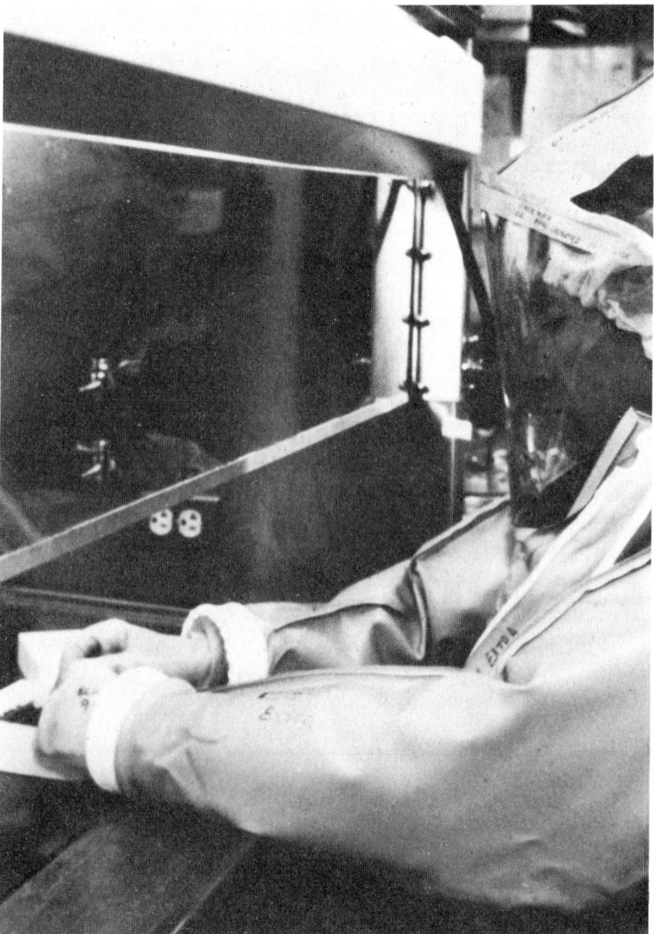

Figure 14-3. Class II laminar-flow biological safety cabinet. (Courtesy CDC)

Because of the inward flow of room air, Class 1 safety cabinets are not suitable for experimental systems that are highly susceptible to airborne contamination. Another factor restricting its use is that interruption in ventilation can lead to direct exposure of the user.

Class II laminar-flow biological safety cabinet (LFBSC). As another partial containment device, this cabinet is developed to better protect both the worker and the experiment from airborne contamination.

Laminar-flow biological safety cabinets provide highly localized control of the environment surrounding critical laboratory operations. Thus, they combine the benefits of high-efficiency filtration and unidirectional laminar airflow without the hazards of operator exposure (Figures 14-3 and 14-4).

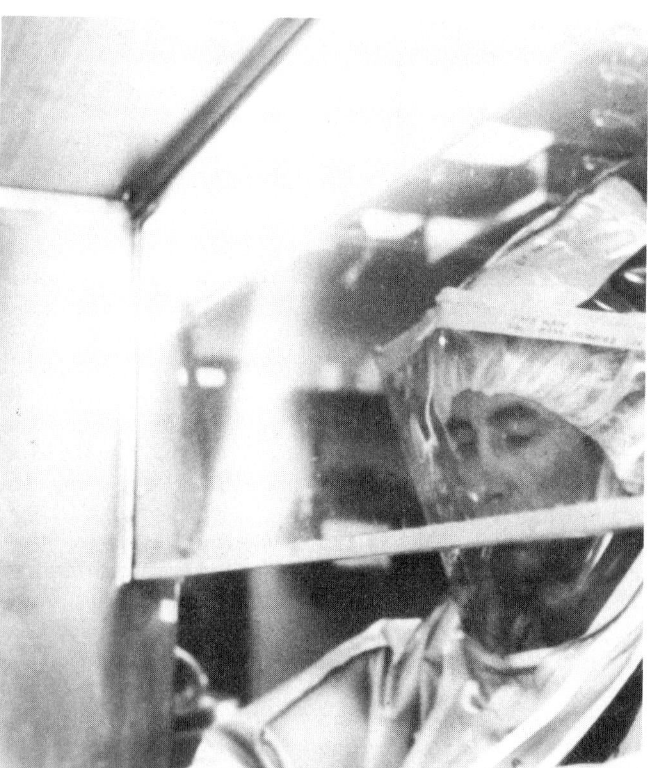

Figure 14-4. A Class II, laminar-flow biological safety cabinet used for activities involving low- or moderate-risk biohazardous material for personnel and experiment protection. (Courtesy CDC)

In Class II cabinets, unlike Class I cabinets, the inward airflow does not cross the work surface; this air is drawn into a front lateral suction grille, which is contiguous with the forward leading edge of the work surface. To protect the product from airborne contamination, the work surface is bathed by a downward flow of filtered air. The appropriate air volume is then recirculated through supply HEPA filters (Figure 14-5).

An optimum air balance between the inward airflow and the filtered cabinet supply air is necessary to achieve both person-

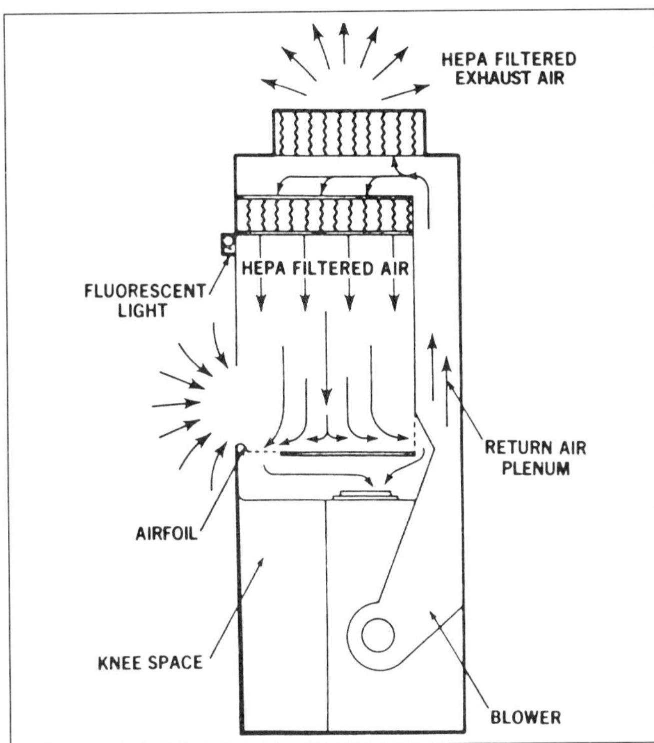

Figure 14-5. Line drawing of a Class II laminar flow cabinet—HEPA filters are standard components of biological safety cabinets.

nel safety and experiment protection. The average velocity of the filtered cabinet supply air should be 0.40 m/sec (80 fpm). The inward air flow average velocity should not be less than 0.38 m/sec (75 fpm). A fixed front access opening will cause a reduction in the average inward airflow velocity and thereby compromise personnel safety. Therefore, vertical laminar-flow cabinets with sliding windows are not recommended for containment of microbiological procedures. For maximum personnel safety, the exhaust air from the laminar flow biological safety cabinets should be ducted out of the facility.

Laminar-flow biological safety cabinets have limitations that must be recognized by the cabinet user. The recirculation of approximately 75 percent of the cabinet supply air precludes the use of the cabinet for containment of flammable solvents. In those installations where the cabinet exhaust air is returned to the room environment, gaseous radioactive materials and highly odorous materials cannot be safely contained. These deficiencies limit the use of the cabinet to the containment of particulate contaminants.

The difficulty of monitoring the performance of the laminar-flow biological safety cabinet is a serious deficiency. Experience has shown that the majority of cabinets that are field tested fail to meet the air balance and filter efficiency requirements and must be corrected to provide adequate safety. Even though the cabinets are certified, after installation the user has no way of knowing whether or not adequate performance is being maintained.

Airflow in the laminar-flow biological safety cabinet can be disrupted by overloading the work space with equipment, altering the size of the front opening, or placing the unit in a location subject to drafts from ventilation or traffic through the laboratory. Disrupting the airflow minimizes the protection of the laminar-flow air curtain across the face of the cabinet and thus compromises safety.

Laminar-flow clean air cabinets are not suitable for work with biohazards, as the positive pressure allows air to flow out of the cabinet exposing workers to contaminated air. Such units are suitable only for use with known "clean" materials; situations in which product protection is the only objective. These devices are often used by microbiological investigators to protect tissue cultures from airborne contamination. The user must recognize that the laminar-flow clean air cabinet affords no personnel protection. The attractiveness of a contamination-free environment for tissue culture and the minimal costs of such units mask the potential hazard of these cabinets.

Class III safety cabinet. This is the only containment cabinet suitable for work with high-risk agents, or for high-risk operations with moderate-risk agents (aerosol generation via centrifuging, blending, homogenizing, etc.). This cabinet provides a physical barrier that isolates the agent from the surrounding environment, and thus provides near absolute protection for operating personnel (Figure 14-6).

Figure 14-6. A Class III gastight biological cabinet, used for all activities involving high-risk biohazards. (Courtesy CDC)

Class III cabinets are gastight modular units maintained under negative air pressure. All manipulations are carried out through arm-length rubber gloves. These devices can be internally equipped with incubators, refrigerators, freezers, and centrifuges for containment during research processes, and can be designed to include all equipment necessary to carry out biological

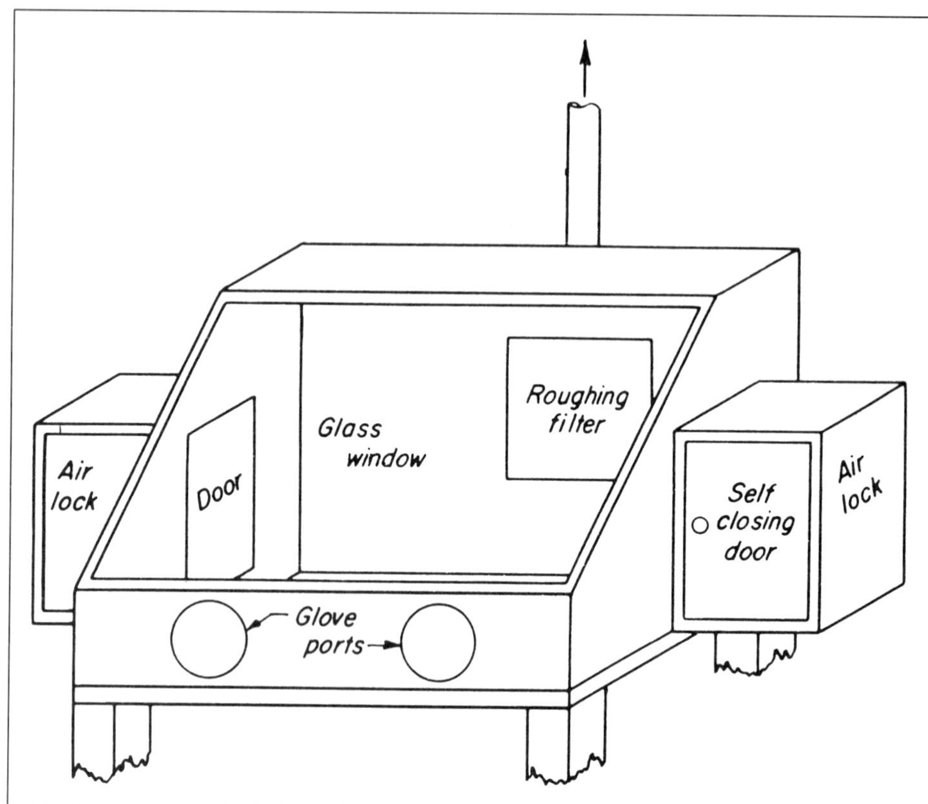

Figure 14-7. Drawing of a chemical fume hood modified with filters and glove ports.

research. These systems, however, are obviously needed only for research of a very high hazard level. Separate exhaust duct systems are required to expel the cabinet air supply that is first filtered through HEPA filters and then incinerated (Figure 14-7).

Since each of the various types of safety cabinet has its advantages and limitations, health and safety professionals and the biohazard control committee must carefully assess the specific research techniques and agents to be used in matching requirements with the appropriate biological safety cabinet. Pertinent factors involved are:

- PROPOSED ACTIVITY. Procedures that can cause aerosols are of particular concern.
- RISK OF THE BIOHAZARDOUS AGENT. All known characteristics of the agent should be evaluated. These include infectivity, concentration of the viable agent to be used, history of known laboratory-acquired human infections, classification of the etiological agent on the basis of hazard, etc.
- CONTROL OBJECTIVES. The protection desired should be determined for the proposed activity; the degree of control used would differ depending upon whether the desired result is for product protection only, for personnel protection only, or for both personnel and product protection.

Ventilation and filtration systems for biosafety cabinets. Ventilation systems serving biological safety cabinets should expel exhaust to the building exterior. Hoods must be equipped with absolute filters, which should be sterilized by a steam formaldehyde purge or an equivalent procedure before changing or handling. A warning system should be available to indicate malfunction. Maintenance workers entering the area after a malfunction should be required to take appropriate precautions, and must wear rubber gloves and masks when handling filters.

Building design

Engineering features commonly used for biohazard control in infectious laboratories include (1) ventilated cabinets and cages that will contain biohazards at their point of use, (2) differential (increasingly negative) air pressures as one moves from clean areas to those of greater biohazardous risk, (3) effective filtration of air from rooms, cabinets, and ventilated cages, (4) ultraviolet air locks and door barriers to separate areas of unequal biohazard risk, (5) treatment of contaminated liquid effluents, (6) change rooms and showers for personnel, and (7) room layout to achieve traffic control along a clean-to-contaminated axis. Choosing the proper engineering safety and health control features in the early stages of planning in new or renovated laboratories is less costly than making changes as a result of problems arising out of work practices.

Control of air contamination in the laboratory is especially important, as the most common source of occupationally acquired infection is the inhalation of accidentally experimentally produced microbial aerosols. Control should begin where the aerosol is formed—at the laboratory bench level. To that end, emphasis should be placed on primary containment devices such as ventilated cabinets.

ACQUIRED IMMUNE DEFICIENCY SYNDROME (AIDS)

Definition

Acquired immune deficiency syndrome (AIDS) is a recent and controversial medical topic. It is caused by human immunodeficiency virus (HIV), a virus that adversely affects the body's natural immunity against diseases (the body's ability to fight infections) (*MMWR* 34:681-685, 691-694, 1985). This virus is known by all of the following names (at the time of this writing): human T-lymphotropic virus, type III (HTLV-III); lymphadenopathy-associated virus (LAV); or AIDS-related retrovirus (ARV). The virus has been given various names, but it appears that the designation HIV (human immunodeficiency virus) is likely to become standard.

History

In the United States, the first AIDS case reported was in 1981; by mid-February 1987, a national cumulative total of 30,839 cases had been reported. Although all 50 states, including the District of Columbia and Puerto Rico have reported cases, according to the U.S. Public Health Service, 33 percent of the cases reported in the United States are from New York and 23 percent are from California. In addition, more than 100 countries have reported cases of AIDS (*Facts About AIDS,* 1986).

Transmission

It is estimated that 1 million people have been infected by the virus, but do not have any symptoms and actually feel well. Not everyone who is exposed to the virus or becomes infected with the virus becomes ill. The incubation period is believed to be six months to five years, or perhaps even longer.

Current research indicates that the virus is *not* transmitted through food, beverages, air, or casual contact, nor is there any evidence that any co-worker, client, customer, or fellow student has contracted AIDS through casual contact with an AIDS patient. Although there have been only a few reported cases of health care workers contracting AIDS from a patient, the severity of the disease warrants strict adherence to the precautions described later in this section.

According to the research literature, AIDS is transmitted via sexual contact, sharing of needles, and transfused blood. This virus can also be transmitted from an infected mother to her infant before, during, or shortly after birth. It appears that the risk of infection increases as the number of sexual partners increases and the number of times needles are shared and used. Note that improved screening methods have all but eliminated the risk of contracting AIDS from routine blood transfusions.

Although AIDS is difficult to contract, the following types of people are most at risk:

1. sexually active homosexual and bisexual men, especially those men who have many sexual partners
2. intravenous drug abusers
3. hemophiliacs and people who have received blood transfusions
4. sexually active heterosexual people who have infected partners
5. infants who have infected mothers.

Symptoms

During the incubation period, many people do not have any symptoms. Those who do develop symptoms experience tiredness, fever, night sweats, loss of appetite and weight, diarrhea, swollen lymph nodes in the neck, armpits, or groin, and/or white spots or unusual blemishes in the mouth.

The two most common diseases developed in AIDS patients are Kaposi's sarcoma (KS), a type of cancer occurring on the surface of the skin or in the mouth, and *Pneumocystis carinii* pneumonia (PCP), a parasitic infection of the lungs. Other infections that may develop include severe infections with yeast, cytomegalovirus (CMV), herpesvirus, and parasites such as *Toxoplasma* or *Cryptosporidia.*

Diagnosis

Currently, there is no single medical/laboratory test to diagnose AIDS. However, there is a test used to detect the presence of antibodies to the HIV virus. Presence of the antibodies indicates that the person had been infected with the virus at one time; the antibodies do not indicate that person is still infected. Therefore, diagnosis also depends upon tests for specific types of white blood cells (WBCs), which indicate impaired immune system, upon the presence of "opportunistic" disease, and the patient's history.

Treatment

Currently, there is not an antiviral drug or medical treatment that will cure AIDS or completely restore the body's natural immune system. Medical treatment for the diseases/infections that are present in the AIDS patients includes interferon, interleukin-2 (a naturally occurring chemical in the body that functions to fight disease), surgery, radiation, and drug treatments. Research is under way in this area.

Prevention

Individuals can decrease their chances of exposure to AIDS by not using intravenous drugs and by controlling their sexual activities, that is, do not have multiple partners nor a partner who has multiple partners. Also, avoid sexual activity with a person who has AIDS. Hospitals prevent the spread of AIDS in several ways. Hospitals and/or blood banks test all donated blood for the HIV antibodies before transfusion. Hospitals use special handling, labeling, and isolation methods when dealing with blood, body fluids, and patient-care equipment that may be contaminated, and tissue samples/cultures from AIDS patients.

Workers

Health care workers. Health care workers include nurses, physicians, dentists, dental workers, optometrists, podiatrists, chiropractors, and blood bank technologists and technicians, phlebotomists, dialysis personnel, paramedics, emergency medical technologists, medical examiners, morticians, housekeepers, laundry workers, and others whose work involves contact with the patients' blood, or other body fluids, or corpses (*MMWR* 34:681–685, 691–694, 1985).

Health care workers should avoid needlestick injuries. In addition, when direct contact with mucous membranes, secretions, or wastes from an AIDS patient is anticipated, the health care worker should wear gloves. The following additional precautions should be taken:

1. Sharp items (needles, scalpel blades, and similar instruments) should be considered potentially infective and be handled with extraordinary care to prevent accidental injuries.
2. Disposable syringes and needles, scalpel blades, and other sharp items should be placed into puncture-resistant containers located as close as practical to the area in which they were used. To prevent needlestick injuries, needles should not be recapped, purposefully bent, broken, removed from disposable syringes, or otherwise manipulated by hand.

3. When the possibility of exposure to blood or other body fluids exists, routinely recommended precautions should be followed. The anticipated exposure may require gloves alone, as in handling items soiled with blood or equipment contaminated with blood or other body fluids, or may also require gowns, masks, and eye-coverings when performing procedures involving more extensive contact with blood or potentially infective body fluids, as in some dental or endoscopic procedures or postmortem examinations. Hands should be washed thoroughly and immediately if they accidentally become contaminated with body fluids.
4. To minimize the need for emergency mouth-to-mouth resuscitation, mouthpieces, resuscitation bags, or other ventilation devices should be strategically located and available for use in areas where the need for resuscitation is predictable.
5. Pregnant health care workers are not known to be at greater risk of contracting HIV infections than health care workers who are not pregnant; however, if a health care worker develops HIV infection during pregnancy, the infant is at increased risk of infection, resulting from perinatal transmission. Because of this risk, pregnant health care workers should be especially familiar with precautions for preventing HIV transmission.

Personal service workers, food service workers, office, school, factories, construction-site workers. Personal service workers include hairdressers, barbers, estheticians, cosmetologists, manicurists, pedicurists, and massage therapists. Food service workers include cooks, caterers, servers, waiters, bartenders, and airline attendants (*MMWR* 34:681-685, 691-694, 1985).

As previously stated, this virus is *not* transmitted through casual contact. Therefore, infected employees should not be restricted from any job site, from working, or from using any equipment, including eating areas and drinking water fountains. In fact, there is no justification to identify infected employees. The same right of privacy applies to food service workers, as in the case of any other illness. Personal service workers should follow the same precautions as those for the health care workers. Food service workers should follow good personal hygiene practices and food sanitation.

Sterilization, disinfection, and housekeeping

The current sterilization and disinfection procedures in health care and dental facilities are adequate to decontaminate instruments including surgical instruments, devices, or other items that were or may have been in contact with blood or other body fluids from infected individuals. Whether decontamination is accomplished by machine- or manual-cleaning operations, appropriate chemical germicides must be used. Only trained personnel wearing appropriate personal protective clothing and/or equipment should perform the decontamination procedures (*MMWR* 34:681-685, 691-694, 1985).

Laundry and dishwashing operations used in hospitals are sufficient to decontaminate linens, dishes, glassware, and utensils. Surfaces exposed to blood and body fluids should be cleaned with a detergent and then decontaminated with an EPA-approved hospital disinfectant that is mycobactericidal. Individuals cleaning up spills should wear such personal protective clothing as disposable gloves (*MMWR* 34:681-685, 691-694, 1985).

Disposal

Blood and other body fluids that can be flushed down the toilet should be. Needles and other sharp items that cannot or should not be decontaminated and then re-used should be placed into puncture-resistant containers and disposed of in accordance with local regulations for biohazardous, solid waste. Other contaminated, disposable items should be wrapped securely in a puncture-resistant plastic bag. It should be placed in a second bag that should be properly labeled as to its contents before being discarded in a manner consistent with local regulations for biohazardous, solid waste disposal. Spills of blood or other body fluids should be cleaned with soap and water or a household detergent. After the spill has been cleaned up, wipe the area with a disinfectant solution or a freshly prepared solution of sodium hypochlorite (household bleach) (*MMWR* 34:681-685, 691-694, 1985).

Future

Extensive research to find a cure for AIDS is continuing. Various U.S. governmental agencies, including the Public Health Service, National Institutes of Health, Centers for Disease Control, the Food and Drug Administration, and the Alcohol, Drug Abuse and Mental Health Administration, are heavily involved in every aspect of AIDS research and education.

LEGIONNAIRES' DISEASE

Definition

The term legionnaires' disease refers to an acute respiratory infection caused by the bacterium *Legionella pneumophila.* This bacterium led to the 1976 outbreak of respiratory illness among persons attending a convention of the American Legion at the Belleview Stratford Hotel in Philadelphia (182 cases and 29 deaths were attributed to the bacteria in the cooling tower) (W. B. Baine, May 1979).

Sources

Organisms of the *Legionella* genus are ubiquitous in the environment and are found in natural fresh water, potable water, as well as in closed-circuit systems, such as cooling tower water, evaporative condensers, humidifiers, recreational whirlpools, and air handling systems.

If the following four conditions are present there is a potential for the presence of *Legionella* bacteria: moisture, temperature (10–60 C [50–140 F]), oxygen, and a source of nourishment (algae/slime) (C. V. Broome, 1983).

Transmission

Legionnaires' disease can be spread through the air. It is not known whether the bacterium can be spread in other ways. In the outbreaks that have been investigated, there has been no evidence of person-to-person spread (*Questions and Answers on Legionnaires' Disease,* CDC No. 28L0343779).

The bacterium has been found in air-conditioning cooling towers and evaporative condensers where outbreaks have occurred. It is not known if this is an important means of spreading of legionnaires' disease. Other outbreaks have occurred in buildings that did not have air-conditioning units.

Target populations

The people at risk for contracting legionnaires' disease include:

1. persons with lowered immunological capacity (for example, cancer patients)
2. cigarette smoking and alcohol abuse are clearly risk factors in acquiring legionnaires' disease.
3. individuals exposed to high concentrations of *Legionella pneumophila*—not well defined. Not all exposed individuals develop the disease.

Symptoms and effects

Legionnaires' disease has most commonly been recognized as a form of pneumonia. Symptoms of this syndrome usually become apparent 2–10 days after known or presumed exposure to airborne legionnaires' disease bacteria. A nonproductive cough is common, but sputum production is sometimes associated with the disease. Within less than a day, the patient can experience rapidly rising fever and the onset of chills. Although physical examination of the patient may reveal unusual sounds on auscultation, fever to 39–41 C (102–105 F), and relatively slow heartbeats, no physical findings are specific to this disease. Associated manifestations can include mental confusion, chest pain, abdominal pain, impaired kidney function, and diarrhea (W.B., Baines, May 1979). Deaths have been reported—about 15 percent of the known cases have been fatal. It is estimated in the United States that around 25,000 people develop legionnaires' disease annually (*Questions and Answers on Legionnaires' Disease,* CDC No. 28L0343779).

Controls

Cooling towers. Procedures used to control bacterial growth vary somewhat on the various regions in a cooling tower system but generally involve a good maintenance program, including repair of damaged components, routine cleaning, and sterilization.

A typical sterilization solution might contain a nominal 50 ppm of residual chlorine combined with a compatible detergent to produce the desired sterilization effect. Special care must be taken in the use of these higher concentrations of biocide, with regard to the wood components of the tower, as these solutions can seriously affect the structural integrity of some of the components being treated. Care must be taken to ensure that the "nominally" inaccessible areas of the inside surfaces of the exterior walls are properly cleaned of algae or slime accumulations.

Air handling units: condensate drain pans. The typical condensate drain pan represents that portion of the air handling unit that offers the most potential for bacteria accumulation. It is important to keep the pans clean and checked for proper gravity drainage of fluid, i.e., to prevent stagnation and the buildup of algae/slime/bacteria. In the event algae or slime have been found to exist in this area, a cleaning and sterilization program will be required.

Other systems that should be reviewed for potential bacterial buildup are as follows: evaporative condensers, humidifiers, stagnant waters, and plumbing systems.

SUMMARY

Before beginning a biohazard control program, management must create and make a firm commitment to a policy about what level of occupational exposure to infectious agents is acceptable, for example, a policy to prevent subclinical infections that can be detected only serologically, or one that prevents those infections resulting in incapacitating illnesses or incurable diseases. Deciding what degree of protection should be provided for those persons working in areas peripheral to the immediate biohazardous work area is essential.

When designing a program, management must first develop the goals of the company in controlling biohazards, develop an understanding of the nature of the biohazards, and write policies to be followed for hazard control. The safety and health program should include:

1. analysis and identification of biohazards;
2. formulation of health and safety regulations and a health and safety plan;
3. training for personnel to recognize and avoid exposures, with extra training for unique hazards;
4. inspections and enforcement of safety and health rules with development of contingency plans for handling emergencies;
5. adequate reporting and investigating of accidents;
6. a medical surveillance program of preplacement and periodic physical examinations;
7. adequate funding to carry out the program in its entirety.

All decisions about safety equipment and safety procedures must be based on the attitude that the work area be made biologically safe before work is begun. This is especially important, since violation of this policy seems to be the source of most infections.

The effects of biohazardous agents are often subtle and slow to develop, thus, people are not alerted to take precautions against exposures. An infecting dose may be odorless, tasteless, and invisible to the eye, thus when an infection does develop, its origin may be difficult to trace. Because of this lack of physical evidence, much educational effort is required to make both employers and employees aware of the need for control of microbiological environments.

The safe handling of potentially hazardous biological agents depends on (1) the laboratory worker's awareness of the situation, and (2) the worker's meticulous attention to safe laboratory practices. Facilities and containment equipment can do much to protect the laboratory worker; however, attitude and concern in complying with all safety rules and procedures is the most important element in an effective biohazard control program. Management must make a firm commitment to the safety and health program and convey this attitude throughout the training, education, and enforcement of safety and health procedures.

Good techniques alone, however, will not prevent infection of laboratory workers. Engineering must design a physical separation of the worker's environment from that of the microorganism. All systems and procedures must be designed to keep the biohazard in its proper container, and to keep the employees out of contact with the biohazards.

The introduction of new biohazardous agents and the execution of unusual operations should be done safely to assure proper biohazards control. A review and analysis of the literature is usually recommended before initiation of new procedures and agents. The principal investigator or supervisor should consult with the safety and health professional and the biohazards committee to obtain an estimate of the degree of risk from the biohazardous material based on established classification schemes and other available references.

Instituting control and safety programs is in everyone's best interest. Injuries and illness cause human suffering and create problems for management. However, effective controls result from a combined effort—all staff involved with biohazardous agents, i.e., technicians, physicians, and administrators—must work together to follow and enforce recommended procedures and guidelines. This will reduce the amount of potential health hazard.

BIBLIOGRAPHY

Acquired immunodeficiency syndrome (AIDS). *MMWR* 34(1985):681–685, 691–694.

American Industrial Hygiene Association (AIHA), AIHA Biohazards Committee. *Biohazards Reference Manual,* Akron, OH: AIHA, 1985.

American Society of Hospital Pharmacists (ASHP). *Safe Handling of Cytotoxic Drugs—Study Guide.* Bethesda, MD: ASHP, 1986.

Baine, W. B. "The Epidemiology of Legionnaires Disease." In *Legionnaires': The Disease, the Bacterium, and Methodology,* HEW Pub. No. (CDC) 79-8375. Atlanta: DHHS, PHS, CDC, May 1979.

Barbeito, M. S., Alg, R. L., and Wedum, A. G. Infectious bacterial aerosol from dropped petri dish cultures. *American Journal Medical Technology* 27 (1961):318-322.

Benenson, A. S. *Control of Communicable Diseases in Man,* 14th ed. Washington, DC: American Public Health Association, 1985.

Brock, T. D. *Biology of Microorganisms,* 4th ed. Englewood Cliffs, NJ: Prentice-Hall Inc., 1984.

Brock, T. D., et al. Basic Microbiology with Application, 3rd ed., Englewood Cliffs, NJ: Prentice-Hall, Inc., 1986.

Broome, C. V. Epidemiologic assessment of methods of transmission of legionellosis. *Zbl. Bakt. Hyg. I., Abt. Orig.* A:225(1983):52-57.

Brown, H. W. *Basic Clinical Parasitology,* 5th ed. New York: Appleton-Century-Crofts, 1983.

Coordinated Framework for Regulation of Biotechnology: Announcement of Policy and Notice for Public Comment. *Federal Register,* June 26, 1986, pp 23, 302–23, 393.

Dickie, H. A., and Murphy, M. E. Laboratory infection with *histoplasma capsulatum. American Review Tuberculosis* 72 (1955):690.

Donham, K. Infectious diseases common to animals and man of occupational significance to agricultural workers. Proceedings of Society for Occupational and Environmental Health Conference on Agricultural Health and Safety, Sept. 4-5, 1974, Iowa City, IA, pp 160-173.

Ehrlich, R., and Henry, M. C. Chronic toxicity of nitrogen dioxide—Effect on resistance to bacteria pneumonia. *Archives of Environmental Health* 17 (1968):860-865.

Ehrenkranz, N. Joel. Statewide hospital infection surveillance. *Archives of Environmental Health* 30 (1975):514-516.

Favero, M. S. Sterilization, disinfection, and antisepsis in the hospital, In: *Manual of Clinical Microbiology,* 4th ed. Washington, DC: American Society for Microbiology, 1985, 129-137.

Fuscalso, A. A., Erlick, B. S., and Hindman, B. *Laboratory Safety: Theory and Practice.* New York: Academic Press, 1980.

Garner, J. S., and Favero, M. S. *Guideline for Handwashing and Hospital Environmental Control, 1985.* Atlanta: CDC, Pub. No. 99-1117, 1985.

Guide for the Care and Use of Laboratory Animals. U.S. Department of Health, Education, and Welfare, Public Health Service, National Institutes of Health, 1972. Pub. No. 73-23.

Hellman, Alfred, ed. *Biohazard Control and Containment in Oncogenic Virus Research.* Washington, DC: U.S. Department of Health, Education, and Welfare, National Institutes of Health, Public Health Service, Pub. No. 73-459.

Hellman, A., Oxman, M. N., and Pollack, R. *Biohazards in Biological Research.* Cold Spring Harbor, NY: Cold Spring Harbor Laboratory, 1973.

Hornibrook, J. W., and Nelson K. R. An institutional outbreak of pneumonitis—Epidemiological and clinical studies. *Public Health Reports* 55 (1940): 1936-1954.

Hubbert, W. T., McCulloch, W. F., and Schnurrenberger, P. R. *Diseases Transmitted from Animals to Man,* 6th ed. Springfield, IL: Charles C Thomas Publ., 1975.

Huddleson, I. F., and Munger, M. "A Study of an Epidemic of Brucellosis Due to Brucella Melitensis." *American Journal of Public Health,* 30 (1940), pp. 944-954.

Hudson, R. P. Lessons from Legionnares' Disease. *Annals of Internal Medicine,* April 1979:704-707.

Huebner, R. J. Report of an outbreak of "Q" fever at National Institutes of Health. *American Journal Public Health,* 37 (1947):431-440.

Hunter, D. *The Diseases of Occupation.* London, U.K.: English Universities Press, Ltd., 1975, chapt. 6, pp 676-717.

Kercher, S. L., and Mortimer, V. D. Before and after: An evaluation of engineering controls for ethylene oxide sterilization in hospitals. *Applied Industrial Hygiene* 2(1)(1987):7-12.

Kneedler, J. A., and Dodge, G. H. *Perioperative Patient Care.* Boston: Blackwell Scientific Publications, 1983:210-217.

Langone, J. AIDS update: Still no reason for hysteria. *Discover,* (Sept. 1986): 28-47.

Loffler, W., and Mooser, H. Mode of transmission of typhus fever: Study based on infection of group of laboratory workers. *Schweizerische Medizinishe Wochenschrift,* 72 (1942): 755-761.

McCoy, G. W. Accidental psittacosis infection among the personnel of the hygienic laboratory, *Public Health Reports,* 45 (1930):843-845.

Merger, C. Hazards associated with the handling of pathogenic bacteria. *Canadian Journal Laboratory Technology,* 18 (1957): 208-210.

Morrison, Robert P. Surveillance of patient areas. *Hospitals* 45 (February 16, 1971).

Murray, J. F., and Howard, D. Laboratory-acquired histoplasmosis. *American Review of Respiratory Diseases* 89 (1964): 631-640.

Osterholm, M. T., et al. A 1957 outbreak of Legionnaires' disease associated with a meat packing plant. *American Journal of Epidemiology* 117(1)(Jan, 1983):60-67.

Oviatt, Vinson, R. Environmental health concerns. *Hospitals* 47 (April 1, 1973).

Phillips, G. B. *Causal Factors in Microbiological Laboratory Accidents and Infections.* Fort Detrick, MD: U.S. Army Biological Laboratories, April 1965.

Phillips, G. B. Microbiological safety in U.S. and foreign laboratories. *Technical Report BL* 35, U.S. Army Chemical Corps Biological Laboratories, Fort Detrick, Frederick, MD (September, 1961).

Phillips, G. B. "Prevention of Laboratory-Acquired Infections." In *Handbook of Laboratory Safety,* 2nd ed, compiled by N. V. Steere. Cleveland: Chemical Rubber Co., 1971, pp 610–617.

Questions and Answers on Legionnaires' Disease, No. 28L0343779. Atlanta: DHHS, CDC.

Proposal for a Coordinated Framework for Regulation of Biotechnology: Notice *Federal Register,* December 31, 1984, pp. 50, 856–50, 907.

OSHA. Proposed Guidelines. *Federal Register,* April 12, 1985, pp. 14, 468–14, 469.

——. *Work Practice Guidelines for Personnel Dealing with Cytotoxic (Antineoplastic) Drugs,* Appendix A, Office of Occupational Medicine, OSHA Instruction Pub. 8-1.1, January 29, 1986.

Robbins, F. C., and Rustigan, R. Q fever in the Mediterranean area: Reports of its occurrence in allied troops, IV. A laboratory outbreak. *American Journal Hygiene,* 44(1946):64-71.

National Cancer Institute, Office of Research Safety. *Safety Standards for Research Involving Oncogenic Viruses.* Washington, DC: U.S. Department of Health, Education, and Welfare (October, 1974).

Smith, C. E. The hazard of acquiring mycotic infections in the laboratory. An address delivered before the Epidemiology and Laboratory Sections, American Public Health Association Meeting, November 2, 1950, St. Louis, MO.

Stark, A. Policy and procedural guidelines for health and safety of workers in virus laboratories. *American Industrial Hygiene Association Journal* 36(1975):234-240.

Steele, J. H. *Zoonoses as Occupational Diseases in Agriculture and Related Industries.* Chicago: American Medical Association, 1977.

Sulkin, S. E., and Pike, R. M. Survey of laboratory infections. *American Journal Public Health* 41(1951):769-781.

Survey of Viral and Rickettsial Infections Contracted in the Laboratory. National Cancer Institute, National Institutes of Health, Bethesda, MD: U.S. Department of Health, Education and Welfare.

USDHEW (now DHHS), CDC. *Biosafety in Microbiological and Biomedical Laboratories.* USDHHS Pub. No. HHS 84-8395. Washington, DC: USHHS, 1984.

USDHEW (now DHHS), CDC, PHS. *Classification of Etiologic Agents on the Basis of Hazard.* Atlanta: CDC, 1974.

USDHEW (now DHHS), NIH. Recombinant DNA research. Proposed revised guidelines. *Federal Register* 42, No. 187.

USDHEW (now DHHS), NIH, PHS. *Biohazards Safety Guide.* Washington, DC: U.S. Government Printing Office, 1974.

Volk, W. A., and Wheeler, M. F. *Basic Microbiology,* 5th ed. Philadelphia: J.B. Lippincott & Co., 1984.

Williams, W. W. Guideline for infection control in hospital personnel. *Infection Control* 1985; 4:326-349.

Wintrobe, M. M., et al., eds. *Harrison's Principles of Internal Medicine,* 10th ed. New York: McGraw-Hill Book Co., 1983.

PART IV

Evaluation of Hazards

15

Industrial Toxicology

by Ralph G. Smith
Julian B. Olishifski, MS, PE, CSP
Revised by Carl Zenz, MD, ScD

Toxicology is the science that deals with the poisonous or toxic properties of substances. Everyone is exposed on and off the job to a variety of chemical substances; most do not present a hazard under ordinary circumstances, but they all have the potential for being injurious at some sufficiently high concentration and level of exposure. How a material is used is the major determinant of its hazard potential. Any substance contacting or entering the body will be injurious at some excessive level of exposure and can be theoretically tolerated without effect at some lower exposure.

DEFINITION

A toxic effect can be defined as any noxious effect on the body—reversible or irreversible; any chemically induced tumor, benign or malignant; any mutagenic or teratogenic effect or death—as a result of contact with a substance via the respiratory tract, skin, eye, mouth, or any other route. Toxic effects are undesirable disturbances of physiological function caused by poisons. Toxic effects can also arise as side effects in response to some medication, vaccines, and exposure to chemicals. As far as medicines are concerned, their acceptability is usually determined by balancing any adverse effects against the therapeutic results achieved.

Toxicity is a physiological property of matter that defines the capacity of a chemical to harm or injure a living organism by other than mechanical means. Toxicity entails a definite dimension—quantity or amount. Then, the toxicity of a chemical depends upon the degree of exposure.

Exposure to or contact with a small amount (the exact amount may have to be very small to be without effect) of a specific material can have no effect. That includes even potentially carcinogenic agents and the most toxic substances; if the concentration is small enough, there may be no effect after exposure. However, if an organism is exposed to or in contact with too much of a specific material, the consequences may be severe and could include death. The extent of the reaction is an aspect of acute toxicity.

There is scientific controversy about whether there actually is a "no effect" level for a carcinogenic substance. The view included in the revision of the Food, Drugs and Cosmetics Act, 1952 (the Delaney Amendment), was that a no-effect level could not be proved. Hence, the permissible level of carcinogens added to foods was set at zero.

The responsibility of the industrial toxicologist is to define how much is too much, and to prescribe precautionary measures and limitations so that normal, recommended use does not result in the intake of too much of that particular material. From a toxicological viewpoint, the industrial hygienist must consider all types of exposure and the subsequent effects on the living organism.

Many chemicals essential for health in small quantities are highly toxic in larger quantities. Small amounts of zinc, manganese, copper, molybdenum, selenium, chromium, nickel, tin, potassium, and many other chemicals are essential for life. However, severe acute and chronic toxicity results from exposure to large amounts of these same materials. For example, nickel and chromium in some of their forms are considered carcinogens.

Toxicity v. hazard

A distinction must be made between toxicity and hazard. Toxicologists generally consider toxicity as the ability of a substance to produce an unwanted effect when the chemical has reached a sufficient concentration at a certain site in the body; hazard is regarded as the probability that this concentration in the body will occur. Many factors contribute to determining the degree of hazard—route of entry, dosage, physiological state, environmental variables, and many others. To some extent, assessing a hazard involves estimating the probability or likelihood that a substance will cause harm. In evaluating a hazard, toxicity is but one factor. Others are chemical and physical properties, including warning properties such as odor. Two liquid materials can possess the same degree of toxicity but present different degrees of hazard. One material may be odorless and not irritating to the eyes and nose while the other may have a pungent or disagreeable odor in minute concentration, or be an eye or respiratory irritant. By comparison, the material with the warning properties presents a lesser degree of hazard. Its presence can be detected in time to avert injury.

Many chemical agents are not selective in their action on tissues or cells; they can exert a harmful effect on all living matter. Other chemical agents may act only on specific cells. Some other agents may be harmful only to certain species; other species may have built-in protective mechanisms.

The term toxicity is commonly used in comparing one chemical agent with another, but is rendered meaningless by omitting data designating the biological species used and the conditions under which the harmful effects were induced.

A chemical stimulus can be considered to have produced a toxic effect when it satisfies the following criteria:

- An observable or measurable physiological deviation has been produced in any organ or organ system. The change may be anatomic in character and may accelerate or inhibit a normal physiological process, or the deviation may consist of a specific biochemical change.
- The observed change can be duplicated from animal to animal even though the dose-effect relationships may vary quantitatively.
- The stimulus has changed normal physiological processes in such a way that a protective mechanism is impaired in its defense against other adverse stimuli.
- The effect is either reversible or at least attenuated when the stimulus is removed.
- The effect does not occur without a stimulus or occurs so infrequently that it indicates generalized or nonspecific response. When high degrees of susceptibility are noted, equally significant degrees of resistance should be apparent.
- The observation must be noted and must be reproducible by other investigators.
- The physiological change reduces the efficiency of an organ or function and impairs physiological reserve in such a way as to interfere with the ability to resist or adapt to other normal stimuli in either a permanent or temporary manner.

Although the toxic effects of many chemical agents used in industry are well known, there are many other commonly used chemicals that are not as well defined. The toxicity of a material is not a physical constant—such as boiling point, melting point, vapor pressure, or temperature—and usually only a general statement can be made concerning the harmful nature of some of the chemical agents.

If the toxicity of a chemical could be predicted from its chemical constitution or structural formula, it would certainly make the job easier. While certain important analogies are apparent between structure and toxicity, important differences exist that require individual study of each compound.

In addition to its toxicity, an evaluation of a chemical hazard involves establishing the amount and duration of exposure, the physical characteristics of the substance, the conditions under which exposure occurs, and the determination of the effects of other substances in a combined exposure. All of these may significantly influence the toxic potency of a substance.

The chemical properties of a compound often can be one of the main factors in its hazard potential. The vapor pressure determines whether or not a given substance has the potential to pose a hazard from inhalation. Many solvents are troublesome because they are quite volatile and vaporize readily into the air to produce high concentrations of vapor in a given set of circumstances. Hence, a low-boiling solvent, other things being equal, would be a greater hazard than a high-boiling solvent simply because it is more volatile and it evaporates quicker.

Chemical injury can be local or systemic. Local injury is the result of direct contact of the irritant with tissue. The skin can be severely burned or surface of the eye can be injured to the extent that vision is impaired. In the respiratory tract, the lining of the trachea and the lungs can be injured as a result of inhaling toxic amounts of vapors, fumes, dusts, or mists. These are all examples of direct chemical contact with tissues and the toxicological reactions can be slight or severe.

ENTRY INTO THE BODY

In discussing toxicity, it is necessary to describe how a material gains entrance into the body and then into the bloodstream. A material cannot produce systemic injury unless it gains entry into the bloodstream. Common routes of entry are ingestion, injection, skin absorption and inhalation. Depending on the substance and its specific properties, however, entry and absorption can occur by more than one route. For instance, inhaling a solvent that can penetrate the skin. Where absorption into the bloodstream occurs, a toxicant may elicit general effects or, more than likely, the critical injury will be localized in specific tissues or organs.

Ingestion

Anything eaten or drunk gets into the intestine and may be absorbed into the blood and thereafter prove to be toxic. In general, the term "toxicity" is associated by most people with things taken by mouth. The problem of ingesting chemicals is not widespread in industry—most workers do not deliberately swallow materials they handle. Occasionally, however, a material must be handled where accidental swallowing of very small but hazardous amounts does occur.

Ingestion of toxic materials may occur as a result of eating lunch in contaminated work areas, and contaminated fingers and hands can lead to accidental oral intake when a worker smokes on the job. Ingestion of inhaled materials can also occur because contaminants deposited in the respiratory tract can be carried out to the throat by the action of the ciliated lining of the respiratory tract. These contaminants are then swallowed and significant absorption of the material may occur from the gastrointestinal tract. Approximately one quart of mucus is produced daily in the lungs of an adult, this constant flow of mucus can carry contaminants out of the lungs into the throat to be swallowed with the saliva or coughed up and expectorated.

Oral toxicity is generally lower than the inhalation toxicity for the same material due to the relatively poor absorption of many materials from the intestines into the bloodstream. Food and liquid mixed with a toxic substance not only provide dilution, but also provide an opportunity to form less soluble substances. After absorption from the gastrointestinal system into the bloodstream, the toxic material goes to the liver, which can further metabolically alter, degrade, or detoxify many substances. This detoxification process is an important body defense mechanism. Basically, detoxification involves a sequence of reactions: deposition in the liver, conversion to a nontoxic compound, transportation to kidney via the bloodstream, and excretion through the kidney and urinary tract.

Injection

A material can be injected into some part of the body. This can be done directly into the bloodstream or into the peritoneal cavity of the abdomen, or the pleural cavity, surrounding the lung. The material can also be injected into the skin, muscle, or just about any other place a needle can be inserted. The effects produced vary with the route of administration. Injection is not too important as a route of worker exposures.

In the laboratory, toxic substances may be injected into animals because it is far more convenient and less costly than establishing blood levels by inhalation exposures. Injection is a rather imperfect way of getting toxic materials into the body if the end result is to gain knowledge concerning the potential industrial hazard. This is because using intravenous injection short-circuits protective mechanisms in the body, which prevent substances from entering the blood.

Skin absorption

An important route of entry in terms of occupational exposure is absorption through either intact or abraded skin. Contact of a substance with skin results in four possible actions: (1) the skin can act as an effective barrier, (2) the substance can react with the skin and cause local irritation, (3) the substance can produce skin sensitization, and (4) the substance can penetrate to the blood vessels under the skin and enter the bloodstream.

For some substances (for example, parathion) the skin is the main portal of entry under typical conditions of occupational exposure. For other substances the amounts absorbed through the skin may be roughly equivalent to inhalatory absorption (for example, aniline, nitrobenzene, phenol). For the majority of other organic compounds the contribution from skin (cutaneous) absorption to the total exposure cannot be neglected or excluded. Results from animal studies using phospho-organic compounds show that toxic effects can occur because of cutaneous penetration.

The cutaneous absorption rate of some organic compounds rises when temperature or perspiration increases. Therefore, absorption is higher in warm climates or seasons. The absorption of liquid organic compounds may follow surface contamination of the skin or clothes; for other compounds it may occur directly from the vapor phase, in which case the rate of absorption is roughly proportional to the air concentration of the vapors. The process may be a combination of absorption of the substances on the skin surface followed by absorption through the skin.

The physiochemical properties of a material mainly determine whether or not a material will be absorbed through the skin. Among the important factors are pH of the skin and the following properties of the chemical: extent of ionization; aqueous and lipid solubilities, and molecular size.

Human skin shows great differences in absorption at different anatomic regions. Using the skin of the forearm as a frame of reference, the skin on the palm of the hand shows approximately the same penetration as that of the forearm for certain organic phosphates. The skin on the back of the hand and the skin of the abdomen have twice the penetration potential of that of the forearm, whereas follicle-rich sites, such as the scalp, forehead, angle of the jaw, postauricular area (behind the ear), and the scrotum show a much greater penetration potential. Temperature elevation can be expected to increase skin absorption by increasing vasodilation. If the skin is damaged by scratching or other abrasion, the normal protective barrier to absorption of chemicals is lessened and penetration can more easily occur. (See Chapter 3, The Skin, and Chapter 8, Industrial Dermatoses, for more information.)

Inhalation

For industrial exposures to chemicals, the most important route of entry is usually inhalation. Nearly all materials that are airborne can be inhaled.

The respiratory system is composed of two main areas: (1) the upper respiratory tract airways—the nose, throat, trachea, and major bronchial tubes leading to the various lobes of the lungs; and (2) the alveoli where the actual transfer of gases across thin cell walls takes place. Only particles smaller than about 5 micrometers in diameter are likely to enter the alveolar sac.

The total amount of a toxic compound absorbed via the respiratory pathways depends upon its concentration in the air, the duration of exposure, and pulmonary ventilation volumes, which increase with higher work loads. If the toxic substance is present in air in the form of an aerosol, deposition and absorption will occur in the respiratory tract. For more details, see Chapter 2, The Lungs.

Gases and vapors having low water solubility but being highly soluble in fats pass through the alveolar lining into the blood and are distributed to organ sites for which they have special affinity. In the course of inhalation exposure at a uniform level, the concentration of the compound in the blood reaches an equilibrium value between absorption, on the one hand, and metabolism and elimination, on the other.

DOSE-RESPONSE RELATIONSHIP

All toxicological considerations are based on the dose-response relationship. A dose is administered to test animals, and, depending on the outcome, is increased or decreased until a range is found where at the upper end all animals die and, at the lower end, all animals survive. The data collected are used to prepare a dose-response curve relating percent mortality to dose administered (Figure 15-1).

The doses given are expressed as the quantity administered per unit body weight, quantity per skin surface area, or quantity per unit volume of the respired air. In addition, the length of time during which the dose was administered should also be listed.

The dose-response relationship can also be expressed as the product of a concentration (C) multiplied by the time duration

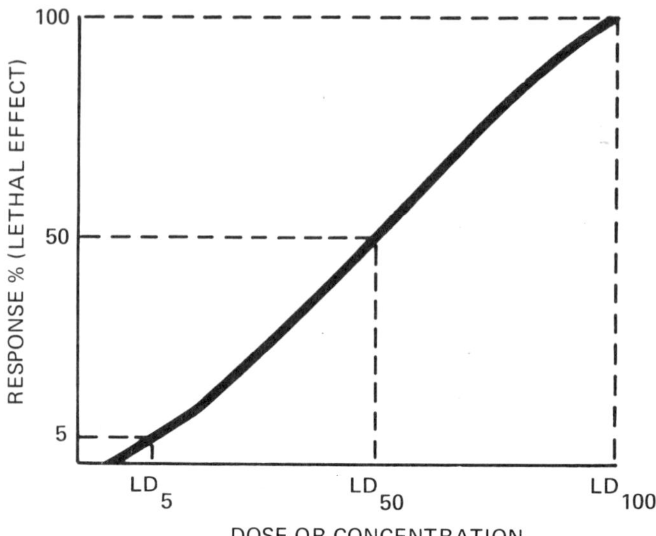

Figure 15-1. Dose-response curves for a chemical agent administered to a uniform population of test animals.

(T) of exposure. This product is proportional more or less to a constant (K); or mathematically, $(C)\times(T)\approx K$. The dose involves two variables—concentration and duration of exposure. For certain chemicals, a high concentration breathed for a short time produces the same effect as a lower concentration breathed for a longer time. The CT value can be used to provide a rough approximation of other combinations of concentration of a chemical and time that would produce similar effects. Although this concept must be used very cautiously and cannot be applied at extreme conditions of either concentration or time, it can be useful in predicting safe limits for airborne contaminants in respect to environmental exposures. Safe limits are set so that the combination of concentrations and time durations are below the levels that will produce injury to exposed individuals.

Exposure to low levels of some chemicals, such as ammonia, causes so much physical irritation that workers will not voluntarily tolerate harmful concentrations. If a person became trapped, however, he would be injured by the excessive exposure.

Threshold concept

A sufficiently small amount of most chemicals is not harmful. This means that there is a threshold of effect or a "no effect" level. The most toxic chemical known, if present in small enough amounts—a few molecules—will produce no measurable effect. It may damage one cell or several cells, but no effect, such as kidney dysfunction, will be measured. As the dose is increased, there is a point when the first measurable effect is noted. The toxic potency of a chemical is defined by the relationship between the dose (the amount) of the chemical and the response that is produced in a biological system. Thus, a high concentration of toxic substance in the target organ will cause a severe reaction, and a low concentration, a less severe reaction (Figure 15-2).

Although many exposures to substances in the industrial community occur by way of the respiratory tract or skin, most

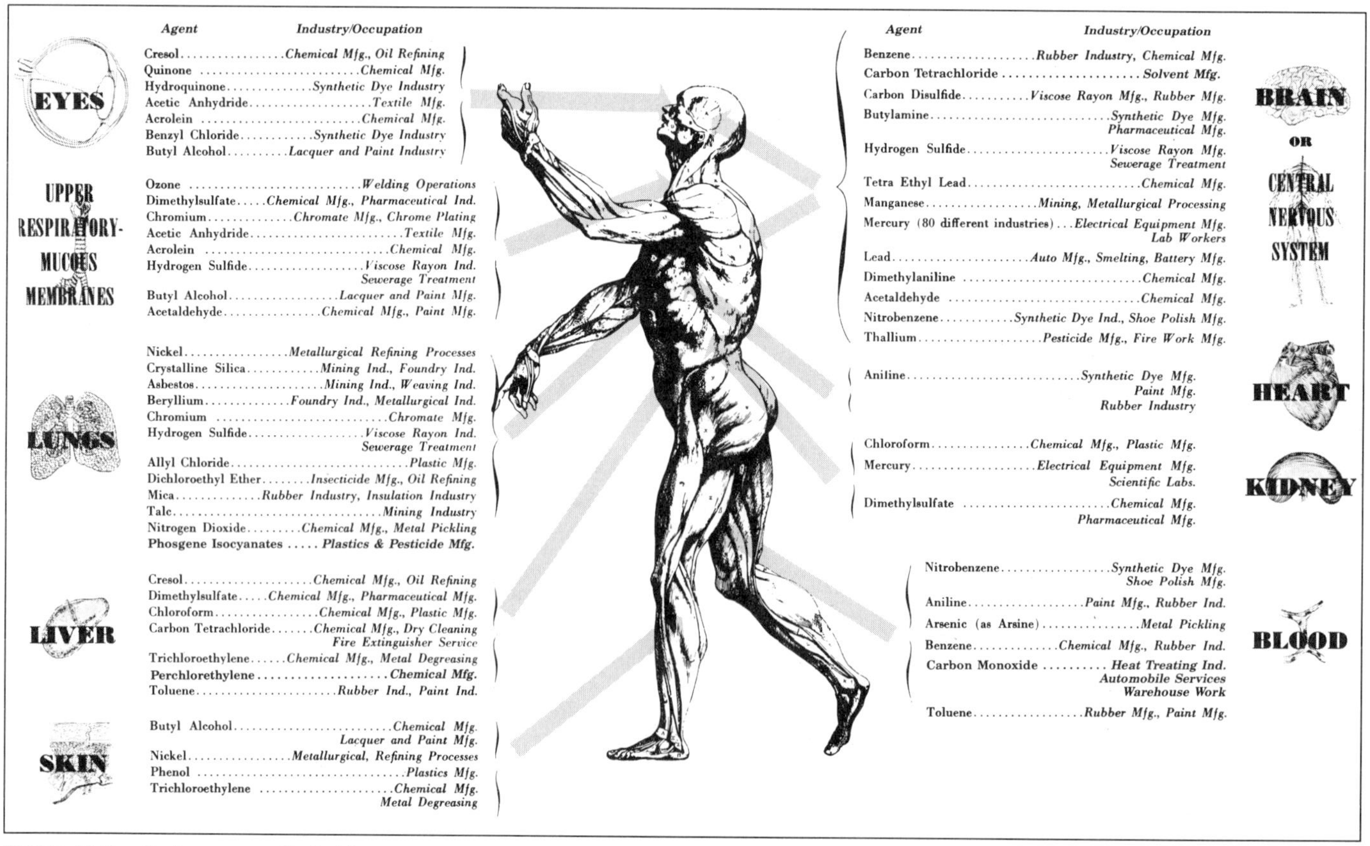

Organ	*Agent*	*Industry/Occupation*
EYES	Cresol	*Chemical Mfg., Oil Refining*
	Quinone	*Chemical Mfg.*
	Hydroquinone	*Synthetic Dye Industry*
	Acetic Anhydride	*Textile Mfg.*
	Acrolein	*Chemical Mfg.*
	Benzyl Chloride	*Synthetic Dye Industry*
	Butyl Alcohol	*Lacquer and Paint Industry*
UPPER RESPIRATORY-MUCOUS MEMBRANES	Ozone	*Welding Operations*
	Dimethylsulfate	*Chemical Mfg., Pharmaceutical Ind.*
	Chromium	*Chromate Mfg., Chrome Plating*
	Acetic Anhydride	*Textile Mfg.*
	Acrolein	*Chemical Mfg.*
	Hydrogen Sulfide	*Viscose Rayon Ind. Sewerage Treatment*
	Butyl Alcohol	*Lacquer and Paint Mfg.*
	Acetaldehyde	*Chemical Mfg., Paint Mfg.*
LUNGS	Nickel	*Metallurgical Refining Processes*
	Crystalline Silica	*Mining Ind., Foundry Ind.*
	Asbestos	*Mining Ind., Weaving Ind.*
	Beryllium	*Foundry Ind., Metallurgical Ind.*
	Chromium	*Chromate Mfg.*
	Hydrogen Sulfide	*Viscose Rayon Ind. Sewerage Treatment*
	Allyl Chloride	*Plastic Mfg.*
	Dichloroethyl Ether	*Insecticide Mfg., Oil Refining*
	Mica	*Rubber Industry, Insulation Industry*
	Talc	*Mining Industry*
	Nitrogen Dioxide	*Chemical Mfg., Metal Pickling*
	Phosgene Isocyanates	*Plastics & Pesticide Mfg.*
LIVER	Cresol	*Chemical Mfg., Oil Refining*
	Dimethylsulfate	*Chemical Mfg., Pharmaceutical Mfg.*
	Chloroform	*Chemical Mfg., Plastic Mfg.*
	Carbon Tetrachloride	*Chemical Mfg., Dry Cleaning Fire Extinguisher Service*
	Trichloroethylene	*Chemical Mfg., Metal Degreasing*
	Perchlorethylene	*Chemical Mfg.*
	Toluene	*Rubber Ind., Paint Ind.*
SKIN	Butyl Alcohol	*Chemical Mfg. Lacquer and Paint Mfg.*
	Nickel	*Metallurgical, Refining Processes*
	Phenol	*Plastics Mfg.*
	Trichloroethylene	*Chemical Mfg. Metal Degreasing*
BRAIN OR CENTRAL NERVOUS SYSTEM	Benzene	*Rubber Industry, Chemical Mfg.*
	Carbon Tetrachloride	*Solvent Mfg.*
	Carbon Disulfide	*Viscose Rayon Mfg., Rubber Mfg.*
	Butylamine	*Synthetic Dye Mfg. Pharmaceutical Mfg.*
	Hydrogen Sulfide	*Viscose Rayon Mfg. Sewerage Treatment*
	Tetra Ethyl Lead	*Chemical Mfg.*
	Manganese	*Mining, Metallurgical Processing*
	Mercury (80 different industries)	*Electrical Equipment Mfg. Lab Workers*
	Lead	*Auto Mfg., Smelting, Battery Mfg.*
	Dimethylaniline	*Chemical Mfg.*
	Acetaldehyde	*Chemical Mfg.*
	Nitrobenzene	*Synthetic Dye Ind., Shoe Polish Mfg.*
	Thallium	*Pesticide Mfg., Fire Work Mfg.*
HEART	Aniline	*Synthetic Dye Mfg. Paint Mfg. Rubber Industry*
KIDNEY	Chloroform	*Chemical Mfg., Plastic Mfg.*
	Mercury	*Electrical Equipment Mfg. Scientific Labs.*
	Dimethylsulfate	*Chemical Mfg. Pharmaceutical Mfg.*
BLOOD	Nitrobenzene	*Synthetic Dye Mfg. Shoe Polish Mfg.*
	Aniline	*Paint Mfg., Rubber Ind.*
	Arsenic (as Arsine)	*Metal Pickling*
	Benzene	*Chemical Mfg., Rubber Ind.*
	Carbon Monoxide	*Heat Treating Ind. Automobile Services Warehouse Work*
	Toluene	*Rubber Mfg., Paint Mfg.*

Figure 15-2. Body organs affected by exposure to some common industrial chemicals are illustrated. (Courtesy *Air Engineering Magazine*)

published reports of exposures concern studies of experimental animals in which the test substances were introduced primarily through the mouth: in food, in drinking water, or by intubation (tube) directly into the stomach.

The harmfulness of a material depends upon its chemical composition, the type and rate or degree of exposure, and the fate of the material in the body. A single large dose of a toxic substance can be expected to produce a greater response than the same total dose administered in small amounts over a long period of time. Each of the small amounts may be detoxified quickly but a large dose may produce its detrimental action before appreciable detoxification occurs. A toxic substance that is detoxified or excreted at a rate that is slower than the rate of intake of the substance becomes a cumulative poison.

Accumulation of a substance in the body is understood as a process in which the level of the substance increases with the duration of exposure and can apply to both continuous and to repeated exposures. Biological tests of exposure show that an accumulation is taking place when rising levels of the substance are seen in the urine, blood, or expired air (Figure 15-3).

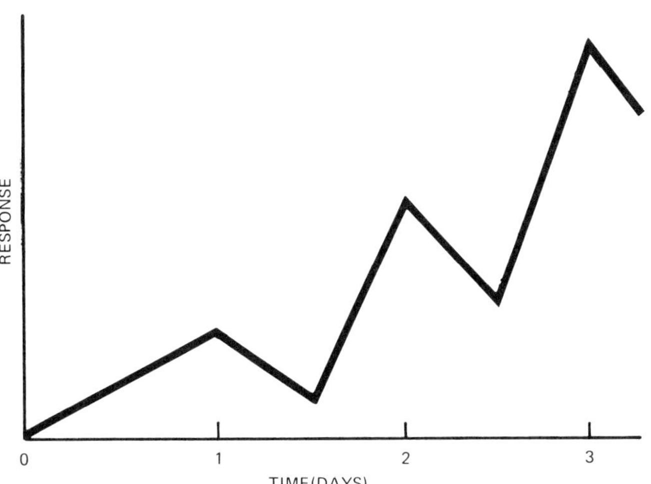

Figure 15-3. Accumulation of a substance in a body is shown in this curve where the level of the substance increases with the duration of exposure.

Lethal dose

If a number of animals are exposed to a toxic substance, when the concentration reaches a certain level, some, but not all, of those animals will be killed. Results of such studies are used to calculate the lethal dose (LD) of toxic substances.

If the only variable being studied is the number of deaths, it is possible to use the concept of the LD. The LD_{50} is the calculated dose of a substance that is expected to cause the death of 50 percent of a defined experimental animal population, as determined from the exposure to the substance, by any route other than inhalation.

Several designations can be used such as LD_{50}, LD_0, LD_{100}, and so on. The designation LD_0, which is not very often used, is the concentration that would produce no deaths in an experimental group and would be the highest concentration that would be tolerated in animals and result in zero deaths; LD_{100} would be the lowest concentration that produces death in 100 percent of the exposed animals. When that dose is administered, there is going to be 100 percent mortality.

The LD_{50} is the concentration that kills half of the exposed animals, but it does not mean that the other half are in good health. They may almost die, and they may or may not recover—that is a separate consideration.

Normally, LD_{50} units are weight of substance per kilogram of animal body weight, usually milligrams per kilogram or micrograms per kilogram, or whatever weight units are appropriate. In small animals, a given dose will have a greater effect than it will in a larger animal simply because it is truly a larger dose to a small animal. The LD_{50} value should be accompanied by an indication of the species of experimental animal used, the route of administration of the compound, the vehicle used to dissolve or suspend the material, if applicable, and the time period during which the animals were observed.

The slope of the dose-response curve provides useful information. It suggests an index of the margin of safety, that is, the magnitude of the range of doses involved in going from a noneffective dose to a lethal dose. If the dose-response curve is very steep, this margin of safety is slight. One compound could be rated as "more toxic" than a second compound because of the shape and slope of the dose-response curve (Figure 15-4).

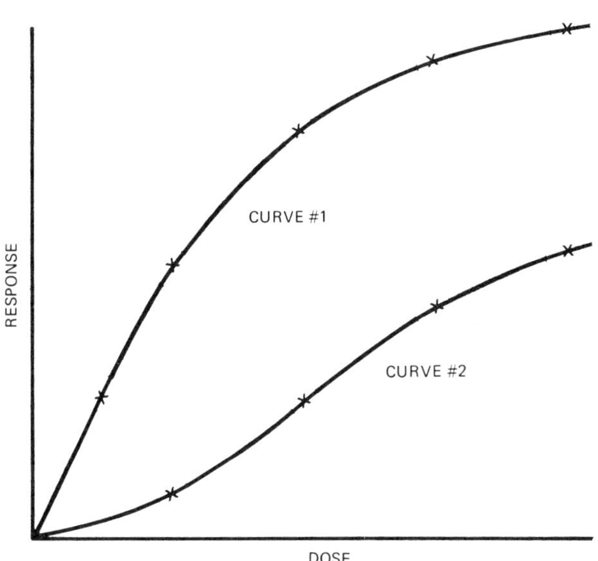

Figure 15-4. The slope of the dose-response curve provides useful information. If the dose-response curve is very steep, the margin of safety for exposure to that material is slight.

Lethal concentration

When considering inhalation exposures, the LD_{50} is not too useful—the dose by inhalation is needed. A similar designation, lethal concentration (LC) is used for airborne materials. An LC_{50} might be 500 parts per million (ppm), which means that when a defined experimental animal population is exposed to a calculated concentration of a substance, for example, a concentration of 500 ppm, that concentration of the substance is expected to cause the death of 50 percent of the animals in a stated length of time.

Time becomes very important because a half-hour exposure

might produce one effect, a few minutes' exposure another, and 24 hours' exposure would produce other effects.

Any publication dealing with LCs should state the species of animals studied, the length of time the exposure was maintained, and the length of time observation was carried out after exposure.

In one study, animals were exposed for a short time to nitrogen dioxide (NO_2). At first there was no observable response, but 36 hours after the exposure, the animals developed a chemical pneumonia, became very sick, and died. If the animals had been observed for only the 24-hour period after the exposure, the significant effects that occurred in the next 24–48 hours would have been missed. Thus, there must be an adequate postexposure time period. The dose should be delivered in a specified length of time with the animals under observation for another specified period of time—this may be 24 hours or 30 days, or even several months when testing for carcinogenesis.

Responses

After the toxic material has been administered, there are various criteria the toxicologist can use to evaluate the response.

Examining the organs removed from the exposed animals reveals the site of action of the toxic agent, the mode of action, and the cause of death. Important pathological changes in tissues can be observed following dose levels below those needed to produce the death of animals. The liver and the kidney are particularly sensitive to the action of many toxic agents.

The effect of the toxic agent on the growth rate of the animals is another criterion of adverse response. Relatively low levels of compounds that do not produce death or signs of serious illness can result in a diminished rate of growth. The food intake must also be measured to learn whether loss of appetite was a cause of diminished growth.

The weight of various organs, or more specifically the ratio of organ weight to body weight can be used as a criterion of adverse response. In some instances, such alterations are specific to the chemical being tested, for example, the increase of lung weight to body weight ratio, as a result of the pulmonary edema produced by irritants such as ozone or oxides of nitrogen.

Physiological function tests also provide useful criteria of response, both in experimental studies and in assessing the response of exposed workers. They can be especially useful in studies of populations with chronic conditions.

Substances can then be rated according to their relative toxicity as shown in animal experiments and the probable LD for humans can be estimated (Table 15-A). These examples are based on the results of short-term (acute) exposures only.

It is quite possible that actual long-term (chronic) exposures to a substance could produce serious tissue damage even though short-term exposure tests indicated a low order of toxicity.

Animal experimental data are sometimes difficult to interpret and apply to human exposure. Such data are valuable as guides for an industrial toxicologist in estimating the likely range of toxicity of a substance as well as in guiding further investigation.

ACTION OF TOXIC SUBSTANCES

The toxic action of a substance can be arbitrarily divided into acute and chronic effects. In addition to acute and chronic toxicity, we can distinguish acute and chronic exposures (Table 15-B). Factors other than immediate effects often determine the type and severity of a chemical's adverse effects. For example, acute benzene poisoning has a different clinical picture than chronic poisoning. And while ethyl alcohol has a somewhat greater systemic toxicity than methyl alcohol, the latter is much more dangerous when ingested, because methyl alcohol can produce a permanent serious effect on the optic nerve.

Acute effects

Acute exposures and acute effects involve short-term high concentrations and immediate results of some kind: illness, irritation, or death. Acute exposures are usually related to an accident. Where people work with toxic substances, the processes should be designed so exposure to large concentrations is not permitted or expected. However, in any plant sometimes materials will escape through a minor leak, a major break in a pipe, or accidental damage of some kind in the equipment.

Acute exposures, typically, are sudden and severe, and are characterized by rapid absorption of the offending material. For example, inhaling high levels of carbon monoxide or swallowing a large quantity of cyanide compound will produce acute poisoning very rapidly. The critical period, during which survival of the victim is uncertain, occurs suddenly. Such incidents generally involve a single exposure in which the chemical is rapidly absorbed and damages one or more of the vital organs. The effect of a chemical hazard is considered acute when it appears within a short time, such as within minutes or hours.

Chronic effects

In contrast to acute effects, chronic effect or illness is characterized by symptoms or disease of long duration or frequent recurrence. Chronic effects often develop slowly. The meaning of each term conforms to its derivation—the Latin *acutus* (sharpened) for acute, and the Greek *chronikos* (time) for chronic.

The term chronic relates to continued exposure to substances presumably throughout a working lifetime. It is impossible to make the workplace absolutely free of toxic substances, and in recognition of that fact, standards-making organizations try to establish limits that will control chronic exposures.

Chronic poisoning assumes that some level of material will be continuously present in the tissues. Chronic poisoning can also be produced by exposure to a harmful material that produces irreversible damage, so the injury accumulates, rather than the poison. The symptoms in chronic poisoning are usually different from those seen in acute poisoning by the same toxic agents, and since the level of contaminant is relatively low, the worker can be unaware of the exposures as they occur.

With most sensitizers, the first few exposures may cause no reaction; but once a person becomes sensitized, reactions can occur from later contact with very small quantities for very short periods of time.

Exposures

Levels of exposure to air contaminants can also be referred to in terms of acute and chronic exposure. Acute exposure generally refers to exposure to very high concentrations during very short time periods; chronic exposure involves repetitive or continuous exposure during long time periods.

Whether or not such chronic exposure is a potential health hazard generally depends upon the material and the worker's time-weighted average (TWA) exposure.

Table 15-A. Combined Tabulation of Toxicity Classes

Commonly Used Term	*LD_{50} Single Oral Dose for Rats (g/kg)*	*4-hr Vapor Exposure Causing 2–4 Deaths in 6-rat Groups (ppm)*	*LD_{50} Skin for Rabbits (g/kg)*	*Probable Lethal Dose for Humans*
Extremely toxic	≤ 0.001	< 10	≤ 0.005	Taste (1 grain)
Highly toxic	0.001–0.05	10–100	0.005–0.043	1 tsp (4 cc)
Moderately toxic	0.05–0.5	100–1,000	0.044–0.340	1 oz (30 gm)
Slightly toxic	0.5–5.0	1,000–10,000	0.35–2.81	1 pint (250 gm)
Practically nontoxic	5.0–15.0	10,000–100,000	2.82–22.6	1 quart
Relatively harmless	> 15.00	$> 100,000$	> 22.6	>1 quart

(Reprinted with permission from Spector, S.W., ed. *Handbook of Toxicology,* Vol. 1, *Acute Toxicities.* Philadelphia, PA: W.B. Saunders Co. 1956.)

Table 15-B. Brief Comparison of Acute, Prolonged, and Chronic Toxicity Tests

	Acute	*Prolonged*	*Chronic*
Exposure lasts. . .	≤ 24 hours, usually single dose	Typically 2, 4, or 6 weeks	≥ 3 months
Typically yields. . .	Single lethal dose, clinical signs of toxicity	Cumulative dose (if any), major metabolic routes, detoxification or excretion	Potential for carcinogenic effect or other delayed effects
Exemplified by. . .	Potassium cyanide rapidly depriving tissues of vital oxygen	Carbon tetrachloride exerting cumulative effect on liver, from repeated exposure over a period of several weeks	Mercury poisoning that may be insidious and slow, as from contaminated food eaten over months or years

Do not assume that chronic effects of air contaminants are less serious than acute effects simply because they result from exposure to lower concentrations of toxic materials. In fact, the opposite may be true—although the onset of damage to health can be slow, the ultimate effect can be quite serious and irreversible.

An effect is local if it harms only that part of the body it comes in contact with, as with an acid burn of the skin. A systemic effect is generalized and changes the normal functioning of related organs operating as a system. An example is the action of carbon monoxide upon the blood, or hydrogen cyanide upon tissue oxidation, and the ultimate effects of both upon the central nervous system.

EFFECTS OF EXPOSURE TO AIR CONTAMINANTS

Air contaminants can be classified on the basis of physiological action into irritants, asphyxiants, and narcotics or anesthetics along with other actions not fitting into these three groups.

The physiological responses to toxic materials depend upon the concentration and duration of exposure. For example, a vapor or gas at one concentration can exert its principal action on the body as an anesthetic, while at a lower concentration for a longer exposure time, the same gas or vapor can injure some internal organ or the blood system.

Irritation

Toxic effects can include simple irritation. Irritation means an aggravation of whatever tissue the material comes in contact with. Contact of some materials with the face and upper respiratory system affects the eyes, the tissues lining the nose, and the mouth.

There are many industrial chemicals that are not truly toxic, because they do not produce irreversible damage to some organ; nonetheless at fairly low concentrations, they will irritate tissues with which they come in contact.

Many irritants are liquids and for many of these the degree of local irritation has no relation to their systemic toxicities. Sometimes differences in viscosity are the determining factors in the type of injury. This applies especially in the lungs where the inhalation hazard from a substance of low viscosity, such as kerosene, is quite different and more severe than the hazard from a higher viscosity substance like medicinal mineral oil.

To a large extent the solubility of an irritant gas influences the part of the respiratory tract that is affected (Table 15-C). Ammonia, which is very soluble in water, irritates the nose and throat, primarily because the moisture on the surface absorbs the ammonia. Nitrogen dioxide, being much less soluble, acts mainly on the tissues in the lungs by traveling deep into the lungs before any significant absorption on moist surfaces occurs.

Some chemicals produce acute pulmonary edema (fluid in lungs), which usually begins as an immediate or intense irritation that is later manifested by coughing, dyspnea, cyanosis, or expectoration of large amounts of mucus. With other chemicals, irritation can be delayed or an immediate reaction can be followed by a period of remission—typically a few hours for phosgene or 24–48 hours for nitrogen oxides. Sensitizing irritants, such as toluene diisocyanate, induce asthmatic bronchitis.

Respiratory irritants can be inhaled in gaseous form, as a mist, or as particles with a coating of absorbed liquid. Irritants are frequently grouped according to their site of action (Table 15-C).

Irritants can be subdivided into primary and secondary irritants. A primary irritant is a material that exerts little systemic toxic action, either because the products formed on the tissues of the respiratory tract are nontoxic or because the irritant action is far in excess of any systemic toxic action.

Table 15-C. Comparison of Several Irritants Affecting Various Areas in Respiratory Tract

Description	*TLV*	*Concentrations Exceeding TLV*
A. Irritants Affecting Upper Respiratory Tract		
Formaldehyde (HCHO) Aldehyde, colorless gas at ordinary temperatures. Soluble in water up to 55%. (Formalin is aqueous solution).	TLV = 1 ppm based on complaints of irritation <1 ppm, constant prickling irritation, disturbed sleep. Substance suspect of carcinogenic potential in humans.	10-20 ppm causes severe difficulty in breathing, intense lacrimation, severe cough. (Allergic asthma may occur at low concentrations, depending on sensitivity of subject.)
Acrolein ($CH_2 = CHCHO$) Aldehyde, colorless or yellowish liquid. Soluble in 2 to 3 parts water.	TLV = 0.1 ppm, low enough to minimize, but not entirely prevent, irritation in exposed individuals.	1 ppm may be strongly irritating to eyes and nose within five minutes or less. 8–10 ppm lethal within four hours or less—100 ppm and above may be lethal within a short time.
Ammonia (NH_3) Alkalai, colorless gas. Soluble in water; pungent odor detected as low as 1 ppm.	TLV = 25 ppm, should protect against irritation to eyes and respiratory tract, minimize complaints of discomfort among unacclimated individuals.	Irritation of respiratory tract and conjunctiva in workers inhaling 100 ppm. Severe eye damage, brain dysfunction at higher concentrations.
Sulfur Dioxide (SO_2) Sulfur dioxide (SO_2), a colorless nonflammable gas with acid odor, pungent taste, one of the commonest contaminants in industrial environment.	TLV $SO_2 = 2$ ppm, expected to prevent irritation and accelerated loss of pulmonary function in most workers.	High acute exposure causes intense irritation, death may follow from suffocation due to respiratory paralysis or pulmonary edema. Industrial poisoning usually chronic—may develop as pulmonary dysfunction progressing to emphysema. No evidence of carcinogenicity, however may act as a promotor.
B. Irritants Affecting Both Upper Respiratory Tract and Lung Tissues		
Chlorine (Cl_2) Halogen, greenish-yellow gas with suffocating odor, which may be noticeable 1–4 ppm. Soluble in water up to 0.8% by weight.	TLV = 1 ppm, to minimize chronic lung changes, accelerated aging, teeth erosion.	30 ppm produces intense coughing.
Ozone (O_3) Bluish or colorless explosive gas or blue liquid. Pleasant characteristic odor ı concentrations of less than 2 ppm. Slightly water-soluble, used as disinfectant.	TLV = 0.1 ppm, which causes no obvious injury but may result in premature aging similar to that from continued exposure to ionizing radiation, if exposure sufficiently prolonged.	Daily intermittent exposure above 5 ppm (reported for arc welders) may cause incapacitating pulmonary congestion.
C. Irritants Affecting Primarily Terminal Respiratory Passages and Air Sacs		
Nitrogen Dioxide (NO_2) Reddish-brown gas with irritating odor. Decomposes in water, nitric acid (HNO_3) and nitric oxide (NO).	TLV = 3 ppm, ceiling, considered sufficiently low to insure against reduced respiratory function.	10–20 ppm may predispose to chronic disease. 100–500 ppm may lead to sudden death, insidious, delayed, and potentially lethal pulmonary edema (most characteristic), delayed inflammatory changes leading to death several weeks after exposure.
Phosgene (Carbonyl chloride) ($COCl_2$) Colorless, non-flammable gas. Suffocating odor when concentrated, otherwise odor suggestive of decaying fruit or moldy hay. Slightly soluble in water and hydrolyzed by it.	TLV = 0.1 ppm because of its irritating effects on the respiratory tract at levels slightly above 0.1 ppm, but from which tolerance develops.	3 ppm causes immediate throat irritation, 50 ppm rapidly lethal. "Tolerance" below 1 ppm in man considered potential triggering mechanism for emphysema and fibrosis.

Table 15-D. Comparison of Some Chemical Asphyxiants

Description	*TLV*	*Concentrations Exceeding TLV*
Hydrogen Cyanide (HCN) Colorless liquid or gas, flammable, protoplasmic poison.	TLV = 10 ppm (ceiling), which may give about a seven or eightfold margin against lethal effects.	18–36 ppm causes slight symptoms after several hours, 135 ppm fatal after 30 minutes, 270 ppm immediately fatal.
Carbon Monoxide (CO) Colorless, odorless gas, sparingly soluble in water, that combines with hemoglobin to form carboxyhenoglobin (COHb), which interferes with oxygen transport to tissues and removal of CO_2 from tissues.	TLV = 50 ppm, based on an air concentration that should not generally result in COHb levels above 10%.* Heavy labor, high temperatures, or altitudes 5,000–8,000 feet above sea level may require 25 ppm TLV.*	Fatal in 1 minute at 1% concentration (= 10,000 ppm), which causes approximately 20% COHb. Severe poisoning from short exposure often followed by complete recovery but neurological, cardiovascular, pulmonary, other complications may occur.
Hydrogen Sulfide (H_2S) Colorless flammable gas, burns to sulfur dioxide. Soluble in water but solutions unstable. Characteristic odor of rotten eggs detectable at concentrations of 0.02 ppm or appreciably less. Higher toxic concentrations can rapidly deaden sense of smell.	TLV = 10 ppm, based primarily on eye effects sometimes reported from slightly lower concentrations.	Concentrations of 500–1,000 ppm cause rapid unconsciousness and death through respiratory paralysis. Associated with an unusual diversity of symptoms—including chronic keratoconjunctivitis, nausea, insomnia, pulmonary edema, balance disorders, polyneuritis, and gray-green discoloration of the teeth.

A secondary irritant produces irritant action on mucous membranes, but this effect is overshadowed by systemic effects resulting from absorption. Examples of materials in this category are many of the aromatic hydrocarbons and other organic compounds. The direct contact of liquid hydrocarbons with the lung can cause chemical pneumonitis. Thus, in the case of accidental ingestion of these materials, inducing vomiting is not recommended because some of the vomited hydrocarbon could be breathed into the lungs.

Normally, irritation is a completely reversible phenomena. Reversible means that if the victim is taken out of the exposure quickly enough the condition of irritation will go away with no residual damage. If a person breathes carbon tetrachloride in a sufficient quantity to produce liver damage, and is then taken out of the exposure, any existing pulmonary irritation may quickly disappear. However, if the liver damage persists even after sufficient time out of the exposure and subsequent treatment, then it is said to be irreversible.

Irritation phenomena are generally reversible after short-term exposures. If a worker goes into a cloud of ammonia, immediate irritation is experienced and unless the worker is greatly overexposed, the sensation of pain and irritation will be largely gone in a very short period after removal from exposure. However, temporary damage to the respiratory epithelium can make the worker susceptible to other irritants that would ordinarily be tolerated.

Asphyxiants

Asphyxiants interfere with oxygenation of the tissues and the affected individual may literally suffocate. This class is generally divided into simple asphyxiants and chemical asphyxiants.

Simple asphyxiants are physiologically inert gases that dilute or displace atmospheric oxygen below that required to maintain blood levels sufficient for normal tissue respiration.

Common examples are carbon dioxide, ethane, helium, hydrogen, methane, and nitrogen. Physiologically inert, as used in this context, may be essentially absolute (nitrogen) or relative (carbon dioxide, which acts as a respiratory stimulant when the level in inspired air is greater than two percent).

Asphyxiants deprive the body of the needed oxygen which must be transported from the lungs via the bloodstream to the cells. With complete deprivation of oxygen, the brain cells perish in 3–5 minutes. Total asphyxiation leads to complete absence of oxygen in the blood (anoxia). Partial asphyxiation induces low levels of oxygen in the blood (hypoxia). If allowed to continue too long, the hypoxia also can result in brain damage or death.

Normal air contains approximately 21 percent oxygen, but a human can get by without too much ill effect if the air contains just a little less oxygen. As the oxygen level goes progressively lower, however, it starts interfering with the life process and, if it gets too low, it can produce death. At higher altitudes, higher percentages of oxygen are required; 100 percent is needed at 33,000 feet. Many substances like nitrogen, which is 80 percent of the air normally breathed, and other chemicals are properly considered simple asphyxiants. Because they are not metabolized or changed to injurious chemicals, they do not cause chemical injury in the body and, hence, are not toxic or irritating. All they do is keep the oxygen level far below what it should be; therefore, they are called simple asphyxiants.

Chemical asphyxiants, through their direct chemical action, either prevent the uptake of oxygen by the blood or interfere with the transporting of oxygen from lungs to the tissues or prevent normal oxygenation of tissues, even though the blood is well oxygenated. Carbon monoxide prevents oxygen transport by preferentially combining with hemoglobin. Hydrogen cyanide inhibits enzyme systems, particularly the cytochrome oxidase needed by practically all aerobic cells to utilize molecular oxygen. Hydrogen sulfide paralyzes the respiratory center of the brain and the olfactory nerve. At sufficiently high levels, all three of these chemical asphyxiants can cause almost instantaneous collapse and unconsciousness (Table 15-D).

The principal action of carbon monoxide is its interference with the delivery of the proper amount of oxygen to the tissues. The concentration of carbon monoxide required to cause death is small compared to the amount of simple asphyxiants. Carbon monoxide combines with hemoglobin to form carboxyhemoglobin. Hemoglobin will combine with carbon monoxide much more readily than it will with oxygen by the ratio of approximately 300 to 1.

Over and above the familiar acute effects of carbon monoxide, there is concern about how low-level exposures will affect performance of such tasks as automobile driving. The blood has a certain oxygen-carrying capacity; this is called percent oxygen saturation. The actual amount of oxygen that can be transported varies with the amount of hemoglobin in a person's blood. Some of the hemoglobin can be tied up by carbon monoxide.

In nonsmokers, a small amount of the hemoglobin is tied up with carbon monoxide at any given time. This comes about because in the normal life process, some carbon monoxide is always being formed in the body (it is a normal process of metabolism). This carbon monoxide as it evolves forms a little bit of carboxyhemoglobin in all of us just due to the life process.

In smokers, depending on the amount smoked, a certain percentage of the hemoglobin is tied up in carboxyhemoglobin. A pack-a-day smoker will frequently be in the 5–10 percent carboxyhemoglobin range. That is important, because studies involving carbon monoxide exposure in industrial operations have to take into account the difference between smokers and nonsmokers. The effects of smoking will usually overshadow and outweigh the environmental effects expected from carbon monoxide as an air pollutant. The smoker's carboxyhemoglobin will be comparable to that produced by exposure to carbon monoxide at the TLV of 50 ppm in air.

Another example of a chemical asphyxiant is hydrogen cyanide. It is transported by the bloodstream to the individual cells of the body where it blocks oxygen uptake at the cellular level by combining with the enzymes that control cellular oxidation. Oxygen uptake at the cellular level is blocked only as long as the cyanide is present. Normal cellular oxygen uptake resumes if death of the cells has not readily occurred.

Cyanide is toxic to body tissues. It is generally referred to as a protoplasmic poison, which is another way of saying that it is incompatible with life. It poisons any enzyme system it comes in contact with. The whole life process is a very delicate balance of thousands of systems working together catalyzed by thousands of organic molecules called enzymes. Many things can happen to interfere with this catalytic action or enzymatic action, and a lot of toxicity effects can be traced to a particular enzyme system being poisoned by a given substance—the result is the same as when that particular enzyme is inhibited.

There are many other chemicals and asphyxiants that interfere with the body's ability to get the amount of oxygen needed for life.

Narcotics and anesthetics

Narcotic substances can produce unconsciousness and many of the same symptons that asphyxiants cause. Narcotics prevent the central nervous system from doing its normal job.

Anesthetics and narcotics exert their principal action by causing simple anesthesia without serious systemic effects, unless the dose is massive. Depending on the concentration present, the depth of anesthesia ranges from mild symptoms to complete loss of consciousness and death. In accidents involving very high concentrations, death may be due to asphyxiation.

Anesthetics and narcotics include: aliphatic alcohols (ethyl, propyl, etc.), aliphatic ketones (acetone, methyl-ethyl-ketone, etc.), acetylene hydrocarbons, and ethers (ethyl, isopropyl, etc.).

Substances like ether, chloroform, and other anesthetics that are very effective in producing anesthesia are selected when the intent is to make someone unconscious. A successful anesthetic is one that produces narcosis, but does it with a lot of room for making a mistake. Some narcotics are quite good but do not have a sufficient margin of safety so they are not intentionally used as anesthetics. In the industrial environment, there are many substances that have narcotic properties—nearly all organic solvents, for example.

Other effects

There are a large number of other substances with a variety of toxicological action that does not fit into any of the three groups described previously.

Cardiac sensitization from inhalation of certain volatile hydrocarbons can make the heart abnormally sensitive to epinephrine. Some persons exposed to these materials can develop cardiac arrhythmias, usually ventricular in origin. Cardiac sensitization has been observed with the anesthetic use of chloroform and cyclopropane, solvent misuse such as with aerosol containers or sniffing glue in a deliberate attempt to get "high," and exposure to some industrial solvents at levels grossly exceeding the recommended TLV.

Neurotoxic agents are materials that produce their main toxic effects on the nervous system. Metals such as manganese and mercury are examples. The central nervous system seems particularly sensitive to organometallic compounds.

In the transfer of energy from one part of the nerve to the next part, the chemical acetylcholine is essential, and the enzyme cholinesterase sees to it that the level of acetylcholine is maintained at the proper levels. As soon as that enzyme is inhibited, then the acetylcholine level starts increasing and reaches the level where it is incompatible with the transfer of nervous energy, and the nervous system undergoes a collapse and fails to work. The triggers do not fire the way they should. Some of the inhibitors of the cholinesterase include chemicals such as organic phosphate and organophosphate pesticides such as parathion.

Dusts. In considering health effects from inhaled dust, primary concern is given to solid material that is small enough to enter the alveoli. There is a certain amount of filtration by the upper respiratory system which prevents the large particles from ever getting into the lung. In the workplace, particles are dispersed in a nonuniform way and in a full spectrum of sizes; only a portion of them are small enough to get into the lung. There is need, therefore, for standards for the respirable fraction of dusts.

Dusts less than one or two micrometers in diameter reach the deep lung readily, and can be expected to exert some effect once they get there. The simplest effect is to be deposited in the lung without damaging tissue. Inert dusts such as calcium carbonate (the principal ingredient in limestone, marble, and chalk) and calcium sulfate (the principal ingredient in gypsum) are considered relatively harmless unless the exposure is severe.

Inert dusts are sometimes called nuisance dusts. The definition of inert is relative because all particulates evoke a tissue response when inhaled in sufficient amount. The TLV for nuisance dusts or nuisance aerosols has been set at 10 mg/m^3 (total particulates) or 5 mg/m^3 when the dust sample is comprised of particulates of respirable diameter, provided that their inhalation:

- does not alter the structure of lung air spaces;
- does not cause collagen or scar formation to any significant extent;
- results in a potentially reversible tissue reaction.

Some inert dusts may cause radiopaque deposits in the pulmonary system that are visible on x-ray films but they produce little or no tissue reactions unless the exposure is overwhelming.

Acute reactions to inhaled dust can be described as irritant, toxic, or allergenic. Chronic exposure to dust is associated with various types of pneumoconiosis (see Chapter 7, Particulates).

Far more serious is the effect of insoluble materials which do damage to the lung. When these get into the lung, they have no place to go. They are either removed by the cleansing mechanisms of the lung, or they are transported from the alveoli to a lymph node.

Pneumoconioses or "dusty lungs" may be differentiated into the simple or nonfibrotic type and the complicated, fibrotic, or fibrogenic type. Potentially reversible pneumoconioses associated with inert dusts are sometimes called benign pneumoconioses.

Fibrotic changes are produced by materials such as free silica which produces the typical silicotic nodule or small area of scarlike tissue. Asbestos also produces typical fibrotic damage to lung tissue (asbestosis) and, in addition, cancerous lung changes. There is newly aroused interest in the possible effects of nonoccupational low-level exposure to asbestos.

More recently, it was found that in combination with smoking, exposure to airborne asbestos particles leads to an excessive incidence of lung cancer. The incidence of these cancers in nonsmokers exposed to asbestos, though lower, is still abnormally high. Asbestos exposure has also been shown to cause mesothelioma, a very rare cancer of the lining of the abdomen or the lung. (See Chapter 6, Particulates, for more information on asbestos-related diseases and other pneumoconioses.)

Soluble dusts

Some inhaled particles that gain entrance to the body through the lung are soluble in body fluids. More specifically, they are soluble in the fluid of the tissues that line the lung. Though they may not damage the lung tissues, they are absorbed into the blood and distributed throughout the body and can damage some other target organ. They have been characterized by studies that identify their target site to be the nervous system, the kidney, the liver, or some other organ.

The effects that result from heavy metals being absorbed through the respiratory tract vary appreciably from substance to substance. Usually there is slow, cumulative, and other irreversible retention of metal in the body; however, the insidious toxic symptoms develop so slowly that the source or cause of the symptoms is often not recognized.

Inorganic lead poisoning differs physiologically from organic lead poisoning. Excessive exposure to the inorganic form usually results from ingestion or inhalation of dust or fume causing physical abnormalities that may be characterized by anemia, headache, anorexia, weakness, and weight loss. With chronic, high exposures, other more serious reactions include peripheral neuritis, colic that can simulate acute appendicitis, and bone marrow changes.

Organic lead, unlike inorganic lead, tends to concentrate in the brain. Some organic lead materials can easily be absorbed through the skin and add to the hazard from ingestion or inhalation. A single exposure to tetrathyl lead can cause symptoms in a few hours and absorption of a relatively small amount can be fatal.

Although mercury poisoning was described by Paracelsus among miners several centuries ago, it has received much more attention recently. Inorganic compounds of mercury are readily absorbed from the intestine, and tend to concentrate primarily in the kidneys, but can also damage the brain. Organic mercury tends to be especially concentrated in blood and brain.

Industrial manganese poisoning, except for its action related to metal fume fever, is primarily a chronic disease resulting from inhalation of fume or dust in the mining or refining of manganese ores. It also can be caused by cutting and welding metals containing high manganese content, while in a confined space, without respiratory protection. Manganese poisoning is noted for its peculiar neurological effects, especially psychomotor instability.

NEOPLASMS AND CONGENITAL MALFORMATIONS

Although any new and abnormal tissue growth may be classed as a neoplasm, this term is most frequently used to describe cancerous or potentially cancerous tissue. The cells of a neoplasm are to some extent out of control. If neoplastic cells invade tissues or spread to new locations in the body (metastasize) the neoplasm has become cancerous or malignant.

Carcinogenesis

It is well established that exposure to some chemicals can produce cancer in laboratory animals and man. In common usage, a carcinogen indicates any agent that can produce or accelerate the development of malignant or potentially malignant tumors or malignant neoplastic proliferation of cells. Carcinogen refers specifically to agents that cause carcinoma, but the current trend is to broaden its usage to indicate an agent that possesses carcinogenic potential. The terms tumorigen, oncogen, and blastomogen have all been used synonymously with carcinogen.

There are a number of factors that have been related to the incidence of cancer—the genetic pattern of the host, viruses, radiation including sunshine, hormone imbalance, along with exposure to certain chemicals. Other factors such as cocarcinogens and tumor accelerators are involved. It is also possible that some combination of factors must be present to induce cancers. There is pretty good clinical evidence that some cancers are virus-related. It may be that a given chemical in some way inactivates a virus, activates one, or acts as a cofactor.

Definitions. A carcinogen can be defined as a substance that will induce a malignant tumor in humans following a reasonable exposure.

A carcinogen has also been defined as a substance that will induce any neoplastic growth in any tissue of any animal at any dose by any method of application applied for as long as the lifetime of the animal. Both of these definitions are useful if we are clear as to which one we are using and use it consistently.

Problems arise when all substances that would fulfill the second definition in the experimental laboratory are classified as carcinogens and it is implied that they will cause malignancies in humans in accordance with the first definition. According to NIOSH, a substance is considered "a suspect carcinogen" to humans if it produces cancers in two or more animal species.

Even if we could extrapolate from a specific strain of a laboratory animal to all species including humans, taking into consideration such factors as weight, surface area, metabolic profiles, genetic susceptibilities, and accounting for drug-induced changes in metabolism, we still might not be able to reach justifiable conclusions as to carcinogenic potential because of the existence of synergism. One of the best examples is the effect of cigarette smoking on the lung: the synergistic action of unknown components increases many times over the carcinogenic activity of trace amounts of polycyclic hydrocarbons in cigarette tars.

Typical carcinogens. The chemicals that induce cancer do it by a mechanism that is still not understood; no one really knows why some chemicals are carcinogenic whereas others are not. Some materials can produce cancer in the lungs after inhalation or they can just use the lung as a route of entry and produce the cancer elsewhere.

A biochemist can look at the structure of an organic chemical and speculate that because of certain functional groups the chemical is carcinogenic but another chemical having functional groups very much like it is not carcinogenic. The theories of why a chemical is carcinogenic are very involved and, as yet, imperfect.

Coal tar and various petroleum products have been identified as skin and subcutaneous carcinogens. Pitch, creosote oil, anthracene oil, soot, lamp black, lignite, asphalt, bitumen, certain cutting oils, waxes, and paraffin oils have also been implicated as potential carcinogens.

Arsenic, also recognized as a carcinogen, is associated with exposures in the manufacture or use of roasting metallic sulfide ores as well as certain paints or enamels, dyes or tints, pesticides, and miscellaneous chemicals.

Inorganic salts of metals such as chromium, and to a lesser extent nickel compounds are associated with cancer of the respiratory tract, usually the lungs. Other metals such as beryllium and cobalt are suspect, but their direct toxic effects in humans can obscure carcinogenic potential.

Leukemia describes a group of diseases characterized by widespread, uncontrolled proliferation of white blood cells, failure of many cells to reach maturity, and abnormal accumulation of such cells. Exposures to x-rays and radioactive substances are principal occupational causes. Benzene exposures are also associated with blood dyscrasias (diseased state of the blood; generally abnormal or deficiently formed cellular elements), which may progress to leukemia or aplastic anemia.

Osteogenic sarcomas (bone tumors) have been detected in workers who used radioactive luminous paint on instrument and watch dials. Angiosarcomas (a relatively rare malignant growth) of the liver have been found to be associated with human exposure to vinyl chloride monomer. Oat cell carcinomas of the lung have been found in workers exposed to bis (chloromethyl) ether (BCME). The BCME can occur as an unsought intermediate in certain reactions involving formaldehyde and hydrochloric acid. The BCME is carcinogenic by inhalation, skin, and subcutaneous routes in animals.

The net results of exposure to chemical carcinogens is pretty clear-cut—people who work around a chemical like benzidine are at increased risk of getting bladder cancer. The incidence of benzidine-induced bladder cancer is known. With many other substances, it is increasingly clear that there is a higher incidence of cancer in certain groups of people who work with carcinogenic materials, perhaps with (general) air pollution as a factor.

Environmental factors. Cancer is considered so insidious and with such a feared end result that carcinogenic chemicals are isolated and looked at differently than everthing else. The statement that 80–90 percent of all cancers are environmentally caused does not mean that 80–90 percent of cancers are caused by industry. The environment includes not only the air we breathe and water we drink but out diet and all the elements of our lifestyle, on and off the job. Major causes of environmental cancer are tobacco smoke and diet.

Mutagenesis

A mutagen is something that affects the genetic system of the exposed people or animals in such a way that it may cause cancer or an undesirable mutation to occur in some later generation. People who work with a certain chemical may not be hurt by it, but their offspring can be. It is conceivable that the damage being done could show up some generations later.

The problem of time lag between exposure and effect is particularly severe for mutagenic agents. Mutations will not show up until the next generation at the earliest, and may not appear for several generations. The long latency makes it difficult to discover the connection between hazard and genetic damage.

Mutagens are chemical or physical agents that cause inheritable changes in the chromosomes. In humans, specific mutagens have not been clearly associated with visible deformities, but radiation has defintely been associated with sterility.

A mutagen can have an effect on somatic cells, and not on germ cells. In this case its effects are not passed on to offspring, but depend on the kind of cell affected. For example, the cells of the bone marrow go on multiplying throughout life and shed the products of division into the blood, where they function for a time as red and white blood cells before they are removed and replaced. Gross interference with the genetic material of such cells may make cell division ineffective.

Another type of interference with the genetic materials of these cells by a virus or chemical can make them capable of more rapid growth and multiplication, so that they are formed far more rapidly than they can be removed from the blood, where they interfere with normal body functions: if the white cells are affected in this way, the outcome is a leukemia.

Similar interference with the genetic material could theoretically start up cell division in cells that do not normally divide during adult life. If the products of such division displace or invade normal tissues, the result is a cancer. In both these instances, the mutagen responsible would have manifested activity as a carcinogen.

Teratogenesis

Since the thalidomide disaster some years ago, everybody has been very much aware that teratogenesis is something to be concerned with. The term literally means monster-making. The terato- prefix refers to the production of abnormal offspring. The exact nature of the deformity however will vary with the substance that causes it.

Teratogenesis (congenital malformation) results from interference with normal embryonic development by an environmental agent. Chemicals administered to a pregnant animal may, under certain conditions, produce malformations of the fetus without inducing damage to the mother or killing the fetus. Such malformations are not hereditary. In contrast, the congenital malformations resulting from changes in the genetic material are mutations and are hereditary.

Typical teratogens. Agents presently identified as human teratogens include infections such as rubella, chemicals such as thalidomide and possibly steroids, and ionizing radiation.

The property of being teratogenic means that if the chemical is administered to or absorbed by pregnant women, it will produce deformed offspring.

Pregnant women in the workplace. A teratogen, by definition, is different from a mutagen in that there has to be a developing fetus to be adversely affected. This is extremely important today because of the very considerable pressure to address the problem of pregnant women in the workplace.

The fetus is protected from toxic chemicals as well as nature knows how, by means of the placenta and other mechanisms that do all they can to filter or remove toxic materials from the fetal blood supply. It is believed that, if a pregnant woman was exposed to sufficient concentrations of industrial chemicals, they would pass the placental barrier. Damage to the fetus (embryo) is most likely to occur in early pregnancy, particularly during the first 8–10 weeks. During much of this critical period, many women are not even aware that they are pregnant. To protect unborn babies during the most crucial stage of their development, some industries have barred all women of childbearing age from jobs with even the slightest degree of hazard; this is a point of social and legal conflict in the United States, as work can be made safe for all. The effects of mutagenic chemicals on men can also be passed on to their offspring.

It can be extremely difficult to establish specific cause-and-effect relationships between a teratogen and the birth defect it can produce. Animal studies need to be supplemented with epidemiological data; and it may be decades before researchers know with certainty what substances hold how much risk for which unborn infants.

The fact that there are women in the workplace, and that they are exposed to teratogens leads to a problem in setting occupational health standards. An embryo of a few weeks or a fetus of a few months should be given consideration and should not be exposed to a toxic environment.

Even today, there are people who believe that one way to solve this problem is to keep women out of the workplace. This may be effective but is not very acceptable to those who believe that the workplace should be such that a pregnant women should be able to work there without any likelihood of harm. Of course, the effects of mutagenic chemicals on men also can be passed to their offspring.

FEDERAL REGULATIONS

OSHA

The Occupational Safety and Health Administration (OSHA) has the regulatory authority to protect workers from hazardous dusts and chemicals in the work environment. Shortly after OSHA was formed, it adopted as federal standards or Permissible Exposure Limits (PELs) the 1968 Threshold Limit Values (TLVs) suggested for industrial chemical exposures by the American Conference of Governmental Industrial Hygienists (ACGIH).

Recently OSHA has enacted the Hazard Communication Standard, 29 *CFR* 1910.1200, which sets standards for worker notification and training for chemicals in the workplace. This is known as right-to-know legislation, and is being followed up by many state and local governments. For more information on the right-to-know legislation, see Chapter 30, Governmental Regulations.

The American National Standards Institute (ANSI) through its former Committee on Acceptable Concentrations of Toxic Dusts and Gases also published standards on levels of materials in the air in work areas. Wherever possible, OSHA adopted ANSI standards because ANSI was considered to be a consensus standards setting group.

The ACGIH TLVs were directed primarily at substances which caused physiological reactions such as poisoning, irritation of eyes or respiratory tract, and skin rashes. The 1968 TLVs were not established on the basis of carcinogenic, teratogenic, or mutagenic properties, and synergistic effects of chemical mixtures were not included in tests used to determine the TLVs. Approximately 400 TLVs were adopted as OSHA standards. Since that time, 24 new standards for carcinogenic and other chemical agents have been promulgated by OSHA.

TOSCA

The Toxic Substances Control Act of 1976 (TOSCA) provides protection from chemical hazards, and it specifically refers to mutagenesis as a health effect for which testing standards may be prescribed. The Act requires that adequate data be developed on the health and environmental effects of chemicals, and that this activity should be the responsibility of the chemical manufacturers and processors. The Environmental Protection Agency (EPA) is required to establish standards for the testing of chemicals.

Companies are required to notify the EPA 90 days before manufacturing any new chemical and to provide test data and other information about the safety of the product. The EPA has the authority to ban or regulate such chemicals if test information is insufficient and if the chemical is to be produced in substantial quantities with wide distribution. The EPA is required to ban or restrict the use of any chemical presenting an unreasonable risk of injury to health or the environment.

Unreasonable risk. The term unreasonable risk is ambiguous and subject to varied interpretation. The definition of unreasonable emerges from regulatory and administrative decisions of EPA, and there are debates around the issue. Consumer advocates and environmentalists might prefer a definition leading to stringent and rigidly enforced control over hazardous chemicals: representatives from industry might prefer a definition that allows for individual discretion and voluntary compliance.

Toxic substances list

The Occupational Safety and Health Act (OSHAct) of 1970, Section 20(a)(6), requires that "the Secretary of Health, Education, and Welfare shall publish within six months of enactment of this act and thereafter as needed, but at least annually, a list of all known toxic substances by generic, family or other useful grouping, and the concentrations at which such toxicity is known to occur." The first such list was prepared in 1971.

The 1983-84 edition, now entitled *Cumulative Supplement to the 1981-82 Edition of the Registry of Toxic Effects of Chemical Substances,* represents a substantial revision and expansion of the earlier list. Under the OSHAct, the Secretary of Labor must issue regulations requiring employers to monitor employee exposure to toxic materials and to keep records of any such employee exposure. This requirement is set forth in Section 8(c)(3) of the act (see Chapter 31, History of OSHA, for more details).

Purpose. The purpose of the Toxic Substances List is to identify all known toxic substances in accordance with defintions that can be used by all sections of our society to describe toxicity. The entry of a substance on the list does not automatically mean that it is to be avoided but that the listed substance has the documented potential of being hazardous if misused. Therefore, care must be exercised to prevent tragic consequences.

The absence of a substance from the list does not necessarily indicate that a substance is not toxic. Some hazardous substances may not qualify for the list, because the dose that causes the toxic effect is not known.

Other chemicals associated with skin sensitization and carcinogenicity may be omitted from the list, because the effects have not been reproduced in experimental animals, or because the human data are not definitive. Thus, the published comments and evaluations of the scientific community are relied on; also, there has been no attempt at an evaluation of the degree of hazard that might be expected from substances on the list—that is a goal of the hazard-evaluation studies.

Hazard evaluation. It is not the purpose of the list to quantify the hazard by way of the toxic concentration or dose that is presented with each of the substances listed. Hazard evaluation involves far more than the recognition of a toxic substance and a knowledge of its relative toxic potency. It involves a measurement of the quantity that is available for absorption by the user, the amount of time that is available for absorption, the frequency with which the exposure occurs, the physical form of the substances, and the presence of other substances, toxic or nontoxic, additives, or contaminants.

Ventilation, appropriate hygienic practices, housekeeping, protective clothing, and pertinent training for safe handling may eliminate or diminish any hazard that might exist.

Hazard evaluation requires, therefore, engineers, chemists, toxicologists, and physicians who have been trained in the fields of toxicology, industrial hygiene, and occupational medicine to recognize, measure, and control these hazards.

NIOSH/OSHA Standards

Since the passage of the OSHAct, both NIOSH and OSHA have been committed to establishing permissible standards for the workplace which are far more complete than the TLVs issued by the ACGIH.

A complete standard should include—the concentration of the substance that has been determined to provide a safe, healthful work environment; the methods for collecting, sampling, and analyzing for the substance; the engineering controls necessary for maintaining a safe environment; appropriate equipment and clothing for the safe handling of the substance; emergency procedures in the event of an accident; medical surveillance procedures necessary for the prevention of illness or injury from inadvertent overexposure; and the use of signs and labels to identify the hazardous substances.

NIOSH Criteria Documents. Except in the case of emergency standards, the normal first step in the standard-setting process is the creation of a criteria document for the substance of concern by NIOSH. These documents are brought into being by NIOSH using NIOSH scientists, or outside contracting organizations which principally conduct an extremely thorough search of the literature, and create a document that summarizes all significant work related to the establishment of the desired standard. Such documents are forwarded to OSHA for consideration as permanent OSHA standards.

Since 1976, the NIOSH criteria documents have incorporated animal and human data, if available, on the effects of reproduction, carcinogenicity, mutagenicity, and teratogenicity. When possible, attempts are made to correlate these adverse reactions with exposures and effects.

Current Intelligence Bulletins (CIBs) have been issued by NIOSH since 1975 for more rapid dissemination of new scientific information about occupational hazards. A CIB may draw attention to a hazard previously unrecognized or may report new data suggesting that a known hazard is either more or less dangerous than was previously thought (see Bibliography).

OSHA standards. The Secretary of Labor has the responsibility for promulgating standards. In some cases a recommended standard is referred to an advisory committee for study and review in accordance with provisions of the Act. OSHA standards are arrived at after extensive activities including public hearings as required, and the selected standard may bear little or no resemblance to that originally proposed by NIOSH. Regardless of the status of the proposed standards in the criteria documents, these documents constitute valuable and readily available sources of information on a variety of subjects, and should be consulted whenever there is interest in a substance for which a criteria document has been written.

Although the standards-setting process just described is an extremely thorough one that allows all interested groups an opportunity to express their opinion, it is also a lengthy and very costly activity that has resulted in the promulgation of very few permanent standards to date.

Emergency standards. From time to time concern with such cancer-producing substances as asbestos, ethylene oxide, vinyl chloride, and other organic chemicals has produced the need for emergency standards that could be established more quickly than the usual standards-setting process. Because relatively few standards (emergency or otherwise) have been established since the passage of the Occupational Safety and Health Act, there is a clear need for some means of accelerating this activity.

BASIS FOR WORKPLACE STANDARDS

Chemical analogy

When dealing with a new chemical, animal or human toxicity data are usually unavailable. Therefore, the nature of response to a chemical may be assumed to be analogous to that produced by contact with a substance with a similar chemical structure. Chemicals that are similar may be assumed initially to produce similar biological responses. As a first approximation, some estimate of toxic potential can then be obtained. As of 1968, 24 percent of all Threshold Limit Values published by the American Conference of Governmental Industrial Hygienists have been based upon analogy (Table 15-E).

Table 15-E. Distribution of Procedures Used to Develop ACGIH TLVs for 414 Substances Through 1968*

Procedure	*No.*	*Percent Total*
Industrial (human) experience	157	38
Human volunteer experiments	45	11
Animal, inhalation—chronic	83	20
Animal, inhalation—acute	8	2
Animal, oral—chronic	18	4.5
Animal, oral—acute	2	0.5
Analogy	101	24

*Exclusive of inert particulates and vapors.

(Reprinted with with permission from Stokinger, H.E. Criteria and procedures for assessing the toxic responses to industrial chemicals. *Permissible Levels of Toxic Substances in the Working Environment, Occupational Safety and Health,* ser. 20. Geneva, Switzerland: International Labor Office, 1970.)

Animal experimentation

Before chemical agents are introduced into the workplace, it is advisable to know their toxic effects. Then preventive measures can be designed to protect workers, and emergency procedures can be put in place, in case of accidental exposure. Since, in the case of new chemicals, there is often little or no information available, an important method of developing such new information uses animal experimentation.

Exposure standards. The toxicological effects of vapors, gases, fumes, or dusts are initially determined in the laboratory by actually exposing animals to known concentrations. Range-finding studies are usually conducted on animals by exposing them to known concentrations for controlled periods of time. This is followed by more definitive studies involving several species of animals and involving the quantitating of exposure both in terms of concentration and duration.

Groups of animals can be exposed to controlled concentrations for 8 hours a day, 5 days a week, for weeks, months, or periods up to a year. In such cases, extreme care must be taken not only in the selection but also in the subsequent care of the animals throughout the experimental period. Elaborate equipment is employed to maintain and record accurately the concentration and duration of exposure.

Animals must be observed daily to ascertain any untoward physiological responses during exposure and postexposure periods. On terminating chronic experiments, all animals are sacrificed and the internal organs are weighed and examined histopathologically. In this manner, the toxicologist gains information regarding no-effect levels as well as those levels which produce systemic injury of various types and degrees.

Screening procedures. Toxicological screening should include both acute toxicity studies and studies of repeated administration at short intervals. Long-term studies performed during the lifespan of the animal, and in some instances—if indicated—over several generations, should be part of the complete test program.

In view of biological variations that influence the reaction of a foreign chemical in different species, it is difficult to duplicate in animal experiments the precise situation to which a person may be exposed. In the development of a specific test program, preliminary studies are necessary to select species that absorb and metabolize related classes of chemicals in ways similar to humans.

A route of administration different from that usually used in people, for example, parenteral (outside the intestine) administration instead of inhalation or ingestion, can give misleading results.

Chronic toxicity studies involve repeated administration of test substances. However, chronic effects can also be expected from a single exposure to a substance if the body stores the material so that it remains in the organism for long periods of time. Repeated administration of the test substance is useful in the investigation of such problems as cumulative toxicity, tolerance, and enzyme-induction phenomena.

Problem areas. It is very difficult to extrapolate an LD_{50} or an LC_{50} to an acceptable Threshold Limit Value. There is a rough relationship and people have used it, but for intelligent standard setting for chronic toxicity, acute toxicity data are almost useless.

Since the primary concern is with the prevention of harm to humans, the limitations inherent in animal-derived data should be recognized. Whether people will respond as the most reactive or least reactive species tested frequently is not known. The question as to whether the most sensitive species has been tested is frequently unanswered. Finally, whether the animal response is an exact parallel to human response cannot always be predicted.

Human epidemiological data

Records of human experience for exposures to many substances are available. This is particularly true for older chemicals such as carbon monoxide, lead, and many others.

Epidemiological data may be: descriptive, retrospective, or prospective.

Descriptive studies can identify a cluster, or a change or difference in incidence of a disease in a subgroup of the population.

Retrospective studies reveal a relationship between a chemical and a certain effect caused by exposure covering months or years prior to the initiation of data collection.

Prospective studies can define more precisely the time relationship and the magnitude of the risk. Prospective sudies are present and future continuing studies that follow along as the exposures are occurring in work areas. The population studies is divided into groups in accordance with the degree and duration of exposure to the material being investigated. The groups compared should, in so far as possible, differ only in exposure to the agent. Follow-up studies are then carried out to determine the occurrence of the effect in the various exposure categories.

The epidemiological analysis can reveal the relationship between time of occurrence of an adverse effect and age at the time of the first exposure, which help to establish or clarify the influence of variables other than the agent under study. For example, cigarette smoking in the study of lung cancer among asbestos workers, is such a variable.

Finally, if a specific chemical is removed from the environment, it should be followed by epidemiological evidence of a decline in the frequency of the effect.

DISCUSSION OF ACGIH THRESHOLD LIMIT VALUES

Many guides for exposure to airborne contaminants have been proposed and some of them have been used throughout the years. The guides that gradually have become the most widely accepted are those issued annually by the American Conference of Governmental Industrial Hygienists, and are termed Threshold Limit Values or (more commonly) TLVs.

Contrary to popular opinion, these TLVs are not the work of a governmental agency, but are instead the product of a committee whose members are associated with a governmental or educational industrial hygiene activity.

Essential features

The essential features of TLVs are described in these selected portions of the preface to the 1986-1987 list.

> **Introduction to the chemical substances.** Threshold limit values refer to airborne concentrations of substances and represent conditions under which it is believed that nearly all workers may be repeatedly exposed day after day without adverse effect. Because of wide variation in individual susceptibility, however, a small percentage of workers may experience discomfort from some substances at concentrations at or below the threshold limit; a smaller percentage may be affected more seriously by aggravation of a preexisting condition or by development of an occupational illness.
>
> Threshold limits are based on the best available information from industrial experience, from experimental human and animal studies, and, when possible, from a combination of the three. The basis on which the values are established may differ from substance to substance; protection against impairment of health may be a guiding factor for some, whereas reasonable freedom from irritation, narcosis, nuisance or other forms of stress may form the basis for others.
>
> **Definitions.** There are three categories of Threshold Limit Values:
>
> a) *The Threshold Limit Value—Time Weighted Average (TLV-TWA)*—the time-weighted average concentration for a normal 8-hour workday and a 40-hour workweek, to which nearly all workers may be repeatedly exposed, day after day, without adverse effect.
>
> b) *Threshold Limit Value—Short Term Exposure Limit (TLV-STEL)*—the concentration to which workers can be exposed continuously for a short period of time without suffering from 1) irritation, 2) chronic or irreversible tissue damage, or 3) narcosis of sufficient degree to increase the likelihood of accidental injury, impair self-rescue or materially reduce work efficiency, and provided that the daily TLV-TWA is not exceeded. It is not a separate independent exposure limit, rather it supplements the time-weighted average (TWA) limit where there are recognized acute effects from a substance whose toxic effects are primarily of a chronic nature. STELs are recommended only where toxic effects have been reported from high short-term exposures in either humans or animals.
>
> A STEL is defined as a 15-minute time-weighted average exposure which should not be exceeded at any time during a work day even if the eight-hour time-weighted average is within the TLV. Exposures at the STEL should not be longer than 15 minutes and should not be repeated more than four times per day. There should be at least 60 minutes between successive exposures at the STEL. An averaging period other than 15 minutes may be recommended when this is warranted by observed biological effects.
>
> c) *Threshold Limit Value-Ceiling (TLV-C)*—the concentration that should not be exceeded during any part of the working exposure.
>
> The Committee holds to the opinion that limits based on physical irritation should be considered no less binding than those based on physical impairment. There is increasing evidence that physical irritation may initiate, promote or accelerate physical impairment through interaction with other chemical or biological agents.
>
> **Excursion limits.** For the vast majority of substances with a TLV-TWA, there is not enough toxicological data available to warrant a STEL. Nevertheless, excursions above the TLV-TWA should be controlled even where the eight-hour TWA is within recommended limits. The excursion recommendations are as follows.
>
> > Short-term exposures should exceed three times the TLV-TWA for no more than a total of 30 minites during a work day and under no circumstances should they exceed five times the TLV-TWA, provided that the TLV-TWA is not exceeded.
>
> **"Skin" notation.** Listed substances followed by the designation "Skin" refer to the potential contribution to the overall exposure by the cutaneous route including mucous membranes and eye, either by airborne, or more particularly, by direct contact with the substance.

Guides

It was the intention of the TLV committee that the TLVs which they issued would be used as guides in the control of health hazards and should not be used as fine lines between safe and dangerous concentrations. Many reasons for the inadequacy of these numbers for such purposes were noted by the committee. In spite of this admonition, however, the TLV list gradually became incorporated into state and federal regulations. The revision of the Walsh-Healey Act incorporated the 1968 TLV list and, later the Occupational Safety and Health Act included virtually all of that list as standards in Section 1910.1000. Although the TLV list issued annually still is not a list of legal standards, it is a fact that the OSHA standards now enforced are for the most part TLVs of 1968 vintage (Table 15-F). For this reason it is worth examining the nature of the TLVs, in some detail.

Time-weighted average

It is implicit in all TLVs that measurements are made in the breathing zone of a worker, and are obtained in such a way that a TWA can be calculated. This concept has proven to be a useful means of estimating the long-term chronic effects of exposure to most substances in the workroom atmosphere. Although the TWA does not necessarily predict the amount of a substance which will be absorbed, it does measure the amount that can be inhaled during a workday and considerations of the extent of absorption will enable a value to be selected which can afford the desired degree of protection.

It is inherent in the definition of a TWA that concentrations higher than the recommended value will be permitted for some

Table 15-F. Distribution of Criteria Used to Develop ACGIH TLVs for 414 Substances Through 1968*

Criteria	*Number*	*Percent*	*Criteria*	*Number*	*Percent***
Organ or organ system affected	201	49	Biochemical changes	8	2
Irritation	165	40	Fever	2	0.5
Narcosis	21	5	Visual changes (halo)	2	0.5
Odor	9	2	Visibility	2	0.5
Organ function changes	8	2	Taste	1	0.25
Allergic sensitivity	6	1.5	Roentgenographic changes	1	0.25
Cancer	6	1.5	Cosmetic effect	1	0.25

*Exclusive of inert particulates and vapors.
**Number of times a criterion was used of total number of substances examined × 100, rounded to nearest 0.25 percent. Total percentages exceed 100 because more than one criterion formed the basis of the TLV of some substances.

Reprinted with permission from Stokinger, H. E. Criteria and procedures for assessing the toxic responses to industrial chemicals. *Permissible Levels of Toxic Substances in the Working Environment, Occupational Safety and Health,* ser. 20. Geneva, Switzerland: International Labor Office, 1970.

period of time provided that these levels are offset by periods of lesser concentration. In some instances, it may be permissible to calculate the average concentration for a workweek rather than for a workday. The degree of permissible excursion is related to the magnitude of the TLV of a particular substance. The relationship between threshold limit and permissible excursion is a rule of thumb and, in certain cases, may not apply (Figure 15-5).

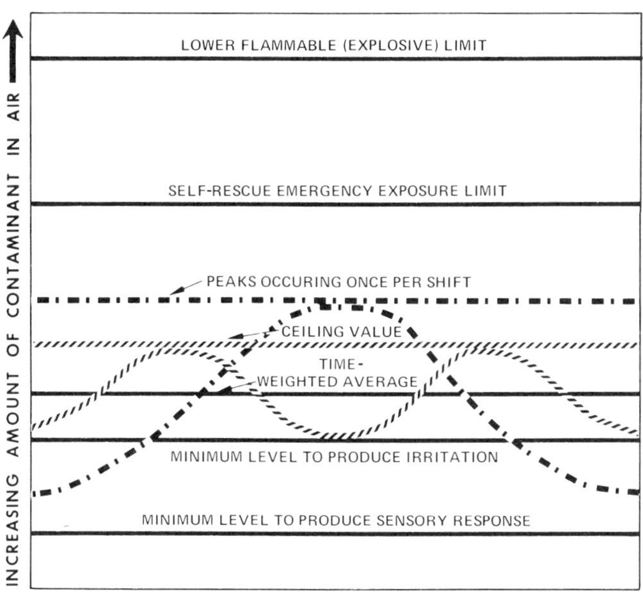

Figure 15-5. Knowledge of the type of injury that would result from exposure to these various contaminant levels is important to health and safety professionals.

Ceiling values

For some substances, it is not advisable to permit concentrations substantially in excess of the recommended TWA, and the practice of the TLV Committee is to designate these substances with the letter C, which stands for ceiling value. Most substances designated with a C value tend to be irritants for which a TLV has been set only slightly below the level where irritation will be noticed by the most sensitive individuals.

It is implicit in these definitions that the manner of sampling to determine if the exposures are within the limits for each group must differ. A single brief sample, that is applicable to a C limit, is not appropriate to calculating a TWA; here, a sufficient number of samples are needed to permit calculation of the TWA concentration throughout the workshift. The ceiling limit places a definite boundary that should not be exceeded.

The 1986-87 TLV list also contains another listing of values for each substance which has been designated short-term exposure limits or STELS. (This term was defined earlier.)

One of the most fundamental tasks confronting the industrial hygienist, therefore, is assessing the degree of exposure likely to be encountered to a bewildering array of chemicals in the work environment. It is almost universally accepted that there is a threshold level for any of these substances below which no impairment of health or other undesirable effects will occur. Because the most common route of entry for a chemical in the workplace is inhalation, the practice for many years has been to sample the air being breathed by the workers, and compare the result with a suitable standard.

Although air standards and guides such as those developed by ACGIH, NIOSH, and OSHA are those most widely used in industrial hygiene practice, there are certain shortcomings inherent to any air standard; this limits their applicability to all situations. Some of the more common recognized problems include:

1. difficulty in acquiring a truly representative breathing zone sample
2. uncertainties concerning the extent of absorption of the amount inhaled
3. nonroutine or nonrepetitive work; air samples can only characterize work operations on the day the sample was taken
4. misleading information resulting from variations in particle size and particle solubility
5. accidental or deliberate contamination of sample.

The TLV is basically an alphabetical listing of substances with the recommended limits expressed either in parts per million by volume or milligrams per cubic meter; see Appendix B-1. It is the practice to give the TLV for all substances expected to be present in the air as particulate suspensions in milligrams per cubic meter only. For those substances expected to be present as gases or vapors, the TLV is expressed in parts per million, and for convenience the equivalence in milligrams per cubic meter is also presented.

In the special case of those substances classified as mineral dusts, formulas are presented for calculating the TLV based upon the quartz or free crystalline silica content of the airborne dust. Several alternative formulas are presented such that evaluation by optical dust counting may be used, with the results expressed in millions of particles per cubic foot (mppcf). Similar formulas are available for so-called "total dust" and the respirable fraction, obtained by means of a suitable size discriminating device. The asbestos TLV is unique, and is expressed in fibers per cubic centimeter of air.

Mixtures

When two or more hazardous substances that act upon the same body organ system are present, their combined effect, rather than that of either component, should be given primary consideration. In the absence of information to the contrary, the effects of the different hazards should be considered as additive. Exceptions may be made when there is a good reason to believe that the chief effects of the different harmful substances are not in fact additive, but independent as when purely local effects on different organs of the body are produced by the various components of the mixture.

Antagonistic action or potentiation may occur with some combinations of atmospheric contaminants. At present such cases must be determined individually. Potentiating or antagonistic agents may not necessarily be harmful by themselves. Potentiating effects of exposure by routes other than that of inhalation is also possible, for example, imbibed alcohol and inhaled narcotic (trichloroethylene). Potentiation is characteristically exhibited at high concentrations, and probably less at low concentrations.

When a given operation or process characteristically emits a number of harmful dusts, fumes, vapors, or gases, it is frequently only feasible to attempt to evaluate the hazard by measurement of a single substance. In such cases, the threshold limit used for this substance should be reduced by a suitable factor, the magnitude of which will depend on the number, toxicity, and relative quantity of the other contaminants ordinarily present.

Examples of processes that are typically associated with two or more harmful atmospheric contaminants are welding, automobile repair, blasting, painting, lacquering, certain foundry operations, reinforced plastic fabrication, and shipbuilding.

Carcinogens

The ACGIH, in its *Threshold Limit Values and Biological Exposure Indices for 1986-1987* (see Appendix B-1) defines a carcinogen as follows:

> The ACGIH considers information from the following kinds of studies to be indicators of a substance's potential to be a carcinogen in humans: epidemiology studies, toxicology studies and, to a lesser extent, case histories. Because of the long latent period for many carcinogens, and for ethical reasons, it is often impossible to base timely risk-management decisions on results from human toxicological studies. In order to recognize the qualitative differences in research results, two categories of carcinogens are designated in the TLV booklet: A1—Confirmed Human Carcinogens; and A2—Suspected Human Carcinogens. All steps must be taken to keep exposures to all A1 carcinogens to a minimum. Workers exposed to A1 carcinogens without a TLV should be properly equipped to insure virtually no contact with the carcinogen.

Physical factors

It is recognized that such physical factors as heat, ultraviolet, and ionizing radiation, work under compressed air or at high altitude may place added stress on the body so that the effects from exposure to chemical agents may be altered. Certain physical stresses may adversely increase the toxic response of a substance. Although most threshold limits have built-in safety factors to guard against moderate deviations from normal environments, the safety factors of most subtances are not of such a magnitude as to take care of gross deviations. For example, continuous work at temperatures above 32 C (90 F) or overtime extending the workweek by more than 25 percent, might be considered gross deviations. In such instances, judgment must be exercised in the proper downward adjustments of the TLVs (see Zielhuis [1987], Bibliography).

Unlisted substances

Many substances present or produced as by-products in industrial processes do not appear on the TLV list. In a number of instances the material is rarely present as a particulate, vapor or other airborne contaminant, and a TLV is not necessary. In other cases, sufficient information to warrant development of a TLV, even on a tentative basis, is not available to the Committee. There are some substances of considerable toxicity that have been omitted primarily because only a limited number of workers, such as employees of a single plant, are known to have potential exposure to possibly harmful concentrations.

Basic data used for TLVs

Whenever there is such data in the published literature, the ACGIH TLV Committee selects a value based on human experience. Good epidemiological studies, complete with environmental data as well as morbidity and mortality data, are perhaps the best possible basis for a TLV; but in most cases, such studies are nonexistent. In the absence of epidemiological studies, individual cases involving human exposures are considered, but the majority of the literature available to justify any given value is based on animal toxicological studies. The preferred studies for determining acceptable exposure limits are those based upon long-term inhalation tests involving several animal species at concentrations both above and below the recommended level.

Frequently, however, scientists must rely upon short-term inhalation data or, in many cases, toxicity studies in which the substance was introduced into the experimental animals by routes other than inhalation. The least useful toxicological data are those based upon short-term oral intake, when the intent is to measure the acute toxicity, or the ability of the substance to kill the exposure animals. Unfortunately for some chemicals listed, it will be found that there is no human exposure data whatsoever, and only meager data based on animal feeding studies is available. It is not surprising, therefore, that the publication of new information frequently results in dramatic changes in the magnitude of some TLVs.

Table 15-G. Classification of Criteria for ACGIH TLVs Applicable to Humans and Animals

Morphologic	*Functional*	*Biochemical*	*Miscellaneous*
	APPLIED CRITERIA		
Systems or organs affected—Lung, liver, kidney, blood, skin, eye, bone, CNS, endocrines, exocrines Carcinogenesis Roentgenographic changes	Changes in organ function—Lung, liver, kidney, etc. Irritation Mucuous membranes Epithelial linings Eye Skin Narcosis Odor	Changes in amounts biochemical constituents including hematologic. Changes in enzyme activity. Immunochemical allergic sensitization.	Nuisance Visibility Cosmetic Comfort Esthetic (Analogy)
	POTENTIALLY USEFUL CRITERIA		
Altered reproduction Body-weight changes Organ/body weight changes Food consumption	Behavioral changes Higher nervous functions Conditioned and unconditioned reflexes—learning Audible and visual responses Endocrine glands Exocrine glands	Changes in isoenzyme patterns Radiomimetic effects Teratogenesis Mutagenesis	

(Reprinted with permission from Stokinger, H. E. Criteria and procedures for assessing the toxic responses to industrial chemicals. *Permissible Levels of Toxic Substances in the Working Environment, Occupational Safety and Health,* ser. 20. Geneva, Switzerland: International Labor Office, 1970)

Documentation

The policy of the TLV Committee is to prepare a justification for each proposed TLV, and from time to time these are published in a document entitled *Documentation of Threshold Limit Values* available from the ACGIH (see Bibliography). In these rather brief discussions, the principal data that the Committee considered significant are reviewed, and references are cited. This document should be consulted whenever a particular TLV is to be applied, for it is important to be aware of the basis for each standard. In most cases, a particular value is selected on the basis of one or more of the consequences of overexposure listed in Table 15-G.

One of the advantages of the TLV list is its timeliness. The list contains a section entitled "Notice of Intended Changes," and, as the name suggests, all changes, including additions, are listed for a period of at least two years. During this period of time, the TLV Committee solicits comments from interested parties concerning the suggested changes. It is unfortunate that the OSHA standards which are for the most part the unchanged 1968 TLVs, cannot be so readily updated.

Threshold Limit Values are intended for use in the practice of industrial hygiene as guidelines or recommendations in the control of potential health hazards and for no other use. The TLVs should be interpreted and applied only by a person trained in industrial hygiene. They are not intended for use, or for modification for use, (1) as a relative index of hazard or toxicity, (2) in the evaluation or control of community air pollution nuisances, (3) in estimating the toxic potential of continuous, uninterrupted exposures, or other extended work periods, (4) as proof or disproof of an existing disease or physical condition, or (5) for adoption by countries in which working conditions differ from those in the United States of America and where substances and processes differ. These limits are *not* fine lines between safe and dangerous concentrations.

BIOLOGICAL STANDARDS

One of the most useful means of assessing occupational exposure to a harmful material is the analysis of biological samples obtained from exposed workers. This analysis may provide an indication of the body burden of the substance, the amount circulating in the blood, or the amount being excreted. Virtually every tissue and fluid in the body can be analyzed, but for practical reasons, most bioassays are confined to specimens of urine or blood. For substances such as carbon monoxide and many solvents, the analysis of exhaled breath samples provides an indication of the level of previous exposure. Occasionally the analysis of samples of hair, nails, feces, or other tissues may be useful.

While air monitoring measures the composition of the external environment surrounding the worker, biological monitoring measures the amount of chemical absorbed into the body. Substances being absorbed through the skin and gastrointestinal tract are accounted for. In addition, the effects of added stress (such as increased work load resulting in a higher respiration rate with increased intake of the air contaminant) will be reflected in the analytical results. The total exposure (both on and off the job) to harmful materials will be accounted for. In the case of substances with a long biological half-life, the concentration in tissues or fluids is more or less independent from variations in concentrations in workroom air. For some chemicals, biological assays, in addition to air measurements, can be much more reliable indicators of health risks than just measurements of air contaminants alone (Table 15-H).

Examples of analyses which can be performed on biological samples are:

- analysis for the unchanged substance (for example, lead, arsenic, mercury) in body fluids and tissues.
- analysis for a metabolite of the substance in body fluids or tissues, for example, phenol in urine resulting from exposure to benzene.
- analysis to determine the variations in the level of a naturally occuring enzyme or other biochemical substance normally present in body fluids or tissues, for example, depression of cholinesterase activity as a result of exposure to organic phosphate compounds.

The rates of absorption, metabolism, and excretion for a par-

Table 15-H. Body Tissues and Fluids Suitable for Biological Analysis

Analysis of **urine** samples should be useful for the following compounds:

Acrylonitrile	Fluoride	Parathion
Aniline	Hydrogen bromide	Selenium
Antimony	Hydrogen cyanide	Selenium hexafluoride
Arsenic	Hydrogen fluoride	Stibine
Arsine	Hydrogen selenide	Sulfuryl fluoride
Benzene	Lead arsenate	Tellurium
Boron trifluoride	Manganese	Tellurium hexafluoride
Cadmium	Mercury	Thallium
Chlorinated benzenes	Molybdenum	Uranium
Chromium (H20-soluble compounds)	Nickel	Vanadium
Cobalt	Nickel carbonyl	Zinc
Cyanide	Nitrobenzene	Zind compounds
	Nitrogen trifluoride	

Analysis of **blood** samples may be useful for the following compounds:

Aluminum	Carbon monoxide	Mercury
Cadmium dust	Lead	Methyl bromide
Cadmium fume	Manganese	Zinc

Breath analysis good for many of the following:

Alcohols	Chlorohydrocarbons	Ketones
Aliphatic hydrocarbons		

NOTE: Analysis of blood, breath, or urine samples can be done in conjunction with air analysis to determine if workers are in danger of injury or intoxication. Ideally, such biological or excretory threshold limits should correspond to the average levels found when workers are exposed to the atmospheric TLV.

(Adapted with permission from Elkins, H. B. Excretory and biologic threshold limits. *American Industrial Hygiene Association Journal* 28 [4] [July-August 1967]: 305-314.)

ticular substance determine when it is most appropriate to analyze samples in relation to duration and time of exposure. For rapidly excreted or exhaled substances, peak concentrations will be found soon after exposure. Peak excretion rates for metabolites of some organic solvents and some inorganic substances may occur 1 to 3 days after exposure. Biological levels of metals with cumulative properties (such as lead or mercury) may reflect the response to several weeks' prior exposure.

Individuals with virtually identical exposure histories can show a wide variation in response, due to subtle differences in their rates of absorption, tissue storage, or metabolism. Greater significance should be given to the variations in an individual's level from period to period than to the variations between different individuals within a group.

Many harmful substances can be stored for long periods of time in various parts of the body. The concentrations are unlikely to have an even distribution throughout the body. In many cases the organ with the highest concentration of the material is the liver or kidney, and in a number of other substances, both the liver and bones. In toxicity studies with radioactive isotopes, the organ that suffers the most severe damage and appears to store most of the toxic material is called the target or critical organ.

Solvents are stored in body fat, the lipid-containing sheaths of nervous tissues, including the brain. Certain substances have been studied in humans, demonstrating storage and excretion long after (days to weeks) exposure has ceased. The "halflives" of various chemicals in the body have been established, and are an important consideration in the total assessment of exposure.

Many materials including organic compounds undergo detoxification in the body. The body converts the material to something else that usually reduces its ability to cause injury. Occasionally, the conversion enhances the toxicity, but in any event, the process helps the body to dispose of the material. The conversion products may appear in the urine or blood as metabolites (see Table 15-I). When the body metabolizes benzene, increased levels of phenols are found in the urine; more specifically, the metabolites are conjugated sulphates (the phenols are hooked up to some sulphur molecules). Either the sulfate or the phenols can be determined as an index of exposure to benzene. For many years, this has been a very useful test to determine the extent of benzene exposure.

The enzyme cholinesterase is inhibited by organic phosphates. The cholinesterase activity in the red blood cells can be measured, and a certain reduction of activity below normal is significant.

Urine tests

Tests for the level of metabolites of toxic agents in the urine have found wide use in industrial toxicology as a means of evaluating exposure of workers. The concentration of the metabolic product is related to the exposure level of the toxic agent. Because normal values of such metabolites have been established, an increase above normal levels indicates that exposure has occurred. This provides a valuable screening mechanism for estimating the hazard from continued or excessive exposure. Because lead, for example, interferes in the porphyrin metabolism, erythrocyte protoporphyrin may be measured and the results are useful in some cases (see Chapter 26, The Occupational Physician).

Table 15-I. Metabolic Products Useful as Indices of Exposure

Product in Urine	*Toxic Agents*
Phenol	Benzene Phenol Aniline
Hippuric acid	Toluene Mandelic acid Styrene Ethyl benzene
Methylhippuric acids	Xylene
Thiocyanate	Cyanate Nitriles
Phenol	Phenol Terpenes
Formic acid	Methyl alcohol
2,6, Dinitro-4-amino toluene	TNT
p-Nitrophenol	Parathion
p-Aminophenol	Aniline

Blood analysis

One of the best documented examples of the effectiveness of biological sampling is that of analysis for exposure to lead and its compounds. In most work operations, there is rather poor correlation between breathing zone levels of lead, and blood-lead levels. It is almost universally agreed that the level of lead in the blood is the best index of the probability of damage resulting from lead exposure.

The purpose of a workplace air-sample is to determine that the lead-in-air concentration has not been exceeded. If it has, then steps can be taken to reduce the blood-lead level to a safe value and prevent possible lead intoxication in workers.

It is possible to have a lead-in-air concentration above the TLV, but still not have a hazardous exposure; the reason is that some lead particles can be very large and they do not get into the alveoli of the lung where they would contribute to any significant intake. Even though large particles may get as far as the nose, they will eventually be excreted without coming in contact with lung tissue.

Galena, a common lead ore, is extremely insoluble; there is very little lead poisoning in lead miners who are mining galena. The worker can be exposed to quite a lot of lead, such as lead sulfide, without adverse effect.

There has been extensive controversy regarding the use of the worker as an integrating air-sampling device. The controversy is not justified, inasmuch as a good industrial hygiene and medical program should require the judicious use of both air and biological samples.

An air-lead program is necessary to define the problem, and a blood-lead program is needed to keep track of each employee's individual exposure. This is the best way to protect the worker. It is very probable that if the air-lead levels are kept below the TLV, nobody is going to have an elevated blood-lead from inhalation exposure.

For many other substances there is no known biological test as useful as the measurement of lead in blood, while for other substances, the correlation between bioassay tests and symptoms may be so poor as to render a biological analysis of little value. The aim of an industrial hygienist or safety professional is to control exposure to harmful materials and it is highly probable that both air sampling and biological monitoring programs will be required.

The industrial hygienist or the safety professional is charged with maintaining a safe, healthful environment and probably will want to do air-sampling and, where appropriate, biological sampling. In the case of a battery plant using lead, a blood-lead program is a necessity. In addition, air-sampling at preselected locations would monitor the controls. Air-sampling can be used to find sources of dispersion.

Inorganic lead is not absorbed to a great extent by the skin, but there are some organic lead compounds, that is, tetraethyl and tetramethyl lead, which can be absorbed quite readily. Although intake by ingestion is secondary to inhalation, it is not negligible. Significant amounts of lead can be absorbed from ingested material. Therefore, personal cleanliness, good sanitation practices, and a prohibition against eating, drinking, or smoking in work areas containing hazardous substances are as important as keeping the air levels down.

Breath analysis

If the inhaled gases and vapors are fat-soluble and are not metabolized, they are cleared from the body primarily through the respiratory system. Examples of these are the volatile halogenated hydrocarbons; the volatile aliphatic, olefinic, and aromatic hydrocarbons; some volatile aliphatic saturated ketones and ethers; esters of low molecular weight; and certain other organic solvents such as carbon disulfide.

For those industrial solvents that continue clearing from the body in the exhaled breath for several hours after exposure, analysis of progressive decrease in the rate of excretion in the breath of the exposed worker offers a laboratory test that may be very helpful in showing not only the nature of the substances to which the worker was exposed, but also the magnitude of the exposure and probable blood levels. By the use of gas chromatography or infrared analysis of the breath samples, the identification of the substance is established, permitting comparison of the exposed workers' breath decay rate with published excretion curves. There is, however, considerable individual variation and it is not easy to set standard values.

Biological limits

There now exists a considerable body of knowledge relative to a large number of substances used in the workplace, and for many of these substances a biological analysis, in addition to air-sampling, will be more useful as a means of evaluating exposure to a toxic substance. Many organic chemicals of high molecular weight and low vapor pressure are not found in the work room air at elevated concentrations under normal conditions of work, but the same substances may be absorbed through the intact skin, giving rise to excessive absorption that cannot be measured by air sampling. In such cases a suitable biological analysis may be an excellent means of detecting the failure of adequate skin-protective measures.

The ACGIH has adopted a set of advisory biological limit values called the Biological Exposure Indices (BEIs) for a limited number of substances. These indices may utilize urine, blood, or expired air sampled under strictly defined conditions. The

Material Safety Data Sheet May be used to comply with OSHA's Hazard Communication Standard, 29 CFR 1910.1200. Standard must be consulted for specific requirements.	**U.S. Department of Labor** Occupational Safety and Health Administration (Non-Mandatory Form) Form Approved OMB No. 1218-0072
IDENTITY *(As Used on Label and List)*	*Note: Blank spaces are not permitted. If any item is not applicable, or no information is available, the space must be marked to indicate that.*

Section I

Manufacturer's Name	Emergency Telephone Number
Address *(Number, Street, City, State, and ZIP Code)*	Telephone Number for Information
	Date Prepared
	Signature of Preparer *(optional)*

Section II — Hazardous Ingredients/Identity Information

Hazardous Components (Specific Chemical Identity; Common Name(s))	OSHA PEL	ACGIH TLV	Other Limits Recommended	% *(optional)*

Section III — Physical/Chemical Characteristics

Boiling Point		Specific Gravity (H_2O = 1)	
Vapor Pressure (mm Hg.)		Melting Point	
Vapor Density (AIR = 1)		Evaporation Rate (Butyl Acetate = 1)	
Solubility in Water			
Appearance and Odor			

Section IV — Fire and Explosion Hazard Data

Flash Point (Method Used)	Flammable Limits	LEL	UEL
Extinguishing Media			
Special Fire Fighting Procedures			
Unusual Fire and Explosion Hazards			

(Reproduce locally) OSHA 174, Sept. 1985

Figure 15-6. Material Safety Data Sheet (Continued).

Section V — Reactivity Data

Stability	Unstable		Conditions to Avoid
	Stable		

Incompatibility (*Materials to Avoid*)

Hazardous Decomposition or Byproducts

Hazardous Polymerization	May Occur		Conditions to Avoid
	Will Not Occur		

Section VI — Health Hazard Data

Route(s) of Entry: Inhalation? Skin? Ingestion?

Health Hazards (*Acute and Chronic*)

Carcinogenicity: NTP? IARC Monographs? OSHA Regulated?

Signs and Symptoms of Exposure

Medical Conditions
Generally Aggravated by Exposure

Emergency and First Aid Procedures

Section VII — Precautions for Safe Handling and Use

Steps to Be Taken in Case Material Is Released or Spilled

Waste Disposal Method

Precautions to Be Taken in Handling and Storing

Other Precautions

Section VIII — Control Measures

Respiratory Protection (*Specify Type*)

Ventilation	Local Exhaust	Special
	Mechanical (*General*)	Other

Protective Gloves	Eye Protection

Other Protective Clothing or Equipment

Work/Hygienic Practices

Page 2

☆ USGPO 1986-491-529/45775

Figure 15-6. Material Safety Data Sheet (Concluded).

user should become familiar with the extensive documentation that accompanies these indices in the "Documentation of TLVs and BEIs, 5th ed." (See Chapter 16, Evaluation, for a more detailed discussion of Biological Exposure Indices.)

It is obvious that the collection of blood samples will require the intervention of medically approved personnel, and in general most programs of biological monitoring become a cooperative effort between safety, industrial hygiene, and medical departments when such are present.

The analysis and interpretation of biological samples is obviously of the greatest importance, and because the quantities involved are almost always very slight, the greatest care must be taken in performing such analyses. Ordinarily, existing plant laboratories are not equipped nor trained to perform these analyses in a satisfactory manner, and it is advisable to use a laboratory which has a proven capability in this area.

Clinical laboratories in general are not equipped to perform such analyses either, although a growing number of such laboratories are entering this field. It is recommended that the services of a laboratory accredited by the American Industrial Hygiene Association be obtained. These laboratories, which now number more than 250, are listed periodically in the *American Industrial Hygiene Association Journal.* This list may be obtained from the American Industrial Hygiene Association.

SOURCES OF TOXICOLOGICAL INFORMATION

The health and safety professional can turn to several sources for information when a question arises about the toxicity and hazard of a material in a certain process or usage. It is important to choose the supplier who can and will provide technical information to guide engineers in designing or controlling the plant or process to minimize exposure.

Material safety data sheet

Material Safety Data Sheets (MSDSs) are a prime source of information on the hazardous properties of chemical products. The OSHA Hazard Communication Standard requires that all chemical manufacturers and importers supply an appropriate MSDS to their customers. The MSDS is usually developed by the chemical manufacturer. Additionally, all users (employers) of the product must have an MSDS for every hazardous chemical used in the workplace.

While OSHA does not specify the format of the MSDS, it does require certain specific information. A sample form approved by OSHA for compliance with the Hazard Communication Standard is shown in Figure 15-6.

There are eight basic required categories of information on the MSDS:

Section I

Manufacturer's Name and Address—This applies to the originator of the MSDS.

Emergency Telephone Number—A number that can be used in an emergency to contact a "responsible party" for information about the product.

Information Telephone Number—To be used in nonemergency cases to contact the manufacturer.

Signature and Date—The signature of the person responsible for the MSDS and the date it was developed or revised.

Section II—Hazardous Ingredients

Common Name—Any identification as used on the label, that is, code name or number, trade, brand, or generic name.

Chemical Name—The scientific designation of a chemical in accordance with the nomenclature systems of the International Union of Pure and Applied Chemistry (IUPAC) or the Chemical Abstracts Service (CAS).

CAS Number—The identification number assigned by the Chemical Abstracts Service, which is unique to a particular chemical.

Section III—Physical and Chemical Characteristics

The physical and chemical data that indicate the potential for vaporization are listed in this section.

Section IV—Fire and Explosion Hazard Data

The data that indicate fire and explosion hazard potentials and special fire-fighting procedures are found in this section.

Section V—Reactivity Data

The stability of the product and the potential for hazardous polymerization and decomposition are outlined in this section. Materials and conditions to avoid during use and storage are also listed.

Section VI—Health Hazards

The most common sensations or symptoms a person might expect to experience from acute and chronic *overexposure* to the material or its components are explained. Emergency and first aid procedures are noted. Any TLVs or PELs are listed. If the chemical is a carcinogen, the source of this designation must be noted.

Section VII—Safe Handling and Use

Designated special handling and disposal methods, and storage and spill precautions are in this section.

Section VIII—Control Measures

The manufacturer recommends the use of ventilation, personal protective equipment, and hygienic practices in this section.

All required sections must be covered. If the required information is not available or not applicable, this must be shown on the form. Additionally, if the ingredients of a chemical mixture are trade secrets, their identity can be withheld, however, their hazardous properties must be given.

These forms are required to be readily available to employees. Training in their use should be included as part of employee training required under the Hazard Communication legislation (see Chapter 30, Governmental Regulations).

The health and safety professional can consult the published literature (see Appendix A, Sources of Help). Such literature includes publications of federal agencies—for example, the National Institute for Occupational Safety and Health—as well as of state agencies. If the material is a newly developed chemical or formulation, very little information may be available.

It is highly advisable that the documentation and/or justification for the various limits or guides be consulted for detailed information concerning mode of action and effect of exposure to a particular toxic material. Such information is given in the reference listed at the end of this chapter. (See Appendices A and B).

SUMMARY

The word toxicity is used to describe a substance that is capable of producing adverse reaction on the health or well-being of an individual. Whether or not any ill effects occur depends on:

(1) the properties of the chemical, (2) the dose (the amount of the chemical acting on the body or system), (3) the route by which the substance enters the body, and (4) the susceptibility or resistance of the exposed individual.

There are four routes of entry or means by which a substance may enter or act on the body: (1) inhalation, (2) ingestion, (3) injection, and (4) contact with or absorption through the skin. Of these, inhalation is the most important insofar as industrial poisioning is concerned.

When a toxic chemical acts on the human body, the nature and extent of the injurious response depends upon the dose received—that is, the amount of the chemical that actually enters the body or system and the time interval during which this dose was administered. Response can vary widely and might be as little as a cough or mild respiratory irritation or as serious as unconsciousness and death.

The practice of industrial hygiene is based on the concept that for each substance there is a safe (or tolerable) level of exposure below which significant injury, illness, or discomfort will seldom occur. The industrial hygienist protects the health of workers by determining the safe limit of exposure for a substance and then controlling the environmental conditions so that exposure does not exceed that limit.

BIBLIOGRAPHY

Ahlborg, G., Jr., Bergström, B., Hogstedt, C., et al. Urinary screening for potentially genotoxic exposures in a chemical industry. *British Journal of Industrial Medicine,* 42(1985):691.

Aitio, A., Pekari, K., and Järvisalo, J. "Skin absorption as a source of error in biological monitoring," *Scandinavian Journal of Work, Environment and Health,* 10(1984):317.

Aitio, A., Riihimäki, V., and Vainio, H., eds. *Biological Monitoring and Surveillance of Workers Exposed to Chemicals,* New York: Hemisphere, 1984.

American Conference of Governmental Industrial Hygienists. *Documentation of Threshold Limit Values,* 4th ed., Cincinnati, OH: ACGIH, 1980.

American Conference of Governmental Industrial Hygienists (ACGIH). *Threshold Limit Values for Chemical Substances and Physical Agents in the Workroom Environment with Intended Changes for 1978.* Cincinnati: ACGIH, 1986-87.

American Industrial Hygiene Association (AIHA), Biohazards Reference Manual. Akron, OH: AIHA, 1985.

AIHA. *Hygienic Guides.* Akron, OH: AIHA.

American National Standards Institute. *Acceptable Concentrations, ser. Z-37.* New York, ANSI.

Anderson, K., and Scott, R. *Fundamentals of Industrial Toxicology.* Ann Arbor, MI: Ann Arbor Science Publishers, 1981.

API Toxicological Reviews. New York: The American Petroleum Institute.

Biological indicators for the assessment of human exposure to industrial chemicals (EUR 8903): Alessio, L., Berlin, A., Boni, M., et al. eds., Office for Official Publications of the European Communities, Luxembourg, 1984.

Boyland, E., and Goulding, R. *Modern Trends in Toxicology,* New York: Appleton-Century-Crofts, 1968.

Browning, E. *Toxicity of Industrial Metals.* London: Butterworths, 1969.

Browning, E. *Toxicity and Metabolism of Industrial Solvents.* New York: Elsevier Publishing Co., 1965.

Calabrese, E. J. *Principles of Animal Extrapolation.* New York: John Wiley & Sons, 1983.

Clayton, G. D., Clayton, F. E., Cralley, L. J., and Cralley, L. V., eds. *Patty's Industrial Hygiene and Toxicology,* 3rd ed., vol. 1 (1978), 2A (1981), 2B (1981), 2C (1982), 3 (1979). New York: John Wiley & Sons, 1978, 1979, 1981, 1982.

Council on Environmental Quality. *Chemical Hazards to Human Reproduction,* Washington, D.C. U.S. Government Printing Office, 1981.

Cralley, L. V., et al., eds. *Industrial Environmental Health.* New York: Academic Press, 1972.

Dreisbach, R. H. *Handbook of Poisoning: Prevention, Diagnosis and Treatment,* 11th ed. Los Altos, CA: Lange Medical Publications, 1983.

Effects of toxic chemicals on the reproductive system, Council on Scientific Affairs, *Journal of the American Medical Association,* 253(1985):23.

Elkins, H. B. *The Chemistry of Industrial Toxicology.* New York: John Wiley & Sons, Inc., 1969.

Elkins, H. B. Excretory and biological threshold limits, *American Industrial Hygiene Association Journal,* 28(1967):305.

Elsevier Monographs on Toxic Agents. New York: Elsevier Publishing Co.

Emergency Exposure Limits. Akron, OH: AIHA.

Fairhall, L. T. *Industrial Toxicology,* 2nd ed. Baltimore, MD.: The Williams & Wilkins, Co., 1969.

Fishbein, L. *Potential Industrial Carcinogens and Mutagens.* New York: Elsevier Scientific Publishing Co., 1979.

Fourth Annual Report on Carcinogens—Summary, 1985. Washington, DC: DHHS/PHS Pub. No. NTP 85-002. Available from the U.S. Government Printing Office.

Friberg, L., Nordberg, G. F., and Vouk, V. B., eds. *Handbook of Toxicology of Metals.* Elsevier, Amsterdam: 1979.

Gerarde, H. W. *Toxicology and Biochemistry of Aromatic Hydrocarbons.* New York: Elsevier Publishing Co., 1960.

Gleason, M. N., et al. *Clinical Toxicology of Commercial Products,* 5th ed. Baltimore, MD: The Williams & Wilkins Co., 1985.

Grant, W. M. *Toxicology of the Eye,* 2nd ed. Springfield, IL: Charles C. Thomas, 1974.

Hamilton, A., and Hardy, H. *Industrial Toxicology.* Edited by Asher J. Finkel, Boston: John Wright—Publishing Sciences Group, 1983.

Hayes, W. J. *Toxicology of Pesticides.* Baltimore: The Williams & Wilkins Company, 1975.

Henderson and Haggard. *Noxious Gases.* New York: Reinhold Publishing Co., 1943.

Irish, D. D. Monitoring the environment and judging its significance, National Safety Congress Transactions, Vol. 12. Chicago, IL: National Safety Council, 1966.

Key, M. M., et al. Occupational Diseases: A Guide to Their Recognition, rev. ed., DHHS (NIOSH) Pub. No. 77-181., Washington, DC: U.S. Department of Health, Education and Welfare, June 1977. (Order from NTIS No.: PB-83-129-528/A99.)

Klassen, C. D., Amdur, M. O., Doull, J., eds. *Casarett and Doull's Toxicology—The Basic Science of Poisons,* 3rd ed. New York: McMillan Publication Co., 1986.

Kurppa, K., ed. Proceedings of the International Symposium of Research on Work-Related Diseases, Espoo, Finland, *Scandinavian Journal of Work, Environment and Health 10,* 6(1984):337.

LaDou, J., ed. The Microelectronics Industry. State Of The Art Reviews, Occupational Medicine—1:1, January-March, 1986, Hanley & Belfus, Inc., Philadelphia, 1986.

LaDou, J. Potential occupational health hazards in the microelectronics industry, *Scandinavian Journal of Work, Environment and Health,* 9(1983):42.

Lauwerys, R. R. *Industrial Chemical Exposure: Guidelines for Biological Monitoring:* Davis, CA: Biomedical Publications, 1983.

Linch, A. L. *Biological Monitoring for Industrial Chemical Exposure Control.* Cleveland: CRC Press, 1973.

Loomis, T. A. *Essentials of Toxicology.* Philadelphia: Lea & Febiger, 1974.

Lundberg, P., ed. Proceedings of the International Conference on Organic Solvent Toxicity, Stockholm, 15-17 October, 1984. *Scandinavian Journal of Environment and Health 11, suppl. 1, 1985.*

McRae, A. et al., eds. *Toxic Substances Control Sourcebook.* Germantown, MD: Aspen Systems Corporation, 1978.

Messite, J. and Bond, M. B. *Reproductive Toxicology and Occupational Exposure.* In *Occupational Medicine: Principles and Practical Applications,* 2nd ed., edited by C. Zenz. Chicago: Year Book Medical Publ., 1988.

NIOSH/OSHA—Occupational Health Guidelines for Chemical Hazards: DHHS (NIOSH) Pub. No. 81-123. U.S. Government Printing Office, Washington, DC: 1981.

National Institute for Occupational Safety and Health: Criteria Documents for Recommended Standards... see: NIOSH Publication Catalog, 6th ed. DHHS (NIOSH) Pub. No. 84-118. Cincinnati, 1984. National Institute for Occupational Safety and Health (Order from Publications Dissemination DSDTT, NIOSH) (Also, see Bibliography, Chapter 26, for list of Current Intelligence Bulletins).

NIOSH: Registry of Toxic Effects of Chemical Substances 1981-82, DHHS (NIOSH) Pub. No. 83-107. Washington, DC: U.S. Government Printing Office, 1983. (Order from NTIS No.: PB-81-154-478/A99.)

Parmeggiani, L., ed. *The Encyclopedia of Occupational Health and Safety,* 3rd ed., 2 vol. Geneva: International Labor Office, 1983.

Peto, R., and Schneiderman, M. *Quantification of Occupational Cancer, Banbury Report No. 9.* Cold Spring Harbor, NY: Cold Spring Harbor Laboratory, 1981.

Piotrowski, J. K. *Exposure Tests for Organic Compounds in Industrial Toxicology.* National Institute for Occupational Safety and Health (NIOSH), Cincinnati, OH: 1977.

Rowe, V. K., Wolf, M. A., Weil, C. S., and Smyth, H. F., Jr. The toxicological basis of threshold limit values, 2. Pathological and biochemical criteria, *AIHA Journal,* 20(1959):346.

Searle, C. E., ed. *Chemical Carcinogens.* Washington, DC: American Chemical Society, 1976.

Scientific Basis for Swedish Occupational Standards, VI: Lundberg, P. (ed.), Arbete öch Hălsa, National Board of Occupational Safety and Health, S-171 84 Solna, Sweden, 1985.

Sittig, M. *Handbook of Toxic and Hazardous Chemicals and Carcinogens,* 2nd ed. Park Ridge, NJ: Noyes Publications, 1985.

Sorsa, M., Hemminki, K., and Vainio, H. Occupational exposure to anticancer drugs—Potential and real hazards. *Mutation Research,* 154(1985):135.

Sunderman, F. W., Jr. Recent advances in metal carcinogenesis. *Annals of Clinical and Laboratory Science,* 14(1984):2.

The Industrial Environment, Its Evaluation and Control: 3rd ed., DHHS/NIOSH Pub. No. 74-117. Rockville, MD: NIOSH, 1973. (Order from GPO No.: 017-001-00396-4.)

The Merck Index, 10th ed. Rahway, NJ: Merck & Co., Inc., 1983.

U.S. Air Force Aerospace Medical Research Laboratory: Proceedings Of The Fourteenth Conference On Environmental Toxicology, November 1983, Dayton, Ohio. Technical Report distributed by Defense Technical Information Center, Defense Logistics Agency, Cameron Station, Alexandria, VA, 1984.

Waldron, H. A., ed. *Metals in the Environment.* London: Academic Press, 1980.

Wexler, P. *Information Resources in Toxicology.* New York: Elsevier Science Publishing Co., 1982.

Your Body at Work—Human Physiology and The Working Environment: Kungsholms Hamnplan 3, S-112 20 Stockholm: The Work Environment Association, 1984.

Zapp, J. A. The toxicological basis of threshold limit values. 3. Physiological criteria, *AIHA,* 20(1959):350.

Zenz, C., ed. *Occupational Medicine—Principles and Practical Applications,* 2nd ed. Chicago: Year Book Medical Publ., Inc., 1988.

Zenz, C. Reproductive risks in the workplace. *National Safety News,* September, 1984.

Zielhuis, R. L. *Occupational Exposure Limits for Chemical Agents. In Occupational Medicine: Principles and Practical Applications,* 2nd ed., edited by C. Zenz. Chicago: Year Book Medical Publ., 1988.

WHO Publications

World Health Organization: Publications Distribution and Sales, CH-1211 *Geneva 27,* Switzerland. (In the United States—WHO Publications Center USA, 49 Sheridan Avenue, Albany, N.Y., 12210). World Health Organization Publication: Analysing and Interpreting Air Monitoring Data, WHO Publication, No. 51, 1980. Evaluation of Airborne Particles in the Work Environment, WHO Publication, No. 80, 1984.

Recommended Health-based Limits in Occupational Exposure to Heavy Metals, Report of a WHO Study Group (Geneva, 1979), Technical Series, No. 647, 1980.

Health Effects of Combined Exposures in the Work Environment, Report of a WHO Expert Committee (Geneva, 1980), Technical Report Series, No. 662, 1981.

Recommended Health-based Limits in Occupational Exposure to Selected Organic Solvents, Report of a WHO Study Group (Geneva, 1980), Technical Report Series, No. 664, 1981.

Recommended Health-based Limits in Occupational Exposure to Pesticides, Report of a WHO Study Group (Geneva, 1981), Technical Report Series, No. 677, 1982.

Recommended Health-based Occupational Exposure Limits for Selected Vegetable Dusts, Report of a WHO Study Group (Geneva, 1982), Technical Report Series, No. 684, 1983.

Methods of Monitoring and Evaluating Airborne Manmade Mineral Fibers, Report on a WHO Consultation (Copenhagen, 1980), EURO Reports and Studies, No. 48, 1981.

Delayed and Chronic Effects of Chemicals in the Workplace, Report on a WHO Meeting (Kiev, USSR, 1980), EURO Reports and Studies, No. 64, 1982.

Women and Occupational Health Risks, Report on a WHO Meeting (Budapest, 1982), EURO Reports and Studies, No. 76, 1983.

Occupational Hazards in Hospitals, Report on a WHO Meeting (The Hague, 1981), EURO Reports and Studies, No. 80, 1983.

Biological Effects of Man-Made Mineral Fibres, Report on a WHO/IARC Meeting (Copenhagen, 1982), EURO Reports and Studies, No. 81, 1983.

Environmental Health Criteria

The *WHO Environmental Health Criteria* series contains assessments of existing information on the relationship between exposure to environmental pollutants or other physical and chemical factors and human health, and provides guidelines for setting exposure limits consistent with health protection.

No. 15 Tin and Organotin Compounds. A preliminary review, 1980.

No. 16 Radiofrequency and Microwaves, 1981.

No. 17 Manganese, 1981.

No. 18 Arsenic, 1981.

No. 19 Hydrogen Sulfide, 1981.

No. 20 Selected Petroleum Products, 1982.

No. 21 Chlorine and Hydrogen Chloride, 1982.

No. 22 Ultrasound, 1982.

No. 23 Lasers and Optical Radiation, 1982.

No. 24 Titanium, 1982.

No. 25 Selected Radionuclides, 1983.

No. 26 Styrene, 1983.

No. 27 Guidelines on Studies in Environmental Epidemiology, 1983.

No. 28 Acrylonitrile, 1983.

No. 29 2,4-Dichlorophenoxyacetic Acid (2,4-D), 1984.

No. 30 Principles for Evaluating Health Risks to Progeny Associated with Exposure to Chemicals during Pregnancy, 1984.

No. 31 Tetrachloroethylene, 1984.

16

Evaluation

by Edward R. Hermann, CE, PhD, PE, CIH
Jack E. Peterson, PhD, PE, CIH

THE BASIC PRINCIPLES used to evaluate occupational health hazards are discussed in this chapter, as well as the philosophical basis for establishing safe levels of exposure to chemical, physical, and biological agents. Evaluating the occupational environment requires knowledge of the etiology of occupational disease, effects of exposure, sampling procedures, biochemical dynamics, Threshold Limit Values (TLVs), and current environmental standards.

MEDICINE AND ENVIRONMENTAL HEALTH

The fields of curative and preventive medicine and environmental protection and health contribute to reduction of disease and promotion of health (Figure 16-1).

- Curative medicine provides treatment to affected persons so that they can overcome the agents of disease and repair and manage injuries to ensure healing.
- Preventive medicine strives to make us more resistant to disease agents, such as pathogenic organisms, in the environment—Developing active immunity to the smallpox virus or *Bordetella pertussis* (the causal agent of whooping cough) is an example of this process.
- Protection of the environment is necessary to protect humans. Although many species besides humans are affected by industrial pollution, lack of environmental protection measures results in decrements in human health.
- The objective of environmental health is to control various biological, chemical, or physical agents in the environment or relevant ecosystem to prevent adverse effects on populations.

These fields of endeavor are ranked in an increasing order of desirability and a decreasing order regarding expenditures of time, energy, and material resources. Since the turn of the century, the tremendous reductions in the mortality and morbidity of infectious diseases are mainly the results of improvements in environmental engineering involving mass sanitation.

The occupational aspects of environmental health can be designated industrial hygiene. This has been defined by the American Industrial Hygiene Association as "the science and art devoted to the recognition, evaluation, and control of those environmental factors and stresses, arising in or from the workplace, which may cause sickness, impaired health and well-being, or significant discomfort and inefficiency among workers along with citizens of the community." (See Chapters 1 and 24 for more details.) The natural environmental diseases of the past have been conquered; today there is a growing concern regarding the culturally spawned environmental diseases of the present and future.

ETIOLOGY OF OCCUPATIONAL DISEASE

In contrast to some of the classical communicable diseases in which a specific agent gives rise to a specific disease, the etiology of occupational diseases is often quite complex. Frequently, there are many chemical and physical causal factors. Although an infectious agent may have a pathognomonic (disease-specifying) characteristic, there may be no apparent unique relationship to the stressing agent when chronic exposure to chemical or physical factors are involved. With most diseases caused by living microorganisms there is usually a short and fairly definite period between invasion of the host and development of the disease. However, the physical and chemical agents usually require long

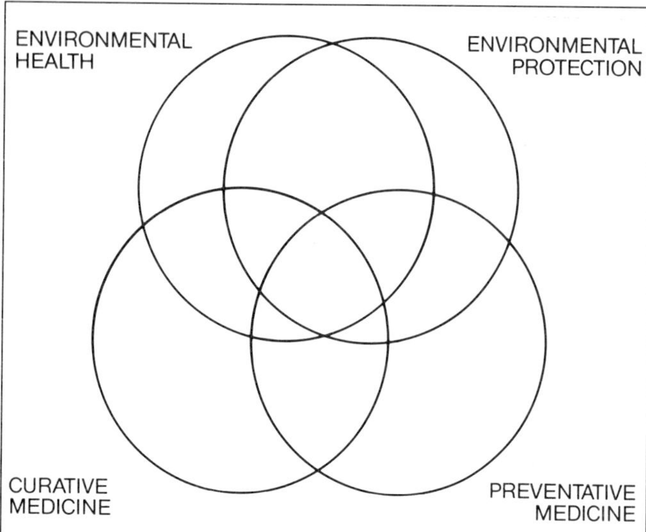

Figure 16-1. Common bases in both pure and derivative arts and science are shared by environmental health, curative medicine, environmental protection, and preventive medicine; however, there are important differences in their professional thrusts.

periods to develop the observable effects of occupational disease. Substances that trigger allergic responses, histamine release, and bronchospasm, however, as well as substances that induce other rapid acute injurious effects are exceptions.

Statistics

Most infectious diseases frequently are diagnosed correctly, and the number of cases and deaths are easily enumerated. In contrast, the excess number of deaths arising from chronic occupational diseases may only be demonstrated by application of statistical methods. The health status of a particular population has classically been measured in terms of absence of disease. In the future, to properly assess health status, more and better measures of positive health attributes of human capacity rather than incapacity must be developed.

The primary objective of industrial hygiene has been to reduce the incidence and mortality of occupational disease. A second objective is to improve the effective quality as well as the length of life. Inasmuch as occupational diseases arise from multiple factors having a complex etiology, evaluation is often difficult and, consequently, environmental control measures evolve slowly.

EFFECTS OF EXPOSURE

Although many biological agents are still very important in the field of occupational health and must not be overlooked, the effects of various chemical and physical agents are a major concern in industrial hygiene. In addition to the various qualities such as concentration, level, and type of matter and energy, the time required for a potentially harmful agent to act on susceptible cells or tissues is of prime importance in assessing its effect.

Dose-response relationship

In general terms, the two items of information needed to evaluate the likelihood of tissue injury are the dose (or total amount) of the agent and the probable response to this dose. Dose can refer to (1) the amount of material inhaled over a period of many years, (2) to the sound energy impinging on cochlear structures in short—say 5-minute—periods or (3) to the integrated flux and duration of other types of energy via other routes of exposure. Probable response relates not only to the dose itself, but also to subsequent periods of insignificant exposure that may or may not allow for recovery of the susceptible tissues of exposed individuals or populations at risk.

Survey

An industrial hygiene survey usually is conducted for evaluation of hazards. The survey consists of estimating a specific parameter of the dose, such as the air concentration of a contaminant, in as much detail as is required by the situation. After this has been accomplished, the probable response can be determined; thus the necesssity, and perhaps urgency, of instituting control measures will become apparent. In the event that control measures are required, the data previously collected may provide clues as to where the controls are most needed and will be most effective.

The results of an industrial hygiene survey must be communicated, both in verbal and written forms, to people with authority to implement action. The best industrial hygiene survey does little good if the necessary remedial measures are not taken.

Dose estimation

Estimating dose actually received consists of integrating the variations in air concentrations, sound pressures, or energy fluxes with those of the absorber. While a dose consists of a quantity of matter or energy and not a flux or concentration, these latter analogs of dose usually are measured in a particular work space or environment. The airborne concentration of a material in the breathing zone of a worker may vary considerably, depending upon many variables acting at the source of emission, the worker's movements, and the amount of incidental or deliberate ventilation.

SAMPLING THE ENVIRONMENT

Environmental sampling whether for gases, vapors, aerosol concentrations, noise, or heat stress can be focused on a particular individual or the general work area. (See Chapters 18 and 19 for more details about air-sampling and instrumentation.)

Personal sampling may be used to determine air concentrations and energy fluxes for one or a number of individuals by sampling their immediate vicinities for durations representative of whole workshift values. If workers whose exposures are determined in this way are truly representative of those in their job classifications, this type of sampling will yield excellent estimates. In general-area sampling, the loci of specific individuals are ignored; such sampling is oriented toward machines, equipment, or areas, rather than persons. General-area sampling, if done correctly, will provide both adequate estimates of exposures and neeeded information about exposure sources. A combination of both personal and general-area sampling is often desirable to get a complete picture of environmental exposure and emission sources.

Biological monitoring is a general method applicable to any stressor that produces on people a quantifiable effect that relates to the absorbed dose (ACGIH, TLVs, and BEIs, 1986) (see

Chapter 26, The Occupational Physician). To be meaningful, this measured effect should be generated at dose levels that are not harmful to the person exposed. Analysis of body fluids and tissues for chemical substances, their metabolites or for shifts in neural responses or enzyme concentrations has long been used for dose quantitation in health evaluation. This same principle may be applied to almost any imaginable stressor—from vaporized heavy metals to allergenic dusts, and from noise to gamma-radiation.

Personal sampling

The first, widely used, personal samplers were film badges and pocket chambers, which estimated the amount of ionizing radiation received by exposed workers. Industrial hygiene is now beginning to approach health physics in this respect through the use of badgelike devices that absorb specific air contaminants at reasonably reproducible rates and can be analyzed to give an estimate of exposure or dose. Noise dosimeters also have been available for some time. To date, however, little attempt has been made to use this approach with nonionizing electromagnetic radiations, heat, cold, ambient pressure or aerosols. (See also Chapters 18 and 19 for more details on sampling.)

Breathing zone. Personal sampling techniques originally were limited to holding the inlet of a sampler in the vicinity of a worker's nose and mouth during the time a particular work function was being performed. This method of obtaining samples produced reasonable results when breathing zones of typical workers were sampled during an appreciable fraction of a workshift. Better results are now obtained by having employees wear battery-powered pumps capable of maintaining adequate airflow through sampling devices for a full workday (Figure 16-2).

Breathing-zone air-sampling of aerosols is usually done using a cyclone or other preselecting device that permits only respirable-sized particulate matter to be collected on a filter. Also, many vapors and gases can be adsorbed onto activated charcoal and later desorbed into carbon disulfide or thermally desorbed for chromatographic analysis.

Stain detector tubes. Detector tubes are probably the most used (and abused) analytical devices for obtaining direct readings of gas or vapor concentrations in work spaces (Figure 16-3). Although methods have been devised for integrating air samples through detector tubes during periods approximating one working day (say, 500 minutes), such devices are not well suited to general area monitoring and integration of exposures. The most valid use of detector tubes is to determine the magnitude of peak gas or vapor concentrations. However, this can only be accomplished when the time and place such peaks occur can be reasonably well estimated from a knowledge of the kinetics of emission and mixing in the workroom. Although gas detector tubes can be used for estimating daily general area concentrations, they are seldom used properly when integrating personal exposures.

Regardless of the duration of the sampling period involved when making a contaminant-gas-in-air measurement, the actual concentration measured is averaged over the specific time period of sampling. What is really needed to produce valid sampling and measurement of gases and vapors in the air for occupational health use is instrumentation that will provide a reading resolution of about 3 seconds (the nominal time of one respiration), and continuous recording of concentration for a full workday of 7–10 hours. With such instrumentation integrated air concentrations could be validly measured on a sound basis for estimating definable exposures. Of course, off-the job exposures to various environmental contaminants are sometimes quite significant and may complicate or even confound the evaluation.

Figure 16-2. Battery-powered pumps can be used to obtain full-shift breathing zone air samples. (Courtesy E. I. du Pont Nemours & Co., Inc.)

Analysis of personal protective equipment. Analysis of protective clothing may be used to evaluate exposures to low volatility materials (such as pesticides), which can be absorbed through the skin in toxic amounts (Wolfe et al, 1967). Use of absorbent pads is another method used to detect penetration of chemicals when the area of contact is small and known in advance (such as with gloves).

A possible extension of breathing-zone sampling is to analyze the filter elements of a respirator to determine the concentrations in the breathing zone of the wearer. Often sampling procedures can be quantified and made somewhat more useful by correlations with biological sampling.

An industrial hygienist must obtain the cooperation of workers if personal sampling is to be valid. In general, when a careful and thorough explanation of the procedure is given, this is not a serious problem.

The principal advantage of personal sampling is that when performed properly it gives a very good indication of a particular person's exposure during an entire workshift. Regardless of how a worker moves about, whether at work or on a break, the sampler collects information concerning the amount of exposure arising from the worker's immediate environment. The main disadvantage of personal sampling is that it gives few clues about the

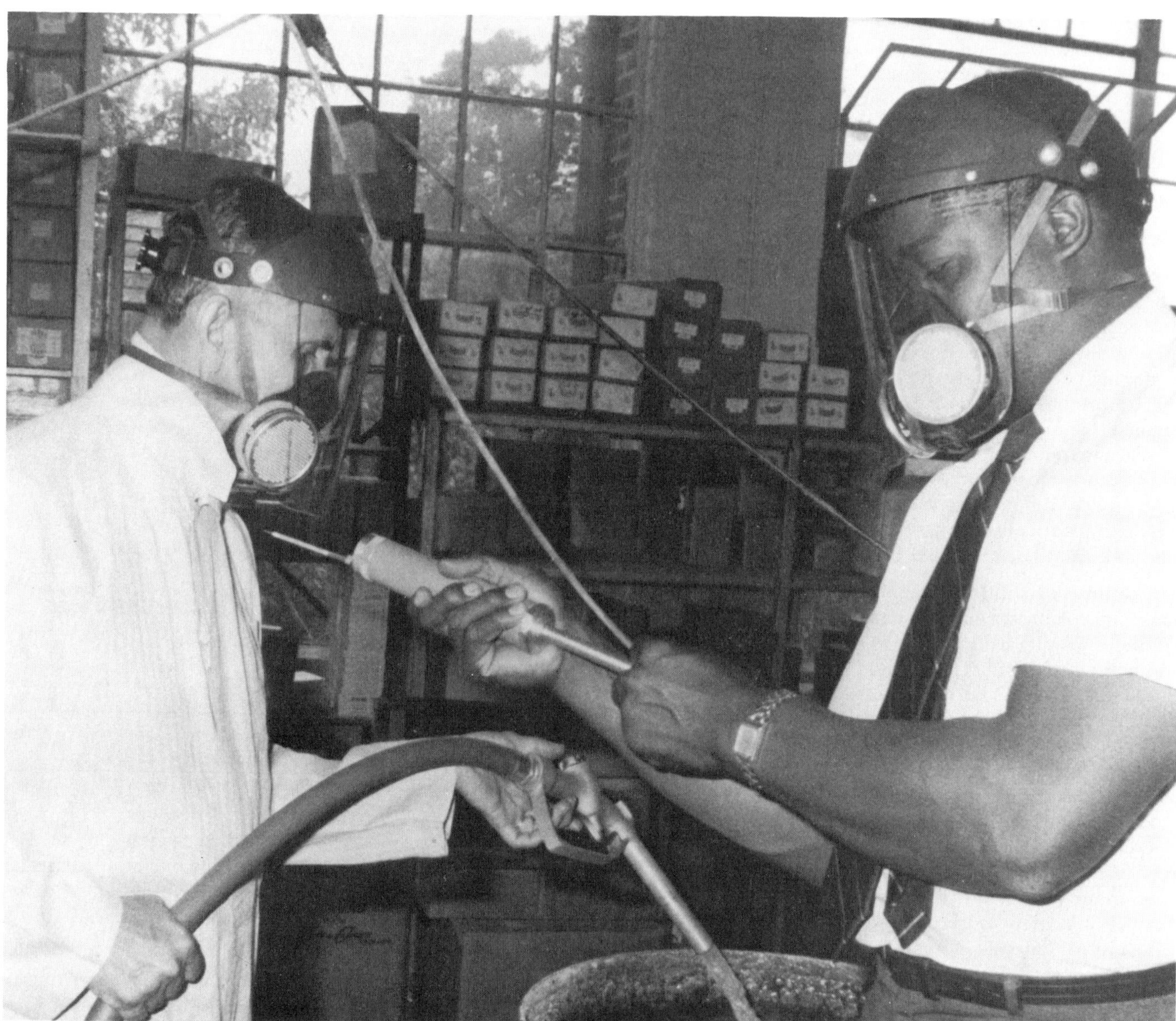

Figure 16-3. The gas detector tubes shown here are useful for obtaining direct readings of the gas or vapor contamination present in the workplace air.

most important sources of exposure, and therefore few hints at how to reduce the specific sources of excessive exposures.

General-area sampling

General-area sampling is used by industrial hygienists to evaluate exposure. Ideally, the sampling would be so thorough and the pattern of potential exposure of workers so well defined that in any given workspace knowledge of a worker's activity would be sufficient to estimate that person's exposure. If, for instance, vapor concentrations and their duration around equipment were known well enough to allow drawing lines of equal concentration (isopleths) superimposed on a floor plan, then a worker's exposure could be determined from observing his/her movements and plotting the frequency and duration spent in each area. The employee's daily exposure could then be found by adding exposure increments to compute the time-weighted average (TWA). This method has several advantages, some of which are not obvious.

In any location where a person may be exposed, the incident flux and/or airborne concentration usually varies with time, and these ranges of variation can sometimes be astonishing. The fluctuations may have a short time base in relation to a full workshift, measurable in seconds or minutes, or they may have a long time base lasting weeks or months. In some situations, both short and long periods of fluctuation are important in the evaluation. These sampling measurements can be correlated with other variables to provide a method for interpolating or extrapolating exposure data to other times and places.

Although only extensive and continuing general-area sampling can provide information about such fluctuations as they pertain to specific locations, the data can provide valuable clues about the main sources of exposure, and hence may provide a means to determine the type and extent of controls.

Another advantage of general-area sampling is that the necessary equipment has few size or shape limitations. The sampling

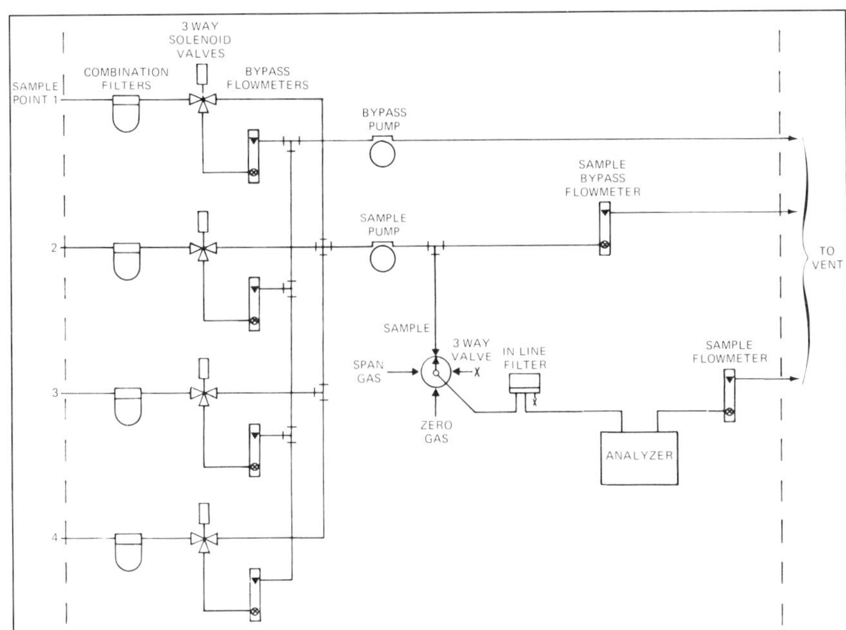

Figure 16-4. Centralized analytical devices can be attached to remote probes to simultaneously acquire data from several areas.

equipment can be made rugged, long-lasting, and even tamper-proof. It can be programmed to provide data for extended periods. Coupled with adequate recording devices, the equipment can operate unattended for days or weeks if its reliability is good. Centralized analytical devices can be attached to remote probes so that data can be acquired from several areas simultaneously (Figure 16-4). An alarm can be sounded if some preset limit is exceeded. General-area sampling equipment may be connected to a strip chart recorder and the result used for historical records of environmental data, and it can be connected to a computer and used to record and retain exposure estimates for any individual in a relatively large work force (Peterson et al, 1966; and Kramer and Mutchler, 1972) (Figure 16-5).

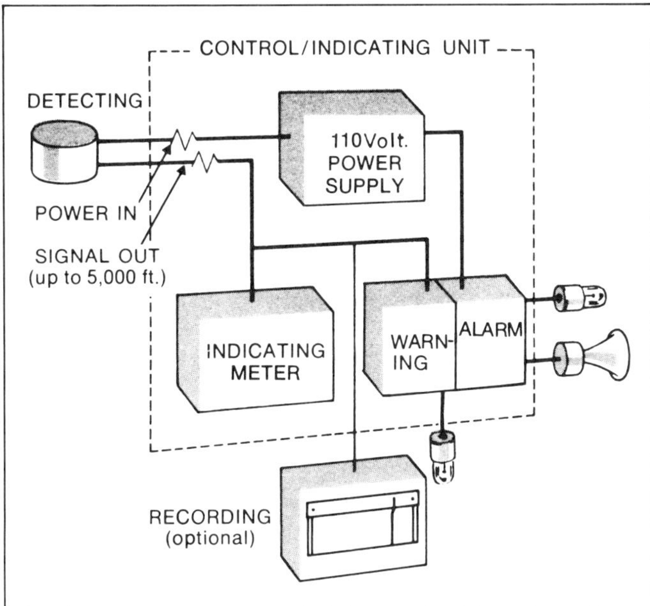

Figure 16-5. General-area sampling equipment can be connected to a strip chart recorder to obtain a permanent record of the contamination levels. (Courtesy MSA)

General-area sampling also has numerous disadvantages. Although sampling equipment can be made rugged and reliable, more often it is not, and as equipment sophistication increases so does the demand for well-trained, highly paid technicians. While these disadvantages are not limited to general-area sampling equipment, the complexity of such equipment is usually greater than that of personal sampling instruments.

Sampling points. General-area sampling can provide a good estimate of exposure at numerous locations; however, the number of sampling stations often is not sufficient to determine the complete description of the actual exposure of a particular person. Even when the equipment is adequate and reliable, other aspects of the study (whether observational or a summation of time and location) frequently are not. The weakest point in the whole process of calculating the exposure from general-area samples is in determining work patterns. Observing work patterns is most frequently conducted during the day shift; but it is very possible that patterns during other shifts are quite different, not only because of different types of supervision but also because of different work schedules and job demands. Because of rapidly changing patterns of work, the difficulty in the determination of work patterns is the major disadvantage in estimating individual exposures with general-area sampling.

Biological monitoring

Urine, blood, and breath are the usual body fluids examined for evidence of past exposure to chemical toxicants. Lead concentrations in urine or blood have long been used as indices of lead exposure. Blood zinc protoporphyrin (ZPP) is a measure of lead exposure during the previous 3–4 months.

Expired breath analysis. Expired breath samples are well accepted for estimating blood alcohol concentrations for legal purposes, and more recently have been used to determine exposures to carbon monoxide and a number of solvents (Kramer and Mutchler, 1972; Breysse and Bovee, 1969; Butt et al, 1974).

(See also Chapters 6, Solvents, and 15, Toxicology.) Biological sampling is especially important for detecting chemical exposures in which a major portion of the dose may have been absorbed through the skin (Linch, 1974).

Saliva and tissues. Saliva is yet another body fluid that may be used for estimating excessive absorption of chemical contaminants. Hair and nail clippings are examples of tissues that have been used for the estimation of exposures to heavy metals. Body fat samples also can be used to determine storage of fat-soluble substances, but obtaining these samples or bone marrow samples is invasive and not suitable for screening.

Many specific biological tests also can be used to indicate past exposure, for example, the direct effects of laser radiation on the retina can be observed with the proper kind of eye examination. Audiometric examinations can be used to determine the extent of temporary or permanent shifts in thresholds of hearing acuity, and grading and analysis of the audiograms yield results making it possible to evaluate occupational noise exposures and prescribe appropriate hearing conservation procedures (Hermann, 1963).

The appearance of lung tissue on x-ray films is characteristic for several pneumoconioses and, in some cases, valid correlations can be made between an x-ray grade and the severity of exposure (Ashford et al, 1965) (Figure 16-6). In another biological sampling procedure, the dynamics of the uptake of rubidium-86 may be useful in estimating the past exposure to x- or gamma-rays (Scott et al, 1973).

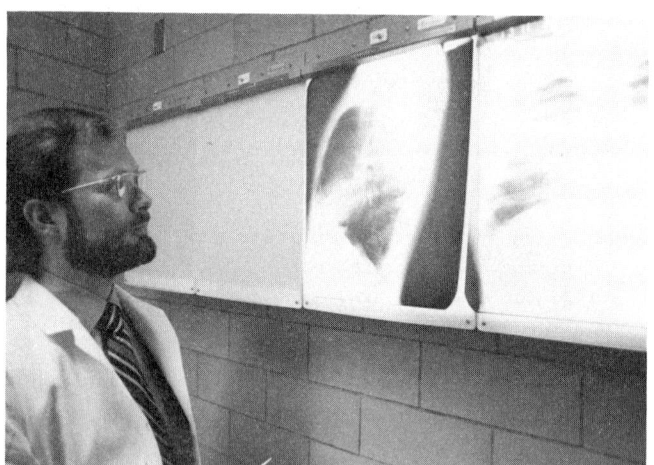

Figure 16-6. X-ray films are used to diagnose pneumoconiosis.

BIOCHEMICAL DYNAMICS

The usefulness and reliability of biological monitoring methods depend on the biochemical dynamics of the contaminant in relation to human life processes. Measurements of biological samples may represent early excretions following recent absorption, or equilibrium levels attained after considerable time. It should not be assumed from this statement, however, that all biological monitoring methods provide adequate correlations with the degree of exposure, or even of absorption.

In an investigation to determine the influence of the size of a dose on the distribution and elimination of inorganic mercury when administered as mercuric nitrate to rats, Cember (1962) found that under his experimental conditions, the dynamics of mercury elimination were strongly influenced by the size of intraperitoneally administered doses, even with much smaller doses than those that caused death. On the other hand, distribution of the mercury to the liver, blood, and kidneys was relatively unaffected by the size of the dose.

From an industrial hygiene viewpoint, note that Cember's data (1962) indicated that mercury excreted through the urine was carried by cells of the tubular epithelium which had sloughed off either as the result of normal cell turnover, or because of desquamation of the tubular epithelium following the accumulation of cytocidal (cell-killing) amounts of mercury in the cells. The cytocidal threshold concentration in this experiment was between 0.2 and 3 micrograms of mercury per gram of kidney. Later, Bell et al (1973) published a study on urinary mercury excretion in humans in relation to inhaled air concentrations; results of this study indicated poor correlation between exposure and excretion.

At the same time, Lauwerys and Buchet (1973) concluded from another study that urinary excretion of mercury by humans did not correlate with occupational exposures to mercury vapors at air concentrations near the TLV. Yet, despite all this information, urinary excretion of mercury is still used as an index of mercury absorption. Obviously, a better measure is needed, and if the uptake and excretion of inhaled mercury by humans resembles that in the rat by intraperitoneal administration, then the mercury content of feces should provide a good if not a much better index of exposure.

Exposure v. dose

Various types of environmental sampling may be used to determine, with varying degrees of precision, the amount of exposure an individual receives. The determination of this exposure level or concentration, however, is not the same as the determination of a dose. With regard to absorbed chemical substances, for example, excretion usually takes place during exposure and dose must represent the manner in which the body absorbs and integrates a contaminant during this exposure.

Biological sampling appears to be one way to reconcile exposure with dose in those cases when reliable information has been developed regarding the uptake, storage, detoxification, and excretion of a chemical in relation to time (ACGIH TLVs and BEIs, 1986; Kramer and Mutchler, 1972; Linch, 1974; Menz et al 1974; Polankoff et al, 1974). Unfortunately, what is frequently labeled a dose-effect relationship is in reality only an exposure-effect relationship, and consequently industrial hygienists are often forced to use the latter correlations to provide health evaluations and prescribe environmental controls. Despite these shortcomings, however, many excellent examples of health evaluation and subsequent control measures have appeared in the literature (Polankoff et al, 1974; Williams, 1970; Elkins et al, 1962; Cholak et al, 1967; Danziger and Possick, 1973; Rosensteel et al, 1974).

Response

Assuming that exposure and dose have been estimated with reasonable accuracy, the next step is to use this information to determine the probable response. What should be a rather straightforward application of known principles is often complicated by concurrent stressors. Exposure to just a single chemical rarely occurs. Frequently, the exposure is to a mixture of several chemicals. (For more details, see Chapter 15, Industrial Toxicology.) Exposure to physical as well as chemical stressors may also occur simultaneously. For example, the process of tunneling may involve simultaneous exposures to high atmospheric pressure, dust, noise, heat, high humidity, and carbon monoxide. An assessment of the strain produced by any one of these stressors may be complicated by the presence of the others.

Combined effects

At present very little is known about how the body integrates two different types of stress and the resultant strain, even if both stressors are chemical. The usual assumption is that chemicals affecting different organs or tissues should be considered independently, while those that affect the same critical organ or tissue should be studied jointly as they may produce additive or even synergistic effects. The present level of knowledge regarding the combined effects of various toxicants on human beings is meager.

Toxicant studies employing the methods used by Wahed (1968) need to be performed on test animals and possibly humans. The studies of Wahed, Frank, and Hermann (1968; 1981) clearly illustrated that the combined effects of several common toxicants including mercuric chloride, nickel chloride, potassium dichromate, copper sulfate, and sodium cyanide were additive, and not synergistic or antagonistic, when tested for their inhibitory effect on oxygen utilization by aerobic microorganisms.

Although studies on aerobic microorganisms may find little direct application to effects on human health, the methods of Wahed et al are quite important inasmuch as it is currently quite popular to claim that exposure to one chemical in combination with another may be synergistic, not merely additive, and that both should, therefore, be banned. Concrete examples of truly synergistic effects, though infrequent, do exist, however.

Another complicating factor is that even for single-substance exposure, well-controlled prospective studies using human subjects are seldom available (Franke et al, 1981). Toxicological data have usually been obtained from animal experiments and applied to people by extrapolation. In some situations, the data are related to a different route of exposure and thus have limited use. Finally, in all too many instances no definitive toxicological data are known and the industrial hygienist must make a judgment based on analogy with another similar form of matter or energy and hope that a safety factor will compensate for any error in evaluation and judgment. The greater the consequence of ignorance, the greater the safety factor that will be required.

If the exposure and related absorbed dose are at or below those known to produce little or no effect, then analysis is uncomplicated, unless the time span of the data base is too short in relation to a human life or unless genetic effects are involved. With most occupational stresses, an upper limit exists, which should not be exceeded even for very brief periods of time, whether or not it is actually recognized. For example, brief exposure to continuous noise exceeding 115 dBA should be avoided. Similar limits exist for abnormal pressure, heat, and nonionizing radiation.

If exposures or doses are not at or below some established control level, then analysis of the situation may indicate a need for information about the expected effects. Often knowledge of the toxic effects of substances on humans has been developed from accumulation of information regarding incidental exposure. Needless to say the validity of such data is often difficult to assess.

WORK SPACE STANDARDS

In the 1920s, lists of personally used maximum allowable concentrations (MACs) for gases, vapors, and dusts began to appear. Because many of the users and developers of such information were industrial hygienists who were affiliated with governmental agencies, the lists were assembled and became the basis for the Threshold Limit Values (TLVs) of the American Conference of Governmental Industrial Hygienists (ACGIH). (See Chapter 15, Industrial Toxicology, for a complete discussion of TLVs.) The TLV lists now contain values for physical as well as chemical agents (see Appendix B-1). Most industrial hygienists refer to this list when they are faced with finding the upper limit of safe exposure for a particular substance or physical agent. The list is reviewed and updated annually by a committee of experts, and then republished by the ACGIH. Although TLVs have generally referred to time-weighted average (TWA) air concentrations, in 1960 the data on chemicals was extended to include a ceiling value (or "C" notation) for certain substances, and a "skin" notation for material if absorption through the intact skin may contribute significantly to the absorbed dose during any particular exposure. In 1976, Short Term Exposure Limits (STELs) were added for many substances.

Short-term exposure limit

The STEL is defined as a 15-minute TWA concentration, which should not be exceeded at any time during a workday, even if the 8-hour TWA is within the TLV. The duration of a STEL exposure should not be repeated more than four times per day. There should be at least 60 minutes between successive exposures at the STEL.

In addition to listing several hundred adopted values, the Threshold Limit Value Committee issues a "Notice of Intended Changes" for the forthcoming year. In the past, this list has largely dealt with the addition of chemical substances to be added to the existing TLV listings, and not only provided an opportunity for specific comment, it also encouraged suggestions for additions. In 1985, however, a major philosophical shift in the collective opinion of the TLV Committee took place, and the Notice of Intended Changes for 1985-86 listed 199 substances for which it was recommended that the TLV-STELs be deleted.

In the 1986-87 TLV booklet, deletion of STELs for 51 of the previously listed 199 chemical substances was reiterated. The rationale underlying this shift in the opinion of the TLV Committee was that toxicological data and documentation were lacking regarding the STELs, and that elimination of the STEL should alert the user that exposure at air concentrations greater than the 8-hour TLV might entail undue risk and should not be condoned.

Even the most astute industrial hygienist should review the Introduction to each set of Threshold Limit Values and refer to their documentation with current supplements (ACGIH, 1986).

Acceptable concentrations

Separate pamphlets designating "acceptable concentrations" for a number of substances have been published by the Z-37 committee of the American National Standards Institute (ANSI). Acceptable concentrations are not single values but, assuming an 8-hour work exposure, contain estimates of an acceptable ceiling concentration, an acceptable TWA, which usually corresponds closely with the ACGIH (TLV), and an acceptable maximum for peaks above the ceiling.

Unfortunately, the number of substances for which the ANSI has published pamphlets is quite limited, and the existing ones are not updated with any regularity.

'Hygienic Guides'

The American Industrial Hygiene Association has published "Hygiene Guides" for more than 150 materials. Each guide contains a summary of toxicological information, including a maximal acceptable concentration which is usually identical to the TLV for the year of publication. The guides provide a summary of significant physical properties of each chemical, as well as concise information on hygiene standards, toxic properties, industrial hygiene practices, medical information, and selected references.

'Criteria Documents'

The National Institute for Occupational Safety and Health (NIOSH) publishes summaries of information concerning chemical and physical stressors in the work environment. The NIOSH publications of most interest are called "Criteria Documents" and contain information from many sources. Although the "Criteria Documents" are more thorough than either the "Acceptable Concentrations" pamphlets or "Hygienic Guides," their contents may become outdated soon after publication.

Permissible Exposure Limits

The U.S. Department of Labor's Occupational Safety and Health Administration (OSHA) adopted the 1968 TLV list in its first attempt to provide legal control of the work environment by making the guides mandatory. These were then called Permissible Exposure Limits (PELs).

As a result of the 1970 Occupational Safety and Health Act, many industrial hygienists are faced with problems of legal compliance as well as protecting the health of the workers. Considerable time has passed since the adoption of official control values, and more recent information may well indicate a need for greater restrictions to assure no injury to susceptible persons, or in some cases it may suggest that a relaxation of restrictions is in order. (Occurrence of the latter situation, however, is rare.)

The industrial health engineer or industrial hygienist has a professional responsibility to provide a good honest evaluation of the hazards of the work environment. Blind adherence to a standard is no substitute for accurate knowledge or even professional judgment. And because legal obligations should be met, when applicable standards are found to be unworkable they should either be rescinded or changed.

Biological Exposure Indices

Control values for other than air concentrations do exist and these may be necessary in addition to environmental monitoring for monitoring worker exposures to industrial air. Some Biological Exposure Indices (BEIs) have been developed and more will be developed in the future by the ACGIH. These values appear in the annually published booklet, *Threshold Limit Values and Biological Exposure Indices.* (See Appendix B-1 for the 1986-87 edition.) These values represent limits on the amount of substances (or their metabolites) to which a worker may be exposed without demonstrable hazard to health or well-being as determined from the worker's tissues, body fluids, or exhaled breath. The biochemical or physiological measurements on which BEIs are based can furnish two kinds of information useful in the control of worker exposure:

- measurement of the individual worker's overall exposure.
- measurement of the worker's individual characteristic response.

Measurement of responses furnishes an estimate of the physiological status of the worker and may consist of (1) changes in the level of some critical biochemical constituent, (2) changes in the activity of a critical enzyme, or (3) changes in some physical function. A Biological Exposure Index limit may be used as an adjunct to its corresponding TLV and in some cases it may alert the health and safety professional to undesirable exposures before they are detected by environmental monitoring. However, under no circumstances should the use of biological monitoring be substituted for environmental monitoring. The BEI is considered supplementary to an airborne TLV (ACGIH, 1986). The development of BEIs is discussed in H.B. Elkin's, *The Chemistry of Industrial Toxicology* (1959). Extensive consideration of biochemical indicators of chemical absorption and body burden has been provided by R.S. Waritz (1979).

ZERO REFERENCE LEVELS

Although the Romans had no number for zero, later cultures did. The ease of calculations afforded by introduction of the Arabic numbering system that includes a zero has contributed immeasurably to the development of mathematics, science, engineering, economics and world trade. However, those who understand the laws of nature realize that achievement of a truly zero concentration (or zero amount of matter in a given quantity of space) is essentially impossible in the real world. When most people think of controlling a hazard, their thoughts center on reduction or elimination rather than control. They try to apply the "zero concept" of mathematical science to the real world, and promulgate regulations demanding zero levels of contaminants in some portion of the environment. That these human-made laws are in conflict with the inviolate laws of nature seems to be beyond the comprehension of some in the body politic and results in continuing controversy and litigation.

Some reflection on a basic difference between biological and chemical agents and disease may be of interest. While technological application of environmental health and medical sciences during the past century has reduced many of the infectious diseases to insignificant incidence rates, only a few have been truly eradicated from sizable geographic areas. For example, in 1976, the World Health Organization reported that there were very few active cases of smallpox in the entire world. Since humans are the sole living reservoir of this viral disease, it was hoped that the infectious agent of smallpox would be eradicated when (1) these cases had run their course, (2) no new cases were reported for a few years, and (3) present stores of smallpox virus used in vaccine manufacture had been destroyed without incident. Although outbreaks in Africa during 1977 temporarily dimmed hope for worldwide eradication, nevertheless, on May 8, 1980, the Thirty-third Assembly of the World Health Organization (WHO) certified that smallpox had been eradicated from the world. Despite this political pronouncement, absolute eradication obviously has not taken place. The United States and the Soviet Union continue to vaccinate their military personnel, each apparently fearing the other will use smallpox virus in biological warfare. Because a stock of variola virus must be maintained to produce the millions of doses (at a cost of millions of dollars) of vaccine provided annually, it is apparent that true eradication from planet Earth has not been achieved.

In dealing with pure or mixed cultures of living organisms in small, well-defined ecosystems, sterility or eradication of a single species may be readily achieved. However, when one considers the absolute elimination of the last molecule of a chemical element or compound from a given volume of space—even such a small volume as one liter or one cubic meter—the problem of eradication of a contaminating substance is hopelessly unachievable inasmuch as this calls for production of an absolute vacuum or perfect cleanout. To remove the last molecule of a contaminant from 22.4 liters of gas would mean scavenging one molecule out of 6×10^{23} molecules (Avogadro's number), or from a cubic meter of air (at 760 mm Hg pressure and 25 C), one molecule out of 2.5×10^{25} molecules.

For an element such as mercury which is monatomic in its vapor phase, eradication of the last molecule in one cubic meter of air would mean reducing the weight-per-volume (W/V) concentration from 3.3×10^{-22} grams per cubic meter to absolute zero. It should be noted by way of comparison that a background concentration of mercury in clean country air has been measured at about 5 nanograms per cubic meter or 1.5×10^{13} mercury molecules per cubic meter. Since the lower limit of our present ability to measure concentrations of chemical substances is about 1 part in 10^{18} parts (1 billionth of one part per billion or 10^{-18}), it is obvious that we cannot even satisfactorily measure air, liquid, or solid concentrations approaching absolute zero for even one cubic meter of air, let alone the one half million or so cubic meters of air that a human being might breathe in an average lifetime. Obviously we cannot use zero in our regulations in the literal sense.

INTERPRETATION OF RESULTS

Despite the impossibility of achieving an absolute zero concentration, a reference point must be established as a guide to the need for control measures. Without something at which to aim, control is unlikely, and achievement of needed reduction lacks a standard of comparison. In addition to an environmental standard or reference point, it is necessary to have workable information regarding the system, the source of contamination, or the process that may need to be controlled. Without appropriate data neither the necessity nor the extent of control can be developed; nor can achievements be verified.

Information about the variable being controlled is called feedback if this information is used to generate an error signal in which intensity is in proportion to the deviation of the variable from its reference point. In order to control anything, including the time-weighted average concentration of a compound in the breathing zone of a worker, the desired control level must be established and information obtained regarding variation in the actual working levels from time to time.

The aim of much of the health activity in the occupational environment is to assure that workers are not exposed to dangerous amounts of anything and, secondarily, when such exposures do occur, to keep them at minimum practical levels. Most important, then, is a reference point somewhat less than the minimum exposure required to produce an unwanted effect. After the desired degree of control has been attained, the reference point can be reduced, if desired, to bring the exposure as near to zero as is consistent with other constraints on the system. Some leeway is often needed in the ongoing body burden for the particular substance of interest to allow for environmental exposures other than those occurring at work. An example of consideration for the total environment is the adjustment of the TLV for lead in order to accommodate absorbed doses from food, drinking water, and community air.

Throughout the monitoring process, continuing information regarding the variable being controlled is essential. Without feedback, control and even reduction may not be possible because the measures taken may have little or no effect. If this lack cannot be sensed, control is unlikely. Thus, in the field of occupational health, ongoing development of analytical systems and devices, with continual improvements in accuracy, precision, sensitivity, selectivity, stability, portability, ruggedness, and cost are constantly sought.

Quantitative measurements

Industrial hygiene and health engineering rest on a base of quantitative measurement which provides the feedback necessary for control. To a large extent, quantitative evaluation of a health hazard is what distinguishes the occupational health engineer or industrial hygienist from other professionals who also seek to reduce occupational diseases and hazards. The reference points used for control purposes in this field have been called MACs, TLVs, STLs, STELs, BEIs, and other names. To the uninitiated, the TLV for a gas in air appears to designate either a line between harmless and toxic concentrations or to be an expression of the relative toxicity of the material. It is in fact, neither! It is a reference point for control purposes, and this should constantly be kept in mind. The rationale for developing the TLV for hydrogen cyanide was different from that behind the TLV for hydrogen chloride, and the criteria underlying the establishment of a TLV for the halocarbon dichlorodifluoromethane is also different. Of these three, only the TLV for hydrogen cyanide has to do with systemic toxicity at the cellular enzyme level; and, whereas the TLVs for hydrogen cyanide and hydrogen chloride are based on related injury hazards, that for dichlorodifluoromethane is not. All TLVs have in common the fact that they are reference points for control purposes; to use them in other ways is to ignore and perhaps pervert their derivation and purpose. Converting TLVs into OSHA standards was viewed by many as unjustified.

Threshold Limit Values for physical agents are established in the same manner as those for chemical agents. They too are not fine lines between "safe" and "dangerous," nor are they necessarily related in any way to the severity of effect that may be experienced upon overexposure. They are hygienic standards or reference points presented to the occupational health community and others as primary control goals. Achievement of these goals will assure that nearly all workers may be repeatedly exposed day after day without adverse effect.

SUMMARY

Knowledge of the concentration, level, type of matter and energy, and length of time during which a harmful agent reacts with susceptible tissues is of prime importance in assessing the effects. The dose or total amount of the agent and the probable response to this dose are needed to accurately evaluate the degree of injury. Estimation of the dose requires integrating the variations in concentration or energy fluxes as a function of time.

Environmental sampling techniques can be classified as personal or general area. Biological sampling of the urine, blood, or breath can provide evidence of past exposure to toxic agents. Expired breath samples have been useful for estimation of blood alcohol levels (AMA, 1972). The usefulness of biological sampling methods depends upon the biochemical dynamics in relation to human life processes and storage in target organs.

Knowledge of environmental standards is required to provide an accurate evaluation of the hazards of the work environment. Blind adherence to a standard is no substitute for scientific knowledge or even professional judgment.

As explained before, rules and regulations demanding zero levels of contaminants are in conflict with the laws of nature. In making occupational health hazard evaluations, it is essential that a reference point be established as a guide for the establishment of control measures.

In addition to the environmental standard or reference point, it is necessary to have information concerning the source of contamination or process that needs to be controlled. Industrial hygiene engineering rests on a base of quantitative measurement that provides the feedback necessary for control.

BIBLIOGRAPHY

American Conference of Governmental Industrial Hygienists. *Threshold Limit Values and Biological Exposure Indices for 1986-87.* Cincinnati: ACGIH, 1986, pp 54-62.

American Conference of Governmental Industrial Hygienists. *Documentation of Threshold Limit Values, Including Biological Exposure Indices and Issue of Supplements through 1989,* 5th ed., Cincinnati: ACGIH, 1986.

American Medical Association. *Alcohol and the Impaired Driver.* Chicago: National Safety Council, 1972.

Ashford, J.R., Fay, J.W.J., and Smith, C.S. The correlation of dust exposure with progression of radiological pneumoconiosis in British coal miners. *American Industrial Hygiene Association Journal* 26 (July-August 1965):347-361.

Baetjer, A. M. Changes—Stress or benefit? *American Industrial Hygiene Association Journal* 25 (May-June 1964):207-212.

Bell, Z.G., Lovejoy, H.B., and Vizena, T.R. Mercury exposure evaluations and their correlation with urine mercury excretions, Pt. 3: Time-weighted average (TWA) mercury exposures and urine mercury levels. *Journal of Occupational Medicine* 15 (June 1973):501-508.

Breysse, P.A., and Bovee, H.H. Use of expired air-carbon monoxide for carboxyhemoglobin determinations in evaluating carbon monoxide exposures resulting from the operation of gasoline fork lift trucks in holds of ships. *American Industrial Hygiene Association Journal* 30 (September-October 1969):477-483.

Butt, J., et al. Carboxyhemoglobin levels in blast furnace workers. *Annals of Occupational Hygiene* 17 (August 1974):57-63.

Cember, H. The influence of the size of the dose on the distribution and elimination of inorganic mercury, $Hg(NO_3)_2$ in the rat. *American Industrial Hygiene Association Journal* 23 (July-August 1962):304-313.

Cholak, J., Schafer, L., and Yeager, D. Exposures to beryllium in a beryllium alloying plant. *American Industrial Hygiene Association Journal* 28 (September-October 1967):399-407.

Danziger, S.J., and Possick, P.A. Metallic mercury exposure in scientific glassware manufacturing plants. *Journal of Occupational Medicine* 15 (January 1973):15-20.

Elkins, H.B. *The Chemistry of Industrial Toxicology.* New York: John Wiley & Sons, 1959.

Elkins, H. B., et al. Massachusetts experience with toluene diisocyanate. *American Industrial Hygiene Association Journal* 23 (July-August 1962): 265-272.

Franke, J.E., Wahed, S.A., and Hermann, E.R. Inhibition of oxygen utilization by substances in combination. *Journal of Applied Toxicology* 1 (March 1981):154-158.

Hermann, E.R. An audiometric approach to noise control. *American Industrial Hygiene Association Journal* 24 (July-August 1963):344-356.

Kramer, C.G., and Mutchler, J.E. The correlation of clinical and environmental measurements for workers exposed to vinyl chloride. *American Industrial Hygiene Association Journal* 33 (January 1972):19-30.

Lauwerys. R.R., and Buchet, J.P. Occupational exposure to mercury vapors and biological action. *Archives of Environmental Health* 27 (August 1973):65-68.

Linch, A.L. Biological monitoring for industrial exposure to cyanogenic aromatic nitro and amino compounds. *American Industrial Hygiene Association Journal* 35 (July 1974): 426-432.

Menz, M., Luetkemeir, H., and Sachsse, K. Long-term exposure of factory workers to dichlorvos (DDVP) insecticide. *Archives of Environmental Health* 28 (February 1974):72-76.

Peterson, J.E., Hoyle, H.R., and Schneider, E.J. The application of computer science to industrial hygiene. *American Industrial Hygiene Association Journal* 27 (March-April 1966): 183-185.

Polankoff, P.L., Busch, K.A., and Okawa, M.T. Urinary fluoride levels in polytetrafluoroethylene fabricators. *American Industrial Hygiene Association Journal* 35 (February 1974):99-106.

Rosensteel, R.E., Shama, S.K., and Flesch, J.P. Occupational health case report: No. 1—Toxic substances: Carbon disulfide; industry; viscose rayon manufacture. *Journal of Occupational Medicine* 16 (January 1974):22-32.

Scott, K.G., et al. Occupational X-ray exposure. *Archives of Environmental Health* 26 (February 1973):64-66.

Stewart, R.D., Hake, C.L., and Peterson, J.E., Use of breath analysis to monitor trichloroethylene exposures. *Archives of Environmental Health* 29 (July 1974):6-13.

Waritz, R.S. "Biological indicators of chemical dosage and burden," Chap. 7 in *Patty's Industrial Hygiene and Toxicology. Theory and Rationale of Industrial Hygiene Practice,* Vol. 3, edited by Lewis J. and Lester V. Cralley. New York: Wiley-Interscience, 1979.

Williams, H.L. A quarter century of industrial hygiene surveys in the fibrous glass industry. *American Industrial Hygiene Association Journal* 31 (May-June 1970):362-367.

Wolfe, H.R., Durham, W.F., and Armstrong, J.F. Exposure of workers to pesticides. *Archives of Environmental Health* 14 (April 1967):622-633.

Acknowledgment

Some of the material in this chapter has appeared in *Industrial Health,* by Jack E. Peterson, Englewood Cliffs, NJ: Prentice-Hall, Inc., 1977.

17

Methods of Evaluation

by Julian B. Olishifski, MS, PE, CSP

EVALUATION CAN BE DEFINED as the decision-making process that results in an opinion as to the degree of risk arising from exposure to chemical, physical, or biological agents. Evaluation involves making a judgment based on observation and measurement of the magnitude of these agents. Evaluation also involves determining (1) the levels of energy or air contaminants arising from a process or work operation and (2) the effectiveness of any control measures used.

GENERAL PRINCIPLES

An industrial hygiene appraisal of the work environment is performed to define and evaluate the exposure of workers to a chemical or biological agent or to a physical energy stress.

After researching the potential health hazard of these stresses, the extent of the hazard must be determined. The nature of the process in which the agent is used or generated, the toxicity and concentration of the chemical or physical agent, the possibility of reaction with other agents (either chemical or physical), the degree of effective ventilation control, the extent of enclosure, the duration of employee exposure, and individual susceptibility all relate to the degree of hazard.

Recognition

The recognition of potential hazards requires familiarity with the processes and work operations involved, the maintenance of an inventory of the physical and chemical agents associated with that process, and a periodic review of the different job activities conducted in that work area. It is also necessary to study the effectiveness of existing control measures (Table 17-A).

New chemical and physical agents are continually being introduced and used in industrial processes. The health and safety professional must be aware of these agents' properties and must ascertain their potentially hazardous nature before they are used.

The health and safety professional who is responsible for the maintenance of a safe, healthful work environment should be thoroughly acquainted with the presence of harmful materials or energies that may be encountered in the industrial environment.

If a plant is going to handle a hazardous material, it is necessary that the health and safety professional also consider all the unexpected events that can occur and the precautions needed to prevent inadvertent atmospheric release of a toxic material.

Raw materials. To recognize hazardous environmental factors or stresses, a health and safety professional must first know about the raw materials used and the nature of the products and by-products manufactured. Sometimes this requires considerable effort. Possible impurities in raw materials, such as benzene in some solvents, should be considered.

By-products. By-products, intermediates, and final products formed from raw materials can be a source of potential hazard. The identity of chemical intermediates formed in the course of an industrial process and the toxicological properties of these intermediates may be difficult to establish. Undesirable chemical by-products, such as carbon monoxide (CO) resulting from the incomplete combustion of organic material, may be formed. Any process involving combustion of organic material is a potential source of CO contamination.

Another example of an undesirable by-product is the lead oxide fumes formed when steel coated with a lead-containing

Table 17-A. Potentially Hazardous Operations and Air Contaminants

Process Types	*Contaminant Type*	*Contaminant Examples*
Hot operations		
Welding	Gases (g)	Chromates (p)
Chemical reactions	Particulates (p)	Zinc and compounds (p)
Soldering	(Dusts, fumes, mists)	Manganese and compounds (p)
Melting		Metal oxides (p)
Molding		Carbon monoxide (g)
Burning		Ozone (g)
		Cadmium oxide (p)
		Fluorides (p)
		Lead (p)
		Vinyl chloride (g)
Liquid operations		
Painting	Vapors (v)	Benzene (v)
Degreasing	Gases (g)	Trichloroethylene (v)
Dipping	Mists (m)	Methylene chloride (v)
Spraying		1,1,1-trichloroethylene (v)
Brushing		Hydrochloric acid (m)
Coating		Sulfuric acid (m)
Etching		Hydrogen chloride (g)
Cleaning		Cyanide salts (m)
Dry cleaning		Chromic acid (m)
Pickling		Hydrogen cyanide (g)
Plating		TDI, MDI (v)
Mixing		Hydrogen sulfide (g)
Galvanizing		Sulfur dioxide (g)
Chemical reactions		Carbon tetrachloride (v)
Solid operations		
Pouring	Dusts (d)	Cement
Mixing		Quartz (free silica)
Separations		Fibrous glass
Extraction		
Crushing		
Conveying		
Loading		
Bagging		
Pressurized spraying		
Cleaning parts	Vapors (v)	Organic solvents (v)
Applying pesticides	Dusts (d)	Chlordane (m)
Degreasing	Mists (m)	Parathion (m)
Sand blasting		Trichloroethylene (v)
Painting		1,1,1-trichloroethane (v)
		Methylene chloride (v)
		Quartz (free silica, d)
Shaping operations		
Cutting	Dusts (d)	Asbestos
Grinding		Beryllium
Filing		Uranium
Milling		Zinc
Molding		Lead
Sawing		
Drilling		

(Reprinted with permission from *Occupational Exposure Sampling Strategy Manual,* NIOSH Pub. No. 77–173)
Abbreviations: d=dusts, g=gases, m=mists, p=particulates, v=vapors.

paint is heated during welding or torch-cutting operations. The high temperatures involved cause the lead to vaporize and combine with oxygen to produce lead oxide.

The inadvertent breakdown of chemical compounds can also occur; this process may produce more toxic or irritating products. Probably the best example of this type of by-product is the generation of phosgene gas when arc-welding operations are conducted near a vapor degreaser that contains a chlorinated hydrocarbon solvent such as trichloroethylene or perchloroethylene.

List of materials. A list should be prepared of all raw materials currently used in the plant, as well as a list of those that may be used in the near future. A chemical inventory is required by the Chemical Hazard Communication Standard published by the Occupational Safety and Health Administration (OSHA). One source of information concerning such materials is the supplier. The supplier must send a material safety data sheet (MSDS) with every hazardous chemical shipped. It is strongly recommended that a close liaison be set up between the supplier of raw materials and the health and safety personnel so that additional information concerning materials in use and those which are to be ordered will be available.

Basic hazard-recognition procedures

Almost any work environment has either potential or actual environmental hazards that need to be recognized, measured, and monitored. The health and safety professional must always be aware of potential environmental hazards. The first consideration is the raw materials and the potential of those materials to do harm. The next consideration is how these raw materials are modified through intermediate steps. Finally, the health and safety professional must determine whether the finished product poses any possible harmful effects to the worker. Each step from raw material to finished product must be evaluated under normal conditions and also under anticipated emergency conditions.

The basic hazard-recognition procedures are similar whether a chemical, physical, or biological agent is involved. As pointed out in previous chapters of this text, there is no substitute for knowledge concerning the process, work operations, raw materials, intermediates, and by-products involved in the work area.

A basic, systematic procedure can be followed in the recognition of environmental health hazards. Answers should be obtained to the following questions:

- What is produced?
- What intermediates are formed in the process?
- What by-products are produced?

For each air contaminant released in an industrial process, determine the threshold limit values (TLVs) and the allowable exposure established by the OSHA or by other exposure guidelines based on the toxicological effect of the material. Then determine where air contaminants are released and where and how long employees are exposed.

If the potential hazard is a physical agent, determine whether it involves electromagnetic radiation, noise, extremes of temperature, humidity, or pressure. New uses for physical agents in industrial processes are increasing at a rapid rate. Examples include the use of lasers and microwaves.

Variations in an industrial process, air currents in a room, changes in work practices, and variations in the emission rate of a contaminant are some of the factors that result in changes in the degree of hazard.

Overtime requirements should be determined so that the effects of 12-hour or double-shift exposure of workers to health hazards can be included in the evaluation.

PROCESS OR OPERATION DESCRIPTION

The degree of hazard depends on both the level or concentration and duration of exposure; therefore, information about the industrial process is needed. Plant engineering personnel should be consulted for information about abnormal operating conditions and other factors that affect exposures.

Unless there is specific information that the process is ade-

Table 17-B. Typical Industrial Operations and Their Associated Health Hazards

Process or Operation: Nature and Description of Hazards

Abrasive blasting. Abrasive blasting equipment may be automatic, or it may be manually operated. Either type may use sand, steel, shot, or artificial abrasives. The dust levels of workroom air should be examined to make sure that the operators are not overexposed.

Abrasive machining. An abrasive machining operation is characterized by the removal of material from a workpiece by the cutting action of abrasive particles contained in or on a machine tool. The workpiece material is removed in the form of small particles and, whenever the operation is performed dry, these particles are projected into the air in the vicinity of the operation.

Assembly operations. Improper positioning of equipment and handling of work parts may present ergonomic hazards due to repeated awkward motion and resulting in excessive stresses.

Bagging and handling of dry materials. The bagging of powdered materials (such as plastic resins, paint pigments, pesticides, cement, and the like) is generally accompanied by the generation of airborne dusts. This occurs as a result of the displacement of air from the bag, spillage, and motions of the bagging machine and the worker. The conveying, sifting, sieving, screening, packaging, or bagging of any dry material may present a dust hazard. The transfer of dry, finely divided powder may result in the formation of considerable quantities of airborne dust. Inhalation and skin contact hazards may be present.

Ceramic coating. Ceramic coating may present the hazard of airborne dispersion of toxic pigments plus hazards of heat stress from the furnaces and hot ware.

Coating operations. Whenever a substance containing volatile constituents is applied to a surface in an industrial environment, there is obviously potential for any vapors evolved to enter the breathing zones of workers. If the volatiles evaporate at a sufficient rate and/or the particular operation is such that workers must remain in the immediate vicinity of the wet coating, these vapors may result in excessive exposures.

Crushing and grinding. Size reduction refers to the mechanical reduction in size of solid particulate material. Two of the principal methods of achieving size reduction are crushing and grinding, but the terms are not synonymous. Crushing generally refers to a relatively slow compressive action on individual pieces of coarse material ranging in size from several feet to under one inch. Grinding is performed on finer pieces and involves an attrition or rubbing action as well as interaction between individual pieces of material. Pulverizing and disintegrating are terms related to grinding. The former applies to an operation that produces a fine powder; the latter indicates the breakdown of relatively weak interparticulate bonds, such as those present in caked powders. Dry grinding operations should be examined for airborne dust, noise, and ergonomic hazards.

Dry mixing. Mixing of dry material may present a dust hazard and should take place in completely enclosed mixers whenever air sampling indicates excessive amounts of airborne dust are present.

Drying ovens. Much of the equipment used for drying purposes is also used for curing, i.e., the application of heat to bring about a physical or chemical change in a substance. The first major category includes direct dryers in which hot gases are in direct contact with the material and carry away any vaporized substances to be exhausted. Limited-use dryers include radiant-heat and dielectric-heat dryers. The operation of the former is based on the generation, transmission, and absorption of infrared rays. The latter rely on heat generation within the solid when it is placed in a high-frequency electric field. Oven vapors (sometimes including carbon monoxide) are often released into the workroom as are a variety of solvents and other substances found in the drying or curing products.

Electron-beam welding. Any process involving an electric discharge in a vacuum may be a source of ionizing radiation. Such processes involve the use of electron beam equipment and similar devices.

Fabric and paper coating. The coating and impregnating of fabric and paper with plastic or rubber solutions may involve evaporation into the workroom air of large quantities of solvents.

Forming and forging. Hot bending, forming, or cutting of metals or nonmetals may involve hazards of lubricant mist, decomposition products of the lubricant, skin contact with the lubricant, heat stress (including radiant heat), noise, and dust.

Gas furnace or oven heating operations (annealing, baking, drying, and so on). Any gas- or oil-fired combustion process should be examined to determine the level of by-products of combustion that may be released into the workroom atmosphere. Noise measurements should also be made to determine the level of burner noise.

Grinding operations. Grinding, crushing, or comminuting of any material may contaminate workroom air as a result of the dust produced from the material being processed or from the grinding wheel.

High temperatures from hot castings, unlagged steam pipes, process equipment, and so on. Any process or operation involving high ambient temperatures (dry-bulb temperature), radiant heat load (globe temperature), or excessive humidity (wet-bulb temperature) should be examined to determine the magnitude of the physical stresses that may be present.

Materials handling, warehousing. Work areas should be checked for levels of carbon monoxide and oxides of nitrogen arising from internal combustion engine fork-lift operations. Operations should also be evaluated for ergonomic hazards.

Metalizing. Uncontrolled coating of parts with molten metals presents hazards of dust and fumes of metals and fluxes in addition to heat and nonionizing radiation.

Microwave and radio-frequency heating operations. Any process or operation involving microwaves or induction heating should be examined to determine the magnitude of heating effects and, in some cases, noise exposure to the employees.

Molten metals. Any process involving the melting and pouring of molten metals should be examined to determine the level of air contaminants of any toxic gas, metal fume, or dust produced in the operation.

Open-surface tanks. Open-surface tanks are used by industry for numerous purposes. Among their applications can be included the common operations of degreasing, electroplating, metal stripping, fur and leather finishing, dyeing, and pickling. An open-surface tank operation is defined as "any operation involving the immersion of materials in liquids, which are contained in pots, tanks, vats, or similar containers." Excluded from consideration in this definition, however, are certain similar operations, such as surface-coating operations and operations involving molten metals for which different engineering control requirements exist.

Paint spraying. Spray-painting operations should be examined for the possibility of hazards from inhalation and skin contact with toxic and irritating solvents and inhalation of toxic pigments. The solvent vapor evaporating from the sprayed surface may also be a source of hazard because ventilation may be provided only for the paint spray booth.

Plating. Electroplating processes involve risk of skin contact with strong chemicals and in addition may present a respiratory hazard if mist or gases from the plating solutions are dispersed into the workroom air.

Pouring stations for liquids. Wherever volatile substances are poured from a spout into a container, some release of contaminants can be expected. In the paint and other coatings industries, solvents may be released when the contents of mills or mixers are poured into portable change cans. While ladles and change cans can be covered while being moved through a workplace, they must of necessity be open at the points they are filled or discharged.

Punch press, press brake, drawing operations, and so on. Cold bending, forming, or cutting of metals or nonmetals should be examined for hazards of contact with lubricant, inhalation of lubricant mist, and excessive noise.

Vapor degreasing. The removal of oil and grease from metal products may present hazards. This operation should be examined to determine that excessive amounts of vapor are not being released into the workroom atmosphere.

Welding—gas or electric arc. Welding operations generally involve melting of a metal in the presence of a flux or a shielding gas by means of a flame or an electric arc. The operation may produce gases or fumes from the metal, the flux, metal surface coatings, or surface contaminants. Certain toxic gases such as ozone or nitrogen dioxide may also be formed by the flame or arc. If there is an arc or spark discharge, the effects of nonionizing radiation and the products of destruction of the electrodes should be investigated. These operations also commonly involve hazards of high potential electric circuits of low internal resistance.

Wet grinding. Wet grinding of any material may produce possible hazards of mist, dust, and noise.

Wet mixing. Mixing of wet materials may present possible hazards of solvent vapors, mists, and possibly dust. The noise levels produced by the associated equipment should be checked.

(Adapted with permission from the OSHA *Compliance Operations Manual,* 1972)

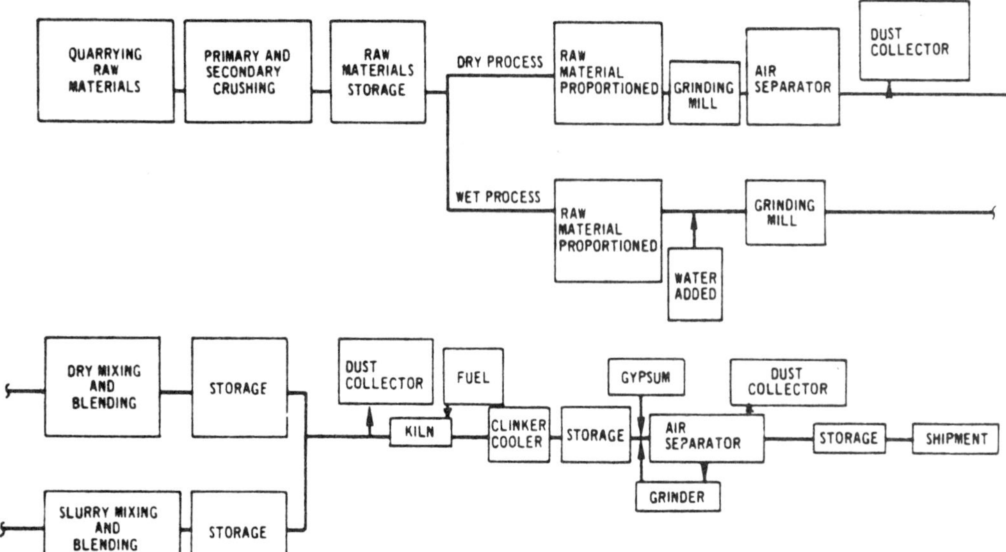

Figure 17-1. Basic flow diagram of portland cement manufacturing process. (Reprinted with permission from the *Compilation of Air Pollutant Emission Factors, 2nd ed., EPA Pub. No. AP-42)*

quately controlled, there are many types of industrial operations that should immediately alert the health and safety professional to a potential health hazard.

The list of industrial operations shown in Table 17-B is adapted from the OSHA *Compliance Operations Manual* (1972), and it may be of help in alerting the health and safety professional to the source and types of possible health hazards. After a list of process operations that can produce harmful air contaminants has been prepared, the professional must determine which operations produce conditions that could result in hazardous exposures and need further study.

Process flow sheet

A simple process flow sheet could be drawn, which will show, stepwise, the introduction of each material and the product of each step (Figure 17-1). Process flow sheets and the standard operating procedures that describe the particular operations involved should be obtained and studied. They not only provide a good description of the general operations involved, but also serve as an excellent source of information for the terminology used in that particular industry. In many industrial operations, many different hazards may exist simultaneously. Therefore, it is necessary to examine carefully the overall process in order not to overlook potentially hazardous conditions.

It is important to identify the air contaminants produced and to pinpoint the location of personnel that might be exposed to these air contaminants. Repetitive operations (where a worker remains in one location and repeats the same task) can be simple to define. The identification process involves measuring the time schedule of each cycle of operation with emphasis on when different contaminants are formed, and then counting the number of cycles per work shift. This procedure is followed for each different job operation or position on the assembly line, if needed as part of the hazard assessment for the entire plant. In many plants, the specific tasks vary from day to day depending on the work assignment, the products being made, and other factors.

Processes in the chemical industry involve chemical or physical changes in the nature of substances and particularly in the chemical structure and composition of the substances. Chemical processes involve crushing, grinding, size separation, filtration, drying, heating, cooling, solvent extraction, absorption, distillation, fractionation, electrolysis, mixing, blending, analysis and process control, packaging, and transport.

Chemical process companies involved with the manufacture of large volumes of chemicals utilize closed systems. Where the chemicals are not routinely released to the atmosphere, we can theorize that the exposure to air contaminants in work areas is primarily due to:

1. releases or leaks from joints, closures, and other system components normally and functionally considered to be a part of a closed system;
2. intentional releases of contaminants from vents or process sampling points;
3. stack gases from combustion processes; and
4. accidental or unintentional releases due to equipment malfunction or failure.

Many valves leak even when they are supposed to be shut, and such leaks can build up the air concentration to the limits set for air contaminants by the OSHA. Purges, minor overpressures, and system breathing into the atmosphere should be contained by collection, scrubbing, reaction, incineration, or other measures that safely dispose of the products or eliminate their release completely. Environmental Protection Agency (EPA) emission requirements may require further control measures.

An analysis of cleaning operations can be followed to determine the degree of hazard. The primary cleaning methods used in industry include:

- manual wiping of parts or equipment with a solvent-soaked rag;
- use of handheld or mechanical brushes;
- use of a scraper;
- wet mopping;
- use of a wet sponge;
- abrasive blasting and hydroblasting;
- steam-cleaning;
- use of compressed air to blow off dust;
- use of vacuum cleaning devices.

The common feature of all these operations is that by some physical and/or chemical action, a contaminant is dislodged from the surface it is adhering to, and the contaminant is collected and removed from the immediate environment for disposal.

When a rag is used with a solvent, the vapors evaporate and may present a problem. A brush used to sweep up dusty substances or a scraper used to dislodge built-up cakes of dry substances can disperse dust into the air. The use of compressed air to blow dust from surfaces will probably produce the greatest concentration of air contaminants. (The OSHA regulations prohibit the use of compressed air to clean except when the pressure is reduced to 30 psi—and then only if an effective chip guard and personal protective equipment are used. The use of compressed air with asbestos dust is forbidden.)

Hydroblasting and steam-cleaning are essentially wet methods that might initially appear to be methods designed to suppress the generation of air contaminants. However, hydroblasting equipment to remove a solid can produce substantial concentrations of air contaminants. Because of the temperatures and forces involved, steam-cleaning can also produce air contaminants under some circumstances.

The use of such vacuum devices as vacuum cleaners appears to be the most satisfactory method of cleaning dry, dusty materials without producing excessive amounts of contaminants. However, a special vacuum with a high-efficiency particulate aerosol (HEPA) filter is needed when dealing with certain highly toxic dusts, such as asbestos.

Checklists

A checklist for evaluating the environmental hazards that arise from industrial operations is presented here. It should be modified to fit each particular situation.

For the overall process or operation. List all hazardous chemical or physical agents used or formed in the process. Carry out or check the following for each chemical agent or material.

- List the conditions necessary for the agent to be released into the workroom atmosphere. Does it normally occur in the process as a dust, mist, gas, fume, vapor, as a low-volatile liquid, or as a solid (Table 17-A)? What process conditions could cause material to be sprayed or discharged into the workroom atmosphere as a liquid aerosol or dust cloud? Have the consequences of the exposure of raw materials or intermediates to adjacent operations been considered? Are unstable materials properly stored? Has the process been laboratory-checked for runaway explosive conditions? Have provisions been made for the rapid disposal of reactants in an emergency? Have provisions been made for the safe disposal of toxic materials?
- List the ambient airborne concentration levels in the workroom atmosphere that would normally be present as a function of time. List the peak airborne concentration levels as a function of event duration. List the appropriate OSHA permissible exposure limits (PELs) and TLVs.
- Are fire extinguishers, respiratory protective equipment, stretchers, and other such equipment required? Are the fire-extinguishing agents compatible with the process materials? Are special emergency procedures and alarms required?
- For all raw materials used in the process, consider chemical reactions that could take place to produce other toxic materials; for example, the inadvertent mixing of acids and plating tank liquids containing cyanides produces hydrogen cyanide (HCN). Is there safe storage space for raw materials and finished products?
- List the levels of those physical agents that may normally be present (electromagnetic radiation, temperature extremes, noise, and so on.) List any appropriate OSHA exposure limits and TLVs.

For process equipment. List those pieces of equipment that contain sufficient hazardous material that a hazard would be produced if their contents were suddenly released to the environment.

- List those pieces of equipment that could produce hazardous levels of physical agents during normal or abnormal plant operations.
- List the machinery and equipment that may produce hazardous concentrations of airborne contaminants. For each item listed here, indicate the control measures installed to minimize the hazard. Is the health and safety control measure adequate, failsafe, and reliable?
- List process equipment with components that are likely to leak hazardous materials—such as valves, pump packing, and tank vents—or equipment with components that are susceptible to leaks due to corrosion, for example. For each item listed, indicate what safeguard has been taken to prevent normal leakage—packless pumps, double rotary seals, vents connected to a scrubber, and so on. Is each safeguard adequate, failsafe, and reliable?
- Are labels provided for all valves and switches? Are all containers of hazardous chemicals labeled according to the OSHA Hazard Communication Standard? Is there some method available that complies with the Hazard Communication Standard and that can be used to identify the contents of pipes, piping systems, and stationary process containers? Is equipment designed to permit lockout procedures? Are emergency disconnect switches properly marked?

Process operations requiring detailed study. For each item on the list for process equipment, consider the following questions.

- What utilities are required? What could happen if one of these utilities were suddenly and permanently interrupted? if temporarily interrupted?
- What would happen if the flow of one or more of the process streams entering or leaving the equipment was interrupted? What if the stream were twice the desired quantity? What would happen if the normal outlet connections from the equipment became plugged? Can flammable fluids feeding production units be shut off from a safe distance in case of fire or other emergency?
- In the case of a small leak, what would happen to the leaking material? Would it produce an environmental health hazard? How long might it leak before being detected? Would the leak hasten the corrosion and failure of the vessel or equipment?
- If the vessel is no longer functional, how would the contents be contained? If the contents are volatile, what would be done to diminish and control their vaporization?
- What would be the effect of overpressure? If the vessel is protected by overpressure release devices, where would the material be vented? What would be the effect of overheating? of overcooling? Would the vessel be surrounded by burning material if other equipment in the area failed? If so, what would happen?
- What instrument failure or operator error could result in the accidental release of vessel contents?

For instrumentation on equipment that handles hazardous materials, consider the following questions.

- What would happen to one instrument or to a group of instruments following a power loss or motive force failure? to all instruments? What position would control valves assume? Is this the best failsafe position to minimize hazard?
- For each instrument, what can happen in case of sensor failure? In case of control valve seizing, would the operator receive warning in time to take corrective action? In an emergency situation, would it be clear to the operator what corrective action is needed, or can the situation be handled with built-in automatic overrides?

For material-collection systems, such as industrial ventilation, or pollution-prevention equipment (scrubbers, hoods, ducts, blowers, dust collectors, condensers, flares, combustion incinerators, and so on), consider the following questions.

- Under normal operating conditions, is the capacity of the system adequate to prevent hazardous concentrations of harmful air contaminants in the workroom air? What abnormal plant conditions can impose the greatest load on this equipment? How long can these abnormal conditions persist? What would be the effect on the system under these conditions of increased or decreased flow? Would the system be adequate?
- What would happen in the case of utility failure or flow stoppage (loss of motive force or line blockage) to the collection device?
- What would happen to the collection device if it were surrounded by fire? Could it suffer an internal fire? What would happen then?

After these considerations have been studied and proper countermeasures have been taken, it is necessary to instruct operating maintenance personnel on the proper operation of the health and safety control measures. Through such instruction, personnel will be alerted to the possible hazards and will understand the reasons for installing certain built-in safety features. Operating and maintenance personnel should set up a routine procedure for testing the industrial hygiene and safety control measures that are not used in normal, ordinary plant operations.

INITIAL JUDGMENT

Estimation of the potential health hazard to employees requires that an initial judgment be made. The initial determination is usually a simple one that is based on factors such as the chemical, physical, and toxic properties of the substance being processed, the size of the workplace, the amount and type of ventilation, and the proximity of the employee to the source of contamination (Table 17-C).

Worst case

When variations in job operations make it difficult to define typical exposure, a "worst case" exposure pattern can be used to represent the longest exposure time to the highest expected air contaminant concentration. This exposure pattern provides a convenient starting point for comparing employee exposures. If the initial judgment or determination indicates that exposure in the worst case is not hazardous to health, then it may be that no problems should be expected to develop with other, lesser exposures. If the worst case exposure pattern does exceed the TLV or OSHA standards, separate initial determination reports should be prepared for each of the individual exposure patterns that may be encountered.

Table 17-C. Initial Determination of Employee Health Hazards

1. Date of report
2. Name and social security number of each employee considered at a work operation
3. Work operations performed by the employee at the time of the report
4. Location of work operations within the worksite
5. Chemical substances to which the employee may be exposed at each work operation
6. Any information, observations, and estimates that may indicate exposure of this employee to a chemical substance; list any exposure measurement data and calculations
7. Federal permissible exposure limits and/or the TLVs established by the American Conference of Governmental Industrial Hygienists (ACGIH) for each chemical
8. Complaints or symptoms that may be attributable to chemical exposure
9. Type and effectiveness of any control measures used; for mechanical ventilation controls, list measurements taken to demonstrate system effectiveness
10. Operating condition ranges for production, process, and control measures for which the determination applies

(Reprinted with permission from the *Occupational Exposure Sampling Strategy Manual,* NIOSH Pub. No. 77-173)

Typical procedures

When investigating a situation in which solvents are used in the workplace, the health and safety professional should ascertain the circumstances of use, the surface area of the exposed solvent, the air and solvent temperatures, the control measures in use, the proximity of workers, and other conditions that might affect employee exposure. The professional should determine the relative volatility and the toxic, irritating, and narcotic properties of the solvent. Based on these facts and on previous experience, a judgment can be made as to whether further investigation is necessary. A great many exposures can be readily classified as nonhazardous or potentially hazardous simply by inspection.

The initial determination ensures that exposure to the contaminant has been investigated in a detailed way. Common sense must be used in the assumptions made in the initial evaluation.

FIELD SURVEY

A survey of work operations is usually performed to identify actual and potential health hazards under normal and abnormal conditions. Surveys and studies are conducted to determine levels of exposure among workers to various atmospheric contaminants and physical agents; to determine the effectiveness of control measures; to investigate complaints; and to determine compliance with federal and state regulations.

Purpose

A walk-through survey is performed to pinpoint the location of existing health hazards so that proper corrective action can be taken. The initial survey is used to establish a baseline from which future conditions can be compared and evaluated. Prior to the walk-through survey, the health and safety professional should review the process to become familiar with the raw materials, equipment, by-products, and products. The process flows, pressure, and temperature conditions should also be reviewed (Table 17-B and Figure 17-1).

Processes and job operations that are run only intermittently may present some of the greatest potential health hazards. Maintenance operations often generate high exposure conditions. If work is done on the night shift or on weekends, a survey should also be carried out at these times to see how air contaminant concentrations at night compare with those during normal daytime operations. Because the production rate may be significantly different on the night shift, conditions could be better or worse.

Information required

Of particular concern during the field survey should be work areas that have the following:

- excessive noise;
- excessive heat;
- inadequate ventilation;
- awkward operator positions or human/machine interface;
- radiation exposure (ionizing and nonionizing);
- excessive air contaminants.

Data are collected to determine the amount of chemicals used or produced, the exposure times and the number of employees exposed. Also, observations are made with regard to handling procedures, housekeeping, and potential skin contact with a contaminant. Although a contaminant's primary mode of entry into the body is through the respiratory tract, contaminants can also enter the body through skin absorption and ingestion.

Sensory perception

Dusty operations can be easily spotted visually, although this does not necessarily mean they are the most hazardous. It must be remembered that respirable dust particles cannot be seen by the unaided eye. Furthermore, respirable-size dust concentrations must reach very high levels before they are readily visible in the air.

The presence of many vapors and gases can be detected by the sense of smell. Trained observers are able to identify a limited number of solvent vapors and gases in the workroom air by their characteristic odor. This detection method is not recommended, however, because many vapors and gases may be present in concentrations considerably in excess of the TLV or of the OSHA permissible exposure limit without being detectable by their odor. One must also consider the differences among individuals. What may be regarded as only a barely perceptible odor by one individual may be considered as strong and objectionable by another. Also, the continuous inhalation of many vapors fatigues the sense of smell, which renders the use of odor as a warning signal unreliable or ineffective.

Sometimes during the field survey, one can judge that air contaminant concentrations are so low (or so high) that quantitative measurements are unnecessary. However, in most cases, the air concentration is determined by sampling the workroom air for the known contaminant.

Equipment capable of producing excessive or abnormal levels of radiant heat, temperatures and humidity, noise, illumination, ultraviolet radiations, microwaves, lasers, x-rays and gamma-rays, and other ionizing and nonionizing radiations should be noted. (See Chapters 10, Ionizing Radiation, and 11, Nonionizing Radiation, for more details.)

Control measures in use

The types of control measures in use and their effectiveness should be appraised. Controls include local exhaust and general ventilation measures, respiratory protective devices, other personnel protective measures, and shielding from radiant or ultraviolet energy.

Control measures in use may be evaluated by a number of simple techniques. Ventilation can be estimated by velocity measurements at hood openings, and air flow patterns can be determined by smoke tubes or visual estimates. Static suction measurements should be made at all operations that are controlled by mechanical ventilation. (See Chapter 21, Industrial Ventilation, for more details.) General guides to check the effectiveness of controls include the presence or absence of dust on floors and ledges, holes in ductwork, nonoperational fans, and the manner in which personal protective measures are treated by the worker.

Particular attention should be paid to the housekeeping and general maintenance measures provided in work areas. Observations should be made as to the deposition of contaminants on the face or on the clothing of workers.

It should be noted whether or not personal protective equipment should be required, whether or not it is in use, whether it is used properly, and what condition it is in. Other points of observation include whether or not local exhaust hoods are used. Particular attention should be paid to process sampling points, since air contaminants are more likely to be released at this location.

SAMPLING METHODS

There are a number of reasons why environmental measurements should be made in the workplace. One reason is to evaluate the degree of employee exposure. Another is to obtain measurements that provide design engineers with the information needed to properly design engineering control measures.

Measurements are also needed to determine the effect of process changes. There are many other reasons to obtain environmental measurements: air pollution control, process control, and so forth. However, measurements that are made for the purpose of evaluating employee exposure are usually of primary interest to the health and safety professional.

Monitoring

Monitoring is a continuing program of observation, measurement, and judgment—all of which are necessary to recognize potential health hazards and to judge the adequacy of protection. Monitoring requires an awareness of the presence of potential health hazards and an assessment on a continuing basis of the adequacy of the control measures in place.

Monitoring is more than simply sampling the air to which an employee is exposed or the medical status of that employee. It is an entire series of actions that permits a judgment to be made relative to the adequacy of protection afforded employees (Figure 17-2).

Four types of monitoring systems are generally used in occupational health surveillance. These are personal, environmental, biological, and medical monitoring systems.

Personal monitoring. Personal monitoring is the measurement of a particular employee's exposure to airborne contaminants. In personal monitoring, the measurement device, or dosimeter, is placed as close as possible to the contaminant's entry portal into the body (Figure 17-3). When monitoring an air contaminant that is toxic if inhalated, the measurement device is

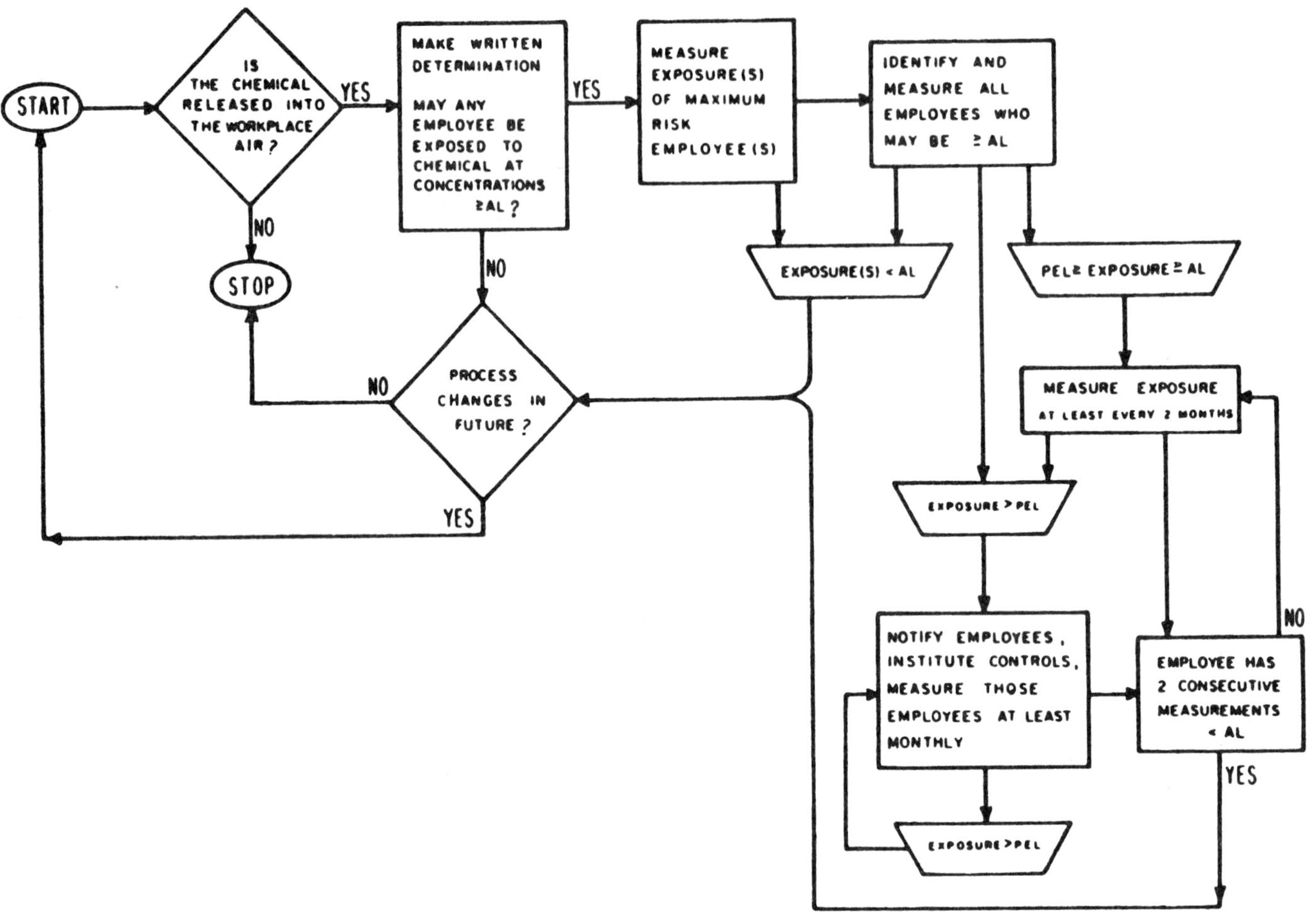

Figure 17-2. The NIOSH-recommended employee exposure determination and measurement strategy. The health standard for each individual substance should be consulted for detailed requirements. In the flow chart, AL = action level and PEL = Permissible Exposure Limit. (Reprinted with permission from the *Occupational Exposure Sampling Strategy Manual,* NIOSH Pub. No. 77-173)

placed close to the breathing zone. When monitoring noise, the device is placed close to the ear. (See Chapter 18, Air-Sampling Instruments, for more information.)

Environmental monitoring. Environmental monitoring is the measurement of contaminant concentrations in the workroom. The measurement device is placed adjacent to the worker's normal workstation (Figure 17-4). The air contaminant concentration or physical energy stress is then measured. (See Chapter 19, Direct-Reading Gas and Vapor Monitors, for more details.)

Biological monitoring. Biological monitoring involves the measurement of changes in the composition of body fluid, tissues, or expired air to determine absorption of a potentially hazardous material. Examples are the measurement of lead, fluoride, cadmium, and mercury in blood or in urine to determine intoxicant absorption. By establishing baseline levels, such monitoring may indicate inadequate environmental controls or improper work methods before excessive absorption occurs. Biological monitoring is *not* a replacement for environmental or personal monitoring but may be used to complement these. (See Chapter 15, Industrial Toxicology, and Chapter 16, Evaluation, for a discussion of Biological Exposure Indices.)

Medical monitoring. Medical monitoring refers to the examination by medical personnel to determine the worker's response to an intoxicant. Biological and medical monitoring provide information only after absorption has occurred. (See Chapter 26, The Occupational Physician, for more details.)

While the concept of air-sampling and the use of air-monitoring devices may appear to be simple, the details of a good monitoring program may be misunderstood unless the person engaged in sampling is adequately trained. Technical guidance by a professional industrial hygienist can help ensure that the technique chosen is the correct one. Errors in the interpretation of sampling data are common when monitoring is conducted by individuals whose training in air sampling techniques is limited.

Major problems arise when an instrument reading is accepted as reliable before its calibration and the reproducibility of its response have been determined (Figure 17-5). The personnel performing these tests must have some experience with and knowledge of the resolution of these problems. The conduct of monitoring programs by untrained or poorly trained persons relegates the program to a numbers game that serves no useful purpose.

After engineering control measures have been put in place,

Figure 17-3. A personal monitoring measuring device should be placed as close to the breathing zone as possible. (Courtesy SKC, Inc.)

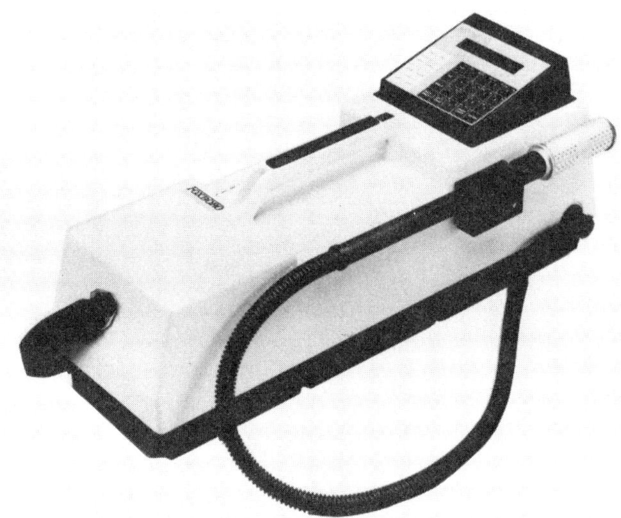

Figure 17-4. This vapor and gas analyzer can be used to measure concentrations at the operator's workstation. (Courtesy Foxboro Co.)

the adequacy of the occupational health program depends on the proper operation of these control measures. The health and safety professional must monitor and assess the efficient performance of the engineering control measures. The health and safety professional can shift from the measurement of employee exposure to the measurement of the continued adequate performance of the ventilation system as a surrogate indicator of

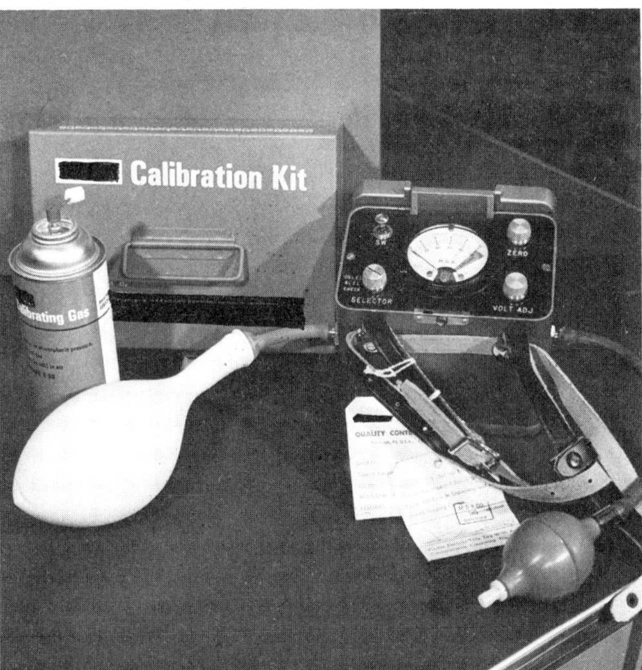

Figure 17-5. Combustible gas indicators must be calibrated before use to guarantee accuracy of the measurements of the instruments. (Courtesy MSA)

employee exposure. This same concept could apply to other kinds of control measures. The adequacy of employee work practices with respect to protection against occupational health and safety hazards must also be monitored, and proper training given in order to comply with the OSHA Hazard Communication Standard.

Sufficient knowledge of the operation or process—particularly its relative magnitude at a given time—is essential for proper evaluation of exposure. For example, samples obtained during a production cutback would yield low results and would not be representative of exposure conditions under normal operation.

Concentrations of atmospheric contaminants and levels of physical agents fluctuate as a process cycles or as leaks or equipment malfunctions occur. Before any test of the workroom air is performed, the health and safety professional must be thoroughly acquainted with the nature of the contaminant in question and with the other substances likely to be present in the atmosphere so that the proper sampling method can be chosen. The sampling method used should yield a valid and reliable measurement of the air contaminant concentration as a function of space and time.

Errors

The word "sampling" is used commonly, but sometimes the full implications of the word are not realized. To sample means to measure only part of the environment and, from the measurements taken, infer things about the totality of the environment.

In the strict statistical sense, a sample consists of several items, each of which has some characteristic that can be measured. In the field of industrial hygiene, however, a sample consists of an airborne contaminant(s) collected on a physical device (such as a filter or charcoal tube), or a sample is a measurement

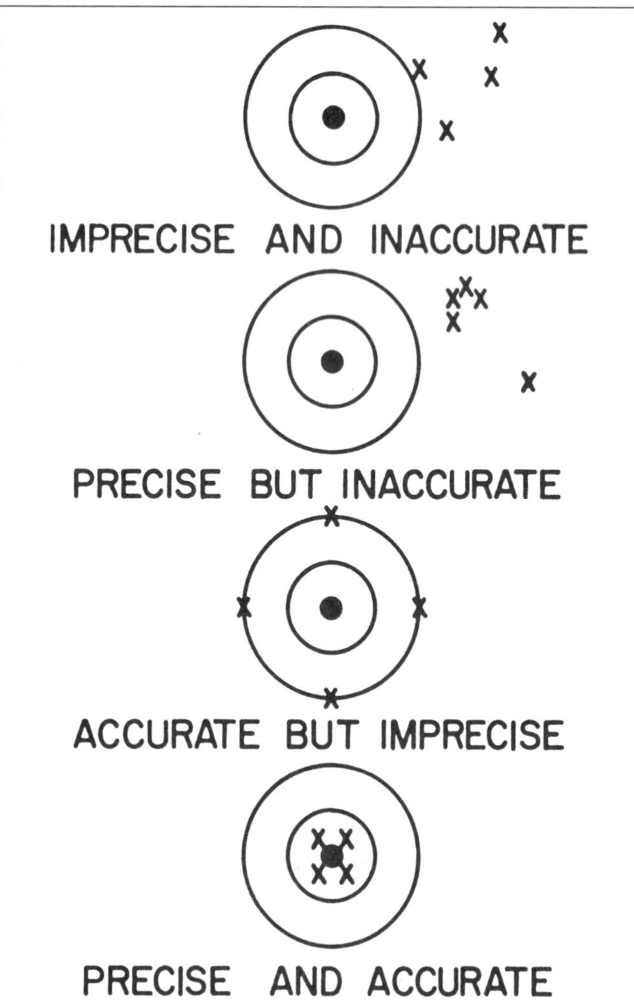

Figure 17-6. It is possible for a measurement to be precise but not accurate—and vice versa. (Reprinted with permission from Powell, C.H., and Hosey, A.D., eds. *The Industrial Environment—Its Evaluation and Control,* 2nd ed. Public Health Services Pub. No. 614, 1965)

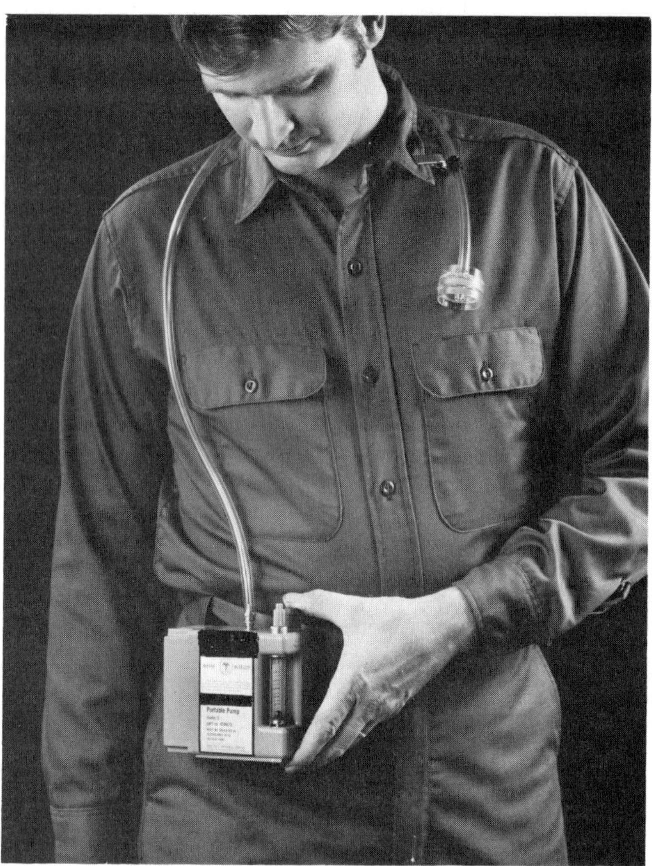

Figure 17-7. Personal monitor for toxic gases, vapors, and dusts.

obtained from a direct-reading instrument. A sample can also refer to measurements of physical energy stresses. Unfortunately, all sampling and analytical methods are subject to error.

Accuracy. Accuracy refers to the relationship between a measured value and the true value. For a measurement to be accurate, it must be close to the true value.

Precision. Precision is the degree of agreement among results obtained by repeated measurements under the same conditions and under a given set of parameters (Figure 17-6). It is possible for a measurement to be precise but not accurate—and vice versa.

Accuracy is affected by controllable sources of error. These errors, called determinate or systematic errors, include method error, personal error, and instrument error.

Personal carelessness, incorrect calculations, poorly calibrated equipment, and use of contaminated reagents are all examples of systematic error. These errors contribute a consistent bias to the results, which renders them inaccurate. These sources of errors must be identified and eliminated or controlled.

Precision is affected by indeterminate or random errors that are not controllable. These errors include:

- intra- or interday concentration fluctuations;
- sampling equipment variations (i.e., random pump flow fluctuations); and
- analytical method fluctuations (i.e., variation in reagent addition or instrument response).

These factors cause variability among the sample results. Statistical techniques are used to account for random error. Increasing the number of samples taken, for example, will minimize the effect of random error.

Strategy

A single sampling strategy cannot be established for all situations. Several factors must be taken into consideration.

Collection techniques. There are three basic types of industrial hygiene sample collection techniques.

- Personal: The sampling device is directly attached to the employee who wears it continuously during all work and rest operations (Figure 17-7).
- Breathing zone: The sampling device is held in the employee's breathing zone. Air that would most likely be inhaled by the employee enters the breathing zone.
- General area: The sampling device is placed in a fixed location in the work area that is generally occupied by employees. This type of collection is also referred to as environmental monitoring (Figure 17-8).

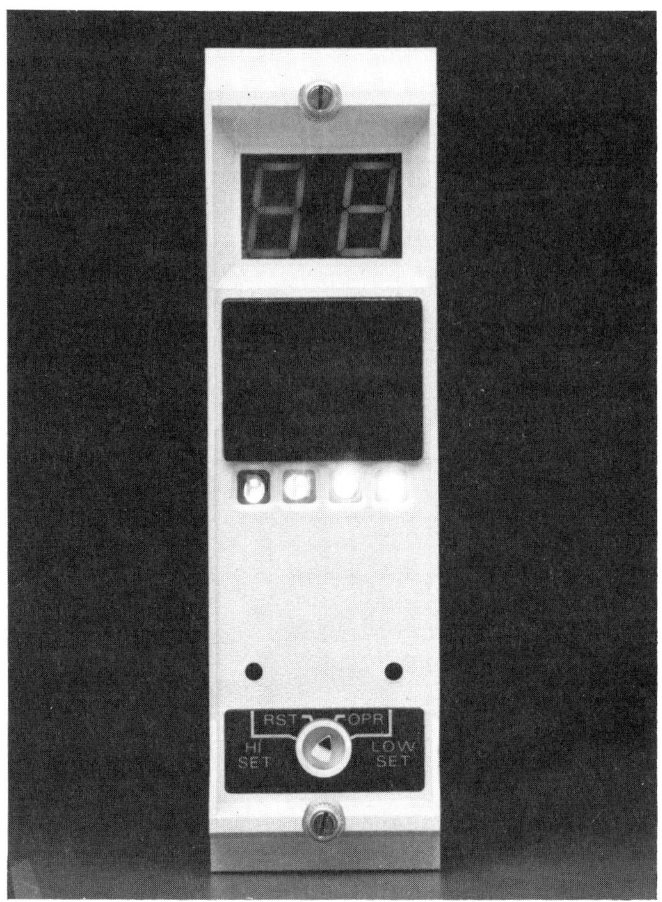

Figure 17-8. The sampling device is placed in a fixed location in the work area to obtain environmental air samples. (Courtesy Sensidyne, Inc.)

Samples taken for the purpose of measuring employee exposure normally should be obtained using personal or breathing zone methods. If samples taken by the general area method are to be used to determine employee exposure, then it is necessary to obtain a comprehensive job time and motion study for each employee.

Where to sample. If the purpose of sampling is to evaluate a worker's exposure, it is necessary to collect samples at or as near as practical to the worker's breathing zone. If the purpose is to define a potential hazard or to obtain data for control purposes, samples would normally be collected in the vicinity of the source (Figure 17-9).

The exposure of a worker can be measured most accurately by determining the concentration of the contaminant in the air that the worker breathes. The instrument should be held or located as close as possible to the employee's nose and mouth without interfering with freedom of movement in the normal conduct of work. Breathing zone samples reflect the worker's actual exposure. Exposure is reduced when the worker moves away from the source. Conversely, exposure is higher when the worker bends over the source. Air-samples that do not take the worker's movements into account do not accurately measure true exposure.

Fixed-position sampling, although not a direct measurement of employee exposure, is used to continuously monitor environmental conditions. Fixed-position samplers are commonly used in parking garages and tunnels to determine the concentrations of carbon monoxide in air. The sampling point should be located in close proximity to the points where the highest concentration of gases may develop (Figure 17-10).

Whom to sample. If the initial determination indicates the possibility of excessive exposure to airborne concentrations of a toxic substance, measurements of the employee's exposure to the substance should be made. Samples are usually obtained for the most highly exposed employee.

In general, the most logical and the simplest means of identifying the maximum-risk employee is to observe and select that employee who is closest to the source of the hazardous air contaminant being generated. Individual differences in the work habits of different workers at a single workstation can have profound effects on the levels of exposure. Even though they may be performing essentially the same jobs with the same materials, their individual method of performing the task could produce varying exposure levels.

Air movement patterns within a workroom must sometimes be analyzed to accurately predict the risk potential to which employees are subjected. Especially in operations or processes involving heating or combustion, the natural air circulation could be such that the maximum-risk employee might be located at considerable distance from the source. The location of ventilation booths, air supply inlets, open doors, and windows and the size and shape of the work area are all factors that affect workroom air flow patterns that can produce higher contaminant concentrations farther away from the source.

When to sample. Another factor to consider is when to sample. Plants located in areas where large temperature differences occur during different seasons of the year should be sampled during summer and winter months. Normally during the summer months, doors and windows are open, and therefore, more dilution of the air contaminant occurs during the summer than in winter. If the work area is air-conditioned, the air contaminant level would be fairly constant throughout the year.

If the plant has more than one shift, samples should be collected during each shift. Airborne concentrations of toxic substances or exposure to physical agents may be different for each shift. Ordinarily, contaminants are not generated at a constant rate, and the concentration can vary considerably from time to time. Air currents within a room, process variations, changes in the work practice performed by an operator, and variation in the emission rate of contaminants are significant factors that result in continual changes in concentration throughout a work shift.

How long to sample. The volume of air sampled and the duration of sampling is based on the sensitivity of the analytical procedure or direct-reading instrument, the estimated air concentration, and the OSHA standard or the TLV for the particular contaminant.

The duration of the sampling period should represent some identifiable period of time—usually a complete cycle of an operation.

The evaluation of a worker's daily time-weighted average (TWA) exposure is best accomplished by having the worker work a full shift with a personal sampler. The OSHA standards require that either a full-shift, 8-hour sample be obtained for

AFTER COMPLETING THE SAMPLING FORM, REMOVE AND RETAIN THE BOTTOM COPY AS YOUR RECORD OF THE SAMPLE TAKEN. SEND THE COMPLETED SAMPLING FORMS TOGETHER WITH THE SAMPLES, PROPERLY PACKAGED TO AVOID DAMAGE TO THE ANALYTICAL (INDUSTRIAL HYGIENE) LABORATORY. ANY BULK SAMPLES SHOULD BE SHIPPED SEPERATLEY FROM THE OTHER SAMPLES TO AVOID CROSS CONTAMI-NATION. PROVIDE ONE OR TWO BLANKS (UNUSED FILTER, CHARCOAL TUBE, IMPINGER SOLUTION, ETC., FROM THE SAME BATCH USED FOR SAMPLING) WITH EACH SET OF SAMPLES SENT TO THE LABORATORY.

SHELL OIL COMPANY
INDUSTRIAL HYGIENE
SAMPLING FORM

SAMPLING FORM NO. XXXXXX

ENTER THE SAMPLING FORM NO. FROM THE UPPER RIGHT CORNER OF OTHER SAMPLING FORMS RELATED — RELATED SAMPLING FORM NO'S.

FILL IN IF MORE THAN ONE FORM IS REQUIRED TO DOCUMENT A SAMPLE OR TO ESTABLISH THE TIME WEIGHTED AVERAGE EXPOSURE. — PAGE ___ OF ___

EMPLOYEE SAMPLED IN BREATHING ZONE | COMPANY NAME (e.g. M.W. KELLOGG) | SAMPLE TYPE: ☐ PERSONAL ☐ AREA ☐ BLANK ☐ ☐ BULK ☐ SOURCE | DATE | ILH NO.

SOCIAL SECURITY NO. | SHELL LOCATION (e.g. DEER PARK M.C.) | ANALYSIS REQUIRED / CONTAMINANTS | SHIFT | SAMPLE NO

EMPLOYEE NO | AREA / UNIT (GIVE SPECIFIC NAME) | SAMPLED BY | PHONE

JOB CLASSIFICATION DESCRIPTION | OPERATING CONDITIONS | POSSIBLE INTERFERENCES | RETURN RESULTS TO (ADDRESS)

WIND | BAROMETER | TEMPERATURE | HUMIDITY | SPECIAL INSTRUCTIONS

TIME	WORK TASKS	*	TIME AT TASK	INSTRUM'T READING	SKETCH AREA (AND/OR REMARKS)

* CHECK (✓) IN THIS COLUMN FOR EACH TASK WHERE PROTECTIVE EQUIPMENT WAS WORN. | TYPE: (GIVE EQUIPMENT USED (e.g. RESPIRATOR, GLOVES, ETC.) MANUFACTURER, MODEL NO., ETC.)

INSTRUMENT TYPE & MODEL NO. | SERIAL NO | CALIBRATED: BY: | DATE | REFERENCE | METHOD

FLOW RATE | SAMPLING TIME | COUNTER READING INITIAL: FINAL: NET: | VOLUME PER COUNT | SAMPLE VOLUME

SAMPLE COLLECTOR: ☐ CHARCOAL TUBE ☐ FILTER (TOTAL) ☐ FILTER (RESPIRABLE) ☐ IMPINGER ☐
DESCRIBE: (MFR/TYPE/LOT NO/COMPOSITION)
IF AN IMPINGER/BUBBLER IS USED GIVE COMPOSITION OF ABSORBING SOLUTION (e.g. 10 ML OF 0.1 N NaOH).

Figure 17-9. The industrial hygiene sampling form shown here can be used to obtain data for control purposes.

an 8-hour TWA exposure limit comparison; or a short-term sample can be obtained when the Permissible Exposure Limit has a ceiling value, or peak concentration limit.

How many samples to take. There is no set rule to determine the number of samples that are necessary to evaluate a worker's exposure. The number of samples to be taken depends on the purpose of the sampling. A single sample measurement may be high or low due to the possible sources of error already mentioned. Several dozen samples may be necessary to accurately define a daily, TWA and short-term exposure for a worker who performs a number of tasks during the shift.

The concentration of contaminant in the ambient air is generally small. Direct-reading instruments and other devices used to collect samples for subsequent analysis must collect a sufficient quantity of the sample so that the chemist can accurately determine the presence of minute amounts (parts per million or milligrams per cubic meter quantities) of the air contaminant.

When a worker's exposure level or the work environment is evaluated, an instrument must be used that will provide the necessary sensitivity, accuracy, reproducibility, and, optimally, rapid results. Detailed discussions of instruments used for sampling for particulates are given in Chapter 7, for gases and vapors in Chapter 18, and for direct-reading instruments in Chapter 19. Instruments used for assessing noise exposure are discussed in Chapter 9, Industrial Noise.

Continuous monitoring devices used to evaluate the working environment are also available (Figure 17-11). Continuous monitors are used in plant areas where hydrogen sulfide is produced or used. Many continuous detecting and recording instruments can be equipped to sample at several remote locations in a plant and record the general air concentration to which workers may be exposed during a shift.

Required accuracy and precision. The accuracy and precision of the sampling method must be accounted for. Not only does this assure that meaningful data are gathered, but compliance with the OSHA regulations is also thereby assured. The Ethylene Oxide Standard, for example, requires that a sampling method with accuracy to a confidence level of 95 percent (within ± 25 percent) be obtained for airborne concentrations of ethylene oxide at the 1.0 parts per million (ppm) TWA and within ± 35 percent for airborne concentrations of ethylene oxide at the action level of 0.5 ppm.

The following include some methods designed to assure accuracy and precision.

- Obtain manufacturer's data. For example, information on detector tubes should be gathered. The National Institute for Occupational Safety and Health (NIOSH) no longer has a certification procedure for these tubes. The Safety Equipment

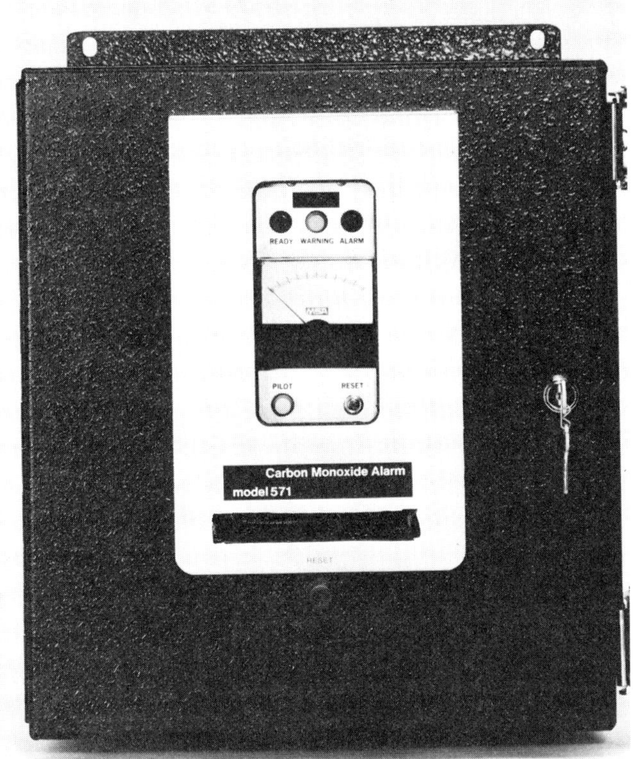

Figure 17-10. Continuous monitors can be placed in work areas to actuate audible alarms. (Courtesy Mine Safety Appliances Co.)

Figure 17-11. The continuous monitoring device shown here is used to measure the concentration of chlorine in the air. (Courtesy Gastech, Inc.)

Institute (SEI), a private organization, has its own certification process. The SEI uses independent laboratories that are certified by the American Industrial Hygiene Association (AIHA) to assure that product models conform to applicable standards. A quality-control audit is then made by an independent expert at the manufacturing facility.

- Establish a calibration schedule for all sampling equipment. Keep a record of results.
- Use a laboratory that participates in an industrial hygiene quality control program, such as the one conducted by the AIHA.

INDUSTRIAL HYGIENE CALCULATIONS

Gases and vapors

Calculations for gas and vapor concentrations are based on the gas laws. Briefly, these are as follows:

- The volume of gas under constant temperature is inversely proportional to the pressure.
- The volume of a gas under constant pressure is directly proportional to the Kelvin temperature. The Kelvin temperature scale is based on absolute zero (0 C = 273 K). The Rankine temperature scale is also used (0 F = 460 R).
- The pressure of a gas of a constant volume is directly proportional to the Kelvin (or Rankine) temperature.

Thus, when measuring contaminant concentrations, it is necessary to know the atmospheric temperature and pressure under which the samples were taken. At standard temperatures and pressure (STP), 1 gm-mol of an ideal gas occupies 22.4 liters (L). The STP is 0 C and 760 mm of mercury (Hg). If the temperature is increased to 25 C (room temperature) and the pressure remains the same, 1 g-mol of gas occupies 24.45 liters.

The concentration of gases and vapors is usually expressed in parts per million. That is:

$$\text{ppm} = \frac{\text{parts of contaminant}}{\text{million parts of air}} \tag{1}$$

The latter is a volume-to-volume relationship. Other similar volumetric relationships are:

$$\frac{\text{liter}}{10^6 \text{ liters}} = \frac{\text{centimeters}^3}{10^6 \text{ centimeters}^3} = \frac{\text{feet}^3}{10^6 \text{ feet}^3} \tag{2}$$

Sometimes it is necessary to convert milligrams per cubic meter (mg/m^3)—a weight-per-unit volume ratio—into a volume-per-unit volume ratio. If it is understood that one gram-mole of an ideal gas at 25 C occupies 24.45 L, the following relationships can be calculated.

$$\text{ppm} = \frac{24.45}{\text{molecular wt}}\ \text{mg/m}^3 \quad (3)$$

$$\text{mg/m}^3 = \frac{\text{molecular wt}}{24.45}\ \text{ppm} \quad (4)$$

The partial pressure method is another method of expressing gas or vapor concentrations. This method uses the vapor pressure exerted by the substance. Vapor pressure is directly convertible to percentage by volume by multiplying by 100 and then dividing by the barometric pressure. Or vapor pressure can be converted to parts per million by multiplying by 1,000,000 (10^6) and then dividing by the barometric pressure.

$$\frac{\text{partial vapor pressure of one constituent}}{\text{total barometric pressure}} \times 10^6 \quad (5)$$

$$= \text{ppm of constituent}$$

EXAMPLE. Given the concentration of a vapor at STP in grams per liter, convert this to parts per million.

The gram-molecular volume at STP=22.4 L/mol, therefore:

Concentration of vapor at STP =

$$\frac{\text{grams of vapor}}{\text{liters of vapor}} = \frac{\text{molecular wt}}{22.4} \quad (6)$$

Rearranging terms:

$$\text{liters of vapor} = \frac{(\text{grams of vapor})(22.4)}{(\text{molecular wt})} \quad (7)$$

$$\text{ppm} = \frac{\text{parts of vapor}}{1{,}000{,}000 \text{ parts of air}} = \frac{\text{liters of vapor}}{10^6\ \text{ L of air}} \quad (8)$$

Substituting liters of vapor from equation (7) into equation (8):

$$\text{ppm} = \frac{\left[\dfrac{(\text{grams of vapor})(22.4)}{(\text{molecular wt, gm})}\right]}{10^6 \text{ L of air}} \quad (9)$$

$$\text{ppm} = \frac{(\text{grams of vapor})(22.4)}{(10^6)(\text{molecular wt of vapor})} = \frac{(\text{milligrams})\ 22.4\ (10^3)}{\text{molecular wt}} \quad (10)$$

Chemical analysis of some atmospheric samples requires the examination of some fixed volume of absorbing or reacting solution. The concentration of contaminant is multiplied by the volume of solution to calculate the total amount of contaminant collected during the sampling period. This is then related to the total volume of air sampled during the sampling period and converted to parts per million. For example, after bubbling 15 L of air at 25 C and 755 mm Hg through 30 mL of an appropriate absorbing solution (100 percent collection efficiency), it was determined that the hydrogen chloride (molecular weight = 36.5) concentration in solution was 15 μg/mL.

The total amount of hydrogen chloride measured was:

$$\frac{15\ \mu\text{g}}{\text{mL}} \times 30 \text{ mL} = 450\ \mu\text{g} \quad (11)$$

The volume of 1 μmol of hydrogen chloride (HCl), corrected for temperature and pressure is as follows:

$$1\ \mu\text{mol} \times 22.4 \times \frac{298}{273} \times \frac{760}{755} = 24.6\mu\text{L} \quad (12)$$

The concentration in parts per million is

$$\frac{450\ \mu\text{g HCl}}{15 \text{ L of air}} \times \frac{\text{moles HCl}}{36.5\ \mu\text{g HCl}} \times \frac{24.6\ \mu\text{L HCl}}{\text{moles HCl}} =$$

$$\frac{11{,}070\ \mu\text{L HCl}}{547.5 \text{ L of air}} = 20.22 \text{ ppm} \quad (13)$$

Dimensional analysis

Occasionally confusion arises when milligrams per cubic meter are converted to parts per million. Dimensional analysis is very useful in avoiding these errors. Milligrams per cubic meter of air must be converted to millimoles per cubic meter and to milliliters per cubic meter or parts per million.

$$\left(\frac{\text{mg}_x}{\text{m}^3 \text{ air}}\right)\left(\frac{\text{mmole}_x}{\text{mg}_x}\right)\left(\frac{22.4 \text{ mL}_x}{\text{mmole}_x}\right)(F_t)(F_p)$$

$$= \frac{\text{mL}_x}{\text{m}^3 \text{ air}} = \text{ppm} \quad (14)$$

In equation (14), F_t and F_p are the pressure and temperature conversion factors, and the subscript x refers to the contaminant.

Conversely:

$$\text{ppm} = \left(\frac{\text{mL}_x}{\text{m}^3 \text{ air}}\right)\left(\frac{\text{mmol}_x}{22.4 \text{ mL}_x}\right)\left(\frac{\text{mg}_x}{\text{mmol}_x}\right)(F_t)(F_p)$$

$$= \frac{\text{mg}_x}{\text{m}^3 \text{ air}} \quad (15)$$

The following conversion formulas are useful for gas or vapor calculations:

milligrams per liter × 1,000 = milligrams per cubic meter
milligrams per liter × 28.32 = milligrams per cubic foot
milligrams per cubic foot × 35.314 = milligrams per cubic meter
milligrams per cubic meter × 0.02832 = milligrams per cubic foot

EXAMPLE. Given the following, derive an equation for the concentration of vapor in a chamber:

V_T = chamber volume in liters
MW = molecular weight of a substance (gm/mol)
T = absolute temperature of apparatus in degrees Kelvin (°K=°C+273)
P = pressure in mm Hg
ρ = density (gm/mL)
V_x = volume of material to be used (in mL)
C = concentration (ppm)

liters (pure vapor) =

$$\frac{(V_x \text{ mL})\ (p \text{ gm/mL})(22.4 \text{ L/mol})}{\text{mol wt of material gm/mol}}\left(\frac{T}{273}\right)\left(\frac{760}{P}\right) \quad (16)$$

$$C\ (\text{ppm}) = \left(\frac{\text{liters of pure vapor}}{V_T \text{ chamber volume}}\right)\left(10^6 \text{ parts of air}\right) \quad (17)$$

C (ppm) =

$$\frac{(V_x\text{mL})\left(p\ \frac{\text{gm}}{\text{mL}}\right)\left(\frac{22.4 \text{ L}}{\text{gm-mole}}\ \frac{\text{gram-mole}}{\text{MW gm}}\right)\left(\frac{T}{273}\right)\left(\frac{760}{P}\right)}{V_T \text{ Liters}} \times 10^6 \text{ parts} \quad (18)$$

$$C\ (\text{ppm}) = \frac{(V_x)(p)\left(\frac{22.4}{MW}\right)\left(\frac{T}{273}\right)\left(\frac{760}{P}\right)}{V_T} \times 10^6 \quad (19)$$

Example. To calculate the volume of a liquid solvent necessary to produce a desired concentration, C(ppm), in air in a given tank volume at room temperature and standard pressure, the following calculation is made:

$$V_x = \frac{C \times MW \times 298 \times P \times V_T}{p \times 24.45 \times T \times 760 \times 10^6} \quad (20)$$

Example. It is desired to generate a concentration of 200 ppm of acetone in a 20.0-L glass container:

molecular weight of acetone=58.08 gm/mol
density of acetone=0.7899 gm/mL
temperature=25 C (77 F)
pressure=740 mm Hg

$$V_x = \frac{(200)(20)(58.08)}{(0.7899)(24.45)} \times \frac{740}{760} \times \frac{298}{273.25} \times \frac{1}{10^6} = 0.0117 \text{ mL} \quad (21)$$

Vapor equivalents

As discussed, the amount of pure vapor formed at sea level by the evaporation of a known volume or weight of liquid can be calculated by one of the following formulas:

$$\frac{\text{cubic feet of vapor}}{\text{pound of liquid}} = \frac{(\text{liters/mole})(\text{grams/pound})}{(\text{liters/cubic foot})(\text{grams/mole})} \quad (22)$$

liters per mole at STP=22.41
grams per pound=453.6
liters per cubic foot=28.32
grams per gram mole=mol wt

$$\frac{\text{cubic feet of vapor}}{\text{pound of liquid}} \text{ at 0 C} = \frac{(22.41)(453.6)}{(28.32)(\text{mol wt})} = \frac{359}{\text{mol wt}} \quad (23)$$

Example. One pound of liquid toluene forms $\left(\frac{359}{92}\right) = 3.9$ cu ft of pure vapor upon complete evaporation at 0 C.

To determine cubic feet of vapor per pound of liquid at 70 F:

$$\frac{\text{cubic feet}}{\text{pound}} \text{ at 70 F} = \frac{(530 \text{ R})(359)}{(492 \text{ R})(\text{mol wt})} = \frac{387}{\text{mol wt}} \quad (24)$$

530 R=Deg F+460=70+460
492 R=32 F+460

At constant pressure, the volume of a given mass of vapor varies directly with the absolute temperature. The factor $\left(\frac{530}{492}\right)$ corrects for the temperature increase of 32 F to 70 F.

At 70 F, 1 lb of liquid toluene would form $\left(\frac{387}{92}\right) = 4.163$ cu ft of vapor.

Example. To determine the cubic feet of vapor per pint of liquid at 70 F:

$$\left[\frac{\text{cubic feet}}{\text{pound}}\right]\left[\frac{\text{pound}}{\text{pint}}\right] = \frac{\text{cubic feet}}{\text{pint}} \quad (25)$$

$$\frac{\text{cubic feet}}{\text{pint}} \text{ at 70 F} = \frac{(387)(1.041)(\text{sp gr})}{\text{mol wt}} = \frac{403(\text{sp gr})}{\text{mol wt}} \quad (26)$$

pounds of water per pint=1.041.
specific gravity (sp gr)=ratio of mass of that liquid to mass of an equal volume of water at 4 C.
sp gr of toluene=0.866

$$\frac{\text{cubic feet of toluene}}{\text{pint of liquid}} = \frac{(403)(0.866)}{92} \quad (27)$$
$$= 3.79 \text{ cu ft/pint at 70 F}$$

Weight-per-unit volume

When the contaminant is dispersed in the atmosphere in solid or liquid form as a mist, dust, or fume, its concentration is usually expressed on a weight-per-volume basis. Outdoor air contaminants and stack effluents are frequently expressed as grams, milligrams, or micrograms per cubic meter, ounces per thousand cubic feet, pounds per thousand pounds of air, and grains per cubic foot. Most measurements are expressed in metric units. However, the use of standard U.S. units is justified for purposes of comparison with existing data, especially those relative to the specifications for air-moving equipment.

Time-weighted average (TWA) exposure

To determine the TWA exposure of an individual to airborne contaminants, a detailed description of the worker's job tasks during the sampling period should be obtained. If personal sampling is not performed, a sufficient number of air-sampling measurements in the breathing zone during these job tasks should be obtained under various plant operating conditions.

In a typical work environment, the employee may be exposed to several different short-term average concentrations during the workshift. Changes in job assignment, process variations, and ventilation conditons are some of the factors that affect the exposure level. The TWA exposure evolved as a method of calculating daily or full-shift average exposure by weighting the different short-term average concentrations by exposure time. It is the equivalent of integrating the concentration values over the total time base of the TWA. It may be determined by the following formula:

$$TWA = \frac{C_1T_1 + C_2T_2 + C_n}{8\ hr} T_n \quad (28)$$

In the formula, TWA=the time-weighted average concentration which is usually expressed in parts per million or in milligrams per cubic meter; C=the concentration of contaminant present during the incremental exposure time; T=time; and T_1, T_2, T_3 . . . T_n are the incremental exposure times at average concentrations C_1, C_2, $+C_n$.

A period of 8 hours is used as the denominator, since the OSHA standards and the TLVs are based on an 8-hour workday.

Or:

$$\frac{\sum_{i=1}^{i=n} (T_i) \cdot (C_i)}{T_{total}\ \text{work time}} = TWA \quad (29)$$

In this formula, the TWA concentration for a particular material is determined: T_i=duration of incremental exposure; C_i=concentration of a specific air contaminant during the incremental time period T_i; and T_{total}=the total work time per shift over an 8-hour workday.

Several samples (of equal or unequal time duration) are obtained during the entire period. The total time covered by the samples must be as close to the total exposure time as possible. For an 8-hour average, for example, the sampling period should be close to 8 hours.

EXAMPLE. The following personal samples for particulates were obtained from a foundry worker:

Sample No.	Time
1	7:00 a.m. (start of shift) to 8:00 p.m.
2	8:00 a.m. to 9:30 a.m.
3	9:30 a.m. to 11:00 a.m.
4	11:00 a.m. to 1:00 p.m. (turned off and covered for 30 minutes during lunch)
5	1:00 p.m. to 3:30 p.m.

The measurement obtained is a full-period consecutive sample measurement because it covers the entire time period appropriate to the TLV and the OSHA standard (8 hours), and the samples were taken consecutively (or serially).

EXAMPLE. A personal sampling pump with a respirable dust sampling head is attached to an employee at the start of the shift at 8:00 a.m., turned off from 11:30 a.m. to 12:00 noon (lunch), and turned on again from 12:00 noon to 4:30 p.m. The sample collected constitutes a full-period sample that can be used for the determination of respirable dust exposure because it covers the entire time period appropriate to the TLV or the OSHA standard (8 hours).

In some cases, because of limitations in measurement methods, such as direct-reading meters or charcoal tubes, it is impossible to collect a series of consecutive samples whose total duration approximates the period for which the standard is defined. In this case, grab samples are obtained over a certain number of short periods that are representative of the entire workshift.

EXAMPLE. It is necessary to obtain a measurement for exposure to benzene using charcoal tubes. Each charcoal tube sample takes 30 minutes to collect. Out of 16 possible 30-minute periods in the 8-hour period, 10 samples are collected. These 10, 30-minute duration samples constitute 10 grab samples of the worker's exposure on the given day. The estimate of the 8-hour TWA exposure would be determined after analyzing the readings of the 10 tubes.

For example, suppose a worker is exposed as detailed in the following:

	Time of exposure (T_i)	Average exposure concentration (ppm) (C_i)
	1 hour	350
	3 hours	200
	4 hours	150
Total T =	8 hours	

Then the TWA for the 8-hour workday will be:

$$TWA = \frac{(1\ hr)(350\ ppm)+(3\ hr)(200\ ppm)+(4\ hr)(150\ ppm)}{8\ hr} \quad (30)$$

$$= \frac{1{,}550\ ppm}{8\ hr} = 194\ ppm$$

In general, when this formula is used, the time spent by a worker at various jobs should be determined to the nearest 30-minute interval. If there is considerable variation in the level of air contamination, shorter time intervals will have to be used.

EXAMPLE. Assume that an employee spends the first 4 hours of an 8-hour shift in the vicinity of a heat-treating operation where the measured carbon monoxide concentration in air at the breathing-zone level remains fairly constant at 50 ppm.

For the remainder of the shift (4 hours), assume that the operator worked in a different part of the plant where the exposure to carbon monoxide was essentially zero.

What is the employee's TWA exposure to carbon monoxide?

Solution:

$$\frac{(4\ hr)(50\ ppm)+(4\ hr)(0\ ppm)}{8\ hr} = \quad (31)$$

$$\frac{200}{8} = 25\ ppm\ (TWA)\ \text{for carbon monoxide.}$$

EXAMPLE. In another case, assume that a machine operator (working the 7:00 a.m. to 4:00 p.m. shift) tending an automatic screw machine was exposed to average levels of oil mists as indicated below.

Time	*Average Level of Oil Mists (mg/m³)*
7:00–8:00	0
8:00–9:00	1.0
9:00–10:00	1.5
10:00–11:00	1.5
11:00–12:00	2.0
12:00–1:00	0.0*
1:00–3:00	4.0
3:00–4:00	5.0

*lunch period, no exposure

What is the TWA exposure to oil mists?

Solution:

$$\frac{\sum_{i=1}^{i=8}(T_i)(C_i)}{T_{total} = (8\text{ hr})} = \text{TWA} \qquad (32)$$

Time (hrs) × Concentration (mg/m³)

$(1)\cdot(0) = 0.0$
$(1)\cdot(1) = 1.0$
$(1)\cdot(1.5) = 1.5$
$(1)\cdot(1.5) = 1.5$
$(1)\cdot(2.) = 2.0$
$(1)\cdot(0.0) = 0.0$
$(2)\cdot(4.0) = 8.0$
$(1)\cdot(5.0) = 5.0$

$$\frac{\sum_{i=1}^{i=8}(T_i)(C_i) = 19.0\text{ hr}\ \frac{\text{mg}}{\text{m}^3}}{8\text{ hr}}$$

$$\frac{19.0\text{ hr}\ \frac{\text{mg}}{\text{m}^3}}{8\text{ hr}} \approx 2.38\text{ mg/m}^3 \text{ TWA exposure to oil mists.}$$

Suppose, as indicated in the previous example, that in addition to exposure to oil mists, this same machine operator is exposed to an average level of 100 ppm of carbon monoxide for a 10-minute period each hour, while forklift materials handling operations are being performed in the vicinity. During the remainder of each hour-long period and during the lunch period, the carbon monoxide level is essentially zero. The calculation for determining the TWA exposure to carbon monoxide for the employee is as follows:

Solution:

(Time in minutes) × (concentration in parts per million) for each hourly period = (10 minutes) × (100 ppm) = 1,000 minutes-ppm. For eight such periods: 8 × (1,000 min-ppm) = 8,000 min-ppm. Total work time per shift = 8 hr = 480 minutes.

$$\text{TWA for CO} = \frac{8{,}000\text{ min-ppm}}{480\text{ minutes}} = 16.7 \approx 17\text{ ppm} \qquad (33)$$

Example. If an employee were exposed to the same material at several work locations or operations during the 8-hour shift, and several grab samples were taken during each of the operations, then the results would be analyzed as follows.

Operation	*Duration*	*Sample*	*Results (ppm) (of each 5-min sample)*
Cleaning room	8:00–11:30 a.m.	A	120
		B	150
		C	170
		D	190
		E	210
Print shop	12:30–4:30 p.m.	F	70
		G	90
		H	110
		I	120

The cleaning room average exposure is:

$$\overline{C_1} = \frac{120+150+170+190+210}{5} = 168\text{ ppm} \qquad (34)$$

The print shop average exposure is:

$$\overline{C_2} = \frac{70+90+110+120}{4} = 98\text{ ppm} \qquad (35)$$

Then the TWA exposure for the 8-hour shift (excluding 60 minutes for lunch) is:

$$\text{TWA} = \frac{(168\text{ ppm})(3.5\text{ hr})+(98\text{ ppm})(4.0\text{ hr})}{8} \qquad (36)$$

$$= 122.5\text{ ppm}$$

Excursions. TWA concentrations imply fluctuations in the level of airborne contaminant. Excursions above the TLV are permissible provided that equivalent excursions below the TLV occur. The TLV booklet stipulates, too, that short-term exposures may exceed three times the TLV for no more than a total of 30 minutes during the workday; under no circumstances should exposures exceed five times the TLV. This stipulation is valid providing the TLV-TWA is not exceeded. In some cases, a specific short-term exposure limit (STEL) has been established.

INTERPRETATION OF RESULTS

The interpretation of results involves the identification of problem areas, the determination of the exposure levels, and the establishment of priorities (i.e., which operations or processes should be controlled first).

Evaluation requires familiarity with the process or operation as well as with the analytical test results. It is important that the analytical results reflect normal working conditions.

Exposure factors

To determine the degree of employee exposure using air samples or measurements of noise levels or radiation, measurements obtained at the worker's actual job station are needed (Figure 17-12). Generally, samples or measurements that cover most of a complete working day should be obtained; measurements that cover several nonconsecutive workdays are even better.

Samples obtained at a stationary point in the work environment (area samples) can indicate possible exposure, but these samples can be very misleading. For example, measuring noise levels a few inches from a noisy machine when the worker is located several feet away may indicate erroneously high exposures. Obtaining air-samples for a solvent at the center of the room would indicate erroneously low exposures if the worker must lean over a solvent tank. Welding fume samples must be obtained in the welder's breathing zone (under the mask) so that these measurements can be related to the worker's actual exposure.

Enough air samples from enough sampling points over a sufficient period of time are required to estimate the probable air contaminant concentration during normal operation, the distribution of peaks under exceptional circumstances, and a reasonably accurate TWA. Not all operations, however, require that such complete information be obtained every time. What is required in any particular operation must be left to the judgment of the people responsible for that operation, and this includes not only the plant superintendent but the chief advisors—the health and safety professional, the industrial hygienist, the plant physician, the nurse, and others. There is no general rule of thumb that can be applied.

When the industrial hygienist encounters potential health hazards, exposure measurements should be obtained. In cases where these measurements clearly demonstrate concentrations above the guidelines (TLVs), the industrial hygienist can then recommend that controls be installed.

The type and extent of controls will depend on the contaminant, the evaluation made of the exposure, and the operation that disperses the contaminant into the work area. If the air concentrations are far below the TLVs, the industrial hygienist can report that conditions are below the recommended standards. If however, some sample measurements are above the TLV concentrations and some below, the industrial hygienist cannot state with certainty that excessive exposures to workers are not occurring. Further monitoring should be conducted and precautionary measures taken.

Comparison with the OSHA standards

Results of the environmental study must be compared with the OSHA standards. It must be emphasized that the samples collected during the study must be representative of the worker's daily TWA exposure before a comparison can be made.

The general industry OSHA standards for air contaminants are listed in the *Code of Federal Regulations (CFR),* Title 29, Part 1910.1000, which contains three tables of permissible exposure limits. More comprehensive standards on specific chemicals begin with 1910.1001 (asbestos).

Harmful effects

The harmful effects of chemical agents are determined by the amount accumulated in the body—by total dosage. The number of milligrams of harmful material deposited per kilogram of body weight determines the toxic effect produced. This is often overlooked when individual exposures are expressed only in terms of the concentration (parts per million) in the air breathed and the length of exposure time. The amount in the inspired air is not necessarily the amount absorbed.

Differences in particle size or solubility, which are practically impossible in many cases to measure accurately, determine the fraction of inhaled impurity that is retained or absorbed. Also, the concentration in the inspired air of materials may be only a partial indication of an individual's true exposure. The absorbed dose also depends on whether or not significant skin absorption or ingestion has occurred. Biological monitoring, which can detect a compound or its metabolite, can be useful in determining the absorbed dose.

Medical surveillance

Medical surveillance procedures can be used to assess the adequacy of protective measures as well as the overall health of employees. Medical surveillance includes the development of a baseline health inventory, followed by periodic reevaluation. The baseline health inventory is obtained by collecting the results of an employee's preplacement medical history and physical examination. It is important to establish a preplacement baseline with regard to organ systems that are known to be affected by materials present in work areas. Preplacement medical examinations are also used to detect medical conditions that may predispose an employee to the toxic effects of the agent.

The preplacement evaluation is used to determine if the applicant's work activities must be restricted because of physical handicap or because of measured susceptibility to specific workplace exposures. Subsequent periodic examinations can help determine if there has been any change in the employee's health that may indicate a need for further evaluation. If the employee has experienced or is experiencing any health problems, continued assignment on the present job may be precluded. Additional tests, appropriate to the toxicities of the materials involved, may be added to this basic examination. Or, if necessary, these or other tests may be performed more frequently so that medical surveillance appropriate to the hazard involved can be performed. Some OSHA standards require that certain types of medical or biological monitoring be performed in addition to environmental monitoring.

SUMMARY

The degree of health hazard to an individual arising from exposure to environmental factors or stresses depends on four factors: (1) the nature of the environmental factor or stress, (2) the level of exposure, (3) the duration of exposure, and (4) individual susceptibility.

It is of the utmost importance to obtain a list of all the chemicals used as raw materials and to determine the nature of the products and by-products arising out of industrial operations.

After the list of chemicals and physical agents to which employees are exposed has been prepared, it is necessary to determine which of the environmental factors or stresses may result in hazardous exposures and need further evaluation.

Valuable information can be obtained by observing the manner in which health hazards are generated, control measures are used, and by determining the number of people involved.

The kind of provisions that must be installed to protect against health hazards will vary from plant to plant, from one toxic material to another, and from process to process.

EMPLOYEE EXPOSURE MEASUREMENT RECORD

Facility ______________________ Area ______________________

Sampled by ______________________ Date ______________________

Temperature ______________________ Altitude ______________________

Sample # ______________________ Employee name ______________________ SS # __________

Operation(s) monitored ______________________

Type of sample: Personal ______________ Breathing zone ______________ Area ______________

Operating conditions and control methods ______________________

____

Time on ______________________ Time off ______________________

Elapsed time (min) ______________ Indicated flow rate (LPM) ______________ Volume (liters) ______________

Calibration location ______________ By ______________ Date ______________

Sampling/analytical method ______________________

Evidence of accuracy ______________________

____

Remarks, possible interferences, action taken, etc. ______________________

____

____

Results of sample analysis or instrument reading ______________________

____

Exposure of employee (indicate 8-hr average or 15 min) and sample numbers it is based on ______________

____

____

Figure 17-12. Employee exposure measurement record. (Reprinted with permission from the *Occupational Exposure Sampling Strategy Manual,* NIOSH Pub. No. 77-173)

The level of exposure is determined by direct-reading instruments or by collecting air samples for laboratory analysis at various times and under different operating conditions. Ordinarily, the industrial hygienist collects the contaminant samples in the field; these samples are later analyzed in the laboratory by the chemist. Fortunately, nearly all physical agents can be evaluated by means of direct-reading instruments.

To determine the duration of exposure for an individual, a detailed job description must be obtained. Then, the overall process or operation must be broken down into a series of successive steps. The equipment, materials, and time necessary for each step should be checked with the operator. The potential health hazards for each step should be identified with regard to time and magnitude. The final step in evaluating an environmental exposure is the interpretation of results obtained in the field survey.

From the information developed in preparing a job description and the measurements made of the magnitude of the environmental factor or stress, time-weighted daily 8-hour exposures and peak exposures can be calculated, and a decision can be made as to the degree to which an individual or a group of individuals is exposed.

Experience, discussion with others in the field, and a thorough knowledge of the effects of exposure to a particular chemical or physical agent arising out of manufacturing operations are required to properly evaluate the hazards arising out of the industrial environment.

BIBLIOGRAPHY

American Conference of Governmental Industrial Hygienists (ACGIH). *Air Sampling Instruments for Evaluation of Atmospheric Contaminants,* 6th ed. Cincinnati: ACGIH, 1983.

——. *Documentation of the Threshold Limit Values for Substances in Workroom Air,* 5th ed. Cincinnati: ACGIH, 1985.

——. *Industrial Ventilation—A Manual of Recommended Practice,* 19th ed. Cincinnati: ACGIH, 1986.

——. *Threshold Limit Values and Biological Exposure Indices for 1986-1987* Cincinnati: ACGIH, 1986.

American Industrial Hygiene Association (AIHA). *Analytic Abstracts.* Akron: AIHA.

——. *Hygienic Guide Series.* Akron: AIHA.

Boggs, R.F. "Environmental Monitoring Requirements for Occupational Health Standards." Paper presented at the National Safety Congress, Chicago: October 19, 1976.

Burgess, W.A. *Recognition of Health Hazards in Industry.* New York: John Wiley & Sons, 1981.

Clayton, G.D., and Clayton, F.E., eds. *Patty's Industrial Hygiene and Toxicology,* 3rd rev. ed., vol I. New York: John Wiley & Sons, 1978.

Crocker, B.B. Preventing hazardous pollution during plant catastrophes. *Chemical Engineering,* May 4, 1970.

Hamilton, A., and Hardy, H.L. *Industrial Toxicology,* 3rd ed. Acton, MA: Publishing Sciences Group, Inc., 1974.

Hutchison, M.K., ed. *A Guide to the Work-Relatedness of Disease.* Washington, DC: U.S. Government Printing Office, 1976. NIOSH Pub. No. 77-123.

Irish, D.D. Monitoring the environment and judging its significance. *National Safety Congress Transactions,* Vol. 12, 1986.

Key, M.M., et al., eds. *Occupational Diseases—A Guide to Their Recognition,* rev. ed. Washington, DC: U.S. Government Printing Office, June 1977.

Kusnetz, H.L. Big industry's environmental monitoring programs. *National Safety News,* April 1976, pp. 87-90.

Leidel, N.A., Busch, K.A., and Crouse, W.E. *Exposure Measurement Action Level and Occupational Environmental Variability.* Cincinnati: NIOSH, December 1975. HEW Pub. No. 76-131.

Leidel, N.A., Busch, K.A., and Lynch, J.R. *Occupational Exposure Sampling Strategy Manual.* Cincinnati: NIOSH, January 1977. NIOSH Pub. No. 77-173.

Linch, A.L. *Biological Monitoring for Industrial Chemical Exposure Control.* Boca Raton, FL: CRC Press, 1974.

Linch, A.L. *Evaluation of Ambient Air Quality by Personnel Monitoring.* Boca Raton, FL: CRC Press, 1974.

Lynch, J.R. Action level (its concept and implications). *National Safety News,* May 1976.

Lynch, J.R. Valid Representative Air Samples—Location, Time, Duration, Work Cycles and Peaks and Time Weighted Average Concentrations. Paper presented at the National Safety Congress, Chicago, October 31, 1972.

McRae, A., and Whelchel, L., eds. *Toxic Substances Control Sourcebook.* Germantown, MD: Aspen Systems Corp., 1978.

National Institute for Occupational Safety and Health. *The Industrial Environment—Its Evaluation and Control.* Washington, DC: U.S. Government Printing Office, 1973.

National Institute for Occupational Safety and Health. *Occupational Medicine Symposia.* Rockville, MD: NIOSH, May 1975, HEW Pub. No. 75-189.

National Institute for Occupational Safety and Health. *Occupational Safety and Health Symposia 1976.* Cincinnati: NIOSH, July 1977. NIOSH Pub. No. 77-179.

Occupational Safety and Health Administration. *Compliance Operations Manual* (OSHA 2006), Chapter XIII. Washington, DC: U.S. Government Printing Office, January 1972.

Peterson, J.E. *Industrial Health.* Englewood Cliffs, NJ: Prentice-Hall, Inc., 1977.

Schilling, R.S.F., ed. *Occupational Health Practice.* Toronto: Butterworths, 1981.

Tasto, D.L., et al. *Health Consequences of Shift Work.* Cincinnati: NIOSH, March 1978. DHEW (NIOSH) Pub. No. 78-154.

United States Steel Corporation. *Environmental Health Monitoring Manual.* Pittsburgh: U.S. Steel Corp., 1973.

Zenz, C. *Occupational Medicine—Principles and Practical Applications,* 2nd ed. Chicago: Year Book Medical Publ, Inc., 1988.

18

Air-Sampling Instruments

by Julian B. Olishifski, MS, PE, CSP
Revised by Maureen A. Kerwin, MPH

HEALTH AND SAFETY PROFESSIONALS are interested in knowing what contaminants employees are exposed to and their concentrations. Accurate determination of the extent of an employee's exposure to one or more contaminants on the job will enable employers to determine compliance with OSHA requirements, Threshold Limit Values (TLVs), and other recommended health standards, or to evaluate the effectiveness of engineering controls.

An employer may choose to conduct air-sampling in-house or rely on outside consultants or insurance company personnel. In either case, a basic understanding of the principles involved in air-sampling is necessary in order to understand, interpret, and respond to the results.

TYPES OF AIR CONTAMINANTS

The type of air contaminants that occur in the workplace depends upon the raw materials used and the processes employed. Air contaminants can be divided into two broad groups, depending upon physical characteristics: (1) gases and vapors and (2) particulates.

Gases and vapors

Gases are formless fluids that expand to occupy the space or enclosure in which they are confined. They can be liquified only by the combined effects of increased pressure and decreased temperature. Vapors result when a solid or liquid is converted by heating into a gaseous state, and is the result of volatilization or sublimation. The term gaseous can be used, in its general sense, to include both gases and vapors.

Gases and vapors follow the normal laws of diffusion and mix freely with the surrounding atmosphere. They are not affected greatly by inertial and electrostatic forces, which may disturb particles. These characteristics make gases and vapors easier to sample than dusts or fumes.

Particulates

Airborne particulates (aerosols) are defined as solid particles or liquid droplets dispersed in the air. Dusts, fumes, and smoke are dispersed solids, while mists and fogs are dispersed liquids. Particulates range in size from visible to submicroscopic. In taking air-samples of fibrogenic dusts (dusts that can cause scarring in lung tissue), the primary concern is with respirable particulates smaller than 10μm in diameter; whereas for systemic poisons, dust particles of all sizes at the breathing level of the worker must be collected.

The aerodynamic diameter of airborne particulate matter is defined as the diameter of a hypothetical sphere of unit density having the same terminal settling velocity as the particle in question, regardless of its geometric size, shape, and true density. It is of special importance for evaluating toxicological effects.

Particles can be sampled to determine the total quantity, or only the respirable fraction. A sample can be collected for chemical analysis to determine substances such as lead, chromium, zinc, etc. The method of sampling for particulates depends upon the reason for sampling and the nature of the particulate.

As most samples are taken to determine the level of concentration of a substance or to analyze the air for one or more specific substances; the procedure usually used is to pass the air through a suitable filter. Fumes and mists can be (1) absorbed and measured in the field, (2) absorbed in the field and evaluated later in the laboratory, or (3) collected on filter media and analyzed later in the laboratory. Dusts can be (1) collected by personal air-sampler, fractionated into respirable size by a cyclone separator, and the fractions weighed to determine the concentration; (2) collected on a filter and weighed; or (3) collected in an appropriate trapping medium and counted in the field or later in the laboratory.

TYPES OF AIR-SAMPLING

The type of air-sampling to be used depends upon a number of factors. The purpose of the sampling, the equipment available, the environmental conditions, and the nature of the toxic contaminant, are some of the considerations. The design of a sampling strategy is covered in Chapter 17, Methods of Evaluation.

Area sampling v. personal sampling

Personal air-sampling is the preferred method of evaluating workers' exposure to air contaminants. It involves the collection of an air-sample by a sampling device worn by the worker. The sampling device is positioned as close as possible to the breathing zone of the worker (usually clipped to the shirt lapel) so the data collected closely approximate the concentration inhaled.

Area- or general-room air samples are taken at fixed locations in the workplace. In general, this type of sampling does not provide a good estimate of worker exposure. For that reason, it is used mainly to pinpoint high exposure areas, indicate flammable or explosive concentrations, or determine if an area should be isolated or restricted to prevent employees from entering a highly contaminated area. Confined spaces, such as sewer lines and reaction vessels, are examples where area monitoring is necessary. In addition, continuous air monitors can be used to detect leaks, ventilation failures, and equipment malfunctions before worker exposure occurs.

Grab sampling v. integrated sampling

Air-sampling can be conducted for long or very short periods, depending on what type of information is needed. Instantaneous or grab sampling is the collection of an air-sample over a short period whereas longer period sampling is called integrated air-sampling. There is no sharp dividing line between the two sampling methods, however, grab samples are usually taken in a period of less than 5 minutes.

Grab samples represent the environmental concentration at a particular point in time. It is ideal for following several phases of a cyclic process and for determining airborne concentrations of brief duration but is seldom used to estimate 8-hour average concentrations. When primary irritants such as ammonia are present, grab-sampling techniques can provide useful information for evaluating peak exposures.

Grab samplers consist of various devices. An evacuated flask or plastic bag can be useful in gas and vapor concentration analysis. After introducing the sample of air into the container, it is sealed to prevent loss or further contamination and sent to a laboratory for analysis. There, trace analysis procedures such as gas chromatography, infrared spectrophotometry, or other methods are used to determine concentrations of gaseous contaminants.

Direct-reading instruments which are described in detail in Chapter 19, Direct-Reading Gas and Vapor Monitors, can also be used. These units are analytical instruments with rapid response times. They are defined as devices in which sampling and analysis are carried out within the instrument and the required information can be read directly from an indicator. Direct-reading devices can be portable instruments or fixed site monitors.

In some cases, it may be necessary to estimate an 8-hour average concentration based upon grab samples. This can occur when the only equipment available for a particular analysis is a direct-reading instrument. In this case, samples are collected at random to ensure that each part of the workday has the same chance of being included in the evaluation. This method, however, must be used with caution. If the atmospheric concentration varies irregularly, this environmental variance will greatly affect the accuracy of the results, and many more samples would have to be taken to determine the average concentration of the contaminant.

In integrated air-sampling, a known volume of air is passed through a collection media to remove the contaminant from the sampled airstream. It is the preferred method of determining time-weighted average (TWA) exposures. Integrated sampling consists of one or a series of samples taken for the full or partial duration of the time averaging period. The time averaging period can be from 15 minutes to 8 hours, depending upon whether a ceiling, short-term, or 8-hour exposure limit is being evaluated.

SELECTION OF EQUIPMENT

The selection of the air-sampling method depends upon a number of factors including:

- the sampling objective (documenting exposures, determining compliance, pinpointing source of exposure)
- the physical and chemical characteristics of the chemical
- presence of other chemicals that may interfere with the collection and/or analysis
- required accuracy and sensitivity
- regulatory requirements
- complexity of method (portability and ease of operation)
- cost
- reliability
- type of sample needed (area vs. personal)
- duration of sampling
- professional judgment

Other than direct-reading devices, many of these factors have already been considered by industrial hygiene analytical laboratories. The first step in selecting an air-sampling device is to consult with the laboratory to ensure that the method of measurement selected meets the sampling objective. Only laboratories certified by the American Industrial Hygiene Association (AIHA) should be used. These laboratories typically publish a manual of analytical methods and are available for consultation.

GRAB SAMPLING

Analytical procedures specify the collection media, sample volume, flow-rate, and chemical analysis. These analytical procedures are based upon (1) the NIOSH *Manual of Analytical*

ACETIC ACID

FORMULA: CH_3COOH; $C_2H_4O_2$

METHOD: 1603
ISSUED: 2/15/84

M.W.: 60.05

OSHA: 10 ppm
NIOSH: no standard
ACGIH: 10 ppm
(1 ppm = 2.46 mg/m³ @ NTP)

PROPERTIES: liquid; d 1.049 g/mL @ 25 °C;
BP 118 °C; MP 17 °C;
VP 1.5 kPa (11 mm Hg; 1.4% v/v) @ 20 °C;
explosive range 5.4 to 16% v/v in air

SYNONYMS: glacial acetic acid, methane carboxylic acid, ethanoic acid, CAS #64-19-7.

SAMPLING	MEASUREMENT
SAMPLER: SOLID SORBENT TUBE (coconut shell charcoal, 100 mg/50 mg)	TECHNIQUE: GAS CHROMATOGRAPHY, FID
FLOW RATE: 0.01 to 1.0 L/min	ANALYTE: acetic acid
VOL-MIN: 20 L @ 25 mg/m³ -MAX: 270 L	DESORPTION: 1 mL formic acid; stand 60 min
SAMPLE STABILITY: at least 7 days @ 25 °C [1,2]	INJECTION VOLUME: 5 µL
BLANKS: 2 to 10 field blanks per set	TEMPERATURE-INJECTION: 230 °C -DETECTOR: 230 °C -COLUMN: 130 to 180 °C, 10°/min or 100 °C isothermal
ACCURACY	CARRIER GASES: N_2 or He, 60 mL/min
RANGE STUDIED: 12.5 to 50 mg/m³ [2] (173-L samples)	COLUMN: 1 m x 4 mm ID glass; Carbopack B 60/80 mesh/3% Carbowax 20M/0.5% H_3PO_4
BIAS: not significant [2]	CALIBRATION: acetic acid in formic acid
OVERALL PRECISION (s_r): 0.058 [2]	RANGE: 0.5 to 10 mg per sample
	ESTIMATED LOD: 0.01 mg per sample [3]
	PRECISION (s_r): 0.007 [1,2]

APPLICABILITY: The working range is 2.5 to 50 mg/m³ for a 200-L air sample. High (90% RH) humidity during sampling did not cause breakthrough at 39 mg/m³ for 4.6 hrs [2].

INTERFERENCES: Formic acid contains a small amount of acetic acid which gives a significant blank value. High purity formic acid must be used to achieve an acceptable detection limit. Alternate columns are 3 m glass, 2 mm ID, 0.3% SP-1000 + 0.3% H_3PO_4 on Carbopack A and 2.4 m x 2 mm ID glass, 0.3% Carbowax 20M/0.1% H_3PO_4 on Carbopack C.

OTHER METHODS: This is Method S169 [1] in a revised format.

Figure 18-1. NIOSH analytical method (No. 1603) for acetic acid.

Methods (2) the AIHA *Analytical Guide Series* (3) the American Public Health Association (APHA) *Methods of Air Sampling and Analysis* (4) OSHA regulations, and (5) the laboratory's own experience.

It may be necessary to collect samples using a number of different methods to determine the concentrations for a mixture of the contaminants. Figure 18-1 is the NIOSH analytical procedure for acetic acid.

For direct-reading devices, the manufacturer should supply information on collection efficiency, sensitivity, and reproducibility. The instrument must be simple to use and require a minimum of manipulation in the field.

Gases and vapors

The detection of hazardous gases in the work environment spans some four areas of measuring magnitude from parts per million to significant percentages, depending on whether toxic or flammable properties are of primary concern.

Many toxic substances encountered in the work environment are also flammable. As a rule the toxic effects of flammable substances occur at much lower concentrations than their lower explosive limit (LEL).

Gases and vapors offer the least difficulty in air-sampling. They follow the normal laws of diffusion, mix freely with the general atmosphere, and can, in a short time, become thoroughly diffused at equilibrium. Direct-reading devices are most often used for grab samples, but a collection of a known volume of air for subsequent laboratory analysis can also be used. In the latter method, a sample of air is collected in an evacuated container, displacement collector, or a flexible plastic bag. It is critical that after introducing the sample of contaminated air into the container, that it is properly sealed to prevent sample loss and properly handled to reduce sample decay. It is also necessary to record the temperature and pressure at which the sample was collected in order to report the concentrations in terms of standard conditions.

Evacuated containers can be used to collect atmospheres containing carbon dioxide, oxygen, methane, carbon monoxide, and nitrogen. They have been used by the Mine Safety and Health Administration (MSHA) to test underground work atmospheres. They are not suitable for reactive gases such as hydrogen sulfide, oxides of nitrogen, or sulfur dioxide since they may become altered due to a reaction with dust particles, moisture, or the walls of the container. Nitrogen oxides, for example, are tenaciously absorbed on glass surfaces.

Evacuated containers are usually constructed of glass and range from 200–1,000 mL in volume. Other containers include 50- to 250-mL glass bulbs, steel evacuated containers lined with a nonabsorbing interior surface, and 10-mL syringes (vacutainer syringe system).

Gas or liquid displacement collectors can be used to collect atmospheres containing oxygen, carbon dioxide, carbon monoxide, nitrogen, and hydrogen. These units are filled with a gas or liquid, which are displaced by the sample gas or vapor. They are primarily 250- to 300-mL glass aspirator bulbs fitted with end tubes that can be closed with greased stopcocks or with rubber tubing and screw clamps. Air is drawn into the container with a source of suction such as a bulb aspirator or a battery-operated pump, until the original content of air is replaced.

In liquid displacement, the liquid is drained from the container, allowing the test atmosphere to enter. This method is restricted to gases that are insoluble or nonreactive in the displacement liquid.

Flexible plastic bags can be used to collect air-samples containing organic or inorganic vapors and gases. They are available in a variety of sizes and materials. Polyester, polyvinylidene chloride, Teflon and fluorocarbons are the most commonly used materials. Due to the possible reaction of the collected gas or vapor with the bag, it is neccesary to know the reactivity, absorbtivity, and diffusive properties of the plastic.

Air is drawn into the bag with a hand- or battery-operated pump (Figure 18-2) or squeeze bulb. Collection bags can be used to collect short- or long-term samples, depending on the size of the bag and the pump flow-rate. A common field application is the collection of a series of air-samples that are analyzed on-site with a portable analytical instrument. This greatly reduces the possibility of gas decomposition before analysis.

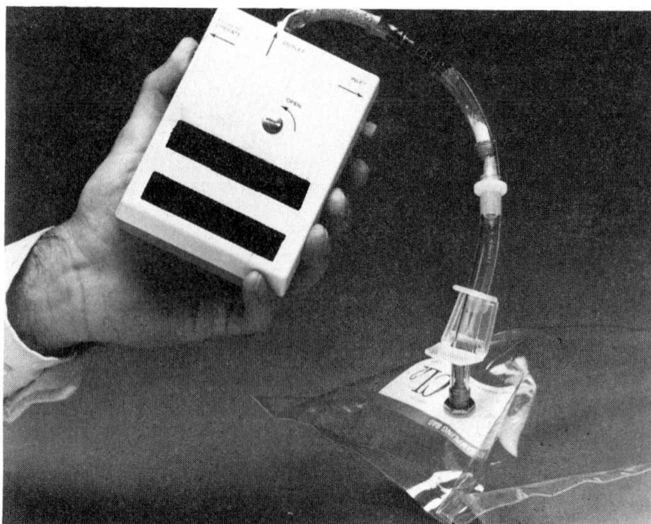
Figure 18-2. Portable, battery-operated pump used to fill flexible plastic gas sampling bag. (Courtesy MSA)

Particulates

Grab samplers for particulates include direct-reading devices, such as fibrous aerosol monitors, respirable mass monitors, and aerosol photometric monitors. These units are nonspecific, they respond to the size, shape, or mass of the particle not to the specific material itself.

INTEGRATED AIR-SAMPLING

Integrated air-sampling is used to determine a worker's 8-hour TWA exposure to a toxic chemical. It is also used if the concentration of the toxic material varies significantly during the work shift or if a large sample volume is necessary to satisfy the sensitivity requirements of the analytical method.

Instrumentation

The entire array of equipment used to perform the air-sampling task is contained in the air-sampling train. An air-sampling train consists of an air-inlet orifice, collection media, airflow meter, flow-rate control valve, and suction pump.

Air-sampling pumps

A critical component of the air-sampling system is the air-sampling pump. The proper selection and calibration of that pump is a critical element of the sampling process. A pump must be selected, based on the desired sampling flow-rate, ease of servicing and calibration, and suitability in a hazardous environment, among other factors. The primary determinant is the desired flow-rate. Personal, battery-powered air-sampling pumps can be classified into these basic categories.

1. Low-flow pumps (approximately 0.5–500 mL/min)
2. High-flow pumps (approximately 500–4,500 mL/min)
3. Dual-range pumps

A standardized air-sampling technique specifies the desired

air-sampling flow-rate and the minimum volume of air that must be sampled. The sample volume is based on the sensitivity of the analytical method. Sensitivity describes the smallest concentration of contaminant that can be detected by the laboratory. To achieve the desired sample volume, the flow-rate is selected based on the desired sampling time (usually 8 hours), and the collection efficiency characteristics of the collection media.

Figure 18-3. Low-flow air-monitoring pump. (Courtesy E.I. DuPont de Nemours & Co., Inc.)

Low-flow pumps (Figure 18-3) are used for gas and vapor sampling. The common flow-rate, for example, for organic vapors is 200 mL/min.

High-flow pumps (Figure 18-4) are used for particulate sampling and gas and vapor sampling. The common flow-rate for fume or dust sampling is 2 L/min. On sampling for respirable crystalline silica, a flow-rate of 1.7 L/min. is usually used.

Hazardous environments can also be a factor when selecting a pump. In a flammable or explosive atmosphere, only explosion-proof electrical devices can be used. Most sampling pumps are intrinsically safe and approved for such environments. They are approved by MSHA and groups such as Underwriters Laboratory (UL) Inc. or Factory Mutual Engineering Corp., Inc. (FM). The UL or FM label should be consulted for the flammable class and group of the approval.

Flow-rate meters

Maintaining a constant pump flow-rate is critical in determining the sample volume based on elapsed time. Sample volume is one of the determining factors in the formula for concentration.

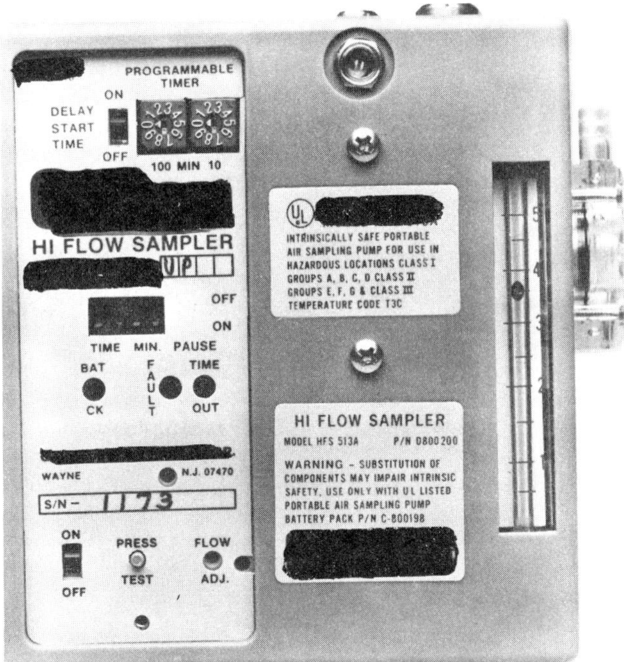

Figure 18-4. High-flow personal air sampler. (Courtesy Gilian Instrument Corp.)

$$\text{Concentration} = \frac{\text{mass}}{\text{volume}}$$

Air-sampling pumps use a variety of devices to assure a constant flow. These include rotameters, critical orifices, and pressure-compensating devices.

Rotameters. Rotameters are commonly attached to high-flow sampling pumps (Figure 18-5). They consist of a "float" or ball that is free to move up or down within a vertical tapered tube which is larger at the top than the bottom. As air is pumped into the bottom of the tube, the ball will rise until an equilibrium is reached between the weight of the ball and the force exerted on the ball by the pressure of air flowing through the annular area between the ball and the wall of the tube. The higher the ball goes, the faster the flow-rate. The height of the ball is measured against a numerical scale attached to the tube and is an indirect measure of flow-rate. The ball is commonly read at the point of maximum diameter (ball center). The flow rate can be adjusted up or down with a calibration screw or knob on the pump. The rotameters found on air-sampling pumps are not very accurate and are crude indications of the flow-rate, at best. For this reason, pumps must be calibrated before and after use with a primary calibration standard, described later in this chapter. (Pump rotameters are not to be confused with precision rotameters. Precision rotameters stand alone and are calibrated against a primary standard of calibration. They are considered to be a secondary calibration standard.)

Critical orifice. Some high- and low-flow pumps use critical or limiting orifices to regulate airflow-rate. A critical orifice is a precision drilled hole in a metal plate through which the airstream being sampled is directed. When certain critical parameters are met, the flow-rate through the orifice remains constant despite changed in inlet conditions (e.g., clogged filter).

Figure 18-5. Rotameter on high-flow sampling pump.

Constant-flow samplers. Constant-flow samplers are designed to overcome the flow-rate variation problems inherent in many sampling situations. Increased collection media resistance, such as a filter loaded with dust or a crimped hose can effect the flow-rate. These pumps have sophisticated flow-rate sensors with feedback mechanisms, which permit maintenance of the preset flow-rate during the sampling period. They are available on both high- and low-flow sampling pumps.

Stroke counters. Some low-flow sampling pumps have stroke counters. Stroke counters provide an indirect measurement of sample volume on pumps with a piston-action motor. Each stroke of the piston is recorded on a dial and relates to the amount of air moved by the pump during that piston stroke. While the flow-rate may change during the sampling period, the stroke counter maintains an accurate record of the sample volume.

Calibration

Calibration of air-sampling pumps is an integral part of the accurate measurement of airflow-rate and volume. There are two categories of calibration methods, primary and secondary. Primary methods are generally direct measurements of volume on the basis of the physical dimensions of an enclosed space. Secondary methods must be calibrated themselves against a primary standard and have been shown to maintain their accuracy with use.

Primary standards. *Soap-bubble meters.* A soap-bubble meter is considered a primary method of calibration and is commonly used to calibrate high- and low-flow air-sampling pumps (method accuracy can be within ±1 percent). It consists of an upturned laboratory burette (a 1,000-mL burette is used to calibrate high-flow pumps, a 100-mL burette is used to calibrate low-flow pumps), a ring stand and clamp, tubing, selected collection media, and the pump (Figure 18-6). The selected collection media to be used in air-sampling must be used during the calibration of the pump. Electronic bubble flow meters for pump calibration (Figure 18-7) are now available from various manufacturers. These electronic models are usually used only for that particular manufacturer's brand of pumps.

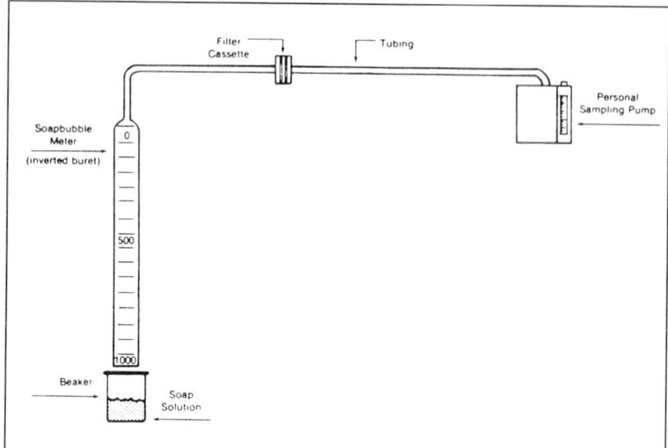

Figure 18-6. Calibration setup for personal sampling pump with filter cassette. (Reprinted from NIOSH *Manual of Analytical Methods,* 1984)

Bubble-meter method

Perform pump calibration using the bubble meter method as follows:

1. Allow sampling pumps to run for 5 min before calibration. Check voltage after 5 min. If the voltage is below the manufacturer's specified level, the pump needs to be recharged. Check the manufacturer's instructions for proper charging procedures.
2. Wet the inside of the burette with the soap solution before setup.
3. Assemble bubble meter and connect sampling pump and the type of collection device intended for the field sampling.
4. Momentarily submerge the opening of the burette to capture a film of soap.
5. Draw two or three bubbles up the burette to ensure that they will reach the top.
6. Visually capture a single bubble and time (with a stopwatch) the bubble from 0–1,000 mL or 0–100 mL, depending on the pump being calibrated.
7. Adjust the pump flow-rate until the desired flow-rate is achieved. For example, for a flow-rate of 2 L/min, the bubble must travel from 0–1,000 mL in 30 seconds. Verify the flow rate at least twice.
8. For pumps with rotameters, mark or record the position of the float (ball) when the pump is running at the desired flow-rate. This will allow the industrial hygienist to adjust the pump flow-rate back to the correct rotameter position if the float moves off the marked setting during sampling.

Secondary standards. Though they rely on periodic recalibration against primary standards, secondary standards can provide an accuracy of ±1.0 percent or better. Secondary standards include wet-test meters, dry-gas meters, and precision rotameters.

Wet-test meter. A typical wet-test meter (Figure 18-8) is a par-

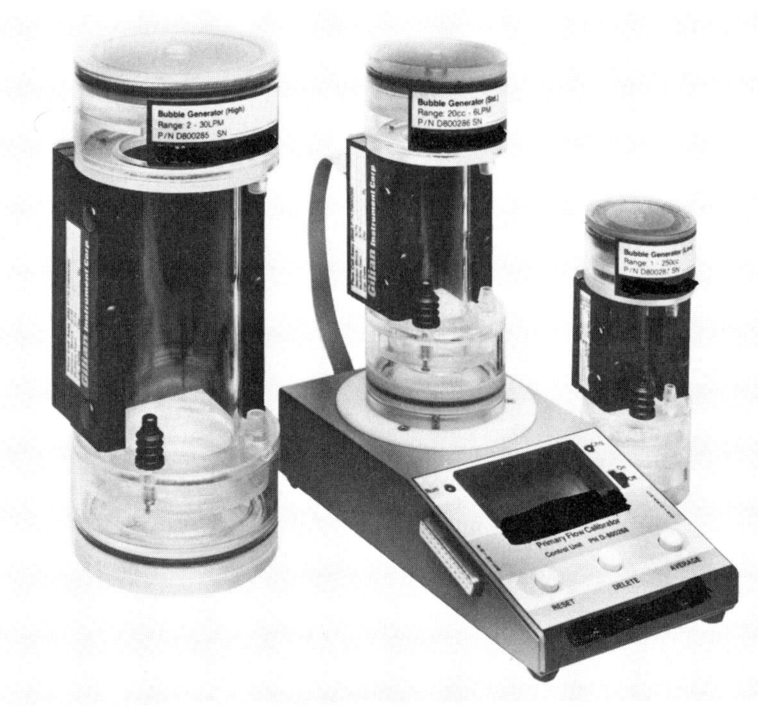

Figure 18-7. Electronic bubble meter calibration system. (Courtesy Gilian Instrument Corp.)

titioned drum, which is half submerged in a liquid (usually water), with openings in the center and periphery of each radial chamber. Air or gas enters at the center and flows into one compartment causing it to raise, thereby causing rotation. The number of revolutions is indicated on a dial on the face of the meter.

Because liquid is being displaced by air, the volume measured depends on the height of the fluid in the meter. A sight gage for determining fluid height is provided along with a gas pressure gage and a thermometer.

Dry-gas meter. A dry-gas meter, similar to a domestic-gas meter, consists of two bellows connected by mechanical valves and a counting device (Figure 18-9). As air or gas fills one bag, it mechanically empties another.

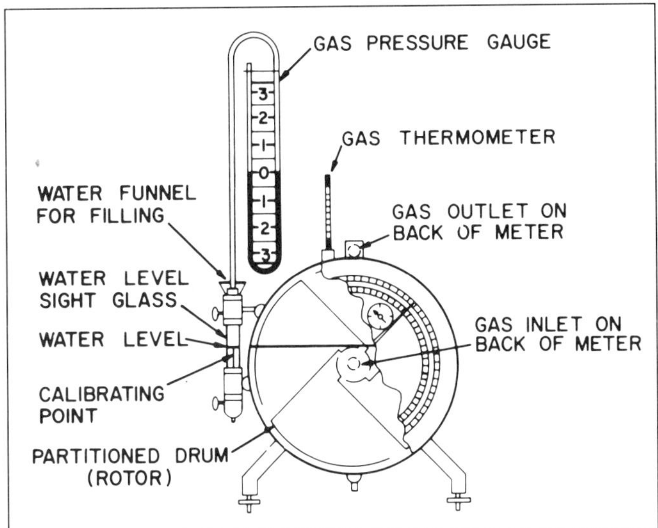

Figure 18-8. The "working parts" of a wet-test meter are shown.

Both wet- and dry-gas meters come in many sizes and should be selected for the flow-rate range of the sampling system to be calibrated. Both types must be periodically recalibrated. The manufacturer usually provides this service.

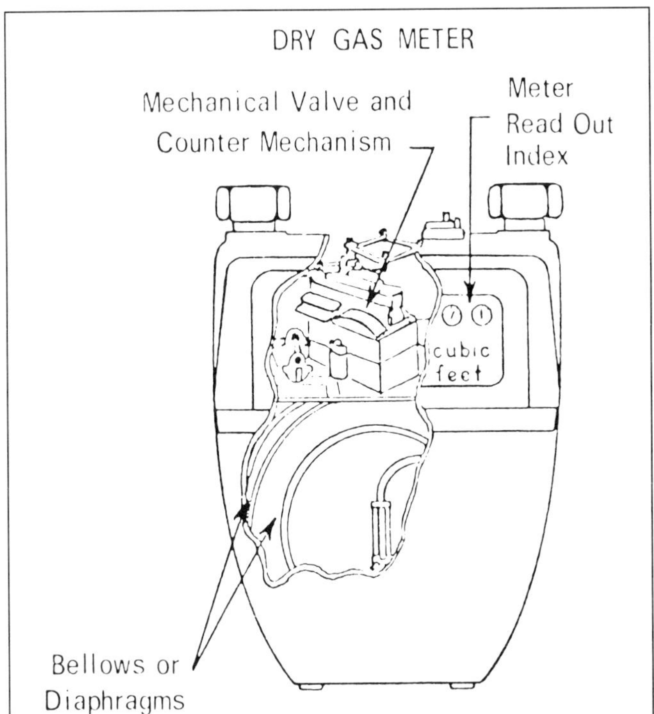

Figure 18-9. A dry-gas meter, consisting of two bags, connected by mechanical valves and a counting device. (Reprinted from *The Industrial Environment—Its Evaluation and Control,* PHS Publication No. 614)

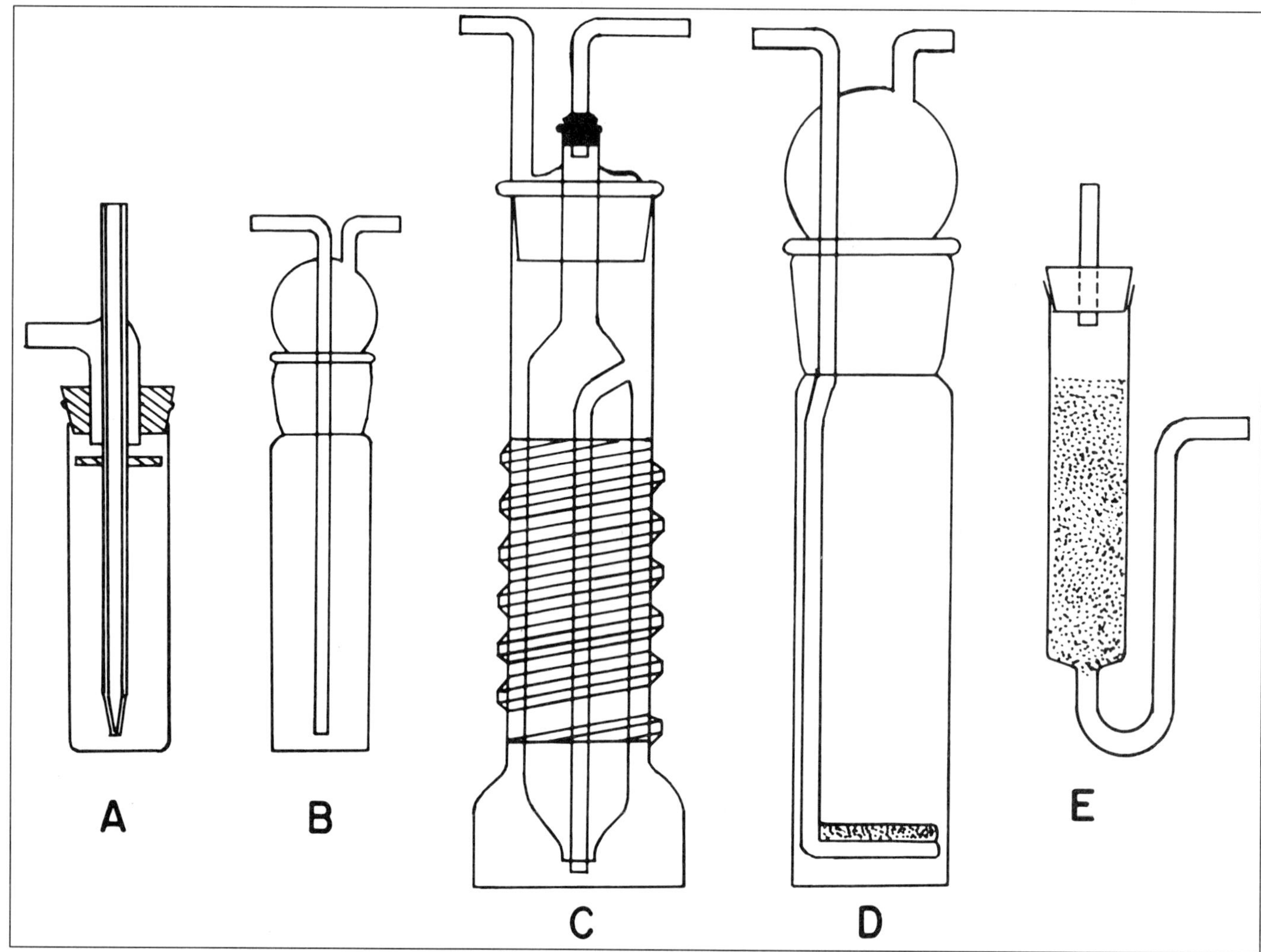

Figure 18-10. Basic absorbers are shown: gas washing (A and B); helical (C), fritted bubbler (D), glass-bead column (E). They provide contact between sampled air and liquid surface for absorption of gaseous contaminants. (Reprinted from *The Industrial Environment—Its Evaluation and Control,* PHS Publication No. 614)

Precision rotameters. Precision rotameters, unlike rotameters incorporated into a sampling pump, can provide an accuracy of ± 1–5 percent. These rotameters are usually 12 in. long and can easily be placed in a sampling train to calibrate the pump before or after the sampling period. It can also be used to periodically check sampler flow-rate during the sampling period.

Precision rotameters must be calibrated against a primary standard at regularly scheduled intervals. Typically, a bubble meter is used as the primary standard. Several points on the rotameter scale are plotted on a graph against selected flow-rates. Then, it is possible to determine flow-rate from the rotameter scale for the particular pump that is being calibrated.

Collection media

The physical form of the contaminant determines the type of medium that the sample is collected on. Thus, there is one set of media for gases and vapors and another grouping for particulates.

Gases and vapors. Gases and vapors are collected using liquid media samplers, solid-sorbent tubes, and passive monitors.

Liquid-media samplers. In these devices, an absorbing liquid is used to capture gaseous substances. When air is passed through liquid, the contaminant is captured by the liquid through solvation or a chemical reaction. Selection of an appropriate absorbing solution, therefore, depends on the chemical nature of the gas or vapor being collected.

There are many sampling devices available, including: gas-wash bottles, spiral absorbers, fritted bubblers, and glass-bead columns (Figure 18-10). The most commonly used gas-wash bottle is the midget impinger. It can be used to collect area or personal samples. Personal samples must be collected with care to prevent accidental spillage of the liquid on the employee wearing the equipment. In addition, sample lines and flow-rates must be checked to assure the liquid media is not sucked into the air-moving pump.

Fritted bubblers are also commonly used. Air is drawn through a sintered or fritted glass bubbler, which is submerged in an absorbing solution or reagent. The fritted glass contains thousands of small holes that disperse the air-sample into tiny bubbles as it is pulled through the solution. This increases the surface area and time of contact between the air-sample and the liquid media, thereby increasing the efficiency of collection.

Solid-sorbent tubes. Air-sampling for insoluble or nonreactive gas and vapor contaminants is commonly conducted using tubes filled with granular sorbents, such as activated charcoal, silica gel, porous polymers, or other materials. Those sorbents have the ability to retain or adsorb gaseous substances chemically unchanged on their surfaces, which can later be desorbed (extracted) and analyzed at a laboratory. Table 18-A provides a listing of some NIOSH-recommended sorbent tubes for different contaminants.

Table 18-A. NIOSH Analytical Methods

Collection Method	*Contaminant*	*NIOSH Reference No.*
Solid-Sorbent Tubes		
Silica gel	Methanol	2000
	Amines, Aromatic	1602
Charcoal tubes	Dioxane	1602
	Ethyl Ether	1610
Coconut-shell charcoal	Hydrocarbons, halogenated	1003
	naphthas	1550
Porapak P	Dimethyl sulfate	2524
Tenax-GC	Nitroglycerin	2507
	Ethylene glycol dinitrate	2507
Ambersorb XE-347	2-Butanone (MEK)	2500
Filters		
Mixed cellulose ester	Arsenic	7900
Polyvinyl chloride (PVC)	Silica, crystalline, respirable	7500
Bubblers		
With 0.1 M HCl	Hydrazine	3503
Solution of 1-(2-methoxyphenyl)-piperzine in toluene	Isocyanate group	5505
Passive Monitor		
Liquid sorbent badge of 0.01 N H_2SO_4	Ammonia	6701
Passive badge	Nitrogen dioxide	6700

Activated charcoal is the most widely used solid sorbent for organic vapor sampling. It is an amorphous form of carbon formed by the burning and steam-heating of wood, nutshells, or other carbonaceous materials. The most commonly used is coconut-shell charcoal.

It has a large surface area, reactive surface, and high adsorptive capacity. It is also electrically nonpolar, meaning it will preferentially adsorb organic vapors rather than polar molecules such as water vapor.

The standard charcoal tube is 7 cm long and 4 mm wide. The first section contains 100 mg of charcoal, which is separated from the second or backup section of 50 mg of charcoal by a fiberglass, glass wool, or urethane foam plug (Figure 18-11). Larger tubes are available and used for specific air-sampling protocols.

Silica gel can be used to sample for air contaminants that are not efficiently collected or desorbed from activated charcoal. Silica gel, an amorphous form of silica, is an efficient collector of inorganic substances and electrically polar molecules. It is currently the NIOSH-recommended sampling media for aromatic and aliphatic amines. A major disadvantage of silica gel is that due to its polar nature, water vapor can seriously interfere with the adsorbtion of the contaminant.

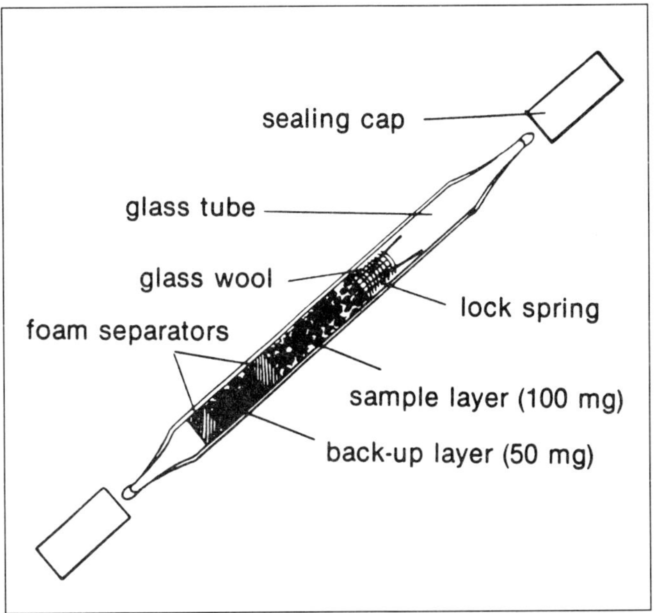

Figure 18-11. Activated charcoal sampling tube. (© SKC Inc., Eighty-Four, PA)

Impregnated solid sorbent tubes are a recent development in sampling for reactive and other gases and vapors. In this method, standard solid sorbants are impregnated with an absorbic reagent, which changes the collected gas or vapor into a more stable ionic form. After the sample is desorbed from the sorbent, it is then followed by ion chromatography analysis.

Passive monitors. Recently, passive monitors have been increasingly used to monitor gases and vapors. They are available for a variety of contaminants, including organic vapors, carbon monoxide, sulfur dioxide, ammonia, and formaldehyde. They are easy to use, have no moving parts to maintain, are small, lightweight and inexpensive, and can be used relatively unattended.

These monitors rely on two methods of collection. The first used primarily for organic vapors is passive diffusion of the contaminated air into a standard collection medium, such as activated charcoal. Diffusion occurs at a specific rate and depends upon the diffusion coefficient of the organic vapor being sampled, the total cross-sectional area of the badge, and the length of the cavity (Fick's Law). Specific information on the calculation of flow-rates and sample volume can be obtained from the manufacturer.

Monitoring begins once the monitor cover is removed. The time is recorded and the monitor is clipped in the breathing zone of the worker usually on the collar (Figure 18-12). When sampling is completed, the monitor is removed and resealed. Again, the time is recorded. The badge is then sent to a laboratory for analysis. Remember that:

$$\text{Concentration} = \frac{\text{mass}}{\text{volume}}.$$

For badges, mass is determined by the amount of contaminant trapped on the badge. Volume is calculated using the manufacturer's formula and the duration time of the sample.

The accuracy of passive monitors has been studies extensively. Most commercially available monitors meet or exceed NIOSH standards—NIOSH recommends an accuracy of ± 25 percent for 95 percent of the samples tested in the range of 0.5–2.0 times the environmental standards. Information on accuracy should be obtained from the manufacturer.

The second passive monitoring method, used to sample for reactive gases, is to react the contaminant with a chemical coating on the collection surface of the monitor. These are available for sulfur dioxide, nitrogen dioxide, and mercury vapor, as well as other gases.

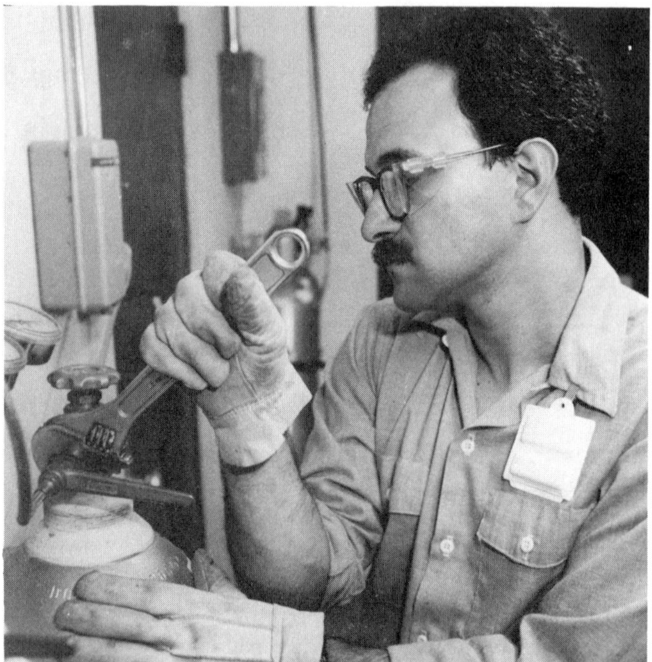

Figure 18-12. Passive badge placed in the worker's breathing zone. (Courtesy E.I. Du Pont de Nemours & Co., Inc.)

Particulates. Particulates are collected on filters, electronic precipitators, cascade impactors, impingers, and elutriators.

Filters. The most common collection method for airborne particulates is a filter. They are easy to use and are available in a wide variety of synthetic materials and pore sizes. Filter materials include glass fiber, mixed fiber, plastic fiber, membrane, and nucleopore. The selection of a filter depends on cost, availability, collection efficiency, requirements of the analytical procedure, and the ability of the filter to retain its filtering properties and physical integrity under sampling conditions. Mixed-fiber filters, for example, are used when a simple gravimetric analysis is to be performed, but cannot be used when a chemical analysis of the material deposited on the filter is needed. Table 18-B shows the filter type for two different NIOSH analytical methods—one for sampling for arsenic and one for respirable silica dust.

Standard filters are 37 mm in diameter and are placed in closed-face cassette, with a backup pad, to avoid contamination (Figure 18-13). The OSHA Asbestos Standard, however, is an exception. It requires a 25-mm filter and open-face (top removed) cassette and a 50-mm extension cowl on the cassette (see NIOSH Analytical Method 7400).

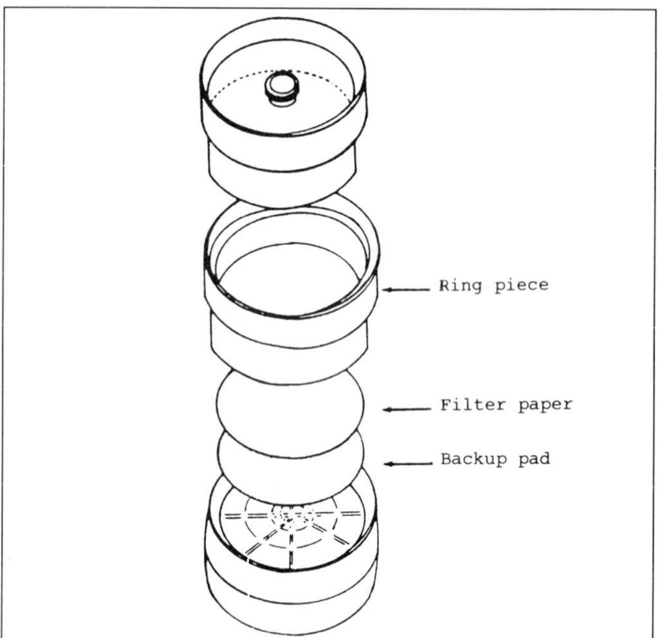

Figure 18-13. Standard filters are 37 mm in diameter, and are placed in closed-face cassettes with a backup pad, to avoid contamination. (Reprinted from OSHA *Industrial Hygiene Technical Manual,* 1984)

Cyclones. It is sometimes necessary to measure only the respirable fraction of dust. These are particles that are 10μ or less in size and penetrate deep into the lungs. Therefore, collection of respirable dust provides the best estimate of a health hazard caused by the inhalation of insoluble dust particles.

A 10-cm cyclone with a preweighed filter is used to collect personal respirable dust samples. The cyclone and filter assembly is usually attached to the workers lapel (Figure 18-14). Air is drawn into the cyclone through a small orifice. The air is then accelerated and whirled causing the heavier (larger) particles to be thrown out to the edge of the airstream and dropped to a removal section at the bottom of the cyclone. The respirable particles remain in the center stream and are drawn up and collected on the filter.

Gravimetric procedures require the careful preweighing and postweighing of the filter. Weighing errors are usually the result of moisture, improper handling, or poor technique. Even filters described as nonhygroscopic can accumulate some moisture. Special techniques are required in handling filters before and during weighing. At a minimum, procedures should include:

1. The same person should weigh the filter before and after sampling using the same balance.
2. Filters should never be touched with the hands, because moisture and oil from the skin will cause weighing errors. Use tweezers or tongs.
3. Before initial (before exposure) weighing, all filters should be stored in a desiccant cabinet to remove any moisture.
4. For sampling, remove filters from the desiccant cabinet one at a time; weigh one at a time.
5. Use a balance with a drying cylinder to prevent moisture errors during weighing.
6. Record preexposed weight of the filter correlating the weight to the filter number. Place filter in its holder (still using

Table 18-B. Sampling Techniques for Collection of Airborne Particulates

Sampling Technique	*Force or Mechanism*	*Examples*
Filters	Combination of inertial impaction, interception, diffusion, electrostatic attraction, and gravitational forces	Various types and sizes of fibrous, membrane, and nucleopore filters with holders
Impactors	Inertial—Impaction on a solid surface	Single and multi-jet cascade impactors and single-stage impactors
Impingers	Inertial—Impingement and capture in liquid media	Greenburg-Smith and midget impingers
Elutriators	Gravitational separation	Horizontal and vertical type elutriators
Electrostatic precipitation	Electrical charging with collection on an electrode of opposite polarity	Tube type, point-to-plane, and plate precipitators
Thermal precipitation	Thermophoresis—particle movement under the influence of a temperature gradient in the direction of decreasing temperature	Various devices have been designed for particulate collection for microscopy analysis
Cyclones	Inertial—Centrifugal separation with collection on a secondary stage	Tangential and axial inlet cyclones in varying sizes

Reprinted from *Occupational Respiratory Diseases.* Pub. No. DHHS (NIOSH) 86-102, 1986.

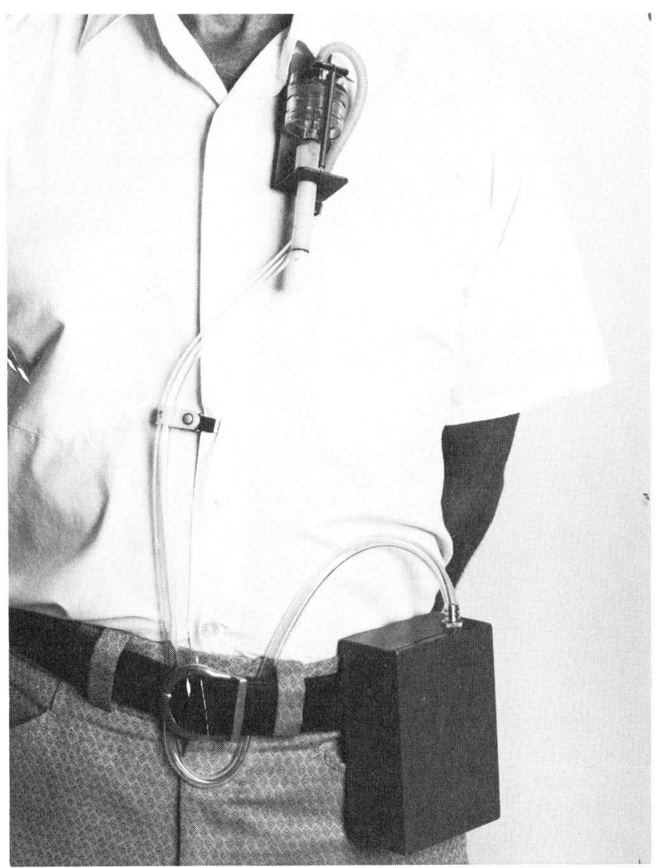

Figure 18-14. Personal sampler clipped to worker's belt pumps air through filter-cyclone assembly attached to worker's collar, to sample dust in the breathing zone.

tweezers or tongs), and seal the holder.

7. After sampling, remove the cover from the filter holder and place the exposed filter in the desiccant cabinet for a period of 16–24 hours to remove excess moisture; filter must be free from the backup disk before drying.
8. After drying, filters should be removed from the drying cabinet one at a time (using tweezers) and placed on the balance for weighing.
9. Record the final weight.
10. Take care not to damage the filter after it has been exposed, because any loss of filter substance will result in erroneous weight calculation.

Electrostatic precipitator. An electrostatic precipitator passes a relatively high airflow through a high-voltage electrical field. It consists of a grounded metal tube that serves as a collecting electrode. Particulates passing through the electrical field are attracted to the cylinder wall where they are collected.

Such a precipitator can be used to collect air-samples for microscopic or chemical analysis. They are particularly well suited to collect particles of submicron size, such as metal fumes. But these units cannot be used in areas containing combustibles or in areas where sources of ignition are prohibited.

Cascade impactors. Cascade impactors (Figure 18-15) can be used to determine the particle size distribution of an air sample. Their operation is based on the principle that when a high velocity dust-laden air stream strikes a flat surface at a 90-degree angle, the sudden change in direction and momentum causes the dust to separate from the air and impact on the deflecting obstacle. There can be one or a series of impactor plates. If a series of impingement plates are used, particles are collected in different size ranges. Thus, the particles on each plate can be analyzed for total weight, particle count, and chemical composition. Airflow requirements are high with this type of instrument.

Impinger. An impinger is an apparatus through which air is drawn at high velocity and impinged onto a plate that is immersed in a liquid collection medium. As the particles are impinged on the plate, they lose their high velocity and are wetted by the liquid. The particles trapped in the liquid are counted using a microscope. Impingers are seldom used today for particulate air-sampling.

Elutriators. Elutriators are used in the front of a sampling train to remove coarse particulate matter enabling a filter or other collection device to collect only the smaller size particles. There are two types: horizontal and vertical. They use normal normal gravitational forces to separate particles. Vertical elutriators are typically used for sampling cotton dust.

Personal Air-Monitoring Methods

Sampling instructions for solid-sorbent tube sampler. Use these instructions for personal sampling for substances that are retained on solid sorbent, such as activated charcoal, silica gel, porous polymers, etc.

1. Calibrate each personal sampling pump at the desired flow-rate with a representative solid-sorbent tube in line, using a bubble meter or equivalent flow measuring device.
2. Break the ends of the solid-sorbent tube immediately before sampling to provide an opening at least one half of the internal diameter at each end.
3. Connect the solid-sorbent tube to a calibrated personal sampling pump with flexible tubing with the smaller sorbent section (backup section) nearer to the pump. Do not pass the air being sampled through any hose or tubing before entering the solid-sorbent tube. Position the solid-sorbent tube vertically during sampling to avoid channeling and premature breakthrough.
4. Field blanks are a necessary quality control measure for the analytical laboratory. Prepare the field blanks at about the same time as sampling is begun. These field blanks should consist of unused solid sorbent tubes *from the same lot* used for sample collection. Handle and ship the field blanks exactly as the samples (e.g., break the ends and seal with plastic caps), but do not draw air through the field blanks. Two field blanks are required for each 10 samples with a maximum of 10 field blanks per sample set.
5. Take the sample at an accurately known flow-rate as specified in the NIOSH analytical method for the substance in order to obtain the specified air volume. Typical flow-rates are in the range 0.01–0.2 L/min. Check the pump during sampling to determine that the flow-rate has not changed. If sampling problems preclude the accurate measurement of air volume, discard the sample.
6. Record pertinent sampling data including location of sample, beginning and ending times of the air sampling, initial and final air temperatures, relative humidity and atmospheric pressure or elevation above sea level, and name and job title of the employee.
7. Seal the ends of the glass tube with plastic caps immediately after sampling. Label each sample and mark clearly with waterproof identification.
8. Pack the tubes tightly for shipment to the laboratory, with adequate padding to minimize breakage.
9. It may be necessary to collect bulk samples of the contaminant to properly identify it. Ship bulk samples in a separate package from the air-samples to avoid contamination of the samples. Suitable containers for bulk samples are glass with a polytetrafluoroethylene (PTFE)-lined cap, e.g., 20-mL glass scintillation vials.

Sampling instructions for filter sampler. Use these instructions for personal sampling of total aerosols. These instructions are not intended for respirable aerosol sampling.

1. Calibrate the personal sampling pump with a representative filter in line using a bubble meter or equivalent flow measuring device.
2. Assemble the filter in the two-piece cassette filter holder. Support the filter by a stainless steel screen or cellulose backup pad. Close firmly to prevent sample leakage around the filter. Seal the filter holder with plastic tape or a shrinkable cellulose band. Connect the filter holder to the personal sampling pump with a piece of flexible tubing.
3. Remove the filter holder plugs and attach the filter holder to the personal sampling pump tubing. Clip the filter holder to the worker's lapel. Air being sampled should not be passed through any hose or tubing before entering the filter holder.
4. Prepare the field blanks at about the same time as sampling is begun. These field blanks should consist of unused filters and filter holders from the same lot used for sample collection. Handle and ship the field blanks exactly as the samples, but do not draw air through the field blanks. Two field blanks are required for each 10 samples with a maximum of 10 field blanks per sample set.
5. Sample at a flow-rate of 1–3 L/min until the recommended sample volume is reached. (Use the flow-rate specified by the analytical method.) Set the flow-rate as accurately as possible (e.g., within ±5 percent) using the personal sampling pump manufacturer's directions.
6. Observe the sampler frequently and terminate sampling at the first evidence of excessive filter loading or change in personal sampling pump flow-rate. (It is possible for a filter to become plugged by heavy particulate loading or by the presence of oil mists or other liquids in the air.)
7. Disconnect the filter after sampling. Cap the inlet and outlet of the filter holder with plugs. Label the sample. Record pertinent sampling data including beginning and ending times of sampling, initial and final air temperatures, relative humidity, and atmospheric pressure or elevation above sea level. Record the type of personal sampling pump used, location of sampler, and name and job title of the employee.
8. Ship the samples to the laboratory as soon as possible in a suitable container designed to prevent damage in transit. Ship bulk material to the laboratory in a glass container with a lined cap. Never store, transport, or mail the bulk sample in the same container as the samples or field blanks.

Sampling instructions for filter and cyclone sampler. Use these instructions for personal sampling of respirable aerosols.

1. Calibrate the pump to 1.7 L/min, with a representative cyclone sampler in line using a bubble meter or a secondary flow measuring device, which has been calibrated against a bubble meter. The calibration of the personal sampling pump should be done close to the same altitude where the sample will be taken.
2. Assemble the preweighed filter in the two-piece cassette filter holder. Support the filter with a stainless steel screen or cellulose backup pad. Close firmly to prevent sample leakage around the filter. Seal the filter holder with plastic tape or a shrinkable cellulose band.

3. Inspect the cyclone interior. If the inside is visibly scored, discard this cyclone since the dust separation characteristics of the cyclone might be altered. Clean the interior of the cyclone to prevent reentrainment of large particles.
4. Assemble the two-piece filter holder, coupler, cyclone, and sampling head. The sampling head rigidly holds together the cyclone and filter holder. Check and adjust the alignment of the filter holder and cyclone in the sampling head to prevent leakage. Connect the sampling head to the sampling pump.
5. Clip the cyclone assembly to the worker's lapel and the personal sampling pump to the belt. Ensure that the cyclone hangs vertically. Explain to the worker why the cyclone must not be inverted.
6. Prepare the field blanks at about the same time as sampling is begun. These field blanks should consist of unused filters and filter holders from the same lot used for sample collection. Handle and ship the field blanks exactly as the samples, but do not draw air through the field blanks. Two field blanks are required for each 10 samples with a maximum of 10 field blanks per sample set.
7. Turn on the pump and begin sample collection. If necessary, reset the flow-rate to the precalibration 1.7 L/min level, using the manufacturer's adjustment procedures. Since it is possible for a filter to become plugged by heavy particulate loading or by the presence of oil mists or other liquids in the air, observe the filter and personal sampling pump frequently to keep the flow-rate within ± 5 percent of 1.7 L/min. The sampling should be terminated at the first evidence of a problem.
8. Disconnect the filter after sampling. Cap the inlet and outlet of the filter holder with plugs. Label the sample. Record pertinent sampling data including beginning and ending times of sampling, initial and final air temperatures, and atmospheric pressure or elevation above sea level. Record the type of personal sampling pump, filter, cyclone used, the location of the sampler, and the name and job title of the employee.
9. Ship the samples and field blanks to the laboratory in a suitable container designed to prevent damage in transit. Ship bulk samples in a separate package.

(This information is adapted from the NIOSH *Manual of Analytical Methods* (1984), see Bibliography.)

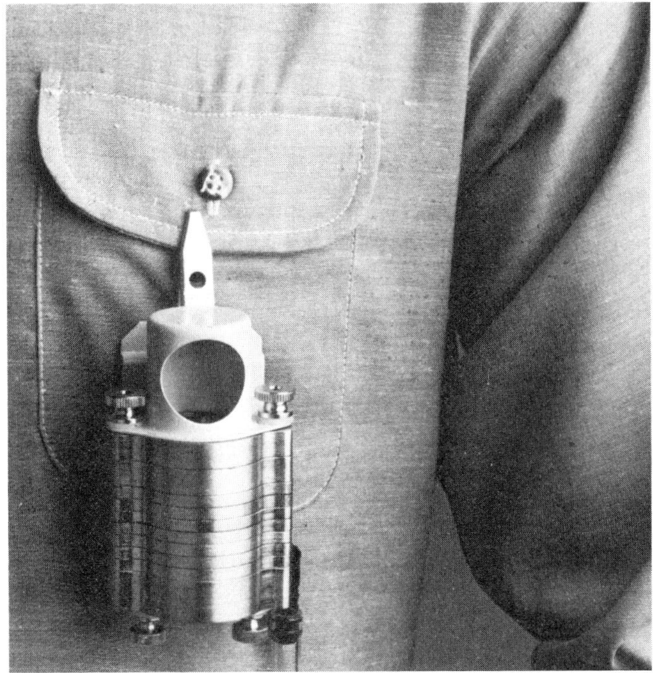

Figure 18-15. Cascade impactor. (Courtesy Anderson Sampler Inc.)

DIRECT-READING INSTRUMENTS

Direct-reading samplers include simple devices, such as colorimetric indicating tubes in which a color change indicates the presence of the contaminant in air passed through the tube, or instruments that are more or less specific for a particular substance. In the latter category are carbon monoxide indicators, combustible gas indicators (explosimeters) and mercury vapor meters.

Some instruments are designed to accomplish another very important task, that of screening, as a means of determining if further and more extensive sampling is required. If process and engineering controls are adequate, it is possible that only simple direct-reading instruments may be needed.

Direct-reading instruments are used for preliminary evaluations for a number of reasons:

- to locate the source of hazardous agents
- to ascertain whether select OSHA air standards may have been violated.

They are also sometimes used as continuous monitors to retain permanently recorded concentrations of contaminants as documentation for use in legal actions, to obtain data needed to inform employees of their exposure, as well as data for epidemiological and other occupational studies.

Colorimetric methods. Direct-reading colorimetric devices use the chemical properties of a contaminant to cause a reaction with a color-producing agent. Liquid reagents and chemically treated papers may be used.

Colorimetric indicator tubes (detector tubes) contain solid granular material, such as silica gel or aluminum oxide, which has been impregnated with an appropriate chemical agent. These tubes are also provided with an appropriate means of drawing sample air through the tube. The length of the stain produced or the ratio of stain length to gel length is compared to a chart to determine the percent concentration of the contaminant.

Unfortunately, not all gaseous substances can be sampled or analyzed by the direct-reading method, so that indirect methods, which require laboratory analysis, must be used.

Other direct-reading instruments rely on physical phenomenon. In general, the sensor portion of the instrument operates on one of the following principles.

Aerosol photometry (Figure 18-16) is used for particulate sampling. The principle employed is the generation of an electrical pulse by a photocell, which detects the light scattered by a particulate. The number of electronic pulses is related to the number of particles counted in the sample gaseous medium.

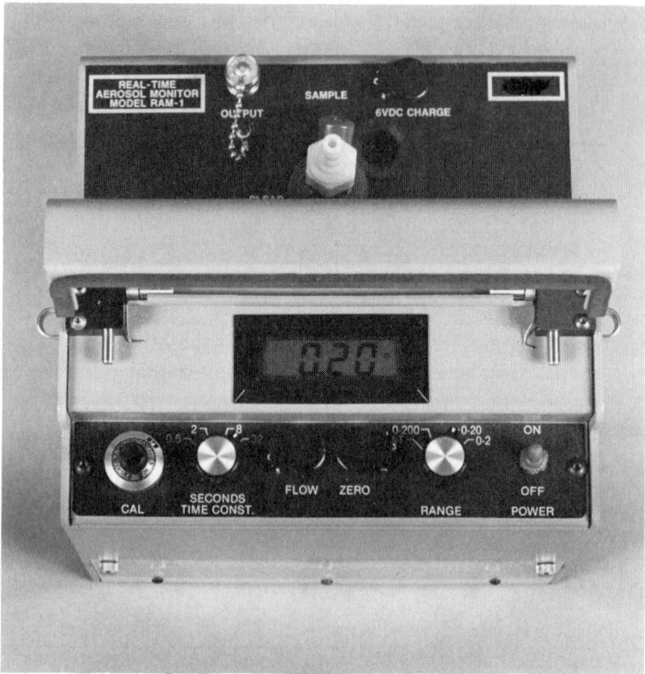

Figure 18-16. Real-time aerosol monitor uses a pulsed light emitting diode in combination with a silicon detector to sense particle concentration. (Courtesy Monitoring Instruments for the Environment, Inc.)

Aerosol photometry can provide only an approximate analysis of particulates according to size. For detailed analysis, the only way, so far, is to examine precollected samples using a microscope or other equally sophisticated method.

Fibrous aerosol monitor. The fibrous aerosol monitor (FAM) is a direct-reading instrument designed to measure airborne concentrations of fibrous materials, such as asbestos and fiberglass with a length-to-diameter ratio greater than three. It uses a helium-neon laser and electro-optical sensors that detect fiber oscillations as the aerosol passes through a rapidly oscillating high-intensity electrical field. Because larger fibers produce narrower pulses than shorter ones, the FAM is able to discriminate between fibers of different length by sensing the sharpness of individual pulses.

A fiber count is digitally displayed in fibers per cubic centimeter during the sampling period. Sample times of 1, 10, 100, or 1,000 minutes can be selected. A standard 37-mm cassette and 0.8-μ mixed cellulose ester filter positioned at the exhaust of the unit can be used for qualitative analysis.

Respirable dust monitor. Ionizing radiation can be used to monitor dust concentrations. The respirable dust monitor uses beta-attenuation. In these instruments, the aerosols are drawn through a cyclone to remove nonrespirable dust and impacted on a suitable surface. The area of impaction is positioned between a beta source and a counter. Attenuation of the beta radiation directly relates to the amount of collected particulates.

Chemiluminescence. Chemiluminescense is a phenomenon whereby some chemical reactions provide a distinct glow, which is analyzed to detect and quantify certain contaminants. It has been used in the monitoring of ozone and oxides of nitrogen. Use of narrow band optical filters enhances the selectivity of such instruments, because interfering contaminants are suppressed.

Combustion. Air containing combustible gas or vapor is passed over a filament heated above the ignition temperature of the contaminant under study. If the filament is part of a Wheatstone bridge circuit, the resulting combustion alters the resistance of the filament. The imbalance caused in the circuit is then used as a measure of concentration. As such, the method is nonspecific, but careful selection of filament temperature may give a degree of specificity.

Combustible gas indicators must be calibrated for their response to the anticipated individual gases or vapors. Since industrial atmospheres rarely have one gaseous contaminant, the meter will respond to all combustible gases present. Other analytical techniques are needed to identify and quantify these gases.

Electrical conductivity. This may be used as a measure of concentration in that some contaminant gases and vapors form electrolytes in aqueous solutions. In others, positively and negatively charged particles are produced in the solution. The resulting increase in conductivity is then used as a measure of concentration. The method is generally nonspecific, but previous knowledge of the environment may help in identifying the contaminant.

Thermal conductivity. In this method the specific heat of conductance of a gas or vapor can be used as a measure of its concentration in a carrier gas, such as air, argon, helium, hydrogen, or nitrogen. This method is quite nonspecific and finds its best use in estimation of gaseous contaminants coming out of a chromatographic column.

Coulometry. This measures the number of electrons in terms of coulombs transferred across an electrode-solution interface. This method is basically nonspecific, but it can be made specific by adjusting concentration, pH, and composition of the electrolyte.

Flame ionization. Flame ionization provides a sensitive method of detection of various hydrocarbons. The great increase in production of ions by introducing a volatile carbon compound into a hydrogen flame is the principle involved. A loop of platinum held above the flame serves as the collector electrode. The current carried across the electrode gap is directly proportional to the number of ions produced. This detector responds to all organic compounds except formic acids, and its response is greatest with hydrocarbons. The method, abbreviated FID, is often used in combination with gas chromatography.

Potentiometry. This is a change in pH value or the hydrogen-ion concentration in a solution caused by a gas reacting with a reagent in the solution as a measure of the concentration of the gas. The resulting potentiometric change—as a result of change in hydrogen ion concentration—is sensed by a galvanic cell known as a pH electrode. This method is also nonspecific, but, in practice, a certain amount of specificity may be introduced through the choice of reagents in the solution.

Air Sampling Worksheet

U.S. Department of Labor
Occupational Safety and Health Administration

1. Reporting ID	2. Inspection Number	3. Sampling Number 0741041

4. Establishment Name	5. Sampling Date	6. Shipping Date

7. Person Performing Sampling (Signature)	8. Print Last Name	9. CSHO ID

10. Employee (Name, Address, Telephone Number)

14. Exposure Information | a. Number | b. Duration

c. Frequency

15. Weather Conditions

16. Photo(s) Y

11. Job Title	12. Occupation Code

13. PPE (Type and effectiveness)

17. Pump Checks and Adjustments

18. Job Description, Operation, Work Location(s), Ventilation, and Controls

Cont'd

19. Pump Number:

Sampling Data

20. Lab Sample Number						
21. Sample Submission Number						
22. Sample Type						
23. Sample Media						
24. Filter/Tube Number						
25. Time On/Off						
26. Total Time *(in minutes)*						
27. Flow Rate ☐ *l/min* ☐ *cc/min*						
28. Volume *(in liters)*						
29. Net Sample Weight *(in mg)*						

30. Analyze Samples for:	31. Indicate Which Samples To Include in TWA, Ceiling, etc. Calculations					

32. Interferences and IH Comments to Lab

33. Supporting Samples

a. Blanks:

b. Bulks:

34. Chain of Custody	Initials	Date
a. Seals Intact?	Y N	
b. Rec'd in Lab		
c. Rec'd by Anal.		
d. Anal. Completed		
e. Calc. Checked		
f. Supr. OK'd		

Case File Page / of

OSHA-91A (Rev. 1/84)

Figure 18-17. OSHA air-sampling form.

Pre-Sampling Calibration Records

Pre					
	35. Pump Mfg. & SN	38. Flow Rate Calculations			
	36. Voltage Checked? ☐ Yes ☐ No				
	37. Location/T & Alt.				
		39. Flow Rate	40. Method ☐ Bubble ☐ PR	41. Initials	42. Date/Time

Post-Sampling Calibration Records

Post			
	43. Location/T & Alt.	44. Flow Rate Calculations	
	45. Flow Rate	46. Initials	47. Date/Time

Sample Weight Calculations

48. Filter No.						
49. Final Weight *(mg)*						
50. Initial Weight *(mg)*						
51. Weight Gained *(mg)*						
52. Blank Adjustment						
53. Net Sample Weight *(mg)*						

54. Calculations and Notes:

☆ U.S. GOVERNMENT PRINTING OFFICE: 1984—448-944

Photometry. Photometry is a measurement of the relative radiant power of a beam of radiant energy in the visible, ultraviolet, or infrared region of the electromagnetic spectrum. The beam will be attenuated as a result of passing through a solution or a gas-air mixture containing a substance such as mercury vapor, ozone, or benzene vapor. The more sophisticated technique is termed spectrophotometry.

Polarographic analysis. This is based on the electrolysis of a sample solution using an easily polarized microelectrode (the indicator electrode) and a large nonpolarizable reference electrode. This method provides both qualitative and quantitative information.

Gas chromatography (GC). This is a physical process for separating the components of complex mixtures, and it is widely used for analysis of air-samples collected in the workplace or in the community. Many plants use a portable gas chromatograph for vinyl chloride detection. Such a device consists of the following:

1. a carrier gas supply with a pressure regulator and flow meter
2. an injection system for introduction of a gas or vaporizable sample.
3. a stainless steel, glass, or copper separation column
4. a heater and oven assembly
5. a detector
6. a recorder for the chromatograms.

The packing materials used in the GC columns may be diatomaceous earth or many other solid materials.

When a sample in a carrier gas is introduced into a GC column, each compound in the mixture, in the course of its travel through the packed column, will be retained by the stationary phase in a unique manner. The result is that each gaseous compound will emerge from the tube at a unique time. The emerging gaseous substance can then be detected using the flame ionization detector (FID), electron capture detector, or other methods.

RECORD-KEEPING

Complete records of all calibration procedures, data, and results must be maintained. This record should include pump calibration, sample data, instrumentation used, temperature and pressure, and the employee's name, job title, and social security number, among other variables. Figure 18-17 shows an OSHA air-sampling form used by compliance officers to collect this information.

The OSHA Access to Employee Exposure and Medical Records (29 CFR 1910.20) Standard requires that employee exposure records must be preserved and maintained for at least 30 years. Background data to environmental (workplace) monitoring or measuring, such as laboratory reports and worksheets, need only be retained for one year, so long as the sampling method (sampling plan), a description of the analytical and mathematical methods used, and a summary of other background data relevant to interpretation of the results obtained, are retained for at least 30 years.

This Standard also gives employees and their representatives the right to these exposure and medical records.

BIBLIOGRAPHY

American Conference of Governmental Industrial Hygienists (ACGIH). *Air Sampling Instruments for Evaluation of Atmospheric Contamination,* 6th ed. Cincinnati: ACGIH, 1983.

American Industrial Hygiene Association. *Analytical Guide Series.* Akron, OH: AIHA, (updated periodically).

APHA Intersociety Committee. *Methods of Air Sampling and Analysis,* 2nd ed., Katz, M., Ed. Washington, DC: American Public Health Association (APHA), 1977.

Beaulieu, H.S., Buchan, R. *Quantitative Industrial Hygiene,* New York: Garland STPM Press, 1981.

Leidel, N.A., Busch, K.A., and Lynch, J.R. *Occupational Exposure Sample Strategy Manual,* Pub. No. PB-274-792. Springfield, VA: National Tech Information Service (NTIS), 1977.

Merchant, J.A., ed. *Occupational Respiratory Diseases,* DHHS (NIOSH) Publication No. 86-102. Washington, DC: Superintendent of Documents, U.S. Government Printing Office, 1986.

National Institute for Occupational Safety and Health (NIOSH). *Manual of Analytical Methods,* vols 1, 2. Springfield, VA: NTIS, 1984.

NIOSH. *The Industrial Environment—Its Evaluation and Control,* Washington, DC: U.S. Government Printing Office, 1973.

OSHA: *OSHA Instruction CPL2-2. 20A* (updated regularly). Washington, DC: OSHA, 1984 (Available from U.S. Government Printing Office).

Patty, F.A., ed. *Industrial Hygiene and Toxicology,* 3rd rev. ed. Vol. I, *General Principles.* New York: Interscience Publishers, Inc., 1978.

Patty's Industrial Hygiene and Toxicology. Vol. III, *Theory and Rationale of Industrial Hygiene Practice,* 2nd ed. 3A. *The Work Environment,* edited by Lewis J. Cralley, and Lester V. Cralley. New York: John Wiley & Sons, 1985.

19

Direct-Reading Gas and Vapor Monitors

by Joseph E. Zatek, CSP, CIH, CHCM, CHM

THE CONCENTRATION OF GASES AND VAPORS IN AIR can be determined readily with the use of direct-reading instruments. Sampling and analysis are carried out within the direct-reading instrument, and the required information can be read directly from a dial or indicator. Many direct-reading instruments also feature digital displays that indicate the percentage of gas or of the vapor in the air on a numerical digital scale (Figure 19-1).

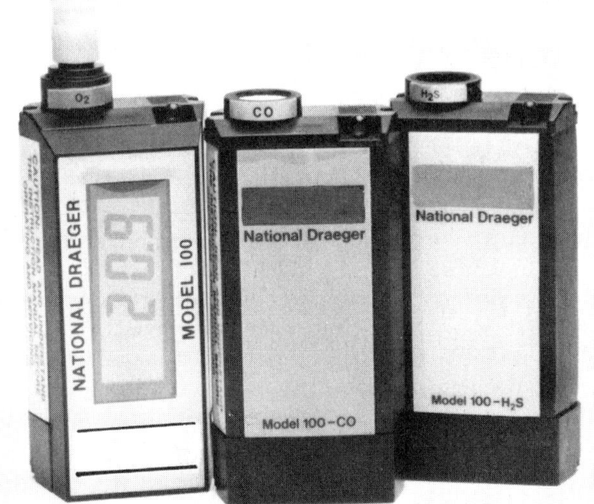

Figure 19-1. Examples of direct-reading instruments that feature digital display are shown. (Courtesy National Draeger, Inc.)

The ideal direct-reading instrument should be capable of sampling air in the breathing zone of the worker and should specify the concentration of the substances under investigation—either as an instantaneous concentration or as a time-weighted average, depending on what information is required. Alternatively, the reading may be in terms of the percentages of an appropriate standard. In most cases, provision for keeping permanent records of the reading is essential.

Direct reading instruments for vapors and gases include the following types:

Colorimetric-type devices.

SOLID. A known volume of air is passed through a small-diameter glass tube containing porous solid granules impregnated with a reagent at a fixed rate. The reagent reacts with the vapor or gas contaminant and then changes color. The degree or shade of color or the length of the color stain is related to the vapor or gas concentration.

PAPER TAPE SAMPLERS. A known volume of air is passed through paper impregnated with a reagent that reacts with vapor or gas and then changes color. The degree of color indicates vapor or gas concentration.

LIQUID. A known volume of air is bubbled through a liquid reagent that reacts with the vapor or gas. An indicator that changes color when the reagent is consumed by a definite quantity of vapor or gas indicates the vapor or gas concentration in the sample.

Thermal.

CONDUCTIVITY. The specific heat of the conductance of vapor or gas is measured to indicate the vapor or gas concentration.

COMBUSTION. The change in electrical resistance of the heated filament in a Wheatstone Bridge is caused by combustion of the vapor or gas. This change is measured to determine vapor or gas concentration.

OTHERS. Other measurements are based on potentiometric and coulometric study, the study of thin film electrochemical cells, infrared analysis, and polarographic study.

Gas chromatography. Various vapors and gases migrate differentially in a porous sorption medium contained in a column. Separated vapors and gases are later desorbed by heat and carried by an inert gas to a detector, such as an ionization or electromagnetic device, for measurement. Each device type will be discussed in turn.

COLORIMETRIC-TYPE DEVICES

Direct-reading colorimetric devices measure the reaction of a contaminant with a color-producing agent. The colorimetric indicator or detector tube has been widely used by industrial hygienists, safety engineers, and others. Its simplicity of operation, low initial cost, and versatility regarding detection of numerous contaminants make this a popular instrument for field use. Nevertheless, like all instruments, these devices are limited with regard to applicability, specificity, and accuracy. The user must be familiar with these critical limitations if proper judgments are to be made.

Detector tube systems

Active. Basically, the active colorimetric detector tube system is composed of two elements: the pump and the colorimetric indicator tubes (Figure 19-2). The bellows and piston-type pumps are designed to draw a fixed volume of air with each full stroke. The indicator tube is an hermetically sealed glass tube that contains solid granular material such as silica gel, alumina, or pumice. This material has been impregnated with an agent that reacts when air containing a specific contaminant or group of contaminants is drawn through the tube (Figure 19-3).

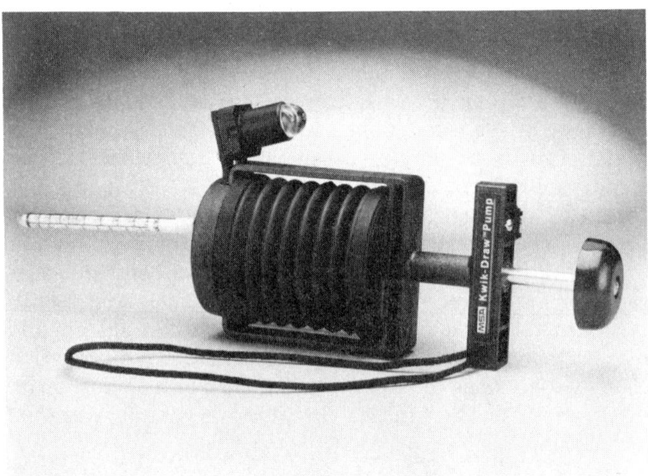

Figure 19-2. Bellows-style sampling pump designed for one-handed operation, shown with detector tube attached to its intake end. (Courtesy MSA)

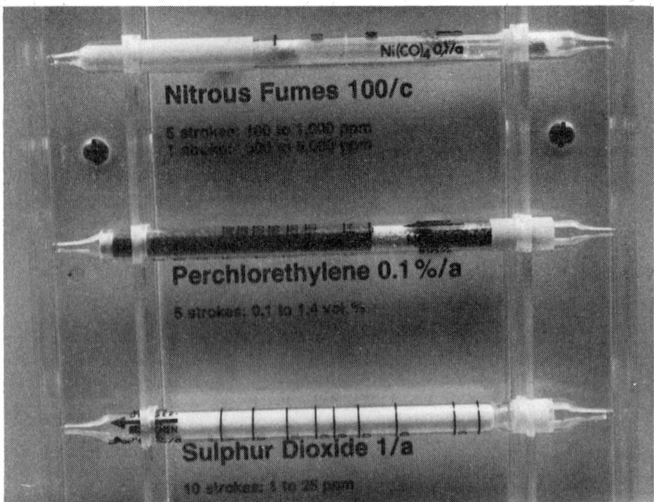

Figure 19-3. Examples of length-of-stain tubes intended for short sampling periods. (Courtesy National Draeger, Inc.)

To conduct a test, the two sealed ends of the indicator tube are broken off, the tube is inserted into the tube holder, and the specified volume of air is drawn through the tube. A specified period of time is permitted for each pump stroke to completely draw its full volume of air.

Passive. Sampling for contaminants can also be performed without using a pump to draw air into the detector tube. The instruments used in this kind of sampling are commonly referred to as passive detectors. Passive colorimetric indicator tubes, which work by diffusion, provide a simple and economical method for measuring the exposure of workers to toxic vapors (Figure 19-4). The user simply breaks off one end of the detector tube, inserts the tube into a holder, and attaches the holder to clothing. No pump is required. Some tubes can be used to sample the atmosphere for several hours, while others are intended for only short sampling periods. Passive colorimetric indicator badges, which work on the same principle as the tubes, are also used as an inexpensive way to quantitatively determine vapor and gas concentrations in ambient air (Figure 19-5).

Sampling for contaminants is more cost-efficient when these types of instruments are used. It is more economical to supply workers with their own detector tube or badge, whenever practical, than it is to supply workers with a tube and a pump.

Length-of-stain type. The length of the stain produced or the ratio of stain length to gel length is compared against a chart to determine the percentage of concentration (Figure 19-3). With some length-of-stain tubes, the stain front may not be sharp, and thus the exact length of stain cannot be readily determined. One can make an approximate estimate by rolling the tube between the fingers to determine the variation of the stain front. The decision to use the visual average, the maximum stain length, or some other means of determining the stain end will depend on the user's experience. It would be helpful to obtain the results of the calibration test performed on known concentrations before the tubes are used in the field.

Color-change type. Colorimetric indicator tubes and badges of the color-change type are very similar to length-of-stain tubes. The primary difference is that when the contaminant gas or

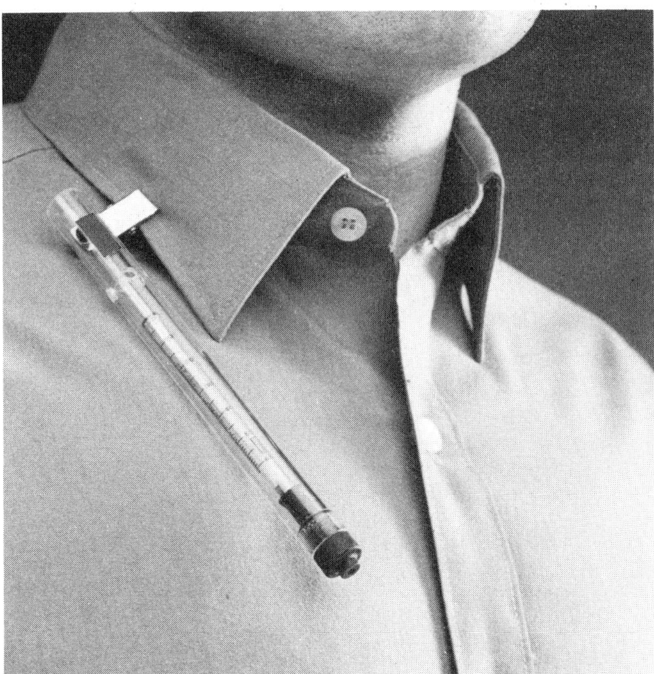

Figure 19-4. Example of a passive length-of-stain tube intended for sampling periods of several hours. (Courtesy MSA)

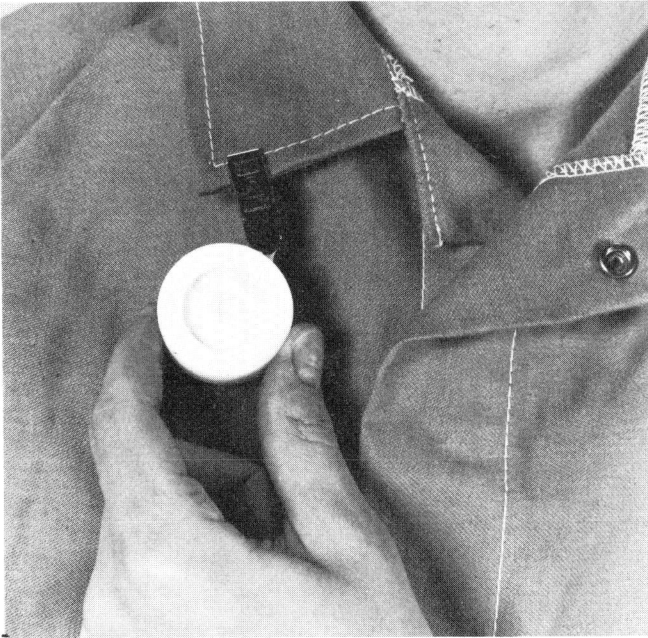

Figure 19-5. An example of one type of passive, color-change dosimeter badge. The badge is clipped to a shirt or lapel and provides an immediate visual indication of vapor concentrations in ambient air. (Courtesy MSA)

vapor reacts with the indicating reagent in a color change device, a relatively uniform color is produced in the reactive section. A color comparison must be made between the indicator tube or badge and the standard color chart that is provided with each instrument of this type. It is also important to recognize that some color stains fade or change with time; thus, the readings should be made as promptly as possible or in accordance with the manufacturer's recommendations.

The visual judgment required to read color-change detector tubes and badges depends to a great degree on the color perception of the observer and the lighting conditions. The exposed detectors should be examined in an area with daylight or incandescent illumination. Fluorescent lighting may not yield a good match for some color systems. Mercury vapor lamps should generally be avoided because the color change may not be visible, and the color stain end is difficult to pinpoint.

Flow-rate

The flow-rates of the pumps used in active color-change and length-of-stain instruments must be checked regularly. Airborne dusts and lint will quickly clog the orifices of these instruments, cause low flow-rates, and generate inaccurate measurements. Flow-rates for color-change and length-of-stain devices must be maintained in accordance with the manufacturer's operating instructions. Proper flow-rate ensures the adequate residence time of the sample in the device and provides sufficient time for the contaminant to react with the chemicals in the detector tube. To obtain meaningful test results, the residence time must be the same as that used to develop the color chart or length-of-stain chart supplied by the manufacturer.

Specificity

Each detector system, both passive and active, is designed to measure a specific gas, such as hydrogen sulfide, chlorine, mercury vapor, nitrogen dioxide, carbon dioxide, or hydrocarbons. Because no device is completely specific for the substances of interest, care must be taken to ensure that interferences do not invalidate the sampling results. Many common gases and vapors react with the same chemicals or have similar physical properties; thus, the instrument may yield falsely high or low readings for the substance being sampled.

Specificity is one of the primary considerations used to select a detector system. However, only a few tubes are limited to the detection of a single gas or vapor. A preconditioning section is employed in a number of these systems to:

1. remove potentially contaminating substances;
2. react with the gas or vapor to convert the compound to a more suitable reacting compound; and
3. react with the gas or vapor with the release of a new gas or vapor that can be measured by the second section.

Interference from other substances in the sample air should always be given consideration. In most cases, the manufacturer has identified interfering substances and conditions and has included this information in the instruction sheet enclosed with the tubes.

Chemical reactions that occur in the detector tubes are temperature-dependent. Ideally, the tubes should be used at room temperature. Any hot or cold extremes will tend to alter the reliability of the detector tube, and false results could be reported. The volume of air going through the detector tube is also temperature-dependent.

Interchanging tubes obtained from various manufacturers will lead to erroneous results. Errors occur because the sampling rate of the various pumps is not the same nor is the reaction rate of the chemical reagents in the indicator tubes. Each manufacturer produces, calibrates, and sells equipment as an integral system and never advocates interchanging tubes or pumps produced by other manufacturers.

Shelf life

The shelf life of detector tubes is a critical consideration since the tubes may not be used very often and, therefore, may not be exhausted within the manufacturer's indicated expiration date. Frequently, the tube life can be extended by storage under refrigeration. Freezing temperatures should not adversely affect a tube's shelf life; however, the tubes must be warmed to room temperature before use. Detector tubes should be stored at temperatures below 30 C (86 F) and never in direct sunlight.

Each tube manufacturer provides specific instructions concerning the operating principles of the sampling kit as well as conversion tables to help interpret the concentration of the contaminant. Some brands of detector tubes are calibrated in terms of milligrams per liter by the manufacturer. Conversion from milligrams per liter to parts per million (ppm) at 25 C (77 F) and 760 mmHg (considered normal atmospheric pressure) can be performed by using the following equation:

$$\text{ppm} = \frac{\text{milligrams per cubic meter}}{\text{molecular weight (grams per mole)}} \times 24.5$$

Long-term detector tubes

The tubes shown in Figure 19-3 are intended for brief sampling periods using hand aspirators to sample the air. The tube shown in Figure 19-4 is a passive detector tube intended for prolonged sampling over periods of several hours. Since the passive tube works by diffusion, no pump or aspirator is necessary. However, in heavy dust environments, passive tubes could be ineffective. In such a case, a miniature air pump, which samples at a rate of 5 mL/min, can be used (Figure 19-6). Chemical detector tubes that are used with a miniature air mover are a comparatively recent development and should correct an important omission in the use of detector tubes; that is, the need to test conformity to the 8-hour time-weighted average (Figure 19-7).

Obviously, performing reliable tests with indicating tubes requires thorough knowledge of their limitations and careful use. Experience has shown that the following measures help to minimize some errors:

1. Test each batch of tubes with a known concentration of the air contaminant that is to be measured.
2. Read the length of stain in a well-lighted area.
3. Read the longest length of stain if stain development is not sharp or even.
4. Observe the manufacturer's expiration date and discard outdated tubes.
5. Store detector tubes in accordance with the manufacturer's recommendations.
6. Refer to the manufacturer's data for a list of interfering materials.

Certification of chemical detector tube systems

Prior to September 30, 1985, many chemical detector tube systems were certified by the National Institute for Occupational Safety and Health (NIOSH) of the U.S. Department of Health and Human Services. The NIOSH program was designed to ensure that commercial detector tube units comply with established performance specifications. However, as a result of budget cuts, the NIOSH prioritized its safety product certification activities, thus eliminating the certification program for detector tubes.

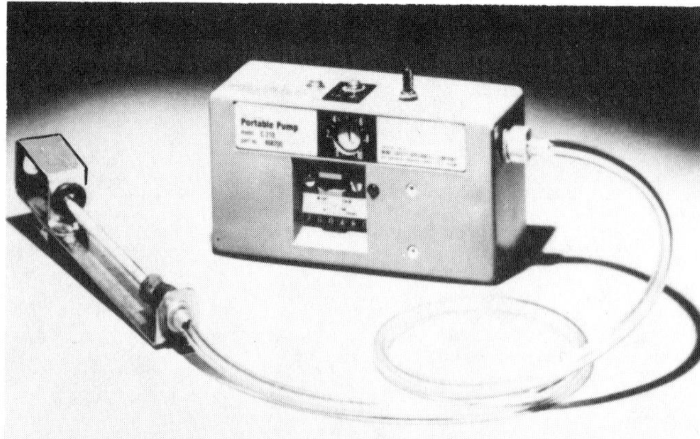

Figure 19-6. Battery-operated pump used for prolonged sampling with detector tube in a holder that attaches to a worker's shirt. (Courtesy MSA)

Figure 19-7. Battery-operated pump used for prolonged sampling periods with detector tube in the instrument. (Courtesy National Draeger, Inc.)

To ensure the program's continuance, a private organization, known as the Safety Equipment Institute (SEI), has filled the void left by the NIOSH. The SEI now certifies detector tube systems through a program similar to that established by the NIOSH, involving product testing and quality assurance audits conducted by designated, third-party independent laboratories. In fact, the SEI program adheres to the same test standard established by the NIOSH program: Title 42 *CFR* Part 84 of the *Code of Federal Regulations.* The SEI offers certification of the manufacturers' product models, and grants the right to use the SEI certification mark if: (1) the testing laboratory has determined that the product models submitted have been tested according to the appropriate standard; and (2) the quality assurance auditor has determined that the manufacturer has complied with SEI quality assurance requirements.

Quality control. The quality control practiced by the manufacturer of the detector tube can critically limit the tube's use. Obviously, improper preparation of the reagents and solid support can cause incorrect results to be obtained. Improper packing within the tube can either be too tight, causing restriction of airflow, or can be too loose, allowing the material to shift while being handled. Variations in the particle size of the detecting material in the tube can lead to striations in the stain and cause a very indefinite stain line.

Calibration. The calibration by the manufacturer is critical as is the preparation of the standard charts. Most manufacturers have comprehensive quality control programs that take these factors into account. However, it is the responsibility of the tube's user to make sure that the entire detector tube system is reliable.

Accuracy requirements. The SEI certifies a manufacturer to produce a gas detector tube unit if it meets the minimum requirements set forth in the regulations (basically ±35 percent accuracy at one half the exposure limit and ±25 percent at one to five times the exposure limit). The quality of future production lots is secured by a quality assurance plan, which the SEI approves as part of the certification process. Adherence to the quality assurance plan is verified by periodic plant inspections and by testing samples obtained from actual inventory.

COLORIMETRIC TAPE SAMPLERS

A chemically impregnated paper tape has been in use since the early 1950s. The first units were developed for use in the detection of hydrogen sulfide (H_2S). A lead acetate reagent system was impregnated into filter paper that produced a dark stain (lead sulfide) when exposed to H_2S. The concentration of H_2S was then determined by measuring the transmission of light energy through the stained paper.

Principle of operation

In modern devices, the chemically treated paper tape is drawn at a constant rate over the sampling orifice, and the contaminant reacts with the chemical to produce a stain. The intensity of the stain can be measured by a reflectometer, and the result can be displayed as a function of concentration.

Figure 19-8 shows a section of the double-track, impregnated paper tape. Only the top track is exposed to the contaminant; the bottom or reference track, is kept unexposed. Therefore, only the top track becomes stained.

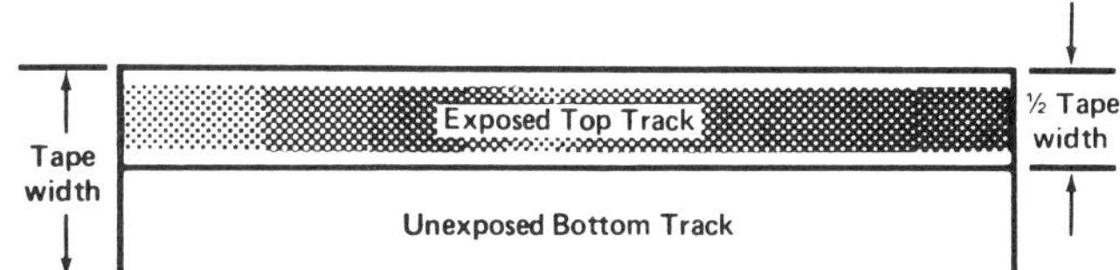

Figure 19-8. A section of the double-track, impregnated paper tape with only the top half exposed to the contaminant is shown. The bottom half, or reference track, is kept unexposed. (Courtesy MDA Scientific, Inc.)

By directing light of equal intensity (from a common source through matched fiber optics) to both the top and bottom track and by mounting a set of matched photoelectric detectors at an angle of 45 degrees, the difference in reflected light can be measured. The system thereby compensates for slight tape variations. This is illustrated in Figure 19-9, which schematically illustrates the general principle of operation.

This diagram also shows the capstan-driven cassette, which constantly moves the impregnated paper tape past the exposure orifice and readout section of the optical block/gate assembly.

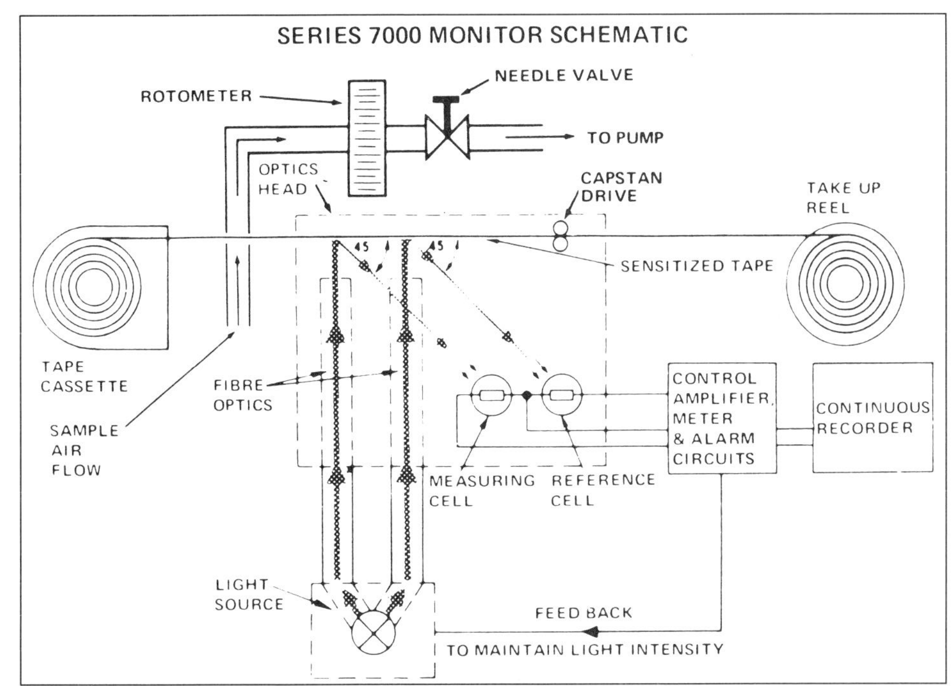

Figure 19-9. A schematic of the Series 7000 monitor, showing the use of reflected light and its measurement. The capstan-driven cassette provides constant movement of the impregnated paper tape. (Courtesy MDA Scientific, Inc.)

While the tape moves, a constant flow of the sample air is aspirated through the porous impregnated tape, which is controlled by means of the self-contained pump and flow controller.

A typical monitor in the series, together with the strip chart recorder is shown in Figure 19-10.

Field calibrations can be accomplished at any time by use of a test strip provided with each monitor. The strip has a calibrated stain, which is equivalent to the stain that would be produced by a known concentration of the contaminant.

Figure 19-10. A continuous colorimetric monitor is depicted. (Courtesy MDA Scientific, Inc.)

ELECTRICAL DIRECT-READING INSTRUMENTS

Dangerous concentrations of flammable gases and vapors can occur in many operations. In some cases, the complete elimination of the hazard is possible, but in many working environments, the presence of flammables may be unavoidable. In these cases, it is essential to determine whether the concentration of the flammable gas or vapor is an explosion hazard. Instruments for both the detection and measurement of combustible gases and vapors are described in this section.

Introduction

Portable direct-reading air-sampling instruments have eliminated much of the guesswork in detecting the presence of flammable gases or vapors. Before the development of these sampling devices, it was necessary to collect a sample of air from the suspected atmosphere and take it to a laboratory for analysis.

This required the services of experienced technicians and a considerable investment in laboratory equipment. By the time the analysis was completed, the concentration of the suspected atmosphere could have changed considerably.

Direct-reading instruments enable the operator to obtain immediate indications of gas or vapor concentration by reading a digital display meter or a meter dial. This does not mean, however, that merely reading a meter constitutes a valid test. On the contrary, the operator must be thoroughly familiar with the calibration, use, and limitations of the instruments and devices.

One of the most useful instruments of the direct-reading type is the hot wire, or combustible gas indicator. Instruments of this type were designed to help detect the presence of explosive or combustible gases in the air.

Combustion theory

Combustion processes are usually explained in terms of the fire pyramid shown in Figure 19-11. Combustion results when oxygen and fuel combine and heat is released. For example, the combustion of methane (CH_4) in oxygen (O_2) with the resultant formation of carbon dioxide (CO_2), water (H_2O), and heat (Δ), may be shown by the chemical equation:

$$CH_4 + 2\ O_2 \rightarrow CO_2 + 2\ H_2O + \Delta$$

fuel + oxygen → carbon dioxide + water + heat

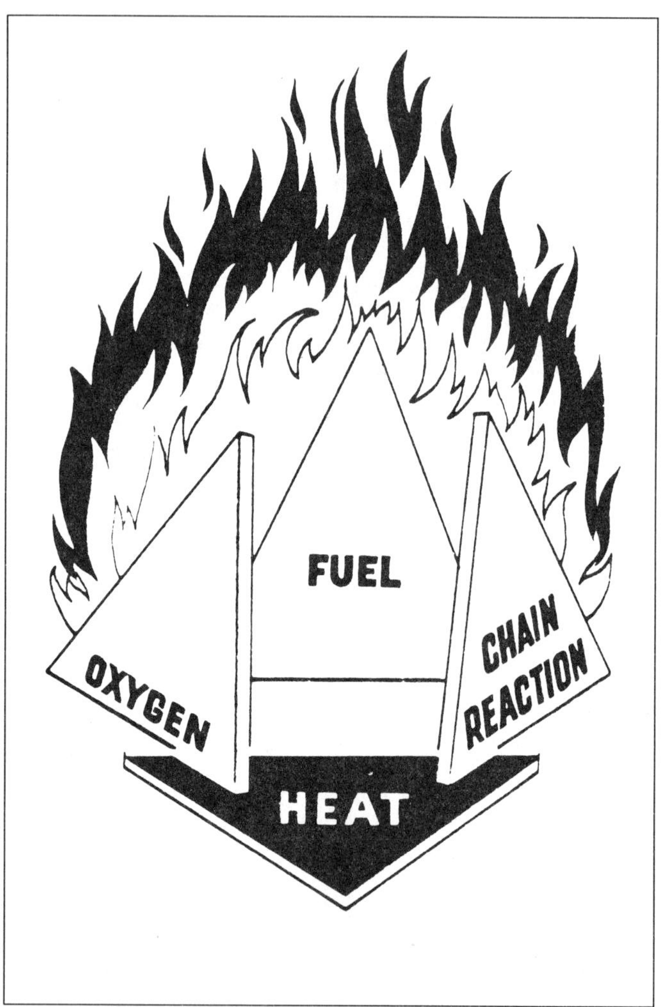

Figure 19-11. The "fire pyramid." Oxygen, heat, fuel, and chain reactions are necessary components of a fire. Speed up the process, and an explosion results.

Reaction mechanism. The mechanism of the combustion reaction involves a number of intermediate steps. The free radicals, or intermediate combustion products, react with each other and with CH_4 and O_2 until they are ultimately converted to CO_2 and water. Heat is released in the process.

As the combustion of methane proceeds, the rate of energy release is balanced by energy dissipation such that a limiting rate of reaction is reached. The example used to demonstrate that the combustion of methane in air produces carbon dioxide and water shows that when methane is burned, the com-

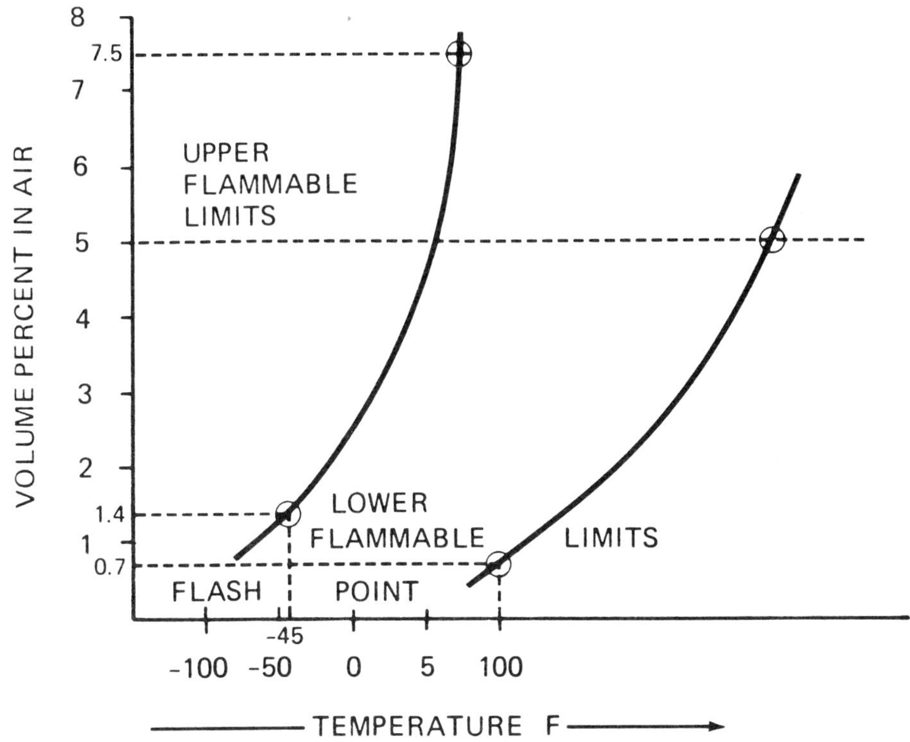

Figure 19-12. Physical steps involved in the combustion behavior of kerosene-air and gasoline-air.

bustion products expand away from the combustion zone taking with them kinetic energy in the form of heat. A limit is placed on the temperature and pressure of the system, and the rate of reaction reaches an equilibrium value.

Explosions. If methane and oxygen are placed in a closed container and the reaction is initiated, an explosion results. In this case, the products of combustion and the heat thereby produced are not removed from the reaction zone. The temperature of this system would continue to rise. Since the system is confined, the pressure rises also. Both effects increase the rate of reaction. There is no mechanism for dissipating the energy; therefore, the rate of reaction increases more rapidly until all of the reactants are consumed.

Rapid oxidations, such as those encountered with gasoline-air mixtures, are explosive if enough air is premixed with the gasoline vapor in a confined space to permit an essentially complete reaction. The reaction becomes explosive because the oxygen atoms in air are in close contact with the carbon and hydrogen atoms and can therefore react instantaneously. In other circumstances, gasoline simply burns because the rate of the reaction is controlled by the diffusion of oxygen to the combustion zone. Many chemical reactions can be potentially explosive if the heat buildup proceeds too rapidly, and the reaction rate escalates.

Flash point. The flash point is the minimum temperature at which the vapor concentration above a liquid is high enough to propagate a flame front when a source of ignition is present. Each material also has a characteristic lower flammable limit, which is the minimum volume percentage of the material in air that can be ignited.

Explosive or flammable range

Each material also has an upper flammable limit, which is the maximum volume percentage of the material in air that can be ignited. For gasoline, this value is 7.6 percent. The significance of the upper flammable limit is not always as obvious as that of the lower flammable limit. If the upper flammable limit is exceeded, the mixture cannot be ignited and sustain combustion. The balanced chemical equation for the combustion of gasoline represented by octane is shown as:

$$2\ C_8H_{10} + 21\ O_2 \rightarrow 16\ CO_2 + 10\ H_2O + \text{heat}$$

The percentage of gasoline in the air under stoichiometric combustion conditions is determined by
(1) calculating the nitrogen content:

$$\frac{21 \text{ moles of } O_2}{21 \text{ percent of } O_2 \text{ in air}} \times 79 \text{ percent nitrogen in air} = 79 \text{ moles of } N_2$$

(2) calculating the percentage of gasoline in the air:

$$\frac{2 \text{ moles of gasoline}}{2 \text{ moles of gasoline} + 21 \text{ moles of } O_2 + 79 \text{ moles of } N_2} \times 100 = 1.96\%$$

If the proportion of gasoline molecules is increased greatly, insufficient oxygen will be present to support combustion. Similarly, if the proportion of gasoline molecules is decreased, insufficient gasoline will be present to sustain combustion. Stable combustion of gasoline is sustained only within a narrow range of concentration limits.

The sequence of events that occur with any fuel-air mixture are the same as those described for the gasoline-air mixture. Fuel-oil-air mixtures are basically the same except that the flash point of kerosene, for example, is approximately 38 C (100 F). The lower flammable limit is 0.7 percent of the volume, and the upper flammable limit is 5.0 percent of the volume.

A comparison of the similarities and differences involved in the interrelationship between physical properties and combustion behavior for No. 1 fuel oil and gasoline is summarized in Figure 19-12.

Lower flammable (explosive) limit (LEL). When certain proportions of combustible vapor are mixed with air and a source of ignition is present, an explosion can occur. The range of concentrations over which this will occur is called the flammable (or explosive) range. This range includes all concentrations in which a flash will occur or a flame will travel if the mixture is ignited. The lowest percentage at which this occurs is the lower flammable (or explosive) limit (LEL), and the highest percentage is the upper flammable (or explosive) limit (UEL). Mixtures below the LEL are too lean to ignite, and mixtures above the UEL are too rich. (Care must be taken, however, when a mixture is too rich because dilution with fresh air could bring the mixture into the flammable or explosive range.)

On the simplest type of instrument (an explosimeter), only one scale is provided, usually with readings from 0–100 percent of the LEL. However, the detectable changes produced by combustion are too small to be measured accurately in the presence of the low concentrations of contaminants usually encountered in the evaluation of potential health hazards. For example, the LEL of even the most explosive gas is approximately 1 percent or 10,000 ppm, which is well in excess of the threshold limit value for any gas.

Instrument design

Several manufacturers make explosimeters or combustible gas indicators. Although the indicators differ somewhat in design and operating features, their operation is based on the fact that a measurable amount of heat is released when a combustible gas or vapor is burned. Most meters contain a battery-operated electrical circuit, known as a Wheatstone Bridge, which is balanced by means of controls on the outside of the instrument. A schematic illustration of the basic flow system and wiring diagram of a Wheatstone Bridge is shown in Figure 19-13.

Wheatstone Bridge circuit. In one part of the bridge, the air being sampled is passed over filaments that have been brought to a high temperature. If the air contains a combustible gas or vapor, the heated filaments cause combustion and additional heat is released, thereby increasing the electrical resistance of the filaments.

Another part of the bridge contains similar filaments that have been sealed. These are heated in identical fashion although not in the air stream. The sealed filaments cancel out all changes in electrical current and resistance due to temperature variations in the wire or to characteristics of the instrument itself. The net effect is that the change in resistance to the electrical current flow in the filaments in the air stream is only due to the presence of combustible gases. These changes in the electrical current are registered as "percent LEL" on the instrument's meter.

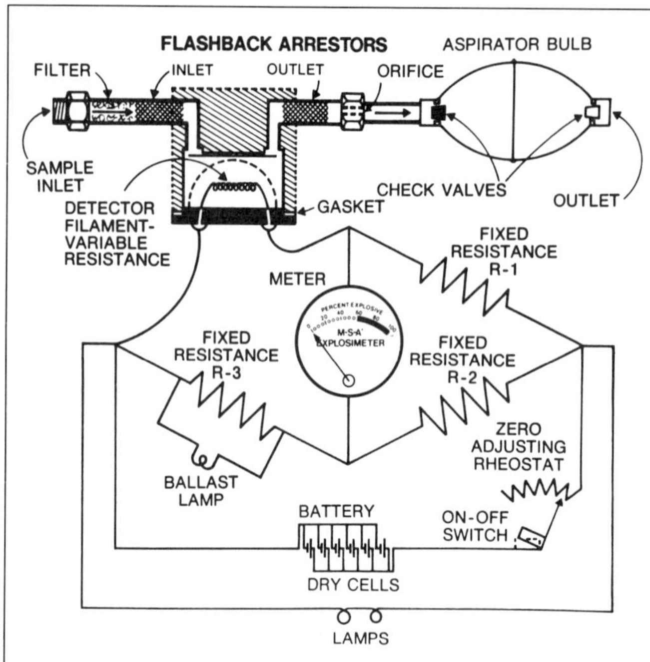

Figure 19-13. A schematic diagram of a typical hot-wire sampling device. (Courtesy MSA)

The manufacturer's instructions for operating a combustible gas indicator should be carefully reviewed before the device is used. In general, all explosimeters require a brief initial warm-up period so that the batteries can heat the filaments.

Most combustible gas indicators are equipped with a length of sampling tubing with a metal probe at the end. The probe is held at the sampling point, and, a few seconds later, the response can be read on the meter (Figure 19-14).

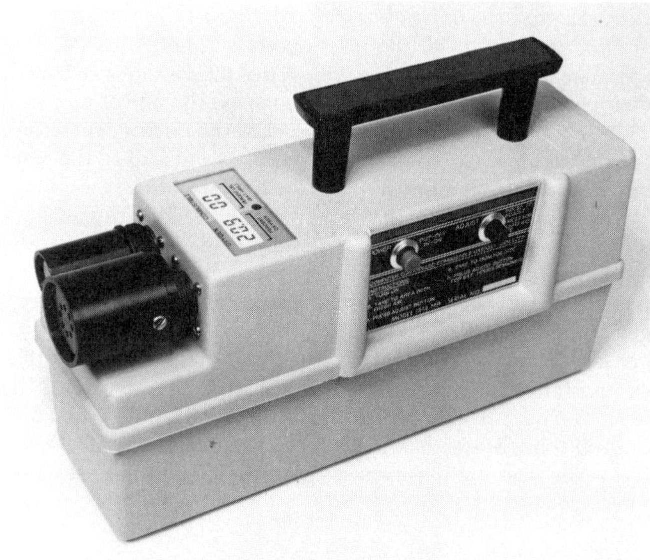

Figure 19-14. A combustible gas and oxygen monitor. (Courtesy Gas Technology, Inc.)

Generally, the air is drawn through the probe and meter by means of a hand-operated rubber squeeze bulb. In some instances, however, a small electrically operated pump in the instrument case is used for this purpose. In most work areas, the concentration of combustible gas or vapor fluctuates constantly, and it is necessary to observe the instrument carefully and to make a judgment concerning average and peak readings.

Zero adjustment. The zero adjustment must be made by taking the instrument to a source of air that does not contain combustible gases or vapors or by passing air into it through an activated carbon filter that will remove all combustible vapors and gases except methane. Since methane is not removed by activated charcoal filters, extra caution is required if the presence of methane is suspected. In addition, the filter should be changed periodically since it becomes saturated during prolonged use and will no longer remove many of the combustible gases and vapors. If the zero adjustments are made in fresh air, care must be taken to ensure that no combustible gas or vapor is present in an amount that would influence the instrument's response.

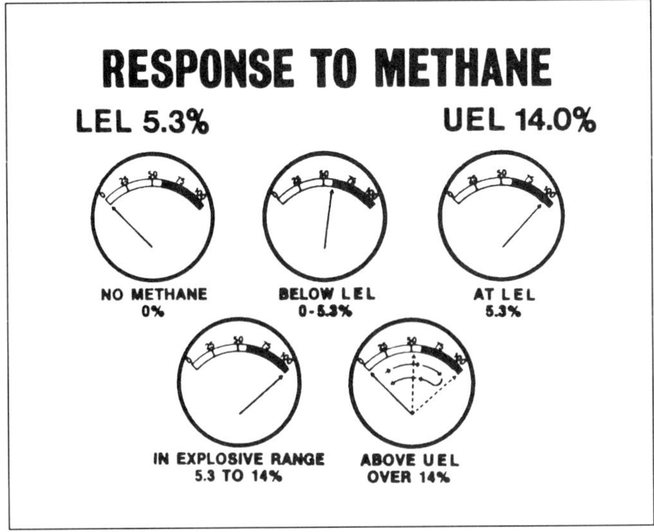

Figure 19-15. Meter readings at various methane-in-air concentrations. (Courtesy MSA)

Interpretation of meter readings

The user of any instrument should be thoroughly familiar with the precautions to be taken during its operation. Users of combustible-gas indicators must also be aware of interfering gases and vapors that could create major aberrations in instrument response. As a precaution, the 0-to-100 percent LEL scale should be used first to determine whether an explosive atmosphere exists and to prevent overloading the 0-to-10 percent LEL scale.

The typical meter responses to methane gas are shown in Figure 19-15 at the LEL, in the explosive range, and above the UEL. If the pointer of the meter travels into the red portion of the scale and remains there, an explosive concentration is present. However, if the pointer climbs rapidly to the red area and then falls back to zero, there is either a concentration above the UEL, or a gas mixture that lacks sufficient oxygen to support combustion is present (Figures 19-15 and 19-16).

Zero reading. Sometimes the probe of a combustible-gas indicator or explosimeter is placed in a manhole (or some other space not normally occupied by people) to determine if there is a potentially explosive or dangerous concentration of gas present. Under this condition, the instrument may show a zero response for several different reasons. Assuming that the batteries are charged and the instrument functions properly, the absence of a continued meter response can mean either that there is little or no combustible gas in the space being tested or that the concentration is so high that it is above the UEL, and combustion cannot occur because of insufficient oxygen. Great care must be exercised to ensure that a reading above the UEL is not misinterpreted as a true zero reading.

A very high concentration of combustible gas can be identified by carefully watching the needle as the probe is moved into and withdrawn from the space being tested. At some point during entry and withdrawal, the probe must pass through the LEL

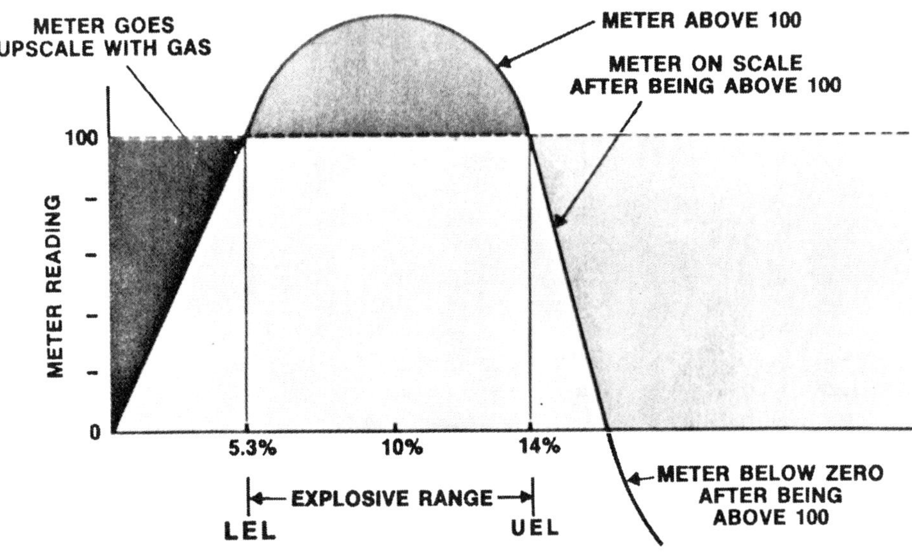

Figure 19-16. The relationship between meter reading and combustible gas concentration. (Courtesy MSA)

concentration and enter the flammable range. At this point, the needle will jump briefly and then settle back to zero. This needle jump is a clear indication that a high concentration is present.

Thermal conductivity. Atmospheres that do not contain oxygen, combustible gas, or vapor, but that do contain pure argon, produce an upscale reading. Argon has lower thermal conductivity than air, which is composed chiefly of nitrogen and oxygen.

When the instrument is brought to zero in a sample of air, it is balanced for a given wire temperature—a function of the electrical heat input on the filament and of the cooling effect of air. If the air-sampling medium is replaced with argon, there will be less dissipation of heat from the filament. This would cause a temperature rise of the filament and a resulting increase in resistance and an upscale meter reading.

High flash point solvents. Although it is relatively easy to operate a combustible gas indicator to detect a flammable gas or vapor, the instrument has some limitations.

The instrument will respond only to those combustibles drawn through the sampling system. If the flash point of a material is higher than the normal ambient room temperature, a relatively low concentration will be indicated. If a closed vessel holding such a contaminant is later heated by welding or cutting, for example, the vapor concentrations will increase, and the atmosphere of the container, which originally showed a low concentration of vapor, may then increase and become explosive.

When testing the atmosphere in drying ovens or other places where the temperature is unusually high, there may be some difficulty in measuring solvents (such as naphthas) that have a relatively high boiling point. The vapors in such samples may condense in the sampling line, which is at a temperature below that of the oven, thus giving a false indication of safety. In some instances, condensation can be prevented by heating the sampling line and the instrument to a temperature equivalent to or above that of the space to be tested.

Combustible gas indicators test only flammable gases and vapors in air. They are, therefore, not applicable for measuring combustibles in steam or inert atmospheres because the oxygen necessary to support combustion on the filament unit is absent.

Dual-scale instruments

To measure high concentrations of combustibles (above 100 percent of the LEL), it is necessary to increase the basic range of the instrument. Until recently, such a range increase was accomplished by using a dilution tube, a dilution valve, or a range multiplier. Any of these instruments essentially caused the sample to be diluted in air at a ratio of 10:1 to 20:1 and produced a new mixture that could be indicated on the scale of the indicator.

Industrial hygienists and health and safety professionals use another instrument to measure toxic as well as fire hazards. Because the Threshold Limit Value of many gases and vapors represents a very small fraction of the LEL concentration (1 percent of any combustible in air is equivalent to 10,000 ppm), it is necessary to use a highly sensitive combustible gas indicator.

Models are equipped with a dual-scale meter that is graduated from 0 to 100 percent and 0 to 10 percent of the LEL. To determine the explosive concentration of a variety of different combustibles, calibration curves are furnished so that meter readings for individual gases can be correlated. Readings are taken on the 0-to-10 percent LEL range to measure toxic concentrations of the gases and vapors that may be flammable (Figure 19-17).

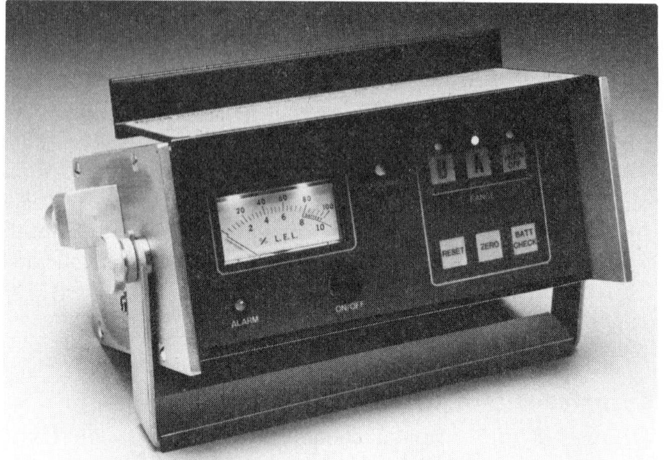

Figure 19-17. A portable continuous combustible gas detector/alarm with a dual-range meter. (Courtesy MSA)

The selector switch on the panel of these instruments changes the metering circuit from one calibration to another. Thus measurements of the percentage of the LEL can be made directly, thereby eliminating the need for calibration curves or charts.

Several types of combustible gas indicators have been designed to be calibrated so that specific combustibles can be measured. One variation of the instrument has adjustable calibration controls and can measure five different gases or vapors in the 0-to-100 percent LEL range for each. Another type has a dual-scale multiple calibration curve in the 0-to-10 percent and 0-to-100 percent range of the LEL.

Leaded gasoline. When a hot-wire type combustible gas indicator is used to test vapors of leaded gasoline, a combustion product of the tetraethyl lead is deposited on the filament unit, thereby reducing the catalytic activity of the filament. To circumvent this poisoning effect, a special version of the standard instrument is available for gasoline service. When such an instrument is used, the voltage is boosted across the detector, a sufficiently high filament temperature is maintained, and contamination is prevented.

Inhibitor filter. Lead poisoning can also be prevented with the use of an inhibitor filter inserted in the filter cavity of the standard instrument. This filter produces a chemical reaction with the tetraethyl lead vapors to produce a more volatile lead product of combustion, thereby preventing contamination of the catalytic platinum filament.

In all hot-wire instruments, flashback arrestors are mounted in the flow system at the inlet and outlet ends of the detector housing. This prevents propagation of a flame outside the filament chamber if there is a flammable mixture of gas or vapor surrounding the detector.

Poisoning the catalyst

Minute concentrations of silicone vapors—even one or two parts

Figure 19-18. A firefighter uses a combustible gas monitor and oxygen alarm to test for the presence of an explosive gas and/or an oxygen deficiency in an underground sewer system. A battery-powered pump inside the unit draws the sampled atmosphere through the probe and into the instrument where it is subsequently analyzed. (Courtesy MSA)

per million—can rapidly poison the catalytic activity of a platinum filament. A hot-wire combustible gas indicator should not be used in areas where silicone vapors are present.

Effective performance requires that the operators know how to correctly use the instruments to detect explosive and toxic concentrations of combustibles. Various refinements and design improvements have been incorporated into the conventional gas indicator to meet occupational safety requirements.

Interferences. Interfering gases and vapors can seriously affect instrument response, and an experienced tester recognizes the indications of their presence. The manufacturer's instructions should be thoroughly understood because high concentrations of chlorinated hydrocarbons (for example, trichloroethylene) or of an acid gas (sulfur dioxide) may cause depressed meter readings in areas where high concentrations of combustibles are present.

Trace amounts of these interferences may not affect the readings directly but can corrode the sensitive detector elements. High molecular weight alcohols in the atmosphere may burn out the filaments, rendering the instrument inoperative. When such limitations are understood, the tester can obtain sufficiently valid results.

Portable combustible gas alarm systems

A portable combustible gas alarm is available to monitor industrial atmospheres for concentrations of flammable gases. This instrument is sensitive enough to detect the presence of a broad range of flammable gases and vapors before they reach their LEL in air. These instruments are well suited for testing the atmosphere for the presence of solvent vapors, volatile fuel vapors (except leaded gasoline), and natural and manufactured gases. An illuminated meter indicates the presence of combustible gas or vapor in concentrations within the 0-to-100 percent LEL range. Some models feature a dual-range meter for measuring both 0-to-100 percent LEL and 0-to-10 percent LEL concentrations. The lower scale allows for the detection of leaks or for the measurement of very low levels of LEL with increased readability.

When the concentration reaches the preset limit, a red alarm light and a loud horn inside the unit are activated, providing both visual and audio warnings of a dangerous concentration.

The horn may then be switched off; in this case, the pilot light will blink until the unit is reset, and the combustible gas concentration falls below the set point. The red alarm light will remain on, even if the concentration falls below the preset danger point, until the reset button is pressed. Figure 19-17 shows an example of a portable combustible gas detector/alarm with a dual-range analog meter.

Combination oxygen and combustible gas monitors

A portable combustible gas monitor and oxygen indicator is designed to simplify the job of inspecting areas for combustible gases or vapors or for an oxygen deficiency by combining two instruments in a single unit (Figure 19-18).

In some models, a battery-powered pump draws the sample across two sensors. In more compact units, such as a belt-worn instrument, the atmosphere is sampled by diffusion (Figure 19-19). Regardless of what type is used, concentration readings are provided within seconds on a meter or digital display.

Figure 19-19. Example of a belt-worn combustible gas and oxygen monitor with LCD digital display. The two sensors located above the digital display sample the atmosphere by diffusion. No pump is required. (Courtesy MSA)

The combination instrument's gas detector operates on the same principle as the combustible gas alarm that has already been described. However, in the combination model, oxygen is sensed directly by a unique galvanic cell inside the oxygen analyzer. The cell, which contains one gold and one lead electrode in an electrolyte, is encapsulated in inert plastic. Oxygen diffusing through the fluorocarbon polymer face of the cell initiates redox reactions, which generate a minute electrical current that is directly proportional to the oxygen partial pressure. A temperature-compensated electronic circuit converts the current to a proportional voltage displayed on the analyzer's meter that represents the concentration of oxygen.

Also included in the combination group are three-channel monitors that sample for such toxic gases as carbon monoxide and hydrogen sulfide as well as oxygen and combustible gases. These units consist of three distinct detection units housed inside the monitor. The user can select single sensor readout or automatic sequential scanning of all three sensors. Similar to the combination units previously described, the multipurpose instruments sample oxygen directly with the use of a galvanic cell. Toxic gases and combustible gases are sensed by electrochemical sensors such as those found in the combustible gas alarm (Figure 19-20).

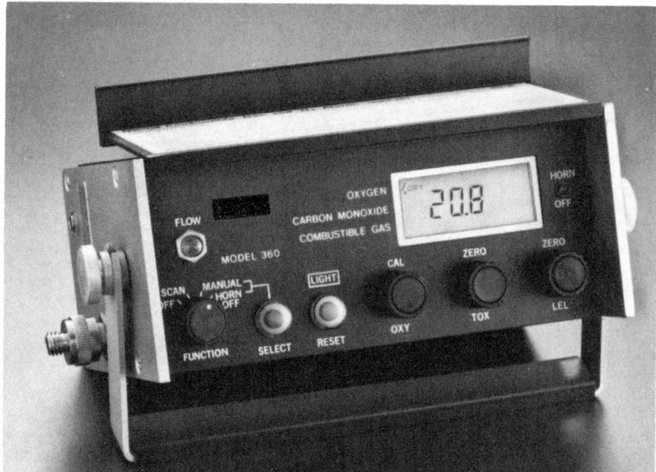

Figure 19-20. Example of a portable indicator and alarm that can detect oxygen, combustible gas, and carbon monoxide; readings are recorded on a digital display. A built-in, battery-powered pump draws samples from the immediate area or from confined spaces when used with the appropriate sampling line. (Courtesy MSA)

OXYGEN MONITORS

Although oxygen does not have a specific threshold limit value, its level in industrial air must often be measured, particularly in enclosed areas where combustion or other processes may use up the available oxygen. Excess oxygen that may result from oxyacetylene or oxyhydrogen operation should also be monitored to prevent a fire hazard.

Air normally contains about 21 percent oxygen by volume. Sixteen percent oxygen is considered the minimum to support life. In some cases, however, air with less than 19.5 percent oxygen may be considered deficient, such as at high altitudes where atmospheric pressures are lower. Federal regulations vary on oxygen requirements. The OSHA regulation 1910.94(d) (9) states that air that contains less than 19.5 percent is oxygen-deficient. The OSHA Maritime Standard (*CFR* 1915.12), however, considers that air containing less than 16.5 percent is oxygen-deficient.

In many locations—such as in mines, manholes, tunnels, or other confined spaces—it is possible for the oxygen content to be low enough to be hazardous to life. In such situations, it is necessary to determine the oxygen content in the air. In addition, it is necessary to take a sample to determine whether combustible gases are present in dangerous concentrations.

Both direct- and indirect-reading instruments are available to sample breathing air for oxygen content. Direct-reading samplers are, for the most part, small, lightweight, and relatively easy to use. Direct-reading instruments include those that are based on the coulometric principle and on colorimetric and paramagnetic analysis.

Galvanic sensing cells

Portable oxygen indicators use a small sealed galvanic sensing cell to measure oxygen on either a 0–25-percent or a 0–100-percent by-volume range (Figure 19-21). A minute electrical current is generated in the cell in proportion to the oxygen content of the atmosphere being sampled.

Coulometric oxygen detectors

The coulometric detector is used to measure oxygen content and to determine oxygen-deficient atmospheres, areas of excess oxygen, and the oxygen concentrations in storage tanks and compartments of vessels. Detection is accomplished with a primary galvanic cell that consists of a zinc and hollow carbon electrode in a special electrolyte.

A gas mixture flows through the interior of the carbon electrode and then diffuses through the carbon to interface with the electrolyte. The oxygen combines with hydrogen that has been conducted to the electrode as hydrogen ions by an electric current generated by the cell itself, thereby causing polarization.

Polarization is counteracted by oxygen, which depolarizes the cell and changes the terminal voltage and meter reading according to the amount of oxygen in the gas being tested. The instrument has a scale ranging between 0 and 25 percent oxygen by volume.

Coulometric instruments are suitable for a wide variety of process applications in which oxygen must be measured in gaseous samples, or dissolved oxygen must be determined in aqueous or nonaqueous solutions. Coulometric instruments can be used to take gaseous measurements of the percentage of oxygen present, measure the millimeters of partial pressure, and take a liquid measurement of the percentage of saturation and the parts per million in a wide variety of process applications in which oxygen must be measured in gaseous samples. These applications include situations involving flue gas, oxygen in hazardous environments, and in tests used to measure the purity of oxygen in respiration studies.

Figure 19-21. A portable, hand-held oxygen indicator with digital display and remote galvanic sensor is shown. The unit operates by diffusion and measures oxygen on a 0–100-percent by-volume range. (Courtesy MSA)

Paramagnetic analysis

Paramagnetic oxygen analyzers are available in standard models that can be used to measure the oxygen content of an atmosphere. Most units operate on the principle that magnetic lines of flux pass through oxygen more easily than through other gases.

Oxygen is paramagnetic (that is, attracted into a magnetic field). The paramagnetic property of oxygen is caused by its atomic and molecular structure and is inversely proportional to its absolute temperature. When oxygen is heated, it loses its paramagnetic property and becomes diamagnetic, or repelled out of a magnetic field.

The paramagnetic analyzer is an accurate, easy-to-use instrument for reliable oxygen measurements. Hand-held units are available in models powered by flashlight batteries.

During the operation of a typical unit, a sample of air passes through the oxygen analysis cell over an electrically heated resistor. Oxygen is attracted to the magnetic field of the active resistor where it is heated and loses its magnetic property.

The cooling effect on the active resistor is proportional to the oxygen content of the gas. The nonactive resistor acts as a reference. The difference in resistance produces a voltage proportional to the amount of oxygen present in the sample. Other units operate simply by passing a gas sample into a test chamber where a suspended magnetized object will rotate in proportion to the oxygen content of the sample.

Other techniques

Oxygen can reduce the intensity of light emitted by fluorescent materials. A device that uses an ultraviolet light source to excite a fluorescent chemical has been developed to determine the oxygen content of a given environment. The fluorescent chemical is absorbed by a porous glass disk, and the degree of fluorescence is monitored by a photoconductive cell. As the oxygen content of the atmosphere in contact with the disk is reduced, the fluorescence increases.

Carbon dioxide monitors, such as those used in breweries, indirectly function as oxygen-deficiency indicators. Such detectors contain a regenerable cartridge of soda lime through which an electrically driven pump draws atmospheric air. Any carbon dioxide in the sample is absorbed by the cartridge, and the difference in the volume of gas sampled and the volume after passage through the cartridge is registered by the levels of liquid in graduated glass tubes on the front of the instrument. Many of these instruments feature alarms that sound when the carbon dioxide concentration becomes hazardous.

Polarographic detectors operate on a current that is generated between two electrodes immersed in a reagent electrolyte which, in turn, is contained in a special cell. A current is generated when gas diffuses through a porous membrane separating the electrolyte liquid from the atmosphere. By varying the electrode materials and the electrolyte, such cells can be made to react to the presence of specific inorganic gases that undergo self-ionization in aqueous solution. Such devices can be used to monitor the depletion of oxygen in respirable atmospheres.

Figure 19-23. A battery-operated carbon monoxide detection instrument using an electrochemical cell. A pump inside the unit draws a sample into the unit where it is subsequently analyzed. (Courtesy MSA)

CARBON MONOXIDE MONITORS

One of the most insidious toxic gas hazards in an industrial atmosphere is carbon monoxide. Odorless, tasteless, and colorless, carbon monoxide can be deadly even in small concentrations. Carbon monoxide can occur in many areas, including gas and utility properties, garages, bus terminals, sewers, vaults, blast furnaces, open-hearth furnaces, and mines.

There are a number of instruments available for measuring carbon monoxide (Figures 19-22 and 19-23). Most of today's instruments combine solid electrolyte technology with thin-film deposition processes. The sensor is an electrochemical polarographic cell. The cell electro-oxidizes CO to produce carbon dioxide in proportion to the partial pressure in the sample area. The resulting electrochemical signal is amplified and the temperature is compensated to drive the meter. Samples are introduced to the sensor by either a battery-operated pump or by diffusion through a gas-porous tetrafluoroethylene membrane (Figure 19-24).

Figure 19-22. This carbon monoxide monitor is small enough to be worn by the user. (Courtesy National Draeger, Inc.)

Figure 19-24. Example of a direct-reading carbon monoxide monitor. Samples are introduced to an electrochemical sensor by diffusion through a gas-porous tetrafluoroethylene membrane. The unit permits spot testing for carbon monoxide concentrations in ambient air or can be used for time-weighted-average testing. (Courtesy MSA)

The portable-type carbon monoxide detector features both visual and audible alarms that alert the user when the danger level is reached. The battery-powered instrument can measure carbon monoxide in the atmosphere in the range of 1–500 ppm by volume.

Chlorine and hydrogen sulfide monitors

There are also several instruments available for the measurement of other toxic gases such as hydrogen sulfide and chlorine gas. Similar to the carbon monoxide monitors, the hydrogen sulfide instruments analyze ambient air using an electrochemical polarographic cell. Many H_2S instruments feature a dual-range meter that provides an analog readout in parts per million of hydrogen sulfide. For general readings, a high range measures 0–100 ppm hydrogen sulfide. For more accurate readings, a low range measures from 0–20 ppm H_2S (Figure 19-25). There are also compact hydrogen sulfide monitors available that are designed for applications in which continual checks for hydrogen sulfide are necessary (Figure 19-26). These instruments are usually hand-held or belt-worn and do not feature a dual-range meter. However, they are ideal for applications in confined spaces since they operate by diffusion and do not require a battery-operated pump.

Chlorine indicators work in a manner that is similar to the operation of hydrogen sulfide monitors. Many measure chlorine on a dual range; the high range measures from 0–10 ppm chlorine, while the lower scale provides a more accurate reading at 0–2 ppm chlorine. Most chlorine indicators feature an electrochemical sensor cell that electroreduces chlorine in proportion to the partial pressure of chlorine in the sample chamber. The resulting electrical signal is amplified to drive an analog meter and an alarm comparator circuit if the instrument features an audio alarm (Figure 19-27).

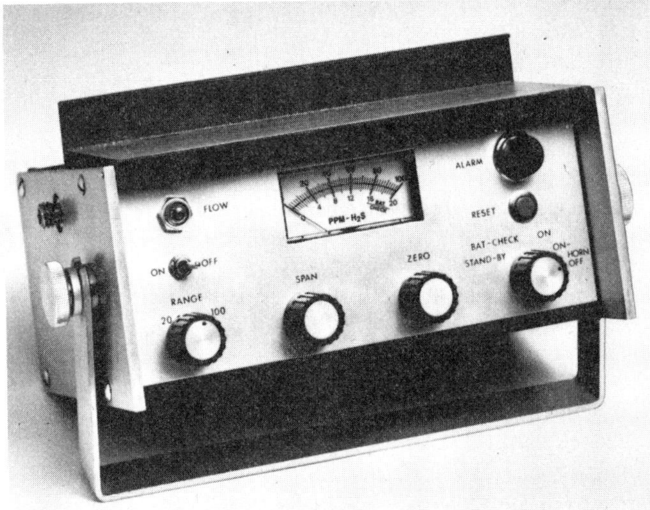

Figure 19-25. A H_2S monitor that features a dual-range meter. Samples are introduced to an electrochemical sensor by a battery-operated pump located inside the unit. (Courtesy MSA)

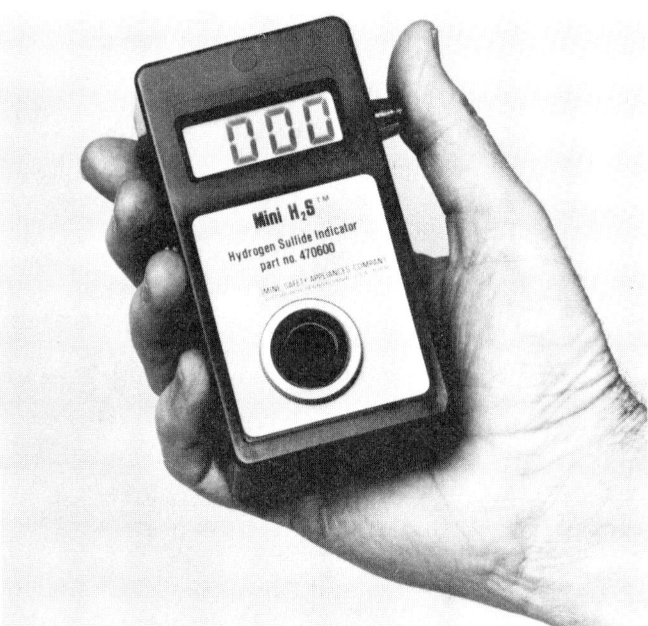

Figure 19-26. This hand-held hydrogen sulfide indicator provides a readout of hydrogen sulfide concentrations in ambient air over the range of 0-200 ppm. Samples are introduced to the electrochemical sensor by diffusion. (Courtesy MSA)

Figure 19-27. Example of chlorine indicator and alarm that features two ranges for monitoring chlorine concentrations. (Courtesy MSA)

Trichloroethylene and methyl bromide monitors

In addition to the hydrogen sulfide and chlorine monitors, recent technological advances have led to the development of instruments that monitor concentrations of trichloroethylene and methyl bromide (Figure 19-28). The trichloroethylene monitor is useful in metal degreasing vats, solvent recovery tanks, and in many other applications in which trichloroethylene is used. The methyl bromide instrument is designed for fumigation monitoring and similar applications. Both of these instruments feature electrochemical sensor cells. Trichloroethylene and methyl bromide are detected via sample conditioning and electrochemical reaction in the sensor cell. The resulting electrical signal is amplified and converted to a digital value.

Figure 19-28. An instrument for detecting toxic concentrations of trichloroethylene. The instrument measures trichloroethylene on a range of 0–100 ppm. Concentrations are indicated on a digital LCD readout. (Courtesy MSA)

OTHER TYPES OF EQUIPMENT

Infrared analyzers

Many gases and vapors, both inorganic and organic, have characteristic infrared spectra that absorb infrared radiation over a spectrum of wavelengths in a manner that can be converted into characteristic graphs. These can be used to detect the presence of air contaminants and also to determine their concentrations in air (Figure 19-29).

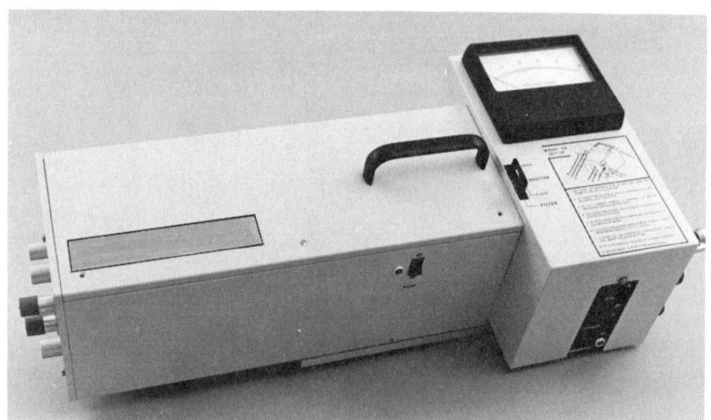

Figure 19-29. An infrared analyzer can be used to monitor gas and vapor exposures. (Courtesy Foxboro Co.)

Formerly requiring permanent installation, recent advances in technology have reduced the power requirements for infrared analyzers. Today these highly sensitive devices can be housed in a portable instrument design.

Infrared analyzers operate on the absorption of infrared energy by certain contaminants. Basically, an infrared generator in the analyzer emits the full frequency range. It is an established fact that various gases and chemicals absorb infrared energy at specific frequencies (bands). Also, the window material in the cell limits the frequencies that can be used. The combination of the window materials and the absorption of energy by gases at specific frequencies determines how the instrument is made selective to a specific chemical or gas.

The instrument's actual operation is really quite simple. At one end there is a source, which emits the infrared energy through two cell paths simultaneously. At the opposite end is a detector, which monitors the energy being transmitted through the two cells. One of the cells is the sample cell. A gas is injected into the sample cell, which is then analyzed. The other cell is sealed with a special mixture inside. If the gas injected into the sample cell contains a gas that absorbs energy at the selected frequency, then the detector will detect less energy coming through the sample cell than the comparison cell. The detector emits an electrical signal to alert the user to this imbalance.

When a single contaminant is present, identification and measurement are achieved with ease. However, when a number of absorbing contaminants are present, separation of the contaminants is not possible. In such cases, the total concentration (that is, of all absorbing contaminants) can be determined, and supporting methods, such as mass spectrometry and gas chromatography, are required.

Photo-ionization and flame-ionization devices are also now available in portable models. These devices operate when ions form in the presence of certain kinds of contaminants.

Gas chromatography

In gas chromatography, the components of a mixture migrate differentially in a porous sorptive medium. Chromatography is primarily a method of resolving complex mixtures. This resolution depends on the differential migration of the components through the porous medium. Differential migration is carried out so that each component separates as a discrete substance. The separated substances appear in a carrier gas as a function of time as the carrier gas passes through the absorption column. Detection of the separated components takes place as the carrier gas emerges from the column.

Analysis for a specific component requires a method (either specific or nonspecific) for the detection and identification of the isolated components of a mixture. The use of particular reference substances and the sorption time sequence technique are suitable methods as is the measurement of the relative migration of carrier gas and components under standardized conditions.

The readout system includes a thermal desorption unit for removal of the collected sample from the charcoal tubes. The thermal desorption unit includes a gas chromatograph with column backflush capability as well as suitable data readout and recording equipment. The readout system can be portable or fixed (see Figure 19-30).

Instrument reliability, ease of operation, and ease of calibration are key considerations when defining the minimum technical skill required to operate the device. Many advantages are obtained by on-site readout systems, such as low cost, rapid data collection, increased flexibility, and greatly reduced sample handling and storage problems.

Another type of monitoring device is shown in Figure 19-31.

CALIBRATION

All instruments that are used for sampling gases or vapors must be calibrated before use, and their limitations and possible sources of error must be fully understood. It is very important

Figure 19-30. A fully portable gas chromatograph for on-site vapor analysis. (Courtesy Sentex Sensing Technology, Inc.)

Figure 19-31. Oxidant monitor provides continuous measuring of oxidant concentrations. Sensor reactions are electrochemically activated and occur at the polarized electrodes, which are immersed in a flowing film of sensing solution. The current produced is directly proportional to the mass per unit time of oxidant entering the sensor. (Courtesy Mass Development Co.)

to establish that an instrument responds properly to the substance it is designed to sample. This is generally carried out by performing calibration procedures with standard concentrations of the substance of interest.

There are two generally accepted methods for calibrating direct-reading, air-sampling instruments: (1) the static method, which involves introducing a known volume of gas into the instrument and sampling for a limited period of time; and (2) the dynamic method, whereby a known concentration of the contaminant is prepared in a test chamber, and the instrument is used to monitor that concentration. The static method is by far the easier, more efficient technique for checking the response of portable gas-detection instruments.

A wide range of static-type calibration kits are in general use today. Formerly two separate kits were required: one for testing combustible gas instruments and one for testing toxic gas instruments. However, with the appropriate container of check gas, today's kits can be used for both types of instruments. These kits generally contain one or more cylinders filled with a known concentration of a specified gas-air mixture, a regulating valve, pressure gauge for measuring the pressure in the container, and a hose adapter that connects the cylinder to the instrument to be checked (Figures 19-32 and 19-33).

Once the container kit is attached to the instrument, a sample of the gas-air mixture from the container is permitted to flow into the device. The meter reading of the instrument is then compared with the known concentration of the sample to verify the proper calibration response.

In contrast, the dynamic method of calibrating gas-detection instruments is somewhat more difficult and requires facilities that are not always available to the average industrial plant.

In dynamic calibration, a specific concentration of the contaminant is prepared in a laboratory and placed in a special test chamber along with the instrument. Then the instrument is activated, and a meter reading is taken to determine the correctness of the response.

The rate of airflow and the rate at which the contaminant is added to the sample stream must be carefully controlled under dynamic calibration methods to produce a known dilution ratio. Dynamic systems offer a continuous supply of contaminant, allow for rapid and predictable concentration changes, and minimize the effect of wall losses as the test substance comes into equilibrium with the interior surfaces of the system.

As discussed earlier, combustible gas detectors are generally calibrated to indicate the percentage of the LEL of the gas or vapor being tested. However, interpreting the results of tests with these instruments requires careful evaluation of all factors involved. For instance, some instruments do not respond in the same manner to different flammable contaminants. If the meter reading of an instrument is plotted on a vertical axis and the percentage of the LEL is plotted on the horizontal axis, different curves will be established for such materials as natural gas, acetylene, gasoline, and carbon disulfide. For this reason, it is important that the manufacturer supplies calibration curves for the air contaminant being monitored and the user understands the limitations of the instrument.

SUMMARY

The ultimate goal of the hazard evaluation process is to determine the exact amount of vapor or gaseous contaminants present in the work environment. Proper operation of various instruments used in hazard analysis is essential to ensure that the information obtained in air quality tests is accurate enough to provide a useful interpretation.

Faulty operation of air-sampling instruments can result in low

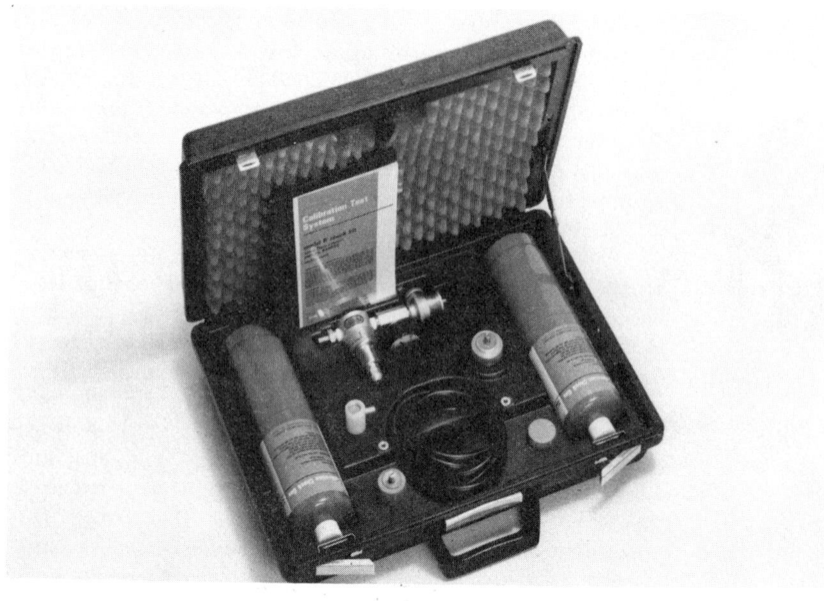

Figure 19-32. A calibration test kit can be used for combustible-gas detection instruments and toxic-gas detection instruments. (Courtesy MSA)

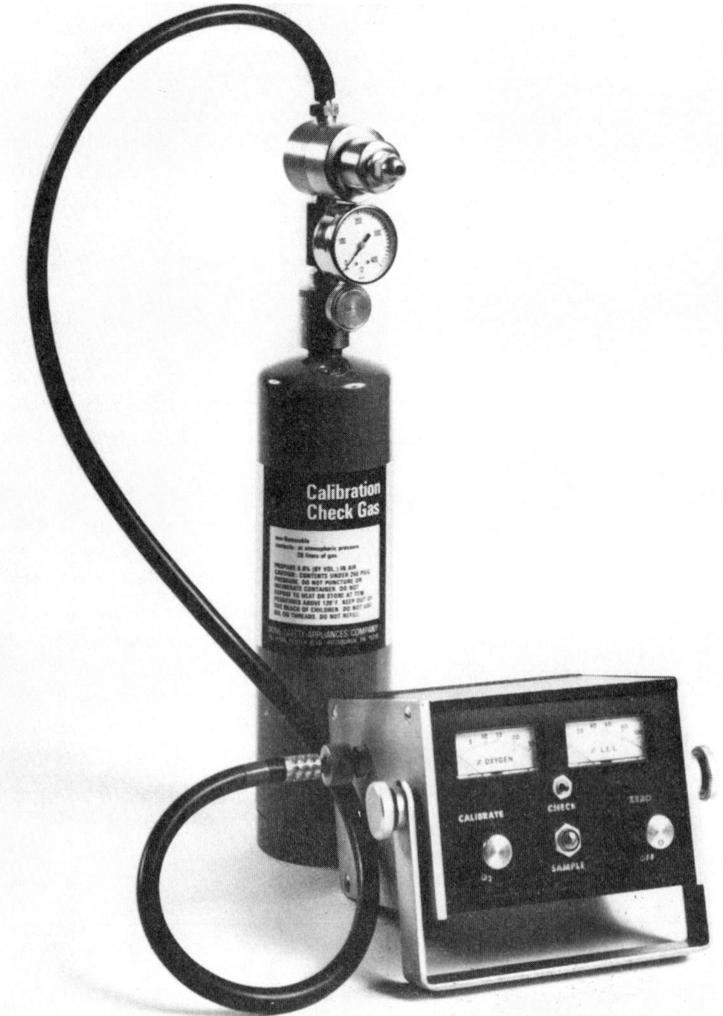

Figure 19-33. The calibration test kit shown in Figure 19-32 can also be used for combination-type instruments, such as the combustible gas and oxygen indicator shown here. (Courtesy MSA)

readings; these can falsely indicate that no hazard is present when, in fact, dangerous conditions might exist. Likewise, high instrument readings and the resulting implementation of a hazard-control procedure may be instituted where none is needed.

A permanent record should be maintained of all calibration procedures, data, and results. The type of information to be kept for this record includes instrument identification, temperature, humidity, trial run results, and final results. It is important that the operator thoroughly understand how to operate the instrument and be aware of the instrument's intended use and the calibration procedures recommended by the manufacturer.

BIBLIOGRAPHY

American Conference of Governmental Industrial Hygienists (ACGIH). *Air Sampling Instruments for Evaluation of Atmospheric Contaminants,* 6th ed., Cincinnati: ACGIH, 1983.

American Industrial Hygiene Association (AIHA). *Direct Reading Colorimetric Indicator Tubes Manual.* Akron, OH: AIHA, 1976.

National Institute for Occupational Safety and Health. *The Industrial Environment—Its Evaluation and Control.* Washington, DC: Superintendent of Documents, U.S. Government Printing Office, 1973.

Patty, F.A., ed. *Industrial Hygiene and Toxicology,* 3rd rev. ed., vol I. New York: Interscience Publ. Inc., 1978.

PART V

Control of Hazards

20

Methods of Control

by Julian B. Olishifski, MS, PE, CSP

THE GENERAL PRINCIPLES AND METHODS involved in controlling occupational health hazards will be discussed in this chapter.

A fundamental engineering principle is that in order to monitor a process or operation, suitable variables must be measured. In the field of industrial hygiene, the control of occupational health hazards requires that an employee's exposure to harmful chemical stresses and physical agents does not exceed permissible levels. The variables or quantities of interest that must be measured are the concentration or intensity of the particular hazard and the duration of exposure.

The types of industrial hygiene control measures to be installed depend on the nature of the harmful substance or agent, and its routes of entry or absorption into the body. An employee's exposure to an airborne substance is related to the amount of contaminants in the breathing zone and the time interval during which an employee is exposed to this concentration. Reducing the amount of contaminant in the employee's breathing zone or the amount of time that an employee spends in the area will reduce the overall exposure.

METHODS OF CONTROL

Various methods of control available to industrial hygienists are broken down into these categories:

- Engineering controls that engineer out the hazard, either by initial design specifications or by applying methods of substitution, isolation, or ventilation.
- Administrative controls that control employees' exposures by scheduling reduced work times in contaminant areas, and/or other work rules.
- Personal protective equipment that should be considered a method of last resort when engineering controls are not sufficient to achieve acceptable limits of exposure. Personal protective equipment can be used in conjunction with engineering controls and other methods.

Built-in protection, inherent in the design of a process, is preferable to a method that depends on continual human implementation or intervention. A complete understanding of the circumstances surrounding the problem is required in choosing methods that will provide adequate control. To lower exposures, determine the contaminant source, the path it travels to the worker, and finally the employee's work pattern and use of protective equipment (Figure 20-1). Hazards can change with time so that health hazard control systems require continuous review and updating.

The health of workers in industry must be protected by controlling exposures to air contaminants, occupational noise, and radiation. Exposures to chemical and physical agents must be controlled, first by the application of engineering control measures that are supplemented when needed by administrative control or the use of personal protective equipment. Engineering controls include local exhaust ventilation to minimize the dispersion of harmful substances into the work environment, and the provision of enclosures or barriers that minimize the dispersion of potentially harmful substances. In some instances, general or dilution-ventilation as a means of controlling health hazards can be used, but only when the toxic effects of the materials in question are considered to be low.

Although administrative control measures can limit the duration of individual exposures, they are not generally favored by

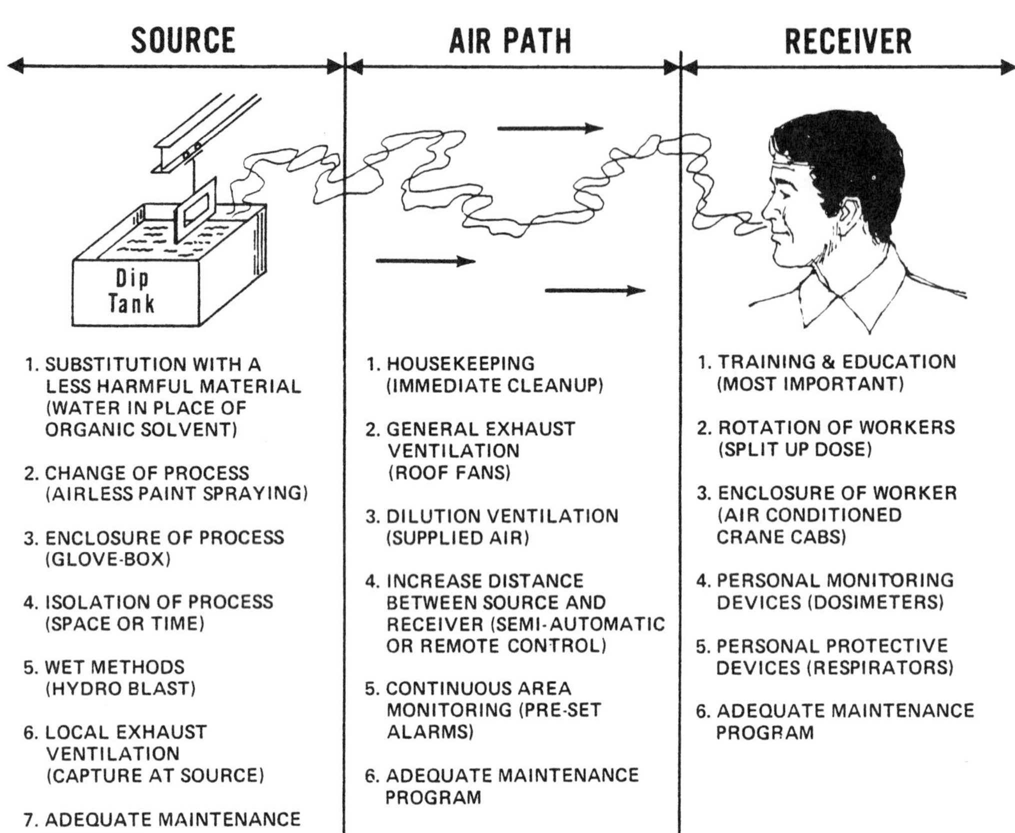

Figure 20-1. To determine the extent of exposure, locate the contaminant source, its path to the employee, and the employee's work pattern and use of protective equipment.

industry because they are difficult to implement and maintain. Control of health hazards by using respirators and other protective devices is usually considered secondary to the use of engineering control methods. Most federal regulations mandate that engineering controls are provided to the extent feasible; if these are insufficient, then the use of personal protective equipment and other corrective measures may be considered to achieve acceptable limits of exposure.

ENGINEERING CONTROLS AT DESIGN STAGE

The best time to introduce engineering controls is when a plant is at blueprint stage. At that time, control measures can be integrated more readily into the design than after the plant has been built or the process gone on-stream.

The systematic layout of the physical plant, and the attainment of compliance with occupational safety and health standards require a knowledge of the vast number of statutorily mandated standards. What is planned must be reconciled with what is permissible or advised. In any particular situation, jurisdiction and applicability of standards may be simple or complex; issues frequently arise that carry unique constraints and resolutions. When more than one agency or standard is involved, the more stringent standard can be assumed to be controlling.

The proposed plant layout must be characterized with respect to construction type, proposed activities in all areas, and possible health hazards. The influence of one area on another and one work activity on another must be assessed as to combined hazards.

It is becoming increasingly common for plant and design engineers to consult with the industrial hygiene engineer at the design stage of a new plant or process. Including industrial hygiene control measures at this point can be less costly than adding them later on in the construction process.

When air contaminants are created, generated, or released in concentrations that can injure the health of workers, ventilation is the usual method of providing protection. However, there are other methods of protection that should be investigated; one example is automatic operations.

Ideally, operations should be conducted in entirely closed systems, but not all industrial processes lend themselves to this approach. When closed systems are used, raw materials can be brought to the manufacturing site in sealed containers and their contents emptied into storage tanks or bins, minimizing employee contact with the material being processed.

Design all systems and components so that airborne contaminants are kept below their acceptable Threshold Limit Values (TLVs). Do not permit leaking of toxic chemicals from process equipment, such as pumps, piping, and tanks, to the working environment to cause a condition in which the TLVs are routinely exceeded in any location where employees may be present. Isolate process equipment and vent to a scrubber,

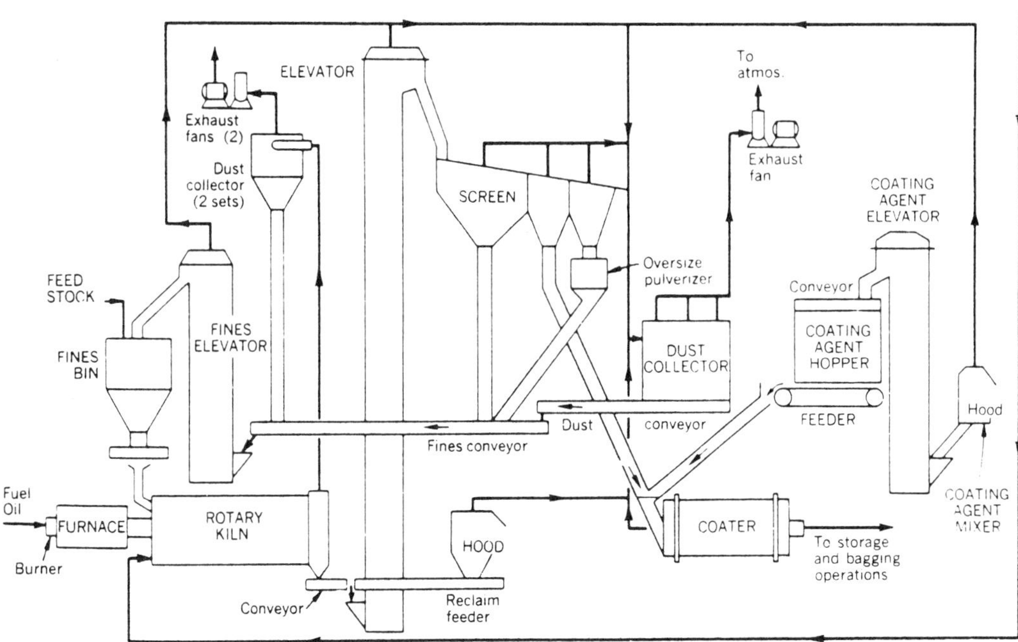

Figure 20-2. A simple process flow sheet showing the stepwise introduction of raw material and the product of each step. The extent of chemical or physical hazards that can occur at any step in the operation should be determined.

absorber, or incinerator. If feasible, remotely control the process from a suitable protected control room.

Some work operations, if conducted separately, do not present a serious hazard; but when combined with other job operations, they can become hazardous in certain situations. Two types of interrelationships can exist.

The first concerns accumulation, as can arise when additional welding stations are provided in a building of fixed general ventilation or when additional noise sources are added to an already noisy work area.

The second type of interrelationship concerns many activities going on in the same area. Activities that by themselves are safe can become hazardous in certain circumstances. For example, vapor degreasing with chlorinated solvents, even when the airborne concentration of the vapors is within permissible limits, may create major hazards when the activity is nearby work areas where ultraviolet (UV) radiation (from welding arcs, bright sunlight, or molten metal) exists. The decomposition of these solvents caused by the UV radiation can produce phosgene gas—a potent and toxic eye and lung irritant. Merely maintaining the concentration of solvent vapor below the TLV is not satisfactory. The most positive control is to prevent the chlorinated solvent vapors from entering the welding area in any detectable concentrations. If vapors cannot be reduced to a minimum, the UV field should be reduced to a minimum by shielding the welding arc. Pyrex® glass is an effective shield.

The problem of considering safety and health with activity and workstation relationships becomes difficult when more than three or four activities must be considered, as in laying out workstations for new or relocated manufacturing operations where as many as 20 or more activities might require consideration. Making appropriate judgments to arrive at either an optimum arrangement or a preferred best compromise is the solution.

Design

Industrial health hazards can best be minimized by process machinery design that controls the contaminant as much as possible. This requires close cooperation between the industrial hygienist and the design engineer. The ideal situation would have the design engineer so thoroughly imbued in the principle of health hazard protection that the health and safety professional need only be a passive reviewer. However, the design engineer needs the help of the health and safety professional during the design process to make sure that a system can be set up so that it does not pose safety and/or health hazards to the operator.

Production processes in chemical plants should be designed so that hazardous materials are not released into the environment. It is important to keep the materials and the by-products and the wastes within the closed system.

To maintain that integrity, a chemical process flow sheet should be reviewed from an overall material balance point of view (Figure 20-2). The material that becomes airborne and gets into the work environment to cause problems can be an insignificant fraction of the total amount of material that is circulating through the system. So much so that in a material balance, the quantity of material that escapes and is released into the workplace that causes the hazard can be insignificant when compared with the total amount present.

Design factors that should be addressed include:

- To what degree is it possible to remove hazardous residues from a piece of equipment before it is opened?
- To what extent can a system be designed so as to be relatively maintenance free?
- Can the system be designed so that the entire operation can be conducted as a closed system?
- Can the process be conducted automatically without worker involvement?

A design engineer not only should have extensive knowledge

of the main aspects of the process being created, but also the finer details, such as health hazard controls and safety devices. Design engineers are usually more familiar with the safety hazards because the effects of their being overlooked are much more obvious than those occurring when health hazards are overlooked.

When health professionals are involved early in the design process, it is possible to plan the development of sampling and analytical methods to yield the health data concurrent with the development of the engineering design. Health hazards monitoring systems can be included as part of the engineering design. Elaborate automated leak-detecting systems designed into the process can yield valuable information for evaluating health hazards in the operating unit (Figure 20-3).

Figure 20-3. This multipoint ambient air monitor is capable of continually measuring up to five gases in up to 24 remote locations. (Courtesy Foxboro Co.)

Neglect of the health professional-engineer interaction in plant design can lead to major management problems. What could have been an easy solution in the design stage can become an extremely difficult problem later on. Changes that might have been readily accomplished during the blueprint stage must now be done as a matter of equipment change and compromise. Worse yet, it may be necessary to shut down production in order to correct a hazard that was overlooked. Consequently, management should consider that for certain processes and materials, the initial design of facilities to cope with the health hazards may be a significant and necessary part of the investment.

Maintenance considerations

It is important to look not only at the mainstream of what is happening, but also at the fine details of what is not supposed to be happening. These untoward events may best be described in two general classes.

First, there may be releases of contaminants into the work environment that are relatively continuous, such as flange leaks, exhaust hoods that are not completely effective, pump seals that have weakened, diffusion that occurs along the valve stems, or noise emission from leaks in acoustic lagging on a machine that it does not fit. This general class of airborne contaminants or furtive emissions may have begun as a low-level background that initially was not high enough to be of serious concern. Coupled with this is another kind of episodic exposure. As equipment becomes worn and starts to leak, the general level of background emissions may eventually result in major worker exposures. A lot of this leakage can be dealt with by continuous, careful, intensive maintenance; however, much of it might have been avoided in the initial design. The degree to which any possibility of leakage is engineered or designed out of a system depends to a great extent on just how much these potential leaks have been anticipated.

The second class of emissions of airborne contaminants arises when a closed system or control process becomes momentarily open or uncontrolled. For instance, the lagging has to be removed from the compressor in order to perform some adjustments or perhaps samples have to be collected or filters replaced. These situations are very common in chemical industries. A filter change operation may occur as infrequently as once every six months; or, when things are going badly, it may have to be done four times a shift. The system has to be designed so that it is possible to get the filter container cleaned and purged so that an employee can perform needed maintenance without hazard.

From time to time, the system as a whole will need to be shut down for cleaning and purging and afterwards opened for maintenance. Under these circumstances, most exposures tend to be brief, but can be quite high and detected by industrial hygiene surveillance only when that surveillance is closely maintained on a day-to-day basis.

Knowledge of the hazards that will be present and the potential for the exposure that may exist in the operating units gives an industrial hygienist an ideal starting point from which to develop the surveillance program of the operating unit. All too often this step is omitted and the industrial hygienist only becomes aware of an engineering project in the advanced stage of development. Waiting to make changes in the design when the system is just about ready to go to construction can turn out to be extremely expensive.

Design specifications

The design specifications are the drawings and documents that enable the engineers to precisely define the process. The industrial hygienist or safety professional should have a clear understanding of where in these drawings significant health hazards may occur in the process.

Review. Before starting a new unit or process, engineering on-site reviews that go over the whole process once again should be made to be sure that nothing was forgotten and everything will proceed as planned. Although these reviews are very detailed and time-consuming, it is worthwhile for an industrial hygienist or safety professional to be involved. Sometimes, last minute changes in the process or equipment are made, and these changes can significantly increase or decrease the health hazard.

Start-up. The industrial hygiene surveillance begins when a unit or process is put into operation and should continue for

Figure 20-4. Loading or unloading of tank trucks can release airborne contaminants.

as long as the operation continues. Problems in handling and operating procedures that were not anticipated during the design stage will now become apparent. Prompt correction of these problems is much easier during the early set-up phase when procedures and people are still somewhat flexible.

Sample-taking. In many industrial operations, such as steel mills and petrochemical plants, taking product samples is a common procedure. The design engineer and the industrial hygienist can choose between a product sampling system that does not provide much control or a system that provides almost total control. Each of these choices probably has some cost increment associated with it. Whichever is chosen should be based upon assessment of the severity of the potential health hazard.

Loading operations. One of the most serious problems in the field of health hazard control is the loading and unloading of tank cars, tank trucks, or barges. Putting a liquid into a space previously occupied by air or vapor quickly saturates that air with vapor. It may become necessary to go to vented systems, enclosed systems, and automatic loading systems that include vapor recovery so that the vapor that is pushed out of the tank will be recovered (Figure 20-4).

Episodic exposures are difficult to control from an engineering point of view. Also, for those infrequent emergency or nonroutine events, personal protection can be the appropriate solution. However, design engineers should recognize that these exposure events will happen, that product samples must be taken, that equipment must be maintained and that filters must be changed. The industrial hygienist working with the designer must consider how these operations can be conducted so that the worker need not be overexposed.

Hazardous materials

Some materials must be very carefully handled because of their toxicity, flammability, or reactivity. The facilities and practices to be used must be consistent with the standards for such things.

Stringent controls regulating mutual proximities, ventilation, sources of ignition, and design are imposed on general industry by federal codes. When potentially photochemically reactive solvents are involved, process controls and discharges to the atmosphere are subject to regulation by air-quality regulatory authorities.

Compressed gas and equipment for its use in industry are extensively referenced in the Compressed Gas Association's standards. Methods of marking, hydrostatic testing of cylinders and vessels, labeling, metering, safety devices, and pipework and outlet and inlet valve-connecting are thoroughly described in pamphlets issued by the Association.

Standards for the design and use of air receivers are promulgated based on the ASME "Boiler and Pressure Vessel Codes." The provision and use of compressed gases in industrial settings must be carefully undertaken; otherwise, catastrophic situations may occur.

INDUSTRIAL HYGIENE CONTROL METHODS

Industrial hygiene control methods for reducing or eliminating environmental factors or stresses include the following:

- substitution of a less hazardous material for one that is harmful to health
- change or alteration of a process to minimize worker exposure
- isolation or enclosure of a process or work operation to reduce the number of employees exposed, or isolation or enclosure of a worker in a control booth or area
- wet methods to reduce generation of dust
- local exhaust ventilation at the point of generation or dispersion of contaminants
- general or dilution ventilation to provide circulation of fresh air without drafts or to control temperature, humidity, or radiant heat load

- personal protective devices, such as special clothing or eye and respiratory protection
- good housekeeping and maintenance, including cleanliness of the workplace, waste disposal, and adequate washing, toilet, and eating facilities
- administrative exposure controls, including adjusting work schedules or rotating job assignments so that no employee receives an overexposure
- special control methods for specific hazards, such as shielding, monitoring devices, and continuous sampling with preset alarms
- medical controls to detect evidence of absorption of toxic materials
- training and education to supplement engineering controls
- emergency response training and education.

A generalized diagram of these methods is shown in Figure 20-1. Each of these industrial hygiene control methods will be discussed in turn.

Substitution

An often-effective industrial hygiene method of control is the substitution of nontoxic or less toxic materials for highly toxic ones. However, an industrial hygienist must exercise extreme caution when substituting one chemical for another, to ensure that some previously unforeseen hazard does not occur along with the substitution. Examples of this include fire hazards, synergistic interactions between chemical exposures, or previously unknown toxicity problems attributed to the "nontoxic" substitute chemical. The classic examples of substitution as an industrial hygiene control measure include replacement of white lead in paint pigments by zinc, barium, or titanium oxides, the use of phosphorus sesquisulfide instead of white phosphorus in match-making, shotblasting instead of sandblasting, Freon instead of methyl bromide as a refrigerant, and less toxic substitutes for asbestos as an insulating material.

When substituting solvents, it is always advisable to experiment on a small scale before making the new solvent part of the operation or process. For example, carbon tetrachloride can be replaced by solvents, such as methyl chloroform, dichloromethane, aliphatic petroleum hydrocarbons, or one of the fluorochlorohydrocarbons. Detergent-and-water-cleaning solutions or a steam-cleaning process should be considered for use in place of organic solvents. Natural rubber cements with aliphatic hydrocarbon solvents can perform virtually the same function as benzene cements.

Synthetic materials rather than sandstone can be used as grinding wheels, and as nonsilica parting compounds in foundry molding operations. Removing beryllium phosphors from formulations for fluorescent lamps eliminated a serious pulmonary hazard to the workers making such lamps.

A change in the physical condition of raw materials received by a plant for further processing may eliminate health hazards. Pelletized or briquetted forms of materials are less dusty and can drastically reduce atmospheric dust contamination in some processes.

However, there are instances when substitution of some toxic materials may be impossible or impractical, as in the manufacture of pesticides, drugs, or solvents, and processes producing ionizing radiations.

Substituting less hazardous materials or process equipment may be the least expensive and most positive method of controlling many occupational health hazards. Reluctance by production engineers to change the status quo can usually be overcome if it can be shown that substitution with less harmful materials can often result in substantial savings.

Changing the process

A change in process offers an ideal chance to concomitantly improve working conditions. Most changes are made to improve quality or reduce cost of production. However, in some cases a process can be modified to reduce the dispersion of dust or fume and thus markedly reduce the hazard. For example, in the automotive industry, the amount of lead dust created by grinding solder seams with small, rotary, high-speed sanding disks was greatly reduced by changing to low-speed, oscillating-type sanders.

Brush-painting or dipping instead of spray-painting can minimize the concentration of airborne contaminants. Other examples of process changes are arc-welding to replace riveting, vapor-degreasing in tanks with adequate ventilation controls to replace hand-washing of parts in open containers or steam-cleaning of the parts, airless paint-spraying techniques to minimize overspray as replacements for compressed-air spraying, and machine application of lead oxide to battery grids, which reduces lead exposure to operators making storage batteries.

Automatic electrostatic paint-spraying instead of manual compressed-air paint-spraying, and mechanical continuous hopper-charging instead of manual batch-charging are additional examples of a change in process to control health hazards.

Isolation

Potentially hazardous operations should be isolated to minimize exposure to employees. The isolation can be a physical barrier, such as acoustic panels, to minimize noise transmission from a whining blower or a screaming ripsaw (Figure 20-5). The isolation can be in terms of time, such as providing remote control semiautomatic equipment so that an operator doesn't have to stay near the noisy machine constantly; or the worker may be isolated or enclosed in a soundproof control booth with a clean source of air supplied to the booth.

Isolation is particularly useful for jobs requiring relatively few workers and when control by other methods is difficult or not feasible. The hazardous job can be isolated from the rest of the work operations, thus eliminating exposures for the majority of the workers. The workers actually at workstations where contaminants are released should be protected by installing ventilation systems, which probably would not have been satisfactory if the job had not been isolated (Figure 20-6).

It may not be feasible to enclose and exhaust all operations. Abrasive blasting operations, such as those found in shipbuilding, are an example. The sandblasting should be done in a specified location, which is as far as practical from other employees. Another way to isolate the sandblasting is to do it when the least number of other employees would be exposed.

In some foundries, the shakeout operation may be performed during the swing shift after employees on the regular shift have gone for the day. The few shakeout workers can be provided with suitable respirators for the short time during which they are exposed to the airborne dust.

Other work that can be scheduled to minimize the number of workers exposed to a hazard includes blasting in mines or

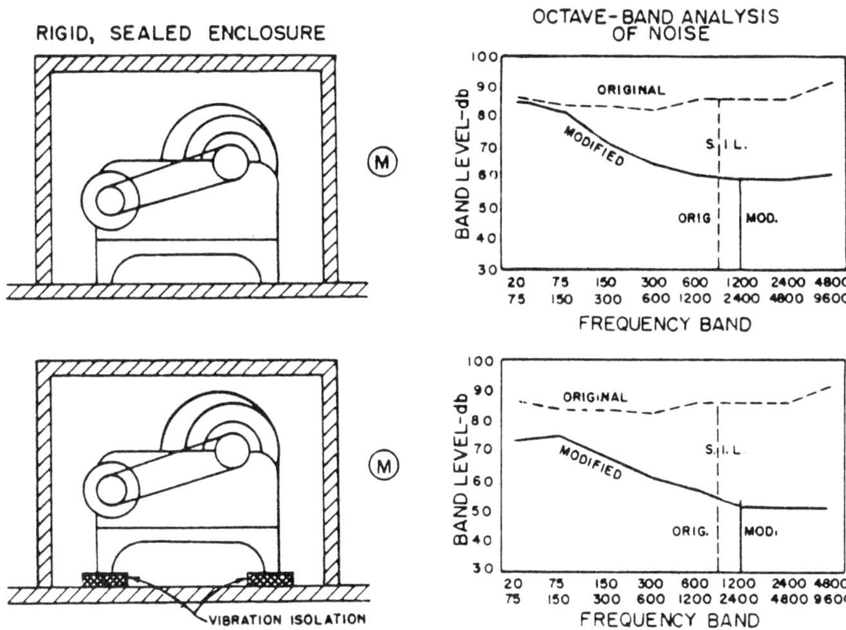

Figure 20-5. Noise can be reduced by enclosing an operation (top); and adding vibration isolators reduces sound transmission even more (bottom).

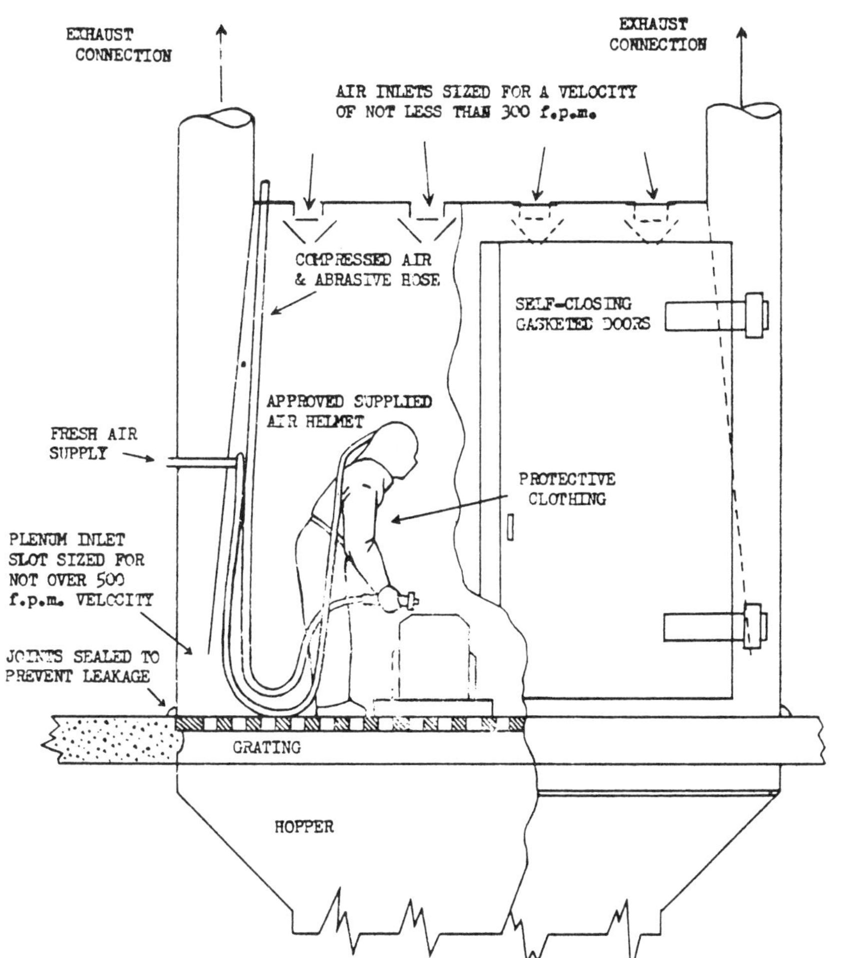

Figure 20-6. Air inlets and exhausts are arranged to sweep contaminated air away from the worker's breathing zone in this enclosed sandblast area. Downdraft averages 80 fpm over the entire floor area. Air should exhaust downward (as shown), or on two sides of the room at the floor line. (Courtesy Connecticut State Department of Health)

Figure 20-7. Some operations require complete enclosure. Here, a technician works with aluminum powder, used in atomic reactor fuel elements, at a glove box. He is wearing a film badge and air sampler on his lapel.

quarries, which can be done at the end of or between shifts, and maintenance procedures, such as cleaning tanks on weekends when few workers are present.

Some operations do not readily lend themselves to other methods of control and should be isolated. They may generate contaminants in large quantities that permeate a workroom or building to expose all workers to a hazard, although only a few of them are actually engaged in the operation.

Equipment isolation can be the easiest method of preventing hazardous physical contact. Insulating a hot water line may not be economical from a strictly heat conservation standpoint, but may be necessary if that line is not sufficiently isolated from people.

When very toxic materials are to be processed, master-slave manipulators can be used to allow handling of equipment from a remote location. The work area can be viewed by remote control television cameras, mirrors, or periscopes. The degree of isolation required depends on the toxicity of the contaminant, the amount released, and work patterns around the process. Often moving a process to another area is sufficient, while in other cases a control room supplied with fresh air may be needed to isolate the process from the operators.

Many modern chemical plants have centralized control rooms with automatic sampling and analysis, remote readout of various sensors, and on-line computer processing of the data and operation of the process. Some operations require complete enclosure and remote control so that nobody is exposed, as in many processes involving nuclear radiation (Figure 20-7).

Total enclosure can be accomplished by mechanization or automation to ensure that workers do not come into contact with toxic materials.

The crane operators in a large foundry, a bulk material storage building, or in a cement clinker shed can be provided with a completely enclosed cab ventilated with filtered air under positive pressure to keep contaminants out. In automatic stone-crushing, grinding, and conveying processes, only periodic or emergency attendance is required by an operator, small ventilated rooms, supplied with filtered air and strategically located within the large workroom, can be occupied by the workers during the major part of the workshift.

Automated plating tanks, paint-dipping operations, and similar processes can be located in separate rooms. When continuous supervision of such operations by a worker is not necessary, general ventilation may be adequate to prevent buildup of air contamination in the workroom. If necessary, an exposed worker can be given a respirator for protection during the brief periods of exposure.

Segregating a hazardous operation or locating one or more such operations together in a separate enclosure or building not only sharply reduces the number of workers exposed but greatly simplifies the necessary control procedures.

Figure 20-8. The fumes arising from lead-melting operations are controlled by local lateral slot exhaust ventilation. (Courtesy Ford Motor Co.)

Enclosing the process or equipment is a desirable method of control, since the enclosure prevents or minimizes the escape of contaminants into the workroom atmosphere. Enclosure should be one of the first control measures attempted, after considering substitution. Additional precautions must be taken when cleaning enclosed equipment or during start-up or shut-down, to avoid high concentrations of the contaminant.

Enclosed equipment is usually tightly sealed and is only opened during cleaning or filling operations. Examples of when this type of control is effective include glove boxes (Figure 20-7), airless-blast or shotblast machines for cleaning castings, and abrasive blasting cabinets.

In the chemical industry, the isolation of hazardous processes in closed systems is a widespread practice. This explains why the initial manufacture of toxic substances is often less hazardous than their subsequent use under less well-controlled conditions at other locations. In the mechanical industries, complete enclosure is frequently the best solution for severe dust or fume hazards, such as those from sandblasting or metal spraying operations (Figure 20-8).

All equipment, whether enclosed or automated, requires maintenance and repair, during which control measures may have to be removed. In such circumstances, safety procedures must be specified, including lockouts and hot work permits to work on such maintenance operations.

Wet methods

Airborne dust hazards can frequently be minimized or greatly reduced by applying water or other suitable liquid. Wetting of floors before sweeping to keep down the dispersion of harmful dust is advisable when better methods, such as vacuum cleaning, cannot be used.

Wetting down is one of the simplest methods for dust control. Its effectiveness, however, depends upon proper wetting of the dust. This may require the addition of a wetting agent to the water and proper disposal of the wetted dust before it dries out and is redispersed.

Significant reductions in airborne dust concentrations have been achieved by the use of water forced through the drill bits used in rock drilling operations. Many foundries successfully use water under high pressure for cleaning castings in place of sandblasting. Airborne dust concentrations can be kept down if molding sand is kept moist, molds with cooled castings can be moistened before shakeout, and the floors are wetted intermittently.

In some instances it may be necessary to blanket the dust source completely. The particles must be thoroughly wetted by means of high-pressure sprays, wetting agents, deluge sprays, or other procedures.

Batch charging of materials that are slightly moistened or that are packaged in paper bags rather than in a dry bulk state may eliminate or reduce the need for dust control in storage bins and batch mixers.

Local exhaust ventilation

Local exhaust ventilation is considered the classical method of control. Local exhaust systems capture or contain contaminants at their source before they escape into the workroom environment. A typical system consists of one or more hoods, ducts, an air cleaner, if needed, and a fan (Figure 20-9).

Figure 20-9. A typical local exhaust ventilation system—a dust collector—traps contaminants near their source, so the worker is not exposed to harmful concentrations.

Local exhaust systems remove air contaminants rather than just dilute them, but removal of the contaminant does not always reach 100 percent. This method should be used when the contaminant cannot be controlled by substitution, changing the process, isolation, or enclosure. Although a process has been isolated, it still may require a local exhaust system.

A major advantage of local exhaust ventilation is that these systems require less airflow than dilution ventilation systems. The total airflow is important for plants that are heated or cooled since heating and air conditioning costs are an important operating expense. Also, local exhaust systems can be used to conserve or reclaim reusable materials.

Two main principles govern the proper use of local exhaust ventilation to control airborne hazards. First, the process or equipment is enclosed as much as possible, and second, air is withdrawn at a rate sufficient to assure that the direction of airflow is into the hood and that the airflow rate will entrain the contaminant into the airstream, and thus draw it into the hood.

The proper design of exhaust ventilation systems depends on many factors, such as the temperature of the process, the physical state of the contaminant (dust, fume, smoke, mist, gas, or vapor), the manner in which it is generated; the velocity and direction with which it is released to the atmosphere; and its toxicity.

Local exhaust systems can be difficult to design. The hoods or pickup points must be properly shaped and located to capture air contaminants, and the fan and ducts must be designed to draw the correct amount of air in through each hood. Hood selection is based on the characteristics of the contaminants and how they are dispersed. (See Chapter 21, Industrial Ventilation, for more details.)

The low-volume, high-velocity exhaust system uses small volumes of air at relatively high velocities to control dust. Control is achieved by exhausting the air directly at the point of dust generation using close-fitting hoods. Capture velocities are relatively high but the exhaust volume is low. For flexibility, small-diameter, lightweight plastic hoses are used with portable tools resulting in very high duct velocities. This method allows the application of local exhaust ventilation to portable tools, which otherwise would require relatively large air volumes and large ductwork when controlled by conventional exhaust methods.

After the local exhaust ventilation system is installed and set in operation, its performance should be checked to see that it meets the engineering specifications—correct rates of airflow and duct velocities. Its peformance should be rechecked periodically as a maintenance measure.

Full details on the design and operation of local exhaust ventilation systems are given in Chapter 21, Industrial Ventilation.

General ventilation

General ventilation systems add or remove air from work areas to keep the concentration of an air contaminant below hazardous levels. This system uses natural convection through open doors or windows, roof ventilators, and chimneys, or air movement produced by fans or blowers. Exhaust fans mounted in roofs, walls, or windows constitute general ventilation.

Only use general ventilation in situations meeting the following criteria:

- small quantities of air contaminants released into the workroom at fairly uniform rates
- sufficient distance between the worker and the contaminant source to allow sufficient air movement to dilute the contaminant to safe levels
- only contaminants of low toxicity are being used
- no need to collect or filter the contaminants before the exhaust air is discharged into the community environment
- no corrosion or other damage to equipment from the diluted contaminants in the workroom air.

The major disadvantage of general or dilution ventilation is that employee exposures can be very difficult to control near the source of contaminant where sufficient dilution has not yet occurred. This is why local exhaust ventilation is most often the proper method to control exposure to toxic contaminants.

When air is exhausted from a work area, consideration must be given to providing makeup, or replacement, air, especially during winter months. Makeup air volumes should be matched to the air being removed; it should be clean and humidified and the temperature regulated as required for comfort.

Care should be taken in selecting the makeup air intake locations so that toxic gases and vapors from discharge stacks or emergency vents are not brought back into work areas. When exhaust stacks and air supply inlets are not separated adequately, the exhaust air may be directed into the air inlet and recirculated to work areas. Inadvertent recirculation of exhausted air contaminants is a growing problem.

Since air-moving, filtering, and tempering equipment is expensive, some engineers attempt to save money by recirculating some exhaust air into the supply system. Adequate monitoring of the recirculated air is necessary to prevent buildup of harmful contaminants. Recirculation of exhaust air may be forbidden in certain instances. Check state and federal regulations.

General ventilation should not be used where there are major, localized sources of air contamination (especially highly toxic dusts and fumes); local exhaust ventilation is more effective and economical in such cases. More information on this subject is presented in Chapter 22, General Ventilation.

Personal protective equipment

When it is not feasible to render the environment completely safe, it may be necessary to protect the worker from the environment with personal protective equipment. This is normally considered to be secondary to the controls mentioned previously and is used when it is not possible to enclose or isolate the process or equipment, provide ventilation, or other adequate control measures; or when there are short exposures to hazardous concentrations or contaminants (for example, where unavoidable spills may occur).

Personal protective devices have one serious drawback—they do nothing to reduce or eliminate the hazard. Their failure means immediate exposure to the hazard. The fact that a protective device may become ineffective without the knowledge of the wearer is particularly serious.

Eye and face protection includes safety glasses, face shields, and similar items used to protect against corrosive solids, liquids and vapors, and foreign bodies (Figure 20-10). Tinted lenses can be used to screen out ultraviolet and infrared radiations. Of the many types of eye and face protection available, there is a correct type for each hazard and the protective device should be worn at all times when the hazard is present.

Hearing protection. Protection against noise-induced hearing impairment, such as earplugs or earmuffs, may often present special difficulties. First, as with air contaminants, the primary objective should be to reduce the noise exposure. But in plants with old machines and processes, and even in new plants, it may not be feasible to lower the noise level. In such cases, wearing hearing protection may be mandatory (see Chapter 9, Industrial Noise, for a discussion of federal regulations on hearing protection).

Protective clothing. Gloves, aprons, boots, coveralls, and other items made of impervious materials should be worn to eliminate prolonged or repeated contact with solvents or chemicals. Care must be taken in choosing the correct type of protective equipment for the specific application.

Manufacturers can provide a large selection of protective garments made of rubber, plastic films, leather, cotton, or synthetic fibers designed for protection against specific hazards. For example, there are now different types of clothing that can protect against acids, alkalis, temperature extremes, moisture, oils, as well as other chemical and physical agents (Figure 20-11).

Figure 20-10. This worker wears safety glasses and a face shield to protect against operational hazards.

Cotton or leather gloves are useful for protecting the hands against friction and dust. Synthetic rubber gloves are used for protection against acid and alkali. Neoprene-dipped cotton gloves will protect against some liquid irritants, and there are specific glove materials that provide protection against chlorinated solvents.

For intermittent protection aganist radiant heat, reflective aluminum clothing is available. Such garments need special care to preserve their essential shiny surface. Air-cooled jackets and suits are available to make endurable a high radiant heat load.

Lead-bearing materials are available for use in garments that protect against ionizing radiation.

Protective creams and lotions. These help minimize skin contact with irritant chemicals. Their effectiveness varies, but if properly selected and correctly used, they can be very helpful. There are instances when a protective cream may be helpful in preventing contact with harmful agents, for example, if the face cannot be covered by a shield or gloves cannot be worn (Figure 20-12).

There is no all-purpose protective cream. Several manufacturers compound a variety of different products, each designed for protection against a certain type of hazard. Some barrier creams are useful for protecting against organic solvents while other types protect against water-soluble materials.

To use a protective cream correctly, it must be applied on clean skin at the beginning of the workshift, removed at lunch, reapplied after lunch, again in the afternoon, and removed at the close of the workshift.

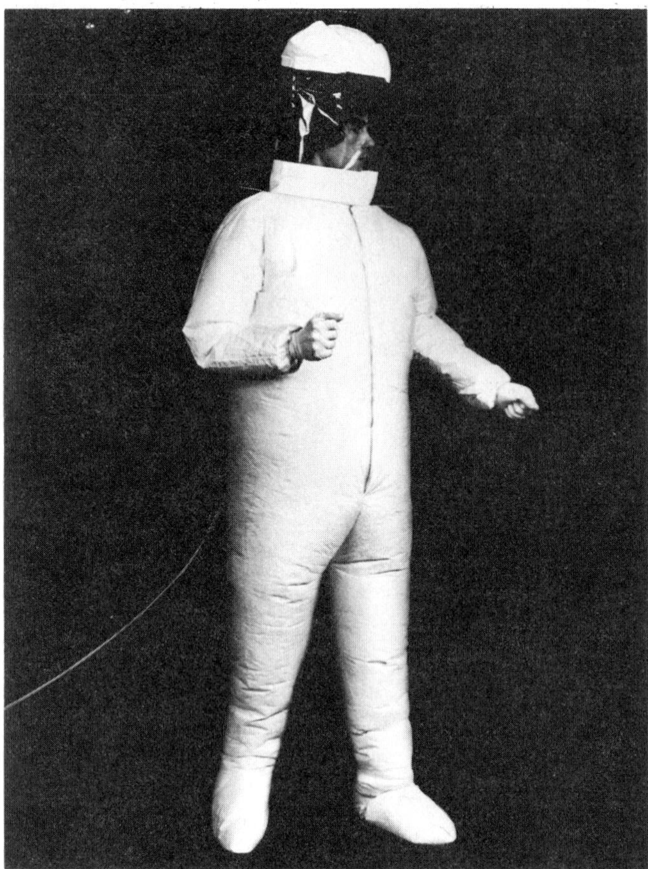

Figure 20-11. This full-coverage suit can be constructed with synthetic material with serged or bound-and-lockstitch seams to provide complete protection, and is ideal for asbestos ripout and other hazardous work.

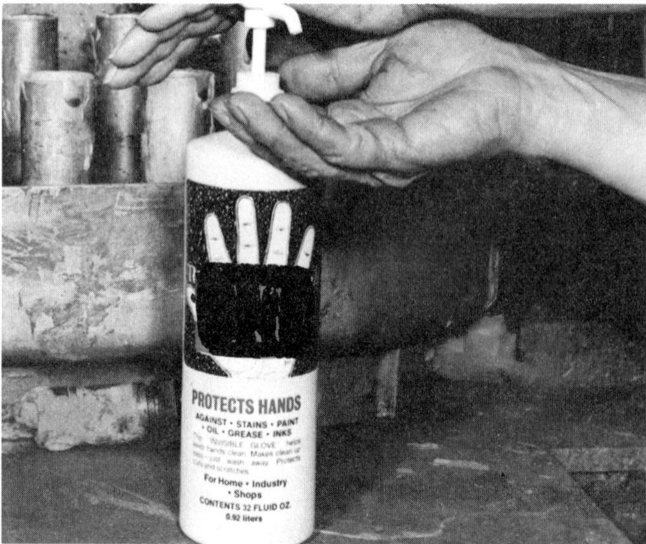

Figure 20-12. There are times when a protective cream may be helpful in preventing contact with harmful agents.

The cream or lotion must be selected on the basis of competent medical advice. The worker must be instructed in the proper type to use and its proper application. (Refer to Protective creams in Chapter 3, The Skin, and Chapter 8, Industrial Dermatoses, for more information.)

Respiratory protective devices are normally restricted for use in intermittent exposures or for operations that are not feasible to control by other methods. Respiratory protection should not be considered a substitute for engineering control methods.

Respiratory protection devices offer emergency or short-term protection. Respirators are a primary protective device for normal operations only when no other method of control is possible (Figure 20-13). Respirators should be used when it is necessary to enter a highly contaminated atmosphere for rescue or emergency repair work; as a means of escape from a suddenly highly contaminated atmosphere; for short-term maintenance or repair of equipment located in a contaminated atmosphere; and for normal operation in conjunction with other control measures when the contaminant is so toxic that other control measures such as ventilation cannot safely be relied on.

An approved respirator must be selected for the particular hazard and environment in which it is to be used (Figure 20-14). The type of air contaminant, its expected maximum concentration, the possibility of oxygen deficiency, the useful life of the respirator, the escape routes available—all these and other factors must be considered in selecting the proper type respirator for emergency use or for standby purposes. When these factors are not known with certainty, the device providing the greatest factor of safety must be used.

There are two general types of respiratory protective devices: (1) air purifiers, which remove the contaminant from the breathing air by filtering or chemical absorption, and (2) air suppliers, which provide clean air from an outside source or from a tank. Full details of types of respirators certified by the National Institute of Occupational Safety and Health (NIOSH) and the Mine Safety and Health Administration (MSHA) should be obtained from the manufacturer. Only NIOSH/MSHA-certified respirators should be used. (See Chapter 23, Respiratory Protective Equipment, for more details.)

Half-mask, cartridge respirators cover the mouth and nose. Full-facepiece respirators also protect the eyes. For dust protection, there are a large number of respirators that have met the requirements established by NIOSH/MSHA, which call for high filtering efficiency and low resistance to breathing. A smaller number of respirators have been certified for protection against metal fumes and mists.

Air-line respirators are usually preferred by the workers to chemical cartridge or mechanical filter respirators, because they are cooler and offer no resistance to breathing; however, they require a proper source of fresh air and a suitable compressor.

Self-contained breathing apparatus—mostly used for emergency and rescue work—have face masks attached by hoses to compressed air cylinders. Such apparatus enables a worker to enter a contaminated or oxygen-deficient atmosphere, up to certain limits specified in the respirator certification.

Selection of the proper type of respiratory protective equipment should be based on the following factors:

1. identification of the substance or substances for which respiratory protection is necessary

Figure 20-13. This operation is provided with clean, respirable air.

2. determination of the hazards of each substance and its significant physical and chemical properties
3. determination of the maximum levels of air contamination expected, probability of oxygen deficiency, and the conditions of exposure
4. determine the capabilities and characteristics essential to the safe use of the respiratory protective device
5. determine what facilities are needed for maintenance.

Since a respirator often becomes uncomfortable after wearing for extended periods, the worker must fully realize the need for protection or he or she will not wear it. To get the worker's cooperation, these factors are important:

1. Prescribe respiratory protective equipment only after every effort has been made to eliminate the hazard.
2. Explain the situation fully to the worker.
3. Fit the respirator carefully according to federal guidelines.
4. Provide for maintenance and cleanliness, including sterilization before reissue.
5. Instruct the worker in the proper use of the respirator.

There is an OSHA respirator program that is required whenever respirators are used. The OSHA requirements for a respiratory protection program are contained in the *Code of Federal Regulations* for General Industry at 29 *CFR* 1910.134. Certain OSHA standards (e.g., Asbestos and Lead Standards) have other specific regulations on respirator use. Check the *Code of Federal Regulations* for this information. (See Chapter 23, Respiratory Protective Equipment, for more details.)

Personal hygiene

Personal hygiene is an important control measure. The worker should be able to wash exposed skin promptly to remove accidental splashes of toxic or irritant materials. If a worker is to minimize contact with harmful chemical agents, there must be easy access to handwashing facilities (Figure 20-15). Inconveniently located washbasins invite such undesirable practices as washing at workstations with solvents, mineral oils, or industrial detergents, none of which is appropriate or intended for skin cleansing.

Many industrial hand cleansers are available as plain soap powders, abrasive soap powders, abrasive soap cakes, liquids, cream soaps, and waterless hand cleaners (Figure 20-16). Powdered soaps provide a feeling of removing soils because of stimulation of the nerve endings in the skin by the abrasives present. Waterless cleaners have become very popular because they remove most soils, such as greases, grimes, tars, and paint, with relative ease. Be aware, however, that some waterless hand cleaners have solvent bases. Soaps may also contribute to industrial dermatitis. Sensitive persons may require pH-neutral soaps or moisturizing agents.

The provision of washing facilities, emergency showers, and eyewash fountains is required when handling hazardous or extremely toxic materials.

When designated or suspected carcinogens are involved, stringent regulation of work areas and activities must be undertaken. The OSHA carcinogen regulations state that the employer must set aside a regulated area where only the particular carcinogen may be produced or used. Only authorized and specially trained personnel with proper personal protection may be allowed to enter that area.

The eating, storage, or drinking of foods and liquids in areas where toxic materials are used should be forbidden. Provide

Figure 20-14. An approved respirator must be selected for the particular hazard and environment.

convenient washing or showering facilities for protection against accidental exposures.

All entrances to the regulated area where biohazards or suspected carcinogens are handled must be properly posted to inform employees of hazards and regular and emergency procedures required. Set aside special areas for employees to change clothing and protective equipment.

Housekeeping and maintenance

Good housekeeping plays a key role in the control of occupational health hazards. Remove dust on overhead ledges and on the floor before it can become airborne by traffic, vibration, and random air currents. Good housekeeping is always important; but where there are toxic materials, it is of paramount importance (Figure 20-17).

Immediate cleanup of any spills of toxic material is a very important control measure. A regular cleanup schedule using vacuum cleaners is an effective method of removing dirt and dust from the work area. Never use an air hose to remove dust from rafters and ledges.

Good housekeeping is essential where solvents are stored, handled, and used. Immediately remedy leaking containers or spigots by transferring the solvent to sound containers or by repairing the spigots. Clean up spills promptly. Deposit all solvent-soaked rags or absorbents in airtight metal receptacles and remove daily to a safe location.

If the thermostat on a vapor degreaser fails, or is accidentally broken, excessive concentrations of trichloroethylene might quickly build up in the work area unless the equipment is shut down immediately and the necessary repairs made.

Continuously monitor airborne contaminants with instrumentation that triggers an alarm when concentrations exceed an established level. The workers or supervisors can then take steps to reduce airborne levels. This approach is also useful to detect abnormal operation conditions.

Maintenance provisions. A key objective should be to provide for periodic shutdown of equipment for maintenance. Provisions should be made for cleaning the equipment and piping systems by flushing them with water, steam, or a neutralizing agent (depending on the particular conditions involved) to render them nonhazardous before dismantling.

Before any equipment is disassembled, it is essential that it be checked for the presence of toxic or hazardous materials. In cases in which this is not possible, employees involved in the disassembling operation should wear proper protective clothing and respirators, if needed. Contaminated equipment, tools and protective clothing must be decontaminated before removing them from the work area.

Waste disposal

Disposal of hazardous materials must be done by highly trained individuals under strict supervision. Procedures should be established in accordance with RCRA regulations (see Chapter 30, Governmental Regulations) for the safe disposal of dangerous chemicals, toxic residues, and other contaminated waste, as well as containers of chemicals that are no longer needed, and containers from which labels have been lost or obliterated.

A competent chemist can determine the best way to neutralize or detoxify chemicals that are no longer needed. In some instances it may be appropriate to perform experimental investigations to determine a means of neutralizing and rendering waste products harmless before full-scale disposal operations are begun. There are a number of methods by which some dangerous chemicals can be rendered safe for disposal. (See the manual, *Hazardous Waste Management—A Manager's Perspective* [National Safety Council, 1986] for detailed information on new RCRA and CERCLA regulations and complete hazardous waste management programs. Chapter 30, Governmental Regulations also contains information on this topic.)

A number of systems and facilities have been suggested for disposal of dangerous chemicals—all of them are expensive and none of them provides a universal means of disposal for all hazardous materials. Deep-well disposal requires geological investigations and the approval of the authorities having jurisdiction, plus a large monetary expenditure. Incinerators require an adequate stack to carry off the products of combustion. Incinerators are also subject to air pollution controls, and may generate off-gases and other products of combustion that may not be acceptable in a populated area.

Reaction with neutralizing agents entails elaborate scrubber facilities, holding tanks, and final treatment facilities, all of which are expensive, require extensive investigation and study, and have limited application. Each disposal method should be considered individually before emergency disposal is imminent.

Special control methods

Many of the general methods mentioned previously (either alone

Figure 20-15. To minimize worker contact with harmful chemical agents, handwashing facilities must be conveniently located.

Figure 20-16. Industrial hand cleansers are available as plain soap powders, abrasive soap cakes, liquids, cream soaps, and waterless hand cleaners.

Figure 20-17. Good housekeeping and maintenance play a key role in the control of occupational hazards.

or in combination) can be used for the control of most occupational health hazards. A few special methods, however, deserve particular mention.

Shielding. This is one of the better control measures used to reduce or eliminate exposures to physical stresses such as heat and ionizing radiation. Lead and concrete are two materials commonly used to shield employees from high-energy ionizing radiation sources, such as particle generators and radioisotopes.

Shielding can also be used to protect employees against exposure to radiant heat sources. Furnaces can be shielded with shiny reflective aluminum panels. Nonreflective metal is not effective for it may act as a "black body," which absorbs heat and then reradiates heat as a secondary source of radiation.

Administrative controls. Reduction of work periods is another method of control in limited areas where engineering control methods at the source are not practical. For example, in the job forge industry, especially in hot weather, a shorter workday and frequent rest periods are used to minimize the effects of exposures to high temperatures, thereby lessening the danger of heat exhaustion or heatstroke.

For workers who must labor in a compressed-air environment, schedules of maximum length of workshift and length of decompression time have been prepared. The higher the pressure, the shorter the workshift and the longer the decompression time period.

However, job rotation when used as a way to reduce employee exposure to toxic chemicals or harmful physical agents must be used with care. Rotation, while it may keep exposure below recommended limits, exposes more workers to the hazard.

Proper illumination of the work area is necessary to prevent eyestrain leading to loss of visual acuity and to eliminate glare that may interfere with vision. Lighting and levels of illumination are discussed in Chapter 11, Nonionizing Radiation.

Medical controls

Medical controls are an important part of an occupational health control program. A medical program can also serve as a verification of the engineering controls. Symptoms of exposure to an occupational health hazard in a group of workers will indicate a failure that must be corrected (Figure 20-18). The extent of the medical controls will depend upon the hazards and seriousness of the risks involved. An industrial hygiene program should parallel the medical program. Both are essential to protect the health of employees.

A physical examination for new employees should include obtaining a thorough detailed history of previous occupational exposures to chemical and physical agents. The preplacement medical examination provides an opportunity to identify persons who are hypersusceptible to specific materials. Of course, those individuals should be restricted from exposure to substances to which they are known to be hypersensitive—even if exposure levels are within permissible levels. The examining physician should decide what physical capabilities are needed to perform the required work.

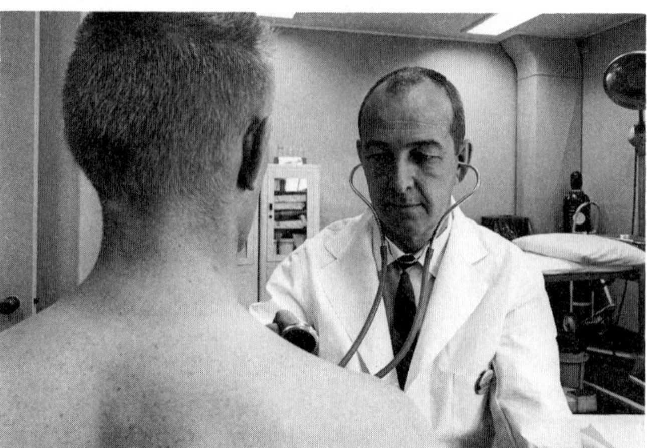

Figure 20-18. Medical controls, including a physical examination for new employees, are an important part of an occupational health control program.

The physician should be aware of the job hazards, and recommend restrictions when this seems warranted. The preplacement examination also provides valuable baseline data which, together with the results of periodic examinations, allow the occupational physician to detect any deleterious effects of work and to assess their severity.

The periodic medical examination is a monitoring procedure supplementary to environmental monitoring. A group of workers with clinical or biological evidence of increased exposure to a contaminant may be the first indication of a breakdown in industrial hygiene control measures. The periodic medical examination is also a means of detecting individual workers who are unduly susceptible to a toxic agent. This provides an opportunity to recommend a change of job for that employee before the onset of serious or permanently disabling disease.

Routine or secondary monitoring of the environment and of people at risk is required in most work situations. A combination of both environmental and medical monitoring is often necessary. Data from the periodic examination of persons and their exposures enable dose-response relationships to be deduced.

Routine periodic analyses of blood and urine samples are practical methods for checking employees exposed to harmful materials. If high levels are found, more stringent environmental controls are needed. For more details, see Chapter 15, Industrial Toxicology.

Medical controls for employees who work with radioactive dusts are more stringent than the medical controls for most industrial materials. An extensive bioassay sampling program is nearly always required.

EDUCATION AND TRAINING

Proper training and education are required to supplement engineering controls. The provision of worker training and information is a requirement under the federal Hazard Communication Standard or applicable state Hazard Communication Standards. In a typical manufacturing plant, the primary responsibility for safe operation and control rests with the line organization of the operations department. This generally would include a first-line supervisor, a shift supervisor, and a plant area manager—all people familiar with every aspect of the day-to-day operation of the plant and the manufacturing process, and readily available when critical decisions must be made.

Supervisors

The education of supervisors usually is process-and-equipment oriented. The aim of the safety and health professional should be to teach them about the safety and health hazards that may be found in their work areas. The supervisors should be told when and under what circumstances to request aid in solving the problems those hazards pose. Supervisors should be knowledgeable and well informed about hazardous processes, operations, and materials for which they are responsible.

Short courses on industrial hygiene can be an easy way to transmit a lot of valuable information with a small expenditure of time. This approach has been used successfully in courses on management techniques and new processes or operations.

Industrial hygiene short courses for managers should identify health hazards in broad areas. The courses should also consider the cost-benefit relationships of controlling health hazards in the work environment.

Workers

The worker must know the proper operating procedures that make engineering controls effective. If performing an operation away from an exhaust hood, the purpose of the control measure will be defeated and the work area may become contaminated. Workers can be alerted to safe operating procedures though booklets, instruction signs, labels, safety meetings, and other educational devices.

The safety and health professional, by persuading a worker to position the exhaust hood properly or to change the manner of weighing a toxic material or of handling a scoop or shovel, can do much to minimize unnecessary exposure to air contaminants. For normal plant operations, a prescribed health hazard evaluation routine should be set up. This should include monitoring the exposures of the personnel involved. This can be accomplished by keeping a record of the exposures to chemical and physical agents in work areas.

In addition to the normal operating instructions that each employee is given when starting a new job, employees assigned to areas where exposures to toxic chemicals can occur must by law be given a special indoctrination program. Such training is required by OSHA under the Chemical Hazard Communication Standard. This program should cover the following types of information:

1. requirements of the standard
2. identity of operations in the work place where hazardous chemicals are present
3. methods and observations used to detect the presence of hazardous chemicals in the work area
4. physical and health hazards of those chemicals
5. hazards associated with chemicals in unlabeled pipes
6. hazards of nonroutine tasks
7. measures that employees can take to protect themselves from these hazards
8. explanation of chemical labeling system
9. explanation of Material Safety Data Sheets (MSDSs)
10. details on the availability and locations of the Hazardous Material Inventory, MSDSs, and other printed Hazard Communication Program materials.

Also, be sure to give employees training in how to respond to emergencies. Information on when NOT to respond is also critical. Many deaths have occurred when untrained workers rushed in to save fallen coworkers and were overcome, themselves.

In order to minimize operator error, employees should be supplied with a detailed instruction manual outlining procedures for all foreseeable situations.

Health hazards affect the workers who are exposed and work directly with materials, process equipment, and processes. These employees should know about the effects of exposure to the materials and energies they work with so that controls can be installed before those problems become severe. A properly informed worker can often anticipate and take steps to control health hazards before they become serious. Once the hazard is known the supervisor or plant engineers can issue work orders to eliminate the problem.

Workers should be given reasons for wearing respirators, protective clothing, and goggles. They also should be informed of the necessity for good housekeeping and maintenance. Since new materials are constantly being marketed and new processes being developed, reeducation and follow-up instruction must also be part of an effective industrial hygiene control program.

SUMMARY

Control of occupational exposures to injurious materials or conditions may be accomplished by means of one or more of the following methods:

- proper design engineering
- substitution of less toxic materials, or change of process
- isolation or enclosure of the source or the employee
- local exhaust ventilation at point of generation or dissemination of the air contaminant
- general ventilation, or dilution with uncontaminated air
- maintenance, housekeeping
- personal protective equipment
- supervisor education
- employee information and training.

One or a combination of these methods may be necessary to prevent excessive exposures to hazardous materials or physical agents.

Education of the workers is important if they are to effectively use the control measures provided for their health and safety. Knowledge by the worker of the proper operating practice is necessary if the engineering controls are to perform the tasks for which they were designed. Knowledge of emergency response procedures is also critical.

Management is responsible for furnishing the facilities and products required to keep the workplace healthful and safe. The worker also has responsibilities in a health hazards control program, including the following: wear protective equipment if it is required; use the local exhaust ventilation system properly, and observe all company rules relating to cleanup and disposal of harmful materials.

BIBLIOGRAPHY

Allen, R.W., Ells, M.D., and Hart, A.W. *Industrial Hygiene.* Englewood Cliffs, NJ: Prentice-Hall, Inc., 1976.

American Conference of Governmental Industrial Hygienists, Committee on Industrial Ventilation. *Industrial Ventilation—A Manual of Recommended Practice,* 19th ed. Lansing, MI: ACGIH, 1986.

Beddows, N.A. Safety and health criteria for plant layout. *National Safety News,* November, 1976.

Burgess, W.A. *Recognition of Health Hazards in Industry—A Review of Materials and Processes.* New York: John Wiley & Sons, 1981.

Cralley, L.V., Cralley, L.J., et al. *Industrial Hygiene Aspects of Plant Operations,* vols. 1-3. New York: MacMillan Publ. Co., 1985.

Cralley, L.E., et al. *Industrial Environmental Health.* New York: Academic Press, 1972.

Hazardous Waste Operations and Emergency Response. 51 *Federal Register* Part 244, Dec. 19, 1986.

Key, M.M., et al., eds. *Occupational Diseases—A Guide to Their Recognition,* rev. ed. DHHS (NIOSH) Publication No. 77-181. Washington, DC: U.S. Government Printing Office, June 1977.

Lynch, J. Industrial hygiene imput into plant design. Paper presented at 65th *National Safety Congress,* Chicago, October 18, 1977.

McDermott, H.J. *Handbook of Ventilation for Contaminant Control (Including OSHA Requirements),* Ann Arbor, MI: Ann Arbor Science Publisher, Inc., 1976.

McRae, A., et al., eds. *Toxic Substances Control Sourcebook.* Germantown, MD: Aspen Systems Corporation, 1978.

NIOSH. *A Guide To Industrial Respiratory Protection.* DHHS (NIOSH) Publication No. 76-189. Washington, DC: NIOSH, June, 1976.

____ *Engineering Control Research Recommendations.* DHHS (NIOSH) Publication No. 76-180. Cincinnati: NIOSH, February, 1976.

____ *The Industrial Environment—Its Evaluation & Control.* Washington, DC: U.S. Government Printing Office, 1973.

National Safety Council. *Hazardous Waste Management—A Manager's Perspective.* Chicago: National Safety Council, 1986.

____ *The Consultant Series: RCRA (Resource Conservation Recovery Act) Compliance Package,* including software and documentation. Chicago: National Safety Council, 1986.

OSHA. Hazard Communication Standard (50 *Federal Register* 51852, December 20, 1985).

Patty, F.A. *Industrial Hygiene and Toxicology, General Principles.* New York: John Wiley & Sons, 1978.

Peterson, J.E. *Industrial Health.* Englewood Cliffs, NJ: Prentice-Hall, Inc., 1977.

Schilling, R.S.F., ed. *Occupational Health Practice.* Toronto, Canada: Butterworth, 1973.

21

Industrial Ventilation

by Willis G. Hazard, AM, CSP
Revised by D. Jeff Burton, PE, CIH

THE PURPOSE OF THIS CHAPTER is to acquaint the health and safety professional with the basic principles of airflow and to use ventilation systems more effectively. Since the prime purpose of industrial ventilation is the protection of health, the health and safety professional has a vital interest in learning the basics.

Regardless of whether a building or plant is large or small, exhaust systems are normally designed by and under the supervision of the plant engineer. However, because the exhaust systems are usually not part of the production equipment of the operating unit, the plant engineer may not take as much time as needed to ensure their satisfactory performance. The health and safety professional can help fill this gap by becoming familiar with ventilation fundamentals. With this book as a guide, and with such assistance as needed from the industrial hygienist or plant engineer, the performance of local exhaust systems can be evaluated. (Note: Publication information on all references given in this chapter can be found in the Bibliography at the end of Chapter 22.)

LOCAL EXHAUST SYSTEMS

A local exhaust system is used to control an air contaminant by trapping it near its source, as contrasted with dilution ventilation which lets the contaminant spread throughout the workroom, later to be diluted by exhausting quantities of air from the workroom.

A local exhaust system is often preferred to ventilation-by-dilution because it provides a cleaner and healthier work environment and because it handles a relatively smaller volume of air, with less attendant heat loss. A local exhaust system also uses a smaller fan and dust arrester. However, before any exhaust sytem is installed, every effort should be made to control the contaminant by isolation or by a change in the process, or by a substitution of less harmful materials. This has already been discussed in Chapter 20, Methods of Control.

A local exhaust system is usually the proper method of contaminant control if:

- air-samples show that the contaminant in the atmosphere constitutes a health, fire, or explosion hazard.
- state or city codes require local exhaust ventilation at that particular process. (For example, most industrial states require ventilation at grinding or buffing wheels and woodworking machines.)
- maintenance of production machinery would otherwise be difficult.
- marked improvement in housekeeping or employee comfort will result.
- emission sources are large, few, fixed and/or widely dispersed.
- emission sources are near the employee breathing zone.
- emission rates vary widely by time.

A local exhaust system consists of five parts (Figure 21-1):

1. *Hoods,* into which the airborne contaminant is drawn
2. *Ducts,* for carrying the contaminated air to a central point
3. *An air-cleaning device,* such as a dust arrester for purifying the air before it is discharged
4. *A fan* and motor to create the required airflow through the system
5. *A stack* to disperse remaining air contaminants.

Some principles of good design relating to each of these are discussed in detail in this chapter.

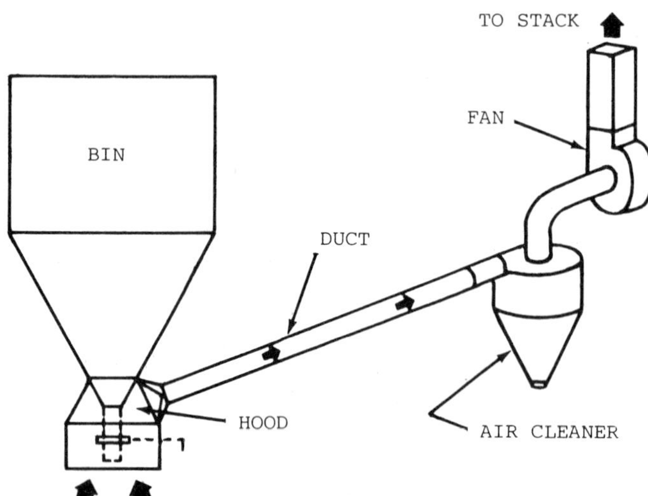

Figure 21-1. A typical local exhaust ventilation system consists of hoods, ducts, air cleaner, fan, and stack. (Courtesy American Conference of Industrial Hygienists [ACGIH])

Hoods

The local exhaust hood is the point of air entry into the duct system. The term "hood" is used in a broad sense to include all suction openings regardless of their shape or mounting arrangement.

No local exhaust system can succeed unless the contaminant is drawn into the hood. Clearly, no matter how well built the ducts and arrester are, or how large the fan is, if the contaminant is not controlled by the hood, the overall value of the installation is nil. The hood is, therefore, a critical component.

There are three basic types of hoods: capture, enclosing, and receiving (canopy) (Figure 21-2).

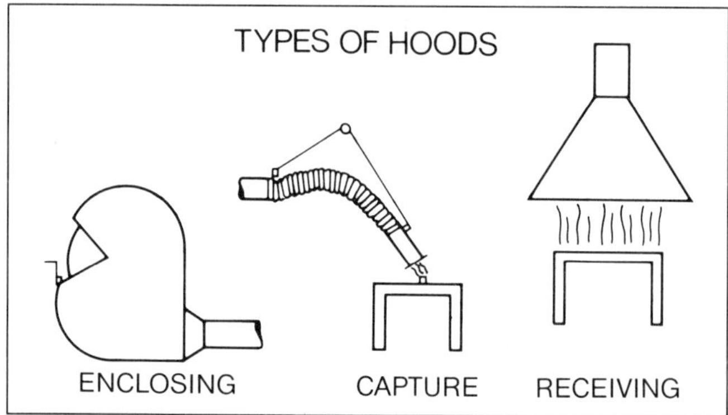

Figure 21-2. Types of hoods include enclosing hoods, capture hoods, and receiving hoods.

Capture hoods. When a duct is connected to the inlet of an exhaust fan, "suction," or an area of low pressure, is set up at the other end of the duct. Air from the room will move toward this low-pressure region. But the air moves in from all directions towards the hood face. Thus, as shown in Figure 21-3, air will move into a freely suspended duct opening from both front and back. The dashed lines going into the duct opening (stream lines) indicate the direction of airflow at that point. The solid curved lines represent spherical contours of equal velocity. What is needed for dust control is an air velocity V, at the point of dust release at a distance x from the duct opening. The air velocity must be high enough to carry the particles into the hood (that is, into the duct). If the amount of air entering the pipe is Q, the velocity at the surface of the sphere (where the dust producer is located) is given by the equation $V = Q/A$, where A is the surface of the sphere. The surface area of any sphere equals:

$$A = 4 \pi x^2 \text{ (ft}^2\text{) or (sq ft)}$$

Then:

$$V = \frac{Q}{4 \pi x^2} \text{ (fpm)} \qquad (1)$$

This relation indicates that the velocity at a point where dust is being released is (1) proportional to volume of air, Q, flowing into the duct cubic feet per minute (cfm) or cubic meters per second (m^3/sec), and (2) inversely proportional to the square of the distance x from the opening.

In practice, the basic equation to measure air velocity (Equation 1) has been modified empirically, and when x is less than 1.5 hood diameters, it has the form:

$$V = \frac{bQ}{x^2 + bA} \text{ (fpm)} \qquad (2)$$

where V = centerline velocity at x *distance from hood*
Q = air flow into duct (cfm),
x = distance outward along hood axis (ft),
A = area of hood opening (sq ft),
b = a constant which depends on the shape of the opening,
fpm = feet per minute.

For circular or square openings, b is essentially 0.1 and the equation becomes:

$$V = \frac{0.1Q}{x^2 + 0.1A} \qquad (3)$$

Suppose the duct is 15 cm (6 in.) in diameter, and the velocity of air passing through is 1,200 m/min (4,000 fpm)—a common situation in dust exhaust systems. Since a 15-cm (6 in.) diameter circle is 180 cm^2 (0.196 sq ft) in area of cross section, Q = 0.196 x 4,000 or 780 cfm (.37 m^3/sec). (In ducts, $Q = AV_{duct}$).

Five centimeters (2 in.) out from the duct end, V has fallen from 500 m/min (4,000 to 1,650 fpm).

At 10 cm (4 in.) from the duct, the velocity is 3 m/sec (600 fpm); at 15 cm (6 in.) away, it is only 88 m/min (290 fpm)—just enough to be felt by the hand.

Where x is very large compared with A, equation (3) becomes:

$$V = \frac{Q}{10x^2} \qquad (4)$$

Flanged hood. If a flange is placed around the duct opening, as shown in Figure 21-3, it will reduce entrance or turbulance loss by preventing the hood from drawing air from behind the hood face. While the same total amount of air is exhausted, a larger portion will come from the front of the duct. This is beneficial, since air that moves from the back of the hood does not help control the contaminant in front. A large flange will increase the useful airflow by 30–40 percent, for the same total

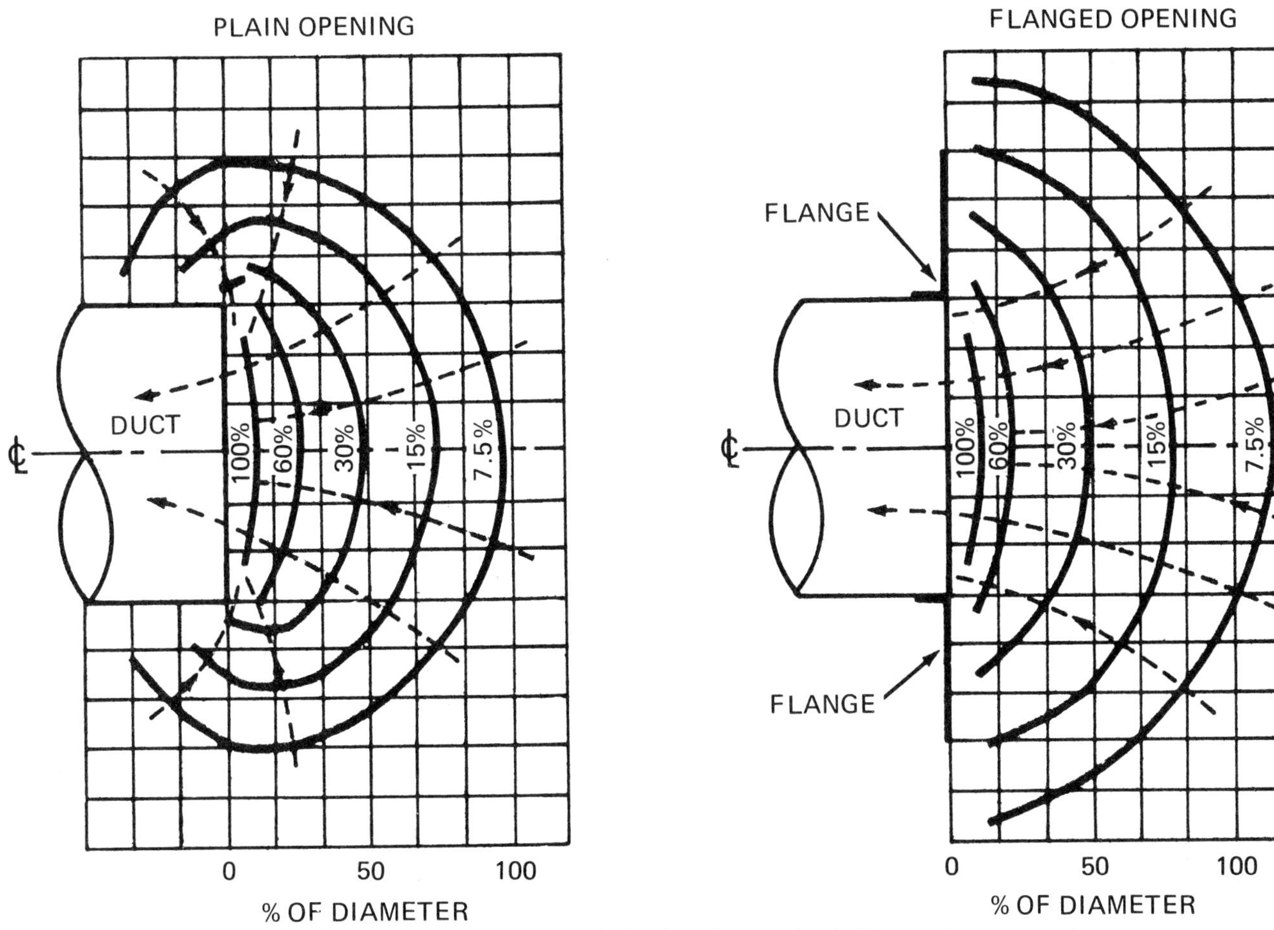

Figure 21-3. Velocity contours expressed as percentages of velocity at the opening (solid curved lines) and stream lines (dashed line) for both plain and flanged circular openings. (Courtesy ACGIH)

volume of air handled. Usually, the flange width equals:

$$W_f = 12x - \tfrac{1}{2}D \tag{5}$$

where:
W_f = flange width (inches)
D = duct diameter (inches) (round ducts)
X = distance outward along hood axis (ft)

Face velocity v. mass air movement. The preceding equations show that hood capture velocity, V, depends on the total airflow entering the hood. This fact is frequently overlooked. A high face velocity or, in the case of a slot-shaped hood, a high slot velocity is not the important factor. The capture of air contaminants depends on mass air movement, Q, not on mere face velocity.

Regardless of face velocity, a source of suction has a woefully poor ability to reach out in a particular direction and induce an inflowing stream of air even a few inches from the usual hood face. Yet, not infrequently, a plant engineer will attempt to improve the exhaust hood by decreasing its size, believing this will raise the velocity. This does no good whatsoever, except in very close proximity to the hood face. It often, in fact, does harm because the total airflow, Q, is then reduced due to increased hood resistance. "Reaching out" can be achieved only from greater mass air movement.

Confusion on this point may occur because common experience shows that air can be blown from a pipe or hose for considerable distances. Air escaping from a compressed air line, for example, will be felt many feet away. Air from a simple desk fan can be felt across the room; a person can blow a match out from several feet away, but try "sucking" or exhausting a match out. It is impossible at any distance beyond a couple of inches.

Figure 21-4 illustrates that air under pressure has a "throw" about 30 times farther than the "pull" on the suction side of the same fan or blower. Air from the pressure side is discharged in a specific restricted direction. Air on the suction side is drawn from all directions into the low-pressure area.

The purpose of capture hoods is to set up air movement at the point or area where the contaminant is released so that a sufficiently large percentage of the contaminant will be drawn to the hood and captured. In selecting the correct control velocity, Table 21-A may serve as a guide.

Baffles and enclosures. Most control velocities at the point of dust or fume emissions are relatively low and can be nullified by drafts from cross breezes or nearby moving equipment. For example, a person walking briskly (at a rate of 1.5 m/sec—300 fpm) or a piece of moving equipment, like a crane or lift truck, often stirs up a significant counter air movement. Unwanted cross drafts can easily blow the contaminant away before it ever comes under the influence of the hood. A cooling or pedestal fan, blowing across an operator working in front of a hood, can defeat the purpose of the hood. Cross drafts from windows, doors, or ventilators can have the same detrimental effect.

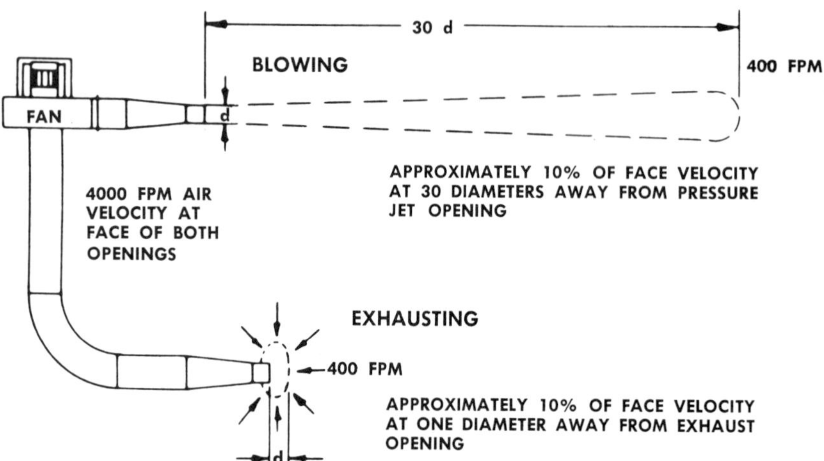

Figure 21-4. The "throw" of a blower can vary greatly, depending on whether it is exhausting or blowing. (Courtesy ACGIH)

Eliminate cross drafts by using baffles, side shields, booths, and other semienclosures and full enclosures wherever possible. All too often such shields are left off, or, if installed originally are left open after maintenance jobs, or for the convenience of the operators. This often destroys the effectiveness of the capture hood.

A full enclosure acts differently and more efficiently than a capture hood because it completely envelops the operation. The only openings in the enclosure are the duct and unavoidable leaks (since most enclosures cannot be completely airtight.) With a capture hood, on the other hand, the air velocity at the point of emission must be high enough to transport the contaminant through free space to the hood and duct, without being diverted by a cross draft.

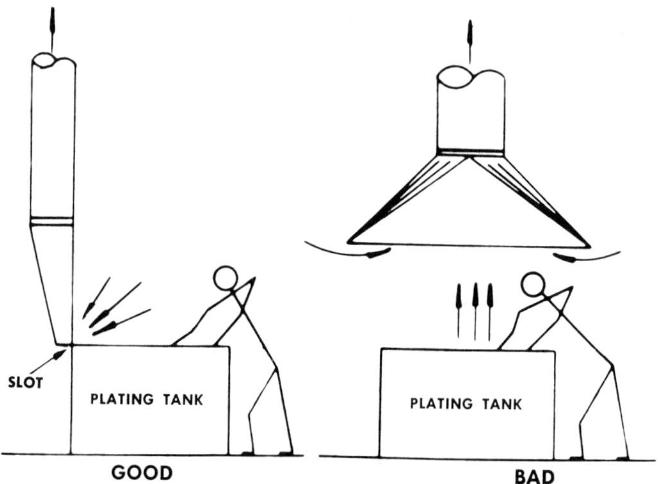

Figure 21-6. Direction of airflow—the hood should be located so the contaminant is removed from the breathing zone of this worker. (Courtesy ACGIH)

Principles of hood design. The following are the more important features of good exhaust hood design:

1. Enclose the operation as much as possible to reduce the rate of airflow needed to control the contaminant, and to prevent cross drafts from blowing the contaminant away from the field of influence of the hood (Figure 21-5).

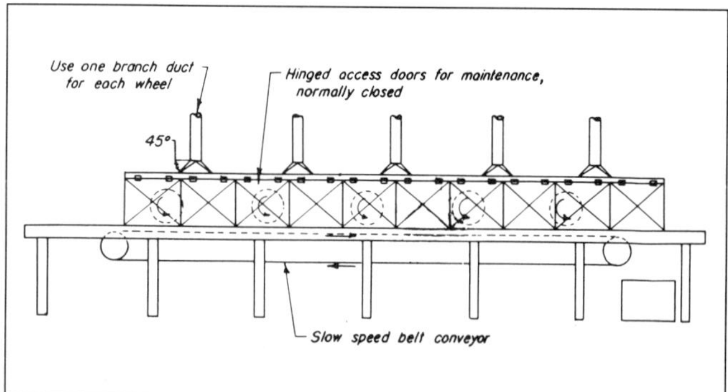

Figure 21-5. Enclosure—the more completely the hood encloses the source, the less air is required for control in this straight-line automatic buffing operation. (Courtesy ACGIH)

2. Locate the hood so the contaminant is moved away from the breathing zone of the operator (Figure 21-6).

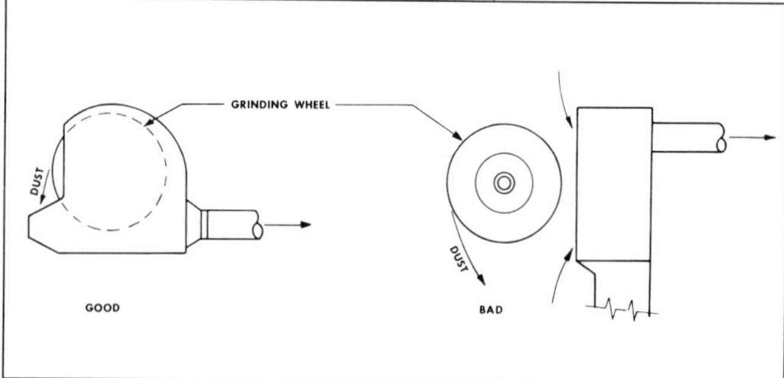

Figure 21-7. The hood should be located and shaped so that the initial velocity of the contaminant will throw it into the hood opening.

3. Locate and shape the hood so the initial velocity of the contaminant will throw it into the hood opening (Figure 21-7).
4. Solvent vapors in health-hazard concentrations are not appreciably heavier than air. Capture them at their source rather than attempt to collect them at the floor level (Figure 21-8).

Table 21-A. Minimum Capture Velocities
Minimum Air Velocities Recommended for the Capture of Dusts, Fumes, Smokes, Mists, Gases, and Vapors Released at Various Types of Operations

Conditions of Generation, Dispersion, or Release of Contaminant	*Minimum Capture Velocity in fpm (m/sec)*	*Examples of Processes or Operations*
Released with no significant velocity into relatively quiet air	50–100 (.25–0.5)	Evaporation or escape of vapors, gases, or fumes from open vessels; degreasing; pickling; plating
Released with low initial velocity into moderately quiet air	100–200 (0.5–1.0)	Spray paint booths, cabinets, and rooms; intermittent dumping of dry materials into containers; welding
Released with considerable velocity or into zone of rapid air movement	200–500 (1.0–2.5)	Some spray painting in small booths and with high pressure; active barrel or container filling; conveyor loading
Released with high velocity or into zone of very rapid air movement	500–2,000 (2.5–10)	Grinding; abrasive blasting; surfacing operations on rock

(Reprinted with permission from ACGIH. *Industrial Ventilation—A Manual of Recommended Practice,* 19th ed. Cincinnati: ACGIH, 1986)

5. Locate the hood as close as possible to source of contaminant (Figure 21-9).
6. Design the hood so it will not interfere with the worker.

Hood design for specific operations. A hood should be shaped and positioned to take advantage of any existing natural movement of the contaminant. For instance, if heat is released, the flow of convection currents should be toward the hood. If heavy particles are thrown off, as in rough grinding, their trajectory should be toward the hood intake. But note this precaution: airborne dust which is of health significance is made up of particles that are extremely small; 0.1 to perhaps 5μ in diameter. (A micrometer is 1/25,000 in.) It has been shown by T.F. Hatch (1935) (see Bibliography) and others that such particles, even if impelled at extremely high initial velocities, travel a very short distance in air—a matter of a few inches at most. Unlike chips or large particles, dust cannot be thrown into a hood which is any considerable distance away (see Figure 21-10).

For even distribution of the airflow through a large hood, flow can be controlled by passing the air through a narrow slot into a plenum chamber. The flow distribution along a hood is fixed by the ratio of the velocity through the slot to the velocity in the chamber. For most purposes, the distribution will be good enough if the velocity through the slot is about twice the average velocity through the plenum. Usually a slot velocity of 10 m/sec (2,000 fpm) is suggested (Figure 21-11).

Recommended designs for hoods for many special operations or applications are given in Section 5 of the ACGIH *Industrial Ventilation—A Manual of Recommended Practice* (see Bibliography). Designs define the size or proportions of hoods and are complete with recommended air quantities, duct sizes, air velocities, and hood entrance losses. This manual should be used for any exhaust hood design problem. The ACGIH also distributes *Industrial Ventilation—A Self-Study Companion to the Ventilation Manual,* a book which explains industrial ventilation in understandable terms (see Bibliography).

Receiving canopy hoods. These seldom succeed as capture hoods. There are several reasons for this—air velocity drops off tremendously (in proportion to the square of the distance) between the canopy and the source of contaminant evolution. A greater volume of airflow is required for contaminant control. Control may be lost due to room air cross currents. Finally, the canopy type of hood frequently forces the employee to work beneath it, that is, between the point of contaminant emission and the hood.

Receiving (canopy) hoods are best used where rising hot air carries the contaminant into the exhaust system. Where this is not vigorous, a fairly even distribution of airflow (within limits) can sometimes be obtained by putting baffles in the center to create a plenum chamber in the upper part of the hood. This is not, however, a safe solution where the operator works under the hood.

The amount of air required for a canopy hood can be approximated by the equation:

$$Q = 1.4 \times 2(L+W)HV \qquad (6)$$

where Q = rate of airflow (cfm)
L = tank length (ft)
W = tank width (ft)
H = height of canopy above tank (ft)
V = desired control velocity (fpm)

V ranges from 0.5-2.5 m/sec (100 fpm-500 fpm), depending on cross drafts.

The effectiveness of receiving hoods can be improved by a vertical baffle or curtain wall (located between the hood and the contaminant source) on one or more sides. If the tank can be enclosed so that only one long side is left open and the hood can be kept at the same height from the tank surface, the air requirement for the same control velocity would be reduced by two-thirds. Even in this instance the operator must not work under the hood.

This reduction in air requirement well illustrates the reason for designing and building hoods that partially or completely enclose the process. If such hoods must be constructed with channels to accommodate monorails or must be made so they will rise or swing to one side for access to the tank or other

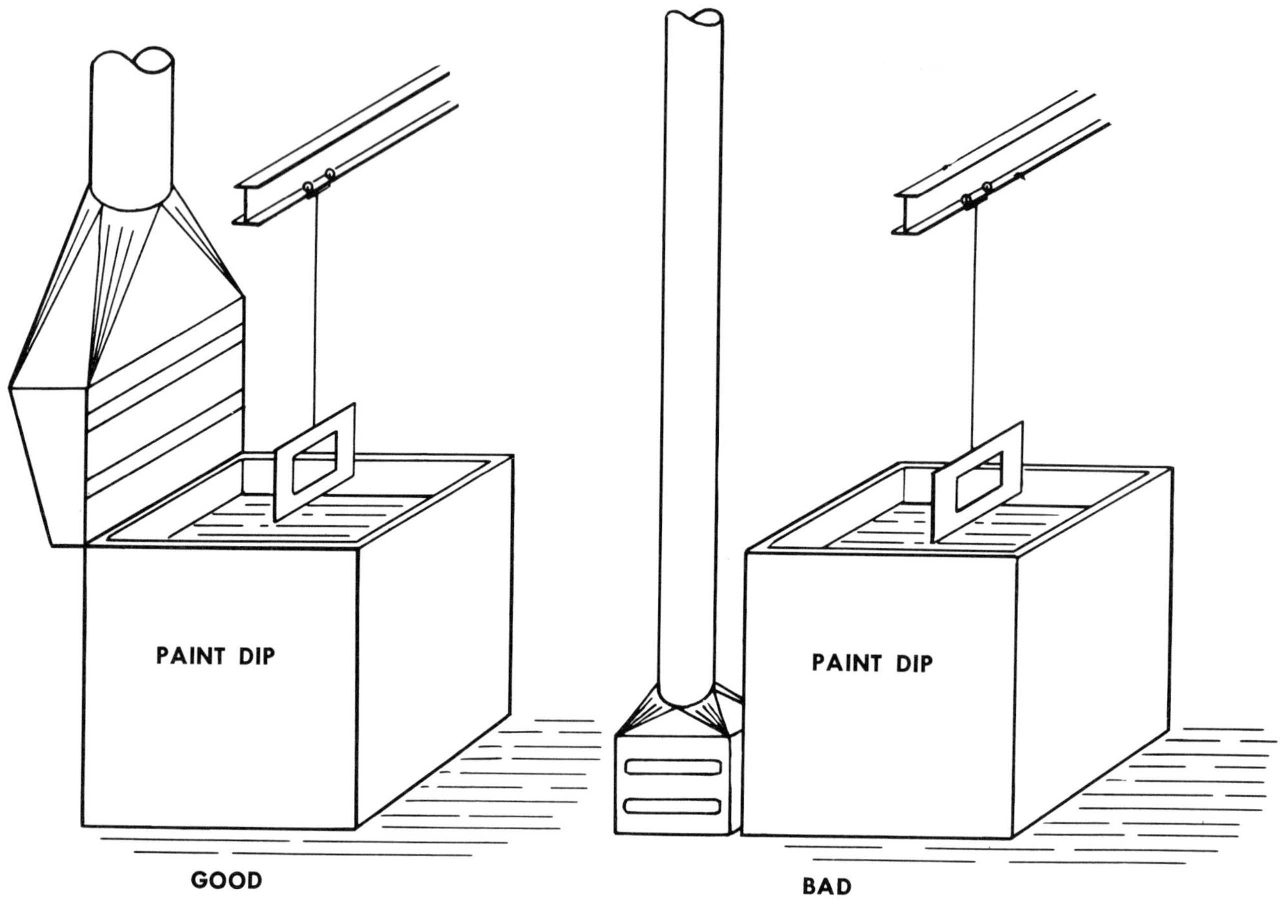

Figure 21-8. Solvent vapors in health-hazard concentration are not appreciably heavier than air. Exhaust from the floor usually gives fire protection only. (Courtesy ACGIH)

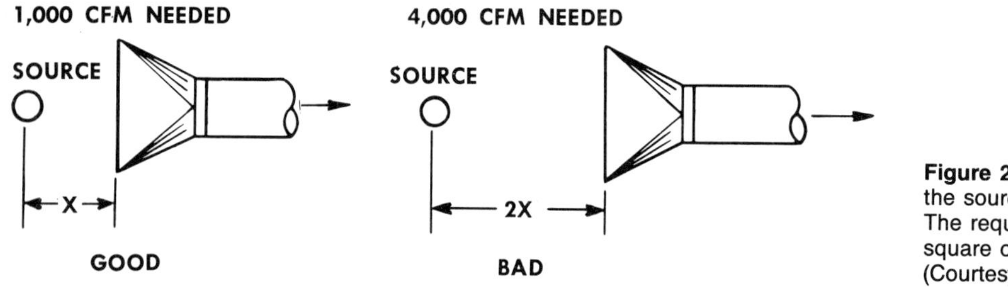

Figure 21-9. Place the hood as close to the source of contaminant as possible. The required volume varies with the square of the distance from the source. (Courtesy ACGIH)

equipment, enclosing or baffling of the contaminant source is useful.

If the canopy cannot be used because of the nature of the process, the most logical compromise is a side hood which overhangs the tank as much as possible.

A lateral exhaust hood, this type of hood shown in Figure 21-11, is used where full access to the top of the tank must be had, as in electroplating. (See the section Open-surface tanks under Special Operations, later in this chapter.)

With tanks up to 0.7 m (24 in.) wide, sufficient control may be obtained with a slot on only one long side. On wider tanks, two sides should have exhaust slots. Total airflow is given by this formula:

$$Q = CLW \qquad (7)$$

where Q = airflow (m^3/sec or cfm)

C = ventilation rate, varying from 50 to 500 cfm per square foot of tank surface, the usual choice being 75–225 fpm

C = cfm/sq ft

L = tank length (ft)

W = tank width (ft)

Each slot would handle ½Q. Since the velocity of air decreases rapidly at increasing distances away from the slot face, lateral exhaust should not be used when the tank is more than 1.2 m (4 ft) wide.

Control velocity over a wider tank can sometimes be obtained by a *push-pull* system, that is, by blowing air from a nozzle at one side into a hood on the other side of the tank. A jet of air under pressure can be confined to a much narrower angle than can be obtained from a suction hood, so that a wider tank can

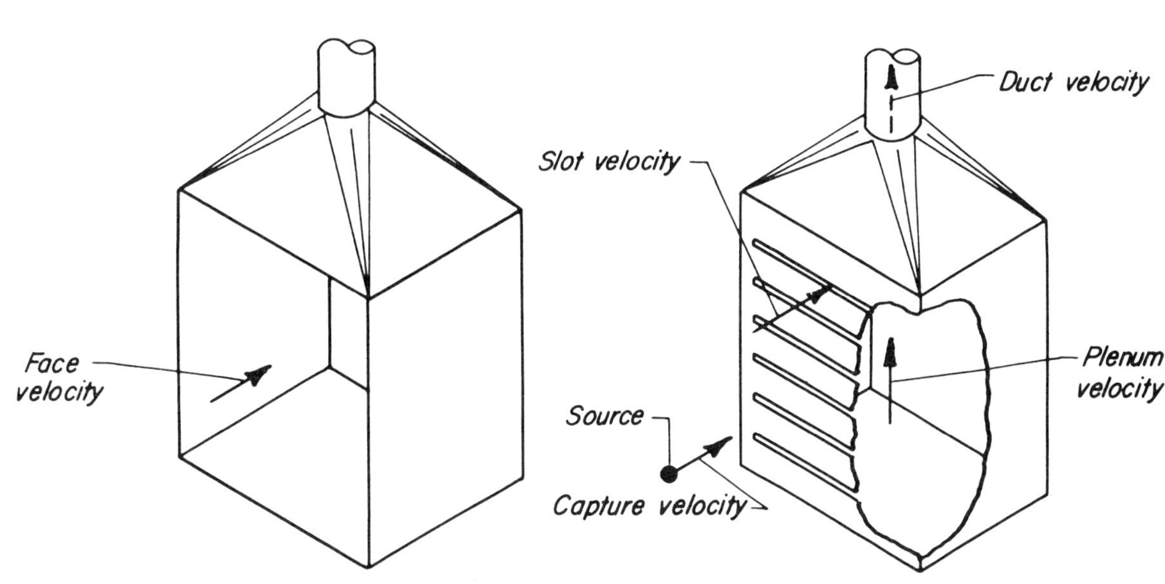

Capture Velocity – *Air velocity at any point in front of the hood or at the hood opening necessary to overcome opposing air currents and to capture the contaminated air at that point by causing it to flow into the hood.*

Face Velocity – *Air velocity at the hood opening.*

Slot Velocity – *Air velocity through the openings in a slot-type hood, fpm. It is used primarily as a means of obtaining uniform air distribution across the face of the hood.*

Plenum Velocity – *Air velocity in the plenum, fpm. For good air distribution with slot-types of hoods, the maximum plenum velocity should be 1/2 of the Slot Velocity or less.*

Duct Velocity – *Air velocity through the duct cross section, fpm. When solid material is present in the air stream, the duct velocity must be equal to the Minimum Design Duct Velocity.*

Minimum Design Duct Velocity – *Minimum air velocity required to move the particulates in the air stream, fpm.*

Figure 21-10. Principles of exhaust hoods. (Courtesy ACGIH)

be served. The jets will entrain material rising from the tank surface and carry it into the hood.

Air jets should be used only with great caution, because it is possible that they can spread the contamination into the working area. For instance, if material is lowered into and raised from the tank and passes through the air jet to dry, the air jet can be broken up and deflected so as to spread the contamination. The lateral exhaust hoods must be adequately sized and exhaust an adequate amount of air. The jet airstream must be pointed in the proper direction so that the contamination can be picked up by the hood.

Employee participation. Finally, a hood should not interfere with the employee's job. The hood is there to *help* the employee, not hinder. If an employee discovers that a hood inhibits the operation, the employee is likely to remove or alter it.

A job must be analyzed thoroughly in advance. The operator's motions must be studied, and process changes must be

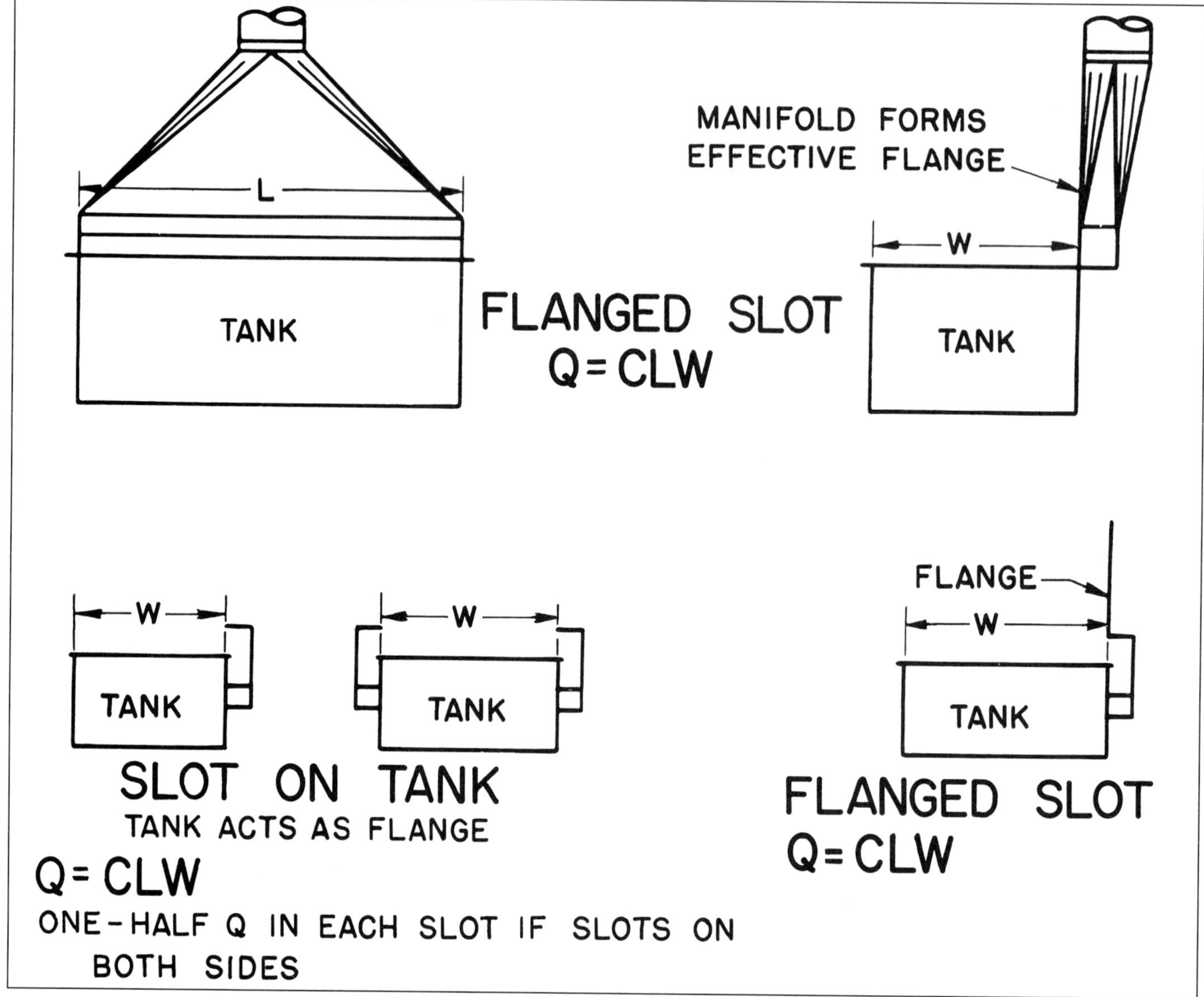

Figure 21-11. Installations that give good results create an airflow past the source to capture the contaminant. In these diagrams Q=required exhaust volume in m^3/s (cfm); L=length of slot in meters (ft); W=width of table or tank in meters (ft); and C is a coefficient that varies from 50–500, with the usual choice being 75–225. (Courtesy ACGIH)

considered. Wherever possible, the person on the job should be consulted and opinions solicited. It may be wise to explain exactly how the proposed hood will work to help. Training of this sort is important to the optimum use of a hood.

After the hood is installed, the employee should use it to the fullest. A cooling or pedestal fan should *not* be placed where it interferes with the effectiveness of an exhaust hood. Nor should any adjustments be made in the exhaust system by the employee. The employee should be instructed to tell the supervisor when there is a decline in exhaust control.

Ducts

After contaminated air has been drawn into a hood, ducts serve the purpose of carrying that air to an air cleaner or to the outdoors. When air passes through any duct or pipe, friction must be overcome. That is, energy must be expended. The amount of this friction loss must be calculated before the system is installed, so the proper size fan and motor can be purchased. However, this is a task for a plant engineer or the ventilation engineer, not for the health and safety professional.

Several excellent references have been published; these remove the guesswork from duct design. (See Bibliography, especially Brandt [1945], Burton [1986], Dalla Valle [1948, 1952, 1953], and Hemeon [1963], and the publications of American Foundrymen's Society and the American Conference of Governmental Industrial Hygienists).

Some general comments on sizing of ducts will be helpful to the health and safety professional in appraising the merits of a given system.

A good starting point in designing a local exhaust system is determining the volume of air per minute that must be handled by each hood to control the contaminant released in the workroom. Based on such data, careful duct design accomplishes the following objectives:

- holds power consumption to a minimum
- maintains proper transport velocity so the contaminant, if it is a dust or fume, will not settle out and plug the pipe
- keeps the system "balanced" at all times.

Multiple ducts. Local exhaust systems with multiple hoods pose problems (Figure 21-12). After deciding the amount of airflow needed at each hood to control the contaminant, the task of the duct designer is to select pipe sizes and fittings (such as elbows, Y's, and reducers) so that air will be distributed from hood to hood as desired. When two branches, coming from two hoods (hood A and hood B, for example) join at a Y to form a single main (or submain or header), the static pressure between this junction point and the face or inlet of hood A is of necessity the same as between this point and the face of hood B. If the same rate of airflow is desired from both hoods, the duct loss in each branch must be equal. But if the branch to hood A is longer than the one to hood B, or has more elbows, its loss, for the same diameter pipe in both branches, will be greater than in branch B. The velocity of air in branch A will be less and, therefore, less air will flow into hood A than hood B.

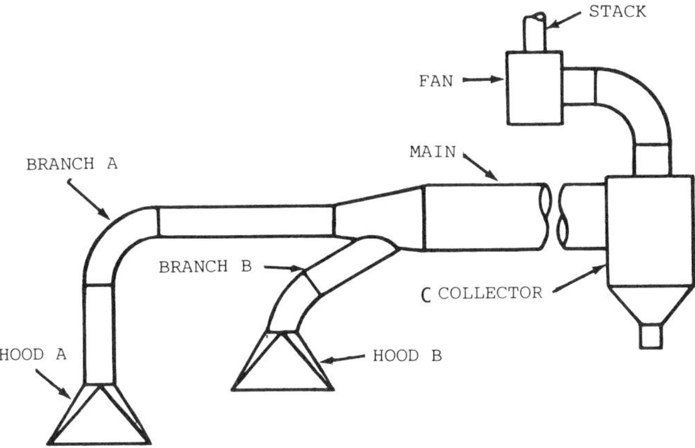

Figure 21-12. A typical local exhaust system: hoods, ducts, collector, fan, and stack.

How can the hoods be made to have equal airflows? There are two solutions, the first presenting the preferred choice.

1. Balance can be achieved. That is, each hood can be constructed to handle the amount of air it is supposed to. This can be achieved by sizing the ductwork and selecting fittings that will give equal pressure drops in the two branches when airflow through hood A and hood B is at the desired rates. If one hood will exhaust more air than it should, smaller diameter piping can be installed. This will increase the resistance, cause air velocity to drop, and allow the air flowing in that duct to be reduced.
2. Air balance may be achieved by use of blast gates in each branch. These blast gates are slide gates or dampers that can be set to partially block the flow of air in order to lower the amount of air entering the hood. However, this is not the preferred method.

Advantages of the first method are (1) if velocities are initially chosen properly, the ductwork will not become plugged; (2) the system cannot be tampered with (tampering may deprive another workstation of the air needed for contaminant control there); and (3) erosion will be reduced and there will be no buildup of dust or linty material caused by blast-gate obstruction. This method is preferred where toxic materials are handled, and is mandatory where explosives, magnesium, and radioactive dusts are exhausted.

The second method gives some leeway for correcting improperly estimated exhaust volumes and also provides some flexibility for future changes or additions. However, a cardinal rule in local exhaust system design is that once a multihood layout is completed and balanced, additional hoods should not be added later. Additional hoods can alter the airflow and make some hoods entirely ineffective.

Whether a system operates under positive (blowing) or negative (suction) pressure makes no difference insofar as duct design features are concerned. Usually, however, exhaust systems have high air velocities to keep dust in suspension within the duct. Positive pressure systems (those that supply ventilating air to a given space) operate with relatively low duct velocities. Table 21-B lists recommended minimum duct velocities for ducting different types of contaminants. Three terms that relate to duct velocity should be understood:

- *Static pressure.* Static pressure is created by the fan and is the energy source of the system. Static pressure is converted to velocity pressure and to heat in the form of friction loss and other "losses."
- *Velocity pressure.* A definite pressure is created by moving air. Velocity pressure is created by converting static pressure to air movement at the hood.
- *Duct friction loss.* For the same velocity of air, narrow-diameter ducts have higher friction loss than large-diameter ducts. As an example, with the same velocity in each, a 10-cm (4-in.) diameter duct 4.2 m (14 ft.) in length has the same friction loss as a 25-cm (10-in.) diameter duct 12 m (40 ft) in length. Narrow ducts create extremely high power demands.
- *Other losses.* Elbows and tapered sections where a branch enters a main, all add their loss. The sharper the bend and the more abrupt the taper in transition pieces, the higher the loss.

Common piping defects. A quick inspection of an exhaust system will reveal a good deal about how well designed it is. For example, square and rectangular ducts, so common in the heating and ventilating field, are seldom used for local exhaust systems. Round piping is used because of its lower friction loss and resistance to collapsing. Duct velocities in the local exhaust systems are much higher to prevent a settling of particles.

Furnace-type elbows and T-fittings that come with "hardware-store" furnace pipe should be avoided. Because their sharp corners waste power, a fan must be sized larger than normal to handle the amount of air necessary to keep dust from dropping out of the air stream. This settling may plug ducts.

Do not be misled by the outdated rule that the sum of the areas of cross section of all branches must at least equal the cross sectional area of the main pipe. This will not result in a balanced system.

Air cleaners

Air cleaners fall into two broad classes according to their use.

- First, there are industrial air cleaners whose purpose is to remove airborne contaminants (dust, fume, mist, vapor, gas,

Table 21-B. Range of Design Velocities

Nature of Contaminant	*Examples*	*Design Velocity*
Vapors, gases, smoke	All vapors, gases, and smokes	Any desired velocity (economic optimum velocity usually 1,000–1,200 fpm)
Fumes	Zinc and aluminum oxide fumes	1,400–2,000
Very fine light dust	Cotton lint, wood flour, litho powder	2,000–2,500
Dry dusts and powders	Fine rubber dust, Bakelite molding powder dust, jute lint, cotton dust, shavings (light), soap dust, leather shavings	2,500–3,500
Average industrial dust	Sawdust (heavy and wet), grinding dust, buffing lint (dry), wool jute dust (shaker waste), coffee beans, shoe dust, granite dust, silica flour, general material handling, brick cutting, clay dust, foundry (general), limestone dust, packaging and weighing asbestos dust in textile industries	3,500–4,000
Heavy dusts	Metal turnings, foundry tumbling barrels and shakeout, sand blast dust, wood blocks, hog waste, brass turnings, cast iron boring dust, lead dust	4,000–4,500
Heavy or moist dusts	Lead dust with small chips, moist cement dust, asbestos chunks from transite pipe cutting machines, buffing lint (sticky), quick-lime dust	4,500 and up

(Reprinted with permission from ACGIH. *Industrial Ventilation—A Manual of Recommended Practice,* 19th ed. Cincinnati: ACGIH, 1986)

or odor) that would otherwise pollute the surroundings, either in-plant, or more generally, the outside plant neighborhood. These air cleaners are designed to function effectively under light to heavy loadings of contaminant. They can be cleaned of collected contaminant. They handle a relatively moderate rate of airflow at relatively high static pressure. This type of air cleaner is the principal concern of this section.

- Second, there are air cleaners that handle relatively high rates of airflow at low static pressure. These are associated with heating, ventilating, and air conditioning systems used in domestic, commerical, office, public, and other such buildings. This type of air cleaner removes particulates from incoming outdoor air and recirculated air to provide "clean" air for the building. This filter, unlike the industrial type, functions usually under a light dust loading (usual range: loadings of up to [70 gm/m³] 0.003 grains/cu ft of air handled). In addition, the filter unit as a whole, or the filter medium, cannot be cleaned. Rather, it is disposable (although there are exceptions, like the electrostatic precipitator).

Industrial-type air cleaners include simple settling chambers, wet and dry centrifugals, wet and dry dynamic precipitators, electrostatic precipitators, wet-packed towers and venturi scrubbers, washers, and fabric filters.

A suitable air cleaning device should be standard equipment in every exhaust system handling a contaminant that might result in a health hazard, a nuisance, or cause air pollution. Increasing community awareness of air pollution and its cost to health and property in the neighborhood is making air cleaning necessary. Many areas have pollution-control ordinances that require a permit. Be sure the collected contaminants are disposed of properly so as not to cause additional pollution (Chapter 20, Methods of Control).

Many types of air-cleaning devices are in use, and the selection of the proper one depends on several factors. The nature of the material to be removed, the quantity of the material, the degree of cleanliness of the outflow that must be achieved, and cost must all be considered. Most air cleaners are housed in unrevealing enclosures, so the lay person generally has no idea what type they are or how to check their performance. General familiarity with types of cleaners and familiarity with their gross shortcomings, in particular may be useful. Before discussing individual cleaners, we will look at how to measure and compare air cleaner efficiency.

Air-cleaner efficiency

Efficiency is the ratio of the amount of dust (or other contaminant) collected by the air cleaner to the amount of the dust that enters the device. Ordinarily, the units of measurement are in terms of weight of dust. The term *weight efficiency* is sometimes used because arresters are often used in the processing industries where the mass of the product handled, shipped, or reclaimed is the important dollars-and-cents consideration.

Other factors enter the efficiency concept when air cleaners are used to arrest a contaminant that is a hazard to health. Such contaminants may include silica, lead, beryllium, cadmium, or zinc fume, arsenic, or other toxic respirable dust. Workers may be exposed to an increased risk of lung disease, which can be caused by small particles—those less than 5μ in size.

Most inertial air cleaners (for example, cyclones) have a variable collection efficiency depending on particle size. These collectors can capture essentially 100 percent of the large particles (say, greater than 100 micrometers), but their efficiency drops off sharply with decreasing particle size. They may be ineffective in collecting particles less than 5 micrometers, the size that is of concern to workers' health.

Because the mass of a particle depends on the cube of its diameter, the weight pyramids enormously as the diameter of the particle increases. For example, a particle of 100μ (0.004 in. or 0.10 mm) weighs as much as one million particles 1μ in size. An air cleaner that passed practically all the dust in the 0.5- to 5-μ range might have a weight efficiency of 99+ percent, if the air contains large numbers of large particles. What the health and safety professional is interested in is not weight efficiency alone, but the number of particles in the respirable size range that will get through the filter (Chapters 1, Overview of Industrial Hygiene, and 7, Particulates).

Dalla Valle states that "efficiency" is merely one aspect on which the customer wants assurance. What the customer really seeks is a "performance rating," that would include not only efficiency, expressed in terms of count instead of mass, but also other important factors. These factors include rate of resistance

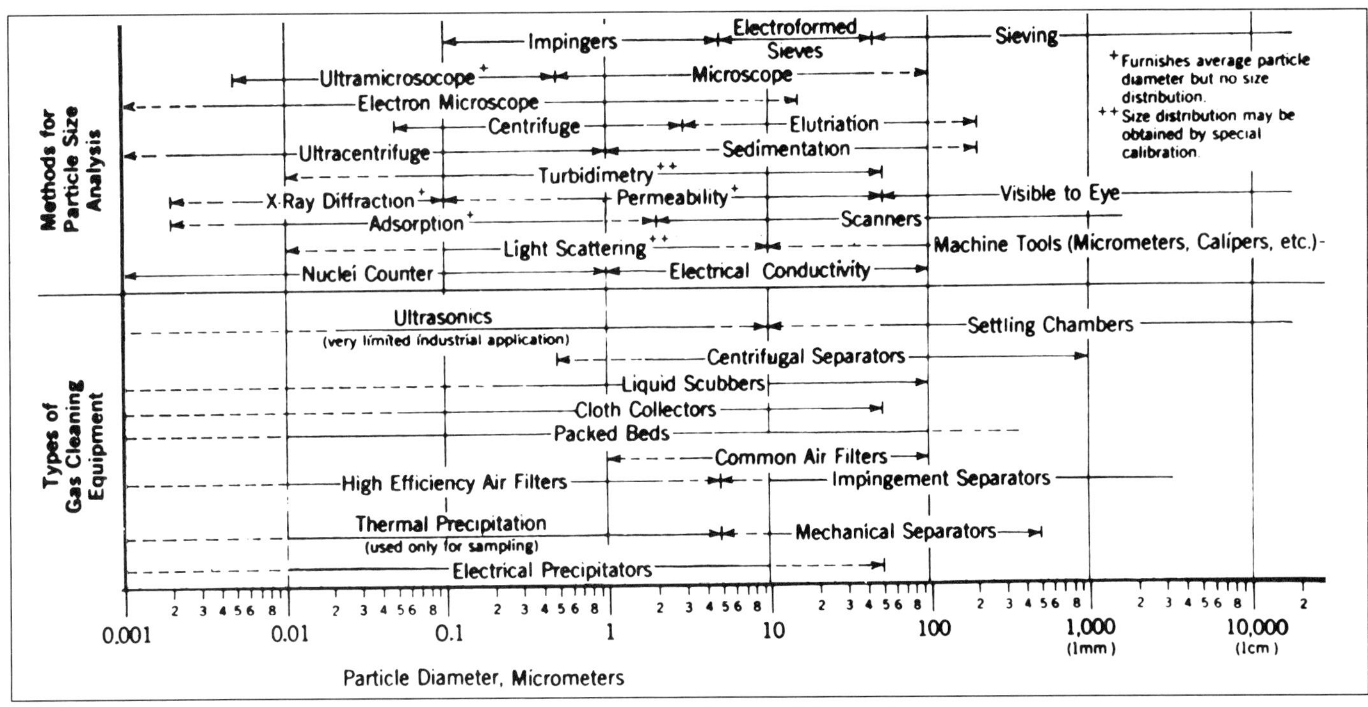

Figure 21-13 Particle size ranges that can be analyzed by various techniques and that can be collected by various types of gas-cleaning equipment. (Reprinted with permission from *Stanford Research Institute Journal* 5 [1961]:95)

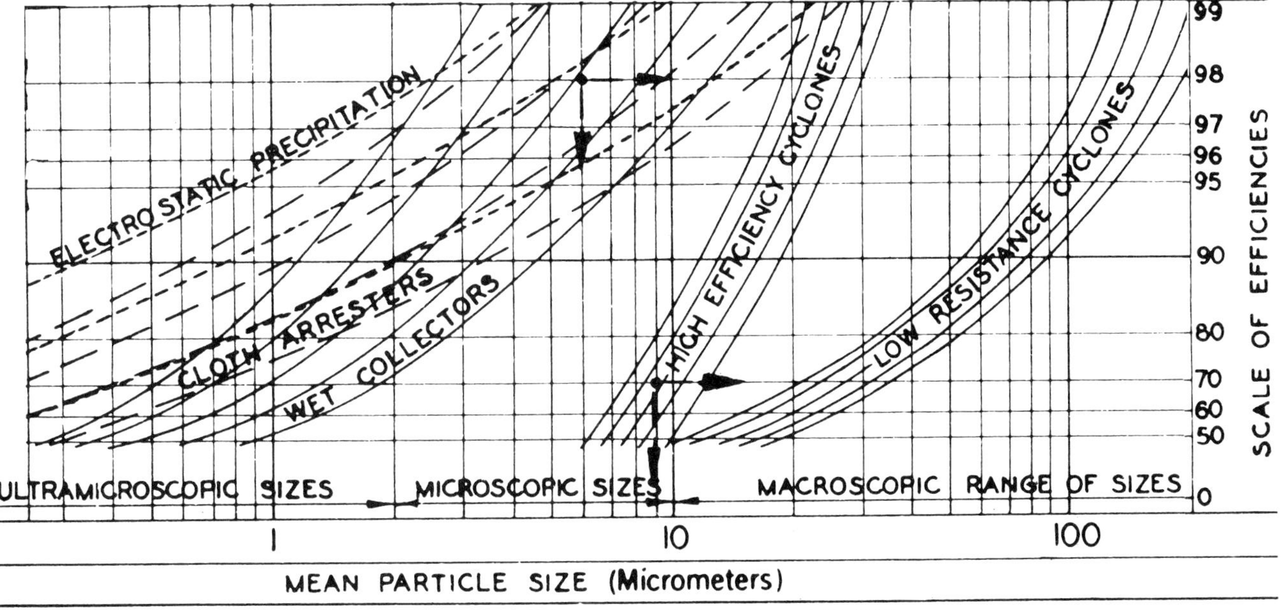

Figure 21-14 Relative performance and efficiencies of various dust collectors. (Copyright 1952, American Air Filter Co., Inc., from ACGIH, *Industrial Ventilation—A Manual of Recommended Practice*)

buildup, dust-holding capacity, and cleanabilty (in permanent filters) or longevity (in throwaway types).

Performance measures vary between manufacturers. The health and safety professional, if involved with the project, should make certain that the vendor agrees to a proper performance guarantee as part of the purchase contract. The health and safety professional must beware of that often misleading word *efficiency* as expressed by percent.

A comparison of important filter characteristics of ventilating system air filters is given in Table 21-C. A comparison of the size range of particles that can be collected by the various types of collectors and filters is shown in Figures 21-13 and 21-14.

Dust arresters

Dust is formed by mechanical actions like grinding, crushing, blasting, and drilling. Dust already formed may be dispersed by the handling, conveying, and dumping of bulk materials, and their weighing, mixing, and packaging. Dust particles range in

Table 21-C. Comparison of Some Important Air Filter Characteristics

Air filters should be used only for supply air systems or other applications where dust loading does not exceed 1 grain per 1,000 cubic feet of air.

Efficiency	Type	Pressure Drop in wg (Notes 1 & 2)		ASHRAE Performance (Note 4)		Face Velocity fpm	Maintenance (Note 5)	
		Initial	Final	Arrestance	Efficiency		Labor	Material
Low/ Medium	1. Glass Throwaway (2″ Deep)	0.1	0.5	77%	NA Note 6	300	High	High
	2. High Velocity (Permanent Units) (2″ Deep)	0.1	0.5	73%	NA Note 6	500	High	Low
	3. Automatic (Viscous)	0.4	0.4	80%	NA Note 6	500	Low	Low
Medium/ High	1. Extended Surface (Dry)	0.15-0.60	0.5-1.25	90-99%	25-95%	300-625	Medium	Medium
	2. Electrostatic:							
	a. Dry Agglomerator/ Roll Media	0.35	0.35	NA Note 7	90%	500	Medium	Low
	b. Dry Agglomerator/ Extended Surface Media	0.55	1.25	NA Note 7	95% +	530	Medium	Medium
	c. Automatic Wash Type	0.25	0.25	NA Note 7	85-95%	400-600	Low	Low
Ultra High	1. HEPA	0.5-1.0	1.0-3.0	Note 3	Note 3	250-500	High	High

Note 1: Pressure drop values shown constitute a range or average, whichever is applicable.

Note 2: Final pressure drop indicates point at which filter or filter media is removed and the media is either cleaned or replaced. All others are cleaned in place, automatically, manually or media renewed automatically. Therefore, pressure drop remains approximately constant.

Note 3: 95-99.97% by particle count, DOP test.

Note 4: ASHRAE Standard 52-76 defines (a) Arrestance as a measure of the ability to remove injected synthetic dust, calculated as a percentage on a weight basis and (b) Efficiency as a measure of the ability to remove atmospheric dust determined on a light-transmission (dust spot) basis.

Note 5: Compared to other types within efficiency category.

Note 6: Too low to be meaningful.

(Reprinted with permission from ACGIH. *Industrial Ventilation—A Manual of Recommended Practice,* 19th ed. Cincinnati: ACGIH, 1986)

size from submicroscopic (less than 0.5 μ), through the range of dusts most likely to cause lung injury (0.5 – 5 μ for free silica, for example), and up into the standard screen mesh sizes (325 mesh = 40 μ, approximately). "Dust," of course, has many meanings. People speak of wood dust, house dust, highway dust, as well as industrial dust. Dust may also loosely suggest dirt.

Centrifugal collectors. The cyclone belongs to a class which relies on centrifugal force to throw the dust out of the airstream.

LOW-PRESSURE CYCLONES are sheet metal cylinders set on top of a cone. Air enters at an angle on the side of the cylinder and swirls around inside, passing downward. It then rises through a center tube and passes out the top of the cylinder. Centrifugal force throws the particles out of the airstream to the bottom of the cone.

There are several advantages and disadvantages of the low-pressure cyclone. The advantages are low cost, low maintenance, and low pressure drop (0.2–0.4 kPa [¾–1½ in.] water gage). The disadvantage is that the low-pressure cyclone is incapable of collecting fine particles. In fact, it is only about 75 percent efficient on particles that are as large as 40 μ. The low-pressure cyclone is worthless against dusts of hygienic significance (5 μ and less).

The primary uses of this cyclone-type are to collect woodworking dust (sawdust, shavings, and chips), paper scraps, and bulk materials after being pneumatically conveyed.

Collecting efficiency increases with an increasing pressure drop. Pressure drop results from making the diameter of the cyclone smaller. Small diameter cyclones cannot carry large volumes of air. This disadvantage, however, is overcome if several are placed in parallel.

HIGH-EFFICIENCY CYCLONES are made by combining many small diameter, high-resistance units. Efficiencies of 75 percent against 10-μ particles, and 90 percent against 15-μ particles can be obtained. However, efficiency may be inadequate for control of dusts of health importance (5 μ and less). Its pressure drop is high: 0.75 –2.0 kPa (3–8 in.) water gage. With abrasive dusts, castings or heavy sheet metal of special alloy must often be used to allow for the expected excessive wear.

DRY DYNAMIC PRECIPITATORS are a combined air cleaner and fan in one unit. They have a specially shaped impeller which precipitates the dust by centrifugal force, while inducing airflow. The collection efficiency is about the same as that of the high-efficiency or high-pressure cyclone.

With abrasive dusts, the impeller blades are subjected to considerable wear. Normally the fan for inducing airflow is placed downstream from the collector to provide maximum protection.

Wet collectors. The principle employed in wet collectors is to get the dust into intimate contact with a liquid. Either the particles are trapped directly because they are in a bath or stream of water, or, because of the increased mass the liquid gives them, they can be better thrown out of the air stream by centrifugal force. Wet collectors can handle gases that are at a high temperature or are laden with moisture. With this type of collector, disposal of collected material presents little problem. However, wet collectors have the following disadvantages:

1. Unless cheap water is handy, settling tanks and accompanying equipment will be needed to reuse the water.
2. Freezing cannot be tolerated.
3. If corrosive chemicals are collected, expensive corrosion-resistant materials must be used.

Although the efficiency of wet collectors is usually only 75 percent or less, it can be as high as 90 percent against 1-μ particles. The efficiency averages less than 50 percent against 0.5-μ particles. The pressure drop of the wet collectors also varies over a wide range.

There are at least seven types of wet collectors.

SPRAY CHAMBERS, consisting of sprays and scrubber plates, are one of the oldest methods of air cleaning. Pressure drop varies, usually being in the range 0.6–1.5 kPa (2½–6 in.) water gage. Water consumption is about 0.32–0.53 liters/min per m^3/min (3–5 gpm per 1,000 cfm) of air handled.

PACKED TOWERS, extensively used by the chemical industry for gas absorption, are also used for collecting toxic dusts. In this type of collector, water usually flows downward over packing, which may be odd-shaped ceramic saddles, or rings, or coke, gravel, or similar material. Water flow is 0.53–1.1 L/min/m^3/min (5–10 gpm/1,000 cfm) of air. Packing depth is usually about 1.2 m (4 ft). Pressure drop is 0.4–0.9 kPa (1½–3½ in.).

WET CENTRIFUGALS use a combination of centrifugal force and water contact to effect collection. Air is introduced tangentially and frequently directed counter-current to flow of water by baffles or directional plates. Pressure losses range from 6.0–0.15 kPa or 2½–6 in. (wg).

WET-DYNAMIC COLLECTORS are similar to the dry dynamic precipitator with the specially shaped impeller, except that water is sprayed on the blades. Water consumption is 1.8–3.8 L/min for 0.47 m^3/sec (0.5–1 gpm for 1,000 cfm) of air.

ORIFICE-TYPE COLLECTORS are arranged so airflow through them is brought into contact with a sheet of water in a restricted passage.

VENTURI-TYPE COLLECTORS make use of the high velocity of air through a venturi throat to "break up" water fed into the throat. These fine water droplets collide with the dust in the air stream, and the dust is wetted. These are high-efficiency collectors being 90–99 percent efficient in the size range 0.1–1 micrometer. Their power consumption is, however, high. The pressure drop is from 2.5 kPa–12.4 kPa (10–15 in. wg).

FOG FILTERS use many small high-pressure nozzles in centrifugal tower-type units to increase the probability of water droplets wetting the dust particles. High water pressure is used to form the droplets of high-energy fog. Good efficiency has been obtained.

Electrostatic precipitators. Air that contains solid or liquid particles is passed between a pair of electrodes—a discharge electrode at high negative potential and an electrically grounded collecting electrode. The potential difference must be high enough to cause a corona discharge to surround the discharge electrode. Gas ions formed there move rapidly under action of the electric field and toward the collecting electrode or plates. They transfer their charge to the particles by colliding with them. The electric field, interacting with the charge on the particles, causes them to move towards and be deposited on the collecting electrode (Figure 21-15).

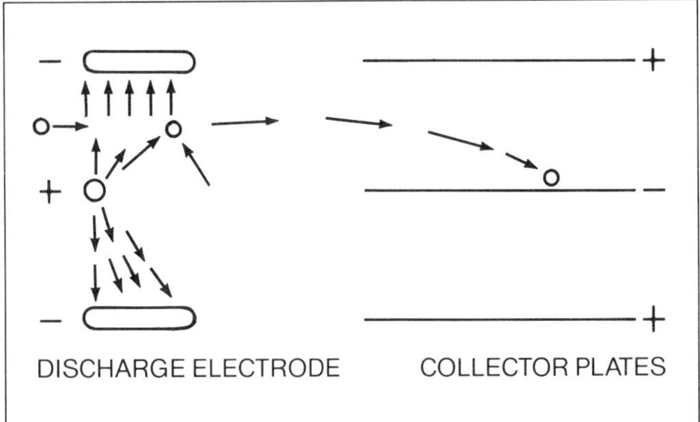

Figure 21-15. In an electrostatic precipitator, particles passing through the electrostatic field (left) become ionized (electrically charged). They are then attracted to collector plates that have the opposite charge. Collector plates are cleaned periodically. (Courtesy ACGIH)

These units are capable of achieving high collection efficiency. This high efficiency is not always profit producing, however, because of cost, worthlessness of collected material, and space limitations.

Electrostatic precipitators have the great advantage of negligible pressure drop. If the dust load is heavy, they are usually preceded by primary collectors that work under a different principle. Although unable to collect nonparticulate gases or vapors, they can collect mists, such as the oil mist given off during the cutting and machining of metals.

Settling chambers. Theoretically, it would be possible to settle out dust in a large chamber by dropping the conveying velocity to a point where the particles would be no longer conveyed. However, space requirements are excessive, and the presence of eddy currents and natural mixing nullifies the effective velocity. Settling chambers can be used only for bulky, coarse materials. Nevertheless, this solution to collection problems is offered perenially in manifold forms such as settling barrels, "dust boxes," and water drums. These appeal to the layperson because of their straightforwardness. A simple calculation will show that dust of health significance (5 μ and less) cannot possibly "settle out" in a reasonable time and space (Chapter 7, Particulates.)

Settling chambers have their place as chip traps and as precleaners for high-efficiency cleaners.

Filters. Two classes of filters are in common use: (1) "throwaway" of furnace-type filters and (2) cloth filters. The first type is not suitable for industrial exhaust air cleaning. Their principal use is in general ventilating systems.

The second filter type, cloth filters, can be made of a wide variety of fabrics such as cotton, wool, Dacron, Orlon, and for high-temperature applications, glass cloth filters. Cloth filters can be sewn together in the shape of tubes (hung vertically) or envelopes that fit over frames to keep the cloth sides from collapsing. In baghouses, the filters are equipped with manual or motor-driven shakers which dislodge the collected dust when the exhaust fan is shut off. This allows the dust to fall into hoppers below. From there it can be removed for disposal.

Cloth filters function by building up a layer of dust on the fabric which then filters additional incoming dust. New cloth filters, therefore, may pass dust for a brief period—perhaps to the consternation of someone who has installed new bags for the first time—before the dust layer accumulates. As the filter operates, its resistance gradually rises. When the filter is cleaned, its resistance falls, but never to as low a point as it was when the filter was new. The residual dust mat necessary for efficient separation always remains in use.

Cloth filters operate at low velocities: 0.47–2.4 L/sec/0.09 m^2 (1–5 cfm/sq ft) of cloth. They have relatively high resistance 0.5–3.7 kPa (2–15 in.) water gage. They have high efficiencies removing 99 percent or more of the incoming dust.

Cloth filters are relatively expensive—at least in the opinion of the heating and ventilating contractor who is accustomed to throwaway filters. Their cost may be $0.50 to $5.00/28.2L/min (per cfm) filtered. Increased filter resistance means higher operating cost. For example, a system handling 4.72 m^3/sec (10,000 cfm) might expend 5.6 kW (7½ horsepower) or more just overcoming the filter pressure drop.

Cloth filters can be used only against dry dust. Sticky or oily material clogs the bags, and when the temperature is below the dew point, condensation results in an exorbitant pressure drop and ineffective cleaning of the filter.

Cloth filters are ordinarily used for intermittent operation (the airflow is shut off every few hours to permit cleaning of the cloth). They may also be equipped for automatic continuous use. Periodically, a section is shut off from the system and the bags in this section are cleaned. This requires a larger cloth area, and additional air dampers and controls.

Regular cleaning is essential, and to assure it, U-tube water manometer should be permanently mounted on the outside of the casing. This will allow the pressure drop to be checked daily. A buildup of filter resistance, as shown by increased pressure drop, is the warning sign that the filter is becoming plugged and needs to be cleaned.

Access doors should be located conveniently so the interior of the casing can be checked periodically for leaks in the bags, bridging of dust in the hopper, or any other problems. Too often filters are located where they are inaccessible—and are then neglected. Filters should be checked weekly.

Cloth filters of the "reverse air jet" type that can be periodically cleaned during filtration (no shutdown) are widely used. They consist of a feltlike tube with the dusty air entering on the inside. On the outside, a slotted ring made of pipe continuously travels up and down the bag surface. Compressed air is blown at high velocity out the slot, which, in passing through the bag, dislodges the dust, so that it falls down in the dust hopper. The rate of airflow delivered by the exhaust system may be as high as 556 L/min/929.03 sq cm (20 cfm/sq ft) of cloth, which is far higher than in the conventional cloth filter. The pressure drop and the efficiency of the two types are comparable.

High-efficiency (HEPA) filters, developed to handle low inlet loadings in special applications, particularly where the air contaminant is radioactive, are available. They are made of various combinations of pads of glass fibers, compressed glass fibers, or resins carded into wool. They are generally disposable units, but are quite different from the common disposable viscous filter used in heating and air conditioning.

Air cleaners for radioactive dusts

Air cleaners for radioactive or highly toxic dusts must meet the following three major requirements.

1. The cleaner must be highly efficient because the amount of these materials permitted in effluents of stacks and discharge pipes is extremely low. In some instances, these materials must be collected because they are valuable or must be accounted for.
2. Low maintenance, that is, infrequent and quick servicing of the collector, is essential to minimize the exposure of the workers.
3. Finally, the radioactive or highly toxic material must be concentrated during collection to the smallest possible volume. Disposal through discharge to air, water, or land in the usual manner is not permitted. Incineration is often used, with the irreducible residue consigned to a burial ground.

Only a few types if collectors meet these three requirements. Dry cyclones are not satisfactory because their collecting efficiency is too low. Any unit with an involved internal structure must be ruled out because of service difficulties. The reverse-jet cloth arrester with wool felt media has become almost a standard providing temperature or corrosion is not a problem. Where such problems do exist, the high-voltage electrostatic collector can be used.

Low-efficiency, low-cost filters are often used as integral parts of laboratory hoods or dry boxes in which radioactive materials are handled. They serve as scalping filters to keep high concentrations of contaminated material from entering the main exhaust system. They must be followed or backed up by an ultra-high-efficiency filter or dust arrester. They have a limited capacity and must be changed frequently. Before they are changed, they can be sprayed while still in position with an acrylic spray to fix any radioactive dust.

Air cleaners for fumes and smokes

Fumes result from reactions such as burning, sublimation, distillation, and especially the condensing of a vapor given off by a molten metal. The composition of fumes may be different from that of the parent material. Lead oxide fume, cadmium fume, zinc oxide, and iron oxide (welding fume) are examples. Fume particles are generally less than 1 μ, so they show active Brownian movement (that is, they are far too small to settle), and they are remarkably uniform in size. Smoke is commonly organic, coming from incomplete combustion of coal, oil, wood, and other fuels. It is usually dark or jet black in color and obscures light. Its particles compare in size to the particles of metallurgical fumes.

The high-efficiency dust collectors already mentioned—the cloth filter, high-efficiency wet collector, and electrostatic precipitator—are best for arresting fumes. For smoke, the answer is improved combustion, or the high-voltage electrostatic precipitator, or both.

Air cleaners for gases and vapors

Vapors can be defined as the gaseous form of a material normally in the liquid or solid state at a room temperature. Gases and vapors are not particles, but individual molecules dispersed among the molecules of the air. Because gases and vapors diffuse, arresters that rely on the straining action of filters or on centrifugal force are not applicable. Methods generally used are as follows:

Absorption. A liquid is used in a packed tower or scrubber which dissolves or reacts chemically with the gas or vapor and removes it from the air. The disposal of this liquid may present problems of waste control and stream pollution.

Adsorption. Many solid particles have an adsorbing action for certain gases and vapors. The action takes place at the surface of the adsorbent where gas and solid come in contact with each other. The most widely used material for removing odors is activated carbon. The activated carbon is placed in granule form in trays, canisters, or perforated cans. Activated carbon can adsorb some vapors and gases up to 50 percent of its own weight. It can then be reactivated (for reuse) by heating. Activated carbon is frequently used for the removal of solvent vapors. This is a particularly appealing method because the solvent can later be reclaimed.

Combustion. If the gas or vapor can be oxidized into harmless or odorless products, combustion can be used for removal. All hydrocarbons can be so removed because on complete oxidation, water and carbon dioxide are the only end products.

The most satisfactory method of consuming odorous or otherwise troublesome organic gases and vapors is by means of direct-flame afterburning. However, proper safety precautions must be observed. These include using flame barriers or operating outside the flammable range of the vapor mixture.

A catalyst can be used to accelerate the reaction in the combustion process. Catalytic combustion solves problems in removal of vapors and odors from paint, varnish, enamel, and foundry core baking ovens; rendering of fish oils and animal fats; roasting coffee; asphalt processing; plastics manufacturing; and kitchen range hood exhausts.

Catalytic units must have essentially pure vapor to give long-term effective control. The catalyst may be poisoned or deactivated by vapors of the metallic group that have a high vapor pressure such as mercury, zinc, or arsenic. If the incoming air has large quantities of a noncombustible dust, it must be removed first.

Condensation. By reducing the temperature of the incoming air, vapor can be changed to the liquid state and removed. Several types of condensers and refrigeration units can be used to chill the incoming vapor.

Fans

We have discussed hoods, ducts, and air-cleaning devices for a local exhaust system. The fourth element in a local exhaust system is the fan and motor. Two groups of fans are in use: centrifugal and axial flow. They are shown schematically in Figure 21-16. The radial blade centrifugal fan can be used to move air that contains particulate material.

Centrifugal fans. Depending on how the blades are pitched, centrifugal fans are all modifications of the basic wheel type. They can be used where static pressure is medium to high, say, 34 kPa (10 or more inches) water gage and up.

- The radial wheel is the workhorse of fans in the industrial ventilation field. Its straight radial blades, made of steel or cast iron, do not readily clog with material passing through. They also withstand considerable abrasion. For decades, this fan has been used in buffing and woodworking shops where lint, chips, and shavings pass through the blades. It has medium tip speed, medium noise factor, and medium mechanical efficiency.
- Backward-curved blades on a centrifugal-type fan permit higher tip speed and thus greater fan efficiency. Material will build up on the blades, so that it often has an air cleaner ahead of it. Although this type of blade has a higher noise factor, its high efficiency and nonoverloading feature makes it the choice for many large volume exhaust systems.
- Forward blades, tipped in the direction of rotation, yield a fan that has low space requirements, low tip speed, and a low noise factor. It is popular in heating and air conditioning work because of these characteristics, and also because under these conditions the static pressures are from low to moderate. Because airborne material will stick to the short curved blades and throw the wheel out of balance, this fan may be preceded by an air cleaner.
- Fans with straight or forward-curved blades demand more power as airflow increases. If the actual duct system resistance is less than the estimate from which the fan was selected, the actual airflow will exceed the estimate, more power will be required, and the driving motor may be overloaded.
- Fans with backward-curved blades, on the other hand, have a power demand that reaches a maximum. If the driving motor is rated to carry this maximum, it cannot be overloaded at the given speed. For this reason, fans with backward-curved blades are called nonoverloading fans.

There are a number of intermediate designs between the extremes of forward and full-backward curved blades that are similar in varying degrees to the performance characteristics of each type.

Axial-flow fans. Fans of this type are modifications of the familiar desk fan or propeller fan. Air leaves in the same straight line in which it enters, whereas in the centrifugal fan air leaves at right angles to the direction in which it enters.

- Propeller fans move large volumes of air against negligible resistance, and are generally mounted on pedestals for cooling of individuals and general circulation. They may also be set in window boards or in walls with no ducts connected. Sometimes the limitation that they cannot operate against the friction that a duct would entail is forgotten.
- A narrow-blade propeller fan is often provided for spray booth exhausts, but duct connections must be held to a minimum. These fans are sensitive to added resistance, and a small increase in resistance will drop the air volume handled markedly.
- Tube-axial fans are an improved version of the propeller fan and are mounted by the manufacturer in a short section of duct to which other piping can be attached. They are effective against only an inch or two of static pressure, water gage.
- Vane-axial fans have guide vanes in the short pipe section which straighten the airflow. They will operate against low

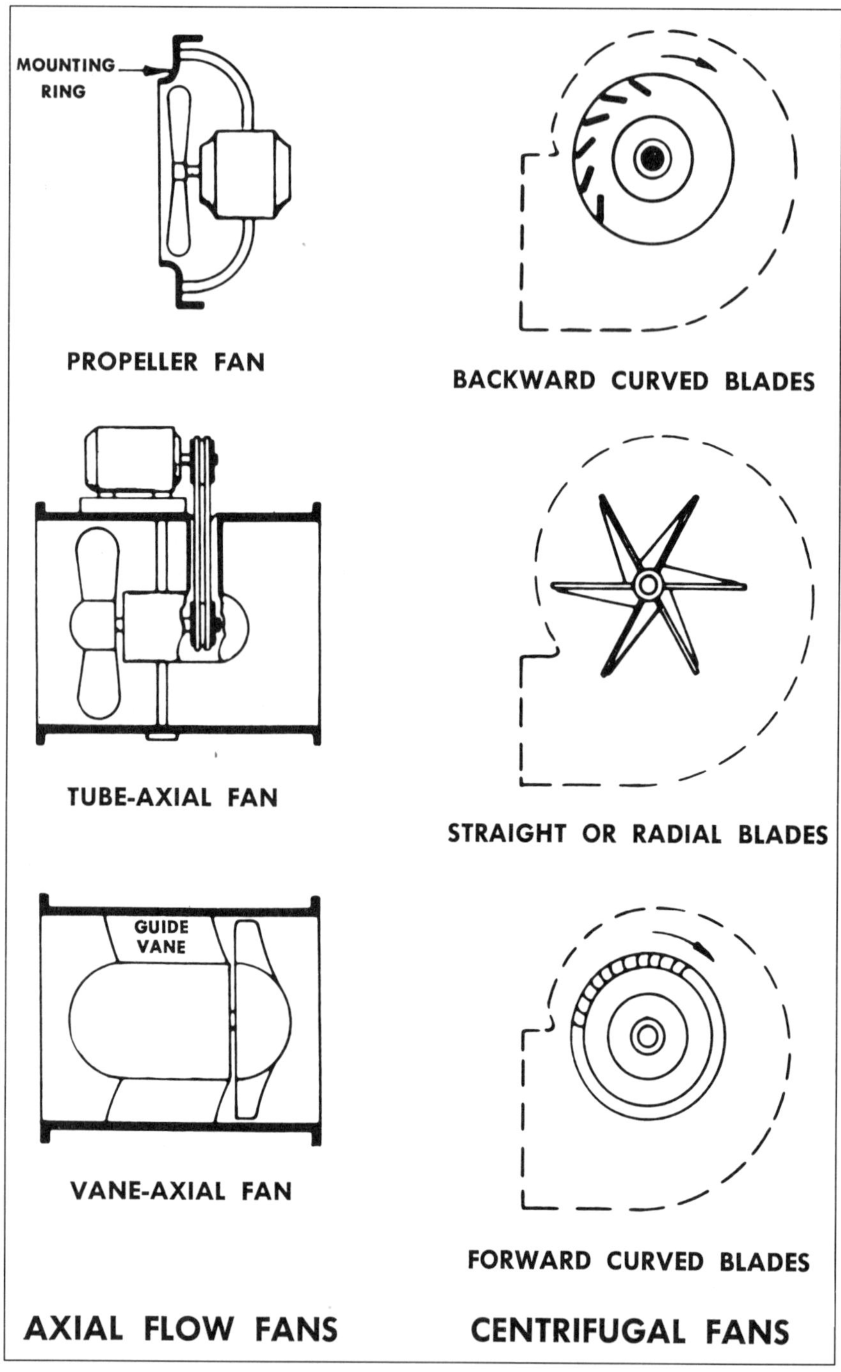

Figure 21-16. Types of exhaust fans. (Courtesy ACGIH)

static pressures, say, up to 1.0 kPa (4 in.) water gage. They must, however, be used only in clean air.

For roof or wall exhaust, there are available direct-discharge fans (no scroll) with backward-curved blades similar to the wheel used in centrifugal fans.

Fan noise. Except for low-speed fan units, fans usually are noisy. This can be distracting, irritating, and/or damaging to the ear. Noise may interfere with speech. Fan noise can be a problem both in the plant area and to neighbors outside.

Fan manufacturers, through technical organizations, such as the Air Moving and Control Association and the American Society of Heating, Refrigerating and Air-Conditioning Engineers, have developed noise ratings for fans based on considerations such as blade-tip speed, brake horsepower, and pressure. (See discussion of these organizations, Appendix A, Sources of Help.)

A large-size fan, operating at slower speed and sometimes even at reduced horsepower, is often better economically in the long term and produces much less bothersome noise. Reasonable quietness can be maintained by keeping the tip or peripheral speed at 30 m/sec or less (6,000 linear feet per minute or less).

When blower noise (inherent in the unit) is disturbing, one solution is to surround the fan by a sound-attenuating enclosure. Depending on the size, the enclosure may be made of masonry, heavy gage sheet metal, or even 19 mm (3/4 in.) plywood, if the unit is small. The enclosure should be lined fully with acoustically-absorbent material.

Fire prevention

It is important to refer to available standards, such as NFPA Standard 91, *Blower and Exhaust Systems for Dust, Stock and Vapor Removal or Conveying* (see Bibliography).

Where fans handle flammable solid materials or vapors, the rotating element must be nonferrous or spark-resistant material, or the casing must consist of or be lined with such material. This requirement also applies to both the rotating element and the casing where solid foreign material passing through the fan could produce a spark.

Fan motors located in rooms or areas in which flammable vapors or flammable dust is being generated and removed should be of the type approved for the particular conditions or hazard. When exhaust systems are used to handle flammable gases or vapors or combustible or flammable dust, stock, or refuse, static electricity must be removed from belts by grounded metal combs or other effective means.

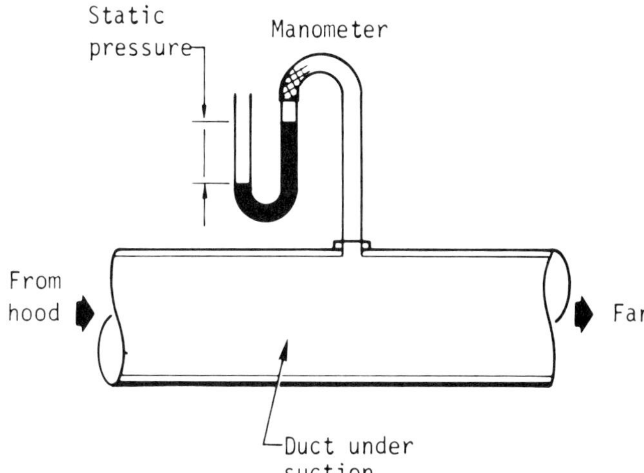

Figure 21-17. A water-filled manometer indicates pressure as "inches of water" column displaced in water gage. (Adapted with permission from McDermott, H.J. *Handbook of Ventilation for Contaminant Control,* 2nd ed. Stoneham, MA: Butterworth, 1985)

CHECKING EXHAUST SYSTEM PERFORMANCE

No local exhaust system is foolproof. Its performance needs to be checked, and maintenance work is required. All too often, a system is well designed and properly installed, but develops problems as time goes on. Perhaps the collectors have not been cleaned regularly, or the ducts have become partially plugged with settled material. Perhaps the hoods have become battered or abraded. It is possible that with plant expansion, additions have been made to the existing system that made the original hood ineffective. This fault is not uncommon and often remains unnoticed.

The system should be tested when it is new and clean, and all readings recorded so they can be used as a level of acceptable performance when checks are made later during regular operation.

Local exhaust systems are primarily used to prevent a health or fire hazard. Their purpose is to keep people safe. Hence, the health and safety professional has a responsibility to see that exhaust systems perform as intended.

The health and safety professional can be equipped to make simple tests that will prove revealing. First, using a simple U-tube or water manometer and permanently positioned test holes in branch ducts, at inlets and outlets of dust arresters, at fan inlets, etc., he or she can measure the static pressure or "suction" (see Figure 21-17). Comparison with earlier readings will tell whether or not anything has gone wrong (Figure 21-18). Second, a simple smoke tube can trace the flow of air into hoods and show visually whether the contaminated air is being removed from the work area.

Records of static pressure tests should be maintained. They may be supplemented by checks on fan speed and on weight of material collected in the dust arrester, as needed (Figure 21-19).

For a full discussion of this subject, see National Safety Council's Industrial Data Sheet 428, *Checking Performance of Local Exhaust Systems.*

Recirculation of "cleaned" air

When the rate of air exhausted by a contaminant control system is high, either in absolute volume of air handled or in comparison to the volume of the room, cleaned air may need to be recirculated. It is often argued that the dollars-and-cents savings in heat will be such that recirculation is attractive; or the existing heating system may have so little reserve capacity that recirculation will be the only way to maintain a comfortable working temperature. There are some basic principles to follow in deciding whether to recirculate air or temper makeup air.

- If the contaminant is harmless, and if after passing through a dust, mist, fume, or vapor separator, the processed air is no more contaminated than the outdoor air, recirculation would be safe.
- Savings in fuel or in heating plant usage are never a valid reason for recirculating air if the original air contaminant is a serious health hazard. For example, if the exhaust system is installed to control carcinogens, silica dust, lead dust, and other toxic dusts, gases, or vapors, then the contaminated air from the air cleaner should be exhausted outdoors. Even if the dust filter or other arrester is operating at high efficiency so that the concentration of contaminant downstream is normally within safe limits, a breakdown can occur that would cause dangerous amounts of material to be blown back into the workroom.
- In every recirculation system, therefore, a Y-connection should be included in the discharge of the arrester. One leg should go to the outdoors, and the other should lead to the recirculation ductwork. Adjustable dampers should be arranged so that when the control is in one position, all of the air will be discharged outdoors, and when in the opposite position, all air will be sent back to the room. Various inbetween ratios can be adjusted depending on the outdoor air temperature.

A further advantage of this Y is that should the arrester break down, the uncollected contaminant can immediately be blown outdoors temporarily, instead of into the workroom.

Present federal and state laws, standards, and rules permit recirculation except in the specific instance of prohibiting recirculation from spray-finishing systems and with car-

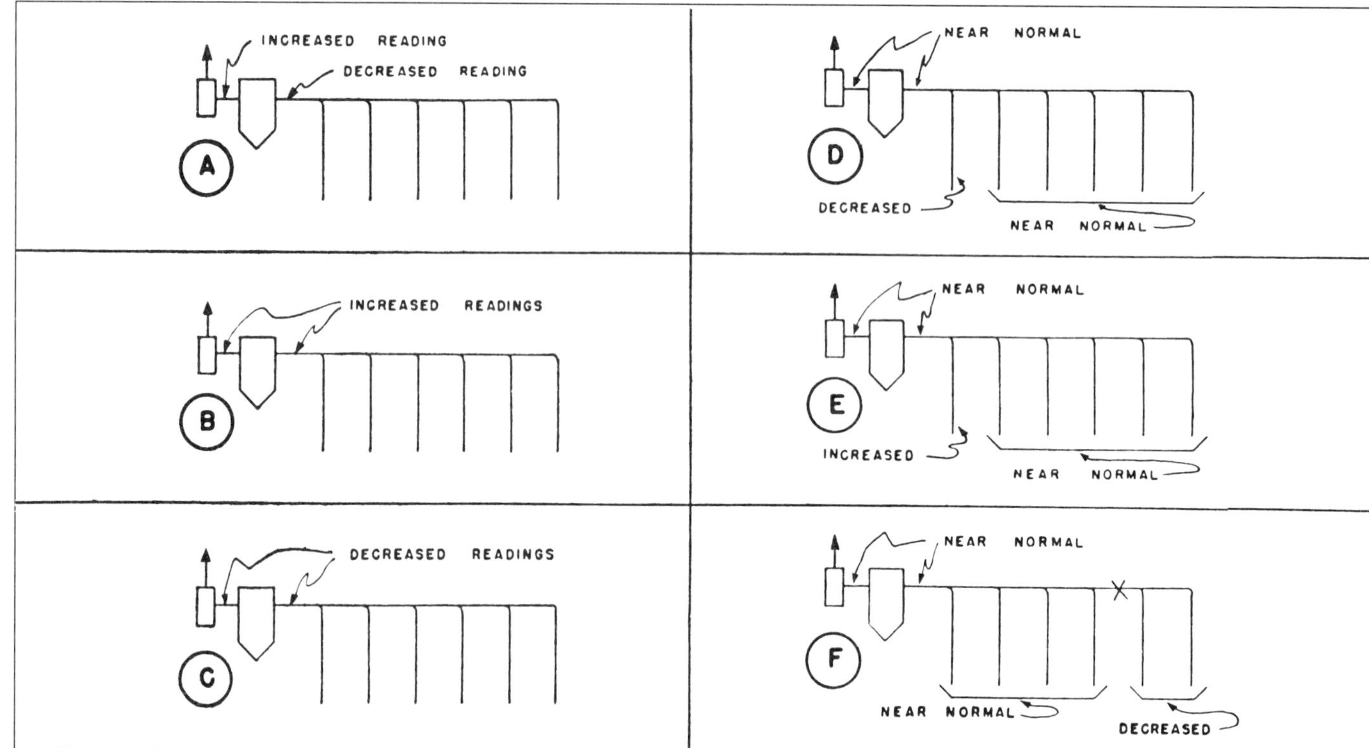

Figure 21-18. U-tube readings diagnose dust arrester troubles: (A) Partially plugged arrester causes increased resistance, needs cleaning; (B) illustrates increased resistance or plugging in header pipe or in branch; (C) indicates decreased fan capacity, open cleanout door, or disjointed pipe; (D) first branch pipe is partially plugged somewhere between header and test hole near hood; (E) first branch pipe is plugged somewhere between test hole and hood face; (F) increased resistance is due to plugging in header pipe at "X". (Courtesy ACGIH)

cinogens. Questions of application, operation, and maintenance of recirculation systems remain to be solved on an individual basis, evaluating each set of circumstances and considering the expected or known contaminants. There cannot be a "cook-book" approach to recirculation except for the simplest systems involving innocuous or "nuisance" substances. The performance of most systems must be tested by a competent health and safety professional to evaluate worker exposure before and after the system is installed. The system must be carefully maintained and the system performance monitored to ensure optimum performance for adequate health protection. All recirculation systems should be equipped with automatic monitoring systems that will alert employees if the system fails.

LOW-VOLUME, HIGH-VELOCITY EXHAUST SYSTEMS

A low-volume, high-velocity exhaust system is designed and operated to produce high capture velocities at points of contaminant release (Figure 21-20). Low-volume, high-velocity hoods may be used to achieve greater effectiveness in contaminant control of portable tools or to control contaminants released with high initial velocities.

The low-volume, high-velocity exhaust system is the unique application of exhaust in which small volumes of air at relatively high velocities are used to control dust from portable hand tools and machining operations. Control is achieved by exhausting the air directly at the point of dust generation using close-fitting, custom-made hoods. Capture velocities are relatively high but the exhaust volume is low due to the small air volume required. For flexibility, small diameter, lightweight plastic hoses are used with portable tools resulting in very high duct velocities. This method allows the application of local exhaust ventilation to portable tools that otherwise would require relatively large air volumes and large ductwork when controlled by conventional exhaust methods.

This technique has found a variety of applications, although its use is not common and many disappointing failures have been reported. Rock-drilling dust has been controlled by using a hollow-core steel drill with suitable exhaust holes in the drill bits. Air is exhausted by one of two methods. A multistage turbine of the size generally used in industrial vacuum cleaners can be used or the exhaust air from the pneumatic tool that operates a venturi that withdraws air from the drill. Application has been made with flexible connections to a central vacuum system to aid in the control of graphite dust at conventional machining operations. A 25- to 50-mm (1–2 in.) diameter flexible hose was used with a simple exhaust hood mounted directly at the cutting tool. In a similar application for the machining of beryllium, described by Chamberlin (1959), a central vacuum system using 1½-in. ID flexible hoses was employed. The exhaust hoods were made of Lucite or other transparent material and were tailor-made to surround the cutting tools and much of the work. Exhaust volumes vary from 0.057–0.071 m^3/sec (120–150 cfm) with inlet velocities of 56–71 m/sec (11,000–14,000 fpm).

Figure 21-20 illustrates one custom-made exhaust hood that is available. (The ACGIH book *Industrial Ventilation—A Man-*

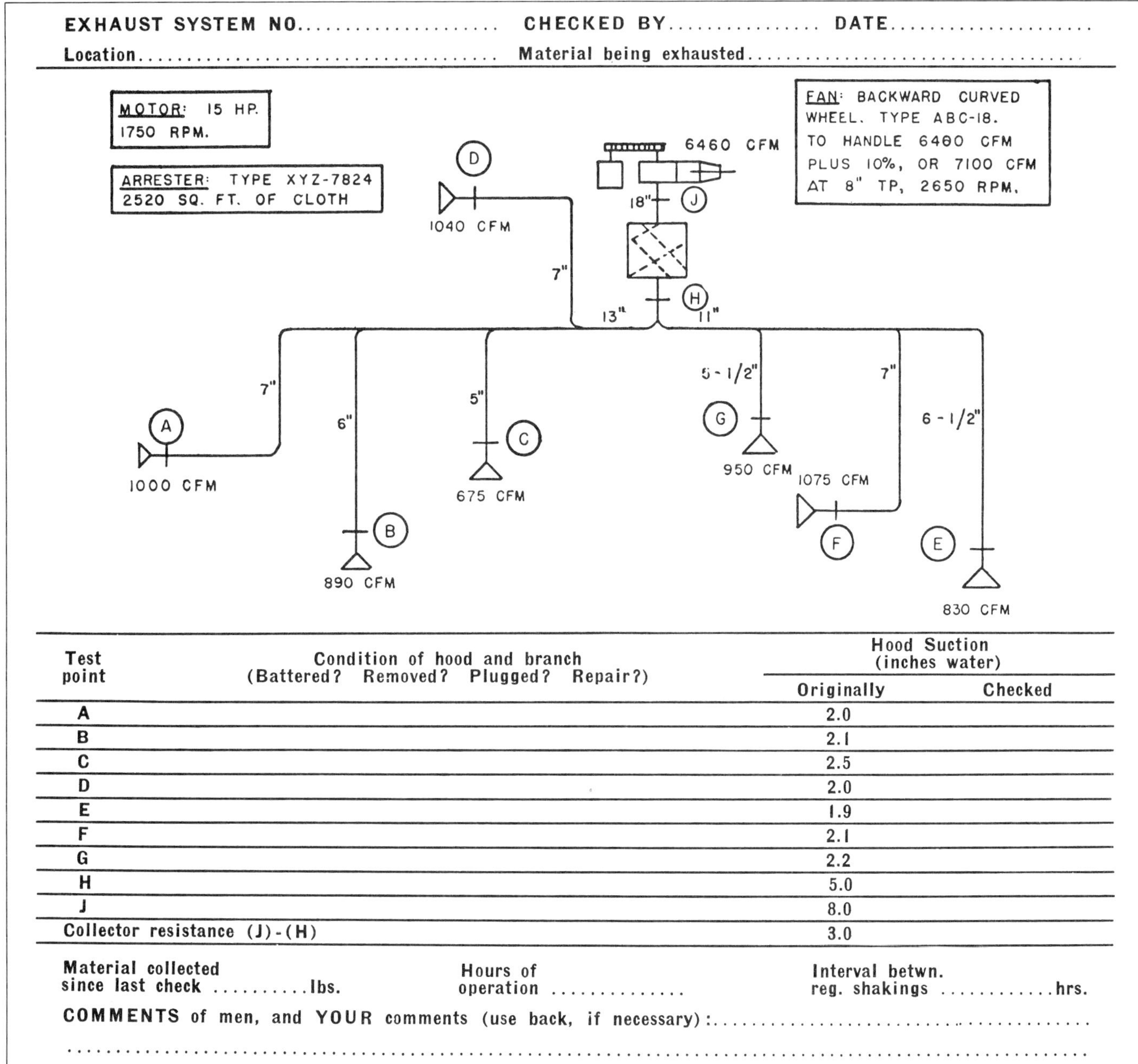

EXHAUST SYSTEM NO.................... CHECKED BY............... DATE....................

Location.................................... Material being exhausted..................................

MOTOR: 15 HP.
1750 RPM.

ARRESTER: TYPE XYZ-7824
2520 SQ. FT. OF CLOTH

FAN: BACKWARD CURVED WHEEL. TYPE ABC-18.
TO HANDLE 6460 CFM PLUS 10%, OR 7100 CFM AT 8" TP, 2650 RPM.

6460 CFM
18"
7"
13"
11"
1040 CFM
1000 CFM
890 CFM
675 CFM
950 CFM
1075 CFM
830 CFM
7"
6"
5"
5-1/2"
7"
6-1/2"
A B C D E F G H J

Test point	Condition of hood and branch (Battered? Removed? Plugged? Repair?)	Hood Suction (inches water) Originally	Checked
A		2.0	
B		2.1	
C		2.5	
D		2.0	
E		1.9	
F		2.1	
G		2.2	
H		5.0	
J		8.0	
Collector resistance (J)-(H)		3.0	

Material collected since last check ....lbs. Hours of operation .... Interval betwn. reg. shakings ....hrs.

COMMENTS of men, and YOUR comments (use back, if necessary):..

....

Figure 21-19. The health and safety professional should obtain or create simple line drawings to show the layout of each main duct and branch. General characteristics of the arrester, fan, and motor should be shown. Suction pressure readings should be made at each test hold once the system has been balanced and dust counts, or air samples, made to show that the air contaminants are under control. These readings are indicated on the drawing by each test hole. A simple drawing should be made for each exhaust system.

ual of Recommended Practice, 19th ed, 1986) illustrates many more applications.) The required air volumes range from 0.038 m³/sec (8 cfm) for pneumatic chisels to 1.4 m³/sec (300 cfm) for swing grinders (Table 21-D). Due to the high entering velocities involved, static pressures are in the range of 5 in. mercury, 17 kPa or 68 in. water gage. This high pressure is necessary to create the high capture velocities at the dust source.

The dust is conveyed at high velocities in small diameter flexible hoses ranging from 10–50 mm (⅜–2 in.) ID. The dust-laden air is passed into primary and secondary cyclone collectors, discharging through a fabric filtering unit. Exhaust is provided by a multistage centrifugal turbine capable of producing static pressure of about 27 kPa below atmospheric pressure or 8 in. mercury (109 in. water). The fabric collector can be cleaned by a simple, manual valve that admits air into the clean side of the fabric, bringing this side of the fabric to atmospheric pressure. The dirty side of the fabric is at a level far below atmospheric pressure. This causes rapid airflow through the fabric, provides reserve cleaning, and drops the collected dust into a hopper.

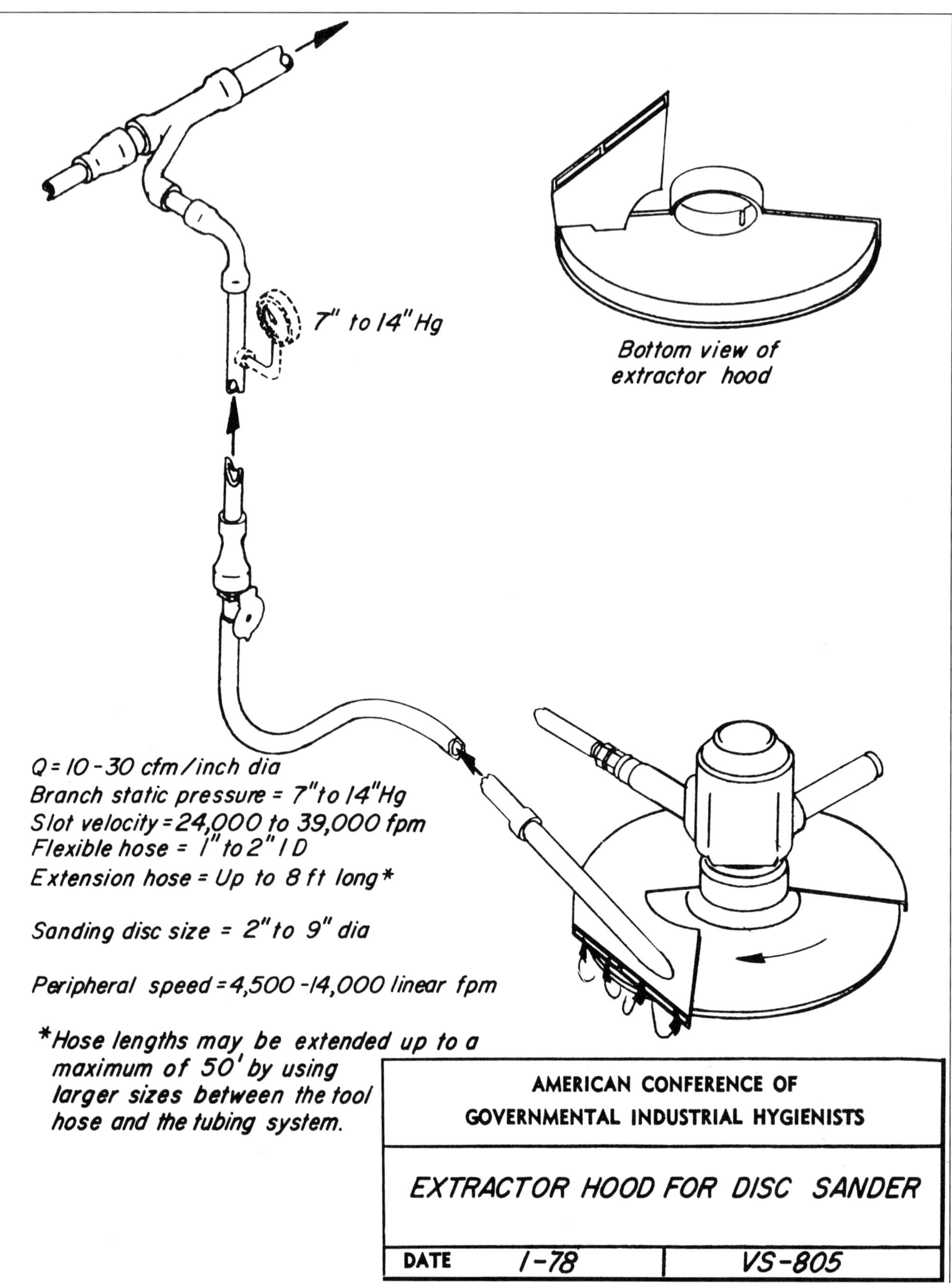

Figure 21-20. Extractor hood for disk sander. (Courtesy ACGIH)

Table 21-D. Exhaust Volumes Required for Low Volume-High Velocity Systems

	cfm	*I.D. Plastic Hose Size*
Disc sanders, 3-9″ diameter	60-175	1″-1½″
Vibratory pad sander-4 x 9″	100	1¼″
Router, ⅛″-1″	80-100	1″-1¼″
Belt sander 3″-4,000 fpm	70	1″
Pneumatic chisel	60	1″
Radial wheel grinder	70	1″
Surface die grinder, ¼″	60	1″
Cone wheel grinder	90	1¼″
Cup stone grinder, 4″	100	1¼″
Cup type brush, 6″	150	1½″
Radial wire brush, 6″	90	1¼″
Hand wire brush 3 x 7″	60	1″
Rip out knife	175	1½″
Rip out cast cutter	150	1½″
Saber saw	120	1½″
Swing frame grinder 2 x 18″	380	2½″
Saw abrasive 3″	100	1¼″

Design calculations

With the exception of custom systems, purchased as a "package," the design calculations for these systems are largely empirical. In normal ventilation practice air is considered to be incompressible since static pressures vary only slightly from atmospheric pressures. However, the extreme pressures required in these systems introduce problems of air density, compressibility, and viscosity which are not easily solved. Also, pressure drop data for small diameter pipe, especially flexible tubing, are not commonly available. The turbine exhauster should be selected for the maximum simultaneous cfm (m^3/sec) exhaust required. Resistance in the pipe should be kept as low as possible; flexible tubing of less than 25 to 38 mm (1–1½ in.) diameter should be limited to 3.0 m (10 ft) or less. In most applications this is not a severe problem.

There are two main considerations in piping for such systems. Firstly, piping must provide smooth internal configuration so as to reduce friction and, hence, pressure loss at the high velocities involved. Secondly, abrasion should be minimized. Ordinarily threaded piping is to be avoided because the lip of the pipe or male fitting, being of smaller diameter than the female thread, presents a discontinuity which increases friction and pressure loss, and may be a point of rapid abrasion.

If a threaded system is to be used, cast-iron drainage fittings and Schedule 40 pipe should be provided. Fittings of this type are recessed so that the inside diameter of the pipe and fitting is the same. Tubing systems of 16-gage wall thickness up to and including 10 cm (4 in.) diameter, and 14-gage for 12.5 cm (5 in.) and above, provide for 10 percent less pressure loss than a Schedule 40 piping system and offer a lower installed cost in most cases. Commercially available tube fittings and clamps or a slip-on flange system may be used so that internal discontinuities are eliminated. In all cases, long-sweep elbows and bends should be provided.

A good collector for dust exhaust systems should be mounted ahead of the exhauster to minimize erosion of the precision blades and subsequent loss in performance. Final balance of the system is achieved by varying the length and diameters of the small flexible hoses.

It must be emphasized that although data are empirical, these systems require the same careful design as the most conventional ones. Abrupt changes of direction, expansions, and contractions must be avoided and care must always be taken to minimize pressure losses.

SPECIAL OPERATIONS

Open-surface tanks

A large variety of industrial operations are conducted in open-surface tanks. For example, surface treatment (such as anodizing and parkerizing), pickling, acid dripping, metal cleaning, plating, etching, and stripping. Since some of these operations involve heat and gassing of the liquid, a simple solution may seem to be a canopy hood over the tank. This has three faults, however: (1) with any canopy hood the worker's head is likely to be between it and the tank surface, so that he or she may have greater exposure with the hood than without it; (2) many of these operations require the work to be raised and lowered by a hoist on a monorail over the tank—the canopy hood would interfere with the operation, (3) air currents caused by open doors and windows, as well as passing traffic, can seriously affect contaminant control (Figure 21-21); and (4) canopy hoods should be used only as receiving hoods over hot processes.

The American National Standards Institute has formulated a code on ventilation of open-surface tanks, *Practices for Ventilation and Operation of Open-Surface Tanks*, Z9.1. This method of design is also included in *Industrial Ventilation—A Manual of Recommended Practice.* Figure 21-22 illustrates how air should flow away from the operator.

Spray booths

A spray booth is a partial enclosure through which a flow of air is drawn by an exhaust fan. The booths may vary in size from small bench type to huge installations large enough for a railroad car. A spray booth is simply an enclosure with sides, a rear, top, and bottom—the front, called the face, being open. The booth must, of course, be large enough to accommodate the work with space around it so the operator has the necessary freedom of movement.

Most booths are used for spraying paint, enamel, or lacquer.

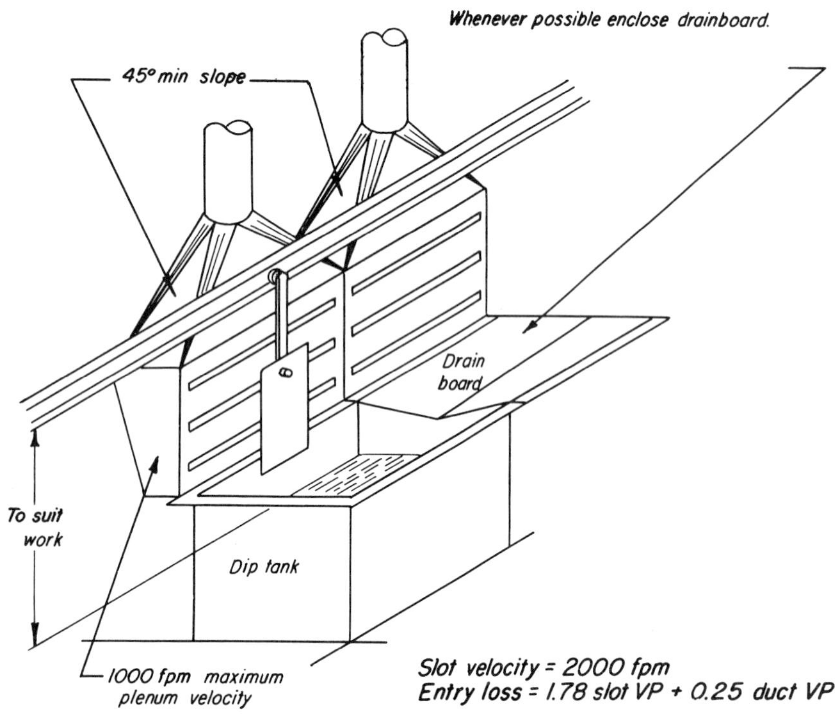

Figure 21-21. Exhaust for open-surface tank draws air and contaminants across the tank, away from any operator. (Courtesy ACGIH)

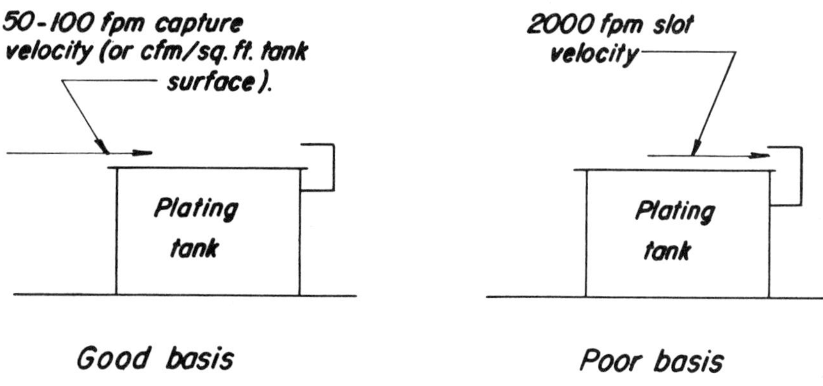

Figure 21-22. To ventilate open-surface tanks, create an airflow past the source, sufficient to capture the contaminant(s). (Courtesy ACGIH)

Fewer booths are used for bleaching, cementing, glazing, metallizing, and cleaning. Another widespread use of booths is for welding.

A paint-spray booth has two main purposes: to protect the health of the painter; and to reduce the fire and explosion hazard. These two interests are served by several regulatory bodies. The requirements of each must be met.

Individual states may have a code administered by its health or labor department designed to control occupational disease exposures, whereas another bureau may prescribe fire control aspects. The federal Occupational Safety and Health Administration (OSHA) has requirements for both. At the same time, the fire insurance carrier may have specific requirements that must be met. It may also have a secondary concern to reduce the total damage resulting from a fire, including water damage. The health and safety professional should be thoroughly aware of all requirements that apply.

When paint-spraying is in progress, the overspray must be carried away from the operator's breathing zone. This can be accomplished usually with an air velocity of 0.51-0.76 m/sec (100–150 fpm) through the booth face. A small shallow booth is usually designed for a face velocity of 1.0 m/sec (200 fpm) or more. In metallizing, when the material is toxic, the velocity should also be at least 10 m/sec.

Tempered makeup air to replace air exhausted by the booth should be supplied. This is needed for any exhaust system, but deserves special attention here because some spray booths remove huge amounts of room air, which in turn should be supplied directly to the booth or to the area from which the air is drawn into the booth.

Routine and scheduled cleaning and maintenance of spray booths are a must. Flammable materials when allowed to accumulate increase the fire risk. In addition, paint in ducts also cuts down needed airflow. There are several products on the market that can be used to cover the inside of the booth that successfully prevent the sticking of the overspray to the booth surfaces.

In some high production painting, much solvent is lost by

evaporation from the wet, painted surfaces. If painted units are stacked out in the shop to dry, vapors may present a health and fire hazard. A separate ventilated area should be provided for preliminary drying of the finished materials before they are sent to the ovens for baking.

When the production is not high, the material can be left in the booth for preliminary drying. In no instance should finished material be stacked in front of the booth so that the vapor-laden airflow drifts from the freshly finished material to the operator.

Dusts from some of the materials used in metallizing are explosive if finely divided and suspended in air in a critical concentration. When collecting dust from metallizing, it is sometimes necessary to provide wet-dust arrester units in the rear of the metallizing booth.

Fume control for welding

Local exhaust hoods are applicable to many indoor welding applications, and are a must when welding potentially toxic materials such as lead-containing or lead-painted metals, and beryllium alloys. Except for production spot welding sometimes done at a fixed jig, a welder usually travels over a wide area. This necessitates some arrangement of flexible or movable ventilation.

The hood itself can have a circular or rectangular face with a flared section to which the duct is connected, thus giving it a streamlined airflow. Normally the hood is surrounded by a flange. The breadth of the face and the amount of the airflow exhausted should be such that fumes will be removed from the welding arc over the length of weld formed. The capacity of the exhaust, therefore, should be sufficient to reach laterally over the length of a weld of this distance.

Flexible metal duct branches or solid sections with swivel joints and counterbalances permit easy relocating of the hood. A duct branch should not be less than 15 cm (6 in.) in diameter and should exhaust sufficient air to control emissions. Face velocities can exceed 7.6 m/sec or 1,500 fpm (average) and duct velocities should assure a sufficiently high airflow into the hood to control fumes at the point of generation (usually several inches away from the hood face), as well as avoid settlement in the duct (Figure 21-23).

Dust control in foundries

Foundries present many difficult problems in dust control, particularly in the shakeout and cleaning rooms. Dust at the shakeout in a hand foundry that processes small castings can be adequately controlled by proper work practices. For example, castings can removed from the sand before the sand dries enough to form a great deal of dust.

In a more mechanized foundry, the shakeout generally can be covered by an enclosing hood, providing the flasks are small enough to be handled on conveyors. The major part of the exhaust should be taken from overhead to utilize the thermal circulation of air over the hot metal. When the flasks are so large that they must be handled by cranes, a side hood on the shakeout grate is about the only possible choice unless the whole operation can be enclosed and isolated. (Chapter 7, Particulates.)

In either case, the sand-conveying and reconditioning equipment should be completely enclosed, and ventilation should be supplied at such dust-producing spots as transfer points on sand conveyor belts, mills, and mullers as well as bin loading and unloading points.

Dust-control measures for foundry cleaning rooms have been well standardized by the American Foundrymen's Society. For all such common equipment as tumbling mills, sandblast cabinets, sandblast rooms, and automatic cleaning machines, the required air volumes and duct sizes are specified in *Recommendations for Control of Occupational Safety and Health in Foundries* (Pub. No. 85-116) NIOSH.

Grinding operations

Health, comfort, and good housekeeping suggests that grinding, buffing, and polishing equipment be equipped with local exhaust systems. Many state or provincial codes require them as does OSHA (whenever its Standard 1910.1000 is exceeded), and where such codes exist, the health and safety professional should obtain copies so that these regulations can be met.

Hoods on grinding and cutting wheels serve a dual purpose: (1) they protect the operator from hazards of bursting wheels, and (2) they provide for removal of dust and dirt generated. Most such wheels today are made of artificial abrasives that contain no free silica and so eliminate the silicosis hazard.

Uncleaned castings, on the other hand, may have mold sand adhering to them. This may be ground to dust, causing a silicosis hazard. In addition, grinding of some alloys, such as high-manganese steel, cemented tungsten-tipped tools, and spark-resistant tools, presents other occupational disease hazards. (Chapter 7, Particulates.)

Furthermore, grinding operations need local exhaust hoods to remove the dust and grindings that otherwise cause poor shop housekeeping.

The exhaust hoods must have sufficient structural strength, and must enclose the wheel sufficiently. Details are given in state codes, industry codes, and codes of the American National Standards Institute.

The hoods for a floor stand, pedestal, or bench grinder should have an adjustable tongue that can be placed so that it is within 3 mm (1/8 in.) of the periphery of the wheel at all times. This is an important adjunct, because it tends to "peel" off the dust as it is carried around the wheel in the airstream set up by the wheel's rotation. No more than 25 percent of the wheel should be exposed (Figure 21-24).

Minimum velocity in the branch ducts should be 23 m/sec (4,500 fpm), and in the main duct, 18 m/sec (3,500 fpm). The branch duct should attach to the hood at a tapered connecting piece. It should be inclined in the direction that material is thrown off the wheel. Chip trays may be provided if desired for heavy grinding.

Local codes should be consulted for hoods on horizontal and vertical spindle grinders, grinding and polishing straps and belts, and miscellaneous grinders.

Portable hand grinding can be done in a booth or on a table with downdraft ventilation (air drawn downward through a grating set in the table top). Shields should be attached to the back and sides of the table to reduce cross drafts and to ensure the ventilation provided is more effective (Figure 21-25). This can also be used for soldering and arc-welding.

Woodworking

The woodworking industry was one of the first to recognize the value of local exhaust systems. They are applied almost universally in producing shops, their benefits being better housekeeping, improved working conditions, and decreased fire haz-

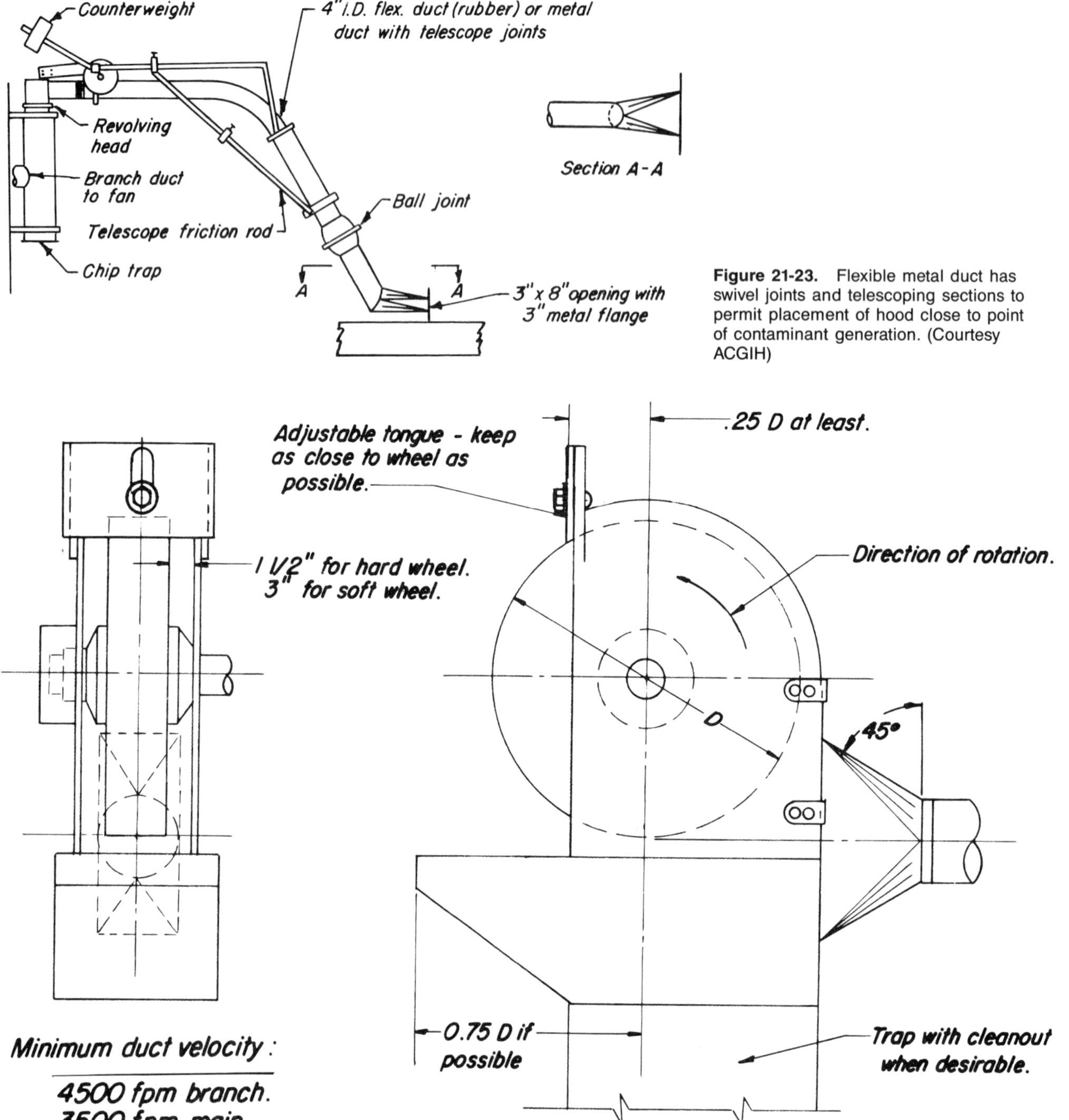

Figure 21-23. Flexible metal duct has swivel joints and telescoping sections to permit placement of hood close to point of contaminant generation. (Courtesy ACGIH)

Figure 21-24. Buffing and polishing hood has adjustable tongue that should be kept as close to the wheel as possible. (Courtesy ACGIH)

ard. The OSHA and many local jurisdictions have specific codes covering this type of ventilation, and these, of course, should be complied with.

Higher cutting speeds developed for woodworking machinery have led to higher production, which in turn means more shaving, chips, and dust. Exhaust ventilation must be increased correspondingly. In fact, with some of the newer machines, more air should be exhausted than is called for by the codes, which often lag behind technological progress.

The ACGIH (1986) (see Bibliography) gives recommended exhaust volume and branch duct diameter for average-sized woodworking machines, with excellent drawings of typical hoods. Exhaust volume will vary with cutting speeds, length of cutter heads, length of sander drums, and diameter of saws. In general, saws require 10- to 15-cm (4 to 6 in.), and occasionally 20-cm (8 in.) branches; most cutting heads, a 12.7-cm (5 in.) branch.

Oil mist

Oil mist is a problem in shops where oil is used as a coolant—

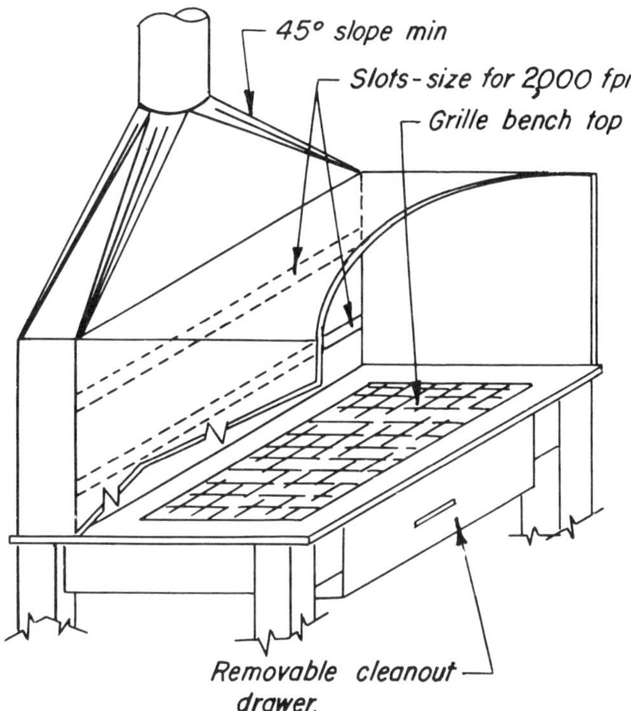

Figure 21-25. Exhaust booth for portable grinding operations, soldering, or arc welding. (Courtesy ACGIH)

rooms can become filled with an oil haze, floors soaked with oil present a slipping hazard, and condensation on walls and ceilings becomes a fire hazard. The International Agency for Research on Cancer (IARC) has stated that there is sufficient evidence that mildly hydrotreated oils are potential carcinogens. Based on this, OSHA requires that such oils be considered as carcinogens under the Hazard Communication Standard (50 *Fed Reg* 51852, Dec 20, 1985).

Hood sides must act as splash guards, because airflow alone will not arrest large drops of oil thrown from the operation. The hood must incorporate such splash guards, but still permit the machine to be serviced and the operation observed when required. A few manufacturers have provided their machine tools with plastic enclosures that meet these requirements.

Hoods attach to machines in a variety of ways and ingenuity is the best instruction manual. Usually, an exhaust volume of 0.2–0.3 m^3/sec (400–600 cfm) will provide the recommended 0.51–0.76 m/sec (100–150 fpm) indraft. Sometimes, when considerable heat is generated or where the hooding fails in being a complete enclosure, 0.5–0.7 m^3/sec (1,000–1,500 cfm) must be exhausted. Duct velocities of 12.7 m/sec (2,500 fpm) are ample. Duct joints should be tight to prevent oil leakage and ducts should be pitched to allow condensed oil to flow back to the machine or to a special collector.

Electrostatic precipitators and wet (oil-filled) centrifugal collectors can be used as arresters for oil mist exhaust systems. To eliminate ductwork, units small enough to mount directly on the machine are available.

Collection and reuse of oil also may offer considerable savings.

Melting furnaces

The design of hoods for metallurgical melting furnaces depends on the type of furnace and the metal being melted. The fumes may constitute a health hazard to the operator or become a source of community air pollution. The tilting crucible furnace is often equipped with a simple overhead, flared hood. Sometimes there is a fan in the duct, but often there is not. Because such hoods are at the mercy of cross drafts, and are at some distance from the metal during pouring, relatively large volumes of air 1.9–2.8 m^3/sec (4,000–6,000 cfm) are needed per furnace to get passing performance. For this reason, the gravity stack without a fan is often unsatisfactory.

By following the basic principle of hood design, enclosing the process, a much more economical airflow and more effective fume control can be obtained. Some furnaces have been completely enclosed, with an access door in the hood rear for charging and a counterweighted panel in front for pouring. Others have a cantilever hood, similar to a movable window awning. Crucible furnaces of the pit type are sometimes placed in booths with sliding doors front and rear to permit removal of the crucible.

Rocker-type furnaces should have overhead hoods with walls enclosing as much of the unit as possible. Hinged doors at the front and back provide access.

Molten cyanide heat treating furnaces would be ventilated by open-surface tank hoods. Usually, a complete cylindrical enclosure can be provided with a small access door or hole in the periphery. An overhead hood can be hinged to provide access for the operation of an overhead crane.

Electric melting furnace fumes have been collected in the past by one of two methods: (1) canopy-type hoods remotely located above the electric melting furnace; and (2) a close-fitting hood designed with a portion of the hood over the charging and pouring spout. This latter type captures the dust close to the source and has cutouts for raising and lowering the electrodes.

Another system is direct evacuation of the furnace shell itself. A hole is tapped into the furnace roof or the shell and connected directly to the local exhaust system.

Materials conveying

Dust that accompanies production processes when granular materials are conveyed, weighed, mixed, packaged, or otherwise handled falls in a different category from the dust generated during the operations considered thus far. Although dust that is generated can be detrimental either to health or comfort, its recovery carries no by-product value.

In materials conveying, the dust does not differ from the product being processed. The more dust removed, the greater the direct loss of usable product. The objective, therefore, is to trap the dust that would otherwise escape to the workroom air, but not to remove useful material being processed.

Often the collected dust can be salvaged. In the handling of sand, for example, or lime, or flour, the dust is chemically the same as the bulk material, and it can be reclaimed without downgrading.

Belt conveyors, bucket elevators, skip hoists, bin fillers and drum fillers, weighers, and mixers are all extremely common

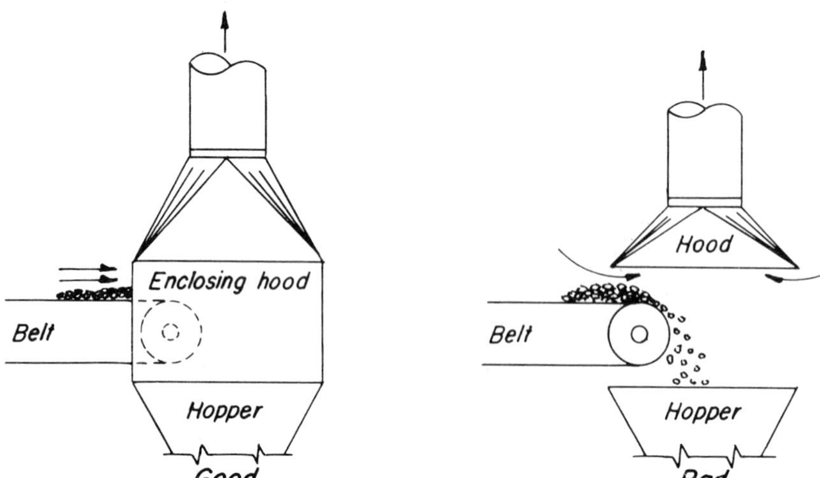

Figure 21-26. Enclose the operation as much as possible—the more completely enclosed the source, the less air is required for control. (Courtesy ACGIH)

operations throughout industry. Yet there is a scarcity of basic design data for related dust control. A good deal of rule of thumb practice is available. Pring et al (1949) and Hemeon (1963), see Bibliography, are among the few who have explored the fundamentals of needed control of these dusts.

No mixer, hopper, bin, silo, tank or other container can be filled with a dry, dusty material without the air that is already inside the container being displaced and carrying with it dust to the outside. One solution to the dust problem is to make the container, chutes, and casing airtight (meaning completely enclosed) and run a hairpin-shaped air vent (breather) or duct from the container being filled to the container from which the dusty bulk material is draining, in order to let the dust-laden air escape back to the feed container. When the hairpin duct is mounted vertically, no dust will settle out in the piping and so the breather cannot become plugged.

A basic fact, often overlooked, is that granular material falling through air (as when discharging from a belt into a storage bin) does not just displace its own volume of air, but induces additional airflow because air is entrained by the falling motion of the individual particles of the material. The aim of the engineer is to control this induced airflow by exhaust ducts.

Because the induced airflow is caused by the fall of particles, air enters with the solids at the top of a circuit; for example, the discharge of material through a chute at the head of a bucket elevator to a conveyor belt. The air, consequently, must be removed from the bottom (in this case, from the bottom of the elevator discharge chute) by properly located exhaust connections. To permit settling out of coarser particles before they are carried into the exhaust system, the volume of air exhausted must be kept at a minimum by:

1. restricting the opening through which incoming air can enter
2. reducing the height of fall as much as possible
3. effectively enclosing the impact zone—the region where the material lands on the belt.

Figure 21-26, shows how the exhausted air can be controlled. The enclosure around the head pulley has a restricting opening which the belt passes, following recommendation (1). The fall of material should be broken by a pocket in the chute, into which the material drops rather than splashing directly on the belt itself. This is in line with (2) and might be further improved if riffles or cleats could be built in, over which the material would cascade, resulting in practically *no free fall.* Requirement (3) can be met by providing a generous enclosure around the skirts on the belt at the foot of the chute. The single exhaust hood should be located well above the belt. This keeps the air velocity at the belt level so low that coarse, useful particles will not be picked up. An exhaust connection is rarely required at the top of the enclosure. However, if the material falls *directly* to the conveyor, a second exhaust connection should be provided at the back of the chute, because dusty air dislodged behind the impact zone cannot be captured by a single hood at the front without dragging coarse particles into the airstream.

Breaking the free fall of material when it is transferred (from belt to belt, elevator to belt, or chute to bin) results in a savings. "Choke feeding," the piling up of material in cone shapes under a discharge point, can often control the dust so that an exhaust hood is not required. For example, a hopper-bottom car might dump material into a hopper opening beneath the tracks, at the bottom of which is a screw conveyor that takes the product to some other point. If the material is discharged from the car at a faster rate than it is removed by the screw conveyor, it will "cone-up" under the car hopper doors. The operation will then be essentially dustless, provided stray cross breezes are screened off. This is a far superior method of dust control than exhaust ventilation.

Internal combustion engine ventilation

Gasoline, diesel, and natural gas fueled engines are used occasionally inside factory buildings. They are operated not only in garages, where maintenance checks and repairs are made, but also in warehouses, storage spaces, loading docks, aisleways, and other indoor areas. Engines are used as drives for lift trucks and for air conditioning and electric generating equipment. Their products of combustion contain such toxic materials as aldehydes, carbon monoxide, and nitrogen oxides.

Aldehydes in air are extremely irritating to the eyes. The Threshold Limit Value for formaldehyde, for example, is only one part per million. (Formaldehyde is designated as a suspect human carcinogen as well.) Some aldehydes may be generated if the engine is in poor operating condition and is burning oil, or has a smoky exhaust. However, they usually are of less importance, from the industrial hygiene standpoint, than the carbon monoxide (CO) released.

For an 8-hour exposure, the Threshold Limit Value of carbon monoxide gas in air is 50 ppm in the breathing zone area of employees. However, a ventilating system should be designed to keep a much lower level of CO. Exhaust gases as released from a gasoline engine contain from 0.1 to more than 10 percent CO (1 percent = 10,000 ppm). If the engine is operating at full-rated horsepower, its exhaust gases will contain about 0.3 percent CO. They may contain more than 10 percent when idling.

Gasoline engines are built in many power levels, ranging from less than one (0.7 kW) to more than 200 hp (150 kW). Lift trucks commonly have a maximum delivery of 35–50 hp (26–37 kW). Since the engine normally discharges about 0.5 L/sec (1 cu ft/min) of exhaust gases per operating horsepower, a lift truck when operating near its maximum rated power would release 0.03 times 50 or 0.7 L/sec (1.5 cfm) CO. The quantity of fresh air needed to dilute this CO output so the concentration would be below the Threshold Limit Value depends on many factors. (See Department of National Health (1959); Michigan Department of Health (1962-1963); and Sheinbaum (1953) in Bibliography, Chapter 22, for more information.)

The distribution of the incoming fresh air throughout the building is just as important as the ventilation rate. A warehouse, for example, may have a central runway with products or goods piled on each side in bays. Deadend alleys or aisles may lead into these bays and it is possible that these contaminant pockets are devoid of ventilation. The CO concentration may also build up to excessive levels in an aisle when a truck is stacking there, although the CO level might still be negligible along the central runway. Such a situation can best be solved by introducing fresh air by a duct having discharge grilles in the areas most likely to have CO concentrations, such as in aisles, rather than along the central runway. The latter generally receives adequate ventilation from the building doors.

Improvement is sometimes effected by extending the tailpipe of a warehouse truck upward so it discharges vertically at a point above the operator's head. This solution is effective, too, in a situation where a dump truck backs up to a pit hopper, at the bottom of which there may be a conveyor. A worker may stand across the pit where the conveyor controls are located and get the full blast of a truck exhaust. A vertical discharge pipe would help prevent this.

There are areas where it may not be practical or economical to ventilate for the prevention of hazardous concentrations of CO, in frozen food storage rooms, for example. Electric trucks are recommended for these areas.

Finally, there are garages with fixed repair stations. Here, a local exhaust system is preferred (Figure 21-27). Seven and one-half centimeter (3 in.) flexible ducts are fitted over the tailpipes. The flexible ducts join 10 cm (4-in.) branches, which in turn are connected to a header located below floor level or overhead. (Exhaust requirements are shown in Table 21-E.)

For general garage ventilation where cars are in motion or are idling outside of the repair stall, the following ventilation guideline is provided for diluting carbon monoxide:

- operating auto engine—2.4 m^3/sec (5,000 cfm)
- operating truck engine—4.7 m^3/sec or more (10,000 cfm)
- per horsepower (0.7 kW) for diesel—100 cfm

Table 21-E. Tailpipe Exhaust Requirements

Type	*Cfm per Tailpipe*	*Diam. (in.) Flexible Duct (Minimum)*
Auto and truck (up to 200 hp)	100	3
Auto and truck (over 200 hp)	200	4
Diesel	See reference below	
On dynamometer test rolls		
Auto and light-duty trucks	2 × cfm above	
Heavy-duty trucks	1,200 cfm min.	

1 cfm = 0.00047 m^3/s

(Reprinted with permission from American Conference of Governmental Industrial Hygienists, Committee on Industrial Ventilation, *Industrial Ventilation Manual*, 19th edition)

Air-sampling. The ultimate effectiveness of any ventilation procedure for controlling CO should be verified by checking air samples for CO content. Several instruments are available that use detector tubes in which a color change is produced if CO is present in the air-sample. Other instruments are available for continuous monitoring if necessary. See Chapter 16, Evaluation.

Fog removal

Hot water tanks are used for steaming purposes in various industries and operations. These may include foundries, veneer mills, plating, and metal cleaning. The steam that is released into a room may lead to excessive moisture condensation, lowered visibility, or discomfort. This is particularly true in winter weather, and at extensive vat operations such as those found in veneer mills. In vat operations, one way to attack fog problems is to install vertical stacks (not elbows) above the vat. If the vat is of the totally enclosed type, the stack inlet should be flush with the cover (centrally located, or proportionally located should there be two or more stacks). If the vat is used for dipping or quenching, it cannot be totally enclosed, but it is almost essential to have three sides baffled or enclosed (above the top edge of the vat) otherwise cross drafts will nullify the purpose of the stacks. The stacks should extend above the high point of the roof, or that of the adjoining building, whichever is higher.

Exhaust fans should be installed in these stacks to ensure a positive outward flow so that airflow does not depend on outdoor weather conditions, thereby making "downdrafts" impossible.

Kitchen range hoods

Some of the most elaborate exhaust hoods, and largest in terms of physical size and airflow, are kitchen range hoods, and yet as a class they receive a minimum of attention and research by health and safety professionals. Today with many company plants supporting their own well-equipped cafeterias, proper ventilation is as necessary in the kitchen as in any other production department. However, greater concern is often shown the aesthetic appearance of kitchen hoods than is given to their performance.

The investment in kitchen hoods may be wasted if installations are underdesigned. Because ceilings, walls, and fixtures become coated with solidified fat, maintenance of sanitary conditions becomes a problem and unappetizing odors pollute the

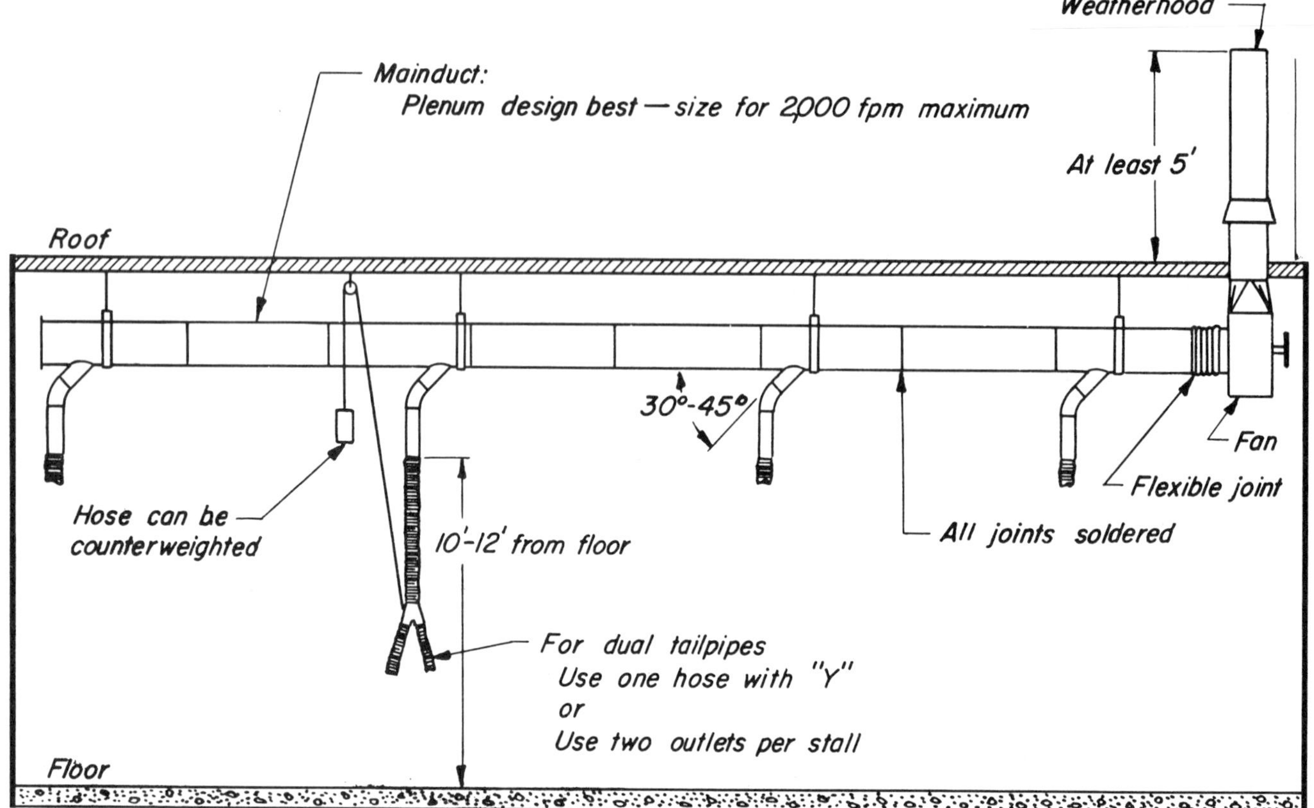

Figure 21-27. Local exhaust ventilation for service station garage. Specifications are given in Table 21-E. (Courtesy ACGIH)

dining area. If hoods are overventilated, original equipment cost is expensive, operating cost is high, and heating of the dining room may be a problem, because kitchen hoods sometimes exhaust a very large volume of air.

Range hoods, in large kitchens at least, are often extremely long for their width, because they must cover a row of ranges, ovens, grills, and deep-fat fryers.

They frequently can be set so the long side is against a wall (although the soundness of doing so is often overlooked), allowing the hood to become a shed-roof type hood. Also, the temperature gradient set up by heat of the stove aids in creating air movement into the hood. Finally, as J.M. Dalla Valle (1952) has written in "Design of Kitchen Range Hoods" (see Bibliography), the control velocities needed at the surface of the range are low compared with those needed, for example, in dust control where particles must be transported into the duct.

Dalla Valle recommends 0.15 m/sec (30 fpm) as the control velocity to be established at the range surface. For different shaped hoods, he has calculated the face velocity. That is, the velocity of air at the hood opening necessary to set up this 30-fpm control velocity. His dimensions are in terms of hood width; "W." Hood shape is expressed as "side ratio." A side ratio of 1:4, for example, means a rectangular hood that is four times as long as it is wide. (Table 21-F.)

The distance that a canopy hood should extend beyond the cooking equipment is regulated in many cities by ordinance. Where there is no code, it is customary to increase the dimension of a canopy hood 0.33–0.41 m/m (4–5 in. for every foot) of distance between the hood face and operating surface (range, fryer, or other heat source). Some designers use a minimum overhang of 30.5 cm (12 in.).

A rectangular exhaust hood placed along the wall at the back of deep fat fryers and grills can be installed closer to the source of smoke and odor. In effect it becomes a lateral exhaust for an "open-surface tank," and may be designed as already outlined in the section under that heading. The smoke and oil vapor are trapped before they rise from the cooking unit, and they have less chance to spread through the room. End baffles further improve the operation.

Grease filters are desirable in all kitchen hoods, including the lateral exhaust type, for two reasons: (1) they prevent fat from entering the duct, where it condenses and causes a fire hazard; and (2) when placed in the hood itself, they serve as excellent diffusers to assure a uniform distribution of air over the hood opening, which is generally large in area. Grease filters, however, add resistance to the system, sometimes as much as 0.12 kPa (0.5 in.) water gage. The exhaust fan should be able to handle the necessary volume of air over the range of filter resistance pressures. Grease filters should be cleaned regularly.

An approved fire damper with a fusible link should be installed in the main exhaust duct or branch adjacent to the range hood. This is required by code in many states.

Codes are often mandatory. These are not to be used as a crutch for poor or inadequate design, however. The designer should try his best to make certain his plans and specifications meet, as a minimum, the requirements of any applicable codes. Additional requirements should be anticipated and met. The health and safety professional, the contractor, or the main-

Table 21-F. Hood Face Velocity (fpm) to Maintain a Control Velocity of 30 fpm at Range

	Distance X from Hood Opening to Working Area					
Side Ratio (W:L)	*0.25W*	*0.5W*	*0.75W*	*W*	*1.25W*	*1.5W*
1:1 Free hanging	40	120	300			
1:2 Shed type		40	77	120	170	240
1:2 Free hanging	24	60	120	215	355	
1:4 Shed type		24	40	60	90	120
1:6 Shed type		16	27	40	60	80

(Reprinted with permission from J.M. Dalla Valle, *Exhaust Hoods,* 2nd ed. New York, The Industrial Press, 1952)

tenance man should not later need to salvage a poor engineering job.

A bibliography related to ventilation for this and the next chapter have all been combined and are listed together at the end of Chapter 22. Particular emphasis is placed on the basic text on ventilation, *Industrial Ventilation—A Manual of Recommended Practice,* of the American Conference of Governmental Industrial Hygienists.

22

General Ventilation

by Willis G. Hazard, AM, CSP
Revised by D. Jeff Burton, PE, CIH

LOCAL EXHAUST SYSTEMS have several inherent advantages, as compared with general ventilation—for example, removal of a contaminant before it spreads throughout the room, economy of airflow, and less heat loss. There are, however, some operations in which local exhaust systems are impractical. These operations can sometimes be controlled by diluting the general room atmosphere with enough fresh air to keep the concentration of toxic material in the room air within recommended limits. Some examples of dilution ventilation are shown in Figure 22-1.

Solvents are used in compounding synthetic varnishes and lacquers, in cements and adhesives, in liquid coatings for fabrics and other materials, in cleaning operations, and in many other processes. In these applications, the basic purpose of solvents is to evaporate into the atmosphere, leaving behind some physically changed substance—the desired end product. Thus, the very heart of the process involves potentially polluting the air with vapor. The aim of the equipment designer or process engineer is to keep this vapor concentration as low as possible, certainly below the toxic limit. See Chapter 6, Solvents, for more details.

The average rate of solvent evaporation can often be ascertained as can the chemical nature of the solvent. It is known that a given volume of vapor will be formed upon evaporation of a given weight of solvent. It is possible, then, to calculate how much fresh air must be mixed with this vapor to hold the air plus vapor concentration down to safe limits. The American Conference of Governmental Industrial Hygienists (ACGIH) has published a table from which general ventilation rates can be estimated (Table 22-A).

This table is based on the formula:

$$\text{Cfm required} = \frac{\text{lb solvent evap.}}{\text{min}} \times \frac{387}{\text{MW solvent}} \times \frac{10^6}{\text{TLV}}$$

where TLV is the ventilation design concentration, and MW stands for molecular weight. For more details, see Appendix C, Chemical Hazards.

The Threshold Limit Value (TLV) should not be used for calculating final air volumes, because vapor dilution in the working space is bound to be uneven and concentrations must always be maintained *below* the TLV in order to provide a factor of safety. In turn, this factor of safety depends on whether the solvent vapor is to be controlled because of its inherent toxicity or its disagreeable odor. (Details were given in Chapter 6, Solvents.) Industrial hygienists often use one half of the TLV as an acceptable exposure level, effectively doubling the dilution rate.

For example, suppose one gallon (3.8 L) of methyl ethyl ketone is evaporated per hour. If one pint of methyl ethyl ketone requires a ventilation rate of 22,500 cu ft of air, one gallon would then require (22,500 × 8) or 180,000 cu ft of air. If this amount is evaporated per hour, the ventilation rate would be 1.42 m^3/sec (3,000 cfm).

It is important to note that this example also assumes perfect mixing of the clean air with the solvent vapor, but in practice this does not occur. The ventilation rate calculated is, therefore, a minimum and should be increased depending upon other factors involved—type and location of air diffusers, location of people in the room, and relative toxicity of the vapor.

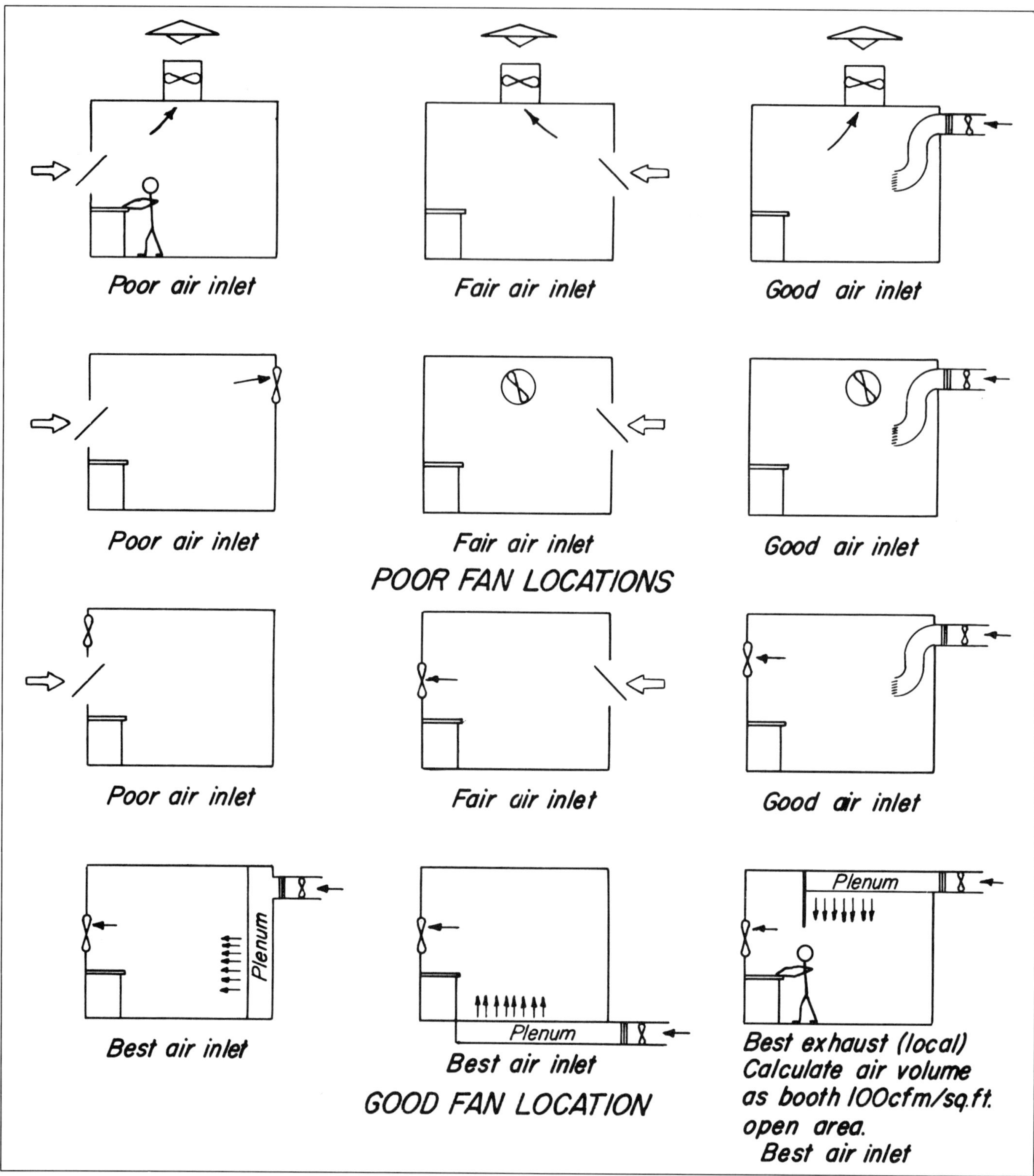

Figure 22-1. Poor, fair, good, and best locations for fans and air inlets for dilution ventilation. Key:⇨= air inlet and airflow direction; ∞ = exhaust fan. (Courtesy American Conference of Governmental Industrial Hygienists [ACGIH])

The size of the workroom does not enter the calculation. This is at variance with the common practice of specifying ventilation requirements in terms of "number of room air changes per minute," which, of course, directly involves room volume. The fact is the rule of thumb based on room air changes per minute, though in widespread use over many years, has been used improperly more often than not. This is especially true when there are solvent vapor contaminants being released within the room.

Table 22-A. Dilution Air Volumes for Vapors

Liquid (TLV in ppm)*	Cu Ft of Air (STP) Required for Dilution to TLV**	
	Per Pint Evaporation	Per Pound Evaporation
Acetone (1000)	5,500	6,650
n-Amyl acetate (100)	27,200	29,800
Benzol (10)	NOT RECOMMENDED	
n-Butanol (butyl alcohol) (50)	88,000	104,000
n-Butyl acetate (150)	20,400	22,200
Butyl cellosolve (2-Butoxyethanol) (25)	NOT RECOMMENDED	
Carbon disulfide (10)	NOT RECOMMENDED	
Carbon tetrachloride (5)	NOT RECOMMENDED	
Cellosolve (2-Ethoxyethanol) (5)	NOT RECOMMENDED	
Cellosolve acetate (2-ethoxyethyl-acetate) (5)	NOT RECOMMENDED	
Chloroform (10)	NOT RECOMMENDED	
1-2 Dichloroethane (10) (ethylene dichloride)	NOT RECOMMENDED	
1-2 Dichloroethylene (200)	26,900	20,000
Dioxane (25)	NOT RECOMMENDED	
Ethyl acetate (400)	10,300	11,000
Ethyl alcohol (1000)	6,900	8,400
Ethyl ether (400)	9,630	13,100
Gasoline (300)	REQUIRES SPECIAL CONSIDERATION	
Isoamyl alcohol (100)	37,200	43,900
Isopropyl alcohol (400)	13,200	16,100
Isopropyl ether (250)	11,400	15,140
Methyl acetate (200)	25,000	26,100
Methyl alcohol (200)	49,100	60,500
Methyl n-butyl ketone (5)	NOT RECOMMENDED	
Methyl cellosolve (2-Methoxyethanol) (5)	NOT RECOMMENDED	
Methyl cellosolve acetate (2-Methoxyethyl acetate) (5)	NOT RECOMMENDED	
Methyl ethyl ketone (200)	22,500	26,900
Methyl isobutyl ketone (50)	64,600	77,400
Methyl propyl ketone (200)	19,900	22,400
Naptha (coal tar)	REQUIRES SPECIAL CONSIDERATION	
Naptha VM & P (300)	REQUIRES SPECIAL CONSIDERATION	
Nitrobenzene (1)	NOT RECOMMENDED	
n-Propyl acetate (200)	17,500	18,900
Stoddard solvent (100)	30,000–35,000	40,000–50,000
1,1,2,2-Tetrachloroethane (1)	NOT RECOMMENDED	
Tetrachloroethylene (50)	79,200	46,800
Toluol (Toluene) (100)	38,000	42,000
Trichloroethylene (50)	90,000	60,000
Xylol (xylene) (100)	33,000	36,400

*See Threshold Limit Values, *Industrial Ventilation Manual,* ACGIH.
**The tabulated dilution air quantities must be multiplied by the selected K value.
(Reprinted with permission from the *Industrial Ventilation Manual,* 19th ed. ACGIH, 1986)

COMFORT VENTILATION

Ventilation is defined as the process of supplying air to, or removing air from, any space by natural or mechanical means. The word implies quantity, not necessarily quality, of air supplied. From the standpoint of comfort and health, however, the problem now is considered to involve both quantity and quality.

The term comfort conditioning (also called air conditioning) implies control of the physical and chemical qualities of the air. The American Society of Heating, Refrigerating and Air Conditioning Engineers (ASHRAE) defines it as "the process of treating air so as to control simultaneously its temperature, humidity, cleanliness, and distribution to meet the requirements of the conditioned space" (ASHRAE *Handbook of Fundamentals;* see Bibliography).

General ventilation of the work place contributes to the comfort and efficiency of the workers. Good general ventilation also makes a definite contribution to health, since working under extreme conditions of temperature and humidity may have an adverse effect on health. It is difficult to say just where the dividing line between health and comfort falls.

DEGRADING OF AIR

There are a number of reasons for air becoming stale or spoiled in a workplace. Some of them are as follows:

Carbon dioxide buildup

Contrary to old theories, the normal changes in carbon dioxide (CO_2) and oxygen in work rooms are of no physiological concern, because they are too small to produce appreciable effects even under the worst conditions of normal human occupancy.

Biological organisms

Various studies indicate that when air is recirculated for building ventilation, the spread of communicable diseases may be accelerated by recirculation of biological agents in contaminated dust and droplets.

The bacterial content of the air in ventilating ducts can be reduced by ultraviolet irradiation, and perhaps by sprays of polyglycols, or by efficient filters. Other biological agents, such as spores and fungi, can be controlled by keeping air distribution systems dry, or chemically treated. (See Chapter 14, Biological Hazards, for more details.)

Odors

Odors may be completely harmless, but they have some significance as an index of air contamination. Disagreeable, though harmless, odors may cause so much discomfort that employees will refuse to work in their presence.

Deodorants merely mask offensive odors and should seldom be used alone. Generally, disinfectants that will destroy the bacteria responsible for the disagreeable odors can be used, or the source of the odors can be eliminated by proper cleanliness or by other comparable means. Many odors, though, are not caused by bacteria.

Some odors can be removed from air by using filters of activated charcoal, a material that absorbs the substance that produces the odor. It is important to filter the air first to remove the particulate matter so that the charcoal filter does not become so clogged that it cannot remove the chemical odor-producing components.

Often, however, the best remedy for odors is to add fresh outdoor air to the recirculated air. In many public buildings, the amount is specified by the local building code and depends upon the proposed occupancy.

Heat

Environmental thermal factors profoundly influence everyday living and comfort, and not infrequently health itself. The body has remarkably complex and delicate mechanisms, which hold body temperature within extremely narrow limits despite a wide range of external air conditions.

Misplaced emphasis has been given the problem because of the idea that a heating system exists to warm the occupants. With body temperature regulated at 37 C (98.6 F), and skin temperature at perhaps 33 C (92 F) in winter and 35 C (95 F) in summer, the task of a heating and ventilating system is *not* to heat the body, but to permit the body heat to escape at a controlled rate. This follows because air temperatures in the mid-20s C (mid-70s F) make up the comfortable range for humans. Whenever air temperatures approach body temperature, an individual is no longer comfortable. (See Chapter 12, Temperature Extremes, for more details.)

The human body has remarkable powers of adaptation. As soon as skin temperature or body tissue temperature rises or falls above or below an optimum, the body sets about to correct matters—sweat is secreted in a hot environment, and blood supply is redistributed between the skin and deeper tissues in a cold environment. Nerves from the skin carry the sense impressions to the brain, and the response comes back over another set of nerves, the motor nerves to the musculature and to all the active tissues in the body, including the endocrine glands. In this way, a two-sided mechanism controls the body temperature by (1) regulation of internal heat production (chemical regulation), and (2) regulation of heat loss through automatic variation in skin circulation and the operation of sweat glands (physical regulation). The reactions involved in cold and hot environments are radically different in nature. The task of the ventilating or air conditioning engineer is always to allow the body to *lose* heat at the desired rate (refer to Chapter 12, Temperature Extremes).

Natural ventilation

Wind, aided by air currents created by heating devices and higher indoor temperatures, provides a certain amount of ventilation even though doors and windows are closed. Air enters and leaves through the pores of building materials as well as through cracks and crevices.

Heating devices, such as ovens, cause the air to expand and rise, and if suitable openings are provided in the roof or near the ceiling, the rising air currents will pass to the outside. Then, if there are suitable openings near the floor, outside air will come in to replace what is lost.

Skylights and louvers satisfactorily control the escape of air in some cases, but they must be adjusted when changes of weather and wind occur, and it is sometimes difficult to prevent downdrafts.

Vertical stacks through roofs usually have tops to exclude rain and snow. The amount of air discharged by a vertical stack and roof ventilator depends on wind velocity, temperature differences, stack height, and the type of ventilator used. The stationary types are generally less effective than the wind-actuated rotary ones, but a great deal depends upon the design of the individual unit. Manfacturers can supply tables showing the effectiveness of each design.

The chimney or stack effect is the tendency of air or gas in a duct or other vertical passage to rise when heated, due to its lower density compared with that of the surrounding air or gas.

Exhaust ventilation

General exhaust ventilation may be achieved by placing a fan at a window or some other opening in the wall, thus removing the air from the room directly. (Windows that are near the fan should be kept closed to contribute to better mixing efficiency.) This type of ventilation involves low initial cost and comparatively low maintenance cost, but is likely to result in drafts near doors and windows.

Local cooling fans

Local fans stir up the air and tend to prevent stagnant accumulations of heat and moisture in limited areas where workers are present. Local fans, however, do not constitute a method of ventilation—the process of supplying or removing air, by natural or mechanical means, to or from any space.

Simple circulation by itself will not remove impurities from the air, nor will it supply pure air. But, power-driven fans placed in the open room may increase the comfort of the worker by increasing the rate of cooling and perspiration evaporation.

Cooling air should come from outdoors, where the temperature may be 5–15 C (10–30 F) cooler than it is in the plant. The duct supplying the air should be kept away from furnace walls, and other hot areas, and even insulated to keep the air temperature down. When the cooling air that blows on the employee is hotter than room air (to the extent of 5 C [10 F] in some installations), then good effects are lost and even the power consumed is largely wasted.

In arid areas outside air passed through an evaporative cooler before being delivered to the workers has been used with great success. When the plant air temperature is high, say 45 C (110–115 F), a 5–10 C (10–15 F) drop in temperature of the air blowing on a person gives extremely welcome relief. Some reference books suggest the velocity of air striking workers should not exceed 1 m/sec (300 fpm). However, if the heat is severe, experience indicates that in order to achieve any degree of relief, the air must be moving at least 5.8 m/sec (1,000 fpm).

Air conditioning

The objective of modern air conditioning is to regulate the temperature, motion, and humidity of the air, and to eliminate dust and dirt. Because normal increases in the amount of carbon dioxide and normal decreases in the amount of oxygen are not important in office and general factory structures, it is not necessary to introduce large volumes of fresh air. Instead, much of the room air can usually be recirculated (to reduce heating or cooling requirements and wasted fuel). Of course, this is not the case, as mentioned previously, when fresh air is needed to dilute concentrations of toxic substances in the workroom air.

In general, air conditioning equipment provides means for moving, distributing, and cleaning air and for controlling temperature and humidity of the air required in the space.

Many factory areas, such as those having ventilated plating or spray painting operations, are under appreciable negative pressure because of the relatively large quantities of air that must be exhausted. In winter, cold makeup air infiltrates through sash, entranceways, and similar places, causing drafts and discomfort. A solution is to blow heated replacement air into the plant at strategic locations.

Table 22-B. Negative Pressures and Corresponding Velocities Through Crack Openings

(Calculated with air at room temperature, standard atmospheric pressure, C=0.6)

Negative Pressure (Inches Water Gage)	*Velocity (fpm)*
0.004	150
0.008	215
0.010	240
0.014	285
0.016	300
0.018	320
0.020	340
0.025	380
0.030	415
0.040	480
0.050	540
0.060	590
0.080	680
0.100	760
0.150	930
0.200	1,030
0.250	1,200
0.300	1,310
0.400	1,510
0.500	1,690
0.600	1,860

1 in. wg = 0.25 kPa
1 fpm = 0.005 m/sec
(Reprinted with permission of the ACGIH)

REPLACEMENT ('MAKEUP') AIR

Whenever air is exhausted from a building, regardless of the method, outdoor air must enter to take its place. The incoming air may enter through random cracks around windows and doors, as general infiltration, or (better yet) a special provision may be made to supply makeup air (Table 22-B and Figure 22-2).

Replacement air is an obvious necessity. For this very reason, it is amazing how often its importance is completely overlooked. The problem is not simple, but often one of considerable magnitude.

Take an installation where 100 percent tempered air should be used at factory doors that are 3 × 3 m (10 × 10 ft) wide on a −12 C (0 F) winter day when air velocity is 5 m/s (1,000 fpm). To heat this incoming air, 2.2 mW (7.5 million Btu) will have to be supplied. (The heat given off by burning one wooden kitchen match is about 1,000 joules [1 Btu].) The weight of this air in itself is 250 tons. Certainly no small unit heater spotted over this doorway will be able to solve the problem.

The following discussion of replacement air is adapted from *Industrial Ventilation—A Manual of Recommended Practice*, published by the ACGIH; see Bibliography.

Necessity for replacement air

Replacement air will enter the building to equal the volume actually exhausted, whether or not provision is made for this replacement. However, the actual exhausted volume may not be the design volume unless an adequate supply of makeup air is provided. Replacement air, when used as a ventilation term, indicates the supply of outdoor replacement air to a building in a controlled manner. In some cases, especially if the exhaust system is small, replacement occurs with no adverse effects even in the absence of a makeup system. However, when exhaust volumes are large relative to the size of the free inlet area of the building, difficulty will be encountered. A relatively old building with large sash areas may have quite pronounced air leakage. On the other hand, a modern windowless plant of masonry construction can be practically airtight and the building will be "air starved" if there is appreciable exhaust ventilation.

When the building is relatively open, air leakage is often undesirable since the influx of cold outside air in the northern climates chills the perimeter of the building. Exposed workers are subjected to drafts, space temperatures are not uniform, and the building heating system is usually overtaxed (Figure 22-3). Although the air may eventually be tempered to acceptable conditions by mixing as it moves to the building interior, this is an ineffective way of transferring heat to the air and usually results in fuel wastage.

Experience has shown that adequate replacement air is necessary for the following reasons:

Assure proper exhaust hood operation. A lack of replacement air creates a negative pressure condition, which increases the static pressure the exhaust fans must overcome. This can cause a reduction in exhaust volume from all fans and is particularly serious with low-pressure fans, such as wall fans and roof exhausters (see Figure 22-4).

Eliminate high-velocity cross drafts. Cross drafts, from windows and doors, not only interfere with the proper operation of exhaust hoods, but may also disperse contaminated air from one section of the building into another; this can interfere with the proper operation of process equipment such as solvent degreasers. In the case of dusty operations, settled material may be dislodged from beams and ledges and result in recontamination of the workroom.

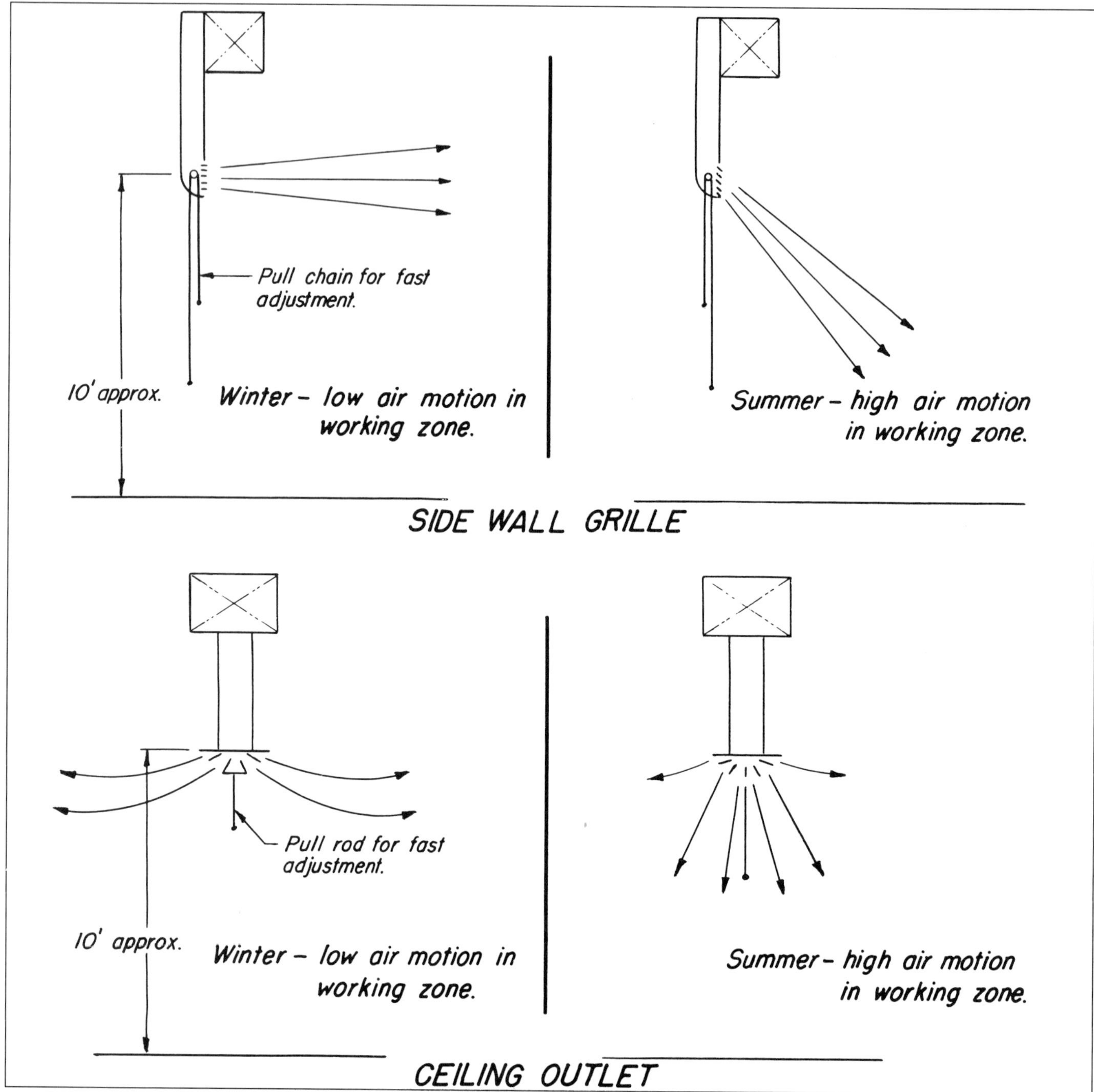

Figure 22-2. Seasonal air control for comfort. (Courtesy ACGIH)

Assure natural-draft stack operation. Even moderate negative pressures can result in backdrafting of flues, which may cause a dangerous health hazard from the release of combustion products (principally carbon monoxide) into the work room. Backdrafting may occur in natural draft stacks at negative pressures as low as 0.01 in. water gage (Table 22-C). Secondary problems include difficulty in maintaining pilot lights in burners, poor operation of temperature controls, and corrosion damage in stacks and heat exchangers due to condensation of the water vapor.

Eliminate cold drafts. Drafts not only cause discomfort and reduce working efficiency but may also result in lower overall ambient temperatures.

Eliminate differential pressure. High differential pressures make doors difficult to open or shut and, in some instances, can cause personnel safety hazards when the doors move in an uncontrolled fashion (Table 22-C and Figure 22-5).

Conserve fuel. Without adequate replacement air, uncomfortable cold conditions near the building perimeter frequently require the installation of more heating equipment in those areas to correct the problem. Unfortunately, these heaters warm the air too late and the overheated air that moves toward the build-

Table 22-C. Negative Pressures That May Cause Unsatisfactory Conditions Within Buildings

Negative Pressure Inches of Water	*Adverse Conditions That May Result*
0.01–0.02	**Worker Draft Complaints.** High velocity drafts through doors and windows.
0.01–0.05	**Natural Draft Stacks Ineffective.** Ventilation through roof exhaust ventilators, flow through stacks with natural draft greatly reduced.
0.02–0.05	**Carbon Monoxide Hazard.** Back drafting will take place in hot water heaters, unit heaters, furnaces and other combustion equipment not provided with induced draft.
0.03–0.10	**General Mechanical Ventilation Reduced.** Airflows reduced in propeller fans and low pressure supply and exhaust systems.
0.05–0.10	**Doors Difficult to Open.** Serious injury may result from non-checked, slamming doors.
0.10–0.25	**Local Exhaust Ventilation Impaired.** Centrifugal fan fume exhaust flow reduced.

1 in. wg=0.25 kPa
(Reprinted with permission of the ACGIH)

ing interior makes those areas uncomfortably warm (Figure 22-3). This in turn leads to the installation of more exhaust fans to remove the excess heat, and further aggravates the problem. Heat is wasted without curing the problem. The fuel consumption with a well-designed replacement air system is usually lower than when attempts are made to achieve comfort without a proper replacement system.

Replacement air volume

In most cases, replacement air volume should equal the total volume of air removed from the building by exhaust ventilation systems, process systems, and combustion processes. Determination of actual volumes of air removed usually requires a simple inventory of air exhaust locations accompanied by any necessary testing under atmospheric pressure conditions.

When conducting the exhaust inventory, it is necessary to determine not only the quantity of air removed but also the need for a particular piece of equipment. At the same time, reasonable projections should be made of the total plant exhaust requirements for the next 1 to 2 years, particularly if process changes or plant expansions are contemplated. In such a case, it can be practical to purchase a replacement air unit that is slightly larger than immediately necessary with the knowledge that the increased capacity will be required within a short time. The additional cost of a larger unit is relatively small and in most cases the fan drive can be regulated to supply only the desired quantity of air.

Having established the minimum air supply quantity necessary for replacement air purposes, many engineers have found that is is wise to provide additional supply air volume to overcome natural ventilation leakage and further minimize drafts at the perimeter of the building.

Environmental control

In addition to toxic contaminants, which are best handled by exhaust ventilation systems, modern industrial processes create additional physical stress agents in the work space—one of the most important of which is heat. Modern automated machining, conveying, and transferring equipment require considerable horsepower. Precision manufacturing and assembling demand increasingly higher light levels in the plant with correspondingly greater heat release. The resulting in-plant heat burden raises indoor temperatures, often beyond the limits of efficient working conditions and, in some cases, beyond the tolerance limits for the product.

Many industrial processes release minor amounts of nuisance contaminants that have no known health effects but which are unpleasant or disagreeable to the workers or harmful to the product. Finally, the desire to provide a clean working environment for both the people and the product often dictates controlled airflow between rooms or entire departments.

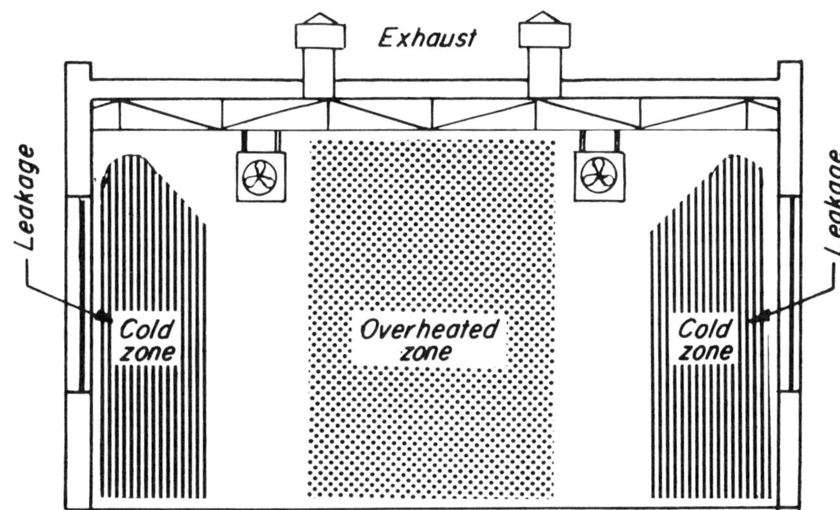

Figure 22-3. Under negative pressure conditions, workers in the cold zones (left and right) turned up the thermostats in an attempt to get heat. Because this did nothing to stop leakage of cold air, they remained cold while center of plant was overheated. (Courtesy ACGIH)

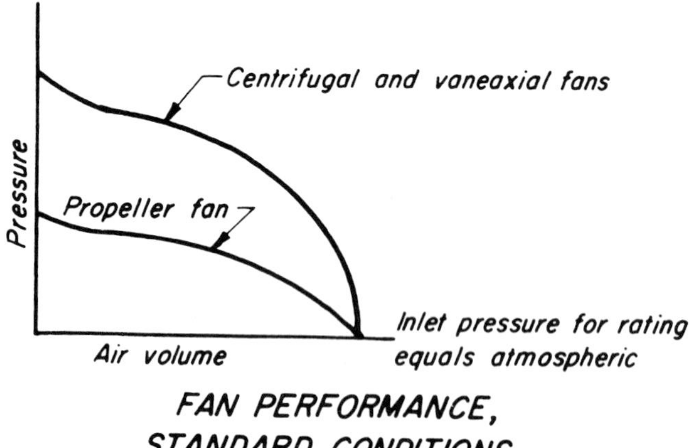

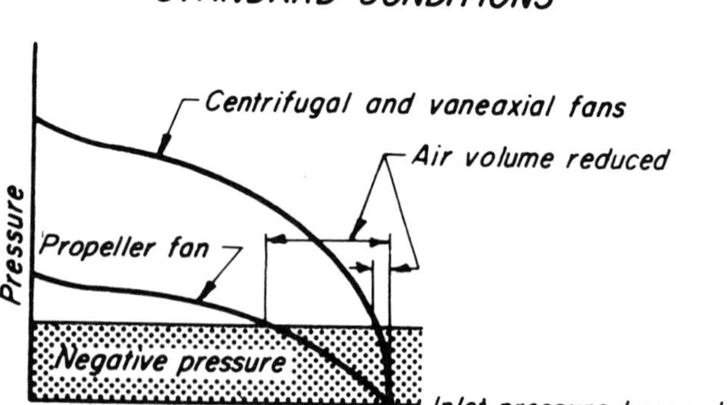

Figure 22-4. Graphic representation shows how fan performance falls off under negative pressure.

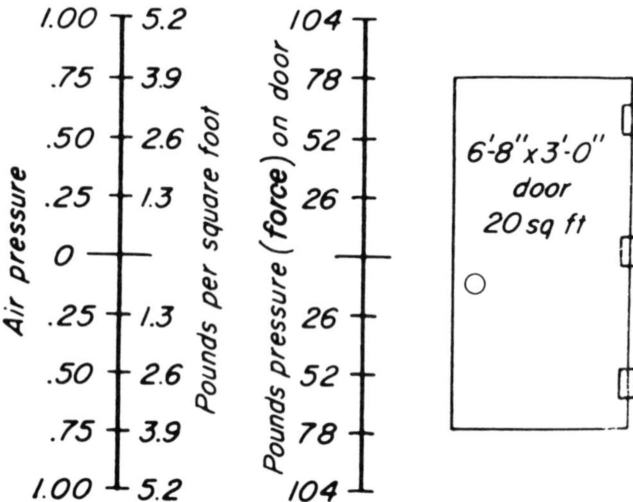

Figure 22-5. Relationship between air pressure and amount of force needed to open or close an average-sized door. (1 psf=990 kPa; 1 lb force=4.4 newtons). (Courtesy ACGIH)

Environmental control of these factors can be accomplished through the careful use of air-supply systems. (It must be noted that radiant heat cannot be controlled by ventilation and methods such as shielding described in Chapter 12, Temperature Extremes, are required.) Sensible and latent heat released by people and the process can be controlled to desired limits by proper use of ventilation.

Air changes

"Air changes per hour" or "air changes per minute" is a poor basis for ventilation criteria when environmental control of hazards, heat, and/or odors is required. The required ventilation depends on the problem, not on the size of the room in which it occurs. For example, let us assume a situation where 11,650 cfm would be required to control solvent vapors by dilution. Assume that the operation can be conducted in either of two rooms, each of different size. Although the air replacement volume would remain constant, the required air changes show quite a difference.

Room size (feet)	Room (ft^3)	Air changes per minute	Air changes per hour
20 × 40 × 12 high	9,600	1.21	73
40 × 40 × 20 high	32,000	0.36	22

Conversely, to maintain the same air change rate, a high ceiling space will require more ventilation than a low ceiling space of a room with the same floor area. There is little relationship between air changes and the required contaminant control.

The air change basis for ventilation does have some applicability for relatively standard situations, such as school rooms, where a correspondingly standard ventilation rate is reasonable. The air change concept is easily understood and reduces the engineering effort required to establish design criteria for ventilation. It is this ease of application, in fact, which in so many cases leads to lack of investigation of the real engineering parameters involved and, consequently, gives correspondingly poor results.

Air-supply temperatures

In most cases, outside air will be supplied in the winter months at (or slightly above) desired space temperatures, and during the summer, at whatever temperature is available out-of-doors. In cases when high internal heat loads are to be controlled, however, the temperature of the air supply can be appreciably below that of the space to be cooled; the amount of heat supplied to the air during the winter months can be reduced and the air in the summer can be cooled.

When large volumes of air are delivered at approximately space temperature or somewhat below, the distribution of the air becomes vitally important in order to maintain satisfactory environmental conditions for the persons in the space capsules. Maximum use of the supply air is achieved when the air is distributed in the living zone of work areas, below the 2.44- to 3.05-m (8- to 10-ft) level (Figure 22-2). When delivered in this manner—where the majority of the people and processes are located—maximum ventilation results with minimum air handling.

During the warm months of the year, large volumes of air at relatively high velocities are welcomed by the workers. During the winter months, however, care must be taken to make sure that air blowing over the person is kept within acceptable values (Table 22-D). To accomplish this, the air can be distributed uniformly in the space or where required for worker comfort. Heavy-duty, adjustable, directional grilles and louvers have proven to be very successful in allowing individual workers to direct the air as needed. Light-gage, stamped grilles intended for commercial use are not satisfactory. Suitable control must be provided to accommodate seasonal and even daily requirements with a minimum of supervision or maintenance

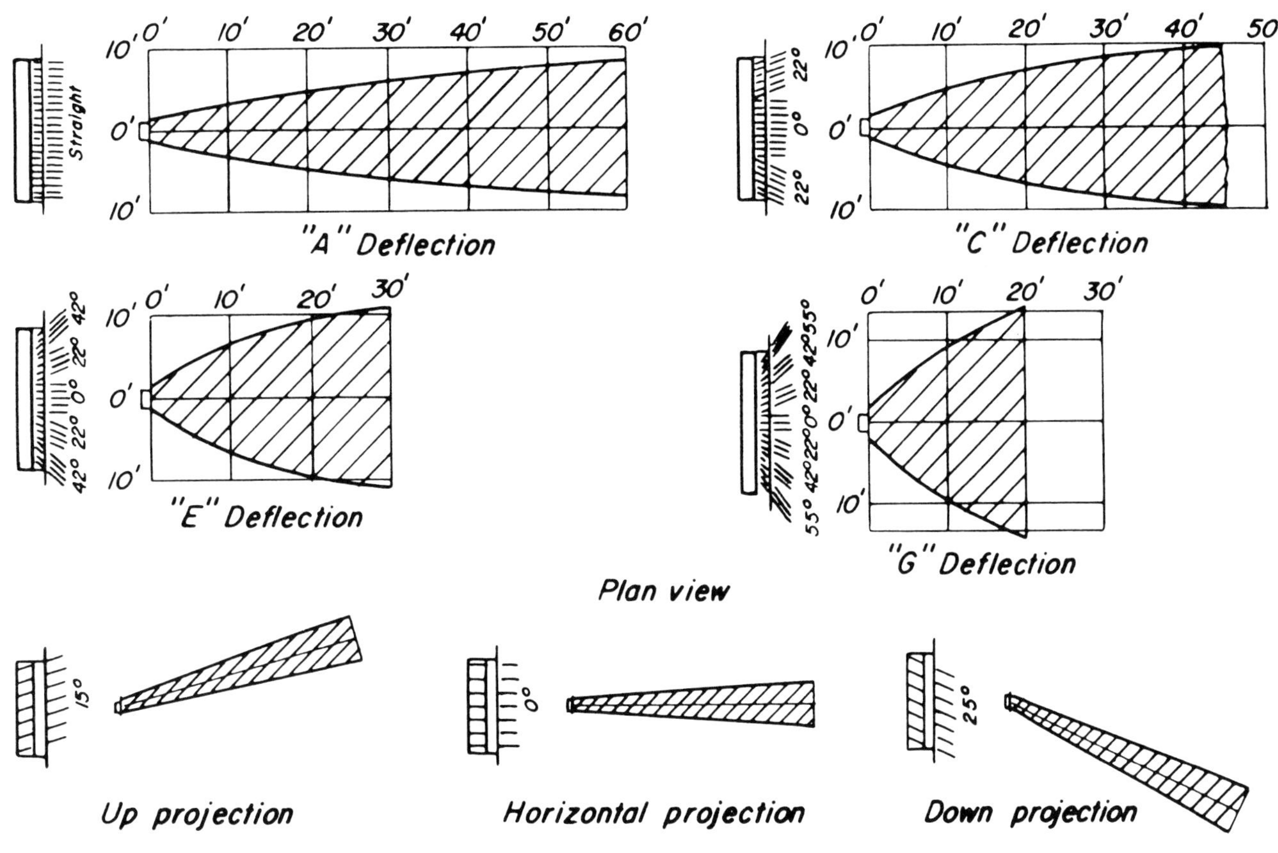

Figure 22-6. Throw patterns and distance form different register adjustments. (Courtesy ACGIH)

Table 22-D. Air Motion Acceptable to a Worker

Continuous Exposure	*Air Velocity (fpm)*
Air conditioned space	50–75
Fixed workstation, general ventilation or spot cooling:	
Sitting	75–125
Standing	100–200
Intermittent Exposure, Spot Cooling or Relief Stations	
Light heat loads and activity	1,000–2,000
Moderate heat loads and activity	2,000–3,000
High heat loads and activity	3,000–4,000

1 fpm = 0.005 m/sec
(Reprinted with permission by the ACGIH)

attention. Ventilation system controls should be maintained by supervision.

The effective temperature charts in Chapter 12, Temperature Extremes, illustrate the effect of air velocity, temperature, and moisture content on the relative comfort that can be derived through adequate airflow control. Published tables of data by register and diffuser manufacturers indicate the amount of throw (projection) and spread that can be achieved with different designs at different flow-rates (Figure 22-6).

Multiple-point distribution is usually best since it provides uniformity of air delivery and minimizes the reentrainment of contaminated air that is observed when large volumes are "dumped" at relatively high velocities. Various distributional layouts are employed depending on the size and shape of the space and the amount of air to be delivered. Single-point distribution can be used; however, it is usually necessary to redirect the large volume of air with a baffle or series of baffles in order to reduce the velocity close to the outlet and minimize reentrainment. In determining the number and types of registers or outlet points, it is also necessary to consider the effect of the terminal air supply velocity on the performance of local exhaust hoods.

When large amounts of sensible heat are to be removed from the space during the winter months, it is most practical to plan for rapid mixing of the cooler air supply with the warmer air in the space. During the summer months the best distribution usually involves laminar flow (minimum mixing) so that the air supply will reach the worker at higher velocities and with a minimum of heat pickup. These results can be obtained by providing horizontal distribution of winter air over the worker's head, allowing for mixing before it reaches the work area; and, for the summer months, directing the air toward the worker through register adjustment (Figure 22-2).

Delivered air temperatures during the winter are usually selected at 18 C (65 F), or even 13 C (55 F) when hard work or significant heat sources are involved. For summer operation,

Table 22-E. Advantages and Disadvantages of Direct-Fired Unvented Replacement Air Heaters and Indirect Exchanger-Type Replacement Air Heaters

Direct-Fired Unvented Replacement Heater	*Indirect Exchanger Replacement Air Heater*
1. Products of combustion in heater air stream (some CO_2, CO, oxides of nitrogen, and water vapor present).	1. No products of combustion, outdoor air only is discharged into building.
2. May be limited in application by state and municipal regulations. Consult local ordinances.	2. Allowable in all types of applications and buildings, if provided with proper safety controls.
3. Better turn down ratio 8:1 in small sizes, 25:1 in large sizes. Better control, lower operating cost.	3. Turn down ratio limited, 3:1 usual, maximum 5:1.
4. No vent stack, flue or chimney necessary. Can be located in side walls of building.	4. Flue or chimney required. Can be only located where flue or chimney is available.
5. Higher efficiency (90 percent). Lower operating cost. (Efficiency based on available sensible heat.)	5. Efficiency lower (80 percent). Higher operating cost.
6. Can heat air over a wide temperature range.	6. Can heat air over a limited range of temperature.
7. Extreme care must be exercised to prevent minute quantities of chlorinated hydrocarbons from entering air intake, or toxic products may be produced in heated air.	7. Small quantities of chlorinated hydrocarbon will normally not break down on exchanger to form toxic products in heated air.
8. Can be used with gas only as a fuel.	8. Can be used with both oil and gas as a fuel.
9. Burner must be tested to assure low CO and oxides of nitrogen content in air stream.	9. No contaminants in air stream from combustion.
10. First cost higher in small size units, and lower in large size units.	10. First cost lower in small size units and higher in large size units.
11. No heat exchanger to corrode or leak. Burner plates are very durable.	11. Heat exchanger subject to severe corrosion condition. Needs to be checked for leaks after a period of use.
12. Can be easily adapted to take all combustion air from outdoors. This is important if corrosive or contaminated work room air is present.	12. Difficult to adapt to take all combustion air from outdoors, unless roof mounted or outdoor mounted.
	13. Recirculation as well as replacement.

(Reprinted with permission of the ACGIH)

the temperature rise in indoor air can be estimated as described in Chapter 12, Temperature Extremes. Evaporative cooling should be considered for summer operation. Although not as effective as mechanical refrigeration under all conditions, evaporative cooling significantly lowers the temperature of the outdoor air being supplied to work areas and improves the ability of the ventilation air to reduce heat stress. In humid climates this could be helpful also, providing that heavy muscular work or heat-producing processes are not involved, since evaporative cooling used in humid climates can significantly impair the workers' evaporative cooling of their bodies.

Air-supply equipment

Air heaters are a specific type of equipment and are usually designed to supply 100 percent outdoor air. The basic requirements for an air heater are that it be capable of continuous operation and delivering constant air volume at a constant preselected discharge temperature. The heater must meet these requirements under varying conditions of service and accommodate outside air temperatures which vary as much as 22 C (40 F) daily.

Standard design heating and ventilating units are usually selected for mixed air applications, that is, partial outdoor air and partial recirculated air. It is rare, however, that the construction and operating capabilities of standard units will meet the requirements of manufacturing industry. Such units are better suited for use in commercial buildings and institutional facilities where the requirements are less severe and where mixed air service is more common.

Air heaters are usually categorized according to their source of heat—steam and hot water units, indirect-fired gas and oil units, and direct-fired natural gas and LP-gas, or LPG, units. Each basic type is capable of meeting the first two requirements—constant operation and constant delivered air volume. Variations occur within each type in relation to the third requirement, that of constant, preselected discharge temperature. One exception to the rule is the direct-fired air heater where the inherent design provides a wide range of temperature control. Each type of air heater (direct-fired and indirect exchanger) has specific advantages and limitations, which must be understood by the designer in making the selection (Table 22-E).

Steam coil units were probably the earliest air heaters applied to general industry as well as commercial and institutional buildings (Figure 22-7). When properly designed, selected, and installed, they are reliable and safe. They require a reliable source of clean steam at dependable pressure. For this reason, they are most widely applied in large installations; smaller industrial plants often do not have sufficient boiler or steam capacity for operating a steam air heater. Principal disadvantages of steam units are potential damage from freezing or water hammer in the coils, the complexity of controls when close temperature limits must be maintained, high cost, and excessive piping.

Freezing and water hammer problems result from poor selection and installation; both can be minimized through careful planning. The coil must be sized to provide the desired heat output at the available steam pressure and flow. The coil preferably should be of the steam-distributing type with vertical tubes. The traps and return piping must be sized for the maximum condensate flow at minimum steam pressure, plus a safety factor. Atmospheric vents must be provided to minimize the dan-

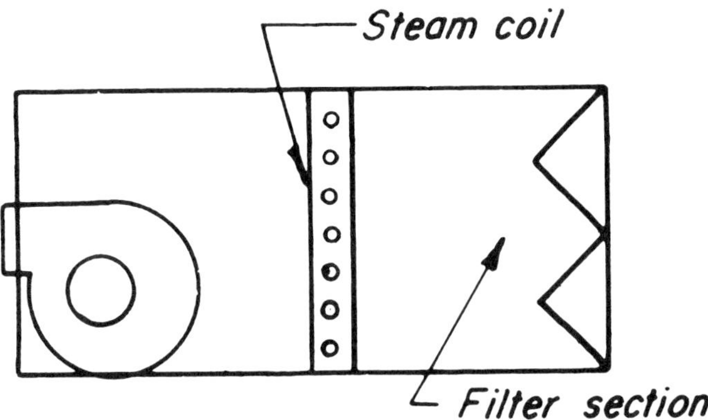

Figure 22-7. Single steam coil air heater. (Courtesy ACGIH)

ger of a vacuum in the coil which would hold up the condensate. Finally, the condensate must never be lifted by steam pressure. The majority of freezeup and water hammer problems relate to the steam-modulating type of unit which relies on throttling of the steam supply to achieve temperature control. When throttling occurs, a vacuum can be created in the coil and, unless adequate venting is provided, the condensate will not drain and can freeze rapidly under the influence of cold outside air. Most freezeups occur when the outdoor air is in the range of −7 to −1 C (20–30 F) and the steam control valve is partially closed, rather than when the outside air is at minimum temperature and full steam supply is on (Figure 22-8).

Safety controls are often used to detect imminent danger from freezeup. A thermostat in the condensate line, or an extended-bulb thermostat on the downstream side of the coil can be connected into the control circuit to shut the unit down when the temperature falls below a safe point. As an alternate, the thermostat can call for full steam flow to the coil, resorting to shutdown if a safe temperature is not maintained. An obvious disadvantage is that the plant air supply is reduced; if the building should be subjected to an appreciable negative pressure, unit freezeup may still occur due to cold air leakage through the fresh air dampers.

The throttling range of a single coil unit can be extended by using two valves—one valve is usually sized for about ⅔ the capacity and the other valve, ⅓. Through suitable control arrangements, both valves will provide 100 percent steam flow when fully opened and various combinations will provide a wide range of temperature control. Controls are complex in this type of unit and care must be taken to make sure that pressure drop through the two valve circuits is essentially equal so as to provide expected steam flow.

Multiple-coil steam units (Figure 22-9) and bypass (Figure 22-10) designs are available to extend the temperature control range and help minimize freezeup. With multiple-coil units, the first coil (preheat) is usually sized to raise the air temperature from design outdoor temperature to at least 5 C (40 F). The coil is controlled with an on-off valve, which is fully open whenever outdoor temperature is below 5 C. The second (reheat) coil is designed to raise the air temperature from 5 C to the desired discharge condition. Temperature control will be satisfactory for most outdoor conditions but overheating can occur when the outside air temperature approaches 5 C (40 F—39 F plus the rise through the preheat coil can give temperatures of 26–32 C [79–89 F] entering the reheat). Refined temperature control can be accomplished by using a second preheat coil to split the preheat load.

Bypass units incorporate dampers to direct the airflow. When maximum temperature rise is required, all air is directed through the coil. As the outdoor temperature rises, more and more air is diverted through the bypass section until finally all air is bypassed. Controls are relatively simple. The principal disadvantage is that the bypass is not always sized for full airflow at the same pressure drop as through the coil, thus the unit may deliver differing volumes of air depending on the damper position. Damper airflow characteristics are also a factor. An additional concern is that in some units, the air coming through the bypass and entering the fan compartment may have a nonuniform flow characteristic, which will affect the fan's ability to deliver air.

A novel type of bypass design called integral face and bypass (Figure 22-11) features alternating sections of coil and bypass. This design is said to promote more uniform mixing of the air stream, minimize any nonuniform flow effect and, through carefully engineered damper design, permit minimum temperature pickup even at full steam flow and full bypass.

Hot water. This is an acceptable heating medium for air heaters. As with steam, there must be a dependable source of water at predetermined temperatures for accurate sizing of the coil. Hot water units are less susceptible to freezing than steam because of the forced convection which ensures that cooler water can be positively removed from the coil. Practical difficulties and pumping requirements thus far have limited the application of hot water to relatively small systems—for a 55-C (100 F) air temperature rise and an allowable 55 C water temperature drop, 3.78 liters/min (1 gallon per minute [gpm]) of water will provide sufficient heat for only 0.227 m^3/sec (460 cfm) of air. This range can be extended with high-temperature hot water systems. Applications to date have been primarily in commercial and institutional buildings with mixed air service.

Hybrid systems using an intermediate exchange fluid such as ethylene glycol have also been installed by industries with critical air-supply problems and a desire to eliminate all freezeup dangers. A primary steam system provides the necessary heat to an exchanger which supplies a secondary closed loop of the selected heat exchange fluid. The added equipment cost is at least partially offset by the less complex control system.

Indirect-fired gas and oil units (Figure 22-12) are widely applied in small industrial and commercial applications. Economics appear to favor their use up to approximately 4.7 m^3/sec (10,000 cfm) above this size the capital cost of direct-fired air heaters is lower. Indirect-fired heaters incorporate a heat exchanger, commonly made of stainless steel, which effectively separates the incoming air stream from the products of combustion of the fuel being burned; thus heated room air can be recirculated. Positive venting of combustion products is usually accomplished with induced-draft fans. These precautions are taken to minimize interior corrosion damage from condensation in the heat exchanger due to the chilling effect of the incoming cold air stream. Another major advantage is that this type of unit is economical in the smaller volume sizes and is widely applied as a package unit in small installations, such as commercial kitchens and laundries.

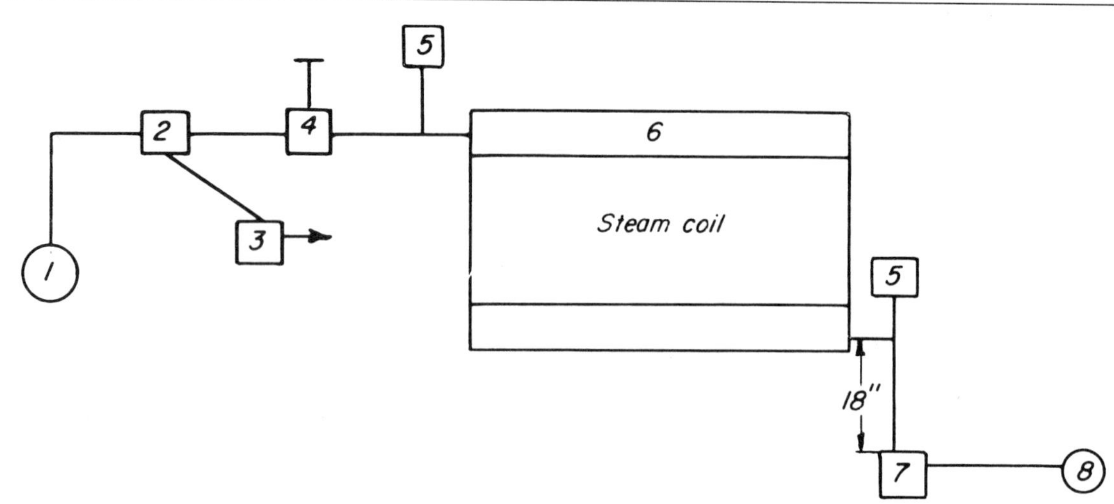

1. Steam supply
   Provide steam from a clean source
   Maintain constant pressure with reducing valves if required
   Provide trapped drips for supply lines
   Size supply piping for full load at available pressure
2. Strainer
   1/32" diameter minimum perforations
3. Drip trap
   Inverted bucket trap preferred
4. Control valve
   Size for maximum steam flow
   Maximum pressure drop equal to 50% inlet steam pressure
5. Vacuum breaker
   1/2" check valve to atmosphere

5'. Alternate vacuum breaker

6. Steam coil
   a. Size for design capacity at inlet steam pressure (supply – valve drop)
   b. Vertical coils preferred
   c. Horizontal coils must be pitched 1/4" per foot toward drain. 6' maximum length recommended
7. Condensate trap
   a. Inverted bucket preferred
   b. Size trap for three times maximum condensate load at pressure drop equal to 50% inlet pressure
   c. Individual trap for each coil
8. Condensate return
   Atmospheric drain only

Figure 22-8. Block diagram of a steam coil makeup air heating unit. (Courtesy ACGIH)

Temperature control (turndown ratio) is limited to about 3:1 or 5:1 due to burner design limitations and the necessity to maintain minimum temperatures in the heat exchanger and flues. Temperature control can be made more versatile through the use of a bypass system similar to that described for single-coil steam air heaters. Bypass units of this design offer the same advantages and disadvantages as the steam bypass units.

A newer design of indirect-fired unit incorporates a rotating heat exchanger. Temperature control is stated to be as high as 20:1. Presently this unit is available for natural and liquefied petroleum gases.

Direct-fired air heaters, wherein the fuel, natural or LPG gas, is burned directly in the air stream and the products of combustion are released in the air supply, have been commercially available for the past 20–25 years (Figure 22-13). These units

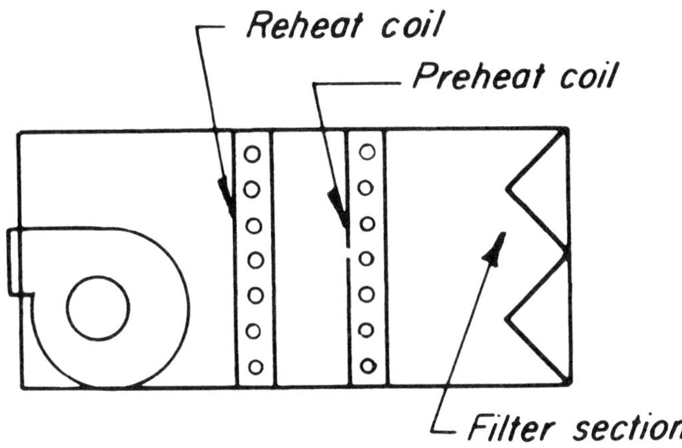

Figure 22-9. Multiple coil steam unit. (Courtesy ACGIH)

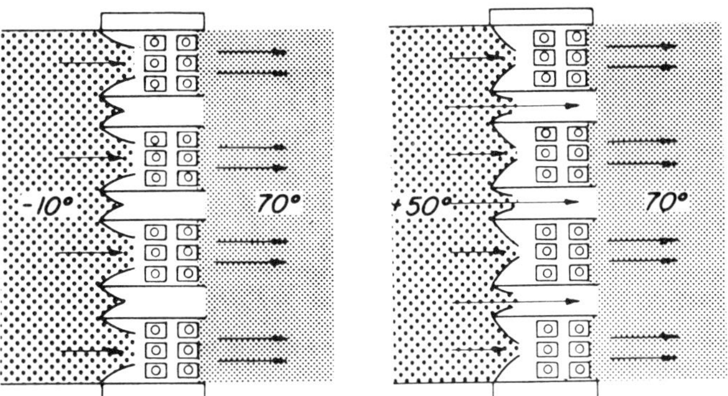

Figure 22-10. Bypass steam unit. (Courtesy ACGIH)

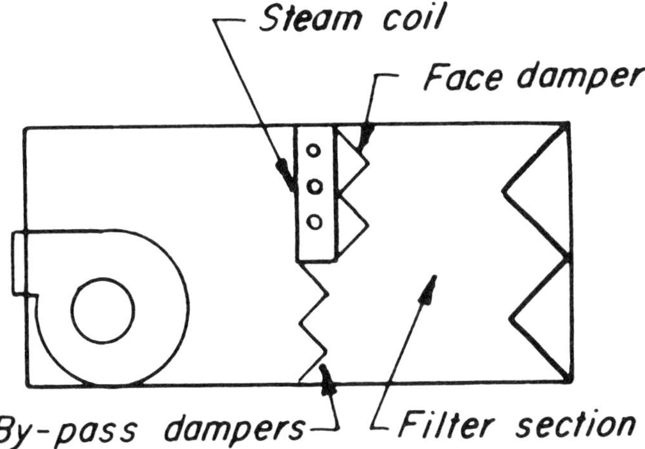

Figure 22-11. Integral face and bypass coil. (Courtesy ACGIH)

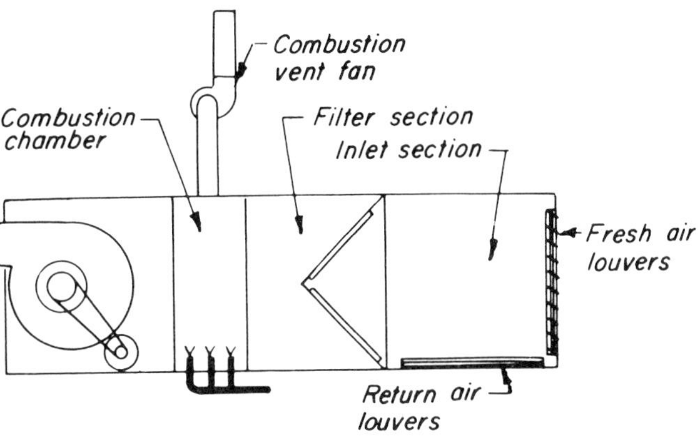

Figure 22-12. Indirect-fired unit. (Courtesy ACGIH)

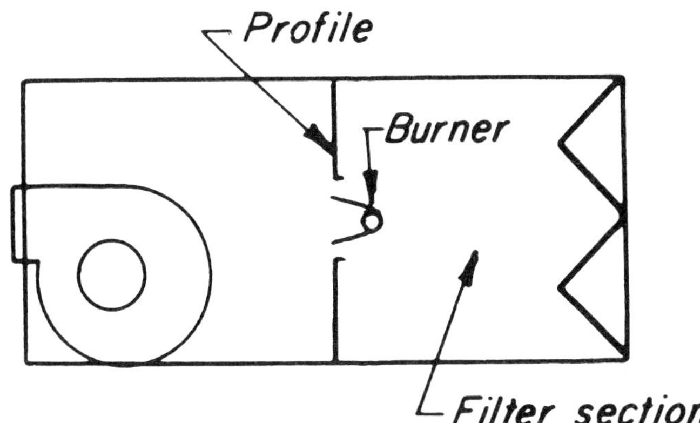

Figure 22-13. Direct-fired unit. (Courtesy ACGIH)

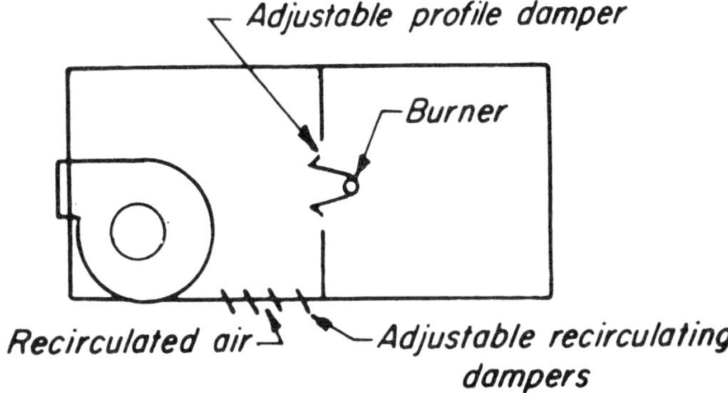

Figure 22-14. Direct-fired bypass unit. (Courtesy ACGIH)

are economical to operate since all of the net heating value of the fuel is available to raise the temperature of the air; this results in a net heating efficiency approaching 100 percent. Commercially available burner designs provide turndown ratios from approximately 25:1 to as high as 45:1, permitting excellent temperature control. In sizes above 4.7 m^3/sec (10,000 cfm), the units are relatively inexpensive on a cost per volume-of-air handled basis; below this capacity, the costs of the additional combustion and safety controls weigh heavily against this design.

A further disadvantage is that state and local codes prohibit the recirculation of room air across the burner. Controls on these units are designed to provide a positive proof of airflow before the burner can ignite, a timed preignition purge to make sure than any leakage gases will be removed from the housing, and constantly supervised flame operation including both flame controls and high-temperature limits.

Concerns are often expressed with respect to potentially toxic concentrations of carbon monoxide, oxides of nitrogen, aldehydes, and other contaminants produced by combustion and released into the supply airstream. Practical field evaluations and detailed studies show that with a properly operated, well-maintained unit, carbon monoxide concentrations will not normally be expected to exceed 5 ppm, and that oxides of nitrogen and aldehydes are well within acceptable limits.

A variation of this unit, known as a bypass design, has gained acceptance in larger buildings where there is a desire to circulate large volumes of air at all times (Figure 22-14). In this design controls are arranged to reduce the flow of outside air across the burner and permit the entry of room air into the fan compartment. In this way, the fan air volume remains constant, maintaining circulation in the space. It is important to note that the bypass air does not cross the burner—100 percent outside air only is allowed to pass through the combustion zone. Controls are arranged to regulate outside air volume and also to make sure that burner profile velocity remains within the limits specified by the burner manufacturer, usually in the range of 10–15 m^3/sec (2,000–3,000 fpm). This is accomplished by providing a variable profile, which changes area as the damper positions change.

Inasmuch as there are advantages and disadvantages to both direct fired and indirect fired makeup air heaters, a careful consideration of characteristics of each heater should be made. A comparison of the heaters is given in Table 22-E.

HEATING OF REPLACEMENT AIR

Replacement air heating should not be confused with space heating, which heats a building whether or not processing lines are functioning within it. However, replacement air can be supplied to an entire building area as well as directly compensating for air being exhausted at one point or being lost through loading dock doors.

General factors which must be considered are:

1. Proper location, size and construction of air intakes.
2. Filtering of replacement air if necessary.
3. Air handling of equipment must be adequate to keep air in the building under positive pressure, even with exhaust systems in use. This will reduce drafts and minimize dispersal of contaminants in the work area.
4. Automatic controls must be provided to assure delivery of replacement air at acceptable temperatures to persons working in the area. They should be located in a place as clean and free of vibration as possible.
5. Proper startup and shutdown procedures must be enforced.
6. Controls and safety devices should be readily accessible to authorized personnel.
7. Air distribution arrangement must provide proper delivery of makeup air.
8. Check building codes to see what special precautions must be taken.

Sources of heat: oil, gas, steam

The three most common heating media for low-temperature industrial heating processes are oil, gas, and steam; all have some application in heating replacement air.

Oil. Precautions that must be taken when using oil-fired heaters include:

1. The possibility of formation of smoke and soot during lightoff make it necessary to transfer the heat through a tubular heat exchanger. Firing must be in closed combustion chambers.
2. Multiple burners and heaters are usually used to obtain turndown, so control system must be adequate.

Gas. Precautions which must be taken when using gas-fired heaters include:

1. Be certain that gas is properly proportioned and mixed with excess air. Use burner systems designed for the operating conditions which will be present.
2. The Threshold Limit Value for CO_2 (carbon dioxide) and CO (carbon monoxide) that is recommended for a continuous 8-hour exposure by the ACGIH is 5,000 ppm for CO_2 and 50 ppm for CO (see Appendix B-1).

In choosing the type of gas combustion system, the following basic requirements should be kept in mind:

1. Heat must be introduced into the airstream to give uniform distribution and uniformity of temperature.
2. The system must be able to throttle smoothly over the total range of turndown (which can be as great as 25:1).
3. The burner must be able to produce complete combustion of the fuel throughout the turndown range.
4. Flame retention and stability must be maintained in high-velocity flow of air.

So-called line burners are almost exclusively used because they give uniform temperature distribution across the airstream. Line burners are sectional burners available in a variety of shapes; they are flanged so they can be bolted together into assemblies to fit the size and shape of the unit. Air/gas proportional and premixing equipment is required, as is a positive flame retention feature. A single flame failure or combustion safeguard device is usually satisfactory as is a single gas pilot for ignition. The wide-range type line burner was developed especially to meet the needs of air heating as it utilizes oxygen from the stream of air being heated to supplement that provided in the primary air/gas mixture in order to extend the range of turndown to the 25:1 figure required. Be sure that heater manufacturer's recommendations are followed in installation, especially with regard to velocity of air flow across burner.

Steam. Where excess steam capacity is available, use of steam coils or finned heaters should be considered carefully as steam heaters are relatively simple, inexpensive, easily maintained, and require no combustion safety equipment.

Precautions which must be taken when using steam include:

1. Avoid long heating coils. These may give uneven temperature distribution from side to side in the duct as the steam supply is throttled over the wide range of turndown required (20:1 to 25:1).
2. Protect against freezeup. Outside air at subfreezing temperatures can be drawn over the heating coils.
3. Consider use of oil- or gas-fired tempering air heaters ahead of steam coils.

It is important that a qualified ventilation engineer be consulted.

The air exhausted from a room must be replaced. The current OSHA standards require that replacement air equal to that exhausted be supplied to each room having exhaust hoods.

The air must enter the room in such a manner that it does not interfere with the operation of any exhaust hood. Air blowing out of a duct carries 30 times as far as the volume of air moving toward a similarly sized exhaust duct. This frequently dictates the use of a diffuser at the end of a supply duct.

The airflow of the replacement system must be measured on installation and rechecked periodically. If replacement air falls below required amount, corrective action must be taken.

Obviously, supply air can be measured inside the ducts as in the case of exhaust air. However, it is also possible to make rather good measurements at the discharge end of supply ducts even when they are equipped with diffusers.

Major diffuser manufacturers have established "K Factors"—effective area factors—for use with the diffuser probe and their various models and sizes of diffusers. Typical instructions will illustrate the location at which the diffuser probe should be placed and for tabulating the K factor for various sizes. Four readings around the circumference should be taken and averaged. (Volume per unit time) = K (average velocity of the air).

SUMMARY

There are some basic things that safety and health professionals should look for in evaluating ventilation systems or when discussing them with the engineering personnel.

- With today's industry replacement air must be supplied if the exhaust equipment is to function properly. Many industrial exhaust systems fail because it is impossible for them to cope with the added resistance required to draw large volumes of air through cracks in the windows and doors when mechanical supply for replacement air is lacking.
- Whenever a building operates in a vacuum, there is an increased flow of air through each and every crack. In fact, air will infiltrate through a brick wall. Whenever a door is opened, large quantities of outside air enter the building. Whenever such indrafts do occur, the people located near the flow of cold air are chilled and cannot be made comfortable regardless of how much heat is added to the building itself.
- Whenever air enters the space and is exhausted, some dilution ventilation takes place. This alone may be sufficient to provide control in an area. If the air is introduced under controlled conditions, the maximum dilution and benefits will occur. Ideally, the best air available is always introduced to the cleanest part of the plant and is allowed to flow across the individuals to the contaminated areas where it is exhausted.
- One of the most successful methods of improving the environment for the personnel in summer is to direct air over their bodies at high velocities. Air-supply systems can, if properly equipped with suitable grilles, provide such air motion. The air must be piped to the lower 3 m (10 ft) of the plant in order to function properly for this purpose. When air is mechanically supplied to a space, it can, if desired, provide pressurization if sufficient volumes of air are introduced. Usually, it is easier to control the environment within a space that is balanced or pressurized.

Most production facilities must be warmed in winter. Sometimes, it is desirable to condition the air in summer. The filtering of the air is now common practice to either provide a cleaner environment for the employees or to protect the equipment. Properly designed air supply systems can fulfill all of these responsibilities thereby eliminating duplication of equipment.

BIBLIOGRAPHY

American Conference of Governmental Industrial Hygienists, (ACGIH), Committee on Industrial Ventilation, Lansing, MI. *Industrial Ventilation—A Manual of Recommended Practice,* 19th ed. Cincinnati: ACGIH, 1986.

American Industrial Hygiene Association (AIHA). *Heating and Cooling for Man in Industry,* 2nd ed. Akron, OH: AIHA, 1975.

AIHA and ACGIH. *Respiratory Protection Monograph.* Akron, OH:AIHA, 1985.

American National Standards Institute, 1430 Broadway, New York, NY: 10018.
Standards:
Fundamentals Governing the Design and Operation of Local Exhaust Systems, Z9.2–1979.
Practices for Ventilation and Operation of Open-Surface Tanks, Z9.1–1977.
Ventilation Control of Grinding, Polishing, and Buffing Operations, Z43.1–1966.

American Society of Heating, Refrigerating and Air-Conditioning Engineers (ASHRAE). *Guide and Data Book—Fundamentals and Equipment.* New York: ASHRAE, latest edition.

Brandt, A.D. A summary of design data for exhaust systems. *Heating and Ventilating* (May 1945).

Burton, J.D. *Industrial Ventilation: A Companion Study Guide to the ACGIH Ventilation Manual,* 3rd ed. Cincinnati: ACGIH, 1986.

Chamberlin, R. I. The control of beryllium machining operations. *AMA Archives of Industrial Health* (Feb. 1959).

Dalla Valle, J.M. Design of kitchen range hoods. *Heating and Ventilating* (Aug. 1953):95–100.

Dalla Valle, J.M. *Exhaust Hoods,* 2nd ed. New York: The Industrial Press, 1952.

Dalla Valle, J.M. *The Industrial Environment and Its Control.* New York: Pitman Pub. Corp., 1948, pp. 133–146.

Department of National Health and Welfare, Ottawa, Canada. The problem of garage ventilation. *Occupational Health Bulletin,* 14 (Jan. 1959):1.

Goodfellow, H.D. *Advanced Design of Ventilation Systems for Contaminant Control.* New York: Elsevier, 1985.

Hama, G.M. How safe are direct-fired units. *Air Engineering* (Sept. 1962):22.

Hatch, T.F. Dust control, present and future design considerations. *Mechanical Engineering,* 57 1935:154.

Hemeon, W.C.L. *Plant and Process Ventilation,* 2nd ed. New York: The Industrial Press, 1963.

Henschel, A., et. al. Assessment of industrial heat stress. *American Industrial Hygiene Association Journal* 27 (1966):13–16.

Jorgensen, R. *Fan Engineering,* 8th ed. Buffalo, NY: Buffalo Forge, 1983.

Lee, D.H.K. *Heat and Cold Effects and Their Control,* Public Health Service Monograph No. 72. Washington, DC: U.S. Dept. of Health and Human Services (DHHS), 1964.

Lee, D.H.K. *Man's Relation to His Thermal Environment: Interactions of Man and His Environment.* New York: Plenum Publishing Corp., 1966.

Lee, D.H.K., and Henschel, A. Effects of physiological and clinical factors on response to heat. *Annals N.Y. Academy of Sciences,* 134 (1966):743–749.

McDermott, H.J. *Handbook of Ventilation for Contaminant Control,* 2nd ed. Stoneham, MA: Butterworth, 1985

McKarnes, J.S., and Brief, R.S. Nomographs give refined estimate of heat stress index. *Heating, Piping and Air Conditioning* (Jan. 1966).

Michigan Department of Health, Lansing, MI. Garage Ventilation. *Michigan Occupational Health* 8:2, Winter 1962–1963.

National Fire Protection Association, Batterymarch Park, Quincy, MA 02269.
NFPA Standards:
Air Conditioning and Ventilating Systems, NFPA 90A.
Blower and Exhaust Systems for Dust, Stock and Vapor Removal or Conveying, NFPA 91.
National Electrical Code. NFPA 70.
Spray Application Using Flammable and Combustible Materials, NFPA 33.
Installation of Warm Air Heating and Air Conditioning Systems, NFPA 90B.

National Safety Council, 444 North Michigan Avenue, Chicago, IL 60611.
Industrial Data Sheets:
Checking Performance of Local Exhaust Systems, 428–1986.
Instruments for Testing Exhaust Ventilation Systems, 431–1981.

NIOSH. *Recommendations for Control of Occupational Safety and Health in Foundries,* Publication No. 85–116. Cincinnati: NIOSH.

Pring, R.T., Knudsen, J.F., and Dennis, R. Design of exhaust ventilation for solid materials handling. *Industrial and Engineering Chemistry,* 41 (1949):442.

Ross, C.R., and Rispler, L. Garage health hazards and ventilation. *Occupational Health Review,* 11 (1960):4.

Sheinbaum, M. Ventilation of factories, garages, and warehouses for products of combustion of gasoline engines. *Monthly Review,* 32 (Oct. 1953):37. New York: State Department of Labor, Division of Industrial Hygiene and Safety Standards.

Stern, A.C., ed. *Air Pollution,* vols 2 and 5. New York: Academic Press, 1977.

Stern, A.C., and O'Neil, D.P. Recirculation from industrial exhaust systems. *Monthly Review,* 30 (May, June, July, 1951). New York State Department of Labor, Division of Industrial Hygiene and Safety Standards.

U.S. Department of Health and Human Services, Public Health Service, Centers for Disease Control, National Institute for Occupational Safety and Health, Division of Physical Sciences and Engineering.
Recirculation of Exhaust Air, Feb. 1976. Available from U.S. Government Printing Office, Washington, DC 20402.
Recommended Industrial Ventilation Guidelines, Jan. 1976. Available from U.S. Government Printing Office, Washington, DC 20402.

23

Respiratory Protective Equipment

by Allen M. Lundin
Revised by Craig E. Colton, CIH

Respiratory protective devices vary in design, equipment specifications, application, and protective capability. Proper selection depends on the toxic substance involved, conditions of exposure, human capabilities, and equipment fit. If, after effective engineering controls have been fully used in reducing exposure to the lowest possible level, the environment is still not completely safe, it will be necessary to protect the worker from contact with airborne contaminants or oxygen-deficient environments.

LEGAL BASIS FOR RESPIRATORY PROTECTION

After enactment of the Occupational Safety and Health Administration Act (OSHAct), the National Institute for Occupational Safety and Health (NIOSH) and the U.S. Bureau of Mines (USBM) jointly promulgated 30 *CFR,* Part 11, which prescribes approval procedures, establishes fees, and extends the requirements for obtaining joint approval of respirators. (The official name for these regulations is Part 11 of Title 30 of the *Code of Federal Regulations,* referred to as *CFR.*) The NIOSH has been named as the testing, approving, and certifying agency for a variety of personal safety devices, including respirators. The approval was issued jointly by NIOSH and the USBM. In 1973 the USBM was reorganized and the regulatory functions, including respirator approval and certification, formed into a new agency, the Mining Enforcement and Safety Administration (MESA). Another reorganization occurred in 1977 in which MESA's functions were transferred to the Mine Safety and Health Administration (MSHA) and henceforth, the approvals were issued jointly between NIOSH-MSHA. This has removed sole responsibility for respirator testing and approval from the MSHA. Figure 23-1 shows these interrelationships.

The NIOSH Testing and Certification Laboratory has the following responsibilities:

1. developing and promulgating certification tests and requirements for personal protective devices and industrial hazard measuring instruments
2. testing and certifying these products
3. preparing lists of certified products
4. surveying manufacturer's plants to check their quality inspection programs
5. periodic testing of certified devices and instruments procured on the open market
6. developing new test methods and requirements for product improvement, when necessary to assure worker protection.

Use of the terms approved and certified reflects the applicable departmental regulations. Regulations issued under authority of the federal Coal Mine Health and Safety Act of 1969 refer to approval of respiratory protective devices. The NIOSH regulations issued under authority of the 1970 OSHAct refer to "certification" of devices. Approved respiratory protection devices display a label similar to that shown in Figure 23-2.

Part 11 of Title 30 *CFR* also placed an end-date of June 30, 1975, on the manufacture of USBM-approved respirators. Users of USBM-approved devices at the time of the cutoff date are permitted to continue to use them under a grandfather clause. Most manufacturers will continue to furnish parts for the devices through the grandfather clause period.

Use of USBM-approved respirators under the grandfather clause is permitted as follows:

- Dust, fume, and mist (particulate) respirators and chemical

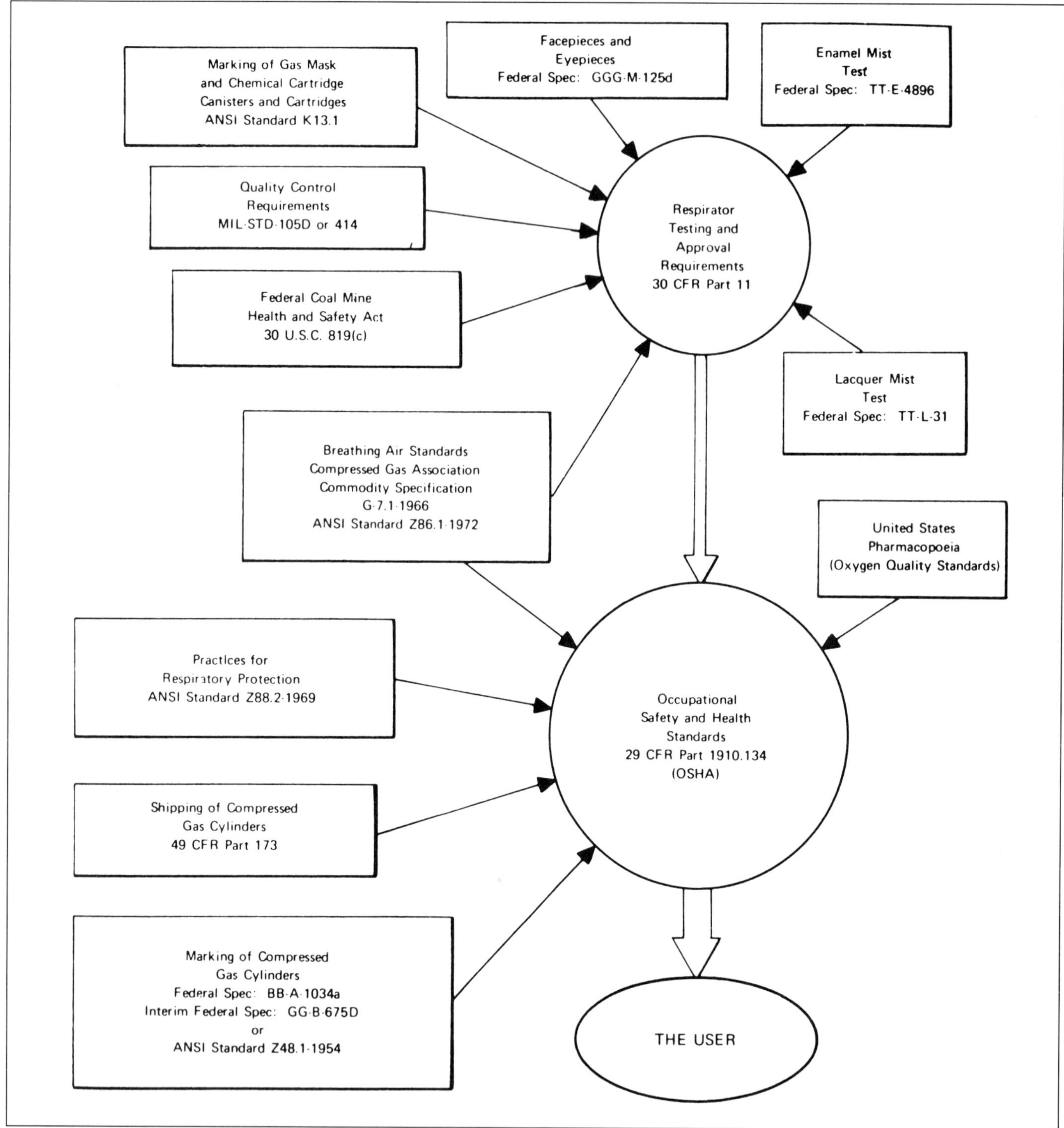

Figure 23-1. Source of OSHA respiratory equipment standards. (Adapted from NIOSH, *Guide to Industrial Respiratory Protection,* HEW Publ. No. (NIOSH) 76-189.)

cartridge (nonemergency gas) respirators (approved under USBM Schedules 21B and 23B, respectively) expired April 1, 1976. Only NIOSH-MSHA-approved respirators can now be accepted as approved devices.

- All gas masks approved under USBM Schedule 14F can continue to be used under the USBM approval until notification by NIOSH. Permission to continue use and production of this type of device under the existing USBM approval was granted in the *Federal Register* of Nov. 22, 1974, provided that they are maintained in an approved condition. (See Addendum C at the end of this chapter.)

  The single exception is the use of gas masks against vinyl chloride monomers and the use of certain chin-type gas masks. The NIOSH acted to approve gas masks and canisters for vinyl chloride on an emergency basis.
- Air-line respirators and hose masks (approved under USBM

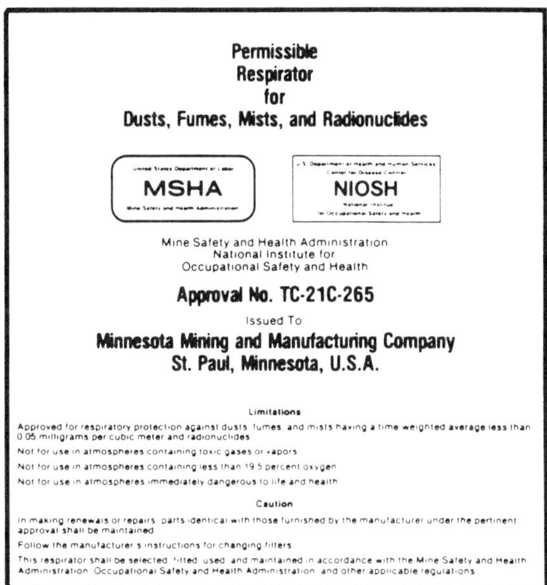

Permissible
Respirator
for
Dusts, Fumes, Mists, and Radionuclides

MSHA

NIOSH

Mine Safety and Health Administration
National Institute for
Occupational Safety and Health

Approval No. TC-21C-265

Issued To

Minnesota Mining and Manufacturing Company
St. Paul, Minnesota, U.S.A.

Limitations

Approved for respiratory protection against dusts, fumes, and mists having a time weighted average less than 0.05 milligrams per cubic meter and radionuclides

Not for use in atmospheres containing toxic gases or vapors

Not for use in atmospheres containing less than 19.5 percent oxygen

Not for use in atmospheres immediately dangerous to life and health

Caution

In making renewals or repairs, parts identical with those furnished by the manufacturer under the pertinent approval shall be maintained

Follow the manufacturer's instructions for changing filters

This respirator shall be selected, fitted, used, and maintained in accordance with the Mine Safety and Health Administration, Occupational Safety and Health Administration, and other applicable regulations

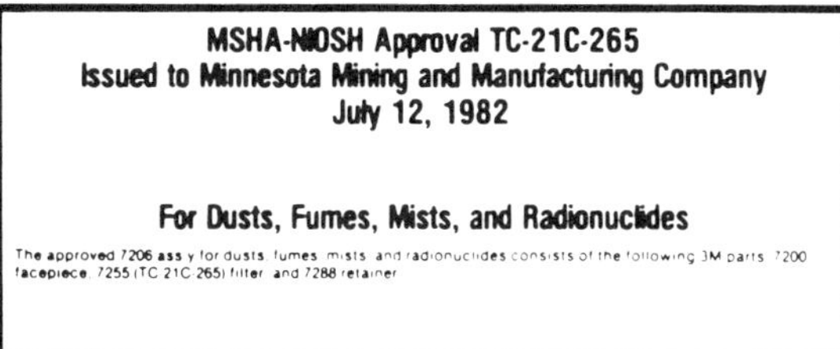

MSHA-NIOSH Approval TC-21C-265
Issued to Minnesota Mining and Manufacturing Company
July 12, 1982

For Dusts, Fumes, Mists, and Radionuclides

The approved 7206 ass'y for dusts, fumes, mists, and radionuclides consists of the following 3M parts: 7200 facepiece, 7255 (TC 21C-265) filter, and 7288 retainer

Figure 23-2. The typical approval label that appears on a NIOSH/MSHA-approved respiratory protection device. The user should make sure that the limitations of the device are understood. (Courtesy 3M Company)

Schedule 19B) expired March 31, 1980; only NIOSH-MSHA-approved devices retain an acceptable approved status.

- Self-contained breathing apparatus (approved under USBM Schedule 13-13E), both open and closed circuit, may continue in service under their existing USBM approval until further notice by NIOSH. However, they must have a low-air warning device.

Voiding an approval

Once approval has been granted to a device by NIOSH-MSHA, the user should become acquainted with the limitations of the device as set forth in the approval (Figure 23-2). The approval will be void if the device is knowingly used in conditions beyond the limitations set by NIOSH or those established by the manufacturer. The user should also guard against any alteration being made to the device. All parts, filters, canisters, cartridges, or anything else not specifically intended to be used on the device by NIOSH-MSHA or the manufacturer, will void the existing approval. If there is any question concerning parts, alteration, or limitation of the device, always check with the manufacturer. The employer should take care so as not to knowingly void the aproval for a piece of equipment.

The NIOSH has the authority to purchase and test respiratory protective devices on the open market as a continuing check on manufacturers' quality control standards and adherence to approvals. Manufacturers may not institute design changes of the device or its components without obtaining an extension of an existing approval or resubmitting the device for a new approval.

OSHA requirements

The primary objective of industrial hygiene programs in industry is the control of airborne contaminants by accepted engineering control measures. When effective controls are not feasible, or while they are being instituted, appropriate respirators shall be used, pursuant to the requirements of Section 1910.134 (Occupational Safety and Health Standards) in the *Code of Federal Regulations* (29 *CFR,* Part 1910, §1910.134, which is reprinted as Addendum A to this chapter.)

The Occupational Safety and Health Review Commission has ruled that, "Providing employees with protective equipment, and requiring them to use it, does not constitute abatement unless the employer demonstrates that administrative or engineering controls are not feasible." The Commission further ruled that "not feasible" is to be interpreted as *technically not feasible* and as *economically* not feasible.

RESPIRATORY PROTECTION PROGRAM

A respiratory protection program must be established when respiratory protection is needed; see Addendum A, 1910.134(a)(2). It should include the minimum requirements listed below; the order of importance may differ for each application:

1. administration
2. knowledge of respiratory hazards
3. assessment of respiratory hazards
4. control of respiratory hazards
5. selection of proper respiratory protective equipment
6. training
7. inspection, maintenance, and repair of equipment
8. medical surveillance.

More requirements are listed in the Addenda to this chapter.

Administration

Responsibility and authority for administration of a respiratory protection program must be assigned to one person who may and probably will have assistance. The need for central authority and responsibility is to make sure that there is coordination and direction. The respiratory protection program will vary widely from company to company, and depends upon many factors; it may involve specialists such as safety personnel, industrial hygienists, health physicists, and physicians. In small plants or companies having no formal industrial hygiene, health physics, or safety engineering department, the respirator program should be administered by an upper-level superintendent, supervisor, or other qualified person responsible to the principal manager. The administrator should have sufficient knowledge of respiratory protection to properly supervise the respirator program. In any case, overall responsibility and authority must reside in a single individual if the program is to achieve optimum results.

Respiratory hazards

Toxic materials can enter the body in three ways: (1) through the gastrointestinal tract, (2) through the skin, and (3) through

Classification of Respiratory Hazards According to Their Properties Which Influence Respirator Selection

Gas and Vapor Contaminants

Inert: Substances that do not react with other substances under most conditions, but create a respiratory hazard by displacing air and producing oxygen deficiency (for example: helium, neon, argon).

Acidic: Substances that are acids or that react with water to produce an acid. In water, they produce positively charged hydrogen ions (H^{+1}) and a pH of less than 7. They taste sour, and many are corrosive to tissues (for example: hydrogen chloride, sulfur dioxide, fluorine, nitrogen dioxide, acetic acid, carbon dioxide, hydrogen sulfide, and hydrogen cyanide).

Alkaline: Substances that are alkalies or that react with water to produce an alkali. In water, they result in the production of negatively charged hydroxyl ions (OH^{-1}) and a pH greater than 7. They taste bitter, and many are corrosive to tissues (for example: ammonia, amines, phosphine, arsine, and stibine).

Organic: The compounds of carbon. Examples are saturated hydrocarbons (methane, ethane, butane), unsaturated hydrocarbons (ethylene, acetylene), alcohols (methyl ether, ethyl ether), aldehydes (formaldehyde), ketones (methyl ketone), organic acids (formic acid, acetic acid), halides (chloroform, carbon tetrachloride), amides (formamide, acetamide), nitriles (acetonitrile), isocyanates (toluene diisocyanate), amines (methylamine), epoxies (epoxyethane, propylene oxide), and aromatics (benzene, toluene, xylene).

Organometallic: Compounds in which metals are chemically bonded to organic groups (for example: ethyl silicate, tetraethyl lead, and organic phosphate).

Hydrides: Compounds in which hydrogen is chemically bonded to metals and certain other elements (for example: diborane and tetraborane).

Particulate Contaminants

Particles are produced by mechanical means by disintegration processes such as grinding, crushing, drilling, blasting, and spraying; or by physiochemical reactions such as combustion, vaporization, distillation, sublimation, calcination, and condensation. Particles are classified as follows:

Dust: A solid, mechanically produced particle with sizes varying from submicroscopic to visible or macroscopic.

Spray: A liquid, mechanically produced particle with sizes generally in the visible or macroscopic range.

Fume: A solid condensation particle of extremely small paricle size, generally less than one micrometer in diameter.

Mist: A liquid condensation particle with sizes ranging from submicroscopic to visible or macroscopic.

Fog: A mist of sufficient concentration to perceptibly obscure vision.

Smoke: A system which includes the products of combustion, pyrolysis, or chemical reaction of substances in the form of visible and invisible solid and liquid particles and gaseous products in air. Smoke is usually of sufficient concentration to perceptibly obscure vision.

From ANSI Z-88.2-1980, Practices for Respiratory Protection

Figure 23-3. Classification of respiratory hazards according to their properties that influence selection of respirator. (Reprinted with permission from ANSI Z88.2-1980, *Practices for Respiratory Protection*)

the lungs. Of these three modes of entry, the human respiratory system presents the quickest and most direct avenue of entry because of its intimate association with the circulatory system and the constant need to oxygenate the tissue cells to sustain life processes.

This section lists and briefly describes various categories of respiratory hazards that might be encountered and which may require use of respirators. The information provides a general background for relating the guidance in subsequent sections to the type of hazard encountered. Management, however, will find it necessary to consult references on industrial hygiene and toxicology and perhaps experts to develop the necessary comprehensive information on specific airborne contaminants (Figure 23-3).

Respiratory hazards can be classified as follows:

1. oxygen deficiency
2. gas and vapor contaminants
   a. Immediately dangerous to life or health (IDLH)
   b. Not immediately dangerous to life or health (nonIDLH)
3. particulate contaminants (aerosols including dust, fog, fume, mist, smoke, and spray)
   a. IDLH
   b. nonIDLH
4. combination of gas, vapor, and particulate contaminants
   a. IDLH
   b. nonIDLH.

Air contaminants include particulate matter in the form of discrete particles of solids or liquids, gaseous material in the form of a true gas or vapor, or a combination of both gaseous and particulate matter.

Control of hazards

Hazard control should start at the process, equipment, and plant design levels where effluents can be effectively controlled at the outset. This would include consideration of process encapsulation or isolation, use of less toxic materials in the process, suitable exhaust ventilation, filters, and scrubbers to control the effluents. Since it is not always practical to provide and maintain engineering controls, proper respiratory protective devices should be made available and used for respiratory protection when required (Table 23-A).

CLASSES OF RESPIRATORY PROTECTIVE DEVICES

Proper respiratory selection shall be made when engineering controls are not feasible, or if they are feasible, while they are being instituted, as noted in the OSHA regulations—see §1910.134(a)(1) and (c), reprinted in the Addendum to this Chapter. The ANSI Standard Z88.2-1969 (updated in 1980) can be of great value for this purpose because much of the OSHA

Table 23-A. Outline for Selecting Respiratory Protective Devices*

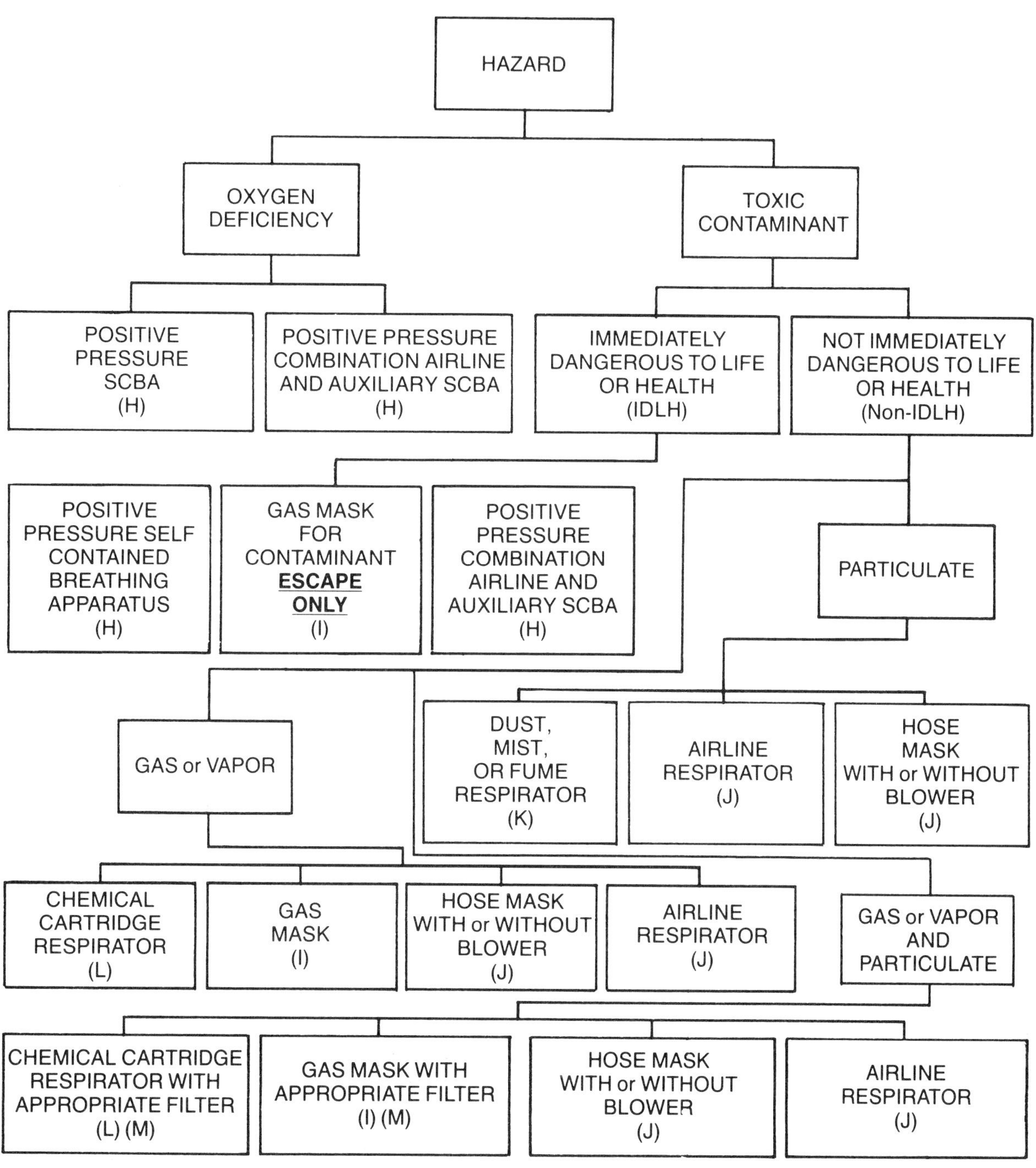

*Letters in parentheses refer to Subpart of 30 *CFR* Part 11.

respiratory protection standard was taken from it. In 1980, ANSI Standard Z88.2 was updated and provides "state of the art" information for respirator selection. In addition, the manufacturer of the respiratory devices should be consulted. Whenever possible, a respiratory device having both NIOSH and MSHA approval should be used.

Base the selection of the proper type of respirator upon (1) the nature of the hazardous operation or process, (2) the type of respiratory hazard (including physical properties, chemical properties, warning properties, effects on the body, concentration of toxic material or airborne radioactivity level, established permissible time weighted-average concentration for toxic material or established maximum permissible airborne concentration for radioactive material, and concentrations IDLH for toxic material), (3) the location of the hazardous area in relation to the nearest area having respirable air, (4) the period of time for which respiratory protection must be provided, (5) the activities of workers in the hazardous area, (6) the functional and physical characteristics of the various types of respirators, and (7) the respirator-protection factors and respirator fit.

There is one limitation applicable to all respiratory protective equipment—certain gaseous contaminants can damage or enter the body by means other than the respiratory tract. For example, ammonia, in concentrations of approximately 3 percent or higher, can cause skin burns (particularly on moist skin). To avoid that possibility, rubberized clothing should be worn in addition to the proper respiratory protection.

Similarly, protective clothing should be worn when appreciable amounts of gases, such as hydrocyanic acid, are present. Hydrocyanic acid—a gas at just above room temperature—is capable of penetrating the skin and causing a systemic poisoning; although to do so, a concentration considerably higher than that required for poisoning through the respiratory tract must be present.

Respiratory protective devices fall into three classes: air-purifying; atmosphere- or air-supplying devices; and combination air-purifying and atmosphere-supplying devices. Each will be discussed in turn.

Class 1. Air-purifying devices

The air-purifying device cleanses the contaminated atmosphere. Chemicals can be used to remove specific gases and vapors and mechanical filters can remove particulate matter. This type of device is limited in its use to those environments where there is sufficient oxygen to sustain life and the air-contaminant level is within the specified concentration limitation of the device. The useful life of an air-purifying device is limited by the concentration of the air contaminants, the breathing demand of the wearer, and the removal capacity of the air-purifying medium.

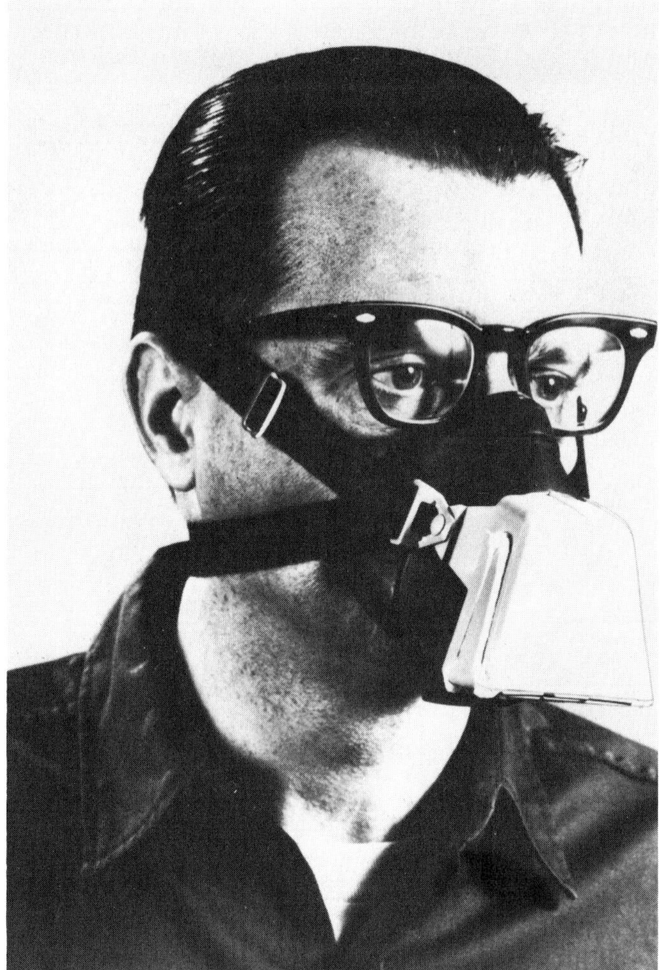

Figure 23-4. A mechanical filter respirator being used here to protect against dust inhalation. (Courtesy MSA)

Mechanical-filter respirators (Figure 23-4). These offer respiratory protection against airborne particulate matter, including dusts, mists, metal fumes, and smokes, but they do not provide protection against gases, vapors, or oxygen deficiency. They consist essentially of a facepiece, either quarter-mask (over the chin), half-mask (under the chin), or full-face design. Directly attached to the facepiece is one of several types of mechanical filters made up of a fibrous material that removes the harmful particles by trapping them as air is inhaled through the material.

The filter must be highly efficient, however, to trap the smaller particles. There are many classes of mechanical filter respirators specifically designed for the various classes of airborne particulate matter. Although a single respirator can be made to provide effective protection against all true particulates, in most cases, it would be too expensive and perhaps too cumbersome for the great majority of users. Therefore, many special-purpose respirators are available to provide the most economical and efficient protection against specific particulate hazards (Figure 23-5).

The NIOSH-MSHA certifies mechanical filter respirators under Subpart K of Title 30 *CFR*, Part 11. The agency approves respirators for one or any combination of particulate hazards—nuisance, fibrosis-producing, and/or toxic dusts, mists, and fumes (Figure 23-6).

The most desirable compromise must be worked out for each classification with respect to filter-surface area, resistance to breathing, efficiency in filtering particulates of specific size ranges, and the time required to clog the filter. For example, in choosing a respirator for protection against particulates more toxic than lead, the user must accept an increase in bulk, weight, and cost in relation to a respirator needed to protect only against so-called nuisance dusts.

Chemical-cartridge respirators (Figure 23-7). These afford protection against concentrations (10 ppm–1,000 ppm by volume, depending upon the contaminant) of certain gases and vapors by using various chemical filters to purify the inhaled air. They differ from mechanical filter respirators in that they use cartridges containing chemicals to remove harmful gases and vapors.

The NIOSH-MSHA Subpart L covers chemical-cartridge respirators. The NIOSH-MSHA will issue certificates of approval to manufacturers whose products meet their performance requirements, and the approval number will be shown on the respirator and/or packing carton. It is good practice and required by OSHA to use only NIOSH-MSHA-approved equipment.

Chemical-cartridge respirators (Figure 23-7) offer protection

Figure 23-5. High-efficiency filter cartridges can be used for protection against asbestos. (Courtesy MSA)

against intermittent exposure to light concentrations of gases and vapors.

Chemical-cartridge respirators are nonemergency respiratory protective devices and should never be used in IDLH atmospheres.

Four major negative rules that apply to chemical-cartridge respirators are as follows:

- Do not use chemical-cartridge respirators for protection against gaseous material that is extremely toxic in very small concentrations.
- Chemical-cartridge respirators should not be used for exposures to harmful gaseous matter that cannot clearly be detected by odor. Example: methyl chloride and hydrogen sulfide. The former is odorless; and the latter, although foul smelling, paralyzes the olfactory nerves so quickly that detection by odor is unreliable.
- Chemical-cartridge respirators should not be used against any gaseous material in concentrations that are highly irritating to the eyes without satisfactory eye protection.
- Chemical-cartridge respirators cannot be used for protection against gaseous material that is not effectively stopped by chemical fills used, regardless of concentrations.

Combination mechanical-filter/chemical-cartridge respirators (Figure 23-8). These use dust, mist, or fume filters with a chemical cartridge for dual or multiple exposure. Respirators with independently replaceable mechanical filters are sometimes used for this type unit because the dust filter normally plugs before the chemical cartridge is exhausted.

One combination mechanical-filter chemical-cartridge respirator uses a backmounted filter element and is especially well suited for spray-painting and welding operations, where the air contaminant is concentrated in front of the worker.

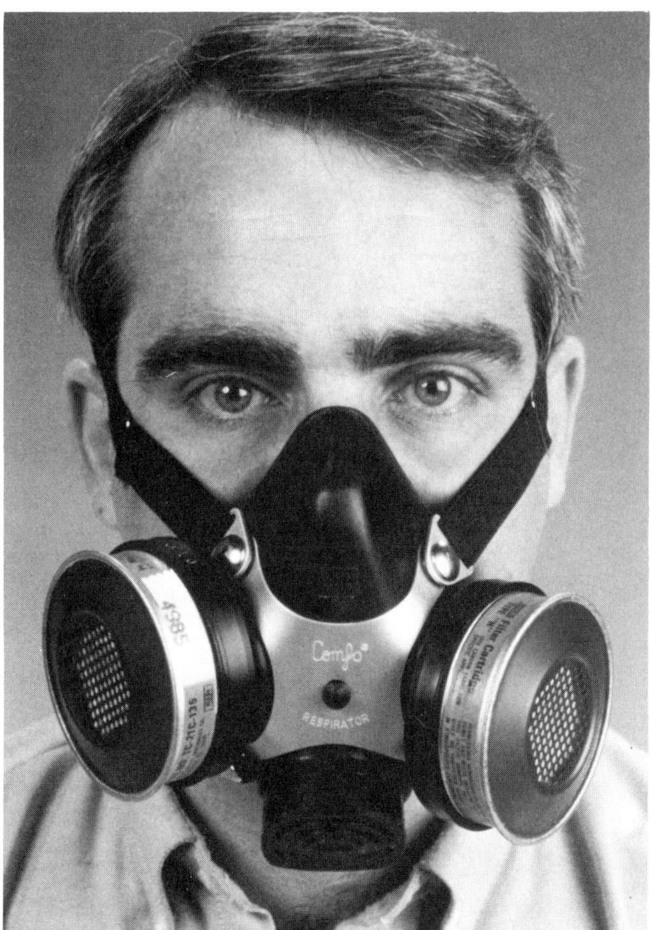

Figure 23-6. Twin-filter, half-mask mechanical filter respirator provides respiratory protection against airborne particulates, such as dusts, mists, and fumes. (Courtesy MSA)

A separate test procedure has been developed for spray-paint respirators by NIOSH-MSHA. This type of respirator has a separate prefilter for mists; it is easily replaced and extends the life of the chemical cartridge (Figure 23-9).

Responsibility for testing respirators to prevent exposure to agricultural chemicals (pesticides) now belongs to NIOSH-MSHA (Title 30, *CFR*, Part 11, Subpart M). Approved combination type units for pesticides are now available (Figure 23-10).

Gas masks (Figures 23-11 and 23-12). These have been used effectively for many years for respiratory protection against certain gases, vapors, and particulate matter that otherwise might be harmful to life or health.

Gas masks are air-purifying devices, designed solely to remove specific contaminants from the air; therefore, it is essential that their use be restricted to atmospheres which contain sufficient oxygen to support life. Gas masks may be used for escape only from IDLH atmospheres, but never for entry into such environments. It is imperative that the user assess the exposure conditions carefully before selecting a specific mask for respiratory protection.

If the specific exposure concentrations are suspected of

Figure 23-7. Full facepiece chemical-cartridge respirator affords protection against light concentrations of certain gases and organic vapors. (Courtesy MSA)

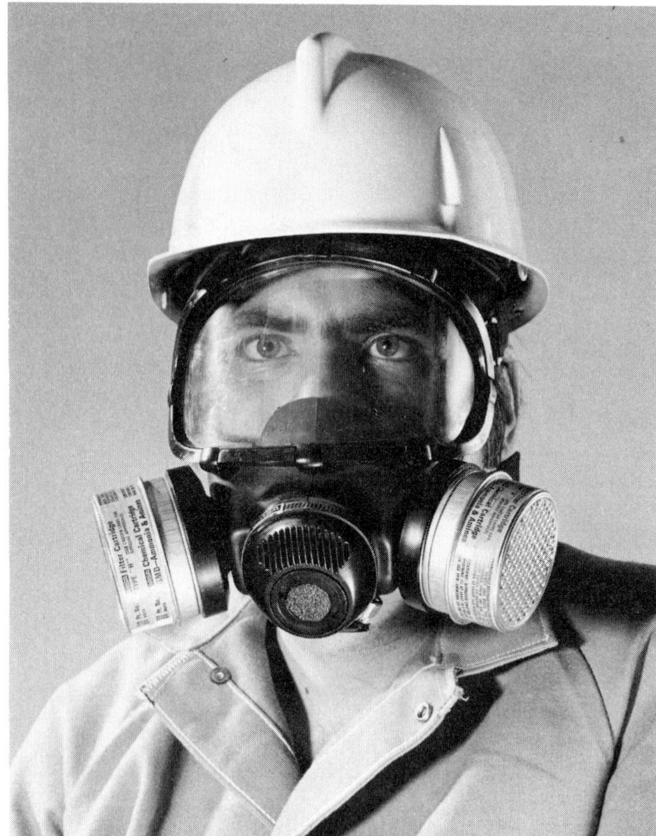

Figure 23-8. Combination mechanical-filter chemical-filter respirators use a mechanical filter with a chemical cartridge for dual or multiple exposure. (Courtesy MSA)

exceeding the specific limitations—only a self-contained breathing apparatus (SCBA) should be used.

From a practical standpoint, gas masks are generally suitable for ventilated areas not subject to rapid change in air-contaminant levels, and should never be used in confined spaces below or above ground where oxygen deficiency and high gas concentrations may occur.

In assessing exposure conditions, remember that oxygen deficiency can occur in a confined space through the displacement of air by other gases or vapors; or by means of processes (such as fire, rusting, and aerobic bacteria) that consume oxygen.

Types of gas masks. Various gas masks with conventional-size canisters were tested and approved by USBM under Approval Schedule 14F for respiratory protection against specific gases and vapors in concentrations up to 2 percent by volume (3 percent for ammonia), or as specified on the canister label. Each canister is specifically labeled and color-coded to indicate the type of protection afforded (see Table 23-B).

The NIOSH-MSHA certification has been granted to three types of canister gas masks; "supersize," "industrial size," and "chin-style" (Figure 23-13). The certificates of approval do not detail the maximum concentration of gaseous contaminants against which the masks should be used except to state that the masks should only be used in concentrations *not* IDLH, or as escape only from IDLH atmospheres and the organic-vapor mask must not be used against vapors with poor warning properties or those that generate high heats of reaction to the absorbent materials in the canister.

Many canisters contain a filter for the removal of dust and other particulate matter, indicated by a purple or orange stripe around the canister. The chin-mask style may also be approved for use only against dusts and mists having a time-weighted average Threshold Limit Value (TWA-TLV) of not less than 0.05 mg/m^3 or 2 million particles per cubic foot.

Service life of an air-purifying device depends on the following factors:

- the design including the quality and amount of chemical fill, packing uniformity, and density.
- exposure conditions, including concentration of contaminants in the air, breathing rate of the wearer, temperature, and humidity. (Generally, high concentrations, a fast breathing rate, and humid conditions adversely affect service life.)

Because exposure conditions are subject to wide variation, it is difficult to estimate the service life of a gas-mask canister even with other conditions being constant. However, for guidance purposes, 30 *CFR* Part 11, Subpart I stipulates the following minimum service requirements at an average breathing

Figure 23-9. The NIOSH-MSHA has approved a separate test procedure for spray-paint respirators. (Courtesy Willson Products Division, ESB Inc.)

rate of 32 liters per minute (L/min) in concentrations of 2 percent for most gases and vapors, or 3 percent for ammonia:

Industrial Size Canisters . . . . 12 minutes
Type-N Canister
Acid Gases . . . . 6 minutes
Organic Vapors . . . . 6 minutes
Ammonia . . . . 6 minutes
Carbon Monoxide . . . . 60 minutes

Supersize canisters, because of their greater volume of chemical fill, will last approximately twice as long as the equivalent industrial-size canister.

Chin-style canisters, because of their small size, should be used in concentrations not in excess of 0.5 percent at which the service life could be as much as 12 minutes as long as this concentration is not IDLH.

Canister replacement. It generally is recommended that gas-mask canisters that have been used for escape should be replaced after each use. The canisters should also be replaced under any one or more of the following conditions:

- if canisters with window indicators show specified color changes
- if leakage is detected by smell, taste, or eye, nose, or throat irritation
- if high resistance to breathing develops
- if the canister shelf life is exceeded.

If a person is wearing a canister that needs replacement, he or she should return to fresh air as quickly as possible. In addition, there are two conditions under which it is imperative that the person return to fresh air:

- Uncomfortable heat in the inhaled air (a properly operating canister will become warm on exposure to certain gases or vapors, but a canister that becomes extremely hot indicates that concentrations greater than the canister limit have been reached.)
- if the wearer has a feeling of nausea, dizziness, or ill-being.

Powered air-purifying respirators. These protect against particulates, gases and vapors, or particulates and gases and vapors. The air-purifying element may be a filter, chemical cartridge, combination filter and chemical cartridge (Figure 23-11), or canister. The powered air-purifying respirator uses a power source (usually a battery pack) to operate a blower that passes air across the air-cleansing element to supply purified air to a respiratory inlet (mouth and nose) covering. To be certified as a powered air-purifying respirator by NIOSH-MSHA, the blower must provide at least 4 cubic feet per minute (cfm) of air to a tight-fitting facepiece (e.g., half-mask or full-face) and at least 6 cfm to a loose-fitting helmet or hood. The great advantage of the powered air-purifying respirator is that it usually supplies air

Figure 23-10. Approved combination-type units for pesticide applicators are now available. (Courtesy 3M Company)

Figure 23-11. Gas masks offer protection against many commonly encountered industrial vapors and gases. (Courtesy Willson Products Division ESB Inc.)

Figure 23-12. Gas masks may be used for escape only from IDLH atmospheres. (Courtesy MSA)

Table 23-B. Color Code for Cartridges and Gas Mask Canisters

Atmospheric Contaminants to be Protected Against	Color Assigned
Acid gases	White
Organic vapors	Black
Ammonia gas	Green
Carbon monoxide gas	Blue
Acid gases and organic vapors	Yellow
Acid gases, ammonia, and organic vapors	Brown
Acid gases, ammonia, carbon monoxide, and organic vapors	Red
Other vapors and gases not listed above	Olive
Radioactive materials (except tritium and noble gases)	Purple
Dusts, fumes, and mists (other than radioactive materials)	Orange

Notes:

(1) A purple stripe shall be used to identify radioactive materials in combination with any vapor or gas.
(2) An orange stripe shall be used to identify dusts, fumes, and mists in combination with any vapor or gas.
(3) Where labels only are colored to conform with this table, the canister or cartridge body shall be gray or a metal canister or cartridge body may be left in its natural metallic color.
(4) The user shall refer to the wording of the label to determine the type and degree of protection the canister or cartridge will afford.

NOTE: See Table 23-H, page 748, for OSHA requirements for color coding of canisters.

(Reprinted with permission from ANSI K13.1-1973, *Identification of Air-Purifying Respirator Canisters and Cartridges*)

at positive pressure so that any leakage is outward from the facepiece. However, it is possible at high work rates to create a negative pressure in the facepiece, thereby increasing facepiece leakage.

Class 2. Atmosphere- or air-supplying devices

Atmosphere- or air-supplying devices are the class of respirators that provide a respirable atmosphere to the wearer, independent of the ambient air. Atmosphere-supplying respirators fall into three groups: supplied-air respirators, self-contained breathing apparatus (SCBA), and combination-SCBA and supplied-air respirators. Each will be discussed in turn.

Supplied-air respirators deliver breathing air through a supply hose connected to the wearer's facepiece or enclosure. The air delivered must be free of contaminants and must be from a source located in clean air. The OSHA requirements for compressed air used for breathing, including monitoring for carbon monoxide are listed in 1910.134(d), (reprinted in Addendum A of this chapter). These devices should be used only in nonIDLH atmospheres. There are three types of supplied-air respirators certified under 30 *CFR* Part 11 Subpart J, Type A, Type B, and Type C.

Type A supplied-air respirators are also known as hose masks with blower. Air is supplied by a motor-driven or hand-operated blower through a strong, large diameter hose. The maximum allowable hose length is 91.5 m (300 ft). The wearer can continue to breathe through the hose if the blower fails.

Type B supplied-air respirators are hose masks as described above without a blower. The wearer draws the air through the hose by breathing. The hose inlet is anchored and fitted with a funnel or similar object with a fine mesh screen to prevent coarse particulate matter. Up to 75 ft of hose can be used with this respirator.

Type C supplied-air respirators are commonly referred to as air-line respirators. The air-line respirator is connected to a suitable compressed air source by a hose of small inside diameter (i.d.) and air is delivered continuously or intermittently in sufficient volume to meet the wearer's breathing requirements. The quality of the compressed air must be assured (see 1910.134(d) and Figure 23-14). Accessory equipment such as pressure regulators, pressure-relief valves and air filters may be necessary to make sure that the air is at the proper pressure and quality for breathing.

An air-line respirator must be supplied with respirable air conforming to Grade D Compressed Gas Association's Standard CGA G-7.1-73, *Commodity Specification for Air,* 1973. (This standard requires air to have the oxygen content normally present in the atmosphere, no more than 5 mg/m^3 of condensed hydrocarbon contamination, no more than 20 ppm carbon monoxide, no pronounced odor, and a maximum of 1,000 ppm of carbon dioxide.) See Figure 23-15.

Air-line respirators must be used only in nonIDLH atmospheres or from which the wearer can escape without the use of the respirator. This limitation is necessary because the air-line respirator is entirely dependent upon an air supply that is not carried by the wearer of the respirator. If this air supply fails, the wearer might not escape immediately from a hazardous atmosphere. Another limitation of air-line respirators is that the air hose limits the wearer to a fixed distance from the air supply source.

Air-line respirators are furnished in many types, but there are three basic classes—constant-flow, demand-flow, and pressure-

Figure 23-13. A chin-style gas mask. (Courtesy MSA)

demand-flow. The respirators are equipped with half masks, full facepiece, and helmets or hoods. If eye protection is required, a full facepiece must be used.

Constant-flow. A constant- or continuous-flow unit (Figure 23-16) has a regulated amount of air fed to the facepiece and is normally used where there is an ample air supply such as that provided by an air compressor.

The NIOSH-MSHA approves air-line (supplied-air) respirators under Title 30, *CFR,* Part 11, Subpart J, which has these significant requirements: (1) the maximum hose length for which approval is granted is 91.5 m (300 ft). (2) The maximum permissible inlet pressure is 683 kPa (125 psig).

The approved pressure range is noted for each approved device on the certification label or instructions supplied with the device.

With a 91.5-m hose assembled to the respirator, and the lowest inlet pressure introduced to the air-supply hose, constant-flow units must deliver at least 115 L/min (4 cfm) measured at the facepiece.

When helmets or hoods are used (Figure 23-17), the same requirements must be met, except that the flow-rate must be at least 170 L/min (6cfm). For both constant-flow and demand-type of assembly with the highest inlet pressure and shortest hose, the maximum flow should not exceed 430 L/min (15 cfm).

Constant-flow, air-line respirators with facepieces only are used where respiratory protection only is needed. A hood can be added to the facepiece for protection against sand or shot blast, and frequently a helmet for this application with a hood or cape fitted to it is used. Similar units are also used for lead grinding.

Demand-flow type. These air-line respirators with half masks or full facepieces deliver airflow only during inhalation. Such respirators are normally used when the air supply is restricted to high-pressure compressed air cylinders. A suitable pressure regulator is required to make sure that the air is reduced to the proper pressure for breathing. The same requirements for approval apply to the demand type as to the constant-flow units.

Pressure-demand flow. For those conditions where the possible inward leakage (caused by the negative pressure during inhalation that is always present in demand systems) is unacceptable and where there cannot be the relatively high air consumption of the constant-flow units, a pressure-demand air-line respirator may be the best choice. It provides a positive pressure during both inhalation and exhalation and must deliver a flow of at least 115 L/min (4 cfm) before a negative pressure can be measured in the facepiece.

Air supply. This is the responsibility of the user, and the air-line respirator is approved for use only when it supplies respirable air at the correct pressure and flow. The compressed air should meet the most recent requirements of Compressed Gas Association's Standard, CGA G-7.1-73, for Type I, Class D, gaseous air described earlier in this chapter (Figure 23-18).

With internally lubricated piston-type compressors, overheating may produce carbon monoxide; therefore, either a constant-monitoring analyzer or a heat-rise alarm, with frequent measurement of effluent carbon monoxide, should be installed.

Some air compressors are manufactured specifically to provide respirable air. They use compression seal liquids (such as water) or diaphragms for delivering the air.

Self-contained breathing apparatus. This provides complete respiratory protection against toxic gases and an oxygen deficiency. The wearer is independent of the surrounding atmosphere because he or she is breathing with a system that is portable and admits no outside air. The oxygen or air supply of the apparatus itself takes care of respiratory requirements.

There are four basic types of self-contained breathing apparatus: (1) oxygen-cylinder rebreathing, (2) self-generating types, (3) demand, and (3) pressure-demand type.

The oxygen cylinder rebreathing type. This includes the "lung-governed" method that automatically compensates for the breathing demand of the user, the constant-flow type, or a combination of the two. The unit has a relatively small cylinder of compressed oxygen, reducing and regulating valves, a breathing bag, facepiece, and chemical container to remove carbon dioxide from the exhaled breath. The types of cylinder rebreathing units now manufactured are approved by NIOSH-MSHA for 45 minutes, 60 minutes, 2-hour, 3-hour or 4-hour durations, and all function in the same basic manner. The following description covers all five types.

The high pressure of the oxygen from the cylinder is lessened by means of a reducing and regulating valve. Some units have a small constant-flow valve plus a lung-controlled valve that adds any required additional flow. Another type delivers a constant flow of oxygen from the breathing bag to the wearer's face.

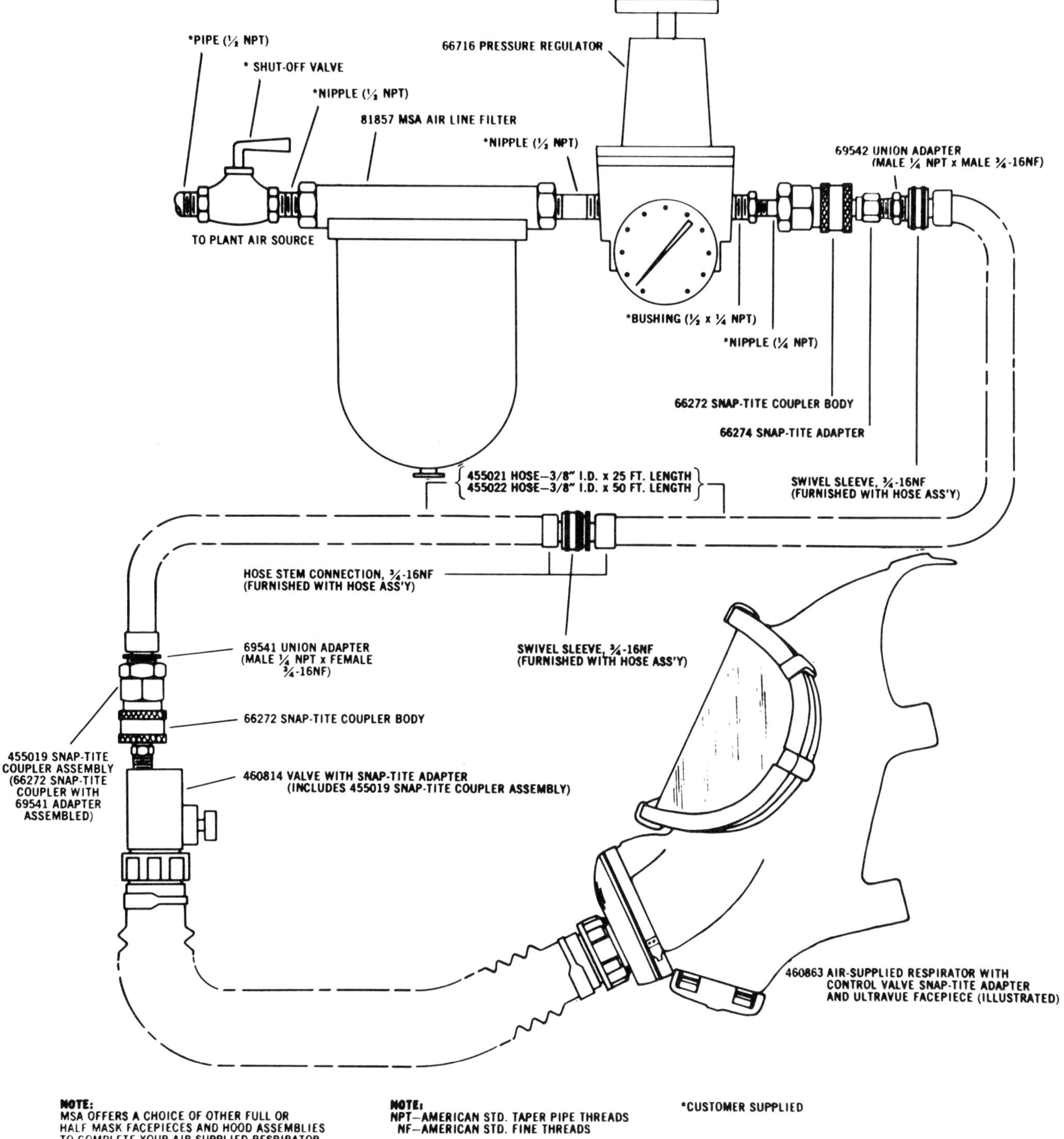

Figure 23-14. Constant-flow air-line respirator—typical assembly as used in conjunction with a compressor system. (Courtesy MSA)

Exhaled breath passes through another tube into a container holding the carbon dioxide-removing chemical and then through a cooler. Finally, the purified air flows into the breathing bag where it mixes with the incoming oxygen from the cylinder.

The rebreathing principle permits most efficient use of the oxygen supply. The exhaled breath contains both oxygen and carbon dioxide as the human body extracts only a small part of the oxygen inhaled. As the user exhales, the carbon dioxide is removed by the chemical and the oxygen which is left is used. This method of operation applies to all oxygen cylinder rebreathing-type apparatus as well as those using liquid oxygen.

The oxygen cylinder must be refilled and the carbon dioxide-removing chemical replaced after each use. As is true of all respiratory protective equipment, training in proper use and maintenance is essential for the most efficient operation.

Demand-type. This apparatus is available in different models for specific applications. All consist of a high-pressure air

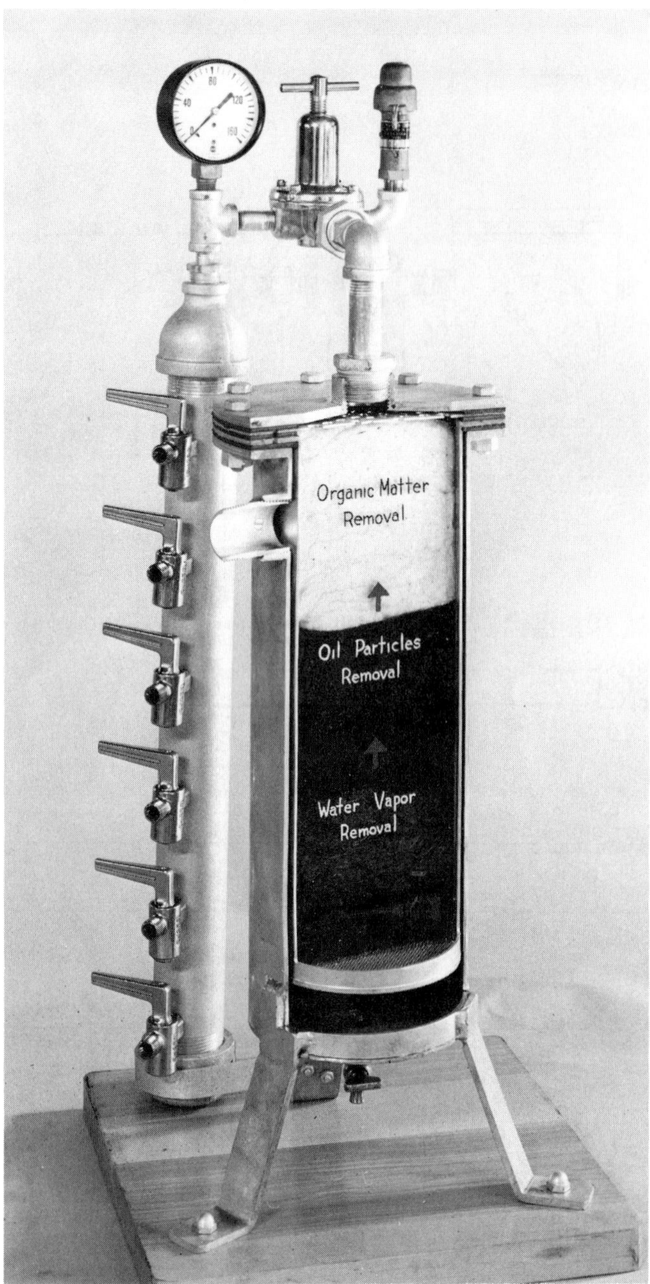

Figure 23-15. Cutaway view shows air path through purifier and the filtering materials for air-line respiratory equipment. Note manifold (left) that permits six air lines to be connected.

cylinder, a demand regulator connected either directly or by a high-pressure tube to the cylinder, a facepiece and tube assembly with an exhalation valve or valves, and a method of mounting the complete apparatus on the body. In use, the wearer opens the cylinder valve after putting on the facepiece and inhales air at breathing pressure. The exhaled air passes through a valve in the facepiece to the surrounding atmosphere.

The term "demand regulator" means that the airflow is on inhalation demand, that is, automatic regulation to the desired level to compensate for variation in breathing needs. All demand apparatus is relatively inefficient when compared with the rebreathing type, because the exhaled air (rich in oxygen) is released into the atmosphere instead of being used again.

Of particular application to some industries is a short duration unit known as an escape self-contained breathing apparatus. It is normally worn when workers are in an area where a potentially toxic atmosphere exists that may be beyond the capacity of a gas mask. Areas in which high-pressure gases are piped fall into this category.

Pressure-demand. This apparatus uses the same principle as the pressure-demand air-line respirators, and is approved and used where the toxicity is such that the potential facepiece leakage of demand apparatus is not tolerable. (Figure 23-19).

Under Title 30 *CFR* Part 11, Subpart H, all demand and pressure-demand type apparatus must function in temperatures as low as 0 C (32 F). Approval at lower temperatures can be obtained. Various components, such as nosecups to reduce facepiece lens fogging, may be added if needed to meet the low-temperature requirements.

Warning device. The wearer of the open-circuit apparatus depends upon watching a pressure gage to learn when air supply has dropped to a point necessitating return to fresh air. A self-actuating warning device is required for NIOSH-MSHA approval. The warning device must activate at 20–25 percent of the full cylinder pressure to alert the wearer that he or she must soon return to fresh air.

Units of 15-, 30-, and 60-minute duration are currently approved by NIOSH-MSHA under the 30 *CFR*, Part 11, Subpart H, for entry into and escape from nonrespirable atmospheres. Three-, 5-, 10-minute duration units are approved for escape only.

The self-generating type. This apparatus (Figure 23-20), previously approved by USBM for 1 hour, differs from the conventional cylinder rebreathing apparatus in that it has a chemical canister that evolves oxygen and removes the exhaled carbon dioxide in accordance with breathing requirements. It eliminates high-pressure cylinders, regulating valves, and other mechanical components.

The canister, which contains potassium superoxide, evolves oxygen when contacted by the moisture and carbon dioxide in the exhaled breath and retains carbon dioxide and moisture. Retaining moisture is important as it aids in preventing lens fogging.

In use, the self-generating unit operates as other rebreathing apparatus except that wearers, using the canister, make their own oxygen instead of drawing it from a compressed gas cylinder or liquid oxygen source. The important features of this type are the simplicity of construction and use, and reduced need for maintenance when compared with high-pressure apparatus.

The combination self-contained breathing apparatus (SCBA) and Type C supplied-air respirators. These are air-line respirators with an auxiliary self-contained air supply. (An auxiliary SCBA is an independent air supply that allows a person to evacuate a contaminated area or enter such an area for a very short period of time where a connection to an outside air supply can be made.) When NIOSH-MSHA approves a device with an auxiliary cylinder, it may be worn in toxic areas having concentrations greater than those permitted for air-line units without the egress cylinder. For this reason, these devices are usable in IDLH atmospheres. The auxiliary air supply can be switched to in the

Figure 23-16. Air-line respirators help provide cool, clean, breathable air for welders and others who must work in close, confined areas. (Courtesy Willson Products Division, ESB Inc.)

event the primary air supply fails to operate. This allows the wearer to escape from the IDLH atmosphere.

The auxiliary self-contained air supply may be NIOSH-MSHA approved for 3, 5, or 10 minutes' service time, or for 15 minutes or greater. If the auxiliary air supply is classified as 3, 5, or 10 minutes in service time, the wearer must use the air line during entry into the hazardous atmosphere and the SCBA portion is used for emergency egress. When the SCBA is rated for service of 15 minutes or longer, the SCBA may be used for emergency entry into a hazardous atmosphere (to connect the air line) provided not more than 20 percent of the air supply's rated capacity is used during entry. This allows for enough air for egress when the warning device indicates a low air supply.

Combination air-line respirators with auxiliary SCBA are designed to operate in three modes: constant-flow, demand-flow, and pressure-demand flow. These devices use the same principles as the respective air-line respirators. Both NIOSH-MSHA approve these devices under 30 *CFR* Part 11, Subpart H.

Class 3. Combination air-purifying and atmosphere-supplying devices

Lately, another type respirator is gaining in popularity. It is a device that is a combination of an air-line respirator with an auxiliary air-purifying attachment, which provides protection in the event the air supply fails. The NIOSH-MSHA has approved combination air-line and air-purifying respirators with the air line operating in either constant flow or pressure-demand flow. The most popular versions are ones in which the air-purifying element is a high-efficiency filter. This filter is approved for dusts, fumes, and mist that have a Permissible Exposure Limit (PEL) less than 0.05 mg/m^3 and radionuclides. The air-line respirator can only be used when the respirator is supplied with respirable air.

The pressure-demand flow air-line/high-efficiency filter respirator has additional limitations. They include the following:

1. Use in the filtering mode is allowed for *escape* only.
2. Not for use in IDLH atmospheres
3. Not for use in atmospheres containing less than 19.5 percent oxygen.
4. Use only the hose lengths and pressure ranges specified on the approval label.
5. When airflow is cut off, switch to the filter and immediately exit to clean air.

These NIOSH-MSHA devices are approved under 30 *CFR* Part 11, Subpart K. Because of the positive-pressure and escape provisions, these respirators have been recommended for asbestos work.

Communication aids. Many full facepiece respirators are furnished with a speaking diaphragm mounted in the facepiece. Without such a speaking diaphragm, audible communication between wearers of respiratory protective equipment is inadequate at best and at times may be impossible. Facepieces with an exhalation valve provide some measure of voice transmission through this valve, and an extremely limited transmission through the facepiece.

Another type of voice transmission equipment that can be used with facepieces containing a speaking diaphragm is the battery-powered amplification unit. These devices are particularly

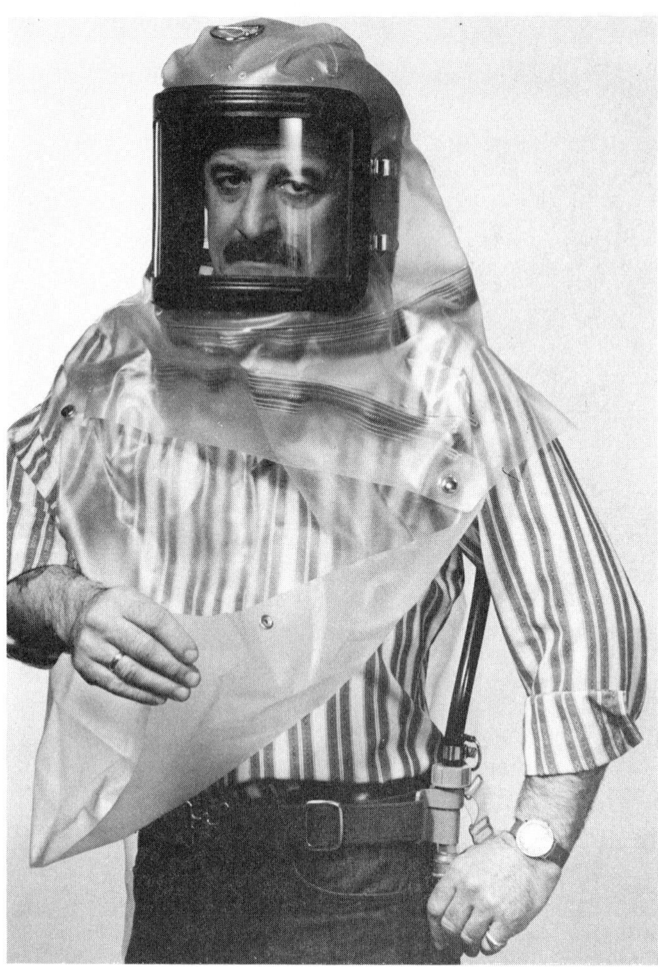

Figure 23-17. When helmets or hoods are used, the flow rate must be at least 170 L/min (6 cfm). (Courtesy American Optical Corp.)

effective in overcoming high background noise levels and enable mask wearers to communicate with other workers.

TRAINING

For safe use of any respiratory protective device, it is essential that the user be properly instructed in its use. Supervisors and the person issuing respirators as well as workers must be so instructed by competent persons.

The OSHA requires that all employees be trained in the proper use of the device assigned to them. (See 1910.134(b)(3) and (e)(5); see Addendum A to this chapter.) Many companies have their employees sign a document attesting to their having completed a training session with the respiratory device.

Each respirator wearer should be given training that would include (1) an explanation of the respiratory hazard and what happens if the respirator is not used properly, (2) a discussion of what engineering and administrative controls are being used and why respirators still are needed for protection, (3) an explanation of why a particular type of respirator has been selected, (4) a discussion of the function, capabilities, and limitations of the selected respirator, (5) instruction in how to don the respirator and to check its fit and operation, (6) instruction in the proper wearing of the respirator, (7) instruction in respirator maintenance, (8) instruction in recognizing and handling emergency situations, (9) instruction as needed for special respirator use and (10) regulations concerning respirator use.

Monitoring respirator use. Supervisory personnel should periodically monitor the use of respirators to insure that they are worn properly.

Respirator fit. Each respirator wearer should be provided with a respirator that fits: to find the respirator that fits, the worker must be fit-tested. This will be discussed later in this chapter. In addition, each respirator wearer should be required

Figure 23-18. Fresh air is first passed through a compressor-purification system for supplying clean, breathable air to workers who must enter confined spaces.

Figure 23-19. A pressure-demand self-contained breathing apparatus (SCBA) shown here is designed for use in oxygen-deficient atmospheres or where dangerous concentrations of toxic gas may be present. (Courtesy MSA)

to check the seal of the respirator by appropriate means before entering a harmful atmosphere (negative or positive-pressure fit-check; see Addendum B for description of this procedure). A respirator equipped with a facepiece must not be worn if facial hair comes between the sealing periphery of the facepiece and the face or if facial hair interferes with valve function. The wearer of a respirator equipped with a full facepiece, helmet, hood, or suit should not be allowed to wear contact lenses. If spectacle or goggles or faceshield must be worn with a facepiece, they should be worn so as not to adversely affect the seal of the facepiece to the face. Certain facepieces enable the wearer to also wear prescription spectacles without disturbing the facepiece seal (Figure 23-21).

INSPECTION, MAINTENANCE, AND REPAIR OF EQUIPMENT

Proper inspection, maintenance, and repair of respiratory protective equipment is mandatory to ensure success of any respiratory protection program. The precise nature of the program will vary widely based on such variables as size of the facility and the equipment involved.

All equipment must be inspected periodically before use and after each use. For equipment used only for emergencies the period between inspections can be as individually desired, but should be no more than one month. A record should be kept of all inspections by date with the results tabulated.

Respirator maintenance should be performed regularly. Maintenance should be carried out on a schedule that insures that each respirator wearer is provided with a respirator that is clean and in good operating condition. Maintenance should include (1) washing, sanitizing, rinsing, and drying, (2) inspection for defects, (3) replacement of worn or deteriorated parts, (4) repair if necessary, and (5) storage to protect against dust, sunlight, excessive heat, extreme cold, excessive moisture, damaging chemicals, and physical damage.

Replacement of other than disposable parts and any repair should be done only by personnel with adequate training to insure the equipment is functionally sound after the work is accomplished.

Medical and bioassay surveillance

Medical surveillance, including bioassay when applicable, of respirator wearers should be carried out periodically to determine if respirator wearers are receiving adequate respiratory protection. A physician should determine the requirements of the

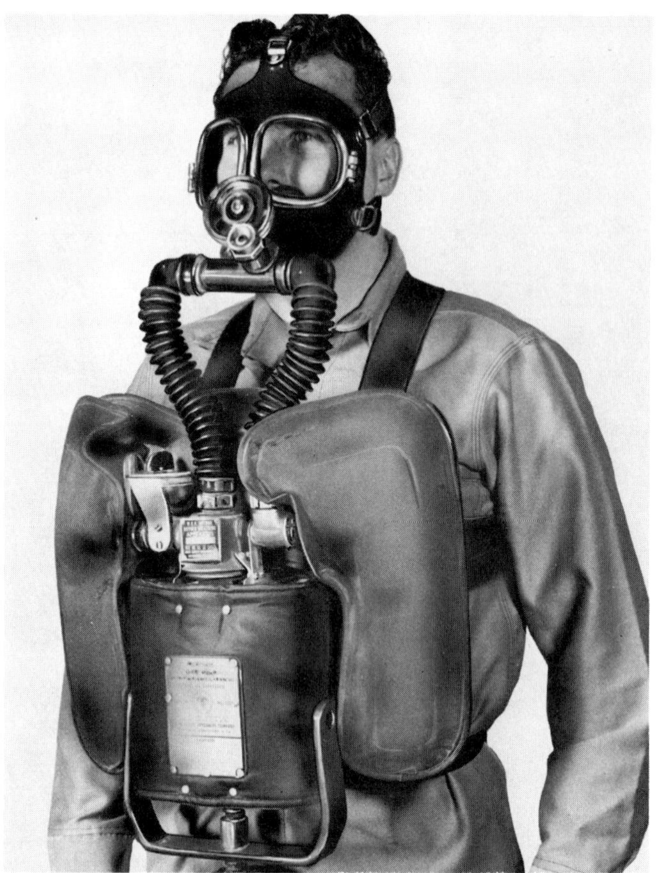

Figure 23-20. Self-generating oxygen breathing apparatus. (Courtesy MSA)

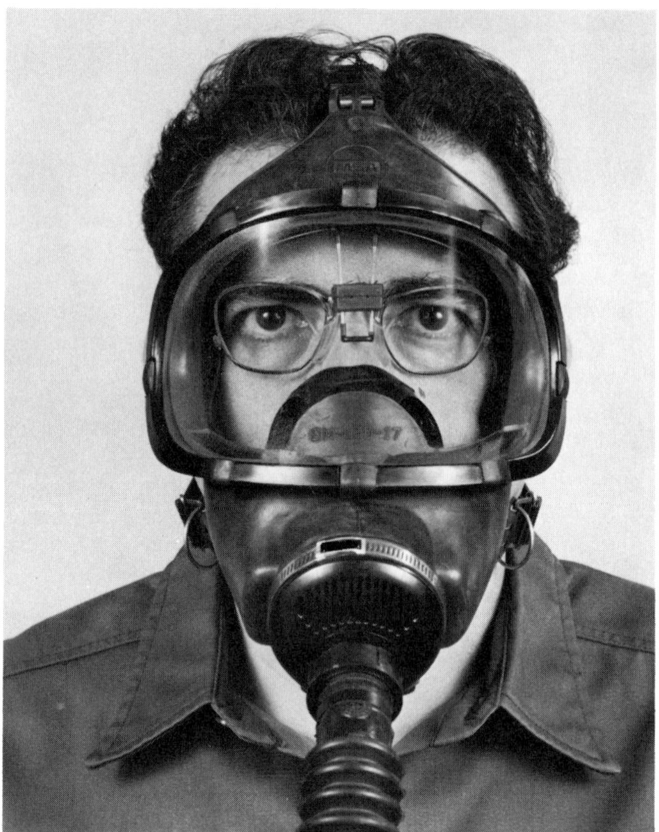

Figure 23-21. Facepiece has provision for prescription spectacles. (Courtesy MSA)

surveillance program. A physician should determine what physical and psychological conditions are pertinent for the wearing of different types of respirators. The respirator program administrator or designee, using guidelines established by a physician, should determine whether or not a person can be assigned to a task requiring the use of a respirator.

OSHA REQUIREMENTS FOR A RESPIRATORY PROTECTION PROGRAM

The OSHA regulation lists several elements of a respiratory protection program. The key elements of any respiratory protection program must start with an asessment of the inhalation hazards present in the workplace.

This initial step involves gathering the necessary toxicological, safety, and research data on the substance or substances in the atmosphere. A simple questionnaire (shown in Figure 23-22) can be used to assist in gathering the pertinent information about the air contaminant and the exposure conditions.

The OSHA methods of compliance are pretty straightforward. The standards require the use of engineering or work-practice controls to reduce exposures to the PEL or below. However, if such controls cannot get exposures down to the permissible limit, they have to be used anyway to reduce the exposure to the lowest level possible. At this point, the controls can be supplemented by respirators.

The OSHA states that respirators are the least satisfactory means of exposure control, because they provide good protection only if they are properly selected, properly fitted, worn by employees, and replaced when their service life is over. In addition, some employees may not be able to wear a respirator.

Despite these difficulties, respirators are the only form of protection available during the interval necessary to install or implement feasible engineering and work practice controls, in work operations such as maintenance and repair activities or other activities for which engineering and work practice controls are not feasible, in work situations where feasible engineering and work practice controls are not yet sufficient to reduce exposure to or below the PEL, and in emergencies.

Respirator selection guide

The OSHA Standard 1910.134 states that respirators shall be selected on the basis of the hazards to which workers are exposed and that ANSI Z88.2-1969 (revised in 1980) shall be used for guidance in their selection. For certain respiratory hazards, specific instructions about respirator use are given in other OSHA regulations (e.g., Asbestos, 1910.1001 and 1926.58; Vinyl Chloride, 1910.1017; and substances regulated after promulgation of Vinyl Chloride). The trend is toward regulations that specify the conditions of respirator use for each substance.

The new OSHA Standards list respiratory protection equipment for various concentrations of a substance. This is called the Respirator Selection Guide (Table 23-C). To provide additional protection, an employer may always select a respirator prescribed for concentrations higher than those found in the workplace. The employer may not, however, use respirators that are not listed.

QUESTIONNAIRE
GAS MASK AND RESPIRATOR RECOMMENDATION

NAME AND ADDRESS OF COMPANY
SEEKING RECOMMENDATION ____________

____

____

NAME OF INDIVIDUAL ____________
HIS PHONE NUMBER ____________

1. Material—
   a. Chemical Name ____________
   b. Trade Name ____________
   c. Formula ____________
   d. TLV or TWA OSHA 1910.1000 ______ Current ACGIH ______
2. Form in which it will be used—
   a. Liquid? ______ b. Solid? ______ c. Gaseous? ______
   d. If gaseous, is it an organic vapor? ______ or acid gas? ______
   other? ____________
3. Maximum expected concentration—
   a. ______ parts per million, or
   b. ______ milligrams per cubic meter
4. Will material be heated? ______
   a. If so, to what temperature? ______ °F.
5. What is the odor threshold of the material? ______
6. At what concentration is the material considered to be immediately dangerous to life or health? ______
7. Can the substance be absorbed through the skin? ______
8. Irritant to eyes? ______ respiratory tract? ______ skin? ______
9. At what concentration is it an irritant? ______
10. If the substance is known to be flammable, what are the lower and upper flammable limits, in per cent by volume? ______
11. What is the vapor pressure of the material? ______
12. Will material be mixed with other chemicals? ______ If so, give details ____________

    ____

    ____
13. Any possibility of oxygen deficiency? ______
14. Can good ventilation of the area be maintained? ______
15. Will exposure be continuous? ______ or intermittent? ______
16. Will the respiratory device be used for routine exposures, or will it be used as an escape device? ______

    ____
17. Provide as much detail as possible concerning exposure conditions.

    ____

    ____

    ____

Figure 23-22. Questionnaire for gas mask and respirator recommendations. (Courtesy MSA)

These Standards also call for a Respiratory Protection Program spelled out in section 1910.134 of the OSHA regulations. Of course the respirators must be approved by NIOSH-MSHA or carry a Bureau of Mines approval still valid under the grandfather clause.

Decision logic

The respirator selection guides use the strict guidelines of a decision logic system. The purpose of this decision logic system is to assure that the Respiratory Protection Selection Guides for the various substances are uniform. This decision logic was originally developed for the joint NIOSH and OSHA Standards Completion Program. It is helpful to use for selection of respirators that will protect against the hazards listed in 29 *CFR* 1910.1000.

The system works by eliminating, in a step-by-step fashion, respirators that are inappropriate until only those types that are acceptable remain.

To do this, the necessary toxicological, safety, and research

Table 23-C. Example of OSHA Respirator Selection Guide: Respiratory Protection for Asbestos, Tremolite, Anthophyllite, and Actinolite Fibers

Airborne concentration of asbestos, tremolite, anthophyllite, actinolite, or a combination of these minerals	*Required respirator*
Not in excess of 2 f/cc (10 × PEL).	1. Half-mask air-purifying respirator equipped with high-efficiency filters.
Not in excess of 10 f/cc (50 × PEL).	1. Full facepiece air-purifying respirator equipped with high-efficiency filters.
Not in excess of 20 f/cc (100 × PEL).	1. Any powered air-purifying respirator equipped with high-efficiency filters. 2. Any supplied-air respirator operated in continuous flow mode.
Not in excess of 200 f/cc (1,000 × PEL).	1. Full facepiece supplied-air respirator operated in pressure demand mode.
Greater than 200 f/cc (>1,000 × PEL) or unknown concentration.	1. Full facepiece supplied-air respirator operated in pressure demand mode equipped with an auxiliary positive pressure self-contained breathing apparatus.

Note: a. Respirators assigned for higher environmental concentrations may be used at lower concentrations.

b. A high-efficiency filter means a filter that is at least 99.97 percent efficient against mono-dispersed particles of 0.3 micrometers or larger. (Reprinted from 29 *CFR* 1910.1001 [g][2])

information on the substances is assembled in order to answer six basic questions.

Skin absorption. Personal protection requirements for splashes or spills that may cause the substance to be absorbed through the skin are not covered, as respirator selection criteria are based primarily on the inhalation hazard. It is possible that a supplied-air suit may provide both skin and respiratory protection from extremely toxic substances that may be absorbed through the skin. Supplied-air suits, however, are not covered in NIOSH-MSHA Approval Regulations, 30 *CFR* Part 11, and the data needed to recommend such suits for all types of exposures simply is not available.

Where information indicates systemic injury or death from absorbing a gas or vapor though the skin, the Standards say:

> Use of supplied-air suits or other impervious coverings may be necessary to prevent skin contact with the substance when the concentration of the substance is unknown or greater than the IDLH concentration. Supplied-air suits should be selected, used, and maintained under the immediate supervision of persons knowledgeable in the limitations and potential life-endangering characteristics of supplied-air suits.

Regardless of the use of supplied-air suits, the entry-and-escape-from-unknown-concentrations category of the Respiratory Protection Guides says:

> Self-contained breathing apparatus with a full facepiece operated in the pressure demand or other positive-pressure mode, or a combination respirator which includes a Type C supplied-air respirator with a full facepiece operated in the pressure demand or other positive-pressure or continu ous flow mode and an auxiliary self-contained breathing apparatus operated in the pressure demand or other positive-pressure mode.

Translated, this section of the respiratory table means a pressure-demand self-contained breathing apparatus or a pressure-demand combination air-line respirator with auxiliary SCBA may be used.

Warning properties. Warning properties such as odor, eye irritation, and respiratory irritation that rely upon human senses are not foolproof. However, they do provide some indication to the wearer that the service life of the cartridge or canister is reaching the end, the facepiece is not fitted properly, or there is some other respirator malfunction. Warning properties may be assumed to be adequate when odor, taste, or irritation effects of the substance can be detected and are persistent at concentrations at or below the PEL.

If the odor or irritation threshold of the substance is many time greater than the PEL, the substance is considered to have poor warning properties, and atmosphere-suppling respirators would be specified (see Addendum B for more details).

Borderline cases are governed by other rules. This same thinking is reflected in NIOSH-MSHA approvals for organic vapor chemical-cartridge respirators and gas masks that prohibit their use against organic vapors with poor warning properties (Table 23-D).

Sorbent efficiency. When evidence shows immediate or less than 3 minutes' breakthrough time at or below the IDLH concentration for a chemical cartridge or canister, these air-purifying devices will not be allowed for any use, including escape. Only atmosphere-supplying devices are permitted.

Eye irritation. For concentrations that produce eye irritation, it will be necessary to provide protection for the eyes. This may best be done by providing full facepiece respirators.

Immediately dangerous to life or health (IDLH). NIOSH-MSHA approval regulations define IDLH as: "Conditions that pose an immediate threat to life or health or conditions that pose an immediate threat of severe exposure to contaminants such as radioactive materials which are likely to have adverse cumulative or delayed effects on health."

Two factors are considered when establishing IDLH concentrations:

1. The worker must be able to escape without losing his or her life or suffering permanent health damage within 30 minutes. Thirty minutes is considered by OSHA as the maximum permissible exposure time for escape.
2. The worker must be able to escape without severe eye or respiratory irritation or other reactions that could inhibit escape.

If the concentration is above the IDLH levels only highly reliable breathing apparatus, such as pressure-demand self-contained breathing apparatus (SCBA), is allowed. Since the IDLH limits are conservative, any approved respirator may be used up to this limit as long as its maximum use concentration, or the limitations on the air-purifying element are not exceeded.

The IDLH limits should be determined from the following sources:

1. specific IDLH values provided in the literature such as *AIHA Hygienic Guides* or *NIOSH Pocket Guide to Chemical Hazards* (see Appendix A, Sources of Information).
2. human exposure data
3. acute animal exposure data.

Table 23-D. Comparison of Odor Tresholds and 1976 TLV's for Selected Compounds

Group 1 - Odor Threshold and TLV approximately the same		
	Odor Threshold (ppm)	**TLV (ppm)**
Acrylonitrile	21.4	20 S
Arsine	0.21	0.05
Cyclohexane	300	300
Cyclohexanol	100	50
Epichlorhydrin	10	5 S
Ethyl benzene	200	100
Ethylene diamine	11.2	10
Hydrogen chloride	10	5
Methyl acetate	200	200
Methylamine	10	10
Methyl chloroform	500	350
Nitrogen dioxide	5	5
Propyl alcohol	200	200
Styrene, monomer	200	100
Turpentine	200	100
Group 2 — Odor threshold from 2 to 10 times the TLV		
Acrolein	0.21	0.1
Allyl alcohol	7	2
Carbon tetrachloride	75	10 S
Chloroform	200	25
Crotonaldehyde	7.32	0.1
1,2 Dichloroethylene	500	200
Dichloroethyl ether	35	5 S
Dimethyl acetamide	46.8	10 S
Hydrogen selenide	0.3	0.05
Isopropyl glycidyl ether (IGE)	300	50
Group 3 - Odor threshold equal to or greater than 10 times TLV		
Bromoform	530	0.5
Camphor (synthetic)	1.6-200	2
Carbon disulfide	(c)	20
αChloroacetophenone	1.34	0.05
Chloropicrin	1.08	0.1
Diglycidyl ether (DGE)	5.0	0.5
Dimethylformamide	100	0 S
Ethylene oxide	500	50
Mercury vapor	(c)	0.5 mg/m^3
Methyl bromide	(c)	15
Methyl chloride	(c)	100
Methyl formate	2000	100
Methanol	2000	200
Methyl cyclohexanol	500	50
Phosgene	1.0	0.1
Phosphine	(c)	0.3
Radioactive gases and vapors	(c)	
Tolune 2,4 diisocyanate (TDI)	2.14	0.2

(Reprinted with permission from Reist, P. C., and Rex, F. Odor detection and respirator cartridge replacement. *American Industrial Hygiene Association Journal,* October 1977)

Table 23-E. Respirator Protection Factors

Type Respirator	Facepiece Pressure	Protection Factor
I. Air-Purifying		
A. Particulate Removing		
Single-Use, Dust	—	5
Quarter-Mask, Dust	—	5
Half-Mask, Dust	—	10
Half- or Quarter-Mask, Fume	—	10
Half- or Quarter-Mask, High-Efficiency	—	10
Full Facepiece, High-Efficiency	—	50
Powered, High-Efficiency, All Enclosures	+	1,000
Powered, Dust or Fume, All Enclosures	+	X
B. Gas and Vapor-Removing		
Half-Mask	—	10
Full Facepiece	—	50
II. Atmosphere-Supplying		
A. Supplied-Air		
Demand, Half-Mask	—	10
Demand, Full Facepiece	—	50
Hose Mask Without Blower, Full Facepiece	—	50
Pressure-Demand, Half-Mask	+	1,000
Pressure-Demand, Full Facepiece	+	2,000
Hose Mask With Blower, Full Facepiece	—	50
Continuous Flow, Half-Mask	+	1,000
Continuous Flow, Full Facepiece	+	2,000
Continuous Flow, Hood, Helmet, or Suit	+	2,000
B. Self-Contained Breathing Apparatus		
Open-Circuit, Demand, Full Facepiece	—	50
Open-Circuit, Pressure-Demand Full Facepiece	+	10,000
Closed-Circuit, Oxygen Tank-Type, Full Facepiece	—	50
III. Combination Respirator		
A. Any Combination of Air-Purifying and Atmosphere-Supplying Respirator	Use Minimum Protection Factor Listed Above for Type and Mode of Operation	
B. Any Combination of Supplied-Air Respirator and an SCBA		

(Reprinted with permission from E. C. Hyatt. "Respirator Protection Factors," LA-6084-MS, Los Alamos National Laboratory, Los Alamos, NM 1976)

Lower flammable limit (LFL) and firefighting. Concentrations in excess of the lower flammable limit (LFL) are considered to be IDLH. At or above the LFL, respirators must provide maximum protection. Such devices include pressure-demand self-contained breathing apparatus and combination positive-pressure supplied-air respirators with egress cylinders (work masks).

The ANSI standard Z88.5 *Practices for Respiratory Protection for the Fire Service,* 1981, defines firefighting as immediately dangerous to life, so for firefighting, the only device providing adequate protection is pressure-demand self-contained breathing apparatus.

Protection factors

Protection factors, a very important part of the decision logic system, are simply a measure of the overall effectiveness of a respirator. These numbers have been assigned to an entire class of respirators. The protection factors being used by NIOSH are based on quantitative fit tests peformed at Los Alamos National Laboratory and elsewhere, and in some instances on professional judgment (Table 23-E). Another source of protection factors for the various types of respirators can be found in ANSI Standard Z88.2-1980 (Table 23-F). Sometimes the protection factors that OSHA uses will be different than either of these sources (Table 23-C).

Protection factors are determined by dividing the ambient airborne concentration by the concentration inside the facepiece. Instead of establishing protection factors based on quantitative fit testing, today's emphasis is upon sampling in the workplace under workplace conditions. In the future these will be referred to as workplace protection factors. As the data are gathered, changes in the protection factors assigned to entire classes of devices are likely to occur.

Protection factors (PF) are used in the selection process to determine the maximum use concentration (MUC) for the respirator. It is determined by multiplying the TLV (Appendix B-1) by the protection factor (Figure 23-23). These recommended respirator protection factors should only be used when the employer has established a minimal acceptable respirator program meeting the requirement of 29 *CFR* 1910.134 and satisfactory fit-testing has been performed.

Impact of proposals on respiratory protection

This basic example shows that the eye irritation, protection factor, and warning property sections of the decision logic system will result in substantial changes in the types of respirators required in the future. Some half-mask respirators will be replaced with full-facepiece respirators due to eye irritation, if it cannot be handled by protective goggles.

The protection factors being applied to respirators will require not only the replacement of any half-facepiece respirators but will also require the wholesale replacement of negative pressure

Table 23-F. ANSI Z88-2-1980 Respirator Protection Factors[a]

Type of Respirator	Permitted for Use in Oxygen-Deficient Atmosphere	Permitted for Use in Immediately-Dangerous-to-Life-or-Health Atmosphere[f]	Respirator Protection Factor	
			Qualitative Test	Quantitative Test
Particulate-filter, quarter-mask or half-mask facepiece[b,c]	No	No	10	As measured on each person with maximum of 100.
Vapor- or gas-removing, quarter-mask or half-mask facepiece[c]	No	No	10, or maximum use limit of cartridge or canister for vapor or gas, whichever is less.	As measured on each person with maximum of 100, or maximum use limit of cartridge or canister for vapor or gas[i,j], whichever is less.
Combination particulate-filter and vapor- or gas-removing, quarter-mask or half-mask facepiece[b,c]	No	No	10, or maximum use limit of cartridge or canister for vapor or gas, whichever is less.	As measured on each person with maximum of 100, or maximum use limit of cartridge or canister for vapor or gas[i,j], whichever is less.
Particulate-filter, full facepiece[b]	No	No	100	As measured on each person with maximum of 100 if dust, fume, or mist filter is used, or maximum of 1000 if high-efficiency filter is used.
Vapor- or gas-removing, full facepiece	No	No	100, or maximum use limit of cartridge or canister for vapor or gas, whichever is less.	As measured on each person with maximum of 1000, or maximum use limit of cartridge or canister for vapor or gas[i,j], whichever is less.
Combination particulate-filter and vapor- or gas-removing, full facepiece[b]	No	No	100, or maximum use limit of cartridge or canister for vapor or gas, whichever is less.	As measured on each person with maximum of 100 if dust, fume, or mist filter is used and maximum of 1000 if high-efficiency filter is used, or maximum use limit of cartridge or canister for vapor or gas[i,j], whichever is less.
Powered particulate-filter, any respiratory-inlet covering[b,c,d]	No	No (yes, if escape provisions are provided[d])	N/A No tests are required due to positive-pressure operation of respirator. The maximum protection factor is 100 if dust, fume, or mist filter is used and 3000 if high-efficiency filter is used.	N/A
Powered vapor- or gas-removing, any respiratory-inlet covering[c,d]	No	No (yes, if escape provisions are provided[d])	N/A No tests are required due to positive pressure operation of respirator. The maximum protection factor is 3000, or maximum use limit of cartridge or canister for vapor or gas[i,j], whichever is less.	N/A
Powered combination particulate-filter and vapor- or gas-removing, any respiratory-inlet covering[b,c,d]	No	No (yes, if escape provisions are provided[d])	N/A No tests are required due to positive-pressure operation of respirator. The maximum protection factor is 100 if dust, fume, or mist filter is used and 3000 if high-efficiency filter is used, or maximum use limit of cartridge or canister for vapor or gas[i,j], whichever is less.	N/A

(Continued)

respirators with positive-pressure respirators. This includes the bulk of the self-contained breathing apparatus and gas masks in use today.

In the category of adequate warning properties, OSHA and NIOSH consider substances with no published odor or irritation data to have poor warning properties. This category eliminates all gas-sorbent air-purifying respirators and will require their replacement with supplied-air respirators or self-contained breathing apparatus.

Respirator fit-testing

After close consideration of all the details pertaining to respirator selection, proper protection will not be provided if the respirator facepiece does not fit the wearer properly. We should not expect one make and model of respirator to fit the entire work force. Because of the great variety in face sizes and shapes encountered in male and female workers, most respirator manufacturers make their models of respirators available in more than one size. In addition, the size and shape of each facepiece varies among the different manufacturers. In other words, the medium-size half-mask facepiece of one manufacturer is not the same shape and size as the medium-size half-mask facepiece from another manufacturer. For these reasons, it is necessary to buy several commercially available respirators and to conduct a respirator fit-testing program. These are also requirements of the OSHA standards.

The OSHA Standard, 29 *CFR* 1910.134, requires that all negative-pressure respirators be fit-tested (Table 23-F) by exposure to a "test atmosphere." This includes disposable respirators. This can be achieved by one of two fitting methods, qualitative or quantitative fit-testing.

Qualitative fit-testing. A qualitative fit-test relies on the wearer's subjective response. The test atmosphere is a substance that typically can be detected by the wearer such as isoamyl acetate (banana oil), irritant smoke, or saccharin. The respirator must be equipped to remove the test atmosphere (Figures 23-24 and 23-25). For example, if using isoamyl acetate, which is an organic chemical that gives off a vapor, an organic-vapor chemical cartridge must be used. Then with a respirator in good repair, if the wearer smells isoamyl acetate, the respirator does not fit well.

These tests are relatively fast, easily performed, and use inexpensive equipment. Because these tests are based on the respirator wearer's subjective response to a test chemical, reproducibility and accuracy may vary. Addendum D of this chapter includes three protocols for qualitative fit-testing. It is

Table 23-F. (Concluded)

Type of Respirator	Permitted for Use in Oxygen-Deficient Atmosphere	Permitted for Use in Immediately-Dangerous-to-Life-or-Health Atmosphere[f]	Respirator Protection Factor	
			Qualitative Test	Quantitative Test
Air-line, demand, quarter-mask or half-mask face-piece, with or without escape provisions[c,e]	Yes[f]	No	10	As measured on each person, but limited to the use of the respirator in concentrations of contaminants below the immediately-dangerous-to-life-or-health (IDLH) values.
Air-line, demand, full face-piece, with or without escape provisions[e]	Yes[f]	No	100	As measured on each person, but limited to the use of the respirator in concentrations of contaminants below the immediately-dangerous-to-life-or-health (IDLH) values.
Air-line, continuous-flow or pressure-demand type, any facepiece, without escape provisions[c]	Yes[f]	No	N/A No tests are required due to positive-pressure operation of respirator. The protection factor provided by the respirator is limited to use of the respirator in concentrations of contaminants below the immediately-dangerous-to-life-or-health (IDLH) values.	N/A
Air-line, continuous-flow or pressure-demand type, any facepiece, with escape provisions[c,e]	Yes[g]	Yes	N/A No tests are required due to positive-pressure operation of respirator. The maximum protection factor is 10 000 plus.[h]	N/A
Air-line, continuous flow, helmet, hood, or suit, without escape provisions	Yes[f]	No	N/A No tests are required due to positive-pressure operation of respirator. The protection factor provided by the respirator is limited to the use of the respirator in concentrations of contaminants below the immediately-dangerous-to-life-or-health (IDLH) values.	N/A
Air-line, continuous-flow, helmet, hood, or suit, with escape provisions[e]	Yes[g]	Yes	N/A No tests are required due to positive-pressure operation of respirator. The maximum protection factor is 10 000 plus.[h]	N/A
Hose mask, with or without blower, full facepiece	Yes[f]	No	10	As measured on each person, but limited to the use of the respirator in concentrations of contaminants below the immediately-dangerous-to-life-or-health (IDLH) values.
Self-contained breathing apparatus, demand-type open-circuit or negative-pressure-type closed-circuit, quarter-mask or half-mask facepiece[c]	Yes[f]	No	10	As measured on each person, but limited to the use of the respirator in concentrations of contaminants below the immediately-dangerous-to-life-or-health (IDLH) values.
Self-contained breathing apparatus, demand-type open-circuit or negative-pressure-type closed-circuit, full facepiece or mouthpiece/nose clamp[c]	Yes[f] (Yes[g], if respirator is used for mine rescue and mine recovery operations.)	No (yes, if respirator is used for mine rescue and mine recovery operations.)	100	As measured on each person, but limited to the use of the respirator in concentrations of contaminants below the immediately-dangerous-to-life-or-health (IDLH) values, except when the respirator is used for mine rescue and mine recovery operations.
Self-contained breathing apparatus, pressure-demand-type open-circuit or positive-pressure-type closed-circuit, quarter-mask or half-mask facepiece, full facepiece, or mouthpiece/nose clamp[c]	Yes[g]	Yes	N/A No tests are required due to positive-pressure operation of respirator. The maximum protection factor is 10 000 plus.[h]	N/A
Combination respirators not listed.	The type and mode of operation having the lowest respirator protection factor shall be applied to the combination respirator.			

N/A means not applicable since a respirator-fitting test is not carried out.

[a]A respirator protection factor is a measure of the degree of protection provided by a respirator to a respirator wearer. Multiplying the permissible time-weighted average concentration or the permissible ceiling concentration, whichever is applicable, for a toxic substance, or the maximum permissible airborne concentration for a radionuclide, by a protection factor assigned to a respirator gives the maximum concentration of the hazardous substance for which the respirator can be used. Limitations of filters, cartridges, and canisters used in air-purifying respirators shall be considered in determining protection factors.

[b]When the respirator is used for protection against airborne particulate matter having a permissible time-weighted average concentration less than 0.05 milligram particulate matter per cubic meter of air or less than 2 million particles per cubic foot of air, or for protection against airborne radionuclide particulate matter, the respirator shall be equipped with a high-efficiency filter(s).

[c]If the air contaminant causes eye irritation, the wearer of a respirator equipped with a quarter-mask or half-mask facepiece or mouthpiece and nose clamp shall be permitted to use a protective goggle or to use a respirator equipped with a full facepiece.

[d]If the powered air-purifying respirator is equipped with a facepiece, the escape provision means that the wearer is able to breathe through the filter, cartridge, or canister and through the pump. If the powered air-purifying respirator is equipped with a helmet, hood, or suit, the escape provision shall be an auxiliary self-contained supply of respirable air.

[e]The escape provision shall be an auxiliary self-contained supply of respirable air.

[f]For definition of "oxygen deficiency – not immediately dangerous to life or health" see Section 2.

[g]For definition of "oxygen deficiency – immediately dangerous to life or health" see Section 2 and A10.

[h]The protection factor measurement exceeds the limit of sensitivity of the test apparatus. Therefore, the respirator has been classified for use in atmospheres having unknown concentrations of contaminants.

[i]The service life of a vapor- or gas-removing cartridge or canister depends on the specific vapor or gas, the concentration of the vapor or gas in air, the temperature and humidity of the air, the type and quantity of the sorbent in the cartridge or canister, and the activity of the respirator wearer. Cartridges and canisters may provide only very short service lives for certain vapors and gases. Vapor/gas service life testing is recommended to ensure that cartridges and canisters provide adequate service lives. Reference should be made to published reports which give vapor/gas life data for cartridges and canisters.

[j]Vapor- and gas-removing respirators are not approved for contaminants that lack adequate warning properties of odor, irritation, or taste at concentrations in air at or above the permissible exposure limits.

NOTE: Respirator protection factors for air-purifying-type respirators equipped with a mouthpiece/nose clamp form of respiratory-inlet covering are not given, since such respirators are approved only for escape purposes.

(Reprinted with permission from ANSI Z88.2-1980, *Respirator Protection Factors*)

Example of Maximum Use Concentration Determination

What is the MUC for a half-mask respirator with dust/mist filters for copper dust?
TLV for copper dust: 1 mg/m^3 PF for half-mask respirator with dust/mist filter: 10 MUC = TLV × PF = 1 mg/m^3 × 10 = 10 mg/m^3
If air-sampling indicates an ambient concentration greater than 10 mg/m^3, this respirator does not provide sufficient protection!

Figure 23-23. Protection factors (PF) are used in the selection process to determine the maximum use concentration (MUC) for the respirator. It is determined by multiplying the TLV (see Appendix B-1) by the protection factor. These recommended respirator protection factors should only be used when the employer has established a minimal acceptable respirator program meeting the requirements of 29 *CFR* 1910.134, and satisfactory fit-testing has been performed.

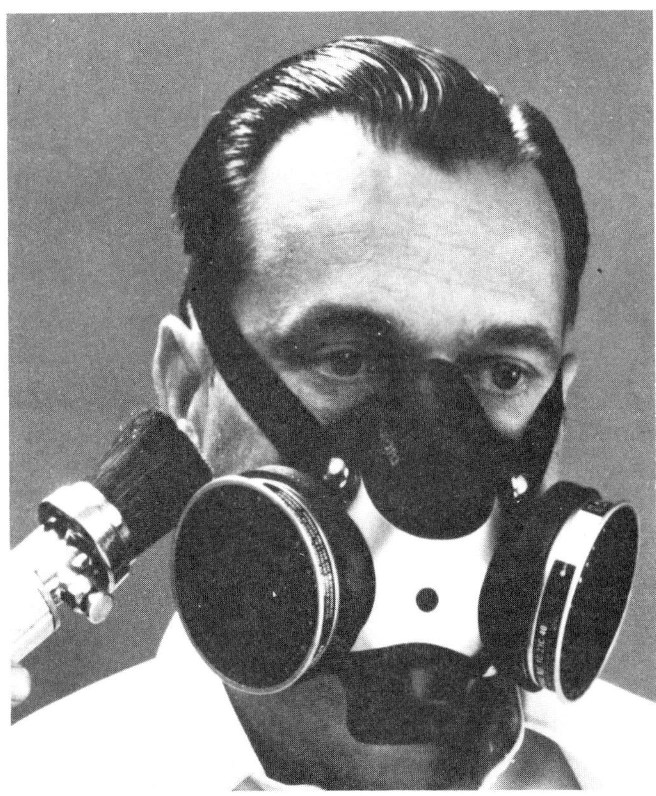

Figure 23-25. The wearer should determine if the respirator leaked by odor of the "banana oil." (Courtesy MSA)

Figure 23-24. The qualitative fit method determines respirator maximum use concentrations (MUC), see Figure 23-23, based on the simplest fit-testing. An irritant smoke tube is used in this test and the wearer determines if the respirator leaks by the irritation that is detected. (Courtesy MSA)

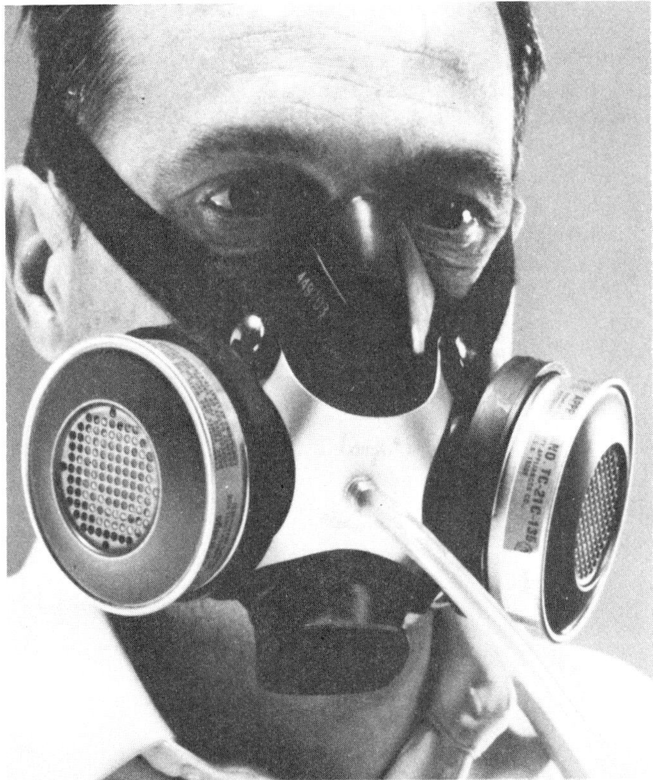

Figure 23-26. The quantitative fit method measures facepiece leakage from inside the facepiece, using a probe, sampling line, and analytical equipment. (Courtesy MSA)

suggested that the reader review these for more specific directions on qualitative fit-testing.

Quantitative fit-testing. Quantitative fit-tests involve exposing the respirator wearer to a test atmosphere containing an easily detectable, nontoxic aerosol, vapor, or gas as the test agent. Instrumentation, which samples the test atmosphere and the air inside the facepiece of the respirator, is used to measure quantitatively the leakage into the respirator. With this information, a quantitative fit factor can be calculated. This is an index that indicates how well the respirator fits the wearer. The higher the number the better the fit.

The advantages to this type of testing is that it does not rely on a subjective response and it is much more accurate. The disadvantages are: cost of instrumentation, need for highly trained personnel to conduct the test, use of special respirators equipped with a sampling probe so the same facepiece cannot be worn in actual service (Figure 23-26), and time consumption as a result of performing three trials.

SUMMARY

The material presented in this chapter is intended for persons concerned with establishing and maintaining a respiratory protection program. It presents certain basic information for guidance purposes. However, it is not intended to be all-inclusive in content or scope.

Simplified interpretations of certain federal regulations pertaining to respiratory protection and monitoring were presented in this chapter. While these interpretations convey background information about the regulations, under no circumstances should they be used as the sole basis of a respiratory protection program. In all cases, the current federal regulations, as published in the *Federal Register* and later collected in the *Code of Federal Regulations,* should be carefully studied, and the rules and procedures in those regulations explicitly followed. Only they define the specific requirements that are in force.

BIBLIOGRAPHY

The American Conference of Governmental Industrial Hygienists (ACGIH). *Threshold Limit Values.* Cincinnati: ACGIH, published annually.

American Industrial Hygiene Association (AIHA). *Respiratory Protection: A Manual and Guideline.* Akron, OH: AIHA, 1980.

American National Standards Institute, 1430 Broadway, New York, NY 10018.
Standards:
Practices for Respiratory Protection, Z88.2, 1969.
Practices for Respiratory Protection, Z88.2, 1980.

Code of Federal Regulations, Title 29, Labor, Parts 1900-1926, and *Code of Federal Regulations, Title 30, Mineral Resources, Parts 11-14A.* Washington, DC: U.S. Government Printing Office.

Compressed Gas Association, 1235 Jefferson Davis Hwy., Arlington, VA 22202.
Pamphlets:
Compressed Air for Human Respiration, G-7, 1976.
Commodity Specification for Air, Pamphlet G-7.1 (also ANSI/CGA G-7.1-73), 1973.

Hyatt, E. C. *Respirator Protection Factors,* LA-6084-MS. Los Alamos, NM: Los Alamos National Laboratory, 1976.

Lundin, A. M. Personal protection—Establishing an effective respiratory program. *National Safety News,* October 1971.

National Fire Protection Association, Batterymarch Park, Quincy, MA 02269.
Self-Contained Breathing Apparatus for Fire Fighters, 1981.

National Institute for Occupational Safety and Health (NIOSH). *A Guide to Industrial Respiratory Protection,* No. 76-189, Cincinnati: NIOSH, 1976.

——. *NIOSH-Certified Equipment List,* DHHS 87-102 (October 1, 1986). Cincinnati: NIOSH, 1986.

Reist, P. C., and Rex, F. Odor detection and respirator cartridge replacement. *American Industrial Hygiene Association Journal* (October 1977): 563-566.

U.S. Dept. of Labor. Respiratory Protection, in *Industrial Hygiene Technical Manual.* U.S. Government Printing Office, Washington DC: March 30, 1984.

Addendum A

§ 1910.134 Respiratory protection.

(a) *Permissible practice.* (1) In the control of those occupational diseases caused by breathing air contaminated with harmful dusts, fogs, fumes, mists, gases, smokes, sprays, or vapors, the primary objective shall be to prevent atmospheric contamination. This shall be accomplished as far as feasible by accepted engineering control measures (for example, enclosure or confinement of the operation, general and local ventilation, and substitution of less toxic materials). When effective engineering controls are not feasible, or while they are being instituted, appropriate respirators shall be used pursuant to the following requirements.

(2) Respirators shall be provided by the employer when such equipment is necessary to protect the health of the employee. The employer shall provide the respirators which are applicable and suitable for the purpose intended. The employer shall be responsible for the establishment and maintenance of a respiratory protective program which shall include the requirements outlined in paragraph (b) of this section.

(3) The employee shall use the provided respiratory protection in accordance with instructions and training received.

(b) *Requirements for a minimal acceptable program.* (1) Written standard operating procedures governing the selection and use of respirators shall be established.

(2) Respirators shall be selected on the basis of hazards to which the worker is exposed.

(3) The user shall be instructed and trained in the proper use of respirators and their limitations.

(4) [Removed]

(5) Respirators shall be regularly cleaned and disinfected. Those used by more than one worker shall be thoroughly cleaned and disinfected after each use.

(6) Respirators shall be stored in a convenient, clean, and sanitary location.

(7) Respirators used routinely shall be inspected during cleaning. Worn or deteriorated parts shall be replaced. Respirators for emergency use such as self-contained devices shall be thoroughly inspected at least once a month and after each use.

(8) Appropriate surveillance of work area conditions and degree of employee exposure or stress shall be maintained.

(9) There shall be regular inspection and evaluation to determine the continued effectiveness of the program.

(10) Persons should not be assigned to tasks requiring use of respirators unless it has been determined that they are physically able to perform the work and use the equipment. The local physician shall determine what health and physical conditions are pertinent. The respirator user's medical status should be reviewed periodically (for instance, annually).

(11) Approved or accepted respirators shall be used when they are available. The respirator furnished shall provide adequate respiratory protection against the particular hazard for which it is designed in accordance with standards established by competent authorities. The U.S. Department of Interior, Bureau of Mines, and the U.S. Department of Agriculture are recognized as such authorities. Although respirators listed by the U.S. Department of Agriculture continue to be acceptable for protection against specified pesticides, the U.S. Department of the Interior, Bureau of Mines, is the agency now responsible for testing and approving pesticide respirators.

(c) *Selection of respirators.* Proper selection of respirators shall be made according to the guidance of American National Standard Practices for Respiratory Protection Z88.2-1969.

(d) *Air quality.* (1) Compressed air, compressed oxygen, liquid air, and liquid oxygen used for respiration shall be of high purity. Oxygen shall meet the requirements of the United States Pharmacopoeia for medical or breathing oxygen. Breathing air shall meet at least the requirements of the specification for Grade D breathing air as described in Compressed Gas Association Commodity Specification G-7.1-1966. Compressed oxygen shall not be used in supplied-air respirators or in open circuit self-contained breathing apparatus that have previously used compressed air. Oxygen must never be used with air line respirators.

(2) Breathing air may be supplied to respirators from cylinders or air compressors.

(i) Cylinders shall be tested and maintained as prescribed in the Shipping Container Specification Regulations of the Department of Transportation (49 CFR Part 178).

(ii) The compressor for supplying air shall be equipped with necessary safety and standby devices. A breathing air-type compressor shall be used. Compressors shall be constructed and situated so as to avoid entry of contaminated air into the system and suitable in-line air purifying sorbent beds and filters installed to further assure breathing air quality. A receiver of sufficient capacity to enable the respirator wearer to escape from a contaminated atmosphere in event of compressor failure, and alarms to indicate compressor failure and overheating shall be installed in the system. If an oil-lubricated compressor is used, it shall have a high-temperature or carbon monoxide alarm, or both. If only a high-temperature alarm is used, the air from the compressor shall be frequently tested for carbon monoxide to insure that it meets the specifications in paragraph (d)(1) of this section.

(3) Air line couplings shall be incompatible with outlets for other gas systems to prevent inadvertent servicing of air line respirators with nonrespirable gases or oxygen.

(4) Breathing gas containers shall be marked in accordance with American National Standard Method of Marking Portable Compressed Gas Containers to Identify the Material Contained, Z48.1-1954; Federal Specification BB-A-1034a, June 21, 1968, Air, Compressed for Breathing Purposes; or Interim Federal Specification GG-B-00675b, April 27, 1965, Breathing Apparatus, Self-Contained.

(e) *Use of respirators.* (1) Standard procedures shall be developed for respirator use. These should include all information and guidance necessary for their proper selection, use, and care. Possible emergency and routine uses of respirators should be anticipated and planned for.

(2) The correct respirator shall be specified for each job. The respirator type is usually specified in the work procedures by a qualified individual supervising the respiratory protective program. The individual issuing them shall be adequately instructed to insure that the correct respirator is issued.

(3) Written procedures shall be prepared covering safe use of respirators in dangerous atmospheres that might be encountered in normal operations or in emergencies. Personnel shall be familiar with these procedures and the available respirators.

(i) In areas where the wearer, with failure of the respirator, could be overcome by a toxic or oxygen-deficient atmosphere, at least one additional man shall be present. Communications (visual, voice, or signal line) shall be maintained between both or all individuals present. Planning shall be such that one individual will be unaffected by any likely incident and have the proper rescue equipment to be able to assist the other(s) in case of emergency.

(ii) When self-contained breathing apparatus or hose masks with blowers are used in atmospheres immediately dangerous to life or health, standby men must be present with suitable rescue equipment.

(iii) Persons using air line respirators in atmospheres immediately hazardous to life or health shall be equipped with safety harnesses and safety lines for lifting or removing persons from hazardous atmospheres or other and equivalent provisions for the rescue of persons from hazardous atmospheres shall be used. A standby man or men with suitable self-contained breathing apparatus shall be at the nearest fresh air base for emergency rescue.

(4) Respiratory protection is no better than the respirator in use, even though it is worn conscientiously. Frequent random inspections shall be conducted by a qualified individual to assure that respirators are properly selected, used, cleaned, and maintained.

(5) For safe use of any respirator, it is essential that the user be properly instructed in its selection, use, and maintenance. Both supervisors and workers shall be so instructed by competent persons. Training shall provide the men an opportunity to handle the respirator, have it fitted properly, test its face-piece-to-face seal, wear it in normal air for a long familiarity period, and, finally, to wear it in a test atmosphere.

(i) Every respirator wearer shall receive fitting instructions including demonstrations and practice in how the respirator should be worn, how to adjust it, and how to determine if it fits properly. Respirators shall not be worn when conditions prevent a good face seal. Such conditions may be a growth of beard, sideburns, a skull cap that projects under the facepiece, or temple pieces on glasses. Also, the absence of one or both dentures can seriously affect the fit of a facepiece. The worker's diligence in observing these factors shall be evaluated by periodic check. To assure proper protection, the facepiece fit shall be checked by the wearer each time he puts on the respirator. This may be done by following the manufacturer's facepiece fitting instructions.

(ii) Providing respiratory protection for individuals wearing corrective glasses is a serious problem. A proper seal cannot be established if the temple bars of eye glasses extend through the sealing edge of the full facepiece. As a temporary measure, glasses with short temple bars or without temple bars may be taped to the wearer's head. Wearing of contact lenses in contaminated atmospheres with a respirator shall not be allowed. Systems have been developed for mounting corrective lenses inside full facepieces. When a workman must wear corrective lenses as part of the facepiece, the facepiece and lenses shall be fitted by qualified individuals to provide good vision, comfort, and a gas-tight seal.

(iii) If corrective spectacles or goggles are required, they shall be worn so as not to affect the fit of the facepiece. Proper selection of equipment will minimize or avoid this problem.

(f) *Maintenance and care of respirators.* (1) A program for maintenance and care of respirators shall be adjusted to the type of plant, working conditions, and hazards involved, and shall include the following basic services:

(i) Inspection for defects (including a leak check),

(ii) Cleaning and disinfecting,

(iii) Repair,

(iv) Storage

Equipment shall be properly maintained to retain its original effectiveness.

(2) (i) All respirators shall be inspected routinely before and after each use. A respirator that is not routinely used but is kept ready for emergency use shall be inspected after each use and at least monthly to assure that it is in satisfactory working condition.

(ii) Self-contained breathing apparatus shall be inspected monthly. Air and oxygen cylinders shall be fully charged according to the manufacturer's instructions. It shall be determined that the regulator and warning devices function properly.

(iii) Respirator inspection shall include a check of the tightness of connections and the condition of the facepiece, headbands, valves, connecting tube, and canisters. Rubber or elastomer parts shall be inspected for pliability and signs of deterioration. Stretching and manipulating rubber or elastomer parts with a massaging action will keep them pliable and flexible and prevent them from taking a set during storage.

(iv) A record shall be kept of inspection dates and findings for respirators maintained for emergency use.

(3) Routinely used respirators shall be collected, cleaned, and disinfected as frequently as necessary to insure that proper protection is provided for the wearer. Respirators maintained for emergency use shall be cleaned and disinfected after each use.

(4) Replacement or repairs shall be done only by experienced persons with parts designed for the respirator. No attempt shall be made to replace components or to make adjustment or repairs beyond the manufacturer's recommendations. Reducing or admission valves or regulators shall be returned to the manufacturer or to a trained technician for adjustment or repair.

(5) (i) After inspection, cleaning, and necessary repair, respirators shall be stored to protect against dust, sunlight, heat, extreme cold, excessive moisture, or damaging chemicals. Respirators placed at stations and work areas for emergency use should be quickly accessible at all times and should be stored in compartments built for the purpose. The compartments should be clearly marked. Routinely used respirators, such as dust respirators, may be placed in plastic bags. Respirators should not be stored in such places as lockers or tool boxes unless they are in carrying cases or cartons.

(ii) Respirators should be packed or stored so that the facepiece and exhalation valve will rest in a normal position and function will not be impaired by the elastomer setting in an abnormal position.

(iii) Instructions for proper storage of emergency respirators, such as gas masks and self-contained breathing apparatus, are found in "use and care" instructions usually mounted inside the carrying case lid.

(g) *Identification of gas mask canisters.* (1) The primary means of identifying a gas mask canister shall be by means of properly worded labels. The secondary means of identifying a gas mask canister shall be by a color code.

(2) All who issue or use gas masks falling within the scope of this section shall see that all gas mask canisters purchased or used by them are properly labeled and colored in accordance with these requirements before they are placed in service and that the labels and colors are properly maintained at all times thereafter until the canisters have completely served their purpose.

(3) On each canister shall appear in bold letters the following:

(i)—

Canister for ——————————————
(Name for atmospheric contaminant)

or

Type N Gas Mask Canister

(ii) In addition, essentially the following wording shall appear beneath the appropriate phrase on the canister label: "For respiratory protection in atmospheres containing not more than ———— percent by volume of ——————————."
(Name of atmospheric contaminant)

(4) Canisters having a special high-efficiency filter for protection against radionuclides and other highly toxic particulates shall be labeled with a statement of the type and degree of protection afforded by the filter. The label shall be affixed to the neck end of, or to the gray stripe which is around and near the top of, the canister. The degree of protection shall be marked as the percent of penetration of the canister by a 0.3-micron-diameter dioctyl phthalate (DOP) smoke at a flow rate of 85 liters per minute.

(5) Each canister shall have a label warning that gas masks should be used only in atmospheres containing sufficient oxygen to support life (at least 16 percent by volume), since gas mask canisters are only designed to neutral-

ize or remove contaminants from the air.

(6) Each gas mask canister shall be painted a distinctive color or combination of colors indicated in Table I-1. All colors used shall be such that they are clearly identifiable by the user and clearly distinguishable from one another. The color coating used shall offer a high degree of resistance to chipping, scaling, peeling, blistering, fading, and the effects of the ordinary atmospheres to which they may be exposed under normal conditions of storage and use. Appropriately colored pressure sensitive tape may be used for the stripes.

TABLE I-1

Atmospheric contaminants to be protected against	Colors assigned [1]
Acid gases	White.
Hydrocyanic acid gas	White with ½-inch green stripe completely around the canister near the bottom.
Chlorine gas	White with ½-inch yellow stripe completely around the canister near the bottom.
Organic vapors	Black.
Ammonia gas	Green.
Acid gases and ammonia gas	Green with ½-inch white stripe completely around the canister near the bottom.
Carbon monoxide	Blue.
Acid gases and organic vapors	Yellow.
Hydrocyanic acid gas and chloropicrin vapor	Yellow with ½-inch blue stripe completely around the canister near the bottom.
Acid gases, organic vapors, and ammonia gases	Brown.
Radioactive materials, excepting tritium and noble gases	Purple (Magenta).
Particulates (dusts, fumes, mists, fogs, or smokes) in combination with any of the above gases or vapors.	Canister color for contaminant, as designated above, with ½-inch gray stripe completely around the canister near the top.
All of the above atmospheric contaminants	Red with ½-inch gray stripe completely around the canister near the top.

[1] Gray shall not be assigned as the main color for a canister designed to remove acids or vapors.

NOTE: Orange shall be used as a complete body, or stripe color to represent gases not included in this table. The user will need to refer to the canister label to determine the degree of protection the canister will afford.

(Approved by the Office of Management and Budget under control number 1218-0099)

(Secs. 4(b)(2), 6(b) and 8(c), 84 Stat. 1592, 1593, 1596, 29 U.S.C. 653, 655, 657; Secretary of Labor's Order No. 8-76 (41 FR 25059); 29 CFR Part 1911); secs. 6, 8, 84 Stat. 1593, 1600 (29 U.S.C. 655, 657), Secretary of Labor's Order No. 9-83 (48 FR 35736), 29 CFR Part 1911)

[39 FR 23502, June 27, 1974, as amended at 43 FR 49748, Oct. 24, 1978; 49 FR 5322, Feb. 10, 1984; 49 FR 18295, Apr. 30, 1984]

Note that additional requirements for respiratory protection are now being included by OSHA in individual standards as, for example, in 1910.1001(d), Asbestos; 1910.1017(g), Vinyl Chloride; 1910.1025(f) Lead; and 1910.1029(g), Coal Tar Pitch Emissions.

Addendum B

OSHA INSTRUCTION CPL 2-2.20A

This OSHA directive is referred to as the Industrial Hygiene Technical Manual and reflects current OSHA industrial hygiene practices and procedures, including those procedures and practices relating to respiratory protective equipment. The following is taken from this directive.

CHAPTER V, RESPIRATORY PROTECTION

CHAPTER V

RESPIRATORY PROTECTION

A. General.

1. Requirements. The OSHA General Industry standard for respiratory protection, 29 CFR 1910.134, requires that a respiratory protection program be established by the employer, and that respirators be provided and be effective when such equipment is necessary to protect the health of the employee. However, OSHA health standards place primary emphasis on engineering, administrative and work practice controls in light of the inherent deficiencies of respirators and respirator programs.

2. Exposure. Exposure to a contaminant is determined by measuring the airborne concentrations of contaminants in the breathing zone of the employee. Procedures in Chapter IV, Evaluation of Exposure Levels for Air Contaminants, establish the criteria to be used for determining whether an overexposure occurs.

3. Definitions. For purposes of this chapter, the terms "exposure" and "overexposure" refer to concentrations in the breathing zone of the employee, outside of the respirator (i.e., the exposure of the employee if no respirator were being worn).

B. Written Operating Procedures.

1. Requirements. The employer must have written standard operating procedures governing the selection and use of respirators. The procedures must include a discussion or explanation of all the items specified in 29 CFR 1910.134 (b). Determining the adequacy of the written procedures is a professional judgment made by the CSHO.

2. Written Procedures. Inspect the written procedures for completeness and accuracy. If possible, obtain a copy of the procedures for the case file. Where the procedures cannot be obtained or do not exist, record pertinent details of the employer's program.

3. Deficiencies. Complete lack of written procedures usually indicates a deficiency in the respiratory protection program. Deficiencies in the written program which do not detract from its overall effectiveness should be brought to the attention of the employer in the closing conference.

C. Respirator Selection. In selecting the correct respirator for a given circumstance, the following factors must be taken into consideration:

1. Nature of the Hazard. In order to make subsequent decisions, the nature of the hazard must be identified to ensure that an overexposure does not occur. The following considerations must be included in this identification:

a. Oxygen Deficiency.

(1) NIOSH/MSHA approval for supplied-air and air-purifying respirators is valid only for atmospheres containing greater than 19.5% oxygen. If it is determined that an oxygen deficient atmosphere may exist, then selection must be made from the two types of appropriate respirators specified in Figure V-1.

(2) If oxygen deficiency is not an issue, then the contaminant(s) and their concentration(s) must be determined. Figure V-1 presents an outline for the selection process based on these criteria.

b. Physical Properties of the Hazard. Physical properties to be considered include:

- Physical state.
- Particle size.
- Molecular weight.
- Vapor pressure.

c. Chemical Properties of the Hazard. Chemical properties include:

- Solubility in water and other liquids.

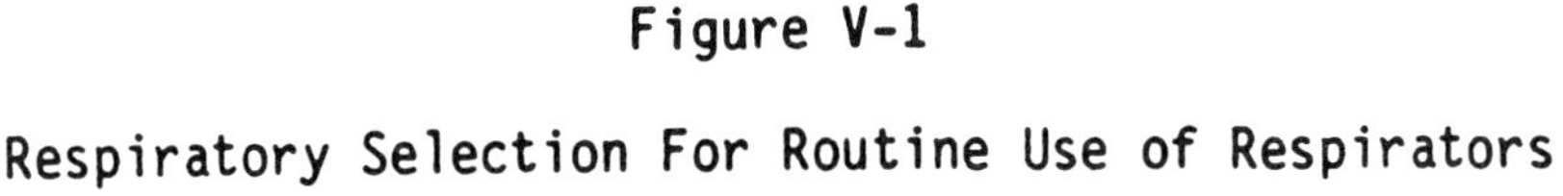

Figure V-1

Respiratory Selection For Routine Use of Respirators

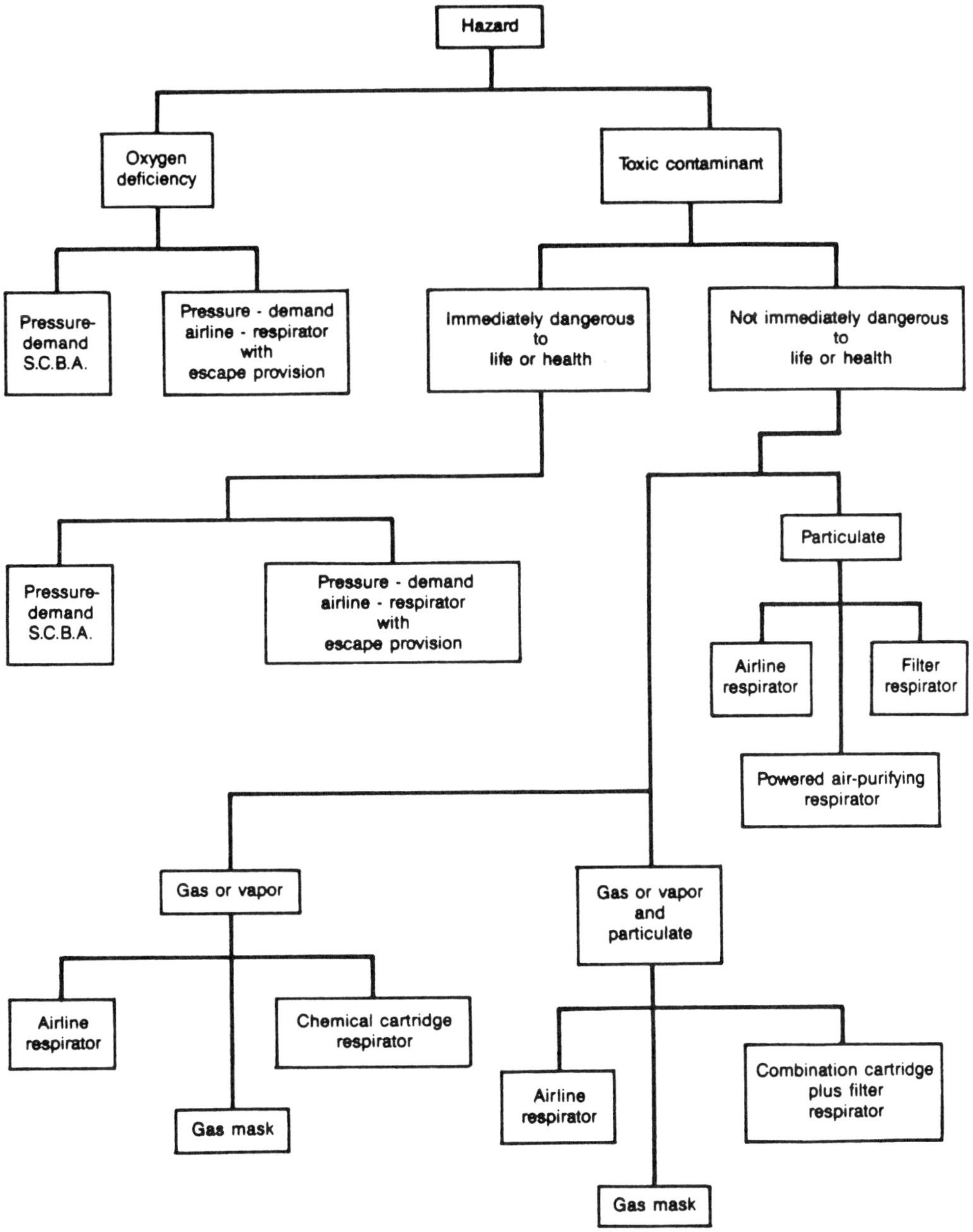

- Hazard
  - Oxygen deficiency
    - Pressure-demand S.C.B.A.
    - Pressure - demand airline - respirator with escape provision
  - Toxic contaminant
    - Immediately dangerous to life or health
      - Pressure-demand S.C.B.A.
      - Pressure - demand airline - respirator with escape provision
    - Not immediately dangerous to life or health
      - Particulate
        - Airline respirator
        - Filter respirator
        - Powered air-purifying respirator
      - Gas or vapor
        - Airline respirator
        - Chemical cartridge respirator
        - Gas mask
      - Gas or vapor and particulate
        - Airline respirator
        - Combination cartridge plus filter respirator
        - Gas mask

"Reprinted by permission from Olin Corporation's Respiratory Protection Manual".

- Reactivity with other chemicals.

- Hazardous decomposition products.

d. Physiologic Effects on the Body. Determine the toxicological (including synergistic) effects on the body in terms of:

- Eye irritation.

- Skin absorption.

- Adverse effects (if any) on olfactory sense.

e. Actual Concentration of a Toxic Compound. If a measurement has been made, then this is extremely useful information because bounds are established for the degree of protection necessary. This is to be used in conjunction with permissible exposure limits to select the correct respirator.

f. Permissible Exposure Limits (PEL). The permissible exposure limits (time-weighted average or ceiling value) may be used to establish proper selection. The concentrations and PEL are compared to protection factors (discussed at C.8. of this chapter) assigned to certain types of respirators.

NOTE: Table V-3 contains a listing of protection factors for various types of respirators.

g. Warning Properties. If an air-purifying respirator is to be used for protection against gas or vapor contaminants, then there must (with limited exceptions) be suitable warning properties of contaminant breakthrough or respirator malfunction.

(1) Adequate warning properties can be assumed when the odor, taste, or irritation effects of the substance are detectable and persistent at concentrations at or below the PEL.

(2) If the odor or irritation threshold of a substance occurs at concentrations greater than three times the PEL, this substance should be considered to have poor warning properties.

(3) If the odor or irritation threshold is somewhat above the PEL (but not in excess of three times the limit) and there is no ceiling limit, determine whether an undetected exposure in this concentration range could cause serious or irreversible health effects. (Refer to Appendix A, Chemical Information Table.) If not, the substance is considered to have adequate warning properties. In such a situation, it is expected that environmental concentrations will vary considerably, and warning of respirator failure would therefore soon be perceived at contaminant concentrations somewhat above the PEL.

(4) It is important to realize that 30 CFR 11, NIOSH/MSHA approvals for respirators, generally do not apply to gases or vapors with poor warning properties except where the device is equipped with an end-of-service life indicator (e.g., carbon monoxide). However, OSHA standard 29 CFR 1910.134(b)(11) may permit such a use for a specific gas or vapor where approved respirators are not available (e.g., vinyl chloride, acrylonitrile).

2. Nature of the Hazardous Operation. For proper respirator selection, it is necessary to know the details of operations which require workers to use respiratory devices. These include:

- Operation or process characteristics.

- Work area characteristics.

- Materials used or produced during the process.

- Workers' duties and actions.

- Abnormal situation characteristics which may necessitate different respirator selection; i.e., upset conditions or emergencies.

3. Location of the Hazardous Area. This is important in the selection process so that backup systems may be planned if necessary. Respirable air locations must be known prior to entry into a hazardous area so escape or emergency operations may be planned.

4. Time Respiratory Protection Is Required. The length of time a respirator will have to be worn by a worker is a factor which must be evaluated. This is most pronounced when using a Self-Contained Breathing Apparatus (SCBA), where, by definition, the air supply is finite. However, time is also a factor during routine use of air-purifying respirators when worker acceptance and comfort are essential to ensure proper use of the device.

5. Employee's Health. Effective usage of a respirator is dependent on an individual's ability to wear a respirator, as determined by a physician. Most respiratory devices increase physical stress on the body, especially the heart and lungs. Care should be taken to ensure that a medical determination has been made that an individual is capable of wearing a respirator for the duration of the work assignment. (See paragraph k. of this chapter, page V-18.)

6. Work Activity. The type of work activities to be performed while wearing a respirator is vitally important in the respirator selection. The proper respirator will be one which is least disruptive to the task being conducted yet providing the desired protection.

7. Respirator Characteristics, Capabilities, and Limitations. Tables V-1 and V-2, reproduced from ANSI Z88.2-1980, provide a description of various respirator characteristics, capabilities, and limitations.

8. Protection Factors. The protection afforded by respirators is dependent upon the seal of the facepiece to the face, leakage around valves, and leakage through or around cartridges or canisters. Depending on these criteria, the degree of protection may be ascertained and a relative safety factor assigned. Protection factors are only applicable if all elements of an effective respirator program are in place and being enforced.

a. The protection factor is a ratio of the air contaminant concentration outside the respirator to the air contaminant concentration inside the respirator facepiece. The higher the protection factor, the greater the degree of protection offered by the respirator.

b. Protection factors are used in conjunction with permissible exposure limits of contaminants to estimate the upper concentration limits to which respirators can be utilized safely. Table V-3, which is reproduced from ANSI Z88.2-1980, provides protection factors and explanations for various types of respirators.

c. Protection factors are invalid when employees remove their respiratory protection for unspecified periods while in the contaminated atmosphere.

NOTE: Field studies of respirator performance have not correlated well with the laboratory test data. Hence, the reported values should only be taken as estimates. For example, recent studies have found that Powered Air-Purifying Respirators (PAPR's) have not achieved the protection factors suggested by laboratory data.

D. Training.

1. Requirements. 29 CFR 1910.134(b)(3) requires that the user "be instructed and trained in the proper use of respirators and their limitations," and 1910.134(e)(5) requires that "the user be properly instructed in its selection, use, and maintenance." 29 CFR 1910.134(e)(5) and (e)(5)(i) also include requirements for demonstrations on "how the respirator should be worn, how to adjust it, and how to determine if it fits properly."

2. Evaluation of Employee's Training. In evaluating if an employer has properly trained employees, consider the following:

a. Whether the employee wears the respirator as it was originally intended. All of the following shall be considered:

(1) The respirator and all functional parts including straps must be in place and worn in the appropriate positions.

(2) All straps must be secure and properly adjusted.

(3) There must be no modification to the respirator or straps; for example, replacing the straps with string or rubber bands indicates an ineffective respirator.

(4) Use of a facelet or a knitted covering over the rubber face seal voids the approval of the respirator.

b. Whether the respirator's capabilities and limitations have been discussed. Essential topics requiring explanation include:

(1) Identification of contaminants or contaminant types against which the respirator is designed to afford protection.

(2) Limitations on the service life of the cartridge, canister, or filter which is used.

(3) Warning properties of the contaminant.

c. How the employee maintains the respirator.

3. Adequacy of Training. The employer and/or employee should be asked to demonstrate familiarity with the fitting technique, respirator capabilities and limitations, and maintenance procedures to determine the adequacy of training.

E. Fit Testing.

1. Requirements. 29 CFR 1910.134 (e)(5) states that respirators shall be fitted properly and shall be tested for their facepiece-to-face seal. 29 CFR 1910 (e)(5)(i) states that respirators shall not be worn when conditions prevent a good face seal. Examples listed in the standard of conditions that may interfere with facial seal are :

a. Sideburns and/or skull caps that project under the facepiece.

b. Temple bars on glasses (especially when wearing full face respirators) and/or the absence of one or both dentures.

2. Quantitative Fit Test. The purpose of the quantitative fit test is to determine the proper fit and degree of integrity of the face fit under actual wearing conditions. It is intended to provide the best method of fitting the respirator to the individual, using sensitive methods of detection for leakage or malfunction.

a. Quantitative respirator fit tests involve exposing the respirator wearer to a test atmosphere containing an easily detectable, relatively nontoxic aerosol, vapor, or gas as the test agent and then measuring the penetration of the test agent into the respirator. There are a number of test atmospheres, test agents, and exercises to perform during the tests. Manufacturers' recommendations should be followed for specific tests.

b. ANSI Z88.2-1980 describes a typical test protocol and exercise for performing quantitative fit testing. The compounds listed in Table V-4 (page V-10) are suitable test agents.

3. Qualitative Fit Test. Qualitative fit tests involve a test subject's responding (either voluntarily or involuntarily) to a chemical outside the respirator facepiece. These tests are fast, easily performed, and use inexpensive equipment. Because they are based on the respirator wearer's subjective response to the test chemical, however, reproducibility and accuracy may vary. Three of the most popular methods are an irritant smoke test, an odorous vapor test, and a taste test. Procedures for some of these test methods are detailed in an addendum to OSHA's "Occupational Exposure to Lead: Respirator Fit Testing," Federal Register (FR 51118). ANSI Z88.2-1980 also includes protocols for the irritant smoke and odorous vapor tests. The following represent a brief summary of each of these tests.

a. Irritant Smoke Test

(1) The irritant smoke test is performed by directing an irritant smoke, usually either stannic chloride or titanium tetracholoride, from a smoke tube towards the respirator being worn. If the wearer cannot detect the irritant smoke, a satisfactory fit is assumed to be achieved.

TABLE V-4

Test Agents Suitable for Carrying Out Quantitative Respirator-Fitting Tests

Test Agent	Concentration of Test Agent in Test Atmosphere	Particle Size if Test Agent is an Aerosol
Polydisperse sodium chloride aerosol	10-20 milligrams particulate matter per cubic meter of air	Mass median aerodynamic diameter of 0.5 to 0.7 micrometer with standard deviation of 2.0 to 2.4
Polydisperse DEHP (diethyl hexylphthalate) aerosol	20-30 milligrams particulate matter per cubic meter of air	Mass median aerodynamic diameter of 0.5 to 0.7 micrometer with standard deviation of 2.0 to 2.4

NOTE: The use of Diethylhexylphthalate (DEHP) is listed as an acceptable test atmosphere in OSHA Instruction CPL 2-2.29 dated October 27, 1980. However, the National Toxicology Program's Third Annual Report (1982) listed DEHP as a suspect carcinogen; therefore, caution is suggested. Corn oil is an acceptable substitute for DEHP.

(2) The respirator wearer will react involuntarily, usually by coughing or sneezing, to leakage around or through the respirator. Since this is a qualitative test, the testor is interested in any response to the smoke. The degree of response is not important.

NOTE: The test substances are irritants to the eyes, skin, and mucous membranes. Therefore, the respirator wearer should keep his/her eyes closed during testing.

(3) When an air-purifying respirator is tested, it has to be equipped with a high efficiency filter.

b. Odorous Vapor Test.

(1) The odorous vapor test relies on the respirator wearer's ability to detect an odorous material, usually isoamyl acetate (banana oil) inside the respirator. The test is performed by passing an isoamyl acetate saturated material around the outside of the respirator, or by introducing the wearer to a concentration of the chemical in a room, chamber, or hood. If the wearer is unable to smell the chemical, then a satisfactory fit is assumed to be achieved.

(2) The use of isoamyl acetate as a test agent has the following limitations:

(a) The odor threshold varies widely among individuals.

(b) Olfactory fatigue may cause a person to fail to detect the odor.

(c) The test is dependent on the wearer's honest response. There is no involuntary reaction.

(3) When an air-purifying respirator is tested, it should be equipped with an organic cartridge or canister which removes the test vapor from the air.

fitting respirator inlet coverings and breathing tubes which can be squeezed or blocked at the inlet to prevent the passage of air.

(1) The inlet opening of the respirator's canister(s), cartridge(s), or filter(s) is closed off by covering with the palm of the hand(s), by replacing the inlet seal on canister(s), or by squeezing a breathing tube or blocking its inlet so that it will not allow the passage of air.

(2) The wearer is instructed to inhale gently and hold his breath for at least 10 seconds.

(3) If a facepiece collapses slightly and no inward leakage of air into the facepiece is detected, it can be reasonably assured that the respirator has been properly donned and the exhalation valve and facepiece are not leaking.

b. Positive Pressure Test. A positive pressure test can be used on respirators equipped with tight fitting respiratory inlet coverings which contain both inhalation and exhalation valves. This test may be impossible to carry out on valveless respirators and on many disposable respirators.

(1) The exhalation valve or breathing tube, or both, is closed off and the wearer is instructed to exhale gently.

(2) The respirator has been properly donned if a slight positive pressure can be built up inside the facepiece without the detection of any outward leakage of air between the sealing surface of the facepiece and the wearer's face.

(3) For some respirators, this test method requires that the respirator wearer first remove the exhalation valve cover from the respirator and then replace it after completion of the test. These tasks often are difficult to carry out without disturbing the fit of the respirator to the wearer.

c. Taste Test.

(1) The taste test relies upon the respirator wearer's ability to detect a chemical substance, usually sodium saccharin, by tasting it inside the respirator. The test is performed by placing an enclosure (hood) over the respirator wearer's head and shoulders and spraying the test agent into the enclosure with a nebulizer. If the wearer is unable to taste the chemical, then a satisfactory fit is assumed to be achieved.

(2) Limitations to the use of sodium saccharin are that taste thresholds for it are highly variable, and the test is totally dependent on the wearer's honest indication of taste. There is no involuntary response. The wearer must not eat, drink (except plain water), chew gum or tobacco for 15 minutes before the test to avoid masking the taste of saccharin.

(3) This test can be used for both air-purifying respira-tors and atmosphere-supplying respirators. When air-purifying respirators are tested, they should be equipped with a particulate filter cartridge.

NOTE: This test is normally used for single-use, disposable type dust respirators.

4. Field Test Measures. There are two tests that can be used in the field to check the seal of the respirator: positive and negative pressure sealing tests. Each should be performed every time a respirator is donned, or else the procedures recommended by the manufacturer should be followed. Neither field test may be substituted for quantitative or qualitative tests. Adequate training of respirator users is essential for satisfactory field tests. The following procedures are recommended by ANSI Z88.2-1980:

a. Negative Pressure Test. This test may be impossible to carry out on valveless respirators and on many disposable (single-use) respirators. However, this test can be used on air-purifying respirators equipped with tight fitting respirator inlet coverings and atmosphere-supplying respirators equipped with tight

F. Cleaning.

1. Requirements. 29 CFR 1910.134(b)(5) states, "Respirators shall be regularly cleaned and disinfected. Those issued for the exclusive use of one worker should be cleaned after each day's use, or more often if necessary." This applies only to those respirators that are routinely used throughout the day. Cleaning at less than daily frequency is acceptable if proper protection is still afforded to the employee.

2. Methods. Respirators are usually cleaned by one of the following methods.

a. Manual Cleaning.

(1) A generalized cleaning procedure is:

(a) Remove canisters, filters, valves, straps, and speaking diaphragms from the facepiece.

(b) Wash facepiece and accessories in warm soapy water. Gently scrub with a brush.

(c) Rinse parts thoroughly in clean water.

(d) Air dry in a clean place or wipe dry with a lintless cloth.

(e) Reassemble.

(2) An alternate method is to use a commercially available cleaner, following the manufacturer's instructions.

b. Machine Cleaning. Machines may be used to expedite the cleaning, sanitizing, rinsing, and drying of large numbers of respirators.

(1) Extreme care must be taken to ensure against excessive tumbling and agitation, or exposure to temperatures above those recommended by the manufacturer (normally 49° C or 120° F maximum), as these conditions are likely to result in damage to the respirators.

(2) Ultrasonic cleaners, clothes-washing machines, dishwashers, and clothes dryers have been specially adapted and successfully used for cleaning and drying respirators.

c. Disinfection. Disinfection is required when the respirator is used by more than one person.

(1) Disinfection procedures recommended by NIOSH are:

(a) Immerse the respirator body for two minutes in a 50 ppm chlorine solution (about 2 ml bleach to 1 liter of water). Rinse thoroughly in clean water and dry.

(b) Immerse the respirator body for two minutes in an aqueous solution of iodine (add 0.8 ml tincture of iodine in 1 liter water). The iodine is about 7 percent ammonium and potassium iodide, 45 percent alcohol and 48 percent water. Rinse thoroughly in clean water and dry.

(2) Immersion times have to be limited to minimize damage to the respirator. The solutions can age rubber and rust metal parts. Caution must be taken to thoroughly rinse the respirator after cleaning and disinfection to prevent dermatitis.

(3) An alternate method is to purchase a commercially prepared solution for disinfection/decontamination and follow the directions recommended by the manufacturer.

G. Storage.

1. Requirements. 29 CFR 1910.134(b)(6) requires that respirators "be stored in a convenient, clean, and sanitary location." The purpose of good respirator storage is to ensure that the respirator will function properly when used.

2. Proper Storage. Care must be taken to ensure that respirators are stored in such a manner as to protect against dust, harmful chemicals, sunlight, excessive heat or cold, and moisture. Storage measures which can be used to protect respirators against dusts, chemicals, and moisture include:

a. Hermetically-sealed plastic bags, or plastic bags capable of being sealed.

b. Plastic containers with tight-fitting lids, such as freezer containers.

c. Cans with tight fitting lids.

3. 29 CFR 1910.134(f)(5)(ii). As recommended by 1910.134(f)(5)(ii), pack or store the respirator so that the facepiece and exhalation valves will rest in a normal position. Do not hang the respirator by its straps. This is to ensure that proper function will not be impaired by the distortion of the respirator or its straps.

4. Emergency Use Respirators. Emergency use respirators should be stored where they are easily accessible and their location clearly marked.

H. Inspection and Maintenance.

1. Requirements. 29 CFR 1910.134(b)(7) and 1910.134(f) provide basic information for the inspection and maintenance of respirators. Supplementary inspection information for the various types of respirators is included below.

2. Disposable Respirators.

a. Check for holes in the filter or damage to sorbent such as loose charcoal granules.

b. Check straps for elasticity and deterioration.

c. Check metal nose clip for rust or deterioration.

3. Air Purifying Respirators.

a. Check rubber facepiece for dirt, pliability of rubber, deterioration, and cracks, tears, or holes.

b. Check straps for breaks, tears, loss of elasticity, broken attachment snaps and proper tightness.

c. Check valves (exhalation and inhalation) for holes, warpage, cracks, and dirt.

d. Check filters, cartridges and canisters for dents, corrosion and expiration dates. Check protection afforded by canister and its limitations.

4. Atmosphere-supplying Respirators.

a. Check appropriate items listed under air purifying respirators.

b. Check hood, helmet, blouse or suit for cracks and tears, torn seams and abrasions; check integrity of headgear suspensions.

c. Check faceshields for cracks or breaks, abrasions or distortions that would interfere with vision.

d. Check abrasive blasting protective screen for integrity and condition. Check that screen fits in designated place.

e. Check air supply system for air quality (see paragraph M., page V-21), breaks or kinks in supply hoses and detachable coupling attachments, tightness of connectors, and manufacturer's recommendations concerning the proper setting of regulators and valves. Check that couplings are compatible with other couplings used at the plant.

f. When an air compressor is used to provide breathable air, check air purifying elements, carbon monoxide and/or high temperature alarm.

5. Self-contained Breathing Apparatus.

a. Check the facepiece and breathing hose for integrity as described above for atmosphere-supplying respirators.

b. Check the integrity and air or oxygen pressure for the cylinder. Also, check integrity of the regulator, harness assembly, and all straps and buckles.

c. Ensure that the regulator and warning devices (end-of-service alarm) function properly.

I. Work Area Surveillance. Although not specifically discussed in 29 CFR 1910.134(b)(8), the standard requires "appropriate surveillance." This should include identification of the contaminant, nature of the hazard, concentration at the breathing zone, and, if appropriate, biological monitoring. The Industrial Hygienist must carefully and fully document any apparent deficiencies in surveillance neccessary to a respirator program.

J. Employee Acceptance.

1. Many factors affect the employee's acceptance of respirators, including comfort, ability to breathe without objectionable effort, adequate visibility under all conditions, provisions for wearing prescription glasses (if necessary), ability to communicate, ability to perform all tasks without undue interference, and confidence in the facepiece fit.

2. Failure to consider these factors is likely to reduce cooperation of the users in promoting a satisfactory program. How well these problems are resolved can be determined by employer observation of respirator users during normal activities and employer solicitation of employee comments.

K. Medical Examination.

1. Requirements. 29 CFR 1910.134(b)(10) states that "persons should not be assigned to tasks requiring the use of respirators unless it has been determined that they are physically able to perform the work and use the equipment."

2. Health and Physical Conditions. A physician shall determine the health and physical conditions that are pertinent for an employee's ability to work while wearing a respirator. The following factors may be pertinent for this determination:

- Emphysema.

- Chronic obstructive pulmonary disease.

- Bronchial asthma.

- X-ray evidence of pneumoconiosis.
- Evidence of reduced pulmonary function.
- Coronary artery disease or cerebral blood vessel disease.
- Severe or progressive hypertension.
- Epilepsy, grand mal or petit mal.
- Anemia.
- Diabetes, insipidus or mellitus.
- Punctured eardrum.
- Pneumomediastinum.
- Communication of sinus through upper jaw to oral cavity.
- Breathing difficulty when wearing a respirator.
- Claustrophobia or anxiety when wearing a respirator.

L. Approved Respirators.

1. Requirements. 29 CFR 1910.134(b)(11) states "approved or accepted respirators shall be used when they are available." A respirator is approved as the whole unit with specific components.

2. Nullification of Approval. A respirator's approval is nullified when:

   a. Components between different types or makes of respirators are mixed.

   b. Nonapproved components are used.

   c. An approved respirator is used in atmospheric concentrations for which it is not approved.

3. MSHA/NIOSH Approval. OSHA recognizes a respirator as approved if it has been jointly approved by NIOSH and the Mine Safety and Health Administration (formerly the U.S. Bureau of Mines and Mine Enforcement Safety Administration) under the provisions of 30 CFR 11.

   a. Gas masks approved in accordance with U.S. Bureau of Mines Schedule 14F are approved until further notice. SCBAs approved by the Bureau of Mines (Schedule 13-13E) which have a low air warning device and which were purchased before June 30, 1975 are also approved. The approval for all other Bureau of Mines respirators has expired and shall not be considered valid.

   b. An "accepted" respirator is one that cannot be tested in accordance with MSHA/NIOSH criteria and which OSHA has evaluated and found acceptable. For example, MSHA/NIOSH has no criteria for testing air-purifying respirators for protection against mercury vapor. Nevertheless, OSHA may determine that in some cases certain models of such respirators are more practicable than alternatives and are effective, leading OSHA to accept their use. Air monitoring should ensure that contaminant concentrations do not exceed the protection afforded by the respirator.

   c. NIOSH has identified approved respirators in a booklet entitled "NIOSH Certified Equipment List," initially published in 1975. There have been periodic updates. Each Area Office should have an updated copy. Additional copies are available from:

      Publications Dissemination, DTS
      National Institute for Occupational Safety and Health
      U.S. Department of Health and Human Services
      4676 Columbia Parkway
      Cincinnati, Ohio 45226

      or

      National Technical Information Service
      5285 Port Royal Road
      Springfield, Virginia 22161

M. Special Procedures for Supplied-Air Respirators. Inspect the sources of the supplied air as follows:

1. Compressed breathing air shall meet the quality specification for grade D breathing air as described in "Compressed Gas Association Commodity Specification G-7.1-1966," as referenced by OSHA standard 29 CFR 1910. 134(d)(1). (CGA revised this pamphlet in 1973).

Oxygen Content(v/v)	19-23% (Atmosphere Air)
Hydrocarbon (Condensed)	5 mg/m^3
Carbon Monoxide	20 ppm
Carbon dioxide	1,000 ppm

2. Where it is necessary to check the quality of air, use appropriate methods to check for carbon monoxide, carbon dioxide, hydrocarbons and oxygen. Record the results.

3. The dewpoint of the air must be at least 10° F below the lowest recorded temperature to assure that supplied air parts do not freeze in cold weather.

N. References. For further information the following references are recommended:

1. American National Standards Institute, American National Standard: Practices for Respiratory Protection, Z88.2-1980; 1430 Broadway, New York, New York 10018.

2. American Industrial Hygiene Association, AIHA Respiratory Protection: A Manual and Guideline, 1980.

3. National Institute for Occupational Safety and Health, A Guide To Industrial Respiratory Protection, June 1976.

NOTE: The following table is reprinted from ANSI Z88.2-1980. However, some of the respirator descriptions and limitations listed may be inconsistent with NIOSH-approved regulations (30 CFR 11). The specific approval label on the device should be read to determine its limitations.

Classification and Description of Respirators by Mode of Operation

Atmosphere-Supplying Respirators

A respirable atmosphere independent of the ambient air is supplied to the wearer.

Self-Contained Breathing Apparatus (SCBA)

A supply of air, oxygen, or oxygen-generating material is carried by the wearer. Normally equipped with full facepiece, but may be equipped with a quarter-mask facepiece, half-mask facepiece, helmet, hood, or mouthpiece and nose clamp.

(1) Closed-Circuit SCBA (oxygen only, negative pressure[a] or positive pressure[b]).

(a) Compressed or liquid oxygen type. Equipped with a facepiece or mouthpiece and nose clamp. High-pressure oxygen from a gas cylinder passes through a high-pressure reducing valve and, in some designs, through a low-pressure admission valve to a breathing bag or container. Liquid oxygen is converted to low-pressure gaseous oxygen and delivered to the breathing bag. The wearer inhales from the bag, through a corrugated tube connected to a mouthpiece or facepiece and a one-way check valve. Exhaled air passes through another check valve and tube into a container of carbon-dioxide removing chemical and reenters the breathing bag. Make-up oxygen enters the bag continuously or as the bag deflates sufficiently to actuate an admission valve. A pressure-relief system is provided, and a manual bypass system and saliva trap may be provided depending upon the design.

(b) Oxygen-generating type. Equipped with a facepiece or mouthpiece and nose clamp. Water vapor in the exhaled breath reacts with chemical in the canister to release oxygen to the breathing bag. The wearer inhales from the bag through a corrugated tube and one-way check valve at the facepiece. Exhaled air passes through a second check valve/breathing tube assembly into the canister. The oxygen-release rate is governed by the volume of exhaled air. Carbon dioxide in the exhaled breath is removed by the canister fill.

(2) Open-Circuit SCBA (compressed air, compressed oxygen, liquid air, liquid oxygen). A bypass system is provided in case of regulator failure except on escape-type units.

(a) Demand type.[c] Equipped with a facepiece or mouthpiece and nose clamp. The demand valve permits oxygen or air flow only during inhalation. Exhaled breath passes to ambient atmosphere through a valve(s) in the facepiece.

(b) Pressure-demand type.[d] Equipped with a facepiece only. Positive pressure is maintained in the facepiece. The apparatus may have provision for the wearer to select the demand or pressure-demand mode of operation, in which case the demand mode should be used only when donning or removing the apparatus.

Supplied-Air Respirators

(1) Hose Mask

Equipped with a facepiece, breathing tube, rugged safety harness, and large-diameter heavy-duty nonkinking air-supply hose. The breathing tube and air-supply hose are securely attached to the harness. The facepiece is equipped with an exhalation valve. The harness has provision for attaching a safety line.

(a) Hose mask with blower. Air is supplied by a motor-driven or hand-operated blower. The wearer can continue to inhale through the hose if the blower fails. Up to 300 feet (91 meters) of hose length is permissible.

(b) Hose mask without blower. The wearer provides motivating force to pull air through the hose. The hose inlet is anchored and fitted with a funnel or like object covered with a fine mesh screen to prevent entrance of coarse particulate matter. Up to 75 feet (23 meters) of hose length is permissible.

(2) Air-Line Respirator

Respirable air is supplied through a small-diameter hose from a compressor or compressed-air cylinder(s). The hose is attached to the wearer by a belt or other suitable means and can be detached rapidly in an emergency. A flow-control valve or orifice is provided to govern the rate of air flow to the wearer. Exhaled air passes to the ambient atmosphere through a valve(s) or opening(s) in the enclosure (facepiece, helmet, hood, or suit). Up to 300 feet (91 meters) of hose length is permissible.

(a) Continuous-flow class. Equipped with a facepiece, hood, helmet, or suit. At least 115 liters (four cubic feet) of air per minute to tight-fitting facepieces and 170 liters (six cubic feet) of air per minute to loose-fitting helmets, hoods, and suits is required. Air is supplied to a suit through a system of internal tubes to the head, trunk, and extremities through valves located in appropriate parts of the suit.

(b) Demand type.[c] Equipped with a facepiece only. The demand valve permits flow of air only during inhalation.

(c) Pressure-demand type.[d] Equipped with a facepiece only. A positive pressure is maintained in the facepiece.

Combination Air-Line Respirators with Auxiliary Self-Contained Air Supply

Include an air-line respirator with an auxiliary self-contained air supply. To escape from a hazardous atmosphere in the event the primary air supply fails to operate, the wearer switches to the auxiliary self-contained air supply. Devices approved for both entry into and escape from dangerous atmospheres have a low-pressure warning alarm and contain at least a 15-minute self-contained air supply.

Air-Purifying Respirators

Ambient air, prior to being inhaled, is passed through a filter, cartridge, or canister which removes particles, vapors, gases, or a combination of these contaminants. The breathing action of the wearer operates the nonpowered type of respirator. The powered type contains a blower – stationary or carried by the wearer – which passes ambient air through an air-purifying component and then supplies purified air to the respirator-inlet covering. The nonpowered type is equipped with a facepiece or mouthpiece and nose clamp. The powered type is equipped with a facepiece, helmet, hood, or suit.

Vapor- and Gas-Removing Respirators

Equipped with cartridge(s) or canister(s) to remove a single vapor or gas (for example: chlorine gas), a single class of vapors or gases (for example: organic vapors), or a combination of two or more classes of vapors or gases (for example: organic vapors and acidic gases) from air.

Particulate-Removing Respirators

Equipped with filter(s) to remove a single type of particulate matter (for example: dust) or a combination of two or more types of particulate matter (for example: dust and fume) from air. Filter may be a replaceable part or a permanent part of the respirator. Filter may be of the single-use or the reusable type.

Combination Particulate- and Vapor- and Gas-Removing Respirators

Equipped with cartridge(s) or canister(s) to remove particulate matter, vapors, and gases from air. The filter may be a permanent part or a replaceable part of a cartridge or canister.

Combination Atmosphere-Supplying and Air-Purifying Respirators

Provide the wearer with the option of using either of two different modes of operation: (1) an atmosphere-supplying respirator with an auxiliary air-purifying attachment which provides protection in the event the air supply fails or (2) an air-purifying respirator with an auxiliary self-contained air supply which is used when the atmosphere may exceed safe conditions for use of an air-purifying respirator.

[a]Device produces negative pressure in respiratory-inlet covering during inhalation.

[b]Device produces positive pressure in respiratory-inlet covering during both inhalation and exhalation.

[c]Equipped with a demand valve that is activated on initiation of inhalation and permits the flow of breathing atmosphere to the facepiece. On exhalation, pressure in the facepiece becomes positive and the demand valve is deactivated.

[d]A positive pressure is maintained in the facepiece by a spring-loaded or balanced regulator and exhalation valve.

Capabilities and Limitations of Respirators

Atmosphere-Supplying Respirators

(See 5.2 for specifications on respirable atmospheres.)

Atmosphere-supplying respirators provide protection against oxygen deficiency and toxic atmospheres. The breathing atmosphere is independent of ambient atmospheric conditions.

General limitations: Except for some air-line suits, no protection is provided against skin irritation by materials such as ammonia and hydrogen chloride, or against sorption of materials such as hydrogen cyanide, tritium, or organic phosphate pesticides through the skin. Facepieces present special problems to individuals required to wear prescription lenses (see 9.1). Use of atmosphere-supplying respirators in atmospheres immediately dangerous to life or health is limited to specific devices under specified conditions (see Table 5 and 9.3 and 9.4).

Self-Contained Breathing Apparatus (SCBA)

The wearer carries his own breathing atmosphere.

Limitations: The period over which the device will provide protection is limited by the amount of air or oxygen in the apparatus, the ambient atmospheric pressure (service life of open-circuit devices is cut in half by a doubling of the atmospheric pressure), and the type of work being performed. Some SCBA devices have a short service life (less than 15 minutes) and are suitable only for escape (self-rescue) from an irrespirable atmosphere.

Chief limitations of SCBA devices are their weight or bulk, or both, limited service life, and the training required for their maintenance and safe use.

(1) Closed-Circuit SCBA. The closed-circuit operation conserves oxygen and permits longer service life at reduced weight. The negative-pressure type produces a negative pressure in the respiratory-inlet covering during inhalation, and this may permit inward leakage of contaminants; whereas the positive-pressure type always maintains a positive pressure in the respiratory-inlet covering and is less apt to permit inward leakage of contaminants.

(2) Open-Circuit SCBA. The demand type produces a negative pressure in the respiratory-inlet covering during inhalation, whereas the pressure-demand type maintains a positive pressure in the respiratory-inlet covering during inhalation and is less apt to permit inward leakage of contaminants.

Supplied-Air Respirators

The respirable air supply is not limited to the quantity the individual can carry, and the devices are lightweight and simple.

Limitations: Limited to use in atmospheres from which the wearer can escape unharmed without the aid of the respirator.

The wearer is restricted in movement by the hose and must return to a respirable atmosphere by retracing his route of entry. The hose is subject to being severed or pinched off.

(1) Hose Mask. The hose inlet or blower must be located and secured in a respirable atmosphere.

(a) Hose mask with blower. If the blower fails, the unit still provides protection, although a negative pressure exists in the facepiece during inhalation.

(b) Hose mask without blower. Maximum hose length may restrict application of device.

(2) Air-Line Respirator (Continuous Flow, Demand, and Pressure-Demand Types)

The demand type produces a negative pressure in the facepiece on inhalation, whereas continuous-flow and pressure-demand types maintain a positive pressure in the respiratory-inlet covering and are less apt to permit inward leakage of contaminants.

Air-line suits may protect against atmospheres that irritate the skin or that may be absorbed through the unbroken skin.

Limitations: Air-line respirators provide no protection if the air supply fails. Some contaminants, such as tritium, may penetrate the material of an air-line suit and limit its effectiveness.

Other contaminants, such as fluorine, may react chemically with the material of an air-line suit and damage it.

Combination Airline Respirators with Auxiliary SC Air Supply

The auxiliary self-contained air supply on this type of device allows the wearer to escape from a dangerous atmosphere. This device with auxiliary self-contained air supply is approved for escape and may be used for entry when it contains at least a 15-minute auxiliary self-contained air supply. (See Table 5).

Air-Purifying Respirators

General limitations: Air-purifying respirators do not protect against oxygen-deficient atmospheres nor against skin irritations by, or sorption through the skin of, airborne contaminants.

The maximum contaminant concentration against which an air-purifying respirator will protect is determined by the design efficiency and capacity of the cartridge, canister, or filter and the facepiece-to-face seal on the user. For gases and vapors, the maximum concentration for which the air-purifying element is designed is specified by the manufacturer or is listed on labels of cartridges and canisters.

Nonpowered air-purifying respirators will not provide the maximum design protection specified unless the facepiece or mouthpiece/nose clamp is carefully fitted to the wearer's face to prevent inward leakage (see 7.4). The time period over which protection is provided is dependent on canister, cartridge, or filter type; concentration of contaminant; humidity levels in the ambient atmosphere; and the wearer's respiratory rate.

The proper type of canister, cartridge, or filter must be selected for the particular atmosphere and conditions. Nonpowered air-purifying respirators may cause discomfort due to a noticeable resistance to inhalation. This problem is minimized in powered respirators. Respirator facepieces present special problems to individuals required to wear prescription lenses (see 9.1). These devices do have the advantage of being small, light, and simple in operation.

Use of air-purifying respirators in atmospheres immediately dangerous to life or health is limited to specific devices under specified conditions (see Table 5 and 9.3 and 9.4).

Vapor- and Gas-Removing Respirators

Limitations: No protection is provided against particulate contaminants. A rise in canister or cartridge temperature indicates that a gas or vapor is being removed from the inspired air.

An uncomfortably high temperature indicates a high concentration of gas or vapor and requires an immediate return to fresh air.

Use should be avoided in atmospheres where the contaminant(s) lacks sufficient warning properties (that is: odor, taste, or irritation at a concentration in air at or above the permissible exposure limit.) (Vapor- and gas-removing respirators are not approved for contaminants that lack adequate warning properties.)

Not for use in atmospheres immediately dangerous to life or health unless the device is a powered-type respirator with escape provisions (see Table 5).

(1) Full Facepiece Respirator. Provides protection against eye irritation in addition to respiratory protection.

(2) Quarter-Mask and Half-Mask Facepiece Respirator. A fabric covering (facelet) available from some manufacturers shall not be used.

(3) Mouthpiece Respirator. Shall be used only for escape applications. Mouth breathing prevents detection of contaminant by odor. Nose clamp must be securely in place to prevent nasal breathing.

A small lightweight device that can be donned quickly.

Particulate-Removing Respirators

Limitations: Protection against nonvolatile particles only. No protection against gases and vapors.

Not for use in atmospheres immediately dangerous to life or health unless the device is a powered-type respirator with escape provisions (see Table 5).

(1) Full Facepiece Respirator. Provides protection against eye irritation in addition to respiratory protection.

(2) Quarter-Mask and Half-Mask Facepiece Respirator. A fabric covering (facelet) available from some manufacturers shall not be used unless approved for use with respirator.

(3) Mouthpiece Respirator. Shall be used only for escape applications. Mouth breathing prevents detection of contaminant by odor. Nose clamp must be securely in place to prevent nasal breathing.

A small, lightweight device that can be donned quickly.

Combination Particulate- and Vapor- and Gas-Removing Respirators

The advantages and disadvantages of the component sections of the combination respirator as described above apply.

Combination Atmosphere-Supplying and Air-Purifying Respirators

The advantages and disadvantages, expressed above, of the mode of operation being used will govern. The mode with the greater limitations (air-purifying mode) will mainly determine the overall capabilities and limitations of the respirator, since the wearer may for some reason fail to change the mode of operation even though conditions would require such a change.

Respirator Protection Factors[a]

Type of Respirator	Permitted for Use in Oxygen-Deficient Atmosphere	Permitted for Use in Immediately-Dangerous-to-Life-or-Health Atmosphere[f]	Respirator Protection Factor	
			Qualitative Test	Quantitative Test
Particulate-filter, quarter-mask or half-mask facepiece[b,c]	No	No	10	As measured on each person with maximum of 100.
Vapor- or gas-removing, quarter-mask or half-mask facepiece[c]	No	No	10, or maximum use limit of cartridge or canister for vapor or gas, whichever is less.	As measured on each person with maximum of 100, or maximum use limit of cartridge or canister for vapor or gas[i,j], whichever is less.
Combination particulate-filter and vapor- or gas-removing, quarter-mask or half-mask facepiece[b,c]	No	No	10, or maximum use limit of cartridge or canister for vapor or gas, whichever is less.	As measured on each person with maximum of 100, or maximum use limit of cartridge or canister for vapor or gas[i,j], whichever is less.
Particulate-filter, full facepiece[b]	No	No	100	As measured on each person with maximum of 100 if dust, fume, or mist filter is used, or maximum of 1000 if high-efficiency filter is used.
Vapor- or gas-removing, full facepiece	No	No	100, or maximum use limit of cartridge or canister for vapor or gas, whichever is less.	As measured on each person with maximum of 1000, or maximum use limit of cartridge or canister for vapor or gas[i,j], whichever is less.
Combination particulate-filter and vapor- or gas-removing, full facepiece[b]	No	No	100, or maximum use limit of cartridge or canister for vapor or gas, whichever is less.	As measured on each person with maximum of 100 if dust, fume, or mist filter is used and maximum of 1000 if high-efficiency filter is used, or maximum use limit of cartridge or canister for vapor or gas[i,j], whichever is less.
Powered particulate-filter, any respiratory-inlet covering[b,c,d]	No	No (yes, if escape provisions are provided[d])	N/A No tests are required due to positive-pressure operation of respirator. The maximum protection factor is 100 if dust, fume, or mist filter is used and 3000 if high-efficiency filter is used.	N/A
Powered vapor- or gas-removing, any respiratory-inlet covering[c,d]	No	No (yes, if escape provisions are provided[d])	N/A No tests are required due to positive pressure operation of respirator. The maximum protection factor is 3000, or maximum use limit of cartridge or canister for vapor or gas[i,j], whichever is less.	N/A
Powered combination particulate-filter and vapor- or gas-removing, any respiratory-inlet covering[b,c,d]	No	No (yes, if escape provisions are provided[d])	N/A No tests are required due to positive-pressure operation of respirator. The maximum protection factor is 100 if dust, fume, or mist filter is used and 3000 if high-efficiency filter is used, or maximum use limit of cartridge or canister for vapor or gas[i,j], whichever is less.	N/A
Air-line, demand, quarter-mask or half-mask facepiece, with or without escape provisions[c,e]	Yes[f]	No	10	As measured on each person, but limited to the use of the respirator in concentrations of contaminants below the immediately-dangerous-to-life-or-health (IDLH) values.
Air-line, demand, full facepiece, with or without escape provisions[e]	Yes[f]	No	100	As measured on each person, but limited to the use of the respirator in concentrations of contaminants below the immediately-dangerous-to-life-or-health (IDLH) values.
Air-line, continuous-flow or pressure-demand type, any facepiece, without escape provisions[c]	Yes[f]	No	N/A No tests are required due to positive-pressure operation of respirator. The protection factor provided by the respirator is limited to use of the respirator in concentrations of contaminants below the immediately-dangerous-to-life-or-health (IDLH) values.	N/A
Air-line, continuous-flow or pressure-demand type, any facepiece, with escape provisions[c,e]	Yes[g]	Yes	N/A No tests are required due to positive-pressure operation of respirator. The maximum protection factor is 10 000 plus.[h]	N/A
Air-line, continuous flow, helmet, hood, or suit, without escape provisions	Yes[f]	No	N/A No tests are required due to positive-pressure operation of respirator. The protection factor provided by the respirator is limited to the use of the respirator in concentrations of contaminants below the immediately-dangerous-to-life-or-health (IDLH) values.	N/A
Air-line, continuous-flow, helmet, hood, or suit, with escape provisions[e]	Yes[g]	Yes	N/A No tests are required due to positive-pressure operation of respirator. The maximum protection factor is 10 000 plus.[h]	N/A
Hose mask, with or without blower, full facepiece	Yes[f]	No	10	As measured on each person, but limited to the use of the respirator in concentrations of contaminants below the immediately-dangerous-to-life-or-health (IDLH) values.
Self-contained breathing apparatus, demand-type open-circuit or negative-pressure-type closed-circuit, quarter-mask or half-mask facepiece[c]	Yes[f]	No	10	As measured on each person, but limited to the use of the respirator in concentrations of contaminants below the immediately-dangerous-to-life-or-health (IDLH) values.
Self-contained breathing apparatus, demand-type open-circuit or negative-pressure-type closed-circuit, full facepiece or mouthpiece/nose clamp[c]	Yes[f] (Yes[g], if respirator is used for mine rescue and mine recovery operations.)	No (yes, if respirator is used for mine rescue and mine recovery operations.)	100	As measured on each person, but limited to the use of the respirator in concentrations of contaminants below the immediately-dangerous-to-life-or-health (IDLH) values, except when the respirator is used for mine rescue and mine recovery operations.
Self-contained breathing apparatus, pressure-demand-type open-circuit or positive-pressure-type closed-circuit, quarter-mask or half-mask facepiece, full facepiece, or mouthpiece/nose clamp[c]	Yes[g]	Yes	N/A No tests are required due to positive-pressure operation of respirator. The maximum protection factor is 10 000 plus.[h]	N/A
Combination respirators not listed.	The type and mode of operation having the lowest respirator protection factor shall be applied to the combination respirator.			

N/A means not applicable since a respirator-fitting test is not carried out.

[a]A respirator protection factor is a measure of the degree of protection provided by a respirator to a respirator wearer. Multiplying the permissible time-weighted average concentration or the permissible ceiling concentration, whichever is applicable, for a toxic substance, or the maximum permissible airborne concentration for a radionuclide, by a protection factor assigned to a respirator gives the maximum concentration of the hazardous substance for which the respirator can be used. Limitations of filters, cartridges, and canisters used in air-purifying respirators shall be considered in determining protection factors.

[b]When the respirator is used for protection against airborne particulate matter having a permissible time-weighted average concentration less than 0.05 milligram particulate matter per cubic meter of air or less than 2 million particles per cubic foot of air, or for protection against airborne radionuclide particulate matter, the respirator shall be equipped with a high-efficiency filter(s).

[c]If the air contaminant causes eye irritation, the wearer of a respirator equipped with a quarter-mask or half-mask facepiece or mouthpiece and nose clamp shall be permitted to use a protective goggle or to use a respirator equipped with a full facepiece.

[d]If the powered air-purifying respirator is equipped with a facepiece, the escape provision means that the wearer is able to breathe through the filter, cartridge, or canister and through the pump. If the powered air-purifying respirator is equipped with a helmet, hood, or suit, the escape provision shall be an auxiliary self-contained supply of respirable air.

[e]The escape provision shall be an auxiliary self-contained supply of respirable air.

[f]For definition of "oxygen deficiency not immediately dangerous to life or health" see Section 2.

[g]For definition of "oxygen deficiency immediately dangerous to life or health" see Section 2 and A10.

[h]The protection factor measurement exceeds the limit of sensitivity of the test apparatus. Therefore, the respirator has been classified for use in atmospheres having unknown concentrations of contaminants.

[i]The service life of a vapor- or gas-removing cartridge or canister depends on the specific vapor or gas, the concentration of the vapor or gas in air, the temperature and humidity of the air, the type and quantity of the sorbent in the cartridge or canister, and the activity of the respirator wearer. Cartridges and canisters may provide only very short service lives for certain vapors and gases. Vapor/gas service life testing is recommended to ensure that cartridges and canisters provide adequate service lives. Reference should be made to published reports which give vapor/gas life data for cartridges and canisters.

[j]Vapor- and gas-removing respirators are not approved for contaminants that lack adequate warning properties of odor, irritation, or taste at concentrations in air at or above the permissible exposure limits.

NOTE: Respirator protection factors for air-purifying-type respirators equipped with a mouthpiece/nose clamp form of respiratory-inlet covering are not given, since such respirators are approved only for escape purposes.

Addendum C

OSHA Statement on the Use of U.S. Bureau of Mines Approved Gas Mask Canisters

The following notice takes precedence over NIOSH's "grandfather" clause and contains information that will be useful to employers using these devices.

U.S. Department of Labor

Occupational Safety and Health Administration
Washington, D.C. 20210

Reply to the Attention of:

NOV 15 1985

MEMORANDUM FOR REGIONAL ADMINISTRATORS

THRU: JOHN B. MILES
Director
Directorate of Field Operations

FROM: STEPHEN J. MALLINGER
Acting Director
Directorate of Technical Support

SUBJECT: Use of Bureau of Mines Approved Gas Mask Canisters

We have received several inquiries concerning the use of Bureau of Mines approved gas mask canisters. These canisters were approved under the Bureau of Mines Schedule 14F for protection against many highly toxic substances such as hydrogen sulfide, hydrogen cyanide and phosphine. All these canisters were approved for concentrations far above their respective immediately dangerous to life or health (IDLH) values and none of these compounds has adequate odor warning properties for the respirator wearer to detect excessive facepiece leakage or sorbent breakthrough.

Although the Mine Safety and Health Administration (MSHA) and the National Institute for Occupational Safety and Health (NIOSH) have extended the expiration date for the Bureau of Mines approved gas mask canisters, NIOSH indicated that they could not conduct quality control testing on these canisters to assure that the performance meets the certification requirements.

In view of the above facts, it is concluded that the Bureau of Mines approved gas mask canisters for protection against hydrogen sulfide, hydrogen cyanide and phosphine may not provide adequate margin of safety to the respirator wearers. Their use for other than emergency escape is not acceptable.

Addendum D

Qualitative Fit Testing Procedures

The following procedures are found in the OSHA Lead Standard, 29 CFR 1910.1025. These procedures are the only allowable qualitative fit test protocols permissible for compliance with paragraph (f)(3)(ii) of this standard. While these exact protocols are not required for other substances, it is good practice to use them. These methods represent validated qualitative fit test protocols for half mask respirators. In addition, OSHA has published very similar protocols in the current asbestos standards, 29 CFR 1910.1001 and 29 CFR 1926.58, as well as in several proposed standards.

Appendix B1. Qualitative Fit Test Procedures

[Note: The following procedures are found in the OSHA Lead Standard (29 CFR 1910.1025) Appendix D.]

This appendix specifies the only allowable qualitative fit test protocols permissible for compliance with paragraph (f)(3)(ii).

I. ISOAMYL ACETATE PROTOCOL

A. Odor Threshold Screening

1. Three 1-liter glass jars with metal lids (e.g. Mason or Bell jars) are required.

2. Odor-free water (e.g. distilled or spring water) at approximately 25°C shall be used for the solution.

3. The isoamyl acetate (IAA)(also known as isopentyl acetate) stock solution is prepared by adding 1 cc of pure IAA to 800 cc of odor free water in a 1-liter jar and shaking for 30 seconds. The solution shall be prepared new at least weekly.

4. The screening test shall be conducted in a room separate from the room used for actual fit testing. The two rooms shall be well ventilated but may not be connected to the same recirculating ventilation system.

5. The odor test solution is prepared in a second jar by placing .4 cc of the stock solution into 500 cc of odor free water using a clean dropper or pipette. Shake for 30 seconds and allow to stand for two to three minutes so that the IAA concentration above the liquid may reach equilibrium. This solution may be used for only one day.

6. A test blank is prepared in a third jar by adding 500 cc of odor free water.

7. The odor test and test blank jars shall be labeled 1 and 2 for jar identification. If the labels are put on the lids they can be periodically *dried off* and switched to avoid people thinking the same jar always has the IAA.

8. The following instructions shall be typed on a card and placed on the table in front of the two test jars (i.e. 1 and 2);

   "The purpose of this test is to determine if you can smell banana oil at a low concentration. The two bottles in front of you contain water. One of these bottles also contains a small amount of banana oil. Be sure the covers are on tight, then shake each bottle for two seconds. Unscrew the lid of each bottle, one at a time, and sniff at the mouth of the bottle. Indicate to the test conductor which bottle contains banana oil."

9. The mixtures used in the IAA odor detection test shall be prepared in an area separate from where the test is performed, in order to prevent olfactory fatique in the subject.

10. If the test subject is unable to correctly identify the jar containing the odor test solution, the IAA QLFT may not be used.

11. If the test subject correctly identifies the jar containing the odor test solution he may proceed to respirator selection and fit testing.

B. Respirator Selection

1. The test subject shall be allowed to select the most comfortable respirator from a large array of various sizes and manufacturers that includes at least three sizes of elastomeric half facepieces and units of at least two manufacturers.

2. The selection process shall be conducted in a room separate from the fit-test chamber to prevent odor fatigue. Prior to the selection process, the test subject shall be shown how to put on a respirator, how it should be positioned on the face, how to set strap tension and how to assess a "comfortable" respirator. A mirror shall be available to assist the subject in evaluating the fit and positioning of the respirator. This may not constitute his formal training on respirator use, only a review.

3. The test subject should understand that he is being asked to select the respirator which provides the most comfortable fit for him. Each respirator represents a different size and shape and, if fit properly, will provide adequate protection.

4. The test subject holds each facepiece up to his face and eliminates those which are obviously not giving a comfortable fit. Normally, selection will begin with a half-mask and if a fit cannot be found here, the subject will be asked to go to the full facepiece respirators. (A small percentage of users will not be able to wear any half-mask.)

5. The more comfortable facepieces are recorded; the most comfortable mask is donned and *worn at least five minutes* to assess comfort. Assistance in assessing comfort can be given by discussing the points in #6 below. If the test subject is not familiar with using a particular respirator, he shall be directed to don the mask several times and to adjust the straps each time, so that he becomes adept at setting proper tension on the straps.

6. Assessment of comfort shall include reviewing the following points with the test subject:

   - Chin properly placed.
   - Positioning of mask on nose.
   - Strap tension.
   - Fit across nose bridge.
   - Room for safety glasses.
   - Distance from nose to chin.
   - Room to talk.
   - Tendency to slip.
   - Cheeks filled out.
   - Self-observation in mirror.
   - Adequate time for assessment.

7. The test subject shall conduct the conventional negative and positive-pressure fit checks (e.g. see ANSI Z88.2-1980). Before conducting the negative- or positive-pressure checks, the subject shall be told to "seat" his mask by rapidly moving the head side-to-side and up and down, taking a few deep breaths.

8. The test subject is now ready for fit testing.

9. After passing the fit test, the test subject shall be questioned again regarding the comfort of the respirator. If it has become uncomfortable, another model of respirator shall be tried.

10. The employee shall be given the opportunity to select a different facepiece and be retested if during the first two weeks of on-the-job wear the chosen facepiece becomes unacceptably uncomfortable.

C. Fit test.

1. The fit test chamber shall be substantially similar to a clear 55 gallon drum liner suspended inverted over a 2 foot diameter frame, so that the top of chamber is about 6 inches above the test subject's head. The inside top center of the chamber shall have a small hook attached.

2. Each respirator used for the fitting and fit testing shall be equipped with organic vapor cartridges or offer protection against organic vapors. The cartridges or masks shall be changed at least weekly.

3. After selecting, donning, and properly adjusting a respirator himself, the test subject shall wear it to the fit testing room. This room shall be separate from the room used for odor threshold screening and respirator selection, and shall be well ventilated, as by an exhaust fan or lab hook, to prevent general room contamination.

4. A copy of the following test exercises and rainbow (or equally effective) passage shall be taped to the inside of the test chamber:

   Test Exercises

   i. Normal breathing.

   ii. Deep breathing. Be certain breaths are *deep* and *regular.*

   iii. Turning head from side-to-side. Be certain movement is complete. Alert the test subject not to bump the respirator on the shoulders. Have the test subject inhale when his head is at either side.

   iv. Nodding head up-and-down. Be certain motions are complete and made about every second. Alert the test subject not to bump the respirator on the chest. Have the test subject inhale when his head is in the fully up position.

   v. Talking. Talk aloud and slowly for several minutes. The following paragraph is called the Rainbow Passage. Reading it will result in a wide range of facial movements, and thus be useful to satisfy this requirement. Alternative passages which serve the same purpose may also be used.

   Rainbow Passage

   When the sunlight strikes raindrops in the air, they act like a prism and form a rainbow. The rainbow is a divison of white light into many beautiful colors. These take the shape of a long round arch, with its path high above, and its two ends apparently beyond the horizon. There is, according to legend, a boiling pot of gold at one end. People look, but no one ever finds it. When a man looks for something beyond reach, his friends say he is looking for the pot of gold at the end of the rainbow.

   vi. Normal breathing.

5. Each test subject shall wear his respirator for at least 10 minutes before starting the fit test.

6. Upon entering the test chamber, the test subject shall be given a 6 inch by 5 inch piece of paper towel or other porous absorbent single ply material, folded in half and wetted with three-quarters of one cc of pure IAA. The test subject shall hang the wet towel on the hook at the top of the chamber.

7. Allow two minutes for the IAA test concentration to be reached before starting the fit-test exercises. This would be an appropriate time to talk with the test subject, to explain the fit test, the importance of his cooperation, the purpose for the head exercises, or to demonstrate some of the exercises.

8. Each exercise described in No. 4 above shall be performed for at least one minute.

9. If at any time during the test, the subject detects the banana-like odor of IAA, he shall quickly exit from the test chamber and leave the test area to avoid olfactory fatigue.

10. Upon returning to the selection room, the subject shall remove the respirator, repeat the odor sensitivity test, select and put on another respirator, return to the test chamber, etc. The process continues until a respirator that fits well has been found. Should the odor sensitivity test be failed, the subject shall wait about 5 minutes before retesting. Odor sensitivity will usually have returned by this time.

11. If a person cannot be fitted with the selection of half-mask respirators, include full facepiece models in the selection process. When a respirator is found that passes the test, its efficiency shall be demonstrated for the subject by having him break the face seal and take a breath before exiting the chamber.

12. When the test subject leaves the chamber he shall remove the saturated towel, returning it to the test conductor. To keep the area from becoming contaminated, the used towels shall be kept in a self-sealing bag. There is no significant IAA concentration buildup in the test chamber from subsequent tests.

13. Persons who have successfully passed this fit test may be assigned the use of the tested respirator in atmospheres with up to 10 times the PEL of airborne lead. In other words this IAA protocol may be used to assign a protection factor no higher than 10.

II. SACCHARIN SOLUTION AEROSOL PROTOCOL

A. Taste threshold screening.

1. Threshold screening as well as fit testing employees shall use an enclosure about the head and shoulders that is approximately 12 inches in diameter by 14 inches tall with at least the front portion clear and that allows free movement of the head when a respirator is worn. An enclosure substantially similar to the 3M hood assembly of part #FT 14 and FT 15 combined is adequate.

2. The test enclosure shall have a three-quarter inch hole in front of the test subject's nose and mouth area to accommodate the nebulizer nozzle.

3. The entire screening and testing procedure shall be explained to the test subject prior to the conduct of the screening test.

4. The test subject shall don the test enclosure. For the threshold screening test, he shall breathe through his open mouth with tongue extended.

5. Using a DeVilbiss Model 40 Inhalation Medication Nebulizer, the test conductor shall spray the threshold check solution into the enclosure. This nebulizer shall be clearly marked to distinguish it from the fit test solution nebulizer or equivalent.

6. The threshold check solution consists of 0.83 grams of sodium saccharin, USP in water. It can be prepared by putting 1 cc of the test solution (see C6 below) in 100 cc of water.

7. To produce the aerosol, the nebulizer bulb is firmly squeezed so that it collapses completely then released and allowed to fully expand.

8. Ten squeezes are repeated rapidly and then the test subject is asked whether the saccharin can be tasted.

9. If the first response is negative, ten more squeezes are repeated rapidly and the test subject is again asked whether the saccharin is tasted.

10. If the second response is negative ten more squeezes are repeated rapidly and the test subject is again asked whether the saccharin is tasted.

11. The test conductor will take note of the number of squeezes required to ellicit a taste response.

12. If the saccharin is not tasted after 30 squeezes (Step 9), the test subject may not perform the saccharin fit test.

13. If a taste response is elicited, the test subject shall be asked to take note of the taste for reference in the fit test.

14. Correct use of the nebulizer means that approximately 1 cc of liquid is used at a time in the nebulizer body.

15. The nebulizer shall be thoroughly rinsed in water, shaken dry, and refilled at least each morning and afternoon or at least every four hours.

B. Respirator selection.

Respirators shall be selected as described in section IB above, except that each respirator shall be equipped with a particular filter cartridge.

C. Fit test.

1. The fit test uses the same enclosure described in B1 and B2 above.

2. Each test subject shall wear his respirator for at least 10 minutes before starting the fit test.

3. The test subject shall don the enclosure while wearing the respirator selected in section A above. This respirator shall be properly adjusted and equipped with a particular filter cartridge.

4. The test subject may not eat, drink (except plain water), or chew gum for 15 minutes before the test.

5. A second DeVilbiss Model 40 Inhalation Medication Nebulizer is used to spray the fit test solution into the enclosure. This nebulizer shall be clearly marked to distinguish it from the screening test solution nebulizer or equivalent.

6. The fit test solution is prepared by adding 83 grams of sodium saccharin to 100 cc of warm water.

7. As before, the test subject shall breathe through the open mouth with tongue extended.

8. The nebulizer is inserted into the hole in the front of the enclosure and the fit test solution is sprayed into the enclosure using the same technique as for the taste threshold screening and the same number of squeezes required to elicit a taste response in the screening. (See B10 above).

9. After generation of the aerosol the test subject shall be instructed to perform the following exercises for one minute each.

   i. Normal breathing.

   ii. Deep breathing. Be certain breaths are *deep* and *regular*.

   iii. Turning head from side-to-side. Be certain movement is complete. Alert the test subject not to bump the respirator on the shoulders. Have the test subject inhale when his head is at either side.

   iv. Nodding head up-and-down. Be certain motions are complete and made about every second. Alert the test subject not to bump the respirator on the chest. Have the test subject inhale when his head is in the fully up position.

   v. Talking. Talk aloud and slowly for several minutes. The folowing paragraph is called the Rainbow Passage. Reading it will result in a wide range of facial movements, and thus be useful to satisfy this requirement. Alternative passages which serve the same purpose may also be used.

   Rainbow Passage

   When the sunlight strikes raindrops in the air, they act like a prism and form a rainbow. The rainbow is a divison of white light into many beautiful colors. These take the shape of a long round arch, with its path high above, and its two ends apparently beyond the horizon. There is, according to legend, a boiling pot of gold at one end. People look, but no one ever finds it. When a man looks for something beyond reach, his friends say he is looking for the pot of gold at the end of the rainbow.

10. Every 30 seconds, the aerosol concentration shall be replenished using one-half the number of squeezes as initially (C8).

11. The test subject shall so indicate to the test conductor if at any time during the fit test the taste of saccharin is detected.

12. If the saccharin is detected the fit is deemed unsatisfactory and a different respirator shall be tried.

13. Successful completion of the test protocol shall allow the use of the tested respirator in contaminated atmospheres up to 10 times the PEL. In other words this protocol may be used to assign protection factors no higher than ten.

III. IRRITANT FUME PROTOCOL

A. Respirator selection.

Respirators shall be selected as described in section IB above, except that each respirator shall be equipped with high efficiency cartridges.

B. Fit test.

1. The test subject shall be allowed to smell a weak concentration of the irritant smoke to familiarize him with its characteristic odor.

2. The test subject shall properly don the respirator selected as above, and wear it for at least 10 minutes before starting the fit test.

3. The test conductor shall review this protocol with the test subject before testing.

4. The test subject shall perform the conventional positive pressure and negative pressure fit checks. Failure of either check shall be cause to select an alternate respirator.

5. Break both ends of a ventilation smoke tube containing stannic oxychloride, such as the MSA part No. 5645, or equivalent. Attach a short length of tubing to one end of the smoke tube. Attach the other end of the smoke tube to a low pressure air pump set to deliver 200 milliliters per minute.

6. Advise the test subject that the smoke can be irritating to the eyes and instruct him to keep his eyes closed while the test is performed.

7. The test conductor shall direct the stream of irritant smoke from the tube towards the faceseal area of the test subject. He shall begin at least 12 inches from the facepiece and gradually move to within one inch, moving around the whole perimeter of the mask.

8. The following exercises shall be performed while the respirator seal is being challenged by the smoke. Each shall be performed for one minute.

   i. Normal breathing.

   ii. Deep breathing. Be certain breaths are *deep* and *regular*.

   iii. Turning head from side-to-side. Be certain movement is complete. Alert the test subject not to bump the respirator on the shoulders. Have test subject inhale when his head is at either side.

   iv. Nodding head up-and-down. Be certain motions are complete. Alert the test subject not to bump the respirator on the chest. Have the test subject inhale when his head is in the fully up position.

   v. Talking—slowly and distinctly, count backwards from 100.

   vi. Normal breathing.

9. If the irritant smoke produces an involuntary reaction (cough) by the test subject, the test conductor shall stop the test. In this case the test respirator is rejected and another respirator shall be selected.

10. Each test subject passing the smoke test without evidence of a response shall be given a sensitivity check of the smoke from the same tube to determine whether he reacts to the smoke. Failure to evoke a response shall void the fit test.

11. Steps B4, B7, B8 of this protocol shall be performed in a location with exhaust ventilation sufficient to prevent general contamination of the testing area by the test agents (IAA, irritant smoke).

12. Respirators successfully tested by the protocol may be used in contaminated atmospheres up to ten times the PEL. In other words this protocol may be used to assign protection factors not exceeding ten. **[appendix D amended at 48 F.R. 9641, 3/8/83.]**

PART VI

Industrial Hygiene Programs

24

The Industrial Hygienist

by Clyde M. Berry
Revised by
Barbara A. Plog, MPH, CIH, CSP

THIS CHAPTER WILL DISCUSS the background and definition of industrial hygiene, its interrelationships with other occupational health groups, the functions and characterization of industrial hygienists, personnel needs, and training programs.

BACKGROUND

The industrial hygienist is trained to recognize, evaluate, and control health hazards in the industrial environment—particularly the chemical stresses, physical agents, biological and ergonomic hazards which can induce injurious effects in humans. The goals of industrial hygiene are reached through competence in a variety of scientific fields—principally chemistry, engineering, physics, biology, and medicine. Trained initially in one of these fields, most industrial hygienists have acquired, by experience and postgraduate study, a knowledge of the other allied disciplines.

Medical diagnoses and remedies and surgical repairs were early needs that were met by the ship doctor, the military surgeon, and the railroad surgeon for specific situations and for specific organizations. From these early beginnings the specialty of industrial or occupational medicine as we know it today has evolved.

Except for instances of rapid acute health effects, the relationship of occupational stress to disease is different. Cause and effect may be widely separated in time, as in industrially induced cancer. Medical diagnoses may be made with certainty only if there is a record of industrial exposure.

In the past, assistance with industrial hygiene needs was sought from individuals in the safety area, but in most instances the level of technical expertise required to perform this type of task could not be found in a single individual.

Interrelationship among occupational health groups

The accompanying schematic diagram (Figure 24-1) illustrates some of the basic relationships in industrial organizations. No area, no person, no profession, no specialized discipline fails to interrelate with another function or responsibility. Rarely would a single individual possess expertise to perform all the functions listed in the accompanying figure.

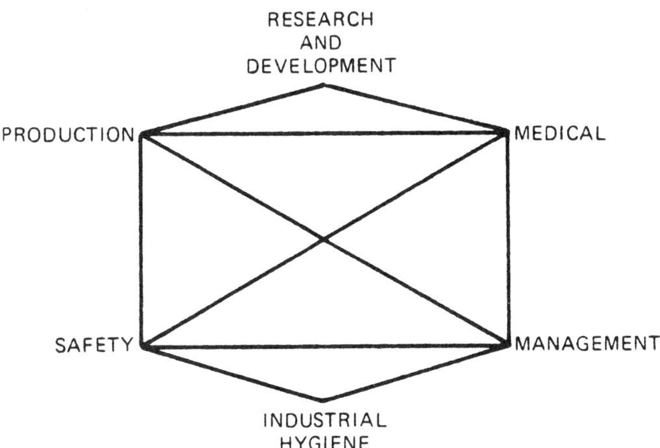

Figure 24-1. The basic relationships of industrial hygiene with other disciplines involved with occupational health hazards are shown in this diagram.

The best interest of the total production enterprise or economic unit can be best served by cooperation in the overlapping areas—not by compartmentalization. If the latter occurs, then conflict and competition are likely. Even worse is the need or task that is ignored because "it isn't my job."

Definition of industrial hygiene

As stated in Chapter 1, the American Industrial Hygiene Association (AIHA) defines industrial hygiene as "that science and art devoted to the recognition, evaluation, and control of those environmental factors or stresses, arising in and from the workplace, which may cause sickness, impaired health and well being, or significant discomfort and inefficiency among workers or among citizens of the community." Perhaps "anticipation" should be added to recognition, evaluation, and control.

An industrial hygienist is a person having a college or university degree or degrees in engineering, chemistry, physics, medicine, or related physical and biological sciences who, by virtue of special studies and training, has acquired competence in industrial hygiene. Such special studies and training must have been sufficient in all of the above cognate sciences to provide the abilities: (a) to recognize the environmental factors and to understand their effect on humans and their well-being; (b) to evaluate, on the basis of experience and with the aid of quantitative measurement techniques, the magnitude of these stresses in terms of ability to impair human health and well-being; and (c) to prescribe methods to eliminate, control, or reduce such stresses when necessary to alleviate their effects.

A person who excels in performing all of these activities would be the ideal. But the capabilities to excel in every field of specialization cannot exist within a single individual. Specialists emerge in the fields of toxicology, epidemiology, chemistry, ergonomics, acoustics, ventilation engineering, and statistics—to name only a few.

Physicians, nurses, and safety professionals can move part or all of the way into industrial hygiene functions. Not to be ignored are overlaps into health physics, air pollution, water pollution, solid waste disposal, and disaster planning.

The industrial hygienist has a contribution to make in employee education and training, law and product liability, sales, labeling, and public information, as well.

JOB DESCRIPTIONS

The job descriptions of industrial hygiene personnel follow those related to safety. The entry-level classification, normally called a safety and/or health technologist, relates to a person who regularly inspects operations using a checklist and a few simple instruments, and investigates minor accidents.

The next higher level, normally called an industrial hygienist is similar, in function, to a safety engineer. This person carries out more detailed studies of accidents, prepares recommendations and other reports, reviews new processes or machinery and layouts from a health or safety engineering viewpoint, promotes health or safety education, and advises all levels of management in safe practices, procedures, and equipment needs.

The industrial hygiene supervisor is similar to the safety director, and manages the total hygiene program with responsibilities equivalent to those of the safety supervisor.

The American Board of Industrial Hygiene (ABIH) has established a level of competence in industrial hygiene, achieved through examination. Successful passage of this examination qualifies the person for membership in the American Academy of Industrial Hygiene; details are given in the Addendum.

Many certified industrial hygienists are also certified safety professionals and vice versa. Proficiency in industrial hygiene, by examination and experience, follows a route roughly comparable to that of occupational safety.

The job descriptions for someone performing in these two areas are grouped together because of their general similarity. Governmental concerns relate to capabilities associated with compliance, research, education and training, program development and management, and standards development. Consultants would be engaged in a wide range of professional activities that would be limited by the capabilities of on-board personnel. Insurance and industry require a range of professional competency from the technologist level to the program manager. Universities require professional capabilities in research, teaching, and program administration, while labor unions require inspection, research, training, program development and administration, and participation in standards development activities.

Industrial hygienist-in-training (IHIT)

This designation is part of the ABIH's certification program. It is awarded to persons having a college or university degree in industrial hygiene, chemistry, engineering, physics, medicine, or related biological sciences, who by virtue of special studies or training have acquired competence in the basic principles of industrial hygiene and have successfully completed the core examination. The candidate for the examination must also have completed at least one full year of industrial hygiene practice acceptable to the Board (see Addendum).

The ABIH instituted the IHIT category in 1972 because it recognized that persons employed in industrial hygiene and recent graduates wished to take the core examination before completing the 5 years of experience required to be eligible to take the comprehensive examination. At the time of this writing, the ABIH has certified 794 IHITs.

The IHIT would perform field and laboratory work that gradually increases in variety as experience and proficiency develop. This person would prepare technical reports that would be reviewed by the next higher level of professional supervision.

This period of apprenticeship is the time for the IHIT to be introduced to the elements of organization and management. The IHIT should also learn to understand a flowchart and a blueprint before progressing to anticipating, recognizing, evaluating, and controlling industrial hygiene hazards. Not much can be accomplished in an anticipatory or remedial way if one cannot interpret flowcharts, blueprints, or an organizational chart.

The IHIT will find that increasing emphasis is placed upon communication skills, and should be encouraged to draft replies to letters, to write reports, or to outline a verbal presentation that must be made—all subject to editing by the supervisor.

If the IHIT has not already done so, this is the time to become deeply involved in the local section or chapter of the professional association most appropriate. The IHIT should attend meetings and work on committees, and should begin to meet other professionals outside the immediate "industrial family."

Occupational health and safety technologist (OHST)

In 1976, the ABIH, in recognition of the growing group of

technologists engaged in industrial hygiene activities, established an Industrial Hygiene Technologist Certification Program.

An Industrial Hygiene Technologist is a person who, by virtue of special studies and training, has acquired proficiency in an aspect or phase of industrial hygiene, that is, air-sampling, monitoring, instrumentation, specialized investigations, or specialized laboratory procedures, and who performs such duties under the supervision of an industrial hygienist. The designation of Certified Industrial Hygiene Technologist (CIHT) was awarded after the applicant demonstrated eligibility to sit for examination and passed the examination.

In 1985, the ABIH and the Board of Certified Safety Professionals (BCSP) began joint sponsorship of a new technologist certification—the Occupational Health and Safety Technologist (OHST), which replaced the Industrial Hygiene Technologist designation.

This certification is *not* intended for persons who are Certified Safety Professionals (CSPs) or Certified Industrial Hygienists (CIHs), nor is it intended for those eligible to take the CSP or CIH examinations. It is a joint technologist certification (see Chapter 25, The Safety Professional).

Candidates for OHST certification must show 5 years of occupational health or safety technologist experience or a combination of acceptable education credit and experience before being allowed to take the OHST examination. The OHST may be able to gain the education and experience to progress to either a CSP or CIH. The OHST certification process is administered by the BCSP (see Chapter 25, The Safety Professional). At the time of this writing, there are 810 OHSTs in the United States.

Industrial hygiene technologists or technicians have acquired knowledge, skills, and field experience, so they can function efficiently in their limited technical area. Industrial hygiene technicians can take samples and make measurements in the plant or community. Their data and observations can be used to provide information for an industrial hygiene plan or program.

The performance of the duties of a technician requires that some very real criteria be met. There are some absolutes involved—such as thoroughness, dependability, and a concern over the accuracy of the data being collected is very important.

Industrial hygiene technicians should be given a detailed and specific outline of their duties. Manuals should be available to technicians for reference. Throughout, however, is the need for technicians to see the relevance and value of their efforts and these should be acknowledged by appropriate manifestations of which salary is only one. For example, the industrial hygiene technician is part of a team, a partner in an effort in which the technician and the industrial hygienist share responsibility. The technician should have accessibility to the industrial hygienist.

Technology changes and adds new problems to the old ones. Rarely are the old hazards totally replaced. New problems call for new approaches, new instruments, new ways of recording, compiling, and reducing data. Concomitantly, the technician must be willing to make appropriate adjustments. Technicians become, therefore, specialists in their own right. Some will be content to remain technicians. Many will move on and become industrial hygienists.

Industrial hygienist

Industrial hygienists are persons who, because of their more generalized skills, should be able to make independent decisions. The industrial hygienist decides what information is available, what additional facts are needed, and how they will be used or acquired.

Functions. According to Radcliffe et al (1959), within their sphere of responsibility, industrial hygienists will:

1. direct the industrial hygiene program.
2. examine the work environment.
   a. study work operations and processes and obtain full details of the nature of the work, materials and equipment used, products and by-products, number and sex of employees, and hours of work
   b. make appropriate measurements to determine the magnitude of exposure or nuisance to workers and the public, devise methods and select instruments suitable for such measurements, personally (or through others under direct supervision) conduct such measurements, and study and test material associated with the work operations
   c. using chemical and physical means, study the results of tests of biological materials, such as blood and urine, when such examination will aid in determining the extent of exposure
3. interpret results of the examination of the environment in terms of its ability to impair health, nature of health impairment, workers' efficiency, and community nuisance or damage, and present specific conclusions to appropriate interested parties such as management, health officials, and employee representatives.
4. make specific decisions as to the need for, or effectiveness of, control measures, and, when necessary, advise as to the procedures that will be suitable and effective for both the work environment and the environs.
5. prepare rules, regulations, standards, and procedures for the healthful conduct of work and the prevention of nuisance in the community.
6. present expert testimony before courts of law, hearing boards, workers' compensation commissions, regulatory agencies, and legally appointed investigative bodies covering all matters pertaining to Industrial Hygiene as described here.
7. prepare appropriate text for labels and precautionary information for materials and products to be used by workers and the public.
8. conduct programs for the education of workers and the public in the prevention of occupational disease and community nuisance.
9. conduct epidemiological studies of workers and industries to discover possibilities of the presence of occupational disease, and establish or improve Threshold Limit Values or standards as guides for the maintenance of health and efficiency.
10. conduct research to advance knowledge concerning the effects of occupation upon health and means of preventing occupational health impairment, community air pollution, noise, nuisance, and related problems.

The industrial hygienist should be able to decide if there are alternative solutions to a problem. Obviously, leadership and management skills are required by this individual if appropriate results are to be attained.

Few problems are so unique that peer acceptance is not

required. Thus, the industrial hygienist must be able to work with other industrial hygienists in the same functional area whether it be industry, government, labor unions, insurance, consulting, or teaching.

An industrial hygienist is an individual who, by virtue of special studies and training, has acquired competence and the ability to recognize and evaluate the hazard potential of environmental factors and stresses associated with work operations and to understand their effect on people and their well-being. The industrial hygienist should have the experience, knowledge, and capability to specify corrective procedures to minimize or control environmental health hazards.

The industrial hygiene manager. In an industry-setting, the industrial hygiene manager would supervise the technical and clerical staff in an industrial hygiene office, prepare budgets and plans, be familiar with governmental agencies related to this operation, relate industrial hygiene operations to research and development, sales, air and water pollution, and other departments or functions within the corporate group, and prepare appropriate weekly, monthly, and annual reports. He or she should be certified by ABIH (see the description of this organization in the Addendum to this chapter).

A difficult route for a company to take is that of "growing its own industrial hygienist"—that is, taking someone from inside the organization, with some scientific background and a knowledge of the firm's products, and exposing them to a crash program in industrial hygiene. A company contemplating an industrial hygiene effort where none existed before must recognize that company knowledge alone is not enough for the optimal solution of industrial hygiene problems in a short period of time. The industrial hygienist must possess or obtain the necessary professional expertise.

The capable industrial hygienist who has made the in-house adjustment to the company's problems should have the versatility and capability to deal with any industrial hygiene problem that may arise. He or she should be able to sit down with research and development people and find out what information will be needed. This might include such items as toxicological information, labeling requirements, assistance to customers, and any special engineering control requirements as the research effort progresses through pilot plant state to full scale-up and commercial production.

With the assistance of a qualified epidemiologist, an existing (or even a suspected) environmental health problem can be studied by the industrial hygienist through epidemiological and biostatistical approaches, in addition to the usual sampling and measuring procedures. The industrial hygienist should know where to go (for example, personnel, purchasing, or process engineering) for the information that might be needed to investigate and solve a problem. If the industrial hygienist knows of another company engaged in making similar products, he or she can exchange information with its industrial hygienist, and, under certain circumstances, exchange visits.

Communicating and working together apply as well to the other professionals, such as physicians, nurses, safety engineers, toxicologists, health physicists, and others, in and out of the company. Industrial hygienists must also communicate with and work very closely with employees. Employees have insights into potential health hazards present in their work area that only those working with the processes every day can possess. They are a primary source of information and suggestions for the industrial hygienist.

Many aspects of industrial hygiene expertise are unique. It makes sense for the industrial hygienist to extend capabilities and sphere of activity by delegating responsibilities to others. This calls for supervisory and planning skills. Not only must the industrial hygienist be capable of planning, directing, and supervising technicians and assistants, but must also be able to plan, program, and budget the activities of the department and staff. As a manager, the decisions will be theirs in many instances where priorities must be established and appropriate corrective action initiated.

Certified industrial hygienist (CIH)

The designation of Certified Industrial Hygienist by the ABIH indicates that a person has received special education, lengthy experience, and proven professional ability in the Comprehensive Practice or in an Aspect of industrial hygiene. The Aspects are Acoustical, Air Pollution, Chemical, Engineering, Radiological, and Toxicological.

The employer, employees, and the public have a right to be reasonably assured that the person to whom their lives are entrusted is professionally capable. The normal manner in which such protection is provided is through licensing; this licensing can be through a governmental agency or through a peer review arrangement, or both. Certification by the ABIH provides this assurance.

For certification by the ABIH, an individual must meet rigorous standards of education and experience prior to proving, by written examination, competency in either the comprehensive practice of industrial hygiene or one of the specialties or Aspects (see Addendum). Diplomates of the American Board of Industrial Hygiene are eligible for membership in the American Academy of Industrial Hygiene.

Certification would provide some assurance that this individual would possess a high level of professional competence. The certified industrial hygienist would be the most likely person to direct an industrial hygiene program capably, to work with other professions and governmental agencies, and to provide the vision and leadership of an industrial hygiene program to make sure that occupational hazards will be kept at a minimum in a rapidly changing technology and society. At the time of this writing, there are about 3,000 CIHs.

All CIHs must actively work to maintain their certification by earning a specified number of certification maintenance points during a 6-year cycle. These points are awarded for working as an industrial hygienist, attending designated technical seminars, courses, and meetings, contributing to publications, participating in professional associations, etc., all designed to keep CIHs abreast of current developments in the field.

Of course, all of the previously described categories of ABIH certifications—CIH, IHIT, and OHST—are open to all industrial hygiene personnel whether they are employed in industry, federal, state, and local governments, labor unions, educational institutions, consulting, etc. as long as they meet the qualifications. However, federally-employed industrial hygienists also have their own unique training programs that reflect the structure and duties of their positions.

INDUSTRIAL HYGIENE, CIVIL SERVICE

For industrial hygiene trainees, assignments are selected and designed to orient the new employee into the field of industrial hygiene, to determine areas of interest and potential, to relieve experienced industrial hygienists of detailed and simple work, and to develop the trainee's knowledge and competence. Specific assignments are carried out under direct supervision of a qualified industrial hygienist, including recognition of hazards, identification of controls, calibration of equipment, collection of samples, and initial preparation of reports with assistance from a qualified industrial hygienist. During inspections, the trainee observes specific safety items when assigned.

Under the general supervision of a senior industrial hygienist, the industrial hygienist will conduct complete industrial hygiene inspections, including selection of sampling methods and locations, evaluation of controls and monitoring procedures and preparation of reports. Completed work is reviewed for overall adequacy and confirmation with policy and precedents. The industrial hygienist determines engineering feasibility, sets periods of abatement, interprets standards and defends appeals under supervision of a senior industrial hygienist.

The senior industrial hygienist performs complete industrial hygiene inspections including the preparation of the final report. He or she determines engineering feasibility, sets periods of abatement, defends appeals, interprets standards, and provides off-site consultation. They receive general assignment of objectives and definition of policy from supervisors. The senior industrial hygienist differs from the industrial hygienist in that he or she receives more complex assignments and may act in place of the industrial hygiene supervisor when the supervisor is absent.

Training plan for entry-level OSHA industrial hygienists

On January 22, 1987, Assistant Secretary for Occupational Safety and Health John A. Pendergrass issued an OSHA Instruction specifying a new Training Operations Program for OSHA Compliance Personnel. The instruction provided policy and guidelines for the implementation of the technical training programs and described a Federal program change which also affects state OSHA programs. This revised training program applies to both newly-hired and experienced compliance personnel.

The training program is designed to provide a series of training courses that are supported and interspersed with on-the-job training and self-instructional activities to ensure that compliance personnel are able to apply technical information skills to their work; however, the elements of the training program are not meant to be prerequisites for advancement.

Objectives. Upon completion of the elements of the developmental training program, the Compliance Safety and Health Officer (CSHO) will have:

a. a working knowledge of the fundamentals of hazard recognition, evaluation and control.
b. adequate knowledge of the implementation of engineering controls, abatement strategies, and the interpretation of data.
c. a reasonable comprehension of basic industrial processes and the ability to make quantitative observations and measurements.
d. field experience in the proper calibration and use of measuring instruments.
e. the ability to perform solo or team inspections in most types of industries.
f. knowledge of regulations and laws that involve safety and health in the workplace.
g. the ability to present inspection data in a legal proceeding efficiently.
h. the ability to make a referral to other appropriate Industrial Hygienists or Safety Officers.

Organizational training responsibilities.

1. *Office of Training and Education.* The mission of the OSHA Office of Training and Education is to provide a program to educate and train employers and employees in the recognition, avoidance and prevention of unsafe and unhealthful working conditions, and to improve the skill and knowledge levels of personnel engaged in work relating to the Occupational Safety and Health Act of 1970.
   a. The Office of Training and Education consists of four components:
      (1) *Division of Training and Educational Programs.* This Division is responsible for planning agency technical training programs and for managing the New Directions grants.
      (2) *Division of Training and Educational Development.* This Division is responsible for developing and updating safety and health training programs and related materials.
      (3) *Division of Administration and Training Information.* This Division is responsible for providing administrative and informational programs for the Office of Training and Education.
      (4) *OSHA Training Institute.* The Training Institute is responsible for the delivery of training to the populations served by the agency.
   b. Specific responsibilities of the OSHA Training Institute include:
      (1) conducting programs of instruction for Federal and State compliance officers, State consultants, other Federal agency personnel and private sector employers, employees, and their representatives.
      (2) participating in the development of course outlines, detailed lesson plans and other educational aids necessary to carry out Institute training programs.

The Regional Training Officer assists the Assistant Regional Administrator for Training, Education and Consultation in coordinating the management of all regionwide training programs. The Regional Training Officer serves as the main focal point in the Regional Office for ensuring the successful implementation of the training program for Regional compliance personnel as outlined in this instruction. Specifically, the Regional Training Officer will assist in providing resource material and current training information to Area Directors and Supervisors concerning the implementation of the objectives of the training program and will evaluate and monitor all records of training.

In OSHA Area Offices, the Area Director has overall responsibility for ensuring and implementing the development and training of newly hired and experienced CSHOs under his or her supervision. The Supervisor, however, serves as the main focal point in the Area Office for ensuring training. The Supervisor provides and coordinates instruction, assistance and guidance to the CSHOs in order to meet the training program objectives. Reviewing and maintaining progress records for each

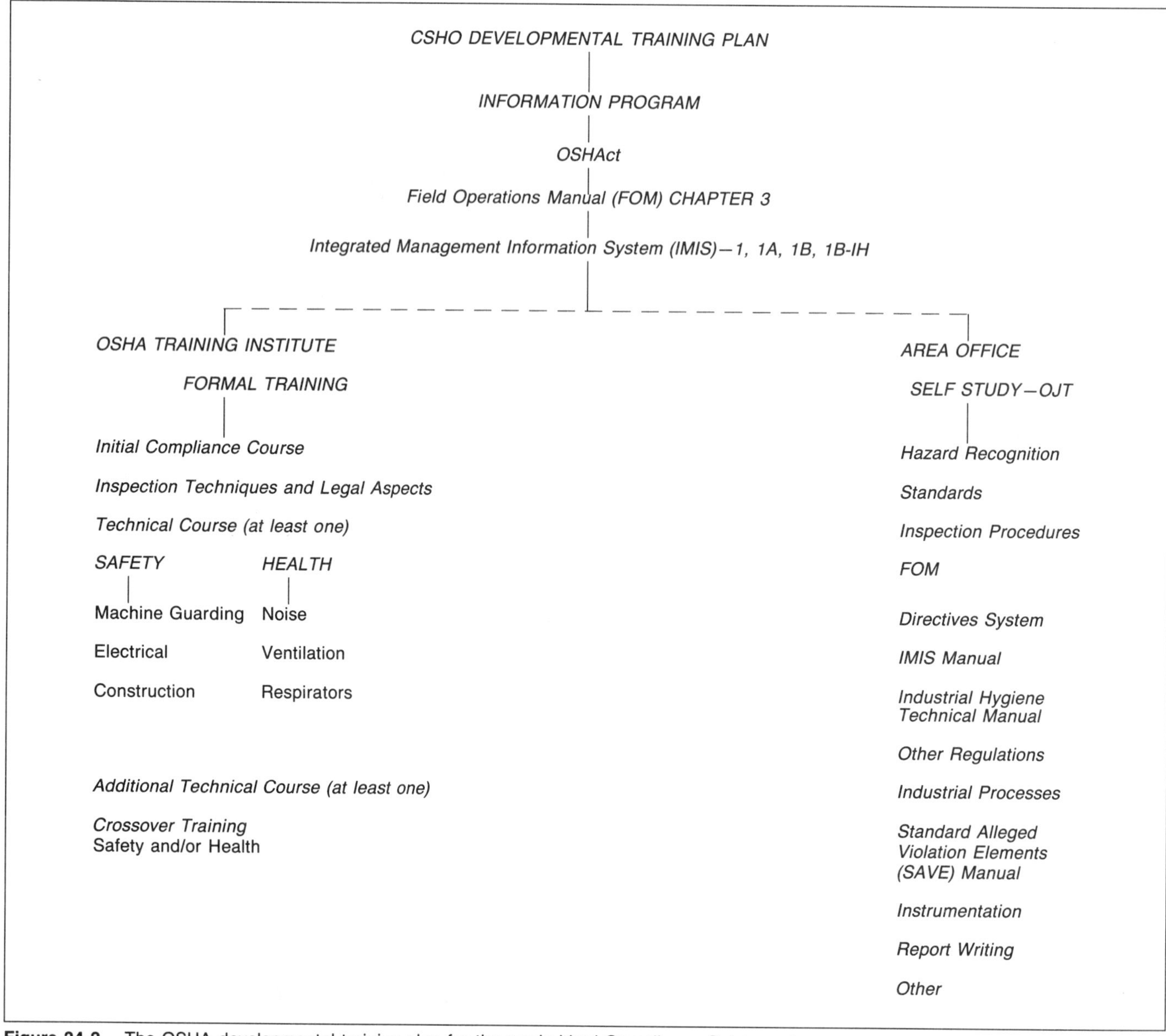

Figure 24-2. The OSHA developmental training plan for the newly hired Compliance Safety and Health Officer. (Adapted from the OSHA Instruction TED, Office of Training and Education)

CSHO and assigning senior CSHOs to assist in on-the-job training of new-hires is also performed by the supervisor.

The program itself provides a well-articulated progression of training requirements for newly-hired personnel. The elements include formal training at the OSHA Training Institute and informal training such as self-study and on-the-job training (OJT). Figure 24-2 illustrates the developmental training plan for the new-hire.

Informational program. The developmental training plan begins with the distribution and self-study of an informational package of materials developed jointly by the National office, the Regional office and the Office of Training and Education. Contents include information on: the U.S. Department of Labor; an introduction, history, purpose, program mix to the OSHA program; the structures of Regional and Area offices, procedures, and libraries; common OSHA acronyms; individual training development programs; and such handout items as organizational charts, *Field Operations Manual* (FOM), standards, directives, personal protective equipment and instruments.

Self-study program. Prior to attending the initial compliance course at the OSHA Training Institute, each CSHO is required to complete three self-study programs on the OSHAct, Chapter III of the Field Operations Manual and on the Integrated Management Information Systems (IMIS) forms 1, 1A, 1B and 1B-IH. During these self-study assignments, the CSHO becomes familiar with the basic OSHAct requirements; studies basic inspection procedures in Chapter III of the FOM, and familiarizes him- or herself with the most commonly used forms.

OSHA Training Institute. After completion of the basic self-study prerequisites, each CSHO is required to complete the following courses at the OSHA Training Institute in Des Plaines, Illinois:

Initial Compliance Course—provides new CSHOs with an understanding of occupational safety and health programs, a working knowledge of the FOM, and a thorough introduction to the organization and content of the standards. (This course may not be waived.)

Inspection Techniques and Legal Aspects—provides new CSHOs with an understanding of basic communication skills, formal requirements and processes of the legal system, and investigative techniques related to OSHA compliance activity.

Technical Courses—(at least two courses required)—provide the CSHO with technical knowledge, skills, and information on hazard recognition as related to OSHA requirements. The specific courses will be determined by the supervisor based on individual need. At least one of the courses shall be selected from the following basic core courses.

Safety

200—Construction Standards
203—Electrical Standards
204—Machinery and Machine Guarding Standards

Health

220—Industrial Noise
221—Principles of Industrial Ventilation
222—Respiratory Protection

Crossover Training—Recognizing the need for CSHOs to be familiar with general concepts of both safety and health, each CSHO is required to complete crossover training during the developmental period.

Area office. The training plan incorporates alternative modes of instruction including self-instructional techniques and on-the-job training (OJT) assignments with supervision. The OJT and self-study programs are designed to reinforce formal training. Self-Study is training that involves independently gained knowledge in the Area Office that will aid in preparation for formal training and course work. On-the-Job Training (OJT) is training that relates principles and theories to work skills which are then taught and applied in the field and office environment.

The OJT and self-study assignments are provided concurrently with formal training to emphasize and complement material covered in formal training courses. Time allowed to accomplish OJT and self-study assignments should be compatible with the new-hire CSHO's current knowledge, skill, and experience levels. Verification of a CSHO's ability to successfully complete OJT and self-study assignments must be documented by the supervisor.

The expertise and judgment of the supervisor will be required when assessing a CSHO's progress during the training program. The supervisor must make certain that the CSHO is ready to perform an assigned task on an individual basis. The program is flexibile enough to also afford the CSHO time for proper sequencing of training. Training assignments may also be supplemented by other task assignments as deemed necessary by the supervisor. Training in the following subject areas, at a minimum, is to be accomplished through both OJT and self-study assignments:

- Hazard Recognition Overview
- Inspection Procedures
- Standards:
  - General Industry
  - Construction
  - Maritime (dependent on geographic location of Area Office)
  - American National Standards Institute (ANSI)
  - National Electrical Code (NEC)
  - National Fire Protection Association (NFPA)
  - American Conference of Governmental Industrial Hygienists (ACGIH)
  - International Agency for Research on Cancer (IARC)
  - National Toxicology Program (NTP)
- Field Operations Manual (FOM)
- IMIS Forms Manual
- OSHA Directives System
- IH Technical Manual (Safety personnel should have basic familiarity, while Industrial Hygienists will have more in-depth knowledge.)
- Other Regulations and Procedures
- Common Industrial Processes
- Standards Alleged Violation Elements (SAFE's) Manuals:
  - Regulatory and General Industry
  - Construction
  - Maritime
- Instrumentation
- Report Writing

Continuing maintenance of skills and knowledge. Once the training developmental period is completed, CSHOs will typically require additional training to keep themselves current in the safety and health field. At a minimum, each CSHO is required to attend a technical course once every 3 years at the OSHA Training Institute. If an Institute course has changed significantly during the years, the CSHO is permitted to repeat the course.

The CSHOs are also encouraged to pursue other training opportunities available both within the Department of Labor and elsewhere.

PERSONNEL NEEDS AND PROBLEMS

The number of practicing industrial hygienists in the United States in 1977 was estimated to be about 5,000 by John Short and associates under a National Institute for Occupational Safety and Health (NIOSH) contract. Hence, the apparent ratio of workforce population to industrial hygiene practitioner is calculated to be 17,100 to 1 by Bernard E. Salzman, PhD, in the April 1982 *American Industrial Hygiene Association Journal* article (see Bibliography).

Although no parallel survey data exists for 1987, membership figures in professional societies are helpful in assessing present trends.

The American Industrial Hygiene Association reports a national membership of 7,300. If local section AIHA members, who are not also national members, are included, the figure rises to approximately 11,000 professionals. Of these there are currently 2,983 CIHs, 794 IHITs and 810 OHSTs.

Short, et al, estimated (see Table 24-A) the 1977 census and future census for industrial hygienists and radiation specialists based on a survey of nonagricultural workplaces of 100 or more employees in which half (3,300) of the workplaces responded, representing 8 percent of the targeted work force. One might suspect that the units which failed to respond were less likely to have had health and safety programs. Thus the sample may have presented a more favorable picture than was truly representative. Indeed, some sources indicate that the 1977 ratio of 1 industrial hygienist to approximately 18,000 employees may in fact be closer to 1 industrial hygienist to 40,000–50,000 employees

Table 24-A. Some Results of 1977 Nationwide Survey Conducted by Short et al.

Item	Census 1977	Projected 1980	Status 1985	Accel 1985	Status 1990	Accel 1990
Industrial Hygiene						
Professional work force	4,810	5,350	5,630	8,160	5,880	8,440
Annual new hires required for interval from preceding date	—	310	160	690	160	170
Annual BS, MS, Ph.D. degrees in industrial hygiene	182	290	—	397	—	454
Occupational Safety						
Professional work force	37,260	41,420	43,690	44,210	45,590	46,030
Annual new hires required for interval from preceding date	—	2,350	1,290	1,440	1,310	1,300
Annual BS, MS, Ph.D. degrees in occupational safety	367	540	—	709	—	776
Annual BS, MS, Ph.D. degrees in occupational safety and health	173	335	—	421	—	477
Annual associate degrees in occupational safety and health	208	212	—	250	—	317
Total of above safety degrees	748	1,087	—	1,380	—	1,570
Radiation						
Professional work force	1,370	1,520	1,650	1,660	1,770	1,780
Annual new hires required for interval from preceding date	—	70	40	50	50	50

(Reprinted with permission from the *American Industrial Hygiene Association Journal* 43 [April 1982])

Table 24-B. OSHA Estimates in 1975: Summary Occupational Safety and Health Manpower Forecast*

Specialty	Present Census	Present Needs	Projected Needs 1980	Projected Needs 1985	Projected Deficit 1980	Projected Deficit 1985	Assumed Annual Growth Rate (%)
Occupational physicians (board certified)	500	1,700	1,971	2,285	1,392	1,614	3
Industrial physicians (short post-graduate course work)	2,700	6,900	8,806	11,239	5,360	6,841	5
Industrial hygienists	500	5,500	8,081	11,874	7,346	10,795	8
Safety professionals	3,000	8,700	14,011	22,566	9,179	14,785	10
Certified occupational nurses	1,000	9,400	11,997	15,312	10,721	13,683	5
Industrial nurses (short-term training)	17,000	36,700	46,840	59,780	25,143	32,089	5
Occupational safety and health specialists	15,000	24,000	38,650	62,246	14,492	23,340	10

*Present needs and census figures from "Plan for Estimating Occupational Safety and Health Needs." Unpublished NIOSH paper January 16, 1973.

Note: The projected increases in manpower are based on the following assumptions:
1. Growth of industry at an accelerated pace.
2. Continued introduction into the workplace of materials and processes that will require increased knowledge and skill to recognize, evaluate and control their attendant hazards.
3. Increased emphasis on creating and maintaining a corps of highly qualified safety and health professionals both in the public and private sectors.

(Reprinted with permission from the *American Industrial Hygiene Association Journal* 43 [April 1982])

especially when one considers individuals who are unable to devote full time to industrial hygiene.

Short's future estimates (accelerated growth) were based on a 50 percent accelerated growth rate allowing for a 2.5 percent retirement figure.

The results of an OSHA Estimate (Table 24-B) prepared in 1975 based upon a 1973 NIOSH estimate showed a census of only 500 industrial hygienists but also showed a census category listing 15,000 Occupational Safety and Health Specialists. The OSHA estimate indicated a then-current need for 5,500 industrial hygienists and 24,000 safety and health specialists. While the projected industrial hygienists needs for 1985 according to OSHA (11,874) and Short—adding accelerated estimates for industrial hygiene and radiation—(10,820) are fairly consistent, OSHA lists a need for 62,246 "Occupational Safety and Health Specialists" as well.

It should be noted that the low estimates of future manpower training needs from the Short survey were based upon the numbers of positions deemed necessary by the managements of the industries surveyed. This demand reflected the regulatory environment, labor relations, business conditions, and many other factors in 1977, which have changed substantially in later years. The allowance of 2.5 percent per year for professional staff turnover and retirements assumes an average stay in the field of 40 years, and that professionals entering the field from other jobs do not need training, both unlikely assumptions. Nor do these estimates consider that most of the current personnel lacks specific professional training in industrial hygiene and safety.

If the prevention of disease and preservation of health and safety are more desirable than having to medically treat occupational diseases or repair industrial injuries, then it is obvious that there is a personnel shortage in industrial hygiene.

The education and training programs for industrial hygiene include: (1) professional school training, (2) graduate curricula, and (3) continuing education (short courses). Professional school curricula in industrial hygiene are generally culminating in a Master of Public Health or Master of Science degree.

Training grants

NIOSH's findings of shortages of trained occupational safety and health graduates were cited in successful efforts to expand training grants programs. One part of this expansion was to introduce the concept of multidisciplinary Educational Resource Centers (ERCs). The other part was growth of single-discipline training grants.

Single-discipline training project grants have been established in 28 universities, and over 100 different academic degree programs were in place in 1980. Because of budget cutbacks, the number of programs was reduced to 60 in 1982. The number of professionals graduating from these and ERC programs each year increased from 217 in 1976 to 747 in 1980.

Educational Resource Centers

Congress authorized creation of up to 20 Educational Resource Centers for occupational safety and health in 1976. Funding increased from $2.9 million in 1977 to $12.9 million in 1980, and the ERCs now number 15. These centers: (1) provide continuing education to occupational health and safety professionals; (2) combine medical, industrial hygiene, safety, and nursing training so that graduates are better able to work effectively in complex and diverse conditions; (3) conduct research; and (4) conduct regional consultation services. All ERCs but one are located in universities.

The centers are distributed as far as possible to give regional representation and to meet training needs for all areas of the Nation. The Federal cost of ERC education is approximately $7,000 for each degree graduate and $70 for each attendee at continuing education courses (102).

Recent budget cuts have reduced the current level of funding for Educational Resource Centers to $5.8 million, 55 percent below the level in fiscal year 1980. According to projections by the Association of University Programs for Occupational Health and Safety, if Federal funding for ERCs had been eliminated as proposed in the President's fiscal year 1984 budget, the number of graduates completing their programs would have decreased to approximately 338 (compared with about 781 in 1981) and only about 25 degree programs of the 112 programs currently in existence could have been expected to survive. Furthermore, there was concern that without Federal funds, multidisciplinary programs would revert to more narrow and limited single-discipline programs.

While the President's 1984 budget for NIOSH contained no request for ERC funding, the Congress added $8.8 million for the centers.

Professional schooling

A program of study leading to a professional degree in industrial hygiene should start with two years of basic arts and sciences, two years of derivative sciences and advanced subjects, and two years of professional courses. Such an advanced degree might appropriately be designated Doctor of Occupational Health, Doctor of Public Health, Doctor of Science, or Doctor of Engineering. Regardless of name, however, it should be clearly understood that such a degree would be a professional scholar's degree.

One could readily "tinker-toy" such a six-year curriculum from existing university programs. For example: four years of undergraduate study in environmental health engineering would satisfy the basic requirements in mathematics, chemistry, physics, biology, economics, liberal arts, and humanities; the first year of study in medical school could fill in the essential medical sciences (there should be some emphasis on health as contrasted to disease); and one year of specialized course work in industrial hygiene would complete the didactic training. With a minimum of two years of "real world" experience, the metamorphosis to a licensable practitioner would be complete.

Graduate curricula

Graduate study programs have generally been developed along the lines of providing in-depth knowledge of a particular subject area and developing scholarly research capabilities. A typical base program for a master of science degree in industrial hygiene could be as follows: course work—50 quarter-hours of credit; thesis research—22 quarter-hours of credit.

An integrated course curriculum would consist of: two epidemiology or biometry courses; three courses in industrial hygiene, occupational safety, and radiological health; three human physiology courses; two principles of occupational health courses; three environmental and occupational laboratory or toxicology courses; and sufficient electives to equal or exceed the required total quarter-hours of course credit.

Specifying a completely rigid core curriculum does not, however, take into account the student's backgrounds. They may be repeating that which they already know in one area while leaving a hiatus in another. Therefore, the curriculum design can be improved by offering individually prescribed programs. Programs should be carefully designed to meet individual student needs. An example of one such specific curriculum from the University of Illinois School of Public Health is shown in the Addendum to this chapter.

For those students whose interests and abilities lie in the area of scholarly research, an additional 16 courses (48 quarter-hours of credit) may be used to satisfy the course requirements for a doctor of philosophy. The research requirement would be open ended and generally takes another two to five years. Most large universities, especially those having colleges of engineering, medicine, and a graduate school, generally have enough suitable courses to round out the curricula. Finding faculty members with the academic preparation, research capability, and a background of real industrial experience may present a problem.

Baccalaureate programs

Educational requirements for baccalaureate programs in occupational health are reviewed by Levine, et al, 1977. The National Institute for Occupational Safety and Health (NIOSH) has offered training grants to academic institutions to develop educational programs in occupational health and safety. It was hoped that these programs might serve as a bridge between present associate arts and graduate degree programs. Students completing the baccalaureate programs would be ideally qualified to function in the workplace or to pursue graduate study and specialization.

In his review, Levine grouped the activities of professional industrial hygienists into the following major areas of responsibility:

1. recognizing and identifying all chemical, physical, and biological agents which may adversely affect the physical, mental, and social well-being of the worker and the community
2. measuring and documenting levels of environmental exposure to specific hazardous agents
3. evaluating the significance of exposures and their relationships to the etiology of occupationally and environmentally induced diseases
4. establishing appropriate controls to prevent hazardous exposures and monitor their effectiveness
5. administering the occupational and environmental hygiene program
6. joining with medical, safety and other members of the occupational health team in developing and presenting a comprehensive approach to prevention programs
7. developing procedures to assure continuing professional development
8. participating in policy-making decision.

Specific tasks were then identified which must be performed to fulfill these responsibilities. The knowledge and the skills necessary for completing these tasks suggested the proposed course contents.

From this activity analysis, courses were formulated to present an integrated approach to the baccalaureate program curriculum (see Addendum). These courses describe only the occupational hygiene content of the program. Students would also be expected to take certain other required courses and units, which may be distributed as follows—20 units of chemistry (inorganic, organic, qualitative and quantitative analysis and biochemistry), 4 units in mathematics (algebra, calculus, or statistics), 4 units in physics, 10 units in biological sciences, 3 units in engineering sciences, 15 units in humanities and behavioral sciences, and 3 units of free electives.

Continuing education

A wide variety of opportunities exist for industrial hygienists who wish to either remain technically current, receive training in previously unfamiliar aspects of industrial hygiene, pursue academic coursework leading to a more advanced degree, or to earn certification maintenance points in order to maintain their CIH certification.

A number of universities offer coursework leading to degrees. Also available at such universities are usually short-courses of days, or week's duration on specific industrial hygiene topics. Summer institutes, concentrating upon a particular area of industrial hygiene of perhaps 3–4 weeks length, are another continuing education opportunity.

The National Institute for Occupational Safety and Health (NIOSH) publishes an annual catalogue of all such training courses nationwide at universities that have been funded as NIOSH Educational Resource Centers (ERCs). Most ERCs contain an industrial hygiene component that includes coursework leading to academic degrees and short courses. The catalogue can be obtained through the NIOSH publications dissemination office.

SUMMARY

The need to control exposures to a rapidly rising number of chemicals and hazardous agents and to comply with and to enforce OSHA regulations has brought about greater demand for industrial hygienists. This demand exists in private industry, labor unions, government and academic organizations.

Individuals practicing industrial hygiene routinely work as a team; thus the physician, the nurse, the safety professional and the industrial hygienist are quite accustomed to working together. Other professions are included as needed; these include toxicologists, health physicists, epidemiologists, statisticians, professional trainers and educators. A team approach, using the knowledge and skills of all these professionals, increases the effectiveness of programs to prevent occupational disease and injuries and helps to anticipate future requirements.

The need continues for industrial hygienists to assume professional responsibility for interpretation of the findings of environmental investigation and for the design and implementation of control measures. The industrial hygienist must, therefore, have the generalist's grasp of varied disciplines in order to interact with divergent groups to develop and maintain the most effective program.

Educational requirements for industrial hygienists will continue to expand with the increasing need to monitor and control hazardous agents and to comply with more stringent governmental regulations. Proposed curricula and course descriptions were included in this chapter. The training program for OSHA CSHOs was also discussed.

Personnel from three professional specialties—industrial hygiene, safety, and fire protection—will be working even more closely together in the future, their responsibilities overlapping in many instances. The separation between these professions has become increasingly blurred, and "melding" may eventually lead to the creation of a single profession whose scope is made up of what is currently recognized today as industrial hygiene and safety.

BIBLIOGRAPHY

Berry C. M. Industrial hygiene manpower. *American Industrial Hygiene Association Journal* 36(1975):433-446.

Berry, C. M. What is an industrial hygienist? *National Safety News* 107 (1973):69-75.

Corn, M., and Heath, E. D. OSHA response to occupational health personnel needs and resources. *American Industrial Hygiene Association Journal* 38(1977):11-17.

Hermann, E. R. Education and training of industrial hygienists. *National Safety Congress Transactions* 12 (1975):64-66.

Levine, M. S., Watfa, N., Hanna, F., et al. A plan for baccalaureate education in occupational hygiene. *American Industrial Hygiene Association Journal* 38(1977):447-455.

Office of Technology Assessment, U.S. Congress. Preventing injury and illness in the workplace, unpublished. New York: InfoSource, 1985.

Radcliffe, J. C.; Clayton, G. D.; Frederick, W. G., et al. Industrial hygiene—definition, scope, function, and organization. *American Industrial Hygiene Association Journal* 20(1959):429-430.

Salzman, B. E. Adequacy of current industrial hygiene and occupational safety professional manpower. *American Industrial Hygiene Association Journal* 43(1982):254-260.

PROFESSIONAL ORGANIZATIONS

American Board of Industrial Hygiene Bulletin and Examination Information, 4600 W. Saginaw, Suite 101, Lansing MI 48917.

American Conference of Governmental Industrial Hygienists, 6500 Glenway Ave., Bldg. D-7, Cincinnati, OH 45211-4438.

American Industrial Hygiene Association, 475 Wolf Ledges Pkwy, Akron, OH 44311-1087.

American Public Health Association, 1015 15th St, NW, Washington, DC 20005.

Board of Certified Safety Professionals, 208 Burwash, Savoy, IL 61874.

ADDENDUM: Description of Societies and Courses of Interest to Industrial Hygienists

The objectives and/or activities and membership qualifications are given for three organizations of interest to industrial hygienists—the American Industrial Hygiene Association, the American Board of Industrial Hygiene, and the American Conference of Governmental Industrial Hygienists. Suggested course curricula for both the bachelor and master of science degrees in industrial hygiene are outlined.

AMERICAN INDUSTRIAL HYGIENE ASSOCIATION

The AIHA is a nonprofit, professional society for persons practicing industrial hygiene in industry, government, labor, academic institutions and independent organizations. Currently, close to 11,000 professionals are affiliated with AIHA. Of these, over 7,300 are members of the National Association. Membership is drawn from the United States, Canada, and 43 other countries.

The AIHA was established in 1939 by a group of industrial hygienists as a result of the need for an association devoted exclusively to industrial hygiene. AIHA is a national professional society of persons engaged in protecting the health and well-being of workers and the general public through the scientific application of knowledge concerning the chemical, engineering, physical, biological, or medical principles to minimize environmental stress and to prevent occupational disease.

The AIHA's purpose is to promote the recognition, evaluation and control of environmental stresses arising in or from the workplace or its products, and to encourage increased knowledge of industrial and environmental health by bringing together specialists in this professional field.

AIHA membership qualifications and types

Membership in the AIHA shall be open to persons who are engaged in industrial hygiene activities and such other persons or organizations as may be provided in the Bylaws. There shall be no limitation or discrimination because of race, creed, sex, or color. Approval by two-thirds of the Board of Directors is required for election to all classes of membership.

The classes of membership are: Student, Affiliate, Associate, Full Member, Emeritus, Honorary, and Organizational. Application for membership may be obtained from the Association Office. (See References.)

A full-time student at the college undergraduate level may become a Student Member upon a yearly application and submission of adequate matriculation documentation to the Association. A Student Member may not serve on committees, vote, or hold office. A Student Member shall have an option to subscribe to the Association *Journal* at a price established by the Board of Directors.

An Affiliate Member shall be a person who is employed full-time in an occupation requiring interaction with and cooperation of Associate Members and Full Members of the Association. An Affiliate Member may not vote, or serve on the Board of Directors, but may serve as a member of committees.

An Associate Member shall be a graduate of an accredited school of college grade with a baccalaureate degree in industrial hygiene, chemistry, physics, engineering, biology or a cognate discipline, who is currently engaged a majority of time in industrial hygiene activities as defined by the Board of Directors, or is a full-time graduate student in industrial hygiene or a cognate discipline. Where an applicant does not have a baccalaureate degree, experience may be substituted on the basis of two years of qualifying experience for one year of undergraduate education. Membership as an Associate Member is limited to a maximum of five (5) years. An Associate Member may serve on committees and vote but may not be elected to the Board of Directors of the Association.

A Full member shall be a graduate of an accredited school of college grade with a baccalaureate degree in industrial hygiene, chemistry, physics, engineering, biology, or a cognate discipline who has been engaged a majority of time for at least three years in industrial hygiene-related activities as defined by the Board of Directors. The Board of Directors will consider, and may accept, any other degree proposed by a candidate who provides evidence of the scientific content of the curriculum. The social sciences are not considered as qualifying sciences. Where an applicant does not have a baccalaureate degree, experience may be substituted on the basis of two years of qualifying experience for one year of undergraduate education. Full-time graduate study in industrial hygiene or a cognate discipline may be accepted on an equivalent time basis for any portion of the required three years of experience. A Full Member may serve on committees, vote and be elected to the Board of Directors of the Association.

Persons particularly distinguished in the general field of industrial hygiene or in a closely related scientific field may be granted Honorary Membership by the Board of Directors. Emeritus membership may be granted by the Board of Directors to a Full Member who has retired from the practice of the industrial hygiene profession.

Any organization may apply for Organizational Membership.

Local section membership

Any person having a professional interest in industrial hygiene may apply for membership in a local section and, if elected, is entitled to all the privileges of membership of the section, except those of the national Association. Application for membership in a local section should be made to the local section.

Address: American Industrial Hygiene Assn., 475 Wolf Ledges Parkway, Akron, Ohio 44311-1087.

AMERICAN BOARD OF INDUSTRIAL HYGIENE

The American Board of Industrial Hygiene (ABIH) was established to improve the practice and educational standards of the profession of industrial hygiene.

The activities presently engaged in for carrying out this purpose are:

1. To receive and process applications for examinations, and to evaluate the education and experience qualifications of the applicants for such examinations.
2. To grant and to issue to qualified persons, who pass the Board's certification examinations, certificates acknowledging their competence in industrial hygiene or aspects thereof, and to revoke for cause certificates so granted or issued.
3. To provide for maintenance of certification by requiring submission of evidence of continued professional qualifications by the holders of certificates in the Comprehensive Practice or an Aspect of industrial hygiene.
4. To maintain a record of holders of certificates granted by the corporation.
5. To furnish to the public, and to interested persons or organizations, a roster of those persons in good standing, having special training, knowledge, and competence in industrial hygiene as evidenced by certification granted by the corporation.

The American Board of Industrial Hygiene issues three categories of certificates.

1. The first certifies that the individual has the required education, experience, and professional ability in the comprehensive practice of industrial hygiene (CIH).
2. The second certifies as to education, experience, and professional ability of the individual in the application and practice of a specialized aspect of industrial hygiene. The specialized aspects are designated by the words acoustical, air pollution, chemical, engineering, radiological, and toxicological (CIH).
3. The third established by the board is the Industrial Hygienist in Training (IHIT) certification. This refers to individuals who are permitted to take the "core" examination before completion of the five-year eligibility requirement.
4. A fourth category—the Occupational Health and Safety Technologist (OHST) designation is a joint certification with the Board of Certified Safety Professionals (BCSP). The OHST examination procedure is administered by the BCSP. (See Bibliography.)

Each applicant for a certificate must meet certain minimum eligibility requirements, and must pass an examination.

The examination for certification consists of two parts.

- The first part, the "core" examination, covers general aspects of industrial hygiene to the degree which, in the opinion of the Board, should be familiar to be candidate.
- The second part of the examination consists of different sets of questions for each category of certificate to be issued. Those seeking certification in the comprehensive practice of industrial hygiene will be given detailed questions covering the comprehensive aspects. Those seeking certification in a specialized aspect will be required to provide answers to questions largely pertinent to, but not confined to, that particular specialty.

Maintenance of certification. The ABIH also administers a certification maintenance program for CIHs. The purpose of the certification maintenance program is to ensure that CIHs continue to develop and enhance their professional industrial hygiene skills for the duration of their active careers.

a. The certificate is granted for a period of six years after which time it expires unless renewed as set out in this section.
b. Certificate holders shall provide evidence to the Board of their continued professional qualifications before the certificate shall be renewed.
c. Activities which will be accepted as such evidence include continuing professional industrial hygiene services; membership in approved professional societies other than the American Academy of Industrial Hygiene; attendance at approved meetings, seminars and short courses; approved extracurricular professional activities; or reexamination.
d. Points for the approved activities will be awarded and publicized by the Board.
e. From time to time the Board will publish a list of approved activities and their point value.
f. A schedule for renewal of certificates is published by the Board.
g. In addition to meeting the aforementioned requirements, maintenance of certification will not be granted if all annual dues during the maintenance cycle period have not been paid.

Besides being entitled to use the "CIH" designation, persons certified in either comprehensive practice or an aspect become members of the American Academy of Industrial Hygiene and their names are published in the annual roster of the Academy.

The names of IHITs are also published in the Academy roster.

Address: American Board of Industrial Hygiene, 4600 W. Saginaw, Suite 101, Lansing, Michigan 48917.

AMERICAN CONFERENCE OF GOVERNMENTAL INDUSTRIAL HYGIENISTS

The American Conference of Governmental Industrial Hygienists was organized in 1938 by a group of governmental industrial hygienists who desired a medium for (1) the free exchange of ideas and experiences and (2) the promotion of standards and techniques in industrial health.

As an organization devoted to the development of administrative and technical aspects of worker health protection, the Conference has contributed substantially to the development and improvement of official industrial health services to industry and labor. The committees on Industrial Ventilation and Threshold Limit Values are recognized throughout the world for their expertise and contributions to industrial hygiene. It is the ACGIH that sets TLVs and annually updates these values.

Membership is comprised of professional and technical personnel in governmental agencies or educational institutions engaged in occupational health activities. The more than 3,200 members from across the United States and from 35 countries give the organization an international scope.

Objectives

The objectives and goals of the Conference are to:

- promote sound industrial hygiene practices
- encourage the coordination of industrial hygiene activities through official federal, state, local, and territorial industrial hygiene agencies
- encourage the free exchange of experiences and ideas among

industrial hygiene personnel in such official agencies and educational institutions

- collect and make available to all governmental industrial hygienists information and data which may be of assistance to them in the fulfillment of their duties
- hold annual and other such meetings as may be necessary to accomplish the purpose of the organization (The ACGIH jointly sponsors the annual American Industrial Hygiene Conference with the AIHA).

Membership

ACGIH is continually seeking new members to perpetuate the exchange of ideas and experiences and application can be made for any of the five membership types discussed here. Annual dues entitle the member to the bimonthly *Bulletin Board* newsletter, yearly TLV booklet, membership directory and the *Applied Industrial Hygiene* journal.

Full member—An occupational health professional who works for the government or an educational institution and who is engaged in occupational health services, consultation, enforcement, research or education on full-time basis. Industrial hygienists, occupational physicians and nurses are examples of those eligible.

Associate member—A professional employed in a full-time activity closely allied to industrial hygiene and employed by a government or an educational institution. Associate members shall have the right to vote on all Conference matters except those concerned with constitutional amendments or adoption of bylaws. They are eligible to serve on any appointive committees and as member-at-large of the Board of Directors, but shall not be eligible to serve as an officer of the Conference. One example of this grade of membership could be a safety professional.

Technical member—An industrial hygiene technician employed full-time by the government or an educational institution who, by virtue of special studies and training, has acquired proficiency in an aspect or phase of industrial hygiene and who performs such duties under the supervision of an industrial hygienist. Technical members have the right to vote on all Conference matters except those concerned with constitutional amendments or adoption of bylaws. They are eligible to serve on any appointive committees and as a member-at-large of the Board of Directors, but are not eligible to serve as an officer of the Conference. Technicians skilled in air sampling, monitoring, and instrumentation are included in this grade.

Affiliate member—A person still employed in the field of industrial hygiene who has been a member of the Conference for at least five years, but is not currently eligible for another class of membership. Affiliate members cannot hold elective office or vote on Conference matters.

Student member—A person engaged in a course of study in industrial hygiene or an aspect or phase thereof. Student Members are eligible to serve on any appointed committee except the nominating committee, but are not eligible to hold elective offices or to vote. ACGIH student members can only qualify for a free subscription to AIH by supplying verification of student status. Dues are $5.00 per year while in student status which must be verified each year at renewal of membership dues.

All persons meeting the eligibility requirements for membership are encouraged to apply. Active participation in ACGIH affairs will enable those individuals in governmental industrial hygiene programs to keep abreast of the rapidly changing field as well as offer the opportunity to express views and contribute professional expertise in solving many of the problems faced today.

Address: American Conference of Governmental Industrial Hygienists, 6500 Glenway Ave., Bldg D-7, Cincinnati, Ohio 45211-4438.

AMERICAN PUBLIC HEALTH ASSOCIATION

The American Public Health Association (APHA), established in 1872, is 50,000 strong in its collective membership, which represents all the disciplines and specialties in the public health spectrum. The APHA is devoted to the protection and promotion of public health. It achieves this goal in several ways:

- sets standards for alleviating health problems
- initiates projects designed for improving health, both nationally and internationally
- researches health problems and offers possible solutions based on that research
- launches public awareness campaigns about special health dangers
- publishes materials reflecting the latest findings and developments in public health.

The APHA has 23 special Sections including an Occupational Health and Safety Section that includes occupational health physicians and nurses, industrial hygienists, and other allied occupational health professionals. Each APHA section has its own professional meetings to provide a forum for the exchange of ideas.

Membership

Three types of membership are available.

Regular membership—Available to all health professionals.

Contributing membership—Provides additional association benefits.

Special memberships—*Student/Trainee:* Persons enrolled full-time in a college or university, or occupied in a formal training program in preparation for entry into a health career.

Retired: APHA members who have retired from active public health practice and no longer derive significant income from professional health-related activities.

Consumer: Persons who do not derive income from health-related activities.

Special health workers: Persons employed in community health whose annual salary is less than $10,000, or its equivalent for foreign nationals.

Address: American Public Health Association, 1015 15th Street, NW, Washington, DC 20005.

BACHELOR OF SCIENCE IN OCCUPATIONAL HYGIENE

Suggested curriculum

First year

FALL	UNITS
Humanities, behavioral and social sciences elective	3
Social sciences elective	2
Mathematics	4
Inorganic chemistry	4
Biology elective	4
	17

SPRING	
Humanities, behavioral and social sciences, elective	3
Inorganic chemistry	4
Biology elective	3
Physics	4
Occupational hygiene and occupational health	2
	16

Second year

FALL	
Physiology	4
Organic chemistry	4
Quantitative and qualitative analysis	4
Humanities elective	2
Fundamentals of occupational hygiene	3
	17

SPRING	
Biology elective	3
Biochemistry	4
Principles of industrial processes	3
Environmental sanitation	4
Humanities elective	3
	17

Third year

FALL	
Human biology	3
Fundamentals of industrial toxicology	3
Principles of occupational hygiene measurements	3
Industrial dust	3
Engineering sciences elective	3
Humanities, social sciences elective	2
	17

SPRING	
Thermal environment: assessment and control	3
Noise: assessment and control	3
Ionizing radiation, lighting and atmospheric pressure	3
Industrial safety and accident prevention	3
Air scrubbing techniques	2
Free elective	3
	17

SUMMER FIELD TRAINING

Fourth year

FALL	
Advanced occupational hygiene instrumentation techniques	3
Laboratory analysis	3
Advanced toxicology	4
Compliance and legal aspects of health and safety	3
Principles of health surveillance and medical monitoring	3
	16

SPRING	
Humanistic and Social Sciences	3
Organization and administration of occupational hygiene problems	3
Approach to problem solving	3
Internship and occupational health project	4–6
	13–15

SUMMER FIELD TRAINING

Faculty supervised training is given in a variety of industrial settings.

MASTER OF SCIENCE DEGREE IN INDUSTRIAL HYGIENE

Suggested curriculum

Required Courses

TITLE	QUARTER HOURS
Principles of Epidemiology	4
Biometrics in Public Health I	4
Biometrics in Public Health II	4
Principles of Environmental Health	4
Environmental Calculations	2
Environmental and Occupational Laboratory	3
Environmental and Occupational Health Seminar	1 (2 quarters required)
Recommended Courses (tailored to student's background)	
Fundamentals of Industrial Hygiene	3
Environmental Accoustics	3
Advanced Laboratory Topics	4
Occupational & Environmental Diseases I	3
Occupational & Environmental Diseases II	3
Radiological Health	4
Air Quality Management I	4
Water Quality Management I	4

Additional courses recommended in industrial ventilation, ergonomics, occupational safety, hazardous waste management, environmental and occupational epidemiology, and human physiology.

Total course credits	50
Research and thesis	22
Total program	72

25

The Safety Professional

by Walter T. McLean, CSP
Revised by Fred A. Manuele, PE, CSP

THIS CHAPTER DEALS WITH THE ROLE of the safety professional in an effective occupational safety and health hazard control program. The duties and function of the safety professional and how he/she works with other professionals are briefly discussed as well as the many ways the safety professional can contribute to the success of the occupational health program.

Safety is a multidisciplinary profession, drawing its workers from many different areas—education, engineering, psychology, medicine, biophysics, and so forth. Interestingly, many workers in the field did not initially choose safety as a profession, but instead became interested in accident prevention and/or loss control while working in other areas. Many made the change into the safety profession through the recognition that a well-defined safety program accomplished management's aim of producing high-quality products at the lowest cost.

The safety profession today is a sophisticated discipline combining engineering, preventive medicine, industrial hygiene, behavioral psychology, and a knowledge about such things as systems safety analysis, human factors engineering, biomechanics, and product safety (Figure 25-1). In addition, the modern safety professional must possess those finer attributes of the old-time "safety expert"—a thorough knowledge of the company's or plant's equipment, facilities, manufacturing process, and its employees. The safety professional must have the ability to work with all types of individuals to get the message across.

The safety professional must be a diagnostician, and a combination civil, mechanical, and industrial engineer, who must display tact, diplomacy, and, at times, even an abrasive aggressiveness, and be ready to back up forcefulness with facts. In short, the safety professional of today must wear many hats and play many roles. The safety professional should serve as a counselor to the company's chief executive. The safety leader must be able to enter the boardroom of a company as an equal—and to do this, must understand the basic technology of the industry.

An increasing number of institutions of higher education offer degrees in safety engineering; a listing of schools and of training courses is given in Appendix A, Sources of Help. Such courses do much toward developing a professional attitude by safety engineers.

Just as all the professional fields change with advancing technology, so does the field of safety. In the early days of safety when someone was designated the "safety man," the responsibility was chiefly to determine where machine guards were needed, conduct a housekeeping inspection, and write reports on any injuries or fatalities. Later, safety programs were organized and the professional qualifications for the "safety man" were tightened. In many of these programs, the testing of the work environment for toxic or harmful gases and dusts that were known to be immediately harmful became a routine safety function. The "safety man" grew to professional status, becoming the "safety professional" without gender reflection.

In its broadest sense, occupational health has come to mean not only freedom from disease but from injury as well. And because of this, the safety professional has become more closely aligned with the industrial hygienist and the field of industrial medicine. Whether a person is disabled because of heart disease, pneumonia, or a mangled hand, the problem is of equal concern to employee, the employee's family, and the employer.

There is no question that accidents are costly to industry and society. Elimination of accidents is vital to the public interest.

The Scope of the Professional Safety Position

Safety professionals are engaged in hazard identification, hazard evaluation and hazard elimination or mitigation to an acceptable level in all aspects of human endeavor. In this context, a hazard is any condition or action which presents an exposure to danger or harm to people, property or the environment.

The safety professional may serve as an engineer, as a manager or as a consultant in every aspect of the planning, organization, direction, operation and control of an endeavor. In performing these functions, the safety professional must draw upon specialized knowledge in the physical and social sciences. Knowledge of engineering, physics, chemistry, statistics, mathematics and principles of measurement and analysis is integrated in the evaluation of safety performance. The safety professional must have a thorough understanding of the causative factors contributing to accident occurrence and combine this with knowledge of motivation, behavior and communication in order to devise methods and procedures to control hazards. Knowledge of management principles, theory of business and government organization is essential to apply the safety professional's specialized knowledge to the work, home or leisure environment.

A safety professional is then defined as "an individual who, by virtue of his specialized knowledge, skill and educational accomplishments, has achieved professional status in the safety field".

The safety professional needs a unique and diversified education and training to meet today's challenges. The recent population explosion, urban expansion and transportation systems development, as well as the complexities of every day life create many problems and extend the safety professional's creativity to its maximum to successfully provide the knowledge and leadership to conserve life, health and property. Areas of specialization in which the safety professional could be involved include occupational safety and health programs, safety engineering, fire protection, traffic and transportation safety, industrial hygiene, product safety, system safety and public safety.

The major functions of the safety professional are contained within four basic areas. Application of all or some of the functions described in this text will depend upon the nature and scope of the hazardous condition or loss problem encountered.

Functions of the Professional Safety Position

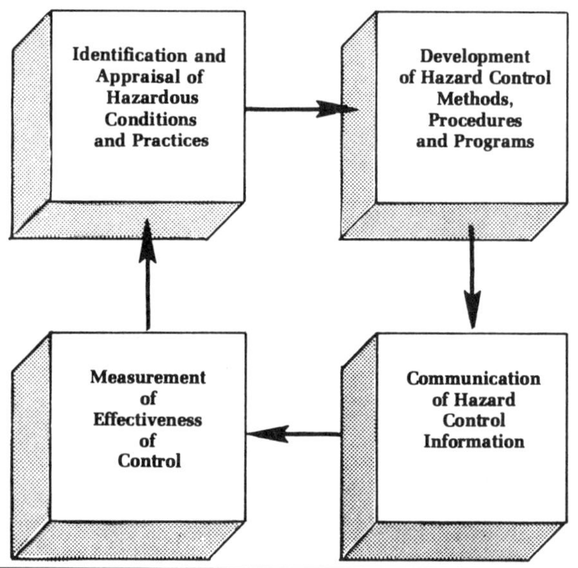

Figure 25-1a. This is taken from a brochure prepared by the American Society of Safety Engineers, and is reprinted through the courtesy of the Society.

Accidents produce economic and social loss, impair individual and group productivity, cause inefficiency, and generally retard progress. (The term "accident" as used here is defined to mean any unexpected happening that interrupts the work sequence or process and that may result in injury, illness, or property damage to the extent that it causes loss.)

Dedicated safety professionals continue to be accident prevention's most valuable asset. Their ranks have grown to the point where membership in the American Society of Safety Engineers (ASSE) is now approaching 20,000. This organization, dedicated both to their interests and to their professional development, has approximately 120 chapters in the United States and Canada. Individual membership is worldwide. (See the discussion in Appendix A.)

Functions of the Professional Safety Position

The major areas are:

A. Identification and appraisal of hazardous conditions and practices and evaluation of the severity of the accident or loss problem.

B. Development of hazard control methods, procedures and programs.

C. Communication of hazard control information to those directly involved, including the management, planning and motivation necessary to integrate safety considerations into operations.

D. Measurement and evaluation of the effectiveness of the hazard control system and development of the modifications needed to achieve optimum results.

A. Identification and appraisal of hazardous conditions and practices and evaluation of the severity of the accident or loss problem

These functions involve:

1. Development of methods of identifying hazards and evaluating the loss producing potential of a given system, operation or process by:
   a. Detailed studies of hazards of planned and proposed facilities, systems, operations, equipment and products.
   b. Hazard analysis of existing facilities, operations and products.
2. Preparation and interpretation of analyses of the total loss resulting from the hazards and exposures under consideration.
3. Review of the entire system in detail to define likely modes of failure, including human error, and their effects on the safety of the system.
   a. Identification of incomplete decision-making, faulty judgment and administrative miscalculation.
   b. Isolation of potential weaknesses found in existing policies, directives, objectives or practices.
   c. Use of specialists for the evaluation of specific problem areas.
4. Review of reports of injuries, property damage, environmental diseases or public liability accidents and the compilation, analysis and interpretation of relevant causative factor information.
   a. Establishment of a classification system which will identify significant causative factors and determine trends.
   b. Establishment of a system to insure the completeness, accuracy and validity of the reported information.
   c. Conduction of a thorough investigation of those situations where specialized knowledge and skill are required.
5. Provision of advice and counsel concerning compliance with applicable laws, codes, regulations and standards.
6. Conducting research studies of technical safety problems.
7. Determination of the need for surveys and appraisals by related specialists such as physicians, health physicists, industrial hygienists, fire protection engineers, design engineers and psychologists to identify conditions or practices affecting the health and safety of individuals.
8. Systematic studies of the various elements of the environment to assure that tasks and exposures of individuals are within psychological and physiological limitations and capacities.

Figure 25-1b.

There are many other qualified safety professionals, in addition to the ASSE members, who, together with thousands of specialists and technicians, carry out a limited scope of activities within the occupational health field.

In 1968, the ASSE was instrumental in forming the Board of Certified Safety Professions (BCSP). Its purpose is to provide the professional status of a Certified Safety Professional (CSP) to qualified safety people by certification after meeting strict education and experience requirements and passing an examination. As of March, 1987, 6,055 CSPs and approximately 1,000 Associate Safety Professionals (ASPs) had been certified by the BCSP. (The ASP designation is awarded to those who pass the Core Examination, the initial exam of the certification process; ASP indicates a recognition of the person's progress towards certification.)

In 1985, the BCSP and the American Board of Industrial Hygiene (ABIH) began joint sponsorship of a certification program for Occupational Health and Safety Technologists (OHSTs). This designation is not intended for persons who are Certified Industrial Hygienists (CIH) or CSPs, nor is it intended for those eligible to take either the CIH or CSP examinations. Its purpose is to recognize technologists in the fields of health and safety.

There has been an orderly development of safety knowledge, which, when applied with sufficient skill and judgment, has produced significant reductions in occupational disease and in many types of accidents and accidental injuries. However, the tremendous increase in scientific knowledge and technological progress has added to the complexities of safety work.

The focus of control of industrial disease and accident prevention efforts has oscillated between environmental control or engineering, and human factors. From this, some important trends

B. Development of hazard control methods, procedures and programs.

In performing this function, the safety professional:

1. Uses the specialized knowledge of accident causation and loss control to prescribe an integrated hazard control system designed to:
   a. Eliminate causative factors associated with the loss problem, preferably before a loss occurs.
   b. Devise mechanisms to reduce the degree of hazard when it is not possible to eliminate the hazard.
   c. Reduce the severity of the results of an accident, harmful exposure or event by prescribing specialized equipment designed to reduce the severity of the injury, illness or loss.
2. Establishes methods to demonstrate the relationship of safety performance to the primary function of the entire operation or any of its components.
3. Develops policies, codes, safety and health standards and procedures that become part of the operational policies of the organization.
4. Incorporates essential safety and health requirements in all purchasing and contracting specifications.
5. As a consultant to personnel engaged in planning, design, development and installation of various parts of a system, advises and consults to insure consideration of all potential hazards.
6. Coordinates the results of hazards analyses to assist in proper selection and placement of personnel, whose capabilities and limitations are suited to the operation involved.
7. Encompasses product safety, including the intended and potential uses of the product as well as its material and construction, through the establishment of general requirements for the application of safety principles throughout planning, design, development, fabrication and testing of various products, to achieve maximum product safety.
8. Systematically reviews technological developments and equipment to keep up-to-date on devices and techniques designed to eliminate or minimize hazards.

C. Communication of hazard control information on those directly involved, including the management, planning and motivation necessary to integrate safety considerations into operations.

In performing this function, the safety professional:

1. Compiles, analyzes and interprets accident, loss and exposure statistical data and prepares reports designed to communicate valid and comprehensive recommendations for corrective action.
2. Communicates recommended hazard control procedures and programs, designed to eliminate or minimize hazard potential, to decision-making personnel.
3. Through appropriate communications media, presents information to those who have ultimate decision-making responsibilities to adopt and utilize those controls which the preponderence of evidence indicates are best suited to achieve the desired results.
4. Directs or assists in the development of specialized education and training materials, and in the conduct of specialized training programs.
5. Provides advice and counsel on the type and channels of communication to insure the timely and efficient transmission of useable hazard control information to those concerned.

D. Measurement and evaluation of the effectiveness of the hazard control system and development of the modifications needed to achieve optimum results.

In performing this function, the safety professional:

1. Establishes measurement techniques, such as cost statistics, cost benefit analysis, loss rate, risk analysis, work sampling or other appropriate means for obtaining periodic and systematic evaluation of the effectiveness of the control system.
2. Develops methods to evaluate the costs of the control system in terms of the effectiveness of each part of the system and its contribution to the control of hazards.
3. Provides feedback information concerning the effectiveness of the control measures to those with ultimate responsibility, with the recommended adjustments or changes as indicated by the analyses.

Figure 25-1c.

in the pattern of the safety professional's development have emerged.

- First, increasing emphasis toward analyzing the loss potential of the activity with which the safety professional is concerned—such analysis requires greater ability (1) to predict where and how loss- and injury-producing events will occur and (2) to find the means of preventing such events.
- Second, increased development of factual, unbiased, and objective information about loss-producing problems and accident causation, so that those who have ultimate decision-making responsibilities can make sound decisions.
- Third, increasing use of the safety professional's help in developing safe products. The application of the principles of accident causation and control to the product being designed or produced has become more important because of product liability cases, legal aspects in the entire field of

HAZARD IDENTIFICATION CHECKLIST
TYPE I------SAFETY

1A. INTERNAL ENERGY RELEASE

1A1	ENERGY SOURCES	POTENTIAL	(PRESSURE VESSEL)
		KINETIC	(CENTRIFUGE)
		CHEMICAL	(FUEL, EXPLOSIVES)

1A2 UNSAFE CONDITIONS (INCLUDES HARDWARE UNRELIABILITIES)

1A3 HUMAN ERROR (CONSIDER PROGRAM OR TRAINING DEFICIENCY)

1B. EXTERNAL ENERGY DAMAGE

1B1 SYSTEM ENVIRONMENTS (SHOCK, TEMPERATURE, CONTAMINATION)

1B2 NATURAL ENVIRONMENTS (LIGHTNING, EARTHQUAKE, HURRICANE)

1C. PHYSIOLOGICAL DAMAGE

1C1 TOXIC SOURCES (POISON GASES, X-RAYS, NOISE)

1C2 DEPRIVATIONS (ANOXIA, STARVATION, DEHYDRATION)

Figure 25-2. Hazard identification checklists are very useful in pinpointing safety and health hazards.

safety and health including negligent design, and the obvious impact that a safer product has on the overall health and safety of the environment.

DEFINITIONS

What, then, is a safety professional? In the broad sense, the safety professional is the person whose basic job function and responsibility is to prevent accidents and other harmful exposures and the personal injury or disease or property damage that may ensue. In the narrower sense, the safety professional is additionally that person who has successfully met the requirements for certification developed by the Board of Certified Safety Professionals of the Americas, Inc., or other certifying groups, or one who passes state certification requirements. This chapter will deal primarily with the broad definition.

Whether called a safety engineer, safety director, loss control manager, or by some other title, the job of the safety professional is to prevent—as nearly as possible—the human suffering brought about by exposure to health hazards and unsafe conditions. A secondary goal is to prevent damage to equipment or materials, to minimize interruptions in operations, and to reduce the costs associated with employee accidents or illness originating in operational processes. The scope of the safety professional's job is illustrated in Figure 25-1, from the ASSE.

The safety professional functions as a specialist whose authority is based on specialized knowledge and soundness of information provided. An ability to gather well-documented facts, based on valid and reliable methods of reducing accidents and losses, determines the validity of that authority. For the most part, the safety professional's efforts are directed at supplying management with accurate decision-making information.

One of the more important skills necessary for attaining a reduction in the accident rate is the ability to see and identify health and safety hazards that others may overlook (Figure 25-2). Because of the safety professional's skill, management is provided with an evaluation of the health and safety problem and methods by which the problem can be resolved.

Based upon the information collected and analyzed, together with recommendations based upon specialized knowledge and experience, the safety professional can propose alternate solutions to those who have ultimate decision-making responsibilities.

The person who is the safety professional frequently has the staff responsibility for fire protection and prevention, may be involved with security, and, in small establishments, may have many additional nonsafety functions.

The safety professional brings together in one person those elements of various disciplines, especially those that are necessary to identify and evaluate the magnitude of the health and safety problem. Concerned with all facets of the problem—personal and environmental, transient and permanent—the safety professional must determine the causes of accidents or the existence of loss-producing conditions, practices, or materials.

The safety professional draws upon specialized knowledge in both the physical and social sciences; and applies the principles of measurement and analysis to evaluate safety performance. Necessary to the job is a fundamental knowledge of statistics, mathematics, physics, chemistry, and the engineering disciplines (including human factors engineering), as well as a familiarity with industrial hygiene and medical applications.

Knowledge of behavior, motivation, and communication is important as is a command of management principles and the theory of business and government organization. This specialized knowledge must include a thorough understanding of accident causation as well as methods and procedures designed to control such events.

Accident prevention activities

The basic accident prevention activities (in descending order of effectiveness and preference) are as follows:

Figure 25-3. Personal protective equipment can be used to shield employees against a hazard.

1. Eliminate the hazard from the machine, method, material, or facility structure.
2. Control or contain the hazard by enclosing or guarding it at its source, or exhausting an airborne hazard away from the operator.
3. Train operating personnel to be aware of the hazard and to follow safe job procedures to avoid it.
4. Prescribe personal protective equipment for personnel to shield them against the hazard (Figure 25-3).

It is beyond the scope of this section to describe completely all accident prevention activities of safety professionals at each operation; however, the primary responsibilities are outlined here:

- Provide advisory services on health and safety problems and other matters related to accident prevention.
- Develop a centralized program to control accident and fire hazards.
- Keep informed of changes in federal, state, and local safety codes and communicate such information to mangement.
- Develop and apply safety standards both for production facilities (equipment, tools, work methods, and safeguarding) and for products, based on applicable legal and voluntary codes, rules, and standards.
- Work closely with the engineering, industrial hygiene, medical, and purchasing departments during the development and construction of new equipment and facilities. Ensure that a procedure is established to assure that only safe tools, equipment, and supplies are purchased; advise the purchasing department on acceptable supplies and materials; review and approve purchase requisitions for personal protective equipment and safety items.
- Develop, plan, and implement the health and safety-inspection program carried out by the operating supervisors and field safety personnel to identify potential hazards, both in the workplace and in the use of the company's products. Inspect all new equipment in conjunction with engineering, operating, and personnel representatives for adequate safeguards and freedom from major safety and health hazards.
- Guide operating supervision in accident investigation to determine its cause and to prevent recurrence. Review nondisabling injury accident reports on a sample basis to check the thoroughness of the accident investigation and corrective actions taken.
- Collect and analyze data on illness and accidents for the purpose of corrective action and to determine accident trends and provide targets for corrective action. Maintain necessary records with regard to accidents, inspections, recommendations, and other accident prevention activities.
- Assure education and training of employees in general health and safety principles and techniques. Maintain supervisory contacts for new instructions, follow-up, and general health and safety motivation (Figure 25-4).
- Cooperate with medical personnel on matters of employee health and fitness to work; and with industrial hygiene or environmental quality control personnel on industrial hygiene problems.

HEALTH AND SAFETY PROGRAMS

Management usually places administration of the accident prevention or safety program in the hands of a safety professional whose title is safety director, manager of safety, or loss control manager.

Full staff responsibility for the safety activities should be assigned to one person. The decision concerning proper placement of responsibility should be based upon the size of the company and the nature of the hazard involved in its operation.

Employment of full-time safety professionals is increasing for the following reasons:

- The passage of the Occupational Safety and Health Act (OSHA) of 1970 requires that certain safety standards be met and maintained.
- A better understanding of the safety professional's services and functions is developing. To administer a safety program effectively, the individual in charge must be highly trained and/or have many years of experience in the safety field.

A health and safety program is not something that is imposed on company operations as an afterthought. Safety—an integral

Figure 25-4. Educating employees about the importance of general health and safety principles and techniques is one of the mainstays of an effective health and safety program.

part of company operations—must be built into every process or product design and into every operation.

The prevention of illness, accidents, and injuries is basically achieved through control of the working environment and control of people's actions. The safety professional can assist management to implement such control.

A company with an effective health and safety program has a working environment in which operations can be conducted safely, economically, and efficiently, with a minimum of employee, customer, and public complaints.

Staff v. line status

In general, the health and safety program is administered by safety professionals or other persons holding line positions in a small company, and staff positions in a large company. In a large corporation, the safety professional and the organization probably have staff status and authority. The exact organizational status of the safety staff is determined by each firm in terms of its own operating policies.

The health and safety program as a staff function should have the following objectives:

- to establish staff credibility to advise and counsel regarding safety or health matters
- to keep all personnel adequately informed regarding safety or health matters
- to assure accountability for safety is properly identified with every staff group and operating management
- to program activities that support harmonious supervisor/employee interaction on safety or health matters
- to establish and reinforce consistent attention to preventive practices and actions.

Sometimes the safety professional is delegated authority that is usually limited to line officials. On fast-moving and rapidly changing operations or those on which delayed action would endanger the lives of workers or others, as in construction and demolition work, fumigation, some phases of manufacture of explosives, chemicals, or other dangerous substances, or emergency work, it is common to find that the safety professional has authority to order immediate changes, including the shutting down of specific equipment or operations.

CODES AND STANDARDS

The safety professional must be familiar with codes and standards applicable to equipment, material, environmental controls, or energy sources. Only by knowing which codes and standards apply, can the safety professional give valid advice regarding company standards for purchasing specifications. While the safety professional must know what meets government agency regulatory requirements (i.e., OSHA, MSHA, EPA), there are many other guides and consensus standards that provide state-of-the-art models. Therefore, the safety professional should be familiar with the following:

- codes and standards approved by the American National Standards Institute (ANSI) and other standards and specifications groups (see Bibliography).
- codes and standards adopted or set by federal, state, and local governmental agencies. This is particularly important where local or state codes are more stringent than federal codes.
- codes, standards, and lists of approved or tested devices published by such recognized agencies as Underwriters Laboratories Inc. and by fire protection organizations.
- safety practice recommendations of such agencies as the National Safety Council, insurance carriers or their associations, and trade and industrial organizations.

Plans and specifications

An important duty of the safety professional should be that of checking plans for new or remodeled facilities; new, rebuilt or rearranged equipment; changes in material used in product or processes; material storage and handling procedures; and for products to be made. This important function must be done early enough to afford an opportunity to discover health and safety hazards and to correct conditions that may otherwise be built into the building and its equipment and that would later result in injuries or other casualty losses. There is also the opportunity, at this planning stage, to build in safety or fire protection features, and to provide adequate space for exit aisles, janitor closets, waste-collection equipment, and other often-overlooked functions.

Many companies will not permit a drawing or specification to be used until it has been approved by the safety professional. Hazards involved in making products should be minimized insofar as possible. Instructions and warnings developed for employee use should be reviewed for safe manufacturing procedures. Plans that in any way affect current applicable hazard communication requirements must be reviewed carefully.

The safety professional should also make sure that company policies and applicable standards are followed in purchase specifications for new materials and equipment and for modification of existing equipment. Some companies have arranged for the purchasing department to notify the safety department when new materials or equipment are to be purchased, or when there is a new supplier of safety-related materials. For instance, when a new chemical is requested, the safety department should see that any applicable Material Safety Data Sheet (MSDS) is obtained from the manufacturer.

The engineering department, with the help of the safety professional, should check with the purchasing department to determine the necessary safety and health measures to be built on or into a machine before it is purchased. Purchasing agents in an industrial plant are necessarily cost-conscious. Consequently, the safety professional must know the occupational disease and accident losses to the company in terms of specific machines, materials, and processes. If the professional is to recommend the expenditure of several thousand dollars for protection of health or additional guarding to be used throughout the plant, for instance, there should be valid evidence that the investment is justified.

Because of highly competitive marketing, manufacturers of machine tools and processing equipment often list safety devices, such as guards or noise enclosures, designed for the protection of operators as separate auxiliary equipment (Figure 25-5). The supplier may not know the ultimate use of the product. The actual needs for guards and automatic controls depend on the proximity of the operator and vary from one installation to another. The safety professional must evaluate each installation and be in a position to satisfy the purchasing agent of the need for health and safety equipment to be included in the original order, or recommend the issuance of additional purchasing orders to provide adequate protection to the operator.

Figure 25-5. This soundproofed enclosure surrounding a swaging machine effectively prevents transmission of sound to adjacent work areas.

In many organizations, particularly where certain personal protective equipment items, such as goggles or safety shoes, are to be reordered from time to time, standard lists have been prepared and purchases are selected only from among the types and from the companies shown on these approved lists.

In many companies, safety and health functions are placed in three coordinate departments:

1. the engineering department, where plans and specifications are prepared for all machinery and equipment to be purchased
2. the safety and health department, where plans and specifications are carefully checked for safety and health

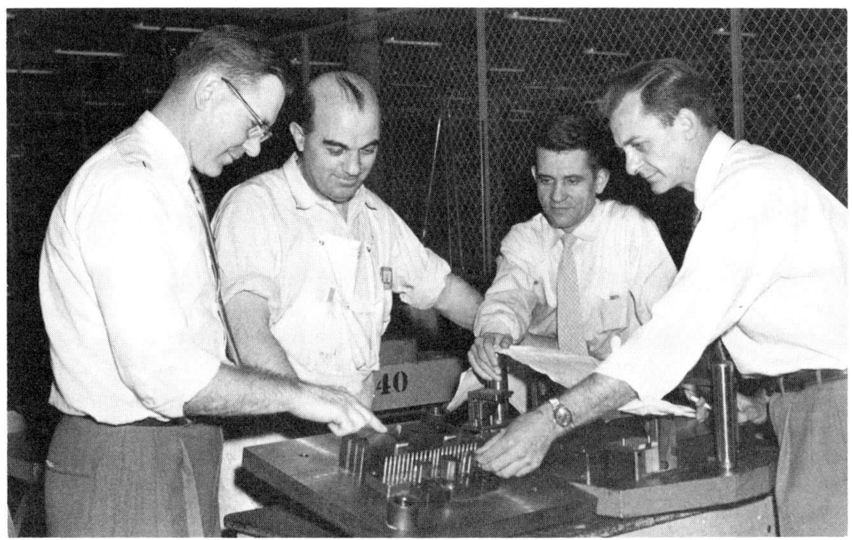

Figure 25-6. As tool is being built, toolmakers and designers discuss production and safety features.

3. the purchasing department, which has much latitude in making selections and determining standards of quality, efficiency, and price.
   Note that even in a smaller organization, someone must be responsible for these three functions for an effective safety and health program.

Machine design

The ultimate objective of a company's engineering program is to design equipment and processes and to plan job procedures so the company produces the best product with the highest quality, at the lowest cost. The customer must be satisfied and the company should make a profit. It is the safety professional's job to see that engineering personnel are acquainted with the particular safety and health hazards involved and to suggest methods of eliminating these hazards.

The safety professional should discuss with all levels of management and supervisors the conditions responsible for accidents or potential accidents. The safety professional should have the necessary knowledge and skills for developing procedures to prevent accidents and be able to make recommendations for effective use of the facilities at hand for safer as well as more efficient production.

The ultimate goal is to design safe and healthful environments and equipment and to set up job procedures so that employee exposure to the hazards of illness and injury will either be eliminated or controlled as completely as possible. This can be accomplished when a high degree of safety and health is incorporated into the design of the equipment or the planning of the process, along with adequate training and supervision.

Company policy should specify that safety and health measures must be designed and built into the job or work instructions before the job is executed. To add health and safety features after work on a job has begun is usually less effective, less efficient, and more costly.

The most efficient time to engineer safety and health hazards out of the plant, product, process, or job is before building or remodeling, while a product is being designed, before a change in a process is put into effect, or before a job is started. Every effort, therefore, should be made to find and remove potential health and safety hazards at the blueprint or planning stage (Figure 25-6).

The machine manufacturer, like any other businessman, wants to have satisfied customers. If the machines cause accidents, customers will be dissatisfied. If the customer's order for a machine specifies that the machine must meet specific regulations of OSHA (or other agency) and have safety built into it, the manufacturer's designers will regard such a specification as a design requirement which they must meet. If only a general statement such as "must meet OSHA standards," is used, the manufacturer will not know which standards apply and the equipment may not be properly guarded. The OSHAct is a "user" law and sets few standards for the manufacturer.

In many instances, guards added to a machine after it has been installed in the plant are easily removed and often are not replaced. If a guard or enclosure is an aid to production and efficiency rather than a hindrance, it is unlikely that the machine would be operated without having the guard in place. Machine safety must be improved without hindering the worker or reducing the efficiency of the machine. (See Bibliography for more information.)

The best solution lies in a basic guard design that eliminates the safety and health hazard and might even increase efficiency. There can be little prospect for safe operation of a machine unless the idea of building safety and health measures into the machine's function is applied right on the drawing board for the establishment that is going to use the equipment. Installation of multiple units usually requires different safety and health specifications than the installation of only a single unit.

Purchasing

The safety professional is responsible for generating and documenting health and safety standards that will guide the purchasing department. These standards should be set up to so that the safety and health hazards involved in a particular kind of equipment or material being purchased are eliminated or, at the very least, substantially reduced.

The purchasing staff, while not concerned closely with educational and enforcement activities, is vitally concerned with many phases of the engineering activities. The purchasing staff selects and purchases the various items of machinery, tools, equipment, and materials used in the organization; and it is their responsibility—to a considerable degree—to see that safety has received adequate attention in the design, manufacture, and particulars of shipment of these items.

The safety professional should be well prepared to advise the purchasing department when required to do so. The purchasing staff can reasonably expect that the safety professional will offer the following:

- give specific information about safety and health hazards that can be eliminated by change in design or application of guarding by the manufacturer
- supply information about equipment, tools, and materials that can cause injuries if misused
- give specific information about health and fire hazards in the workplaces
- provide information on federal and state safety and health rules and regulations
- supply, on request, information on accident experience with machines, equipment, or materials when such articles are about to be reordered.

Health and safety considerations. In purchasing, items such as lifting devices, automatic packaging equipment, chemical processing, or storage equipment, may have a great bearing on health and safety; for example, extreme caution must be observed in the purchase of personal protective equipment, including eye protection, respirators, gloves, and the like; of equipment for the movement of suspended loads, such as ropes, chains, slings, and cables; of equipment for the movement and storage of materials; of miscellaneous substances and fluids for cleaning and other purposes that might constitute or aggravate a fire or health hazard. Adequate labeling that identifies contents and calls attention to safety and health hazards should be specified. This labeling must comply with state or federal hazard communication ("right-to-know") standards. With the changing rules and regulations by federal and state agencies, the safety professional must keep up-to-date on both employee and community right-to-know regulations.

Many unsuspected safety and health hazards must be considered when very ordinary items, such as common hand tools, reflectors, tool racks, cleaning rags, and paint for shop walls and machinery are purchased. Among the factors to be considered are, for example, maximum load strength; long life without deterioration; sharp, rough, or pointed characteristics of articles; need for frequent adjustment; ease of maintenance; ergonomic factors that may result in production of excessive fatigue. Where toxic chemicals are involved, disposal of residue, scrap, and shipping containers, must be considered. Safety professionals who are in day-to-day contact with the operating problems must give such information to the purchasing agent.

Price considerations. The purchasing staff must constantly reconcile quality, work efficiency, and health and safety with the price of an item.

Sometimes the cost of an adequately protected machine may seem out of all proportion to the cost of an unguarded machine. As previously stated, the experience of many industrial facilities has proved that the best time to safeguard a machine or process is in the design stage. Safeguards planned and built as integral parts of a machine are usually the most efficient and durable. Thus, the purchase of machinery or equipment involves a system whereby definite specifications, including drawings, are first prepared by engineers. These plans and specifications should then be carefully checked by the safety professional before bids and cost estimates are solicited. The purchasing agent should have these plans and specifications at hand when prices are requested.

HEALTH AND SAFETY INSPECTIONS

Many people may be involved in health and safety inspections. Their duties are outlined in this section.

Safety inspectors or technicians

Inspectors should know how to locate health and safety hazards and have the authority to act and make recommendations. A good safety inspector must know the company's accident experience, familiarity with accident potentials, ability to make intelligent recommendations for corrective action, and diplomacy in handling situations and personnel.

The safety inspector must be equipped with the proper personal protective equipment, protective clothing, and other required equipment to carry out duties. It would be difficult for the safety inspector to persuade an employee to wear eye protection or safety shoes when the inspector does not wear them, or to require workers to use respirators unless the inspector sets the example and uses one in a hazardous environment. It is essential that the inspector practice what he or she preaches.

Safety professionals

The safety professional has a very productive role during safety inspections, coordinates the safety and health program, and teaches by firsthand contact and on-the-spot examples.

The number of safety professionals and inspectors needed for adequate safety inspection activities depends a great deal on the size and complexity of the facility and the type of industry. Large companies with well-organized safety and health programs usually employ a staff of full-time safety professionals and inspectors who work directly under the safety director or safety supervisor. Large companies also can have specially designated employees who spend part of their time on inspections; and there may be employee inspection committees.

The safety professional should be in full charge of developing health and safety inspection activities and receive reports of all inspectors. Special departmental inspectors should either make safety and health inspections personally or supervise the inspectors in their work. Although the safety professional may have considerable desk work, he or she should get out into the production and maintenance areas as often as possible and make general as well as specific safety and health inspections. If there is more than one facility involved, there should be a plan to make at least an annual inspection survey of each plant.

Purpose of a health and safety inspection program

Health and safety inspections are one of the principal means of locating potential causes of accidents and illness and help determine what safeguarding is necessary to protect against safety and health hazards before accidents and personal injuries occur (Figure 25-7).

As inspections of the manufacturing process are important functions in quality control, health and safety inspections are likewise important in accident control.

Inspections should not be limited to a search for unsafe physical conditions, but should also try to detect unsafe or unhealthful work practices.

Finding unsafe conditions and work practices and promptly

Figure 25-7. Weekly safety inspections are conducted to spot potential health and safety hazards before an accident occurs.

correcting them is one of the most effective methods to prevent accidents and safeguard employees. Management can also show employees its interest and sincerity in accident prevention by correcting unsafe conditions or work practices immediately. Inspections help to sell the health and safety program to employees. Each time a safety professional or an inspection committee passes through the work area, management's interest in health and safety is advertised. Regular plant inspections encourage individual employees to inspect their immediate work areas.

In addition, inspections facilitate the safety professional's contact with individual workers and obtaining their help in eliminating accidents and illnesses. Frequently, the workers can point out unsafe conditions that might otherwise go unnoticed and uncorrected. When employee suggestions are acted upon, all employees are helped to realize their cooperation is essential and appreciated.

Health and safety inspections should not be conducted primarily to find how many things are wrong, but rather to determine if everything is satisfactory. Their purpose should be to discover conditions that, if corrected, will bring the plant up to accepted and approved health and safety standards, and result in making it a safer and more healthful place in which to work. When observed, inspectors should tactfully point out any unsafe work procedures to the employees involved. They should be certain to indicate the hazards. Inspectors may need to recommend new or continuing health and safety training for supervisors and employees.

Inspection of work areas

It is advisable to review the reports containing analyses of all accidents (including noninjury accidents and near misses, if possible) for the previous several years, so that special attention can be given to those conditions and locations known to be scenes of accidents.

The inspection made at irregular intervals is the type of inspection used in most plants. It can include an unannounced inspection of a particular department, piece of equipment, or small work area. Such inspections made by the safety department tend to keep the supervisory staff alert to find and correct unsafe conditions before they are found by the safety inspector.

The need for intermittent inspections is frequently indicated by accident tabulations and analysis. If the analysis shows an unusual number of accidents for a particular department or location, or an increase in certain types of injuries, inspection should be made to determine the reasons for the increase and to find out what is necessary to make corrections. All results of inspections must be discussed with operating supervision if any gain is to be made.

Supervisors should continuously be sure that tools, machines, and other department equipment are maintained properly and are safe to use. To do this effectively, they should use systematic inspection procedures and can delegate authority to others in a department.

Inspection programs should be set up for new equipment, materials, and processes. A process should not be put into regular operation until it has been checked for hazards, additional safeguards installed (if necessary), and safety instructions or procedures developed. Serious injury has occurred because this routine was not followed. This is also a good time to make a complete job safety analysis (JSA) of the operation and to train employees in safe operations. It takes less time and effort now than if done later.

Third-party inspections or audits

The value of a third-party inspection of policies, procedures, and practices, as well as an inspection of the physical facility and equipment is increasingly evident. The advantages of such audits are as follows:

- Objectiveness of the inspecting party is less likely to bias the findings or reporting of findings.
- Results of the audit usually are directed at a higher level of decision-making authority and thus are more likely to be acted upon promptly.
- Performance of the audit does not have to depend on the time/convenience of organization staff.
- Professionals contracted for such audits usually have much expertise in a given industry.

Many businesses currently find an annual audit and inspection of their facilities to assess the state of their safety, health, and environmental affairs to be as important as the traditional financial audit. Results of these third-party audits often are included in the company's annual report. More information on third-party audit services is often available from insurance carriers, independent safety/health consulting firms, or the National Safety Council.

ACCIDENT AND OCCUPATIONAL ILLNESS INVESTIGATIONS

This section deals with both noninjury accidents and injury accidents.

Purpose

Investigation and analysis is used by safety professionals to prevent accidents and illnesses. The investigation or analysis can produce information that leads to countermeasures that prevent or reduce the number of accidents and illnesses. The more complete the information, the easier it will be for the safety professional to design effective countermeasures. For example, knowing that 40 percent of a plant's accidents involve ladders is useful, but it is general and not as useful as knowing that 80 percent of the plant's ladder accidents involve broken rungs.

An investigation of at least every disabling injury or illness (or every OSHA lost workday case) should be made. Accidents resulting in nondisabling injuries, or no injuries, and also "near-accidents" should be investigated to evaluate the cause in relation to injury-producing accidents or breakdowns, especially if there is frequent recurrence of certain types of nondisabling injuries, or if the frequency of accidents is high in certain areas of operations.

The consequences of certain types of accidents are so devastating, that any hint of conditions that might lead to their occurrence warrants an investigation. In such cases, any change from standard safety specification calls for a thorough investigation.

For purposes of accident prevention, investigations must be fact-finding, not fault-finding; otherwise, they may do more harm than good. This is not to say that responsibility may not be fixed where personal failure has caused injury, or that such persons should be excused from the consequences of their actions.

Types of investigations

There are many accident investigation and analysis techniques available. Some of these techniques are more complicated than others. The choice of a particular method depends upon the purpose and orientation of the investigation.

The accident investigation and analysis procedure focuses primarily on unsafe circumstances surrounding the occurrence of the accident, and is the most often used technique. Other similar techniques involve investigation within the framework of defects in man, machine, media, and management (the "4 M's"); or education, enforcement, and engineering (the "3 E's of Safety").

For analysis, these techniques involve classifying the data about a group of accidents into various categories. This has been referred to as the statistical method of analysis. Countermeasures are designed on the basis of most frequent patterns of occurrence.

Other techniques are discussed later in this chapter under the systems approach to safety. Systems safety stresses an enlarged viewpoint that takes into account the interrelationships between the various events that could lead to an accident. Because accidents rarely have a single cause, the systems approach to safety can point to more than one place in a system where effective countermeasures can be introduced. This allows the safety professional to choose the countermeasure that best meets the criteria for effectiveness, speed of installment, etc. Systems safety techniques also have the advantage of application before accidents or illnesses occur and can be applied to new procedures and operations.

Who makes the investigation?

Depending on the nature of the accident and other conditions, the investigation can be made by the supervisor, the safety engineer or inspector, the workers' health and safety committee, the general safety committee, the safety professional, or a loss control specialist from the insurance company or other external source. Regardless of who conducts the initial investigation, a representative of the company safety department should verify the findings and a written report should be made to the proper official or to the general safety committee.

The safety professional's value and ability are best shown in the investigation of an accident. Specialized training and analytical experience enable the professional to search for all the facts, both apparent and hidden, and to submit an unbiased, unprejudiced report. The safety professional should have no interest in the investigation other than to get information that can be used to prevent a similar accident.

During an investigation, methods to prevent a recurrence can occur, but decisions about the specific course to take are best made after all facts are well established. There are usually several alternatives; all need full understanding to allow the most effective decision. The safety professional should present every valid, feasible alternative to operating management for their consideration. At this stage, input from employees is most beneficial in determining the best corrective measure.

RECORDS

The Williams-Steiger OSHAct of 1970 requires employers to maintain records of work-related employee injuries and illnesses, plus many inspection reports of high injury potential equipment. In addition, many employers are also required to make reports to state compensation authorities.

Safety professionals are faced with two tasks: (1) maintaining those records required by law and by their management, and (2) maintaining records that are useful to an effective safety program. Unfortunately, the two are not always synonymous. A good record-keeping system necessitates more data than that contained in almost all OSHA-required forms.

Records of accidents and injuries and the training experience of the persons involved are essential to efficient and successful safety programs, just as records of production, costs, sales, and profits and losses are essential to efficient and successful operation of a business. Records supply the information necessary to transform haphazard, costly, ineffective safety and training efforts into a planned safety and health program that controls both conditions and acts that contribute to accidents. Good record-keeping is the foundation of a scientific approach to occupational safety.

Uses of records

A good record-keeping system can help the safety professional in the following ways:

- provide safety personnel (1) with the means for an objective evaluation of the magnitude of occupational illness and accident problems and (2) with a measurement of the overall progress and effectiveness of the health and safety program.
- identify high-hazard units, plants, or departments and problem areas so that extra effort can be made in those areas.
- provide data for an analysis of accidents and illnesses that can point to specific circumstances of occurrence, which can then be attacked by specific countermeasures.
- create interest in health and safety among supervisors by furnishing them with information about the accident and illness experience of their own departments.
- provide supervisors and safety committees with hard facts about their safety and health problems so that their efforts can be concentrated.
- measure the effectiveness of individual countermeasures and determine if specific programs are doing the job that they were designed to do.
- establish the need for, and the content of, employee and management development training programs that can be tailor-made to fit the particular needs of that company or plant.

Accident reports and illness records

To be effective, preventive measures must be based on complete and unbiased knowledge of the causes of accidents, and the knowledge of the supervisor and employee about the operation. The primary purpose of an accident report—like the inspection—is to obtain information, not to fix blame. Since the completeness and accuracy of the entire accident record system depend upon information in the individual accident reports and the employee training history, it is important that the forms and their purpose are understood by those who must fill them out. Necessary training or instruction by the safety professional should be given to those who are responsible for generating the information. (Illustrations of typical forms are given in the latest editions of the National Safety Council's *Accident Prevention Manual for Industrial Operations: Administration and Programs,* and the *Supervisors Safety Manual*—see Bibliography).

The first aid report. Collecting injury or illness data generally begins in the first aid department. The first aid attendant or nurse fills out a first aid report for each new case. Copies are sent to the safety department or safety committee, the worker's first-line supervisor, and other departments as management may wish.

The first aid attendant or the nurse should know enough about accident analysis and illness investigation to be able to record the principal facts about each case. Note that the questioning of the injured or sick person must be complete enough to establish whether the incident is or is not work-related. Current emphasis on chemical air contaminants makes it necessary to include or exclude exposure to known health hazards. First aid reports can be very helpful to the safety or industrial hygiene personnel. The employer's physician who treats injured employees also should be informed of the basic rules for classifying cases since, at times, the physician's opinion of the severity of an injury may be necessary to record the case accurately.

The supervisor's accident report form. This should be completed as soon as possible after an accident occurs, and copies sent to the safety department and to other designated persons. Information concerning unsafe or unhealthful work conditions and improper work procedures is important in the prevention of accidents, but information that shows why the unsafe or unhealthful conditions existed can be even more important. This type of information is particularly difficult to get unless it is obtained promptly after the accident occurs. If the information is based on opinion, not on proven facts, it is still important, but should be so identified.

Generally, analyses of accidents are made only periodically, and often long after the accidents have occurred. Because it is often impossible to accurately recall the details of an accident, this information must be recorded accurately and completely at once or they may be lost forever.

Injury and illness record of an employee. The first aid report and the supervisor's report contain information about the agency of injury (type of machine, tool, or material), and the type of accident or other factor that will facilitate use of the reports for accident prevention. Another form, therefore, must be used to record the injury experience of individual employees.

The employee training record card should have space to record injury information such as the date, classification, days charged, and costs.

Because of the importance of the personal factor in accidents, much can be learned about accident causes from studying employee injury records. If certain employees or job classifications have frequent injuries or illnesses, a study of the work environment, job training, health and safety training, work practices, and the instructions and supervision given them may reveal more than a study of accident locations, agencies, or other factors.

EDUCATION AND TRAINING

Health and safety training begins at the time of employment, before the employee actually starts work. An effective safety and health training program will include a carefully prepared and presented introduction to the company (Figure 25-4).

New employees immediately begin to learn and form attitudes about the company, job, boss, and co-workers, whether or not the employer offers a training program. To encourage a new employee to form positive attitudes, it is important for the employer to provide a sound basis—providing health and safety information is vital. Training about exposure to chemical hazards in the workplace is now mandated by state and federal Hazard Communication Standards (right-to-know laws); see Chapter 30, Governmental Regulations, for more information on these and other relevant regulations and standards. (The National Safety Council offers many audiovisual programs concerned with on- and off-the-job health and safety; and these presentations can be modified to suit each company's needs.)

An effective accident prevention and occupational health hazard control program is based on proper job performance. When people are properly trained to do their jobs, they will do them safely. This, in turn, means that supervisors must (1) know what employee training needs to be given, which means knowing the training requirements of the job; (2) know how to train an employee in the safe way of doing a job; as well as (3) know how to supervise. It also means the safety professional should be familiar with good training techniques. Although the professional is not always directly involved in the training effort, he or she should be able to recognize the elements of a practical training program.

A training program is needed (1) for new employees, (2) when new equipment or processes are introduced, (3) when procedures have been revised or updated, (4) when new information must be made available, (5) when employee performance needs to be improved, and (6) on a periodic basis to refresh the material. Older employees also need training so that they have the same information about new equipment, products, or company policies that new employees are receiving. Unless the training programs are planned to upgrade skills of both management and employees, they may be an excess cost item.

Many supervisors acquired their present positions in organizations where some sort of safety and health program already existed, and their understanding of the program is firmly established. However, a safety professional undertaking the safety training of supervisors will almost invariable find that the first major job is to get supervisors at all levels to understand and accept their role in accident and illness prevention. This job cannot be done in a single meeting or by a single communication.

Simply getting supervisors to agree in theory that responsibility for safety and health is one of their duties is not enough. They must come to understand the many ways in which they can prevent illness and accidents, and they must become interested in improving their safety performance. For a safety and health program to be effective, all levels of management must be firmly committed to the program, and express that commitment by action and example. Management is ultimately responsible for the safety and health of the employees. Much of the effort put into an industrial safety and health program by a safety professional is, therefore, directed toward educating and influencing people.

Employee training

The training of the employee begins the day the employee starts the job. As observed earlier in this chapter, whether or not the firm has a formal safety and health indoctrination program, the employee starts to learn about the job and to form attitudes about many things—including safety and health—on the first day.

The safety professional assists supervisors in instructing employees in the safe way of doing each job. But accidents can be prevented only when these recommended procedures are based upon a thorough analysis of the job and when these procedures are followed. This is why a complete Job Safety Analysis is so valuable (Figure 25-8). It provides a consistent baseline against which to compare and it details all necessary safety elements of the various job tasks.

The safety professional can provide methods for observing all workers in the performance of their tasks to establish the job safety requirements. The National Safety Council's training program on the implementation of Job Safety Analyses can be very beneficial. The safety professional should participate in follow-up observations to reinforce the supervisor's training. In this way, supervisors are informed of any weaknesses in the company's safety and health program, and will have a common reference point for monitoring these occurrences.

Maintaining interest

A prime objective of a safety and health program is to maintain interest in safety to prevent accidents. It is, however, as difficult to determine the degree of success achieved by an interest-maintaining effort as part of a safety program as it is to isolate the effectiveness of an advertising campaign separate from an entire marketing program. The reason is that, generally, companies with sound basic safety and health programs have working conditions that are safe, employees who are well trained and safety minded, and supervision that is of a high caliber.

Health and safety rule enforcement

Obeying health and safety rules is actually a matter of education; employees must understand the rules and the importance of following them. In helping an employee understand, the possibility of language barriers should be considered. However, language barriers not only are caused by national origins but more frequently by the jargon of a particular profession or industry. A considerable amount of confusion can occur when a new employee comes from a different industry or field of work. To avoid problems of that nature and to set a good example for employee education, both top management and supervisors must know and believe in safety and health rules, and must conscientiously follow them.

Role of the supervisor

Supervisors are the "key persons" in any program designed to create and maintain interest in safety and health, because they are responsible for translating management's policies into action and for promoting safe and healthful work practices directly among the employees. The supervisor's attitude toward safety and health is a significant factor in the success, not only of specific promotional activities, but also of the entire safety and health program, because his or her views will be reflected by the employees in the department.

How well the supervisor meets this responsibility will be determined to a large extent by how well the supervisor has been trained. And training and educating the supervisor in matters of safety and health is the responsibility of the safety professional.

The supervisor who is sincere and enthusiastic about accident prevention can do much to maintain interest because of a direct connection with the worker. Conversely, if the supervisor pays only lip service to the program or ridicules any part of it, this attitude offsets any good that might be accomplished by the safety professional.

Some supervisors are reluctant to change their mode of operation and are slow to accept new ideas. It is the safety professional's task to sell these supervisors on the benefits of accident prevention, to convince them that promotional activities are not "frills," but rather projects that can help them do their job easier and prevent illness and injuries, and to persuade them that their wholehearted cooperation is essential to the success of the entire safety and health program.

Setting a good example by wearing safety glasses and other personal protective equipment whenever it is required is an excellent way in which the supervisor can promote the use of personal protective equipment and demonstrate interest in safety. Teaching safety and health principles to supervisors is an important function of the safety professional; safety posters, a few warning signs, or even general rules are not enough to do this job.

JOB SAFETY ANALYSIS *INSTRUCTIONS ON REVERSE SIDE*	JOB TITLE (and number if applicable): PAGE____ OF____ JSA NO.____		DATE:	☐ NEW ☐ REVISED
	TITLE OF PERSON WHO DOES JOB:	SUPERVISOR:	ANALYSIS BY:	
COMPANY/ORGANIZATION:	PLANT/LOCATION:	DEPARTMENT:	REVIEWED BY:	
REQUIRED AND/OR RECOMMENDED PERSONAL PROTECTIVE EQUIPMENT:			APPROVED BY:	

SEQUENCE OF BASIC JOB STEPS	POTENTIAL HAZARDS	RECOMMENDED ACTION OR PROCEDURE

80M385 Printed in U.S.A. Product No. 156.15

INSTRUCTIONS FOR COMPLETING JOB SAFETY ANALYSIS FORM

Job Safety Analysis (JSA) is an important accident prevention tool that works by finding hazards and eliminating or minimizing them *before* the job is performed, and *before* they have a chance to become accidents. Use your JSA for job clarification and hazard awareness, as a guide in new employee training, for periodic contacts and for retraining of senior employees, as a refresher on jobs which run infrequently, as an accident investigation tool, and for informing employees of specific job hazards and protective measures.

Set priorities for doing JSAs: jobs that have a history of many accidents, jobs that have produced disabling injuries, jobs with high potential for disabling injury or death, and new jobs with no accident history.

Here's how to do each of the three parts of a Job Safety Analysis:

SEQUENCE OF BASIC JOB STEPS

Break the job down into steps. Each of the steps of a job should accomplish some major task. The task will consist of a *set* of movements. Look at the first *set* of movements used to perform a task, and then determine the next logical *set* of movements. For example, the job might be to move a box from a conveyor in the receiving area to a shelf in the storage area. How does that break down into job steps? Picking up the box from the conveyor and putting it on a handtruck is one logical set of movments, so it is one job step. Everything related to that one logical set of movements is part of that job step.

The next logical *set* of movements might be pushing the loaded handtruck to the storeroom. Removing the boxes from the truck and placing them on the shelf is another logical set of movements. And finally, returning the handtruck to the receiving area might be the final step in this type of job.

Be sure to list *all* the steps in a job. Some steps might not be done each time—checking the casters on a handtruck, for example. However, that task is a part of the job as a whole, and should be listed and analyzed.

POTENTIAL HAZARDS

Identify the hazards associated with each step. Examine each step to find and identify hazards—actions, conditions and possibilities that could lead to an accident.

It's not enough to look at the obvious hazards. It's also important to look at the entire environment and discover every conceivable hazard that might exist.

Be sure to list health hazards as well, even though the harmful effect may not be immediate. A good example is the harmful effect of inhaling a solvent or chemical dust over a long period of time.

It's important to list *all* hazards. Hazards contribute to accidents, injuries and occupational illnesses.

In order to do part three of a JSA effectively, you must identify potential and existing *hazards*. That's why it's important to distinguish between a hazard, an accident and an injury. Each of these terms has a specific meaning:

HAZARD—A potential danger. Oil on the floor is a *hazard*.

ACCIDENT—An unintended happening that may result in injury, loss or damage. Slipping on the oil is an *accident*.

INJURY—The *result* of an accident. A sprained wrist from the fall would be an injury.

Some people find it easier to identify possible accidents and illnesses and work back from them to the hazards. If you do that, you can list the accident and illness types in parentheses following the hazard. But be sure you focus on the *hazard* for developing recommended actions and safe work procedures.

RECOMMENDED ACTION OR PROCEDURE

Using the first two columns as a guide, decide what actions are necessary to eliminate or minimize the hazards that could lead to an accident, injury, or occupational illness.

Among the actions that can be taken are: 1) engineering the hazard out; 2) providing personal protective equipment; 3) job instruction training; 4) good housekeeping; and 5) good ergonomics (positioning the person in relation to the machine or other elements in the environment in such a way as to eliminate stresses and strains).

List recommended safe operating procedures on the form, and also list required or recommended personal protective equipment for each step of the job.

Be specific. Say *exactly* what needs to be done to correct the hazard, such as, "lift, using your leg muscles." Avoid general statements like "be careful."

Give a recommended action or procedure for *every* hazard.

If the hazard is a serious one, it should be corrected immediately. The JSA should then be changed to reflect the new conditions.

Figure 25-8. A Job Safety Analysis form is used to record information that will be used as a consistent baseline against which to compare, and includes information on all necessary safety elements of the various job tasks.

The safety professional should educate supervisors so that working conditions are kept as safe and healthful as possible and that the workers follow safe procedures consistently, simply as part of good job performance. The supervisor is entitled to all of the help the safety professional can give through supplies of educational material for distribution and frequent visits as circumstances permit. Supervisors should also receive adequate recognition for independent and original safety activity.

Supervisors can be most effective in giving facts and personal reminders on safety and health to employees as part of their daily work instructions. This procedure is particularly necessary in the transportation and utility industries where the work crews are on their own.

In any case, supervisors should be encouraged to take every opportunity to exchange ideas on accident prevention with workers, to commend them for their efforts to do the job safely, and to invite them to submit suggestions for better ways to do the job that will prevent injuries or illness.

Job safety and health analysis

Job safety and health analysis (Figure 25-8) is a procedure used by safety professionals and supervisors to review job methods and uncover hazards (1) that may have been overlooked in the layout of the plant or building and in the design of the machinery, equipment, and processes, (2) that may have developed after production started, and (3) resulted from changes in available work force.

Once the safety and health hazards are known, the proper solution can be developed. Some solutions may be physical changes that control the hazard, such as enclosures to contain an air contaminant or placing a guard over exposed moving machine parts. Others may be job procedures that eliminate or minimize the hazard, for example, safe piling of materials. These will require training and supervision.

A job safety and health analysis is a procedure to make a job safe by:

1. identifying the hazards or potential injuries or illnesses associated with each step of a job
2. developing a solution for each hazard that will either eliminate or control the exposure.

Benefits. The principal benefits that arise from job safety analysis are as follows:

- giving individual training in safe, efficient procedures
- instructing the new employees on health and safety procedures
- preparing for planned safety and health observations
- giving "prejob" instructions on irregular jobs
- reviewing job procedures after accidents occur.

New employees must be trained in the basic job steps. They must be taught to recognize the safety and health hazards associated with each job step and must learn the necessary precautions. There is no better guide for this training than a well-prepared job safety analysis (Figure 25-8) used by the job instruction training method.

All supervisors are concerned with improving job methods to increase safety, reduce costs, and step up production. The job safety analysis is an excellent starting point for questioning the established way of doing a job.

TOTAL LOSS AND DAMAGE CONTROL

An important, recent development by safety professionals is the concept of total loss and damage control. Many progressive safety professionals have investigated and applied a concept of total accident and illness control based on studies of "near misses" (noninjury accidents), and on detailed analyses of hidden as well as direct accident costs.

Total loss control

Cost-conscious companies find they must control accident costs if they are to continue to do business in a highly competitive market. They do not consider safety a nuisance, but a necessity—not a fringe benefit, but a prerequisite.

Health and safety activities related to safeguarding corporate investment and continuity of operation need coordination with the safety professional. For example, related functions, as personnel safety and fire protection, must be closely integrated. Those activities that are related to loss control must be functionally integrated.

One large company uses an approach consisting of five closely related, logically ordered steps for a coordinated program.

Hazard identification. To prevent accidents and control losses, first identify all safety and health hazards—to determine those areas or activities in an operation where losses can occur. This requires studying processes at the research stage, reviewing design during engineering, checking pilot plant operations and startup, and regularly monitoring normal production.

Hazard elimination. Toxic, flammable, or corrosive chemicals can sometimes be replaced by safer materials. Machines can be redesigned to remove danger points. Plant layouts can be improved by eliminating blind corners or limited visibility crossings, for example.

Hazard protection. Hazards that cannot be removed must be protected against. Familiar examples include mechanical guards to keep fingers from pinchpoints, safety shoes to safeguard toes against dropped objects, and ventilation systems to control the buildup of air contaminants. Industry is concerned with all losses, injury to personnel, damage to products, and destruction of property.

Maximum possible loss. This step involves the determination of the maximum loss that could occur if everything went wrong. Entire buildings or areas can be lost by fires or explosions; thus the amount that a company could lose under the most adverse conditions can be estimated.

Loss retention. Having some idea of the amount that could be lost under a combination of unfavorable circumstances, one can then determine what portion of such a loss a company is willing to bear itself. Industrial companies can afford to retain a portion of each loss. The remaining loss potential is then insured through the company's insurance carrier. This proves a good incentive for strong safety and health programs. These activities can be consolidated in one department; such as a loss prevention department, bringing together the safety professional, the fire protection manager, the security and plant protection manager, the industrial physician, the industrial hygienist, and the insurance manager. The administrator of a total loss control program does not need to know all the details of each function but should be able to develop an atmosphere in which there is harmonious cooperation and mutual understanding. Primary concerns are the control of occupational disease and personnel safety.

Damage control

There is a damage control program for investigating all accidents, not just those that produce injuries. This approach of studying accidents instead of injuries recognizes that a so-called "no-injury accident" if repeated in the future could result in some personal injury, property damage, or both.

Ferreting out the causes of accidents determines what unsafe conditions and/or work practices were responsible for the accident.

Three basic steps are successfully used to reduce property damage (and injuries): (1) spot-checking, (2) reporting by repair control centers, and (3) auditing.

Spot-checking. Spot-checking involves observing and taking notes to permit damage estimates, by comparing total costs for a repair period with those found during sample observations.

Reporting by repair control centers. This step involves developing a system in which the repair or cost control center records property damage. The system should be designed with the least amount of paperwork. No one system will work in all companies because repair cost accounting methods vary greatly from company to company, and even from plant to plant.

Auditing. An effective reporting program requires complete auditing. Safety personnel should receive a copy of every original work order processed through the maintenance, planning, and cost control center. Safety professionals make on-the-spot checks to see if accidental damage was involved.

SYSTEMS SAFETY

Recently, safety professionals are increasingly exploring system approaches to industrial accident prevention. Safety professionals are asked to find ways of implementing system safety techniques. And although complete system safety analysis requires specially trained engineers and rather sophisticated mathematics, safety professionals find that some knowledge of these techniques can directly benefit helping codify and direct their safety and health programs.

A system analysis can clarify a complex process by devising a chart or model that provides a comprehensive, overall view of the process by showing its principal elements and the ways in which they are interrelated (Figure 25-9).

Having established the concept of a system, the next step is the analysis of systems. Progress in the analysis of complex systems enables application by safety professionals to solving problems in the control of occupational illness and accident prevention.

Methods of analysis

There are four principal methods of analysis: (1) failure mode and effect, (2) fault tree, (3) THERP, and (4) cost-effectiveness. Each has variations and more than one may be combined in a single analysis.

Failure mode and effect. In this method the failure or malfunction of each component is considered, including the mode of failure (such as, a switch jammed in the ON position), the effects of the failure are traced through the system, and the ultimate effect on the task performance is evaluated.

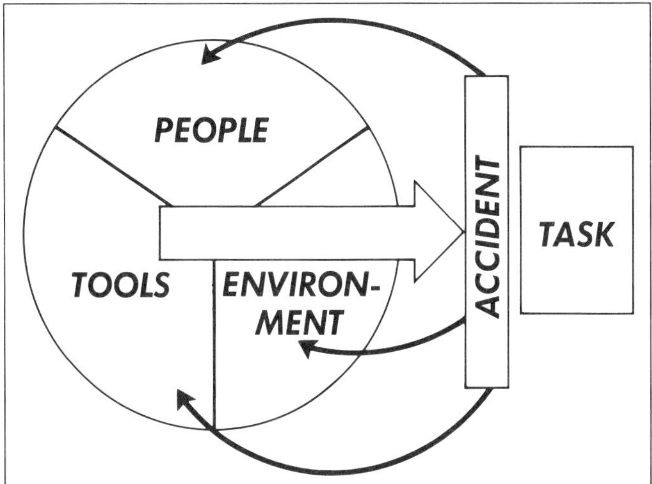

Figure 25-9. A system analysis can show how the interrelationship of people, tools, and the environment can be combined to produce an accident.

Fault tree. In this method an undesired event is selected and all the possible occurrences that can contribute to the event are diagrammed in the form of a tree. The branches of the tree are continued until independent events are reached. Probabilities are determined for the independent events and after simplifying the tree, both the probability of the undesired event and the most likely chain of events leading up to it can be computed.

THERP. This is a technique for human error prediction, developed by Scandia Corporation, which provides a means for quantitatively evaluating the contribution of human error to the degradation of product quality. It can be used for human components in systems and thus can be combined either with the failure mode and effect or the fault-tree method.

Cost-effectiveness. In the cost-effectiveness method, the cost of system changes made to increase safety and health measures is compared with either the decreased costs of fewer serious failures, or with the increased effectiveness of the system to perform its task, to determine the relative value of these changes. Ultimately all system changes have to be evaluated, but this method makes such cost comparisons explicit. Moreover, cost-effectiveness is frequently used to help make decisions concerning the choice of one of several systems that can perform the same task.

In all of these analytical methods, the main point is to measure quantitatively the effects of various failures within a system. In each case, probability theory is an important element.

The systems approach to safety can help to change the safety profession from an art to a science by codifying much of safety and health knowledge. It can help change the application of safety measures from piecemeal problem solving (such as putting a pan under a leak) to a safely designed operation (preventing the leak itself).

The safety professional determines what can happen if a component fails or the effects of malfunction in the various elements of the system and provides answers before the accident occurs instead of after the damage has been done.

Board of Certified Safety Professionals
208 Burwash Avenue
Savoy, Illinois 61874
Phone: (217) 359-9263

PROCEDURES FOR CERTIFICATION

A. REQUISITES

Section 1. General Requirements

To be eligible for the designation of "Certified Safety Professional," an applicant must be of good character and reputation, submit an application on the proper forms, satisfy the educational and experience requirements, complete the examination requirements, and pay the required fees.

Section 2. Definitions

a. "Certified Safety Professional" — An individual who utilizes the expertise derived from a knowledge of the various sciences and professional experience, to create or develop procedures, processes, standards, specifications and systems to achieve an optimal control or reduction of the hazards and exposures which are detrimental to people and/or property by the utilization of analysis, synthesis, investigation, evaluation, research, planning, design, and consultation and who has met all of the requirements for certification established by the Board of Certified Safety Professionals.

b. "Associate Safety Professional" — An individual who has successfully completed the Core Examination but who has not successfully completed one of the specialty examinations.

c. Professional Safety Experience — Professional safety experience differs from non-professional safety experience in the degree of responsible charge and ability to defend analytical approaches and engineering/administrative control recommendations. The safety professional must be able to demonstrate to the satisfaction of his peers, employer, and clients the ability to use analysis, synthesis, design, investigation, planning and communication to optimally control or reduce the hazards and exposures that would be detrimental to people, property and environment. For a position to be accepted as qualifying professional safety experience, professional safety work must account for at least 50% of the position's responsibilities. Additionally, the first six months of an applicant's first safety position is considered trainee experience and is not acceptable as a professional safety experience.

Section 3. Academic Requirement

The academic requirement for certification is an accredited baccalaureate degree in safety meeting the Board's current requirements or an acceptable combination of other education and professional safety experience.

a. An applicant's academic background is evaluated on an individual basis and units of credit are assigned up to a maximum of 48 units for a fully acceptable baccalaureate degree in safety.

b. For applicants who receive less than the maximum credit for their education, professional safety experience may be substituted for the remaining credit needed at the rate of one month of experience for each unit of credit needed.

Section 4. Experience Requirement

A minimum of four years of professional safety experience as defined in Section 2 is required in addition to the academic requirement. Professional Safety experience used to meet the academic requirement may not be used to meet the experience requirement.

Section 5. Experience Equivalents

An earned graduate degree from an accredited institution may be accepted in lieu of a portion of the required professional safety experience. The specific credit allowed will depend upon the degree and the major area of study.

a. A master's degree is acceptable for credit ranging from a minimum of three months of experience to a maximum of twelve months.

b. A doctorate is acceptable for credit ranging from a minimum of six months of experience to a maximum of twenty-four months.

c. Only one graduate degree is acceptable for credit. For applicants with more than one graduate degree, the Board will use the degree which will receive the most credit.

Section 6. Examination Requirements

a. The examination requirements applicable to these requisites consist of a *Core Examination* and a *Specialty Examination*. These examinations are selected by the Board to effectively determine the applicant's knowledge and ability to perform the functions of a professional safety position.

b. Examination Eligibility:

(1) **Core Examination.** Applicants who are otherwise qualified and who have completed the minimum academic requirements of Section 3 are eligible to sit for the Core Examination. Applicants who have passed the Core Examination must pass a Specialty Examination within a period of eight years after passing the Core Examination.

(2) **Specialty Examinations.** Applicants who are otherwise qualified, meet the minimum academic requirements of Section 3, the experience requirements of Section 4, and who have received passing scores on the Core Examination are eligible to sit for a Specialty Examination. Applicants who meet the minimum academic and experience requirements and who hold one of the following current certifications or registrations may sit for a Specialty Examination without passing the Core Examination:

(a) Registration as a Professional Engineer in any U.S. state, district or territory.

(b) Certification as an Industrial Hygienist by the American Board of Industrial Hygiene.

(c) Certification as a Health Physicist by the American Board of Health Physics.

Section 7. Application for Certification

a. Application for certification shall be on forms prescribed and furnished by the Board. All applications and supporting documentation submitted to the Board become the property of the Board.

b. Each application will be evaluated by an Application Review Team composed of current members of the Board of Directors. The Review Team will determine the applicant's eligibility for the examinations and eligibility for certification upon successfully completing the required examinations. The results of the application review will be communicated to the applicant in writing.

c. An applicant who disagrees with the results of the Board's evaluation of his application has the right to appeal to the Board in accordance with the following procedures:

(1) Request for Re-evaluation

If an applicant disagrees with the Board's initial evaluation, he may request, *in writing*, that his file be re-evaluated. This appeal should include the reasons for appealing and provide any necessary clarification of previously submitted material which may have a bearing on the Board's evaluation.

(2) Request for Adjudication Committee Review

If an applicant is not satisfied with the results of a re-evaluation of his application, he may request, *in writing*, a second appeal to be submitted for Adjudication Committee Review. The Adjudication Committee consists of three senior members of the Board of Directors and the results of its review are considered final and binding.

(3) Time Limits on Appeals

A request for re-evaluation must be made within one year of the receipt of the results of the Board's initial evaluation of the application. A request for a review by the Adjudication Committee must be made within one [illegible] of the receipt of the results of a re-evaluation.

(4) Special Re-evaluations

Cases in which applicants are requesting re-evaluations on the basis of the attainment of additional qualifications since their initial evaluations will not be considered as appeals and will not be subject to the time con-constraints of an appeal. Examples of such re-evaluations are those resulting from the attainment of additional academic degrees and additional registrations or certifications.

Figure 25-10. Current Procedures for Certification from the Board of Certified Safety Professionals. (Continued)

d. Eligible applicants who do not sit for an examination within a period of three years after the notarized date of application will be required to reapply. Applicants, other than current Associate Safety Professionals and Certified Safety Professionals, who have failed an examination and who have not retaken the examination within a period of three years after the examination was taken will be required to reapply. The application files will be destroyed for applicants who do not sit for an examination within the time limits described above. Applicants are responsible for keeping the Board informed of any changes in mailing addresses.

Section 8. Fees

a. The application fee for certification shall be 50 U.S. Dollars. It shall accompany all applications and is not refundable.

b. The fee for each examination shall be established by the Board. If, after paying an examination fee, an applicant is unable to take the scheduled examination for a reason acceptable to the Board, the fee will be refunded. An applicant taking a re-examination shall pay another examination fee.

c. Upon notification that the Board of Examiners has authorized the issuance of a certificate, such applicant shall forward to the Board an initial fee equivalent to the appropriate annual renewal fee.

d. Should the Board deny the issuance of a certificate of qualification to any applicant, the fee(s) paid shall be retained by the Board.

e. The annual renewal fees shall be as follows:

(1) "Associate Safety Professional" — The annual renewal fee for applicants receiving the designation of Associate Safety Professional shall be 20 U.S. Dollars.

(2) "Certified Safety Professional" — The annual renewal fee for applicants receiving the designation of Certified Safety Professional shall be 30 U.S. Dollars.

Section 9. Certificates and Designations

a. The Board shall issue a "Certificate of Qualification" to any applicant who, in the opinion of the Board, has met the requirements contained in the requisites for the designation of Certified Safety Professional and who has paid the required fee(s). The Board shall also issue a certificate and the designation "Associate Safety Professional" to any applicant who has passed the Core Examination in recognition of the progress made toward certification.

b. The certificates issued by the Board shall be as follows:

(1) Certified Safety Professional — The certificate of qualification issued to applicants meeting the requirements for the designation "Certified Safety Professional" shall carry the designation "Certified Safety Professional", shall show the full name of the applicant without any titles, shall show the area of specialization, shall have a serial number and shall be signed by the President and the Secretary under the Seal of the Board.

(2) Associate Safety Professional — The certificate of recognition issued to applicants meeting the requirements for the designation "Associate Safety Professional" shall carry the designation "Associate Safety Professional", shall show the full name of the applicant without any titles, shall have a serial number, and shall be signed by the President and the Secretary under the Seal of the Board. This designation may be maintained for a period not to exceed eight (8) years. At the end of the eight-year period, the designation shall not be renewed except when authorized by petition to the Board.

c. The Board shall, upon payment of the annual renewal fee, issue a wallet-sized "Certificate" showing the expiration date of said "Certificate".

d. Each "Certified Safety Professional" may obtain a seal and/or stamp of the specific design, authorized by the Board, bearing the certificate-holder's name, serial number, and the legend, "Certified Safety Professional". The seal or stamp may be used by the individual to identify his work except where such use is prohibited by law.

e. The seal and/or stamp as authorized above shall be purchased from the Board at its registered address and the cost shall be borne by the designee.

f. An applicant issued the designation "Certified Safety Professional" may use the initials "CSP" after his name. The designation "CSP" will not be further modified when used in such a manner.

B. EXPIRATIONS AND RENEWALS

Section 1. Expirations

Each certificate issued by the Board shall expire annually on a date set by the Board and shall become invalid after that date unless renewed. The Secretary of the Board shall notify every person holding a valid certificate of the amount of the fee required for its renewal. Such notice shall be mailed to the last known address of each certificate holder at least thirty days in advance of the expiration date.

Section 2. Renewals

Payment of the renewal fee is due on or before the expiration date of the certificate. If payment is not received within six (6) months after the expiration of the certificate, the renewal fee will not be accepted, and a new certificate will be issued only by re-application and re-examination, except when timely submission of the fee is prevented by circumstances beyond the control of the designee.

C. DISCIPLINARY ACTION

Section 1. Grounds

The Board shall have the power to reprimand a certificate holder, suspend, refuse to renew or revoke the certificate of any certificate holder who is found guilty of the following:

a. Obtaining a certificate through the use of fraud or deceit.

b. Misconduct, incompetence or gross negligence in the practice for which a certificate was issued.

c. Any felony or any crime involving moral turpitude.

Section 2. Procedure

a. Any person may bring charges on the grounds listed in Section 1 against someone holding a certificate issued by the Board. Such charges shall be in writing and shall be sworn to before a Notary Public by the person or persons making them and shall be filed with the Secretary of the Board.

b. All charges, unless dismissed by the Board of Directors, as unfounded or trivial, shall be heard by the Board within six months of the date on which they shall have been received by the Secretary.

c. The time and place of said hearing shall be fixed by the Board, and a copy of the charges, together with a notice of the time and place of hearing, shall be personally served on or mailed to the last known address of the person charged at least thirty days before the date fixed for said hearing.

d. At any hearing, the person against whom the charges were filed may appear personally or be represented or both and shall have the right to cross-examine witnesses in his defense and to produce evidence and witnesses in his own defense. If the person charged fails or refuses to appear, the Board may proceed to hear and determine the validity of the charges.

If after such hearing, a majority of the Board votes in favor of sustaining the charges, the Board shall reprimand, suspend, refuse to renew, or revoke the certificate issued to the individual charged.

e. The Board may, for reasons it may deem sufficient, reissue a certificate to any person whose certificate has been revoked.

f. Any certificate holder receiving a reprimand or whose certificate has been suspended or revoked shall have the right to a personal appeal. Such appeal shall be directed to the President who shall call a meeting of the Board within ninety days of the filing of the appeal with the Secretary of the Board. Action taken by the Board at such appeal hearing shall be final.

RESPONSIBILITIES OF SAFETY PROFESSIONALS certified by the BOARD OF CERTIFIED SAFETY PROFESSIONALS OF THE AMERICAS

The Certified Safety Professional in achieving certification, recognizes and assumes responsibility to the safety profession by maintaining:

Professional integrity in his relations with clients, associates and the public that reflects the highest standard of ethics.

Professional competence by remaining abreast of the technical administrative and regulatory developments in his chosen field.

Figure 25-10. (Concluded.)

CERTIFIED SAFETY PROFESSIONAL

Employers, employees, and the public deserve some assurance that the individuals practicing safety are professionals and are able to provide the safety expertise that, in turn, should provide adequate protection.

Usually a candidate for professional status must complete a specified course of study followed by practical experience in that field. The applicant must pass an examination to prove mastery of a specific body of knowledge. Finally, a board composed of members of the profession reviews that candidate's qualifications and grants professional certification.

Professional regulation usually centers on the need to protect the public from potential harm at the hands of unqualified persons. Clearly, there was a need for professional regulation, and the Board of Certified Safety Professionals was created to fill this need.

The Board of Certified Safety Professionals of the Americas was incorporated in Illinois to establish criteria for professional certification, accept applications, evaluate the credentials of candidates, and issue certificates to those who met the requisites.

One method of determining professional abilities is to compare education and experience against a predetermined set of requirements. Once these criteria have been developed, each application showing the candidate's education and experience is evaluated against that base. The applicant may be found to be eligible to go to the next step—to take the certification examinations; upon successful completion of the examinations, the candidate is granted certification as a Certified Safety Professional or CSP.

Current requirements

Currently, a candidate must take six specific steps before being designated a CSP (Figure 25-10):

1. complete the application for certification by filling out the Board's application form
2. pay the application fee
3. be of good character and reputation (in part, this is determined by three reference forms from individuals selected by the applicant)
4. meet the educational requisite (or have acceptable experience in lieu of education)
5. meet the experience requisite
6. successfully pass the professional safety examinations.

Academic and experience requisites

The educational requirement states that the candidate must be a graduate of an accredited college or university with a bachelor's degree program in safety, which meets the Board's requirements, in order to receive full credit toward the academic requisites.

An additional paragraph on education states that applicants not meeting the academic requirements can substitute one year of professional safety experience of each academic year needed by the applicant to satisfy the requirements. The experience requirement calls for four years or more of "professional safety experience." Graduate degrees can be recognized as professional experience. A doctoral degree may be recognized in lieu of up to two years, and a master's degree may be recognized in lieu of up to one year. These experience equivalents shall be acceptable at the highest level, and shall not be cumulative.

The Board recognized that there were many persons presently in safety work who were not college graduates, which is the reason for permitting one year of practical safety experience to be substituted for each year of needed college work. Without a degree, to be eligible for certification, an applicant would need a total of four years of professional safety experience (educational requisite) plus four more years of professional safety experience and passing scores on the examinations. In contrast, a person with an acceptable bachelor's degreee in safety plus a doctoral degree would need only two years of professional safety experience and passing scores on the examinations to be certified.

To be eligible to sit for the first examination or "core" examination, the applicant need only meet the educational requirement. The examination can be taken at any time after meeting this requisite. Then eligibility criteria for the second examination require that both the academic and the experience requisites be met.

THE FUTURE OF SAFETY AS A PROFESSION

The future remains uncertain. Problems, predictable and unpredictable, can be expected to impact upon the safety professional. Some of these problems will call for reapplication of established safety techniques. Others will call for radical departures and the creation of new methods and new organizational forms. To be able to discriminate between the two situations will, perhaps, be the safety professional's greatest test.

The field of industrial safety continues to progress and improve, largely through the continued application of techniques and knowledge slowly and painfully acquired through the years. There appears to be no limit to the progress possible through the application of the universally accepted safety techniques of education, engineering, and enforcement.

Yet large and serious problems remain unsolved. A number of industries still have high accident rates. There are still far too many instances where management and labor are not working together or have different goals for the safety program.

The resources of the safety movement are great and strong—an impressive body of knowledge, a corps of able professional safety people, a high level of prestige, and strong organizations for cooperation and exchange of information.

Well-trained workers in practically all phases of safety are needed. This trend is expected to continue with the extension of legislation in occupational safety, product safety, and environmental health. Growth in the trade and service industries and the expanding safety needs of educational institutions, construction, transportation, insurance, and governmental groups should further accentuate the demand for safety workers.

Obviously, there is a need in safety work for people with varying degrees of education and experience. The range of opportunities extends from what could be considered paraprofessional to the highly trained and skilled professional at the corporate management level, and includes safety educators and government safety inspectors and researchers.

The safety professional will also need diversified education and training to meet the challenges of the future. Growth in the population, the communication and information explosion, problems of urban areas and future transportation systems, and the increasing complexities of everyday life will create many problems and may extend the safety professional's creativity to the maximum to successfully provide the knowledge and leadership needed to conserve life, health, and property.

Training of the future safety professional can no longer be confined to the on-the-job, one-on-one type, but must include specialized undergraduate level training, leading to a bachelor's degree or higher degree.

The type of training needed will depend on the individual job requirements. This presents some difficulties for those preparing to enter the health and safety occupations. Some authorities view the health and safety specialist as a behaviorist and therefore would direct training toward the behavioral sciences, such as psychology. Others see the specialist as a technician able to handle the technical problems of hazard control, and emphasize a heavy background in engineering. Still others believe the safety worker's background should include both the engineering and behavioral aspects.

Therefore, future application of this knowledge in all aspects of our civilization—whether to industry, to transportation, at home, or in recreation—makes it imperative that those in this field be trained to use scientific principles and methods to achieve adequate results. The knowledge, skill, and ability to integrate machines, equipment, and environments with humans and their capabilities will be of prime importance.

SUMMARY

The work of the safety professional follows a pattern. Before taking any steps in the containment of illness or accidents, the safety professional first identifies and appraises all existing safety and health hazards, both immediate and potential. Once having identified the hazards, the necessary accident-prevention procedures are developed and put into operation. However, this is not enough; safety and health information must be communicated to both management and workers. Finally, the safety professional must evaluate the effectiveness of safety and health control measures after they have been put into practice. If the conditions warrant the safety professional can recommend changes in materials or operational procedures or possibly that additional enclosures or safety equipment be added to machinery.

Accurate records are essential in the search for the cause of an illness or an accident, and can help find the means to prevent future similar incidents. When studying records to determine the cause or causes of accidents, the records of other companies with similar operations should not be overlooked. Upon determining the cause, the safety professional will have a firm basis on which to propose preventive measures.

Preventive measures are obviously better than *corrective measures* after an accident. This can be one of most valuable functions of the safety professional who can examine the specifications for materials, job procedures, new machinery and equipment, and for new structures from the standpoint of health and safety well before installation or construction. In some cases, the safety professional can even help draft the necessary specifications.

As part of the overall safety and health program, the safety professional should recommend policies, codes, safety standards, and procedures that should become part of the operational policies of the organization.

The safety professional draws upon specialized knowledge in both the physical and social sciences; and applies the principles of measurement and analysis to evaluate safety performance. The safety professional should have a fundamental knowledge of statistics, mathematics, physics, chemistry, and engineering.

The safety professional should be a well-informed specialist, coordinating the safety and health program and supplying the ideas and inspiration, while enlisting the wholehearted support of management, supervision, and employees.

To summarize—successful prevention of occupational illness and accidents requires the participation by the safety professional in five fundamental activities:

1. a study of all working areas to detect and eliminate or control physical or environmental hazards that contribute to accidents or illness
2. a study of all operating methods and practices
3. education, instruction, training, and discipline to minimize hazards that contribute to accidents
4. a thorough investigation of at least every accident that results in a disabling injury
5. association with peers to maintain a balanced view toward the entire field of safety.

BIBLIOGRAPHY

American National Standards Institute, 1430 Broadway, New York, NY 10018. Standards:

Method of Recording and Measuring Work Injury Experience, ANSI Z16.3-1967(R1973).

Method of Recording Basic Facts Relating to Nature and Occurrence of Work Injuries, ANSI Z16.2-1962(R1969).

Uniform Recordkeeping for Occupational Injuries and Illnesses, ANSI Z16.4-1977.

American Society of Safety Engineers (ASSE). *Scope and Function of the Professional Safety Position.* Des Plaines, IL, ASSE, 1982.

Bird, F. E., and Loftus, R. G. *Loss Control Management.* Santa Monica, CA: Institute Press, 1976.

DeReamer, R. *Modern Safety and Health Technology.* New York: John Wiley & Sons, 1980.

Firenze, R. J. *The Process of Hazard Control.* Dubuque, IA: Kendall/Hunt Publishing Co., 1978.

Grandjean, E. *Fitting the Task to the Man.* London: Taylor & Francis Ltd., 1980.

Gypsum Association. *Industrial Safety Manual,* 4th ed. Evanston, IL: Gypsum Assoc., 1985.

Hammer, W. *Occupational Safety Management & Engineering,* 3rd ed. Englewood Cliffs, NJ: Prentice-Hall, 1985.

Harner, R. E. Safety review: A system of program development and evaluation. *Professional Safety* 27 (Oct. 1982):27-29.

Heinrich, H. W., Petersen, D., and Roos, N. *Industrial Accident Prevention.* New York: McGraw-Hill Book Co., 1980.

Hughes, L. M., and LeBlanc, J. The audit: A vital force in system safety. *Hazard Prevention* (Sep./Oct. 1980): 13-15.

Moore, C. J., and Alloh, R. V. *Industrial Safety.* Portsmouth, NH: Heineman Educational Books, Inc., 1981.

National Safety Council, 444 N. Michigan Ave., Chicago, IL 60611.

Management Safety Policies, Industrial Data Sheet. No. 585, R(1985).

Accident Facts. Chicago: National Safety Council, published annually.

Accident Prevention Manual for Industrial Operations: Vol. 1: *Administration and Programs;* Vol. 2: *Engineering and Technology.* Chicago: National Safety Council, 1988.

Safeguarding Concepts Illustrated, 5th ed. Chicago: National Safety Council, 1987.

Supervisors Safety Manual, 6th ed. Chicago: National Safety Council, 1985.

Partlow, H. A. Case history of a safety sampling program. *ASSE Journal* (Aug. 1971):22-26.

Pollina, V. Safety sampling—Technique in accident control. *Environmental Control & Safety Management* (Sept. 1970).

Simonds, R. H., and Grimaldi, J. V. *Safety Management.* Homewood, IL: Richard D. Irwin, 1984.

Stone, J. R. Safety inspections—A safety catalyst. *National Safety Congress Transactions* 12(1964).

Strong, M. E. *Accident Prevention Manual for Training Programs.* American Technical Society, 1975.

Surrey, J. *Industrial Accident Research—A Human Engineering Appraisal.* Toronto: Toronto University Press, 1968.

Tarrants, W. E. *The Measurement of Safety Performance.* New York: Garland Press, 1980.

Tyler, W. W. Measuring unsafe behavior. *Professional Safety* 31(Nov. 1986): 22-24.

U.S. Dept. of Labor, OSHA. *Recordkeeping Requirements Under the Williams-Steiger Occupational Safety and Health Act of 1970.* Washington, DC: OSHA.

Workman, J. Safety sampling. *National Safety Congress Transactions* 10(1974).

26

The Occupational Physician

by Carl Zenz, MD

Since the 1940s, occupational medicine has progressed to an important medical specialty. In 1955 occupational medicine was classified by the American Board of Preventive Medicine as a specialty. This includes public health and aerospace medicine specialties.

As a subspecialty field of preventive medicine, occupational medicine is concerned with the following:

- appraisal, maintenance, restoration, and improvement of the workers' health through application of the principles of preventive medicine, emergency medical care, rehabilitation, and environmental medicine.
- promotion of a productive and fulfilling interaction of the worker and the job, via application of principles of human behavior.
- active appreciation of the social, economic, and administrative needs and responsibilities of both the worker and work community.
- team approach to health and safety, involving cooperation of the physician with occupational or industrial hygienists, occupational health nurses, safety personnel, and other specialties.

Requirements to qualify as a specialist in occupational medicine include completion of postgraduate courses in the following areas: biostatistics and epidemiology, industrial toxicology, work physiology, radiation (ionizing and nonionizing), noise and hearing conservation, effects of certain environmental conditions such as high altitude and high pressures (hyperbaric and hypobaric factors), principles of occupational safety, fundamentals of industrial hygiene, occupational aspects of dermatology, psychiatric and psychological factors, occupational respiratory diseases, biological monitoring, ergonomics, basic personnel management functions, record and data collection, governmental regulations, general environmental health (air, water, ground pollution, and waste management control).

Since 1911, workers' compensation laws have been enacted in all states. These laws require employers to (1) provide medical care for occupationally injured employees and (2) compensate employees or their heirs for occupational disability or death. In addition, most of these laws require employers to provide medical care for employees with occupational diseases.

These and other laws, especially those promulgated by the U.S. Department of Labor's Occupational Safety and Health Administration *(OSHA),* have given employers a real incentive, as well as an obligation, to maintain safe and healthful working environments. The problems associated with increasingly complex technology and with ever-new, potentially hazardous physical and chemical agents serve as an important stimulus to the development of occupational health programs (Figure 26-1). These developments caused a broadening of the earlier concept of curative occupational medicine. The field now includes and emphasizes preventive medicine and health maintenance. This new emphasis led to creation of the kind of occupational health program described in this chapter.

GOVERNMENTAL PROGRAMS

The passage of the Occupational Safety and Health Act (OSHAct) of 1970 has given new and long-needed impetus to health and safety programs as well as research in job-related hazards. The Act seeks ". . . to assure so far as possible every working man and woman in the nation safe and healthful working

Figure 26-1. Worker chipping on a large iron casting using an air-powered hammer/chisel. Some of the factors that can influence the person on the job are indicated. (Reprinted with permission from Zenz, C. *Occupational Medicine: Principles and Practical Applications,* 2nd ed. Chicago: Year Book Medical Publ., 1988)

conditions." The law became effective on April 28, 1971, and involves approximately 60 million workers and more than 4 million places of employment.

Medical regulations first promulgated in 1972 under the OSHAct were applicable to all employers. Subpart K, §1910.151, Medical services and first aid (*Federal Register (FR)* 39:125. p.23682, June 27, 1974 (revised), of these regulations states:

> (a) The employer shall ensure the ready availability of medical personnel for advice and consultation on matters of plant health.
>
> (b) In the absence of an infirmity, clinic or hospital in near proximity to the workplace which is used for the treatment of all injured employees, a person or persons shall be adequately trained to render first aid. First aid supplies approved by the consulting physicians shall be readily available.
>
> (c) Where the eyes or body of any person may be exposed to injurious corrosive materials, suitable facilities for quick drenching or flushing of the eyes and body shall be provided within the work area for immediate emergency use.

Later, 1972 Regulations for Construction included everything in (a) above, and substituted the following portion of (b). (See Subpart D, Occupational Health and Environmental Controls; §1926.50, Medical services and first aid [*FR* 37:243.p.27510, Dec 16, 1972]).

> . . . a person or persons who have a valid certificate in first aid training from the U.S. Bureau of Mines, the American National Red Cross, or equivalent training that can be verified by documentary evidence, shall be available at the worksite to render first aid.

The regulations continue:

> The first aid kit shall consist of materials approved by the consulting physician. The materials shall be in a weatherproof container with individual sealed packages for each type of item. The contents of the first aid kit shall be checked at least weekly to ensure that the expended items are replaced.
>
> Proper equipment for prompt transportation of the injured person to a physician or hospital, and a communication system for contacting an ambulance service shall be provided.
>
> The telephone numbers of the physicians, hospitals, and ambulances shall be conspicuously posted.

Occupational medical standards are also promulgated by the

Secretary of Labor based on recommendations from the National Institute of Occupational Safety and Health (NIOSH), from advisory committees, or other groups; usually hearings are held. An example of such a standard is §1910.1001, Asbestos. Paragraph (l) of the Asbestos Standard, Medical examinations. The Standard specifies requirements for preplacement, annual, and termination of employment examinations, and contains provisions for maintenance and access to medical records.

Using the Asbestos Standard as an example, the regulations call for annual comprehensive medical examinations of all employees exposed to airborne concentrations of asbestos fibers above the action level of 0.1 fiber per cubic centimeter. Minimum requirements are a 35 × 43 cm (14 × 17 in.) chest roentgenogram, a "history to elicit symptomatology of respiratory disease, and pulmonary function tests to include forced vital capacity (FVC) and forced expiratory volume at one second (FEV_1)." Examinations shall be at the employer's cost. Medical records shall be retained by employers for at least 30 years, and be accessible to OSHA, NIOSH, and certain physicians.

Responsibilities for health standards are divided between two government agencies—the Department of Labor (DOL) and the Department of Health and Human Services (HHS) (formerly the Department of Health, Education, and Welfare). The DOL has the responsibility for promulgating and enforcing new, mandatory occupational safety and health standards and has the authority to enter factories and workplaces to conduct inspections, to issue citations, and to impose penalties.

The DOL regulations require that employers maintain accurate records and reports of work-related injuries, illnesses, and deaths. It initiated programs for educating employees and employers in the recognition, avoidance, and prevention of unsafe or unhealthy working conditions covered by the OSHAct.

The NIOSH was created within the Department of Health and Human Services when the OSHAct was enacted. The NIOSH is required to develop the criteria for new safety and health standards and to transmit them to the DOL for promulgation and enforcement. Related to the development of criteria for recommended standards are the requirements to conduct ". . . research into the motivational and behavioral factors relating to the field of occupational safety and health . . ." and to ". . . conduct and publish industrywide studies of the effect of chronic or low-level exposure to industrial materials, processes, and stresses on the potential for illness, disease, or loss of functional capacity in aging adults." NIOSH will also ". . . conduct special research, experiments, and demonstrations relating to occupational safety and health as are necessary to explore new problems, including those created by new technology. . ."

The OSHA adopts standards and conducts inspections of workplaces to determine whether the standards are being met (see Chapter 31, History of the Occupational Safety and Health Administration). The OSHA requires each employer to provide a workplace free from safety and health hazards and to comply with the standards.

A safety or health standard is a legally enforceable regulation governing conditions, practices, or operations meant to assure a safe and healthy workplace; these standards are published in the *Federal Register* from time to time. The OSHA standards are divided into three major categories—general industry, maritime, and construction.

The OSHA requires employers of 10 or more employees to maintain certain records of job-related fatalities, injuries, and illnesses. The OSHA requires that only two simple forms be maintained. On January 1, 1978, the OSHA Recordkeeping and Reporting System was revised to streamline and simplify the forms. The new system is a comprehensive redesign around a new form—OSHA no. 200, the basic *Log and Summary of Occupational Injuries and Illnesses* (Figure 26-2). In addition, every recordable injury or illness entered on the log must be recorded on a supplementary record (OSHA form no. 101) or similar form, which contains the same information.

On the OSHA form no. 200 (Figure 26-2), employers can use a checkoff procedure instead of assigning code numbers for distinguishing the type of case that occurred in the establishment. Also, the employer is able to have a running summary of injuries and illnesses at any time during the year. This new format makes it easier for the employer, the employee, and the compliance officer to spot injury and illness problems and trends.

The basic objectives of an occupational health program as set forth in the revision of the American Medical Association's (AMA's) *Scope, Objectives and Functions of Occupational Health Programs,* 1972, follow:

1. to protect employees against health and safety hazards in their work situation.
2. insofar as practical and feasible, to protect the general environment of the community.
3. to facilitate the placement of workers according to their physical, mental, and emotional capacities in work which they can perform with an acceptable degree of efficiency and without endangering their own health and safety or that of others.
4. to assure adequate medical care and rehabilitation of the occupationally ill and injured.
5. to encourage and assist in measures for personal health maintenance, including the acquisition of a personal physician whenever possible.

Achievement of these objectives benefits both employees and employers by improving employee health, morale, and productivity.

Despite the comprehensive approach of modern occupational health programs, the primary duty of any in-house medical department is to operate a dispensary for the treatment of occupational injuries and illnesses.

TREATING OCCUPATIONAL INJURIES

This brief discussion assumes certain principles that depend on the available treatment facilities and medical staff. In all work settings, injured workers should be segregated from those who need preplacement examinations or other physical examinations, assuring the immediate treatment of the injured workers and privacy. An on-site medical facility should be central to most workers. Personnel trained in first aid should be available in all remote areas, and appropriate litters or other means of transportation, available, to transport ill or injured workers to the centralized, on-site medical facility or to a nearby hospital.

A physician should serve as medical director or medical advisor. A physician's services are required to properly conduct the employee health program and to provide medical direction for the occupational health nurse (a registered nurse [RN] working in the occupational health field, see Chapter 27, The Occupational Health Nurse). These principles were first stated in 1932, when the State Medical Society of Wisconsin published

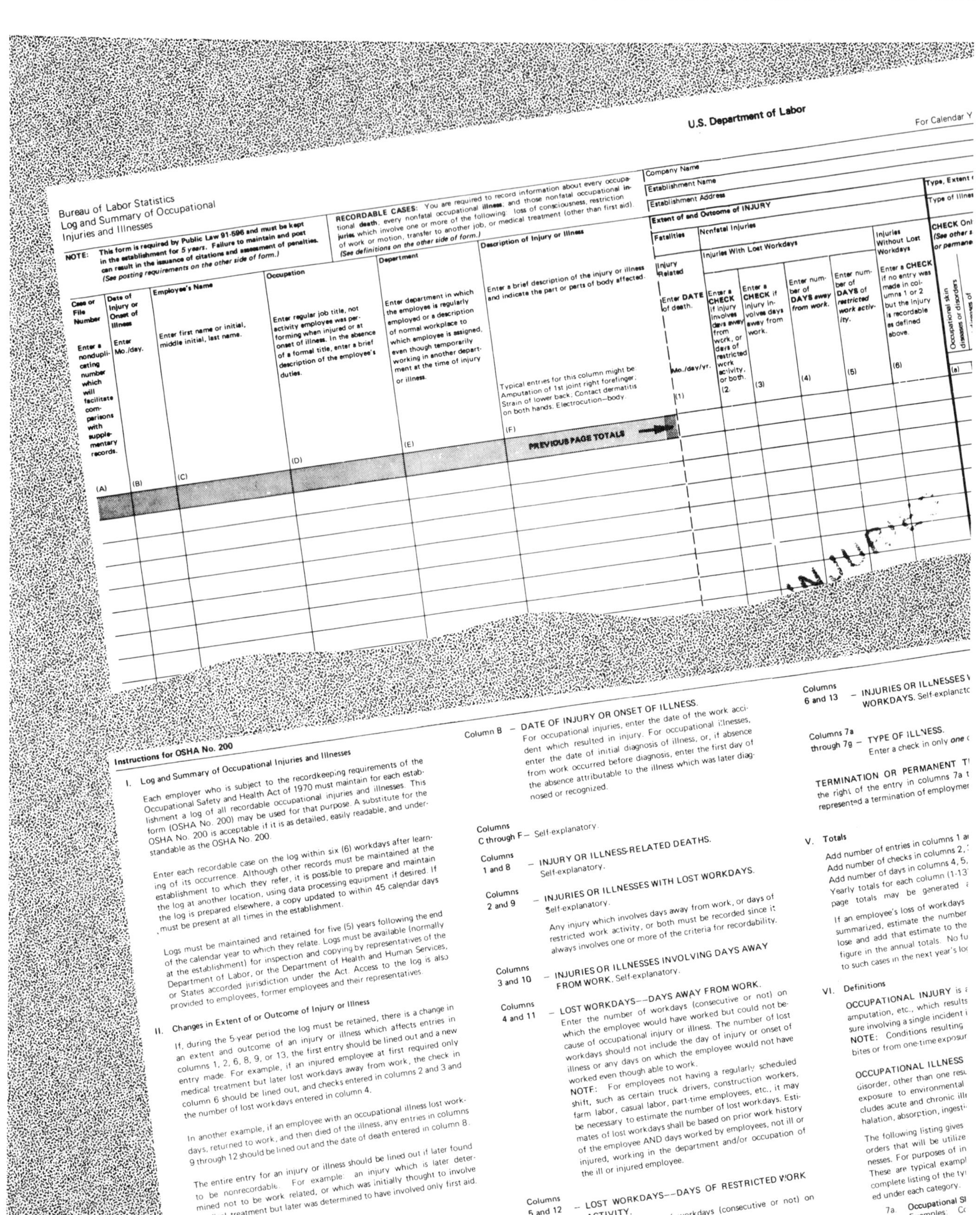

U.S. Department of Labor

For Calendar Y

Bureau of Labor Statistics
Log and Summary of Occupational
Injuries and Illnesses

NOTE: This form is required by Public Law 91-596 and must be kept in the establishment for *5 years*. Failure to maintain and post can result in the issuance of citations and assessment of penalties. *(See posting requirements on the other side of form.)*

RECORDABLE CASES: You are required to record information about every occupational **death**; every nonfatal occupational **illness**; and those nonfatal occupational **injuries** which involve one or more of the following: loss of consciousness, restriction of work or motion, transfer to another job, or medical treatment (other than first aid). *(See definitions on the other side of form.)*

Company Name

Establishment Name

Establishment Address

Case or File Number	Date of Injury or Onset of Illness	Employee's Name	Occupation	Department	Description of Injury or Illness	Extent of and Outcome of INJURY: Fatalities — Injury Related	Nonfatal Injuries — Injuries With Lost Workdays				Injuries Without Lost Workdays	Type, Extent of … Type of Illness — CHECK Only … *(See other s… or permane…*
Enter a nonduplicating number which will facilitate comparisons with supplementary records.	Enter Mo./day.	Enter first name or initial, middle initial, last name.	Enter regular job title, not activity employee was performing when injured or at onset of illness. In the absence of a formal title, enter a brief description of the employee's duties.	Enter department in which the employee is regularly employed or a description of normal workplace to which employee is assigned, even though temporarily working in another department at the time of injury or illness.	Enter a brief description of the injury or illness and indicate the part or parts of body affected. Typical entries for this column might be: Amputation of 1st joint right forefinger; Strain of lower back; Contact dermatitis on both hands; Electrocution—body.	Enter DATE of death. Mo./day/yr.	Enter a CHECK if injury involves days away from work, or days of restricted work activity, or both.	Enter a CHECK if injury involves days away from work.	Enter number of DAYS away from work.	Enter number of DAYS of restricted work activity.	Enter a CHECK if no entry was made in columns 1 or 2 but the injury is recordable as defined above.	Occupational skin diseases or disorders
(A)	(B)	(C)	(D)	(E)	(F)	(1)	(2)	(3)	(4)	(5)	(6)	(a)
					PREVIOUS PAGE TOTALS →							

Instructions for OSHA No. 200

I. **Log and Summary of Occupational Injuries and Illnesses**

Each employer who is subject to the recordkeeping requirements of the Occupational Safety and Health Act of 1970 must maintain for each establishment a log of all recordable occupational injuries and illnesses. This form (OSHA No. 200) may be used for that purpose. A substitute for the OSHA No. 200 is acceptable if it is as detailed, easily readable, and understandable as the OSHA No. 200.

Enter each recordable case on the log within six (6) workdays after learning of its occurrence. Although other records must be maintained at the establishment to which they refer, it is possible to prepare and maintain the log at another location, using data processing equipment if desired. If the log is prepared elsewhere, a copy updated to within 45 calendar days must be present at all times in the establishment.

Logs must be maintained and retained for five (5) years following the end of the calendar year to which they relate. Logs must be available (normally at the establishment) for inspection and copying by representatives of the Department of Labor, or the Department of Health and Human Services, or States accorded jurisdiction under the Act. Access to the log is also provided to employees, former employees and their representatives.

II. **Changes in Extent of or Outcome of Injury or Illness**

If, during the 5-year period the log must be retained, there is a change in an extent and outcome of an injury or illness which affects entries in columns 1, 2, 6, 8, 9, or 13, the first entry should be lined out and a new entry made. For example, if an injured employee at first required only medical treatment but later lost workdays away from work, the check in column 6 should be lined out, and checks entered in columns 2 and 3 and the number of lost workdays entered in column 4.

In another example, if an employee with an occupational illness lost workdays, returned to work, and then died of the illness, any entries in columns 9 through 12 should be lined out and the date of death entered in column 8.

The entire entry for an injury or illness should be lined out if later found to be nonrecordable. For example: an injury which is later determined not to be work related, or which was initially thought to involve medical treatment but later was determined to have involved only first aid.

III. **Posting Requirements**

A copy of the totals and information following the fold line of the last page for the year must be posted at each establishment in the place or places where notices to employees are customarily posted. This copy must be posted no later than *February 1 and must remain in place until March 1.*

…ough there were no injuries or illnesses during the year, zeros must … and the form posted.

… that the

Column B — DATE OF INJURY OR ONSET OF ILLNESS.
For occupational injuries, enter the date of the work accident which resulted in injury. For occupational illnesses, enter the date of initial diagnosis of illness, or, if absence from work occurred before diagnosis, enter the first day of the absence attributable to the illness which was later diagnosed or recognized.

Columns C through F — Self-explanatory.

Columns 1 and 8 — INJURY OR ILLNESS-RELATED DEATHS.
Self-explanatory.

Columns 2 and 9 — INJURIES OR ILLNESSES WITH LOST WORKDAYS.
Self-explanatory.

Any injury which involves days away from work, or days of restricted work activity, or both must be recorded since it always involves one or more of the criteria for recordability.

Columns 3 and 10 — INJURIES OR ILLNESSES INVOLVING DAYS AWAY FROM WORK. Self-explanatory.

Columns 4 and 11 — LOST WORKDAYS—DAYS AWAY FROM WORK.
Enter the number of workdays (consecutive or not) on which the employee would have worked but could not because of occupational injury or illness. The number of lost workdays should not include the day of injury or onset of illness or any days on which the employee would not have worked even though able to work.
NOTE: For employees not having a regularly scheduled shift, such as certain truck drivers, construction workers, farm labor, casual labor, part-time employees, etc., it may be necessary to estimate the number of lost workdays. Estimates of lost workdays shall be based on prior work history of the employee AND days worked by employees, not ill or injured, working in the department and/or occupation of the ill or injured employee.

Columns 5 and 12 — LOST WORKDAYS—DAYS OF RESTRICTED WORK ACTIVITY.
Enter the number of workdays (consecutive or not) on which because of injury or illness:
(1) the employee was assigned to another job on a temporary basis, or
(2) the employee worked at a permanent job less than full time, or
(3) the employee worked at a permanently assigned job but could not perform all duties normally connected with it.

…clude the day of

Columns 6 and 13 — INJURIES OR ILLNESSES W… WORKDAYS. Self-explanato…

Columns 7a through 7g — TYPE OF ILLNESS.
Enter a check in only *one* …

TERMINATION OR PERMANENT T… the right of the entry in columns 7a t… represented a termination of employme…

V. **Totals**

Add number of entries in columns 1 a…
Add number of checks in columns 2, …
Add number of days in columns 4, 5, …
Yearly totals for each column (1-13) …
page totals may be generated …

If an employee's loss of workdays … summarized, estimate the number … lose and add that estimate to the … figure in the annual totals. No fu… to such cases in the next year's lo…

VI. **Definitions**

OCCUPATIONAL INJURY is … amputation, etc., which results … sure involving a single incident i…
NOTE: Conditions resulting … bites or from one-time exposur…

OCCUPATIONAL ILLNESS … disorder, other than one resu… exposure to environmental … cludes acute and chronic ill… halation, absorption, ingesti…

The following listing gives … orders that will be utilize… nesses. For purposes of in… These are typical exampl… complete listing of the ty… ed under each category.

7a. Occupational Sk… Examples: Co… mary irritants … chrome ulcers; …

7b. Dust Diseases … Examples: … eases, coal … other pneum…

7c Respiratory … Examples: … due t…

Figure 26-2. The OSHA form no. 200, Log and Summary of Occupational Injuries and Illnesses.

Suggestions for the Guidance of the Nurse in Industry. This was the first U.S. medical society to formally acknowledge the need for a well-defined physician-nurse relationship in occupational health programs (Figure 26-3).

MEDICAL AUTHORIZATION

Under my supervision and direction, Mrs. Mary Jones, RN, is authorized to draw blood samples from employees of any department for any medical monitoring program at Wisconsin Chemical Corporation or from prospective employees, and also to remove sutures from minor wounds.

________________ ________________

Date Approved Date Approved

____

Date

Figure 26-3. Sample of a medical authorization is indicative of a well-defined relationship between the occupational physician and the occupational nurse.

All physicians concerned with an occupational health program should initiate written medical directives to guide the occupational health nurse in making a provisional diagnosis and determining the next step needed to aid an ill or injured employee. Although preventive health services are important, one of the nurse's major functions is the emergency care of ill and injured employees.

Examples of company policies

Policies range from concise to elaborate.

Concise policy. The American Medical Association (AMA) has a concise statement:

> The aim of the plant medical service is to provide a healthy, effective working force by encouraging personal health improvement, preventing illness and injury, and treating those industrial diseases and injuries that do occur.

Elaborate policy. An elaborate personnel statement is exemplified by the Volvo Company of Sweden that has issued a policy statement, distributed to all employees concerning the physical environment.

> In the design of products, workplaces, machines and equipment, attention must also be paid to the physical and mental conditions of the employees and the need for a good working environment. The working conditions to be aimed at are those in which the employees can carry out their jobs without any detrimental stress; systematic efforts are to be made to identify and eliminate health risks; and the working environment is to be designed in such a way that it provides opportunities for new forms of work and encourages cooperation and satisfaction at the workplace (Figure 26-4 gives another example of a company policy statement.)

Multiservice program. Activities of a fully developed multiservice program in occupational medicine are outlined in the following seven sections quoted from the AMA's *Scope, Objectives, and Functions of Occupational Health Programs.* This document is the most widely used guide for physicians who are concerned with planning and administering such programs. It also is used by many government agencies as a reference for recommendations.

1. Maintenance of a Healthful Work Environment—This requires that personnel skilled in industrial hygiene (environmental control) perform periodic inspections on the premises, including all facilities used by employees, and evaluate the work environment in order to detect and appraise health hazards, mental as well as physical. Such inspections and appraisals, together with the knowledge of processes and materials used, provide current information on health aspects of the work environment. This information will serve as the basis for appropriate recommendations to management for preventive and corrective measures.
2. Preplacement Examinations and/or Screening—As an aid to suitable placement, some form of preplacement health assessment is desirable. The scope of such assessments will be influenced by such factors as size, nature, and location of the industry as well as by the availability of physician and nurse services. They may range from medical examination by a physician, through medical questionnaires and screening procedures by a nurse, to more simple screening by allied health personnel. The type of assessment to be performed as a routine, as well as the specific tests to be included in each category, should be determined by the physician in charge of the health program. In general, the assessment should include: (a) personal and family history, (b) occupational history, (c) such physical examinations and/or laboratory and appropriate screening tests as seem advisable for the particular industry and availability of professional personnel.

   Unrealistic and needlessly stringent standards of physical fitness for employment defeat the purpose of health examination and of maximum utilization of the available work force.
3. Periodic Health Appraisals—These health evaluations are performed at appropriate intervals to determine whether the employee's health remains compatible with his job assignment and to detect any evidence of ill health that might be attributable to his employment. Certain employees and groups may require more frequent examinations as well as additional procedures and tests, depending on their age, their physical condition, the nature of their work, and any special hazards involved. Health examinations and appraisals should be conducted or supervised by a physician with the assistance of such qualified allied health personnel as may be indicated and available. The examination may be made in any properly equipped medical facility, at the work place.

   The individual to be examined should be informed by appropriate means of the purpose and value of the examination. The physician should discuss the findings of the examination with the individual; he should explain the importance of further medical attention for any significant health defects found.

Environmental Health & Hygiene Policy

As a responsible manufacturer of a wide variety of specialty chemicals and pesticides, this company's plants, laboratories and offices shall be maintained as clean and healthful places of employment.

At all times, this company's facilities shall be designed and operated in compliance with the spirit and letter of federal, state and local occupational health and hygiene regulations.

This company acknowledges and shall satisfy its repsonsibility to promptly provide current, comprehensive information on potential adverse health effects and appropriate handling procedures for chemicals handled by both our employees and our customers.

It is a basic responsibility of all this company's employees to make the health and safety of fellow human beings a part of their daily, hourly concern. This responsibility must be accepted by each one who conducts the affairs of the corporation, no matter in what capacity he may function.

A goal of this company is to achieve and maintain a corporate occupational health and hygiene program which is a model for the chemical industry.

Chairman of the Board & Chief Executive Officer

President & Chief Operating Officer

________________ ________________

Figure 26-4. This statement is a good example of a health and hygiene policy. (Courtesy Velsicol Chemical Corporation)

4. Diagnosis and Treatment:
   (a) Occupational injury and disease—Diagnosis and treatment in occupational injury and disease cases should be prompt and should be directed toward rehabilitation. Workers' compensation laws, insurance coverage arrangements, and policies of medical societies usually govern the provision of medical services for such cases.
   (b) Nonoccupational injury and illness—Every employee should be encouraged to use the services of a personal physician or medical service where these are available for care of off-the-job illness or injuries. Treatment of nonoccupational injury and illnesses never has been and is not now ordinarily considered to be a routine responsibility of an occupational health program with these limited exceptions:

In an emergency, of course, the employee should be given the attention required to prevent loss of life or limb or to relieve suffering until placed under the care of his personal physician. For minor disorders, first aid or palliative treatment may be given if the condition is one for which the employee would not reasonably by expected to seek the attention of his personal physician, or if it will enable the employee to complete his current work shift.

Due to the shortage of physicians in some communities, it is recognized that some employees on occasion may find it impossible to locate or to obtain the services of a personal physician or health service. In such circumstances, limited to where treatment is otherwise unavailable, the occupational physician may undertake additional and continuing treatment of an employee's nonoccupational condition if requested to do so by the employee or his family. If such services might become ongoing within the occupational health program, approval of the employer should be obtained. In order to help assure high-quality medical care, discussion with the local medical society in developing such projects, including the methods of payment for services, is urged.

5. Immunization Programs—An employer may properly make immunization procedures available to his employees under the principles set forth in the AMA's Guide for Industrial Immunization Program.
6. Medical Records—The maintenance of accurate and complete medical records of each employee from the time of his first examination or treatment is a basic requirement. The confidential character of these records, including the results of health examinations, should be rigidly observed by all members of the occupational health staff. Such records should remain in the exclusive custody and control of the medical personnel. Disclosure of information from an employee's health record should not be made without his consent, except as required by law.
7. Health, Education and Counseling—Occupational health personnel should educate employees in personal hygiene and health maintenance. The most favorable opportunity for reaching an employee with health education and counseling arises when he visits a health facility.

   Health education appropriately goes hand in hand with safety education. The occupational health and safety personnel, therefore, should work cooperatively with supervisory personnel in imparting appropriate health and safety information to employees. Health and safety education should (a) encourage habits of cleanliness, orderliness, and safety, and (b) teach safe work practices, the use and maintenance of available protective clothing and equipment, and the use of available health services and facilities. Experience has shown that health education is most effective when the employer demonstrates his sincere and continuing interest in the health of his employees and when employees are encouraged to participate in the planning and conduct of health education activities.

SMALL UNITS OF LARGE ORGANIZATIONS

Small operations belonging to large companies are the main users of partial in-house medical programs. At an AMA Congress on Occupational Health in 1971, Marcus B. Bond, M.D., gave an excellent summary on this complex subject. The remainder of this section is quoted from his report.

If small employee groups are part of a large company with headquarters elsewhere, advice and instructions can be supplied from the corporate offices where staffs of specialists exist. Physicians, nurses, industrial hygienists, safety experts, and labor relations representatives can combine their knowledge and provide specific instructions to supervisors and local physicians in any location covering any type work operation. Such instructions should, of course, be consistent with general company policies, insurance and benefit provisions, labor contracts, and the laws of the particular state. (See Figure 26-5.)

Selection of a physician

Selection of a physician obviously is a first step because a local physician, licensed in the state, is required for examinations, treatment, or any other professional activities. It is best if local management and the company medical director share the responsibility for obtaining a doctor's services. Usually local managers are acquainted with several physicians in the community through social, business, and civic activities and they can recommend one or more physicians to the medical director. The medical director can check the physician's credentials by consulting with the *American Medical Directory,* published by the American Medical Association. Often the medical director will be personally acquainted with one or more doctors in the particular community or may know officials in the state or local medical society who can provide information about available physicians. When the choice of a physician has been agreed upon by the medical director and local management, the physician should be contacted to see if he or she is interested in serving the organization. If so, then the medical director should visit him personally. If this is not possible, the director could write a letter and provide the physician's manual or other written policies, standards for applicants, description of jobs, and the necessary periodic tests or examinations for employees.

Work site visits

The physician should visit the work sites and become familiar with all phases of the different jobs. The local manager should arrange for the physician to visit the plant or place of work and actually see the working conditions and become acquainted with at least some of the supervisors. The frequency of visits thereafter should be determined by the nature of the work and major changes in materials or processes. If significant hazards exist in the work environment, the physician should tour the work area every 1 or 2 months.

Often, raw materials, by-products, wastes (effluents), solvents, and related materials in the processing or manufacturing can be more hazardous than the final product. Indeed, emphasis was made in this regard in a chapter introduction in *Developments in Occupational Medicine* (see Bibliography):

> With expanding uses of new materials such as are found in adhesives, coatings (paints and finishes), a multitude of plastics and other synthetic entities, more workers are potentially exposed, at various levels of risk, often unintentionally. *Those concerned with occupational health practices* (preventive programs) *should and must be aware of the more commonly encountered substances and have some knowledge of what is going on in the factories near them.* The immense quantities of certain hazardous materials used are awesome. Indeed, some chemicals being produced and used for other processes range into and beyond millions to billions of kilograms each year; e.g., phosgene, used as an intermediate to make isocyanates, has an annual production in the United States in excess of 300 million

Procedure for Handling First Aid/Medical Cases Outside Normal Working Hours

General First Aid

In the event of any injury/illness, follow normal first aid practice to control emergency.

If resuscitation is required, act immediately; commence mouth-to-mouth resuscitation until the ______ (name of equipment) can be utilized.

During the first normal shift following the incident, notify the personnel manager or other designated person as to the location, nature, and time of injury. This information may be needed should complications arise.

More Serious Injuries/Illnesses Requiring Medical Treatment

- Immediate transport to City Hospital
- One of the following ambulance services may have to be called:
  City Hospital 234-2500
  Police Department 233-1234
  Gold Cross Ambulance 234-4000
  Mid-South Ambulance 235-8800
- Give the hospital the names of the physicians, Drs. ______ or ______ Phone number ______
- The supervisor under whom the employee was working should begin investigation of the cause of the accident.
- Personnel should be notified as soon as practical.

Less-Serious Injuries/Illnesses Requiring Medical Attention

- Saturdays, 8:00 a.m. to 12:00 noon, refer to doctor at the Industrial Clinic, 2345 Smith Highway, after calling 234-4567 to confirm that one of the doctors is there.
- Otherwise, refer to City Hospital, and follow the procedure given above for more serious injuries.
- In either case, an accident investigation should be conducted by the supervisor, and personnel should be notified at the next normal shift.

Charges for medical treatment of work-related illnesses/injuries should be billed to XYZ Insurance Company.

Figure 26-5. This company bulletin was posted throughout the facility.

kg; sodium hydroxide produced each year in the United States is more than 10 billion kg. In the smaller workshops particularly, points of concern may be found in processing and formulating, packaging, storage, shipping, reformulating, applications, maintenance work, laboratory testing, etc. Often overlooked is that these varied activities occur over geographically widely scattered areas and mainly involve small numbers or groups of workers. Despite technologic advances in the highly industrialized countries, some occupational exposures continue and are troublesome: exposures to carbon monoxide, hydrogen sulfide, other irritating agents such as acids and caustics, oxides of nitrogen, etc., all found in varying amounts in certain occupational settings.

Remuneration of the physician may be on a fee-for-service basis, including hourly charges for visits to the workplace. If the services provided are frequent, then a retainer or salary may be desirable for both parties—it increases the physician's attention to the employer's business and may provide certain fringe benefits associated with employment such as health and life insurance, pensions, Social Security, and others.

During the past decade, more physicians have become active in the private practice of occupational medicine, either as solo practitioners, in partnerships, or from clinic groups. Often, several employers may receive professional guidance, or consultations on a regular basis from one physician. In such circumstances, this externalization of services does not lend itself to accept employment status. Rather, these physicians find billing on a monthly basis more suitable and further enhancing neutrality, a vital aspect, often difficult to maintain in a salaried position, whether full- or part-time.

The practicing physician who agrees to provide service to a company or other organization should honor this commitment. A small amount of effort initially will save much time and confusion of all concerned, and will particularly save the physician's time. As mentioned above, a visit to the worksite should be made for an observation of the working environment and a frank discussion of what the employer thinks the employee problems are. The wise physician will do more listening than talking, and recognize that each employee group is unique in terms of employee policy and details of getting the job done. Employers are most often delighted to conduct the physician through the worksite and share any information about the business.

This may permit recognizing the occasional employer who would misuse the physician by asking him to "cover up" or deny injuries or illnesses from work exposure, or who would not reveal to the physician the actual composition of certain materials used at work under the guise of protecting trade secrets. There can be no secrets kept from the physician if he or she is to perform his function properly. Employers who would misuse physicians should be advised there can be no compromise with honesty in dealing with the health of people. Further, competent physicians need not continue their professional relationship with a manipulative organization, regardless of payments received.

Medical examinations

After the physician has become familiar with the work environment and company policies, he should advise the employer how and when to send employees to his office for nonemergency conditions such as applicant examinations, return-to-work checkups, and periodic tests and examinations such as blood counts, chest x-rays, etc. Instructions for these and for emergencies should be written and provided to the employer so the latter may post them on bulletin boards, distribute them to supervisors, place them in first aid kits and in company trucks, etc. Emergency instructions should also include where to send patients at different times during the day and on special days such as weekends and days off; to be listed are the name, address, and phone number of hospital emergency room, plus the name of associate or colleague that should be contacted if the physician is not available.

The physician must then instruct his or her own nurses and office assistant as to what commitments have been made to the employer and provide them with the company forms or instructions. Reference books or articles pertaining to the hazards in the particular working environment should be obtained. The company medical director should furnish these or can advise on them.

At times, the local physician can have the advantage of working with . . . . the medical director—on problem cases. These can be reassuring when an unusual illness or reaction occurs that could possibly be related to work exposure. The medical director should be familiar with products, materials, and processes in the company and be able to provide guidance, even if at long distance, Dr. Bond concludes.

The physician should be given a list of all materials used throughout the workplace, including catalysts. The industrial hygienist and the responsible safety person will find the physician to be valuable professionally. They should ensure that the physician knows of all processes, intermediate products, effluents, and methods for handling or dealing with hazardous situations; and that all relevant data and documentation pertaining to this information be provided (e.g., the material safety data sheets [MSDSs], etc.)

The physician who is new to occupational health practice should acquire certain reference literature, including this book, *Fundamentals of Industrial Hygiene,* and *Introduction to Occupational Safety and Health.* Both texts are published by the National Safety Council, and are valuable review and refresher sources for physicians. See the Bibliography for more information.

LOCATION OF EMPLOYEE HEALTH SERVICES

Although good occupational medicine can be practiced in any logical facility location that is clean and private, experience indicates that the medical unit commands respect only if careful attention is paid to suitable and efficient housing, appearance, and equipment. The entire unit should be painted in light colors and kept spotlessly clean. The dispensary should have hot and cold running water and be adequately heated, ventilated, and illuminated. Toilet facilities are necessary. Suitable provision should be made for examining men and women if both are employed. Cleanliness and privacy are essential.

The medical department should be easily accessible and near the greatest number of employees so that distance does not become a barrier against the immediate reporting for treatment of even minor injuries. If possible, it should also be connected with the employment and safety departments; this facilitates prompt physical examinations of job applicants, the mutual use of clerical service, and the interchange of ideas and plans relative to employment, accident, and health problems.

Another location consideration is to have the facilities near the entrance so an ambulance can be brought to the door, if necessary. Also, injured workers, who are off duty, but under treatment, may be admitted through a separate entrance.

The medical department should be in a place of greatest safety in case of a major disaster that might otherwise destroy first aid or dispensary supplies or facilities.

Health service office (dispensary)

A minimum of three rooms, consisting of a waiting room, a treatment room, and a room for consultation or for making physical examinations is recommended. Rooms for special purposes can be added according to the needs and size of the company. As mentioned earlier, the layout of the dispensary should be such that applicants for employment waiting for physical examinations do not mingle with any injured workers.

The surgical treatment room should be large enough to treat more than one person at a time—small dressing booths can be arranged to give some degree of privacy.

First aid room

It is always advisable to set aside a room at a convenient location for the sole purpose of administering first aid. It upsets morale to administer treatments in public as injured persons prefer privacy. Furthermore, the person administering first aid should have a proper place to do his or her work.

The environment of a good first aid room should be similar to that of the dispensary. At a minimum, the room should be equipped with the following items:

1. desk and chair
2. examining table
3. bed for emergency cases, enclosed by a movable curtain, hospital-type bed preferred—the typical cot or sofa style beds are too low and too narrow
4. treatment table and containers for first aid supplies
5. small table at bedside
6. chair with arms and one without arms
7. Mayo stand
8. switchable lighting fixtures.

Emergency oxygen is of great benefit in the treatment of many first aid cases. When oxygen is being administered, smoking should be prohibited because of the danger of fire or explosion. Any type of resuscitating device should only be used by trained persons.

First aid kits

There are many types of emergency first aid kits; they are designed to fill every need, depending on the special hazards that might occur. Commercial or cabinet-type first aid kits, as well as unit-type kits, must meet OSHA requirements (Figure 26-6).

Kits vary in size from a pocket-size model to what amounts to a portable first aid room. The size and content depend on the intended use and the types of injuries that could occur (Figure 26-7). For example, personal kits contain only essential articles for the immediate treatment of injuries. Departmental kits are larger—they are planned to cover a group of workers and the quantity of material depends on the size of the working force. Truck kits are the most complete—they can be carried easily to the accident site, or can be stored near working areas that are distant from well-equipped emergency first aid rooms. Although they include such bulky items as a wash basin, blankets, splints, and stretchers, they conveniently can be carried by two persons.

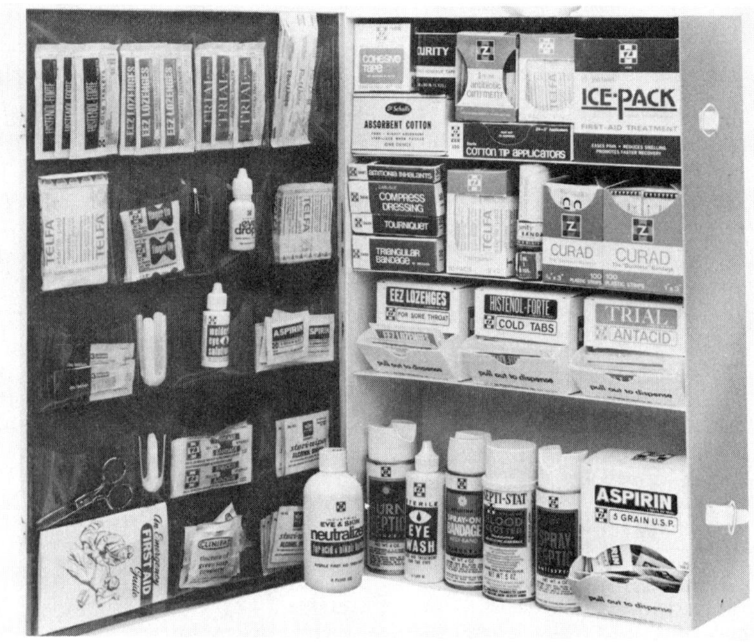

Figure 26-6(a) (b). Typical unit-type first aid kit meets the OSHA requirements.

Distribution of first aid kits throughout the plant or job site seems to work to best advantage when supervised properly. Each kit must be the responsibility of one trained individual who should understand that he or she is to care for the most trivial injuries only or to render only temporary treatment for more serious cases (Figure 26-8). It is believed that with this system many slight scratches and cuts are given attention that they would not otherwise receive.

In industrial organizations such as mining companies and public utilities, where activities are widely scattered, the use of first aid kits and some self-medication may be necessary. However, it is better if first aid service can be controlled by having the attendant in charge properly instructed in first aid and by seeing that the service as a whole has proper medical supervision.

The maintenance and use of all of these first aid kits should be under the supervision of medical personnel, and they should contain the materials approved by the consulting physician. A member of the medical services group should be assigned to regularly inspect all first aid materials, and to submit a report of their content level and serviceability.

Maintenance of quantities of materials in the first aid kit is easier if each kit lists the original contents and the quantities at which new materials should be ordered. All bottles or other containers should be clearly labeled.

Recommended materials for first aid kits are listed in textbooks from the American National Red Cross and the U.S. Mine Safety and Health Administration. Suggestions also are available from the American Medical Association and manufacturers of first aid materials. The physician or nurse routinely should inspect first aid supplies, stretchers, and, especially, oxygen tanks.

Stretchers

It is essential to have adequate means of transporting a seriously injured person from the scene of the accident to a first aid room or a hospital. Unnecessary delays and inadequate transportation can add to the seriousness of the injury and may be the determining factor between life and death.

The stretcher provides the most acceptable method of hand transportation and it also can be used as a temporary cot at the scene of the accident, during transit in a vehicle, and in the first aid room or dispensary (Figure 26-9).

There are several types of stretchers. The army-type is commonly used and is satisfactory. However, when it is necessary

Figure 26-7. First aid kits vary in size, depending on the specific requirements. This first aid station is for use in mines, remote work sites, and places far from the dispensary in large plants. It contains a clam-type (scoop) stretcher, blankets, splints, and a first aid kit with assorted emergency treatment supplies.

to hoist or lower the injured person out of awkward places, it is better to use specially shaped stretchers with straps to hold the patient in an immovable position.

Stretchers should be conveniently located in all areas where employees are exposed to serious hazards. It is customary to keep stretchers, blankets, and splints in well-marked, conspicuous cabinets. Stretchers should be kept clean, not be exposed to destructive fumes, dust, or other substances, protected against mechanical damage, and ready for use at all times. If the stretcher is of a type that will deteriorate, it should be tested periodically for durability and for strength and cleanliness.

Planning for emergency care in the occupational setting

The following policies and procedures are helpful in planning for emergency care in the occupational setting. Some obviously are independent nursing functions and responsibilities, while others are not and must be carried out in cooperation with the physician, management, safety personnel, industrial hygienists, first aid teams, and other appropriate individuals and groups (see Chapter 27, The Occupational Health Nurse) (Figure 26-5). The names and phone numbers should be changed to fit local conditions.)

Planning is necessary to assure and maintain a safe working environment, proper health protection, prompt and definitive emergency care for injured and ill employees, and safe handling and transportation of injured and ill employees.

Planning should include items such as:

1. *Assessment of hazards in the working environment.* Hazards must be identified and assessed for accident and corrosiveness, irritative capability, toxic or otherwise injurious materials, flammability, and explosiveness; their location in the plant and their use; the symptoms that could be produced by exposure to these; and treatment to counteract affects. Treatment procedures should be posted and necessary medical supplies readily available in the employee health services areas.

   Since May, 1986, employers have been required to label in-plant containers of hazardous chemicals, to inform employees of workplace hazards, to make MSDSs available to employees, and to train workers in protective measures for specific chemical hazards. Employers will have to develop written hazard communication programs outlining their plans to accomplish these objectives. (See the discussion on hazard communication laws in Chapter 30, Governmental Regulations.)
2. *Occupational and nonoccupational health emergencies.* A plan and procedure should be developed for meeting personal health emergencies of both minor and major magnitude.
3. *Medical directives and nursing procedures* should be formulated to cover all emergency situations.
4. *Emergency care equipment.* Plans should be drawn up for providing stationary and portable emergency equipment. Portable emergency equipment should be clearly labeled and stored in appropriate areas throughout the plant, as well as in the employee health services.
5. *Transportation of injured and ill employees.* There should be a written policy approved by management outlining the transportation of ill and injured employees.
6. *Training first aid teams and auxiliary personnel.* It is essential to have plans for providing well-trained first aid teams with clearly identified responsibilities in emergency situations.
7. *Cardiopulmonary resuscitation (CPR).* When a physician is available, he should take charge of the resuscitation of a cardiac (or) pulmonary arrest case. In the absence of a physician, a professional nurse or another person specially trained and certified in the recognition of cardiopulmonary arrest and the technique of resuscitation may apply the appropriate emergency procedures (Figure 26-10).

   A policy must be clearly stated to cover this emergency and the requisite special training should be made available to the staff.
8. *Critical illness and death.* A written policy should be prepared in collaboration with all concerned persons, and approved by management. This policy should cover the procedure to be followed in case of critical illness or death.
9. *Employee health record.* Emergency medical identification information should also be recorded on individual employee records. These records should be kept in the employee health service so the nurse giving emergency care has them available. (An example is shown in Figure 26-11.)

   It is especially important to have up-to-date information on each individual employee's immunization status against tetanus.

Figure 26-8. First aid kits should be distributed throughout the job site or facility. Each should be the responsibility of one trained person, who should understand that he or she is to care for the most trivial injuries only or to give only temporary treatment for more serious cases.

10. *Anticipatory orders from employee's personal physician.* Special information and orders should be on file from an employee's personal physician covering emergency or routine care for special health problems, for example, diabetes, asthma, and cardiovascular disease. These anticipatory orders may be obtained either by the employee or by the nurse with the employee's permission.
11. *Personal emergency medical identification.* All employees with special health problems should be encouraged to wear or carry emergency medical identification information both on and off the job. Wallet cards, for instance, should include:
    - medical conditions—i.e., diabetes
    - allergies—i.e., horse serum, penicillin
    - medication regularly used—i.e., heart patients receiving anticoagulants
    - immunization status—i.e., date of tetanus booster
    - name, phone number, and address of personal physician and next of kin.

Durable devices should be worn on wrist or around neck to identify employees with diabetes, epilepsy, and other problems for which emergency medical care may be needed. Further information on emergency medical identification is available from the American Medical Association, 535 North Dearborn St., Chicago, IL 60610. Also refer to Medic Alert Foundation, Turlock, CA.

12. *Disaster planning and community resources.* It is recommended that all planning for disaster and emergency care be correlated with community health resources.

All factors listed above must be regularly reviewed by all appropriate personnel; never less than once yearly. For further details and sample directives, refer to the 4th edition of the *Occupational Health Guide for Medical and Nursing Personnel,* published by the State Medical Society of Wisconsin (see Bibliography). The objectives, principles, and directives found in this guide have worldwide acceptance and are adaptable to most jurisdictions.

MEDICAL PLACEMENT

Preplacement examination

A preplacement examination is performed for the following reasons:

- to determine the individual's physical and emotional capacity to perform a particular job
- to assess the individual's general health
- to establish a baseline record of physical condition for the personal needs of the employee as well as the employer. Not only is this important in the event of job changes, but it is also valuable for medical-legal purposes.

Determination of the physical and emotional capacity for job performance is the main purpose for a preplacement physical examination. Theoretically, this is how individuals can be matched to specific jobs according to their physical and emotional capabilities, as well as aptitudes. With a good match, a happy, healthy employee can result; if not, the employee may not succeed.

Medical evaluation of physical capacity arose from the need to determine which individuals could do work that required various levels of work effort without doing harm to themselves or others. Although employers have a fear of an employee aggravating a preexisting condition with a resultant workers' compen-

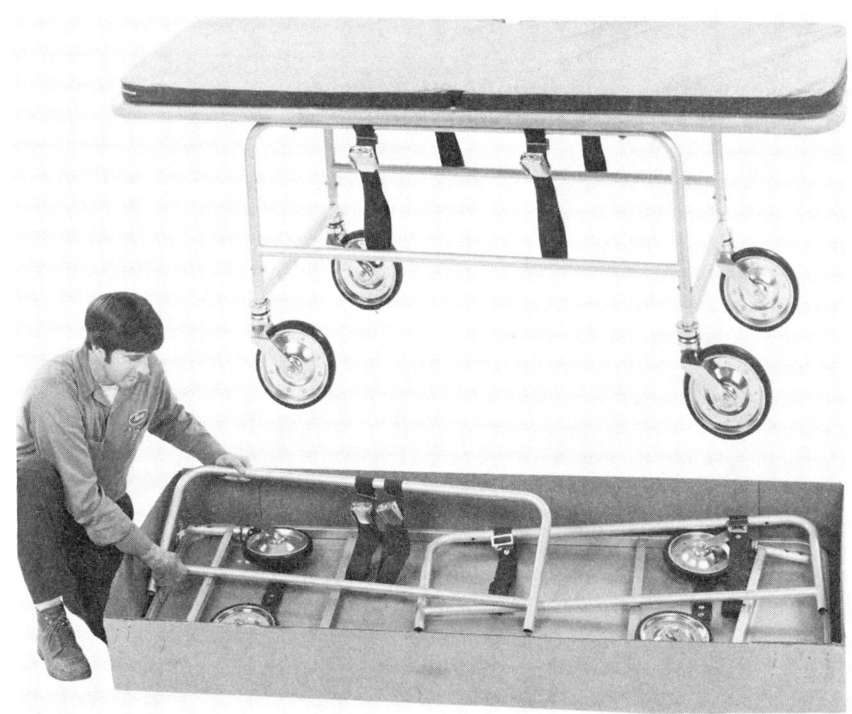

Figure 26-9. The stretcher provides the most acceptable method of hand transportation to transport a seriously injured person from the scene of the accident to a first aid room or hospital.

sation case and attendant costs, this has never been borne out as valid; in the author's experience, this is an erroneous concept.

Unfortunately, many jobs involving considerable exposure to health hazards still remain. Industrial hygiene surveys and engineering controls have resulted in a gradual reduction of the physical demands of many jobs. Automation and high employee turnover rates have diminished the physical fitness aspect of the preplacement examination (Table 26-A).

Nevertheless, jobs causing exposure to heat, noise, vibrations, chemicals, dusts, and other adverse conditions still exist (Figure 26-12). Some jobs also require heavy labor. As described earlier in this chapter, with the advent of OSHA standards delineating physical requirements for certain jobs, especially those characterized by exposures to chemical and physical agents, it appears that the preplacement examination will assume a new and increasingly important role.

These examinations should not be undertaken with the intent to exclude a person from work, although that may have to be the case in some positions, such as air pilot, some military duties, and commercial diver.

Evaluating physical capacity

In evaluating physical capacity for a particular job, the physician can categorize individuals with significant impairment as follows.

Stable impairments that cause some limitations but are unlikely to progress or be worsened by job activity if the employee is properly placed. Amputations, ankylosis (fixed joints), spasticity, blindness, deafness, cerebral palsy, and residual effects of polio are examples of these. Under these conditions, the major consideration is whether the employee can perform the job with reasonable efficiency and safety and without undue stress or harm to himself and others.

Impairments, stable or unstable, that are likely to be made worse by the work exposure. Examples of these are (1) chronic skin diseases, such as psoriasis or certain eczemas, in an employee to be placed in a job that may involve solvent exposure and (2) liver disease in an employee being placed where there might be excessive and uncontrolled exposure to solvents. Under these job conditions, a careful evaluation should be made of the prospective employee's medical condition and the job requirements to be sure the person's condition would not be made worse by the work.

Progressive conditions. Chronic obstructive lung disease (such as emphysema), congestive heart failure, and certain arthritic diseases of the spine are examples of progressive conditions. Although placement in other than sedentary work may be possible, progression of the condition may make necessary a future placement in less demanding work.

Conditions resulting in intermittent impairment which may result in incapacity. Uncontrolled or inadequately treated epilepsy, and poorly controlled diabetes subject to intermittent reactions such as dizziness or insulin shock would be examples. An employee with these conditions suddenly could become unable to perform his job and become a grave risk to himself and co-workers.

Guidelines for the employment of individuals who are variously incapacitated have been outlined by several professional organizations (see Bibliography).

It is not the function of the physician, however, to inform the applicant whether or not he is to be employed. This is the prerogative and duty of management because there are many other factors in addition to physical qualification that bear upon suitability for employment.

SCOPE OF THE EXAMINATION

It is impossible to define what constitutes a complete examina-

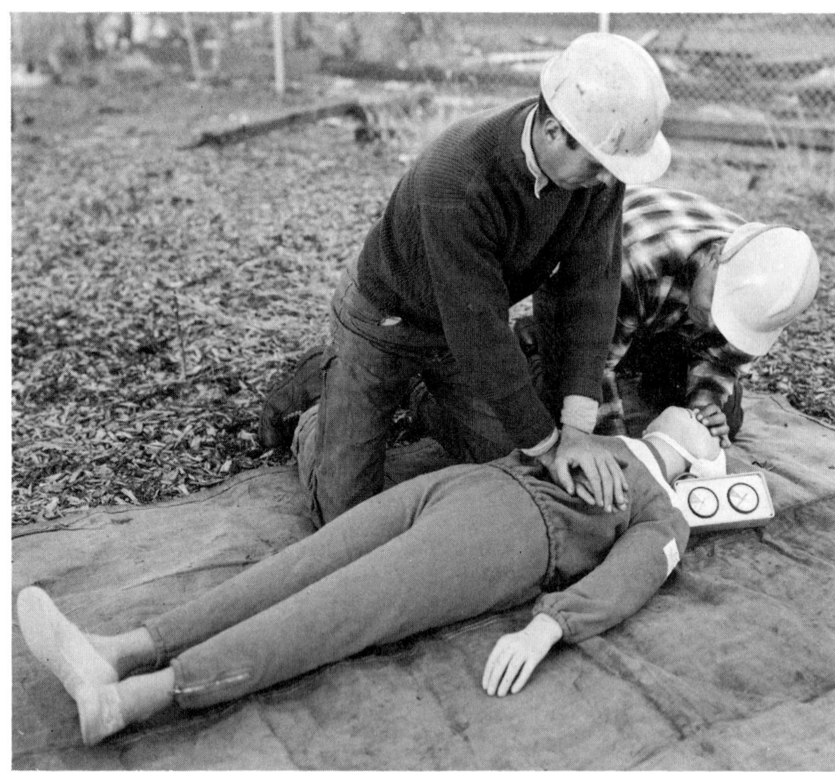

Figure 26-10. These linemen are participating in cardiopulmonary resuscitation (CPR) training in the field, using a doll as the "victim." In the absence of a physician, a professional nurse or another person trained and certified in CPR can apply appropriate emergency procedures.

tion or even a suitable examination because physicians have different opinions regarding the relative values of various test procedures. Therefore, the scope of physical examinations should be determined by the physician who is familiar with the job and working conditions. The nature of the industry, its inherent hazards, the variations in jobs, in physical demands and in health exposures are determinants. The values of different test procedures and their cost in time and dollars must be assayed. Perhaps examinations should be different in scope for different jobs. For example, the physical condition of an ironworker who will help build a multistory building is a far different problem than that of the sedentary seamstress. There are, however, basic physical examination considerations applicable to each.

It is estimated that between 4 and 15 percent of diseases encountered by a practicing physician may be caused by the occupational environment or are naturally occurring diseases aggravated by occupational circumstances. If such statistics are applied to the number of patients seen by physicians, it would suggest that there are a larger number of occupationally induced medical problems than reported by physicians each year. The missed condition could better be diagnosed by increasing awareness of the large number of undiagnosed medical problems, developing a more adequate data base, especially an occupational medical history, and acquiring more information about the diseases of people at their workplaces.

The most common error made by the clinician who attempts to diagnose an occupational disease is to relate an ill-defined medical problem to a work environment about which no environmental information is available. More correctly, the first step should be to make a diagnosis of the disease or observe the symptoms, signs, and abnormal laboratory parameters. As much as possible this evaluation should be quantitative, both for symptoms and signs and for the laboratory data. The second step is to obtain quantitative information about the work environment (such as opportunity for exposure, the materials used, and environmental air measurements taken by the company, a governmental agency, or insurance carrier). This step often requires obtaining MSDSs from the employer, eliciting the employer's help in determining just what exposures really occur to the patient, investigating trade names as possible toxic substances through the use of the poison index at the local poison control center, and often writing or calling manufacturers. Finally, experience and published reports in the world's literature allow the investigation of any possible relationships between the medical problem and the occupational environmental circumstances.

The occupational history plays an important role in fully investigating the possibilities of relationship between disease or symptom complex and the work environment. At the same time other environmental exposures outside of the workplace, such as those experienced in hobbies, sports, and the home, also need to be considered. The State Medical Society of Wisconsin's Committee on Environmental and Occupational Health suggests consideration for use of the occupational history format as a guideline to help the practicing physician in acquiring the appropriate occupational medical data base (Figure 26-4). This suggested form can be adapted to meet clinical needs. It has been designed so that the worker-patient may complete the form, and the physician should review the form with the patient to clarify and avoid incompleteness by helping the patient to understand parts of the form if necessary (Figures 26-13 and 26-14).

The various kinds of examinations may be classified as follows—preplacement (described earlier), periodic, transfer, promotion, special, retirement, and termination. In general, medical examinations should be voluntary. There are, however,

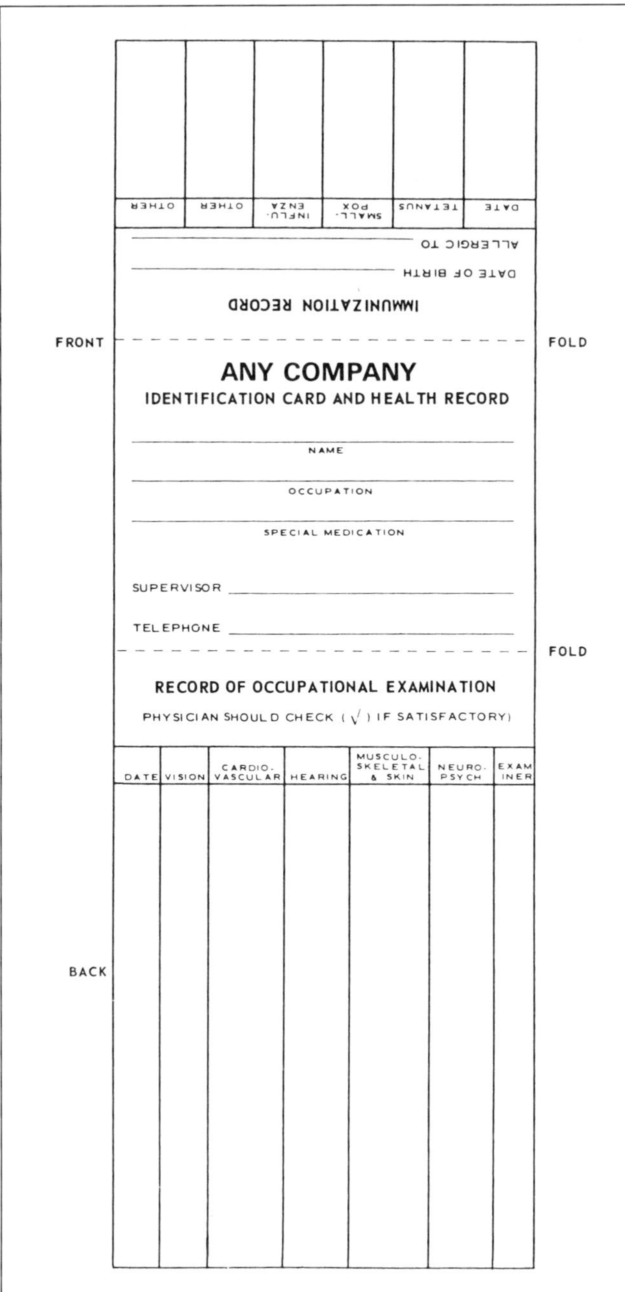

Figure 26-11. Illustration of an employee health record. (Courtesy M.B. Bond, MD)

exceptions—the flight physical for the FAA, the school bus driver, the interstate truck driver, the overhead crane operator, and others. As noted and strongly emphasized, the physical examination cannot be used in place of credible environmental monitoring.

Fitness or periodic examination

Fitness examination is the term used for a health evaluation made any time during the career of an employee. These occupational evaluations of the fitness of an employee for his or her work may be done periodically or only when there is (or appears to be) a health problem that might interfere with attendance or performance.

For certain jobs, periodic fitness examinations (1) ensure that the employee still has the special skills and abilities to safely continue his or her job or (2) evaluate whether the employee has suffered potential harm from work exposures (noise, toxic substances in the environment, and the like). The latter examinations are covered in other chapters.

It is recommended that all employees in potentially hazardous work be given a fitness examination at intervals, varying from 1 or 2–4 years, depending on age and type of work. Requirements for such examinations covering certain jobs are included in the OSHA regulations—and this list likely will be expanded. Federal regulations require operators of passenger or cargo vehicles to be periodically examined.

Fitness examinations requested by the employer also are important. These would include those requested by the physician and nurse based on their knowledge of a particular health problem, those requested by the insurance or benefit department and, most important, those requested by the supervisor.

Personality change, increased absence from work, tardiness, or other factors in the employee's behavior and appearance can help the supervisor identify a health problem early, when treatment may be most effective. When a supervisor refers an

Table 26-A. Medical Surveillance Program

Employee identification (personnel)	Prior to work with chemicals: Identify each individual employee who will work with ______. Complete the information requested on program cover (pink color).
Identify preplacement risk (physician)	Employee completed preplacement risk assessment form for predisposition, (re: skin, respiratory, allergies, dermatitis, etc.)
Review last physical examination record (physician)	Determine physical fitness for work and use of protective equipment—respirator, hats, gloves, protective clothing.
Complete physician's assessment form (physician)	Based upon record review of the above and the physician's professional judgment.
Counsel the employee (physician)	1. Review possible health hazards from work with ______. 2. Provide employee information requested. 3. Identify the need for early and prompt problem notification to health personnel or supervisor followed by prompt referral to the plant physician.
Record retention (personnel)	All records pertaining to the employee's work with ______ are to be retained in the individual's locked medical record file.
Medical records access (personnel)	Refer to *CFR* 1910: 20 (OSHA) Right and Opportunity to Examine and Copy Employee Medical Records. Inform each employee of this right and obtain a completed authorization form prior to record release.
Corporate awareness (personnel)	NOTIFY HEALTH/SAFETY/HYGIENE SERVICES OF ANY IDENTIFIED ______ HEALTH RELATED PROBLEMS.

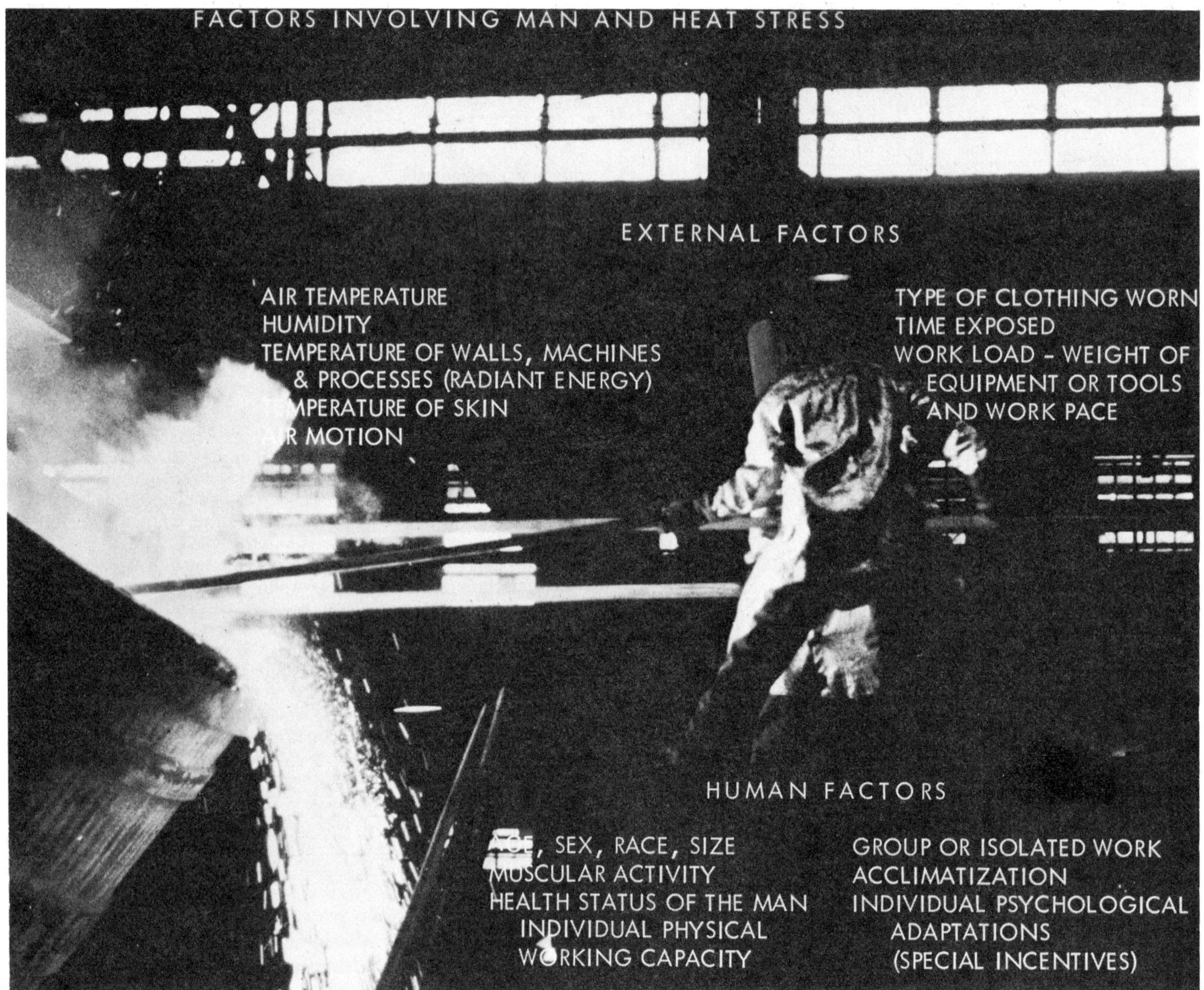

Figure 26-12. Two workers on a hot job—skimming slag from the surface of a large ladle of molten iron (30 tons) before it is poured into a mold. This work can take up to 20 minutes of sustained moderate effort, together with the high radiant heat stress. (Reprinted with permission from Zenz, C. *Occupational Medicine: Principles and Practical Applications.* Chicago: Year Book Medical Publ., 1988)

employee for a fitness examination, he or she should communicate objective observations and the reasons for concern to the medical department if the physician is to properly evaluate a possible health problem. The physician must correlate the supervisor's statement with the employee's health and exposure history if he or she is to uncover a condition that the employee may wish to conceal, such as alcoholism or mental illness.

When the physician has prior medical history of the particular employee, he is better prepared to conduct a fitness examination. Therefore, the examination need only evaluate the specific problem; this might be completed during an office visit or could require a very thorough examination, and even include referral to other specialists.

A report to the supervisor is necessary, and should include the following information—the employee is (or is not) able to perform specific tasks; employee does or does not have a health problem; if so, advice should be given to the employee for proper medical care, along with a general statement concerning the prognosis. No diagnosis or other detailed information should be revealed to the supervisor; this protects the confidentiality of information concerning the employee and contributes much to obtaining the employee's respect and cooperation.

Fitness examinations should be given on return to work after a sickness disability of 1 week or longer. At this time, it is determined if the employee actually is able to return to work or not. Return to lighter work or shortened hours often is helpful in rehabilitation.

A fitness examination during an absence due to illness sometimes is indicated if return to work is delayed beyond that expected. In this instance, it may be discovered that the diagnosis or treatment is improper and helpful recommendations can be made, especially if present care is not working. It also may permit detection of the employee who would abuse the sickness disability payments plan by not returning to work when adequate recovery has been achieved.

See Table 26-B for sample guidelines for required physical examinations and other special tests based on occupation.

It must be emphasized that periodic examinations and annual physicals cannot be expected to reveal responses to occupational exposures if the working conditions do not exceed the recommended ACGIH Threshold Limit Values for chemical or physical agents; indeed, if exposures reach these levels, it is expecting too much to have a physician detect any chemical changes. Even with the use of multiple biochemistry blood tests, work exposures must be excessively high to ordinarily cause overt physiological response (Chapter 15, Industrial Toxicology, and other related chapters).

Table 26-B. Suggested Examples of Recommended Physical Examinations and Other Special Tests Based on Occupation

Employees in the following occupations or located in the listed work areas require periodic physical examination and/or special laboratory and x-ray studies.

Occupation	*Type of Examination*	*Frequency*
Climbing—10 ft or more (includes hitchers, electricians, maintenance and repair, power house, plant engineering workers, and steamfitters)	Examination by physician, orthorater, audiometer test, urinalysis, electrocardiogram (ECG) if over 40 years old or if otherwise medically indicated	Annually, if over 40 Every 3 yr, if 20-30 Every 2 yr, if 30-40
Crane operators	As noted above	
Locomotive engineer and in train crew	As noted above	As noted above
Truck driver and vehicle test drivers	As noted above	As noted above

Note: Not intended to be complete or to be regarded as official procedures; merely represented as a guide. (Physical examinations, laboratory and x-ray studies are not to be used as substitutes for workplace hygienic studies, such as air-monitoring.)

Vision

Visual tasks vary greatly and are influenced by illumination, reflectance, color, object size, and many other factors that are beyond the scope of this brief section (Chapter 5, The Eyes). For most jobs, it is adequate to consider simply near and far visual acuity, field of vision, and color vision. Employees who operate commercial passenger or cargo vehicles, including aircraft, must conform to federal standards, copies of which should be on hand for ready reference.

Vision testing of applicants and employees is best accomplished with one of the vision-screening devices that screens acuity (far and near for each eye separately), color vision, deviations from normal vision (phorias), and depth perception (stereopsis). (The depth-perception test is not considered accurate enough to use as the only basis for work restrictions.)

The results of such vision screening are entered on a card that indicates those abnormalities requiring referral to an ophthalmologist for detailed evaluation. The ophthalmologist is an important consultant to the occupational physician, and such ophthalmological services frequently are needed.

Industrial safety glasses, conforming to American National Standard Institute Z87.1, *Practice for Occupational and Educational Eye and Face Protection,* should be required where eye hazards exist in the working environment. The term eye hazard must be interpreted with judgment after observing the work site to consider not only routine potential hazards such as dust and smoke but also the possible effects of a sudden disruption of the work process. For example, molten metal or irritating chemicals may splash into the eye because of an accident, and although the risk based on frequency of occurrence may appear small, the potential seriousness of the injury may be sufficient to recommend routine wearing of protective eye equipment.

The ANSI Standard Z87.1 states that spectacles must have nonflammable frames and lenses of plastic or heat-treated glass that are at least 3 mm thick in the center and able to withstand the impact of a 25.4-mm (1 in.) steel ball dropped from a height of 1.27 m (50 in.).

Prescription glasses. There is another standard for prescription glasses, ANSI Z80.1, *Requirements for First-Quality Prescription Ophthalmic Lenses,* which provides that lenses must be at least 2 mm thick in the center and able to withstand the force of a 15.9-mm (⅝ in.) steel ball dropped 1.27 m (50 in.). Plastic lenses must meet other stringent requirements.

Contact lenses. Contact lenses both of the hard corneal type and of the soft, large hydrophilic type are quite common, and the occupational physician must understand the potential problems related to such lenses. First, it is important to know whether the individual can tolerate them comfortably for 10 hours or more each day. Second, where there are appreciable amounts of dust, smoke, or irritating fumes or liquid irritants that could splash into the eyes, contact lenses are not recommended by most ophthalmologists. Often a judgment must be made about the degree of potential hazard, because some persons obtain better visual correction with contact lenses and do need them, but these cases are rare. These persons should wear the proper safety spectacles or goggles over their contacts when in an area where any degree of potential eye hazard exists. The National Society to Prevent Blindness gave a position statement concerning use of contact lenses.

The use of contact lenses by employees may constitute a special eye hazard in certain job categories. The nurse should take special care to record prominently on the medical record when an employee wears contact lenses. One little-noticed regulation under the OSHAct prohibits the wearing of contact lenses if a face-type respirator is to be used (Addendum A to Chapter 23, Respiratory Protective Equipment, and Chapter 5, The Eyes).

Monocular vision. Persons with longstanding monocular vision usually learn a degree of depth judgment that makes them able to work at any job except those covered by federal regulations for operating commercial vehicles, trains, or airplanes. However, individuals who have recently become monocular may be at increased risk for a period of time when working with moving machinery or work at heights, and may need temporary or even permanent work restrictions.

SUGGESTED OCCUPATIONAL EXPOSURE HISTORY
Part I

Please answer the following questions. Begin with your present job and list all jobs or military service you have held in order of date whether full or part time.

TODAY'S DATE NAME SOCIAL SECURITY NUMBER JOB CLASSIFICATION(S)	**List potential hazards exposed to: (examples)**				Work related illnesses or injuries	
	Physical	**Chemical**	**Biological**	**Psychological**		
	Noise Radiation Vibration Electrical shock Temperature extremes Repetitive motion Heavy lifting	Mercury Lead Dust Gases Fumes Acids Solvents Caustics	Viruses Bacteria Parasites Fungus Animal bites Etc	Boredom Work shift fatigue Risk of falling Risk of being buried Repetition	YES	NO
COMPANY NAME CITY, STATE JOB TITLE FROM: TO: AVERAGE HR/WK	HAZARDS: COMMENTS:					
COMPANY NAME CITY, STATE JOB TITLE FROM: TO: AVERAGE HR/WK	HAZARDS: COMMENTS:					
COMPANY NAME CITY, STATE JOB TITLE FROM: TO: AVERAGE HR/WK	HAZARDS: COMMENTS:					
COMPANY NAME CITY, STATE JOB TITLE FROM: TO: AVERAGE HR/WK	HAZARDS: COMMENTS:					

Figure 26-13. An exposure history should be kept in the employee's medical file. (Reprinted with permission from the *Wisconsin Medical Journal* 80[1981]) (Continued.)

SUGGESTED OCCUPATIONAL EXPOSURE HISTORY
Part II

SECONDARY WORK (examples) Firefighting, Civil defense, Farming, Gardening, Civic activities, Etc	List potential hazards exposed to: (examples)				Work related illnesses or injuries	
	Physical	Chemical	Biological	Psychological	YES	NO
	Noise Radiation Vibration Electrical shock Temperature extremes Etc	Mercury Lead Dust Gases Fumes Acids Solvents Caustics	Viruses Bacteria Parasites Fungus Animal bites Etc	Boredom Work shift fatigue Risk of falling Risk of being buried Repetition Etc		
ORGANIZATION CITY, STATE JOB TITLE FROM: TO: AVERAGE HR/WK	HAZARDS: COMMENTS:					
ORGANIZATION CITY, STATE JOB TITLE FROM: TO: AVERAGE HR/WK	HAZARDS: COMMENTS:					
ORGANIZATION CITY, STATE JOB TITLE FROM: TO: AVERAGE HR/WK	HAZARDS: COMMENTS:					
HOBBIES & ACTIVE SPORTS						
ACTIVITY CITY, STATE FROM: TO: AVERAGE HR/WK	HAZARDS: COMMENTS:					
ACTIVITY CITY, STATE FROM: TO: AVERAGE HR/WK	COMMENTS:					

Some chemicals have effects on the reproductive system. Have you or your present or former spouse had any problems with reproduction? If so, please indicate circumstances (e.g., stillborn, deformed, miscarriage, infertility).

EXPLAIN: ______________________________

____

Figure 26-13. (Concluded.)

WISCONSIN CHEMICAL CORPORATION
EMPLOYEE OCCUPATIONAL HEALTH RECORD

Name (Last First Middle)	Sex	SS#	Date of Birth	Location of Plant

TO THE EXAMINING PHYSICIAN:

In order that Wisconsin Chemical Corporation may protect and maintain the occupational health of our employees, the attached examining physician's job information card is to appraise you of the demands of the job under consideration or now being performed.

Please complete the physical examination and test data evaluation and indicate the employee's fitness for work.

DATE	JOB TITLE	FITNESS YES	FITNESS NO	PHYSICIAN'S SIGNATURE	PHYSICIAN'S COMMENTS

Figure 26-14. Employee occupational health record. (Courtesy Wisconsin Chemical Corporation)

Vision standards are used by the military and other federal government agencies (such as the Department of Transportation [DOT] and Federal Aviation Administration [FAA]) and most companies. Although there is reasonable agreement among these standards, there is no single set of visual abilities that could be considered a universal consensus standard. For example, the DOT requirements for interstate truck drivers are 20/40 (Snellen) in each eye and with both eyes, with or without correction, and a field of vision of at least 70 degrees in the horizontal meridian in each eye, plus the ability to recognize traffic signal colors—red, green, and amber.

The AMA's Ad Hoc Committee on Medical Aspects of Automotive Safety published its recommendations in 1969 and these are listed in the AMA publication "Medical Conditions Affecting Drivers—1986." This Committee recognized three classes of drivers, namely, *Class 1*—Professional drivers of ambulances and the like; *Class 2*—commercial taxi drivers and professional drivers of freight trucks; and *Class 3*—operators of personal vehicles. Visual acuity of 20/25 (Snellen) in each eye with spectacle correction of less than 10 diopters is required for Class 1. For Class 2, central visual acuity of 20/40 in the better eye and at least 20/60 in the worse eye is required, with a special notation if spectacle correction is 10 diopters or more. Class 3 drivers must have 20/40 or better in one eye. These figures are given for quick reference, but the entire AMA publication should be consulted.

Low-back problems

Perhaps the greatest hope for detecting the high-risk individual with reference to the low-back problem has been based on the use of preplacement lumbar x-rays and is mentioned because of the magnitude of the problem. Many physicians have claimed such procedures are effective in reducing the frequency and severity of low-back problem cases in industry, particularly during the 1950s and early 1960s. More controlled studies by experts have not justified the validity of low-back x-rays for employment screening and placement purposes *when used as the sole criterion for employability.* The indiscriminate use of x-ray procedures is improper and such techniques must only be used for selected occupations posing the greatest risk to the prospective worker, and must be taken into account with the medical history and physical examination, and knowledge of the job assignments.

Low-back medical histories. The alternative to x-ray-based screening seems to be to acquire both a better history of low-back health and a better examination and functional evaluation of the person's back. However, a person could easily distort his or her own medical history, intentionally or unintentionally, to acquire a job.

Selection of people for materials handling jobs based on their heights or weights is not well justified according to the statistical evidence showing reduced low-back pain incidence rates. There is, however, the ever-present need to specifically consider a person's height and weight in relation to the physical characteristics of the prospective job. This concern goes far beyond simple height and weight. All jobs that do not allow for a large range of variation in the population, as stated in various reference books, should be identified and those specific limit dimensions should be stated in job descriptions. The medical personnel then should select people within the stated dimensional categories. Obviously, continued efforts should be made to eliminate the restrictions through job design.

Chest x-rays. The routine use of preplacement chest x-rays (or hospital admission x-rays) has been proven to be redundant; indeed many states have prohibited the taking of routine x-rays unless clinically indicated and justified by the physician. Likewise, annual or other periodic chest x-rays are not to be taken routinely.

Pulmonary function testing. The routine taking of pulmonary function tests (spirometry) is frequently overdone, especially in healthy, younger workers; under age 40. The equipment used must be of an approved type, as recommended by The American Thoracic Society, proper calibration must be maintained and the person performing the tests must have approved training, as well as be supervised by a physician, in order to produce credible test results (see Horvath, *Pulmonary Function Testing* in Industry, Chapter 18, in Zenz, 1988).

Emotional abnormalities

The mental or emotional state of employees has enormous consequence in our working environment and it is well known that emotional problems account for the largest single source of lost work time. The usual applicant examination will not detect most persons with common emotional difficulties such as psychoneuroses unless current or prior treatment is admitted. Perhaps this is proper, because psychoneurosis is so common and many persons with the disorder are good employees. The physician's role is, after all, to evaluate for employment and not to play God. Yet, it is known that many neurotic employees have excessive lost time from work, and occupational physicians commonly see work schedules and production disrupted because of personality clashes between neurotic workers and other employees.

History of personality disorders. Those applicants with a history of treatment for mental disease should be carefully investigated. Before placement, written reports should be obtained from attending physicians and institutions. Persons with symptomatic phobias or with certain psychophysiological reactions, especially cardiac, may require special placement and may need psychiatric treatment either before or after employment. Appropriate, written permission must be obtained before requesting medical information. The signed release form should be kept in the professional's files (Figure 26-15).

Applicants with potentially serious personality disorders should be referred for evaluation by a psychiatrist and/or psychologist and special attention given to their histories. In many of these cases, it is necessary to recommend against their working in job situations that might involve too much freedom and responsibility—in other words, these persons often do better in a small group with limited contact and responsibility with the public. Periodic follow-up visits are recommended and this may permit many with mental abnormalities to remain useful and productive citizens.

Applicants with a history of presence of a psychosis (mental derangement characterized by defective or lost contact with reality) often are unsatisfactory in the average work setting. Very careful investigation must be made of the background health history. If all other factors appear favorable, such persons may be able to work successfully if placed in positions with limited responsibility, at least initially, and with an understanding supervisor. Follow-up of these persons on the job is indicated to make sure they pursue proper treatment and to detect recurrence of their disease.

HEALTH EDUCATION AND HEALTH COUNSELING

Clinical experience has demonstrated that early detection, referral, and psychiatric care for emotionally troubled employees are essential in reducing prolonged disability. Psychiatrists and physicians associated with industry have long known that the work situation is unique in its possibility for early detection of mental illness. There is a complex interplay involving the patient—employee, family, supervisor, and co-workers. Some or all of these relationships can have a strong influence on reducing lost time and in successful rehabilitation.

Treatment delays, lack of follow-up and outright disinterest are seen far too often. Medical departments, management personnel, and employees and their families are equally guilty. In an occupational injury, for example, it is routine practice to refer the injured employee to a physician or a specialist for prompt attention, especially when the employee elects to be directed by the medical department of the plant. Frequent referrals are made to ophthalmologists, dermatologists, plastic surgeons, and other specialists, but rarely to psychiatrists. With emotional disorders, on the other hand, busy practitioners usually are not eager to become involved in the care of persons with depressive reactions, severe psychoneuroses, and alcoholism. The personal physician may appreciate the problem, but, because of the pace of his or her practice, often does not take the time to convince a sick person to seek the required social, psychological, and psychiatric casework.

A resultant concern in this situation is the high cost of absences. Delays in obtaining treatment add a financial burden, since a person's absence from the job may cost a company from 4 to 6 times the actual wages.

The traditional patient-physician relationship is well known and understood. Physicians must appreciate the employee-employer relationship, which may extend throughout the adult life of the individual, sometimes up to 40 to 50 years. During this period, the employee may have seen many physicians. Today, most employers pay nearly the entire cost of illness and disability insurance, and they recognize that maintenance of good

EMPLOYEE AUTHORIZATION FOR MEDICAL RECORD RELEASE TO HIS/HER DESIGNATED REPRESENTATIVE

I ______________________ hereby authorize Wisconsin Chemical Corporation ______________________
(worker's full name) (location)

to release to ______________________ the following medical information from my personal medical records.
(authorized representative)

I give my permission for this medical information to be used for the following purpose.

I *do not* give my permission for any other use or re-disclosure of this information.

Full name of Employee/Legal Representative: ______________________

Signature of Employee/Legal Representative: ______________________

Date of Signature: ______________________

Figure 26-15. Employee release of personal medical information form. (Courtesy Wisconsin Chemical Corporation)

health and the sustained capability to remain on the job with optimal performance can be attained and maintained only with the assistance of experts, such as physicians.

Access to medical records

Part 1910:20 of the *CFR* is the standard covering an employee's right to have access to personal medical records. Its purpose is as follows:

> To provide employees and their designated representatives a right of access to relevant medical records, and provide a right of access to the representatives of the Assistant Secretary to these records in order to fulfill responsibilities under the OSHAct.
>
> Access by employees, their representatives, and the Assistant Secretary will yield direct and indirect improvement in detection, treatment and prevention of occupational diseases.
>
> It is the employer's responsibility to assure compliance with this section. However, access activities to medical records, may be performed on the employer's behalf by physician or other health care personnel.
>
> Except, as expressly provided nothing is intended to affect existing legal or ethical obligations which concern maintenance and confidentiality of the employee's medical information/duty; to disclose information to an employee or any aspect of medical care relationship or affect legal obligations which exist; to protect trade secret information.

Substance abuse

Alcoholism and drug abuse problems are exceedingly troublesome, and it is essential that the physician as well as the employer have a good understanding of the potential for helping many of these persons in the work setting. Employee assistance programs (EAPs) are increasingly available in many companies. They have been designed to offer in-house services to troubled employees or to refer the employees to the best source of treatment. Alcoholism and drug abuse continue to be among the nation's leading illnesses. Dependence on alcohol or other drugs is a major contributor to deterioration of family life, impaired job performance, morale and disciplinary problems, increased insurance rates, occupational accidents, increased absenteeism, and the rising crime rate. These illnesses know no boundaries. There is no generation gap among abusers—all races are susceptible, and socioeconomic status is no barrier.

It is estimated that one out of every 10 U.S. workers may have a drinking problem, but fewer than 10 percent of those who have drinking problems actually receive treatment.

Every time an applicant or employee visits a physician or nurse, there should be some health education and/or counseling given. Industrial medical departments provide an ideal opportunity for these services, which constitute an asset for both the employee and the employer. Employers are paying a large part of the costs for health care for employees and their dependents. It is economically important for the employer that his employees have an entry into the health care system when needed and that quality care is obtained.

Both physicians and nurses play an important role in guiding employees to the proper health care resources. An almost equally important role is played when the need is for counseling rather than for pills or surgery, and physicians and nurses experienced in occupational health programs generally are very skilled in the art of counseling. The education and counseling functions often are overlooked when assessing costs versus benefits of occupational health programs. This is because of the difficulties in determining the units of service that can be defined. This is unfortunate because experience and common sense indicate these activities are among the most valuable to both employer and employees. And these benefits are available to all employees, not just those exposed to potential hazards. This is an important point in justifying the costs of a health program in any sizable employee group.

Drug screening

The initiation of an employer-sponsored drug screening program is a delicate legal and social problem. If instituted, the direct involvement of medical and nursing personnel is inappropriate. As repeatedly stated, "Policing for absenteeism, insurance fraud, and socially deviant behavior is not a proper role for the physician." The company drug-screening program is a legal decision, and the physician must not become an unwitting tool.

MEDICAL ASPECTS OF USE OF RESPIRATORY PROTECTIVE EQUIPMENT

The problem of providing properly fitting respirators is complicated by the wide range of facial sizes and shapes that must be accommodated. Differences in facial sizes and shapes result from a wide variety of factors, the most significant of which are age, sex, and race.

Facial hair, such as beards or sideburns, makes it impossible to achieve an airtight seal between the facepiece and the face, particularly with the half-mask respirator. Even the stubble resulting from failure to shave daily can cause serious inward leakage of contaminated air. Therefore, workers who wear respirators with half or full facepieces should be cleanly shaven.

The wearer's comfort and his acceptance of the distress caused by wearing a respirator are no less important than the device's effectiveness. Some factors that can alter a respirator's acceptability are: (1) improperly fitted respirators; (2) uncomfortable resistance to breathing (which may result from increased physical effort demanding *respiratory air movement upwards of 50–60 liters per minute,* and can become intolerable if continued for more than a few minutes); and (3) limitations of vision and speech transmission.

A three-part program to effectively assess the employee's ability to use respiratory protective equipment should be initiated. Assessment should be done on a case-by-case basis. Such a program consists of three parts:

1. Employee evaluation
   a. Health history
   b. Health screening records
   c. Employee counseling interview
2. Respirator-fitness testing performed by the company industrial hygienist
3. Physician's review of data and employee fitness assessment. (See Chapter 23, Respiratory Protective Equipment, for more information.)

Eyeglasses

Providing respiratory protection for individuals wearing corrective glasses is also a serious problem. The ability to wear corrective glasses with a half-mask depends on the face fit. For a full-face mask, a proper face seal cannot be established if the temple bars of the eyeglasses extend through the sealing edge. Some full-facepiece designs provide for the mounting of special corrective lenses within the facepiece.

Any respirator affects the wearer's ability to see. The half-mask and the attached elements can restrict normal downward vision appreciably. Diminished vision in the full-face mask may be caused not only by the facepiece, but also by the design and placement of the eyepieces (see Chapter 23, Respiratory Protective Equipment).

Speech transmission

Speech transmission through a respirator can be difficult, annoying, and even fatiguing. Moving the jaws to speak can cause leakage between the facepiece and face, especially with the half-mask respirator.

WOMEN IN THE WORK FORCE

In the United States, women now make up nearly 50 percent of the labor force. Because of social legislation and government regulations, women are entitled to apply and they should be considered for any job. Nevertheless, there are some extremely important physiological elements of concern for the safety and health status for the women of childbearing capability.

Exposure to the unborn child

Of immediate concern is inadvertent exposure to a number of potentially very toxic materials that could prove hazardous to a fetus during its early development. Radiation exposure has long been recognized as an area of grave concern needing strict controls. Some other potentially hazardous substances, suspected or known to adversely affect the embryo or fetus, are: alcohol, aniline, arsenic, benzene, carbon disulfide, carbon monoxide, captan, formaldehyde, hexacholorobutadiene, hydrogen sulfide, lead (see later in this chapter for a discussion), mercury, methyl mercury, nicotine, nitrates (and other chemicals capable of causing methemoglobinemia), nitrobenzene, polychlorinated biphenyls (PCBs), selenium compounds, vinyl chloride, and xylene.

The AMA's Council on Scientific Affairs recently issued a report, Effects of Toxic Chemicals on the Reproductive System.

Spontaneous abortions

Concern about the possible chronic effects of anesthetic gases has been growing since 1967 reports of spontaneous abortions among Russian anesthetists working with ether in poorly ventilated operating rooms in the Soviet Union. In the United States, many operating rooms lack equipment for venting overflow gases to the outside. Other possible significant exposures in these occupational groups include ethyl chloride, ethyl ether, nitrous oxide, trichloroethylene, ionizing radiation, and infectious diseases.

Teratogenic effects

Götel and Stahl reported teratogenic effects, liver damage, and addiction upon exposure to halothane (2-bromo-2-chloro-1, 1, 1-trifluoroethane). Fatigue, headache, tiredness, and irritability are common symptoms of such exposure. Based on studies at the Regional Hospital in Örebro, Sweden, they stated that an anesthetist may be exposed to about two patient doses of halothane per year. It is easy to reduce exposure to halothane by connecting an evacuation unit to the anesthetic machine. With such equipment at work, the authors found that the gas in the anesthetist's breathing zone could be reduced 90 percent or more. Indeed, many women at work may be unaware of an underlying pregnancy, and the growing embryo or fetus is extremely susceptible to harm from agents external to the mother.

Since neither the mother nor anyone else can know of the pregnancy until at least several days and usually a few weeks after conception, harm to their fertilized ovum or fetus can occur from work exposures that are considered acceptable for the nonpregnant worker. The question that follows then is, "Are there work exposures acceptable to the nonpregnant person but from which the risks of fetal harm in very early pregnancy are so great as to make it wise to forbid the fertile woman to experience them?" The answers to such questions are social and not merely medical.

It is proper to attempt to define these risks, however, but information on which such definitions can be based is only partially available at this time. Whatever information is available to permit quantifying risks to at least the more common work exposures, such information seems appropriate for inclusion in this chapter as a guide to employers. There will never be a complete definition of the range of risks because of the myriad substances and conditions that can potentially cause harm to the fertilized ovum or growing fetus. In addition there are new substances being developed, which complicates the immense problem of determining whether an external agent will actually cause harm.

The following instructions should be accepted only as guidelines, and at the same time the variables of the individual worker, the pregnancy, and the job should be recognized. Jobs must be considered from the standpoint of physical demands and potential hazards in the work environment. (The efficiency of a woman in various stages of pregnancy as compared to a nonpregnant woman will not be considered as this is held to be a matter of employee-employer relations and not strictly a medical subject.)

Guidelines for the working pregnant woman

The following guidelines are recommendations set forth by joint efforts of the American College of Obstetricians and Gynecologists and NIOSH.

Normal woman—normal pregnancy. A normal woman who is experiencing a normal pregnancy should be given work that is neither strenuous nor potentially hazardous (the latter term refers both to the physical hazards and external agents in the work environment that constitute a significant risk). The most common jobs in this class are clerical and administrative, but include many craft, trade, and professional positions. The usual limiting factor on the work that is "strenuous and/or potentially hazardous" to a pregnant woman, fetus, or both is the pregnancy itself as acceptable limits have been established in most positions for the nonpregnant worker. (See discussion on lead exposure later in this chapter.)

Normal woman—abnormal pregnancy. If an abnormal pregnancy is known to exist, a list of the abnormalities of the pregnancy should be developed and comments made on the ability to work in each instance, identifying the specific risk. Included in the evaluation should be the appropriate intervals for the follow-up examinations while working.

Abnormal woman—normal pregnancy. An abnormal woman is used here as one who has a health condition such as heart disease, diabetes, thyroid disease, or multiple sclerosis. Guidelines should be developed for proper evaluation and monitoring during pregnancy while working.

Abnormal woman—abnormal pregnancy. For this condition, the guidelines for the last two conditions should be used.

After childbirth "disability." A normal woman who had a normal delivery should work only in nonstrenuous and nonhazardous jobs beginning about 4–6 weeks after delivery. If the woman is abnormal and/or the delivery is complicated, a list of complications should be developed with an explanation of how each affects the ability to work.

Hazardous exposure. Without question, a pregnant woman should avoid unnecessary exposure to any of the following:

- ionizing radiation
- chemical substances that are mutagenic, teratogenic, or abortifacient
- biological agents of potential harm.

("Unnecessary exposures" may be construed in this context to be regularly at, or above, the "action level.")

BIOMECHANICAL AND ERGONOMIC FACTORS

Using height and weight alone when selecting people for materials-handling jobs is not justified according to the statistical evidence showing low-back pain incidence rates. Although a person's height and weight in relation to the physical requirements of the prospective job is an important consideration, it is not the only one.

All jobs that do not allow for a large range of physical variation in the population should be identified. Specific physical limits should be stated in job descriptions. The medical department then should be prepared to select people within the stated physical limit categories. Obviously, continued efforts should be made to eliminate the restrictions wherever possible by job redesign.

Strength testing

Some occupational physicians have adopted informal methods of strength testing to assess what a person can handle safely. One test is to simply ask the person to lift a box of a weight similar to that of the maximal lift on the job. If the person appears to be capable of lifting the load easily, he or she obtains the job. Some limitations of this particular procedure are worth noting, however; this technique still relies on a subjective estimate of how well the person handled the load. This subjective judgment may be difficult to defend. Such a test also has an inherent danger of subjecting a person to a potentially high impulse stress that they cannot readily sense and control due to the box-lifting dynamics.

Many investigators have determined that the American female worker has a mean strength of 70 percent of her male counterpart. But more recently, studies have indicated that, depending on the past physical activities of the population, the woman's strength may average as low as 43 percent of the strength of the male.

It is reasonable to expect that some women are stronger because they have held jobs requiring considerable physical effort, and if tested, their mean strength values probably would be closer to those generally reported for men (that is, 70 percent). One could characterize that as a natural selection of people who accept certain types of work. On the other hand, if an industry is selecting women from existing jobs that require little physical effort, their resulting strength values undoubtedly will be lower. A person's past performance is an important factor in predicting strength capability in future jobs. Another important factor in comparing male and female strength capacities is the work history of the men: whether they have been doing sedentary or heavy work or are the test subjects a mix of both types?

Of all the factors affecting population strengths, certainly the sex of a group of people is one of the most important factors. Generally, in overall terms, a woman has about 60 percent of the muscle capacity of a man of similar physical build. Consequently, separate standards for men and women are required, or the strength requirements must be engineered to include a mixed male-female population.

LEAD EXPOSURE—CLINICAL AND ENVIRONMENTAL CONTROL CONCEPTS

Lead (and its compounds) is an important industrial material. The medical, engineering, and hygienic principles of control outlined here for lead compounds apply to most other potentially toxic exposures. The annual industrial consumption of lead, in the United States, alone, is well over one million tons and because it is used in more than 100 occupations, lead can affect countless workers.

Inorganic lead absorption can occur from inhalation of finely divided particles of metallic or soluble lead compounds when exposure occurs in excess of 50 μg of lead per cubic meter of air per 8 hours. The lead dust or fumes (for example, lead oxide) from heated metal is the typical exposure hazard. Potential sources of exposure when lead poisoning is most frequently reported are as follows:

1. storage battery manufacture—particularly in oxide mixing, plate pasting and assembly operations with inadequate controls
2. grinding lead solder smooth (common in the auto industry)
3. smelting lead scrap metal and old battery plates
4. brass foundries, where 1–30 percent lead alloys are melted at 205 C (400 F) and become volatile at 540 C (1,000 F), along with copper, which melts at about 1,100 C (2,000 F), thus causing intense fuming of lead in its most dangerous form
5. cutting "red-lead" painted steel with oxyacetylene torches volatizes the finely divided lead
6. making, mixing, and bagging paint pigments; sanding or burning off.

The most important bodily route of entry is through the respiratory tract. (Swallowed lead is absorbed relatively less than inhaled lead but is still dangerous.) *Lead absorption, not lead intoxication* may be ascertained by: (1) history of health by systemic review, by physical examination including a neurological examination; (2) a blood lead level over 45 micrograms (μg)/100 mL blood; (3) hemoglobin below 13 gm, for a male—for a female below 12 gm; and (4) there may be stippling of the red cells.

Lead intoxication (for adults)

Clinical types of lead intoxication include:

1. clinical lead intoxication characterized by anemia, elevated urinary porphyrins and delta aminolevulinic acid (ALA)
2. colic—a smooth muscle spasm, causing abdominal cramping pain
3. palsy—a neuromuscular deficit, which causes either a wrist drop or foot drop
4. blood lead levels consistently greater than 40 μg/100 mL blood (verified by repeated, approved [accredited] laboratory tests).

The diagnosis of acute lead colic or lead intoxication is made by medical history and physical examination and the blood level (>50–60 μg/100 mL blood).

Estimated duration of the intoxication:

1. Chronic anemia or "lead colic" may take 3–6 months to resolve with no residual findings.
2. Palsy may take 2 months–2 years, with occasional partial disability.
3. The blood lead levels take several months to fall to normal, depending on extent of total exposure, the magnitude of past exposure, and how high the blood level was found to be. Also, older, sedentary, and less physically active persons require longer periods; 1–2 years is not unusual even with no further exposure.

Treatment: Reduce or eliminate exposure.

Chelating therapy

"Deleading" of persons with lead exposure is commonly reported. Chelating agents must not be given to employees who are still exposed to lead. Chelating agents must be used with caution as they can cause kidney damage, which could be irreversible. Other trace metals, normally found in the body, may be forced out during chelation.

There are only two reasons for treating lead toxicity of patients with a chelating agent such as calcium disodium ethylenediaminetetraacetic acid (EDTA) or penicillamine:

1. The primary reason for chelation therapy is to reduce acute symptoms of lead colic (substantiated by high blood lead levels and other findings previously described). Occasionally, lead colic will not respond to intravenous calcium

gluconate and the abdominal pain is not relieved. After treatment with EDTA, the abdominal colic will be relieved, usually within a period of 6 hours, and always by the end of 24 hours. If one is treating a patient with a chelating agent, 2 or more 24-hour urinary lead level analyses should be obtained prior to EDTA therapy and several 24-hour specimens should be obtained following EDTA therapy.

2. The second reason for using chelating therapy is for occasional cases of marked and unresponsive chronic anemia. On occasion, dramatic recovery of chronic anemia due to lead occurs after EDTA therapy.

Female workers exposed to lead compounds have elicited tremendous interest in recent years, but there is lack of understanding, and conflicting opinions on this subject. Hernberg, the Scientific Director of the Institute of Occupational Health in Helsinki, Finland, has investigated whether women of reproductive age should be employed at all in lead works, because of the risk for the embryo in case of pregnancy. There is sufficient evidence to prove that lead passes the placenta, so the levels in the embryo will reflect those of the mother. On the other hand, there is not yet conclusive evidence of any particular sensitivity of the embryo, although common sense makes one likely to suspect such a hypersensitivity.

In the book, *Lead: Airborne Lead in Perspective,* the National Academy of Sciences concludes its review in the following way: "Clinical responses to moderate increases in soft-tissue lead are still ill defined, particularly in very young children. Nevertheless, a rapidly growing child's response to moderately increased lead content may well differ from that of an adult, despite the current inability to perceive the response." Thus, it seems prudent to regard developing embryos as a risk group.

Tolerance limit

Many experts now believe that a blood concentration of 40 μg/100 mL is the upper tolerance limit for the general adult population. However, under no circumstances can commonly used occupational norms be applied for pregnant women. Accordingly, any exposure resulting in 30 μg/100 mL in the mother's blood should be regarded unacceptable, and pregnant women should not be allowed to be exposed to such intensities.

Although potential reproductive problems related to excess lead exposure are vexing, it is well known that lead at high levels can be harmful to all adults. What is considered safe for the female worker? According to OSHA, a maximum of 50 $\mu g/m^3$ in air and 40 μg/100 mL blood. In adult men, it has been estimated that chronic exposure to 50 $\mu g/m^3$ in air can produce blood lead levels of 35–45 μg/100 mL. Messite and Bond (1987) have concluded that blood lead levels in women should be no higher than 30 μg, implying chronic airborne (and respirable) exposures be no greater than 50 $\mu g/m^3$, and without other sources of exposure.

In this situation, the decision is simple. But most women start their work without being pregnant, and the problem does not arise until the pregnancy starts. Should the use of female workers in the reproductive age in lead work be completely prohibited only because of the possibility of later pregnancy? It is hard to believe that such a prohibition would work in practice.

Thus, preventing an embryo from becoming overexposed must rely on intensive health education, stressing the possible dangers should the woman become pregnant, and the demand for reporting immediately to the health service when this happens. The prospective mother also should be assured that another job with the same economic and social conditions will be arranged for her in case of pregnancy to encourage her to report without fear of losing her job.

Properly, occupational working conditions should be so that exposures to lead and other harmful substances are controlled.

Preplacement examination

All workers who are applying for jobs where there is exposure to lead should be submitted to preplacement examinations. In the United States and several other countries, such examinations are required by law. The aim of the examination is to prevent workers with dispositions suggestive of higher-than-normal sensitivity to lead from being exposed.

Symptoms indicating such dispositions are nail-biting, anemia, porphyrinuria, cardiovascular disease, recurrent peptic ulcer or gastritis, psychiatric and neurological disease, neurasthenic disorders, renal and hepatic disease, and all other conditions that significantly impair health. In many countries, the employment of workers below 18 years of age is prohibited.

The preplacement examination should thus include a carefully compiled history, a physical examination, a complete blood count (CBC), blood lead, urine analysis (including microscopic), and other tests necessary to make sure that the worker is free from significant disorders. Medical records should be kept and should contain all relevant data.

In summary, then, with respect to lead, if all potential hazards of toxic work exposures were properly controlled to minimize or eliminate exposure of employees, and if, for example, the high-volume exhaust system is functioning adequately, there would be less concern about lead absorption by employees on the job, and setting up and maintaining lead screening programs would be unnecessary. But factory conditions are rarely ideal; thus, exposure to, absorption of, and intoxication from lead and its compounds pose a never-ending problem in industry.

Common expressions, control and supervision of the worker exposed to various toxic materials, ought to give way to redirection to and emphasis on designing, installing, and maintaining clean working areas. If hygiene studies of workplaces reveal lead levels greater than 30 $\mu g/m^3$ during an 8-hour shift and engineering controls that are less than desirable, with dirty work areas and poor personal hygiene, then appropriate action is necessary.

First, respirable lead levels must be reduced. Other cleanup measures, including restrictions on eating, smoking, and drinking in the work areas should be enforced. To make all this workable, *decent* eating areas must be designated; kept clean with equipment to store food and liquids—the air must be of known cleanliness. Also, the employer must provide suitable washup facilities, (including showers); separate lockers for each employee, one for street clothes and personal shoes and one for work clothing and work shoes (work clothes should be provided by the employer along with appropriate laundry arrangements—no work clothes and shoes permitted to be taken home). Medical screening should be instituted to include occupational and medical history, physical examination, and a venous blood sample for lead analysis.

1978 NIOSH *Criteria Document on Lead*

This NIOSH *Criterial Document* states that, "unacceptable absorption of lead posing a risk of lead poisoning (in adults) is demonstrated at lead levels of 60 μg/100 mL of whole blood or greater." Excellent as they are, criteria documents are not substitutes for good clinical judgment. Furthermore, laboratories that perform blood and urine analyses for lead must demonstrate competence and provide the limits of variation and error of their results; this variation in current laboratory techniques can be large (± 20 μg). If the blood level of lead is found to be elevated by a few micrograms above the established upper level and the worker is asymptomatic, take additional blood samples. A worker should not be removed from the job without good clinical cause, and since laboratory results vary widely, this decision must not be based on a single test. Repeated blood lead determinations after 1–2 weeks will be useful for confirming absorption and following the course of possible toxicity. See OSHA regulation for medical removal requirements.

Chelating agents. "Again, it must be emphasized that treatment with chelating agents should not be instituted in the absence of positive clinical findings. Chelating agents must be used with caution, for side effects such as kidney damage can occur. Deleading of a worker must not be undertaken while the worker remains exposed to lead. Sound occupational health practice is based on two major fundamentals, (1) control of contaminants at the worksite by engineering, and hygienic measures (including environmental monitoring) and (2) good medical surveillance. Neither of these should be substituted for the other."

BIOLOGICAL MONITORING

The use of biological monitoring in assessing workplace exposures to chemicals and other substances has gained in importance, particularly for observing workers in lead-using jobs, where blood lead level determinations predominate as example of biological monitoring. Other examples include measurement of carbon monoxide in the blood (as carboxyhemoglobin), various solvents detectable in the air expired from lungs (alcohol concentrations used in traffic law enforcement), phenol in urine, and mandelic acid in urine to detect styrene and many other substances. Even the analysis for arsenic in hair, nails, and other tissues is a method of biological monitoring. Another example of biological monitoring is the study of seminal fluid in cases of exposure to the pesticide dibromochloropropane (DBCP), which proved a valuable indicator for site-specific toxicity in workers making this product.

These measurements can indicate that there was an exposure and absorption of a chemical or substance into the body, but do not indicate the quantity that the person was exposed to and absorbed nor the source of the exposure. In very specific circumstances, fat biopsies can be taken to ascertain quantification and storage time of certain substances, e.g., DDT, PCBs, solvents, etc. Lung, liver, and other organ tissue samples can be used to augment or verify clinical findings.

Biological monitoring is based on the relationship between a measure of some absorbed dose and the environmental exposure, or as a measure of a health effect. Monitoring can offer information useful in studying a worker's individual response and in measuring overall exposure. Many factors can affect results: the metabolism and toxicokinetics of the chemical or substance, individual factors, such as the person's state of health, body size, workload, and life-style. Biological monitoring takes into account the uptake of a chemical or substance by all routes of exposure, including respiratory, cutaneous, oral, and nonoccupational sources (background levels). The main advantage of biological monitoring is that assessment of exposure by all routes can be made. A limitation is that these measurements are estimates of uptake and may not distinguish between occupational and nonoccupational exposures. Also, there is wide interpersonal variation due to factors outside the workplace, e.g., diet, alcohol usage, smoking, and drug usage; all potentially can affect the levels of a chemical or metabolite in a biological specimen, making interpretation difficult.

Explanation of "skin" notation

The "Skin" notation used in the American Conference of Governmental Industrial Hygienists' (ACGIH) *Threshold Limit Values and Biological Exposure Indices* (see Appendix B-1) deals with the potential contribution of the cutaneous route of exposure to the overall bodily exposure. Although there is scant data available from animal or human studies describing quantitative skin penetration, routing, and absorption into tissues and/or systemic uptake of materials, such as dusts, vapors, and gases, the ACGIH substance listing describes the impact that skin exposures play to the overall exposure. The cutaneous route includes the mucous membranes and eye, either by airborne or direct contact with substances(s) that have a "skin" notation in the TLV booklet.

The rate of absorption is a function of the concentration of the chemical or substance and the duration to which the skin is exposed. Certain "vehicles" can alter this potential absorption. Substances having a "skin" notation and a low TLV can be problematic at high airborne concentrations, particularly if significant areas of skin are exposed for long periods. Respiratory tract protection, while the rest of the body surface is exposed to a high concentration, may produce this type of situation.

Biological monitoring should be considered to determine the relative contribution of skin exposure to the total dose. The "skin" designation is intended to suggest appropriate measures to prevent skin absorption, so that the respiratory threshold limit is not underestimated, misinterpreted, or otherwise misused (so as not to provide an accurate assessment of the degree of potential worker exposure and consequent uptake, influencing substance body burden and certain physiological functions (ACGIH).

ACGIH Biological Exposure Indices (BEIs)

The ACGIH uses the term Biological Exposure Index (BEI) (Appendix B-1) for an "index" chemical that appears in a biological fluid or in expired air after an exposure to a workplace chemical or substance. It serves as a warning of exposure, either by the appearance of the chemical or its metabolite, or by the appearance of a physiological response associated to exposure. The BEIs are intended as guidelines to assess total exposure; and compliance with BEIs is not a substitute for controlling the workplace environment. The proper role of biological monitoring is to complement environmental monitoring. The OSHA regulations do require environmental monitoring, but, with some exceptions, do not require biological monitoring. *Environmental monitoring should be used to assess workplace levels of chemicals and substances, while biological monitoring can be used to more accurately determine uptake of workplace chemicals by an employee.*

SUMMARY

Sound occupational health practice is based on two major fundamentals—control of contaminants at the worksite by engineering and hygienic measures (including environmental monitoring), and good medical surveillance (including biological monitoring). Neither of these should be considered as a replacement for the other.

The elements of occupational environmental control are of utmost importance for protecting employee health in all work areas. This broad professional, occupational, or environmental control is most commonly the field of concern of the industrial hygienist who most recognize, study, and recommend appropriate measures to prevent illness or impaired health of the worker or workers.

The occupational health physician depends on the knowledge and techniques of the hygienist who provides insight on the type and magnitude of potentially hazardous or stressful occupational environment factors, be they chemical, physical, or a combination of these. Without adequate industrial hygiene information, it may be difficult, if not impossible, for a physician to determine whether or not a relationship exists between symptoms and findings of occupational or nonoccupational disorders or diseases.

The mere history of exposure to a substance does not mean that occupational disease has occurred or that some form of poisoning to the worker is present. These especially complicated and difficult finer points constitute the heart of the practice of occupational medicine and industrial hygiene activities.

BIBLIOGRAPHY

Adams, R. M. *Occupational Dermatology.* New York: Grune & Stratton, 1983.

Aitio, A.; Riihimaki, V.; and Vainio, H., eds. *Biological Monitoring and Surveillance of Workers Exposed to Chemicals.* New York: Hemisphere, 1984.

American Conference of Governmental Industrial Hygienists (ACGIH). *Documentation of Threshold Limit Values,* 5th ed. Cincinnati: ACGIH, 1985.

——. *Threshold Limit Values and Biological Exposure Indices for 1986-1987.* Cincinnati: ACGIH, 1986.

American Industrial Hygiene Association (AIHA). *Biohazards Reference Manual.* Akron, OH: AIHA, 1985.

American Medical Association (AMA). *Guide to Developing Small Plant Occupational Health Programs.* Chicago: AMA, 1983.

——. *Guide to the Development of an Occupational Medical Records System,* 2nd ed. Chicago: AMA, 1984.

——. *Guides to the Evaluation of Permanent Impairment,* 2nd ed. Chicago: AMA, 1984.

——. *Role of the Family Physician in Occupational Health Care.* Chicago: AMA, 1984.

——. *Traveler's Health Abroad: A Guide for Physicians.* Chicago: AMA, 1982.

——. *Medical Conditions Affecting Drivers.* Chicago: AMA, 1986.

American Petroleum, Inc. (API). *Medical Management of Chemical Exposures in the Petroleum Industry,* Washington, DC: API, 1982.

Bond, M. B., and Messite, J. "Preplacement Medical Evaluations and Recommendations." In *Occupational Medicine: Principles and Practical Applications,* 2nd ed., edited by C. Zenz. Chicago: Year Book Medical Publ., 1988.

Brown, M. L. *Occupational Health Nursing.* New York: Springer Publishing Co., 1981.

Chapman, W. P. *Alcoholism: Forty Questions and Answers on its Recognition, Treatment and Natural History.* Boston: Boston Globe, 1982.

Clayton, G. D., et al. *Patty's Industrial Hygiene and Toxicology,* vols. 1, 2a, 2b, 2c, and 3. New York: John Wiley & Sons, 1978-82.

Dreiabach, R. H. *Handbook of Poisoning: Prevention, Diagnosis and Treatment.* Los Altos, CA: Lange Medical Publications, 1983.

——. Effects of toxic chemicals on the reproductive system, Council on Scientific Affairs. *JAMA* 253(June 21, 1985):23.

——. *Emergency Care and Transportation of the Sick and Injured.* Menasha, WI: George Banta, Inc., 1971.

Friberg, L.; Nordberg, G. F.; and Vouk, V. B.; eds. *Handbook of Toxicology of Metals.* Amsterdam: Elsevier, 1979.

Frostman, T. O., and Horvath, E. P. *NIOSH Spirometry Workbook.* Cincinnati: NIOSH, DHHS, PHS, CDC, May 1980.

Gleason, M. N., et al. *Clinical Toxicology of Commercial Products,* 5th ed. Baltimore: Williams & Wilkins Co., 1985.

Guidelines for infection control in hospital personnel. *Infection Control* 4(1983):327.

Hayes, W. J. *Toxicology of Pesticides.* Baltimore: Williams & Wilkins, Co., 1975.

Hernberg, S. *Epidemiology in Occupational Medicine,* 2nd ed. Edited by C. Zenz. Chicago: Year Book Medical Publ., 1988.

Horvath, E. P. *Manual of Spirometry in Occupational Medicine.* Cincinnati: NIOSH, DHHS, PHS, CDC, November 1981.

Horvath, E. P. Jr.; Andronian, J. J.; and Rowe, D. M. Attending physician's return-to-work recommendations record. *Wisconsin Medical Journal* 83(June 1984):202.

Key, M. M., et al., eds. *Occupational Diseases—A Guide to Their Recognition,* Cincinnati: NIOSH, June 1977.

Kurppa, K., ed. Proceedings of the International Symposium of research on work-related diseases: Espoo, Finland, *Scandinavian Journal of Work and Environmental Health* 10 (June 1984):337.

LaDou, J., ed. *Introduction to Occupational Health and Safety.* Chicago: National Safety Council, 1986.

LaDou, J., ed. The microelectronics industry, State of the Art Reviews. *Occupational Medicine,* 1(January-March 1986):1.

LaDou, J., ed. *Occupational Health Law: A Guide for Industry.* New York: Marcel Dekker, 1981.

LaDou, J. Potential occupational health hazards in the microelectronics industry, *Scandinavian Journal of Work and Environmental Health* 9(1983):42.

Maibach, H. I., and Gellin, G. A. *Occupational and Industrial Dermatology.* Chicago: Year Book Medical Publ., 1982.

——. *Merck Index,* 10th ed. Rahway, NJ: Merck & Co., 1983.

——. *Merck Manual of Diagnosis and Therapy,* 14th ed. Rahway, NJ: Merck Sharp & Dohme Research Laboratories, 1982.

Messite, J., and Bond, M. B. *Occupational Health Considerations for Women at Work,* edited by C. Zenz. Chicago: Year Book Medical Publ., 1987.

Messite, J., and Bond, M. B. *Reproductive Toxicology and Occupational Exposure in Occupational Medicine,* 2nd ed., edited by C. Zenz. Chicago: Year Book Medical Publ., 1987.

National Council on Alcoholism, Inc. *The EAP Manual—A Practical Step-by-Step Guide to Establishing an Effective Employee Alcoholism/Assistance Program.* New York: National Council on Alcoholism, Inc., 1982.

National Institute for Occupational Safety and Health (NIOSH). *NIOSH Publications Catalog,* 6th ed., DHHS No. 84-118. Cincinnati: NIOSH, 1984.

Murray, F., Schwetz, B., Nitschke, K., et al. Teratogenicity of acrylonitrile given to rats by gavage or inhalation. *Food Cosmetics Toxicology,* 16(1978):547-551.

National Safety Council, 444 N. Michigan Ave., Chicago, IL 60611.

———. *Cutting Oils, Emulsions, Drawing Compounds,* Data Sheet No. 719, 1986. *Pocket Guide to Occupational Health,* 1982.

Olishifski, J. B., and Harford, E. R. *Industrial Noise and Hearing Conservation.* Chicago: National Safety Council, 1975.

Parmeggiani, L., ed. *The Encyclopedia of Occupational Health and Safety,* 3rd ed., 2 vols. Geneva: International Labor Office, 1983.

Physicians' Desk Reference (PDR). Oradell, N.J.: Medical Economics Company, Inc., published annually.

Sauter, S., et al. *Well-Being of Video Display Terminal Users.* Cincinnati: NIOSH, Div. of Biomedical and Behavioral Science, 1983.

Skjei, E., and Whorton, D. *Of Mice and Molecules, Technology and Human Survival.* New York: Dial Press, 1983.

Solomayer, J. A., and Boardman, T. L. *Get the Health Care You Deserve—A Manual for Managers in Industry and Commerce.* Euclid, OH: VME of the Americas, 1986.

State Medical Society of Wisconsin, Committee on Occupational Health. *Occupational Health Guide for Medical and Nursing Personnel,* edited by C. Zenz. Madison, WI: State Medical Society of Wisconsin, 1985.

Theiss, A., and Fleig, I. "Analysis of Chromosomes of Workers Exposed to Acrylonitrile." In *Reproductive Hazards of Industrial Chemicals,* edited by S. Barlow, and F. Sullivan. New York: Academic Press Inc., 1982, pp. 52-53.

U.S. Department of Labor. *All about OSHA,* No. 2056. Washington, DC: U.S. Government Printing Office, 1985.

U.S. Department of Transportation. *Federal Aviation Regulations,* Part 67, Chapter III. Washington, DC: U.S. Government Printing Office, 1969.

U.S. Department of Transportation. *Motor Carrier Regulations,* Sections 391.41 and 391.43. Washington, DC: U.S. Government Printing Office, 1971.

U.S. Environmental Protection Agency. *Recognition and Management of Pesticide Poisonings,* 3rd ed. Washington, DC: U.S. Government Printing Office, 1982.

———. *Health Information for International Travel,* DHHS No. 83-8280, 1983.

———. Industrial Environment, Its Evaluation and Control, DHHS (NIOSH) No. 74-117, 1974.

———. Occupational Health Guidelines for Chemical Hazards, DHHS (NIOSH) No. 81-123, 1981.

———. *Registry of Toxic Effects of Chemical Substances,* DHHS (NIOSH) No. 83-107, 1983.

Voelz, G. L. "Ionizing Radiation." In *Occupational Medicine: Principles and Practical Applications,* 2nd ed., edited by C. Zenz. Chicago: Year Book Medical Publ., 1988.

Waldron, H. A., ed. *Metals in the Environment.* London: Academic Press, 1980.

Whorton, M. D., and Davis, M. E. "Perspective on Health Intervention Programs in Industry," in *Managing Health Promotion in the Workplace.* Palo Alto, CA: Mayfield Publishing, 1982.

Whyte, A. A. A guide to comprehensive occupational health and safety information management systems, *Occupational Health & Safety,* 53(1984):4.

Zenz, C., ed. *Occupational Medicine: Principles and Practical Applications,* 2nd ed. Chicago: Year Book Medical Publ., 1988.

Zenz, C. Reproductive risks in the workplace. *National Safety News,* September 1984.

Suggested reading

American College of Obstetricians and Gynecologists. *Guidelines on Pregnancy and Work.* Rockville, MD: DHHS, 1978.

American Hospital Association. *A Hospital-wide Infection Approach to AIDS. Recommendations of the Advisory Committee on Infection within Hospitals.* Chicago: AHA, December 1983.

Ayoub, M. M., et al. Development of strength and capacity norms for manual materials handling activities, State of the Art. *Human Factors* 22(3)(1980):2.

Cady, L. D., et al. Strength and fitness and subsequent back injuries in firefighters, *Journal of Occupational Medicine,* 21(1979):269-272.

Chaffin, D.B., and Andersson, G.B.J. *Occupational Biomechanics.* New York: J. Wiley & Sons, Inc., 1984.

Chaffin, D. B., et al. *Preemployment Strength Testing, NIOSH Technical Report.* Cincinnati: NIOSH, Physiology and Ergonomics Branch, 1977.

Davies, B. T. "Training in Manual Materials Handling," in *Safety in Manual Materials Handling.* Cincinnati: NIOSH, DHHS No. 78-185, 1978.

Dodson, V. N., et al. Diagnosing occupationally-induced diseases, *Wisconsin Medical Journal* 80(December 1901):199.

Horvath, E. P. "Occupational Health Programs in Clinics and Hospitals." In *Occupational Medicine: Principles and Practical Applications,* 2nd ed., edited by C. Zenz. Chicago: Year Book Medical Publ., 1988.

Keyserling, W. M., Herrin, G. D., and Chaffin, D. B. Isometric strength testing as a means of controlling medical incidents on strenuous jobs. *Journal of Occupational Medicine* 22(1980):332-336.

NIOSH. *Work Practices Guide to Manual Lifting,* NIOSH Technical Report No. 81-122. Cincinnati: NIOSH, 1981, and Akron, OH: American Industrial Hygiene Association, 1983.

Novak, J. F., and Paul, R. W. Contact lenses in industry, *Journal Occupational Medicine,* 13(1971):175.

———. Recommendations for preventing transmission of infection with HTLV-III/LAV in the workplace, *MMWR* 34(1985):681.

Snook, S. H. The design of manual handling tasks, *Ergonomics,* 21(1978):963-985.

Snook, S. H., Campanelli, R. A., and Hart, J. W. A study of

preventive approaches to low back injury, *Journal of Occupational Medicine* (1978):478-481.

Solomayer, J. A., and Boardman, T. L. *Get the Health Care You Deserve—A Manual for Managers in Industry and Commerce.* St. Clair, OH: VME of the Americas, 1986.

Tichauer, E. R. *The Biomechanical Basis of Ergonomics.* New York: Wiley-Intersciences, 1978.

White, A. H. *Your Aching Back.* New York: Bantam Books, 1983.

Yu, T., et al. Low back pain in industry, *Journal of Occupational Medicine* 26(7)(1984):517-524.

Cumulative list of NIOSH current intelligence bulletins

1. Chloroprene, 1975.
2. Trichloroethylene (TCE), 1975.
3. Ethylene Dibromide (EDB), 1975.
4. Chrome Pigment, 1975, 1976.
5. Asbestos—Asbestos Exposure during Servicing of Motor Vehicle Brake and Clutch Assemblies, 1975.
6. Hexamethylphosphoric Triamide (HMPA), 1975.
7. Polychlorinated Biphenyls (PCB's), 1975, 1976.
8. 4,4'—Diaminodiphenylmethane (DDM), 1976.
9. Chloroform, 1976.
10. Radon Daughters, 1976.
11. Dimethylcarbamoyl Chloride (DMCC), Revised, 1976.
12. Diethylcarbamoyl Chloride (DECC), 1976.
13. Explosive Azide Hazard, 1976.
14. Inorganic Arsenic—Respiratory Protection, 1976.
15. Nitrosamines in Cutting Fluids, 1976.
16. Metabolic Precursors of a Known Human Carcinogen, Beta-Naphthylamine, 1976.
17. 2-Nitropropane, 1977.
18. Acrylonitrile, 1977.
19. 2,4-Diaminoanisole in Hair and Fur Dyes, 1978.
20. Tetrachloroethylene (Perchloroethylene), 1978.
21. Trimellitic Anhydride (TMA), 1978.
22. Ethylene Thiourea (ETU), 1978.
23. Ethylene Dibromide and Disulfiram Toxic Interaction, 1978.
24. Direct Black 38, Direct Blue 6, and Direct Brown 95 Benzidine Derived Dyes, 1978.
25. Ethylene Dichloride (1,2-Dichloroethane), 1978.
26. NIAX Catalyst ESN, 1978.
27. Chloroethanes—Review of Toxicity, 1978.
28. Vinyl Halides—Carcinogenicity, 1978.
29. Glycidyl Ethers, 1978.
30. Epichlorohydrin, 1978.
31. Adverse Health Effects of Smoking and the Occupational Environment, 1979.
32. Arsine (Arsenic Hydride) Poisoning in the Workplace, 1979.
33. Radiofrequency (RF) Sealers and Heaters: Potential Health Hazards and Their Prevention, 1979.
34. Formaldehyde: Evidence of Carcinogenicity, 1981.
35. Ethylene Oxide (EtO): Evidence of Carcinogenicity, 1981.
36. Silica Flour: Silicosis, 1981.
37. Ethylene Dibromide (EDB) Revised, 1981.
38. Vibration Syndrome, 1983.
39. The Glycol Ethers, with Particular Reference to 2-Methoxyethanol and 2-Ethoxyethanol: Evidence of Adverse Reproductive Effects, 1983.
40. 2,3,7,8-Tetrachlorodibenzo-p-dioxin (TCDD, Dioxin), 1984.
41. 1,3-Butadiene, 1984.
42. Cadmium, 1984.
43. Monohalomethanes: Methyl Chloride, Methyl Bromide, Methyl Iodide, 1984.
44. Dinitrotoluenes (DNT), 1985.
45. Polychlorinated Biphenyls (PCBs): Potential Health Hazards from Electrical Equipment Fires or Failures, 1986.
46. Methylene Chloride, 1986.
47. 4,4′ Methylenedianiline (MDA) Revised, 1986.
48. Organic Solvent Neurotoxicity, 1987.

For the convenience of those who desire a complete series of Current Intelligence Bulletins, #1 through #30, have been reprinted as NIOSH publications, #78-127 and #79-146, respectively. These publications and single copies of Bulletins #31 and higher are available from Publications Dissemination, DSDTT; National Institute for Occupational Safety and Health, 4676 Columbia Parkway, Cincinnati, OH 45226.

The American Medical Association (AMA) and the American Occupational Medical Association (AOMA) provide educational publications for occupational medicine practitioners. AMA publications cover topics such as the physician's role in workers' compensation; guides for developing small plant occupational health programs, industrial medical records systems, and for conducting medical examinations in industry; and the role of nurses in occupational health. A list can be obtained from:

Environmental & Occupational Health,
American Medical Association,
535 North Dearborn Street,
Chicago, Ill. 60610

The AOMA has developed a basic curriculum for practicing physicians who are interested in expanding their careers into occupational medicine. Additionally, AOMA has a list of over 200 publications on matters of occupational health, including articles on specific industrial exposures and diseases, cancer and mutagenicity, occupational medical practice, multiphasic screening, and ethics. A list can be obtained from:

American Occupational Medical Association,
2340 South Arlington Heights Road,
Arlington Heights, Ill. 60005

27

The Occupational Health Nurse

by Jeanette M. Cornyn, RN
Revised by Larry Hannigan, RN

An occupational health nurse applies current principles of nursing to the needs of clients and employees. This practice requires handling issues that arise within the work environment and also many off-the-job safety and health concerns. Managing programs in safety and health education, on subjects such as nutrition, life-style changes and health promotion, also is part of the occupational health nurse's overall responsibility.

PRIMARY GOAL OF THE NURSE

The American Association of Occupational Health Nurses (AAOHN) defines occupational health nursing as the application of nursing principles in conserving the health of workers in all occupations. It involves prevention, recognition, and treatment of illness and injury, and requires special skills and knowledge in the fields of health education, counseling, environmental health, rehabilitation, and human relations (*Occupational Health,* 1985) (Figure 27-1). In *Occupational Health Nursing,* Mary Louise Brown (1981) says, "Occupational health nursing is the easiest job in the world to do badly and the most demanding to do well." Mutual support, cooperation, and communication are required between nursing and other occupational health and safety disciplines are vital for an effective program.

Scope

The work of an occupational health nurse can be complicated because health and safety programs greatly vary from company to company, even among those in the same industry. A plant's geographic location, number of employees and their age distribution, type of products and processes, chemicals, and machinery needed to produce them, and management philosophy combine to make each assignment unique in its problems and solutions.

COMPANY PROGRAMS

Companies that maintain an in-house occupational health program usually provide services for their workers in one of the following patterns:

- In-house programs staffed by one or more full-time occupational health nurses and one or more full-time physicians. Other professional services are provided as needed by consultants or part-time employees.
- A full-time nurse and a part-time physician provide in-house occupational health services for workers. The number of full-time staff nurses, and their areas of specialization, depends on the type of industry and the number of workers.
- Part-time in-house programs. All members of the occupational health team are on a part-time or consulting basis.

Health service staffing

A nurse is frequently the only health professional in an industrial complex, and is a full- or part-time employee or has been retained on contract through a nursing service agency. A physician's services are retained in much the same way or on a fee-for-service basis. Similarly, on retainer or contract, the company provides supporting health services to their nurse and physician, such as those offered by industrial hygienists, rehabilitation counselors, nursing consultants, occupational therapists, x-ray technologists, psychologists, and optometrists.

DEFINITION, PURPOSE, AND OBJECTIVES OF AN OCCUPATIONAL HEALTH NURSING SERVICE

Occupational health nursing is the application of nursing principles in conserving the health of workers in all occupations. It involves prevention, recognition and treatment of illness and injury, and requires special skills and knowledge in the areas of health education and counseling, environmental health, rehabilitation and human relations.

The major objectives of an occupational health nursing service are health maintenance and disease prevention of employees. Contributory objectives are:

- To adapt the nursing program to serve the health needs of the company with respect to employee population, type of industry and location
- To give competent nursing care and treatment
- To have and maintain adequate space, equipment and supplies
- To obtain medical direction where none exists
- To establish and maintain an adequate record system
- To promote health education among the employees
- To maintain good working relations with all departments of the company
- To maintain cooperative relations with local physicians, allied health services and other community agencies
- To upgrade the quality of the nursing service through continual personal and professional educational growth
- To evaluate the health program and nursing service periodically and revise accordingly

Figure 27-1. The goals of the occupational health nurse are shown. (Reprinted with permission from American Association of Occupational Health Nurses [AAOHN]. *A Guide for Establishing an Occupational Health Nursing Service.* New York: AAOHN, 1977)

Regardless of the condition of employment, the physician normally has ultimate authority for the occupational health program. In cooperation with the nurse, the physician develops procedures for ordering and stocking first aid materials and medications and treating injuries and illnesses in his or her absence. These guidelines must periodically be reviewed, updated, and authorized by both the physician and the nurse. When necessary, a company provides additional personnel to help administrate the occupational health program, such as clerks, receptionists, secretaries, and data processors.

The majority of nurses working in occupational health are employed in one-nurse units. The major requirements for nurses pursuing this specialty are autonomy, responsibility, and accountability.

Responsibilities

Different responsibilities and position subtitles of an occupational nurse—communicator, screener, therapist, counselor, educator—are described in detail by the American Association of Occupational Health Nurses (AAOHN) (Figure 27-2). In large companies, supervisory categories include charge nurse, staff nurse, junior supervisor, administrator of nursing services, administrator of a health service, consultant, and other titles not included in the AAOHN definitions.

If medical direction is provided to a nurse by a physician who rarely or never visits the workplace, the nurse may be considered as having responsibility for a one-nurse service; in this case the physician should provide written orders covering the development and administration of the first aid and nursing service in the occupational health service.

Duties

The nurse may have some or all of the following assignments:

- responsibility for operating the in-house dispensary, administering first aid to sick or injured employees, determining severity of accidents, and arranging for treatment of employees by a physician when required
- developing all relevant information regarding work injuries, circumstances, and causal factors; the nurse establishes, maintains, and processes detailed and accurate records of all visits to the dispensary, and follows up on the status of employees unable to work due to lost-time work injuries
- adminstering preplacement physical examinations and developing of medical histories. The nurse periodically examines or arranges for examinations of employees
- maintaining adequate stocks of dispensary supplies, pamphlets and health bulletins, and providing first aid supplies when needed
- maintaining a nursing policy and procedure manual
- maintaining a system of clinical records from which departmental statistical summaries and periodic reports can be developed
- participating with the physician in administering health examinations
- coordinating and managing health education programs for employees and their families
- counseling employees on matters of physical or emotional health and when necessary referring workers to a personal physician, dentist, or community health or welfare agency
- serving on facility committees charged with accident prevention, disaster preparedness, and first aid training

OCCUPATIONAL HEALTH NURSING RESPONSIBILITIES

- The occupational health nurse shall have professional preparation commensurate with job responsibilities.
- The occupational health nurse shall be licensed and registered in the state in which the nurse practices.
- Background in emergency room and/or community health is desirable.

Clinical and administrative functions to be considered in establishing an occupational health nurse service should include:

A. Administrative

The organization, the administration, and the policies of the nursing program shall be established in accordance with the general company policies that govern the health service.

The occupational health nurse plans, organizes, and develops an efficient occupational health nursing service consistent with the objectives and scope of the company health programs.

Provides leadership and self-example, which motivate employees to function effectively and efficiently.

Recommends budget needs as related to the nursing service provided.

Assesses the health program, nursing procedures, and responsibilities.

Recommends methods that will increase effectiveness.

Initiates, participates, and cooperates in research and studies for the improvement and delivery of health care.

Participates in preparing and devising programs to be used when medical personnel are unavailable.

Takes an active role in the formulation of policy for an in-plant disaster.

Serves as an active member of the safety committee.

The nurse must conform to the legal and ethical principles of the profession.

In views of the fact that each nurse is responsible for his or her own acts, it is recommended that adequate professsional liability coverage be obtained; it is recommended that the nurse investigate the adequacy of the professional liability provided by his or her employer.

1. Communications

An adequate communication system is essential to efficient health service functioning.

The occupational health nurse develops and maintains a communication system with the physician, management, and labor to promote understanding and acceptance of health policies and practices.

Assists management in the interpretation of company medical related policies and procedures.

taining a safe and healthful work environment.

b. Establishes communication with and makes use of services from government, insurance companies, and other agencies concerned with environmental health.

c. Visits work areas to identify environmental hazards and maintains a record of visits and conditions observed. The nurse recommends preventive and corrective measures to the responsible person.

d. The nurse develops and maintains knowledge of OSHA, NIOSH, and worker's compensation relative to the state in which the nurse practices.

5. Rehabilitation: The nurse develops and implements a nursing-care plan which provides for continued care and treatment, rehabilitation, and return to work. The rehabilitation process is started at the earliest knowledge of illness or injury.

6. Communicable disease: The nurse must have an understanding of epidemiology, and control and prevention of communicable disease.

7. Interviewing and counseling

Interviewing and counseling are essential in providing effective nursing care.

The occupational health nurse performs interviews and counsels employees to identify physical and emotional problems and institutes referrals as necessary.

8. Health Education

Health education is an essential part of all

Communicates and relates to employees in terms meaningful to them.

2. Records, reports, and directives.

Health records and directives are essential in providing effective nursing care.

The occupational health nurse develops, reviews, and maintains a nursing policy and procedure manual.

Participates in the formulation and maintenance of signed medical directives within the limits of the state's Nurse Practice Act.

Develops and maintains a record and reporting system which meets legal requirements, and ensures continuity of care and confidentiality.

3. Physical facilities and equipment

The facilities and equipment to provide an efficient nursing service shall be determined by management with the physician and the nurse.

The occupational health nurse recommends physical facilities, space, layouts, equipment, and supplies.

Establishes and periodically reviews facilities and equipment to be used for in-plant disaster.

Aids in the planning and establishment of new medical facilities and changes to existing facilities for which the nurse has responsibility.

The name chosen for the department should reflect the routine of the program. It should not be designated "medical department," nor "dispensary," (without adequate medical supervision), and certainly not "first aid station." Employee Health Service seems most suitable.

B. Clinical

1. Health assessment

Major health assessment topics to be considered include:

Health examinations of employees are the responsibilities of the physician to which the nurse shall make meaningful contributions.

The occupational health nurse develops and implements a nursing care plan which provides for continued care and treatment, rehabilitation, and return to work.

2. Health screening: Observes, examines, and tests by the use of selective diagnostic procedures. The nurse selects appropriate actions and referral.

3. Acute illness intervention: Recognizes and assesses symptoms of acute illness to ascertain a nursing diagnosis in order to render appropriate treatment within the scope of the state's Nurse Practice Act.

4. Occupational injury or disease

a. Cooperates with all other company personnel involved in providing and main-

health services. The occupational health nurse provides:

a. Education, support, and motivation in all areas of health and safety.

b. Teaching and/or coordinating instruction in first aid, cardiopulmonary resuscitation, and other health-related topics utilizing various media, such as audiovisuals, company publications, and bulletin boards.

c. Health teaching for employees is done in both formal and informal settings.

9. Community resources and referral

The health education program shall be correlated with community health activities.

The occcupational health nurse cooperates with community agencies in case-finding projects and maintains an up-to-date knowledge of availability of health and social agencies in the community.

Serves as a consultant to the community and encourages development of community resources.

Acts as a liaison with the community for establishment of a disaster program.

C. Continuing education

It is recommended that the occupational health nurse seek out continuing education programs and read professional journals to increase personal and professional development. Additionally, that the nurse maintain active membership and participation in the activities of professional nursing organizations.

Figure 27-2. The responsibilities of an occupational health nurse. (Reprinted with permission from the AAOHN. *A Guide for Establishing an Occupational Health Nursing Service.* New York: AAOHN, 1977)

- working with management and supervisory personnel to provide understanding and support of such related activities as alcoholism rehabilitation and employment of physically disabled job applicants
- updating and maintaining skills by continuing education through attendance at professional meetings and through coursework
- being informed about pertinent OSHA regulations and standards. Helping to keep the company informed of pertinent federal, state, and local standards and requirements related to occupational health
- being aware of the environmental hazards located in the plant; being aware of new processes, chemicals, and ways of handling material in the plant
- being familiar with employees' work areas. Touring the plant helps pinpoint air contamination areas, noise, heat stress situations, skin hazards, housekeeping, sanitation, improper storage of hazardous chemicals and the apparent effectiveness of control measures in use.

Expanded duties

Increasingly, the occupational health nurse assumes new responsibilities, and must be prepared to provide emergency care for workers suffering major injuries or acute medical conditions. At other times, the nurse functions within the occupational health service's policy and the medical directives of the physician and gives definitive care to workers whose injuries and illnesses are within his or her authority and capability to treat.

The occupational health nurse specialist must be able to identify health problems and give immediate nursing care. The nurse must recognize when to refer a problem and how to involve other members of the occupational health team when their specialized skills are required.

The safety professional, the industrial hygienist, and the physician must be ready to receive and respond to the nurse's referral. Management must recognize that the nurse can be effective only if sufficient health service backup is available. The nurse also must have sufficient help with processing insurance, compensation, and government forms to provide quality nursing care.

Objectives

The plant nurse can assist in providing a safe and healthful environment. Because management is interested in work efficiency and preventing employees' lost work time, the objectives of an occupational health nursing program are as follows:

- to protect the employee against health hazards in the work environment;
- to facilitate placement and ensure the suitability of the individual according to physical capabilities, mental abilities, and emotional makeup;
- to assure adequate medical care and rehabilitation of the occupationally ill and injured;
- to encourage personal health maintenance.

Nursing is concerned with people, well and sick, at work and at play. Nursing knowledge brings dimensions to human safety and human service. Occupational health nurses can safely and competently make complex judgments requiring scientific knowledge and intellectual skill.

Obviously, the nurse should be given the opportunity to know about plant processes, hazardous exposures, and community medical facilities. The nurse must have qualities of tactfulness and be able to interpret (1) the results of physical examination findings and laboratory test results and to make them understandable to the worker, and (2) benefits under insurance coverage. Ideally, the nurse should remain professionally neutral and not be unduly swayed by the safety professional or others for the sake of the record.

Relationship to first aid workers

The first aid worker's purpose is to render first aid only. He or she is required to have completed an approved course in first aid and should have the skills necessary to meet the legal requirements. Where there are unusual risks, chemicals or gases, for example, the first aid worker should be given added instruction by the physician or nurse in special procedures or equipment use.

Auxiliary workers employed to assist the nurse in the health unit should have first aid training and in-service instruction by the nurse. Sometimes these workers have served in medical units in the armed forces and can serve a useful purpose within the health unit. They should work under the direction of a nurse or physician.

Every company should have one or more persons trained to give first aid. When the full-time occupational health nurse is given the opportunity to work with the first aid worker, their effectiveness will increase through a developing team attitude and feelings of mutual respect. The first aid worker should understand the extent to which he can and is expected to participate in handling emergencies.

When there is an established occupational health unit staffed by an occupational health professional, the first aid worker should not provide health care for workers. Injured or ill employees should be required to go to the occupational health unit for care. There are two major exceptions: a chemical injury (especially to the eye), and a major medical emergency, when, for example, a person is not breathing. Both require immediate and specific attention.

Standard procedures should be developed and used as the basis for the emergency care provided by the first aid worker. These procedures should: (1) outline the treatment, (2) state when the employee is to be seen by the nurse, (3) explain how and when referrals are to be made by the first aid worker to the physician, and (4) determine when the injured or ill person is to be taken to the emergency room of the hospital.

All care provided by the first aid worker should be recorded. But before the information is put into the employee's record, the nurse should review and discuss each situation with the first aider.

PRIMARY CARE

Primary health care is the employee's most common contact with the occupational health program. The first step in primary care is detecting, evaluating, and solving an immediate problem. It includes making a referral for appropriate medical care. The second step is the maintenance of health and disease prevention.

Major responsibility

Management of occupational and nonoccupational injuries and illnesses will always be a major function of the occupational health nurse, who has the responsibility to initiate prompt and skillful emergency care, consistent with the degree of professional training and knowledge of first aid techniques. The nurse's interest in and concern for the welfare of the ill or injured combined with prompt attention, good judgment, and sound management are necessary for prompt and effective rehabilitation.

The first care usually has been preordered and authorized by the physician. However, in most instances it is implemented by the nurse in the physician's absence. A skilled nursing assessment or judgment must be made in each case before action can be taken and medical intervention occurs. This, of course, is the professional skill for which the nurse is hired. Follow-up care, medical referral for further treatment, transmittal to a hospital or physician's office, returning home, notifying the family, and all other such matters are usually occupational health nursing duties.

The nature and extent of the nursing care provided for nonoccupational injuries and illnesses should be determined by the company physician. Generally accepted practice allows for the simple treatment of minor complaints that are not expected to require the services of a physician. The more serious nonoccupational illnesses should be referred to the worker's family physician.

Emotional problems

The nurse should recognize the possibility that an employee's visit to the health service may have deeper significance than medical attention. The chronic complainer, the employee who suffers frequent minor injuries, or even the daily visitor may be an individual crying for help.

The occupational health nurse can help determine if there is a nonoccupational physical disability or emotional problem. While treating an employee or dressing a wound, the nurse is constantly observing the patient. At this time signs and symptoms of an emotional disturbance can be discerned. Emotions may be predisposing factors in many industrial accidents. Close attention on the part of the nurse may help to uncover an underlying emotional problem and avoid a major tragedy. For this reason, enlightened management has learned not to discourage nonoccupational visits to the medical department, no matter how simple the complaint.

Emergency care

Provision of emergency care is a chief responsibility of the occupational health nurse, and can be the primary reason for employing health professionals in industry. Although some emergencies may not be serious, others may require life-saving care.

First aid team. Many facilities develop an in-house emergency first aid team, and the occupational health nurse should contribute to the development of an emergency first aid program for the team; team training; and maintain continued contact with team members. The nurse should hold frequent review sessions, alerting team members to changes in policies and procedures affecting emergency care.

Emergency guidelines. Both nurse and physician must develop written protocols for dealing with emergencies. The nurse, who may be faced with an emeregency situation without the presence of a physician, should have the skills required in the guidelines (Figure 27-2). These nursing skills should meet or exceed the requirements of the nurse practice act of the state in which the nurse is employed.

Community emergency facilities. The nurse should become familiar with the location and staff of community emergency treatment facilities. In order to make an appropriate referral for definitive care, knowledge of specially designated facilities is necessary, such as trauma and burn centers. The nurse should encourage a cooperative and close relationship with local private, volunteer, and municipal ambulance services. Ambulance personnel can be oriented to the work site through frequent dry runs of the emergency plans. During these practices, the ambulance personnel should note entrance and exit locations, the layout of the facility, elevator availability, equipment compatibility, and other pertinent items. These practices are necessary to test the effectiveness of emergency plans.

Nursing evaluation

A nurse's evaluation of a patient's condition must meet the standard of learning, skill, and care to which nurses practicing in the community are held responsible. A nurse, in order to properly and effectively administer first aid to employees, must make a sufficient diagnosis before applying a remedy (Figure 27-3).

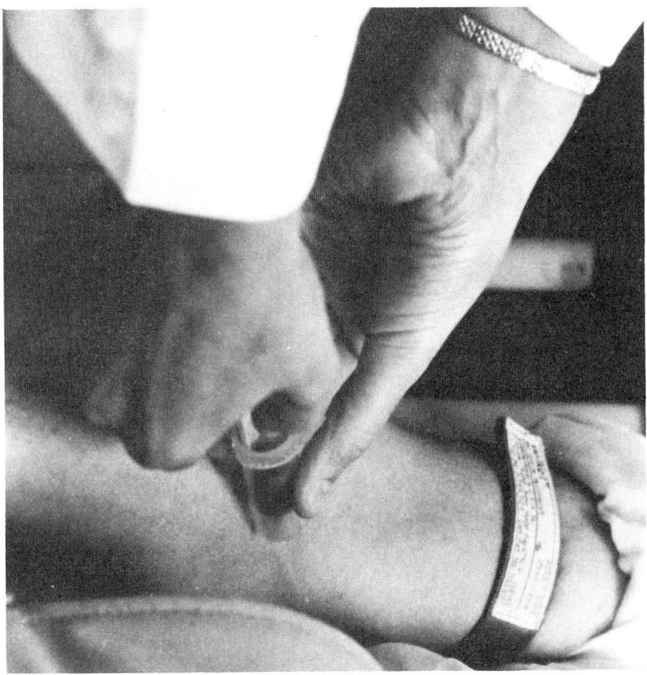

Figure 27-3. In emergency cases, the occupational health nurse takes such measures as the situation demands until additional medical services are available.

In emergencies, the primary significance in the nursing diagnosis is that the nurse sees the injured or ill worker before the physician does and this evaluation often determines the initial care.

Nursing and medical practice are interrelated and sometimes indistinguishable from each other. The same act may be clearly the practice of medicine when performed by the physician and likewise the practice of nursing when performed by the nurse.

Figure 27-4. The ideal situation is having the occupational health nurse and the consulting physician work together until they are familiar with each other's ways of functioning.

Medical directives

Medical directives give the occupational health nurse a broader scope in administering the department and caring for the employees. The nurse should participate with management and the medical director in formulating policies and directives affecting his or her work.

The services performed by the occupational health nurse are unique in the field of nursing. About one third of all occupational health nurses work without direct supervision by a physician. In hospitals, standing orders are written by a physician for a particular patient after the physician has made a diagnosis.

The nurse in industry performs under a set of standard procedures or medical directives for routine situations before the ill or injured person is seen by a physician. A nurse is often pressured by management, who, without realizing it, asks the nurse to perform duties beyond the ethical or legal limits of the profession. This is when established and definite policies and directives should prevent many problems.

Relationship with physician

The nurse, physician, and employer must recognize it takes time for a trusting and collaborative relationship to develop. The physician needs to know how the nurse arrives at the nursing evaluation and the nurse needs to know what the physician would do under certain circumstances. Respect for the ability of the nurse and recognition of any limitations are required.

The ideal situation is having the occupational health nurse and the consulting physician work together until they know each other's ways of functioning (Figure 27-4). What frequently happens, however, is the nurse is employed full-time and the physician is on call. This leads, in some instances, to less than adequate guidance for the nurse.

Ideally, the best working relationship permits the physician to be responsible for the medical aspects of the program and the nurse to be responsible for providing most of the services.

MEDICAL EXAMINATIONS

Medical examinations are an important component of a sound occupational health program. They are valuable for providing criteria for safe job placement, uncovering early physical and emotional changes, and detecting the effects of harmful working conditions. All these factors contribute to the maintenance of a safe, healthy, and productive employee population.

The occupational nurse, an important participant in the company examination program, conducts the health interview and performs the preliminary evaluations designated by the company physician (Figure 27-5). The health history elicited during the interview is necessary to the physician's comprehensive evaluation. To secure the best possible personal and occupational health information, the nurse should know the individual's job assignment and be familiar with its requirements, stresses, and hazards.

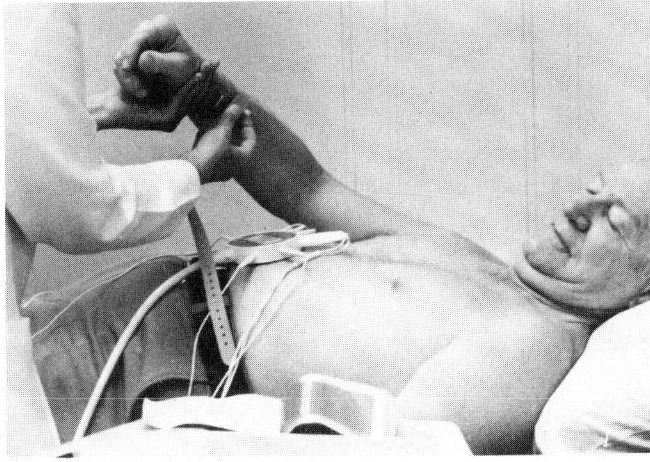

Figure 27-5. The occupational nurse conducts the health interview and performs the preliminary evaluations designated by the company physician.

Health evaluation

A health evaluation includes checking the vision of employees, preferably with a vision tester, to determine near and distance vision, visual acuity, depth perception, and color blindness. Sometimes it also includes baseline audiograms. To perform legal and acceptable audiograms, the nurse eventually will have to be a certified industrial audiometric technician.

In conducting a health evaluation, the nurse records the height and weight of all employees. This gives a frame of reference for future comparisons.

Blood pressure is another component of a health evaluation. The employer can use blood pressure information in properly placing new workers.

Hypertension, obesity, and high cholesterol level are some of the risk factors for heart attack and stroke. Urinalyses as part of a health evaluation can determine kidney, bladder, circulatory, and certain metabolic problems such as diabetes. The level of lead, toluene, or other toxic substances in the urine can be used as an index of exposure to these substances.

Health examinations

Companies providing health examinations probably use one of the following methods:

1. An examination conducted in the facility's health unit—supplementary services such as chest x-rays, special laboratory tests, electrocardiograms, audiograms, and other screening tests are conducted at outside resources if in-plant equipment or technology is not available.
2. An examination conducted in the physician's private office or clinic—additional supplementary screening procedures are conducted elsewhere.

When examinations are made outside the facility, the nurse should receive a copy of the physician's report for the employee's medical file. The nurse may explain the findings of the examination to the employee and to management to ensure proper job placement. Where the examinations are performed at the facility, the nurse can assist with the examination in the following ways:

- planning and scheduling examinations;
- taking health and work histories;
- performing the various screening activities, such as blood pressure, audiograms, and visual acuity tests. If equipment is not available, the nurse usually is responsible for scheduling the various screening activities at outside resources;
- explaining positive findings to the employer; if a physical finding is shown that would make a difference in the individual's job placement, the nurse should consult with the personnel office, in a confidential manner and preserving the employees' rights to privacy;
- the occupational health nurse should be responsible for keeping all medical findings and updating them in the medical files.

Types of examinations

A company's program may include one or all of the following types of examinations:

Preplacement examinations. Initial examination and health assessment are completed to be sure proper job placement is made from a health standpoint. The nurse should be familiar with the tasks, stresses, and hazards associated with a particular assignment. The type of medical examination given depends on the company philosophy, as well as legal requirements and medical staff recommendations. Typical components are a comprehensive health history and a method for recording physical findings (both are discussed later in this chapter). Tests and measurements should include height, weight, blood pressure, vision and hearing acuity, and urinalysis. In addition, blood tests, x-rays, pulmonary function, cytology and other tests may be necessary for particular exposures; these are determined by the medical staff or are mandated by regulatory agencies, such as OSHA. This type of screening exemplifies an occupational health program and the nurse must be responsible for providing most of the services (Figure 27-6).

Placement of a female worker of childbearing age presents special problems. When taking a medical history from a female employee, the nurse must determine if the employee is pregnant. For example, in its recommended standard for occupational exposure to vinyl chloride, the NIOSH recommendation to OSHA is that a woman who is pregnant or who reasonably expects to be pregnant should not be directly employed in vinyl chloride monomer operations. (The OSHA standard states that "any employee who would be materially impaired shall be withdrawn from possible contact with vinyl chloride.") This is because the placenta is a limited barrier and many substances can cross it by simple diffusion. Many chemicals, such as

I.

APPLICANT'S HEALTH STATEMENT

Date ____________

Name ______________________ Date of Birth __________ Sex __________

Address ______________________ Marital Status ________ No. of Dependents ________

______________________ Job Applied for ______________________

Previous Occupations: ______________________

Personal Physician ______________________

Name Address

HISTORY

A. Have you ever:

Yes	No	
☐	☐	1. Received worker's compensation for injury, or occupational disease?
☐	☐	2. Received government pension?
☐	☐	3. Been hospitalized?
☐	☐	4. Been operated on?
☐	☐	5. Been rejected in any medical examination?
☐	☐	6. Given up job because of back trouble or other health problems?
☐	☐	7. Had back operation?

Explain above answers: (Give dates, places, name of physician and his address)

____

____

____

B. Have you ever had (or suffered disease of, or injury to:)

Yes	No	
☐	☐	1. Back injury or pain in back, bones, joints, or skeletal muscles?
☐	☐	2. Heart, lungs, kidneys, intestines, spine?
☐	☐	3. Rheumatism, diabetes, fits, fainting spells?
☐	☐	4. Asthma, TB, high blood pressure, skin irritations?
☐	☐	5. Head, eyes, ears, nose, throat?
☐	☐	6. Shortness of breath, pain in chest, or abdomen, dizziness or fainting, indigestion at rest or following exercise?
☐	☐	7. Nervous or mental conditions?
☐	☐	8. Colds, headaches, infections, other upsets?
IF FEMALE:		
☐	☐	Do you have female disorders?
☐	☐	Do you lose time from work because of period?
		Number of miscarriages ______________________
		Ages of living children ______________________
		Date of last menstrual period ______________ 19 ________
☐	☐	9. Other — explain: ______________________

Explain above answers (Give dates, places, name of physician and address)

____

____

____

____

The above answers are true and correct to the best of my knowledge and belief. I understand that falsification is grounds for termination of employment. I agree to a Company physical examination and authorize release to the Company of any medical information concerning my past or present condition by any practitioner or hospital.

______________________ ______________________

Signature of Witness Signature of Applicant

Figure 27-6. Initial examination and health assessment are completed to ensure that proper job placement is made from a health standpoint. Completion of these forms is necessary for a comprehensive evaluation. (Continued.)

II. **EXAMINING PHYSICIAN'S REPORT**

1. Temp. ________ Height ________ Weight ________
GENERAL APPEARANCE ________
2. VISUAL ACUITY: (If applicant wears glasses, test and record acuity with and without glasses.)
Without glasses R 20/ ____ L 20/ ____ Depth perception ________
With glasses R 20/ ____ L 20/ ____ Color perception ________
Corneal scars or capacities ________
Pupils: Equal ________ Reaction ________
Eye grounds ________
Form fields of vision Rt. eye ________ Lt. eye ________
(Record degrees of temporal fields in spaces above)
Coordination:
Lateral Exo° ________ Eso° ________
Vertical R. Hyper. ________ L. Hyper. ________
Fusion
Excellent ____ Good ____ Poor ____ None ____
Evidence of suppression ________
3. Ears: Hearing 20 ft: Rt. Ear ________/20 Left ear ________/20
Disease or injury ________

Yes	No		Yes	No	
☐	☐	4. Deformities	☐	☐	7. Hydrocele
☐	☐	5. Hernia	☐	☐	8. Venereal disease
☐	☐	6. Varicocele	☐	☐	9. Enlarged nodes

BLOOD PRESSURE: Sitting: Systolic ________ Diastolic ________ Pulse ________
2 minutes after
exercise: Systolic ________ Diastolic ________ Pulse ________

Abnormal	Normal	
☐	☐	10. Skin
☐	☐	11. Hygiene
☐	☐	12. EENT
☐	☐	13. Head, Neck
☐	☐	14. Breasts
☐	☐	15. Lungs
☐	☐	16. Heart
☐	☐	17. Abdomen
☐	☐	18. Genitalia
☐	☐	19. Nervous system
☐	☐	20. Reflexes
☐	☐	21. Extremities (include range of motion)
☐	☐	22. Varicosities
☐	☐	23. Back (include range of motion)
☐	☐	24. Rectal examination
☐	☐	25. Pelvic examination (optional)
☐	☐	26. Evidence of infectious disease or mental disability.
☐	☐	27. Hands (include range of motion)
		Deformities ________
		Amputations ________
		Scars ________
☐	☐	28. Any chronic disease or injury.

REMARKS: (to include physician's review of history and elaboration of findings of physical examination)
____
____
____

LABORATORY WORK:
Kline or VDRL ________
Urinalysis: Sp. Gr. ____ React. ____ Alb. ____ Sug. ____ Micro. ____
Agglutination for Brucella ________

PHYSICAL CLASSIFICATION
☐ 1. Employable ________
☐ 2. Rejected ________

____
M.D. signature

Figure 27-6. (Concluded.)

solvents, lead fumes, gases, and vapors may be absorbed by the pregnant woman through inhalation, skin contact, or ingestion and ultimately be carried to the fetus.

Female workers should be encouraged to report pregnancy early. The first 3 months of pregnancy are particularly important because that is when the developing fetus is most vulnerable. Consideration should be given to reassigning a pregnant woman to an area where she will not be exposed to potentially toxic chemicals.

Periodic medical surveillance. Examinations and health assessments may be required because of certain job exposures, as in the case of employees who work with chemicals or toxic substances, or those who handle food. The frequency and content of the examinations are based on the type of exposure and amount, duration, and frequency of exposure. Examinations can include a comprehensive health evaluation or be limited to assessing specific target organs or body burden, such as audiometric tests for high noise exposure or blood lead measurements. Periodic examinations may also be offered to employees on a voluntary basis depending upon the resources of the provider. Recommendations for the frequency of routine examination of healthy individuals have been made by various medical organizations. One scheme is every 5 years up to age 30; three times in the 40s and every 2 years thereafter. These examinations are beneficial because they can detect changes in an individual's health at an early stage and thus make intervention possible. Hence, health problems can be corrected or their progression contained or retarded (Figure 27-7).

International travel. Health assessments are performed for individuals and possibly dependents before and/or upon completion of an assignment to or from another country. The purpose of such examinations is to be sure that there is no health contraindication to those persons undertaking foreign assignments, and that those returning from such an assignment are in good health. As with many examinations, a comprehensive examination is performed with particular attention paid to issues that are relevant to the person's health history and to the country to be visited (for example, status of control in diabetics or malaria prophylaxis and education for those going to an endemic area).

Return to work or fitness for duty examinations are usually required for employees who have been temporarily off work due to illness or injury.

Design of examinations

The occupational health nurse, in collaboration with the physician, plans and designs various types of health examinations along with reporting and record forms. He or she is responsible for:

- knowing the physical demands for specific jobs and keeping up to date on job requirements, and physical and chemical accident hazards as well as severities of exposure of each job;
- interviewing employees and recording their personal health, or illness, and occupational history;
- collecting biological samples and completing other parts of the examination delegated by the physician;
- assisting the physician throughout the examination, if there is an in-plant doctor;
- recording medical findings on the employee's permanent record.

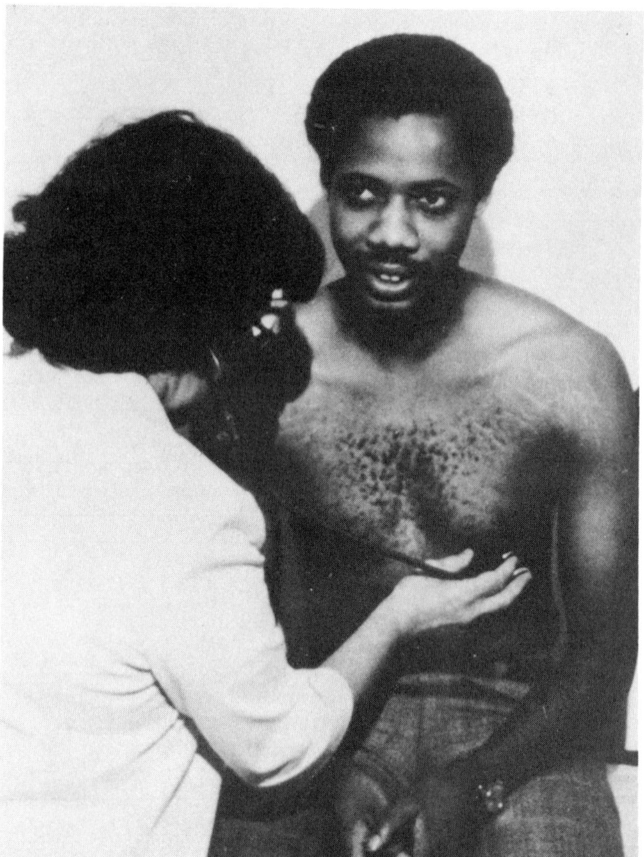

Figure 27-7. Physical examinations can be required due to certain job-related exposures to toxic chemicals and substances; the frequency and content of the examination are based on the type of exposure and amount of the substance, duration, and frequency of exposure. Examinations can also be offered to employees on a voluntary basis. These can detect changes in a person's health, and many health problems can be corrected before they progress to a more serious stage of development.

The nurse also evaluates physical, mental, and emotional fitness for the job and makes recommendations to management. However, the final appraisal is that of the medical director.

Interpreting medical findings and recommendations, arranging employee referrals to outside medical care or special services, follow-up care, rehabilitation services, and supervising the physically handicapped are some duties that may be assigned to the occupational health nurse.

Medical controls

Medical controls must be exerted on certain exposed employees. In some health standards, OSHA requires physical examinations as a check against engineering control measures. Included are examinations such as an electrocardiogram, x-ray of the chest, and spirometry (pulmonary functions) tests.

Biological monitoring procedure varies between companies and industrial processes. A health examination program is contingent on many factors; the most important is the assessment of the work environment.

This assessment requires the expertise of both the safety professional and industrial hygienist. The information compiled from their monitoring and surveys is essential and guides the firm's occupational safety and health program. Sharing this vital information is very important to the effectiveness of the medical and nursing program.

Follow-up. The nurse's responsibility includes checking up on employees with known health defects or physical handicaps. These periodic health checks help determine if the employee's disabilities are adversely affected by work and also assure continued medical assistance where needed. When the person's physical abilities and job demands are not in harmony, the nurse will notify the physician who, in turn, will work with those responsible for job assignments.

Rehabilitation

Assuming an active role in employee rehabilitation is a key responsibility of the occupational health nurse whether the rehabilitation is for an occupational injury or nonoccupational injury or illness. From the employer's point of view, regardless of the reason, the employee's time away from work is costly in benefits, as well as lost production, impact on employee morale, and selecting and training a replacement.

The rehabilitation process commences with the onset of injury or illness. Therefore, the quality of care provided initially, as well as subsequently, will have an influence on the duration and quality of the rehabilitation process. The nurse must know community, state, regional, and federal rehabilitation services in certain circumstances in order to provide appropriate referrals. A networking system should provide a means of learning about and sharing various resources with others. A close working relationship with the company's worker's compensation carrier is imperative. Often follow-up of occupational cases by the company nurse and the carrier, especially the carrier's rehabilitation nurse, can expedite an employee's return to work and prepare the worker for any special work arrangements.

The nurses can be a valuable asset to management in dealing with applicants who are handicapped and may require special accommodations. Other sensitive areas such as alcoholism, drug abuse, and psychiatric problems often can be helped by the nursing professional.

MEDICAL RECORDS

A good medical record system is basic to an effective occupational health program. The supervision and maintenance of employee health records is a responsibility of both the physician and the nurse. When the services of the company physician are limited to an on-call or part-time basis, the nurse's responsibilities are, of necessity, increased (Figure 27-8).

Figure 27-8. A good medical record system is basic to an effective occupational health program. When the services of the company physician are limited to an on-call or part-time basis, the nurse's responsibilities are increased.

The information the records contain is confidential, except when otherwise provided by law. Only members of the medical department should have direct access to the records. The nurse may, however, discuss specific but limited information from the medical record that might relate to job safety with the personnel director, safety director, and supervisory personnel.

Medical recommendations, such as work restrictions, are given to management based upon a health evaluation or opinion to ensure a proper work environment. However, the nature of the illness need not and should not be released. It is generally accepted that confidentiality is waived for conditions covered by workers' compensation. Personnel staff who administer benefits may also require access to confidential records.

Most institutions maintain original, written or typed medical records, sometimes called hard copies. Microfilming of some records, especially for long-term retention or storage, is used to reduce both storage space and costs. There are many computerized medical records systems commercially available, which not only save time and space but offer unique capability for storage and retrievability of huge amounts of data.

Medical record system

The record system for an occupational health service should provide:

- health information in a readily usable form
- competent evidence when matters have legal importance
- data for health program planning and evaluation.

The types, complexity, and sophistication of a record system will be determined by the size and goals of the health service, by the size of the work force, and the type of company.

Medical records and reports are an important data base, and the occupational health nurse can use this data to:

- evaluate the employee health service
- audit nursing care
- determine employee educational and training needs

- prepare statistical reports used to compute incidence rates, lost time due to illness and injury, absenteeism, and costs and benefits
- identify hazardous or problem areas
- support research activities
- demonstrate trends
- recommend program changes and expansion
- propose policy.

Analysis

A periodic analysis of the work-induced injuries should be made in every occupational health program. This provides a continuous search for trouble areas and gives information about unsafe working practices and conditions. A careful study of small, nonreportable injuries can prevent more serious problems and injuries in the same area.

Records will show the comparison of injury experience with other companies in the same field. This is an excellent opportunity for the nurse to evaluate his or her accomplishments by showing a realistic aim for the future and demonstrating the results to management.

Retention of records

Frequently, it is not practical to maintain records or portions of records in-house because of the lack of use and cost of maintenance. For efficient medical office operation, medical records, log books, appointment schedules, records of services rendered, reports, and other documents should periodically be purged and the unwanted data put into separate files. When such records are placed in storage, a systematic method should be employed to expedite retrieval in case the data is needed. As always, the confidentiality of such records should be maintained even when they are not physically located in the medical unit.

OSHA record-keeping requirements

The Occupational Safety and Health Act (OSHAct) of 1970 requires that the employer maintain accurate records of work-related deaths, injuries, and illnesses. The employer is required to record those injuries that involved medical treatment, loss of consciousness, restriction of work or motion, or transfer to another job. Also, records of employees exposed to potentially toxic materials or harmful agents must be maintained. Requirements for specific substances are listed in *CFR* 29.

The OSHA requires the use of its form no. 200, Log and Summary of Occupational Injuries and Illnesses (Figure 27-9). This form records, classifies, and describes each recordable case of injury or illnesses. The OSHA form no. 101, Supplementary Record of Occupational Injuries and Illnesses, records additional information on each case listed in the log. It describes how the accident occurred, the causative agents, the nature of the accident or illness, and the body parts involved. The employer can substitute other forms for OSHA forms 200 and 101 as long as they meet statutory requirements. Proper maintenance of these records is the legal responsibility of the employer. The OSHA records must be retained for at least 5 years.

A guide to using OSHA 200 and 101 is available from the Bureau of Labor Statistics and state labor agencies. This guide helps the employer determine which cases are reportable.

Absenteeism

Absenteeism is a leading concern of American industry due to costs, lost production, and loss of human resource from the workstation. The nurse can be instrumental in preventing or reducing absenteeism. Early diagnosis, timely intervention, proper referral, and periodic follow-up by the nurse can aid in a prompt return to work. Employees with known chronic medical conditions, such as diabetes, cardiovascular disorders, and respiratory problems, when regularly followed, can often avoid complications and extended lost time.

Return to work

Employees who have been on extended absence should be evaluated before returning to work, to assess their readiness to resume normal work schedules and activities. Health professionals working in the occupational health environment must remember that community physicians or health personnel in agencies or other facilities may not fully be aware of the individual's work environment and the actual demands placed on the worker. Temporary work restrictions, such as limited lifting or work hours, may be appropriate, based on the employee's health status upon release to work. Should an employee be out of work due to communicable disease, make certain he or she is no longer infectious at the time of return. When necessary, report the communicable disease to the designated health department.

HEALTH EDUCATION

Providing health education is a major responsibility of most nurse-practitioners, as charged in many nurse practice acts. The occupational setting is a very conducive environment for the nurse to provide health education, either on a one-to-one basis or through a group process. Subjects such as self breast examination, nutrition, and cardiovascular and other risk factors may be presented. Films and handout literature and guest speakers from the community or various agencies can be used. Programs on wellness and preventive aspects may have long-term gains for both employee and employer and are often supported by senior management and employee participation. Emergency care training is usually included in an occupational health education program, to include first aid and cardiopulmonary resuscitation (CPR). No setting other than an industrial environment, except perhaps a school itself, offers a better opportunity to provide health education to the adult.

Training programs

By taking advantage of times when employees are assembled, the occupational health nurse can provide much information to employees on health- and safety-related matters. Contact with community agencies for handout literature, speakers, and audiovisuals usually can be easily arranged. Such activity fosters a good relationship between industry and agencies within the area. Usually, group health education programs that are held during company time are most readily received, but require the support of management. Without commitment from management and a positive philosophy regarding health and safety, the success of such programs will be in doubt. Occasionally, after-hours programs allow family members or friends to attend, and in some situations may be more profitable.

On a continuing basis, the occupational health nurse usually finds that health teaching is done on a one-on-one basis, geared to individual needs. This activity is usually triggered by an employee's comments or problems shared with the nurse during

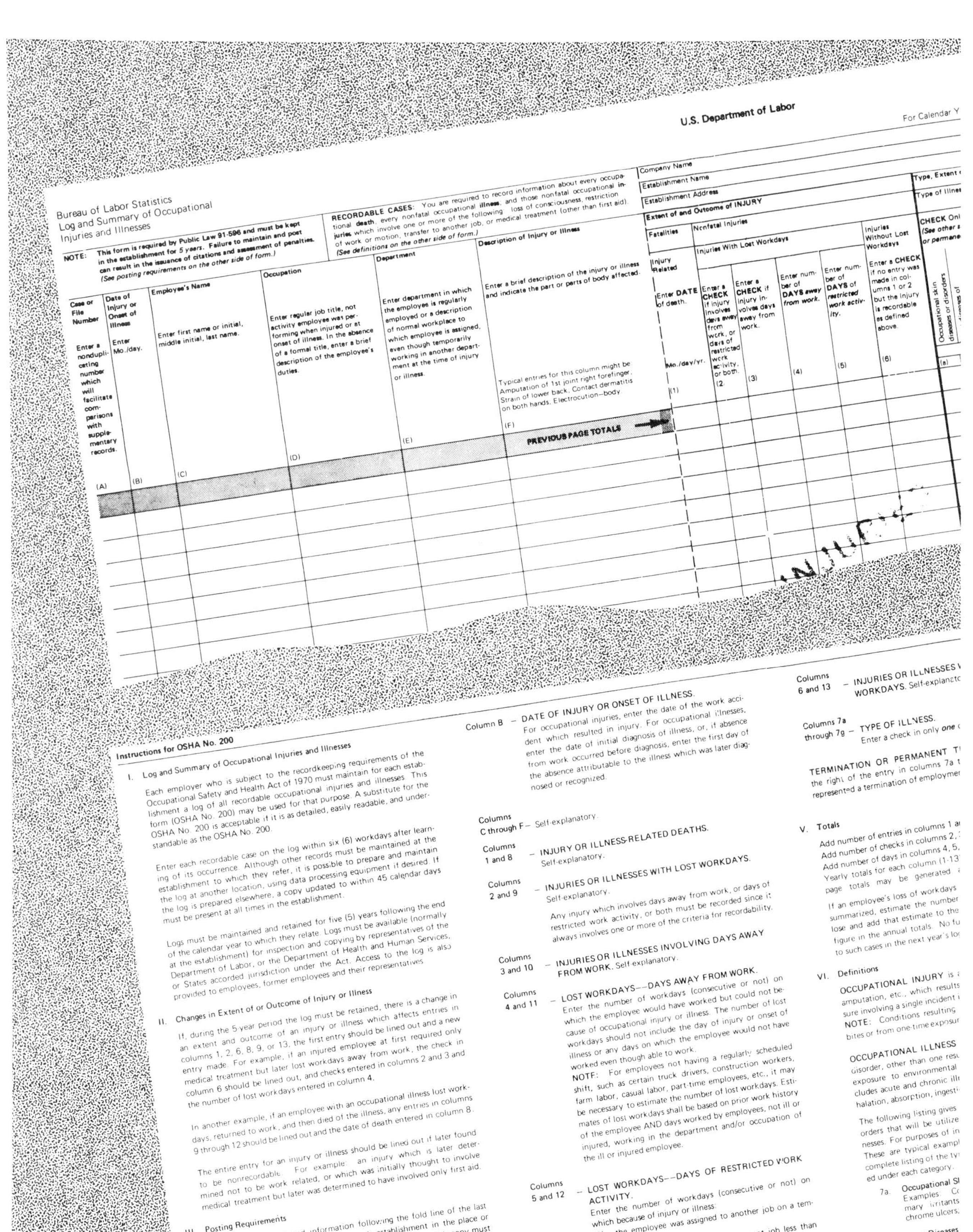

U.S. Department of Labor

For Calendar Y

Bureau of Labor Statistics
Log and Summary of Occupational Injuries and Illnesses

NOTE: **This form is required by Public Law 91-596 and must be kept in the establishment for *5 years*. Failure to maintain and post can result in the issuance of citations and assessment of penalties. *(See posting requirements on the other side of form.)***

RECORDABLE CASES: You are required to record information about every occupational **death**, every nonfatal occupational **illness**; and those nonfatal occupational **injuries** which involve one or more of the following: loss of consciousness, restriction of work or motion, transfer to another job, or medical treatment (other than first aid). ***(See definitions on the other side of form.)***

Company Name

Establishment Name

Establishment Address

Case or File Number	Date of Injury or Onset of Illness	Employee's Name	Occupation	Department	Description of Injury or Illness
Enter a nonduplicating number which will facilitate comparisons with supplementary records.	Enter Mo./day.	Enter first name or initial, middle initial, last name.	Enter regular job title, not activity employee was performing when injured or at onset of illness. In the absence of a formal title, enter a brief description of the employee's duties.	Enter department in which the employee is regularly employed or a description of normal workplace to which employee is assigned, even though temporarily working in another department at the time of injury or illness.	Enter a brief description of the injury or illness and indicate the part or parts of body affected. Typical entries for this column might be: Amputation of 1st joint right forefinger. Strain of lower back. Contact dermatitis on both hands. Electrocution—body.
(A)	(B)	(C)	(D)	(E)	(F)
					PREVIOUS PAGE TOTALS

Extent of and Outcome of INJURY

Fatalities	Nonfatal Injuries				
Injury Related	Injuries With Lost Workdays				Injuries Without Lost Workdays
Enter **DATE** of death. Mo./day/yr.	Enter a **CHECK** if injury involves days away from work, or days of restricted work activity, or both.	Enter a **CHECK** if injury involves days away from work.	Enter number of **DAYS away from work.**	Enter number of **DAYS** of ***restricted work activity.***	Enter a **CHECK** if no entry was made in columns 1 or 2 but the injury is recordable as defined above.
(1)	(2)	(3)	(4)	(5)	(6)

Type, Extent

Type of Illne

CHECK Onl
(See other s or permane

Occupational skin diseases or disorders

(a)

Instructions for OSHA No. 200

I. **Log and Summary of Occupational Injuries and Illnesses**

Each employer who is subject to the recordkeeping requirements of the Occupational Safety and Health Act of 1970 must maintain for each establishment a log of all recordable occupational injuries and illnesses. This form (OSHA No. 200) may be used for that purpose. A substitute for the OSHA No. 200 is acceptable if it is as detailed, easily readable, and understandable as the OSHA No. 200.

Enter each recordable case on the log within six (6) workdays after learning of its occurrence. Although other records must be maintained at the establishment to which they refer, it is possible to prepare and maintain the log at another location, using data processing equipment if desired. If the log is prepared elsewhere, a copy updated to within 45 calendar days must be present at all times in the establishment.

Logs must be maintained and retained for five (5) years following the end of the calendar year to which they relate. Logs must be available (normally at the establishment) for inspection and copying by representatives of the Department of Labor, or the Department of Health and Human Services, or States accorded jurisdiction under the Act. Access to the log is also provided to employees, former employees and their representatives.

II. **Changes in Extent of or Outcome of Injury or Illness**

If, during the 5-year period the log must be retained, there is a change in an extent and outcome of an injury or illness which affects entries in columns 1, 2, 6, 8, 9, or 13, the first entry should be lined out and a new entry made. For example, if an injured employee at first required only medical treatment but later lost workdays away from work, the check in column 6 should be lined out, and checks entered in columns 2 and 3 and the number of lost workdays entered in column 4.

In another example, if an employee with an occupational illness lost workdays, returned to work, and then died of the illness, any entries in columns 9 through 12 should be lined out and the date of death entered in column 8.

The entire entry for an injury or illness should be lined out if later found to be nonrecordable. For example: an injury which is later determined not to be work related, or which was initially thought to involve medical treatment but later was determined to have involved only first aid.

III. **Posting Requirements**

A copy of the totals and information following the fold line of the last page for the year must be posted at each establishment in the place or places where notices to employees are customarily posted. This copy must be posted no later than ***February 1 and must remain in place until March 1.***

... were no injuries or illnesses during the year, zeros must ... form posted.

Column B – DATE OF INJURY OR ONSET OF ILLNESS.
For occupational injuries, enter the date of the work accident which resulted in injury. For occupational illnesses, enter the date of initial diagnosis of illness, or, if absence from work occurred before diagnosis, enter the first day of the absence attributable to the illness which was later diagnosed or recognized.

Columns C through F – Self-explanatory.

Columns 1 and 8 – INJURY OR ILLNESS-RELATED DEATHS.
Self-explanatory.

Columns 2 and 9 – INJURIES OR ILLNESSES WITH LOST WORKDAYS.
Self-explanatory.

Any injury which involves days away from work, or days of restricted work activity, or both must be recorded since it always involves one or more of the criteria for recordability.

Columns 3 and 10 – INJURIES OR ILLNESSES INVOLVING DAYS AWAY FROM WORK. Self-explanatory.

Columns 4 and 11 – LOST WORKDAYS––DAYS AWAY FROM WORK.
Enter the number of workdays (consecutive or not) on which the employee would have worked but could not because of occupational injury or illness. The number of lost workdays should not include the day of injury or onset of illness or any days on which the employee would not have worked even though able to work.
NOTE: For employees not having a regularly scheduled shift, such as certain truck drivers, construction workers, farm labor, casual labor, part-time employees, etc., it may be necessary to estimate the number of lost workdays. Estimates of lost workdays shall be based on prior work history of the employee AND days worked by employees, not ill or injured, working in the department and/or occupation of the ill or injured employee.

Columns 5 and 12 – LOST WORKDAYS––DAYS OF RESTRICTED WORK ACTIVITY.
Enter the number of workdays (consecutive or not) on which because of injury or illness:
(1) the employee was assigned to another job on a temporary basis, or
(2) the employee worked at a permanent job less than full time, or
(3) the employee worked at a permanently assigned job but could not perform all duties normally connected

Columns 6 and 13 – INJURIES OR ILLNESSES W... WORKDAYS. Self-explanato...

Columns 7a through 7g – TYPE OF ILLNESS.
Enter a check in only *one* ...

TERMINATION OR PERMANENT T... the right of the entry in columns 7a t... represented a termination of employmen...

V. **Totals**

Add number of entries in columns 1 a...
Add number of checks in columns 2, ...
Add number of days in columns 4, 5, ...
Yearly totals for each column (1-13) ...
page totals may be generated ...

If an employee's loss of workdays ... summarized, estimate the number ... lose and add that estimate to the ... figure in the annual totals. No fu... to such cases in the next year's lo...

VI. **Definitions**

OCCUPATIONAL INJURY is ... amputation, etc., which results ... sure involving a single incident i... NOTE: Conditions resulting ... bites or from one-time exposur...

OCCUPATIONAL ILLNESS ... disorder, other than one resu... exposure to environmental ... cludes acute and chronic ill... halation, absorption, ingesti...

The following listing gives ... orders that will be utilize... nesses. For purposes of in... These are typical exampl... complete listing of the ty... ed under each category.

7a. **Occupational S...** Examples: C... mary irritants ... chrome ulcers; ...

7b. **Dust Diseases** Examples ... eases, coal ... other pneum...

7c. **Respiratory** ...

Figure 27-9. The OSHA form no. 200, Log and Summary of Occupational Injuries and Illnesses.

routine visits, or in response to a direct request for information. Knowledge of the theories and practices of adult education helps the nurse to more adequately respond to client needs for information.

Counseling

Advice and counsel to employees and management on health and related concerns are major responsibilities of the occupational health nurse. Being a good listener and knowing about community and internal company resources are key elements for the nurse. Frequently, finding the resources is a major problem, especially if they are located some distance away. Often, networking among professional peers can unearth these resources. Community social service agencies, such as the Community Chest can provide reference books listing area agencies and the services they provide.

One of the most rewarding experiences any nurse can have is to counsel a client and to see behavioral change, therapeutic intervention, or other action which improves physical or mental well-being. Counseling requires skillful listening, the ability to allow the client self-expression and discovery of his concerns, and being able to provide a choice of solutions.

The nurse must be aware of the limitations of practice and personal abilities. It should not be considered a failure to refer a client to someone more qualified. The welfare and interest of the client and the quality of care must be the first priority.

Employee personal health problems

The nurse often is the employee's primary source of health information. During this contact with the nurse problems are defined—a nonoccupational health problem (physical or mental), a family problem, or a job-related problem. The nurse frequently can help employees resolve their problems by providing them with guidance and direction. In some cases, the nurse may refer the employee to agencies or persons who can provide special services.

Here the nurse can help executives and supervisors recognize problems in human relations which may have medical implications and recommend adjustments in job requirements or relationships. This is a very sensitive area and requires diplomacy and a sincere commitment to confidentiality.

Community health resources. Community health agencies often can assist an occupational health program. The occupational health nurse should be familiar with health agencies in the community and establish good working relationships with them. Employees frequently have problems which are outside the scope of an occupational health program. These agencies are often in a position to offer the specific assistance needed.

The nurse, who is aware of the services and is familiar with specific referral procedures, can direct employees to the appropriate source with ease and efficiency. Also, the workers who have unsolved problems can be a hazard to both themselves and their fellow employees.

The occupational health nurse interprets the objectives of the health programs to other community resources and establishes a mutual exchange of information. It is as important for the community to know about a company's health facility as it is for the nurse to know the community's resources.

Being a good listener

The occupational health nurse has learned to be a good listener. Being aware of the problems and stresses of everyday living, the nurse can help an employee understand a problem and help him continue on the job. Some companies offer employee assistance programs (EAPs) to workers. If possible, employees with problems can be referred to the facility's EAP for assistance.

The employee's supervisor may become involved by recognizing problems on the job and allowing time for the employee to visit the nurse. If the employee's problem is not recognized and resolved early, it can become a crisis. It is better to have a company policy encouraging good communication and resolve these situations early, than have problems later with employee absenteeism.

Crisis intervention

The nurse intervenes to help a person influence the course of a crisis towards a more rewarding outcome. The nurse encourages the worker to examine the problem—whether this is a misuse of drugs or alcohol or a concern about a sick child or unpaid bills. The nurse can help the employee plan how to handle his problem.

The nurse also gives encouragement and support and, at times, anticipatory guidance. The nurse uses the anxiety of the moment to help the person accept his need for care, or for changing his behavior. The nurse helps the worker to identify alternative ways to handle his problem and to select the most satisfactory one.

The men and women to whom the nurse frequently gives care take their home and family problems to work with them. The occupational health nurse can help these people who often are in a crisis situation. Such upsets may be strong, but usually they are temporary. The worker's response to stress represents a response to a difficult situation that cannot be understood or handled alone. The nurse looks for and tries to understand nonverbal communications. The nurse makes a conscious effort to understand what the problem or situation means to the worker.

Drugs and alcoholism

Substance abuse is difficult to cope with. When these problems arise, the nurse is probably the first to suspect. It is doubtful that anyone will positively know until the employee is obviously affected.

Management makes a mistake if it allows the health services group to be used as a disciplinary arm. Discipline should come from supervision in the employee's operating unit. If it results in absenteeism or sloppy or indifferent work, then it should be dealt with on that basis until it is certain the underlying cause is drugs or alcoholism.

The occupational health nurse should not be used as a shield by an employee who has confided he or she has a problem. If the employee is willing to admit it, then the nurse should make it known she can only help with the aid of other members of the occupational health team.

SAFETY COMMITTEES

Nurse's role

Plant safety and health committees are essential. The nurse

should become a member, attend committee meetings, advise in the selection and care of protective equipment, report unsafe conditions, and interpret the health aspects of the safety program to employees. Further, the nurse can participate in the prevention and control of occupational disease hazards and advise on sanitation standards and safe food handling.

The occupational health nurse should be familiar with safety hazard recognition and prevention, use of personal protective equipment, good housekeeping and maintenance, company standards, and employee-employer responsibilities.

The nurse should understand what constitutes a safe working environment and know the plant's safety practices and safety devices. General knowledge of the principles of industrial hygiene and toxicology and basic measuring and sampling techniques also are required.

Familiarity with the workplace and its operations can enable the nurse to seek consultation with other specialists—safety professionals, industrial hygienists, and medical specialists. Health hazards such as asbestos, lead, silica, cotton dust, and carbon monoxide have been given special attention by federal agencies. The biological effects of these substances should be thoroughly understood by the nurse.

Touring the facility

Awareness of the mechanical hazards on assembly lines, in the machine shops, and even in the general offices, adds greater effectiveness to the nurse's participation as a member of the safety team. Frequent plant tours can provide clearer insight of industrial processes.

These tours keep the nurse informed of changes in the plant environment, its operations, and existing or potential health hazards. Equipped with this information, the nurse can compile accurate accident histories and contribute to the prevention of similar injuries. Through such tours the nurse can recognize the relationship between nonoccupational complaints and a work-related exposure. The nurse should bring findings to the attention of the plant physician and safety professional, so proper investigation can be initiated and appropriate preventive measures instituted.

Facility operations

An occupational nurse must have a working knowledge of facility operations. If a welder gets a burn or laceration, the nurse should know the metals in use. Wounds which heal slowly, ulcerate, and incapacitate may be caused by exposures to certain metals.

The nurse can motivate employees to work safely. Areas in which the nurse can be of assistance include—using and caring for protective equipment; stressing the need of meticulous skin care when exposed to materials such as epoxy resin hardeners, solvents and acids; providing for immediate copious rinsing following chemical splash contact to the skin or eye; and motivating workers to use proper bending and lifting procedures. Further, as a member of the safety committee, the nurse collaborates with the safety and health professionals and other members of the committee to plan and arrange educational activities for groups of workers.

THE PROFESSION

Ethical-moral issues

The occupational health nurse must be keenly aware of the ethical issues facing nursing, medicine, employees, management, and society in general. Sensitivity to such issues as mental health, venereal disease, family discord, crisis intervention, finances, confidentiality, and individual rights must be displayed in a professional manner in order to establish needed rapport. In some circumstances it may be difficult to maintain confidentiality in the occupational environment. Management may request information that is either confidential or for which there is no business need. In such situations, it would be improper for the nurse to provide the data—and would breach nurse-client confidentiality. The nurse is obligated to provide the necessary information to ensure a safe and healthy work environment, but this obligation does not necessarily require the disclosure of the medical reason or diagnosis behind the recommendations given to management.

The nurse must be aware of personal bias and attempt to maintain a nonjudgmental position when interacting with employees. One's personal values and feelings should not interfere with intervention, especially in areas concerning drugs, alcohol, sex, life-style, and religious beliefs.

Occasionally the nurse may be in conflict over an ethical or moral issue. At these times, consulting with peers and other professionals can supply support, understanding, and clarification. Private or community resources on ethical policy also can help the nurse resolve this type of conflict.

Nurse-physician relationship

Without question, a positive and cooperative relationship should exist between the nurse and staff physician. The most successful occupational health programs encourage an active teamwork approach among nursing and physician groups. However, it must be recognized that the disciplines of nursing and medicine are separate and distinct, even though they greatly overlap in theory and practice. Responsibilities and limitations are established for both disciplines by the state practice acts. These regulations are particularly important when developing a nursing guidelines and principles statement. Care must be taken not to describe the nurse acting in a manner that could be interpreted as practicing medicine, rather than nursing. The physician too must be cognizant of the responsibilities that appropriately may be delegated to nursing personnel. Through a cooperative effort and professionalism, the two disciplines can provide a health program of which they both can be proud.

Professional organizations

In recent years, the importance of professional nursing organizations has increased because of changes in regulations, legislation governing the scope of practice, and internal and external pressures on the nursing profession. As with other professional specialities, organizations have developed to meet the educational and professional needs of practicing nurses. The American Association of Occupational Health Nurses, previously known as the American Association of Industrial Nurses, was founded in 1942, and has since been constantly changing as it meets the needs of its membership. Based in Atlanta, Georgia, the organization's primary mission is to provide education to nurses employed in the occupational setting. In addition, it carefully monitors legislation affecting public workers and health care, especially that related to industry.

Nurses employed in occupational health are encouraged to join and become active members in their nursing specialty organization on local, state, and national levels. Contacts with other nurses provide an excellent network or resource group for exchange of ideas, policies, and methods. An annual conference provides continuing educational offerings and up-to-date information relative to practice and concern in the field. Membership is not only beneficial for the nurse, but also for the company and its employees.

Membership and participation in other nursing, professional, or work-related organizations also can be advantageous, such as the American Nurses Association, Public Health Association, National Safety Council, Critical Care Nurses Association, and Emergency Department and Nurses Association. Health-related community agencies also provide benefits to the nurse: American Red Cross and Heart, Lung, and Cancer Associations. The company often will fund the nurse's professional organization activities and support and encourage participation.

Certification

The American Board for Occupational Health Nursing was established in 1972 in response to the need for certification in the occupational nursing specialty. As with most such processes, certification as a Certified Occupational Health Nurse (COHN) is encouraged to maintain the quality and professionalism of the practitioner in this field.

Certification is evidence of expertise in the nursing specialty of occupational health. Application requirements for the examination and recertification are available through the Executive Director of the American Board, located in Santa Monica, California.

Continuing education

Maintenance of excellent nursing skills and keeping abreast of new developments in the field are the personal and professional responsibilities of every nurse. Although evidence of continuing education is required for recertification as an Occupational Health Nurse and in some states for relicensing, most nurses voluntarily spend time each year on the educational process. Course offerings are presented by many nursing organizations, private enterprises, and universities with nursing and health-related curriculums.

SUMMARY

Occupational health nurses are specialists—members of a team contributing to the health and well-being of workers. They practice the principles and concepts of nursing: physical assessment, preventive medicine, chronic disease control, counseling, health education, and emergency care. Occupational health nurses should have knowledge of safety practices, injury compensation regulations, labor laws, OSHA regulations, toxicology, and industrial hygiene.

The occupational health program should be under the direction of the company physician, who may be employed on a full-time or part-time basis. The occupational health nurse should be directly responsible to the physician for all professional matters and, when the physician is employed full-time, for defined administrative matters.

If there is no physician in charge of the service the nurse should be responsible to management and will advise on health problems in collaboration with their physician consultant. When the company physician is employed on a part-time basis, he or she may authorize the nurse to discuss the placement of workers. Where no company physician is employed the nurse advises management on the placement of workers when a health problem has been detected or a disability is known to exist.

Where an industrial physician is in charge of the occupational health service, principles for the nurse should be drawn up by him or her in consultation with the nurse and approved by the employer.

Management expects the occupational health nurse to:

- be technically qualified
- be aware of the health needs and problems of others
- be able to establish and maintain a clear and positive line of communication
- have the ability to develop and maintain the confidence of management, physicians, and employees.

The occupational health nurse is usually responsible for the daily administration of the health unit.

The nurse assists the company physician with the preemployment and routine medical examination of workers. Following up the health progress of workers who are under the physician's care or orders also is the nurse's responsibility.

The nurse should interview all workers on return to work after an injury or illness, and seek the advice of the physician if there is any doubt about the worker's wellness.

Maintenance of records is a very important duty undertaken by the nurse. Any personal medical information received by the nurse must be regarded as confidential and not disclosed with any person other than the company physician. There are certain records that always are maintained by the nurse while others may be needed to meet any legal requirements.

Health counseling is another important aspect of the occupational health nurse's work, and many of the minor ailments reported may be found to be due to underlying stress and anxiety. At times no advice is requested by the worker, the nurse should be a willing listener.

The nurse is a member of the safety team and, where appropriate, assists the safety professional with the implementation of the safety program. The nurse should have access to all workplaces and should conform to regulations relating to confidential technical information.

The OSHAct increasingly compels the occupational health nurse to give more comprehensive service. This necessitates a close working relationship with the industrial hygienist and the safety officer and requires a knowledge of toxic materials and harmful agents.

The occupational health nurse and the industrial hygienist have much in common. Each must recognize the value of the other and work together for the good of the employee. The nurse's competency, interest, and motivation, and the demands of the job determine the nursing role. Every occupational health nurse should have a thorough knowledge of the fundamentals of industrial hygiene. This knowledge can be acquired in several ways: through practical experience, by attendance at professional meetings and short courses, by participation in academic training, and through personal research.

Briefly, these are the major functions and responsibilities. Although the occupational health nurse's primary responsibility concerns guarding employee health, he or she is strategically placed and can promote good will and facilitate a better understanding between management and workers.

BIBLIOGRAPHY

American Association of Occupational Health Nurses, Inc. (AAOHN). *Guide for the Development of Functions and Responsibilities in Occupational Health Nursing,* 3rd ed. Atlanta: AAOHN, 1977.

——. *Guide for the Preparation of Manual of Policies and Procedures.* Atlanta, 1969.

——. *Standards for Evaluating an Occupational Health Nursing Service,* 3rd ed. Atlanta: AAOHN, 1977.

——. *The Nurse in Industry.* Atlanta: AAOHN, 1976.

American College of Surgeons, Committee on Trauma. *Emergency Care of the Sick and Injured.* Philadelphia: W. B. Saunders Co.

American Conference of Governmental Industrial Hygienists. *Guide to Health Records for Health Services in Small Industries,* Cincinnati, OH: ACGIH.

American Medical Association (AMA), Dept. of Occupational Medicine. *Guide to Developing an Industrial Disaster Medical Service.* Chicago: AMA.

——. *Guide to the Development of an Industrial Medical Records System.* Chicago: AMA.

——. *Guide to Small Plan Occupational Health Programs.* Chicago: AMA.

——. *Guiding Principles of Medical Examinations in Industry.* Chicago: AMA.

——. *The Legal Scope of Industrial Nursing Practice.* Chicago: AMA.

——. *Occupational Health Service for Women Employees.* Chicago: AMA.

——. *Scope, Objectives, and Functions of Occupational Health Programs.* Chicago: AMA.

Brown, M. L. "The Occupational Health Nurse: A New Perspective." In *Occupational Medicine—Principles and Practical Applications* 2nd ed., edited by C. Zenz. Chicago: Year Book Medical Publ., Inc., 1988.

——. *Occupational Health Nursing.* New York: Springer Publishing Co., Inc., 1981.

Burkeen, O. E. The nurse and industrial hygiene. *Occupational Health Nursing,* 4(1976):7-10.

Draffin, W. What management expects of the occupational health nurse. *Occupational Health Nursing,* 1(1973):7-8.

French, M. The nurse's role in prevention and treatment. *Occupational Health Nursing,* 2(1973):15-17.

A Guide for Establishing an Occupational Health Nursing Service, New York: American Association of Occupational Health Nurses, Inc., 1977.

Hannigan, L. Nurse, health assessment: The new 'physical' in American industry. *Occupational Health Nursing,* (August 1982).

Lee, J. A. *The New Nurse in Industry.* Cincinnati: DHHS, PHS, CDC, NIOSH Division of Technical Services, 1978.

Murphy, A. J. The identity of the nurse in an industrial hearing conservation program. *Occupational Health Nursing,* 5(1969):32-36.

Nelson, E. Let's get some answers on nurse role expansion. *Occupational Health and Safety,* 2(1976):28-31.

Nelson, E. The nurse's role in toxicology. *Safety Newsletter.* Chicago: National Safety Council, 1974.

Occupational Health, 4(July/August 1985). Kenilworth, NJ: Schering Corp., 1985.

Onyett, H. P. The nurse's role in industry. *Safety Newsletter.* Chicago: National Safety Council, 1974.

Popiel, E. S. Principles and concepts of occupational health nursing. *Occupational Health Nursing,* 9(1973):23-25.

Tuohey, P. M. The role of the industrial nurse in the electronic and electrical equipment industry. *National Safety Congress Transactions.* Chicago: National Safety Council, 1968.

Wolter, M. J. Evolving role of the occupational health nurse. Occupational Health and Safety, 2(1975):28-31.

——. The nurse's expanding nursing role. *Occupational Health & Safety,* 3(1975):22-25.

U.S. Department of Labor Statistics. *A Brief Guide for Recordkeeping Requirements for Occupational Injuries and Illnesses.* Washington, DC: U.S. Government Printing Office, 1986.

28

The Industrial Hygiene Program

by Edward J. Largent
Julian Olishifski, MS, PE, CSP
Revised by Maureen Kerwin, MPH

INTRODUCTION

INDUSTRIAL HYGIENE is an essential component of the occupational health program. Its chief goal is to prevent occupational disease or injury through the recognition, evaluation, and control of occupational health hazards. Providing a safe and healthful workplace benefits both employees and employers by improving health, morale, and productivity.

There are no set rules for developing and implementing a program—its form depends upon a variety of factors. The type and size of the industry, management philosophy, and organization are some factors that play a role in forming the program. For example, small companies may not have the resources for even a rudimentary occupational health program. These firms must rely upon private or governmental consulting agencies for their industrial hygiene needs. On the other hand, large corporations are increasingly integrating comprehensive industrial hygiene programs into their overall occupational health programs. In 1965, industrial hygiene was a major activity in only 9 percent of the companies surveyed by the Conference Board, a management organization. By 1975, that figure had increased to 27 percent.

Though structure and responsibilities can vary, each industrial hygiene program has five fundamental components: (1) Written policy, (2) Hazard recognition, (3) Hazard evaluation, (4) Hazard control, and (5) Employee training.

BENEFITS AND GOALS OF THE INDUSTRIAL HYGIENE PROGRAM

Senior management's commitment to the industrial hygiene program is imperative for its success. To achieve this, the benefits—both social and economic—must be made apparent. The following description of the benefits of an industrial hygiene program is adapted from a brochure prepared by the American Industrial Hygiene Association.

> The industrial hygienist is a specialist. A scientist concerned with solving industrial health problems, the industrial hygienist is trained to recognize, evaluate, and control health hazards in the industrial environment—particularly the chemical and physical agents that can induce injurious effects in humans.

The goals of industrial hygiene are reached through competence in a variety of scientific fields—principally, chemistry, engineering, physics, biology, and medicine. Initially trained in one of these fields, most industrial hygienists have acquired, by experience and postgraduate study, knowledge of the other allied disciplines.

As will be discussed later in this chapter, industrial hygienists are expected to become involved with the health problems of the workplace and deal with many different aspects of these problems—work practices, the cleanup of spilled chemicals, labeling and placarding needs, housekeeping, reactor vessel entry, respiratory protection, air monitoring, medical surveillance, measurements of physical stresses, employee training, and record-keeping.

Many larger corporations use industrial hygiene teams composed of analytical chemists, physicists, engineers, toxicologists, nurses, and industrial physicians, each applying a specialty to combat industrial health problems (Figure 28-1). Smaller corporations rely on the comprehensive knowledge of an individual industrial hygienist.

Figure 28-1. The industrial hygienist uses analytical techniques ranging from classical wet chemical procedures to this mass spectrometer. (Courtesy American Industrial Hygiene Association)

To bridge a gap

The industrial hygienist bridges the gap between manufacturing areas and the medical department by applying chemical, engineering, physical, biological, and/or physiological principles to prevent occupational disease, which may range from deafness to heatstroke, from simple skin rash to cancer of the bladder. Special techniques and knowledge enable the industrial hygienist to assess plant environmental health problems and to provide the plant physician with the health background of the employees' jobs.

To inform the plant physician

Many occupational diseases cause symptoms similar to those of diseases that are not occupational in origin. The industrial hygienist can focus the attention of the physician on hazardous tasks within the plant. Such guidance enables the physician to correlate the patient's condition and complaints with the potential health hazards of the job. And when physicians know what effects to look for, they can ask for appropriate, specific, clinical tests to find out whether the normal operation of the patient's bodily functions has been altered due to any hazard at work (Figure 28-2).

To protect

The industrial hygienist's aim is to devise controls in the work area so hazardous agents or materials are confined or otherwise kept away from employees. The hygienist can contribute to the design of health-protecting measures that can be incorporated into a process at low cost and with maximum effectiveness. Periodic checks of the control methods, the environmental conditions, and the physician's clinical evaluation of the health of the employees are used to verify and justify judgment concerning the adequacy of environmental controls.

Other prime benefits

Industrial hygiene operates in areas related to those jointly

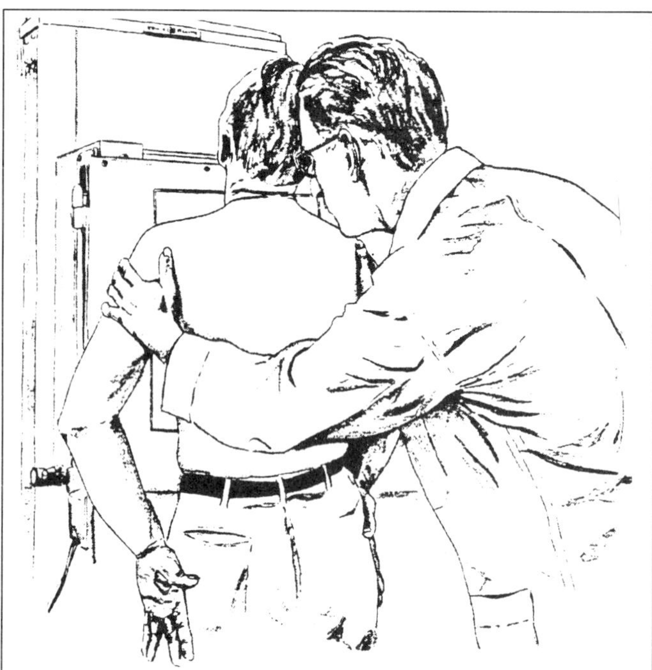

Figure 28-2. Guided by the industrial hygienist's evaluation of job hazards, the physician is able to use to best advantage such procedures as radiography to detect illness. (Courtesy American Industrial Hygiene Association)

covered by medical and safety programs. The industrial hygienist can provide a unique perspective to the company operation and can offer specialized solutions to many problems.

Reduced compensation costs. In 1982, the average compensation paid per closed workers' compensation case was about $4,350 according to data on 121,028 cases from the New York Compensation Board. Lower insurance premiums and a decrease in medical expenses related to job-induced illness should follow the favorable loss-control experience resulting from the industrial hygienist's activities.

Increased productivity. Reducing discomfort and disabilities makes increased production possible. Example: A glass producer eliminated summer slowdowns when the industrial hygienist reduced exposure to process heat and improved working conditions.

Increased employee efficiency. Production depends on people and machines. Management has done a great deal to install the correct machinery, maintain it, and keep it trouble-free. The industrial hygienist, on the other hand, is trained to protect the human component. The hygienist can organize the overall maintenance of employee workplace health so production can continue as smoothly on the human side as it does on the mechanical side. Management needs an industrial hygienist to preserve the health and efficiency of employees, just as it requires a millwright to preserve the optimum efficiency of machinery or a financial expert to preserve a sound financial structure (Figure 28-3).

Better product design. Products and equipment with built-in protection against health problems are welcomed by a company's

Figure 28-3. Three principal sources of heat—radiation, convection, and metabolism—affect this worker's health and efficiency. The industrial hygienist can measure the effects of each source and plan effective controls. (Courtesy American Industrial Hygiene Association)

customers. Sales can be lost if design engineers overlook the possible health hazards of a product. Consumer items are coming increasingly under government scrutiny with regard to hazards associated with a product's presence or use. With each promulgation for additional federal safety controls for manufactured products, the tasks of industrial hygienists have increased. These increasing requirements for safety in products are described in:

1. *Title 16, Federal Hazardous Substances Act (FHSA), Parts 1500-1512.* The Act allows the Consumer Product Safety Commission (CPSC) to ban or regulate hazardous substances produced for consumer use. Under the Act, the CPSC has labeling authority over consumer products that meet definitions as being toxic, corrosive, flammable, irritant, or radioactive.
2. *Title 16, Consumer Product Safety Act, Parts 1000-1406.* Standards promulgated under the Act require specific labeling design, packaging, or composition of products intended for sale to the public.
3. *Title 16, Poison Prevention Packaging Act (PPPA).* The Act authorizes the CPSC to establish packaging standards for household substances to prevent children from ingesting or using hazardous substances.
4. *Title 21, Federal Food, Drug and Cosmetic Act (FFDCA), Parts 1-1300.* The Act enables the Food and Drug Administration (FDA) to establish standards for chemical content for various covered products, specify level of pesticides, chemicals, and naturally occurring poisons that may be in food products. The Delaney Clause, Section 409 (c) (3) of the Act, provides that no food additive can be considered safe by the FDA if it produces cancer in animals.
5. *Title 40, Toxic Substances Control Act (TSCA), Parts 700-799.* The Act, under the regulatory authority of the Environmental Protection Agency (EPA), has the following provisions:
   1) a premanufacture notification program under which the EPA assesses the safety of new chemicals before manufacture; (40*CFR* 720-723)
   2) the EPA can require industry to test a chemical on the market to show if the substance poses a health or environmental risk;
   3) the EPA can require a manufacturer to collect, record, and submit information on certain chemicals. Information includes data on chemical production, use, exposure and disposal, records of significant adverse reactions to health or the environment, any health and safety studies, and notification of previously unknown risks;
   4) allows the EPA to control or ban chemicals that the EPA finds pose an unreasonable risk to health or the environment.
6. *Title 40, Federal Insecticide, Fungicide and Rodenticide Act (FIFRA), Parts 162-180.* The Act provides the EPA with regulatory authority for the registration and use of pesticides and herbicides.
7. *29 CFR 1910.1200. Hazard Communication Standard.* This Standard requires that chemical manufacturers assess the physical and health hazards posed by their products. The chemical product must be labeled with this information when leaving the manufacturer and a material safety data sheet (MSDS) must be supplied to the purchaser. Employers within covered SIC codes who purchase these products must label all in-plant containers, provide information and training to all exposed employees, make MSDSs available and create a written hazard communication program. (State and community hazard communication requirements may differ from the federal Standard.)

Process design. It saves delays and makes sense to recognize potential health hazards when a process is in the design phase. Industrial hygiene-oriented engineering avoids, or provides control for, such potential problems. This function can be accomplished by having industrial hygiene review plans for new processes at an early stage. Industrial hygienists have spearheaded the drive to formulate, standardize, and publish adequate engineering design criteria for exhaust ventilation systems (Figure 28-4).

Toxicity and labeling data. The industrial hygienist skilled in toxicology can supply data required by the FDA, the Department of Transportation (DOT), the Department of Labor (DOL), and EPA. Large firms frequently maintain toxicology laboratories to provide such information. This information is now included with sales literature and MSDSs.

Liaison with governmental agencies. The industrial hygienist works with agencies charged with the responsibility for safeguarding the health of workers. Administrators of federal, state, and local regulations are authorized to inspect and control unhealthful operations. In addition, the industrial hygienist is frequently involved with agencies charged with environmental consumer protection. Regulatory compliance is a major concern for industry. The large increase in the number of industrial

Figure 28-4. The industrial hygienist is always concerned with the supply, movement, and cleaning of air. In industry, harmful materials most commonly enter the body by inhalation of contaminated air. (Courtesy American Industrial Hygiene Association)

hygiene programs is largely seen as a response to the Occupational Safety and Health Administration Act (OSHAct) of 1970 (see Chapter 31).

Information and training for employees. The industrial hygienist can be an important source of information and training for employees. Many OSHA regulations require the dissemination of occupational health information and the training of employees in hazard recognition and proper work practices. Chemical- or hazard-specific regulations such as the noise regulation or the Hazard Communication Standard require employee training and access to information.

Reduced labor turnover. Good, safe, and healthful working conditions lessen labor turnover. When management obviously strives to improve the working environment, employees are less likely to look elsewhere for work.

Better labor relations. Since the industrial hygienist is clearly concerned with the health and welfare of employees, a good working relationship—benefiting all concerned—should be developed between shop stewards, who often receive complaints about work conditions, and the industrial hygienist. Union officials are well aware of the necessity for proper industrial hygiene practices, and such considerations frequently enter into collective bargaining procedures.

To solve problems

Here are a few examples of occupational health problems an industrial hygienist can help solve:

Noise. Noise is not an inevitable and unavoidable consequence of modern industrial operations. Compliance with the noise provisions of the OSHAct, compensation awards for hearing loss, interference with communication, inability to hear safety signals, and the annoyance to employees as well as to the neighborhood have required management to expend increasing amounts of effort and money to prevent and control unwanted noise (Figure 28-5).

With sound-level meters, octave-band analyzers, and many other specialized instruments, the industrial hygienist is able to identify and characterize noise sources. Armed with this information and guided by experience and the published work of fellow professionals, the industrial hygienist is able to recommend the most effective and economical measures for noise abatement. In addition, the industrial hygienist is able to recommend the necessary personal protective equipment and provide employee training (see Chapter 9, Industrial Noise, for more information).

Figure 28-5. Preplacement audiometric testing, with routine retesting during working years, provides the necessary background required to evaluate the effect of a worker's exposure to on-the-job noise. (Courtesy American Industrial Hygiene Association)

Dermatitis. In terms of cost to industry, dermatitis resulting from contact with skin irritants and sensitizing chemicals has been far and away the greatest offender. The industrial hygienist can suggest alternative chemicals, appropriate personal protective equipment, or personal hygiene programs to alleviate the problem (see Chapters 3, The Skin, and 8, Industrial Dermatoses).

Of course, the industrial hygienist is available to help solve the entire range of concerns coming under the definition of the field, including chemical exposures in the forms of dusts, mists, fumes, gases and vapors (see Chapters 6, Solvents, and 7, Particulates); any biological hazard control problems (see Chapter 14, Biological Hazards); the whole range of physical hazards including noise (see Chapter 9, Industrial Noise); heat and cold extremes (see Chapter 12, Temperature Extremes); ionizing and nonionizing radiation and vibration (see Chapters 10, Radiation, and 11, Nonionizing Radiation); and ergonomic hazards (see Chapter 13, Ergonomics).

Other plant problems. In the plant, there are many ways in which specialized industrial hygiene procedures can be applied toward solving problems unrelated to employee health.

- *Control of waste process heat.* The industrial hygienist's familiarity with the physics of radiant and convective heat may prove valuable. This is especially important concerning waste heat recovery in relation to the increasing need to conserve fuel.
- *Product contamination.* Airborne contamination can ruin products requiring strict control of purity. If this is a problem, an industrial hygienist can be of help (Figure 28-6).
- *Control of solvent vapor losses.* Air-sampling and analysis help chart the extent of losses and reveal the routes by which dollars are leaking from processes.

Figure 28-6. Techniques pioneered in industrial hygiene proved valuable in the design of "white rooms," required to make and assemble space-age components free from airborne contamination. (Courtesy American Industrial Hygiene Association)

SETTING UP AN INDUSTRIAL HYGIENE PROGRAM

Every industrial hygiene program must cover four basic areas:

- health hazard recognition
- health hazard evaluation
- health hazard control
- employee education and training.

How the program proposes to address these four areas should be contained in a written plan. The written plan must clearly delineate scope, responsibilities, and authority of the program. It should be written in coordination with other related departments and have the support of senior management.

Recognition

The industrial hygienist is responsible for the recognition of occupational health stressors. Several program elements are used to achieve this goal. Walk-through surveys, chemical inventories, and process and equipment reviews are all used to identify employee exposures to potential health hazards.

An essential element of any program is an initial and periodic walk-through survey. During the walk-through, the industrial hygienist becomes familiar with the process, potential chemical and physical hazards, numbers of employees exposed, condition and effectiveness of engineering and/or administrative controls, selection, use, and care of personal protective equipment, level of hazard awareness of employees and management staff, the adequacy of emergency response procedures, and specific employee complaints regarding occupational health hazards.

An important tool of the survey is the chemical inventory. The inventory, which includes chemical intermediaries, is very useful in evaluating potential health hazards. It is also a required part of the written OSHA Hazard Communication Program (29*CFR* 1910.1200). This list can be compiled from purchase orders, material safety data sheets, and a physical inventory.

The written program should state the frequency with which surveys will be accomplished and who should perform them. Surveys can be done annually, every six months, or in response to an employee complaint. Staff for the survey can range from the corporate industrial hygienist to supervisory personnel, who have received specialized training in industrial hygiene, and members of the joint labor/management health and safety committee. Survey frequency and staff make-up depend upon the severity of the hazards and the needs of the company.

Evaluation

Evaluation is the process where the industrial hygienist gives a professional opinion on the degree of risk present in the workplace. The risk associated with a given health stressor depends upon these factors:

- nature of the hazard
- magnitude of exposure
- duration of the exposure
- individual's susceptibility.

The magnitude and duration of exposure constitute the dose to the individual. Sampling—or the quantitative evaluation of the dose—constitutes a major element in the industrial hygiene program.

A basic premise of industrial hygiene is that occupational health stressors can be quantitatively evaluated and the results compared with recommended standards. Program elements include environmental (area and personal) monitoring, sample analysis, statistical evaluation of the data, and biological monitoring.

A written program must detail how each program element will be accomplished. For example, environmental monitoring must include maintenance and calibration of monitoring equipment, restocking expendable items (i.e., filters, charcoal tubes), and sampling procedure manual.

Record-keeping is an essential program element. Records of equipment used, equipment calibration, and sample results must be kept in an easily accessible system. The OSHA Access to Employee Exposure and Medical Records (29*CFR* 1910.20) requires the following:

1. Employee exposure records must be maintained for 30 years except that
   a) *Background data.* Background data to environmental (workplace) monitoring or measuring such as laboratory reports and worksheets need only be retained for one year so long as the sampling results, the collection methods, description of the analytical and mathematical methods used, and a summary of other relevant interpretation of the results obtained, are retained for at least 30 years.
   b) *Employee medical records (biological monitoring).* Each employee's medical record shall be preserved and maintained for at least the duration of employment plus 30 years.
   c) *Analysis using exposure or medical records.* Each analysis using exposure or medical records shall be maintained for at least 30 years.

Some OSHA standards have their own record-keeping requirements (e.g., Lead or Asbestos Standards). In all cases, provisions are made for employee access to these records.

Control

When a health stressor is judged to pose a health hazard, control measures must be instituted. Program elements can include recommendations for chemical or machine substitution, design and recommendations for engineering and/or administrative controls in proposed or existing systems, and establishment of a personal protective equipment program.

A written program must ensure that the industrial hygienist is involved in the planning stages of any new processes or facilities. It is much easier to implement a control during the planning stage, and spares workers unnecessary exposures. The program should also specify how necessary controls in existing equipment will be implemented.

The industrial hygienist should have primary responsibility for respiratory protection (see Chapter 23, Respiratory Protective Equipment) and hearing protection programs (see Chapter 9, Industrial Noise). The written program should specify that the industrial hygienist will select and/or approve all purchases in this area. In addition, protocol for fit-testing and employee training should be specified.

NIOSH occupational health guidelines. The National Institute for Occupational Safety and Health (NIOSH) has published *Occupational Health Guidelines for Chemical Hazards* (Pub. No. 81-123) with guidance on the evaluation and control of specific chemical hazards. These are excellent sources of information for beginning the necessary program elements for each component of the hygiene program. This publication provides information on the following:

- health hazards
- monitoring and measurement procedures
- respiratory protection guidelines
- medical surveillance
- personal protective equipment
- emergency first aid procedures.

The NIOSH guideline for carbon monoxide is shown in Figure 28-7.

Training

Employee information and training is a critical part of the industrial hygiene program. Program elements include new employee orientation, periodic information and training sessions, written safety and health guidelines (standard operating procedures), posting of dangerous areas, and chemical labeling. The OSHA Hazard Communication Standard and state right-to-know laws mandate training of employees exposed to hazardous chemicals.

There have been many guidelines published on the organization and content of training programs. The OSHA has published a series of publications providing guidance on the development of a training program (49 *CFR* 30290, July 27, 1984). These guidelines, while not mandatory, are useful in establishing an effective, meaningful program. The following summary of these guidelines was written by Margaret Samway, 1986-87 chairman of the AIHA Employee Education and Training Committee.

> **If training is needed.** The first step assists Industrial Hygienists in avoiding the trap of using training as a band-aid for all situations when employees are not performing their job properly. Problems that can be addressed effectively by training are defined as those in which there is a deficit of knowledge or skill, either in an existing or a new task. Training is not effective when there is an underlying engineering or administrative deficiency, such as lack of engineering controls or precise job definition.
>
> **Identifying training needs.** Once it is determined that a problem can be addressed by training, the next step is to identify what training is needed. This step requires some homework; for example, for a new program on an unfamiliar procedure, MSDSs and engineering data should be examined. Other needs assessment techniques include a study of company accident and injury records, observations of employees on the job, and the examination of all pertinent standards and regulations. By focusing on precisely what is needed, this process has the additional advantage of eliminating what is not needed. Unnecessary training is expensive both in direct costs and because it tends to detract from the impact of the necessary training.
>
> **Identifying goals and objectives.** Clear and measurable instructional objectives will tell employers and employees exactly what must be achieved by the end of training. This includes what employees will be able to do, to do better, or to stop doing. The objectives should define how competence will be demonstrated after training, under what workplace conditions, and to what level of performance.

Detailed statements of objectives will allow other qualified persons to recognize whether the desired knowledge or skills have been achieved, that is, to evaluate the effectiveness of training.

Developing learning activities. The fourth step is identifying and sequencing learning activities, in such a way that they simulate real life as closely as possible. It involves an analysis of what resources are available, such as a training room and audiovisual equipment; what type of learning activity is appropriate for the defined objectives; and what constraints, such as language problems, may be anticipated in the trainees. Whatever methods and media are selected, the learning activities should be developed in such a way that employees can clearly demonstrate that they have acquired the target skills or knowledge.

Conducting the training. This step, usually performed as the first and only training activity, should follow the completion of all previous steps. It is a positive approach that involves explaining the objectives, relating to the trainees' skills and interests, overviewing the main training points, encouraging participation and questions, and giving employees a chance to demonstrate their new knowledge and skills.

Evaluating program effectiveness. The training process does not stop with the delivery of training; the industrial hygienist needs to know if it has been effective. An evaluation plan, drawn up at the same time as the instructional objectives are written, will yield information on whether those objectives have been achieved. Methods of evaluation range from short-range observations of immediate retention to the use of long-range measures of changes in illness or injury rates, clinic visits, downtime and other indices.

Improving the program. Evaluation results will reveal deficiencies that can be remedied in a continuous process of improving and upgrading the training program. This step closes the loop; evaluation data point to new and refresher training needs.

The *Guidelines* close with some advice on how to prioritize candidate groups of employees for training by identifying employee populations at high levels of occupational risk. Some variables identified by research are listed as selection tools; these variables are factors that relate to a disproportionate share of worksite illnesses and injuries on the part of employees. Further tools recommended for determining training content are Job Safety Analysis (see Chapter 25, The Safety Professional) and information found in the MSDS.

Finally, an important step omitted from the *Guidelines* is documentation of training. Documentation is essential for tracking the training status of each employee, and also establishes a record for regulatory compliance reasons and for possible protection from legal liability.

The model outlined in the *Voluntary Training Guidelines* is sound and helpful. However, most industrial hygienists will need further assistance to put "flesh on the bones." In addition to assistance that is offered by many trade associations, insurance companies, and governmental agencies, there are some standard references that give practical help (see Bibliography).

EVALUATING THE INDUSTRIAL HYGIENE PROGRAM

The effectiveness of the industrial hygiene program must be periodically evaluated. The evaluation serves two purposes:

1. determines the extent to which goals have been attained as a result of program activities.
2. generates information useful for the continued success of the program.

While the goal of any industrial hygiene program is the prevention of occupational disease, data on the incidence rates of occupational disease cannot serve as the only measure of performance. Incidence rates frequently reflect only acute effects (such as exposure to an irritating chemical or acid burns) and neglect chronic effects. Occupational cancer, for example, may not occur for 20 years after exposure to a carcinogen.

Each program element, therefore, must be evaluated. Table 28-A, which was written by Frederick M. Toca, PhD (see Bibliography) summarizes suggested measurement criteria for each program element. Hygienists must develop criteria which fit their particular program elements and give a true indication of program efficiency.

COORDINATING WITH OTHER ORGANIZATIONS

The success of safety and industrial hygiene programs requires the cooperation of many organizations. Medical, engineering, safety, and purchasing organizations, the supervisor and the individual employee are all needed for a successful program.

Medical organizations

Medical programs can range from the elaborate to the bare minimum required by OSHA. One establishment may have a full-time staff of physicians, nurses, and technicians, housed in a model dispensary; another may have only the required first aid kit with an adequately trained person to render first aid.

Modern occupational health programs, regardless of size, ideally are composed of elements and services designed to maintain the health of the work force, to prevent or control occupational and nonoccupational diseases and accidents, as well as to prevent and lower disability and the resulting lost time. They should provide for the following elements:

- health examinations
- diagnosis and treatment
- medical records
- health education and counseling.

Supervision of the health status of workers by qualified medical personnel is essential if any occupational health program is to obtain maximum benefits for both employer and employee. Therefore, a program of physical examinations should be established. The examining physician should discuss all significant findings with the worker while emphasizing the importance of obtaining adequate personal medical care. A transcript of the data could be supplied to the worker's personal physician with the consent of the employee. Certainly, the confidential character of health examination records must be observed. The OSHA Standard, Access to Medical Records, specifies the regulatory means employees or their representatives can use to obtain copies of their medical records.

Management must be informed of a potentially harmful work

Occupational Health Guidelines for Carbon Monoxide

Introduction

This guideline is intended as a source of information for employees, employers, physicians, industrial hygienists, and other occupational health professionals who may have a need for such information. It does not attempt to present all data; rather, it presents pertinent information and data in summary form.

Substance identification

- Formula: CO
- Synonyms: Monoxide
- Appearance and odor: Colorless, odorless gas

Permissible Exposure Limit (PEL)

The current OSHA standard for carbon monoxide is 50 parts of carbon monoxide per million parts of air (ppm) averaged over an 8-hour workshift. This may also be expressed as 55 milligrams of carbon monoxide per cubic meter of air (mg/m^3). The NIOSH has recommended that the PEL be reduced to 35 ppm averaged over a workshift of up to 10 hours per day, 40 hours per week, with a ceiling of 200 ppm. The *NIOSH Criteria Document for Carbon Monoxide* should be consulted for more detailed information.

Health hazard information

- **Routes of exposure.** Carbon monoxide can affect the body if it is inhaled or if liquid carbon monoxide comes in contact with the eyes or skin.
- **Effects of overexposure.** Exposure to carbon monoxide decreases the ability of the blood to carry oxygen to the tissues. Inhalation of carbon monoxide may cause headache, nausea, dizziness, weakness, rapid breathing, unconsciousness, and death. High concentrations may be rapidly fatal without producing significant warning symptoms. Exposure to this gas may aggravate heart disease and artery disease and may cause chest pain in those with preexisting heart disease. Pregnant women are more susceptible to the effects of carbon monoxide exposure. The effects are also more severe in people who are working in places where the temperature is high or at altitudes above 2,000 feet. Skin exposure to liquid carbon monoxide may cause frostbite-type burns.
- **Reporting signs and symptoms.** A physician should be contacted if anyone develops any signs or symptoms and suspects that they are caused by exposure to carbon monoxide.
- **Recommended medical surveillance.** The following medical procedures should be made available to each employee who is exposed to carbon monoxide at potentially hazardous levels:

1. *Initial medical examination:*
   —A complete history and physical examination: The purpose is to detect preexisting conditions that might place the exposed employee at increased risk, and to establish a baseline for future health monitoring.
   Persons with a history of coronary heart disease, anemia, pulmonary heart disease, cerebrovascular disease, thyrotoxicosis (condition due to excessive thyroid hormone), and smokers would be expected to be at increased risk from exposure. Pregnant women have an increased sensitivity to the effects of carbon monoxide. Examination of the cardiovascular system, the pulmonary system, the blood, the central nervous system should be stressed.
   —A complete blood count: Carbon monoxide affects the ability of the blood to carry oxygen. A complete blood count should be performed including a red cell count, a white cell count, a differential count of a stained smear, as well as hemoglobin and hematocrit.
2. *Periodic medical examination:* The aforementioned medical examinations should be repeated on an annual basis, with the exception that a carboxyhemoglobin determination should be performed at any time overexposure is suspected or signs or symptoms of toxicity occur.

- **Summary of toxicology.** Carbon monoxide gas causes tissue hypoxia by preventing the blood from carrying sufficient oxygen. Carbon monoxide combines reversibly with the oxygen-carrying sites on the hemoglobin molecule with an affinity ranging from 210–240 times greater than that of oxygen; the carboxyhemoglobin thus formed is unavailable to carry oxygen. In addition, carboxyhemoglobin interferes with the release of oxygen carried by unaltered hemoglobin. With exposure to high concentrations, such as 4,000 ppm and above, transient weakness and dizziness may be the only premonitory warnings before coma supervenes; the most common early aftermath of severe intoxication is cerebral edema. Exposure to concentrations of 500–1,000 ppm causes the development of headache, tachypnea (rapid breathing), nausea, weakness, dizziness, mental confusion, and in some instances hallucinations, and may result in brain damage. The affected person is commonly cyanotic. Concentrations as low as 50 ppm result in blood carbon monoxide hemoglobin levels up to 10 percent in an 8-hour day. This greatly increases the risk of angina pectoris and coronary infarctions by decreasing the oxygen supply in the blood and also in the myoglobin of the heart muscle. These effects are aggravated by heavy work, high ambient temperatures, and high altitudes. Pregnant women are especially susceptible to the effects of increased carbon monoxide levels. Smoking also increases the risk: cigarette smoke contains 4 percent carbon monoxide, which results in 5.9 percent carbon monoxide hemoglobin level if a pack a day is smoked. The blood of persons not exposed to carbon monoxide contains about 1 percent carbon monoxide, probably as a result of normal heme metabolism. The diagnosis of carbon monoxide intoxication depends primarily on the demonstration of significantly increased carboxyhemoglobin in the blood. Levels over 60 percent are usually fatal; 40 percent is associated with collapse and syncope; above 25 percent there may be electrocardiographic evidence of a depression of the S-T segment; between 15 percent and 25 percent there may be headache and nausea. The reaction to a given blood level of carboxyhemoglobin is extremely variable: some persons may be in coma with a carboxyhemoglobin level of 38 percent while others may maintain an apparently clear sensorium with levels as high as 55 percent.

Chemical and physical properties

- **Physical data**
  molecular weight: 28
  boiling point (760 mmHg): −191.5 C (−313 F)
  specify gravity (water =1): 0.79 (liquid at boiling point)
  vapor density (air = 1 at boiling point of carbon monoxide): 0.97
  melting point: −199 C (−326 F)
  vapor pressure at 20 C (68 F): Greater than 1 atmosphere
  solubility in water, g/100 g water at 20 C (68 F): 0.004
  evaporation rate (butyl acetate = 1): Not applicable.
- **Reactivity**
  conditions contributing to instability: elevated temperatures may cause cylinders to explode.
  incompatibilities: contact with strong oxidizers may cause fires and explosions.
  hazardous decomposition products: none
  special precautions: none.
- **Flammability**
  flash point: not applicable
  autoignition temperature: 609 C (1128 F)
  flammable limits in air, percent by volume: lower: 12.5; upper: 74
  extinguishant: dry chemical—if flow of gas cannot be stopped, let fire burn.
- **Warning properties**
  odor threshold: the *AIHA Hygienic Guide* points out that carbon monoxide is odorless.
  Eye irritation level: Grant states that carbon monoxide is a nonirritating gas.

Figure 28-7. NIOSH guidelines for carbon monoxide. (Continued.)

Occupational Health Guidelines for Carbon Monoxide

Evaluation of warning properties: carbon monoxide is an odorless, nonirritating gas; it has no warning properties.

Monitoring and measurement procedures

- **8-hour exposure evaluation**
  Measurements to determine employee exposure are best taken so that the average 8-hour exposure is based on a single 8-hour sample or on two 4-hour samples. Several short-time interval samples (up to 30 minutes) may also be used to determine the average exposure level. Air samples should be taken in the employee's breathing zone (air that would most nearly represent that inhaled by the employee).
- **Ceiling evaluation**
  Measurements to determine employee ceiling exposure are best taken during periods of maximum expected airborne carbon monoxide concentrations. Each measurement should consist of a 15-minute sample or series of consecutive samples totaling 15 minutes in the employee's breathing zone (air that would most nearly represent that inhaled by the employee). A minimum of three measurements should be taken on one workshift and the highest of all measurements taken is an estimate of the employee's exposure.
- **Method**
  Sampling and analysis may be performed by collection of carbon monoxide vapors using an adsorption tube with a subsequent chemical analysis of the adsorption tube. Also, detector tubes or other direct-reading devices calibrated to measure carbon monoxide may be used (see Chapter 19, Direct-Reading Gas and Vapor Instruments).

Respirators

- Good industrial hygiene practices recommend that engineering controls be used to reduce environmental concentrations to the PEL. However, there are some exceptions when respirators may be used to control exposure. Respirators may be used when engineering and work practice controls are not technically feasible, when such controls are in the process of being installed, or when they fail and need to be supplemented. Respirators may also be used for operations which require entry into tanks or closed vessels, and in emergency situations. If the use of respirators is necessary, the only respirators permitted are those that have been approved by the Mine Safety and Health Administration (MSHA) (formerly Mining Enforcement and Safety Administration) or by NIOSH.
- In addition to respirator selection, a complete respiratory protection program should be instituted, which includes regular training, maintenance, inspection, cleaning, and evaluation.

Personal protective equipment

- Employees should be provided with and required to use impervious clothing, gloves, face shields (8-inch minimum), and other appropriate protective clothing necessary to prevent the skin from becoming frozen from contact with liquid carbon monoxide or from contact with vessels containing liquid carbon monoxide.
- Any clothing which becomes wet with liquid carbon monoxide should be removed immediately and not reworn until the carbon monoxide has evaporated.
- Employees should be provided with and required to use splashproof safety goggles where liquid carbon monoxide may contact the eyes.

Common operations and controls

The following list includes some common operations in which exposure to carbon monoxide may occur and control methods which may be effective in each case:

Operation	Controls
Liberation from emissions in enclosed places from exhaust fumes of internal combustion engines; from metallurgic industry and foundries; from chemical industry for synthesis and emission as results of incomplete combustion	Local exhaust ventilation; respiratory protective device
Liberation during acetylene welding; from enclosed areas as mines or tunnels; from fire-damp explosions	Local exhaust ventilation; respiratory protective device
Liberation from industrial heating	Local exhaust ventilation

Emergency first aid procedures

In the event of an emergency, institute first aid procedures and send for first aid or medical assistance.

- **Breathing**
  If a person breathes in large amounts of carbon monoxide, move the exposed person to fresh air at once. If breathing has stopped, perform artificial respiration. Keep the affected person warm and at rest. Get medical attention as soon as possible.
- **Rescue**
  Move the affected person from the hazardous exposure. If the exposed person has been overcome, notify someone else and put into effect the established emergency rescue procedures. Do not become a casualty. Understand the facility's emergency rescue procedures and know the locations of rescue equipment before the need arises.

Leak procedures

- Persons not wearing protective equipment and clothing should be restricted from areas of leaks or releases until cleanup has been completed.
- If carbon monoxide is leaked or released in hazardous concentrations, the following steps should be taken:
  1. Ventilate area of leak or release to disperse gas.
  2. Stop flow of gas. If source of leak is a cylinder and the leak cannot be stopped inplace, remove the leaking cylinder to a safe place in the open air, and repair the leak or allow the cylinder to empty.

Respiratory protection for carbon monoxide*

Condition	Minimum Respiratory Protection Required above 50 PPM
Gas Concentration	
500 ppm or less	Any supplied-air respirator Any self-contained breathing apparatus
1,500 ppm or less	Any supplied-air respirator with a full facepiece, helmet, or hood Any self-contained breathing apparatus with full facepiece A Type C supplied-air respirator operated in pressure-demand or other positive-pressure or continuous-flow mode
Greater than 1,500 ppm or entry and escape from unknown concentrations	Self-contained breathing apparatus with a full facepiece operated in pressure-demand or other positive-pressure mode A combination respirator which includes a Type C supplied-air respirator with a full facepiece operated in pressure-demand or other positive pressure or continuous-flow mode and an auxiliary self-contained breathing apparatus operated in pressure-demand or other positive-pressure mode
Fire fighting	Self-contained breathing apparatus with a full facepiece operated in pressure-demand or other positive-pressure mode
Escape	Any gas mask providing protection against carbon monoxide Any escape self-contained breathing apparatus

*Only NIOSH-approved or MSHA-approved equipment should be used.

Figure 28-7. (Concluded.)

Table 28-A. Summary of Criteria v. Activities

Program Element Activity	*Activity*	*Measurement Criteria*	*Goal*
Policy	Write, prepare/present for management acceptance	Is policy complete? Is policy understood and supported by management/employees? Does policy carry authority needed for implementation?	An accepted and working policy that clearly states the scope, responsibilities and authority of the program.
Education	New employee orientation Periodic information and education sessions Written safety and health guidelines Posting of dangerous areas Labeling of materials handled by employees	No. educational materials produced and distributed Increase in employee knowledge of safety and health issues Employee avoidance of hazards	Increased employee awareness of health and safety in the workplace.
Health hazard recognition	Plant survey Chemical inventory Process and equipment review Health hazard review procedures Process change review procedures	No. surveys Completion and procedure update Procedures and staff in place for review, etc.	Identify all present and potential hazards in the workplace.
Health hazard evaluation	Environmental monitoring (area, personal) Sample analysis Statistical analysis of data Biological monitoring Records of data Establishment of criteria	No. samples collected No. analyses performed Statistical significance of sample data Well-documented record-keeping system Established criteria for each stress	Measure and quantitatively evaluate stresses and hazards, determine their impact upon the work environment.
Health hazard control	Design and/or recommend administrative and engineering controls Procedural mechanism for implementing controls Procedural mechanism for including controls as a part of planning for new processes and changes in existing processes Administrative review of rejected procedures	Controls implemented and working Administrative procedures in place	Control or reduce to the lowest level all potential workplace hazards.
OSHA compliance	Review all present and future regulations, standards Determine level of compliance obtained via compliance inspections	No violations present Program positioned to comply with regulations	Complete compliance with all laws, regulations, standards, etc.

(Printed with permission from Toca, F.M. Program evaluation: Industrial Hygiene. *American Industrial Hygiene Association Journal* 42 [March 1981]:213-216.)

environment detected through examination of persons subjected to it.

Physicians have different opinions regarding the relative values of various test procedures, based on their own training and experience. Therefore, the scope of physical examinations should be determined by the occupational physician who is familiar with the operations and regulations involved.

In summary, the responsibilities of the medical organization in the industrial hygiene program are as follows:

- to cooperate in developing adequate, effective measures to prevent exposure to harmful agents
- to examine periodically those employees who are working with or exposed to hazardous agents or materials
- to restrict employees from further potential exposure, whenever warranted by findings of such periodic examinations.

In order to do the job effectively, the physician must be fully cognizant of what is made at the plant, how it is made, what raw materials are used, the potential and actual health hazards associated with these manufacturing processes, and the physical requirements of the various types of jobs. The physician must have this information to adequately carry out the preplacement health appraisals, periodic health examinations, and the health education programs.

Engineering organizations

The introduction of new plant processes and operations begins with the engineering organization. For this reason, the engineering group plays a most important role in the control of occupational health hazards. Its responsibilities are as follows:

- to plan all operations using established engineering procedures to prevent unnecessary exposure to harmful environmental factors or stresses. The industrial hygienist should be consulted at the planning stage
- to notify the medical, industrial hygiene, and safety organi-

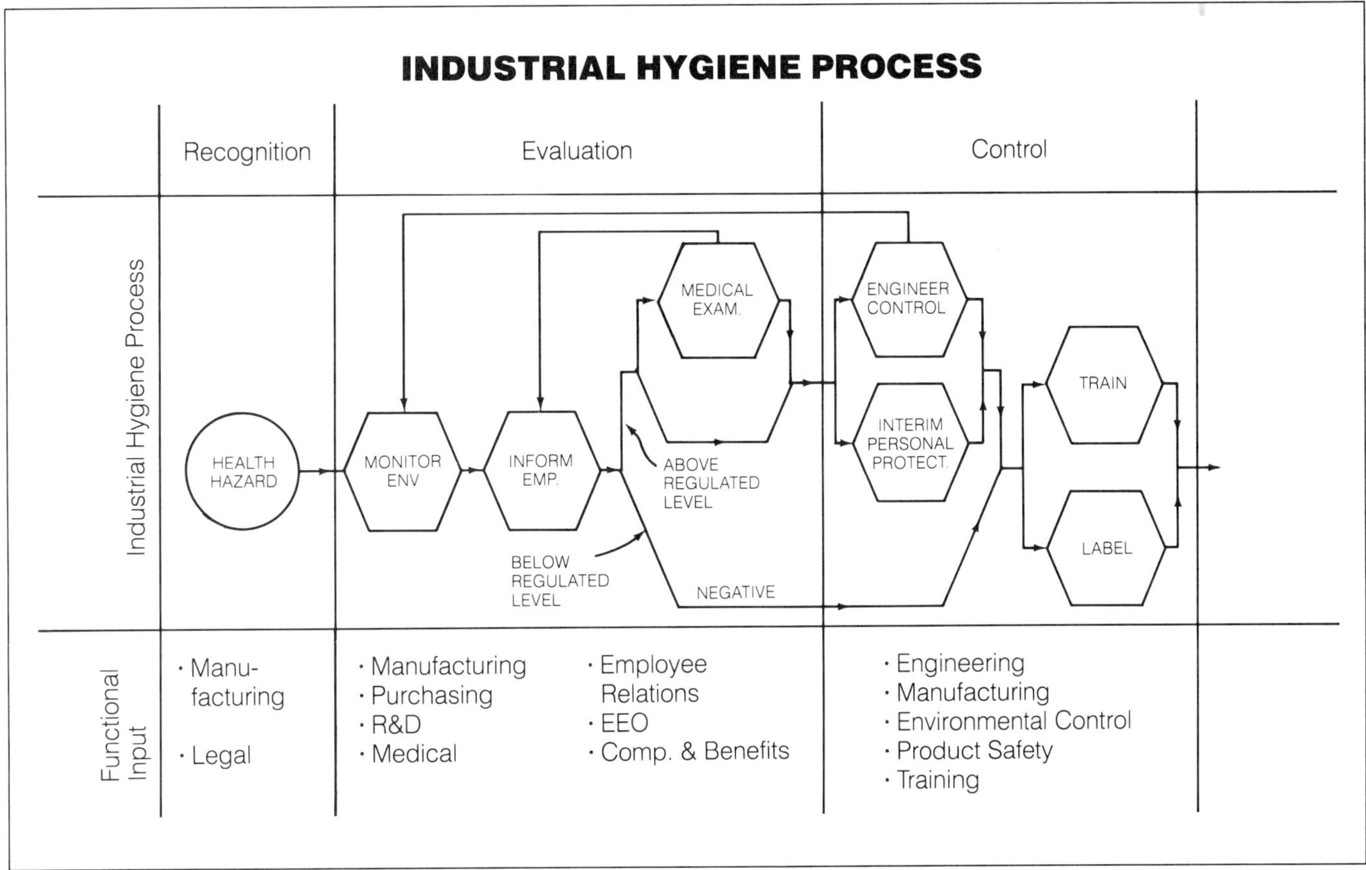

Figure 28-8. (Adapted with permission from Bridge, D. P. Developing and implementing an industrial hygiene and safety program in industry. *American Industrial Hygiene Association Journal* 40[1979]:255-263)

zations, whenever the company is planning to introduce new operations or processes
- to request an industrial hygiene survey of new installations before permitting shop personnel to operate the equipment.

Safety organizations

The safety organization plays an integral part in the overall industrial hygiene program. Its responsibilities are as follows:
- to conduct an effective safety program by coordinating the educational, engineering, supervisory, and enforcement activities related to the safety program
- to provide educational material for any safety training program for personnel working with hazardous materials
- to assist the supervisor in teaching employees safety rules, regulations, and procedures
- to conduct safety surveys to ensure that proper practices and procedures are being followed
- to recommend changes in safety rules, regulations, and procedures to keep pace with technological advancements.

Purchasing organizations

The purchasing organization's responsibilities are to ensure that only equipment and material approved by the industrial hygiene, safety, medical, and other appropriate departments are purchased for use in the company. Purchasing may also be made responsible for ensuring that MSDSs are provided along with all chemical purchases.

Supervisor's responsibilities

Each supervisor is responsible (1) for maintaining safe working conditions within his or her sphere of responsibility and (2) for directly implementing the safety program. Their responsibilities are as follows:
- to maintain a work environment that assures the maximum safety for employees
- to make certain that the employees (newly hired or transferred) have been examined by the medical organization and approved for the job before being assigned to work
- to instruct employees periodically on precautions, procedures, and practices to be followed to minimize exposure to harmful agents
- to make sure that meticulous housekeeping practices are developed and used at all times
- to make sure that food, candy, beverages, or other edibles are not stored or consumed in work areas where toxic materials are being used
- to inform promptly the engineering, industrial hygiene, and safety organizations of any operation or condition, which appears to present a hazard to employees
- to inform the medical organization promptly in case of accidental exposure to harmful agents, and to send the employee(s) involved to the medical department for examination
- to furnish employees with the proper personal protective equipment, instruct them in its proper use, and enforce the wearing of such equipment

- to observe all work restrictions imposed by the medical organization
- to consult with safety, industrial hygiene, engineering, and medical personnel for aid in fulfilling his or her responsibilities
- to administer appropriate action when health and safety rules are violated
- to teach health and safety procedures by example.

Employee's responsibilities

Each employee is responsible for contributing his or her part towards the success of the industrial hygiene program. This includes the following:

- to notify the supervisor immediately when conditions or practices can cause personal injury or property damage
- to observe all safety rules, and to make maximum use of all prescribed personal protective equipment, and to follow practices and procedures established to maintain health and safety
- to report immediately to the supervisor an accidental exposure to harmful agents
- to practice good habits of personal hygiene and housekeeping.

Several other organizations can be involved to some degree with the industrial hygiene program. For example, the legal department can become involved in OSHA regulatory proceedings or labor relation matters as they relate to safety and health. Personnel functions such as Equal Employment Opportunities (EEO) can become involved when employees are exposed to teratogenic chemicals. Compensation and benefit functions can become involved with workers' compensation.

Communication among all concerned groups is essential for a smoothly functioning industrial hygiene program. Figure 28-8 graphically demonstrates how each group is involved in the integrated response to the recognition, evaluation, and control of occupational health hazards.

SUMMARY

The industrial hygienist is trained to interrelate a wide variety of scientific disciplines needed to bring about a healthful working environment. The industrial hygienist is skilled in guiding health-related activities in relation to complex legal requirements aimed at protecting workers, consumers, and the environment.

The form and content of industrial hygiene programs vary throughout industry. However, every program covers four basic points—the recognition, evaluation, and control of health hazards and employee training. In reaching these objectives, the program must depend upon many different managerial groups. Communication and coordination among all concerned is essential for the success of the program.

BIBLIOGRAPHY

Bridge, D.P. Developing and implementing an industrial hygiene and safety program in industry. *American Industrial Hygiene Association Journal* 40(1979):255-263.

Cralley, L. J., and Cralley, L. V., eds. *Patty's Industrial Hygiene and Toxicology.* Vol. 3: *Theory and Rationale of Industrial Hygiene Practice:* 3B *Biological Responses,* 2nd ed. New York: John Wiley & Sons, Inc., 1985.

National Institute for Occupational Safety and Health (NIOSH). *The Industrial Environment—Its Evaluation and Control.* Washington, DC: U.S. Government Printing Office, 1973.

Toca, F. M. Program evaluation: Industrial hygiene. *American Industrial Hygiene Association Journal* 42 (March 1981): 213-216.

29

Computerizing an Industrial Hygiene Program

by Adrienne A. Whyte, PhD

INFORMATION SYSTEMS

THE COMPUTER HAS COME OF AGE in health and safety departments. Once the exclusive domain of personnel trained in data processing, computers have been "demystified," especially since the widespread use of personal computers. Not too many years ago, health and safety professionals had to expend considerable effort to justify the need for computer systems to support their organizations. Today, computer systems are an integral part of the business environment, and health and safety professionals are using them to help prevent work-related deaths, injuries, and illnesses; manage risks; and control losses (Figure 29-1).

Computers are used by safety and health professionals to:

- compile, store, analyze, and produce corporate summary reports of occupational injuries and illnesses—reports that would otherwise be time-consuming and expensive to manually prepare each month, quarter, or year.
- compile, store, analyze, and produce summary reports of workplace sampling results for chemical, physical, and biological hazards.
- produce follow-up reports pointing to hazards that were identified during inspections, but are not yet abated.
- store and correlate employee health records with data on workplace exposures to chemicals.
- maintain records needed to document regulatory compliance, such as training required by the OSHA Hazard Communications Standard.
- Using computers, health and safety professionals can access growing numbers of data bases, which contain volumes of information needed by industrial, health, safety, and environmental professionals. Within minutes, a user of a data base can obtain lists of needed references, abstracts of specific articles, or properties of a substance used in a manufacturing process, to name only some of the information available on-line. A detailed listing of currently available data bases is included in Appendix A, Sources of Information.

Computers are tools to help you obtain the information necessary to make decisions; computers make your job easier, your efforts more productive, and help protect your employees and organization.

The importance of information management to occupational health and safety programs did not become critical until the 1970s. The passage of the Occupational Safety and Health Act (OSHAct) in 1970, followed by passage of environmental laws such as the Toxic Substances Control Act and the Resource Conservation and Recovery Act, introduced record-keeping requirements that strained the resources of some occupational health and safety programs. In the late 1970s and early 1980s, right-to-know laws and regulations added to regulatory requirements with complex procedural, reporting, and record-keeping requirements for industry and, by executive order, for government agencies. As safety, industrial hygiene, and medical surveillance requirements increased, so did the requirements for documentation, reporting, and analysis.

These factors, combined with the need for better internal records management, led many organizations to look for ways to improve their record-keeping systems. Most of them chose a common solution—computerized occupational health and safety information systems. Today, it is known that computers can solve many of the information management problems faced by health and safety professionals.

Figure 29-1. The microcomputer already is in use by many safety and health professionals for word processing and data base management.

ADVANTAGES OF AUTOMATED INFORMATION SYSTEMS

Improved availability of data

A well-designed computer system can store a huge amount of data and retrieve it easily and quickly for reporting and analysis. For example, an employer might want to know if any employees exposed to particular solvents have had abnormal hematological results. With a manual system, a time-consuming search through industrial hygiene and medical files for this information would be necessary. With an interactive computer system, this search should be completed within minutes. In addition, interactive systems can be used to provide emergency information, such as first aid and clean-up procedures for chemical spills.

Elimination of duplication

The time spent maintaining duplicate records within an organization can be reduced with an automated system. For example, with a manual system, it could be necessary to keep records of occupational illnesses and injuries in four offices: health and safety, medical, human resources, and area supervision. With a shared information system, one comprehensive file on each illness or injury can be kept in the computer.

Improved communications

With manual systems, important data can be buried in unfinished piles of paperwork. Often, statistics generated by one of the company's offices do not match those generated by another. Again take occupational illness and injuries as an example; illness and injury reports from one facility may not reach other facilities, decreasing the chance that all groups within the organization can benefit from the "lesson learned." A good data base management system makes data available to all authorized users. In addition, electronic mail systems can facilitate communication within and between facilities.

Data standardization and accuracy

Standardizing the data input and the way in which it is organized is particularly important in organizations with many system users. When the same data is being input in the same way by each user, the analysis and decision-making capabilities of the data base are enlarged. For instance, accident and incident rates can be analyzed for all locations at the same time, rather than analyzing each location separately and then standardizing and entering the collective data for final analysis.

Accuracy of data can be vastly improved by customized entry screens, editing routines, and checks for the completeness of records. For instance, a computer system can alert the medical department that part of an OSHA-required physical examination was not conducted.

Improved analytical capabilities

With most manual systems, compilation and analysis of data from various records consumes professional time that could be better spent on program management activities. A good automated system performs analytical tasks quickly and formats the results for reports to management. For example, the ability to correlate workplace exposure and medical records leads to improved medical surveillance. Trends in employee health can become apparent in time to act to prevent injuries and illnesses. Analyses not practical with manual systems routinely can be done using an automated system.

Cost savings

This advantage is probably the hardest to document, but automated systems can save organizations money above and beyond the costs of system development and operation. Increased employee productivity, decreased incident rates with attendant savings, and more effective program management can lead to cost savings.

HEALTH AND SAFETY FUNCTIONS

Today's occupational health and safety systems support single functions, such as management of training records, or a comprehensive set of functions. This section is an overview of the system functions and tools in use by health and safety professionals today.

Incident management

Many organizations are using automated systems to store, analyze, and report on incident data. Data for all types of incidents is entered, such as occupational injury, property damage, near miss, and transportation. This data usually parallels the organization's first report and incident investigation forms. The system can produce: (1) individual incident reports; (2) summary reports that categorize incidents by location, type, rate, severity, loss, and other factors; (3) OSHA 200 logs; and (4) analytical reports that pinpoint major causes and types of accidents within organizational subsets. For example, an organization might determine through use of its automated system that a disproportionate number of accidents is occurring in a particular operation, and then arrange for a health and safety inspection and evaluation of the operation.

Workplace conditions

Many systems manage safety, industrial hygiene, health physics, environmental, and other sampling, audit, and inspection data capable of describing and quantifying workplace conditions.

Industrial hygiene sampling results can be entered and used to produce exposure profiles for each workplace. Also, if health and safety inspection findings, recommendations, target dates for completion, and completion dates are automated, a system can produce follow-up reports highlighting identified deficiencies that have not been corrected.

In comprehensive systems, workplace data can be correlated with employee health data to support health surveillance programs.

Environmental agents

Automated systems greatly facilitate environmental agent monitoring, including the preparation and maintenance of facility and corporate inventories of environmental agents, toxicology data, and material safety information on toxic agents within the workplace. These agents include chemical, biological, radiological, and physical hazards. Inventories of agents can be maintained by department, area, process, job, or any combination of these.

Right-to-know and hazard communications requirements make it mandatory that organizations identify agents in each workplace, train employees in their safe use, maintain material safety data sheets, and promptly respond to requests for information. Automated systems can facilitate compliance by maintaining agent inventories, training records, material safety data sheets, and information for right-to-know requests.

Protective measures

Computer systems can maintain information on measures taken to protect employee health and safety through: (1) engineering controls such as hoods for ventilation; (2) administrative controls for health hazards; (3) personal protective equipment assignment and fit-testing; and (4) individual employee training and group training programs. These functions allow an organization to maintain the information needed to facilitate regulatory compliance and document the measures taken to protect employees.

Employee health

Monitoring employee health is a key function of comprehensive occupational health and safety systems. The protection of employees is the major goal of health and safety programs. This function primarily is used by occupational medicine specialists to manage and store records related to employee health surveillance. These records, shown in Table 29-A, result from both scheduled and unscheduled events. Scheduled health events include all physical examinations (preplacement, periodic, certification, and termination examinations and biological monitoring). Unscheduled health events are those that cannot be anticipated. It is necessary to maintain a record of both scheduled and unscheduled health events to compile a historical profile of an employee's health.

As with other system functions, individual components of employee health programs can be automated. Hearing conservation program records are a frequent candidate. Noise monitoring, audiogram, and employee history data can be entered and stored. These data can be used to produce monitoring schedules, threshold shift evaluations, notices to employees, and summary reports for program management.

Employee demographics and job histories

In addition to maintaining health records, this function (or an independent but related function) must also maintain data that identify and describe individual employees and maintain a history of job assignments and locations. The identification of employees is usually accomplished through use of one or more of the following—name, social security number, employee number, and badge number. These normally are found in personnel information systems and can be downloaded to a health and safety system.

Job assignments and location tell where an employee works and what he or she does. This data is a necessary bridge between the employee health records and the workplace and environmental records. They allow health professionals to correlate medical records with workplace records and are an integral part of comprehensive occupational health and safety surveillance systems.

Table 29-A. Occupational Medicine Records

- Physical Examination Records
  - Personal and Family Medical History
  - Occupational History
  - Physical Measurements (such as vision tests and blood pressure levels)
  - Blood and Urine Laboratory Tests
  - Clinical Evaluations and Diagnoses
  - Audiometry Results
  - Spirometry Results
  - Electrocardiogram Results
  - X-Ray Results
  - Other or Special Test Results
- Immunization Records
- Clinic Visit Records
- Occupational Injury and Illness Records
- Disability Data
- Workers' Compensation Records
- Biological Monitoring Results
- Sickness and Absence Records
- Mortality Records

REGULATORY, ADMINISTRATIVE, AND ACTION ITEMS

Another function can provide information on regulatory requirements, including OSHA Permissible Exposure Limits (PELs) or action levels, or internally adopted standards, or such advisory guidelines as the American Conference of Governmental Industrial Hygienists (ACGIH) Threshold Limit Values (TLVs). This function also can be used to track ongoing events or problems requiring follow-up, for example, hazard abatement plans and schedules.

Scheduling

The computerized system can be used to schedule physical examinations, workplace monitoring or inspections, fit testing of personal protective equipment, training, or any other event documented in an occupational health and safety system. The schedules can be generated by system logic, or they can be generated by users and maintained by the system.

Standard and ad hoc reporting

Reporting is the most important function of an automated system. Standard reports are predefined and routinely used, and they are usually preformatted. They can be prepared with one or few instructions (often a choice of a menu of reports). Examples of typical standard reports are audit reports and lists of occupational illnesses and injuries by workplace. All systems should have a standard reporting function.

Ad hoc reports cannot be predefined or preformatted, but are designed by users for special analyses and reporting. There are several methodologies for producing ad hoc reports. Two of them, the question-and-answer and fill-in-the-blanks approaches, prompt the system user to name the variables for study, indicate how they are to be manipulated, and specify the format for the report. Another method requires the use of a simple computer language for queries. The method chosen should be easy to learn and simple to use. Adequate query capability is one of the most important features of an occupational health and safety system, since nothing is more frustrating than to find out after years of data entry that some data cannot be retrieved in the manner required.

Statistical analysis

Statistical analyses are necessary for many health and safety studies. Commercial statistical software packages are available for this function, but they often are separate from the occupational health and safety software. These are available on personal computers and mini and mainframe computers. It is usually an easy task to pass data from newer data base management systems to these packages for analysis. In addition, many packages are available for preparation of graphics from health and safety data systems.

DEVELOPMENT AND IMPLEMENTATION OF A COMPUTER SYSTEM

There are five major steps to the successful development and implementation of an automated system:

1. Understand the operations and needs
2. Identify and evaluate software and hardware meeting those needs
3. Purchase and customize a system or develop a system
4. Implement the system
5. Evaluate the system.

Understanding needs

When considering buying a computer system, conduct a requirements study, which is a careful assessment of the needs that a system will fulfill. The requirements study must say *why* a system is needed, based on current and foreseen conditions. It must say *what* system features will satisfy the needs. It must also say *how* the system is to be constructed.

To do this, the requirements analysis must address three subjects:

1. *Needs analysis.* The reasons *why* the system should be created. How the system will solve a problem. Why certain technical, operational, and economic feasibilities are the criteria for the system.
2. *Functional specification.* A description of *what* the system will be in terms of the functions it must accomplish. How the system will be in terms of the functions it must accomplish. How the system will fulfill its role. Why certain design components should be considered.

To determine the functional specifications you need to look at the operations of the organization, record-keeping and reporting requirements, and information management problems. Then estimate the time spent on each activity, quantify the activities, make a "wish list," and answer questions such as:

- How much of what types of work is performed?
- How many employees are covered by the health and safety office?
- How many accident cases or employee-health related problems occur per year?
- How many on-site chemicals are tracked for employee exposure?
- How many Material Safety Data Sheets (MSDSs) are maintained?
- How is employee health and safety training tracked?
- How are Permissible Exposure Limits measured and tracked?
- How are safety equipment distribution and maintenance tracked?
- How are employee medical records, such as audiograms tracked?

3. *Design constraints.* A summary of conditions specifying *how* the system is to be constructed and implemented. What will compose the system. Why particular designs are feasible. Design constraints could include:

- Computer hardware—If the department already has IBM PCs the system must operate on them.
- Computer software standards—Does your company's data processing group support only certain data base management systems?
- Personnel—Must the system not require additional staff for data entry or operation?

Each of these subjects should be fully documented during the requirements analysis. Collectively, these analysis components should provide all of the information needed to design or evaluate a system.

Identify and evaluate systems

After the requirements and specifications for a system have been defined, there should be enough information to identify the computer systems capable of meeting the needs of the organization. There are two alternatives: (1) purchase commercially available software and customize it, if necessary, or (2) develop a system.

Since 1977, occupational health and safety software has been commercially available. Today, many good packages are available, and any group planning to implement a system should find out if any of these packages meet their requirements. The advantage of purchasing software is the time and expense that can be saved. The disadvantages are that modifications and work practice adaptations are sometimes necessary when generalized packages are used. So carefully study the software features before purchasing. See the Bibliography and Appendix A, Sources of Help, for additional information.

The Consultant Series: RCRA Compliance Package, Hazard Communication System Software, Ergonomic Lifting Calculator, A Wellness and Health Promotion Package, Accident/Illness Analysis System Software, and *k-Graph* are some software packages published by the National Safety Council that have many applications for health and safety professionals.

The major advantage of developing customized software is the exact fit between the requirements and the system. Little, if any, accommodation should be required by users. Today, new hardware and software technologies make custom software development a viable alternative.

When comparing the available alternatives, remember two cardinal rules. First, the alternative should be judged on the basis of criteria established by the requirements analysis. A commercially available system cannot be evaluated for use in an organization until the requirements for the system have thoroughly been defined. Second, life-cycle costs should be used to compare the costs of the considered alternatives. Life-cycle costs are all costs associated with system design, development, implementation, and operation over the life of the system. The life of a system is the amount of time the software-hardware configuration can be expected to perform without major modification.

Purchase and customization

A software system may have to be modified before it will meet all requirements. Naturally, when evaluating systems, look at the ease with which they can be changed. Once purchased, treat the package as a prototype. Let system users work with it and define its deficiencies. Then make any required changes. There may have to be several rounds of modifications before all users are completely happy with the system.

System development

If developing a system, participate in the design and development process. At this stage, obtain assistance from a systems analyst. The intricacies of system development will depend on the size and complexity of the system and the hardware and software environment in which it will be built. Naturally, a small PC-based system for management of one or more unrelated functions will require less effort and systems expertise than a comprehensive health and safety system.

The first step is to establish the system structure. This activity, often called general design, consists of defining the subparts of the system and the interfaces between them. In the detailed design step that follows, the precise algorithms or system processes and data structures are defined. Detailed design may involve several iterations. Then, system production begins; this involves programming or use of system tools for defining file structures and report formats, testing of individual pieces, integrating various pieces, system testing, documentation, and, finally, the initial performance evaluation.

Many organizations build and implement large systems in phases. This can work well *if* an overall plan for total system design is prepared and followed during the phased development. Too often, groups have built system modules without considering total system requirements and necessary connections between modules, thus leading to project failure.

System implementation

Implementating the new information system should be a joint effort between system developers and users. While there are many activities associated with implementation, the most important one is user training. This training should establish realistic expectations among users about the system's capabilities and requirements. Any changes required in record-keeping procedures and potential transitional difficulties should be thoroughly explained.

User training should not be left to programmers or systems analysts. Ideally, it should be carried out by user representatives. If a system has been purchased, some training will, by necessity, have to be conducted by the supplier. However, a core of system users should be trained by the supplier, and then, if possible, these users should train others.

Training should be conducted as part of the everyday work routine. Naturally, some general training sessions must be held, but the most effective type of training is on-the-job training, with health and safety professionals learning to use terminals for data input and output in the context of their actual work.

Training should be timed so there is no gap between training and actual use of the system. When something entirely new is learned, most people need to practice it to remember it. The longer the gap between training and system use, the less learning retention there will be.

Follow-up to training also is important. Periodically, the system manager should return to the users to determine if their expectations for the system are realistic and if they are using the system correctly. Error rates should be studied. If they are too high, the reasons should be identified and the users retrained. System success depends just as much on follow-up training as it does on initial training.

An effective user's manual is important to user training. During the development process for any information system, system documentation is written. Much of this documentation is for the use of systems analysts and programmers. However, a user's manual is also prepared, and great care must be taken in its preparation. Both the system developers and user representatives must participate in writing the user's manual. It must be simple to understand and direct in its instructions. If it is properly written, it will be a permanent, valued reference for all users. If it is not well prepared, it either will not be used or will promote errors.

Every system should have a manager or data base administrator. This person has ultimate responsibility for the day-to-day operation of the system, including: (1) management of system security; (2) supervision of the data base content; (3) problem solving; (4) coordination of changes to the system; (5) archiving of data; (6) data quality; and (7) planning for future needs and applications. The system manager should know the application well and work closely with users to make sure the system meets user expectations.

Evaluating the system

After an information system has been implemented, it should be evaluated annually. Estimating an information system's value to an occupational health and safety program can be difficult for the same reasons that conducting a cost-benefit analysis is difficult. However, evaluation is an important follow-up, and the following factors should be considered when evaluating the system:

- completeness
- reliability
- user acceptance
- costs
- improved availability of information
- new capabilities.

If problems are identified in any but the last area, corrective steps should be taken. Problems can result from human, hardware, or software elements of the system.

SYSTEM SOFTWARE

Advancements in microcomputer technologies have made personal computers a part of our personal and professional lives. Personal computers are ideal tools for information management because they are easy to use and relatively inexpensive. Many companies are making personal computers available to managers for information management, word processing, spreadsheet manipulation, and electronic mail. They are part of an overall strategy for office automation.

Word processing, one of the most favored software packages because of its ease-of-use and range of office applications, is the ability to create and change text on-screen before printing it out. The text is saved, and then retrieved at a later time for possible editing before reprinting. The computer does indeed become a typewriter, but with additional features that make word processors save about 50 percent of the time it would take a typist to accomplish the same amount of work.

Data base management systems are an integral component of most contemporary occupational health and safety systems. It is important to understand the distinction between applications and data base management software. Today, most applications, including health and safety systems, are written with data base management systems. All data base management systems allow the storage, sorting, and retrieval of information in useful ways. They have tools for building files and reports, and they offer programming languages or commands for building complex programs. A data base management system looks like a generic computer facility until its tools are used to build a specific application.

Applications are programs or sets of programs that provide tailored menus, data entry screens, and reports to support specific functions—like occupational health and safety recordkeeping and analysis. The commercially available packages for health and safety are applications software, and most of them were built with data base management systems.

Many different data base management systems are available. Some health and safety professionals are using them to develop automated functions on personal computers. There are data base management systems available for all types of computers—some are easier to use than others. If an organization plans to develop its own applications software, the resources in the organization's information center or data processing department should be investigated to determine what is available and if there is any development support.

SYSTEM HARDWARE

The types and placement of computers to be used for an occupational health and safety information system must be determined during the requirements study. Today, there are many alternatives for computers, ranging from large mainframe computers to personal computers, and information processing can be accomplished on one or many of these.

Many health and safety departments are using personal computers for information management. Personal computer technologies are evolving at a rapid pace. Every year, vendors offer greater speed, more memory, and better connectivity. New operating systems permit users to run multiple tasks simultaneously, and new networking capabilities let several PCs share disks, programs, and printers.

Because rapid advances in personal computer technologies have occurred and prices have decreased, many companies have developed occupational health and safety systems relying solely on personal computers, and some use local area networks to link the PCs.

Personal computers also can be used in system configurations that dedicate the PCs to local information management and tie them into central mainframe or minicomputers for long-term storage, cross-functional analyses, and corporate data base management. This type of system configuration distributes system tools and mimics the decentralized organizational structure of many corporations. It is ideally suited to the seemingly opposing needs for local records management and control and corporate-wide data base management.

Other configurations also are being used in the development of health and safety systems. They include: (1) use of a mainframe or minicomputer for all system functions with terminals distributed to users and communication over telephone lines and (2) use of mini or maxiframe computers for regional data base management with terminals or PCs used for data input, and offering communication among the larger machines to allow corporate-wide analyses.

There are many factors that will influence an organization's choice and location of computer hardware for its information system, including currently available hardware, corporate standards for computer hardware, telecommunications, and software, and the information requirements of system users. All system options and constraints should be studied during the requirements study so an economical, efficient, and integrated system can be developed.

SYSTEM COMMUNICATIONS

After deciding the roles of the mainframe, minicomputer, and microcomputers in the organization, as well as what the information sharing ideal would be, consider the communication tools available including: modems, mainframe-mini-micro links, and networking.

When considering communication tools, modems usually are mentioned first. The modem is an excellent communication link, especially if departments are located at considerable distances from each other, perhaps in another city. With a modem, two computers are always as close as the phone lines.

Most communications software packages not only provide a way to transmit information, but make possible transfers between different brands of incompatible computers. Advanced terminal emulation features of some communications software enable communications between mainframes and minicomputers, and other microcomputers.

The hardware and software to make mainframes and micros talk to each other is available, but is dependent on the hardware and software you use.

Networking is a communication form gaining rapid popularity. Local area networking (LAN) is a team of personal computers tied together, so they can talk to each other and easily share information. The network is local in the sense that the computers are generally in the same building or within 305 m (1,000 ft) of each other.

There are several species of LANs made by a number of different manufacturers. These varieties have names like *star, token ring,* and *bus,* which describe how the computers are connected.

The LAN can be connected to a minicomputer or even a mainframe so information can be transferred back and forth, thus enabling the company or department to communicate with the main company system. This way department can talk to department, and departments can talk to the main company system as well.

Because a LAN can connect two or more microcomputers, each of these computers has the power to work on its own but in addition has the ability to use information from the other computers connected to the network. Each department, then, can operate its microcomputers independently, but still remain in communication with the other computers on the network.

In addition to the benefits of information sharing, networking microcomputers permit the sharing of devices. For example: Several different stations can be connected to a printer. A laser printer could serve several different locations; the same hard disk could accommodate the storage needs. This multiple use of equipment represents some savings, especially with expensive printers.

Similarly, a network enables a number of users to access the same software. This is useful when network versions of popular programs are used by several different workstations. This capability is even more valuable when there is customized in-house software in use by a number of different people in separate locations. The LAN permits the establishment of an information network, integrating departments in useful ways, while effortlessly increasing communication.

Networking does have some additional costs associated with it. Yet, because of the associated benefits, more companies are turning away from pure mainframe systems, and are incorporating microcomputers in a networking environment. With an increased emphasis on cost control, the LAN offers a way to get operating information quickly down to the plant level to facilitate effective management and produce competitive results.

Networks have unique properties that make costs justification easier:

- *First,* it can start small and expand. A simple four-unit network can be expanded to a complex 20-terminal LAN. Later, if needed, the LAN can be connected to another LAN. This means initial authorization for a modest investment, and after the value is shown, propose expanding the system and the functions performed by the system, thereby meeting all your system communications needs.
- *Second,* the initial costs of a LAN can be substantially less than a comparable minicomputer. This is especially important when trying to justify a capital expenditure without the customary return on investment calculations.

What kind of LAN installation is best?

There are several choices, and the choice depends upon the performance required and your budget. In any case, the answer requires some skilled professional assistance from your management information department, information center, or outside consultant. It is very important that your resource have actual operating experience with networks.

Initially, you will need a good deal of assistance because networks are complicated to set up and do require some training in their use. Once in operation, however, they are easy to use and have abundant advantages.

It is a good idea to assign someone the job of network administrator. This person will assign passwords, access rights, write the log-in scripts, train people to use the network, and make sure the network is up and ready to serve the users.

BIBLIOGRAPHY

American Conference of Governmental Industrial Hygienists. *Microcomputer Applications in Occupational Health and Safety.* Chelsea, MI: Lewis Publ. Inc., 1987.

——. *Best's Loss Control Engineering Manual.* Oldwick, NJ: A.M. Best Co., published annually.

Helander, M.G. (ed.) *Handbook of Human/Computer Interaction.* New York: Elsevier, 1987.

Klonicke, D.W. Loss control in the computer room. *Professional Safety* (April 1983)17-20.

Miller, E., and O'Hern, C. Microcomputers can make it work for you. *Safety & Health,* 35(Jan. 1987)28-32.

National Fire Protection Association, Batterymarch Park, Quincy, Mass. 02269.

Fire Protection Handbook, 15th ed.

Standards:

Electronic/Data Processing Equipment, NFPA 75.

Halogenated Fire Extinguishing Agent Systems—HALON 1301, Fire Extinguishing System, NFPA 12A.

Protection of Records, NFPA 232.

Ross, D.T., and Schoman, K.E. Structural Analysis for Requirements Definition. In *Software Design Techniques,* 3rd ed. Edited by P. Freeman and A.I. Wasserman. Long Beach, CA: IEEE Computer Society, 1980.

Whyte, A. Occupational health and safety information management systems. In *Introduction to Occupational Health and Safety.* Edited by Joseph La Dou. Chicago: National Safety Council, 1986.

PART VII

Governmental Regulations and Their Impact

30

Governmental Regulations

by M. Chain Robbins, BS, MPH, CSP, PE

BETWEEN 1970 AND 1977, governmental regulations of safety and health matters had been largely the concern of state agencies. There was little uniformity of application of codes and standards from one state to another and almost no enforcement proceedings were undertaken against violators of those standards. Some states adopted as guidelines the Threshold Limit Values (TLVs) for exposure to toxic materials as recommended by the American Conference of Governmental Industrial Hygienists (ACGIH). Enforcement of those guidelines, however, was minimal.

The federal government had some safety and health standards for its contractors and for the stevedoring industry. Enforcement of those standards rested with the Bureau of Labor Standards in the U.S. Department of Labor. Although there were thousands of federal contractors, the inspection and enforcement activities were restricted by the U.S. Department of Labor's limited budget and staff.

Between 1970 and 1977, Congress enacted four new safety and health laws. These legislative efforts continue to have a significant impact on industrial hygiene activities in the United States. These laws are:

- Public Law 91-596, December 29, 1970, the Occupational Safety and Health Act of 1970, popularly known as the OSHAct.
- Public Law 91-173, December 30, 1969, Federal Coal Mine Health and Safety Act.
- Public Law 89-577, September 16, 1966, Federal Metal and Nonmetallic Mine Safety Act.
- Public Law 91-173, November 9, 1977, Federal Mine Safety and Health Act of 1977.

Each of these laws will briefly be discussed.

THE OCCUPATIONAL SAFETY AND HEALTH ADMINISTRATION

The Occupational Safety and Health Administration (OSHA) came into official existence on April 28, 1971, the date the OSHAct became effective. This organization was created by the Department of Labor to discharge the responsibilities assigned to it by the Act. (See Chapter 31, History of the Occupational Safety and Health Administration.)

Major authorities, functional areas, and responsibilities

The Act grants the Secretary of Labor the authority, among other things, to promulgate, modify, and revoke safety and health standards; to conduct inspections and investigations and to issue citations, including proposed penalties; to require employers to keep records of safety and health data; to petition the courts to restrain imminent danger situations; and to approve or reject state plans for programs under the Act. The Secretary's authority includes right of access to the records of other federal agencies, and a shared responsibility with other federal agency heads for the adequacy of programs in the organizations reporting to them.

The Act authorizes the Secretary to have the Department of Labor conduct training of personnel involved in performance of duties related to their responsibilities under the Act, and, in consultation with the U.S. Department of Health and Human Services (DHHS) (formerly the Department of Health, Education and Welfare), to provide training and education to employers and employees. The Secretary and his designees are authorized to consult with employers, employees, and organizations regarding prevention of injuries and illnesses. The Secretary, after consultation with the Secretary of the DHHS, may grant funds to the states for identification of program needs and plan development, experiments, demonstrations, administration and operation of programs. In conjunction with the Secretary of the DHHS, the Secretary is charged with developing and maintaining a statistics program for occupational safety and health.

Major duties delegated by the Secretary of Labor

In establishing the Occupational Safety and Health Administration, the Secretary of Labor delegated to the Assistant Secretary for Occupational Safety and Health the authority and responsibility for safety and health programs and activities of the Department of Labor, including responsibilities derived from:

1. Occupational Safety and Health Act of 1970
2. Walsh-Healey Public Contracts Act of 1936, as amended
3. Service Contract Act of 1965
4. Public Law 91-54 of 1969 (construction safety amendments)
5. Public Law 85-742 of 1958 (maritime safety amendments)
6. National Foundation on the Arts and Humanities Act of 1965
7. Federal safety program under 5 U.S.C. §7902.

The delegated authority includes responsibility for organizational changes, for coordination with other officials and agencies having responsibilities in the occupational safety and health area, and for contracting.

At the same time, the Commissioner of the Bureau of Labor Statistics was delegated the authority and given the responsibility for developing and maintaining an effective program for collection, compilation, and analysis of occupational safety and health statistics, providing grants to the states to assist in developing and administering the statistics programs, and coordinating functions with the Assistant Secretary for Occupational Safety and Health.

The Solicitor of Labor is assigned responsibility for providing legal advice and assistance to the Secretary and all officers of the Department in the administration of statutes and Executive orders relating to occupational safety and health. In enforcing the Act's requirements, the Solicitor of Labor also has the responsibility for representing the Secretary in litigation before the Occupational Safety and Health Review Commission, and, subject to the control and direction of the Attorney General, before the federal courts.

The Labor Department regulations dealing with OSHA are published in Title 29 of the *Code of Federal Regulations* (CFR) as:

29 CFR Part 1910—General Industry Standards
29 CFR Part 1915—Maritime Standards
29 CFR Part 1926—Construction Standards.

General duty and obligations

The OSHAct sets out two duties for employers and one for employees. The general duty provisions are:

- Each employer shall furnish to each employee a place of employment which is free from recognized hazards that are causing or are likely to cause death or serious physical harm to his employees.
- Each employer shall comply with occupational safety and health standards under the Act.
- Each employee shall comply with occupational safety and health standards and all rules, regulations, and orders issued pursuant to the Act which are applicable to his own actions and conduct.

The greatest potential impact of these provisions may be in the area of health. Determination of excessive exposures requires tests with proper instrumentation. Evaluation of the test data requires individuals who by experience and judgment are qualified in the field of industrial hygiene.

The interpretation of the general duty clause for providing a safe and healthful working environment adds new dimensions to the protection of employee health. Control of health exposures oftentimes involves costly and sophisticated engineering systems.

Key provisions

Some of the key provisions of the Act are to:

- Assure, insofar as possible, that every employee has safe and healthful working conditions.
- Require employers to maintain accurate records of exposures to potentially toxic materials or harmful physical agents that are required in the various safety and health standards to be monitored or measured, and inform employees of the monitoring results.
- Provide for employee walkaround or interview of employees during the inspection process.
- Provide procedures for investigating alleged violations, at the request of any employee or employee representative, issuing citations and assessing monetary penalties against employers.
- Empower the Secretary of Labor (through the Occupational Safety and Health Administration) to issue safety and health regulations and standards that have the force and effect of law.
- Provide for establishment of new rules and regulations for new or anticipated hazards to health and safety (Section 6(b) OSHAct).
- Establish a National Institute for Occupational Safety and Health (NIOSH), with the same "right of entry" as OSHA representatives, to undertake health studies of alleged hazardous conditions, and to develop criteria to support revisions of, or recommendations to OSHA for new health standards.
- Provide for 50/50 funding with states that wish to establish state programs, which are at least as effective as the federal program in providing safe and healthful employment (Section 18 OSHAct). There are 24 state programs at the time of this writing.

OSHA STANDARDS

Health standards are promulgated under the OSHAct by the Labor Department with technical advice from the National Institute for Occupational Safety and Health (NIOSH) in the DHHS. A review of OSHA's standards-setting process will be helpful in understanding how regulations are derived.

Most of the safety and health standards now in force under OSHAct for general industry were promulgated 30 days after the law went into effect on April 28, 1971, as 29 CFR Part 1910 of Labor Department regulations (Title 29). They represented a compilation of material authorized by the Act from existing federal standards and national consensus standards (ANSI and

NFPA). These, with some amendments, deletions, and additions, remain the body of standards under OSHAct.

The Act prescribes procedures for use by the Secretary of Labor in promulgating regulations. It is of special interest that the 1968 ACGIH Threshold Limit Values for exposures to toxic materials and harmful agents have been adopted in the regulations, and have the effect of law. Although procedures are given for measuring exposure levels to specific materials and agents in the standards promulgated by the Department of Labor, professional skills and judgments are still required in applying the intent of the many aspects of the Act.

Categories

The OSHA standards consist of the four following categories:

Design standards. Examples of these detailed design criteria are the ventilation design details contained in Section 1910.94 of the initial standards.

Performance standards. Such standards are the Threshold Limit Values (TLVs) of the American Conference of Governmental Industrial Hygienists which are contained in Section 1910-1000. A performance standard states the objective that must be obtained and leaves the method for achieving it up to the employer.

Vertical standards. A vertical standard applies to a particular industry, with specifications that relate to individual operations. Section 1910.261 (Subpart R) of the initial standards is in this category—it applies only to pulp, paper, and paperboard mills.

Horizontal standards. A horizontal standard is one which applies to all workplaces and relates to broad areas, such as Sanitation (§1910.141) or Walking and Working Surfaces (Part 1910 Subpart D).

Standards development

The development of standards is a continuing process. The National Institute for Occupational Safety and Health (NIOSH) of the DHHS provides information and data about health hazards, but the final authority for promulgation of the standards remains with the Secretary of Labor.

Section 6 of the OSHAct defines how safety and health standards are to be set. The Secretary of Labor promulgates standards ". . . based upon research, demonstrations, experiments, and such other information as may be appropriate. In addition to the attainment of the highest degree of health and safety protection for the employee, other considerations shall be the latest scientific data in the field, the feasibility of the standards, and experience gained in this and other health and safety laws. Whenever practicable, the standard promulgated shall be expressed in terms of objective criteria and of the performance desired" (Section 6(b) (5)).

There is a mechanism in the Act (Section 6(c)) by which the Labor Secretary can promulgate emergency standards if he or she believes the evidence supports it. Several emergency health standards have been promulgated—for asbestos, carcinogens, and acrylonitrile. Following the time period required by the Act, these have then been followed by the public rulemaking process with promulgation of final standards.

As mentioned earlier, the first health standards were the 8-hour time-weighted average (TWA) values of air contaminants from the 1968 ACGIH TLV list; they now have the force of legal requirements. Guidance for specific sampling strategies, medical surveillance, and protective measures were lacking. However, the Act in Section 8(c)(3) called out requirements for employers to measure contaminants, maintain records, and notify employees of overexposures and corrective action to be taken for all future health standards.

Because there has been a significant effort in the Labor Department to increase the rate at which health standards are to be promulgated, NIOSH (the advisory group) has prepared a number of criteria for health standards. In turn OSHA has proposed some health standards that include the areas of action described in Section 8(c)(3).

For reference purposes, the indices of OSHA Health Standards 1910, Subpart G—Occupational Health and Environmental Control, and Subpart Z—Toxic and Hazardous Substances, are given in Table 30-A.

Table 30-A. Part 1910—Occupational Safety and Health Standards

Section	Title
Subpart G—Occupational Health and Environmental Control	
1910.94	Ventilation
1910.95	Occupational noise exposure
1910.96	Ionizing radiation
1910.97	Nonionizing radiation
1910.98	Additional delay in effective date
1910.99	Sources of standards
1910.100	Standards organizations
Subpart Z—Toxic and Hazardous Substances	
1910.1000	Air contaminants
1910.1001	Asbestos, tremolite, anthophyllite and actinolite
1910.1002	Coal tar pitch volatiles; interpretation of term
1910.1003	4-Nitrobiphenyl
1910.1004	alpha-Naphthylamine
1910.1005	4, 4'-Methylene bis (2-chloroaniline)
1910.1006	Methyl chloromethyl ether
1910.1007	3,3'-Dichlorobenzidine (and its salts)
1910.1008	bis-Chloromethyl ether
1910.1009	beta-Naphthylamine
1910.1010	Benzidine
1910.1011	4-Aminodiphenyl
1910.1012	Ethyleneimine
1910.1013	beta-Propiolactone
1910.1014	2-Acetylaminofluorene
1910.1015	4-Dimethylaminoazobenzene
1910.1016	N-Nitrosodimethylamine
1910.1017	Vinyl chloride
1910.1018	Inorganic Arsenic
1910.1025	Lead
1910.1029	Coke oven emissions
1910.1143	Cotton dust
1910.1044	1,2-dibromo-3-chloropropane
1910.1045	Acrylonitrile
1910.1047	Ethylene oxide
1910.1200	Hazard communication

Action level. Of interest to the industrial hygienist is the "action level" concept. In 1976, OSHA defined the action level as one-half the Permissible Exposure Limit (PEL). Where exposures reach or exceed the action level, additional requirements apply, including medical surveillance and a full air-monitoring program. Exposures to an airborne concentration in excess of the PEL trigger still further requirements, including reduction of exposures to (or below) the PEL by means of engineering controls supplemented by work practice controls, use of specified respirators, and use of other appropriate protective clothing and equipment.

Employee protection. The OSHA decided as a policy matter that the action level, which triggers the measurement requirements, be set at some level below the PEL to better protect employees from overexposure. The OSHA reasoned that this method was the most reasonable approach to a recurring problem, that is, how to provide the maximum employee protection necessary with the minimum burden to the employer. Thus where the results of employee exposure measurements demonstrate that no employee is exposed to airborne concentrations of a substance in excess of the action level, employers are, in effect, exempted from major provisions of the particular standard.

A duty to measure employee exposure and provide medical surveillance only when the employee exposure was equal to or greater than the PEL was rejected by OSHA as not providing sufficient protection for the exposed employees. Among other things, such a scheme would not protect the employee from overexposure because the employer would have no way of knowing when airborne concentrations of a regulated substance approached the TWA. It is not possible to assure that all exposures are within the permissible limits simply because sampling was done when an employee's exposure was at the PEL.

Other requirements. It has been determined, therefore, that three key duties should be triggered when an action level is reached—exposure measurement, medical surveillance, and employee training. All three actions are considered necessary by OSHA before employee exposure reaches the PEL or higher. It is important to begin measurement procedures when approaching the Permissible Exposure Limit to assure that employee exposure does not exceed it. Similarly, employees should be screened for preexisting medical conditions and trained so they learn suitable precautions against dangerous properties of the substance when there is some chance that their exposure will become significant.

The OSHA has claimed that an alternative to having the action level would be to require the medical and measurement procedures at any level of exposure, no matter how low. This alternative, however, would burden employers unnecessarily because they would be required to implement the medical and measurement provisions even where concentrations were so low that they presented no health problem.

The action level concept will, no doubt, continue to be a subject of discussion between the Labor Department and practicing industrial hygienists. There are those hygienists who argue that a *new* permissible exposure level is created by arbitrarily setting 50 percent of the PEL as the action level. Some maintain that the blanket concept should not be imposed unless there is clear evidence of toxicity and there should be a differentiation between toxic substances and irritating substances. Other criticisms have been that the action level approach used by OSHA is overly conservative because it assumes a single lognormal distribution for all cases and uses only one measured 8-hour TWA to estimate the mean of the distribution of other 8-hour TWAs.

Exposures at or just above the action level are not citable by OSHA, but failure to take the specified actions that should be triggered when reaching the action level are citable.

HAZARD COMMUNICATION

Affecting worker health and the industrial hygiene profession, the Hazard Communication Standard, 29 *CFR* 1910.1200, is one of OSHA's most significant standards. This Standard was promulgated as a final rule on November 25, 1983, and contains three important compliance dates. The first, November 25, 1985, is the date on which all chemical manufacturers, importers, and distributors became required to label chemical shipping containers, assess chemical hazards, and provide material safety data sheets (MSDSs) to recipients of their chemicals.

The second compliance date, May 26, 1986, affected all employers covered by the Standard under manufacturing division Standard Industrial Classification (SIC) codes 20 through 39. Beginning on this date, all relevant employers must have had a written and operating hazard communication program assuring the necessary information and training is provided to affected employees, including chemical container warnings, MSDSs, and other forms of warnings.

On August 24, 1987, OSHA expanded the scope of the Hazard Communication Standard to include the nonmanufacturing sector. As of the third compliance date, May 24, 1988, *all* employers must be in compliance with the Standard.

The intent of the Hazard Communication Standard is to provide employees with information about the potential health hazards from exposure to workplace chemicals. The objective of the training is to provide employees with enough information to allow them to make more knowledgeable decisions with respect to any personal risks of their work and to impress the need for safe work practices. The Standard requires that employee training includes:

- explanations of the requirements of the standard
- identification of workplace operations where hazardous chemicals are present
- knowledge of the methods and observations used to detect the presence of hazardous workplace chemicals
- assessment of the physical and health hazards of those chemicals
- warnings about hazards associated with chemicals in unlabeled pipes
- descriptions of hazards associated with nonroutine tasks
- details about the measures employees can take to protect themselves against these hazards, including specific procedures
- explanation of the labeling system
- instructions on location and use of material safety data sheets (MSDSs)
- details on the availability and location of the hazardous material inventory, MSDSs, and other written hazard communication material.

The Standard applies to any chemical known to be present in the workplace that employees may be exposed to under normal conditions of use, or may be exposed to in a foreseeable emergency. Pesticides, foods, food additives, cosmetics, distilled spirits, certain consumer products, and hazardous wastes are all covered under other federal legislation and, therefore, are exempt from this OSHA Standard.

Numerous state and local governments have promulgated similar legislation, also known as right-to-know laws. These laws differ in some ways from the Hazard Communication Standard.

In states with a federal OSHA enforcement program, OSHA

says the federal Hazard Communication Standard preempts the state right-to-know law as it applies to the manufacturing sector. This question, however, is currently being clarified through litigation in various states and cities.

The important role of the industrial hygienist in assuring compliance with hazard communication legislation is obvious—from initial risk assessment through employee training and MSDS interpretation.

ENFORCEMENT OF OSHAct

The Secretary of Labor is the principal administering officer of the OSHAct. OSHA is authorized to conduct inspections and, when alleged violations of safety and health standards are found, to issue citations, and, when the situation calls for them, to assess penalties.

Highlights

- The OSHA schedules inspections on a priority system: First, in response to fatalities and multiple (five or more) hospitalization incidents and imminent danger situations; second, in response to employee complaints; third, random inspections of high hazard industries; fourth, follow-up inspections.
- The OSHA compliance officers may enter the employer's premises without delay to conduct inspections and usually without advance notice (Section 8(a) OSHAct); however, if the employer refuses entry, a search warrant may be requested.
- The OSHA's right to inspect includes records of injuries and illnesses, including certain medical records.
- The OSHA compliance officers who find conditions of imminent danger can only request, not demand, shutdown of an operation. If shutdown is refused, the compliance officer notifies employees of the hazard, and the Department of Labor may seek court authority to shut down the operation.
- Criminal penalties can be invoked only by court action and in extreme cases (usually willful violations leading to death).
- An appeal system has been set up under which employers and employees can appeal certain OSHA actions to the independent Occupational Safety and Health Review Commission (OSHRC) (Section 12 OSHAct).

OSHA FIELD OPERATIONS

Field Operations Manual (FOM)

The *Field Operations Manual (FOM)* was issued by OSHA in order to contain inspection policy. This manual contains general instructions and policies on field compliance operations, with special emphasis on the assessment of health hazards found in the workplace. The OSHA *Industrial Hygiene Technical Manual* sets forth the technical industrial hygiene practices and procedures used by OSHA personnel. This section summarizes these procedures and gives background about OSHA functions at the regional and local levels.

Health compliance operations involve several technical and professional disciplines possessed by industrial hygienists, safety engineers, and safety specialists. Therefore, the processing of all health inspections and citations requires close coordination between industrial hygienists, engineers, and local, regional, and national staff. In addition, nonagency assistance, either from outside consultants or from other agencies, such as NIOSH, may be required. (NIOSH is discussed later in this chapter.)

Responsibilities. The national OSHA office, through its technical and analytical units, coordinates the technical aspects of health programming among the regions. An industrial hygienist in each regional office is responsible for coordinating the technical aspects of the health program within the region. This responsibility includes, but is not limited to, providing guidelines for inspections, evaluating and assisting in contested cases, and guidance in using technical equipment in accordance with criteria provided by the national office.

The Area Director. This official administers the field compliance program within the designated geographic area. Each area director designates an industrial hygiene supervisor who is responsible for the technical aspects of the health compliance program, and for recommending health inspection priorities in his or her area.

Health inspections are conducted in industries in accordance with priorities outlined earlier. A health inspection can be either a complete survey of a particular workplace for all health hazards, or a special survey such as an accident investigation. Some inspections require a team effort because they involve more than one specialty. In such cases, the area director designates a team leader to coordinate the efforts. The industrial hygiene supervisor provides guidance, especially on health inspection procedures.

Safety Specialist-Compliance Officers. The officers doing full-time inspection work have taken course work in industrial hygiene so they can recognize potential health hazards.

After being trained to recognize and evaluate health hazards, the officer collects health-hazard information for possible referral to an industrial hygienist. Health referrals are incorporated into the regular inspection schedule.

The OSHA instructs the safety specialist-compliance officer that a health hazard can be suspected when any of the following occur: (1) eye irritation is felt when entering the work area, (2) a strong odor is noticed when entering the work area, (3) visible fume or dust clouds are observed coming from the operation into the workplace or visible clouds are observed coming from poorly maintained ventilation systems.

If, during a routine safety inspection, the safety specialist-compliance officer observed a potential health hazard, he or she makes a written notation that consists of the following (as applicable): date and time, name of plant or establishment, suspected health hazard, description of the operation, description of the location of the operation in the plant, reason for suspecting the health hazard, worker comments regarding the health hazards, the number of employees affected, and the results of any spot samples that were taken.

The industrial hygiene supervisor carefully screens all referrals to determine their appropriateness and inspection priority.

Health complaints are investigated by an industrial hygiene compliance officer who has been trained to recognize health hazards and evaluate conditions. Complaints involving the appropriateness of unusual medical testing or questionable results of medical findings are discussed with the regional industrial hygienist and the technical and analytical assistance unit.

OSHA INDUSTRIAL HYGIENE INSPECTIONS

An industrial hygiene inspection is conducted by an OSHA

industrial hygienist and often is complex and time consuming. The essential elements of a visit are preinspection planning, opening conference, walk-through inspection, collecting samples, and closing conference. These guidelines are given to all OSHA industrial hygienists.

Preinspection planning

The OSHA industrial hygienist should become familiar with the particular industry and general process information and size of the plant or establishment. The industrial hygienist should also review appropriate standards and select sampling methods. If the inspection is a referral visit to a workplace previously visited by a compliance officer, the industrial hygienist should review all information contained in the previous inspection reports. Based both on his experiences and specific study, the necessary field instrumentation should be selected. The hygienist then can prepare instruments and equipment according to standard methods of sampling and calibration.

Opening conference

The instructions state that upon entering the establishment, the OSHA industrial hygienist presents identification credentials. An opening conference with plant management is arranged to discuss the purpose and scope of the inspection. If the employees are represented by a labor union, they must be notified of OSHA's presence and invited to participate in the inspection. At the beginning of inspection, the industrial hygienist usually will request a complete process flow diagram or plant layout from the plant personnel, or, if the plant layout chart is not available, he can ask that a sketch be made to identify the operations, distribution of equipment including engineering controls, and approximate layout of the plant.

A brief examination is made of all required records kept at the establishment, such as the nature of any injuries or illnesses shown on OSHA form no. 200, Recordkeeping Log. Recordkeeping requirements can be discussed at the closing conference after the inspection has been made or after sampling results have been analyzed. At this time, preliminary information is gathered about the occupational health program.

Walk-through inspection

A walk-through inspection is required for all health inspections regardless of whether the establishment was previously inspected. The main purpose of the walk-through inspection is to identify potential workplace health hazards, because during the walk-through, the industrial hygienist becomes familiar with plant processes, collects information on chemical and physical agents, and observes worker's activities. As the walk-through inspection proceeds, information is obtained concerning raw materials used, intermediates (if any), and final products. Estimated amounts of substances present in the plant and a complete inventory are also obtained. In addition, a request is made for a list of raw materials received at the loading dock. The industrial hygienist also checks for hazardous physical agents present in the plant, such as noise and excessive heat. He or she observes work activity throughout the plant, but concentrates particularly on potential health hazard areas.

The approximate number of workers in each area is written on the sketch of the plant. The industrial hygienist observes and records the general mobility of the workers and indicates whether they are engaged in stationary or transient activities. Existing engineering controls are marked on the plant layout or sketch. Ventilation measurements are made at strategic locations in the duct system and recorded on the plant sketch.

Information usually is requested from the plant manager concerning a preventive maintenance program for engineering controls. The industrial hygienist keeps alert for any imminent dangers during the walk-through inspections and takes appropriate action if necessary. Photographs can be taken to document the survey. Employee interviews are encouraged at this phase of the inspection.

In a small establishment or plant, health hazards sampling can be initiated after the opening conference. A final sampling schedule can be prepared using the information collected.

Collecting samples

Representative jobs should be selected and personal sampling devices prepared. Operations with the highest expected exposure should be monitored first. The sampling program should be planned to follow the particular industrial processes as closely as possible in order to keep a logical sequence to any information obtained.

The OSHA industrial hygienist determines compliance with air quality and noise standards based upon one or more days of full-shift, TWA concentration measurements. These findings are tempered by the hygienist's professional judgment of the data and conditions that exist at the plant.

For 8-hour exposures, the TWA concentration must be determined. Sampling devices monitoring full-shift exposures must operate for a minimum of 7 hours. This implies that the concentration is calculated using the air volume sampled during the shift for a time exceeding 7 hours for some chemicals. Spot samples must be taken throughout the work shift to represent at least 7 hours of exposure and to represent periods of exposure for TWA calculations. When the actual work shift exceeds 8 hours, the TWA can be calculated using the results of 7- or 8-hour sampling period and separate samples taken to determine any additional exposure. This value is compared with the standard for compliance determination.

Materials in Table Z-1 of the OSHA Standard 1910.1000 preceded by a "C" have maximum peak ceiling limits to which an employee can be exposed; these values are never to be exceeded. Generally a 15-minute sampling period should be applied to ceiling measurements (except for imminent danger situations where immediate escape from the atmosphere is necessary).

A minimum number of samples must be taken. These can be a single 7–8-hour sample for full-shift assessment, several spot samples that represent on a TWA a full-shift assessment, or one 15-minute sample for a ceiling assessment. The measurement(s) must exceed the sum of the allowable limit and a calculated margin of error before a citation can be issued.

The factors to be applied to full-shift TWA concentrations have been developed because legal considerations require a degree of certainty that the standard is violated when a citation is issued. A citation shall not be issued unless the measured level exceeds the calculated upper confidence level of the permissible level based upon a single day's sampling results. Measurements below the confidence limit indicate that overexposure may have occurred.

All sampling equipment must be checked and calibrated (in accordance with standard procedures described in the *Indus-*

trial Hygiene Technical Manual) prior to taking any samples. The sampling for both physical and chemical contaminants in the environment must relate to the worker's exposure, unless otherwise specified in a standard. So the measurements will represent, insofar as possible, actual exposure of the employee, the measurements for air pollutants should be taken at close proximity to the employee's breathing zone, or taken near the ear for noise level exposure.

Closing conference

Because the industrial hygienist may not have the results of the environmental measurements at the end of the inspection or during the closing conference, a second closing conference may be held; for this, a telephone call or a letter can be used in place of a personal visit. Employees or their representatives also will be informed of the inspection results at that time. If the results indicate noncompliance, the alleged violations can be discussed at the second conference along with abatement procedures and methods of control.

Employer's occupational health programs

Information on the following aspects of the employer's occupational health program is gathered during the inspection for evaluation. These aspects may be discussed during the closing conference and later considered in relation to the standard requirements and also as evidence of good faith when penalties are being proposed.

Monitoring program. Detailed information concerning the industrial hygiene program will be obtained. Especially valuable data that will be asked for include number and qualifications of personnel, and availability of necessary sampling and calibration equipment, ventilation-measuring equipment, and laboratory services.

Medical program. The industrial hygienist should determine whether the employer provides employees with preemployment and regular medical examinations as required by certain OSHA standards. The medical examination protocol is reviewed to determine the extent of the medical examination.

Education and training program. A special check will be made to evaluate the company's efforts to comply with the Hazard Communication Standard, 1910.1200.

Record-keeping program. The employer's record-keeping program will be checked, including types of records, how long they are maintained, and the accessibility of these records to the employees.

Compliance programs

The industrial hygienist determines the employer's implementation of engineering and administrative controls, and checks appropriate equipment and preventive maintenance. A specific engineering control should be evaluated for effectiveness in its present application.

Work-practice and administrative controls. Control techniques include isolation of hazardous operations, rotation of employee job assignments, and sanitation and housekeeping practices. A detailed description of such controls should be obtained. It is essential that work practice controls and the education program be implemented simultaneously because the overall effectiveness of such practices is enhanced by employees' knowledge of their exposures.

Protective devices. The industrial hygienist determines whether protective devices are effectively used in the plant or establishment. A detailed investigation of the personal protection program will include a determination of compliance with 29 *CFR* 1910.134, Respiratory Protection. Special emphasis is given to the sanitary condition of the equipment, the maintenance of the equipment, and the use of training programs on selection and use of protective devices. It is required that the program be available as a written procedure. See Chapter 23, Respiratory Protective Equipment, for more information.

Regulated areas. Some standards require the establishment of regulated areas. Sampling is used to determine the limits of areas that must be regulated because employee exposure is expected to be greater than prescribed levels. The industrial hygienist will make sure regulated areas meet certain standards such as:

- Being clearly identified and known to affected employees.
- Maintaining the regulated area designations according to the prescribed criteria of the standard.
- Maintaining daily rosters of authorized personnel entering and leaving the area. Summaries of such rosters are acceptable.

Emergency procedures. Some standards provide for specific emergency procedures to be followed when handling certain hazardous substances. Company procedures should be checked for the inclusion of potential emergency conditions in the written plan, explanation of such emergency conditions to the employees, training schemes for the protection of affected employees, and delegation of authority for the implementation of the operational plan in emergency situations.

Evaluation of sampling data. The OSHA industrial hygienist will use professional judgment when evaluating the data and conditions to confirm the sampling results. Sampling results should reasonably correlate with each other. For example, an occupation that is expected to be dusty should have higher dust concentrations measured than those expected to be less dusty. Sampling results must reasonably correlate with previous measurements made by OSHA, the company, and others (taking into account variations due to weather, production, breakdowns, and the like). The sampling results must take into account accuracy of the instrument. The industrial hygienist's record must state that all sampling was performed according to OSHA standard methods.

Issuance of citations

The industrial hygienist uses OSHA guidelines when classifying violations. Here are some examples.

The workplace may be found to be in compliance with OSHA standards. In this case, no citations are issued or penalties proposed.

Violations may be found in the establishment. In that case, citations may be issued, and civil penalties may be proposed. In order of significance, these are the types of standard violations normally considered on a first inspection.

Imminent danger. A condition where there is reasonable certainty a hazard exists that can be expected to cause death or serious physical harm immediately or before the hazard can be eliminated through regular procedures. If the employer fails to abate such conditions immediately, the compliance officer, through his area director, can go directly to the nearest Federal District Court for legal action as necessary.

Serious violations. A serious violation has a substantial probability that death or serious physical harm could result and that the employer knew, or should have known, of the hazard. For example, absence of point-of-operation guards on punch presses or saws. A proposed penalty of up to $1,000 is mandatory. A serious penalty may be adjusted downward based on the employer's good faith, history of previous violations, and size of the business.

Other violations. Other violations are those that have a direct relationship to job safety and health but probably would not cause death or serious physical harm. For example, a tripping hazard. A nonserious penalty may be adjusted downward depending on the severity of the hazard, employer's good faith, his history of previous violations, and size of business. Congress has told OSHA that exclusive of serious violations it must find more than 10 other violations before any penalty can be imposed.

De minimis. A condition that has no direct or immediate relationship to job safety and health.

If respirators and other personal protective equipment are not properly fitted or are not worn and the affected employee is exposed to a toxic agent in excess of the Permissible Exposure Limits, a citation will be issued with classification as serious.

Employers will be cited for an other-than-serious violation where they (1) have not established written operating procedures governing the use of respirators, (2) have not trained and instructed employees in their proper use, or (3) have not regularly cleaned and disinfected the respirators even though such respirators are properly fitted and worn.

Sometimes, when appropriate personal protective equipment is being properly used, citations are issued for failure to use administrative or engineering controls if the industrial hygienist believes that such controls are feasible. Generally, such citations will be other-than-serious unless available data indicates the personal protective equipment, even though properly fitted and worn, is not effective in fully reducing exposure to acceptable limits as required by the standard.

To explain this a bit further, the Occupational Safety and Health Administration has a technical support group for establishing and maintaining uniform compliance activities across the nation so that different items are not cited differently by different inspectors. This support group will help determine which engineering controls are feasible, and it will give this information to OSHA industrial hygienists to help them judge whether an item is serious or nonserious.

During the course of an inspection, the industrial hygienist will carefully investigate the source or cause of observed hazards to determine if some type of engineering or administrative control (or combination) may be applied that would significantly reduce employee exposure. In order to issue a citation, an OSHA representative need not show that the controls would reduce exposure to the limits prescribed by the applicable standard.

Engineering controls

Engineering controls include any procedure, other than administrative controls or use of personal protective equipment, that reduces exposure at its source or close to the employee's breathing or hearing zone. Proper work practices and personal hygiene facilities are defined as engineering controls. To be feasible, a general technical knowledge must exist about materials or methods that are available or adaptable to specific circumstances, and there must be a reasonable possibility that employee exposure to violative conditions of noise, dust, or other substances will be reduced.

Contest of OSHA citations

Suppose the employer disagrees with the citation and/or proposed penalty. What can be done? First, the employer can request an informal meeting with the area director to discuss the case.

If the employer decides to contest the citation, the Act contains a specific appeal procedure, guaranteeing full review of the case by an agency separate from the Labor Department. That agency is the independent Occupational Safety and Health Review Commission (OSHRC) which has no connection with the U.S. Department of Labor.

The Review Commission was created to adjudicate enforcement actions initiated under the act when they are contested by employers, employees, or representatives of employees.

The Commission's functions are strictly adjudicatory; it is, however, more of a court system than a simple tribunal, for within the Review Commission there are two levels of adjudication. All cases which require a hearing are assigned to a Review Commission Judge who will decide the case. Each such decision is subject to the discretionary review by the three members of the Review Commission upon the motion of any one of the three.

Employers may contest a citation, a proposed penalty, a notice of failure to correct a violation, the time allotted for abatement of a violation, or any combination of these. Employees or employee representatives can contest the time allotted for abatement.

A notice of contest must be filed by certified mail, the OSHA Area Office initiating the action being contested within 15 federal working days of the receipt of the notice of the enforcement action. If this step is not taken within 15 working days, the proposed penalties and citations are final and not subject to review by any court or agency.

There is no prescribed form for the notice of contest. The notice must, however, clearly indicate what is being contested—whether it is the citation, the proposed penalty, or the time for abatement.

Employees working on the site of the alleged violation must be notified of a contest filed by their employer. If they are members of a union, the union must be given a copy of the notice of contest. Nonunion employees must be notified either by posting a copy of the notice of contest where they will see it or by serving each a copy.

The notice of contest must contain the names of those to whom it has been served or the address of the place it was posted. The posted or individually served copy of the notice of contest must be accompanied by a warning that they may not be allowed to participate in the case unless they identify themselves before the hearings begin to the commission or hearing examiner. This is called filing for "party status." To file for party status, a notice of intent should be sent to the OSHRC.

If an employer contests an alleged violation in good faith and not solely for delay or variance of penalties, the abatement period does not begin until the entry of the final order by the OSHRC.

When a notice of contest reaches the area director, the executive secretary of the OSHRC is notified of the contest and its details. The latter gives the case a docket number. In time a hearing will be held by an OSHRC judge. OSHA first presents its case, subject to cross examination by the other parties. The defendant then presents the case, subject to cross examination by the other parties. Employees or their representatives may participate in this hearing if they have filed for party status. The judge is allowed to consider only what is on the record. Any statements that go unchallenged will therefore be considered to be fact.

The hearing will ordinarily be held in the community where the alleged violation occurred or close thereto. At the hearing, the Secretary of Labor will have the burden of proving the case.

After the hearing, the judge must issue a report, based on findings of fact, that affirms, modifies, or vacates the Secretary's citation or proposed penalty, or that directs other appropriate relief. The report will become a final order of the Commission 30 days thereafter unless, within such period, any Commission member directs that such report shall be reviewed by the Commission itself. When this occurs, the members of the Commission will thereafter issue their own decisions on the case.

Once a case is decided, any person adversely affected or aggrieved thereby, may obtain a review of such decision in the United States Court of Appeals.

NATIONAL INSTITUTE FOR OCCUPATIONAL SAFETY AND HEALTH

The National Institute for Occupational Safety and Health (NIOSH) is the principal federal agency engaged in research to eliminate on-the-job hazards to the health and safety of America's working men and women. It was established within the Department of Health, Education, and Welfare (now the DHHS) under the provision of Public Law 91-596—the Occupational Safety and Health Act of 1970. Administratively, NIOSH is located within DHHS's Centers for Disease Control of the Public Health Service.

Responsibilities

The NIOSH is responsible for identifying occupational safety and health hazards and for recommending changes in the regulations limiting them. It also has obligations for training occupational health manpower.

The institute's main research laboratories are in Cincinnati, Ohio, where studies include not only the effects of exposure to hazardous substances used in the workplace, but also the psychological, motivational, and behavioral factors involved in occupational safety and health. Much of the institute's research centers on specific hazards, such as asbestos and other fibers, beryllium, coal tar pitch volatiles, silica, noise, and stress.

At the NIOSH Appalachian Laboratory for Occupational Safety and Health (ALOSH), in Morgantown, West Virginia, research has primarily focused on coal workers' pneumoconiosis—black lung disease—but the program is being expanded to include other occupational respiratory diseases. Also located in Morgantown is the NIOSH Testing and Certification Branch, which evaluates and certifies the performance of workers' personal safety equipment.

Testing and certification

Certification tests are performed under the authority of the Federal Coal Mine Health and Safety Act of 1969, the Occupational Safety and Health Act of 1970, and the numerous regulations and standards issued by the Mine Safety and Health Administration (MSHA) and the Occupational Safety and Health Administration (OSHA).

The purposes of these tests are threefold. First, to determine if currently available protective devices conform or fail to conform with existing performance standards for such devices. Second, to encourage manufacturers of such devices and instruments to improve their performance and quality when NIOSH finds them to be out of conformance, or of marginal quality. And third, the results of the tests are used by NIOSH in determining (1) the need for future NIOSH certification projects and (2) priorities for research efforts directed toward product improvements through increased quality and scope of performance testing.

The NIOSH personnel, in addition to doing respirator research, serve as respirator consultants to OSHA and industry on the broad subject of respirator selection, use, and maintenance. NIOSH personnel regularly participate in OSHA hearings and submit new respirator information to OSHA and industry on a routine basis.

Research

In addition to conducting its own research program, NIOSH also funds supportive research activities at a number of colleges and universities as well as at private facilities.

Not all of NIOSH's work is done in the laboratories. A legislatively mandated activity involves Health Hazard Evaluations (HHE); these are on-the-job investigations of reported worker exposures to toxic or potentially toxic substances. Made as a direct response to requests by management and/or authorized representatives of employees, HHEs are usually initiated through NIOSH's representatives, although scientists from other Institute facilities often are involved.

Under the authority of the Occupational Safety and Health Act, NIOSH conducts research for new occupational safety and health standards. Its recommended standards are transmitted to the Department of Labor, which has the responsibility for development, promulgation, and enforcement of the standards.

Training

The NIOSH has a training grant program to develop two-year, baccalaureate and graduate programs in colleges and universities across the nation. NIOSH also offers a spectrum of short-term training courses for upgrading the knowledge and skills of present occupational health practitioners.

The NIOSH also maintains staff in Regional Offices throughout the United States. The Regional Offices are focal points for special surveys and evaluations of existing occupational safety and health problems, consultative services to the states, and other activities. Headquarters for NIOSH are in Atlanta.

Recommendations for standards

One of the Institute's most important responsibilities under this act is to transmit recommended standards to the Occupational

Safety and Health Administration (OSHA) in the Department of Labor. The NIOSH recommendations are intended to serve as the basis, along with other available information, for assisting OSHA in developing new standards and in revising the approximately 400 consensus health standards (most of them consist only of a numerical exposure limit) that were promulgated when the act was passed.

The NIOSH recommendations, termed *"Criteria Documents,"* include an environmental limit for workplace exposure, as well as recommendations on (1) the use of labels and other forms of warning, (2) type and frequency of medical examinations to be provided by the employer, (3) sampling and analytical methods, (4) procedures for technological controls of hazards, and (5) suitable personal protective equipment. In addition to the *Criteria Documents,* NIOSH developed, under the standards completion program, technical standards for most of the consensus health standards. These standards supplement the existing environmental limits with procedures for informing employees of hazards, monitoring techniques, engineering and control mechanisms, and medical surveillance programs. These recommendations should protect workers from many of the more serious occupational exposures. These recommendations are based on laboratory and epidemiologic research conducted by NIOSH and others.

Field investigations

The NIOSH has promulgated regulations governing field investigations (42 CFR Part 85). Under these regulations, it is the practice to meet with company management and with employee representatives, before initiating a study, in order to explain its purpose and scope. Before conducting medical examinations, investigators must receive specific approval from the NIOSH Human Subjects Review Board and obtain the informed consent of each employee examined. All participating employees and their designated physicians are given the results of these medical examinations. Before releasing final reports on group data (with individual identifiers removed), draft copies are provided to employers and employee representatives for their comments on technical accuracy. The results of the epidemiologic studies are then presented in NIOSH criteria documents and technical reports, in scientific journals, at scientific meetings, and at OSHA hearings on workplace standards.

Workplace investigations also are conducted in a health hazard evaluation program. Under this program, NIOSH responds to requests from employers and employee representatives to investigate a workplace, collect environmental samples, make toxicity determinations, and provide medical examinations for workers. The results of these investigations, including recommendations for work practices, personal protective equipment, and engineering controls, are reported back to company or plant management, employee representatives, and OSHA.

OTHER U.S. GOVERNMENT REGULATORY AGENCIES

In addition to OSHA and NIOSH, already described, there are other federal government regulatory agencies and commissions.

Mining Enforcement and Safety Administration

The Mining Enforcement and Safety Administration (MESA) was established on May 7, 1973, by the Secretary of the Interior; the Administration became operative on July 16, 1973. The Secretary's Order assigned to the Administrator, MESA, the responsibility for administering the enforcement provisions of the Federal Coal Mine Health and Safety Act of 1969 and the Federal Metal and Nonmetallic Mine Safety Act.

The Mining Enforcement and Safety Administration administers the enforcement provisions of the public laws and related standards and training programs in a manner that guards the health and safety of American miners.

The Secretary's order of May 7, 1973, designated the Bureau of Mines of the Department to continue its research functions for mine health and safety.

Federal Mine Safety and Health Amendments Act of 1977

On November 9, 1977, President Carter signed the Federal Mine Safety and Health Amendments Act of 1977. The Act transfers authority for enforcement of mining safety and health from the Department of Interior to the Department of Labor. Most of the provisions of the Act became effective March 9, 1978.

The Federal Mine Safety and Health Act of 1977 repeals the Metal and Nonmetallic Mine Safety Act, and establishes a single mine safety and health law for all mining operations under an amended Coal Mine Health and Safety Act of 1969.

The Mine Safety and Health Administration (MSHA) is created to replace the Interior Department's Mining Enforcement and Safety Administration. This agency is headed by an Assistant Secretary of Labor for Mine Safety and Health.

The new agency is separate from the Occupational Safety and Health Administration and the Secretary of Labor is authorized to settle jurisdictional disputes between the two agencies.

The 1977 Act defines "mine" broadly to include all underground or surface areas from which a mineral is extracted; all surface facilities used in preparing or processing minerals; and roads, structures, dams, impoundments, tailing ponds, and similar facilities related to mining activity. Included within the coverage of the Act is the protection of miners from radiation hazards connected with the milling of certain radioactive materials. Mine construction activity on the surface is included in the scope of the Act.

The Act does not amend the section of the Coal Act dealing with benefits for victims of black lung.

Provisions. The Federal Mine Safety and Health Act of 1977 provides, in part, as follows: joins all mines, both coal and noncoal, under one statute; transfers administration from the Interior to the Labor Department; provides that existing and new standards applicable to metal and nonmetal mining remain separate from existing and new standards for coal mining; and establishes statutory timetables for each step of the standard-setting process.

This Act vests in the Secretary of Labor all authority for developing safety and health standards; it authorizes NIOSH to prepare criteria documents for the development of health standards, with the Secretary of Labor required to act within 60 days of receipt of criteria.

Federal Mine Safety and Health Review Commission. This Act established an independent Federal Mine Safety and Health Review Commission. It requires each mine operator to have a safety and health training program that is approved by the Secretary of Labor (with such training to be provided at the operator's expense and during normal working hours) and authorizes miners' representatives to participate in inspections not only during the actual inspection, but also in pre- and post-inspection conferences held at the mine. While more than one miner representative may participate, only one representative, who is also an employee of the operator, is to be paid by the operator for participation in the inspection and conferences.

The Act authorizes a withdrawal order based on the existence of a pattern of violations of standards that could "significantly or substantially contribute to the cause and effect of a mine hazard and health risk"; authorizes miners or their representatives to make written requests for inspection based on suspected violations of standards or conditions of imminent danger; and mandates a minimum of two inspections each year of all surface mines.

The Secretary of Health and Human Services is required to appoint an advisory committee on coal or other mine health research composed of representatives of the Director of the Bureau of Mines, the Director of the National Science Foundation, and the Director of NIH, as well as other persons knowledgeable in the field of mine health research. The Secretary of HHS also designates the committee chairman.

The purpose of the advisory committee is to consult with and make recommendations to the Secretary of HHS on matters relating to mine health research. The Secretary is required to consider the recommendations of the committee in the conduct of such research and the awarding of any research grants and contracts. The chairman and majority of committee members are not permitted to have economic interests in mining, or be miners, mine operators or government employees. In effect, the new mining law expands the existing Coal Mine Health Research Advisory Committee to cover all mine health research.

Under the existing Federal Coal Mine Health and Safety Act of 1969, NIOSH conducted research on occupational diseases of coal miners, established coal mine health standards, and assured the availability of medical examinations for underground coal miners. The MESA in the Department of Interior established coal mine safety standards and enforced both health and safety standards. In addition, NIOSH has established joint regulations with MESA to test and certify respirators and coal mine dust personal samplers.

In the past NIOSH has not had specific legislative authority to conduct research in noncoal mines but has done so under the general authority of the Public Health Service Act and as designees of the Interior Department. Under the Occupational Safety and Health Act, NIOSH has developed recommended standards that could be applicable to some of the most important health hazards facing workers in metal and nonmetal mines, such as noise, silica, asbestos, beryllium, and arsenic.

The NIOSH research authority into occupational diseases of coal miners is changed significantly by the new legislation. The law does provide specific authority to conduct health hazard evaluations and establish a list of toxic substances and hazardous physical agents found in mines. Significantly, however, instead of actually setting coal mine health standards, as is provided by the Coal Act, NIOSH would be submitting recommended standards to the Department of Labor, as is done under the Occupational Safety and Health Act.

Research on health hazards in metal and nonmetallic mines can be expected to expand considerably since NIOSH has not previously had specific authority to investigate conditions in such workplaces and develop recommended standards.

Environmental Protection Agency

Toxic Substances Control Act. In 1976, Congress enacted the Toxic Substances Control Act (TSCA) PL 94-469. The Act provides the Environmental Protection Agency (EPA) with the authority to require testing of chemical substances entering the environment and to regulate them when necessary. The regulatory actions include toxicity testing and environmental monitoring. This authority supplements and closes the loop of already existing hazardous substances laws in the EPA and other federal agencies.

Resource Conservation and Recovery Act. Also in 1976, Congress enacted the Resource Conservation and Recovery Act (RCRA) PL 94-580. This Act amended the Solid Waste Disposal Act of 1970 that, unlike the Clean Air and Water Acts, did not contain standards and timetables for compliance. It reauthorized the two Acts and put into place a universal hazardous waste management program applying not only to federal agencies but also to the private sector. RCRA greatly expanded the federal government's role in solid waste disposal management with emphasis on hazardous waste disposal. RCRA continued the federal facilities guidelines under the program established by the 1970 Solid Waste Disposal Act, created a major, new hazardous waste regulatory program, and prohibited the practice of open dumping. In order to accomplish this, the Act provides for extensive federal aid through grants to state and regional agencies for solid waste planning and information programs.

In the "cradle to the grave" program of control established by the Resource Conservation and Recovery Act (RCRA), custody and responsibility moves with a waste material from the generator and the transporter to its final disposal site. The generator, however, never loses liability for the waste created. While the original RCRA regulations (1980-1984) exempted small quantity generators if they produced less than 1,000 kilograms of hazardous waste per calendar month, newer RCRA regulations require generators producing between 100 and 1,000 kilograms per month to meet certain procedural standards.

Since 1980, general waste management requirements of RCRA have included proper notification and recording of hazardous waste activities, along with adequate packaging, labeling, and manifesting of wastes for shipment off-site. A RCRA permit is required for storage, treatment, or disposal of hazardous waste on-site or off-site. Standards for treatment, storage, and disposal (TSD) facilities include rigorous facility management plans, preparedness and prevention of emergencies and releases, contingency plans, operating records and reports, groundwater protection for land disposal facilities, and closure and postclosure plans with financial responsibility assurance.

The new RCRA requirements, following the amendments of 1984, make TSD facilities responsible for assessing human exposure to current and past waste management operations and for corrective action needed to remedy releases of hazardous constituents to the environment.

The RCRA regulations also require worker training, development of safe handling procedures, and emergency response measures. Documentation of training is required and inspectors may require a review of the documentation.

Comprehensive Environmental Response, Compensation and Liability Act. In 1980, Congress enacted the Comprehensive Environmental Response, Compensation and Liability Act

(CERCLA) PL 96-510. This Act established the Superfund program to handle emergencies at uncontrolled waste sites, to clean up the sites, and to deal with related problems. In 1986, CERCLA was reauthorized by the Congress to provide additional funding and additional provisions. The new authorities and programs included in this reauthorization, which are of considerable interest to the industrial hygienist, include underground storage tanks, emergency planning, risk assessment, community right-to-know, research, development, demonstrations and training.

These regulations include requirements that owners-operators of leaking, underground storage tanks undertake corrective action to protect human health and the environment.

Superfund program. Companies covered under the OSHA Hazard Communication Standard also are subject to EPA Hazardous Chemical Reporting Rules under Title III of Superfund.

Covered facilities are required to submit copies of either the same MSDSs they must prepare for OSHA compliance or a list of all chemicals for which MSDSs are required to the state emergency response commission, the local emergency planning committee and the local fire department. The first MSDS submission date: October 17, 1987.

Covered facilities must also submit emergency and hazardous chemical inventory forms to the same state and local authorities. Information on the maximum daily amounts and chemical locations (designated Tier I information) submittal date is March 1, 1988 and annually thereafter. The more detailed Tier II information would be submitted upon request.

Reportable quantities or reporting thresholds determine when a submittal is necessary. In the first year, a reporting threshold of 1,000 pounds would be used. In the second year, the reporting threshold drops to 500 pounds and by the third year, any quantity generates the MSDS and Tier I information reporting requirements.

CERCLA Amendment. The reauthorization amendment to CERCLA also establishes a comprehensive federal program to promote various research, development, demonstration, and training activities, including:

- techniques for detecting, assessing, and evaluating health effects of hazardous substances
- methods to assess human health risks
- methods and technologies to detect hazardous substances and to reduce volume and toxicity.

With the RCRA amendments and the requirements for cleanup under the Superfund (CERCLA) legislation, the time of total accountability for current and past waste management practices has arrived. Quantitative liabilities for waste management practices are directly in proportion to how much a facility is affected by these requirements and remedial actions.

Nuclear Regulatory Commission

The Nuclear Regulatory Commission (NRC) licenses and regulates the uses of nuclear energy to protect public health and safety and the environment. It does this by licensing persons and companies to build and operate nuclear reactors and to own and use nuclear materials. The NRC makes rules and sets standards for these types of licenses. The NRC also carefully inspects the activities of the persons and companies licensed to make sure that they do not violate the safety rules of the Commission.

The Nuclear Regulatory Commission was established as an independent regulatory agency under the provisions of the Energy Reorganization Act of 1974. Transferred to the NRC were all licensing and related regulatory functions formerly assigned to the Atomic Energy Commission, which was established by the Atomic Energy Act of 1946.

The major program components of the NRC are the Office of Nuclear Reactor Regulation, the Office of Nuclear Material Safety and Safeguards, and the Office of Nuclear Regulatory Research, which were created by the Energy Reorganization Act of 1974; plus the Commission-created Office of Standards Development and Office of Inspection and Enforcement. Headquarters offices are located in Bethesda, Maryland; and regional offices are located in five domestic areas.

The Nuclear Regulatory Commission's purpose is to make certain that the civilian uses of nuclear materials and facilities are conducted in a manner consistent with the public health and safety, environmental quality, national security, and antitrust laws. The major share of the Commission effort is focused on the use of nuclear energy to generate electric power.

Programs and activities. NRC fulfills its responsibilities through a system of licensing and regulation that includes the following: licensing the construction and operation of nuclear reactors and other nuclear facilities and the possession, use, processing, transport, handling and disposal of nuclear materials; regulation of licensed activities including assurance that measures are taken for the physical protection of facilities and materials; development and implementation of rules and regulations governing licensed nuclear activities; conducting public hearings on radiological safety, environmental, common defense and security and antitrust matters; and the development of effective working relationships with the states regarding the regulation of nuclear materials. This relationship includes the assurance that adequate regulatory programs are maintained by those states that exercise, by agreement with the Commission, regulatory control over certain nuclear materials within their respective borders. Inspection of NRC-licensed activities is carried out from five regional offices.

The Commission also contracts for research deemed necessary for performing licensing and related regulatory functions and is conducting surveys dealing with the feasibility of a NRC security agency for safeguards and safeguard aspects in using mixed oxide fuels in light water reactors.

Food and Drug Administration

The Food and Drug Administration was created by the Agriculture Appropriation Act of 1931. The Food and Drug Administration's (FDA) activities are directed toward protecting the health of the nation against impure and unsafe foods, drugs and cosmetics, and other potential hazards.

The National Center for Toxicological Research. This Center conducts research programs to study the biological effects of potentially toxic chemical substances found in the human environment. Its studies determine probable health effects resulting from long-term, low-level exposure to chemical toxicants; the basic biological processes for chemical toxicants in animal organisms; help develop improved methodologies and test protocols for evaluating the safety of chemical toxicants and the data that will facilitate the extrapolation of toxicological data from laboratory animals to man.

The Bureau of Drugs. This Bureau develops FDA policy with regard to the safety, effectiveness, and labeling of all drugs for human use, evaluates new drug applications and notices of claimed investigational exemption for new drugs; develops standards for the safety and effectiveness of all over-the-counter drugs; monitors the quality of marketed drugs through product testing, surveillance, and compliance programs; develops guidelines on good manufacturing practices; conducts research and develops scientific standards on the composition, quality, safety, and efficacy of human drugs; and disseminates toxicity and treatment information on household products and medicines.

The Bureau of Foods. This Bureau conducts research and develops standards on the composition, quality, nutrition, and safety of foods, food additives, colors, and cosmetics; conducts research designed to improve the detection, prevention, and control of contamination that may be responsible for illness or injury conveyed by foods, colors, and cosmetics.

The Bureau of Radiological Health. This Bureau carries out programs designed to reduce human exposure to hazardous ionizing and nonionizing radiation, develops standards for safe limits of radiation exposure, develops methodology for controlling radiation exposures, conducts research on the health effects of radiation exposure, and conducts an electronic product radiation control program to protect public health and safety. The Bureau's programs include development and administration of performance standards to control the emission of radiation from electronic products and the undertaking by public and private organizations of research and investigation into the effects and control of such radiation emissions.

BIBLIOGRAPHY

A Guide to Voluntary Compliance. OSHA Publication No. 2088. Out-of-print.

National Safety Council. *Hazardous Waste Management—A Manager's Perspective.* Chicago: National Safety Council, 1986.

The Occupational Safety and Health Act of 1970. U.S. Congress (91st), S.2193. Public Law 91-596. Washington, DC: U.S. Government Printing Office, 1977.

OSHA Industrial Hygiene Technical Manual, OSHA Instruction CPL 2-2, 20A, 1984. Washington, DC: U.S. Government Printing Office, 1984.

U.S. Department of Labor, Occupational Safety and Health Administration. *All About OSHA,* OSHA Publication No. 2056. Washington, DC: U.S. Government Printing Office, 1985.

31

Occupational Safety and Health: The Federal Regulatory Program— A History

by Benjamin W. Mintz

INTRODUCTION

Testifying before a congressional subcommittee in 1968, Willard Wirtz, Secretary of Labor, spoke dramatically of the industrial casualty rate. "Each day," he stated, "there will be 55 dead, 8,500 disabled and 27,200 hurt in America's workplaces." Secretary Wirtz was pressing the subcommittee to approve the comprehensive federal occupational safety and health bill which had been sponsored by the Administration of President Lyndon B. Johnson. The issue, Secretary Wirtz asserted, is "simply whether the Congress will act to stop a carnage which continues only because people don't realize its magnitude, and can't see the blood on the things they buy, on the food they eat, and the services they get." Dr. Irving Selikoff, a pioneer in the field of occupational medicine, testified that it was an "unhappy reflection" on the country that in the 1960s in the United States, 7 percent of all insulation workers would die of asbestosis, a "completely preventable disease." (*Hearing on S.2864 Before the Subcomm. on Labor of the Senate Comm. on Labor and Public Welfare,* 90th Cong., 2nd Sess. 69 [1968]).

Despite this testimony, Congress did not pass occupational safety and health legislation that year. Not until 1970, after a difficult legislative battle, an OSHA law was enacted. It was signed into law by Richard Nixon on December 29, 1970, and became effective 120 days later.

Fifteen years later, in April 1985, a major report on OSHA, entitled *Preventing Illness and Injury in the Workplace,* was issued, by the Office of Technology Assessment (OTA) of the Congress. The OTA estimated that there were between 2.5–11.3 million nonfatal occupational injuries each year and 6,000 deaths annually due to injuries. Noting the great disagreement about the number of workplace illnesses, OTA refused even to estimate the correct number. (*OTA Report,* 1985, pp. 29-38).

This chapter covers the history of OSHA from 1970–1987, in a framework that details the OSHA program's achievements and failures during its brief 17-year history.

1968 TO 1970: THE AGENCY IS CREATED

The legislative battle

The 91st Congress is remembered as the occupational safety and health Congress. In 1969, it passed two landmark statutes: the Coal Mine Safety Act and the Construction Safety Act. In December 1970, the most comprehensive statute of all, the OSHAct, was adopted by overwhelming votes in the Senate and the House of Representatives, (OSHAct 29 U.S.C. §651 and following). Earlier in 1968, a federal OSHA bill was introduced, but did not reach a vote, largely because of strong opposition from the business community. A representative of the American Iron and Steel Institute told a subcommittee that voluntary employer efforts, with state activity, and federally-sponsored research, would accomplish far more in reducing workplace injuries, "than would a program of federal penalties and other attributes of overwhelming federal authority reaching into hundreds of thousands of large and small business operations" (*Hearing on S. 2864 Before the Subcomm. on Labor of the Senate Comm. on Labor and Public Welfare,* 90th Cong., 2nd Sess. 347, 349–352 [1968]).

In 1969, with support from newly-elected President Nixon, a broad consensus had emerged on the need for federal legislation to prevent illnesses and injuries in the workplace. Several important factors contributed to this consensus.

- The workers' compensation system in the past had not provided sufficient financial incentives for employers to undertake efforts to improve workplace safety and health.
- The accident rate was rising.
- Illness in the workplace was a serious and rapidly increasing problem.
- State efforts had proven inadequate (MacLaury, 1981, p. 18).

The main issue was: what would be the substance of the legislation. There was basic agreement on several points; for example, broad coverage by the new law; the need for occupational safety and health standards prescribing the employer conduct necessary to achieve safety and health; government enforcement of these standards; and a role for the states in the program. The disagreement centered on the extent of the powers to be assigned to the Secretary of Labor. The unions, generally supported by Democratic members of Congress, favored a strong role for the Secretary of Labor. They argued in favor of that office's authority to issue standards, to enforce the standards, and to adjudicate violations of the standards, as well as administrate closing down employer operations in the event of imminent dangers. With considerable support from the Republican party, the business community vigorously objected. In the words of Senator Dominick of Colorado, who favored the legislation but supported a "division of responsibilities," the "concentration of authority" advocated by the Democrats was not "balanced," and was "objectionable because concentration of power gives rise to a great potential for abuse" (116 CONG. REC. 37,336 [1970]).

The Democratic view was embodied in the Williams bill which was adopted by the Senate early in 1970. The Republican approach was taken in the bill sponsored by Congressman William A. Steiger of Wisconsin, also an advocate of OSHA legislation, which passed the House of Representatives. The two bills went to a congressional conference committee late in 1970, and its version became the OSHAct. The conference OSHA bill was a compromise, but the thrust of the legislation was much closer to the stringent Senate bill than to the House version. The Democratic Chairman of the House Labor Committee selected representatives to the conference committee who favored the Senate bill, and who no doubt often joined their Senate counterparts in voting on critical issues (Page and O'Brien, 1972, pp. 176-179). The OSHA law as passed has generally been viewed as a stringent regulatory statute; its subsequent history describes the process by which the courts, the Congress, and the Department of Labor joined in an often uncomfortable alliance to make this new and historic legislation—borrowing the words of Senator Dominick—"workable and effective."

OSHA program structure

The statutory structure of the OSHAct is articulated frequently; therefore, only the basic policy principles underlying the new law will be stated.

Universal coverage. The OSHAct applies to all private employers, without exception. Federal employees and state and local employees in states with approved plans are covered by separate occupational safety and health programs. A limited number of particular working conditions subject to the enforcement of occupational safety and health standards of other federal agencies—such as the Coast Guard—are also not covered by the OSHAct.

Employer obligations. The employer's obligation to provide safe and healthful working conditions is defined primarily by means of standards promulgated by the Secretary. The standards are generally promulgated after public rulemaking proceedings designed to elicit data and views upon which these standards are based. The standards define employer obligations prospectively and in considerable detail. To the extent that a hazard is not covered by a standard, the employer must comply with the general duty obligation to provide a workplace "free from recognized hazards likely to cause death or serious physical harm."

Enforcement. Enforcement of the OSHAct through workplace inspections, citations, and assessment of civil penalties is designed to achieve safe and healthful workplaces in two significant ways.

1. With respect to workplaces actually inspected, OSHA imposes legally enforceable abatement requirements and conducts follow-up inspections. These are expected to bring about neutralization of hazards at *that* workplace. The OSHA system for determining priorities for inspection—usually called "worst-first"—should thus result in substantial elimination of hazards in the most dangerous workplaces.
2. As for other workplaces, the enforcement program is intended to constitute an incentive to employers to abate hazards without regard as to whether an inspection takes place. Two elements in the statutory structure are particularly designed to achieve this result: sanctions are imposed when violations are disclosed at the first inspection and no advance notice is given of workplace inspections.

Employee participation. The OSHA is preeminently designed to protect workers, and the statute gives them a crucial role in virtually every aspect of the program. For example, an employee representative has the right to request a workplace inspection, to participate in the walkaround inspection, to participate in adjudicatory and court review proceedings and rulemaking hearings, and, significantly, to be free from employer reprisal for the exercise of rights protected under the Act. Regulatory requirements have also been imposed on employers to assure that employees are informed of workplace hazards; this has been called worker right-to-know.

Checks and balances. Reflecting the concern of Congress that the OSHA program be balanced, the Act contains both broad agency authority and constraints on that authority. Discretion is given to the Secretary to issue occupational safety and health standards; however, these standards may be issued, generally, only if the Agency follows detailed rulemaking proceedings. Specific criteria are set forth in the statute, notably in section 6(b) (5) and (7), for the content of OSHA health standards, which must be based on the best available evidence. All standards are subject to court of appeals review based on the record in the rulemaking proceeding. Similarly, OSHA enforcement actions may be contested and are adjudicated before a neutral and independent commission; employees have the right to intervene and protect their interest in these adjudicatory proceedings. Other constraints on the Agency, though not explicitly in the Act, are implicit in our governmental system. Noteworthy is the right of an employer to refuse a workplace inspection unless preceded by a warrant based on probable cause. Congressional and White House scrutiny of OSHA implementation of the Act, particularly in the area of standards development, has been pervasive.

Role of the states. The OSHA program is primarily a federal program. After a brief transition period, state occupational safety and health enforcement was preempted. However, the Act assigns a role to the states to develop their individual OSHA programs, which are supported by 50 percent federal funding; the state programs must be "at least as effective as" the federal program and are subject to continuing federal monitoring and evaluation. The history of OSHA has been replete with sharp controversies over the proper role of the states in the enforcement of the OSHAct.

Information and data. Congress was acutely aware of the inadequacy of the then available fund of information on occupational injuries and illnesses. In order to more acurately determine injury and illness rates, the Act mandates that employers maintain accurate records on all but minor work-related deaths, injuries, and illnesses and that they periodically report that information. Further, OSHA was required to develop "an effective program of collection, compilation, and analysis" of occupational safety and health statistics. This responsibility was assigned to the Bureau of Labor Statistics (BLS) in the Department of Labor, which publishes annually the results of its statistical survey of occupational related injuries and illnesses.

Training and education. In addition to its power to impose sanctions, OSHA has a wide array of authority that includes education, training, and consultation, which are designed to elicit voluntary employer and employee cooperation in bringing about safe and healthful workplaces. These voluntary programs have been undertaken by OSHA throughout its history and have received special emphasis during the administration of President Reagan. The program of enforcement-free workplace consultation visits, which has been financed mostly by OSHA and operated by the states, has been particularly important.

1971 TO 1973: EARLY AGENCY ACTIVITIES

The new agency

The Secretary of Labor first created an agency within the Department—the Occupational Safety and Health Administration—to implement the new program (Figure 31-1). Its head, OSHA's first Assistant Secretary, was George C. Guenther, who served until January 1973. Guenther had been head of the Department of Labor's Bureau of Labor Standards, which had responsibility for administering the pre-OSHA occupational safety and health programs under the Walsh-Healey Act, the Longshoremen and Harbor Workers Act, and related legislation. An administrative structure was established for enforcing the Act: this included 10 regional offices and 49 area offices, staffed with compliance officers assigned to the area offices who were responsible for conducting workplace inspections. At that time, most of the OSHA inspectors had expertise in the area of safety. Thus, in June 1973, OSHA had 456 safety inspectors and 68 industrial hygienists. Not unexpectedly, the bulk of OSHA inspection activity, 93.4 percent in fiscal year 1973, was safety-related (*OTA Report,* 1985, p. 336). This reflected the primary emphasis that had been placed in the past, both in and out of government, on industrial accidents, about which a great deal more was known. It would be another 5 years, when Assistant Secretaries Corn and Bingham shifted OSHA's emphasis to health hazards, before industrial hygiene inspectors constituted a significant portion of OSHA's compliance staff.

Compliance activity commences

A regulatory framework for OSHA activity was also established in the first two years of OSHA's existence. Two major regulations were issued in 1972 after public comment. The first, known as Part 1903, dealt with Agency inspection procedures (29 *CFR* Part 1903). The OSHA statute sets forth the basic outline for OSHA enforcement activity. The inspection regulations fleshed out inspection procedures in numerous ways, and provided interpretations of key provisions. The regulations in Part 1903 have remained substantially intact throughout the history of OSHA.

Another major element in OSHA's regulatory framework was the first *Compliance Manual* orginally issued in April 1971, and periodically revised thereafter (OSHA *Compliance Manual,* Jan. 1972, OSHA 2006). Although intended as instructions to field staff for implementing the OSHA program, the *Compliance Manual* from the start was a public document and, in practice, has constituted the primary means by which the Agency advised interested parties of its policies in administering the statute. Many interpretations of statutory terms—what is a serious, willful and repeated violation—are included in the manual; a general definition of imminent danger is also included.

The *Field Operations Manual* (FOM), much expanded, is currently in two volumes, the *OSHA Field Operations Manual* and the *Industrial Hygiene Technical Manual.*

A particularly crucial component of OSHA's *Compliance Manual* was the establishment of inspection priorities. Under the Act, OSHA has general inspection authority; however, inspections are mandated only in response to certain written employee complaints. It has therefore been essential for OSHA to define the criteria for its selecting for inspection by its limited compliance resources among the approximately 5 million workplaces. In its first *Compliance Manual,* OSHA gave instructions on what it called "compliance programming." The first priority (after, of course, dealing with imminent dangers) was to investigate workplaces where fatalities or catastrophes occurred. The second priority was response to employee complaints; the third priority was the "special hazard elimination programs" which included target industries and health hazards. These priorities brought OSHA inspectors to the most hazardous workplaces, "hazardous" as determined in part in the annual statistical surveys of the BLS.

Finally, to the extent resources are available, OSHA said it would conduct inspections of all other workplaces in order to make clear that the Act's obligations are applicable to *all* private employers. (The latter two categories are usually called "programmed inspections" [*OSHA Compliance Manual*, Jan. 1972, chapt. 4]). While these broad priority categories have remained substantially the same since 1972, they have been modified in detail in many respects, particularly respecting Agency response to complaints. Additionally, considerable controversy later developed, in Congress and elsewhere, on whether employers with above-average safety records should be exempt from inspection activity.

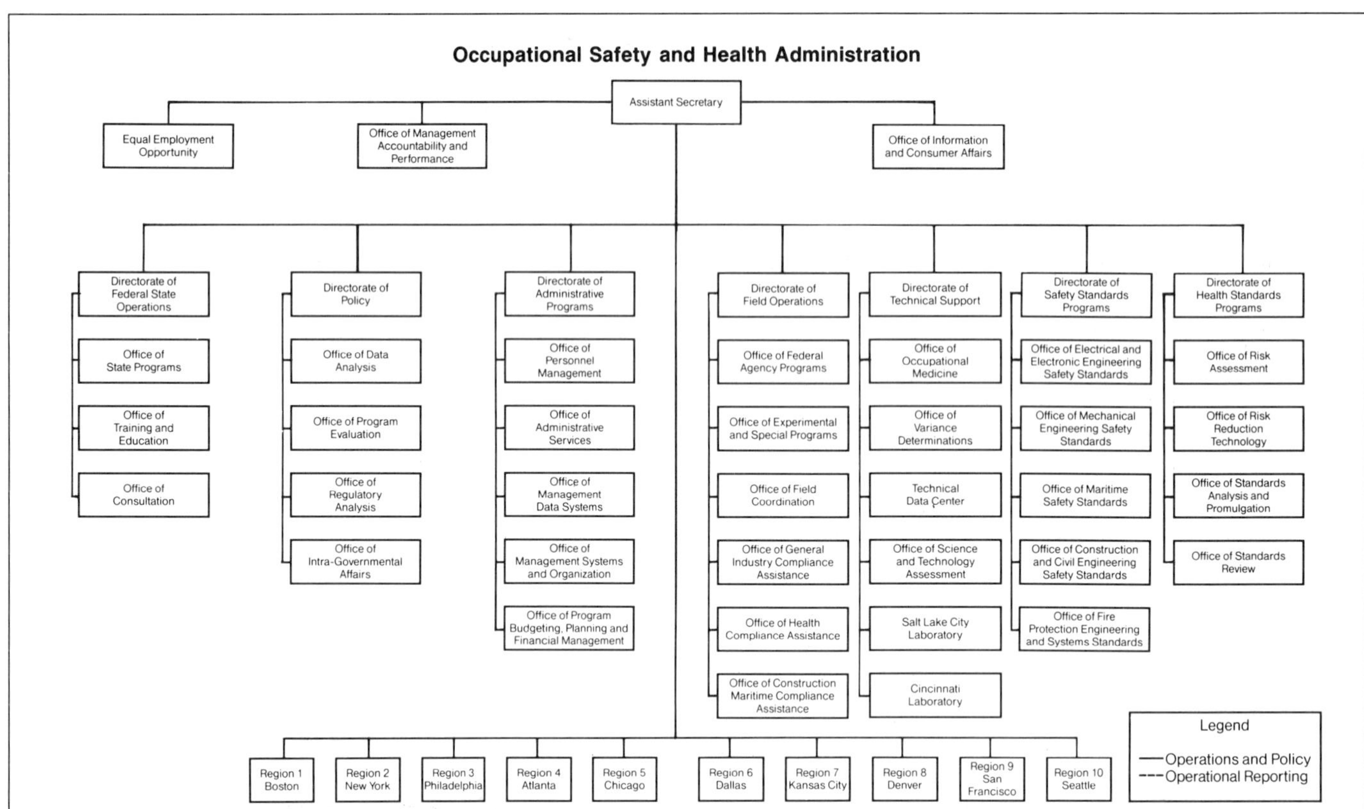

Figure 31-1. Current organizational chart of OSHA. (Adapted from *Preventing Illness and Injury in the Workplace.* Washington, DC: U.S. Congress, Office of Technology Assessment, OTA-H-256, April 1985)

Field staff were required to report regularly on enforcement activity to the OSHA National Office in Washington. One of the required statistical reports is the number and type of OSHA inspections. In fiscal year 1973, for example, OSHA conducted about 48,000 inspections; 5.1 percent of these were fatality/catastrophe inspections; 13.7 percent were complaint inspections; 66.5 percent "programmed" inspections—target industries and general industry inspections; follow-up inspections in workplaces where violations had been found constituted 14.7 percent. Since then, the total number of inspections has grown to a high of about 90,000 in fiscal year 1974, leveling to about 60,000 or fewer after 1977, which may reflect the new emphasis on more time-consuming health inspections. While the number of fatality/catastrophe inspections has remained fairly constant over the years, there have been significant variations in the number of complaint inspections, resulting from important Agency policy changes either emphasizing or deemphasizing complaint inspections. A significant deemphasis in follow-up inspections occurred after 1981 (*OTA Report,* 1985, p. 368).

The enforcement program began at the end of August 1971, when OSHA's newly adopted standards became generally effective. The first OSHA citation was issued at the beginning of May; it was a general duty violation based on excessive employee exposure to mercury. During the remainder of the calendar year, 9,507 citations were issued with proposed penalties amounting to $737,486. In 1972, the number of issued citations increased to 23,900 and the amount of the proposed penalties was more than $3 million (The President's Report on Occupational Safety and Health, May 1972, p. 87; Dec. 1973, p. 36).

Start-up standards

One of OSHA's most important tasks after the Act became effective was to issue standards that would provide the Agency with a basis for promptly commencing its enforcement program. Congress gave the Agency authority under section 6(a) to promulgate certain standards without rulemaking—that is, without the delays inherent in public comment proceedings. These have been referred to as start-up standards and included national consensus standards and established federal standards, both of which had gone through at least some public comment process before they were issued. The OSHA quickly—many have said too quickly—promulgated these standards, and on May 29, 1971, barely a month after the Act's effective date, OSHA issued a large body of these start-up standards. Most were in the safety area, taken from standards issued by the American National Standards Institute (ANSI) and the National Fire Protection Association (NFPA), both determined by Congress to be national consensus organizations; others were pre-OSHA standards issued under the Walsh-Healey Act, the Longshoremen's and Harbor Workers' Compensation Act, and the Construction Safety Act, which already had gone through rulemaking proceedings. Some health standards were issued in 1971. Among them were the Threshold Limit Value (TLV) levels of the American Conference of Governmental Industrial Hygienists (ACGIH) for about 400 substances, which had become established federal standards under the Walsh-Healey Act. These contained no more than the Permissible Exposure Limit (PEL) and a requirement for implementation of engineering and protective equipment controls to reach the limit. But these limits were often based on inadequate information and in 1970, Congress recognized that the standards "may not be as effective and up-to-date as desirable" and that they would provide only a "minimum level of health and safety" (S. REP. No. 1282, 91st Cong. 2nd Sess. 6 [1970]).

There is no question that OSHA's adoption of start-up standards in 1971 got enforcement off to a rapid start. Even today, the bulk of OSHA safety standards appearing in the *Code of Federal Regulations* is the original material that was issued at the start of the program. But OSHA's hasty action—as it was called—led to no end of controversy and criticism of the young regulatory agency. Litigation ensued when OSHA began to enforce its national consensus safety standards, and many citations and penalties were reversed by the Review Commission and the courts. Indeed, one of the most outspoken critics of the OSHA's start-up standards was Robert D. Moran, Chairman of the Review Commission, who argued that these standards "violate the spirit and purpose of the Act" (Moran, 1976, pp. 19-20). Even the President of the AFL-CIO, testifying many years later, said: "Because of the hue and cry over these 'nitpicking' aspects of the program, the more serious and important goals of the OSHA program became lost in a morass of largely unintelligible debate and political animosity" (*Oversight on the Administration of the Occupational Safety and Health Act, 1980: Hearing Before the Senate Comm. on Labor and Human Resources,* 96th Cong., 2nd Sess. 730-731 [1980] [Testimony of Lane Kirkland, President of the AFL-CIO]). Largely in response to the criticism, a Presidential Task Force was organized in 1976 to develop a "model approach to safety standards." The Task Force reported a year later that OSHA should avoid design or specification requirements and recommended a "performance" approach, by means of which employers would have had considerable flexibility to achieve the safety goals of the standard "by any appropriate means" (MacAvoy, 1977, pp. 17-21). The OSHA Appropriations Act in 1976 directed the Agency to undertake a "review and simplification" of existing standards and to eliminate "nuisance" standards. (Department of Labor and Health, Education and Welfare Appropriations Act of 1976, Pub. L. No. 94-206, 90 Stat. 3 [1976]).

The pressure continued to mount and in 1978, after receiving public comment, OSHA deleted about 600 safety standards, using seven criteria for determining which should be revoked; among those deleted were "obsolete" or "inconsequential" standards and those "encumbered by unnecessary detail" (43 *Fed. Reg.* 49, 726 [1978]). The OSHA also undertook a broad revision of all its national consensus standards, a project that has moved slowly, with only the Standards on Fire Protection and Electrical Hazards revised by 1985. Other issues arising from the initial promulgation continued to receive OSHA's attention. One of these was whether so-called "should" standards—those stating that the employers "should" take a particular action rather than "shall" do so—are mandatory in effect. This issue was litigated extensively and was not resolved until February 1984 when OSHA revoked a group of its "should" standards (49 *Fed. Reg.* 5318 [1984]).

Regulating asbestos

In enacting OSHA, Congress was well aware of the dangers of employee exposure to asbestos fibers. The Senate Labor Committee in its 1970 report referred to asbestos as "another material which continues to destroy lives of workers," noting that as many as 3.5 million workers are at risk of asbestosis, pulmonary cancer and mesothelioma as a result of asbestos exposure (S. REP. No. 1282, 91st Cong. 2nd Sess. 2-4 [1970]). In 1971, OSHA adopted, as an established federal standard, the Walsh-Healey Asbestos Standard, requiring the implementation of engineering controls to achieve a PEL of 12 fibers per cubic centimeter of air. However, it was immediately apparent that this standard provided inadequate protection. Several months after the effective date of the Asbestos Start-up Standard, on December 7, 1971, OSHA invoked its emergency authority to issue an Emergency Temporary Standard (ETS) for asbestos. The ETS mandated, effective immediately, a 5-fiber per cubic centimeter of air 8-hour time weighted average (TWA), with a 10-fiber per cubic centimeter of air ceiling for any 15-minute period (36 *Fed. Reg.* 23,207 [1971]).

The Standard was based on OSHA's finding that there was a "grave danger" to employees from asbestos exposure. Unlike many later OSHA emergency standards which were challenged, and the challenges sustained, the Asbestos ETS was not challenged and remained in effect for the entire 6-month period. In June, 1972, after receiving written comment and testimony at a public hearing, OSHA issued a permanent Asbestos Standard. In its preamble to the Standard—which explains the basis for the Standard—OSHA asserted that there was no dispute that exposure to asbestos is "causally related" to cancers and asbestosis; the only issue, it said, was the specific level below which exposure is safe. This was an issue, because the Agency did not have "accurate measures" of the levels of exposure occurring 20–30 years ago, which caused the disease. However, OSHA concluded that in view of the undisputed "grave consequences" from exposure to asbestos fibers, it is essential that the exposure be regulated "on the basis of the best evidence available now, even though it may not be as good as scientifically desirable" (37 *Fed. Reg.* 11,318 [1972]). The Agency therefore determined that the PEL be set at a 2-fiber per cubic centimeter of air level. However, concluding that "many work operations" will meet "varying degrees of difficulty," such as the necessity for extensive redesign and relocation of equipment, in complying with the 2-fiber per cubic centimeter of air standard, OSHA delayed the effective date for 4 years, applying a less strict 5-fiber standard in the interim.

The AFL-CIO's Industrial Union Department challenged the standard in the Court of Appeals for the District of Columbia Circuit. Among other provisions, the union attacked the delay in applying the stricter PEL. The Court decision in the case was a landmark in the development of OSHA law. It affirmed OSHA's decision to adopt a 2-fiber per cubic centimeter standard. In much quoted language, the Court, in an opinion by Judge Carl McGowan, recognized that some of the questions involved in OSHA standards development are "on the frontiers of scientific knowledge," and therefore not susceptible to precise factual determination. Court review, therefore, cannot be factual-type review which is typically undertaken by the courts but rather must be deferential to the Agency, calculated only "to negate the dangers of arbitrariness and irrationality." In Asbestos, the Court said, the choice of a lower level was "doubtless sound" inasmuch "as the protection of the health of employees is the overriding concern of OSHA" (*Industrial Union Department v. Hodgson,* 499 F. 2nd 467, 478 [D.C. Cir. 1974]).

But, while affirming the Agency's 2-fiber per cubic centimeter of air level, the court rejected OSHA's delay of the effective date.

The Court granted that "feasibility" considerations—both economic and technical—may play a role in OSHA decision-making. However, OSHA had not shown in the record why it was necessary to impose an across-the-board delay in needed protection for workers in all industries. The case was remanded to OSHA for further development of a record on the feasibility of the lower PELs in specific industries. The language of the court on the feasibility issue has also served as a significant precedent for later decisions on OSHA standards. "Practical considerations can temper protective requirements," Judge McGowan said. At the same time, standards may be economically feasible "even though, from the standpoint of employers, they are financially burdensome and affect profit margins adversely." However, it determined that a standard was not feasible if it required "protective devices unavailable under existing technology or by making financial viability generally impossible" (*Industrial Union Department v. Hodgson,* 499 F. 2nd 467, 478 [D.C. Cir. 1974]).

Involving the states

The dissatisfaction of organized labor with OSHA's Asbestos Standard was exceeded in intensity by the controversy engendered by the Agency's initial policy in the area of state plans. The Act contains a detailed scheme defining the OSHA relationship between the federal government and the states. Primary authority was assigned to the federal agency; the states were preempted except if they submitted "at least as effective" plans, which are approved by OSHA and found in practice to be equally effective. Finally, in order to prevent preemption of ongoing state programs while the states were still developing plans for submissions, section 18(h) authorized 2-year agreements permitting continued state enforcement.

The OSHA acted quickly to encourage state participation in the OSHA program. The Secretary of Labor quickly wrote to the Governors urging the states to enter into section 18(h) agreements and to submit "at least as effective" plans. All but three states accepted the Secretary's invitation, and entered into section 18(h) agreements. By the end of 1972, 44 states, the District of Columbia and four territories had submitted plans. It was apparent, however, that as submitted, these plans did not meet the statutory criteria for equal effectiveness. Thus, many states lacked occupational safety and health enabling statutes which could provide authority for the state OSHA program. To meet this problem, apparently unanticipated by Congress, and "to allow the transition that states must undergo to upgrade an existing program," OSHA developed a new concept—the "developmental" plan—in its State Plan Regulations (Part 1902). This authorized approval of state plans—triggering 50 percent federal funding—on the basis of commitments by the states to effectively meet goals in the future, generally within 3 years, even though, on submission, the plans were concededly not equally effective (29 *CFR,* §1902.2).

Organized labor was skeptical from the start about state efforts, arguing before the committees considering OSHA bills that state OSHA programs were largely understaffed and ineffective. When the developmental concept appeared by means of OSHA interpretation in 1971, it was vigorously opposed by the unions. The main forum for debate was NACOSH—the National Advisory Committee on Occupational Safety and Health—created by the Act, where Jack J. Sheehan, legislative representative of the United Steelworkers of America, said: "On the developmental plan, it's a non-plan as far as I can understand it" (Proc. NACOSH, Sept. 24, 1971, pp. 119–126).

The debate on developmental plans led to further debate on OSHA's attempt to extend the effective period of section 18(h) agreements by means of so-called "temporary orders." These orders were even more "temporary" than OSHA planned, because in January 1973 a federal district court found them beyond OSHA authority. By the end of 1972, only three plans—those of South Carolina, Oregon, and Montana—had been approved. In explaining the need for temporary orders, Assistant Secretary George Guenther argued that with an "absolutely strict interpretation" after 1972 there would be a "protection gap" for employees in states with plans pending approval. The temporary orders would close the gap by allowing state enforcement for an additional 6 months (Proc. NACOSH, Nov. 16, 1972, pp. 44–46). Judge Barrington Parker did not find this policy-imperative sufficiently compelling and refused to approve OSHA's expansive interpretation of its authority. Noting congressional and union objections to temporary orders, Judge Parker relied particularly on the "express language of the Act, which mandated exclusive federal jurisdiction after December 28, 1972"; he also took the position that even without temporary orders, state OSHA jurisdictions would not be "seriously" disrupted, which turned out to be the case (*AFL-CIO v. Hodgson,* 1971–1973 OSHD [CCH] ¶15,353 [D.D.C., 1973]). The OSHA did not pursue an appeal from Judge Parker's decision and temporary orders were abandoned.

Involving workers

The Act pervasively provides for the participation of employees in the OSHA program. The right of employees or their representatives to request an inspection and their right to participate in the physical inspection of the workplace is foremost. Under section 8(f)(1), OSHA must conduct an inspection if an employee or representative files a written complaint alleging that a hazard is threatening physical harm or is an imminent danger and OSHA determines that there are "reasonable grounds" to believe there is merit to the complaint. Early, the issue arose on OSHA's response to complaints not meeting the formality requirements of the Act. These, usually known as "informal" complaints, include oral communications by employees—telephone calls to an area office—alleging hazardous workplace conditions. The OSHA emphasizing its limited inspection resources, determined in its first *Compliance Manual* that, as a general rule, informal complaints would not trigger workplace inspections except in situations apparently involving imminent dangers (*OSHA Compliance Manual,* Jan. 1972, chapt. 6). This inspection policy continued essentially unchanged for the next 4 years, until the traumatic "Kepone incident," to be discussed later in this chapter, forced OSHA to a major reversal of its policy on informal complaints.

The right of employees to participate in the physical inspection is known as the walkaround right. In reporting favorably on the OSHA bill in 1970, the Senate Labor Committee observed that under pre-OSHA laws, "workers tend[ed] to be cynical regarding the thoroughness and efficacy" of inspections because they were usually not advised that an inspection was taking place. The walkaround right was therefore added to "provide an appropriate degree of involvement of employees themselves in the physical inspections of their own places of employment" (S. Rep. No. 1281, 91st Cong., 2nd Sess., 11–12 [1970]).

The question arose almost immediately—are employees selected as representatives in the walkaround entitled to the wages they would otherwise have received if they continued actual work. In a proceeding involving an inspection of the Mobil Oil Company, the Oil, Chemical and Atomic Workers Union complained that pay was withheld from its members, thus interfering with a basic statutory right. The Solicitor of Labor rejected the claim, deciding that time spent in accompanying an inspector was not hours worked and therefore no discrimination had taken place. The Court of Appeals for the District of Columbia upheld the Solicitor's view in a related case involving the Fair Labor Standards Act (*Leone v. Mobil Oil Corp.*, 523 F. 2nd 1153 [D.C. Cir. 1975]). The OSHA policy remained the same until 1977 when it was reexamined and reversed by Assistant Secretary Bingham.

Many statutes, both federal and state, contain whistleblower provisions protecting the exercise by individuals of their protected rights against reprisal and retaliation. Section 11(c) of the OSHAct is a whistleblower provision: it prohibits discrimination in conditions against employees "because of the exercise of any rights afforded by the Act." Specifically mentioned are the protected right to file complaints, to institute OSHA proceedings, and to testify at these proceedings. In its initial interpretation of the provision, OSHA said that the provision also protected "other rights [which] exist by necessary implication." That interpretation confronted a particularly difficult issue: whether an employer could discharge or discipline an employee for refusing to work in particularly dangerous circumstances. The OSHA expansively interpreted the Act and determined that an employee was protected against discipline if "with no reasonable alternative" he or she "in good faith" refuses to expose himself to imminently dangerous conditions (29 *CFR* 1977.12).

Injury and illness rates

One of Congress' important goals in enacting OSHA legislation in 1970 was to improve the reliability of the work-injury and work-illness data. Although the BLS had been collecting and publishing this data for more than 30 years, there were serious limitations on the usefulness of the statistics that were based on the Z.16.1 ANSI standard, the *American Standard Method of Measuring and Recording Work Injury Experience.* Only disabling injuries had been counted in the injury rates; occupational illnesses were seldom, if ever, recorded, except in the most obvious and extreme cases; and the information was limited by its dependence on voluntary reports from employers.

The OSHA statute required the Secretary of Labor to issue regulations requiring employers "to maintain accurate records of, and to make periodic reports on, work-related deaths, injuries and illnesses other than minor injuries requiring only first aid treatment and which do not involve medical treatment, loss of consciousness, restriction of work or motion or transfer to another job." The Secretary was also required "to develop and maintain an effective program of collection, compilation, and analysis of occupational safety and health statistics." Soon after the Act was passed, the Secretary directed the BLS to continue to collect and publish work-injury statistics under the new law. Mandatory record-keeping regulations (Part 1906) were adopted by OSHA and published on July 2, 1971, following public written comment, advice and consultation by an interagency group and a public hearing held by the Office of Management and Budget (OMB). Under the regulations, employers were required to maintain a log of occupational injuries and illnesses, a supplemental record containing more detailed information on each injury and illness, and a yearly statistical summary of injuries and illnesses. These documents were maintained by all employers at individual establishments; but only a statistical sample of employers was required to report to BLS on injury and illness experience.

The BLS's first full-year survey in 1972 included data for all employments outside of farms and government (which were separately surveyed) and showed almost 5.7 million recordable work-related injuries and illnesses; that is, one out of every 10 workers experienced a job-related injury or illness. Then, as it has in each survey since 1972, BLS emphasized that "underreporting of occupational illnesses is prevalent due to problems of identification and measurement." These problems include "lack of facilities and trained medical personnel for proper diagnosis; long latency periods thwarting timely detection; questions of occupational illness coverage under workers' compensation; and factors outside the work environment that cloud the work relationship concept." In addition to reporting on numbers of injuries and illnesses, BLS also reports on incidence rates—that is, injuries and illnesses per 100 full-time employees. In 1972, the incidence rate for all recordable cases was 10.9. The incidence rate for lost-workday cases was 3.3; and for nonfatal cases without lost workdays was 7.6 (The President's Report, 1973, pp. 61–74).

Additional OSHA activities during this period included establishing OSHA's Training Institute in Rosemont, IL (now located in Des Plaines, IL). The Institute opened in January 1972 to provide training primarily for federal and state inspection staff and employees in the private sector. On July 28, 1971, President Nixon signed Executive Order 11,612, implementing Section 19 of the Act and setting up the framework for the safety and health program for federal employees. The OSHA's role in the Federal Safety Program was essentially advisory; primary responsibility for carrying out the responsibilities under the law and the Executive Order was assigned to the individual federal agency heads. Finally, OSHA embarked on a program of cooperation with 28 other federal agencies that had authority relating to occupational safety and health in the private sector. Despite an agreement with the Department of Transportation in May 1972, which was designed to avoid duplication in the handling of complaints in the railroad industry, it became clear very quickly that major jurisdictional disagreements among the various federal agencies were emerging, which would be difficult to resolve.

1973 TO 1976: THE AGENCY CONTINUES TO GROW

OSHA's first transition

George C. Guenther left office in January 1973, at the beginning of President Nixon's second term. After a brief hiatus, John H. Stender, a former union official and a Republican legislator from the State of Washington, became Assistant Secretary of OSHA. He served until July 1975 in an often stormy term of office. There was no Assistant Secretary from June until December 2, 1975, when Dr. Morton Corn, who had been Professor of Industrial Hygiene and Engineering at the University of Pittsburgh took office. As the first health professional who headed

OSHA, Dr. Corn brought major changes, emphasizing professionalism and bringing about a major reorientation towards health regulation and enforcement. Dr. Corn left office in January 1977 when the newly elected President Jimmy Carter chose Dr. Eula Bingham, also a health professional, as Assistant Secretary.

During the period between 1973–1976, OSHA continued to grow, both in size and in the magnitude of the controversy it engendered. The OSHA budget in fiscal year 1972 in current dollars was 33.9 million; by fiscal 1977 it was $130.2 million. (As adjusted for inflation, the growth rate, of course, was less marked: $33.9 million in 1972 and $90.0 million in 1977.) The OSHA inspection staff also grew. At the end of 1972, the field enforcement staff included 456 compliance officers and 68 industrial hygienists. Almost all compliance officer positions authorized for fiscal year 1976 were industrial hygienists, so that in December 1976 there were 358 hygienists in OSHA, or 27 percent of the total staff (The President's Report, 1976, pp. 28–31; 36–39).

There were parallel increases in the number of OSHA inspections; in particular Assistant Secretary Stender emphasized the importance of numbers of inspections. In fiscal year 1973, there were 48,409 establishment inspections; by fiscal 1976 the number had grown to 90,482 inspections.

The total number of OSHA inspections was reduced in fiscal 1977, largely as a result of the new emphasis on health enforcement, which was introduced by Dr. Morton Corn. In fiscal 1976, 8.4 percent of OSHA inspections were listed as health inspections; in fiscal 1977, 15.2 percent and in fiscal 1978 (the beginning of the administration of Dr. Bingham) the number of health inspections grew further to 18.6 percent. Health inspections are typically far more time-consuming, requiring, among other things, calibration of equipment, monitoring of workplace atmospheres, and collection and analysis of monitoring results. In 1977, OSHA separated from its *Compliance Manual,* which had come to be known as the *Field Operations Manual* (FOM), a detailed instructions manual for health inspections—then named the *Industrial Hygiene Field Operations Manual* (IHFOM). This IHFOM specifies the technical procedures for making an industrial hygiene inspection to insure uniformity. OSHA also embarked on a 3-year apprenticeship program for entry level health compliance officers.

The Kepone incident

The basic OSHA priorities for selecting workplaces for inspection remained basically the same during the Stender and Corn administrations. Special emphasis programs, emphasizing inspections of specific hazardous industries, were adopted. In March 1973, the emphasis was placed on trenching and excavation inspections; a new program was instituted in 1975, called the National Emphasis Program (NEP), targeting foundries for inspection. The purpose of the program was to combine a variety of OSHA resources including training and education, consultation and enforcement in concentrating on a single high-hazard area. The most important change in OSHA's inspection priority system was its major modification in response to employee complaints which took place in 1976 as a result of the "Kepone incident." In September 1974, OSHA had received a complaint from a former employee of Life Science Products Company of Hopewell, VA, a chemical manufacturing company, alleging his exposure to pesticide fumes and dust. The OSHA Area Office did not conduct an inspection of the plant because of the complaint's informality, treating it instead as a discrimination complaint under section 11(c). The matter rested there for about 10 months when it became known that Life Science employees had been massively exposed to a pesticide, Kepone, which caused serious illness in seven of these employees. The plant was quickly closed down by the State of Virginia, and it further became known that pervasive ecological damage had been caused by the company's irresponsible and unlawful disposal of Kepone into the James River.

The great human and environmental damage, caused—on the surface, at least—by OSHA's failure to inspect, led to great public outrage. In March 1976 a subcommittee of the House Labor Committee held a hearing in Hopewell, VA, on the Kepone incident. Dr. Corn had just assumed office and he was thoroughly and angrily questioned on handling of the Kepone complaint, incidents which preceded his association with OSHA. Although OSHA sometimes was blamed for workplace accidents with little or no justification, the Agency's performance in the Kepone matter required much explanation. Dr. Corn wrote to Chairman Gaydos, saying that the "episode has pointed up a distinct need for improvements in OSHA's response to employee complaints of hazardous working conditions" (Corn letter, 1976). Soon thereafter OSHA revised its field instructions, instructing staff to conduct a workplace inspection "whenever information comes to the attention of the Area Director without regard to its source and without regard to whether it meets the formality requirements of Section 8(f), indicating that safety or health hazards exist at a workplace." The directive also shortened the time frame for complaint responses: no more than 24 hours for imminent dangers, 3 days for serious violations, and 7 working days for nonserious violations (OSHA Field Information Memo No. 76-9, 1976). The Kepone incident also solidified Dr. Corn's intention to focus a greater portion of OSHA's resources on health regulation.

It was expected that the new complaint policy would substantially increase the number of these inspections. In fiscal year 1976, complaints were 10.2 percent of total inspections; 1 year later, complaint inspections constituted 32.4 percent (19,415 out of 60,004 inspections.) There were similar high percentages of complaint inspections for the next 2 years. The impact that this new emphasis had on OSHA enforcement activity was not entirely anticipated, however. In December 1977 OSHA said that the number of complaints received and investigated "overtaxed the resources available. . . introduced complaint backlogs, reduced inspection activity at some field offices in several important safety and health programs, and severely decimated planned regional inspection programs" (OSHA Prog. Dir. No. 200-69 [1977]). The complaint policy was criticized in a 1979 report of the Comptroller General, as will be discussed later, and was significantly modified by Dr. Bingham in 1979.

With the increase in OSHA inspections, there were substantial increases in the number of OSHA citations issued, the numbers of violations of all kinds, and the rate of employer contest of citations and penalties. Thus, for example, in fiscal year 1973, 3.2 percent of inspections resulted in serious violations (1,535 violations), by fiscal year 1977 this increased to 18.5 percent of inspections (11,092 violations). The increase in the number of serious citations resulted in part from a series of issue papers by the General Accounting Office (GAO) at the request of the Senate Committee of Labor and Human Resources, which criticized OSHA for its citation policy (*OSHA Review, 1974: Hearing Before the Subcomm. on Labor of the Senate Comm. on Labor and Public Welfare,* 93rd Cong. 2nd Sess., apps., 941-1,238 [1974]). The percentages of OSHA inspections with willful and repeated violations, which are subject to penalties of up to $10,000, were substantially smaller (in fiscal 1977, 0.3 percent willful violations and 3.9 percent repeated violations) but they too increased over the 4-year period. The contest rate almost tripled between 1973–1977, reflecting the growing adversary quality of the OSHA program. During 1975, OSHA implemented a comprehensive field performance evaluation system, establishing qualitative and quantitative criteria for federal field activities (The President's Report, 1975, pp. 49-53).

OSHA in court

With the increased litigation over citations and penalties, the OSHA enforcement case law evolved at a rapid rate. Some of the proceedings challenged the constitutionality of the enforcement scheme. The issue was decided favorably to OSHA in 1977 in *Atlas Roofing Company Inc. v. OSHRC,* 430 U.S. 442 (1977). In this major constitutional decision, the U.S. Supreme Court held that Congress could constitutionally establish a flexible administrative procedure for the imposition of civil penalties, even though no jury trial was provided.

The general duty clause also resulted in numerous court decisions. In a major case decided in 1974, the Court of Appeals for the Eighth Circuit, *American Smelting and Refining Company v. OSHRC,* 501 F. 2nd 504 (8th Cir. 1974), held that high levels of airborne concentrations of lead were a "recognized" hazard under section 5(a)(1), even though they could not be detected except through the use of monitoring equipment. Although a lead standard was soon to become effective, and the general duty clause no longer applied to lead hazards, the basic principle had a major impact. In *National Realty and Construction Co. v. OSHRC,* 489 F. 2nd 1257 (D.C. Cir. 1973), decided a year before by the Court of Appeals for the District of Columbia Circuit, several principles of general duty law were set forth: (1) In determining whether a hazard was "recognized," the standard "would be the common knowledge of safety experts who were familiar with the circumstances of the industry or activity in question"; (2) a hazard was "likely to cause" death or serious physical harm if the result "could eventuate" "upon other than a freakish or utterly implausible concurrence of circumstances"; and (3) the general duty clause imposes sanctions only on "preventable hazards"; hazardous conduct by employees is not "preventable" "if it is so idiosyncratic in motive or means that conscientious employers familiar with the industry could not take it into account in prescribing a safety program." This last issue—known as the employee misconduct defense—followed OSHA in its commission and court litigation throughout the Agency's history.

The increased court litigation, both involving citations and penalties and, as we shall see, litigation on the review of OSHA standards, raised a critical question of which group of attorneys would be responsible for this litigation. The Act stated that OSHA litigation (except for litigation in the U.S. Supreme Court) would be handled by Department of Labor attorneys in the Office of the Solicitor, "subject to the direction and control of the Attorney General." Almost immediately after the effective date of the Act, a dispute arose as to the extent of this Department of Justice "direction and control." The Labor Department claimed that its attorneys who were experienced and expert in OSHA affairs were more qualified to handle its litigation with minimum supervision by the Department of Justice. Justice argued, on the other hand, that they were more experienced litigators and that they should handle OSHA litigation, just as they handle most other Executive Department litigation. The Department of Justice's view prevailed at first, causing considerable bad feeling and morale deterioration on the part of OSHA and its attorneys. However, in 1975, an agreement was reached between the Departments of Labor and Justice relating to OSHA and two other enforcement programs, assigning primary litigating authority to the Office of the Solicitor, and establishing procedures to determine the rare circumstance when the Department of Justice would become involved. This arrangement has continued to the present time on court of appeals litigation; Supreme Court litigation continues to be handled by the Solicitor General and procedures for handling federal district court litigation are normally worked out on a local level between the U.S. attorneys and the regional solicitor's offices.

Disputes over jurisdiction

The OSHA is not the only federal agency with authority on occupational safety and health. Other agencies, notably the Department of Transportation (DOT), also enforce statutes which, although less comprehensive in scope than OSHA, concern worker protection. To avoid duplication of effort of agencies as well as to ensure that there is no hiatus in protection, section 4(b)(1) of the Act provides that OSHA will not apply to "working conditions" with respect to which another federal agency "exercise[s] statutory authority to prescribe or enforce standards or regulations affecting occupational safety or health." The meaning of the provision, and the delineation of jurisdiction among agencies, has been yet another source of continuing controversy and litigation from the start of the program.

The OSHA has construed the section to mean that the exemption from OSHA applies only to specific working conditions as to which another agency exercised its authority, that is, only when the agency issues standards affecting occupational safety and health. The railroad industry, among others, often supported by federal regulatory agencies eager to protect their jurisdiction, argued that section 4(b)(1) provides an "industry exemption" for any industry subject to *any* occupational safety or health standards. The Review Commission in 1974 and several courts of appeal in 1976 rejected the "industry exemption" argument (e.g., *Southern Railway v. OSHRC,* 539 F. 2nd 335 [4th Cir.], *cert. denied,* 429 U.S. 999 [1976]); however, litigation continued on a variety of related issues under section 4(b)(1). One of these issues was the meaning of "working condition." The courts split on whether it means "physical surroundings" or "hazards." Another issue was the definition of "exercise"; how much "exercise" does the other federal agency have to undertake to preempt OSHA (e.g., *Northwest Airlines,* 8 OSHC 1982 [Review Comm'n., 1980]).

Litigation, it was generally thought, was a poor and expensive way to decide jurisdictional issues and considerable effort was then invested by the affected agencies to negotiate jurisdictional agreements intended to clarify to employers and employees which agency was responsible for occupational safety and health enforcement activity. These agreements were useful in principle, but seldom succeeded in clarifying any jurisdictional issues.

In 1972, OSHA's memorandum with the Federal Railroad Administration provided an expeditious method of handling employee complaints in the railroad industry, without attempting to decide jurisdictional questions. However, the agreement was cancelled in 1974 by Assistant Secretary Stender, with no reasons given, and was never renewed. Various agreements were negotiated between OSHA and the U.S. Coast Guard in the DOT in respect to jurisdiction during activities and in the nation's waterways and on the Outer Continental Shelf. In 1983, OSHA conceded authority to the Coast Guard respecting "working conditions of seamen aboard inspected vessels" (48 *Fed. Reg.* 11,365 [1983]). The most successful interagency agreements have been between OSHA and the Department of Interior regarding mine safety, with any remaining jurisdictional issues regarding mines largely resolved when Congress passed the Mine Safety and Health Act of 1977, broadening the protection of mine employees and transferring enforcement authority to the newly-created Mine Safety and Health Administration in the Department of Labor 30 U.S.C. §801 and following.

Anticipating the possibility of disputes in this sensitive area, Congress in the OSHA Act directed the Agency to submit a report to the Congress within 3 years of the Act's effective date with recommendations to avoid "unnecessary duplication" and to achieve coordination between agencies. The report was filed 6 years late, and only after OSHA was sued by members of Congress in a federal court for not submitting a report. The report stated that it was satisfied with the *status quo,* concluding that court opinions will give "an even clearer picture" of jurisdiction, that cooperative efforts among agencies "will continue to expand," and, therefore, that no new legislation was needed (Mintz, 1984, p. 485).

Amendments, riders, and oversight

The involvement of Congress with the OSH statute did not end with its enactment in 1970. One of the main techniques available to the Congress for monitoring Agency performance is the oversight function; this activity includes public hearings and, less formally, contacts, letters, speeches, and telephone calls, between members of Congress and staff and the Agency, all designed to monitor and influence Agency action. The source of Congress' influence is its ultimate authority to amend the Act, and, on a continuing basis, its power to control the Agency's annual appropriations.

The OSHAct has not been amended substantively since 1970, but as will be discussed, appropriations bills have often been accompanied by legislative riders. Oversight hearings have frequently taken place, often with notable impact on Agency activity. While the Senate and House Labor Committees have primary jurisdiction over OSHA issues, and frequently conducted oversight hearings, other committees have also held oversight hearings on OSHA. Among these are the small business committees, concerned with the impact of OSHA enforcement on small business, and the Government Operations Committees, which have dealt particularly with regulation of toxic chemicals, hazard communication, and the safety of federal employees.

One of the earliest oversight hearings took place in 1972. A House subcommittee on agricultural labor critically questioned Assistant Secretary Guenther on the absence of OSHA standards to protect farmworkers and the lack of enforcement activity in agriculture. At the close of that hearing, Chairman O'Hara said to Mr. Guenther in an exercise in understatement: "I want you to know that it is not my intention that you leave this hearing with the impression that we want you to slow down on your enforcement of OSHA in agriculture" (*Farm Workers Occupational Safety and Health: Oversight Hearings Before the Subcomm. on Agricultural Labor of the House Comm. on Education and Labor,* 92nd Cong., 2nd Sess., 15-20 [1972]). At an oversight hearing before the Senate Labor Committee in 1974, Assistant Secretary Stender was pressed on OSHA's delay in setting up a Standards Advisory Committee for Coke Oven Emissions. Mr. Stender emphasized his frustration with "bureaucratic delays," an explanation that has often been heard at oversight hearings (*OSHA Review 1974: Hearing Before the Subcomm. on Labor of the Senate Comm. on Labor and Public Welfare,* 93rd Cong., 2nd Sess. 221-250 [1974]). Oversight hearings sometimes lead to committee reports, but frequently they do not. But the absence of a report does not mean that the hearing was without impact. After the Kepone hearing, at which Assistant Secretary Corn appeared, OSHA changed its complaint policy, even though no committee report was issued.

Sometimes Congress makes its views known to the Agency without a hearing. In 1974, the Senate Labor Committee, with the assistance of the GAO, sent 17 "issue" papers to OSHA, critical of various aspects of its enforcement activity. One of these—on classification of violations—suggested that the "number of total serious violations should be far greater than that which have been reported." Shortly thereafter, OSHA issued a "major clarification" of policy on serious violations to regional and area offices in order to "focus attention" of field staff on detecting during inspections hazards involving a significant probability of serious harm (*OSHA Review, 1974: Hearing Before Subcomm. on Labor of the Senate Comm. on Labor and Public Welfare,* 93rd Cong., 2nd Sess., apps. 941-1,238 [1974]). In the next 3 years, the percentage of serious violations cited by OSHA increased from 2.1 percent of all violations in fiscal year 1976 to 29 percent in fiscal year 1979. Other factors, including new and stricter enforcement policies already discussed, played a part; but congressional pressure, it seems clear, was a crucial factor in making sure that field staff would not overlook the need for serious citations.

The appropriation process is another way that Congress influences the policy of administrative agencies. Frequently, an appropriations committee report gives "instructions" to the Agency on how the money should or should not be spent. An example discussed previously was Congress' direction to OSHA in the 1976 appropriations act to undertake a "review and simplification of existing [national consensus] standards" and to eliminate "nuisance" standards.

Congressional interest in expanding OSHA's on-site consultation program was also transmitted to OSHA through the appropriations process. Section 21(c) provides specific authorization for OSHA to provide for the education and training of employers and employees in the recognition, avoidance, and prevention of unsafe and unhealthful working conditions. From the beginning, OSHA undertook programs to educate and train individuals in the private sector in OSHA matters. Some of these were carried out by OSHA at its Des Plaines, IL, training institute or by regional area offices; others were implemented by OSHA contractors, such as the National Safety Council and schools for workers at universities. Numerous educational materials, books, pamphlets, and audiovisual materials were distributed as part of this activity. However, an insistent demand continued for an additional component in employer education—on-site consultation. This involved OSHA providing information on hazards and controls *at the worksite* without threat of citation or penalty. Early in its history, OSHA determined that when OSHA personnel enter a workplace and observe hazards, the statute requires that a citation must be issued. The determination elicited criticism but attempts to amend the OSHAct to authorize on-site consultation were opposed by unions and were not enacted.

States with approved plans generally included on-site consultation programs in their plans (the legal interpretation didn't apply to the states) and in 1974 Congressman William A. Steiger, one of the main sponsors of the Act in 1970, sponsored an amendment to the OSHA appropriations bill which would authorize additional funds for the specific purpose of financing agreements between OSHA and states *without* approved plans for on-site consultation. The funds were appropriated and OSHA entered agreements with 12 states, providing 50 percent federal financing for the on-site consultation. The idea was politically attractive and the Senate Appropriations Committee later directed OSHA to increase the level of funding; OSHA responded, amending its regulations to provide 90 percent financing of these agreements. By 1980, on-site consultation had been made available to employers in all states, either through state plans, agreements between OSHA and states without plans, or in limited instances, by private consultants (*Oversight on the Administration of the OSHAct: Hearing Before the Senate Comm. on Labor and Human Resources,* 96th Cong., 2nd Sess., Pt. 1 at 24 [1980] [Testimony of Basil Whiting, Deputy Assistant Secretary of OSHA]).

More commonly, the appropriations process is used by Congress as a check on the Agency, through the enactment of limitations on the expenditure of funds—so called limitations riders. While the rules of the Congress restrict use of appropriations riders to some extent, in practice they frequently are introduced and enacted and provide a potent weapon for members of Congress to control Agency action without resort to the formal amendment process. This means that limitations riders are not preceded by committee hearings held by the standing committee with jurisdiction over the statute (in the case of OSHA, the labor committees), which are typically opposed to limitation of Agency authority. Often, even the Appropriations Committee does not consider the rider, which is introduced on the floor. Finally, the pressure to enact appropriations laws often is a strong impetus for Congress to accept the riders. As early as 1972, riders limiting OSHA enforcement activity were passed by one or both Houses of Congress but for a variety of reasons never became law. In 1974, a rider was enacted exempting employers with 10 or fewer employees from OSHA's recordkeeping requirements. The provision was incorporated into OSHA regulations and dropped from the appropriation law (29 *CFR* §1904.15).

The breakthrough in OSHA riders took place in 1976 when two major riders were passed. The first exempted small farms (with 10 or fewer employees) from OSHA enforcement. This rider was Congress' reaction to OSHA's proposal to regulate field sanitation (the history of OSHA's regulation of field sanitation will be discussed later in this chapter) and Congress' irritation at what it viewed to be a patronizing educational pamphlet published by OSHA on farm safety. The debate in the House was remarkable in its extreme hostility to the Agency. For example, one congressman from a farm state, a sponsor of the rider, responded to an accusation that he wanted to "castrate OSHA" by saying: "Believe me, my colleagues, I do not want to castrate OSHA because if I do it might grow more rapidly. . . But if castration is the only solution I would sooner castrate the zealots who are drawing up regulations at OSHA than let them destroy the smaller farmers of America" (122 Cong. Rec. 20,366-20,372 [1976]). The other rider which was enacted that year eliminated OSHA authority to impose penalties which are cited for fewer than 10 nonserious violations during an inspection.

Since 1976, five more riders were added to the OSHA appropriations acts and, with minor changes, have remained in effect to the present. These riders relate for the most part to enforcement against small businesses; one deals with state plan monitoring and one with OSHA-Coast Guard jurisdiction; another, demonstrating the legislative power of special small interest groups, limits OSHA enforcement in recreational hunting and fishing (Pub. L. No. 98-619, 98 Statutes at Large 3305).

The state plan framework and the benchmarks controversy

With the approval of the Virginia Plan in 1976, 23 states and one territory, the Virgin Islands, had received initial approval of their plans. Eleven states that had originally submitted plans had withdrawn them by 1976; six of these (New Jersey, New York, Illinois, Wisconsin, Montana, and North Dakota) had already received OSHA approval when their plans were withdrawn. Five states never submitted plans. Under the statutory scheme, the state first receives "initial" approval of its plan. After initial approval, federal OSHA has concurrent enforcement authority with the state. This concurrent jurisdiction was designed to assure that workers in the state would continue to receive occupational safety and health protection during this transition period. This concurrent federal authority is discretionary, so that federal OSHA legally could discontinue its enforcement either in whole or in part at any time between initial and final approval. The transition period ends when the state's plan receives final approval, which is granted only after the state demonstrates that *in fact* implementation of its occupational safety and health program was at least as effective as the federal program. After "final" approval, federal authority must end, unless formal proceedings are undertaken. However, because OSHA had decided to give initial approval to plans on the basis of the promises later to meet their effectiveness goal (developmental plans), in 1972 OSHA committed itself not to exercise its discretionary authority to terminate federal enforcement during the developmental stage.

This commitment was reexamined and overriden by Assistant Secretary Stender in 1973, who expressed his special concern to avoid "redundant" state and federal enforcement. He decided to enter into "operational" agreements with states, thus ending federal enforcement. In order to achieve this sought-after operational status, the state was required to have in place legislation, standards which are at least as effective, a procedure to review enforcement actions and sufficient number of enforcement personnel. This new policy drew strong criticism, not only from organized labor, which from the start had distrusted OSHA for what the unions believed was its abdication of federal enforcement responsibility to the states, but, as well, from public interest groups and NACOSH. One of NACOSH's subcommittees issued a report objecting to the new policy and urging that federal enforcement continue until there had been evaluation by OSHA of the "in fact" effectiveness of the state plan (Report of the NACOSH Subcomm. on State Programs). The NACOSH criticism elicited a strong negative response from Mr. Stender, who recommended that NACOSH abolish its subcommittees and concentrate on training and education issues. Congressman Steiger, whose relationships with Assistant Secretary Stender were strained, said this step in restricting NACOSH's role would be a "tragic mistake." Mr. Stender's "operational" policy was carried out despite the opposition, and by the end of 1975 there were 13 "operational" agreements in effect (The President's Report, 1975, p. 57).

Once a state has met all of its developments, it is certified by OSHA. After a period of further concentrated monitoring, the state plan is eligible for final approval, which can take place no sooner than 1 year after certification and 3 years after initial approval. The determination of whether a state in practice was at least as effective—and therefore entitled to final approval—or deficient, and therefore subject to withdrawal proceedings is based on OSHA's monitoring of the plan. The OSHA continuing monitoring of state plan effectiveness is required by section 18 of the Act; and has included collection of statistical data, the examination of state case files and investigation of complaints about state plan performance—known as CASPAs. On the basis of this monitoring, OSHA prepares reports with recommendations to the states for changes to improve effectiveness. Major problems with state plans in the past have been the lack of sufficient health enforcement personnel and lack of thoroughness in inspection activity.

By the end of 1976, five states had received certification, South Carolina, Iowa, Minnesota, North Carolina and Utah. Fourteen states had operational status agreements (The President's Report, 1976, pp. 44-50). However, it would be 8 more years before OSHA would give final approval to a state plan. The primary reason for this was the benchmarks litigation, which began in 1974 and did not end until 1983.

The benchmarks litigation involved the validity of OSHA's numerical requirements for state compliance personnel under state plans. The OSHA's interpretations of the "at least as effective" requirements were contained in the state plan regulations—Part 1902—issued in 1972. Under OSHA's interpretation of the Act, states were required to achieve staffs and budgets equal to those federal OSHA would have provided in that state in the absence of a state plan. Because of budget restrictions and resource limitations, it was conceded that the federal OSHA compliance staff was not as large as optimally needed. This meant that OSHA set state staffing requirements—or "benchmarks," as they were called—at a relatively low level. Thus, a state could receive initial approval, certification, and final approval without providing—in the words of section 18(c)(4) of the Act, as understood by organized labor—"satisfactory assurances" that the state ultimately (that is, at the end of the developmental period) will have "qualified personnel necessary for the enforcement" of the state program. The AFL-CIO challenged the OSHA interpretation, urging that state programs be required to provide adequate funds and staff "to ensure that normative standards are in fact enforced"; in other words, a judgment would be made as to whether the "necessary" staff had been hired by the state.

In 1978, the Court of Appeals for the District of Columbia decided *AFL-CIO v. Marshall*, 570 F. 2nd 1030; it rejected the notion that "at least as effective" means "at least as ineffective" and agreed with the AFL-CIO that personnel and funding benchmarks must be "part of a coherent program to realize a fully effective enforcement effort at some point in the foreseeable future." The Court remanded the case to OSHA for the establishing of criteria in accordance with the decision. In 1980, Assistant Secretary Bingham submitted benchmarks increasing substantially the personnel requirements for the states. Particularly in the health area, these became the basis for continuing disagreement among OSHA, the states, and the AFL-CIO (U.S. Department of Labor News Release, April 25, 1980).

Standards for construction, agriculture, and others

The Stender and Corn administrations devoted much attention to dealing with criticism of the so-called "nuisance" safety standards adopted in 1971. Many of these were revoked in 1978; in addition, revisions of all other national consensus standards were undertaken and a complete revision of subparts on Fire Protection and Electrical Hazards was issued in 1980 and 1981.

A number of safety standards projects were completed by OSHA between 1973–1976; some of these related to the construction industry. While, generally, OSHA standards were vertical—that is, applied across-the-board to all industries—in the construction and maritime industries OSHA's standards were horizontal, applying only to those industries. This special treatment was largely due to historical reasons, OSHA's start-up standards in these areas were adopted as "established federal standards" from pre-OSHA statutes applying to these specific industries. Another unique feature in the construction industry's standards was OSHA's obligation to consult with the Construction Safety and Health Advisory Committee, which had been established under the Construction Safety Act, to advise OSHA on standards for the construction industry. The Construction Committee was a standing committee: most other standards advisory committees were ad hoc, meaning that OSHA had discretion whether to seek their advice on specific standards.

In 1972 OSHA issued a major addition to the construction standards requiring rollover protection structures on construction equipment. Also in 1972, a subpart on power transmission and distribution lines was added to the construction standards. In addition in 1976, OSHA required Ground Fault Circuit Interrupters (GFCIs) on electrical circuits in the construction industry.

Another area which OSHA treated vertically was agriculture. The OSHA start-up standards contained only four limited standards applicable to agriculture. In 1972, the Migrant Legal Action Program and other groups petitioned OSHA to promulgate additional standards for agriculture. Shortly after the petition, OSHA in 1973 issued an ETS protecting farm workers against pesticide exposure. There was protest from Congress and from agricultural employers against OSHA's finding that pesticides create a "grave danger" to workers. The Chairman of the Pesticide Subcommittee of OSHA's Agriculture Advisory Committee wrote to the Assistant Secretary that he was "shocked" by the action since there was "no disagreement" in the subcommittee "regarding the absence of any need" for emergency action (Arant-Stender letter, May 9, 1973). The Court of Appeals for the Fifth Circuit in 1974 vacated the Standard, finding that the "easily curable and fleeting effects" of pesticide exposure on health did not meet the statutory requirements for a finding of "grave danger" (*Florida Peach Growers Association v. Department of Labor*, 489 F. 2nd 120 [5th Cir. 1974]).

Although OSHA started rulemaking for a permanent Pesticide Standard, imposing field reentry times for pesticide exposure, sharp disagreement arose with EPA which claimed jurisdiction over pesticide regulation under the Federal Insecticide, Fungicide and Rodenticide Act (FIFRA). The OSHA conceded EPA authority and ultimately EPA jurisdiction was upheld by the Court of Appeals for the District of Columbia in a suit brought by a migrant worker public interest group against OSHA for abdicating its regulatory responsibility (*Organized Migrants in Community Action v. Brennan,* 520 F. 2nd 1161 [D.C. Cir. 1975]). The OSHA issued two major agriculture standards, which were not challanged in court. The first requiring rollover protection for agricultural tractors was issued in 1975, and the other requiring guarding for farm equipment was issued in early 1976.

Two other OSHA safety standards led to court decisions important in the evolution of the law governing OSHA standards. In partially vacating OSHA's action in reducing the number of lavatories required for office employees, the Court of Appeals for the Second Circuit faulted OSHA for the lack of substantial evidence and because of the inadequacy of the statement of reasons. The Court said that, where the public opposes a provision on "substantial" grounds, OSHA "has the burden of offering *some* reasoned explanation" (*Associated Industries v. Department of Labor* 487 F. 2nd 342 [2nd Cir. 1973]).

In *AFL-CIO v. Brennan* (530 F. 2nd 109 [3rd Cir. 1975]), the Court of Appeals for the Third Circuit followed the precedent of the *Asbestos* case in upholding OSHA's elimination of the so-called "no-hands-in-dies" machines-guarding requirement. The court agreed with OSHA that the Standard was not feasible as originally written and said: "[a]n economically impossible standard would in all likelihood prove unenforceable, inducing employers faced with going out of business to evade rather than to comply with the regulation."

Health standards: An overview

A number of important actions on occupational health standards were taken by OSHA during this period, with a mixed record of success in the courts (Table 31-A). Four emergency standards were promulgated—on vinyl chloride, pesticides, 14 carcinogens, and diving. Several were challenged and all which were challenged were vacated or stayed by a court of appeals. The OSHA issued four "permanent," or final, Standards on Vinyl Chloride, 14 Carcinogens, Diving, and Coke Oven Emissions. These Standards, issued after rulemaking, were affirmed in court of appeals decisions, except the Standards for one of the carcinogenic substances, known as MOCA (4,4-methylene(bis)-2-chloroanaline), and one provision—medical surveillance—in the Diving Standard. Several rulemakings were begun but not completed until later; these included proposed revisions of OSHA Standards on Lead, Hearing Conservation, Arsenic, and Asbestos. A number of rulemakings were begun and abandoned; among these were for trichloroethylene, beryllium, and sulfur dioxide (The President's Report, 1975, pp. 22-23).

A number of general observations should be made regarding developments in health standards rulemaking and in court review of these proceedings.

Issues: All standards rulemaking, particularly on health standards, had become much more lengthy, complex and controversial. (In the 1974 Annual Report for OSHA, 22 steps were listed in the standards development process [The President's Report, 1974, pp. 9-11]). Two overriding issues, typically, were argued and resolved in standards proceedings: the PEL necessary to protect employees and the feasibility, economic and technological, of reaching that level through engineering controls. The question whether OSHA should require engineering controls as the primary method of achieving the permissible limit recurred in each proceeding. Although the details of the argument varied, the thrust of the business community's contention was that adequate protection from toxic substances could be achieved through the much less costly means of protective equipment. However, OSHA argued and continues to maintain that engineering controls are the preferred method of compliance and that protective equipment should be used only when the preferred engineering controls were not feasible or inadequate. The basis for the argument is that protective equipment is unreliable because of the uncertainty as to whether it will be worn, and whether it will afford complete protection.

Other issues relating to the nature of the protective equipment, medical surveillance, and monitoring and other requirements, were frequently raised in standards proceedings. Affected parties uniformly presented witnesses who gave public testimony on the issues in the proceedings. Examination of witnesses, under the direction of the administrative law judge, routinely took place. At first OSHA allowed public comment but did not itself present witnesses or ask questions, but the Agency soon found it necessary to offer expert witnesses and to question other witnesses to better defend the standard in court. Beginning with the Vinyl Chloride proceeding, OSHA contracted for feasibility studies that became part of the record; often, as in the Coke Oven Emissions and Cotton Dust proceedings, additional feasibility

Table 31-A. OSHA Health Standards

OSHA regulation	Final standard
1. Fourteen carcinogens[a]	1/29/74
2. Vinyl chloride[a]	10/04/74
3. Coke oven emissions	10/22/76
4. Benzene[a]	2/10/78
5. DBCP[ac]	3/17/78
6. Inorganic arsenic	5/05/78
7. Cotton dust/cotton gins	6/23/78
8. Acrylonitrile[a]	10/03/78
9. Lead	11/14/78
10. Cancer policy	1/22/80
11. Access to employee exposure and medical records	5/23/80
12. Occupational noise exposure/hearing conservation	1/16/81
13. Lead—reconsideration of respirator fit-testing requirements	11/12/82
14. Coal tar pitch volatiles—modification of interpretation	1/21/83
15. Hearing conservation—reconsideration	3/08/83
16. Hazard communication (labeling)	1/25/83
17. Ethylene oxide	6/22/84
18. Asbestos[ab]	6/20/86
19. Hazardous Waste Sites (Interim Final Standard)	12/19/86
20. Field Sanitation Standard	4/28/87

(Adapted from *Preventing Illness and Injury in the Workplace.* Washington, DC: U.S. Congress, Office of Technology Assessment, OTA 1-1-256, April 1985)

[a]Subject of an Emergency Temporary Standard.

[b]Emergency standards were issued for asbestos in 1971 and 1983.

[c]1,2-dibromo-3-chloropropane.

studies were introduced by employer associations. Presidential orders issued beginning with the Administration of President Gerald Ford, who insisted on inflationary impact statements (Baram, 1980), required economic or regulatory analyses; most recently, President Reagan's Executive Order 12,291 required an agency to prepare a proposed and final regulatory impact analysis (RIA) for all "major" actions.

Preambles: As the issues in the health standards proceedings became more complex, and the records more lengthy, the preambles to OSHA's proposed and final standards became considerably more detailed, including a detailed analysis of the record, a summary of the contentions of the parties, and OSHA's resolution of each issue, with a section-by-section analysis, discussing the basis of each provision and presenting preliminary interpretations of these provisions. Some recent standard preambles have been longer than 100 three-columned, printed pages in the *Federal Register.* The significant trend toward very detailed statements of reasons was also made necessary by the increasing scrutiny that the courts of appeals were giving to the OSHA "statement of reasons" for the standard. The Court of Appeals for the Third Circuit, for example, vacated the challenged portion of OSHA's ETS on 14 carcinogens because of the inadequacy of the statement of reasons (*Dry Color Manufacturers Association v. Department of Labor,* 486 F. 2nd 98 [3rd Cir. 1973]).

Requirements: The basic content of a health standard remained similar to the content of OSHA's first health Standard, Asbestos. The key provision is the appropriate PEL: the level above which the employer is not permitted to expose employees. The PEL is usually expressed as an 8-hour TWA. Sometimes a ceiling, or short-term exposure level (STEL), is added; the issue whether to include a more protective STEL issue has been critical in the recent Ethylene Oxide proceeding. In addition, an action level, usually one-half the PEL, is often included, and is defined as the point where certain provisions of the standard, such as medical surveillance, are mandated. Other requirements—medical surveillance, monitoring, training, record-keeping, for example—continued to be included, with refinements and elaborations reflecting Agency experience and knowledge. In the area of medical surveillance, for example, OSHA evolved toward a statement of policy that employers were required to make medical examinations available to employees, but that OSHA would not require that employees take the examination. (An employer could, of course, make the examination a condition of employment.) Questions on the type and frequency of monitoring, and the extent and the availability of employer records were often significant in the rulemaking proceedings, both during this period and continuing into the administration of Dr. Bingham. As will be discussed, the issues of mandatory transfer of employees who were at increased risk from exposure and wage retention became major issues after 1976.

Pace of promulgation: Because of resource limitations, and the increasing length of the proceedings, OSHA was falling further and further behind in its regulations of toxic substances. A variety of strategies were discussed and tried, with little noticeable impact. The so-called "standards completion" project, to fill out the bare-boned health standards adopted in 1971, was initiated by OSHA and NIOSH as a cooperative venture in 1975 but never completed. Later, during the administration of Dr. Bingham, the Carcinogens Policy was issued for the purpose of facilitating the issuance of standards on carcinogens; for a variety of reasons, the policy was never implemented. The OSHA policy for determining standards' priorities was therefore extremely critical in its overall regulatory effort. Regulation of carcinogens was almost always OSHA's first priority, although other toxic substances, though not carcinogens, but causing serious health effects and pervasively present in workplaces, were also regulated. Examples of the latter efforts were the Cotton Dust and Lead Standards. Court suits requiring OSHA to rearrange its priorities and to initiate rulemaking were undertaken, usually by unions or public interest groups. The field sanitation suit in 1973 is an example; later, suits were brought to require OSHA to start rulemaking on ethylene oxide and formaldehye. By 1987, court involvement in decisions on OSHA standards priorities were a regular feature of standard activity.

Court precedents: The courts of appeals, with only one exception, followed the lead of the Court of Appeals for the District of Columbia in the *Asbestos* case, generally deferring to OSHA's policy judgments, so long as they were within the bounds of rationality, particularly when the Agency was acting to afford greater protection for workers. Thus, in the *Vinyl Chloride* case, the Court of Appeals for the Second Circuit upheld OSHA's one part per million PEL. Even though the Court said, "The factual finger points, [but] it does not conclude" it decided that "under the command of OSHA, it remains the duty of the Secretary to act to protect the working man, and to act even in circumstances where existing methodology or research is deficient." In that case, and in other standards review cases, the court upheld OSHA's policy judgment that evidence of carcinogenicity from animal studies should be extrapolated "from mouse to man." The court also applied the doctrine of "technology forcing," in *Vinyl Chloride* in deciding that, despite lack of substantial evidence establishing the technological feasibility of the one part per million (ppm) level, the Standard was feasible because employers "simply need more faith in their own technological potentialities" (*Society of Plastics Industry, Inc. v. OSHA,* 509 F. 2nd 1301 [2nd Cir.] *cert. denied sub nom. Firestone Plastics Co. v. Department of Labor,* 421 U.S. 992 [1975]). On the other hand, in the Court of Appeals for the Fifth Circuit, the view was evolving towards closer scrutiny of OSHA's actions; this appeared first in that Circuit's decisions on the Pesticide and Diving Emergency Standards and was proclaimed fully in 1978 when the Court set aside OSHA's Benzene Standard, disagreeing with the deferential view of other courts of appeals (*American Petroleum Inst. v. OSHA,* 581 F. 2nd 493 [5th Cir. 1978]). This decision was later upheld by the Supreme Court, but on somewhat more narrow grounds.

Procedural scrutiny: While giving the Agency substantial deference on policy judgments, the courts nevertheless insisted on rigorous adherence to the procedural requirements on rulemaking as stated in the OSHAct and the Administrative Procedure Act. As the court of appeals later said in the *Cotton Dust* case, the courts' role in the partnership is to ensure that the regulations resulted from a process of reasoned decision-making including, "notice to the interested parties of issues presented in the proposed rule," "opportunities for these parties to offer contrary evidence and arguments" and assurance that the Agency has "explicated" the basis for its decision (*AFL-CIO v. Marshall* 617 F. 2nd 636 [D.C. Cir. 1979]). A number of OSHA standards, particularly during the early years, were vacated because of procedural defects.

Emergency standards: Throughout OSHA's history the courts of appeals applied particularly rigorous scrutiny to ETS. The view was expressed by the Court of Appeals for the Fifth Circuit in the *Pesticides* case as follows: "Extraordinary power is delivered to the Secretary under the emergency provisions of the Occupational Safety and Health Act. That power should be delicately exercised, and only in those emergency situations which require it" (*Florida Peach Growers Association v. Department of Labor,* 489 F. 2nd 120 [5th Cir. 1974]). Other courts agreed with the Fifth Circuit Court of Appeals, at least on principle, for example, the Third Circuit (*Dry Color Manufacturers Association v. Department of Labor,* 486 F. 2nd 98 [3rd Cir. 1973]). While it is clear that some emergency standards, such as the Carcinogens Standards, were vacated because OSHA had failed to follow proper procedures; increasingly, there were questions on whether any challenged emergency standard could be upheld, particularly after the Fifth Circuit Court of Appeals vacated OSHA's second Asbestos Emergency Standard (*Asbestos Information Association v. OSHA,* 727 F. 2nd 415 [5th Cir. 1984]). Indeed, throughout the history of OSHA, only in one case—*Acrylonitrile*—when the judicial challenge was withdrawn after the Court of Appeals refused a stay of the Standard (*Visitron v. OSHA,* 6 OSHC 1483 [6th Cir. 1978]), did OSHA prevail in court in a proceeding on an emergency standard.

Health standards proceedings

In March 1974, OSHA issued an ETS on Vinyl Chloride, based on recently discovered evidence, both in animal studies and in humans, that the substance causes an unusual form of liver cancer. The ETS was preceded by a brief fact-finding hearing held by the Agency, though not required by law. The Emergency Standard was not challenged and, after rulemaking, a "permanent" Standard was issued in October 1974—6 months after the ETS, that is, within the statutory period. After a stay which lasted for a brief period, the court of appeals affirmed the Standard and dissolved the stay. The Supreme Court refused to hear the case. This proceeding was generally recognized as successful both in terms of result and in speed in achieving protection for employees from a life-threatening hazard. In the words of Dr. Irving Selikoff, the regulation was "a success for science in having defined the problem; success for labor in rapid mobilization of concern; success for government in urgently collecting data, evaluating it, and translating it into necessary regulations; and success for industry in preparing the necessary engineering controls to minimize or eliminate the hazard" (*OTA Report,* pp. 230–231).

Some other OSHA health standard proceedings were not quite as successful. The regulation of field reentry time for pesticides, which has already been discussed, did not result in any OSHA regulation. The OSHA Emergency Standard on 14 Carcinogens was set aside in part; after rulemaking, OSHA issued 14 permanent Standards. The Court of Appeals for the Third Circuit vacated the Standard for one of the carcinogens—MOCA—for procedural reasons: the Court held that OSHA had failed to use the proper sequence of procedures in obtaining the recommendation of an advisory committee. The Court upheld the Standard as it applied to another carcinogen, ethyleneimine, deferring to OSHA's policy judgments on the interpretation of scientific evidence, and no challenge was filed as to the standards for other substances. The OSHA has not issued another standard on MOCA (*Synthetic Organic Chemical Manufacturers Association v. Brennan,* 503 F. 2nd 1155; 506 F. 2nd 385 [3rd Cir. 1974]).

In 1977, OSHA completed rulemaking on the "permanent" Diving Standard, and on challenge from the Diving Contractors Association the Court of Appeals for the Fifth Circuit set aside the medical surveillance provisions. The Court held that a portion of the medical requirements was beyond the Agency's authority because the purpose was to protect the jobs of workers rather than to protect their occupational safety and health (*Taylor Diving & Salvage Co. v. Department of Labor,* 599 F. 2nd 622 [5th Cir. 1979]).

The widely praised Coke Oven Emission proceeding was completed during Dr. Corn's term. The United Steelworkers of America petitioned OSHA for rulemaking based on strong epidemiological evidence demonstrating that coke oven emissions were carcinogenic. A Standards Advisory Committee was formed in November 1974, under the chairmanship of Dr. Eula Bingham, who later became Assistant Secretary. (Dr. Bingham's statements favoring medical removal protection at a Coke Oven Advisory Committee meeting became an issue in the Lead Standard court proceeding in 1980, which will be discussed later in this chapter). Based on the Advisory Committee recommendations, but differing substantially from them, OSHA issued a proposed Coke Oven Emissions Standard in July 1975. After a comment period and a public hearing OSHA issued a final Standard in October 1976. The Standard regulated the benzene-soluble fraction of total particulate matter present during the coking process, establishing a PEL of 150 $\mu g/m^3$. It also specified the engineering controls required: most health standards are "performance" in this respect, mandating a level and permitting the employer to select the specific controls. The Standard was upheld by the Court of Appeals of the Third Circuit against an industry challange; the Court found the PEL necessary for the protection of employees and economically and technologically feasible. The Court also upheld OSHA's authority to prescribe specific engineering controls (*American Iron and Steel Institute v. OSHA,* 577 F. 2nd 825 [3rd Cir. 1978]). Industry appealed the case to the Supreme Court, and although the Court agreed to hear the appeal, the petition was withdrawn before decision, and the Court of Appeals ruling thus became final.

Federal safety: Trends in injury rates

In an effort to strengthen the federal Agency safety and health program, a new Executive Order No. 11,807 was issued, effective September 28, 1974, superseding the Order that had been issued in 1971. At the end of 1974, OSHA published guideline regulations implementing the new Executive Order and specifying the responsibilities of the federal agencies and OSHA's Office of Federal Agency Safety Programs (29 *CFR,* Part 1960). The order was based on section 19 of the Act, which requires the head of each federal agency, after consulting with employee representatives, to establish and maintain an effective and comprehensive occupational safety and health program, "consistent" with the standards promulgated by OSHA for the private sector.

The underlying premise of the Federal Agency Program was

that employees of the federal government need and are also entitled to protection. In addition, federal efforts to require private sector compliance would be severely hampered if the government were shown to have failed to provide a model by keeping its own house in order. However, despite the good intentions, criticism of the federal government's effort continued. In March 1973, the GAO published a report showing the need for improvements in the Federal Agency Program (GAO Report, March 15, 1973). In 1975, hearings were held in Charleston, SC, dealing with the U.S. Navy Shipyard and in Washington, DC, with testimony received from various government executive departments, unions, and the GAO (*Safety in the Federal Workplace: Hearings Before a Subcomm. of the House Comm. on Government Operations,* 94th Cong., 1st Sess. [1975]). The Committee on Government Operations' report, entitled "Safety in the Federal Workplace", was issued in 1976 (H.R. REP. No. 94-784 [1976]). The report was sharply critical of the federal effort, pointing to the fact that agency policy directives were "vague and ambiguous," "deficiencies" existed in consultation with employee representatives and the Department of Labor was "not staffed to perform the evaluations" of agency programs mandated by the law. The report noted that agencies had made commitments to improve their programs but that because of "bureaucratic infighting," little progress had been made in meeting these commitments. The Committee made numerous recommendations, including a suggestion that OSHA accelerate its efforts to improve the accident and illness reporting system for federal employees. Congressional hearings and committee reports criticizing the federal government's internal safety and health effort have been a recurring component of the federal government program.

Meanwhile, the BLS continued to make its annual surveys of injuries and illnesses in the private sector. Based on reports submitted by private employers in 1976, the 1975 statistical survey showed that the overall incidence rate (the number of injuries and illnesses per 100 full-time workers) dropped from 10.4 in 1974 to 9.1 in 1975; however, OSHA said that the reduction could be explained due to the disproportionate decline in manufacturing and contract construction employment from 1974–1975. Despite the improvements, on the average, one out of every 11 workers experienced a job-related injury or illness, and the report itself again recognized the deficiency of the illness statistics. Thus, four out of every 10 recorded illnesses were for skin diseases or disorders (The President's Report, 1976, pp. 86-105). The decline in incidence rates did not continue, a fact leading to renewed and bitter controversy over the effectiveness of the OSHA program.

1977 TO 1981: GIVING TEETH TO THE TIGER

A new assistant secretary

At an oversight hearing on OSHA held early in the Bingham administration, Senator Harrison A. Williams, Chairman of the Senate Labor Committee, observed that while OSHA originally enjoyed "broad support among legislators and the public," it was now perceived either as a "meddling, mischievous intrusion by the Government into the affairs of our Nation's businesses" or by others as a "paper tiger" (*Oversight on the Administration of the OSHAct: 1978 Hearings Before the Subcomm. on Labor of the Senate Comm. on Human Resources,* 95th Cong., 2nd Sess. [1978]). Jimmy Carter, elected President in 1976, gave high priority to improving the deteriorated image of OSHA. He selected Ray Marshall, a labor economist and professor, as Secretary of Labor and Dr. Eula Bingham, Professor of Toxicology at the University of Cincinnati, as Assistant Secretary. Dr. Bingham was no stranger either to occupational safety and health or to OSHA; she had served in 1973 as a member of the Carcinogens Advisory Committee and in 1974 as Chairperson of the Coke Oven Emissions Advisory Committee. The new Assistant Secretary sought to give teeth to the OSHA tiger and, at the same time, to eliminate the Agency as an irritant to business; her basic policy was a "shift to common sense priorities." The reorganization of the national office staff, begun in 1976 and designed to improve coordination of field activity and to improve the Agency's technical support activities, was fully implemented. The executive staff of the Agency was almost completely changed. Dr. Bingham remained Assistant Secretary during President Carter's entire term of office. This was the longest administration in the history of OSHA. During this 4-year period, there were numerous major shifts in OSHA policy and new initiatives. However, it is by no means clear that in 1981 OSHA enjoyed wider support than at the start of the administration.

Health standards: The new priority

From the start, Dr. Bingham emphasized the importance of standards activity, particularly standards regulating health hazards. In April 1977, soon after she came to office, Dr. Bingham told a subcommittee of the House Committee on Government Operations, "Quite honestly, I plan to stretch the resources of the agency in putting out health standards, and I intend to use the ETS authority whenever employees are exposed to grave danger." She added, "All I can say is watch the *Federal Register"* (*Performance of the OSH Administration: Hearings Before the Subcomm. on Manpower and Housing of the House Comm. on Government Operations,* 95th Cong., 1st Sess., pp. 77–78, 92[1977]). Dr. Bingham's first standards effort was to lower the PEL for benzene, and while ultimately unsuccessful in terms of practical protective results, to many, the Benzene proceeding was a turning point in the history of OSHA standards rulemaking.

One of OSHA's start-up standards regulated benzene exposure; the Standard established a permissible level of 10 ppm. Increasing epidemiological evidence demonstrating that benzene causes leukemia had become available, and soon after reaching office, Dr. Bingham's attention was directed to the pressing need for increased protection to workers from benzene hazards. On April 29, 1977, Dr. Bingham signed an ETS lowering the PEL of benzene to 1 ppm, with a ceiling level of 5 ppm.

Both the Industrial Union Department of the AFL-CIO (IUD) and the American Petroleum Institute (API) filed petitions to review the Emergency Standard. Courts of appeals are divided into separate circuits, and both the IUD and the API sought review in a court that each thought would be favorable to its view, the IUD in the District of Columbia Circuit and API in the Fifth Circuit. A federal statute provides that when petitions are filed in different circuits, the court of appeals that decides the case is the one in which the first petition for review is filed. As a result, interested parties rush to file the first petition, in order to win what has been called the "race to the courthouse." The "race," at best, is unseemly and wasteful; in the *Benzene* litigation, the litigation was particularly complex because of factual and legal disputes creating the difficulty in determining which was the first-filed petition. The litigation over the venue (as it is called) of the *Benzene* appeal lasted for 5 months, during which time the Standard was stayed and therefore not in effect. In September 1977, the Court of Appeals for the District of Columbia issued a decision transferring the proceeding to the Court of Appeals for the Fifth Circuit. While the judges agreed that the case should be transferred, they disagreed on the theory and three separate opinions were written, each presenting a different legal approach. In any event, since an emergency standard remains in effect no longer than 6 months, OSHA decided that it was not worthwhile to pursue the litigation any further and the ETS expired without ever being effective. The Agency decided instead to concentrate its efforts on a permanent standard (*Industrial Union Department v. Bingham,* 570 F. 2nd 965 [D.C. Cir. 1977]).

After full rulemaking OSHA issued a "permanent" Benzene Standard, modifying the PEL to 1 ppm, a TWA, and a 5-ppm ceiling. There was no animal data on the carcinogenic effect of benzene; and the human data was at exposure levels considerably higher than the prior PEL of 10 ppm. In its preamble to the final Benzene Standard, OSHA asserted that the conclusions to be derived from the available data were that higher exposures to a toxic substance carry a greater risk, and since a determination of the precise level of benzene that presents no hazard cannot be made, the question of whether there is a "safe level" "cannot be answered" on the basis of present knowledge. Prudent health policy, the Agency said, requires that the level be set at the "lowest feasible level" which is found to be 1 ppm (43 *Fed. Reg.* 5918 [1978]). Thus, the lowest feasible level policy for carcinogens—as it was called—was the culmination of the regulatory theory, first expressed by OSHA in regulating asbestos, to resolve all doubts in favor of workers' protection.

The permanent Standard was challenged primarily by API in the Court of Appeals for the Fifth Circuit, this time without any race to the courthouse. The Standard was immediately stayed and in October 1978 the Court of Appeals vacated the Standard on two grounds:

1. The Agency had failed to provide an estimate, supported by substantial evidence, of the expected benefits from reducing the PEL.
2. The OSHA did not assess the reasonableness of the relationship between expected costs and benefits.

In more familiar terminology, the Court of Appeals decided that OSHA must do a cost-benefit analysis. The Court's decision was noteworthy because the Fifth Circuit departed markedly from the decisions of other courts of appeals which gave great deference to OSHA decisions on health standards. The Fifth Circuit, however, defined its partnership with OSHA in a completely different way, insisting that OSHA "regulate on the basis of more knowledge and fewer assumptions" (*American Petroleum Insititue v. OSHA,* 581 F. 2nd [5th Cir. 1978]).

Both OSHA and the IUD sought review, and the Supreme Court agreed to hear the case. On July 2, 1980, in one of the major regulatory decisions of the decade, the Supreme Court, in a sharply divided vote, invalidated the Benzene Standard. The many lengthy opinions of the Justices of the Supreme Court, including the opinions of the Justices concurring in the results but disagreeing on rationale, are almost as significant as the decision itself. There was no majority opinion. The plurality—four Justices—ruled that the Act addresses only "significant risks" and does not seek to provide a "risk free" workplace; and that in developing standards, OSHA has the burden of showing by substantial evidence that it is addressing a "significant risk" of harm in the workplace and the proposed standard would eliminate or reduce that significant risk. Since OSHA did not meet that burden (indeed, since OSHA did not know of the requirement, it did not even try to do so), the Benzene Standard was vacated. The practical impact of the decision was that, in the future, OSHA would have to establish the extent of risk from the toxic substance, even of carcinogens, quantitatively—usually by means of quantitative risk assessments—and then find that the risk is "significant." The OSHA "lowest feasible level" policy for carcinogens was rejected outright by the Supreme Court, and quantitative risk assessments are now a routine part of OSHA health standards development.

Since the Supreme Court decided the case on other grounds, it did not have to reach the cost-benefit issue, leaving it for the next case. Both the plurality and dissenting opinions were sharply worded and partly ideological in their thrust. Justice John Paul Stevens who wrote the plurality opinion asserted, for example, that it would be "unreasonable to assume that Congress intended to give the Secretary the unprecedented power over industry that would result from the Government's view"; he refused to agree with OSHA that the "mere possibility that some employee somewhere in the country may confront some risk" as a basis for the Secretary's requiring "the expenditure of hundreds of millions of dollars to minimize that risk." The dissent written by Justice Marshall for four Justices was equally strident, accusing the majority of deciding the case not on the basis of congressional intent but rather "in line with the plurality's own view of proper regulatory policy" (*Industrial Union Department v. American Petroleum Institute,* 448 U.S. 607 [1980]). The OSHA did not issue a new proposal on benzene until December 1985.

The cost-benefit issue was not resolved by the Supreme Court until the Cotton Dust proceeding. A proposed Cotton Dust Standard had been published at the end of December 1976, shortly before Dr. Corn left. The rulemaking was completed, and after a major controversy with economists in the White House, who exerted pressure for a less costly final Standard, Dr. Bingham issued the Standard in June 1978 with only minor changes to accommodate the economists' view. The Standard established a PEL of 200 $\mu g/m^3$ for yarn manufacturing and cotton washing, 750 $\mu g/m^3$ for slashing and weaving in the textile industry, and 500 $\mu g/m^3$, for textile mill waste house operations or for exposure to dust from "lower grade washed cotton" in yarn manufacturing. Rejecting industry arguments during the rulemaking, OSHA refused to perform a cost-benefit analysis, saying that prior attempts to quantify benefits as an aid to decision-making had not proven "fruitful"; it based the PEL in the textile industry on an excellent epidemiological study performed by Dr. James Merchant and found that the Standard was technologically and economically feasible. The Court of Appeals for the District of Columbia Circuit affirmed the Standard for the most part, rejecting the argument that the OSHAct requires cost-benefit analysis. The Court said: "Especially where a policy aims to protect the health and lives of thousands of people, the difficulties in comparing widely dispersed benefits with more concentrated and calculable costs may overwhelm the advantages of such analysis" (*AFL v. Marshall,* 617 F. 2nd 636 [D.C. Cir. 1979]).

Industry appealed to the Supreme Court; and it withdrew its

appeal to the Supreme Court in the *Coke Oven Emissions* case, apparently so that the Supreme Court could direct its full attention to the *Cotton Dust* case. By this point, cost-benefit had emerged as *the* major OSHA regulatory issue: on the one hand, industry argued that cost-benefit would provide a "potential legislative check on what might otherwise amount to the exercise of virtually untrammeled authority" and allow the "correction, through the political process, of actions that are deemed by the Congress to be extreme, unwarranted, and inconsistent with congressional intent" (Brief, American Textile Manufacturers Institute in *American Textile Manufacturers' Institute v. Donovan,* 452 U.S. 490 [1981]). To other sectors of the public, unions and public interest groups, in particular, cost-benefit analysis was an anathema because it "places a monetary value on human life, thereby obliterating the moral purpose which led Congress to pass the OSHAct." (Comments of the United Steelworkers of America on the Advance Notice of Proposed Rulemaking for Cotton Dust, Docket No. 052B, May 29, 1981). The Supreme Court in June 1981 agreed with OSHA and the unions, and upheld the Cotton Dust Standard as it applied to the textile industry. Relying in particular on the fact that neither the Act nor the legislative history mentions cost-benefit analysis, the Court held that the Act did not require that it be used. In his majority opinion, Justice William Brennan looked to the legislative history and concluded that Congress viewed the costs of safety as a cost of doing business. He quoted Senator Yarborough's statement, which was still relevant to the OSHA program: "We are talking about people's lives, not the indifference of some cost accountants" (*American Textile Manufacturers Institute v. Donovan,* 452 U.S. 490 [1981]). The Supreme Court opinion addressed only the issue in the case: whether the Act requires cost-benefit. The decision was generally understood to mean that cost-benefit was prohibited. The Supreme Court, however, opened the regulatory door somewhat, saying that "cost-effectiveness" analysis could be used; that is, once the Agency decided what level of protection was necessary, it could adopt the least expensive means to achieve that level. As we shall see, cost-effectiveness analysis in the 1980s became one of the keystones in OSHA standards development.

Dr. Bingham issued two other emergency standards during the first years of her administration:

1. The first in September 1977 was for 1.2-Dibromo-3-Chloropropane (DBCP); this was based on findings that the substance was a carcinogen and caused sterility, and therefore created a "grave danger". The Standard was not challenged and was replaced by a permanent Standard in March 1978.
2. In January 1978, OSHA issued another ETS, this time for Acrylonitrile, also a carcinogen. The Standard was challenged in the Court of Appeals for the Sixth Circuit, but the Court refused a stay, and the court challenge was withdrawn. The ETS was superseded by a permanent Standard in November 1978, which was not challenged.

Although Dr. Bingham had urged Congress to watch the *Federal Register* for emergency standards, none were published by her after January 1978. Indeed, in 1980, OSHA asserted a different policy approach. It said that there may be other occasions when the "level of the Agency's resources including compliance, legal and technical personnel, at a given time, may suggest that employee health may be more effectively protected by concentrating those resources in work on permanent standards" (Carcinogens Policy, 45 *Fed. Reg.* 5002, 5215–5216 [1980]). Undoubtedly, this judgment was also predicated on the legal vulnerability of emergency standards.

That statement was made in OSHA's Carcinogens Policy, issued in 1980. Its purpose was to provide a framework for the regulation of carcinogens in a "timely and efficient manner." The Carcinogens Policy contained policy determinations on the issues in the regulation of carcinogens which could be questioned only in specified circumstances in later substance-specific rulemaking proceedings. This, the Agency said, was to avoid reargument of the same policy issues in each rulemaking proceeding. While the lengthy preamble to the Carcinogens Policy was an important contribution to the principles of identification, classification and regulation of occupational carcinogens, mainly because of the change of administration, the Policy was never used and never resulted in the promulgation of a carcinogen standard; crucial portions of the Policy were stayed by the Agency in 1981. Thus, the considerable resources expended on the promulgation of the Policy have failed at least for the present to achieve the resource-saving purposes for which it was designed.

Several other health standards were issued during Dr. Bingham's administration. An Arsenic Standard was issued in May 1978, lowering the PEL for the carcinogenic substance to 10 $\mu g/m^3$. In 1985, after reconsideration by the Agency in light of the intervening Supreme Court decisions in the *Benzene* case and the reaffirmation of the original PEL, the Standard was affirmed by the Court of Appeals for the Ninth Circuit (*ASARCO v. OSHA,* 746 F. 2nd 483, [9th Cir. 1984]).

Another major proceeding was OSHA's regulation of lead hazards. Although not an occupational carcinogen, lead exposure affects numerous employees and industries and long had been known to result in serious illness. A proposed revised Lead Standard was issued in November 1975, during the Corn administration; it proposed reducing the PEL for lead from 200 $\mu g/m^3$ to 100 $\mu g/m^3$. A final Lead Standard was issued in November 1978 by Dr. Bingham, further reducing the PEL to 50 $\mu g/m^3$, requiring the use of engineering controls but affording affected industries periods of up to 10 years to comply with these requirements. The Standard also contained a novel provision, requiring employers to transfer employees who were at excess risk from lead exposure to lower exposure jobs, and to maintain their wage levels and seniority generally for a period of up to 18 months while the employees were on the other jobs or laid off. This program came to be known as medical removal protection (MRP). It was designed to protect employees by encouraging participation in the medical surveillance program; the MRP provision had not been a subject of the original proposal and a reopened hearing on MRP was held before the issuance of the final Standard.

The Standard was challenged in the court of appeals by the Steelworkers Union and the Lead Industries Association (LIA); two petitions for review were filed simultaneously in two different circuit courts of appeal. After a preliminary round of litigation on venue, the case was transferred to the Court of Appeals for the District of Columbia Circuit. In a lengthy opinion in August 1980, the Court, with a vigorous dissent by Judge McKinnon on some issues, rejected numerous procedural challenges raised by LIA—based on lack of adequate notice by OSHA on the permissible level, *ex parte* communications, and bias of Dr. Bingham in prejudging the issue of MRP—and found that substantial evidence supported the new PEL and that it

Table 31-B. Federal OSHA Safety and Health Inspections

	Total Establishment Inspections (number)	*Safety Inspections (number)*	*Safety Inspections (percent)*	*Health Inspections (number)*	*Health Inspections (percent)*
1973	48,409	45,225	93.4	3,184	6.6
1974	77,142	73,189	94.9	3,953	5.1
1975	80,978	75,459	93.2	5,519	6.8
1976	90,482	82,885	91.6	7,597	8.4
1977	60,004	50,892	84.8	9,112	15.2
1978	57,278	46,621	81.4	10,657	18.6
1979	57,734	46,657	80.8	11,077	19.2
1980	63,404	51,565	81.3	11,839	18.7
1981	56,994	46,236	81.1	10,758	18.9
1982	52,818[a]	43,609	82.6	9,209	17.4
1983	58,516[b]	48,269	82.5	10,247	17.5
1984	62,118[c]	51,038	82.2	11,081	17.8
1985	64,167[d]	53,421	83.2	10,746	16.7
1986	59,452[e]	49,696	83.6	9,756	16.4
1987 (1st 6 months)	24,988[f]	20,932	83.8	4,056	16.2
TOTALS	914,484	785,694	86.0	128,791	14.0

(Adapted from *Preventing Illness and Injury in the Workplace.* Washington, DC: U.S. Congress, Office of Technology Assessment, OTA H-256, April 1985)

[a]Does not include 8,444 "Records Review" inspections.
[b]Does not include 10,402 "Records Review" inspections.
[c]Does not include 9,253 "Records Review" inspections.
[d]Does not include 7,136 "Records Review" inspections.
[e]Does not include 4,619 "Records Review" inspections.
[f]Does not include 1,280 "Records Review" inspections.

was feasible for the major industries affected. The Court also interpreted OSHA's authority expansively and ruled that it could require wage-guarantees under the MRP provision. The Court, however, directed the Agency to determine the feasibility of engineering controls for 38 other industries. The Supreme Court refused certiorari (*United Steelworkers v. Marshall,* 647 F. 2nd 1189, [D.C. Cir. 1980] *cert. denied* 453, U.S. 913 [1981]).

On the last day of her term, Dr. Bingham issued another final health standard regulating Occupational Noise. The original proposal was issued in 1974 by Mr. Stender; after numerous studies and reopenings of the record, OSHA decided on January 16, 1980, to retain the 90-decibel level to be achieved by engineering controls but to require a "Hearing Conservation" Program for employees exposed above 85 decibels. Mr. Auchter later suspended the Standard, modified it, and extended litigation ensued. The Court of Appeals for the Fourth Circuit, *en banc*, upheld the Standard in 1985. This litigation will be discussed later in this chapter.

Deletion of de minimis standards

One of the major accomplishments during the Bingham Administration was the deletion of approximately 600 safety standards in November 1978 (Revocation of Selected General Industry Safety and Health Standards, 43 *Fed. Reg.* 49,726–49,727, 1978). These Standards were determined by OSHA to be unsuitable for regulatory purposes for a variety of reasons. For example, the Standards were obsolete, inconsequential, or directed to public safety. The broad review of OSHA national consensus standards promulgated in 1971 began in 1977, and two complete subparts were revised and simplified: the Fire Protection Standards in 1980 and the Electrical Standards in 1981. In addition to eliminating provisions unrelated to worker safety and health, lengthy provisions of the reference materials were removed and placed in nonmandatory appendices. The OSHA emphasized the "performance" approach, that is, giving employers the "flexibility" of selecting from among a variety of methods to provide the required protection. A Standard regulating Commercial Diving Operations was issued in 1977. A provision on medical surveillance in the Standard was vacated by the Court of Appeals for the Fifth Circuit (*Taylor Diving and Salvage Co. v. Department of Labor,* 599 F. 2nd 622 [5th Cir. 1979]), and OSHA's enforcement of the Diving Standard was largely superseded when the Coast Guard issued a parallel standard in 1978. Two other safety standards were issued during the Bingham administration:

1. A Standard on Servicing of Multipiece Wheel Rims
2. In response to a series of unfavorable court decisions on the issue of whether OSHA's Perimeter Guarding Standard covered "roofs" (e.g., *Diamond Roofing Co., Inc. v. OSHRC,* 528 F. 2nd 645 [5th Cir. 1976]), a Standard on the Guarding of Low-Pitched Roof Perimeters.

"Common sense" enforcement policy

The OSHA program of "common sense priorities" was intended to focus workplace inspections on health hazards, larger workplaces and the more serious health hazards (Table 31-B). From fiscal year 1976–1977, the percentage of total inspections that were health inspections rose from 8.4 percent to 15.2 percent, and then to 18.6 percent and 19.2 percent in 1978 and 1979. The focus on health hazards was demonstrated by the continuing growth of the industrial hygiene staff. In 1976, OSHA had 967 safety inspectors and 314 industrial hygienists, but by the end of 1980, there were 972 safety inspectors and 548 hygienists. A critical aspect of Dr. Bingham's common sense priorities was her emphasis on serious hazards. Under the program, 95 percent of OSHA's programmed inspections were targeted to the industries with the most serious health and safety hazards; the determination of hazards was based on data from the annual

BLS survey. Similar targeting was initiated for health hazards, but the lack of adequate illness data limited the effectiveness of this targeting. Finally, OSHA directed that the field compliance officers spend the bulk of their time on actual inspection activity with only 30 percent of the time permitted for support work. While the National Emphasis Program continued to be implemented at least through 1977, new crises brought forth new emphases. A series of grain elevator explosions at the end of 1977, leading to more than 50 employee deaths, resulted in OSHA's targeting inspection activity to that industry and initiating a review of standards in that industry (The President's Report, 1977, p. 18). The new OSHA targeting resulted in an increase in inspections in the more hazardous manufacturing sector (43.7 percent in fiscal year 1976 and 52.1 percent in fiscal 1977, and 52.3 percent in fiscal 1978). The percentage of inspections in the hazardous construction sector also grew, but more slowly, rising from 25.9 percent in fiscal 1977 to 45.5 percent in fiscal 1981. By 1983, the construction percentage had grown even more, to 58.1 percent. The percentage of inspections in "other industries" continued to drop as the number in the targeted industries increased.

During this period OSHA continued to give priority to inspections in response to employee complaints; however, a major change was made in 1979 on the issue of informal complaints. Because a great proportion of OSHA inspection resources was being expended on inspecting in response to nonwritten, sometimes nonemployee, complaints, many of which did not result in the discovery of workplace hazards, OSHA revised its "Kepone" Policy and decided that inspections would be conducted in response to informal complaints only where they appeared to involve imminent dangers or "extremely serious hazards." Otherwise, OSHA said, it would send a letter to the employer, advising him or her of the complaint "and of the action required." If the employer response was satisfactory, no inspection would be conducted; if unsatisfactory, or no response was received, an inspection would be conducted. In addition, as a check, random inspections would be conducted in the case of every 10th informal complaint. As a result of this change, the percentage of complaint inspections dropped significantly, from a high of 37.6 percent in 1978 to 23.4 percent in fiscal 1981. This made possible substantial increases in the numbers of targeted inspections, which, of course, was the major reason for the new Policy. The issue of response to employee complaints was again a major issue—this time legislative—in 1980, with the introduction of the Schweiker bill, as will be discussed in the section relating to OSHA and Congress.

During Dr. Bingham's administration, OSHA also embarked on a serious effort to give added credibility to the sanctions imposed for violations disclosed during workplace inspections. In the past, no sanctions were being imposed for nonserious violations—largely because of congressional action in appropriations riders—and even when penalties were imposed for more serious hazards, they were rarely of sufficient amount to constitute a deterrence to future violation. The charge the OSHA was a "paper tiger" was based in major part on the lack of meaningful sanctions. Under Dr. Bingham, the amounts of proposed penalties for serious, willful, repeated, and failure to abate violations all rose sharply. For example, in fiscal 1977, the proposed penalties for serious violations were approximately $6 million; in fiscal 1980, $11,301,487. In February 1980, OSHA proposed its "record" penalty (as of then $786,190) against Newport News Shipbuilding and Dry Dock Company for 551 alleged safety and 66 alleged health violations. A major factor in the increase in total penalties was the parallel increases in the percentages of serious and willful violations being cited. Increased emphasis was also placed by OSHA on criminal cases to the U.S. Justice Department. Under the Act, criminal penalties may be imposed for willful violations of a standard causing the death of an employee. Instructions were sent to field staff, emphasizing the importance of the criminal sanctions and in 1980, for example, nine cases were referred to the Justice Department for possible criminal action, bringing the total since OSHA began to 27 referrals (The President's Report, 1980, pp. 41-48). However, case convictions were few and far between and the effectiveness of the criminal provisions was seriously questioned (*Yale Law Journal* 91 (1982):1446).

Not surprisingly, this more vigorous enforcement brought with it a sharp rise in the contest rate, and the number of contests. In fiscal 1973, the number of contests was 2.7 percent, with 1315 cases being contested; in fiscal 1980 the rate was 11.7 percent and there were 7,391 contested cases. This meant, first, that the backlog of cases for decision by the Review Commission, which had always been a problem, increased at an alarming rate. In addition, regional solicitors, the attorneys for OSHA, found it increasingly difficult to handle the greater litigation load, particularly since many of the administrative proceedings were complex and time-consuming, involving expert testimony on the feasibility of engineering controls. Attempts by regional solicitors to settle cases with reduced penalties and sometimes by reducing the nature of the violation were often criticized by unions that were parties to the proceedings and even by OSHA officials. Increasing tension between lawyers and clients resulted, until 1980, when Dr. Bingham issued a directive to field staff authorizing area directors themselves to adjust citations and penalties before the contest period. This "Informal Settlement" Policy had the desired effect and the contest rate dropped sharply in fiscal 1981 to 6.3 percent. The Informal Settlement Policy was well received and particularly attractive to the next OSHA administration of Mr. Auchter, which gave added emphasis to the Policy, leading to an even greater reduction in the contest rate, to 1.9 percent in fiscal 1983.

A major legal development during the Bingham administration affecting enforcement activity was the Supreme Court decision that the Fourth Amendment protection against unreasonable searches and seizures applied to OSHA inspections. The OSHA argued that Congress had intended prompt and unannounced inspections and that a warrant requirement would delay entry to the workplace, defeat the legislative purpose and "significantly impede the implementation of OSHA." The Supreme Court in *Marshall v. Barlow's Inc.*, 436 U.S. 307 (1978) (decided in May 1978) rejected OSHA's arguments, saying that it was "unconvinced" that the warrant requirement would impose "serious burdens on the inspection system or the courts." In the first place, the Supreme Court said, the "great majority" of businessmen could be expected to consent to inspection. Further, "probable cause in the criminal law sense is not required." According to the Court, OSHA could obtain the warrant, based on a showing that the establishment was selected on an "administrative plan for the enforcement of the Act derived from neutral sources"; the Court referred specifically to OSHA's targeting plan based on accident experience and dispersal of employees. Finally, the Court said, there was no reason why

OSHA could not change its regulations to authorize its obtaining *ex parte* warrants; that is, without first notifying the employer or holding a hearing held before the district court. Since *Barlow's,* about 3 percent of OSHA inspections have resulted in warrant proceedings.

The OSHA had consistently emphasized that the "compliance assistance" aspects of its program—education, consultation and informational assistance—were an integral part of the total OSHA program. In April 1978, OSHA launched its "New Directions" Grants Program; its purpose was to utilize labor unions, trade associations, educational institutions, and nonprofit organizations to provide to employers and employees job safety and health education and training, including assistance in hazard recognition and control and training in employer and worker rights. The amounts directed by OSHA in the "New Directions" Program increased during the Bingham administration, and in August 1980, OSHA awarded $3.5 million to 66 organizations for training and education in hazard abatement. This was in addition to 82 continuing grants refunded at $13.4 million (The President's Report, 1980, pp. 70-71). At 1980 oversight hearings before the Senate Labor Committee, Basil Whiting, Deputy Assistant Secretary, testified on three "success stories" involving the New Directions Program: health and safety seminars for high-level foundry management sponsored by the Pennsylvania Foundrymen's Association; formal instruction in safety and health to employees of a large electronics manufacturing company by Indiana University; resolution of a problem causing job illnesses by the Machinists Union, in cooperation with management. At the same oversight hearing, Mr. Whiting testified on the continued expansion of the on-site consultation program. He noted that the employer demand for those services had grown and that in 1979 one out of every six OSHA-related visits to a worksite was a consultative visit and not an inspection (*Oversight on the Administration of the Occupational Safety and Health Act: Hearing Before the Senate Comm. on Labor and Human Resources,* 96th Cong., 2nd Sess., 24 [1980] [referred to hereafter as *Oversight Hearings,* 1980]).

Increasing worker participation

Secretary Ray Marshall said in 1980: "During its first 7 years, the Agency dealt almost exclusively" with the rights of the employer "with little attention to the role of the workers in recognition and abatement of hazards." This, he said, had been remedied by the Bingham administration, which concentrated equally on the contributions "all parties can make" (*Oversight Hearings,* 1980, pp. 1034–1044). One of OSHA's first steps in this direction was the revision of instructions in the FOM in 1978 to "assure" and to "encourage" employee participation in the opening, closing and informal conferences, "when practical." If it was not "practical" to hold a joint conference with employers and employees, the instructions required separate conferences to be held (Program Directive No. 200-82, Aug. 15, 1978). The issue of the payment of employee wages for walkaround time reemerged at the start of Dr. Bingham's term of office. Early in the history of OSHA, a legal determination was made that employees were not entitled to pay for walkaround time. One of Dr. Bingham's first actions in 1977 was to reverse the decision; the Solicitor of Labor issued a new interpretation that the employer's refusal to pay for walkaround was discrimination because of the exercise of a protected right and therefore illegal. The U.S. Chamber of Commerce challenged this interpretation and in 1980 the Court of Appeals for the District of Columbia upset the walkaround pay rule on the grounds that it had been issued by OSHA without OSHA first having gone through public notice and comment proceedings. Sharply criticizing OSHA for its "high handed" action, and for treating the procedural obligations as "meaningless ritual," Judge Tamm for the Court noted that public comment serves the practical purposes of reducing the risk of factual errors, arbitrary actions, and unforeseen detrimental actions (*Chamber of Commerce of the United States v. OSHA,* 636 F. 2nd 464 [D.C. Cir. 1980]). At once OSHA embarked upon notice-and-comment rulemaking in January 1981, shortly before the end of Dr. Bingham's administration. The 1981 regulation was short-lived, however. On assuming office Assistant Secretary Auchter delayed the effective date of the rule, and, following a period of public comment, revoked it. At present, there is no legal requirement on walkaround pay, although payment is required by some collective bargaining agreements.

A particularly troublesome issue was employee participation in Commission and court litigation. Although the Act expressly says that workers have a right "to participate as parties to hearings," major disagreements arose on the role of employee-parties with respect to the withdrawal of a citation and penalty by OSHA, and employees' right to participate in the settlement of contested citations and penalties. Employee groups urged that they have the right to object to withdrawals and prejudicial settlements, which derives from their statutory right to participate as parties in Commission proceedings. They argued further that it is essential that they—for whose protection the law was passed—have the right to prevent OSHA's lawyers from reducing or eliminating citations and penalties for reasons often unrelated to worker participation, such as resource limitations. The Solicitor of Labor argued, on the other hand, that it has "prosecutorial discretion" to decide not only whether to issue citations but also whether to prosecute them or settle them. The controversy became particularly sensitive because the OSHA view was being advanced by its lawyers, who were committed to the view that they should control litigation, while OSHA—the agency—did not uniformly side with its attorneys on the issue. Attempts to settle the controversy by providing for informal consultation between unions and the Solicitor's office were not successful, and the issue was litigated through the Commission and the courts. In 1985 the Supreme Court sided with the Solicitor, giving the Agency exclusive control over withdrawal and settlement of citations (*Cuyahoga Valley Ry. Co. v. United Transportation Union,* 106 S. Ct. 286 [1985]). At the same time, the courts have held that employees have the right to appeal adverse Commission decisions to the court of appeals, even if OSHA does not wish to pursue the case (*Oil, Chemical & Atomic Workers International Union v. OSHRC [American Cyanamid],* 671 F. 2nd 643 [D.C. Cir. 1982], *cert. denied sub nom. American Cyanamid Co. v. OCAW,* 456 U.S. 969 [1982]).

Section 11(c) of the Act, like analogous provisions in other regulatory statutes, protects employees against reprisal in the exercise of their statutory rights. One section 11(c) case reached the Supreme Court, *Whirlpool Corp. v. Marshall,* involving the question of whether employees have the right to refuse to work under conditions that are reasonably believed to be imminently dangerous. The Supreme Court unanimously held that the Secretary's regulation affirming this right was valid, saying that the regulation "on its face appears to further the overriding purpose of the Act, and rationally to complement its remedial scheme" (445 U.S. 1 [1980]). The *Whirlpool* case involved a safety hazard—the fall from potentially dangerous heights. The application of the refusal to work principle to health hazards, with long latency periods, is less clear and has not been definitively resolved. The issue is analogous to the question of whether health hazards would constitute imminent dangers. This, too, has not been resolved; indeed, throughout the history of OSHA, the number of imminent danger proceedings, either safety or health, have been extremely limited. It is not entirely clear that, as has sometimes been claimed, voluntary action by employers in apparent imminent danger situations is the reason that resort to court proceedings has been unnecessary.

Criticism of OSHA's implementation of the section 11(c) program has been continuous. One of the major issues has been the delays in the processing of employee section 11(c) complaints by OSHA. Another is the delays in cases reaching trial, even if OSHA decides to prosecute the case. In 1980, a representative of the Oil, Chemical and Atomic Workers International Union told a Senate Committee "without exaggeration" that section 11(c) "no longer works" (*Oversight Hearings,* 1980, pp. 873–879). To help remedy the situation, an attempt was made to imply the right of individual employees to sue an employer under section 11(c), at least where OSHA refuses to act, but the Court of Appeals for the Sixth Circuit in 1980 decided that the Act means what it says: that only the Secretary of Labor can sue in court to vindicate section 11(c) rights (*Taylor v. Brighton,* 616 F. 2nd 256 [6th Cir. 1980]).

The right of employees to know of conditions in the workplace was addressed in the Act and amplified considerably during the Administration of Dr. Bingham. Under the statute, an employer who is cited must post the citation prominently "at or near" the place where the violation occurred. Also, under the mandates of section 6(b)(7) and section 8(b)(3), all OSHA substance-specific health standards include provisions requiring labels or other forms of warning on toxic substances, access of employees to their medical records, and provisions requiring that employees have an opportunity to observe workplace monitoring and have access to monitoring records. In July 1978, OSHA amended its recordkeeping regulations, issued originally in 1971, to give employees, former employees and their representatives access to the employer's injury and illness log and to the summary of recorded occupational injuries and illnesses (29 *CFR* §1904.7[b]).

In 1980, another major step in employees' right-to-know occurred with the issuance of a regulation providing for employee access to existing employer monitoring and medical records. The new rule did not mandate the creation of new records; it applied only to those which employers had already developed under the employer's own ongoing programs. However, the new rule applied to records relating to a broad range of toxic substances and not only to the limited number of substances that OSHA had regulated with specific health standards. In 1982, a Federal District Court upheld the access rule in all respects, rejecting, among others, arguments based on employer trade secret rights and employee rights of privacy. In May 1984, the Court of Appeals for the Fifth Circuit summarily affirmed the rule without opinion (*Louisiana Chemical Association v. Bingham,* II OSHC 1922 [5th Cir. 1984]).

Probably the most important area of employees' right-to-know addressed by OSHA was hazard communication. The Agency's involvement with requirements for employer identification and communication of hazards in the workplace began almost at the beginning of the program. An advisory committee recommended a standard in 1975, as did the National Institute for Occupational Safety and Health, and the House Government Operations Committee held a number of hearings aimed, at first unsuccessfully, at pressing OSHA to issue a hazard communication standard. Dr. Bingham published a proposed Standard on "Hazards Identification," as it was then called, just prior to the end of her administration. The Standard would have required employers to assess the hazards in their workplace; and labels containing extensive information about hazards would have been required on all containers, including pipes. The proposal was withdrawn just after the beginning of Mr. Auchter's administration, as part of the regulatory reevaluation which was undertaken under Executive Order 12,291. As will be discussed, a significantly revised Hazards Communication Standard was issued in 1983, and in 1985 largely upheld by the Court of Appeals for the Third Circuit.

Continuing congressional oversight

Many amendments have been proposed throughout the history of OSHA, almost all curtailing the Agency's authority, but none has passed. The OSHA has invariably opposed these weakening amendments, a position that has been generally supported by the two labor committees, which, as strong proponents of the OSHA program, have consistently refused to report out "anti-OSHA" bills. The pressure for amendment of the Act has been significantly relieved through the device of appropriation riders, limiting OSHA authority in various respects, which have functioned as a catharsis for congressional frustration with OSHA and opposition to certain of its policies.

The greatest legislative threat to OSHA was the bill introduced by former Senator Richard Schweiker of Pennsylvania in 1979. The bill primarily would have completely revamped OSHA inspection priorities; it sought to reduce OSHA safety inspection activity in "safe" workplaces, "safe" being determined by establishment injury data for past years in that workplace. OSHA's targeting programs had been based on industry-wide, rather than individual establishment injury inspection rates. In introducing the bill, which would also have limited OSHA response to employee complaints—even those meeting formality requirements—in "safe" workplaces, Senator Schweiker said: "The bottom line is this: After 9 years under the Act's present safety regulatory scheme, we are left with no demonstrable record that it works and with a bad taste all around from the experience" (125 CONG. REC. 37,135–37,137 [1979]).

Senator Schweiker's assertion that OSHA does not "work" seemed to be borne out by the statistical record of injuries. According to BLS surveys, beginning with 1976 and continuing to 1980, there were increases in the lost workday incidence rates and in the number of lost workday cases. Thus, in the 1975 survey, the incidence rate per 100 full-time workers for lost workday cases was 3.3; four years later, the same incidence rate was 4.3. Similarly, the incidence rate for lost workdays in 1975 was 56.1 and in 1975 it was 67.7. While the incidence rate for cases without lost workdays went down somewhat during that period, the statistics far from established the marked improvement in workplace safety that had been anticipated by the OSHAct's sponsors. Dr. Bingham and others countered by saying that statistics "provide only a partial picture of the true state of safety and health in the workplace" (U.S. Department of Labor News Release, Nov. 20, 1980). The testimony of Lloyd McBride, President of the United Steelworkers of America, at the Senate oversight hearing, also argued for the wider benefits of OSHA, not measurable through statistics (*Oversight Hearing,* 1980, 698–699, 745–751).

Assistant Secretary Bingham testified vigorously in opposition to the Schweiker bill on both practical and philosophical grounds. The targeting system would decrease protection, she said. Is an air carrier not inspected for safety because it had no accidents the prior year, she asked. Dr. Bingham also sharply criticized the bill for its change in complaint policy in "overturning" OSHA's "fundamental policy judgment," that "if an employee actually working in the plant and exposed on a daily basis to hazards cares about his safety and health sufficiently to write the Secretary of Labor about it, and is courageous enough to ask for direct help from OSHA by signing his name to the complaint, that employee deserves and should receive an on-site inspection if the complaint, in OSHA's judgment, appears to have merit" (*Hearings on S. 2153 Before the Senate Comm. on Labor and Human Resources,* 96th Cong., 2nd Sess., 33-36 [1980]). Organized labor joined OSHA's major effort to defeat the bill, which was never reported by the Senate Labor Committee. In 1979, Senator Schweiker introduced an appropriations rider, that was enacted, which served a similar purpose limiting OSHA inspections to "safe" employers with 10 or fewer workers. An appropriations rider does not require the approval of the Agency's standing committee—for OSHA, the labor committees—thus avoiding a major stumbling block to passage. The Schweiker Rider has remained in effect.

Simultaneous with its legislative hearings on the Schweiker Bill, the Senate Labor Committee held also general oversight hearings on the administration of the OSHAct. As is usually the case in oversight hearings, testimony covering many topics related to the program was presented by representatives of a variety of interests: unions, employers associations, academicians, state officials, professional organizations, public interest groups, and many others. Representatives of OSHA—the Agency—typically also appear at oversight hearings; their testimony constitutes a report on the "state-of-the-Agency," often anticipating issues of concern to the committee. Questioning by members of the Committee, particularly of the OSHA witnesses, is frequent, and often sharp.

In 1979, for example, OSHA testified at six congressional hearings, including subcommittee on investigations of the House Post Office and Civil Service Committee held a hearing on the Federal Agency Program; a hearing was held by a subcommittee of the Home Labor and Education Committee on the effectiveness of the OSHA enforcement program in the Philadelphia area. In July, OSHA testified on the concerns of small businesses before the Judiciary Committee's Subcommittee on Administrative Practice and Procedure. The OSHA representatives also testified before three different subcommittees on cost-benefit analysis, on the issue of extending OSHA coverage to legislative and executive employees and on OSHA enforcement of standards affecting migrant workers (The President's Report, 1979, pp 10–11). Tragedies causing multiple deaths of employees frequently lead to congressional oversight hearings to determine the cause of the accident and whether action was needed to avoid recurrences. Examples were the hearings after a series of grain elevator explosions causing many employee deaths in December 1977 and January 1978, and the hearing held by a subcommittee of the House Labor and Education Committee in June 1978 in St. Mary's, WV, following the Willow Island cooling tower collapse, which caused the death of 78 employees (*OSHA Oversight—Willow Island, West Virginia Cooling Tower Collapse: Hearings Before the Subcomm. on Compensation, Health and Safety of the House Comm. on Education and Labor,* 95th Cong., 2nd Sess. [1978]).

The 1980 oversight hearing elicited wide interest, particularly because of the parallel legislative hearing. Many of the witnesses were prominent. Secretary of Labor Marshall appeared in addition to Dr. Bingham and Basil Whiting; Lane Kirkland, President of the AFL-CIO, Howard Samuel, President of the IUD, and Lloyd McBride, President of the United Steelworkers of America, testified for organized labor. (The witnesses representing business interests had appeared for the most part in the prior legislative hearings.) Mr. McBride's testimony, for example, was a broad-ranging evaluation of OSHA activity over a period of 10 years. He commented: "Probably the major impact of the first decade of OSHA has been the development of an occupational safety and health infrastructure, which, in turn, generates its own ameliorating influence upon the hard conditions of work and the workplace." He cited, among other things, safety and health clauses in bargaining agreements and the increase in the number of professionals in the area (*Oversight Hearings, 1980, 698–699, 745–751*). Of course, not all testimony was favorable to the Agency; many groups and interests, including organized labor, criticized various facets of OSHA activity. Mike McKevitt for the National Federation of Independent Business claimed that OSHA would continue to have little positive impact unless it gave up its "steadfast adherence" to specification standards. In sum, oversight and legislative activity, including consideration of proposed amendments and the appropriations process, have served as the major means for Congress' continuous monitoring of the administration of regulatory programs.

Interagency cooperation and controversy

The OSHA continued to work closely with other federal agencies with occupational safety and health responsibilities, but disputes continued and it was often necessary for parties to resort to litigation in order to settle jurisdictional issues. An example is *Northwest Airlines,* 8 OSHC 1982 (Review Comm. 1980), involving safety responsibility for maintenance work on airplanes, the Review Commission held that the Federal Aviation Administration had exercised authority and preempted OSHA. Some jurisdictional agreements were reached notably with the Coast Guard in the Department of Transportation (45 *Fed. Reg.* 9142 [1980]) on jurisdiction over employees working on the Outer Continental Shelf and the Mine Safety and Health Administration in the Department of Labor (44 *Fed. Reg.* 22, 827 [1979]). Dr. Bingham was particularly committed to the cooperative governmental efforts of the Interagency Regulatory Liaison Group (IRLG). Established in 1977, the IRLG comprised five major social regulatory agencies, OSHA, EPA, Consumer Product Safety Commission, the Food and Drug Administration and the Food Safety and Quality Service of the Department of Agriculture. Since these agencies have common regulatory responsibilities, their heads determined that it would be beneficial for them to share their research facilities' knowledge and personnel. The IRLG established seven work groups with the responsibility of developing consistent approaches among the agencies in such areas as compliance and enforcement, epidemiological activity, risk assessments, and testing standards (The President's Report, 1978, pp. 10–14). One of the major products of IRLG activity was its publication of a major policy report on procedures for the determination of whether chemicals were carcinogenic and the extent of the risk. The report was prepared by the Risk Assessment Group of IRLG, with the assistance of scientists from the National Cancer Institute and the National Institute of Environmental Health Sciences and was published in February 1979. (The President's Report, 1979, pp. 15-17).

OSHA: Leader or partner in state programs?

During the administration of Dr. Bingham, the tone of the relationship between OSHA and the states changed markedly. Between 1971 and 1976, the states were greeted with federal encouragement of their developing their individual programs. Interpretations of the Act, for example, the novel concept of developmental plans, were designed to facilitate the approval of state plans, allowing states to receive 50 percent funding to administer their own programs. With Dr. Bingham, however, OSHA emphasized that the Act set up a national program and that the Agency had a crucial responsibility to monitor and evaluate the effectiveness of state programs. A withdrawal proceeding was initiated for the first time during the Bingham administration in 1980 against Indiana, because of OSHA's view of the inadequacy of its program, particularly the health component. Consideration was also given to initiation of proceedings against the Virginia Plan in 1979. An earlier withdrawal proceeding against the Wyoming Plan, initiated in 1978, was withdrawn after the plan deficiencies were corrected by the state legislature. Dr. Bingham testified in 1980 at state plan oversight hearings that she had a responsibility to "take some positive action to deal with those states that were clearly operating deficient programs." She added that Federal OSHA had a "leadership role which does not lend itself to a traditional partnership of equality as I believe a number of States desire" (*Oversight Hearings on OSHA—Occupational Safety and Health for Federal Employees, Part 4: State Plans: Hearings Before the Subcomm. on Health Safety of the House Comm. on Education and Labor,* 96th Cong., 2nd Sess. 476–497 [1980]). State displeasure with OSHA policy in this area was strong and often expressed. The House Labor Committee held an oversight hearing in 1980 that featured state complaints against OSHA, notably about federal monitoring activity. At the end of the hearing, Chairman Gaydos summarized the testimony by saying that state representatives had charged that the Agency was "insensitive to the states' needs," that its requirements were "prohibitive" and "self-serving" and that the Agency wishes to "preserve complete federal jurisdiction," contrary to the statutory policy (*Oversight Hearings on OSHA—Occupational Safety and Health for Federal Employees, Part 4: State Plans: Hearings Before the Subcomm. on Health Safety of the House Comm. on Education and Labor,* 96th Cong., 2nd Sess., 515–516 [1980]).

The most controversial issues at that time were the State Plan benchmarks. As already discussed, the court of appeals rejected OSHA's view on state staff requirements and required OSHA to submit new staffing benchmarks to achieve a "fully effective" program. In 1980, with the agreement of the AFL-CIO, the plaintiffs in the litigation, OSHA submitted these benchmarks to the court; the benchmarks were based on recommendations of the National Advisory Committee and panels of federal and state experts, who considered the question of what is a "fully effective" program and the resource implications of the program. Under the benchmarks, for state safety inspections, OSHA would have required an increase from 849–1,154 compliance officers for all approved plans over a 5-year period; for health inspectors, OSHA would have required an increase from 332 to 1,683, over a longer 10-year period. The states bitterly opposed the benchmarks, particularly those for health inspectors, arguing that they demanded staffing far above the level in the federal sector, that they were unachievable, were not based on the "best available" information and techniques, and were procedurally deficient because they were not adopted after notice and comment. Dr. Bingham admitted that achievement of the benchmarks would be a "costly undertaking" but insisted that the "workplace penetration rates on which the benchmarks are based do not reflect an excessive or unreasonable amount of coverage"—for example, health inspections of textile mills "only" once every 3.9 years (*Oversight Hearings on OSHA—Occupational Safety and Health for Federal Employees, Part 4: State Plans: Hearings Before the Subcomm. on Health Safety of the House Comm. on Education and Labor,* 96th Cong., 2nd Sess. 492–493 [1980]).

At the end of 1980, 24 states and jurisdictions had received initial approval from OSHA for their occupational safety and health plans. Of these 13 states had been certified as having completed their developmental steps. However, as of 1980, there were no final approvals of state plans. The Bingham administration had been extremely reluctant to grant final approval since resumption of federal enforcement at that point would require a full administrative proceeding. The "Benchmarks" decision, with the stringent benchmarks, virtually assured that final approvals would not be granted for most states, since under the court of appeals decision, a state would have to meet the stringent personnel benchmarks in order to be eligible for final approval. A major portion of the state plan activity during the ensuing Reagan Administration was designed to remove the barrier posed by the benchmarks to final approval.

A new Executive Order

In February 1980, President Jimmy Carter signed Executive Order No. 12,196, adding many new features to the Federal Employee Safety and Health Program (45 *Fed. Reg.* 45,235 [1980]). Among the provisions in the order, which greatly strengthened the Program, federal agency heads are required to comply with OSHA standards applicable to the private sector, unless the Secretary of Labor approves compliance with alternate standards. For the first time OSHA was given authority to conduct on-site inspections of federal agency facilities in specified circumstances and to recommend abatement measures. Agency heads are authorized to establish safety and health committees comprising an equal number of management and employee representatives. The committee, by majority vote, may request an OSHA inspection if it is not satisfied with the agency response to a report of hazardous working conditions. The Order tracked the OSHAct applicable to private employers in prohibiting discrimination against employees for the exercise of protected rights and providing for employee walkaround rights. The Order also authorizes "official time" for employees participating in activities under the Order. A substantial revision of the Department of Labor's regulations, newly entitled "Basic Program Elements for Federal Employee Occupational Safety and Health Programs," reflecting the new Executive Order, was also issued by OSHA in 1980 (29 *CFR* Part 1960).

The Bingham administration ends

Dr. Bingham's administration ended as it began—with a flurry of activity. During the last several weeks, a number of standards and regulations were issued: Walkaround Pay, Hearing Conservation, Hazard Communication, and several others of somewhat less major consequence involving supplemental decisions in Lead and the Carcinogens Policy. Not all were destined to be long-lived, but the Assistant Secretary's commitment to vigorous standards and enforcement action continued to the last day of her term of office.

1981 TO THE PRESENT: OSHA'S BALANCED APPROACH

"A balanced approach"

The election of Ronald Reagan in 1980 inaugurated a new era in the history of OSHA. President Reagan appointed Ray Donovan as Secretary of Labor and Thorne Auchter as Assistant Secretary for OSHA. Mr. Auchter, who had been a construction executive, served as Assistant Secretary from March 1981 to March 1984. After Mr. Auchter's resignation, Robert Rowland, a Texas attorney, who was previously Chairman of the Occupational Safety and Health Review Commission, served as Assistant Secretary under a recess appointment for about nine months. Mr. Rowland's term was controversial, and he resigned in mid-1985, soon after William Brock became Secretary of Labor. Patrick Tyson, a former attorney in the Office of the Solicitor, headed OSHA in the absence of an Assistant Secretary until early 1986. John Pendergrass, a certified industrial hygienist, took office on May 22, 1986, and is currently Assistant Secretary.

Assistant Secretary Auchter summarized his approach to OSHA to a subcommittee of the House Labor Committee in 1982:

> Only by working together can business, labor and government achieve their common goal of safe and healthful workplaces. The varied authorities granted under the Act allow a wide range of agency activities and programs that involve the government and the private sector in cooperative efforts. We have accordingly developed a balanced program mix that focuses not only on standards-setting and enforcement, but also on ways of helping employers and employees solve safety and health problems in the workplace.

He listed as examples of "self-help approaches": on-site consultation, education and training, and other voluntary protection methods (*OSHA Oversight—Agency Report by Assistant Secretary of Labor for OSHA: Hearing Before the Subcomm. on Health Safety of the House Comm. on Education and Labor,* 97th Cong., 2nd Sess., 2–4 [1982]).

A new era in compliance

Consistent with his philosophy of "working together," Mr. Auchter instituted a number of changes in enforcement priorities and procedures. In the past, follow-up inspections—that is, OSHA reinspections of establishments under a Commission order to abate hazards to determine if abatement in fact had taken place—were more-or-less routine. This policy was revised by Assistant Secretary Auchter; he stated that since experience showed that "almost all" firms visited in follow-up inspections were in compliance, the Agency was deemphasizing follow-ups, keeping them to an "essential minimum." This would permit OSHA inspection resources to be utilized in higher priority areas. Sharp reductions in the number of follow-up inspections took place: in fiscal year 1980, follow-ups were 18.4 percent of all inspections; in 1981, 9.5 percent and in 1982, 2.5 percent. In fiscal year 1986, 3 percent of OSHA inspections were a follow-up (16 OSHR 911, 1987). Also noting the high contest rate, Mr. Auchter underscored the existing policy of informal conferences between OSHA regional staff and employers, so that, "whenever possible," settlement agreements under which the employer agrees to comply would be reached. The result, as anticipated, was in Mr. Auchter's words, a "dramatic drop" in contested cases: from one fourth of all OSHA inspections in fiscal 1980, to one twelfth in fiscal 1981, and 2.8 in 1982. In fiscal 1986, the contest rate was 3.6 percent (16 OSHR 912, 1987). The OSHA also issued guidelines on the use of the general duty clause to avoid its unwarranted use. In early 1982, one of Auchter's earliest actions was to withdraw the walkaround pay regulation after notice and comment.

Another innovation by OSHA was to target inspections in the manufacturing sector to high hazard establishments. Under the new procedures, applicable to programmed inspections for general industry in the safety area, compliance officers continue to visit establishments based on the high hazard industry list. At the beginning of the inspection, however, the compliance officer would inspect the firm's injury and illness records and calculate the firm's lost workday injury rate (the number of lost workdays per 100 workers). If the particular firm's rate was below the most recently published national lost workday rate for manufacturing, the inspector would not walk through the workplace and conduct a full scale safety inspection. The purpose of this system, according to Mr. Auchter, was to identify and inspect "only those workplaces where there is a high likelihood of finding serious problems." The program did not apply to health inspections, nor did it affect OSHA's response to complaints. In fiscal 1983 there were 71,303 inspections and 8,444 "records inspections"—a new category, inspections of records and not of workplaces. In fiscal 1986, there were 64,071 inspections, of which 4,619 were "records inspections" (16 OSHR 911, 1987).

Responding to criticism that excluding companies with low lost workday injury rates from the threat of inspection eliminates the incentive for compliance, OSHA announced in 1986 that it was allocating 5 percent, or about 700, of its programmed safety inspections to manufacturing establishments with below average lost workday rates. In addition, OSHA announced its intention to undertake a comprehensive inspection of every 10th high hazard industry manufacturer, which had undergone only a records review by the Agency (15 OSHR 867, 1986).

The new "balanced" approach of OSHA was viewed differently by all groups. Shortly after the program was instituted, the AFL-CIO stated that the system was "ill-conceived" and "unsound," removing numerous manufacturing employers from "one of OSHA's most effective compliance tools. . .the threat of general scheduled safety inspections" (*Oversight on the Administration of the Occupational Safety and Health Act: Joint Hearings Before the Subcomm. on Investigations and General Oversight and Subcomm. on Labor of the Subcomm.*

on Labor and Human Resources, 97th Cong., 1st Sess. 171 [1981]).

The OSHA has also continued to implement special emphasis targeting programs, in response to new information showing particular hazards to employees. The grain elevator inspection, in effect since 1977, was renewed. In August 1985, OSHA instituted a targeting program for the fireworks industry after 30 employee deaths in fireworks plants (16 OSHR 133-134, 1986). Following the Bhopal (India) catastrophe, OSHA instituted a pilot Special Emphasis Inspection Effort in the chemical industry in 1985. Congress called for a report on the Program, and on July 15, 1985, OSHA preliminarily noted instances disclosed in its inspections of chemical plants where hazardous conditions were not addressed by OSHA standards (16 OSHR 147-148, 1986).

Consultation, education, and training were major components of the Auchter compliance effort. On-site consultation activity was emphasized and expanded. By 1985, 52 states and jurisdictions provided consultation services to employers. Under revised consultation regulations issued in 1984, an employer is given a 1-year exemption from OSHA general schedule inspections (but not complaint inspections or accident investigations) if it undergoes a comprehensive consultation visit, corrects all identified hazards, and demonstrates that it has an effective safety and health program in operation (49 *Fed. Reg.* 25, 082, 1984). The regulations also broadened the scope of consultation to include advice on the "effectiveness of the employer's total management system" to ensure safety and health at the workplace. In 1985, the Program's first full year of operation, OSHA granted a total of 382 exemptions to participating employers (The President's Report, 1985, pp. 57-58). In light of the diminished prospects that an employer would receive an OSHA programmed inspection, however, some have questioned whether there is any "large" incentive for employers to participate in the consultation program (*OTA Report,* p. 238).

Three new Voluntary Protection Programs-Star, Praise, and Try—were instituted in 1982. (The Praise Program has since been discontinued.) Each of the Programs was directed at a different category of employers but all were "based on the premise that [an employer] having a comprehensive safety and health program which operates effectively can provide greater worker protection than the chance of an OSHA enforcement inspection." For employers accepted under one of these Programs, OSHA programmed inspections are discontinued, but complaint inspections and accident investigations are handled in accordance with regular procedures (The President's Report, 1985, pp 60-61). In early 1986, there were 26 general industry and three construction employers in Star, and six general industry employers in Try. A construction company with injury rates nearly four times the injury average was removed from the Try Program in November 1985, the first time, according to OSHA, that an employer was removed (15 OSHR 845, 1985).

The New Directions grant program, which was designed to develop competence in nonprofit organizations for safety and health training and education for employers and employees, continued. Significant changes were made, however. The program funding was cut down substantially, from $13.9 million in fiscal year 1981 to $6.8 million in fiscal years 1982 and 1983, and $5.6 million in 1985 (*OTA Report,* p. 238; The President's Report, 1985, p. 61). In 1981 OSHA stopped using a peer review process under which persons affiliated with grant recipients evaluated applicants for new grants. The reason given was that peer review was "too costly and resulted in a possible conflict of interest" because those who had received grants evaluated applicants (How OSHA Controls and Monitors Its New Directions Program, Draft Report, GAO No. HRD, 85-29, p. 7 [1984]).

In 1983, OSHA made a major change in New Directions Program requirements, making educational and other nonprofit organizations ineligible to receive grants unless they had previously been approved for a planning grant. Educational and other nonprofit organizations may be members of a consortium eligible for a grant, but there must be a labor or employer organization in the consortium which assumes responsibility for submitting the proposal and administering the grant (The President's Report, 1985, p. 64). The COSH groups (local safety and health coalitions) were no longer funded, and a 1986 report issued by the public interest group, Public Citizen, sharply criticized OSHA for eliminating its funding of these occupational safety and health coalitions (*Working in the Dark—Reagan and the "Right to Know" About Occupational Hazards,* 1986, pp. 40-42).

In order to better focus the activity of New Direction grantees in high priority areas, in November 1985 OSHA announced that in the future it would award grants only to organizations that proposed to develop education and training programs in one of four specific areas: chemical industry, chemical and toxic substances, hazardous waste sites, and new OSHA standards (Notice of Grant Program, 50 *Fed. Reg.* 47,294 [1985]; The President's Report, 1985, p. 65).

The debate over the wisdom and success of OSHA's new orientation continued. In 1983 the Center for Responsive Law, a public interest group, published a report critical of the Auchter administration, asserting that the terms "voluntary," "cooperative," and "nonadversarial" are "clear code words for regulatory abdication." Much the same view was expressed in another public interest report, issued in 1984, *Retreat from Safety,* which concluded that Mr. Auchter's "legacy to the American workers of today and tomorrow was a 'sorry' one indeed." (Claybrook, 1984, p. 113). Mr. Auchter responded to the earlier report of the Center for Responsive Law, saying it was "flawed and biased," that it relied on "inaccurate and misleading statements, undocumented opinions, and unrepresentative anecdotes to support its preconceived conclusions" (13 OSHR 408, 1983). And, when the BLS injury statistics showed improvement during the early years of Mr. Auchter's administration, Mr. Auchter referred to this fact as demonstrating that the administration's new approach to enforcement had succeeded while the prior "tough" enforcement approaches had failed, as evidenced by the growing injury rate before 1981.

The states: A partnership once again

Assistant Secretary Auchter also took a radically different approach to state programs. He said in 1981 that it was his "firm intention" to "resolve differences" that have existed in the past between federal and state OSHA, and "to develop a management and policy framework that in the future will integrate the states into the overall OSHA program." In the last analysis, he said, "local problems are best addressed by those closest to them" (*State Implementation of Federal Standards: Hearings Before the Subcomm. on Intergovernmental Regulations of the Senate Comm. on Governmental Affairs,* 97th Cong., 1st Sess. 14-25 [1981]). Among the steps taken by Mr. Auchter to implement this partnership was to terminate the withdrawal proceedings against Indiana, deny the petition for withdrawal of the

Virginia Plan, and enter into operational agreements with the remaining states with approved plans, thus ending discretionary federal enforcement in those states. One of the major initiatives of the Auchter administration was to radically reverse Dr. Bingham's "benchmarks," which Mr. Auchter viewed as the major impediment to "final" approval because of their "stringent requirements." The OSHA supported a rider to its appropriations bill that would preclude the expenditure of Agency funds to implement the court of appeals benchmarks decision; this was an unusual response, since the Agency had traditionally opposed riders as a means of legislating. The rider was passed, but elicited opposition and resentment and was deleted by Congress in December 1982, never to be renewed.

At this point OSHA embarked a broad reconsideration of Bingham benchmarks. At the request of OSHA, a State Plan Task Group was established in August 1983 to work with OSHA in reviewing and revising the benchmarks. The Task Group decided that the original benchmark formula used in 1980 was "conceptually sound" but that modifications in input were necessary to incorporate, where available, state-specific data and "to build flexibility into the formula to accommodate differences among states" (*Kentucky, Final Approval Notice,* 50 *Fed. Reg.* 24,884, 24,886 [1985]). Twelve states completed revision of their benchmarks. During 1985, eight states (Arizona, Iowa, Kentucky, Maryland, Minnesota, Tennessee, Utah and Wyoming) obtained approval of their revised benchmarks and final approval of their state plans, making 11 final approvals in all (The President's Report, 1985, pp. 43-47). In January 1986, OSHA approved revised benchmarks for four additional states (Indiana, North Carolina, South Carolina and Virginia) and final approval was granted to the Indiana State Plan in September 1986 (51 *Fed. Reg.* 34,215). The AFL-CIO voiced strong opposition to the new benchmarks because of the substantial reductions in the number of inspectors required for final approval. In the 12 states, for example, 171 health inspectors would make the states eligible, instead of 746, as in 1980. The unions argued that the decrease conflicted with the court of appeals decision and that the number of inspectors required failed to provide, among other things, for enforcement of new standards (15 OSHR 903-4, 1986). In commenting on the revised benchmarks, the United Steelworkers of America complained that the revision enables each state "to manipulate" the number of workplaces to be covered by the inspection program in order to justify current staff levels (O'Brien letter to OSHA, 1985, p. 1).

Assistant Secretary Auchter also changed OSHA's state monitoring system. Section 18(f) of the Act requires OSHA to make a "continuing evaluation" of how the state is "carrying out" its approved plan which would be the basis for determining whether the state should receive final approval. The main components of OSHA's monitoring system had been: (1) case file reviews; (2) spot check visits by federal inspectors of establishments previously inspected by state enforcement personnel; and (3) accompanied visits by federal inspectors of actual state inspection and consultation visits. An appropriations rider in 1980 radically limited OSHA's spot-check monitoring authority; during that year state representatives testified at House oversight hearings, strongly criticizing OSHA monitoring activity.

Under Mr. Auchter's new monitoring system, first implemented in August 1983, there was a shift in emphasis in state plan evaluation from "intrusive on-site monitoring to analysis of the state-submitted statistical data" (The President's Report, 1983, p. 41). The monitoring system compared state statistical information in 11 major program areas—such as standards, variances, consultation, and enforcement—to federal performance, also as statistically measured, although states were "not necessarily" expected to equal federal performance respecting each measure. A primary component of the system was the identification and analysis of "outliers," that is, "state performance on a particular performance measure that falls outside the established level or range of performance" required in comparison to the federal program (*OSHA Instruction,* STP 2.22A "State Plan Policies and Procedures," 1986, chapt. 3). An "outlier" is not necessarily a deficiency, but requires "further explanation." Most data used to evaluate state performance is obtained through state participation in the Integrated Management Information System (IMIS). As of 1986, 22 states were participating in IMIS. These states provide data to the system either by forms for OSHA entry or by direct entry using OSHA-provided equipment.

On July 1, 1987, the California state program covering private employers ceased operations when funding ended because of a budget dispute (17 OSHR 199, 1987).

Health standards—Continued

Mr. Auchter early directed his attention to OSHA's health standards activity. The administration attempt to modify OSHA's historic position against cost-benefit analysis was rejected by the U.S. Supreme Court in the *Cotton Dust* case, which held that the Act prohibited cost-benefit analysis. This was only the beginning; the health standards area continued to be characterized by controversy, including some major litigation, throughout the Reagan Administration.

Among the major trends in standards development during this period were the following:

1. In 1986, OSHA stated that its approach to standards encompassed the following elements:
   - adopting "less rigid, performance-oriented standards"
   - addressing "existing, significant risks" and adopting requirements that will "significantly" reduce such risks
   - adopting requirements that are technologically and economically feasible
   - using the most "cost-effective" approach (The President's Report, 1985, pp. 7-8).

   In determining "significant risk" for health standards, OSHA typically would prepare one or more quantitative risk assessments, by means of which OSHA would quantify the excess risk from the disease for various PELs. Based on these numerical calculations, OSHA would determine whether the risk was "significant" under the *Benzene* case guidelines.
2. On February 17, 1981, President Reagan issued Executive Order 12,291 "to reduce the burdens of existing and future regulations, increase agency accountability for regulatory actions, provide for presidential oversight of the regulatory process, minimize duplication and conflict of regulations, and insure well-reasoned regulations" (E.O. 12,291). Significant requirements of the Order were the preparation by agencies of a preliminary and final Regulatory Impact Analysis (RIA) for each "major" regulatory action and the review of these analyses by officials in the OMB. As a result of Executive Order 12,219, the OMB was assigned a major role in reviewing OSHA standard activity. In several standards proceedings, OSHA standards actions were questioned and often delayed by OMB. In the Ethylene Oxide proceeding,

the legality of OMB involvement was sharply challenged by the Public Citizen Group (a public interest group). As will be discussed, the court of appeals vacated OSHA's Standard partly, but without reaching the OMB question.

3. In this period OSHA undertook only limited new health standard actions and, in some cases it was pressured to do so by litigation. Two new final Standards were issued, Asbestos and Ethylene Oxide, and several rulemaking proceedings are under way at this writing. At the same time, OSHA's attempts to significantly cut back the stringency of previously issued standards were mostly unsuccessful. A notable example is the revision of the Cotton Dust Standard, which, although revised, was retained substantially as originally promulgated in 1979.
4. Somewhat reluctantly, the courts of appeals have been thrust into deciding cases raising the question of whether OSHA should be ordered to commence rulemaking, or complete previously delayed rulemaking. Two major decisions were issued by the Court of Appeals for the District of Columbia in this area. The first directed OSHA to promulgate an Ethylene Oxide Proposal within 30 days. Ultimately the Agency issued a final Standard, which was further challenged in the court of appeals. Early in 1987, the Court of Appeals sharply criticized OSHA for its 14-year delay in failing to issue a Field Sanitation Standard, and directed that a final standard be promulgated within 30 days. The Standard was issued in April 1987 (52 *Fed. Reg.* 16,050).
5. The OSHA's overall past success in defending its standards in court continued throughout this period. However, this success, as before, did not extend to the Court of Appeals for the Fifth Circuit, which in 1983 set aside OSHA's Emergency Standard on Asbestos. The OSHA issued no other emergency standard during the Reagan Administration.
6. A major area of new litigation was the preemption of state standards activity; this arose particularly in connection with OSHA's Hazard Communication Standard.
7. The OSHA sought to use "negotiated" rulemaking procedures in developing a revised Benzene Standard. This cooperative approach earlier had been attempted briefly and unsuccessfully with the Coke Ovens Standard. While negotiations ultimately failed in the Benzene proceeding (OSHA later proposed its own revised Standard); the "negotiated" approach later succeeded in developing a proposed Standard for MDA (4,4'-methylene dianiline) (16 OSHR 1451, 1987). However "skepticism" continued to be expressed about the usefulness, or even legality, of these efforts (15 OSHR 942, 1986).
8. The Assistant Secretary and many others continued to express concern over OSHA's slow pace in standards development. Early in 1987, the Administrative Conference published a report noting "severe management problems" in OSHA's standards-setting process (16 OSHR 995, 1987). Congress made clear its concern about OSHA's promptness in issuing standards when it required in the Superfund Amendments and Reauthorization Act of 1986 (SARA) that OSHA issue a protective standard for hazardous waste site operations within 60 days of the date of enactment. The law was enacted on October 17, 1986; the OSHA interim Standard, effective immediately, was published on December 19, 1986 (51 *Fed. Reg.* 45,654, [1986]).

The major individual OSHA rulemakings during the Reagan Administration related to cotton dust, hearing conservation, hazard communication, asbestos, field sanitation, and ethylene oxide.

Cotton Dust

One of the requirements of Executive Order 12,291 related to regulations pending when the Order was issued in 1981. These had to be reviewed by the Agency to determine their consistency with the policies of the new Executive Order. The *Cotton Dust* case, involving the issue of cost-benefit analysis, was then pending before the U.S. Supreme Court, after having been briefed and oral arguments held. In March 1981, OSHA, through the Solicitor General, filed a supplemental memorandum with the Supreme Court, asking the Court to refrain from deciding the case, so that the Agency would be able to "reconsider the Cotton Dust Standard and the role of cost-benefit analysis under the Act" in light of the new order. The Supreme Court, in deciding the case and prohibiting cost-benefit analysis, rejected OSHA's request for a second chance in a footnote, without giving reasons. Although the Supreme Court thus affirmed the Cotton Dust Standard as it applied to the textile industry, the Agency proceeded with its reevaluation of the Cotton Dust Standard, publishing a notice of proposed revisions in June 1983. Prior to the issuance of the proposal, there was an extended dispute between OSHA and OMB over the scope of the reconsideration—particularly whether engineering controls should continue to be required—but ultimately in its proposal, OSHA adhered to its established policy in this instance of requiring engineering controls. Revisions in the Cotton Dust Standard were eventually issued in December 1985, following, according to reports, another dispute between OSHA and OMB over the scope of the medical surveillance requirements in the Standard. The revised Standard, according to OSHA, contained substantial cost savings while maintaining health protection for workers (50 *Fed. Reg.* 51,120 [1985]). Two challenges to the Cotton Dust Standard are pending in the Court of Appeals for the District of Columbia, at this writing.

Hearing Conservation

The OSHA Hearing Conservation Amendment, issued by Dr. Bingham in January 1981, was also significantly affected by the change of administration. Mr. Auchter quickly stayed the effective date of the Amendment, and then stayed it a second time; there followed a suit by the AFL-CIO challenging the stays because they were issued without notice and comment. In the meantime, Mr. Auchter issued an interim revised Hearing Conservation Amendment in August 1981 and, after rulemaking, a final revised Amendment in March 1983. A variety of changes were made in the Bingham Standard, mostly emphasizing a more flexible "performance" approach. In other words, employers were given "flexibility" in complying, and compliance would therefore be encouraged "in the manner that is easiest under the circumstances present in the particular work environment" (48 *Fed. Reg.* 9738 [1983]).

The Standard was challenged by an employer association, and in November 1984, in an unexpected and remarkable decision, the court of appeals, with one dissent, vacated the Hearing Conservation Amendment on the ground that the Agency failed to distinguish between hearing losses caused by workplace noise and those caused by nonworkplace noise, and therefore "clearly imposes responsibilities on employers based on nonwork-related hazards." This, the court said, was "not a problem that

Congress delegated to OSHA to remedy." OSHA and the AFL-CIO moved for a hearing by the full court; the union emphasized the impact of the court's decision not only on the Noise Standard but on all other OSHA health standards where it is frequently impossible to separate between workplace-caused illness and illness caused outside the workplace. The full court reconsidered the case and in October 1985 reversed the original decision, completely rejecting its reasoning and finding the Standard feasible, supported by substantial evidence (*Forging Industry v. Secretary,* 773 F. 2nd 1436, [4th Cir. 1985]).

Hazard Communication

The proposed Hazard Communication Standard was also published at the end of Dr. Bingham's administration. The Standard was quickly withdrawn by the new administration in February 1981 "for further consideration of regulatory alternatives" and in March 1982 a revised Hazard Communications Proposal was published, limited in its application to employers in manufacturing classifications. In November 1983, a final Standard was published, also limited to the manufacturing sector. The Standard also "accommodated" the health interest and the economic interest in trade secret protection by "narrowly defining the circumstances under which specific chemical identity must be disclosed." The Standard was challenged, significantly only by a union—the United Steelworkers of America—a public interest group—Public Citizen—and by several states. Employers did not challenge the Standard; indeed, they generally supported a federal OSHA Hazard Communications Standard; employers hoped its effect would be to preempt the "multiplicity of differing and conflicting State and local hazard communications laws" which "impose... an undue burden on products moving in interstate commerce and on multistate employers" (*Final Hazard Communication Standard,* 48 *Fed. Reg.* 53,280 [1983]).

In May 1985, the Court of Appeals for the Third Circuit issued an important decision ruling in favor of the challenging parties on several critical issues. First, the Court ruled that OSHA had failed adequately to justify its limiting the hazard communication requirements to the manufacturing sector, thus ignoring the recorded evidence that workers in sectors outside the manufacturing sector are also exposed to toxic materials hazards. Secondly, the Court found that OSHA afforded unduly broad trade secret protection. Suggesting that a rule which protected only "formula and process information but [required] disclosure of hazardous ingredients" would adequately protect trade secrets, the Court of Appeals directed OSHA to reconsider this issue as well as the scope of the Standard. Finally, the Court rejected OSHA's limitation of access to certain confidential information to health professionals (*United Steelworkers of America v. Auchter,* 763 F. 2nd 728 [3rd Cir. 1985]).

On September 30, 1986, OSHA issued a final rule amending the Hazard Communication Standard to provide wider access to trade secrets in nonemergency situations and to eliminate trade secret protection for chemical information that can be discovered readily through reverse engineering processes (51 *Fed. Reg.* 34,590 [1986]). However, OSHA delayed issuing even a proposal on extending the scope of the Hazard Communication Standard for more than 18 months and issued a second decision ordering OSHA to broaden the Standard within 60 days, or to explain why it was not feasible to do so (*United Steelworkers v. Pendergrass,* 13 OSHC 1305 [3rd Cir. 1987]).

In the meantime, in October 1985, the Court of Appeals for the Third Circuit decided another case involving New Jersey's Right-to-Know Law. (New Jersey does not have an OSHA State Plan.) A variety of business groups claimed that the New Jersey Law was preempted by the OSHA Hazard Communication Rule. Under the OSHAct, the laws and regulations relating to occupational safety and health in states without approved OSHA plans are preempted as to "issues" covered by the Federal Program. The court made two distinctions: between the manufacturing and nonmanufacturing sectors and between workplace and environmental hazards. It decided that the New Jersey Law was preempted by the OSHA Standard *only* respecting the identification and disclosure of workplace hazardous substances in the manufacturing area—that is, on the "issue" that was covered by OSHA's standard. There was no preemption on nonmanufacturing activities, the Court held, since the OSHA Standard did not apply to those activities. Finally, the Court ruled that there were unresolved factual questions on whether there was preemption as to state labeling requirements for environmental hazards. (*New Jersey State Chamber of Commerce v. Hughey,* 774 F. 2nd 587 [3rd Cir. 1985]). In September 1986, courts of appeals handed down two additional decisions further outlining the complex rules on the effect of the Hazard Communication Standard in preempting state and municipal requirements. The first related to the Pennsylvania Right-to-Know Law, the second the City of Akron Right-to-Know Ordinance (*Manufacturers Association of Tri-County v. Krepper,* 801 F. 2nd 130 [3rd Cir. 1986]; *Ohio Manufacturers' Association v. City of Akron,* 801 F. 2nd 824 [6th Cir. 1986]). Note: Effective May 4, 1988, *all* employers will be required to follow the rules of the Hazard Communication Standard.

Asbestos

Assistant Secretary Auchter took a major regulatory action in November 1983, issuing an ETS lowering the PEL for asbestos to 0.5 fibers per cubic centimeter of air. The existing PEL was 2 fibers per cubic centimeter, which was established in 1976 at the expiration of the 4-year delayed effective date of OSHA's first Asbestos Standard, issued in 1972. In 1975, meanwhile, OSHA proposed a 0.5 PEL for asbestos but no further action was taken on the proposal. In 1983, OSHA performed a risk assessment and concluded that for all workers exposed to 0.5 fibers per cubic centimeter, there would be an estimated 196 excess cancer deaths per 1,000 workers for 45 years of exposure, 139 deaths for 20 years, 10 for one year, and 6 excess deaths per 1,000 workers for 6 months of exposure. On the basis of these statistics, OSHA said: "The overall extraordinary degree of risk, the extent that very high risk is found in many asbestos using industries, and the unusually high quality of the data utilized to make these assessments present a very strong evidentiary basis for a 'grave danger' finding" (*ETS on Asbestos,* 48 *Fed. Reg.* 51,086 [1983]). The ETS was quickly challenged by a trade association representing 47 employers and by a number of individual employers. On November 23, the Court of Appeals for the Fifth Circuit granted the stay and in March 1984 held that the Standard was invalid.

The decision of the Court of Appeals was significant not only because it vacated the particular ETS, but because of its negative implications for any OSHA emergency standards, at least in the Fifth Circuit. This Standard, unlike some preceding ones that were vacated, would have regulated a substance that is known to cause death and serious physical harm; OSHA performed extensive and careful risk assessments, statistically demonstrating the excess risk at various levels of asbestos exposure, and the preamble set forth the Agency data, its reasoning and findings in detail. Yet, the Court of Appeals, once again warning that the emergency authority must be "delicately exercised," faulted OSHA for its mistake in calculating the number of deaths which could be avoided over a 6-month period—"substantially less than 80," the Court said—and held that "evidence based on risk assessment analysis is precisely the type of data that may be more uncritically accepted after public scrutiny, through notice-and-comment rulemaking especially when the conclusions it suggests are controversial or subject to different interpretations" (*Asbestos Information Association v. OSHA,* 727 F. 2nd 415 [5th Cir. 1984]).

After rulemaking, including a public hearing, which lasted from June 19 to July 10, 1984, and a printed record of 55,000 pages, on June 20, 1986, OSHA issued two final Standards regulating exposure to asbestos and related materials; one Standard applied to general industry (including maritime) and the second to the construction industry. The Standards lowered the PEL for asbestos to 0.2 fibers per cubic centimeter (OSHA's 1972 Standard was 12 fibers per cubic centimeter). The requirements of the Construction Standard were tailored to the unique characteristics of the industry, notably, the fact that construction industry worksites are nonfixed and are temporary in nature (*Final Asbestos Standard,* 51 *Fed. Reg.* 22,612 [1986] [to be codified at 29 *CFR* §1910.1001]). Challenges to the new Asbestos Standard were filed quickly. Two departments of the AFL-CIO sought review, claiming that OSHA's regulation was not stringent enough. The Asbestos Information Association, an employer organization, sought review, because it had "questions about the feasibility, particularly of monitoring, at the new Permissible Exposure Limit." Petitions for review were also filed by R. T. Vanderbilt and the National Stone Association, challenging OSHA's regulation of nonasbestiform tremolite, actinolite, and anthophyllite and in July 1986, OSHA stayed the Standard insofar as it applied to these substances, pending reconsideration. The review proceedings have all been transferred to the Court of Appeals for the District of Columbia (16 OSHR 885 1986).

Field Sanitation

Throughout the history of OSHA, courts have been petitioned usually to review standards already issued by OSHA. However, more recently the courts have been pressed into another role: deciding suits brought to compel OSHA (or other regulatory agencies) to initiate rulemaking actions for the issuance of a standard, or to issue standards when the completion of rulemaking is unduly delayed. In a period when OSHA standards development, in the view of many groups, has slowed down, it is not surprising that this type of proceeding has become more common.

The court's authority to review a standard that has been issued is explicitly granted in the OSHAct. It is less clear that a court can or should decide whether a standards proceeding should be initiated. A decision to undertake rulemaking typically involves the weighing of competing demands for Agency resources, the importance of a variety of projects, as well as considerations of general policy, priorities, and politics. In such situations, the court obviously has preferred to give the Agency great discretion in demanding what actions to take. Also, in a suit to compel Agency action, there is typically no record before the court on the basis of which a decision could reasonably be made. Despite these strong considerations against review, courts on occasion have found the Agency actions so arbitrary as to justify a finding of abuse of Agency discretion and have directed that action be taken. As an alternative, the court may require the Agency to make a decision—whether to publish a proposal or not—without determining the particular action that should be undertaken.

The Field Sanitation proceeding raised the question of OSHA's failure to promulgate standards requiring toilet and drinking facilities for agricultural workers, parallel to the sanitation requirements for nonagricultural workers that were imposed by OSHA in 1971. A suit was brought in December 1974 by the Migrant Legal Action Program to require OSHA to initiate rulemaking. First OSHA published a proposed Standard, but angry congressional pressure forced it to drop the rulemaking. The case reached the Court of Appeals for the District of Columbia for the first time, in 1977. The Court was willing to give broad deference to OSHA's discretion on rulemaking priorities. The Court said: "With its broader perspective, and access to a broad range of undertakings, and not merely the program before the Court, the Agency has a better capacity than the Court to make the comparative judgments involved in determining priorities and allocating resources." However, the Court of Appeals insisted that OSHA, which had never said that a Field Sanitation Standard should not be issued, develop a timetable on when it would be in a position to complete rulemaking (*National Congress of Hispanic American Citizens [El Congreso] v. Usery,* 554 F. 2nd 1196 [D.C. Cir. 1977]). The Agency developed timetables and on several occasions in 1981 said that it could not work on field sanitation for the next 2 years because of higher priorities and that it would be between 58 and 63 months before the Standard would be issued. The district judge, who had been sympathetic to the suit of the migrant workers from the outset, rejected the timetable, saying that the existence of other OSHA work doesn't justify "relegating a simple Standard of Field Sanitation to the dust bin" (*National Congress of Hispanic American Citizens v. Donovan,* 2142-73 [D.D.C., 1981]). The case was heading for the Court of Appeals again, but was settled, with OSHA promising that it would make a "good faith" effort to issue a final Standard in 31 months.

Consistent with its commitments, OSHA commenced rulemaking and public hearings were held in Washington DC, Florida, Texas, Ohio, and California. Over 200 witnesses were heard, and 4,000 pages of transcript were received. On April 16, 1985, with Robert Rowland under a recess appointment, OSHA published a *Federal Register* notice containing a determination that no final Field Sanitation Standard would be issued. The OSHA gave three reasons: (1) other enforcement priorities would make it difficult for the Agency to enforce a Field Sanitation Standard, and therefore future expenditure of resources on the Standard would serve no useful purpose; (2) it would be more appropriate under principles of federalism for the field sanitation issue to be regulated by the states; and, (3) finally, OSHA was reluctant to preempt field sanitation standards already issued by states with approved plans (*Decision,* 50 *Fed. Reg.* 15,086 [1985]). (Since OSHA up to now had no standard on the "issue," there was no preemption.) There was a storm of angry criticism of OSHA, including congressional oversight hearings, and litigation was renewed in the Court of

Appeals in a new proceeding brought by the Farmworker Justice Fund. The public interest brief to the Court begins by saying, "It is difficult to imagine an agency action less supported by substantial evidence in the record than OSHA's refusal to adopt the Field Sanitation Standard" (Brief of Migrant Legal Action Program in *Farmworker Justice Fund Inc. v. Brock,* D.C. Cir. No. 85-1349, p. 27 [1985]).

Meanwhile, William E. Brock became Secretary of Labor and Robert Rowland left the Agency. In October 1985, in a notice signed by the Deputy Assistant Secretary Patrick R. Tyson, OSHA set aside its original determination not to issue a standard. The new decision again concludes that state regulation of field sanitation "would be preferable to, and more effective than, federal actions"; but this time around OSHA gave the states 18 months (until April 1987) to develop and implement field sanitation standards of their own; if the state response was "inadequate," as measured by certain criteria established by OSHA, within 6 months OSHA would issue a federal Field Sanitation Standard (*Comment Period Reopened, 50 Fed. Reg.* 42,660 [1985]). The OSHA stated that the state field sanitation standards would have to provide "protection equivalent to the Federal Field Sanitation Proposal of 1984. "At the same time," according to OSHA, "specific requirements may vary from the Federal Proposal." Finally, OSHA insisted that the state must have "adequate enforcement programs."

Testifying on November 6, 1985 before a subcommittee of the House Government Operations Committee, Department of Labor representatives stated confidently that the Secretary of Labor "has ended, not prolonged, the 13-year debate over field sanitation because he has made an unequivocal commitment to provide additional needed protections to American agricultural field workers" (*OSHA's Failure to Establish a Farmworker Field Sanitation Standard: Hearing Before a Subcomm. of the House Comm. on Government Operations,* 99th Cong., 1st Sess. 79 [1985]).

The OSHA statement was premature, for on February 6, 1987, the Court of Appeals for the District of Columbia decided that OSHA must issue a final Field Sanitation Standard within 30 days. The Court angrily stated that the Agency had utilized an "arsenal of administrative law doctrines" as a justification for "ricocheting" the case between OSHA and the courts for over a decade. It expressed the hope that its decision will "bring to an end this disgraceful chapter of legal neglect." In particular, the Court concluded that OSHA was legally wrong in relying on state action, and had acted unreasonably in delaying the Standard for yet another 2 years in the "unsupported and unrealistic" hope that the states would move *en masse* to issue field sanitation standards (*Farmworker Justice Fund v. Brock,* 811 F.2d 613, [D.C. Cir. 1987]). The OSHA issued the Field Sanitation Standard on April 28, and the Court then vacated its decision as "moot."

Ethylene Oxide

The Ethylene Oxide (ETO) proceeding raised several important issues on OSHA rulemaking. There, on petition from Public Citizen, the public interest group, the court of appeals in 1983 directed OSHA to initiate rulemaking on the carcinogenic substance within 30 days of the decision. The court said: "Three years from announced intent to regulate to final rule is simply too long given the significant risk of grave danger that ETO poses. . . ." *Public Citizen Health Research Group v. Auchter,* 702 F. 2nd 1150 [D.C. Cir., 1983]). The OSHA conducted a rulemaking and at the end of the proceeding was prepared to issue a Standard containing both a TWA of 1 ppm and a STEL of 5 ppm. As required by Executive Order 12,291, OSHA submitted the draft Standard for OMB review. Following OMB review, OSHA deleted the STEL and held further rulemaking on the issue. After the rulemaking, a final determination was made to delete the STEL, Public Citizen challenged the Standard in the court of appeals, and argued, among other things, that OMB interference with OSHA rulemaking violated the OSH statute which gives authority to issue standards to the Agency and not to OMB, and that the off-the-record communications between OSHA and OMB violate the procedural requirements for rulemaking by denying parties their right of rebuttal. Significantly, a number of chairpersons of House committees filed a brief as a friend of the court, also arguing against OMB "control" of health and safety regulation. The Justice Department filed a brief for OSHA, defending the right, indeed, the obligation of the President to oversee the Executive Department's regulatory actions, to assure its consistency with national policy. The proper role of OMB as representative of the President in monitoring the Executive Department Agencies has been widely debated, and in the OSHA context, has also been the subject of congressional hearings. For example, in 1983, the House Government Operations Committee, following hearings, issued a report entitled "OMB Interference with OSHA Rulemaking." The majority report said that Executive Order 12,291 had been "used as a backdoor, unpublicized channel of access to the highest levels of political authority in the Administration for industry alone." Twelve members of the Committee filed a dissent arguing in defense of the OMB role (*OMB Interference with OSHA Rulemaking, House Government Operations Comm.*, H.R. REP. No. 98, 98th Cong., 1st Sess. [1983]).

On July 25, 1986, the Court of Appeals for the District of Columbia issued a major decision, *Public Citizen Health Research Group v. Tyson,* 796 F.2d 1479 (D.C. Cir. 1986), affirming OSHA's long-term exposure limit for ethylene oxide, but concluding that OSHA's decision not to set a STEL was not supported by the record and remanding the issue to OSHA for further proceedings. In view of this disposition, the Court found it unnecessary to decide the "difficult Constitutional questions concerning the Executive's proper role in administrative proceedings" and the proper scope of power delegated by Congress to certain Executive Agencies. Meanwhile, OMB's activity in the rulemaking brought consideration in Congress in the form of legislation proposed by Senators Levin, Durenburger, and Rudman. A bill, entitled "The Rulemaking Information Act of 1986" (S. 2023), would require the disclosure to the public of a number of documents exchanged between OMB and Executive Agencies in the process of review by OMB of agency rules under Executive Order 12,291. At the same time, the House Energy and Commerce Subcommittee on Oversight and Investigation was investigating allegations that OMB interfered with cancer risk assessment guidelines being prepared by OSHA and other regulatory agencies. In response to these actions, OMB agreed in June 1986 publicly to disclose certain information pertaining to its review process.

On July 21, 1987, the court of appeals directed OSHA to issue a final rule on the short-term exposure issue by March 1988, or face a contempt citation (17 OSHR 237–238, 1987).

Safety standards

A number of actions were taken on safety standards since 1981. Some were the promulgation of new standards: Marine Terminals in July 1983, Servicing Single and Multipiece Wheel Rims in February 1984, and Electrical Standards for the Construction Industry in 1986 (the Electrical Standard has been challenged in court). Other actions were amendments limiting earlier standards, in particular, exemptions to the Diving Standards for Educational Diving issued in November 1982, and revocation of "should" standards in February 1984. In September 1986, OSHA amended its accident tag requirements to provide employers with more "flexibility" in meeting the Standard (51 *Fed. Reg.* 33,251 [1986]). Finally, rulemaking continued in other safety areas, but disputes with OMB led to extensive delays and no final standards were promulgated; these included oil and gas drilling, and grain elevators, the latter an issue that was high on OSHA's agenda since the grain elevator explosions in 1975 (*Oversight of the OMB Regulatory Review and Planning Process: Hearing Before the Subcomm. on Inter-Governmental Relations of the Senate Governmental Affairs Comm.,* 99th Cong., 2nd Sess. 1-58 [1986]: 16 OSHR 260, 15 OSHR 557 1985).

The federal program

The OSHA recently has undertaken a number of new initiatives in the area of federal employee safety and health, primarily on the basis of its expanded authority in 1980 Presidential Executive Order. In 1983, OSHA revised its staff manual, outlining three types of federal agency inspections which would be conducted by OSHA: unannounced inspections in response to employee reports of hazardous conditions; fatality and catastrophe inspections; and inspections of targeted high hazard workplaces. The number of OSHA inspections has continued to rise: in fiscal 1985, it conducted 1,883 inspections (*OSHA Oversight, Status of Federal Agency, Health and Safety Programs: Hearing Before Subcomm. on Health and Safety of the House Comm. on Education and Labor,* 99th Cong., 1st Sess. 2–3 [1985] [referred to hereafter as *OSHA Oversight, Status of Federal Agency* [1985]], The President's Report, 1985, p. 38). The OSHA not only inspects federal agency workplaces but evaluates the "complete safety and health programs" of federal agencies. Thus, for example, in 1985 OSHA conducted a full evaluation of the Agriculture and Interior Departments and conducted a follow-up evaluation of the programs of the Tennessee Valley Authority, the Postal Service, and the Navy Department (*OSHA Oversight, Status of Federal Agency* 8–9 [1985]).

In 1983 important changes were made to OSHA's practice in reporting on federal injury and illness rates. Previously, the annual statistical report was based on data submitted to OSHA by the agencies. Beginning in 1984, however, OSHA no longer required the submission of data and used as a basis for its surveys compensation claims data that had already been submitted to the Department of Labor by the agencies. Using this new method, OSHA concluded that the incidence rate for all federal civilian employee injuries and illnesses stayed essentially the same from fiscal year 1984–1985 (5.77 in 1984; 5.65 in 1985). However, the lost workday case incidence rate dropped 8 percent, from 2.91 to 2.6, continuing a downward trend, which began in 1980. The OSHA also reported on related statistics on employee compensation charge back costs. In 1985, the rate of growth increased substantially to 10 percent, although OSHA asserted that this was not due to an increase in the number of claims filed (Occupational Safety and Health Statistics, Federal Government, pp. 1-5, 1986).

Criticism of OSHA's efforts continued, with Michael Urquhart, President, Local 12, American Federation of Government Employees, telling a House Subcommittee in 1985 that OSHA "has abandoned the goal of reducing worker injuries and illnesses and instead is pursuing the goal of reducing workers' compensation costs." He asserted that the reduction of workplace risks involves commitment, resources, and leadership, "three things we find sorely lacking in OSHA's federal agency program today" (*OSHA Oversight, Status of Federal Agency,* etc., p. 17, [1985]).

BLS statistical survey

In November 1986, BLS reported that its 1985 injury and illness survey showed virtually no change in the injury and illness rates between 1984 and 1985. Thus, in 1985, there were 7.9 injuries and illnesses for every 100 full-time workers as compared to 8 in 1984. The actual number of injuries and illnesses increased by 1.6 percent but the rate did not increase because of the increased number of workers and hours of work. Assistant Secretary Pendergrass saw the data as affording encouragement to OSHA on the "course we have mapped out" for the Agency. The AFL-CIO, on the other hand, termed the results "disappointing," focusing on increases in the service industries, where employment growth had occurred (16 OSHR 628, 640-646, 1986).

In the meantime, serious questions have arisen over the accuracy of the data of injuries and illnesses that are being submitted by employers to BLS. An expert panel under the National Academy of Sciences has been asked to report at the end of 1987 on a number of issues related to the record-keeping and reporting system. Labor unions have charged that employers seriously underreport injuries and illnesses, which not only distort BLS statistics but also result in inappropriate exemptions from OSHA inspection activity. The unions argue that its assertions are confirmed by a number of major OSHA citations and penalties for willful violations of the record-keeping regulations; in January 1987, for example, a large automobile company agreed to pay a $295,000 penalty for record-keeping violations, reduced from $910,000, the second largest penalty ever imposed by OSHA. In July, OSHA proposed a record $2.59 million fine against a meat-packing company for alleged intentional failure to record 1,000 worker injuries (17 OSHR 235, 1987).

In an effort to study the accuracy of BLS statistics, OSHA has begun a pilot program to examine the injury and illness records of 200 employers and compare them with a "reconstructed" picture of the actual number of injuries and illnesses, based on interviews and other records (16 OSHR 880-82, 1987).

CONCLUSION

There has been considerable activity during the 16-year history of the Occupational Safety and Health Act. The OSHA has performed almost 1 million workplace inspections, and has proposed over $150.5 million in civil penalties. Its staff has grown substantially, as has its budget, although both have leveled off in the recent austerity years. (The total number of compliance offers in 1986 was 897.) A regulatory structure has been established covering such areas as enforcement, state plans, standards-setting, and statistics. In addition to the 400 PELs issued in

1971, OSHA has promulgated detailed health standards for 24 toxic substances, mostly carcinogens, as well as major hazard communication and employee access rules. About 25 safety standards, and numerous revisions of existing standards, have also been issued. With some important exceptions, these standards have either not been challenged or have been upheld by the courts. Twenty-two states with plans approved by OSHA have ongoing Occupational Safety and Health Programs, financed at 50 percent by OSHA and subject to OSHA's continuing monitoring. Eleven of these state plans have now received final approval. In fiscal 1985, states with plans conducted 114,215 safety and health instructions (The President's Report, 1985, p. 49). Numerous programs for training for employers and employees in safety and health, and voluntary compliance programs, notably on-site consultation, have been successfully undertaken by OSHA throughout its history. Annual statistical surveys on job-related injuries and illnesses have been published by the BLS, and these have provided substantially more complete data particularly with respect to the prevalence of occupational injuries. The National Institute for Occupational Safety and Health (NIOSH), charged with responsibility for research in the identification, evaluation and control of work-related illnesses and injuries, the dissemination of information to workers, employees and health professionals, and the training of occupational safety and health professionals, has engaged in numerous programs in fulfilling its statutory mandate and supporting OSHA regulatory activity.

The OSHA activity has been accompanied often by wide-ranging criticism, emanating from virtually every interest group affected by the Agency's program. Employers and their associations have accused OSHA, among other things, of imposing extremely burdensome and largely unnecessary costs; of "nit-picking" enforcement; of denials of due process and basic fairness; of confused priorities; of inefficiency and of excessive zeal. Unions and public interest groups, on the other hand, have been equally critical, usually at different times than employers, for OSHA's failure to act promptly to regulate carcinogens; for its abdication of responsibility to the states; for its failure to marshall sufficient staff and resources effectively to enforce the Act's requirements; and for inefficiency and lack of zeal. Members of Congress and congressional committees have voiced many of these same criticisms and the White House and the OMB have frequently exercised their executive supervisory authority in constraining and limiting OSHA rulemaking activity.

In this controversial context, some general conclusions may be suggested:

1. The original Act remains intact. It has not been amended or weakened, although a variety of appropriations riders have been passed, attenuating OSHA enforcement responsibility in certain respects. By means of its flexibility and responsiveness to criticism—sometimes amounting to excessive responsiveness—OSHA has managed to avoid any wholesale dismantling of its regulatory structure.
2. Some have taken the position that the "bottom line" in evaluating OSHA is whether OSHA has reduced workplace injuries and illnesses. After an extensive study, the OTA of the U.S. Congress in 1981 stated that it was unable to make any judgment of OSHA's impact in the area of health because of the inadequacy of the data and that it attributed "most" of the workplace injury rate changes since 1972 to changes in business activity, and not to OSHA (*OTA Report,* pp. 34-36). Thus, judged by this limited yardstick, OSHA's value is unproven. At the same time, OTA noted studies showing significant declines in employee exposures to major toxic hazards, specifically, asbestos, cotton dust and lead resulting from the Agency's recent stringent standards covering these substances (*OTA Report,* p. 268). Similar reductions in employee exposure to vinyl chloride, one of the earliest carcinogens regulated by the Agency, have also been demonstrated (*OTA Report,* pp. 230-231).
3. Testifying in 1980, the late Lloyd McBride, President of the Steelworkers Union, presented a different measure for evaluating OSHA success. He asserted that most of the importance of OSHA as a regulatory agency lies in its bringing about the evolution of the "OSHA movement." The 1970 Act, he said, was a necessary ingredient to bring about a "rise in the national consciousness and the development of other private sector institutional responses" *(Oversight Hearings,* 1980, 698-699). Much the same point was advanced by OTA, which emphasized the "increased attention" to safety and health on the part of managers and employees which has been engendered by OSHA activity (*OTA Report,* pp. 268-269).

Some of these nonregulatory aspects should be mentioned. There has been the significant increase in collective bargaining activities relating to occupational safety and health. The number of OSHA-type provisions in collective bargaining agreements has grown (in 1986, they were found in 84 percent of the agreements), and numerous mechanisms for continuing dialogue and joint-activity between unions and management, notably joint labor-management safety and health committees, exist in many plants (*OTA Report,* pp. 314-320; BNA, *Collective Bargaining Negotiations and Contracts,* 95.1, p. 169 [1986]). Voluntary efforts by employers, unions, and private organizations in the fields of workplace safety and health are extensive. Voluntary standards developed by various groups, while not unenforceable, have provided a valuable source of technical information to regulatory agencies and to employers and employees. The *OTA Report* describes the "effective company-run safety and health organization" which is being implemented at E.I. du Pont de Nemours and Co. DuPont has an extensive corporate structure to carry out "top management's commitment to environmental quality, including safety and health" (*OTA Report,* p. 299).

Unions also have established important safety and health programs to support and expand upon government activity. The United Steelworkers of America, for example, has an International Safety and Health Department, which includes industrial hygienists, whose major functions include assistance to local unions' education and training programs for union committee members and staff, assistance in the negotiation of safety and health clauses in collective bargaining agreements, and efforts to bring about more effective safety and health standards and enforcement by the federal and state OSHA. The Steelworkers, whose OSHA program has been one of the most vigorous, played a key role in the development of OSHA's 1976 Coke Oven Emission Standard.

A number of cities have organized Coalitions for Occupational Safety and Health (COSH), which function to coordinate workers' efforts in the OSHA area. One of the more successful of these, MASSCOSH, recently celebrated its 10th anniversary. There are about 25 of these coalitions throughout the country, the largest representing over 150,000 workers. Among their most important activities are health and safety training for workers and unions, particularly in the area of "right to know"; technical assistance in recognition and control of hazards, legal assistance to employees in asserting their rights under the Act, and political action campaigns to create the best safety and health laws and regulations and strengthen their enforcement (16 OSHR 62, 1986).

Activity by states without approved plans in the occupational safety and health field has increased particularly since 1981. State right-to-know laws have been a major factor in requiring the provision of information to workers on workplace hazards, although legal issues relating to federal preemption have complicated greatly the effectiveness of these programs. Another significant development has been the increasingly active role being taken by local officials in prosecuting cases under the criminal laws involving worker deaths (15 OSHR 1132-1135, 1986; 15 OSHR 1208, 1986; 16 OSHR 886, 1987). A major impetus to this new trend took place in July 1985 when four executives of Film Recovery System Inc., an Illinois silver extraction facility, were sentenced to 25 years in prison on a murder charge for bringing about the death of an employee by exposure to cyanide (16 OSHR, 150, 1986). However, on June 29, 1987, the Illinois Appellate Court, First Judicial District, ruled that the federal OSHAct preempted state enforcement of its criminal statutes applied to workplace conditions regulated by OSHA (*Illinois v. Chicago Magnet Wire Corp.*, 13 OSHC 1337 [1987]).

BIBLIOGRAPHY

Arant, F. S., Chairman of the Subcommittee on Pesticides of OSHA's Agricultural Advisory Committee, letter, dated May 9, 1973. Reprinted in part, in Mintz, B. *History, Law, and Policy.* Washington, DC: Bureau of National Affairs, 1984.

Corn, M., Assistant Secretary, to Chairman Daniels of Subcommittee on Manpower, Compensation and Health and Safety of the House Education and Labor Committee, letter, dated, March 25, 1976. Reprinted in part in Mintz, B. *OSHA History, Law, and Policy.* Washington, DC: Bureau of National Affairs, 1984, pp. 411-414.

MacAvoy, P., ed. *Report of the Presidential Task Force, OSHA Safety Regulation.* Washington, DC, 1977.

MacLaury, J. The Job Safety Law of 1970: Its passage was perilous. *Monthly Labor Review* (March 1981):18.

Mintz, B. *OSHA: History, Law, and Policy.* Washington, DC: Bureau of National Affairs, 1984.

Moran, R. *Cite OSHA For Violations, Occupational Safety and Health* (March-April, 1976):19-20.

More Concentrated Effort Needed by the Federal Government on Occupational Safety and Health Programs for Federal Employees. General Accounting Office, Report No. B-163375, March 15, 1973.

O'Brien, M. W., Assistant General Counsel, United Steelworkers of America, to OSHA, letter on state programs, dated March 15, 1985.

Office of Technology Assessment (OTA), U.S. Congress. *Preventing Illness and Injury in the Workplace.* Washington, DC: OTA-H-256, April 1985.

Page, J., and O'Brien, M. W. *Bitter Wages.* Viking-Penguin Inc., 1973.

*Report of the NACOSH Subcommittee on State Plans,*1974. Reprinted in part in Mintz, B. *OSHA: History, Law and Policy.* Washington, DC: Bureau of National Affairs, 1984, pp. 630-632.

Working in the Dark: Reagan and the "Right to Know" About Occupational Hazards. Washington, DC: Public Citizen, June 1986.

APPENDICES

Sources of Help

by Julian B. Olishifski, MS, PE, CSP
Revised by Ruth Hammersmith

THE SAFETY PROFESSIONAL in routine accident prevention activities is frequently required to make a decision as to the degree of health hazard arising from a process or operation. In emergency situations and in the absence of an industrial hygienist, it becomes the safety professional's duty to see that immediate corrective action is taken toward the recognition, evaluation, and control of occupational health hazards. If the unit is part of a multiplant corporation, the safety professional should consult with the home office. The supplier of materials that are suspected of being toxic can also be consulted for information on the potential hazards.

Many companies or plants, especially those of medium or small size, do not employ an industrial hygienist, chemical engineer, or fire protection engineer, although they may have a part-time or full-time safety professional. Of necessity, these safety professionals must acquire some understanding of potential health hazards to personnel and to the community that may result from their company or plant operations.

Specialized help is available from a number of sources.

- Most employers are likely to have trained professional or scientific personnel who can provide some technical assistance or guidance, even though their primary interest is in a field other than industrial hygiene.
- Many insurance companies that carry worker's compensation insurance provide industrial hygiene service, just as they provide, for example, periodic safety audits.
- Professional consultants and privately owned or endowed laboratories are available on a fee basis for concentrated studies of a specific problem or for a plant-wide or company-wide survey, which may be undertaken to identify and catalog individual environmental exposures.
- Many states have excellent industrial hygiene departments, which are fertile sources of help. Some state laboratories are extremely well equipped and have sophisticated sampling and analyzing equipment that an individual company could not possibly justify purchasing because of large initial cost, specialized operator training, or infrequent or intermittent use.

ASSOCIATIONS AND ORGANIZATIONS

Many associations are concerned with industrial hygiene problems. For example, many of the industrial sections of the National Safety Council have an occupational health or industrial hygiene subcommittee that can be called upon for assistance.

Technical societies

Scientific and technical societies that can help with health conservation or with a specific problem area are listed in this section. Some are prepared to provide consultation service to nonmembers as well as members; they all have a wealth of available technical information.

Air Pollution Control Association
Three Gateway Center, Four W
Pittsburgh, PA 15222
Industrialists, researchers, equipment manufacturers, governmental control personnel, educators, meteorologists, and others seeking economical answers to the problem of air pollution. Sponsors continuing education courses and maintains a library.

American Academy of Industrial Hygiene
302 S Waverly Rd
Lansing, MI 48917
Professional society of industrial hygienists.

American Academy of Occupational Medicine
2340 Arlington Heights Rd, Ste 400
Arlington Heights, IL 60005
Physicians who devote full time to some phase of occupational medicine. Promotes maintenance and improvement of the health of industrial workers.

American Academy of Ophthalmology
655 Beach St
San Francisco, CA 94120
Ophthalmologists concerned with high quality eye care and the continuing education of members.

American Academy of Optometry
5530 Wisconsin Ave, NW, Ste 745
Washington, DC 20815
Professional society of optometrists, educators, and scientists interested in conserving human vision, clinical and experimental research in visual problems.

American Academy of Otolaryngology—Head and Neck Surgery
1101 Vermont Ave, NW, Ste 302
Washington, DC 20005
Professional society of medical doctors specializing in otolaryngology, disease of the ear, nose and throat, and head and neck surgery.

American Association of Occupational Health Nurses
3500 Piedmont Rd, NE
Atlanta, GA 30305
Registered professional nurses employed by business and industrial firms, nurse educators, nurse editors, nurse writers and others interested in industrial nursing.

American Board of Industrial Hygiene
302 S Waverly Rd
Lansing, MI 48917
This specialty board is authorized to certify properly qualified industrial hygienists. The overall objectives are to encourage the study, improve the practice, elevate the standards, and issue certificates to qualified applicants.

American Board of Preventive Medicine
Department of Community Medicine
Wright State University
Dayton, OH 45401
The Medical Specialty Board is authorized to certify properly qualified specialists. Overall purposes are to encourage the study, improve the practice, elevate the standards, and advance the cause of preventive medicine. Issues certificates of special knowledge to duly licensed physicians specializing in the fields of public health, aviation medicine, occupational medicine, and general preventive medicine.

American Chemical Society
1155 16th St, NW
Washington, DC 20036
Scientific, educational, and professional society of chemists and chemical engineers.

American Conference of Governmental Industrial Hygienists
6500 Glenway Ave, Bldg D-5
Cincinnati, OH 45211
Professional society of persons responsible for full-time programs of industrial hygiene, who are employed by official governmental units. Primary function is to encourage the interchange of experience among governmental industrial hygienists, and to collect and make available information of value to them. Also promotes standards and techniques in industrial hygiene, and coordinates governmental activities with community agencies.

American Industrial Hygiene Association
475 Wolf Ledges Pkwy
Akron, OH 44311
Organization of professionals trained in the recognition and control of health hazards and the prevention of illness related thereto. Industries are eligible for associate membership. Promotes the study and control of environmental factors affecting the health of industrial workers; provides information and communication services pertaining to industrial hygiene.

American Institute of Chemical Engineers
345 47th St
New York, NY 10017
Professional society of chemical engineers. Establishes standards for chemical engineering curricula.

American Institute of Chemists
7315 Wisconsin Ave, NW
Bethesda, MD 20814
Chemists and chemical engineers. To elevate the professional and economic status of chemists and chemical engineers.

American Medical Association
535 N Dearborn St
Chicago, IL 60610
Professional association of persons holding either a medical degree or an unrestricted license to practice medicine. Its purpose is to promote the science and art of medicine and the betterment of public health. Consequently, it provides information on medical and health topics; represents the medical profession before Congress and governmental agencies; cooperates in setting standards for medical schools and training programs; keeps members informed on significant medical and health legislation.

American Occupational Medical Association
2340 S Arlington Heights Rd
Arlington Heights, IL 60005
Professional society of medical directors and plant physicians specializing in industrial medicine and surgery, established to foster the study of the problems peculiar to the practice of industrial medicine and surgery; to encourage the development of methods of conserving and improving the health of workers; and to promote a more general understanding of the purpose and results of the medical care of these workers.

American Optometric Association
243 N Lindbergh Blvd
St. Louis, MO 63141
Professional society of optometrists. Supports a library on ophthalmic and related sciences, with emphasis on optometry, its history and socioeconomic aspects.

American Public Health Association
1015 18th St, NW
Washington, DC 20036
Professional association of physicians, nurses, educators, engineers, environmentalists, social workers, industrial hygienists, and others who have an interest in personal and environmental health. Its services include—promulgation of standards; development of the etiology of communicable diseases; establishment of minimum educational qualifications for public health workers; accreditation of schools of public health; and research in public health.

American Society of Heating, Refrigerating and Air Conditioning Engineers
1791 Tullie Circle, NE
Atlanta, GA 30329
Professional society of heating, ventilating, refrigeration, and air conditioning engineers carries out a number of research programs in cooperation with universities and research laboratories.

American Society of Safety Engineers
1800 E Oakton St
Des Plaines, IL 60016
Professional society of safety engineers, safety directors and others concerned with accident prevention and safety programs. Compiles statistics.

American Society for Testing and Materials
1916 Race St
Philadelphia, PA 19103
Engineers, scientists, and skilled technicians holding membership as individuals or as representatives of business firms, government agencies, educational institutions, and laboratories. Establishes voluntary consensus standards for materials, products, systems, and services.

American Society for Training and Development
1630 Duke St
Alexandria, VA 22313
Professional society of persons engaged in the training and development of business, industrial, and government personnel.

Board of Certified Safety Professionals
208 Burwash Ave
Savoy, IL 61874
The principal purpose of the board is to establish minimum academic and experience attainments necessary to qualify as a certified safety professional, and to determine the competence and issue certificates to qualified applicants.

Health Physics Society
1340 Old Chain Bridge Rd
McLean, VA 22101
Professional society of persons active in the field of health physics—the profession devoted to the protection of man and his environment from radiation hazards. Aids research in the field, improves dissemination of information to individuals in the profession and in related fields, works to improve public understanding of the problems which exist in matters of radiation protection, and promotes the profession of health physics. Sponsors the American Board of Health Physics for the voluntary certification of health physicists.

Human Factors Society
PO Box 1369
Santa Monica, CA 90406
Professional society of psychologists, engineers, physiologists, and other related scientists who are concerned with the use of human factors in the development of systems and devices of all kinds.

Illuminating Engineering Society of North America
345 E 47th St
New York, NY 10017
Professional society whose members include engineers, architects, designers, educators, students, contractors, distributors, utility personnel, scientists, and manufacturers dealing with the art or science of illumination. Provides assistance with technical problems, reference help, speakers, training aids. Maintains liaison with schools and colleges.

Institute of Electrical and Electronics Engineers
345 E 47th St
New York, NY 10017
Engineers and scientists in electrical engineering, electronics, and allied fields. Holds numerous meetings and special technical conferences. Conducts lecture courses at the local level on topics of current engineering and scientific interest. Assists student groups.

Institute of Environmental Sciences
940 E Northwest Hwy
Mount Prospect, IL 60056
Engineers, scientists, and management people engaged in the simulation of the natural environment and the environments induced by equipment operation and the testing of men, materials, and equipment in the simulated environments.

Society of Automotive Engineers
400 Commonwealth Dr
Warrendale, PA 15096
Professional society of engineers in field or self-propelled ground, flight, and space vehicles; engineering students are enrolled in special affiliation. To promote the arts, sciences, standards, and engineering practices related to the design, construction, and use of self-propelled mechanisms, prime movers, components thereof, and related equipment.

Society of Manufacturing Engineers
One SME Drive
Dearborn, MI 48121
Professional society of manufacturing engineers and management executives concerned with manufacturing techniques. To advance the science of manufacturing through the continuing education of manufacturing engineers and management.

Society of Toxicology
1133 I St, NW, Ste 800
Washington, DC 20005
Persons who have conducted and published original investigations in some phase of toxicology and who have a continuing professional interest in this field.

System Safety Society
14252 Culver Dr, Ste A-261
Irvine, CA 92714
The System Safety Society, Inc., is a nonprofit organization

dedicated to the advancement of system safety principles, techniques and methodology in all technical endeavors where potential of injury damage or loss is present. Additional information on the society and membership requirements may be obtained by writing to the society.

Trade associations

Another group of associations come under the broad heading of "trade associations." They are concerned with furthering the aims of their field of productive enterprise, including health preservation of employees and the public. These associations have trained personnel, cooperating committees, and publications that can be extremely helpful.

Alliance of American Insurers
1501 Woodfield Rd
Schaumburg, IL 60195
An organization of leading mutual property and casualty insurance companies that promotes loss prevention principles amongst its members as well as disseminates safety information.

American Foundrymen's Society
Golf and Wolf Rds
Des Plaines, IL 60016
Technical society of foundrymen, patternmakers, technologists, and educators. Maintains a Technical Information Center that provides literature searching and document retrieval service.

American Insurance Association
85 John St
New York, NY 10038
Represents companies providing property and liability insurance. Seeks to promote the economic, legislative, and public standing of its participating companies through a broad spectrum of activities.

American Iron and Steel Institute
1000 16th St, NW
Washington, DC 20036
Basic manufacturers and individuals in the steel industry. Conducts extensive research programs and workshops.

American Petroleum Institute
1220 L St, NW
Washington, DC 20005
Producers, refiners, marketers, and transporters of petroleum and allied products. Seeks to maintain cooperation between government and industry, fosters foreign and domestic trade in American petroleum products, promotes the interests of the petroleum industry, and provides extensive publication and information services.

Compressed Gas Association
1235 Jefferson Davis Hwy
Arlington, VA 22202
Firms producing and distributing compressed, liquefied, and cryogenic gases; also manufacturers of related equipment. Submits recommendations to appropriate government agencies to improve safety standards and methods of handling, transporting, and storing gases; acts as advisor to regulatory authorities and other agencies concerned with safe handling of compressed gases; collaborates with national organizations to develop specifications and standards of safety.

Lead Industries Association
292 Madison Ave
New York, NY 10017
An association of mining companies, smelters, and refiners; and manufacturers of lead products of which lead is a component. Provides technical service and information to consumers; gathers statistical information.

National Machine Tool Builders' Association
7901 Westpark Dr
McLean, VA 22101
Manufacturers of power-driven machines that are used to shape or form metal. Seeks to improve methods of producing and marketing machine tools; promotes research and development in the industry.

Soap and Detergent Association
475 Park Ave
New York, NY 10016
Manufacturers of soap, synthetic detergents, fatty acids, and glycerine, and raw materials suppliers. Activities include cleanliness promotion, consumer information, environmental and human safety research, government liaison.

Technical Association of the Pulp and Paper Industry
PO Box 105113
Atlanta, GA 30348
Executives, managers, engineers, research scientists, superintendents, and technologists in the pulp, paper, and allied industries. Conducts conferences and develops testing procedures for laboratory analyses and process control.

Scientific and service organizations

These organizations have a high interest in industrial hygiene.

American National Standards Institute
1430 Broadway
New York, NY 10018
This federation of industrial, trade, technical, labor, and professional organizations; government agencies; and consumer groups coordinates development of standards in multiple subject areas, and oversees their publication.

Factory Mutual System
1151 Boston-Providence Turnpike
Norwood, MA 02062
Association of mutual fire insurance companies which insure large industrial and commercial properties in the U.S. and Canada. Its object is to minimize fire losses and resultant interruption of production, and to provide insurance at actual cost.

Industrial Health Foundation, Inc
34 Penn Circle, W
Pittsburgh, PA 15206
Nonprofit organization for the advancement of healthful working conditions in industry. Members are industrial companies or organizations of employer firms or corporations, such as trade associations. Provides engineering, occupational medicine, toxicological, and information services; these include plant visits and surveys, special studies, and availability of highly qualified personnel to supplement members' staffs on a consultant basis at cost.

International Atomic Energy Agency
Wagramstrasse 5
A-1400 Vienna, Austria
Established in 1956 as an autonomous intergovernmental organization under the aegis of the United Nations, for the purpose of accelerating and enlarging the contribution of atomic energy to peace, health, and prosperity. Required to ensure that assistance provided by it or under its suggestion, supervision or control, is not used to further any military purpose. Provides technical assistance and advice on developments in the use of nuclear power in electricity generating, and the use of radiation and radioisotopes in medicine, agriculture and industry; on health and safety aspects of its use; and on the management of radioactive wastes.

International Radiation Protection Association
c/o G. Bresson
PO Box 33
F-92269 Fontenay aux Roses, France
Organization of individuals and national affiliated societies. Provides a medium for contacts and cooperation among scientists engaged in radiation protection work; encourages establishment of radiation protection societies throughout the world; encourages establishment of standards and recommendations; promotes research and education.

National Council on Radiation Protection and Measurements
7910 Woodmont Ave, Ste 1016
Bethesda, MD 20014
Congressionally chartered, nonprofit organization. Collects, analyzes, and disseminates information about radiation protection and measurements. Members are nationally recognized scientists who volunteer their services to the Council's program.

National Fire Protection Association
Batterymarch Park
Quincy, MA 02269
Membership drawn from fire service centers, business and industry, health care, educational and other institutions, insurance companies, government at all levels, architects and engineers, and others. Serves as a clearinghouse of information; compiles annual statistics on causes and occupancies of all fires.

National Safety Council
444 N Michigan Ave
Chicago, IL 60611
Independent nonprofit organization with the goal of reducing the number and severity of all kinds of accidents and industrial illnesses by collecting and distributing information about the causes of accidents and illnesses and ways to prevent them. Gathers and analyzes statistics, performs research in various areas of accident prevention and safety and health program effectiveness, sponsors special-interest conferences and committees, and provides research consultant services.

Underwriters Laboratories Inc.
333 Pfingsten Rd
Northbrook, IL 60062
Independent, nonprofit organization for public safety testing. Operates laboratories for examination and testing of devices, systems, and materials. Product services include listing, classification, recognition, certification, and inspection. Fact-finding and research services are also conducted on a contract basis for manufacturers whose products meet UL safety requirements.

Emergency information service

Emergency response system that can help.

CHEMTREC
Emergency information about hazardous chemicals involved in transportation accidents can now be obtained 24 hours a day. It is the Chemical Transportation Emergency Center (CHEMTREC), and it can be reached by a nation-wide telephone number—800: 424-9300. The Area Code 800 WATS line permits the caller to dial the station-to-station number without charge. CHEMTREC will provide the caller with response/action information for the product or products and tell what to do in case of spills, leaks, fires, and exposures. This informs the caller of the hazards, if any, and provides sufficient information to take immediate first steps in controlling the emergency. CHEMTREC is strictly an emergency operation provided for fire, police, and other emergency services. It is not a source of general chemical information of a nonemergency nature.

Other organizations

Information concerning hazardous materials may also be acquired from such organizations as:

American Association of Poison Control Centers
c/o Dr. Gary Oderda
Maryland Poison Center
20 N Pine
Baltimore, MD 21201
This association keeps up-to-date information on the ingredients and potential acute toxicity of substances that may cause accidental poisoning and of the proper management of such poisonings. It has established standards for poison information and control centers. Publishes a listing on Poison Centers which is updated annually.

ICES (Information Center for Energy Safety)
Oak Ridge National Laboratory
PO Box Y
Oak Ridge, TN 37830
The Information Center for Energy Safety (ICES) was established at Oak Ridge National Laboratory (ORNL) by the Energy Research and Development Administration (ERDA) as a national center for collecting, storing, evaluating, and disseminating safety information related to the development and use of several nonnuclear forms of energy.

The Center, established in 1976, has a staff that will analyze current information, prepare state-of-the-art reviews of the safety of the various energy systems, and answer technical inquiries. Within these categories, the environmental, physical science, and engineering aspects of energy conversion, storage, and use are searchable. Requests for information may be made by correspondence, telephone, or a visit to the Center. Arrangements for a visit may be initiated by contacting the Center, preferably one week in advance of the date of arrival.

Telephone calls can be made by direct dialing of the number (615) 483-8611, extension 3-5453.

National Response Center
2100 Second St, SW, Room 2611
Washington, DC 20593

U.S. Coast Guard
2100 Second St, SW
Washington, DC 20593

Chemical Manufacturers Association
2501 M St, NW
Washington, DC 20037

GOVERNMENTAL ORGANIZATIONS

There is an overwhelming amount of information available from the federal government that concerns all aspects of safety and health, environmental problems, pollution, statistical data, and other industry problems.

Federal agencies

Because of constant changes in government agency activities and frequent reorganizations within the government, it is recommended that the reader consult the latest issue of the *United States Government Organization Manual,* published annually by the Government Printing Office, Washington, DC 20402. This paperbound book can be found in most libraries.

Bureau of Mines
Department of the Interior
2401 E St, NW
Washington, DC 20241
The Bureau of Mines is primarily a research and fact-finding agency. Its goal is to stimulate private industry to produce a substantial share of the Nation's mineral needs in ways that best protect the public interest. Applied and basic research are conducted to develop the technology for the extraction, processing, use, and recycling of the Nation's mineral resources at a reasonable cost without harm to the environment or the workers involved. Typical areas of research are mine health and safety, recycling of solid wastes, improvement of coal production technology, abatement of pollution and land damage caused by mineral extraction and processing operations, and development of ways to use domestic low-grade ores as alternative sources of critical minerals that must currently be imported.

Center for Devices and Radiological Health
Food and Drug Administration
8757 Georgia Ave
Silver Springs, MD 20910
The center (1) carries out programs designed to reduce the exposure of man to hazardous ionizing and nonionizing radiation, (2) develops standards for safe limits of radiation exposure, (3) develops methodology for controlling radiation exposures, (4) conducts research on the health effects of radiation exposure, and (5) conducts an electronic product radiation control program to protect public health and safety, including the development and administration of performance standards to control the emission of radiation from electronic products and the undertaking by public and private organizations of research and investigation into the effects and control of such radiation emissions.

Consumer Product Safety Commission
1111 Eighteenth St, NW
Washington, DC 20207
The purpose of the Consumer Product Safety Commission is to protect the public against unreasonable risks of injury from consumer products, to assist consumers to evaluate the comparative safety of consumer products, to develop uniform safety standards for consumer products and minimize conflicting State and local regulations, and to promote research and investigation into the causes and prevention of product-related deaths, illnesses, and injuries.

Department of Labor
200 Constitution Ave, NW
Washington, DC 20210
The purpose of the Department of Labor is to foster, promote, and develop the welfare of the wage earners of the United States, to improve their working conditions, and to advance their opportunities for profitable employment. In carrying out this mission, the Department administers more than 130 federal labor laws guaranteeing workers' rights to safe and healthful working conditions, a minimum hourly wage and overtime pay scale, freedom from employment discrimination, unemployment insurance, and workers' compensation. The Department also protects workers' pension rights; sponsors job training programs; helps workers find jobs; works to strengthen free collective bargaining; and keeps track of changes in employment, prices, and other national economic measurements. As the Department seeks to assist all Americans who need and want to work, special efforts are made to meet the unique job market problems of older workers, youths, minority group members, women, the handicapped, and other groups.

Environmental Protection Agency
401 M St, SW
Washington, DC 20460
The purpose of the Environmental Protection Agency (EPA) is to protect and enhance our environment today and for future generations to the fullest extent possible under the laws enacted by Congress. The Agency's mission is to control and abate pollution in the areas of water, air, solid waste, pesticides, noise, and radiation. EPA's mandate is to mount an integrated, coordinated attack on environmental pollution in cooperation with state and local governments.

Mine Safety and Health Administration
Department of Labor
4015 Wilson Blvd
Arlington, VA 22203
The Mine Safety and Health Administration (MSHA) conducts programs to control health hazards and reduce fatalities and injuries in the mineral industries through inspection, investigation, and enforcement; assessment of penalties for violations; technical support; and education, training, and safety motivation. Mandatory health and safety standards and regulations are developed or revised as warranted by new technology or by changing conditions. MSHA replaces the Mining Enforcement and Safety Administration.

National Bureau of Standards
Headquarters:
Route I-270 and Quince Orchard Rd
Gaithersburg, MD 20899
The Bureau provides the basis for the Nation's measurement standards. These standards are the means through which people

and nations buy and sell goods, develop products, judge the quality of their environment, and provide guidelines for the protection of health and safety. The Bureau's overall goal is to strengthen and advance the Nation's science and technology and facilitate their effective application for public benefit. NBS is involved in over 1,500 projects aimed at dealing with such national concerns as energy conservation and research, fire protection and prevention, and consumer product safety.

National Center for Toxicological Research
Food and Drug Administration
5600 Fishers Lane
Rockville, MD 20857
The National Center for Toxicological Research conducts research programs to study the biological effects of potentially toxic chemical substances found in man's environment emphasizing the determination of the health effects resulting from long-term low-level exposure to chemical toxicants and the basic biological processes for chemical toxicants in animal organisms, and the development of improved methodologies and test protocols for evaluating the safety of chemical toxicants and the data that will facilitate the extrapolation of toxicological data from laboratory animals to man.

National Technical Information Service
Department of Commerce
5285 Port Royal Rd
Springfield, VA 22161
NTIS was established to simplify and improve public access to Department of Commerce publications and to data files and scientific and technical reports sponsored by federal agencies. NTIS is the central point in the United States for the public sale of government-funded research and development reports and other analyses prepared by federal agencies, their contractors, or grantees.

Nuclear Regulatory Commission
1717 H St, NW
Washington, DC 20555
The Nuclear Regulatory Commission (NRC) licenses and regulates the uses of nuclear energy to protect the public health and safety and the environment. It does this by licensing persons and companies to build and operate nuclear reactors and to own and use nuclear materials. The NRC makes rules and sets standards for these types of licenses. The NRC also carefully inspects the activities of the persons and companies licensed to make sure that they do not violate the safety rules of the Commission.

Occupational Safety and Health Administration
U.S. Department of Labor
200 Constitution Ave, NW
Washington, DC 20210
The Assistant Secretary of Labor for Occupational Safety and Health has responsibility for occupational safety and health activities.

The Occupational Safety and Health Administration, established pursuant to the Occupational Safety and Health Act of 1970 (84 Stat. 1590), develops and promulgates occupational safety and health standards; develops and issues regulations; conducts investigations and inspections to determine the status of compliance with safety and health standards and regulations; and issues citations and proposes penalties for noncompliance with safety and health standards and regulations.

Occupational Safety and Health Review Commission
1825 K St, NW
Washington, DC 20006
The Occupational Safety and Health Review Commission (OSHRC) is concerned with providing safe and healthful working conditions for both the employer and the employee. It adjudicates cases forwarded to it by the Department of Labor when disagreements arise over the results of safety and health inspections performed by the Department.

Office of Energy Research
Department of Energy
1000 Independence Ave, SW
Washington, DC 20585
The Office of Energy Research advises the Secretary of Energy on the physical and energy research and development programs of the department. It is responsible for conducting basic research in (1) basic energy (physical) sciences; (2) high energy and nuclear physics; (3) biological and environmental research; (4) magnetic fusion energy.

Office of Hazardous Materials Transportation
Department of Transportation
400 Seventh St, SW
Washington, DC 20590
The Office of Hazardous Materials Transportation develops and issues regulations for the safe transportation of hazardous materials by all modes, excluding bulk transportation by water. The regulations cover shipping and carrier operations, packaging and container specifications, and hazardous materials definitions. The Office is also responsible for the enforcement of regulations other than those applicable to a single mode of transportation. It reviews and analyzes reports made by the industry and by field staffs bearing upon compliance with the regulations, and conducts training and education programs. The Office is the national focal point for coordination and control of the Department's multimodal hazardous materials regulatory programs, ensuring uniformity of approach and action by all modal administrations.

U.S. Fire Adminstration
Federal Emergency Management Agency
16825 S Seton Ave
Emmitsburg, MD 21727
The primary mission of the U.S. Fire Administration is to reduce the loss of life and property through better fire prevention and control with a program coordinated to support and reinforce the fire prevention and control activities of state and local governments.

State agencies

Many states have excellent industrial hygiene departments; these are a fertile source of help.

Some state laboratories are extremely well equipped and have numerous devices for sampling and analyzing that an individual company could not possibly justify purchasing because of large initial cost, specialized operator training, or infrequent or intermediate use. Not only are state services helpful in solving day-to-day problems, but they can also assist in unusual or complex problems.

Other agencies

Centers for Disease Control
Department of Health and Human Resources
Public Health Service
1600 Clifton Rd, NE
Atlanta, GA 30333

Health Resources and Service Administration
Food and Drug Administration
5600 Fishers Lane
Rockville, MD 20857

Library of Congress
Science and Technology Division
10 First St, SE
Washington, DC 20540

National Aeronautical and Space Administration
600 Independence Ave, SW
Washington, DC 20546

National Cancer Institute
National Institutes of Health
9000 Rockville Pike
Bethesda, MD 20892

National Center for Health Statistics
Public Health Service
3700 East-West Highway
Hyattsville, MD 20782

National Council on Radiation Protection and Measurement (NCRP)
7910 Woodmont Ave, Ste 1016
Bethesda, MD 20814

National Heart, Lung, and Blood Institute
National Institutes of Health
9000 Rockville Pike
Bethesda, MD 20892

National Institute of Environmental Health Sciences (NIEHS)
PO Box 12233
Research Triangle Park, NC 27709

National Institute of Mental Health
Alcohol, Drug Abuse, and Mental Health Administration
Department of Health and Human Services
5600 Fishers Lane
Rockville, MD 20857

National Institute for Occupational Safety and Health (NIOSH)
1600 Clifton Rd, NE
Atlanta, GA 30333
—also located at
4676 Columbia Pkwy
Cincinnati, OH 45226,
and
944 Chestnut Ridge Rd
Morgantown, WV 26505

National Library of Medicine
8600 Rockville Pike
Bethesda, MD 20894
Note: See details under MEDLARS On-line Network in the next section.

U.S. Government Printing Office
Washington, DC 20402

DATA BASES

General

Books in Print. Information source on U.S. book publishing; from R. R. Bowker Co., 1180 Avenue of the Americas, New York, NY 10036.

International Dissertation Abstracts. Doctoral dissertations from accredited universities (predominantly in the U.S.); from University Microfilm International, 300 North Zeeb Rd, Ann Arbor, MI 48106.

GPO Monthly Catalog. The machine equivalent of the printed "Monthly Catalog of U.S. Government Publications," from U.S. Government Printing Office, Washington, DC 20402.

National Newspaper Index. An index to *The Christian Science Monitor, The New York Times,* and *The Wall Street Journal;* from Information Access Corporation, 11 Davis Dr, Belmont, CA 94002.

National Technical Information Service. Citations to government-sponsored research, mostly federal, but also contains some state and local reports; from National Technical Information Service, U.S. Dept. of Commerce, Springfield, VA 22161.

Pharmaceutical News Index. Indexing records for seven major drug industry newsletters; from Data Courier, Inc, 620 S. Fifth, Lexington, KY 40202.

TSCA Initial Inventory. A nonbibliographic dictionary listing of chemical substances in commercial use in the United States as of June 1, 1979. It is not a list of toxic chemicals. From Dialog Information Retrieval Service, 3460 Hillview Avenue, Palo Alto, CA 94304.

Ulrich's International Periodicals and Irregular Serials. Information on periodicals and serials; from R. R. Bowker Co, 1180 Avenue of the Americas, New York, NY 10036, 1983.

MEDLARS on-line network

Health and safety professionals will find the data bases in the MEDLARS on-line network most useful. MEDLARS is an on-line network of approximately 20 bibliographic data bases covering worldwide literature in the health sciences. References may be retrieved by searching one or a combination of the 14,000 designated Medical Subject Headings used by NLM in indexing and cataloging materials. Requestors may obtain a complete record for each reference retrieved—including subject headings and abstracts—or a less detailed format including only the elements necessary to locate the item: author, title and publication source. (A description of each MEDLARS data base begins on page, following the explanation of MEDLARS services).

Records: MEDLARS contains more than 20 data bases covering in excess of six million documents published after 1965. The system is updated continuously.

Services: On-line search service is available directly from the National Library of Medicine (cost is $22.00 per hour during prime time), any of its seven Regional Medical Libraries, or through a nationwide NLM network at more than 2,000 universities, medical schools, hospitals, government agencies, and commercial organizations. To find your nearest on-line center, or how your institution can become a center, write the Regional Library for your area. At many on-line centers, the librarian will do the search for you; at others you may be encouraged to do your own searching after some preliminary instructions. The charge for services varies among centers. Some absorb all or most of the costs, others levy a modest cost-recovery fee.

Individuals can obtain direct access to MEDLARS. To do so, simply contact NLM or a Regional Library to obtain an application form. Any subscriber with a home computer and telephone link-up can dial in. Currently, cost is $15.00 minimum per month and MEDLARS offers a free three-day training session to subscribers.

AVLINE. AVLINE (Audio Visuals on-Line) is one of the data bases maintained by the National Library of Medicine.

Subject: Educational aids in the health sciences.
Sources: From audiovisuals acquired from producers from all over the U.S.
Content: The data base contains citations to over 11,000 audiovisual teaching packages. Subject coverage includes: dentistry, nursing, allied health, and other disciplines. Descriptive review information regarding review data are included in some cases. This information includes: rating, audience levels, instructional design, specialties, and abstracts. Procurement information on titles is provided. Most items have been produced within the last ten years.
Producer: National Library of Medicine
MEDLARS Management Section
8600 Rockville Pike
Bethesda, MD 20894
(301) 496-6193
Available on: National Library of Medicine (NML) and BlaiseLink (in the U.K. and Ireland); and on on-line systems in several other countries abroad.

CANCERLIT. CANCERLIT (Cancer Literature) contains more than 475,000 citations, with abstracts, to the worldwide literature on oncological epidemiology, pathology, treatment, and research. Before 1976, the data base corresponded to Cancer Therapy Abstracts from 1967, and Carcinogenesis Abstracts from 1963.

Subject: Biomedicine
Source: NIH's National Cancer Institute (NCI), more than 300,000 U.S. and foreign journals, books, reports and meeting abstracts.
Content: The data base contains English abstract references dealing with various aspects of cancer. The data base abstracts over 3,000 U.S. and foreign journals, as well as books, reports and meeting abstracts.
Date of information: 1963 to present.
Producer: U.S. National Institutes of Health
National Cancer Institute
International Cancer Research Data Bank Program
9000 Rockville Pike, Building 82
Bethesda, MD 20894
(301) 496-7403
Available on: National Library of Medicine.

CANCERNET (International Database on Oncology)

Type: Reference (Bibliographic)
Subject: Biomedicine
Producer: CANCERNET/Centre National de la Recherche Scientifique (CNRS)
On-line Service: University of Tsukuba
Content: Contains citations, with abstracts, to literature on oncology. Covers clinical and experimental carcinology, epidemiology, public health, and fundamental sciences (e.g., immunology, virology). Corresponds to *Cancer/Oncology, Bulletin Signaletique 251.*
Language: English and French
Coverage: International
Time Span: 1968 to date; abstracts from some articles, from 1981 to date.
Updating: About 1500 records a month.

CANCERPROJ (Cancer Research Projects)

Subject: Biomedicine research projects.
Sources: Cancer researchers in many countries.
Content: This data base contains summaries of ongoing and recently completed cancer research projects. It includes projects that are solicited through various international collaborating organizations by CCRESPAC for ICRDB.
Date of information: Most recent two to three fiscal years.
Producer: National Institutes of Health
National Cancer Institute
9000 Rockville Pike, Building 82
Bethesda, MD 20894
(301) 496-7403
Available on: National Library of Medicine.

CHEMLINE. CHEMLINE (Chemical Dictionary on-Line) is the National Library of Medicine's on-line, interactive chemical dictionary file created by the Specialized Information Services in collaboration with Chemical Abstracts Service (CAS). It provides a mechanism whereby thousands of chemical substance names can be searched and retrieved on-line. This file contains CAS Registry Numbers; molecular formulas; preferred chemical index nomenclature; generic and trivial names derived from the CAS Registry Nomenclature File; and a locator designation which points to other files in the NLM system containing information on that particular chemical substance. For a limited number of records in the file, there are Medical Subject Headings (MeSH) terms and Wiswesser Line Notations (WLN).

In addition, where applicable, each Registry Number record in CHEMLINE contains ring information including—number of rings within a ring system, ring sizes, ring elemental composition, and component line formulas.

Subject: Chemical dictionary.

Sources: TOXLINE, TOXBACK 65, TOXBACK 74, RTECS, MEDLINE, and TDB data bases; also the EPA Toxic Substances Control Act Inventory.

Content: Over 500,000 records on chemical substances found in the data bases mentioned above. This data base helps the user in searching the other MEDLARS data base by providing synonyms and CAS Registry Numbers, the use of which can increase retrieval in those data bases. Can also be searched to locate classes of chemical substances. Information is from 1965 to present.

Producer: U.S. National Institutes of Health
National Cancer Institute
International Cancer Research Data Bank Program
5333 Westbard Ave
Westbard Building, Room 10A-18
Bethesda, MD 20894
(301) 496-7403

Available on: National Library of Medicine (CHEMLINE), DIALOG (CHEMNAME), SDC Information Service (CHEMDEX).

MEDLINE. MEDLINE (Medical Literature Analysis and Retrieval System on-Line) contains references to citations from 3,000 biomedical journals. It is designed to help health professionals find out easily and quickly what has been published recently on any specific biomedical subject.

Subject: References to biomedical literature.

Sources: Indexed articles from over 3,000 international journals. 40% of the records added since 1975 include author abstracts taken from published articles. Each year over 250,000 records are added.

Content: Approximately 600,000 references to biomedical journal articles published in the present year and two years before. MEDLINE corresponds to three printed indexes: Index Medicus, Index to Dental Literature, and International Nursing Index. Indexes using NLM's (National Library of Medicine) controlled vocabulary MeSh (Medical Subject Headings). Can be used to update a search periodically. Information from 1966 to present totals more than 3.8 million references.

Producer: Office of Inquiries and Publications Management
National Library of Medicine
8600 Rockville Pike
Bethesda, MD 20894

Available on: DIALOG, BRS.

RTECS. RTECS (Registry of Toxic Effects of Chemical Substances) is an on-line, interacting version of the National Institute of Occupational Safety and Health (NIOSH) publication, *Registry of Toxic Effects of Chemical Substances.* It contains basic acute and chronic toxicity data for more than 57,000 potentially toxic chemical identifiers, exposure standards, and status under various Federal regulations and programs. The file can be searched by chemical identifiers, type of effect, or other criteria.

Subject: Toxicology

Sources: Based on the National Institute of Occupational Safety and Health (NIOSH) publication.

Content: The direct search on this data base allows the user to display, for a given list of compounds, the CAS Registry Number, and the details of each published toxicity measurement for each compound, including literature references. The user may ask for all entries relating to specific end effects on specific classes of animals for specific means of application, having dosage within a given range. The result of a query such as this is a list of Registry Numbers which can then be used to display RTECS data or to obtain information from other modules of the CIS.

Producer: U.S. National Institutes of Health
National Institute for Occupational Safety and Health
1600 Clifton Road
Atlanta, GA 30333
(404) 329-3771

Available on: National Library of Medicine, and NIH/EPA Chemical Information System.

TDB. TDB (Toxicology Data Bank) is composed of approximately 4,000 comprehensive, peer-reviewed chemical records. Compounds selected for TDB include highly regulated chemicals, high-volume production/exposure chemicals, and drugs and pesticides exhibiting high toxicity potential.

Subject: Toxicology.

Sources: Approximately 4,000 chemical records.

Content: This data base contains toxicological, pharmacological, environmental, occupational, manufacturing, and use information as well as chemical and physical properties.

Producer: National Library of Medicine
Toxicology Information Program
8600 Rockville Pike
Bethesda, MD 20894
(301) 496-6193

Available on: National Library of Medicine.

TOXLINE. TOXLINE (Toxicology Information on-Line) is the National Library of Medicine's extensive collection of computerized toxicology references to published human and animal toxicity studies, effects of environmental chemicals and pollutants, adverse drug reactions, and analytical method.

Subject: Toxicology.

Sources: Selected chemical abstracts Toxbib, which is selected from Medline; International Pharmaceutical Abstracts; HEEP, which is from BioSciences Information Services; EMIC and ETIC.

Content: This is a bibliographic data base which covers the pharmacological, biochemical, physiological, environmental, and toxicological effects of drugs and other chemicals. Most of the references in the data base have abstracts and/or indexing terms and Chemical Abstracts Service (CAS) Registry Numbers. Over 1.4 million references to toxicology literature are provided. Recent references are in the data base with older information in the TOXLINE backfiles, TOXBACK74 and TOXBACK65. The date of information is from 1965 to present with some data going back further:

Producer: National Library of Medicine
Toxicology Information Program
8600 Rockville Pike
Bethesda, MD 20894
(301) 496-6193

Available on: National Library of Medicine.

NLM regional locations. These are the NLM regional locations at the time of this writing:

MEDLARS National Library of Medicine (NLM) Computer Literature Retrieval Services can be accessed through either of these two means.

1) Office of Inquiries and Publications Management
National Library of Medicine (NLM)
8600 Rockville Pike
Bethesda, MD 20894
(301) 496-6308

2) The MEDLAR's Regional Library in your area for information or referral to your nearest MEDLAR's search service:

Region 1: Greater Northeastern Regional Medical Library Program (Connecticut, Delaware, Maine, Massachusetts, New Hampshire, New Jersey, New York, Pennsylvania, Rhode Island, Vermont, and Puerto Rico)
The New York Academy of Medicine
2 East 103rd St
New York, NY 10029
(212) 876-1232

Region 2: Southeastern/Atlantic Regional Medical Library Services (Alabama, Florida, Georgia, Maryland, Mississippi, North Carolina, South Carolina, Tennessee, Virginia, West Virginia, and District of Columbia)
University of Maryland
Health Sciences Library
111 South Greene St
Baltimore, MD 21201
(301) 528-7545

Region 3: Greater Midwest Regional Medical Library Network (Iowa, Illinois, Indiana, Kentucky, Michigan, Minnesota, North Dakota, Ohio, South Dakota, and Wisconsin)
University of Illinois at Chicago
Library of the Health Sciences
Health Sciences Center
P.O. Box 7509
Chicago, IL 60680
(312) 996-8974

Region 4: Midcontinental Regional Medical Library Program (Colorado, Kansas, Missouri, Nebraska, Utah, and Wyoming)
University of Nebraska
Medical Center Library
42nd and Dewey Avenue
Omaha, NE 68105
(402) 559-5346

Region 5: South Central Regional Medical Library Program (Arkansas, Louisiana, New Mexico, Oklahoma, and Texas)
University of Texas
Health Science Center at Dallas
5323 Harry Hines Blvd
Dallas, TX 75235
(214) 688-2626

Region 6: Pacific Northwest Regional Health Sciences Library Service (Alaska, Idaho, Montana, Oregon, and Washington)
Health Sciences Library
University of Washington
Seattle, WA 98195
(206) 543-8262

Region 7: Pacific Southwest Regional Medical Library Service (Arizona, California, Hawaii, and Nevada)
UCLA Biomedical Library
Center for the Health Sciences
Los Angeles, CA 90024
(213) 825-5781

BASIC REFERENCE BOOKS

The basic reference books listed in this section supplement the specific references appended at the end of the other chapters in this manual. The reference material cited in this bibliography was selected to provide safety professionals with sources of information which are likely to prove most useful in coping with problems of worker health protection and hazard assessment. This compilation is not to be viewed as a comprehensive coverage of the abundant literature on this subject, nor is any endorsement implied. The reference books are listed according to the following outline:

A. General principles
B. Risk assessment
C. Sampling methods
D. Toxicology
E. Medical
F. Dermatitis
G. Physical stresses
H. Ergonomics
I. Biological
J. Chemical
K. Pollution and hazardous waste
L. Control
M. Encyclopedias and handbooks

The intent of this section is to provide health and safety professionals with brief descriptions of the basic reference books and publications in the field of occupational health and industrial hygiene. Out-of-print books can usually be obtained from a library or by inter-library exchange.

A. General principles

Accident Prevention Manual for Industrial Operations, 9th ed. Vol. I: *Administration and Programs;* Vol. II: *Engineering and Technology.* Chicago: National Safety Council, 1988.

The *Administration and Programs* volume contains information on hazard control, organizing a program, governmental regulations and workers' compensation, computers, and information management systems, training, motivation, maintaining interest, publicizing safety, environmental concepts, emergency planning, personal protective equipment, industrial sanitation, handicapped workers, and product safety.

The *Engineering and Technology* volume is more nuts and bolts oriented, covering heavy and light construction (including tools and machines), material handling of all types, safeguarding of both woodworking and metalworking machinery, automated processes, welding and cutting, electrical hazards, flammable liquids hazards, and fire protection.

Allen, R. W., Ells, M. D., and Hart, A. W. *Industrial Hygiene,* Englewood Cliffs, NJ: Prentice-Hall, Inc., 1976.

This comprehensive book contains detailed information on safety and medical programs in industry for both large and small companies and for those who wish to revamp existing programs or instigate new ones. The book also contains important information on federal regulations such as the Occupational Safety and Health Act standards, and on procedures and forms that have been tested and proven in actual working conditions.

American Conference of Governmental Industrial Hygienists. *Microcomputer Applications in Occupational Health and Safety.* Chelsea, MI, Lewis Publishers, Inc., 1986.

Derived from a symposium sponsored by the ACGIH in March of 1986, scope of this work includes information systems and communications, integration of data, use of electronic spreadsheets, materials inventory and MSDS information, data base management, toxic substances research, expenditure prediction, and health and safety computer-aided design.

Annino, R., and Driver, R. *Scientific and Engineering Applications with Personal Computers.* New York, NY: John Wiley & Sons, Inc., 1986.

How to use the personal computer in the laboratory; both hardware and software are covered, as are a number of examples for the Apple, IBM-PC, and CPM. Chapters are devoted to programming, storage and program management, computer graphics, numerical analysis and modeling, and analysis of experimental data.

Baetjer, A. M. *Women in Industry: Their Health and Efficiency.* (Reprint of 1961 edition). Salem, MA: Ayers Co., Publishers, 1977.

Box, G. E. R., Hunter, W. G., and Hunter, J. S. *Statistics for Experimenters.* New York, NY: John Wiley & Sons, Inc., 1978.

As an introduction to design, data analysis, and model building, this text focuses on applications in physical, engineering, biological, and social sciences. It shows step-by-step how to set up experimental programs of high statistical and engineering efficiency. Useful to experimenters in all fields.

Brief, R. S. *Basic Industrial Hygiene Manual.* Akron, OH: American Industrial Hygiene Assoc., 1975.

Burgess, W. A. *Recognition of Health Hazards in Industry.* New York, NY: John Wiley & Sons, Inc., 1981.

Discusses how industrial operations can affect worker health. An example of a plant survey is given, along with the information needed for evaluating common unit operations. Materials and equipment are described, the physical form and origin of air contaminants are identified, as well as physical stresses encountered in the process or industry.

Clayton, G. D., and Clayton, F. E., eds. *Patty's Industrial Hygiene and Toxicology.* Vol. 1: *General Principles.* New York: John Wiley & Sons, Inc., 1978.

This authoritative handbook and reference provides a complete guide to methods of evaluation, record-keeping, and control, as well as historical background. Written more for the industrial hygienist.

Cohen, B., ed. *Human Aspects in Office Automation.* New York: Elsevier Science Publishing Co., Inc., 1984.

This book illustrates the relationship of employee health and job satisfaction to the creation and maintenance of proper working conditions.

Computer Systems for Occupational Safety and Health Management. New York: Marcel Dekker, Inc., 1984.

Book provides a clear step-by-step explanation for building and using a computerized safety data system.

Cralley, L. J., and Cralley, L. V., eds. *Patty's Industrial Hygiene and Toxicology.* Vol. 3: *Theory and Rationale of Industrial Hygiene Practice,* 2nd ed. (in two parts: A and B). New York: John Wiley & Sons, Inc., 1985.

Part 3A covers health promotion in the workplace, health surveillance programs in industry, occupational exposure limits, pharmacokinetics, and the effects of unusual work schedules.

Part 3B discusses the biological responses of the body to chemical and environmental hazards likely found in industrial workplaces. Text tells how the safety of chemicals, biological agents, ionizing radiation, and noise levels is evaluated.

Daubenspeck, W. G. *Occupational Health Hazards.* Hicksville, NY: Exposition Press, 1974.

A complete and updated rewrite of a Bureau of Labor Standards Bulletin, this book should be of help to safety professionals and others who are concerned with the recognition, evaluation, and control of occupational health hazards.

DeReamer, R. *Modern Safety and Health Technology.* New York: John Wiley & Sons, Inc., 1980.

Emphasis is on designing safety and health programs that conform to modern management principles and practices.

Esmen, N., and Mehlman, M. A., eds. *Occupational and Industrial Hygiene: Concepts and Methods.* Princeton, NJ: Princeton Scientific Publishers, Inc., 1984.

This volume gives a comprehensive picture of the discipline that Theodore F. Hatch helped to create, including discussions of the conceptual and methodological basis for the contemporary practice of industrial and occupational hygiene.

Firenze, R. H. *The Process of Hazard Control.* Dubuque, IA: Kendall/Hunt Publishing Co., 1978.

This guide has been specifically prepared to be used by both instructor and student in courses and seminars in order to guide discussion, stimulate interest and direct the study of occupational safety, occupational health, and industrial hazard control. It will explore areas of engineering, management, occupational health, hazard analysis, and fire protection as they relate to effective hazard reduction. The term *hazard control* is used throughout to familiarize management more thoroughly with the full dimension of hazards occurring from failures in techniques, equipment, systems, and operations that are responsible for dollar and manpower losses, repair or replacement of tools, equipment, litigation expenses, and the like.

Gunn, A., and Vesilind, P. A. *Environmental Ethics for Engineers.* Chelsea, MI: Lewis Publishers, Inc., 1986.
Environmental ethics apply to interactions of engineers with clients or employers, and to interactions between engineers and nature; that is, the effect of their work on the natural environment. Book discusses the issues of environmental ethics in-depth.

Johnson, W. G. *MORT Safety Assurance Systems.* New York: Marcel Dekker, Inc., 1980.
MORT (management oversight and risk tree) is a proven guide to the generic causal factors in an accident, a formal disciplined logic that integrates a wide variety of safety concepts systematically.

Key, M. M., et al., eds., *Occupational Diseases: A Guide to Their Recognition,* Revised. Washington, DC: U.S. Dept. of Health, Education, and Welfare, June 1977.
Occupational diseases are discussed in terms of occupational health hazards as a means to recognition of the disease. The text covers routes of entry and modes of action, chemical hazards, physical hazards, biological hazards, dermatoses, airway diseases, wood and plant hazards, chemical carcinogens, and pesticides. Sources of consultation and a list of references are included.

LaDou, J. L., ed. *Introduction to Occupational Health and Safety.* Chicago: National Safety Council, 1986.
Written for the safety professional who must start the *health* portion of the phrase "safety and health program"; for the manager who must administer and back them; and for medical professionals who want to learn more about occupational health.

Lee, J. S., and Rom, W. N. *Legal and Ethical Dilemmas in Occupational Health.* Stoneham, MA: Butterworth Publishers, 1982.
This book identifies what is being done, what is needed, and what lies ahead. Leading authorities present illuminating and frequently controversial views on the difficult legal and ethical issues in the occupational health field.
This book does not attempt to solve all the problems or answer all the questions. It does, however, provide the latest cases and current science as well as the experienced conclusions of professionals from many disciplines.

Lowry, G. G., and Lowry, R. C. *Handbook of Hazard Communication and OSHA Requirements.* Chelsea, MI: Lewis Publishers, Inc., 1986.
Explains how to meet OSHA's right-to-know requirements. Chapters include discussion of physical and health hazards characteristics, written hazard communication program, training, label design, consequences, and legal

...tional Safety and *Health.* New York: Free Press, Division of MacMillan Pub. Co., 1981.
This guide to OSHA regulations and compliance also covers NIOSH, labor relations law, civil liability, and Title VII.

O'Donnell, M. P., and Ainsworth, T. H., eds. *Health Promotion in the Workplace.* Somerset, NJ: John Wiley & Sons, Inc., 1983.
Book covers the past, current, and future status of the evolving health promotion movement. A major portion is devoted to in-depth analyses of program design, management, evaluation, facility design, and an assessment of general health.

Otway, H., and Peltu, M. *Regulating Industrial Risks Science, Hazards and Public Protection.* Stoneham, MA: Butterworth Publishers, 1986.
Regulations covering industrial plant licensing, operation, and work conditions define to a great extent the risks to which society is exposed. This book deals systematically with regulatory processes and the factors that influence them.
The book looks at the influence of national styles, organizational forces, international aspects, and problems of implementation, and examines the roles played by public policy groups, expert testimony and analysis, and the communications media. This book will interest all concerned with risks to the environment and to public health and safety.

Parmeggiani, L., ed. *Encyclopedia of Occupational Health and Safety,* 3rd rev. ed. New York: McGraw-Hill Book Co., 1983.
Reference work containing 900 articles prepared by 700 specialists in more than 70 countries and 10 international organizations. Covers all aspects of occupational safety and health. Emphasizes the safety precautions to be taken against the main hazards encountered in each branch of industry. Examples are quoted from international standards rather than national legislation. Articles are arranged alphabetically and each article includes bibliographic references. The second volume contains nine appendixes, a list of authors, and a comprehensive analytic index.

Peterson, J. E. *Industrial Health.* Englewood Cliffs, NJ: Prentice Hall, Inc., 1977 (out-of-print).
Focusing on immediate concerns of the field, this text stresses occupational hazards ranging from chemical toxins to those of various energy forms, and considers many of today's most pressing environmental and ecological problems. *Industrial Health* stresses principles that underlie various facets of the field—and highlights the generalizations that emerge with practical examples.

Schilling, R.S.F., ed. *Occupational Health Practice,* 2nd. ed. London, England: Butterworths, 1981.

Scott, R. *Muscle and Blood, The Massive, Hidden Agony of Industrial Slaughter in America.* Fresh Meadows, NY: Alsyl/Alexander, 1974.
A popular work written in the style of an exposé.

Stellman, J. M., and Daum, S. M. *Work Is Dangerous to Your Health: A Handbook of Health Hazards in the Workplace and What You Can Do About Them.* New York: Vintage, 1973.

This is a book written in the popular vein and addresses itself to the worker.

B. Risk assessment

Andelman, J. B., and Underhill, D. W. *Evaluation of Health Effects from Hazardous Waste Sites.* Chelsea, MI: Lewis Publishers, Inc., 1986.

The information and data for evaluating health effects from hazardous waste sites are the result of efforts of specialists representing research centers, hospitals, and government agencies. Consultant, as well as corporate viewpoints are presented. The work evolved from the Fourth Annual Symposium on Environmental Epidemiology.

Blakeslee, H. W., and Grabowski, T. M. *A Practical Guide to Plant Environmental Audits.* New York: Van Nostrand Reinhold, 1985.

This guide gives step-by-step instructions based on real-life situations and issues for conducting environmental audits at both small shops and large manufacturing plants. The whys and hows of environmental audit programs are detailed.

Conway, R. A., ed. *Environmental Risk Analysis for Chemicals.* New York: Van Nostrand Reinhold, 1981.

This book provides information for preparing an overall plan for basic testing. Step-by-step procedures are provided for determining the environmental risk of any chemical. The best methods for determining environmentally safe entry quantities of chemicals are presented.

Hallenbeck, W. H., and Cunningham, K. M. *Quantitative Risk Assessment for Environmental and Occupational Health.* Chelsea, MI: Lewis Publishers, Inc. 1986.

Of value wherever chemicals are manufactured or used, this book treats a complicated subject in a straightforward and understandable way. The reader needs only a basic knowledge of toxicology, epidemiology, and statistics to understand and perform the calculations presented. Sophisticated computer programs are not required.

C. Sampling methods

American Conference of Governmental Industrial Hygienists. *Air Sampling Instruments—For Evaluation of Atmospheric Contaminants,* 6th ed. Cincinnati: ACGIH, 1986.

Describes uses, principles, physical and performance data, operating and maintenance instructions, and commercial sources for air-sampling instructions. Includes a technical discussion of the principles of air sampling and the use of instruments for the evaluation of airborne contaminants.

Dux, J. P. *Handbook of Quality Assurance for the Analytical Chemical Laboratory.* New York: Van Nostrand Reinhold, 1985.

This handbook pays special attention to documenting procedures carried out by both degreed and non-degreed personnel. Specific problems of laboratory accreditation are explained in detail. With the information provided here, supervisors will be prepared to prove to management and outsiders alike that proper quality control is exercised in their labs.

Hesketh, H. E. *Fine Particles in Gaseous Media.* Chelsea, MI: Lewis Publishers, Inc., 1986.

Both theory and practice on the behavior and control of particles, their collection and measurement, are presented. The understanding of particle behavior becomes increasingly important with the development of hazardous emissions regulations. A useful book for environmental, mechanical, chemical, and civil engineers as well as regulatory officials in source testing, research, designing and using control equipment.

Hinds, W. C. *Aerosol Technology.* New York: John Wiley & Sons, Inc., 1982.

This book gives a complete exposition of the basic principles of aerosol science, including particle motion, forces on particles, the interaction of particles with the suspending gas, other particles, and electromagnetic radiation, and the application of these principles to aerosol measurement.

Intersociety Committee. *Methods of Air Sampling and Analysis.* 2nd ed. Washington, DC: American Public Health Association, 1977.

Represents the methods adopted by The Committee for a Manual of Methods of Air Sampling and Analysis, according to its established procedures.

Jungreis, E. *Spot Test Analysis.* New York: John Wiley & Sons, Inc., 1985.

This book provides procedures for conducting quick tests for a wide range of investigations, including biochemical components, forensic substances, geochemical composition, air and water pollutants, soil chemistry, and food adulteration and composition.

Katz, M. *Measurement of Air Pollutants: Guide to Selection of Methods.* Geneva, Switzerland: World Health Organization, 1969.

Laing, W. R. *Analytical Chemical Instrumentation.* Chelsea, MI: Lewis Publishers, Inc., 1986.

A new book of particular value to environmental and energy scientists as well as analytical chemists, with special emphasis on instrumentation. The contents of the volume are from the 28th ORNL–DOE Conference, October 1–3, 1985, Knoxville, TN.

Lee, S. D., Schneider, T., Grant, L. D., and Verkerk, P. J. *Aerosols: Research, Risk Assessment and Control Strategies.* Chelsea, MI: Lewis Publishers, Inc., 1985.

The volume on aerosols is the compilation of papers from a U.S.–Dutch Symposium held in Williamsburg, VA, May 198[illegible]. The topics include aerosol characterization and distribution and methodologies for sampling and analysis; transportation and deposition; exposure-effect relationships; epidemiological studies; risk assessment and economic impact; and policy on aerosol control strategies.

Leith, W. *The Analysis of Air Pollutants.* Philadelphia: Coronet Books, 1970.

——. *Evaluation of Ambient Air* [illegible] *ing,* 2nd ed. Boca Ra[illegible]

Personnel mo[illegible]

tion of the inhaled dose of an airborne toxic material or of an air-mediated hazardous physical force by the continuous collection of samples in the breathing or auditory zone, or other appropriate exposed body area, over a finite period of exposure time. A personnel monitor is a self-powered device worn by the monitored individual to collect a representative sample for laboratory analysis, or to provide accumulated dose or instantaneous warning of immediately hazardous conditions by visible or auditory means while being worn.

Mercer, T. T. *Aerosol Technology in Hazard Evaluation.* New York: Academic Press, 1973.

Produced by the American Industrial Hygiene Association under contract with the former U.S. Atomic Energy Commission. Theoretical and practical reference text relating to the science of aerosol technology. Discusses the nature, behavior, and properties of aerosols that predict whether they will become a biological hazard. Includes collecting methods, methods of establishing toxic level criteria, and monitoring atmospheres.

NIOSH Manual of Analytical Methods. In seven volumes, 77-157A–C, 78–175, 79–141, 80–125, 82–100. Cincinnati: National Institute for Occupational Safety and Health, 1977–1982.

A compilation of procedures covering many different chemicals that chemists in the Physical and Chemical Analysis Branch of NIOSH have used for industrial hygiene analyses.

NIOSH Manual of Sampling Data Sheets and Supplement. Cincinnati: National Institute for Occupational Safety and Health, 1977, 1978.

This edition includes sampling data sheets proposing methods to sample the industrial environment for contaminants.

OSHA Analytical Methods Manual. Salt Lake City: OSHA Salt Lake City Analytical Laboratory (SLCAL), 1985.

This manual was prepared by the OSHA Salt Lake City Analytical Laboratory to meet the requests of other governmental agencies as well as the private sector for sampling and analytical procedures of toxic substances. There are over 75 different organic methods and 20 inorganic methods presented in the manual. The manual does not replace the comprehensive NIOSH manual.

D. Toxicology

Albert, R. E. *Thorium: Its Industrial Hygiene Aspects.* San Diego, CA: Academic Press, 1966.

Summarizes the major technical uses of thorium, the hazards common to the various industrial processes, the techniques and objectives for the control of those hazards, and the biological and medical foundations on which the hazard controls are based. Also gives the physical, chemical, and radioactive properties of thorium and thoron.

American Conference of Governmental Industrial Hygienists. *Documentation of the Threshold Limit Values for Substances in Workroom Air,* 5th rev. ed. Cincinnati: ACGIH, 1985.

Contains the basis of TLVs for more than 475 substances. Includes discussions, limitations, and cautions for understanding and application of TLVs.

Arena, J. M. *Poisoning: Toxicology, Symptoms, Treatments,* 5th ed. Springfield, IL: Charles C Thomas, 1986.

Topics include general considerations of poisoning; insecticides, rodenticides, and herbicides; industrial hazards; occupational hazards; drugs; soaps and detergents; poisonous plants, insects, and fish; and miscellaneous compounds and topics, including radioactive isotope poisoning, rocket fuels, and welding hazards. Appendix of normal laboratory values used in the diagnosis and treatment of poisoning.

Caldwell, J., ed. *Amphetamines and Related Stimulants: Chemical, Biological, Clinical, and Sociological Aspects.* Boca Raton, FL: CRC Press, Inc., 1978.

The aim of this volume is to explore historical, chemical, biological, clinical, and sociological aspects of the amphetamines and related stimulants with reference both to legitimate medical use and to their abuse.

There is at the present time an enormous literature on the amphetamines and related stimulants, particularly in the area of neuropsychopharmacology, but it is extremely difficult to distill from this the information of relevance to the problem of the abuse of these compounds. The aim of this volume is to draw together those aspects of the chemical, biological, clinical, and sociological studies that have the maximum impact on the abuse problem.

Carson, B. L., Ellis, H. V. III, and McCann, J. L. *Toxicology and Biological Monitoring of Metals in Humans Including Feasibility and Need.* Chelsea, MI: Lewis Publishers, Inc., 1986.

Persons in many disciplines besides toxicology need information about toxic effects of substances related to exposure and how to monitor exposure. This book contains toxicological, exposure, and monitoring information about metals in a brief, uniform format.

Casarett, L. J., and Doull, J., eds. *Toxicology, The Basic Science of Poisons,* 2nd ed. New York: Macmillan Publishing Co., 1980. (Now called *Casarett and Doull's Toxicology, The Basic Science of Poisons.* Edited by J. Doull, C. D. Klaassen, and M. O. Amdur.)

Christian, M. S., et al., Assessment of Reproductive and Teratogenic Hazards. Princeton, NJ: Princeton Scientific Publishers, Inc., 1983.

Human survival depends on our ability to carry out normal reproduction from generation to generation in this highly complex century and we are continually striving to improve the quality of life through the development of new drugs and chemicals. The potential adverse effects of these substances is the concern of reproductive toxicology. The information presented in this book within the limitations imposed by toxicology's heavy reliance on animal testing, should be of value to toxicologists, teratologists and biological scientists in government and industry, who are concerned with the evaluation and assessment of reproductive hazards to men and women.

Clayson, D. B., Krewski, D., and Munro, I. *Toxicological Risk Assessment.* Boca Raton, FL: CRC Press, Inc., 1985.

This book discusses measurements of certain environmental risks and their application to the relatively low levels to which humans are generally exposed.

Clayton, G. D., and Clayton, F. E., eds. *Patty's Industrial Hygiene and Toxicology.* Volume 2A, 2B, 2C: Toxicology. New York: John Wiley & Sons, Inc., 1981, 1982.

Volume 2 of *Patty's Industrial Hygiene and Toxicology* is divided into three parts. Part A contains a historical summary of toxicology, followed by a thorough description of the handling of 40 metals. Part B encompasses the hundreds of newly released data appearing since the previous edition. Topics discussed include occupational carcinogenesis, the halogens, and the nonmetals. Part C gives a thorough treatment of glycols, alcohols, and ketones.

Cohen, Gerald M., Ph.D. *Target Organ Toxicity.* Boca Raton, FL: CRC Press, Inc., 1986.

The volume provides essential information on the general principles of target organ toxicity. The general principles are then illustrated using specific examples of toxicity in different target organs and systems. Modification of DNA and repair in tumor induction, and specificity in tumor initiation are also explained. This book is of primary interest to toxicologists, pharmacologists, biochemists, and environmental toxicologists.

Committee on Fire Research, Commission on Sociotechnical Systems, National Research Council. *Physiological and Toxicological Aspects of Combustion Products: International Symposium.* Washington, DC: National Academy of Sciences, 1976.

The project that is the subject of this report was approved by the Governing Board of the National Research Council, whose members are drawn from the National Academy of Sciences, the National Academy of Engineering, and the Institute of Medicine. The members of the Committee responsible for the report were chosen for their special competences and with regard for appropriate balance.

Committee on Medical and Biologic Effects of Environmental Pollutants. *Nickel.* Washington, DC: National Academy of Sciences, 1975.

This document was written by the Panel on Nickel under the chairmanship of Dr. F. William Sunderman, Jr. Although each section was prepared initially by a member of the panel or an invited contributor, some material was later combined, and the total document was reviewed and approved by the entire panel and thus represents its cooperative effort. Dr. Sunderman was responsible for the introduction, large parts of the sections on nickel metabolism in man and animals and on nickel toxicity, and most of the sections on nickel carcinogenesis and nickel in the reproductive system.

Dominguez, G. S., ed. *Guidebook: Toxic Substances Control Act.* Boca Raton, FL: CRC Press, Inc., 1977.

The book is not only a complete guide to the law, but it is also a source on how to prepare for and respond to the many newly instituted federal requirements. The book provides practical and clear advice on anticipated compliance approaches, as well as suggestions for early organizational preparation and planning.

Dreisbach, R. H. *Handbook of Poisoning: Diagnosis and Treatment,* 11th ed. Palo Alto, CA: Lange Medical Publications, 1983.

A concise summary of the diagnosis and treatment of clinically important poisons including those encountered in industry and agriculture.

Dunnom, D. *Health Effects of Synthetic Silica Particulates.* Philadelphia: American Society for Testing and Materials, 1981.

Focused on health problems caused by synthetic silica dust, this book gives the picture of current understanding of the physiological effects of silica particulates.

Dutka, B. J., and Bitton, G. *Toxicology Testing Using Microorganisms.* Boca Raton, FL: CRC Press, 1986.

A compendium of new and traditional technology for microbiological toxicity testing procedures. Procedures, apparatus, degree of reliability, advantages, and pitfalls of each technology are outlined, with references to original literature.

Essays in Toxicology. San Diego, CA: Academic Press, 1969–1976.

Seven volumes have thus far been issued in this continuing series treating a wide range of subjects in the field of toxicology.

Flamm, W. G., and Lorentzen, R. J., eds. *Mechanisms and Toxicity of Chemical Carcinogens and Mutagens.* Princeton, NJ: Princeton Scientific Publishers, Inc., 1985.

This volume provides both a status report and an indication of future directions of research on the genetic mechanisms of carcinogenesis. The introduction reviews two etiological factors involved in cancer causation and shows, where possible, how various theories on the mechanism of carcinogenesis relate to what is known about the etiology of cancer. The experimental evidence supporting the somatic mutation theory of cancer induction is discussed.

Friberg, L., Elinder, C-G., Kjellstrom, T., and Nordberg, G. F. *Cadmium and Health: A Toxicological and Epidemiological Appraisal.* Boca Raton, FL: CRC Press, Inc., 1986.

A comprehensive review and critical evaluation of studies conducted in many nations, including important research in Japan. The most significant studies are discussed and analyzed at length.

Gleason, M. N., Gosselin, R. E., Hodge, H. C., and Smith, R. P. *Clinical Toxicology of Commercial Products,* 5th ed. Baltimore: The Williams & Wilkins Co., 1986.

This book assists physicians in dealing quickly and effectively with acute chemical poisonings in the home and on the farm, arising through misuse of commercial products. It provides a list of trade-name products together with their ingredients when these have been revealed, addresses and telephone numbers of companies for use when ingredients are not listed, sample formulas of many types of products with an estimate of the toxicity of each formula, toxicological information including an estimate of the toxicity of individual ingredients, recommendations for treatment, names and addresses of manufacturers, and a system of standard nomenclature for the clarification of poisonings. Medical libraries, pharmacies, industrial medical departments, public health nursing centers, and any other agency frequently called upon for emergency help should also find it helpful as a quick source of information on first aid, treatment procedures, and other questions.

Goldsmith, J. R. *Environmental Epidemiology: Epidemiological Investigation of Community Environmental Health Problems.* Boca Raton, FL: CRC Press, Inc., 1985.

The experiences of practicing epidemiologists in solving worldwide community environmental health problems are discussed. Emphasis is placed on problems facing the community, methods of analysis, and means and results of action. Actual case histories of various complexity provide exercises in solving community health problems using applicable elementary concepts of statistics. Selected tables offer quantitation of problems and authoritative references direct interested students to further reading.

Grant, W. M. *Toxicology of the Eye,* 2nd ed. Springfield, IL: Charles C Thomas, 1974.

Grover, P. L., ed. *Chemical Carcinogens and DNA.* (2 vols.) Boca Raton, FL: CRC Press, Inc., 1979.

Deals with the chemical modification of the DNA molecule and the ways these modifications can be detected.

Halpern, S. *Drug Abuse and Your Company.* Ann Arbor, MI: University Microfilms International, Books-on-Demand, 1972.

Hamilton, A., and Hardy, H. L. *Industrial Toxicology,* 3rd ed. Littleton, MA: Publishing Science Group, Inc., 1974.

Hardin, J. W., and Arena, J. M. *Human Poisoning From Native and Cultivated Plants.* Durham, NC: Duke University Press, 1973.

Most of the existing literature on poisonous plants deals with those that are poisonous to livestock. A real need exists for a source of information on just those plants poisonous to humans—particularly children. Physicians, health officers, nurses, scout leaders, camp counselors, teachers, parents, and many others should not only know the dangerous plants of their area but have a ready reference in case of emergencies. This book has been written with these people in mind and has grown out of a number of years' experience with poisonous plants accumulated by both of the authors in the field, laboratory, and clinic.

Kopfler, F., and Crawn, G., eds. *Environmental Epidemiology.* Chelsea, MI: Lewis Publishers, Inc., 1986.

This book is valuable to a broad spectrum of individuals active in the environmental and health sciences as well as those involved in the measurement and effects of numerous kinds of drinking water contamination and other indoor and ambient air pollution. Environmental researchers involved in human exposure to toxic substances, regulators and administrators also will find this book of value.

MacFarland, H. N., et al., ed. *Applied Toxicology of Petroleum Hydrocarbons.* Princeton, NJ: Princeton Scientific Publishers, Inc., 1984.

Studies on the toxicology of hydrocarbons consist of interdisciplinary approaches aiming to evaluate the toxicity of these compounds from an applied, as well as from a mechanistic, point of view. These investigations use biological systems composed of human subjects, animal models and in vitro cellular systems of mammalian and bacterial origin. The symposium focused on key problems and approaches in the area of toxicology of petrochemicals.

Matheson Chemicals Company. *Effects of Exposure to Toxic Gases First Aid and Medical Treatment.* Lyndhurst, NJ: Matheson Co., Div. of Searles Medical Products, 1983.

This edition includes a chapter on industrial hygiene in an effort to stress the necessity of incorporating good safety and health control in the design of work areas where hazardous gases will be used. Workers' attention to the need for good work practices, special emergency measures, and equipment that should be available to minimize the risk and consequences of accidents are also discussed.

Milman, H. A., and Weisburger, E. K., eds. *Handbook of Carcinogen Testing.* Park Ridge, NJ: Noyes Data Corp., 1985.

This book considers short-term and long-term testing for carcinogens, and all other facets of the operation. The volume affords a total view of a bioassay from initial phases to application.

Monson, Richard R. *Occupational Epidemiology.* Boca Raton, FL: CRC Press, Inc., 1980.

The detection and control of long-term effects, and a review of general epidemiologic methods and their application in the context of occupational settings are included. The book discusses how data from occupational populations can be used to prevent illness, and principles to be followed in setting up and running comprehensive programs in occupational epidemiology are included.

Neely, W. B., and Blau, G. E. *Environmental Exposure in Chemicals.* Boca Raton, FL: CRC Press, Inc., 1985.

This book includes estimation of physical properties, pollution sorption in environmental systems, air/soil exchange coefficients, air/water exchange coefficients, biogradation, hydrolysis, photochemical transformations, and equilibrium models for initial integration of physical and chemical properties.

Nordberg, G. F., ed. *Effects and Dose-Response Relationships of Toxic Metals.* Amsterdam, The Netherlands: Elsevier, 1976.

O'Donoghue, J. L., ed. *Neurotoxicity of Industrial and Commercial Chemicals.* Boca Raton, FL: CRC Press, Inc., 1985.

A collection of up-to-date information on the neurotoxicity of chemicals used in industry or having commercial value. Chemicals reported to cause a variety of effects on the nervous system are thoroughly reviewed. Exposure data, clinical manifestations, pathology, experimental neurology, metabolism, and structure activity correlates are integrated and presented by the anatomical and functional areas of the nervous systems affected, and by chemical classes with neurotoxic effects. Much of the information is presented in tabular format.

Ottoboni, M. A. *The Dose Makes the Poison.* Berkeley, CA: Vincente Books, 1984.

This book explains, in nontechnical language, what makes chemicals harmful or harmless. Dr. Ottoboni states that all living organisms have to deal with exposure to many noxious substances; but only when we overwhelm our bodily natural defense mechanisms by taking in too much too often, do we get into trouble.

Plunkett, E. R. *Handbook of Industrial Toxicology,* 2nd ed. New York: Chemical Publishing Co., 1976.

Randolph, T. G. *Human Ecology and Susceptibility to the Chemical Environment.* Springfield, IL: Charles C Thomas, 1981.

Most illnesses were originally believed to have arisen within the body—only recently has this age-old concept been challenged. The importance of the outside environment as a cause of sickness was first demonstrated in respect to infectious diseases approximately eighty years ago and to allergic diseases approximately fifty years ago. Although the general principles of infectious disease are now fully accepted and applied, the medical profession has been slow in learning and applying the necessary techniques to demonstrate cause-and-effect relationships between the nonmicrobial environment and ill health.

Reeves, A. L., ed. *Toxicology: Principles and Practice,* Volume 1. New York: John Wiley & Sons, Inc., 1981.

This volume addresses toxicology as the basic science and as the applied doctrine, with the practical purpose of uncovering and eliminating potential sources of poisoning from the environment. It will aid physicians and public health officials in the recognition, treatment, and prevention of poisonings.

Registry of Toxic Effects of Chemical Substances, vol. 2. Edited by E. J. Fairchild et al. Cincinnati, OH: NIOSH, published annually.

Annual list of known toxic substances that may exist in the environment, or that are manufactured, processed, or synthesized, such as drugs, food additives, preservatives, ores, pesticides, dyes, detergents, lubricants, soaps, or plastics. Information on each substance includes prime name of chemical substance, Chemical Abstracts Service registry number, molecular weight and formula, Wiswesser Line Notation, synonyms, toxic dose data, units of dose measurement, notations descriptive of the toxicology, cited reference, U.S. Occupational Standards, and NIOSH Criteria Documents.

Scher, J. M. *Drug Abuse in Industry: Growing Corporate Dilemma.* Springfield, IL: Charles C Thomas, 1973.

Searle, C. E., ed. *Chemical Carcinogens,* 2nd rev. ed. Washington, DC: American Chemical Society, 1984.

This volume contains information which will be invaluable to scientists working on all aspects of cancer research and occupational health. Recent advances in cancer research are presented in over 800 pages with more than 750 structural formulas.

Cancer-causing agents are now known to exist throughout the environment—in polluted air and tobacco smoke, in various plants and foods, and in many chemicals that are used in industry and laboratories.

This timely monograph contains comprehensive accounts of the theories of cancer chemistry and biology and of major hazards identified. Authorities from the United States, United Kingdom, France, and West Germany have contributed 16 chapters.

Sperling, F. *Toxicology: Principles and Practice,* vol. 2. New York: John Wiley & Sons, Inc., 1984.

Toxicology is discussed, as a basic science and as an applied doctrine, with the practical purpose of uncovering and eliminating potential sources of poisoning from the environment. This volume covers such topics as toxicity, respiratory toxicity studies, immune response in toxicology, eye and skin toxicity testing, and renal toxicology.

Stich, H. F., ed. *Carcinogens and Mutagens in the Environment, vol 4: The Workplace: Monitoring and Prevention of Occupational Hazards.* Boca Raton, FL: CRC Press, Inc., 1985.

This volume includes chapters on cancer as an occupational hazard; tissues in cancer prevention; improvement for worker protection; government, employers, and labor relations through information and participation; risk assessment; exposed populations; hazards and safety of the modern office environment.

Stokinger, H. E., ed. *Beryllium: Its Industrial Hygiene Aspects.* San Diego, CA: Academic Press, 1966.

Detailed account of how beryllium and its compounds can be handled safely and the engineering controls needed. Treats the chemical, biological, and medical aspects of beryllium, including sampling and analysis of atmospheric and biological specimens. Also reviews its toxicology and pathology.

Sunshine, I. *Methodology for Analytical Toxicology,* 3 volumes. Boca Raton, FL: CRC Press, Inc., 1979, 1982, 1985.

Thoma, J. J., and Bondo, P. B., eds. *Guidelines for Analytical Toxicology Programs.* Boca Raton, FL: CRC Press, Inc., 1978.

This two-volume work provides physicians and laboratory directors with practical guidelines for development of a reliable toxicology service for their community. It also features detailed information on instrumentation for various toxicological testing procedures. Emphasis is placed on defining proper technique in the use of instrument and establishing routine maintenance schedules to assure optimum performance. This work is prepared by 25 leading toxicologists, physicians, and pharmacologists.

Threshold Limit Values and Biological Exposure Indices. Cincinnati: American Conference of Government Industrial Hygienists. Published annually.

This pocket-size booklet contains a listing of adopted and proposed Threshold Limit Values (TLVs) for more than 700 of the most common and widely found toxic substances and physical agents. The booklet also includes Biological Exposure Indices (BEIs) and an extensive preface on the intended role of these values.

Williams, P. L., and Burson, J. L., eds. *Industrial Toxicology Safety and Health Applications in the Workplace.* Belmont, CA: Lifetime Learning Publications, 1984.

This book was written for those health professionals who need toxicological information and assistance beyond that of an introductory text in general toxicology, yet more practical than that in advanced scientific works on toxicology. It is of particular interest to industrial hygienists, occupational physicians, safety engineers, sanitarians, occupational health nurses, safety directors, and environmental scientists.

Zimmerman, F. K., and Taylor-Mayer, R. E. *Mutagenicity*

Testing in Environmental Pollution Control. New York: John Wiley & Sons, Inc., 1985.

In the past decade, mutagens have been demonstrated to be carcinogens. This book introduces the problems of mutagenicity testing, and provides examples of how mutagenic activity can be detected in the environment.

E. Medical

Alderman, M. H., and Hanley, M. J., eds. *Clinical Medicine for the Occupational Physician.* New York: Marcel Dekker, Inc., 1982.

This book covers the interdisciplinary topics related to personal health and medical care. It presents the workplace as a site for health promotion; stressing the management of total personal health by occupational physicians and nurses. The book provides focused examinations of the special characteristics of women, elderly and disabled workers, and employees with chronic illnesses; and features in-depth coverage of toxicology and both chemical and physical hazards, emphasizing occupational cancer and problems in dermatology.

Alderson, M. *Occupational Cancer.* Stoneham, MA: Butterworth Publishers, 1986.

The increased risks of cancer resulting from exposure to particular agents in the workplace or from working in a particular industry have now been extensively studied. The book reviews the current information on the occurrence and causes of cancer, concentrating on material drawn from epidemiological studies.

The greater portion of the book discusses each chemical and physical agent associated with an increased risk of cancer, and is an essential reference for those involved with the health-care of workers. Chapters include: epidemiological methods; agents which cause occupationally-induced cancers; occupations associated with increased risk of cancer; aetiology of cancer; and towards control of occupational cancer.

Brown, M. L. *Occupational Health Nursing Principles and Practices.* New York: Springer Publishing Co., Inc., 1981.

This book provides comprehensive information in areas such as roles, objectives, legislation, and ecology for those in occupational health and safety. Emphasis is placed on the nurse's role on the occupational health team, and the need to access and control the impact of the work environment on the health and safety of the worker.

Cataldo, M. F., and Coates, T. J. *Health and Industry.* New York: John Wiley & Sons, Inc., 1986.

A behavioral medicine perspective, this book covers the principles of behavioral medicine in the industrial setting, plus the day-to-day problems affecting workers including obesity, stress, hypertension, cardiovascular disease, smoking, dental care, cancer, and accidents.

Craun, G. F. *Waterborne Diseases in the United States.* Boca Raton, FL: CRC Press, Inc., 1986.

Water-related illness is examined, emphasizing transmission of infectious diseases through contaminated drinking water supplies and deficiencies in water supply systems that allow waterborne outbreaks. Also included are important etiologic agents, surveillance activities, regulations, preventive measures, procedures for investigating disease outbreaks, and laboratory methods for identifying pathogens, and diseases that result from ingesting, bathing or wading, and inhalation. This volume is a good reference for public health investigators, environmental engineers, and sanitarians.

Guides to the Evaluation of Permanent Impairment, 2nd ed. Chicago: American Medical Association, Committee on Rating of Mental and Physical Impairment, 1984.

The AMA has provided authoritative material to assist physicians and others in determining levels of impairment of patients, clients, or applicants who are seeking benefits from the various agencies and programs serving the disabled.

Hoover, H. C., and Hoover, L. H. *Georgius Agricola: De Re Metallica,* translated from the First Latin Edition of 1556. New York: Dover Publications, Inc., 1950.

Hunter, D. *The Diseases of Occupations,* 6th ed. London, England: The English Universities Press, 1978.

Includes historical outline of occupational diseases and their treatment. Discusses hazardous materials, such as metals and noxious gases, describes the processes that led to their being recognized as hazards, symptoms, preventive methods, treatment, and some case histories.

Johnstone, R. T., and Miller, S. E. *Occupational Diseases and Industrial Medicine.* Ann Arbor, MI: University Microfilms International, Books-on-Demand, 1960.

Textbook discusses diseases caused by gases, vapors, and dusts, physical agents, such as vibration, noise, and extremes of temperature and pressure; photoactinic diseases, other hazards and protective measures and devices.

Jones, D. M., and Chapman, A. J. *Occupational Lung Disease: Research Approaches and Methods.* New York: Marcel Dekker, Inc., 1981.

Investigations into occupational lung diseases now yield information on pathogenic processes. This book covers how this occurs.

Key, M. M., et al. *Occupational Diseases: A Guide to Their Recognition,* rev. ed. Washington, DC: U.S. Department of Health, Education, and Welfare, Public Health Service, 1977.

LaDou, J., ed. *Occupational Health Law: A Guide for Industry.* New York: Marcel Dekker, Inc., 1981.

This book investigates the complex legal and regulatory issues faced by health and safety administrators who work in industry. With the information provided, the occupational health specialist is better equipped to understand the benefits and shortcomings of occupational health legislation.

Lee, D. H. K., and Kotin, P., eds. *Multiple Factors in the Causation of Environmentally Induced Disease,* Fogarty International Center Proceedings, No. 12. New York: Academic Press, 1972.

Lefevre, M. J. *First Aid Manual for Chemical Accidents: For Use with Nonpharmaceutical Chemicals.* New York: Van Nostrand Reinhold, 1980.

First aid for chemical accidents requires specific, quick, correct action to obviate and minimize potentially serious, harmful effects. This first aid manual is dedicated totally to emergency care at the workplace in which industrial chemicals are used.

A color-indexing scheme makes proper treatment information easily accessible. Each compound of the nearly 500 industrial products is referred to a color-coded section: toxicology; symptoms of overexposure; and first aid procedures in cases of inhalation, ingestion, skin contact, and eye contact.

This quick-reference manual for emergencies can be equally valuable for the prevention of accidents in the training of anyone who works with industrial solvents, insecticides, commercial chemicals, fertilizers, or chemical intermediates.

Magi, S. Z. *Disability and Rehabilitation: Legal, Clinical, Self-Concepts, and Measurements.* Columbus, OH: Ohio State University Press, 1978.

McLean, A., ed. *Occupational Stress.* Springfield, IL: Charles C Thomas, 1974.

Nichols, P. J. *Rehabilitation Medicine,* 2nd ed. Stoneham, MA: Butterworth Publishers, 1980.

Parkes, W. R. *Occupational Lung Disorders,* 2nd ed. Stoneham, MA: Butterworth Publishers, 1981.

This new edition of an internationally acclaimed reference provides an up-to-date and thorough overview of occupational skin disorders. Dr. Parkes presents an excellent medical appraisal of the damaging ways in which industrial processes affect the lung. This single volume offers a clear picture of the physical and chemical nature of a wide range of agents, the industrial process involved, and the clinical and pathological features seen.

Preventing Illness and Injury in the Workplace. Washington, DC: Office of Technology Assessment, 1985.

Workers, employers, health and safety professionals, and government officials have all contributed to progress in this field. But improvements can still be made. More concerted effort and better use of existing methods would enhance hazard identification. Further research could improve health and safety technologies and contribute to their incorporation in U.S. workplaces. Employers' decisions to control hazards might be fostered by changing the incentives and imperatives that affect those decisions.

Ramazzini, B. *De Morbis Artificum (Diseases of Workers,* text of 1713, revised), translation from the Latin with notes by W. C. Wright. University of Chicago Press, 1940. Reprint: Halner Publishing Company, New York, 1964.

The historical work by the "Father of occupational medicine."

Rom, W. N., ed. *Environmental and Occupational Medicine.* Boston: Little, Brown & Co., 1982.

The book contains many areas of information, including recognition and evaluation of occupational health problems, asbestos and related fibers, and occupational and environmental health standards.

Safety Guide for Health Care Institutions, 3rd ed. Chicago: National Safety Council and the American Hospital Association, 1983.

This guide is intended to recognize and identify hazards in health care facilities, provide information for their elimination, and stimulate each hospital's personnel to improve its safety program. To be most effective, this book should be used with other safety books from the NFPA, National Safety Council, and the AIHA.

Schilling, R. S. F., ed. *Occupational Health Practice,* 2nd ed. London, England: Butterworths, 1981.

Shepard, W. P. *The Physician in Industry.* (reprint of 1961 ed) Salem, MA: Ayer Co. Publishers, 1977.

A guide to industrial medicine, for the practicing physician. Topics include the role of the physician and the nurse, environmental effects and control, industrial toxicology, radiation hazards, accidents, care for disabled workers, and mental health in industry.

Sunderman, E. W., and Sunderman, F. W., Jr., eds. *Laboratory Diagnosis of Diseases Caused by Toxic Agents.* St. Louis: Warren H. Green, 1970.

Zenz, C., ed. *Occupational Medicine: Principles and Practical Applications,* 2nd ed. Chicago: Year Book Medical Publishers, 1988.

F. Dermatological

Adams, R. M. *Occupational Skin Disease.* Orlando, FL: Grune & Stratton, Inc., 1982.

This book guides dermatologists, allergists, and industrial physicians in the successful management of occupational dermatoses. The information on prevention and treatment will enable workers, industrial hygienists, and plant superintendents to plan and implement methods that should lower the incidence of such disease.

Gellin, G. A., and Maibach, H. I. *Occupational Industrial Dermatology.* Chicago, IL: Year Book Medical Publishers, 1982.

Rees, R. B. *Dermatoses Due to Environmental and Physical Factors.* Springfield, IL: Charles C Thomas, 1962.

G. Physical stresses

Attix, F. H., and Roesch, W. C., eds. *Radiation Dosimetry,* 2nd ed. San Diego, CA: Academic Press, 1969.

This second edition fills the need for a comprehensive treatise, written primarily as a reference work for radiation workers. Many useful tables, curves, illustrations, formulas, and references to the literature have been included. Every effort was made to present the material as clearly as possible, for those just entering the field.

Brodsky, A. *CRC Handbook of Radiation Measurement and Protection.* Boca Raton, FL: CRC Press, Inc., 1982.

This handbook is a comprehensive guide to methods and data used for measuring or estimating radiation doses, and provides information on radiation protection for industrial, research, and medical installations. This book will assist in solving problems of radiation dose estimation, biological risk evaluation, and facility design.

Cember, H. *Introduction to Health Physics,* 2nd ed. New York: Pergamon, 1983.
Contents deal with a review of physical principles, atomic and nuclear structure, radioactivity, interaction of radiation with matter, radiation dosimetry, biological effects of radiation, radiation protection guides, health physics instrumentation, external and internal protection, criticality, and evaluation of protective measures.

Cheremisinoff, P. N. and P. P., eds. *Industrial Noise Control Handbook.* Ann Arbor, MI: Ann Arbor Science Publishers, Inc., 1977.
This volume was designed for use by engineers faced with industrial noise problems. It should also be of use to consultants and planners, as well as the student. It is written in general technical language aimed to facilitate its use and stresses the practical rather than the theoretical. The Occupational Safety and Health Act requires that employers provide a noise control program whenever employees work in an environment exposing them to such hazards. The authors have tried to give information necessary to meet OSHA requirements, thus making possible a safer and more productive work environment.

Eichholz, G. G. *Environmental Aspects of Nuclear Power.* Chelsea, MI: Lewis Publishers, Inc., 1985.
This volume reviews all design and chemical features of nuclear power plants and associated fuel cycle operations that may be considered relevant to environmental considerations. It describes the engineering solutions available to solve such problems and assesses the relative importance of factors involved.
The book presents a complete description of the whole nuclear power cycle and supplies a fair and objective treatment of many problem areas that have come under public scrutiny. The work brings together many details previously available in report form only.

Eichholz, G. G., and Poston, J. W. *Principles of Nuclear Radiation Detection.* Chelsea, MI: Lewis Publishers, Inc., 1985.
This book provides discussion of the principles of those radiation detectors most widely used in nuclear technology, medical practice, and radiation protection. It stresses the alternative detectors available and discusses practical considerations in choosing and setting up detector systems for actual use. The book covers traditional materials, including semiconductors, TLDs, and modern data handling.

Harris, C. M., ed. *Handbook of Noise Control,* 2nd ed. New York: McGraw-Hill Book Co., 1979.
In general, the material presented relates to properties of sound, effects of noise on man, vibration control, instrumentation and noise measurement, techniques of noise control, noise control in buildings, sources of noise and examples of noise control of machinery and electrical equipment, noise control in transportation, community noise, and the legal aspects of noise problems.
This manual is a handy reference source because of the large number of references given. It should prove helpful to those concerned with almost any kind of noise problem, legal or technical.

Heating and Cooling for Man in Industry, 2nd ed. Akron, OH: AIHA, 1975.
Written for the working industrial hygienist and heating and ventilation engineer, this manual contains information to obviate the need for extensive research when attempting to solve a problem. It has been written with one objective—describing the means of controlling the working environment to conduct a variety of operations with fluctuating outdoor conditions. Methods of varying temperature, air motion, and humidities within the work space are described. A worker's space is the primary area of interest.

Industrial Noise Manual, 3rd ed. Akron, OH: AIHA, 1975.
Physics of sound, instruments for sound measurement, technique of sound measurement, noise surveys, vibration, anatomy and physiology of the ear, effects of noise on man, hearing measurement, medical aspects of industrial hearing conservation, personal protection, engineering control and legal aspects of the industrial noise problem are covered in this edition.

International Atomic Energy Agency. *Radiation Protection Procedures.* Vienna, Austria: 1973. Available from UNIPUB, Inc., P.O. Box 433, New York, NY 10016. Order No. STI/PUB-257.
Text reviews the fundamentals of nuclear physics and interactions of ionizing radiations with matter and living cells; the basic concepts governing the formulation of units for the measurement of radiations; methods used for measurements of radiations; selection, calibration, and maintenance of instruments used for monitoring; shielding; protective clothing; decontamination measures; radioactive waste management; the transport of radioactive materials; and emergency procedures for radiation accidents.
Discusses various administrative and technical measures which could form the basis for establishing a successful radiation protection program.

Kryter, K. D. *The Effects of Noise on Man,* 2nd ed. San Diego, CA: Academic Press, 1985.
Treats auditory system responses, subjective responses, and nonauditory system responses to noise; includes an extensive set of references.

LeBlanc, J. *Man in the Cold.* Springfield, IL: Charles C Thomas, 1975.
This volume serves as a summary statement of our present understanding of human functional responses to cold exposure. It helps the reader tie into the overall pattern of human responses to cold, such observations as local changes in fat composition, changes in amounts and distribution of isoenzymes, and modifications of gluconeogenesis.

Miller, D. G. *Radioactivity and Radiation Detection.* New York: Gordon & Breach, 1972.

Miller, K. L., and Weidner, W. A. *CRC Handbook of Management of Radiation Protection Programs.* Boca Raton, FL: CRC Press, Inc., 1986.
This guidebook organizes the profusion of rules and regulations surrounding radiation protection into an easy-to-use, single-volume reference. Employee and public protection, acci-

dent prevention, and emergency preparedness are included in this comprehensive coverage. Whenever possible, information is presented in convenient checklists, tables, or outlines that enable readers to locate information quickly.

The book is ideal as a starting point for organizations that are establishing a program, and as a self-evaluation tool for existing programs.

Morgan, K. Z., and Turner, J. E., eds. *Principles of Radiation Protection: A Textbook of Health Physics.* New York: John Wiley & Sons, Inc., 1967 (Rep. 1973).

Contents include history of health physics; passage of heavy charged particles, gamma-rays, and x-rays through matter; radiation quantities and units; physical basis of dosimetry; detection and measurement of ionization; dose from electrons and beta-rays; dose from external sources; internal exposure; and radiation biology and biophysics.

National Council on Radiation Protection and Measurements. *Basic Radiation Protection Criteria.* (NCRP Report No. 39.) Bethesda, MD, 1971.

Includes chapters on radiation and man, radiation exposure conditions that may require consideration, basic biological factors, specific radiation effects, manifestations of overexposure in adults, bases for radiation protection standards, specific protection concepts or standards, dose limiting recommendations, and guidance for special cases.

Norwood, W. D. *Health Protection of Radiation Workers.* Springfield, IL: Charles C Thomas, 1975.

Okress, E. C., ed. *Microwave Power Engineering.* San Diego, CA: Academic Press, 1968.

This book introduces the electronics technology of microwave power and its applications. This technology emphasizes microwave (and eventually quantum) electronics for direct power utilization and transmission purposes rather than exclusively for information and communications applications. Essentially, microwave power can be divided into microwave heating, microwave processing, microwave dynamics, and microwave power transmission involving generation and power amplification, direct power utilization, and closed waveguide or radiation beam propagation for remote utilization and rectification.

Pearce, B. ed. *Health Hazards of VDTs.* New York: John Wiley & Sons, Inc., 1984.

The author takes an objective look at the highly emotional debate of the health, safety, and quality of working life of those using the new VDT technology. It examines the evidence for the alleged direct health hazards, such as facial rashes, radiation emissions, and cataracts, and indirect hazards, such as vision and lighting problems and emotional stress. The authors also consider some of the ways by which the working conditions of VDT users might be improved.

Permissible Levels of Toxic Substances in the Working Environment, sixth session of the Joint ILO/WHO Committee on Occupational Health, Geneva, 4-10 June, 1968; Occupational Safety and Health Series, Geneva 22, Switzerland: International Labour Office, Occupational Safety and Health Branch, 1970.

The Committee's report on reaching an international agreement on the basic principles for defining permissible limits; lists various countries; MACs.

Polk, C. *CRC Handbook of Biological Effects of Electromagnetic Fields.* Boca Raton, FL: CRC Press, Inc., 1986.

This book presents the current knowledge about the effects of electromagnetic fields on living matter. The three-part format covers dielectric permittivity and electrical conductivity of biological materials; effect of direct current and low frequency fields; and effects of radio frequency (including microwave) fields. The parts are designed to be consulted independently or in sequence, depending upon the needs of the reader. Useful appendices on measurement units and safety standards are also included.

Rahn, F. J. *A Guide to Nuclear Power Technology.* New York: John Wiley & Sons, Inc., 1984.

A resource for decision-making, the entire field is brought into focus, from the mining and production of reactor fuel materials, fuel utilization, and waste disposal, to safety procedures, regulation, and proliferation.

Sataloff, J., and Michael, P. L. *Hearing Conservation.* Springfield, IL: Charles C Thomas, 1973.

Shapiro, J. *Radiation Protection: A Guide for Scientists and Physicians,* 2nd ed. Cambridge, MA: Harvard University Press, 1981.

Provides the radiation user with information on protection, and compliance with governmental and institutional regulations regarding the use of radionuclides and radiation sources. Designed to obviate the need for reviews of atomic and radiation physics; the mathematics is limited to elementary arithmetical and algebraic operations.

Sliney, D., and Wolbarsht, M. L. *Safety with Lasers and Other Optical Sources.* New York: Plenum Publishing Corp., 1980.

This book thoroughly reviews current knowledge of biological hazards from optical radiation and lasers, presents current exposure limits, and provides a wealth of information required for the control of health hazards.

Stewart, D. C. *Data for Radioactive Waste Management and Nuclear Applications.* New York: John Wiley & Sons, Inc., 1985.

This comprehensive reference provides essential information for solutions to radioactive waste disposal problems. It provides the data needed for safely disposing of high-level liquid wastes, intermediate-level wastes, and packaged radioactive wastes.

Tables and charts provide data on waste container sizes, nuclear migration through rock, leachability, and exposure rates. There is information on nuclear reactor operational data concerning shielding materials, radiation damages, and decontamination.

Taylor, L. S. *Radiation Protection Standards.* Boca Raton, FL: CRC Press, Inc., 1971.

Summarizes and gives background information for various standards, including those of the ICRP, and NCRP, and the National Bureau of Standards. Arranged chronologically, 1928-1970.

Wasserman, D. E. *Vibration and the Workers' Health and Safety,* Technical Report No. 77. Cincinnati, Ohio: NIOSH, Division of Laboratories and Criteria Development, 1973.

Waxler, M., and Hitchens, V. M. *Optical Radiation and Visual Health.* Boca Raton, FL: CRC Press, Inc., 1986.
A focus on the parameters of ultraviolet, visible, and infrared radiations that could cause long-term visual health problems in humans. It reviews early research on radiation effects on the eye, and gives detailed attention to the hazardous effects of optical radiation on the retinal pigment epithelium and the photoreceptors. These data are further analyzed with regard to a number of long-term visual health problems. Epidemiological principles for studying the relationships between optical radiation and long-term visual health problems are reviewed, concluding with the implications for future research and radiation protection.

H. Ergonomics

Alexander, D. C., and Babur, M. P. *Industrial Ergonomics: A Practitioner's Guide.* Atlanta: Industrial Engineering & Management Press, 1985.
A very practical, job-oriented approach.

Astrand, P. O., and Rodahl, K. *Textbook of Work Physiology,* 3rd ed. New York: McGraw-Hill Book Co., 1986.

Barnes, R. *Motion and Time Study,* 7th ed. New York: John Wiley & Sons, 1980.

Bioastronautics Data Book, 2nd ed. NASA SP 3006, Washington, DC, 1973 (out-of-print).

Brammer, A. J., and Taylor, W. *Vibration Effects on the Hand and Arm in Industry.* New York: John Wiley & Sons, Inc., 1983.
This collection of papers explores the medical, acoustical, engineering, and legal aspects of exposure of the arm and hand to industrial vibration; and examines various methods of measuring exposure, the effects of such exposure, including the development of white fingers and dead hand, and the resulting dose-response relationships.

Chaffin, D. B., and Andersson, J. B. *Occupational Biomechanics.* New York: John Wiley & Sons, Inc., 1984.
The book provides an understanding of musculoskeletal mechanics to assist in modifying potentially debilitating conditions in the workplace. The book contains more than 200 information-packed diagrams demonstrating the mechanics of the musculoskeletal system as well as solutions to mechanical workload stress.

Chapanis, A. *Research Techniques in Human Engineering.* Ann Arbor, MI: University Microfilms International, Books-on-Demand, 1959.

Damon, A., Stoudt, H. W., and McFarland, R. A. *The Human Body in Equipment Design.* Cambridge, MA: Harvard University Press, 1966.
Applies principles of physical anthropology to the design of equipment for human use. Major areas of discussion include anthropometry and human engineering; biomechanics and equipment design; human body compositon and tolerance to physical and mechanical force; and design recommendations.

Eastman Kodak Company. *Ergonomic Design for People at Work,* vol. 1. Belmont, CA: Lifetime Learning Publications, 1983.
This book offers a practical discussion of workplace, equipment, environmental design and of the transfer of information in the workplace. It summarizes current data, experience, and thoughts assembled from the published literature, internal research, and observation by the Human Factors Section of the Eastman Kodak Company. The guidelines and examples are drawn from case studies. These principles have been successfully applied in the workplace to reduce the potential for occupational injury, increase the number of people who can perform a job, and improve performance on the job, thereby increasing productivity and quality.

Eastman Kodak Company. *Ergonomic Design for People at Work,* vol. 2. New York: Van Nostrand Reinhold, 1986.
Effective ergonomic design must take every aspect of human physiology into account. This volume shows how to design a work environment that is in conformance with basic human physiology. It goes well beyond factors such as height and reach and into the effects of muscular contractions, biological rhythms, and heart rate on worker capabilities. Human mechanics and job demand; the effects of work patterns on worker efficiency; and manual materials handling are discussed.
The book also provides methods to measure maximum human capacities, interpret heart rate, and analyze timed activities. Practical explanations of basic psychological factors help in taking steps to reduce unnecessary stress in the workplace.

Fitts, P. M., and Posner, M. J. *Human Performance.* Westport, CT: Greenwood Press, 1979.

Geldard, F. *The Human Senses,* 2nd ed. New York: John Wiley & Sons, 1973.

Grandjean, E. *Fitting the Task to the Man—An Ergonomic Approach,* 3rd ed. London, England: Taylor & Francis, Ltd., 1982.
This book contains chapters on man-machine systems and the questionnaire for controlling working conditions; a number of topical factors such as seating at work, heart rate as an indication of physical stress, monotony, daytime lighting, environmental climate in offices, and some recent advances in the assessment of heat stress are also considered.

Helander, M. G., ed. *Handbook of Human-Computer Interaction.* New York: Elsevier Science Pubishing Co., Inc., 1987.

Jacob, S. W., and Francone, C. A. *Structure and Function in Man,* 5th ed. Philadelphia: W. B. Saunders Co., 1982.

Konz, S. *Work Design: Industrial Ergonomics,* 2nd ed. Columbus, OH: Grid Publishing Co. (subs. of John Wiley & Sons), 1983.

Kroemer, K. H. E., Kroemer, H. J., and Kroemer-Elbert, K. E. *Engineering Physiology: Physiologic Bases of Human Factors/Ergonomics.* New York: Elsevier Science Publishing Co., Inc., 1987.

Kvalseth, T. *Ergonomics of Workstation Design.* Stoneham, MA: Butterworth Publishers, 1983.

This up-to-date and informative book is the combined effort of more than 30 experts from the United States, Europe, and Japan. It offers a clear presentation of the ergonomic aspects of the working environment in a way that can be put to immediate use by those responsible for the well-being and efficiency of the workforce in business and industry.

Problems associated with the factory, office, hospital, and transport have been the subject of intensive observation and research of the contributors, and their results and conclusions are presented clearly, with ample use of tables and diagrams. Industrial accidents and their prevention, and the avoidance of injury feature prominently. There is additional reference data provided.

Morgan, C. T., Cook, J. S., III, Chapanis, A., and Lund, M. W. *Human Engineering Guide to Equipment Design.* New York: McGraw-Hill Book Co., 1963.

A guide in human engineering that the designer can use in the same manner as handbooks in other areas to assist in solving design problems as they arise; the primary emphasis in the guide is on recommended design principles and practices in relation to general design problems rather than on the compilation of research data. However, research data can, if necessary, be included as a means of supporting or clarifying the design recommendations.

Murrel, K. F. H. *Ergonomics—Man in His Working Environment.* London, England: Chapman & Hall, 1980.

Parson, H. M. *Man-Machine System Experiments.* Baltimore: Johns Hopkins Press, 1972.

Proshansky, H. M. *Environmental Psychology: People and Their Physical Settings, 2nd ed.* New York: Holt, Rinehart & Winston, 1976.

Salvendy, G., ed. *Handbook of Human Factors.* New York: John Wiley & Sons, Inc., 1987.

Sanders, M. S., and McCormick, E. J. *Human Factors in Engineering and Design,* 6th ed. New York: McGraw-Hill Book Co., 1987.

Shephard, R. J. *Men at Work: Applications of Ergonomics to Performance and Design.* Springfield, IL: Charles C Thomas, 1973.

Survey of the physiology and psychology of work, biomechanics, and human factors engineering, with problems from industry, the home, teaching, and urban planning.

Singleton, W. T., Fox, J. F., and Whitfield, D., eds. *Measurement of Man at Work: Papers.* Philadelphia: Taylor & Francis, 1973.

Steindler, A. *Kinesiology: Of the Human Body under Normal and Of Pathological Conditions.* Springfield, IL: Charles C Thomas, 1977.

Thompson, C. W. *Kranz Manual of Structural Kinesiology,* 10th ed. St. Louis: C. V. Mosby Co., 1984.

Tichauer, E. R. *The Biomechanical Basis of Ergonomics Anatomy Applied to the Design.* New York: John Wiley & Sons, Inc., 1978.

Now professionals in manufacturing and service industries concerned with the health, welfare, and performance of people at work can apply functional anatomy to increase productivity while reducing on-the-job hazards.

Tichauer, E. R. "Human Factors Engineering," in *1984 McGraw-Hill Yearbook of Science and Technology.* New York: McGraw-Hill Book Co., 1983.

Van Cott, H., and Kinkade, R., eds. *Human Engineering Guide to Equipment Design,* rev. ed. New York: John Wiley & Sons, 1984.

Woodson, W. E. *Human Factors Design Handbook.* New York: McGraw-Hill Book Co., 1981.

Woodson, W. E., and Conover, D. W. *Human Engineering Guide for Equipment Designers,* 2nd rev. ed. Berkeley, CA: University of California Press, 1966.

The greatest expansion in this new revision has occurred in the first parts of the Guide. The first chapter, "Design Philosophy," is entirely new, having replaced the former introductory section. Chapter 2 is a considerably expanded version of the original material; however, an attempt has been made to retain the original direct format, which seems to have been appreciated by most designers.

The chapter on "Body Measurement" has been revised appreciably and made more practical from the designer's point of view. This change is a reflection of the application experience of the writers in working very closely with aerospace and weapon system designers since the beginning of the Jet Age. Revisions in the remaining parts of the book are less extensive, but reflect many of the changes brought about by more recent research—especially in the area of man-in-space and in industrial applications.

I. Biological

Hall, T. G. *Diseases Transmitted from Animals to Man,* 6th ed. Springfield, IL: Charles C Thomas, 1975.

Laskin, A. L., and Lechevalier, H., eds. *CRC Handbook of Microbiology,* 2nd ed: Vol. 1, *Bacteria;* Vol. 2, *Fungi, Algae, Protozoa, and Viruses.* Boca Raton, FL: CRC Press., Inc., 1978.

J. Chemical

Alliance of American Insurers. *Handbook of Organic Industrial Solvents* (Technical Guide No. 6), 5th ed. Schaumburg, IL: AAI, 1980.

This handbook was developed to assist the safety professional in his analysis of problems involving the use of solvents. A list of common solvents was compiled along with pertinent data needed in evaluating hazards.

Bretherick, L. *Handbook of Reactive Chemical Hazards,* 3rd ed. Stoneham, MA: Butterworth Publishers, 1985.

The majority of the book is devoted to specific information on the stability of the listed compounds, or the reactivity of mixtures of two or more under various circumstances. Each description of an incident or violent reaction gives references to the original literature. Each chemical is classified on the basis of similarity in structure or reactivity and each class is described in a separate section. Many cross references throughout both sections emphasize similarity in compounds or incidents not obviously related, and an introductory chapter identifies the underlying principles which govern the complex subject of reactive chemical hazards. In many cases, quantitative thermodynamic data on energies of decomposition or reaction are included. An appendix lists the fire-related properties of the higher risk materials and there are comprehensive indexes and a glossary of specialized terms.

Castleman, B. I. *Asbestos: Medical and Legal Aspects.* New York: Harcourt Brace Jovanovich, Inc., 1984.

This volume might well be the most complete account in print of the genesis of today's asbestos disease epidemic. It draws upon published literature of all kinds, supplemented by unpublished documentation. The social and industrial responses to the hazard are analyzed, and lessons in prevention are drawn that apply to a wide range of hazards.

Chissick, S. S., and Derricott, R., eds. *Asbestos: Properties, Applications, and Hazards.* New York: John Wiley & Sons, Inc., 1983.

This volume is a comprehensive compendium of published information on asbestos from world sources. It is a useful sourcebook on the literature of the subject for both experts and nonexperts.

Clary, J. J., Gibson, J. E., and Warity, R. S., eds. *Formaldehyde: Toxicology, Epidemiology, Mechanisms.* New York: Marcel Dekker, Inc., 1983.

Advances in the state-of-the-art in toxicology, the study of toxic effects, mechanisms of toxicity, and hazards from exposure are described. The book spotlights the critical problem of extrapolating results obtained in animal experiments to humans, and presents major information needed for sensible regulation as well as giving biological and toxicological principles.

Cold Cleaning with Halogenated Solvents (STP 403A). Philadelphia: American Society for Testing and Materials, 1981.

This book has important information of flammability, health hazards (such as inhalation, skin contact, and ingestion), and container handling and storage. The manual would be useful to industries, such as electrical, electronic, electromechanical, automotive and equipment maintenance; metal and metal fabrication; aerospace and nuclear.

Collings, A. J., and Luxon, S. G., *Safe Use of Solvents.* Orlando, FL: Academic Press, Inc., 1982.

Solvents have varying uses in different areas of modern industry, ranging from the extraction of petroleum products, foods, and natural products, to their use as coupling agents in adhesives and surface coatings, and as carriers for perfumes, essences, and insecticides. Thus, government, industry, and trade unions have an active interest in how solvents are manufactured, stored, transported, used, and disposed of in a safe manner.

Compressed Gas Association. *Handbook of Compressed Gases,* 2nd ed. New York: Van Nostrand Reinhold, 1981.

Discusses 49 widely used compressed gases in terms of their properties, methods of manufacture, commercial uses, and physiological effects. Includes data relative to the materials of construction required for all types of compressed gas installations, equipment, and containers. Also includes a chapter on safe handling of compressed gases, as well as information on hazardous materials regulations.

Englund, A., Ringen, K., and Mehlman, M. A., eds. *Occupational Health Hazards of Solvents.* Princeton, NJ: Princeton Scientific Publishers, Inc., 1982.

Western European and North American representatives from labor, industry, government, and academia present data on the toxicity of the various substances used in solvent-based materials, as well as expert scientific discussions of the health hazards presented by these substances.

Fawcett, H. H., and Wood, W. S. *Safety and Accident Prevention in Chemical Operations,* 2nd ed. New York: Interscience Publishers, 1982.

An excellent description of the hazards and respiratory protective devices available is presented. This volume is a source book or guide, rather than a mere tabulation of hazardous materials. Comprehensive treatment is enhanced by many photographs and authoritative references. It should be included in the libraries of schools and colleges where chemistry and chemical engineering are taught. In addition, the small user of chemicals, who frequently has had no orientation in chemical safety, should welcome the availability of this knowledge.

Fawcett, H. H., ed. *Hazardous and Toxic Materials: Safe Handling and Disposal.* New York: John Wiley & Sons, Inc., 1984.

The book presents a balanced view of the latest scientific information about hazardous and toxic materials, their containment, and their availability to man, animals, and plants. It takes a close look at the laboratory where most of these materials originate and then proceeds into fire and explosion hazards and their detection. It details the personal protection that is necessary in the waste site investigation and clean-up environments. Assuming a positive problem-solving approach the book emphasizes the importance of alternative disposal methods and shows how to control and prevent future environmental disasters.

Fire Protection Guide on Hazardous Materials, Quincy, MA: National Fire Protection Association, 1978.

Green, A. E., ed. *High Risk Safety Technology.* New York: John Wiley & Sons, Inc., 1982.

The techniques for evaluating the safety of technological systems are described in a logical and quantitative form. Risk evaluation and criteria are treated in such a way as to make this a sourcebook for developing ideas in a scientific and analytical manner in those areas of technology where high risks of any nature are involved.

Hazard Assessment of the Electronic Component Manufacturing Industry. Cincinnati: National Institute for Occupational Safety and Health (NIOSH), 1985.

The hazard assessment is the first step in the NIOSH process of evaluating the occupational environment of industry operations. This type of investigation is based on health and process data collected from the entire industry, including trade organizations, labor groups, and management. Tripartite meetings with industry, government, and trade unions followed by a literature review and facility surveys provided the information required to perform the assessment.

Hoffman, J. M., and Maser, D. C., eds. *Chemical Process Hazard Review.* ACS Symposium Series No. 274. Washington, DC, American Chemical Society, 1987.

James, D. *Fire Prevention Handbook.* Stoneham, MA: Butterworth Publishers, 1986.

This practical handbook explains clearly and without technical jargon the causes of fire and the processes by which it spreads. It sets down the basic rules of fire prevention and shows how good housekeeping, training, and general motivation can contribute to the prevention of fire.

LeFeure, M. J. *First Aid Manual for Chemical Accidents, for Use with Pharmaceutical Chemicals.* New York: Van Nostrand Reinhold Co., Inc., 1980.

This manual deals entirely with first aid at the workplace, in which industrial chemicals are used.

Levadic, B., ed. *Definitions for Asbestos and Other Health-Related Silicates.* Philadelphia: American Society for Testing and Materials, 1984.

This book addresses the need for clarification of terminology used to refer to groups of minerals; namely, asbestos, silica, and talc. The volume includes a comprehensive glossary for health-related silicates defining 46 terms. A practical handbook for the scientific community, regulatory agencies, medical and legal professions, industry management, and the media.

McKinnon, G. P., and Tower, K., eds. *Fire Protection Handbook,* 16th ed. Quincy, MA: National Fire Protection Association, 1986.

Much new material has been added, recognizing the many advances made in fire protection technology since the previous edition was published. New fire problems, and the solutions to them, that in the last decade were only then beginning to make themselves known, are now deserving of extensive attention. (Highrise buildings, for example, received only passing mention in the last edition with no direct reference to their potential as a source of hazard to life.)

O'Connor, C. J., and Lirtzman, S. I. *Handbook of Chemical Industry Labeling.* Park Ridge, NJ: Noyes Data Corp., 1984.

The need for informative labeling in workplace, transportation, distribution and disposal operations has been recognized by all levels of government agencies. Society-at-large has demanded increased information on chemical products, and organized labor has actively pursued an improved hazardous label communication program.

Powers, P. W. *How to Dispose of Toxic Substances and Industrial Wastes.* Park Ridge, NJ: Noyes Data Corp., 1976 (out-of-print).

This book discusses all recognized and allowed ultimate disposal methods in detail and contains a long list of specific recommendations for specific substances plus alternative disposal or recovery methods.

In this book are condensed vital data that are scattered and often difficult to assemble. Important techniques are interpreted and explained by actual case histories. This condensed information will enable you to establish a sound background for action towards disposal of toxic and hazardous materials with safety.

Rappe, C., Choudhary, G., and Keith, L. H. *Chlorinated Dioxins and Dibenzofurans in Perspective.* Chelsea, MI: Lewis Publishers, Inc., 1986.

This book provides the latest human exposure data and the most advanced analytical techniques, developed in the continuing effort against contamination by chlorinated dioxins and dibenzofurans.

Skoog, R. F., and Twombly, R. C. *Asbestos Abatement Training Manual.* Washington, DC: Asbestos Abatement Council/AWCI, 1985.

This manual offers a comprehensive overview of the procedures necessary to correct those conditions that promote an asbestos hazard to the environment and to the population at large.

Turk, A., Johnson, J. W., Jr., and Moulton, D. G., eds. *Human Responses to Environmental Odors.* New York: Academic Press, 1974.

The purpose of this volume is to bring together some of the more recent approaches to the study of the human olfactory response in which both sensory and physico-chemical aspects are presented.

Zabetakis, M. G. *Safety with Cryogenic Fluids.* Ann Arbor, MI: University Microfilms International, Books-on-Demand, 1967.

This monograph was prepared in an effort to present in concise form the principles of safety that are applicable to the field of cryogenics. Thus, while it includes safety rules, design data, first aid and hazard control procedures, emphasis has been placed on basic principles. An appreciation of these principles permits an individual to conduct a safe operation under a wider variety of conditions than is possible if he is familiar only with a list of safety rules.

K. Pollution and hazardous waste

Bhatt, H. G., Sykes, R. M., and Sweeney, T. R. *Management of Toxic and Hazardous Wastes.* Chelsea, MI: Lewis Publishers, Inc., 1985.

The demand for cleaning of hazardous waste disposal sites has grown since the passage of Superfund. This book presents the important aspects of hazardous waste management. Attention is focused on waste treatment and recycling, risk assessment, public participation and land disposal. A section on legal considerations provides pointers on precautions to be taken to minimize legal liabilities.

Calvert, S. and Englund, H. M. *Handbook of Air Pollution Technology.* New York: John Wiley & Sons, Inc., 1984.
The text provides an up-to-date guide for defining, analyzing, and controlling a wide variety of air pollution problems. Experts in the field discuss methods for the control of gases, gaseous pollutant characteristics, the effects of air pollutants on the atmosphere and materials. It contains valuable material for the professional concerned with reducing and controlling the often invisible pollutants existing in our atmosphere.

Canter, L. W. *Acid Rain and Dry Deposition: Sourcebook.* Chelsea, MI: Lewis Publishers, Inc., 1986.
This book deals with the terrestrial and aquatic impacts of both acid rain and the dry deposition of metals and organics.

Canter, L. W. *Environmental Impacts of Agricultural Production Activities.* Chelsea, MI: Lewis Publishers, Inc., 1986.
A valuable resource and research tool, this book summarizes and analyzes the environmental impacts of agricultural practices relative to water and soil, air quality, noise, and solid wastes. An evaluation scheme for examining environmental impacts is included along with comparative information on the impacts of various tillage practices. The book includes a bibliography of 300 abstracts.

Canter, L. W. *River Water Quality Monitoring.* Chelsea, MI: Lewis Publishers, Inc., 1985.
This practical guide gives the information required to plan and conduct river water quality studies. These studies are necessary to etablish baseline conditions, set water quality criteria and standards, monitor temporal changes, and determine the impacts of specific projects and developments.

Canter, L. W., and Knox, R. C. *Ground Water Pollution Control.* Chelsea, MI: Lewis Publishers, Inc., 1986.
This comprehensive work covers thoroughly technologies for ground pollution in part one and deals in depth with aquifer restoration decision-making in part two. Part three gives detailed abstracts of 225 selected references and a range of case studies.

Canter, L. W., Knox, R. C., and Fairchild, D. M. *Ground Water Quality Protection,* Chelsea, MI: Lewis Publishers, Inc., 1986.
Considered by the EPA to be one of the "major environmental issues of the 1980s," groundwater supplies a large majority of the water we use. The book deals with this critical problem and the action to be taken to prevent despoliation of the aquifers where this water is now found, because once contaminated, an aquifer is difficult to decontaminate.

Dawson, G. W., and Mercer, B. W. *Hazardous Waste Management.* New York: John Wiley & Sons, Inc., 1986.
This reference work deals with hazardous waste materials. The book gives cost-effective practical approaches as well as technological and policy considerations of facilities design, waste site reclamation treatment processes, incineration alternatives, land fill, salt dome disposal, as well as survey techniques.

Gammage, R. B., and Kaye, S. V. *Indoor Air and Human Health.* Chelsea, MI: Lewis Publishers, Inc., 1985.
This book deals with five principal classes of indoor pollutants: radon, microorganisms, passive cigarette smoke, combustion products, and organics. It examines each pollutant from the viewpoint of measurement and source; characteristics; habitat studies; health effects; risk analysis and future need.

Godish, T. *Air Quality.* Chelsea, MI: Lewis Publishers, Inc., 1985.
This book on air quality provides comprehensive coverage of the subject with special treatment of atmospheric effects, effects on humans, plants, and materials, as well as pollution regulation and motor vehicle emission control.

Levine, S. P., and Martin, W. F. *Protecting Personnel at Hazardous Waste Sites.* Stoneham, MA: Butterworth Publishers, 1984.
The procedures and programs outlined in the text can be applied to disposal operations already under way, and will also be useful in correcting old problems due to improperly managed hazardous chemicals. Everyone responsible for the protection of workers at hazardous waste sites should have a copy of this book for standard reference in the field.

Lioy, P. J., and Daisey, J. M., eds. *Toxic Air Pollution.* Chelsea, MI: Lewis Publishers, Inc., 1986.
The book deals with the characteristics and dynamics of noncriteria pollutants. It stems from the work of the authors on the Airborne Toxic Element and Organics Substances (ATEOS) project, which is the first successful attempt to develop an understanding of the subject. The text reports the results and major conclusions from a two-year study in three different urban environments.

Martin, E. J., and Johnson, J. H. *Hazardous Waste Management Engineering.* New York: Van Nostrand Reinhold, 1987.
The book includes discussions on the following topics: hazardous waste and chemical substances; exposure and risk assessment; chemical, physical, and biological treatment of hazardous waste; incineration of hazardous waste; storage of hazardous waste; land disposal of hazardous waste; hazardous waste leachate management; hazardous waste facility siting.

Mudrack, K., and Kunst, S. *Biology of Sewage Treatment and Water Pollution Control.* New York: John Wiley & Sons, Inc., 1986.
Biological methods for treating pollution in industrial-use water and in natural bodies of water receiving sewage are presented. Clear explanations of the microbiological principles applicable to sewage treatment are presented. The book also contains the step by step process of the biological treatment of sludge and the biotechnological processes for treating concentrated effluents to recover useful materials.

Payne, J. R., and Phillips, C. R. *Petroleum Spills in the Marine Environment.* Chelsea, MI: Lewis Publishers, Inc., 1985.
Subtitled "The Chemistry and Formation of Water-in-Oil Emulsions and Tar Balls," this book covers research completed between 1981 and 1985. It includes reviews of recently completed studies, sitings and investigations of spills-of-opportunity,

results of arctic and subarctic oil weathering experiments and observations of the oil weathering experiments and observations of the behavior of crude oil in the presence of ice.

Robinson, W. D. *The Solid Waste Handbook, A Practical Guide.* New York: John Wiley & Sons, Inc., 1986.

The handbook covers all aspects of solid waste management, including public policy, implementation of hazardous waste administration guidelines. Changes in waste management practices mandated by RCRA through 1984 are covered. It also examines the state-of-the-art practices for land disposal, resource recovery, and decision making in both the public and private sectors.

Wadden, R. A., and Scheff, P. A. *Indoor Air Pollution.* New York: John Wiley & Sons, Inc., 1982.

This book provides guidance to evaluate and control indoor air pollution caused by reduced ventilation and energy saving measures. The emphasis is on the environment of domestic and public building, although the material is equally applicable to many indoor spaces. The text also includes state-of-the-art information on indoor pollution hazards, methods for measuring and predictive models.

L. Control

Air Pollution Reference Library. Cincinnati: ACGIH, latest edition.

This is a compilation of references on air pollution and includes lists of books, handbooks, journals, periodicals, and other references.

Alden, J. L., and Kane, J. M. *Design of Industrial Ventilation Systems,* 5th ed. New York: Industrial Press, Inc., 1982.

This edition covers the inter-related areas of general exhaust ventilation and air makeup supply. The manual contains the information required to design, purchase, and operate an exhaust system that will comply with government regulations for the protection of employees and that will minimize atmospheric pollution. Like the earlier editions, the book also deals with local exhaust systems.

American Foundrymen's Society (AFS). *Engineering Manual for Control of In-Plant Environment in Foundries.* Des Plaines, IL: AFS, 1956.

An extremely well-organized and informative book covering foundry ventilation and foundry hygiene problems. Section 2, "Exhaust Hoods and Exhaust System Design," contains some excellent information on designing exhaust systems. A step-by-step explanation of the nomograph methods of designing exhaust system is covered; also a step-by-step procedure for using the widely accepted friction chart for designing exhaust systems is covered.

American Industrial Hygiene Association. *Heating and Cooling for Man in Industry,* 2nd ed. Akron, OH: AIHA, 1975.

A wealth of information on this subject is presented here on methods of evaluation and control of heat and cold stress conditions.

Borup, B. *Pollution Control for the Petrochemicals Industry.* Chelsea, MI: Lewis Publishers, Inc., 1986.

A concise presentation of all aspects of pollution control in the petrochemicals industry is given. Solutions to water, air, solid, and hazardous wastes problems are summarized. Energy use and conservation are also discussed.

Buonicore, A. J., and Theodore, L. *Industrial Control Equipment for Gaseous Pollutants,* Vol. 1. Boca Raton, FL: CRC Press, Inc., 1975.

It is the intent of this book to offer the reader the fundamentals and principles of control equipment for gaseous pollutants with appropriate practical applications and to serve as an introduction to the specialized and more sophisticated texts in this area.

Burton, J. D. *Industrial Ventilation, A Self Study Companion,* 3rd ed. Cincinnati: American Conference of Government Industrial Hygienists, 1986.

This book is the text for the American Board of Industrial Hygiene-approved Industrial Ventilation Home Study Course.

Constance, J. D. *Controlling In-Plant Airborne Contaminants.* New York: Marcel Dekker, Inc., 1983.

This book covers the control of each of the various types of in-plant airborne contaminants. Emphasizing the interaction of theory with practical considerations, the book details proven control methods, allowing for the evaluation, selection, design, and implementation of the most economic and effective option.

Cralley, Lewis J., and Cralley, L. V., eds. *Industrial Hygiene Aspects of Plant Operations* (in three volumes). New York: Macmillan Publishing Co., 1982, 1984, 1986.

Volume I: *Process Flows* details all of the basic operations involved in the process flow of the material from inception through production in the manufacture of a cross section of representative industrial products.

Volume 2: *Unit Operation and Products Fabrication* fills an especially important and urgent need with its flowsheet style of presentation. Contributors discuss unit operations as distinct entities along an industry-wide concept and then cover the operations and procedures for assembling parts and materials into final products.

Volume 3: *Engineering Considerations in Equipment Selection, Layout, and Building Design* provides a comprehensive, up-to-date, and useful reference on recognizing, measuring, and controlling job-related health hazards.

Deisler, Paul F., Jr. *Reducing the Carcinogenic Risks in Industry.* New York: Marcel Dekker, Inc., 1984.

This work examines various private and government policy approaches to risk reduction. It discusses government regulatory and industry self-regulatory methods in the U.S. and Europe, spotlighting specific examples and accomplishments.

Deitz, V. R., ed. *Removal of Trace Contaminants from the Air.* (Symposium Series No. 17). Washington, DC: American Chemical Society, 1975.

Sixteen chapters provide critical and in-depth coverage of air pollution characterization and removal. This compendium stresses interaction among particulates and gas phase contaminants, pesticides, occupational contaminants, cigarette smoke and aerosol filtration, sulfur dioxide, trace gas adsorption, nitrogen oxides, and high ozone concentrations.

Fullman, J. B. *Construction Safety, Security, and Loss Prevention.* New York: John Wiley & Sons, Inc., 1984.

Actual examples illustrate successful solutions for each phase of construction. The author takes you step by step from preconstruction safety preparations to site preparation, excavation, and mechanical systems inspection. He also tells you how to avoid structural collapse, reduce insurance premiums, and handle public relations.

Heating, Ventilating, and Air Conditioning Guide. New York: Society of Heating, Ventilating, and Air-Conditioning Engineers, Inc., published annually.

Hemeon, W., ed. *Plant and Process Ventilation,* 2nd ed. Ann Arbor, MI: University Microfilms International, Books-on-Demand.

This book was written to assist those individuals charged with the responsibility of designing exhaust ventilation systems. The first half is concerned with methods for analyzing a factory ventilation problem and explains the dynamics of the air-polluting process to determine in what manner the air is to be channeled through the space in question.

Industrial Ventilation, A Manual of Recommended Practice, 19th ed. Lansing, MI: Committee on Industrial Ventilation, ACGIH, 1986.

A practical reference for use by designers, contractors, engineers, industrial hygienists, and anyone else concerned with industrial ventilation. Basic ventilation principles and sample calculations are presented in a clear and simplified manner, enabling many users to "do-it-yourself." Text provides practical solutions to the entire range of ventilation problems.

LaDou, J. L., ed. *Introduction to Occupational Health and Safety.* Chicago: National Safety Council, 1986.

Written for the safety professional who must start the health portion of a health and safety program; for the manager who must administer and support them; and for the medical professionals who want to learn more about occupational health.

Lees, R., and Smith, A. F., eds. *Design, Construction, and Refurbishment of Laboratories.* New York: John Wiley & Sons, Inc., 1984.

Laboratory design specialists solve many of the problems of designing new laboratories and refurbishing older ones. They analyze the specific difficulties facing research facilities for chemistry, microbiology, engineering, and other fields. Full consideration is given to improved health and safety standards as well as to the effective and efficient use of space. This reference examines design flexibility, special features required in tropical climates and the Third World, and furniture design, enabling the reader to compare modernization and new construction costs.

McDermott, H. *Handbook of Ventilation for Contaminant Control,* 2nd ed. Stoneham, MA: Butterworth Publishers, 1985.

The volume covers the following topics: indoor air pollution; OSHA ventilation standards; hazard assessment; how local exhaust systems work; hood selection and design; air cleaner selection; ventilation system design; fans; ventilation for high toxicity or high-nuisance contaminants; saving ventilation dollars; testing; and solving ventilation system problems.

Noll, K. E., Haas, C. N., Schmidt, C., and Kodukula, P. *Recovery, Recycle and Reuse of Industrial Wastes.* Chelsea, MI: Lewis Publishers, Inc., 1985.

Designed to explain underlying concepts, advantages, and, in certain instances, disadvantages of recovery, recycle, and reuse. Examples of recovery technology applications are provided in addition to discussions of economic implications and nontechnical overviews.

Obayashi, A. W., and Gorgan, J. M. *Management of Industrial Pollutants by Anaerobic Processes.* Chelsea, MI: Lewis Publishers, Inc., 1985.

This book will interest those industrial and environmental engineers who are familiar with fermentation and similar anaerobic process technology and its economic potential. The text incorporates a broad spectrum of information, including an overview of anaerobic technology, as well as applications for industrial pollutants and associated toxicity effects. An innovative recovery application to sulfur waste management is included as an example.

Patterson, J. W., ed. *Metals Speciation, Separation and Recovery.* Chelsea, MI: Lewis Publishers, Inc., 1986.

By and for engineers and scientists, this book covers all aspects of metals chemistry, separation chemistry, and metals separation processes. State-of-the-art papers give new and recent developments and future research needs.

Pfeiffer, J. B., ed. *Sulfur Removal and Recovery from Industrial Processes.* (Advances in Chemistry Series No. 139.) Washington, DC: American Chemical Society, 1975.

Sixteen chapters form a consolidated reference source of sulfur removal and recovery methods concentrating on recovery techniques from sources other than power plant stacks. Emissions from smelter gas streams and Claus units are discussed, and seven scrubbing processes are described. Companion volume is No. 140 *New Uses of Sulfur.*

Pipitone, D. A., ed. *Safe Storage of Laboratory Chemicals.* New York: John Wiley & Sons, Inc., 1984.

This volume provides full information on fundamental principles of chemical storage. Readers will learn the variety and type of chemical storage hazard, degrees of hazards, and practical guidelines for analyzing and correcting deficiencies. Problems related to hazardous, flammable, unstable, and incompatible chemicals are discussed. Information on facts about surveys and inspections, storage systems, computerized warehousing functions, and choosing a computer for its information retrieval functions are also included.

Pitt, M. J., and Pitt, E. *Handbook of Laboratory Waste Disposal.* New York: John Wiley & Sons, Inc., 1985.

This safety-first guide shows approved procedures for fume extraction, laboratory drainage, burning and incineration, and clearing out a dead store. Italicized cautionary notices alert readers to the most frequent mistakes made in handling dangerous wastes. Special attention is given to materials recycling, emergency procedures, and the disposal of carcinogens, and cytotoxic agents. Handy chemical tables list hazardous waste specifications.

Rajhans, G., and Blackwell, D. S. *Practical Guide to Respirator Usage in Industry.* Stoneham, MA: Butterworth Publishers, 1985.

This practical reference will greatly simplify the task of anyone responsible for worker protection in industry. The table of contents includes—overview of respiratory hazards and evaluation; respirator types and limitations; criteria for selection and fitting; administration and training; maintenance and care; medical supervision; criteria for respiratory protection program; industrial applications; research needs.

One chapter of particular interest makes use of case studies to point out common mistakes in judgment or choice which can lead to severe health problems or even death. Each study is followed by an explanation of how the specific situation could have been avoided or remedied.

Rice, R. G., Bollyky, L. J., and Lacy, W. J. *Analytical Aspects of Ozone Treatment for Water and Wastewater.* Chelsea, MI: Lewis Publishers, Inc., 1986.

This book, written by experts in the field of ozone technology, is divided into the following sections: 1. Fundamental Parameters; 2. Analysis of Ozone in Aqueous Solution; 3. Analysis of Ozone in Gaseous Phase; 4. Instrumental Methods of Analysis and Monitoring of Ozone; 5. Automated Procedures for Controlling Ozonation Processes.

Ruch, W. E., and Held, B. J. *Respiratory Protection—OSHA and the Small Businessman.* Ann Arbor, MI: Ann Arbor Science, 1975.

This volume was designed for use by the small businessman to provide a respiratory-protection program for his employees. The Occupational Safety and Health Act requires that the employer provide a respiratory protection program for the employees whenever they must work in a hazardous atmosphere which cannot be controlled by engineering methods. This volume should provide the necessary information to aid the employer in meeting OSHA requirements, thereby providing a safer workplace for the employee.

Salvato, J. A. *Environmental Engineering and Sanitation,* 3rd ed. New York: John Wiley & Sons, Inc., 1982.

This guide covers virtually every problem encountered in sanitary and environmental engineering and adminstration. The third edition includes updated and expanded coverage of alternate on-site sewage disposal, water reclamation and reuse; protection of groundwater quality; control and management of hazardous waste; resource recovery and energy conservation. There are also sections on food sanitation and integrated pest management.

Schifftner, K. C. *Wet Scrubbers: A Practical Handbook.* Chelsea, MI: Lewis Publishers, Inc., 1983.

This handbook gives a straightforward and concise guide to the selection, engineering basics, applications, and maintenance of wet scrubbers, especially valuable in product recovery and air pollution control. Of special interest are sections on process design, mechanical design of scrubbers and related subsystems, system operations, control and maintenance descriptions, and how regulations work.

Schwope, A. D., Costas, P. P., Jackson, J. D., and Weitzman, D. J. *Guidelines for Selection of Chemical Protective Clothing,* 2nd ed. Cincinnati: American Conference of Governmental Industrial Hygienists (ACGIH), 1983.

Chemical protective clothing (CPC) is a key element in minimizing the potential for worker exposure to chemicals. Volume 1 contains CPC performance information, and provides the basic information required to select, order, and intelligently use CPC. The sources are identified in volume 2. The second volume also includes recommendations for actions that would benefit the CPC selection and use.

Sittig, M. *How to Remove Pollutants and Toxic Materials from Air and Water: A Practical Guide.* Park Ridge, NJ: Noyes Data Corporation, 1977.

Stoner, D. L. *et al. Engineering a Safe Hospital Environment.* New York: John Wiley & Sons, Inc., 1982.

The book focuses on safety: electrical, general and building, mechanical and laboratory, and radiation; control: infection, environmental; fire protection; and safe use and operation of medical equipment. The book provides summaries of standards, codes, and regulations as well as safety checklists, which enable readers to assimilate the information easily and install routine practices essential to an adequate safety program.

Tavlarides, L. L. *Process Modifications for Industrial Pollution Source Reduction.* Chelsea, MI: Lewis Publishers, Inc., 1985.

This text describes one of the field's most innovative and potentially advantageous modern concepts, process modification. In many industries process modification is a technically feasible and economically attractive approach to industrial waste management.

M. Encyclopedias and handbooks

Agricultural Health and Safety Resource Directory. Cincinnati: ACGIH, 1984.

This directory was developed through a questionnaire mailed to individuals and organizations. It is considered comprehensive but not exhaustive. The topics chosen were those affecting producers rather than processors and other agriculturally related industries.

Calvert, S., and Englund, H. M., eds. *Handbook of Air Pollution Technology.* New York: John Wiley & Sons, Inc., 1984.

Essential technical principles, design ideas and methods, practical examples, and other valuable information are included. This book is of interest to all concerned with reducing and controlling the often invisible pollutants that exist in our atmosphere.

Considine, D. M., ed. *Encyclopedia of Chemistry,* 4th ed. New York: Van Nostrand Reinhold, 1984.

The revised and enlarged edition incorporates the advancements in chemistry achieved during the past decade and fully updates the traditional aspects of the field. Nearly 90 percent of the text is new—approximately 1,300 alphabetically arranged entries, each prepared by an expert in the field, give you valuable facts on processing and use of natural and synthetic chemical solids, liquids, and gases.

Biomass energy, chromatography, exomosis, freeze-concentrating, graphite structures, hypobaric systems, immunochemistry, macromolecules, recombinant DNA, and many other vital topics are covered.

Encyclopedia of Occupational Health and Safety. Geneva, Switzerland: International Labour Office, 1983.

The encyclopedia is a reference tool providing information about causes of accidents, resultant illnesses, descriptions of occupational terms, and recommendations for preventive measures. This information is presented in a practical manner within the context of economic and social conditions, covering specific hazards encountered in every working environment and their prevention. The encyclopedia is in two volumes.

Handbook of Hazardous Materials, 2nd ed. (Technical Guide No. 7.) Schaumburg, IL: Alliance of American Insurers, 1983.

Handbook of Hazardous Waste Regulation. Vol. 2: *How To Protect Employees During Environmental Incident Response—Official EPA Health and Safety Guidance.* Madison, CT: Bureau of Law and Business, Inc., 1985.

This guidance manual provides needed information for establishing and maintaining health and safety protection for response personnel.

Handbook of Organic Industrial Solvents, 5th ed. (Technical Guide No. 6.) Schaumburg, IL: Alliance of American Insurers, 1980.

Hawley, G. *The Condensed Chemical Dictionary,* 11th ed. New York: Van Nostrand Reinhold, 1987. (Now called *Hawley's Condensed Chemical Dictionary.* Revised by: N. Irving Sax and Richard J. Lewis, Jr.)

This edition defines many new terms covering thousands of chemicals and chemical phenomena. It provides information on an abundance of topics, including pollution and waste control, chemical manufacturing equipment, energy sources and their potential.

Three distinct types of information are provided—(1) technical descriptions of chemicals, raw materials, and processes, (2) Expanded definitions of chemical entities, phenomena, and terminology, (3) Descriptions or identifications of a broad range of trademarked products used in the chemical industries. Aspects of hazardous materials, such as toxicity and flammability; explosion risks; radioactive hazards; oxidizing and corrosive properties; and tissue irritants are described.

Hazardous Waste Compliance Checklists for Supervisors. Madison, CT: Bureau of Law and Business, Inc., 1985.

This booklet contains checklists on a number of different areas associated with hazardous waste. There are over 35 checklists provided, including: hazardous spills checklist; ethylene oxide and toluene checklist; benzene checklist; decontamination checklist.

King, R., and Hudson, R. *Construction Hazard and Safety Handbook.* Stoneham, MA: Butterworth Publishers, 1985.

From the earliest times, construction has been one of the most dangerous activities undertaken by man. Despite efforts to make construction work safer, real improvements have lagged behind those of other industries. This book was written in an attempt to identify and analyze the hazards of construction, not only in buildings, but also industrial plant construction, offshore work, and civil engineering projects. It covers hazards of health, accidental injury, fire and explosion, and much more. The authors also offer general guidance in improving safety, training, and hazard monitoring.

Kirk-Othmer Concise Encyclopedia of Chemical Technology. New York: John Wiley & Sons, Inc., 1984.

This book is an authoritative abridgement of the 26-volume third edition of Kirk-Othmer. This concise encyclopedia covers all the topics of the original set, and features the contributions of the original contributors or their colleagues. Complete indexing and cross-referencing makes it simple to retrieve exactly what you need.

Lederer, W. H. *Regulatory Chemicals of Health and Environmental Concern.* New York: Van Nostrand Reinhold Co., 1985.

This sourcebook contains detailed guidance to important regulations, standards, and other related information regarding chemicals, of health and environmental concern. The book covers, in alphabetical order, approximately 2,000 chemicals that are listed in federal regulations, voluntary standards, and consent decrees.

Lowry, G. G., and Lowry, R. C. *Handbook of Hazard Communication and OSHA Requirements.* Chelsea, MI: Lewis Publishing Co., Inc., 1985.

This book provides guidance, explanation, and critical evaluation for the thousands of companies required to comply with the OSHA Hazard Communication Standard. The volume will help companies of all sizes to comply with the standard, which was created to protect employee health and provide safe working conditions for individuals who may come in contact with hazardous substances.

McGraw-Hill Encyclopedia of Science and Technology, 5th ed. 15 vols. New York: McGraw-Hill Book Co., 1982.

The Merck Index. An Encyclopedia of Chemicals and Drugs. 10th ed. Rahway, NJ: Merck & Co., 1983.

An essential work which includes information on toxicology.

Miller, D. E., ed. *Occupational Safety, Health and Fire Index,* vol. 1. Ann Arbor, MI: University Microfilms International, Books-on-Demand.

This is a comprehensive and unified reference listing of the many safety, health and fire codes, standards, guides, and publications in the field.

Parker, J. F., Jr., and West, V. R. *Bioastronautics Data Book,* 2nd ed. Washington, DC: National Aeronautics and Space Administration, 1973.

This revision of the *Bioastronautics Data Book* was prepared in order to bring together the large body of human research information generated in recent years and to present it in a form suitable for engineers and others concerned with the development and evaluation of modern systems.

Perry, J. H., et al., eds. *Chemical Engineers' Handbook,* 5th ed. New York: McGraw-Hill Book Co., 1973.

Ridley, J., ed. *Safety at Work,* 2nd ed. Stoneham, MA: Butterworth Publishers, 1986.

This encyclopedic volume contains 90 chapters on safety management, occupational health and hygiene, and behavioral science. More than 20 specialists have combined their expertise to produce this valuable reference. This new edition has been completely updated to provide the most up-to-date information.

The book will serve as a useful general reference source for safety managers as well as those who are faced with specific problems such as company liability or compensation claims. The practicing engineer will find chapters on subjects vital to the understanding of safety in industry, but are outside of his or her routine experience, such as psychology, legislation, and industrial hygiene.

Sax, N. I. *Dangerous Properties of Industrial Materials,* 6th ed. New York: Van Nostrand Reinhold Co., 1984.

This handbook includes single-figure toxic hazard ratings for all the chemicals discussed. The edition includes a 50,000-synonym index, and NIOSH and CAS numbers for nearly all entries.

Sax, N. I., and Lewis, R. J., Sr., eds. *Rapid Guide to Hazardous Chemicals in the Workplace.* New York: Van Nostrand Reinhold Co., 1986.

This guide is designed to afford easy access to information on the adverse properties of commonly encountered industrial materials. The information provided allows for a quick assessment of the relative hazards of the material and the types and nature of the hazards likely to be encountered. Reference codes included with each entry refer to sources of additional information.

Schleien, B., and Terpilak, M. *The Health Physical and Radiological Health Handbook.* Olney, MD: Necleon Lectern Associates, 1984.

The handbook was developed for health physics practitioners, technicians, and students as an easy-to-use, practical handbook containing health physics and radiological health data. While briefer and more specific data sources are available on single subject areas as are multivolume compendia, there is no other current up-to-date compilation of information useful on a daily basis for the health physicist.

Sittig, M. *Handbook of Toxic and Hazardous Chemicals and Carcinogens,* 2nd ed. Park Ridge, NJ: Noyes Data Corp., 1985.

This handbook presents concise chemical, health, and safety information on nearly 800 toxic and hazardous chemicals so that responsible decisions can be made by those who may have contact with or interest in these chemicals due to their own or third-party exposure.

Stedman's Medical Dictionary, 24th ed. Baltimore, MD: Williams & Wilkins Co., 1982.

Steere, N. *Handbook of Laboratory Safety,* 2nd ed. Boca Raton, FL: CRC Press, Inc., 1971.

This book provides useful and accurate information for preventing or controlling accidents, injuries, fires, and losses in laboratories. Some of the topics included in the handbook are responsibility for laboratory safety, protective equipment, legal liability for accidents, working alone, fire hazards, and ventilation and exhaust systems.

Weast, R. C., ed. *Handbook of Chemistry and Physics.* Boca Raton, FL: CRC Press, Inc., latest edition.

Weast, R. C. *Handbook of Data on Organic Compounds.* Boca Raton, FL: CRC Press, Inc., 1985.

This handbook presents chemical and physical data on 24,000 organic compounds encountered in academic, medical and industrial research, manufacturing, and processing, and environmental control. Compounds are listed in alphabetical sequence by IUPAC name and accompanied by other information including CAS number, boiling and melting points, and synonyms.

NIOSH PUBLICATIONS

The National Institute for Occupational Safety and Health (NIOSH) has published many useful publications in the field of industrial hygiene. Consult the current "Publications Catalog," listing all NIOSH publications in print and their prices, for further information. The NIOSH publications can be obtained by requesting single copies from:

National Institute for Occupational Safety and Health
Division of Technical Services
Publications Dissemination
4676 Columbia Parkway
Cincinnati, OH 45226

Many of these publications are also available from:

Superintendent of Documents
U.S. Government Printing Office
Washington, DC 20402

Some NIOSH publications can also be obtained from:

National Technical Information Service (NTIS)
Springfield, VA 22161

'Criteria Documents'

The NIOSH is responsible for providing relevant data from which valid criteria for effective standards can be derived. Recommended standards for occupational exposure, which are the result of this work, are based on the health effects of exposure.

The single most comprehensive source of information on a particular material will probably be found in the NIOSH "Criteria Document" for that substance. The Table of Contents for a Criteria Document is as follows:

I. Recommendations for an Occupational Exposure Standard
   - Section 1—Environmental (workplace air)
   - Section 2—Medical
   - Section 3—Labeling and posting
   - Section 4—Personal protective equipment and clothing

Section 5—Informing employees of hazards
Section 6—Work practices
Section 7—Sanitation practices
Section 8—Monitoring and recordkeeping

II. Introduction

III. Biological Effects of Exposure
Extent of exposure
Historical reports
Effects on humans
Epidemiological studies
Animal toxicity
Correlation of exposure and effect
Carcinogenicity, mutagenicity, teratogenicity, and effects on reproduction
Summary tables of exposure and effect

IV. Environmental Data
Environmental concentrations
Sampling and analysis
Engineering controls

V. Work Practices

DATA SHEETS AND GUIDES

AIHA Ergonomics Guide Series. Akron, OH: American Industrial Hygiene Association.
Contents: No. 1—Guide to Manual Lifting; No. 2—Guide to Assessment of Metabolic and Cardiac Costs of Physical Work. Each provides basic information on new concepts.

AIHA Hygienic Guide Series. Akron, OH: American Industrial Hygiene Association.
Separate data sheets on specific substances giving hygienic standards, properties, industrial hygiene practice, specific procedures, and references.

ANSI Standards, Z37 Series, "Acceptable Concentrations of Toxic Dusts and Gases." New York: American National Standards Institute.
These guides represent a concensus of expertise concerning minimum safety requirements for the storage, transportation, and handling of toxic substances and are intended to aid the manufacturer, the consumer, and the general public.

ASTM Standards with Related Material. Phildelphia: American Society for Testing and Materials.

CIS Information Sheets. Geneva 20, Switzerland: International Occupational Safety and Health Centre (CIS), International Labour Office.
Pamphlets on general or specific occupational safety and health topics; published irregularly.

Industrial Data Sheets. Chicago, IL: National Safety Council.
Information and recommendations regarding health and safety hazards arising in the occupational environment.

Technical Bulletins: Chemical-Toxicological Series. Pittsburgh: Industrial Health Foundation, Inc.
No. 1. Industrial Air Sampling and Analysis (1974). No. 2. Hygienic Guides (1963). No. 3. Threshold Limit Values: recommended revisions and additions for 1966 (1965). No. 4. Emergency Exposure Limits and Hygienic Guides (1965). No. 6. Range Finding Toxicity Data for 43 compounds. No. 7. Threshold Limit Values; a report of progress in 1967 and annnouncement of 1968 workshop on TLVs (1967). No. 8. Suggested Principles and Procedures for Developing Data for Threshold Limit Values for air (1969).

Technical Bulletins: Medical Series. Pittsburgh: Industrial Health Foundation, Inc.
Reports on various topics in occupational health and medicine. Representative titles include: Emphysema in Industry; Asbestos Bioeffects Research for Industry; the Pneumoconioses; Acute Radiation Syndrome; and Proteolytic Enzymes in Cotton Mill Dust—A Possible Cause of Byssinosis.

Threshold Limit Values and Biological Exposure Indices for 1986-1987. Cincinnati: American Conference of Governmental Industrial Hygienists.
Threshold limits and exposure indices based on information from industrial experience and experimental human and animal studies; they are intended for use in the practice of industrial hygiene.

NEWSLETTERS AND REPORTS

AOMA Report. American Occupational Medical Association. 1845 W. Morse Ave., Chicago, IL 60626.

Environmental Health Letter. Semimonthly. Environews, Inc., 1331 Pennsylvania Ave. NW, Washington, DC 20004.

Environmental Health and Safety News. Department of Environmental Health, School of Public Health and Community Medicine, University of Washington, F-461 Health Sciences Building, Seattle, WA 98195.

Environmental and Public Health News. Monthly. Department of Environmental, Public, and Occupational Health, American Medical Association, 535 N. Dearborn St., Chicago, IL 60610.

Industrial Hygiene News Report. Monthly. Flournoy and Associates, 1845 W. Morse Ave., Chicago, IL 60626.

Industrial Section Newsletters. Bimonthly. National Safety Council, 444 N. Michigan Ave., Chicago, IL 60611.

Occupational Health Reporter. Division of Occupational Health Bureau of Retirement, Insurance, and Occupational Health, U.S. Civil Service Commission, 1900 E Street, NW, Washington, DC 20415.

Occupational Health and Safety Letter. Semimonthly. Environews, Inc., 1331 Pennsylvania Ave, NW, Washington, DC 20004.

Occupational Safety and Health Reporter. Bureau of National Affairs, 1231 25th St., NW, Washington, DC 20037.

OSHA Up-to-Date. Monthly. National Safety Council, 444 N. Michigan Ave., Chicago, IL 60611.

Publications on Occupational Safety and Health. International Labour Office, CH-1211, Geneva 22, Switzerland.

World Health Organization Technical Report Series. Available from WHO, PO Box 5284, Church Street Station, New York, NY 10249.

JOURNALS AND MAGAZINES

Many articles on industrial hygiene can be found in the following journals and magazines.

Across the Board (formerly *Conference Board Report).* Conference Board, 845 Third Ave., New York, NY 10022.

American Industrial Hygiene Association Journal. American Industrial Hygiene Association, 475 Wolf Ledges Pkwy., Akron, OH: 44311. Monthly.

American Journal of Epidemiology. Johns Hopkins University, School of Hygiene and Public Health, 624 N. Broadway, Baltimore, MD 21205.

American Journal of Medicine. York Medical Group, Magazine Division, Technical Publishing Co., 875 Third Ave., New York, NY 10022.

American Journal of Public Health. American Public Health Association, 1015 Fifteenth St., Washington, DC 20005. Monthly.

American Medical News. American Medical Association, 535 N. Dearborn St., Chicago, IL 60610.

American Review of Respiratory Disease. American Lung Association, 1740 Broadway, New York, NY 10019.

Analytical Chemistry. American Chemical Society, 1155 16th St. NW, Washington, DC 20036.

Annals of Occupational Hygiene. Pergamon Press, Maxwell House, Fairview Park, Elmsford, NY 10523. Published quarterly for the British Occupational Hygiene Society.

Applied Industrial Hygiene. American Conference of Governmental Industrial Hygienists, 6500 Glenway Avenue, Bldg. D-7, Cincinnati, OH: 45211.

Archives of Environmental Health. Heldref Publications, 4000 Albemarle St., NW, Washington, DC 20016. Monthly.

Archives of Internal Medicine. American Medical Association, 535 N. Dearborn St., Chicago, IL 60610.

Archives of Pathology and Laboratory Medicine, American Medical Association, 535 N. Dearborn St., Chicago, IL 60610.

ASHRAE Journal. American Society of Heating, Refrigerating and Air-Conditioning Engineers, Inc., 1791 Tullie Circle, NE, Atlanta, GA 30329.

British Journal of Industrial Medicine. British Medical Association, Tavistock Sq., London WC1 H9JR, England. Monthly.

British Medical Journal. Tavistock Sq., London WC1 H9JR, England.

Chemical and Engineering News. American Chemical Society, 1155 16th St., NW, Washington, DC 20036.

Environment. Scientists' Institute for Public Information, Heldref Publications, 9000 Albemarle St., NW, Washington, DC 20016.

Environmental Health and Safety News (formerly: *Occupational Health Newsletter).* Department of Environmental Health SC-34, Scool of Public Health and Community Medicine, University of Washington, Seattle, WA 98195. Monthly.

Environmental Health Perspectives. National Institute of Environmental Health Sciences. PO Box 12233, Research Triangle Park, NC 27709.

Environmental Research. Academic Press, Inc., Journal Division, 1250 Sixth Ave., San Diego, CA 92101.

Environmental Science and Technology. American Chemical Society, 1155 16th St., NW, Washington, DC 20036. Monthly.

Ergonomics. Taylor and Francis, Ltd., 242 Cherry St., Philadelphia PA 19106.

Excerpta Medica. Section 35: "Occupational Health and Industrial Medicine." Excerpta Medica, Inc., Princeton, NJ 08540. Monthly.

Health Physics. Pergamon Press. Maxwell House, Fairview Park, Elmsford, NY 10523. Sponsored by the Health Physics Society. Monthly.

Industrial Health & Toxicology. BioSciences Information Services, Biological Abstracts, 2100 Arch St., Philadelphia, PA 19103. Monthly.

Industrial Hygiene Digest. Industrial Health Foundation, Inc., 34 Penn Circle W, Pittsburgh, PA 15206. Monthly.

Industrial Hygiene News Report. Flournoy and Associates, 1845 W. Morse Ave., Chicago, IL 60626.

International Archives of Occupational and Environmental Health. Springer-Verlag, 175 Fifth Avenue, New York, NY 10010. Quarterly.

Job Safety and Health Report. Business Publishers, Inc., PO Box 1067. Blair Station, Silver Spring, MD 20910. Biweekly.

Journal of the Acoustical Society of America. Acoustical Society of America, 335 E. 45th St., New York, NY 10017.

Journal of the Air Pollution Control Association. Air Pollution Control Association, Three Gateway Center, Four West, Pittsburgh, PA 15222. Monthly.

Journal of American Insurance. Alliance of American Insurers. 1501 Woodfield Dr., Schaumburg, IL 60195.

Journal of the American Medical Association. American Medical Association, 535 N. Dearborn St., Chicago, IL 60610.

Journal of Auditory Research. The C. W. Skilling Auditory Research Center, Inc., Box M, Groton, CT 06340.

Journal of Aviation, Space and Environmental Medicine (formerly: *Aerospace Medicine*). Aerospace Medical Association, Washington National Airport, Washington, DC 20001. Monthly.

Journal of Toxicology: Clinical Toxicology. Marcel Dekker, Inc., 305 E. 45th St., New York, NY 10017. Bimonthly.

Journal of Occupational Medicine. American Occupational Medicine Association, 1845 W. Morse Ave., Chicago, IL 60626.

Journal of the Society of Occupational Medicine (formerly: *Transactions of the Society of Occupational Medicine*). Royal College of Physicians, 11 St. Andrew's Place, London NW1 4LE, England. Quarterly.

Journal of Toxicology & Environmental Health. Hemisphere Publishing Corporation, 79 Madison Ave., New York, NY 10016.

Lancet. Little, Brown & Company, 34 Beacon St., Boston, MA 02106.

Medical Bulletin. Exxon Corporation and Affiliated Companies, 1251 Avenue of the Americas, New York, NY 10020.

New England Journal of Medicine, New England Journal of Medicine, 1440 Main St., Waltham, MA 02254.

Noise Control Engineering Journal. Institute of Noise Control Engineering, Department of Mechanical Engineering, Auburn University, Auburn, AL 36849.

Noise Control Report. Business Publishers, Inc., 951 Pershing Drive, Blair Station, Silver Spring, MD 20910. Biweekly.

Noise/News. Institute of Noise Control Engineering, 3469 Arlington Branch, Poughkeepsie, NY 12603.

Occupational Hazards. Occupational Hazards, Penton-IPC, 1100 Superior Ave., Cleveland, OH 44114.

Occupational Health, Macmillan Journals, Ltd., Little Essex Street, London WC2, England. Monthly.

Occupational Health Nursing. American Association of Industrial Nurses, 79 Madison Avenue, New York, NY 10016.

Occupational Health & Safety. Stevens Publishing Corp., 5002 Lakeland Circle, Box 7573, Waco, TX 76710.

Occupational Health and Safety Letter. Gershon W. Fishbein, Publisher, Environews, Inc., 1331 Pennsylvania Ave., NW, Washington, DC 20004.

Occupational Safety and Health, Royal Society for the Prevention of Accidents, Cannon House, The Priory, Queensway, Birmingham B46B5, U.K. Monthly.

Occupational Safety & Health Reporter. Bureau of National Affairs, Inc., 1231 25th Street, NW, Washington, DC 20037. Weekly.

Occupational Safety and Health Subscription Service. Standards, Interpretations, Regulations, and Procedures. U.S. Occupational Safety and Health Administration, Washington, DC 5 vols. (OSH 01 through OSH 05). Available from Government Printing Office. Catalog No. L35.6/3-1 through -5.

OSHA Up-to-Date. National Safety Council, 444 N. Michigan Ave., Chicago, IL 60611. Monthly.

Plant Engineering. Technical Publishing Company, 875 Third Ave., New York, NY 10022.

Professional Safety (formerly *American Society of Safety Engineers Journal*). ASSE, 1800 E. Oakton St., Des Plaines, IL 60018.

Public Health Reports. Department of Health and Human Services, Hubert Humphrey Bldg., 200 Independence Ave., SW, Washington, DC 20201.

Safety and Health. National Safety Council, 444 N. Michigan Ave., Chicago, IL 60611.

Scandinavian Journal of Work, Environment & Health. Haartmanin Katu 1, SF-00290, Helsinki 29, Finland. Monthly.

Skin and Allergy News. American Medical News Service, Inc., 12230 Wilkins Ave., Rockville, MD 20852.

Sound and Vibration. Acoustical Publications, Inc., 27101 E. Oviatt Rd., Bay Village, OH 44140.

State of the Art Reviews: Occupational Medicine. Hanley & Belfus, Inc., 210 S. 13th St., Philadelphia, PA 19107. Quarterly.

Toxicology and Applied Pharmacology. Academic Press, Journal Division, 1250 Sixth Ave., San Diego, CA 92101.

STATES WITH ON-SITE OSHA CONSULTATION AGREEMENTS

Rules were promulgated under section 7(c)(1) of the Occupational Safety and Health Act of 1970 (84 Stat. 1590), which authorizes OSHA to reimburse state agencies for use of their personnel to provide on-site consultation services to employers. The service will be made available at no cost to employers to assist them in providing their employees a place of employment which is safe and healthful. Consultants will identify specific hazards in the workplace and provide advice on their elimination. On-site consultation will be conducted independently of any OSHA enforcement activity, and the discovery of hazards will not mandate citation or penalties.

Here is a listing by state to request on-site consultation:

OSHA Onsite Consultation Project Directory, June 1986

State/Region, Address, and Telephone:

Alabama (IV)
7(c)(1) Onsite Consultation Program
P.O. Box 6005
University, Alabama 35486
(205) 348-7136

Alaska (X)
Division of Consultation & Training LS&S/OSH
Alaska Department of Labor
3301 Eagle Street, Suite 303
Pouch 7-022
Anchorage, Alaska 99510
(907) 264-2599

Arizona (IX)
Consultation and Training
Division of Occupational Safety & Health
Industrial Commission of Arizona
P.O. Box 19070
800 West Washington
Phoenix, Arizona 85005
(602) 255-5795

Arkansas (VI)
OSHA Consultation
Arkansas Department of Labor
1022 High Street
Little Rock, Arkansas 72202
(501) 375-8442

California (IX)
CAL/OSHA Consultation Service
525 Golden Gate Avenue, 2nd Floor
San Francisco, California 94102
(415) 557-2870

Colorado (VIII)
Occupational Safety & Health Section
Institute of Rural Environmental Health
Colorado State University
110 Veterinary Science Building
Fort Collins, Colorado 80523
(303) 491-6151

Connecticut (I)
Division of Occupational Safety & Health
Connecticut Department of Labor
200 Folly Brook Boulevard
Wethersfield, Connecticut 06109
(203) 566-4550

Delaware (III)
Occupational Safety and Health
Division of Industrial Affairs
Delaware Department of Labor
820 North French Street, 6th Floor
Wilmington, Delaware 19801
(302) 571-3908

District of Columbia (III)
Office of Occupational Safety & Health
District of Columbia Department of Employment Services
950 Upshur Street, NW
Washington, DC 20011
(202) 576-6339

Florida (IV)
7(c)(1) Onsite Consultation Program
Bureau of Industrial Safety & Health
Florida Department of Labor & Employment Security
LaFayette Building, Room 204
2551 Executive Center Circle, West
Tallahassee, Florida 32301
(904) 488-3044

Georgia (IV)
7(c)(1) Onsite Consultation Program
Georgia Institute of Technology
O'Keefe Bldg., Rm. 23
Atlanta, Georgia 30332
(404) 894-3806

Guam (IX)
OSHA Onsite Consultation
Government of Guam
P.O. Box 23548, Guam Main Facility
Agana, Guam 96921-0318
9-011 (671) 646-9446

Hawaii (IX)
ATTN: Consultation Program Manager
Division of Occupational Safety & Health
677 Ala Moana Blvd., Suite 910
Honolulu, Hawaii 96813
(808) 548-2511

Idaho (X)
Safety & Health Consultation Program
Boise State University
Dept. of Comm. & Env. Health
Boise, Idaho 83725
(208) 385-3283

Illinois (V)
Division of Industrial Services
Illinois Department of Commerce & Community Affairs
100 West Randolph St., Suite 3-400
Chicago, Illinois 60601
(312) 917-2337

Indiana (V)
Bureau of Safety, Education & Training
Indiana Division of Labor
1013 State Office Building
Indianapolis, Indiana 46204
(317) 232-2688

Iowa (VII)
7(c)(1) Consultation Program
Iowa Bureau of Labor
307 East Seventh Street
Des Moines, Iowa 50319
(515) 281-5352

Kansas (VII)
7(c)(1) Consultation Program
Kansas Department of Human Resources
512 West 6th Street
Topeka, Kansas 66603
(913) 296-4386

Kentucky (IV)
Education and Training
Occupational Safety & Health Program
Kentucky Department of Labor
U.S. Highway 127 South
Frankfort, Kentucky 40601
(502) 564-6895

Louisiana (VI)
No services available

Maine (I)
Division of Industrial Safety
Maine Department of Labor
Labor Station 82
283 State Street
Augusta, Maine 04333
(207) 289-2591

Maryland (III)
7(c)(1) Consultation Program
Division of Labor & Industry
501 Saint Paul Place
Baltimore, Maryland 21202
(301) 659-4218

Massachusetts (I)
7(c)(1) Consultation Program
Division of Industrial Safety
Massachusetts Department of Labor and Industries
100 Cambridge Street
Boston, Massachusetts 02202
(617) 727-3463

Michigan (Health) (V)
Special Programs Section
Division of Occupational Health
Michigan Department of Public Health
3500 North Logan
P.O. Box 30035
Lansing, Michigan 48909
(517) 335-8250

Michigan (Safety) (V)
Safety Education & Training Division
Bureau of Safety and Regulation
Michigan Department of Labor
7150 Harris Drive
P.O. Box 30015
Lansing, Michigan 48909
(517) 322-1809

Minnesota (Safety) (V)
Consultation Division
Minnesota Department of Labor and Industry
444 Lafayette Road, 4th Floor
St. Paul, Minnesota 55101
(612) 297-2393

Minnesota (Health) (V)
Consultation Unit
Department of Public Health
717 Delaware, S.E.
Minneapolis, Minnesota 55440
(612) 623-5100

Mississippi (IV)
7(c)(1) Onsite Consultation Program
Division of Occupational Safety & Health
Mississippi State Board of Health
P.O. Box 1700
Jackson, Mississippi 39205
(601) 982-6315

Missouri (VII)
Onsite Consultation Program
Division of Labor Standards
Missouri Department of Labor and Industrial Relations
1001-D Southwest Blvd.
P.O. Box 449
Jefferson City, Missouri 65102
(314) 751-3403

Montana (VIII)
Montana Bureau of Safety & Health
Division of Workers' Compensation
5 South Last Chance Gulch
Helena, Montana 59601
(406) 444-6401

Nebraska (VII)
Division of Safety, Labor and Safety Standards
Nebraska Department of Labor
State Office Building
301 Centennial Mall, South
Lincoln, Nebraska 68509-5024
(402) 471-2239

Nevada (IX)
Training and Consultation
Division of Occupational Safety & Health
4600 Kietzke Lane, Bldg. D-139
Reno, Nevada 89502
(702) 789-0546

New Hampshire (I)
Onsite Consultation Program
New Hampshire Department of Labor
19 Pillsbury Street
Concord, New Hampshire 03301
(603) 271-3170

New Jersey (II)
Division of Workplace Standards
New Jersey Department of Labor
110 East Front Street
Trenton, New Jersey 08625-0054
(609) 292-2313

New Mexico (VI)
OSHA Consultation
Occupational Health & Safety Bureau
1190 St. Francis Drive, Rm. 2200 No.
P.O. Box 968
Santa Fe, New Mexico 87504-0968
(505) 827-8949

New York (II)
Division of Safety and Health
New York State Department of Labor
One Main Street
Brooklyn, New York 11201
(718) 797-7645

North Carolina (IV)
North Carolina Consultative Services
North Carolina Department of Labor
Shore Memorial Building
214 West Jones Street
Raleigh, North Carolina 27603
(919) 733-4880

North Dakota (VIII)
North Dakota State Department of Health
Div. of Environmental Engineering
1200 Missouri Ave., Rm. 304
Bismarck, North Dakota 58502-5520
(701) 224-2348

Ohio (V)
Division of Onsite Consultation
Ohio Department of Industrial Relations
P.O.B. 825
2323 West 5th Avenue
Columbus, Ohio 43216
(800) 282-1425
(Toll-free in State)
(614) 466-7485

Oklahoma (VI)
OSHA Division
Oklahoma Department of Labor
1315 Broadway Place, Rm. 301
Oklahoma City, Oklahoma 73103
(405) 235-0530 X240

Oregon (X)
Voluntary Compliance, Consultative Sec.
Accident Prevention Division
Oregon Dept. of Workers' Compensation
Labor and Industries Bldg., Rm. 115
Salem, Oregon 97310
(503) 378-2890

Pennsylvania (III)
Indiana University of Pennsylvania
Safety Sciences Department
Uhler Hall
Indiana, Pennsylvania 15705
(800) 382-1241
(Toll-free in State)
(412) 357-2561/2396

Puerto Rico (II)
Occupational Safety & Health Office
Puerto Rico Department of Labor and Human Resources
505 Munoz Rivera Avenue, 21st Floor
Hato Rey, Puerto Rico 00918
(809) 754-2134/2171

Rhode Island (I)
Division of Occupational Health
Rhode Island Department of Health
206 Cannon Building
75 Davis Street
Providence, Rhode Island 02908
(401) 277-2438

South Carolina (IV)
7(c)(1) Onsite Consultation Program
Consultation and Monitoring
South Carolina Department of Labor
3600 Forest Drive
P.O. Box 11329
Columbia, South Carolina 29211
(803) 734-9599

South Dakota (VIII)
S.T.A.T.E. Engineering Extension
Onsite Technical Division
South Dakota State University
Box 2218
Brookings, South Dakota 57007
(605) 688-4101

Tennessee (IV)
OSHA Consultative Services
Tennessee Department of Labor
501 Union Building, 6th Floor
Nashville, Tennessee 37219
(615) 741-2793

Texas (VI)
Division of Occupational Safety and State Safety Engineer
Texas Department of Health
1100 West 49th Street
Austin, Texas 78756
(512) 458-7287

Utah (VIII)
Utah Safety & Health
Consultation Service
P.O. Box 45580
Salt Lake City, Utah 84145-0580
(801) 530-6868

Vermont (I)
Division of Occupational Safety & Health
Vermont Department of Labor and Industry
120 State Street
Montpelier, Vermont 05602
(802) 828-2765

Virginia (III)
Virginia Department of Labor and Industry
P.O. Box 12064
205 N. 4th Street
Richmond, Virginia 23241
(804) 786-5875

Virgin Islands (II)
Division of Occupational Safety & Health
Virgin Islands Department of Labor
Lagoon Street
Frederiksted
Virgin Islands 00840
(809) 772-1315

Washington (X)
Washington Department of Labor and Industries
P.O. Box 207
814 East 4th
Olympia, Washington 98504
(206) 753-6500

West Virginia (III)
West Virginia Department of Labor
State Capitol, Bldg. 3 Rm. 319
1800 E. Washington Street
Charleston, West Virginia 25305
(304) 348-7890

Wisconsin (Health) (V)
Section of Occupational Health
Wisconsin Department of Health and Social Services
1414 E. Washington Ave., Rm. 112
P.O. Box 309
Madison, Wisconsin 53701
(608) 266-0417

Wisconsin (Safety) (V)
Division of Safety and Buildings
Wisconsin Department of Industry, Labor and Human Relations
1570 East Moreland Boulevard
Waukesha, Wisconsin 53186
(414) 521-5063

Wyoming (VIII)
Occupational Health and Safety
State of Wyoming
604 East 25th Street
Cheyenne, Wyoming 82002
(307) 777-7786

1908 Consultation Training Coordination

Occupational Safety and Health Administration
Training Institute
1555 Times Drive
Des Plaines, Illinois 60018
(312) 297-4810

Service Agreement

Wisconsin Laboratory
Wisconsin Occupational Health Laboratory
979 Jonathon Drive
Madison, Wisconsin 53713
(608) 263-8807

SELECTING AND USING A HYGIENE CONSULTANT

At times, an industrial hygiene problem will defy your best efforts to find a solution: those are the times to seek the advice of an expert. Someone who has studied industrial hygiene problems for many years can often find a more economical solution than would occur to a newcomer or inexperienced individual.

This section covers what a consultant does; thus enabling you to make better use of the industrial hygiene consultant's services.

Many companies selling industrial hygiene instruments and products are staffed by engineers who are knowledgeable in industrial hygiene measurements and controls. They can advise you on industrial hygiene measurement techniques at no cost.

An industrial hygiene consultant, however, is an independent professional, group of professionals, a private consultant, or state government agency performs services for clients on a fee basis. Consulting services can be as broad or narrow as needed. Using a consultant usually guarantees that the project, whether long-term or occasional, will be solved with maximum skill and economy.

A consultant can offer the following:

- Creativity born of diverse experience with a wide range of problem-solving applications.
- Professional independence. An independent consultant is in a position to divorce judgment of methods and materials from all secondary interests and keep a purely objective viewpoint.
- The consultant strives to provide successful solutions at a reasonable cost.
- Organizational stability. The consultant lessens the need for the organization, administration and eventual disbandment of large internal staffs formed to meet peak load projects or specialized problems.

- Complete accreditation. Consultants are usually qualified by education and experience to give complete consulting services.
- Flexibility and mobility. The consultant is available whenever special counsel is required.

Before a decision is made about hiring a consultant, question whether the consultant can:

- Do the job faster?
- Do it better?
- Do it at less overall cost?

Typical situations when a consultant can be used are as follows:

- When you do not know exactly what to do
- When you do not know how to do it
- When you want outside advice
- When you need an unbiased viewpoint
- When you need access to special facilities. Special facilities or equipment may be needed for one project and cannot be justified as a purchase, or acquired in the time available.
- When you need to pursue alternate solutions. Staff may not be available internally to analyze many possible solutions to a problem.
- When you need to convince management to take a particular course of action. An in-company solution to a problem may exist, but management may refuse to believe the recommendations of its own people.

In the final analysis, the decision to hire an industrial hygiene consultant is determined by the economics of the situation.

Much of the material in this following section is adapted from "Occupational Exposure Sampling Strategy Manual," NIOSH Publication No. 77-173, available through the U.S. Government Printing Office.

Industrial hygiene consultants are hired primarily to accomplish two major objectives: (1) to identify and evaluate potential health and safety hazards to workers in the occupational environment, and/or (2) to design effective controls to protect the workers. Competent industrial hygiene consultants should be able to perform both of these tasks because of their training and experience. Usually, consultants can evaluate the extent of employee exposures more efficiently because of their detailed knowledge of the proper sampling equipment and analytical procedures.

Industrial hygiene consultants can recommend whether or not control measures are required and the alternatives available. They can design, supervise the installation of, and evaluate the effectiveness of control measures.

Consultants can be used to keep management aware of both current and proposed federal and state regulations in the area of occupational health and safety. They can inform management when medical examinations of employees may be recommended or required by regulation. They should be able to recommend appropriate physicians or clinics specializing in occupational medicine. An industrial hygiene consultant can play a valuable role in providing the examining physician with information on the occupational exposures of employees.

Consultants can design employee training programs. A consultant can serve as an expert witness if your company is involved in a lawsuit and data must be obtained, interpreted, and presented by a disinterested third party.

To maximize use of consulting services, the purpose of the investigation or study must be defined. In any problem-solving situation, it is important to find out what the problem really is—before you try to solve it. This is the essence of what a good consultant does. Company personnel may not always be able to describe the problem clearly.

The consultant is obligated to pinpoint the problem, regardless of what the client thinks it is. This can be difficult, to say the least, but it must be done. Nothing is so disconcerting as solving the wrong problem or a nonexistent problem.

Define the problem

First, try to formulate and define the problem; then list the specific qualifications and experience a consultant should have to solve this problem.

Currently any person can legally offer services as an industrial hygiene consultant; consequently, it is important to avoid hiring those who are unsuitable because of lack of training, inexperience, or incompetence. Individuals or firms billing themselves as industrial hygiene consultants can be broadly classified according to (1) whether they recommend a particular procedure, product, service, or control process, or (2) whether they are independent consultants.

Product-oriented individuals or firms vary in their backgrounds from nontechnical product sales personnel to experienced industrial hygiene professionals. In this case, "consulting" consists mainly of recommending appropriate equipment and facilities. This type of consultation may include assistance in soliciting proposals for the design and installation of control equipment, such as ventilation control systems or respirators.

The advantage of using this group directly is that you avoid consultant costs and pay only for the product or service. The disadvantage in dealing with a product-oriented consultant is that these consultants may not consider all options available. Thousands of dollars could be spent in purchasing a particular type of monitoring equipment or in implementing a particular control system, only to discover later that the desired results cannot be obtained or that another solution could have been obtained for less money.

If there are any doubts as to the proper method for solving a problem, then an independent consultant (one free from ties to a particular service or line of products) should be called in. It is this type of industrial hygiene consultant that will be discussed in the remainder of this section.

Sources

There are several sources one can go to for information and names of consultants. One source or information is from professional associations and public-service organizations related to occupational safety and health. Three national groups are the American Industrial Hygiene Association (AIHA), American Society of Safety Engineers (ASSE), and the National Safety Council (NSC). These three have local chapters, sections, or offices in major cities that can be a source of information and assistance. The AIHA published a nationwide list of industrial hygiene consultants.

Many insurance companies now have loss prevention programs that employ industrial hygienists. Inquiries should be made of your present insurer; you may want to compare the services they offer with those of other insurance companies. Finally, there may be a university or college in your area that has an environmental health program (see the listing earlier in this section).

Selection

Selection of a consultant should be guided by one primary consideration—the qualification of the consulting staff for the project to be undertaken. The size of a consulting firm is seldom a reliable single determinant, nor is the length of time in practice a major factor for consideration.

A good line of action to follow is to consider the qualifications of a number of individuals or firms that appear to be capable of meeting the requirements of the project to be undertaken.

Select a limited number of individuals or firms that appear to be best qualified for the particular project. Write each of them individual letters—describe briefly the project and inquire as to their interest in it. On receipt of affirmative answers, invite the companies to come in for separate personal interviews. At the interview, go over the qualifications and record of each firm. Have the firms submit up-to-date data and available staff information, a brief description of work on hand that might possibly conflict with your project, and the qualifications of specific key personnel who will be assigned to your company's project.

A series of questions for the consultant to answer is given here. They should not be given equal weight since some are minor in importance. (The list is organized roughly in descending order of importance.)

EXPERIENCE

1. For how many years have you been professionally active in industrial hygiene?
2. Please supply a list of recent clients that you have served, preferably in this geographical area and on problems similar to those in which I am interested.
3. What teaching have you done or training have you had in industrial hygiene? What groups were involved—university, industry, trade associations, civic groups, engineers, symposia?

CONSULTATION STATUS

1. Are you now an independent consultant? For how many years? Full time or part time?
2. If part time:
   a. Who is your chief employer or in what other business ventures are you involved?
   b. Is your employer aware and does he approve of your part-time activity as an industrial hygiene consultant?

EDUCATION

1. What schools did you attend and what courses did you take related to industrial hygiene?
2. What degrees did you receive and when?
3. What special conferences, seminars, symposia, or short courses have you attended (especially recently) to stay current with industrial hygiene technical information and governmental regulations?
4. What other sources of information do you use to stay current with the field of industrial hygiene?

PROFESSIONAL AFFILIATIONS

1. What professional associations do you belong to? (Representative ones are the American Industrial Hygiene Association, American Conference of Governmental Industrial Hygienists, and the American Society of Safety Engineers.) What is your present grade of membership and length of time in that grade for each association?
2. Are you certified by any of the following?
   a. American Board of Industrial Hygiene (specify area of certification)
   b. Board of Certified Safety Professionals
3. Are you a registered professional engineer? In what states and disciplines?
4. Of what professional engineer associations are you or your firm a member?
5. Of what trade associations, chambers of commerce, or similar business groups are you or your firm a member?

SPECIAL CAPABILITIES

1. In what areas of industrial hygiene do you specialize?
2. What equipment do you have for conducting industrial hygiene evaluations?
3. What laboratories do you use for the analysis of your exposure measurement samples?
4. Can you serve as an expert witness, either for your client or as a friend of the court?
5. What experience have you had as an expert witness?

BUSINESS PRACTICES

1. Please indicate your fee structure. Do you work by hourly charges, estimates for the total job, retainer charges, or any combination of these?
2. In your charges, how do you treat such expenses as travel, subsistence, shipping, report reproduction, and computer time?
3. Can you supply a list of typical laboratory analytical fees?

Compensation

Compensation for consultant services may be calculated and established by a variety of methods.

- Fixed lump sum
- Cost plus a fixed amount
- Salary cost times a factor, plus incurred or out-of-pocket expenses
- Per diem

As with medical, legal, and other professional services, consulting services should never be secured purely on the basis of a price comparison. Competitive proposals for professional consulting services are undesirable because there is no direct cost basis for comparison of services that involve judgment and creative thinking. These factors cannot be evaluated precisely in advance of performance.

The proposal

Once you have selected a consultant, you can arrange to obtain services in several ways. A verbal commitment is sometimes all that is necessary. However, you may wish to request a written proposal that spells out the steps to be taken in the solution of your problem.

Aside from background qualifications of the consultant, the proposal should answer the questions:

1. How much is the service going to cost? Smaller jobs are often bid on an hourly basis, with a minimum of one-half day's work, plus direct expenses commonly specified. Larger jobs are usually bid at a fixed amount.
2. What is the consultant going to do? The answer to this question may range all the way from a simple agreement to study the problem to a comprehensive step-by-step plan to solve it.
3. What will be the end result? The answer to this question

is all too often not clearly understood; the result is usually a report that specifies the consultant's recommendation. If you do not want to pay for the preparation of a written report because a verbal one will do, specify this in advance.

Since recommendations often call for construction or other operations to be carried out by others whose work is not subject to the consultant's control, results can usually not be guaranteed by the consultant. Rather, an estimate of the exposure control to be attained is all that can be expected.

If the consultant is to provide drawings from which the contractor will work, the proposal must specify sketches or finished drawings. If special materials are required, the consultant should agree to specify alternative selections, if possible. If you want a guaranteed result, experimental work will usually be necessary and will have to be paid for.

The consultant can also monitor construction to determine compliance with specifications. The consultant can also measure after installation to confirm predictions and supply oral briefings as needed.

Even if your consultant is to serve as an expert witness for you, the consultant is not automatically on your side. Rather, the consultant is more like a friend of the court, devoted to bringing out the facts, with careful separation of fact from expert opinion.

Working with the consultant

Once an industrial hygiene consultant is hired, the problem should be defined as exactly as possible. There should be no guesswork unless there is no other choice. All the data bearing on the problem should be provided at the beginning of the study or investigation—it does no good to leave out embarrassing or unpleasant facts. (The consultant will find them anyway, and the extra time spent will cost you money.) The consultant should be introduced to key members in your organization, and appropriate personnel should be available when needed.

Before hiring a consultant, consider the following items. Be sure to have any necessary equipment available and a place for the consultant to work. Do not let the consultant wander around alone trying to find information or things. Do not hold back information from the consultant.

Monitor the consultant's progress regularly. Check to see that corporate personnel are cooperating fully, that there is no personality conflict, and that the consultant is getting the help desired.

Review particularly the beginning phases including the formulation of the problem and the analysis of the problem. Review at the end of each phase.

The three phases of consulting work. Most consulting work can be divided into three phases—problem definition, problem analysis, and the solution phase.

The problem definition phase lays the groundwork for subsequent stages. The need for thoroughness and accuracy at this stage is obvious.

The problem analysis phase deals with the facts. The consultant defines the problem and identifies the opportunities for improvement, determines the causes of the problem, determines the objectives to be met by the solution, and develops alternative solutions.

The solution phase involves selecting the most effective, workable, timely, and practical solution. The details of the solution must be worked out carefully. The solution should be the equivalent of a blueprint describing what needs to be done, how it is to be done, by whom, and in what sequence the actions are to take place.

Problems and pitfalls. The successful consulting job is always a team effort between company and the consultant. Too often a consultant is literally challenged to do a successful job as though someone were saying, "We couldn't do it and neither can you."

Some typical problems and pitfalls that you may run into are described here. Some of these items are negative versions of items on the checklist. Some of the failures listed are the company's fault, some are the consultant's, and some involve both.

These pitfalls include the following:

- Failure to clearly define the problem.
- Failure to set specific objectives.
- Failure to establish financial arrangements in the beginning.
- Failure to establish realistic time requirements.
- Failure to select the best qualified consultant.
- Failure to make all pertinent information available to the consultant.
- Failure to obtain a clear and complete proposal from the consultant.
- Failure to review the oral or written proposal for clarity, completeness, creative approaches and qualification.
- Failure to make a cost/benefit analysis.
- Failure to review progress periodically.
- Failure because of exceeding the scope of the assignment.

Achieving the greatest value from a consultant is largely a matter of providing full information, defining responsibilities clearly, and establishing workable lines of communication. The consultant should have full information regarding the project from the outset. The where, when, how, and why of the task are the very tools needed to solve the problem.

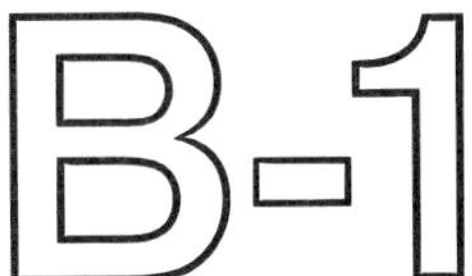

Threshold Limit Values and Biological Exposure Indices

THIS APPENDIX GIVES THE Threshold Limit Values (TLVs) for chemical substances and Physical Agents in the Work Environment and the Biological Exposure Indices (BEIs) that were adopted by the American Conference of Governmental Industrial Hygienists (ACGIH). This material is reprinted with permission from the ACGIH.

INTRODUCTION TO THE CHEMICAL SUBSTANCES

Threshold limit values refer to airborne concentrations of substances and represent conditions under which it is believed that nearly all workers may be repeatedly exposed day after day without adverse effect. Because of wide variation in individual susceptibility, however, a small percentage of workers may experience discomfort from some substances at concentrations at or below the threshold limit; a smaller percentage may be affected more seriously by aggravation of a pre-existing condition or by development of an occupational illness.

Threshold limits are based on the best available information from industrial experience, from experimental human and animal studies, and, when possible, from a combination of the three. The basis on which the values are established may differ from substance to substance; protection against impairment of health may be a guiding factor for some, whereas reasonable freedom from irritation, narcosis, nuisance or other forms of stress may form the basis for others.

The amount and nature of the information available for establishing a TLV varies from substance to substance; consequently, the precision of the estimated TLV is also subject to variation and the latest *Documentation* should be consulted in order to assess the extent of the data available for a given substance.

These limits are intended for use in the practice of industrial hygiene as guidelines or recommendations in the control of potential health hazards and for no other use, e.g., in the evaluation or control of community air pollution nuisances, in estimating the toxic potential of continuous, uninterrupted exposures or other extended work periods, as proof or disproof of an existing disease or physical condition, or adoption by countries whose working conditions differ from those in the United States of America and where substances and processes differ. These limits *are not* fine lines between safe and dangerous concentration and *should not* be used by anyone untrained in the discipline of industrial hygiene.

The Threshold Limit Values, as issued by ACGIH, are recommendations and should be used as guidelines for good practices. In spite of the fact that serious injury is not believed likely as a result of exposure to the threshold limit concentrations, the best practice is to maintain concentrations of all atmospheric contaminants as low as is practical.

The ACGIH disclaims liability with respect to the use of TLVs in a manner inconsistent with their intended use as stated herein.

Notice of Intent. At the beginning of each year, proposed actions of the Committee for the forthcoming year are issued in the form of a "Notice of Intended Changes." This Notice provides not only an opportunity for comment, *but solicits suggestions of substances to be added to the list. The suggestions should be accompanied by substantiating evidence.* The list of Intended Changes follows the Adopted Values in the TLV booklet. Values listed in parenthesis in the "Adopted" list are to be used during the period in which a proposed change for that Value is listed in the Notice of Intended Changes.

Definitions. Three categories of Threshold Limit Values (TLVs) are specified herein, as follows:

***a)** The Threshold Limit Value-Time Weighted Average (TLV-TWA)*—the time-weighted average concentration for a normal 8-hour workday and a 40-hour workweek, to which nearly all workers may be repeatedly exposed, day after day, without adverse effect.

***b)** Threshold Limit Value-Short Term Exposure Limit (TLV-STEL)*—the concentration to which workers can be exposed continuously for a short period of time without suffering from 1) irritation, 2) chronic or irreversible tissue damage, or 3) narcosis of sufficient degree to increase the likelihood of accidental injury, impair self-rescue or materially reduce work efficiency, and provided that the daily TLV-TWA is not exceeded. It is not a separate independent exposure limit, rather it supplements the time-weighted average (TWA) limit where there are recognized acute effects from a substance whose toxic effects are primarily of a chronic nature. STELs are recommended only where toxic effects have been reported from high short-term exposures in either humans or animals.

A STEL is defined as a 15-minute time-weighted average exposure which should not be exceeded at any time during a work day even if the eight-hour time-weighted average is within the TLV. Exposures at the STEL should not be longer than 15 minutes and should not be repeated more than four times per day. There should be at least 60 minutes between successive exposures at the STEL. An averaging period other than 15 minutes may be recommended when this is warranted by observed biological effects.

***c)** Threshold Limit Value-Ceiling (TLV-C)*—the concentration that should not be exceeded during any part of the working exposure.

In conventional industrial hygiene practice if instantaneous monitoring is not feasible, then the TLV-C can be assessed by sampling over a 15-minute period except for those substances which may cause immediate irritation with exceedingly short exposures.

For some substances, e.g., irritant gases, only one category, the TLV-Ceiling, may be relevant. For other substances, either two or three categories may be relevant, depending upon their physiologic action. It is important to observe that if any one of these three TLVs is exceeded, a potential hazard from that substance is presumed to exist.

The Committee holds to the opinion that limits based on physical irritation should be considered no less binding than those based on physical impairment. There is increasing evidence that physical irritation may initiate, promote or accelerate physical impairment through interaction with other chemical or biologic agents.

Time-Weighted Average vs Ceiling Limits. Time-weighted averages permit excursions above the limit provided they are compensated by equivalent excursions below the limit during the workday. In some instances it may be permissible to calculate the average concentration for a workweek rather than for a workday. The relationship between threshold limit and permissible excursion is a rule of thumb and in certain cases may not apply. The amount by which threshold limits may be exceeded for short periods without injury to health depends upon a number of factors such as the nature of the contaminant, whether very high concentrations—even for short periods—produce acute poisoning, whether the effects are cumulative, the frequency with which high concentrations occur, and the duration of such periods. All factors must be taken into consideration in arriving at a decision as to whether a hazardous condition exists.

Although the time-weighted average concentration provides the most satisfactory, practical way of monitoring airborne agents for compliance with the limits, there are certain substances for which it is inappropriate. In the latter group are substances which are predominantly fast acting and whose threshold limit is more appropriately based on this particular response. Sustances with this type of response are best controlled by a ceiling "C" limit that should not be exceeded. It is implicit in these definitions that the manner of sampling to determine noncompliance with the limits for each group must differ; a single brief sample, that is applicable to a "C" limit, is not appropriate to the time-weighted limit; here, a sufficient number of samples are needed to permit a time-weighted average concentration throughout a complete cycle of operations or throughout the work shift.

Whereas the ceiling limit places a definite boundary which concentrations should not be permitted to exceed, the time-weighted average limit requires an explicit limit to the excursions that are permissible above the listed values. It should be noted that the same factors are used by the Committee in determining the magnitude of the value of the STELs, or whether to include or exclude a substance for a "C" listing.

Excursion Limits. For the vast majority of substances with a TLV-TWA, there is not enough toxicological data available to warrant a STEL. Nevertheless, excursions above the TLV-TWA should be controlled even where the eight-hour TWA is within recommended limits. Earlier editions of the TLV list included such limits whose values depended on the TLV-TWAs of the substance in question.

While no rigorous rationale was provided for these particular values, the basic concept was intuitive: in a well controlled process exposure, excursions should be held within some reasonable limits.

Unfortunately, neither toxicology nor collective industrial hygiene experience provide a solid basis for quantifying what those limits should be. The approach here is that the maximum recommended excursion should be related to variability generally observed in actual industrial processes. Leidel, Busch and Crouse,* in reviewing large numbers of industrial hygiene surveys conducted by NIOSH, found that short-term exposure measurements were generally log normally distributed with geometric standard deviation mostly in the range of 1.5 to 2.0.

While a complete discussion of the theory and properties of the log normal distribution is beyond the scope of this section, a brief description of some important terms is presented. The measure of central tendency in a log normal description is the antilog of the mean logarithm of the sample values. The distribution is skewed and the geometric mean is always smaller than the arithmetic mean by an amount which depends on the geometric standard deviation. In the log normal distribution, the geometric standard deviation (sd_g) is the antilog of the standard deviation of the sample value logarithms and 68.26% of all values lie between m_g/sd_g and $m_g \times sd_g$.

If the short-term exposure values in a given situation have a geometric standard deviation of 2.0, 5% of all values exceed 3.13 times the geometric mean. If a process displays a variability greater than this, it is not under good control and efforts should be made to restore control. This concept is the basis for the new excursion limit recommendations which are as follows:

> *Short-term exposures should exceed three times the TLV-TWA for no more than a total of 30 minutes during a work day and under no circumstances should they exceed five times the TLV-TWA, provided that the TLV-TWA is not exceeded.*

The approach is a considerable simplification of the idea of the log normal concentration distribution but is considered more convenient to use by the practicing industrial hygienist. If exposure excursions are maintained within the recommended limits, the geometric standard deviation of the concentration measurements will be near 2.0 and the goal of the recommendations will be accomplished.

When the toxicological data for a specific substance are available to establish a STEL, this value takes precedence over the excursion limit regardless of whether it is more or less stringent.

"Skin" Notation. Listed substances followed by the designation "Skin" refer to the potential contribution to the overall exposure by the cutaneous route including mucous membranes and eye, either by airborne, or more particularly, by direct contact with the substance. Vehicles can alter skin absorption.

Little quantitative data are available describing absorption of vapors and gases through the skin. The rate of absorption is a function of the concentration to which the skin is exposed.

Substances having a skin notation and a low TLV may present a problem at high airborne concentrations, particularly if a significant area of the skin is exposed for a long period of time. Protection of the respiratory tract, while the rest of the body surface is exposed to a high concentration, may present such a situation.

Biological monitoring should be considered to determine the relative contribution of dermal exposure to the total dose.

This attention-calling designation is intended to suggest appropriate measures for the prevention of cutaneous absorption so that the threshold limit is not invalidated.

Mixtures. Special consideration should be given also to the application of the TLVs in assessing the health hazards which may be associated with exposure to mixtures of two or more substances. A brief discussion of basic considerations involved in developing threshold limit values for mixtures, and methods for their development, amplified by specific examples, are given in Appendix C.

Nuisance Particulates. In contrast to fibrogenic dusts which cause scar tissue to be formed in lungs when inhaled in excessive amounts, so-called "nuisance" dusts have a long history of little adverse effect on lungs and do not produce significant organic disease or toxic effect when exposures are kept under reasonable control. The nuisance dusts have also been called (biologically) "inert" dusts, but the latter term is inappropriate to the extent that there is no dust which does not evoke some cellular response in the lung when inhaled in sufficient amount. However, the lung-tissue reaction caused by inhalation of nuisance dusts has the following characteristics: 1) the architecture of the air spaces remains intact; 2) collagen (scar tissue) is not formed to a significant extent; and 3) the tissue reaction is potentially reversible.

Excessive concentrations of nuisance dusts in the workroom air may seriously reduce visibility, may cause unpleasant deposits in the eyes, ears and nasal passages (Portland cement dust), or cause injury to the skin or mucous membranes by chemical or mechanical action *per se* or by the rigorous skin cleansing procedures necessary for their removal.

A threshold limit of 10 mg/m³ of total dust < 1% quartz is recommended for substances in these categories and for which no specific threshold limits have been assigned. This limit, for a normal workday, does not apply to brief exposures at higher concentrations. Neither does it apply to those substances which may cause physiologic impairment at lower concentrations but for which a threshold limit has not yet been adopted. Some nuisance particulates are given in Appendix D. This list is not meant to be all inclusive; the substances serve only as examples.

Simple Asphyxiants — 'Inert" Gases or Vapors. A number of gases and vapors, when present in high concentrations in air, act primarily as simple asphyxiants without other significant physiologic effects. A TLV may not be recommended for each simple asphyxiant because the limiting factor is the available oxygen. The minimal oxygen content should be 18 percent by volume under normal atmospheric pressure (equivalent to a partial pressure, pO_2 of 135 mm Hg). Atmospheres deficient in O_2 do not provide adequate warning and most simple asphyxiants are odorless. Several simple asphyxiants present an explosion hazard. Account should be taken of this factor in limiting the concentration of the asphyxiant. Specific examples are listed in Appendix E. This list is not meant to be all inclusive; the substances serve only as examples.

Physical Factors. It is recognized that such physical factors as heat, ultraviolet and ionizing radiation, humidity, abnormal pressure (altitude), and the like may place added stress on the body so that the effects from exposure at a threshold limit may be altered. Most of these stresses act adversely to increase the toxic response of a substance. *Although most threshold limits have built-in safety factors to guard against adverse effects to moderate deviations* from normal environments, the safety factors of most substances are not of such a magnitude as to take care of gross deviations. For example, continuous work at temperatures above 90°F, or overtime extending the workweek more than 25%, might be considered gross deviations. In such instances judgment must be exercised in the proper adjustments of the Threshold Limit Values.

Hypersusceptibility. Tests are available (*J. Occup. Med. 15*:564, 1973; *Ann. N.Y. Acad. Sci. 151, Art. 2*:968, 1968) that may be used to detect those individuals hypersusceptible to a variety of industrial chemicals (respiratory irritants, hemolytic chemicals, organic isocyanates, carbon disulfide).

Unlisted Substances. Many substances present or handled in industrial processes do not appear on the TLV list. In a number of instances the material is rarely present as a particulate, vapor or other airborne contaminant, and a TLV is not necessary. In other cases sufficient information to warrant development of a TLV, even on a tentative basis, is not available to the Committee. Other substances, of low toxicity, could be included in Appendix D pertaining to nuisance particulates. This list (as well as Appendix E) is not meant to be all inclusive: the substances serve only as examples.

In addition there are some substances of not inconsiderable toxicity, which have been omitted primarily because only a limited number of workers (e.g., employees of a single plant) are known to have potential exposure to possibly harmful concentrations.

Trade names. Because many chemical substances are marketed under several trade names, the trade names have been replaced with their generic equivalent in the alphabetical listing. Appendix G was created to ease this transition and the CAS number appears with the generic name to aid identification.

Operational Guidelines. The ACGIH Board of Directors has adopted operational guidelines for Chemical Substances TLV Committee. These guidelines prescribe: charge, authority, policies,

* Leidel, N.A., K.A. Busch and W.E. Crouse: *Exposure Measurement, Action Level and Occupational Environmental Variability.* NIOSH Pub. No. 76-131 (December 1975).

membership, organization, and operating procedures. The policies include the appeals procedures. Copies of the guidelines document are available from the Publications Office at a cost of $5 per copy.

Substance [CAS #]	ADOPTED VALUES			
	TWA		STEL	
	ppm[a]	mg/m³[b]	ppm[a]	mg/m³[b]
Acetaldehyde [75-07-0]	100	180	150	270
Acetic acid [64-19-7]	10	25	15	37
Acetic anhydride [108-24-7]	C 5	C 20	—	—
Acetone [67-64-1]	750	1,780	1,000	2,375
Acetonitrile [75-05-8]—Skin	40	70	60	105
Acetylene [74-86-2]	E	—	—	—
Acetylene dichloride, *see* 1,2-Dichloroethylene				
*Acetylene tetrabromide [79-27-6]	1	15	—	—
Acetylsalicylic acid (Aspirin) [50-78-2]	—	5	—	—
Acrolein [107-02-8]	0.1	0.25	0.3	0.8
‡Acrylamide [79-06-1]—Skin	—	(0.3)	—	(0.6)
Acrylic acid [79-10-7]	10	30	—	—
Acrylonitrile [107-13-1]—Skin	2,A2	4.5,A2	—	—
*Aldrin [309-00-2] — Skin	—	0.25	—	—
Allyl alcohol [107-18-6]—Skin	2	5	4	10
Allyl chloride [107-5-1]	1	3	2	6
Allyl glycidyl ether (AGE) [106-92-3]—Skin	5	22	10	44
Allyl propyl disulfide [2179-59-1]	2	12	3	18
*α-Alumina [1344-28-1]	—	D	—	—
Aluminum [7429-90-5]				
* Metal & oxide	—	10	—	—
Pyro powders	—	5	—	—
Welding fumes	—	5	—	—
Soluble salts	—	2	—	—
Alkyls (NOC†)	—	2	—	—
4-Aminodiphenyl [92-67-1]—Skin	—	A1b	—	—
2-Aminoethanol, *see* Ethanolamine				
*2-Aminopyridine [504-29-0]	0.5	2	—	—
3-Amino 1,2,4-triazole, *see* Amitrole				
*Amitrole [61-82-5]	—	0.2	—	—
Ammonia [7664-41-7]	25	18	35	27
Ammonium chloride fume [12125-02-9]	—	10	—	20
*Ammonium sulfamate [7773-06-0]	—	10	—	—
‡n-Amyl acetate [628-63-7]	100	530	(150)	(800)
‡sec-Amyl acetate [626-38-0]	125	665	(150)	(800)
*Aniline [62-53-3] & homologues—Skin	2	10	—	—
Anisidine [29191-52-4] (o-, p-isomers)—Skin	0.1	0.5	—	—
Antimony [7440-36-0] & compounds, as Sb	—	0.5	—	—
Antimony trioxide [1309-64-4]				
Handling and use, as Sb	—	0.5	—	—
Production	—	A2	—	—
*ANTU [86-88-4]	—	0.3	—	—
Argon [7440-37-1]	E	—	—	—

(a) Parts of vapor or gas per million parts of contaminated air by volume at 25°C and 760 torr.
(b) Approximate milligrams of substance per cubic meter of air.
‡ See Notice of Intended Changes.
* 1986-87 Addition.
Capital letters A, B, D & E refer to Appendices; C denotes ceiling limit.

Substance [CAS #]	ADOPTED VALUES			
	TWA		STEL	
	ppm[a]	mg/m³[b]	ppm[a]	mg/m³[b]
Arsenic [7440-38-2] & soluble compounds, as As	—	0.2	—	—
Arsenic trioxide production [1327-53-3]	—	A2	—	—
Arsine [7784-42-1]	0.05	0.2	—	—
Asbestos [1332-21-4], *see* DUSTS	—	A1a	—	—
‡Asphalt (petroleum) fumes [8052-42-4]	—	5	—	(10)
Atrazine [1912-24-9]	—	5	—	—
*Azinphos-methyl [86-50-0]—Skin	—	0.2	—	—
Barium [7440-39-3], soluble compounds, as Ba	—	0.5	—	—
*Benomyl [17804-35-2]	0.8	10	—	—
‡Benzene [71-43-2]	10,A2	30,A2	(25,A2)	(75,A2)
Benzidine [92-87-5]—Skin	—	A1b	—	—
p-Benzoquinone, *see* Quinone				
Benzoyl peroxide [94-36-0]	—	5	—	—
Benzo(a)pyrene [50-32-8]	—	A2	—	—
Benzyl chloride [100-44-7]	1	5	—	—
Beryllium [7440-41-7] & compounds, as Be	—	0.002,A2	—	—
‡Biphenyl [92-52-4]	0.2	1.5	(0.6)	(4)
*Bismuth telluride [1304-82-1]	—	10	—	—
* Se-doped	—	5	—	—
Borates, tetra, sodium salts [1303-96-4]				
Anhydrous	—	1	—	—
Decahydrate	—	5	—	—
Pentahydrate	—	1	—	—
*Boron oxide [1303-86-2]	—	10	—	—
*Boron tribromide [10294-33-4]	C 1	C 10	—	—
Boron trifluoride [7637-07-2]	C 1	C 3	—	—
*Bromacil [314-40-9]	1	10	—	—
Bromine [7726-95-6]	0.1	0.7	0.3	2
*Bromine pentafluoride [7789-30-2]	0.1	0.7	—	—
Bromochloromethane, *see* Chlorobromomethane				
Bromoform [75-25-2]—Skin	0.5	5	—	—
*1,3-Butadiene [106-99-0]	10,A2	22,A2	—	—
Butane [106-97-8]	800	1,900	—	—
Butanethiol, *see* Butyl mercaptan				
2-Butanone, *see* Methyl ethyl ketone (MEK)				
‡2-Butoxyethanol [111-76-2]—Skin	25	120	(75)	(360)
n-Butyl acetate [123-86-4]	150	710	200	950
‡sec-Butyl acetate [105-46-4]	200	950	(250)	(1,190)
‡tert-Butyl acetate [540-88-5]	200	950	(250)	(1,190)
Butyl acrylate [141-32-2]	10	55	—	—
n-Butyl alcohol [71-36-3]—Skin	C 50	C 150	—	—
sec-Butyl alcohol [78-92-2]	100	305	150	455
tert-Butyl alcohol [75-65-0]	100	300	150	450
Butylamine [109-73-9]—Skin	C 5	C 15	—	—
tert-Butyl chromate, as CrO_3 [1189-85-1]—Skin	—	C 0.1	—	—
n-Butyl glycidyl ether (BGE) [2426-08-6]	25	135	—	—
n-Butyl lactate [138-22-7]	5	25	—	—
Butyl mercaptan [109-79-5]	0.5	1.5	—	—
o-sec-Butylphenol [89-72-5]—Skin	5	30	—	—
p-tert-Butyltoluene [98-51-1]	10	60	20	120

Capital letters A, B, D & E refer to Appendices; C denotes ceiling limit.
* 1986-87 Addition.

Substance [CAS #]	ADOPTED VALUES TWA ppm[a]	TWA mg/m³[b]	STEL ppm[a]	STEL mg/m³[b]
*Cadmium [7440-43-9] Dusts & salts, as Cd	—	0.05	—	—
Cadmium oxide [1306-19-0]				
Fume, as Cd	—	C 0.05	—	—
Production	—	0.05	—	—
*Calcium carbonate/marble [1317-65-3]	--	D	—	—
*Calcium cyanamide [156-62-7]	—	0.5	—	—
Calcium hydroxide [1305-62-0]	—	5	—	—
Calcium oxide [1305-78-8]	—	2	—	—
Calcium silicate [1344-95-2]	—	D	—	—
Camphor, synthetic [76-22-2]	2	12	3	18
‡Caprolactam [105-60-2]				
Dust	—	(1)	—	(3)
Vapor	(5)	(20)	(10)	(40)
Captafol [2425-06-1]—Skin	—	0.1	—	—
*Captan [133-06-2]	—	5	—	—
*Carbaryl [63-25-2]	—	5	—	—
Carbofuran [1563-66-2]	—	0.1	—	—
*Carbon black [1333-86-4]	—	3.5	—	—
*Carbon dioxide [124-38-9]	5,000	9,000	30,000	54,000
Carbon disulfide [75-15-0]—Skin	10	30	—	—
Carbon monoxide [630-08-0]	50	55	400	440
Carbon tetrabromide [558-13-4]	0.1	1.4	0.3	4
*Carbon tetrachloride [56-23-5]—Skin	5,A2	30,A2	—	—
Carbonyl chloride, *see* Phosgene				
Carbonyl fluoride [353-50-4]	2	5	5	15
Catechol [120-80-9]	5	20	—	—
*Cellulose (paper fiber) [9004-34-6]	—	D	—	—
Cesium hydroxide [21351-79-1]	—	2	—	—
Chlordane [57-74-9]—Skin	—	0.5	—	2
Chlorinated camphene [8001-35-2]—Skin	—	0.5	—	1
Chlorinated diphenyl oxide [55720-99-5]	—	0.5	—	2
Chlorine [7782-50-5]	1	3	3	9
Chlorine dioxide [10049-04-4]	0.1	0.3	0.3	0.9
Chlorine trifluoride [7790-91-2]	C 0.1	C 0.4	—	—
Chloroacetaldehyde [107-20-0]	C 1	C 3	—	—
α-Chloroacetophenone [532-27-4]	0.05	0.3	—	—
Chloroacetyl chloride [79-04-9]	0.05	0.2	—	—
Chlorobenzene [108-90-7]	75	350	—	—
o-Chlorobenzylidene malononitrile [2698-41-1]—Skin	C 0.05	C 0.4	—	—
Chlorobromomethane [74-97-5]	200	1,050	250	1,300
2-Chloro-1,3-butadiene, *see* β-Chloroprene				
Chlorodifluoromethane [75-45-6]	1,000	3,500	1,250	4,375
Chlorodiphenyl (42% Chlorine) [53469-21-9]—Skin	—	1	—	2
Chlorodiphenyl (54% Chlorine) [11097-69-1]—Skin	—	0.5	—	1
1-Chloro,2,3-epoxy-propane, *see* Epichlorohydrin				
2-Chloroethanol, *see* Ethylene chlorohydrin				
Chloroethylene, *see* Vinyl chloride				
*Chloroform [67-66-3]	10,A2	50,A2	—	—
bis(Chloromethyl) ether [542-88-1]	0.001, A1a	0.005, A1a	—	—
Chloromethyl methyl ether [107-30-2]	A2	A2	—	—
1-Chloro-1-nitropropane [600-25-9]	2	10	—	—
Chloropentafluoroethane [76-15-3]	1,000	6,320	—	—
Chloropicrin [76-06-2]	0.1	0.7	0.3	2
β-Chloroprene [126-99-8]—Skin	10	35	—	—
o-Chlorostyrene [2039-87-4]	50	285	75	430
o-Chlorotoluene [95-49-8]	50	250	75	375
2-Chloro-6-(trichloromethyl) pyridine, *see* Nitrapyrin				
Chlorpyrifos [2921-88-2]—Skin	—	0.2	—	0.6
Chromite ore processing (Chromate), as Cr	—	0.05,A1a	—	—
Chromium [7440-47-3] Metal	—	0.5	—	—
Chromium (II) compounds, as Cr	—	0.5	—	—
Chromium (III) compounds, as Cr	—	0.5	—	—
Chromium (VI) compounds, as Cr				
Water soluble	—	0.05	—	—
Certain water insoluble	—	0.05,A1a	—	—
Chromyl chloride [14977-61-8]	0.025	0.15	—	—
Chrysene [218-01-9]	A2	A2	—	—
Clopidol [2971-90-6]	—	10	—	20
Coal tar pitch volatiles [8007-45-2], as benzene solubles	—	0.2,A1a	—	—
‡Cobalt [7440-48-4], as Co Metal, dust & fume	—	(0.1)	—	—
Cobalt carbonyl [10210-68-1], as Co	—	0.1	—	—
Cobalt hydrocarbonyl [16842-03-8], as Co	—	0.1	—	—
Copper [7440-50-8]				
Fume	—	0.2	—	—
* Dusts & mists, as Cu	—	1	—	—
*Cotton dust, raw	—	0.2[d]	—	—
Cresol [1319-77-3], all isomers—Skin	5	22	—	—
‡Crotonaldehyde [123-73-9]	2	6	(6)	(18)
Crufomate [299-86-5]	—	5	—	20
‡Cumene [98-82-8]—Skin	50	245	(75)	(365)
Cyanamide [420-04-2]	—	2	—	—
Cyanides [151-50-8; 143-33-9], as CN—Skin	—	5	—	—
Cyanogen [460-19-5]	10	20	—	—
Cyanogen chloride [506-77-4]	C 0.3	C 0.6	—	—
‡Cyclohexane [110-82-7]	300	1,050	(375)	(1,300)
Cyclohexanol [108-93-0]—Skin	50	200	—	—
‡Cyclohexanone [108-94-1]—Skin	25	100	(100)	(400)

Capital letters A, B, D & E refer to Appendices; C denotes ceiling limit
‡ See Notice of Intended Changes.
* 1986-1987 Addition.

Substance [CAS #]	ADOPTED VALUES			
	TWA		STEL	
	ppm[a]	mg/m³[b]	ppm[a]	mg/m³[b]
Cyclohexene [110-83-8]	300	1,015	—	—
Cyclohexylamine [108-91-8]	10	40	—	—
Cyclonite [121-82-4]—Skin	—	1.5	—	3
‡Cyclopentadiene [542-92-7]	75	200	(150)	(400)
‡Cyclopentane [287-92-3]	600	1,720	(900)	(2,580)
*Cyhexatin [13121-70-5]	—	5	—	—
*2,4-D [94-75-7]	—	10	—	—
*DDT (Dichlorodiphenyl-trichloroethane) [50-29-3]	—	1	—	—
Decaborane [17702-41-9]—Skin	0.05	0.3	0.15	0.9
*Demeton [8065-48-3]—Skin	0.01	0.1	—	—
Diacetone alcohol [123-42-2]	50	240	75	360
1,2-Diaminoethane, *see* Ethylenediamine				
*Diazinon [333-41-5]—Skin	—	0.1	—	—
Diazomethane [334-88-3]	0.2	0.4	—	—
Diborane [19287-45-7]	0.1	0.1	—	—
1,2-Dibromoethane, *see* Ethylene dibromide				
*2-N-Dibutylaminoethanol [102-81-8]—Skin	2	14	—	—
Dibutyl phosphate [107-66-4]	1	5	2	10
‡Dibutyl phthalate [84-74-2]	—	5	—	(10)
Dichloroacetylene [7572-29-4]	C 0.1	C 0.4	—	—
o-Dichlorobenzene [95-50-1]	C 50	C 300	—	—
p-Dichlorobenzene [106-46-7]	75	450	110	675
3,3'-Dichlorobenzidine [91-94-1]—Skin	—	A2	—	—
*Dichlorodifluoromethane [75-71-8]	1,000	4,950	—	—
1,3-Dichloro-5,5-dimethyl hydantoin [118-52-5]	—	0.2	—	0.4
1,1-Dichloroethane [75-34-3]	200	810	250	1,010
1,2-Dichloroethane, *see* Ethylene dichloride				
1,1-Dichloroethylene, *see* Vinylidene chloride				
1,2-Dichloroethylene [540-59-0]	200	790	250	1,000
Dichloroethyl ether [111-44-4]—Skin	5	30	10	60
Dichlorofluoromethane [75-43-4]	10	40	—	—
Dichloromethane, *see* Methylene chloride				
*1,1-Dichloro-1-nitroethane [594-72-9]	2	10	—	—
1,2-Dichloropropane, *see* Propylene dichloride				
*Dichloropropene [542-75-6]—Skin	1	5	—	—
2,2-Dichloropropionic acid [75-99-0]	1	6	—	—
*Dichlorotetrafluoroethane [76-14-2]	1,000	7,000	—	—
*Dichlorvos [62-73-7]—Skin	0.1	1	—	—
Dicrotophos [141-66-2]—Skin	—	0.25	—	—
Dicyclopentadiene [77-73-6]	5	30	—	—
*Dicyclopentadienyl iron [102-54-5]	—	10	—	—
*Dieldrin [60-57-1]—Skin	—	0.25	—	—
Diethanolamine [111-42-2]	3	15	—	—
Diethylamine [109-89-7]	10	30	25	75
2-Diethylaminoethanol [100-37-8]—Skin	10	50	—	—

Capital letters A, B, D & E refer to Appendices; C denotes ceiling limit
‡ See Notice of Intended Changes.
* 1986-1987 Addition.
(d) Lint-free dust as measured by the vertical elutriator cotton-dust sampler described in the *Transactions of the National Conference on Cotton Dust*, p. 33, by J.R. Lynch (May 2, 1970).

Substance [CAS #]	ADOPTED VALUES			
	TWA		STEL	
	ppm[a]	mg/m³[b]	ppm[a]	mg/m³[b]
Diethylene triamine [111-40-0]—Skin	1	4	—	—
Diethyl ether, *see* Ethyl ether				
Di-2-ethylhexylphthalate, *see* Di-sec, octyl phthalate				
Diethyl ketone [96-22-0]	200	705	—	—
‡Diethyl phthalate [84-66-2]	—	5	—	(10)
*Difluorodibromomethane [75-61-6]	100	860	—	—
Diglycidyl ether (DGE) [2238-07-5]	0.1	0.5	—	—
Dihydroxybenzene, *see* Hydroquinone				
Diisobutyl ketone [108-83-8]	25	250	—	—
Diisopropylamine [108-18-9]—Skin	5	20	—	—
Dimethoxymethane, *see* Methylal				
*Dimethyl acetamide [127-19-5]—Skin	10	35	—	—
Dimethylamine [124-40-3]	10	18	—	—
Dimethylaminobenzene, *see* Xylidene				
Dimethylaniline [121-69-7] (N,N-Dimethylaniline)—Skin	5	25	10	50
Dimethylbenzene, *see* Xylene				
Dimethyl carbamoyl chloride [79-44-7]	A2	A2	—	—
Dimethyl-1,2-dibromo-2-dichloroethyl phospate, *see* Naled				
*Dimethylformamide [68-12-2]—Skin	10	30	—	—
2,6-Dimethyl-4-heptanone, *see* Diisobutyl ketone				
*1,1-Dimethylhydrazine [57-14-7]—Skin	0.5,A2	1,A2	—	—
Dimethylnitrosoamine, *see* N-Nitrosodimethylamine				
*Dimethylphthalate [131-11-3]	—	5	—	—
Dimethyl sulfate [77-78-1]—Skin	0.1,A2	0.5,A2	—	—
Dinitolmide [148-01-6]	—	5	—	—
*Dinitrobenzene [528-29-0; 99-65-0; 100-25-4] (all isomers)—Skin	0.15	1	—	—
*Dinitro-o-cresol [534-52-1]—Skin	—	0.2	—	—
3,5-Dinitro-o-toluamide, *see* Dinitolmide				
*Dinitrotoluene [121-14-2]—Skin	—	1.5	—	—
*Dioxane [123-91-1]—Skin	25	90	—	—
Dioxathion [78-34-2]—Skin	—	0.2	—	—
Diphenyl, *see* Biphenyl				
*Diphenylamine [122-39-4]	—	10	—	—
Diphenylmethane diisocyanate, *see* Methylene bisphenyl isocyanate				
Dipropylene glycol methyl ether [34590-94-8]	100	600	150	900
Dipropyl ketone [123-19-3]	50	235	—	—
*Diquat [85-00-7]	—	0.5	—	—
Di-sec, octyl phthalate [117-81-7]	—	5	—	10
*Disulfiram [97-77-8]	—	2	—	—
*Disulfoton [298-04-4]	—	0.1	—	—
‡2,6-Ditert. butyl-p-cresol [128-37-0]	—	10	—	(20)
Diuron [330-54-1]	—	10	—	—
Divinyl benzene [108-57-6]	10	50	—	—
*Emery [112-62-9]	—	D	—	—

Capital letters A, B, D & E refer to Appendices; C denotes ceiling limit
‡ See Notice of Intended Changes.
* 1986-1987 Addition.

Substance [CAS #]	ADOPTED VALUES			
	TWA		STEL	
	ppm[a]	mg/m^3[b]	ppm[a]	mg/m^3[b]
*Endosulfan [115-29-7] — Skin	—	0.1	—	—
*Endrin [72-20-8] — Skin	—	0.1	—	—
Enzymes, *see* Subtilisins				
*Epichlorohydrin [106-89-8] — Skin	2	10	(5)	—
*EPN [2104-64-5] — Skin	—	0.5	—	—
1,2-Epoxypropane, *see* Propylene oxide				
2,3-Epoxy-1-propanol, *see* Glycidol				
Ethane [74-84-0]	E	—	—	—
Ethanethiol, *see* Ethyl mercaptan				
Ethanol, *see* Ethyl alcohol				
Ethanolamine [141-43-5]	3	8	6	15
Ethion [563-12-2] — Skin	—	0.4	—	—
2-Ethoxyethanol [110-80-5] — Skin	5	19	—	—
2-Ethoxyethyl acetate [111-15-9] — Skin	5	27	—	—
Ethyl acetate [141-78-6]	400	1,400	—	—
‡Ethyl acrylate [140-88-5] — Skin	5	20	(25)	(100)
Ethyl alcohol [64-17-5]	1,000	1,900	—	—
Ethylamine [75-04-7]	10	18	—	—
Ethyl amyl ketone [541-85-5]	25	130	—	—
Ethyl benzene [100-41-4]	100	435	125	545
Ethyl bromide [74-96-4]	200	890	250	1,110
‡Ethyl butyl ketone [106-35-4]	50	230	(75)	(345)
*Ethyl chloride [75-00-3]	1,000	2,600	—	—
Ethylene [74-85-1]	E	—	—	—
Ethylene chlorohydrin [107-07-3] — Skin	C 1	C 3	—	—
Ethylenediamine [107-15-3]	10	25	—	—
Ethylene dibromide [106-93-4] — Skin	A2	A2	—	—
*Ethylene dichloride [107-06-2]	10	40	—	—
Ethylene glycol [107-21-1] Vapor	C 50	C 125	—	—
Ethylene glycol dinitrate [628-96-6] — Skin	0.05	0.3	—	—
Ethylene glycol methyl ether acetate, *see* 2-Methoxyethyl acetate				
Ethylene oxide [75-21-8]	1,A2	2,A2	—	—
Ethylenimine [151-56-4] — Skin	0.5	1	—	—
Ethyl ether [60-29-7]	400	1,200	500	1,500
‡Ethyl formate [109-94-4]	100	300	(150)	(450)
Ethylidene chloride, *see* 1,1-Dichloroethane				
Ethylidene norbornene [16219-75-3]	C 5	C 25	—	—
*Ethyl mercaptan [75-08-1]	0.5	1	—	—
*N-Ethylmorpholine [100-74-3] — Skin	5	23	—	—
*Ethyl silicate [78-10-4]	10	85	—	—
Fenamiphos [22224-92-6] — Skin	—	0.1	—	—
Fensulfothion [115-90-2]	—	0.1	—	—
Fenthion [55-38-9] — Skin	—	0.2	—	—
*Ferbam [14484-64-1]	—	10	—	—
Ferrovanadium dust [12604-58-9]	—	1	—	3
Fibrous glass dust	—	10	—	—
Fluorides, as F	—	2.5	—	—
Fluorine [7782-41-4]	1	2	2	4
Fluorotrichloromethane, *see* Trichlorofluoromethane				
Fonofos [944-22-9] — Skin	—	0.1	—	—
Formaldehyde [50-00-0]	1,A2	1.5,A2	2,A2	3,A2
‡Formamide [75-12-7]	(20)	(30)	(30)	(45)
Formic acid [64-18-6]	5	9	—	—
‡Furfural [98-01-1] — Skin	2	8	(10)	(40)

Substance [CAS #]	ADOPTED VALUES			
	TWA		STEL	
	ppm[a]	mg/m^3[b]	ppm[a]	mg/m^3[b]
Furfuryl alcohol [98-00-0] — Skin	10	40	15	60
Gasoline [8006-61-9]	300	900	500	1,500
*Germanium tetrahydride [7782-65-2]	0.2	0.6	—	—
Glass, fibrous or dust, *see* Fibrous glass dust				
Glutaraldehyde [111-30-8]	C 0.2	C 0.7	—	—
Glycerin mist [56-81-5]	—	D	—	—
‡Glycidol [556-52-5]	25	75	(100)	(300)
Glycol monoethyl ether, *see* 2-Ethoxyethanol				
‡Graphite (Natural) [7782-42-5], *see* DUSTS				
‡Graphite (Synthetic)	—	(D)	—	—
*Gypsum [10101-4-4]	—	D	—	—
*Hafnium [7440-58-6]	—	0.5	—	—
Helium [7440-59-7]	E	—	—	—
*Heptachlor [76-44-8] — Skin	—	0.5	—	—
Heptane [142-82-5] (n-Heptane)	400	1,600	500	2,000
2-Heptanone, *see* Methyl n-amyl ketone				
3-Heptanone, *see* Ethyl butyl ketone				
Hexachlorobutadiene [87-68-3] — Skin	0.02,A2	0.24,A2	—	—
*Hexachlorocyclopentadiene [77-47-4]	0.01	0.1	—	—
Hexachloroethane [67-72-1]	10	100	—	—
*Hexachloronaphthalene [1335-87-1] — Skin	—	0.2	—	—
*Hexafluoroacetone [684-16-2] — Skin	0.1	0.7	—	—
Hexamethyl phosphoramide [680-31-9] — Skin	A2	A2	—	—
Hexane (n-Hexane) [110-54-3]	50	180	—	—
Other isomers	500	1,800	1,000	3,600
2-Hexanone, *see* Methyl n-butyl ketone				
Hexone, *see* Methyl isobutyl ketone				
sec-Hexyl acetate [108-84-9]	50	300	—	—
Hexylene glycol [107-41-5]	C 25	C 125	—	—
Hydrazine [302-01-2] — Skin	0.1,A2	0.1,A2	—	—
Hydrogen [1333-74-0]	E	—	—	—
Hydrogenated terphenyls [61788-32-7]	0.5	5	—	—
*Hydrogen bromide [10035-10-6]	C 3	C 10	—	—
Hydrogen chloride [7647-01-0]	C 5	C 7	—	—
Hydrogen cyanide [74-90-8] — Skin	C 10	C 10	—	—
*Hydrogen fluoride [7664-39-3], as F	C 3	C 2.5	—	—
*Hydrogen peroxide [7722-84-1]	1	1.5	—	—
Hydrogen selenide [7783-07-5], as Se	0.05	0.2	—	—
Hydrogen sulfide [7783-06-4]	10	14	15	21
‡Hydroquinone [123-31-9]	—	2	—	(4)
4-Hydroxy-4-methyl-2-pentanone, *see* Diacetone alcohol				
2-Hydroxypropyl acrylate [999-61-1] — Skin	0.5	3	—	—
‡Indene [95-13-6]	10	45	(15)	(70)
*Indium [7440-74-6] & compounds, as In	—	0.1	—	—
Iodine [7553-56-2]	C 0.1	C 1	—	—
*Iodoform [75-47-8]	0.6	10	—	—

Capital letters A, B, D & E refer to Appendices; C denotes ceiling limit
‡ See Notice of Intended Changes.
* 1986-1987 Addition.

Substance [CAS #]	TWA ppm[a]	TWA mg/m³[b]	STEL ppm[a]	STEL mg/m³[b]
*Iron oxide fume (Fe_2O_3) [1309-37-1], as Fe	B2	5	—	—
Iron pentacarbonyl [13463-40-6], as Fe	0.1	0.8	0.2	1.6
*Iron salts, soluble, as Fe	—	1	—	—
‡Isoamyl acetate [123-92-2]	100	525	(125)	(655)
Isoamyl alcohol [123-51-3]	100	360	125	450
Isobutyl acetate [110-19-0]	150	700	187	875
‡Isobutyl alcohol [78-83-1]	50	150	(75)	(225)
Isoocytl alcohol [26952-21-6]—Skin	50	270	—	—
Isophorone [78-59-1]	C 5	C 25	—	—
‡Isophorone diisocyanate [4098-71-9] — Skin	(0.01)	(0.09)	—	—
‡Isopropoxyethanol [109-59-1]	25	105	(75)	(320)
Isopropyl acetate [108-21-4]	250	950	310	1,185
Isopropyl alcohol [67-63-0]	400	980	500	1,225
Isopropylamine [75-31-0]	5	12	10	24
*N-Isopropylaniline [768-52-5] — Skin	2	10	—	—
Isopropyl ether [108-20-3]	250	1,050	310	1,320
Isopropyl glycidyl ether (IGE) [4016-14-2]	50	240	75	360
*Kaolin	—	D	—	—
Ketene [463-51-4]	0.5	0.9	1.5	3
*Lead [7439-92-1], inorg. dusts & fumes, as Pb	—	0.15	—	—
Lead arsenate [10102-48-4], as $Pb_3(AsO_4)_2$	—	0.15	—	—
Lead chromate [7758-97-6], as Cr	—	0.05,A2	—	—
*Limestone [1317-65-3]	—	D	—	—
*Lindane [58-89-9] — Skin	—	0.5	—	—
Lithium hydride [7580-67-8]	—	0.025	—	—
‡L.P.G. (Liquified petroleum gas)	1,000	1,800	(1,250)	(2,250)
*Magnesite [546-93-0]	—	D	—	—
Magnesium oxide fume [1309-48-4]	—	10	—	—
Malathion [121-75-5] — Skin	—	10	—	—
Maleic anhydride [108-31-6]	0.25	1	—	—
‡Manganese [7439-96-5], as Mn				
‡ Dust & compounds	—	(C 5)	—	—
Fume	—	1	—	3
*Manganese cyclopentadienyl tricarbonyl [12079-65-1], as Mn — Skin	—	0.1	—	—
Manganese tetroxide [1317-35-7]	—	1	—	—
*Marble/calcium carbonate [1317-65-3]	—	D	—	—
Mercury [7439-97-6], as Hg — Skin				
Alkyl compounds	—	0.01	—	0.03
All forms expect alkyl				
Vapor	—	0.05	—	—
Aryl & inorganic compounds	—	0.1	—	—
Mesityl oxide [141-79-7]	15	60	25	100
Methacrylic acid [79-41-4]	20	70	—	—
Methane [74-82-8]	E	—	—	—
Methanethiol, *see* Methyl mercaptan				
Methanol, *see* Methyl alcohol				
Methomyl [16752-77-5]	—	2.5	—	—

Capital letters A, B, D & E refer to Appendices; C denotes ceiling limit
‡ See Notice of Intended Changes.
* 1986-1987 Addition.

Substance [CAS #]	TWA ppm[a]	TWA mg/m³[b]	STEL ppm[a]	STEL mg/m³[b]
Methoxychlor [72-43-5]	—	10	—	—
2-Methoxyethanol [109-86-4] — Skin	5	16	—	—
2-Methoxyethyl acetate [110-49-6] — Skin	5	24	—	—
4-Methoxyphenol [150-76-5]	—	5	—	—
Methyl acetate [79-20-9]	200	610	250	760
Methyl acetylene [74-99-7]	1,000	1,650	1,250	2,040
Methyl acetylene-propadiene mixture (MAPP)	1,000	1,800	1,250	2,250
Methyl acrylate [96-33-3] — Skin	10	35	—	—
*Methylacrylonitrile [126-98-7] — Skin	1	3	—	—
‡Methylal [109-87-5]	1,000	3,100	(1,250)	(3,875)
Methyl alcohol [67-56-1] — Skin	200	260	250	310
Methylamine [74-89-5]	10	12	—	—
Methyl amyl alcohol, *see* Methyl isobutyl carbinol				
‡Methyl n-amyl ketone [110-43-0]	50	235	(100)	(465)
*N-Methyl aniline [100-61-8] — Skin	0.5	2	—	—
*Methyl bromide [74-83-9] — Skin	5	20	—	—
Methyl n-butyl ketone [591-78-6]	5	20	—	—
Methyl chloride [74-87-3]	50	105	100	205
Methyl chloroform [71-55-6]	350	1,900	450	2,450
Methyl 2-cyanoacrylate [137-05-3]	2	8	4	16
‡Methylcyclohexane [108-87-2]	400	1,600	(500)	(2,000)
‡Methylcyclohexanol [25639-42-3]	50	235	(75)	(350)
o-Methylcyclohexanone [583-60-8] — Skin	50	230	75	345
*Methylcyclopentadienyl manganese tricarbonyl [12108-13-3] — Skin, as Mn	—	0.2	—	—
*Methyl demeton [8022-00-2] — Skin	—	0.5	—	—
‡Methylene bisphenyl isocyanate (MDI) [101-68-8]	(C 0.02)	(C 0.2)	—	—
‡Methylene chloride [75-09-2]	(100)	(350)	(500)	(1,740)
4,4'-Methylene bis(2-chloroaniline) [101-14-4] — Skin	0.02,A2	0.22,A2	—	—
‡Methylene bis(4-cyclohexylisocyanate) [5124-30-1]	(C 0.01)	(C0.11)	—	—
*4,4'-Methylene dianiline [107-77-9] — Skin	0.1,A2	0.8,A2	—	—
Methyl ethyl ketone (MEK) [78-93-3]	200	590	300	885
Methyl ethyl ketone peroxide [1338-23-4]	C 0.2	C 1.5	—	—
Methyl formate [107-31-3]	100	250	150	375
5-Methyl-3-heptanone, *see* Ethyl amyl ketone				
Methyl hydrazine [60-34-4] — Skin	C 0.2,A2	C 0.35,A2	—	—
*Methyl iodide [74-88-4] — Skin	2,A2	10,A2	—	—

Capital letters A, B, D & E refer to Appendices; C denotes ceiling limit
‡ See Notice of Intended Changes.
* 1986-1987 Addition.

Substance [CAS #]	ADOPTED VALUES TWA ppm[a)]	TWA mg/m³[b)]	STEL ppm[a)]	STEL mg/m³[b)]
Methyl isoamyl ketone [110-12-3]	50	240	—	—
Methyl isobutyl carbinol [108-11-2]—Skin	25	100	40	165
Methyl isobutyl ketone [108-10-1]	50	205	75	300
Methyl isocyanate [624-83-9]—Skin	0.02	0.05	—	—
Methyl isopropyl ketone [563-80-4]	200	705	—	—
Methyl mercaptan [74-93-1]	0.5	1	—	—
‡Methyl methacrylate [80-62-6]	100	410	(125)	(510)
*Methyl parathion [298-00-0]—Skin	—	0.2	—	—
Methyl propyl ketone [107-87-9]	200	700	250	875
*Methyl silicate [681-84-5]	1	6	—	—
α-Methyl styrene [98-83-9]	50	240	100	485
Metribuzin [21087-64-9]	—	5	—	—
Mevinphos [7786-34-7]—Skin	0.01	0.1	0.03	0.3
Molybdenum [7439-98-7], as Mo				
* Soluble compounds	—	5	—	—
* Insoluble compounds	—	10	—	—
Monochlorobenzene, *see* Chlorobenzene				
Monocrotophos [6923-22-4]	—	0.25	—	—
Morpholine [110-91-8]—Skin	20	70	30	105
*Naled [300-76-5]—Skin	—	3	—	—
Naphthalene [91-20-3]	10	50	15	75
β-Naphthylamine [91-59-8]	—	A1b	—	—
Neon [7440-01-9]	E	—	—	—
Nickel [7440-02-0]				
Metal	—	1	—	—
* Soluble compounds, as Ni	—	0.1	—	—
Nickel carbonyl [13463-39-3], as Ni	0.05	0.35	—	—
Nickel sulfide roasting, fume & dust, as Ni	—	1,A1a	—	—
*Nicotine [54-11-5]—Skin	—	0.5	—	—
Nitrapyrin [1929-82-4]	—	10	—	20
Nitric acid [7697-37-2]	2	5	4	10
*Nitric oxide [10102-43-9]	25	30	—	—
p-Nitroaniline [100-01-6]—Skin	—	3	—	—
*Nitrobenzene [98-95-3]—Skin	1	5	—	—
‡p-Nitrochlorobenzene [100-00-5]—Skin	(0.5)	(3)	—	—
4-Nitrodiphenyl [92-93-3]	—	A1b	—	—
*Nitroethane [79-24-3]	100	310	—	—
Nitrogen dioxide [10102-44-0]	3	6	5	10
*Nitrogen trifluoride [7783-54-2]	10	30	—	—
Nitroglycerin (NG) [55-63-0]—Skin	0.05	0.5	—	—
*Nitromethane [75-52-5]	100	250	—	—
*1-Nitropropane [108-03-2]	25	90	—	—
‡2-Nitropropane [79-46-9]	(C 25, A2)	(C 90 A2)	—	—

Capital letters A, B, D & E refer to Appendices; C denotes ceiling limit
‡ See Notice of Intended Changes.
* 1986-1987 Addition.

Substance [CAS #]	ADOPTED VALUES TWA ppm[a)]	TWA mg/m³[b)]	STEL ppm[a)]	STEL mg/m³[b)]
N-Nitrosodimethylamine [62-75-9]—Skin	—	A2	—	—
Nitrotoluene [99-08-1]—Skin	2	11	—	—
Nitrotrichloromethane, *see* Chloropicrin				
‡Nonane [111-84-2]	200	1,050	(250)	(1,300)
Octachloronaphthalene [2234-13-1]—Skin	—	0.1	—	0.3
Octane [111-65-9]	300	1,450	375	1,800
Oil mist, mineral [8012-95-1]	—	5[(e)]	—	10
Osmium tetroxide [20816-12-0], as Os	0.0002	0.002	0.0006	0.006
Oxalic acid [144-62-7]	—	1	—	2
*Oxygen difluoride [7783-41-7]	C 0.05	C 0.1	—	—
Ozone [10028-15-6]	0.1	0.2	0.3	0.6
‡Paraffin wax fume [8002-74-2]	—	2	—	(6)
Paraquat [4685-14-7], respirable sizes	—	0.1	—	—
*Parathion [56-38-2]—Skin	—	0.1	—	—
Particulate polycyclic aromatic hydrocarbons (PPAH), *see* Coal tar pitch volatiles				
Pentaborane [19624-22-7]	0.005	0.01	0.015	0.03
*Pentachloronaphthalene [1321-64-8]	—	0.5	—	—
*Pentachlorophenol [87-86-5]—Skin	—	0.5	—	—
*Pentaerythritol [115-77-5]	—	D	—	—
Pentane [109-66-0]	600	1,800	750	2,250
2-Pentanone, *see* Methyl propyl ketone				
Perchloroethylene [127-18-4]	50	335	200	1,340
Perchloromethyl mercaptan [594-42-3]	0.1	0.8	—	—
Perchloryl fluoride [7616-94-6]	3	14	6	28
Phenacyl chloride, *see* α-Chloroacetophenone				
‡Phenol [108-95-2]—Skin	5	19	(10)	(38)
*Phenothiazine [92-84-2]—Skin	—	5	—	—
N-Phenyl-beta-naphthylamine [135-88-6]	A2	A2	—	—
p-Phenylene diamine [106-50-3]—Skin	—	0.1	—	—
Phenyl ether [101-84-8], vapor	1	7	2	14
Phenylethylene, *see* Styrene, monomer				
Phenyl glycidyl ether (PGE) [122-60-1]	1	6	—	—
Phenylhydrazine [100-63-0]—Skin	5,A2	20,A2	10,A2	45,A2
Phenyl mercaptan [108-98-5]	0.5	2	—	—
Phenylphosphine [638-21-1]	C 0.05	C 0.25	—	—
Phorate [298-02-2]—Skin	—	0.05	—	0.2
Phosdrin, *see* Mevinphos				
Phosgene [75-44-5]	0.1	0.4	—	—
Phosphine [7803-51-2]	0.3	0.4	1	1
Phosphoric acid [7664-38-2]	—	1	—	3
*Phosphorus (yellow) [7723-14-0]	—	0.1	—	—
Phosphorus oxychloride [10025-87-3]	0.1	0.6	0.5	3

Capital letters A, B, D & E refer to Appendices; C denotes ceiling limit
‡ See Notice of Intended Changes.
* 1986-1987 Addition.
(e) As Sampled by method that does not collect vapor.

	ADOPTED VALUES			
	TWA		STEL	
Substance [CAS #]	ppm[a]	mg/m^3[b]	ppm[a]	mg/m^3[b]
Phosphorus pentachloride [10026-13-8]	0.1	1	—	—
Phosphorus pentasulfide [1314-80-3]	—	1	—	3
Phosphorus trichloride [7719-12-2]	0.2	1.5	0.5	3
‡Phthalic anhydride [85-44-9]	1	6	(4)	(24)
m-Phthalodinitrile [626-17-5]	—	5	—	—
Picloram [1918-02-1]	—	10	—	20
Picric acid [88-89-1]—Skin	—	0.1	—	0.3
‡Pindone [83-26-1]	—	0.1	—	(0.3)
Piperazine dihydrochloride [142-64-3]	—	5	—	—
2-Pivalyl-1,3-indandione, *see* Pindone				
*Plaster of Paris	—	D	—	—
Platinum [7440-06-4]				
Metal	—	1	—	—
Soluble salts, as Pt	—	0.002	—	—
Polychlorobiphenyls, *see* Chlorodiphenyls				
Polytetrafluoroethylene decomposition products	—	B1	—	—
Portland cement	—	D	—	—
Potassium hydroxide [1310-58-3]	—	C 2	—	—
Propane [74-98-6]	E	—	—	—
Propane sultone [1120-71-4]	A2	A2	—	—
‡Propargyl alcohol [107-19-7]—Skin	1	2	(3)	(6)
‡β-Propiolactone [57-57-8]	0.5,A2	1.5,A2	(1,A2)	(3,A2)
‡Propionic acid [79-09-4]	10	30	(15)	(45)
‡Propoxur [114-26-1]	—	0.5	—	(2)
n-Propyl acetate [109-60-4]	200	840	250	1,050
Propyl alcohol [71-23-8]—Skin	200	500	250	625
Propylene [115-07-1]	E	—	—	—
Propylene dichloride [78-87-5]	75	350	110	510
Propylene glycol dinitrate [6423-43-4]—Skin	0.05	0.3	—	—
Propylene glycol monomethyl ether [107-98-2]	100	360	150	540
Propylene imine [75-55-8]—Skin	2,A2	5,A2	—	—
Propylene oxide [75-56-9]	20	50	—	—
n-Propyl nitrate [627-13-4]	25	105	40	170
Propyne, *see* Methyl acetylene				
‡Pyrethrum [8003-34-7]	—	5	—	(10)
‡Pyridine [110-86-1]	5	15	(10)	(30)
Pyrocatechol, *see* Catechol				
‡Quinone [106-51-4]	0.1	0.4	(0.3)	(1)
RDX, *see* Cyclonite				
Resorcinol [108-46-3]	10	45	20	90
Rhodium [7440-16-6]				
Metal	—	1	—	—
Insoluble compounds, as Rh	—	1	—	—
Soluble compounds, as Rh	—	0.01	—	—
Ronnel [299-84-3]	—	10	—	—
‡Rosin core solder pyrolysis products, as formaldehyde	—	0.1	—	(0.3)
‡Rotenone (commercial) [83-79-4]	—	5	—	(10)
*Rouge	—	D	—	—
Rubber solvent (Naphtha)	400	1,600	—	—
Selenium compounds [7782-49-2], as Se	—	0.2	—	—
Selenium hexafluoride [7783-79-1], as Se	0.05	0.2	—	—
*Sesone [136-78-7]	—	10	—	—
Silane, *see* Silicon tetrahydride				
*Silicon [7440-21-3]	—	D	—	—
*Silicon carbide [409-21-2]	—	D	—	—
Silicon tetrahydride [7803-62-5]	5	7	—	—
Silver [7440-22-4]				
Metal	—	0.1	—	—
Soluble compounds, as Ag	—	0.01	—	—
Sodium azide [26628-22-8]	C 0.1	C 0.3	—	—
Sodium bisulfite [7631-90-5]	—	5	—	—
Sodium 2,4-dichloro-phenoxyethyl sulfate, *see* Sesone				
Sodium fluoroacetate [62-74-8] — Skin	—	0.05	—	0.15
Sodium hydroxide [1310-73-2]	—	C 2	—	—
Sodium metabisulfite [7681-57-4]	—	5	—	—
*Starch [9005-25-8]	—	D	—	—
*Stibine [7803-52-3]	0.1	0.5	—	—
‡Stoddard solvent [8052-41-3]	100	525	(200)	(1,050)
*Strychnine [57-24-9]	—	0.15	—	—
Styrene, monomer [100-42-5]	50	215	100	425
Subtilisins [1395-21-7] (Proteolytic enzymes as 100% pure crystalline enzyme)	—	C 0.00006[f]	—	—
*Sucrose [57-50-1]	—	D	—	—
*Sulfotep [3689-24-5]—Skin	—	0.2	—	—
*Sulfur dioxide [7446-09-5]	2	5	5	10
*Sulfur hexafluoride [2551-62-4]	1,000	6,000	—	—
Sulfuric acid [7664-93-9]	—	1	—	—
*Sulfur monochloride [10025-67-9]	C 1	C 6	—	—
*Sulfur pentafluoride [5714-22-7]	C 0.01	C 0.1	—	—
*Sulfur tetrafluoride [7783-60-0]	C 0.1	C 0.4	—	—
Sulfuryl fluoride [2699-79-8]	5	20	10	40
Sulprofos [35400-43-2]	—	1	—	—
Systox, *see* Demeton				
*2,4,5-T [93-76-5]	—	10	—	—
‡Tantalum [7440-25-7]	—	(5)	—	(10)
TEDP, *see* Sulfotep				
Tellurium & compounds [13494-80-9], as Te	—	0.1	—	—
Tellurium hexafluoride [7783-80-4], as Te	0.02	0.2	—	—
*Temephos [3383-96-8]	—	10	—	—
*TEPP [107-49-3] — Skin	0.004	0.05	—	—
Terphenyls [26140-60-3]	C 0.5	C 5	—	—
*1,1,1,2-Tetrachloro-2,2-difluoroethane [76-11-9]	500	4,170	—	—
*1,1,2,2-Tetrochloro-1,2-difluoroethane [76-12-0]	500	4,170	—	—
*1,1,2,2-Tetrachloroethane [79-34-5]—Skin	1	7	—	—
Tetrachloroethylene, *see* Perchloroethylene				
Tetrachloromethane, *see* Carbon tetrachloride				
*Tetrachloronaphthalene [1335-88-2]	—	2	—	—

Capital letters A, B, D & E refer to Appendices; C denotes ceiling limit
‡ See Notice of Intended Changes.
* 1986-1987 Addition.

Substance [CAS #]	ADOPTED VALUES TWA ppm[a]	TWA mg/m^3[b]	STEL ppm[a]	STEL mg/m^3[b]
*Tetraethyl lead [78-00-2], as Pb—Skin	—	0.1(g)	—	—
Tetrahydrofuran [109-99-9]	200	590	250	735
*Tetramethyl lead [75-74-1], as Pb —Skin	—	0.15(g)	—	—
*Tetramethyl succinonitrile [3333-52-6]—Skin	0.5	3	—	—
Tetranitromethane [509-14-8]	1	8	—	—
Tetrasodium pyrophosphate [7722-88-5]	—	5	—	—
*Tetryl [479-45-8]—Skin	—	1.5	—	—
Thallium [7440-28-0] Soluble compounds, as Tl—Skin	—	0.1	—	—
*4,4'-Thiobis(6-tert, butyl-m-cresol) [96-69-5]	—	10	—	—
Thioglycolic acid [68-11-1]—Skin	1	4	—	—
*Thionyl chloride [7719-09-7]	C 1	C 5	—	—
*Thiram [137-26-8]	—	5	—	—
Tin [7440-31-5]				
* Metal	—	2	—	—
* Oxide & inorganic compounds, except SnH_4, as Sn	—	2	—	—
* Organic compounds, as Sn—Skin	—	0.1	—	—
*Titanium dioxide [13463-67-7]	—	D	—	—
o-Tolidine [119-93-7]—Skin	A2	A2	—	—
Toluene (toluol) [108-88-3]	100	375	150	560
Toluene-2,4-diisocyanate (TDI) [584-84-9]	0.005	0.04	0.02	0.15
o-Toluidine [95-53-4]—Skin	2,A2	9,A2	—	—
*m-Toluidine [108-44-1]—Skin	2	9	—	—
*p-Toluidine [106-49-0]—Skin	2,A2	9,A2	—	—
Toxaphene, *see* Chlorinated camphene				
*Tributyl phosphate [126-73-8]	0.2	2.5	—	—
Trichloroacetic acid [76-03-9]	1	7	—	—
1,2,4-Trichlorobenzene [120-82-1]	C 5	C 40	—	—
1,1,1-Trichloroethane, *see* Methyl chloroform				
*1,1,2-Trichloroethane [79-00-5]—Skin	10	45	—	—
Trichloroethylene [79-01-6]	50	270	200	1,080
Trichlorofluoromethane [75-69-4]	C 1,000	C 5,600	—	—
Trichloromethane, *see* Chloroform				
*Trichloronaphthalene [1321-65-9]—Skin	—	5	—	—
Trichloronitromethane, *see* Chloropicrin				
‡1,2,3-Trichloropropane [96-18-4]—Skin	(50)	(300)	(75)	(450)
1,1,2-Trichloro-1,2,2-trifluoroethane [76-13-1]	1,000	7,600	1,250	9,500
Tricyclohexyltin hydroxide, *see* Cyhexatin				
Triethylamine [121-44-8]	10	40	15	60

Capital letters A, B, D & E refer to Appendices; C denotes ceiling limit
‡ See Notice of Intended Changes.
* 1986-1987 Addition.
(f) Based on "high Volume" sampling.
(g) For control of general room air, biologic monitoring is essential for personnel control.

Substance [CAS #]	ADOPTED VALUES TWA ppm[a]	TWA mg/m^3[b]	STEL ppm[a]	STEL mg/m^3[b]
*Trifluorobromomethane [75-63-8]	1,000	6,100	—	—
Trimellitic anhydride [552-30-7]	0.005	0.04	—	—
Trimethylamine [75-50-3]	10	24	15	36
‡Trimethyl benzene [2551-13-7]	25	125	(35)	(170)
*Trimethyl phosphite [121-45-9]	2	10	—	—
2,4,6-Trinitrophenol, *see* Picric acid				
2,4,6-Trinitrophenylmethylnitramine, *see* Tetrvl				
*2,4,6-Trinitrotoluene (TNT) [118-96-7]—Skin	—	0.5	—	—
*Triorthocresyl phosphate [78-30-8]—Skin	—	0.1	—	—
Triphenyl amine [603-34-9]	—	5	—	—
*Triphenyl phosphate [115-86-6]	—	3	—	—
Tungsten [7440-33-7], as W				
Insoluble compounds	—	5	—	10
Soluble compounds	—	1	—	3
‡Turpentine [8006-64-2]	100	560	(150)	(840)
Uranium (natural) [7440-61-1]				
Soluble & insoluble compounds, as U	—	0.2	—	0.6
n-Valeraldehyde [110-62-3]	50	175	—	—
Vanadium, as V_2O_5 [1314-62-1]				
Respirable dust & fume	—	0.05	—	—
Vegetable oil mists	—	D	—	—
Vinyl acetate [108-05-4]	10	30	20	60
Vinyl benzene, *see* Styrene				
Vinyl bromide [593-60-2]	5,A2	20,A2	—	—
Vinyl chloride [75-01-4]	5,A1a	10,A1a	—	—
Vinyl cyanide, *see* Acrylonitrile				
Vinyl cyclohexene dioxide [106-87-6]—Skin	10,A2	60,A2	—	—
Vinylidene chloride [75-35-4]	5	20	20	80
Vinyl toluene [25013-15-4]	50	240	100	485
‡VM & P Naphtha [8032-32-4]	300	1,350	(400)	(1,800)
‡Warfarin [81-81-2]	—	0.1	—	(0.3)
Welding fumes (NOC†)	—	5,B2	—	—
Wood dust (certain hard woods as beech & oak)	—	1	—	—
Soft wood	—	5	—	10
Xylene [1330-20-7] (o-, m-, p-isomers)	100	435	150	655
m-Xylene α,α'-diamine [1477-55-0]—Skin	—	C 0.1	—	—
Xylidine [1300-73-8]—Skin	2	10	—	—
‡Yttrium [7440-65-5]	—	(1)	—	(3)
‡Zinc chloride fume [7646-85-7]	—	1	—	2
‡Zinc chromate [13530-65-9], as Cr	—	(0.05,A2)	—	—
Zinc oxide [1314-13-2]				
Fume	—	5	—	10
Dust	—	D	—	—
‡Zinc stearate [557-05-1]	—	(D)	—	(20)
Zirconium compounds [7440-67-2], as Zr	—	5	—	10

Capital letters A, B, D & E refer to Appendices; C denotes ceiling limit
‡ See Notice of Intended Changes.
* 1986-1987 Addition.
† NOC = Not otherwise classified.

Radioactivity: See Physical Agents section on Ionizing Radiation.

DUSTS

Substance	TLV-TWA
SILICA, SiO_2	
Crystalline	
* Quartz[h] [14808-60-7]	0.1 mg/m³, Respirable dust
* Cristobalite [14464-46-1]	0.05 mg/m³, Respirable dust
Silica, fused [60676-86-0]	0.1 mg/m³, Respirable dust
* Tridymite [15468-32-3]	0.05 mg/m³, Respirable dust
Tripoli [1317-95-9]	0.1 mg/³ of contained respirable quartz dust.
Amorphous	
*Diatomaceous earth (uncalcined)[i] [68855-54-9]	10 mg/m³, Total dust
‡ Precipitated silica	(5mg/m³, Respirable dust) 10 mg/m³, Total dust
‡ Silica gel	(5 mg/m³, Respirable dust) 10 mg/m³, Total dust
* *SILCATES[i]*	
Asbestos[j]	
Amosite [12172-73-5]	0.5 fiber/cc,[k] A1a
Chrysotile [12001-29-5]	2 fibers/cc,[k] A1a
Crocidolite [12001-28-4]	0.2 fiber/cc,[k] A1a
Other forms	2 fibers/cc,[k] A1a
‡Graphite (natural) [7782-42-5]	2.5 mg/m³, Respirable dust (5 mg/m³, Total dust)
*Mica [12001-25-2]	3 mg/m³, Respirable dust
Mineral wool fiber	10 mg/m³
*Perlite	10 mg/m³, Total dust
*Portland cement	10 mg/m³, Total dust
Soapstone	3 mg/m³, Respirable dust 6 mg/m³, Total dust
Talc (containing no asbestos fibers) [14807-96-6]	2 mg/m³, Respirable dust
Talc (containing asbestos fibers)	Use asbestos TLV-TWA. However, should not exceed 2 mg/m³ respirable dust.
OTHER DUSTS	
*Barium sulfate [7727-43-7]	10 mg/m³, Total dust
*Grain dust (oats, wheat, barley)	4 mg/m³, Total particulate
*Graphite, synthetic	10 mg/m³, Total dust
*Nuisance particulates (see Appendix D)	10 mg/m³, Total dust
‡*COAL DUST*	

‡ (2 mg/m³ (respirable dust fraction < 5% quartz). If > 5% quartz, use respirable quartz value.

FOOTNOTES FOR DUSTS

(h) Both concentration and percent quartz (if applicable) for respirable dust for the application of this limit are to be determined from the fraction passing a size-selector with the characteristics defined in Appendix F.

(i) For silicates, the values are for dust containing no asbestos and <1% crystalline silica in the total dust. For coal dust, the value is for coal dust containing <5% crystalline silica in the respirable fraction. For materials containing more than these percentages of crystalline silica, the environment should be evaluated against the TLV-TWA of 0.1 mg/m³ for respirable quartz. Even where the respirable quartz concentration is less than 0.1 mg/m³, the level of the major component should not exceed its TLV.

(j) As determined by the membrane filter method at 400-450 x magnification (4 mm objective) phase contrast illumination.

(k) Fibers longer than 5 μm and with an aspect ratio equal to or greater than 3:1.

NOTICE OF INTENDED CHANGES (for 1986-87)

These substances, with their corresponding values, comprise those for which either a limit has been proposed for the first time, or for which a change in the "Adopted" listing has been proposed. In both cases, the proposed limits should be considered trial limits that will remain in the listing for a period of at least two years. If, after two years no evidence comes to light that questions the appropriateness of the values herein, the values will be reconsidered for the "Adopted" list. Documentation is available for each of these substances.

Substance [CAS #]	TWA ppm[a]	TWA mg/m³[b]	STEL ppm[a]	STEL mg/m³[b]
Acrylamide [79-06-1]—Skin	—	0.03,A2	—	—
†Ammonium perfluoro-octanoate [3825-26-1]	—	0.1	—	—
†Caprolactam [105-60-2] Vapor & aerosol	0.25	1	—	—
Cobalt metal, dust & fume [7440-48-4], as Co	—	0.05	—	0.1
Enflurane [13838-16-9]	75	575	—	—
†Formamide [75-12-7]—Skin	10	15	—	—
Halothane [151-67-7]	50	400	—	—
†Hexamethylene diisocyanate [822-06-0]	0.005	0.035	—	—
†Isophorone diisocyanate [4098-71-9]	0.005	0.045	—	—
†Manganese dust & compounds [739-96-5]	—	5	—	—
†Methylene bisphenyl isocyanate [101-68-8]	0.005	0.055	—	—
†Methylene chloride [75-09-2]	50,A2	175,A2	—	—
†Methylene bis-(4-cyclohexylisocyanate [5124-30-1]	0.005	0.055	—	—
†p-Nitrochlorobenzene [100-00-5]	0.1	0.6	—	—
2-Nitropropane [79-46-9]	10,A2	35,A2	(20,A2)	(70,A2)
†Stearates	—	D	—	—
†Tantalum [7440-25-7], metal and oxide, *see* DUSTS				
1,2,3-Trichloropropane [96-18-4]—Skin	10	60	—	—
†Yttrium [7440-65-5] metal and compounds, and Y	—	1	—	—
†Zinc chromates [13530-65-9; 1103-86-9; 37300-23-5], as Cr	—	0.01,A1	—	—

DELETE THE SHORT-TERM EXPOSURE LIMITS (TLV-STELs) FOR THE FOLLOWING SUBSTANCES:

n-Amyl acetate
sec-Amyl acetate
Asphalt (petroleum) fumes
Benzene
Biphenyl
L.P.G. (liquified petroleum gas)
Methylal
Methyl n-amyl ketone
Methylcyclohexane
Methylcyclohexanol

2-Butoxyethanol
sec-Butyl acetate
tert-Butyl acetate
Crotonaldehyde
Cumene
Cyclohexane
Cyclohexanone
Cyclopentadiene
Cyclopentane
Dibutyl phthalate
Diethyl phthalate
2,6-Ditert. butyl-p-cresol
Ethyl acrylate
Ethylbutyl ketone
Ethyl formate
Furfural
Glycidol
Hydroquinone
Indene
Isoamyl acetate
Isobutyl alcohol
Isopropoxyethanol
Methyl methacrylate
Nonane
Parafffin wax fume
Phenol
Phthalic anhydride
Pindone
Propargyl alcohol
β-Propiolactone
Propionic acid
Propoxur
Pyrethrum
Pyridine
Quinone
Rosin core solder pyrolysis products
Rotenone (commercial)
Stoddard solvent
Trimethyl benzene
Turpentine
VM & P naphtha
Warfarin

NOTICE OF INTENDED CHANGES

DUSTS

Substance	TLV-TWA
SILICA, SiO_2	
Amorphous	
Precipitated silica[i]	10 mg/m³, Total dust
Silica gel[i]	10 mg/m³, Total dust
SILICATES[i]	
Coal dust	2 mg/m³ respirable dust fraction[i]
Graphite (natural) [7782-42-5]	2.5 mg/m³, Respirable dust
OTHER DUSTS	
†Stearates (Appendix D—Nuisance Particulates)	10 mg/m³, Total dust
†Tantalum [7440-25-7], metal and oxide	10 mg/m³, Total dust

NOTICE OF INTENT TO CHANGE

APPENDIX A

Carcinogens

The guidelines explain how the Chemical Substances Threshold Limit Values Committee classifies substances found in the occupational environment as either carcinogenic in man or experimental animals. Scientific debate over the existence of biological thresholds for carcinogens is unlikely to be resolved in the near future. For most substances determined to be carcinogenic by the Committee, a value is given to provide practical guidelines for the industrial hygienist to control exposures in the workplace.

The Chemical Substances TLV Committee considers information from the following kinds of studies to be indicators of a substance's potential to be a carcinogen in humans: epidemiology studies, toxicology studies and, to a lesser extent, case histories. Because of the long latent period for many carcinogens, and for ethical reasons, it is often impossible to base timely risk-management decisions on results from human toxicological studies. In order to recognize the qualitative differences in research results, two categories of carcinogens are designated in this booklet: A1—Confirmed Human Carcinogens; and A2 — Suspected Human Carcinogens. All steps must be taken to keep exposures to all A1 carcinogens to a minimum. Workers exposed to A1 carcinogens without a TLV should be properly equipped to insure virtually no contact with the carcinogen. Please see the *Documentation of the Threshold Limit Values* for a more complete description and derivation of these designations.

(a) Parts of vapor or gas per million parts of contaminated air by volume at 25°C and 760 torr.
(b) Approximate milligrams of substance per cubic meter of air.
† 1986-1987 Revision or Addition.
Capital letters A, B, D & E refer to Appendices; C denotes ceiling limit.

ADOPTED APPENDICES

APPENDIX A

Carcinogens

The Committee lists below those substances in industrial use that have proven carcinogenic in man, or have induced cancer in animals under appropriate experimental conditions. Present listing of those substances carcinogenic for man takes two forms: Those for which a TLV has been assigned (1a) and those for which environmental conditions have not been sufficiently defined to assign a TLV (1b).

A1a. *Human Carcinogens.* Substances, or substances associated with industrial processes, recognized to have carcinogenic or cocarcinogenic potential, with an assigned TLV:

Substance	TLV
Asbestos	
Amosite	0.5 fiber[k]
Chrysotile	2 fibers[k]
Crocidolite	0.2 fiber[k]
Other forms	2 fibers[k]
bis-(Chloromethyl) ether	0.001 ppm
Chromite ore processing (chromate)	0.05 mg/m³, as Cr
Chromium (VI), certain water insoluble compounds	0.05 mg/m³, as Cr
Coal tar pitch volatiles	0.2 mg/m³, as benzene solubles
Nickel sulfide roasting, fume and dust	1 mg/m³, as Ni
Vinyl chloride	5 ppm
†Zinc chromates	0.01 mg/m³, as Cr

A1b. *Human Carcinogens.* Substances, or substances associated with industrial processes, recognized to have carcinogenic potential without an assigned TLV:

4-Aminodiphenyl—Skin
Benzidine—Skin
β-Naphthylamine
4-Nitrodiphenyl—Skin

For the substances in 1b, no exposure or contact by any route—respiratory, skin or oral, as detected by the most sensitive methods—shall be permitted. The worker should be properly equipped to insure virtually no contact with the carcinogen.

A2. *Industrial Substances Suspect of Carcinogenic Potential for MAN.* Chemical substances or substances associated with industrial processes, which are suspect of inducing cancer, based on either 1) limited epidemiologic evidence, exclusive of clinical reports of single cases, or 2) demonstration of carcinogenesis in one or more animal species by appropriate methods.

Acrylamide—Skin	0.03 mg/m³
Acrylonitrile—Skin	2 ppm
Antimony trioxide production	—

† 1986-1987 Revision or Addition.

Arsenic trioxide production	—
Benzene	10 ppm
Benzo(a)pyrene	—
Beryllium	2 μg/m³
‡1,3-Butadiene	10 ppm
Carbon tetrachloride — Skin	5 ppm
Chloroform	10 ppm
Chlormethyl methyl ether	—
**Chromates of lead and zinc, as Cr	0.05 mg/m³
Chrysene	—
3,3'-Dichlorobenzidine — Skin	—
Dimethyl carbamoyl chloride	—
1,1-Dimethylhydrazine — Skin	0.5 ppm
Dimethyl sulfate — Skin	0.1 ppm
Ethylene dibromide — Skin	—
Ethylene oxide	1 ppm
Formaldehyde	1 ppm
Hexachlorobutadiene	0.02 ppm
Hexamethyl phosphoramide — Skin	—
Hydrazine — Skin	0.1 ppm
4,4'-Methylene bis(2-chloroaniline) — Skin	0.02 ppm
†Methylene chloride	50 ppm
‡4,4'-Methylene dianiline	0.1 ppm
Methyl hydrazine — Skin	C 0.2 ppm
Methyl iodide — Skin	2 ppm
2-Nitropropane	10 ppm
N-Nitrosodimethylamine — Skin	—
N-Phenyl-beta-naphthylamine	—
Phenylhydrazine — Skin	5 ppm
Propane sultone	—
β-Propiolactone	0.5 ppm
Propylene imine — Skin	2 ppm
o-Tolidine — Skin	—
o-Toluidine — Skin	2 ppm
‡p-Toluidine — Skin	2 ppm
Vinyl bromide	5 ppm
Vinyl cyclohexene dioxide — Skin	10 ppm

For the above, worker exposure by all routes should be carefully controlled to levels consistent with the animal and human experience data (*see* Documentation), including those substances with a listed TLV.

* * * * *

The Committee Guidelines for Classification of Experimental Animal Carcinogens

The following guidelines are offered in the present state of knowledge as an aid in classifying substances in the occupational environment found to be carcinogenic in experimental animals. A need was felt by the Threshold Limits Committee for such a classification in order to take the first step in developing an appropriate TLV for occupational exposure.

Determination of Approximate Threshold Response Requirement. In order to determine in which category to classify an experimental carcinogen for the purpose of assigning an industrial air limit (TLV), an approximate threshold of neoplastic response must be determined. Because of practical experimental difficulties, a precisely defined threshold cannot be attained. For the purpose of standard-setting, this is of little moment, as an appropriate risk, or safety, factor can be applied to the approximate threshold, the magnitude of which is dependent on the degree of potency of the carcinogenic response.

To obtain the best "practical" threshold of neoplastic response, dosage decrements should be less than logarithmic. This becomes particularly important at levels greater than 10 ppm (or corresponding mg/m³). Accordingly, after a range-finding determination has been made by logarithmic decreases, two additional dosage levels are required within the levels of "effect" and "no effect" to approximate the true threshold of neoplastic response.

The second step should attempt to establish a metabolic relationship between animal and man for the particular substance found carcinogenic in animals. If the metabolic pathways are found comparable, the substance should be classed highly suspect as a carcinogen for man. If no such relation is found, the substance should remain listed as an experimental animal carcinogen until evidence to the contrary is found.

Proposed Classification of Experimental Animal Carcinogens. Substances occurring in the occupational environment found carcinogenic for animals may be grouped into three classes, those of high, intermediate and low potency. In evaluating the incidence of animal cancers, significant incidence of cancer is defined as a neoplastic response which represents, in the judgment of the Committee, a significant excess of cancers above that occurring in negative controls.

EXCEPTIONS: No substance is to considered an occupational carcinogen of any practical significance which reacts by the respiratory route at or above 1000 mg/m³ for the mouse, 2000 mg/m³ for the rat; by the dermal route, at or above 1500 mg/kg for the mouse, 3000 mg/kg for the rat; by the gastrointestinal route at or above 500 mg/kg/d for a lifetime, equivalent to about 100 g T.D. for the rat, 10 g T.D. for the mouse. These dosage limitations exclude such substances as dioxane and trichlorethylene from consideration as carcinogens.

Examples: **1)** Dioxane — rats, hepatocellular and nasal tumors from 1015 mg/kg/d, oral; **2)** Trichloroethylene — female mice, tumors (30/98 at 900 mg/kg/d), oral.

A. *Industrial Substances of High Carcinogenic Potency in Experiemental Animals.*

1. A substance to qualify as a carcinogen of high potency must fulfill *one* of the three following conditions (a, b, or c) in two animal species:

a. *Respiratory.* Elicit cancer from 1) dosages below 1 mg/m³ (or equivalent ppm) via the respiratory tract in 6- to 7-hour daily repeated inhalation exposures throughout lifetime; or 2) from a single intratracheally administered dose not exceeding 1 mg of particulate, or liquid, per 100 ml or less of animal minute respiratory volume.

Examples: **1)** bis(chloromethyl) ether — malignant tumors, rats, at 0.47 mg/m³ (0.1 ppm) in 2 years; **2)** Hexamethyl phosphoramide — nasal squamous cell carcinoma, rats at 0.05 ppm, in 13 months.

b. *Dermal.* Elicit cancer within 20 weeks by skin-painting, twice weekly at 2 mg/kg body or less per application for a total dose equal to or less than 1.5 mg, in a biologically inert vehicle.

Examples: **1)** 7,12-Dimethylbenz(a)anthracene — skin tumors at 0.12-0.8 mg T.D. in four weeks; **2)** Benzo(a)pyrene — mice 12 μg, 3 x /wk for 18 mos. T.D. 2.6 mg. 90.9% skin tumors

c. *Gastrointestinal.* Elicit cancer by daily intake via the gastrointestinal tract, within six months, with a six-month holding period, at a dosage below 1 mg/kg body weight per day; total dose, rat, ⩽ 50 mg; mouse, ⩽ 3.5 mg.

Examples: **1)** 7,12-Dimethylbenz(a)anthracene — mammary tumors from 10 mg 1X; **2)** 3-Methylcholanthrene — tumors at 3 sites from 8 mg in 89 weeks; **3)** Benzo(a)pyrene — mice, 3.9% leukemias, from 30 mg T.D. 198 days.

2. Elicit cancer by all three routes in at least two animal species at dose levels prescribed for high or intermediate potency.

B. *Industrial Substances of Intermediate Carcinogenic Potency in Experimental Animals.*

To qualify as a carcinogen of intermediate potency, a substance should elicit cancer in two animal species at dosages intermediate between those described in A and C by two routes of administration.

Example: Carbamic acid ethyl ester — dermal, mammary tumors, mice, 100%, 63 weeks, 500-1400 mg T.D. Gastrointestinal, vari-

** See Notice of Intended Changes.
† 1986-1987 Addition.
‡ 1986-1987 Adoption.

ous type tumors, mice 42 weeks, 320 mg T.D. Gastrointestinal, various type tumors, rats, 60 weeks, 110-930 mg T.D.

C. *Industrial Substances of Low Carcinogenic Potency in Experimental Animals.*

To qualify as a carcinogen of low potency, a substance should elicit cancer in one animal species by any *one* of the three routes of administration at the following prescribed dosages and conditions:

1. ***Respiratory.*** Elicit cancer from *a)* dosages greater than 10 mg/m³ (or equivalent ppm) via the respiratory tract in 6- to 7-hour, daily repeated inhalation exposures, for 12 months' observation period; or *b)* from intratracheally administered dosages totaling more than 10 mg of particulate or liquid per 100 ml or more of animal minute respiratory volume.

Examples: **1)** Beryl (beryllium aluminum silicate) — malignant lung tumors, rats, at 15 mg/m³ at 17 months; **2)** Benzidine — various tumors, rats, 10-20 mg/m³ at > 13 months.

2. ***Dermal.*** Elicit cancer by skin-painting of mice in twice weekly dosages of > 10 mg/kg body weight in a biologically inert vehicle for at least 75 weeks, i.e., ⩾ 1.5 g T.D.

Examples: **1)** Shale tar — mouse, 0.1 ml × 50 = 5 g T.D. 59/60 skin tumors; **2)** Arsenic trioxide — man, dose unknown, but estimated to be high.

3. ***Gastrointestinal.*** Elicit cancer from daily oral dosages of 50 mg/kg/day or greater during the lifetime of the animal.

APPENDIX B
Substances of Variable Composition

B1. *Polytetrafluoroethylene* decomposition products.* Thermal decomposition of the fluorocarbon chain in air leads to the formation of oxidized products containing carbon, fluorine and oxygen. Because these products decompose in part by hydrolysis in alkaline solution, they can be quantitatively determined in air as fluoride to provide an index of exposure. No TLV is recommended pending determination of the toxicity of the products, but air concentration should be minimal.

B2. *Welding Fumes — Total Particulate (NOC)*†
TLV-TWA, 5 mg/m³

Welding fumes cannot be classified simply. The composition and quantity of both are dependent on the alloy being welded and the process and electrodes used. Reliable analysis of fumes cannot be made without considering the nature of the welding process and system being examined; reactive metals and alloys such as aluminum and titanium are arc-welded in a protective, inert atmosphere such as argon. These arcs create relatively little fume, but an intense radiation which can produce ozone. Similar processes are used to arc-weld steels, also creating a relatively low level of fumes. Ferrous alloys also are arc-welded in oxidizing environments which generate considerable fume, and can produce carbon monoxide instead of ozone. Such fumes generally are composed of discreet particles of amorphous slags containing iron, manganese, silicon and other metallic constituents depending on the alloy system involved. Chromium and nickel compounds are found in fumes when stainless steels are arc-welded. Some coated and flux-cored electrodes are formulated with fluorides and the fumes associated with them can contain significantly more fluorides than oxides. Because of the above factors, arc-welding fumes frequently must be tested for individual constituents which are likely to be present to determine whether specific TLVs are exceeded. Conclusions based on total fume concentration are generally adequate if no toxic elements are present in welding rod, metal, or metal coating and conditions are not conducive to the formation of toxic gases.

Most welding, even with primitive ventilation, does not produce exposures inside the welding helmet above 5 mg/m³. That which does, should be controlled.

APPENDIX C
Threshold Limit Values for Mixtures

When two or more hazardous substances, which act upon the same organ system, are present, their combined effect, rather than that of either individually, should be given primary consideration. In the absence of information to the contrary, the effects of the different hazards should be considered as additive. That is, if the sum of the following fractions,

$$\frac{C_1}{T_1} + \frac{C_2}{T_2} + \ldots \frac{C_n}{T_n}$$

exceeds unity, then the threshold limit of the mixture should be considered as being exceeded. C_1 indicates the observed atmospheric concentration, and T_1 the corresponding threshold limit (*see* Example A.1 and B.1).

Exceptions to the above rule may be made when there is a good reason to believe that the chief effects of the different harmful substances are not in fact additive, but *independent* as when purely local effects or different organs of the body are produced by the various components of the mixture. In such cases the threshold limit ordinarily is exceeded only when at least one member of the series (C_1/T_1 + or + C_2/T_2, etc.) itself has a value exceeding unity (*see* Example B.1).

Synergistic action or potentiation may occur with some combinations of atmospheric contaminants. Such cases at present must be determined individually. Potentiating or synergistic agents are not necessarily harmful by themselves. Potentiating effects of exposure to such agents by routes other than that of inhalation is also possible, e.g., imbibed alcohol and inhaled narcotic (trichloroethylene). Potentiation is characteristically exhibited at high concentrations, less probably at low.

When a given operation or process characteristically emits a number of harmful dusts, fumes, vapors or gases, it will frequently be only feasible to attempt to evaluate the hazard by measurement of a single substance. In such cases, the threshold limit used for this substance should be reduced by a suitable factor, the magnitude of which will depend on the number, toxicity and relative quantity of the other contaminants ordinarily present.

Examples of processes which are typically associated with two or more harmful atmospheric contaminants are welding, automobile repair, blasting, painting, lacquering, certain foundry operations, diesel exhausts, etc.

Examples of TLVs for Mixtures

A. *Additive effects.* The following formulae apply only when the components in a mixture have similar toxicologic effects; they should not be used for mixtures with widely differing reactivities, e.g., hydrogen cyanide and sulfur dioxide. In such case the formula for Independent Effects (B) should be used.

1. General case, where air is analyzed for each component, the TLV of mixture =

$$\frac{C_1}{T_1} + \frac{C_2}{T_2} + \frac{C_3}{T_3} + \ldots = 1$$

Note: It is essential that the atmosphere be analyzed both qualitatively and quantitatively for each component present, in order to evaluate compliance or non-compliance with this calculated TLV.

Example A.1: Air contains 400 ppm of acetone (TLV, 750 ppm), 150 ppm of sec-butyl acetate (TLV, 200 ppm) and 100 ppm of methyl ethel ketone (TLV, 200 ppm).

Atmospheric concentration of mixture = 400 + 150 + 100 = 650 ppm of mixture.

$$\frac{400}{750} + \frac{150}{200} + \frac{100}{200} = 0.53 + 0.75 + 0.5 = 1.78$$

Threshold Limit is exceeded.

* Trade Names: Algoflon, Fluon, Teflon, Tetran.

† Not otherwise classified (NOC).

2. Special case when the source of contaminant is a liquid mixture and the atmospheric composition is *assumed* to be similar to that of the original material, e.g., on a time-weighted average exposure basis, all of the liquid (solvent) mixture eventually evaporates. When the percent composition (by weight) of the liquid mixture is known, the TLVs of the constituents must be listed in mg/m³. TLV of mixture =

$$\frac{1}{\frac{f_a}{TLV_a} + \frac{f_b}{TLV_b} + \frac{f_c}{TLV_c} + \ldots \frac{f_n}{TLV_n}}$$

Note: In order to evaluate compliance with this TLV, field sampling instruments should be calibrated, in the laboratory, for response to this specific quantitative and qualitative air-vapor mixture, and also to fractional concentrations of this mixture, e.g., 1/2 the TLV; 1/10 the TLV; 2 × the TLV; 10 × the TLV; etc.)

Example A.2: Liquid contains (by weight):

50% heptane: TLV = 400 ppm or 1600 mg/m³
1 mg/m³ ≡ 0.25 ppm
30% methyl chloroform: TLV = 350 ppm or 1900 mg/m³
1 mg/m³ ≡ 0.18 ppm
20% perchloroethylene: TLV = 50 ppm or 335 mg/m³
1 mg/m³ ≡ 0.15 ppm

$$\text{TLV of Mixture} = \frac{1}{\frac{0.5}{1600} + \frac{0.3}{1900} + \frac{0.2}{335}}$$

$$= \frac{1}{0.00031 + 0.00016 + 0.0006}$$

$$= \frac{1}{0.00107} = 935 \text{ mg/m}^3$$

of this mixture
50% or (935)(0.5) = 468 mg/m³ is heptane
30% or (935)(0.3) = 281 mg/m³ is methyl chloroform
20% or (935)(0.2) = 187 mg/m³ is perchloroethylene

These values can be converted to ppm as follows:

heptane: 468 mg/m³ × 0.25 = 117 ppm
methyl chloroform: 281 mg/m³ × 0.18 = 51 ppm
perchloroethylene: 187 mg/m³ × 0.15 = 29 ppm

TLV of mixture = 117 + 51 + 29 = 197 ppm, or 935 mg/m³

B. *Independent effects.* TLV for mixture =

$$\frac{C_1}{T_1} = 1; \quad \frac{C_2}{T_2} = 1 \quad \frac{C_3}{T_3} = 1; \text{ etc.}$$

Example B.1: Air contains 0.15 mg/m³ of lead (TLV, 0.15) and 0.7 mg/m³ of sulfuric acid (TLV, 1).

$$\frac{0.15}{0.15} = 1; \qquad \frac{0.7}{1} = 0.7$$

Threshold limit is not exceeded.

C. *TLV for mixtures of mineral dusts.* For mixtures of biologically active mineral dusts the general formula for mixtures given in A.2 may be used.

APPENDIX D
Some Nuisance Particulates(I)

*TLV-TWA, 10 mg/m³ of total dust(i)

α-Alumina (Al_2O_3)
Calcium carbonate
Calcium silicate
Cellulose (paper fiber)
Emery
Glycerin mist
Gypsum
Kaolin
Limestone
Magnesite
Marble
Vegetable oil mists (except castor oil, cashew nut or similar irritant oils)

Mineral wool fiber
Pentaerythritol
Plaster of Paris
Portland cement
Rouge
Silicon
Silicon carbide
Starch
†Stearates
Sucrose
Titanium dioxide
‡Zinc stearate
Zinc oxide dust

(i) When toxic impurities are not present, e.g., quartz < 1%.
(I) As defined in the Introduction.
‡ See Notice of Intended Changes.

APPENDIX E
Some Simple Asphyxiants(I)

Acetylene
Argon
Ethane
Ethylene
Helium

Hydrogen
Methane
Neon
Propane
Propylene

(I) As defined in the Introduction.

APPENDIX F
Chemical Substances and Other Issues Under Study[A]

Chemical Substances

Acetomethylchloride
Acetophenone
Acetylacetone
Acrylic acid
Allyl chloride
Bromodichloromethane
Ceramic fibers
Dibutyl phenyl phosphate
Dichlorvos
Dinitrotoluene
Epichlorohydrin
Ethylamines
Gasoline (unleaded)
Graphite fibers
Hexachlorocyclopentadiene
Hydrazine
Jet, petroleum and diesel fuels
Malathion

Methyl bromide
Methyl hydrazines
Mineral wool fibers
Naled
Nitrous oxide
Pentachlorophenol
Perchloroethylene
Persulfates
Petroleum solvents
o-Phenylenediamine
Propylene dichloride
Rosin core solder pyrolysis p
Skydrol hydraulic fluid
1,1,1,2-Tetrachloro-2,2-difluor
1,1,2,2-Tetrachloro-1,2-difluor
1,1,2,2-Tetrachloroethane
Thiram
Trichloroethylene

Other Issues

1. Particle Size-Selective Sampling Criteria for Airborne Particulate Matter[B]

   For chemical substances present in inhaled air as suspensions of solid particles or droplets, the potential hazard depends on particle size as well as mass concentration because of: 1) effects of particle size on deposition site within the respiratory tract, and 2) the tendency for many occupational diseases to be associated with material deposited in particular regions of the respiratory tract.

   ACGIH has recommended particle size-selective TLVs for crystalline silica for many years in recognition of the well established association between silicosis and respirable mass concentrations. It now has embarked on a re-examination of other chemical substances encountered in particulate form in occupational environments with the objective of defining: 1) the size-fraction most closely associated for each substance with the health effect of concern, and 2) the mass concentration within that size fraction which should represent the TLV.

The Particle Size-Selective TLVs (PSS-TLVs) will be expressed in three forms, e.g.,

a. *Inspirable Particulate Mass TLVs (IPM-TLVs)* for those materials which are hazardous when deposited anywhere in the respiratory tract.
b. *Thoracic Particulate Mass TLVs (TPM-TLVs)* for those materials which are hazardous when deposited anywhere within the lung airways and the gas-exchange region.
c. *Respirable Particulate Mass TLVs (RPM-TLVs)* for those materials which are hazardous when deposited in the gas-exchange region.

The three particulate mass fractions described above are defined in quantitative terms as follows:

a. Inspirable Particulate Mass consists of those particles that are captured according to the following collection efficiency regardless of sampler orientation with respect to wind direction:

$$E = 50(1 + \exp[-0.06\ d_a]) \pm 10;$$
$$\text{for } 0 < d_a \leqslant E\ 100\ \mu m$$

Collection characteristics for $d_a > 100$ μm are presently unknown. E is collection efficiency in percent and d_a is aerodynamic diameter in μm.

b. Thoracic Particulate Mass consists of those particles that penetrate a separator whose size collection efficiency is described by a cumulative lognormal function with a median aerodynamic diameter of 10 μm $\pm$1.0 μm and with a geometric standard deviation of 1.5 ($\pm$0.1).
c. Respirable Particulate Mass consists of those particles that penetrate a separator whose size collection efficiency is described by a cumulative lognormal function with a median aerodynamic diameter of 3.5 μm $\pm$0.3 μm and with a geometric standard deviation of 1.5 ($\pm$0.1). This incorporates and clarifies the previous ACGIH Respirable Dust Sampling Criteria.

   These definitions provide a range of acceptable performance for each type of size-selective sampler. Further information is available on the background and performance criteria for these particle size-selective sampling recommendations.[1-3]

References

1. **ACGIH:** *Particle Size-Selective Sampling in the Workplace*, 80 pp. Cincinnati, Ohio (1984).
2. Particle Size-Selective Sampling in the Workplace. *Ann. Am. Conf. Govt. Ind. Hyg. 11*:23-100 (1984).
3. Chapter 7, Performance Considerations for Size-Selective Samplers (revised). Submitted to *Ann. Am. Conf. Govt. Ind. Hyg.*(1986).

2. Should the TLVs currently expressed as "total dust"[C] be changed to "inspirable particulate mass" (as defined in the above criteria) without changing the numerical value?
3. Applications of TLVs to altered work schedules.

[A] Information, data especially, and comments are solicited to assist the Committee in its deliberations and in the development of draft documents. Draft documentations are used by the Committee to decide what action, if any, to recommend on a given question.

[B] Includes redefinition of respirable dust and notice of additional size-selective concentrations to be used in TLVs for particulate matter under consideration for revision.

[C] As used for the mineral dusts.

APPENDIX G
Registered Trade Names

Trade Name	Generic Name	CAS No.
Abate	Temephos	3383-96-8
Ammate	Ammonium sulfamate	7773-06-0
Azodrin	Monocrotophos	6923-22-4
Baygon	Propoxur	114-26-1
Baytex	Fenthion	55-38-9
Bidrin	Dicrotophos	141-66-2
Bolstar	Sulprofos	35400-43-2
Butyl Cellosolve	2-Butoxyethanol	111-76-2
Cellosolve acetate	2-Ethoxyethyl acetate	111-15-9
Coyden	Clopidol	2971-90-6
Crag herbicide	Sesone	136-78-7
Dasanit	Fensulfothion	115-90-2
Delnav	Dioxathion	78-34-2
Dibrom	Naled	300-76-5
Difolatan	Captafol	2425-06-1
Disyston	Disulfoton	298-04-4
Dursban	Chlorpyrifos	2921-88-2
Dyfonate	Fonofos	944-22-9
Furadan	Carbofuran	1563-66-2
Guthion	Azinphos-methyl	86-50-0
Lannate	Methomyl	16752-77-5
Methyl Cellosolve	2-Methoxyethanol	109-84-4
Methyl Cellosolve acetate	2-Methoxyethyl acetate	110-49-6
Nemacur	Fenamiphos	22224-92-6
Nialate	Ethion	563-12-2
N-Serve	Nitrapyrin	1929-82-4
Pival	Pindone	83-26-1
Plictran	Cyhexatin	1312 1-70-5
Sencor	Metribuzin	21087-64-9
Sevin	Carbaryl	63-25-2
Teflon	Polytetrafluoro-ethylene	9002-84-0
Thimet	Phorate	298-02-2
Thiodan	Endosulfan	115-29-7
Tordon	Picloram	1918-02-1
Zoalene	Dinitolmide	148-01-6

INTRODUCTION
BIOLOGICAL EXPOSURE INDICES

Biological Exposure Indices (BEIs) represent warning levels of biological response to the chemical, or warning levels of the chemical or its metabolic product(s) in tissues, fluids, or exhaled air of exposed workers, regardless of whether the chemical was inhaled, ingested, or absorbed via skin. Introduction of the BEI is a step in the evolution of the concept of TLVs. The BEI provides the health personnel with an additional tool to provide protection for the worker. Use of body fluids and appendages such as hair or nails for measuring the absorbed amount of a substance has long been a standard practice for certain substances. Lead is a classical example of a substance for which blood concentrations have long been considered the critical value in determining "safe" versus "unsafe" exposures. Two problems hindered the wider use of biological measurements as indicators of "safe" environmental exposures: 1) the relatively wide range in individual response to a substance and the wide range of "normal" that has to be considered; and 2) the lack of simple specific analytical methods of sufficient sensitivity. Both problems are capable of solution, and we believe that sufficient progress has been made to begin utilizing selected BEIs which can be used as a guide to "safe" exposures to toxic chemicals. The BEI is considered supplementary to an airborne TLV.

TLVs are intended to provide the industrial hygienist with an additional measure to aid in the design of engineering controls or for temporary use of personnel protective equipment which will protect almost all exposed workers from untoward effects of chemical exposure. In principle, TWA-TLVs are designed to prevent exposures which may cause acute or chronic adverse effects. They are also intended to avoid attendant deterioration of normal physiological function. This approach is based on the assumption that for nearly all workers there is a tolerable exposure limit and a tolerable body burden of airborne material. If this assumption is valid, there should be a range of safe biologically insignificant changes of various measures of body function.

TLVs are a measure of the composition of the external environment surrounding the worker. BEIs are a measure of the amount of chemical *absorbed* into the body. The concept of the BEI is particularly useful in evaluating exposures to substances with significant absorption through the skin.

The biological determinant on which the BEIs are based can furnish two kinds of information useful in the control of worker exposure; 1) measure of the worker's individual response, and 2) measure of the worker's individual overall exposure. Measurements of response furnish an estimate of the physiological status of the worker and can be made by, *a)* determining changes in the amount of a critical biochemical constituent, *b)* determining changes in activity of a critical enzyme, and *c)* determining changes in a physiological function. Measurements of exposure can be made by, *a)* determining the chemical in exhaled air, urine, blood, hair, nails, body tissues and fluids, *b)* determining the metabolite(s) of the chemical in tissues and fluids, and *c)* determining the extent of specific biochemical and physiological changes induced by the chemical.

Recommended values of BEIs are based on data obtained in epidemiological and field studies or determined as bioequivalent to a TLV by means of pharmacokinetic analysis of data from controlled human studies. Most chemicals (including organic solvents) are initially absorbed and eliminated fairly rapidly—usually with initial half-life values measured in a few hours or even minutes. Rapidly changing concentrations in body fluids complicate the interpretation of data and the average body burden of a chemical attained during a work shift can easily be over-predicted or underpredicted. Furthermore, biological measurements fail in most instances to detect transient periods of over-exposure during the work shift. Because elimination of chemicals and their metabolic products, as well as biological changes induced by exposure to the chemical, are kinetic events, the listed BEIs are strictly related to 8-hour exposures and to the specified timing for the collection of biological samples.

Factors to be considered. There are other factors to be considered when the BEI is applied. Among the factors which must be considered in using BEIs are: *a)* changes induced by strenuous physical activity; *b)* changes induced by environmental conditions (altitude, heat, diet, etc.); *c)* changes induced by water intake; *d)* changes in physiological functions induced by preexisting disease or congenital variation; *e)* changes in metabolism induced by congenital variation of metabolic pathways; and *f)* changes in metabolic pathway induced by simultaneous administration of another chemical (induction or inhibition of activity of a critical enzyme by medication or by preexposure or coexposure to another chemical).

For BEIs based on urine analysis, a simple measurements of concentrations can provide sufficient information on exposure, but in many instances, measurements of elimination rates provide more precise information. Urinary concentrations related to creatinine represent a reasonable compromise between the accuracy of the information and the technical means of obtaining the data.

Some BEIs are not protective of an identified population, or are nonspecific. The correlation between the exposure and biological determinant is weakened by variables introduced by large interindividual variation in response to the chemical or by time factors and fluctuation of exposure concentration. In such cases the BEIs carry the following notations:

"R" Notation. Indicates that an identifiable population group might have an increased susceptibility to the effect of the chemical which leaves it unprotected by the recommended BEI. The specific documentation should be consulted for detailed information.

"*" Notation. Some determinants are nonspecific, since different chemicals may bear the same biological response. Such BEIs carry "*" notation. These nonspecific tests are preferred because they are easy to use and, in many instances, offer a better correlation between exposure and response than specific tests. In such instances, a BEI for a specific less quantitative biological determinant is recommended as a confirmatory test. The documentation should be consulted for information on factors affecting interpretation of such a BEI.

"**" Notation. Indicates that the biological determinant is a specific indicator of exposure to the chemical but the quantitative interpretation of the measurements is very ambiguous, and that the relationship between the TLV and BEI is markedly weakened by variables or time factors, fluctuation of exposure concentrations, and by other circumstantial variables. Such biological determinants should be used as confirmatory tests; and their BEIs should be applied cautiously, mainly for confirmation of exposures indicated by nonspecific BEIs.

"†" Notation. The determinant is usually present in the biological specimens collected from subjects without occupational exposure. For information on background levels consult the documentation.

"G" Notation. Because of the wide interindividual variation in response to some chemicals, the BEI is in some instances recommended as a mean value of a group test for workers subjected to a similar level of occupational exposure, rather than as an index for an individual. Such BEIs carry "G" notation.

The table includes BEIs for which a sufficient data base is available and on which the committee took action. Some BEIs are more suitable for correlation with TLV-TWAs than others. Other BEIs are preferable for evaluating recent exposure or for confirmation of exposure. For specific instances the documentation should be consulted.

Workers are not expected to suffer any ill effects as long as the described measurement of the determinants are maintained within limits of the recommended BEIs. Measurements outside these limits are not necessarily indicators of a disease process. However, if these deviate measurements persist, it is an indication that the individual should be examined by a physician to determine whether there is any health effect. The workplace and work practices should be investigated further.

NOTE: *It is strongly advisable to consult the specific documentation published in the* Documentation of Threshold Limit Values and Biological Exposure Indices, 5th edition, *before invoking the BEIs listed in the the following table.*

ADOPTED BIOLOGICAL EXPOSURE INDICES

Airborne Chemical [CAS #] *Indices*	*Timing*	*BEI*	*Additional Notation*
• CARBON MONOXIDE [630-08-0]			R
* Carboxyhemoglobin in blood	End of shift	less than 8%	†
* CO in end-exhaled air	End of shift	less than 40 ppm	†
• ETHYL BENZENE [100-41-4]			
* Mandelic acid in urine	End of shift and end of workweek	2 g/L 1.5 g/g creat.	G G
** Ethyl benzene in end-exhaled air	Prior to shift	2 ppm	
• STYRENE [100-42-5]			
* Mandelic acid in urine	End of shift	1 g/L 0.8 g/g creat.	G G
** Styrene in mixed-exhaled air	Prior to shift	40 ppb	
* Phenylglyoxylic acid in urine	End of shift	250 mg/L 240 mg/g creat.	† G G
** Styrene in mixed-exhaled air	During shift	18 ppm	
** Styrene in blood	End of shift Prior to shift	0.55 mg/L 0.02 mg/L	
• TOLUENE [108-88-3]			
* Hippuric acid in urine	End of shift Last 4 hrs of shift	2.5 g/g creat. 3 mg/min.	† G G
** Toluene in venous blood	End of shift	1 mg/L	
** Toluene in end-exhaled air	During shift	20 ppm	
• TRICHLOROETHYLENE [79-01-6]			
* Trichloroacetic acid in urine	End of workweek	100 mg/L	G
* Trichloroacetic acid and trichloroethanol in urine	End of workweek and end of shift	300 mg/L 320 mg/g creat.	G G
* Free trichloroethanol in blood	End of shift and end of workweek	4 mg/L	
** Trichloroethylene in end-exhaled air	Prior to shift and end of workweek	0.5 ppm	
• XYLENES [1330-20-7]			
Methylhippuric acids in urine	End of shift Last 4 hrs of shift	1.5 g/g creat. 2 mg/min	

• 1986-1987 Adoption.

NOTICE OF INTENT TO ESTABLISH

Airborne Chemical [CAS #] *Indices*	*Timing*	*BEI*	*Additional Notation*
‡ANILINE [62-53-3]			
*Total p-aminophenol in urine	End of shift	50 mg/L	G
BENZENE [71-43-2]			
*Total phenol in urine	End of shift	50 mg/L	† G
**Benzene in exhaled air	Prior to next shift		
mixed-exhaled:		0.08 ppm	
end-exhaled:		0 12 ppm	
‡CADMIUM [7440-43-9]			
Cadmium in urine	Not critical	10 μg/g creat.	†
Cadmium in blood	Not critical	10 μg/L	†
‡CARBON DISULFIDE [75-15-0]			
2-Thiothiazolidine-4-carboxylic acid (=TTCA) in urine	End of shift and	5 mg/g creat.	
‡DIMETHYLFORMAMIDE [68-12-2]			
N-Methylformamide in urine	End of shift	40 mg/g creat.	
n-HEXANE [110-54-3]			
*2,5-Hexanedione in urine	End of shift	5 mg/L	G
**n-Hexane in end-exhaled air	During shift	40 ppm	
LEAD [7439-92-1]			R
Lead in blood	Not critical	50 μg/100 ml	†
Lead in urine	Not critical	150 μg/g creat.	† G
Zinc protoporphyrin in blood	After 1 month of exposure	250 μg/100 ml erythrocytes or 100 μg/100 ml blood	† G
‡METHYL ETHYL KETONE (MEK) [78-93-3]			
MEK in urine	End of shift	2 mg/L	G
‡PENTACHLOROPHENOL (PCP) [87-86-5]			
Total PCP In urine	Prior to the last shift of workweek	2 mg/L	†
Free PCP in plasma	End of shift	5 mg/L	†
PHENOL [108-95-2]			
*Total phenol in urine	End of shift	250 mg/g creat.	† G
	Last 2 hrs of shift	15 mg/hr	† G

‡1986-1987 Addition or Revision.

Chemical Substances Under Study to Establish BEIs

Acetone
Chromium
Fluoride
Malathion
Mercury
Nitrobenzene
Parathion
Perchloroethylene
Polychlorinated biphenyls

INTRODUCTION

PHYSICAL AGENTS

These threshold limit values refer to levels of physical agents and represent conditions under which it is believed that nearly all workers may be repeatedly exposed day after day without adverse effect. Because of wide variations in individual susceptibility, exposure of an occasional individual at, or even below, the threshold limit may not prevent annoyance, aggravation of a pre-existing condition, or physiological damage.

These threshold limits are based on the best available information from industrial experience, from experimental human and animal studies, and when possible, from a combination of the three.

These limits are intended for use in the practice of industrial hygiene and should be interpreted and applied only by a person trained in this discipline. They are not intended for use, or for modification for use, 1) in the evaluation or control of the levels of physical agents in the community, 2) as proof or disproof of an existing physical disability, or 3) for adoption by countries whose working conditions differ from those in the United States of America.

These values are reviewed annually by the Committee on Threshold Limits for Physical Agents for revision or additions, as further information becomes available.

The ACGIH disclaims liability with respect to the use of TLVs in a manner inconsistent with their intended use as stated herein.

Notice of Intent—At the beginning of each year, proposed actions of the Committee for the forthcoming year are issued in the form of a "Notice of Intent." This notice provides not only an opportunity for comment, but also solicits suggestions of physical agents to be added to the list. The suggestions should be accompanied by substantiating evidence.

Definitions—Two categories of Threshold Limit Values (TLVs) are specified herein, as follows:

a) *Threshold Limit Value-Time Weighted Average (TLV-TWA)*—the time-weighted average concentration for a normal 8-hour workday and a 40-hour workweek, to which nearly all workers may be repeatedly exposed day after day, without adverse effect. Examples of their use can be found in the TLVs for Heat and Noise.

b) *Threshold Limit Value-Ceiling (TLV-C)*—the concentration that should not be exceeded even instantaneously, as in the case of 115 dBA limit for noise.

Physical Factors. It is recognized that combinations of such physical factors as heat, ultraviolet and ionizing radiation, humidity, abnormal pressure (altitude), and the like may place added stress on the body so that the effects from exposure at a threshold limit may be altered. Also, most of these stresses may act adversely to increase the toxic response of a foreign substance. Although most threshold limits have built-in safety factors to guard against adverse effects to moderate deviations from normal environments, the safety factors of most substances are not of such a magnitude as to take care of gross deviations. For example, continuous work at WBGT temperatures above 30°C (86°F), or overtime extending the workweek more than 25%, might be considered gross deviations. In such instances judgment must be exercised in the proper adjustments of the Threshold Limit Values.

ADOPTED

THRESHOLD LIMIT VALUES

HEAT STRESS

These Threshold Limit Values (TLVs) refer to heat stress conditions under which it is believed that nearly all workers may be repeatedly exposed without adverse health effects. The TLVs shown in Table 1 are based on the assumption that nearly all acclimatized, fully clothed workers with adequate water and salt intake should be able to function effectively under the given working conditions without exceeding a deep body temperature of 38°C.[1,2]

Since measurement of deep body temperature is impractical for monitoring the workers' heat load, the measurement of environmental factors is required which most nearly correlate with deep body temperature and other physiological responses to heat. At the present time Wet Bulb Globe Temperature Index (WBGT) is the simplest and most suitable technique to measure the environmental factors. WBGT values are calculated by the following equations:

1. Outdoors with solar load:
   $WBGT = 0.7\ NWB + 0.2\ GT + 0.1\ DB$
2. Indoors or Outdoors with no solar load:
   $WBGT = 0.7\ NWB + 0.3\ GT$

where:

WBGT = Wet Bulb Globe Temperature Index
NWB = Natural Wet-Bulb Temperature
DB = Dry-Bulb Temperature
GT = Globe Temperature

TABLE 1
Permissible Heat Exposure Threshold Limit Values
(Values are given in °C WBGT)

	Work Load		
Work—Rest Regimen	**Light**	**Moderate**	**Heavy**
Continuous work	30.0	26.7	25.0
75% Work — 25% Rest, each hour	30.6	28.0	25.9
50% Work — 50% Rest, each hour	31.4	29.4	27.9
25% Work — 75% Rest, each hour	32.2	31.1	30.0

The determination of WBGT requires the use of a black globe thermometer, a natural (static) wet-bulb thermometer, and a dry-bulb thermometer.

Higher heat exposures than shown in Table 1 are permissible if the workers have been undergoing medical surveillance and it has been established that they are more tolerant to work in heat than the average worker. Workers should not be permitted to continue their work when their deep body temperature exceeds 38.0°C.

Evaluation and Control

I. *Measurement of the Environment*

The instruments required are a dry-bulb, a natural wet-bulb, a globe thermometer, and a stand. The measurement of the environmental factors shall be performed as follows:

A. The range of the dry and the natural wet bulb thermometer shall be -5°C to 50°C with an accuracy of ± 0.5°C. The dry bulb thermometer must be shielded from the sun and the other radiant surfaces of the environment without restricting the airflow around the bulb. The wick of the natural wet-bulb thermometer shall be kept wet with distilled water for at least 1/2 hour before the temperature reading is made. It is not enough to immerse the other end of the wick into a reservoir of distilled water and wait until the whole wick becomes wet by capillarity. The wick shall be wetted by direct application of water from a syringe 1/2 hour before each reading. The wick shall extend over the bulb of the thermometer, covering the stem about one additional bulb length. The wick should always be clean and new wicks should be washed before using.

B. A globe thermometer, consisting of a 15 cm (6-inch) diameter hollow copper sphere painted on the outside with a matte black finish or equivalent, shall be used. The bulb or sensor of a thermometer (range –5° to +100°C with an accuracy of ± 0.5°C) must be fixed in the center of the sphere. The globe thermometer shall be exposed at least 25 minutes before it is read.

C. A stand shall be used to suspend the three thermometers so that they do not restrict free air flow around the bulbs, and the wet-bulb and globe thermometer are not shaded.

D. It is permissible to use any other type of temperature sensor that gives identical reading as that of a mercury thermometer under the same conditions.

E. The thermometers must be so placed that the readings are representative of the condition where the men work or rest, respectively.

TABLE 2
Assessment of Work Load[9]

Average values of metabolic rate during different activities.

A. Body position and movement	kcal/min
Sitting	0.3
Standing	0.6
Walking	2.0-3.0
Walking up hill	add 0.8 per meter (yard) rise

B. Type of Work		Average kal/min	Range kcal/min
Hand work	*light*	0.4	0.2-1.2
	heavy	0.9	
Work with one arm	*light*	1.0	0.7-2.5
	heavy	1.7	
Work with both arms	*light*	1.5	1.0-3.5
	heavy	2.5	
Work with body	*light*	3.5	2.5-15.0
	moderate	5.0	
	heavy	7.0	
	very heavy	9.0	

TABLE 3
Activity Examples[9]

- Light hand work: writing, hand knitting
- Heavy hand work: typewriting
- Heavy work with one arm: hammering in nails (shoemaker, upholsterer)
- Light work with two arms: filing metal, planing wood, raking of a garden
- Moderate work with the body: cleaning a floor, beating a carpet
- Heavy work with the body: railroad track laying, digging, barking trees

Sample Calculation

Assembly line work using a heavy hand tool.

A.	Walking along	2.0 kcal/min
B.	Intermediate value between heavy work with two arms and light work with the body	3.0 kcal/min
	Subtotal:	5.0 kcal/min
C.	Add for basal metabolism	1.0 kcal/min
	Total:	6.0 kcal/min

The methodology outlined above is more fully explained by Minard.[3,4]

II. *Work Load Categories*

Heat produced by the body and the environmental heat together determine the total heat load. Therefore, if work is to be performed under hot environmental conditions, the workload category of each job shall be established and the heat exposure limit pertinent to the workload evaluated against the applicable standard in order to protect the worker exposure beyond the permissible limit.

A. The work load category may be established by ranking each job into light, medium, and heavy categories on the basis of type of operation. Where the work load is ranked into one of said three categories, i.e.,

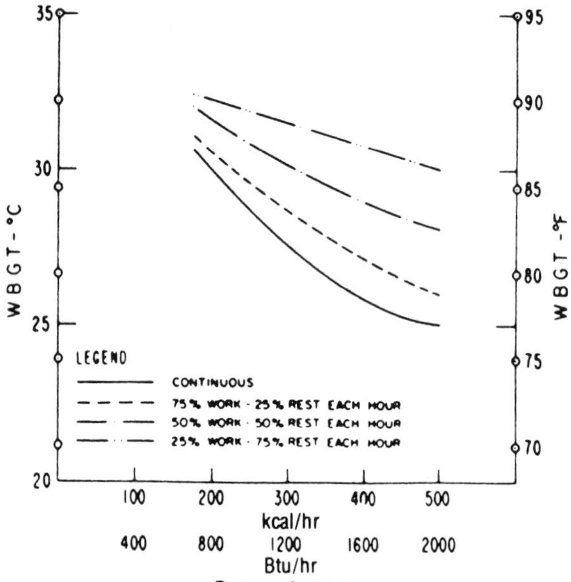

Figure 1—Permissible Heat Exposure Threshold Limit Values.

(1) light work (up to 200 kcal/hr or 800 Btu/hr): e.g., sitting or standing to control machines, performing light hand or arm work,

(2) moderate work (200-350 kcal/hr or 800-1400 Btu/hr): e.g., walking about with moderate lifting and pushing, or

(3) heavy work (359-500 kcal/hr or 1400-2000 Btu/hr): e.g., pick and shovel work,

The permissible heat exposure limit for that workload shall be determined from Table 1.

B. The ranking of the job may be performed either by measuring the worker's metabolic rate while performing his job or by estimating his metabolic rate with the use of Tables 2 and 3. Additional tables available in the literature[5-8] may be utilized also. When this method is used the permissible heat exposure limit can be determined by Figure 1.

III. *Work-Rest Regimen*

The permissible exposure limits specified in Table 1 and Figure 1 are based on the assumption that the WBGT value of the resting place is the same or very close to that of the workplace. Where the WBGT of the work area is different from that of the rest area a time-weighted average value should be used for both environmental and metabolic heat. When time-weighted average values are used, the appropriate curve on Figure 1 is the solid line labeled "continuous."

The time-weighted average metabolic rate (M) shall be determined by the equation:

$$\text{Av. M} = \frac{M_1 \times t_1 + M_2 \times t_2 + \ldots + M_n \times t_n}{t_1 + t_2 + \ldots + t_n}$$

where M_1, M_2...and M_n are estimated or measured metabolic rates for the various activities and rest periods of the worker during the time periods t_1, t_2...and t_n (in minutes) as determined by a time study.

The time-weighted average WBGT shall be determined by the equation:

$$\text{Av. WGBT} = \frac{\text{WBGT}_1 \times t_1 + \text{WBGT}_2 \times t_2 + \ldots + \text{WBGT}_n \times t_n}{t_1 + t_2 + \ldots + t_n}$$

where WBGT_1, WBGT_2 ... and WBGT_n are calculated values of WBGT for the various work and rest areas occupied during total time periods t_1, t_2 ... and t_n are the elapsed times in minutes spent in the corresponding areas which are determined by a time study. Where exposure to hot environmental conditions is continuous for several hours or the entire work day, the time-weighted averages shall be calculated as hourly time-weighted average, i.e., $t_1 + t_2 + \ldots t_n = 60$ minutes. Where the exposure is intermittent, the time-weighted averages shall be calculated as two-hour time-weighted averages, i.e., $t_1 + t_2 + \ldots + t_n = 120$ minutes.

The permissible exposure limits for continuous work are applicable where there is a work-rest regimen of a 5-day work week and an 8-hour work day with a short morning and afternoon break (approximately 15 minutes) and a longer lunch break (approximately 30 minutes). Higher exposure limits are permitted if additional resting time is allowed. All breaks, including unscheduled pauses and administrative or operational waiting periods during work, may be counted as rest time when additional rest allowance must be given because of high environmental temperatures.

IV. *Water and Salt Supplementation*

During the hot season or when the worker is exposed to artificially generated heat, drinking water shall be made available to the workers in such a way that they are stimulated to frequently drink small amounts, i.e., one cup every 15-20 minutes (about 150 ml or 1/4 pint).

The water shall be kept reasonably cool (10°-15°C or 50.0°-60.0°F) and shall be placed close to the workplace so that the worker can reach it without abandoning the work area.

The workers should be encouraged to salt their food abundantly during the hot season and particularly during hot spells. If the workers are unacclimatized, salted drinking water shall be made available in a concentration of 0.1% (1 g NaCl to 1.0 liter or 1 level tablespoon of salt to 15 quarts of water). The added salt shall be completely dissolved before the water is distributed, and the water shall be kept reasonably cool.

V. *Other Considerations*

A. *Clothing:* The permissible heat exposure TLVs are valid for light summer clothing as customarily worn by workers when working under hot environmental conditions. If special clothing is required for performing a particular job and this clothing is heavier or it impedes sweat evaporation or has higher insulation value, the worker's heat tolerance is reduced, and the permissible heat exposure limits indicated in Table 1 and Figure 1 are not applicable. For each job category where special clothing is required, the permissible heat exposure limit shall be established by an expert.

B. *Acclimatization and Fitness:* Acclimatization to heat involves a series of physiological and psychological adjustments that occur in an individual during this first week of exposure to hot environmental conditions. The recommended heat stress TLVs are valid for acclimated workers who are physically fit. Extra caution must be employed when unacclimated or physically unfit workers must be exposed to heat stress conditions.

References

1. *Health Factors Involved in Working Under Conditions of Heat Stress.* WHO Technical Report Series No. 412 (1969).
2. **Dukes-Dobos, F.N. and A. Henschel:** Development of Permissible Heat Exposure Limits for Occupational Work. *ASHRAE Journal 15(9)*:57-62 (Sept. 1973).
3. **Minard, D.:** *Prevention of Heat Casualities in Marine Corps Recruits, Period of 1955-60, with Comparative Incidence Rates and Climatic Heat Stresses in Other Training Categories.* Research Report No. 4, Contract No. MR 005.01-0001.01, Naval Medical Research Institute, Bethesda, MD (Feb. 21 1961). Published in *Military Medicine 126(44)*:261-272 (April 1961).
4. **Minard, D. and R.L. O'Brien:** *Heat Casualties in the Navy and the Marine Corps 1959-1962 with Appendices on the Field Use of the Wet-Bulb Globe Temperature Index.* Research Report No. 7, Contract No. MR 005.01-0001.01, Naval Medical Research Institute, Bethesda, MD (March 12, 1964).
5. **Astrand, Per-Olof and Kaare Rodahl:** *Textbook of Work Physiology.* McGraw-Hill Book Co., New York, San Francisco (1970).
6. Ergonomics Guide to Assessment of Metabolic and Cardiac Costs of Physical Work. *Am. Ind. Hyg. Assoc. J. 32*:560 (1971).
7. *Energy Requirements for Physical Work.* Research Progress Report No. 30. Purdue Farm Cardiac Project, Agricultural Experiment Station, West Lafayette, IN (1961).
8. **Durnin, J.V.G.A. and R. Passmore:** *Energy, Work and Leisure.* Heinemann Educational Books, Ltd., London (1967).
9. **Lehmann, G.E., A. Muller and H. Spitzer:** Der Kalorienbedarb bie Gewerblicher Arbeit. *Arbeitsphysiol. 14*:166 (1950).

*COLD STRESS

These Threshold Limit Values (TLVs) are intended to protect workers from the severest effects of cold stress (hypothermia) and cold injury and to describe exposures to cold working conditions under which it is believed that nearly all workers can be repeatedly exposed without adverse health effects. The TLV objective is to prevent the deep body core temperature from falling below 36°C and to prevent cold injury to body extremities. Deep body temperature is the core temperature of the body as determined by rectal temperature measurements. For a single, occasional exposure to a cold environment a drop in core temperature to no lower that 35°C should be permitted. In addition to provisions for total body protection, the TLV objective is to protect all parts of the body with emphasis on hands, feet and head from cold injury.

Introduction

Fatal exposures to cold among workers have almost always resulted from accidental exposures involving failure to escape from low environmental air temperatures or from immersion in low temperature water. The single most important aspect of life-threatening hypothermia is the fall in the deep core temperature of the body. The clinical presentations of victims of hypothermia are shown in Table 4 (taken from Dembert in *AFP*, January 1982). Workmen should be protected from exposure to cold so that the deep core temperature does not fall below 36°C (96.8°F); lower body temperatures will very likely result in reduced mental alertness, reduction in rational decision making, or loss of consciousness with the threat of fatal consequences.

Pain in the extremities may be the first early warning of danger to cold stress. During exposure to cold, maximum severe shivering develops when the body temperature has fallen to 35°C (95°F). This must be taken as a sign of danger to the workers and exposure to cold should be immediately terminated for any workers when severe shivering becomes evident. Useful physical or mental work is limited when severe shivering occurs.

Since prolonged exposure to cold air, or to immersion in cold water, at temperatures well above freezing can lead to dangerous hypothermia, whole body protection must be provided.

---

* 1986-1987 Adoption.

[A] Wind chill factor is a unit of heat loss from a body defined in watts per meter squared per hour being a function of the air temperature and wind velocity upon the exposed body.

1. Adequate insulating clothing to maintain core temperatures above 36°C must be provided to workers if work is performed in air temperatures below 4°C (40°F). Wind chill factor[A] or the cooling power of the air is a critical factor. The higher the wind speed and the lower the temperature in the work area,

TABLE 4
Progressive Clinical Presentations of Hypothermia*

Core Temperature °C	°F	Clinical Signs
37.6	99.6	"Normal" rectal temperature
37	98.6	"Normal" oral temperature
36	96.8	Metabolic rate increases in an attempt to compensate for heat loss
35	95.0	Maximum shivering
34	93.2	Victim conscious and responsive, with normal blood pressure
33	91.4	Severe hypothermia below this temperature
32 } 31	89.6 } 87.8	Consciousness clouded; blood pressure becomes difficult to obtain; pupils dilated but react to light; shivering ceases
30 } 29	86.0 } 84.2	Progressive loss of consciousness; muscular rigidity increases; pulse and blood pressure difficult to obtain; respiratory rate decreases
28	82.4	Ventricular fibrillation possible with myocardial irritability
27	80.6	Voluntary motion ceases; pupils nonreactive to light; deep tendon and superficial reflexes absent
26	78.8	Victim seldom conscious
25	77.0	Ventricular fibrillation may occur spontaneously
24	75.2	Pulmonary edema
22 } 21	71.6 } 69.8	Maximum risk of ventricular fibrilation
20	68.0	Cardiac standstill
18	64.4	Lowest accidental hypothermia victim to recover
17	62.6	Isoelectric electroencephalogram
9	48.2	Lowest artificially cooled hypothermia patient to recover

the greater the insulation value of the protective clothing required. An equivalent chill temperature chart relating the actual dry bulb air temperature and the wind velocity is presented in Table 5. The equivalent chill temperature should be used when estimating the combined cooling effect of wind and low air temperatures on exposed skin or when determining clothing insulation requirements to maintain the deep body core temperature.

2. Unless there are unusual or extenuating circumstances cold injury to other than hands, feet, and head is not likely to occur without the development of the initial signs of hypothermia. Older workers or workers with circulatory problems require special precautionary protection against cold injury. The use of extra insulating clothing and/or a reduction in the duration of the exposure period are among the special precautions which should be considered. The precautionary actions to be taken will depend upon the physical condition of the worker and should be determined with the advice of a physician with knowledge of the cold stress factors and the medical condition of the worker.

* Presentations approximately related to core temperature. Reprinted from the January 1982 issue of *American Family Physician*, published by the American Academy of Family Physicians.

Evaluation and Control

For exposed skin, continuous exposure should not be permitted when the air speed and temperature results in an equivalent chill temperature of −32°C (−25°F). Superficial or deep local tissue freezing will occur only at temperatures below −1°C regardless of wind speed.

At air temperatures of 2°C (35.6°F) or less it is imperative that workers who become immersed in water or whose clothing becomes wet be immediately provided a change of clothing and be treated for hypothermia.

Recommended limits for properly clothed workers for periods of work at temperatures below freezing are shown in Table 6.

Special protection of the hands is required to maintain manual dexterity for the prevention accidents:

1. If fine work is to be performed with bare hands for more than 10-20 minutes in an environment below 16°C (60°F), special provisions should be established for keeping the workers' hands warm. For this purpose, warm air jets, radiant heaters (fuel burner or electric radiator), or contact warm plates may be utilized. Metal handles of tools and control bars shall be covered by thermal insulating material at temperatures below −1°C (30°F).
2. If the air temperature falls below 16°C (60°F) for sedentary, 4°C (40°F) for light, −7°C (20°F) for moderate work and fine manual dexterity is not required then gloves shall be used by the workers.

To prevent contact frostbite, the workers should wear anti-contact gloves.

1. When cold surfaces below −7°C (20°F) are within reach, a warning should be given to each worker by his supervisor to prevent inadvertent contact by bare skin.
2. If the air temperature is −17.5°C (0°F) or less, the hands should be protected by mittens. Machine controls and tools for use in cold conditions should be designed so that they can be handled without removing the mittens.

Provisions for additional total body protection is required if work is performed in an environment at or below 4°C (40°F). The workers shall wear cold protective clothing appropriate for the level of cold and physical activity:

1. If the air velocity at the job site is increased by wind, draft, or artificial ventilating equipment, the cooling effect of the wind shall be reduced by shielding the work area, or by wearing an easily removable outer windbreak layer garment. Wind chill cooling rates are illustrated in Figure 2 and Table 7.
2. If only light work is involved and if the clothing on the worker may become wet on the job site, the outer layer of the clothing in use may be of a type impermeable to water. With more severe work under such conditions the outer layer should be water repellent, and the outerwear should be changed as it becomes wetted. The outer garments must include provisions for easy ventilation in order to prevent wetting of inner layers by sweat. If work is done at normal temperatures or in a hot environment before entering the cold area, the employee shall make sure that his clothing is not wet as a consequence of sweating. If his clothing is wet, the employee shall change into dry clothes before entering the cold area. The workers shall change socks and any removable felt insoles at regular daily intervals or use vapor barrier boots. The optimal frequency of change shall be determined empirically and will vary individually and according to the type of shoe worn and how much the individual's feet sweat.
3. If extremities, ears, toes and nose, cannot be protected sufficiently to prevent sensation of excessive cold or frostbite by handware, footwear and face masks, these protective items shall be supplied in auxiliary heated versions.

4. If the available clothing does not give adequate protection to prevent hypothermia or frostbite, work shall be modified or suspended until adequate clothing is made available or until weather conditons improve.
5. Workers handling evaporative liquid (gasoline, alcohol or cleaning fluids) at air temperatures below 4°C (40°F) shall take special precautions to avoid soaking of clothing or gloves with the liquids because of the added danger of cold injury due to evaporative cooling. Special note should be taken of the particularly acute effects of splashes of "cryogenic fluids" or those liquids with a boiling point only just above ambient temperatures.

TABLE 5
Wind Chill Factor Chart*

Estimated Wind Speed (in mph)	Actual Temperature Reading (°F) 50	40	30	20	10	0	−10	−20	−30	−40	−50	−60
	Equivalent Chill Temperature (°F)											
calm	50	40	30	20	10	0	−10	−20	−30	−40	−50	−60
5	48	37	27	16	6	−5	−15	−26	−36	−47	−57	−68
10	40	28	16	4	−9	−24	−33	−46	−58	−70	−83	−95
15	36	22	9	−5	−18	−32	−45	−58	−72	−85	−99	−112
20	32	18	4	−10	−25	−39	−53	−67	−82	−96	−110	−121
25	30	16	0	−15	−29	−44	−59	−74	−88	−104	−118	−133
30	28	13	−2	−18	−33	−48	−63	−79	−94	−109	−125	−140
35	27	11	−4	−20	−35	−51	−67	−82	−98	−113	−129	−145
40	26	10	−6	−21	−37	−53	−69	−85	−100	−116	−132	−148
(Wind speeds greater than 40 mph have little additional effect.)	*LITTLE DANGER* In < hr with dry skin. Maximum danger of false sense of security				*INCREASING DANGER* Danger from freezing of exposed flesh within one minute.				*GREAT DANGER* Flesh may freeze within 30 seconds.			
	Trenchfoot and immersion foot may occur at any point on this chart.											

* Developed by U.S. Army Research Institute of Environmental Medicine, Natick, MA.

TABLE 6
Threshold Limit Values Work/Warm-up Schedule for Four-Hour Shift*

Air Temperature—Sunny Sky		No Noticeable Wind		5 mph Wind		10 mph Wind		15 mph Wind		20 mph Wind	
°C (approx.)	°F	Max. Work Period	No. of Breaks	Max. Work Period	No. of Breaks	Max. Work Period	No. of Breaks	Max. Work Period	No. of Breaks	Max. Work Period	No. of Breaks
1. -26° to -28°	-15° to -19°	(Norm. Breaks)	1	(Norm. Breaks)	1	75 min	2	55 min	3	40 min	4
2. -29° to -31°	-20° to -24°	(Norm. Breaks)	1	75min	2	55 min	3	40 min	4	30 min	5
3. -32° to -34°	-25° to -29°	75 min	2	55 min	3	40 min	4	30 min	5	Non-emergency work should cease	
4. -35° to -37°	-30° to -34°	55 min	3	40 min	4	30 min	5	Non-emergency work should cease		↓	
5. -38° to -39°	-35° to -39°	40 min	4	30 min	5	Non-emergency work should cease		↓		↓	
6. -40° to -42°	-40° to -44°	30 min	5	Non-emergency work should cease		↓		↓		↓	
7. -43° & below	-45° & below	Non-emergency work should cease		↓		↓		↓		↓	

Notes for Table 6:

1. Schedule applies to moderate to heavy work activity with warm-up breaks of ten (10) minutes in a warm location. For Light-to-Moderate Work (limited physical movement): apply the schedule one step lower. For example, at -30°F with no noticeable wind (Step 4), a worker at a job with little physical movement should have a maximum work period of 40 minutes with 4 breaks in a 4-hour period (Step 5).
2. The following is suggested as a guide for estimating wind velocity if accurate information is not available: 5 mph: light flag moves; 10 mph: light flag fully extended; 15 mph: raises newspaper sheet; 20 mph: blowing and drifting snow.
3. If only the wind chill cooling rate is available, a rough rule of thumb for applying it rather than the temperature and wind velocity factors given above would be: 1) special warm-up breaks should be initiated at a wind chill of about 1750 W/m²; 2) all non-emergency work should have ceased at or before a wind chill of 2250 W/m². In general the warm-up schedule provided above slightly under-compensates for the wind at the warmer temperatures, assuming acclimatization and clothing appropriate for winter work. On the other hand, the chart slightly over-compensates for the actual temperatures in the colder ranges, since windy conditions rarely prevail at extremely low temperatures.

* Adopted from Occupational Health & Safety Division, Saskatchewan Department of Labour.

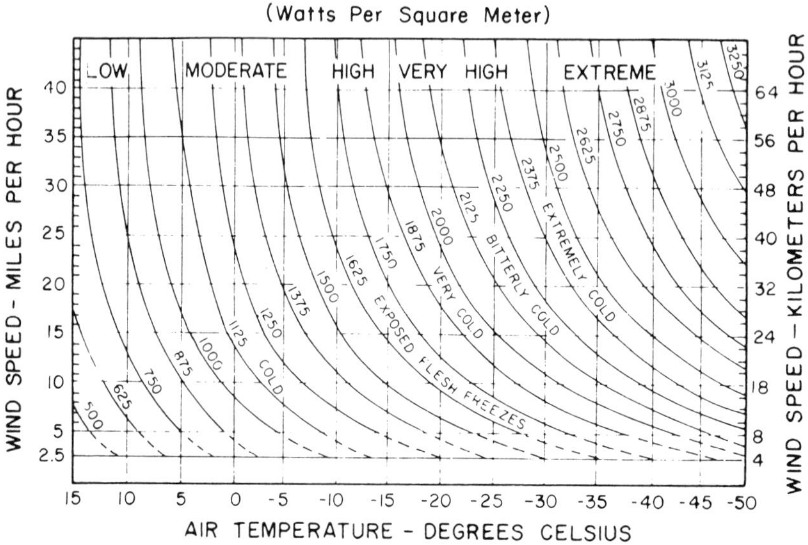

Figure 2—Wind chill cooling rates. Adapted from Canadian Department of the Environment, Atmospheric Environment Service.

Work-Warming Regimen

If work is performed continuously in the cold at an equivalent chill temperature (ECT) or below −7°C (20°F) heated warming shelters (tents, cabins, rest rooms, etc.) shall be made available nearby and the workers should be encouraged to use these shelters at regular intervals, the frequency depending on the severity of the environmental exposure. The onset of heavy shivering, frostnip, the feeling of excessive fatique, drowsiness, irritability, or euphoria, are indications for immediate return to the shelter. When entering the heated shelter the outerlayer of clothing shall be removed and the remainder of the clothing loosened to permit sweat evaporation or a change of dry work clothing provided. A change of dry work clothing shall be provided as necessary to prevent workers from returning to their work with wet clothing. Dehydration, or the loss of body fluids, occurs insidiously in the cold environment and may increase the susceptibility of the worker to cold injury due to a significant change in blood flow to the extremities. Warm sweet drinks and soups should be provided at the work site to provide caloric intake and fluid volume. The intake of coffee should be limited because of a diuretic and circulatory effect.

For work practices at or below −12°C (10°F) ECT the following shall apply:

1. The worker shall be under constant protective observation (buddy system or supervision).
2. The work rate should not be so high as to cause heavy sweating that will result in wet clothing; if heavy work must be done, rest periods must be taken in heated shelters and opportunity for changing into dry clothing shall be provided.
3. New employees shall not be required to work full-time in cold in the first days until they become accustomed to the working conditions and required protective clothing.
4. The weight and bulkiness of clothing shall be included in estimating the required work performance and weights to be lifted by the worker.
5. The work shall be arranged in such a way that sitting still or standing still for long periods is minimized. Unprotected metal chair seats shall not be used. The worker should be protected from drafts to the greatest extent possible.
6. The workers shall be instructed in safety and health procedures. The training program shall include as a minimum instruction in:

TABLE 7
Wind Chill Cooling Rate Effects*

Wind Chill Rates (Watts/m²/hr)	Comments/Effects
700	Conditions considered comfortable when dressed for skiing.
1200	Conditions no longer pleasant for outdoor activities on overcast days.
1400	Conditions no longer pleasant for outdoor activities on sunny days.
1600	Freezing of exposed skin begins for most people depending on the degree of activity and the amount of sunshine.
2300	Conditions for outdoor travel such as walking become dangerous. Exposed areas of the face freeze in less than 1 minute for the average person.
2700	Exposed flesh will freeze within half a minute for the average person.

* From Canadian Department of the Environment, Atmospheric Environment Service.

a. Proper rewarming procedures and appropriate first aid treatment.
b. Proper clothing practices.
c. Proper eating and drinking habits.
d. Recognition of impending frostbite.
e. Recognition signs and symptoms of impending hypothermia or excessive cooling of the body even when shivering does not occur.
f. Safe work practices.

Special Workplace Recommendations

Special design requirements for refrigerator rooms include the following:

1. In refrigerator rooms, the air velocity should be minimized as much as possible and should not exceed 1 meter/sec (200 fpm) at the job site. This can be achieved by properly designed air distribution systems.
2. Special wind protective clothing shall be provided based upon existing air velocities to which workers are exposed.

Special caution shall be exercised when working with toxic substances and when workers are exposed to vibration. Cold exposure may require reduced exposure limits.

Eye protection for workers employed out-of-doors in a snow and/or ice-covered terrain shall be supplied. Special safety goggles to protect against ultraviolet light and glare (which can produce temporary conjuntivitis and/or temporary loss of vision) and blowing ice crystals are required when there is an expanse of snow coverage causing a potential eye exposure hazard.

Workplace monitoring is required as follows:

1. Suitable thermometry should be arranged at any workplace where the environmental temperature is below 16°C (60°F) to enable overall compliance with the requirements of the TLV to be maintained.
2. Whenever the air temperature at a workplace falls below −1°C (30°F), the dry bulb temperature should be measured and recorded at least every 4 hours.
3. In indoor workplaces, the wind speed should also be recorded at least every 4 hours whenever the rate of air movement exceeds 2 meters per second (5 mph).
4. In outdoor work situations, the windspeed should be measured and recorded together with the air temperature whenever the air temperature is below −1°C (30°F).
5. The equivalent chill temperature shall be obtained from Table 16 in all cases where air movement measurements are required, and shall be recorded with the other data whenever the equivalent chill temperature is below −7°C (20°F).

Employees shall be excluded from work in cold at −1°C (30°F) or below if they are suffering from diseases or taking medication which interferes with normal body temperature regulation or reduces tolerance to work in cold environments. Workers who are routinely exposed to temperatures below −24°C (−10°F) with wind speeds less than five miles per hour, or air tempeatures below −18°C (0°F) with wind speeds above five miles per hour should be medically certified as suitable for such exposures.

Trauma sustained in freezing or subzero conditions requires special attention because an injured worker is predisposed to secondary cold injury. Special provisions must be made to prevent hypothermia and secondary freezing of damaged tissues in addition to providing for first aid treatment.

*HAND-ARM (SEGMENTAL) VIBRATION

These Threshold Limit Values (Table 8) refer to component accelerations levels and durations of exposure that represent conditions under which it is believed that most workers may be exposed repeatedly without progressing beyond Stage 3 of the Taylor-Pelmear Classification System for Vibration-induced White Finger (VWF), also known as Raynaud's Phenomenon of Occupational Origin). Since there is a paucity of dose-response relationships for VWF, these recommendations have been derived from epidemiological data from forestry, mining, and metal working. These values should be used as guides in the control of hand-arm vibration exposure and because of individual susceptibility, should not be regarded as defining a boundary between safe and dangerous levels.

It should be recognized that the application of the TLV alone for hand-arm vibration will not protect all workers from the adverse effects of hand-arm vibration exposure. The use of: 1) antivibration tools, 2) antivibration gloves, 3) proper work practices which keep the worker's hands and remaining body warm and also minimize the vibration coupling between the worker and the vibration tool are necessary to minimize vibration exposure, and 4) a conscienciously applied medical surveillance program are ALL necessary to rid VWF from the workplace.

Continuous, Intermittent, Impulsive, or Impact Hand-arm Vibration

The measurement of vibration should be performed in accordance with the procedures and instrumentation specified by the Second Draft International Standard ISO/DIS 5349 (1984), *Guide for the Measurement and the Assessment of Human Exposure to Vibration Transmitted to the Hand,* and summarized below:

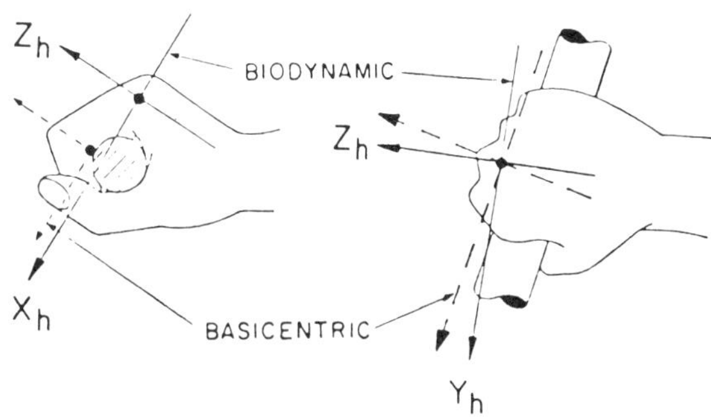

Figure 3—Biodynamic and basicentric coordinate systems for the hand, showing the directions of the acceleration components (ISO 5349).

* 1986-1987 Adoption.

The acceleration of a vibration handle or work piece should be determined in three mutually orthogonal directions at a point close to where vibration enters the hand. The directions shall preferably be those forming the ISO biodynamic coordinate system, but may be a closely related basicentric system with its origin at the interface between the hand and the vibrating surface (see Figure 3) to accommodate different handle or work piece configurations. A small and lightweight transducer shall be mounted so as to record accurately one or more orthogonal components of the source vibration in the frequency range from 5 to 1500 Hz. Each component should be frequency-weighted by a filter network with gain characteristics specified by the ISO for human-response vibration measuring instrumentation, to account for the change in vibration hazard with frequency (see Figure 4).

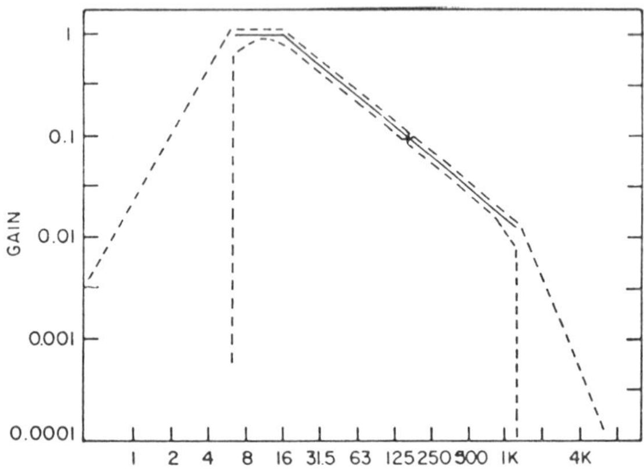

Figure 4—Gain characteristics of the filter network used to frequency-weight acceleration components (continuous line). The filter tolerances (dashed lines) are provisional, and are those contained in ISO 5349.

Assessment of vibration exposure should be made for EACH applicable direction (X_h, Y_h, Z_h) since vibration is a vector quantity (magnitude and direction). In each direction, the magnitude of the vibration during normal operation of the power tool, machine or work piece shall be expressed by the root-mean-square (rms) value of the frequency-weighted component accelerations, in units of meters per second squared (m/s²), or gravitational units (g), the largest of which, a_K, forms the basis for exposure assessment.

For each direction being measured, linear integration shall be employed for vibrations that are of extremely short duration or vary substantially in time. If the total daily vibration exposure in a given direction is composed of several exposures at different rms accelerations, then the equivalent, frequency-weighted component acceleration in that direction shall be determined in accordance with the following equation:

$$\left(a_{K_{eq}}\right) = \left[\frac{1}{T}\sum_{i=1}^{n}\left(a_{K_i}\right)^2 T_i\right]^{1/2}$$

$$= \sqrt{\left(a_{K_1}\right)^2 \frac{T_1}{T} + \left(a_{K_2}\right)^2 \frac{T_2}{T} + \ldots \left(a_{K_n}\right)^2 \frac{T_n}{T}}$$

where: $T = \sum_{i=1}^{n} T_i$

T = total daily exposure duration

a_{K_i} = ith frequency-weighted, rms acceleration component with duration T_i

These computations may be performed by commercially available human-response vibration measuring instruments.

TABLE 8
Threshold Limit Values for Exposure of the Hand to Vibration in Either X_h, Y_h, Z_h Directions

Total Daily Exposure Duration[a]	Values of the Dominant,[b] Frequency-Weighted, rms, Component Acceleration Which Shall not be Exceeded a_K,($a_{K_{eq}}$)	
	m/s²	g[c]
4 hours and less than 8	4	0.40
2 hours and less than 4	6	0.61
1 hour and less than 2	8	0.81
less than 1 hour	12	1.22

[a] The total time vibration enters the hand per day, whether continuously or intermittently.
[b] Usually one axis of vibration is dominant over the remaining two axis. If one or more vibration axis exceeds the Total Daily Exposure then the TLV has been exceeded.
[c] g = 9.81 m/s².

Notes: Table 8:

1. Hardly any person exposed at or below the TLVs for vibration contained in Table 8 has progressed to *Stage 3* Vibration White Finger, in the Taylor-Pelmear classification, i.e., the point at which extension blanching of all fingers has occurred and there is definite interference at work, home, and restricted social activities.(2-7)
2. Acute exposures to frequency-weighted, rms, component accelerations in excess of the TLVs for infrequent periods of time (e.g., 1 day per week, or several days over a two-week period) are not necessarily more harmful.(2-4)
3. Acute exposures to frequency-weighted, rms, component accelerations of three times the magnitude of the TLVs are expected to result in the same health effects after between 5 and 6 years of exposure.(2-4)
4. Preventive measures, including specialized preemployment and annual medical examinations to identify persons susceptible to vibration, should be implemented in situations in which workers are or will be exposed to hand-arm vibration.(3-7)
5. To moderate the adverse effects of vibration exposure, workers should be advised to avoid *continuous* vibration exposure by cessation of vibration exposure for approximately 10 minutes per continuous vibration hour.
6. Good work practices should be used, and should include instructing workers to employ a minimum hand grip force consistent with safe operation of the power tool or process, keep their body and hands warm and dry, and avoid smoking.(2,3)
7. A transducer and its device for attachment to the vibrating source suitable for measurement purposes together should weigh less than 15 grams, and should possess a cross-axis sensitivity of less than 10%.
8. The measurement by many (mechanically underdamped) piezoelectric accelerometers of repetitive, large displacement, impulsive vibrations, such as those produced by percussive pneumatic tools, is subject to error. The insertion of a suitable, low-pass, mechanical filter between the accelerometer and the source of vibration with a cut-off frequency of 1500 Hz or greater (and cross-axis sensitivity of less than 10%) can help eliminate incorrect readings.(3,4)
9. The manufacturer and type number of all apparatus used to measure vibration should be reported, as well as the value of the dominant direction and frequency-weighted, rms, component acceleration.

References

1. **Pyykko I.:** Vibration Syndrome. A Review. *Vibration and Work,* pp. 1-24. O. Dorhonen, Ed. Institute of Occupational Health, Helsinki (1976).
2. *Vibration White Finger in Industry,* W. Taylor and P.L. Pelmear, Eds. Academic Press, London (1975).
3. **NIOSH:** *Proceedings of the International Occupational Hand-Arm Vibration Conference,* D E. Wasserman and W. Taylor, Eds. DHEW NIOSH Pub. No. 77-170 (1977).
4. **Brammar, J.J.:** Threshold Limit for Hand-Arm Vibration Exposure Throughout the Workday. *Vibration Effects on the Hand and Arm in Industry,* pp. 291-301. A.J. Brammer and W. Taylor, Eds. John Wiley & Sons, New York (1982).
5. **Wasserman, D.E. and W. Taylor:** *Environmental and Occupational Medicine,* Chap. 68, *Occupational Vibration,* pp. 743-749. W.N. Rom, Ed. Little, Brown and Co., Boston (1982).
6. **NIOSH:** *Current Intelligence Bulletin #38: Vibration Syndrome.* DHHS (NIOSH) Pub. No. 83-110 (1983).
7. **NIOSH:** *Vibration Syndrome.* NIOSH Videotape #177 (27 minutes). Cincinnati, OH.
8. **International Organization for Standardization:** *Guide for the Measurement and the Assessment of Human Exposure to Vibration Transmitted to the Hand.* Second DIS 5349. International Organization for Standardization, Geneva (in press, 1983).
9. **International Organization for Standardization:** *Human-Response Vibration Measuring Instrumentation.* Second Draft Proposal DP 8041. ISO/TC 108/SC 3 n 99. International Organization for Standardization, Geneva (unpublished, 1982).

IONIZING RADIATION

The Committee accepts the philosophy and recommendations of the National Council on Radiation Protection and Measurements (NCRP) for the ionizing radiation TLV. The NCRP is charted by Congress to, in part, collect analyze, develop and disseminate information and recommendations about protection against radiation and about radiation measurements, quantities and units, including development of basic concepts in these areas. NCRP Report No. 39 provides basic philosophy and concepts leading to protection criteria established in the same report.[1] Other NCRP reports address specific areas of radiation protection and, collectively, provide an excellent basis for establishing a sound program for radiation control. The Committee recommends the listed references as substantative documentation of a sound basis for ionizing radiation protection. The Committee also strongly recommends that all exposure to ionizing radiation be kept as low as reasonably achievable within the stated guidance.

References

1. *Basic Radiation Protection Criteria.* NCRP Report No. 39 (January 15, 1971).
2. *Maximum Permissible Body Burdens and Maximum Permissible Concentrations of Radionuclides in Air and in Water for Occupational Exposure.* National Bureau of Standards Handbook 69, (June 5, 1959), with *Addendum 1* (August 1963). Available as NCRP Report No. 22.

The above documents, as well as information on numerous other NCRP Reports addressing specific subjects in ionizing radiation protection, are available from: NCRP Publications, 7910 Woodmont Ave., Suite 1016, Bethesda, MD 20814.

LASERS

The Threshold Limit Values (TLVs) are for exposure to laser radiation under conditions to which nearly all workers may be exposed without adverse effects. The values should be used as guides in the control of exposures and should not be regarded as fine lines between safe and dangerous levels. They are based on the best available information from experimental studies.

Limiting Apertures

The TLVs expressed as radiant exposure or irradiance in this section may be averaged over an aperture of 1 mm except for TLVs for the eye in the spectral range of 400-1400 nm, which should be averaged over a 7 mm limiting aperture (pupil); and except for all TLVs for wavelengths between 0.1-1 mm where the limiting aperture is 10 mm. No modification of the TLVs is permitted for pupil sizes less than 7 mm.

The TLVs for "extended sources" apply to sources which subtend an angle greater than α (Table 9) which varies with exposure time. This angle is *not* the beam divergence of the source.

TABLE 9
Limiting Angle to Extended Source Which May Be Used For Applying Extended Source TLVs

Exposure Duration(s)	Angle α (mrad)	Exposure Duration(s)	Angle α (mrad)
10^{-9}	8.0	10^{-2}	5.7
10^{-8}	5.4	10^{-1}	9.2
10^{-7}	3.7	1.0	15
10^{-6}	2.5	10	24
10^{-5}	1.7	10^{2}	24
10^{-4}	2.2	10^{3}	24
10^{-3}	3.6	10^{4}	24

Correction Factors A and B (C_A and C_B)

The TLVs for ocular exposure in Tables 10 and 11 are to be used as given for all wavelength ranges. The TLVs for wavelengths between 700 nm and 1049 nm are to be increased by a uniformly extrapolated factor (C_A) as shown in Figure 5. Between 1049 nm and 1400 nm, the TLV has been increased by a factor (C_A) of five. For certain exposure times at wavelengths between 550 nm and 700 nm, correction factor (C_B) must be applied.

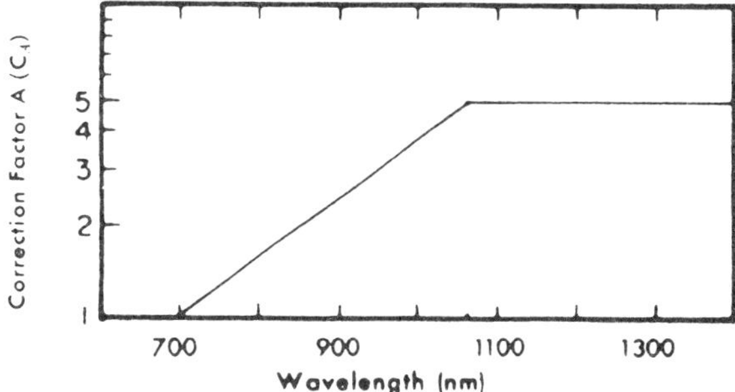

Figure 5—TLV correction factor for λ = 700–1400 nm.* (* For λ = 700–1049 nm, $C_A = 10^{[0.002(\lambda-700)]}$; For λ = 1050–1400 nm, C_A= 5.)

The TLVs for skin exposure are given in Table 12. The TLVs are to be increased by a factor (C_A) as shown in Figure 5 for wavelengths between 700 nm and 1400 nm. To aid in the determination of TLVs for exposure durations requiring calculations of fractional powers Figures 6, 7 and 8 may be used.

* *Repetitively Pulsed Exposures*

Scanned CW lasers or repetitively pulsed lasers can both produce repetitively pulsed exposure conditions. The TLV for intrabeam viewing which is applicable to wavelengths between 400 and 1400 nm and a single-pulse exposure (of pulse duration *t*) is modified in this instance by a correction factor determined by the number of pulses in the exposure. First, calculate the number of pulses (n) in an expected exposure situation; this is the pulse repetition frequency (PRF in Hz) multiplied by the duration of exposure. Normally, realistic exposures may range from 0.25s for a bright visible source to 10s for an infrared source. The corrected TLV on a per-pulse basis is:

$$TLV = (n^{-1/4})\,(\text{TLV for single-pulse}) \qquad (1)$$

This approach applies only to thermal-injury conditions, i.e., all exposures at wavelengths greater than 700 nm, and for many exposures at shorter wavelengths. For wavelengths less than or equal to 700 nm, the corrected TLV from equation 1 above applies if the average irradiance does not exceed the TLV for continuous exposure. The average irradiance (i.e., the total accumulated exposure for *nt* seconds) shall not exceed the radiant exposure given in Table 10 for exposure durations of 10 seconds to T_1.

* 1986-1987 Addition.

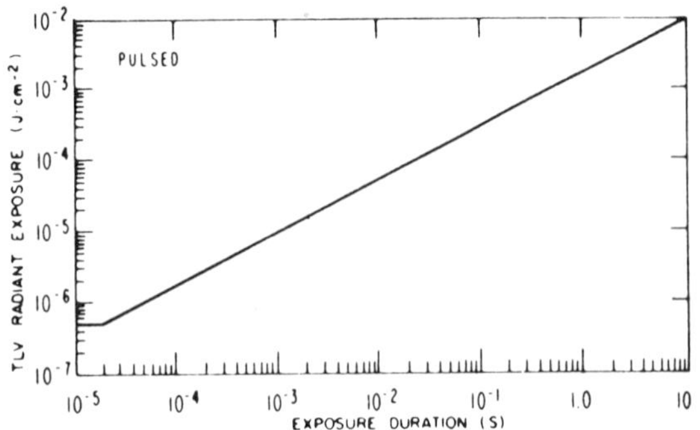

Figure 6a — TLV for intrabeam (direct) viewing of laser beam (400-700 nm).

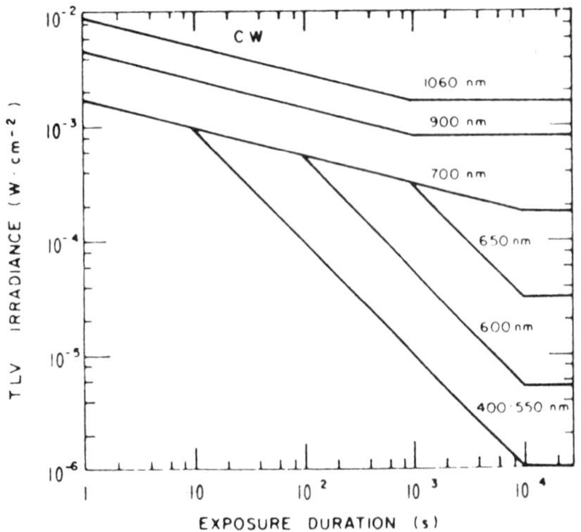

Figure 6b — TLV for intrabeam (direct) viewing of CW laser beam (400-1400 nm).

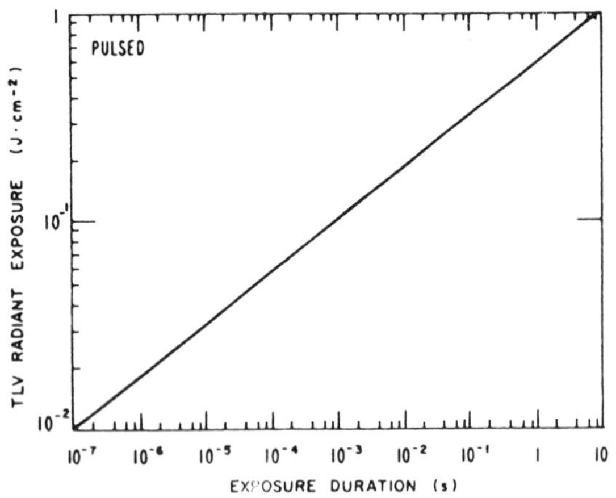

Figure 7a — TLV for laser exposure of skin and eyes for far-infrared radiation (wave-lengths greater than 1.4 μm).

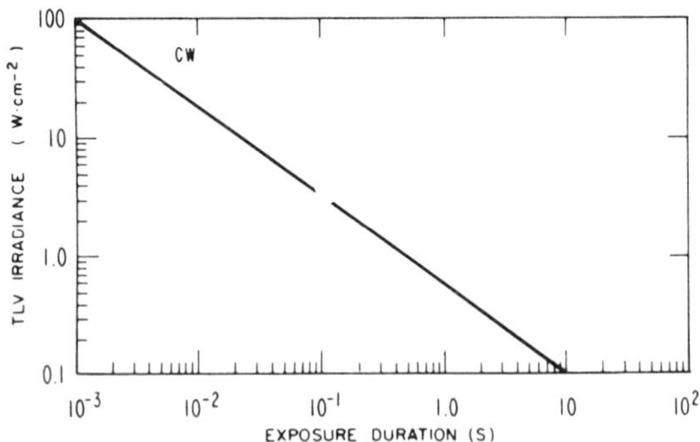

Figure 7b — TLV for CW laser exposure of skin and eyes for far-infrared radiation (wave-lengths greater than 1.4 μm).

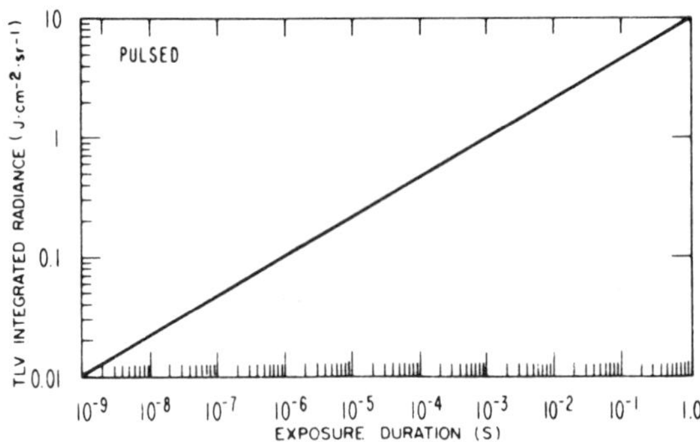

Figure 8a — TLV for extended sources or diffuse reflections of laser radiation (400-700 nm).

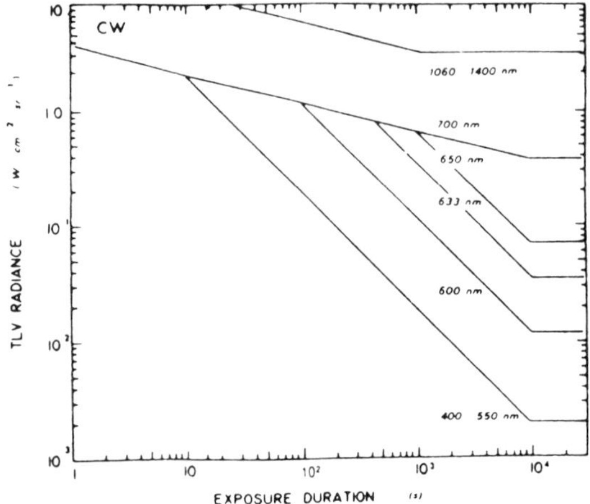

Figure 8b — TLV for extended sources or diffuse reflections of laser radiations (400-1400 nm).

TABLE 10

Threshold Limit Value for Direct Ocular Exposures (Intrabeam Viewing) from a Laser Beam

Spectral Region	Wave Length	Exposure Time, (t) Seconds	TLV	
UVC	200 nm to 280 nm	10^{-9} to 3×10^{4}	3 mJ • cm^{-2}	
UVB	280 nm to 302 nm	"	3 "	*not to exceed $0.56\ t^{1/4}$ J • cm^{-2} for t ≦ 10 s.
	303 nm	"	4 "	
	304 nm	"	6 "	
	305 nm	"	10 "	
	306 nm	"	16 "	
	307 nm	"	25 "	
	308 nm	"	40 "	
	309 nm	"	63 "	
	310 nm	"	100 "	
	311 nm	"	160 "	
	312 nm	"	250 "	
	313 nm	"	400 "	
	314 nm	"	630 "	
UVA	315 nm to 400 nm	10^{-9} to 10	$.56\ t^{1/4}$ J • cm^{-2}	
	" "	10 to 10^{3}	1.0 J • cm^{-2}	
	" "	10^{3} to 3×10^{4}	1.0 mW • cm^{-2}	
Light	400 nm to 700 nm	10^{-9} to 1.8×10^{-5}	5×10^{-7}J • cm^{-2}	
	400 nm to 700 nm	1.8×10^{-5} to 10	1.8 $(t/\sqrt[4]{t})$ mJ • cm^{-2}	
	400 nm to 549 nm	10 to 10^{4}	10 mJ • cm^{-2}	
	550 nm to 700 nm	10 to T_1	1.8 $(t/\sqrt[4]{t})$ mJ • cm^{-2}	
	550 nm to 700 nm	T_1 to 10^{4}	10 C_B mJ • cm^{-2}	
	400 nm to 700 nm	10^{4} to 3×10^{4}	C_B μW • cm^{-2}	
IR-A	700 nm to 1049 nm	10^{-9} to 1.8×10^{-5}	5 $C_A \times 10^{-7}$ J • cm^{-2}	
	700 nm to 1049 nm	1.8×10^{-5} to 10^{3}	1.8 C_A $(t/\sqrt[4]{t})$ mJ • cm^{-2}	
	1050 nm to 1400 nm	10^{-9} to 10^{-4}	5×10^{-6} J • cm^{-2}	
	1050 nm to 1400 nm	10^{-4} to 10^{3}	$9(t/\sqrt[4]{t})$ mJ • cm^{-2}	
	700 nm to 1400 nm	10^{3} to 3×10^{4}	320 C_A μW • cm^{-2}	
IR-B & C	1.4 μm to 10^{3} μm	10^{-9} to 10^{-7}	10^{-2} J • cm^{2}	
	" "	10^{-7} to 10	0.56 $\sqrt[4]{t}$ J • cm^{-2}	
	" "	10 to 3×10^{4}	0.1 W • cm^{-2}	

C_A – See Fig. 5; $C_B = 1$ for $\lambda = 400$ to 549 nm; $C_B = 10^{[(0.015\ (\lambda - 550)]}$ for $\lambda = 550$ to 700 nm; $T_1 = 10$ s for $\lambda = 400$ to 549 nm; $T_1 = 10 \times 10^{[0.02\ (l - 550)]}$ for $\lambda = 550$ to 700 n.

At wavelengths greater than 1400 nm, for beam cross-sectional areas exceeding 100 cm^2 the TLV for exposure durations exceeding 10 seconds is: TLV = $(10{,}000/A_s)$ mW/cm^2 (2), where A_s is the irradiated skin area for 100 to 1000 cm^2, and the TLV for irradiated skin areas exceeding 1000 cm^2 is 10 mW/cm^2 and for irradiated skin areas less than 100 cm^2 is 100 mW/cm^2.

TABLE 11
Threshold Limit Values for Viewing a Diffuse Reflection of a Laser Beam or an Extended Source Laser

Spectral Region	Wave Length	Exposure Time, (t) Seconds	TLV
UV	200 nm to 400 nm	10^{-9} to 3×10^{4}	Same as Table 10
Light	400 nm to 700 nm	10^{-9} to 10	$10 \sqrt[3]{t}$ J • cm^{-2} • sr^{-1}
	400 nm to 549 nm	10 to 10^{4}	21 J • cm^{-2} • sr^{-1}
	550 nm to 700 nm	10 to T_1	3.83 $(t/\sqrt[4]{t})$ J • cm^{-2} • sr^{-1}
	550 nm to 700 nm	T_1 to 10^{4}	21 C_B J • cm^{-2} • sr^{-1}
	400 nm to 700 nm	10^{4} to 3×10^{4}	2.1 $C_B t \times 10^{-3}$ W • cm^{-2} • sr^{-1}
IR-A	700 nm to 1400 nm	10^{-9} to 10	10 $C_A \sqrt[3]{t}$ J • cm^{-2} • sr^{-1}
	700 nm to 1400 nm	10 to 10^{3}	3.83 C_A $(t/\sqrt[4]{t})$ J • cm^{-2} • sr^{-1}
	700 nm to 1400 nm	10^{3} to 3×10^{4}	0.64 C_A W • cm^{-2} • sr^{-1}
IR-B & C	1.4 μm to 10^{3} μm	10^{-9} to 3×10^{4}	Same as Table 10

C_A, C_B, and T_1 are the same as in footnote to Table 10.

TABLE 12
Threshold Limit Value for Skin Exposure from a Laser Beam

Spectral Region	Wave Length	Exposure Time, (t) Seconds	TLV
UV	200 nm to 400 nm	10^{-9} to 3×10^{4}	Same as Table 10
Light &	400 nm to 1400 nm	10^{-9} to 10^{-7}	2 $C_A \times 10^{-2}$ J • cm^{-2}
IR-A	" "	10^{-7} to 10	1.1 $C_A \sqrt[4]{t}$ J • cm^{-2}
IR-A	" "	10 to 3×10^{4}	0.2 C_A W • cm^{-2}
IR-B & C	1.4 μm to 10^{3} μm	10^{-9} to 3×10^{4}	Same as Table 10

C_A = 1.0 for λ = 400-700 nm; see Figure 5 for λ = 700 to 14 nm.

NOISE

These Threshold Limit Values (TLVs) refer to sound pressure levels and durations of exposure that represent conditions under which it is believed that nearly all workers may be repeatedly exposed without adverse effect on their ability to hear and understand normal speech. Prior to 1979, the medical profession had defined hearing impairment as an average hearing threshold level in excess of 25 decibels (ANSI-S3.6-1969) at 500, 1000, and 2000 Hz, and the limits which are given have been established to prevent a hearing loss in excess of this level.[a] The values should be used as guides in the control of noise exposure and, due to individual susceptibility, should not be regarded as fine lines between safe and dangerous levels.

It should be recognized that the application of the TLV for noise will not protect all workers from the adverse effects of noise exposure. A hearing conservation program with audiometric testing is necessary when workers are exposed to noise at or above the TLV levels.

Continuous or Intermittent

The sound level shall be determined by a sound level meter, conforming as a minimum to the requirements of the American National Standard Specification for Sound Level Meters, S1.4 (1971) Type S2A, and set to use the A-weighted network with slow meter response. Duration of exposure shall not exceed that shown in Table 13.

These values apply to total duration of exposure per working day regardless of whether this is one continuous exposure or a number of short-term exposures and does include the impact and impulsive type of noise that contributes to the sound level meter reading at slow response.

When the daily noise exposure is composed of two or more periods of noise exposure of different levels, their combined effect should be considered, rather than the individual effect of each. If the sum of the following fractions:

$$\frac{C_1}{T_1} + \frac{C_2}{T_2} + \ldots \frac{C_n}{T_n}$$

exceeds unity, then, the mixed exposure should be considered to exceed the threshold limit value, C_1 indicates the total duration of exposure at a specific noise level, and T_1 indicates the total duration of exposure permitted at that level. All on-the-job noise exposures of 80 dBA or greater shall be used in the above calculations.

Impulsive or Impact

It is recommended that exposure to impulsive or impact noise shall not exceed the limits listed in Table 14 or taken from Figure 9. No exposures in excess of 140 decibels peak sound pressure level are permitted. Impulsive or impact noise is considered to be those variations in noise levels that involve maxima at intervals of greater than one per second. Where the intervals are less than one second, it should be considered continuous.

---

[a] In 1979 the American Academy of Ophthalmology and Otolaryngology (AAOO) included 3000 Hz in their hearing impairment formula.

TABLE 13
Threshold Limit Values for Noise

Duration per Day Hours	Sound Level dBA†
16	80
8	85
4	90
2	95
1	100
1/2	105
1/4	110
1/8	115*

† Sound level in decibels are measured on a sound level meter, conforming as a minimum to the requirements of the American National Standard Specification for Sound Level Meters, S1.4 (1971) Type S2A, and set to use the A-weighted network with slow meter response.

* No exposure to continuous or intermittent in excess of 115 dBA.

TABLE 14
Threshold Limit Values Impulsive or Impact Noise

Sound Level dB*	Permitted Number of Impulses or Impacts per day
140	100
130	1000
120	10,000

* Decibels peak sound pressure level; re 20 μPa.

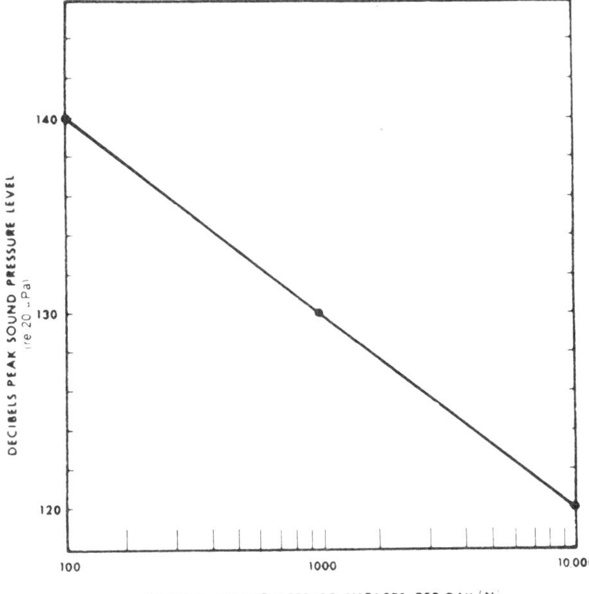

Figure 9—Threshold Limit Values for Impulse/Impact Noise.

RADIOFREQUENCY/MICROWAVE RADIATION

These Threshold Limit Values (TLVs) refer to radiofrequency (RF) and microwave radiation in the frequency range from 10 kHz to 300 GHz, and represent conditions under which it is believed workers may be repeatedly exposed without adverse health effects. The TLVs shown in Table 15 are selected to limit the average whole body specific absorption rate (SAR) to 0.4 W/kg in any six-minute (0.1 hr) period for 3 MHz to 300 GHz, see Figure 10. Between 10 kHz and 3 MHz the average whole body SAR is still limited to 0.4 W/kg, but the plateau at 100 mW/cm² was set to protect against shock and burn hazards.

Since it is usually impractical to measure the SAR, the TLVs are expressed in units that are measurable, viz, squares of the electric and magnetic field strength, averaged over any 0.1 hour period. This can be expressed in units of equivalent plane wave power density for convenience. The electric field strength (E) squared, magnetic field strength (H) squared, and power density (PD) values are shown in Table 15. For near field exposures PD cannot be measured directly, but equivalent plane wave power density can be calculated from the field strength measurement data as follows:

$$\text{PD in mW/cm}^2 = \frac{E^2}{3770}$$

where:

E^2 is in volts squared (V^2) per meter squared (m^2).

$$\text{PD in mW/cm}^2 = 37.7\ H^2$$

where:

H^2 is in amperes squared (A^2) per meter squared (m^2).

These values should be used as guides in the evaluation and control of exposure to radiofrequency/microwave radiation, and should not be regarded as a fine line between safe and dangerous levels.

Notes:

1. Needless exposure to all Radiofrequency Radiation (RFR) exposures should be avoided given the current state of knowledge on human effects, particularly non-thermal effects.
2. For fields consisting of a number of frequencies, the fraction of the protection guide incurred within each frequency level should be determined and the sum of all fractions should not exceed unity.
3. For pulsed and continuous wave fields, the power density is averaged over the six-minute period.
4. For partial body exposures at frequencies between 10 kHz and 1.0 GHz, the protection guides in Table 15 may be exceeded if the output power of a radiating device is 7 watts or less. For example, if a hand held transmitter operating at 27 MHz has a maximum output of 5 watts, it would be excluded from any further field measurements.
5. The TLVs in Table 15 may be exceeded if the exposure conditions can be demonstrated to produce a SAR of less than 0.4 W/kg as averaged over the whole body and spatial peak SAR values less than 8.0 W/kg as averaged over any 1.0 gram of tissue. For example, for frequencies from 3 to 30 MHz, the equivalent power density can be increased by a factor of 10 up to a limit of 100 mW/cm², if it can be assured that exposed individuals are not in contact with the ground plate.
6. At frequencies below 30 MHz, ungrounded objects such as vehicles, fences, etc., can strongly couple to RF fields. For field strengths near the TLV, shock and burn hazards can exist. Care should be taken to eliminate ungrounded objects, to ground such objects, or use insulated gloves when ungrounded objects must be handled.
7. No measurement should be made within 5 cm of any object.
8. All exposures should be limited to a maximum (peak) electric field intensity of 100 kV/m.

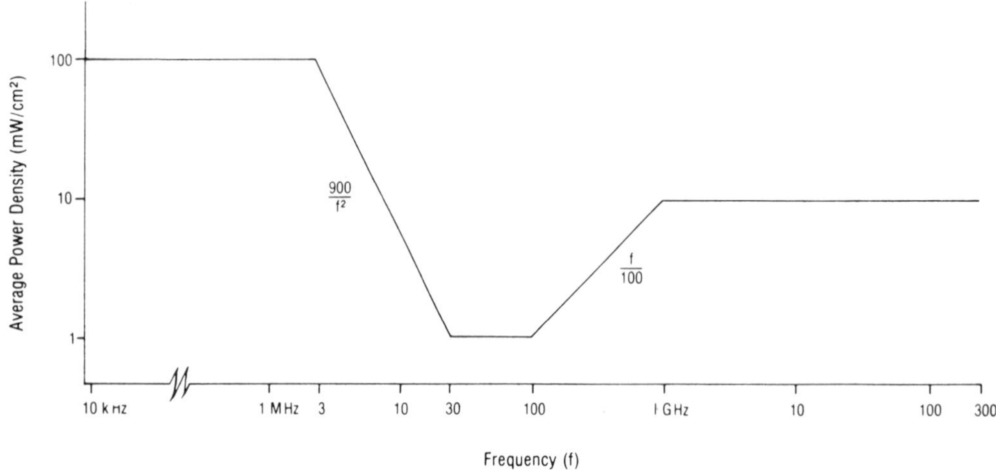

Figure 10—Threshold Limit Values (TLV) for Radiofrequency/Microwave Radiation in the workplace (whole body SAR less than 0.4 W/kg).

TABLE 15
Radiofrequency/Microwave Threshold Limit Values

Frequency	Power Density (mW/cm^2)	Electric Field Strength Squared (V^2/m^2)	Magnetic Field Strength Squared (A^2/m^2)
10 KHz to 3 MHz	100	377,000	2.65
3 MHz to 30 MHz	$900/f^2$*	$3770 \times 900/f^2$	$900/(37.7 \times f^2)$
30 MHz to 100 MHz	1	3770	0.027
100 MHz to 1000 MHz	f/100	$3770 \times f/100$	$f/37.7 \times 100$
1 GHz to 300 GHz	10	37,700	0.265

*f = frequency in MHz.

ULTRAVIOLET RADIATION*

These Threshold Limit Values (TLVs) refer to ultraviolet radiation in the spectral region between 200 and 400 nm and represent conditions under which it is believed that nearly all workers may be repeatedly exposed without adverse effect. These values for exposure of the eye or the skin apply to ultraviolet radiation from arcs, gas and vapor discharges, fluorescent and incandescent sources, and solar radiation, but do not apply to ultraviolet lasers.* These values do not apply to ultraviolet radiation exposure of photosenitive individuals or of individuals concomitantly exposed to photosensitizing agents.[1] These values should be used as guides in the control of exposure to continuous sources where the exposure duration shall not be less than 0.1 sec.

These values should be used as guides in the control of exposure to ultraviolet sources and should not be regarded as a fine line between safe and dangerous levels.

Recommended Values

The threshold limit value for occupational exposure to ultraviolet radiation incident upon skin or eye where irradiance values are known and exposure time is controlled are as follows:

1. For the near ultraviolet spectral region (320 to 400 nm) total irradiance incident upon the unprotected skin or eye should not exceed 1 mW/cm^2 for periods greater than 10^3 seconds (approximately 16 minutes) and for exposure times less than 10^3 seconds should not exceed one J/cm^2.
2. For the actinic ultraviolet spectral region (200-315 nm), radiant exposure incident upon the unprotected skin or eye should not exceed the values given in Table 16 within an 8-hour period.
3. To determine the effective irradiance of a broadband source weighted against the peak of the spectral effectiveness curve (270 nm), the following weighting formula should be used:

$$E_{eff} = \Sigma E_\lambda S_\lambda \Delta\lambda$$

where:

E_{eff} = effective irradiance relative to a monochromatic source at 270 nm in W/cm^2 ($J/s/cm^2$)
E_λ = spectral irradiance in $W/cm^2/nm$
S_λ = relative spectral effectiveness (unitless)
$\Delta\lambda$ = band width in nanometers

4. Permissible exposure time in seconds for exposure to actinic ultraviolet radiation incident upon the unprotected skin or eye may be computed by dividing 0.003 J/cm^2 by E_{eff} in W/cm^2.

* See Laser TLVs.

TABLE 16
Relative Spectral Effectiveness by Wavelength*

Wavelength (nm)	TLV (mJ/cm²)	Relative Spectral Effectiveness S_λ
200	100	0.03
210	40	0.075
220	25	0.12
230	16	0.19
240	10	0.30
250	7.0	0.43
254	6.0	0.5
260	4.6	0.65
270	3.0	1.0
280	3.4	0.88
290	4.7	0.64
300	10	0.30
305	50	0.06
310	200	0.015
315	1000	0.003

* See Laser TLVs.

TABLE 17
Permissible Ultraviolet Exposures

Duration of Exposure Per Day	Effective Irradiance, E_{eff} (μW/cm²)
8 hrs	0.1
4 hr	0.2
2 hrs	0.4
1 hr	0.8
30 min	1.7
15 min	3.3
10 min	5
5 min	10
1 min	50
30 sec	100
10 sec	300
1 sec	3,000
0.5 sec	6,000
0.1 sec	30,000

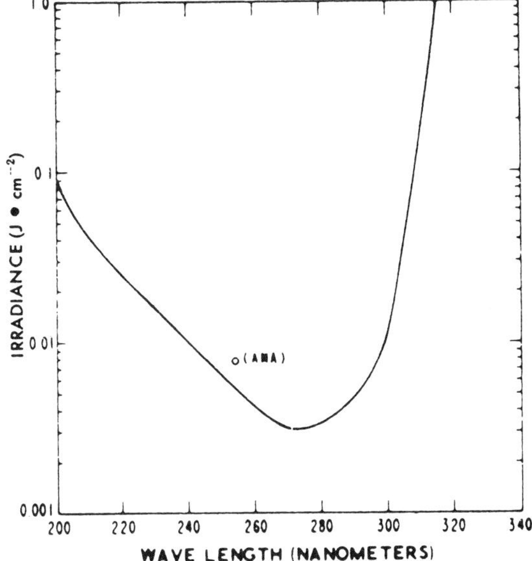

Figure 11—Threshold Limit Values for Ultraviolet Radiation.

The exposure time may also be determined using Table 17 which provides exposure times corresponding to effective irradiances in μW/cm².

5. All the preceding TLVs for ultraviolet energy apply to sources which subtend an angle less than 80°. Sources which subtend a greater angle need to be measured only over an angle of 80°.

Conditioned (tanned) individuals can tolerate skin exposure in excess of the TLV without erythemal effects. However, such conditioning may not protect persons against cancer.

Reference

1. *Sunlight and Man.* Fitzpatrick *et al*, Eds. Univ. of Tokyo Press, Tokyo, Japan (1974).

NOTICE OF INTENDED CHANGES (for 1986-87)

These physical agents, with their corresponding values, comprise those for which either a limit has been proposed for the first time, or for which a change in the "Adopted" listing has been proposed. In both cases, the proposed limits should be considered trial limits that will remain in the listing for a period of at least one year. If after one year no evidence comes to light that questions the appropriateness of the values herein the values will be reconsidered for the the "Adopted" list.

NOTICE OF INTENT TO ESTABLISH THRESHOLD LIMIT VALUES

LIGHT AND NEAR-INFRARED RADIATION

These Threshold Limit Values (TLVs) refer to visible and near-infrared radiaton in the wavelength range of 400 nm to 1400 nm and represent conditions under which it is believed that nearly all workers may be exposed without adverse effect. These values should be used as guides in the control of exposure to light and should not be regarded as a fine line between safe and dangerous levels.

Recommended Values

The Threshold Limit Value for occupational exposure to broad-band light and near-infrared radiation for the eye apply to exposure in any eight-hour workday and require knowledge of the spectral radiance (L_λ) and total irradiance (E) of the source as measured at the position(s) of the eye of the worker. Such detailed spectral data of a white light source is generally only required if the luminance of the source exceeds 1 cd cm^{-2}. At luminances less than this value the TLV would not be exceeded.

The TLVs are:

1. To protect against retinal thermal injury, the spectral radiance of the lamp weighted against the function R (Table 18) should not exceed:

$$\sum_{400}^{1400} L_\lambda \bullet R_\lambda \bullet \Delta\lambda \leqslant 1/\alpha t^{1/2} \qquad (1)^*$$

where L_λ is in W cm^{-2} sr^{-1} nm^{-1} and t is the viewing duration (or pulse duration if the lamp is pulsed) limited to 1 μs to 10 s, and α is the angular substence of the source in radians. If the lamp is oblong, α refers to the longest dimension that can be viewed. For instance, at a viewing distance r = 100 cm from a tubular lamp of length l = 50 cm, the viewing angle is:

* Formulae (1) and (7) are empirical and are not, strictly speaking, dimensionally correct. To make the formulae dimensionally correct, one would have to insert a dimensional correction factor *k* in the right hand numerator in each formula. For formula (1) this would be k_1 = 1 W • rad • s$^{1/2}$/(cm² • sr), and for formula (7) K_2 = 1 W • rad/(cm² • sr).

$$\alpha = l/r = 50/100 = 0.5 \text{ rad} \tag{2}$$

2. To protect against retinal photochemical injury from chronic blue-light exposure the integrated spectral radiance of a light source weighted against the blue-light hazard function B_λ (Table 18) should not exceed:

$$\sum_{400}^{1400} L_\lambda \bullet t \bullet B_\lambda \bullet \Delta\lambda \leqslant 100 \text{ J cm}^{-2} \text{ sr}^{-1} \ (t \leqslant 10^4 \text{ s}) \tag{3a}$$

$$\sum_{400}^{1400} L_\lambda \bullet B_\lambda \bullet \Delta\lambda \leqslant 10^{-2} \text{ sr}^{-1} \ (t > 10^4 \text{ s}) \tag{3b}$$

The weighted product of L_λ and B_λ is termed L (blue). For a source radiance L weighted against the blue-light hazard function (L [blue]) which exceeds 10 mW cm^{-2} sr^{-1} in the blue spectral region, the permissible exposure duration t_{max} in seconds is simply:

$$t_{max} = 100 \text{ J cm}^{-2} \text{ sr}^{-1}/\text{L (blue)} \tag{4}$$

The latter limits are greater than the maximum permissible exposure limits for 440 nm laser radiation (*see* Laser TLV) because of a 2-3 mm pupil is assumed rather than a 7 mm pupil for the Laser TLV. For a light source subtending an angle α less than 11 mrd (0.011 radian) the above limits are relaxed such that the spectral irradiance weighted against the blue-light hazard function B_λ should not exceed E (blue).

$$\sum_{400}^{1400} E_\lambda \bullet t \bullet B_\lambda \bullet \Delta\lambda \leqslant 10 \text{ mJ} \bullet \text{cm}^{-2} \ (t \leqslant 10^4 \text{ s}) \tag{5a}$$

$$\sum_{400}^{1400} E_\lambda \bullet B_\lambda \bullet \Delta\lambda \leqslant 1 \ \mu\text{W} \bullet \text{cm}^2 \ (t \geqslant 10^4 \text{ s}) \tag{5b}$$

For a source where the blue light weighted irradiance E (blue) exceeds 1 μW • cm^{-2} the maximum permissible exposure duration t_{max} in seconds is:

$$t_{max} = 10 \text{ mJ} \bullet \text{cm}^{-2}/\text{E (blue)} \tag{6}$$

3. *Infrared radiation:* To avoid possible delayed effects upon the lens of the eye (cataractogenesis), the infrared radiation (λ = 770 nm) should be limited to 10 mW cm^{-2}. For an infrared heat lamp or any near-infrared source where a strong visual stimulus is absent, the near infrared (770-1400 nm) radiance as viewed by the eye should be limited to:

$$\sum_{770}^{1400} L_\lambda \bullet \Delta\lambda \leqslant 0.6/\alpha \tag{7*}$$

for extended duration viewing conditions. This limit is based upon a 7 mm pupil diameter.

TABLE 18
Spectral Weighting Functions for Assessing Retinal Hazards from Broad-Band Optical Sources

Wavelength (nm)	Blue-Light Hazard-Function B_λ	Burn Hazard Function R_λ
400	0.10	1.0
405	0.20	2.0
410	0.40	4.0
415	0.80	8.0
420	0.90	9.0
425	0.95	9.5
430	0.98	9.8
435	1.0	10.0
440	1.0	10.0
445	0.97	9.7
450	0.94	9.4
455	0.90	9.0
460	0.80	8.0
465	0.70	7.0
470	0.62	6.2
475	0.55	5.5
480	0.45	4.5
485	0.40	4.0
490	0.22	2.2
495	0.16	1.6
500-600	$10^{[(450-\lambda)/50]}$	1.0
600-700	0.001	1.0
700-770	0.001	$10^{[(700-\lambda)/505]}$
770-1400	0.001	0.2

AIRBORNE UPPER SONIC AND ULTRASONIC ACOUSTIC RADIATION

These Threshold Limit Values (TLVs) refer to sound pressure levels that represent conditions under which it is believed that nearly all workers may be repeatedly exposed without adverse effect. The values listed in Table 19 should be used as guides in the control of noise exposure and, due to individual susceptibility, should not be regarded as fine lines between safe and dangerous levels. The levels for the third-octave bands centered below 20 kHz are below those which cause subjective effects. Those levels for 1/3 octaves above 20 kHz are for prevention of possible hearing losses from subharmonics of these frequencies.

TABLE 19
Permissible Ultrasound Exposure Levels

Mid-Frequency of Third-Octave Band kHz	One-Third Octave—Band Level in dB re 20 μPa
10	80
12.5	80
16	80
20	105
25	110
31.5	115
40	115
50	115

PHYSICAL AGENTS UNDER STUDY

The Physical Agents TLV Committee of ACGIH has examined the current literature and has not found sufficient information to propose a TLV. However, these agents will remain under study during the coming year to examine new evidence indicating the need and feasibility for establishing a proposed TLV. Comments and suggestions, accompanied by substantive documentation are solicited and should be forwarded to the Executive Secretary, ACGIH. Documentation summarizing the current status of the biological effects literature is available on those agents preceded by an asterisk (*).

1. **Extremely Low Frequency (ELF) Radiation.* Specifically, that portion of the spectrum from 0 to 300 Hz.
2. *Magnetic Fields.* Both pulsed and *continuous.
3. *Laser Radiation.* Specifically laser exposures of less than one (1) nanosecond.
4. *Vibration.* Whole-body.
5. *Pressure Variations.*

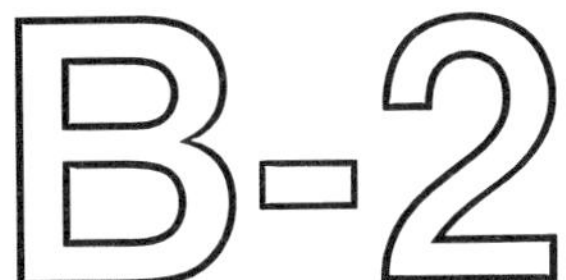

Permissible Exposure Limits

THIS APPENDIX GIVES THE Permissible Exposure Limits (PELs) as set forth in the Occupational Safety and Health Administration (OSHA) standards (Title 29, *Code of Federal Regulations*, Part 1910, Subpart Z, Toxic and Hazardous Substances, Section 1910.1000, Air Contaminants, revised 1989).

Most OSHA PELs were promulgated initially from the 1968 ACGIH list of Threshold Limit Values (TLVs). The 1989 revision updated the PELs on the basis of current research. However, in July 1992, the U.S. Court of Appeals for the Eleventh Circuit vacated these 1989 standards, ruling that OSHA presented insufficient evidence of significant health risk for individual permissible exposure limits. At the time of this printing, OSHA is appealing this decision while continuing to enforce the 1989 PELs.

Subpart Z—Toxic and Hazardous Substances

SOURCE: 39 FR 23502, June 27, 1974, unless otherwise noted. Redesignated at 40 FR 23073, May 28, 1975.

§ 1910.1000 Air contaminants.

An employee's exposure to any material listed in table Z-1, Z-2, or Z-3 of this section shall be limited in accordance with the requirements of the following paragraphs of this section.

(a) Table Z-1:

(1) *Materials with names preceded by "C"—Ceiling Values.* An employee's exposure to any material in table Z-1, the name of which is preceded by a "C" (e.g., C Boron trifluoride), shall at no time exceed the ceiling value given for that material in the table.

(2) *Other materials—8-hour time weighted averages.* An employee's exposure to any material in table Z-1, the name of which is not preceded by "C", in any 8-hour work shift of a 40-hour work week, shall not exceed the 8-hour time weighted average given for that material in the table.

(b) Table Z-2: (1) *8-hour time weighted averages.* An employee's exposure to any material listed in table Z-2, in any 8-hour work shift of a 40-hour work week, shall not exceed the 8-hour time weighted average limit given for that material in the table.

(2) *Acceptable ceiling concentrations.* An employee's exposure to a material listed in table Z-2 shall not exceed at any time during an 8-hour shift the acceptable ceiling concentration limit given for the material in the table, except for a time period, and up to a concentration not exceeding the maximum duration and concentration allowed in the column under "acceptable maximum peak above the acceptable ceiling concentration for an 8-hour shift".

(3) *Example.* During an 8-hour work shift, an employee may be exposed to a concentration of Benzene above 25 p.p.m. (but never above 50 p.p.m.) only for a maximum period of 10 minutes. Such exposure must be compensated by exposures to concentrations less than 10 p/m so that the cumulative exposure for the entire 8-hour work shift does not exceed a weighted average of 10 p/m

(c) Table Z-3: An employee's exposure to any material listed in table Z-3, in any 8-hour work shift of a 40-hour work week, shall not exceed the 8-hour time weighted average limit given for that material in the table.

(d) Computation formulae:

(1) (i) The cumulative exposure for an 8-hour work shift shall be computed as follows:

$$(E = C_a T_a + C_b T_b + \ldots C^n T^n) \div 8$$

Where:

E is the equivalent exposure for the working shift.

C is the concentration during any period of time *T* where the concentration remains constant.

T is the duration in hours of the exposure at the concentration C.

The value of *E* shall not exceed the 8-hour time weighted average limit in table Z-1, Z-2, or Z-3 for the material involved.

(ii) To illustrate the formula prescribed in paragraph (d)(1)(i) of this section, note that isoamyl acetate has an 8-hour time weighted average limit of 100 p.p.m. (table Z-1). Assume that an employee is subject to the following exposure:

Two hours exposure at 150 p/m
Two hours exposure at 75 p/m
Four hours exposure at 50 p/m

Substituting this information in the formula, we have

$$(2 \times 150 + 2 \times 75 + 4 \times 50) \div 8 = 81.25 \text{ p/m}$$

Since 81.25 p.p.m. is less than 100 p.p.m., the 8-hour time weighted average limit, the exposure is acceptable.

(2) (i) In case of a mixture of air contaminants an employer shall compute the equivalent exposure as follows:

$$E_m = (C_1 \div L_1 + C_2 \div L_2) + \ldots (C_n \div L_n)$$

Where:

Em is the equivalent exposure for the mixture.

C is the concentration of a particular contaminant.

L is the exposure limit for that contaminant, from table Z-1, Z-2, or Z-3.

The value of E*m* shall not exceed unity (1).

(ii) To illustrate the formula prescribed in paragraph (d)(2)(i) of this section, consider the following exposures:

Material	Actual concentration of 8-hour exposure	8-hour time weighted average exposure limit
Acetone (Table Z-1)	500 p/m	1,000 p/m
2-Butanone (Table Z-1)	45 p/m	200 p/m
Toluene (Table Z-2)	40 p/m	200 p/m

Substituting in the formula, we have:

$$Em = 500 \div 1{,}000 + 45 \div 200 + 40 \div 200$$
$$Em = 0.500 + 0.225 + 0.200$$
$$Em = 0.925$$

Since E*m* is less than unity (1), the exposure combination is within acceptable limits.

(e) To achieve compliance with paragraph (a) through (d) of this section, administrative or engineering controls must first be determined and implemented whenever feasible. When such controls are not feasible to achieve full compliance, protective equipment or any other protective measures shall be used to keep the exposure of employees to air contaminants within the limits prescribed in this section. Any equipment and/or technical measures used for this purpose must be approved for each particular use by a competent industrial hygienist or other technically qualified person. Whenever respirators are used, their use shall comply with § 1910.134.

TABLE Z-1

Substance	p/m[a]	mg./M³[b]
Acetaldehyde	200	360
Acetic acid	10	25
Acetic anhydride	5	20
Acetone	1,000	2,400
Acetonitrile	40	70
Acetylene dichloride, see 1, 2-Dichloroethylene		
Acetylene tetrabromide	1	14
Acrolein	0.1	0.25
Acrylamide—Skin		0.3
Aldrin—Skin		0.25
Allyl alcohol—Skin	2	5
Allyl chloride	1	3
C Allylglycidyl ether (AGE)	10	45
Allyl propyl disulfide	2	12
2-Aminoethanol, see Ethanolamine		
2-Aminopyridine	0.5	2
Ammonia	50	35
Ammonium sulfamate (Ammate)		15
n-Amyl acetate	100	525
sec-Amyl acetate	125	650
Aniline—Skin	5	19
Anisidine (o, p-isomers)—Skin		0.5
Antimony and compounds (as Sb)		0.5
ANTU (alpha naphthyl thiourea)		0.3
Arsenic organic compounds (as As)		0.5
Arsine	0.05	0.2
Azinphos-methyl—Skin		0.2
Barium (soluble compounds)		0.5
p-Benzoquinone, see Quinone		
Benzoyl peroxide		5
Benzyl chloride	1	5
Biphenyl, see Diphenyl		
Boron oxide		15
C Boron trifluoride	1	3
Bromine	0.1	0.7
Bromoform—Skin	0.5	5
Butadiene (1, 3-butadiene)	1,000	2,200
Butanethiol, see Butyl mercaptan		
2-Butanone	200	590
2-Butoxy ethanol (Butyl Cellosolve)—Skin	50	240
Butyl acetate (n-butyl acetate)	150	710
sec-Butyl acetate	200	950
tert-Butyl acetate	200	950
Butyl alcohol	100	300
sec-Butyl alcohol	150	450
tert-Butyl alcohol	100	300
C Butylamine—Skin	5	15

TABLE Z-1—Continued

Substance	p/m[a]	mg./M[3b]
C tert-Butyl chromate (as CrO_3)—Skin		0.1
n-Butyl glycidyl ether (BGE)	50	270
Butyl mercaptan	10	35
p-tert-Butyltoluene	10	60
Calcium oxide		5
Camphor		2
Carbaryl (Sevin[R])		5
Carbon black		3.5
Carbon dioxide	5,000	9,000
Carbon monoxide	50	55
Chlordane—Skin		0.5
Chlorinated camphene—Skin		0.5
Chlorinated diphenyl oxide		0.5
C Chlorine	1	3
Chlorine dioxide	0.1	0.3
C Chlorine trifluoride	0.1	0.4
C Chloroacetaldehyde	1	3
a-Chloroacetophenone (phenacylchloride)	0.05	0.3
Chlorobenzene (monochlorobenzene)	75	350
o-Chlorobenzylidene malononitrile (OCBM)	0.05	0.4
Chlorobromomethane	200	1,050
2-Chloro-1,3-butadiene, see Chloroprene		
Chlorodiphenyl (42 percent Chlorine)—Skin		1
Chlorodiphenyl (54 percent Chlorine)—Skin		0.5
1-Chloro, 2,3-epoxypropane, see Epichlorhydrin		
2-Chloroethanol, see Ethylene chlorohydrin		
Chloroethylene, see Vinyl chloride		
C Chloroform (trichloromethane)	50	240
1-Chloro-1-nitropropane	20	100
Chloropicrin	0.1	0.7
Chloroprene (2-chloro-1,3- butadiene)—Skin	25	90
Chromium, sol. chromic, chromous salts as Cr		0.5
Metal and insol. salts		1
Coal tar pitch volatiles (benzene soluble fraction) anthracene, BaP, phenanthrene, acridine, chrysene, pyrene		0.2
Cobalt, metal fume and dust		0.1
Copper fume		0.1
Dusts and Mists		1
Cotton dust (raw)		1†
Crag[R] herbicide		15
Cresol (all isomers)—Skin	5	22
Crotonaldehyde	2	6
Cumene—Skin	50	245
Cyanide (as CN)—Skin		5
Cyclohexane	300	1,050
Cyclohexanol	50	200
Cyclohexanone	50	200
Cyclohexene	300	1,015
Cyclopentadiene	75	200
2,4-D		10
DDT—Skin		1
DDVP—Skin		1
Decaborane—Skin	0.05	0.3
Demeton[R]—Skin		0.1
Diacetone alcohol (4-hydroxy-4-methyl-2-pentanone)	50	240
1,2-diaminoethane, see Ethylenediamine		
Diazomethane	0.2	0.4
Diborane	0.1	0.1
Dibutyl phosphate	1	5
Dibutylphthalate		5
C o-Dichlorobenzene	50	300
p-Dichlorobenzene	75	450
Dichlorodifluoromethane	1,000	4,950
1,3-Dichloro-5,5-dimethyl hydantoin		0.2
1,1-Dichloroethane	100	400
1,2-Dichloroethylene	200	790
C Dichloroethyl ether—Skin	15	90
Dichloromethane, see Methylenechloride		
Dichloromonofluoromethane	1,000	4,200
C 1,1-Dichloro-1-nitroethane	10	60
1,2-Dichloropropane, see Propylenedichloride		
Dichlorotetrafluoroethane	1,000	7,000
Dieldrin—Skin		0.25
Diethylamine	25	75
Diethylamino ethanol—Skin	10	50
Diethylether, see Ethyl ether		
Difluorodibromomethane	100	860
C Diglycidyl ether (DGE)	0.5	2.8
Dihydroxybenzene, see Hydroquinone		
Diisobutyl ketone	50	290
Diisopropylamine—Skin	5	20
Dimethoxymethane, see Methylal		
Dimethyl acetamide—Skin	10	35
Dimethylamine	10	18
Dimethylaminobenzene, see Xylidene		
Dimethylaniline (N-dimethyl- aniline)—Skin	5	25
Dimethylbenzene, see Xylene		
Dimethyl 1,2-dibromo-2,2-dichloroetnyl phosphate, (Dibrom)		3
Dimethylformamide—Skin	10	30
2,6-Dimethylheptanone, see Diisobutyl ketone		
1,1-Dimethylhydrazine—Skin	0.5	1
Dimethylphthalate		5
Dimethylsulfate—Skin	1	5
Dinitrobenzene (all isomers)—Skin		1
Dinitro-o-cresol—Skin		0.2
Dinitrotoluene—Skin		1.5
Dioxane (Diethylene dioxide)—Skin	100	360
Diphenyl	0.2	1
Diphenylmethane diisocyanate (see Methylene bisphenyl isocyanate (MDI))		
Dipropylene glycol methyl ether—Skin	100	600
Di-sec, octyl phthalate (Di-2-ethylhexylphthalate)		5
Endrin—Skin		0.1
Epichlorhydrin—Skin	5	19
EPN—Skin		0.5
1,2-Epoxypropane, see Propyleneoxide		
2,3-Epoxy-1-propanol, see Glycidol		
Ethanethiol, see Ethylmercaptan		
Ethanolamine	3	6
2-Ethoxyethanol—Skin	200	740
2-Ethoxyethylacetate (Cello-solve acetate)—Skin	100	540
Ethyl acetate	400	1,400
Ethyl acrylate—Skin	25	100
Ethyl alcohol (ethanol)	1,000	1,900
Ethylamine	10	18
Ethyl sec-amyl ketone (5- methyl-3-heptanone)	25	130
Ethyl benzene	100	435
Ethyl bromide	200	890
Ethyl butyl ketone (3- Heptanone)	50	230
Ethyl chloride	1,000	2,600
Ethyl ether	400	1,200
Ethyl formate	100	300
C Ethyl mercaptan	10	25
Ethyl silicate	100	850
Ethylene chlorohydrin—Skin	5	16
Ethylenediamine	10	25
C Ethylene glycol dinitrate and/or Nitroglycerin—Skin	[d] 0.2	1
Ethylene glycol monomethyl ether acetate, see Methyl cellosolve acetate		
Ethylene imine—Skin	0.5	1
Ethylidine chloride, see 1,1- Dichloroethane		
N-Ethylmorpholine—Skin	20	94
Ferbam		15
Ferrovanadium dust		1
Fluoride (as F)		2.5
Fluorine	0.1	0.2
Fluorotrichloromethane	1,000	5,600
Formic acid	5	9
Furfural—Skin	5	20
Furfuryl alcohol	50	200
Glycidol (2,3-Epoxy-1- propanol)	50	150
Glycol monoethyl ether, see 2-Ethoxyethanol		
Guthion[R], see Azinphosmethyl		
Hafnium		0.5
Heptachlor—Skin		0.5
Heptane (n-heptane)	500	2,000
Hexachloroethane—Skin	1	10
Hexachloronaphthalene—Skin		0.2
Hexane (n-hexane)	500	1,800
2-Hexanone	100	410
Hexone (Methyl isobutyl ketone)	100	410
sec-Hexyl acetate	50	300
Hydrazine—Skin	1	1.3
Hydrogen bromide	3	10
C Hydrogen chloride	5	7
Hydrogen cyanide—Skin	10	11
Hydrogen peroxide (90%)	1	1.4
Hydrogen selenide	0.05	0.2
Hydroquinone		2
C Iodine	0.1	1
Iron oxide fume		10
Isoamyl acetate	100	525
Isoamyl alcohol	100	360
Isobutyl acetate	150	700
Isobutyl alcohol	100	300
Isophorone	25	140
Isopropyl acetate	250	950
Isopropyl alcohol	400	980
Isopropylamine	5	12
Isopropylether	500	2,100
Isopropyl glycidyl ether (IGE)	50	240
Ketene	0.5	0.9
Lindane—Skin		0.5
Lithium hydride		0.025
L.P.G. (liquified petroleum gas)	1,000	1,800
Magnesium oxide fume		15
Malathion—Skin		15
Maleic anhydride	0.25	1
C Manganese		5
Mesityl oxide	25	100
Methanethiol, see Methyl mercaptan		
Methoxychlor		15
2-Methoxyethanol, see Methyl cellosolve		
Methyl acetate	200	610
Methyl acetylene (propyne)	1,000	1,650
Methyl acetylene-propadiene mixture (MAPP)	1,000	1,800
Methyl acrylate—Skin	10	35
Methylal (dimethoxymethane)	1,000	3,100
Methyl alcohol (methanol)	200	260
Methylamine	10	12
Methyl amyl alcohol, see Methyl isobutyl carbinol		
Methyl (n-amyl) ketone (2- Heptanone)	100	465
C Methyl bromide—Skin	20	80
Methyl butyl ketone, see 2- Hexanone		
Methyl cellosolve—Skin	25	80
Methyl cellosolve acetate—Skin	25	120
Methyl chloroform	350	1,900
Methylcyclohexane	500	2,000
Methylcyclohexanol	100	470
o-Methylcyclohexanone—Skin	100	460
Methyl ethyl ketone (MEK), see 2-Butanone		
Methyl formate	100	250
Methyl iodide—Skin	5	28
Methyl isobutyl carbinol—Skin	25	100
Methyl isobutyl ketone, see Hexone		
Methyl isocyanate—Skin	0.02	0.05
C Methyl mercaptan	10	20
Methyl methacrylate	100	410
Methyl propyl ketone, see 2- Pentanone		
C# Methyl styrene	100	480
C Methylene bisphenyl isocyanate (MDI)	0.02	0.2
Molybdenum:		
Soluble compounds		5
Insoluble compounds		15
Monomethyl aniline—Skin	2	9
C Monomethyl hydrazine— Skin	0.2	0.35
Morpholine—Skin	20	70
Naphtha (coaltar)	100	400
Naphthalene	10	50
Nickel carbonyl	0.001	0.007
Nickel, metal and soluble cmpds, as Ni		1
Nicotine—Skin		0.5
Nitric acid	2	5
Nitric oxide	25	30
p-Nitroaniline—Skin	1	6
Nitrobenzene—Skin	1	5
p-Nitrochlorobenzene—Skin		1
Nitroethane	100	310
C Nitrogen dioxide	5	9
Nitrogen trifluoride	10	29
C Nitroglycerin—Skin	0.2	2

Table Z-1—Continued

Substance	p/m[a]	mg./M^{3}[b]
Nitromethane	100	250
1-Nitropropane	25	90
2-Nitropropane	25	90
Nitrotoluene—Skin	5	30
Nitrotrichloromethane, see Chloropicrin		
Octachloronaphthalene—Skin		0.1
Octane	500	2,350
Oil mist, mineral		5
Osmium tetroxide		0.002
Oxalic acid		1
Oxygen difluoride	0.05	0.1
Ozone	0.1	0.2
Paraquat—Skin		0.5
Parathion—Skin		0.1
Pentaborane	0.005	0.01
Pentachloronaphthalene—Skin		0.5
Pentachlorophenol—Skin		0.5
Pentane	1,000	2,950
2-Pentanone	200	700
Perchloromethyl mercaptan	0.1	0.8
Perchloryl fluoride	3	13.5
Petroleum distillates (naphtha)	500	2,000
Phenol—Skin	5	19
p-Phenylene diamine—Skin		0.1
Phenyl ether (vapor)	1	7
Phenyl ether-biphenyl mixture (vapor)	1	7
Phenylethylene, see Styrene		
Phenyl glycidyl ether (PGE)	10	60
Phenylhydrazine—Skin	5	22
Phosdrin (Mevinphos[R])— Skin		0.1
Phosgene (carbonyl chloride)	0.1	0.4
Phosphine	0.3	0.4
Phosphoric acid		1
Phosphorus (yellow)		0.1
Phosphorus pentachloride		1
Phosphorus pentasulfide		1
Phosphorus trichloride	0.5	3
Phthalic anhydride	2	12
Picric acid—Skin		0.1
Pival[R] (2-Pivalyl-1,3- indandione)		0.1
Platinum (Soluble salts) as Pt		0.002
Propane	1,000	1,800
n-Propyl acetate	200	840
Propyl alcohol	200	500
n-Propyl nitrate	25	110
Propylene dichloride	75	350
Propylene imine—Skin	2	5
Propylene oxide	100	240
Propyne, see Methylacetylene		
Pyrethrum		5
Pyridine	5	15
Quinone	0.1	0.4
Rhodium, Metal fume and dusts, as Rh		0.1
Soluble salts		0.001
Ronnel		15
Rotenone (commercial)		5
Selenium compounds (as Se)		0.2
Selenium hexafluoride	0.05	0.4
Silver, metal and soluble compounds		0.01
Sodium fluoroacetate (1080)—Skin		0.05
Sodium hydroxide		2
Stibine	0.1	0.5
Stoddard solvent	500	2,900
Strychnine		0.15
Sulfur dioxide	5	13
Sulfur hexafluoride	1,000	6,000
Sulfuric acid		1
Sulfur monochloride	1	6
Sulfur pentafluoride	0.025	0.25
Sulfuryl fluoride	5	20
Systox, see Demeton[R]		
2,4,5T		10
Tantalum		5
TEDP—Skin		0.2
Tellurium		0.1
Tellurium hexafluoride	0.02	0.2
TEPP—Skin		0.05
C Terphenyls	1	9
1,1,1,2-Tetrachloro-2,2-difluoroethane	500	4,170
1,1,2,2-Tetrachloro-1,2-difluoroethane	500	4,170
1,1,2,2-Tetrachloroethane—Skin	5	35
Tetrachloromethane, see Carbon tetrachloride		
Tetrachloronaphthalene—Skin		2
Tetraethyl lead (as Pb)—Skin		0.075
Tetrahydrofuran	200	590
Tetramethyl lead (as Pb)— Skin		0.075
Tetramethyl succinonitrile— Skin	0.5	3
Tetranitromethane	1	8
Tetryl (2,4,6-trinitrophenyl- methyl-nitramine)—Skin		1.5
Thallium (soluble compounds)—Skin as T1		0.1
Thiram		5
Tin (inorganic cmpds, except oxides		2
Tin (organic cmpds)		0.1
C Toluene-2,4-diisocyanate	0.02	0.14
o-Toluidine—Skin	5	22
Toxaphene, see Chlorinated camphene		
Tributyl phosphate		5
1,1,1-Trichloroethane, see Methyl chloroform		
1,1,2-Trichloroethane—Skin	10	45
Titaniumdioxide		15
Trichloromethane, see Chloroform		
Trichloronaphthalene—Skin		5
1,2,3-Trichloropropane	50	300
1,1,2-Trichloro 1,2,2-trifluoroethane	1,000	7,600
Triethylamine	25	100
Trifluoromonobromomethane	1,000	6,100
2,4,6-Trinitrophenol, see Picric acid		
2,4,6-Trinitrophenylmethyl- nitramine, see Tetryl		
Trinitrotoluene—Skin		1.5
Triorthocresyl phosphate		0.1
Triphenyl phosphate		3
Turpentine	100	560
Uranium (soluble compounds)		0.05
Uranium (insoluble compounds)		0.25
C Vanadium:		
V_2O_5 dust		0.5
V_2O_5 fume		0.1
Vinyl benzene, see Styrene		
Vinylcyanide, see Acrylonitrile		
Vinyl toluene	100	480
Warfarin		0.1
Xylene (xylol)	100	435
Xylidine—Skin	5	25
Yttrium		1
Zinc chloride fume		1
Zinc oxide fume		5
Zirconium compounds (as Zr)		5

†This standard applies in cotton yarn manufacturing until compliance with § 1910.1043 (c) and (e) is achieved.

[a]Parts of vapor or gas per million parts of contaminated air by volume at 25° C. and 760 mm. Hg pressure.

[b]Approximate milligrams of particulate per cubic meter of air.

(No footnote "c" is used to avoid confusion with ceiling value notations.)

[d]An atmospheric concentration of not more than 0.02 p.p.m., or personal protection may be necessary to avoid headache.

Table Z-2

Material	8-hour time weighted average	Acceptable ceiling concentration	Acceptable maximum peak above the acceptance ceiling concentration for an 8-hour shift	
			Concentration	Maximum duration
Benzene (Z37.40-1969)	10 p.p.m	25 p.p.m	50 p.p.m	10 minutes.
Beryllium and beryllium compounds (Z37.29-1970).	2 μg./M^{3}	5 μg./M^{3}	25 μg./M^{3}	30 minutes.
Cadmium fume (Z37.5-1970)	0.1 mg./M^{3}	0.3 mg./M^{3}		
Cadmium dust (Z37.5-1970)	0.2 mg./M^{3}	0.6 mg./M^{3}		
Carbon disulfide (Z37.3-1968)	20 p.p.m	30 p.p.m	100 p.p.m	30 minutes.
Carbon tetrachloride (Z37.17-1967).	10 p.p.m	25 p.p.m	200 p.p.m	5 minutes in any 4 hours.
Chromic acid and chromates (Z37.7-1971).		1 mg./10M^{3}		
Ethylene dibromide (Z[illegible] 1970).	20 p.p.m	30 p.p.m	50 p.p.m	5 minutes.
Ethylene dichloride (Z37.[illegible] 1969).	50 p.p.m	100 p.p.m	200 p.p.m	5 minutes in any 3 hours.
Formaldehyde (Z37.16-1967)	3 p.p.m	5 p.p.m	10 p.p.m	30 minutes.
Hydrogen fluoride (Z37.28-1969).	3 p.p.m			
Hydrogen sulfide (Z37.2-1966)		20 p.p.m	50 p.p.m	10 minutes once only if no other measurable exposure occurs.
Fluoride as dust (Z37.28-1969)	2.5 mg./M^{3}			
Mercury (Z37.8-1971)		1 mg./10M^{3}		
Methyl chloride (Z37.18-1969)	100 p.p.m	200 p.p.m	300 p.p.m	5 minutes in any 3 hours.

TABLE Z-2—Continued

Material	8-hour time weighted average	Acceptable ceiling concentration	Acceptable maximum peak above the acceptance ceiling concentration for an 8-hour shift	
			Concentration	Maximum duration
Methylene chloride (Z37.23-1969).	500 p.p.m	1,000 p.p.m	2,000 p.p.m	5 minutes in any 2 hours.
Organo (alkyl) mercury (Z37.30-1969).	0.01 mg./M³	0.04 mg./M³		
Styrene (Z37.15-1969)	100 p.p.m	200 p.p.m	600 p.p.m	5 minutes in any 3 hours.
Tetrachloroethylene (Z37.22-1967).	......do	......do	300 p.p.m	5 minutes in any 3 hours.
Toluene (Z37.12-1967)	200 p.p.m	300 p.p.m	500 p.p.m	10 minutes.
Trichloroethylene (Z37.19-1967)	100 p.p.m	200 p.p.m	300 p.p.m	5 minutes in any 2 hours.

TABLE Z-3—MINERAL DUSTS

Substance	Mppcf[e]	Mg/M³
Silica:		
Crystalline:		
Quartz (respirable)	250 [f] / $\%SiO_2+5$	10mg/M³ [m] / $\%SiO_2+2$
Quartz (total dust)		30mg/M³ / $\%S_2O_2+2$
Cristobalite: Use ½ the value calculated from the count or mass formulae for quartz		
Tridymite: Use ½ the value calculated from the formulae for quartz		
Amorphous, including natural diatomaceous earth	20	80mg/M³ / $\%SiO_2$
Silicates (less than 1% crystalline silica):		
Mica	20	
Soapstone	20	
Talc (non-asbestos-form)	20[n]	
Talc (fibrous). Use asbestos limit		
Tremolite (see talc, fibrous)		
Portland cement	50	
Graphite (natural)	15	
Coal dust (respirable fraction less than 5% SiO_2)		2.4mg/M³ or
For more than 5% SiO_2		10mg/M³

TABLE Z-3—MINERAL DUSTS—Continued

Substance	Mppcf[e]	Mg/M³
		$\%SiO_2+2$
Inert or Nuisance Dust:		
Respirable fraction	15	5mg/M³
Total dust	50	15mg/M³

NOTE: Conversion factors—mppcf × 35.3 = million particles per cubic meter = particles per c.c.

[e] Millions of particles per cubic foot of air, based on impinger samples counted by light-field technics.

[f] The percentage of crystalline silica in the formula is the amount determined from air-borne samples, except in those instances in which other methods have been shown to be applicable.

[m] Both concentration and percent quartz for the application of this limit are to be determined from the fraction passing a size-selector with the following characteristics:

[n] Containing <1% quartz; if 1% quartz, use quartz limit.

Aerodynamic diameter (unit density sphere)	Percent passing selector
2	90
2.5	75
3.5	50
5.0	25
10	0

The measurements under this note refer to the use of an AEC instrument. The respirable fraction of coal dust is determined with a MRE; the figure corresponding to that of 2.4 Mg/M³ in the table for coal dust is 4.5 Mg/M³.

Chemical Hazards

DEFINITIONS

Acute. Short-term.

Asphyxiant. A chemical (gas or vapor) that can cause death or unconsciousness by suffocation. Simple asphyxiants such as nitrogen either use up or displace oxygen in the air. They become especially dangerous in confined or enclosed spaces. Chemical asphyxiants, such as carbon monoxide and hydrogen sulfide, interfere with the body's ability to absorb or transport oxygen to the tissues.

Cancer. A substance that causes cancer. Regulated as a carcinogen by OSHA.

Chronic. Persistent, prolonged or repeated conditions; long-term.

CNS. Central nervous system.

Mutagen. Anything that can cause a change (or mutation) in the genetic material of a living cell.

Narcosis. Stupor or unconsciousness caused by exposure to a chemical.

Nuisance particulate. Particulate that has a long history of little adverse effect on lungs and does not produce significant organic disease or toxic effect when exposures are kept under reasonable control.

Dermatitis. An inflammation of the skin.

Sensitizer. A substance that may cause no reaction in a person during initial exposures, but afterwards, further exposures will cause an allergic response to the substance.

Suspect carcinogen. A chemical known or suspected to cause cancer in animals or humans.

Systemic. Spread throughout the body: affecting many or all body systems or organs, not localized in one spot or area.

Teratogen. An agent or substance that may cause physical defects in the developing embryo or fetus when a pregnant female is exposed to that substance.

One of the most complex issues the health and safety professional must face is how to recognize, evaluate, and control a chemical hazard. The table in this appendix is an aid to doing just that; it lists physical constants of many common industrial chemicals along with their relative fire and health hazards.

HOW TO USE THE TABLE

Recognize

The table can be used to compare the safety of various chemicals. For example, you can compare the vapor volumes and flash points of all the solvents consistent with process needs, then choose the one that combines the lowest volatility with minimal fire and health hazards.

The more you use the table, the more familiar you will become with the properties of chemicals encountered in the work environment.

Evaluate

The table contains the information necessary to make a

preliminary evaluation of a chemical hazard. Obviously, complex and high risk situations require the services of an industrial hygienist.

The following example shows how to use the table when evaluating a chemical hazard.

Problem: Toluene is being used in a workroom 20×10×20 ft. In an 8-hr workday, 5 gal of the solvent is lost through evaporation. There are two air changes per hour. Is there a potential health hazard?

Solution: Assuming there is complete mixing of the toluene throughout the workroom, then:

Volume of room=20×10×20=4000 cu ft.
Total volume of air supplied to the work room in 8 hr–4000 cu ft×8 hr×2 changes/hr=64,000 cu ft/8-hr day.
Vapor volume of toluene=31 cu ft/gal (from column 11 of Table).
Total vapor volume=31 cu ft/gal×5 gal=155 cu ft.
Threshold Limit Value of toluene=100 ppm.
Total volume of air required to dilute toluene to TLV= 155×1,000,000÷100=1,555,000 cu ft/8-hr day.

Because natural ventilation is supplying only 64,000 cu ft of air per 8-hr day, there is a potential health hazard and additional ventilation is required to maintain the solvent concentration within safe limits.

This type of calculation is valid only when the air contaminant is uniformly distributed at a relatively low concentration. Where the air contaminant is localized in high concentrations, such as in a paint-spraying operation, more complex means of evaluating the hazard must be used.

Control

The amount of exhaust ventilation needed to control an explosion hazard can be estimated by using the Table.

Problem: An average of 1 gal of N-heptane per hour is being evaporated in a drying oven. What is the rate of exhaust ventilation needed to prevent flammable (explosive) concentrations?

Solution: Assuming complete mixing of the N-heptane throughout the oven, the rate of ventilation depends on the volume of evaporating solvent and is independent of the volume of the oven.

From the Table, Columns 8 and 11:

Lower flammable limit for N-heptane=1.2 percent.
Vapor volume=22 cu ft/gal.
Exhaust rate required to dilute vapors to the lower flammable limit=22×100÷1.2=1833 cu ft/hr=30.6 cfm.

This rate is normally multiplied by a safety factor of about 5 to allow for imperfect mixing. Therefore 30.6×5=153 cfm exhaust rate. For more precise calculations, see NFPA 86, *Ovens and Furnaces: Design, Location, and Equipment.*

Listing of substances

The Table of Chemical Hazards lists the physical properties associated with fire and health hazards of common industrial chemicals.

Data in the Table have been taken from the following principal references. Be sure to check latest editions of these standards and books.

Documentation of Threshold Limit Values. Cincinnati: American Conference of Governmental Industrial Hygienists.

Handbook of Chemistry and Physics. Boca Raton, FL: CRC Press, Inc.

Handbook of Compressed Gases. New York, NY: Compressed Gas Association.

Handbook of Environmental Control, vol. I: Air Pollution. Boca Raton, FL: CRC Press, Inc.

Handbook of Laboratory Safety. Boca Raton, FL: CRC Press, Inc.

Handbook of Organic Industrial Solvents, 15th ed. Schaumburg, IL: Alliance of American Insurers.

Industrial Hygiene Technical Manual. Washington, DC: U.S. Department of Labor.

NFPA 49, *Hazardous Chemicals Data,* and 325M, *Fire Hazard Properties of Flammable Liquids, Gases, and Volatile Solids.* Quincy, MA: National Fire Protection Association.

Occupational Safety and Health Standards (Title 29, *CFR,* Part 1910). Washington, DC: U.S. Government Printing Office.

Pocket Guide to Chemical Hazards (Publication No. 85-114) NIOSH 5th printing (1985).

Rose, A. and E. *The Condensed Chemical Dictionary.* New York, NY: Van Nostrand-Reinhold Co.

Column 1

The first column lists the chemical names of substance as cited by the Occupational Safety and Health Administration (OSHA) (29 CFR 1910.1000, Tables Z-1, Z-2, or Z-3) and the American Conference of Governmental Industrial Hygienists (ACGIH) 1986 Threshold Limit Values. See the chemical name cross index in this appendix for chemicals and compounds that may be listed under another name.

Column 2

Physical state of substance:

Sol.=Solid
Liq.=Liquid
L/S=Liquid/Solid
G/L=Gas/Liquid
G=Gas
Fume=Fume
Mix=Mixture
Vap=Vapor
Var=Varies with individual compounds (Var.)

Column 3

The OSHA standard permissible exposure level (PEL) is described in 29 CFR 1910.1000, Tables Z-1, Z-2, or Z-3.

ppm=parts of substance per million parts of air
mppcf=million of particles per cubic foot of air
mg/m^3=milligrams of substance per cubic meter of air
C=Ceiling limit
None=No OSHA standard
Std.=Refers to the standard number

For carcinogens, refer to the appropriate OSHA standard listed in this column.

Column 4

The odor threshold is the minimum concentration, expressed as parts per million, at which the odor can be recognized.

A number of studies have been made on odor thresholds, but they have been conducted in a variety of ways and as a result have produced a number of values. Different studies of the same chemical compound by test panels have produced odor threshold values varying as much as a thousandfold.

Substances with molecular weights equal to or exceeding that of air (about 29) and with appreciable vapor pressures at ordinary temperatures are generally odorous. Odorous gases and vapors may be adsorbed on solid particles or dissolved in liquid droplets in the atmosphere; such aerosols may then be odorous by virture of the gases or vapors desorbed or evaporated from them. Odor measurements used to detect and trace odors and for evaluating intensity and acceptability are limited to the sensory act of smelling.

Techniques and equipment used to reduce the gas and vapor emissions are usually able to eliminate odor problems. However, the concentration of the odorous compound may be considerably lower than gas or vapor emissions, so modifications of these techniques may be necessary.

Column 5

This column gives a summary of the health hazards of a substance. No attempt was made to include an evaluation of the industrial processes for substances or the likelihood that any given material could exceed the OSHA permissible exposure level (PEL).

"Skin" notation. Listed substances that are designated "skin" refers to the potential contribution to the overall exposure by the cutaneous route; this includes mucous membranes, and eye, either by airborne, or, more particularly, by direct contact with the substance. Vehicles can alter skin absorption. This attention-calling designation is intended to suggest appropriate measures for the prevention of cutaneous absorption so that the threshold limit is not invalidated.

Column 6

NFPA hazard classifications. Fires and other emergency situations may involve chemicals that have varying degrees of toxicity, flammability, and reactivity (instability and water reactivity). The National Fire Protection Association grading of these relative hazards (under fire conditions) is given in the columns marked "NFPA Health," "NFPA Flammability," and "NFPA Reactivity."

For a full description of the NFPA classifications, see NFPA 704, *Standard System for the Identification of the Fire Hazards of Materials.* A complete listing of chemicals is given in NFPA 49, *Hazardous Chemicals Data,* and in NFPA 325M, *Fire Hazard Properties of Flammable Liquids, Gases, and Volatile Solids.* An explanation of the degrees of hazard follows.

NFPA health hazards. In general, the health hazard in firefighting is that of a single exposure which may vary from a few seconds up to an hour. The physical exertion caused by firefighting or other emergency may intensify the effects of any exposure.

Health hazards arise from two sources: (1) the inherent properties of the material, and (2) from the toxic products of combustion or decomposition of the material. (Common hazards from burning of ordinary combustible materials are not included.)

The degree of hazard should indicate (1) that people can work safely only with specialized protective equipment, (2) that they can work safely with suitable respiratory protective equipment, or (3) that they can work safely in the area with ordinary clothing.

A health hazard, as defined by the NFPA, is any property of a material that either directly or indirectly can cause injury or incapacitation, either temporary or permanent, for exposure by contact, inhalation, or ingestion.

The degrees of hazard under fire conditions are ranked according to the probable severity of hazard to personnel, as follows:

4—Materials that on very short exposure could cause death or major residual injury even though prompt medical treatment was given, including those that are too dangerous to be approached without specialized protective equipment. This degree should include:

- Materials that can penetrate ordinary rubber or synthetic protective clothing;
- Materials that under normal conditions or under fire conditions give off gases that are extremely hazardous (i.e., toxic or corrosive) through inhalation or through contact with or absorption through the skin.

3—Materials that on short exposure could cause serious temporary or residual injury even though prompt medical treatment was given, including those requiring protection from all bodily contact. This degree should include:

- Materials that give off highly toxic combustion products;
- Materials corrosive to living tissue or toxic by skin absorption.

2—Materials that on intense or continued exposure could cause temporary incapacitation or possible residual injury unless prompt medical treatment is given, including those requiring use of respiratory protective equipment with independent air supply. The degree should include:

- Materials that give off toxic combustion products;
- Materials that give off highly irritating combustion products;
- Materials that either under normal conditions or under fire conditions give off toxic vapors lacking warning properties.

1—Materials that on exposure would cause irritation but only minor residual injury even if no treatment is given, including those that require use of an approved canister type gas mask. This degree should include:

- Materials which under fire conditions would give off irritating combustion products;
- Materials that on the skin could cause irritation without destruction of tissue.

0—Materials that on exposure under fire conditions would offer no hazard beyond that of ordinary combustible material.

NFPA flammability hazards deal with the degree of susceptibility of materials to burning, even though some materials that burn under one set of conditions will not burn under others. The form or condition of material, as well as its properties, affects the hazard.

The degrees of hazard are ranked according to the susceptibility of materials to burning, as follows:

4—Materials that will rapidly or completely vaporize at atmospheric pressure and normal ambient temperature or which are readily dispersed in air, and which will burn readily. This degree should include:

- Gaseous materials;
- Cyrogenic materials;
- Any liquid or gaseous material that is a liquid while under pressure and having a flash point below 22.8 C (73 F) and having a boiling point below 37.8 C (100 F), Class IA flammable liquids.
- Materials that because of their physical form or environmental conditions can form explosive mixtures with air and that are readily dispersed in air, such as dusts of combustible solids and mists of flammable or combustible liquid droplets.

3—Liquids and solids that can be ignited under almost all ambient temperature conditions. Materials in this degree produce hazardous atmospheres with air under almost all ambient temperatures or, though unaffected by ambient temperatures, are readily ignited under almost all conditions. This degree should include:

- Liquids having a flash point below 22.8 C (73 F) and having a boiling point at or above 37.8 C (100 F) and those liquids having a flash point at or above 22.8 C and below 37.8 C, Class IB and Class IC flammable liquids;
- Solid materials in the form of coarse dusts that may burn rapidly but that generally do not form explosive atmospheres with air;
- Solid materials in a fibrous or shredded form that may burn rapidly and create flash fire hazards, such as cotton, sisal and hemp;
- Solids that burn with extreme rapidity, usually by reason of self-contained oxygen (for example, dry nitrocellulose);
- Materials that ignite spontaneously when exposed to air.

2—Materials that must be moderately heated or exposed to relatively high ambient temperatures before ignition can occur. Materials in this degree would not under normal conditions form hazardous atmospheres with air, but under high ambient temperatures or under moderate heating may release vapor in sufficient quantities to produce hazardous atmospheres with air. This degree should include:

- Liquids having a flash point above 37.8 C, but not exceeding 99.3 C (200 F).
- Solids and semisolids that readily give off flammable vapors.

1—Materials that must be preheated before ignition can occur. Materials in this degree require considerable preheating, under all ambient temperature conditions, before ignition and combustion can occur. This degree should include:

- Materials that will burn in air when exposed to a temperature of 815.5 C (1,500 F) for a period of 5 minutes or less;
- Liquids, solids, and semisolids having a flash point above 99.3 C;
- Includes most ordinary combustibles.

0—Materials that will not burn. This degree should include any material which will not burn in air when exposed to a temperature of 815 C (1,500 F) for a period of 5 minutes.

NFPA reactivity (instability) hazards deal with the degree or susceptibility of materials to release energy. Some materials are capable of rapid release of energy by themselves (as by self-reaction or polymerization), or they can undergo violent eruptive or explosive reaction if contacted with water or other extinguishing agents or with certain other materials.

The violence of reaction or decomposition of materials may be increased by heat or pressure, by mixture with certain other materials to form fuel-oxidizer combinations, or by contact with incompatible substances, sensitizing contaminants, or catalysts.

Because of the wide variations of accidental combinations possible in fire emergencies, these extraneous hazard factors (except for the effect of water) cannot be applied in a general numerical scaling of hazards. Such extraneous factors must be considered individually in order to establish appropriate safety factors such as separation or segregation. Such individual consideration is particularly important where significant amounts of materials are to be stored or handled. Guidance for this consideration is provided in NFPA 49, *Hazardous Chemicals Data.*

The degree of hazard should indicate to firefighting personnel that the area should be evacuated, that the fire may be fought from a protected location, that caution must be used in approaching the fire and applying extinguishing agents, or that the fire may be fought using normal procedures.

The relative reactivity of a material is defined as follows.

Reactive materials are those that can enter into a chemical reaction with other stable or unstable materials. For purposes of this guide, the other material to be considered is water and only if its reaction releases energy. While it is recognized that reactions with common materials, other than water, may release energy violently, and that such reactions must be considered in individual cases, inclusion of these reactions is beyond the scope of this identification system.

Unstable materials are those that in the pure state or as commercially produced will vigorously polymerize, decompose or condense, or become self-reactive and undergo other violent chemical changes.

Stable materials are those that normally have the capacity to resist changes in their chemical composition, despite exposure to air, water, and heat as encountered in fire emergencies.

The degrees of hazard are ranked according to ease, rate and quantity of energy release as follows:

4—Materials that are readily capable of detonation or of explosive decomposition or explosive reaction at normal temperatures and pressures. This degree should include materials that are sensitive to mechanical or localized thermal shock at normal temperatures and pressures.

3—Materials that are capable of detonation or of explosive decomposition or explosive reaction but that require a strong initiating source or that must be heated under confinement before initiation. This degree should include materials that are sensitive to thermal or mechanical shock at elevated temperatures and pressures or that react explosively with water without requiring heat or confinement.

2—Materials that are normally unstable and readily undergo violent chemical change but do not detonate. This degree should

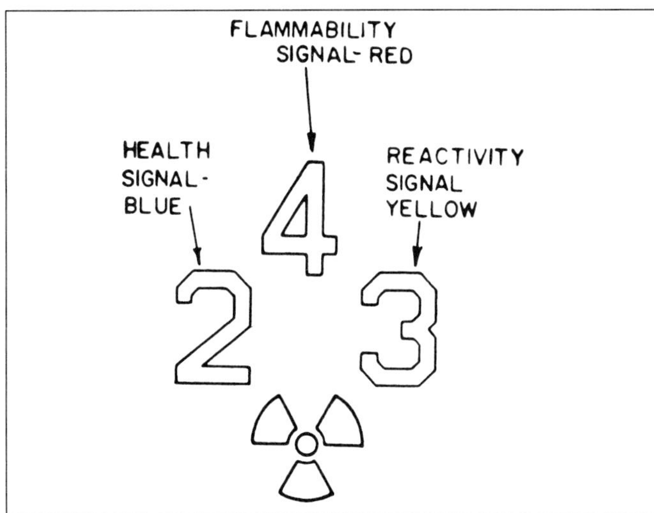

Figure C-1. Correct arrangement and order of signals used on equipment for quick identification of material hazards. See NFPA 704, for details.

include materials that can undergo chemical change with rapid release of energy at normal temperatures and pressures or which can undergo violent chemical change at elevated temperatures and pressures. It should also include those materials that may react violently with water or that may form potentially explosive mixtures with water.

1—Materials that are normally stable, but which can become unstable at elevated temperatures and pressures or which may react with water with some release of energy but not violently.

0—Materials that are normally stable, even under fire exposure conditions, and which are not reactive with water.

Column 7

The flash point is the lowest temperature at which the liquid gives off sufficient vapor to form an ignitable mixture with air and produce a flame when an ignition source is brought near the surface of the liquid. The flash point is used by the NFPA to define and classify the fire hazard of liquids.

A standard closed container is used to determine the closed-cup flash point of a liquid and a standard open-surface dish is used for the open-cup flash point determination, as specified by the standards of the American Society for Testing and Materials (ASTM).

Unless otherwise specified, the flash point values in the Table are closed-cup determinations.

Column 8

Flammable (or explosive) limits are those concentrations of a vapor or gas in air below or above which propagation of a flame does not occur on contact with a source of ignition. The lower limit is the minimum concentration below which the vapor-air mixture is too "lean" to burn or explode. The upper limit is the maximum concentration above which the vapor-air mixture is too "rich" to burn or explode.

Flammable or explosive limits are given in the Table in terms of percentage by volume of gas or vapor in air, and, unless otherwise noted, at normal atmospheric pressures and temperatures. Increasing the temperature or pressure lowers the lower limit and raises the upper limit; decreasing the temperature or pressure has the opposite effect.

Column 9

Boiling point is the temperature at which the vapor pressure of a liquid is just slightly greater than the total pressures of the surrounding atmosphere. This temperature may be approximately determined in many cases by noting the temperature at which ebullition first occurs, that is, when bubbles of vapor are formed within the body of the liquid as its temperature is gradually raised. Precise determination of the boiling point is more complicated, and requires special methods because of superheating of the liquid, formation of bubbles of air or other dissolved gases, and for other reasons.

Solvents with low boiling points will volatilize readily at room temperature, whereas those with high boiling points will usually volatilize more slowly. Volatility is a factor that affects the toxic vapor hazard. A relatively high-boiling, non-volatile solvent would be less dangerous than a volatile one of equal toxicity. Substances boiling above 150 C are ordinarily not considered to be a serious hazard for this reason.

Column 10

Vapor pressure is the pressure (usually expressed in millimeters of mercury) at any given temperature of the vapor of the substance in equilibrium with its liquid or solid form. A method of expressing vapor concentrations is the partial pressure method, using the unit of the pressure exerted by one millimeter of mercury. This is convertible directly to percent by volume by multiplying the vapor pressure by 100 and dividing by the barometric pressure; or to ppm by multiplying by 1,000,000 and dividing by the barometric pressure:

$$\frac{\text{Vapor pressure of one constituent}}{\text{Total barometric pressure}} \times 1{,}000{,}000 = \text{ppm of constituent}$$

When the vapor of a single-component liquid is exposed to the atmosphere, there is a continuous movement of the molecules of the solvent from the surface of the liquid into the free work space. The vapor pressure of a solvent has to be taken into consideration in judging its potential as a fire and health hazard. Knowledge of the vapor pressure also enables one to know what pressure a solvent will exert in a closed vessel at a given temperature.

Vapor pressures of single-component liquids are independent of the quantity of the liquid and vapor which are in equilibrium. Vapor pressure, however, varies with the temperature and for every temperature below the critical pressure a definite pressure is exerted by a vapor in equilibrium with its liquid precursor.

Column 11

The vapor volume, as given in the Table, is the number of cubic feet of solvent vapor formed by the evaporation of one gallon of solvent. An example of the method used for calculating vapor volume is as follows.

The weight of a gallon of solvent can readily be computed by multiplying the specific gravity (SG) of the solvent by the weight of a gallon of water. This formula can be used:

$$SG \times 8.31 = \text{weight of a gallon of the solvent at 23.8 C (75 F)}$$

Molecules of the various solvents have different relative weights and these may be found in any reference tables showing the molecular weight (MW). Dividing the weight of the gallon of solvent by the molecular weight of the solvent gives the number of pound-moles in the gallon.

$$\frac{SG \times 8.31}{MW} = \text{pound-moles in a gallon of solvent}$$

One pound-mole of a solvent contains the same number of molecules as a pound-mole of any other solvent or a pound-mole of a gas. A pound-mole of a vaporized solvent will occupy 392 cubic feet at a temperature of 23.8 C. Knowing the number of pound-moles in a gallon of solvent, the volume occupied can readily be determined by multiplying by 392 to obtain the number of cubic feet. This can be expressed in an equation:

$$\frac{SG \times 8.31 \times 392}{MW} = \text{vapor volume in cubic feet per gallon}$$

Vapor volume can be used advantageously as explained in the preface to this appendix.

For those who work in the metric system, to convert to cubic meters of vapor volume per liter of liquid at 24 C, multiply the figure found in the last equation by 0.00748.

Column 12

Specific gravity is the ratio of the mass of a given volume of liquid to the mass of an equal volume of water at a given temperature. At 23.8 C, water has a specific weight of 8.31 pounds per gallon. The specific weight of a solvent such as benzene at 23.8 C (specific gravity = 0.88) would be 0.88 times 8.31 or 7.31 pounds per gallon.

In the metric system, specific gravity and density are the same number. The SI unit for density (mass) is kilograms per cubic meter.

Column 13

References to publications on health and chemical safety have been included in the Table as an aid to the health and safety professional who is seeking more information on safe handling of a particular chemical. A brief description of the format and content of each of these publications follows. Copies of the publications and further information can be obtained from the organizations listed.

- *NSC Data Sheet*—National Safety Council, 444 N. Michigan Ave., Chicago, IL 60611.

The "Industrial Data Sheets, Chemical Series," cover the major hazards associated with a single substance or family of substances, discuss methods of controlling or eliminating the hazards, and concisely outline safe and efficient procedures. They are written for the safety professional, first-line supervisor, and other management representatives.

A typical Chemical Series Data Sheet is composed of the following topics:

1. Properties
2. Uses
3. Containers
4. Shipping regulations
5. Storage
6. Personnel hazards
7. Handling
8. Personal protective equipment
9. Ventilation
10. Fire and explosion hazards
11. Electrical equipment
12. Symptoms of poisoning
13. First aid
14. Treatment of burns
15. Toxicity
16. Threshold Limit Values
17. Medical examinations
18. Waste disposal

Table of Chemical Hazards

1	2	3		4	5	6			7	8		9	10	11	12	13
Substance Name	Physical State (Sol., Liq., Gas)	OSHA P.E.L Permissible Exposure Levels ppm	mg/m³	Odor Level Thresholds (approximate ppm)	Health Hazard	N.F.P.A. Health	Flammability	Reactivity	Flash Point (Deg. F)	Flammable or Explosive Limits (% by volume) lower	upper	Boiling Point (Deg. F)	Vapor Pressure (mm Hg at S.T.P.)	Vapor Volume (cu ft/gal)	Specific Gravity	References NSC Data Sheet
Abate	Sol.	none	none	–	Marked irritation–eye, nose, throat, skin; mutagen; "skin"	–	–	–	–	–	–	–	–	–	–	–
Acetaldehyde	Liq.	200	360	0.5	Marked irritation–eye, nose, throat, skin; narcosis; kidney damage	2	4	2	–36	4.0	60	70	740	58	0.782	–
Acetic acid (glacial)	Liq.	10	25	2.0	Marked irritation–eye, nose, throat, skin	2	2	1	109	5.4	16.0 (212 F)	245	17	57	1.05	410
Acetic anhydride	Liq.	5	20	–	Marked irritation–eye, nose, throat, skin; acute lung toxicity	2	2	1	129	2.7	10.0	284	10	35	1.08	–
Acetone	Liq.	1000	2400	2.0	Mild irritation–eye, nose, throat; narcosis	1	3	0	0	2.6	12.8	134	227	44	0.791	398
Acetonitrile	Liq.	40	70	–	Mild irritation–eye, nose, throat; acute toxicity (cyanosis); "skin"	2	3	1	42	4.4	16.0	179	75	62	0.783	683
2-Acetylaminofluorene	Sol.	Std. 1910.1014		–	Cancer; "skin"	–	–	–	–	–	–	–	–	–	–	–
Acetylene	Gas	none	none	–	Asphyxiation	1	4	3	–	2.5	81	–	–	–	–	494
Acetylene tetrabromide	Liq.	1	14	–	Cumulative liver and lung damage	3	0	1	–	–	–	275	0.1	–	–	–
Acrolein	Liq.	0.1	0.25	2.0	Marked irritation–eye, nose, throat, lungs, skin; mutagen	3	3	2	–15	2.8	31	125	214	–	0.843	722
Acrylamide	Sol.	–	0.3	–	Polyneuropathy; dermatitis; skin irritation; suspect carcinogen; "skin"	–	–	–	–	–	–	–	–	–	–	–
Acrylonitrile	Liq.	2	–	20.0	Suspect carcinogen; reproductive hazards; CNS depression; "skin"; mutagen	4	3	2	32	3.0	17.0	171	110-115	–	0.800	–
Aldrin	Sol.	–	0.25	–	Cumulative liver damage; suspect carcinogen; "skin"; CNS depression	2	0	0	–	–	–	–	–	–	–	–
Aluminum Oxide	Sol.	none	none	–	"Inert" particulate		–		–	–	–	–	–	–	–	–
Allyl alcohol	Liq.	2	5	2.0	Marked irritation–eye, nose, throat, skin; cumulative eye damage; "skin"; mutagen	3	3	1	70	2.5	18.0	206	24	48	0.854	–
Allyl chloride	Liq.	1	3	0.2	Marked irritation–eye, nose, throat, skin; cumulative liver damage; mutagen; "skin"	3	3	1	–25	2.9	11.1	113	360	40	0.938	–
Allyl glycidyl ether (AGE)	Liq.	C10	45	10.0	Marked irritation–eye, nose, throat, skin; contact skin allergy; "skin"	–	–	–	–	–	–	309	5	–	–	–
Allyl propyl disulfide	Liq.	2	12	–	Marked irritation–eye, nose, throat	–	–	–	–	–	–	–	–	–	–	–
alpha-Chloroaceto-phenone	L/S	0.05	0.3	–	Marked irritation–eye, nose, throat, lungs, skin; lung edema	–	–	–	–	–	–	523	–	–	–	–
alpha-Methyl styrene	Liq.	C 100	480	–	Mild irritation–eye, nose, throat; CNS effects, narcosis	1	2	1	129	1.9	6.1	330	2	–	–	–
alpha-Naphthylamine	Sol.	Std. 1910.1004		–	Cancer; "skin"	2	1	0	315	–	–	572	–	–	–	–
4-Aminodiphenyl	Sol.	Std. 1910.1011		–	Cancer; "skin"	–	–	–	–	–	–	–	–	–	–	–
2-Aminopyridine	Sol.	0.5	2	–	Increased blood pressure; headache, CNS stimulation; "skin"	–	–	–	–	–	–	–	–	–	–	–
Ammonia	Gas	50	35	20.0	Marked irritation–eye, nose, throat, lungs; acute lung damage	3	1	0	Gas	16	25	28	–	–	–	251
Ammonium chloride	Fume	none	none	–	Mild irritation–eye, nose, throat	1	0	0	–	–	–	–	–	–	–	–
Ammonium sulfamate	Sol.	–	15	–	Mild irritation–eye, nose, throat	–	–	–	–	–	–	–	–	–	–	–
n-Amyl acetate	Liq.	100	525	5.0	CNS depression; moderate irritation–eye, nose, throat	1	3	0	77	1.1	7.5	300	5	22	0.879	208
Amyl acetate-sec	Liq.	125	650	–	Moderate irritation–eye, nose, throat; CNS depression	1	3	0	–	–	–	273	9	–	0.861	208
Aniline	Liq.	5	19	1.0	Methemoglobin formation; acute toxic effects; "skin"; suspect carcinogen	3	2	0	158	1.3	–	364	15	3.6	1.022	409
Anisidine (O, P isomers)	Liq.	–	0.5	–	Methemoglobin formation; cumulative CNS effects; suspect carcinogen; "skin"	–	–	–	–	–	–	455	0.01	–	1.097	–
Antimony and compounds as Sb	Sol.	–	0.5	–	Cumulative lung and heart damage; suspect carcinogen	3	1	1	–	–	–	–	–	–	–	408
ANTU (alpha Naphthyl thiourea)	Sol.	–	0.3	–	Cumulative endocrine (thyroid and adrenal) damage; "skin"	–	–	–	–	–	–	–	–	–	–	–
Argon	Gas	none	none	–	Asphyxiation	–	–	–	–	–	–	–	–	–	–	–
Arsenic and compounds, as As	Sol.	–	0.5	–	Cumulative systemic poison; suspect carcinogen	3	1	0	–	–	–	–	–	–	–	499
Arsine	Gas	0.05	0.2	0.5	Acute systemic toxicity	–	–	–	Gas	–	–	–	–	–	–	499
Asbestos (all forms)	Sol.	Std. 1910.1001		–	Asbestosis; cancer	–	–	–	–	–	–	–	–	–	–	–
Asphalt (petroleum) fumes	Fume	none	none	–	Suspect carcinogen; skin irritant	0	2	0	<50	–	–	–	–	–	–	215, 582

(Continued)

Table of Chemical Hazards (continued)

1 Substance Name	2 Physical State (Sol., Liq., Gas)	3 OSHA P.E.L Permissible Exposure Levels ppm	 mg/m3	4 Odor Level Thresholds (approximate ppm)	5 Health Hazard	6 N.F.P.A. Health	 Flammability	 Reactivity	7 Flash Point (Deg. F)	8 Flammable or Explosive Limits (% by volume) lower	 upper	9 Boiling Point (Deg. F)	10 Vapor Pressure (mm Hg at S.T.P.)	11 Vapor Volume (cu ft/gal)	12 Specific Gravity	13 References NSC Data Sheet
Azinphos methyl	Sol.	–	0.2	–	Nervous system disturbances; "skin"	–	–	–	–	–	–	–	–	–	–	–
Barium (soluble compounds)	Sol.	–	0.5	–	Acute lung and gastrointestinal effects; baritosis	1	0	2	–	–	–	–	–	–	–	–
Baygon (propoxur)	Sol.	none	none	–	Nervous system disturbances; "skin"	–	–	–	–	–	–	–	–	–	–	–
Benzene	Liq.	10	30	2.0	Cumulative bone marrow damage; suspect leukemogen; "skin"	2	3	0	12	1.3	7.1	176	100	37	0.880	308
Benzidine	Sol.	Std. 1910.1010		–	Cancer; "skin"	–	–	–	–	–	–	–	–	–	–	–
Benzoyl peroxide	Sol.	–	5	–	Moderate irritation—eye, nose, throat	1	4	4	–	–	–	–	–	–	–	–
Benzyl chloride	Liq.	1	5	0.01	Marked irritation—eye, nose, throat; suspect carcinogen; acute lung edema	2	2	1	153	1.1	–	354	–	28	1.10	–
Beryllium and compounds	Sol.	–	0.002	–	Cumulative lung damage (berylliosis); suspect carcinogen	4	4	1	–	–	–	–	–	–	–	–
beta-Naphthylamine	Sol.	Std. 1910.1009		–	Cancer; "skin"	–	–	–	–	–	–	–	–	–	–	–
beta-Propiolactone	Liq.	Std. 1910.1013		–	Cancer	–	–	–	–	–	–	–	–	–	1.146	–
Biphenyl (diphenyl)	Sol.	–	1	–	Moderate irritation—eye, nose, throat, lungs; CNS effects	2	1	0	–	–	–	489	–	–	–	–
bis-Chloromethyl ether	Liq.	Std. 1910.1008		–	Cancer	2	2	0	–	–	–	–	–	–	–	–
Bismuth telluride	Sol.	none	none	–	Nuisance particulate	–	–	–	–	–	–	–	–	–	–	–
Bismuth telluride (Se-doped)	Sol.	none	none	–	Cumulative lung damage	–	–	–	–	–	–	–	–	–	–	–
Borates, tetra (sodium salt), anhydrate	Sol.	none	none	–	Moderate irritation—eye, nose, throat, skin	–	–	–	–	–	–	–	–	–	–	–
Borates, tetra (sodium salt), decahydrate	Sol.	none	none	–	Moderate irritation—eye, nose, throat, skin	–	–	–	–	–	–	–	–	–	–	–
Borates, tetra (sodium salt), pentahydrate	Sol.	none	none	–	Moderate irritation—eye, nose, throat, skin	–	–	–	–	–	–	–	–	–	–	–
Boron oxide	Sol.	–	15	–	Mild irritation—eyes, nose, throat, skin	–	–	–	–	–	–	–	–	–	–	–
Boron tribromide	Liq.	none	none	–	Marked irritation—eye, nose, throat, lungs	–	–	–	–	–	–	196	72		2.69	–
Boron trifluoride	Gas	C 1	3	–	Acute lung damage; marked irritation— eye, nose, skin	3	0	1	–	–	–	–	–	–	–	–
Bromine	Liq.	0.1	0.7	0.05	Marked irritation—eye, nose, throat, lungs; acute lung damage	4	0	0	–	–	–	138	77	–	3.11	313
Bromine pentafluoride	Liq.	none	none	–	Marked irritation—eye, nose, throat, lungs; acute lung damage	4	0	3	–	–	–	136	441	–	2.46	–
Bromoform	Liq.	0.5	5	–	Marked irritation—eye, nose, throat; cumulative liver damage; "skin"; narcosis	–	–	–	–	–	–	194	5.6	–	2.88	–
Butadiene (1,3-butadiene)	Gas	1000	2200	–	Moderate irritation—eye, nose, throat	2	4	2	Gas	2.0	11.5	–	–	–	–	–
Butane	Gas	none	none	5000	Narcosis; asphyxiant	1	4	0	Gas	1.9	8.5	31	–	–	–	–
2-Butanone (MEK)	Liq.	200	590	5.0	Mild irritation—eye, nose, throat; narcosis	1	3	0	21	1.8	10	175	70	–	0.805	–
2-Butoxyethanol	Liq.	50	240	–	Mild irritation—eye, nose, throat; anemia; "skin"	1	2	0	–	–	–	345	1	–	0.902	–
n-Butyl acetate	Liq.	150	710	10.0	Moderate irritation—eye, nose, throat; narcosis	1	3	0	72	1.7	7.6	260	15	25	0.883	–
sec-Butyl acetate	Liq.	200	950	–	Moderate irritation—eye, nose, throat; narcosis	1	3	0	88	1.7	–	234	10	–	0.890	–
Butyl acetate-tert	Liq.	200	950	–	Moderate irritation—eye, nose, throat; narcosis	1	3	0	–	–	–	208	–	–	0.896	–
sec-Butyl alcohol	Liq.	150	450	20.0	Moderate irritation—eye, nose, throat; narcosis	1	3	0	75	1.7	9.8	201	24	–	0.808	–
n-Butyl alcohol	Liq.	100	300	<25.0	Moderate irritation—eye, nose, throat; "skin"; narcosis	1	3	0	84	1.4	11.2 (212 F)	243	6	36	0.811	–
Butyl alcohol-tert	Sol.	100	300	–	Moderate irritation—eye, nose throat	1	3	0	52	2.4	8.0	181	42	–	–	–
Butyl chromate-tert, as CrO_3	Sol.	–	C 0.1	–	Marked irritation—eye, nose, throat, skin; suspect carcinogen; "skin"	–	–	–	–	–	–	–	–	–	–	–
r.-Butyl glycidyl ether (BGE)	Liq.	50	270	–	Mild irritation—eye, nose, throat, skin; mutagen	–	–	–	–	–	–	327	3	–	–	–

(Continued)

Table of Chemical Hazards (continued)

1 Substance Name	2 Physical State (Sol., Liq., Gas)	3 OSHA P.E.L Permissible Exposure Levels ppm	mg/m³	4 Odor Level Thresholds (approximate ppm)	5 Health Hazard	6 N.F.P.A. Health	Flammability	Reactivity	7 Flash Point (Deg. F)	8 Flammable or Explosive Limits (% by volume) lower	upper	9 Boiling Point (Deg. F)	10 Vapor Pressure (mm Hg at S.T.P.)	11 Vapor Volume (cu ft/gal)	12 Specific Gravity	13 References NSC Data Sheet
Butyl lactate	Liq.	none	none	–	Moderate irritation–eye, nose, throat, lungs; headache	–	–	–	–	–	–	320	0.4	–	0.974	–
Butyl mercaptan	Liq.	10	35	.05	Offensive odor; moderate irritation–eye, nose, throat	2	3	0	3.5	–	–	208	?	–	0.841	–
p-tert-Butyl toluene	Liq.	10	60	–	Cumulative liver, kidney, cardiovascular; CNS damage; nausea	–	–	–	–	–	–	379	0.7	–	–	–
Butylamine	Liq.	C 5	15	–	Marked irritation–eye, nose, throat, lungs, skin; "skin"; suspect carcinogen	2	3	0	10	1.7	9.8	172	72	33	0.738	–
Cadmium (metal dust)	Sol.	–	0.2	–	Cumulative kidney and lung damage; suspect carcinogen	–	–	–	–	–	–	–	–	–	–	726
Cadmium oxide (fume)	Fume	–	0.1	–	Cumulative kidney and lung damage; suspect carcinogen	–	–	–		–	–	–	–	–	–	–
Calcium arsenate, as As	Sol.	–	1	–	Cumulative systemic poisoning; suspect carcinogen	–	–	–	–	–	–	–	–	–	–	–
Calcium carbonate	Sol.	none	none	–	Nuisance particulate	–	–	–	–	–	–	–	–	–	–	–
Calcium cyanamide	Sol.	none	none	–	Moderate irritation–eye, nose, throat, skin; suspect carcinogen	–	–	–	–	–	–	–	–	–	–	–
Calcium hydroxide	Sol.	none	none	–	Marked irritation–eye, nose, throat, skin	–	–	–	–	–	–	–	–	–	–	–
Calcium oxide	Sol.	–	5	–	Marked irritation–eye, nose, throat, skin	1	0	1	–	–	–	–	–	–	–	241
Camphor (synthetic)	Sol.	2	12	–	Moderate irritation–eye, nose, throat; acute toxicity; CNS effects	2	2	0	–	–	–	405	0.2	–	–	–
Caprolactam dust	Sol.	none	none	–	Mild irritation–eye, nose, throat, skin; CNS effects	–	–	–	–	–	–	–	–	–	–	–
Caprolactam (vapor)	Vap.	none	none	–	Mild irritation–eye, nose, throat, skin; CNS effects	–	–	–	–	–	–	–	–	–	–	–
Captafol (Difolatan)	Sol.	none	none	–	Respiratory sensitization (asthma); phototoxic dermatitis; "skin"; suspect teratogen; mutagen	–	–	–	–	–	–	–	–	–	–	–
Captan	Sol.	none	none	–	Suspect carcinogen and mutagen; teratogen	–	–	–	–	–	–	–	–	–	–	–
Carbaryl (Sevin)	Sol.	–	5	–	Teratogen; suspect carcinogen; nervous system disturbances	–	–	–	–	–	–	–	–	–	–	–
Carbofuran	Sol.	none	none	–	Nervous system disturbances	–	–	–	–	–	–	–	–	–	–	–
Carbon black	Sol.	–	3.5	–	Cumulative heart and lung damage	–	–	–	–	–	–	–	–	–	–	–
Carbon dioxide	Gas	5000	9000	none	Asphyxiation	–	–	–	–	Nonflammable		–	–	–	–	682
Carbon disulfide	Liq.	20	–	0.1	Cumulative CNS damage; "skin"; reproductive impairment; chronic effects on heart, kidney, liver, skin	2	3	0	−22	1.3	50	115	360	54	1.26	341
Carbon monoxide	Gas	50	55	none	Chemical anoxia and asphyxiation	2	4	0	–	12.5	74	–	–	–	–	415
Carbon tetrabromide	Sol.	none	none	–	Toxic effects on liver; potent lachrymator	3	0	1	–	–	–	–	–	–	–	–
Carbon tetrachloride	Liq.	10	–	70.0	Cumulative liver damage; suspect carcinogen; "skin"; teratogen	3	0	0	–	Nonflammable		170	117	34	1.58	–
Catechol (Pyrocatechol)	Sol.	none	none	–	Marked irritation eyes & skin; kidney damage	–	–	–	–	–	–	–	–	–	–	–
Cellulose (paper fiber)	Sol.	none	none	–	Nuisance particulate	–	–	–	–	–	–	–	–	–	–	–
Cesium hydroxide	Sol.	none	none	–	Mild irritation–eyes, nose, throat, skin	–	–	–	–	–	–	–	–	–	–	–
Chlordane	Liq.	–	0.5	–	Cumulative liver damage; suspect carcinogen; "skin"	–	–	–	–	–	–	347	–	–	1.57	–
Chlorinated camphene	Sol.	–	0.5	–	Cumulative liver damage; "skin" suspect carcinogen	–	–	–	–	–	–	–	–	–	–	–
Chlorinated diphenyl oxide	Sol.	–	0.5	–	Cumulative liver damage; dermatitis	–	–	–	–	–	–	–	–	–	–	–
Chlorine	Gas	C 1	3	3.0	Lung damage; marked irritation of eyes, nose, throat, bronchi	3	0	1	–	–	–	–	–	–	–	207
Chlorine dioxide	Gas	0.1	0.3	<3.0	Marked irritation–eye, nose, throat, lungs; lung damage	–	–	–	–	>10	–	–	–	–	–	–
Chlorine trifluoride	G/L	C 0.1	0.4	<3.0	Marked irritation–eye, nose, throat, lungs; lung damage	4	0	3	–	–	–	–	–	–	–	–
Chloroacetaldehyde	Liq.	C 1	3	–	Marked irritation–eyes, nose, throat, lungs, skin; mutagen	–	–	–	–	–	–	185	100	–	1.19	–

(Continued)

Table of Chemical Hazards (continued)

1 Substance Name	2 Physical State (Sol., Liq., Gas)	3 OSHA P.E.L Permissible Exposure Levels ppm	 mg/m³	4 Odor Level Thresholds (approximate ppm)	5 Health Hazard	6 N.F.P.A. Health	 Flammability	 Reactivity	7 Flash Point (Deg. F)	8 Flammable or Explosive Limits (% by volume) lower	 upper	9 Boiling Point (Deg. F)	10 Vapor Pressure (mm Hg at S.T.P.)	11 Vapor Volume (cu ft/gal)	12 Specific Gravity	13 References NSC Data Sheet
Chlorobenzene	Liq.	75	350	0.2	Narcosis; cumulative systemic toxicity; moderate irritation of eyes, nose, skin	2	3	0	84	1.3	7.1	270	12	32	1.105	–
0-Chlorobenzylidene malononitrile	Sol.	0.05	0.4	–	Marked irritation–eyes, nose, throat, skin; "skin"	–	–	–	–	–	–	–	–	–	–	–
Chlorobromomethane	Liq.	200	1050	–	Cumulative liver damage; narcosis; mild irritation of eyes & throat	–	–	–	–	–	–	155	160	–	–	–
Chlorodifluoromethane (Refrigerant 22)	Gas	none	none	–	Asphyxiant	–	–	–		–	–	–	–	–	–	–
Chlorodiphenyl (42% Cl)	Liq.	–	1	–	Cumulative liver damage; chloracne; "skin"; suspect carcinogen	–	–	–	–	–	–	–	0.006	–	–	–
Chlorodiphenyl (54% Cl)	Liq.	–	0.5	–	Cumulative liver damage; chloracne; suspect carcinogen; "skin"	–	–	–	–	–	–	–	–	–	–	–
Chloroform	Liq.	C 50	240	100.0	Cumulative liver and kidney damage; suspect carcinogen; narcosis	2	0	0	–	Nonflammable		143	200	41	1.485	–
Chloropicrin	Liq.	0.1	0.7	1.0	Marked irritation–eyes, nose, throat, lungs, skin; acute lung damage	4	0	3	–	–	–	233	17	–	–	–
Chloroprene	Liq.	25	90	–	Systemic toxicity; reproductive hazard; mutagen; suspect carcinogen; "skin"	2	3	0	–4	4.0	20	138	215	–	–	–
Chloropyrifos (Dursban)	Sol.	none	none	–	Nervous system disturbances; "skin"	–	–	–	–	–	–	–	–	–	–	–
o-Chlorostyrene	Liq.	none	none	–	Cumulative liver and kidney damage	–	–	–	–	–	–	390	0.3	–	–	–
o-Chlorotoluene	Liq.	none	none	–	Suspect carcinogen; mild irritation; eye & skin	2	2	0	126	–	–	319	179	–	1.077	–
Chromium, metal & Insoluble salts	Sol.	–	1	–	Cumulative lung damage; suspect carcinogen	–	–	–	–	–	–	–	–	–	–	–
Chromic acid and chromates	Liq.	–	0.1	–	Cumulative lung damage; suspect carcinogen; nasal perforation and ulceration	3	0	1	–	–	–	–	–	–	–	–
Chromium, sol. (chromate salts)	Sol.	–	0.5	–	Cumulative lung damage; dermatitis	–	–	–	–	–	–	–	–	–	–	–
Clopidol (Coyden)	Sol.	none	none	–	Nuisance particulate	–	–	–	–	–	–	–	–	–	–	–
Coal tar pitch volatiles	Mix.	–	0.2	–	Cumulative lung changes; suspect carcinogen	–	–	–	–	–	–	–	–	–	–	–
Cobalt (metal, fume, and dust)	Sol.	–	0.1	–	Cumulative lung changes; dermatitis	–	–	–	–	–	–	–	–	–	–	–
Copper dusts and mists	Sol.	–	1	–	Mild irritation–eyes, nose, throat, skin	–	–	–	–	–	–	–	–	–	–	–
Copper fume	Sol.	–	0.1	–	Mild irritation–eye, nose, throat; acute lung damage	–	–	–	–	–	–	–	–	–	–	–
Corundum (Al_2O_3)	Sol.	none	none	–	Nuisance particulate	–	–	–	–	–	–	–	–	–	–	–
Cotton dust (raw)	Sol.	–	1	–	Cumulative lung damage (byssinosis)	–	–	–	–	–	–	–	–	–	–	–
Crag herbicide (Sesone)	Sol.	–	15	–	Cumulative liver damage	–	–	–	–	–	–	–	–	–	–	–
Cresol (all isomers)	S/L	5	22	0.005	Marked irritation–eye and skin; acute CNS toxicity; cumulative liver, cardiovascular, kidney damage	3	2	0	190	1.3	–	383	4	32	1.03	–
Crotonaldehyde	Liq.	2	6	7.0	Marked irritation–eye, nose, throat; lung edema	3	3	2	55	2.1	15.5	216	19	–	0.858	–
Crufomate (Ruelene)	Sol.	none	none	–	Nervous system disturbances	–	–	–	–	–	–	–	–	–	–	–
Cumene	Liq.	50	245	<10.0	Narcosis; "skin"; moderate irritation of eyes & skin	2	3	0	111	0.9	6.5	307	10	23	0.862	–
Cyanamide	Sol.	none	none	–	Marked irritation–eye, nose, throat, skin; acute toxicity	–	–	–	–	–	–	–	–	–	–	–
Cyanides as Cn	Sol.	–	5	–	Mild irritation–eye, nose, throat; "skin"; acute toxicity (cyanosis)	3	2	–	–	–	–	–	–	–	–	–
Cyanogen	Gas	none	none	–	Moderate irritation–eye, nose, throat; acute toxicity (cyanosis)	4	4	3	–	6.6	32	–6	–	–	–	–
Cyclohexane	Liq.	300	1050	300.0	Moderate irritation–eye, nose, throat; narcosis	1	3	0	–4	1.3	8.0	179	99	30	0.779	–
Cyclohexanol	Sol.	50	200	–	Mild irritation–eye, nose, throat; cumulative liver and lung damage; narcosis	1	2	0	154	–	–	320	3.5	31	–	–
Cyclohexanone	Liq.	50	200	–	Moderate irritation–eye, nose, throat; cumulative liver and kidney damage; narcosis; "skin"	1	2	0	111	1.1 (212 F)	–	313	5	32	0.948	–
Cyclohexene	Liq.	300	1015	–	Moderate irritation–eye, nose, throat; cumulative systemic toxicity; narcosis	1	3	0	<20	–	–	181	?	–	0.810	–
Cyclohexylamine	Liq.	none	none	–	Marked irritation–eye, nose, throat, skin; mutagen; "skin"; CNS disturbances	2	3	0	90	–	–	274	?	–	0.870	–
Cyclopentadiene	Liq.	75	200	–	Moderate irritation–eye, nose, throat	–	–	–	–	–	–	105	?	–	–	–

(Continued)

Table of Chemical Hazards (continued)

1 Substance Name	2 Physical State (Sol., Liq., Gas)	3 OSHA P.E.L Permissible Exposure Levels ppm	mg/m³	4 Odor Level Thresholds (approximate ppm)	5 Health Hazard	6 N.F.P.A. Health	Flammability	Reactivity	7 Flash Point (Deg. F)	8 Flammable or Explosive Limits (% by volume) lower	upper	9 Boiling Point (Deg. F)	10 Vapor Pressure (mm Hg at S.T.P.)	11 Vapor Volume (cu ft/gal)	12 Specific Gravity	13 References NSC Data Sheet
p-Cymene	Liq.	none	none	—	A primary skin irritant	2	2	0	117	0.7 (212 F)	—	349	2	—	0.875	—
2,4-D (2,4-Dicholoro-phenoxyacetic acid)	Sol.	—	10	—	Suspect teratogen, mutagen, and carcinogen, CNS effects	—	—	—	—	—	—	—	—	—	—	—
DDT	Sol.	—	1	—	Chronic and cumulative toxicity; mutagen, suspect carcinogen; "skin"	—	—	—	—	—	—	—	—	—	—	—
Decaborane	Sol.	0.05	0.3	< 2.0	Acute and chronic CNS toxicity; "skin"	3	2	1	176	0.8	98 (approximate)	−66	—	—	—	508
Demeton (Systox)	Liq.	—	0.1	—	Suspect teratogen; mutagen; nervous system disturbances	—	—	—	—	—	—	—	—	—	—	—
Di-sec. octyl phthalate	Sol.	—	5	—	Mild irritation eyes, nose, throat; suspect carcinogen	—	—	—	—	—	—	—	—	—	—	—
Diacetone alcohol	Liq.	50	240	—	Moderate irritation—eye, nose, throat; cumulative kidney damage	1	2	0	148	1.8	6.9	328	1	26	0.940	—
Diazinon	Liq.	none	none	—	"Skin"; suspect teratogen; mutagen; nervous system disturbances	—	—	—	—	—	—	?	—	—	—	—
Diazomethane	Gas	0.2	0.4	—	Suspect carcinogen; acute lung damage; marked irritation—eyes	—	—	—	—	—	—	—	—	—	—	—
Diborane	Gas	0.1	0.1	<2.0	Acute respiratory damage; nervous system damage; marked irritation-eyes	3	4	3	—	—	—	—	—	—	—	—
Dibrom	Sol.	none	none	—	Mutagen; nervous system disturbances	—	—	—	—	—	—	—	—	—	—	—
1,2-Dibromoethane (ethylene dibromide)	Liq.	20	145	25.0	Suspect carcinogen and mutagen; "skin"; cumulative kidney damage; teratogen	3	0	0	—	—	—	268	12	—	—	—
Dibutyl phosphate	Liq.	1	5	—	Mild irritation—eye, nose, throat, lungs (upper); headaches	—	—	—	—	—	—	644	2	—	—	—
2-n-Dibutylamino-ethanol	Liq.	none	none	—	"Skin" moderate irritation eyes, nose, throat, lungs, skin; cumulative liver damage; "skin"	3	2	0	—	—	—	—	?	—	0.859	—
Dibutylphthalate	Liq.	—	5	—	Suspect teratogen; mild irritation eyes & throat	0	1	0	315	—	—	690	2	—	1.048	—
1,1-Dichloro-1-nitroethane	Liq.	C 10	60	—	Acute systemic toxicity (lungs, heart, liver)	2	2	3	168	—	—	255	16	—	—	—
Dichloroacetylene	Liq.	none	none	—	Cumulative CNS toxin; disabling nausea, headaches; acute lung edema	—	—	—	—	—	—	140	270	—	—	—
o-Dichlorobenzene	Liq.	C 50	300	—	Marked irritation—eye, nose, throat; liver damage	2	2	0	151	2.2	9.2	356	1.56	29	1.305	—
p-Dichlorobenzene	Sol.	75	450	—	Cumulative systemic toxicity; suspect cause of cataracts	2	2	0	150	—	—	345	1	29	—	—
3,3-Dichlorobenzidene	Sol.	Std. 1910.1007		—	Cancer; "skin"	—	—	—	—	—	—	—	—	—	—	—
Dichlorodifluoromethane (Refrigerant 12; Freon 12)	Gas	1000	4950	—	Cumulative liver damage	—	—	—	—	—	—	—	—	—	—	—
1,3-Dichloro-5,5-dimethyl hydantoin	Sol.	—	0.2	—	Marked irritation—eye, nose, throat, lungs	—	—	—	—	—	—	—	—	—	—	—
1,1-Dichloroethane	Liq.	100	400	5.0	Cumulative liver damage; mild irritation eyes, throat, bronchi	2	3	0	22	5.6	—	136	234	—	—	—
1,2-Dichloroethane (Ethylene dichloride)	Liq.	50	200	100.0	Cumulative liver and kidney damage; CNS effects; suspect carcinogen	2	3	0	56	6.2	16.0	183	87	49	1.255	—
Dichloroethyl ether	Liq.	C 15	90	0.1	Marked irritation—eye, nose, throat, lungs; "skin"; suspect carcinogen; lung edema	3	2	0	131	—	—	352	0.1	28	1.222	—
1,2-Dichloroethylene	Liq.	200	790	1.0	Narcosis; CNS effects	2	3	2	43	9.7	12.8	141	260	43	1.28	—
Dichloromonofluoro-methane (Refrigerant 21, Freon 21)	Gas	1000	4200	—	Cumulative liver damage; asphyxiant	—	—	—	—	—	—	—	Gas	—	—	—
Dichlorotetra-fluoroethane	Gas	1000	7000	—	Respiratory irritant; acute cardiovascular effects	—	—	—	—	—	—	—	—	—	—	—
Dichlorvos (DDVP)	Liq.	—	1	—	Nervous system disturbances	—	—	—	—	—	—	184	0.01	—	—	—
Dicrotophos (Bidrin)	Liq.	none	none	—	Nervous system disturbances	—	—	—	—	—	—	—	?	—	—	—
Dicyclopentadiene	Liq.	none	none	—	Mild irritation—eye, nose, throat; cumulative liver and kidney damage	1	3	1	90	—	—	342	10	—	0.979	—
Dicyclopentadienyl iron	Sol.	none	none	—	Nuisance dust	—	—	—	—	—	—	—	—	—	—	—
Dieldrin	Sol.	—	0.25	—	Cumulative liver damage; suspect carcinogen; "skin"; suspect teratogen	—	—	—	—	—	—	—	—	—	—	—
Diethylamine	Liq.	25	75	—	Marked irritation—eye, nose, throat, lungs, skin; myocardial degeneration	2	3	0	<0	1.8	10.1	134	195	32	0.706	—
Diethylaminoethanol	Liq.	10	50	—	Marked irritation—eye, nose, throat; "skin"	3	2	0	—	—	—	324	21	—	0.88	—

(Continued)

Table of Chemical Hazards (continued)

1 Substance Name	2 Physical State (Sol., Liq., Gas)	3 OSHA P.E.L Permissible Exposure Levels ppm	 mg/m3	4 Odor Level Thresholds (approximate ppm)	5 Health Hazard	6 N.F.P.A. Health	 Flammability	 Reactivity	7 Flash Point (Deg. F)	8 Flammable or Explosive Limits (% by volume) lower	 upper	9 Boiling Point (Deg. F)	10 Vapor Pressure (mm Hg at S.T.P.)	11 Vapor Volume (cu ft/gal)	12 Specific Gravity	13 References NSC Data Sheet
Diethylene triamine	Liq.	none	none	–	Marked irritation–eye, nose, throat, lungs, skin; "skin"; pulmonary sensitization	3	1	0	215	–	–	405	0.4	–	0.954	–
Diethylphthalate	Liq.	none	none	–	Mild irritation–eye, nose, throat	0	1	0	325	–	–	565	0.5	–	1.12	–
Difluorodibromomethane	Liq.	100	860	–	Cumulative liver damage; narcosis; mild respiratory irritation	–	–	–	–	–	–	76	Gas	–	–	–
Diglycidyl ether (DGE)	Liq.	C 0.5	2.8	–	Marked irritation–eye, nose, throat, lungs, skin; cumulative systemic toxicity; mutagen	–	–	–	–	–	–	500	0.1	–	–	–
Diisobutyl ketone	Liq.	50	290	–	Moderate irritation–eye, nose, throat; narcosis	1	2	0	140	0.8 (212 F)	6.2 (212 F)	335	2	–	0.809	–
Diisopropylamine	Liq.	5	20	–	Moderate irritation–eye, nose, throat, lungs; CNS effects	3	3	0	30	–	–	183	70	–	0.718	–
Dimethyl acetamide	Liq.	10	35	20.0	Cumulative liver damage; "skin"; suspect teratogen; CNS effects	–	–	–	–	–	–	329	9	–	0.937	–
4-Dimethyl aminoazobenzene	Sol.	Std. 1910.1015		–	Cancer	–	–	–	–	–	–	–	–	–	–	–
Dimethyl sulfate	Liq.	1	5	–	Acute lung, eye and skin effects; suspect carcinogen; "skin"; mutagen	4	2	0	182	–	–	370	0.1	33	1.352	–
Dimethylamine	Gas	10	18	0.02	Marked irritation–eye, nose, throat, skin; cumulative liver and testicular damage	3	4	0	–	–	–	–	–	–	–	–
Dimethylaniline	Liq.	5	25	–	Methemoglobinemia; "skin"; CNS effects	3	1	0	206	–	–	382	1	–	0.954	–
Dimethylformamide	Liq.	10	30	20.0	Cumulative liver damage; "skin"; CNS effects; mutagen	1	2	0	136	2.2 (212 F)	15.2	307	4	–	0.954	–
1,1-Dimethylhydrazine	Liq.	0.5	1	<10.0	Acute CNS toxicity and anemia; suspect carcinogen; "skin"; mutagen	3	3	1	5	2	95	145	157	–	0.782	–
Dimethylphthalate	Liq.	–	5	–	Mild irritation of nose & throat	0	1	0	295	–	–	540	6	–	1.19	–
Dinitro-o-cresol	Sol.	–	0.2	–	Cumulative systemic (metabolic) toxin; "skin"; mutagen	–	–	–	–	–	–	–	–	–	–	–
3,5-Dinitro-o-toluamide (Zoalene)	Sol.	none	none	–	Cumulative liver damage; mutagen	–	–	–	–	–	–	–	–	–	–	–
Dinitrobenzene (all isomers)	Sol.	–	1	–	"Skin"; liver & kidney damage; methemoglobinemia	3	1	4	302	–	–	604	–	30	–	–
Dinitrotoluene	Sol.	–	1.5	–	Cumulative liver damage; mutagen; methemoglobinemia	3	1	3	404	–	–	572	–	–	–	–
Dioxane	Liq.	100	360	150.0	Cumulative kidney and liver damage; suspect carcinogen; "skin"; mild irritation of eyes, nose, throat	2	3	1	54	2.0	22	214	37	39	1.036	–
Dioxathion (Delvar)	Liq.	none	none	–	"Skin"; nervous system disturbances	–	–	–	–	–	–	–	–	–	–	–
Diphenylamine	Sol.	none	none	–	Cumulative liver, kidney, bladder damage; suspect teratogen	3	1	0	–	–	–	–	–	–	–	–
Dipropylene glycol methyl ether	Liq.	100	600	–	Moderate irritation–eye, nose, throat; "skin"; slight narcosis	2	0	2	185	–	–	374	0.4	–	0.95	–
Diquat	Sol.	none	none	–	Cumulative effects on eyes (cataracts); mutagen; suspect teratogen	–	–	–	–	–	–	–	–	–	–	–
Disulfiram	Sol.	none	none	–	Acute toxicity (antabuse effects) with alcohol; suspect carcinogen	–	–	–	–	–	–	–	–	–	–	–
Disyston	Sol.	none	none	–	"Skin"; nervous system disturbances	–	–	–	–	–	–	–	–	–	–	–
2,6-Ditert-butyl-p-cresol	Sol.	none	none	–	Nuisance dust	–	–	–	–	–	–	–	–	–	–	–
Dyfonate	Sol.	none	none	–	"Skin"; nervous system disturbances	–	–	–	–	–	–	–	–	–	–	–
Emery	Sol.	none	none	–	Nuisance dust	–	–	–	–	–	–	–	–	–	–	–
Endosulfan (Thiodan)	Sol.	none	none	–	Acute CNS toxin; cumulative kidney damage; "skin"; suspect carcinogen; cumulative kidney damage	–	–	–	–	–	–	–	–	–	–	–
Endrin	Sol.	–	0.1	–	Suspect carcinogen; acute CNS toxicity	3	1	0	–	–	–	–	–	–	–	–
Epichlorohydrin	Liq.	5	19	–	Cumulative kidney and liver damage; marked skin irritation and sensitization; suspect carcinogen & mutagen	3	3	2	105	–	–	239	13	42	1.18	–
EPN	Liq.	–	0.5	–	"Skin"; nervous system disturbances	–	–	–	–	–	–	97	0.003	–	1.598	–
Ethane	Gas	none	none	–	Asphyxiation	1	4	2	–	–	–	–	–	–	–	–
Ethanolamine	Liq.	3	6	–	Cumulative liver, lung, and kidney damage; narcosis; marked irritation eyes, throat, skin	2	2	0	185	–	–	342	0.4	33	1.018	–
Ethion (Nialate)	Liq.	none	none	–	Nervous system disturbances	–	–	–	–	–	–	–	–	–	–	–
2-Ethoxyethanol (Cellosolve)	Liq.	200	740	–	Cumulative blood damage; moderate irritation of eyes and nose	2	2	0	115	2.5	14.0	192	5.3	–	0.931	–

(Continued)

Table of Chemical Hazards (continued)

1 Substance Name	2 Physical State (Sol., Liq., Gas)	3 OSHA P.E.L Permissible Exposure Levels ppm	mg/m³	4 Odor Level Thresholds (approximate ppm)	5 Health Hazard	6 N.F.P.A. Health	Flammability	Reactivity	7 Flash Point (Deg. F)	8 Flammable or Explosive Limits (% by volume) lower	upper	9 Boiling Point (Deg. F)	10 Vapor Pressure (mm Hg at S.T.P.)	11 Vapor Volume (cu ft/gal)	12 Specific Gravity	13 References NSC Data Sheet
2-Ethoxyethylacetate	Liq.	100	540	–	Cumulative kidney & liver damage; mild irritation eyes, nose, throat; "skin"	2	2	0	117	1.7	–	313	2	–	–	–
Ethyl acetate	Liq.	400	1400	10.0	Mild irritation—eye, nose, throat, lungs; objectionable odor; mild narcosis	1	3	0	24	2.2	11.0	171	100	33	0.899	–
Ethyl acrylate	Liq.	25	100	0.0001	Marked irritation—eye, nose, throat, lungs; "skin"; lung edema	2	3	2	60	1.8	–	211	30	–	0.923	–
Ethyl alcohol	Liq.	1000	1900	5.0	Moderate irritation—eye, nose, throat; narcosis; reproductive impairment	0	3	0	55	3.3	19.0	173	50	56	0.816	–
Ethyl benzene	Liq.	100	435	2.0	Moderate irritation—eye, nose, throat; narcosis	2	3	0	59	1.0	6.7	277	9	27	0.867	–
Ethyl bromide	Liq.	200	890	200.0	Narcosis; cumulative liver, kidney, and heart damage; mild irritation eyes, lung, skin	2	3	0	−4	6.7	11.3	100	475	–	1.43	–
Ethyl butyl ketone	Liq.	50	230	–	Mild irritation—eye, nose, throat; narcosis	1	2	0	115	–	–	299	4	–	0.819	–
Ethyl chloride	Gas	1000	2600	–	Narcosis	2	4	0	−58	3.8	15.4	54	539	46	–	–
Ethyl ether	Liq.	400	1200	1.0	Mild irritation—eye, nose, throat; narcosis	2	4	0	−49	1.9	36.0	95	440	31	0.71	–
Ethyl formate	Liq.	100	300	–	Mild irritation—eye, nose, throat; narcosis	2	3	0	−4	2.8	16.0	130	194	41	0.923	–
Ethyl mercaptan	Liq.	C 10	25	0.0005	Objectionable odor; mild irritation nose & throat	2	4	0	<80	2.8	18.1	95	–	–	0.839	–
Ethyl sec-amyl ketone	Liq.	25	130	–	Moderate irritation—eye, nose, throat	–	–	–	–	–	–	321	2	–	0.82	–
Ethyl silicate	Liq.	100	850	–	Mild irritation—eye, nose, throat; cumulative kidney, liver and lung damage	2	2	0	125	–	–	334	1	–	0.936	–
Ethylamine	Liq.	10	18	–	Marked irritation—eye, nose, throat, lungs; corneal injury	3	4	0	<0	3.5	14.0	62		–	0.689	–
Ethylene	Gas	none	none	–	Asphyxiation	1	4	2	Gas	2.7	36.0	−155	–	–	–	–
Ethylene chlorohydrin	Liq.	5	16	–	Acute toxicity, chronic toxicity liver, kidney, skin and lungs; mutagen; "skin"	3	2	0	140	4.9	15.9	265	4.9	49	1.205	–
Ethylene diamine	Liq.	10	25	–	Moderate irritation—eye, nose, throat, skin; contact dermatitis; asthma	3	2	0	110	–	–	241	10	40	0.899	–
Ethylene glycol dinitrate	Liq.	C 0.2	1.0	–	Cumulative effect on blood pressure (hypotension), "skin"; headache	2	3	4	50	4.0	–	190	0.1	–	–	–
Ethylene glycol monomethyl etheracetate (methyl cellosolve acetate)	Liq.	25	120	–	Cumulative CNS and blood (anemia) effects; kidney damage	–	–	–	140	–	–	290	4	–	0.975	–
Ethylene glycol (vapor)	Vap.	none	none	–	Moderate irritation—eye, nose, throat; CNS depression	1	1	0	232	3.2	–	387	0.06	–	1.116	–
Ethylene oxide	Gas	1	1.8	1.0	Cumulative lung, liver damage; marked eye and lung; skin irritation; suspect carcinogen, reproductive effects; mutagen	2	4	3	<0	3.6	100	51	625	–	–	–
Ethyleneimine	Liq.	Std. 1910.1012		–	Cancer	3	3	3	12	3.6	46	132	160	–	0.832	–
Ethylidene norbornene	Sol.	none	none	–	Moderate irritation—eye, nose, throat; cumulative liver and testicular damage; reproductive damage	–	–	–	–	–	–	–	–	–	–	–
n-Ethylmorpholine	Liq.	20	94	–	Moderate irritation—eye, nose, throat; acute visual effect	2	3	0	90	–	–	280	6.1	–	0.916	–
Fensulfothion (Dasanit)	Sol.	none	none	–	Nervous system effects	–	–	–	–	–	–	–	–	–	–	–
Ferbam	Sol.	–	15	–	Mild irritation—eye, nose, upper respiratory; suspect carcinogen; suspect teratogen	–	–	–	–	–	–	–	–	–	–	–
Ferro-vanadium dust	Sol.	–	1	–	Moderate irritation—upper respiratory		–		–	–	–	–	–	–	–	–
Fluoride (as F)	Sol.	–	2.5	–	Marked irritation—eye, nose, throat; cumulative bone damage	–	–	–	–	–	–	–	–	–	–	442
Fluorine	Gas	0.1	0.2	–	Lung edema; marked irritation—eyes, nose, throat, skin, lungs; cumulative liver and kidney damage	4	0	3	–	–	–	–	–	–	–	–
Fluorotrichloromethane (Refrigerant 11)	Liq.	1000	5600	–	Acute CNS and cardiovascular effects	–	–	–	–	–	–	75	–	–	–	–
Formaldehyde	Liq.	3	–	1.0	Marked irritation—eyes, lungs, skin; suspect carcinogen & mutagen	2	4	0	–	7.0	73	−3	–	–	–	342

(Continued)

Table of Chemical Hazards (continued)

1 Substance Name	2 Physical State (Sol., Liq., Gas)	3 OSHA P.E.L Permissible Exposure Levels ppm	mg/m³	4 Odor Level Thresholds (approximate ppm)	5 Health Hazard	6 N.F.P.A. Health	Flammability	Reactivity	7 Flash Point (Deg. F)	8 Flammable or Explosive Limits (% by volume) lower	upper	9 Boiling Point (Deg. F)	10 Vapor Pressure (mm Hg at S.T.P.)	11 Vapor Volume (cu ft/gal)	12 Specific Gravity	13 References NSC Data Sheet
Formamide	Liq.	none	none	–	Cumulative systemic toxicity; suspect reproductive effects; mild irritation eyes & skin	–	–	–	–	–	–	223	10	–	1.14	–
Formic acid	Liq.	5	9	–	Marked irritation–eyes, nose, throat, lungs; mutagen	3	2	0	156	–	–	213	43	–	1.22	–
Furfural	Liq.	5	20	0.2	Moderate irritation–eye, nose, throat; "skin"; CNS effects	2	2	0	140	2.1	19.3	322	2	39	1.16	–
Furfuryl alcohol	Liq.	50	200	–	Moderate irritation–eyes, lungs; narcosis	1	2	1	167	1.8	16.3	340	1	–	1.13	–
Gasoline	Liq.	none	none	<100.0	Mild irritation; CNS effects	1	3	0	−45	1.4	7.6	100-400	–	24 to 32	0.75	–
Germanium tetrahydride	Gas	none	none	–	Acute systemic toxicity (hemolysis)	–	–	–	–	–	–	–	–	–	–	–
Glass, fibrous or dust	Sol.	none	none	–	Moderate irritation–nose, throat, skin	–	–	–	–	–	–	–	–	–	–	–
Glutaraldehyde	Liq.	none	none	–	Marked irritation–eye, nose, throat, skin	–	–	–	–	–	–	–	–	–	–	–
Glutaraldehyde (alkaline activated)	Liq.	none	none	–	Marked irritation–eye, nose, throat, skin; allergy	–	–	–	–	–	–	–	–	–	–	–
Glycerin mist	Liq.	none	none	–	Nuisance particulate	1	1	0	320	–	–	554	–	45	1.26	–
Glycidol (2,3 Epoxy-1-propanol)	Liq.	50	150	–	Moderate irritation–eye, nose, throat, skin; CNS effects; mutagen	–	–	–	–	–	–	133	0.9	–	–	–
Graphite (natural)	Sol.	–	15 mppcf	–	Cumulative lung damage (pneumoconiosis)	–	–	–	–	–	–	–	–	–	–	–
Graphite (synthetic)	Sol.	none	none	–	Cumulative lung damage (pneumoconiosis)	–	–	–	–	–	–	–	–	–	–	–
Gypsum	Sol.	none	none	–	Nuisance particulate	–	–	–	–	–	–	–	–	–	–	–
Hafnium	Sol.	–	0.5	–	Cumulative liver damage; mild irritation eyes, throat, skin	–	–	–	–	–	–	–	–	–	–	–
Helium	Gas	none	none	–	Simple asphyxiation	–	–	–	–	–	–	–	–	–	–	–
Heptachlor	Sol.	–	0.5	–	Cumulative liver damage; suspect carcinogen; "skin"; CNS effects	–	–	–	–	–	–	–	–	–	–	–
n-Heptane	Liq.	500	2000	200.0	Moderate irritation–eye, nose, lungs; CNS effects; narcosis	1	3	0	25	1.1	6.7	209	150	22	0.684	–
Hexachlorocyclo-pentadiene	Liq.	none	none	–	Cumulative (varied) organ damage; marked irritation eyes, throat, lungs; lung edema	–	–	–	–	–	–	–	–	–	–	–
Hexachloroethane	Sol.	1	10	–	Cumulative organ damage; "skin"; CNS effects; suspect carcinogen	–	–	–	–	–	–	–	–	–	–	–
Hexachloronapthalene	Sol.	–	0.2	–	Cumulative liver damage; chloracne; "skin"	–	–	–	–	–	–	–	–	–	–	–
Hexafluoroacetone	Liq.	none	none	–	Cumulative organ damage; suspect teratogen	–	–	–	–	–	–	–	–	–	–	–
n-Hexane	Liq.	500	1800	–	Narcosis; peripheral neuropathy	1	3	0	−7.0	1.1	7.5	156	150	25	0.660	–
2-Hexanone (MBK)	Liq.	100	410	–	Moderate irritation–eye, nose, throat; peripheral neuropathy; "skin"	2	3	0	95	1.2	8.0	262	4	–	0.83	–
Hexone (MIBK)	Liq.	100	410	0.5	Moderate irritation–eye, nose, throat; narcosis	2	3	0	73	1.4	7.5	246	8	–	0.803	–
sec-Hexylacetate	Liq.	50	300	–	Mild irritation–eye, nose, throat	1	2	0	113	–	–	285	4	–	0.89	–
Hexylene glycol	Liq.	none	none	–	Mild irritation–eye, nose, throat, skin; narcosis	1	1	0	215	–	–	–	–	31	0.922	–
Hydrazine	Liq.	1	1.3	3.0	Marked irritation; cumulative organ damage; suspect carcinogen and mutagen; "skin"	3	3	2	100	4.7	100	236	14.4	–	1.004	–
Hydrogen	Gas	none	none	–	Asphyxiation	0	4	0	–	4.0	75	–	–	–	–	–
Hydrogen bromide	Gas	3	10	<2.0	Marked irritation–eye, nose, throat; acute lung damage	3	0	0	–	–	–	–	–	–	–	–
Hydrogen chloride	Gas	C 5	7	10.0	Marked irritation–eye, nose, throat; lung edema, dental erosion	3	0	0	–	–	–	–	–	–	–	–
Hydrogen cyanide	Gas	10	11	1.0	Acute and cumulative systemic toxicity (cyanosis)	4	4	2	–	5.6	40	77	–	–	–	–
Hydrogen fluoride	Gas	3	2	0.3	Marked irritation–eye, nose, throat, acute lung damage; cumulative bone damage	4	0	0	–	–	–	–	–	–	–	459
Hydrogen peroxide (90%)	Liq.	1	1.4	none	Marked irritation–eye, nose, throat, skin; acute lung damage; mutagen	2	0	3	–	Nonflammable		284	5	–	1.46	–
Hydrogen selenide	Gas	0.05	0.2	0.5	Acute lung damage; cumulative CNS effects; liver damage; marked irritation eyes, nose, throat	–	–	–	–	–	–	–	–	–	–	–

(Continued)

Table of Chemical Hazards (continued)

1 Substance Name	2 Physical State (Sol., Liq., Gas)	3 OSHA P.E.L Permissible Exposure Levels ppm	mg/m³	4 Odor Level Thresholds (approximate ppm)	5 Health Hazard	6 N.F.P.A. Health	Flammability	Reactivity	7 Flash Point (Deg. F)	8 Flammable or Explosive Limits (% by volume) lower	upper	9 Boiling Point (Deg. F)	10 Vapor Pressure (mm Hg at S.T.P.)	11 Vapor Volume (cu ft/gal)	12 Specific Gravity	13 References NSC Data Sheet
Hydrogen sulfide	Gas	C 20	30	0.0002	Moderate irritation—eye (conjunctivitis), lungs; acute systemic toxicity; CNS effects	3	4	0	—	4.0	44	−76	—	—	—	284
Hydroquinone	Sol.	—	2	—	Cumulative corneal (eye) damage; mutagen; CNS effects; suspect teratogen	—	—	—	—	—	—	—	—	—	—	—
Indene	Liq.	none	none	—	Marked irritation—eye, nose, throat; cumulative liver and kidney damage	—	—	—	—	—	—	264	—	—	1.006	—
Indium and compounds, as In	Sol.	none	none	—	Cumulative organ damage; suspect teratogen	—	—	—	—	—	—	—	—	—	—	—
Iodine	Sol.	C 0.1	1	—	Marked irritation—eye, nose, throat, lungs; lung edema	—	—	—	—	—	—	—	2	—	—	—
Iodoform	Sol.	none	none	0.002	Moderate irritation—eye, nose, throat, lungs; acute CNS effects	—	—	—	—	—	—	—	—	—	—	—
Iron oxide fume	Fume	—	10	—	Lung changes (siderosis)	—	—	—	—	—	—	—	—	—	—	—
Iron pentacarbonyl	Liq.	none	none	—	Acute toxicity (CNS); acute lung damage	2	3	1	5	—	—	221	—	—	1.446	—
Iron salts, soluble, as Fe	Sol.	none	none	—	Moderate irritation—upper respiratory tract, skin	—	—	—	—	—	—	—	—	—	—	—
Isoamyl acetate	Liq.	100	525	0.003	Moderate irritation—upper respiratory tract; CNS effects	1	3	0	77	1.0	7.5	290	6	—	0.876	—
Isoamyl alcohol	Liq.	100	360	0.002	Mild irritation—eye, nose, throat; narcosis; suspect carcinogen	1	3	0	67	1.2	9.0	215	4	—	0.813	—
Isobutyl acetate	Liq.	150	700	—	Mild irritation—eye, nose, throat; CNS effects	1	3	0	64	2.4	10.5	244	20	—	0.871	—
Isobutyl alcohol	Liq.	100	300	0.003	Mild irritation—eye, nose, throat; suspect carcinogen	1	3	0	82	1.2	10.9 (212 F)	225	12.2	—	0.806	—
Isophorone	Liq.	25	140	—	Marked irritation—eye, nose, throat; acute and chronic CNS effects	2	2	0	184	0.8	3.8	419	0.4	22	0.923	—
Isophorone diisocyanate	Liq.	none	none	—	Marked irritation—eye, nose, throat, lungs, skin; "skin"; respiratory sensitization	—	—	—	—	—	—	418	0.5	—	—	—
Isopropyl acetate	Liq.	250	950	—	Mild irritation—eye, nose, throat	1	3	0	40	1.8	8	184	73	—	0.877	—
Isopropyl alcohol	Liq.	400	980	50.0	Mild irritation—eye, nose, throat; narcosis	1	3	0	53	2.0	12	181	44	—	0.785	—
Isopropyl amine	Liq.	5	12	—	Marked irritation—eye, nose, throat, lungs	3	4	0	−35	—	—	89	460	—	0.69	—
Isopropyl ether	Liq.	500	2100	—	Mild irritation—eye, nose, throat	2	3	1	−18	1.4	7.9	156	119	—	0.723	—
Isopropyl glycidyl ether (IGE)	Liq.	50	240	—	Moderate irritation—eye, nose, throat; skin sensitization	—	—	—	—	—	—	49	9	—	—	—
Kaolin	Sol.	none	none	—	Nuisance particulate	—	—	—	—	—	—	—	—	—	—	—
Ketene	Gas	0.5	0.9	—	Marked irritation eyes, throat, nose, lungs	—	—	—	—	—	—	—	—	—	—	—
Lead arsenate, as Pb	Sol.	—	0.05	—	Cumulative organ toxicity; suspect carcinogen	2	0	0	—	—	—	—	—	—	—	—
Lead, inorg. (fumes and dusts), as Pb	Sol.	—	0.05	—	Cumulative blood effects; cumulative neurologic effects; reproductive hazards	—	—	—	—	—	—	—	—	—	—	—
Limestone	Sol.	none	none	—	Nuisance particulate	1	0	1	—	—	—	—	—	—	—	—
Lindane	Sol.	—	0.5	—	Cumulative liver and CNS damage; suspect carcinogen and mutagen; "skin"	2	1	0	—	—	—	—	—	—	—	—
Lithium hydride	Sol.	—	0.025	—	Marked irritation—eye, nose, throat; lung damage; CNS effects	3	4	2	—	—	—	—	—	—	—	566
Liquified petroleum gas, LP gas	Gas	1000	1800	—	Narcosis; asphyxiant	3	4	1	—	—	—	—	—	—	—	479
Magnesite	Sol.	none	none	—	Nuisance particulate	—	—	—	—	—	—	—	—	—	—	—
Magnesium oxide fume	Sol	—	15	—	Fume fever	—	—	—	—	—	—	—	—	—	—	426
Malathion	Liq.	—	15	—	Mutagen; nervous system disturbances; "skin"	—	—	—	—	—	—	315	0.001	—	1.23	—
Maleic anhydride	Sol.	0.25	1	—	Marked irritation—eye, nose, throat, lungs, skin; asthma; suspect carcinogen and mutagen	3	1	1	215	1.4	7.1	—	—	—	—	—
Manganese and compounds	Sol.	—	C 5	—	Cumulative CNS damage; lung damage	—	—	—	—	—	—	—	—	—	—	306
Manganese cyclopentadienyl tricarbonyl	Sol.	none	none	—	Cumulative kidney; acute CNS effects; acute blood effects	—	—	—	—	—	—	—	—	—	—	—

(Continued)

Table of Chemical Hazards (continued)

1 Substance Name	2 Physical State (Sol., Liq., Gas)	3 OSHA P.E.L Permissible Exposure Levels ppm	mg/m³	4 Odor Level Thresholds (approximate ppm)	5 Health Hazard	6 N.F.P.A. Health	Flammability	Reactivity	7 Flash Point (Deg. F)	8 Flammable or Explosive Limits (% by volume) lower	upper	9 Boiling Point (Deg. F)	10 Vapor Pressure (mm Hg at S.T.P.)	11 Vapor Volume (cu ft/gal)	12 Specific Gravity	13 References NSC Data Sheet
Marble	Sol.	none	none	—	Nuisance particulate	—	—	—	—	—	—	—	—	—	—	—
Mercury (organo) alkyl compounds (as Hg)	Var.	—	0.01	—	Acute cumulative CNS damage; "skin"; and marked irritation of skin	—	—	—	—	—	—	—	—	—	—	—
Mercury (inorganic), as Hg	Var.	—	0.1	—	Acute and cumulative CNS damage; "skin"; gastrointestinal effects; gingivitis; suspect carcinogen	—	—	—	—	—	—	675	0.002	—	—	—
Mesityl oxide	Liq.	25	100	10.0	Marked irritation—eye, nose, throat	3	3	0	87	—	—	266	10	28	0.860	—
Methane	Gas	none	none	none	Simple asphyxiation	1	4	0	—	5.3	14.0	—	—	—	—	—
Methomyl (Lannate)	Sol.	none	none	—	Mutagen; nervous system disturbances	—	—	—	—	—	—	—	—	—	—	—
Methoxychlor	Sol.	—	15	—	Cumulative kidney damage	—	—	—	—	—	—	—	—	—	—	—
2-Methoxyethanol (Methyl cellulose)	Liq.	25	80	—	Cumulative CNS; "skin"; suspect reproductive effects; blood disorders	—	—	—	—	—	—	125	10	—	0.966	—
Methyl 2-cyanoacrylate	Liq.	none	none	—	Moderate irritation—eye, nose, throat	—	—	—	—	—	—	—	—	—	—	—
Methyl acetate	Liq.	200	610	1.0	Mild irritation—eye, nose, throat, lungs; narcosis; CNS effects	1	3	0	14	3.1	16	140	230	41	0.927	—
Methyl acetylene	Gas	1000	1650	—	Narcosis	—	—	—	—	—	—	—	—	—	—	—
Methyl acetylene-propadiene mix (MAPP)	G/L	1000	1800	—	Narcosis		—		—	—	—	—	—	—	—	—
Methyl acrylate	Liq.	10	35	—	Marked irritation—eye, nose, throat, skin; "skin"; acute lung damage; cumulative lung, liver & kidney damage	2	3	2	27	2.8	25	176	65	—	0.957	—
Methyl acrylonitrile	Liq.	none	none	—	Cumulative CNS effects; mild irritation eyes and skin	—	—	—	—	—	—	—	—	—	—	—
Methyl alcohol	Liq.	200	260	10.0	Cumulative CNS effects; narcosis; irritation eyes, nose, throat; "skin"	1	3	0	52	7.3	36	147	95	80	0.792	407
Methyl amine	Gas	10	12	0.02	Marked irritation—eye, nose, throat, skin	3	4	0	—	4.9	20.7	21	—	—	—	—
Methyl n-amyl ketone	Liq.	100	465	—	Moderate irritation—eye, nose, throat; narcosis	1	2	0	120	—	—	302	2	—	0.813	—
Methyl bromide	Gas	C 20	80	none	"Skin"; cumulative CNS & organ damage; acute lung damage	3	1	0	non-flammable	10	15	40	—	—	—	—
Methyl chloride	Gas	100	210	10.0	Acute and chronic CNS effects; cumulative liver and kidney damage	2	4	0	—	10.7	17.4	−11	—	—	—	—
Methyl chloroform	Liq.	350	1900	20.0	Mild irritation—eye, nose, throat; narcosis; suspect carcinogen	2	1	0	—	—	—	165	125	32	1.32	—
Methyl chloromethyl ether	Liq.	Std. 1910.1006		—	Cancer	—	—	—	—	—	—	—	—	—	1.06	—
Methyl cyclohexane	Liq.	500	2000	—	Narcosis	2	3	0	25	1.2	6.7	214	43	26	0.770	—
Methyl cyclohexanol	Liq.	100	470	—	Narcosis; cumulative liver and kidney damage; mild irritation eyes & respiratory tract	—	2	0	149	—	—	329	1.5	27	0.92	—
o-Methyl cyclohexanone	Liq.	100	460	—	Moderate irritation—eye, nose, throat; narcosis; "skin"	—	2	0	118	—	—	325	10	27	0.925	—
Methyl ethyl ketone peroxide	Liq.	none	none	—	Marked irritation—eye, nose, throat, lungs; cumulative liver and kidney damage; suspect carcinogen	—	—	—	—	—	—	—	—	—	—	—
Methyl formate	Liq.	100	250	2.0	Moderate irritation—eye, nose, throat, lungs; CNS effects; narcosis	2	4	0	−2	5.9	23	90	600	53	0.981	—
Methyl iodide	Liq.	5	28	—	Cumulative CNS damage; "skin"; suspect carcinogen and mutagen	—	—	—	—	—	—	108	—	—	2.25	—
Methyl isoamyl ketone	Liq.	none	none	—	Moderate irritation—eye, nose, throat; CNS effects	1	2	0	110	—	—	294	—	—	0.813	—
Methyl isobutyl-carbinol	Liq.	25	100	—	Moderate irritation—eye, nose, throat; "skin"; narcosis	2	2	0	106	1.0	5.5	270	10	—	0.813	—
Methyl isocyanate	Liq.	0.02	0.05	—	Marked irritation—eye, nose, throat, skin, lungs; respiratory sensitization	—	—	—	—	—	—	140	—	—	—	—
Methyl mercaptan	Liq.	C 10	20	0.001	Odor; moderate irritation—eye, nose, throat; CNS effects	2	4	0	—	3.9	21.8	42	—	—	0.87	—
Methyl methacrylate	Liq.	100	410	0.2	Mild irritation—eye, nose; throat; suspect carcinogen and mutagen; suspect teratogen	2	3	2	50	1.7	8.2	212	13	—	0.940	—
Methyl parathion	Sol.	none	none	—	"Skin"; suspect teratogen; mutagen; nervous system disturbances	4	3	2	—	—	—	—	—	—	—	—

(Continued)

Table of Chemical Hazards (continued)

1 Substance Name	2 Physical State (Sol., Liq., Gas)	3 OSHA P.E.L. Permissible Exposure Levels ppm	mg/m³	4 Odor Level Thresholds (approximate ppm)	5 Health Hazard	6 N.F.P.A. Health	Flammability	Reactivity	7 Flash Point (Deg. F)	8 Flammable or Explosive Limits (% by volume) lower	upper	9 Boiling Point (Deg. F)	10 Vapor Pressure (mm Hg at S.T.P.)	11 Vapor Volume (cu ft/gal)	12 Specific Gravity	13 References NSC Data Sheet
Methyl silicate	Sol.	none	none	—	Marked irritation—eye, nose, throat, lungs; kidney damage; severe eye damage	—	—	—	—	—	—	250	12	—	—	—
Methyl styrene	Liq.	C 100	480	—	Moderate irritation—eye, nose, throat; CNS effects; narcosis	1	2	1	129	1.9	6.1	330	2	—	—	—
Methylal (Dimethoxymethane)	Liq.	1000	3100	—	Cumulative systemic toxicity; moderate irritation eyes, nose, throat	2	3	2	0	—	—	111	400	37	0.856	—
Monomethylaniline	Liq.	2	9	—	Methemoglobinemia; "skin"	3	2	0	185	—	—	392	200	—	0.991	—
Methylcyclopentadienyl manganese tricarbonyl (as Mn)	Sol.	none	none	—	Cumulative kidney and liver damage; acute CNS effects; moderate eye irritation	—	—	—	—	—	—	—	—	—	—	—
Methyldemeton	Sol.	none	none	—	"Skin"; nervous system disturbances	—	—	—	—	—	—	—	—	—	—	—
Methylene bis (4-cyclohexylisocyanate)	Sol.	none	none	—	Marked irritation of skin; respiratory and skin sensitization	—	—	—	—	—	—	—	—	—	—	—
4,4-Methylene bis (2-chloroaniline) (MOCA)	Sol.	Std. 1910.1005		—	Cancer	—	—	—	—	—	—	—	—	—	—	—
Methylene bisphenyl isocyanate (MDI)	Sol.	C 0.02	0.2	—	Marked irritation—eye, nose, throat, skin; respiratory sensitization	—	—	—	—	—	—	—	—	—	—	—
Methylene chloride	Liq.	500	1800	200.0	Chemical anoxia (metabolic conversion to CO); cumulative liver damage; CNS effects/narcosis; suspect carcinogen and mutagen	2	0	0	non-flammable (in oxygen)	15.5	66	104	390	51	1.34	474
Monomethyl hydrazine	Liq.	C 0.2	0.35	—	Acute lung, CNS and blood damage; suspect carcinogen; "skin"; suspect teratogen	3	3	1	<80	—	—	190	50	—	0.874	—
Mica (<1% quartz)	Sol.	—	20 mppcf	—	Pneumoconiosis	—	—	—	—	—	—	—	—	—	—	—
Mineral wool fiber	Sol.	none	none	—	Moderate irritation nose, throat, skin	—	—	—	—	—	—	—	—	—	—	—
Molybdenum, as Mo (insolubles)	Sol.	—	15	—	Cumulative liver and kidney damage; blood disorders; mild irritation eyes, nose, throat, lungs	—	—	—	—	—	—	—	—	—	—	—
Molybdenum, as Mo (solubles)	Sol.	—	5	—	Cumulative liver and kidney damage; blood disorders; mild irritation of eyes, nose, throat, lungs	—	—	—	—	—	—	—	—	—	—	—
Monocrotophos (Azodrin)	Sol.	none	none	—	Nervous system disturbances	—	—	—	—	—	—	—	—	—	—	—
Morpholine	Liq.	20	70	—	Moderate irritation—eye, nose, throat; "skin"; cumulative liver and kidney damage, suspect carcinogen	2	3	0	100	—	—	262	8	9	1.00	—
Naphtha (coal tar)	Liq.	100	400	<300.0	Moderate irritation—eye, nose, throat; narcosis	2	2	0	100-110	—	—	360	—	37	0.85	—
Naphthalene	Sol.	10	50	0.3	Marked irritation—eye, nose, throat; anemia; occular damage; CNS damage, suspect carcinogen	2	2	0	174	0.9	5.9	420	0.1	—	—	370
Neon	Gas	none	none	—	Asphyxiation	—	—	—	—	—	—	—	—	—	—	—
Nickel, metal and soluble compounds	Sol.	—	1	—	Cumulative lung damage; suspect carcinogen; dermatitis	—	—	—	—	—	—	—	—	—	—	—
Nickel carbonyl	Liq.	0.001	0.007	1.0	Suspect carcinogen; acute lung edema and CNS effects; suspect teratogen	4	3	3	—	—	—	109	261	—	1.32	—
Nicotine	Liq.	—	0.5	—	Acute systemic toxicity; CNS damage; suspect teratogen	4	1	0	—	0.7	4.0	475	1	—	1.01	—
Nitric acid	Liq.	2	5	<5.0	Marked irritation—eye, nose, throat, skin; acute lung damage	3	0	1	—	—	—	188	10	—	—	—
Nitric oxide	Gas	25	30	—	Methemoglobinemia; CNS effects; delayed lung damage	—	—	—	—	—	—	—	—	—	—	—
p-Nitroaniline	Sol.	1	6	—	Methemoglobinema; cumulative liver damage; "skin"	3	1	3	390	—	—	630	0.1	—	—	—
Nitrobenzene	Liq.	1	5	0.005	Methemoglobinemia; "skin"; CNS effects	3	2	0	190	1.8 (200 F)	—	412	3	32	1.20	—
4-Nitrobiphenyl	Sol.	Std. 1910.1003		—	Cancer	—	—	—	—	—	—	—	—	—	—	—
p-Nitrochlorobenzene	Sol.	—	1	—	Methemoglobinema; "skin"	2	1	3	261	—	—	—	—	—	—	—
Nitroethane	Liq.	100	310	—	Moderate irritation; narcosis	1	3	3	82	3.4	—	237	16	46	1.05	693
Nitrogen	Gas	none	none	—	Asphyxiation	3	0	0	—	—	—	—	—	—	—	—
Nitrogen dioxide	Gas	C 5	9	1.0	Cumulative lung damage (bronchitis and emphysema); moderate irritation eyes and nose	3	0	0	—	—	—	—	—	—	—	—

(Continued)

Table of Chemical Hazards (continued)

1 Substance Name	2 Physical State (Sol., Liq., Gas)	3 OSHA P.E.L Permissible Exposure Levels ppm	mg/m³	4 Odor Level Thresholds (approximate ppm)	5 Health Hazard	6 N.F.P.A. Health	Flammability	Reactivity	7 Flash Point (Deg. F)	8 Flammable or Explosive Limits (% by volume) lower	upper	9 Boiling Point (Deg. F)	10 Vapor Pressure (mm Hg at S.T.P.)	11 Vapor Volume (cu ft/gal)	12 Specific Gravity	13 References NSC Data Sheet
Nitrogen trifluoride	Gas	10	29	–	Methemoglobinemia; cumulative liver damage	–	–	–	–	–	–	–	–	–	–	–
Nitroglycerin	Liq.	C 0.2	2	–	Cumulative effect on lowering blood pressure; "skin"; CNS effects	–	–	–	–	–	–	–	–	–	–	–
Nitromethane	Liq.	100	250	–	Mild irritation–eye, nose, throat; narcosis; cumulative liver and lung damage	1	3	4	95	7.3	–	214	28	61	1.13	–
1-Nitropropane	Liq.	25	90	–	Moderate irritation–eye, nose, throat; cumulative kidney and liver damage	1	3	1	120	2.2	–	268	8	37	1.00	693
2-Nitropropane	Liq.	25	90	–	Moderate irritation–eye, nose, throat; cumulative liver damage; suspect carcinogen and mutagen	1	3	1	103	2.6	–	248	13	36	0.992	–
n-Nitrosodimethylamine	Liq.	Std. 1910.1016		–	Cancer; suspect teratogen	–	–	–	–	–	–	–	–	–	–	–
Nitrotoluene	Liq.	5	30	–	Methemoglobinemia; "skin"; skin sensitization	3	1	0	223	–	–	460	1.5	–	1.16	–
Nitrous oxide	Gas	none	none	–	Reproductive hazard; CNS effects	–	–	–	–	–	–	–	–	–	–	–
Nonane	Liq.	none	none	–	Odor; unknown toxic potential (octane analogy)	0	3	0	–	–	–	290	–	–	0.722	–
Octachloro naphthalene	Sol.	–	0.1	–	Cumulative liver damage; "skin"; chloracne	–	–	–	–	–	–	–	–	–	–	–
Octane	Liq.	500	2350	150.0	Mild irritation eyes, nose, skin; narcosis	0	3	0	56	1.0	6.5	258	11	20	0.704	–
Oil mist (mineral)	Sol.	–	5	–	Accumulation in lungs (pneumonitis)	0	1	0	–	–	–	–	–	–	–	–
Osmium tetroxide (as Os)	Sol.	–	0.002	–	Marked irritation–eye, nose, throat, lungs; lung edema	–	–	–	–	–	–	–	–	–	–	–
Oxalic acid	Sol.	–	1	–	Marked irritation–eye, nose, throat, skin; acute systemic effects	2	1	0	–	–	–	–	–	–	–	406
Oxygen difluoride	Gas	0.05	0.1	–	Marked irritation–lungs; cumulative kidney damage; CNS effects (headache)	–	–	–	–	–	–	–	–	–	–	–
Ozone	Gas	0.1	0.2	0.01	Marked irritation–lungs; lung edema	–	–	–	–	–	–	–	–	–	–	–
Paraffin wax fume	Sol.	none	none	–	Mild irritation–eye, nose, throat	0	1	0	–	–	–	–	–	–	–	–
Paraquat	Sol.	–	0.5	–	Cumulative lung damage; "skin"; mild irritation eyes, nose, throat; suspect teratogen	–	–	–	–	–	–	–	–	–	–	–
Parathion	Liq.	–	0.1	–	"Skin"; suspect teratogen; nervous system disturbances	4	1	2	–	–	–	320	.003	–	1.26	–
Pentaborane	Liq.	0.005	0.01	1.0	Acute and cumulative CNS damage	3	3	2	spontaneous	0.4	–	140	200	–	0.61	508
Pentachloro-naphthalene	Sol.	–	0.5	–	Cumulative liver damage, chloracne; "skin"; marked irritation eyes, nose, skin; acute vascular and CNS damage	–	–	–	–	–	–	–	–	–	–	–
Pentachlorophenol	Sol.	–	0.5	–	Acute systemic toxicity; vascular and CNS damage, "skin"; chloracne; suspect teratogen	3	2	0	–	–	–	–	–	–	–	–
Pentaerythritol	Sol.	none	none	–	Nuisance particulate	–	–	–	–	–	–	–	–	–	–	–
Pentane	Liq.	1000	2950	–	Narcosis	1	4	0	−40	1.5	7.8	97	500	29	0.631	–
2-Pentanone	Liq.	200	700	–	Moderate irritation–eye, nose, throat; narcosis	2	3	0	45	1.5	8.2	216	27	–	0.809	–
Perchloroethylene (tetrachloroethylene)	Liq.	100	670	5.0	Cumulative liver and CNS damage; "skin"; narcosis; suspect carcinogen and mutagen.	2	0	0	–	–	–	250	19	–	1.62	673
Perchloromethyl-mercaptan	Liq.	0.1	0.8	–	Marked irritation–eye, nose, throat; suspect carcinogen	–	–	–	–	–	–	–	–	–	1.72	–
Perchloryl fluoride	Gas	3	13.5	–	Moderate irritation–eye, nose, throat; methemoglobinemia	–	–	–	–	–	–	–	–	–	–	–
Perlite	Sol.	none	none	–	Nuisance particulate	–	–	–	–	–	–	–	–	–	–	–
Petroleum distillates (naphtha)	Liq.	500	2000	–	Moderate irritation; narcosis	1	4	0	<0	1.1	5.9	149	–	–	–	–
Phenol	Sol.	5	19	0.3	Marked irritation–eye, nose, throat, lungs; cumulative liver, and kidney damage; "skin"; suspect carcinogen	3	2	0	175	–	–	358	0.4	–	–	–
Phenothiazine	Sol.	none	none	–	Moderate irritation–skin; photosensitization–skin; "skin"	–	–	–	–	–	–	–	–	–	–	–
Phenyl ether (vapor)	Liq.	1	7	–	Moderate irritation, eye, nose, throat; cumulative liver and kidney damage	–	–	–	–	–	–	500	0.02	–	1.07	–

(Continued)

Table of Chemical Hazards (continued)

1 Substance Name	2 Physical State (Sol., Liq., Gas)	3 OSHA P.E.L Permissible Exposure Levels ppm	mg/m³	4 Odor Level Thresholds (approximate ppm)	5 Health Hazard	6 N.F.P.A. Health	Flammability	Reactivity	7 Flash Point (Deg. F)	8 Flammable or Explosive Limits (% by volume) lower	upper	9 Boiling Point (Deg. F)	10 Vapor Pressure (mm Hg at S.T.P.)	11 Vapor Volume (cu ft/gal)	12 Specific Gravity	13 References NSC Data Sheet
Phenyl ether-diphenyl mix (vapor)	Liq.	1	7	—	Moderate irritation—eye, nose, throat; nausea; cumulative liver and kidney damage				—	—	—	495	0.1	—	—	—
p-Phenylene diamine	Sol.	—	0.1	—	Respiratory and skin sensitization; "skin"	—	—	—	—	—	—	—	—	—	—	—
Phenylglycidyl ether (PGE)	—	10	60	—	Moderate irritation—eye, nose, throat; narcosis; skin sensitization; mutagen	—	—	—	—	—	—	475	0.01	—	—	—
Phenylhydrazine	Liq.	5	22	—	Hemolytic anemia; skin irritation; "skin"; skin sensitization	3	2	0	192	—	—	470	—	—	1.098	—
Phenylphosphine	Liq.	none	none	—	Hemolytic anemia; CNS effects; testicular damage	—	—	—	—	—	—	—	—	—	—	—
Phorate (Thimet)	Liq.	none	none	—	"Skin"; nervous system disturbances	—	—	—	—	—	—	—	—	—	—	—
Phosdrin (Mevinphos)	Liq.	—	0.1	—	"Skin"; nervous system disturbances	—	—	—	—	—	—	—	—	—	—	—
Phosgene	Gas	0.1	0.4	0.5	Marked lung edema; chronic lung disease; moderate irritation eyes and skin	4	0	0	—	—	—	—	—	—	—	—
Phosphine	Gas	0.3	0.4	0.02	Acute and chronic systemic toxicity; CNS effects; lung edema; anemia	—	—	—	—	—	—	—	—	—	—	—
Phosphoric acid	Liq.	—	1	—	Marked irritation—eye, nose, throat	2	0	0	—	—	—	—	2	—	1.88	674
Phosphorus (yellow)	Sol.	—	0.1	—	Cumulative bone and liver damage; acute lung damage	3	3	1	—	—	—	—	—	—	—	282
Phosphorus pentachloride	Sol.	—	1	—	Marked irritation and damage to lungs	3	0	2	—	—	—	—	—	—	—	—
Phosphorus pentasulfide	Sol.	—	1	—	Marked irritation and H_2S hazard	3	1	2	—	—	—	—	—	—	—	—
Phosphorus trichloride	Liq.	0.5	3	0.7	Marked irritation—eye, nose, throat, lungs; bronchial pneumonia, pulmonary edema	3	0	2	—	Nonflammable		168	—	—	1.57	—
Phthalic anhydride	Sol.	2	12	—	Marked irritation—eye, nose, throat, lungs, skin; "skin"; and respiratory sensitization	2	1	0	305	1.7	10.5	543	—	—	—	—
m-Phthalodinitrile	Sol.	none	none	—	Nuisance particulate	—	—	—	—	—	—	—	—	—	—	—
Picloram (Tordon)	Sol.	none	none	—	Suspect carcinogen and mutagen	—	—	—	—	—	—	—	—	—	—	
Picric acid	Sol.	—	0.1	—	Cumulative kidney, liver and red blood cell damage; dermatitis; "skin"	2	4	4	Explodes	—	—	—	—	—	—	351
Pival	Sol.	—	0.1	—	Cumulative anticoagulant effects (warfarin analogy)	—	—	—	—	—	—	—	—	—	—	—
Plaster of Paris	Sol.	none	none	—	Nuisance particulate	—	—	—	—	—	—	—	—	—	—	—
Platinum (soluble salts), as Pt	Sol.	—	0.002	—	Respiratory sensitization (asthma); moderate irritant eyes, nose, throat; dermatitis	—	—	—	—	—	—	—	—	—	—	—
Polytetrafluoroethylene decomposition products	Var.	none	none	—	Acute toxic effects (polymer fume fever)	3	4	3	—	—	—	—	—	—	—	—
Portland cement	Sol.	—	50 mppcf	—	Nuisance particulate mild irritation eyes and nose	—	—	—	—	—	—	—	—	—	—	—
Potassium hydroxide	Sol.	none	none	—	Marked irritation—eye, nose, throat, lungs, skin	3	0	1	—	—	—	—	—	—	—	—
Propane	Gas	1000	1800	—	Narcosis; asphyxiation	1	4	0	—	—	—	—	—	—	—	—
Propargyl alcohol	Liq.	none	none	—	Marked irritation—eye, nose, throat, skin; "skin"	3	3	3	97	—	—	239	—	—	0.922	—
n-Propyl acetate	Liq.	200	840	15.0	Mild irritation—eye, nose, throat; narcosis	1	3	0	58	2.0	8	215	35	28	0.887	—
Propyl alcohol	Liq.	200	500	2.0	Mild irritation—eye, nose, throat; "skin"; narcosis; suspect carcinogen	1	3	0	77	2.1	13.5	207	21	44	0.804	—
n-Propyl nitrate	Liq.	25	110	—	Cumulative systemic effects; methemoglobinemia; moderate skin irritation	—	—	—	—	—	—	231	16	—	—	—
Propylene dichloride	Liq.	75	350	—	Cumulative liver damage; moderate irritation eyes, nose, throat; narcosis	2	3	0	60	3.4	14.5	205	50	33	1.16	—
Propylene glycol monomethyl ether	Liq.	none	none	—	Moderate irritation eyes, nose, throat; narcosis	0	3	0	100	—	—	248	8	45	0.919	—
Propylene imine	Liq.	2	5	—	Moderate irritation—eye, nose, throat; acute kidney and lung damage; "skin"; suspect carcinogen and mutagen	—	—	—	—	—	—	—	—	—	—	—

(Continued)

Table of Chemical Hazards (continued)

1 Substance Name	2 Physical State (Sol., Liq., Gas)	3 OSHA P.E.L Permissible Exposure Levels ppm	mg/m³	4 Odor Level Thresholds (approximate ppm)	5 Health Hazard	6 N.F.P.A. Health	Flammability	Reactivity	7 Flash Point (Deg. F)	8 Flammable or Explosive Limits (% by volume) lower	upper	9 Boiling Point (Deg. F)	10 Vapor Pressure (mm Hg at S.T.P.)	11 Vapor Volume (cu ft/gal)	12 Specific Gravity	13 References NSC Data Sheet
Propylene oxide	Liq.	100	240	200.0	Cumulative liver damage; CNS depression; suspect carcinogen and mutagen	2	4	2	−35	2.8	37	95	445	–	0.83	–
Pyrethrum	Sol.	–	5	–	Mild irritation; contact and allergic dermatitis	–	–	–	–	–	–	–	–	–	–	–
Pyridine	Liq.	5	15	0.01	Cumulative liver, kidney and bone marrow damage; CNS effects	2	3	0	68	1.8	12.4	239	20	41	0.982	310
Quinone	Sol.	0.1	0.4	–	Acute & cumulative eye (cornea) damage; suspect carcinogen; mild irritant	–	–	–	–	–	–	–	–	–	–	–
RDX (Cyclotrimethylene trinitramine)	Sol.	–	–	–	Acute CNS effects (nausea, convulsions); "skin"	–	–	–	–	–	–	–	–	–	–	–
Resorcinol	Sol.	none	none	–	Moderate irritation—eye, nose, throat, skin; cumulative systemic toxicity; mutagen	–	–	–	–	–	–	–	–	–	–	–
Rhodium, metal fume and dusts, as Rh	Sol.	–	0.1	–	Low toxicity	–	–	–	–	–	–	–	–	–	–	–
Rhodium, soluble salts	Sol.	–	0.001	–	Respiratory sensitization; suspect carcinogen	–	–	–	–	–	–	–	–	–	–	–
Ronnel	Sol.	–	15	–	Nervous system disturbances	–	–	–	–	–	–	–	–	–	–	–
Rosin-core solder pyrolysis products	Var.	none	none	–	Marked irritation—eye, nose, throat	–	–	–	–	–	–	–	–	–	–	–
Rotenone (commercial)	Sol.	–	5	–	Cumulative systemic toxicity; suspect carcinogen; mild irritant	–	–	–	–	–	–	–	–	–	–	–
Rouge	Sol.	none	none	–	Nuisance particulate	–	–	–	–	–	–	–	–	–	–	–
Rubber solvent	Liq.	none	none	–	Mild irritant; narcosis	–	–	–	–	–	–	–	–	–	–	–
Selenium compounds (as Se)	Sol.	–	0.2	–	Moderate irritation—eye, nose, throat; cumulative lung, liver, kidney damage; suspect carcinogen	–	–	–	–	–	–	–	–	–	–	578
Selenium hexafluoride, (as Se)	Gas	0.05	0.4	–	Acute lung damage (edema)	–	–	–	–	–	–	–	–	–	–	–
Silica (amorphous)	Sol.	–	20 mppcf	–	Possible pneumoconiosis	–	–	–	–	–	–	–	–	–	–	–
Silica (fused)	Sol.	(Use quartz formula)		–	Pneumoconiosis	–	–	–	–	–	–	–	–	–	–	–
Silica (quartz) respirable	Sol.	(Use quartz formula)		–	Pneumoconiosis (silicosis)	–	–	–	–	–	–	–	–	–	–	–
Silicon	Sol.	none	none	–	Nuisance particulate	–	–	–	–	–	–	–	–	–	–	–
Silicon carbide	Sol.	none	none	–	Nuisance particulate	–	–	–	–	–	–	–	–	–	–	–
Silicon tetrahydride (Silane)	Gas	none	none	–	Acute systemic toxicity by analogy with other metal hydrides	–	–	–	–	–	–	–	–	–	–	–
Silver, metal and soluble compounds, as Ag	Sol.	–	0.01	–	Cumulative skin pigmentation and organ accumulation	–	–	–	–	–	–	–	–	–	–	–
Soapstone	Sol.	–	20 mppcf	–	Pneumoconiosis; suspect carcinogen	–	–	–	–	–	–	–	–	–	–	–
Sodium azide	Sol.	none	none	–	Cumulative CNS and blood pressure damage; mild irritant; mutagen	–	–	–	–	–	–	–	–	–	–	–
Sodium fluoroacetate (1080)	Sol.	–	0.05	–	Acute systemic toxicity (metabolic pathway inhibitor); "skin"	–	–	–	–	–	–	–	–	–	–	–
Sodium hydroxide	Sol.	–	2	–	Marked irritation—eye, nose, throat, lungs, skin	3	0	1	–	–	–	–	–	–	–	214, 373
Starch	Sol.	none	none	–	Nuisance particulate	–	–	–	–	–	–	–	–	–	–	–
Stibine	Gas	0.1	0.5	0.05	Acute systemic toxicity (rbc hemolysis); CNS effects	–	–	–	–	–	–	–	–	–	–	–
Stoddard solvent	Liq.	500	2900	<300.0	Narcosis; mild irritant	0	2	0	100	0.7	5	–	–	–	0.78	–
Strychnine	Sol.	–	0.15	–	Acute systemic toxicity (CNS, paralysis)	–	–	–	–	–	–	–	–	–	–	–
Styrene, monomer	Liq.	100	420	0.05	Moderate irritation—eye, nose, throat; narcosis; CNS effects, mutagen	2	3	2	90	1.1	6.1	295	5	28	0.905	627
Subtilisins (Proteolytic enzymes)	Sol.	none	none	–	Respiratory allergy (asthma and lung damage); mild irritant	–	–	–	–	–	–	–	–	–	–	–
Sucrose	Sol.	none	none	–	Nuisance particulate	–	–	–	–	–	–	–	–	–	–	–

(Continued)

Table of Chemical Hazards (continued)

1 Substance Name	2 Physical State (Sol., Liq., Gas)	3 OSHA P.E.L Permissible Exposure Levels ppm	mg/m3	4 Odor Level Thresholds (approximate ppm)	5 Health Hazard	6 N.F.P.A. Health	Flammability	Reactivity	7 Flash Point (Deg. F)	8 Flammable or Explosive Limits (% by volume) lower	upper	9 Boiling Point (Deg. F)	10 Vapor Pressure (mm Hg at S.T.P.)	11 Vapor Volume (cu ft/gal)	12 Specific Gravity	13 References NSC Data Sheet
Sulfur dioxide	Gas	5	13	0.5	Marked irritation—eye, nose, throat, lungs; bronchoconstriction; mutagen, suspect reproductive effects	3	0	0	—	—	—	—	—	—	—	—
Sulfur hexafluoride	Gas	1000	6000	—	Asphyxiant	3	0	1	—	—	—	—	—	—	—	—
Sulfur monochloride	Liq.	1	6	—	Marked irritation—eye, nose, throat, lung	2	1	1	—	—	—	275	7	—	1.69	—
Sulfur pentafluoride	Gas	0.025	0.25	—	Marked irritation—lungs	—	—	—	—	—	—	—	—	—	—	—
Sulfur tetrafluoride	Gas	none	none	—	Marked irritation—lungs	—	—	—	—	—	—	—	—	—	—	—
Sulfuric acid	Liq.	—	1	—	Marked irritation—eye, nose, throat, skin; cumulative lung damage; dermal erosion	3	0	1	—	—	—	640	—	—	1.84	325
Sulfuryl fluoride	Gas	5	20	—	Cumulative liver damage; cumulative lung damage; acute CNS effects	—	—	—	—	—	—	—	—	—	—	—
2,4,5-T	Sol.	—	10	—	Suspect teratogen; suspect carcinogen & mutagen	—	—	—	—	—	—	—	—	—	—	—
Talc (nonasbestiform)	Sol.	—	20 mppcf	—	Pneumoconiosis (Talcosis)	—	—	—	—	—	—	—	—	—	—	—
Tantalum	Sol.	—	5	—	Low toxicity	—	—	—	—	—	—	—	—	—	—	—
TEDP	Sol.	—	0.2	—	"Skin"; nervous system disturbances	—	—	—	—	—	—	—	—	—	—	—
Teflon decomposition products	Var.	none	none	—	Acute systemic toxicity (polymer fume fever); suspect carcinogen	3	4	3	—	—	—	—	—	—	—	—
Tellurium	Sol.	—	0.1	—	Garlic breath; acute CNS effects; cumulative organ damage	—	—	—	—	—	—	—	—	—	—	—
Tellurium hexafluoride (as Te)	Gas	0.02	0.2	—	Lung edema	—	—	—	—	—	—	—	—	—	—	—
TEPP	Liq.	—	0.05	—	"Skin"; nervous system disturbances	—	—	—	—	—	—	—	—	—	1.20	—
Terphenyls	Sol.	C 1	9	—	Moderate irritation—eye, nose, throat, lungs; cumulative liver and kidney damage; cumulative lung damage	—	—	—	—	—	—	—	—	—	—	—
Tetra methyl lead (as Pb)	Liq.	—	0.075	—	Cumulative liver, CNS, kidney damage; acute CNS effects	3	3	3	100	—	—	212	—	—	1.99	—
Tetra methyl succinonitrile	Sol.	0.5	3	—	Acute systemic toxicity (CNS); "skin"	—	—	—	—	—	—	—	—	—	—	—
1,1,2,2-Tetrachloro-1,2-difluoroethane (Refrigerant 112)	Liq.	500	4170	—	Mild irritant; narcosis; cumulative liver damage; "skin"	—	—	—	—	—	—	199	—	—	—	—
1,1,1,2-Tetrachloro-2,2 difluoroethane (Refrigerant 112a)	Liq.	500	4170	—	Narcosis; mild irritant	—	—	—	—	—	—	197	—	—	—	—
1,1,2,2-Tetrachloroethane	Liq.	5	35	50.0	Cumulative liver damage and other organ damage; suspect carcinogen and mutagen; "skin"	—	—	—	—	Nonflammable		290	6	31	1.59	—
Tetrachloronaphthalene	Sol.	—	2	—	Cumulative liver damage; chloracne; "skin"	—	—	—	—	—	—	—	—	—	—	—
Tetraethyl lead (as Pb)	Liq.	—	0.075	—	Cumulative liver, CNS, kidney damage; "skin"; acute CNS effects; suspect teratogen	3	2	3	200	—	—	230	0.5	—	1.65	—
Tetrahydrofuran	Liq.	200	590	—	Moderate irritation—eye, nose, throat; narcosis; mutagen	2	3	1	6	2	11.8	151	142	40	0.888	—
Tetranitromethane	Liq.	1	8	—	Marked irritation—eye, nose, throat, cumulative systemic damage; acute CNS and lung effects; methemoglobinemia	—	—	—	—	—	—	255	13	—	1.65	—
Tetryl	Sol.	—	1.5	—	Cumulative systemic toxicity; contact dermatitis; "skin"; CNS effects; mutagen	—	—	—	—	—	—	—	—	—	—	—
Thalium (soluble compounds)	Sol.	—	0.1	—	"Skin"; cumulative systemic toxicity; CNS effects	—	—	—	—	—	—	—	—	—	—	—
4,4-Thiobis (6 tert-Butyl-m-cresol)	Sol.	none	none	—	Low toxicity	—	—	—	—	—	—	—	—	—	—	—
Thiram	Sol.	—	5	—	Acute systemic toxicity ("antabuse"-like effects); suspect teratogen, mutagen	—	—	—	—	—	—	—	—	—	—	—
Tin (inorganics), except SnH_4 and SnO_2	Sol.	—	2	—	Mild irritant; acute and chronic systemic toxicity	—	—	—	—	—	—	—	—	—	—	—
Tin (organics), as Sn	Var.	—	0.1	—	Cumulative systemic toxicity; "skin"; CNS effects; moderate irritant	—	—	—	—	—	—	—	—	—	—	—
Tin oxide	Sol.	none	none	—	Pneumoconiosis (stannosis)	—	—	—	—	—	—	—	—	—	—	—

(Continued)

Table of Chemical Hazards (continued)

1 Substance Name	2 Physical State (Sol., Liq., Gas)	3 OSHA P.E.L Permissible Exposure Levels ppm	mg/m³	4 Odor Level Thresholds (approximate ppm)	5 Health Hazard	6 N.F.P.A. Health	Flammability	Reactivity	7 Flash Point (Deg. F)	8 Flammable or Explosive Limits (% by volume) lower	upper	9 Boiling Point (Deg. F)	10 Vapor Pressure (mm Hg at S.T.P.)	11 Vapor Volume (cu ft/gal)	12 Specific Gravity	13 References NSC Data Sheet
Titanium dioxide	Sol.	—	15	—	Nuisance particulate	—	—	—	—	—	—	—	—	—	—	—
Toluene	Liq.	200	750	2.0	Moderate irritation—eye, nose, throat; narcosis; "skin"; suspect teratogen; mutagen	2	3	0	40	1.2	7.1	231	30	31	0.866	—
Toluene-2,4-diisocyanate (TDI)	Liq.	C 0.02	0.14	0.2	Marked irritation—eye, nose, throat, lungs; respiratory and skin sensitization	2	1	2	270	0.9	9.5	484	0.01	—	1.22	—
o-Toluidine	Liq.	5	22	—	Methemoglobinemia; suspect carcinogen; "skin"; acute systemic effects	3	2	0	185	—	—	392	1	—	1.004	—
Tributyl phosphate	Liq.	—	5	—	Moderate irritation—eye, nose, throat, lungs	2	1	0	—	—	—	560	—	—	0.973	—
1,1,2-Trichloro-1,2,2-trifluoroethane	Liq.	1000	7600	—	Narcosis; mild irritant	—	—	—	—	—	—	118	—	—	1.57	—
1,2,4-Trichlorobenzene	Liq.	none	none	3	Moderate irritant	—	—	—	—	—	—	415	—	—	1.46	—
1,1,2-Trichloroethane	Liq.	10	45	—	Narcosis; cumulative liver damage; "skin"; suspect carcinogen	2	1	0	—	—	—	235	17	—	1.44	—
Trichloroethylene	Liq.	100	535	20.0	Narcosis; suspect carcinogen and mutagen; teratogen; cumulative systemic toxicity	2	1	0	99	12.5	90	188	77	36	1.46	—
Trichloronaphthalene	Sol.	—	5	<1.0	Cumulative liver damage; "skin"; chloracne	—	—	—	—	—	—	—	—	—	—	—
1,2,3-Trichloropropane	Liq.	50	300	—	Marked irritation; cumulative liver damage; mutagen; narcosis	3	2	0	—	—	—	313	—	—	1.39	—
Tricyclohexylin hydroxide (Plictran)	Sol.	none	none	—	Low toxicity	—	—	—	—	—	—	—	—	—	—	—
Triethylamine	Liq.	25	100	—	Marked irritation—lungs, skin; lung edema; corneal damage	2	3	0	20	1.2	8.0	193	400	—	0.723	—
Trifluoromono-bromomethane	Gas	1000	6100	—	Low toxicity	—	—	—	—	—	—	—	—	—	—	—
Trimethylbenzene	Liq.	none	none	—	Marked irritation—lungs, skin	0	2	0	—	—	—	349	2	—	0.863	—
Trinitrotoluene	Sol.	—	1.5	—	Methemoglobinemia; aplastic anemia; cumulative eye (cataracts) and liver damage; "skin"	2	4	4	—	—	—	Explodes	0.5	—	—	314
Triorthocresyl phosphate	Liq.	—	0.1	—	Cumulative neuromuscular damage (paralysis)	2	1	0	—	—	—	770	10	—	—	—
Triphenyl phosphate	Sol.	—	3	—	Nervous system disturbances	2	1	0	—	—	—	—	—	—	—	—
Tungsten and compounds, as W (insoluble)	Sol.	none	none	—	Pneumoconiosis	—	—	—	—	—	—	—	—	—	—	—
Tungsten and compounds, as W (soluble)	Sol.	none	none	—	Acute systemic toxicity (CNS, anoxia)	—	—	—	—	—	—	—	—	—	—	—
Turpentine	Liq.	100	560	—	Moderate irritation—eye, nose, throat; cumulative kidney damage; CNS effects	1	3	0	95	0.8	—	300	5	18	0.857	—
Uranium (natural), insoluble	Sol.	—	0.25	—	Cumulative kidney damage	—	—	—	—	—	—	—	—	—	—	—
Uranium (natural), soluble	Sol.	—	0.05	—	Cumulative kidney damage	—	—	—	—	—	—	—	—	—	—	—
Vanadium (V_2O_5), as dust	Sol.	—	C 0.5	—	Marked irritation—eye, nose, throat, lungs; acute and chronic bronchial damage	—	—	—	—	—	—	—	—	—	—	—
Vanadium (V_2O_5), as fume	Sol.	—	C 0.1	—	Marked irritation—eye, nose, throat, lungs; acute and chronic bronchial damage	—	—	—	—	—	—	—	—	—	—	—
Vinyl acetate	Liq.	none	none	—	Mild irritation—eye, nose, throat	2	3	2	18	2.6	13.4	161	115	35	0.932	—
Vinyl bromide	Gas	none	none	—	Cumulative bromide intoxication (CNS effects); suspect carcinogen and mutagen	—	—	—	—	—	—	—	—	—	—	—
Vinyl chloride	Gas	Std. 1910.1017		—	Cancer	2	4	2	—	4	22	—	—	—	—	—
Vinyl cyclohexene dioxide	Liq.	none	none	—	Marked irritation—skin; suspect carcinogen and mutagen	—	—	—	—	—	—	—	—	—	—	—
Vinyl toluene	Liq.	100	480	25.0	Moderate irritation—eye, nose, throat; CNS effects	2	2	1	—	—	—	340	1	—	0.890	—
Vinylidene chloride	Liq.	none	none	—	Cumulative liver and kidney damage; suspect carcinogen	2	4	2	0	7.3	16	99	—	—	—	—

(Continued)

Table of Chemical Hazards (concluded)

1 Substance Name	2 Physical State (Sol., Liq., Gas)	3 OSHA P.E.L Permissible Exposure Levels ppm	mg/m³	4 Odor Level Thresholds (approximate ppm)	5 Health Hazard	6 N.F.P.A. Health	Flammability	Reactivity	7 Flash Point (Deg. F)	8 Flammable or Explosive Limits (% by volume) lower	upper	9 Boiling Point (Deg. F)	10 Vapor Pressure (mm Hg at S.T.P.)	11 Vapor Volume (cu ft/gal)	12 Specific Gravity	13 References NSC Data Sheet
VM and P naphtha	Liq.	none	none	—	Narcosis	1	3	0	28	0.9	6.0	212-320	—	—	—	—
Warfarin	Sol.	—	0.1	—	Cumulative anticoagulant effect; suspect teratogen	—	—	—	—	—	—	—	—	—	—	—
Welding fumes (total particulate)	Fume	See specific metal oxides		—	Moderate irritation; acute toxicity from metal oxides	—	—	—	—	—	—	—	—	—	—	—
Wood dust (nonallergenic)	Sol.	none	none	—	Lung damage; suspect carcinogen; dermatitis	—	—	—	—	—	—	—	—	—	—	—
Xylene (o-, m-, p-, isomers)	Liq.	100	435	0.5	Moderate irritation—eye, nose, throat; narcosis; "skin"	2	3	0	85	1.1	7.0	284	10	27	0.87	—
m-Xylene, alpha, alpha, diamine	Liq.	none	none	—	Moderate irritation—eye, nose, throat, skin; skin sensitization; "skin"	—	—	—	—	—	—	—	—	—	—	—
Xylidene	Liq.	5	25	—	Metheglobinemia; acute systemic toxicity; "skin"	3	1	0	—	—	—	435	—	—	0.97	—
Yttrium	Sol.	—	1	—	Pneumoconiosis (diffuse fibrosis)	—	—	—	—	—	—	—	—	—	—	—
Zinc chloride fume	Sol.	—	1	—	Marked irritation—eye, nose, throat, lungs; acute lung damage; suspect carcinogen and mutagen	2	0	2	—	—	—	—	—	—	—	—
Zinc chromate, as CrO_3	Sol.	none	none	—	See chromates; suspect carcinogen	1	0	1	—	—	—	—	—	—	—	—
Zinc oxide fume	Sol.	—	5	—	Acute systemic toxicity (metal fume fever); mutagen	—	—	—	—	—	—	—	—	—	—	267
Zinc stearate	Sol.	none	none	—	Nuisance particulate	—	—	—	—	—	—	—	—	—	—	—
Zirconium compounds, as Zr	Sol.	—	5	—	Pneumoconiosis; lung and skin granulomas	1	4	1	—	—	—	—	—	—	—	—

(Concluded)

CROSS INDEX

A

B

C

D

E

ethanoic anhydride . . . acetic anhydride
ethanol . . . ethyl alcohol
2-ethanoxyethylacetate . . . ethylene glycol monoethyl ether acetate
ethenyl ethanoate . . . vinyl acetate
ether, petroleum . . . naphtha (petroleum)
ether . . . ethyl ether
ethinyl trichloride . . . trichloroethylene
2-ethoxy ethanol . . . ethylene glycol monoethyl ether
2-ethoxy ethylacetate . . . ethylene glycol monoethyl ether acetate
ethyl acrylate . . . acrylate, methyl
ethyl aldehyde . . . acetaldehyde
ethylcaproaldehyde . . . 2-ethyl hexanol
ethyl Cellosolve . . . ethylene glycol monoethyl ether
ethyl dimethyl methane . . . pentane
ethyl ethanoate . . . ethyl acetate
ethylene dibromide . . . 1,2-dibromoethylene
2,2-ethylene dioxydiethanol . . . triethylene glycol
ethylene glycol monoethyl ether . . . 2-ethoxyethanol
ethylene, tetrachloro . . . tetrachloroethylene
ethylene trichloride . . . trichloroethylene
ethyl methanoate . . . ethyl formate
ethyl methyl ketone . . . methyl ethyl ketone
ethyl oxide . . . ethyl ether

F

formal . . . methylal
formic acid ethyl ester . . . ethyl formate
formic acid methyl ester . . . methyl formate
formic ether . . . ethyl formate
formyl trichloride . . . chloroform
fuel oil No. 1 . . . kerosene
fulminate of mercury . . . mercury fulminate
fuming sulfuric acid . . . oleum
furfuraldehyde . . . furfural
2-furaldehyde . . . furfural

G

glycerol . . . glycerine
glycol . . . ethylene glycol
glycol chlorohydrin . . . ethylene chlorohydrin
glycol dichloride . . . ethylene dichloride
grain alcohol . . . ethyl alcohol

H

hexahydrobenzene . . . cyclohexane
hexahydrotoluene . . . methyl cyclohexane
hexalin . . . cyclohexanol
hexamethylene . . . cyclohexane
hexane diacid . . . adipic acid
hexanedioic acid . . . adipic acid
2-hexanone . . . methyl butyl ketone
hexone . . . methyl butyl ketone
hexyl hydride . . . hexane
hydralin . . . cyclohexanol
hydrochloric ether . . . ethyl chloride
hydrogen chloride . . . hydrochloric acid
hydrogen cyanide . . . hydrocyanic acid
hydrogen fluoride . . . hydrofluoric acid
4-hydroxy-4-methyl-2-pentanone . . . diacetone alcohol
hydroxytoluenes . . . cresols
2-hydroxytriethylamine . . . diethylamino ethanol

I

iso-butyl carbinol . . . i-amyl alcohol
iso-propanol . . . iso-propyl alcohol
iso-propyl benzene . . . cumene
iso-propyl carbinal . . . 1-butanol

L

ligroin . . . naphtha (petroleum)
liquefied petroleum gas . . . LP-gas
LOX . . . oxygen, liquid
lye . . . sodium hydroxide

M

MBK . . . methyl butyl ketone
MDI . . . isocyanates
MEK . . . methyl ethyl ketone
methanal . . . formaldehyde
methanethiol . . . methyl mercaptan
3-methoxy-1-butanol . . . amyl alcohols
2-methoxy ethanol . . . ethylene glycol monomethyl ether
methyl alcohol . . . methanol
methyl Cellosolve acetate . . . ethylene glycol monomethyl ether acetate
methyl acetic ester . . . methyl acetate
methyl acetylene . . . propyne
methyl acrylate . . . acrylate, methyl
methyl benzene . . . toluene
2-methyl butane . . . pentane
3-methyl-1-butanol . . . i-amyl alcohol
3-methyl-1-butanol acetate . . . amyl acetate
2-methyl butyl ethanoate . . . amyl acetate
methyl n-butyl methane . . . hexane
methyl Cellosolve . . . ethylene glycol monomethyl ether
methyl Cellosolve acetate . . . ethylene glycol monomethyl ether acetate
methyl chloride . . . methylene chloride
methyl cyanide . . . acetonitrile
methylenebis (4-phenyl isocyanate) . . . isocyanates
methylene chlorobromide . . . bromochloromethane
methylene dimethyl ether . . . methylal
methyl ethyl carbinol . . . butanol
methyl ethylene glycol . . . propylene glycol
2-methyl hexane . . . heptane
methyl iso-butyl carbinol . . . methyl amyl alcohol
methyl methacrylate . . . acrylate, methyl
methyl methanoate . . . methyl formate
4-methyl-2-pentanol . . . methyl amyl alcohol
4-methyl-2-pentanone . . . methyl iso-butyl ketone
4-methyl-3-pentene-2-one . . . mesityl oxide
methylphenols . . . cresols
methyl propanol . . . butanol

mineral spirits . . . naphtha (petroleum)
monochlorobenzene . . . chlorobenzene
monoethanolamine . . . ethanolamine
monoethyl amine . . . ethylamine
monomethyl hydrazine . . . methyl hydrazine
monomethyl aniline . . . methyl aniline
motor fuel antiknock compound . . . tetraethyl lead
muriatic acid . . . hydrochloric acid
muriatic ether . . . ethyl chloride

N

naphtha 76° . . . naphtha (petroleum)
nitrating acid . . . acids, mixed

O

oil, fuel . . . fuel oil
oil of mirbane . . . nitrobenzene
oil of vitriol . . . sulfuric acid

P

pentanol acetate . . . amyl acetate
pentanols . . . amyl alcohols
2-pentanone . . . methyl propyl ketone
perchloroethylene . . . tetrachloroethylene
petroleum ether . . . naphtha (petroleum)
phenyl amine . . . aniline
phenyl chloride . . . chlorobenzene
phenyl ethane . . . ethylbenzene
phenyl ethylene . . . styrene monomer
phenyl methane . . . toluene
2-phenyl propane . . . cumene
phorone, iso . . . isophorone
pimelic ketone . . . cyclohexanone
polonium . . . radon
propane . . . LP-gas
1,2-propanediol . . . propylene glycol
1,2,3-propane triol . . . glycerine
propanols . . . propyl alcohols
2-propanone . . . acetone
propene oxide . . . propylene oxide
2-propen-1-ol . . . allyl alcohol
propyl acetone . . . methyl butyl ketone
propyl carbinol . . . butanol
propylene dichloride . . . 1,2-dichloropropane
propyl methanol . . . butanol
prussic acid . . . hydrocyanic acid

Q

quick lime . . . lime (quick)

R

red fuming nitric acid . . . nitric acid (red fuming)
refrigerants . . . *see chemical name*

S

sodium cyanide . . . cyanides
sodium pentachlorophenate . . . pentachlorophenol
sulfuric ether . . . ethyl ether

T

TDI . . . isocyanates
Teflon . . . polytetrafluoroethylene
TEL . . . tetraethyl lead
tetrabromoethane . . . acetylene tetrabromide
tetrachloromethane . . . carbon tetrachloride
tetrahydronaphthalene . . . Tetralin
tetramethylene oxide . . . tetrahydrofuran
TML . . . tetramethyl lead
TNT . . . trinitrotoluene
toluene diisocyanate . . . isocyanates
toluidine . . . methyl aniline
toluol . . . toluene
tolylene diisocyanate . . . isocyanates
1,1,1-trichloroethane . . . methyl chloroform
trichloromethane . . . chloroform
trimethylamine . . . methylamine
trimethyl carbinol . . . butanol
2,4,6-trinitrophenylmethylnitramine . . . tetryl
2,4,6-trinitrophenol . . . picric acid
triorthocresyl phosphate . . . tricresyl phosphate

U

UDMH . . . 1,1-dimethylhydrazine
unsymmetrical dimethyl hydrazine . . . 1,1-dimethylhydrazine

V

vinegar acid . . . acetic acid
vinyl benzene . . . styrene monomer
vinylcyanide . . . acrylonitrile
vinyl trichloride . . . methyl chloroform
vinylidene chloride monomer . . . dichloroethylene

W

wood alcohol . . . methyl alcohol
wood spirit . . . methyl alcohol

X

xylols . . . xylenes

Conversion of Units

All physical units of measurement can be reduced to one or more of three dimensions—mass, length, and time. Reducing units to basic dimensions simplifies problem-solving and makes comparison between operations, or operations and standards, easier and more accurate.

For example, three airflows could be measured: the first in liters per second, the second in cubic meters per second, and the third in cubic feet per minute. Then the total volume of air in each of the three samplings could be converted to cubic meters or cubic feet, and the airflows could be compared. In another situation, the results of atmospheric pollution studies and stack sampling surveys are often reported as grains per cubic foot, grams per cubic foot, or pounds per cubic foot. The degree of contamination is usually reported in the standard unit of parts or contaminant per million parts of air.

If physical measurements are made, or reported in different units, they must be converted to the standard units if any comparisons are to be meaningful.

In order to achieve a uniform system of measurement, governments representing 98 percent of the world's population have committed to using the *Système international d'Unités* (SI) version of the metric system (McQueen MJ. Conversion to SI units; The Canadian experience. *JAMA* 256[1986]:3001-3002.) Although, in 1975, Congress passed the Metric Conversion Act, which endorsed a voluntary conversion to SI, the English system is still in popular use in the United States. The SI system, however, is the standard for the international scientific community.

FUNDAMENTAL UNITS

Because of the need to conserve time and space when reporting data, universally accepted abbreviations are often used in place of unit names. This appendix will show the abbreviations used throughout this book, and those generally agreed upon by industrial hygiene practitioners. Conversion factors are provided when data is reported in nonstandard units.

Each measurement unit, for example length, area, and flow, has a table of conversion factors. To use the table to find the numerical value of the quantity desired, locate the unit to be converted in the first column. Then multiply this value by the number appearing at the intersection of the row and the column containing the desired unit. The answer will be the numerical value in the desired unit. Various English system and metric system units are given for your convenience.

An explanation of the SI system and official conversion factors are given to a 6- or 7-place accuracy in ASTM standard E380-76 (ANSI Z210.1-1976). (This standard is available, although not listed in the ANSI Catalog.)

Table D-A. Base Système International (SI) Units

Physical Quantity	*Base Units*	*SI Symbol*
Length	Meter	m
Mass	Kilogram	kg
Time	Second	s
Amount of substance	Mole	mol
Thermodynamic temperature	Kelvin	K
Electric current	Ampere	A
Luminous intensity	Candela	cd

Table D-B. Units Derived From Combinations of Base Units

Derived Unit	*Name and Symbol*	*Expressed as SI Base Derived Unit*
Area	Square meter	m^2
Volume	Cubic meter	m^3
Force	Newton (N)	$kg \cdot m \cdot s^{-2}$ ($kg \cdot m/s^2$)
Frequency	Hertz (Hz)	s^{-1}
Work, energy, heat	Joule (J)	$N \cdot m$
Power	Watt (W)	$J \cdot s^{-1}$(J/s)
Pressure	Pascal (Pa)	$kg \cdot m^{-1} \cdot s^{-2}$($N/m^2$)
Electric potential	Volt (V)	$W \cdot A^{-1}$(W/A)
Electric charge	Coulomb (C)	$A \cdot s$
Electric capacitance	Farad (F)	$A \cdot sV^{-1}$($A \cdot s/V$ or C/V)
Inductance	Henry (H)	$V \cdot s \cdot A^{-1}$($V \cdot s/A$)

Table D-C. Multiples and Submultiples of SI Units

Factor	*Prefix*	*Symbol*
10^{12}	tetra	T
10^{9}	giga	G
10^{6}	mega	M
10^{3}	kilo	k
10^{-3}	milli	m
10^{-6}	micro	μ
10^{-9}	nano	n
10^{-12}	pico	p
10^{-15}	femto	f
10^{-18}	atto	a

Table D-D. Area

To Obtain → *Multiply Number of ↓* *By ↘*	*square meter (m^2)*	*square inch (sq in.)*	*square foot (sq ft)*	*square centimeter (cm^2)*	*square millimeter (mm^2)*
square meter	1	1,550	10.76	10,000	10^6
square inch	6.452×10^{-3}	1	6.94×10^{-3}	6.452	645.2
square foot	0.0929	144	1	929.0	92,903
square centimeter	0.0001	0.155	0.001	1	100
square millimeter	10^{-6}	0.00155	0.00001	0.01	1

Table D-E. Length

To Obtain → *Multiply Number of ↓* *By ↘*	*meter (m)*	*centimeter (cm)*	*millimeter (mm)*	*micron (μ) or micrometer*	*angstrom unit (Å)*	*inch (in.)*	*foot (ft)*
meter	1	100	1,000	10^6	10^{10}	39.37	3.28
centimeter	0.01	1	10	10^4	10^8	0.394	0.0328
millimeter	0.001	0.1	1	10^3	10^7	0.0394	0.00328
micron	10^{-6}	10^{-4}	10^{-3}	1	10^4	3.94×10^{-5}	3.28×10^{-6}
angstrom	10^{-10}	10^{-8}	10^{-7}	10^{-4}	1	3.94×10^{-9}	3.28×10^{-10}
inch	0.0254	2.540	25.40	2.54×10^4	2.54×10^8	1	0.0833
foot	0.305	30.48	304.8	304,800	3.048×10^9	12	1

Table D-F. Density

To Obtain → Multiply Number of ↓ By ↘	gm/cm^3	lb/cu ft	lb/gal
gram/cubic centimeter	1	62.43	8.345
pound/cubic foot	0.01602	1	0.1337
pound/gallon (U.S.)	0.1198	7.481	1

1 grain/cu ft = 2.28 mg/m^3

Table D-G. Force

To Obtain → Multiply Number of ↓ By ↘	dyne	newton (N)	kilogram-force	pound-force (lbf)
dyne	1	1.0×10^{-5}	1.02×10^{4}	2.248×10^{4}
newton	1.0×10^{5}	1	0.1020	0.2248
kilogram-force	9.807×10^{-5}	9.807	1	2.205
pound-force	4.448×10^{-5}	4.448	0.4536	1

Table D-H. Mass

To Obtain → Multiply Number of ↓ By ↘	gram (gm)	kilogram (kg)	grains (gr)	ounce (avoir) (oz)	pound (avoir) (lb)
gram	1	0.001	15.432	0.03527	0.00220
kilogram	1,000	1	15,432	35.27	2.205
grain	0.0648	6.480×10^{-5}	1	2.286×10^{-3}	1.429×10^{-4}
ounce	28.35	0.02835	437.5	1	0.0625
pound	453.59	0.4536	7,000	16	1

Table D-I. Volume

To Obtain → Multiply Number of ↓ By ↘	cu ft	gallon (U.S. liquid)	liters	cm^3	m^3
cubic foot	1	7.481	28.32	28,320	0.0283
gallon (U.S. liquid)	0.1337	1	3.785	3,785	3.79×10^{-3}
liter	0.03531	0.2642	1	1,000	1×10^{-3}
cubic centimeters	3.531×10^{-5}	2.64×10^{-4}	0.001	1	10^{-6}
cubic meters	35.31	264.2	1,000	10^{6}	1

Table D-J. Velocity

To Obtain → Multiply Number of ↓ By ↘	cm/s	m/s	km/hr	ft/s	ft/min	mph
centimeter/second	1	0.01	0.036	0.0328	1.968	0.02237
meter/second	100	1	3.6	3.281	196.85	2.237
kilometer/hour	27.78	0.2778	1	0.9113	54.68	0.6214
foot/second	30.48	0.3048	18.29	1	60	0.6818
foot/minute	0.5080	0.00508	0.0183	0.0166	1	0.01136
mile per hour	44.70	0.4470	1.609	1.467	88	1

Table D-K. Flow rates

To Obtain → *Multiply Number of ↓* *By ↘*	*liters/min*	*m^3/s*	*m^3hr*	*gal/min*	*cu ft/min*	*cu ft/sec*
liter/minute	1	1.67×10^{-5}	0.06	0.2640	0.0353	5.89×10^{-4}
cubic meter/second	4.63×10^{-3}	1	2.77×10^{-4}	1.22×10^{-3}	1.63×10^{-4}	2.7×10^{-6}
cubic meter/hour	16.67	2.78×10^{-4}	1	4.4	0.588	9.89×10^{-3}
gallon (U.S.)/minute	3.78	6.3×10^{-5}	0.227	1	0.1338	2.23×10^{-3}
cubic foot/minute	28.32	4.71×10^{-4}	1.699	7.50	1	0.01667
cubic foot/second	1.69×10^{3}	2.83×10^{-3}	1.02×10^{2}	448.8	60	1

Table D-L. Heat, energy, or work

To Obtain → *Multiply Number of ↓* *By ↘*	*joule*	*ft-lb*	*kwh*	*hp-hour*	*kcal*	*cal*	*Btu*
joules	1	0.737	2.773×10^{-7}	3.725×10^{-7}	2.39×10^{-4}	0.2390	9.478×10^{-4}
foot-pound	1,356	1	3.766×10^{-7}	5.05×10^{-7}	3.24×10^{-4}	0.3241	1.285×10^{-3}
kilowatt-hour	3.6×10^{6}	2.66×10^{6}	1	1.341	860.57	860,565	3,412
hp-hour	2.68×10^{6}	1.98×10^{6}	0.7455	1	641.62	641,615	2,545
kilocalorie	4,184	3,086	1.162×10^{-3}	1.558×10^{-3}	1	1,000	3.9657
calorie	4.184	3.086	1.162×10^{-6}	1.558×10^{-6}	0.001	1	0.00397
British thermal unit	1,055	778.16	2.930×10^{-4}	3.93×10^{-4}	0.252	252	1

Table D-M. Emission rates

To Obtain → *Multiply Number of ↓* *By ↘*	*gm/s*	*gm/min*	*kg/hr*	*kg/day*	*lb/min*	*lb/hr*	*lb/day*
gram/second	1.0	60.0	3.6	86.40	0.13228	7.9367	190.48
gram/minute	0.016667	1.0	0.06	1.4400	2.2046×10^{-3}	0.13228	3.1747
kilogram/hour	0.27778	16.667	1.0	24.000	0.036744	2.2046	52.911
kilogram/day	0.011574	0.69444	0.041667	1.0	1.5310×10^{-3}	9.1860×10^{-2}	2.2046
pound/minute	7.5598	453.59	27.215	653.17	1.0	60.0	1440.
pound/hour	0.12600	7.5598	0.45359	10.886	1.6667×10^{-2}	1.0	24.0
pound/day	5.2499×10^{-3}	0.31499	1.8900×10^{-2}	0.45359	6.9444×10^{-4}	4.1667×1^{-2}	1.0

Table D-N. Pressure

To Obtain → *Multiply Number of ↓* *By ↘*	*lb/sq in. (psi)*	*atm*	*in. (Hg) 32 F 0 C*	*mm (Hg) 32 F 0 C*	*k Pa (k N/m^2)*	*ft (H_2O) 60 F 15 C*	*in. (H_2O)*	*lb/sq ft*
pound/square inch	1	0.068	2.036	51.71	6.895	2.309	27.71	144
atmospheres	14.696	1	29.92	760.0	101.32	33.93	407.2	2,116
inch (Hg)	0.4912	0.033	1	25.40	3.386	1.134	13.61	70.73
millimeter (Hg)	0.01934	0.0013	0.039	1	0.1333	0.04464	0.5357	2.785
kilopascals	0.1450	9.87×10^{-3}	0.2953	7.502	1	0.3460*	4.019	20.89
foot (H_2O)(15 C)	0.4332	0.0294	0.8819	22.40	2.989*	1	12.00	62.37
inch (H_2O)	0.03609	0.0024	0.073	1.867	0.2488	0.0833	1	5.197
pound/square foot	0.0069	4.72×10^{-4}	0.014	0.359	0.04788	0.016	0.193	1

*at 4 C

Table D-O. Radiant energy units

To Obtain → *Multiply ↓* *By ↘*	*erg*	*joule (J)*	*W-s*	*μW-s*	*g-cal*
erg	1	10^{-7}	10^{-7}	0.1	2.39×10^{-8}
joule	10^{7}	1	1	10^{6}	0.239
watt-second	10^{7}	1	1	10^{6}	0.239
micro-watt second	10	10^{-6}	10^{-6}	1	2.39×10^{-7}
gram-calorie	4.19×10^{7}	4.19	4.19	4.19×10^{6}	1

Table D-P. Energy/unit area (dose units)

To Obtain → *Multiply ↓* *By ↘*	*erg/cm²*	*J/cm²*	*W-s/cm²*	*μW-s/cm²*	*g-cal/m²*
erg/square centimeter	1	10^{-7}	10^{-7}	0.1	2.39×10^{-8}
joule/square centimeter	10^{7}	1	1	10^{6}	0.239
watt-second/square centimeter	10^{7}	1	1	10^{6}	0.239
microwatt-second/square centimeter	10	10^{-6}	10^{-6}	1	2.39×10^{-7}
gram-calorie/square centimeter	4.19×10^{7}	4.19	4.19	4.19×10^{6}	1

Table D-Q. Temperature equivalents

Scale	*Symbol*	*Freezing point of water (1 atm)*	*Boiling point of water (1 atm)*
Celsius	C	0	100 deg
Fahrenheit	F	32	212
Thermodynamic Kelvin Absolute Celsius	K, A	273.16 ± 0.01*	373.16 ± 0.01*
Approximate Absolute	AA	273	373
Rankine Absolute Fahrenheit	R	491.69	671.69

Conversion formulae

C = (5/9) (F − 32) = K − 273.16 = AA − 273
F = (9/5)C + 32 = (9/5) (K − 273.16) + 32
K = C + 273.16 = AA + 0.16 = (5/9) (F − 32) + 273.16
AA = C + 273 = K − 0.16 = (5/9) (F − 32) + 273
Rankine = F + 459.69

R.T. Birge, *Rev. Mod. Phys.*, 13:233. 1941.

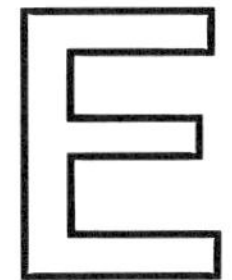

Review of Mathematics

SIGNIFICANT FIGURES

Measurements often result in what are called *approximate numbers* in contrast to *discrete counts*. For example, the dimensions of a table can be reported as 75.2 cm (29.6 in.) by 127.5 cm (50.2 in.). This implies that the measurement is to the nearest tenth of a cm and that the table is less than 127.6 cm (50.25 in.) and more than 127.4 cm (50.15 in.) in length. One can show the same thing for the width, using the symbolic notations:

75.0 cm (29.55 in.) $<$ width $<$ 75.3 cm (29.65 in.)

If, on the other hand, one knows the degree of precision of the measurement (say 0.03 cm or $\pm$ 0.08 cm), one may write:

$50.2 \pm .03$ or $50.2 \pm .08$

to indicate the degree of accuracy of the measurement of the length.

In reporting results, the number of significant digits that can be recorded is determined by the precision of the instruments used.

Rules

- In any approximate number, the significant digits include the digit that determines the degree of precision of the number and all digits to the left of it, except for zeros used to place the decimal.
- All digits from 1 to 9 are significant.
- All zeros that are between significant digits are significant.
- Final zeros of decimal numbers are significant. For example:

Number	*Number of Significant Digits*
0.0702	3
0.07020	4
70.20	4
7,002.	4
7,020.	3

Scientific notation

One case where it is difficult to determine the number of significant digits is the figure, 7000. In general, it is considered to have only one significant digit. It is better to use scientific notation.

In standard scientific notation, the number is written as a number between 1 and 10, in which only the significant digits are shown, multiplied by an exponential number to the base 10. For example:

Number	*Number of Significant Digits*
$5{,}320{,}000 = 5.32 \times 10^6$	3
$= 5.320 \times 10^6$	4
$= 5.3200 \times 10^6$	5
$0.00000532 = 5.32 \times 10^{-6}$	3

Addition and subtraction

The result must not have more decimal places than the number with the fewest decimal places. For example:

21.262	should be	21.3
23.74	should be	23.7
139.6	should be	139.6
184.602	should be	184.6

Multiplication and division

The result must not have more significant places than are possessed by the number with the fewest significant digits. For example:

$$(50.20)(29.6) = 1485.92$$
$$= 1490$$
$$= 1.49 \times 10^3$$

LOGARITHMS

Logarithms are exponents. The logarithm of any number is the power to which a selected base must be raised to produce the number. The laws of exponents apply to logarithms.

The two equations:

$a^x = y$ and
$x = \log_a y$

are two ways of expressing the same thing, that is, the exponent applied to "a" to give "y" is equal to "x." The number, "a" is called the base of the system of logarithms.

Although any positive number greater than 1 can be used as the base of some system of logarithms, there are two systems in general use. These are the *common* (or Briggs') system and the *natural* or Napierian system. In the common system the base is 10; while in the natural system the base is a certain irrational number e = 2.71828. . .

Common logarithms

Common logarithms use the base 10 and are identified by the notation "log." The common logarithm of a number consists of a characteristic, which locates the decimal point in the number, and a mantissa, which defines the numerical arrangement of the number.

A bar over a characteristic indicates a negative characteristic and a positive mantissa. The log may be written $\bar{4}.7$ or 6.7–10 or –3.3. The form –3.3 does not contain a characteristic and mantissa.

The integral part of a logarithm is called the *characteristic* and the decimal part is called the *mantissa.* In log 824, the characteristic is 2 and the mantissa is 0.9162. For convenience in constructing tables, it is desirable to select the mantissa as positive even if the logarithm is a negative number. For example, log ½ = –0.3010; but since –0.3010 = 9.6990 – 10, this may be written log ½ = 9.6990 – 10 with a positive mantissa. This is also the log of 0.5, which we should have looked up in the first place. The following illustration shows the method of writing the characteristic and mantissa:

log	8245	= 3.9162
log	824.5	= 2.9162
log	82.45	= 1.9162
log	8.245	= 0.9162
log	0.8245	= 9.9162 – 10
log	0.08245	= 8.9162 – 10

By using scientific notation, logarithm characteristics can easily be found, as shown in the table at the top of the next page.

How to use logarithms

If the laws of exponents are rewritten in terms of logarithms, they become the *laws of logarithms:*

$$\log_a(x^n) = n \log_a x$$

$$\log_a\left(\frac{x}{y}\right) = \log_a x - \log_a y$$

$$\log_a(x^a) = n \log_a x$$

Logarithms derive their main usefulness in computation from these laws, because they allow multiplication, division, and exponentiation to be replaced by the simpler operations of addition, subtraction, and multiplication, respectively.

Number	*Exponential Form*	*Common Logarithmic Form* *Characteristic*	*Mantissa*	*Complete Log*
0.0005	5×10^{-4}	–4	0.7	$\bar{4}.7$
0.05	5×10^{-2}	–2	0.7	$\bar{2}.7$
5.0	5×10^{0}	0	0.7	0.7
500.0	5×10^{2}	2	0.7	2.7
50000.0	5×10^{4}	4	0.7	4.7

How to use logarithm tables

In this appendix is a "four-place" table of logarithms. In this table, the mantissas of the logarithms of all integers from 1 to 999 are recorded correct to four decimal places, which is all one needs—for instance—to work with decibels, which at best have three significant digits.

To find the logarithm of a given number, use the table as follows.

For example, to find the logarithm of 63.5, glance down the column headed *N* for the first two significant digits (63), and then along the top of the table for the third figure (5). In a row with *63* and under the column with *5* is found *8028.* This is the mantissa. Adding the proper characteristic *1,* the logarithm (or log) of 63.5 is *1.8028.*

Conversely, one can find the number that corresponds to a given logarithm (the antilogarithm).

For example, find the number whose logarithm is 1.6355. The mantissa 6355 corresponds to the number in the table that is under the column with *2* and in the row with *43.* Thus, the mantissa corresponds to the number *432.* Since the characteristic is *1,* the number whose logarithm is 1.6355 is *43.2.*

Because in measuring sound we are concerned only with three significant digits, the number whose logarithm is 1.6360 would also be *43.2*. The number whose logarithm is 1.6361 would be *43.3*.

Decibel notation

Again, using the measurement of sound as an example, if two sound intensities P_1 and P_2 are to be compared according to the ability of the ear to detect intensity differences, we may determine the number of decibels which expresses the relative value of the two intensities by

$$N_{dB} = 10 \log_{10}\frac{P_1}{P_2}$$

where P_1 is greater than P_2.

The factor 10 comes into this picture because the original unit devised was the *bel,* which is the logarithm of 10 to the base 10 and which represents 10 times as many decibels in any expression involving the relation between two sound intensities as there are bels.

The decibel is a logarithmic unit. Each time the amount of power is increased by a factor of 10, we have added 10 decibels (abbreviated dBA).

To determine the number of decibels by which two powers differ, we must *first determine the ratio of the two powers,* then we look up this ratio in a table of logarithms to the base 10 and then we multiply the figure obtained by a factor of 10.

If we want to find the relative loudness of 10,000 people who can shout louder than 100 people can, we use the following reasoning.

The logarithm (to the base 10) of any number is merely the number of times 10 must be multiplied by itself to be equal to the number. In the example here, 100 represents 10 multiplied by itself, and the logarithm of 100 to the base 10, therefore, is 2. For example, the number of decibels expressing the relative loudness of 10,000 people shouting compared with 100 is

$$\begin{aligned} N_{dB} &= 10 \log_{10} (10{,}000 \div 100) \\ &= 10 \log_{10} 100 \\ &= 10 \times 2.0 \\ &= 20 \end{aligned}$$

Now let us see what happens if we double the number of people to 20,000.

$$\begin{aligned} N_{dB} &= 10 \log_{10} (20{,}000 \div 100) \\ &= 10 \log_{10} 200 \\ &= 10 \times 2.3010 \\ &= 23 \text{ (rounded to significant digits)} \end{aligned}$$

It can be seen, therefore, that decibels are logarithm ratios. In their use in sound measurement, P (the usual reference level) is 20 micropascals or 0.0002 dynes/square centimeter, which approximates the "threshold of hearing," the sound that can just be heard by a young person with excellent hearing.

NORMAL AND LOGNORMAL FREQUENCY DISTRIBUTIONS

The statistical methods discussed here assume that measured concentrations of random occupational environmental samples are lognormally and independently distributed within one 8-hour period and over many daily exposure averages.

Before sample data can be statistically analyzed, we must have knowledge of the frequency distribution of the measurements or some assumptions must be made. Most community air pollution environmental data can be described by a lognormal distribution. That is, the logarithms (either base e or base 10) of the data are approximately normally distributed.

What are the differences between normally and lognormally distributed data? A "normal" distribution is completely determined by the parameters: the arithmetic mean (μ); the standard deviation (σ) of the distribution. On the other hand, a "lognormal" distribution is competely determined by the median or geometric mean (GM), and the geometric standard deviation (GSD). For lognormally distributed data, a logarithmic transformation of the original data is normally distributed. The GM and GSD of the lognormal distribution are the antilogs of the mean and standard deviation of the logarithmic transformation. Normally distributed data have a symmetrical distribution curve while lognormally distributed environmental data are generally positively skewed (long "tail" to the right indicating a larger probability of very large concentrations than for normally distributed data.)

Variability

The variability of occupational environmental data (differences between repeated measurements at the same site) can usually be broken into three major components: random errors of the sampling method; random errors of the analytical method; variability of the environment with time. The first two components of the variability are known in advance and are approximately normally distributed. However the environmental fluctuations of a contaminant in a plant usually greatly exceed the variability of known instruments (often by factors of 10 or 20).

When several samples are taken in a plant to determine the average concentration of the contaminant to estimate the average exposure of an employee, then the lognormal distribution should be assumed. However, the normal distribution may be used in the special cases of taking a sample to check compliance with a ceiling standard, and when a sample (or samples) is taken for the entire time period for which the standard is defined (be it 15 minutes or 8 hours). In these cases the entire time interval of interest is represented in the sample, and only sampling and analytical errors are present.

Coefficient of variation

The relative variability of a normal distribution (such as the random errors of the sampling and analytical procedures) is commonly measured by the coefficient of variation (CV). The CV is also known as the *relative standard deviation.* The CV is a useful index of dispersion in that limits consisting of the true mean of a set of data plus or minus twice the CV will contain about 95 percent of the data measurements.

Thus if an analytical procedure with a CV of 10 percent is used to repeatedly measure some nonvarying physical property (as the concentration of a chemical in a beaker of solution) then about 95 percent of the measurements will fall within plus or minus 20 percent (two times the CV) of the true concentration.

EXPOSURE CONCENTRATION

Unfortunately the property we are trying to measure, the employee's exposure concentration, is not a fixed, nonvarying physical property. The exposure concentrations are fluctuating in a lognormal manner. First, the exposure concentrations are fluctuating over the 8-hour period of the time-weighted average (TWA) exposure measurement. Breathing zone grab samples (samples of less than about 30 minutes' duration, typically only a few minutes) tend to reflect this intraday environmental variability so that grab sample results have relatively high variability.

Intraday variability in the sample results can be eliminated from measurement variability by going to a full-period sampling strategy. The day to day (interday) variability of the true 8-hour TWA eposures is also lognormally distributed. It is this interday variability that creates a need for an action level where only one day's exposure measurement is used to draw conclusions regarding compliance on unmeasured days.

GEOMETRIC STANDARD DEVIATION

The parameter often used to express either the intraday or interday environmental variability is the *geometric standard deviation* (GSD). A GSD of 1.0 represents absolutely no variability in the environment. GSDs of 2.0 and above represent relatively high variability.

The shape of lognormal distributions with low variabilities, such as those with GSDs less than about 1.4, roughly approximate normal distribution shapes. For this range of GSDs there is a rough equivalence between the GSD and CV as follows:

GSD	*Approximate CV*
1.40	35 percent
1.30	27 percent
1.20	18 percent
1.10	9.6 percent
1.05	4.9 percent

COMMON LOGARITHMS

N	0	1	2	3	4	5	6	7	8	9
0	....	0000	3010	4771	6021	6990	7782	8451	9031	9542
1	0000	0414	0792	1139	1461	1761	2041	2304	2553	2788
2	3010	3222	3424	3617	3802	3979	4150	4314	4472	4624
3	4771	4914	5051	5185	5315	5441	5563	5682	5798	5911
4	6021	6128	6232	6335	6435	6532	6628	6721	6812	6902
5	6990	7076	7160	7243	7324	7404	7482	7559	7634	7709
6	7782	7853	7924	7993	8062	8129	8195	8261	8325	8388
7	8451	8513	8573	8633	8692	8751	8808	8865	8921	8976
8	9031	9085	9138	9191	9243	9294	9345	9395	9445	9494
9	9542	9590	9638	9685	9731	9777	9823	9868	9912	9956
10	0000	0043	0086	0128	0170	0212	0253	0294	0334	0374
11	0414	0453	0492	0531	0569	0607	0645	0682	0719	0755
12	0792	0828	0864	0899	0934	0969	1004	1038	1072	1106
13	1139	1173	1206	1239	1271	1303	1335	1367	1399	1430
14	1461	1492	1523	1553	1584	1614	1644	1673	1703	1732
15	1761	1790	1818	1847	1875	1903	1931	1959	1987	2014
16	2041	2068	2095	2122	2148	2175	2201	2227	2253	2279
17	2304	2330	2355	2380	2405	2430	2455	2480	2504	2529
18	2553	2577	2601	2625	2648	2672	2695	2718	2742	2765
19	2788	2810	2833	2856	2878	2900	2923	2945	2967	2989
20	3010	3032	3054	3075	3096	3118	3139	3160	3181	3201
21	3222	3243	3263	3284	3304	3324	3345	3365	3385	3404
22	3424	3444	3464	3483	3502	3522	3541	3560	3579	3598
23	3617	3636	3655	3674	3692	3711	3729	3747	3766	3784
24	3802	3820	3838	3856	3874	3892	3909	3927	3945	3962
25	3979	3997	4014	4031	4048	4065	4082	4099	4116	4133
26	4150	4166	4183	4200	4216	4232	4249	4265	4281	4298
27	4314	4330	4346	4362	4378	4393	4409	4425	4440	4456
28	4472	4487	4502	4518	4533	4548	4564	4579	4594	4609
29	4624	4639	4654	4669	4683	4698	4713	4728	4742	4757
30	4771	4786	4800	4814	4829	4843	4857	4871	4886	4900
31	4914	4928	4942	4955	4969	4983	4997	5011	5024	5038
32	5051	5065	5079	5092	5105	5119	5132	5145	5159	5172
33	5185	5198	5211	5224	5237	5250	5263	5276	5289	5302
34	5315	5328	5340	5353	5366	5378	5391	5403	5416	5428
35	5441	5453	5465	5478	5490	5502	5514	5527	5539	5551
36	5563	5575	5587	5599	5611	5623	5635	5647	5658	5670
37	5682	5694	5705	5717	5729	5740	5752	5763	5775	5786
38	5798	5809	5821	5832	5843	5855	5866	5877	5888	5899
39	5911	5922	5933	5944	5955	5966	5977	5988	5999	6010
40	6021	6031	6042	6053	6064	6075	6085	6096	6107	6117
41	6128	6138	6149	6160	6170	6180	6191	6201	6212	6222
42	6232	6243	6253	6263	6274	6284	6294	6304	6314	6325
43	6335	6345	6355	6365	6375	6385	6395	6405	6415	6425
44	6435	6444	6454	6464	6474	6484	6493	6503	6513	6522
45	6532	6542	6551	6561	6571	6580	6590	6599	6609	6618
46	6628	6637	6646	6656	6665	6675	6684	6693	6702	6712
47	6721	6730	6739	6749	6758	6767	6776	6785	6794	6803
48	6812	6821	6830	6839	6848	6857	6866	6875	6884	6893
49	6902	6911	6920	6928	6937	6946	6955	6964	6972	6981
50	6990	6998	7007	7016	7024	7033	7042	7050	7059	7067
N	0	1	2	3	4	5	6	7	8	9

N	0	1	2	3	4	5	6	7	8	9
50	6990	6998	7007	7016	7024	7033	7042	7050	7059	7067
51	7076	7084	7093	7101	7110	7118	7126	7135	7143	7152
52	7160	7168	7177	7185	7193	7202	7210	7218	7226	7235
53	7243	7251	7259	7267	7275	7284	7292	7300	7308	7316
54	7324	7332	7340	7348	7356	7364	7372	7380	7388	7396
55	7404	7412	7419	7427	7435	7443	7451	7459	7466	7474
56	7482	7490	7497	7505	7513	7520	7528	7536	7543	7551
57	7559	7566	7574	7582	7589	7597	7604	7612	7619	7627
58	7634	7642	7649	7657	7664	7672	7679	7686	7694	7701
59	7709	7716	7723	7731	7738	7745	7752	7760	7767	7774
60	7782	7789	7796	7803	7810	7818	7825	7832	7839	7846
61	7853	7860	7868	7875	7882	7889	7896	7903	7910	7917
62	7924	7931	7938	7945	7952	7959	7966	7973	7980	7987
63	7993	8000	8007	8014	8021	8028	8035	8041	8048	8055
64	8062	8069	8075	8082	8089	8096	8102	8109	8116	8122
65	8129	8136	8142	8149	8156	8162	8169	8176	8182	8189
66	8195	8202	8209	8215	8222	8228	8235	8241	8248	8254
67	8261	8267	8274	8280	8287	8293	8299	8306	8312	8319
68	8325	8331	8338	8344	8351	8357	8363	8370	8376	8382
69	8388	8395	8401	8407	8414	8420	8426	8432	8439	8445
70	8451	8457	8463	8470	8476	8482	8488	8494	8500	8506
71	8513	8519	8525	8531	8537	8543	8549	8555	8561	8567
72	8573	8579	8585	8591	8597	8603	8609	8615	8621	8627
73	8633	8639	8645	8651	8657	8663	8669	8675	8681	8686
74	8692	8698	8704	8710	8716	8722	8727	8733	8739	8745
75	8751	8756	8762	8768	8774	8779	8785	8791	8797	8802
76	8808	8814	8820	8825	8831	8837	8842	8848	8854	8859
77	8865	8871	8876	8882	8887	8893	8899	8904	8910	8915
78	8921	8927	8932	8938	8943	8949	8954	8960	8965	8971
79	8976	8982	8987	8993	8998	9004	9009	9015	9020	9025
80	9031	9036	9042	9047	9053	9058	9063	9069	9074	9079
81	9085	9090	9096	9101	9106	9112	9117	9122	9128	9133
82	9138	9143	9149	9154	9159	9165	9170	9175	9180	9186
83	9191	9196	9201	9206	9212	9217	9222	9227	9232	9238
84	9243	9248	9253	9258	9263	9269	9274	9279	9284	9289
85	9294	9299	9304	9309	9315	9320	9325	9330	9335	9340
86	9345	9350	9355	9360	9365	9370	9375	9380	9385	9390
87	9395	9400	9405	9410	9415	9420	9425	9430	9435	9440
88	9445	9450	9455	9460	9465	9469	9474	9479	9484	9489
89	9494	9499	9504	9509	9513	9518	9523	9528	9533	9538
90	9542	9547	9552	9557	9562	9566	9571	9576	9581	9586
91	9590	9595	9600	9605	9609	9614	9619	9624	9628	9633
92	9638	9643	9647	9652	9657	9661	9666	9671	9675	9680
93	9685	9689	9694	9699	9703	9708	9713	9717	9722	9727
94	9731	9736	9741	9745	9750	9754	9759	9763	9768	9773
95	9777	9782	9786	9791	9795	9800	9805	9809	9814	9818
96	9823	9827	9832	9836	9841	9845	9850	9854	9859	9863
97	9868	9872	9877	9881	9886	9890	9894	9899	9903	9908
98	9912	9917	9921	9926	9930	9934	9939	9943	9948	9952
99	9956	9961	9965	9969	9974	9978	9983	9987	9991	9996
100	0000	0004	0009	0013	0017	0022	0026	0030	0035	0039
N	0	1	2	3	4	5	6	7	8	9

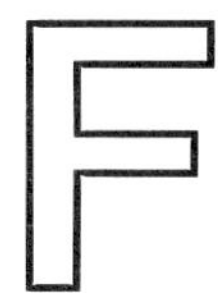

Glossary

EVERY INDUSTRY HAS ITS OWN TERMINOLOGY. The health and safety professional must be aware of the precise meanings of certain words commonly used in industrial hygiene, occupational health, and chemistry, to effectively communicate with other professionals in this area.

A fume respirator, for instance, is worthless as protection against gases or vapors. Too frequently, these terms are used interchangeably. Each term, however, has a definite meaning and describes a certain state of matter that can be achieved only by certain physical changes to the substance itself.

This glossary defines words and terms, some of which are peculiar to a single industry and others common to many industries. Terms were taken or adapted from the latest editions of the following: *The Chemical Industry Facts Book,* published by the Manufacturing Chemists Association, Inc., Washington, DC; *Occupational Diseases and Industrial Medicine,* by R.T. Johnstone, and S.E. Miller, published by W.B. Saunders Co., Philadelphia; *Guide for Industrial Audiometric Technicians,* published by the Safety and Health Services, Employers Insurance of Wausau, WI; American National Standards S1.1 *Acoustical Terminology,* and Z88.2 *Practices for Respiratory Protection; 101 Atomic Terms and What They Mean,* by the Esso Research and Engineering Company, Linden, NJ; *Paramedical Dictionary,* by J. E. Schmidt, M.D., published by Charles C. Thomas, Springfield, IL; *The Condensed Chemical Dictionary,* Published by Reinhold Publishing Corporation, New York; and *Stedman's Medical Dictionary,* 24th ed., published by the Williams & Wilkins Company, Baltimore.

A

A-, an- (prefix). Absent, lacking, deficient, without. Anemia, deficient in blood.

Abrasive blasting. A process for cleaning surfaces by means of such materials as sand, alumina, or steel grit in a stream of high-pressure air.

Sound absorption coefficient. The ratio of the sound energy absorbed by a surface of a medium (or material) exposed to a sound field (or to sound radiation), to the sound energy incident on the surface.

Accelerator. A device for imparting very high velocity to charged particles such as electrons or protons. Also, a chemical additive which increases the speed of a chemical reaction.

Acclimatization. The process of becoming accustomed to new conditions (i.e., heat).

Accommodation. The ability of the eye to adjust focus for various distances.

Accuracy (instrument). Quite often used incorrectly as precision (see Precision). Accuracy refers to the agreement of a reading or observation obtained from an instrument or a technique with the true value.

ACGIH. American Conference of Governmental Industrial Hygienists.

Acid. A proton donor.

Acid pickling. A bath treatment to remove scale and other impurities from metal surfaces prior to plating or other surface treatment. Sulfuric acid commonly is used.

Acne. See Oil dermatitis.

Acoustic, Acoustical. Containing, producing, arising from, actuated by, related to, or associated with sound.

Acoustic trauma. Hearing loss caused by sudden loud noise in one ear, or by sudden blow to head. In most cases, hearing loss is temporary, although there may be some permanent loss.

Acro- (prefix). Topmost; outer end. An extremity of the body. Acro-osteolysis is degeneration of terminal or distal end of bone tissue.

Acrylic. A family of synthetic resins made by polymerizing esters of acrylic acids.

Action level. Term used by OSHA and NIOSH to express the level of toxicant which requires medical surveillance, usually one half the PEL.

Activated charcoal. Charcoal is an amorphous form of carbon formed by burning wood, nutshells, animal bones, and other carbonaceous materials. Charcoal becomes activated by heating it with steam to 800-900 C. During this treatment, a porous, submicroscopic internal structure is formed which gives it an extensive internal surface area. Activated charcoal is commonly used as a gas or vapor adsorbant in air-purifying respirators and as a solid sorbant in air-sampling.

Activation. Making a substance artificially radioactive in an accelerator or by bombarding it with protons or neutrons in a reactor.

Activity. Frequently used as a shortened form of radioactivity, it refers to the radiating power of a radioactive substance. Activity may be given in terms of atoms disintegrating per second.

Acuity. This sense pertains to the sensitivity of receptors used in hearing or vision.

Acute. Health effects which show up a short length of time after exposure. An acute exposure runs a comparatively short course.

Additives. An inclusive name for a wide range of chemical substances which are added in low percentage to stabilize certain end products, such as antioxidants in rubber.

Aden- (prefix). Pertaining to a gland. Adenoma is a tumor of glandlike tissue.

Adenoma. An epithelial tumor, usually benign, with a glandlike structure (the cells lining glandlike depressions or cavities in the stroma).

Adhesion. The ability of one substance to stick to another. There are two types of adhesion: mechanical, which depends on the penetration of the surface, and molecular or polar adhesion, in which adhesion to a smooth surface is obtained because of polar groups such as carboxyl groups.

Administrative controls. Methods of controlling employee exposures by job rotation, work assignment, or time periods away from the hazard.

Adsorption. The condensation of gases, liquids, or dissolved substances on the surfaces of solids.

AEC. Atomic Energy Commission. Now called Nuclear Regulatory Commission in the U.S. Department of Energy.

Aerobe. Microorganisms that require the presence of oxygen.

Aerodynamic diameter. The diameter of a unit density sphere having the same settling velocity as the particle in question of whatever shape and density.

Aerodynamic forces. The forces exerted on a particle in suspension by either the movement of air or gases around the particle or the resistance of the gas or air to movement of the particle through the medium.

Aerosols. Liquid droplets or solid particles dispersed in air, that are of fine enough particle size (0.01 to 100 micrometers) to remain so dispersed for a period of time.

Agglomeration. Implies consolidation of solid particles into larger shapes by means of agitation alone, i.e., without application of mechanical pressure in molds, or between rolls, or through dies. Industrial agglomeration usually is implemented in balling devices such as rotating discs, drums, or cones; but it can occur in a simple mixer. On occasion, however, the word agglomeration has been used to describe the entire field of particulate consolidation.

AIHA. American Industrial Hygiene Association.

Air. The mixture of gases that surrounds the earth; its major components are as follows: 78.08 percent nitrogen, 20.95 percent oxygen, 0.03 percent carbon dioxide, and 0.93 percent argon. Water vapor (humidity) varies. See Standard air.

Air bone gap. The difference in decibels between the hearing levels for a particular frequency as determined by air conduction and bone conduction.

Airborne microorganisms. Biologically active contaminants suspended in the air either as free-floating particles surrounded by a film or organic or inorganic material, or attached to the surface of other suspended particulates.

Air cleaner. A device designed to remove atmospheric airborne impurities, such as dusts, gases, vapors, fumes, and smokes.

Air conditioning. The process of treating air so as to control simultaneously its temperature, humidity, cleanliness, and distribution to meet requirements of the conditioned space.

Air conduction. Air conduction is the process by which sound is conducted to the inner ear through air in the outer ear canal as part of the pathway.

Air filter. An air-cleaning device to remove light particulate matter from normal atmospheric air.

Air hammer. A percussion-type pnuematic tool, fitted with a handle at one end of the shank and a tool chuck at the other, into which a variety of tools may be inserted.

Air horsepower. The theoretical horsepower required to drive a fan if there were no losses in the fan; that is, if it were 100 percent efficient.

Air monitoring. The sampling for and measuring of pollutants in the atmosphere.

Air mover. Any device which is capable of causing air to be moved from one space to another. Such devices are generally used to exhaust, force, or draw gases through specific assemblies.

Air quality criteria. The amounts of pollution and lengths of exposure at which specific adverse effects to health and welfare take place.

Air-regulating valve. An adjustable valve used to regulate airflow to the facepiece, helmet, or hood of an air-line respirator.

Air standard. See Standard air.

Air-purifying respirator. Respirators that use filters or sorbents to remove harmful substances from the air.

Air-supplied respirator. Respirator that provides a supply of breathable air from a clean source outside of the contaminated work area.

Albumin. A protein material found in animal and vegetable fluids, characterized by being soluble in pure water.

Albuminuria. The presence of albumin or other protein substance, such as serum globulin, in the urine.

-algia (suffix). Pain. A prefix such as neur- tells where the pain is (neuralgia).

Algorithm. A precisely stated procedure or set of instructions that can be applied stepwise to solve a problem.

Aliphatic. (Derived from Greek word for oil.) Pertaining to an open-chain carbon compound. Usually applied to petroleum products derived from a paraffin base and having a straight or branched chain, saturated, or unsaturated molecular structure. Substances, such as methane and ethane, are typical aliphatic hydrocarbons. See Aromatic.

Alkali. A compound that has the ability to neutralize an acid and form a salt. Example: sodium hydroxide, referred to as caustic soda or lye. Used in soap manufacture and many other applications. Turns litmus paper blue. See Base.

Alkaline earths. Usually considered to be the oxides of alkaline earth metals: barium, calcium, strontium, beryllium, and radium. Some authorities also include magnesium oxide.

Alkyd. A synthetic resin which is the condensation product of a polybasic acid such as phthalic, a polyhydric alcohol such as glycerin, and an oil fatty acid.

Alkylation. The process of introducing one or more alkyl radicals by addition or substitution into an organic compound.

Allergy. An abnormal response of a hypersensitive person to chemical and physical stimuli. Allergic manifestations of major importance occur in about 10 percent of the population.

Alloy. A mixture of metals such as in brass and in some instances and a nonmetal.

Alpha-emitter. A radioactive substance which gives off alpha particles.

Alpha particle (alpha-ray, alpha-radiation). A small electrically charged particle of very high velocity thrown off by many radioactive materials, including uranium and radium. It is made up of two neutrons and two protons. Its electric charge is positive.

Aluminosis. A form of pneumoconiosis due to the presence of aluminum-bearing dust in the lungs, especially that of alum, bauxite, or clay.

Alveoli. Tiny air sacs of the lungs, formed at the ends of bronchioles; through the thin walls of the alveoli, the blood takes in oxygen and gives up its carbon dioxide in the process of respiration.

Alveolus. A general term used in anatomical nomenclature to designate a small saclike dilation.

Amalgamation. The process of alloying metals with mercury. This is one process used in extracting gold and silver from their ores.

Ambient noise. The all-encompassing noise associated with a given environment, being usually a composite of sounds from many sources near and far.

Amorphous. Noncrystalline.

Anaerobe. Microorganism that grows without oxygen (air). Facultative anaerobe is able to grow with or without oxygen; obligate anaerobe will grow only in the absence of oxygen.

Anaerobic bacteria. Any bacteria that can survive in partial or complete absence of air.

Anaphylaxis. Hypersensitivity resulting from sensitization following prior contact with a chemical or protein.

Andro- (prefix). Man, male. An androgen is an agent which produces masculinizing effects.

Anechoic room (free-field room). One whose boundaries absorb effectively all the sound incident thereon, thereby affording essentially free-field conditions.

Anemias. Deficiency in the hemoglobin and erythrocyte content of the blood. Term refers to a number of pathological states that may be due to a large variety of causes and appear in many different forms.

Anemometer. A device to measure air speed.

Anesthesia. Loss of sensation; in particular, the temporary loss of feeling induced by certain chemical agents.

Angi-, angio- (prefix). Blood or lymph vessel. Angitis is an inflammation of a blood vessel.

Angle of abduction. Angle between the longitudinal axis of a limb and a sagittal plane.

Angstrom (A). Unit of measure of wavelength equal to 10^{-10} meters or 0.1 nanometers.

Anneal. To treat by heat with subsequent cooling for drawing the temper of metels, that is, to soften and render them less brittle. See Temper.

Anode. The positive electrode.

Anorexia. Lack or loss of the appetite for food.

ANSI. The American National Standards Institute is a voluntary membership organization (run with private funding) that develops consensus standards nationally for a wide variety of devices and procedures.

Antagonist. A muscle opposing the action of another muscle. An active antagonist is essential for control and stability of action by a prime mover.

Anthracosilicosis. A complex form of pneumoconiosis; a chronic disease caused by breathing air containing dust that has free silica as one of its components and that is generated in the various processes in mining and preparing anthracite (hard) coal, and, to a lesser degree, bituminous coal.

Anthracosis. A disease of the lungs caused by prolonged inhalation of dust that contains particles of carbon and coal.

Anthrax. A highly virulent bacterial infection picked up from infected animals and animal products.

Anthropometry. The part of the anthropology having to do with measurement of the human body to determine differences in individuals or groups of individuals.

Anti- (prefix). Against. An antibiotic is "against life"—in the case of a drug, against the life of disease germs.

Antibiotic. A substance produced by a microorganism that in dilute solutions kills other organisms, or retards or completely represses their growth, normally in doses that do not harm higher orders of life.

Antibody. Any of the body globulins that combine specifically with antigens and neutralize toxins, agglutinate bacteria or cells, and precipitate soluble antigens. It is found naturally in the body or produced by the body in response to the introduction into its tissues of a foreign substance.

Antigen. A substance that, when introduced into the body, stimulates the production of an antibody.

Antioxidant. A compound which retards deterioration by oxidation. Antioxidants for human food and animal feeds, sometimes referred to as freshness preservers, retard rancidity of fats and lessen loss of fat-soluble vitamins (A, D, E, K). Antioxidants also are added to rubber, motor lubricants, and other materials to inhibit deterioration.

Antiparticle. A particle which interacts with its counterpart of the same mass but opposite electric charge and magnetic properties (e.g., proton and antiproton), with complete annihilation of both and production of an equivalent amount of radiation energy. The positron and its antiparticle, the electron, annihilate each other upon interaction and produce gamma-rays.

Antiseptic. A substance that prevents or inhibits the growth of microorganisms; a substance used to kill microorganisms on animate surfaces, such as skin.

Aplastic anemia. A condition in which the bone marrow fails to produce an adequate number of red blood corpuscles.

Approved. Tested and listed as satisfactory by an authority having jurisdiction, such as U.S. Department of HHS, NIOSH-MSHA; or U.S. Department of Agriculture.

Aqueous humor. Fluid in the anterior chamber of the eye.

Arc welding. One form of electrical welding using either uncoated or coated rods.

Arc-welding electrode. A component of the welding circuit through which current is conducted between the electrode holder and the arc.

Argyria. A slate-gray or bluish discoloration of the skin and deep tissues, due to the deposit of insoluble albuminate of silver, occurring after the medicinal administration for a long period of a soluble silver salt; formerly fairly common after the use of insufflations of silver-containing materials into the nose and sinuses. Also seen with occupational exposure to silver-containing chemicals.

Aromatic. Applied to a group of hydrocarbons and their derivatives characterized by presence of the benzene nucleus (molecular ring structure). See Aliphatic.

Arthr- (prefix). Joint. Arthropathy is disease affecting a joint.

Artificial abrasive. Materials, such as carborundum or emery, substituted for natural abrasive such as sandstone.

Artificial radioactivity. That produced by bombardment of a target element with nuclear particles. Iodine-131 is an artificially produced radioactive substance.

Asbestos. A hydrated magnesium silicate in fibrous form.

Asbestosis. A disease of the lungs caused by the inhalation of fine airborne fibers of asbestos.

Asepsis. Clean and free of microorganisms.

Aseptic technique. The performance of a procedure or operation in a manner that prevents the introduction of septic material.

Aspect ratio. Length to width ratio.

Asphyxia. Suffocation from lack of oxygen. Chemical asphyxia is produced by a substance, such as carbon monoxide, that combines with hemoglobin to reduce the blood's capacity to transport oxygen. Simple asphyxia is the result of exposure to a substance, such as methane, that displaces oxygen.

Asthma. Constriction of the bronchial tubes, in response to irritation, allergy, or other stimulus.

Ataxia. Lack of muscular coordination due to any of several nervous system diseases.

Atmospheric pressure. The pressure exerted in all directions by the atmosphere. At sea level, mean atmospheric pressure is 29.92 inches Hg, 14.7 psi, or 407 inches w.g.

Atmospheric tank. A storage tank which has been designed to operate at pressures from atmopheric through 0.5 psig (3.5 kPa).

Atom. All materials are made of atoms. The elements, such as iron, lead and sulfur, differ from each other because their atomic structures are different. The word "atom" comes from the Greek work meaning indivisible. Now we know it can be split and consists of an inner core (nucleus) surrounded by electrons which rotate around the nucleus. As a chemical unit, it remains unchanged during any chemical reaction, yet may undergo nuclear changes (transmutations) to other atoms as in atomic fission.

Atom smasher. Accelerator that speeds up atomic and subatomic particles so that they can be used as projectiles to literally blast apart the nuclei of other atoms.

Atomic energy. Energy released in nuclear reactions. Of particular interest is the energy released when a neutron splits an atom's nucleus into smaller pieces (fission) or when two nuclei are joined together under millions of degrees of heat (fusion). "Atomic energy" is really a popular misnomer. It is more correctly called "nuclear energy."

Atomic hydrogen welding. A shielded gas-electric welding process using hydrogen as the reducing atmosphere.

Atomic number. The number of protons (positively charged particles) found in the nucleus of an atom. All elements have different atomic numbers. The atomic number of hydrogen is 1, that of oxygen 8, iron 26, lead 82, uranium 92. The atomic number is also called charge number and is usually denoted by Z.

Atomic power. The name given to the production of thermal power in a nuclear reactor or power plant.

Atomic waste. The radioactive ash produced by the splitting of uranium (nuclear) fuel, as in a nuclear reactor. It may include products made radioactive in such a device.

Atomic weight. The atomic weight is approximately the sum of the number of protons and neutrons found in the nucleus of an atom. This sum is also called mass number. The atomic weight of oxygen, for example, is approximately 16, with most oxygen atoms containing 8 neutrons plus 8 protons. Aluminum is 27, it contains 14 neutrons and 13 protons.

Atrophy. Arrested development or wasting away of cells and tissue.

Attenuate. To reduce in amount. Usually refers to noise or ionizing radiation.

Attenuation. The reduction of the intensity at a designated first location as compared with intensity at a second location, which is farther from the source.

Attenuation block. A block or stack, having dimensions 20 cm by 20 cm by 3.8 cm, of Type 1100 aluminum alloy or aluminum alloy having equivalent attenuation.

Audible range. The frequency range over which normal ears hear—approximately 20 Hz through 20,000 Hz. Above the range of 20,000 Hz, the term ultrasonic is used. Below 20 Hz, the term subsonic is used.

Audible sound. Sound containing frequency components lying between 20 and 20,000 Hz.

Audiogram. A record of hearing loss of hearing level measured at several different frequencies—usually 500 to 6,000 Hz. The audiogram may be presented graphically or numerically. Hearing level is shown as a function of frequency.

Audiologist. A person with graduate training in the specialized problems of hearing and deafness.

Audiometer. A signal generator or instrument for measuring objectively the sensitivity of hearing in decibels referred to audiometric zero. Pure tone audiometers are standard instruments for industrial use for audiometric testing.

Audiometric technician. A person who is trained and qualified to administer audiometric examinations.

Audiometric zero. The threshold of hearing—0.0002 microbars of sound pressure. See Decibel.

Auditory. Pertaining to, or involving, the organs of hearing or the sense of hearing.

Auricle. Part of the ear that projects from the head; medically, the pinna. Also, one of the two upper chambers of the heart.

Autoclave. An apparatus using pressurized steam for sterilization.

Autoignition temperature. The lowest temperature at which a flammable gas- or vapor-air mixture will ignite from its own heat source or a contacted heated surface without necessity of spark or flame. Vapors and gases will spontaneously ignite at a lower temperature in oxygen than in air, and their autoignition temperature may be influenced by the presence of catalytic substances.

Avogadro's number. The number of molecules in a mole of any substance; it equals 6.02217×10^{23}. At 0 C and 29.92 inches Hg, 1 mole of any gas occupies 22.414 liters of volume.

Axis of rotation. The true line about which angular motion takes place at any instant. Not necessarily identical with anatomical axis of symmetry of a limb, nor necessarily fixed. Thus, forearm rotates about an axis which extends obliquely from lateral side of elbow to a point between the little finger and ring finger. The elbow joint has a fixed axis maintained by circular joint surfaces, but the knee has a moving axis as its cam-shaped surfaces articulate. Axis of rotation of tools should be aligned with true limb axis of rotation. System of rotation of tools should be aligned with true limb axis of rotation. Systems of predetermined motion times often specify such axis incorrectly.

Axis of thrust. The line along which thrust can be transmitted safely. In the forearm, it cooincides with the longitudinal axis of the radius. Tools should be designed to align with this axis.

B

Babbit. An alloy of tin, antimony, copper, and lead used as a bearing metal.

Babbitting. Process of applying babbitt to a bearing.

Bacillus. A rod-shaped bacterium.

Background radiation. The radiation coming from sources other than the radioactive material to be measured. This "background" is primarily due to cosmic rays which constantly bombard the earth from outer space.

Background noise. Noise coming from sources other than the particular noise source being monitored.

Bacteria. Microscopic organisms living in soil, water, organic matter, or the bodies of plants and animals characterized by lack of a distinct nucleus and lack of ability to photosynthesize. Singular: Bacterium.

Bactericide. Any agent that destroys bacteria.

Bacteriophage. Viruses that infect bacteria and lyse the bacterial cell.

Bacteriostat. An agent that stops the growth and multiplication of bacteria but does not necessarily kill them. Usually growth resumes when the bacteriostat is removed.

Bag house. Many different trade meanings. Term commonly used for the housing containing bag filters for recovery of fumes of arsenic, lead, sulfur, etc., from flues of smelters.

Bagasse. Sugar cane pulp residues.

Bagassosis. Respiratory disorder believed to be caused by breathing fungi found in bagasse.

Balancing by dampers. Method for designing local exhaust system ducts using adjustable dampers to distribute airflow after installation.

Balancing by static pressure. Method for designing local exhaust system ducts by selecting the duct diameters that generate the static pressure to distribute airflow without dampers.

Ball mill. A grinding device using balls usually of steel or stone in a revolving container.

Banbury mixer. A mixing machine which permits control over the temperature of the batch; commonly used in the rubber industry.

Band-pass filter. A wave filter that has a single transmission band extending from a lower cutoff frequency greater than zero to a finite upper cutoff frequency.

Band pressure level of a sound for a specified frequency band is the sound pressure level for the sound contained within the restricted band. The reference pressure must be specified.

Bandwidth. When applied to a band-pass filter, bandwidth is determined by the interval of transmitted waves between the low and high cutoff frequencies.

Baritosis. An inert pneumoconiosis produced by the inhalation of insoluble barium compounds.

Barotrauma. An injury to the ear caused by a sudden alteration in barometric (atmospheric) pressure, Aerotitis.

Basal metabolism. A measure of the amount of energy required by the body at rest.

Base. A compound that reacts with an acid to form a salt. It is another term for alkali. It turns litmus paper blue.

Basilar. Of, relating to, or situated at the base.

Bauxite. Impure mixture of aluminum oxides and hydroxides; the principal source of aluminum.

Bauxite pneumoconiosis. Shaver's disease. Found in workers exposed to fumes containing aluminum oxide and minute silica particles arising from smelting bauxite in the manufacture of corundum.

Beam axis. A line from the source through the centers of the x-ray fields.

Beam divergence. Angle of beam spread measured in milliradians (1 milliradian—3.4 minutes of arc).

Beam-limiting device. A device which provides a means to restrict the dimensions of the x-ray field.

Beat elbow. Bursitis of the elbow; occurs from use of heavy vibrating tools.

Beat knee. Bursitis of the knee joints due to friction or vibration; common in mining.

Beehive kiln. A kiln shaped like a large beehive, used usually for calcining ceramics.

Bel. A unit of sound level based on a logarithmic scale.

Belding-Hatch index. Estimate of the body heat stress of an average man (standard man) for various degrees of activity; also relates to his sweating capcity.

Benign. Not malignant. A benign tumor is one which does not metastasize or invade tissue. Benign tumors may still be lethal, due to pressure on vital organs.

Benzene, C_6H_6. A major organic intermediate and solvent derived from coal or petroleum. The simplest member of the aromatic series of hydrocarbons.

Beryl. A silicate of beryllium and aluminum.

Berylliosis. Chronic beryllium intoxication.

Beta decay. The process whereby some radioactive emitters give off a beta particle. Also called beta disintegration.

Beta particle (beta-radiation). A small electrically charged particle thrown off by many radioactive materials. It is identical with the electron. Beta particles emerge from radioactive material at high speeds.

Betatron. A large doughnut-shaped accelerator in which electrons (beta particles) are whirled through a changing magnetic field gaining speed with each trip and emerging with high energies. Energies of the order of 100 million electron volts have been achieved. The betatron produces artificial beta radiation.

Biceps brachii muscle. The large muscle in the front of the upper arm. Supinates (turns the palm up) the forearm.

Bicipital tuberosity. A protuberance on the medial surface of the radius to which the biceps brachii attaches.

Billet. A piece of semifinished iron or steel, nearly square in section, made by rolling and cutting an ingot.

Binder. The nonvolatile portion of a coating vehicle which is the film-forming ingredient used to bind the paint pigment particles together.

Binding energy. The energy which holds the neutrons and protons of an atomic nucleus together. Represents the difference between the mass of an atom and the sum of the masses of protons and neutrons that make up its nucleus.

Biohazard. A combination of the words biological hazard. Organisms or products of organisms that present a risk to humans.

Biohazard area. Any area (a complete operating complex, a single facility, a room within a facility, etc.) in which work has been or is being performed with biohazardous agents or materials.

Biohazard control. Any set of equipment and procedures utilized to prevent or minimize the exposure of humans and their environment to biohazardous agents or materials.

Biological half-life. The time required to reduce the amount of an exogenous substance in the body by half.

Biological oxygen demand (BOD). Quantity of oxygen required for the biological and chemical oxidation of waterborne substances under test condition.

Biomechanics. The study of the human body as a system operating under two sets of law: the laws of Newtonian mechanics and the biological laws of life.

Biopsy. Careful removal of small bits of living tissue from the body for further study and examination, usually under the microscope.

Black light. Ultraviolet (UV) light radiation between 3,000 and 4,000 angstroms (0.3-0.4 micrometers); it is responsible for pigmentation of the skin following exposure to UV light.

Black liquor. A liquor composed of alkaline and organic matter resulting from digestion of wood pulp and cooking acid during the manufacture of paper.

Bleaching bath. Chemical solution used to bleach colors from a garment preparatory to dyeing it, a solution of chlorine or sodium hypochorite commonly being used.

Bleph- (prefix). Pertaining to the eyelid.

Blind spot. Normal defect in visual field due to position at which optic nerve enters the eye.

Blood count. A count of the number of corpuscles per cubic millimeter of blood. Separate counts may be made for red and white corpuscles.

Body burden. The amount of noxious material in the body at a given time.

Body burden, maximum permissible. That body burden of a radionuclide which, if maintained at a constant level, would produce the maximum permissible dose equivalent in the critical organ.

Boiling point. The temperature at which the vapor pressure of a liquid equals atmospheric pressure.

Bombardment. Shooting neutrons, alpha particles, and other high-energy particles at atomic nuclei usually in an attempt to split the nucleus or to form a new element.

Bone conduction test. A special test conducted by placing an oscillator on the mastoid process to determine the nerve-carrying capacity of the cochlea and the eighth cranial (auditory) nerve.

Bone marrow. A soft tissue which constitutes the central filling of many bones and serves as a producer of blood corpuscles.

Bone-seeker. Any element or radioactive species that predominantly lodges in the bone when introduced into the body.

Brachialis muscle. Short, strong muscles originating at lower end of the humerus and inserting into ulna. Powerful flexor of forearm, employed when lifting.

Brady- (prefix). Slow. Bradycardia is slow heartbeat.

Bradycardia. Abnormal slowness of the heartbeat, as evidenced by slowing of the pulse rate to 50 or less.

Brake horsepower. The horsepower required to drive a unit; it includes the energy losses in the unit and can be determined only by actual test. It does not include drive losses between motor and unit.

Branch (or path) of greatest resistance. The path from a hood to the fan and exhaust stack in a ventilation system that causes the most pressure loss.

Brass. An alloy of copper and zinc and may contain a small portion of lead.

Brattice. A partition constructed in underground passageways to control ventilation in mines.

Braze. To solder with any alloy that is relatively infusible.

Brazing furnace. Used for heating metals to be joined by brazing. Requires a high temperature.

Breathing tube. A tube through which air or oxygen flows to the facepiece, helmet, or hood.

Breathing zone. Imaginary globe of two foot radius surrounding the head.

Breathing zone sample. An air-sample collected in the breathing zone of workers to assess their exposure to airborne contaminants.

Bremsstrahlung. Secondary x-radiation produced when a beta particle is slowed down or stopped by a high-density surface.

Briquette. Coal or ore dust pressed into oval or brick-shaped blocks.

Broach. A cutting tool for cutting nonround holes.

Bronch-, broncho- (prefix). Pertaining to the air tubes of the lung.

Bronchial tubes. Branches or subdivisions of the trachea (windpipe). A bronchiole is a branch of a bronchus, which is a branch of the windpipe.

Bronchiectasis. A chronic dilation of the bronchi or bronchioles marked by fetid breath and paroxysmal coughing, with the expectoration of mucopurulent matter. It may affect the tube uniformly, or may occur in irregular pockets, or the dilated tubes may have terminal bulbous enlargements.

Bronchiole. The slenderest of the many tubes that carry air into and out of the lungs.

Bronchiolitis. Bronchopneumonia. (Fibrosa obliterans—bronchiolitis marked by ingrowth of connective tissue from the wall of the terminal bronchi with occlusion of their lumina.)

Bronchitis. Inflammation of the bronchi or bronchial tubes.

Bronchoalveolitis. Bronchopneumonia.

Bronchopneumonia. A name given to an inflammation of the lungs which usually begins in the terminal bronchioles. These become clogged with a mucopurulent exudate forming consolidated patches in adjacent lobules. The disease is essentially secondary in character, following infections of the upper respiratory tract, specific infectious fevers, and debilitating diseases.

Bronzing. Act or art of imparting a bronze appearance as by powders, painting or chemical process.

Brownian motion. The irregular movement of particles suspended in a fluid as a result of bombardment by atoms and molecules.

Brucella. A genus of short, rod-shaped to coccoid, encapsulated, gram-negative, parasitic, and pathogenic bacteria.

Brucellosis. A group of diseases caused by an organism of the *Brucella* genus. Undulant fever. One source is unpasturized milk from cows suffering from Bang's disease (infectious abortion).

Bubble chamber. A chamber containing a liquefied gas such as liquid hydrogen, under conditions such that a charged particle passing through the liquid forms bubbles to make its path visible.

Bubble tube. A device used to calibrate air-sampling pumps.

Buffer. Any substance in a fluid which tends to resist the change in pH when acid or alkali is added.

Bulk plant shall mean that portion of a property where flammable or combustible liquids are received by tank vessel, pipelines, tank car, or tank vehicle, and are sorted or blended in bulk for the purpose of distributing such liquids by tank vessel, pipeline, tank car, tank vehicle, or container.

Burn-up. The extent to which the nuclear fuel in a fuel element has been consumed by fission, as in a nuclear reactor.

Burns. Result from the application of too much heat to the skins. First degree burns show redness of the unbroken skin; second degree, skin blisters and some breaking of the skin; third degree, skin blisters and destruction of the skin and underlying tissues, which can include charring and blackening.

Burr. The thin rough edges of a machined piece of metal.

Bursa. A synovial lined sac that facilitates the motion of tendons, usually near a joint.

Bursitis. Inflammation of a bursa.

Byssinosis. Disease occurring to those who experience prolonged exposure to heavy air concentrations of cotton or flax dust.

C

Calcination. The heat treatment of solid material to bring about thermal decomposition, to lose moisture or other volatile material, or be oxidized or reduced.

Calender. An assembly of rollers for producing a desired finish on paper, rubber, artificial leather, plastic, or other sheet material.

Calking. The process or material used to fill seams of boats, cracks in tile, etc.

Calorimeter. A device for measuring the total amount of energy absorbed from a source of electromagnetic radiation.

Cancer. A cellular tumor the natural course of which is fatal and usually associated with formation of secondary tumors.

Capitulum of humerus. A smooth hemispherical protuberance at the distal end of the humerus articulating with the head of the radius. Irritation caused by pressure between the capitulum and head of the radius may be a cause of tennis elbow.

Capture velocity. Air velocity at any point in front of the hood necessary to overcome opposing air currents and to capture the contaminated air by causing it to flow into the exhaust hood.

Carbohydrate. An abundant class of organic compounds, serving as food reserves or structural elements for plants and animals. Compounded primarily of carbon, hydrogen, and oxygen, they constitute about two thirds of the average daily adult caloric intake. Sugar, starches, and plant components (cellulose) are all carbohydrates.

Carbon black. Essentially a pure carbon, best known as common soot. Commercial carbon black is produced by making soot under controlled conditions. It is sometimes called channel black, furnace black, acetylene black, or thermal black.

Carbon monoxide. A colorless, odorless toxic gas produced by any process that involves the incomplete combustion of carbon-containing substances. It is emitted through the exhaust of gasoline powered vehicles.

Carbonizing. The immersion in sulfuric acid of semiprocessed felt to remove any vegetable matter present; other applications.

Carborundum. A trade name for silicon carbide, widely used as an abrasive.

Carboy. A large glass bottle, usually protected by a crate.

Carboxyhemoglobin. The reversible combination of carbon monoxide with hemoglobin.

Carcinogenic. Cancer producing.

Carcinoma. Malignant tumors derived from epithelial tissues, that is, the outer skin, the membranes lining the body cavities, and certain glands.

Cardi-, cardio- (prefix). Denoting the heart.

Cardiac. (a) Pertaining to the heart; (b) a cordial or restorative medicine; (c) a person with heart disorder.

Carding. The process of combing or untangling wool, cotton, etc.

Carding machine. In the textile industry, prepares wool, cotton or other fiber for spinning.

Cardiovascular. Relating to the heart and the blood vessels or the circulation.

Carp- (prefix). The wrist.

Carpal tunnel. A passage in the wrist through which the median nerve and many tendons pass to the hand from the forearm.

Carpal tunnel syndrome. A common affliction caused by compression of the median nerve in the carpal tunnel. Often associated with tingling, pain, or numbness, in the thumb and first three fingers—may be job-related.

Carrier. A person in apparent good health who harbors a pathogenic microorganism.

Carrier gas. The mixture of gases which contains and moves the contaminant material. Components of the carrier gas are not considered to cause air pollution or react with the contaminant material.

CAS number. Identifies a particular chemical by the Chemical Abstract Service, a service of the American Chemical Society that indexes and compiles abstracts of worldwide chemical literature called "Chemical Abstracts."

Case-hardening. A process of surface-hardening metals by raising the carbon or nitrogen content of the outer surface.

Cask (or coffin). A thick-walled container (usually lead) used for transporting radioactive materials.

Casting. The pouring of a molten material into a mold and permitting it to solidify to the desired shape.

Catalyst. A substance which changes the speed of a chemical reaction but undergoes no permanent change itself. In respirator use, a substance which converts a toxic gas (or vapor) into a less-toxic gas (or vapor). Usually catalysts greatly increase the reaction rate, as in conversion of petroleum to gasoline by cracking. In paint manufacture, catalysts, which hasten the film-forming, generally become part of the final product. In most uses, however, they do not, and can often be used over again.

Cataract. Opacity in the lens of the eye that may obscure vision.

Cathode. The negative electrode.

Catwalk. A narrow suspended footway usually for inspection or maintenance purposes.

Caustic. Something that strongly irritates, burns, corrodes, or destroys living tissue. See Alkali.

Ceiling Limit (C). An airborne concentration of a toxic substance in the work environment, which should never be exceeded.

Cell. A structural unit of which tissues are made. There are many types—nerve cells, muscle cells, blood cells, connective tissues cells, fat cells, and other. Each has a special form to serve a particular function.

Cellulose ($C_6H_{10}O_5$). A carbohydrate which makes up the structural material of vegetable tissues and fibers. Purest forms: chemical cotton and chemical pulp. Basis of rayon, acetate, and cellophane.

Celsius. The Celsius temperature scale is a designation of the scale previously known as the centigrade scale.

-cele (suffix). Swelling of herniation of a part, as rectocele, prolapse of the rectum.

Cement, portland. Portland cement commonly consists of hydraulic calcium silicates to which the addition of certain materials in limited amounts is permitted. Ordinarily, the mixture consists of calcareous materials such as limestone, chalk, shells, marl, and clay, shale, blast furnace slag, etc. In some specifications, iron ore and limestone are added. The mixture is fused by calcining at temperatures usually up to 1,500 C.

Centrifuge. An apparatus that uses centrifugal force to separate or remove particulate matter suspended in a liquid.

Cephal- (prefix). Pertaining to the head. Encelphal-, "within the head," pertains to the brain.

Ceramic. A term applied to pottery, brick and the tile products molded from clay and subsequently calcined.

Cerumen. Earwax.

Cervi- (prefix). A neck.

CFR. See *Code of Federal Regulations.*

Chain reaction. When a fissionable nucleus is split by a neutron it releases energy and one or more neutrons. These neutrons split other fissionable nuclei releasing more energy and more neutrons making the reaction self-sustaining, as long as there are enough fissionable nuclei present.

Charged particles. A particle which possesses at least a unit electrical charge and which will not disintegrate upon loss of charge. Charged particles are characterized by particle size, number and sign of unit charges and mobility. Also see Ion.

Chelating agent or chelate. (Derived from Greek word kelos for claw.) Any compound which will inactivate a metallic ion with the formation of an inner ring structure in the molecule, the metal ion becoming a member of the ring. The original ion, thus chelated, is effectively out of action.

Chemical cartridge. The type of absorption unit used with a respirator for removal of low concentrations of specific vapors and gases.

Chemical engineering. That branch of engineering concerned with the development and application of manufacturing processes in which chemical or certain physical changes of materials are involved. These processes may usually be resolved into a coordinated series of unit physical operations and unit chemical processes. The work of the chemical engineer is concerned primarily with the design, construction, and operation of equipment and plants in which these unit operations and processes are applied.

Chemical burns. Generally similar to those caused by heat. After emergency first aid, their treatment is the same as that for thermal burns. In certain instances, such as with hydrofluoric acid, special treatment is indicated.

Chemical reaction. A change in the arrangement of atoms or molecules to yield substances of different composition and properties. Common types of reaction are combination, decomposition, double decomposition, replacement, and double replacement.

Chemotherapy. Use of chemicals of particular molecular structure in the treatment of specific disorders on the assumption that known structures exhibit an affinity for certain parts of malignant cells or infectious organisms, and thereby tend to destroy or inactivate them.

Chert. A microcrystalline form of silica. An impure form of flint used in abrasives.

Cheyne-Stokes respiration. The peculiar kind of breathing usually observed with unconscious or sleeping individuals, who seem to stop breathing altogether for 5–40 seconds, then start up again with gradually increasing intensity, stop breathing once more, and then repeat the performance. (Common in healthy infants.)

Chloracne. Caused by chlorinated naphthalenes and polyphenyls acting on sebaceous glands.

Chol-, chole- (prefix). Relating to bile. Cholesterol is a substance found in bile.

Chon-, chondro- (prefix). Cartilage.

Chromatograph. An instrument which can separate and analyze mixtures of chemical substances.

Chromosome. Important rod-shaped constituent of all cells. Chromosomes contain the genes or heredity-determining units, and are made up of deoxyribonucleic acids (DNA).

Chronic. Persistent, prolonged, repeated.

Cilia. Tiny hairlike "whips" in the bronchi and other respiratory passages that aid in the removal of dust trapped on these moist surfaces.

Ciliary. Pertaining to the cilium (pl. cilia), a minute vibratile hairlike process attached to the free surface of a cell.

Clays. A great variety of aluminum-silicate-bearing rocks, plastic when wet, hard when dry. Used in pottery, stoneware, tile, bricks, cements, fillers, and abrasives. Kaolin is one type of clay. Some clay deposits may include appreciable quartz. Commercial grades of clays may contain up to 20 percent quartz.

Clostridium botulinum. A human pathogenic bacteria that produce an exotoxin, botulinin, which causes botulism.

Cloud chamber. A glass-domed chamber filled with moist vapor. When certain types of atomic particles pass through the chamber they leave a cloudlike track much like the vapor trail of a jet plane. This permits scientists to "see" these particles and study their motion. The cloud chamber and the bubble chamber serve the same purpose.

CNS. Central nervous system.

Coagulase. An enzyme produced by pathogenic staphylococci; causes coagulation of blood plasma.

Coagulation. Formation of a clot or gelatinous mass.

Coalesce. To unite into a whole; to fuse; to grow together.

Coated welding rods. The coatings of welding rods vary. For the welding of iron and most steel, the rods contain manganese, titanium and a silicate.

Coccus. A spherical bacterium. Plural: cocci.

Cochlea. The auditory part of the internal ear, shaped like a snail shell. It contains the basilar membrane on which the end organs of the auditory nerve are distributed.

Code of Federal Regulations. The rules promulgated under U.S. law and published in the *Federal Register* and actually in force at the end of a calendar year are incorporated in this code (CFR).

Coefficient of discharge. A factor used in figuring flow through an orifice. The coefficient takes into account the facts that a fluid flowing through an orifice will contract to a cross-sectional area which is smaller than that of the orifice, and there is some dissipation of energy due to turbulence.

Coefficient of entry. The actual rate of flow caused by a given hood static pressure compared to the theoretical flow that would result if the static pressure could be converted to velocity pressure with 100 percent efficiency; it is the ratio of actual to theoretical flow.

Coefficent of variation. The ratio of the standard deviation to the mean value of a population of observations.

Coffin. A thick-walled container (usually lead) used for transporting radioactive materials.

Cohesion. Molecular forces of attraction between particles of like compositions.

Colic. A severe, cramping, gripping pain in or referred to the abdomen.

Collagen. An albuminoid, the main supportive protein of skin, tendon, bone, cartilage, and connective tissue.

Collection efficiency. The percentage of a specific substance removed and retained from air by an air cleaning or sampling device. A measure of the cleaner or sampler performance.

Collimated beam. A beam of light with parallel waves.

Colloid. Generally a liquid mixture or suspension in which the particles of suspended liquid or solid are very finely divided. Colloids do not settle out of suspension appreciably.

Colloid mill. A machine that grinds materials into very fine state of suspension, often simultaneously placing this in suspension in a liquid.

Colorimetry (colorimetric). The term applied to all chemical analysis techniques involving reactions in which a color is developed when a particular contaminant is present in the sample and reacts with the collection medium. The resultant color intensity is measured to determine the contaminant concentration.

Coma. A level of unconsciousness from which a patient cannot be aroused.

Combustible liquids. Combustible liquids are those having a flash point at or above 37.8 C (100F).

Comfort ventilation. Airflow intended to maintain comfort of room occupants (heat, humidity, and odor).

Comfort zone. Average—the range of effective temperatures over which the majority (50 percent or more) of adults feels comfortable.

Communicable. Refers to a disease whose causative agent is readily transferred from one person to another.

Compaction. The consolidation of solid particles between rolls, or by tamp, piston, screw, or other means of applying mechanical pressure.

Compound. A substance composed of two or more elements joined according to the laws of chemical combination. Each compound has its own characteristic properties different from those of its constituent elements.

Compressible flow. Flow of high-pressure gas or air which undergoes a pressure drop sufficient to result in a significant reduction of its density.

Compton effect. The glancing collision of a gamma-ray with an electron. The gamma-ray gives up part of its energy to the electron.

Concentration. The amount of a given substance in a stated unit of measure. Common methods of stating concentration are percent by weight or by volume, weight per unit volume, normality, etc.

Condensate. The liquid resulting from the process of condensation. In sampling, the term is generally applied to the material that is removed from a gas sample by means of cooling.

Condensation. Act or process of reducing from one form to another denser form such as steam to water.

Condensoid. A dispersoid consisting of liquid or solid particles which have been formed by the process of condensation. The dispersoid is commonly referred to as a condensation aerosol.

Conductive hearing loss. Type of hearing loss; not caused by noise exposure, but due to any disorder in the middle or external ear that prevents sound from reaching the inner ear.

Confined space. An enclosure that is difficult to get out of and limited or no ventilation. Examples are storage tanks, boilers, sewers, and tank cars.

Congenital. Some problem that originates before birth.

Conjunctiva. The delicate mucous membrane that lines the eyelids and covers the exposed surface of the eyeball.

Conjunctivitis. Inflammation of the conjunctiva.

Contact dermatitis. Dermatitis caused by contact with a substance—gaseous, liquid, or solid. May be due to primary irritation or an allergy.

Control rod. A rod (containing an element such as boron) used to control the power of a nuclear reactor. The control rod absorbs neutrons that would normally split the fuel nuclei. Pushing the rod in reduces the release of atomic power; pulling out the rod increases it.

Controlled areas. A specified area in which exposure of personnel to radiation or radioactive material is controlled and which is under the supervision of a person who has knowledge of the appropriate radiation protection practices, including pertinent regulations, and who has responsibility for applying them.

Convection. The motions in fluids resulting from the differences in density and the action of gravity.

Converter. A nuclear reactor that uses one kind of fuel and produces another. For example, a converter charged with uranium isotopes might consume uranium-235 and produce plutonium from uranium-238. A breeder reactor produces more atomic fuel than it consumes. A converter does not.

Coolants. Coolants are transfer agents used in a flow system to convey heat from its source.

Copolymers. Mixed polymers or heteropolymers. Products of the polymerization of two or more substances at the same time.

Core. (1) The heart of a nuclear reactor where the nuclei of the fuel fission (split) and release energy. The core is usually surrounded by a reflecting material which bounces stray neutrons back to the fuel. It is usually made up of fuel elements and a moderator. (2) Also, a shaped, hard-baked cake of sand with suitable compounds that is placed within a mold, forming a cavity in the casting when it solidifies. (3) The vital centers of the body—heart, viscera, brain—as opposed to the shell—the limbs and integument.

Corium. The deeper skin layer containing the fine endings of the nerves and the finest divisions of the blood vessels, the capillaries. Also called the derma.

Cornea. Transparent membrane covering the anterior portion of the eye.

Corpuscle. A red or white blood cell.

Corrective lens. A lens ground to the wearer's individual prescription.

Corrosion. Physical change, usually deterioration or destruction, brought about through chemical or electrochemical action as contrasted with erosion caused by mechanical action.

Corundum. An impure form of aluminum oxide.

Cosmic rays. High-energy rays which bombard the earth from outer space. Some penetrate to the earth's surface and others may go deep into the ground. Although each ray is energetic, the number bombarding the planet is so small that the total energy reaching the earth is about the same as that from starlight.

Corrosive. A substance that causes visible destruction or permanent changes in human skin tissue at the site of contact.

Costo-, costal (prefix). Pertaining to the ribs.

Cottrell precipitator. A device for dust collection using high voltage electrodes.

Coulometry. The measurements of the number of electrons that are transferred across an electrode solution interface when a reaction in the solution is created and carried to completion. The reaction is usually caused by a contaminant which is in a sample gas which is drawn through or onto the surface of the solution. The number of electrons transferred in terms of coulombs is an indication of the contaminant concentrations.

Count. May be a click in a Geiger counter or it may refer to the numerical value for the activity of a radioactive specimen.

Counter. A device for counting. See Geiger counter and Scintillation counter.

Count median size. The numerical size of the particle in a sample of particulate matter of which there are contained an equal number of particles larger and smaller than the stated size.

Covered electrode. A composite filler metal electrode consisting of a core of bare electrode or metal-cored electrode to which a covering (sufficient to provide a slag layer on the weld metal) has been applied—the covering may contain materials providing such functions as shielding from the atmosphere, deoxidation, and arc stabilization and can serve as a source of metallic addtions to the weld.

Cps. Cycles per second, now called hertz.

Cracking. Used almost exclusively in the petroleum industry, cracking is thermal or catalytic decomposition of organic compounds usually for the manufacture of gasoline. Petroleum constituents are also cracked for the purpose of manufacturing chemicals.

Cramps. Painful muscular contractions that may affect almost any voluntary or involuntary muscle.

Cranio- (prefix). Skull. As in craniotomy, incision through a skull bone.

Cristobalite. A crystalline form of free silica, extremely hard and inert chemically; very resistant to heat. Quartz in refractory bricks and amorphous silica in diatomaceous earth are altered to cristobalite when exposed to high temperatures (calcined).

Critical mass. The amount of nuclear fuel necessary to sustain a chain reaction. If too little fuel is present, too many neutrons will stray and the reaction will die out.

Critical pressure. The pressure under which a substance may exist as a gas in equilibrium with the liquid at the critical temperature.

Critical temperature. The temperature above which a gas cannot be liquefied by pressure alone.

Crucible. A heat-resistant barrel-shaped pot used to hold metal during melting in a furnace. Other applications.

Crude petroleum. Hydrocarbon mixtures that have a flash point below 65.6 C (150 F), and that have not been processed in a refinery.

Cry-, cryo- (prefix). Very cold.

Cryogenics. The field of science dealing with the behavior of matter at very low temperatures.

Cubic centimeter (cc). Cubic centimeter, a volumetric measurement that is also equal to one milliliter (mL).

Culture (biology). A population of microorganisms or tissue cells cultivated in a medium.

Culture medium. Any substance or preparation suitable for and used for the growth of cultures and cultivation of microorganisms. Selective medium is a medium composed of nutrients designed to allow growth of a particular type of microorganism; broth medium, a liquid medium; agar medium, solid culture medium.

Cubic meter (m^3). A measure of volume in the metric system.

Curie. A measure of the rate at which a radioactive material decays. The radioactivity of one gram of radium is a curie. It is named for Pierre and Marie Curie, pioneers in radioactivity and discoverers of the elements radium, radon, and polonium. One curie corresponds to 37 billion disintegrations per second.

Cutaneous. Pertaining to or affecting the skin.

Cuticle. The superficial scarfskin or upper strata of skin.

Cutie-pie. A portable instrument equipped with a direct reading meter used to determine the level of ionizing radiation in an area.

Cutting fluids (oils). The cutting fluids used in industry today are usually an oil or an oil-water emulsion used to cool and lubricate a cutting tool. Cutting oils are usually light or heavy petroleum fractions.

CW laser. Continuous wave laser.

Cyan- (prefix). Blue.

Cyanide (as CN). Cyanides inhibit tissue oxidation upon inhalation or ingestion and cause death.

Cyanosis. Blue appearance of the skin, especially on the face and extremities, indicating a lack of sufficient oxygen in the arterial blood.

Cyclone separator. A dust-collecting device which has the ability to separate particles by size. Typically used to collect respirable dust samples.

Cyclotron. A particle accelerator. In this atomic "merry-go-round" atomic particles are whirled around in a spiral between the ends of a huge magnet gaining speed with each rotation in preparation for their assault on the target material.

Cyst- (prefix). Pertaining to a bladder or sac, normal or abnormal, filled with gas, liquid, or semisolid material. The term appears in many words concerning the urinary bladder (cystocele, cystitis).

Cyto- (prefix). Cell. Leukocytes are white blood cells.

Cytoplasm. Cell plasma (protoplasm) which does not include the cell's nucleus.

Cytotoxin. A substance, developed in the blood serum, having a toxic effect upon cells.

D

Damage risk criterion. The suggested base line of noise tolerance, which if not exceeded, should result in no hearing loss due to noise. A damage risk criterion may include in its statement a specification of such factors as time of exposure, noise level, and frequency, amount of hearing loss that is considered significant, percentage of the population to be protected, and method of measuring the noise.

Damp. A harmful gas or mixture of gases occurring in coal mining.

Dampers. Adjustable sources of airflow resistance used to distribute airflow in a ventilation system.

Dangerous to life or health, immediately (IDLH). Used to describe very hazardous atmospheres where employee exposure can cause serious injury or death within a short time or serious delayed effects.

Daughter. As used in radioactivity, this refers to the product nucleus or atom resulting from decay of the precursor or parent.

dBA. Sound level in decibels read on the A-scale of a sound-level meter. The A scale discriminates against very low frequencies (as does the human ear) and is, therefore, better for measuring general sound levels. See also Decibel.

dBC. Sound level in decibels read on the C-scale of a sound-level meter. The C scale discriminates very little against very low frequencies. See also Decibel.

Decay. When a radioactive atom disintegrates it is said to decay. What remains is a different element. An atom of polonium decays to form lead, ejecting an alpha particle in the process.

Decibel (dB). A unit used to express sound power level (L_W). Sound power is the total acoustic output of a sound source in watts (W). By definition, sound power level, in decibels, is: $L_W = 10 \log W \div W_0$ where W is the sound power of the source and W_0 is the reference sound power.

Decomposition. The breakdown of a chemical or substance into different parts or simpler compounds. Decomposition can occur due to heat, chemical reaction, decay, etc.

Decontaminate. To make safe by eliminating poisonous or otherwise harmful substances, such as noxious chemicals or radioactive material.

Deltoid muscle. The muscle of the shoulder responsible for abducting the arm sideways, and for swinging the arm at the shoulder. Overuse of the deltoid muscle may cause fatigue and pain in the shoulder.

Density. The ratio of the mass to volume.

Dent-, dento- (prefix). Pertaining to a tooth or teeth; from the Latin.

Derma. The corium or true skin.

Dermatitis. Inflammation of the skin from any cause.

Dermatophytosis. Athlete's foot.

Dermatosis. A broader term than dermatitis; it includes any cutaneous abnormality, thus encompassing folliculitis, acne, pigmentary changes, and nodules and tumors.

Desiccant. Material that absorbs moisture.

Deuterium. Heavy hydrogen. The nucleus of heavy hydrogen is a deuteron. It is called heavy hydrogen because it weighs twice as much as ordinary hydrogen.

Deuteron. The nucleus of an atom of heavy hydrogen containing one proton and one neutron. Deuterons are often used for the bombardment of other nuclei.

Diagnostic x-ray system. An x-ray system designed for irradiation of any part of the human body for the purpose of diagnosis or visualization.

Diaphragm. (1) The musculomembranous partition separating the abdominal and thoracic cavities; (2) any separating membrane or structure; (3) a disk with one or more openings in it, or with an adjustable opening, mounted in relation to a lens, by which part of the light may be excluded from the area.

Diatomaceous earth. A soft, gritty amorphous silica composed of minute siliceous skeletons of small aquatic plants. Used in filtration and decolorization of liquids, insulation, filler in dynamite, wax, textiles, plastics, paint, and rubber. Calcined and flux-calcined diatomaceous earth contains appreciable amounts of cristobalite, and dust levels should be controlled the same as for cristobalite.

Die. A hard metal or plastic form used to shape material to a particular contour or section.

Differential pressure. The difference in static pressure between two locations.

Diffuse sound field. One in which the time average of the mean-square sound pressure is everywhere the same and the flow of energy in all directions is equally probable.

Diffusion, molecular. A process of spontaneous intermixing of different substances attributable to molecular motion and tending to produce uniformity of concentration.

Diffusion rate. A measure of the tendency of one gas or vapor to disperse into or mix with another gas or vapor. This rate depends on the density of the vapor or gas as compared with that of air, which is given a value of 1.

Diluent. A liquid which is blended with a mixture to reduce concentration of the active agents.

Dilution. The process of increasing the proportion of solvent or diluent (liquid) to solute or particulate matter (solid).

Dilution ventilation. See General ventilation.

Diopters. A measure of the power of a lens or prism, equal to the reciprocal of its focal length in meters.

Direct-reading instrumentation. Those instruments that give an immediate indication of the concentration of aerosols, gases, or vapors or magnitude of physical hazard by some means such as a dial or meter.

Disease. A departure from a state of health, usually recognized by a sequence of signs and symptoms.

Disinfectant. An agent that frees from infection by killing the vegetative cells of microorganisms.

Disintegration. A nuclear transformation or decay process that results in the release of energy in the form of radiation.

Dispersion. The general term describing systems consisting of particulate matter suspended in air or other fluid; also, the mixing and dilution of contaminant in the ambient environment.

Distal. Away from the central axis of the body.

Distal phalanx. The last bony segment of a toe or finger.

Distillery. A plant or that portion of a plant where flammable or combustible liquids produced by fermentation are concentrated, and where the concentrated products may also be mixed, stored, or packaged.

Diuretic. Anything that promotes excretion of urine.

DNA. Deoxyribonucleic acid. The genetic material within the cell.

DOP. Dioctyl phthalate, a powdered chemical that can be

aerosolized to an extremely uniform size, i.e., 0.3 μm for a major portion of any sample.

Dose. A term used (1) to express the amount of a chemical or of ionizing radiation energy absorbed in a unit volume or an organ or individual. Dose rate is the dose delivered per unit of time. (See also Roentgen, Rad, Rem.) (2) Used to express amount of exposure to a chemical substance.

Dose, absorbed. The energy imparted to matter in a volume element by ionizing radiation divided by the mass of irradiated material in that volume element.

Dose equivalent. The product of absorbed dose, quality factor, and other modifying factors necessary to express on a common scale, for all ionizing radiations, the irradiation incurred by exposed persons.

Dose equivalent, maximum permissible (MPD). The largest equivalent received within a specified period which is permitted by a regulatory agency or other authoritative group on the assumption that receipt of such dose equivalent creates no appreciable somatic or genetic injury. Different levels of MPD may be set for different groups within a population. (By popular usage, "dose, maximum permissible," is an accepted synonym.)

Dose-response relationship. Correlation between the amount of exposure to an agent or toxic chemical and the resulting effect on the body.

Dosimeter (dose meter). An instrument used to determine the full-shift exposure a person has received to a physical hazard.

DOT. Department of Transporation.

Drier. Any catalytic material which when added to a drying oil, accelerates drying or hardening of the film.

Drop forge. To forge between dies by a drop hammer or drop press.

Droplet. A liquid particle suspended in a gas. The liquid particle is generally of such size and density that it settles rapidly and only remains airborne for an appreciable length of time in a turbulent atmosphere.

Dross. The scum forming on the surface of molten metals, largely oxides and impurities.

Dry-bulb thermometer. An ordinary thermometer, especially one with an unmoistened bulb; not dependent on atmospheric humidity. Reading is the dry-bulb temperature.

Duct. A conduit used for conveying air at low pressures.

Duct velocity. Air velocity through the duct cross section. When solid particulate material is present in the air stream, the duct velocity must exceed the minimum transport velocity.

Ductile. Capable of being molded or worked, as metals.

Dust collector. An air-cleaning device to remove heavy particulate loadings from exhaust systems before discharge to outdoors; usual range is loadings of 0.003 gr/cu ft (0.007 mg/m^3) and higher.

Dusts. Solid particles generated by handling, crushing, grinding, rapid impact, detonation, and decrepitation of organic or inorganic materials, such as rock, ore, metal, coal, wood, and grain. Dusts do not tend to flocculate, except under electrostatic forces; they do not diffuse in air but settle under the influence of gravity.

Dynometer. Apparatus for measuring force or work output external to a subject. Often used to compare external output with associated physiological phenomena to assess physiological work efficiency.

Dyne. The force that would give a free mass of one gram an acceleration of one centimeter per second per second.

Dys- (prefix). Difficult, bad. This prefix occurs in large numbers of medical words, since it is attachable to any organ or process that is not functioning as well as it should.

Dysfunction. Disturbance, impairment, or abnormality of the functioning of an organ.

Dyspnea. Shortness of breath, difficult or labored breathing. More strictly, the sensation of shortness of breath.

Dysuria. Difficulty or pain in urination.

E

EAP. Employee Assistance Program.

Ear. The entire hearing apparatus, consisting of three parts: external ear, the middle ear or tympanic cavity, and the inner ear or labyrinth. Sometimes the pinna is called the ear.

Ecology. The science of the relationships between living organisms and their environments.

-ectomy (suffix). A cutting out; surgical removal. Denotes any operation in which all or part of a named organ is cut out of the body.

Eczema. A skin disease or disorder. Dermatitis.

Edema. A swelling of body tissues as a result of being waterlogged with fluid.

Effective temperature. An arbitrary index that combines into a single value the effect of temperature, humidity, and air movement on the sensation of warmth and cold on the human body.

Effective temperature index. An empirically determined index of the degree of warmth perceived on exposure to different combinations of temperature, humidity, and air movement. The determination of effective temperature requires simultaneous determinations of dry bulb and wet bulb temperatures.

Efficiency, fractional. The percentage of particles of a specified size which are removed and retained by a particular type collector or sampler. A plot of fractional efficiency values versus the respective sized particles yields a fractional efficiency curve which may be related to the total collecting efficiency of air-cleaning or -sampling equipment.

Efflorescence. A phenomenon whereby a whitish crust of fine crystals forms on a surface. These are usually sodium salts that diffuse from the substrate.

Effluent. Generally something that flows out or forth, like a stream flowing out into a lake. In terms of pollution, an outflow of a sewer, storage tank, canal, or other channel.

Ejector. An air mover consisting of a two flow system wherein a primary source of compressed gas is passed through a Venturi and the vacuum which is developed at the throat of the Venturi is used to create a secondary flow of fluid. In the case of air movers for sampling applications, the secondary flow is the sample gas.

Elastomer. In a chemical industry sense, a synthetic polymer with rubber-like characteristics; a synthetic or natural rubber, or a soft, rubbery plastic with some degree of elasticity at room temperature.

Electrical precipitators. A device which removes particles from an airstream by charging the particles and collecting the charged particles on a suitable surface.

Electrolysis. The process of conduction of an electric current by means of a chemical solution.

Electromagnetic radiation. The propagation of varying electric and magnetic fields through space at the speed of light, exhibiting the characteristics of wave motion.

Electron. A minute atomic particle possessing a negative electric charge. In an atom the electrons rotate around a small nucleus. The weight of an electron is so infinitesimal that it would take 500 octillions (500 followed by 27 zeros) of them to make a pound. It is only about a two-thousandth of the mass of a proton or neutron.

Electronvolt (ev). A small unit of energy. An electron gains this much energy when it is acted upon by one volt. See Bev.

Electroplate. To cover with a metal coating (plate) by means of electrolysis.

Element. Solid, liquid, or gaseous matter that cannot be further decomposed into simpler substances by chemical means. The atoms of an element may differ physically but do not differ chemically. All atoms of an element contain a definite number of protons and thus have the same atomic number.

Elutriator. A device used to separate particles according to mass and aerodynamic size by maintaining a laminar flow system at a rate which permits the particles of greatest mass to settle rapidly while the smaller particles are kept airborne by the resistance force of the flowing air for longer times and distances. The various times and distances of deposit may be used to determine representative fractions of particle mass and size.

Embryo. The name for early stage of development of an organism. In humans, the period from conception to the end of the second month.

Emergent beam diameter. Diameter of the laser beam at the exit aperture of the system.

Emery. Aluminum oxide, natural and synthetic abrasive.

Emission factor. Statistical average of the amount of a specific pollutant emitted from each type of polluting source in relation to a unit quality of material handled, processed or burned.

Emission inventory. A list of primary air pollutants emitted into a given community's atmosphere, in amounts per day, by type of source.

Emission standards. The maximum amount of pollutant permitted to be discharged from a single polluting source.

Emmetropia. A state of perfect vision.

Emphysema. A lung disease, in which the walls of the air sacs (aveoli) have been stretched too thin and broken down.

Emulsifier or Emulsifying agent. A chemical that holds one insoluble liquid in suspension in another. Casein, for example, is a natural emulsifier in milk, keeping butterfat droplets dispersed.

Emulsion. A suspension, each in the other, of two or more unlike liquids which usually will not dissolve in each other.

Enamel. A paintlike oily substance that produces a glossy finish to a surface to which it is applied. Often contains various synthetic resins. It is lead free. In contrast is the ceramic enamel, that is, porcelain enamel, which contains lead.

Endemic. (1) Present in a community or among a group of people; usually a disease prevailing continually in a region (2) The continuing prevalence of a disease, as distinguished from an epidemic.

Endo- (prefix). Within, inside of, internal. The endometrium is the lining membrane of the uterus.

Endocrine. Secreting without the means of a duct or tube. The term is applied to certain glands which produce secretions that enter the blood stream or the lymph directly and are then carried to the particular gland or tissue whose function they regulate.

Endothermic. Characterized by or formed with absorption of heat.

Endotoxin. A toxin which is part of the wall of a microorganism and is released when that organism dies.

Energy density. The intensity of electromagnetic radiation per unit area per pulse expressed as joules per square centimeter.

Engineering controls. Methods of controlling employee exposures by modifying the source or reducing the quantity of contaminants released into the workroom environment.

Entero- (prefix). Pertaining to the intestines. Gastroenteritis is an inflammation of the intestines as well as the stomach.

Enteric. Pertaining to the intestines.

Enterotoxin. A toxin specific for cells of the intestine; gives rise to symptoms of food poisoning.

Entrainment velocity. The gas flow velocity, which tends to keep particles suspended and cause deposited particles to become airborne.

Entrance loss. The loss in static pressure of a fluid that flows from an area into and through a hood or duct opening. The loss in static pressure is due to friction and turbulence resulting from the increased gas velocity and configuration of the entrance area.

Entry loss. Loss in pressure caused by air flowing into a duct or hood.

Enzymes. Delicate chemical substances, mostly proteins, that enter into and bring about chemical reactions in living organisms.

EPA. Environmental Protection Agency.

EPA number. The number assigned to chemicals regulated by the Environmental Protection Agency.

Epicondylitis. Inflammation of certain bony prominences in the area of the elbow., e.g., tennis elbow.

Epidemiology. The study of disease in human populations.

Epidermis. The superficial scarfskin or upper (outer) strata of skin.

Epilation. Temporary or permanent loss of body hair.

Epithelioma. Carcinoma of the epithelial cells of the skin and other epithelial surfaces.

Epithelium. The purely cellular, avascular layer covering all the free surfaces—cutaneous, mucous, and serous, including the glands and other structures derived therefrom, e.g., the epidermis.

Erg. The force of one dyne acting through a distance of one centimeter. It would be equivalent to the work done by a "June bug" climbing over a 1 cm (½ in.) high stone; or the energy required to ionize about 20 billion molecules of air.

Ergonomics. A multidisciplinary activity dealing with interactions between man and his total working environment plus stresses related to such environmental elements as atmosphere, heat, light, and sound as well as all tools and equipment of the workplace.

Erysipeloid. A bacterial infection affecting slaughterhouse workers and fish handlers.

Eryth-, erythro- (prefix). Redness. Erythemia is indicated by redness of the skin (including a deep blush). An erythrocyte is a red blood cell.

Erythema. Reddening of the skin.

Erythemal region. Ultraviolet light radiation between 2,800 and 3,200 angstroms (280-320 millimicrons); it is absorbed by the cornea of the eye.

Erythrocyte. A type of red blood corpuscle.

Eschar. The crust formed after injury by a caustic chemical or heat.

Essential oil. Any of a class of volatile, odoriferous oils found in plants and imparting to the plants odor and often other characteristic properties. Used in essence, perfumery, etc.

Esters. Organic compunds that may be made by interaction between an alcohol and an acid, and by other means. Esters are nonionic compounds, including solvents and natural fats.

Etch. To cut or eat away material with acid or other corrosive substance.

Etiologic agent. Refers to organisms, substances, or objects associated with the cause of disease or injury.

Etiology. The study or knowledge of the causes of disease.

Eu- (prefix). Well and good. A euthyroid person has a thyroid gland that couldn't be working better. A euphoric one has a tremendous sense of well-being.

Eustachian tube. A structure about 6 cm (2½ in.) long leading from the back of the nasal cavity to middle ear. It equalizes the pressure of air in the middle ear with that outside the eardrum.

Evaporation. The process by which a liquid is changed into the vapor state.

Evaporation rate. The ratio of the time required to evaporate a measured volume of a liquid to the time required to evaporate the same volume of a reference liquid (ethyl ether) under ideal test conditions. The higher the ratio, the slower the evaporation rate.

Exhalation valve. A device that allows exhaled air to leave a respirator and prevents outside air to leave a respirator and prevents outside air from entering through the valve.

Exhaust ventilation. The removal of air usually by mechanical means from any space. The flow of air between two points is due to the occurrence of a pressure difference between the two points. This pressure difference will cause air to flow from the high pressure to the low pressure zone.

Exothermic, Exothermal. Characterized by or formed with evolution of heat.

Exotoxin. A toxin excreted by a microorganism into the surrounding medium.

Explosive limit. See Flammable limit.

Exposure. Contact with a chemical, biological, or physical hazard.

Extension. Movement whereby the angle between the bones connected by a joint is increased. Motions of this type are produced by contraction of extensor muscles.

Extensor muscles. A muscle that, when active, increases the angle between limb segments, e.g., the muscles that straighten the knee or elbow, open the hand, or straighten the back.

Extensor tendon. Connecting structure between an extensor muscle and the bone into which it inserts. Examples are the hard, longitudinal tendons found on the back of the hand when the fingers are fully extended.

External mechanical environment. The synthetic physical environment; e.g., equipment, tools, machine controls, clothing. Antonym: internal (bio)mechanical environment.

Extrusion. The forcing of raw material through a die or a form in either a hot or cold state, in a solid state or in partial solution. Long used with metals and clays, it is now extensively used in the plastic industry.

Eyepiece. Gastight, transparent window(s) in a full facepiece through which the wearer may see.

F

Face velocity. Average air velocity into the exhaust system measured at the opening into the hood or booth.

Facepiece. That portion of a respirator that covers the wearer's nose and mouth in a half-mask facepiece or nose, mouth, and eyes in a full facepiece. It is designed to make a gastight or dust-tight fit with the face and includes the headbands, exhalation valve(s), and connections for air-purifying device or respirable-gas source or both.

Facing. In foundry work, the final touch-up work of the mold surface to come in contact with metal is called the facing operation and the fine powdered material used is called the facing.

Fainting. Technically called syncope, is a temporary loss of consciousness as a result of a diminished supply of blood to the brain.

Fallout. Dust particles that contain radioactive fission products resulting from a nuclear exposion. The wind can carry fallout particles many miles.

Fan laws. Statements and equations that describe the relationship between fan volume, pressure, brake horsepower, size, and rotating speed.

Fan rating curve or table. Data that describe the volumetric output of a fan at different static pressures.

Fan static pressure. The pressure added to the system by the fan. It equals the sum of pressure losses in the system minus the velocity pressure in the air at the fan inlet.

Far field (free field). In noise measurement, this refers to the distance from the noise source where the sound pressure level decreases 6 dBA for each doubling of distance (inverse square law).

Farmer's lung. Fungus infection and ensuing hypersensitivity from grain dust.

Federal Register. Publication of U.S. government documents officially promulgated under the law, documents whose validity depends upon such publication. It is published on each day following a government working day. It is, in effect, the daily supplement to the *Code of Federal Regulations, CFR.*

Feral animal. A wild animal, or domestic animal which has reverted to the wild state.

Fertilizer. Plant food usually sold in mixed formula containing basic plant nutrients: compounds of nitrogen, potassium, phosphorus, sulfur, and sometimes other minerals.

Fetus. The term used to describe the developing organisms (human) from the third month after conception to birth.

FEV. Forced expired volume.

Fever. A condition in which the body temperature is above its regular or normal level.

Fibrillation. Very rapid irregular contractions of the muscle fibers of the heart resulting in a lack of synchronism of the heartbeat.

Fibrosis. A condition marked by increase of interstitial fibrous tissue.

Film badge. A piece of masked photographic film worn by nuclear workers. It is darkened by nuclear radiation, and radiation exposure can be checked by inspecting the film.

Filter. (1) A device for separating components of a signal on the basis of its frequency. It allows components in one or more frequency bands to pass relatively unattenuated, and it attenuates greatly components in other frequency bands, (2) a fibrous media used in respirators to remove solid or liquid particles from the airstream entering the respirator, (3) a sheet of material that is interposed between patient and the source of x-rays to absorb a selective part of the x-rays, (4) a fibrous or membrane media used to collect dust, fume, or mist air samples.

Filter efficiency. The efficiency of various filters can be established on the basis of entrapped particles; i.e., collection efficiency, or on the basis of particles passed through the filter, i.e., penetration efficiency.

Filter, HEPA. High-efficiency particulate air filter that is at least 99.97 percent efficient in removing thermally generated monodisperse dioctylphthalate smoke particles with a diameter of 0.3μ.

Firebrick. A special clay that is capable of resisting high temperatures without melting or crumbling.

Fire damp. In mining the accumulation of an explosive gas, chiefly methane gas. Miners refer to all dangerous underground gases as "damps."

Fire point. The lowest temperature at which a material can evolve vapors to support continuous combustion.

Fission. The splitting of an atomic nucleus into two parts accompanied by the release of a large amount of radioactivity and heat. Fission reactions occur only with heavy isotopes, such as uranium-233, uranium-235, and plutonium-239.

Fissionable. A nucleus that undergoes fission under the influence of neutrons, even of very slow neutrons.

Fission product. The highly radioactive nuclei into which a fissionable nucleus splits ("fissions") under the influence of neutron bombardment.

Flagellum. A flexible, whiplike appendage on cells used as an organ of locomotion.

Flame propagation. See Propagation of flame.

Flameproofing material. Chemicals that catalytically control the decomposition of cellulose material at flaming temperature. Substances used as fire retardants are borax-boric acid, borax-boric acid diammonium phosphate, ammonium bromide, stannic acid, antimony oxide, and combinations containing formaldehyde.

Flammable aerosol. An aerosol that is required to be labeled "Flammable" under the Federal Hazardous Substances Labeling Act (15 USC 1261).

Flammable limits. Flammables have a minimum concentration below which propagation of flame does not occur on contact with a source of ignition. This is known as the lower flammable explosive limit (LEL). There is also a maximum concentration of vapor or gas in air above which propagation of flame does not occur. This is known as the upper flammable explosive limit (UEL). These units are expressed in percent of gas or vapor in air by volume.

Flammable liquid. Any liquid having a flash point below 37.8 C (100 F).

Flammable range. The difference between the lower and upper flammable limits, expressed in terms of percentage of vapor or gas in air by volume, and is also often referred to as the "explosive range."

Flange. A rim or edge added to a hood to reduce the quantity of air entering the hood from behind the hood.

Flash blindness. Temporary visual disturbance resulting from viewing an intense light source.

Flash point. The lowest temperature at which a liquid gives off enough vapor to form an ignitable mixture with air and produce a flame when a source of ignition is present. Two tests are used—open cup and closed cup.

Flask. In foundry work, the assembly of the cope and the drag constitutes the flask. It is the wooden or iron frame containing sand into which molten metal is poured. Some flasks may have three or four parts.

Flexion. Movement whereby the angle between two bones connected by a joint is reduced. Motions of this type are produced by contraction of flexor muscles.

Flexor muscles. A muscle which, when contracting, decreases the angle between limb segments. The principal flexor of the elbow is the brachialis muscle. Flexors of the fingers and the wrist are the large muscles of the forearm originating at the elbow. See Extensor muscles.

Flocculation. The process of forming a very fluffy mass of material held together by weak forces of adhesion.

Flocculator. A device for aggregating fine particles.

Flora; microflora. Microorganisms present in a given situation (such as intestinal flora, soil flora).

Flotation. A method of ore concentration in which the mineral is caused to float due to chemical frothing agents while the impurities sink.

Flotation reagent. Chemical used in flotation separation of minerals. Added to pulverized mixture of solids and water and oil, it causes preferential nonwetting by water of certain solid particles, making possible the flotation and separation of unwet particles.

Flow coefficient. A correction factor used for figuring volume flow rate of a fluid through an orifice. This factor includes the effects of contraction and turbulence loss (covered by the coefficient of discharge), plus the compressibility effect, and the effect of an upstream velocity other than zero. Since the latter two effects are negligible in many instances, the flow Coefficient is often equal to the coefficient of discharge (see the Coefficient of discharge).

Flow meter. An instrument for measuring the rate of flow of a fluid or gas.

Flow, turbulent. Fluid flow in which the fluid moves transversely as well as in the direction of the tube or pipe axis, as opposed to stream line or viscous flow.

Fluid. A substance tending to flow or conform to the outline of its container. It may be liquid, vapor, gas, or solid (like raw rubber).

Fluorescence. Emission of light from a crystal, after the absorption of energy.

Fluorescent screen. A screen coated with a fluorescent substance so that it emits light when irradiated with x-rays.

Fluoroscope. A fluorescent screen mounted in front of an x-ray tube so that internal organs may be examined through their shadow cast by x-rays. It may also be used for inspection of inanimate objects.

Fluoroscopy. The practice of examining through the use of an x-ray fluoroscope.

Flux. Usually refers to a substance used to clean surfaces and promote fusion in soldering. However, fluxes of various chemical nature are used in the smelting of ores, in the ceramic industry, in assaying silver and gold ores and in other endeavors. The most common fluxes are silica, various silicates, lime, sodium, and potassium carbonate and litharge and red lead in the ceramic industry. See Soldering, Galvanizing, and Luminous flux.

Fly ash. Finely divided particles of ash entrained in flue gases arising from the combustion of fuel.

Foci (focus). A center or site of a disease process.

Follicle. A small anatomical cavity or deep, narrow-mouthed depression; a small lymph node.

Folliculitis. Infection of a hair follicle, often caused by obstruction by natural or industrial oils.

Foot candle. A unit of illumination. The illumination at a point on a surface which is one foot from, and perpendicular to, a uniform point source of one candle.

Foot-pounds of torque. A measurement of the physiological stress exerted upon any joint during the performance of a task. The product of the force exerted and the distance from the point of application to the point of stress. Physiologically, torque that does not produce motion nonetheless causes work stress, the severity of which depends on the duration and magnitude of the torque. In lifting an object or holding it elevated, torque is exerted and applied to the lumbar vertebrae.

Force. That which changes the state of rest or motion in matter. The SI (International System) unit of measurement is the newton (N).

Fovea. A depression or pit in the center of the macula of the eye; it is the area of clearest vision.

Fractionation. Separation of a mixture into different portions or fractions, usually by distillation.

Free sound field (free field). A field in a homogeneous, isotropic medium free from boundaries. In practice it is a field in which the effects of the boundaries are negligible over the region of interest. (see Far field.)

Frequency (in hertz or Hz). Rate at which pressure oscillations are produced. One hertz is equivalent to one cycle per second. A subjective characteristic of sound related to frequency is pitch.

Friction factor. A factor used in calculating loss of pressure due to friction of a fluid flowing through a pipe or duct.

Friction loss. The pressure loss due to friction.

Fuller's earth. A hydrated silica-alumina compound, associated with ferric oxide. Used as a filter medium and as a catalyst and catalyst carrier and in cosmetics and insecticides.

Fume. Airborne particulate formed by the evaporation of solid materials, e.g., metal fume emitted during welding. Usually less than one micron in diameter.

Fume fever. Metal fume fever is an acute condition caused by a brief high exposure to the freshly generated fumes of metals, such as zinc or magnesium, or their oxides.

Functional anatomy. Study of the body and its component parts taking into account those structural features which are directly related to physiological function.

Fundamental frequency. Fundamental frequency is the lowest component frequency of a periodic quantity.

Fundus. The interior surface of a hollow organ, such as the retina of the eye.

Fungus (pl. fungi). Any of a major group of lower plants that lack chlorophyll and live on dead or other living organisms. Fungi include molds, rusts, mildews, smuts, and mushrooms.

Fusion. The joining of atomic nuclei to form a heavier nucleus, accomplished under conditions of extreme heat (millions of degrees). If two nuclei of light atoms fuse, the fusion is accompanied by the release of a great deal of energy. The energy of the sun is believed to be derived from the fusion of hydrogen atoms to form helium.

Fusion. (In welding) the melting together of filler metal and base metal (substrate), or of base metal only, which results in coalesce.

FVC. Forced vital capacity.

G

Gage pressure. Pressure measured with respect to atmospheric pressure.

Galvanizing. An old but still used method of providing a protective coating for metals by dipping them in a bath of molten zinc.

Gamete. A mature germ cell. An unfertilized ovum or spermatozoon.

Gamma-rays (gamma radiation). The most penetrating of all radiation. Gamma-rays are very high-energy x-rays.

Ganglion (pl. ganglia). A knot, or knotlike mass; used as a general term to designate a group of nerve cell bodies located outside of the central nervous system. The term is also applied to certain nuclear groups within the brain or spinal cord.

Gangue. In mining or quarrying, useless chipped rock.

Gas. A state of matter in which the material has very low density and viscosity; can expand and contract greatly in response to changes in temperature and pressure; easily diffuses into other gases; readily and uniformly distributes itself throughout any container. A gas can be changed to the liquid or solid state only by the combined effect of increased pressure and decreased temperature (below the critical temperature).

Gas chromatography. A gaseous detection technique which involves the separation of mixtures by passing them through a column that will enable the components to be held up for varying periods of time before they are detected and recorded.

Gas metal arc-welding (GMAW). An arc-welding process which produces coalescence of metals by heating them with an arc between a continuous filler metal (consumable) electrode and the work—shielding is obtained entirely from an external supplied gas or gas mixture; some methods of this process are called MIG or CO_2 welding.

Gas tungsten arc-welding (GTAW). An arc-welding process that produces coalescence of metals by heating them with an arc between a tungsten (nonconsumable) electrode and the work, shielding is obtained from a gas or gas mixture. Pressure may or may not be used and filler metal may or may not be used. (This process has sometimes been called TIG welding.)

Gastr-, gastro- (prefix). Pertaining to the stomach.

Gastritis. Inflammation of the stomach.

Gate. A groove in a mold to act as a passage for molten metal.

Geiger counter. A gas-filled electrical device which counts the presence of an atomic particle or ray by detecting the ions produced. (Sometimes called a "Geiger-Mueller" counter.)

Genes. The ultimate biological units of heredity.

General ventilation. System of ventilation consisting of either natural or mechanically induced fresh air movements to mix with and dilute contaminants in the workroom air. This is *not* the recomended type of ventilation to control contaminants that are toxic.

Genetic effects. Mutations or other changes which are produced by irradiation of the germ plasm.

Genetically significant dose (GSD). The dose which, if received by every member of the population, would be expected to produce the same total genetic injury to the population as do the actual doses received by the various individuals.

Germ. A microorganism; a microbe usually thought of as a pathogenic organism.

Germicide. An agent capable of killing germs.

GI. Gastrointestinal.

Gingival. Pertaining to the gingivae—gums. (Gingivae—the mucous membrane, with the supporting fibrous tissue, which overlies the crowns of unerupted teeth and encircles the necks of those that have erupted).

Gingivitis. Inflammation of the gums.

Gland. Any body organ that manufactures some liquid product that it secretes from its cells.

Globe thermometer. A thermometer set in the center of a metal sphere that has been painted black in order to measure radiant heat.

Globulin. General name for a group of proteins which are soluble in saline solutions, but not in pure water.

Glossa- (prefix). Pertaining to the tongue.

Glove box. A sealed enclosure in which all handling of items inside the box is carried out through long impervious gloves sealed to ports in the walls of the enclosure.

Gob. "Gob pile" is waste mineral material such as from coal mines but containing sufficient coal that gob fires may arise from spontaneous combustion.

Gonads. The male (testes) and female (ovaries) sex glands.

Grab sample. A sample which is taken within a very short time period. The sample is taken to determine the constituents at a specific time.

Gram (g). A metric unit of weight. One ounce equals 28.4 grams.

Grams per kilogram (g/kg). This indicates the dose of a substance given to test animals in toxicity studies.

Granuloma. A mass or nodule of chronically inflamed tissue with granulations; usually associated with an infective process.

Graticule. See Reticle.

Gravimetric. Of or pertaining to measurement by weight.

Gravimetric method. A procedure dependent upon the formation or use of a precipitate or residue, which is weighed to determine the concentration of a specific contaminant in a previously collected sample.

Gravitation. The universal attraction existing between all material bodies. The gravitational attraction of the earth's mass for bodies at or near its surface is called gravity.

Gravity, specific. The ratio of the mass of a unit volume of a substance to the mass of the same volume of a standard substance at a standard temperature. Water at 4 C (39.2 F) is the standard substance usually referred to. For gases, dry air, at the same temperature and pressure as the gas, is often taken as the standard substance.

Gravity, standard. A gravitational force which will produce an acceleration equal to 9.8 m (32.17 ft) per second. The actual force of gravity varies slightly with altitude and latitude. The standard was arbitrarily established as that at sea level and 45-degree latitude.

Gray iron. The same as cast iron and in general any iron containing high carbon.

Grooving. Designing a tool with grooves on the handle to accommodate the fingers of the user. A bad practice because of the great variation in the size of workers' hands. Grooving interferes with sensory feedback. Intense pain may be caused by the grooves to the arthritic hand.

Gyn-, gyne- (prefix). Women, female. Gynecology. The medical specialty concerned with diseases of women.

Gyratory crusher. A device for crushing rock by means of a heavy steel pestle rotating in a steel cone, with the rock being fed in at the top and passing out of the bottom.

H

Half-life, radioactive. For a single radioactive decay process, the time required for the activity to decrease to half its value by that process.

Half-thickness. The thickness of a specified absorbing material which reduces the dose rate to one half its original value.

Half-value layer (HVL). The thickness of a substance necessary to reduce the intensity of a beam of gamma- or x-rays to one-half of its original value. Also known as half-thickness.

Halogenated hydrocarbon. A chemical material that has carbon plus one or more of these elements: chlorine, fluorine, bromine, or iodine.

Hammer mill. A machine for reducing the size of stone or other bulk material by means of hammers usually placed on a rotating axle inside a steel cylinder.

Hardness. A relative term to describe the penetrating quality of radiation. The higher the energy of the radiation, the more penetrating (harder) is the radiation.

Hardness of water. A "degree" of hardness is the equivalent of one grain of calcium carbonate, $CaCO_3$, in one gallon of water.

Hazardous material. Any substance or compound that has the capability of producing adverse effects on the health and safety of humans.

Heading. In mining, a horizontal passage or drift of a tunnel, also the end of a drift or gallery. In tanning, a layer of ground bark over the tanning liquor.

Health physicist. A professional person especially trained in radiation physics and concerned with problems of radiation damage and protection.

Hearing conservation. The prevention or minimizing of noise-induced deafness through the use of hearing protection devices, the control of noise through engineering methods, annual audiometric tests, and employee training.

Hearing level. The deviation in decibels of an individual's threshold from the zero reference of the audiometer.

Heat cramps. Painful muscle spasms as a result of exposure to excess heat.

Heat exhaustion. A condition usually caused by loss of body water because of exposure to excess heat. Symptoms include headache, tiredness, nausea, and sometimes fainting.

Heat, latent. The quantity of heat absorbed or given off per unit weight of material during a change of state, such as ice to water or water to steam.

Heat of fusion. The heat given off by a liquid freezing to a solid or gained by a solid melting to a liquid, without a change in temperature.

Heat of vaporization. The heat given off by a vapor, condensing to a liquid or gained by a liquid evaporating to a vapor, without a change in temperature.

Heat, sensible. Heat associated with a change in temperature; specific heat exchange with environment; in contrast to a heat interchange in which (only) a change of state (phase) occurs.

Heat, specific. The ratio of the quantity of heat required to raise the temperature of a given mass of any substance one degree to the quantity required to raise the temperature of an equal mass of a standard substance (usually water at 15 C [59 F]) one degree.

Heat stress. Relative amount of thermal strain from the environment.

Heat stress index. Index which combines the environmental heat and metabolic heat into an expression of stress in terms of requirement for evaporation of sweat.

Heatstroke. A serious disorder resulting from exposure to excess heat. It results from sweat suppression and increased storage of body heat. Symptoms include hot dry skin, high temperature, mental confusion, convulsions, and coma. Heatstroke is fatal if not treated promptly.

Heat treatment. Any of several processes of metal modification such as annealing.

Heavy hydrogen. Same as deuterium.

Heavy metals. Metallic elements with high molecular weights.

Heavy water. Water containing heavy hydrogen (deuterium) instead of ordinary hydrogen. It is widely used in reactors to slow down neutrons.

Helmet. A device that shields the eyes, face, neck, and other parts of the head.

Hem-, Hemato-, -em- (prefix). Pertaining to blood. Hematuria means blood in the urine. When the roots occur internally in a word, the "h" is often dropped for the sake of pronunciation, leaving -em- to denote blood as in anoxemia (deficiency of oxygen in the blood).

Hematology. Study of the blood and the blood-forming organs.

Hematuria. Blood in the urine.

Hemi- (prefix). Half. The prefix is plain enough in hemiplegia, "half paralysis," affecting one side of the body. It is not so plain in migraine (one-sided headache), a word which shows how language changes through the centuries. The original word was hemicrania, "half-head."

Hemoglobin. The red coloring matter of the blood that carries the oxygen.

Hemolysis. Breakdown of red blood cells with liberation of hemoglobin.

Hemoptysis. Bleeding from the lungs, spitting blood, or blood-stained sputum.

Hemorrhage. Bleeding; especially, profuse bleeding, as from a ruptured or cut blood vessel (artery or vein).

Hemorrhagic. Pertaining to or characterized by hemorrhage.

HEPA filter. (High Efficiency Particulate Air Filter) A disposable, extended medium, dry type filter with a particle removal efficiency of no less than 99.97 percent for 0.3μ particles.

Hepatitis. Inflammation of the liver.

Hepatotoxin. Chemicals that produce liver damage.

Herpes. An acute inflammation of the skin or mucous membranes, characterized by the development of groups of vesicles on an inflammatory base.

Hertz. The frequency measured in cycles per second. 1 cps = 1 Hz.

High frequency loss. Refers to a hearing deficit starting with 2000 Hz and beyond.

Homogenizer. A machine that forces liquids under high pressure through a perforated shield against a hard surface to blend or emulsify the mixture.

Homoiotherm. Uniform body temperature, a warm-blooded creature remaining so regardless of environment.

Hood. (1) Enclosure, part of a local exhaust system; (2) a device that completely covers the head, neck, and portions of the shoulders.

Hood entry loss. The pressure loss from turbulence and friction as air enters the ventilation system.

Hood, slot. A hood consisting of a narrow slot leading into a plenum chamber under suction to distribute air velocity along the length of the slot.

Hood static pressure. The suction or static pressure in a duct near a hood. It represents the suction that is available to draw air into the hood.

Hormones. Chemical substances secreted by the endocrine glands, exerting influence over practically all body activities.

Horsepower. A unit of power, equivalent to 33,000 foot-pounds per minute (746 W). See Brake horsepower.

Host. The plant or animal harboring another as a parasite or as an infectious agent.

Hot. In addition to meaning "having a relatively high temperature," this is a colloquial term meaning highly radioactive.

Human-equipment interface. Areas of physical or perceptual contact between man and equipment. The design characteristics of the human-equipment interface determine the quality of information. Poorly designed interfaces may lead to excessive fatigue or localized trauma, e.g., calluses.

Humerus. The bone of the upper arm that starts at the shoulder joint and ends at the elbow. Muscles that move the upper arm, forearm and hand are attached to this bone.

Humidify. To add water vapor to the atmosphere, to add water vapor or moisture to any material.

Humidity. (1) Absolute humidity is the weight of water vapor per unit volume, pounds per cubic foot or grams per cubic centimeter; (2) relative humidity is the ratio of the actual partial vapor pressure of the water vapor in a space to the saturation pressure of pure water at the same temperature.

Humidity, specific. The weight of water vapor per unit weight of dry air.

Hyalinization. Conversion into a substance resembling glass.

Hydration. The process of converting raw material into pulp by prolonged beating in water, to combine with water or the elements of water.

Hydrocarbons. Organic compounds, composed solely of carbon and hydrogen. Several hundred thousand molecular combinations of C and H are known to exist. Basic building blocks of all organic chemicals. Main chemical industry sources of hydrocarbons are petroleum, natural gas, and coal.

Hydrogenation. A reaction of molecular hydrogen with numerous organic compounds. An example is the hydrogenation of olefins to paraffins or of the aromatics to the naphthenes or the reduction of aldehydes and ketones to alcohols.

Hydrolysis. The interaction of water with a material resulting in decomposition.

Hydrometallurgy. Science of metal recovery by a process involving treatment of ores in an aqueous medium, such as acid or cyanide solution.

Hydrophobic. Repelled by water or "water hating."

Hygroscopic. Readily absorbing or retaining moisture.

Hyper- (prefix). Over, above, increased. The usual implication is overactivity or excessive production, as in hyperthyroidism.

Hyperkeratosis. Hypertrophy of the horny layer of the skin.

Hypertension. Abnormally high tension; especially high blood pressure.

Hypertrophy. Increase in cell size causing an increase in the size of the organ or tissue.

Hypnotic. Anything that induces sleep, or that produces the effects ascribed to hypnotism.

Hypo- (prefix). Under, below; less, decreased. The two different meanings of this common prefix can be tricky. Hypodermic might reasonably be interpreted to mean that some unfortunate patient has too little skin. The actual meaning is under or beneath the skin, a proper site for an injection. The majority of "hypo" words, however, denote an insufficiency, lessening, reduction from the norm, as in hypoglycemia, too little glucose in the blood.

Hypothermia. Condition of reduced body temperature.

Hyster-, hystero- (prefix). Denoting the womb. Hysteria perpetuates a Greek notion that violent emotional behavior originated in the uterus; when it occurred in men, it must have been called something else.

Hysteresis. A retardation of the effect when the forces acting upon a body are changed (as if from viscosity or internal friction). Specifically, the magnetization of a sample of iron or steel actually lags behind the magnetic field that induced it, when the field varies.

I

IARC. International Agency for Research on Cancer.

Iatro- (prefix). Pertaining to a doctor. A related root, -iatrist, denotes a specialist—pediatrician, obstetrician.

Iatrogenic. Caused by the doctor.

ICC. Interstate Commerce Commission.

Idio- (prefix). Peculiar to, private, distinctive. As in idiosyncrasy.

Idiopathic. Disease that originates in itself.

Idiosyncrasy. A special susceptibility to a particular substance introduced into the body.

Iliac crest. The upper rounded border of the hip bone. No muscles cross the iliac crest and it lies immediately below the skin. It is an important anatomical reference point because it can be felt through the skin. Seat backrests should clear the iliac crest.

Image. As used in this glossary the fluorescent picture produced by x-rays hitting a fluoroscopic screen.

Image receptor. Any device, such as a fluorescent screen or radiographic film, which transforms incident x-ray photons either into a visible image or into another form which can be made into a visible image by further transformations.

IDLH. Immediately dangerous to life or health.

Immiscible. Not miscible. Any liquid that will not mix with another liquid, in which case it forms two separate layers or exhibits cloudiness or turbidity.

Immune. Resistant to disease.

Immunity. The power of the body to resist successfully infection and the effects of toxins. This resistance comes through the possession by the body of certain "fighting substances," called antibodies. To immunize is to confer immunity. Immunization is the process of acquiring or conferring immunity.

Impaction. The forcible contact of particles of matter, a term often used synonymously with impingement, but generally reserved for the case where particles are contacting a dry surface.

Impingement. As used in air-sampling, impingement refers to a process for the collection of particulate matter in which a particle containing gas is directed against a wetted glass plate and the particles are retained by the liquid.

Inches of mercury column. A unit used in measuring pressures. One inch of mercury column equals a pressure of 1.66 kPa (0.491 lb per sq in.).

Inches of water column. A unit used in measuring pressures. One inch of water column equals a pressure of 0.25 kPa (0.036 lb per sq in.).

Incompatible. This term is applied to liquid and solid systems to indicate that one material cannot be mixed with another specified material without the possibility of a dangerous reaction.

Incubation. Holding cultures of microorganisms under conditions favorable to their growth.

Incubation time. The elapsed time between exposure to infection and the appearance of disease symptoms, or the time period during which microorganisms inoculated into a medium are allowed to grow.

Induration. Heat hardening that may involve little more than thermal dehydration.

Inert (chemical). Not having active properties.

Inert gas welding. An electric welding operation utilizing an inert gas such as helium to flush away the air to prevent oxidation of the metal being welded.

Inert gas. A gas that does not normally combine chemically with the base metal or filler metal.

Inertial moment. Related to biomechanics, that moment of force-time caused by sudden accelerations or decelerations. Whiplash of the neck is caused by an inertial moment. In an industrial setting, sidestepping causes application of a lateral inertial moment on the lumbosacral joint, which may cause trauma, pain, and in any case will lower performance efficiency. The inertial moment is one of the seven elements of a lifting task.

Infection. Entrance into the body or its tissues of disease-causing organisms and the causation of damage to the body as a whole or to tissues or organs. It also refers to the entrance into the body of parasites, like certain worms. On the other hand, parasites like mites and ticks that attack the surface of the body are said to infest, not infect.

Infectious. Capable of invading a susceptible host, replicating and causing an altered host reaction, commonly referred to as a disease.

Infestation. Invasion of the body surface by parasites. See Infection.

Inflammation. The reaction of body tissue to injury, whether by infection or trauma. The inflamed area is red, swollen, hot, and usually painful.

Infrared. Those wavelengths of the electromagnetic spectrum longer than those of visible light and shorter than radio waves. 10^{-4} cm to 10^{-1} cm wavelength.

Infrared radiation. Electromagnetic energy with wavelengths from 770 nm to 12,000 nm.

Ingestion. (1) The process of taking substances into the stomach, as food, drink, medicine, etc. (2) With regard to certain cells, the act of engulfing or taking up bacteria and other foreign matter.

Ingot. A block of iron or steel cast in a mold for ease in handling before processing.

Inguinal region. The abdominal area on each side of the body occurring as a depression between the abdomen and the thigh, the groin.

Inhalation valve. A device that allows respirable air to enter the facepiece and prevents exhaled air from leaving the facepiece through the intake opening.

Inhibition. Prevention of growth or multiplication of microorganisms.

Inhibitor. An agent that arrests or slows chemical action or a material used to prevent or retard rust or corrosion.

Injury. Damage or harm to the body, as the result of violence, infection, or anything else that produces a lesion.

Innocuous. Harmless.

Inoculate. The artificial introduction of microorganisms into a system.

Inorganic. Term used to designate compounds that generally do not contain carbon. Source: matter, other than vegetable or animal. Examples: sulfuric acid and salt. Exceptions are carbon monoxide, carbon dioxide.

Insomnia. Inability to sleep; abnormal wakefulness.

Instantaneous radiation. The radiation emitted during the fission process. These instantaneous radiations are frequently called "prompt" gamma-rays or "prompt" neutrons. Most of the fission products continue to emit radiation after the fission process.

Inter- (prefix). Between.

Intermediate. A chemical formed as a "middle-step" in a series of chemical reactions, especially in the manufacture of organic dyes and pigments. In many cases, it may be isolated and used to form a variety of desired products. In other cases, the intermediate may be unstable or used up at once.

Internal biomechanical environment. The muscles, bones and tissues of the body, all of which are subject to the same Newtonian force as external objects in their interaction with other bodies and natural forces. When designing for the body, one must consider the forces that the internal mechanical environment must withstand.

Interphalangeal joints. The finger or the toe joints. The thumb has one interphalangeal joint, the fingers have two interphalangeal joints each.

Interstitial. (1) Pertaining to the small spaces between cells or structures; (2) occupying the interstices of a tissue or organ; (3) designating connective tissue occupying spaces between the functional units of an organ or a structure.

Intoxication. Means either drunkenness or poisoning.

Intra- (prefix). Within.

Intraperitoneal. Inside the space formed by the membrane that lines the interior wall of the abdomen and covers the abdominal organs.

Intravenous. Into or inside the vein.

Inverse square law. The propagation of energy through space is inversely proportional to the square of the distance it must travel. An object 3 m (ft) away from an energy source receives 1/9 as much energy as an object 1 m (ft) away.

Inversion. Phenomenon of a layer of cool air trapped by a layer of warmer air above it so that the bottom layer cannot rise. This is a special problem in polluted areas because the contaminating substances cannot be dispersed.

Investment casting. There are numerous types of investment casting and the materials include fire clay, silicon dioxide, silica flour, stillimanite, cristobalite, aluminum oxide, zirconium oxide, and others. The Mercast process utilizes mercury poured into a steel die. A ceramic shell mold is built around the pattern and then the pattern is frozen. The mercury is subsequently recovered at room temperature. The potential harm from exposure to mercury often is unrecognized.

Ion. An electrically charged atom. An atom that has lost one or more of its electrons is left with a positive electrical charge. Those that have gained one or more extra electrons are left with a negative charge.

Ion-exchange resin. Synthetic resins containing active groups that give the resin the property of combining with or exchanging ions between the resin and a solution.

Ion pair. A positively charged atom (ion) and an electron formed by the action of radiation upon a neutral atom.

Ionization. The process whereby one or more electrons is removed from a neutral atom by the action of radiation. Specific ionization is the number of ion pairs per unit distance in matter, usually air.

Ionization chamber. A device roughly similar to a Geiger counter and used to measure radioactivity.

Ionizing radiation. Refers to (1) electrically charged or neutral particles, or (2) electromagnetic radiation which will interact with gases, liquids, or solids to produce ions. There are five major types: alpha, beta, x (or x-ray), gamma, and neutrons.

Irradiation. The exposure of something to radiation.

Irritant. A substance that produces an irritating effect when it contacts skin, eyes, nose, or respiratory system.

Ischemia. Loss of blood supply to a particular part of the body.

Ischial tuberosity. A rounded projection on the ischium. It is a point of attachment for several muscles involved in moving the femur and the knee. It can be affected by improper designing of chairs and by situations involving trauma to the pelvic region. When seated, pressure is borne at the site of the ischial tuberosities. Chair design should provide support to the pressure projection of the ischial tuberosity through the skin of the buttocks.

Isometric work. Referring to a state of muscular contraction without movement. Although no work in the "physics" sense is done, physiologic work (energy utilization and heat production) occurs. In isometric exercise, muscles are tightened against immovable objects. In work measurements, isometric muscular contractions must be considered as major factor of task severity.

Isotope. One of two or more atomic species of an element differing in atomic weight but having the same atomic number. Each contains the same number of protons but a different number of neutrons. Uranium-238 contians 92 protons and 146 neutrons while the isotope U-235 contains 92 protons and 143 neutrons. Thus the atomic weight (atomic mass) of U-238 is three higher than that of U-235. See also Radioisotope.

Isotropic. Exhibiting properties with the same values when measured along axes in all directions.

-itis (suffix). Inflamation.

J

Jaundice. Icterus. A serious symptom of disease that causes the skin, the whites of the eyes, and even the mucous membranes to turn yellow.

Jigs and fixtures. Often used interchangeably; precisely a jig holds work in position and guides the tools acting on the work, while a fixture holds but does not guide.

Joint. Articulation between two bones that may permit motion in one or more planes. They may become the sites for work-induced trauma (such as tennis elbow, arthritis) or other disorders.

Joule. Unit of energy used in describing a single pulsed output of a laser. It is equal to one wattsecond or 0.239 calories. It equals 1×10^7 ergs.

Joule/cm^2 (j/cm^2). Unit of energy density used in measuring the amount of energy per area of absorbing surface, or per area of a laser beam. It is a unit for predicting damage potential of a laser beam.

K

Kaolin. A type of clay composed of mixed silicates and used for refractories, ceramics, tile, and stoneware. In some deposits, free silica may be present as an impurity.

Kaolinosis. A condition induced by inhalation of the dust released in the grinding and handling of kaolin (china clay).

Kelvin scale. The fundamental temperature scale, also called the absolute or thermodynamic scale, in which the temperature measure is based on the average kinetic energy per molecule of a perfect gas. The zero of the Kelvin scale is -273.18 degrees Celsius.

Keratin. Sulfur-containing proteins that form the chemical basis for epidermis tissues; found in nails, hair, feathers.

Keratitis. Inflammation of the cornea.

Kev. Kilo electron volts or 1,000 electron volts is a kev. A unit of energy.

Kilocurie. 1,000 curies. A unit of radioactivity.

Kilogram (kg). A unit of weight in the metric system equal to 2.2 lb.

Kinesiology. The study of human movement in terms of functional anatomy.

Kinetic chain. A combination of body segments connected by joints which, operating together, provide a wide range of motion for the distal element. A single joint only allows rotation, but kinetic chains, by combining joints enable translatory motion to result from the rotary motions of the limb segments. Familiarity with the separate rotary motions and their limitations is necessary for comprehension of the characteristics of the resultant motion. By combining joints whose axes are not parallel, the kinetic chain enables a person to reach every point within his span of reach.

Kinetic energy. Energy due to motion. See Work.

Kyphosis. Abnormal curvature of the spine of the upper back in the anteroposterior plane.

L

Laboratory acquired infection. Any infection resulting from exposure to biohazardous materials in a laboratory environment. Exposure may be the result of a specific accident or inadequate biohazard control procedure or equipment.

Lacquer. A collodial dispersion or solution of nitrocellulose, or similar film forming compounds, resins and plasticizers in solvents and diluents used as a protective and decorative coating for various surfaces.

Laminar air flow. Streamlined airflow in which the entire body of air within a designated space moves with uniform velocity in one direction along parallel flow lines.

Lapping. The operation of polishing or sanding surfaces such as metal or glass to a precise dimension.

Laryngitis. Inflammation of the larynx.

Larynx. The organ by which the voice is produced. It is situated at the upper part of the trachea.

Laser. The acronym for Light Amplification by Stimulated Emission of Radiation. The gas laser is a type in which laser action takes place in a gas medium, usually a continuous wave (CW) laser, which has a continuous output—with an off time of less than 1 percent of the pulse duration time.

Laser light region. A portion of the electromagnetic spectrum, which includes ultraviolet, visible, and infrared light.

Laser system. An assembly of electrical, mechanical, and optical components, which includes a laser.

Latent heat. The amount of heat absorbed or evolved by 1 mole, or a unit mass, of a substance during a change of state (such as fusion, sublimation, or vaporization) at constant temperature and pressure.

Latent period. The time that elapses between exposure and the first manifestation of damage.

Latex. Original meaning: milky extract from rubber tree, containing about 35 percent rubber hydrocarbon; the remainder is water, proteins and sugars. This word also is applied to water emulsions of synthetic rubbers or resins. In emulsion paints, the film-forming resin is in the form of latex.

Lathe. A machine tool used to perform cutting operations on wood or metal by the rotation of the work piece.

Latissimus dorsi. A large flat muscle of the back that originates from the lower back and inserts into the humerus near the armpit. It adducts the upper arm, and when the elbow is abducted, it rotates the arm medially. It is actively used in operating equipment such as the drill press where a downward pull by the arm is required.

LC_{50}. Lethal concentration that will kill 50 percent of the test animals within a specified time. See LD_{50}.

LD_{50}. The dose required to produce the death in 50 percent of the exposed species within a specified time.

Lead poisoning. Lead compounds can produce poisoning when they are swallowed or inhaled. Inorganic lead compounds commonly cause symptoms of lead colic and lead anemia. Organic lead compounds can attack the nervous system.

Leakage radiation. Radiation emanating from the diagnostic source assembly except for the useful beam and radiation produced when the exposure switch or timer is not activated.

Lens, crystalline. Lens of the eye—a transparent biconvex body situated between the anterior chamber (aqueous) and the posterior chamber (vitreous) through which the light rays are further focused on the retina. The cornea provides most of the refractive power of the eye.

Lesion. An injury, damage, or abnormal change in a tissue or organ.

Lethal. Capable of causing death.

Leuk-, leuko- (prefix). White.

Leukemia. A group of malignent blood diseases distinguished by overproduction of white blood cells.

Leukocyte. White blood cell.

Leukocytosis. An abnormal increase in the number of white blood cells.

Leukopenia. A serious reduction in the number of white blood cells.

Lig- (prefix). Binding. A ligament ties two or more bones together.

Linear accelerator. A machine for speeding up charged particles such as protons. If differs from other accelerators in that the particles move in a straight line at all times instead of in circles or spirals.

Line-voltage regulation. The difference between the no-load and the load-line potentials expressed as a percent of the load-line potential.

Lipo- (prefix). Fat, fatty.

Liquefied petroleum gas. A compressed or liquefied gas usually composed of propane, some butane and lesser quantities of other light hydrocarbons and impurities; obtained as a by-product in petroleum refining. Used chiefly as a fuel, and in chemical synthesis.

Liquid. A state of matter in which the substance is a formless fluid that flows in accord with the law of gravity.

Liter (L). A measure of capacity—one quart equals 0.9 L.

Live room. A reverberant room that is characterized by an unusually small amount of sound absorption.

Liver. The largest gland or organ in the body, situated on the right side of the upper part of the abdomen. It has many important functions, including: regulating the amino acids in the blood; storing iron and copper for the body; forming and secreting bile, which aids in absorption and digestion of fats; transforming glucose into glycogen; and detoxifying exogenous substances.

Local exhaust ventilation. A ventilation system that captures and removes the contaminants at the point they are being produced before they escape into the workroom air.

Localized. Restricted to one spot or area in the body and not spread all through it—contrast with systemic.

Lordosis. The curvature of the lower back in the antero-posterior plane.

Loudness. The intensive attribute of an auditory sensation, in terms of which sounds may be ordered on a scale extending from soft to loud. Loudness depends primarily upon the sound pressure of the stimulus, but it also depends upon the frequency and wave form of the stimulus.

Loudness level. A sound, in phons, is numerically equal to the median sound pressure level, in decibels, relative to 0.0002 microbar, of a free progressive wave of frequency 1,000 Hz presented to listeners facing the source, which in a number of trials is judged by the listeners to be equally loud.

Louver. A slanted panel.

Low-pressure tank. A storage tank designed to operate at pressures at more than 0.5 psig but not more than 15 psig (3.5 to 103 kPa).

Lower explosive limit (LEL). The lower limit of flammability of a gas or vapor at ordinary ambient temperatures expressed in percent of the gas or vapor in air by volume. This limit is assumed constant for temperatures up to 120 C (250 F). Above this, it should be decreased by a factor of 0.7 because explosibility increases with higher temperatures.

LP-gas. See Liquefied petroleum gas.

Lumbar spine. The section of the lower spinal column or vertebral column immediately above the sacrum. Located in the small of the back and consisting of five large lumbar vetebrae, it is a highly stressed area in work situations and in supporting the body structure.

Lumbosacral joint. The joint between the fifth lumbar vertebrae and the sacrum. Often the site of spinal trauma because of large moments imposed by lifting tasks.

Lumen. The flux on one square foot of a sphere, one foot in radius, with a light source of one candle at the center that radiates uniformly in all directions.

Luminous flux. The rate of flow of light, measured in lumens.

Lymph. A pale, coagulable fluid that consists of a liquid portion resembling blood plasma and containing white blood cells (lymphocytes).

Lymph node. Small oval bodies having a glandlike structure and scattered widely throughout the body in the course of the vessels which carry the lymph. Also known as lymphatic nodes, lymph glands, and lymphatic glands.

Lymphoid. Resembling lymph.

Lysis. The distribution or breaking up of cells by either internal or external means.

M

MAC. Maximum allowable concentration.

Macroscopic. Visible without the aid of a microscope.

Macula. An oval area in the center of the retina devoid of blood vessels; the area most responsible for color vision.

Magnification. The number of times the apparent size of an object has been increased by the lens system of a microscope.

Makeup air. Clean, tempered outdoor air supplied to a work space to replace air removed by exhaust ventilation or some industrial process.

Malaise. Vague feeling of bodily discomfort.

Malignant. As applied to a tumor, cancerous and capable of undergoing metastasis, or invasion of surrounding tissue.

Malingerer. One who pretends illness or other abnormalities, or inability to perform one's duties or service.

Manometer. Instrument for measuring pressure; essentially a U-tube partially filled with a liquid (usually water, mercury, or a light oil), so constructed that the amount of displacement of the liquid indicates the pressure being exerted on the instrument.

Maser. Microwave Amplification by Stimulated Emission of Radiation. When used in the term optical maser, it is often interpreted as molecular amplification by stimulated emission of radiation.

Masking. The stimulation of one ear of a subject by controlled noise to prevent the person from hearing with that ear the tone or signal given to the other ear. This procedure is used where there is at least 15 to 20 dBA difference in the hearing level of the two ears.

Mass. Quantity of matter. The units are the gram and the pound.

Matter. Anything that has mass or occupies space.

Maximum evaporative capacity. The amount of evaporating sweat from a human being that the environment can accept.

Maximum line current. The rms current in the supply line of an x-ray machine operating at its maximum rating.

Maximum permissible concentration (MPC). These concentrations are set by the National Committee on Radiation Protection (NCRP). They are recommended maximum average concentrations of radionuclides to which a worker may be exposed, assuming that he works 8 hours a day, 5 days a week, and 50 weeks a year.

Maximum permissible dose (MPD). Currently, a permissible dose is defined as the dose of ionizing radiation that, in the light of present knowledge, is not expected to cause appreciable bodily injury to a person at any time during their lifetime.

Maximum permissible power or energy density. The intensity (power density or energy density) of laser radiation that, in light of present medical knowledge, is not expected to cause detectable bodily injury to a person at any time during their lifetime.

Maximum use concentration (MUC). The product of the protection factor of the respiratory protection equipment and the permissible exposure limit (PEL).

Mechanical efficiency curve. A graphical representation of a fan's relative efficiency in moving air at different airflow rates and static pressures.

Mechanotactic stress. Stress caused by contact with a mechanical environment.

Mechanotaxis. Contact with a mechanical environment consisting of forces (pressure, moment), vibration, etc. One of the ecological stress vectors. Improper design of the mechanotactic interface may lead to instantaneous trauma, cumulative pathogenesis, or death.

Median nerve. A major nerve controlling the flexor muscles of the wrist and hand. Tool handles and other objects to be grasped should make good contact with the sensory feedback area of this nerve located in the palmar surface of the thumb, index, middle, and part of the ring finger.

Medicament. A substance used in therapy.

Medium. See Culture medium.

Mega. One million—for example, megacurie = 1 million curies.

Mega-, megalo- (prefix). Large, huge. The prefix macro- has the same meaning.

Meiosis. The process whereby chromosome pairs undergo nuclear division as the germ cell matures.

Melan- (prefix). Black. The root usually refers in some way to cells that produce melanin, the dark pigment mobilized by a suntan. But it also endures in melancholy, "black bile," a gloomy humor anciently supposed to be the cause of wretchedness.

Melanderma. Abnormal darkening of the skin.

Melanocyte. A cell containing dark pigments.

Melt. In glass industry, the total batch of ingedients that may be introduced into pots or furnace.

Melting point. The transition point between the solid and the liquid state. Expressed as temperature at which this change occurs.

Membrane. A thin, pliable layer or sheet of animal tissue that serves to cover a surface, line the interior of a cavity or organ, or divide a space.

Membrane filter. A filter medium made from various polymeric materials such as cellulose, polyethylene, or tetrapolyethylene. Membrane filters usually exhibit narrow ranges of effective pore diameters and, therefore, are very useful in collecting and sizing microscopic and submicroscopic particles and in sterilizing liquids.

Men.-, meno- (prefix). Pertaining to menstruation, from a Greek word for "month."

Meniere's disease. The combination of deafness, tinnitus, and vertigo.

Meson. A particle that weighs more that the electron but generally less than the proton. Mesons cay be produced artificially. They are also produced by cosmic radiation (natural radiation coming from outer space). Mesons are not stable—they disintegrate in a fraction of a second.

Mesothelioma. Cancer of the membranes that line the chest and abdomen.

Metabolism. Refers to the flow of energy and the associated physical and chemical changes that are constantly taking place in the billions of cells that make up the living body.

Metal fume fever. A flulike condition caused from inhaling of the fumes of heated metal.

Metalizing. An operation involving the melting of wire by means of a flame in a special device that sprays the atomized metal onto a surface to be coated. The metal can be steel, lead, or other metal or alloy.

Metastasis. Transfer of the causal agent (cell or microorganism) of a disease from a primary focus to a distant one through the blood or lymphatic vessels. Also, spread of malignancy from site of primary cancer to secondary sites.

Methemoglobinemia. The presence of methemoglobin in the blood. (Methemoglobin—a compound formed when the iron moiety of hemoglobin is oxidized from the ferrous to the ferric state.) This protein inactivates the hemoglobin as an oxygen carrier.

Mev. Million electron volts is abbreviated as Mev.

Mica. A large group of silicates of varying composition, but similar in physical properties. All have excellent cleavage and can be split into very thin sheets. Used in electrical insulation.

Microbar. A unit of pressure, commonly used in acoutics, equals one dyne per square centimeter. A reference point for the decibel, which is accepted as 0.0002 dyne per square centimeter.

Microbe. A microscopic organism.

Microcurie. One-millionth of a curie (μc). A still smaller unit is the micromicrocurie ($\mu\mu c$).

Micron (micrometer). A unit of length equal to 10^{-4} centimeter, approximately 1/25,000 of an inch.

Microorganism. A minute organism; microbes, bacteria, cocci, viruses, molds, etc., are microorganisms.

Microphone. An electroacoustic transducer that responds to sound waves and delivers essentially equivalent electric waves.

Midsagittal plane. A reference plane formed by bisecting the human anatomy so as to have a right and left aspect. Human motor function can be described in terms of movement relative to the midsaggital plane.

Miliary. Characterized or accompanied by seedlike blisters or inflamed raised portions on tissue.

Milligram (mg). A unit of weight in the metric system. One thousand milligrams equal one gram.

Milligrams per cubic meter (mg/m^3). Unit used to measure air concentrations of dusts, gases, mists, and fumes.

Milliliter (mL). A metric unit used to measure volume. One milliliter equals one cubic centimeter.

Millimeter of mercury (mmHg). The unit of pressure equal to the pressure exerted by a column of liquid mercury one millimeter high at a standard temperature.

Milliroentgen. One one-thousandth of a roentgen. A roentgen is a unit of radioactive dose.

Millwright. A mechanic engaged in the erection and maintenence of machinery.

Mineral pitch. Tar from petroleum or coal in distinction to wood tar.

Mineral spirits. A petroleum fraction with boiling range between 149 and 240 C (300 and 400 F).

Miosis. Excessive smallness or contraction of the pupil of the eye.

Mists. Suspended liquid droplets generated by condensation from the gaseous to the liquid state or by breaking up a liquid into a dispersed state, such as by splashing, foaming, or atomizing. Mist is formed when a finely divided liquid is suspended in air.

Mitosis. Nuclear cell division in which resulting nuclei have the same number and kinds of chromosomes as the original cell.

Mixture. A combination of two or more substances that may be separated by mechanical means. The components may not be uniformly dispersed. See also Solution.

Moderator. A material used to slow neutrons in a reactor. These slow neutrons are particularly effective in causing fission. Neutrons are slowed down when they collide with atoms of light elements such as hydrogen, deuterium, and carbon, three common moderators.

Mold. (1) A growth of fungi forming a furry patch, as on stale bread or cheese. See Spore. (2) A hollow form or matrix into which molten material is poured to produce a cast.

Molecule. A chemical unit composed of one or more atoms.

Moment. Magnitude of the force times distance of application.

Moment concept. The concept based on theoretical and experimental bases that lifting stress depends on the bending moment exerted at susceptible points of the vertebral column rather than depending on weight alone.

Monaural hearing. Refers to hearing with one ear only.

Monochromatic. Single fixed wavelength.

Monomer. A compound of relatively low molecular weight which, under certain conditions, either alone or with another monomer, forms various types and lengths of molecular chains called polymers or copolymers of high molecular weight. Example: Styrene is a monomer which polymerizes readily to form polystyrene. See Polymer.

Morphology. The branch of biological science that deals with the study of the structure and form of living organisms.

Motile. Capable of spontaneous movement.

MPE. Maximum permissible exposure.

MPL. May be either maximum permissible level or limit, or dose. It refers to the tolerable dose rate for humans exposed to nuclear radiation.

Mppcf. Million particles per cubic foot.

mr. Millirem.

mR. Milliroentgen.

MSHA. The Mine Safety and Health Administration; a federal agency that regulates the mining industry in the safety and health area.

MSDS. Material Safety Data Sheet.

Mucous membranes. Lining of the hollow organs of the body, notably the nose, mouth, stomach, intestines, bronchial tubes, and urinary tract.

Musculoskeletal system. The combined system of muscles and bones which comprise the internal biomechanical environment.

Mutagen. Anything that can cause a change (mutation) in the genetic material of a living cell.

Mutation. A transformation of the gene that may alter characteristics of the offspring.

MWD. Megawatt days, usually per ton. The amount of energy obtained from one megawatt power in one day, normally used to measure the extent of burnup of nuclear fuel. 10,000 MWD per ton is about 1 percent burnup.

My-, myo- (prefix). Pertaining to muscle. Myocardium is the heart muscle.

Myelo- (prefix). Pertaining to marrow.

N

Nanometer. A unit of length equal to 10^{-7} centimeter.

Naphthas. Hydrocarbons of the petroleum type that contain substantial portions of paraffins and naphthalenes.

Narcosis. Stupor or unconsciousness produced by chemical substances.

Narcotics. Chemical agents that put a person to sleep, completely or partially.

Narrow band. Applies to a narrow band of transmitted waves, with neither of the critical or cutoff frequencies of the filter being zero or infinite.

Nascent. Just forming, as from a chemical or biological reaction.

Natural gas. A combustible gas composed largely of methane and other hydrocarbons with variable amounts of nitrogen and noncombustible gases; obtained from natural earth fissures or from driven wells. Used as a fuel in the manufacture of carbon black, and in chemical synthesis of many products. Major source of hydrogen for manufacture of ammonia.

Natural radioactivity. The radioactive background or, more properly, the radioactivity which is associated with the heavy naturally occurring elements.

Natural uranium. That which is purified from the naturally occurring ore, as opposed to that which may be enriched in fissionable content by processing at separation plants.

Nausea. An unpleasant sensation, vaguely referred to the epigastrium and abdomen, and often precedes vomiting.

NCRP. National Committee on Radiation Protection, an advisory group of scientists and professional people which makes recommendations for radiation protection in the United States.

Near field. In noise measurement this refers to a field in the immediate vicinity of the noise source where the sound pressure level does not follow the inverse square law.

Necro- (prefix). Dead.

Necrosis. Destruction of body tissue.

Neoplasm. A cellular outgrowth characterized by rapid cell multiplication. It may be benign (semicontrolled and restricted) or malignant.

Nephr-, nephro- (prefix). From the Greek for "kidney." Also see ren- .

Nephrotoxins. Chemicals that produce kidney damage.

Nephritis. Inflammation in the kidneys.

Neur-, neuro- (prefix). Denoting nerves.

Neural loss. Hearing loss; See Sensorineural.

Neuritis. Inflammation of a nerve.

Neurodermatitis. A chronic skin ailment of uncertain origin, possibly neurotic.

Neurological (neurology). That branch of medical science that deals with the nervous system, both normal and in disease.

Neurotoxin. Chemicals which produce their primary effect on the nervous system.

Neutrino. A particle resulting from nuclear reactions that carries energy away from the system but has no mass or charge, and is absorbed only with extreme difficulty.

Neutron. A constituent of the atomic nucleus. The neutron weighs about the same as the proton, and, as its name implies, has no electric charge. Neutrons make effective atomic projectiles for the bombardment of nuclei.

NFPA. The National Fire Protection Association is a voluntary membership organization whose aim is to promote and improve fire protection and prevention. The NFPA publishes 16 volumes of codes known as the National Fire Codes.

NIOSH. The National Institute for Occupational Safety and Health is a federal agency. It conducts research on health and safety concerns, tests, and certifies respirators, and trains occupational health and safety professionals.

Nitrogen fixation. Chemical combination or fixation of atmospheric nitrogen with hydrogen as in synthesis of ammonia. Fixation of nitrogen in soil is done by bacteria. Provides industrial and agricultural source of nitrogen.

Node. (1) A point, line, or surface in a standing wave where some characteristic of the wave field has essentially zero amplitude. (2) A small, round or oval mass of tissue; a collection of cells. (3) One of several constrictions occurring at regular intervals in a structure.

Nodule. A small mass of rounded or irregularly shaped cells or tissue; a small node.

Nodulizing. May be defined as simultaneous sintering and drum balling, usually in a rotary kiln.

Noise. Any unwanted sound.

Noise-induced hearing loss. The terminology used to refer to the slowly progressive inner ear hearing loss that results from exposure to continuous noise over a long period of time as contrasted to acoustic trauma or physical injury to the ear.

Nonauditory effects of noise. Refers to stress, fatigue, health, work efficiency, and performance effects of loud noise that is continuous.

Nonferrous metal. Metal (such as nickel, brass, or bronze) that does not include any appreciable amount of iron.

Nonionizing radiation. Electromagnetic radiation that does not cause ionization. Includes ultraviolet, laser, infrared, microwave, and radiofrequency radiation.

Nonpolar solvents. The aromatic and petroleum hydrocarbon groups characterized by low dielectric constants are referred to as nonpolar solvents.

Nonvolatile matter. The portion of a material that does not evaporate at ordinary temperatures.

Normal pulse (conventional pulse). Heart beat; also a single output event whose pulse duration is between 200 microseconds and one millisecond.

Nosocomial. (1) Pertaining to a hospital. (2) Of disease, caused or aggravated by hospital life.

NRC. Nuclear Regulatory Commission of the U.S. Department of Energy.

NTP. National Toxicology Program.

N-unit (or n-unit). A measure of radiation dose due to fast neutrons.

Nuclear battery. A device in which the energy emitted by decay of a radioisotope is converted first to heat and then directly to electricity.

Nuclear bombardment. The shooting of atomic projectiles at nuclei usually in an attempt to split the atom or to form a new element.

Nuclear energy. The energy released in a nuclear reaction, such as fission or fusion. Nuclear energy is popularly, though mistakenly, called "atomic energy."

Nuclear explosion. The rapid fissioning of a large amount of fissionable material. It creates an intense heat and light flash, a heavy blast and a large amount of radioactive fission products. These may be attached to dust and debris forming fallout. Nuclear explosions also result from nuclear fusion which does not give radioactive fission products.

Nuclear reaction. Result of the bombardment of a nucleus with atomic or subatomic particles or very high energy radiation. Possible reactions are emission of other particles or the splitting of the nucleus (fission). The decay of a radioactive material is also a nuclear reaction. Fusion is also a nuclear reaction.

Nuclear reactor. A machine for producing a controlled chain reaction in fissionable material. It is the heart of nuclear power plants where it serves as a source of heat. See Reactor.

Nucleonics. The application of nuclear science and techniques in physics, chemistry, astronomy, biology, industry, and other fields.

Nucleus. The inner core of the atom. It consists of neutrons and protons tightly locked together.

Nuclide. A species of atom characterized by its mass number, atomic number, and energy state of the nucleus, provided that the mean life in that state is long enough to be observable.

Nuisance dust. Have a long history of little adverse effect on the lungs and do not produce significant organic disease or toxic effect when exposures are kept under reasonable control.

Null point. The distance from a contaminant source that the initial energy or velocity of the contaminants is dissipated, and the material can be captured by a hood.

Nutrient. A substance that can be used for food.

O

Octave. The interval between two sounds having a basic frequency ratio of two.

Octave band. An arbitrary spread of frequencies. The top frequency in an octave band is always twice the bottom one. The octave band may be referred to by a center frequency.

Ocul-, oculo- (prefix) (Latin): and **ophthalmo-** (Greek). Both roots refer to the eye, "ophth" words refer more often to diseases.

Odor. That property of a substance that affects the sense of smell.

Odor threshold. The minimum concentration of a substance at which a majority of test subjects can detect and identify the characteristic odor of a substance.

Oil dermatitis. Blackheads and acne caused by oils and waxes that plug the hair follicles and sweat ducts.

Olecranon fossa. A depression in the back of the lower end of the humerus in which the ulna bone rests when the arm is straight.

Olefins. A class of unsaturated hydrocarbons characterized by relatively great chemical activity. Obtained from petroleum and natural gas. Examples: butene, ethylene and propylene. Generalized formula: C_nH_{2n}.

Olfactory. Has to do with the sense of smell.

Olig-, oligo- (prefix). Scanty, few, little. Oliguria, means scanty urination.

Oncogenic. Tumor generation.

Oncology. Study of causes, development, characteristics, and treatment of tumors.

Opacity. The condition of being nontransparent; a cataract.

Ophthalmologist. A physician who specializes in the structure, function, and diseases of the eye.

Optical density (OD). A logarithmic expression of the attenuation afforded by a filter.

Optically pumped laser. A type of laser which derives its energy from a noncoherent light source such as a xenon flash lamp. This laser is usually pulsed, and is commonly called solid-state laser.

Organ. An organized collection of tissues that have a special and recognized function.

Organ of Corti. An aggregation of nerve cells in the ear lying on the basilar membrane that picks up vibrations and converts them to electrical energy, which is sent to the brain and interpreted as sound. It is the "heart" of the hearing mechanism.

Organic. Term used to designate chemicals that contain carbon. To date nearly one million organic compounds have been synthesized or isolated. Many occur in nature; others are produced by chemical synthesis. See also Inorganic.

Organic disease. Disease in which some change in the structure of body tissue could either be visualized or positively inferred from indirect evidence.

Organic matter. Compounds containing carbon.

Organism. A living thing, as a human being, an animal, germ, plant, etc., especially one consisting of several parts, each specializing in a particular function.

Orifice. The opening that serves as an entrance and/or outlet of a body cavity or organ, especially the opening of a canal or a passage.

Orifice meter. A flowmeter, employing as the measure of flow rate the difference between the pressures measured on the upstream and downstream sides of a restriction within a pipe or duct.

Ortho- (prefix). Straight, correct, normal. Orthopsychiatry is the specialty concerned with "straightening out" behavior disorders.

Orthoaxi. The true anatomical axis about which a limb rotates as opposed to the assumed axis. The assumed axis is usually the most obvious or geometric one, while the orthoaxis is less evident and can only be referenced by the use of anatomical landmarks.

Os-, oste-, osteo- (prefix). Pertaining to bone. The Latin os- is most often associated with anatomical structures, the Greek osteo- with conditions involving bone. Osteogenesis means formation of bone.

Oscillation. The variation, usually with time, of the magnitude of a quantity with respect to a specified reference when the magnitude is alternately greater and smaller than the reference.

OSHA. U.S. Occupational Safety and Health Administration.

Osmosis. The passage of fluid through a semipermeable membrane as a result of osmotic pressure.

Osseous. Pertaining to bone.

Ossicle. A small bone, any member of a chain of three bones from the outer membrane of the tympanum (eardrum) to the membrane covering the oval window of the inner ear.

ot-, oto- (prefix). Pertaining to the ear. Otorrhea means a discharge from the ear.

Otitis media. An inflammation and infection of the middle ear.

Otologist. A physician who has specialized in surgery and diseases of the ear.

Otosclerosis. A condition of the ear caused by a growth of body tissue about the foot plate of the stapes and oval window of the inner ear. It results in a gradual loss of hearing. Surgery can often correct this.

Output power and output energy. Power is used primarily to rate CW lasers since the energy delivered per unit time remains relatively constant (output measured in watts). In contrast, pulsed lasers deliver their energy impulses in pulses and their effects may be best categorized by energy output per pulse. The power output level of CW lasers is usually expressed in milliwatts (mw = 1/1,000 watts), or watts pulsed lasers in kilowatts (kw = 1,000 watts), and q-switch pulsed lasers in megawatts (Mw = million watts) or gigawatts (gw = billion watts). Pulsed energy output is usually expressed as joules per pulse.

Overexposure. Exposure beyond the specified limits.

Oxidation. Process of combining oxygen with some other substance; technically, a chemical change in which an atom loses one or more electrons whether or not oxygen is involved. Opposite of reduction.

Oxygen deficiency. An atmosphere having less than the percentage of oxygen found in normal air. Normally, air contains about 21 percent oxygen at sea level. When the oxygen concentration in air is reduced to approximately 16 percent, many individuals become dizzy, experience a buzzing in the ears, and have a rapid heartbeat. More precisely, the deficiency occurs when the partial pressure of oxygen falls below 120 mmHg.

P

PAH. Polynuclear aromatic hydrocarbons.

Pair production. The conversion of a gamma ray into a pair of particles—an electron and a positron. This is an example of direct conversion of energy into matter according to Einstein's famous formula: $E = mc^2$; (energy) = (mass) × (velocity of light)^2.

Palmar arch. Blood vessels in the palm of the hand from which the arteries supplying blood to the fingers are branched. Pressure against the palmar arch by poorly designed tool handles may cause ischemia of the fingers and loss of tactile sensation and precision of movement.

Palpitation. Rapid heartbeat of which a person is acutely aware.

Papilloma. A small growth or tumor of the skin or mucous membrane. Warts and polyps are papillomas.

Papule. A small solid, usually conical elevation of the skin.

Papulovesicular. Characterized by the presence of papules and vesicles.

Para- (prefix). Alongside, near, abnormal. As in paraproctitis, inflammation of tissues near the rectum. A Latin suffix with the same spelling, -para, denotes bearing, giving birth, as multipara, a woman who has given birth to two or more children.

Paraffins, Paraffin series. (From parum affinis—small affinity.) Those straight- or branched-chain hydrocarbon components of crude oil and natural gas whose molecules are saturated (i.e., carbon atoms attached to each other by single bonds) and therefore very stable. Examples: methane and ethane. Generalized formula: C_nH_{2n+2}.

Parasite. An organism that derives its nourishment from a living plant or animal host. Does not necessarily cause disease.

Parenchyma. The distinguishing or specific (working) tissue of a bodily gland or organ, contained in and supported by the connective tissue framework, or stroma.

Parent. Also called precursor, this is the name given to a radioactive nucleus that disintegrates to form a radioactive product or daughter.

Partial barrier. An enclosure constructed so that sound transmission between its interior and its surroundings is minimized.

Particle. A small discrete mass of solid or liquid matter.

Particle concentration. Concentration expressed in terms of number of particles per unit volume of air or other gas. Note: When expressing particle concentrations the method of determining the concentration should be stated.

Particle size. The measured dimension of liquid or solid particles usually in units of microns.

Particle size distribution. The statistical distribution of the sizes or ranges of size of a population of particles.

Particulate. A particle of solid or liquid matter.

Particulate matter. A suspension of fine solid or liquid particles in air, such as dust, fog, fume, mist, smoke or sprays. Particulate matter suspended in air is commonly known as an aerosol.

Path-, patho-, pathy- (prefix). Feeling, suffering, disease. Pathogenic, producing disease; enteropathy, disease of the intestines; pathology, the medical specialty concerned with all aspects of disease. The root appears in the everyday word sympathy ("to feel with").

Pathogen. Any microorganism capable of causing disease.

Pathogenesis. Describes how a disease takes hold on the body, spreads, and what it does to the body.

Pathogenic. Producing or capable of producing disease.

Pathognomonic. Specifically distinctive or characteristic of a disease or pathologic condition; a sign or symptom on which a diagnosis can be made.

Pathological. Abnormal or diseased.

Pathology. The study of disease processes.

Pelleting. In various industries the material involved in powder form may be made into pellets or briquettes for convenience. The pellet is a distinctly small briquette. See Pelletizing.

Pelletizing. Refers primarily to extrusion by pellet mills. The word "pellet," however, carries its lay meaning in that it is also applied to other small extrusions and to some balled products. Pellets are generally regarded as being larger than grains, but smaller than briquettes.

Percent impairment of hearing. Percent impairment of hearing (percent hearing loss) is an estimate of a person's ability to hear correctly. It is usually based by means of an arbitrary rule, on the pure tone audiogram. The specific rule for calculating this quantity from the audiogram varies from state to state according to rule or law.

Percutaneous. Performed through the unbroken skin, as by application of an ointment to the skin and allowing the medicinal ingredients to be absorbed.

Peri- (prefix). Denoting around, about, surrounding. Periodontium is a word for tissues which surround and support the teeth.

Periodic table. Systematic classification of the elements according to atomic numbers (nearly the same order as by atomic weights) and by physical and chemical properties.

Permissible dose. See MPC, MPL.

Permissible exposure limit (PEL). An exposure limit that is published and enforced by OSHA as a legal standard.

Personal protective equipment. Devices worn by the worker to protect against hazards in the environment. Respirators, gloves, and hearing protectors are examples.

Pesticides. General term for that group of chemicals used to control or kill such pests as rats, insects, fungi, bacteria, weeds, etc., that prey on man or agricultural products. Among these are insecticides, herbicides, fungicides, rodenticides, miticides, fumigants, and repellents.

Petrochemical. A term applied to chemical substances produced from petroleum products and natural gas.

Pink noise. Noise that has been weighted, especially at the low end of the spectrum, so that the energy per band (usually octave band) is about constant over the spectrum.

pH. Means used to express the degree of acidity or alkalinity of a solution with neutrality indicated as seven.

Phagocyte. A cell in the body that characteristically engulfs foreign material and consumes debris and foreign bodies.

Phalanx (pl. phalanges). Any of the bones of the fingers or toes. Frequently used as anatomical reference points in ergonomic work analysis.

Pharmaceuticals. That group of drugs and related chemicals reaching the public primarily through drug suppliers. In government reports, category includes not only such medicinals as aspirin and antibiotics but also such nutriments as vitamins and amino acids for both human and animal use.

Pharyngeal. Pertaining to the pharynx. (Pharynx—the musculo-membranous sac between the mouth and nares and esophagus.) It is continuous below with the esophagus, and above it communicates with the mouth, nasal passages, and auditory tubes.

Phenol, C_6H_5OH. Popularly known as carbolic acid. Important chemical intermediate and base for plastics, pharmaceuticals, explosives, antiseptics, many other end products.

Phenolic resins. A class of resins produced as the condensation product of phenol or substituted phenol and formaldehyde or other aldehydes.

Phosphors. Materials capable of absorbing energy from suitable sources, such as visible light, cathode rays, or ultraviolet radiation and then emitting a portion of the energy in the ultraviolet, visible, or infrared region of the electromagnetic spectrum. In short, they are fluorescent or luminescent.

Photochemical process. Chemical changes brought about by radiant energy acting upon various chemical substances. See Photosynthesis.

Photoelectric effect. Occurs when an electron is thrown out of an atom by a light-ray or gamma-ray. This effect is used in an "electric eye." Light falls on a sensitive surface throwing out electrons which can then be detected.

Photomultiplier tube. A vacuum tube that multiplies the electron input.

Photon. A bundle (quantum) of radiation. Constitutes, for example, x-rays and light. Gamma-rays are also photons.

Photophobia. Abnormal sensitivity to and discomfort from light.

Photosynthesis. The process by which plants produce carbohydrates from carbon dioxide and water when the green tissues (chlorophyll) are exposed to sunlight. In reducing the carbon dioxide, oxygen is released. Were it not for this process, life on the earth would be impossible.

Physiology. The science and study of the functions or actions of living organisms.

Physiopathology. The science of functions in disease, or as modified by the disease.

Pig. (1) A container (usually lead) used to ship or store radioactive materials. The thick walls protect the person handling the container from radiation. (2) In metal refining, a small ingot from the casting of blast furnace metal.

Pigment. A finely divided, insoluble substance which imparts color to the material to which it is added.

Pile. A nuclear reactor. Called a pile because earliest reactors were "piles" of graphite blocks and uranium slugs.

Pilot plant. Small scale operation preliminary to major enterprises. Common in the chemical industry.

Pinna. Ear flap (part of ear that projects from the head); also known as the auricle.

Pitch. That attribute of auditory sensation in terms of which sounds may be ordered on a scale extending from low to high. Pitch depends primarily upon the frequency of the sound stimulus, but it also depends upon the sound pressure and wave form of the stimulus.

Pitot tube. A device consisting of two concentric tubes, one serving to measure the total or impact pressure existing in the airstream, the other to measure the static pressure only. When the annular space between the tubes and the interior of the center tube are connected across a pressure measuring device, the pressure difference automatically nullifies the static pressure, and the velocity pressure alone is registered.

Plasma. (1) The fluid part of the blood in which the blood cells are suspended. Also protoplasm. (2) A gas that has been heated to an at least partially ionized condition, enabling it to conduct an electric current.

Plasma arc welding (PAW). An arc welding process that produces coalescence of metals by heating them with a constricted arc between an electrode and the workpiece (transferred arc) or the electrode and the constricting nozzle (nontransferred arc). Shielding is obtained by the hot, ionized gas issuing from the orifice, which may be supplemented by an auxiliary source of shielding gas. Shielding gas can be an inert gas or a mixture of gases. Pressure may or may not be used, and filler metal may or may not be supplied.

Plastics. Officially defined as any one of a large group of materials that contains as an essential ingredient an organic substance of large molecular weight. Two basic types: thermosetting (irreversibly rigid) and thermoplastic (reversibly rigid). Before compounding and processing, plastics often are referred to as (synthetic) resins. Final form may be as film, sheet, solid, or foam; flexible or rigid.

Plasticizers. Organic chemicals used in modifying plastics, synthetic rubber, and similar materials to facilitate compounding and processing, and to impart flexibility to the end product.

Plenum. Pressure-equalizing chamber.

Plenum chamber. An air compartment connected to one or more ducts or connected to a slot in a hood used for air distribution.

Pleura. The thin membrane investing the lungs and lining the thoracic cavity, completely enclosing a potential space known as the pleural cavity. There are two pleurae, right and left, entirely distinct from each other. The pleura is moistened with a secretion that facilitates the movements of the lungs in the chest.

Plumbism. One name for lead intoxication.

Pleurisy. Caused when the outer lung lining (visceral pleura) and the chest cavity's inner lining (parietal pleura) lose their lubricating properties; the resultant friction causes irritation and pain.

Plume trap. An exhaust ventilation hood designed to capture and remove the plume given off the target upon impact of the laser beam.

Plutonium. A heavy element that undergoes fission under the impact of neutrons. It is a useful fuel in nuclear reactors. Plutonium cannot be found in nature, but can be produced and "burned" in reactors.

Pneumo- (Greek) **and pulmo-** (Latin) (prefix). Both terms pertain to the lungs.

Pneumoconiosis. Dusty lungs, as a result of the continued inhalation of various kinds of dusts or other particles.

Pneumoconiosis-producing dust. Dust, which when inhaled, deposited, and retained in the lungs, may produce signs, symptoms and findings of pulmonary disease.

Pneumonitis. Inflammation of the lungs—refers to certain forms of pneumonitis.

Point source. A source of radiation whose dimensions are small enough compared with distance between source and receptor for those dimensions to be neglected in calculations.

Poison. A material introduced into the reactor core to absorb neutrons. Any substance which when taken into the body, is injurious to health.

Polarography. A physical analysis method for the determination of certain atmospheric pollutants which are electroreducible or electrooxidizable and are in true solution and stable for the duration of the measurement.

Polar solvents. Solvents (such as alcohols and ketones) that contain oxygen. These have high dielectric constants.

Pollution. Synthetic contamination of soil, water, or atmosphere beyond that which is natural.

Poly- (prefix). Many.

Polycythemia. A condition marked by an excess in the number of red corpuscles in the blood.

Polymer. A high molecular weight material formed by the joining together of many simple molecules (monomers). There may be hundreds or even thousands of the original molecules linked end to end and often crosslinked. Rubber and cellulose are naturally occurring polymers. Most resins are chemically produced polymers.

Polymerization. A chemical reaction in which two or more small molecules (monomers) combine to form larger molecules (polymers) that contain repeating structural units of the original molecules. A hazardous polymerization is the above reaction, with an uncontrolled release of energy.

Polystyrene resins. Synthetic resins formed by polymerization of styrene.

Popliteal clearance. Distance between the front of the seating surface and the popliteal crease. This should be about 5 in. in good seat design to prevent pressure on the popliteal artery.

Popliteal crease (or line). The crease in the back of the leg in the hollow of the knee when lower leg is flexed. Important anatomical landmark.

Popliteal height of chair. The height of the highest part of the seating surface above the floor.

Popliteal height of individual. The height from the crease in the hollow of the knee to the floor is called the "popliteal height" of the individual concerned.

Porphyrin. One of a group of complex chemical substances that form the basis of the respiratory pigments of animals and plants, as hemoglobin and chlorophyll.

Portal. Place of entrance.

Portland cement. See Cement, portland.

Positive displacement pump. Any type of air mover pump in which leakage is negligible, so that the pump delivers a constant volume of fluid, building up to any pressure necessary to deliver that volume (unless, of course, the motor stalls or the pump breaks).

Positron. A particle that has the same weight and charge as an electron but is electrically positive rather than negative. The positron's existence was predicted in theory years before it was actually detected. It is not stable in matter because it reacts readily with an electron to give two gamma-rays.

Potential energy. Energy due to position of one body with respect to another or to the relative parts of the same body.

Power. Time rate at which work is done; units are the watt (one joule per second) and the horsepower (33,000 foot-pounds per minute). One horsepower equals 746 watts.

Power density. The intensity of electromagnetic radiation per unit area expressed as watts/cm^2.

Power level. The level, in decibels, is 10 times the logarithm to the base 10, of the ratio of a given power to a reference power. The reference power must be indicated.

ppb. Parts per billion.

PPE, Personal Protective Equipment. Includes items such as gloves, goggles, respirators, and protective clothing.

ppm. Parts per million parts of air by volume of vapor or gas or other contaminant.

Precision. The degree of agreement of repeated measurements of the same property, expressed in terms of dispersion of test results about the mean result obtained by repetitive testing of a homogeneous sample under specified conditions. The precision of a method is expressed quantitatively as the standard deviation computed from the results of a series of controlled determinations.

Presby- (prefix). Old. As in presbyopia, eye changes associated with aging.

Presbycusis. The hearing loss due to age. It is believed by some to be the degeneration of the nerve cells due to the ordinary wear and tear of the aging process.

Pressure. Force applied to, or distributed over a surface; measured as force per unit area. See Absolute pressure, Atmospheric pressure, Gage pressure, Standard temperature and pressure, Static pressure, Total pressure, Vapor pressure, and Velocity pressure.

Pressure drop. The difference in static pressure measured at two locations in a ventilation system due to friction or turbulence.

Pressure loss. Energy lost from a pipe or duct system through friction or turbulence.

Pressure, static. The normal force per unit area that would be exerted by a moving fluid on a small body immersed in it if the body were carried along with the fluid. Practically, it is the normal force per unit area at a small hole in a wall of the duct through which the fluid flows or on the surface of a stationary tube at a point where the disturbances, created by inserting the tube, cancel. The potential pressure exerted in all directions by a fluid at rest. It is the tendency to either burst or collapse the pipe, usually expressed in "inches water gage" when dealing with air.

Pressure, total. In the theory of the flow of fluids, the sum of the static pressure and the velocity pressure at the point of measurement. Also called dynamic pressure.

Pressure, vapor. The pressure exerted by a vapor. If a vapor is kept in confinement over its liquid so that the vapor can accumulate above the liquid, the temperature being held constant, the vapor pressure approaches a fixed limit called the maximum, or saturated, vapor pressure, dependent only on the temperature and the liquid.

Pressure vessel. A storage tank or vessel designed to operate at pressures greater than 15 psig (103 kPa).

PRF laser. A pulsed recurrence frequency laser, which is a pulsed-typed laser with properties similar to a CW laser if the frequency is very high.

Probe. A tube used for sampling or for measuring pressures at a distance from the actual collection or measuring apparatus. It is commonly used for reaching inside stacks or ducts.

Proct-, procto- (prefix). Pertaining to the anus or rectum. Proctoplasty, reparative or reconstructive surgery of the rectum.

Proliferation. The reproduction or multiplication of similar forms, especially of cells and morbid cysts.

Pronation. Rotation of the forearm in a direction to face the palm downward when the forearm is horizontal, and backward when the forearm is in a vertical position.

Propagation of flame. The spread of flame through the entire volume of the flammable vapor-air mixture from a single source of ignition. A vapor-air mixture below the lower flammable limit may burn at the point of ignition without propagating (spreading away) from the ignition source.

Prophylactic. Preventive treatment for protection against disease.

Protection factor (PF). With respiratory protective equipment—the ratio of the ambient airborne concentration of the contaminant to the concentration inside the facepiece.

Protective atmosphere. A gas envelope surrounding the part to be brazed, welded, or thermal sprayed, with the gas composition controlled with respect to chemical composition, dew point, pressure, flow rate, etc.

Protective coating. A thin layer of metal or organic material, as paint applied to a surface primarily to protect it from oxidation, weathering and corrosion.

Proteins. Large molecules found in the cells of all animal and vegetable matter and containing carbon, hydrogen, nitrogen, and oxygen, sometimes sulfur and phosphorus. Proteins are essential to life and growth. The fundamental structural units of proteins are amino acids.

Proteolytic. Capable of splitting or digesting proteins into simpler compounds.

Proton. A fundamental unit of matter having a positive charge and a mass number of one.

Protoplasm. The basic material from which all living tissue is made. Physically it is a viscous, translucent, semifluid colloid, composed mainly of proteins, carbohydrates, fats, salts, and water.

Protozoa. Single-celled microorganisms belonging to the animal kingdom.

Protozoan. Any of the single-celled, usually microscopic organisms of the phylum or subkingdom Protozoa, which includes the most primitive forms of animal life.

Proximal. Describing that part of a limb that is closest to the point of attachment. The elbow is proximal to the wrist, which is proximal to the fingers.

Psych-, psycho- (prefix). Pertaining to the mind, from the Greek word for "soul."

Psychogenic deafness. Loss originating in or produced by the mental reaction of an individual to their physical or social environment. It is sometimes called functional deafness or feigned deafness.

Psychoneuroses. Mental disorders that are of psychogenic origin, but present the essential symptoms of functional nervous diseases.

Psychrometer. An instrument consisting of wet and dry bulb thermometers for measuring relative humidity.

Psychrometric chart. A graphical representation of the thermodynamic properties of moist air.

Pterygium. A growth of the conjunctiva considered to be due to a degenerative process caused by long continued irritation (as from exposure to wind, dust, and possibly to ultraviolet radiation).

Pulmonary. Pertaining to the lungs.

Pulse length. Duration of a pulsed laser flash, it may be measured in milliseconds (msec. = 10^{-3} second) or microseconds ($\mu s = 10^{-6}$ second) or nanoseconds (ns = 10^{-9} second).

Pulsed laser. A class of laser characterized by operation in a pulsed mode, i.e., emission occurs in one or more flashes of short duration (pulse length).

Pumice. A natural silicate being volcanic ash or lava. Used as an abrasive.

Pupil. The variable aperature in the iris through which light travels toward the interior regions of the eye. The pupil size varies from 2 mm to 8 mm.

Pur-, pus- (Latin) and **pyo-** (Greek) (prefix). Indicates pus, as in purulent, suppurative, pustulent, and pyoderma, suppurative disease of the skin.

Pure tone. The sound wave characterized by its singleness of frequency.

Purpura. Extensive hemorrhage into the skin or mucous membrane.

Push-pull hood. A hood consisting of an air supply system on one side of the contaminant source blowing across the source and into an exhaust hood on the other side.

Putrefaction. Decomposition of proteins by microorganisms, producing disagreeable odors.

Pyloric stenosis. Obstruction of the pyloric opening of the stomach due to hypertrophy of the pyloric sphincter.

Pylorus. The orifice of the stomach communicating with the small intestine.

Pyel-, pyelo- (prefix). Pertaining to the urine-collecting chamber (pelvis) of the kidney.

Pyr, pyret- (prefix). Indicates fever.

Pyrethrum. A pesticide obtained from the dried, powdered flowers of the plant of the same name; mixed with petroleum distillates, it is used as an insecticide.

Pyrolysis. The breaking apart of complex molecules into simpler units by the use of heat, as in the pyrolysis of heavy oil into gasoline.

Q

QF. Quality factor for relating biological effects to the different types of radiation. This multiplication factor was once referred to as RBE (relative biological effectiveness).

Q fever. Disease caused by rickettsial organism that infects meat and livestock handlers; similar but not identical to tick fever.

Q-switched laser. (Also known as Q-spoiled). A pulsed laser capable of extremely high peak powers for very short durations (pulse length of several nanoseconds).

Q-value. The energy liberated or absorbed in a nuclear reaction is its Q-value.

Quality. A term used to describe the penetrating power of x-rays or gamma-rays.

Quality factor. A linear energy transfer dependent factor by which absorbed radiation doses are to be multiplied to obtain the dose equivalent.

Quantum. "Bundle of energy"; discrete particle of radiation. Plural: quanta.

Quartz. Vitreous, hard, chemically resistant, free silica, the most common form in nature. The main constituent in sandstone, igneous rocks, and common sands.

Quenching. A heat treating operation in which metal raised to the desired temperature is quickly cooled by immersion in oil bath.

R

Rabbit. A capsule that carries samples in and out of an atomic reactor through a pneumatic tube. Purpose is to permit study of the effect of intense radiation upon various materials.

rad. Radiation absorbed dose.

Radial deviation. Flexion of the hand which decreases the angle between its longitudinal axis and the radius. Tool design should minimize radial deviation. Strength of grasp is diminished in radial deviation.

Radian. An arc of the circle equal in length to the radius.

Radiant temperature. The temperature resulting from the body absorbing radiant energy.

Radiation (nuclear). The emission of atomic particles or electromagnetic radiation from the nucleus of an atom.

Radiation protection guide (RPG). The radiation dose that should not be exceeded without careful consideration of the reasons for doing so; every effort should be made to encourage the maintenance of radiation doses as far below this guide as practicable.

Radiation (radioactivity). See Ionizing radiation.

Radiation source. An apparatus or a material emitting or capable of emitting ionizing radiation.

Radiation (thermal). The transmission of energy by means of electromagnetic waves longer than visible light. Radiant energy of any wavelength may, when absorbed, become thermal energy and result in the increase in the temperature of the absorbing body.

Radiator. That which is capable of emitting energy in wave form.

Radioactive. The property of an isotope or element that is characterized by spontaneous decay to emit radiation.

Radioactivity. Emission of energy in the form of alpha-, beta-, or gamma-radiation from the nucleus of an atom. Always involves change of one kind of atom into a different kind. A few elements, such as radium, are naturally radioactive. Other radioactive forms are induced. See Radioisotope.

Radioactivity concentration guide (RCG). The concentration of radioactivity in the environment that is determined to result in organ doses equal to the radiation protection guide (RPG).

Radiochemical. Any compound or mixture containing a sufficient portion of radioactive elements to be detected by a Geiger counter.

Radiochemistry. That phase of chemistry concerned with the properties and behavior of radioactive materials.

Radiodiagnosis. A method of diagnosis that involves X-ray examination.

Radiohumeral joint. Part of the elbow. Not truly a joint, but a thrust bearing.

Radioisotope. A radioactive isotope of an element. A radioisotope can be produced by placing material in a nuclear reactor and bombarding it with neutrons. Many of the fission products are radioisotopes. Radioisotopes are sometimes used as tracers, as energy sources for chemical processing or food pasteurization, or as heat sources for nuclear batteries. Radioisotopes are at present the most widely used outgrowth of atomic research and are one of the most important peacetime contributions of nuclear energy.

Radionuclide. A radioactive nuclide; one that has the capability of spontaneously emitting radiation.

Radiopoison. A radioactive poison, such as plutonium, radium, or strontium-90.

Radioresistant. Relatively invulnerable to the effects of radiation.

Radiosensitive. Applies to tissues that are more easily damaged by the action of nuclear radiation.

Radiotherapy. Treatment of human ailments with the application of relatively high roentgen dosages.

Radium. One of the earliest known naturally radioactive elements. It is far more radioactive than uranium and is found in the same ores.

Radius. The long bone of the forearm in line with the thumb. It is the active element in the forearm during pronation (inward rotation) and supination (outward rotation). It also provides the forearm connection to the wrist joint.

Rale. Any sound or noise in the lungs that should not be there.

Random noise. A sound or electrical wave whose instantaneous amplitudes occur as a function of time, according to a normal (Gaussian) distribution curve. Random noise is an oscillation whose instantaneous magnitude is not specified for any given instant of time. The instantaneous magnitudes of a random noise are specified only by probability functions giving the fraction of the total time that the magnitude, or some sequence of the magnitudes, lies within a specific range.

Rare earths. Originally those elements in the periodic table with atomic numbers 57 through 71. Often included are numbers 39 and less frequently 21 and 90. Variety of emerging uses include manufacture of special steels and glasses.

Rash. Abnormal reddish coloring or blotch on some part of the skin.

Rated line voltage. The range of potentials in volts, of the supply line specified by the manufacturer at which the x-ray machine is designed to operate.

Rated output current. The maximum allowable lead current of the x-ray high-voltage generator.

Rated output voltage. The allowable peak potential, in volts, at the output terminals of the x-ray high-voltage generator.

Raynaud's syndrome, phenomenon. Abnormal constriction of the blood vessels of the fingers on exposure to cold temperature.

RBE. Relative biological effectiveness, the relative effectiveness of the same absorbed dose of two ionizing radiations in producing a measurable biological response.

Reactivity (chemical). A substance's susceptibility to undergoing a chemical reaction or change that may result in dangerous side effects, such as an explosion, burning, and corrosive or toxic emissions.

Reactor. An atomic "furnace," or nuclear reactor. In a reactor, nuclei of the fuel undergo controlled fission under the influence of neutrons. The fission produces new neutrons, in a chain reaction. This releases large amounts of energy. This energy is removed as heat that can be used to make steam for driving steam engines and to produce electricity. The moderator for the first reactor was piled-up blocks of graphite. Thus, a nuclear reactor was formerly referred to as a pile. Reactors are usually classified now as research, test, process heat, and power, depending on their principal function. No workable design for a controlled fusion reactor has yet been devised.

Reagent. Any substance used in a chemical reaction to produce, measure, examine, or detect another substance.

Recoil energy. When a nucleus undergoes a nuclear reaction such as fission or radioactive decay, the energy emitted is shared by the reaction products and is called recoil energy.

Reduction. Addition of one or more electrons to an atom through chemical change.

Refractories. A refractory material is one especially resistant to the action of heat and hence used for lining furnaces, etc., as fire clay, magnesite, graphite, and silica.

Regenerative process. Replacement of damaged cells by new cells.

Regimen. A regulation of the mode of living, diet, sleep, exercise, etc., for a hygienic or therapeutic purpose; sometimes mistakenly called regime.

Reid method. Method of determining the vapor pressure of a volatile hydrocarbon by the "Standard Method of Test for Vapor Pressure of Petroleum Products," ASTM D323.

Relative humidity. The ratio of the quantity of water vapor present in the air to the quantity that would saturate it at any specific temperature.

Reliability. The degree to which an instrument, component, or system retains its performance characteristics over a period of time.

Rem. Roentgen equivalent man, a dose unit that equals the dose in rads multiplied by the appropriate value of RBE for the particular radiation.

Renal. Having to do with the kidneys.

Replication. A fold or folding back, the act or process of duplicating or reproducing something.

Resin. A solid or semisolid amorphous (noncrystalline) organic compound or mixture of such compounds with no definite melting point and no tendency to crystallize. May be of vegetable (gum arabic), animal (shellac), or synthetic origin (celluloid). Some resins may be molded, cast, or extruded. Others are used as adhesives, in the treatment of textiles and paper, and as protective coatings.

Resistance. (1) Opposition to the flow of air, as through a canister, cartridge, particulate filter, or orifice. (2) A property of conductors depending on their dimensions, material, and temperature which determines the current produced by a given difference in electrical potential.

Resonance. Each object or volume of air will "resonate" or strengthen a sound at one or more particular frequencies. The frequency depends on the size and construction of the object or air volume.

Respirable size particulates. Particulates in the size range that permits them to penetrate deep into the lungs upon inhalation.

Respirator. A device to protect the wearer from inhalation of harmful contaminants.

Respiratory system. Consists of (in descending order)—the nose, mouth, nasal passages, nasal pharynx, pharynx, larynx, trachea, bronchi, bronchioles, air sacs (alveoli) of the lungs, and muscles of respiration.

Reticle. A scale or grid or other pattern located in the focus of the eyepiece of the microscope.

Retina. The light-sensitive inner surface of the eye that receives and transmits the image formed by the lens.

Retro- (prefix). Backward or behind.

Reverbatory furnace. A furnace in which heat is supplied by burning of fuel in a space between the charge and the low roof.

Rheumatoid. Resembling rheumatism that is a disease marked by inflammation of the connective tissue structures of the body, especially the membranous linings of the joints, and by pain in these parts; eventually the joints become stiff and deformed.

Rhin-, rhino- (prefix). Pertaining to the nose.

Rhinitis. Inflammation of the mucous membrane lining in the nasal passages.

Rickettsia. Rod-shaped microorganisms characterized by growing within the cells of animals. These human pathogens are often carried by arthropods.

Riser. In metal casting, a channel in a mold to permit escape of gases.

Roasting of ores. A refining operation in which ore is heated to a high temperature, some with catalytic agents, to drive off certain impurities such as roasting of copper ore to remove sulfur.

Roentgen (R). A unit of radioactive dose, or exposure was called a roentgen (pronounced rentgen). See rad.

Roentgenogram. A film produced by exposing x-ray film to x-rays.

Roentgenography. Photography by means of roentgen rays. Special techniques for roentgenography of different areas of the body have been given specific names.

Route of entry. The path by which chemicals can enter the body. There are three main routes of entry; inhalation, ingestion, and skin absorption.

Rosin. Specifically applies to the resin of the pine tree and chiefly derives from the manufacture of turpentine. Widely used in manufacture of soap, flux.

Rotameter. A flow meter, consisting of a precision-bored, tapered, transparent tube with a solid float inside.

Rotary kiln. Any of several types of kilns used to heat material, such as in the portland cement industry.

Rouge. A finely powdered form of iron oxide used as a polishing agent.

S

Safety can. An approved container, of not more than 19 L (5 gal) capacity, having a spring-closing lid and spout cover, and so designed that it will safely relieve internal pressure when subjected to fire exposure.

Sagittal plane. A plane from back to front vertically dividing the body into the right and left portions. Important in anthropometric definitions. Midsagittal plane is a sagittal plane, symetrically dividing the body.

Salamander. A small furnace usually cylindrical in shape, without grates, used for heating.

Salivation. An excessive discharge of saliva; ptyalism.

Salmonella. A genus of gram-negative, rod-shaped pathogenic bacteria.

Salt. A product of the reaction between an acid and a base. Example: Table salt is a compound of sodium and chlorine. It can be made by reacting sodium hydroxide with hydrochloric acid.

Sampling. A process consisting of the withdrawal or isolation of a fractional part of a whole. In air analysis, the separation of a portion of an ambient atmosphere with subsequent analysis to determine concentration.

Sandblasting. A process for cleaning metal castings and other surfaces with sand by a high-pressure air stream.

Sandhog. Any worker doing tunneling work requiring atmospheric pressure control.

Sanitize. To reduce the microbial flora in or on articles, such as eating utensils, to levels judged safe by public health authorities.

Saprophyte. An organism living on dead organic matter.

Sarcoma. Malignant tumors that arise in connective tissue.

Saturated vapor pressure. See Vapor pressure.

Saturation. The point at which the maximum amount of matter can be held dissolved at a given temperature in a solution.

Saturation pressure. The pressure at which a vapor confined above its liquid will be in stable equilibrium with it. Below saturation pressure, some of the liquid will change to vapor, and above saturation pressure, some of the vapor will condense to liquid.

Scarfskin, superficial. The outer strata of skin, also known as the epithelium, epidermis, or cuticle.

Scattered radiation. Radiation that is scattered in direction by interaction with objects or within tissue.

Scintillation counter. A device for counting atomic particles by means of tiny flashes of light (scintillations) which the particles produce when they strike certain crystals or liquids.

Scler- (prefix). Indicates hard, hardness. Arteriosclerosis is a condition of hardening of the arteries.

Sclera. The tough white outer coat of the eyeball.

Scleroderma. Hardening of the skin.

Scotoma. A blind or partially blind area in the visual field.

Sealed source. A radioactive source sealed in a container or having a bonded cover, where the container or cover has sufficient mechanical strength to prevent contact with and dispersion of the radioactive material under the conditions of use and wear for which it was designed.

Sebaceous. Of, related to, or being fatty material.

Seborrhea. An oily skin condition caused by an excess output of sebum from the sebaceous glands of the skin.

SCBA. Self contained breathing apparatus.

Semicircular canals. The special organs of balance that are closely associated with the hearing mechanism and the eighth cranial nerve.

Semiconductor or junction laser. A class of laser that normally produces relatively low CW power outputs. This class of laser can be "tuned" in wavelength and has the greatest efficiency.

Sensation. The translation into consciousness of the effects of a stimulus exciting one of the organs of the senses.

Sensible. Capable of being perceived by the sense organs.

Sensitivity. The minimum amount of contaminant that can repeatedly be detected by an instrument.

Sensitization. The process of rendering an individual sensitive to the action of a chemical.

Sensitizer. A material that can cause allergic reaction of the skin or respiratory system.

Sensorineural. Type of hearing loss that affects millions of people. If damage to the inner ear is at fault, the hearing loss is sensory; if the fibers of the eighth nerve are affected, it is a neural loss. Since the pattern of hearing loss is the same in either case, the term "sensorineural" is used.

Sensory end organs. Receptor organs of the sensory nerves located in the skin. Each end organ can sense only a specific type of stimulus. Primary stimuli are heat, cold, or pressure, each requiring different end organs.

Sensory feedback. Use of external signals perceived by sense organs (e.g., eye, ear) to indicate quality or level of performance of an event triggered by voluntary action. On the basis of sensory feedback information, decisions may be made; e.g., permitting or not permitting an event to run its course; enhancing or decreasing activity levels.

Septa (septum). A dividing wall or partition; used as a general term in anatomical nomenclature.

Septicemia. Blood poisoning, actually growth of infectious organisms in the blood.

Sequela (Pl. sequelae). Any lesion or affection following or caused by an attack of disease.

Sequestrants. Chelates used to deactivate undesirable properties of metal ions without the necessity for removing these ions from solution. Sequestrants find many uses, including application as antigumming agents in gasoline, antioxidants in rubber, and as rancidity retardants in edible fats and oils.

Serum. (1) The clear fluid that separates from the blood when blood clots. (2) Blood-serum—containing antibodies.

Shakes. Workers' name for "metal fume fever."

Shakeout. In the foundry industry, the separation of the solid, but still not cold, casting from its molding sand.

Shale. Many meanings in industry, but in geology a common fossil rock formed from clay, mud, or silt somewhat stratified but without characteristic cleavage.

Shale oil. Some shale is bituminous and on distillation yields a tarry oil.

Shaver's disease. Bauxite pneumoconiosis.

Shell. The electrons around the nucleus are arranged in shells, that is, spheres centered on the nucleus. The innermost shell, closest to the nucleus, is called K-shell, the next one L-shell and so on to the Q-shell. The nucleus itself may have a shell-type structure.

Shield, shielding. Interposed material (like a wall) that protects workers from harmful radiations released by radioactive materials.

Shielded metal arc welding (SMAW). An arc-welding process that produces coalescence of metals by heating them with an arc between a covered metal electrode and the work. Shielding is obtained from decomposition of the electrode covering. Pressure is not used and filler metal is obtained from the electrode.

Shock. Primarily, the rapid fall in blood pressure following injury, operation, or the administration of anesthesia.

Shotblasting. A process for cleaning of metal castings or other surfaces by small steel shot in a high-pressure air stream. This process is a substitute for sandblasting to avoid silicosis.

SI. The International System of Units, the metric system that is being adopted throughout the world. It is a modern version of the MKSA (meter, kilogram, second, ampere) system whose details are published and controlled by an international treaty organization financed by the Member States of the Metre Convention, of which the United States is a member.

Siderosis. The deposition of iron pigments in the lung—can be associated with disease.

Silica gel. A regenerative absorbent consisting of the amorphous silica manufactured by the action of HCl on sodium silicate. Hard, glossy, quartzlike in appearance. Used in dehydrating and in drying and as a catalyst carrier.

Silicates. Compounds of silicon, oxygen, and one or more metals with or without hydrogen. These dusts cause nonspecific dust reactions, but generally do not interfere with pulmonary function or result in disability.

Silicon. A nonmetallic element being, next to oxygen, the chief elementary constitutent of the earth's crust.

Silicones. Unique group of compounds made by molecular combination of the elements silicon or certain of its compounds with organic chemicals. Produced in variety of forms, including silicone fluids, resins and rubber. Silicones have special properties, such as water repellency, wide temperature resistance, high durability, and great dielectric strength.

Silicosis. A disease of the lungs caused by the inhalation of silica dust.

Silicotuberculosis. Tuberculous infection of the silicotic lung, infective silicosis.

Silver solder. A solder of varying components but usually containing an appreciable amount of cadmium.

Simple tone (pure tone). (1) A simple tone is a sound wave, the instantaneous sound pressure of which is a simple sinusoidal function of time; (2) a simple tone is a sound sensation characterized by its singleness of pitch.

Sintering. Process of making coherent powder of earthy substances by heating, but without melting.

Skin dose. A special instance of tissue dose, referring to the dose immediately on the surface of the skin.

Slag. The dross of flux and impurities that rise to the surface of molten metal during melting and refining.

Slot velocity. Linear flow rate through the opening in a slot-type hood (plating, degreasing operations, etc.).

Short-term exposure limit (STEL). ACGIH-recommended exposure limit. Maximum concentration to which workers can be exposed for a short period of time (15 minutes) for only four times throughout the day with at least one hour between exposures.

Sludge. In general, any muddy or slushy mass. As to specific applications, mud from a drill hole in boring; muddy sediment in the steam boiler; precipitated solid matter arising from sewage treatment processes.

Slug. A "fuel element" for a nuclear reactor, a piece of fissionable material. The slugs in large reactors consist of uranium metal coated with aluminum to prevent corrosion.

Slurry. A thick, creamy liquid resulting from the mixing and grinding of limestone, clay, and other raw materials with water.

Smelting. One step in the procurement of metals from ores—hence to reduce, to refine, to flux or to scorify.

Smog. Irritating haze resulting from the sun's effect on certain pollutants in the air, notably those from automobile and industrial exhaust.

Smoke. An air suspension (aerosol) of particles, originating from combustion or sublimation. Carbon or soot particles less than 0.1μ in size result from incomplete combustion of carbonaceous materials such as coal or oil. Smoke generally contains droplets as well as dry particles. Tobacco, for instance, produces a wet smoke composed of minute tarry droplets.

Soaking pit. A device in steel manufacturing in which ingots with still molten interiors stand in a heated upright chamber until solidification is complete.

Soap. Ordinarily a metal salt of a fatty acid, usually sodium stearate, sodium oleate, sodium palmitate, or some combination of these.

Soapstone. Complex silicate of varied composition similar to some talcs with wide industrial application such as in rubber manufacture.

Solder. A material used for joining metal surfaces together by filling a joint or covering a junction. The most commonly used solder is one containing lead and tin. Silver solder may contain cadmium. Zinc chloride and fluorides are commonly used as fluxes to clean the surfaces.

Solid-state laser. A type of laser that utilizes a solid crystal such as ruby, or glass. This type is most commonly used in pulsed lasers.

Solution. Mixture in which the components lose their identities and are uniformly dispersed. All solutions are composed of a solvent (water or other fluid) and the substance dissolved called the "solute." A true solution is homogeneous as salt in water.

Solvent. A substance that dissolves another substance. Usually refers to organic solvents.

Soma. Body, as distinct from psyche, mind.

Somatic. Pertaining to all tissue other than reproductive cells.

Somnolence. Sleepiness; also unnatural drowsiness.

Soot. Agglomerations of particles of carbon impregnated with "tar," formed in the incomplete combustion of carbonaceous material.

Sorbent. (1) A material that removes toxic gases and vapors from air inhaled through a canister or cartridge. (2) Material used to collect gases and vapors during air-sampling.

Sound. An oscillation in pressure, stress, particle displacement, particle velocity, etc., which is propagated in an elastic material, in a medium with internal forces (e.g., elastic, viscous), or the superposition of such propagated oscillations. Sound is also the sensation produced through the organs of hearing—usually by vibrations transmitted in a material medium, commonly air.

Sound absorption. The change of sound energy into some other form, usually heat, in passing through a medium or on striking a surface. In addition, sound absorption is the property possessed by materials and objects, including air, of absorbing sound energy.

Sound analyzer. A device for measuring the band pressure level or pressure-spectrum level of a sound as a function of frequency.

Sound level. A weighted sound pressure level, obtained by the use of metering characteristics and the weighting A, B, or C specified in ANSI S1.4.

Sound-level meter and octave-band analyzer. Instruments for measuring sound pressure levels in decibels referenced to 0.0002 microbars. Readings can also be made in specific octave bands, usually beginning at 75 Hz and continuing through 10,000 Hz.

Sound pressure level, SPL. The level, in decibels, of a sound is 20 times the logarithm to the base 10 of the ratio of the pressure of this sound to the reference pressure. The reference pressure must be explicitly stated.

Sound transmission. The word "sound" usually means sound waves traveling in air. However, sound waves also travel in solids and liquids. These sound waves may be transmitted to air to make sound we can hear.

Sound transmission loss. A barrier's ability to block transmission is indicated by its transmission loss (TL) rating, measured in decibels.

Sour gas. Slang for either natural gas or a gasoline contaminated with odor-causing sulfur compunds. In natural gas, the contaminant is usually hydrogen sulfide; in gasolines, usually mercaptans.

Source. Any substance that emits radiation. Usually refers to a piece of radioactive material conveniently packaged for scientific or industrial use.

Spasm. Tightening or contraction of any set(s) of muscles.

Specific gravity. The ratio of the mass of a unit volume of a substance to the mass of the same volume of a standard substance at a standard temperature. Water at 4 C (39.2 F) is the standard usually referred to for liquids; for gases, dry air (at the same temperature and pressure as the gas) is often taken as the standard substance. See Density.

Specific ionization. See Ionization.

Specific volume. The volume occupied by a unit mass of a substance under any specified conditions of temperature and pressure.

Specific weight. The weight per unit volume of a substance, same as density.

Specificity. The degree to which an instrument or detection method is capable of accurately detecting or measuring the concentration of a single contaminant in the presence of other contaminants.

Spectrography—spectral emission. An instrumental method for detecting trace contaminants utilizing the formation of a spectrum by exciting the contaminants under study by various means, causing characteristic radiation to be formed, which is dispersed by a grating or a prism and photographed.

Spectrophotometer. An instrument used for comparing the relative intensities of the corresponding colors produced by chemical reactions.

Spectroscopy. Observation of the wavelength and intensity of light, or other electromagnetic waves, absorbed or emitted by various materials. When excited by an arc or spark, each element emits light of certain well-defined wavelengths.

Spectrum. The distribution in frequency of the magnitudes (and sometimes phases) of the components of the wave. Spectrum also is used to signify a continuous range of frequencies, usually wide in extent, within which waves have some specified common characteristics. Also, the pattern of red-to-blue light observed when a beam of sunlight passes through a prism and then projects upon a surface.

Speech interference level (SIL). The speech interference level of a noise is the average, in decibels, of the sound pressure levels of the noise in the three octave bands of frequency 600-1,200, 1,200-2,400, and 2,400-4,800 Hz.

Speech perception test. A measurement of hearing acuity by the administration of a carefully controlled list of words. The identification of correct responses is evaluated in terms of norms established by the average performance of normal listeners.

Speech reading. Also called lip reading or visual hearing.

Sphincter. A muscle that surrounds an orifice and functions to close it.

Sphygmomanometer. Apparatus for measuring blood pressure (and a good word for testing spelling ability).

Spore. A resistant body formed by certain microorganisms; resistant resting cells. Mold spores: unicellular reproductive bodies.

Spot size. Cross-sectional area of laser beam at the target.

Spot welding. One form of electrical-resistance welding in which the current and pressure are restricted to the spots of metal surfaces directly in contact.

Spray coating painting. The result of the application of a spray in painting as a substitute for brush painting or dipping.

Squamous. Covered with or consisting of scales; scaley.

Stain. A dye used to color microorganisms as an aid to visual inspection.

Stamping. Many different usages in industry, but a common one is the crushing of ores by pulverizing.

Standard air. Air at standard temperature and pressure. The most common values are 21.1 C (70 F) and 101.3 kPa (29.92 in. Hg). Also, air with a density of 1.2 kg/m^3 (0.075 lb/cu ft) is substantially equivalent to dry air at 70 F and 29.92 in. Hg barometer.

Standard air density. The density of air. 0.075 lb/cu ft (1.2 kg/m^3), at standard conditions.

Standard conditions. In industrial ventilation, 21.1 C (70 F), 50 percent relative humidity, and 101.3 kPa (29.92 in. of mercury) atmospheric pressure.

Standard gravity. Standard accepted value for the force of gravity. It is equal to the force that will produce an acceleration of 9.8 m/s^2 (32.17 ft per sec).

Standard Industrial Classification (SIC) Code. Classification system for places of employment according to major type of activity.

Standard man. A theoretical physically fit man of standard (average) height, weight, dimensions, and other parameters (blood composition, percentage of water, mass of salivary glands, to name a few), used in studies of how heat or ionizing radiation affects humans.

Standard temperature and pressure. See Standard air.

Standing wave. A standing wave is a periodic wave having a fixed distribution in space which is the result of interference of progressive waves of the same frequency and kind. Such waves are characterized by the existence of nodes or partial nodes and antinodes that are fixed in space.

Stannosis. A form of pneumoconiosis caused by the inhalation of tin-bearing dusts.

Static pressure. The potential pressure exerted in all directions by a fluid at rest. For a fluid in motion, it is measured in a direction normal (at right angles) to the direction of flow, thus it shows the tendency to burst or collapse the pipe. When added to velocity pressure, it gives total pressure.

Static pressure curve. A graphical representation of the volumetric output and fan static pressure relationship for a fan operating at a specific rotating speed.

Static pressure regain. The increase in static pressure in a system as air velocity decreases and velocity pressure is converted into static pressure according to Bernoulli's theorem.

Sterile. Free of living microorganisms.

Sterility. Inability to reproduce.

Sterilization. The process of making sterile; the killing of all forms of life.

Sterilize. To perform any act that results in the absence of all life on or in an object.

Sternomastoid muscles. A pair of muscles connecting the breastbone and lower skull behind the ears, which flex or rotate the head.

Stink damp. In mining, hydrogen sulfide.

Stom-, stomato- (prefix). Pertaining to the mouth or a mouth.

Stp flow rate. The rate of flow of fluid, by volume, corrected to standard temperature and pressure.

Stp volume. The volume that a quantity of gas or air would occupy at standard temperature and pressure.

Stress. A physical, chemical, or emotional factor that causes bodily or mental tension and may be a factor in disease causation or fatigue.

Stressor. Any agent or thing causing a condition of stress.

Strip mine. A mine in which coal or ore is extracted from the earth's surface after removal of overlayers of soil, clay and rock.

Stupor. Partial unconsciousness or nearly complete unconciousness.

Sublimation. A process in which a material passes directly from a solid into a gaseous state and condenses to form solid crystals, without liquefying.

Sulcus (pl. sulci). A groove, trench, or furrow; used in anatomical nomenclature as a general term to designate such a depression, especially one of those on the surface of the brain, separating the gyri; also, a linear depression in the surface of a tooth, the sloping sides of which meet at an angle.

Supination. Rotation of the forearm about its own longitudinal axis. Supination turns the palm upward when the forearm is horizontal, and forward when the body is in anatomical position. Supination is an important element of available motions inventory for industrial application, particularly where tools such as screwdrivers are used. Efficiency in supination depends on arm position. Workplace design should provide for elbow flexion at 90 degrees.

Supplied-air suit. A one- or two-piece suit that is impermeable to most particulate and gaseous contaminants and is provided with an adequate supply of respirable air.

Supra- (prefix). Above, upon.

Surface-active agent or **Surfactant.** Any of a group of compounds added to a liquid to modify surface or interfacial tension. In synthetic detergents, which is the best known use of surface active agents, reduction of interfacial tension provides cleansing action.

Surface coating. Term used to include paint, lacquer, varnish, and other chemical compositions used for protecting and/or decorating surfaces. See Protective coating.

Suspect carcinogen. A material which is believed to be capable of causing cancer but for which there is limited scientific evidence.

Sweating. (1) Visible perspiration; (2) the process of uniting metal parts by heating solder so that it runs between the parts.

Sweetening. The process by which petroleum products are improved in odor by chemically changing certain sulfur compounds of objectionable odor into compounds having little or no odor.

Swing grinder. A large power-driven grinding wheel mounted on a counterbalanced swivel-supported arm guided by two handles.

Symptom. Any bit of evidence from a patient indicating illness; the subjective feelings or the patient.

Syncope. Fainting spell.

Syndrome. A collection, constellation, or concurrence of signs and symptoms, usually of disease.

Synergism. Cooperative action of substances whose total effect is greater than the sum of their separate effects.

Synergistic. Pertaining to an action of two or more substances, organs or organisms to achieve an effect of which is greater than the additive effects of each alone.

Synonym. Another name by which the same chemical may be known.

Synthesis. The reaction or series of reactions by which a complex compound is obtained from simpler compounds or elements.

Synthetic. (From Greek work synthetikos—that which is put together.) "Man-made 'synthetic' should not be thought of as a substitute for the natural," states *Encyclopedia of the Chemical Process Industries;* it adds, "Synthetic chemicals are frequently more pure and uniform than those obtained naturally." Classic example: synthetic indigo.

Synthetic detergents. Chemically tailored cleaning agents soluble in water or other solvents. Originally developed as soap substitutes. Because they do not form insoluble precipitates, they are especially valuable in hard water. They may be composed of surface-active agents alone, but generally are combinations of surface-active agents and other substances, such as complex phosphates, to enhance detergency.

Synthetic rubber. Artifically made polymer with rubberlike properties. Various types have varying composition and properties. Major types designated as S-type, butyl, neoprene (chloroprene polymers) and N-type. Several synthetics duplicate the chemical structure of natural rubber.

Systemic. Spread throughout the body, affecting all body systems and organs, not localized in one spot or area.

T

Tachy- (prefix). Indicates fast, speedy, as in tachycardia, abnormally rapid heartbeat.

Tailings. In mining or metal recovery processes, the gangue rock residue from which all or most of the metal has been extracted.

Talc. A hydrous magnesium silicate used in ceramics, cosmetics, paint, and pharmaceuticals, and as a filler in soap, putty, and plaster.

Tall oil. (Name derived from Swedish word tallolja; material first investigated in Sweeden—not synonymous with U.S. pine oil.) Natural mixture of rosin acids, fatty acids, sterols, high-molecular weight alcohols, and other materials, derived primarily from waste liquors of sulfate wood pulp manufacture. Dark brown, viscous, oily liquid often called liquid rosin.

Tar. A loose term embracing wood, coal, or petroleum exudations. In general represents complex mixture of chemicals of top fractional distillation systems.

Tar crude. Organic raw material derived from distillation of coal tar and used for chemicals.

Tare. A deduction of weight, made in allowance for the weight of a container or medium. The initial weight of a filter, for example.

Target. The material into which the laser beam is fired or at which electrons are fired in an x-ray tube.

Temper. To relieve the internal stresses in metal or glass and to increase ductility by heating the material to a point below its critical temperature and cooling slowly. See Anneal.

Temperature. The condition of a body which determines the transfer of heat to or from other bodies. Specifically, it is a manifestation of the average translational kinetic energy of the molecules of a substance due to heat agitation. See Celsius and Kelvin scale.

Temperature, dry-bulb. The temperature of a gas or mixture of gases indicated by an accurate thermometer after correction for radiation.

Temperature, effective. An arbitrary index that combines into a single value the effect of temperature, humidity, and air movement on the sensation of warmth or cold felt by the human body. The numerical value is that of the temperature of still, saturated air that would induce an identical sensation.

Temperature, mean radiant (MRT). The temperature of a uniform black enclosure in which a solid body or occupant would exchange the same amount of radiant heat as in the existing nonuniform environment.

Temperature, wet-bulb. Thermodynamic wet-bulb temperature is the temperature at which liquid or solid water, by evaporating into air, can bring the air to saturation adiabatically at the same temperature. Wet-bulb temperature (without qualification) is the temperature indicated by a wet-bulb psychrometer.

Tempering. The process of heating or cooling makeup air to the proper temperature.

Temporary threshold shift (TTS). The hearing loss suffered as the result of noise exposure, all or part of which is recovered during an arbitrary period of time when one is removed from the noise. It accounts for the necessity of checking hearing acuity at least 16 hours after a noise exposure.

Tendon. Fibrous component of a "muscle." It frequently attaches to bone at the area of application of tensile force. When its cross section is small, stresses in the tendon are high, particularly because the total force of many muscle fibers is applied at the single terminal tendon. See Tenosynovitis.

Tennis elbow. Sometimes called lateral epicondylitis. An inflammatory reaction of tissues in the lateral elbow region.

Tenosynovitis. Inflammation of the connective tissue sheath of a tendon.

Teratogen. An agent or substance that may cause physical defects in the developing embryo or fetus when a pregnant female is exposed to that substance.

Terminal velocity. The terminal rate of fall of a particle through a fluid as induced by gravity or other external force; the rate at which frictional drag balances the accelerating force (or the external force).

Tetanus. A disease of sudden onset caused by the toxin of the bacterium called *Clostridium tetani*. It is characterized by muscle spasms. Also called "lockjaw."

Therm. A quantity of heat equivalent to 100,000 Btu.

Thermal pollution. Discharge of heat into bodies of water to the point that increased warmth activates all sewage, depletes the oxygen the water needs to cleanse itself, and eventually destroys some of the fish and other organisms in the water. Eventually thermal pollution makes the water smell bad and taste bad.

Thermonuclear reaction. A fusion reaction, that is, a reaction in which two light nuclei combine to form a heavier atom, releasing a large amount of energy. This is believed to be the sun's source of energy. It is called thermonuclear because it occurs only at a very high temperature.

Thermoplastic. Capable of being repeatedly softened by heat.

Thermoplastic plastics. Those that can repeatedly melt or that soften with heat and harden on cooling. Examples: vinyls, acrylics, and polyethylene.

Thermosetting. Capable of undergoing a chemical change from a soft to a hardened substance when heated.

Thermosetting plastics. Those that are heat-set in their final processing to a permanently hard state. Examples: phenolics, ureas, and melamines.

Thermostable. Resistant to changes by heat.

Thimble ionization chamber. A small cylindrical or spherical ionization chamber, usually with walls of organic materials.

Thinner. A liquid used to increase the fluidity of paints, varnishes, and shellac.

Threshold. The level where the first effects occur; also the point at which a person just begins to notice the tone is becoming audible.

Thromb- (prefix). Pertaining to a blood clot.

Thyrotoxicosis. A morbid condition resulting from overactivity of the thyroid gland.

Timbre. The quality given to a sound by its overtones; the tone distinctive of a singing voice or a musical instrument. Pronounced tam′ber or tim′ber.

Time-weighted average concentration (TWA). Refers to concentrations of airborne toxic materials which have been weighted for a certain time duration, usually 8 hours.

Tinning. Any work with tin such as tin roofing, but in particular in soldering, the primary coating with solder of the two surfaces to be united.

Tinnitus. A ringing sound in the ears.

Tissue. A large group of similar cells, bound together to form a structural component. An organ is composed of several kinds of tissue, and in this respect it differs from a tissue as a machine differs from its parts.

TLV. Threshold Limit Value. A time-weighted average concentration under which most people can work consistently for 8 hours a day, day after day, with no harmful effects. A table of these values and accompanying precautions is published annually by the American Conference of Governmental Industrial Hygienists. See Appendix B-1.

Tolerance. (1) The ability of the living organism to resist the usually anticipated stress. (2) Also, the limits of permissible inaccuracy in the fabrication of an article above and below its design specifications.

Tolerance dose. See Maximum permissible concentration and MPL.

Toluene, $C_6H_5CH_3$. Hydrocarbon derived mainly from petroleum but also from coal. Source of TNT, lacquers, saccharin, and many other chemicals.

Tone deafness. The inability to make a close discrimination between fundamental tones close together in pitch.

Topography. Configuration of a surface, including its relief and the position of its natural and man-made features.

Total pressure. The algebraic sum of the velocity pressure and the static pressure (with due regard to sign).

Toxemia. Poisoning by the way of the bloodstream.

Toxicant. A poison or poisonous agent.

Toxicity. A relative property of a chemical agent and refers to a harmful effect on some biologic mechanism and the condition under which this effect occurs.

Toxin. A poisonous substance that is derived from an organism.

Tracer. A radioisotope mixed with a stable material. The radioisotope enables scientists to trace the material as it undergoes chemical and physical changes. Tracers are being used widely in science, industry and agriculture today. When radioactive phosphorus, for example, is mixed with a chemical fertilizer the radioactive substance can be traced through the plant as it grows.

Trachea. The windpipe or tube that conducts air to and from the lungs. It extends between the larynx above and the point where it divides into two bronchi, below.

Trade name. The commercial name or trademark by which a chemical is known. One chemical may have a variety of trade names depending on the manufacturing or distributors involved.

Transducer. Any device or element which converts an input signal into an output signal of a different form; examples include the microphone, phonograph pickup, loudspeaker, barometer, photoelectric cell, automobile horn, doorbell, and underwater sound transducer.

Transmission loss. The ratio, expressed in decibels, of the sound energy incident on a structure to the sound energy that is transmitted. The term is applied both to building structures (walls, floors, etc.) and to air passages (muffler, ducts, etc.).

Transmutation. Any nuclear process which involves a change in energy or identity of the nucleus.

Transport (conveying) velocity. Minimum air velocity required to move the suspended particulates in the air stream.

Trauma. An injury or wound brought about by an outside force.

Tremor. Involuntary shaking, trembling, or quivering.

Triceps. The large muscle at the back of the upper arm that extends the forearm when contracted.

Tridymite. Vitreous, colorless form of free silica formed when quartz is heated to 1,598 F (870 C).

Trigger finger. Also known as snapping finger. A condition of partial obstruction in flexion or extension of a finger. Once past the point of obstruction, movement is eased. Caused by constriction of the tendon sheath.

Tripoli (Rottenstone). A porous, siliceous rock, resulting from the decomposition of chert or siliceous limestone. Used as a base in soap and scouring powders, in metal polishing, as a filtering agent, and in wood and paint fillers. A cryptocrystalline form of free silica.

Tritium. Often called hydrogen-three, extra-heavy hydrogen whose nucleus contains two neutrons and one proton. It is three times as heavy as ordinary hydrogen and is radioactive.

Tuberculosis. A contageous disease caused by infection with the bacterium *Mycobacterium tuberculosis.* It usually affects the lung, but bone, lymph glands, and other tissues may be affected.

Tularemia. A bacterial infection of wild rodents, such as rabbits. It may be generalized, or localized in the eyes, skin, or lymph nodes, or in the respiratory tract. It can be transmitted to humans.

Tumbling. An industrial process, such as in founding, in which small castings are cleaned by friction in revolving drum (tumbling mill, tumbling barrel), which may contain sand, sawdust, stone, etc.

Tumor. A swelling.

Turbid. Cloudy.

Turbidity. Cloudiness; disturbances of solids (sediment) in a solution, so that it is not clear.

Turbulence loss. The pressure or energy lost from a ventilation system through air turbulence.

Turning vanes. Curved pieces added to elbows or fan inlet boxes to direct air and so reduce turbulence losses.

TWA. Time-weighted average.

Tympanic cavity. Another name for the chamber of the middle ear.

U

Ulcer. The destruction of an area of skin or mucous membrane.

Ulceration. The formation or development of an ulcer.

Ulna. One of the two bones of the forearm. It forms the hinge joint at the elbow and does not rotate about its longitudinal axis. It terminates at the wrist on the same side as the little finger. Task design should not impose thrust loads through the ulna.

Ulnar deviation. A position of the hand in which the angle on the little finger side of the hand with the corresponding side of the forearm is decreased. Ulnar deviation is a poor working position for the hand and may cause nerve and tendon damage.

Ultrasonics. The technology of sound at frequencies above the audio range.

Ultraviolet. Those wavelengths of the electromagnetic spectrum which are shorter than those of visible light and longer than x-rays, 10^{-5}cm to 10^{-6}cm wavelength.

Unstable. All radioactive elements are unstable since they emit particles and decay to form other elements.

Unstable (reactive) liquid. A liquid which in the pure state or as commercially produced or transported will vigorously polymerize, decompose, condense, or will become self-reactive under conditions of shocks, pressure, or temperature.

Upper explosive limit (UEL). The highest concentration (expressed in percent vapor or gas in the air by volume) of a substance that will burn or explode when an ignition source is present.

Uranium. A heavy metal. The two principal isotopes of natural uranium are U-235 and U-238. U-235 has the only readily fissionable nucleus, which occurs in appreciable quantities in nature, hence its importance as nuclear fuel. Only one part in 140 of natural uranium is U-235. Highly toxic and a radiation hazard that requires special consideration.

Urethr-, urethro- (prefix). Relating to the urethra, the canal leading from the bladder for discharge of urine.

Urticaria. Hives.

USC. *United States Code* is the offical compilation of federal statutes. New editions are issued approximately every 6 years. Cumulative supplements are issued annually.

V

Vaccine. A suspension of disease-producing microorganisms modified by killing or attenuation so that it will not cause disease and can facilitate the formation of antibodies upon inoculation into man or animals.

Valence. A number indicating the capacity of an atom and certain groups of atoms to hold others in combination. The term also is used in more complex senses.

Valve (air oxygen). A device which controls the direction of air or fluid flow or the rate and pressure at which air or fluid is delivered, or both.

Van de Graaff accelerator. An electrostatic generator—a particle accelerator. To obtain the voltage, static electricity is picked up at one end of the machine by a rubber belt and carried to the other end where it is stored.

Vapor pressure. Pressure (measured in pounds per square inch absolute—psia) exerted by a vapor. If a vapor is kept in confinement over its liquid so that the vapor can accumulate above the liquid (the temperature being held constant), the vapor pressure approaches a fixed limit called the maximum (or saturated) vapor pressure, dependent only on the temperature and the liquid.

Vapor volume. The number of cubic feet of pure solvent vapor formed by the evaporation of one gallon of liquid at 24 C (75 F).

Vapors. The gaseous form of substances that are normally in the solid or liquid state (at room temperature and pressure). The vapor can be changed back to the solid or liquid state either by increasing the pressure or decreasing the temperature alone. Vapors also diffuse. Evaporation is the process by which a liquid is changed into the vapor state and mixed with the surrounding air. Solvents with low boiling points will volatilize readily.

Vasoconstriction. Decrease in the cross-sectional area of blood vessels. This may result from contraction of a muscle layer within the walls of the vessels or may be the result of mechanical pressure. Reduction in bloodflow results.

Vat dyes. Water insoluble, complex coal tar dyes that can be chemically reduced in a heated solution to a soluble form that will impregnate fibers. Subsequent oxidation then produces insoluble color dyestuffs which are remarkably fast to washing, light and chemicals.

Vector. (1) Term applied to an insect or any living carrier which transports a pathogenic microorganism from the sick to the well, inoculating the latter; the organism may or may not pass through any developmental cycle. (2) Anything (e.g., velocity, mechanical force, electromotive force) having magnitude, direction, and sense that can be represented by a straight line of appropriate length and direction.

Velocity. A vector that specifies the time rate of change of displacement with respect to a reference.

Velocity, capture. The air velocity needed to draw contaminants into the hood.

Velocity, face. The inward air velocity in the plane of openings into an enclosure.

Velocity pressure. The kinetic pressure in the direction of flow necessary to cause a fluid at rest to flow at a given velocity. When added to static pressure, it gives total pressure.

Velometer. A device for measuring air velocity.

Vena contracta. The reduction in the diameter of a flowing airstream at hood entries and other locations.

Veni-, veno- (prefix). Relating to the veins.

Ventilation. One of the principal methods to control health hazards, may be defined as "causing fresh air to circulate to replace foul air simultaneously removed."

Ventilation, dilution. Airflow designed to dilute contaminants to acceptable levels.

Ventilation, mechanical. Air movement caused by a fan or other air moving device.

Ventilation, natural. Air movement caused by wind, temperature difference or other nonmechanical factors.

Vermiculite. An expanded mica (hydrated magnesium-aluminum-iron silicate) used in lightweight aggregates, insulation, fertilizer, and soil conditioners, as a filler in rubber and paints, and as a catalyst carrier.

Vertigo. Dizziness; more exactly, the sensation that the environment is revolving around you.

Vesicant. Anything that produces blisters on the skin.

Vesicle. A small blister on the skin.

Viable. Living.

Vibration. An oscillation motion about an equilibrium position produced by a disturbing force.

Vinyl. A general term applied to a class of resins such as polyvinyl chloride, acetate, butyral, etc.

Viruses. A group of pathogens consisting mostly of nucleic acids and lack of cellular structure.

Virulence. The capacity of a microorganism to produce disease.

Virulent. Extremely poisonous or venomous; capable of overcoming bodily defensive mechanisms.

Viscera. Internal organs of the abdomen.

Viscose. Term applied to viscous liquid composed of cellulose xanthate.

Viscose rayon. The type of rayon produced from the reaction of carbon disulfide with cellulose and the hardening of the resulting viscous fluid by passing it through dilute sulfuric acid, this final operation causing the evolution of hydrogen sulfide gas.

Viscosity. The property of a fluid that resists internal flow by releasing counteracting forces.

Viscosity, absolute. A measure of a fluid's tendency to resist flow, without regard to density. The product of a fluid's kinematic viscosity times its density, and is expressed in dyne-seconds per cm^2 or poises (or pascal seconds).

Viscosity, kinematic. The relative tendency of fluids to resist flow. The value of the kinematic viscosity is equal to the absolute viscosity of the fluid divided by the fluid density and is expressed in units of stoke (or square meters per second).

Visible radiation. The wavelengths of the electromagnetic spectrum between 10^{-4} cm and 10^{-5}cm.

Vision, photopic. Vision attributed to cone function characterized by the ability to discriminate colors and small detail; daylight vision.

Vision, scotopic. Vision attributed to rod function characterized by the lack of ability to discriminate colors and small detail and effective primarily in the detection of movement and low luminous intensities; night vision.

Visual acuity. Ability of the eye to sharply perceive the shape of objects in the direct line of vision.

Volatility. The tendency or ability of a liquid to vaporize. Such liquids as alcohol and gasoline, because of their well-known tendency to evaporate rapidly, are called volatile liquids.

Volume flow rate. The quantity (measured in units of volume) of a fluid flowing per unit of time, as cubic feet per minute, gallons per hour, or cubic meters per second.

Volume, specific. The volume occupied by one pound of a substance under any specified conditions of temperature and pressure.

Volumetric analysis. A statement of the various components of a substance (usually applied to gases only), expressed in percentages by volume.

Vulcanization. Process of combining rubber (natural, synthetic or latex) with sulfur and accelerators in presence of zinc oxide under heat and usually pressure in order to change the material permanently, from a thermoplastic to a thermosetting compositon, or from a plastic to an elastic condition. Strength, elasticity and abrasion resistance also are improved.

Vulcanizer. A machine in which raw rubber that has been mixed with chemicals is cured by heat and pressure to render it less plastic and more durable.

W

Wart. A characteristic growth on the skin, appearing most frequently on the fingers; generally regarded as a result of a virus infection. Synonym: verruca.

Water column. A unit used in measuring pressure. See also Inches of water column.

Water curtain or waterfall booth. Many different meanings in industry, but in spray painting the water running down a wall into which the excess paint spray is drawn or blown by fans and the water carries the paint downward to a collecting point.

Waterproofing agents. These usually are formulations of three distinct materials: (1) a coating material, (2) a solvent, and (3) a plasticizer. Among the materials used in waterproofing are cellulose esters and ether, polyvinyl cloride resins or acetates, and variations of vinyl chloride-vinylidine chloride polymers.

Watt (w). A unit of power, equal to one joule per second. See Erg.

Watt/cm². A unit of power density used in measuring the amount of power per area of absorbing surface, or per area of a CW laser beam.

Wavelength. The distance in the line of advance of a wave from any point to a like point on the next wave. It is usually measured in angstroms, microns, micrometers, or nanometers.

Weight. The force with which a body is attracted toward the earth. Although the weight of a body varies with its location, the weights of various standards of mass are often used as units of force. See Force.

Weighting network (sound). Electrical networks (A, B, C) associated with sound level meters. The C network provides a flat response over the frequency range 20-10,000 Hz of interest while the B and A networks selectively discriminate against low (less than 1 kHz) frequencies.

Weld. A localized coalescence of metals or nonmetals produced either by heating the materials to suitable temperatures, with or without the application of pressure, or by the application of pressure alone, and with or without the use of filler material.

Welding. The several types of welding are electric arc-welding, oxyacetylene welding, spot welding, and inert or shielded gas welding utilizing helium or argon. The hazards involved in welding stem from (1) the fumes from the weld metal such as lead or cadmium metal, or (2) the gases created by the process, or (3) the fumes or gases arising from the flux.

Welding process. A materials joining process that produces coalescence of materials by heating them to suitable temperatures, with or without the application of pressure or by the application of pressure alone, and with or without the use of filler metal.

Welding rod. A rod or heavy wire that is melted and fused into metals in arc welding.

Wet-bulb globe temperature index. An index of the heat stress in humans when work is being performed in a hot environment.

Wet-bulb temperature. Temperature as determined by the wet bulb thermometer or a standard sling psychrometer or its equivalent. This temperature is influenced by the evaporation rate of the water, which in turn depends on the humidity (amount of water vapor) in the air.

Wet-bulb thermometer. A thermometer having the bulb covered with a cloth saturated with water.

White damp. In mining, carbon monoxide.

White noise. A noise whose spectrum density (or spectrum level) is substantially independent of frequency over a specified range.

Wide band. Applied to a wide band of transmitted waves, with neither of the critical or cutoff frequencies of the filter being zero or infinite.

Work. When a force acts against resistance to produce motion in a body, the force is said to do work. Work is measured by the product of the force acting and the distance moved through against the resistance. The units of measurement are the erg (the joule is 1×10^7 ergs) and the foot-pound.

Work hardening. The property of metal to become harder and more brittle on being "worked," that is, bent repeatedly or drawn.

Work strain. The natural physiological response reaction of the body to the application of work stress. The locus of the reaction may often be remote from the point of application of work stress. Work strain is not necessarily traumatic, but may appear as trauma when excessive, either directly or cumulatively, and must be considered by the industrial engineer in equipment and task design. Thus a moderate increase of heart rate is non-traumatic work strain resulting from physical work strain if resulting from undue work stress on the wrists.

Work stress. Biomechanically, any external force acting on the body during the performance of a task. It always produces work strain. Application of work stress to the human body is the inevitable consequence of performance of any task, and is, therefore, only synonymous with "stressful work conditions" when excessive. Work stress analysis is an integral part of task design.

Working level (WL). Any combination of radon daughters in one liter of air that will result in the ultimate emission of 1.3×10^5 MeV of alpha energy.

X

Xanth- (prefix). Yellow.

X-ray. Highly penetrating radiation similar to gamma-rays. Unlike gamma-rays, x-rays do not come from the nucleus of the atom but from the surrounding electrons. They are produced by electron bombardment. When these rays pass through an object they give a shadow picture of the denser portions.

X-ray diffraction. Since all crystals act as three-dimensional gratings for x-rays, the pattern of diffracted rays is characteristic for each crystalline material. This method is of particular value in determining the presence or absence of crystalline silica in an industrial dust.

X-ray tube. Any electron tube designed for the conversion of electrical energy into x-ray energy.

Xero- (prefix). Indicated dryness, as xerostomia, dryness of the mouth.

Xeroderma. Dry skin; may be rough as well as dry.

Z

Z. Symbol for atomic number. An elements's atomic number is the same as the number of protons found in one of its nuclei. All isotopes of a given element have the same Z number.

Zoonoses. Diseases biologically adapted to and normally found in lower animals, but which under some conditions also infect humans.

Zygote. Cell produced by the joining of two gametes (sex or germ cells).

Index

A

D

E

F

I

J

K

L

M

N

O

P

Q

R

S

T

U

V

W

X

Y

Z